AF753704

BMW 5 Series
(E60, E61)

Service Manual
525i, 525xi, 528i, 528xi
530i, 530xi, 535i, 535xi
545i, 550i
2004, 2005, 2006, 2007, 2008, 2009, 2010

BentleyPublishers®
.com

[B]® BENTLEY PUBLISHERS® | Automotive Reference™

Bentley Publishers, a division of Robert Bentley, Inc.
1734 Massachusetts Avenue
Cambridge, MA 02138 USA
800-423-4595 / 617-547-4170

Information that makes the difference®

BentleyPublishers®.com

Technical contact information

We welcome your feedback. Please submit corrections and additions to our BMW technical discussion forum at:

`http://www.BentleyPublishers.com`

Updates and corrections

We will evaluate submissions and post appropriate editorial changes online as updates or tech discussion. Appropriate updates will be incorporated with the manual text in future printings. Check for updates and corrections for this book before beginning work on your vehicle. See the following web address for additional information:

`http://www.BentleyPublishers.com/updates/`

WARNING—Important Safety Notice

Do not use this manual for repairs unless you are familiar with automotive repair procedures and safe workshop practices. This manual illustrates the workshop procedures for some maintenance and service work. It is not a substitute for full and up-to-date information from the vehicle manufacturer or for proper training as an automotive technician. Note that it is not possible for us to anticipate all of the ways or conditions under which vehicles may be serviced or to provide cautions as to all of the possible hazards that may result.

We have endeavored to ensure the accuracy of the information in this manual. Please note, however, that considering the vast quantity and the complexity of the service information involved, we cannot warrant the accuracy or completeness of the information contained in this manual.

FOR THESE REASONS, NEITHER THE PUBLISHER NOR THE AUTHOR MAKES ANY WARRANTIES, EXPRESS OR IMPLIED, THAT THE INFORMATION IN THIS MANUAL IS FREE OF ERRORS OR OMISSIONS, AND WE EXPRESSLY DISCLAIM THE IMPLIED WARRANTIES OF MERCHANTABILITY AND OF FITNESS FOR A PARTICULAR PURPOSE, EVEN IF THE PUBLISHER OR AUTHOR HAVE BEEN ADVISED OF A PARTICULAR PURPOSE, AND EVEN IF A PARTICULAR PURPOSE IS INDICATED IN THE MANUAL. THE PUBLISHER AND AUTHOR ALSO DISCLAIM ALL LIABILITY FOR DIRECT, INDIRECT, INCIDENTAL OR CONSEQUENTIAL DAMAGES THAT RESULT FROM ANY USE OF THE EXAMPLES, INSTRUCTIONS OR OTHER INFORMATION IN THIS MANUAL. IN NO EVENT SHALL OUR LIABILITY, WHETHER IN TORT, CONTRACT OR OTHERWISE, EXCEED THE COST OF THIS MANUAL.

Before attempting any work on your BMW, read **001 Warnings and Cautions** and any **WARNING** or **CAUTION** that accompanies a procedure in the manual. Review the **WARNINGS** and **CAUTIONS** each time you prepare to work on your BMW.

Your common sense and good judgment are crucial to safe and successful service work. Read procedures through before starting them. Think about whether the condition of your car, your level of mechanical skill or your level of reading comprehension might result in or contribute in some way to an occurrence which might cause you injury, damage your car or result in an unsafe repair. If you have doubts for these or other reasons about your ability to perform safe repair work on your car, have the work done at an authorized BMW dealer or other qualified shop.

Part numbers listed in this manual are for identification purposes only, not for ordering. Always check with your authorized BMW dealer to verify part numbers and availability before beginning service work that may require new parts.

Special tools required to perform certain service operations are identified in the manual and are recommended for use. Use of improper tools may be detrimental to the car's safe operation as well as the safety of the person servicing the car.

The vehicle manufacturer will continue to issue service information updates and parts retrofits after the editorial closing of this manual. Some of these updates and retrofits will apply to procedures and specifications in this manual. We regret that we cannot supply updates to purchasers of this manual.

This manual is prepared, published and distributed by Bentley Publishers, 1734 Massachusetts Avenue, Cambridge, Massachusetts 02138 USA. All information contained in this manual is based on the information available to the publisher at the time of editorial closing. BMW has not reviewed and does not vouch for the accuracy or completeness of the technical specifications and work procedures described and given in this manual.

© 2010 Robert Bentley, Inc. Bentley Publishers is a trademark of Robert Bentley, Inc.

[B]® is a registered trademark of Bentley Publishers and Robert Bentley, Inc.

All rights reserved. The right is reserved to make changes at any time without notice. No part of this publication may be reproduced, stored in a retrieval system, or transmitted in any form or by any means, electronic, mechanical, photocopying, recording, or otherwise, without the prior written consent of the publisher. This includes text, figures, and tables. All rights reserved under Berne and Pan-American Copyright conventions.

ISBN 978-0-8376-1689-6 (2-volume set)
ISBN 978-0-8376-1690-2 (volume 1)
ISBN 978-0-8376-1691-9 (volume 2) **Job Code: B510-05** **Editorial closing 04 / 2010**

Library of Congress Cataloging-in-Publication Data for the 2010 edition
BMW 5 series, (E60, E61) : service manual : 525i, 525xi, 528i, 528xi, 530i, 530xi, 535i, 535xi, 545i, 550i, 2004, 2005, 2006, 2007, 2008, 2009, 2010
p. cm.
Includes bibliographical references and index.
ISBN 978-0-8376-1621-6 (alk. paper)
1. BMW 5 series automobiles--Maintenance and repair--Handbooks, manuals, etc. I. Robert Bentley, inc.

TL215.B25B57 2010
629.28'722--dc22

2010012318

Cover design and use of blue band on spine and back cover are trade dress and trademarks of Bentley Publishers. All rights reserved.

The paper used in this publication is acid free and meets the requirements of the National Standard for Information Sciences–Permanence of Paper for Printed Library Materials. ∞

Manufactured in the United States of America

202411R001

You are now in **Volume 2**

Bentley Publishers
BentleyPublishers.com–All Rights Reserved

Foreword

For the BMW owner with basic mechanical skills and for independent auto service professionals, this manual includes many of the specifications and procedures that were available in an authorized BMW dealer service department as this manual went to press. The BMW owner with no intention of working on his or her car will find that owning and referring to this manual will make it possible to be better informed and to more knowledgeably discuss repairs with a professional automotive technician.

For those intending to do maintenance and repair on their BMW, it is essential that safety equipment be used and safety precautions observed when working on the vehicle. A minimum safety equipment list includes hand protection, eye protection and a fire extinguisher. A selection of good quality hand tools is also needed. This includes a torque wrench to ensure that fasteners are tightened in accordance with specifications. In some cases, the text refers to special tools that are recommended or required to accomplish adjustments or repairs. These tools are often identified by their BMW special tool number and illustrated.

Disclaimer

We have endeavored to ensure the accuracy of the information in this manual. When the vast array of data presented in the manual is taken into account, however, no claim to infallibility can be made. We therefore cannot be responsible for the result of any errors that may have crept into the text. Please also read the **WARNING—Important Safety Notice** on the copyright page at the beginning of this book.

Prior to starting a repair procedure, read the procedure, **001 Warnings and Cautions** and the warnings and cautions that accompany the procedure. Reading a procedure before beginning work helps you determine in advance the need for specific skills, identify hazards, prepare for appropriate capture and handling of hazardous materials, and the need for particular tools and replacement parts such as gaskets.

Bentley Publishers encourages comments from the readers of this manual with regard to errors, and suggestions for improvement of our product. These communications have been and will be carefully considered in the preparation of this and other manuals. If you identify inconsistencies in the manual, you may have found an error. Please contact the publisher and we will endeavor to post applicable corrections on our website. Posted corrections (errata) should be reviewed before beginning work. Please see the following web address:

`http://www.BentleyPublishers.com/updates/`

BMW will continue to issue service information and parts retrofits after the editorial closing of this manual. Some of this updated information may apply to procedures and specifications in this manual. For the latest information, please see the following web address:

`http://www.bmwtechinfo.com/`

BMW offers extensive warranties, especially on components of the fuel delivery and emission control systems. Therefore, before deciding to repair a BMW that may be covered wholly or in part by any warranties issued by BMW of North America, LLC, consult your authorized BMW dealer. You may find that the dealer can make the repair either free or at minimum cost. Regardless of its age, or whether it is under warranty, your BMW is both an easy car to service and an easy car to get serviced. So if at any time a repair is needed that you feel is too difficult to do yourself, a trained BMW technician is ready to do the job for you.

Bentley Publishers

600 Electrical System–General

TABLES

GENERAL

This section describes the principal parts of the electrical system. Also covered here are basic electrical system troubleshooting tips.

See also:

- **121 Battery, Starter, Alternator**
- **ECL Electrical Component Locations** for common BMW acronyms as well as component location information
- **ELE Electrical Wiring Diagrams**
- **OBD On-Board Diagnostics**

Electrical overview

The vehicle operates on a 12-volt direct current (DC) negative-ground system. Power is supplied by a lead-acid battery in right side of the trunk or cargo compartment.

A voltage regulator controls system voltage at approximately 12 volts. All circuits are grounded by direct or indirect connection to the negative (–) terminal of the battery.

A number of ground connections throughout the vehicle connect the wiring harness to chassis ground. These circuits are completed by the battery cable or ground strap between the body and the battery negative (–) terminal.

BentleyPublishers.com
BentleyPublishers.com—All Rights Reserved

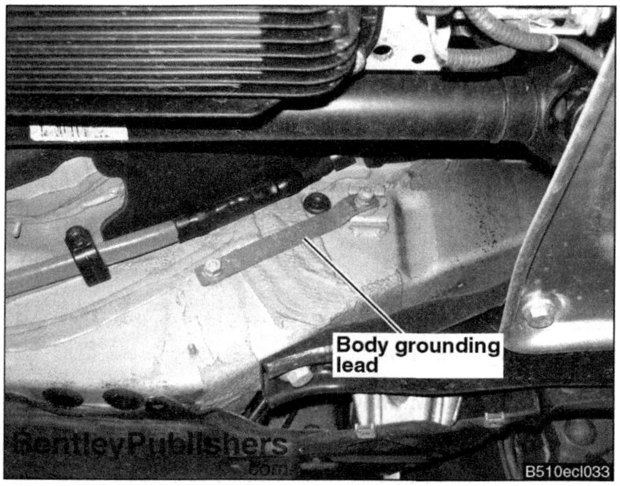

Due to the extensive use of aluminum in 5 Series front end components, grounding leads are used to electrically bridge body components to each other.

Fuses are color coded to indicate current capacities.

Electrical components connect using one of the following:

- Heavy cables with lug-type connectors (battery, starter, alternator).
- Electrical harnesses with keyed, push-on connectors that lock into place.
- Buses with modular connectors.

Relays and control modules are mounted in various places throughout the vehicle. See **ECL Electrical Component Locations**.

Bus basics

To handle electrical complexity, 5 Series vehicles are equipped extensively with bus systems. Signals are shared digitally among electrical components on a bus, eliminating the need for separate connections for each pair of components. The use of bus communication for controls and accessories reduces wiring complexity and improves system response time.

Data transfer over a bus is similar to a telephone conference. A component on the bus transmits a stream of data (data telegrams) which other components receive at the same time. Each component is then free to use or ignore this telegram.

The benefits of the bus method of data transfer are as follows:

- As data and programs are modified and extended, only software modifications are necessary.
- Continuous verification of transmitted data leads to low error rate.
- Sensors and signal wires can be simplified or eliminated due to the transmittal of multiplexed digital data.
- Control modules transfer data at a high rate.
- Control module sizes and connector sizes are smaller.
- Bus architecture conforms to international standards. This facilitates data interchange between components of different manufacture.

Electrical test equipment

Many electrical tests described in this manual call for measuring voltage, current or resistance using a digital multimeter. A digital meter is preferred for precise measurements and for electronics work because it is generally more accurate than an analog meter (swing-needle). An analog meter may draw enough current to damage sensitive electronic components.

An LED test light is a safe, inexpensive tool that can be used to perform many simple electrical tests that would otherwise require a digital multimeter. The LED indicates when voltage is present between any two test-points in a circuit.

The integrated safety, comfort, security and handling systems on 5 Series vehicles are designed with self-diagnostic capabilities. The

BentleyPublishers.com
BentleyPublishers.com—All Rights Reserved

quickest way to diagnose many problems is to start with a scan tool read-out of diagnostic trouble codes (DTCs). See **Scan tool basics** in this repair group.

Static damage prevention

Electronic circuits can be easily damaged by static electricity held by the human body and vehicle electrical components. To avoid static electricity damage, take precautions to ground the static electricity before touching any internal components of a circuit.

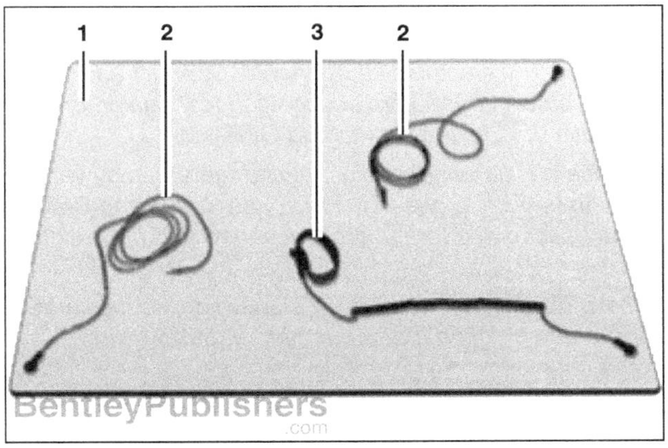

Wearing an electrostatic discharge (ESD) wrist strap or working on an anti-static mat prevents static electricity damage.

1. ESD mat
2. Grounding lead
3. ESD wrist strap

Warnings and Cautions

WARNING—

- *Airbags and pyrotechnic seat belt tensioners utilize explosive devices. Handle with extreme care. See warnings and cautions in* **721 Airbag System (SRS)**.
- *The ignition system operates at lethal voltages. If you have a weak heart or wear a pacemaker, do not expose yourself to ignition system electric currents. Take extra precautions when working on the ignition system or when servicing the engine while it is running or the key is ON. See* **120 Ignition System** *for additional ignition system warnings and cautions.*
- *Keep hands, clothing and other objects clear of the electric engine cooling fan when working on a warm engine. The fan may start at any time, even when the ignition is switched OFF.*

Bentley Publishers
BentleyPublishers.com–All Rights Reserved

CAUTION—
- *Do not disconnect the battery with the engine running.*
- *Prior to disconnecting the battery, read the battery disconnection cautions given in* **001 Warnings and Cautions**.
- *Switch the ignition OFF and remove the negative battery cable before removing any electrical components. Connect and disconnect electrical connectors and ignition test equipment leads only while the ignition is switched OFF.*
- *Relay and fuse positions are subject to change and may vary from vehicle to vehicle. If questions arise, an authorized BMW dealer is the best source for the most accurate and up-to-date information.*
- *Use a digital multimeter for electrical tests. Switch the multimeter to the appropriate function and range before making test connections.*
- *Many control modules are static sensitive. Static discharge damages them permanently. Handle the modules using static prevention equipment and techniques. See* **Static damage prevention** *in this repair group.*
- *To avoid damaging harness connectors or relay panel sockets, use jumper wires with flat-blade connectors that are the same size as the connector or relay terminals.*
- *Do not try to start the engine of a vehicle which has been heated above 176°F (80°C) (for example, in a paint drying booth). Allow it to cool to normal temperature.*
- *Disconnect the battery before doing any electric welding on the vehicle.*
- *Do not wash the engine while it is running, or any time the ignition is ON.*
- *Choose test equipment carefully. Use a digital multimeter with at least 10 MΩ input impedance or an LED test light. An analog meter (swing-needle) or a test light with a normal incandescent bulb may draw enough current to damage sensitive electronic components.*
- *Do not use an ohmmeter to measure resistance on solid state components such as control modules.*
- *Disconnect the battery before making resistance (ohm) measurements on a circuit.*

BentleyPublishers.com
BentleyPublishers.com–All Rights Reserved

VEHICLE ELECTRICAL SYSTEM (2004 - 2005)

Power supply (2004 - 2005)

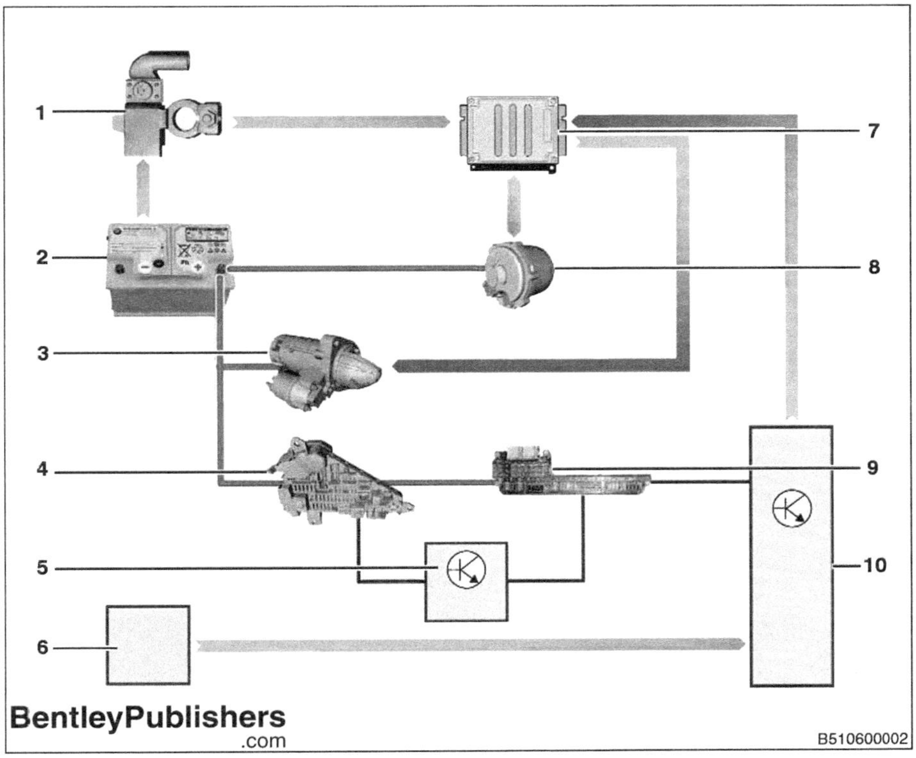

1. Intelligent battery sensor (IBS)
   - Monitors battery state of charge, state of health.
   - See **121 Battery, Starter, Alternator** for additional information.
2. Battery.
3. Starter.
4. Rear power distribution and fuse panel (A46).
   - Houses terminal 30g relay for electric load cut-off function.
   - See **ECL Electrical Component Locations** for fuse and relay positions.
5. Micro-power module (MPM).
   - In trunk, underneath spare tire.
   - Houses bistable (latching) relay for electric load cut-off function.
   - Switches power supply to audio and communications equipment.
6. Ignition and starter switch (ZAS).
7. Engine control module (ECM).
8. Alternator.
9. Front fuse and relay panel (A41).
   - See **ECL Electrical Component Locations** for fuse and relay positions.
10. Car access system (CAS), version 1.
    - Underneath left side dashboard.
    - See **121 Battery, Starter, Alternator** for replacement procedure.

600

BentleyPublishers.com
BentleyPublishers.com—All Rights Reserved

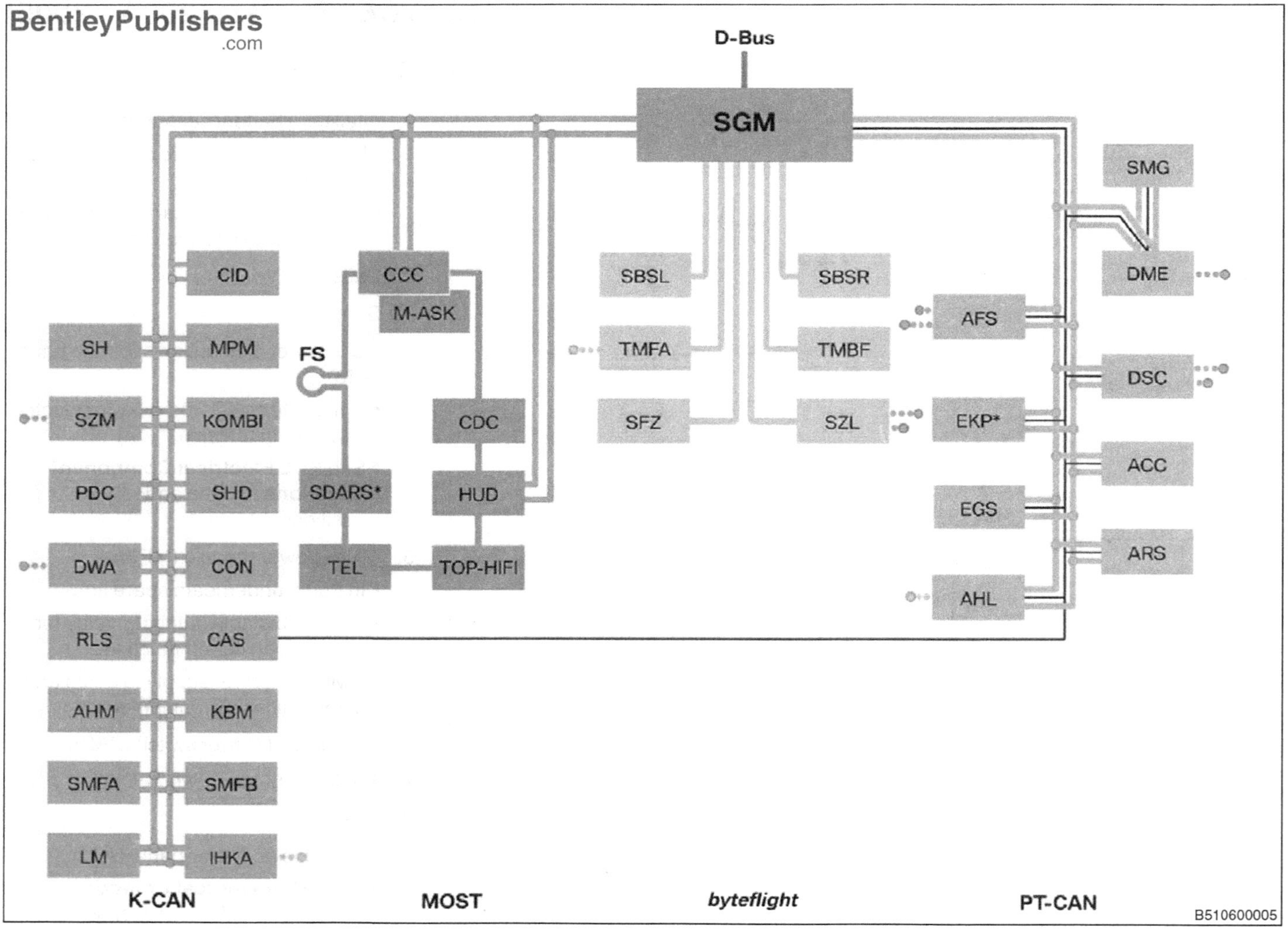

Bus systems (2004 - 2006)

ACC Active cruise control

AFS Active steering

AHL Adaptive headlights

AHM Hazard warning

ARS Acceleration sensor

CAS Car access system

CCC Car communications computer

CDC CD changer

CID Central information display

CON iDrive controller

DME Engine control module (ECM)

DSC Dynamic stability control module

DWA anti-theft alarm

EGS Automatic transmission control module

EKP Fuel pump control module

FS MOST direct access

IHKA Automatic climate control

KBM Body basic module

KOMBI Instrument cluster

LM Light module

M-ASK Multi-audio system controller

MPM Micro-power module

PDC Park distance control

RLS Rain and light sensor

SBSL B-pillar satellite crash sensor, left

SBSR B-pillar satellite crash sensor, right

SDARS Satellite radio

SFZ Center satellite crash sensor

SGM Safety and gateway module

SH Seat heater

SHD Sunroof module

SMBF Passenger seat module

SMG Sequential manual gearbox

SMFA Driver seat module

SZL Steering column switch cluster

SZM Center console switch center

TMBF Passenger door module

TMFA Driver door module

TEL Telematics control module

TOP-HIFI Amplifier

BentleyPublishers.com
BentleyPublishers.com—All Rights Reserved

Bus functions and types (2004 - 2005)

Main buses

Main bus systems exchange data between control modules throughout the vehicle.

D-bus (diagnostic -bus) connects SGM to diagnostic plug (OBD plug).

K-CAN (body CAN) connects the instrument cluster, central information display (CID) and iDrive controller.

MOST (media oriented systems transport fiber optic bus) connects satellite radio, telephone and top HiFi system.

byteflight (fiber optic bus) connects satellite crash sensors, door modules and driver airbag to SGM (primary airbag module).

PT-CAN (powertrain CAN) links fuel injection, powertrain and traction control modules. The change from CAN-bus (twisted pair of wires) is that PT-CAN has three wires, with the third wire used as the wake-up line for associated modules.

Sub-buses

Sub-bus systems exchange relatively small quantities of data within one system.

LIN-bus (local interconnect network), a bidirectional single-wire bus system, includes a main controller and several server units. The 5 Series LIN-bus connects control modules with blower motor, tire pressure monitoring wheel housing antennas and door modules.

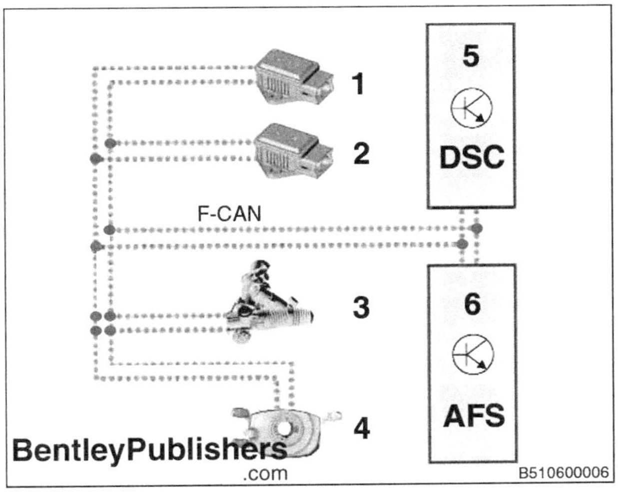

F-CAN (chassis CAN) enables fast data transfer among components in DSC and active steering.

1. DSC acceleration sensor 1
2. DSC acceleration sensor 2
3. Active steering actuator
4. Steering column switch center (SZL)
5. DSC control module
6. Active steering control module

600

BentleyPublishers.com
BentleyPublishers.com–All Rights Reserved

Bus functions and types (2004 - 2005)

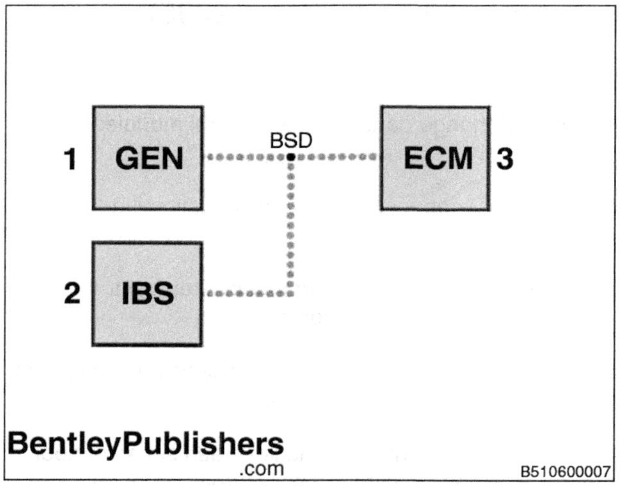

BSD (bit-serial data interface) connects intelligent battery sensor (IBS) with alternator and engine control module (ECM).

1. Alternator
2. Intelligent battery sensor (IBS)
3. Engine control module (ECM)

Table a summarizes bus structure and data speeds.

Table a. Bus system parameters

Bus system	Data rate	Structure
BSD	9.6 kBd	Linear, single wire
byteflight	10 MBd	Star, fiber optic
D-bus	10.5 - 115 kBd	Linear, single wire
DWA-bus	9.6 kBd	Linear, single wire
F-CAN	100 kBd	Linear, 2-wire
K-bus seats	9.6 kBd	Linear, single wire
K-CAN	100 kBd	Linear, 2-wire
LIN-bus adaptive headlights	19.2 kBd	Linear, single wire
LIN-bus doors	19.2 kBd	Linear, single wire
LIN-bus IHKA	9.6 kBd	Linear, single wire
LIN-bus tire pressure monitor	9.6 kBd	Linear, single wire
MOST	22.5 Mbd	Ring, fiber optic
PT-CAN	500 kBd	Linear, 2-wire

BentleyPublishers.com
BentleyPublishers.com—All Rights Reserved

Vehicle Electrical System (2006 - 2007)

System overview (2006 - 2007)

Beginning with 09 / 2005 production (2006 model year), 5 Series electrical system was modified as follows:

System network:

- byteflight bus eliminated.
- Body-gateway module (KGM) mounted in front fuse and relay panel behind glove compartment.
- Door module functions transferred to KGM.

Energy management: Micro-power module (MPM) eliminated. MPM function, including bistable relay, integrated in KGM.

Comfort access (optional):

- Keyless access.
- Electronic outer door handle modules (TAGE)

Car access system (CAS), version 2:

- Start / stop button.
- Key slot.
- Electric steering column lock (ELV).

Expanded diagnostic functions:

- Monitoring of control modules to detect authorized bus system wake-up.
- Monitoring of control modules to detect sleep mode preventer (control unit that prevents bus system from switching to sleep mode).

Steering column switch cluster (SZL): New connection to PT-CAN.

Passive safety: Airbag control system changed from advanced safety electronics (ASE) to advanced crash and safety management (ACSM).

Audio systems: MP3 function in radio module.

600

Bus system (2006 - 2007)

The 2004 - 2005 bus system is modified as follows:

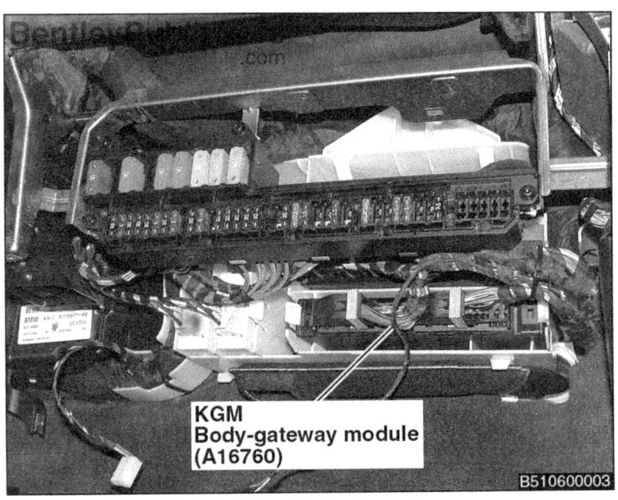

Body-gateway module (KGM) performs the gateway interconnect function for the following buses:

- Diagnostic-bus (D-bus)
- PT-CAN (powertrain CAN)
- K-CAN (body CAN)
- LIN-bus (local interconnect network)

Data transmission rates and data telegram structure vary among different buses. KGM is responsible for buffering data telegrams and forwarding them according to priority.

KGM is the interface to the diagnostic-bus for the purpose of diagnosing or programming control units connected to the bus system.

The KGM additionally features a link to LIN-bus for the purpose of reading information from the driver's door switch cluster.

Main airbag control module functions are transferred to the advanced crash and safety management (ACSM) control module, above center tunnel, connected to sensors and airbags via K-CAN.

Comfort access components, including the control module, door handle modules (TAGE) and electric steering column lock (ELV) interconnect via K-CAN.

BentleyPublishers.com
BentleyPublishers.com—All Rights Reserved

Vehicle Electrical System (2008 - 2010)

Electrical system (2008 - 2010)

Beginning with 03 / 2007 production (2008 model year) new electrical features are either standard or optional on 5 Series:

S-CAN (sensor CAN), a new bus system based on PT-CAN, transmits data at the same speed as PT-CAN and uses PT-CAN protocol.

D-CAN (diagnostics-CAN), also a new bus, is connected to KGM.

Car access system (CAS), version 3 with electronic immobilizer 3 or 4 (EWS 3 or EWS 4), depending on the engine. With EWS 4, engine electronics are connected to CAS by an additional data connection. The data connection is redundant to PT-CAN and uses K-bus protocol.

Rain / light sensor connected by LIN-bus to wiper and lights systems.

Welcome light.

Daytime driving light with corona rings.

Adaptive headlights with turn-off light.

Turn signal LED and additional rear side-marker lights standard.

Diagnostics-CAN on body-gateway module (KGM).

Lane departure warning (TLC) generates a vibration warning in the steering wheel when the driver accidentally changes lane, prompting the driver to counter-steer.

Active cruise control with stop and go function makes it easier to drive in heavy traffic. The system supports repeated pulling away and braking in slow traffic queues.

Longitudinal dynamics management (LDM) is used, among other things, for active cruise control stop and go function.

Long and short range radar are for active cruise control with stop and go function.

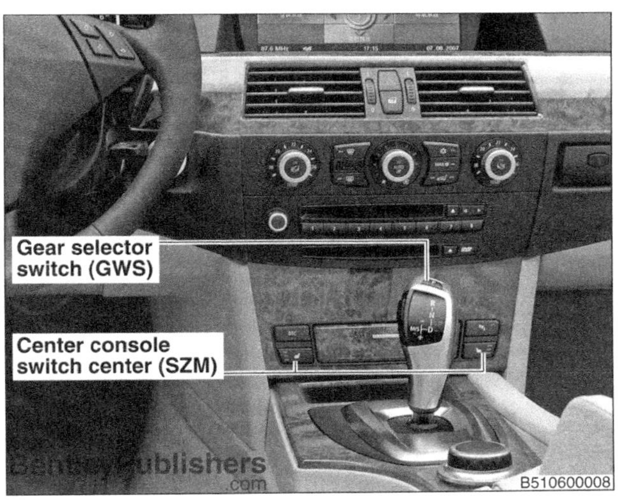

Gear selector switch (GWS) control module replaces automatic transmission mechanical gear shift.

Center console switch cluster (SZM) consists of two columns of buttons on two sides of center dashboard ashtray.

600

BentleyPublishers.com
BentleyPublishers.com—All Rights Reserved

Power Management

Power management basics

Power management strategies in 5 Series vehicles are designed to monitor battery power, minimize power loss (dead battery) and diagnose battery state of charge problems. Intelligent battery sensor (IBS) data is continuously monitored and evaluated for this purpose.

During vehicle operation, engine mechanical energy is converted by the alternator into electrical energy and made available to electric loads in the vehicle. Most electric loads receive power primarily via switched terminal 30g and terminal 30g_f relays. Certain electric loads are supplied unswitched battery power directly by terminal 30 or terminal R. *Example:* Anti-theft alarm system is active when the vehicle is parked and engine is shut OFF. Therefore anti-theft power is from an unswitched terminal.

The power management portion of the engine control module (ECM) operating program is able to:

- Regulate idle speed and charging voltage while the engine is running.
- Regulate power consumption of electric loads with relatively high power demands, or switch OFF those loads if power is needed elsewhere.
- Switch OFF electric loads when engine is stationary either as a timed function (via CAS and terminal 30g relay) or in response to electrical faults (via ECM and terminal 30g_f relay).
- Generate and store fault codes to help with vehicle or battery diagnosis.

Additional information on IBS is in **121 Battery, Starter, Alternator**.

BentleyPublishers.com
BentleyPublishers.com—All Rights Reserved

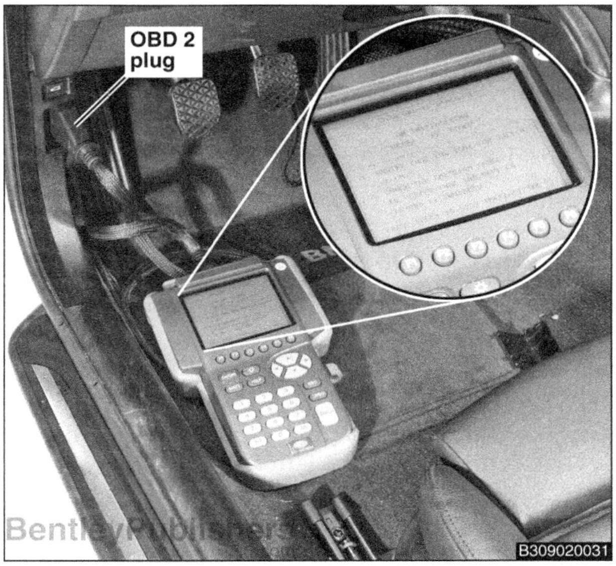

CODING, PROGRAMMING, ADAPTATION

Scan tool basics

The scan tool, shown here connected to the OBD 2 plug, is an electronic device used for the following:

- Displaying vehicle data such as engine, engine management and transmission versions
- Viewing and clearing fault codes (diagnostic trouble codes or DTCs)
- Coding and programming new control modules and other electronic equipment
- Clearing and resetting adaptations

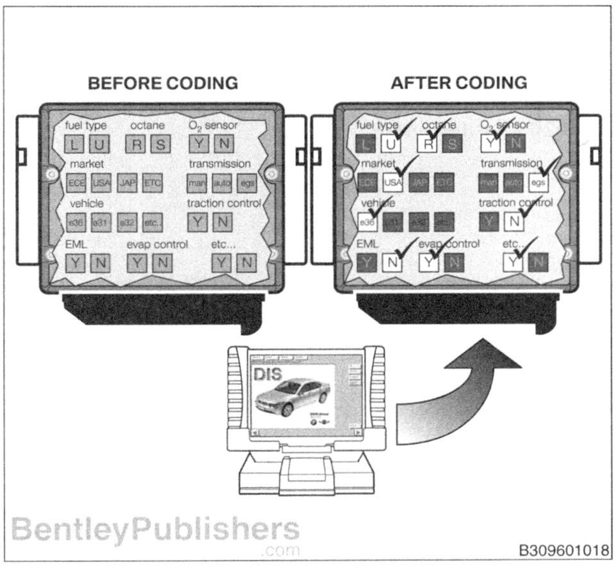

Coding

The procedure of assigning one labeled group of data to the operating program of a control module or component is referred to as coding. A codable control module has a basic operating program already installed along with several variations of operating data. The coding process allows a specific set of operating data to be assigned to the basic operating program of that module or component to match a specific application.

Here are some types of operating requirements:

- Nominal values of device input signals (0.25 - 2.5 volts, 5 - 25 watts, etc.)
- Type of device input signal (PWM, square wave, analog)
- Operational parameters (device activation or deactivation time)
- Market specific operations (oxygen sensors, fuel type, emission standards)
- Country specific regulations (USA, Canada, Japan, UK, EU)
- Powertrain configurations (manual transmission, automatic transmission, TLEV, ULEV, SULEV, Diesel)

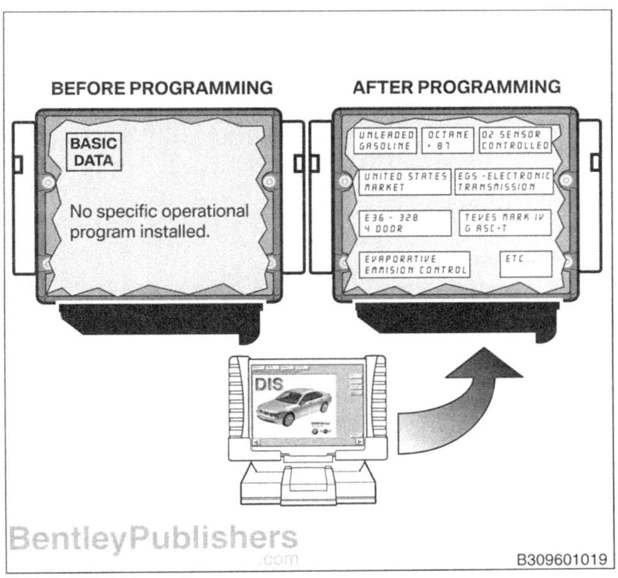

Programming

Programming loads application and system-specific operating instructions (program) into a coded module. Programming is also a means of updating data and operating instructions previously installed in a control module.

600

BentleyPublishers.com
BentleyPublishers.com—All Rights Reserved

Adaptation

The engine control model (ECM), automatic transmission control module (EGS), dynamic stability control (DSC) module and other modules and components in the vehicle are adaptive. The software in each module learns from use and becomes adapted to other accompanying modules.

When installing a new module or component, be sure to use the scan tool to clear old adaptations in order to allow the components to readapt.

Coding and programming procedure

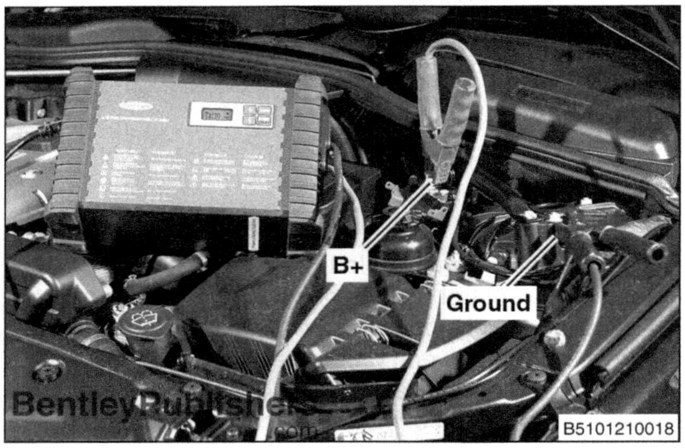

Before starting the coding and programming procedure, use a vehicle power supply to maintain correct voltage throughout the electrical system. See **121 Battery, Starter, Alternator**.

For the most up-to-date data and program information, make sure the scan tool has the current version of CIP (coding, individualization and programming) installed.

Connect BMW scan tool or equivalent to OBD II plug under dashboard. See **Scan tool basics** in this repair group. In addition to BMW scan tools, there are a number of aftermarket tools available. See **Tools** in **020 Maintenance**.

BentleyPublishers.com
BentleyPublishers.com–All Rights Reserved

Electrical Troubleshooting

Four things are required for current to flow in any electrical circuit:

- Voltage source.
- Wires or connections to transport voltage.
- Load or device that uses electricity.
- Connection to ground.

Most problems can be found using a digital multimeter (volt / ohm / ammeter) to check the following:

- Voltage supply.
- Breaks in the wiring (infinite resistance / no continuity).
- A path to ground that completes the circuit.

Electric current is logical in its flow, always moving from the voltage source toward ground. Electrical faults can usually be located through a process of elimination. When troubleshooting a complex circuit, separate the circuit into smaller parts. General tests outlined below may be helpful in finding electrical problems. The information is most helpful when used with wiring diagrams.

Be sure to analyze the problem. Use wiring diagrams to determine the most likely cause. Get an understanding of how the circuit works by following the circuit from ground back to the power source.

When making test connections at connectors and components, use care to avoid spreading or damaging the connectors or terminals. Some tests may require jumper wires to bypass components or connections in the wiring harness. When connecting jumper wires, use blade connectors at the wire ends that match the size of the terminal being tested. The small internal contacts are easily spread apart, and this can cause intermittent or faulty connections that can lead to more problems.

Voltage and voltage drop

Wires, connectors, and switches that carry current are designed with very low resistance so that current flows with a minimum loss of voltage. A voltage drop is caused by higher than normal resistance in a circuit. This additional resistance actually decreases or stops the flow of current. Excessive voltage drop can be noticed by problems ranging from dim headlights to sluggish wipers. Some common sources of voltage drops are corroded or dirty switches, dirty or corroded connections or contacts, and loose or corroded ground wires and ground connections.

A voltage drop test is a good test to perform if current is flowing through the circuit but the circuit is not operating correctly. A voltage drop test helps pinpoint a corroded ground strap or a faulty switch. Normally, there should be less than 1 volt drop across most wires or closed switches. Voltage drop across a connector or short cable varies but should not exceed 0.3 volt.

A voltage drop test is generally more accurate than a simple resistance check because the resistances involved are often too small to measure with most ohmmeters. For example, a resistance as small as 0.02 Ω results in a 3 volt drop in a typical 150 amp starter circuit. (150 amps x 0.02 Ω = 3 volts).

BentleyPublishers.com
BentleyPublishers.com—All Rights Reserved

Keep in mind that voltage with ignition key ON and voltage with engine running are not the same. With ignition ON and engine OFF, fully charged battery voltage is approximately 12.6 volts. With engine running (charging voltage), normal voltage is approximately 14.0 volts. Measure voltage at battery with ignition ON and then with engine running to get exact measurements.

Voltage, measuring

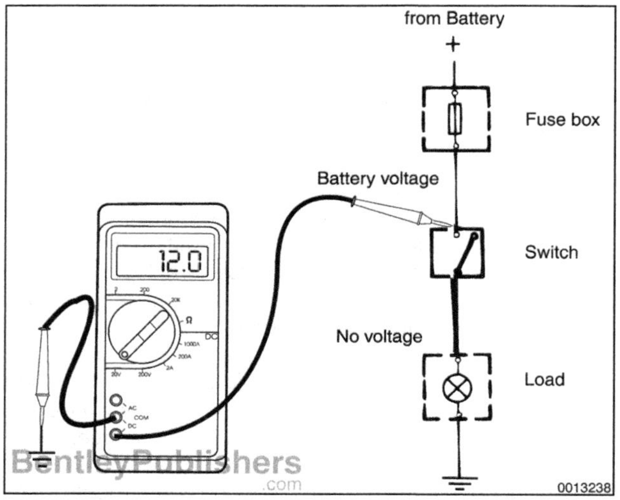

- Connect digital multimeter negative lead to a reliable ground point on vehicle.

NOTE—
- *The negative (-) battery terminal is always a good ground point.*

- Connect digital multimeter positive lead to point in circuit you wish to measure.
- Check that voltage reading does not deviate more than 1 volt from voltage at battery. If voltage drop is more than this, check for a corroded connector or loose ground wire.

Voltage drop, testing

Check voltage drop only when there is a load on the circuit, such as when operating the starter motor or turning on the headlights. Use a digital multimeter to ensure accurate readings.

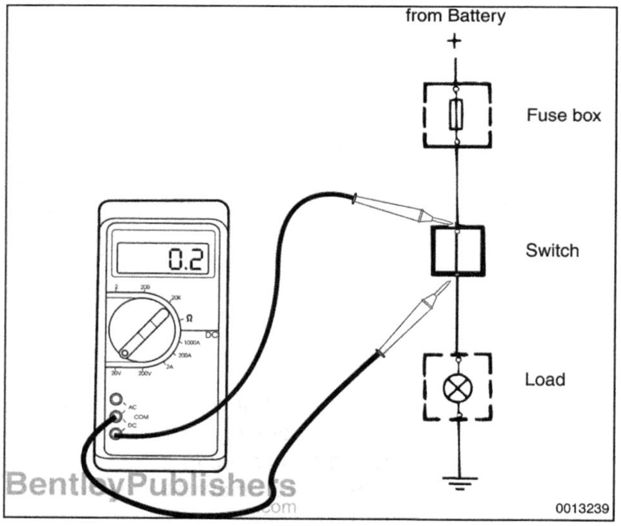

- Connect digital multimeter positive lead to positive (+) battery terminal or a positive power supply close to battery source.
- Connect digital multimeter negative lead to other end of cable or switch being tested.
- With power switched ON and circuit working, meter shows voltage drop (difference between two points). This value should not exceed 1 volt.
- Maximum voltage drop in an automotive circuit, as recommended by the Society of Automotive Engineers (SAE), is as follows:
  - 0 volt for small wire connections
  - 0.1 volt for high current connections
  - 0.2 volt for high current cables
  - 0.3 volt for switch or solenoid contacts
- On longer wires or cables, the drop may be slightly higher. In any case, a voltage drop of more than 1.0 volt usually indicates a problem.

BentleyPublishers.com
BentleyPublishers.com—All Rights Reserved

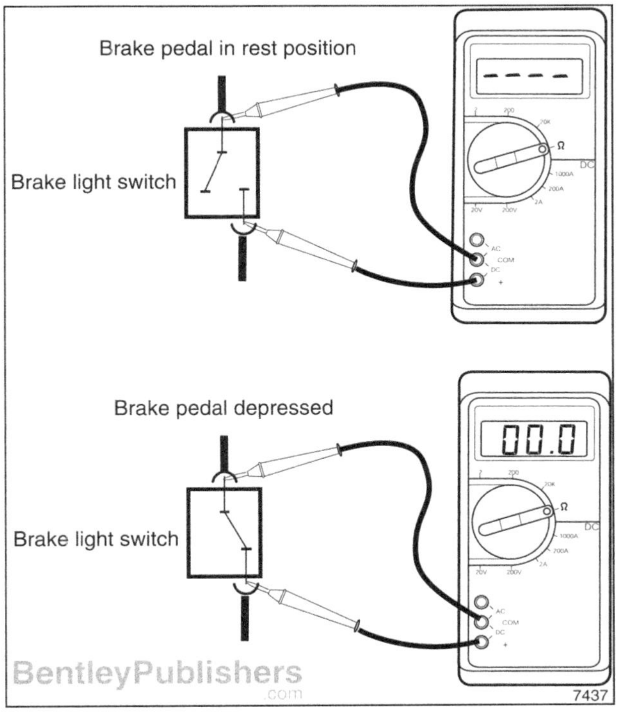

Continuity, checking

Use continuity test to check a circuit or switch. Because most automotive circuits are designed to have little or no resistance, a circuit or part of a circuit can be easily checked for faults using an ohmmeter. An open circuit or a circuit with high resistance does not allow current to flow. A circuit with little or no resistance allows current to flow easily.

When checking continuity, switch ignition OFF. On circuits that are powered at all times, disconnect battery. Using the appropriate wiring diagram, test circuit for faulty connections, wires, switches, relays and engine sensors by checking for continuity.

Example: Test brake light switch for continuity:

- With brake pedal in rest position (switch open) there is no continuity (infinite Ω).
- With pedal depressed (switch closed) there is continuity (0 Ω).

Short circuits, testing

Short circuits are exactly what the name implies. Current in the circuit takes a shorter path than it was designed to take. The most common short that causes problems is a short to ground where the insulation on a positive (+) wire wears away and the metal wire is exposed. When the wire rubs against a metal part of the vehicle or other ground source, the circuit is shorted to ground. If the exposed wire is live (positive battery voltage), a fuse blows or the circuit may be damaged.

Short circuits vary in nature and are often difficult to locate. They can be found using a logical approach based on knowledge of the current path.

Use a digital multimeter to locate short circuits.

> ***CAUTION—***
> - *In circuits protected with high rating fuses (25 amp and greater), wires or circuit components may be damaged before the fuse blows. Check for wiring damage before replacing fuses of this rating. Also, check for correct fuse rating.*

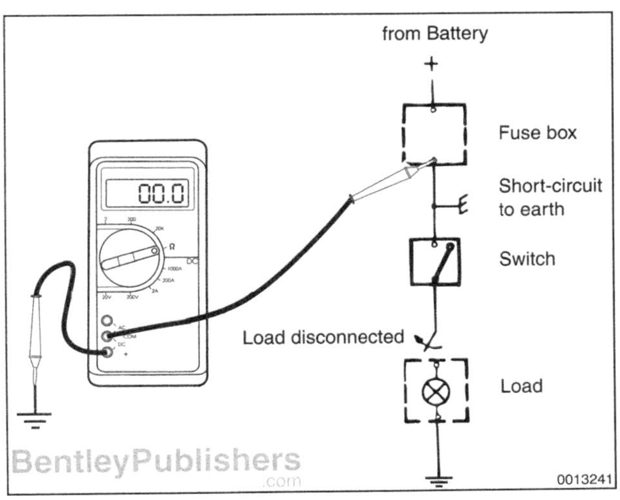

Testing with ohmmeter

– Remove blown fuse from circuit and disconnect cables from battery. Disconnect harness connector from circuit load or consumer.

Using an ohmmeter, connect one test lead to load side of fuse terminal (terminal leading to circuit) and other test lead to ground.

– If there is continuity to ground, there is a short to ground.

600

BentleyPublishers.com
BentleyPublishers.com—All Rights Reserved

– If there is no continuity, work from wire harness nearest to fuse and relay panel and move or wiggle wires while observing meter. Continue to move down harness until meter displays a reading. This is the location of the short to ground.

– Visually inspect wire harness at this point for any faults. If no faults are visible, carefully slice open harness cover or wire insulation for further inspection. Repair any faults found.

Testing with voltmeter

– Remove blown fuse from circuit. Disconnect harness connector from circuit load or consumer.

NOTE—
- *Most fuses power more than one consumer. Be sure all consumers are disconnected when checking for a short circuit.*

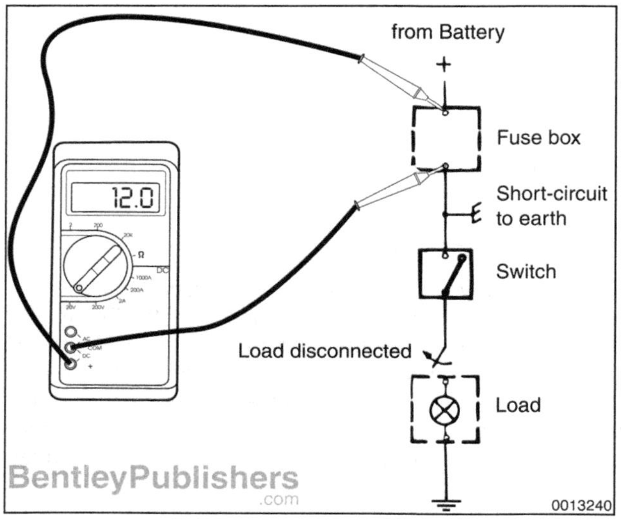

◄ Using a digital multimeter, connect test leads across fuse terminals. Make sure power is present in circuit. If necessary switch ignition ON.

– If voltage is present at voltmeter, there is a short to ground.

– If voltage is not present, work from wire harness nearest to fuse and relay panel and move or wiggle wires while observing meter. Continue to move down harness until meter displays a reading. This is the location of the short to ground.

Visually inspect wire harness at this point for any faults. If no faults are visible, carefully slice open harness cover or wire insulation for further inspection. Repair any faults found.

BentleyPublishers.com
BentleyPublishers.com–All Rights Reserved

611 Wipers and Washers

TABLES

General

This repair group covers repair information for windshield, headlight and rear window wiper and washer systems.

See also:

- **020 Maintenance** for cabin microfilter housing removal
- **612 Switches** for steering column switch cluster replacement
- **ECL Electrical Component Locations** for wiper and washer system fuse and relay locations and applications
- **ELE Electrical Wiring Diagrams**

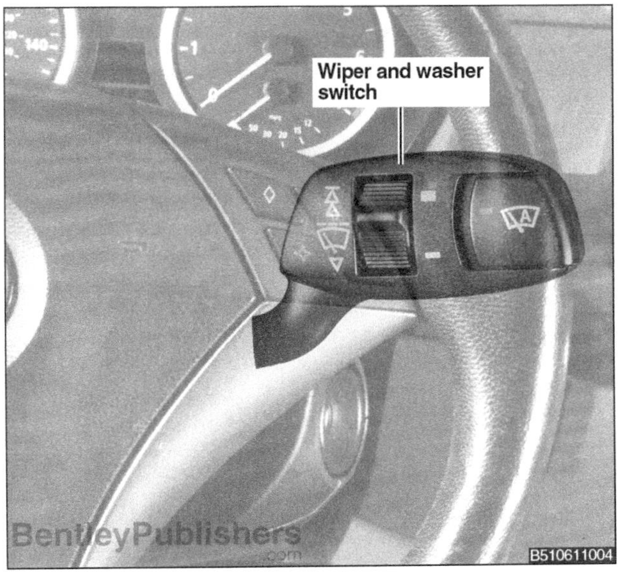

Wiper and washer system operation

Driver input to wiper and washer system is via multi-function stalk switch to right of steering column. Wiper and washer system functions are capable of activation once ignition is switched ON.

611

BentleyPublishers.com
BentleyPublishers.com—All Rights Reserved

Body base module (KBM) is the control hub for wiper and washer operation. Based on driver input, rain and light sensor signals and vehicle speed, KBM activates the following components:

- Windshield wiper motor dual relay in E-box.
- Rear window wiper relay in spare tire well.
- Headlight washer relay.
- Washer pumps in washer reservoir behind left front wheel housing liner.

Windshield washer nozzle heating is activated by IHKA.

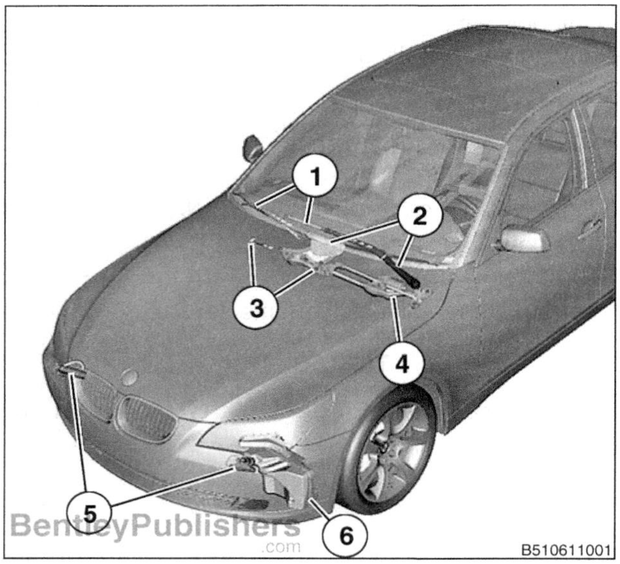

Wiper and washer system components (Sedan)

1. Wiper blades
2. Wiper arms
3. Heated windshield washer jets
4. Wiper motor assembly
5. Headlight washer jets
6. Washer fluid reservoir

Wiper and washer system components (Sports Wagon)

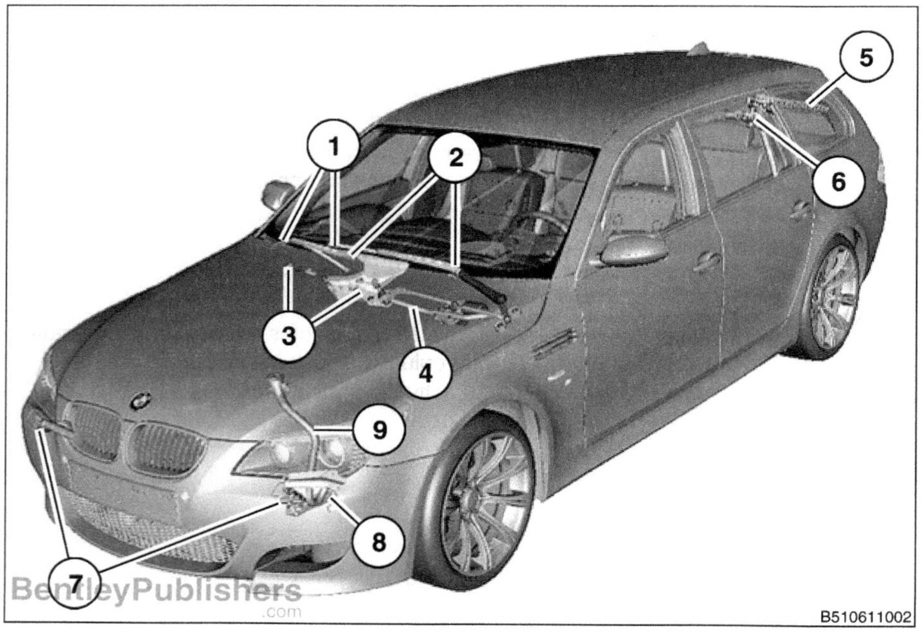

1. Windshield wiper blades
2. Windshield wiper arms
3. Heated windshield washer jets
4. Windshield wiper motor assembly
5. Rear wiper blade
6. Rear wiper motor assembly
7. Headlight washer jets
8. Washer fluid reservoir
9. Washer reservoir filler neck

BentleyPublishers.com
BentleyPublishers.com—All Rights Reserved

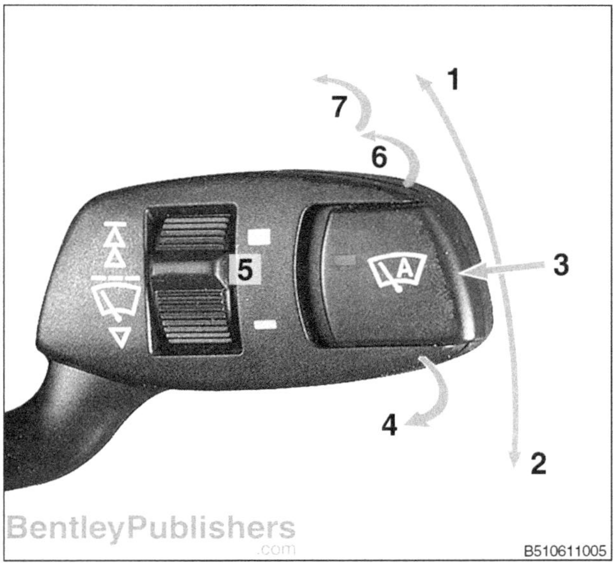

Wiper and washer switch operation

1. Switch wipers ON.
   - Flick UP once for slow speed. Intermittent when vehicle stops.
   - Flick UP twice or push past detent for fast speed. Slows when vehicle stops.
2. Flick DOWN to switch wipers OFF or for single wipe.
3. Push IN to activate or deactivate rain sensor. Wiper operation controlled automatically by rain intensity.
4. PULL toward driver to activate windshield and headlight washers.
5. Sensitivity adjustment thumb wheel for interval control.
6. PUSH away from driver to activate rear wiper in intermittent mode. Continuous operation when in reverse gear.
7. PUSH past detent to activate rear window washer.

WIPER BLADES AND ARMS

Wiper blade cleaning problems

Common problems with the wipers include streaking or sheeting, water drops after wiping and blade chatter.

Streaking is usually caused when wiper blades are coated with road film or car wash wax.

- Clean blades using soapy water. If cleaning does not cure problem, replace blades.

BMW recommends replacing the wiper blades twice a year, before and after the cold season.

Drops that remain behind after wiping are usually caused by oil, road film or diesel exhaust residue on the glass.

- Use an alcohol or ammonia solution or other non-abrasive cleaner to clean glass.

Chatter may be caused by dirty or worn blades or by incorrect wiper blade contact angle.

- Clean blades and glass as described above.
- Check and adjust wiper arm twist. See **Windshield wiper blades and arms, adjusting position** in this repair group.
- If problems persist, replace wiper blades and arms. See **Windshield wiper blade, replacing** and **Windshield wiper arms, removing and installing**, in this repair group.

BentleyPublishers.com
BentleyPublishers.com–All Rights Reserved

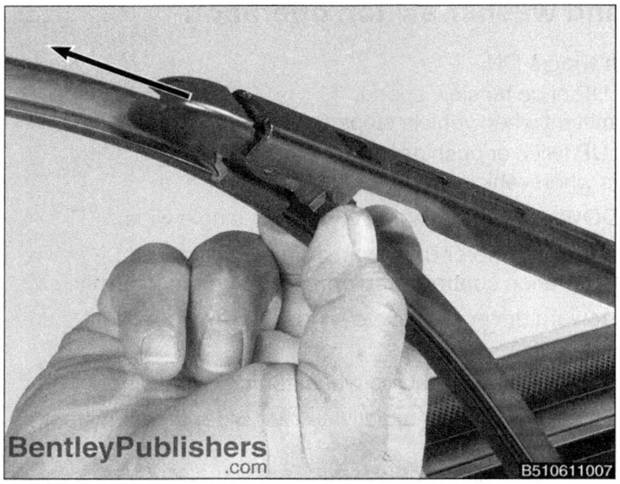

Windshield wiper blade, replacing

– Pivot wiper arm off windshield. Note that arm only pivots approx. 45°.

◄ Working at wiper blade pivot:

- Press on locking clip to release wiper blade.
- Slide wiper blade forward (**arrow**) to detach from arm.

> ***CAUTION—***
> - *Do not allow wiper arm to snap back against windshield when wiper blade is off.*

– Installation is reverse of removal.

Windshield wiper arms, removing and installing

– Open engine hood.

Left wiper arm

◄ Spread retaining tabs (**arrows**) and lift off wiper pivot nut plastic cover (**curved arrow**).

> ***CAUTION—***
> - *Do not shear off or otherwise damage retaining tabs.*

◄ Remove wiper arm pivot nut (**arrow**). Lift arm off pivot.

Note that tightening torque is stamped on wiper arm.

BentleyPublishers.com
BentleyPublishers.com—All Rights Reserved

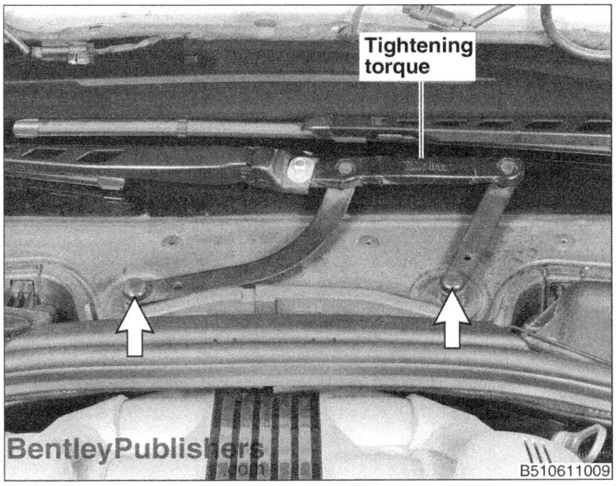

Right wiper arm

◀ Pry off wiper pivot nut plastic caps (**arrows**).

Note that tightening torque is stamped on wiper arm link. (Plastic cover removed for illustration.)

◀ Remove wiper arm pivot nuts (**arrows**). Lift linkage off pivots.

◀ If necessary, use puller to remove wiper arm from pivot.

> ***CAUTION—***
> *• Do not bend wiper arm while removing.*

– Before installing wiper arms, make sure wipers are in PARK position:
  - Switch ignition ON.
  - Switch wipers ON and allow two cycles to complete.
  - Switch wipers OFF. Wait for wiper motor to come to rest, then switch ignition OFF.

611

BentleyPublishers.com
BentleyPublishers.com—All Rights Reserved

– Attach wiper arms and tighten pivot nuts.
  - Left wiper arm: Temporarily tighten pivot nut, then check and adjust wiper blade position. See **Windshield wiper blades and arms, adjusting position** in this repair group.
  - Right wiper arm: Fully tighten pivot nuts.
  - If applicable, tighten pivot nuts to torques stamped on wiper arms.
  - After approximately 50 minutes settling time, recheck torques.

Tightening torque	
Front wiper arm to wiper pivot • M8 nut • M10 nut	 17 - 20 Nm (13 - 15 ft-lb) 30 - 35 Nm (22 - 26 ft-lb)

– Right wiper arm: Check and adjust wiper blade position. See **Windshield wiper blades and arms, adjusting position** in this repair group.

Windshield wiper blades and arms, adjusting position

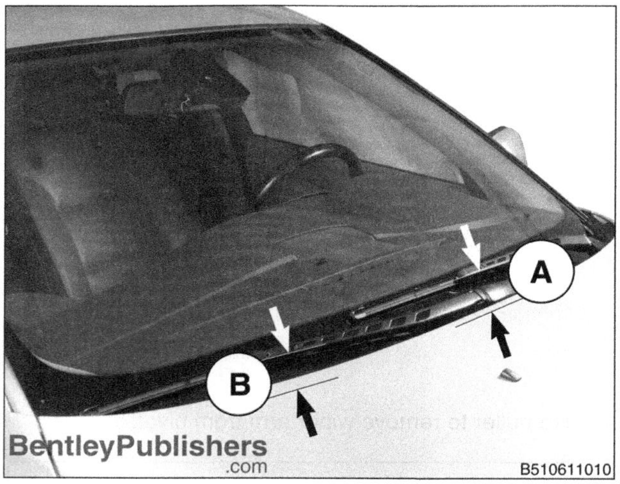

Wiper blade position is set in relation to top of cowl cover at base of windshield.
- Make sure wipers are in PARK position.
- Measure from bottom edge of wiper blade to base of windshield. See **Table a**.

Table a. Wiper blade position in PARK	
Left side (**A**)	69 ± 3 mm (2.7 ± 0.1 in)
Right side (**B**)	87 ± 3 mm (3.4 ± 0.1 in)

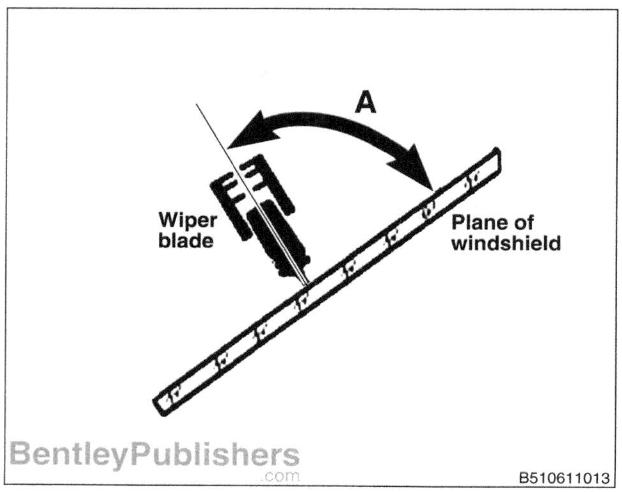

Wiping action is improved by correctly setting contact angle of wiper blade (**A** to plane of windshield. Both wiper arms are designed with a tilt adjustment for setting wiper blade contact angle.

BentleyPublishers.com
BentleyPublishers.com—All Rights Reserved

BMW special tool 61 6 100 or equivalent protractor may be used to measure wiper blade contact angle. Correct wiper blade contact angle is in **Table b**

Table b. Wiper blade contact angle	
Wiper blade angle to windshield (left or right)	86° ± 1°

Left wiper arm, adjusting

Left wiper arm attachment components:

1. Pivot nut, M10
   - Tighten to 30 - 35 Nm (22 - 26 ft-lb)
2. Eccentric washer
3. Wiper arm
   - Slide on slotted hole to adjust wiper blade angle
4. Wiper pivot
5. Domed adjusting disc
   - Use puller to remove from pivot

Curved bottom of wiper arm contact surface and domed adjusting disc allow wiper blade angle adjustment.

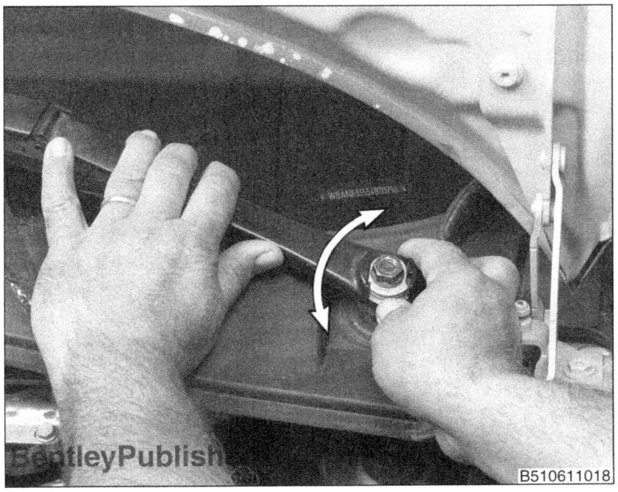

With pivot nut finger tight, slide wiper arm up or down (**arrow**) until blade position and angle are correct. See **Table a** and **Table b**.

– Tighten pivot nut and recheck wiper blade position.
- If applicable, tighten pivot nut to torque stamped on wiper arm.
- After approximately 50 minutes settling time, recheck torque.

Tightening torque	
Front wiper arm to wiper pivot (M10 nut)	30 - 35 Nm (22 - 26 ft-lb)

BentleyPublishers.com
BentleyPublishers.com–All Rights Reserved

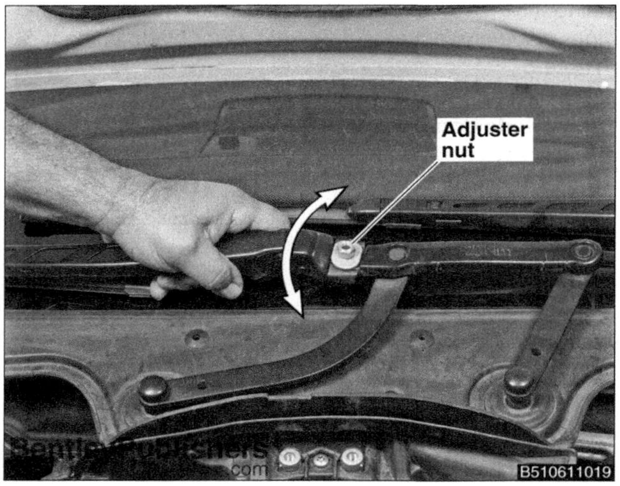

Right wiper arm, adjusting

With right wiper arm pivot nuts fully tightened:

- Remove right wiper arm plastic trim.
- Loosen adjuster nut.
- With adjuster nut finger tight, slide wiper arm up or down (**arrow**) until blade position and angle are correct. See **Table a** and **Table b**.

– Tighten adjuster nut and recheck wiper blade position.

Tightening torque	
Right wiper arm adjusting nut	20 Nm (15 ft-lb)

Rear wiper blade, replacing

– Pivot wiper arm off rear window.

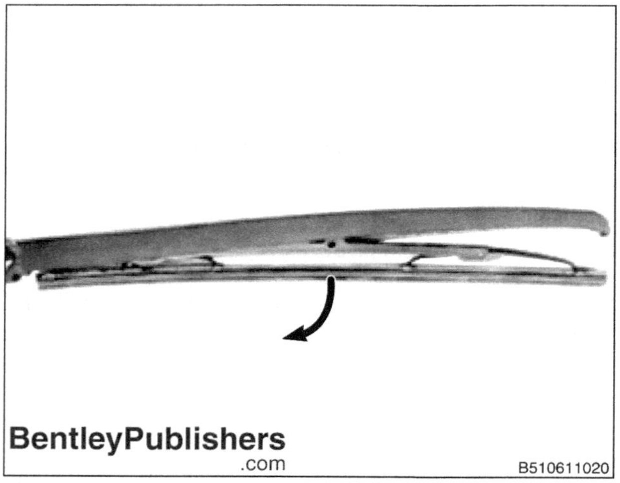

Pull wiper blade off wiper arm (**arrows**).

When installing, make sure wiper blade mounting tab clips firmly in wiper arm slot (**arrow**).

BentleyPublishers.com
BentleyPublishers.com—All Rights Reserved

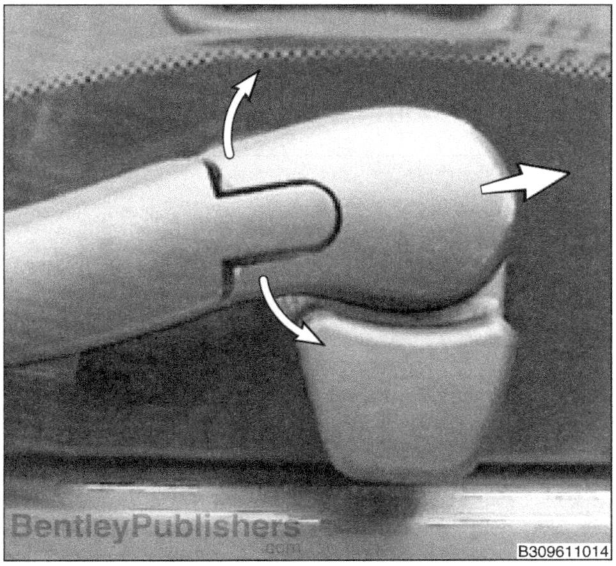

Rear wiper arm, removing and installing

◄ Spread wiper pivot plastic cap catches (**curved arrows**) and remove cap (**flared arrow**).

◄ Pivot wiper arm off rear window, then remove mounting nut (**arrow**).

– Remove wiper arm off pivot. If necessary, use puller to remove wiper arm from pivot.

– Before installing wiper arm, make sure wiper assembly is in PARK position:

- Switch ignition ON.
- Switch rear wiper ON and allow two cycles to complete.
- Switch wiper OFF. Wait for wiper motor to come to rest, then switch ignition OFF.

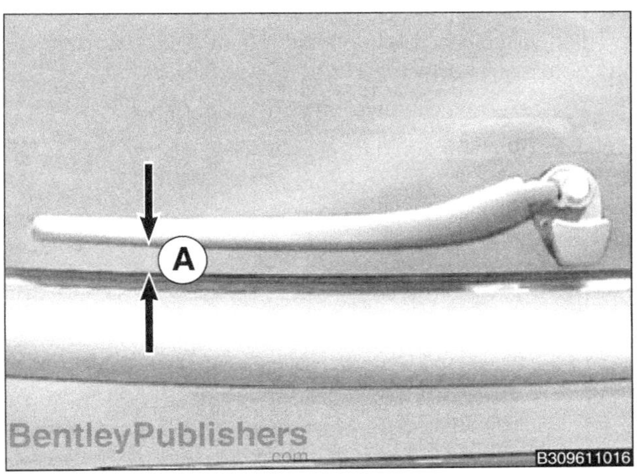

◄ Attach wiper arm and tighten down mounting nut temporarily, then check and adjust wiper blade position.

Rear wiper blade position	
Measurement **A** (distance between bottom of wiper arm and rear window lower edge	30 mm (1.2 in)

– Tighten arm to final torque.

Tightening torque	
Rear wiper arm to wiper pivot (M8)	13 Nm (10 ft-lb)

BentleyPublishers.com
BentleyPublishers.com–All Rights Reserved

Wiper Assemblies

Windshield wiper assembly, removing and installing

> ***CAUTION—***
> - *Make sure wipers are parked (stalk switch in OFF position) and the ignition is also OFF.*
> - *Use fender cover to protect windshield.*
> - *To avoid damaging the wiper arms and pivots, do not manually slide or force the wiper arms across the windshield.*

– Remove wiper arms. See **Windshield wiper arms, removing and installing** in this repair group.

– Remove cabin microfilter housings on left and right. See **020 Maintenance**.

– Remove front strut tower brace. See **310 Front Suspension**.

< Use suitable puller to remove domed adjusting disc from left wiper pivot.

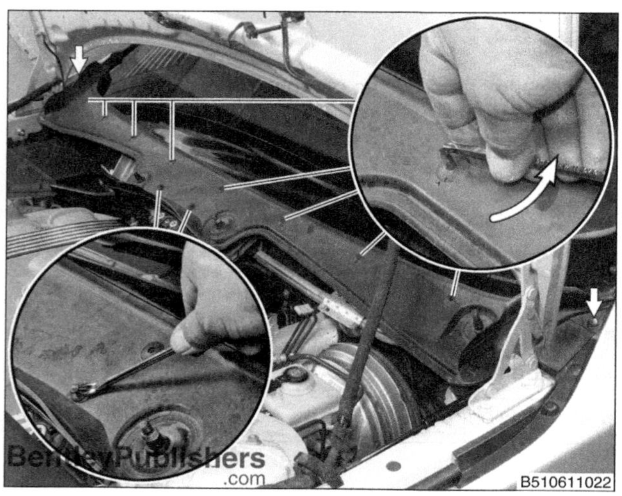

< Remove cowl cover panel:
- Pry out plastic rivets (**lower inset**).
- Use 5 mm Allen wrench to rotate plastic retainers 90° (**upper inset**), thus detaching panel from base of windshield.
- Remove plastic fasteners (**arrows**) at corners of panel.
- Lift panel off base of windshield.

BentleyPublishers.com
BentleyPublishers.com–All Rights Reserved

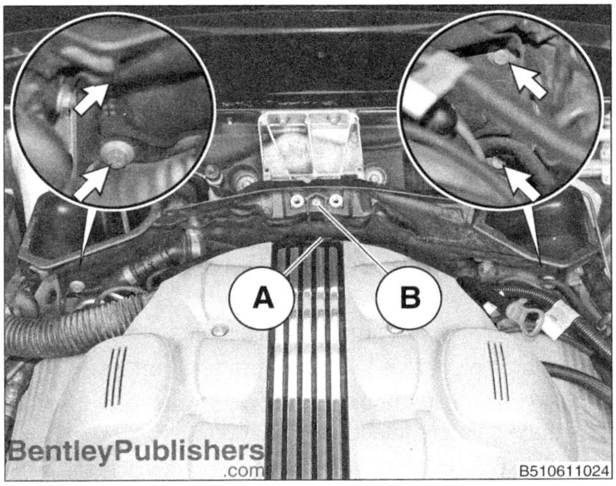

◀ Working at front air duct:

- Remove 4 side duct mounting screws (**arrows**).
- Rotate center quick-release plastic fastener (**A**) approx. 90°.
- Remove Torx screw (**B**) at strut tower brace center anchor.
- Push air duct forward to expose strut tower brace mounting bracket bolts.

◀ Remove strut tower brace mounting bracket bolts (**arrows**) and lift out bracket.

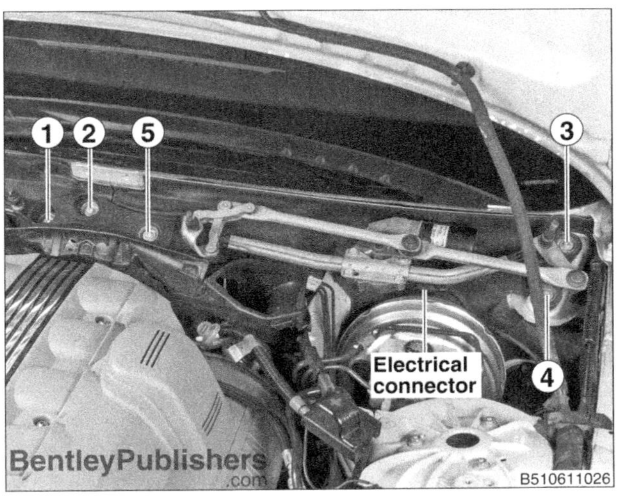

◀ Working at wiper motor assembly:

- Detach wiper motor electrical connector
- Loosen and remove Torx screws in order **1** - **5**.
- Lift wiper motor out of cowl opening.

– When installing:

- Install mounting screws hand tight at first.
- Tighten screws in order **5** - **1**.
- If wiper linkage interferes with installation of screw **4**, attach electrical connector, switch ignition ON, then switch wipers ON long enough to move linkage to right. Switch ignition OFF midway through wipe cycle.

Tightening torque	
Wiper motor assembly to body (M6 Torx bolts)	10 Nm (7 ft-lb)

611

BentleyPublishers
BentleyPublishers.com–All Rights Reserved

– Remainder of installation is reverse of removal. Before installing wiper arms, make sure wipers are in PARK position:
  - Switch ignition ON.
  - Switch wipers ON and allow two cycles to complete.
  - Switch wipers OFF. Wait for wiper motor to come to rest, then switch ignition OFF.
  - Left wiper arm: Install domed adjusting disc and wiper arm. Temporarily tighten pivot nut, then check and adjust wiper blade position. See **Windshield wiper blades and arms, adjusting position** in this repair group.
  - Right wiper arm: Install wiper arm and fully tighten pivot nuts.
  - If applicable, tighten pivot nuts to torques stamped on wiper arms.
  - When reinstalling microfilter housings, to avoid water leaks into cabin, check that seals are intact and seat correctly.
  - After approximately 50 minutes settling time, recheck wiper arm torques.

Tightening torque	
Air duct to side mountings (M6 bolts)	4.5 Nm (40 in-lb)
Air duct to strut tower brace center anchor (Torx screw)	3.5 Nm (31 in-lb)
Brace to strut tower or bulkhead bracket (use new M8 Torx bolts)	30.5 Nm (22 ft-lb)
Front wiper arm to wiper pivot • M8 nut • M10 nut	 17 - 20 Nm (13 - 15 ft-lb) 30 - 35 Nm (22 - 26 ft-lb)
Strut tower brace mounting bracket to bulkhead (Torx bolts)	24 Nm (18 ft-lb)

BentleyPublishers.com
BentleyPublishers.com—All Rights Reserved

Rear wiper assembly

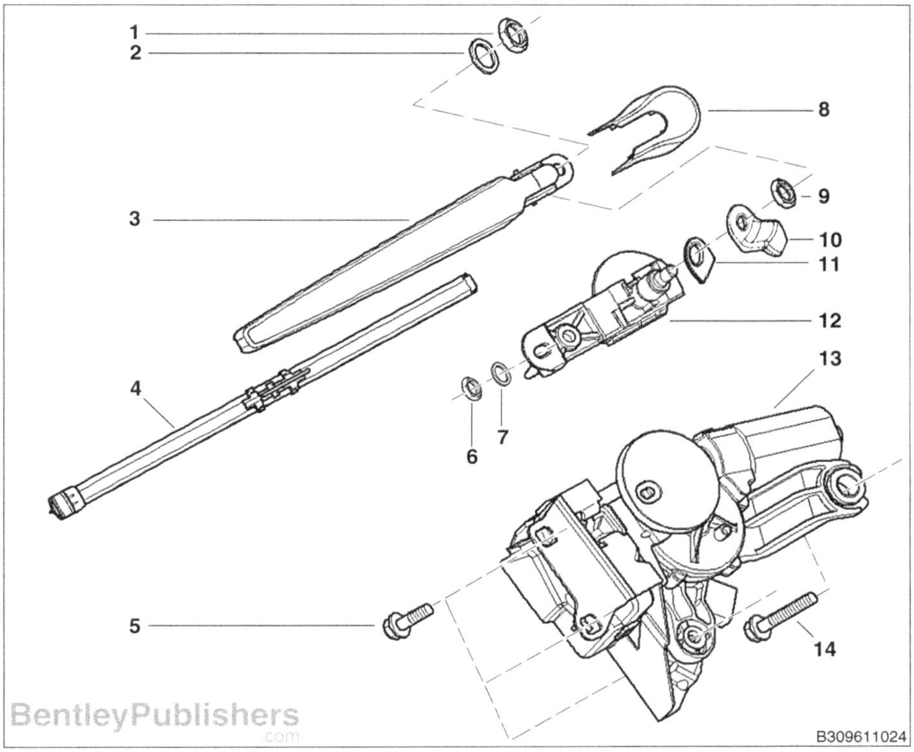

1. Wiper arm nut (M8)
    - Tighten to 10 Nm (7 ft-lb)
2. Wave washer
3. Wiper arm
4. Wiper blade
5. Torx bolt (M6 x 9 mm)
    - Tighten to 10 Nm (7 ft-lb)
6. Wiper pivot nut (M6)
    - Tighten to 8 Nm (6 ft-lb)
7. Gasket
8. Cover
9. Lock nut
10. Rear window lock release button
11. Gasket
12. Wiper pivot assembly
13. Wiper motor assembly
14. Torx bolt (M6 x 22 mm)
    - Tighten to 10 Nm (7 ft-lb)

Rear wiper motor, removing and installing

The rear wiper motor is bolted to the Sports Wagon tailgate. The wiper arm and blade are mounted to the rear window which is hinged at the top. The motor drives the rear wiper arm via a pivot assembly mounted on the rear window. For the window to close correctly and rear wiper to operate, align the pivot assembly on the glass with the drive on the wiper motor using BMW special tool 61 1 330.

The rear window lock is bolted to the rear wiper motor assembly.

– Remove tailgate inner trim panel. See **513 Interior Trim**.

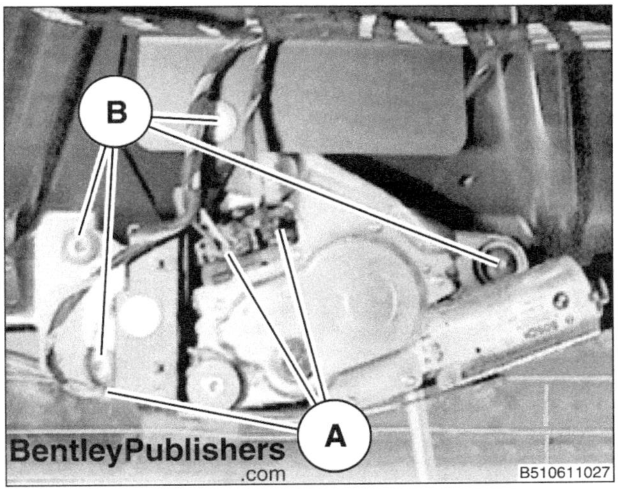

Working at rear wiper motor:
- Detach electrical connectors (**A**).
- Remove mounting screws (**B**).
- Lift off wiper motor.

– If replacing motor: Separate rear window lock from rear wiper motor assembly and bolt to new unit.

– Reinstall wiper motor but tighten mounting bolts finger tight at first.

– Attach wiper motor electrical harness. Switch ignition ON. Cycle rear wiper twice to make sure wiper motor is in PARK position. Switch ignition OFF.

611

BentleyPublishers.com
BentleyPublishers.com—All Rights Reserved

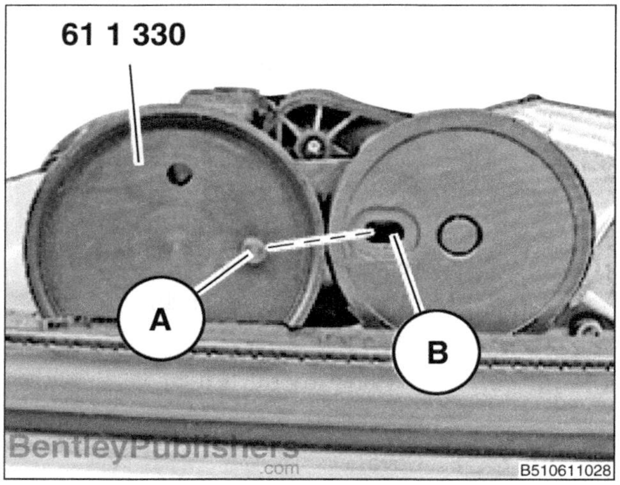

Align assembly using BMW special tool 61 1 330: Position special tool so that pin (**A**) on tool inserts into slot (**B**) in wiper motor drive.

With rear wiper arm vertical to glass, close rear window and allow journal of wiper pivot assembly to snap into bore (**arrow**). Close window completely.

- Tighten rear wiper motor mounting bolts.

Tightening torque	
Rear wiper motor to tailgate	10 Nm (7 ft-lb)

- Open window and remove special tool.
- Reinstall tailgate trim.

Rear window wiper pivot, removing and installing

- Remove rear wiper arm. See **Rear wiper arm, removing and installing** in this repair group.
- Use duct tape to hold window lock release button in place.

Remove nut (**arrow**) securing pivot to window lock release button.

BentleyPublishers.com
BentleyPublishers.com—All Rights Reserved

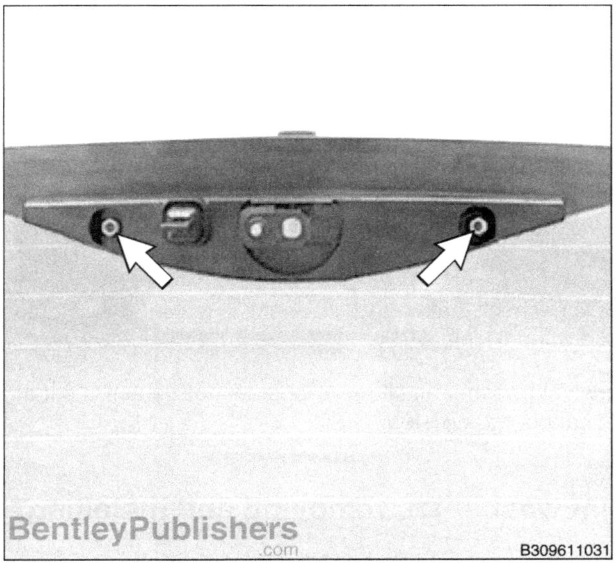

With window open:

- Carefully pry out wiper pivot assembly trim covers.
- Remove trim cover mounting nuts (**arrows**) and pull off cover.

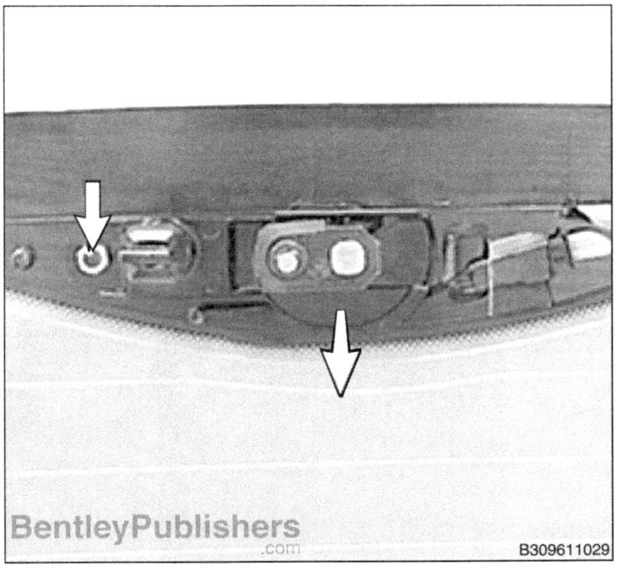

With window open:

- Detach window lock release button electrical connector.
- Remove pivot assembly mounting nut (**arrow**) and remove pivot (**flared arrow**).

– Installation is reverse of removal.

Tightening torques	
Window lock release button to glass	10 Nm (7 ft-lb)
Wiper pivot to glass (M6)	8 Nm (6 ft-lb)

– Synchronize wiper pivot and wiper motor drive. See **Rear wiper motor, removing and installing** in this repair group.

Washer System

The windshield washer system includes electrically heated spray jets under the engine hood, the washer fluid pump, and washer fluid reservoir in front of the left front wheel housing.

The headlight washer system (optional) consists of front washer fluid reservoir (shared with windshield washer system), a separate washer pump, and washer jets in the front bumper.

The rear window washer consists of the fluid reservoir (shared with windshield washer system), a separate washer pump, and a washer jet at the top of the rear window.

611

BentleyPublishers
BentleyPublishers.com—All Rights Reserved

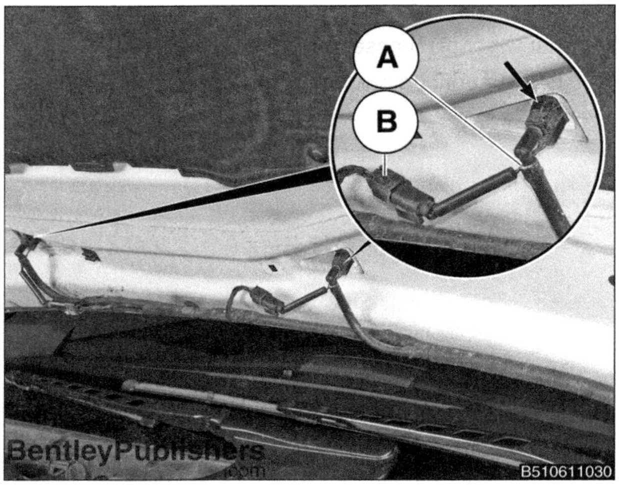

Windshield washer jet, removing and installing

– Open engine hood.

– Working on underside of hood:
  - Pinch off washer hose, then detach washer fluid hose (**A**) from jet.
  - Press on lock clip (**B**) and separate washer jet electrical connector.
  - Press on washer jet locking clip (**arrow**) and push jet up through hood to remove.

– When installing, check that washer jet snaps in place firmly. Route washer hose without kinks.

– Check windshield washer function. Washer jet nozzle aim is set at factory and cannot be adjusted.

Headlight washer jet, removing and installing

– Remove front bumper cover. See **510 Exterior Trim, Bumpers**.

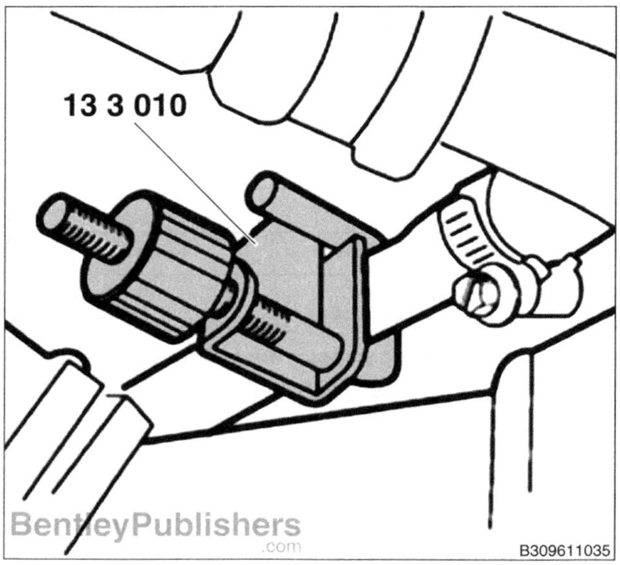

– To avoid washer fluid spillage, pinch off washer hose with hose pinch tool 13 3 010 or equivalent.

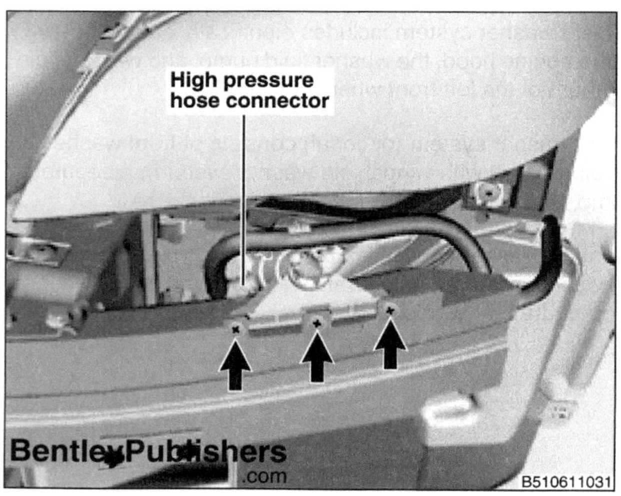

– Working at headlight washer jet:
  - Detach high pressure hose connector.
  - Remove washer jet mounting screws (**arrow**). Lift washer jet out of bumper opening.

– Installation is reverse of removal. Check headlight washer function. Nozzle aim is set at factory and cannot be adjusted.

BentleyPublishers.com
BentleyPublishers.com–All Rights Reserved

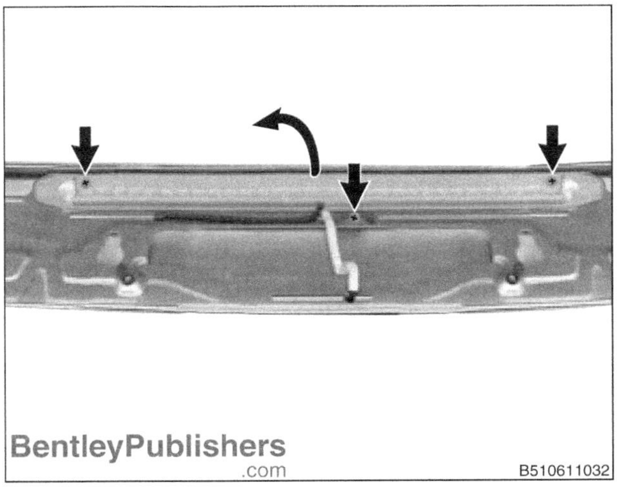

Rear window washer jet, removing and installing

– Remove rear spoiler. See **412 Trunk Lid, Tailgate**.

< Remove center brake light screws (**arrows**) and pull brake light unit and rear window washer nozzle away (**curved arrow**) from spoiler.

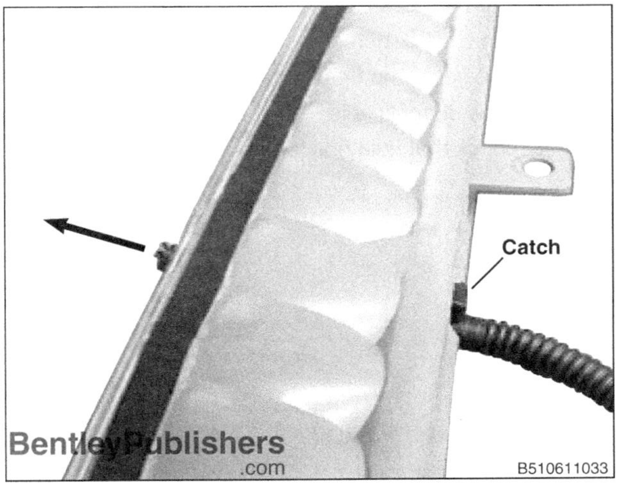

< Unhook washer hose from catch and pull nozzle away from center brake light assembly (**arrow**).

– Installation is reverse of removal. Route washer hose as before, making sure it is not kinked.

BentleyPublishers
BentleyPublishers.com–All Rights Reserved

Washer fluid reservoir components service

– Washer fluid reservoir is behind left front wheel housing liner. See **410 Fenders, Engine Hood** for wheel housing liner removal.

◄ Depending on model, washer reservoir components are as follows:

1. Headlight washer pump
2. Rear window washer pump (Sports Wagon)
3. Windshield washer pump
4. Washer fluid level switch

– To replace any component:

- Detach electrical connector.
- Separate high pressure fluid hose, if applicable.
- Rotate component clockwise and pry out of reservoir. Be prepared to catch dripping fluid.

– Installation is reverse of removal, noting the following:

- Replace strainers on washer pumps.
- Coat grommet at base of pump or sensor with anti-friction agent.
- Use new hose clamps, if applicable.
- After installation, check that fluid hoses are not kinked.

– Fill reservoir and check for leaks.

BentleyPublishers.com
BentleyPublishers.com–All Rights Reserved

612 Switches

General

This repair group covers replacement of electrical switches at the steering wheel, steering column, dashboard, pedal cluster, center console and door panels.

See also the following sections:

- **210 Clutch** for clutch switch
- **230 Manual Transmission** for manual transmission back-up light switch
- **240 Automatic Transmission** for automatic transmission back-up light switch function
- **250 Gearshift Linkage** for automatic shiftlock
- **320 Steering and Wheel Alignment** for steering wheel and steering column components
- **520 Seats** for seat position control switches
- **ECL Electrical Component Locations**
- **ELE Electrical Wiring Diagrams**

612

BentleyPublishers.com
BentleyPublishers.com–All Rights Reserved

Warnings and Cautions

CAUTION—
- *When working on electrical switches or lights, disconnect the negative (–) cable from the battery and insulate the cable end to prevent accidental reconnection.*
- *Prior to disconnecting the battery, read the battery disconnection cautions given in* **001 Warning and Cautions**.
- *To prevent marring interior trim, work with plastic prying tools or wrap the tips of screwdrivers and pliers with tape before prying out switches or trim.*

WARNING—
- *The steering wheel mounted airbag is an explosive device. Treat with extreme caution. Follow the airbag removal procedure in* **721 Airbag System (SRS)**.
- *Airbag removal may set SRS fault codes. The SRS indicator light remains ON until problems are corrected and fault memory cleared.*

STEERING WHEEL SWITCHES

The steering wheel of vehicles covered by this manual are equipped with the following:

- SRS airbag and horn button
- Horn contact
- Cellular telephone controls
- Radio controls

To replace the steering wheel switches and buttons, remove the airbag first. The horn button is integrated with the airbag and is not replaceable separately.

Multifunction steering wheel switches

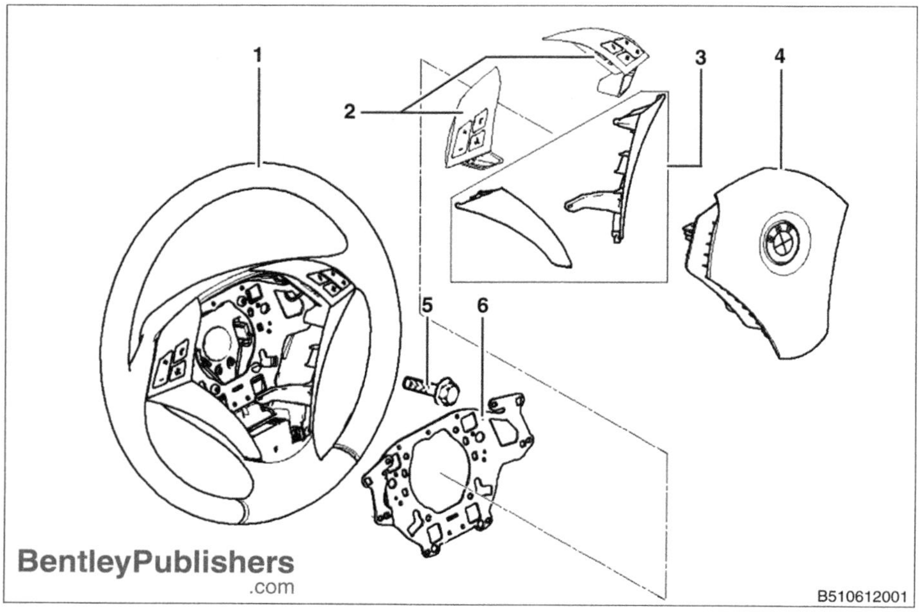

1. Multifunction steering wheel
2. Multifunction switches
3. Steering wheel trim
4. Airbag unit and horn button
5. Mounting bolt (M12)
   - Tighten to 62.5 Nm (46 ft-lb)
6. Multifunction switch mounting plate

BentleyPublishers.com
BentleyPublishers.com—All Rights Reserved

Sport steering wheel switches

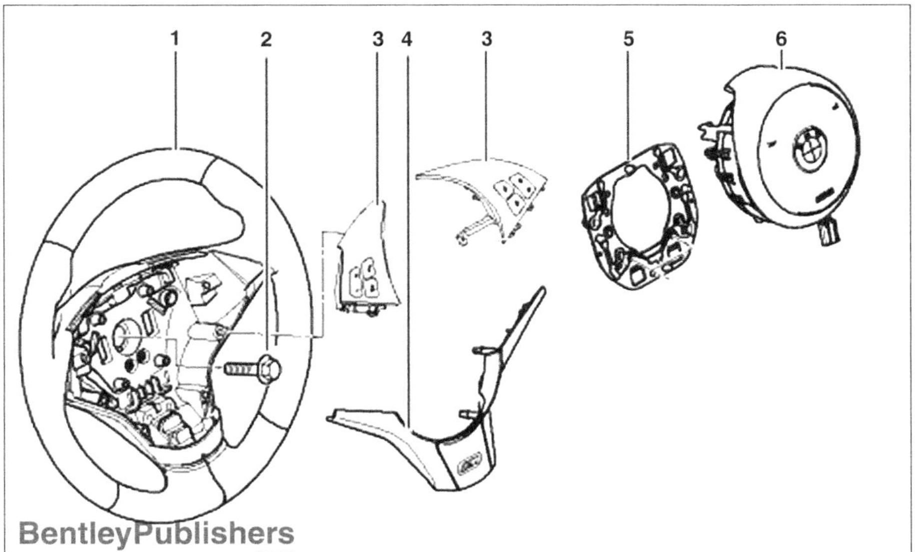

B510612002

1. Sport steering wheel
2. Mounting bolt (M12)
   - Tighten to 62.5 Nm (46 ft-lb)
3. Multifunction switches
4. Steering wheel trim
5. Multifunction switch mounting plate
6. Airbag unit and horn button

Steering wheel switches, removing and installing

– Disconnect negative (-) battery cable and cover battery terminal to keep cable from accidentally contacting terminal.

> ***CAUTION—***
> - *Prior to disconnecting the battery, read the battery disconnection cautions given in* **001 Warnings and Cautions**.

– Remove driver airbag in steering wheel. See **721 Airbag System (SRS)**.

> ***WARNING —***
> - *The steering wheel mounted airbag is an explosive device. Treat with extreme caution. Follow the airbag removal procedure in* **721 Airbag System (SRS)**.
> - *Airbag removal may set SRS fault codes. The SRS indicator light remains ON until problems are corrected and fault memory cleared.*

Multifunction steering wheel

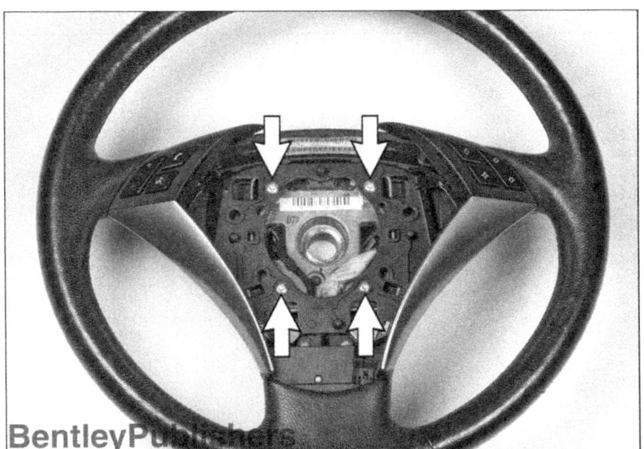

B510612003

◄ Remove multifunction switch mounting bracket fasteners (**arrows**).

612

BentleyPublishers.com
BentleyPublishers.com—All Rights Reserved

Steering wheel switches, removing and installing

◄ Disconnect electrical connectors (**arrows**).

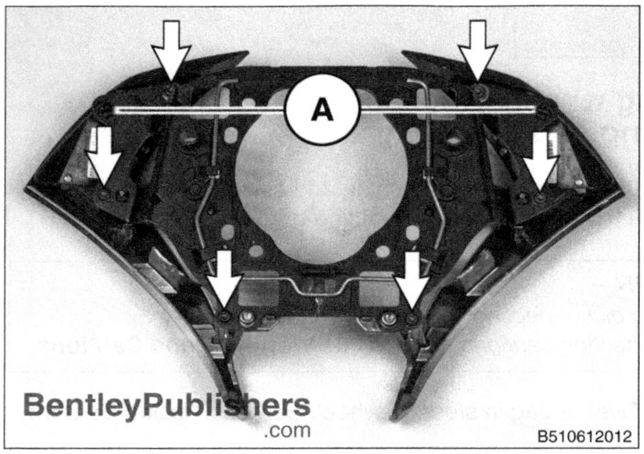

◄ Remove multifunction switch fasteners (**arrows**). Unclip multifunction switch in area **A** and remove from bracket.

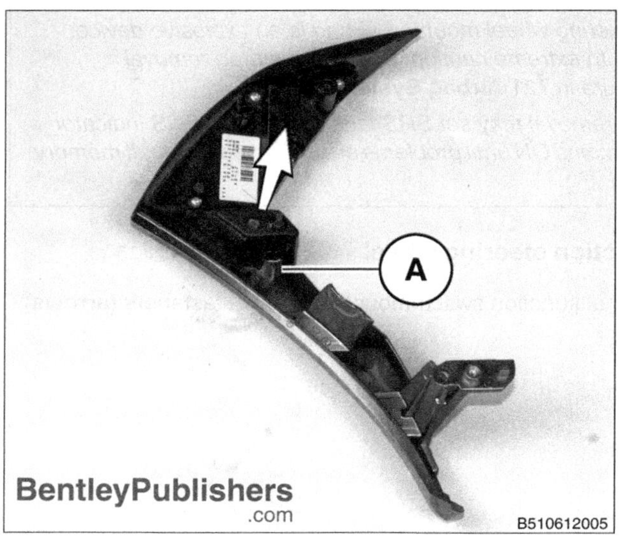

◄ Unclip (**A**) multifunction switch from steering wheel trim and remove in direction of **arrow**.

– Installation is reverse of removal.

Tightening torques	
Mounting plate to steering wheel	6 Nm (4.4 ft-lb)

BentleyPublishers.com
BentleyPublishers.com–All Rights Reserved

Sport steering wheel

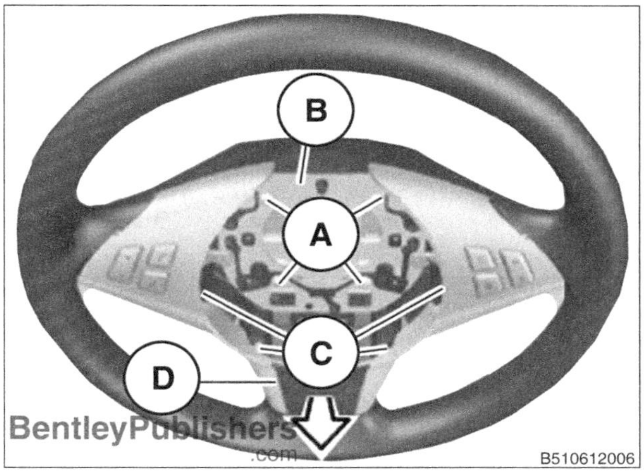

- Remove multifunction switch mounting plate fasteners (**A**).
- Remove mounting plate (**B**), multifunction switches and trim panel (**D**) from mounts (**C**) in direction of **arrow**.
- Disconnect electrical connectors and remove from steering wheel.
- Remove multifunction switch fasteners and separate switches from mounting plate.
- Installation is reverse of removal.

Tightening torques	
Mounting plate to steering wheel	6 Nm (4.4 ft-lb)

STEERING COLUMN SWITCH CLUSTER (SZL)

Steering column switch cluster (SZL) basics

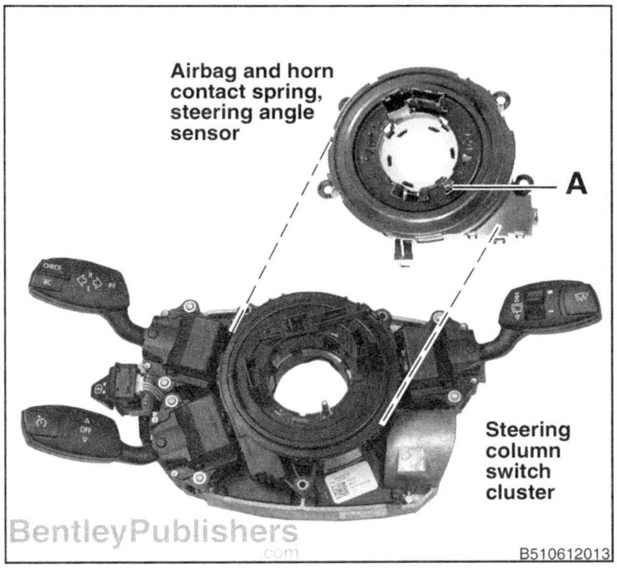

The steering column switch cluster (SZL) integrates the following functions:

- Wiper and washer switch
- Headlight beam switch
- Turn signal switch
- Cruise control
- Airbag and horn contact spring, steering angle sensor with stop pin (**A**)
- On-board computer (BC) switch

SZL switches are optical devices.

Steering column switch cluster (SZL), removing and installing

- Disconnect negative (-) battery cable and cover battery terminal to keep cable from accidentally contacting terminal.

> ***CAUTION—***
> - *Prior to disconnecting the battery, read the battery disconnection cautions given in* **001 Warnings and Cautions**.

612

BentleyPublishers.com
BentleyPublishers.com–All Rights Reserved

Steering column switch cluster (SZL), removing and installing

– Remove driver airbag in steering wheel. See **721 Airbag System (SRS).**

> ***WARNING—***
> - *The steering wheel mounted airbag is an explosive device. Treat with extreme caution. Follow the airbag removal procedure in* **721 Airbag System (SRS)***.*
> - *Airbag removal may set SRS fault codes. The SRS indicator light remains ON until problems are corrected and fault memory cleared.*

– Extend steering column as far back and as low as possible.

– Remove steering wheel. See **320 Steering and Wheel Alignment**.

> ***CAUTION—***
> - *Do not move front wheels from straight-ahead position while steering wheel is off.*
> - *While steering wheel is off, do not turn or twist the airbag and horn contact spring.*

– Remove steering column trim. See **513 Interior Trim**.

B510612007

< Remove SZL mounting fasteners (**arrows**) and pull cluster back.

– Detach electrical connectors and remove SZL.

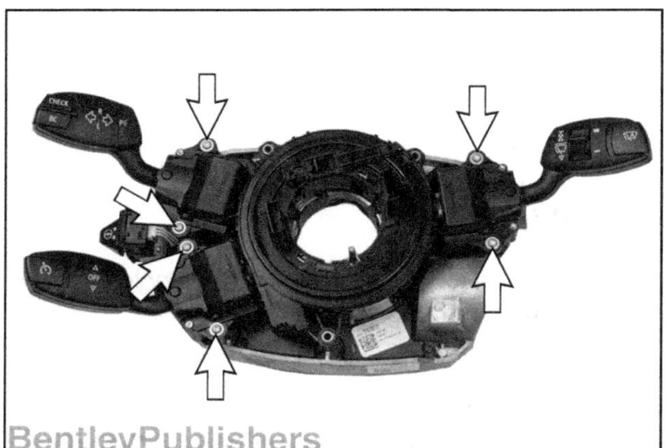

B510612014

< If replacing SZL remember to:

- Remove turn signal, cruise control and windshield wiper stalk fasteners (**arrows**) and swap stalks to new part.
- Remove airbag and horn contact spring. See **Airbag and horn contact spring, steering angle sensor, removing and installing** in this repair group.

– Installation is reverse of removal. Use BMW scan tool to code, program and carry out steering angle sensor adjustment.

BentleyPublishers.com
BentleyPublishers.com—All Rights Reserved

Airbag and horn contact spring, steering angle sensor, removing and installing

– Remove SZL. See **Steering column switch cluster (SZL), removing and installing** in this repair group.

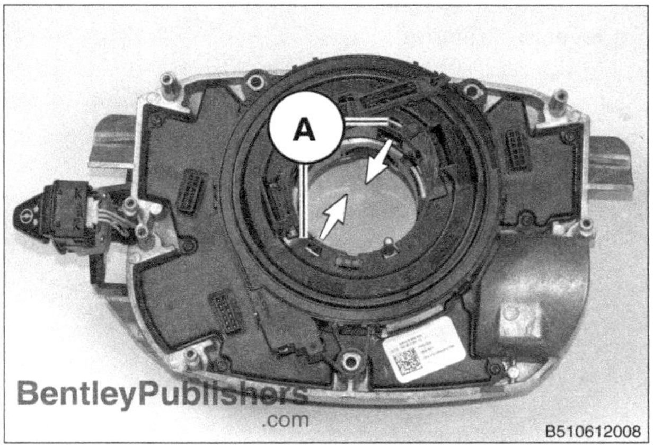

◄ Working at contact spring module:

- Unlock catches **A** in direction of **arrows**.
- Lift contact spring off SZL.

> ***CAUTION—***
> - *Do not rotate inner ring of contact spring. If necessary, secure ring using adhesive tape.*

◄ When installing contact spring, make sure electrical connectors (**arrows**) are properly aligned.

DASHBOARD SWITCHES

Light switch, removing and installing

◄ Use plastic prying tool to lever out light switch bezel with switch.

- Note that switch bezel snaps into dashboard using lug in areas **A**.
- Unclip switch electrical connector.

– Detach switch from bezel and replace.

– Installation is reverse of removal.

BentleyPublishers
BentleyPublishers.com—All Rights Reserved

START STOP switch, removing and installing

- Using a plastic prying tool, lever START STOP switch out of dashboard in direction of **arrow**.
- Disconnect electrical connector and remove.
- Installation is reverse of removal.

Hazard warning, central locking switch assembly, removing and installing

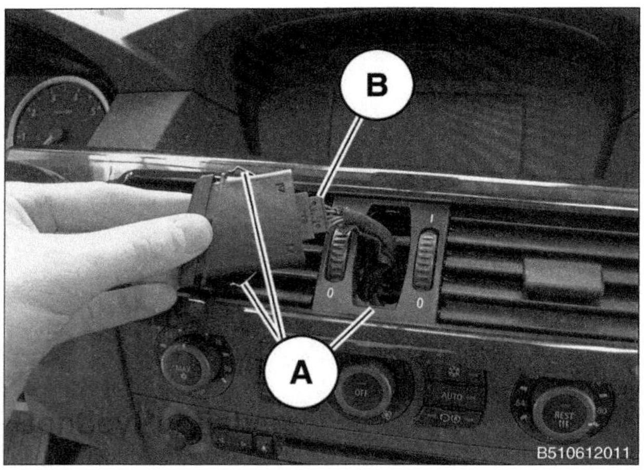

- Using a plastic prying tool, release hazard warning switch retaining tabs (**A**) and pull switch out of dashboard. Disconnect electrical connector (**B**) and remove switch.
- Installation is reverse of removal.

OTHER SWITCHES

Back-up light switch

For manual transmission back-up light switch replacement, see **230 Manual Transmission**.

Automatic transmission electronics combines the following functions:

- Neutral safety switch.
- Back-up light switch.

BentleyPublishers.com
BentleyPublishers.com—All Rights Reserved

Brake light switch, removing and installing

– Remove left lower dashboard trim (above pedals). See **513 Interior Trim**.

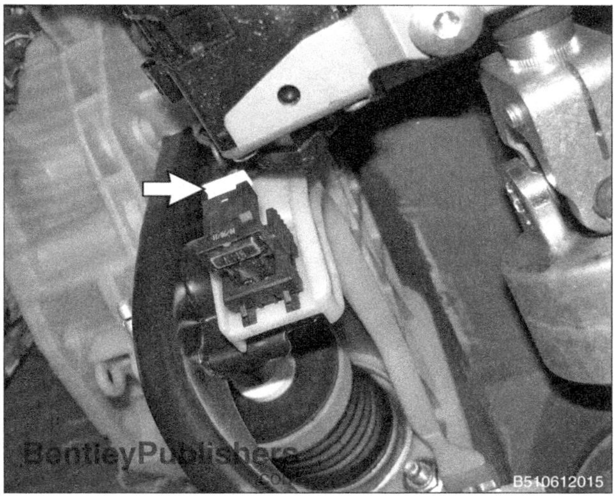

◄ Working at pedal cluster, disconnect electrical connector (**arrow**) from brake light switch.

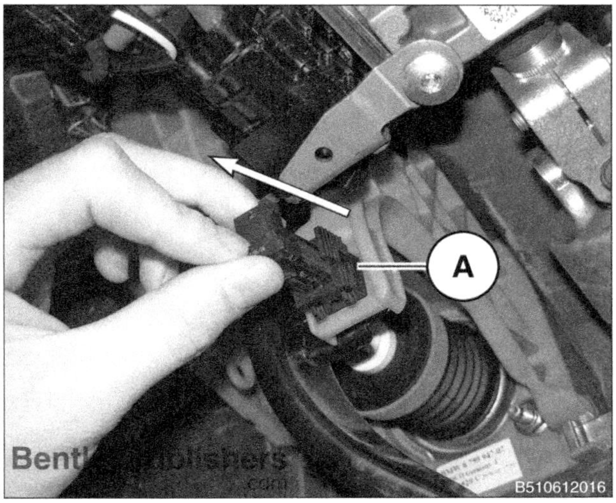

◄ Pull brake light switch (**A**) from bracket mounting in direction of **arrow**.

– To install, depress brake pedal and hold down with brake pedal prop tool.

– Slide brake light switch as far as it will go into brake light switch holder.

– Attach harness connector.

– Grip brake light switch holder, slowly return brake pedal to rest position and pull switch back to stop.

– Check function of brake light switch with assistant.

Trunk lid or tailgate release button, replacing

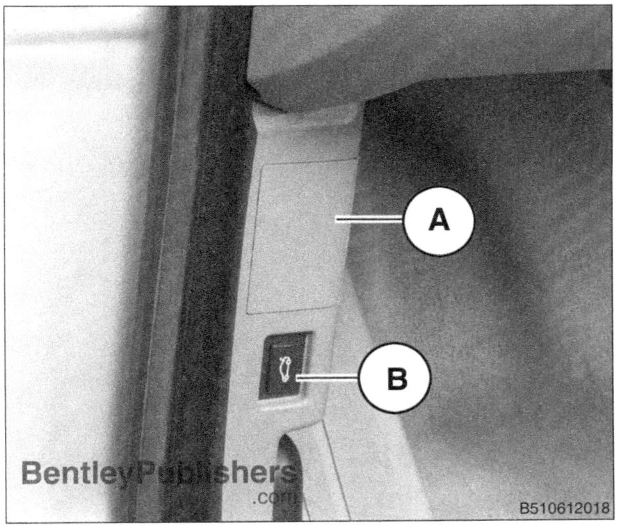

◄ Working at left door jamb:

- Open OBD II plug cover (**A**).
- Reach through opening and push out trunk lid or tailgate switch (**B**).
- Detach electrical connector.

– Installation is reverse of removal.

BentleyPublishers.com–All Rights Reserved

iDrive controller, removing and installing

– Remove center console. See **513 Interior Trim**.

< Remove iDrive controller fasteners (**arrows**) and remove from center console.

– If installing new controller use BMW scan tool to code and program new unit.

Driver door switch, removing and installing

The master front and rear window switches are ganged with child safety switch and outside mirror control switch on the driver door arm rest.

– Remove door trim panel. See **513 Interior Trim**.

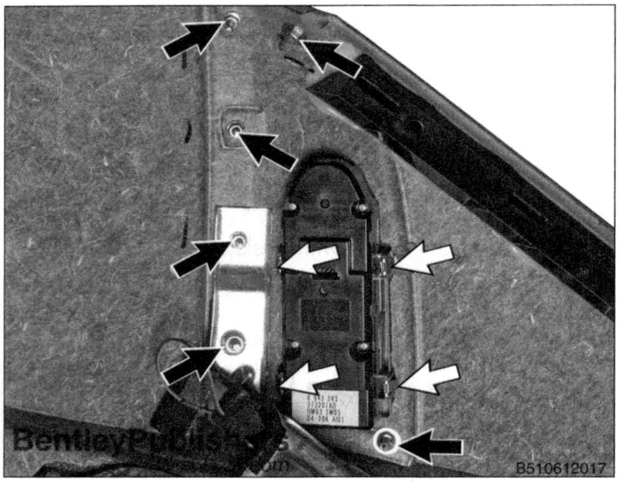

< Remove fasteners (**black arrows**). Unlock retaining clips (**white arrows**) and remove switch from door panel.

– Installation is reverse of removal.

BentleyPublishers.com
BentleyPublishers.com—All Rights Reserved

620 Instruments

GENERAL

This repair group covers removal and installation of the instrument cluster. The cluster does not contain any replaceable components. If necessary, replace the entire unit.

See **020 Maintenance** for an explanation of condition based service (CBS) and Check Control, both of which use instrument cluster displays.

INSTRUMENT CLUSTER

Instrument cluster layout

1. Speedometer
2. Indicator and warning lights
3. Tachometer
4. Trip odometer reset button
5. Fuel gauge
6. Clock
   Outside temperature
   Check Control
   Indicator and warning lights
7. Automatic transmission gear selected
   On-board computer (BC)
   Next scheduled service
   Odometer
   Trip odometer
   Flat tire monitor (FTM) initialization
   Engine oil level
   Settings and information
8. Fuel mileage gauge

620

BentleyPublishers.com—All Rights Reserved

Instrument Cluster Service

Instrument cluster replacement

If replacing the instrument cluster, keep in mind that vehicle odometer reading and other data is redundantly stored in the instrument cluster, light module (LM) and car access system (CAS) module.

If LM, CAS and the instrument cluster are replaced simultaneously, the odometer reading may be lost as a result.

Instrument cluster, removing and installing

– Extend steering column as far back and down as it will go.

< Remove instrument cluster fasteners (**arrows**). Tilt cluster backward (**curved arrow**) out of dashboard.

– Unlock and disconnect harness connectors from back of cluster. Slide cluster sideways to clear steering wheel.

– Installation is reverse of removal, noting the following:
  - Make sure instrument cluster harness connector locking levers are in fully unlocked position before installing connector.
  - Code replacement cluster using BMW scan tool.

Tightening torque	
Instrument cluster to dashboard	1.5 Nm (8.5 in-lb)

BentleyPublishers.com
BentleyPublishers.com—All Rights Reserved

630 Lights

TABLES

GENERAL

This repair group covers interior and exterior lights.

See also:

- **600 Electrical System–General** for bus system basics
- **611 Wipers and Washers** for headlight washer system
- **612 Switches** for light switch replacement
- **620 Instruments** for instrument cluster removal
- **ECL Electrical Component Locations**
- **ELE Electrical Wiring Diagrams**

630

BentleyPublishers.com
BentleyPublishers.com–All Rights Reserved

Bulb applications

Bulb applications for Sedan and Sports Wagon models

Table a. Bulb applications: Sedan, Sports Wagon	
Location	**Type and wattage**
Headlights • Low beam (halogen) • Low beam (xenon) 6 cylinder V8 to 9 / 2005 V8 from 9 / 2005 • High beam • Daytime running from 03 /2007 (xenon only)	 H7 55 w D1S D2S D1S H7 55 w H8
Back-up lights	921
Brake lights • Sides • Center	 7506 21 w LEDs
Licence plate lights	6418 5 w
Rear foglight	7506 21 w
Taillights	7506 21 w
Parking lights to 03 / 2007 from 03 / 2007	 2825 3457
Turn signals • Front to 03 / 2007 from 03 / 2007 • Rear to 03 / 2007 from 03 / 2007 • Side to 03 / 2007 from 03 / 2007	 7507 21 w 3457 21 w 7507 21w 7506 21 w 2827 LED
Interior lights	
• Cargo area light (D-pillar)	6411
• Dome and map lights, front or rear to 03 / 2007 from 03 / 2007	 6486 2886
• Door courtesy lights to 03 / 2007 from 03 / 2007	 6411, 6418, 6486 6411, 6418, 2886x
• Footwell lights	Festoon C10 w
• Glove compartment	6418
• Trunk light	Festoon C10 w
• Vanity mirror	Festoon C10 w

BentleyPublishers.com
BentleyPublishers.com—All Rights Reserved

Exterior lights overview

The exterior lighting system consists of:

- Headlights with low beams, high beams and parking lights
- Turn signal lights
- Brake lights including center brake light
- Side turn signals
- Taillights
- Back-up lights
- Front and rear foglights
- Rain and light sensor with automatic driving light control
- Light module (LM)

The optional bi-xenon headlights are available in conjunction with adaptive headlights.

Light module

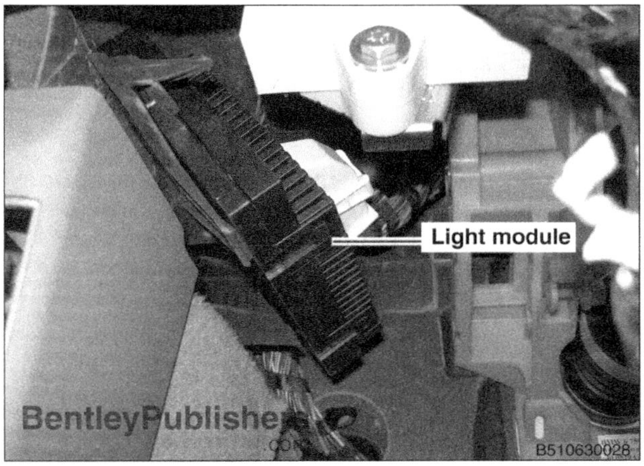

Light module (LM) controls and monitors exterior lights as well as adaptive headlight function. It also controls instrument panel dimming and includes emergency functions in case of a fault.

Exterior lighting functions

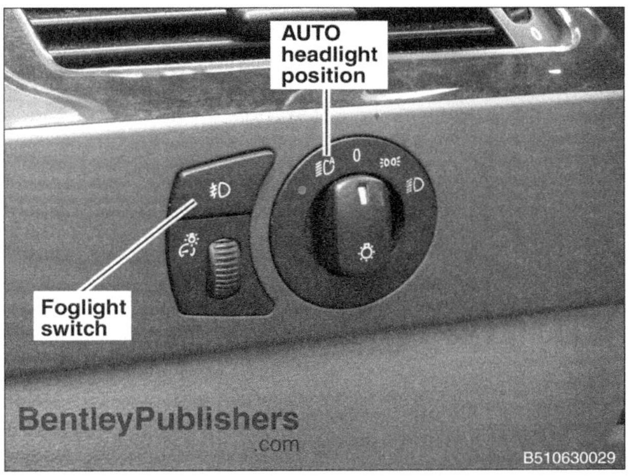

Automatic headlight control is activated by turning the light switch to the AUTO headlight position. Driving lights are switched ON together with the side lights by the automatic driving lights control system. Lights activation is based on input from the rain and light sensor (RLS), depending on ambient lighting conditions (tunnel, twilight, rain or snow).

In case of RLS malfunction, light module (LM) illuminates the headlights.

> ***WARNING*** —
> - *The automatic headlight control cannot serve as a substitute for personal judgement in determining when to switch the lights ON. For example: The system cannot detect fog.*

On vehicles equipped with iDrive, the sensitivity of the lighting activation point can be adjusted by the driver. Refer to the owner's manual.

Foglights are activated by pressing the foglight button; switch parking lights or low beam headlights ON first.

Turn signals are activated via the turn signal stalk on the steering column switch cluster (SZL), an optical switch. A defective bulb in the front or rear turn signals results in the remaining lights flashing at double the normal rate.

The one- touch function flashes the turn signals three times. The function can be modified using car / key memory (personal profile). See **515 Central Locking and Anti-theft**.

Hazard lights are activated by depressing the hazard light switch ganged with the central locking switch and located in the center of the dashboard.

630

BentleyPublishers
BentleyPublishers.com—All Rights Reserved

Turn signals have priority if the hazard flashers are activated first. Hazard flashers have priority if the turn signal is activated first.

Vehicle with alarm system: Hazard lights flash for up to 6 minutes if alarm is triggered and not switched off. LM receives the flasher request directly from hazard switch.

When central locking is activated, car access system (CAS) supplies a signal to LM via K-CAN to flash the hazard lights indicating that the vehicle is locked. This visual feedback can be deactivated using car / key memory (personal profile). See **515 Central Locking and Anti-theft**.

Taillights. In case of a taillight bulb failure or other malfunction, the brake light bulbs are used as a substitute. The bulbs are dimmed by a pulsed signal to provide substitute taillights.

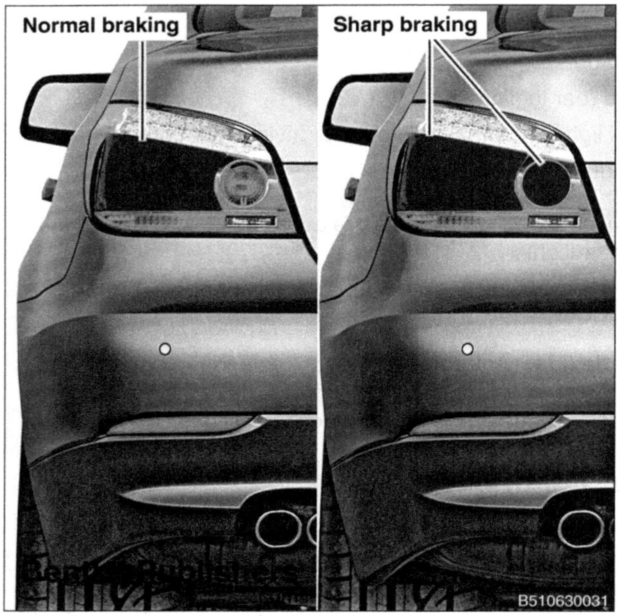

Brake lights, brake force display. Brake lights illuminate in two stages, depending on how sharply the brakes are applied:

- Normal braking: Outboard and center brake lights illuminate
- Sharp braking: additional brake light bulb illuminates. Note that this bulb also functions as the rear foglight. Foglights ON overrides the brake force display.

Brake force display, standard on 5 Series vehicles, is activated at speeds above 5 kph (3 mph). The threshold for triggering brake force display is deceleration above 5 m / s^2 (0.003 mach).

The brake light switch is powered by CAS with terminal R ON.

Back-up lights. Automatic transmission signal for activating back-up lights is communicated from automatic transmission control (EGS) module via K-CAN to LM.

Manual transmission back-up light signal is communicated by reverse gear switch to LM.

Bulb monitoring. LM monitors the exterior lighting system with the lights switched ON or OFF if terminal 15 is active (ignition ON).

Cold monitoring with lights OFF: Brief current pulses, so short that bulbs are not illuminated, are used for measurement purposes. LM evaluates individual bulb outputs to establish whether there is a line break or a bulb is defective.

The number of current pulses is large during the first 4 seconds after terminal 15 ON to check whether bulbs are in working order before setting off. This function is referred to as the predrive check. The number of pulses is then reduced after the predrive check. Pulsed power is then applied to the bulbs every 1.5 minutes.

The center brake light is not included in the cold monitoring system because LEDs react too fast to the current pulse and would consequently light up. The bi-xenon headlight is also not included in the cold monitoring because of legal restrictions.

Hot monitoring with lights ON is based on measuring the current of individual bulbs, used to detect an overload or interruption (break).

Pathway / follow me home light is activated by the headlight flasher position on the SZL stalk.

BentleyPublishers.com
BentleyPublishers.com–All Rights Reserved

Xenon headlights

To 3 / 2007: Vehicles equipped with the optional xenon headlights (also known as high intensity discharge or HID headlights) utilize a special high voltage bulb and control unit for the low beam headlights. High beam bulbs are conventional halogens.

Bi-xenon headlights

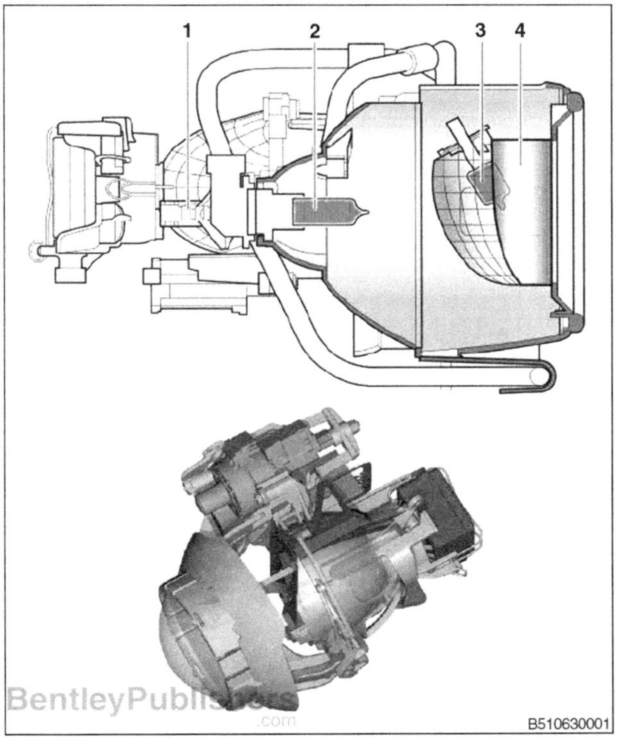

From 3 / 2007: The optional bi-xenon headlight is equipped with an electrically activated shutter which directs the low beam cone of light in the same direction as the high beam.

1. Xenon bulb
2. Bulb for side light and daytime running light
3. Bulb for cornering light
4. Mirror for cornering light

Adaptive headlight control

Bi-xenon headlights are equipped with an adaptive headlight control system.

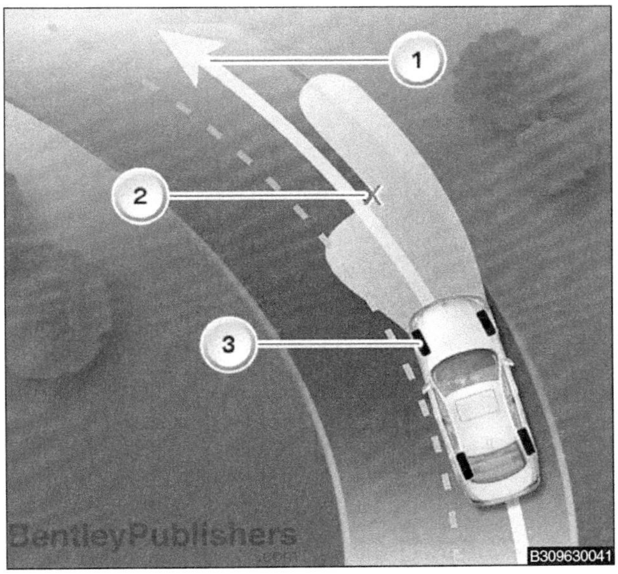

Adaptive headlights rotate slightly as the front wheels (**3**) turn, providing an enhanced lighting path as the vehicle travels through a turn. The light / dark transition point (**2**) is designed to be to the right of the anticipated curve of the vehicle (**1**) in order to reduce the risk of glare to oncoming traffic.

Headlights are rotated via a stepper motor in each headlight assembly. LM controls the amount of headlight rotation based on the following information:

- Steering angle (from steering angle sensor)
- Road speed (from vehicle speed sensors)
- Yaw rate (from DSC acceleration sensor)

BentleyPublishers.com
BentleyPublishers.com—All Rights Reserved

Exterior light troubleshooting

The following table used in conjuction with a BMW scan tool and a digital volt / ohm meter (DVOM) can speed diagnosis of exterior light faults.

Table b. Exterior light troubleshooting

Possible fault or light failure	Faulty, missing signal to light module (LM)	Signal transmitting control module
No permanent high beam, emergency function while lever remains pressed.	Steering_column_lever_operation	Steering column switch center (SZL)
Turn signal is not automatically canceled.	Steering wheel angle (physical angle)	Steering column switch center (SZL)
Turn signal is not automatically canceled.	Steering_wheel_angle_fault (steering angle sensor (SAS))	Steering column switch center (SZL)
Restricted monitoring of back up lights.	Trailer_status	Trailer module (AHM)
Flashing frequency not doubled.	Trailer_indicator _fault_status	Trainler module (AHM)
Lights permanently ON.	Headlamp_status	Rain and light sensor (AIC)
No back up lights.	Transmission_gear_status	Electronic transmission control (EGS)
No headlight flasher, brake lights, turn signals at Terminal R.	Terminal_R_status	Car access system (CAS)
Position light remains ON.	Key_in_ignition_status	Car access system (CAS)
No brake lights via active cruise control (ACC).	Brake_light_control	Dynamic stability control (DSC)
No stop position for turn signals, emergency function while lever remains pressed.	Steering_column_lever_operation	Steering column switch center (SZL)
No hazard warning lights with antitheft alarm.	Anti-theft_alarm_control	Anti-theft alarm (DWA)
No alarm acknowledgment active.	Alarm_acknowledgment_control	Anti-theft alarm (DWA)

BentleyPublishers.com
BentleyPublishers.com–All Rights Reserved

Interior lights overview

Interior lights consist of:

- Front dome lights with switches
- Rear dome lights with switches
- Reading lights with switches
- Trunk light (Sedan)
- Cargo compartment light (Sports Wagon)
- D-pillar lights (Sports Wagon)
- Vanity mirror lights, driver and passenger
- Footwell lights, driver and passenger
- Door courtesy lights
- Door handle courtesy lights

Interior lighting modules and functions

Interior lighting function is distributed over several control module that communicate with each other via K-CAN.

Basic body module (KBM)

The basic body module (KBM) controls interior and rear door lighting. Front door lighting is controlled by:

- **to 09 / 2005** Front door modules (TMFA, TMBF).
- **from 09 / 2005** Body gateway module (KGM).

The interior lights are pulse-width modulated (PWM) and controlled by various inputs. PWM adjusts brightness in relation to vehicle battery voltage. Above 12.8 volts PWM is adjusted to provide a power output of 100%. Below 12.8 volts the interior lights dim accordingly.

Electric load shutdown

Some interior lights such as map, glovebox and trunk can be switched on when the vehicle is OFF. To prevent battery drain, interior lighting is switched off after 16 minutes.

Instrument lights

Instrument lighting via terminal 58g and K-CAN is pulse-width modulated and features two brightness levels:

- Brightness level for instrument lighting is toggled via switch on turn signal lever (lights ON).
- Brightness level for function lighting is not dimmed and is switched on at full brightness.

630

BentleyPublishers.com
BentleyPublishers.com—All Rights Reserved

Warnings and Cautions

WARNING—
- *There is high pressure inside a halogen bulb. Bulb temperature may exceed 700° Celsius. Wear safety glasses and gloves when removing and installing a halogen bulb.*
- *Do not look directly at an operating xenon bulb. The UV emissions are approximately 2.5 times that of a comparable halogen bulb.*
- *When working on electrical systems, remove the fuse protecting the circuit under repair. See* **ECL Electrical Component Locations** *for fuse locations.*

CAUTION—
- *Xenon bulbs and igniter modules are electrically sensitive. Use an antistatic mat and work with caution. See* **600 Electrical System–General**.
- *Do not operate the xenon control module unless a bulb is connected.*
- *Before servicing the headlight system, switch electrical consumers OFF. Switch ignition OFF and remove ignition key.*
- *Do not handle bulb glass with bare fingers. Dirt and skin oils cause a bulb to fail prematurely. If necessary, wipe bulb using a clean cloth dampened with rubbing alcohol.*
- *Use only original equipment replacement bulbs. Non-original bulbs may cause false failure readings in Check Control.*
- *To avoid marring car paint or trim, work with plastic prying tools or wrap the tips of tools with tape.*

BentleyPublishers.com
BentleyPublishers.com—All Rights Reserved

HEADLIGHTS

Halogen headlight components

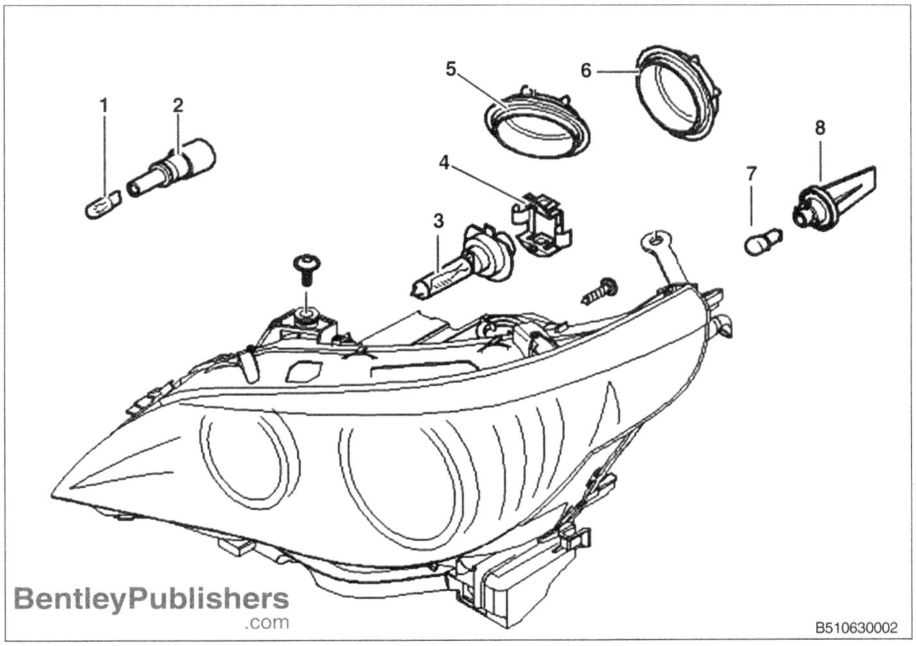

1. Parking light bulb
   - 2825 5 w
2. Parking light bulb socket
3. High and low beam bulb
   - H7 55 w
4. Bulb retaining spring
5. High beam bulb cover
6. Low beam bulb cover
7. Turn signal bulb
   - 7507 21 w (amber)
8. Turn signal socket

Xenon headlight components (to 03 / 2007)

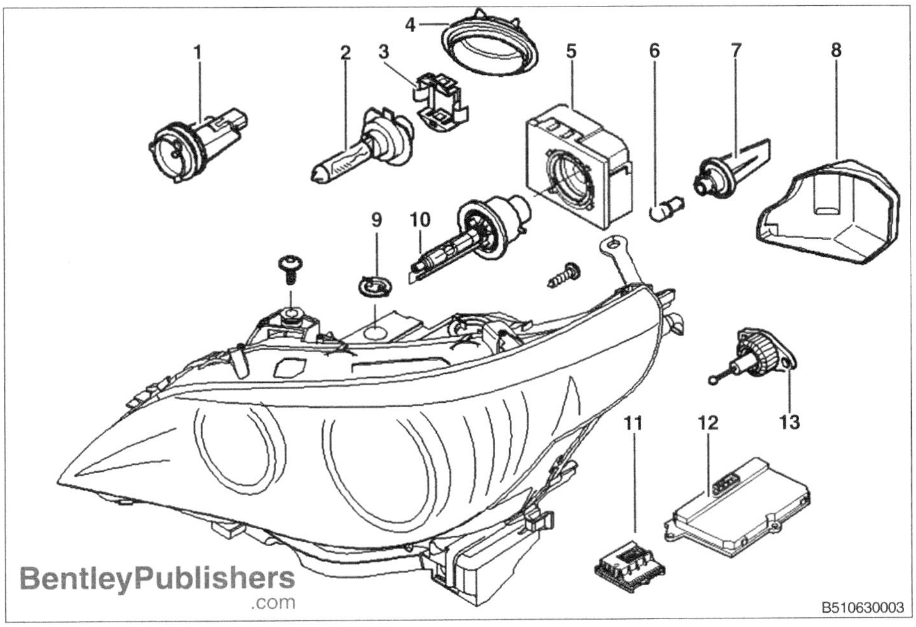

1. Parking light socket with bulb
2. Halogen high beam bulb
   - H7 55 w
3. Bulb retaining spring
4. High beam bulb cover
5. Xenon bulb igniter
6. Turn signal bulb
   - 7507 21 w
7. Turn signal socket
8. Xenon bulb cover
9. Cover
10. Xenon low beam bulb
    - 6 cylinder: D1S
    - V8 to 9 / 2005: D2S
11. Adaptive headlight control module
12. Xenon bulb ignition unit
13. Headlight vertical aim stepper motor

BentleyPublishers.com
BentleyPublishers.com–All Rights Reserved

Xenon headlight components (from 03 / 2007)

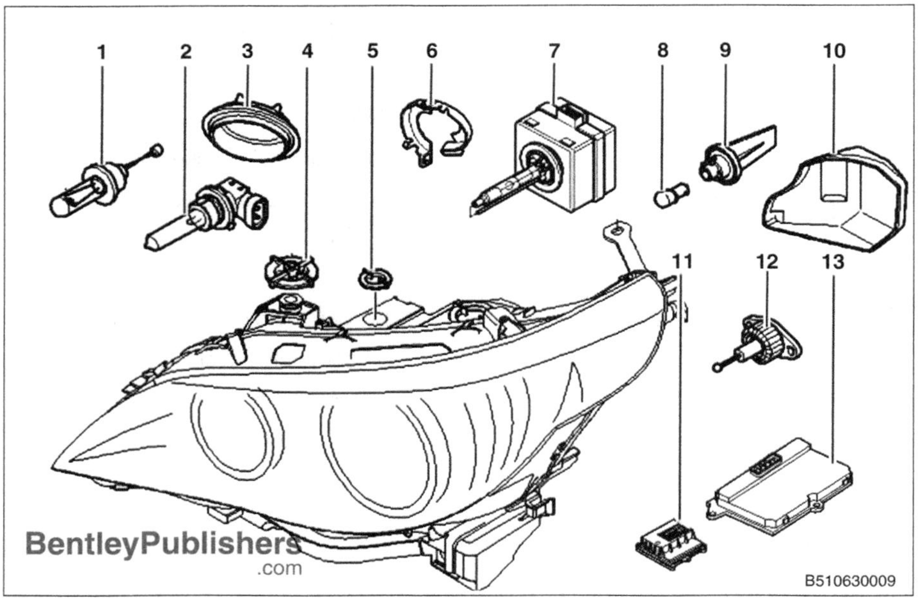

1. Parking light bulb
   - H3 55 w
2. Daytime running light bulb
   - H8 35 w
3. High beam bulb cover
4. Cover
5. Cover
6. Xenon bulb retainer
7. Xenon low beam bulb
   - D1S
8. Turn signal bulb
   - 3457 21 w
9. Turn signal socket
10. Xenon bulb cover
11. Adaptive headlight control module
12. Headlight vertical aim stepper motor
13. Xenon bulb ignition unit

Headlight aim, adjusting

Adjust headlight aim with correct tire pressures, fuel tank full, and weight of one person (approximately 75 kg or 165 lb) in driver seat.

When adjusting xenon headlights, switch lights ON, then wait 80 seconds before proceeding with adjustment. Do not move or vibrate vehicle during that time.

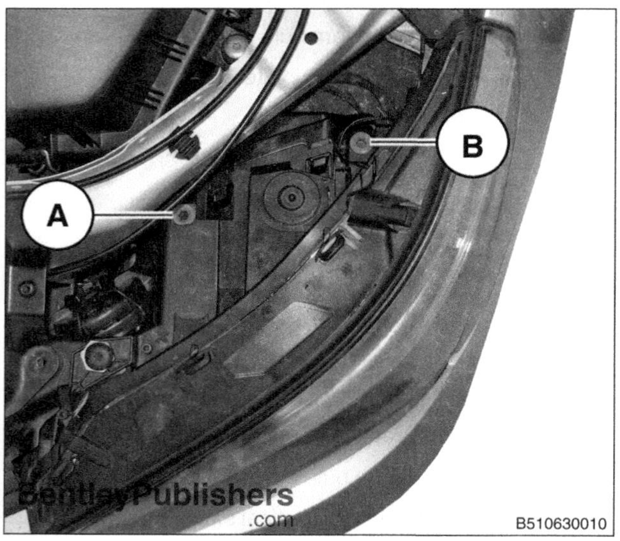

Working above and in rear of headlight housing, use screws to adjust headlight aim:

- **A**: Primarily vertical aim adjustment
- **B**: Primarily lateral aim adjustment

Note that turning each adjuster affects both planes of adjustment.

BentleyPublishers.com
BentleyPublishers.com—All Rights Reserved

Headlight bulb, replacing (halogen headlight assembly)

- Switch lights OFF.
- Left headlight: Remove air filter housing. See **020 Maintenance**.

Low beam bulb

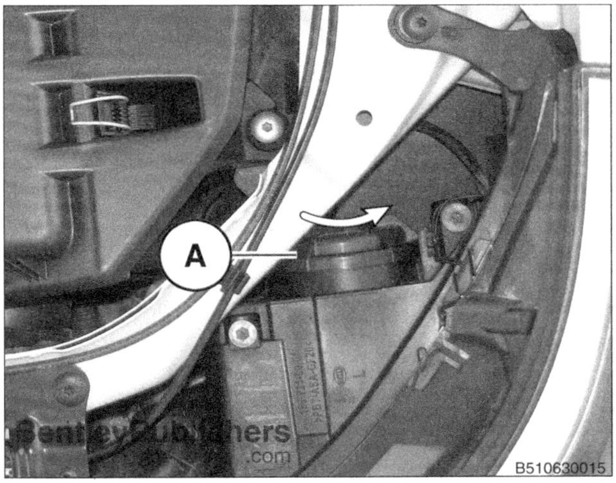

◄ **To 03 / 2007**: Working in engine compartment behind headlight, rotate cap (**A**) approx. 30° counterclockwise and remove from headlight.

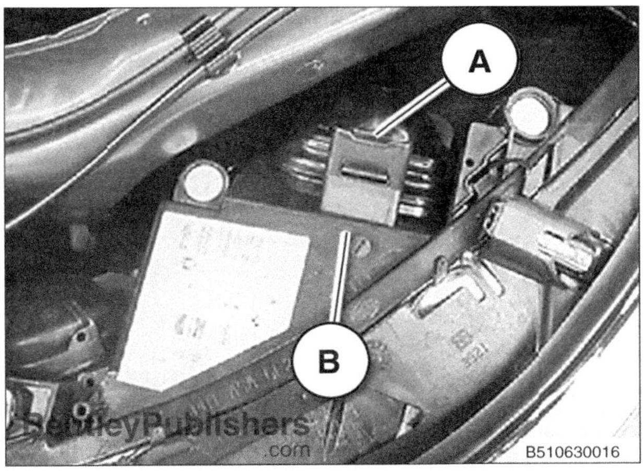

◄ **From 03 / 2007**: Unclip cover cap (**A**) and remove from headlight (**B**) towards rear of vehicle.

◄ Rotate bulb socket (**A**) counterclockwise and remove from headlight.

BentleyPublishers
BentleyPublishers.com–All Rights Reserved

Headlight bulb, replacing (halogen headlight assembly)

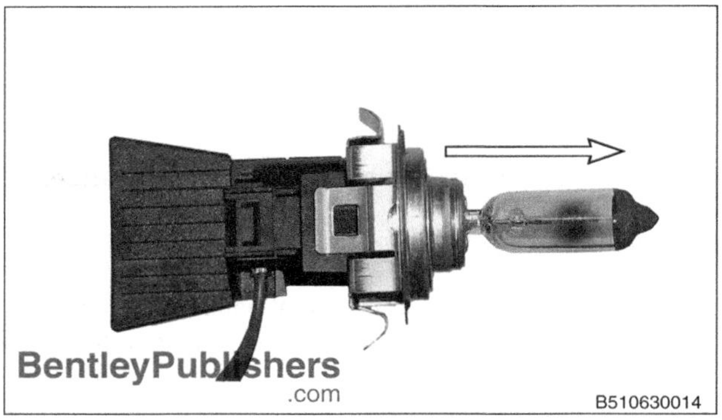

◄ Remove bulb from socket in direction of **arrow**.

– Replace bulb. Do not touch bulb surface with bare finger. Oil from skin can cause bulb to fail prematurely.

– Installation is reverse of removal. Make sure cover is firmly seated on headlight.

High beam bulb

◄ Working in engine compartment behind headlight, rotate cap (**A**) approx. 30° counterclockwise and remove from headlight.

◄ Rotate bulb socket (**A**) counterclockwise and remove from headlight.

BentleyPublishers.com
BentleyPublishers.com—All Rights Reserved

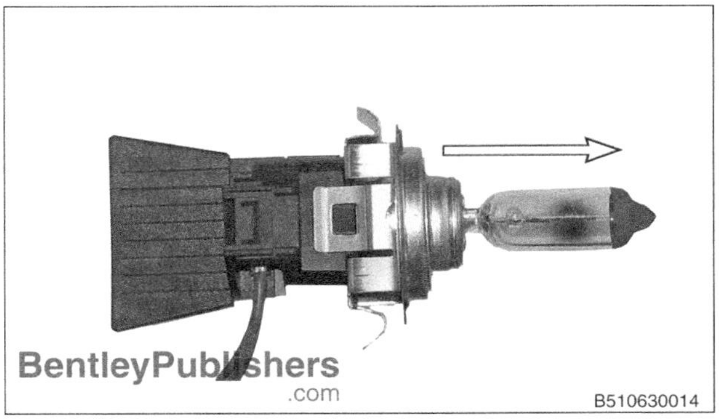

◀ Remove bulb from socket in direction of **arrow**.

- Replace bulb. Do not touch bulb surface with bare finger. Oil from skin can cause bulb to fail prematurely.

- Installation is reverse of removal. Make sure cover is firmly seated on headlight.

Headlight bulb, replacing (xenon headlight assembly)

To **3 / 2007:** Vehicles equipped with the optional xenon headlights (also known as high intensity discharge or HID headlights) utilize a special high voltage bulb and control unit for the low beam headlights. High beam bulbs are conventional halogens.

From **3 / 2007:** The optional bi-xenon headlights can be switched over from low-beam to high-beam. When the low beams are ON, the lower part of the light beam is blocked by a shutter directly in front of the reflector. When high beams are switched ON, an electro-magnet is energized by light module (LM). The electro-magnet flips the shutter open, allowing the entire beam of light generated by the xenon bulb to be the high-beam. The halogen high beam bulb in earlier models is now used for daytime running lights.

- Remove headlight fuse(s). See **ECL Electrical Component Locations**.

> ***WARNING —***
> - *High voltage in xenon headlights is hazardous. Disconnect xenon headlight components from power supply before removing.*
> - *The physical and mechanical properties of xenon bulb and igniter module are very sensitive. Use an antistatic mat and work with caution.*

- Remove headlight assembly. See **Headlight assembly, removing and installing** in this repair group.

Xenon low beam bulb, replacing

◀ Unlock headlight bulb cover retaining clips (**small arrows**) and remove cover (**arrow**).

630

BentleyPublishers.com
BentleyPublishers.com—All Rights Reserved

Headlight bulb, replacing (xenon headlight assembly)

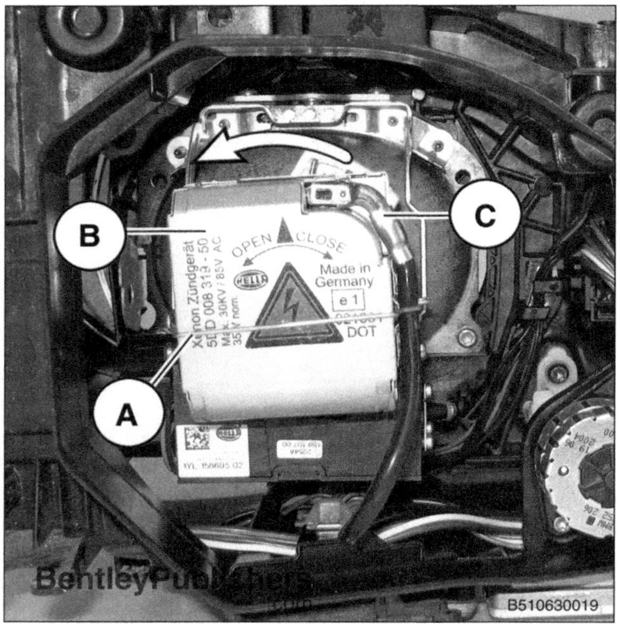

To 02 / 2005: Remove retaining spring (**A**). Rotate igniter (**B**) approx. 30° counterclockwise and remove. Electrical connector (**C**) is automatically released.

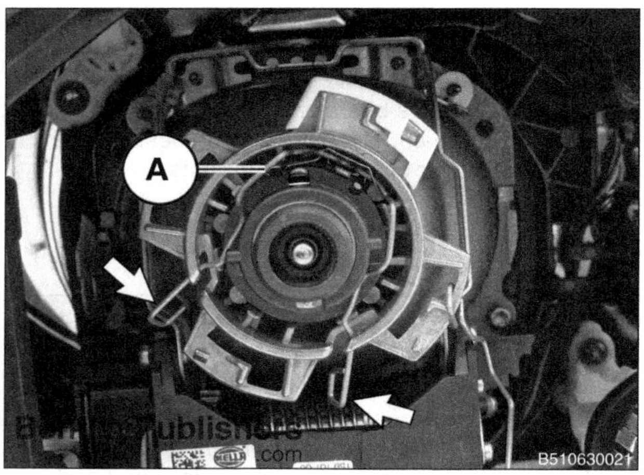

To 02 / 2005: Unlock retaining ring (**A**) and remove bulb from headlight assembly.

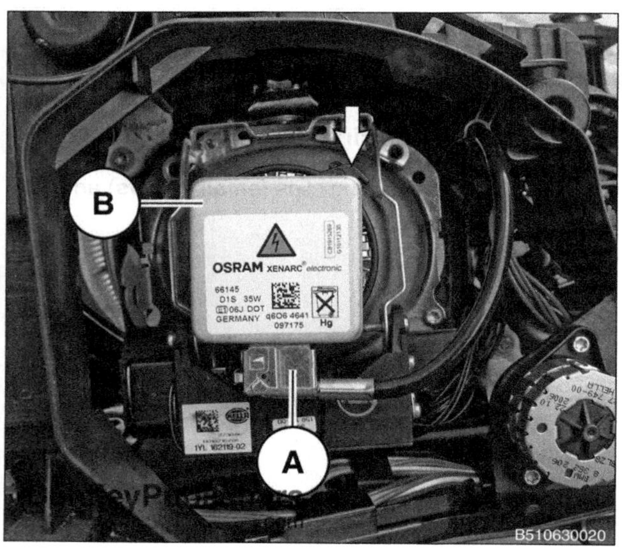

From 02 / 2005: Disconnect electrical connector (**A**). Press tab (**arrow**) and rotate igniter (**B**) approx. 30° counterclockwise and remove.

– Replace bulb. Do not touch bulb surface with bare finger. Oil from skin can cause bulb to fail prematurely.

– Installation is reverse of removal. Make sure cover is firmly seated on headlight.

BentleyPublishers.com
BentleyPublishers.com—All Rights Reserved

High beam bulb: to 3 / 2007

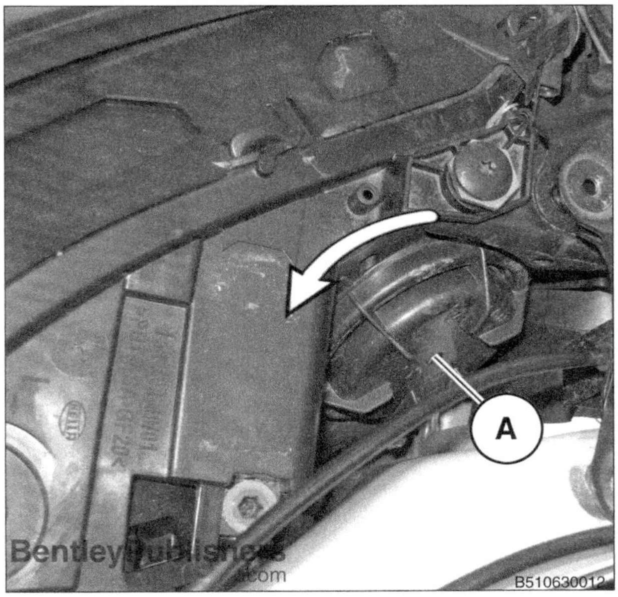

◄ Working in engine compartment behind headlight, rotate cap (**A**) approx. 30° counterclockwise and remove from headlight.

◄ Rotate bulb socket (**A**) counterclockwise and remove from headlight.

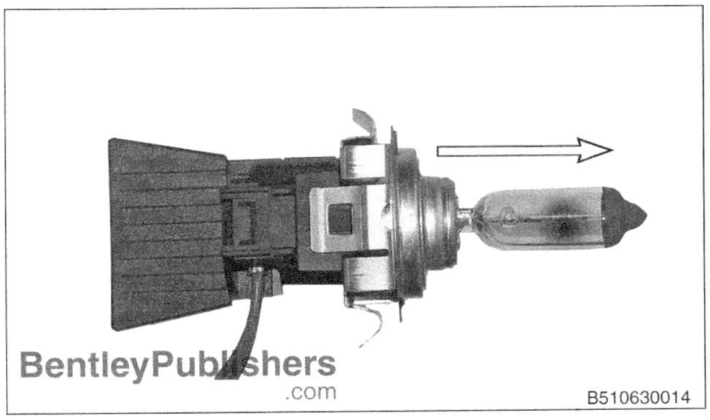

◄ Remove bulb from socket in direction of **arrow**.

- Replace bulb. Do not touch bulb surface with bare finger. Oil from skin can cause bulb to fail prematurely.

- Installation is reverse of removal. Make sure cover is firmly seated on headlight.

BentleyPublishers.com
BentleyPublishers.com–All Rights Reserved

Daytime running light bulb, replacing from 3 / 2007

◀ Working in engine compartment behind headlight, rotate cap (**A**) approx. 30° counterclockwise and remove from headlight.

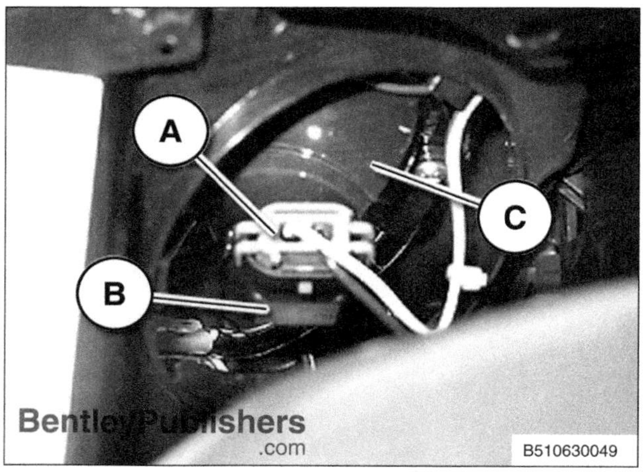

◀ Disconnect electrical connector (**A**). Rotate bulb (**B**) counterclockwise and remove from headlight (**C**).

– Replace bulb. Do not touch bulb surface with bare finger. Oil from skin can cause bulb to fail prematurely.

– Installation is reverse of removal. Make sure cover is firmly seated on headlight.

Headlight assembly, removing and installing

– **Xenon headlight**: Remove headlight fuse(s). See **ECL Electrical Component Locations**.

> ***WARNING —***
> - *High voltage in xenon headlights is hazardous. Disconnect xenon headlight components from power supply before removing.*
> - *The physical and mechanical properties of xenon bulb and igniter module are very sensitive. Use an antistatic mat and work with caution.*

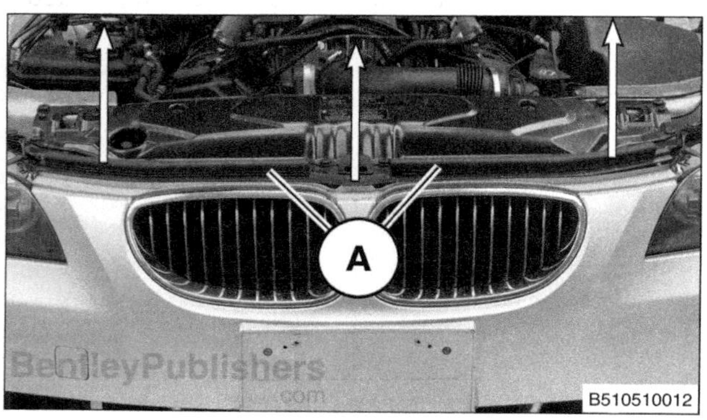

◀ Lift off trim strip with seal (**A**).

BentleyPublishers.com
BentleyPublishers.com–All Rights Reserved

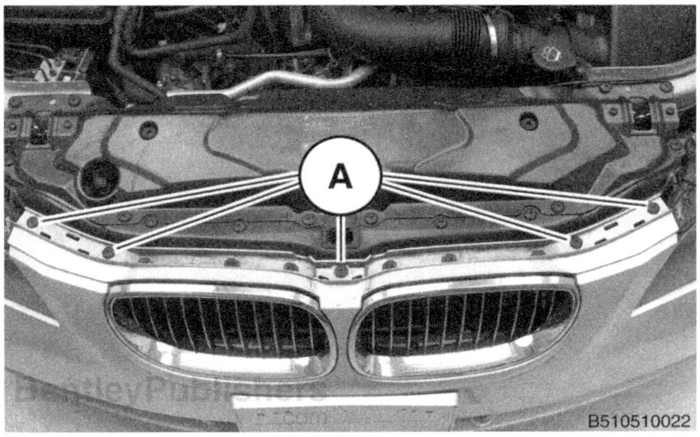

Remove bumper cover fasteners (**A**).

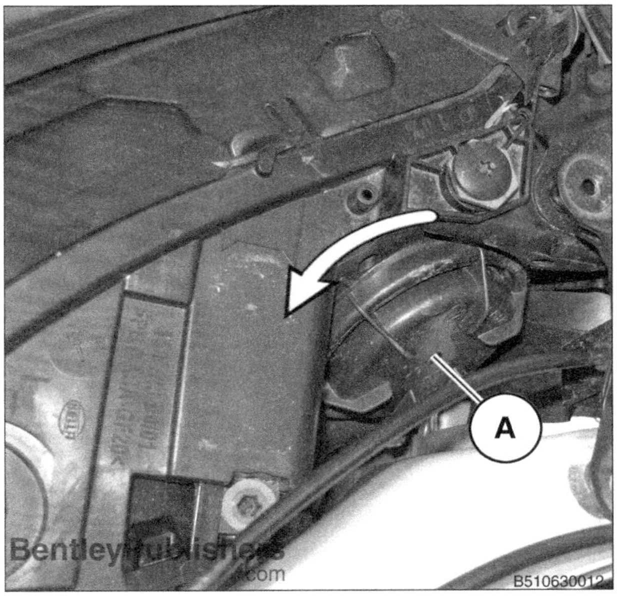

Working in engine compartment behind headlight, rotate cap (**A**) approx. 30° counterclockwise and remove from headlight.

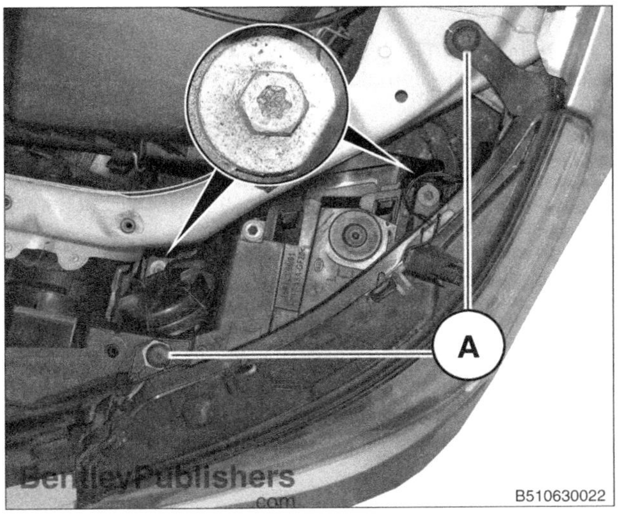

Working above headlight:

- Remove upper headlight assembly fasteners (**A**).
- Loosen headlight assembly fasteners (**inset**) just enough to remove headlight.

BentleyPublishers.com
BentleyPublishers.com—All Rights Reserved

Pull bumper cover (**A**) forward then pull out headlight slightly and disconnect electrical connector in back. Remove headlight in direction of **arrow**.

When installing, set gaps from top (below fender) and bottom (above bumper cover) according to specifications.

Headlight installation gaps	
Dimension **A** (below fender edge)	4.0 ± 1.0 mm (0.16 ± 0.04 in)
Dimension **B** (above bumper cover)	3.0 ± 1.3 mm (0.12 ± 0.05 in)

- Adjust headlight aim when done. See **Headlight aim, adjusting** in this repair group.

Front turn signal bulb, replacing

Twist turn signal bulb socket (**A**) to remove.

- Replace bulb. Do not touch bulb surface with bare finger. Oil from skin can cause bulb to fail prematurely.
- Installation is reverse of removal.

BentleyPublishers.com
BentleyPublishers.com—All Rights Reserved

Side turn signal bulb, replacing

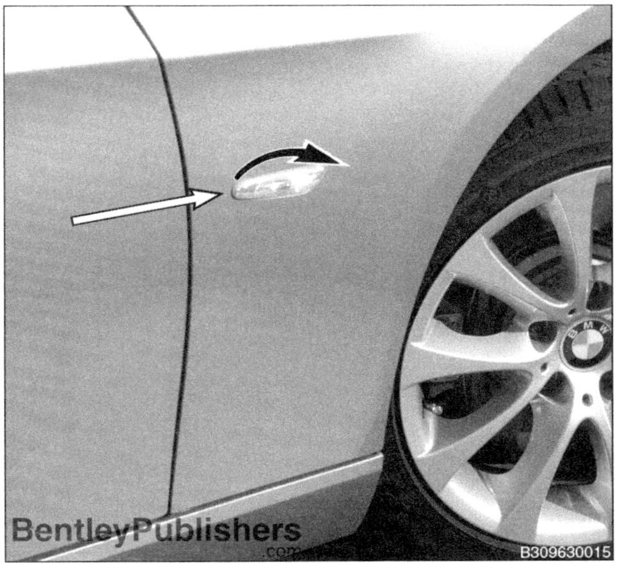

- Use plastic prying tool or finger to push side turn signal housing forward (**white arrow**). Pivot housing away from fender (**curved arrow**) and remove.
- Rotate bulb socket and remove. Replace bulb. Do not touch bulb surface with bare finger. Oil from skin can cause bulb to fail prematurely.
- Installation is reverse of removal.

Front parking light bulb, replacing

- Remove headlight high beam bulb. See **Headlight bulb replacing**, for your model 5 Series in this repair group.

- Rotate socket (**A**) approx. 30° counterclockwise and remove from headlight assembly. Disconnect electrical connector and remove socket from vehicle.
- Replace bulb. Do not touch bulb surface with bare finger. Oil from skin can cause bulb to fail prematurely.
- Installation is reverse of removal.

BentleyPublishers.com
BentleyPublishers.com–All Rights Reserved

Front foglight bulb, replacing

– Remove engine lower covers. See **020 Maintenance**.

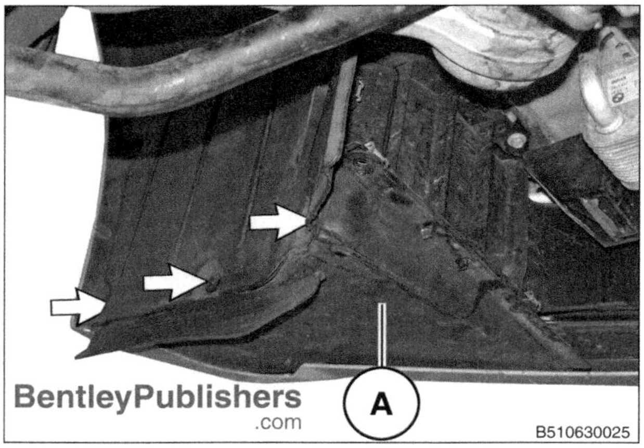

◄ Remove fasteners (**arrows**) and remove foglight splash shield (**A**).

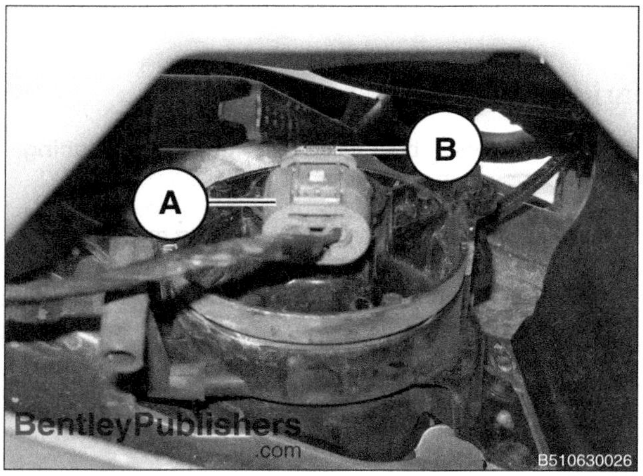

◄ Working behind end of front bumper cover:

- Detach foglight harness connector (**A**).
- Twist foglight socket (**B**) to remove.

– Replace bulb. Do not touch bulb surface with bare finger. Oil from skin can cause bulb to fail prematurely.

– Installation is reverse of removal.

– Adjust foglight aim through slot in front bumper cover. See **Front foglight adjusting** in this repair group.

Front foglight, adjusting

Adjust foglight aim with correct tire pressures, fuel tank full, and weight of one person (approximately 75 kg or 165 lb) in driver seat.

◄ Working above and on inside of foglight, use screw (**A**) to adjust foglight aim.

BentleyPublishers.com
BentleyPublishers.com—All Rights Reserved

Taillights

Taillight assembly (Sedan)

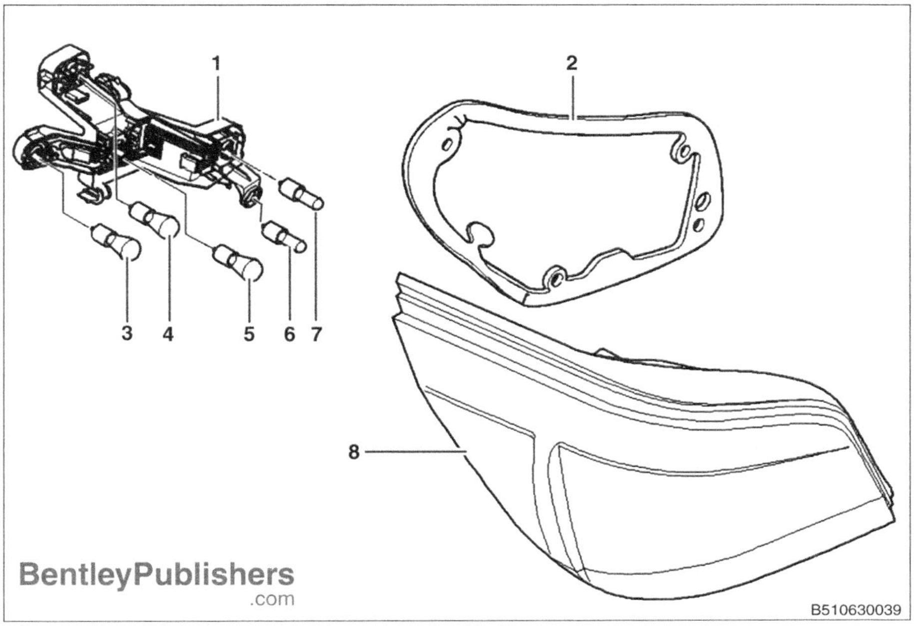

1. Taillight socket
2. Sealing gasket
3. Brake light bulb
   - 7506
4. Turn signal bulb
   - to 03 / 2007: 7507
   - from 03 / 2007: 7506
5. Taillight and parking light bulb
   - 7506
6. Back-up light bulb
   - 921
7. Rear foglight, brake force display light
   - 7506
8. Taillight lens

Taillight assembly (Sports Wagon)

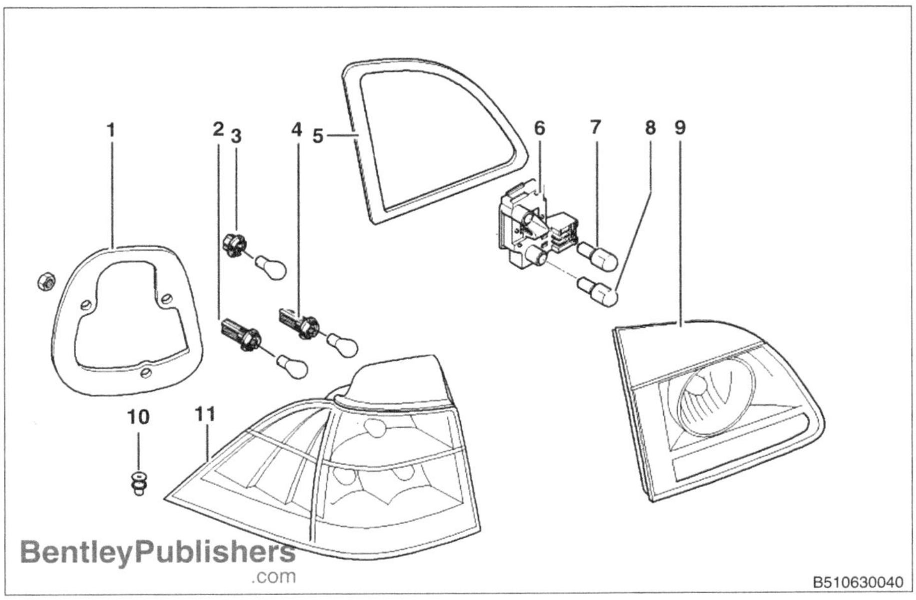

1. Taillight sealing gasket, outboard
2. Taillight and parking light bulb
   - 7506
3. Turn signal bulb
   - 7507
4. Brake light bulb
   - 7506
5. Taillight sealing gasket, inboard
6. Taillight socket, inboard
7. Rear foglight, brake force display light
   - 921
8. Back-up light bulb
   - 921
9. Taillight lens, inboard
10. Mounting clip
11. Taillight lens, outboard

630

BentleyPublishers.com
BentleyPublishers.com—All Rights Reserved

Taillights service

Sedan taillight bulbs

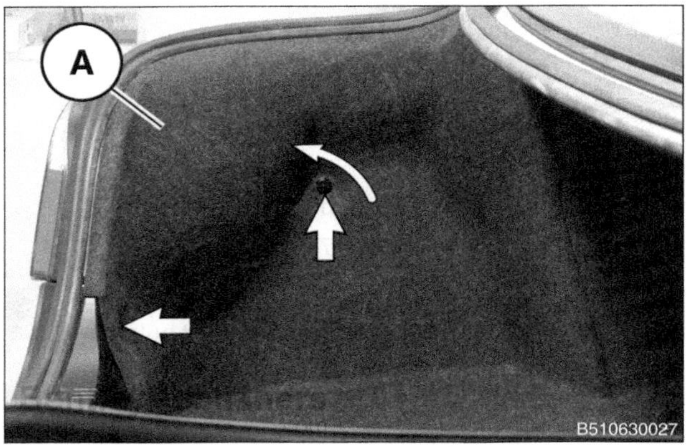

Working in trunk, remove trunk interior trim panel (**A**) by rotating fasteners (**arrows**) 90° and pulling outside edge toward trunk interior. Left side shown, right side is similar.

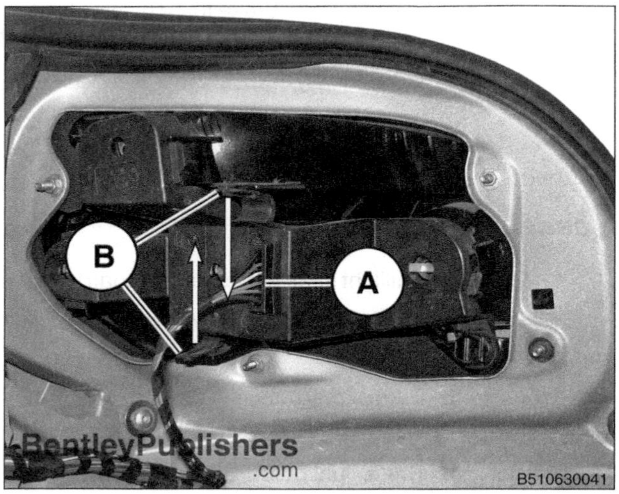

Working at taillight socket:
- Detach electrical connector (**A**).
- Press locking clips (**B**) to unlock socket.
- Remove socket to replace bulb(s). Do not touch bulb surface with bare finger. Oil from skin can cause bulb to fail prematurely.

– Installation is reverse of removal.

Sports Wagon taillight bulbs

Outboard taillight bulbs

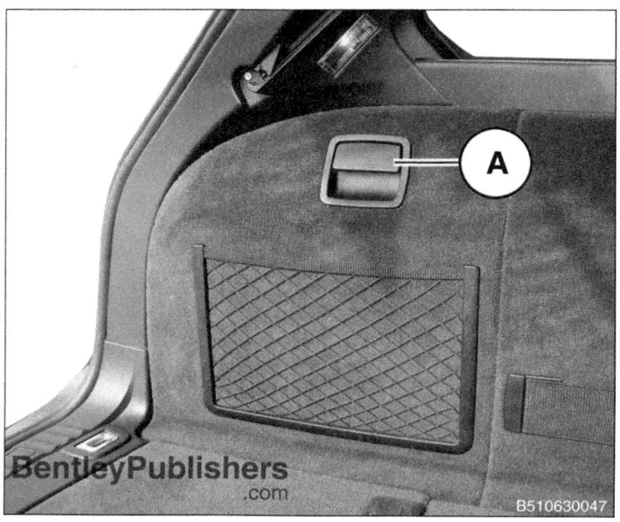

Open tailgate. Open access panel (**A**).

BentleyPublishers.com
BentleyPublishers.com—All Rights Reserved

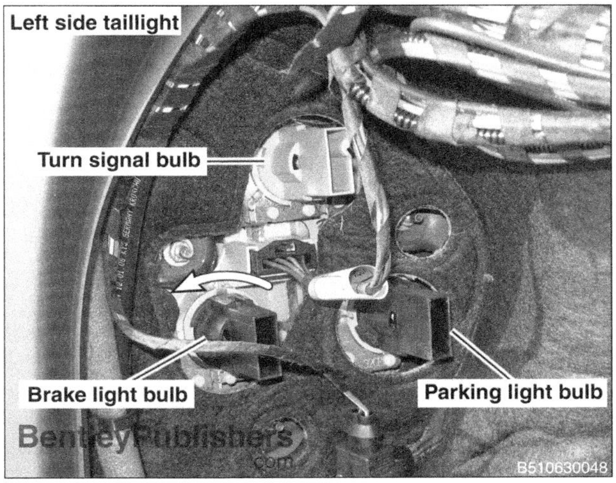

Working at taillight socket:

- Remove socket to replace bulb(s). Do not touch bulb surface with bare finger. Oil from skin can cause bulb to fail prematurely.
- Installation is reverse of removal.

Inboard taillight bulbs

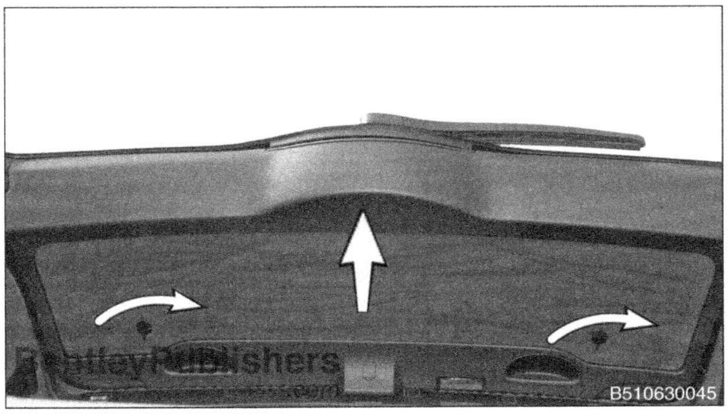

Rotate locks (**curved arrows**) and open tailgate trim panel.

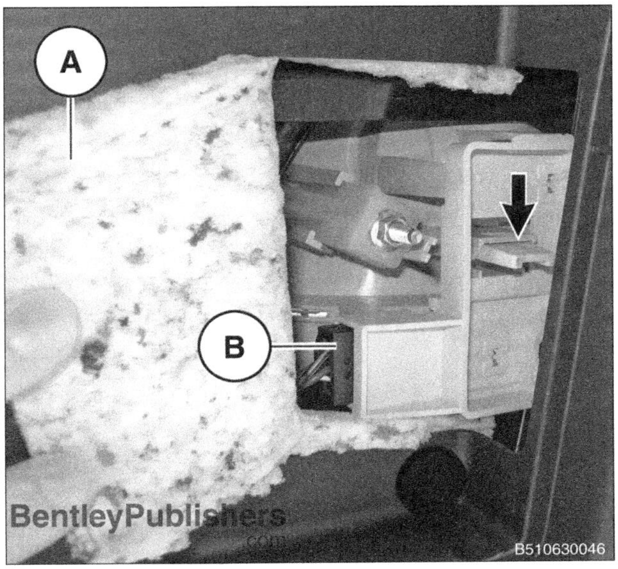

Working at taillight socket:

- Peel back foam access cover (**A**).
- Disconnect electrical connector (**B**).
- Press (**arrow**) locking clip to unlock socket.
- Remove socket to replace bulb(s). Do not touch bulb surface with bare finger. Oil from skin can cause bulb to fail prematurely.
- Installation is reverse of removal.

BentleyPublishers.com
BentleyPublishers.com—All Rights Reserved

Center brake light service

Center brake light bulbs consist of varying numbers of LEDs. In case of bulb failure, replace the entire LED strip.

Sedan, center brake light

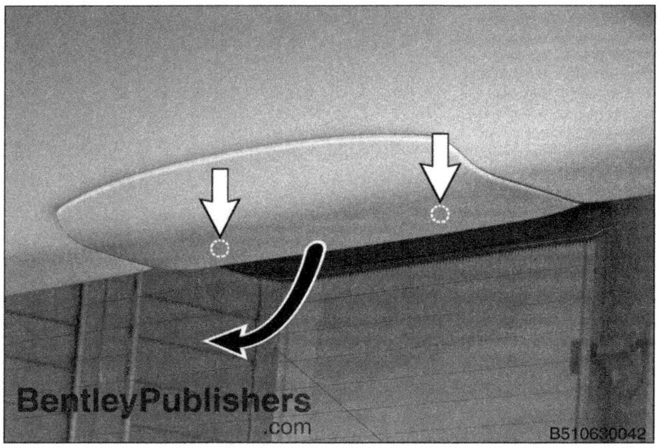

- Working at top of rear window, pull off center brake light cover. Cover is held by plastic clips (**white arrows**).

- Working at center brake light assembly:
  - Detach electrical connector (**A**).
  - Press (**arrows**) locking clips to unlock assembly.
  - Remove assembly to replace LED strip.
- Installation is reverse of removal.

Sports Wagon, center brake light

- Remove rear spoiler. See **412 Trunk Lid, Tailgate**.

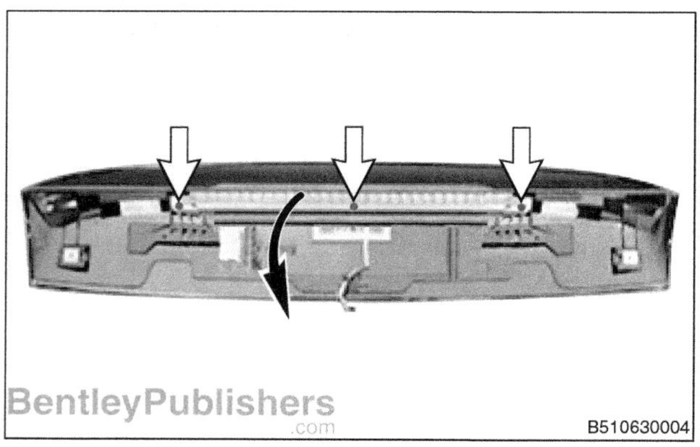

- Remove screws (**white arrows**) and remove brake light assembly from spoiler.
- Replace LED strip.
- Installation is reverse of removal.

BentleyPublishers.com
BentleyPublishers.com–All Rights Reserved

License plate light bulb, replacing

B510630005

◄ Use small screwdriver to carefully pry out license plate light. Detach electrical connector.

– Replace bulb. Do not touch bulb surface with bare finger. Oil from skin can cause bulb to fail prematurely.

– Installation is reverse of removal.

INTERIOR LIGHTS

Interior light bulb replacement

Dome light bulbs

This procedure applies to front dome lights and rear dome light assembly.

B510630006

◄ Use plastic prying tool to lever off dome light from roof switch center.

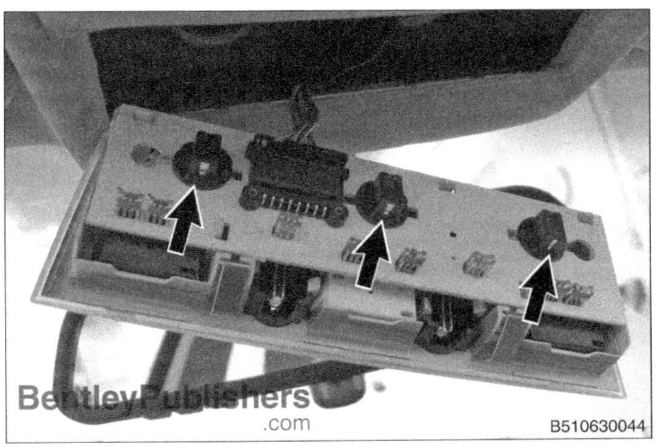

B510630044

◄ Light bulb sockets (**arrows**) are accessible from back of unit. Outer bulbs are reading lights, Inner bulb is dome light. Replace as necessary

– Installation is reverse of removal.

BentleyPublishers.com
BentleyPublishers.com–All Rights Reserved

Glove compartment light bulb

◄ Open glove compartment and use plastic prying tool to lever off light fixture.

– Replace light bulb.

– Installation is reverse of removal.

Door courtesy light bulb

◄ Working at bottom of door trim panel, use plastic prying tool to remove courtesy light.

– Replace bulb.

– Installation is reverse of removal.

Trunk light bulb

◄ Open trunk and use plastic prying tool to lever off light fixture.

– Replace light bulb.

– Installation is reverse of removal.

BentleyPublishers.com
BentleyPublishers.com—All Rights Reserved

Cargo compartment light bulb

- Open tailgate and use plastic prying tool to lever off light fixture.
- Replace light bulb.
- Installation is reverse of removal.

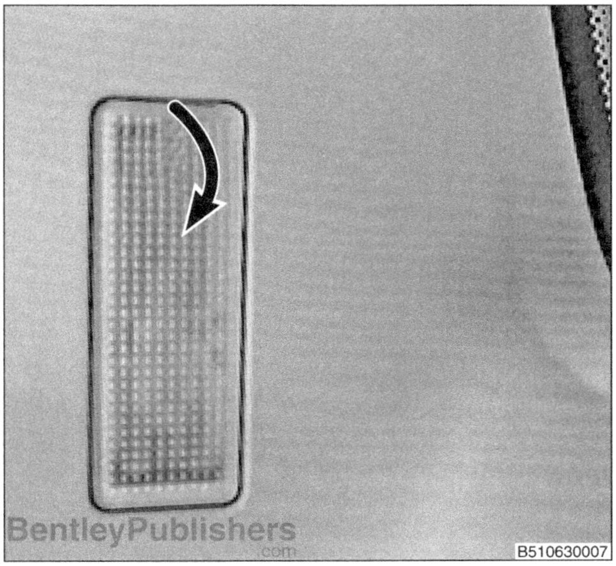

D-pillar light bulb

- Open tailgate and use plastic prying tool to lever off light fixture.
- Replace light bulb.
- Installation is reverse of removal.

BentleyPublishers.com
BentleyPublishers.com—All Rights Reserved

BentleyPublishers.com
BentleyPublishers.com—All Rights Reserved

640 Heating and Air-conditioning

GENERAL

This repair group covers heating and air-conditioning operation and component replacement.

If the air-conditioning system requires evacuation and recharging of refrigerant, use equipment specially designed for A/C system service and follow the equipment manufacturer's instructions.

See also:

- **020 Maintenance** for cabin microfilter service
- **170 Radiator and Cooling System** for cooling system service and electric cooling fan removal and installation
- **600 Electrical System–General**

A/C system fluids

The air-conditioning refrigerant used in modern automobiles is R134a, known chemically as *tetrafluoroethane*. Strict regulations govern the handling and disposal of automotive refrigerant. Be sure to read **Warnings and Cautions** in this repair group if working with A/C and heating components.

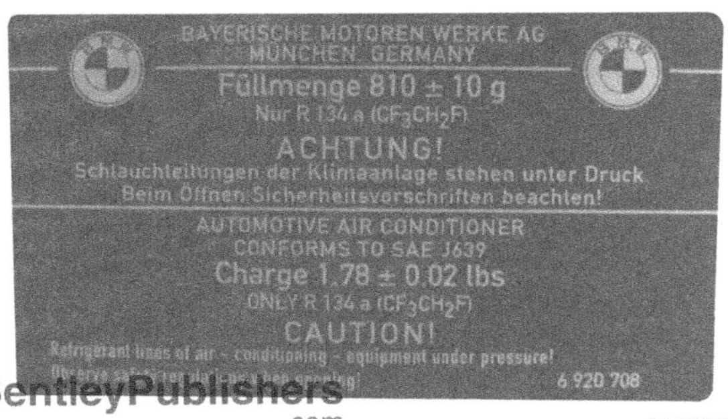

BentleyPublishers.com
B510640002

A sticker in the engine compartment gives the manufacturer's recommendation for refrigerant capacity.

Table a. A/C system refrigerant capacity	
R134a	810 ± 10 grams (28.5 ± 0.3 oz)

A synthetic oil, *polyalkylene glycol* (PAG), is used to lubricate the A/C compressor. This type of oil is highly hygroscopic (absorbs water). Be sure to immediately reseal an opened container after use.

640

BentleyPublishers.com
BentleyPublishers.com–All Rights Reserved

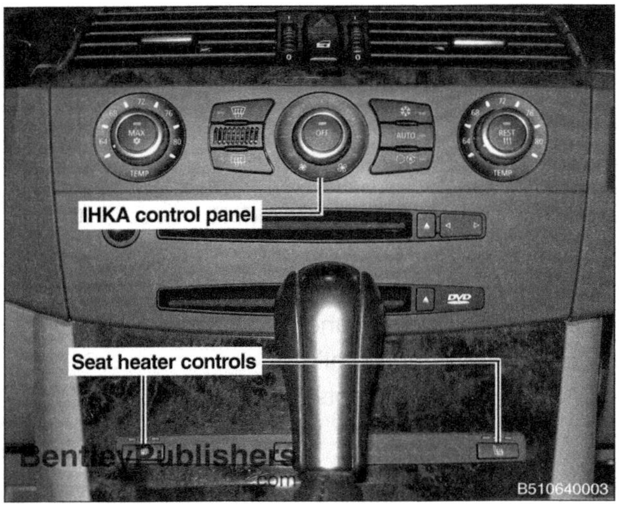

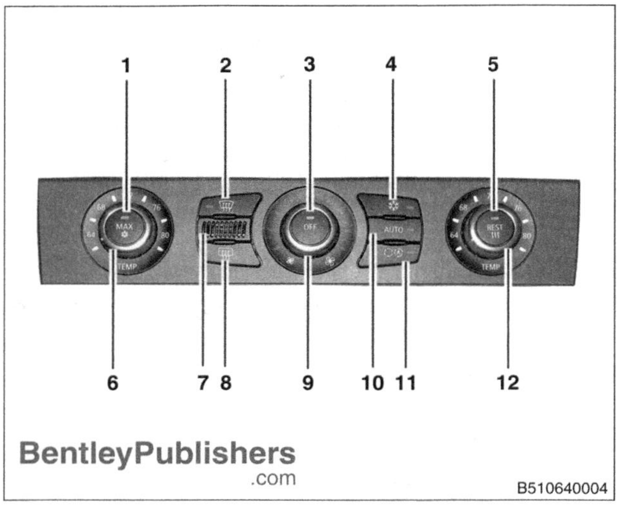

IHKA basics

5 Series models come equipped with an integrated automatic climate control system (IHKA). The dashboard mounted control panel shares functions with iDrive.

IHKA includes the following features:

- Dual zone control system allows driver and passenger to control temperature settings separately.
- Residual heat (REST) function allows brief periods of cabin-heating with the engine OFF.
- Condensation sensor at base of interior rear view mirror.
- Nine air control flap stepper motors (seven are identical).
- Stepper motors control via LIN-bus.
- Seat heater control.
- Rear window defogger control.
- Clutchless A/C compressor.

IHKA control panel

1. Maximum A/C setting
2. Windshield air distribution
3. OFF button
4. A/C button
5. Residual heat (REST) button
6. Driver temperature setting display
7. Interior air temperature sensor intake vent
8. Rear window defogger button
9. Blower speed
10. Automatic program button
11. Recirculation control button
12. Passenger temperature setting

IHKA control panel functions

Climate control sensor data is processed in the IHKA control panel. Driver and passenger climate control settings and requests are also entered in the panel. However, the panel does not directly control all functions and components but rather makes use of other control modules for this purpose. The IHKA control panel directly monitors the following:

- Central information display (CID) and iDrive controls.
- Solar sensor input.
- Condensation sensor.
- Rear air distribution knob (rheostat) signal.
- Ventilation, footwell, and evaporator temperature sensor signals.
- Condensation sensor signal.

BentleyPublishers.com
BentleyPublishers.com—All Rights Reserved

IHKA CID control

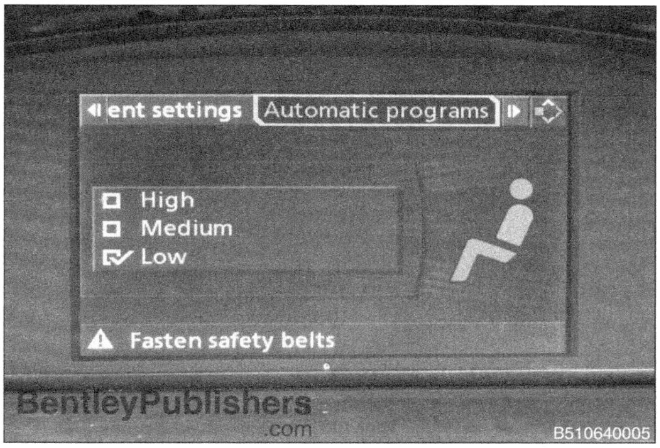

Extended HVAC functions such as air stratification are controlled with the iDrive controller and CID.

Car communication computer (CCC) processes IHKA menu and submenu signals and displays them in the dashboard mounted central information display (CID).

Central information display (CID). Use iDrive controller to activate the following functions and display selections in CID:

- Air distribution setting. Positions for defrost flaps, ventilation flaps and footwell flaps may be individually set in air distribution submenu.
- Automatic IHKA mode. Climate control intensity (low, medium or high) may be set in automatic mode submenu.

IHKA additional controls

Engine control module (ECM) climate control functions are:

- Actuates engine cooling fan for cooling of A/C condenser.
- Interfaces with IHKA control panel via bus network for compressor ON signal.
- Interfaces with IHKA control panel for operation of electric coolant pump for residual heat feature, (N52, N54 engine).

Car access system (CAS) assigns a personal identification code to each remote control key. Personal identification codes are transmitted to IHKA control panel via K-CAN. When vehicle enters sleep mode, current climate control settings are stored in IHKA control panel for remote control key in use at that time.

When vehicle is unlocked via remote control key, settings for this key are called up and executed.

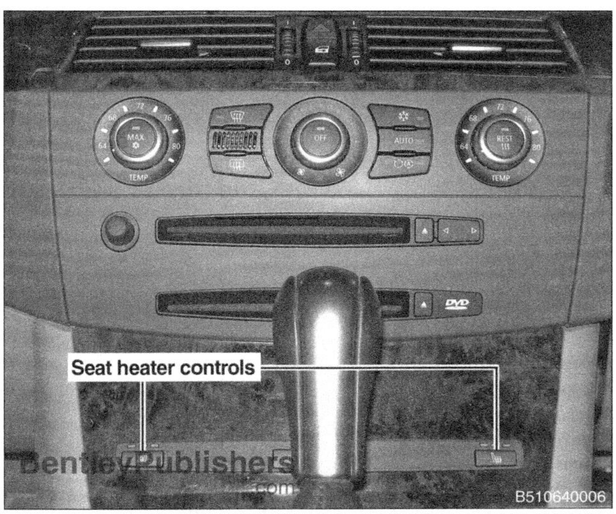

Center console switch cluster (SZM). Seat heater switches in SZM are connected directly to CID and both seat control modules.

Central information display (CID). Use iDrive controller to choose seat cushion and backrest heat distribution.

640

BentleyPublishers.com
BentleyPublishers.com—All Rights Reserved

IHKA sensors

Interior temperature sensor, mounted on IHKA control panel, samples cabin air by using a small electrical fan (non-replaceable) to suck air through the grill on the front face of the control panel. Sensor output is used by the IHKA panel to modulate interior temperature.

Evaporator temperature sensor, mounted close to the evaporator, transfers data directly via cable to IHKA control panel.

Two heater core temperature sensors, mounted close to the heater core, transfer data directly via cable to IHKA control panel.

Outside temperature sensor is hard wired to the instrument cluster which then sends a signal via K-CAN to the IHKA control panel.

Vent flap temperature sensors are 3 NTC type sensors wired directly to the IHKA control panel:

- Stratified temperature sensor
- Ventilation temperature sensor
- Footwell temperature sensor

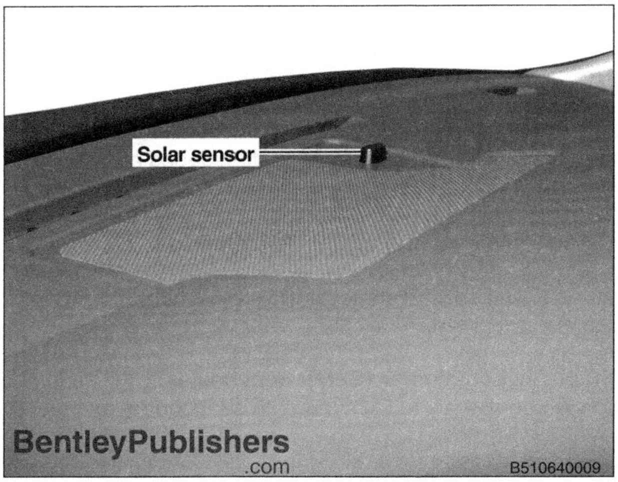

◄ **Solar sensor**, located at the front of the dashboard under the windshield, is hard wired to the IHKA control panel

Refrigerant pressure sensor is in the pressure line between the A/C condenser and the evaporator. In case of excessively high or low refrigerant pressure, A/C compressor output is reduced via signal from IHKA control panel to ECM. Sensor power is provided by IHKA.

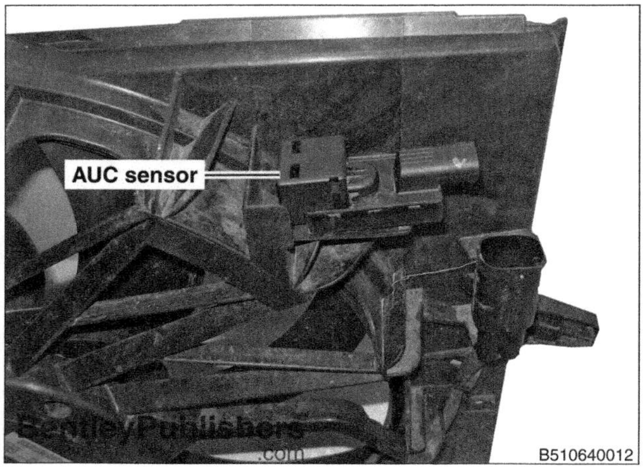

◄ **Air quality (AUC) sensor**:

- To 2006: mounted to the engine cooling fan housing,
- From 2006: mounted to cabin microfilter housing,

is controlled and powered by IHKA. AUC sensor detects the following:

- Carbon monoxide (CO)
- Hydrocarbons (HC)
- Oxides of nitrogen (NO_X).

BentleyPublishers.com
BentleyPublishers.com—All Rights Reserved

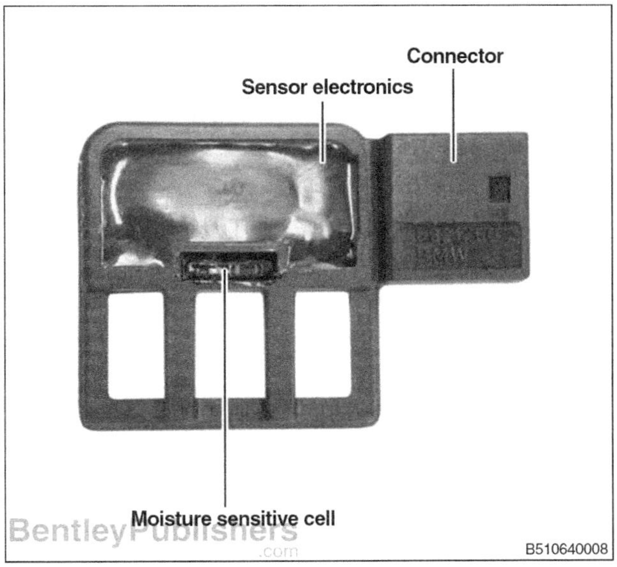

Condensation sensor, mounted under the rain and light sensor (RLS) on interior mirror base at top of windshield.

When moisture is detected on the windshield by the condensation sensor, the following occurs if IHKA system is set on automatic operation:

- Defroster flaps open further.
- Fresh air flaps open 100%.
- Blower speed is increased.
- Footwell flaps are closed.
- Temperature setting increases.
- Evaporator temperature threshold goes to minimum.

IHKA outputs

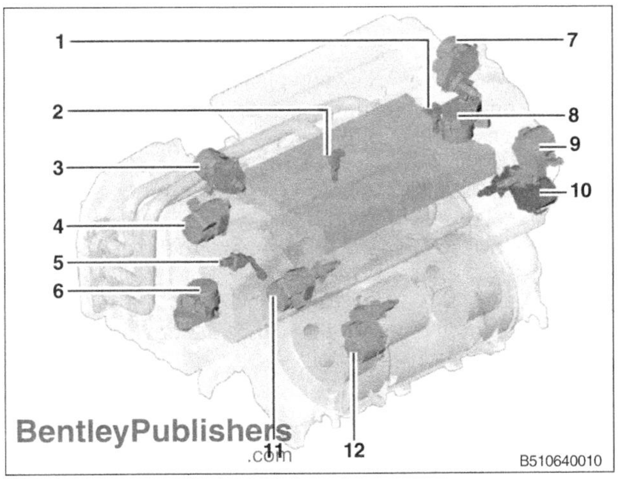

Stepper motors in climate control housing under the dashboard:

1. Right heater core temperature sensor
2. Left heater core temperature sensor
3. Left footwell flap stepper motor
4. Defroster flaps stepper motor
5. Evaporator temperature sensor
6. Left fresh air – recirculation flap stepper motor
7. Right footwell flap stepper motor
8. Right fresh air – recirculation flap stepper motor
9. Right dashboard ventilation flaps stepper motor
10. Front air distribution flap stepper motor
11. Left dashboard ventilation flaps stepper motor
12. Center compartment stepper motor

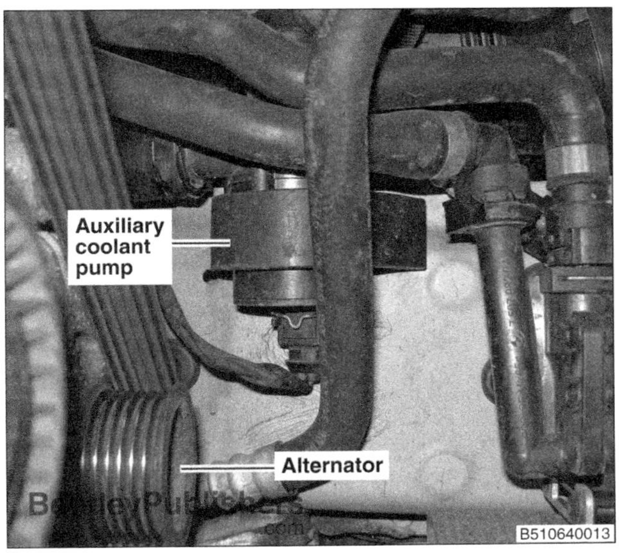

Auxiliary coolant pump is mounted to the driver side frame rail near the alternator. It supplies hot coolant to heater core when engine is idling.

BentleyPublishers.com—All Rights Reserved

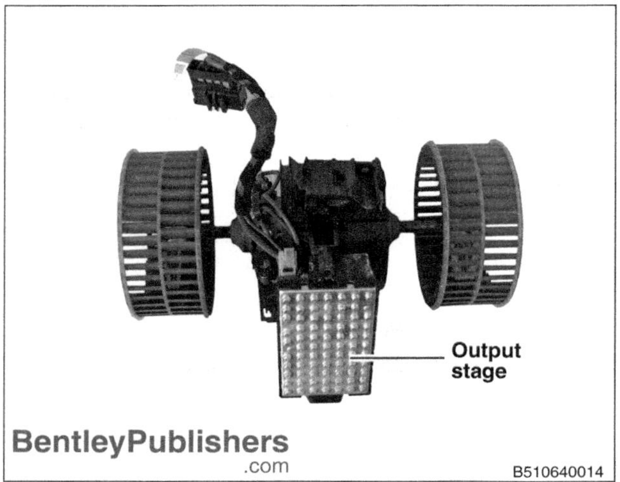

B510640014

Blower and **blower output stage** are mounted in the climate control housing center dashboard, behind IHKA control panel.

Blower output stage is activated via pulse-width modulation by IHKA control panel. Operating power for the blower motor is supplied by fuse 3 (40A), located in the front fuse panel.

Heater valve controls amount of heated engine coolant sent to the heater core. Blend flaps are used to control the temperature settings on the two sides of the dual zone climate control system.

Heater valve is opened via a spring. It is operated via band pulse width modulated signal from IHKA.

A/C compressor is clutchless. A swash plate in the compressor housing varies the amount of compression on the refrigerant. The compressor valve is controlled and powered by IHKA.

Seat heating is controlled and powered by seat control module of the respective seat. Request for seat heating activation is initiated from center console switch cluster (SZM) directly to the seat control module via K-CAN.

Rear window defogger is controlled by IHKA. Defogging request activates a relay in the rear fuse panel.

B309640015

N52, N54 engine: Electric coolant pump is electronically controlled by the engine control module (ECM). When residual heat (REST) function is activated, IHKA control panel interfaces with ECM to activate the coolant pump, circulating hot or warm coolant through the heater core to warm up the passenger cabin.

BentleyPublishers.com
BentleyPublishers.com—All Rights Reserved

IHKA non-electrical components

Refrigerant circuit consists of A/C compressor, expansion valve, evaporator and condenser. The system features:

- Aluminum refrigerant lines.
- Receiver-drier (with replaceable desiccant) integrated in the condenser.
- Clutchless compressor controlled by IHKA logic.
- Expansion valve accessible from engine compartment.

Heating circuit. Heated engine coolant is pumped by coolant pump(s) through heater valve to heater core (heat exchanger). From here, coolant is routed back to engine and coolant thermostat.

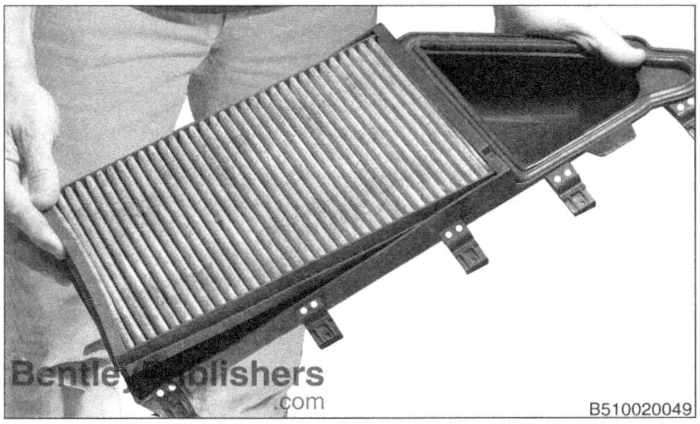

Cabin microfilters. The microfilter with activated carbon is accessible from the engine compartment. For microfilter replacement, see **020 Maintenance**.

To 2008: Filter service is indicated by condition based service (CBS) system. Microfilter condition is monitored by the IHKA control panel using a calculation model (algorithm) based on the following factors:

- Outside temperature.
- Signal from rain and light sensor.
- Signal from solar sensor.
- Blower voltage.
- Air quality, recognized from frequency of recirculation mode usage.
- Average vehicle road speed.
- Service interval display (SIA) timer.

IHKA panel communicates with instrument cluster via K-CAN to detect and report the following:

- Odometer reading.
- Microfilter remaining life in percent.
- Time remaining until next microfilter service.

From 2008: Microfilter replacement is recommended every second oil service.

BentleyPublishers.com
BentleyPublishers.com–All Rights Reserved

Warnings and Cautions

WARNING—
- *Do not discharge or charge the A/C system without proper equipment and training. Damage to the vehicle and personal injury may result.*
- *Wear hand and eye protection (gloves and goggles) when working around the A/C system.*
- *If refrigerant comes in contact with your skin or eyes:*
  -Do not rub skin or eyes.
  -Immediately flush skin or eyes with cool water for 15 minutes.
  -Rush to a doctor or hospital.
  -Do not attempt to treat yourself.
- *Work in a well ventilated area. Switch on exhaust / ventilation systems when working on the refrigerant system.*
- *Do not expose any component of the A/C system to high temperatures (above 80°C or 176°F) or open flames. Excessive heat causes a pressure increase which could burst the system.*
- *Keep refrigerant away from open flames. Poisonous gas is produced if it burns. Do not smoke near refrigerant gases for the same reason.*
- *The A/C system is filled with refrigerant gas which is under pressure. Pressurized refrigerant in the presence of oxygen may form a combustible mixture. Do not introduce compressed air into any refrigerant container (full or empty).*
- *Electric welding near refrigerant hoses causes R-134a to decompose. Discharge system before welding.*
- *At normal operating temperature the cooling system is pressurized. Allow the system to cool as long as possible before opening (a minimum of one hour), then release the cap slowly to allow safe release of pressure.*

CAUTION—
- *In the United States, any person who services a motor vehicle air-conditioner must, by law, be properly trained and certified, and use approved refrigerant recycling equipment. Technicians must complete an EPA-approved recycling course to be certified.*
- *Comply with state and local governments which may have additional requirements regarding air-conditioning servicing.*
- *It is recommended that all A/C service be left to an authorized BMW dealer or other qualified A/C service facility.*
- *Do not top off a partially charged refrigerant system. Discharge system, evacuate and then recharge system.*
- *The mixture of refrigerant oil (PAG oil) and refrigerant R-134a attacks some metals and alloys (for example, copper) and breaks down certain hose materials. Use only hoses and lines that are identified with a green mark (stripe) or marked with R-134a.*
- *Immediately plug open connections on A/C components and lines to prevent dirt and moisture contamination.*
- *Do not steam clean the A/C condenser or evaporator. Use only cold water or compressed air.*
- *To avoid damaging plastic interior trim, use a plastic prying tool or a screwdriver with the tip wrapped with masking tape.*

BentleyPublishers.com
BentleyPublishers.com—All Rights Reserved

IHKA Component Replacement

IHKA control panel, removing and installing

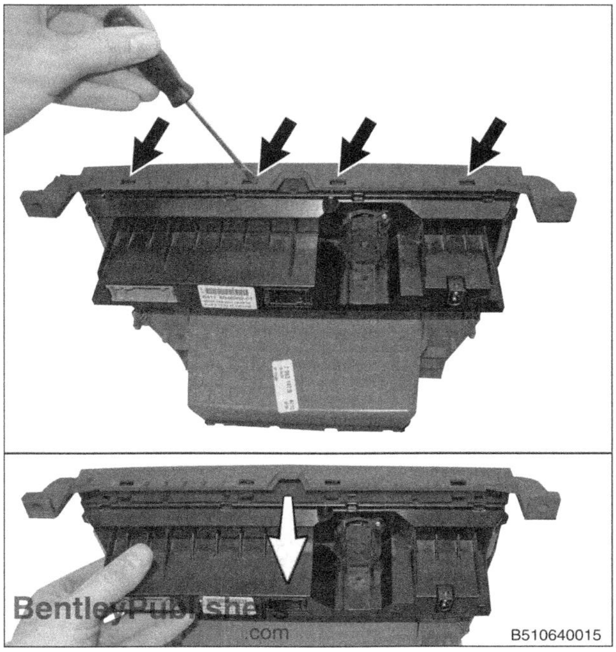

- Remove center dashboard trim panel. See **513 Interior Trim**.
- Release control panel retaining tabs (**arrows**) and remove control panel from center dashboard trim.
- Installation is reverse of removal. Use BMW scan tool to program and code new panel. See **600 Electrical System–General**.

Stepper motors

The IHKA system utilizes nine stepper motors that are wired in series for climate control operation. IHKA control panel operates stepper motors via LIN-bus.

Stepper motors can be replaced without removing the dashboard or climate control housing.

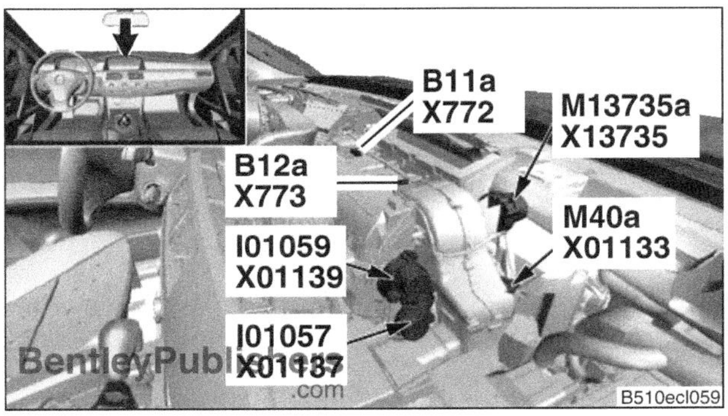

Top of IHKA housing:

- **B11a** Heater core temperature sensor, left side
- **B12a** Heater core temperature sensor, right side
- **I01057** Fresh air flap stepper motor
- **I01059** Right ventilation flap stepper motor
- **M40a** Right fresh-recirculated air flap stepper motor
- **M13735a** Right footwell flap stepper motor
- **X772** Connector
- **X773** Connector
- **X01133** Connector
- **X01137** Connector
- **X01139** Connector

640

BentleyPublishers.com
BentleyPublishers.com–All Rights Reserved

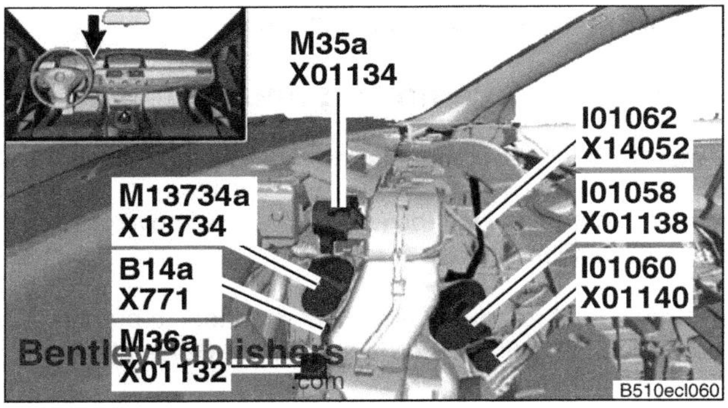

Top of IHKA housing:

- **B14a** Evaporator temperature sensor
- **I01058** Left ventilation flap stepper motor
- **I01060** Rear compartment mixing flap stepper motor
- **I01062** Electric auxiliary heater
- **M35a** Defroster flap stepper motor
- **M36a** Left fresh-recirculated air flap stepper motor
- **M13734a** Left footwell flap stepper motor
- **X771** Connector
- **X01132** Connector
- **X01134** Connector
- **X01138** Connector
- **X01140** Connector
- **X13734** Connector
- **X14052** Connector

Stepper motor, removing and installing

– To gain access to left side stepper motors, remove left lower dashboard trim (pedal cluster trim). See **513 Interior Trim**.

– To gain access to right side stepper motors:
- Remove right lower dashboard trim. See **513 Interior Trim**.
- Remove glove compartment. See **513 Interior Trim**.

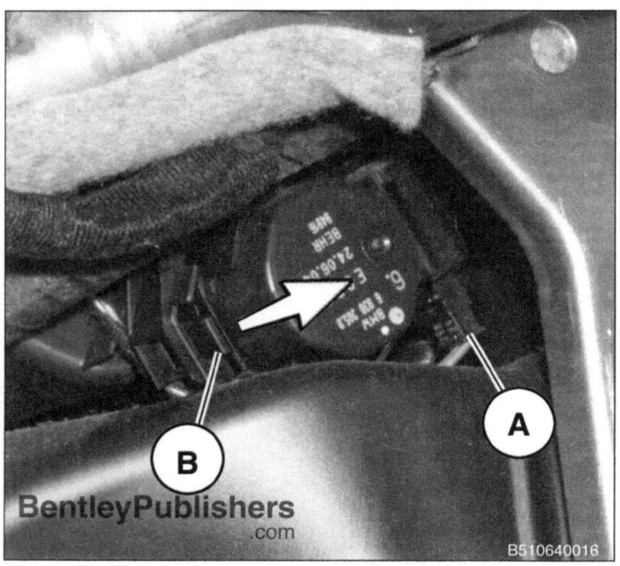

To remove stepper motor:

- Detach electrical connector (**A**).
- Press locking tab (**B**) and remove motor from housing in direction of **arrow**.

BentleyPublishers.com
BentleyPublishers.com—All Rights Reserved

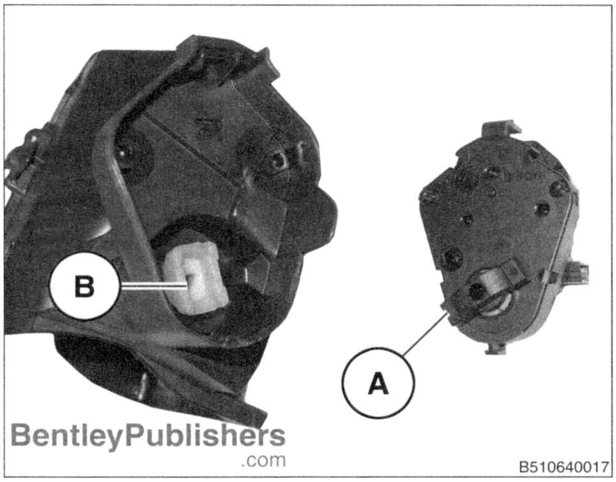

During installation:
- Do not twist motor drive (**A**). Instead align shaft of flap (**B**) with stepper motor.
- Listen for stepper motor to snap into place.

– Reassemble vehicle. Check operation of stepper motor.

Blower, removing and installing

– Remove center console. See **513 Interior Trim**.

– Remove left side lower dashboard trim. See **513 Interior Trim**.

– Remove glove compartment. See **513 Interior Trim**.

– Remove radio. See **650 Radio**.

– Remove CID display. See **650 Radio**.

Remove center air duct in direction of **arrow**.

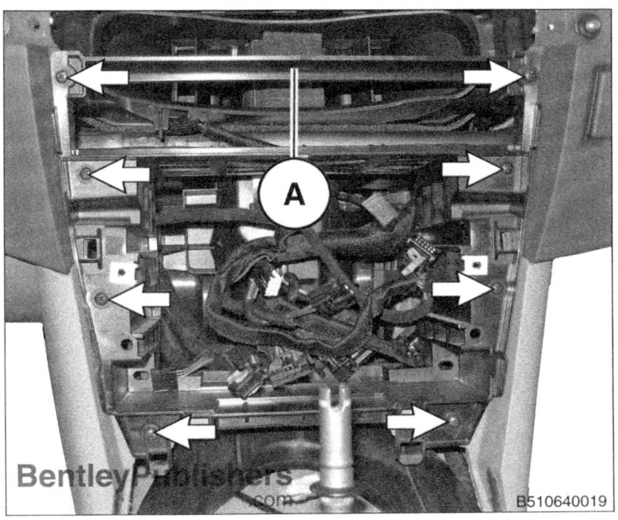

Remove fasteners (**arrows**) then remove bracket (**A**) from dashboard.

BentleyPublishers.com
BentleyPublishers.com–All Rights Reserved

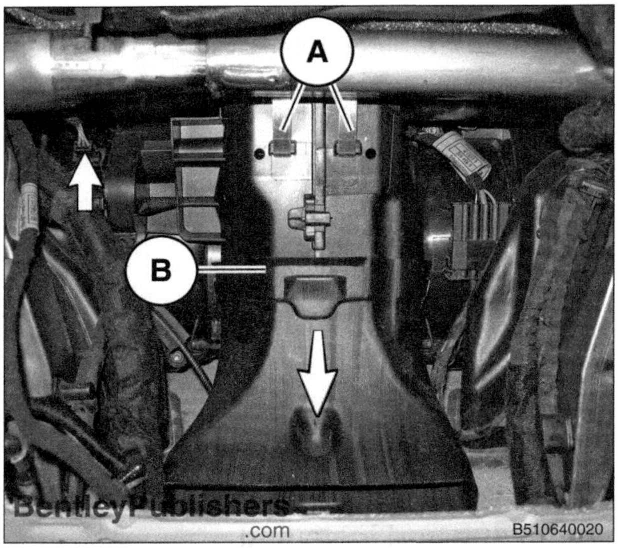

Disconnect electrical connectors (**arrow**). Remove clips (**A**) and remove air duct (**B**) toward rear of vehicle.

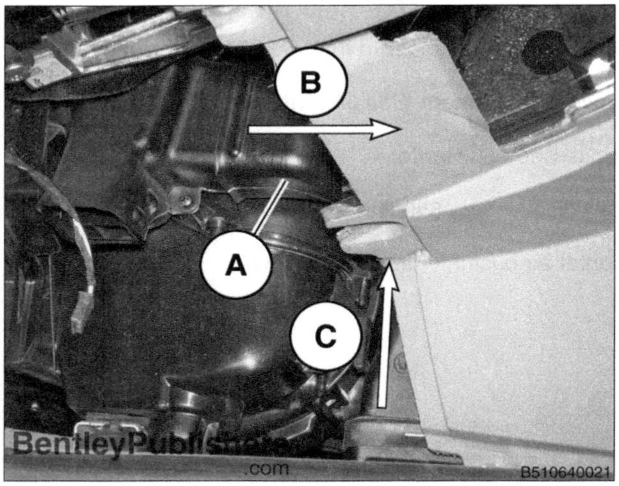

Disconnect left side air duct (**A**) in area **B** then in area **C**.

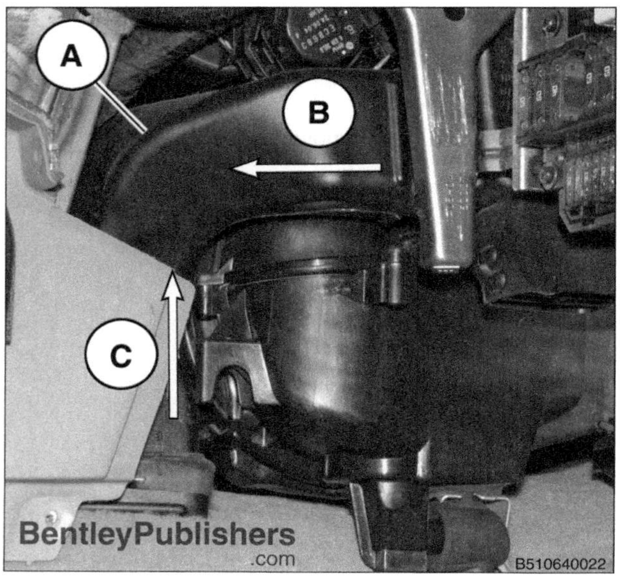

Disconnect right side air duct (**A**) in area **B** then in area **C**.

BentleyPublishers.com
BentleyPublishers.com—All Rights Reserved

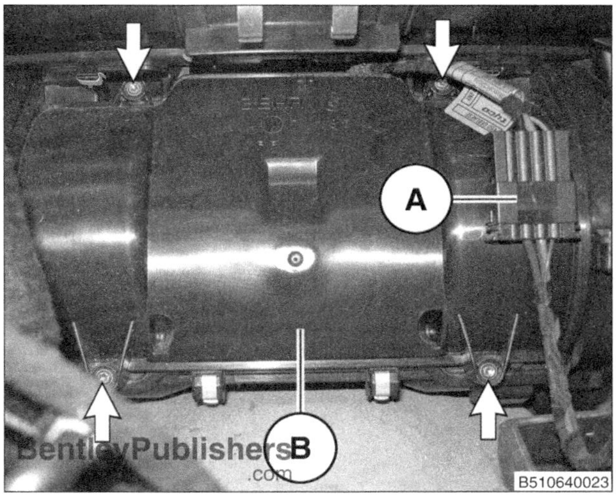

◄ Disconnect electrical connector (**A**) and remove from blow motor cover. Remove fasteners (**arrows**) and remove blow motor cover (**B**).

◄ Remove blower motor fasteners (**arrows**) and remove blower motor from heater housing.

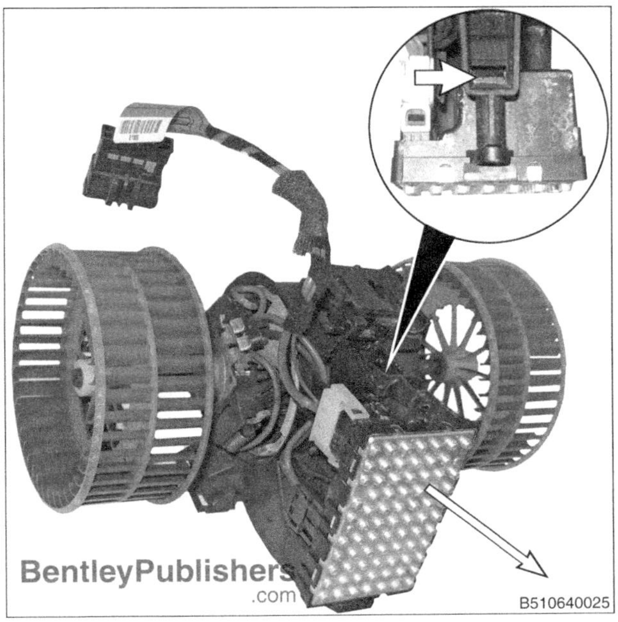

◄ If necessary, separate blower final stage using a flat blade screwdriver.

– Installation is reverse of removal. Remember to:
  - Properly seat final stage.
  - Check that blower motor cages do not rub on heater housing.
  - Properly install all housing covers.
  - Seat air ducts correctly.

BentleyPublishers
BentleyPublishers.com–All Rights Reserved

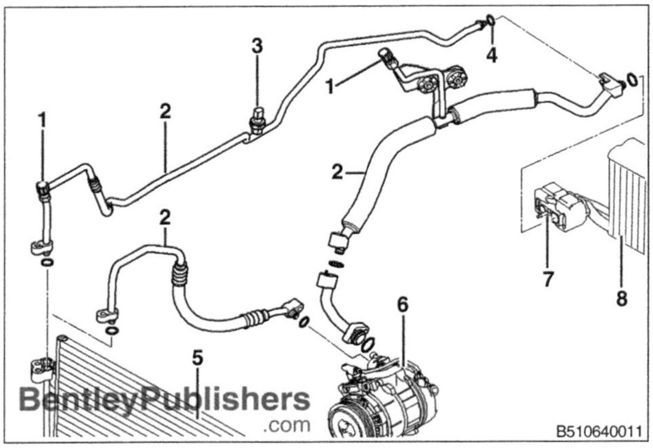

Air-conditioning components

1. Refrigerant service port
2. A/C line
3. Pressure sensor
4. O-ring
5. Condenser
6. A/C compressor
7. Expansion valve
8. Evaporator

BentleyPublishers.com
BentleyPublishers.com—All Rights Reserved

650 Radio

GENERAL

This section covers BMW factory-installed sound and navigation systems.

Radio

The two radio versions supplied with 5 Series models are both on the MOST (optical cable) bus and have CD playing capabilities.

The base radio is referred to as M-ASK. It is the central control module of the information and communication system. Up to four control modules can be combined into M-ASK. The basic 180 watt radio may be upgraded to Top-HiFi Logic7 with 420 watts of power.

The Top-HiFi Logic7 system comes with 420 watts of power standard. In addition to playing standard music CDs, MP3 audio files can be played due to an upgarded CD-ROM drive.

The car communication computer (CCC) serves as the navigation system. The CCC is controlled by the iDrive controller and utilizes the navigation video screen or central information display (CID) as a visual monitor.

MOST-bus (media oriented systems transport)

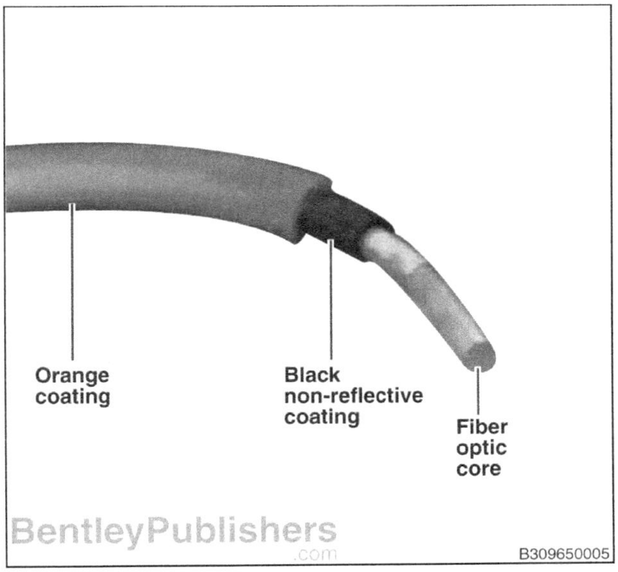

Sound system and navigation components are connected using MOST-bus, a fiber optic network.

650

BentleyPublishers.com—All Rights Reserved

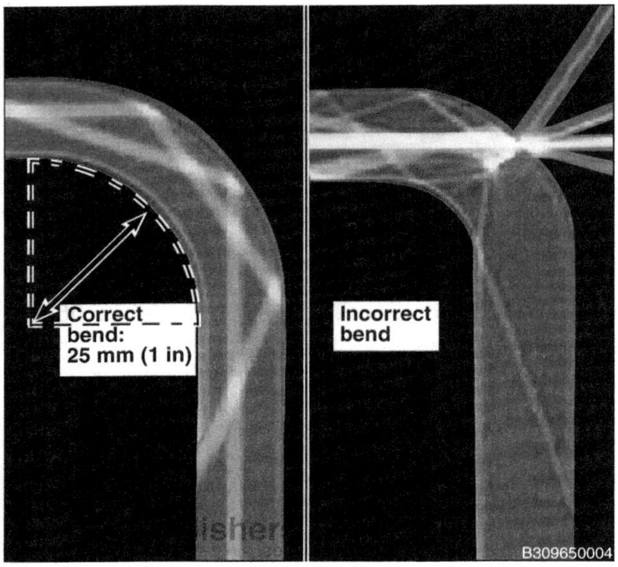

Make sure any bend in MOST-bus fiber optic harness has a radius of 25 mm (1 in) or greater.

> **CAUTION—**
> - *Signal transmission through a fiber optic harness that is bent excessively is impaired.*

Bluetooth radio standard

Bluetooth is a short-range radio standard which allows up to 8 bluetooth-compatible devices within the space of up to 10 meters to be networked together. Up to 4 Bluetooth mobile phones can be coupled to the telematics control unit (TCU).

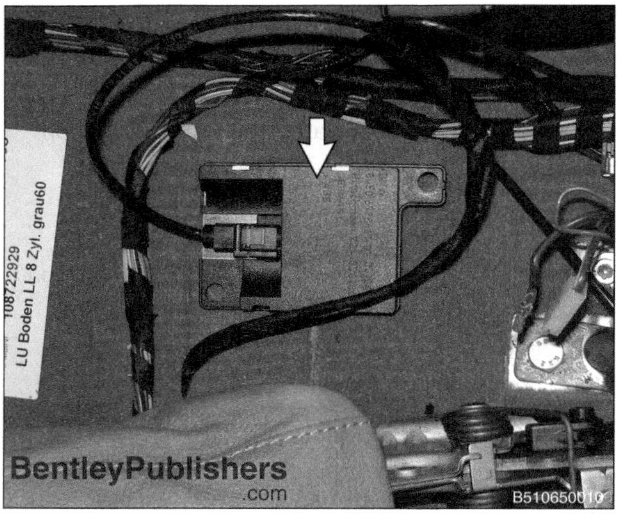

The Bluetooth antenna (**arrow**) is located beneath the center console next to parking brake handle.

The Bluetooth system automatically connects a previously paired phone each time you turn the ignition ON. The first phone found in the search will be connected and one phone can be used at a time.

Bluetooth phone pairing

To use a Bluetooth phone in your vehicle, an initial one-time-only pairing procedure is required.

– Turn ON Bluetooth search mode using iDrive control menu.

– Turn ON Bluetooth search mode in the mobile phone.

– Enter Bluetooth Passkey for verification.
  - A random passkey can be chosen.
  - Enter it on the phone or vehicle, and the other device will verify it.

– Vehicle and device will pair and conect.

See owners manual for additona information on address book integration and hands-free phone operation.

BentleyPublishers.com
BentleyPublishers.com—All Rights Reserved

Satellite digital audio radio service (SDARS)

The satellite tuner converts digital signals from the SDARS antennas into light signals for use on the MOST-bus.

Some benefits of satellite radio are a low degree of interference from outside sources, CD quality sound and national radio stations.

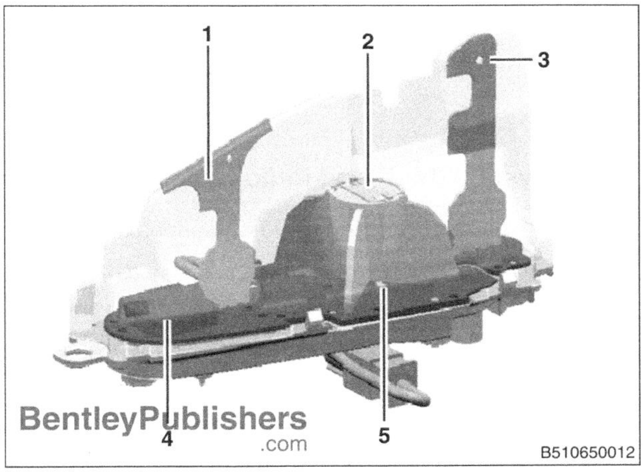

High-frequency digital signals from satellites and stationary antennas are picked up by two SDARS antennas. Separate antenna cables route the signals to the satellite tuner.

1. Telephone antenna 1
2. Telephone antenna 2
3. GPS antenna
4. SDARS for terrestrial reception
5. SDARS for satellite reception

Cautions

CAUTION—
- *Before beginning work on the radio or sound system, verify that the radio is an original equipment BMW radio and that the wiring harness is not modified.*
- *Refer servicing of aftermarket sound equipment to an authorized agent of the equipment manufacturer.*
- *When handling electronic equipment (monitors, control modules, relays), use anti-static tools and techniques to prevent static discharge damage. See* **600 Electrical System–General**.
- *To avoid damaging plastic interior trim, use a plastic prying tool or a screwdriver with the tip wrapped with masking tape.*

650

BentleyPublishers.com
BentleyPublishers.com—All Rights Reserved

Component Locations

Sedan audio equipment

HiFi audio system (basic radio)

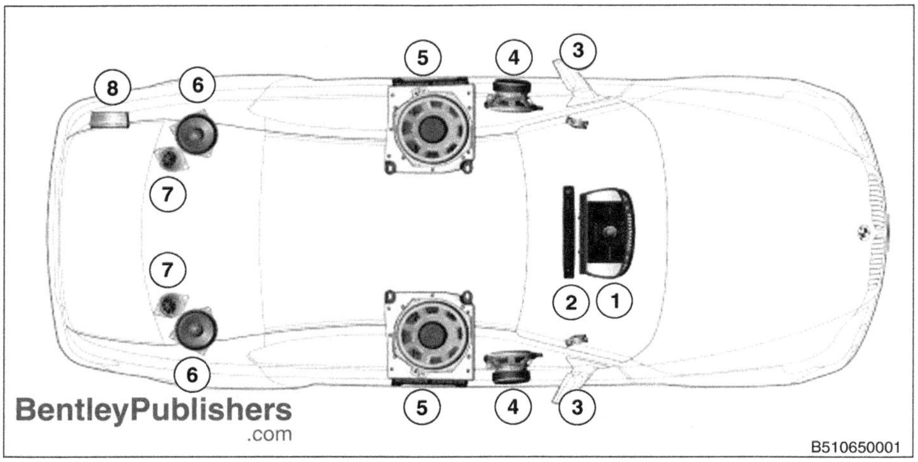

1. Central information display (CID)
2. Radio (M-ASK)
3. Front tweeters
4. Front midrange speakers
5. Floor-mounted woofers
6. Rear midrange speakers
7. Rear tweeters
8. Amplifier

Top-HiFi audio system (high-end radio)

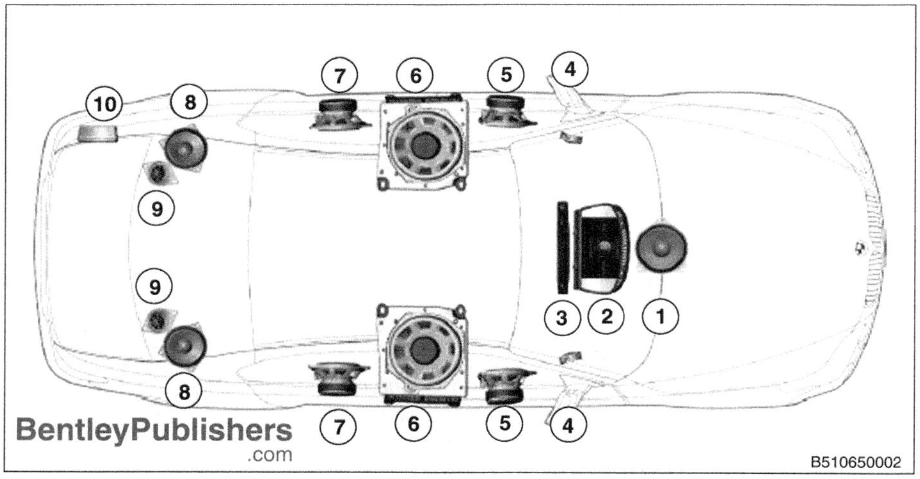

1. Front center midrange speaker
2. Central information display (CID)
3. Radio (M-ASK)
4. Front tweeters
5. Front midrange speakers
6. Floor-mounted woofers
7. Rear door midrange speakers
8. Rear midrange speakers
9. Rear tweeters
10. Amplifier

BentleyPublishers.com
BentleyPublishers.com—All Rights Reserved

Antennas

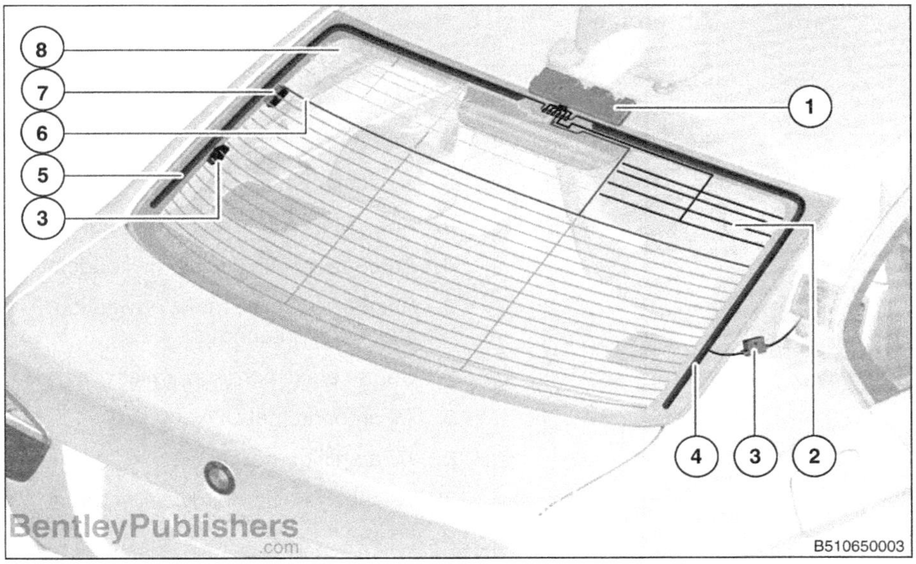

1. Antenna amplifier, diversity module
2. Remote key antenna (FBD)
3. Rear window defogger suppressor filter
4. FM3, TV1 antenna
5. FM2, TV2 antenna
6. FM1 antenna
7. Center brake light filter
8. AM antenna

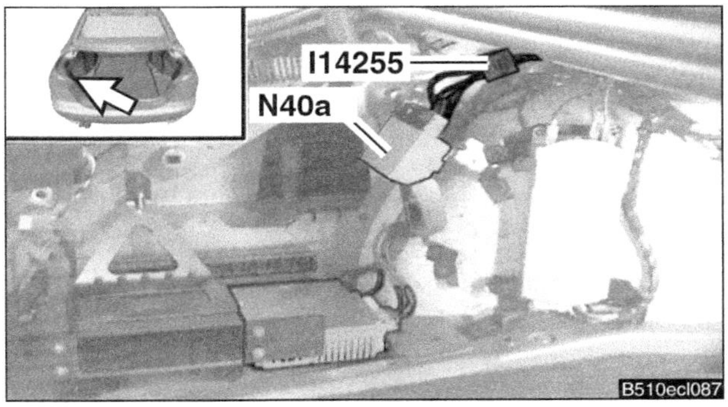

Audio equipment

Left side trunk (Sedan) behind trim:

- **I14255** MOST-bus connector
- **N40a** Sound system amplifier

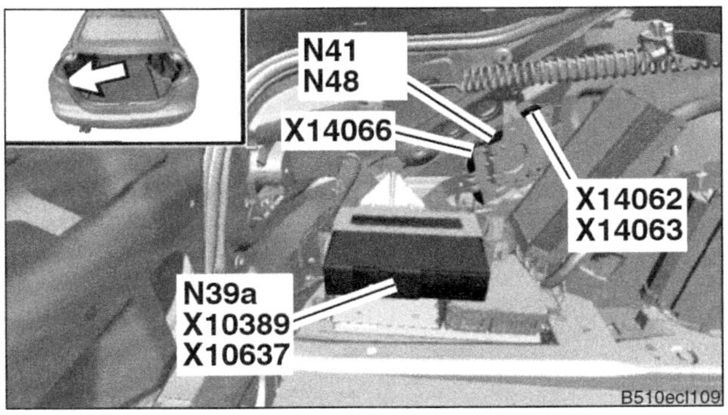

Left side trunk (Sedan) behind trim:

- **N39a** DVD changer
- Options:
  N41 or **N48** Digital tuner
- **X10389** Connector
- **X10637** Connector
- **X14962** Connector
- **X14063** Connector
- **X14066** Connector

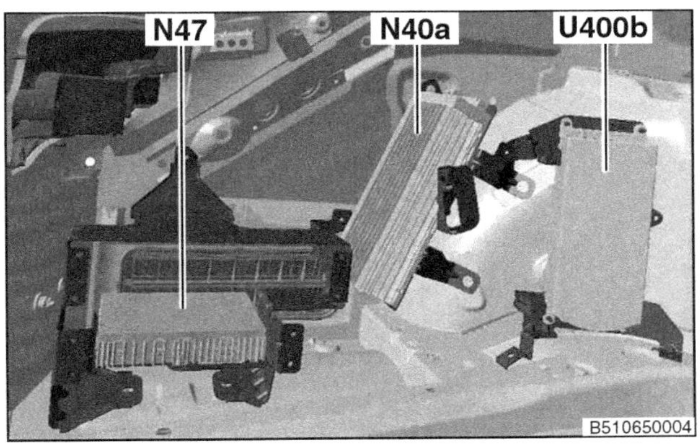

Left side trunk (Sedan) behind trim:

- **N40a** Sound system amplifier
- **N47** Satellite radio
- **U400b** Telematics control module (TCU)

650

BentleyPublishers
BentleyPublishers.com—All Rights Reserved

Sports Wagon audio equipment

Antennas

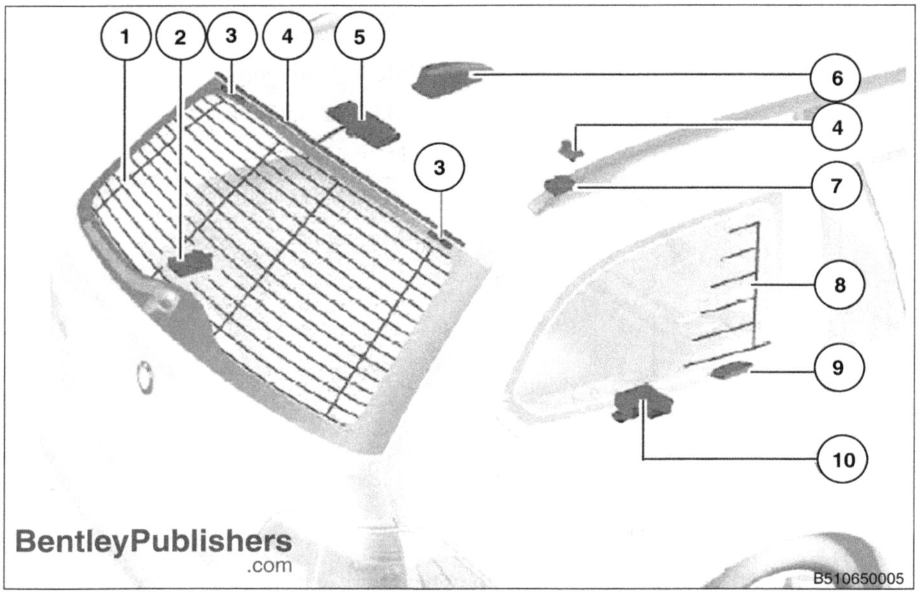

1. Rear window defogger with inegrated FM antennas 1-3
2. Compensator
3. Rejector circuit, ground filter
4. AM, FM4, remote key (FBD) antenna
5. Antenna amplifier, diversity module
6. Sharkfin (satellite radio, navigation, telephone antenna)
7. Center brake light suppressor filter
8. TV antenna (not US)
9. TV amplifier (not US)
10. Telephone emergency call antenna

Sports Wagon audio equipment

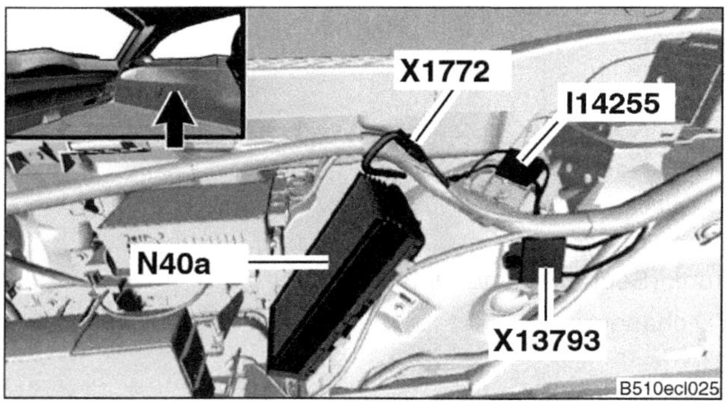

Left side cargo compartment (Sports Wagon) behind trim:

- **I14255** MOST-bus connector
- **N40a** Sound system amplifier
- **X1772** Connector
- **X13793** Ground connection

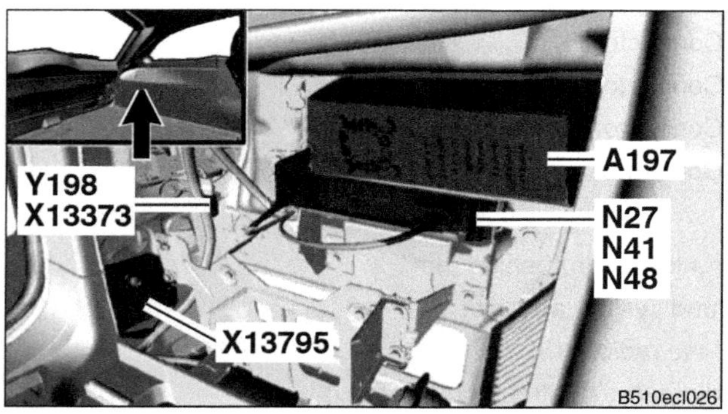

Left rear cargo compartment (Sports Wagon) behind trim:

- **A197** Video module
- Options:
  N27 Mobile phone compensator
  N41 Digital tuner
  N48 Digital tuner
- **X13373** Connector
- **X13793** Ground connection
- **Y198** Tailpipe flap actuator

BentleyPublishers.com
BentleyPublishers.com—All Rights Reserved

Sound System

Radio service mode

Use service mode to check radio functions directly at the radio.

CCC service mode

Information available through the CCC service mode are software, hardware, and system status numbers. This information is primarily intended for equipment developers.

– Using iDrive controller, open Start menu.

– Press and hold iDrive controller for at least 10 seconds.

– Move controller 3 stops to right. Read and follow menus.

– Move controller 3 stops to left. Read and follow menus.

– Move controller 1 stop to right. Read and follow menus.

– Move controller 1 stop to left. Read and follow menus.

– Move controller 1 stop to right. Read and follow menus.

– Press controller once.

– Press Menu button to exit service mode.

CCC reset

– Simultaneously press and hold eject buttons on DVD or CD player and iDrive controller for approximately 10 seconds.
  - Central information display (CID) becomes blank.
  - CCC then restarts.

BentleyPublishers.com
BentleyPublishers.com–All Rights Reserved

Radio, removing and installing

– Disconnect negative (-) battery cable and cover battery terminal to keep cable from accidentally contacting terminal.

> ***CAUTION—***
> • *Prior to disconnecting the battery, read the battery disconnection cautions given in* **001 Warnings and Cautions**.

– Remove center dashboard trim panel. See **513 Interior Trim**.

B510650006

◄ Working below radio face plate, remove mounting screws (**arrows**) and pull radio out of center dashboard.

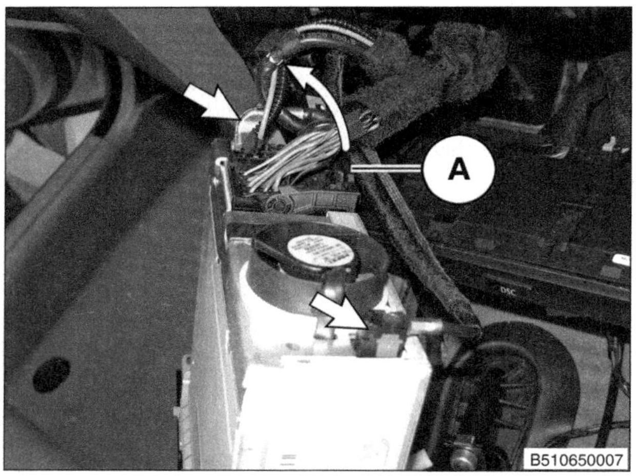

B510650007

◄ Working at rear of radio, remove antenna connectors (**arrows**). Squeeze tab (**A**) to unlock to release remaining electrical connectors in direction of **arrow**.

– Installation is reverse of removal. Check to make sure MOST-bus is not kinked. See **MOST-bus (media oriented systems transport)** in this repair group.

– Using BMW scan tool, code and program new radio. See **600 Electrical System–General**.

BentleyPublishers.com
BentleyPublishers.com—All Rights Reserved

TELEMATICS

Telematics combines information and communications technology such as GPS and telephone to send and receive data via telecommunications devices. Telematics can assist the driver in a wide range of situations and deliver useful information directly to the vehicle. The system consists of several different services which are provided via subscription.

BMW Assist features include the following:

- Vehicle position (can be transmitted if an emergency request is sent).
- Automatic collision notification.
- Enhanced roadside assistance.
- Customer relations.
- Vehicle service status or required inspections.
- Remote door unlock (in case of lost remote key).
- Stolen vehicle recovery.
- Critical calling (allows a limited number of telephone calls when the driver's mobile phone is not available).

BMW Online features include the following:

- Location information for restaurants, hotels, theaters, etc.
- Traffic news, enabling the navigation system to calculate a route that avoids traffic jams.

BMW TeleServices monitors the condition of key vehicle parts and, if required, contacts a BMW service center automatically to arrange for service.

NAVIGATION SYSTEM

Navigation basics

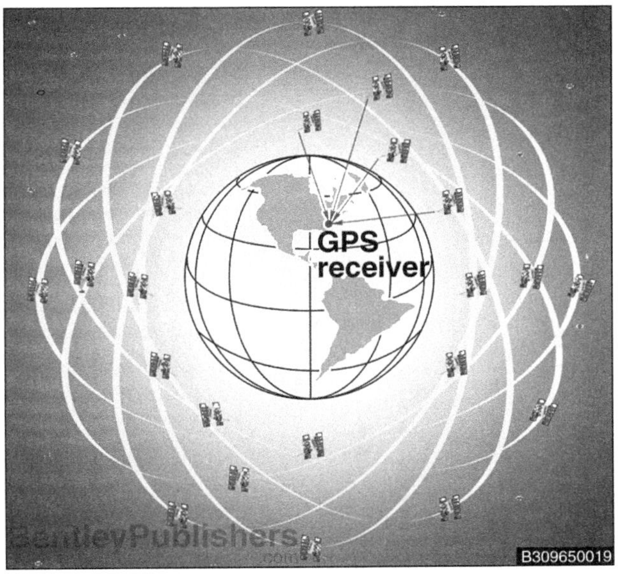

Vehicle navigation systems utilize signals from global positioning system (GPS) satellites.

The 5 Series navigation system operates via the car communication computer (CCC) and features the following range of functions:

- Navigation with map and arrow mode.
- Display on 8.8" color central information display (CID).
- Split-screen view.
- Perspective map presentation.
- Input via iDrive controller with electrically controlled feedback.
- Input optionally controlled by voice commands (SVS).

BentleyPublishers.com
BentleyPublishers.com–All Rights Reserved

Central information display (CID), removing and installing

– Make sure ignition is OFF.

◄ Working at CID screen:

- Remove mounting screws (**arrows**) at top.
- Tilt screen backward (**curved arrow**) and remove.

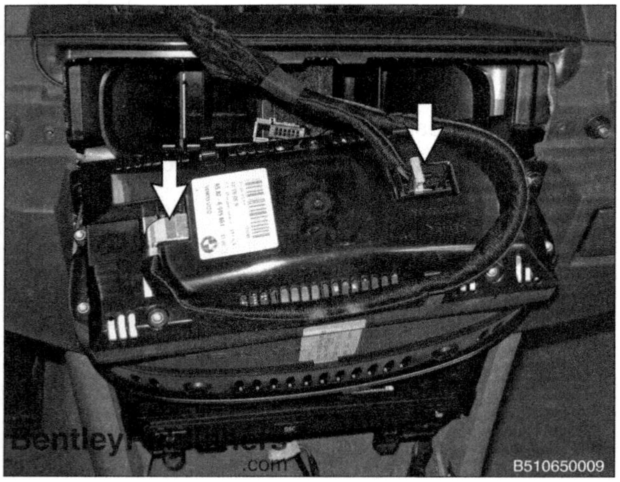

◄ Detach CID screen electrical connectors (**arrows**).

– When installing, route cables without kinking.

– Use BMW scan tool to code and program new CID. See **600 Electrical System–General**.

BentleyPublishers.com
BentleyPublishers.com—All Rights Reserved

720 Seat Belts

TABLE

General

This section covers mechanical repairs to seat belts and seat belt tensioner components. Airbag system electronics integrate seat belt operations and airbag deployment. Airbag system information is covered in **721 Airbag System (SRS)**.

One of two systems are installed in the E60 / E61 cars covered by this manual, depending on model year.

Passive restraint system application	
Year	**Restraint system**
2004 - 2005	Advanced Safety Electronics (ASE)
2006 - 2010	Advanced Crash and Safety Management (ACSM)

See also:

- **513 Interior Trim**
- **520 Seats**

Troubleshooting

Check that seat belt reel locks when driving quickly through curves and during severe braking. The automatic reel does not require any servicing. Replace if it is found faulty.

In case of an accident, follow the seat belt troubleshooting procedures presented in **Table a**.

BentleyPublishers.com
BentleyPublishers.com–All Rights Reserved

Table a. Seat belt troubleshooting		
Problem	**Probable cause**	**Repairs**
Damage to seat belt.	Accident in which bumper impact absorbers were permanently deformed.	Replace complete automatic seat belt. Also check and replace, if necessary, seat belt mounting on car body and seat belt mounting on seat runner.
Belt creased, unraveled, pinched, cut or melted. Belt buckle or belt lock plastic casing worn, damaged or missing.	Seat belts aged or worn.	Replace complete automatic seat belt.
Seat belt buckle not ejected with spring pressure when red button on seat belt lock pressed.	Seat belt lock mechanism worn or damaged.	Replace seat belt buckle.
Seat belt automatic reel does not lock when pulled out suddenly.	Automatic reel defective.	Replace reel assembly.
Seat belt automatic reel jams when pulled out.	Automatic reel loose. Return spring broken inside reel.	Tighten reel mounting bolt. Replace reel assembly.
Seat belt does not retract automatically (see note below).	Automatic reel loose. Return spring broken inside reel.	Tighten reel mounting bolt. Replace reel assembly.
Automatic belt squeaks when fastened or unfastened.	Excessive friction in belt guides. Automatic reel loose. Return spring broken inside reel.	Replace reel assembly. Tighten reel mounting bolt. Replace reel assembly.
Seat belt pyrotechnic tensioner triggered.	Accident triggered pyrotechnic deployment.	Replace complete automatic seat belt. Also check retaining bracket of belt tensioner for twist.

NOTE—

- *The rear bench seat belt strap may not retract fully due to increased friction between strap and seat cover. A small remaining loop in belt strap when fully retracted is acceptable.*

BentleyPublishers.com

BentleyPublishers.com—All Rights Reserved

Warnings

WARNING—
- *For maximum protection from injury, replace seat belts and hardware if subjected to occupant loading in a collision.*
- *Do not modify or repair seat belts or seat belt mounting points.*
- *Do not bleach or dye seat belt webbing. Webbing that is severely faded or redyed does not meet the strength requirements of a collision. Replace it.*
- *Clean belts with a luke-warm soap solution only.*
- *Periodically inspect seat belts for webbing defects such as cuts or pulled threads.*
- *Immediately after replacing a damaged or worn seat belt, destroy the old belt to prevent it from being used again.*
- *Pyrotechnic seat belt tensioners are powerful devices. Handle with extreme care. Incorrect handling can trigger the tensioner.*
- *BMW recommends that repair or replacement work on pyrotechnic devices be carried out by a qualified BMW technician.*
- *Be sure to disconnect the battery and wait 1 minute before attempting to work on pyrotechnic devices.*
- *Pyrotechnic devices cannot be repaired. Always replace them.*
- *Do not treat pyrotechnic components with cleaning agents or grease.*
- *Do not expose pyrotechnic components to temperatures above 75°C (167°F).*
- *Do not fire a pyrotechnic gas generator prior to disposal. It must be fired by a special disposal company or shipped back to BMW in the packaging of the new component.*
- *During body straightening and welding with an electric arc welder, disconnect the battery and the connection to the pyrotechnic gas generators.*

BentleyPublishers.com
BentleyPublishers.com–All Rights Reserved

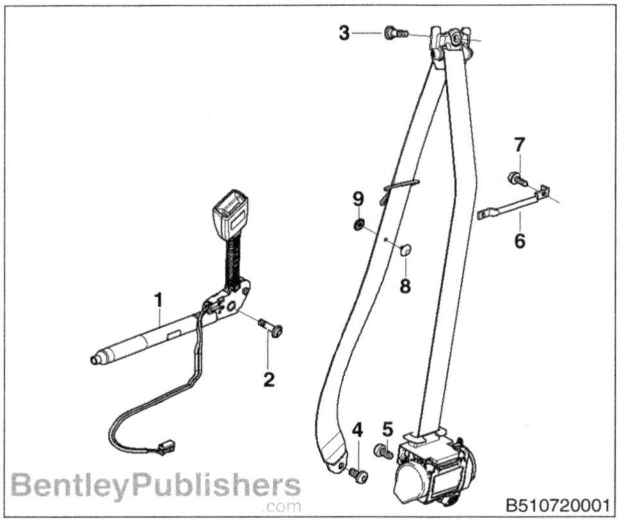

Front Seat Belts

Front seat belt components

1. Seat belt buckle assembly with pyrotechnic (explosive) tensioner
2. Self-locking fastener, replace with new
   Tighten to 36 Nm (27 ft-lb)
3. B-pillar fastener
   Tighten to 36 Nm (27 ft-lb)
4. M10 x 30 mm seat belt anchor self-locking fastener, replace with new
   Tighten to 36 Nm (27 ft-lb)
5. M10 x 26 mm seat belt reel self-locking fastener, replace with new
   Tighten to 36 Nm (27 ft-lb)
6. Belt guide
7. ST5.5 x 13 mm sheet metal screw
   Tighten to 3.5 Nm (31 in-lb)
8. Plastic button
9. Plastic button

Front seat belt buckle and tensioner, removing and installing

The front seat belt buckle is equipped with pyrotechnic (explosive charge) automatic tensioner designed to automatically retract and tension the seat belt by 55 mm (approximately 2 inches) in case of an accident.

> ***WARNING—***
> - *The side impact airbag and pyrotechnic seat belt tensioner attached to the front seat are powerful and potentially dangerous devices. Handle with extreme care. Incorrect handling can trigger off the airbag or tensioner and cause injury.*

– Disconnect battery negative (–) cable in trunk.

> ***CAUTION—***
> - *Prior to disconnecting the battery, read the battery disconnection cautions given in* **001 Warnings and Cautions**.

– Remove front seat. See **520 Seats**.

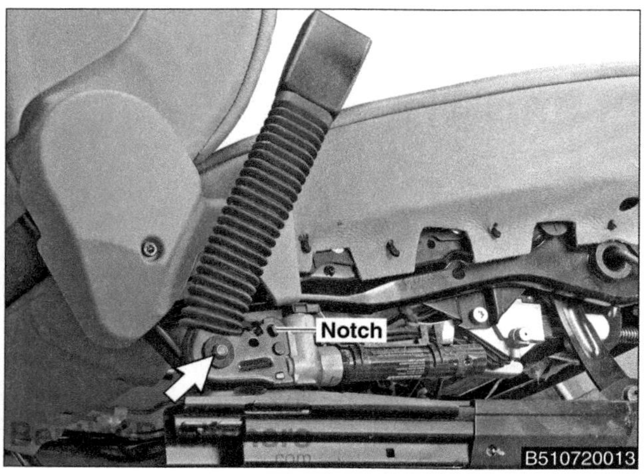

◄ Working at lower side of seat, remove belt buckle mounting fastener (**arrow**) and discard.

BentleyPublishers.com
BentleyPublishers.com—All Rights Reserved

– When installing:
- Secure base of belt buckle to seat using aligning notch.
- Use new self-locking belt buckle fastener.

Tightening torque	
Belt buckle to base of seat (self-locking fastener, replace with new)	36 Nm (27 ft-lb)

– Replace wire ties and reconnect harness connectors before reinstalling seat.

Tightening torque	
Front seat to floor (replace fasteners)	44 Nm (32 ft-lb)

Front seat belt reel, removing and installing

– Remove B-pillar trim. See **513 Interior Trim**.

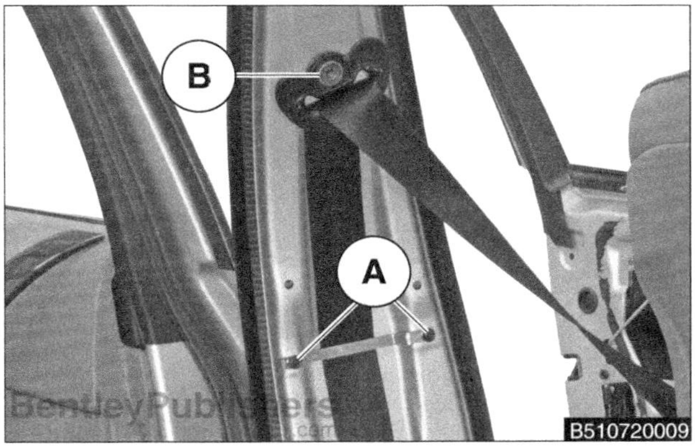

◄ Working at upper B-pillar:
- Remove belt guide mounting fasteners (**A**). Remove guide.
- Remove upper seat belt fastener (**B**).

– Set upper part of seat belt aside temporarily.

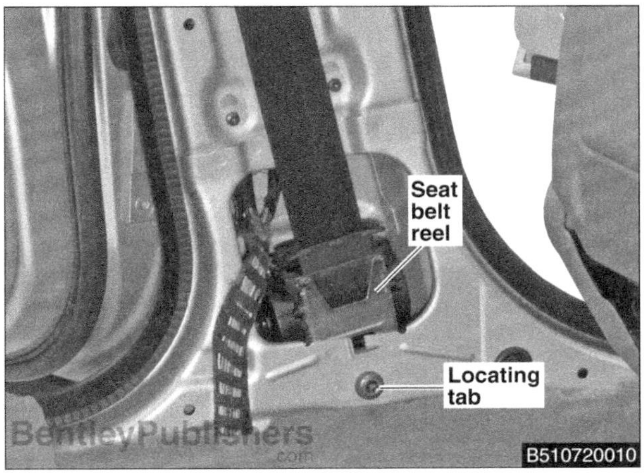

◄ Working at lower B-pillar, remove automatic seat belt reel mounting fastener (**arrow**) and lift out reel with seat belt.

– When installing:
- Make sure reel locating tab fits B-pillar slot.
- Replace self-locking fasteners with new.
- Position trim retaining clips back into trim panels before installing.
- Replace damaged trim retaining clips or trim panels.

Tightening torques	
Seat belt anchor to seat (M10 x 15 mm self-locking fastener, replace with new)	44 Nm (32 ft-lb)
Seat belt guide to B-pillar (ST5.5 x 13 mm sheet metal screw)	3.5 Nm (31 in-lb)
Seat belt reel to lower B-pillar (M10 x 22 mm self-locking fastener, replace with new)	36 Nm (27 ft-lb)
Seat belt to upper B-pillar	36 Nm (27 ft-lb)

720

BentleyPublishers.com
BentleyPublishers.com–All Rights Reserved

Rear Seat Belts

Rear seat belt assembly (Sedan)

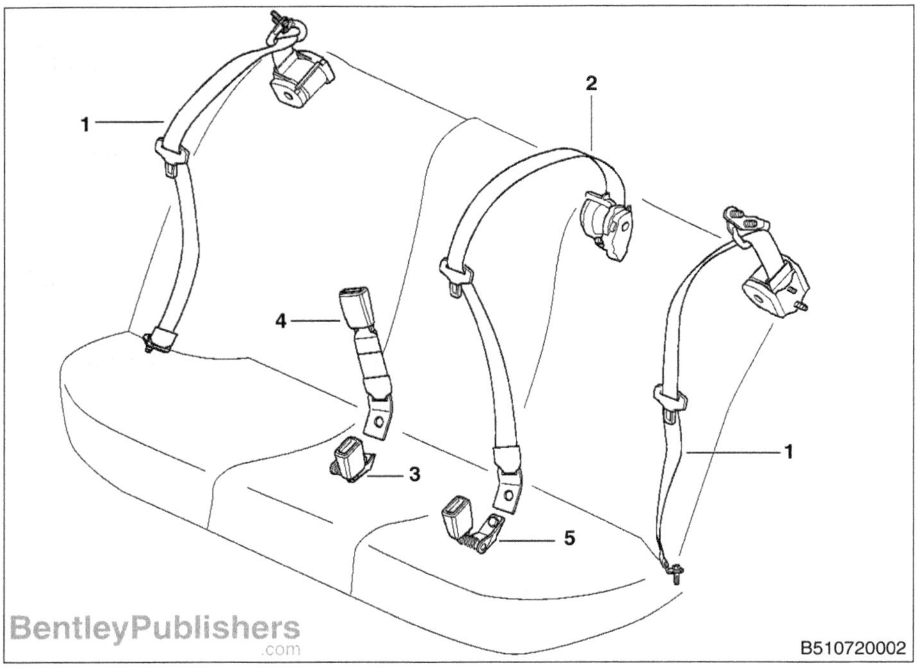

1. Rear seat belt, outboard
2. Rear seat belt, center
3. Center rear seat belt buckle
4. Right seat belt buckle
5. Left seat belt buckle

Rear seat belt assembly (Sports Wagon)

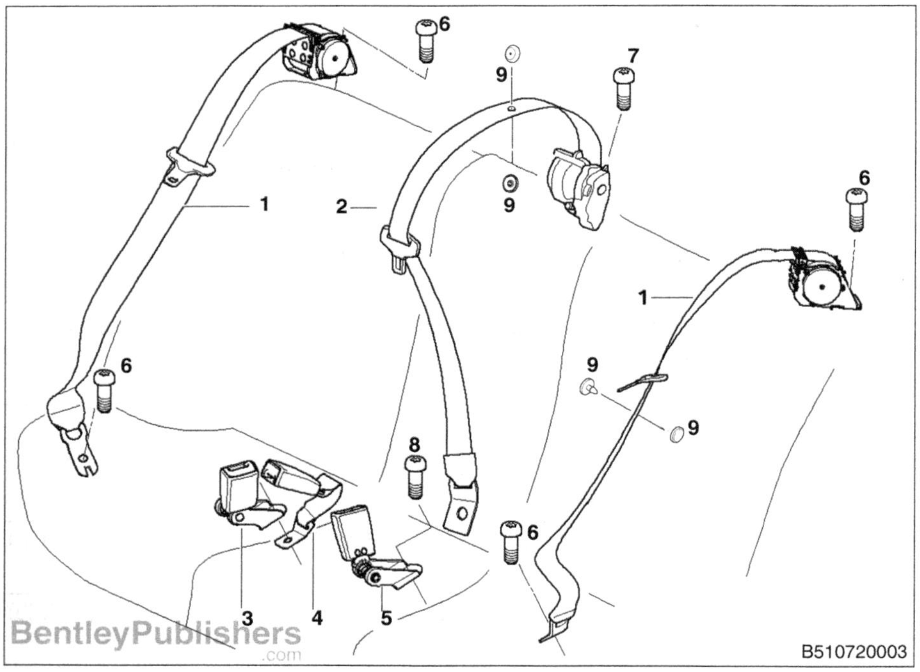

1. Rear seat belt, outboard
2. Rear seat belt, center
3. Right rear seat belt buckle
4. Center rear seat belt buckle
5. Left seat belt buckle
6. M10 x 26 mm self-locking fastener, replace with new
7. M10 x 16 mm self-locking fastener, replace with new
8. Self-locking fastener
9. Plastic button

BentleyPublishers.com
BentleyPublishers.com—All Rights Reserved

Rear seat belt, removing and installing

In Sports Wagon model with split rear seat backrest, the center seat belt is integrated with the left backrest.

– Sedan: Remove rear parcel shelf

– Sports Wagon:
  - Remove C-pillar and D-pillar trim. See **513 Interior Trim**.
  - Remove other trim panels as necessary.

– Remove rear seat. See **520 Seats**.

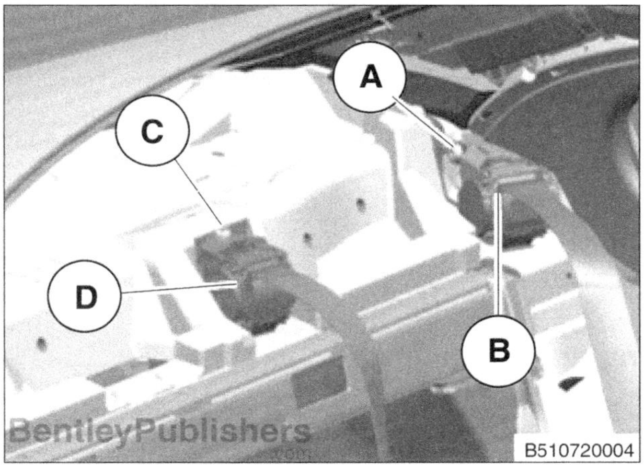

◄ Remove outboard seat belt reel mounting fastener (**A**) and seat belt reel (**B**).

– Sedan: Remove center seat belt reel fastener (**C**) and seat belt reel (**D**).

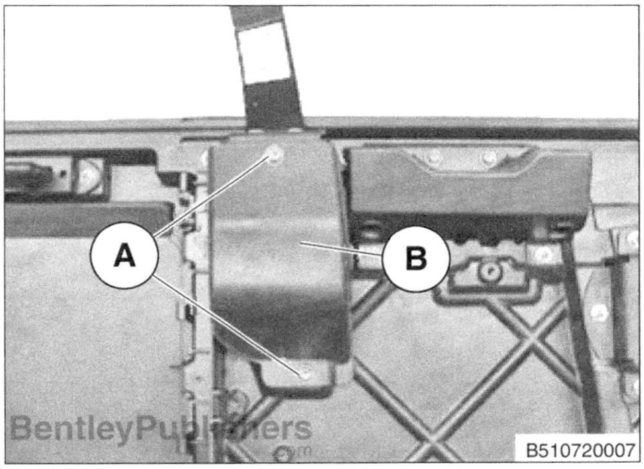

◄ Sports Wagon: Remove fasteners (**A**) and center seat belt reel cover (**B**).

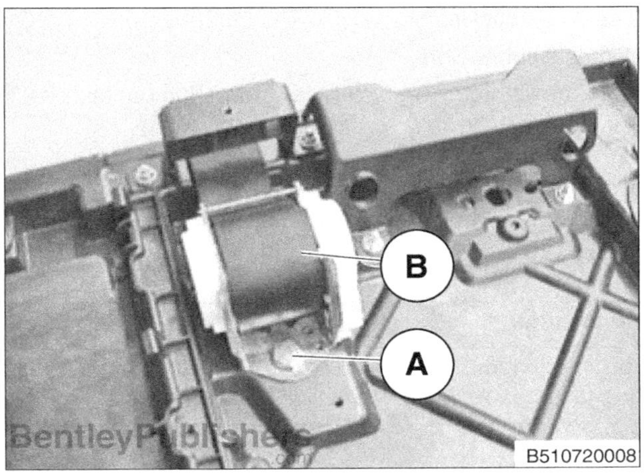

◄ Sports Wagon: Remove center seat belt reel mounting fastener (**A**) and seat belt reel (**B**).

720

BentleyPublishers
BentleyPublishers.com–All Rights Reserved

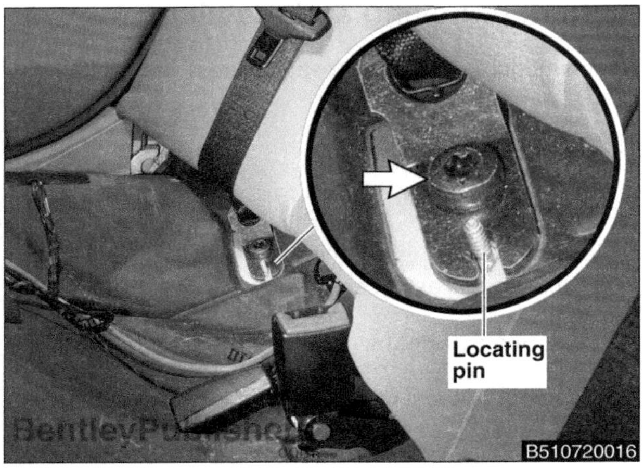

- Remove seat belt anchor fastener (**arrow**) at base of backrest and lift out seat belt.

– Installation is reverse of removal.
  - Place seat belt anchor correctly on locating pin.
  - Make sure tab on seat belt reel engages body slot.
  - Use new self-locking fasteners.

Tightening torques	
Seat belt anchor to body (self-locking fastener, replace with new)	36 Nm (27 ft-lb)
Seat belt reel to body (self-locking fastener, replace with new)	36 Nm (27 ft-lb)

Rear seat belt buckles and tensioners, removing and installing

In vehicles equipped with optional rear side airbags, rear outboard seat belt buckles are equipped with pyrotechnic (explosive charge) automatic tensioners designed to automatically retract and tension the seat belts by 55 mm (approximately 2 inches) in case of an accident.

> ***WARNING—***
> - *Pyrotechnic seat belt tensioners are powerful and potentially dangerous devices. Handle with extreme care. Incorrect handling can trigger off the tensioner and cause injury.*

– Disconnect battery negative (–) cable in trunk.

> ***CAUTION—***
> - *Prior to disconnecting the battery, read the battery disconnection cautions given in* **001 Warnings and Cautions**.

– Remove rear seat. See **520 Seats**.

- Working at seat belt buckles:
  - Detach electrical connectors.
  - Loosen and remove mounting fasteners (**inset**). Lift out buckles and store in a safe place.

– When installing:
  - Make sure tab on seat belt buckle engages body slot.
  - Use new self-locking fasteners.

Tightening torques	
Seat belt buckle to body (Self-locking fastener, replace with new)	36 Nm (27 ft-lb)

BentleyPublishers.com
BentleyPublishers.com—All Rights Reserved

721 Airbag System (SRS)

General

Th is repair group covers airbag systems and airbag system component replacement. Airbag system problems can only be diagnosed using BMW scan tool or equivalent.

One of two systems is installed in the E60 / E61 cars covered by this manual, depending on model year.

Passive restraint system application	
Year	**Restraint system**
2004 - 2005	Advanced Safety Electronics (ASE)
2006 - 2010	Advanced Crash and Safety Management (ACSM)

See also:

- **121 Battery, Starter, Alternator** for battery safety terminal
- **411 Doors** for door panel removal
- **513 Interior Trim** for center console and glove compartment removal
- **520 Seats** for seat removal
- **612 Switches** for steering column switch cluster
- **720 Seat Belts**
- **ECL Electrical Component Locations**

BentleyPublishers.com
BentleyPublishers.com—All Rights Reserved

Warnings and Cautions

WARNING —
- *The airbag is inflated by an explosive device. Handled improperly or without adequate safeguards, it can be very dangerous. Observe special precautions prior to any work at or near airbags.*
- *Serious injury may result if system service is attempted by persons unfamiliar with the BMW airbag system and its approved service procedures. BMW advises that airbag system inspection and service be performed by an authorized BMW dealer.*
- *Special test equipment is required to retrieve airbag fault codes, diagnose system faults and switch the airbag warning light OFF. The warning light remains ON until problems are corrected and fault memory cleared.*
- *If the airbag warning light is ON, there is a risk that airbags will not trigger in case of an accident. Be sure to have the system inspected and repaired immediately.*
- *Disconnect the battery and cover the negative (–) battery terminal with an insulator before starting diagnostic, troubleshooting or service work associated with the airbags, and before doing any welding on the car.*
- *After disconnecting the battery, wait 1 minute before beginning work on airbag components.*
- *If an airbag deploys due to an accident, BMW specifies that airbag and airbag related components be replaced. For more information on post-collision airbag service, contact an authorized BMW dealer.*
- *Do not fire an airbag unit prior to disposal. It must be fired by a special disposal company or shipped back to BMW in the packaging of the new components.*
- *Wear gloves and avoid skin contact when removing a deployed airbag unit. In case of skin contact, wash with water.*
- *Do not allow airbag system components to come in contact with cleaning solutions or grease. Do not subject airbag components to temperatures above 75°C (167°F). When reconnecting the battery, make sure no person is inside the vehicle.*
- *Place a removed airbag unit with the padded side facing upward. Do not leave an airbag unit unattended.*
- *If the airbag unit or airbag control module is dropped from a height of ½ meter (1½ feet) or more, do not use it.*

CAUTION—
- *To avoid damaging plastic interior trim, use a plastic prying tool or a screwdriver with the tip wrapped with masking tape.*

BentleyPublishers.com
BentleyPublishers.com—All Rights Reserved

Advanced Safety Electronics (ASE)

On 2004 and 2005 models, the BMW airbag system is called the Advanced Safety Electronics system (ASE). (The industry name is supplemental restraint system or SRS.)

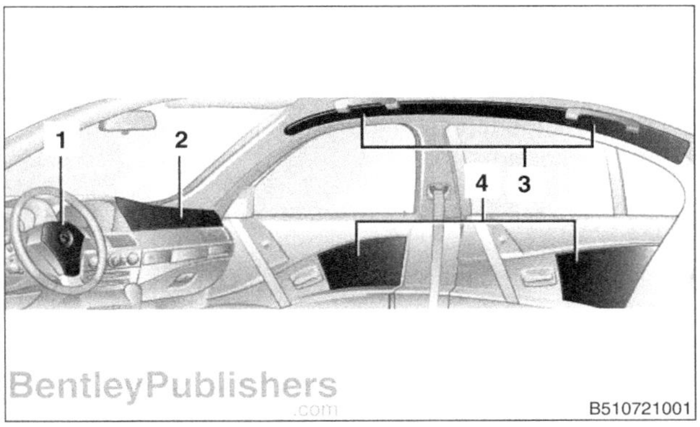

1. Driver airbag
2. Passenger airbag
3. Curtain side airbag
4. Door-mounted side airbag (rear optional)

ASE components

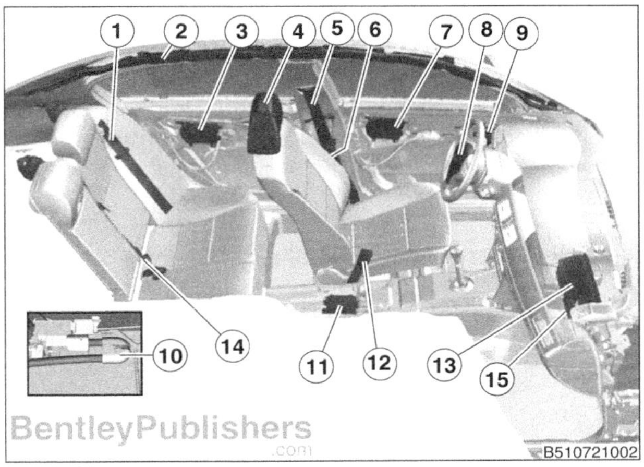

The ASE system consists of the following components:

1. Rear seat belt
2. Curtain side airbag
3. Rear side airbag (optional)
4. Active headrest
5. Front seat belt
6. B-pillar satellite
7. Front side airbag
8. Driver airbag
9. Front door module
10. Battery safety terminal (BST)
11. Vehicle center satellite
12. Front seat belt pyrotechnic tensioner
13. Front passenger airbag
14. Rear seat belt pyrotechnic tensioner
15. Safety and gateway module (SGM)

Safety and gateway module (SGM)

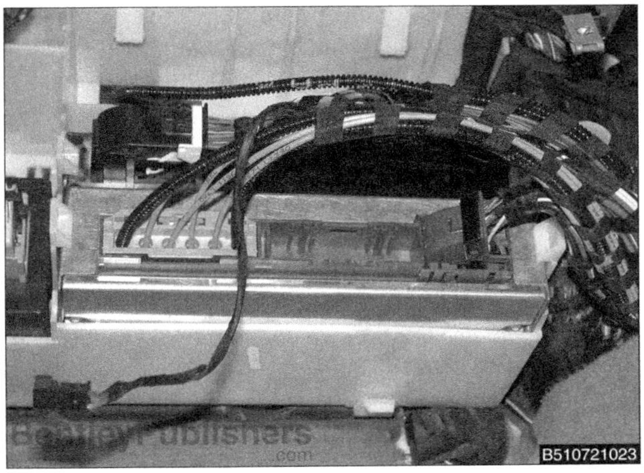

The ASE system is controlled by the safety and gateway module (SGM), located in the module carrier behind the glove compartment. The SGM contains a separate transmitter/receiver for each satellite and connects to the satellites via a fiber optic star bus (byteflight). Data communication is bidirectional, achieved at a rate of 10 Mbit/s.

BentleyPublishers.com
BentleyPublishers.com–All Rights Reserved

Vehicle center satellite

The vehicle center satellite is located under the center console. It connects to other ASE system components and vehicle electronics. The module contains two crash sensors.

Communication with other control modules in the vehicle network takes place via byteflight.

For a description of vehicle bus systems, see **600 Electrical System–General**.

The red ASE warning LED in the instrument cluster is illuminated for a short period after switching the ignition ON.

If a fault is stored in the SGM, the LED illuminates and remains ON until the ignition is switched OFF. A check control icon also appears in either yellow or red depending on the system fault along with a brief text message.

With the warning LED illuminated, ASE may not operate in case of a collision.

Crash sensors

Satellite sensors are installed in the B-pillars, frame rails, doors and center console.The longitudinal and transverse acceleration sensors and door pressure sensors serve the purpose of detecting side crashes.

The longitudinal acceleration sensors detect front end and rear end crashes. Acceleration sensor data is converted to digital signals and transmitted to the SGM via byteflight through the star coupler as data telegrams every 250 microseconds (μs).

B-pillar satellite, a combined longitudinal and transverse acceleration sensor, detects frontal, side and rear end crashes. The left and right sensors are identical and are mechanically coded to side during installation.

BentleyPublishers.com
BentleyPublishers.com—All Rights Reserved

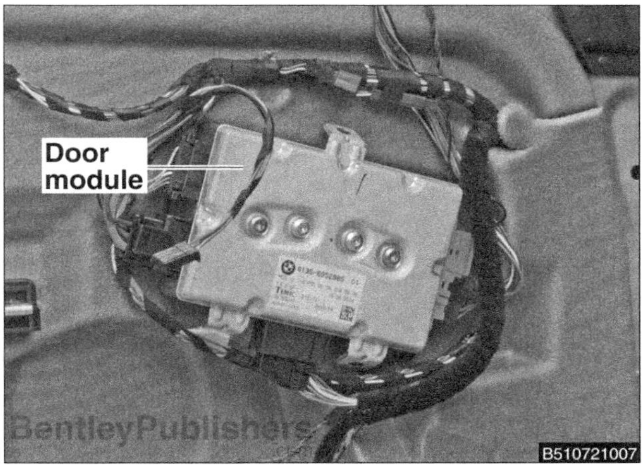

Door pressure sensor is integrated into the front door module and measures the pressure inside the door, providing supplementary side impact detection. In case of a side crash, the outer door panel is pushed inward, thus compressing the space inside the door and increasing the pressure, detected by the door pressure sensor.

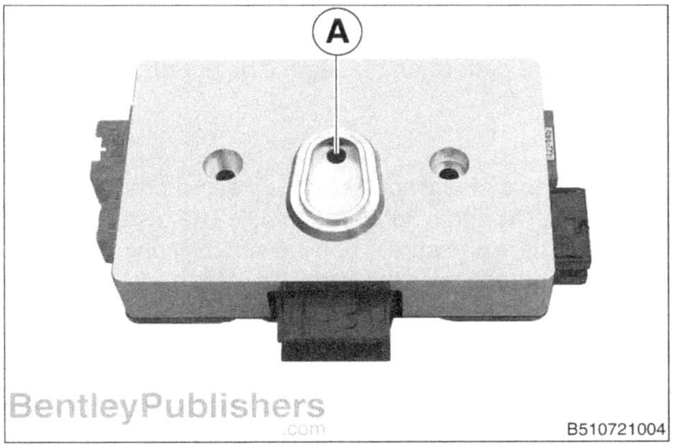

The front door module contains a pressure sensor (**A**) and an electronic module which digitizes and transmits data cyclically to the SGM. At the same time, side impact is detected by the B-pillar satellite sensor. The SGM processes the two signals and triggers the restraint system on the basis of the information provided.

Driver airbag and horn contact spring

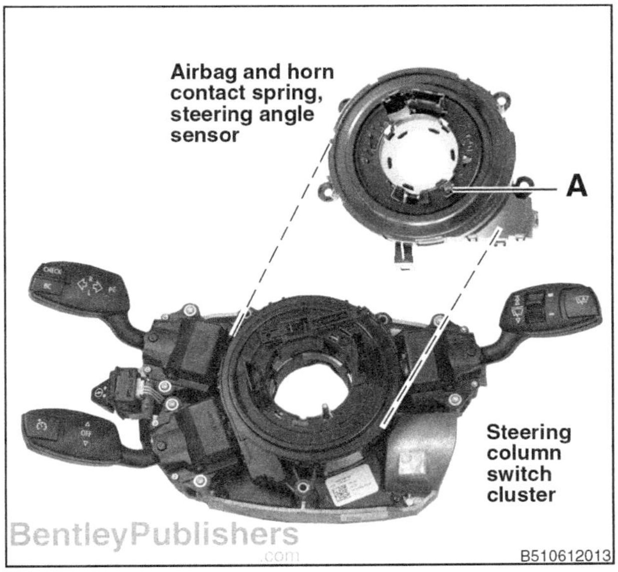

The contact spring on the steering column switch cluster (SZL) behind the steering wheel provides the electrical interface between the fixed wiring harness and the moveable driver airbag. SZL service is covered in **612 Switches**.

The contact spring cassette consists of outer and inner housings with integral connectors and contains a ribbon type flexible cable.

BentleyPublishers.com
BentleyPublishers.com–All Rights Reserved

Passenger seat occupancy sensor (OC3 mat)

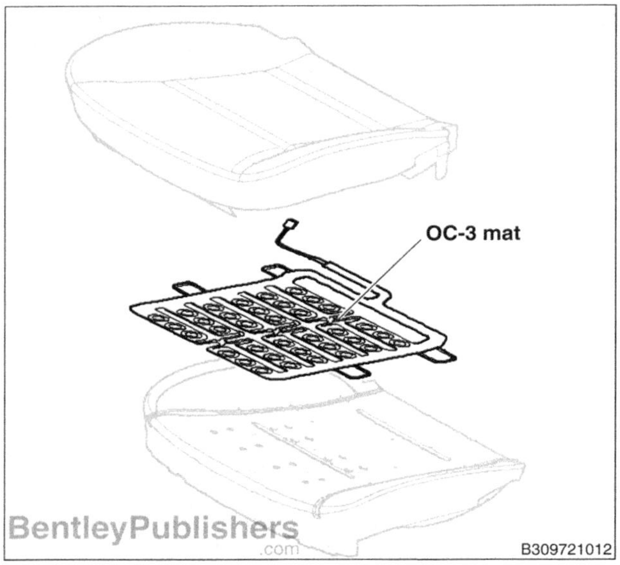

The passenger seat occupancy sensor (occupant classifier or OC3 mat) is integrated into the seat area of the passenger seat. The sensor consists of conductors with pressure-dependent resistor (force sensitive resistance or FSR) elements connected to an electronic analyzer.

The FSR elements are wired in such a way that they can be sampled individually. When the mechanical load on a sensor element increases, its electrical resistance decreases and the current changes accordingly. By analyzing the signals from different sections of the mat, the analyzer maps the occupancy of the seat surface and can identify local concentrations of weight.

The purpose of the OC3 mat is to monitor the presence of a person or object placed on the front passenger seat and determine whether to deactivate passenger side airbags. The mat is capable of detecting a child seat that conforms to the NHTSA FMVSS 208 standard by the weight pattern on the seat. If detected, passenger airbags are deactivated.

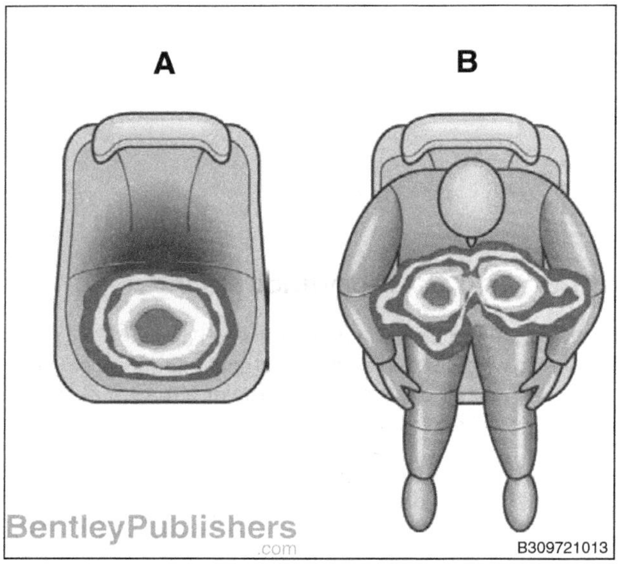

The OC3 mat is capable of distinguishing between a one-year-old child in a child's seat (**A**) and a light person (**B**). The distances between the areas where pressure is applied and the concentrations of pressure reveal who occupies the seat.

The OC3 mat analyzer transmits a message to the right B-pillar satellite which in turn transmits a message to the SGM. If the system detects that the seat is unoccupied or that a child seat for a child up to one year old is fitted, the airbags on the passenger side are deactivated. If a child seat is detected, the passenger airbag status light illuminates, indicating that passenger side airbags are deactivated.

Airbag status light

The airbag status light in the dome light housing glows yellow when the front passenger airbag and the side passenger airbag are deactivated.

BentleyPublishers.com
BentleyPublishers.com—All Rights Reserved

Pyrotechnic devices

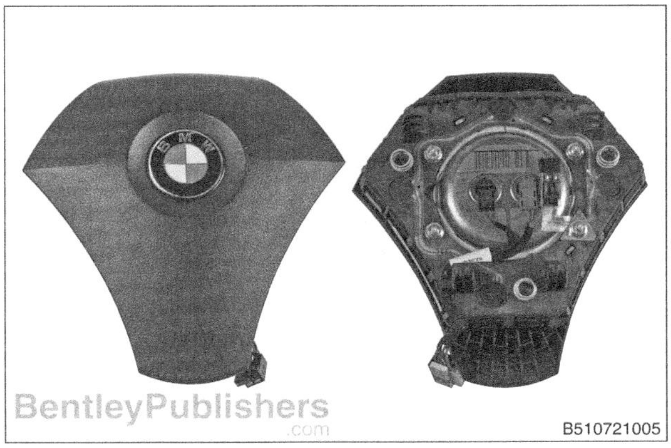

Driver airbag, of two-stage design, is attached to the steering wheel by spring clips. Electrical connection to the SGM is provided via the driver airbag and horn contact spring in the steering column.

Once the airbag is fully inflated, vents in the airbag prevent further pressure build-up, so that progressive deceleration is provided as the driver contacts the cushion and injury due to sudden impact forces is prevented.

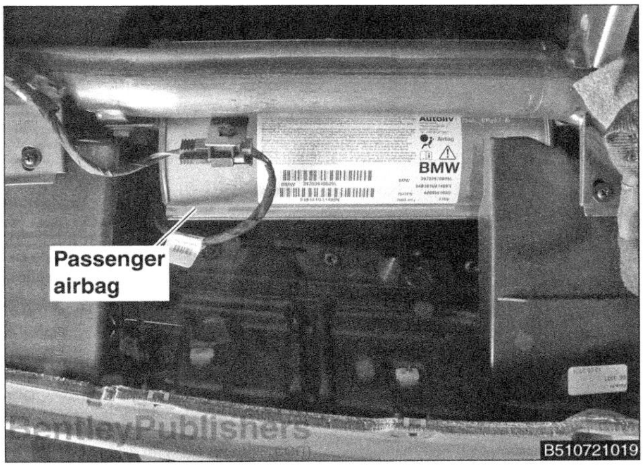

Passenger airbag, of two-stage design, is above the glove compartment, directly in front of the passenger seat, mounted to the body crossbeam.

Upon deployment, the airbag tears the dashboard cover through several predetermined shear points. Once free of the housing and fascia, the airbag inflates to its full extent to provide a protective cushion between the front seat passenger and the windshield.

Vents in the airbag prevent excess pressure bursting the bag and, as soon as the material in the gas generator is exhausted, the airbag deflates.

Thorax (side) airbag, mounted to the front doors behind door trim panel (rear side airbag is optional), is designed to protect the front seat occupant during a side impact. Side airbags are activated by a signal from the SGM in case of a side impact or a front angled impact of sufficient severity to cause both front and side airbag deployment.

The side airbag module consists of an aluminum housing with a plastic cover, a folded nylon fabric bag, gas generating capsules and an igniter.

The side airbag module is marked either “VORNE / FRONT” or “HINTEN / REAR”.

BentleyPublishers.com
BentleyPublishers.com–All Rights Reserved

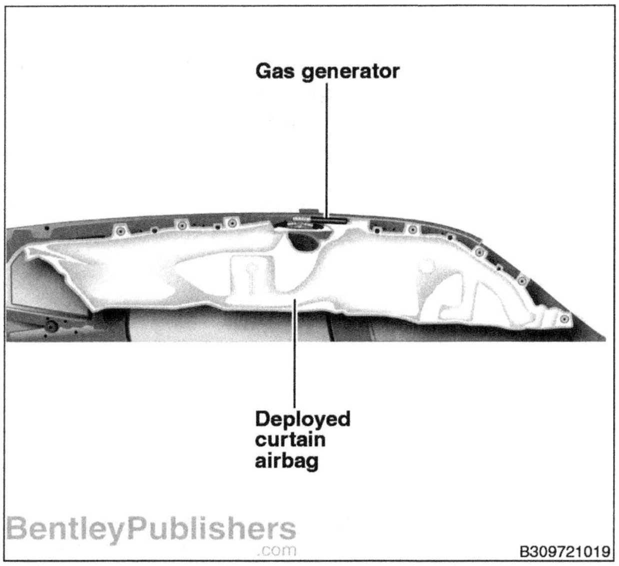

Curtain airbag, extending all the way from the A-pillar to the C-pillar, covering the entire side-window area, inflates between the vehicle occupants and the side windows and pillar trims.

The curtain airbag reduces the risk of occupant's head or other extremities protruding through the windows in a sideways collision. Advantages of the system are:

- Extended covered area for side windows, front and rear.
- Protection against glass splinters and penetrating objects.
- Optimized protective area offering protection for occupants of differing sizes.

The curtain airbag is positioned along the line of the roof side member, folded up. It consists of a gas generator, two gas nozzles and curtain. In the event of a severe enough side collision, the generator is detonated and the gas flows through the two gas nozzles into the curtain. Simultaneous inflation of the curtain at the front and back achieves more even deployment.

Because the curtain generators and curtain are a sealed system, the curtain airbag retains its shape and strength for several seconds.

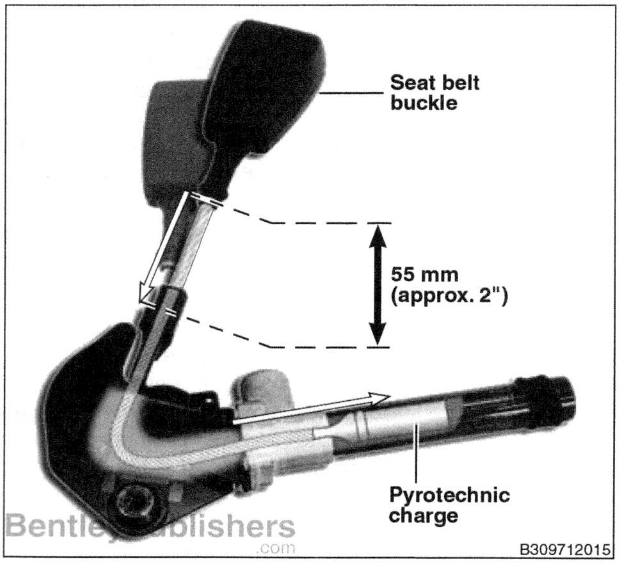

Seat belt tensioner. Each seat belt buckle is equipped with a pyrotechnic (explosive charge) tensioner capable of cinching the belt an additional 55 mm (approx. 2 inches) in case of an accident. The belt buckle is attached via steel cable to a tube containing explosive propellant and a piston. An igniter in the base of the tube provides an ignition source when triggered by a signal from the SGM. A 2-pin connector located on a bracket underneath the seat frame links the igniter to SGM wiring.

The SGM deploys the pyrotechnic seat belt tensioners at a slightly lower impact threshold level than the front airbags.

Belt buckle Hall microswitch in the front seat belt buckle is used to detect whether the seat belt is fastened or not. The detection signal is used by the SGM as a criterion for selective triggering of airbags and seat belt tensioners in case of a crash.

The switch is supplied current by the SGM via a current-signal interface. The current draw of the switch is the signal for switch position. The belt buckle switch is monitored while the ignition is ON.

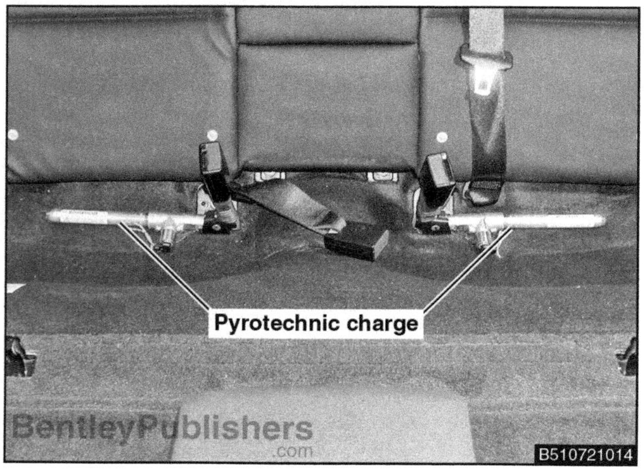

Rear belt tensioners are also equipped with pyrotechnic devices in vehicles equipped with rear side airbags.

Seat belt tensioner replacement is covered in **720 Seat Belts**.

BentleyPublishers.com
BentleyPublishers.com—All Rights Reserved

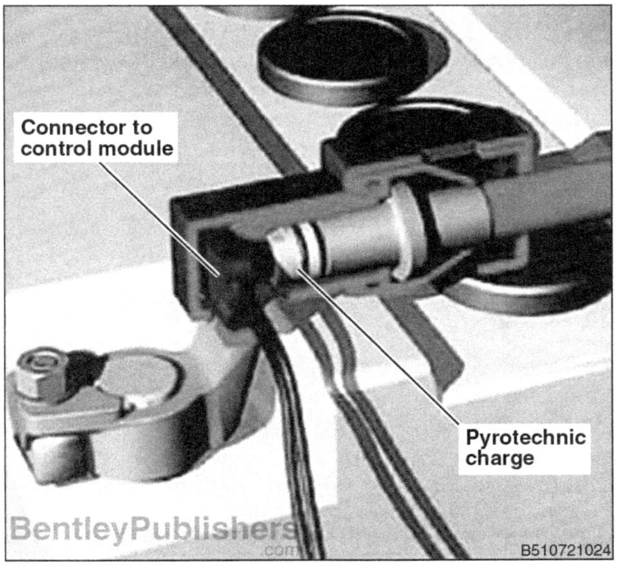

Battery safety terminal (BST) is used to minimize the possibility of a short circuit at the heavy gauge B+ wire that runs the length of the vehicle in case of an accident.

The BST consists of a conventional clamp that is screwed to the battery positive terminal from above and connected with a hollow cylinder. A propellant charge is stored in this hollow cylinder. As with airbags and seat belt tensioners, this pyrotechnic device is controlled and ignited by the SGM. The unit is enclosed in a plastic shell that captures the cable if it is forced out via BST deployment and locks it so that a renewed contact is no longer possible.

See also **121 Battery, Starter, Alternator**.

Principles of operation

The SGM performs the following functions:

- Crash detection and determining triggering point.
- Triggering ignition output stages.
- Documentation of crash data.
- Cyclic monitoring.
- Display of system standby status.
- Fault indication and fault code storage.
- Fault output (diagnosis).
- Output of a crash telegram for other users in communication system network.
- Activation of airbag indicator light when passenger airbag is deactivated.

Crash detection

The SGM determines whether a crash has occurred based on the values detected by crash sensors and calculated by a triggering algorithm.

Signals from different sensors help determine crash severity and direction. Based on this information, the stored algorithm determines the triggering (ignition) points of the restraint systems to be activated.

Documentation of actuator triggering

The triggering (ignition) commands are recorded in case of a crash where one or several actuators are triggered. The most important crash event data are stored in the form of a crash telegram in the non-volatile memory of the SGM. This data is only available to the manufacturer.

BentleyPublishers.com
BentleyPublishers.com—All Rights Reserved

Cyclic monitoring

Once system self-test is successfully concluded and the system is ready for operation, cyclic fault monitoring is performed as long as ignition is ON.

Indication of system operability

System operability is indicated by the airbag warning light switching OFF in the instrument cluster.

Fault code storage

A fault in the system is indicated by the airbag warning light and stored in nonvolatile control module memory. A distinction is made between internal and external faults.

Fault output (diagnosis)

Using BMW scan tool or equivalent, fault codes can be read out via the diagnosis interface. After rectifying faults, the fault codes can be cleared by means of "clear fault code memory" diagnosis command.

Crash telegram output

In case of a crash which triggers restraint system components, the SGM sends a crash telegram via bus system. The control module then activates or deactivates various components depending on crash severity.

Deactivation of passenger airbags

The SGM utilizes the OC-3 mat signal to deactivate the front and side passenger airbags when a child seat is mounted on the passenger side. Airbag deactivation is indicated by a yellow airbag indicator light.

Seat belt reminder

The seat belt reminder (SBR) function monitors whether the driver or passenger has fastened his or her seat belt. If the seat belt is not worn or is unbuckled while driving a distance of approx. 100 meters (328 feet), a reminder is illuminated and a chime is sounded for up to 90 seconds unless belt is buckled in the meantime.

Left and right front seat belt buckle microswitches are monitored separately.

BentleyPublishers.com
BentleyPublishers.com—All Rights Reserved

Advanced Crash and Safety Management (ACSM)

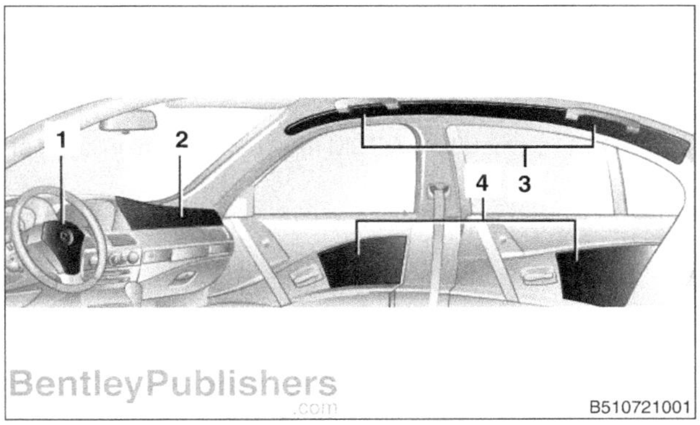

On models produced in 2006 and later, the BMW airbag system is called the advanced crash and safety management system (ACSM). (The industry name is supplemental restraint system or SRS.)

1. Driver airbag
2. Passenger airbag
3. Curtain side airbag
4. Side airbag (rear optional)

ACSM components

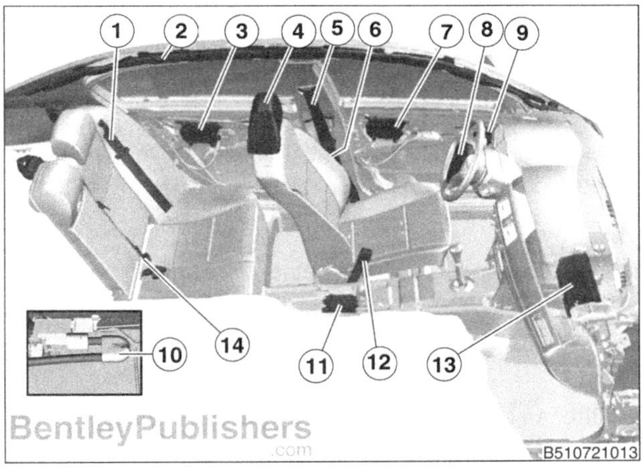

The ACSM system consists of the following components:

1. Rear seat belt
2. Curtain side airbag
3. Rear side airbag (optional)
4. Active headrest
5. Front seat belt
6. B-pillar satellite
7. Front side airbag
8. Driver airbag
9. Door pressure sensor
10. Battery safety terminal (BST)
11. Crash safety (airbag) module
12. Front seat belt pyrotechnic tensioner
13. Front passenger airbag
14. Rear seat belt pyrotechnic tensioner

Crash safety module (airbag control module)

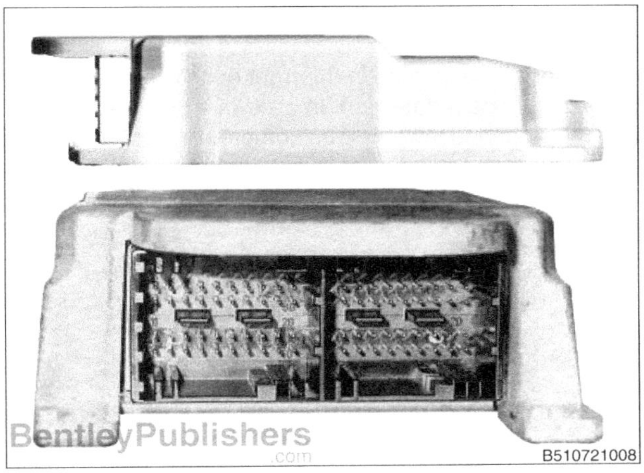

The crash safety (airbag) module is located under the center console. It connects to other ACSM system components and vehicle electronics. The module contains two crash sensors.

Communication with other control modules in the vehicle network takes place via K-CAN.

For a description of vehicle bus systems, see **600 Electrical System–General**.

BentleyPublishers.com
BentleyPublishers.com—All Rights Reserved

B309721005

The red ACSM warning LED in the instrument cluster is illuminated for a short period after switching the ignition ON.

If a fault is stored in the crash safety module, the LED illuminates and remains ON until the ignition is switched OFF. A check control icon also appears in either yellow or red depending on the system fault along with a brief text message.

With the warning LED illuminated, ACSM may not operate in case of a collision.

Crash sensors

Satellite sensors are installed in the B-pillars, frame rails, doors and center console.The longitudinal and transverse acceleration sensors and door pressure sensors serve the purpose of detecting side crashes.

The longitudinal acceleration sensors detect front end and rear end crashes. Acceleration sensor data is converted to digital signals in an ASIC (Application Specific Integrated Circuit) and transmitted to the crash safety module.

B309721006

B-pillar satellite, a combined longitudinal and transverse acceleration sensor, detects frontal, side and rear end crashes. The left and right sensors are identical and are mechanically coded to side during installation.

B510721009

Door pressure sensor is located in the front door, providing supplementary side impact detection. In case of a side crash, the outer door panel is pushed inward, thus compressing the space inside the door and increasing the pressure, detected by the door pressure sensor.

BentleyPublishers.com
BentleyPublishers.com–All Rights Reserved

Up-front sensors are located on the left and right sides of the front end. Each sensor contains an acceleration sensor, a signal conditioner, and an ASIC for data transmission. Information is sent via a data telegram to the crash safety module.

Driver airbag and horn contact spring

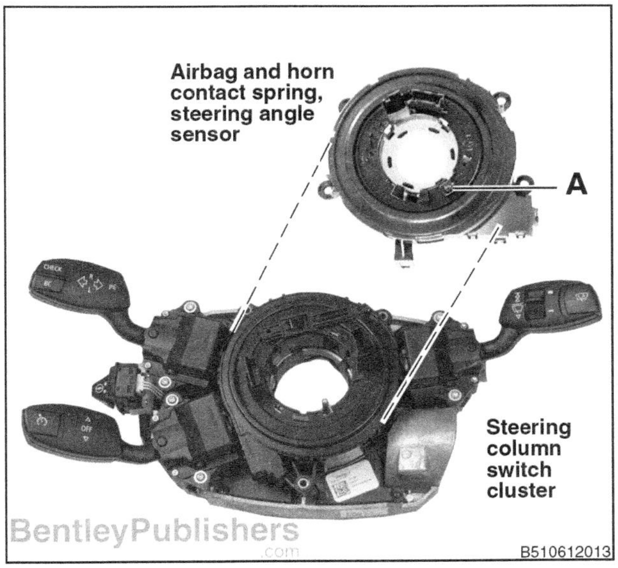

The contact spring on the steering column switch cluster (SZL) behind the steering wheel provides the electrical interface between the fixed wiring harness and the moveable driver airbag. SZL service is covered in **612 Switches**.

The contact spring cassette consists of outer and inner housings with integral connectors and contains a ribbon type flexible cable.

Passenger seat occupancy sensor (OC3 mat)

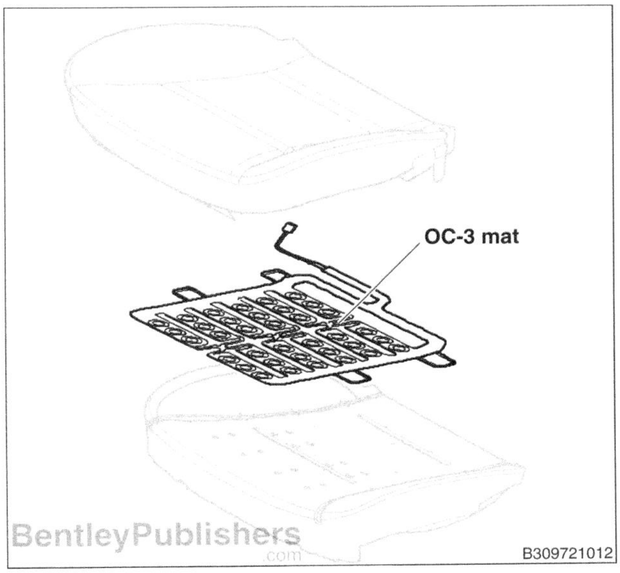

The passenger seat occupancy sensor (occupant classifier or OC3 mat) is integrated into the seat area of the passenger seat. The sensor consists of conductors with pressure-dependent resistor (force sensitive resistance or FSR) elements connected to an electronic analyzer.

The FSR elements are wired in such a way that they can be sampled individually. When the mechanical load on a sensor element increases, its electrical resistance decreases and the current changes accordingly. By analyzing the signals from different sections of the mat, the analyzer maps the occupancy of the seat surface and can identify local concentrations of weight.

The purpose of the OC3 mat is to monitor the presence of a person or object placed on the front passenger seat and determine whether to deactivate passenger side airbags. The mat is capable of detecting a child seat that conforms to the NHTSA FMVSS 208 standard by the weight pattern on the seat. If detected, passenger airbags are deactivated.

721

BentleyPublishers.com
BentleyPublishers.com—All Rights Reserved

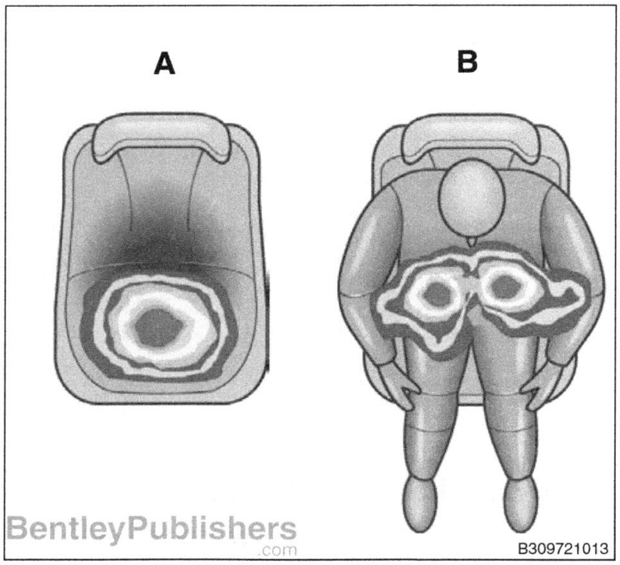

The OC3 mat is capable of distinguishing between a one-year-old child in a child's seat (**A**) and a light person (**B**). The distances between the areas where pressure is applied and the concentrations of pressure reveal who occupies the seat.

The OC3 mat analyzer transmits a message to the right B-pillar satellite which in turn transmits a message to the crash safety module. If the system detects that the seat is unoccupied or that a child seat for a child up to one year old is fitted, the airbags on the passenger side are deactivated. If a child seat is detected, the passenger airbag status light illuminates, indicating that passenger side airbags are deactivated.

Airbag status light

The airbag status light in the dome light housing glows yellow when the front passenger airbag and the side passenger airbag are deactivated.

Pyrotechnic devices

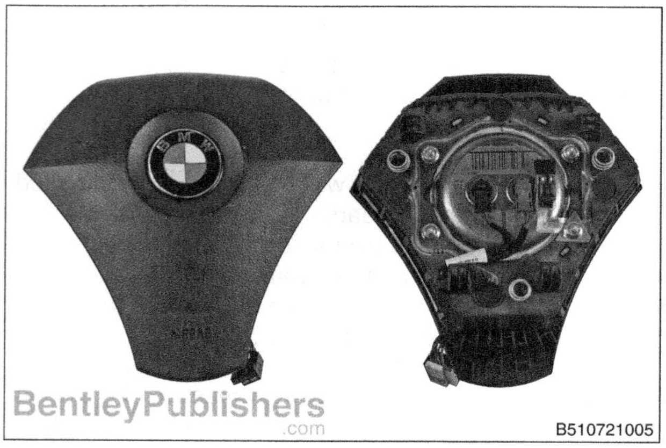

Driver airbag, of two-stage design, is attached to the steering wheel by spring clips. Electrical connection to the crash safety module is provided via the driver airbag and horn contact spring in the steering column.

Once the airbag is fully inflated, vents in the airbag prevent further pressure build-up, so that progressive deceleration is provided as the driver contacts the cushion and injury due to sudden impact forces is prevented.

BentleyPublishers.com
BentleyPublishers.com—All Rights Reserved

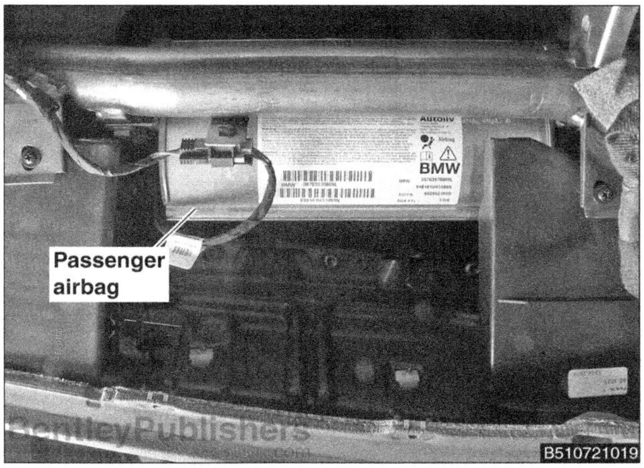

Passenger airbag, of two-stage design, is above the glove compartment, directly in front of the passenger seat, mounted to the body crossbeam.

Upon deployment, the airbag tears the dashboard cover through several predetermined shear points. Once free of the housing and fascia, the airbag inflates to its full extent to provide a protective cushion between the front seat passenger and the windshield.

Vents in the airbag prevent excess pressure bursting the bag and, as soon as the material in the gas generator is exhausted, the airbag deflates.

Thorax (side) airbag, mounted to the front doors behind door trim panel (rear side airbag is optional), is designed to protect the front seat occupant during a side impact. Side airbags are activated by a signal from the crash safety module in case of a side impact or a front angled impact of sufficient severity to cause both front and side airbag deployment.

The side airbag module consists of an aluminum housing with a plastic cover, a folded nylon fabric bag, gas generating capsules and an igniter.

The side airbag module is marked either “VORNE / FRONT” or “HINTEN / REAR”.

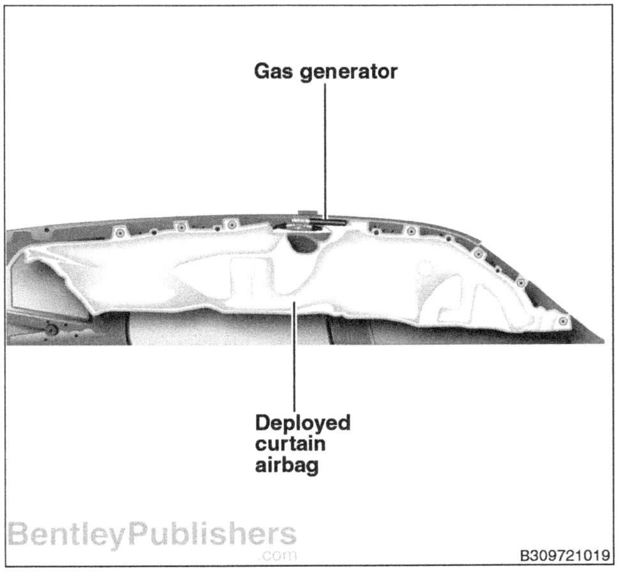

Curtain airbag, extending all the way from the A-pillar to the C-pillar, covering the entire side-window area, inflates between the vehicle occupants and the side windows and pillar trims.

The curtain airbag reduces the risk of occupant’s head or other extremities protruding through the windows in a sideways collision. Advantages of the system are:

- Extended covered area for side windows, front and rear.
- Protection against glass splinters and penetrating objects.
- Optimized protective area offering protection for occupants of differing sizes.

The curtain airbag is positioned along the line of the roof side member, folded up. It consists of a gas generator, two gas nozzles and curtain. In the event of a severe enough side collision, the generator is detonated and the gas flows through the two gas nozzles into the curtain. Simultaneous inflation of the curtain at the front and back achieves more even deployment.

Because the curtain generators and curtain are a sealed system, the curtain airbag retains its shape and strength for several seconds.

BentleyPublishers.com
BentleyPublishers.com—All Rights Reserved

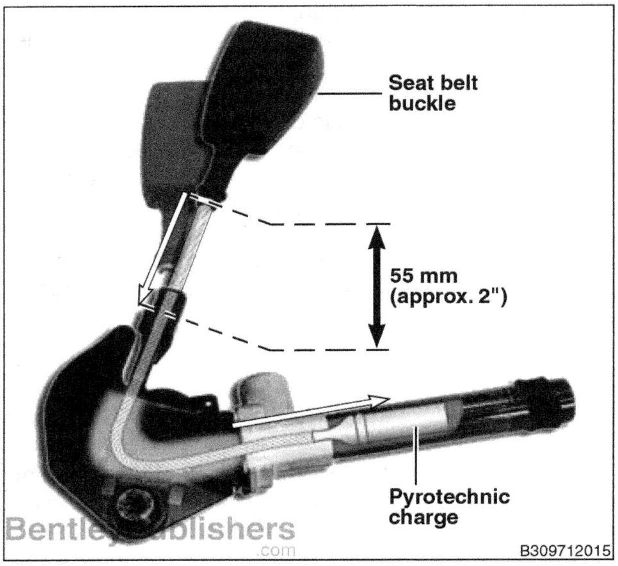

Seat belt tensioner. Each seat belt buckle is equipped with a pyrotechnic (explosive charge) tensioner capable of cinching the belt an additional 55 mm (approx. 2 inches) in case of an accident. The belt buckle is attached via steel cable to a tube containing explosive propellant and a piston. An igniter in the base of the tube provides an ignition source when triggered by a signal from the crash safety module. A 2-pin connector located on a bracket underneath the seat frame links the igniter to crash safety module wiring.

The crash safety module deploys the pyrotechnic seat belt tensioners at a slightly lower impact threshold level than the front airbags.

Belt buckle Hall microswitch in the front seat belt buckle is used to detect whether the seat belt is fastened or not. The detection signal is used by the crash safety module as a criterion for selective triggering of airbags and seat belt tensioners in case of a crash.

The switch is supplied current by the crash safety module via a current-signal interface. The current draw of the switch is the signal for switch position. The belt buckle switch is monitored while the ignition is ON.

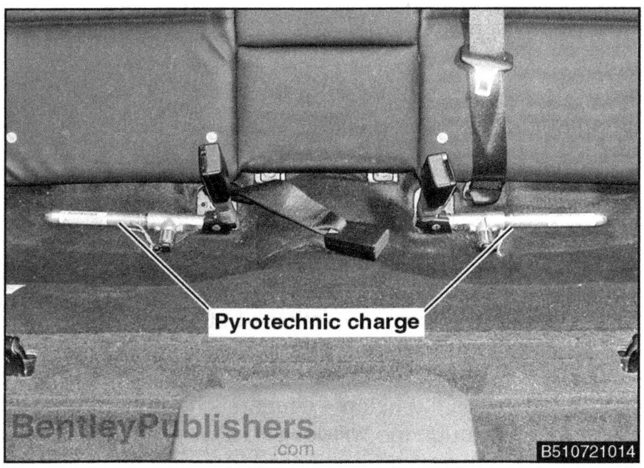

Rear belt tensioners are also equipped with pyrotechnic devices in vehicles equipped with rear side airbags.

Seat belt tensioner replacement is covered in **720 Seat Belts**.

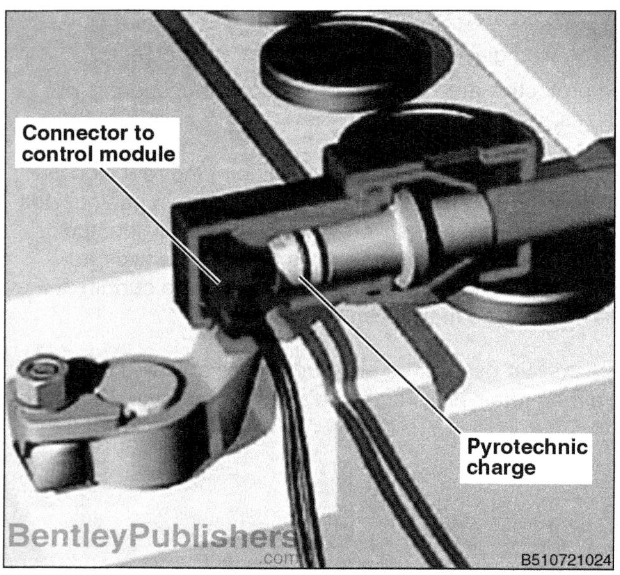

Battery safety terminal (BST) is used to minimize the possibility of a short circuit at the heavy gauge B+ wire that runs the length of the vehicle in case of an accident.

The BST consists of a conventional clamp that is screwed to the battery positive terminal from above and connected with a hollow cylinder. A propellant charge is stored in this hollow cylinder. As with airbags and seat belt tensioners, this pyrotechnic device is controlled and ignited by the crash safety module. The unit is enclosed in a plastic shell that captures the cable if it is forced out via BST deployment and locks it so that a renewed contact is no longer possible.

See also **121 Battery, Starter, Alternator**.

BentleyPublishers.com
BentleyPublishers.com—All Rights Reserved

Principles of operation

The crash safety (airbag) module performs the following functions:

- Crash detection and determining triggering point.
- Triggering ignition output stages.
- Documentation of crash data.
- System self-test.
- Cyclic monitoring.
- Display of system standby status.
- Fault indication and fault code storage.
- Fault output (diagnosis).
- Output of a crash telegram for other users in communication system network.
- Activation of airbag indicator light when passenger airbag is deactivated.

Crash detection

The crash safety module determines whether a crash has occurred based on the values detected by crash sensors and calculated by a triggering algorithm. Deceleration signals from two independent sensors are needed to indicate a crash. In the case of a frontal crash for example, the acceleration values from the B-pillar satellite and from the longitudinal acceleration sensor (in the control module) are needed.

Signals from different sensors help determine crash severity and direction. Based on this information, the stored algorithm determines the triggering (ignition) points of the restraint systems to be activated.

Triggering output stages

Signals from two different sensors, B-pillar satellite and crash safety module sensor, triggers airbag ignition output stages.

With ignition ON, voltage is supplied to the crash safety module, which becomes operational once it completes a system self-test. In case of a crash, a capacitor provides the energy reserve for airbag ignition even if electrical power is interrupted.

Documentation of actuator triggering

The triggering (ignition) commands are recorded in case of a crash where one or several actuators are triggered. The most important crash event data are stored in the form of a crash telegram in the non-volatile memory of the crash safety module. This data is only available to the manufacturer.

A maximum of 3 crash telegrams can be stored, after which the control module needs to be replaced.

System self-test

The control module performs a system self-test when the ignition is switched ON. The airbag warning light is illuminated for 3 - 5 seconds during system self-test.

BentleyPublishers.com
BentleyPublishers.com—All Rights Reserved

If system self-test finds no faults, the airbag warning light switches OFF and the system is ready for operation.

Cyclic monitoring

Once system self-test is successfully concluded and the system is ready for operation, cyclic fault monitoring is performed as long as ignition is ON.

Indication of system operability

System operability is indicated by the airbag warning light switching OFF in the instrument cluster.

Fault code storage

A fault in the system is indicated by the airbag warning light and stored in nonvolatile control module memory. A distinction is made between internal and external faults.

Fault output (diagnosis)

Using BMW scan tool or equivalent, fault codes can be read out via the diagnosis interface. After rectifying faults, the fault codes can be cleared by means of "clear fault code memory" diagnosis command.

Crash telegram output

In case of a crash which triggers restraint system components, the crash safety module sends a crash telegram via bus system. The following control modules then perform the following functions depending on crash severity:

Crash telegram responses	
Fuel pump control module (EKP)	Switches fuel pump OFF
Engine control module (ECM)	Switches alternator OFF
Body gateway module (KGM)	Releases central locking
Light module (LM)	Activates hazard warning lights
Light module (LM)	Activates interior lights
Telematic control module (TCU)	Places emergency call

Deactivation of passenger airbags

The crash safety module utilizes the OC-3 mat signal to deactivate the front and side passenger airbags when a child seat is mounted on the passenger side. Airbag deactivation is indicated by a yellow airbag indicator light.

Seat belt reminder

The seat belt reminder (SBR) function monitors whether the driver or passenger has fastened his or her seat belt. If the seat belt is not worn or is unbuckled a reminder is illuminated and a chime is sounded.

Left and right front seat belt buckle microswitches are monitored separately.

BentleyPublishers.com
BentleyPublishers.com—All Rights Reserved

Component Replacement

Vehicle center satellite (ASE), replacing

- Disconnect negative (–) cable from battery and cover negative terminal with insulating material.

> ***CAUTION—***
> - *Prior to disconnecting the battery, read the battery disconnection cautions given in* **001 Warnings and Cautions**.

- Remove center console between front seats. See **513 Interior Trim**.

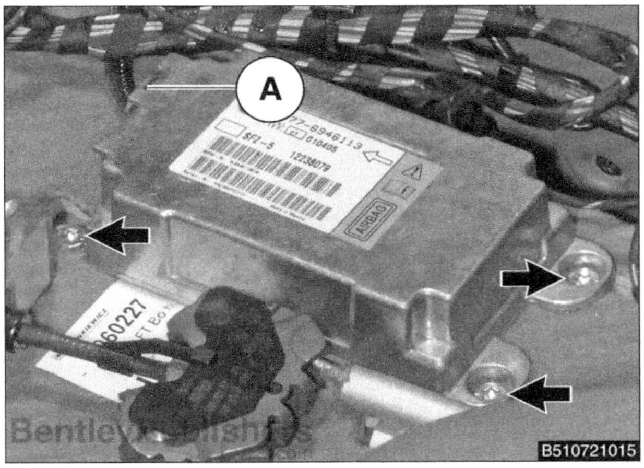

- Disconnect electrical harness connector (**A**). Remove vehicle center satellite mounting fasteners (**arrows**) and lift up module.
- Installation is reverse of removal.

Tightening torque	
Vehicle center satellite to center tunnel	8 Nm (6 ft-lb)

- Use BMW scan tool or equivalent to code new module.

Crash safety (airbag) module (ACSM), replacing

- Disconnect negative (–) cable from battery and cover negative terminal with insulating material.

> ***CAUTION—***
> - *Prior to disconnecting the battery, read the battery disconnection cautions given in* **001 Warnings and Cautions**.

- Remove center console between front seats. See **513 Interior Trim**.

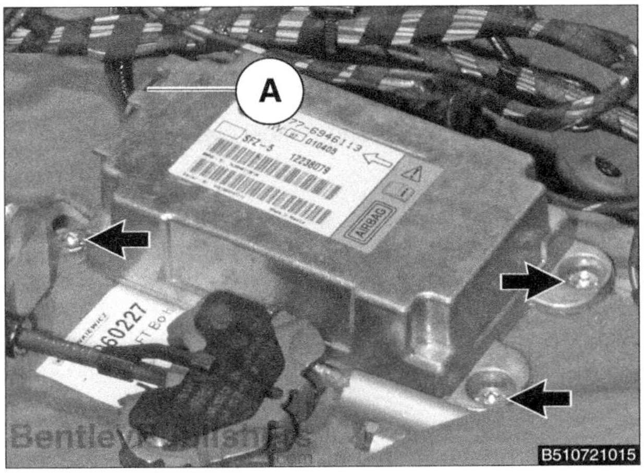

- Disconnect electrical harness connector (**A**). Remove vehicle center satellite mounting fasteners (**arrows**) and lift up module.
- Installation is reverse of removal.

Tightening torque	
Vehicle center satellite to center tunnel	8 Nm (6 ft-lb)

- Use BMW scan tool or equivalent to code new module.

BentleyPublishers.com—All Rights Reserved

Driver airbag, removing and installing

– Disconnect negative (–) cable from battery and cover negative terminal with insulating material.

> ***WARNING —***
> • *After disconnecting the battery, wait 1 minute before beginning work on airbag components.*

> ***CAUTION—***
> • *Prior to disconnecting the battery, read the battery disconnection cautions given in* **001 Warnings and Cautions**.

◄ Insert Torx screwdriver (T25 or T27) into concealed opening on back side of steering wheel until spring resistance is felt. Increase pressure on spring until airbag unit is unlocked.

– Tilt back airbag.

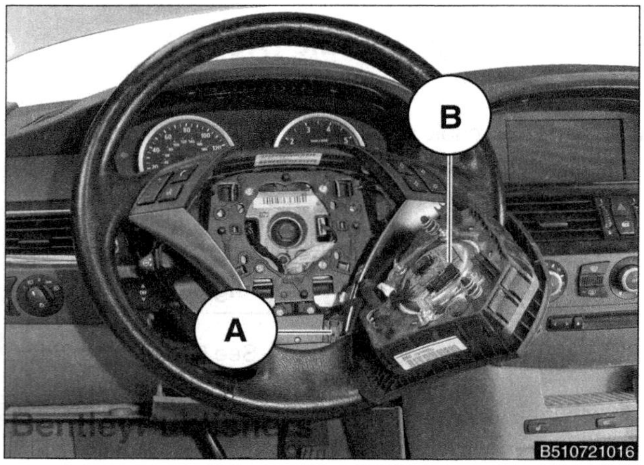

◄ Disconnect electrical connector (**A**) and remove airbag unit (**B**).

> ***WARNING —***
> • *Place a removed airbag unit with the padded side facing upward. Do not leave an airbag unit unattended.*
> • *Do not pinch airbag harness in center of steering wheel when removing or installing airbag.*

– Installation is reverse of removal:
  - Make sure electrical lead is positioned correctly.
  - Press airbag directly down on steering wheel until it snaps in place.

BentleyPublishers.com
BentleyPublishers.com–All Rights Reserved

Passenger airbag, removing and installing

If the passenger airbag deploys, it tears a hole through the dashboard surface. Be prepared to replace the dashboard.

– Disconnect negative (–) cable from battery and cover negative terminal with insulating material.

> ***WARNING —***
> - *After disconnecting the battery, wait 1 minute before beginning work on airbag components.*

> ***CAUTION—***
> - *Prior to disconnecting the battery, read the battery disconnection cautions given in* **001 Warnings and Cautions**.

– Remove glove compartment. See **513 Interior Trim**.

– **Vehicle with front cupholders:** Remove front outer cupholder carrier and release front inner cupholder carrier.

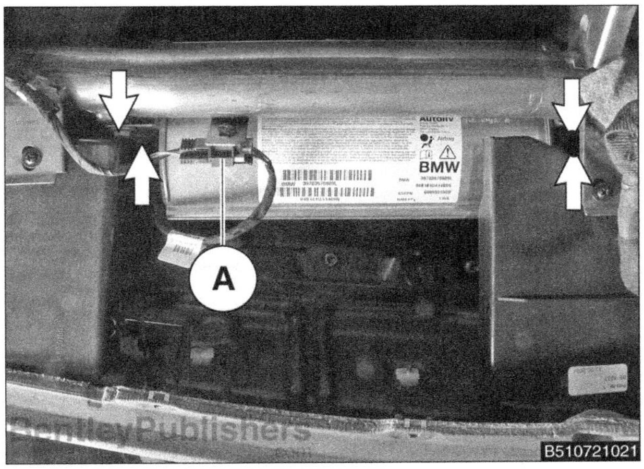

– Working from below right side of dashboard:
- Detach electrical connector (**A**).
- Move front inner cupholder carrier to side if applicable.
- While supporting airbag unit, remove mounting fasteners (**arrows**), then lower airbag.

> ***WARNING —***
> - *Place a removed airbag unit with the padded side facing upward. Do not leave an airbag unit unattended.*

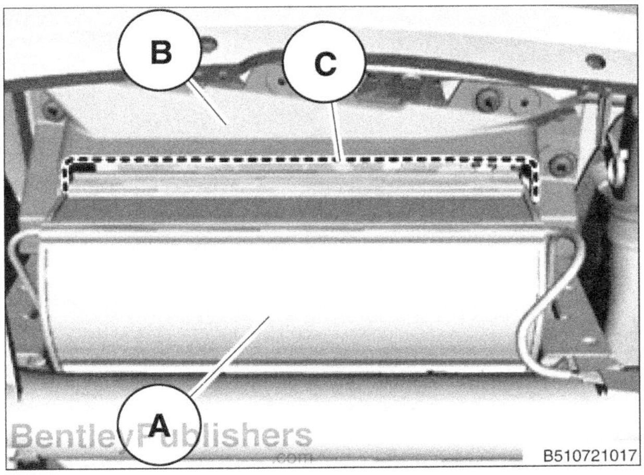

– Install airbag (**A**) in channel (**B**), making sure channel surrounds airbag module. Pull channel apart at sides if necessary.

– Remainder of installation is reverse of removal.

Tightening torque	
Passenger airbag to body crossbeam	8 Nm (6 ft-lb)

BentleyPublishers.com
BentleyPublishers.com—All Rights Reserved

Side airbag (front or rear), removing and installing

– Disconnect negative (–) cable from battery and cover negative terminal with insulating material.

> ***WARNING —***
> - *After disconnecting the battery, wait 1 minute before beginning work on airbag components.*

> ***CAUTION—***
> - *Prior to disconnecting the battery, read the battery disconnection cautions given in* **001 Warnings and Cautions**.
> - *Front and rear side airbags are not interchangeable. The side airbag module is marked either "VORNE / FRONT" or "HINTEN / REAR"*

– Remove door trim panel. See **411 Doors**.

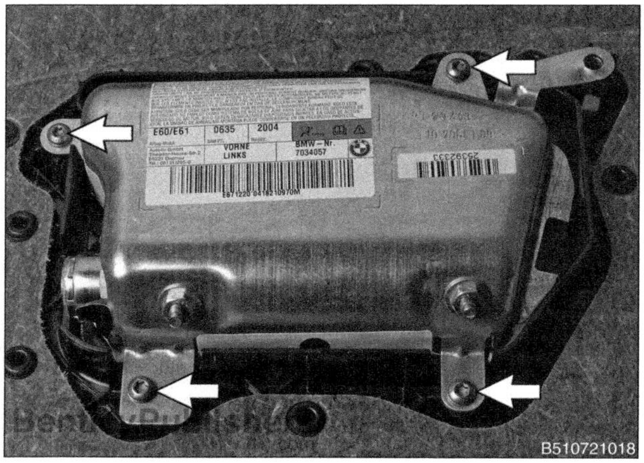

◀ Remove fasteners (**arrows**) and remove airbag module.

– Prior to installation, be sure to remove all foreign objects (fasteners, etc.) from airbag installation recess in door panel.

– Remainder of installation is reverse of removal.

BentleyPublishers.com
BentleyPublishers.com—All Rights Reserved

ECL Electrical Component Locations

TABLES

GENERAL

This repair group covers electrical component locations, fuse locations and ground points.

Electrical equipment and accessories included vary with model and model year. Confirm component identification whenever you begin work on the electrical system.

See also:
- **600 Electrical System–General** for bus systems
- **ELE Electrical Wiring Diagrams** for electrical schematics

BMW acronyms

BMW uses many abbreviations and acronyms in the technical literature, service bulletins, wiring diagrams and parts bulletins it distributes. See **Table a** for common BMW and industry acronyms.

BentleyPublishers.com
BentleyPublishers.com–All Rights Reserved

BMW acronyms

Table a. Common BMW acronyms	
Acronym	**Component or system**
A/C	air-conditioning
ABS	antilock brakes
ACC	active cruise control
AFS	active front steering
AGS	adaptive gearbox system
ARS	active stabilizer (dynamic drive)
AS	active steering
AUC	automatic air recirculation control
BST	battery safety terminal
CA	comfort access
CAN	controller area network (bus)
CAS	car access system (module)
CBS	condition-based service
CCC/M-ASK	car communications computer / audio system controller
CID	central information display
DISA	variable intake manifold system
DME	digital motor electronics
DMTL	diagnosis module—tank leakage
DSC	dynamic stability control
DWA	anti-theft alarm
ECM	engine control module
ECO	electronically controlled orifice
ECT	engine coolant temperature (sensor)
EDK	electronic throttle
EEPROM	flash programmable read-only memory
EGS	electronic transmission control
EHC	electronic height control (air suspension)
EKP	fuel pump module
ELV	electronic steering lock
EPROM	erasable / programmable read-only memory
EVAP	fuel tank evaporative control system
EWS	electronic immobilizer
FH	power windows
FTM	flat tire monitoring
GPS	global positioning system
HPS	head protection airbag
IAT	intake air temperature (sensor)
IBS	intelligent battery sensor
IHKA	automatic heating and air-conditioning
IKE	integrated instrument cluster module
ITS	inflatable tubular structure (head protection airbag)
KBM	basic body module

Table a. Common BMW acronyms	
Acronym	**Component or system**
KGM	body gateway module
KL 15	battery positive, ignition switch on RUN
KL 30	battery positive (B+), power
KL 31	battery / chassis ground
KL 50	ignition start position, power
KL R	battery positive, ignition on accessories
LDP	fuel tank leak diagnosis pump
LED	light emitting diode
LEV	low emissions vehicle
LIN	local interconnected network (bus)
MAP	manifold absolute pressure
MIL	malfunction indicator light
MPM	micro-power module
MRS	multiple restraint system (airbags)
NAV	navigation system
NLEV	national low emissions vehicle
NTC	negative temperature coefficient (resistor)
OBD II	second generation on-board diagnostics
PDC	park distance control
PWG	accelerator pedal module
RAM	random access memory
RDC	tire pressure control
RDW	tire pressure warning
RXD	receive data line
SGM	safety and gateway module
SHD	sunroof module
SMG	sequential manual gearbox
SRA	headlight cleaning system
SRS	supplemental restraint system (airbags)
SZL	steering column switch cluster
SZM	center console switch cluster
TAGE	electronic outer door handle module
TCM	automatic transmission control module
TCU	telematics control module
TDC	top dead center
TI	injection signal
TLEV	transitional low emissions vehicle
TPM	tire pressure monitoring
TXD	transmit data line
ULEV	ultra low emissions vehicle
VANOS	variable camshaft timing
VVT	Valvetronic

BentleyPublishers.com
BentleyPublishers.com – All Rights Reserved

Warnings and Cautions

WARNING—
- *The battery safety terminal, pyrotechnic seat belt tensioners and airbags utilize explosive devices. Handle with extreme care. See warnings and cautions in* **121 Battery, Alternator, Starter**, **720 Seat Belts**, *and* **721 Airbag System (SRS)**.

CAUTION—
- *Prior to disconnecting the battery, read the battery disconnection precautions in* **001 Warnings and Cautions**.
- *Relay and fuse positions vary from car to car. If questions arise, an authorized BMW dealer is the best source for the most accurate and up-to-date information.*
- *A good way to verify a relay position is to compare the wiring colors at the relay socket to the colors indicated on the wiring diagrams in* **ELE Electrical Wiring Diagrams**.
- *Switch the ignition OFF and remove the negative (-) battery cable before removing any electrical components. Connect and disconnect ignition system wires, multiple connectors, and ignition test equipment leads while the ignition is switched OFF.*
- *Use a digital multimeter for electrical tests.*

ELECTRICAL COMPONENTS

Electrical components table

Table b is a cross-referenced listing of electrical components.
- **Column 1**: Components alphabetized by names in current usage
- **Column 2**: BMW alphanumeric code (corresponds to code in electrical wiring diagrams)
- **Column 3**: Location in vehicle
- **Column 4**: Figure number in this repair group, if illustrated
- **Column 5**: Page number for illustration

Table b. Component locations

Component	Code	Location	Fig #	Page #
12v outlet, center console	A839	Rear of center console between front seats		
12v socket, rear	A14112	Right side trunk or cargo compartment		
A/C		*see also* IHKA entries		
A/C and heating control panel (module)	A11a	Center dashboard		
A/C compressor control valve	Y2a	Right side of engine on A/C compressor	**Fig 58**	**Pg 55**
A/C refrigerant pressure sensor	B97	Right side engine compartment underneath coolant reservoir	**Fig 30**	**Pg 45**
ABS		*see also* DSC entries		
ABS / DSC control module	A65	Right front engine compartment	**Fig 38**	**Pg 48**
ABS wheel speed sensor	B1, B2 B3, B4	Wheel hub, left front, right front, left rear, or right rear wheel	**Fig 136**	**Pg 74**
Acceleration sensor, DSC	B75	*from 09 / 2005:* Left front floor, underneath seat		
Acceleration sensor, DSC	B75	*to 09 / 2005:* Right front floor, underneath seat	**Fig 87**	**Pg 62**
Acceleration sensor, left front	B6249	Behind left front wheel housing, behind wheel	**Fig 130**	**Pg 73**

ECL

BentleyPublishers.com
BentleyPublishers.com—All Rights Reserved

Electrical components table

Table b. Component locations				
Component	**Code**	**Location**	**Fig #**	**Page #**
Acceleration sensor, right front	B49	Right front floor, underneath rug	**Fig 87**	**Pg 62**
Acceleration sensor, right front (AFS)	B41	Underneath right front fender	**Fig 134**	**Pg 74**
Acceleration sensor, right rear	B48	Trunk or cargo compartment, right rear behind trim	**Fig 115**	**Pg 69**
Accelerator pedal sensor	B10	Integrated with accelerator pedal	**Fig 78**	**Pg 60**
Active backrest		*see* Backrest adjustment (active backrest) module, driver, passenger seat		
Active cruise control distance sensor	B155	Center right, front bumper		
Active cruise control module	A144a	Center of front bumper, underneath right kidney grille		
Active cruise control proximity sensors	B153 B154	Sides of front grill, left, right		
Active stabilizer		*see* Dynamic drive (active stabilizer) entries		
Active steering control module	A467	Passenger seat floor, underneath rug	**Fig 89**	**Pg 63**
Active steering electronically controlled orifice (ECO) valve	Y14040a	Behind power steering pump, left lower engine	**Fig 27**	**Pg 44**
Active steering motor	M179	Steering shaft at steering rack		
Active steering yaw sensor	B105	*from 09 / 2005:* Underneath passenger seat		
Active steering yaw sensor	B105	*to 09 / 2005:* Underneath rear of driver seat, under center console	**Fig 82**	**Pg 61**
Adaptive headlights control module	A214	Behind glove compartment		
Air suspension		*see* EHC (air suspension) entries		
Airbag		*see also* Crash entries		
Airbag and horn contact (volute spring)	I01086	Underneath steering wheel, at steering column switch cluster (SZL)		
Airbag deactivation switch, passenger airbag	S559	Right end of dashboard		
Airbag gas generator, head protection, left, right	G17 G18	A-pillar next to base of windshield, left, right		
Airbag sensor, rear side airbag	A173a A174a	Lower B-pillar under trim		
Airbag, driver	G5	In steering wheel under horn button		
Airbag, knee, driver	G30	Dashboard lower trim, underneath steering column		
Airbag, knee, passenger	G31	Underneath glove compartment		
Airbag, passenger	G6	Under right end of dashboard	**Fig 77**	**Pg 60**
Airbag, side, driver, passenger	G14b G15b	Front door, under trim, left, right	**Fig 93**	**Pg 64**
Airbag, side, rear	G20a G21a	Left rear, right rear door, under trim panel	**Fig 94**	**Pg 64**
Alarm		*see* Anti-theft		
Alternator (generator)	G6524	Left front of engine	**Fig 38** **Fig 48**	**Pg 48** **Pg 51**
Amplifier, sound system	N40a	Trunk or cargo compartment, left rear behind trim	**Fig 112** **Fig 114** **Fig 120**	**Pg 68** **Pg 69** **Pg 70**
Antenna		*see* **Antenna components** in this repair group *see also* Comfort access (CA) antennas *see also* Tire pressure control (RDC) antenna		**Pg 39**

BentleyPublishers.com
BentleyPublishers.com—All Rights Reserved

Table b. Component locations

Component	Code	Location	Fig #	Page #
Antenna amplifier and diversity module	N8a	*Sedan:* Top right of rear window	**Fig 16**	**Pg 40**
Antenna amplifier and diversity module	N8a	*Sports Wagon:* Top of rear window, behind spoiler	**Fig 19**	**Pg 40**
Antenna noise suppression filter	Z13	*Sedan:* Rear parcel shelf behind trim, left	**Fig 14**	**Pg 39**
Antenna noise suppression filter	Z13 Z402	*Sports Wagon:* Right rear headliner, behind trim	**Fig 18** **Fig 19**	**Pg 40** **Pg 40**
Antenna, SDRS (satellite radio)	I01106	Roof top, center rear	**Fig 19**	**Pg 40**
Antenna, VICS	A457	Center top windshield	**Fig 12**	**Pg 39**
Anti-theft alarm system control module (DWA)	A121a	Integrated with anti-theft alarm system control module (DWA)		
Anti-theft siren and tilt sensor (DWA)	H1b	Behind right rear bumper cover	**Fig 141**	**Pg 76**
AUC sensor		*see* Automatic recirculation sensor		
Audio system controller / car communications computer (CCC/M-ASK)	N38a	Center dashboard		
Audio-visual socket	A18807	Rear of center console between front seats		
Automatic recirculation sensor	B414	*to 09 / 2005:* Right front engine compartment on radiator	**Fig 29**	**Pg 45**
Automatic recirculation sensor	B414	*from 09 / 2005:* Right cabin microfilter housing cover, above E-box		
Automatic transmission control module (TCM)	A7000a	In transmission housing, rear *see also* SMG control module		
Automatic transmission gear selector switch (GWS)	S227a	Center console, front		
Automatic transmission selector lever module	A134	Left of selector lever on center console		
Automatic transmission steptronic switch	S224	Under selector lever bezel, center console		
Auxiliary heater	I01062	IHKA housing, underneath top dashboard trim	**Fig 21**	**Pg 41**
Auxiliary sound system connector	A267a	Rear of center console between front seats		
B+ jump start terminal	G6430	Left rear engine compartment	**Fig 38** **Fig 53**	**Pg 48** **Pg 53**
B+ power distribution	A2076	E-box, right rear engine compartment	**Fig 5** **Fig 8**	**Pg 29** **Pg 30**
Backrest adjustment (active backrest) module, front seat	A853 A854	Underneath front seat	**Fig 23** **Fig 98**	**Pg 43** **Pg 65**
Back-up light switch		*Automatic transmission:* Transmission range switch		
Back-up light switch	S8511	*Manual transmission:* Right rear transmission housing		
Basic body module (KBM)	A1a	Behind glove compartment	**Fig 22** **Fig 23** **Fig 85**	**Pg 42** **Pg 43** **Pg 62**
Battery	G1	Right side trunk or cargo compartment, behind trim	**Fig 115**	**Pg 69**
Battery safety terminal (BST)	G19a	Positive battery terminal, in trunk or cargo compartment	**Fig 115**	**Pg 69**
Battery sensor	B895	Battery negative post, top of battery, in trunk or cargo compartment		
Bistable relay		*2004 - 2005:* In micro-power module (MPM) *2006 - 2010:* In body-gateway module (KGM)		
Blower	M30	Under right side dashboard		
Blower output stage	N2	On blower under right side dashboard		
Body-gateway module (KGM)	A16760	Behind glove compartment	**Fig 85**	**Pg 62**

ECL

BentleyPublishers.com
BentleyPublishers.com—All Rights Reserved

Electrical components table

Table b. Component locations				
Component	**Code**	**Location**	**Fig #**	**Page #**
Brake disc ventilation flap sensor, right, left	I01224 I01225	Behind right, left edge of front of bumper		
Brake fluid level switch	B18a	Brake fluid reservoir, left rear engine compartment	**Fig 38** **Fig 53**	**Pg 48** **Pg 53**
Brake light switch	S29	Above brake pedal		
Brake pad wear sensor, left front	B16a	Left front brake caliper	**Fig 128**	**Pg 72**
Brake pad wear sensor, right front	B118	Right front brake caliper		
Brake pad wear sensor, right rear	B17a	Right rear brake caliper		
BST		*see* Battery safety terminal (BST)		
CA		*see* Comfort access (CA) entries		
Camshaft sensor, exhaust	B6224	*M54 engine:* Front of cylinder head, left	**Fig 37**	**Pg 47**
Camshaft sensor, intake	B6214	*M54 engine:* Front of cylinder head, left	**Fig 36**	**Pg 47**
Camshaft sensor, intake, exhaust	B6214 B6224	*N52 engine:* Front of cylinder head, left, right	**Fig 42**	**Pg 49**
Camshaft sensor, intake, exhaust	B6214 B6224	*N54 engine:* Front of cylinder head, left, right	**Fig 51**	**Pg 52**
Camshaft sensors, intake, exhaust	B62141 B62142 B62241 B62242	*V8 engine:* Rear of cylinder heads	**Fig 54** **Fig 56** **Fig 61** **Fig 62**	**Pg 54** **Pg 54** **Pg 56** **Pg 56**
Car access system (CAS) module	A149a	Underneath left dashboard, left of steering column	**Fig 22** **Fig 23** **Fig 79**	**Pg 42** **Pg 43** **Pg 60**
Car communications computer		*see* Audio system controller / car communications computer (CCC/M-ASK)		
Cargo compartment roller cover motor	M237	Left rear cargo compartment behind trim	**Fig 122**	**Pg 71**
Cargo compartment roller cover release	M300 M301	D-pillar behind quarter glass, left, right	**Fig 122**	**Pg 71**
CAS		*see* Car access system (CAS) module		
CCC/M-ASK		*see* Audio system controller / car communications computer (CCC/M-ASK)		
CD changer	N22	Right side dashboard behind glove compartment *see also* DVD changer	**Fig 84**	**Pg 61**
Center brake light	H34a	Above rear window	**Fig 16**	**Pg 40**
Center console switch center (SZM)	A169	Above center console in lower center dashboard		
Center console switch center electronics	A862	Behind glove compartment		
Central information display	A165	Center dashboard		
Charging and hands-free module	U400c	*Sedan:* Left rear trunk, behind trim		
Charging and hands-free module	U400c	*Sports Wagon:* Floor of cargo compartment		
Charging and hands-free module auxiliary battery	G402	*Sedan:* Left rear trunk, behind trim		
Charging and hands-free module auxiliary battery	G402	*Sports Wagon:* Floor of cargo compartment		
Classification resistor		*see* xDrive classification resistor		
Clutch sensor (SMG)	B6574	Left side SMG transmission	**Fig 110**	**Pg 68**
Clutch switch	S805a	Above clutch pedal		

BentleyPublishers.com
BentleyPublishers.com—All Rights Reserved

Table b. Component locations

Component	Code	Location	Fig #	Page #
Comfort access (CA) antennas	I01039 W10a - W15	Throughout interior *see also* **Antenna components**	**Fig 22** **Fig 23**	**Pg 42** **Pg 43** **Pg 39**
Comfort access (CA) control module	A215	Behind glove compartment	**Fig 22** **Fig 86**	**Pg 42** **Pg 62**
Compressor		*see* A/C compressor		
Condensation sensor	B100	Top center of windshield		
Coolant		*see* Engine coolant entries		
Cooler box	A445	Rear seat armrest		
Crankcase breather heater	E65390	*N52 engine:* Intake manifold	**Fig 39**	**Pg 49**
Crankcase breather heater	E65390	*N54 engine:* Intake manifold	**Fig 49**	**Pg 49**
Crankcase breather heater	E65393 - E65395	*N52 engine:* Behind left rear of cylinder head at intake manifold	**Fig 47**	**Pg 51**
Crankcase breather heater relay	K6539	E-box, right rear engine compartment		
Crankshaft sensor	B6203	*V8 engine:* Right side bell housing	**Fig 125**	**Pg 72**
Crankshaft sensor	B6203	*6-cylinder engine:* Left side bell housing	**Fig 49**	**Pg 51**
Crash module, center satellite	A161	Under center console, between front seats		
Crash modules, left, right satellite	A157 A158	Behind B-pillar trim, left, right	**Fig 90**	**Pg 63**
Crash safety module	A12a	Between front seats, under center console		
Crash sensor, front left, front right	B10508, B10509	Engine compartment front, underneath headlight assemblies	**Fig 26**	**Pg 44**
Crash sensor, front side airbag	B78a B79a	Front of front door behind trim, left, right	**Fig 93**	**Pg 64**
Digital tuner	N41 N48	Trunk or cargo compartment, left rear behind trim	**Fig 113** **Fig 121**	**Pg 69** **Pg 71**
DISA		*see* Variable intake manifold (DISA) controllers		
Diversity		*see* Antenna amplifier and diversity module		
DLC		*see* OBD II socket		
DME main relay	K6300	E-box, right rear engine compartment	**Fig 6** **Fig 8**	**Pg 29** **Pg 30**
Door handle control module, front, rear	S263 S264 S265a S266a	Outer door handle, front, rear, right, left door		
Door lock motor, soft close, right, left, front, rear door	M408 - M411	Rear edge, lower door cavity	**Fig 93**	**Pg 64**
Door locks, driver door, passenger door	S246 S254	Rear of door in door cavity, left, right	**Fig 93**	**Pg 64**
Door locks, right rear, left rear	S255a S256a	Rear edge of door inside door cavity, right, left,	**Fig 94**	**Pg 64**
Door module, driver door, passenger door	A23a A24a	Inside driver door cavity, upper front		
Door switch cluster, driver door	S126	Driver door arm rest		
DSC		*see also* ABS entries		
DSC / DTC switch	S6	Center console switch center (SZM)		
DSC control module		*see* ABS / DSC control module		

BentleyPublishers.com—All Rights Reserved

Table b. Component locations

Component	Code	Location	Fig #	Page #
DSC hydraulic unit		On DSC control module		
DSC steering angle sensor		Steering column switch center (SZL)		
DVD changer	N39a	Trunk or cargo compartment, left rear behind trim *see also* CD changer	**Fig 113**	**Pg 69**
DVD changer cooling fan	M304	*Sedan:* Left rear trunk, behind trim		
DWA		*see* Anti-theft entries		
Dynamic drive (active stabilizer) control module	A148a	Behind right kick panel trim	**Fig 86**	**Pg 62**
Dynamic drive (active stabilizer) lateral acceleration sensor	A139	Right front floor underneath rug	**Fig 87**	**Pg 62**
Dynamic drive (active stabilizer) valve block	Y40	Behind right front wheel, behind wheel housing liner		
Dynamic stability control		*see* ABS / DSC *see also* DSC		
E-box		Right rear engine compartment under cover *see* **E-box fuses, relays and modules**	**Fig 38** **Fig 53**	**Pg 48** **Pg 53**
E-box cooling fan	M6506	Base of E-box, right rear engine compartment		
ECM		*see* Engine control module (ECM)		
ECO valve		*see* Active steering electronically controlled orifice (ECO) valve		
EHC (air suspension) compressor	Y22	Underneath vehicle, between spare tire well and battery box	**Fig 143**	**Pg 76**
EHC (air suspension) compressor relay	K126	*Sports Wagon:* Spare tire well	**Fig 24** **Fig 25**	**Pg 43** **Pg 44**
EHC (air suspension) module	A118a	*Sports Wagon:* Spare tire well	**Fig 24** **Fig 25**	**Pg 43** **Pg 44**
EHC (air suspension) sensor, front	B42	Right front suspension	**Fig 137**	**Pg 75**
EHC (air suspension) sensor, front	B69 B70	Rear suspension, right, left	**Fig 140**	**Pg 76**
EHC (air suspension) valves, rear	Y403 Y404	Rear air spring towers, behind rear seat backrest	**Fig 118**	**Pg 70**
Electric cooling fan	M135	Between radiator and engine	**Fig 29**	**Pg 45**
Electronic height control		*see* EHC (air suspension) entries		
Electronic power steering		*see* Active steering entries		
Electronics box		*see* E-box		
Emergency call button	S253a	Front dome light unit	**Fig 76**	**Pg 59**
Engine control module (ECM)	A6000	E-box, right rear engine compartment	**Fig 6** **Fig 8** **Fig 10** **Fig 38** **Fig 53**	**Pg 29** **Pg 30** **Pg 32** **Pg 48** **Pg 53***
Engine control module relay		*see* DME main relay		
Engine coolant level switch	S63b	Bottom of coolant reservoir, right front engine compartment		
Engine coolant pump, electric	M6035	*N52 engine:* Right front lower engine compartment	**Fig 46**	**Pg 50**
Engine coolant pump, electric	M6035	*N54 engine:* Right front lower engine compartment	**Fig 52**	**Pg 52**
Engine coolant temperature (ECT) sensor	B6236	*M54 engine:* Cylinder head, underneath intake manifold	**Fig 32**	**Pg 46**
Engine coolant temperature (ECT) sensor	B6236	*N54 engine:* Left front of cylinder head, on thermostat housing cover	**Fig 51**	**Pg 52**

BentleyPublishers.com
BentleyPublishers.com—All Rights Reserved

Table b. Component locations				
Component	**Code**	**Location**	**Fig #**	**Page #**
Engine coolant temperature (ECT) sensor	B6236	*N62 engine:* Coolant manifold, front of engine	**Fig 57**	**Pg 55**
Engine coolant temperature (ECT) sensor	B6236	*N62TU engine:* Coolant manifold, front of engine	**Fig 65**	**Pg 57**
Engine coolant temperature (ECT) sensor	B6236a	*N52 engine:* Front of engine on coolant outlet	**Fig 38** **Fig 43**	**Pg 48** **Pg 50**
Engine coolant temperature sensor, radiator outlet	B604	Lower radiator hose, right front engine compartment	**Fig 29**	**Pg 45**
Engine coolant thermostat	Y6279	*M54 engine:* Front of cylinder head on coolant manifold	**Fig 37**	**Pg 47**
Engine coolant thermostat	Y6279	*N52 engine:* Right front lower engine compartment on electric coolant pump	**Fig 46**	**Pg 50**
Engine coolant thermostat	Y6279	*N54 engine:* Right front lower engine compartment on electric coolant pump	**Fig 52**	**Pg 52**
Engine electronics fuse carriers	A8680 A8681 A8684 A8685	*see* **Fuse Locations and Ratings** in this repair group *see also* **E-box fuses, relays and modules** in this repair group		**Pg 21** **Pg 29**
Engine hood contact switch	S19	E-box, right rear engine compartment		
Engine hood lock microswitch, right, left	A137 A138	Radiator upper support, right, left		
Entertainment center, rear compartment	N554	Rear of center console between front seats		
Evaporative emissions valve	Y6120	*M54 engine:* Left front of cylinder head	**Fig 35**	**Pg 46**
Evaporative emissions valve	Y6120	*N52 engine:* Left front of cylinder head	**Fig 40**	**Pg 49**
Evaporative emissions valve	Y6120	*N54 engine:* Behind cylinder head, left side	**Fig 49**	**Pg 51**
Evaporative emissions valve	Y6120	*N62 engine:* Front of left cylinder head	**Fig 55**	**Pg 54**
Evaporative emissions valve	Y6120	*N62TU engine:* Front of left cylinder head	**Fig 66**	**Pg 57**
Exhaust camshaft sensor		*see* Camshaft sensor		
Fan cutoff relay	K9137	Trunk or cargo compartment, right rear behind trim	**Fig 116**	**Pg 69**
Final stage unit		*see* Blower final stage		
Flashlight charging socket, glove compartment	I01100	Left edge of glove compartment		
Fresh air temperature sensor	B13729	Left side dashboard ducts		
Front fuse and relay panel	A41a	Behind glove compartment	**Fig 84** **Fig 85**	**Pg 61** **Pg 62**
Fuel filler flap lock motor	M16	Trunk or cargo compartment, right rear behind trim		
Fuel injector relay	K6327	E-box, right rear engine compartment	**Fig 6**	**Pg 29**
Fuel injectors	Y6101 - Y6108	Non-turbo: Intake manifold Turbo: Top of cylinder head		
Fuel level sensor	B25a	Top of fuel tank, underneath rear seat		
Fuel pressure sensor, low pressure	B6125a	*N54 engine:* Left side of engine underneath intake manifold	**Fig 48**	**Pg 51**
Fuel pump	M2	Underneath right rear seat cushion, top of fuel tank		
Fuel pump control module (EKPS)	A13663	Behind right rear shock absorber tower, right side trunk or cargo compartment, behind trim	**Fig 123**	**Pg 71**
Fuel pump relay	K96	*to 09 / 2005:* Behind glove compartment	**Fig 84**	**Pg 61**
Fuel rail pressure sensor	B2261	*N54 engine:* Left side of engine underneath intake manifold	**Fig 48**	**Pg 51**
Fuel tank flap emergency release cable		Trunk or cargo compartment, behind right trim		
Fuel tank leakage diagnostic module	M119a	Underneath vehicle, behind left rear wheel housing	**Fig 142**	**Pg 76**

Bentley Publishers
BentleyPublishers.com—All Rights Reserved

Electrical components table

Table b. Component locations				
Component	**Code**	**Location**	**Fig #**	**Page #**
Fuel volume control valve	Y2381	*N54 engine:* Left side of engine underneath intake manifold, on high pressure fuel pump	**Fig 48**	**Pg 51**
Fuse carrier, engine electronics		*see* **E-box fuses, relays and modules** in this repair group		**Pg 29**
Fuses		*see* **Fuse Locations and Ratings** in this repair group		**Pg 21**
Gear indicator light, automatic transmission	E82	Under selector lever bezel		
Gearshift paddles	I01091 I01092	On steering wheel spokes		
Gearshift sensor (SMG)	B6572	Top of SMG transmission	**Fig 109**	**Pg 68**
Gearshift solenoids (SMG)	Y6561 - Y6564	Underneath rear of SMG transmission	**Fig 108**	**Pg 67**
Generator		*see* Alternator (generator)		
Glove compartment light switch	S37a	Left of glove compartment, underneath dashboard		
GPS		*see* Audio system controller / car communications computer (CCC/M-ASK) *see also* Navigation		
Ground bus	I01053	Right of E-box, right rear engine compartment	**Fig 5**	**Pg 29**
Grounds		*see* **Grounds** in this repair group		**Pg 34**
GWS		*see* Automatic transmission gear selector switch (GWS)		
Hall effect camshaft position sensor		*see* Camshaft sensor		
Hand-free microphone	B429a	Front dome light unit, left	**Fig 76**	**Pg 59**
Hazard warning switch	S18	Center dashboard		
Head restraint horizontal extender, driver, passenger	G43 G44	Head restraint, driver, passenger seat		
Head restraint vertical extender, driver, passenger	G34a G35a	Top of seat backrest, driver, passenger seat	**Fig 107**	**Pg 67**
Head up display projector	A503	Left front dashboard		
Headlight washer pump		*see* Washer pumps		
Headlight washer relay	K6	*to 09 / 2005:* Behind glove compartment	**Fig 84**	**Pg 61**
Headlight washer relay	K6	*from 09 / 2005:* Rear power distribution and fuse panel, right rear trunk or cargo compartment, behind trim	**Fig 3**	**Pg 23**
Headset socket	A468	Rear of center console between front seats		
Heated oxygen sensor		*see* Oxygen sensor		
Heater valve	Y4	Left rear engine compartment	**Fig 28**	**Pg 45**
Heating		*see* **Climate control components** in this repair group		**Pg 41**
Horn button	S4 S4a	Steering wheel center		
Horn contact spring		*see* Airbag and horn contact (volute spring)		
Horns	H2 H3	Behind front bumper, right side	**Fig 133**	**Pg 74**
Hot-film mass air flow sensor		*see* Mass air flow sensor		
Hydraulic pressure sensor (SMG)	B6571	Underneath rear of SMG transmission	**Fig 108**	**Pg 67**
Hydraulic pump (SMG)	Y6975	Underneath SMG transmission	**Fig 108**	**Pg 67**
IBS		*see* Battery sensor		

BentleyPublishers.com
BentleyPublishers.com—All Rights Reserved

Table b. Component locations				
Component	**Code**	**Location**	**Fig #**	**Page #**
Idle speed control valve	Y6130	*M54 engine:* Underneath intake manifold, above throttle housing	**Fig 33**	**Pg 49**
iDrive controller	A167a S14099	Center console between front seats		
Ignition coils	T6151 - T6158	Top of cylinder head cover(s) above spark plugs	**Fig 38** **Fig 53**	**Pg 48** **Pg 53**
Ignition coils suppression capacitor	I01046	*N52 engine:* Between ignition coil connectors 2 and 3	**Fig 44**	**Pg 50**
Ignition coils suppression capacitor	I01046	*V8 engine:* Next to ignition coil 8 connector	**Fig 60**	**Pg 55**
Ignition key slot	S2 S14028	Right side of steering column		
IHKA		*see* **Climate control components** in this repair group		**Pg 41**
IHKA control panel	A11a	Center dashboard		
Infrared interface, rear seat entertainment	I01174	Rear of center console between front seats		
Instrument cluster module	A2a	Dashboard		
Intake camshaft sensor		*see* Camshaft sensor		
Intake pressure (MAP) sensor	B6239	*N52 engine:* Left rear of intake manifold	**Fig 39**	**Pg 49**
Intake pressure (MAP) sensor	B6239	*N54 engine:* Air intake duct, left rear engine compartment	**Fig 49**	**Pg 51**
Intake pressure (MAP) sensor	B6239	*N62 engine:* Rear of right cylinder head	**Fig 59**	**Pg 55**
Intake pressure (MAP) sensor	B6239	*N62 TU engine:* Rear of intake manifold	**Fig 63**	**Pg 56**
Intake resonance valve		*see* Variable intake manifold (DISA) controllers		
Intake temperature and boost pressure sensor	B6583	*N54 engine:* Rear of intake manifold	**Fig 49**	**Pg 51**
Integrated voltage supply module (IVM)	A6009	E-box, right rear engine compartment	**Fig 10**	**Pg 32**
Intelligent battery sensor		*see* Battery sensor		
Interior movement sensor (ultrasonic detector)		Rear dome light unit		
IVM		*see* Integrated voltage supply module (IVM)		
KBM		*see* Basic body module (KBM)		
KGM		*see* Body-gateway module (KGM)		
Knock sensor	B6257 B6258	*V8 engine:* Underneath intake manifold on engine block		
Knock sensors	B62400	*6-cylinder engine:* Left side of engine underneath intake manifold	**Fig 48**	**Pg 51**
Level sensor		*see* EHC (air suspension) entries		
Light module	A3	Underneath left dashboard, left of steering column	**Fig 22** **Fig 23** **Fig 79**	**Pg 42** **Pg 43** **Pg 60**
Light sensor		*see* Rain and light sensor		
Light switch	S8	Left of steering column		
Load reduction (load shedding) relay terminal 15	K9	*from 09 / 2005:* Behind glove compartment	**Fig 84**	**Pg 61**
Longitudinal dynamics management control module	A144b	Behind glove compartment		
Low fuel pressure sensor		*see* Fuel rail pressure sensor		
MAP sensor		*see* Intake pressure (MAP) sensor entries		

ECL

BentleyPublishers.com
BentleyPublishers.com—All Rights Reserved

Electrical components table

Table b. Component locations				
Component	**Code**	**Location**	**Fig #**	**Page #**
Mass air flow sensor	B6207 B6207b	Air intake duct between air filter housing and throttle housing	**Fig 38** **Fig 68**	**Pg 48** **Pg 57**
Microphone	B400	Front dome light unit, right		
Micro-power module (MPM)	A16021	*Sports Wagon:* Spare tire well	**Fig 24** **Fig 25**	**Pg 43** **Pg 44**
Mirror, electrochromic (inside)	A22	Top center of windshield		
Mirror, outside, left, right	Y5 Y6	Driver door, passenger door		
Mixing flap control, rear compartment	R012	Rear of center console between front seats		
MOST-bus connector	I14255	Trunk or cargo compartment, left rear behind trim	**Fig 111** **Fig 120**	**Pg 68** **Pg 70**
MPM		*see* Micro-power module (MPM)		
MRS		*see* Crash entries		
Multifunction switch assemblies	I01032 I01033	On steering wheel spokes		
Navigation module	A112a	Left side trunk or cargo compartment behind trim		
Neutral safety switch	Y19	Under selector lever bezel, center console		
Night vision camera	A860	Front bumper, center left		
Night vision module	A859	Behind glove compartment		
O_2 sensor		*see* Oxygen sensor		
OBD II socket	X19527	Edge of driver door post		
Oil condition sensor (OZS)	B62540	Oil pan	**Fig 125**	**Pg 72**
Oil level sensor	B6254	Oil pan		
Oil pressure control valve	Y6384	*N52K engine:* Right lower crankcase, toward front	**Fig 45**	**Pg 50**
Oil pressure switch	B6231	*M54 engine:* Left front of cylinder head	**Fig 35**	**Pg 46**
Oil pressure switch	B6231	*N52 engine:* Left front of cylinder head	**Fig 38**	**Pg 48**
Oil pressure switch	B6231	*N54 engine:* Left side of engine underneath intake manifold	**Fig 48**	**Pg 51**
OIl pressure switch	B6231	*N62 engine:* Right front of right cylinder head	**Fig 58**	**Pg 55**
OIl pressure switch	B6231	*N62TU engine:* Right front of right cylinder head	**Fig 65**	**Pg 57**
OIl temperature sensor	B6238	*M54 engine:* Left front of cylinder head	**Fig 35**	**Pg 46**
Outside temperature sensor	B21a	Underneath engine on lower splash shield	**Fig 126**	**Pg 72**
Oxygen sensor		*see* **Oxygen sensors** in this repair group		**Pg 57**
Oxygen sensor connectors	X62101 X62102 X62201 X62202	Top or sides of engine	**Fig 38** **Fig 67** **Fig 69** **Fig 74** **Fig 75**	**Pg 48** **Pg 57** **Pg 58** **Pg 59** **Pg 59**
Panorama		*see* Sunroof		
Park / neutral position switch		*see* Neutral safety switch		
Park distance control (PDC) module	A81a	Spare tire well	**Fig 22** **Fig 25**	**Pg 42** **Pg 44**
Park distance control sensor, rear	B34a - B37a	Rear bumper		
Park distance control sensors, front	B30a - B33a	Front bumper		

BentleyPublishers.com
BentleyPublishers.com–All Rights Reserved

Table b. Component locations

Component	Code	Location	Fig #	Page #
Parking brake warning switch	S31	Base of parking brake handle, under center console		
Passenger airbag warning light	H23	Front dome light unit	**Fig 76**	**Pg 59**
PDC		*see* Park distance control (PDC) entries		
Power steering		*see* Steering *see also* Active steering		
Power steering fluid level switch	S65	Power steering fluid reservoir		
Radiator outlet coolant temperature sensor		*see* Engine coolant temperature sensor, radiator outlet		
Radiator shutter drive motor	M1997	Below center of radiator		
Radiator shutter solenoid	Y13864	Underneath center of radiator behind front bumper		
Rain and light sensor	B57b	Top center of windshield	**Fig 22** **Fig 23** **Fig 76**	**Pg 42** **Pg 43** **Pg 59**
RDC		*see* Tire pressure control (RDC) entries		
Rear power distribution and fuse panel	A46	Above battery, right rear trunk or cargo compartment, behind trim	**Fig 2** **Fig 3**	**Pg 23** **Pg 23**
Rear window defogger	E9	Rear window		
Rear window defogger relay	I01061	Rear power distribution and fuse panel, right rear trunk or cargo compartment, behind trim	**Fig 2** **Fig 3**	**Pg 23** **Pg 23**
Rear window lock motor	M96	*Sports Wagon:* Rear window lower edge, integrated with rear window wiper assembly		
Rear window release button	S147a	*Sports Wagon:* Base of rear window		
Relay, terminal 15	I01069	Rear power distribution and fuse panel, right rear trunk or cargo compartment, behind trim	**Fig 2** **Fig 3**	**Pg 23** **Pg 23**
Relay, terminal 30g	I01068	Rear power distribution and fuse panel, right rear trunk or cargo compartment, behind trim	**Fig 2** **Fig 3**	**Pg 23** **Pg 23**
Resonance valve		*see* Variable intake manifold (DISA) controllers		
Ride height control (EHC)		*see* EHC (air suspension) entries		
Roller cover position switch	S562	*Sports Wagon:* Right D-pillar		
rpm sensor	Y8616	Right side SMG transmission	**Fig 109**	**Pg 68**
Run flat indicator switch	I01137	Center console switch center (SZM)		
Safety and gateway module (SGM)	A220	Behind glove compartment	**Fig 85**	**Pg 62**
Satellite radio	N47	Trunk or cargo compartment, left rear behind trim	**Fig 114**	**Pg 69**
Seat belt pretensioner, driver	G39	Base of driver seat backrest, left		
Seat belt tensioner, pyrotechnic, left, right, front, rear	G12a G13a G25 G26	Base of seat belt buckle, left, right, front or rear seats		
Seat components		*see* **Seat components** in this repair group		**Pg 64**
Seat control module, front seat	A186 A187	Underneath front seat	**Fig 22** **Fig 23**	**Pg 42** **Pg 43**
Seat control switch, front	S50 S56	Outboard edge of front seat, left, right		
Seat heater coil, left, right, front, rear	E56a - E59a E65a - E68a	Seat cushion, backrest, left, right, front or rear seats		

Bentley Publishers
BentleyPublishers.com—All Rights Reserved

Electrical components table

Table b. Component locations

Component	Code	Location	Fig #	Page #
Seat heating module, front seat	A186a A187a	Underneath front seat	**Fig 22** **Fig 23**	**Pg 42** **Pg 43**
Seat heating switch, front, left, right	S838 S840	Center console switch center (SZM)		
Seat heating switch, rear, left, right	S68 S69	Rear of center console between front seats		
Seat occupancy detection module, front seats	A221 A222	Front seat cushion	**Fig 105**	**Pg 67**
Seat occupancy sensor pad	I01023	Seat cushion	**Fig 105**	**Pg 67**
Seatbelt buckle contact microswitch, front	S58a S59a	Front seat belt buckle, left, right		
Secondary air injection pump	M63	Behind front bumper, right side	**Fig 133**	**Pg 74**
Secondary air injection pump relay	K6304a	E-box, right rear engine compartment	**Fig 5** **Fig 10**	**Pg 29** **Pg 32**
Selector angle sensor (SMG)	B6573	Top of SMG transmission *see also* Crankshaft sensor	**Fig 109**	**Pg 68**
Sequential manual gearbox components		*see* **SMG components** in this repair group.		**Pg 67**
Servotronic valve	B15a	Underneath front subframe at steering shaft	**Fig 127**	**Pg 72**
SGM		*see* Safety and gateway module (SGM)		
Slide-in compartment		*see* Ignition key slot		
SMG components		*see* **SMG components** in this repair group.		**Pg 67**
SMG control module	A70010	E-box, right rear engine compartment	**Fig 9**	**Pg 30**
SMG hydraulic pump relay	K6318	E-box, right rear engine compartment	**Fig 11**	**Pg 33**
SMG paddle shifter buttons	I01028 I01030	Above steering column spokes		
Solar sensor	B66	Front center of dashboard		
Sound system control panel cooling fan	M455	Sound system control panel		
Speaker, center dashboard	H77a	Center dashboard		
Speaker, emergency	H82	Underneath steering column		
Speaker, high-range (tweeter), front right, left	H50a H59a	Front of window frame, front door, right, left	**Fig 93**	**Pg 64**
Speaker, high-range (tweeter), rear right, left	H53 H56	*Sedan:* Rear parcel self behind trim, right, left	**Fig 14**	**Pg 39**
Speaker, high-range (tweeter), rear right, left	H53 H56	*Sports Wagon:* Rear headliner behind trim, right, left	**Fig 18**	**Pg 40**
Speaker, low-range (subwoofer), left, right	H75 H76	Underneath rug under front seat, left, right	**Fig 90**	**Pg 63**
Speaker, midrange, front door, left, right	H68a H69a	Center of front door behind trim, left, right		
Speaker, mid-range, rear door, left, right	H78a H79a	Front of rear door behind trim, left, right	**Fig 94**	**Pg 64**
Speaker, midrange, rear left, right	H70a H71a	*Sedan:* Rear parcel shelf behind trim, left, right	**Fig 14**	**Pg 39**
Speaker, midrange, rear left, right	H70a H71a	*Sports Wagon:* Rear headliner behind trim, right, left	**Fig 18**	**Pg 40**
Speed (rpm) sensor		*see* Crankshaft sensor		

BentleyPublishers.com
BentleyPublishers.com–All Rights Reserved

Table b. Component locations

Component	Code	Location	Fig #	Page #
SRS		*see* Airbag entries *see also* Crash safety module		
START - STOP button	S2a	Dashboard, right of steering column		
Starter	M6510a	*V8 engine:* Right side of engine under exhaust manifold		
Starter	M6510 M6510a	*6-cylinder engine:* Left side of engine under intake manifold	**Fig 31** **Fig 48**	**Pg 45** **Pg 51**
Steering angle sensor		Steering column switch center (SZL)		
Steering column adjustment motor	M137	Left side of steering column	**Fig 79**	**Pg 60**
Steering column switch center (SZL)	A72	Behind steering wheel on steering column		
Steering lock	M137a	Underneath steering column		
Steering module		*see* Active steering control module		
Steering wheel heater	I01118	Steering wheel rim		
Steering wheel heater switch	I01122	Stalk switch, left of steering column		
Steering wheel module	I01004	Center spoke, steering wheel		
Stepper motor		*see* **Climate control components** in this repair group		**Pg 41**
Steptronic switch		*see* Automatic transmission steptronic switch		
Sunroof module	A33a A33b	*to 03 / 2007:* Behind glove compartment	**Fig 86**	**Pg 62**
Sunroof module	A33b	*from 03 / 2007:* Spare tire well	**Fig 25**	**Pg 44**
Sunroof motor	M14105	*Sports Wagon:* Above center of windshield, behind headliner		
Sunroof switch	S38a	Front dome light unit	**Fig 76**	**Pg 59**
Sunshade motor, rear window	M74	Underneath rear parcel shelf		
Supplemental restraint system		*see* Crash entries		
Tailgate inner release button	S237	Bottom of tailgate, right of latch		
Tailgate lift module	A204a	*Sports Wagon:* Spare tire well	**Fig 24** **Fig 25**	**Pg 43** **Pg 44**
Tailgate lift motor	M10	Cargo compartment, left rear behind trim	**Fig 119**	**Pg 70**
Tailgate lock motor	M17b	Tailgate		
Tailgate lock motor, soft close	M407	Tailgate lock striker		
Tailgate lock switch	S172	Left side tailgate		
Tailgate opening angle sensor	S236	Tailgate lift strut, left	**Fig 119**	**Pg 70**
Tailgate release	S117	Center tailgate above license plate		
Tailgate release button, interior	S204	Edge of driver door post		
Tailgate soft-close relay	K207	*Sports Wagon:* Spare tire well	**Fig 24** **Fig 25**	**Pg 43** **Pg 44**
Tailpipe flap actuator	Y198	Trunk or cargo compartment, left rear behind trim	**Fig 111** **Fig 121**	**Pg 68** **Pg 71**
TCM		*see* Automatic transmission control module (TCM)		
Telematics control module (TCU)	U400b	*Sports Wagon:* Floor of cargo compartment		
Telematics control module (TCU)	U400b	*Sedan:* Left rear trunk, behind trim	**Fig 114**	**Pg 69**
Telephone		*see* **650 Radio**		
Telephone compensator	N27	Trunk or cargo compartment, left rear behind trim	**Fig 121**	**Pg 71**
Telephone eject box	A117a	Center console, between front seats		
Telephone transceiver	U400	*Sedan:* Left rear trunk, behind trim		

BentleyPublishers.com
BentleyPublishers.com—All Rights Reserved

Electrical components table

Table b. Component locations				
Component	**Code**	**Location**	**Fig #**	**Page #**
Telephone transceiver	U400a	*Sports Wagon:* Floor of cargo compartment		
Temperature sensor, evaporator	B14a	IHKA housing, underneath top dashboard trim	**Fig 21**	**Pg 41**
Temperature sensors, heater core	B11a B12a	Top of IHKA housing, underneath top dashboard trim	**Fig 20**	**Pg 41**
Thermostat, coolant	B6279 Y6279	*M54 engine:* Coolant manifold, front of engine	**Fig 37**	**Pg 47**
Thermostat, engine coolant	Y6279	*N62 engine:* Coolant manifold, front of engine	**Fig 57**	**Pg 55**
Thermostat, engine coolant	Y6279	*N62TU engine:* Coolant manifold, front of engine	**Fig 65**	**Pg 57**
Throttle valve	Y6390	*M54 engine:* Underneath intake manifold	**Fig 33**	**Pg 49**
Throttle valve	Y6390	*N52 engine:* Underneath intake manifold	**Fig 40**	**Pg 49**
Throttle valve	Y6390	*N62 engine:* Front of intake manifold	**Fig 57**	**Pg 55**
Throttle valve	Y6390	*N62TU engine:* Front of intake manifold	**Fig 66**	**Pg 57**
Throttle valve	Y63900	*N52K engine:* Underneath intake manifold	**Fig 41**	**Pg 49**
Throttle valve	Y63900	*N54 engine:* Underneath intake manifold	**Fig 50**	**Pg 51**
Tilt sensor		*see* Anti-theft entries		
Tire pressure control (RDC) antenna	B46a	Underneath vehicle, below driver seat	**Fig 138**	**Pg 75**
Tire pressure control (RDC) module	A85a	Trunk or cargo compartment, underneath spare tire	**Fig 24**	**Pg 43**
Tire pressure control (RDC) sensor	B43b B44b B45b B47b	Inside each wheel	**Fig 132**	**Pg 73**
Tire pressure control (RDC) transmitter	B43a B44a B45a B47a	Left front, right front, left rear, right rear fender housing	**Fig 129** **Fig 141**	**Pg 73** **Pg 76**
Towing hitch motor	M8540a	Under rear bumper		
Trailer hitch module	A195a	*Sports Wagon:* Spare tire well	**Fig 25**	**Pg 44**
Trailer module	A6	*Sports Wagon:* Spare tire well	**Fig 25**	**Pg 44**
Transfer case control module	A70006	Underneath carpet in front of passenger seat	**Fig 88**	**Pg 62**
Transmission		*see* Automatic transmission entries *see also* SMG entries		
Trunk lid lock motor	M17b	Trunk lid	**Fig 117**	**Pg 70**
Trunk lid lock switch	S172	Left side trunk lid	**Fig 117**	**Pg 70**
Trunk lid release	S117	Center trunk lid above license plate		
Trunk lid release button, interior	S204	Edge of driver door post		
Turn signal / high beam switch		Steering column switch center (SZL), right side		
Turn signal, front left, right	H4a H7a	Headlight assembly, left, right		
Turn signal, side, right, left	H8a H9a	Side of front fender, right, left		
Ultrasonic detector		*see* Interior movement sensor (ultrasonic detector)		
USB hub	A873	Trunk or cargo compartment, left rear behind trim		
Vacuum pump relay	K213	Behind right side dashboard	**Fig 77**	**Pg 60**
Valvetronic control module	A60210	E-box, right rear engine compartment	**Fig 53**	**Pg 53**

BentleyPublishers.com
BentleyPublishers.com—All Rights Reserved

Table b. Component locations

Component	Code	Location	Fig #	Page #
Valvetronic actuator	M6351 M6352	*N62 engine:* Top center of cylinder head cover	**Fig 53 Fig 54 Fig 56**	**Pg 53 Pg 54 Pg 54**
Valvetronic actuator	M6353	*N52 engine:* Top center of cylinder head cover	**Fig 38 Fig 44**	**Pg 48 Pg 50**
Valvetronic eccentric shaft sensor	B60213	*N52 engine:* Top front of cylinder head cover	**Fig 44**	**Pg 50**
Valvetronic eccentric shaft sensors, left, right	B60213 B60223	*V8 engine:* Rear of cylinder heads	**Fig 54 Fig 56 Fig 61 Fig 62**	**Pg 54 Pg 54 Pg 56 Pg 56**
Valvetronic relay	K6319	E-box, right rear engine compartment	**Fig 8**	**Pg 30**
Valvetronic sensor		*see* Eccentric shaft sensor		
Vanity mirror switch, left, right	S77 S78	Behind sun visor, left, right		
VANOS solenoid, exhaust	Y6276	*M54 engine:* Front of cylinder head	**Fig 37**	**Pg 47**
VANOS solenoid, intake	Y6275	*M54 engine:* Front of cylinder head	**Fig 36**	**Pg 47**
VANOS solenoid, intake, exhaust	Y6275 Y6276	*N52 engine:* Front of cylinder head	**Fig 42**	**Pg 49**
VANOS solenoid, intake, exhaust	Y6275 Y6276	*N54 engine:* Front of cylinder head	**Fig 51**	**Pg 52**
VANOS solenoids, left cylinder head	Y6283 Y6284	*N62 engine:* Front of left cylinder head	**Fig 55**	**Pg 54**
VANOS solenoids, left cylinder head	Y6283 Y6284	*N62TU engine:* Front of left cylinder head	**Fig 64**	**Pg 56**
VANOS solenoids, right cylinder head	Y6281 Y6282	*N62 engine:* Front of right cylinder head	**Fig 57**	**Pg 55**
VANOS solenoids, right cylinder head	Y6281 Y6282	*N62TU engine:* Front of right cylinder head	**Fig 65**	**Pg 57**
Variable intake manifold (DISA) solenoid	Y6167	*M54 engine:* Rear of engine on intake manifold	**Fig 34**	**Pg 46**
Variable intake manifold (DISA) solenoid	Y6167	*N62 engine:* Rear of engine on intake manifold	**Fig 59**	**Pg 55**
Variable intake manifold (DISA) solenoid	Y6540 Y6541	*N52 engine:* Intake manifold	**Fig 38 Fig 39**	**Pg 48 Pg 49**
Variable intake manifold (DISA) solenoid	Y6540 Y6541	*N62TU engine:* Rear of engine on intake manifold, right, left	**Fig 63**	**Pg 56**
Video module	A197	Left side trunk or cargo compartment behind trim	**Fig 121**	**Pg 71**
Visor motor	M14104	*Sports Wagon:* Behind rear dome light unit		
Volute spring		*see* Airbag and horn contact (volute spring)		
VTG actuator		*see* xDrive clutch (VTG) actuator		
Washer fluid level switch	S136	Washer fluid reservoir in front of left front wheel, behind wheel housing liner	**Fig 131**	**Pg 73**
Washer pumps	M4 M7 M95	Washer fluid reservoir in front of left front wheel, behind wheel housing liner	**Fig 131**	**Pg 73**
Washer switch		Steering column switch center (SZL), right side		
Wastegate solenoid	Y6039 Y6723	*N54 engine:* Right side of cylinder head	**Fig 72**	**Pg 58**
Wheel speed sensor		*see* ABS wheel speed sensor *see also* **340 Brakes**		

BentleyPublishers.com—All Rights Reserved

Table b. Component locations				
Component	**Code**	**Location**	**Fig #**	**Page #**
Window regulator motor, left front, right front	M21 M23	Center of front door inside door cavity, left, right	**Fig 93**	**Pg 64**
Window regulator motors, left rear, right rear	M20a M22a	Center of rear door inside door cavity, left, right	**Fig 94**	**Pg 64**
Window switch, driver		*see* Door switch cluster, driver door		
Window switch, passenger	S127	Passenger door armrest		
Window switch, rear	S128a S129a	Rear door armrest, left, right		
Windshield washer nozzle heater, left, right	E51a E52a	Base of windshield on engine hood		
Windshield wiper relay	K11	E-box, right rear engine compartment	**Fig 5** **Fig 8** **Fig 11**	**Pg 29** **Pg 30** **Pg 33**
Wiper motor, rear window	M94	*Sports Wagon:* Rear window lower edge, integrated with rear window lock		
Wiper motor, windshield	M3a	Underneath plenum chamber cover, left rear engine compartment	**Fig 38** **Fig 53**	**Pg 48** **Pg 53**
Wiper relay, rear window	K91	*Sports Wagon:* Spare tire well	**Fig 24** **Fig 25**	**Pg 43** **Pg 44**
Wiper switch		Steering column switch center (SZL), right side		
X connector		*see* **Harness connectors, wire splices (X connectors)** in this repair group.		
X ground		*see* **Grounds** in this repair group		
xDrive classification resistor	R8554	Rear of transmission housing on transfer case	**Fig 139**	**Pg 75**
xDrive clutch (VTG) actuator	M8533	Rear of transmission housing on transfer case	**Fig 139**	**Pg 75**
xDrive control module		*see* Transfer case control module		
Yaw sensor, active steering		*see* Active steering yaw sensor		

Harness connectors, wire splices (X connectors)

BMW electrical components connect to the electrical harness via X connectors. Most connectors bear the same numerical designation as the component they attach to. For example, the connector for Y6279 (coolant thermostat) is designated X6279.

Note that grounds are also designated with X numbers. See **Grounds** in this repair group.

Table c lists harness connectors and wire splices not attached directly to components and gives brief location information for each.

Table c. X connectors				
Code	**Description and application**	**Location**	**Fig #**	**Page #**
X10116	Wire splice	Behind right side dashboard		
X10183	Wire splice	Behind right side dashboard		
X10184	Wire splice	Behind right side dashboard		
X1019	Wire splice	Behind right side dashboard		
X10547	Wire splice	Behind right side dashboard		

BentleyPublishers.com
BentleyPublishers.com—All Rights Reserved

Table c. X connectors				
Code	**Description and application**	**Location**	**Fig #**	**Page #**
X10550	Wire splice	Behind right side dashboard, base of A-pillar		
X10551 X10552	PT-CAN bus wire splice	*Sports Wagon:* Right side cargo compartment behind trim	**Fig 123**	**Pg 71**
X1091	Wire splice	Right side center console		
X1101	Wire splice	Behind instrument cluster		
X11212	12v power wire splice	Behind right side dashboard		
X11273	Ground splice	Trunk lid behind trim		
X113	Connector, right front wheel speed sensor	Behind right front wheel housing liner		
X114	Connector, left front wheel speed sensor	Behind left front wheel housing liner	**Fig 128**	**Pg 72**
X11544	Wire splice	Right side center console		
X13125	Wire splice	Left side center console		
X13276	Ground splice	Tailgate behind trim		
X13502	Wire splice	Behind right side dashboard		
X13637	Wire splice	Behind instrument cluster		
X13796	Wire splice	Behind right side dashboard		
X13887	Wire splice	Behind right side dashboard		
X13895	Ground splice	Behind right taillight	**Fig 115**	**Pg 69**
X14023 X14024	F-CAN bus wire splice	Right front floor underneath rug	**Fig 87**	**Pg 62**
X14174	Wire splice	Right side center console		
X14197	Ground splice	Behind right side dashboard		
X14252	Wire splice	Right side center console		
X14276	Wire splice	Right side center console		
X15009 X15010	F-CAN bus wire splice	Right C-pillar behind trim	**Fig 92**	**Pg 63**
X1662	Ground splice	Behind right side dashboard		
X1663	Ground splice	Behind right side dashboard		
X16678 X16679	Wire splice	Passenger seat rail, underneath rug	**Fig 89**	**Pg 63**
X1680 X1681	Wire splice	*to 09 / 2004:* Behind instrument cluster from 03 / 2005: Underneath left side dashboard		
X1720	Wire splice	Right side center console		
X1727 X1728	Speaker wire splice	Behind right side dashboard		
X18063	Ground splice	Behind right side dashboard		
X181	Wire splice	Behind right side dashboard		
X18121	Wire splice, basic body module	Behind right side dashboard		
X18122	Wire splice, basic body module	*Sedan:* Behind right side dashboard *Sport Wagon from 03 / 2005:* Right D-pillar behind trim		
X18132	Ground splice	Behind instrument cluster		
X18139	Ground splice	Behind right side dashboard		
X18145	Connector, left front brake pad sensor	Behind left front wheel housing liner	**Fig 128**	**Pg 72**
X18146	Connector, right rear brake pad sensor	Behind right rear wheel housing liner		

BentleyPublishers.com
BentleyPublishers.com—All Rights Reserved

Harness connectors, wire splices (X connectors)

Table c. X connectors

Code	Description and application	Location	Fig #	Page #
X1816	Ground splice	*to 03 / 2005:* Behind right side dashboard *from 03 / 2005:* Right C-pillar behind trim		
X186	Wire splice	Behind right side dashboard		
X18726	Ground splice	Behind instrument cluster		
X18740	Ground splice	Behind right side dashboard		
X18749	Wire splice	Behind right side dashboard		
X18750	Ground splice	Behind right side dashboard		
X18751	Ground splice	*to 09 / 2004:* Behind instrument cluster *from 03 / 2005:* Underneath left side dashboard		
X2350	Terminal 31 support	E-box, right rear engine compartment	**Fig 5**	**Pg 29**
X243	Wire splice	Behind right side dashboard		
X376	Connector, right rear wheel speed sensor	Behind right rear wheel housing liner		
X377	Connector, left rear wheel speed sensor	Behind left rear wheel housing liner		
X60531 X60541	Connector, engine electronics	E-box, right rear engine compartment	**Fig 5**	**Pg 29**
X62101 X62102 X62201 X62202	Oxygen sensor connectors	Top or sides of engine	**Fig 38** **Fig 67** **Fig 69** **Fig 74** **Fig 75**	**Pg 48** **Pg 57** **Pg 58** **Pg 59** **Pg 59**
X6450 X6460	Wire splice, engine management components	Top right center of engine, between valve cover and intake manifold.		
X64541 X64542	Ground splices	Right of E-box, right rear engine compartment	**Fig 5**	**Pg 29**
X793	Wire splice	Right C-pillar behind trim	**Fig 92**	**Pg 63**
X806	Wire splice	*to 09 / 2004:* Behind instrument cluster *from 03 / 2005:* Underneath left side dashboard		
X834	Ground splice, left rear door	Behind left rear door panel		
X835	Ground splice, right rear door	Behind right rear door panel		
X849	Ground splice, driver door	Behind driver door panel		
X875	Wire splice	Behind right side dashboard		
X891	Ground splice, passenger door	Behind passenger door panel		
X908	Wire splice	Behind right side dashboard		
X913	Ground splice	Behind and below left headlight		
X923	Ground splice	Behind and below right headlight		
X273 - X274	Rear door electronics harness connector, left, right	Front of door, B-pillar, left, right	**Fig 94**	**Pg 64**
X256 - X257	Front door electronics harness connector, right, left	Front of door, A-pillar, right, left	**Fig 93**	**Pg 64**
X942	Ground splice	Behind right end of front bumper		
X943	Ground splice	Behind right side dashboard		

BentleyPublishers.com
BentleyPublishers.com—All Rights Reserved

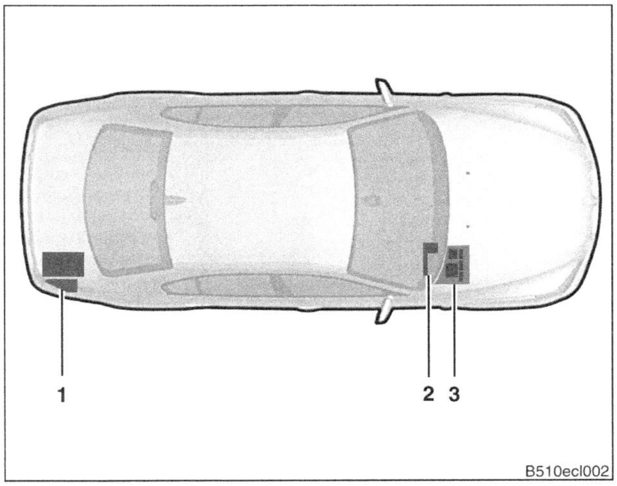

Fuse Locations and Ratings

Fuse panels

For a description of the power supply system, see **600 Electrical System–General**.

Fuses and fusible links in 5 Series vehicle are located as follows:

1. **A46** Rear power distribution and fuse panel in trunk or cargo compartment:
   - High amperage fuses on back of panel. See **Rear power distribution and fuse panel (A46)** in this repair group.
   - Rear fuses. See **Rear fuses** in this repair group.
2. **A41** Front fuse panel under right side dashboard, behind glove compartment. See **Front fuses** in this repair group.
3. E-box fuse carriers, right rear engine compartment: See **E-box fuses, relays and modules** in this repair group.

Rear power distribution and fuse panel (A46)

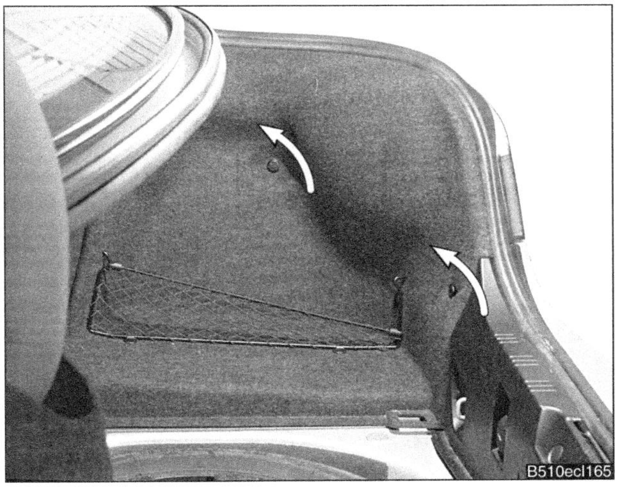

Sedan:

- Remove trunk floor liner.
- Twist trunk right side trim plastic retainers (**arrows**) 90° to remove.
- Slide right side trunk trim backward and remove to expose rear fuse panel.

Sport Wagon: Open right cargo compartment cover (**arrow**). Peel rear fuse panel cover up to expose panel.

BentleyPublishers.com
BentleyPublishers.com—All Rights Reserved

Rear power distribution and fuse panel (A46)

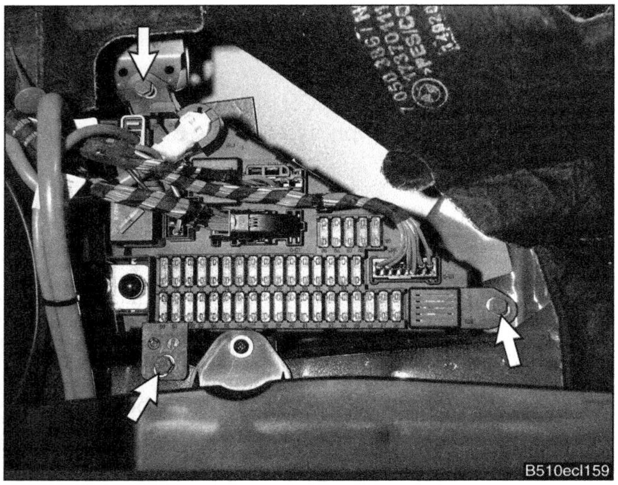

Remove panel mounting bolts (**arrows**) to gain access to high amperage fuses (fusible links) in back of panel.

Using thin-bladed screwdriver, pry back lock tabs (**arrows**) to remove fusible link cover.

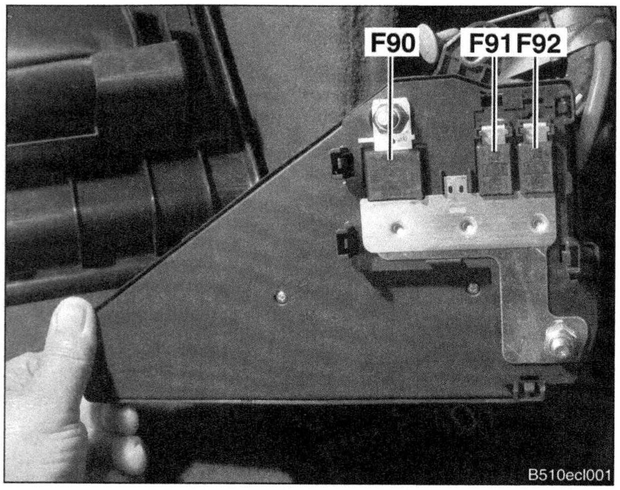

Fig 1 Right rear trunk or cargo compartment

A46 Back of fuse panel.

Ratings for fusible links (high amperage fuses) are in **Table d**

Table d. High amperage fuses in rear power distribution and fuse panel (A46)		
Fuse	**Rating in amps**	**Protected circuits**
F90	200	Front fuse panel (A41)
F91	100	N52: Valvetronic power supply V8: Integrated supply module (IVM) (A6009)
F92	100	Auxiliary heater (not USA)

BentleyPublishers.com
BentleyPublishers.com–All Rights Reserved

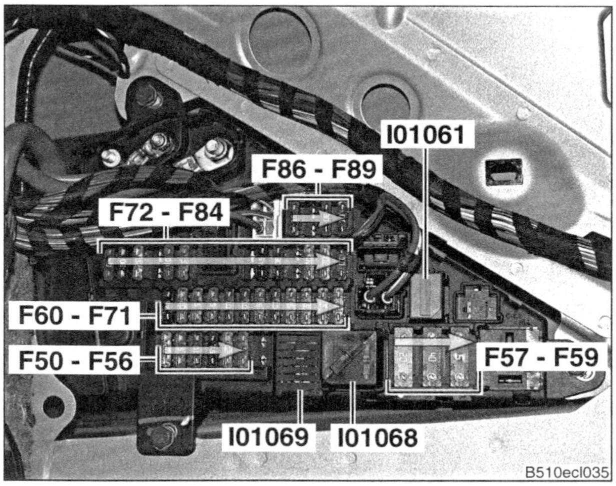

Fig 2 Right rear trunk or cargo compartment (to 09 / 2005)

A46 Front of fuse panel.

- **F50 - F89** Fuses
  See fuse ratings in **Table e**.
- **I01061** Rear window defogger relay
- **I01068** Terminal 30g relay
- **I01069** Terminal 15 relay

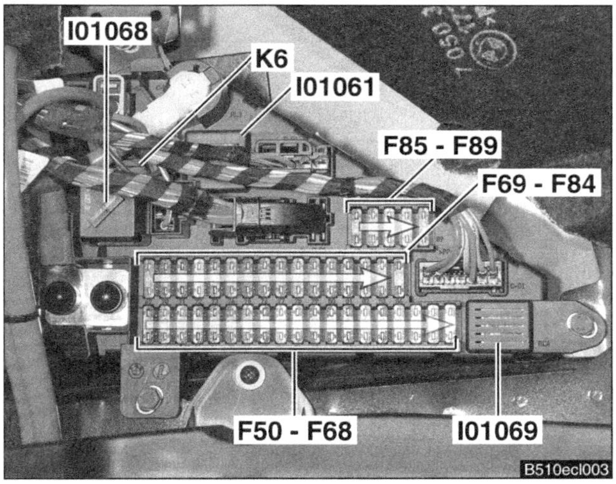

Fig 3 Right rear trunk or cargo compartment (from 09 / 2005)

A46 Front of fuse panel.

- **F50 - F89** Fuses
  See fuse ratings in **Table f**.
- **I01061** Rear window defogger relay
- **I01068** Terminal 30g relay
- **I01069** Terminal 15 relay
- **K6** Headlight washer relay

BentleyPublishers.com—All Rights Reserved

Rear fuses

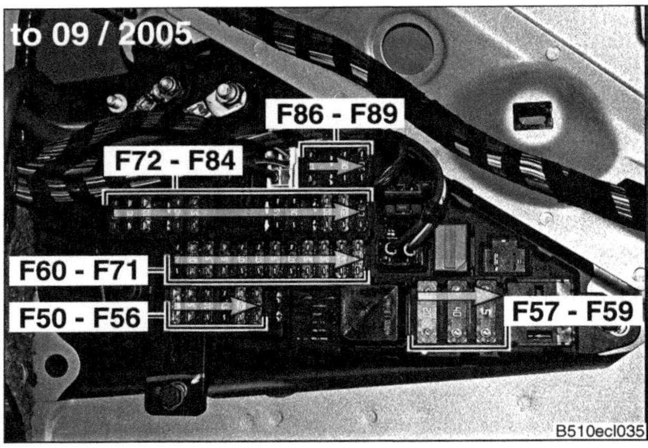

Rear fuses

Table e. Rear fuses to 09 / 2005

Fuse	Rating in amps	Protected circuits
F50	20	Fuel pump
F51	5	Anti-theft alarm Comfort access door Electrochromic mirror
F52	10	Micro-power module
F53	7.5	*to 09 / 2004:* Trunk lid or tailgate soft-close relay
		09 / 2004 - 09 / 2005: Comfort access Outer door handle electronic modules
F54	20	Load-shedding (unloader) relay terminal 15
F55	5	Rain and light sensor
F56		not used
F57	20	Micro-power module
F58	40	Rear window defogger
F59	5	Antenna tuner
F60		not used
F61	20	Front cigar lighter Glove compartment 12v utility socket
F62	5	Load-shedding (unloader) relay terminal 15
F63	5	Electrochromic mirror Park distance control (PDC)
F64	10	Central information display (CID) iDrive controller
F65	10	Head-up display Headset module
F66	5	DSC Transfer box
F67	10	*to 03 / 2006:* Satellite radio Video module
F68	5	*to 03 / 2006:* Seats
F69	5	Park distance control (PDC)
F70	5	Adaptive headlights
F71	20	Rear wiper
F72	40	EHC (air suspension) compressor
F73	30	Trailer
F74	20	Trailer
F75	30	Transfer box
F76	40	Trunk lid lift
F77	5	Antenna
F78		not used
F79	7.5	EHC (air suspension)
F80	30	Amplifier
F81		not used
F82	20	Rear cigar lighter Rear 12v utility socket
F83	20	Sunroof
F84	10	Active cruise control
F85		not used
F86	15	Transmission
F87		not used
F88	30	Center console switch center (SZM)
F89	15	Gearshift lighting Telephone eject box

BentleyPublishers.com
BentleyPublishers.com—All Rights Reserved

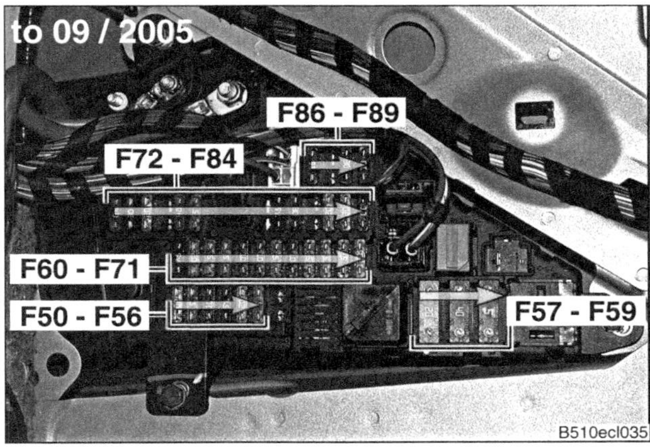

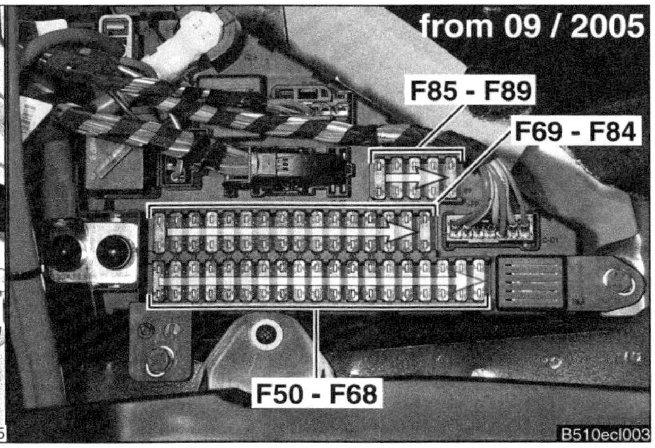

Table f. Rear fuses from 09 / 2005

Fuse	Rating in amps	Protected circuits
F50	30	Headlight washer pump
F51	5	Anti-theft alarm Comfort access Electrochromic mirror
F52	40	EHC (air suspension) compressor
F53	30	Active seat backrest
F54	30	Rear window defogger
F55	40	Trunk lid lift
F56	5	Rain and light sensor
F57	7.5	Climate control Washer system
F58	20	Rear wiper
F59	5	Antenna
F60	10	DVD changer
F61	7.5	Rear arm-rest cooler box
F62	30	Window control
F63	20	Climate control
F64	15	Climate control
F65	20	Trailer connection
F66	20	Window control
F67	20	Window control
F68	20	*from 03 / 2006:* Left doors soft-close
F69	5	Park distance control (PDC)
F70	10	Active cruise control
F71	30	Center console switch center (SZM)
F72	20	Fuel pump

Table f. Rear fuses from 09 / 2005

Fuse	Rating in amps	Protected circuits
F73	30	*09 / 2005 - 09 / 2007:* Amplifier
	40	*from 09 / 2007:* Amplifier
F74	5	*09 / 2005 - 09 / 2007:* DSC Seat control
	7.5	*from 09 / 2007:* Seats
F75	10	Radio Satellite radio Video module
F76	10	Dynamic drive (active stabilizer)
F77	10	Cooling fan Head-up display
F78	5	DSC Transfer case
F79	10	Central information display (CID) iDrive
F80	10	Gearshift light Telephone
F81	7.5	EHC (air suspension)
F82	7.5	Tire pressure control (RDC)
F83	30	Active backrest
F84	15	Transmission
F85	7.5	Transmission
F86	40	Active steering
F87	20	Front cigar lighter Glove compartment 12v utility socket
F88	20	Rear 12v utility socket
F89	5	Electrochromic mirror Load-shedding (unloader) relay terminal 15

BentleyPublishers.com
BentleyPublishers.com—All Rights Reserved

Front fuses

Open glove compartment and twist plastic retainers (**arrows**) 90° to release fuse panel cover.

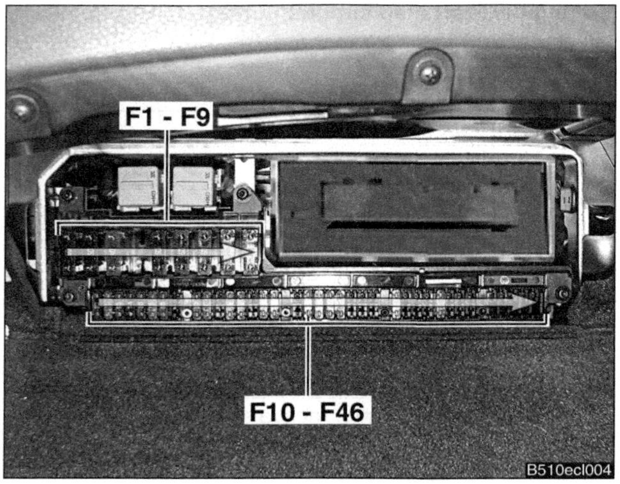

Fig 4 In glove compartment

A41 Front fuse panel.

Ratings for front fuses are in **Table g** and **Table h**.

BentleyPublishers.com
BentleyPublishers.com—All Rights Reserved

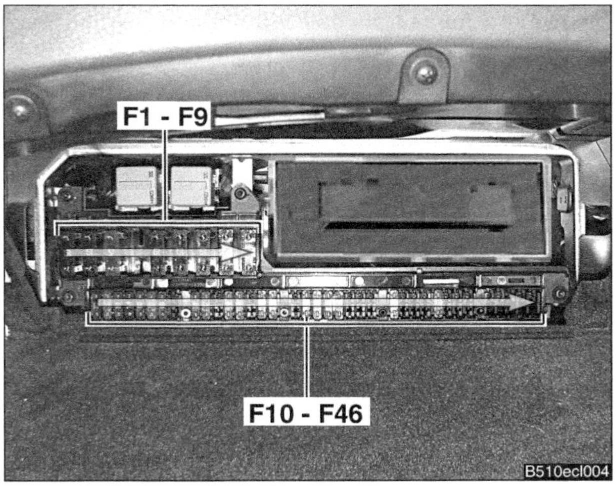

Table g. Front fuses to 09 / 2005

Fuse	Rating in amps	Protected circuits
F1	50	DSC
F2	60	Secondary air injection pump relay
F3	40	Blower output stage
F4	40	Active steering
F5	50	Light module
F6	50	Light module
F7	30	Car access system (CAS)
F8	60	*6-cylinder engines:* A2076 B+ power distribution Engine electronics fuse carriers DME main relay Engine control module (ECM)
		V8 engine, no SMG transmission: Integrated supply module (IVM)
		V8 engine, SMG transmission: A8683 Engine electronics fuse carrier SMG transmission hydraulic pump relay
F9	60	Electric cooling fan
F10	30	Driver door module
F11	5	Basic body module
F12	30	Passenger door module
F13	7.5	Instrument cluster
F14	30	Driver seat control
F15	5	Car access system (CAS)
F16	30	Windshield wiper relay
F17		not used
F18	30	Headlight washer relay
F19	5	Gear indicator lighting
F20		not used
F21	30	Passenger seat control
F22	30	Basic body module
F23		not used
F24	30	Basic body module
F25	30	DSC
F26	7.5	Climate control
F27	30	Basic body module
F28	20	Steering column switch cluster
F29	10	OBD II plug Safety and gateway module
F30	15	Climate control
F31	30	Driver seat module
F32	10	Dynamic drive
F33	30	Passenger seat module
F34	30	Radio (CCC / M - ASK)
F35	5	Navigation
F36		not used
F37	10	TCU Telephone eject box ULF
F38	10	CD changer
F39	5	Steering column switch cluster
F40	10	DVD changer
F41	7.5	*from 03 / 2005:* Instrument cluster
F42		not used
F43		
F44		
F45		
F46		

ECL

BentleyPublishers.com
BentleyPublishers.com—All Rights Reserved

Front fuses

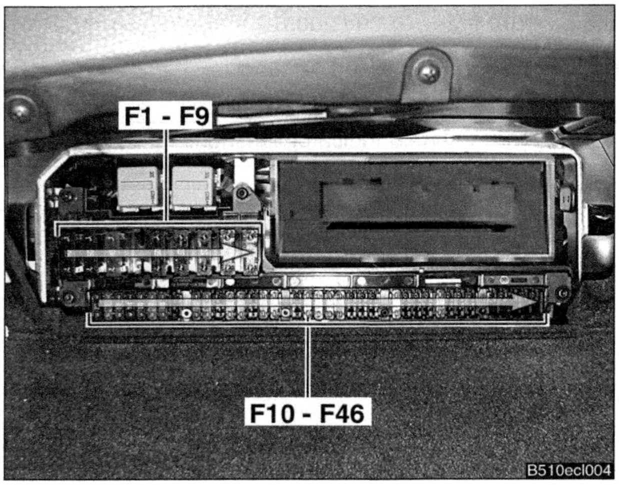

Table h. Front fuses from 09 / 2005

Fuse	Rating in amps	Protected circuits
F1	50	DSC
F2		not used
F3	40	Blower output stage
F4		not used
F5	40	Light module
F6	40	Light module
F7	30	Car access system (CAS)
F8	60	*6-cylinder engines:* A2076 B+ power distribution Engine electronics fuse carriers DME main relay Engine control module (ECM)
		V8 engine, no SMG transmission: Integrated supply module (IVM)
		V8 engine, SMG transmission: A8683 Engine electronics fuse carrier SMG transmission hydraulic pump relay
F9	60	Electric cooling fan
F10	30	Body gateway module
F11	5	Basic body module
F12	30	Body gateway module
F13	30	Transfer case control module
F14	30	Driver seat control
F15	5	Car access system (CAS)
F16	30	Windshield wiper relay
F17	5	Steering column switch cluster
F18	30	Body gateway module
F19	5	Gear indicator lighting
F20	20	On-board diagnostics
F21	30	Passenger seat control
F22	30	Basic body module
F23	30	Engine control module
F24	30	Basic body module
F25	30	DSC
F26	20	Load-shedding (unloader) relay
F27	30	Basic body module
F28	20	Steering column switch cluster
F29	10	Body gateway module OBD II plug
F30	20	Fuel pump relay
F31	10	Engine control module
F32	10	Driver door switch cluster Outside mirror control
F33	30	Instrument cluster Seat heating
F34	20	Radio (CCC / M - ASK)
F35	5	Navigation
F36	10	Comfort access control module Outer door handle electronic modules
F37	5	TCU Telephone eject box ULF
F38	5	CD changer
F39	5	Steering column switch cluster
F40	10	DVD changer
F41	7.5	Instrument cluster
F42		not used
F43		
F44		
F45		
F46		

BentleyPublishers.com
BentleyPublishers.com–All Rights Reserved

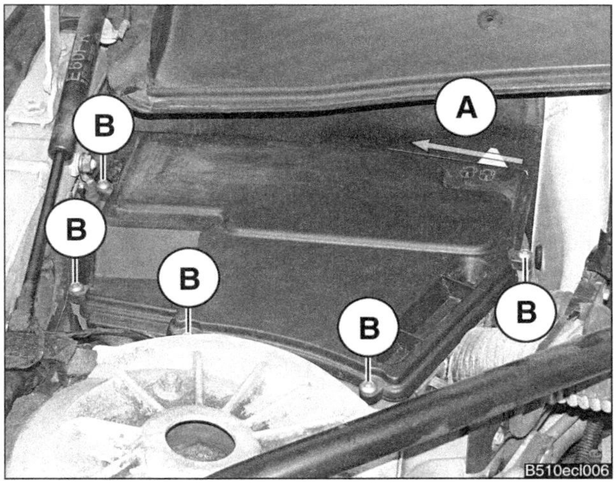

B510ecl006

E-box cover, removing

– Working at right rear of engine compartment: Remove right side cabin microfilter housing and plenum cover. See **Engine Covers** in **020 Maintenance**.

Working at E-box cover:

- Slide lock (**A**) in direction of **arrow** to unlock cover.
- Loosen cover hold-down screws (**B**) approx. ½ in each and remove cover. Note that these are captured screws. Loosening more than
  ½ in may allow them to fall out of cover and become lost.

E-box fuses, relays and modules

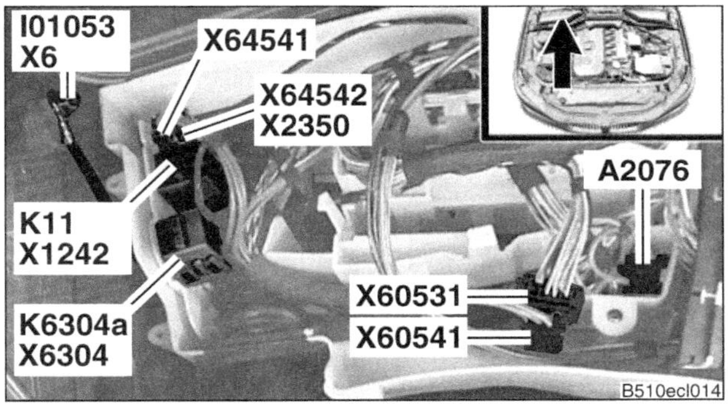

B510ecl014

Fig 5 E-box components (M54 engine)

E-box cover removed (see **E-box cover, removing** in this repair group):

- **A2076** B+ power distribution
- **I01053** Ground bus
- **K11** Windshield wiper relay
- **K6304a** Secondary air injection pump relay
- **X6** Ground connection
  See **Grounds** in this repair group.
- **X1242** Connector
- **X2350** Connector
- **X6304** Connector
- **X60531** Connector
- **X60541** Connector
- **X64541** Ground splice to X6
- **X64542** Ground splice to X6

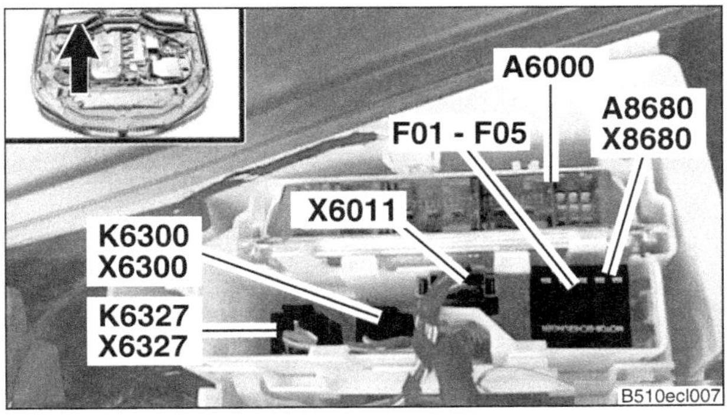

B510ecl007

Fig 6 E-box components (M54 engine)

E-box cover removed (see **E-box cover, removing** in this repair group):

- **A6000** Engine control module (ECM)
- **A8680** Engine electronics fuse carrier
- **F01 - F05** Engine electronics fuses
- **K6300** DME main relay
- **K6327** Fuel injector relay
- **X6011** Connector
- **X6300** Connector
- **X6327** Connector

Engine electronics fuse ratings are in **Table i**.

BentleyPublishers.com
BentleyPublishers.com–All Rights Reserved

E-box fuses, relays and modules

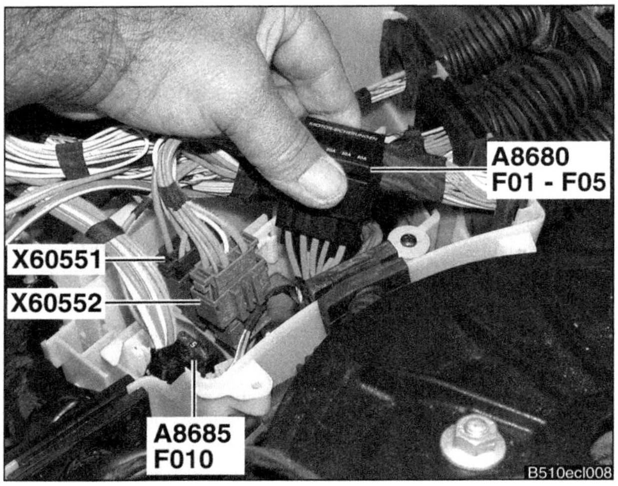

Fig 7 E-box components (N52 engine)

E-box cover removed (see **E-box cover, removing** in this repair group):

- **A8680** Engine electronics fuse carrier
- **A8685** Engine electronics fuse carrier
- **F01 - F05** Engine electronics fuses
- **F010** Engine electronics fuse
- **X60551** Connector
- **X60552** Connector

Engine electronics fuse ratings are in **Table j**.

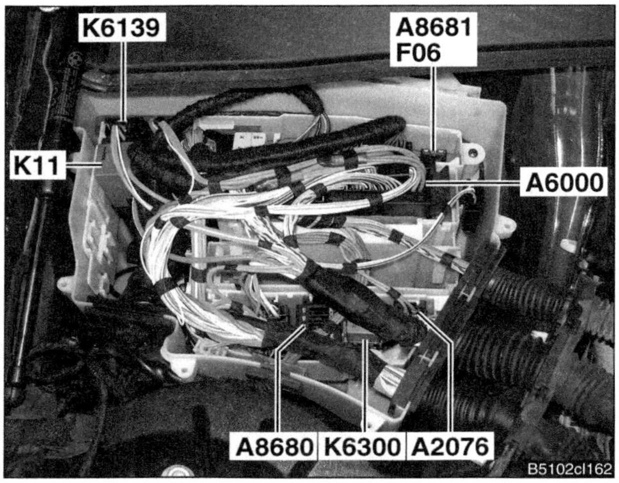

Fig 8 E-box components (N52 engine)

E-box cover removed (see **E-box cover, removing** in this repair group):

- **A2076** B+ power distribution
- **A6000** Engine control module (ECM)
- **A8680** Engine electronics fuse carrier
- **A8681** Engine electronics fuse carrier
- **F06** Engine electronics fuse
- **K11** Windshield wiper relay
- **K6300** DME main relay
- **K6139** Valvetronic relay

Engine electronics fuse ratings are in **Table j**.

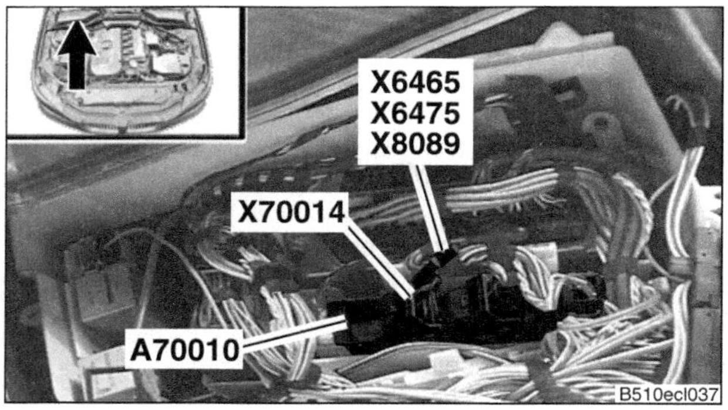

Fig 9 E-box components (N52 engine)

E-box cover removed (see **E-box cover, removing** in this repair group):

- **A70010** SMG control module
- **X6465** Connector
- **X6475v**
- **X8089** Connector
- **X70014** Connector

BentleyPublishers.com
BentleyPublishers.com—All Rights Reserved

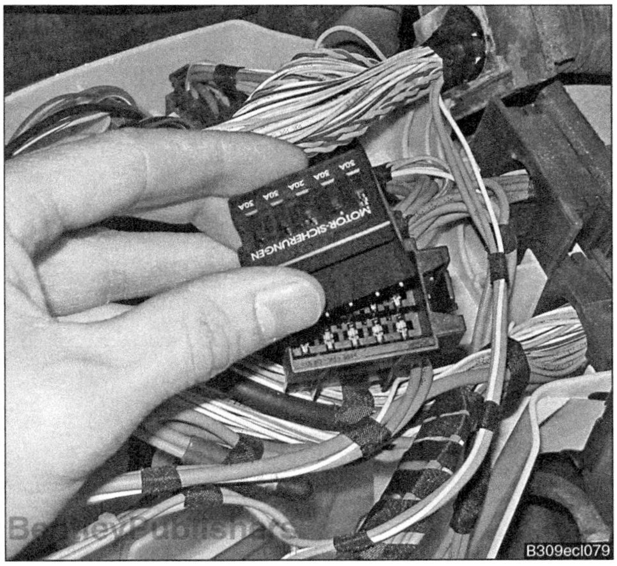

If replacing fuses F01 - F05, unplug fuse carrier A8680.

Pry open A8680 cover using sharp tools. Fuse ratings are in **Table i** and **Table j**.

Table i. Engine electronics fuses in A8680, A8681 (M54 engine)

Fuse	Rating in amps	Protected circuits
F001	30	SMG hydraulic pump relay
F01	30	Ignition coils
F02	30	Engine control module (ECM) Fuel pump relay Fuel tank leakage diagnostic module (DMTL) Fuel tank vent valve Idle actuator Mass air flow sensor Radiator shutter solenoid Secondary air injection pump relay Secondary air injection pump valve VANOS solenoids Variable intake manifold solenoid
F03	20	Camshaft sensors Coolant thermostat Crankshaft sensor Mass air flow sensor
F04	30	Oxygen sensor heater SMG control module
F05	30	Fuel injectors relay

BentleyPublishers.com
BentleyPublishers.com—All Rights Reserved

E-box fuses, relays and modules

Table j. Engine electronics fuses in A8680, A8681, A8682 (N52 engine)		
Fuse	**Rating in amps**	**Protected circuits**
F01	30	Ignition coils Interference suppression capacitor
F02	30	Camshaft sensors Coolant thermostat Electric coolant pump VANOS solenoids
F03	20	Crankshaft sensor Engine control module (ECM) Fuel tank vent valve Mass air flow sensor Oil condition sensor (OZS) Variable intake manifold (DISA) solenoids
F04	30	Crankcase breather heater Oxygen sensor heaters SMG transmission control module
F05	30	Fuel injectors relay
F06	10	AUC sensor Brake air flap sensors E-box fan Fuel tank leakage diagnostic module (DMTL) Nitrogen oxide module Radiator shutter solenoid Secondary air injection mass air flow sensor Secondary air injection pump relay Tailpipe flap
F07	40	Valvetronic relay
F010	5	Engine breather heater

Fig 10 E-box components (V8 engine)

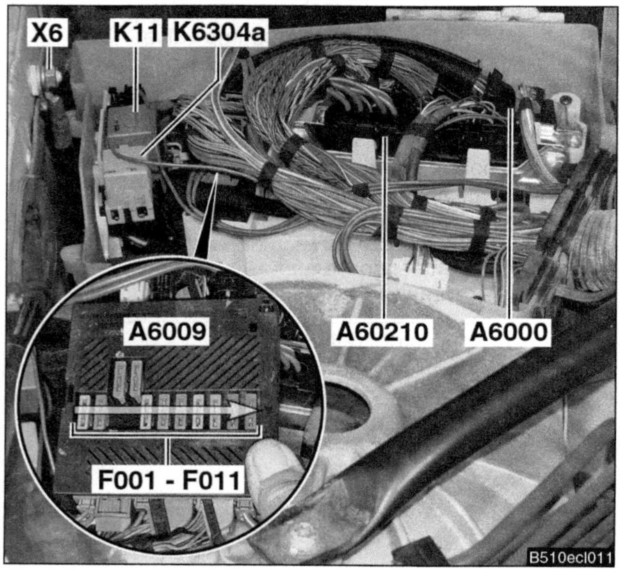

E-box cover removed (see **E-box cover, removing** in this repair group):

- **A6000** Engine control module (ECM)
- **A6009** Integrated voltage supply module (IVM)
  —To gain access to fuses, slide upward out of E-box.
  —Holds engine electronics fuses **F001 - F011.**
  —For fuse ratings, see **Table k**
- **A60210** Valvetronic control module
- **K11** Wiper relay
- **K6304a** Secondary air injection relay
- **X6** Ground post
  See **Ground applications** in this repair group

BentleyPublishers.com
BentleyPublishers.com–All Rights Reserved

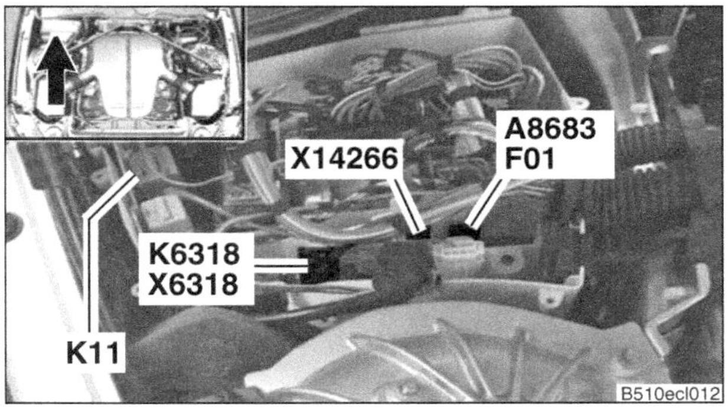

Fig 11 E-box components (V8 engine)

E-box cover removed (see **E-box cover, removing** in this repair group):

- **A8683** Engine electronics fuse carrier
- **F01** Engine electronics fuse
- **K11** Windshield wiper relay
- **K6318** SMG hydraulic pump relay
- **X6318** Connector
- **X14266** Connector

Ratings for engine electronics fuses in IVM and A8683 fuse holder are in **Table k**.

Table k. Engine electronics fuses in IVM, A8683 fuse holder (V8 engine)

Fuse	Rating in amps	Protected circuits
F01	30	SMG transmission hydraulic pump relay
F001	30	Engine control module (ECM) Fuel injectors 5 - 8 SMG transmission control module Valvetronic control module
F002	20	Camshaft sensors (left cylinder head) Fuel tank vent valve VANOS solenoids (left cylinder head)
F003	20	Ignition coils 1 - 4
F004	20	Ignition coils 5 - 8
F005	30	Camshaft sensors (right cylinder head) Coolant thermostat Crankshaft sensor Mass air flow sensor VANOS solenoids (right cylinder head)
F006	20	Fuel injectors 1 - 4
F007	20	Engine control module (ECM) SMG control module
F008	30	Engine control module (ECM) Oil condition sensor (OZS) Oxygen sensor heaters
F009	10	AUC sensor Brake air flap sensors E-box fan Fuel pump relay Fuel tank leakage diagnostic module (DMTL) Radiator shutter solenoid Secondary air injection pump relay Tailpipe flap
F010	40	Valvetronic control module
F011	40	Valvetronic control module

BentleyPublishers.com—All Rights Reserved

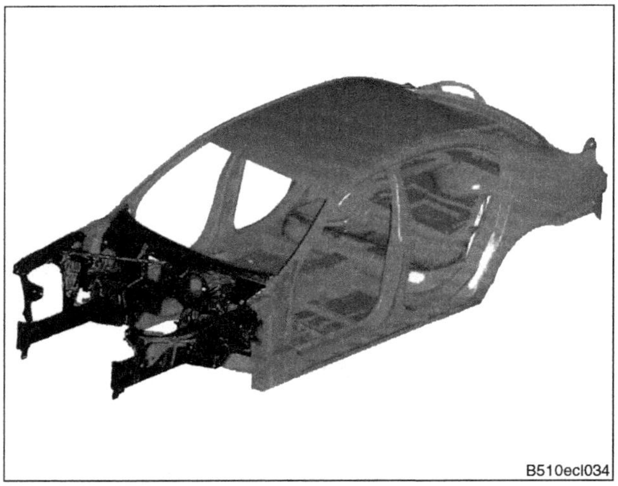

GROUNDS

Body panel grounding

In order to reduce vehicle weight, the front end of the E60 vehicle is built almost entirely of aluminum panels. This reduced-weight aluminum front end or GRAV is bonded and riveted to the remainder of the vehicle, which is steel, in the front bulkhead area. In the accompanying illustration, aluminum components are darker.

Riveted and bonded aluminum components do not necessarily provide a reliable circuit to ground. Therefore, individual front end panels are connected to each other and to steel components with small EMC safety weld seams. When repairing or replacing body components, be sure to use grounding straps (**arrow**) are reinstalled.

Ground distribution

Grounds are distributed throughout the vehicle body. Many are found under the interior carpets or behind trim panels. Several components grounds are often ganged. Ground positions vary among models. Lugs and connectors attached to ground are susceptible to damage and corrosion. Clean or renew as necessary.

Most ground positions are illustrated in photos in this repair group. See figure and page references in **Table l**. Ground applications are detailed in **Table m**.

BentleyPublishers.com
BentleyPublishers.com—All Rights Reserved

Table I. Ground locations

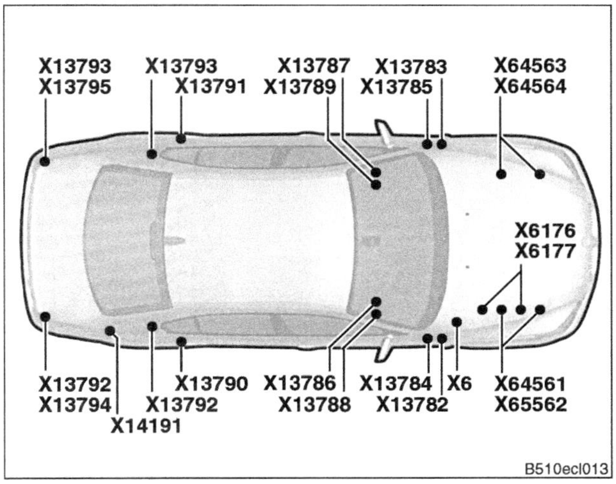

Code	Location, application	Fig #	Page #
X13782	Behind right kick panel, near door opening	Fig 86	Pg 62
X13783	Behind left kick panel, near door opening	Fig 81	Pg 61
X13784	Behind right kick panel, near door opening	Fig 86	Pg 62
X13785	Behind left kick panel, near door opening	Fig 81	Pg 61
X13786	Passenger seat rail, under rug	Fig 89	Pg 63
X13787	Driver seat rail, under rug	Fig 83	Pg 61
X13788	Passenger seat rail, under rug	Fig 89	Pg 63
X13789	Driver seat rail, under rug	Fig 83	Pg 61
X13790	Right C-pillar behind trim	Fig 92	Pg 63
X13791	Left C-pillar behind trim	Fig 91	Pg 63
X13792	*Sports Wagon:* Behind right side cargo compartment trim	Fig 123	Pg 71
X13792	*Sedan:* Inside trunk, underneath right taillight, behind trim	Fig 115	Pg 69
X13793	*Sports Wagon:* Behind left side cargo compartment trim	Fig 120	Pg 70
X13793	*Sedan:* Inside trunk, underneath left taillight, behind trim	Fig 111	Pg 68
X13794	Inside trunk, underneath right taillight, behind trim	Fig 115 Fig 123	Pg 69 Pg 71
X13795	Inside trunk, underneath left taillight, behind trim	Fig 111 Fig 121	Pg 68 Pg 71
X14234	Passenger side floor, under rug	Fig 89	Pg 63
X6176	Ignition coils 1 - 3 ground *M54 engine:* Between ignition coils 2 and 3 *N52 engine:* Right of ignition coil 2		
X6177	Ignition coils 4 - 6 ground *M54 engine:* Between ignition coils 4 and 5 *N52 engine:* Right of ignition coil 5		
X6	Right rear engine compartment, to right of E-box	Fig 5 Fig 38 Fig 53	Pg 29 Pg 48 Pg 53
X64561 X64562	Ignition coils 1 - 4 ground: *V8 engine:* Right side cylinder head cover, top of engine	Fig 56	Pg 54
X64563 X64564	Ignition coils 5 -8: *V8 engine:* Left side cylinder head cover, top of engine	Fig 54 Fig 61	Pg 54 Pg 56

ECL

BentleyPublishers
BentleyPublishers.com—All Rights Reserved

Ground applications

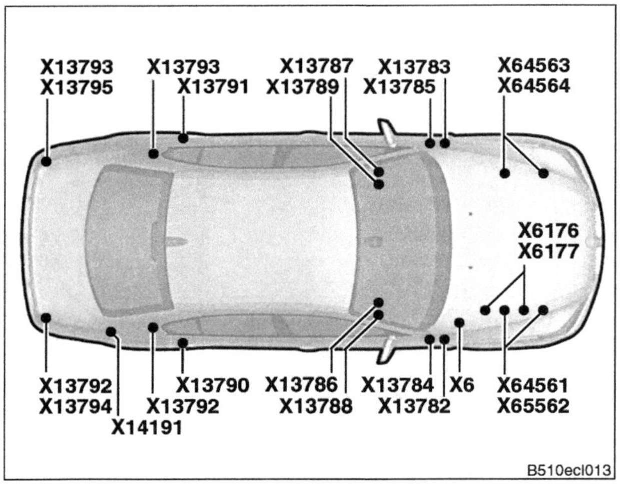

Ground applications

Table m. Ground applications			
Circuit	**Ground**	**Fig #**	**Pg #**
Adaptive headlights	X13782	**Fig 86**	**Pg 62**
Basic body module	X13782	**Fig 86**	**Pg 62**
Cigar lighter, front (from 03 / 2005)	X13782	**Fig 86**	**Pg 62**
Door sill light, right front	X13782	**Fig 86**	**Pg 62**
DSC	X13782	**Fig 86**	**Pg 62**
Dynamic drive (active stabilizer)	X13782	**Fig 86**	**Pg 62**
Foglight, right front	X13782	**Fig 86**	**Pg 62**
Footwell light, right front	X13782	**Fig 86**	**Pg 62**
Headlight, right (to 03 /2005)	X13782	**Fig 86**	**Pg 62**
Hood contact switch (from 03 / 2005)	X13782	**Fig 86**	**Pg 62**
Horns	X13782	**Fig 86**	**Pg 62**
Oil level switch	X13782	**Fig 86**	**Pg 62**
Secondary air pump	X13782	**Fig 86**	**Pg 62**
Turn signal, right front (to 03 / 2005)	X13782	**Fig 86**	**Pg 62**
Windshield wiper motor	X13782	**Fig 86**	**Pg 62**
Coolant level sensor	X13783	**Fig 81**	**Pg 61**
Door sill light, left front	X13783	**Fig 81**	**Pg 61**
Foglight, left front	X13783	**Fig 81**	**Pg 61**
Footwell light, left front	X13783	**Fig 81**	**Pg 61**
Headlight, left	X13783	**Fig 81**	**Pg 61**
Heater valve	X13783	**Fig 81**	**Pg 61**
Seat adjustment switch, driver	X13783	**Fig 81**	**Pg 61**
Turn signal, left front	X13783	**Fig 81**	**Pg 61**
Washer fluid level sensor	X13783	**Fig 81**	**Pg 61**
Washer pumps	X13783	**Fig 81**	**Pg 61**
12v utility outlet, glove compartment	X13784	**Fig 86**	**Pg 62**
AUC sensor	X13784	**Fig 86**	**Pg 62**

Table m. Ground applications			
Circuit	**Ground**	**Fig #**	**Pg #**
Center console switch center (SZM) (to 03 / 2005)	X13784	**Fig 86**	**Pg 62**
Central locking switch	X13784	**Fig 86**	**Pg 62**
Climate control panel	X13784	**Fig 86**	**Pg 62**
Comfort access control module	X13784	**Fig 86**	**Pg 62**
Condensation sensor (to 03 / 2005)	X13784	**Fig 86**	**Pg 62**
Fresh air temperature sensor	X13784	**Fig 86**	**Pg 62**
Gear indicator lighting	X13784	**Fig 86**	**Pg 62**
Glove compartment light	X13784	**Fig 86**	**Pg 62**
Hazard warning switch	X13784	**Fig 86**	**Pg 62**
Hood contact switch (to 03 / 2005)	X13784	**Fig 86**	**Pg 62**
iDrive controller	X13784	**Fig 86**	**Pg 62**
Integrated supply module (IVM)	X13784	**Fig 86**	**Pg 62**
Passenger seat control	X13784	**Fig 86**	**Pg 62**
Secondary air pump relay	X13784	**Fig 86**	**Pg 62**
Shift lock	X13784	**Fig 86**	**Pg 62**
SMG transmission selector lever	X13784	**Fig 86**	**Pg 62**
Steptronic switch	X13784	**Fig 86**	**Pg 62**
Turn signal, right front (from 03 / 2005)	X13784	**Fig 86**	**Pg 62**
Windshield wiper relay	X13784	**Fig 86**	**Pg 62**
12v utility outlet, rear	X13785	**Fig 81**	**Pg 61**
Car access system (CAS) (to 03 / 2005)	X13785	**Fig 81**	**Pg 61**
Cigar lighter, rear	X13785	**Fig 81**	**Pg 61**
Head-up display	X13785	**Fig 81**	**Pg 61**
Light module (to 03 / 2005)	X13785	**Fig 81**	**Pg 61**
Seat control, driver seat	X13785	**Fig 81**	**Pg 61**
Seat heating, rear seat	X13785	**Fig 81**	**Pg 61**
Steering column switch cluster (to 03 / 2005)	X13785	**Fig 81**	**Pg 61**
Washer jet heater	X13785	**Fig 81**	**Pg 61**
Active cruise control	X13786	**Fig 86**	**Pg 62**
Airbag indicator light	X13786	**Fig 86**	**Pg 62**
Automatic transmission control	X13786	**Fig 86**	**Pg 62**
Crash sensor, center console	X13786	**Fig 86**	**Pg 62**
Crash sensor, right B-pillar	X13786	**Fig 86**	**Pg 62**
DSC acceleration sensor	X13786	**Fig 86**	**Pg 62**
Passenger seat occupancy sensor	X13786	**Fig 86**	**Pg 62**
Safety and gateway module	X13786	**Fig 86**	**Pg 62**
Brake fluid level sensor	X13787	**Fig 83**	**Pg 61**
Brake light switch	X13787	**Fig 83**	**Pg 61**
Car access system (CAS) (from 03 / 2005)	X13787	**Fig 83**	**Pg 61**

BentleyPublishers.com
BentleyPublishers.com—All Rights Reserved

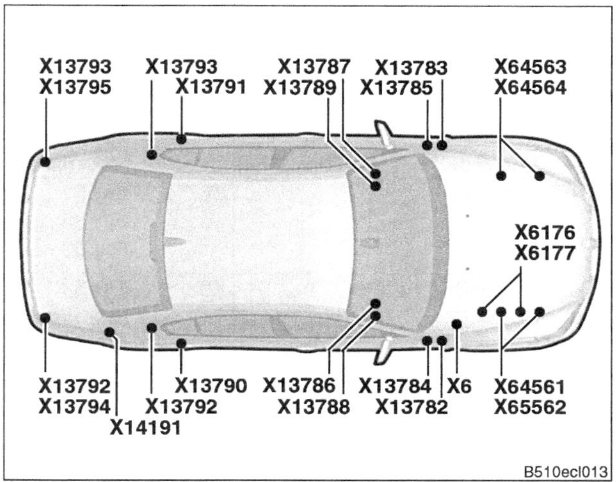

Table m. Ground applications			
Circuit	**Ground**	**Fig #**	**Pg #**
Center console switch center (SZM) (from 03 / 2005)	X13787	**Fig 83**	**Pg 61**
Central information display (CID)	X13787	**Fig 83**	**Pg 61**
Clutch switch module	X13787	**Fig 83**	**Pg 61**
Crash sensor, left B-pillar	X13787	**Fig 83**	**Pg 61**
Instrument cluster	X13787	**Fig 83**	**Pg 61**
Light module (from 03 / 2005)	X13787	**Fig 83**	**Pg 61**
OBD II plug	X13787	**Fig 83**	**Pg 61**
Outer door handle electronic modules	X13787	**Fig 83**	**Pg 61**
Rear compartment display	X13787	**Fig 83**	**Pg 61**
Steering column switch cluster (from 03 / 2005)	X13787	**Fig 83**	**Pg 61**
Steering rack	X13787	**Fig 83**	**Pg 61**
Active steering	X13788	**Fig 89**	**‘Pg 63**
Blower output stage	X13788	**Fig 89**	**‘Pg 63**
Door electrical components, passenger door	X13788	**Fig 89**	**‘Pg 63**
EDC	X13788	**Fig 89**	**‘Pg 63**
Headlight washer relay (from 03 / 2005)	X13788	**Fig 89**	**Pg 63**
12v utility outlet, rear	X13789	**Fig 83**	**Pg 61**
Cigar lighter, rear	X13789	**Fig 83**	**Pg 61**
Door electrical components, driver door	X13789	**Fig 83**	**Pg 61**
Head-up display	X13789	**Fig 83**	**Pg 61**
OBD II plug	X13789	**Fig 83**	**Pg 61**
Seat control, driver seat	X13789	**Fig 83**	**Pg 61**
Trunk lid or tailgate inside release switch	X13789	**Fig 83**	**Pg 61**
Windshield wiper relay	X13789	**Fig 83**	**Pg 61**
Anti-theft alarm	X13790	**Fig 92**	**Pg 63**

Table m. Ground applications			
Circuit	**Ground**	**Fig #**	**Pg #**
Cigar lighter, front (to 03 / 2005)	X13790	**Fig 92**	**Pg 63**
Condensation sensor (from 03 / 2005)	X13790	**Fig 92**	**Pg 63**
Door electrical components, right rear	X13790	**Fig 92**	**Pg 63**
Headlight washer relay (to 03 / 2005)	X13790	**Fig 92**	**Pg 63**
Headlight, right (from 03 / 2005)	X13790	**Fig 92**	**Pg 63**
Micro-power module	X13790	**Fig 92**	**Pg 63**
Park distance control	X13790	**Fig 92**	**Pg 63**
Rain and light sensor	X13790	**Fig 92**	**Pg 63**
Rear fuse and relay panel (A46)	X13790	**Fig 92**	**Pg 63**
Seat heating, rear seat	X13790	**Fig 92**	**Pg 63**
Trunk lid or tailgate soft-close	X13790	**Fig 92**	**Pg 63**
Antenna noise suppression filter	X13791	**Fig 91**	**Pg 63**
Door electrical components, left rear	X13791	**Fig 91**	**Pg 63**
Emergency call button	X13791	**Fig 91**	**Pg 63**
Interior light, left rear	X13791	**Fig 91**	**Pg 63**
Roller cover motor release, cargo compartment, left	X13791	**Fig 91**	**Pg 63**
Seat heating, rear seat	X13791	**Fig 91**	**Pg 63**
Vanity mirrors	X13791	**Fig 91**	**Pg 63**
12v utility outlet, luggage compartment	X13792	**Fig 123**	**Pg 71**
Dome light, rear	X13792	**Fig 123**	**Pg 71**
Electrochromic mirror, inside	X13792	**Fig 115** **Fig 123**	**Pg 69** **Pg 71**
Fuel pump	X13792	**Fig 115** **Fig 123**	**Pg 69** **Pg 71**
Fuel pump module	X13792	**Fig 115** **Fig 123**	**Pg 69** **Pg 71**
License plate lights	X13792	**Fig 115** **Fig 123**	**Pg 69** **Pg 71**
Rear window button	X13792	**Fig 123**	**Pg 71**
Sunroof	X13792	**Fig 123**	**Pg 71**
Taillights, right	X13792	**Fig 115** **Fig 123**	**Pg 69** **Pg 71**
Trailer socket	X13792	**Fig 123**	**Pg 71**
Trunk lid or tailgate lock	X13792	**Fig 115** **Fig 123**	**Pg 69** **Pg 71**
Trunk lid or tailgate lock release button	X13792	**Fig 115** **Fig 123**	**Pg 69** **Pg 71**
Antenna	X13793	**Fig 111** **Fig 120**	**Pg 68** **Pg 70**
Rear window defogger	X13793	**Fig 111** **Fig 120**	**Pg 68** **Pg 70**
Sound system components	X13793	**Fig 111** **Fig 120**	**Pg 68** **Pg 70**

BentleyPublishers
BentleyPublishers.com – All Rights Reserved

Ground applications

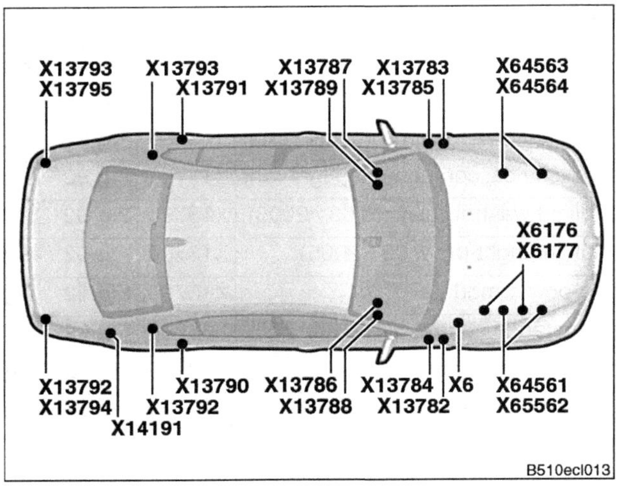

Table m. Ground applications

Circuit	Ground	Fig #	Pg #
Telematics	X13793	**Fig 111** **Fig 120**	**Pg 68** **Pg 70**
Telephone	X13793	**Fig 111** **Fig 120**	**Pg 68** **Pg 70**
Video module	X13793	**Fig 111** **Fig 120**	**Pg 68** **Pg 70**
Antenna noise suppression filter	X13794	**Fig 123**	**Pg 71**
Load-shedding (unloader) relay terminal 15	X13794	**Fig 115** **Fig 123**	**Pg 69** **Pg 71**
Roller cover motor release, cargo compartment, right	X13794	**Fig 123**	**Pg 71**
Spare wheel well fan	X13794	**Fig 115** **Fig 123**	**Pg 69** **Pg 71**
Tailgate lift motor	X13794	**Fig 123**	**Pg 71**
Trailer module	X13794	**Fig 115** **Fig 123**	**Pg 69** **Pg 71**
Trunk lid or tailgate soft-close	X13794	**Fig 115** **Fig 123**	**Pg 69** **Pg 71**
Wiper, rear window	X13794	**Fig 123**	**Pg 71**
CD changer cooling fan	X13795	**Fig 111** **Fig 121**	**Pg 68** **Pg 71**
Rear window lock	X13795	**Fig 111** **Fig 121**	**Pg 68** **Pg 71**
Taillight, left	X13795	**Fig 111** **Fig 121**	**Pg 68** **Pg 71**
Trunk lid or tailgate lift motor	X13795	**Fig 111** **Fig 121**	**Pg 68** **Pg 71**
Trunk lid or tailgate soft close	X13795	**Fig 111** **Fig 121**	**Pg 68** **Pg 71**
Trunk lid or tailgate lock	X13795	**Fig 111** **Fig 121**	**Pg 68** **Pg 71**
Active steering control module	X14234	**Fig 89**	**Pg 63**
A/C compressor	X6	**Fig 5** **Fig 38** **Fig 53**	**Pg 29** **Pg 48** **Pg 53**
Back-up light switch	X6	**Fig 5** **Fig 38** **Fig 53**	**Pg 29** **Pg 48** **Pg 53**
Crankcase breather heater	X6	**Fig 5** **Fig 38** **Fig 53**	**Pg 29** **Pg 48** **Pg 53**
DISA controller	X6	**Fig 5** **Fig 38** **Fig 53**	**Pg 29** **Pg 48** **Pg 53**
Engine control module (ECM)	X6	**Fig 5** **Fig 38** **Fig 53**	**Pg 29** **Pg 48** **Pg 53**
Fuel injector relay	X6	**Fig 5** **Fig 38** **Fig 53**	**Pg 29** **Pg 48** **Pg 53**
Oil level sensor (OZS)	X6	**Fig 5** **Fig 38** **Fig 53**	**Pg 29** **Pg 48** **Pg 53**
SMG transmission control	X6	**Fig 5** **Fig 38** **Fig 53**	**Pg 29** **Pg 48** **Pg 53**
Ignition coils (6-cylinder engine)	X6176 X6177		
Ignition coils (V8 engine)	X64561 X64562 X64563 X64564		

BentleyPublishers.com
BentleyPublishers.com—All Rights Reserved

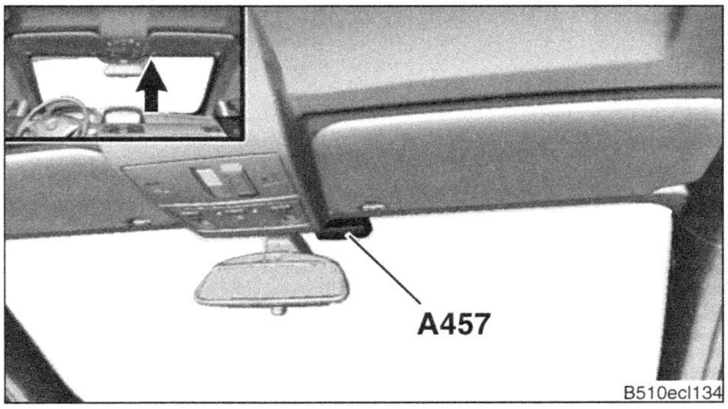

Component Location Photos

Antenna components

For comfort access (CA) antennas, see **Control modules, relays** in this repair group.

Fig 12 Top center windshield

Right of rear view mirror:

- **A457** Antenna, VICS

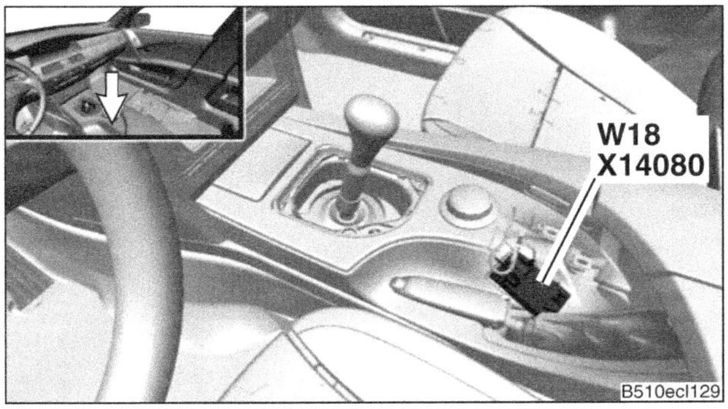

Fig 13 Center of cabin

Underneath center console:

- **W18** Antenna, bluetooth
- **X14080** Connector

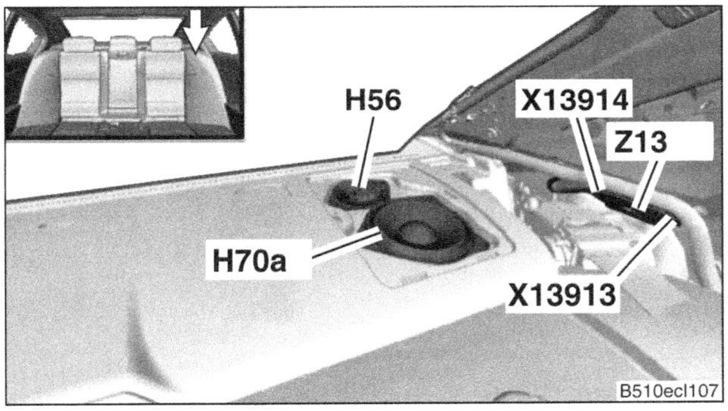

Fig 14 Parcel shelf, left side (Sedan)

Behind trim:

- **H56** Left rear tweeter
- **H70a** Left rear midrange speaker
- **X13913** Connector
- **X13914** Connector
- **Z13** Antenna noise suppression filter

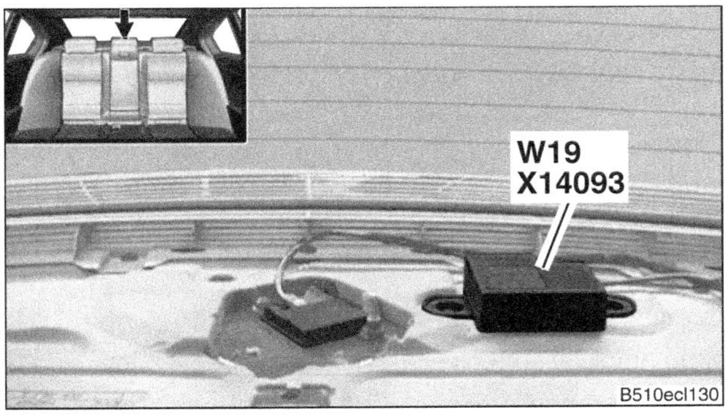

Fig 15 Parcel shelf, center (Sedan)

Behind trim:

- **W19** Antenna, emergency call
- **X14093** Connector

BentleyPublishers.com
BentleyPublishers.com—All Rights Reserved

Antenna components

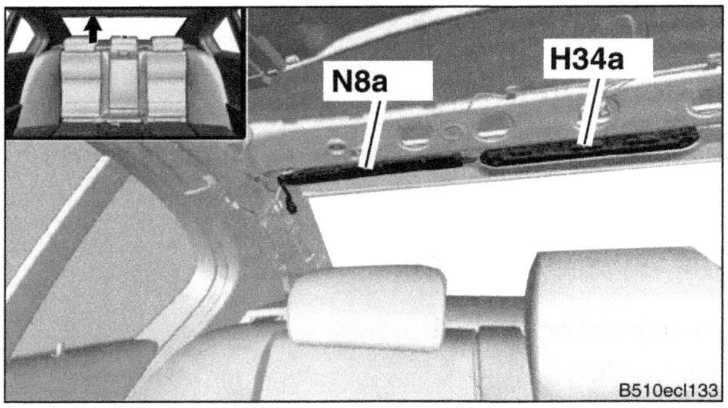

Fig 16 Above rear window (Sedan)

Behind trim:

- **H34a** Center brake light
- **N8a** Antenna amplifier with diversity module

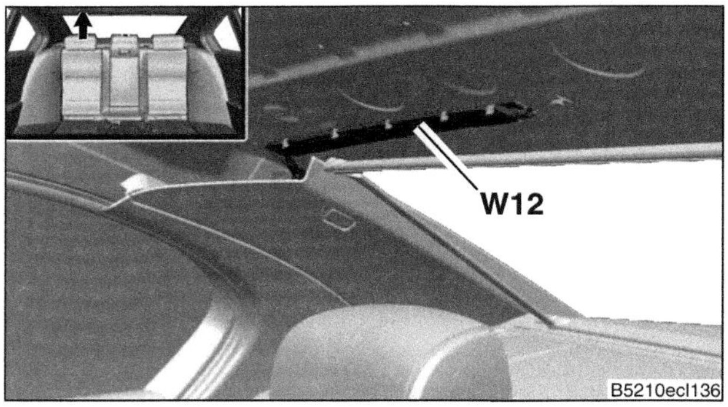

Fig 17 Right rear roof (Sedan)

Behind headliner:

- **W12** Antenna, interior

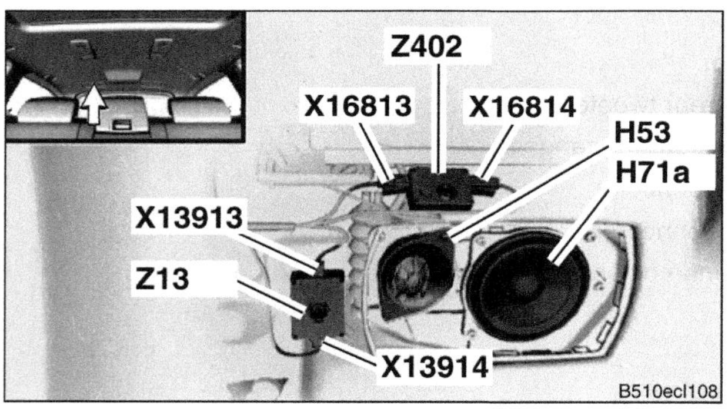

Fig 18 Right rear headliner (Sport Wagon)

Behind trim:

- **H53** Right rear tweeter
- **H71a** Right rear midrange speaker
- **X13913** Connector
- **X13914** Connection
- **X16813** Connector
- **X16814** Connector
- **Z13** Antenna noise suppression filter (center brake light)
- **Z402** Antenna noise suppression filter

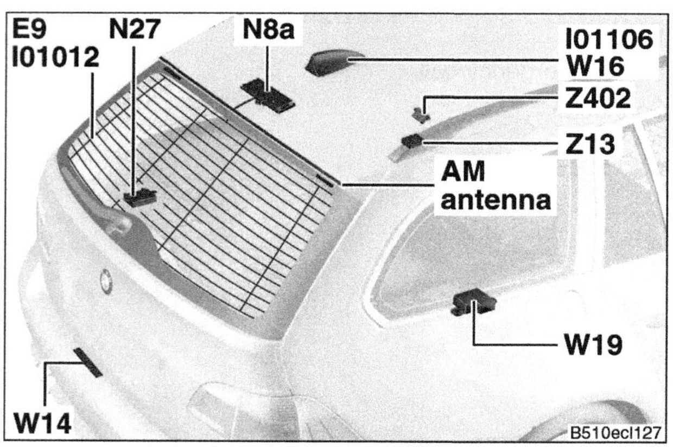

Fig 19 Body rear (Sport Wagon)

Tailgate, headliner:

- **AM antenna** under spoiler
- **E9** Rear window defogger
- **I01012** Antennas, FM1 - FM3
- **I01106** Antenna, satellite (SDARS)
- **N8a** Antenna amplifier with diversity module under spoiler
- **N27** Telephone compensator
- **W14** Antenna, rear bumper
- **W16** Antenna, telematics (TCU)
- **W19** Antenna, emergency call
- **Z13** Noise suppression filter (center brake light)
- **Z402** Noise suppression filter

BentleyPublishers.com
BentleyPublishers.com—All Rights Reserved

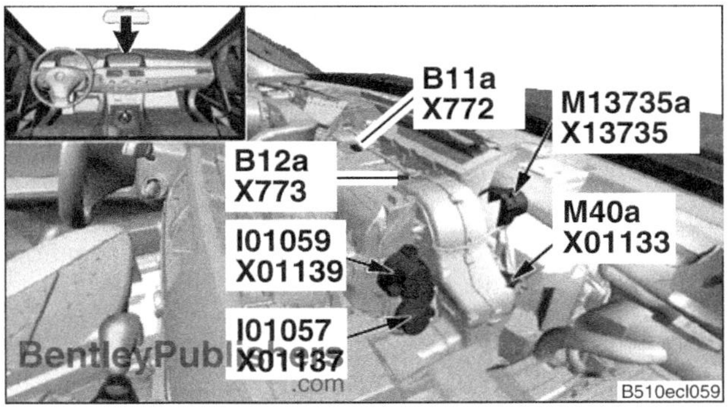

Climate control components

Fig 20 Underneath top dashboard trim

Top of IHKA housing:

- **B11a** Heater core temperature sensor, left side
- **B12a** Heater core temperature sensor, right side
- **I01057** Fresh air flap stepper motor
- **I01059** Right ventilation flap stepper motor
- **M40a** Right fresh-recirculated air flap stepper motor
- **M13735a** Right footwell flap stepper motor
- **X772** Connector
- **X773** Connector
- **X01133** Connector
- **X01137** Connector
- **X01139** Connector

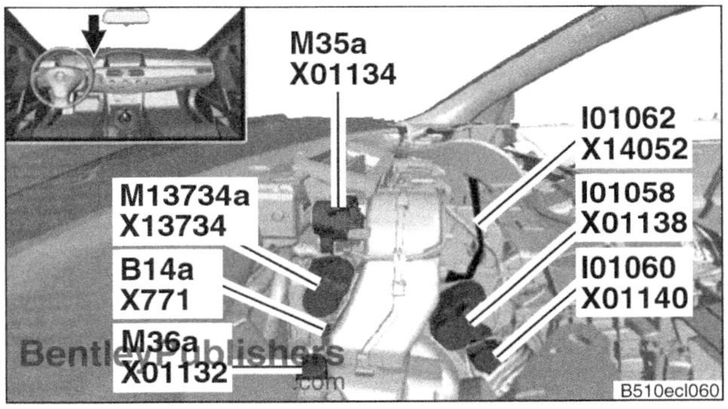

Fig 21 Underneath top dashboard trim

Top of IHKA housing:

- **B14a** Evaporator temperature sensor
- **I01058** Left ventilation flap stepper motor
- **I01060** Rear compartment mixing flap stepper motor
- **I01062** Electric auxiliary heater
- **M35a** Defroster flap stepper motor
- **M36a** Left fresh-recirculated air flap stepper motor
- **M13734a** Left footwell flap stepper motor
- **X771** Connector
- **X01132** Connector
- **X01134** Connector
- **X01138** Connector
- **X01140** Connector
- **X13734** Connector
- **X14052** Connector

BentleyPublishers.com—All Rights Reserved

Control modules, relays

Fig 22 Sedan control modules

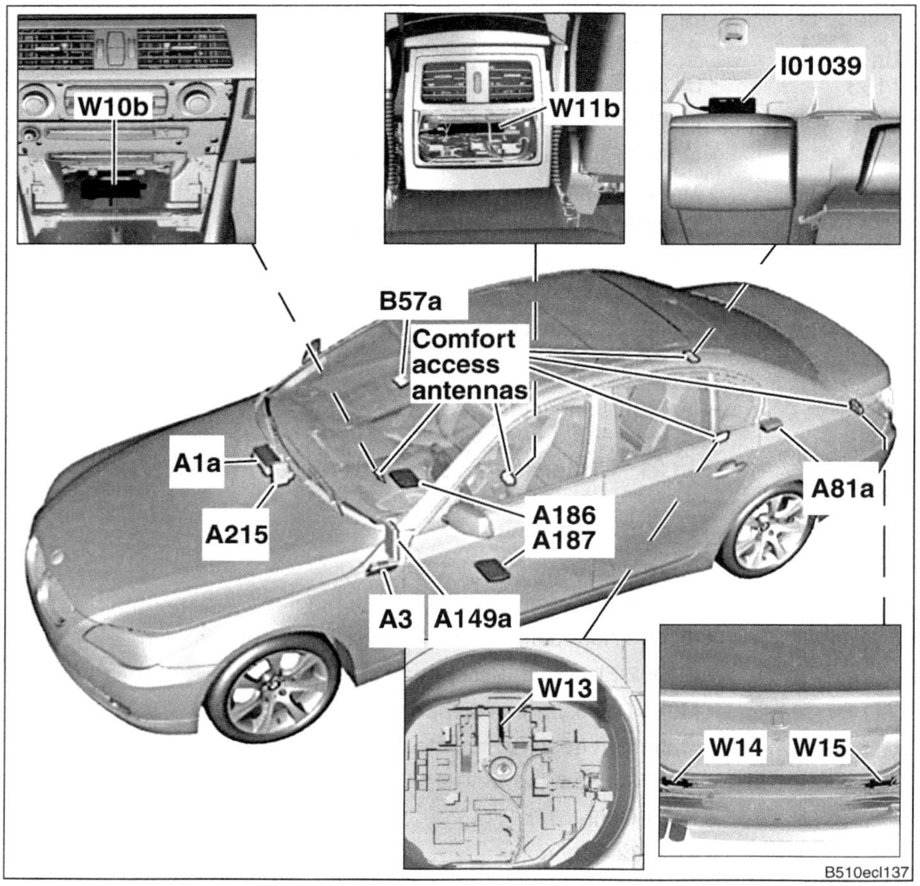

- **A1a** Basic body module (KBM)
- **A3** Light module
- **A81a** Park distance control (PDC) control module
- **A149a** Car access system (CAS)
- **A186 - A187** Seat control modules
- **A215** Comfort access control module
- **B57b** Rain and light sensor
- Comfort access antennas
  I01039 Rear parcel shelf
  W10a Center dashboard
  W11b Rear of center console
  W13 Trunk, spare tire well
  W14 - W15 Rear bumper

BentleyPublishers.com
BentleyPublishers.com—All Rights Reserved

Fig 23 Sport Wagon control modules

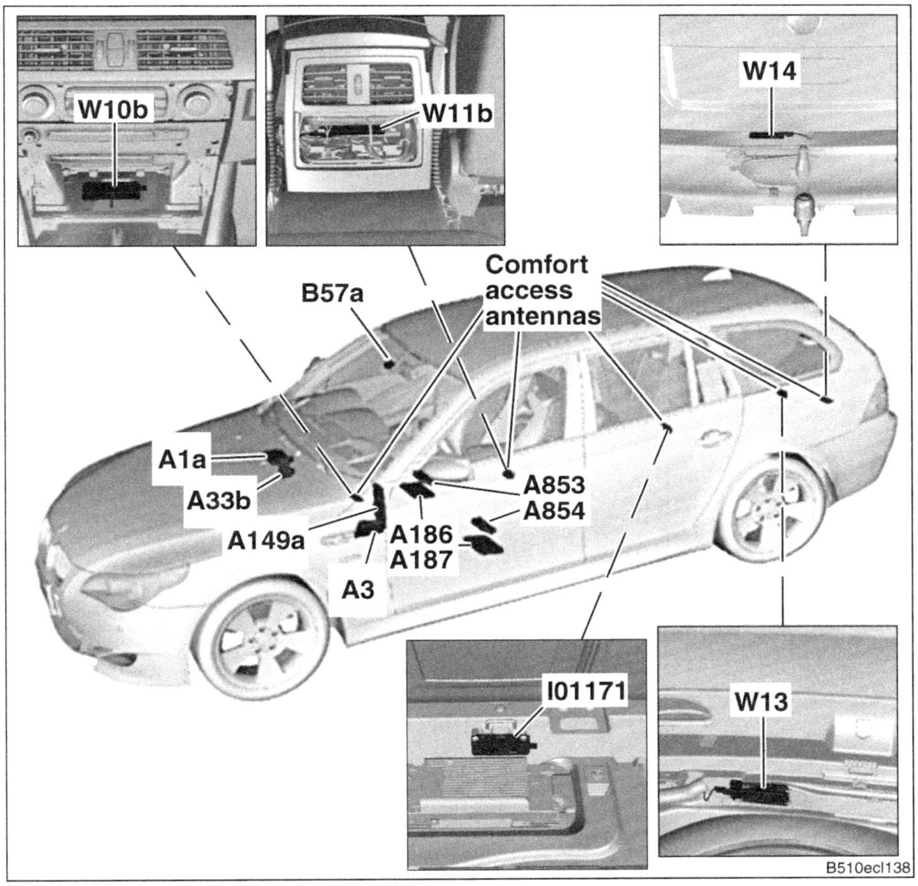

- **A1a** Basic body module (KBM)
- **A3** Light module
- **A33b** Sunroof module (to 03 / 2007)
- **A149a** Car access system (CAS)
- **A186 - A187** Seat control modules
- **A853 - A854** Backrest adjustment (active backrest) control modules
- **B57b** Rain and light sensor
- Comfort access antennas
  I01171 Front, spare tire well
  W10a Center dashboard
  W11b Rear of center console
  W13 Rear, spare tire well
  W14 Rear bumper

Fig 24 Spare tire well, version 1 (Sport Wagon)

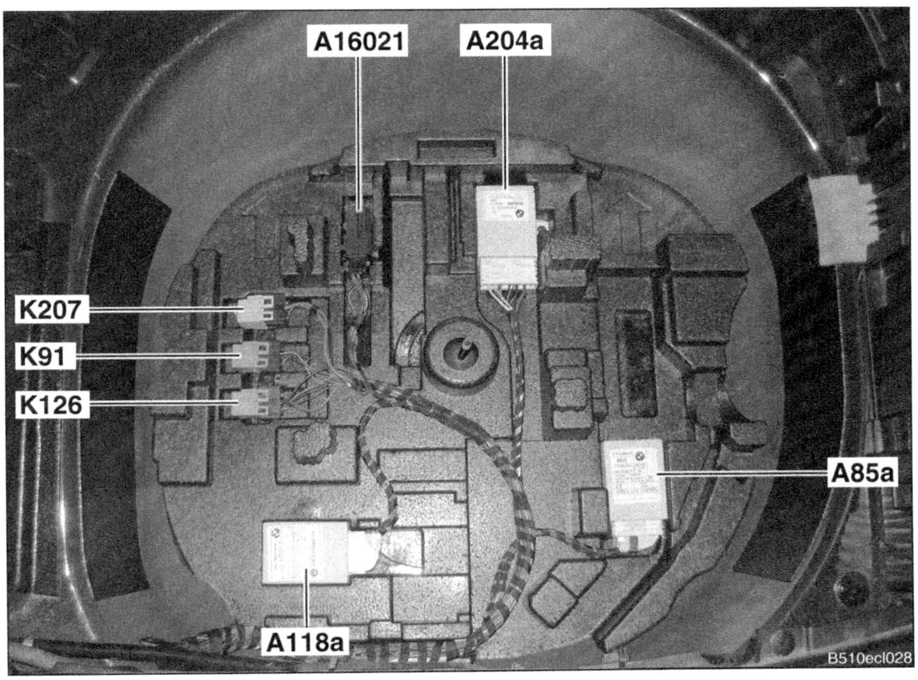

- **A85a** Tire pressure control (RDC) module
- **A118a** EHC (air suspension) module
- **A204a** Tailgate lift module
- **A16021** Micro-power module (MPM)
- **K91** Rear wiper relay
- **K126** EHC (air suspension) compressor relay
- **K207** Tailgate soft-close relay

BentleyPublishers.com–All Rights Reserved

Fig 25 Spare tire well, version 2 (Sport Wagon)

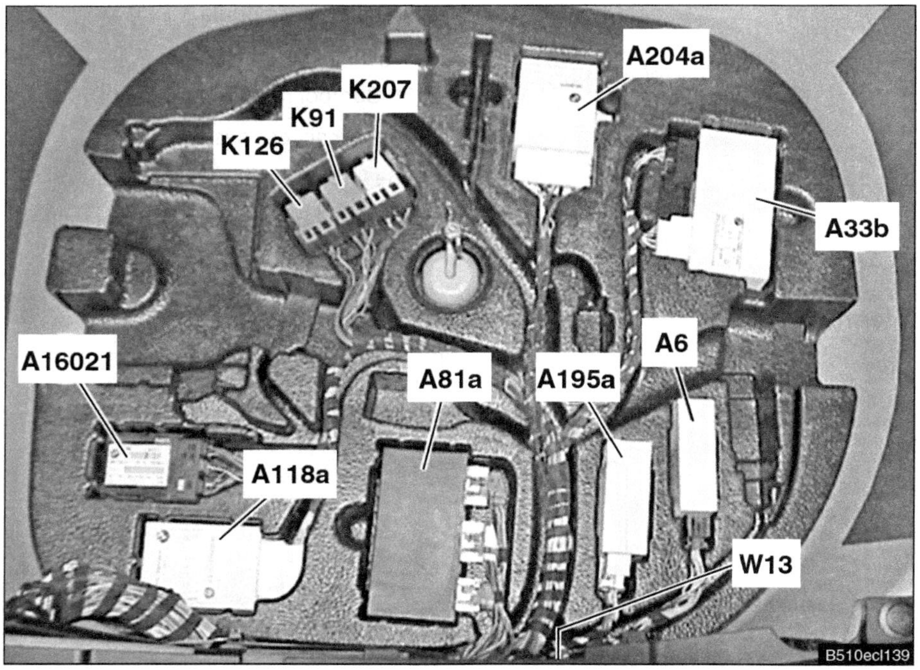

- **A6** Trailer module
- **A33b** Sunroof module (from 03/ 2007)
- **A81a** Park distance control (PDC) module
- **A118a** EHC (air suspension) module
- **A195a** Trailer hitch module
- **A204a** Tailgate lift module
- **A16021** Micro-power module (MPM)
- **K91** Rear wiper relay
- **K126** EHC (air suspension) compressor relay
- **K207** Tailgate soft-close relay
- **W13** Comfort access (CA) antenna

Engine compartment

Fig 26 Left front engine compartment

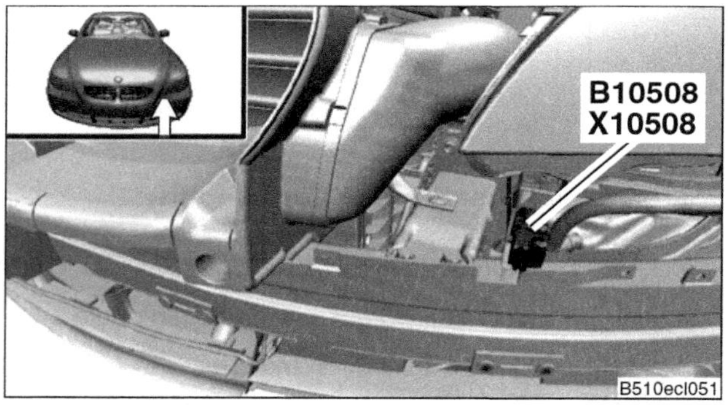

Underneath left headlight assembly:

- **B10508** Left front crash sensor
- **X10508** Connector

Fig 27 Left front engine

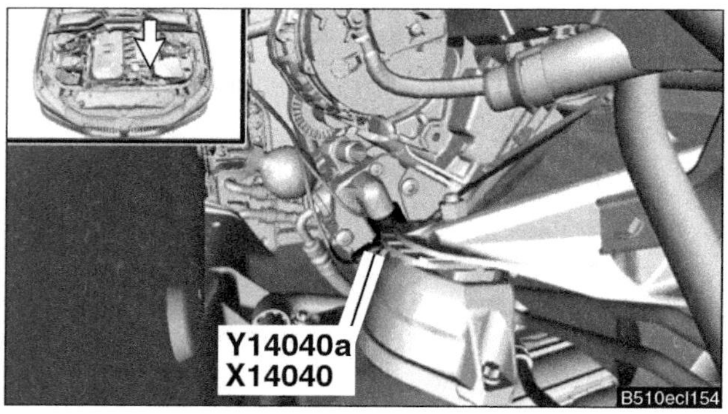

Behind power steering pump:

- **Y14040a** Active steering ECO valve
- **X14040** Connector

BentleyPublishers.com
BentleyPublishers.com—All Rights Reserved

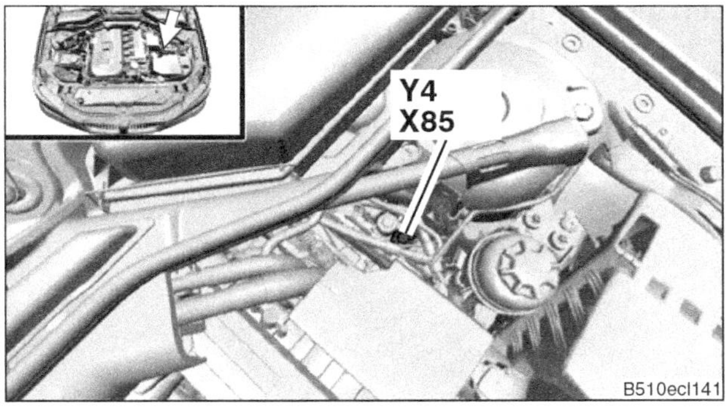

Fig 28 Left rear engine compartment

At strut tower:

- **X85** Connector
- **Y4** Heater valve

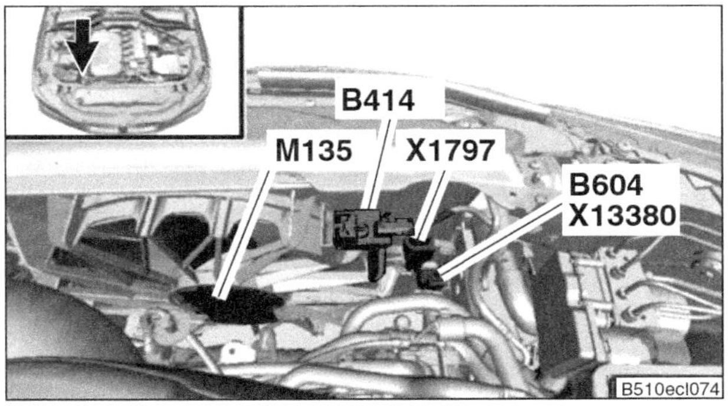

Fig 29 Right front engine compartment

At radiator:

- **B414** Automatic recirculation (AUC) sensor
- **B604** Engine coolant temperature sensor, radiator outlet
- **M135** Engine cooling fan
- **X1797** Connector
- **X13380** Connector

Fig 30 Right side engine compartment

Underneath coolant overflow reservoir:

- B97 A/C refrigerant pressure sensor

Engine compartment (M54 engine)

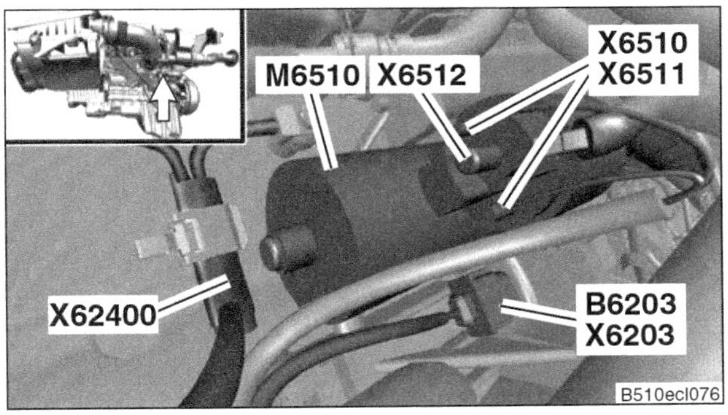

Fig 31 Left rear engine

At transmission bell housing, underneath intake manifold:

- **B6203** Crankshaft sensor
- **M6510** Starter
- **X6203** Connector
- **X6510 - X6512** Starter connectors
- **X62400** Connector

BentleyPublishers.com—All Rights Reserved

Engine compartment (M54 engine)

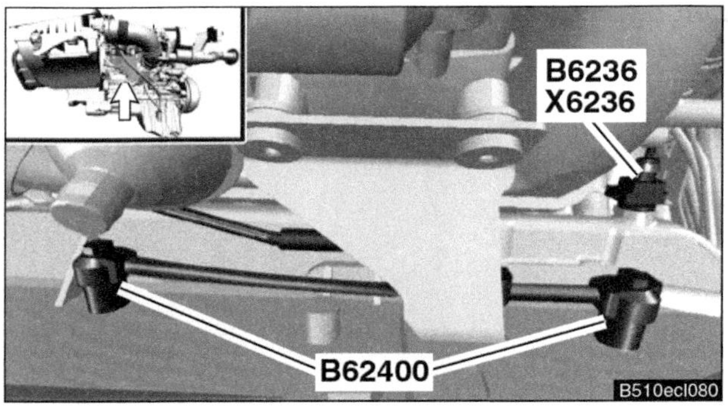

Fig 32 Left side engine

Underneath intake manifold:

- **B6236** Engine coolant temperature (ECT) sensor
- **B62400** Knock sensors
- **X6236** Connector

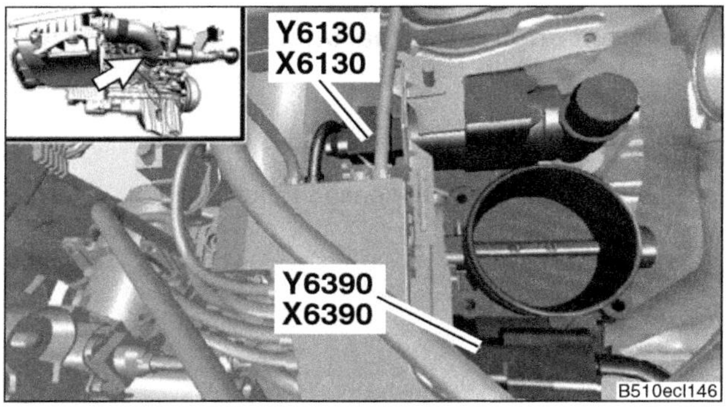

Fig 33 Left side engine

Underneath intake manifold:

- **X6130** Connector
- **X6390** Connector
- **Y6130** Idle speed control valve
- **Y6390** Throttle valve

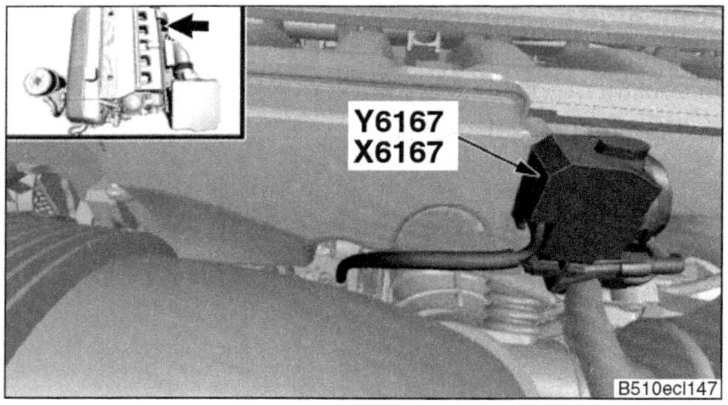

Fig 34 Left side engine

On intake manifold:

- **X6167** Connector
- **Y6167** Variable intake manifold (DISA) solenoid

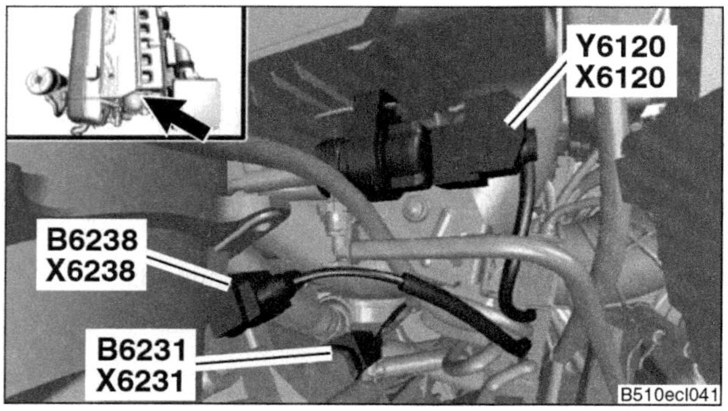

Fig 35 Left front of engine

Behind and below oil filter housing:

- **B6231** Oil pressure switch
- **B6238** Oil temperature sensor
- **X6120** Connector
- **X6231** Connector
- **X6238** Connector
- **Y6120** Fuel tank evaporative emissions control valve

BentleyPublishers.com
BentleyPublishers.com–All Rights Reserved

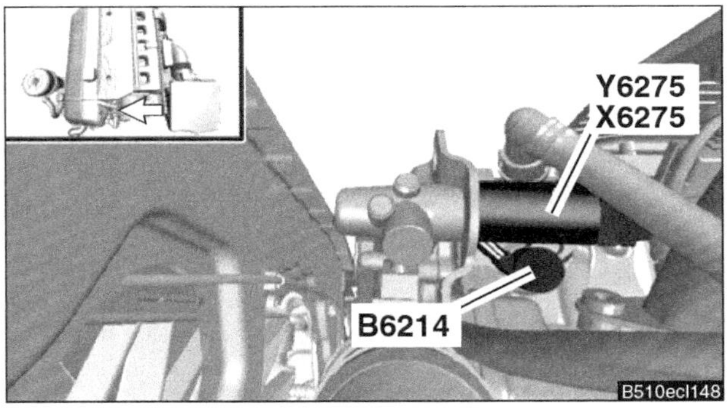

Fig 36 Left front of cylinder head

Above crankshaft vibration damper:

- **B6214** Intake camshaft sensor
- **X6275** Connector
- **Y6275** Intake VANOS solenoid

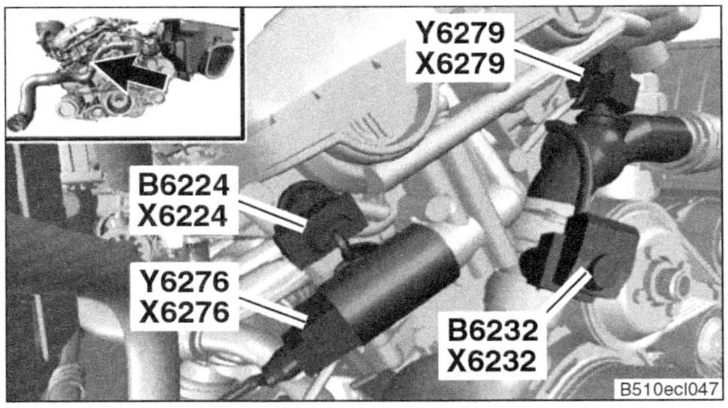

Fig 37 Front of cylinder head

Above crankshaft vibration damper:

- **B6224** Exhaust camshaft sensor
- **B6232** Engine coolant temperature (ECT) sensor
- **B6279** Coolant thermostat
- **X6224** Connector
- **X6232** Connector
- **X6276** Connector
- **X6279** Connector
- **Y6276** Exhaust VANOS solenoid

BentleyPublishers.com
BentleyPublishers.com—All Rights Reserved

Engine compartment (N52 engine)

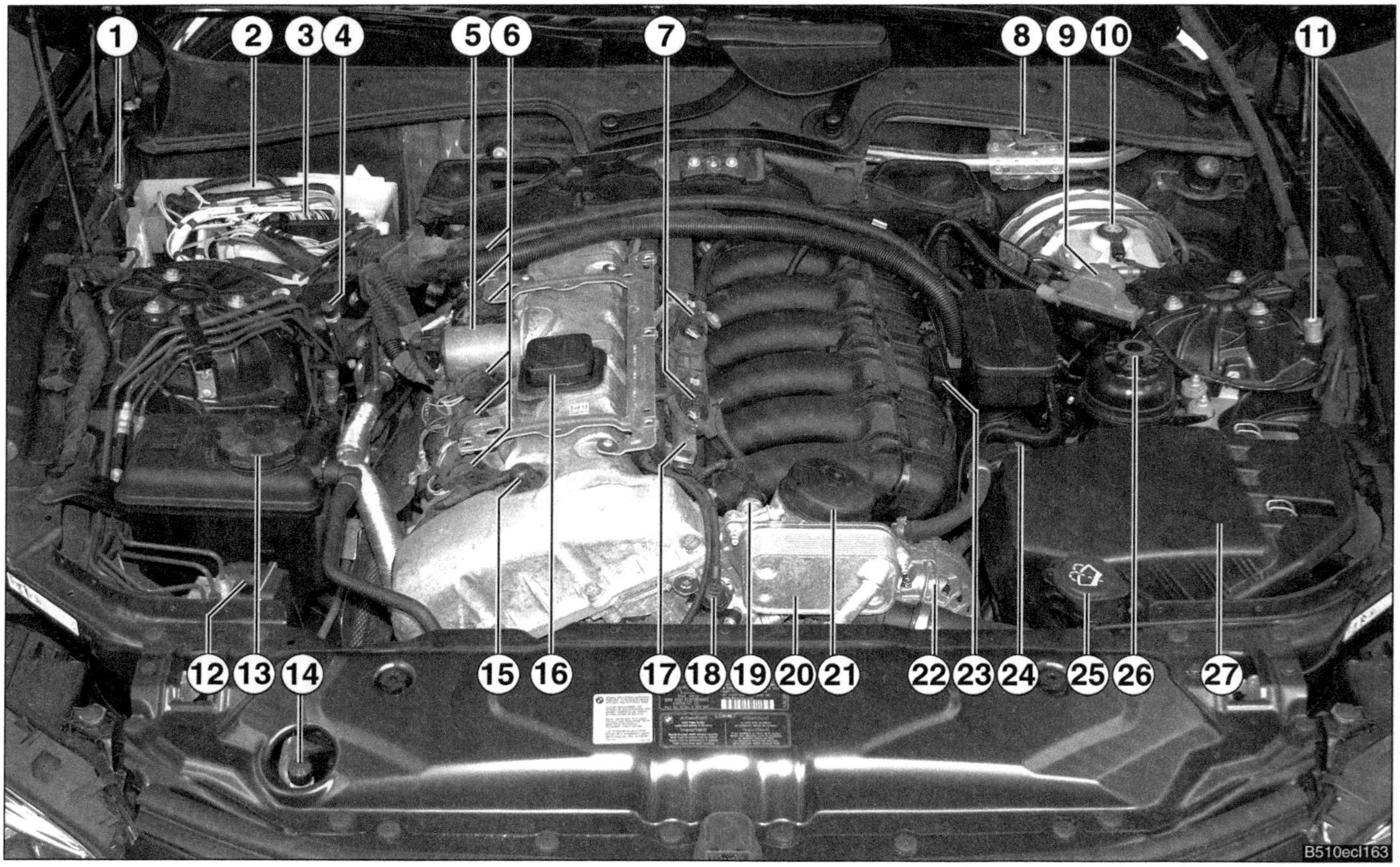

Fig 38 Engine compartment, covers removed (N52 engine)

1. **X6** Ground lug
2. E-box
3. **A6000** Engine control module (ECM)
4. A/C refrigerant port
5. **M6353** Valvetronic actuator
6. **T6151 -T6156** Ignition coils
7. **X62102 - X62202** Post-catalyst oxygen sensor connectors
8. **M3a** Windshield wiper motor
9. **G6430** B+ jump start terminal
10. **B18a** Brake fluid level switch
11. Ground jumper lug
12. **A65** ABS / DSC control module
13. Engine coolant overflow reservoir
14. A/C refrigerant port
15. Cylinder head cover breather port
16. Oil filler cap
17. Fuel rail
18. **B6236a** Engine coolant temperature sensor
19. **B6231** Oil pressure switch
20. Engine oil cooler
21. Oil filter housing cap
22. **G6524** Alternator (generator)
23. **Y6540** Variable intake manifold (DISA) solenoid
24. **B6207** Mass air flow sensor
25. Washer fluid reservoir cap
26. Power steering fluid reservoir cap
27. Air filter housing

BentleyPublishers.com
BentleyPublishers.com–All Rights Reserved

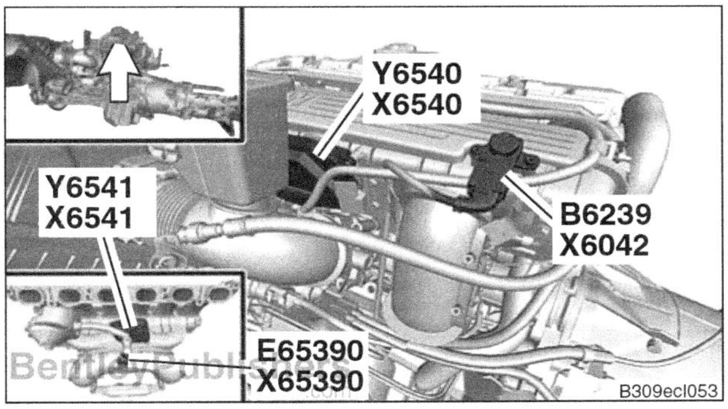

Fig 39 Left of cylinder head

Top of intake manifold:

- **B6239** Intake pressure (MAP) sensor
- **E65390** Crankcase breather heater
- **X6042** Connector
- **X6540** Connector
- **X6541** Connector
- **X65390** Connector
- **Y6540** Variable intake manifold (DISA) controller 1
- **Y6541** Variable intake manifold (DISA) controller 2

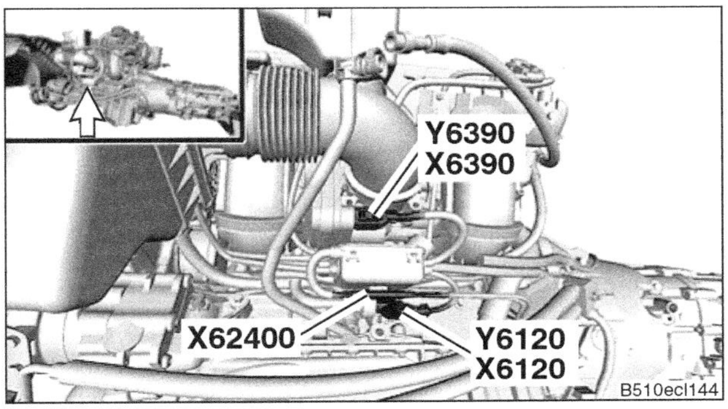

Fig 40 Left of cylinder head (N52 engine)

Underneath intake manifold:

- **X6120** Connector
- **X6390** Connector
- **X62400** Connector
- **Y6120** Fuel tank evaporative emissions control valve
- **Y6390** Throttle valve

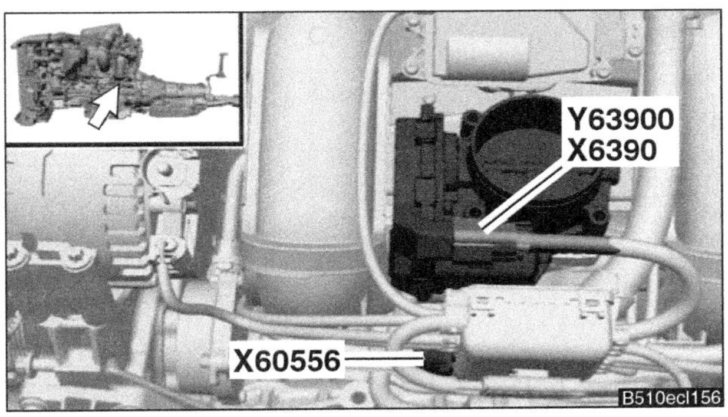

Fig 41 Left of cylinder head (N52K engine)

Underneath intake manifold:

- **X6390** Connector
- **X60556** Connector
- **Y63900** Throttle valve

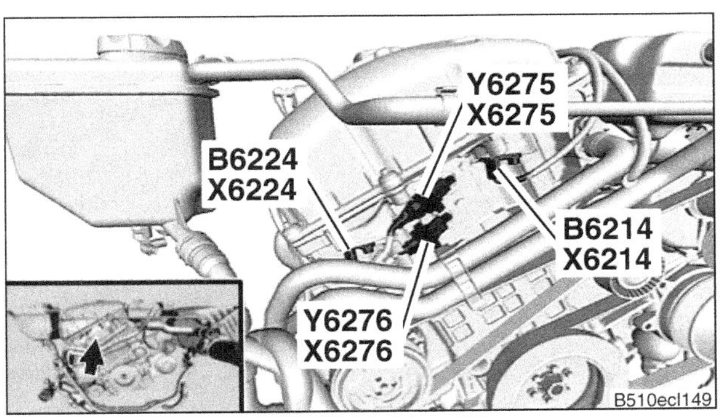

Fig 42 Front of cylinder head

On coolant outlet:

- **B6214** Intake camshaft sensor
- **B6224** Exhaust camshaft sensor
- **X6214** Connector
- **X6224** Connector
- **X6275** Connector
- **X6276** Connector
- **Y6275** Intake VANOS solenoid
- **Y6276** Exhaust VANOS solenoid

ECL

BentleyPublishers.com
BentleyPublishers.com—All Rights Reserved

Engine compartment (N52 engine)

Fig 43 Front of cylinder head

On coolant outlet:

- **B6236a** Engine coolant temperature (ECT) sensor
- **X6236** Connector

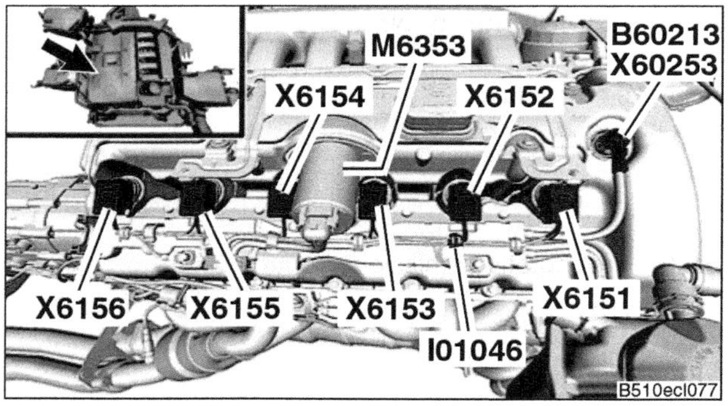

Fig 44 Right of cylinder head

Underneath ignition coil plastic cover:

- **B60213** Valvetronic eccentric shaft sensor
- **M6353** Valvetronic actuator
- **I01046** Interference suppression capacitor, ignition coils
- **X6151 - X6156** Ignition coil connectors
- **X60253** Connector

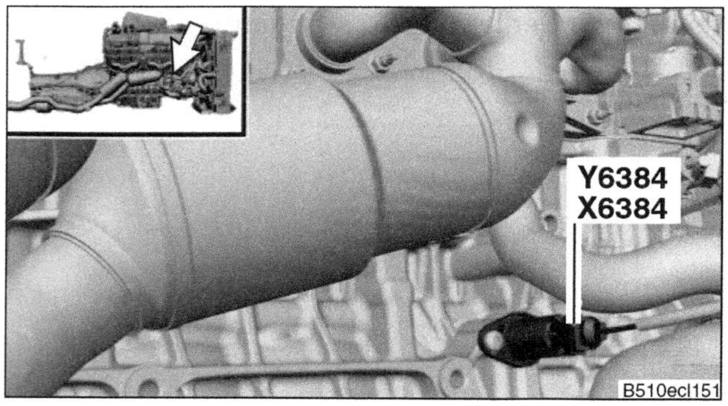

Fig 45 Right front engine compartment

right front lower engine block:

- **X6384** Connector
- **Y6384** Oil pressure control valve

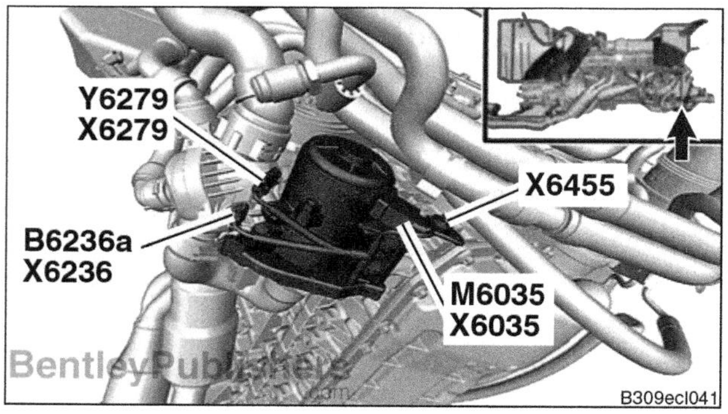

Fig 46 Right front engine compartment

Bolted to lower front right of engine:

- **B6236a** Engine coolant temperature (ECT) sensor
- **M6035** Electric coolant pump
- **X6035** Connector
- **X6236** Connector
- **X6279** Connector
- **X6455** Ground
  Electric coolant pump
- **Y6279** Coolant thermostat

BentleyPublishers.com
BentleyPublishers.com—All Rights Reserved

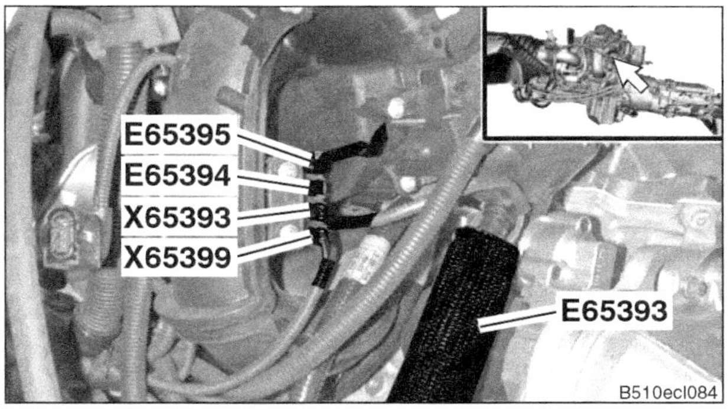

Fig 47 Rear of engine

Rear of intake manifold:

- **E65393 - E65395** Crankcase breather heater
- **X65393** Connector
- **X65399** Connector

Engine compartment (N54 engine)

Fig 48 Left side of engine

Intake manifold removed:

- **B2261** Fuel rail pressure sensor
- **B6125a** Low fuel pressure sensor
- **B6231** Oil pressure switch
- **B62400** Knock sensors
- **G6524** Alternator
- **M6510a** Starter
- **Y2381** Fuel volume control valve

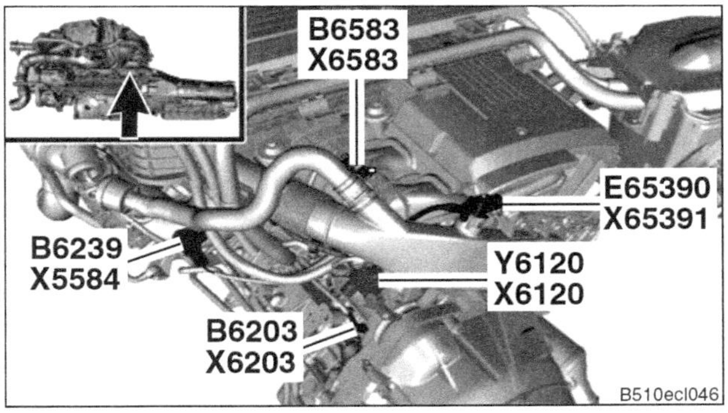

Fig 49 Left side of engine

Left rear of engine:

- **B6203** Crankshaft sensor
- **B6239** Intake pressure (MAP) sensor
- **B6583** Intake temperature and boost pressure sensor
- **E65390** Crankcase breather heater
- **X6120** Connector
- **X6203** Connector
- **X6239** Connector
- **X6583** Connector
- **X65391** Connector
- **Y6120** Evaporative emissions control valve

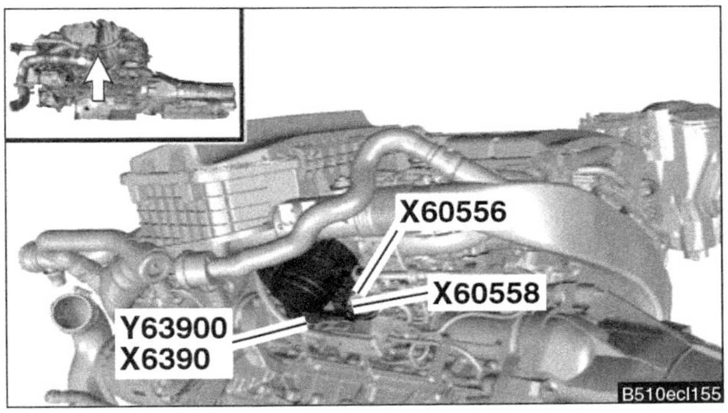

Fig 50 Left side of engine

Underneath intake manifold:

- **X6390** Connector
- **X60556** Connector
- **X60558** Connector
- **Y63900** Throttle valve

BentleyPublishers.com
BentleyPublishers.com—All Rights Reserved

Engine compartment (N54 engine)

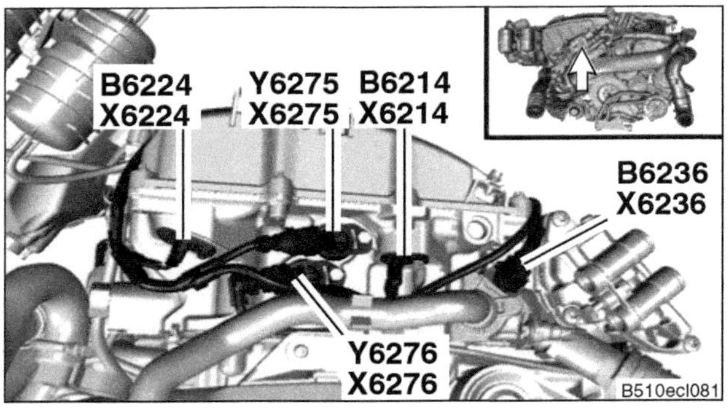

Fig 51 Front of engine

Front of cylinder head:

- **B6214** Intake camshaft sensor
- **B6224** Exhaust camshaft sensor
- **X6214** Connector
- **X6224** Connector
- **X6236** Connector
- **X6275** Connector
- **X6276** Connector
- **Y6275** Intake VANOS solenoid
- **Y6276** Exhaust VANOS solenoid

Fig 52 Right front engine compartment

Bolted to lower front right of engine (**arrow**):

- **M6035** Electric coolant pump
- **X6035** Connector
- **X6455** Ground
  Electric coolant pump
- **X6279** Connector
- **Y6279** Coolant thermostat

BentleyPublishers.com
BentleyPublishers.com—All Rights Reserved

Engine compartment (N62 engine)

Fig 53 Engine compartment, covers removed (N62 engine)

1. **X6** Ground lug
2. E-box
3. **A60210** Valvetronic control module
4. **A6000** Engine control module (ECM)
5. A/C refrigerant port
6. **T6151 - T6154** Ignition coils, right bank
7. **M6351** Valvetronic actuator. right bank
8. **M6352** Valvetronic actuator, left bank
9. Engine oil dipstick
10. **T6155 -T6158** Ignition coils, left bank
11. **M3a** Windshield wiper motor
12. **B18a** Brake fluid level switch
13. **G6430** B+ jump start terminal
14. Ground jumper lug
15. **A65** ABS / DSC control module
16. A/C refrigerant port
17. Engine coolant overflow reservoir
18. Power steering fluid reservoir cap
19. Vacuum pump
20. Oil filler cap
21. **Y6390** Throttle valve
22. **B6236** Engine coolant temperature sensor
23. **B6207** Mass air flow sensor
24. Washer fluid reservoir cap
25. Air filter housing

Bentley Publishers.com
BentleyPublishers.com—All Rights Reserved

Engine compartment (N62 engine)

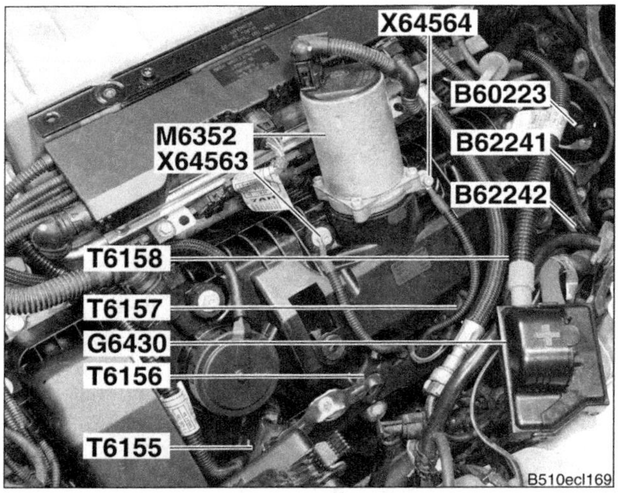

Fig 54 Left side of engine

Cylinder head cover, left bank:

- **B60223** Valvetronic eccentric shaft sensor, left bank
- **B62241** Intake camshaft sensor, left bank
- **B62242** Exhaust camshaft sensor, left bank
- **G6430** B+ jump start terminal
- **M6352** Valvetronic actuator, left bank
- **T6155 - T6158** Ignition coils 5 - 8
- **X64563 - X64564** Ignition coil grounds

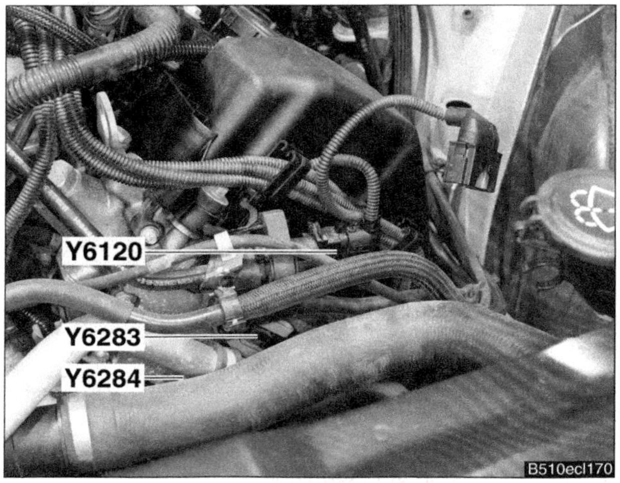

Fig 55 Left front of engine

Front of left cylinder head, air filter housing and duct removed:

- **Y6120** Fuel tank evaporative emissions control valve
- **Y6283** VANOS intake solenoid, left cylinder head
- **Y6284** VANOS exhaust solenoid, left cylinder head

Fig 56 Right side of engine

Cylinder head, right bank:

- **B60213** Valvetronic eccentric shaft sensor, right bank
- **B62141** Intake camshaft sensor, right bank
- **B62142** Exhaust camshaft sensor, right bank
- **M6351** Valvetronic actuator, right bank
- **T6151 - T6154** Ignition coils 1 - 4

BentleyPublishers.com
BentleyPublishers.com—All Rights Reserved

Fig 57 Front of engine

Air intake duct removed:

- **Y6236** Engine coolant temperature (ECT) sensor
- **Y6279** Engine coolant thermostat (characteristic cooling map)
- **Y6281** VANOS intake solenoid, right cylinder head
- **Y6282** VANOS exhaust solenoid, right cylinder head
- **Y6390** Throttle valve actuator

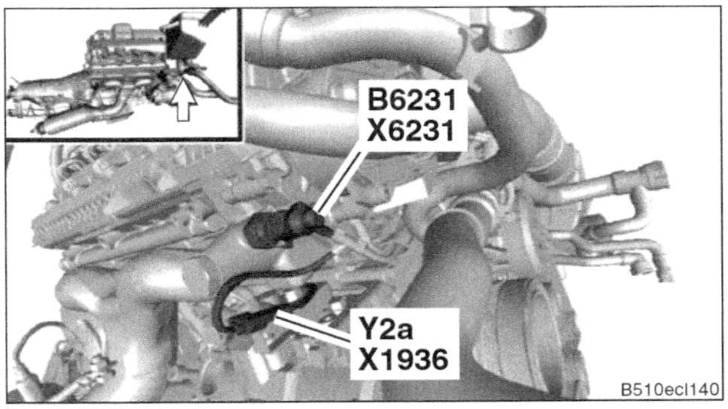

Fig 58 Right front of engine

Front of right cylinder head:

- **B6231** Oil pressure switch
- **X1936** Connector
- **X6231** Connector
- **Y2a** A/C compressor control valve

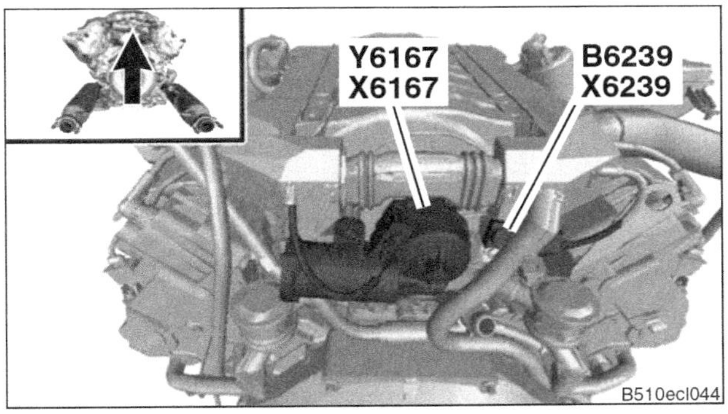

Fig 59 Rear of engine

On intake manifold:

- **B6239** Intake pressure (MAP) sensor
- **X6167** Connector
- **X6239** Connector
- **Y6167** Variable intake manifold (DISA) solenoid

Engine compartment (N62TU engine)

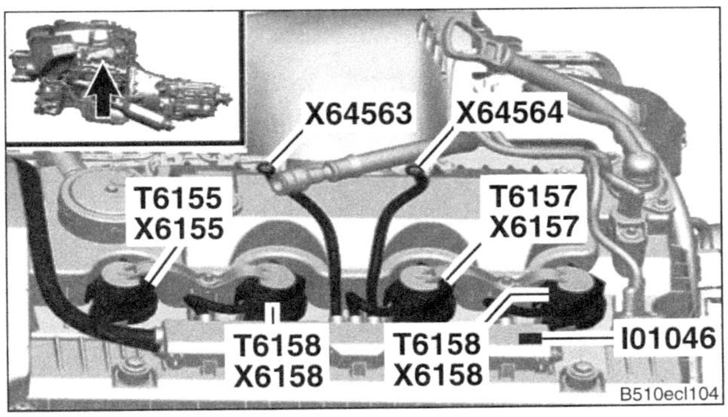

Fig 60 Left side of engine

Cylinder head, left bank:

- **I01046** Interference suppression capacitor, ignition coils
- **T6155 - T6158** Ignition coils 5 - 8
- **X64563 - X64564** Ignition coil grounds

Bentley Publishers
BentleyPublishers.com—All Rights Reserved

Engine compartment (N62TU engine)

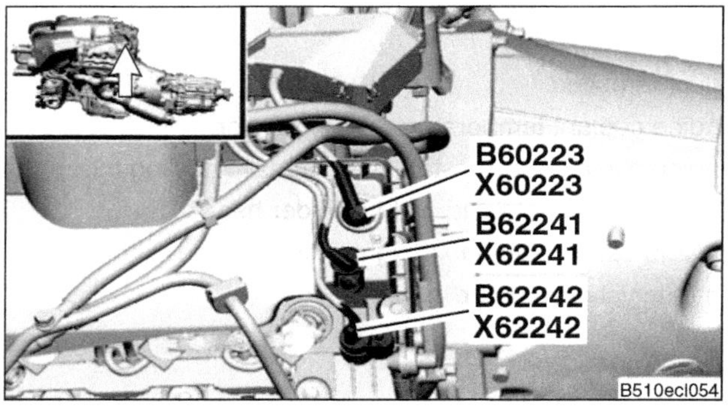

Fig 61 Left side of engine

◀ Cylinder head, left bank, rear:
- **B60223** Eccentric shaft sensor, left bank
- **B62241** Intake camshaft sensor, left bank
- **B62242** Exhaust camshaft sensor, left bank
- **X60223** Connector
- **X62241** Connector
- **X62242** Connector

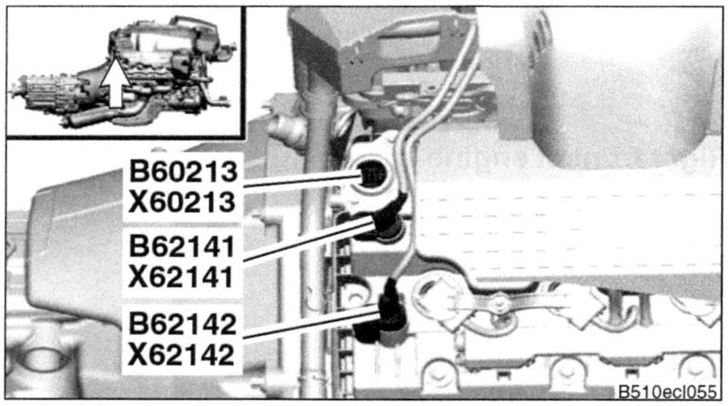

Fig 62 Right side of engine

◀ Cylinder head, left bank, rear:
- **B60213** Eccentric shaft sensor, right bank
- **B62141** Intake camshaft sensor, right bank
- **B62142** Exhaust camshaft sensor, right bank
- **X60213** Connector
- **X62141** Connector
- **X62142** Connector

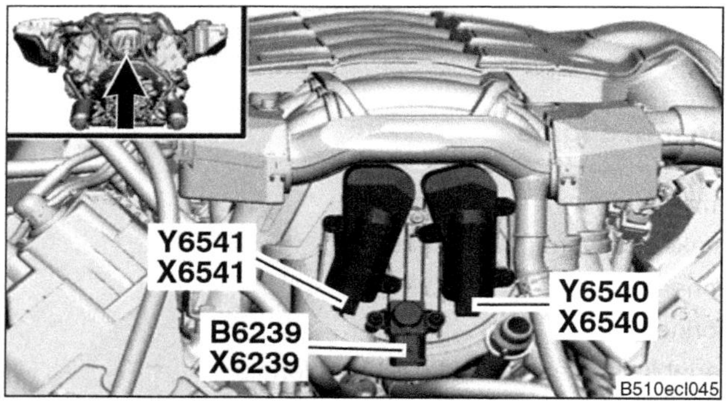

Fig 63 Rear of engine

◀ On intake manifold:
- **B6239** Intake pressure (MAP) sensor
- **X6239** Connector
- **X6540** Connector
- **X6541** Connector
- **Y6540** Variable intake manifold (DISA) controller, right cylinder head
- **Y6541** Variable intake manifold (DISA) controller, left cylinder head

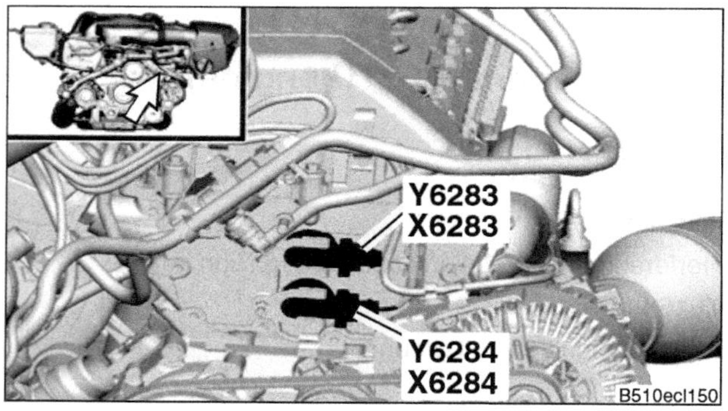

Fig 64 Front of engine

◀ Front of left cylinder head:
- **X6283** Connector
- **X6284** Connector
- **Y6283** VANOS intake solenoid, left cylinder head
- **Y6284** VANOS exhaust solenoid, left cylinder head

BentleyPublishers.com
BentleyPublishers.com–All Rights Reserved

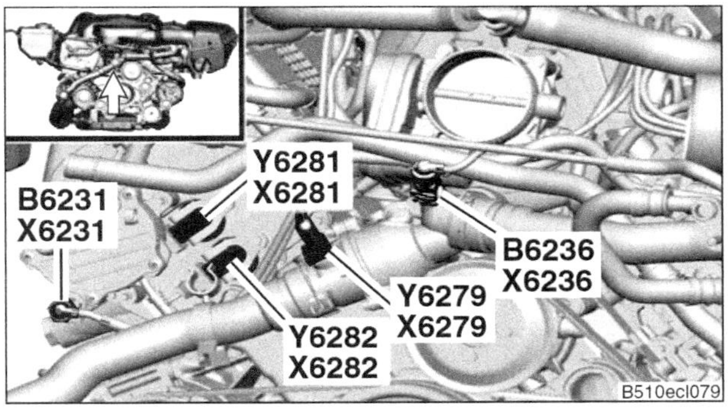

Fig 65 Front of engine

Front of right cylinder head:
- **B6231** Oil pressure switch
- **B6236** Engine coolant temperature (ECT) sensor
- **X6231** Connector
- **X6236** Connector
- **X6279** Connector
- **X6281** Connector
- **X6282** Connector
- **Y6279** Engine coolant thermostat
- **Y6281** VANOS intake solenoid, right cylinder head
- **Y6282** VANOS exhaust solenoid, right cylinder head

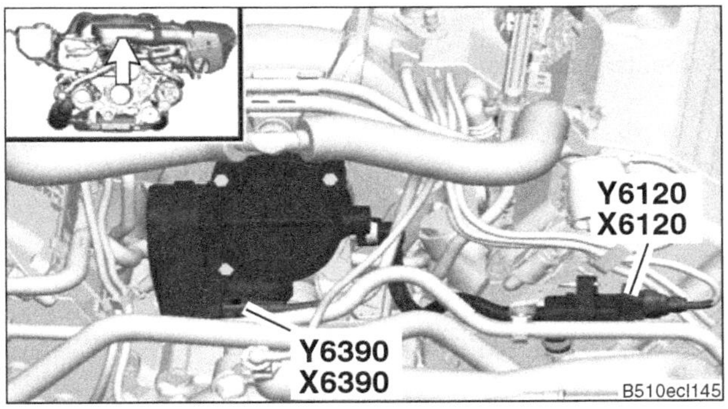

Fig 66 Front of engine

At intake manifold:
- **X6120** Connector
- **X6390** Connector
- **Y6120** Fuel tank evaporative emissions control valve
- **Y6390** Throttle valve

Oxygen sensors

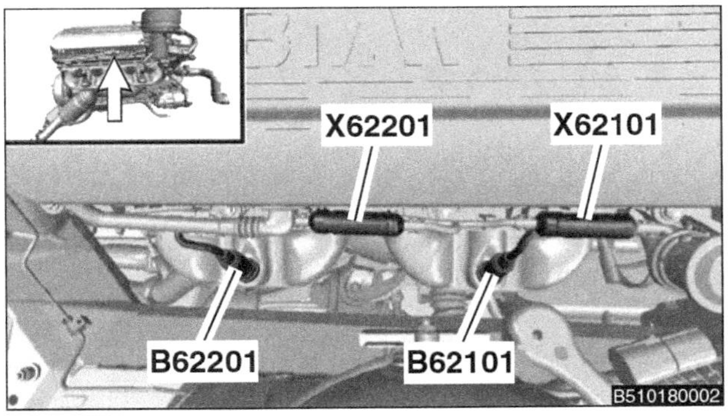

Fig 67 Exhaust manifold (M54 engine)

Right side engine compartment:
- **B62101** Pre-catalyst oxygen sensor, cyl 1 - 3
- **B62201** Pre-catalyst oxygen sensor, cyl 4 - 6
- **X62101** Oxygen sensor connector
- **X62102** Oxygen sensor connector

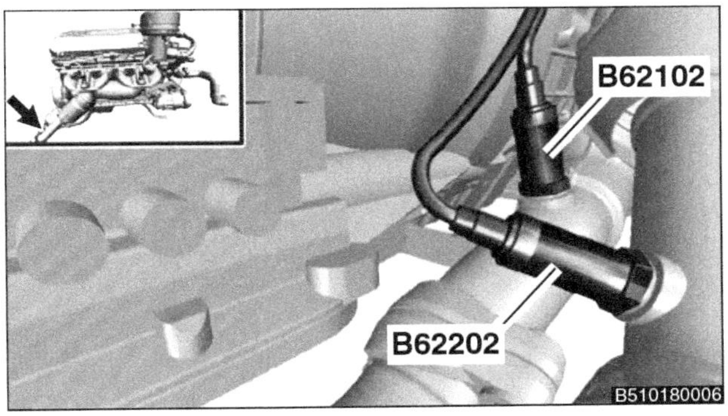

Fig 68 Exhaust pipe (M54 engine)

Underneath right side engine, below catalytic converters:
- **B62102** Post-catalyst oxygen sensor, cyl 1 - 3
- **B62202** Post-catalyst oxygen sensor, cyl 4 - 6

BentleyPublishers
BentleyPublishers.com—All Rights Reserved

Oxygen sensors

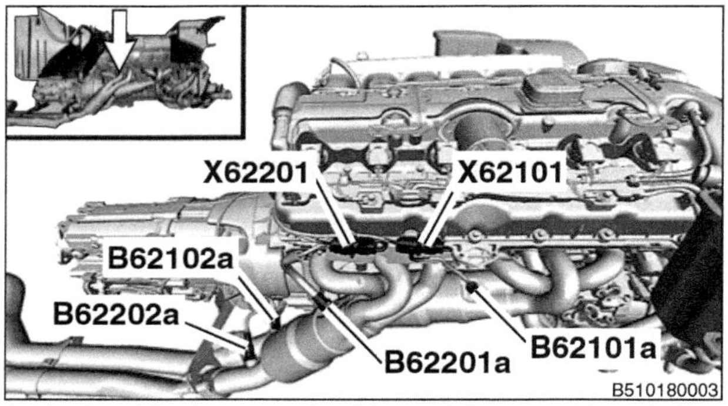

Fig 69 Exhaust manifold (N52 engine)

Right side engine compartment:

- **B62101a** Pre-catalyst oxygen sensor, cyl 1 - 3
- **B62102a** Post-catalyst oxygen sensor, cyl 1 - 3
- **B62201a** Pre-catalyst oxygen sensor, cyl 4 - 6
- **B62202a** Post-catalyst oxygen sensor, cyl 4 - 6
- **X62101** Oxygen sensor connector
- **X62201** Oxygen sensor connector

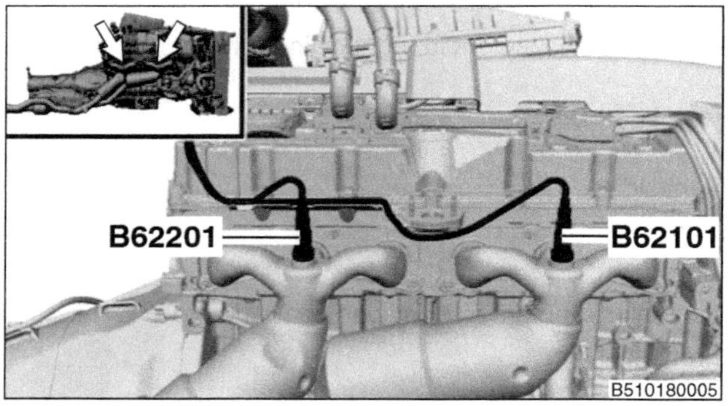

Fig 70 Exhaust manifold (N52K engine)

Right side engine compartment:

- **B62101** Pre-catalyst oxygen sensor, cyl 1 - 3
- **B62202** Pre-catalyst oxygen sensor, cyl 4 - 5

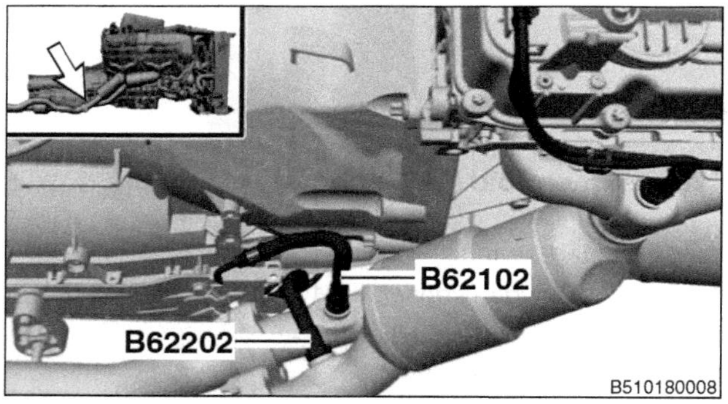

Fig 71 Exhaust pipe (N52K engine)

Underneath right side engine, below catalytic converters:

- **B62102** Post-catalyst oxygen sensor, cyl 1 - 3
- **B62202** Post-catalyst oxygen sensor, cyl 4 - 6

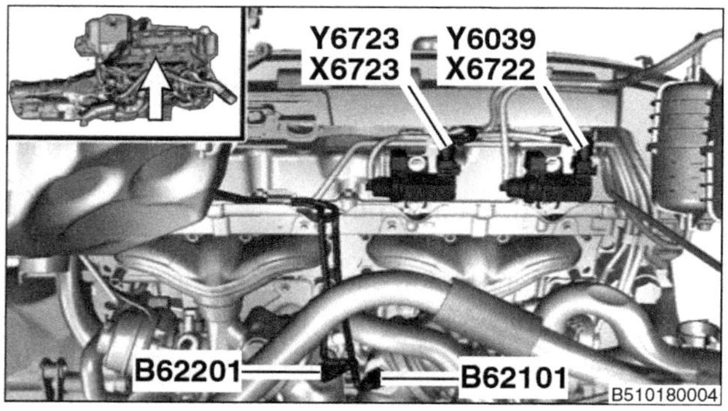

Fig 72 Exhaust manifold (N54 engine)

Right side engine compartment:

- **B62101** Pre-catalyst oxygen sensor, cyl 1 - 3
- **B62201** Pre-catalyst oxygen sensor, cyl 4 - 6
- **X6722** Connector
- **X6723** Connector
- **Y6039** Wastegate solenoid 1
- **Y6723** Wastegate solenoid 2

BentleyPublishers.com
BentleyPublishers.com–All Rights Reserved

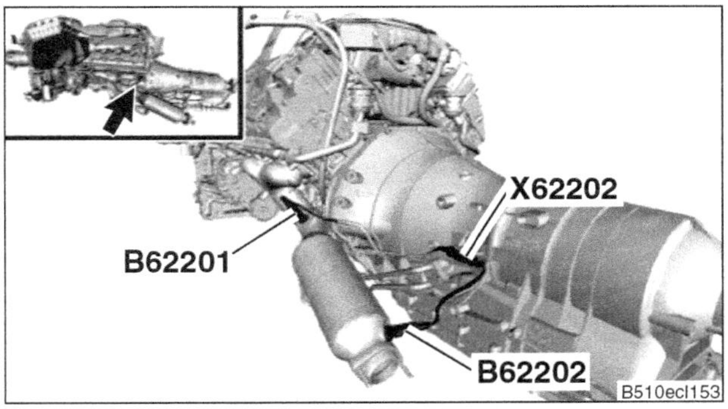

Fig 73 Left bank front pipe and catalytic converter (N62 engine)

Left bank catalytic converter:

- **B62201** Pre-catalyst oxygen sensor, left bank
- **B62202** Post-catalyst oxygen sensor, left bank
- **X62202** Oxygen sensor connector

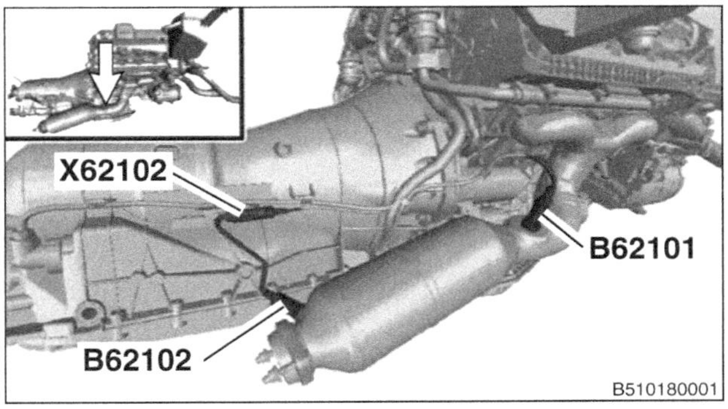

Fig 74 Right bank front pipe and catalytic converter (N62 engine)

Right bank catalytic converter:

- **B62101** Pre-catalyst oxygen sensor, right bank
- **B62102** Post-catalyst oxygen sensor, right bank
- **X62102** Oxygen sensor connector

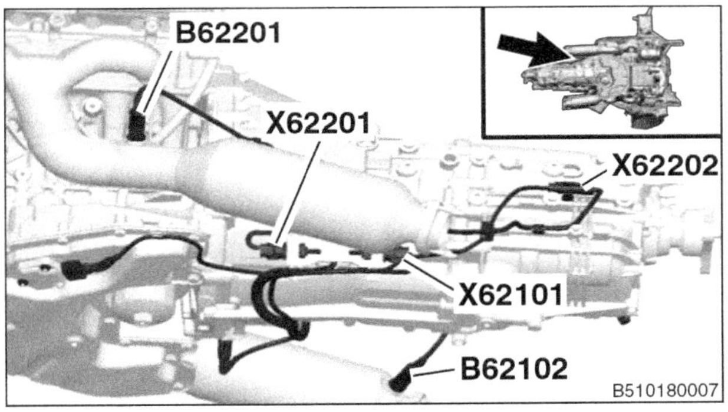

Fig 75 front pipes and catalytic converters (N62TU engine)

Left bank catalytic converter:

- **B62102** Post-catalyst oxygen sensor, right bank
- **B62201** Pre-catalyst oxygen sensor, left bank
- **X62102** Oxygen sensor connector
- **X62201** Oxygen sensor connector
- **X62202** Oxygen sensor connector

Passenger compartment components

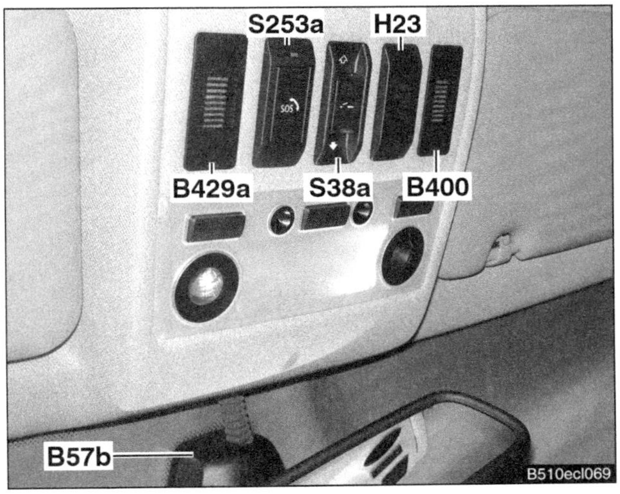

Fig 76 Top of windshield

In headliner:

- **B57b** Rain and light sensor
- **B400** Microphone
- **B429a** Hands-free microphone
- **H23** Passenger airbag deactivated warning light
- **S38a** Sunroof switch
- **S253a** Emergency call button

Bentley Publishers
BentleyPublishers.com—All Rights Reserved

Passenger compartment components

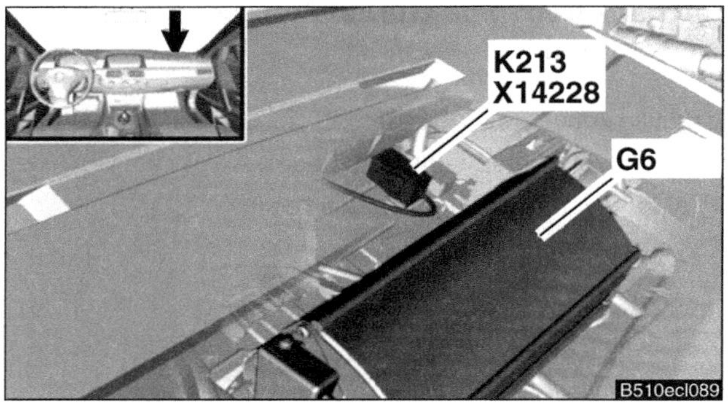

Fig 77 Behind right side dashboard

Dashboard trim removed:
- **G6** Passenger airbag
- **K213** Vacuum pump relay
- **X14228** Connector

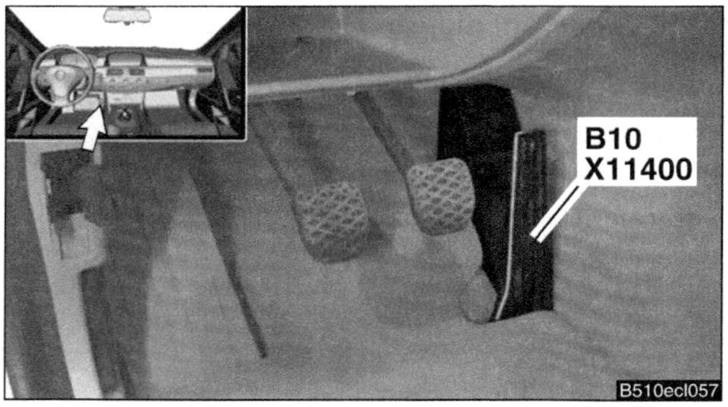

Fig 78 Underneath left side dashboard

At accelerator pedal:
- **B10** Accelerator pedal module
- **X11400** Connector

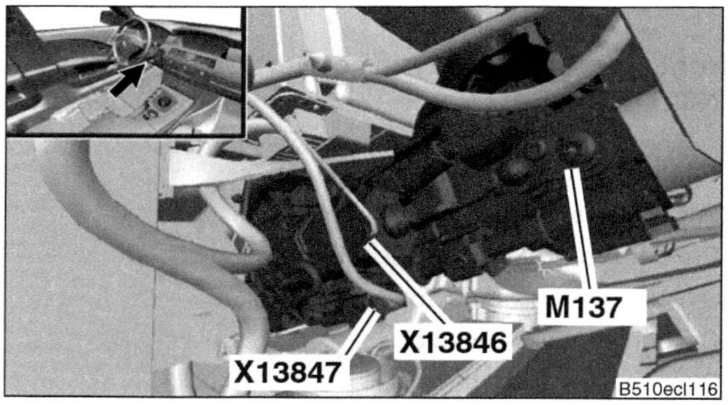

Fig 79 Underneath left side dashboard

On steering column:
- **M137** Steering column adjustment motor
- **X13846** Connector
- **X13847** Connector

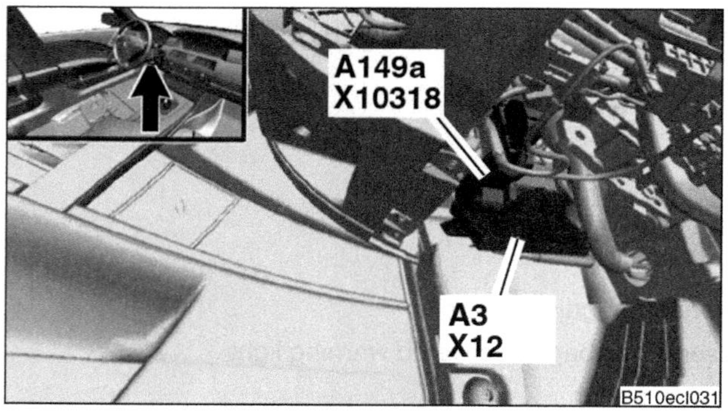

Fig 80 Underneath left side dashboard

Left of steering column:
- **A3** Light module
- **A149a** Car access system (CAS) module
- **X12** Connector
- **X10318** Connector

BentleyPublishers.com
BentleyPublishers.com—All Rights Reserved

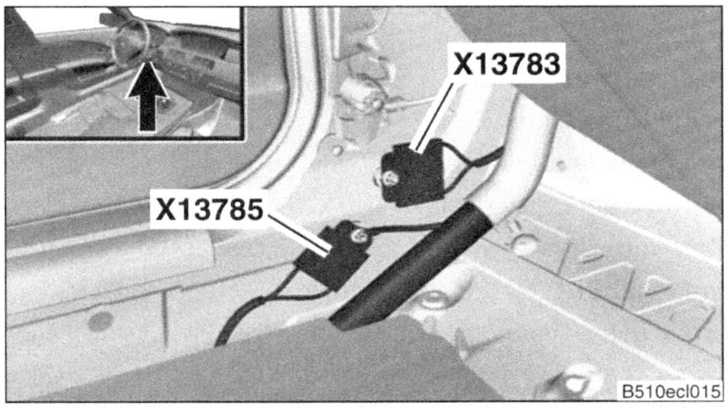

Fig 81 Left front

Left kick panel trim removed:

- **X13783** Ground connection
  See **Grounds** in this repair group.
- **X13785** Ground connection
  See **Grounds** in this repair group.

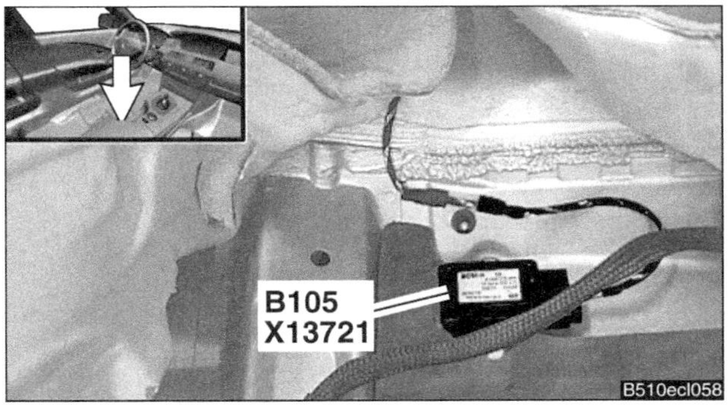

Fig 82 Left front floor

Underneath rear of driver seat:

- **B105** Accelerator sensor, DSC auxiliary
- **X13721** Connector

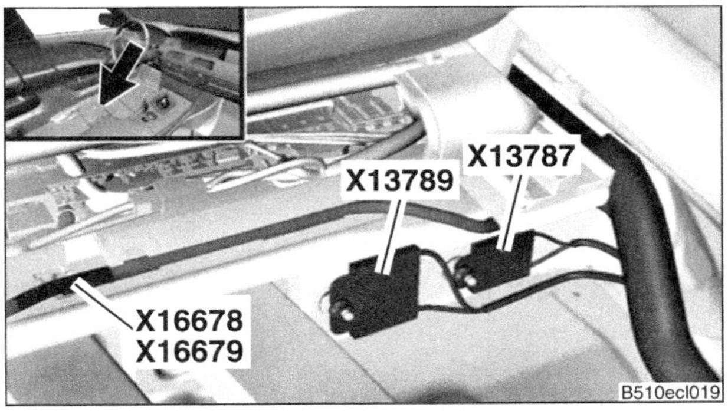

Fig 83 Driver seat rail

Underneath rug:

- **X13787** Ground connection
  See **Grounds** in this repair group.
- **X13789** Ground connection
  See **Grounds** in this repair group.
- **X16678** Connector
- **X16679** Connector

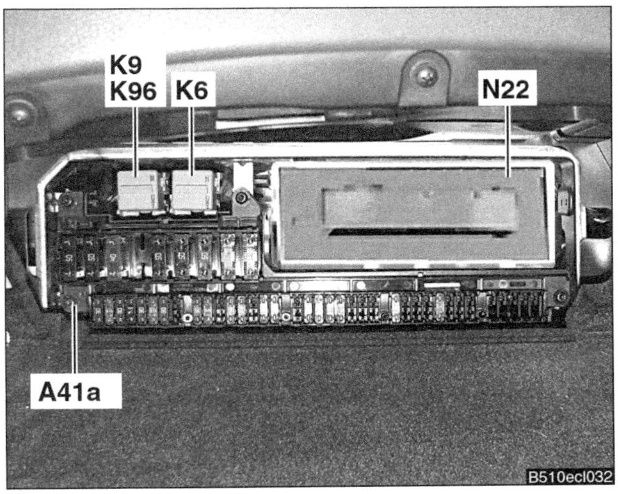

Fig 84 Underneath right side dashboard

Behind glove compartment:

- **A41a** Front fuse and relay panel
- **K6** Headlight washer relay (to 09 / 2005)
- **K9** Load reduction (load shedding) relay terminal 15 (from 09 / 2005)
- **K96** Fuel pump relay (to 09 / 2005)
- **N22** CD changer

A41a fuse applications are in **Front fuses** in this repair group.

BentleyPublishers.com
BentleyPublishers.com—All Rights Reserved

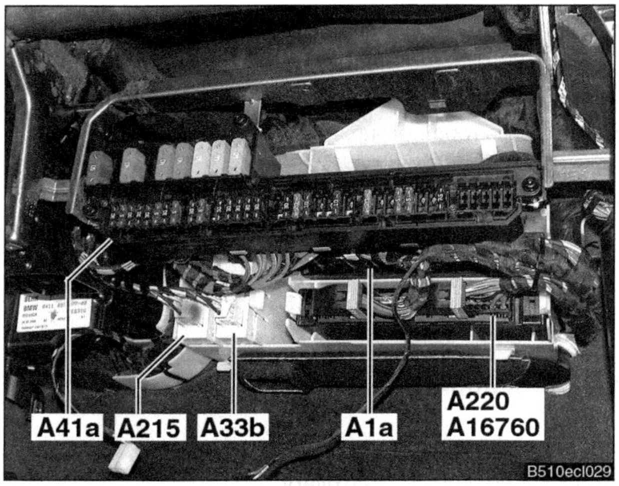

Fig 85 Underneath right side dashboard

Glove compartment removed:

- **A1a** Basic body module (KBM)
- **A33b** Sunroof module (panorama)
- **A41a** Front fuse and relay panel
- **A220** Safety and gateway module (SGM) (to 09 / 2005)
- **A215** Comfort access (CA) control module
- **A16760** Body gateway module (KGM) (from 09 / 2005)

A41a fuse applications are in **Front fuses** in this repair group.

Glove compartment removal is covered in **513 Interior Trim**.

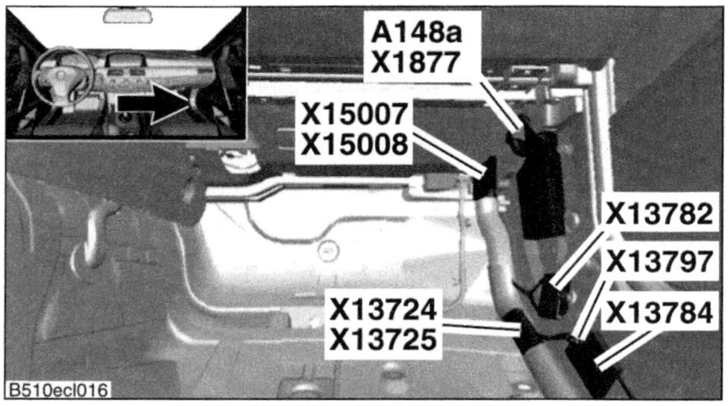

Fig 86 Right front

Right kick panel trim removed:

- **A148a** Dynamic drive (active stabilizer) control module
- **X1877** Connector
- **X13724** F-CAN bus wire splice
- **X13725** F-CAN bus wire splice
- **X13782** Ground connection
  See **Grounds** in this repair group.
- **X13784** Ground connection
  See **Grounds** in this repair group.
- **X13797** Connector
- **X15007** F-CAN bus wire splice
- **X15008** F-CAN bus wire splice

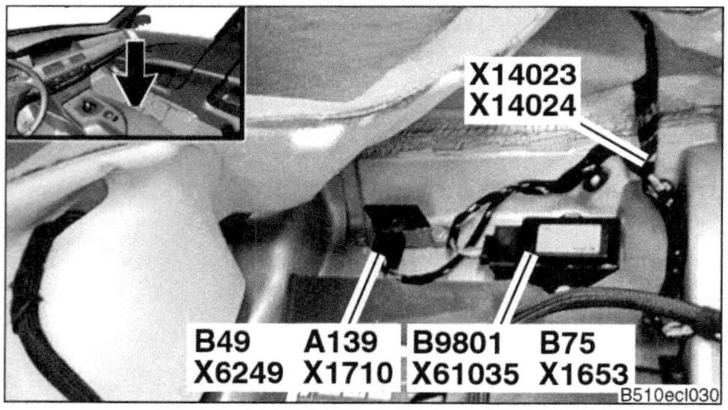

Fig 87 Right front floor

Underneath rug:

- **A139** Lateral acceleration sensor, dynamic drive (active stabilizer)
- **B49** Acceleration sensor
- **B75** DSC acceleration sensor
- **B9801** DSC acceleration sensor
- **X1653** Connector
- **X1710** Connector
- **X6249** Connector
- **X14023** F-CAN bus wire splice
- **X14024** F-CAN bus wire splice
- **X61035** Connector

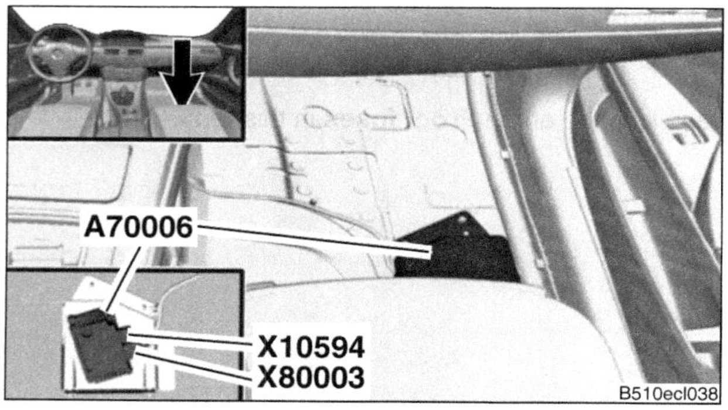

Fig 88 Right front floor

Underneath rug:

- **A70006** Transfer case control module
- **X10594** Connector
- **X80003** Connector

BentleyPublishers.com
BentleyPublishers.com—All Rights Reserved

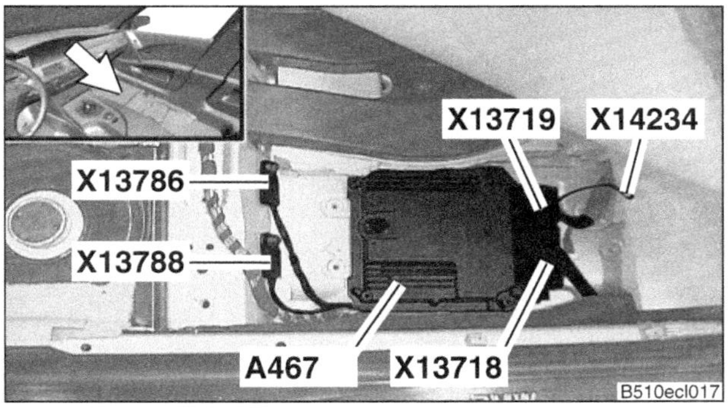

Fig 89 Passenger seat rail

Underneath rug:

- **A467** Active steering control module
- **X13718** Connector
- **X13719** Connector
- **X13786** Ground connection
  See **Grounds** in this repair group.
- **X13788** Ground connection
  See **Grounds** in this repair group.
- **X14234** Ground connection
  See **Grounds** in this repair group.

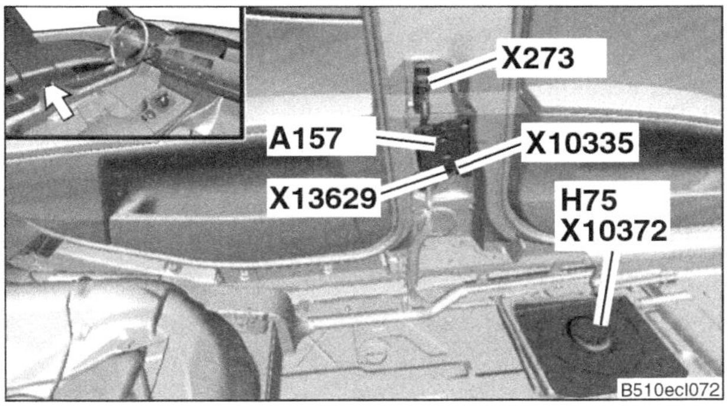

Fig 90 Left B- pillar

Behind trim, driver seat removed:

- **A157** Crash sensor, satellite
- **H75** Subwoofer (underneath rug under driver seat)
- **X273** Left rear door electronics harness connector
- **X10335** Connector
- **X10372** Connector
- **X13629** Connector

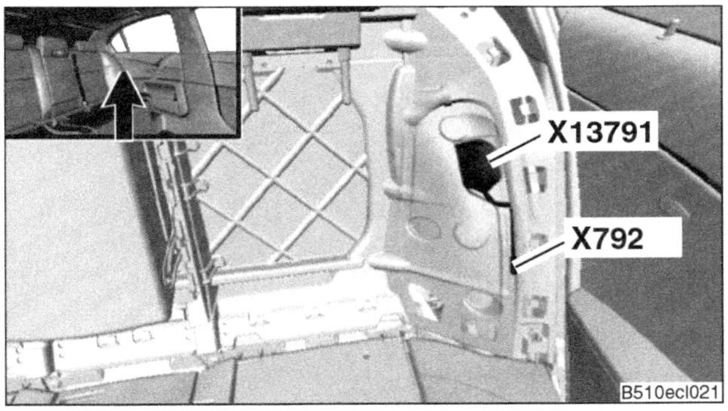

Fig 91 Left C- pillar

Behind trim:

- **X792** Connector
- **X13791** Ground connection
  See **Grounds** in this repair group.

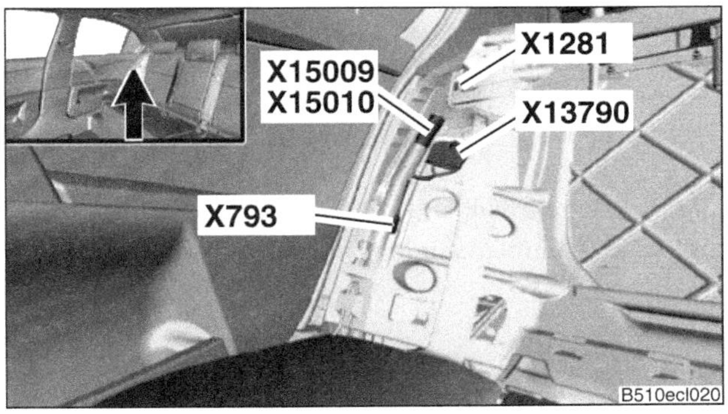

Fig 92 Right C- pillar

Behind trim:

- **X793** Connector
- **X1281** Connector
- **X13790** Ground connection
  See **Grounds** in this repair group.
- **X15009** F-CAN bus wire splice
- **X15010** F-CAN bus wire splice

BentleyPublishers.com
BentleyPublishers.com—All Rights Reserved

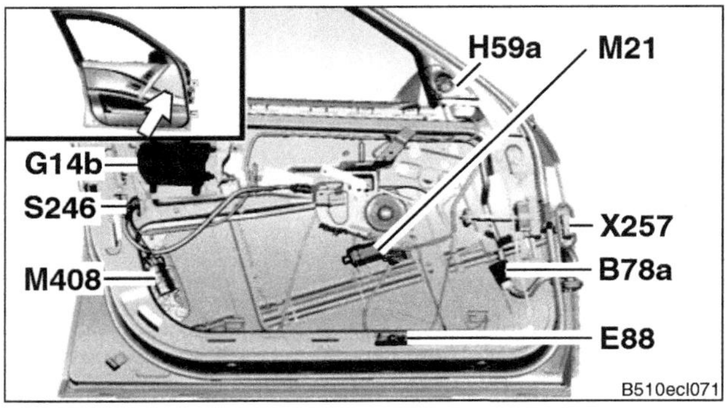

Fig 93 Driver door

Behind trim:

- **B78a** Airbag crash sensor
- **E88** Door courtesy light
- **G14b** Side airbag
- **H59a** Tweeter
- **M21** Window regulator motor
- **M408** Soft-close lock motor
- **S246** Door lock motor
- **X257** Driver door electronics harness connector

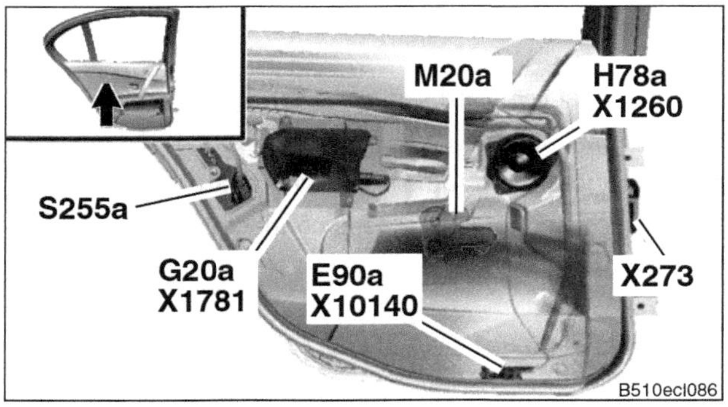

Fig 94 Left rear door

Behind trim:

- **E90a** Door courtesy light
- **G20a** Rear side airbag
- **H78a** Midrange speaker
- **M20a** Window regulator motor
- **S255a** Door lock motor
- **X273** Left rear door electronics harness connector
- **X1260** Connector
- **X1781** Connector
- **X10140** Connector

Seat components

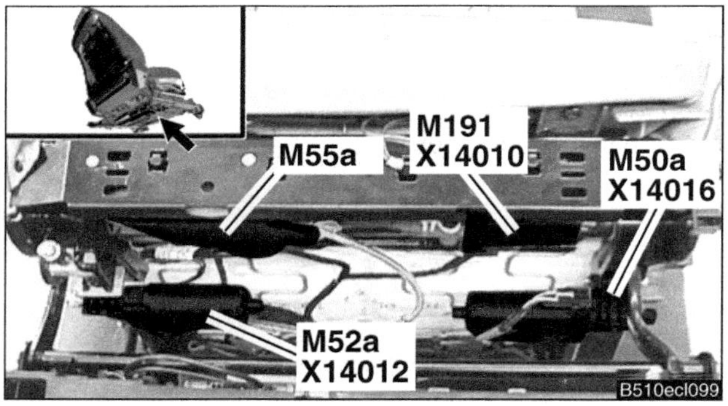

Fig 95 Driver seat

Seat control motors, underneath seat:

- **M50a** Seat angle adjustment motor
- **M52a** Seat height adjustment motor
- **M55a** Lumbar support adjustment motor
- **M191** Seat depth adjustment motor
- **X14010** Connector
- **X14012** Connector
- **X14016** Connector

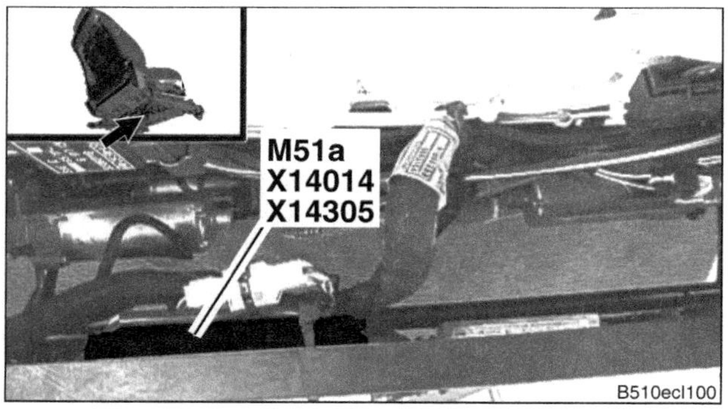

Fig 96 Driver seat

Seat control motor, underneath seat:

- **M51a** Seat fore-aft control motor
- **X14014** Connector
- **X14305** Connector

BentleyPublishers.com
BentleyPublishers.com—All Rights Reserved

Fig 97 Driver seat

Seat control motor, underneath seat:
- **M55c** Lumbar support drive unit
- **X729** Connector
- **X1503** Connector
- **Y193** Valve block, lumbar support

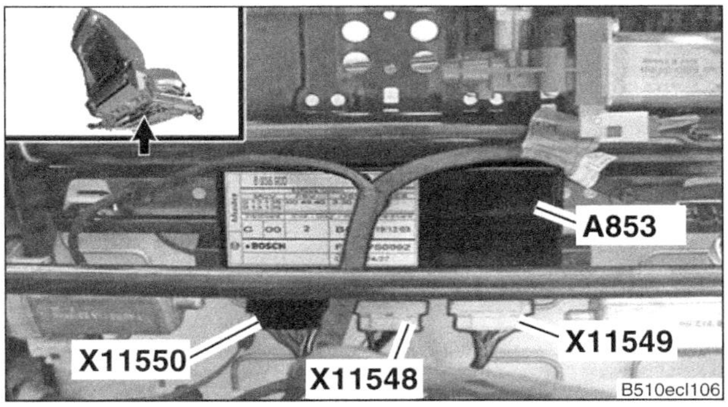

Fig 98 Driver seat

Underneath seat:
- **A853** Backrest width adjustment (active backrest) module
- **X11550** Connector
- **X11548** Connector
- **X11549** Connector

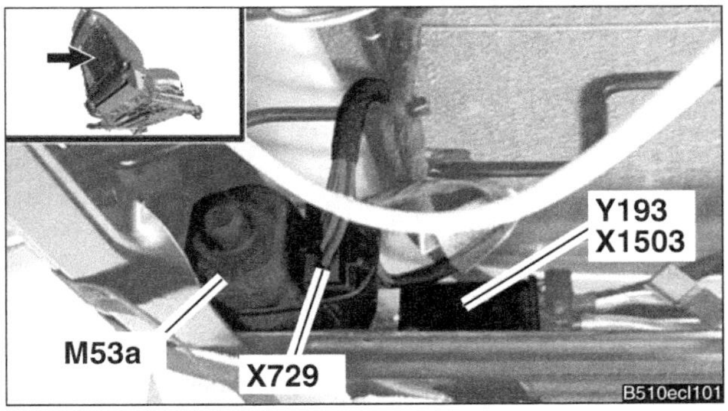

Fig 99 Driver seat backrest

Seat rear panel removed:
- **M53a** Backrest angle adjustment motor
- **X729** Connector
- **X1503** Connector
- **Y193** Valve block, lumbar support

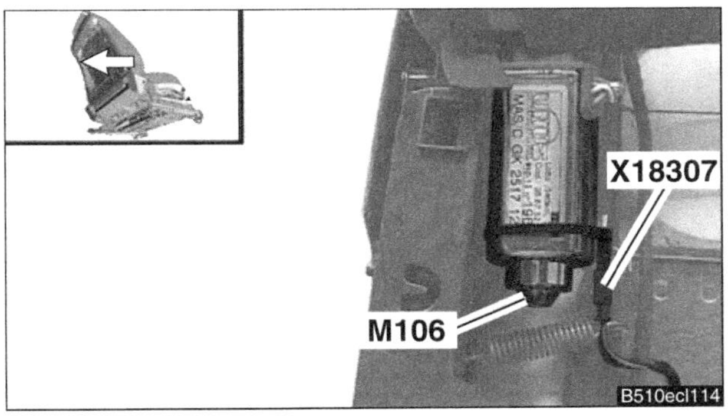

Fig 100 Driver seat backrest

Seat rear panel removed:
- **M106** Upper backrest angle adjustment motor
- **X18307** Connector

ECL

BentleyPublishers.com—All Rights Reserved

Seat components

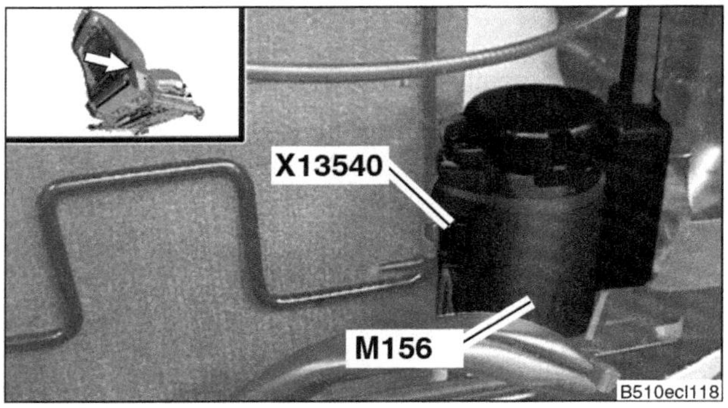

Fig 101 Driver seat backrest

Seat rear panel removed:

- **M156** Backrest width adjustment motor
- **X13540** Connector

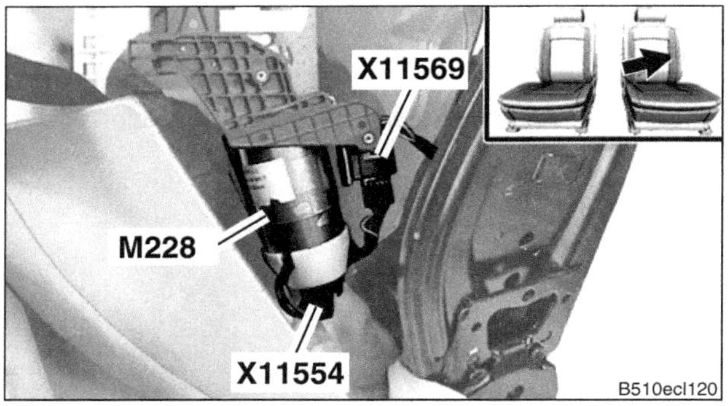

Fig 102 Driver seat backrest

In backrest left side bolster:

- **M228** Active seat backrest width motor
- **X18307** Connector

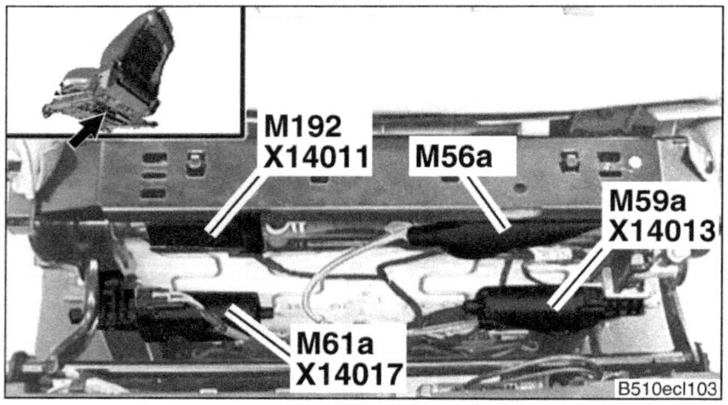

Fig 103 Passenger seat

Seat control motors, underneath seat:

- **M56a** Lumbar support adjustment motor
- **M59a** Seat height adjustment motor
- **M61a** Seat angle adjustment motor
- **M192** Seat depth adjustment motor
- **X14011** Connector
- **X14013** Connector
- **X14017** Connector

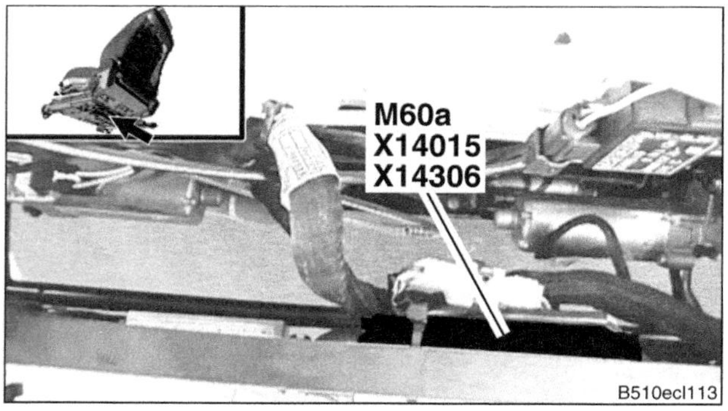

Fig 104 Passenger seat

Seat control motor, underneath seat:

- **M60a** Seat fore-aft control motor
- **X14015** Connector
- **X14306** Connector

BentleyPublishers.com
BentleyPublishers.com—All Rights Reserved

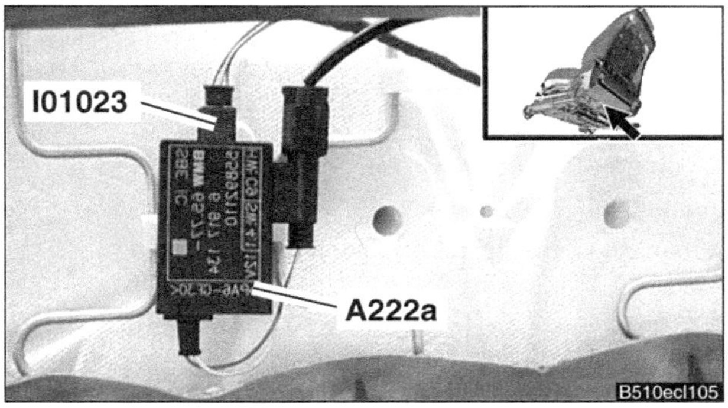

Fig 105 Passenger seat

In seat cushion:

- **A221** Seat occupancy detection module
- **I01023** Seat occupancy sensor mat (connector)

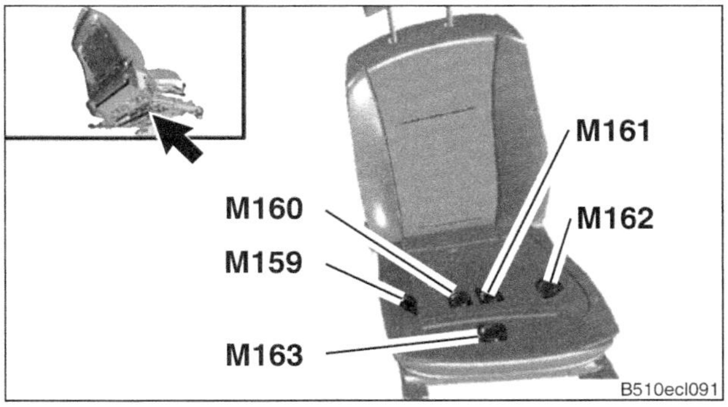

Fig 106 Driver seat

In seat cushion:

- **M159 - M162** Fans

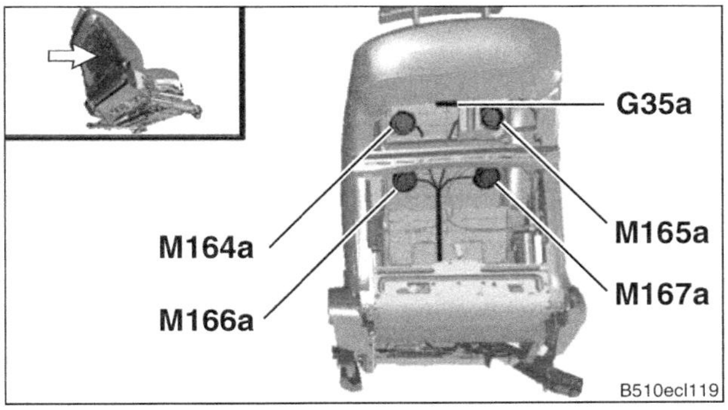

Fig 107 Driver seat

In seat backrest:

- **G35a** Head restraint vertical extender
- **M164a - M167a** Fans

SMG components

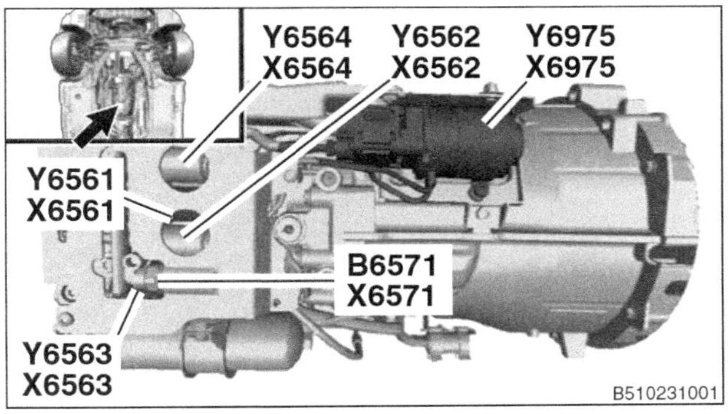

Fig 108 SMG transmission

Underneath transmission:

- **B6571** Hydraulic pressure sensor
- **X6561 - X6564** Connectors
- **X6571** Connector
- **X6975** Connector
- **Y6561** Solenoid 1, gear
- **Y6562** Solenoid 2, gear
- **Y6563** Solenoid, brake
- **Y6564** Solenoid, clutch
- **Y6975** Hydraulic pump

BentleyPublishers
BentleyPublishers.com—All Rights Reserved

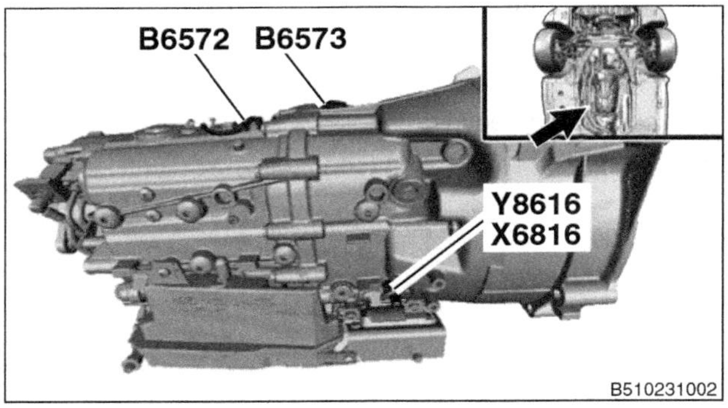

Fig 109 SMG transmission

Right side of transmission:
- **B6572** Gearshift sensor
- **B6573** Selector angle sensor
- **X6816** Connector
- **Y8616** Transmission main shaft rpm sensor

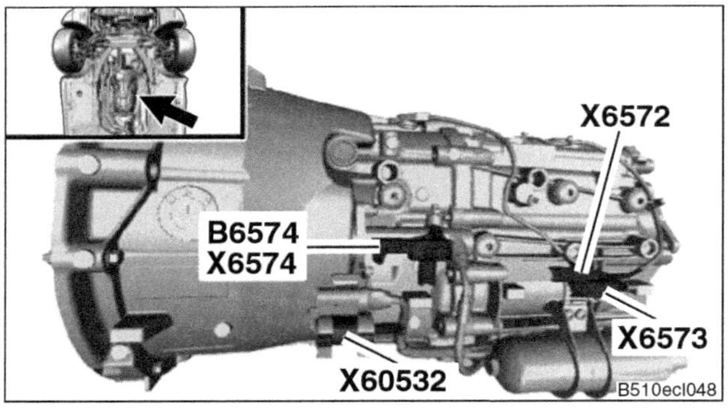

Fig 110 SMG transmission

Left side of transmission:
- **B6574** Clutch sensor
- **X6572 - X6574** Connectors
- **X60532** Connector

Trunk or cargo compartment components

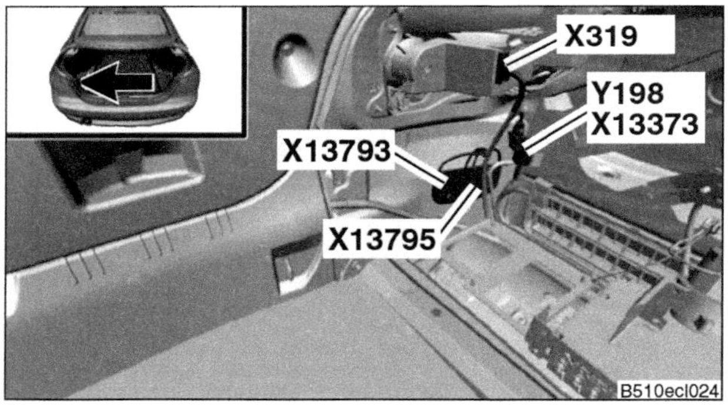

Fig 111 Left rear trunk (Sedan)

Behind trim:
- **X319** Connector
- **X13373** Connector
- **X13793** Ground connection
  See **Grounds** in this repair group.
- **X13795** Ground connection
  See **Grounds** in this repair group.
- **Y198** Tailpipe flap actuator

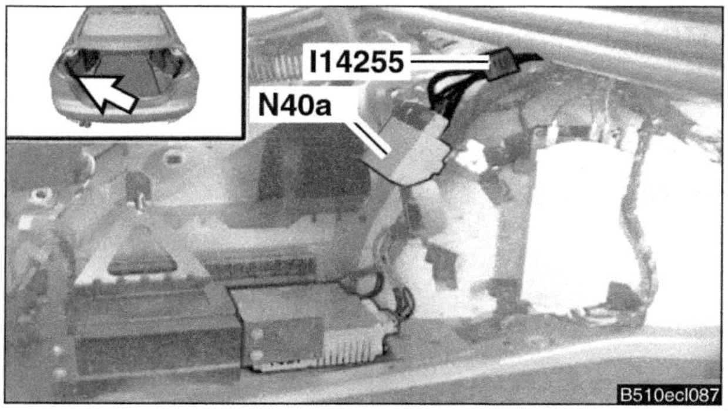

Fig 112 Left side trunk (Sedan)

Behind trim:
- **I14255** MOST-bus connector
- **N40a** Sound system amplifier

BentleyPublishers.com
BentleyPublishers.com—All Rights Reserved

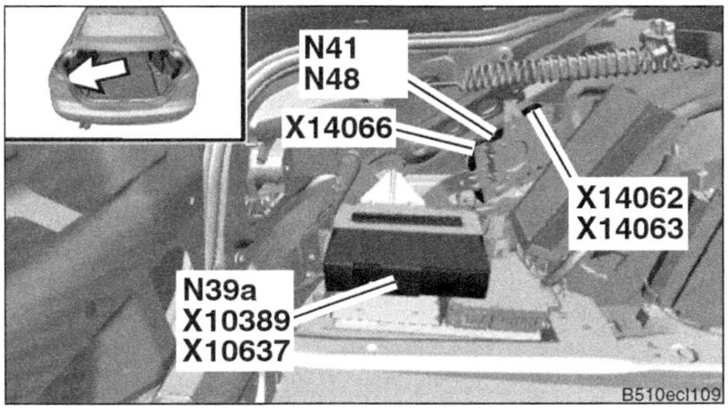

Fig 113 Left side trunk (Sedan)

Behind trim:

- **N39a** DVD changer
- Options:
  N41 or **N48** Digital tuner
- **X10389** Connector
- **X10637** Connector
- **X14962** Connector
- **X14063** Connector
- **X14066** Connector

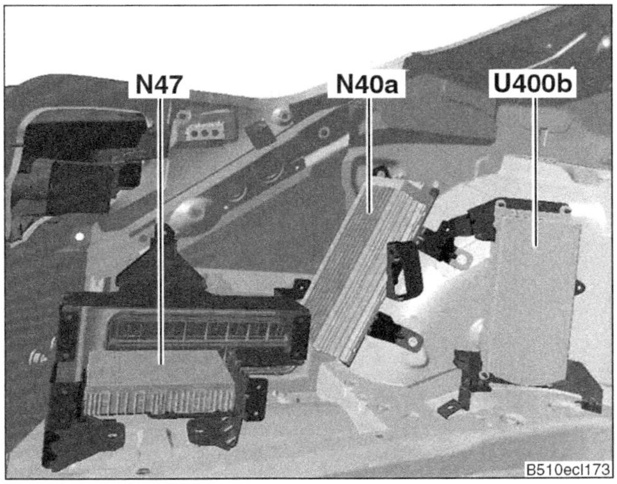

Fig 114 Left side trunk (Sedan)

Behind trim:

- **N40a** Sound system amplifier
- **N47** Satellite radio
- **U400b** Telematics control module (TCU)

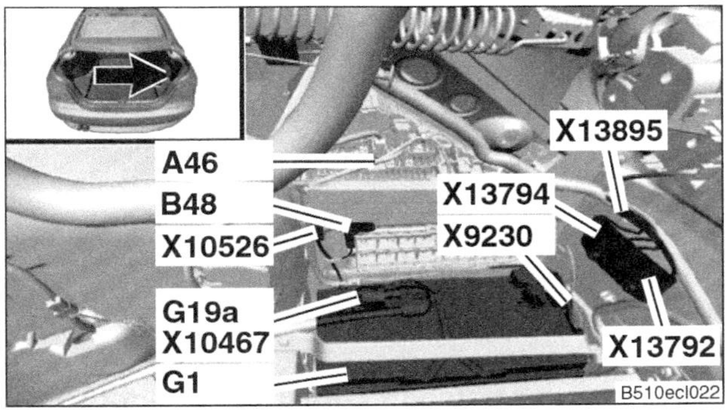

Fig 115 Right side trunk (Sedan)

Behind trim:

- **A46** Rear power distribution and relay panel
- **B48** Acceleration sensor
- **G1** Battery
- **G19a** Battery safety terminal (BST)
- **X9230** Connector
- **X10467** Connector
- **X10526** Connector
- **X13792** Ground connection
  See **Grounds** in this repair group.
- **X13794** Ground connection
  See **Grounds** in this repair group.
- **X13895** Connector

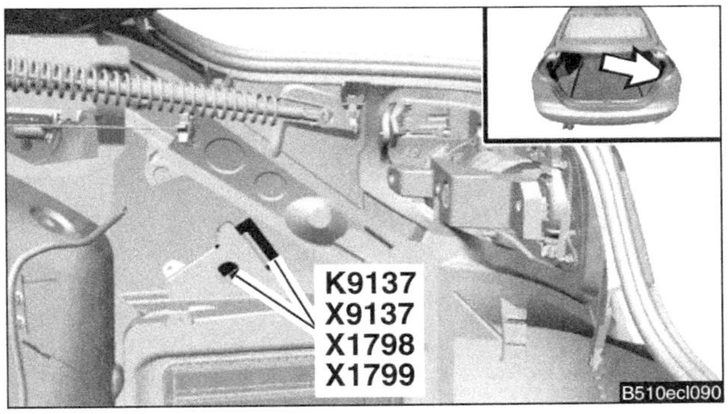

Fig 116 Right side trunk (Sedan)

Behind trim:

- **K9137** Fan cutoff relay
- **X1798** Connector
- **X1799** Connector
- **X9137** Connector

BentleyPublishers.com
BentleyPublishers.com—All Rights Reserved

Trunk or cargo compartment components

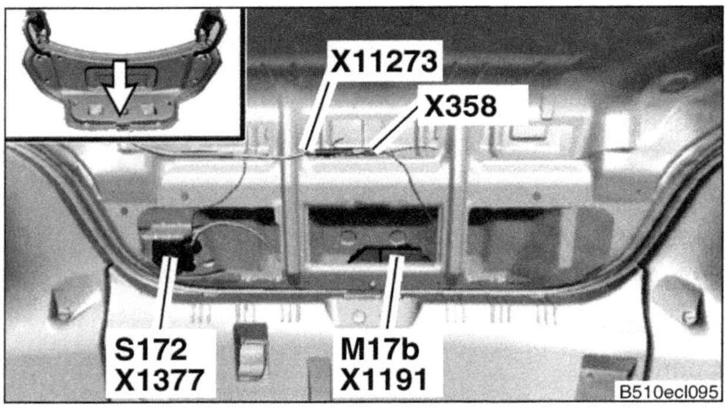

Fig 117 Trunk lid (Sedan)

Behind trim:

- **M17b** Trunk lock motor
- **S172** Trunk lock switch
- **X358** Connector
- **X1191** Connector
- **X1377** Connector
- **X11273** Connector

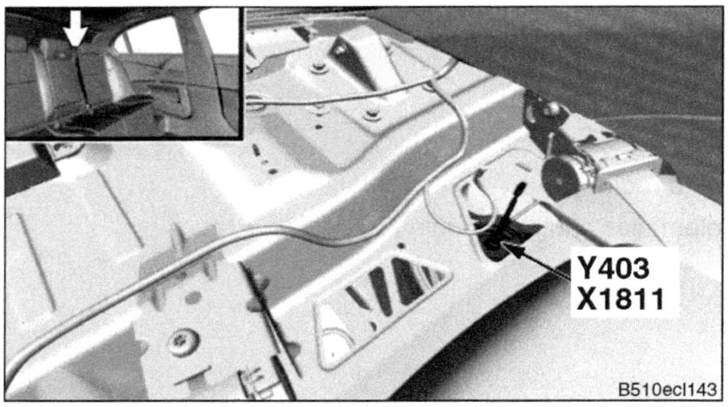

Fig 118 Left front cargo compartment (Sport Wagon)

Behind rear seat backrest:

- **Y403** EHC (air suspension) shock absorber valve
- **X1811** Connector

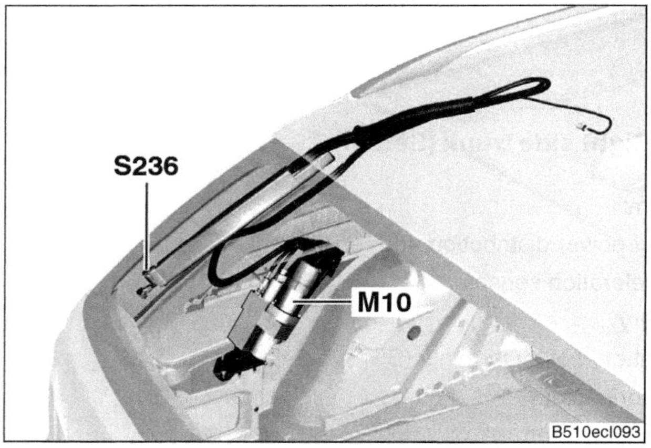

Fig 119 Left side cargo compartment (Sport Wagon)

Automatic tailgate lift components:

- **M10** Hydraulic motor, behind trim
- **S236** Tailgate opening angle sensor, on tailgate lift strut

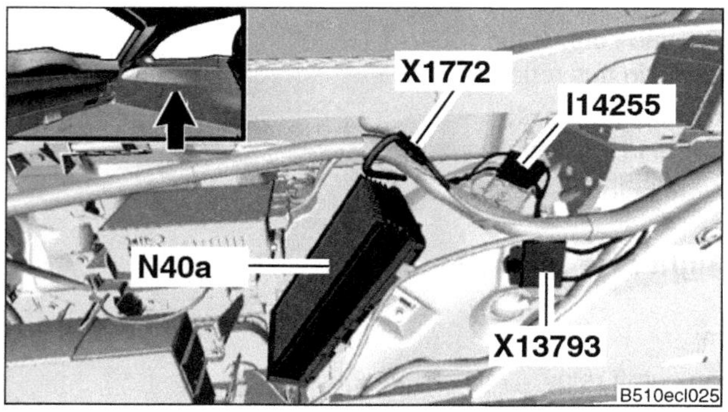

Fig 120 Left side cargo compartment (Sport Wagon)

Behind trim:

- **I14255** MOST-bus connector
- **N40a** Sound system amplifier
- **X1772** Connector
- **X13793** Ground connection
  See **Grounds** in this repair group.

BentleyPublishers.com
BentleyPublishers.com—All Rights Reserved

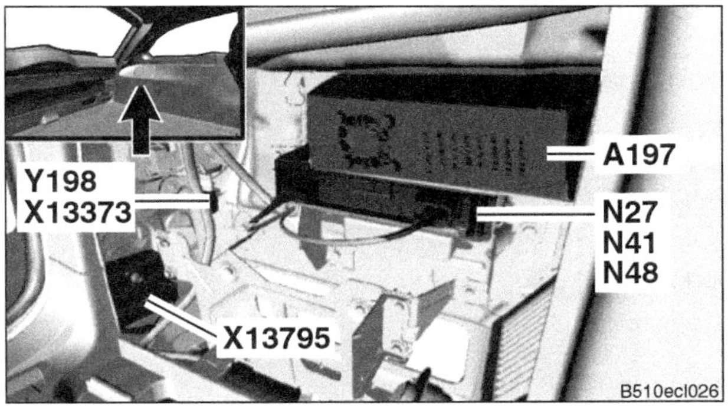

Fig 121 Left rear cargo compartment (Sport Wagon)

Behind trim:

- **A197** Video module
- Options:
  N27 Mobile phone compensator
  N41 Digital tuner
  N48 Digital tuner
- **X13373** Connector
- **X13793** Ground connection
  See **Grounds** in this repair group.
- **Y198** Tailpipe flap actuator

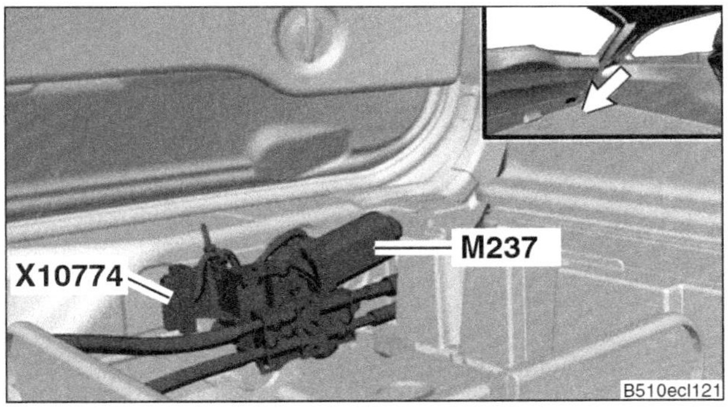

Fig 122 Left rear cargo compartment (Sport Wagon)

Behind trim:

- **M237** Roller cover motor
- **X10774** Connector

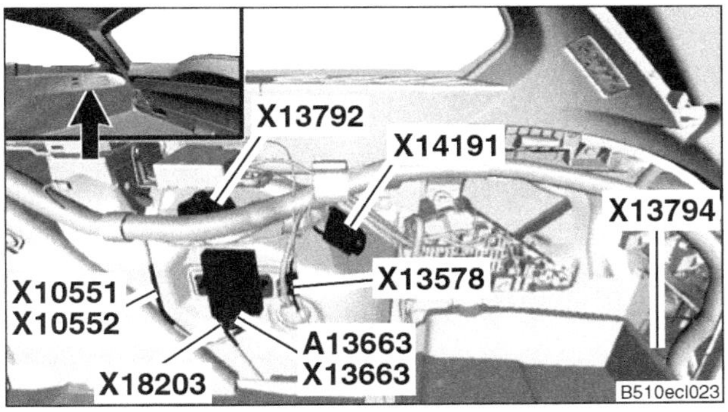

Fig 123 Right side cargo compartment (Sport Wagon)

Behind trim:

- **A13663** Fuel pump control module (EKPS)
- **X10551** PT-CAN bus wire splice
- **X10552** PT-CAN bus wire splice
- **X13578** Wire splice
- **X13663** Connector
- **X13792** Ground connection
  See **Grounds** in this repair group.
- **X13794** Ground connection
  See **Grounds** in this repair group.
- **X14191** Connector
- **X18203** Connector

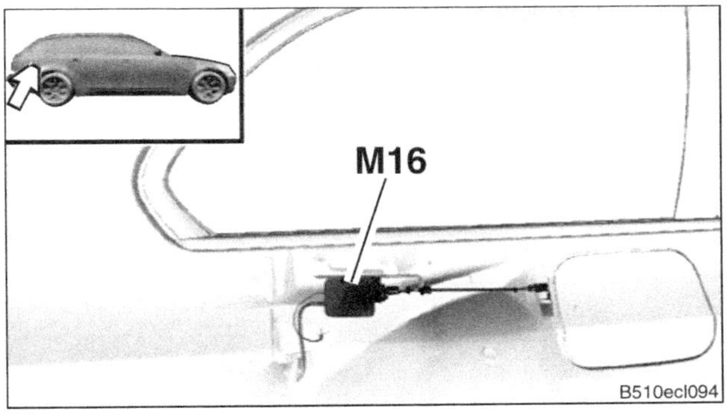

Fig 124 Right side cargo compartment (Sport Wagon)

Behind trim:

- **M16** Fuel filler flap lock motor
- **X312** Connector

Underneath vehicle

Fig 125 Underneath engine

V8 engine:

- **B6203** Crankshaft sensor
- **B62540** Oil quality sensor (OZS)
- **X6203** Connector
- **X62540** Connector

Fig 126 Underneath engine

On lower splash shield:

- **B21a** Outside temperature sensor
- **X770** Connector

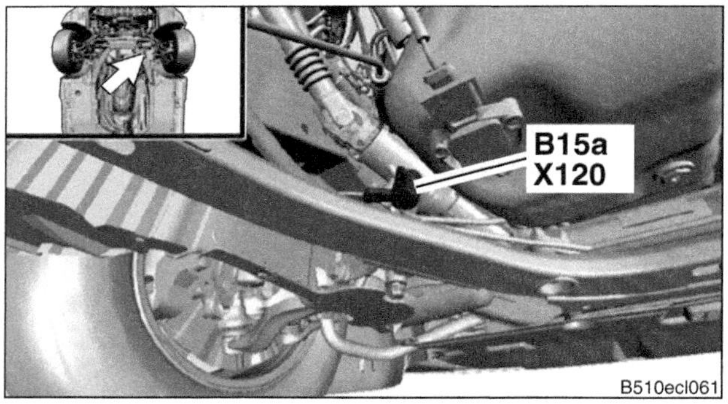

Fig 127 Underneath front subframe

At steering column:

- **B15a** Servotronic valve
- **X120** Connector

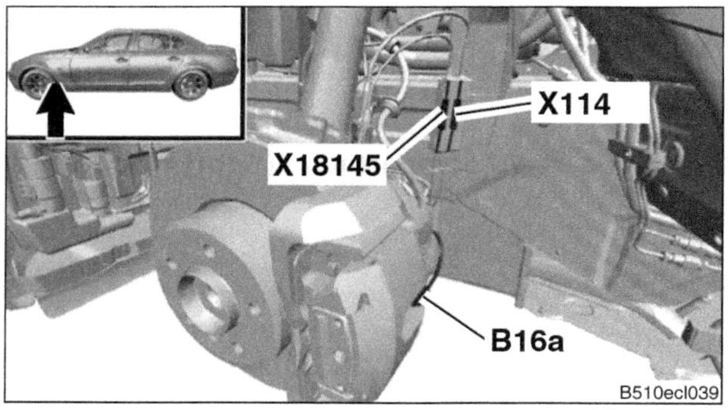

Fig 128 Left front wheel housing

Behind wheel housing liner:

- **B16a** Brake pad sensor
- **X114** Wheel speed sensor connector
- **X18145** Brake pad sensor connector

BentleyPublishers.com
BentleyPublishers.com—All Rights Reserved

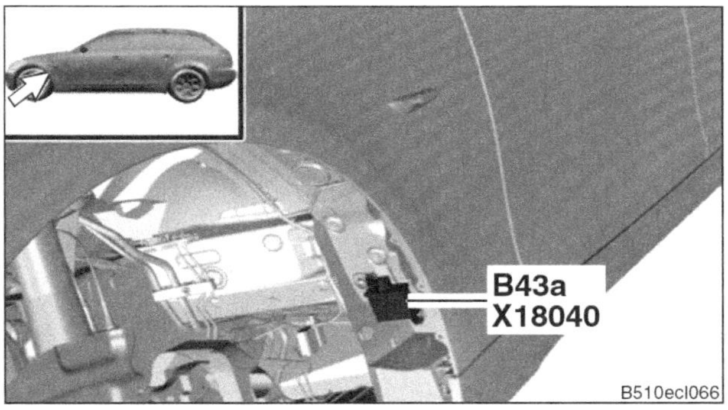

Fig 129 Left front wheel housing

Underneath left front fender behind liner:

- **B43a** Tire pressure control (RDC) transmitter
- **X18040** Connector

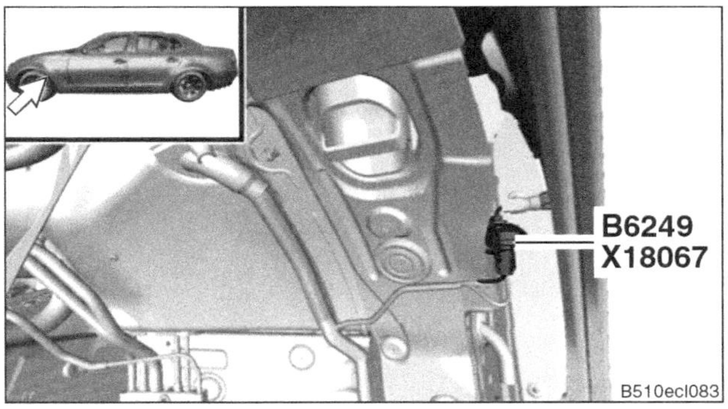

Fig 130 Left front wheel housing

Underneath left front fender:

- **B6249** Acceleration sensor
- **X18067** Connector

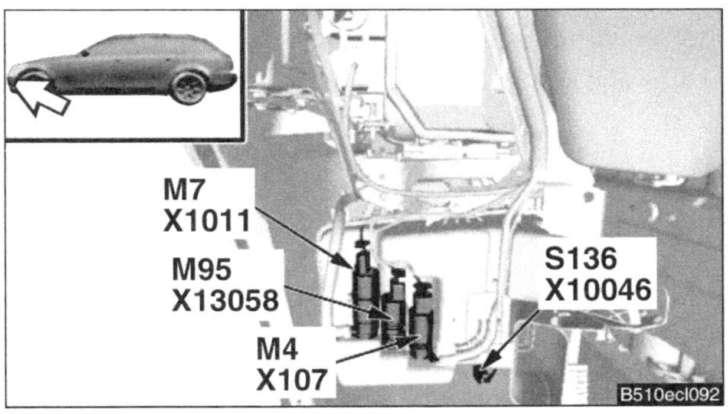

Fig 131 Left front wheel housing

Front of wheel housing, underneath liner:

- **M4** Windshield washer pump
- **M7** Headlight washer pump
- **M95** Rear window washer pump
- **S136** Washer fluid level switch
- **X107** Connector
- **X1011** Connector
- **X10046** Connector
- **X13058** Connector

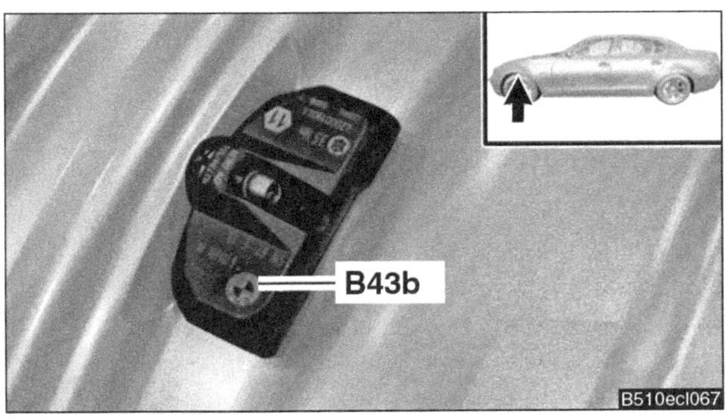

Fig 132 Inside wheel

Attached to tire valve stem:

- **B43b** Tire pressure control (RDC) sensor

BentleyPublishers.com
BentleyPublishers.com—All Rights Reserved

Underneath vehicle

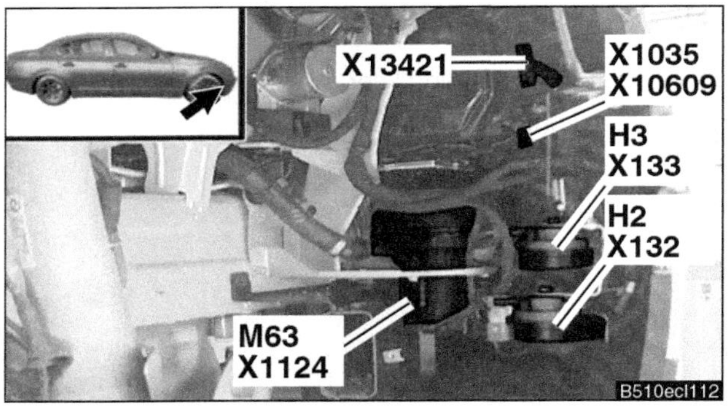

Fig 133 Underneath right front

Behind front bumper, right side:

- **H2 - H3** Horns
- **M63** Secondary air injection pump (M54, N62 engines)
- **X132** Connector
- **X133** Connector
- **X1035** Connector
- **X1124** Connector
- **X13421** Connector

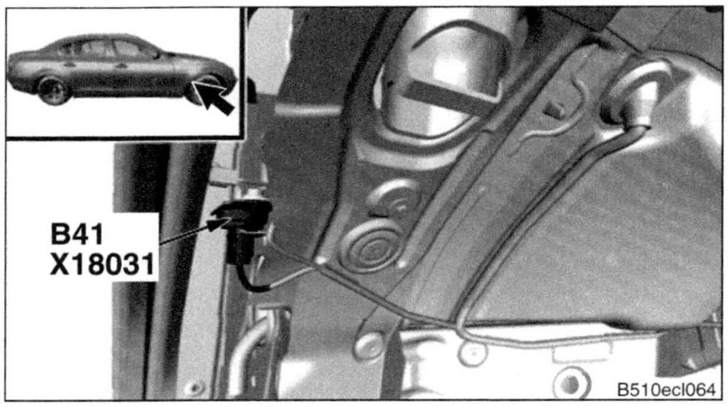

Fig 134 Right front wheel housing

Underneath right front fender:

- **B41** Acceleration sensor

Fig 135 Right front wheel housing

Behind wheel, under wheel housing liner:

- **B47a** Tire pressure control (RDC) transmitter

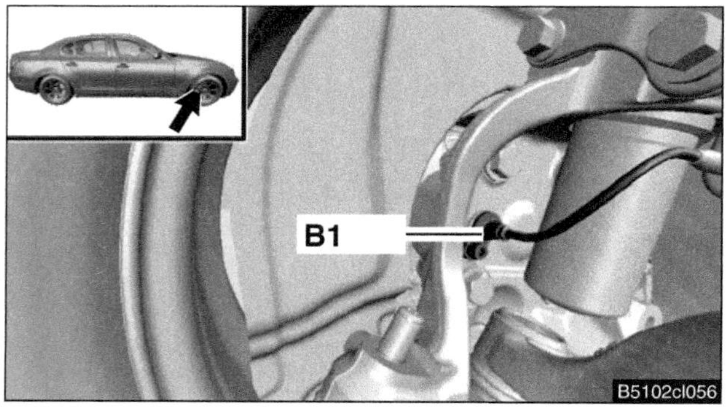

Fig 136 Right front wheel hub

Behind brake backing plate:

- **B1** ABS wheel speed sensor

BentleyPublishers.com
BentleyPublishers.com—All Rights Reserved

Fig 137 Underneath right front

At control arm:

- **B42** Front ride height sensor

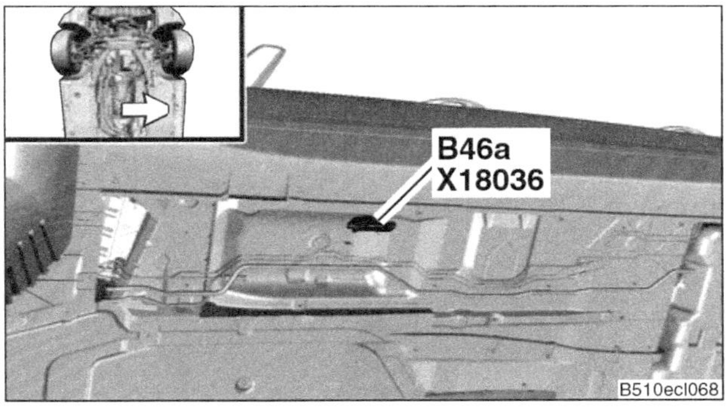

Fig 138 Underneath center of vehicle

Below driver seat:

- **B46a** Tire pressure control (RDC) antenna
- **X18036** Connector

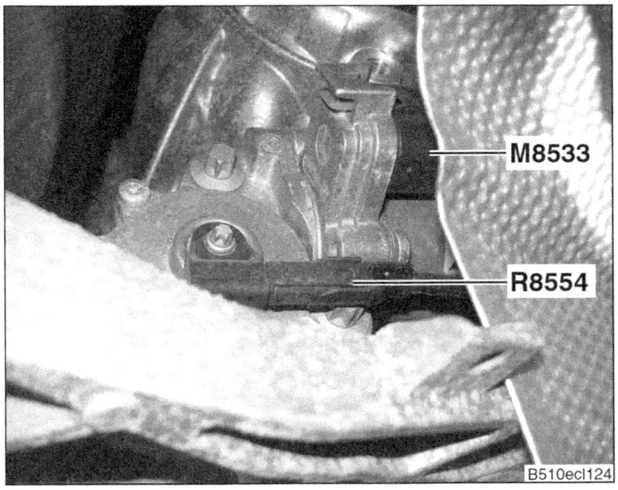

Fig 139 Underneath center of vehicle

Rear end of transmission housing on transfer case:

- **M8553** xDrive clutch (VTG) actuator
- **R8554** xDrive classification resistor

BentleyPublishers.com
BentleyPublishers.com—All Rights Reserved

Underneath vehicle

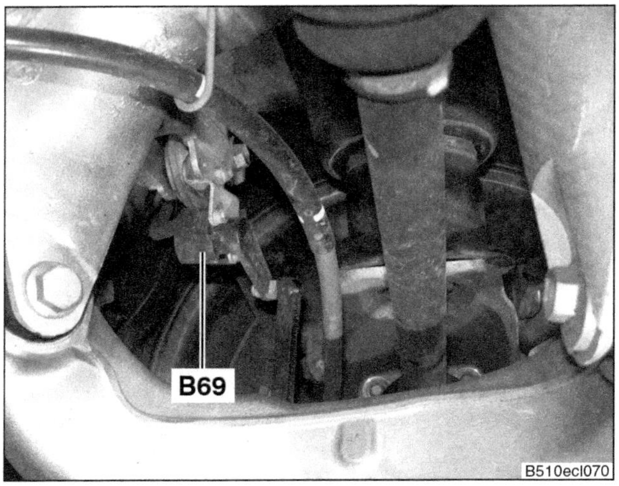

Fig 140 Underneath rear suspension

Right rear swing arm:

- **B69** Right rear ride height sensor

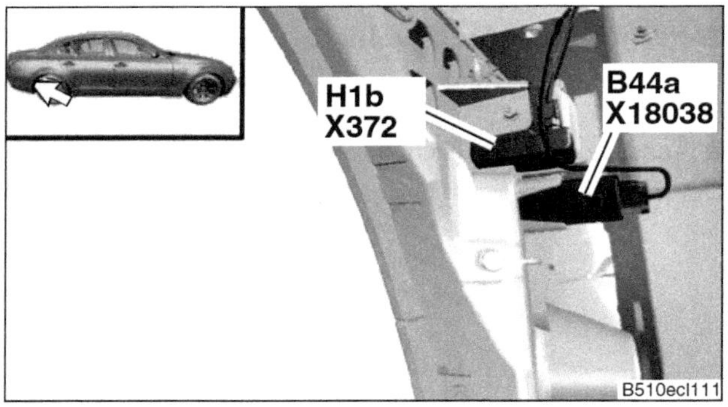

Fig 141 Right rear wheel housing

Underneath right rear fender behind liner:

- **B44a** Tire pressure control (RDC) transmitter
- **H1b** Anti-theft (alarm) siren, tilt sensor
- **X372** Connector
- **X18038** Connector

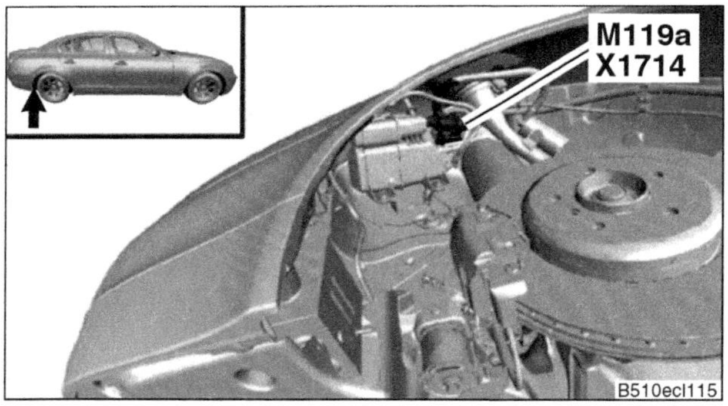

Fig 142 Right rear wheel housing

Underneath right rear fender behind liner:

- **M119a** Fuel tank leakage diagnostic module
- **X1714** Connector

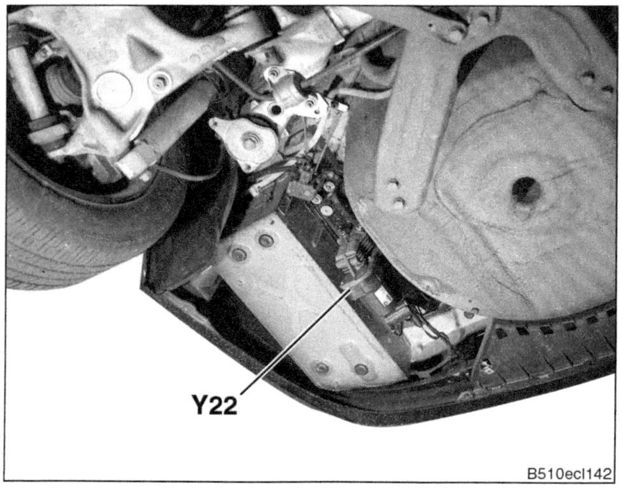

Fig 143 Underneath rear of vehicle (Sport Wagon)

Between spare tire well and battery box:

- **Y22** EHC (air suspension) compressor

BentleyPublishers.com
BentleyPublishers.com—All Rights Reserved

ELE Electrical Wiring Diagrams

ELECTRICAL COMPONENT LOCATIONS (relay and fuse positions, ground locations) see **Repair Group ECL**

GENERAL

This section contains selected electrical wiring diagrams for E90 3 Series models. A detailed index of the diagrams and electrical component is listed in the opening pages.

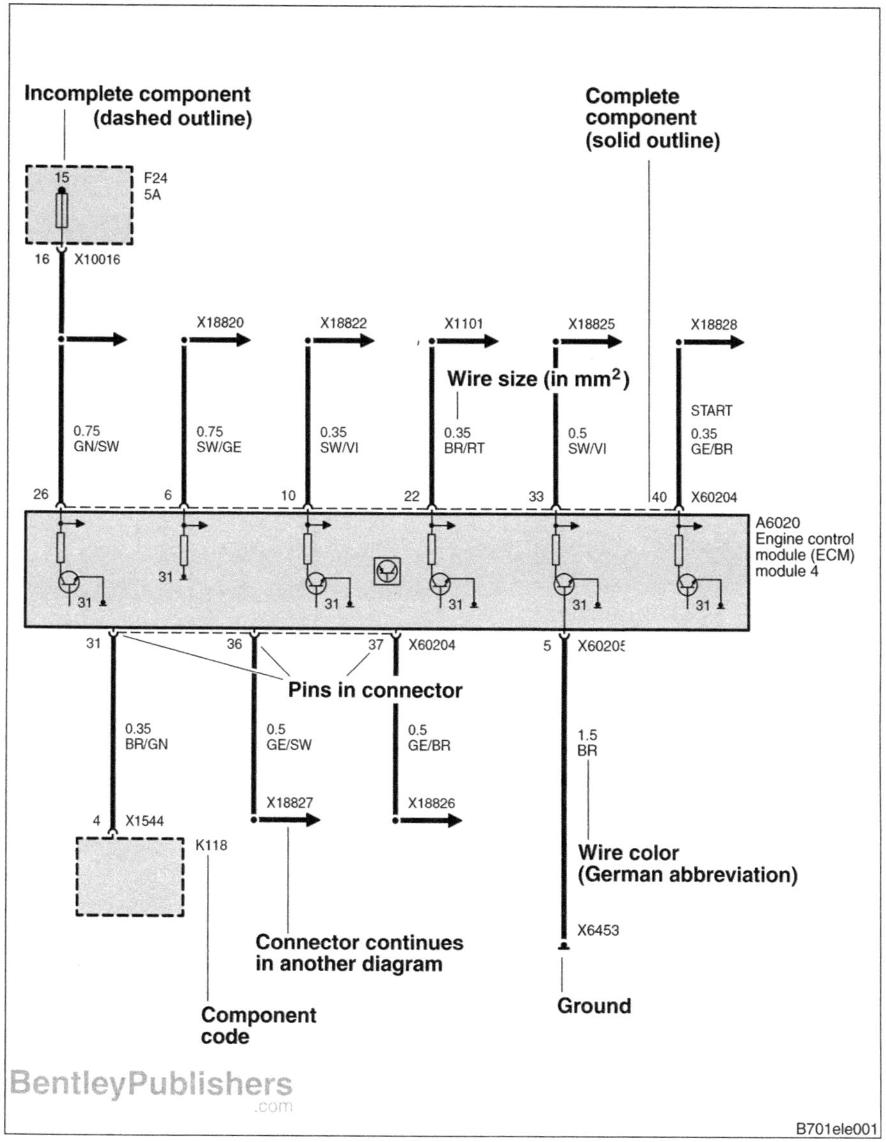

Schematic conventions

The schematics (wiring diagrams) divide the vehicle electrical system into individual circuits. Interacting electrical components are shown on one schematic.

Electrical components are represented in such a way that their general layout and function are self-explanatory. They are usually arranged in the diagrams so that the current path can be followed from positive at the top to negative at the bottom.

- If a schematic spans several pages, this fact is clearly indicated at the top of each page.
- A component (or connector) which is completely represented in the schematic is shown as a solid box.
- A component (or connector) which has other connectors in addition to the ones shown in the schematic is shown with a dashed line.
- Switches and relays are always shown in rest position (generally OFF).

Symbols

The schematics utilize simplified electrical symbols.

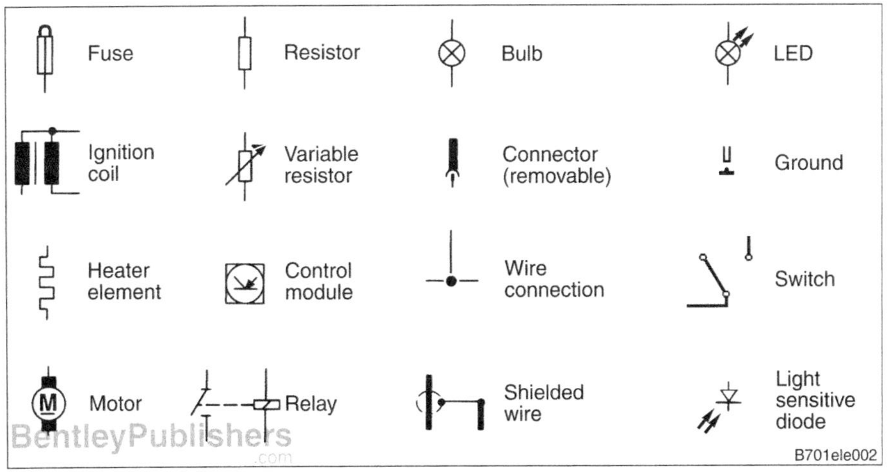

Wire insulation colors in this section are given with German color abbreviations.

Wire sizes follow the DIN (European) convention.

Example: 0.5 wire is ½ mm^2 in cross-section area. This corresponds to approx. SAE 16 gauge wire.

BMW identifies many electrical circuits with unique designations which follow the DIN standard. Also, BMW designates electrical components, junctions and grounds with a unique alphanumeric designation, most of which also follow the DIN standard.

Wiring color code	
BL	= blue
BR	= brown
GE	= yellow
GN	= green
GR	= grey
RT	= red
SW	= black
WS	= white
VI	= violet

B701ele003

BentleyPublishers.com

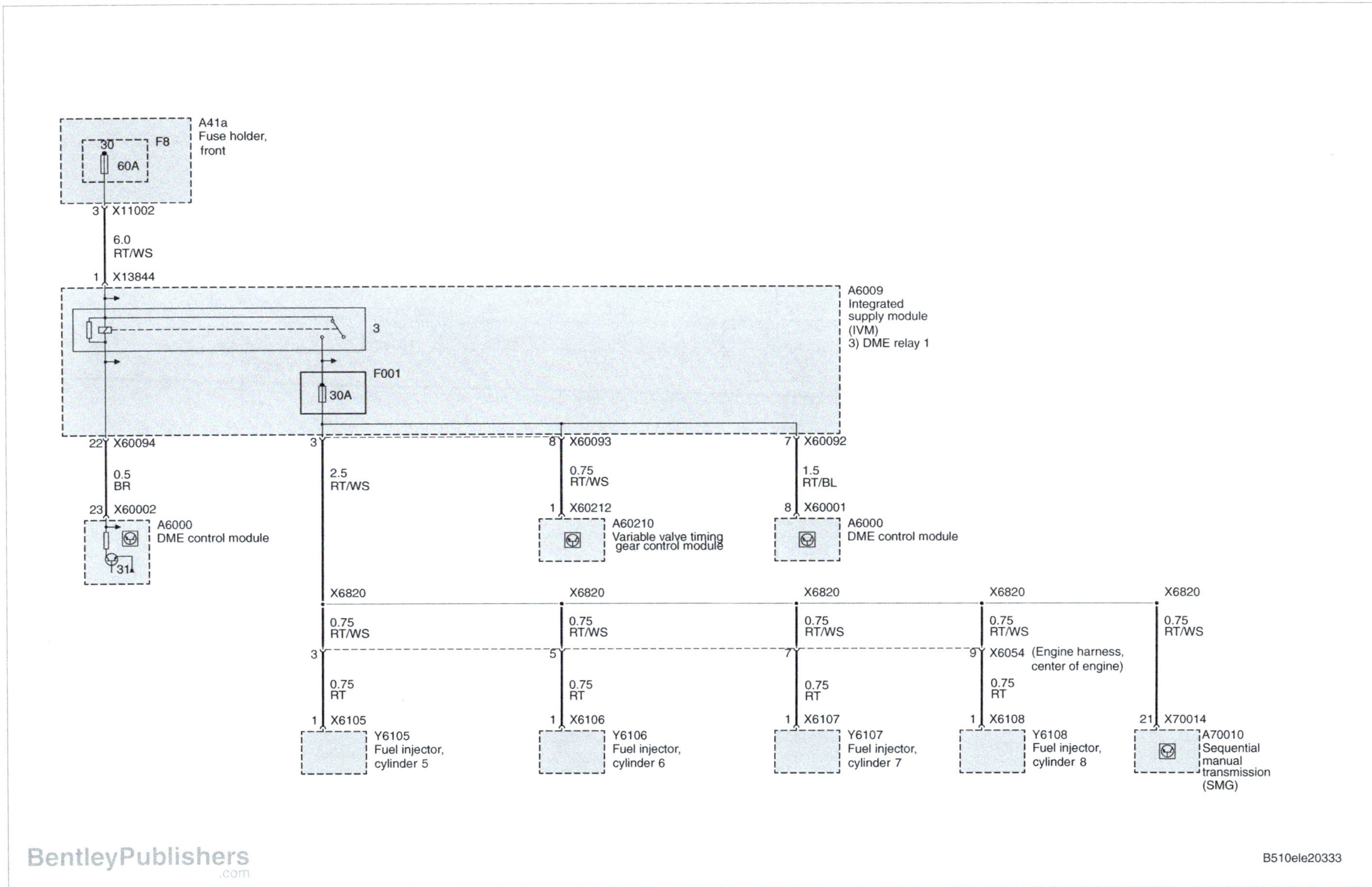
A41a
Fuse holder,
front
F8
30
60A
3 X11002
6.0
RT/WS
1 X13844
A6009
Integrated
supply module
(IVM)
3) DME relay 1
3
F001
30A
22 X60094
0.5
BR
23 X60002
A6000
DME control module
31
3
2.5
RT/WS
8 X60093
0.75
RT/WS
1 X60212
A60210
Variable valve timing
gear control module
7 X60092
1.5
RT/BL
8 X60001
A6000
DME control module
X6820
0.75
RT/WS
3
5
7
9 X6054 (Engine harness,
center of engine)
0.75
RT
1 X6105
Y6105
Fuel injector,
cylinder 5
1 X6106
Y6106
Fuel injector,
cylinder 6
1 X6107
Y6107
Fuel injector,
cylinder 7
1 X6108
Y6108
Fuel injector,
cylinder 8
21 X70014
A70010
Sequential
manual
transmission
(SMG)
BentleyPublishers
.com
B510ele20333

Fuse F001
M54 engine with SMG Sequential Manual Transmission

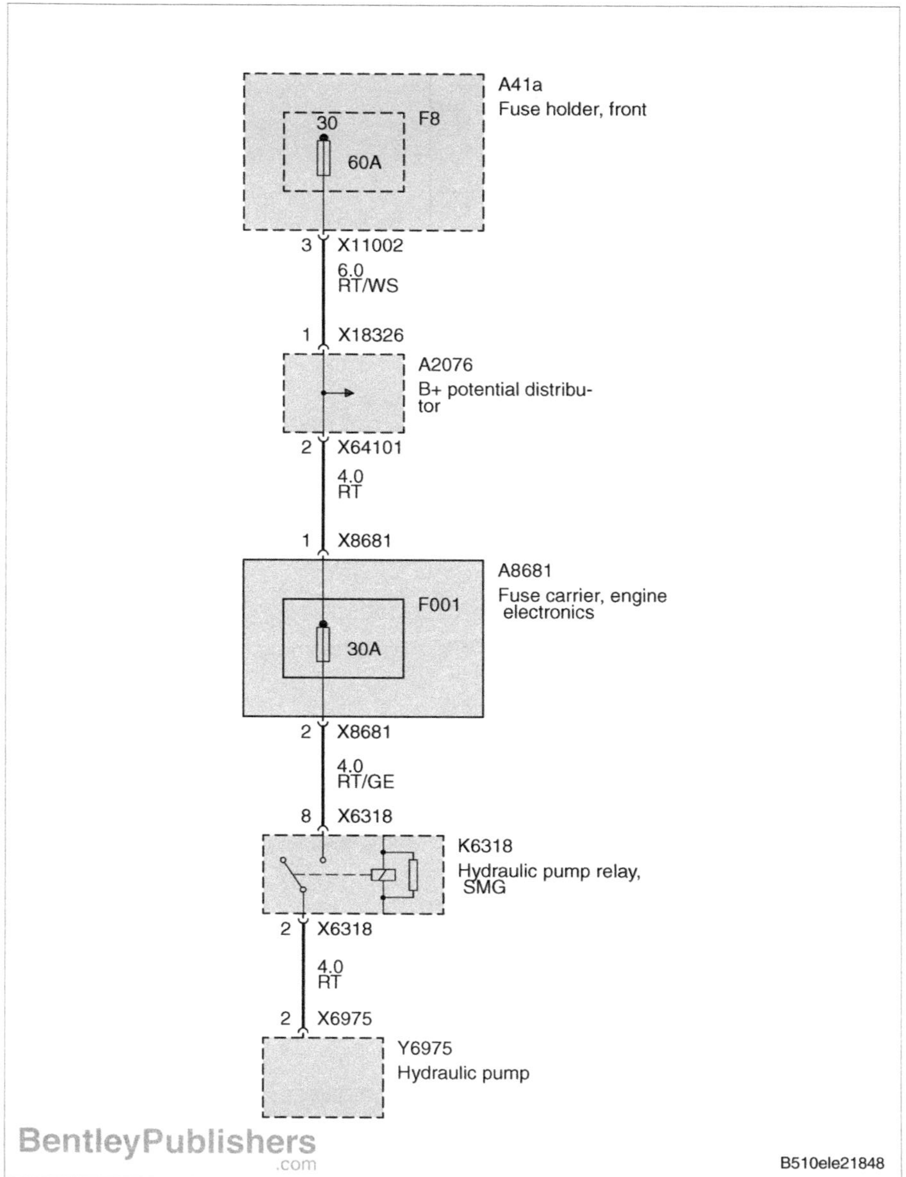

Fuse F002
N62 engine

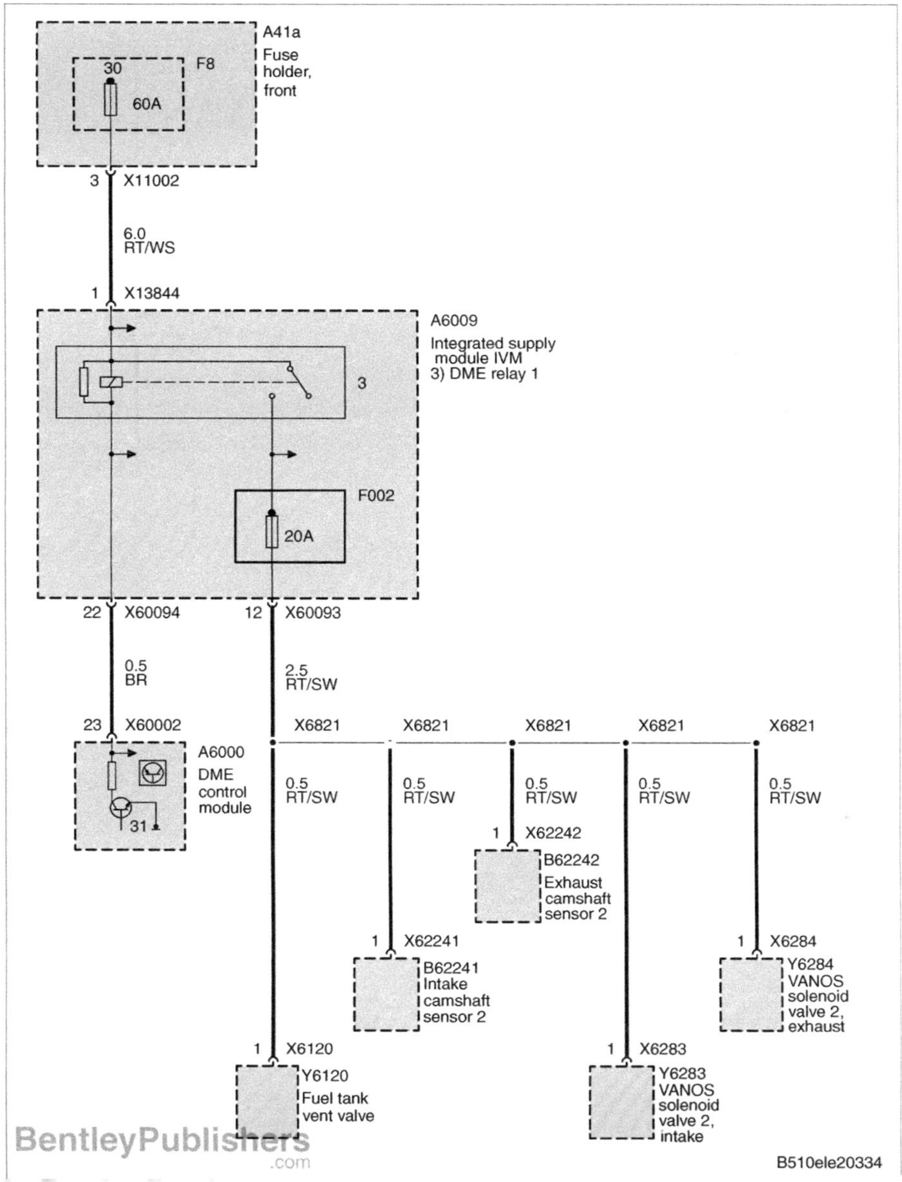

BentleyPublishers.com
BentleyPublishers.com

Fuse F002
N62 TU engine

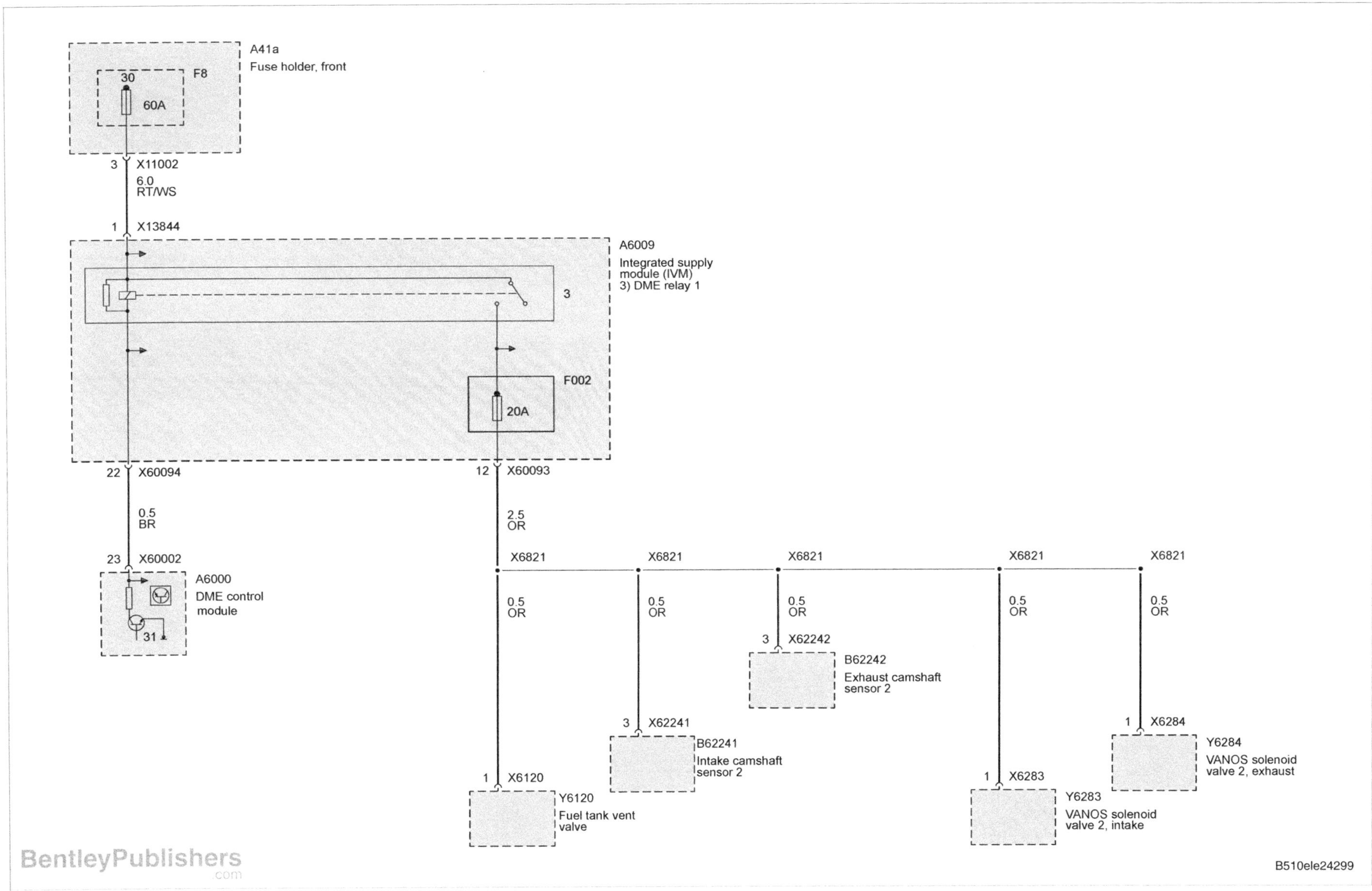

Fuse F003
N62 engine

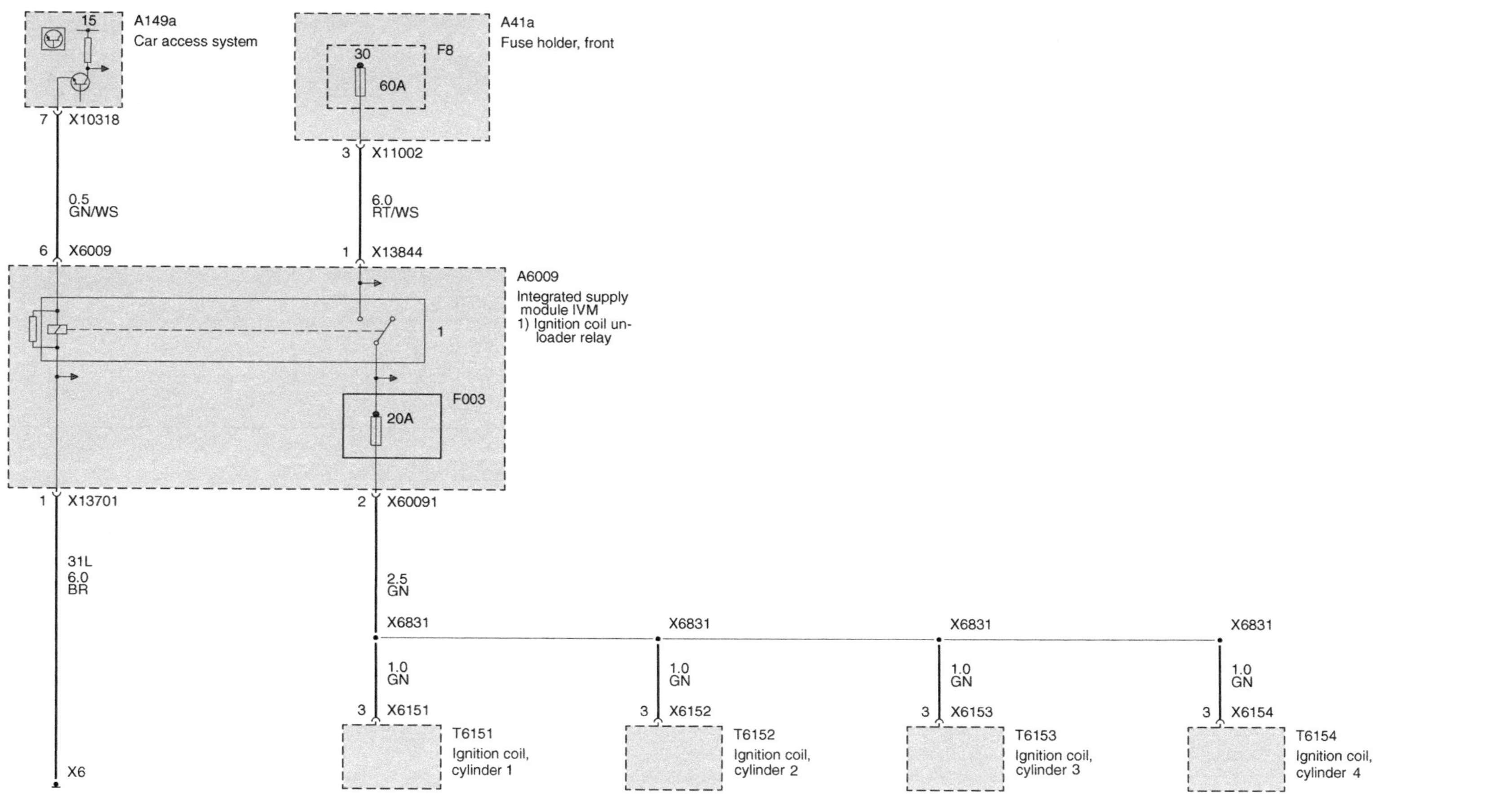

BentleyPublishers.com

Fuse F004
N62 engine

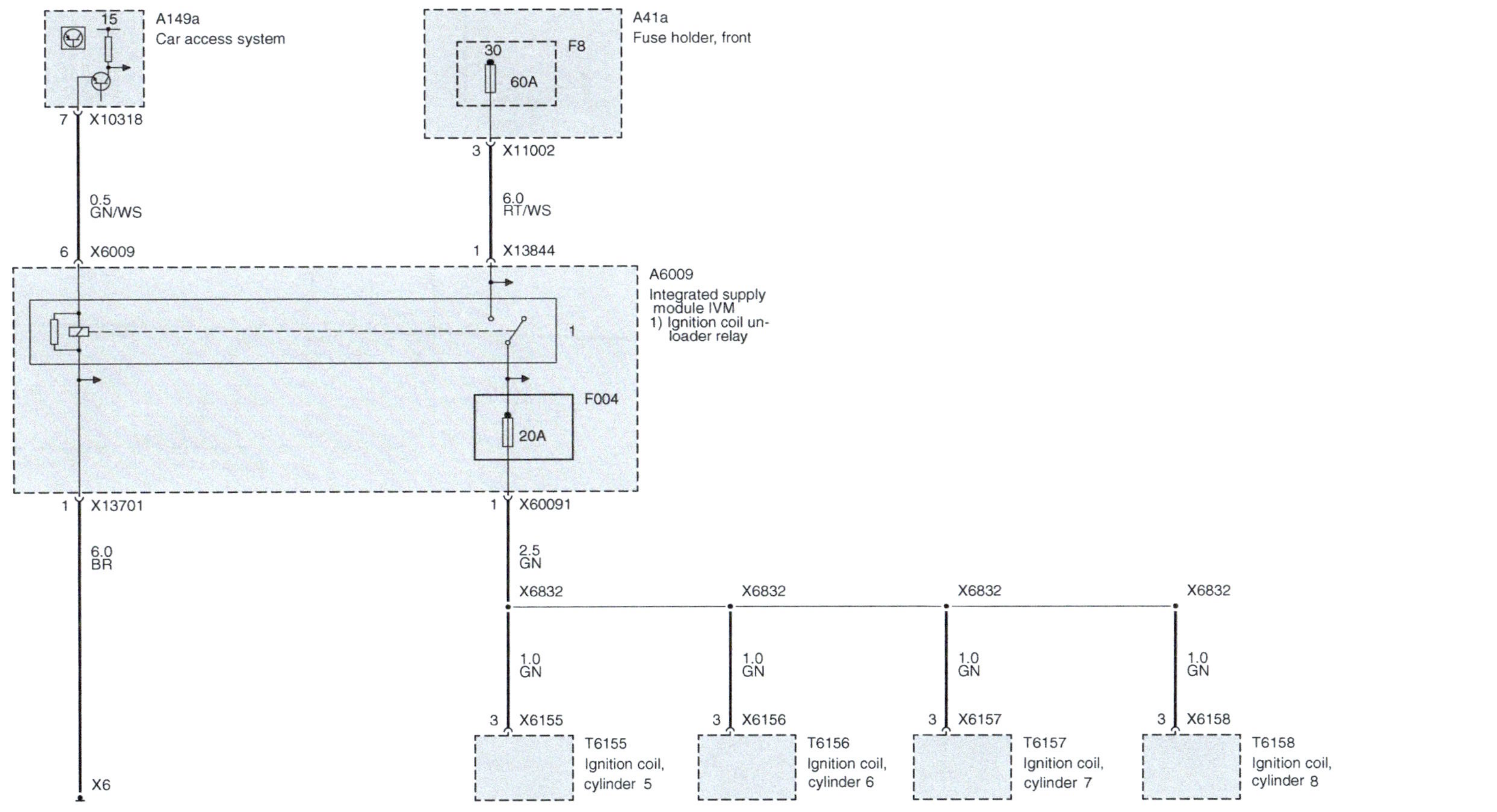

BentleyPublishers.com

Fuse F005
N62 engine

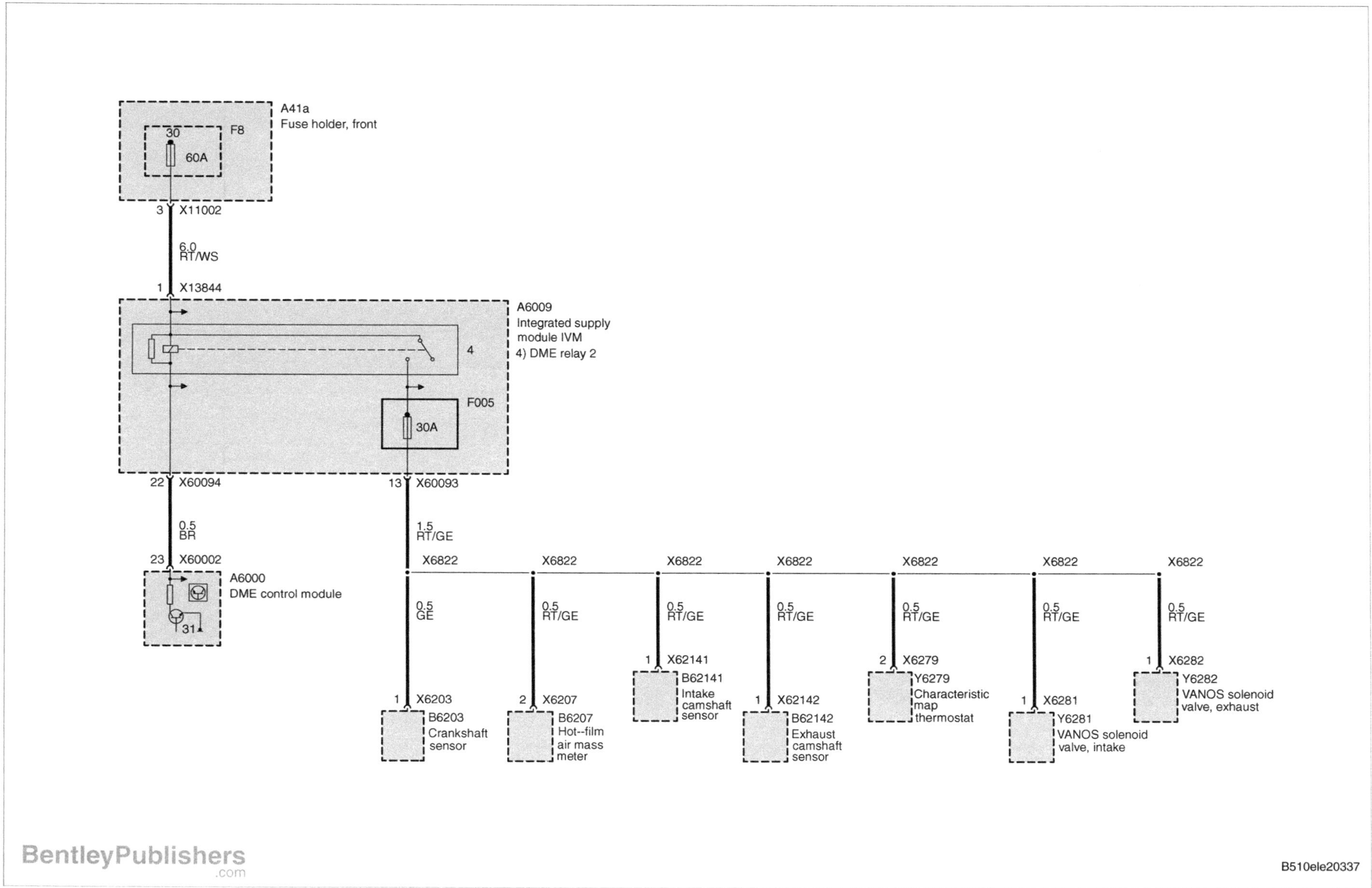

BentleyPublishers.com

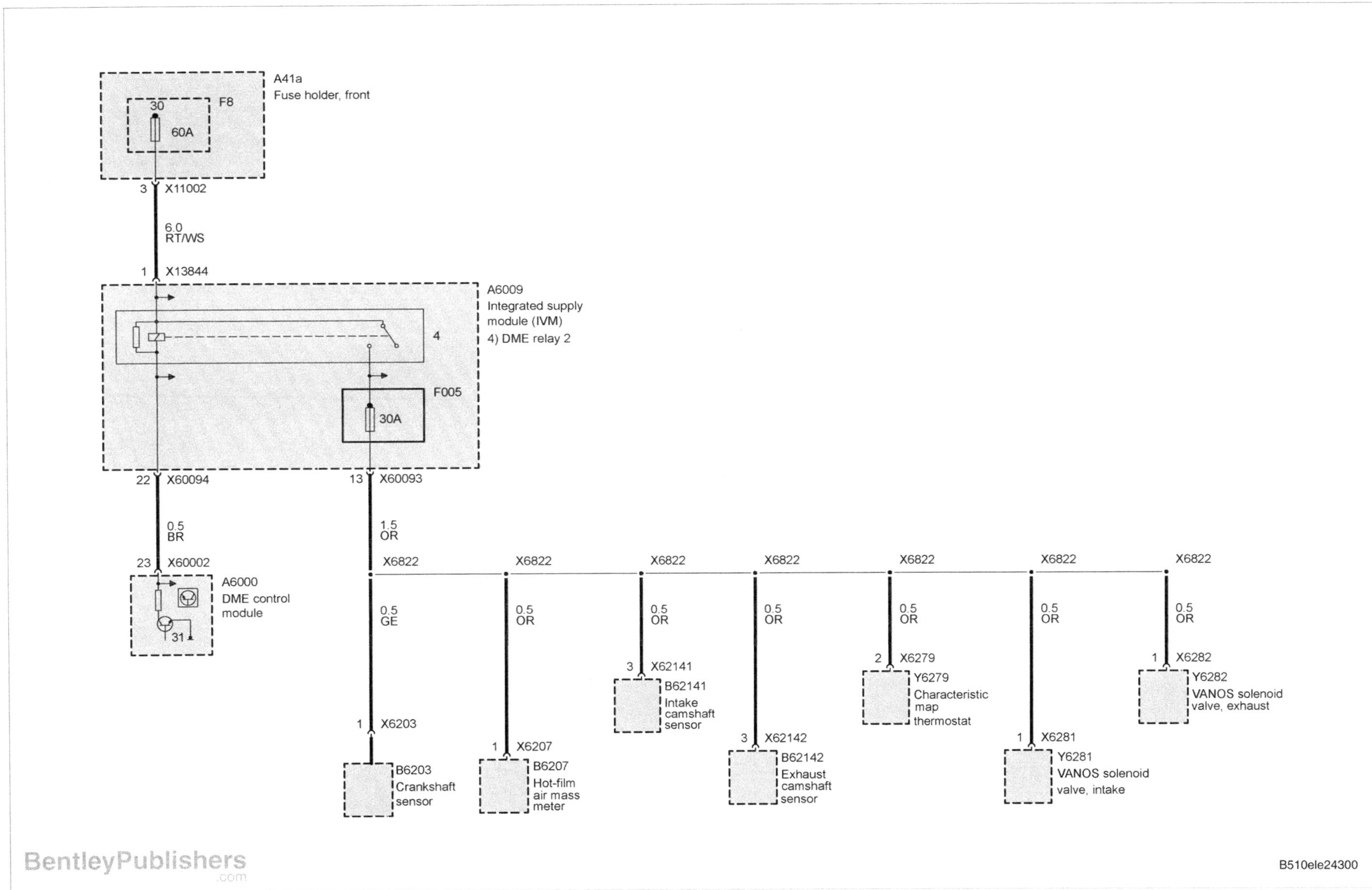
A41a
Fuse holder, front
F8
30
60A
3 X11002
6.0
RT/WS
1 X13844
A6009
Integrated supply
module (IVM)
4) DME relay 2
4
F005
30A
22 X60094
13 X60093
0.5
BR
1.5
OR
23 X60002
A6000
DME control
module
31
X6822
0.5
GE
0.5
OR
1 X6203
B6203
Crankshaft
sensor
1 X6207
B6207
Hot-film
air mass
meter
3 X62141
B62141
Intake
camshaft
sensor
3 X62142
B62142
Exhaust
camshaft
sensor
2 X6279
Y6279
Characteristic
map
thermostat
1 X6281
Y6281
VANOS solenoid
valve, intake
1 X6282
Y6282
VANOS solenoid
valve, exhaust
BentleyPublishers
.com
B510ele24300

Fuse F006
N62 engine

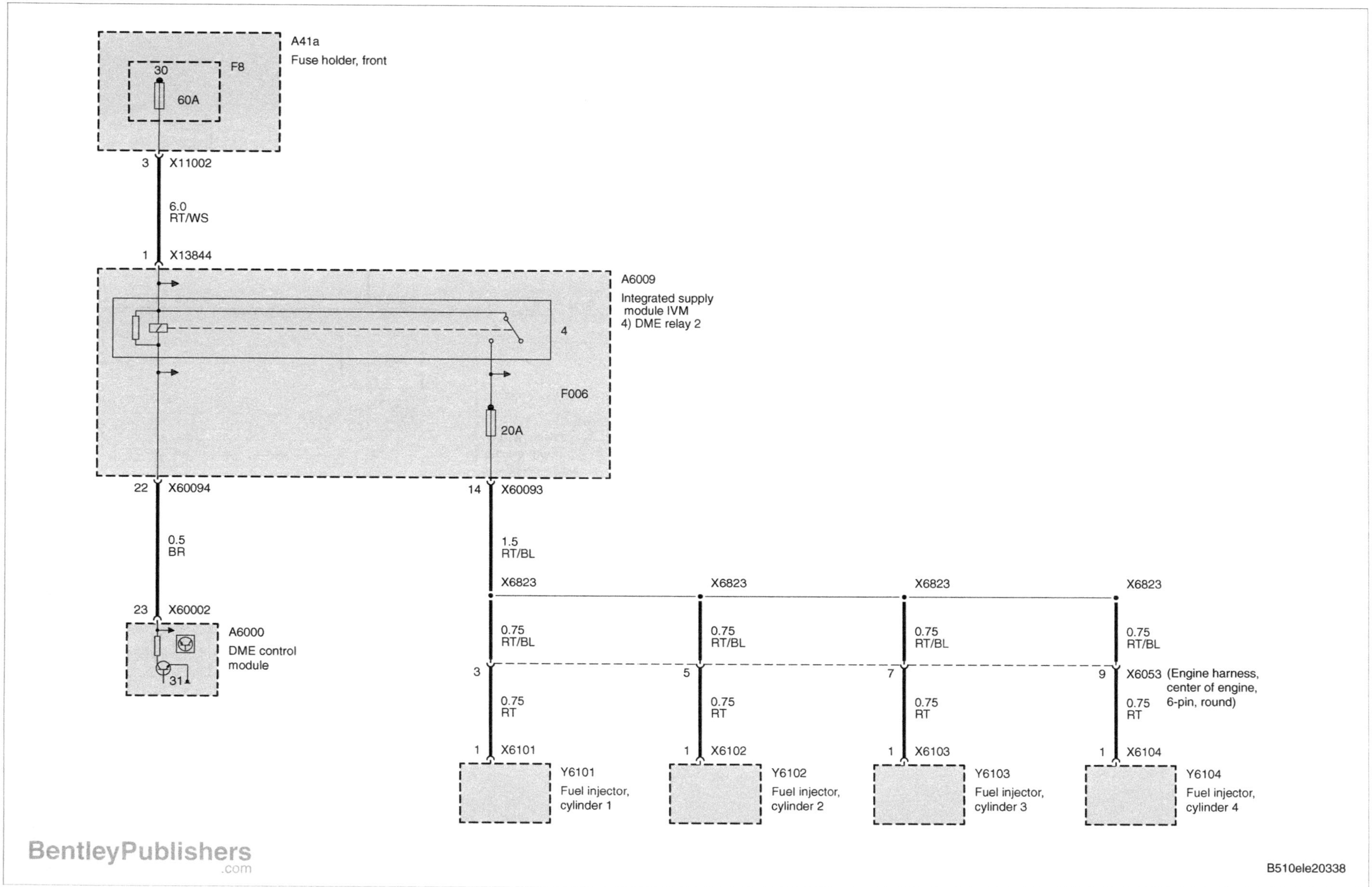

BentleyPublishers
.com

Fuse F007
N62 engine

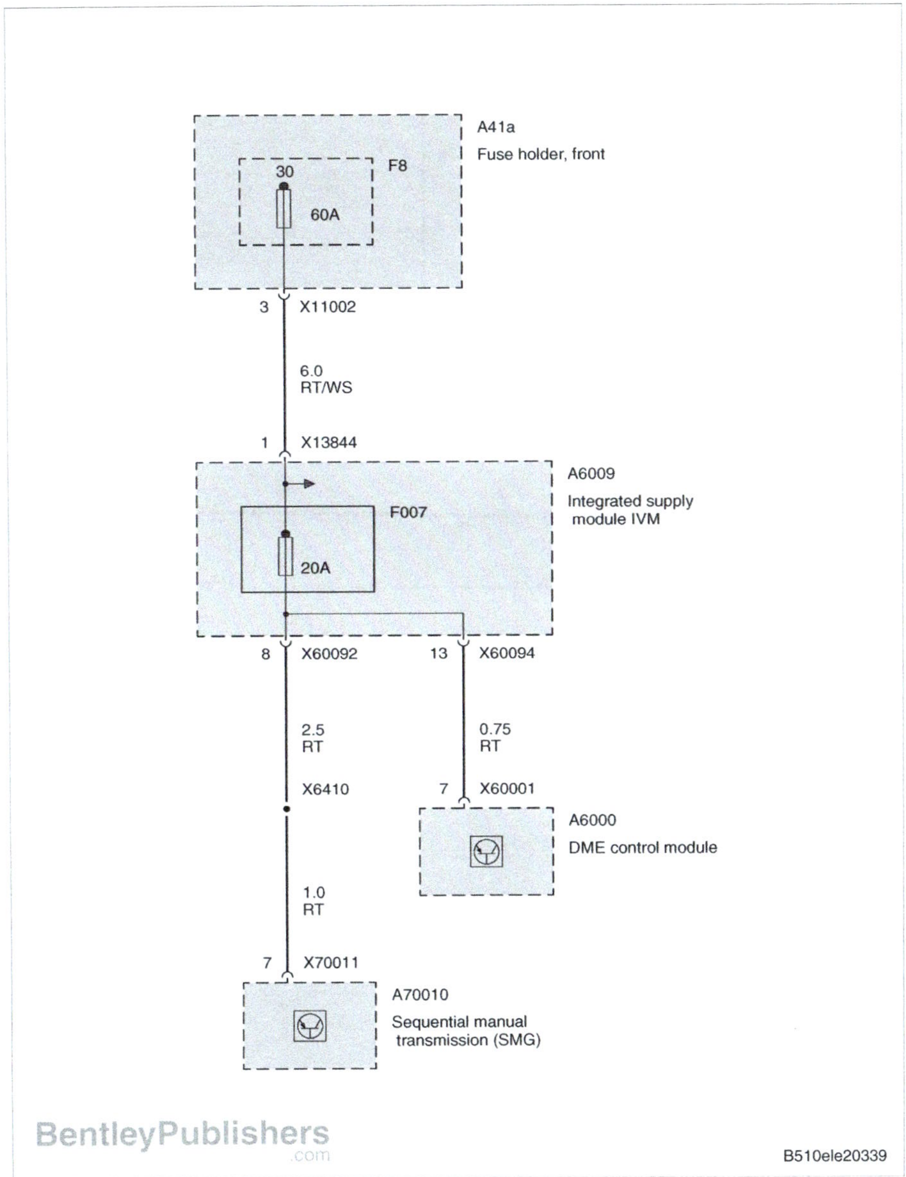

Fuse F008
N62 engine

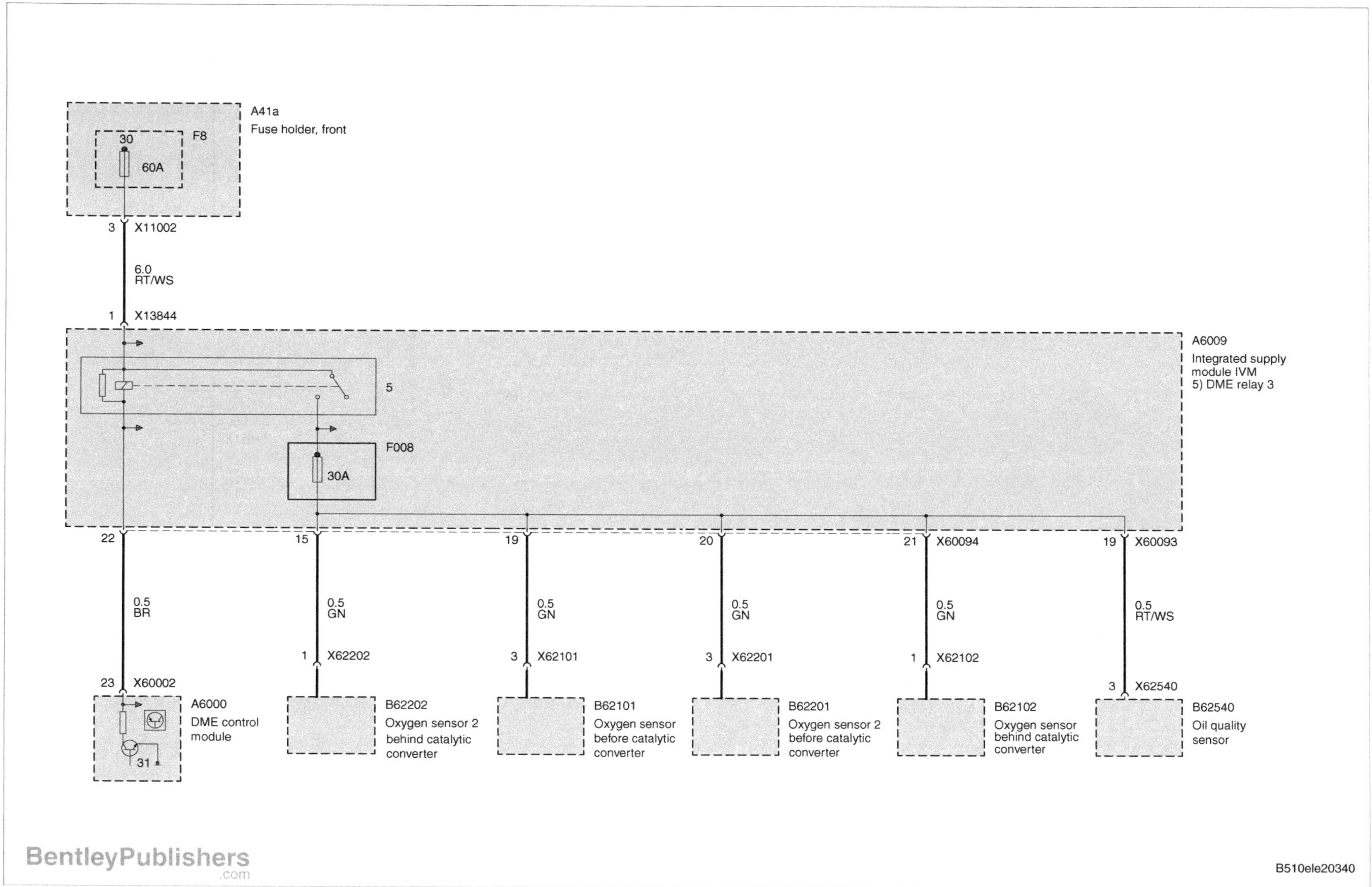

BentleyPublishers.com

Fuse F009
(up to 03 / 2007)

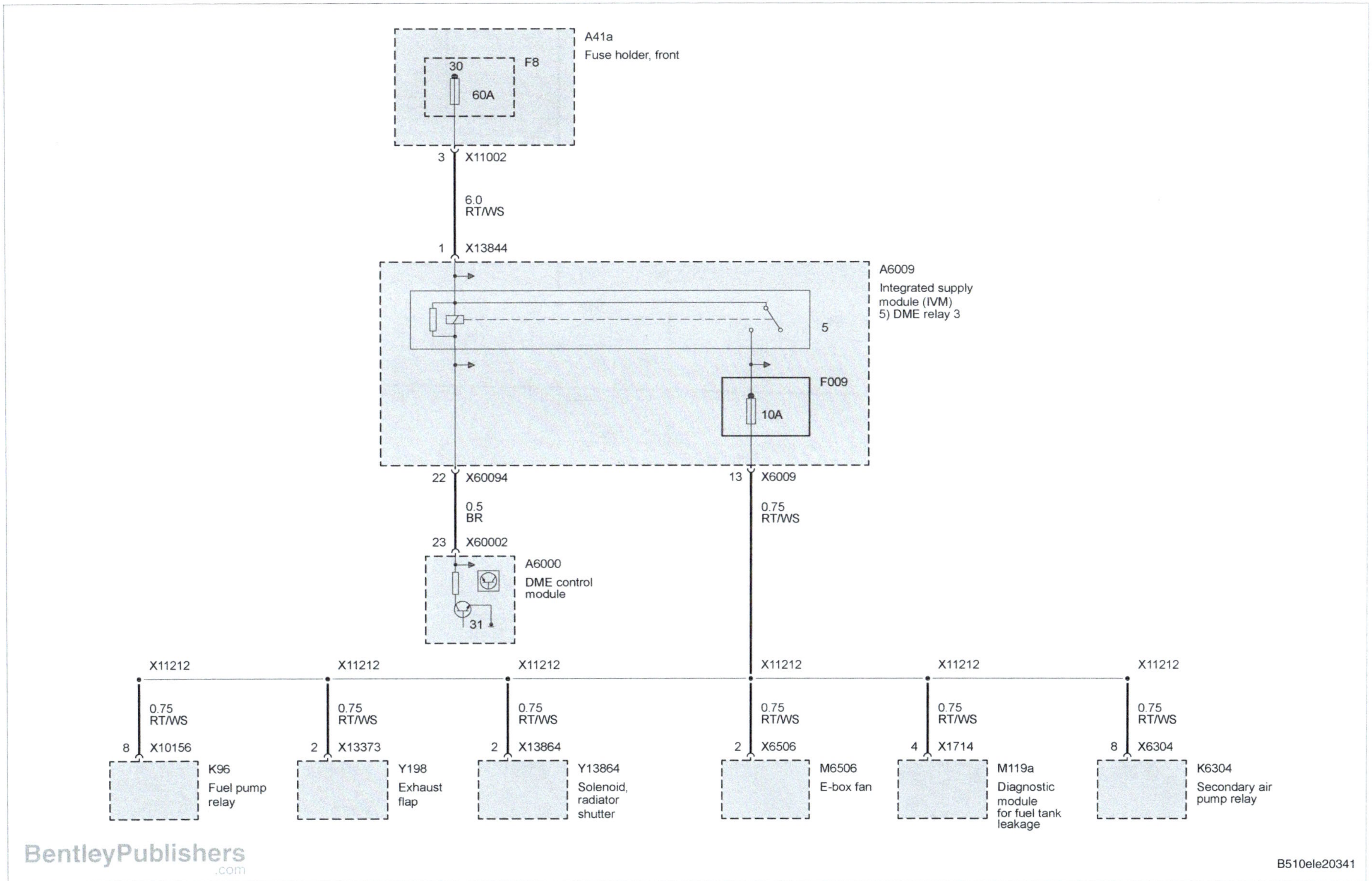

Fuse F009
(as of 03 / 2007)

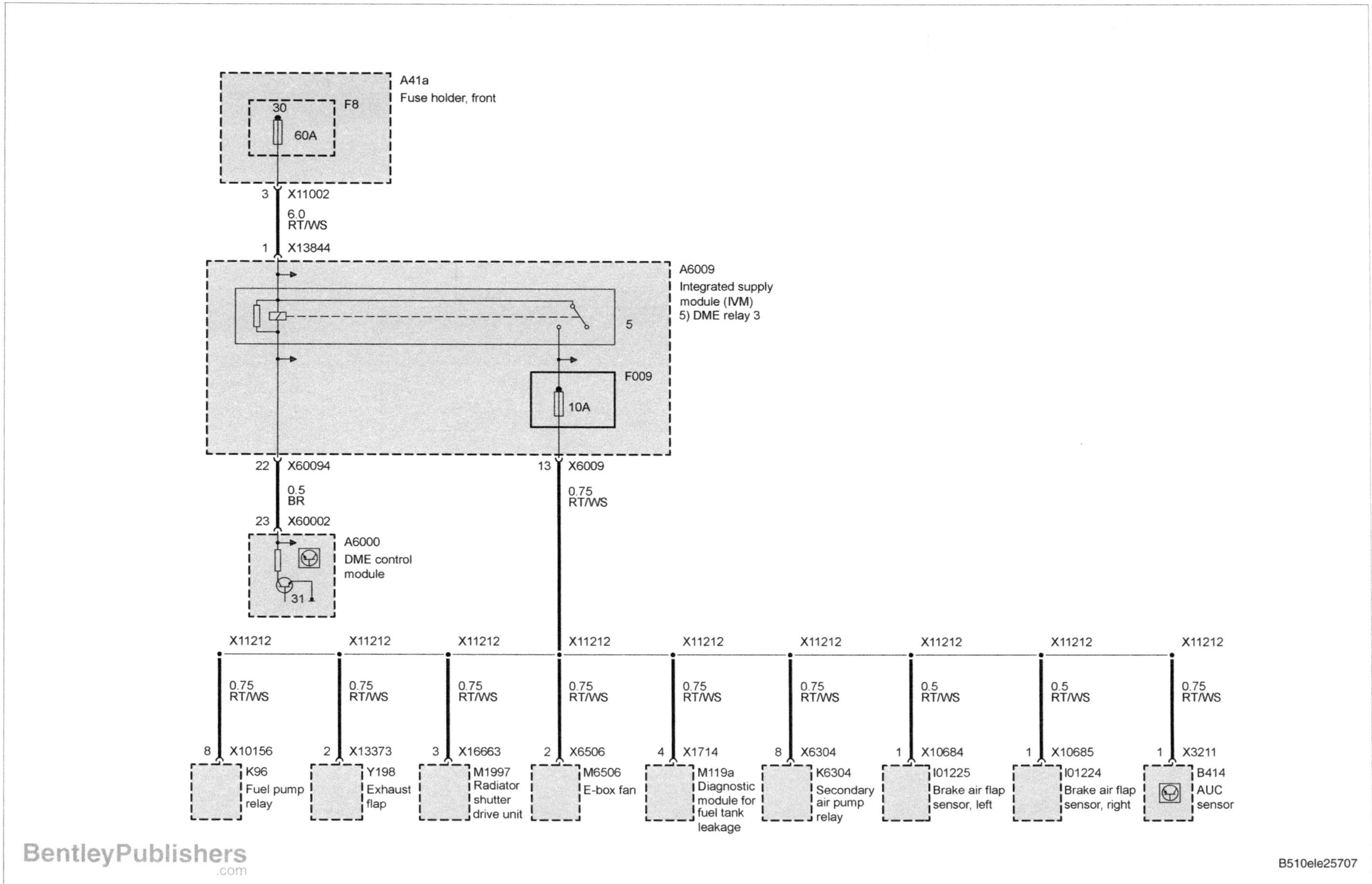

BentleyPublishers.com

Fuse F010

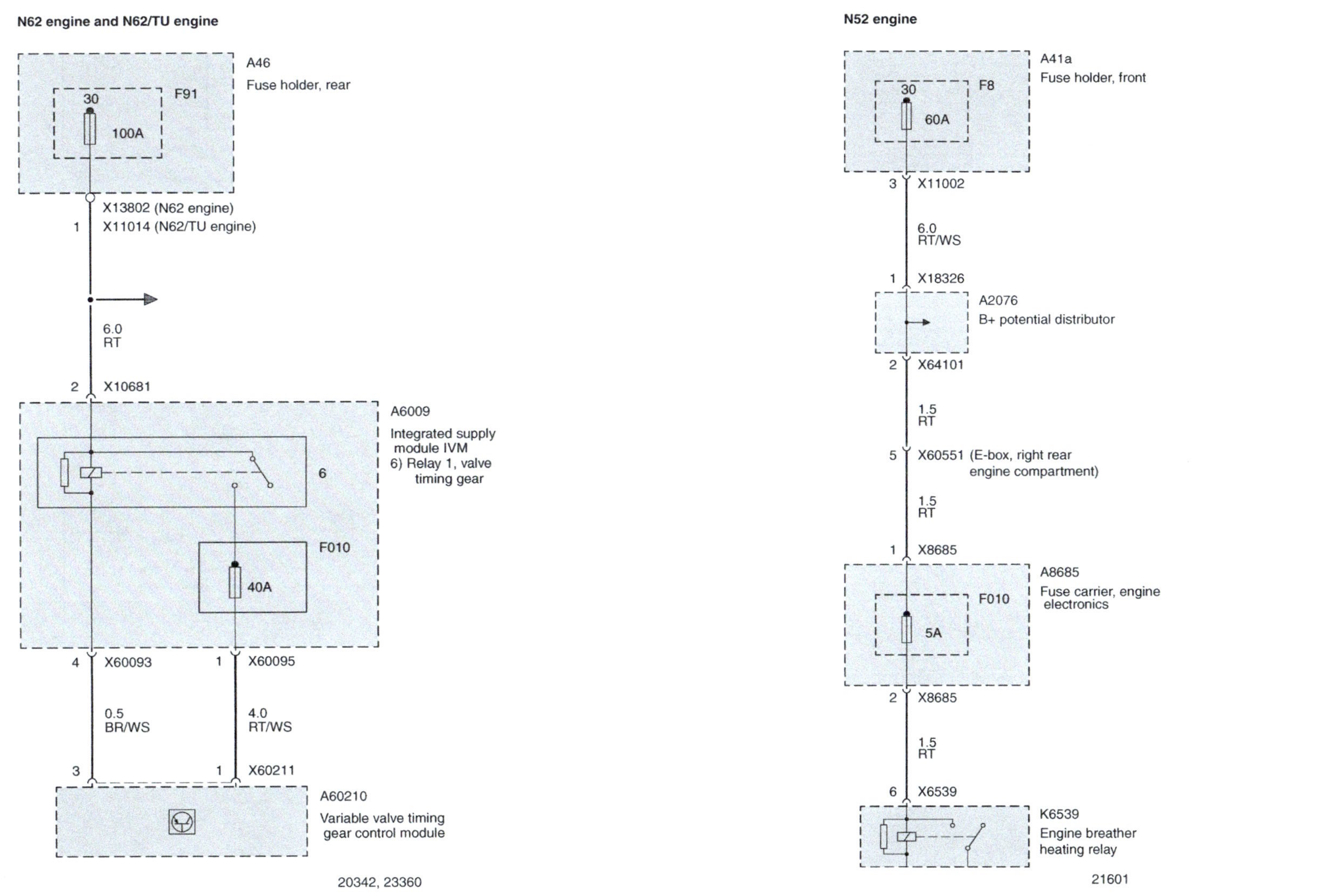

BentleyPublishers.com

Fuse F011
N62 engine and N62 TU engine

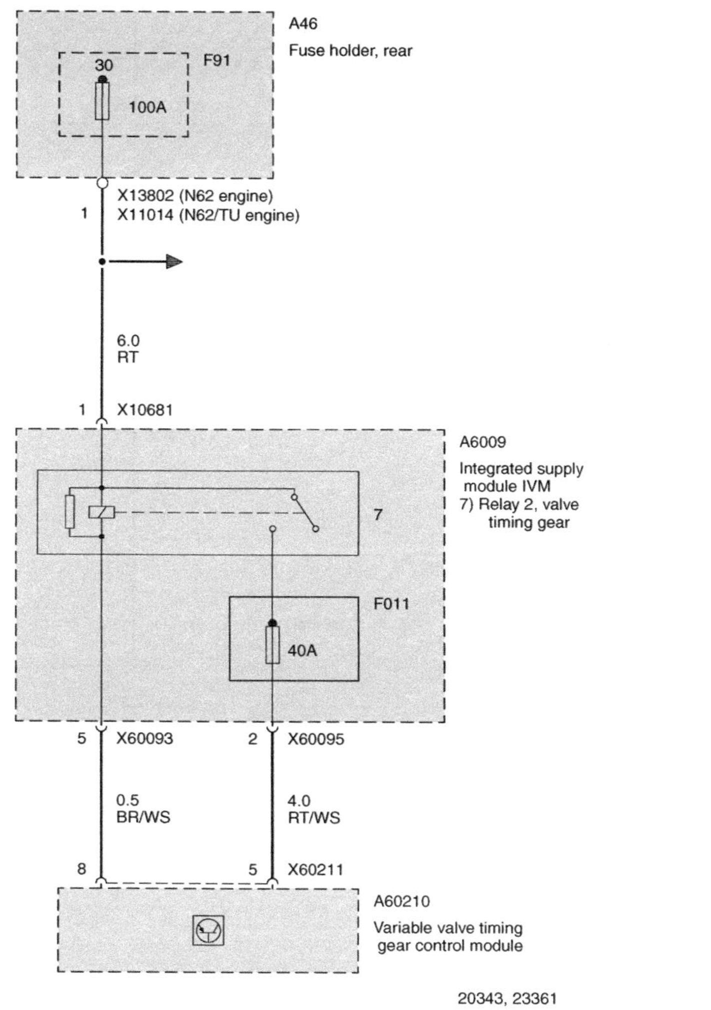

BentleyPublishers.com

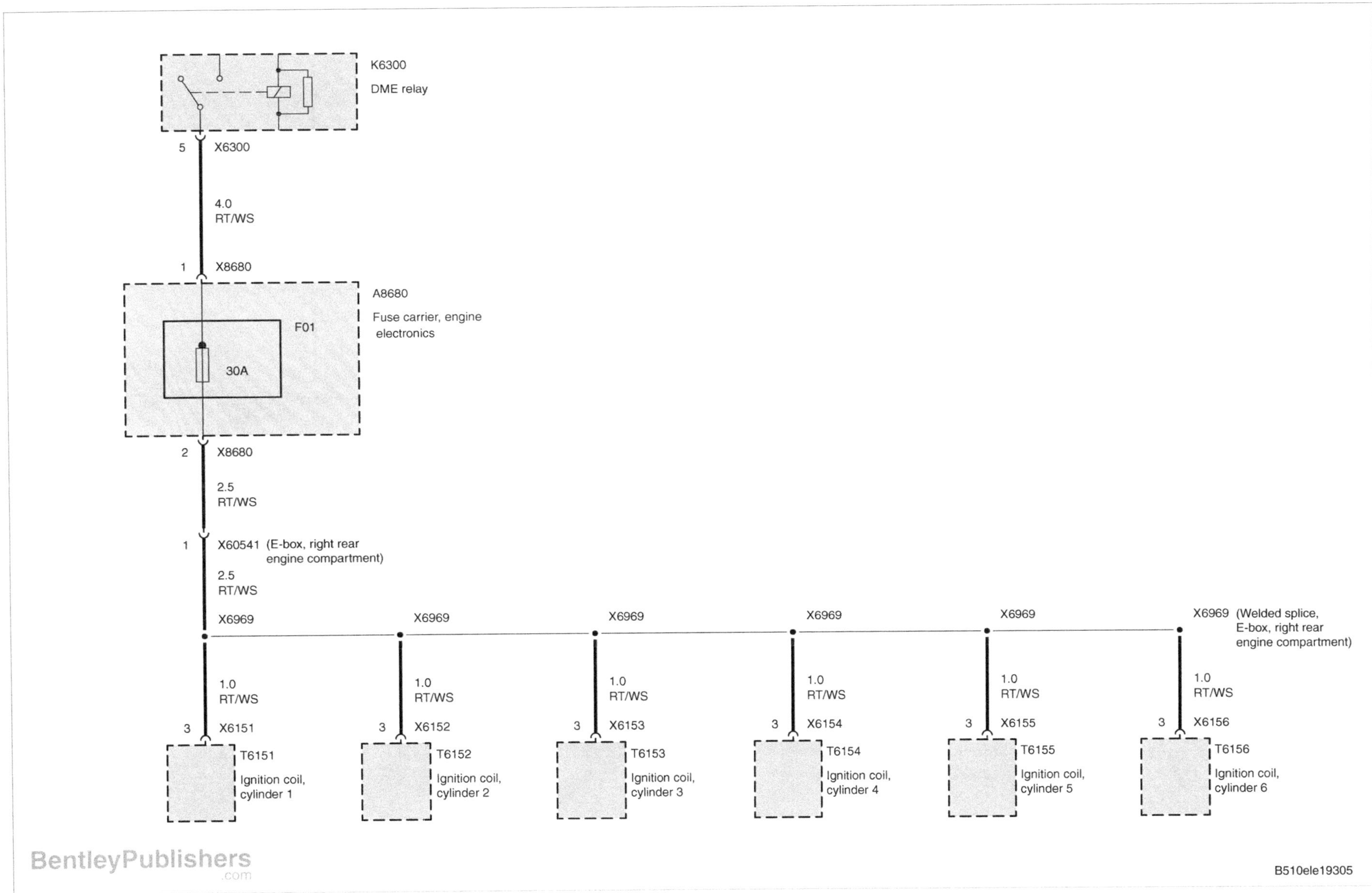

BentleyPublishers.com

Fuse F01
N52 engine

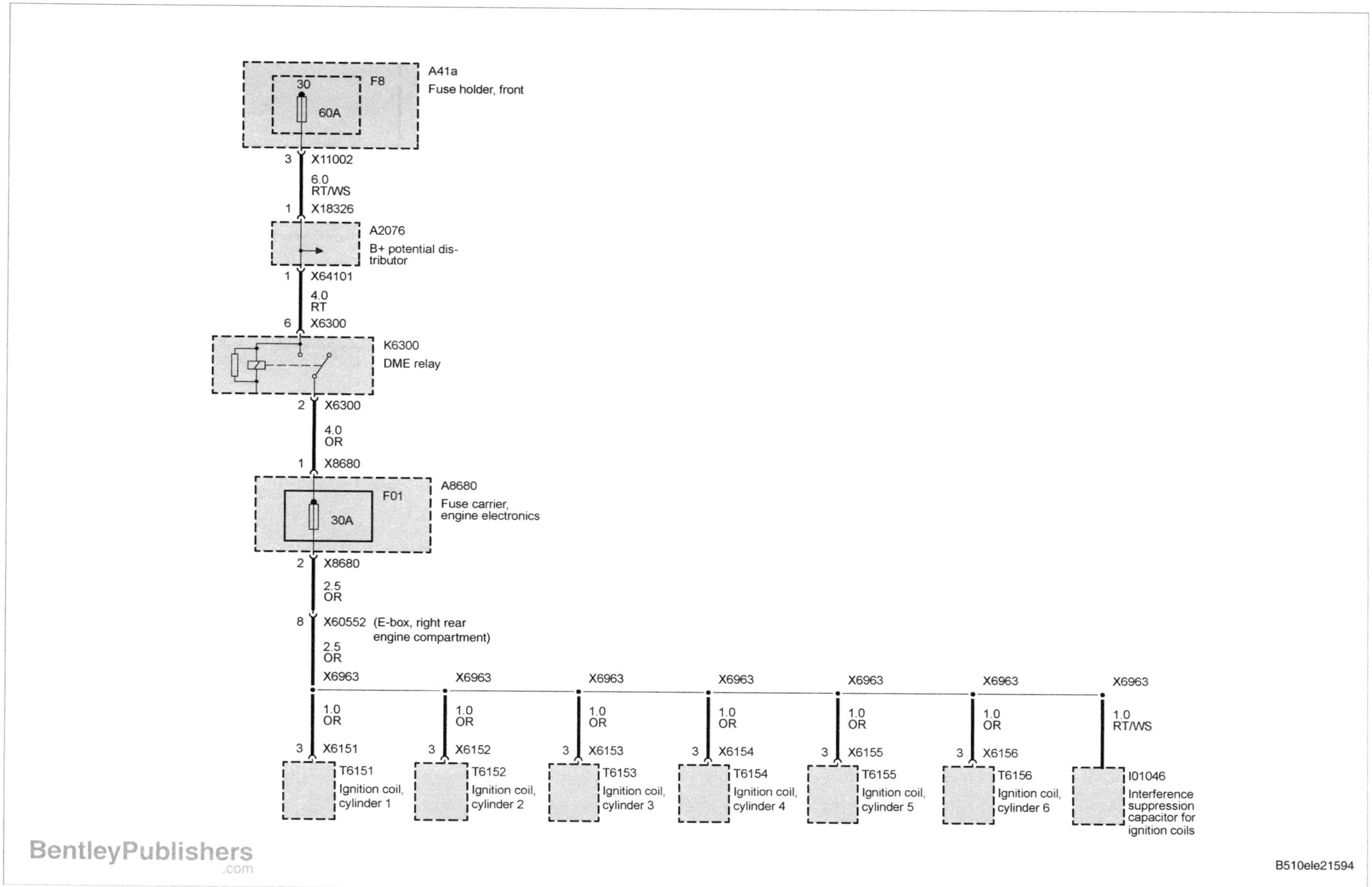

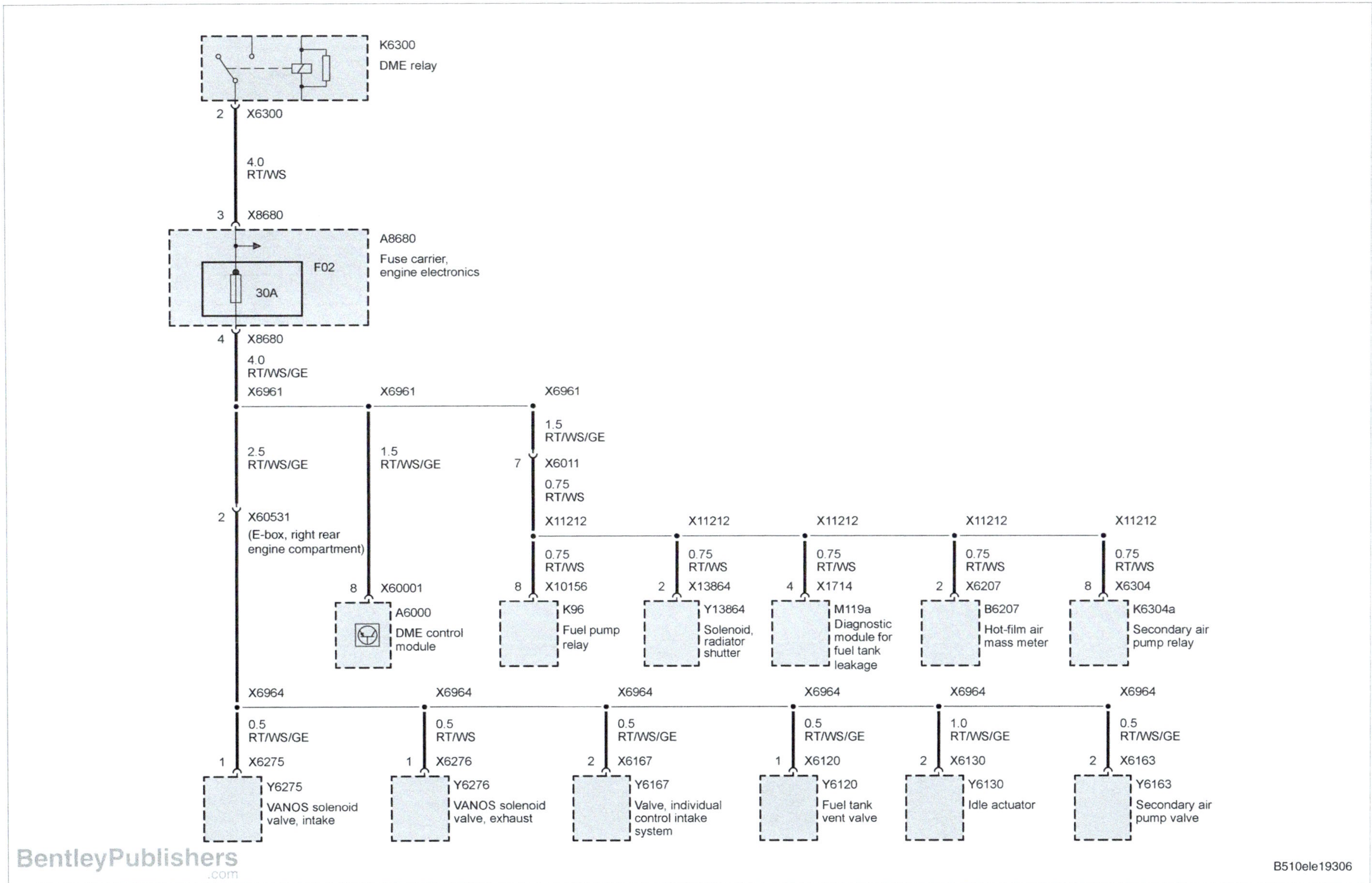

BentleyPublishers.com

Fuse F02
N52 engine

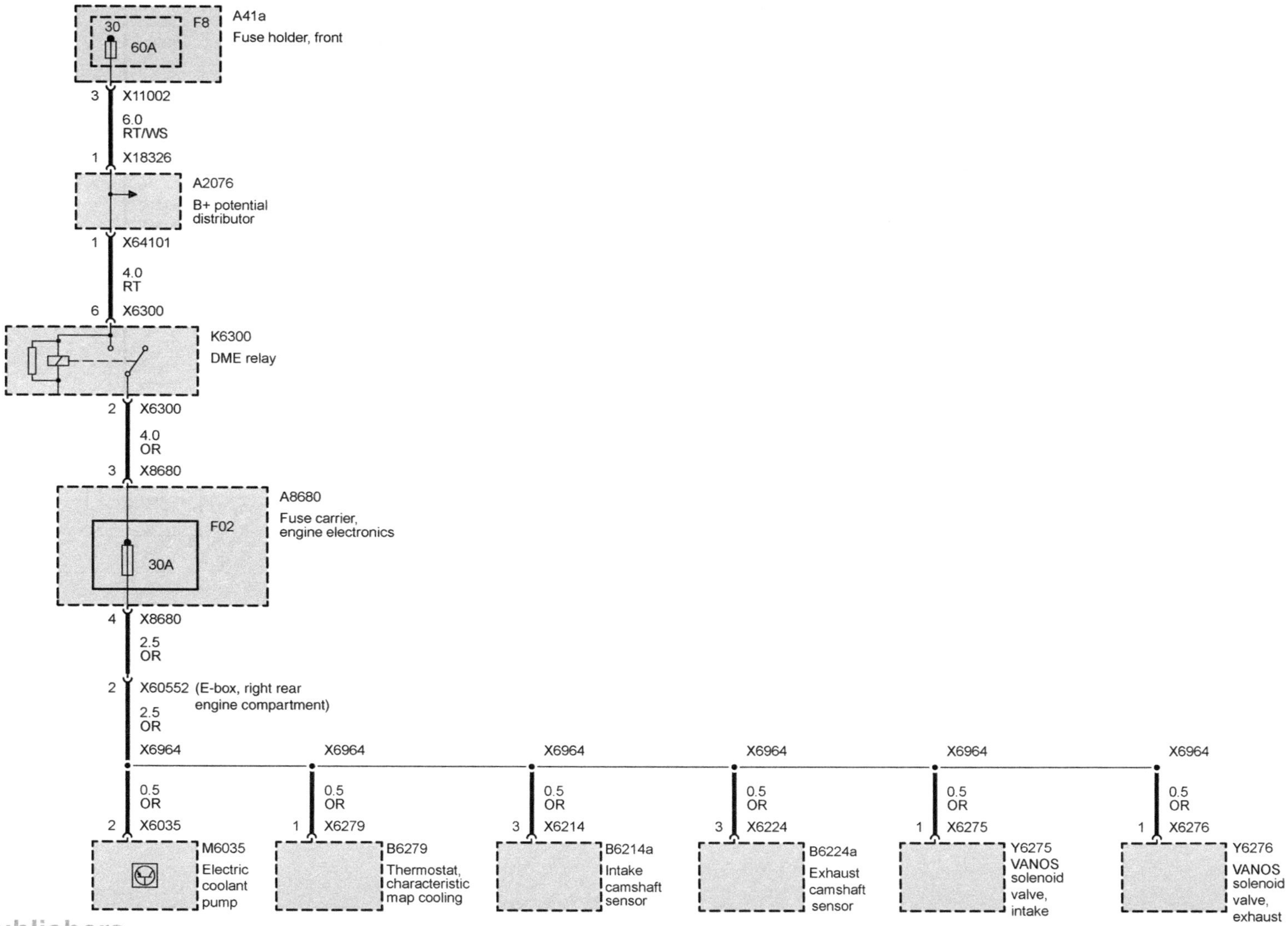

BentleyPublishers.com

B510ele21595

Fuse F03
M54 engine

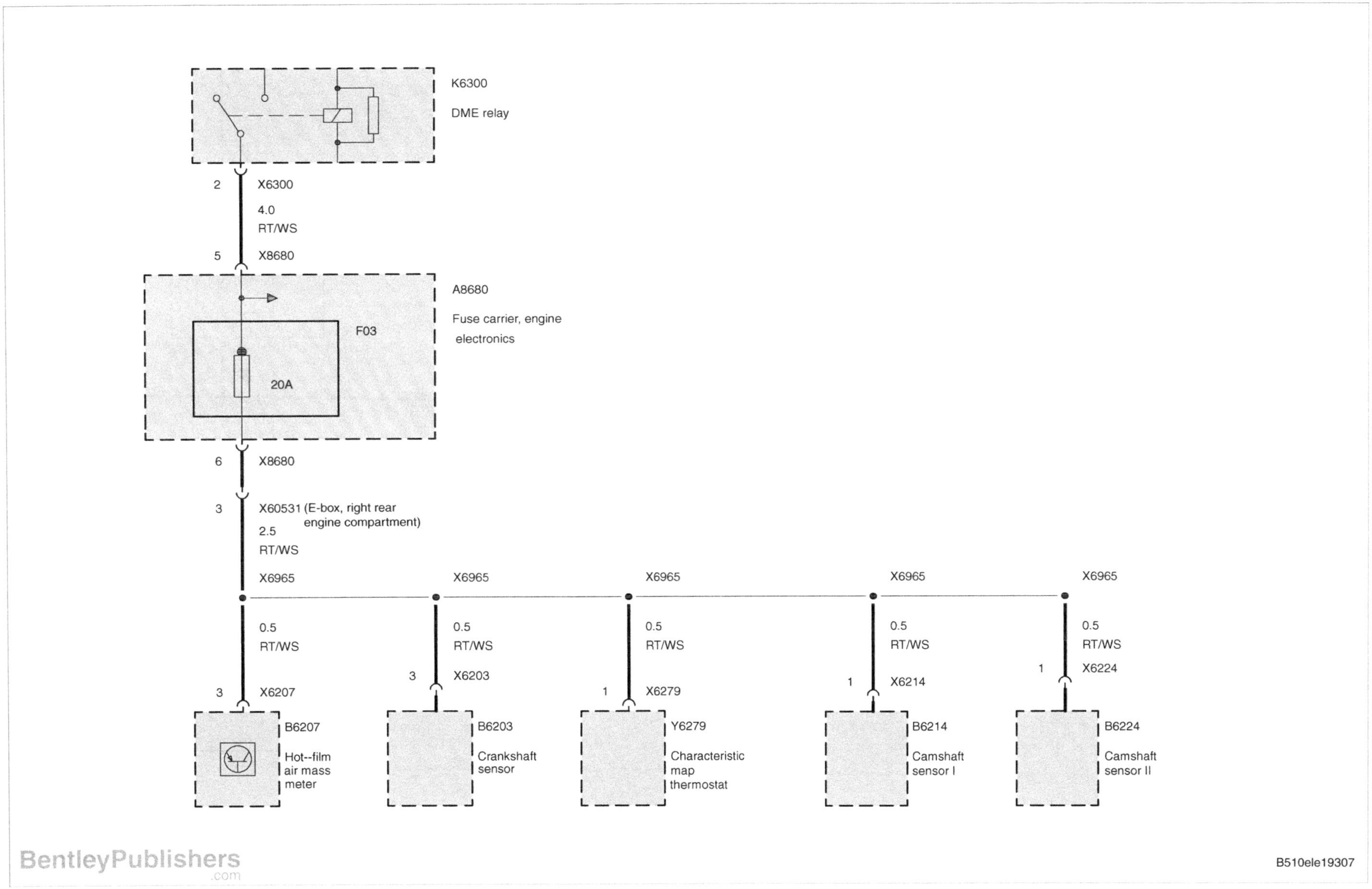

Fuse F03
N52 engine

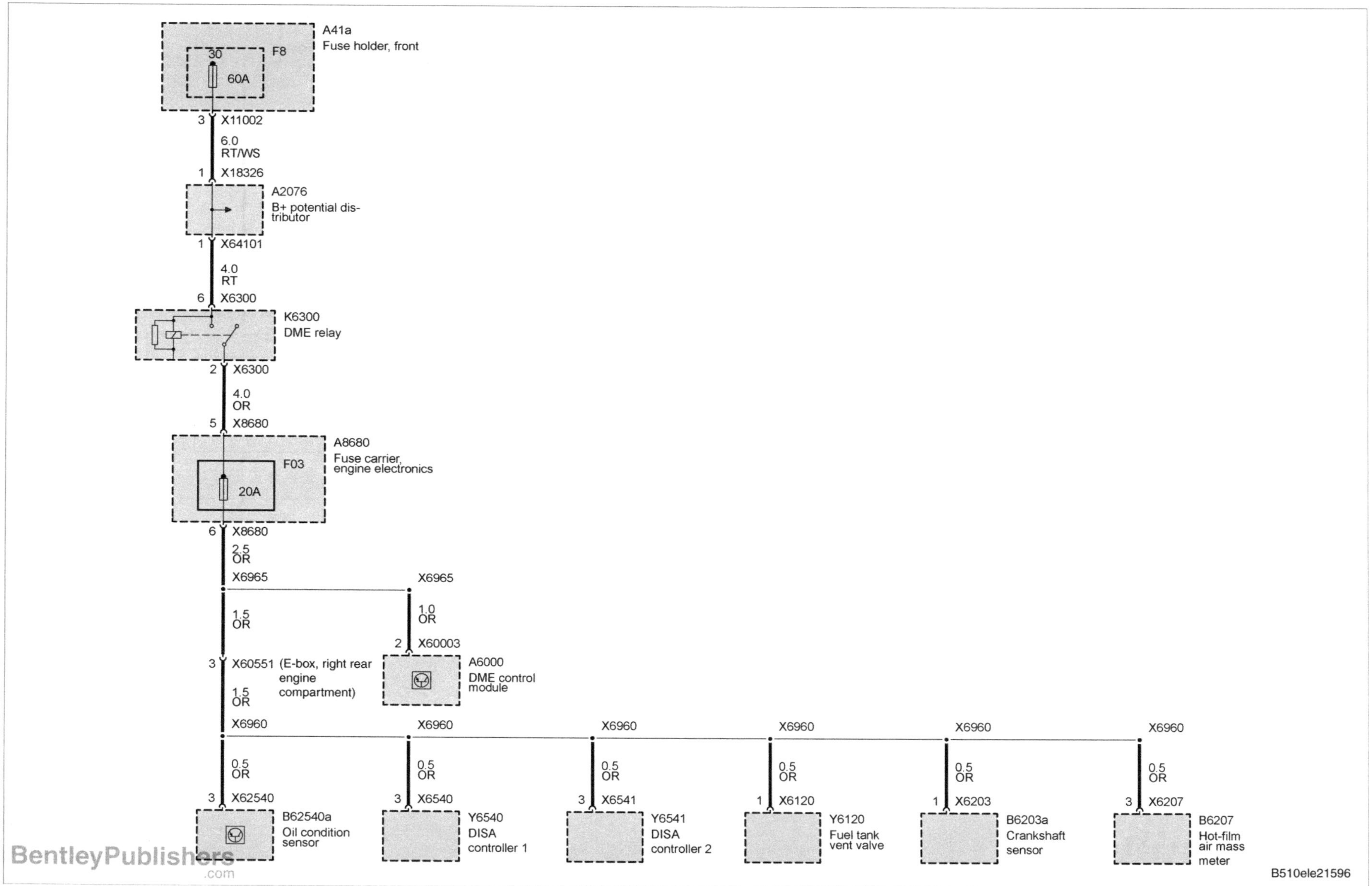

BentleyPublishers.com

Fuse F04
M54 engine

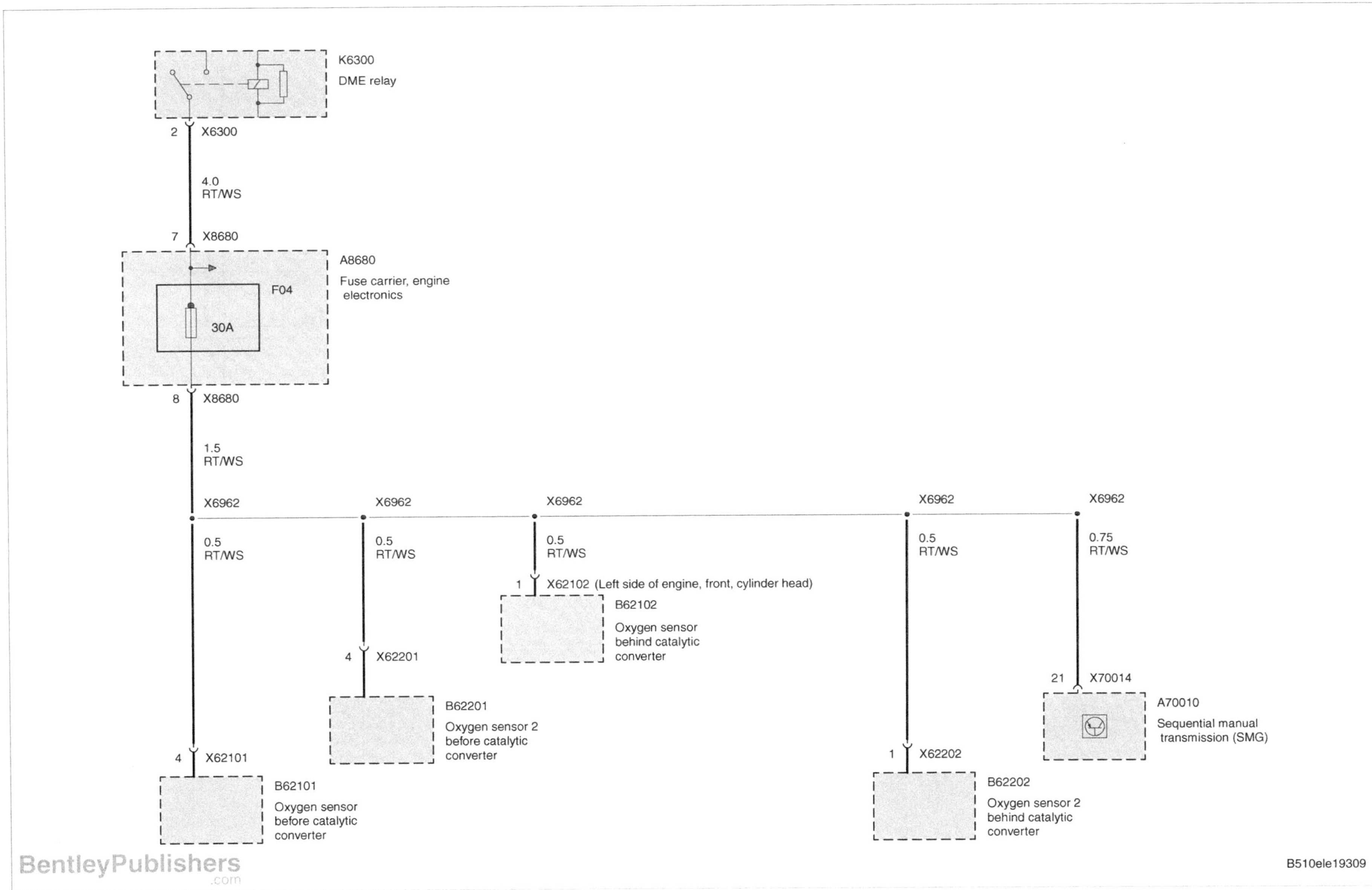

Fuse F04
N52 engine

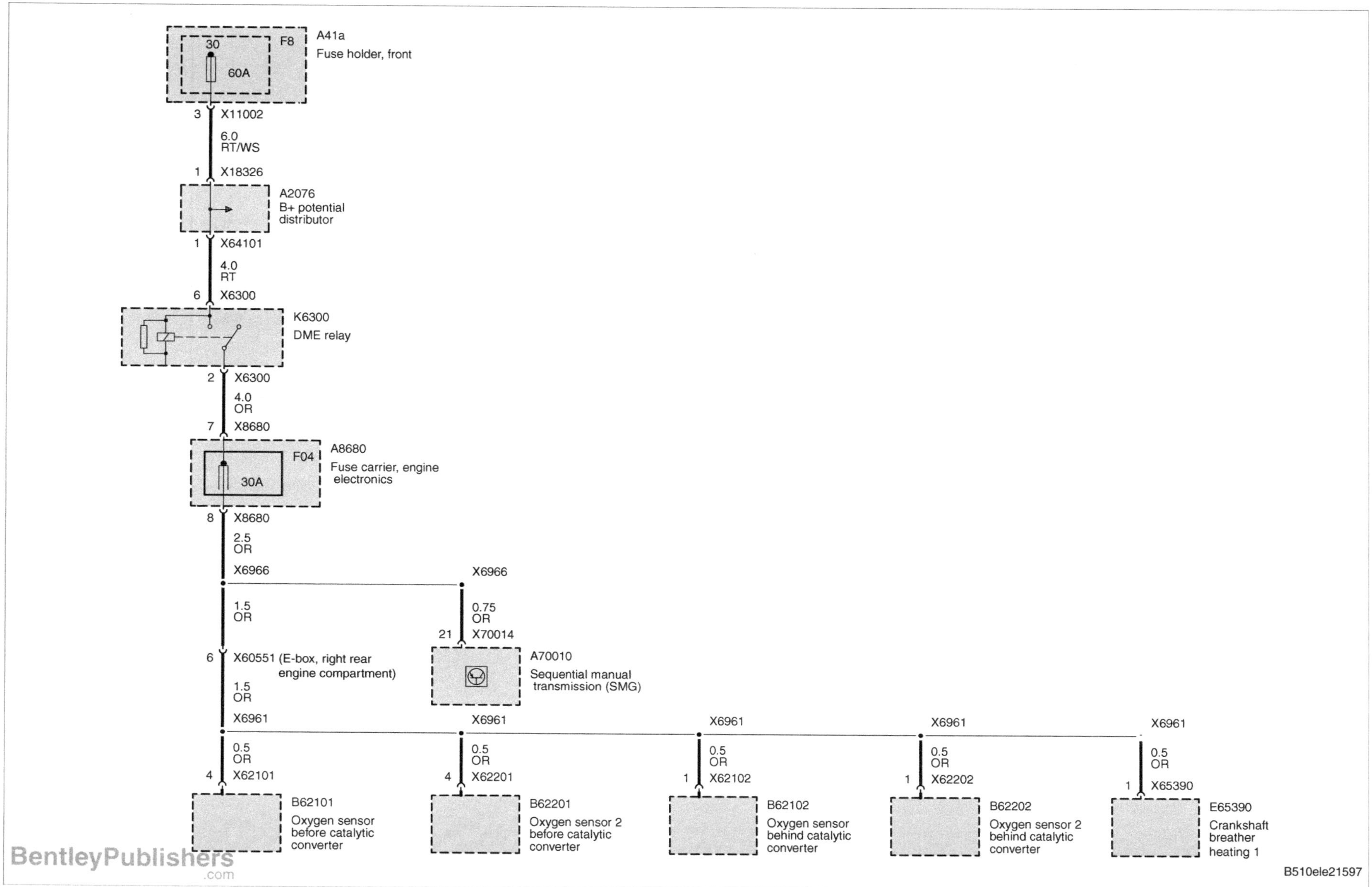

BentleyPublishers.com

Fuse F05
M54 engine or N52 engine

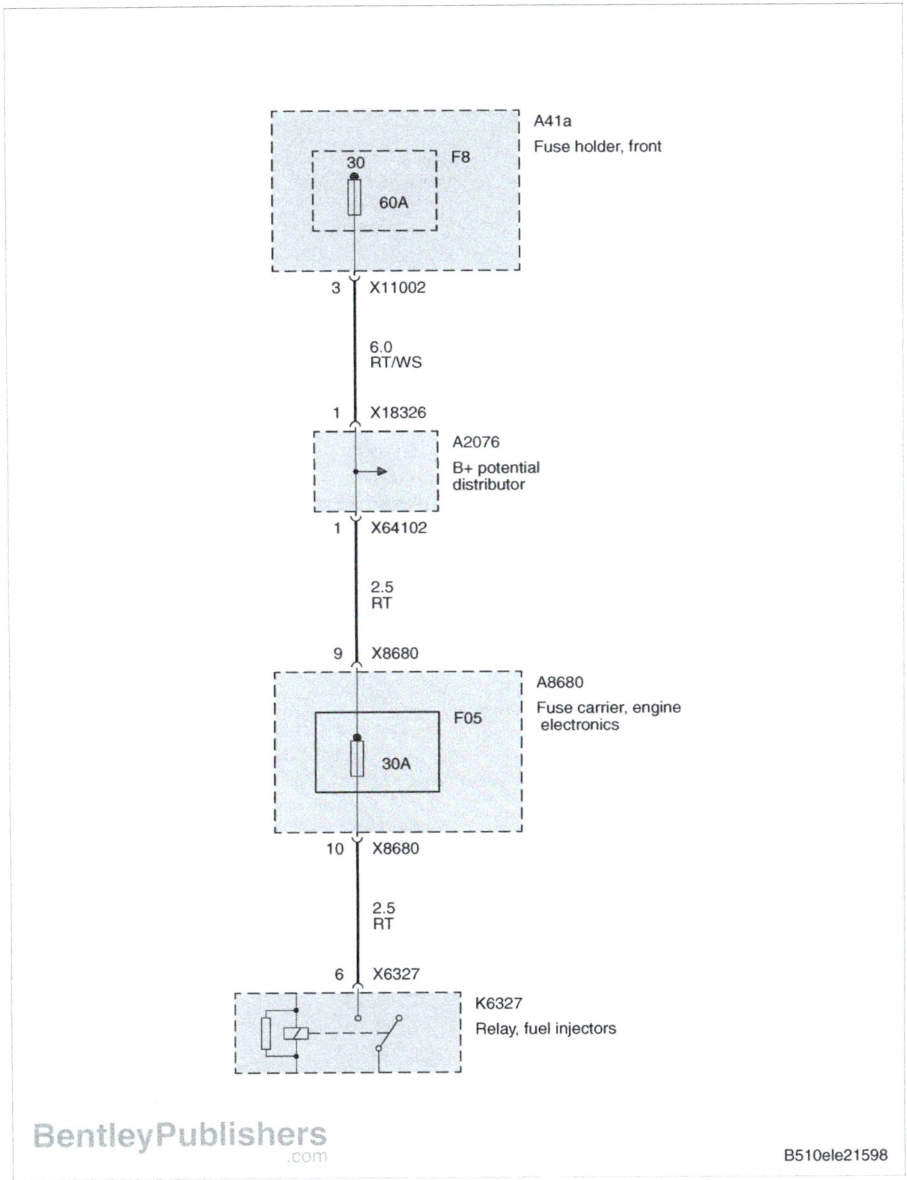

Fuse F06
N52 engine up to 03 / 2007

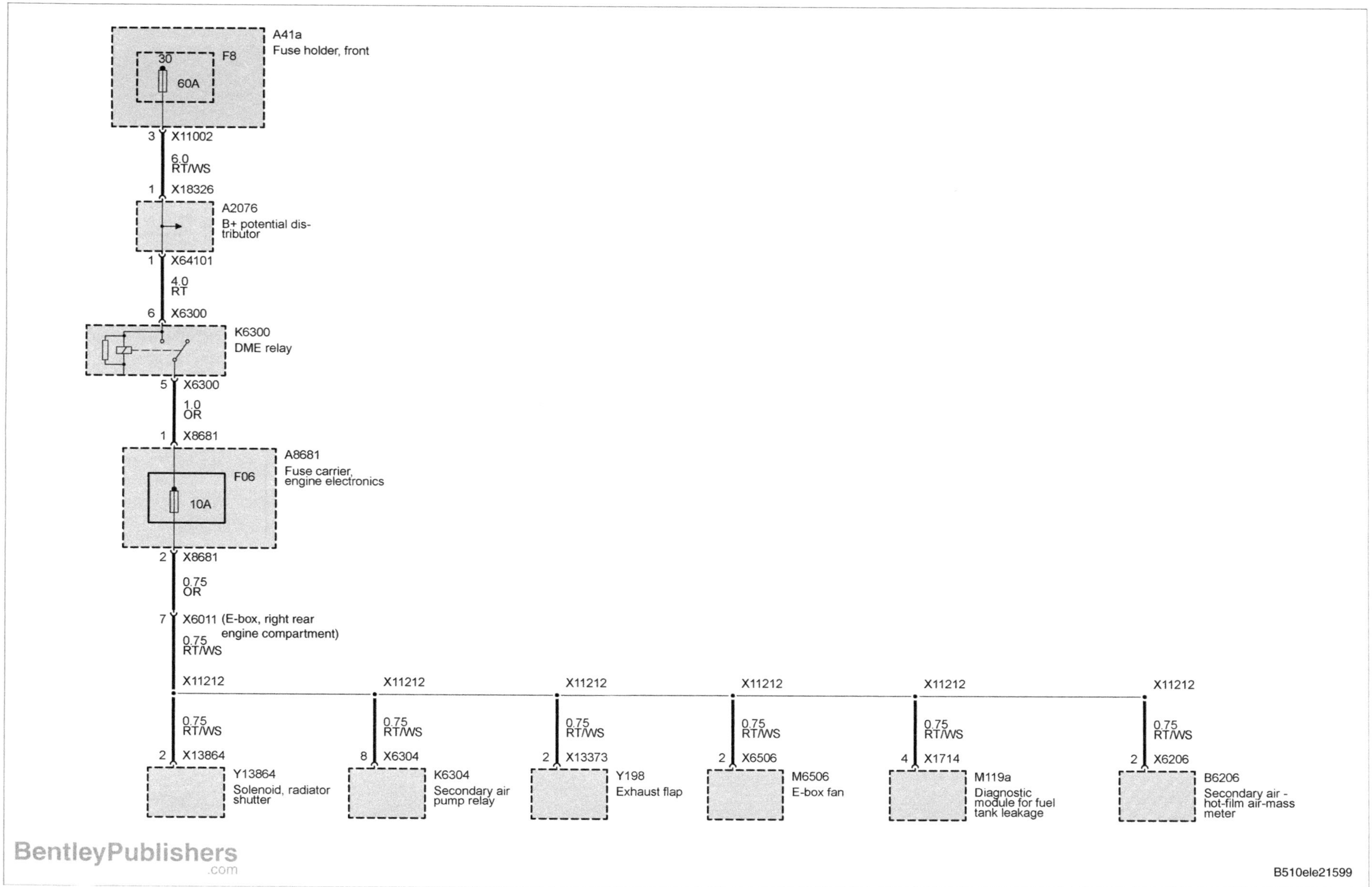

Fuse F06
as of 03 / 2007

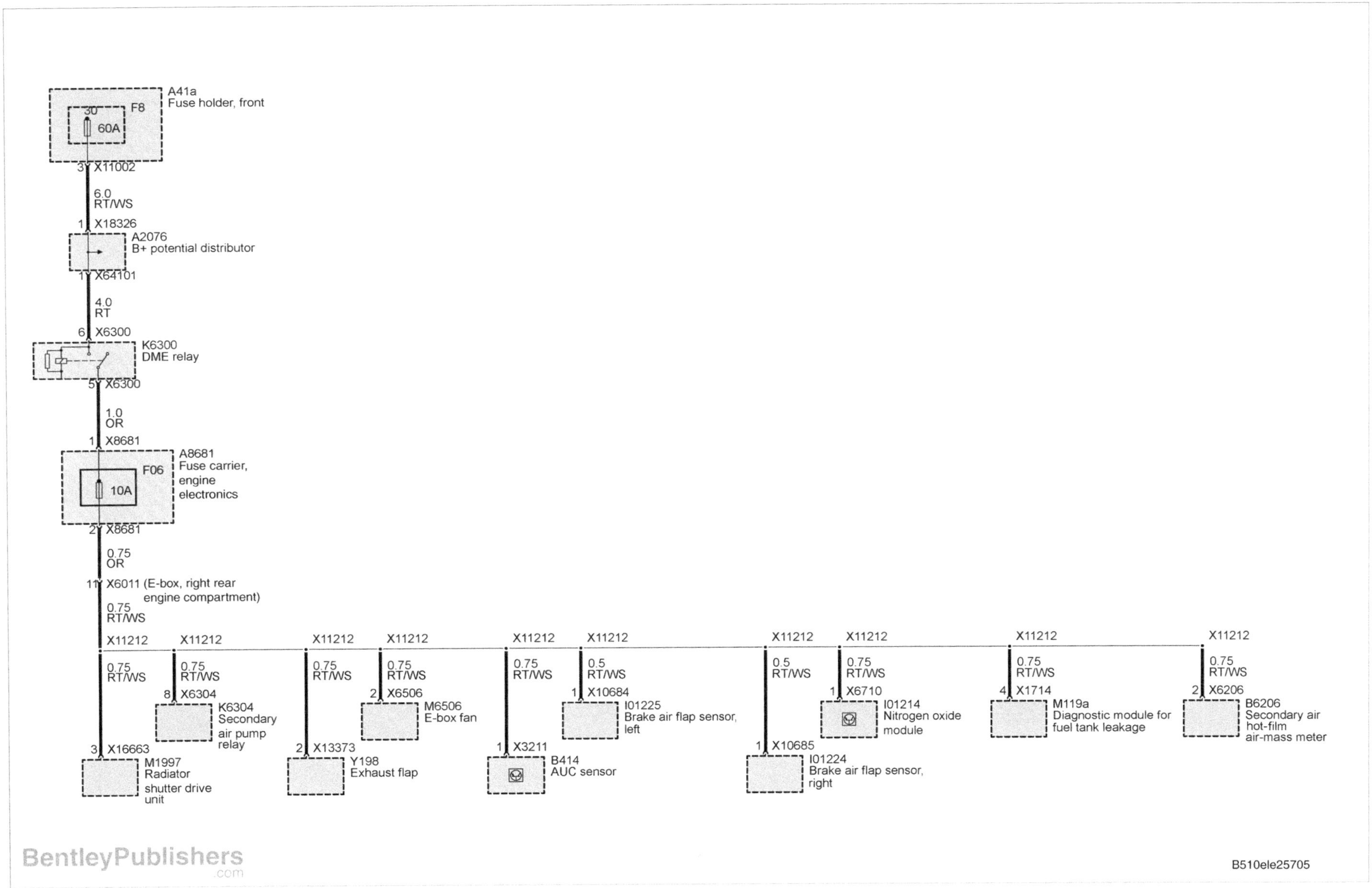

Fuse F07
N52 engine

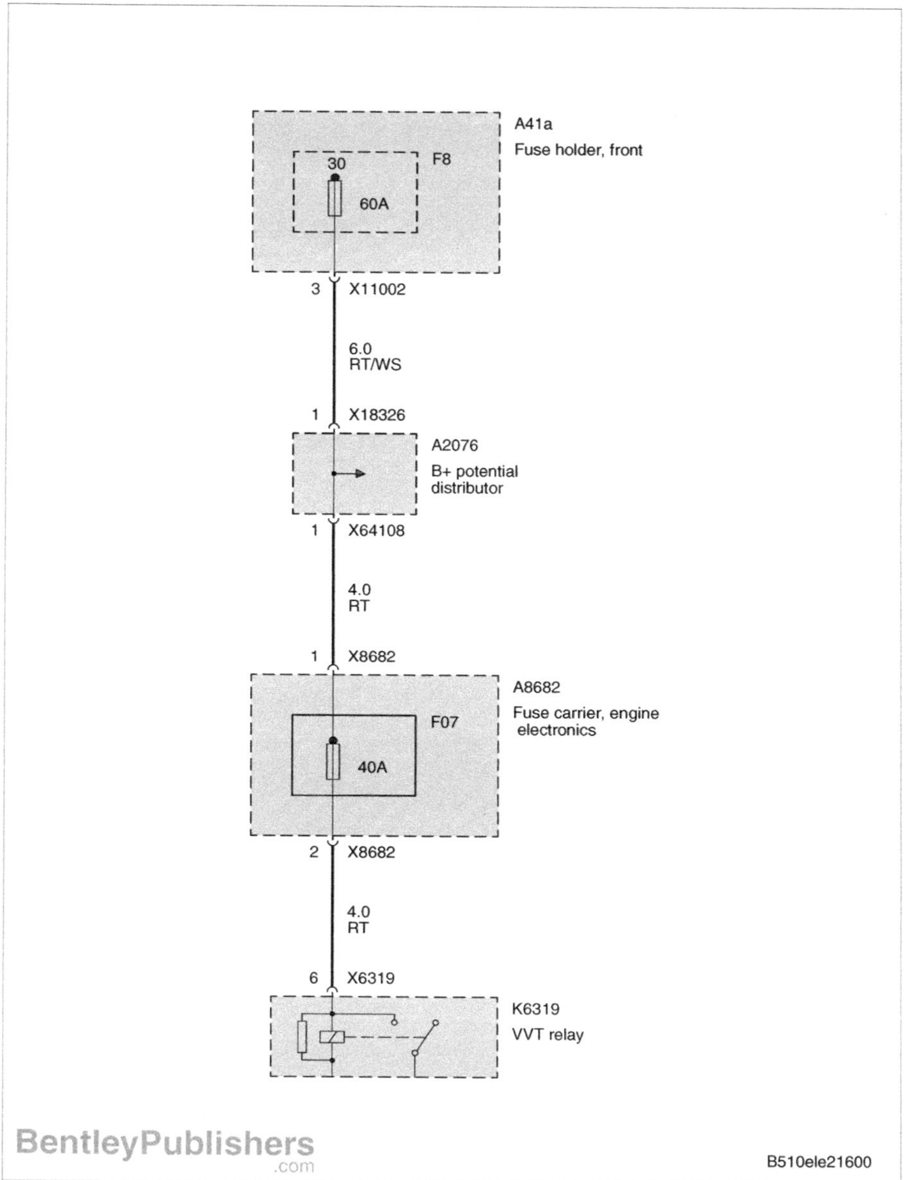

Fuse F08
N52 engine with SMG Sequential Manual Transmission

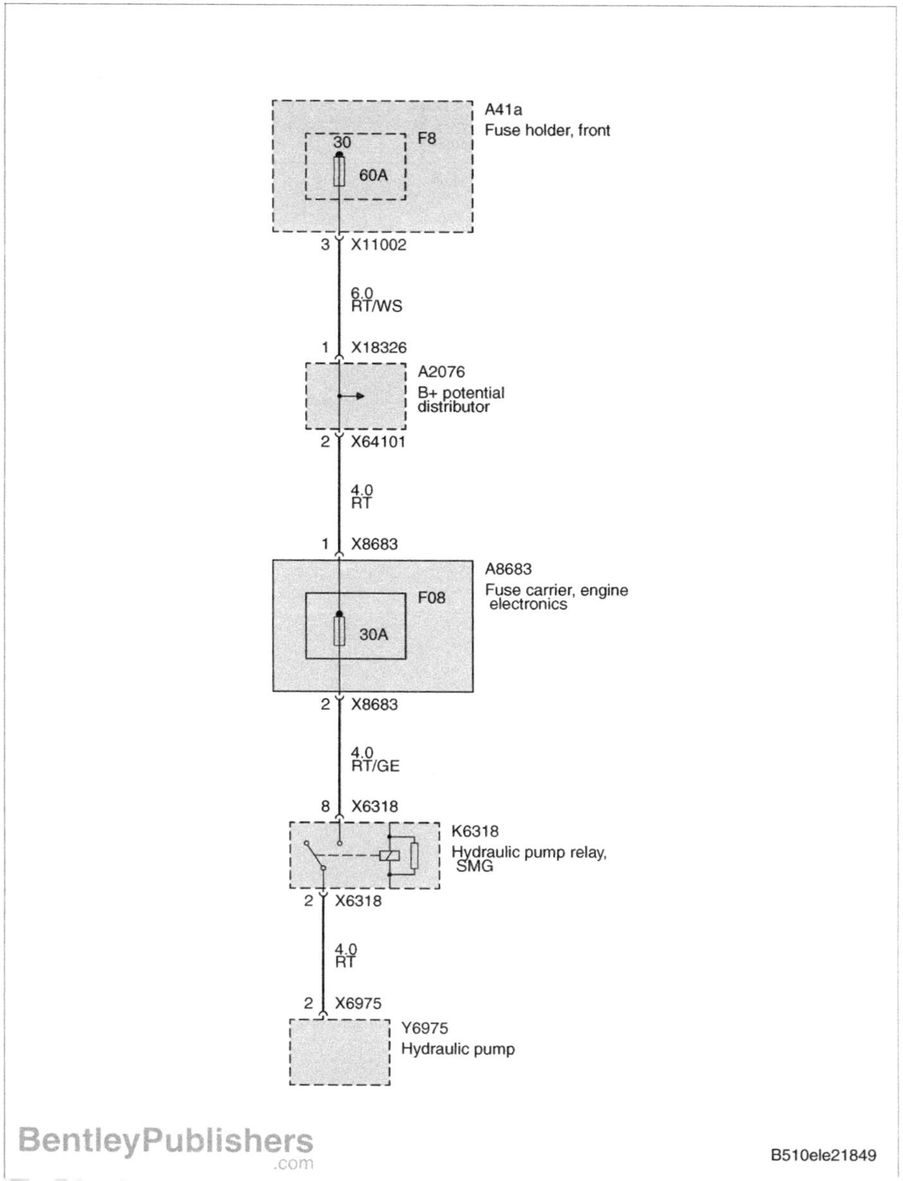

Fuse F1

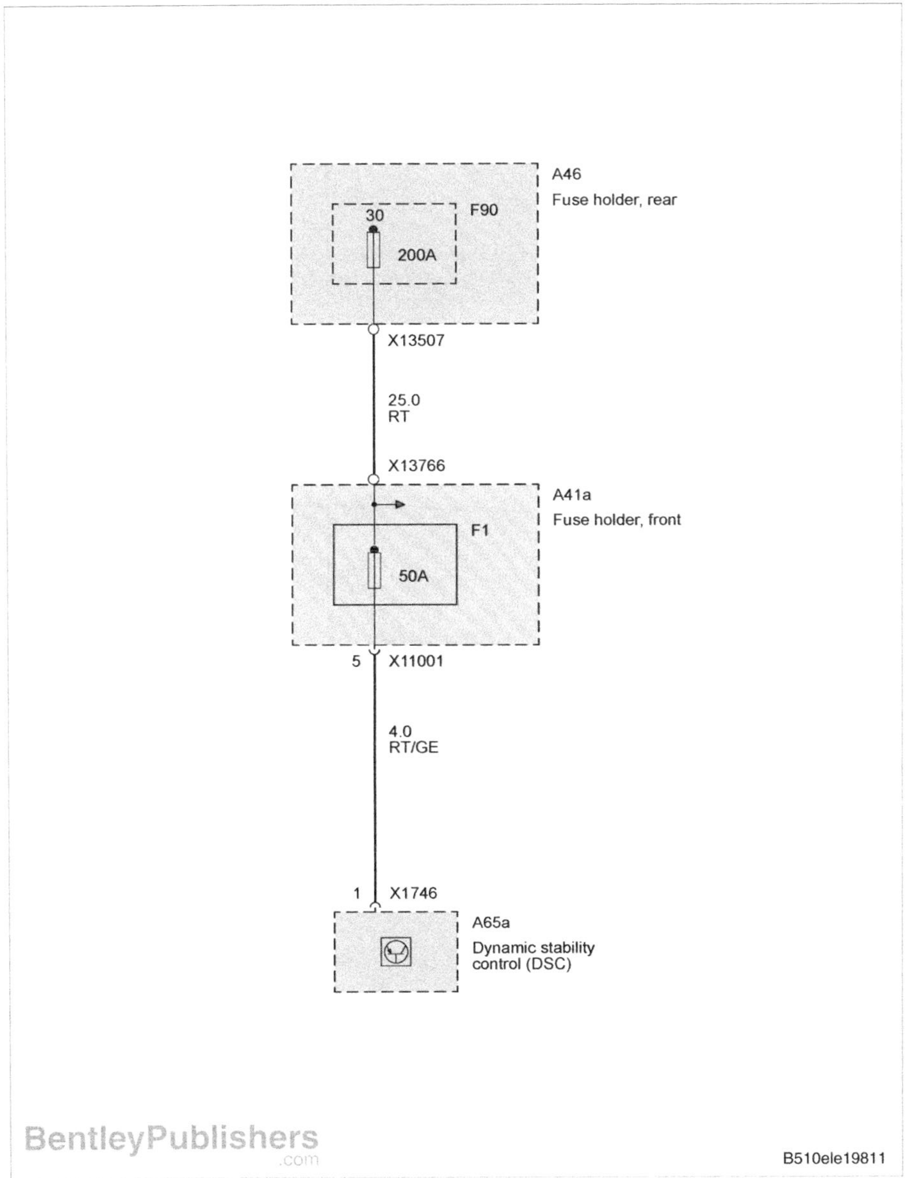

Fuse F2

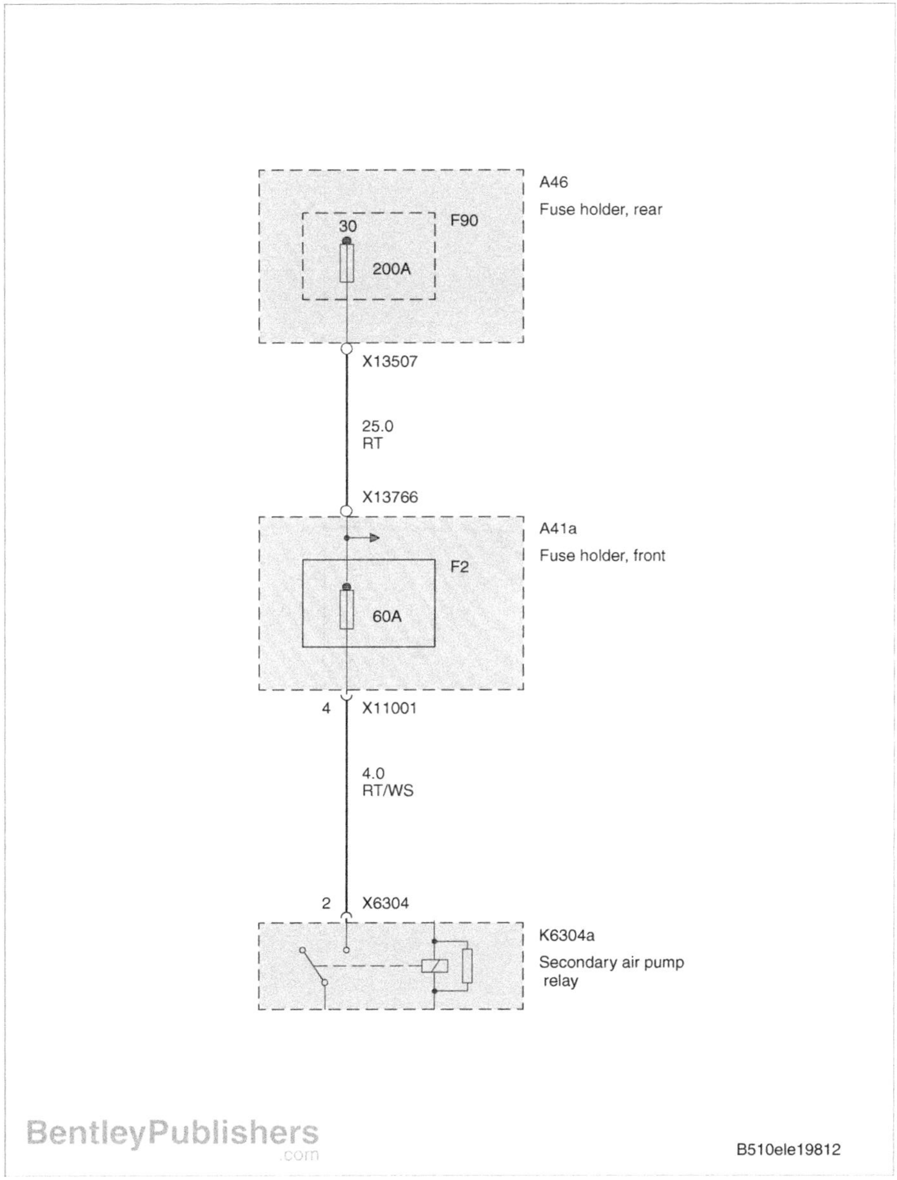

Fuse F3

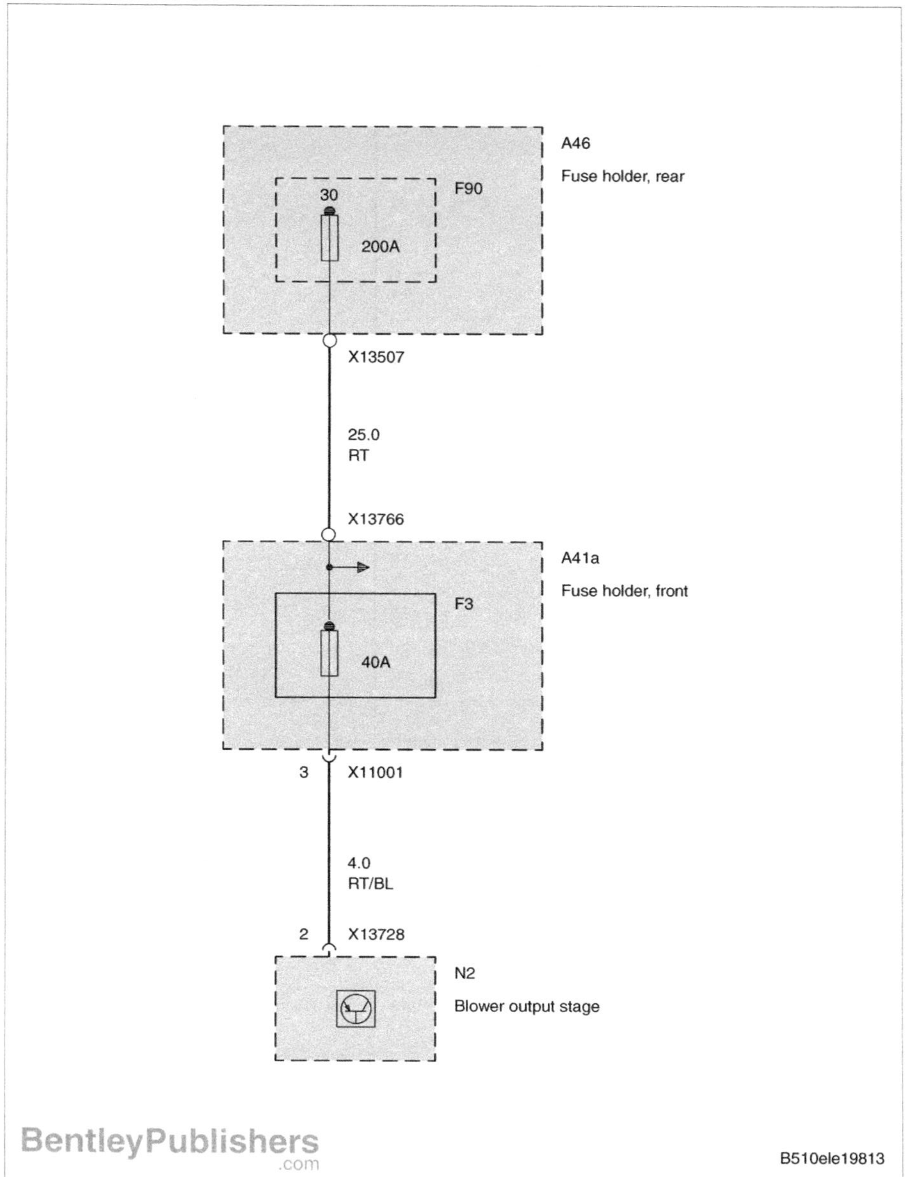

Fuse F4 up to 09 / 2005

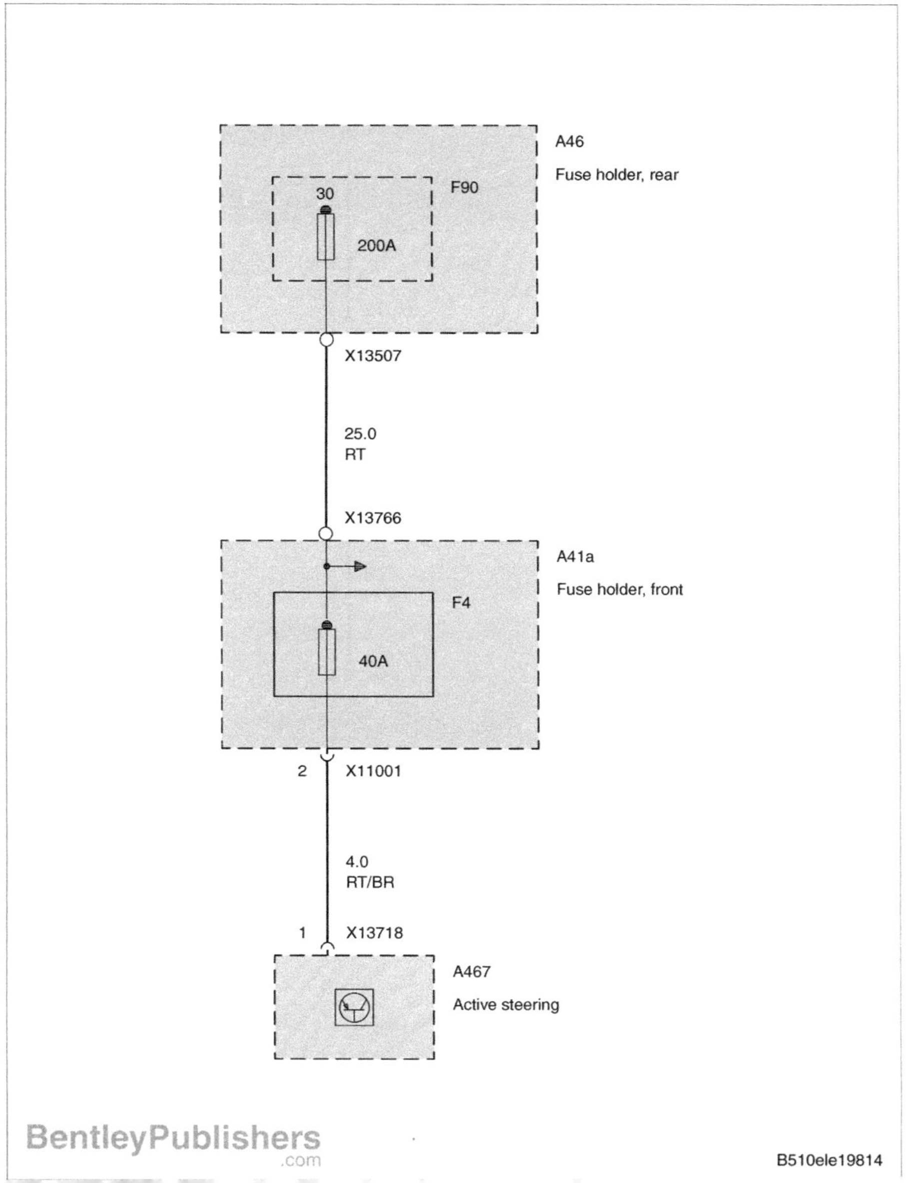

Fuse F4
as of 09 / 2005

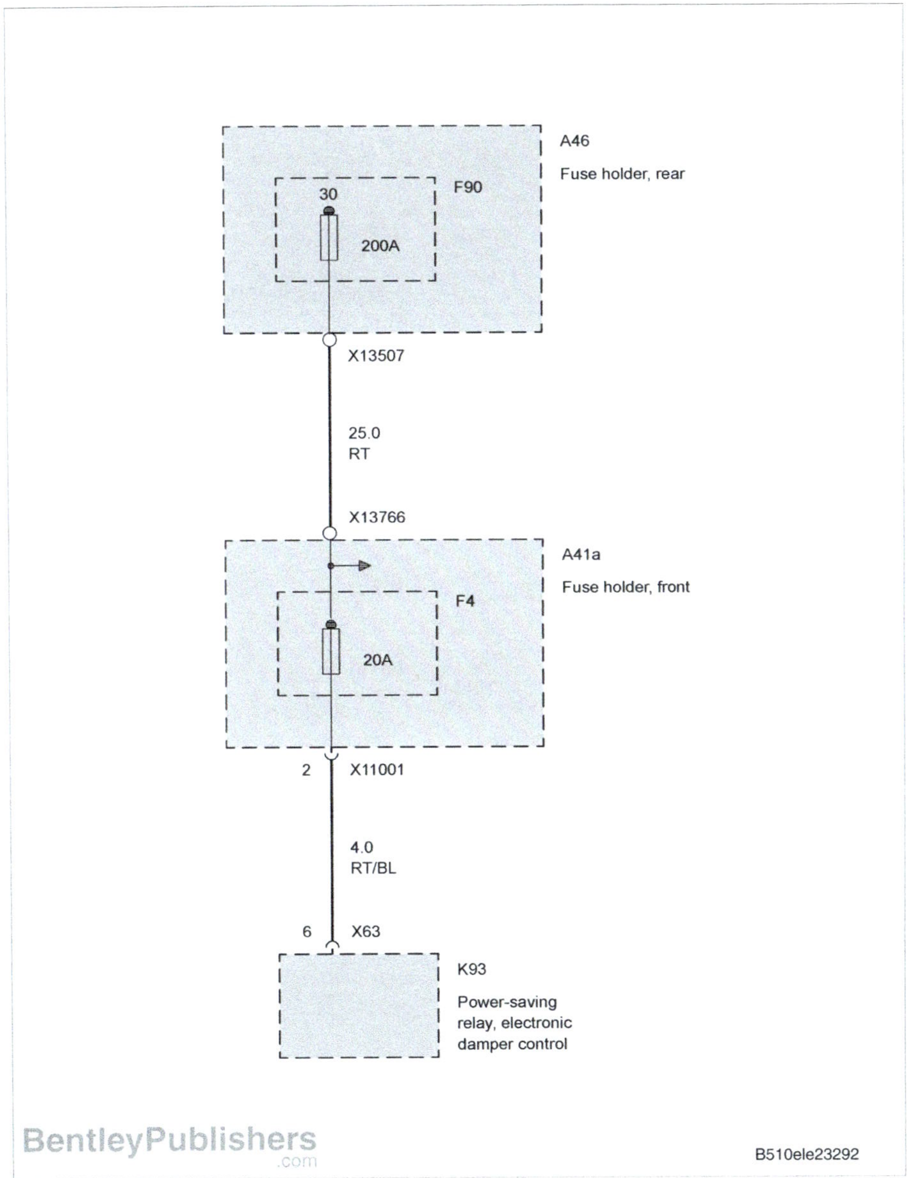

Fuse F5

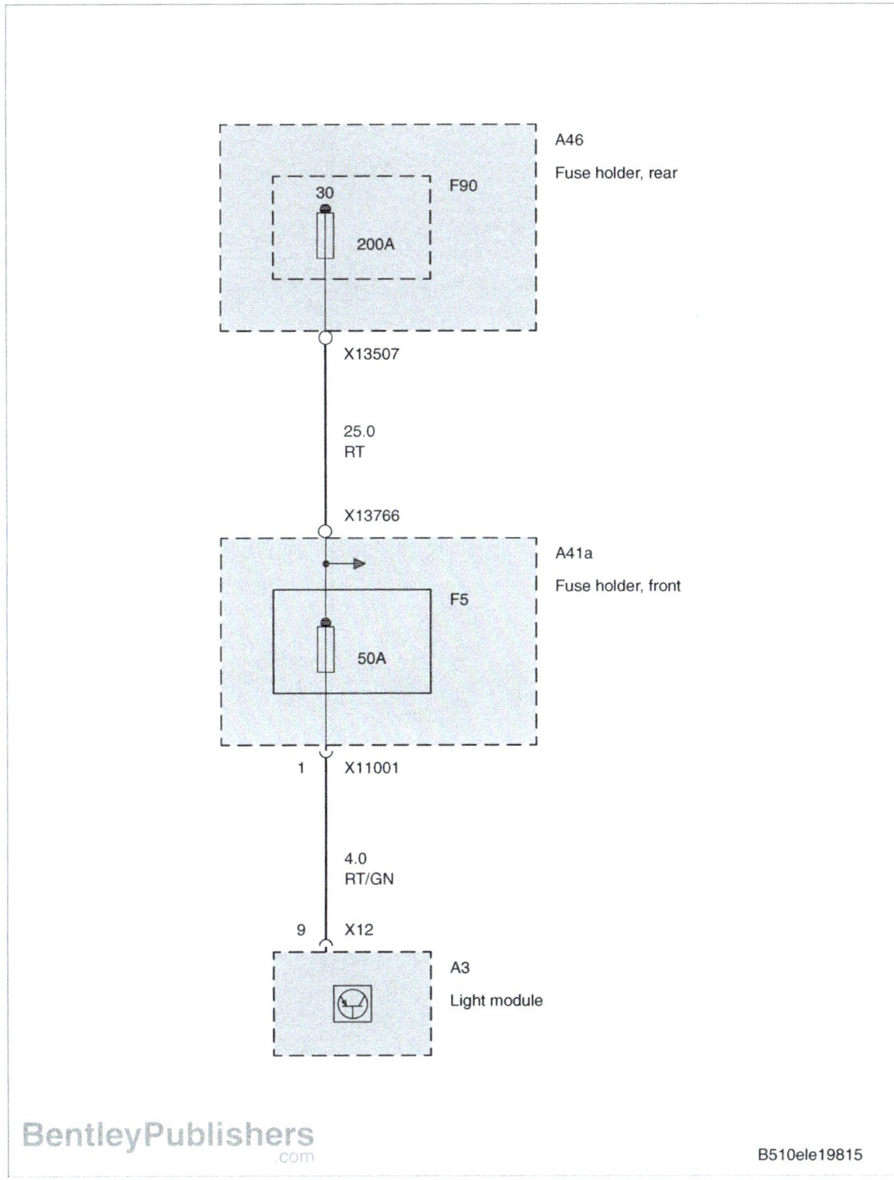

Fuse F6

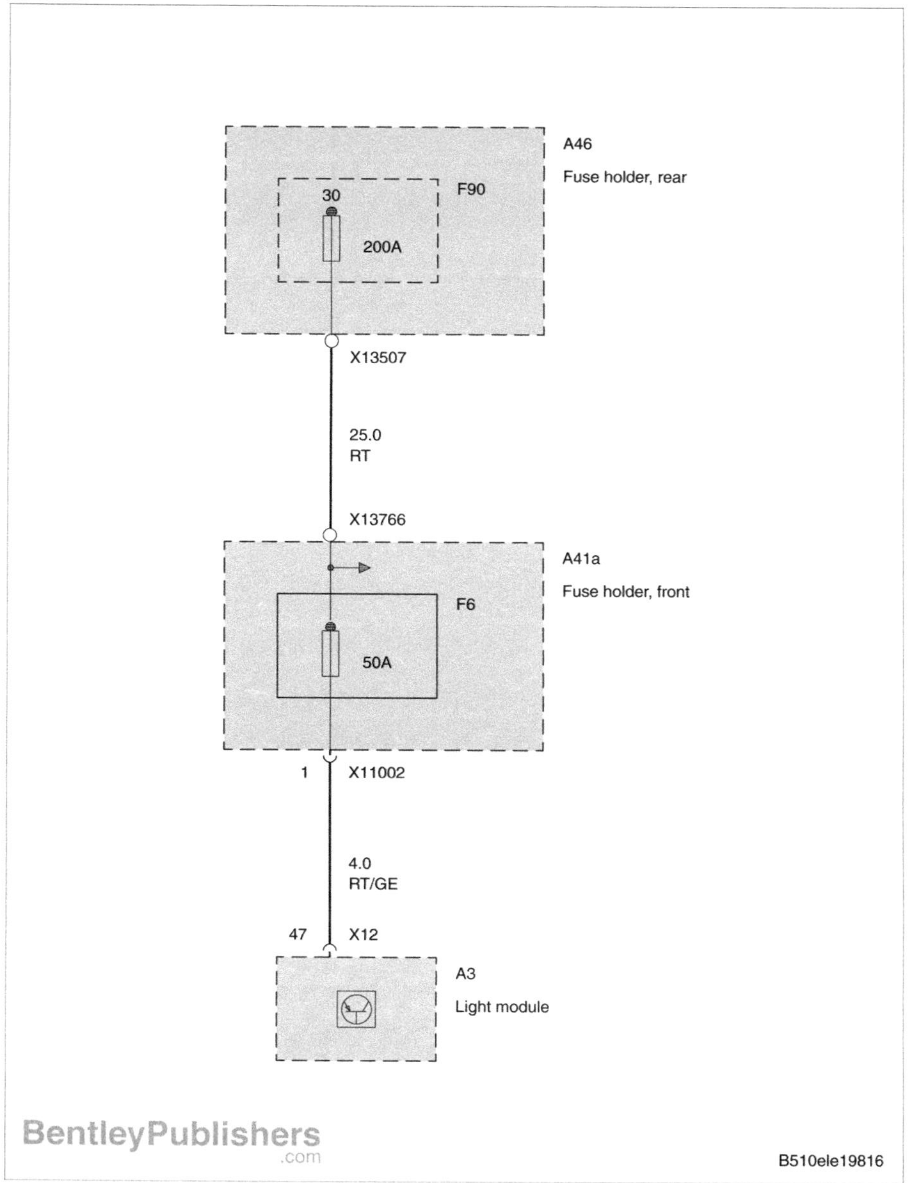

Fuse F7

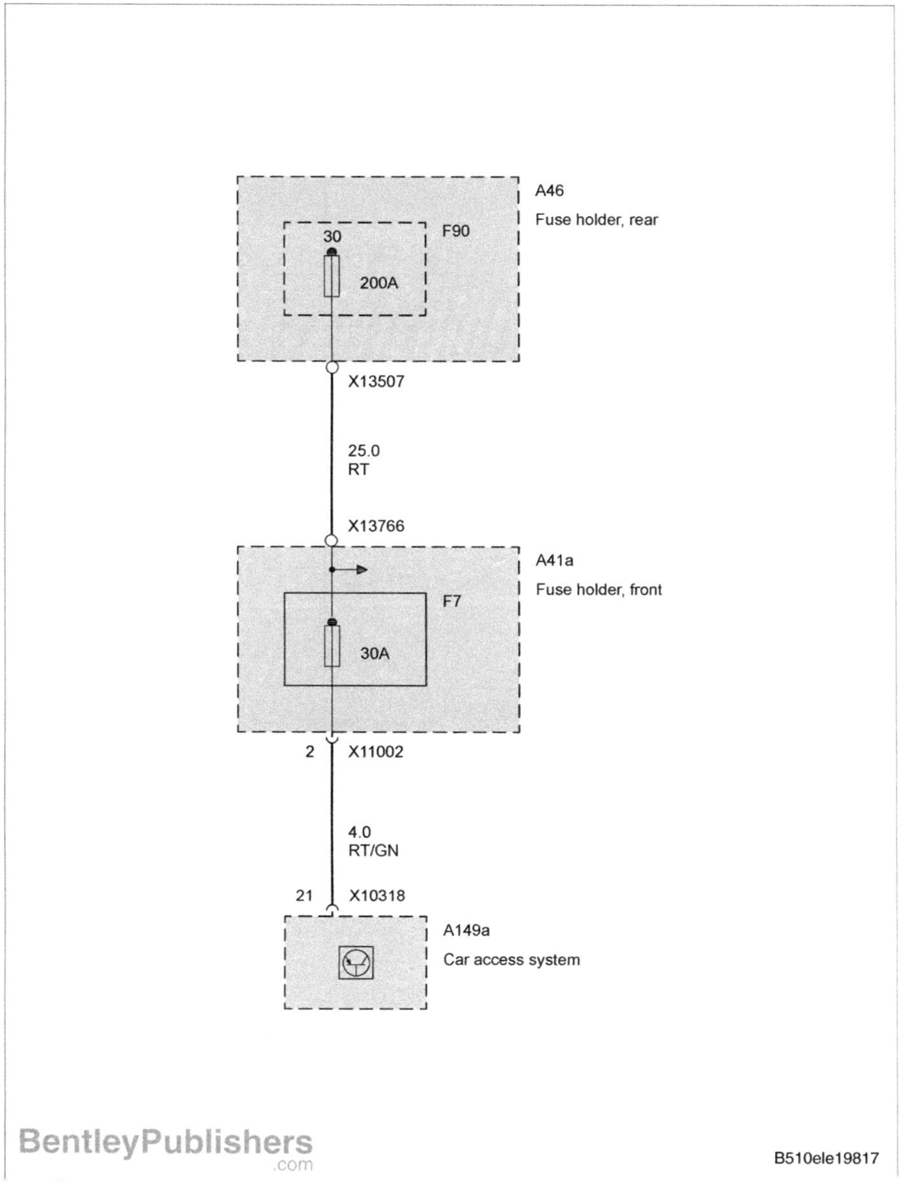

Fuse F8
M54 engine

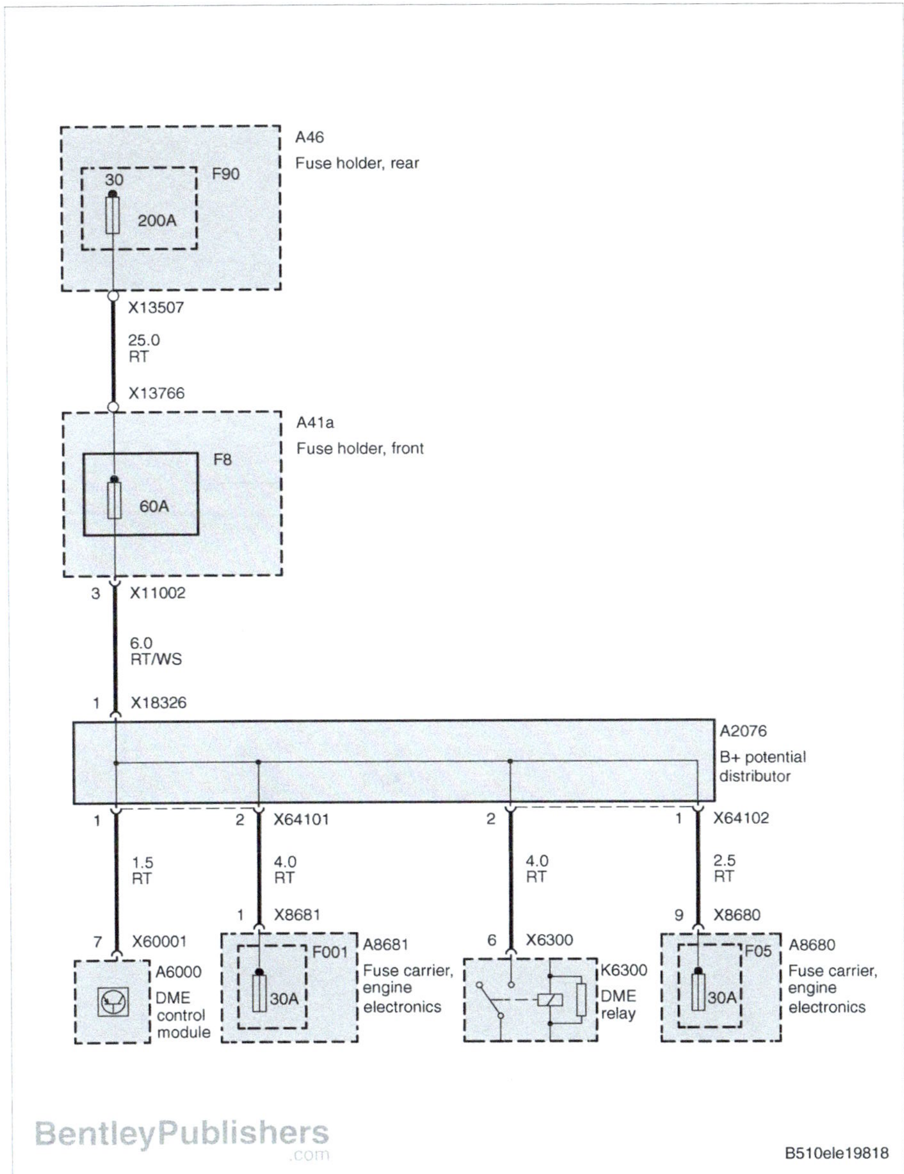

Fuse F8
N52 engine

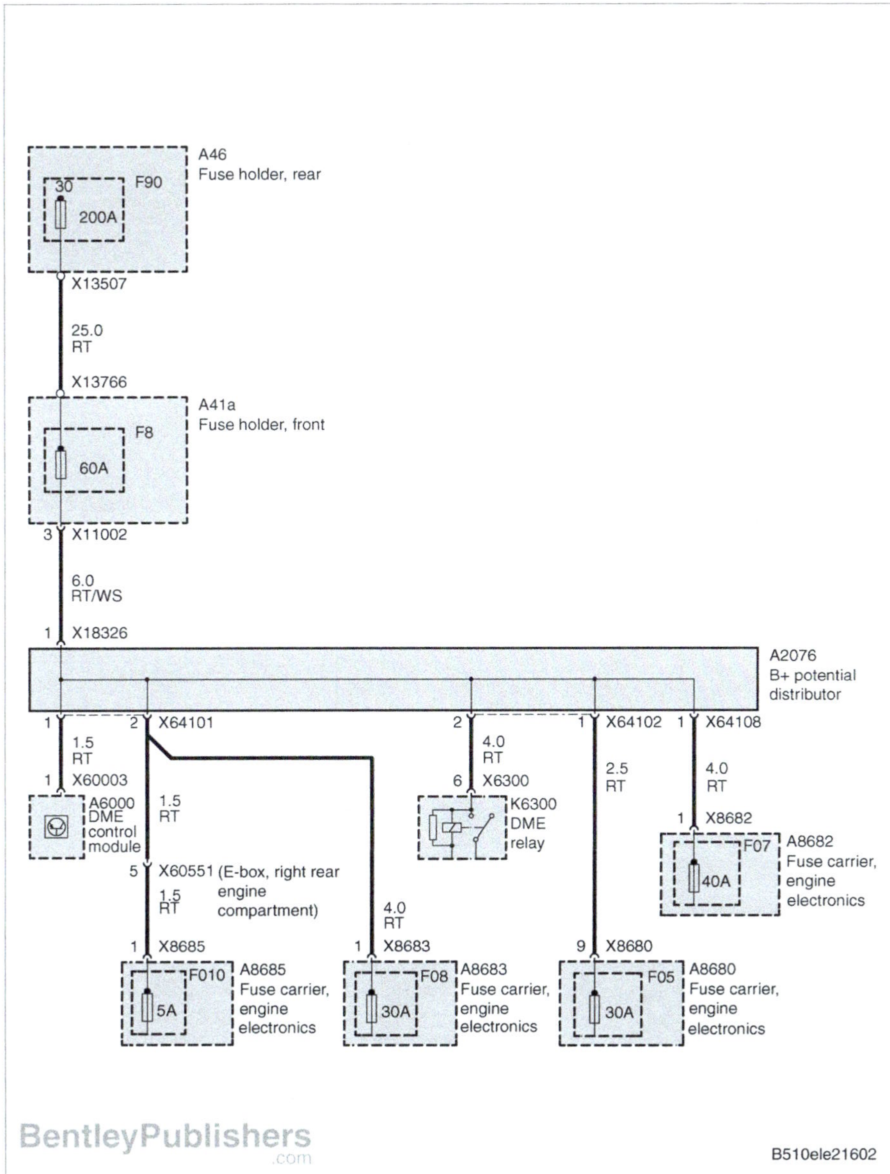

Fuse F8
N62 engine without SMG Sequential Manual Transmission

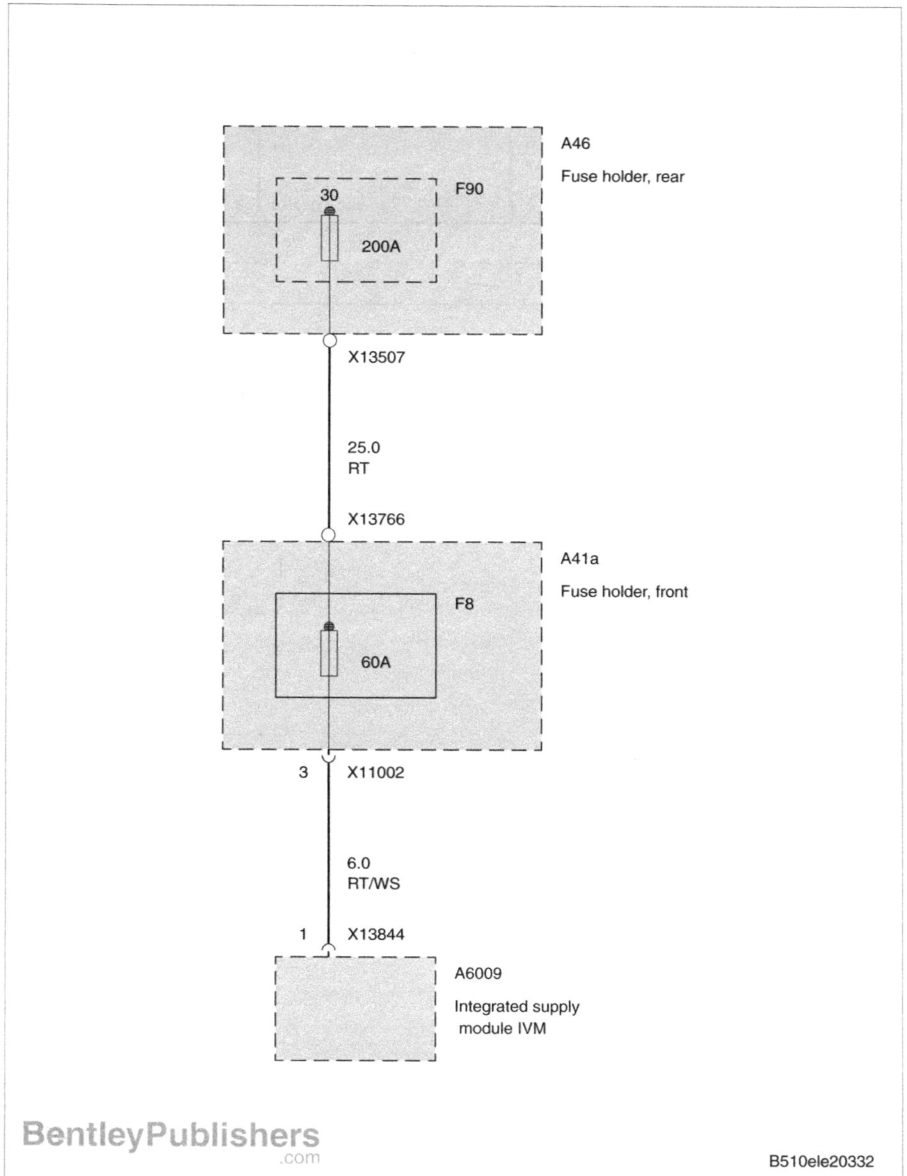

Fuse F8
N62 engine with SMG Sequential Manual Transmission

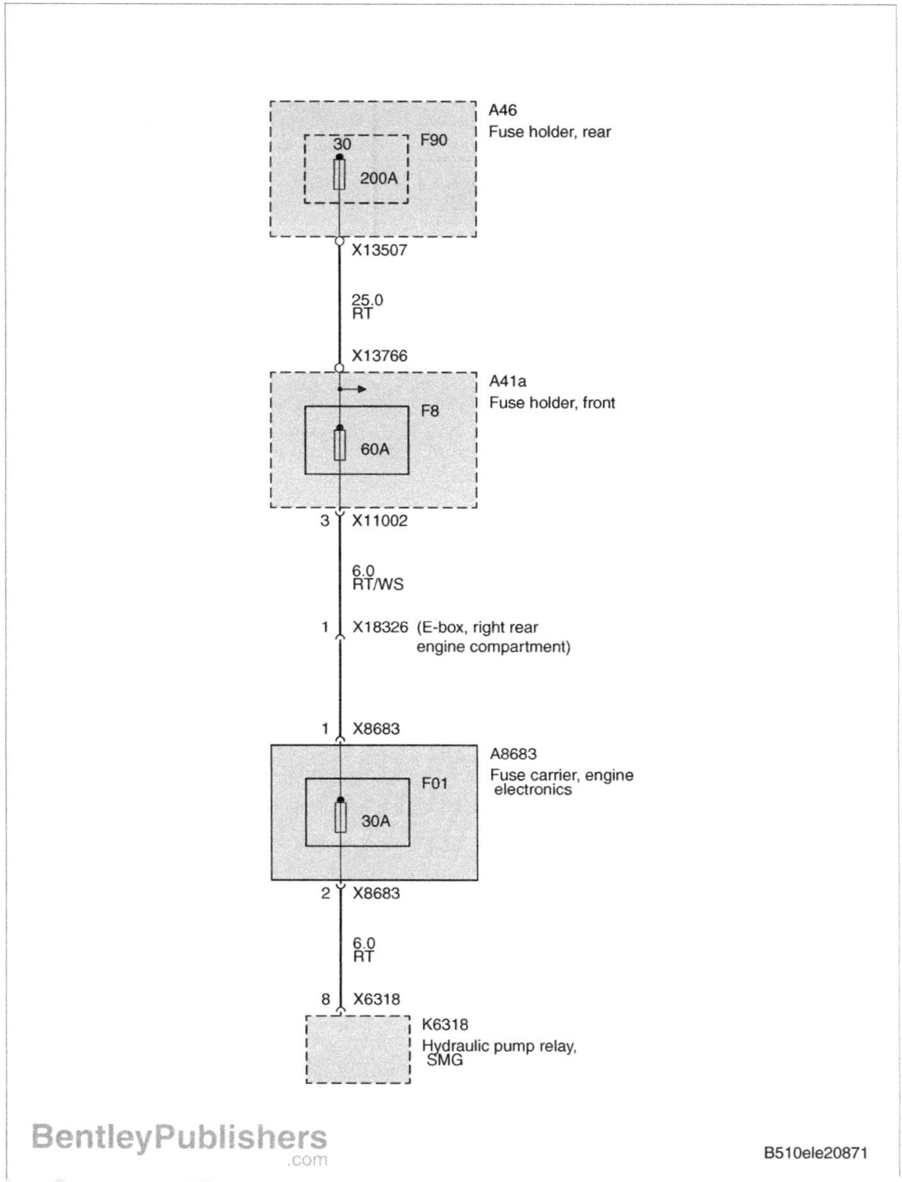

Fuse F9

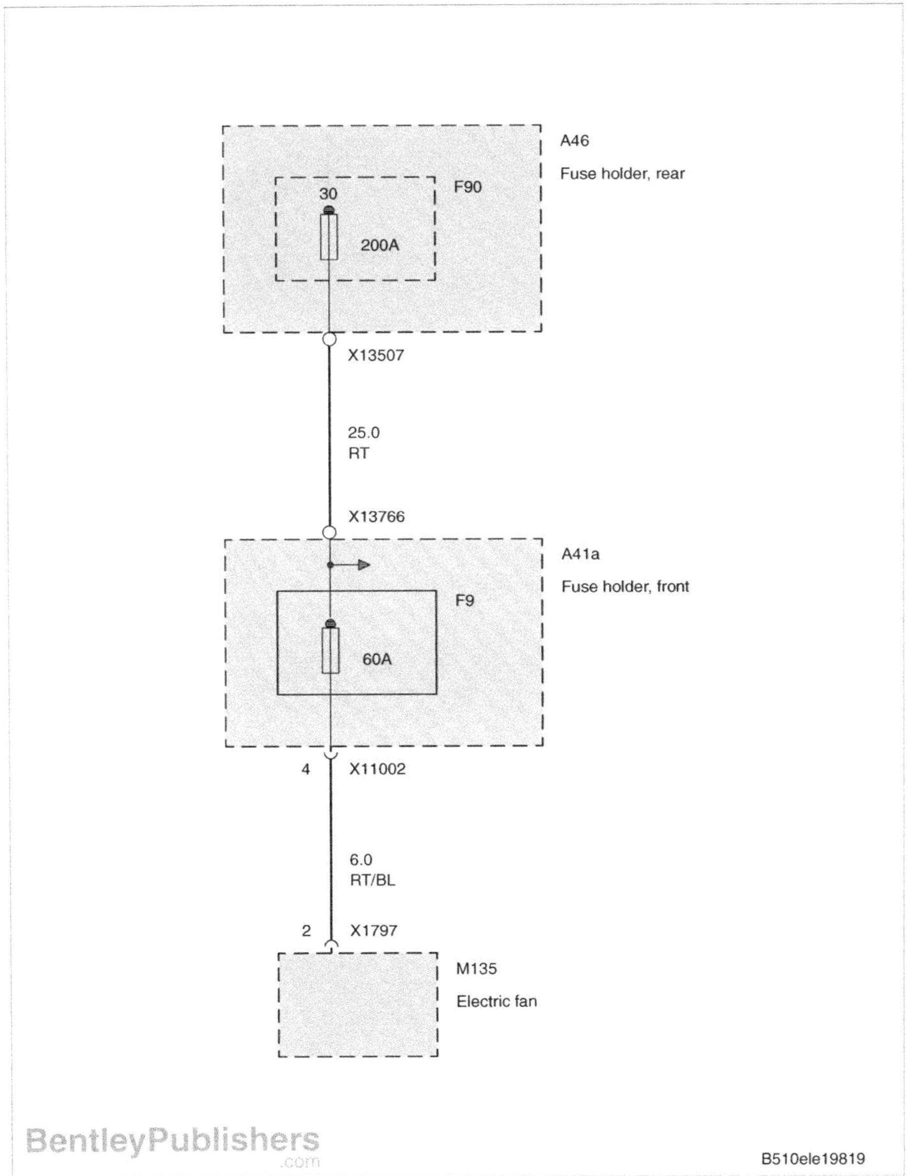

Fuse F10 up to 09 / 2005

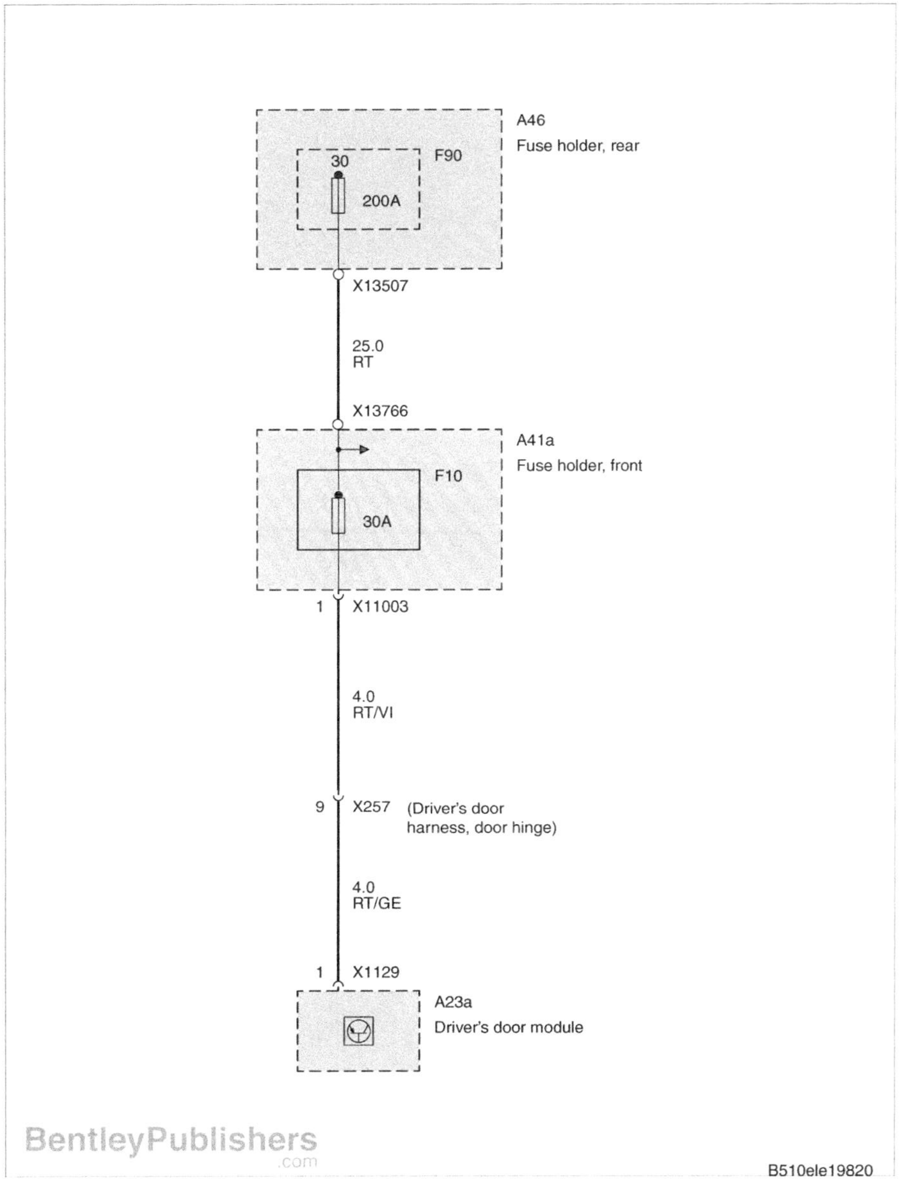

Fuse F10
as of 09 / 2005

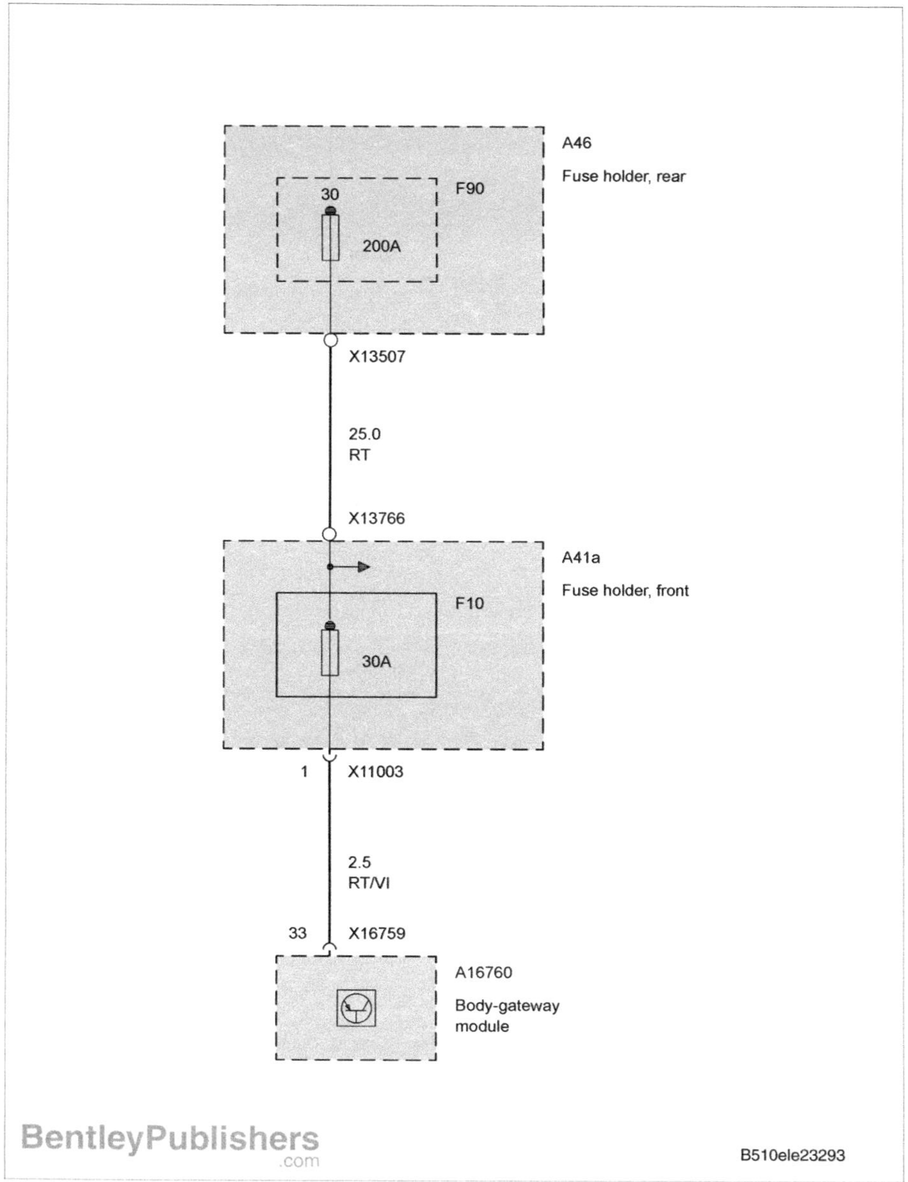

Fuse F11

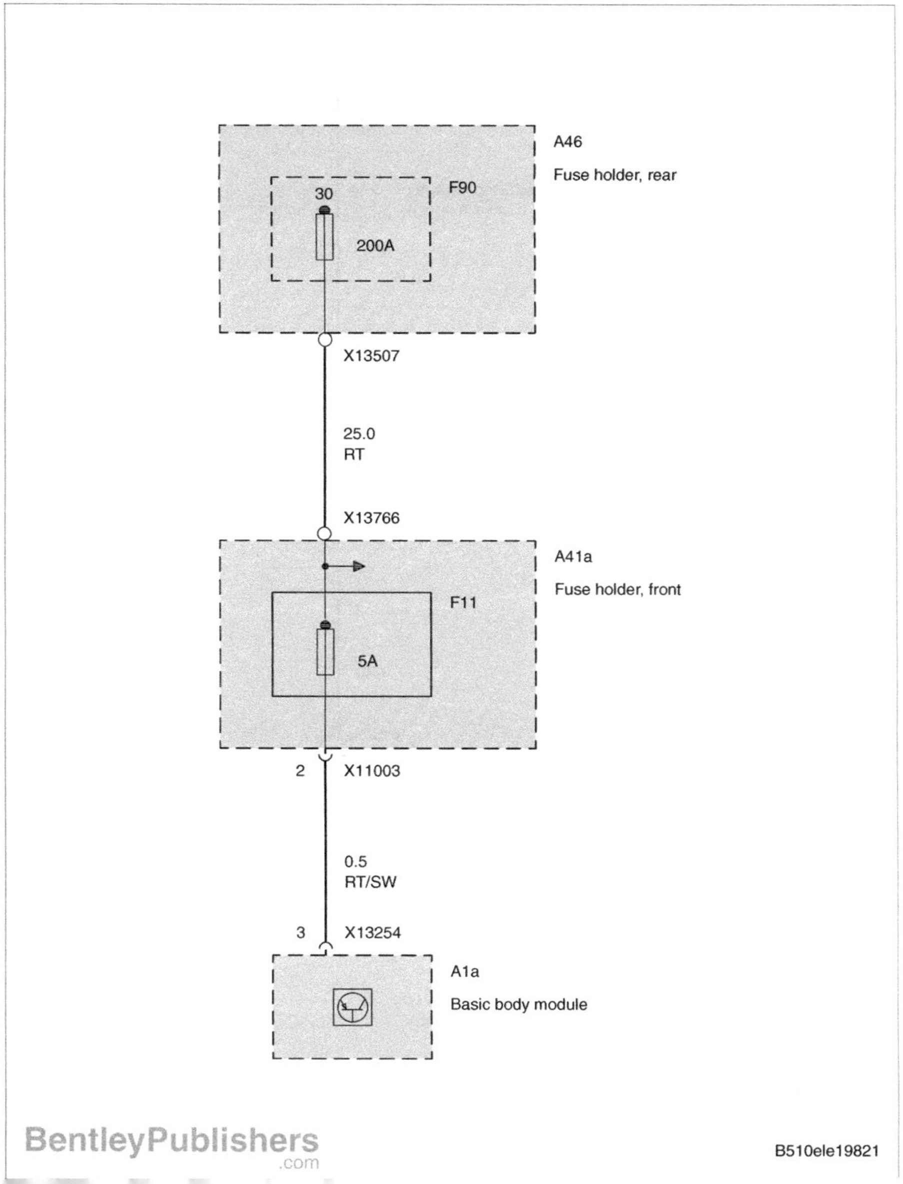

Fuse F12
up to 09 / 2005

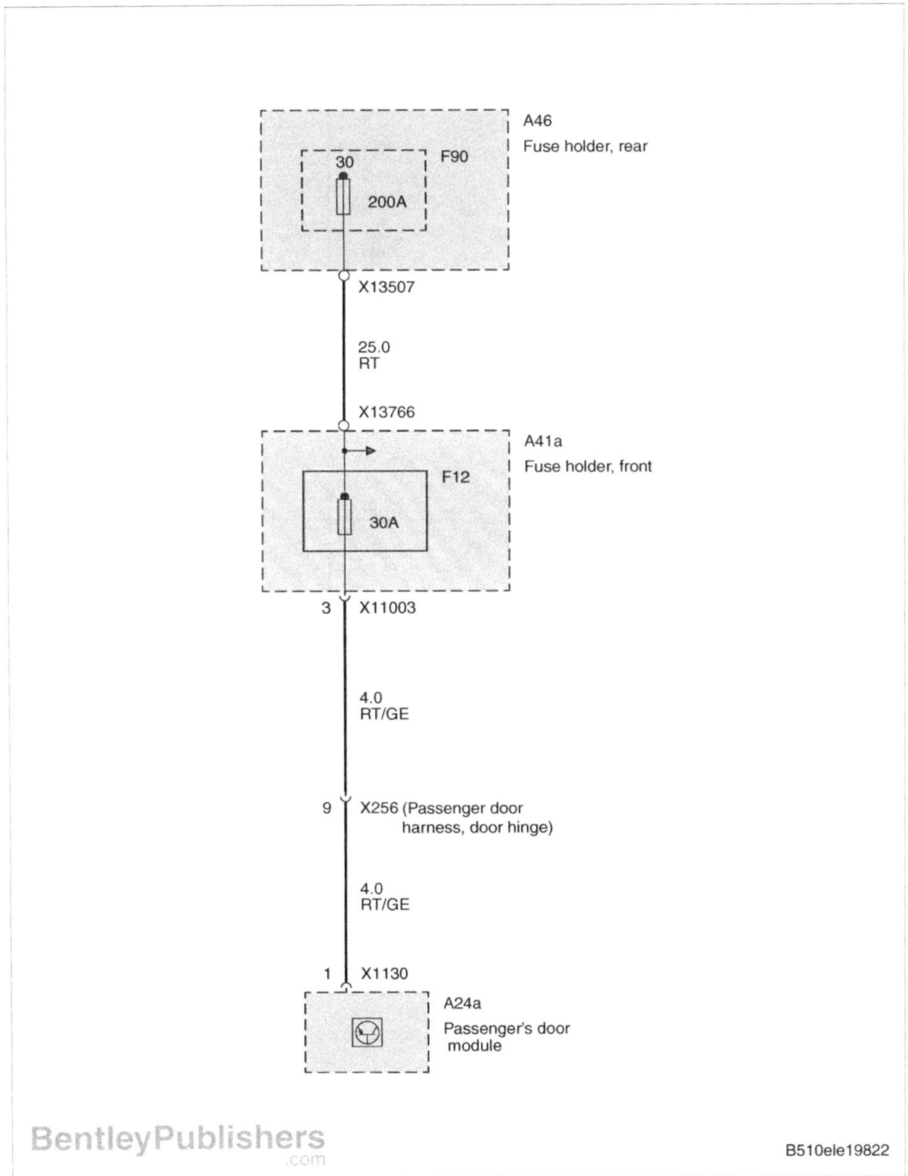

Fuse F12
as of 09 / 2005

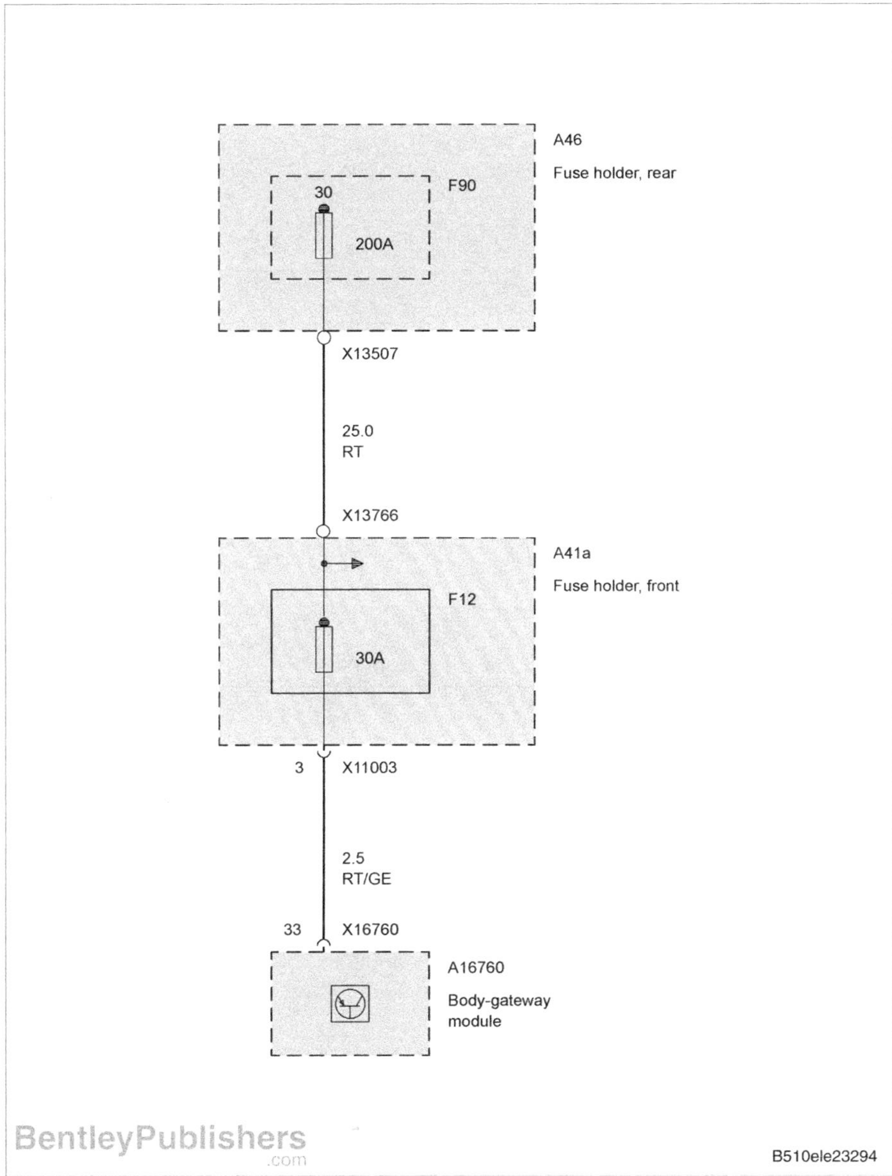

Fuse F13
up to 09 / 2005

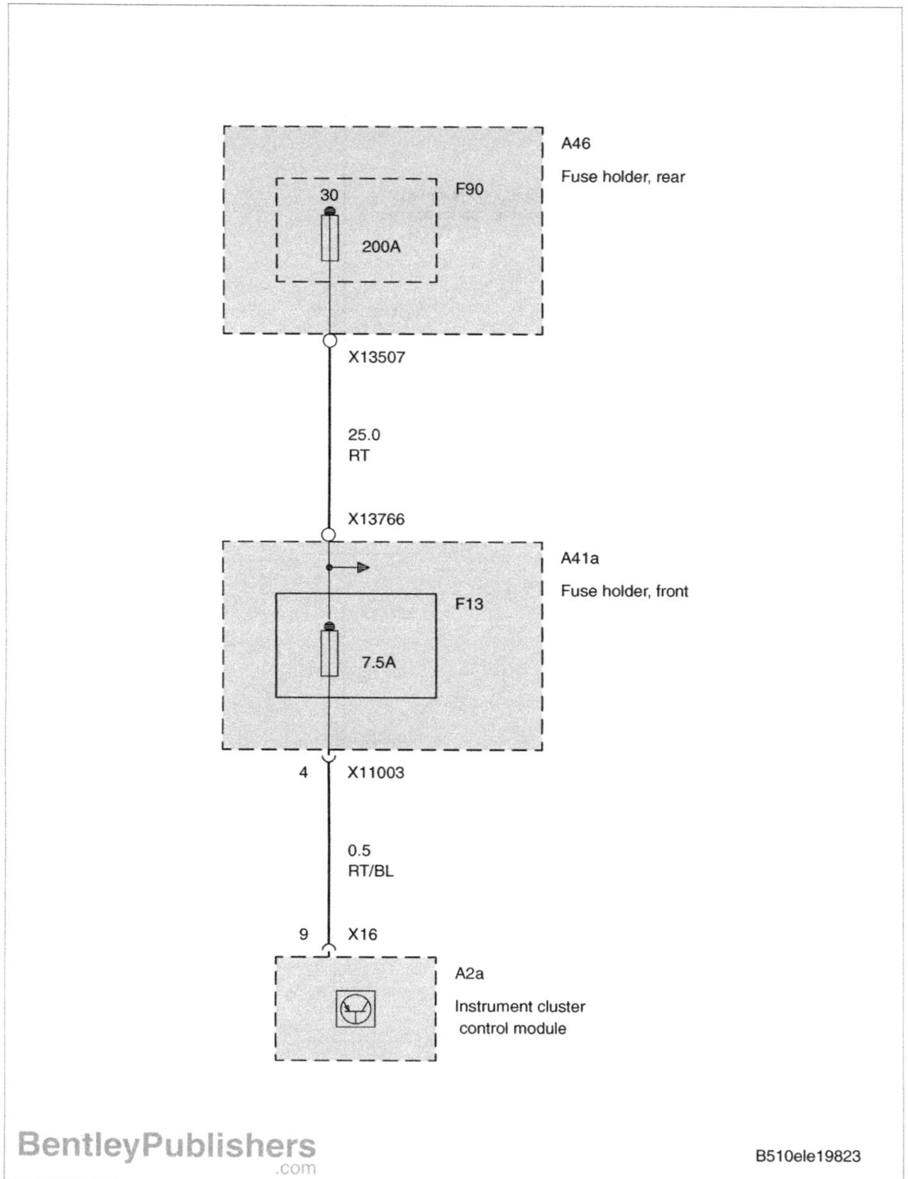

Fuse F13
VTG Transfer case as of 09 / 2005

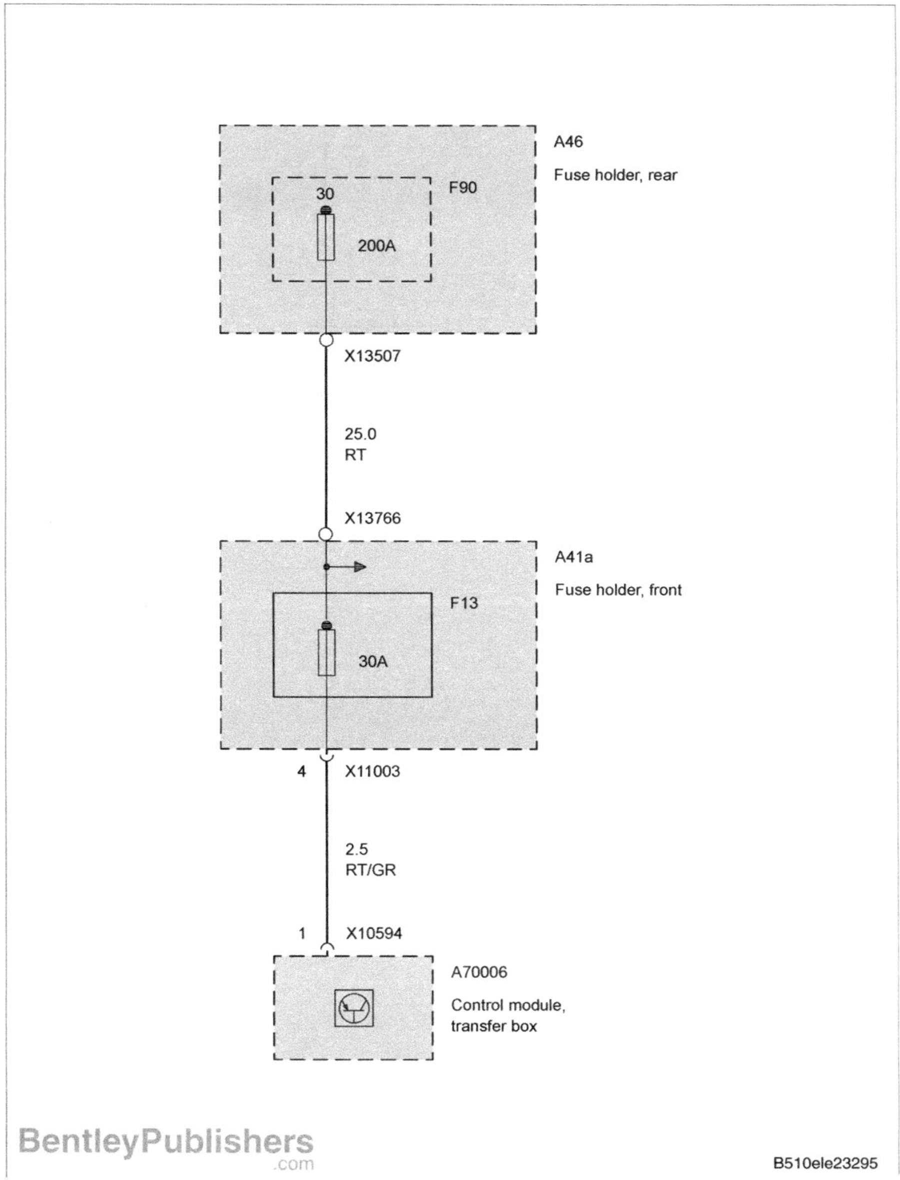

Fuse F14

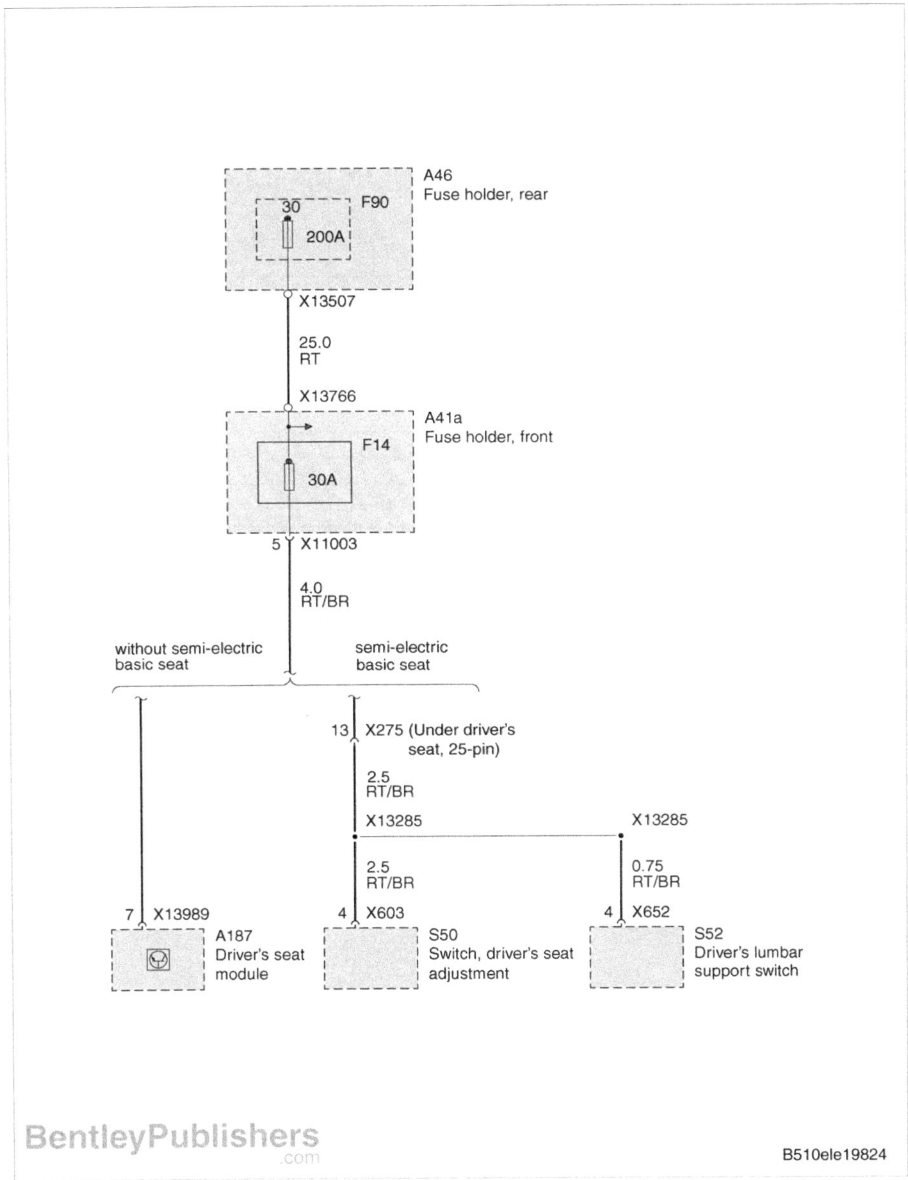

Fuse F15

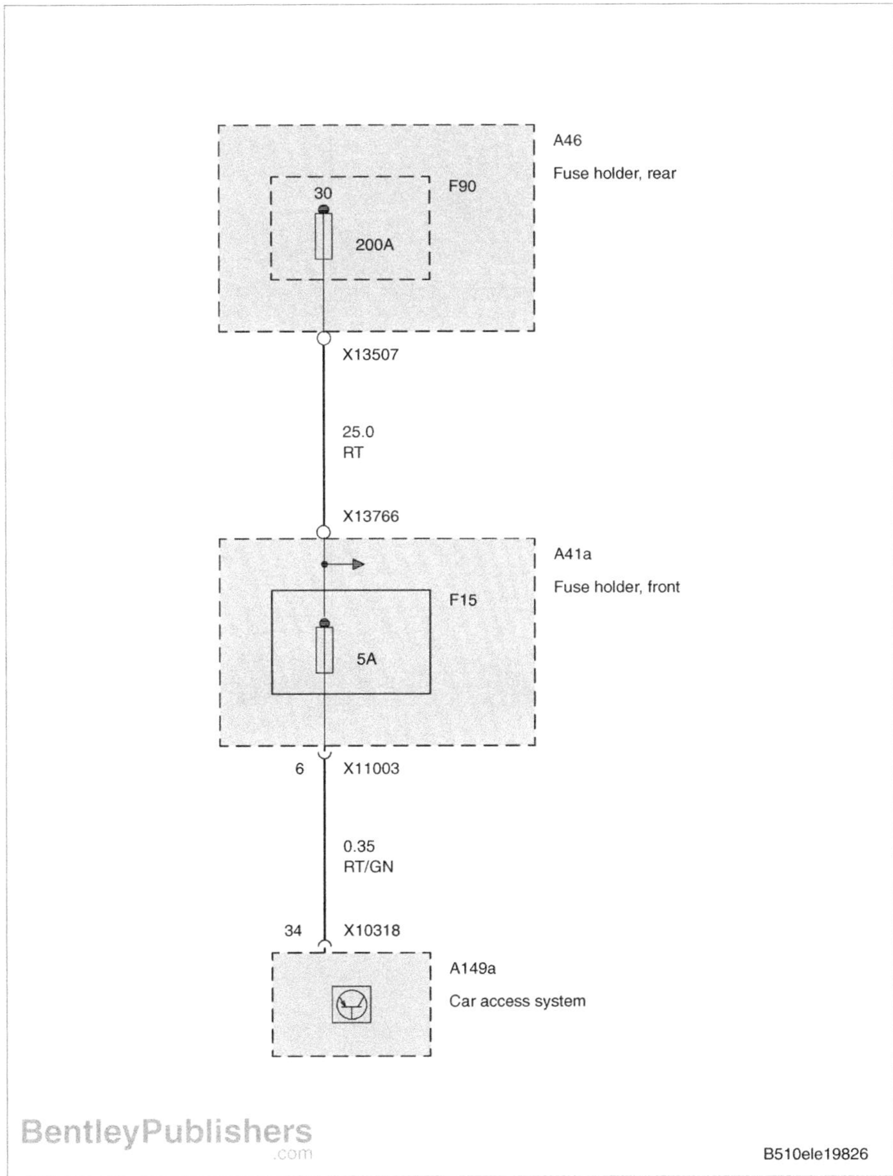

Fuse F16

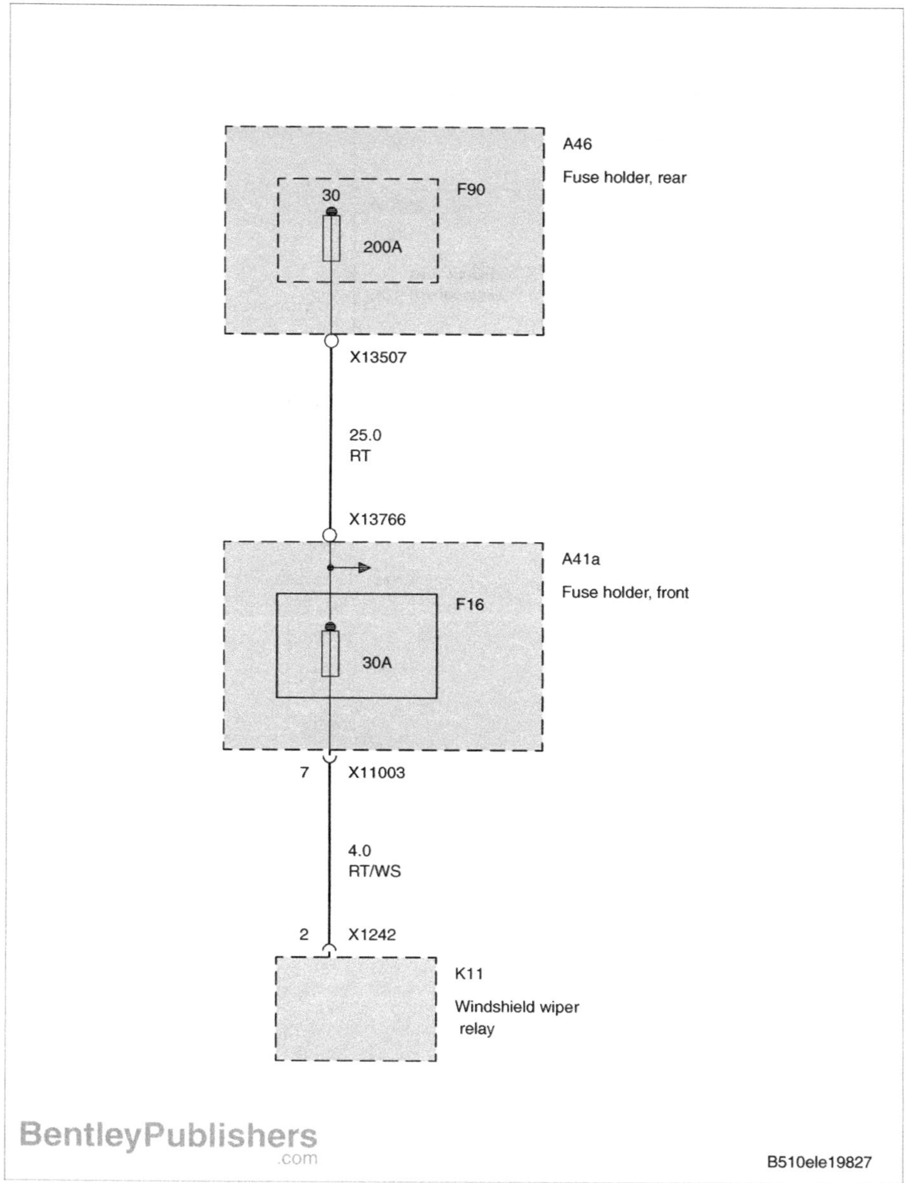

Fuse F17
as of 09 / 2005 up to 09 / 2007

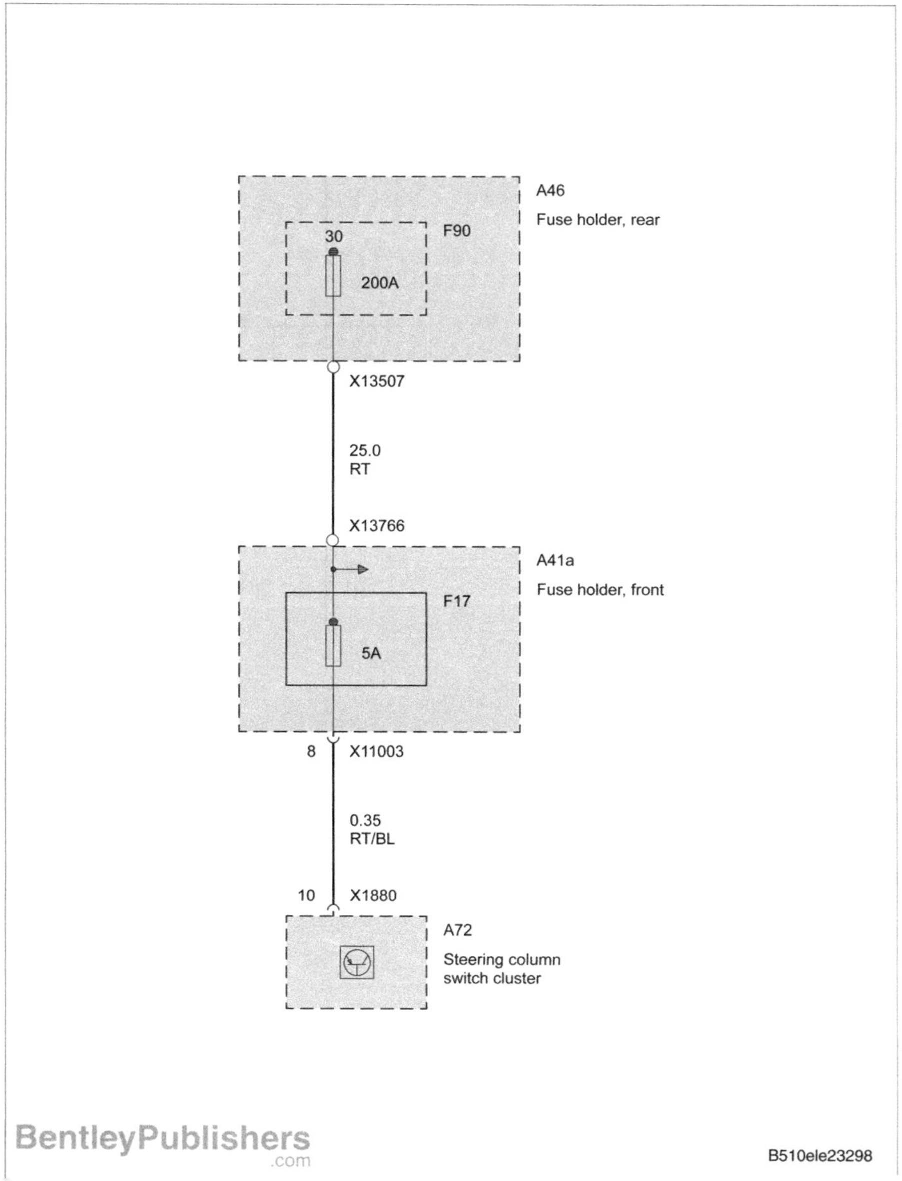

Fuse F17
as of 09 / 2007

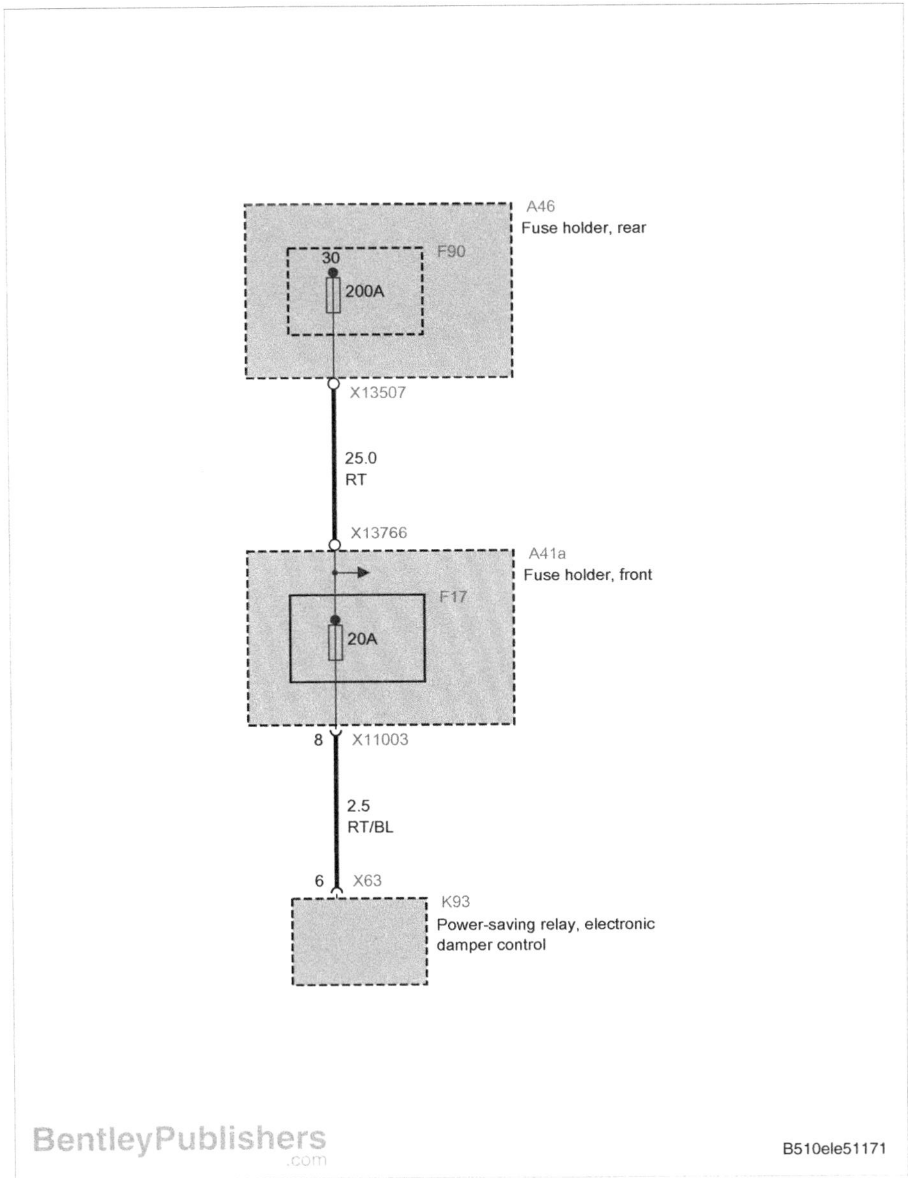

Fuse F18
up to 09 / 2005

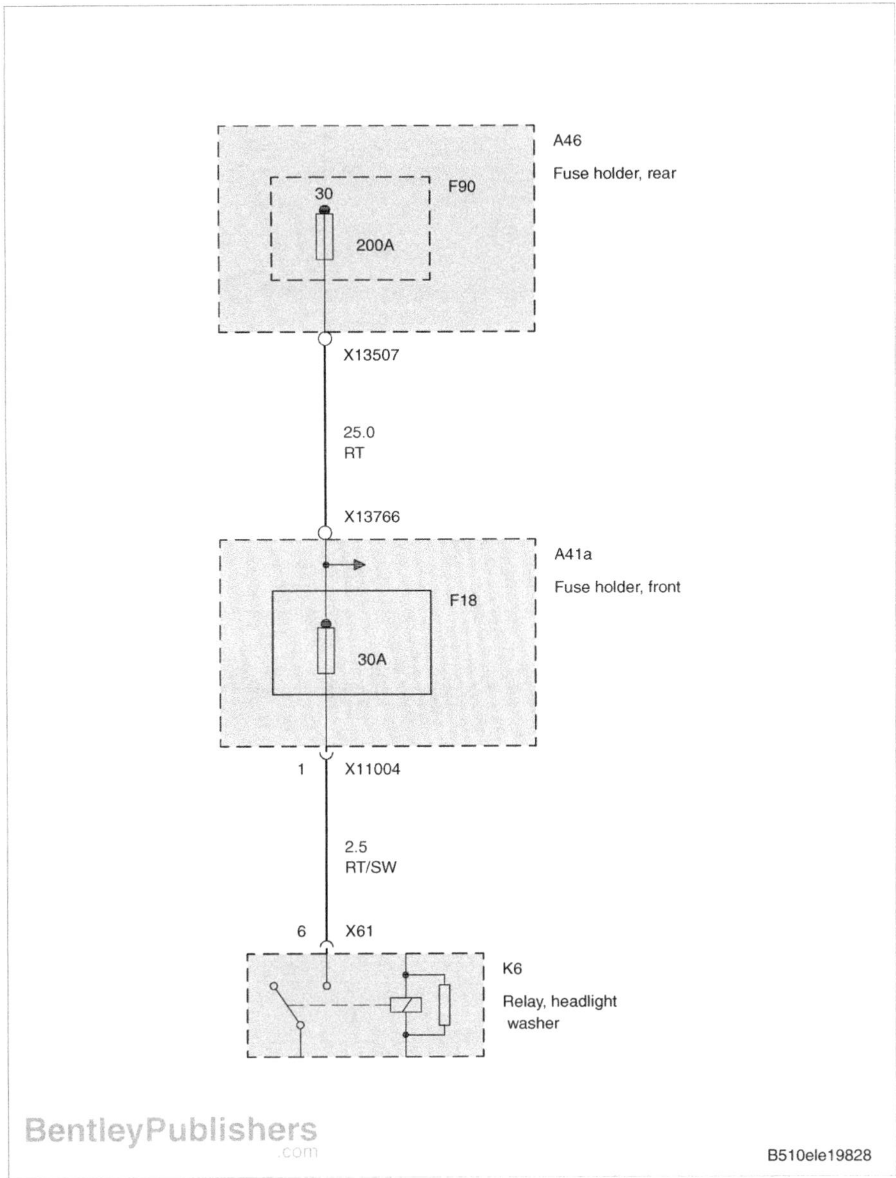

Fuse F18
as of 09 / 2005

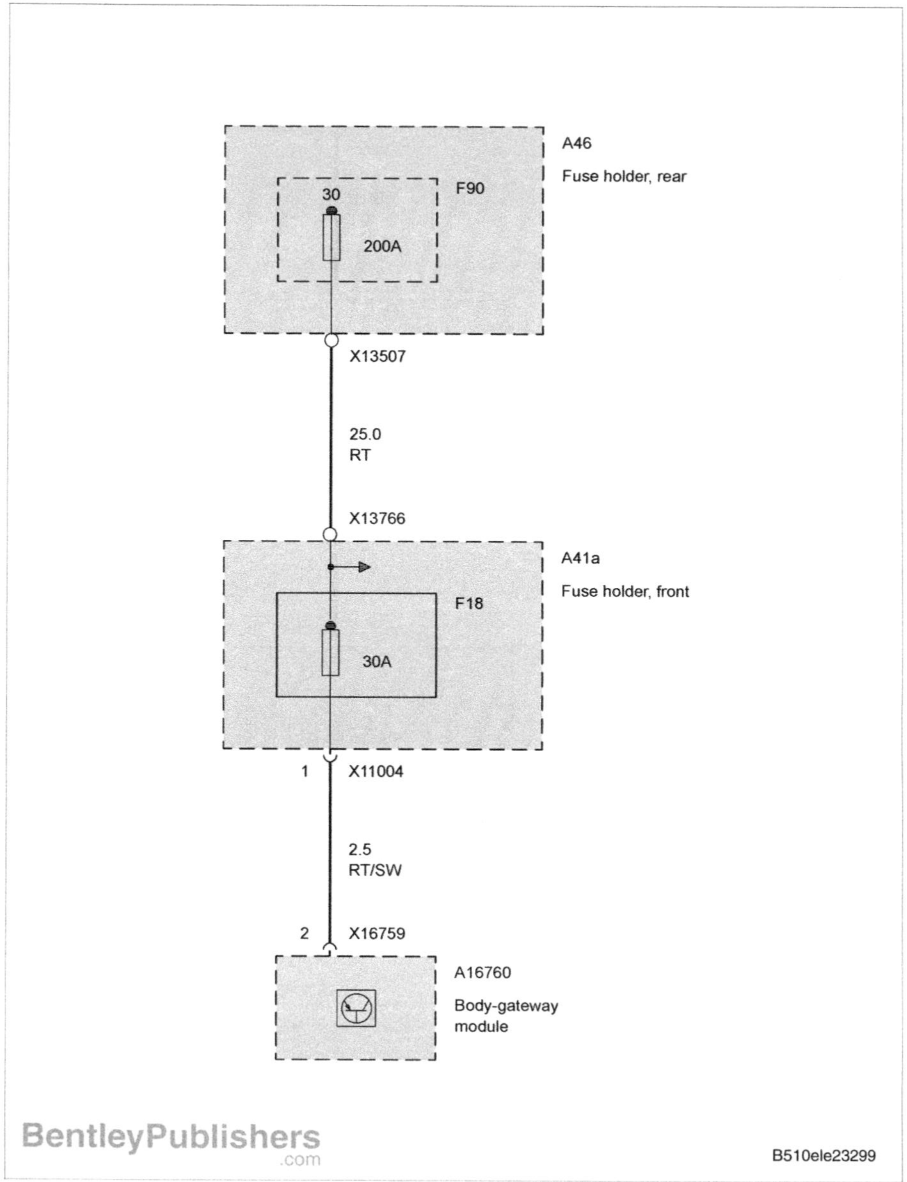

Fuse F19

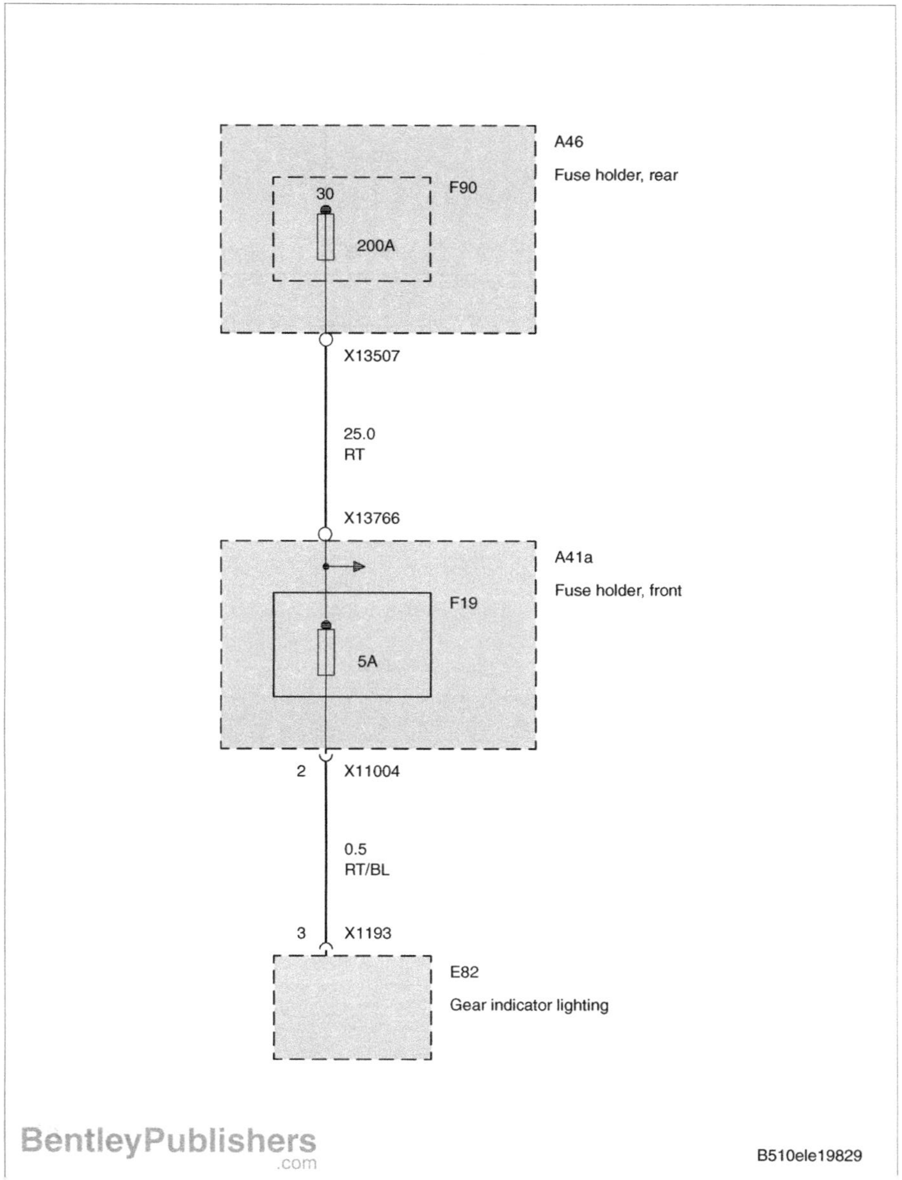

Fuse F20
SHZH Independent / auxiliary heater

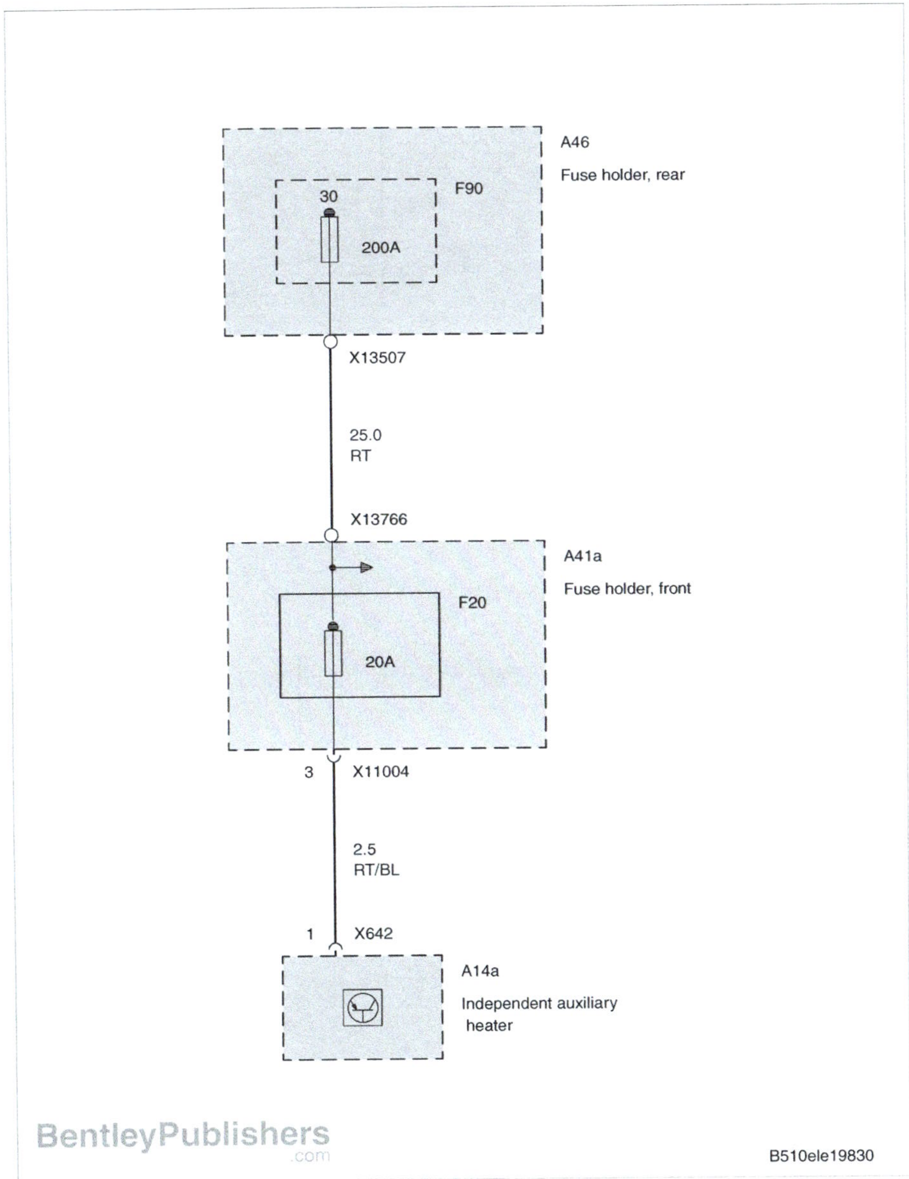

Fuse F20
EDC Electronic Damper Control

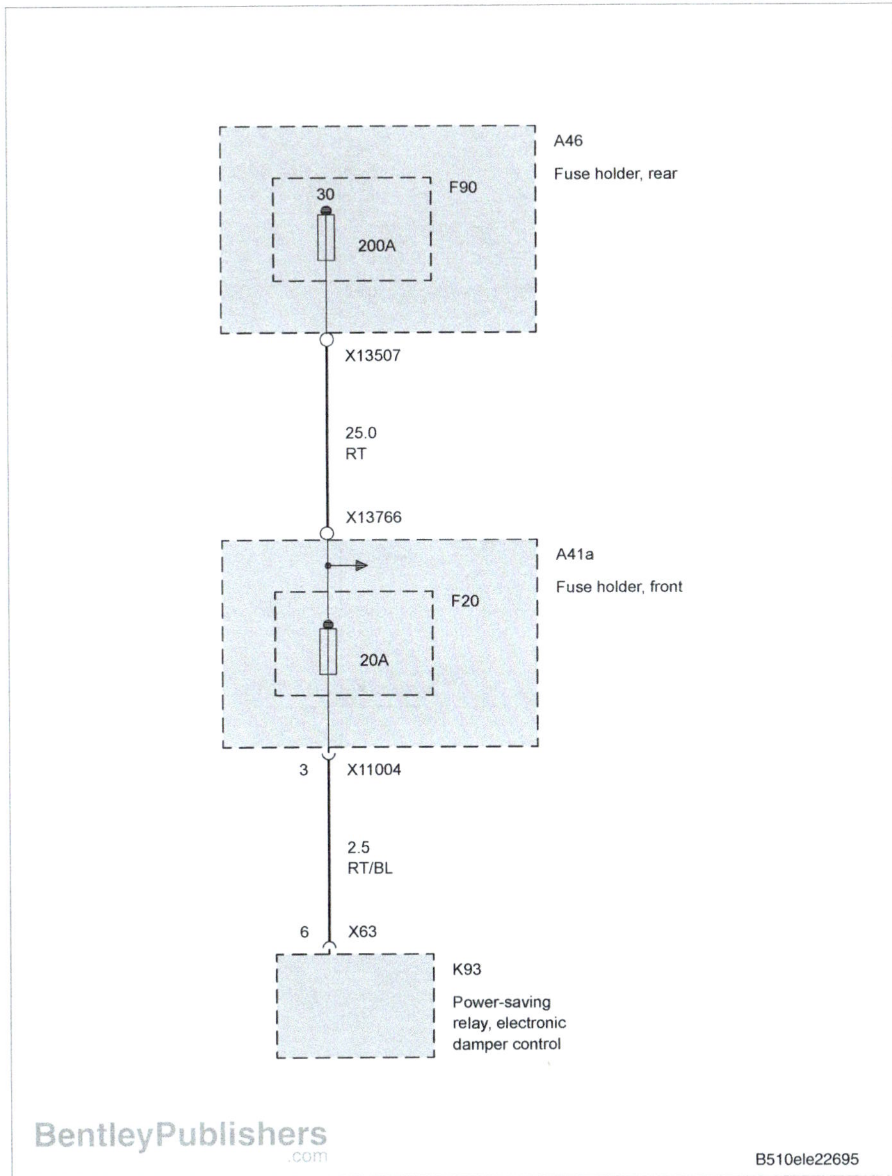

Fuse F21

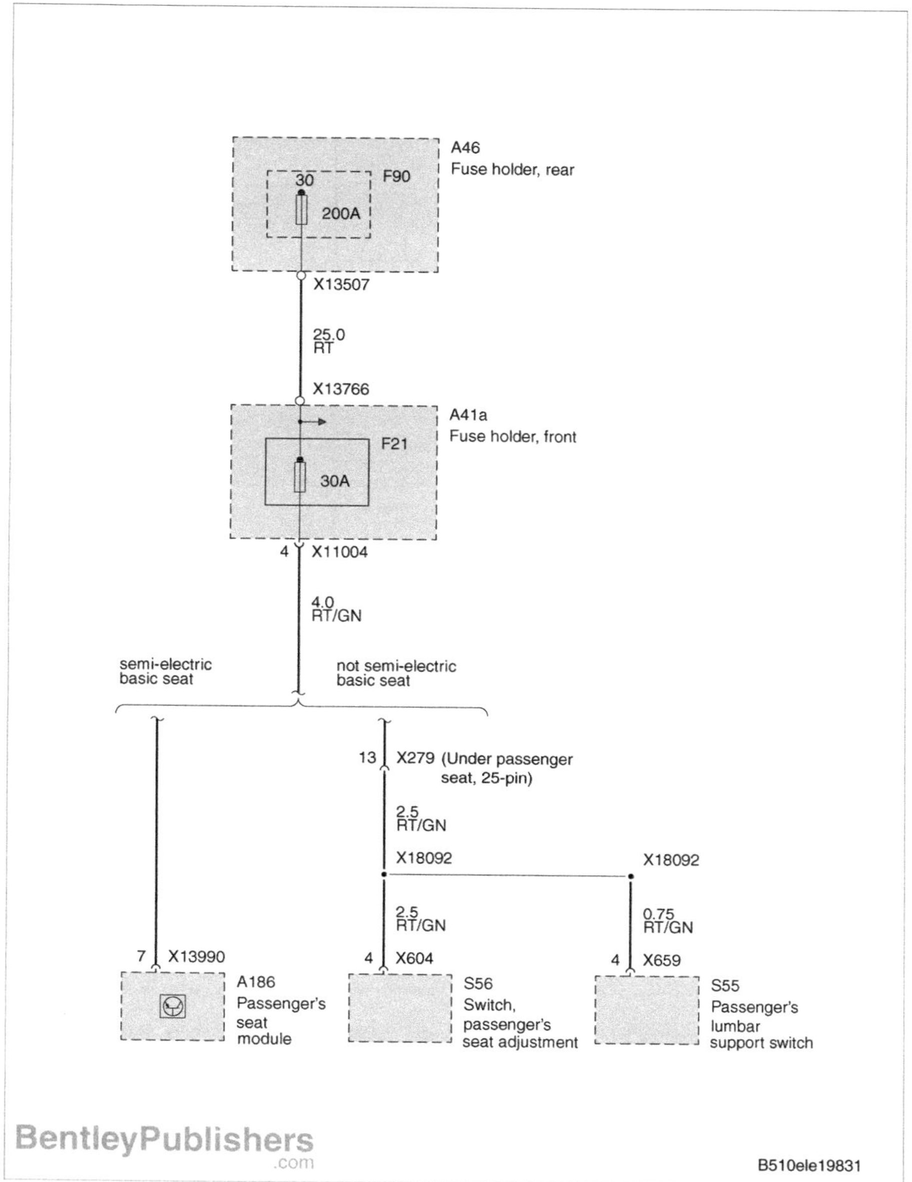

Fuse F22

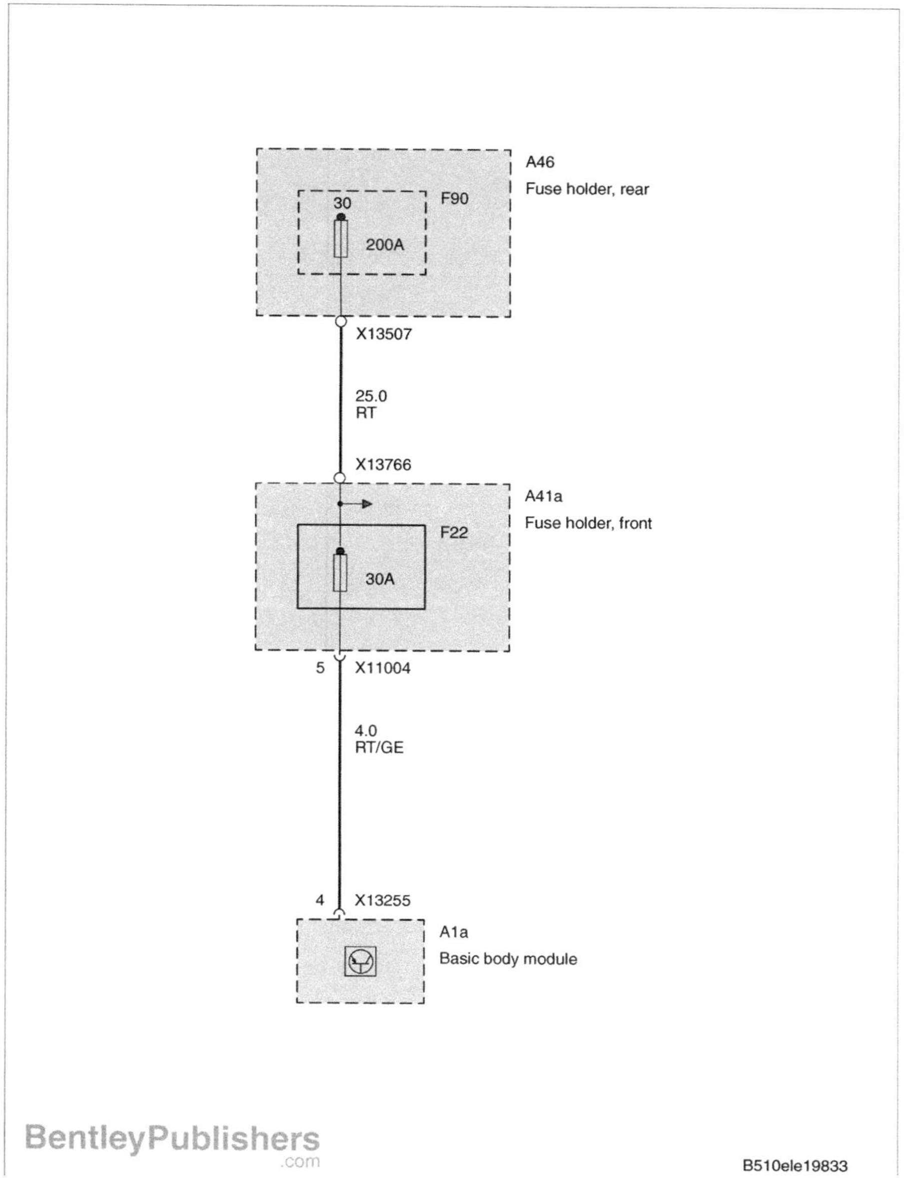

Fuse F24

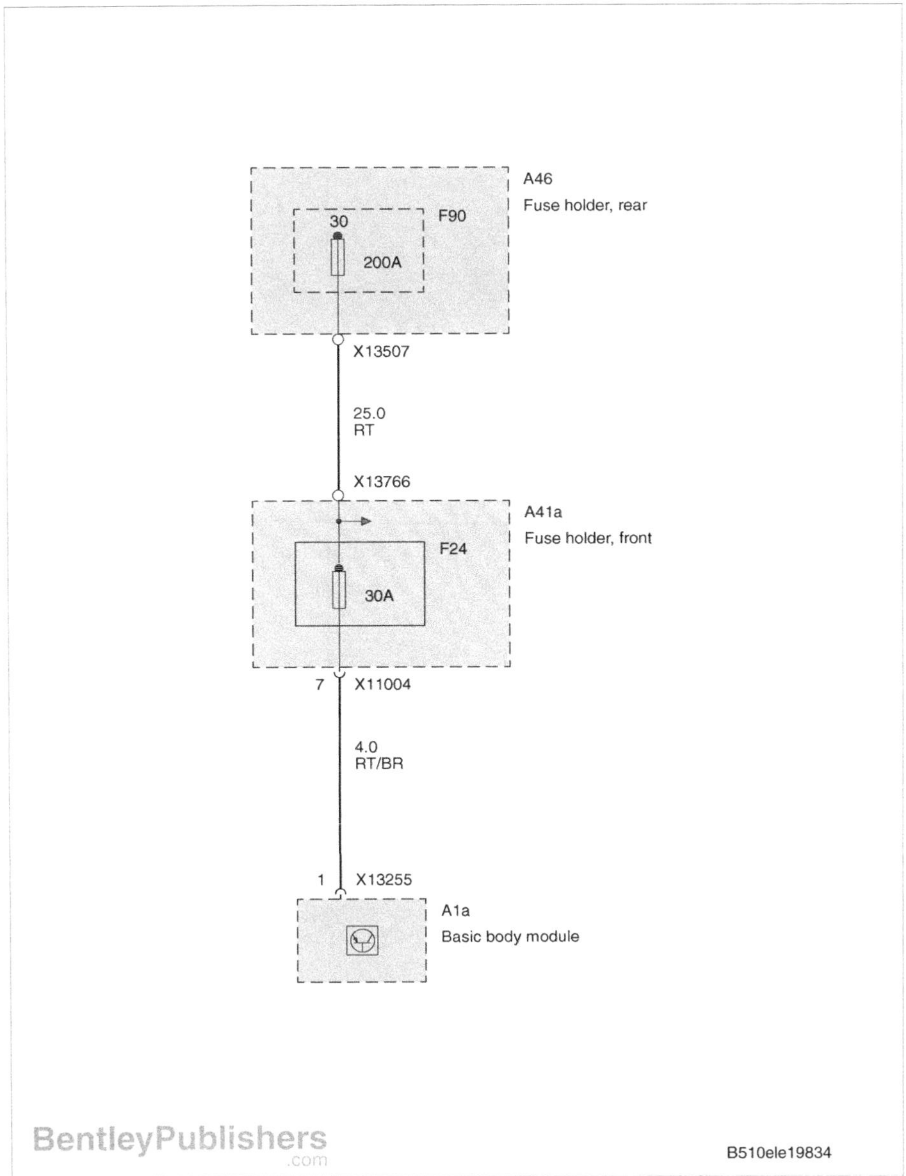

Fuse F25

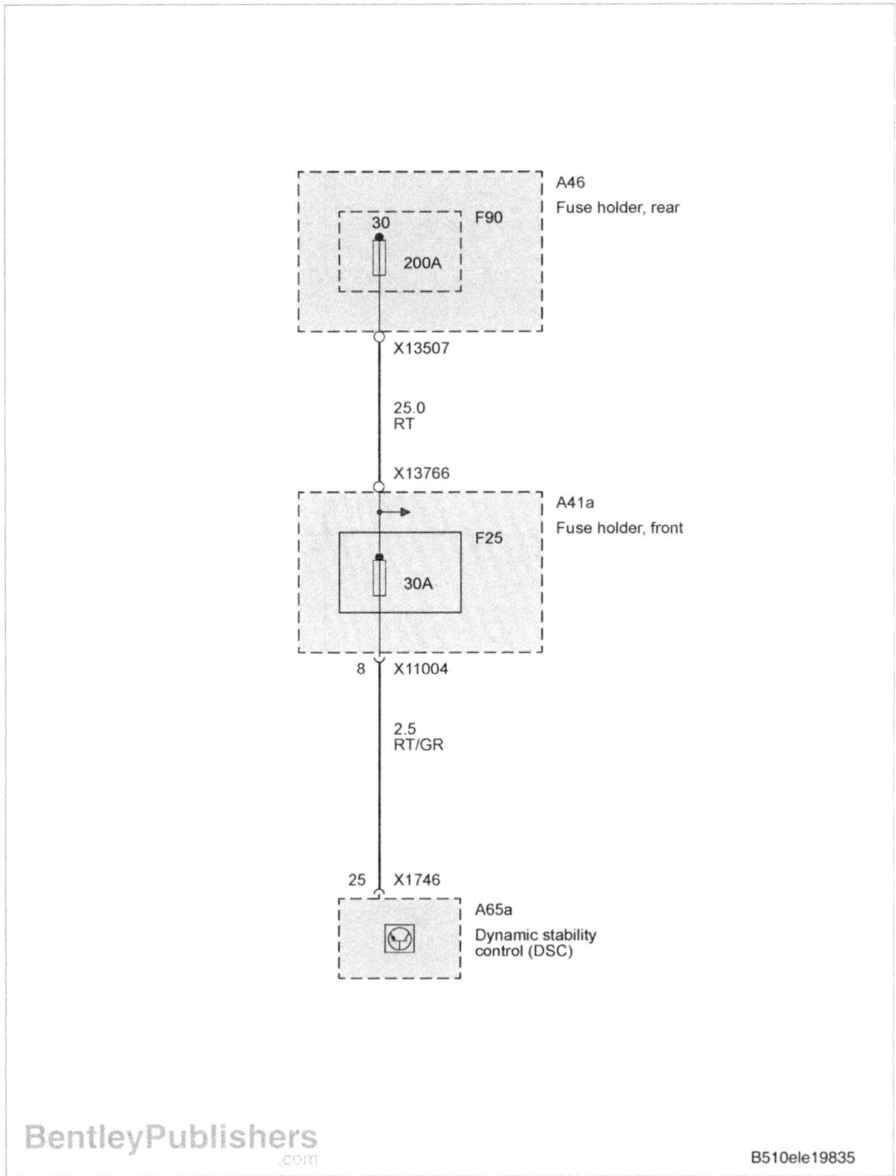

Fuse F26
IHKA Basic up to 09 / 2005

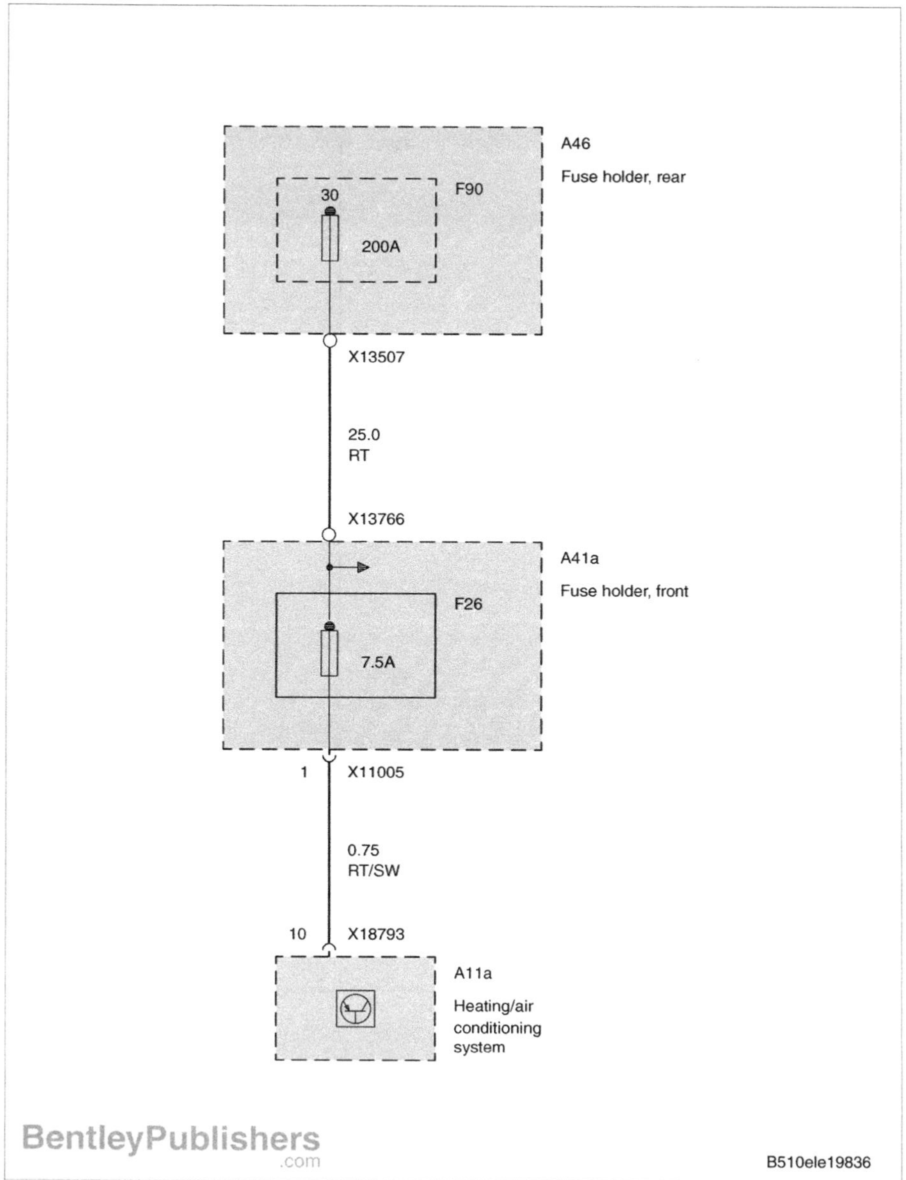

Fuse F26
as of 09 / 2005

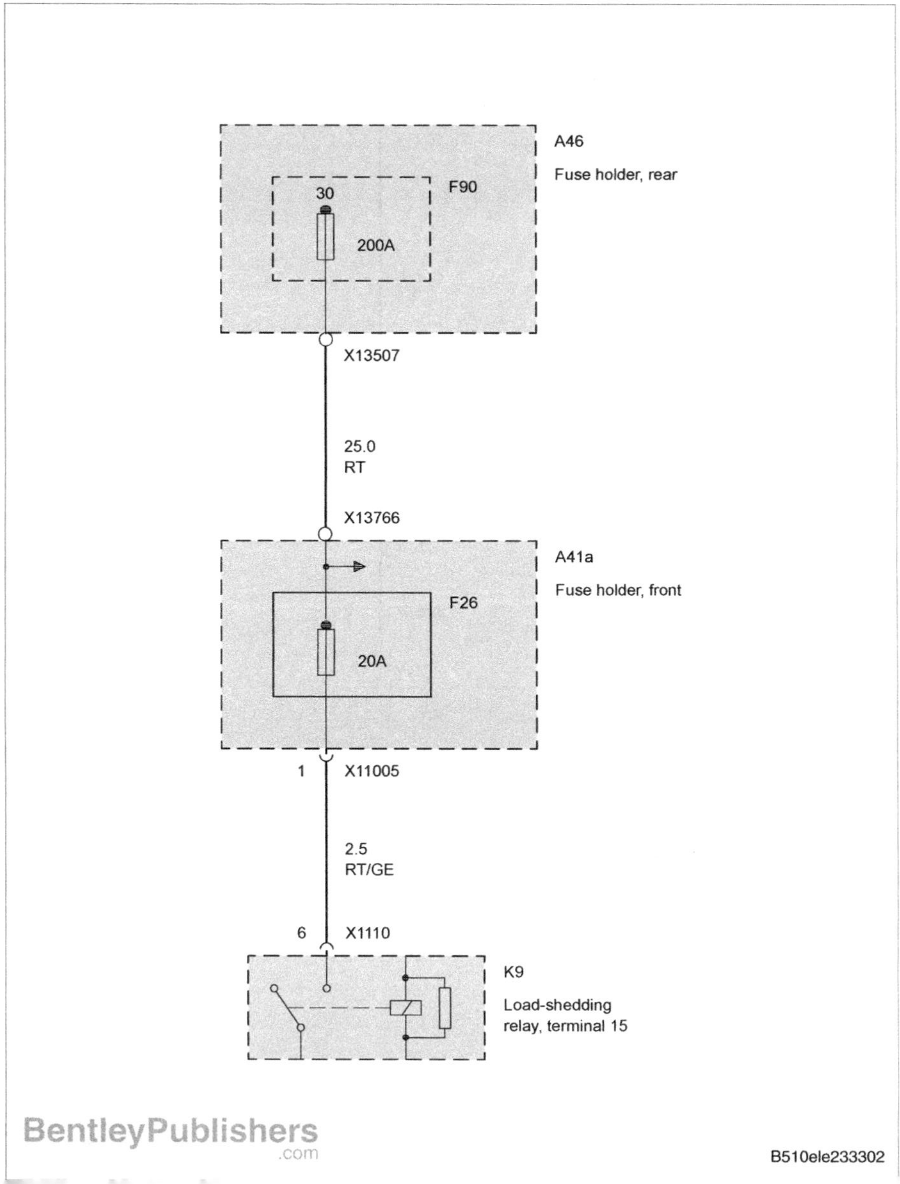

Fuse F27

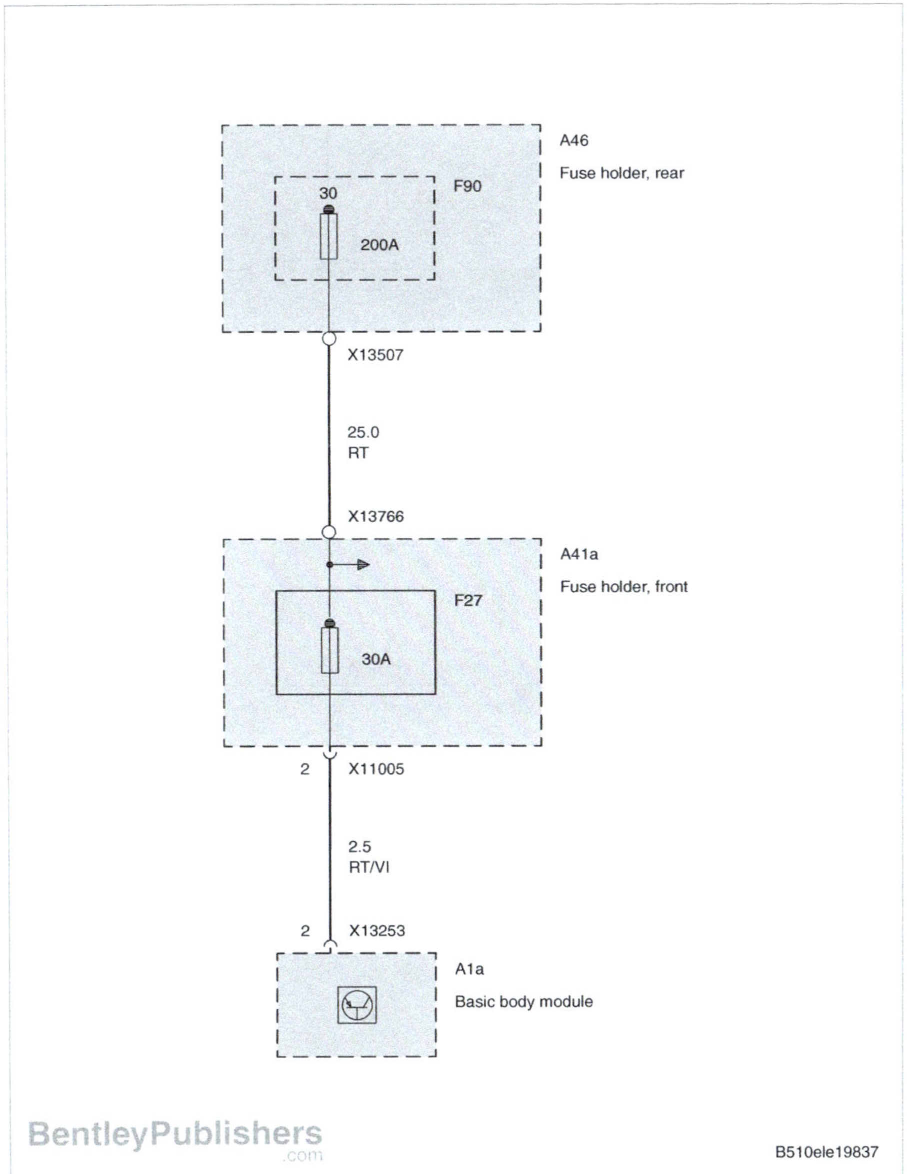

Fuse F28 up to 09 / 2005

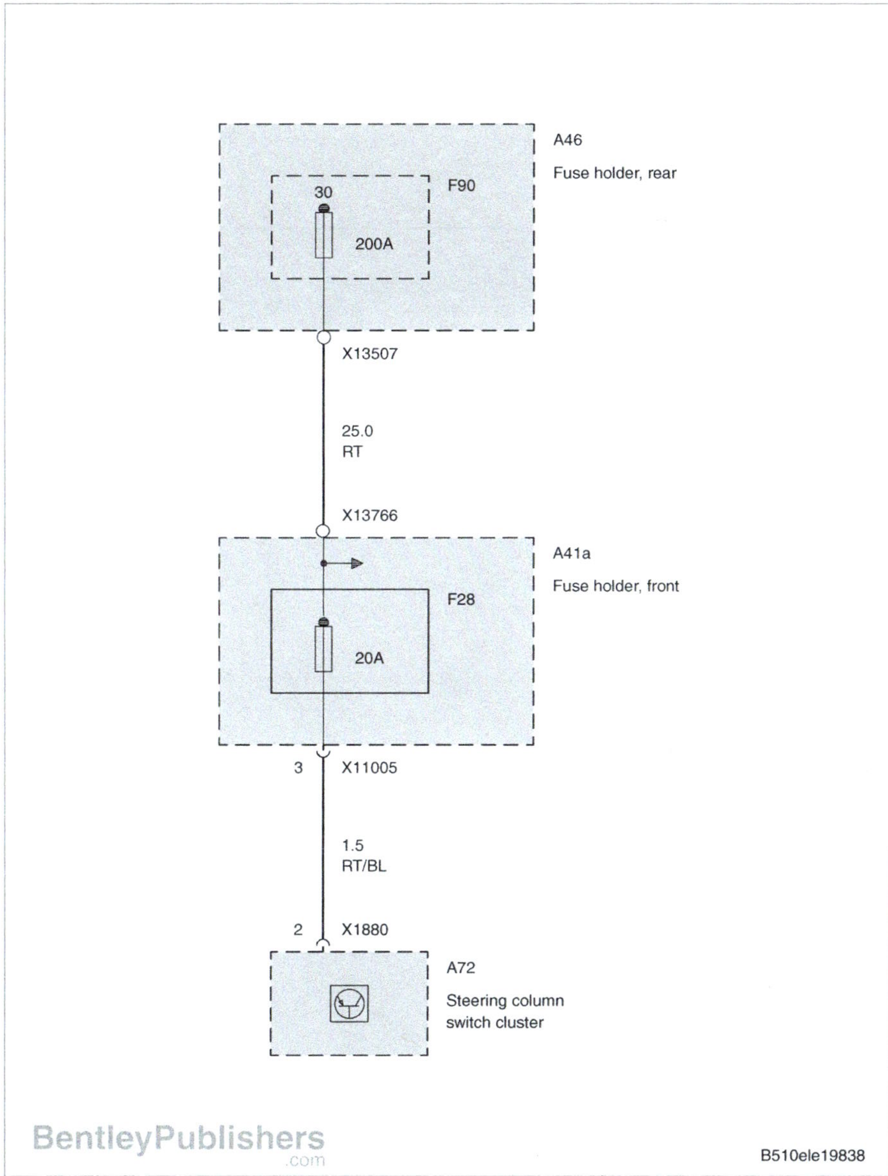

Fuse F28
as of 09 / 2005

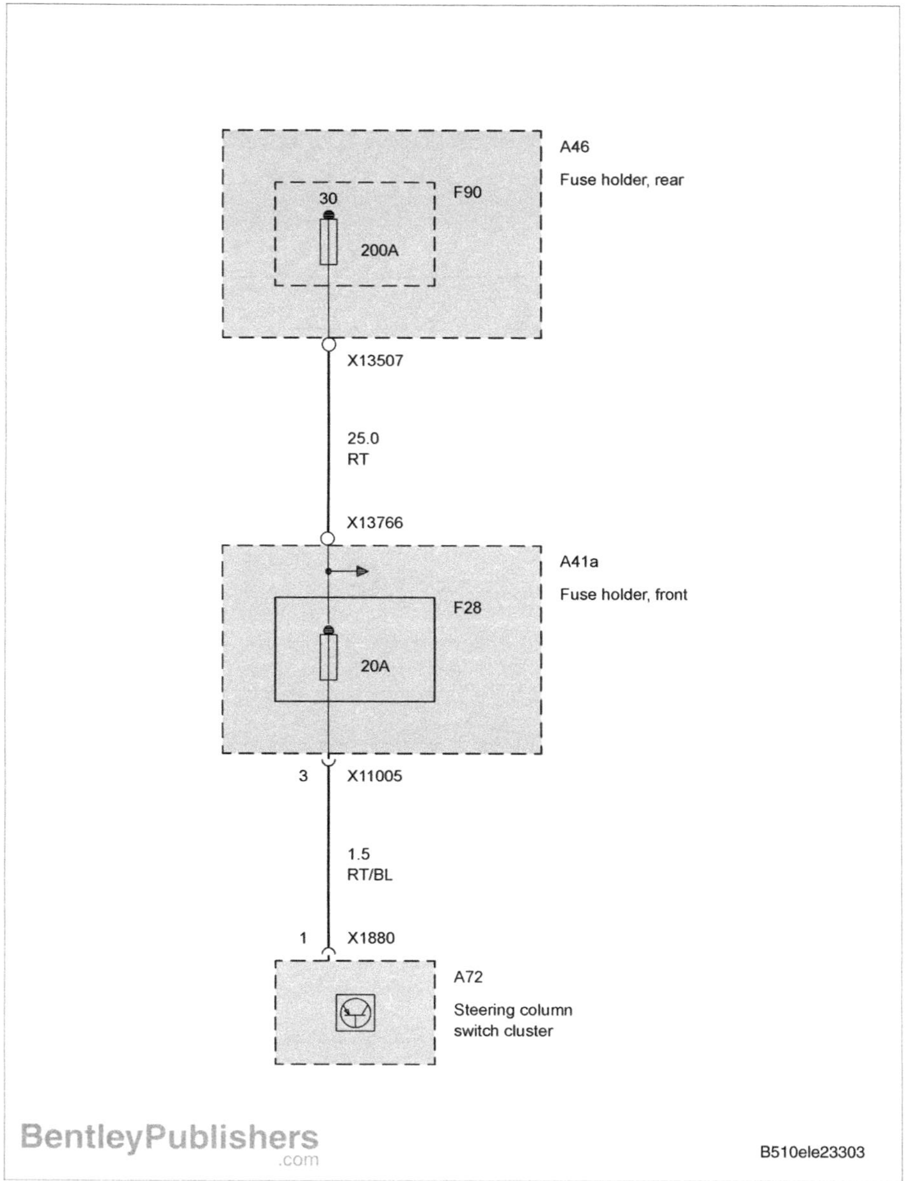

Fuse F29
up to 09 / 2005

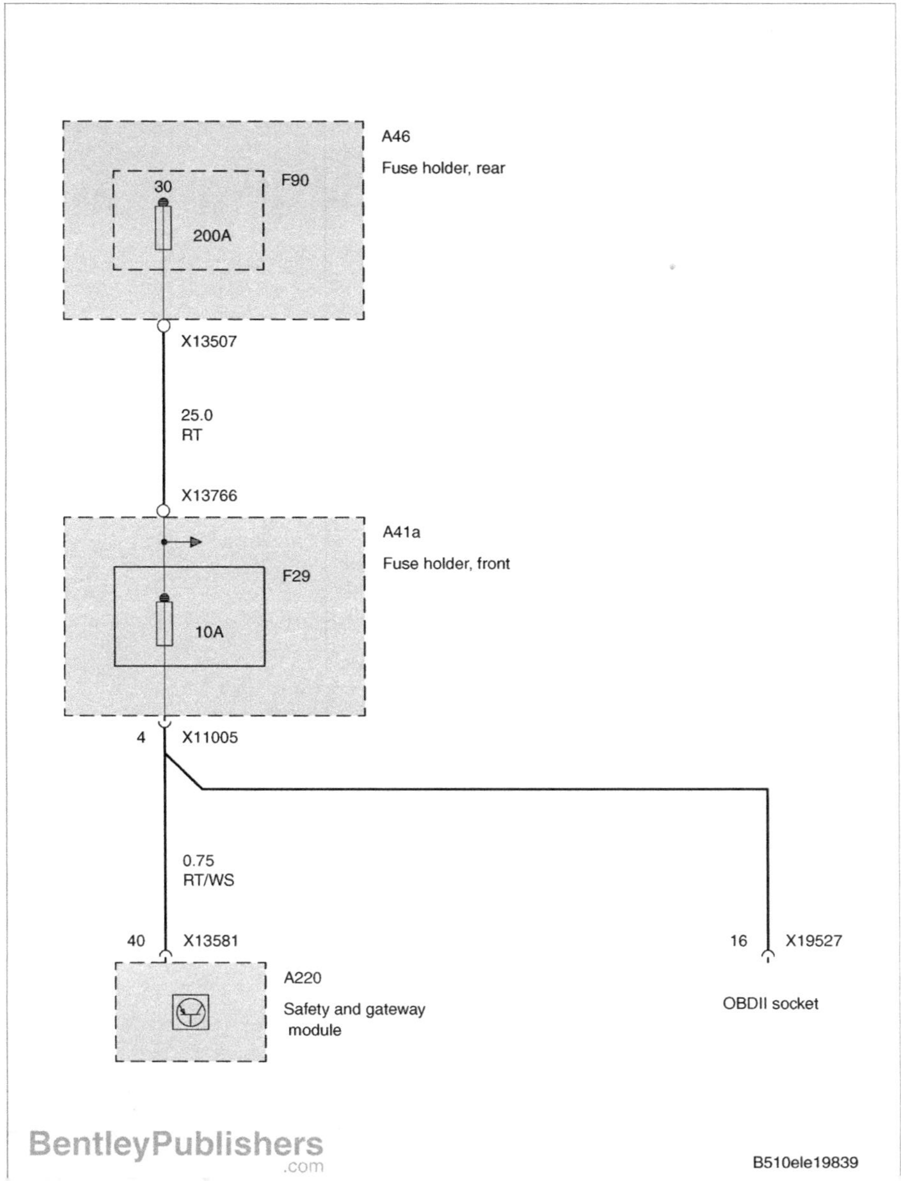

BentleyPublishers.com
BentleyPublishers.com

Fuse F29
as of 09 / 2005

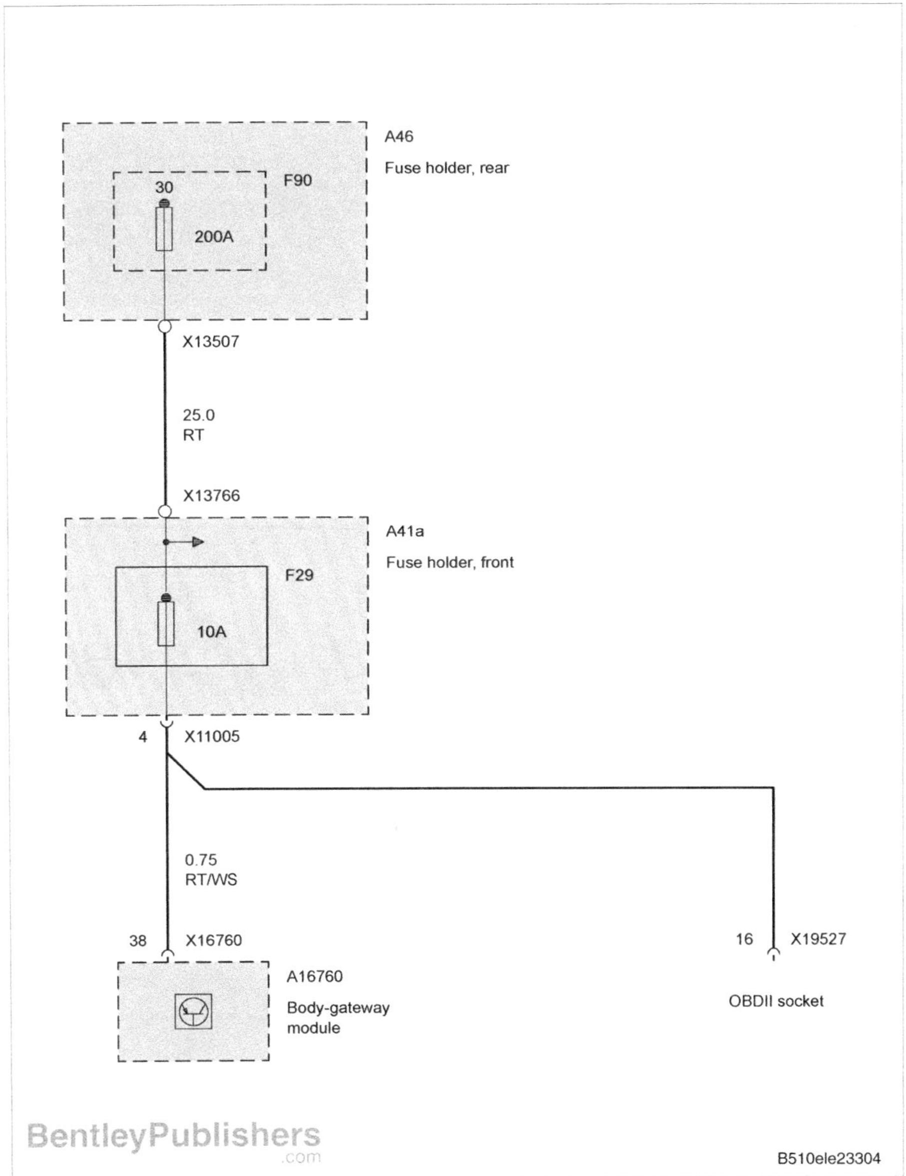

Fuse F30
IHKA High up to 09 / 2005

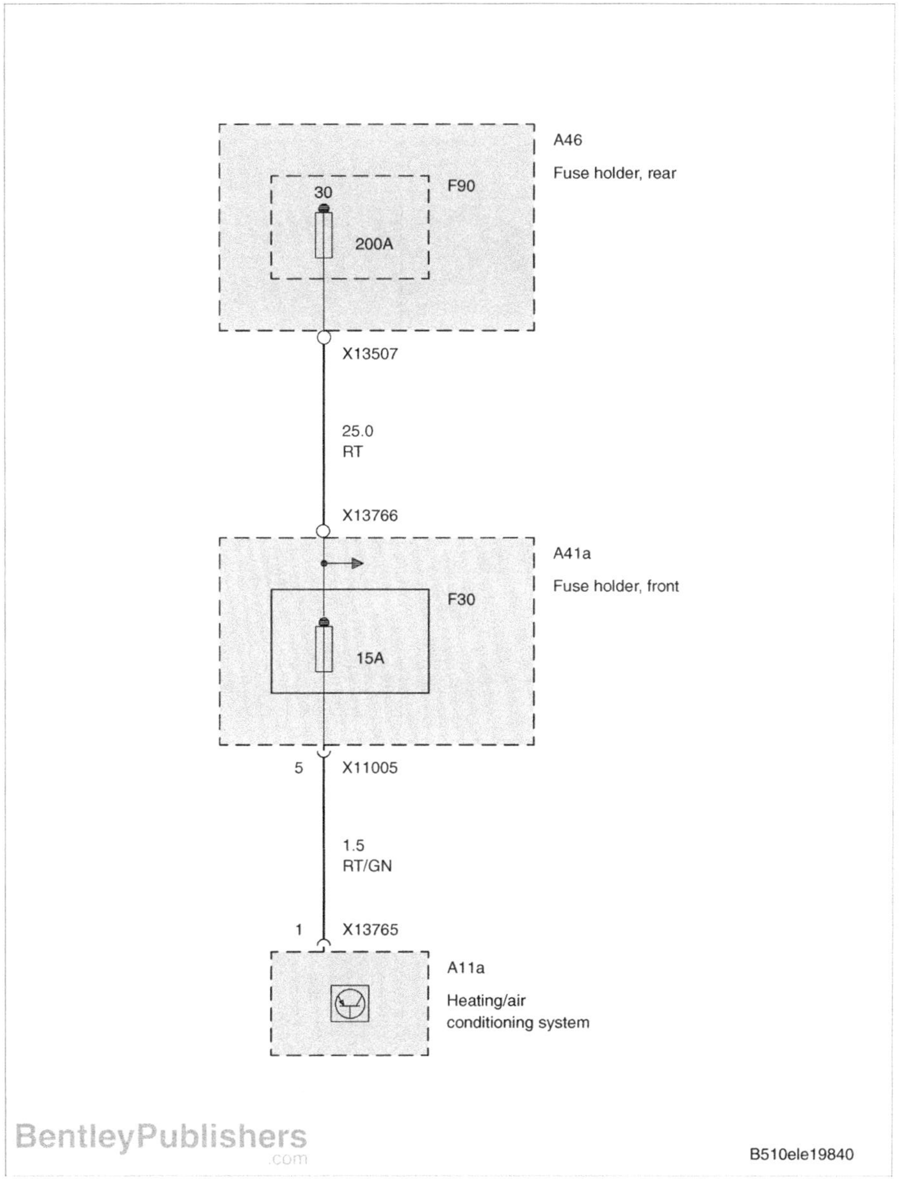

Fuse F30
as of 09 / 2005

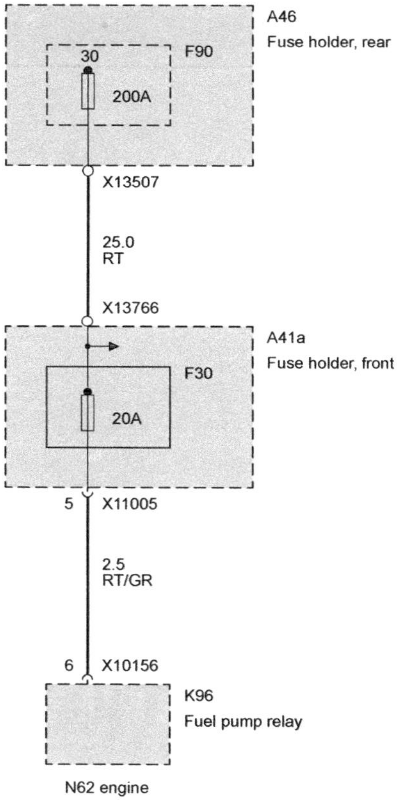

BentleyPublishers.com

Fuse F31
SMFA Seat module, driver

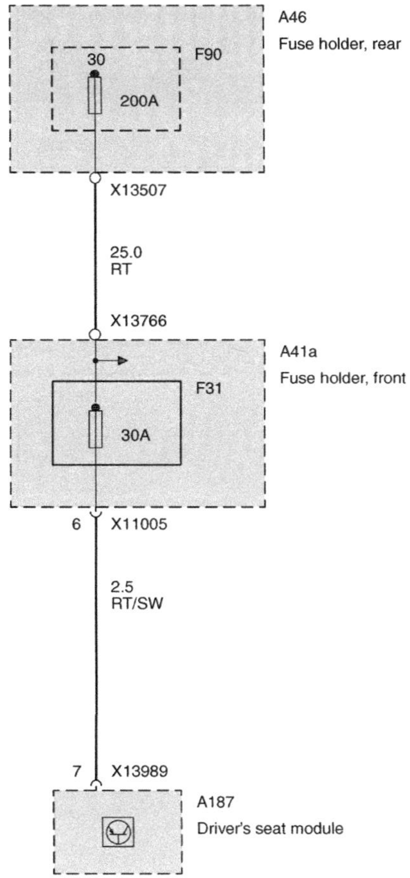

BentleyPublishers.com

Fuse F32
up to 09 / 2005

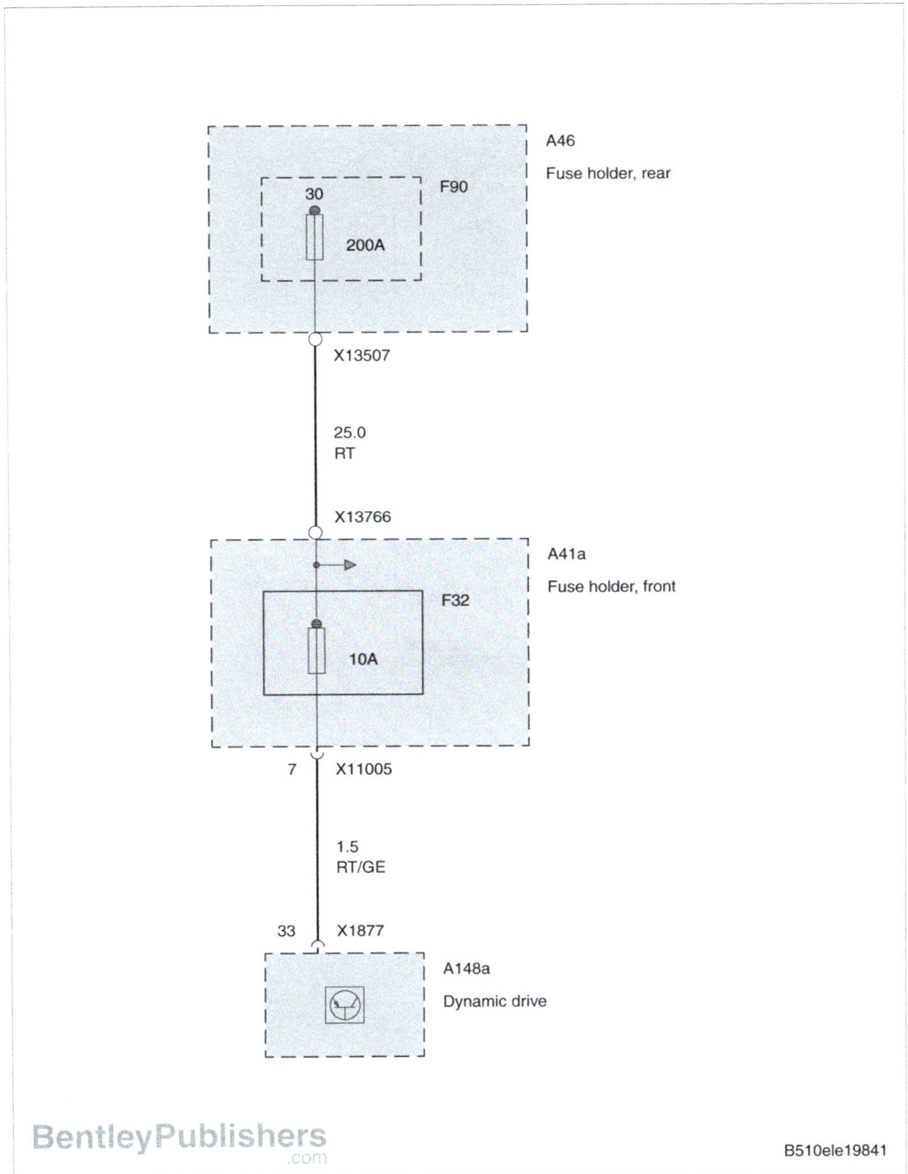

Fuse F32
as of 09 / 2005

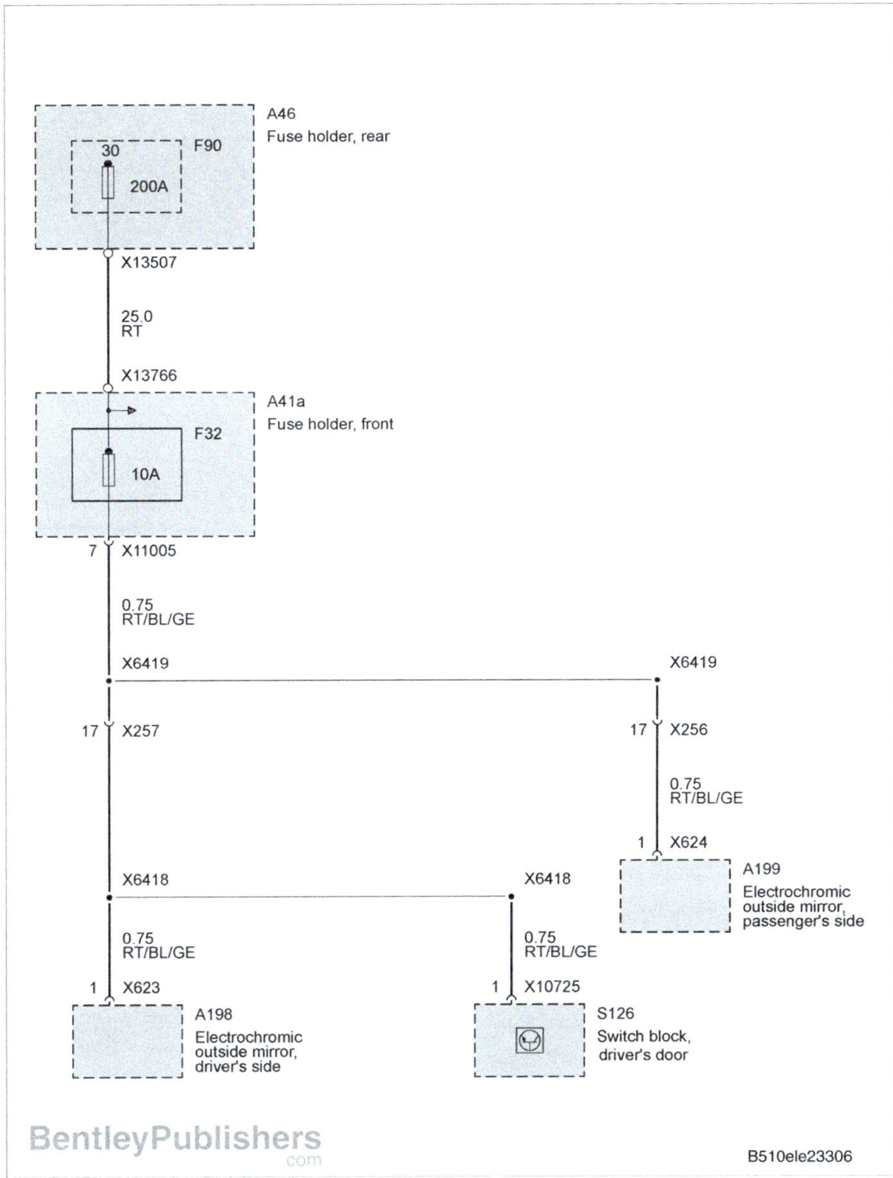

Fuse F33
SMFA Seat module, driver

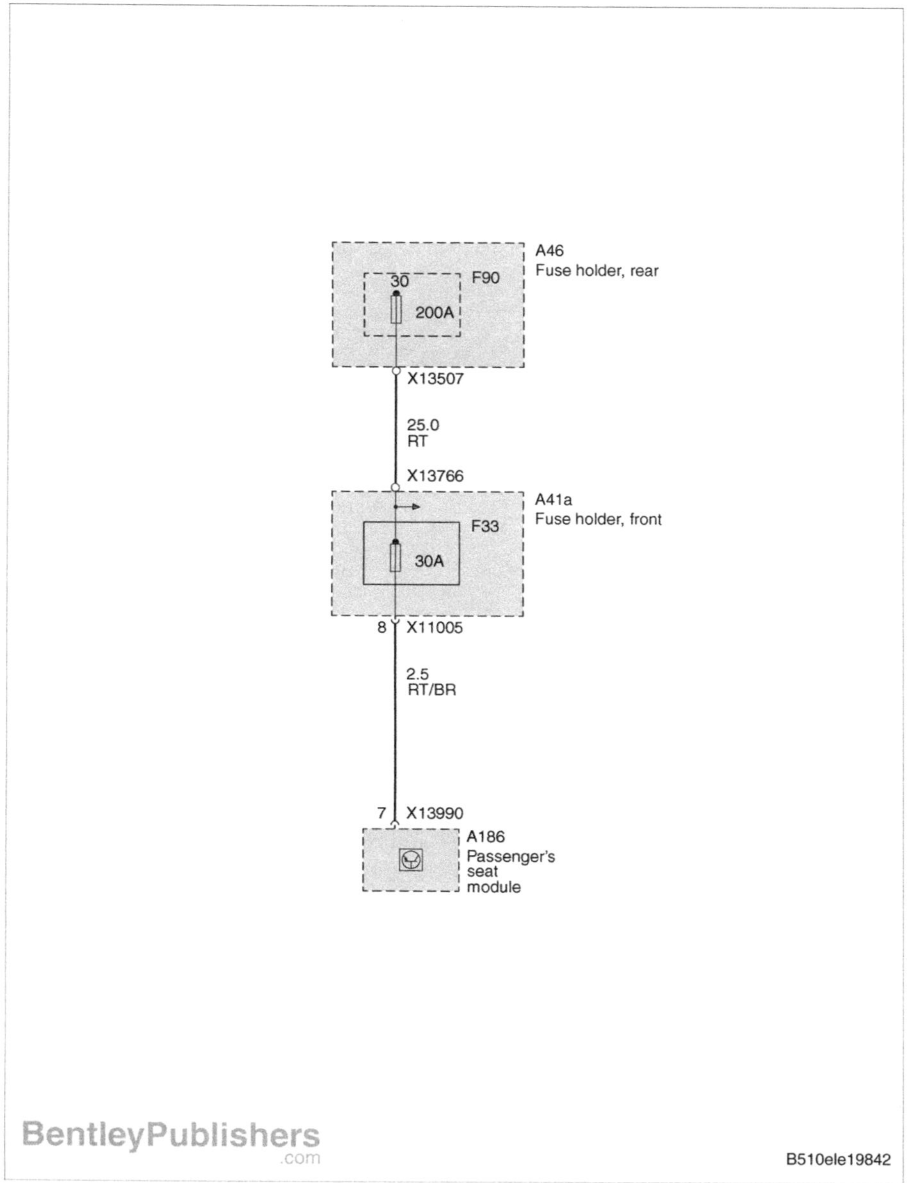

Fuse F34
up to 09 / 2005

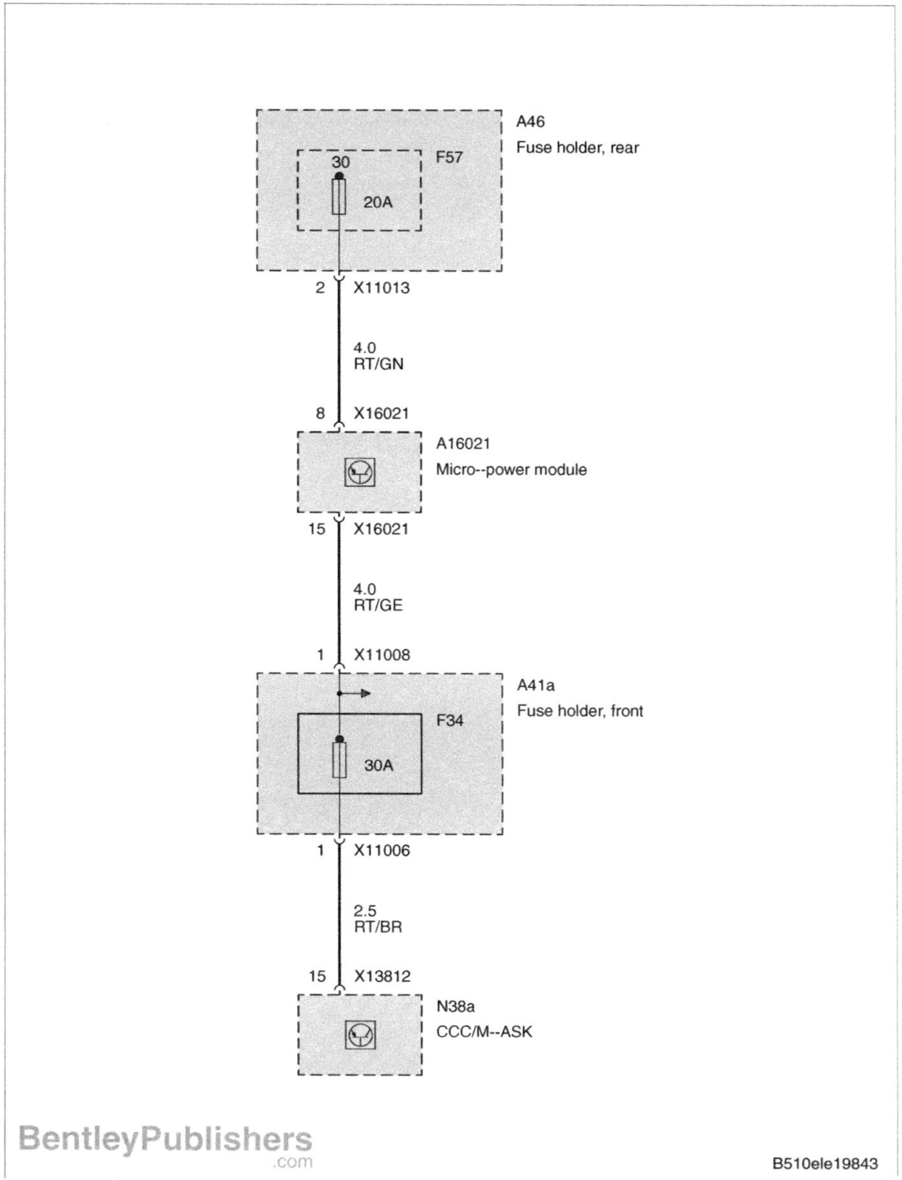

Fuse F34 without CIC Car Information Computer as of 09 / 2005

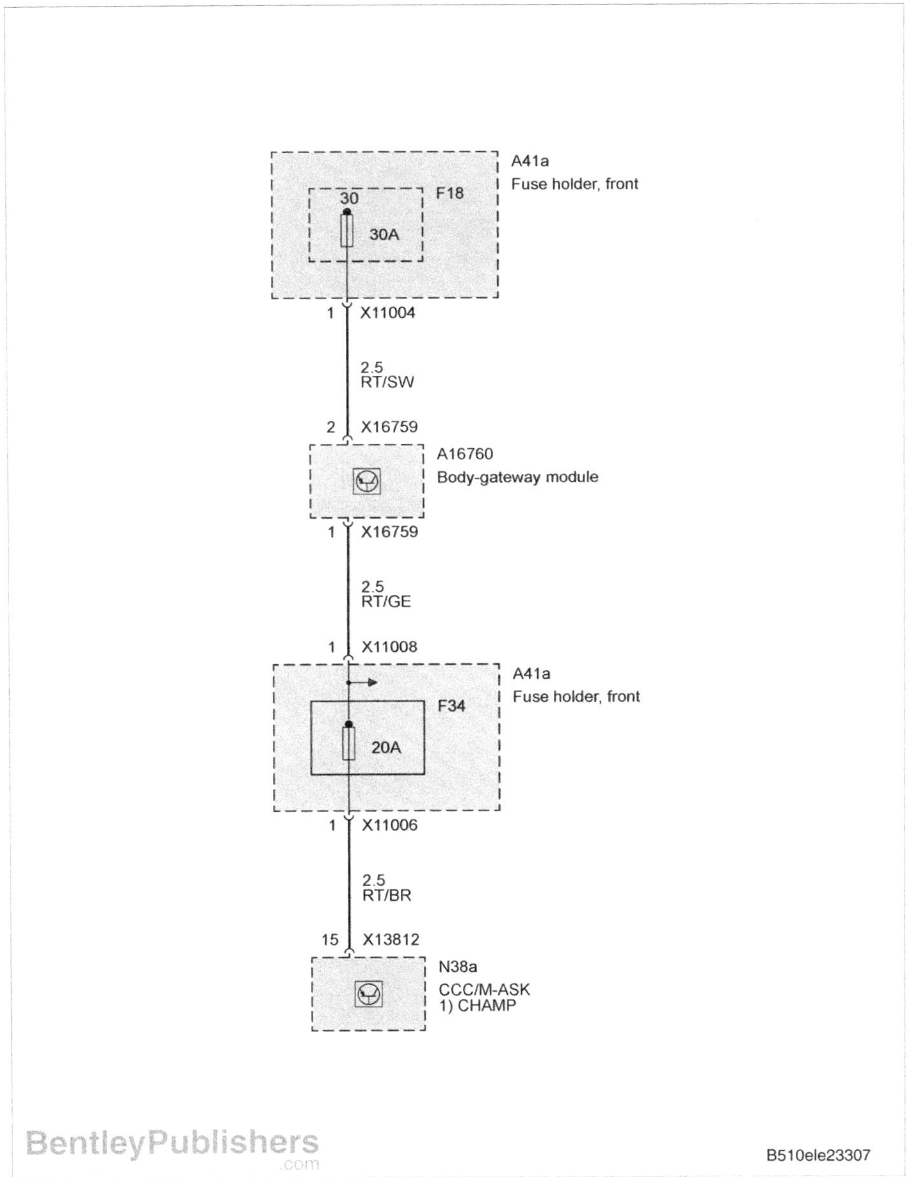

Fuse F34 with CIC Car Information Computer

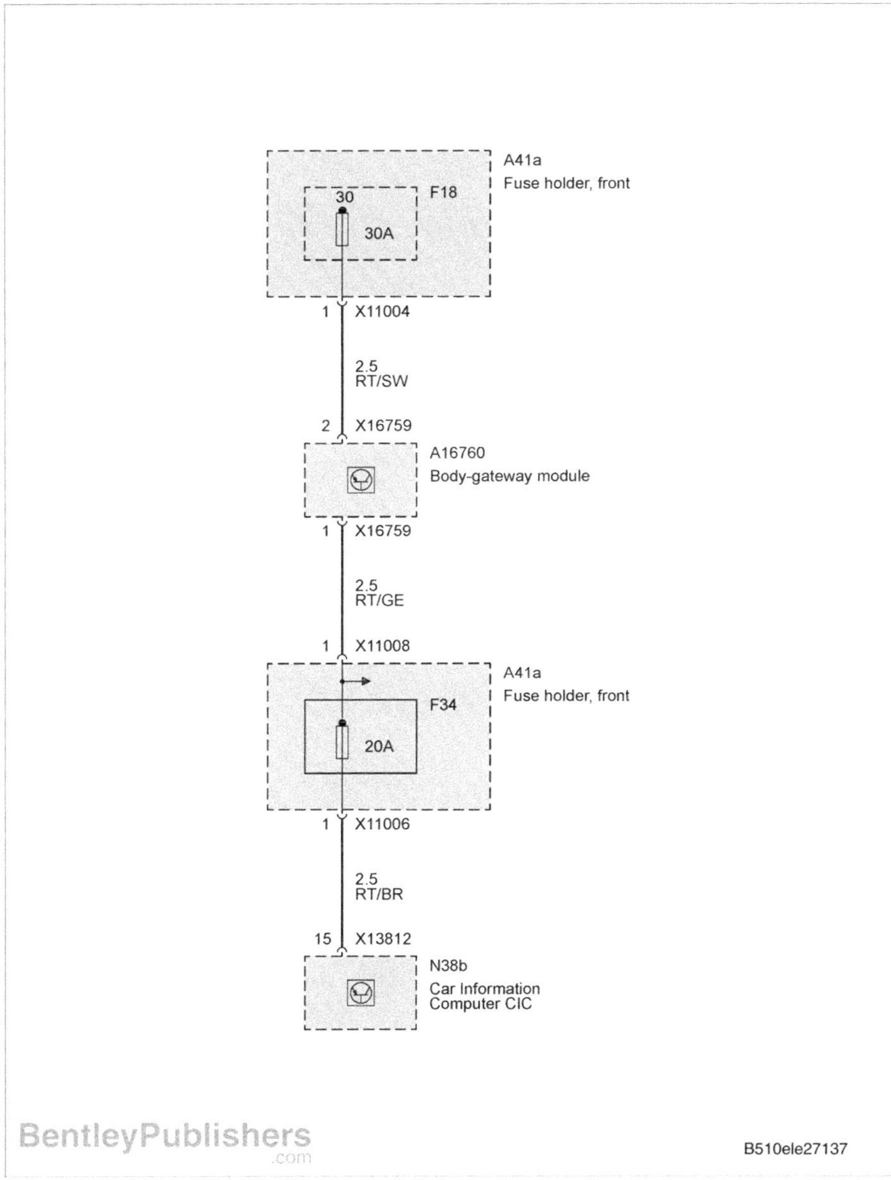

Fuse F35
as of 09 / 2005

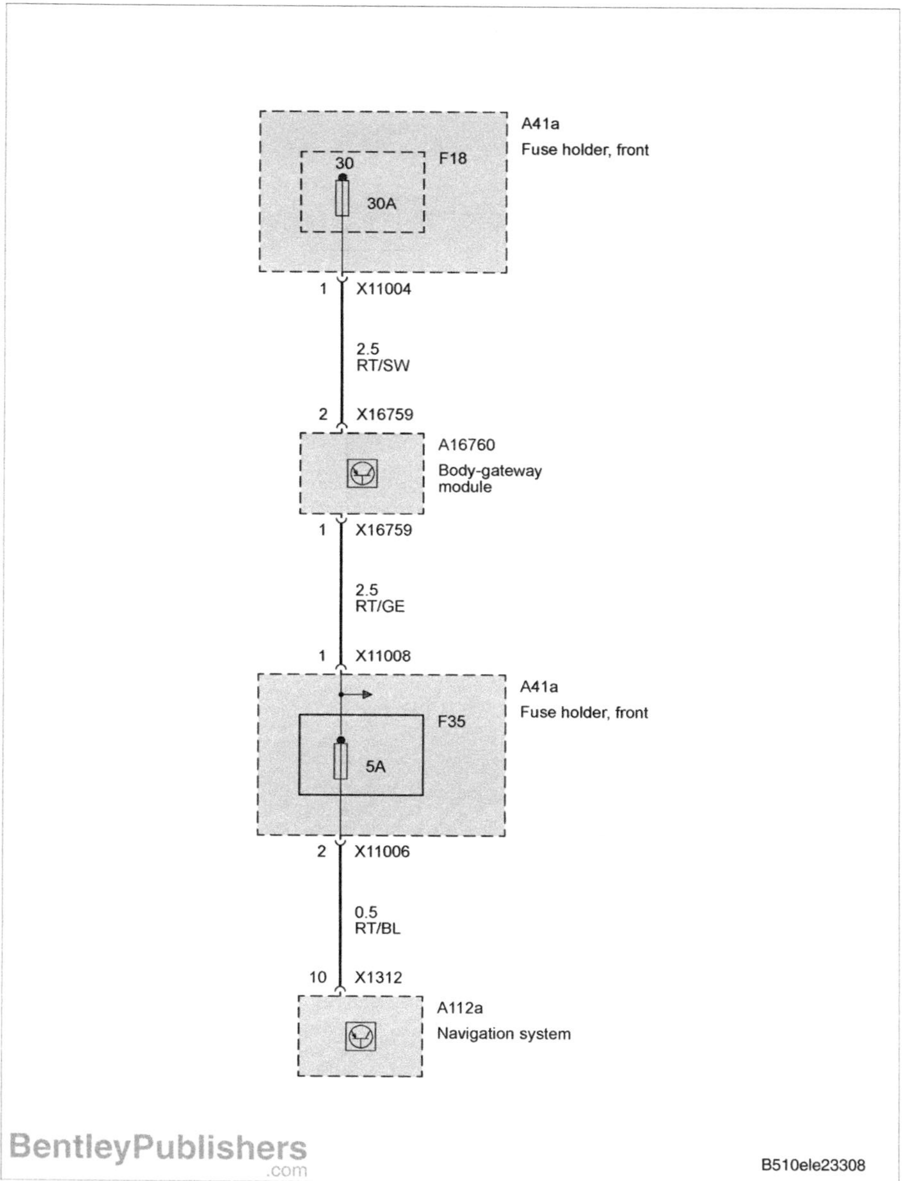

BentleyPublishers.com

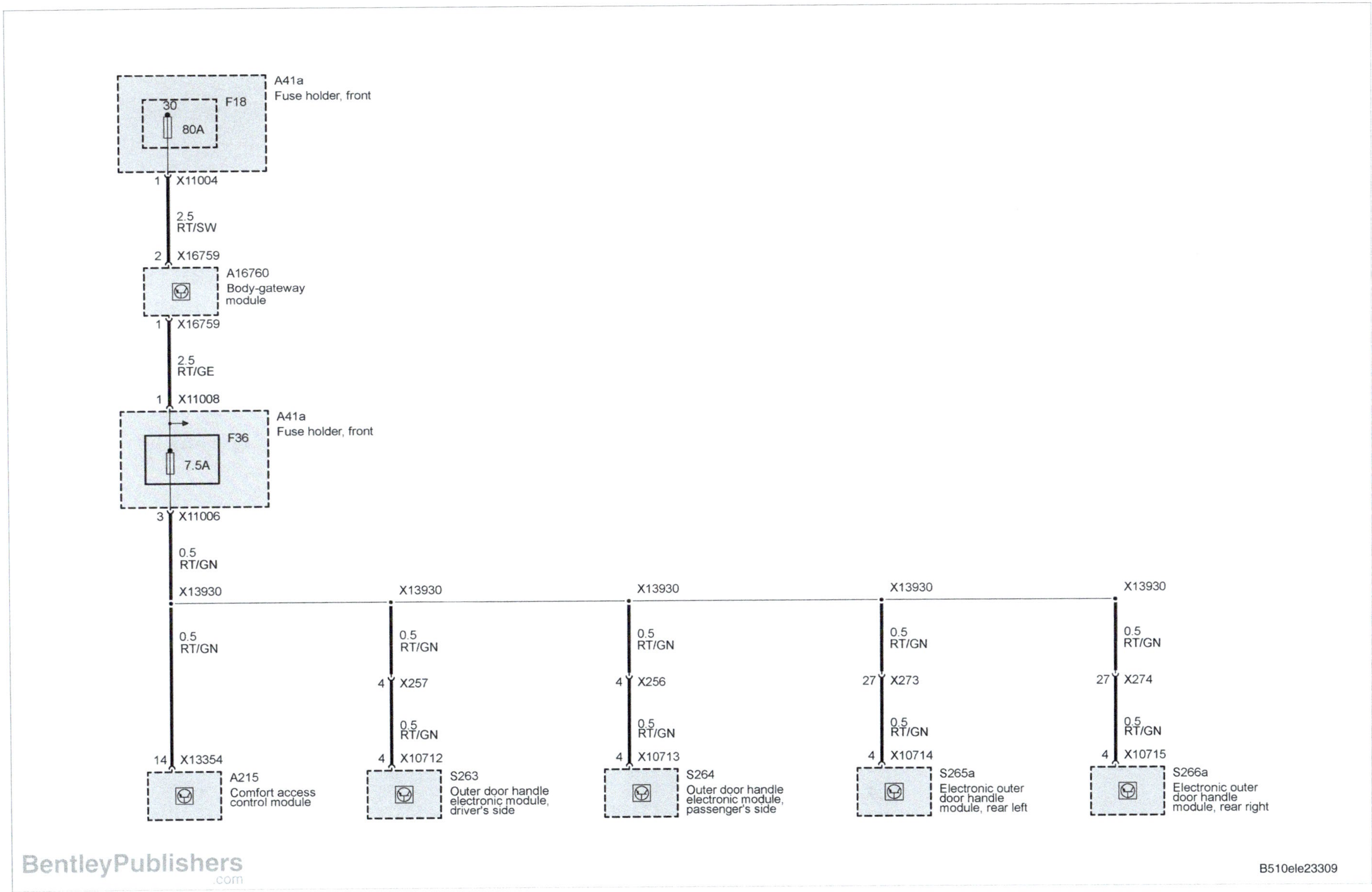
A41a
Fuse holder, front
F18
30
80A
1 X11004
2.5
RT/SW
2 X16759
A16760
Body-gateway module
1 X16759
2.5
RT/GE
1 X11008
A41a
Fuse holder, front
F36
7.5A
3 X11006
0.5
RT/GN
X13930
0.5
RT/GN
14 X13354
A215
Comfort access control module
X13930
0.5
RT/GN
4 X257
0.5
RT/GN
4 X10712
S263
Outer door handle electronic module, driver's side
X13930
0.5
RT/GN
4 X256
0.5
RT/GN
4 X10713
S264
Outer door handle electronic module, passenger's side
X13930
0.5
RT/GN
27 X273
0.5
RT/GN
4 X10714
S265a
Electronic outer door handle module, rear left
X13930
0.5
RT/GN
27 X274
0.5
RT/GN
4 X10715
S266a
Electronic outer door handle module, rear right
BentleyPublishers.com
B510ele23309

Fuse F37
up to 09 / 2005

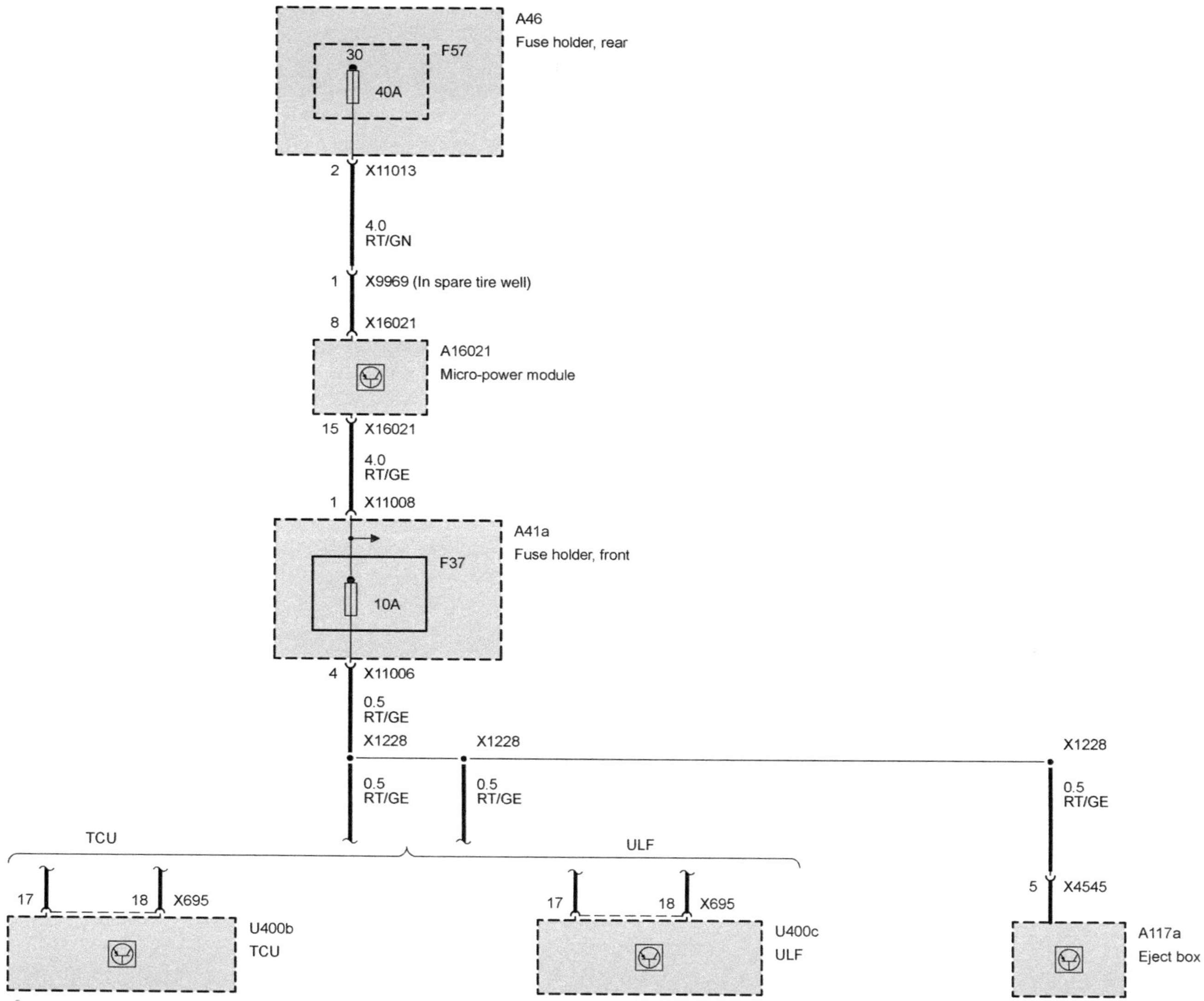

BentleyPublishers.com

B510ele21845

Fuse F37
as of 09 / 2005 up to 03 / 2007

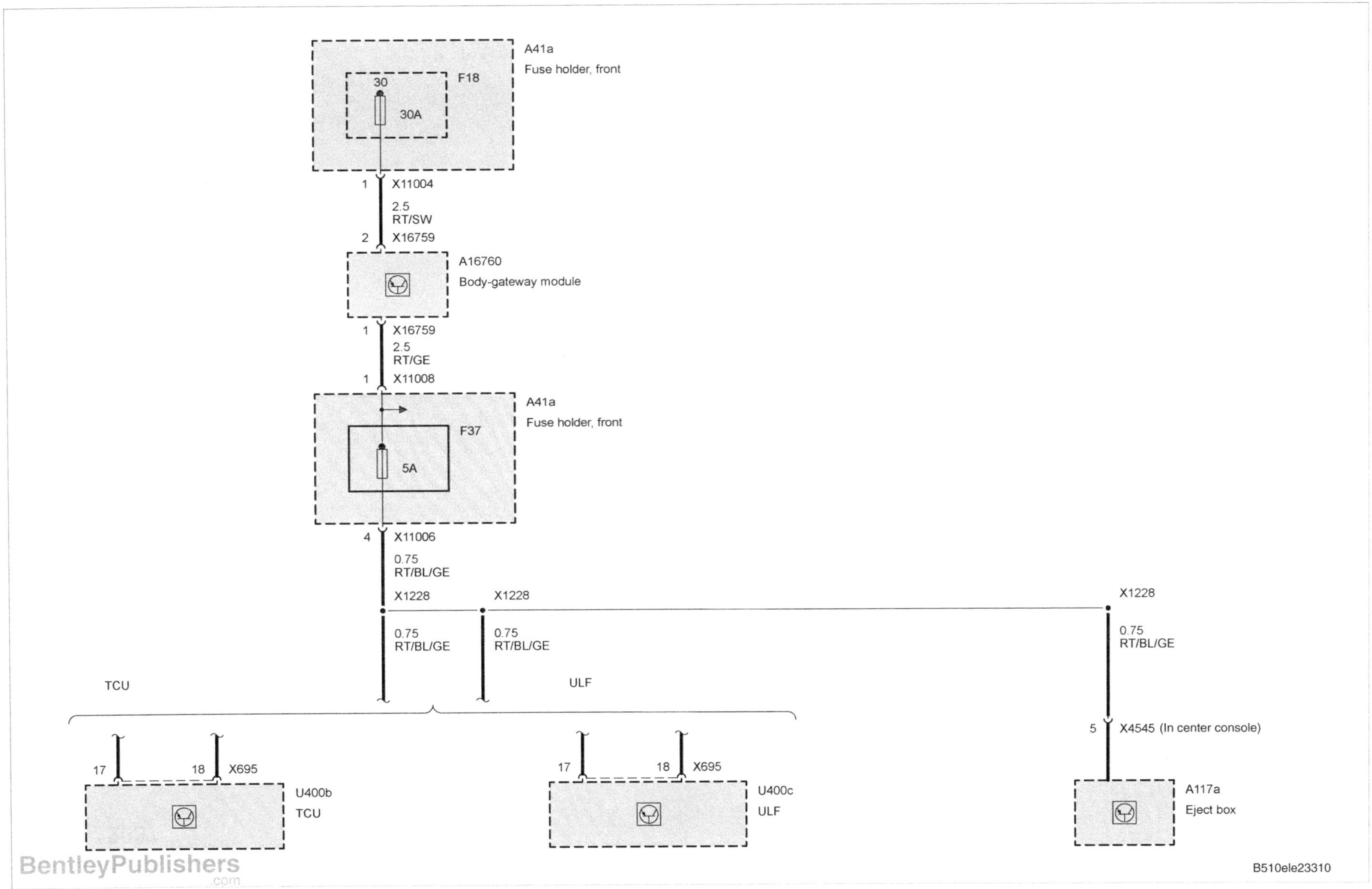

BentleyPublishers.com

Fuse F37
as of 03 / 2007

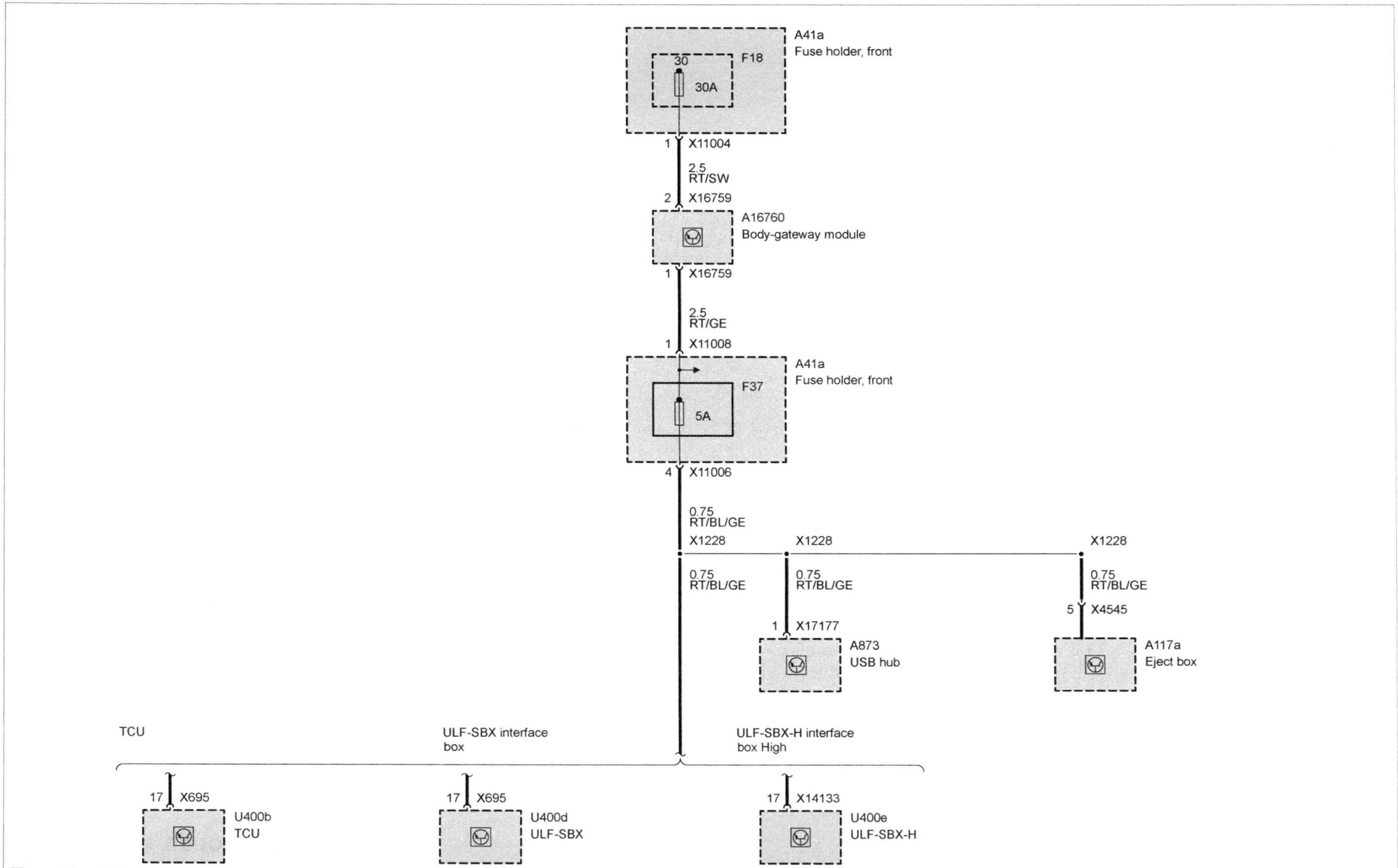

BentleyPublishers.com

Fuse F38
up to 09 / 2005

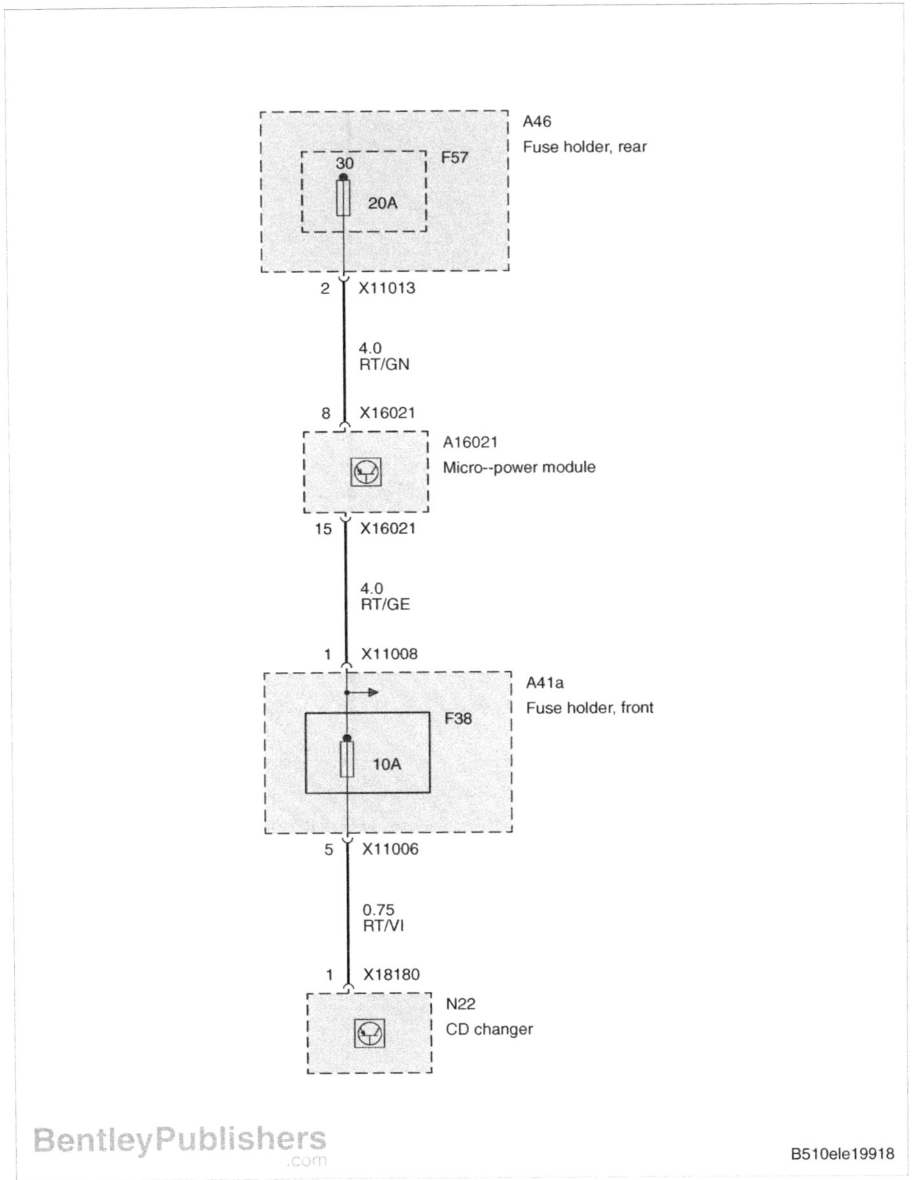

Fuse F38
as of 09 / 2005

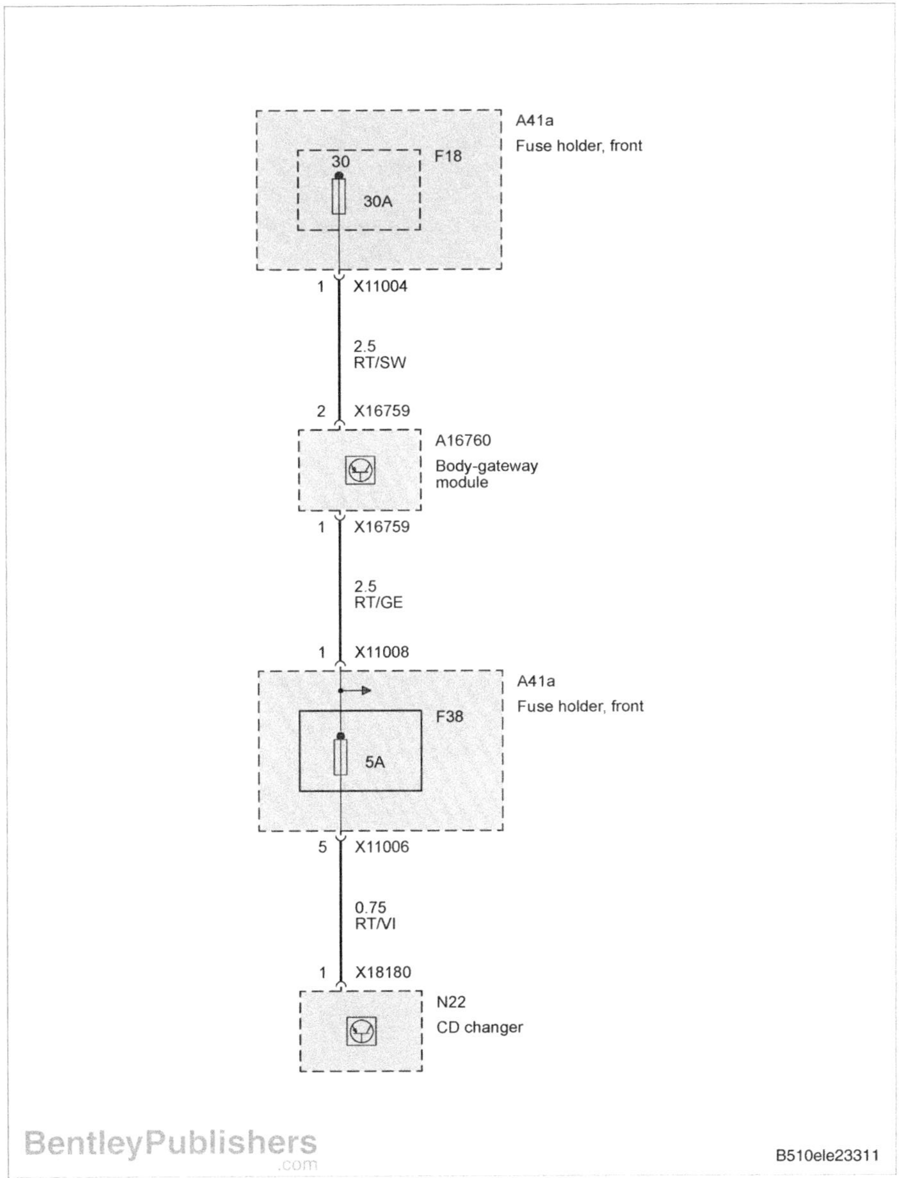

Fuse F39

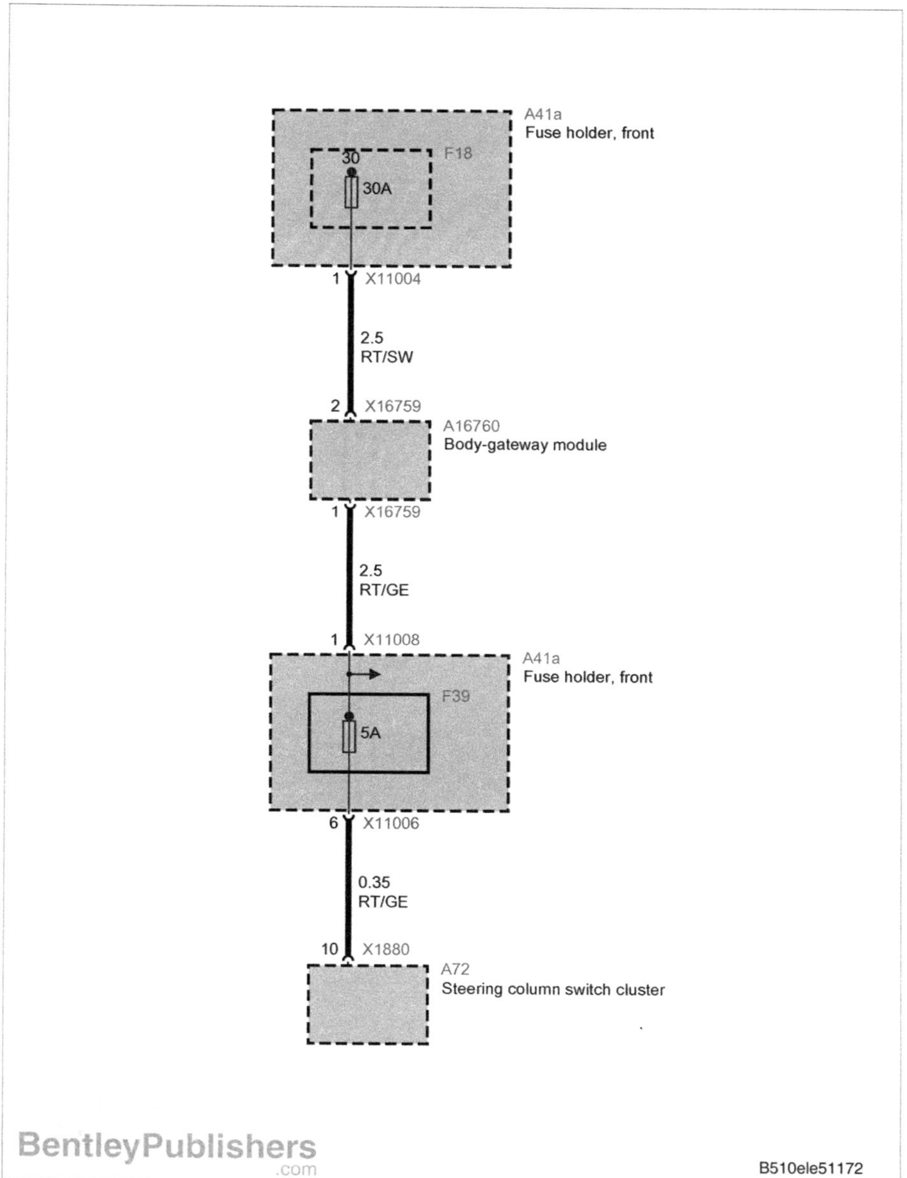

Fuse F40 up to 09 / 2005

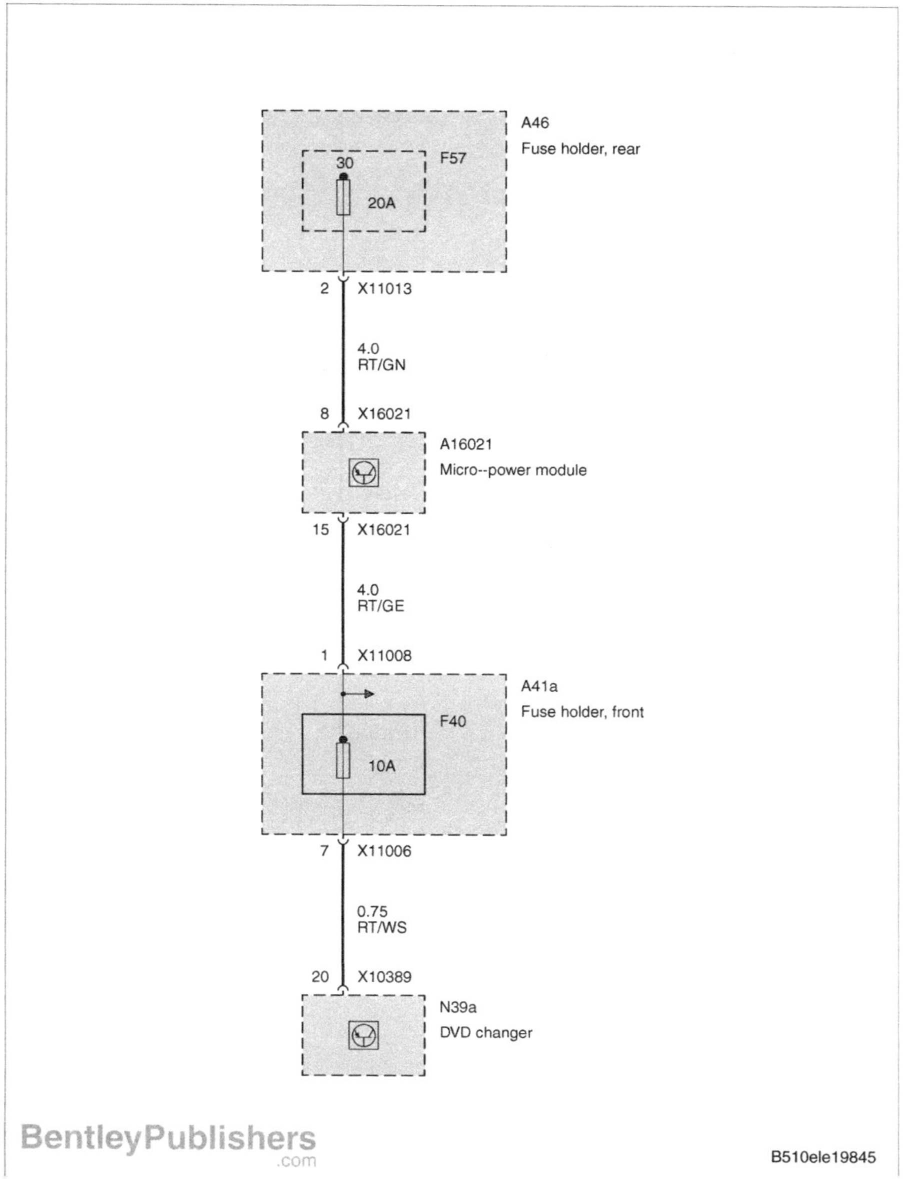

Fuse F40
as of 09 / 2005

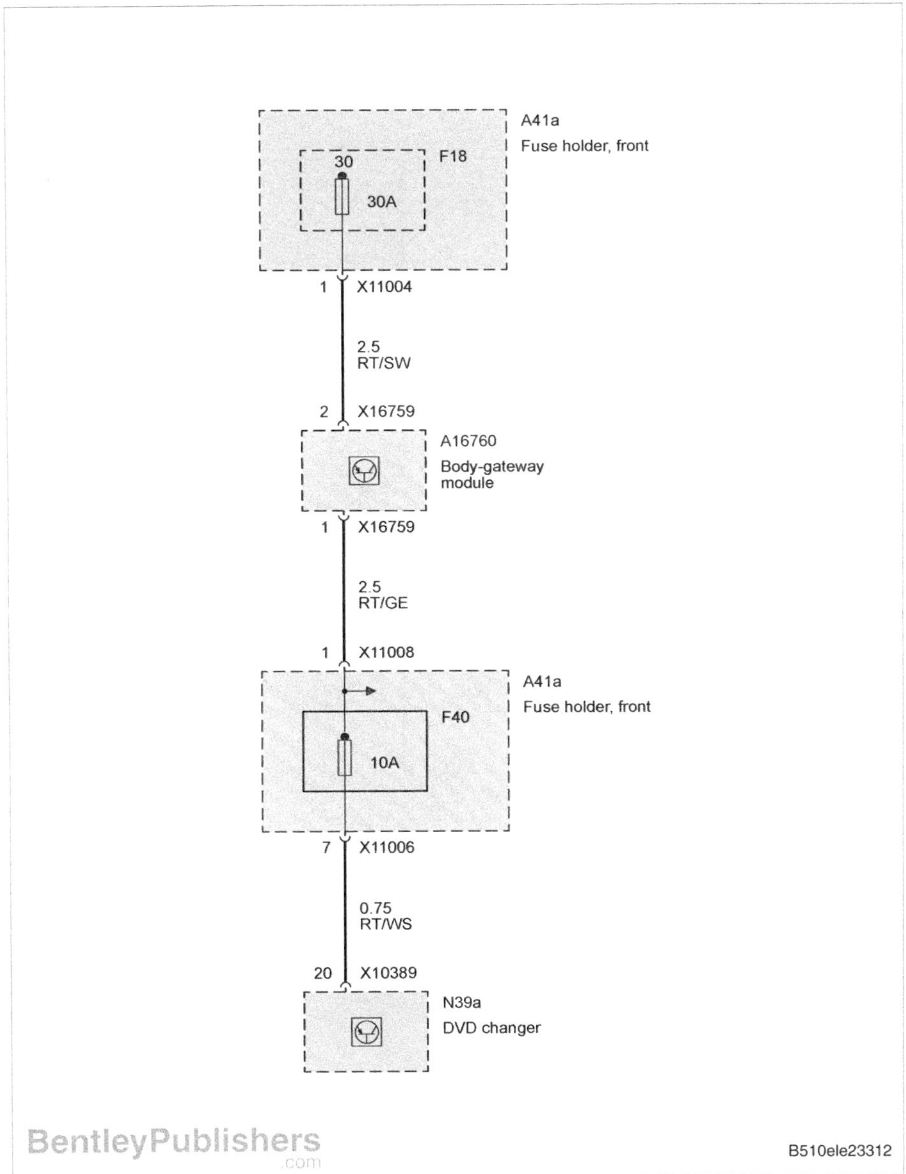

Fuse F41
as of 09 / 2005

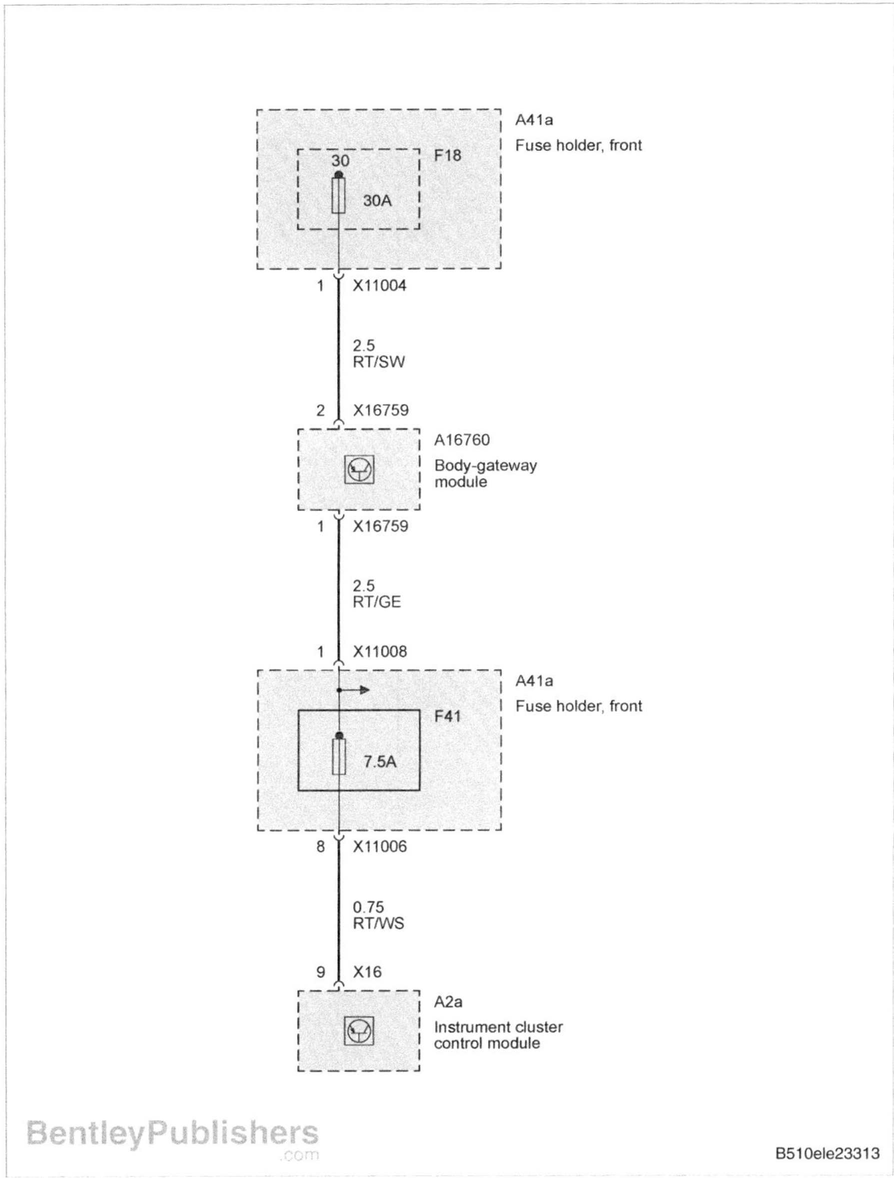

Fuse F50
without S85 engine

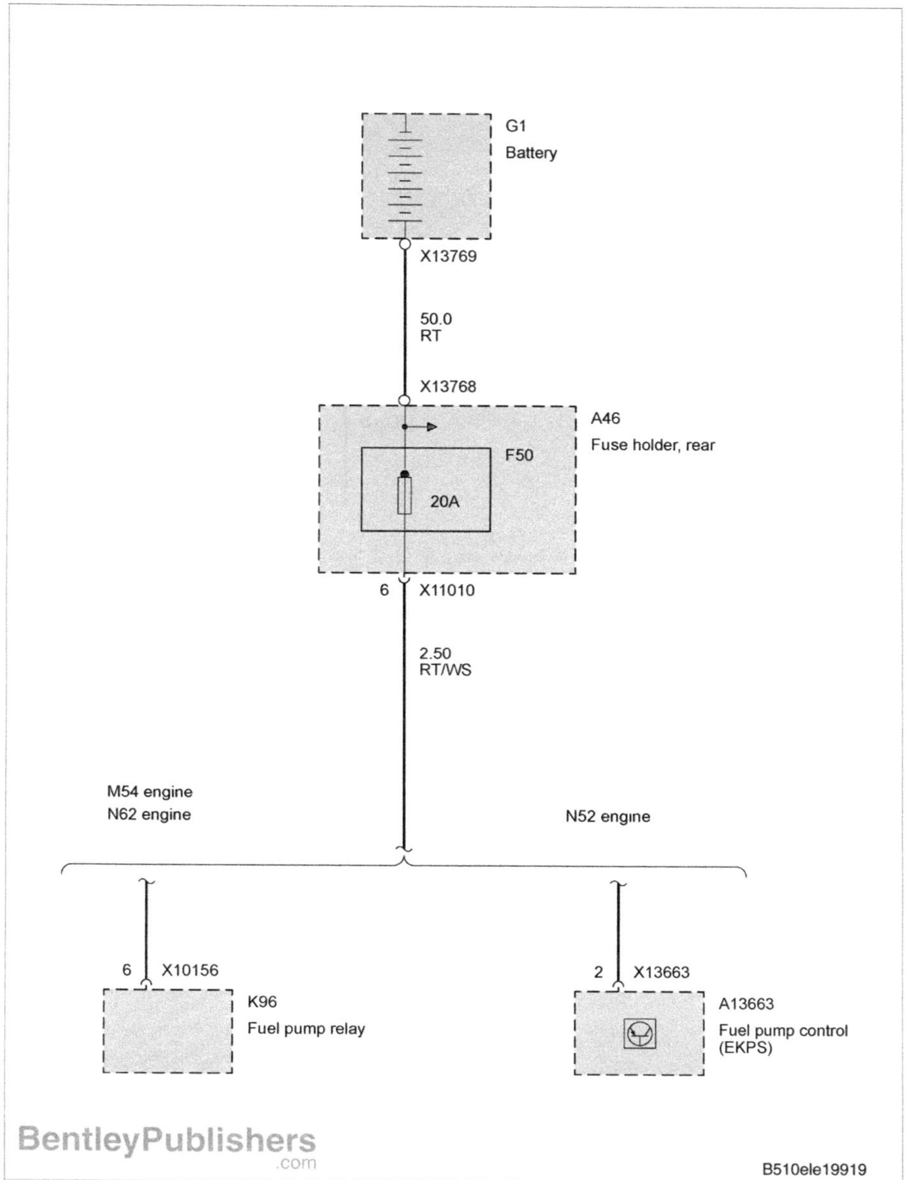

Fuse F50
up to 09 / 2005

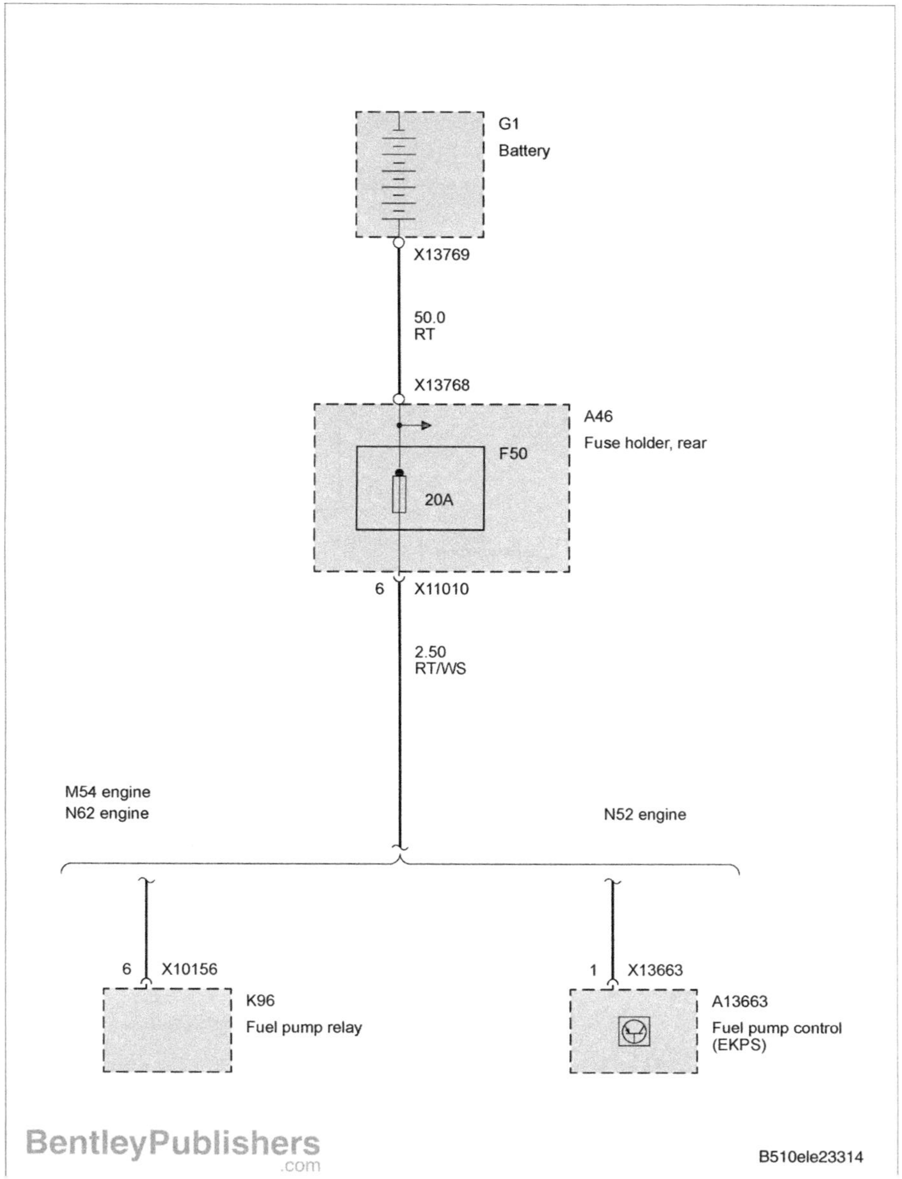

Fuse F50
as of 09 / 2005

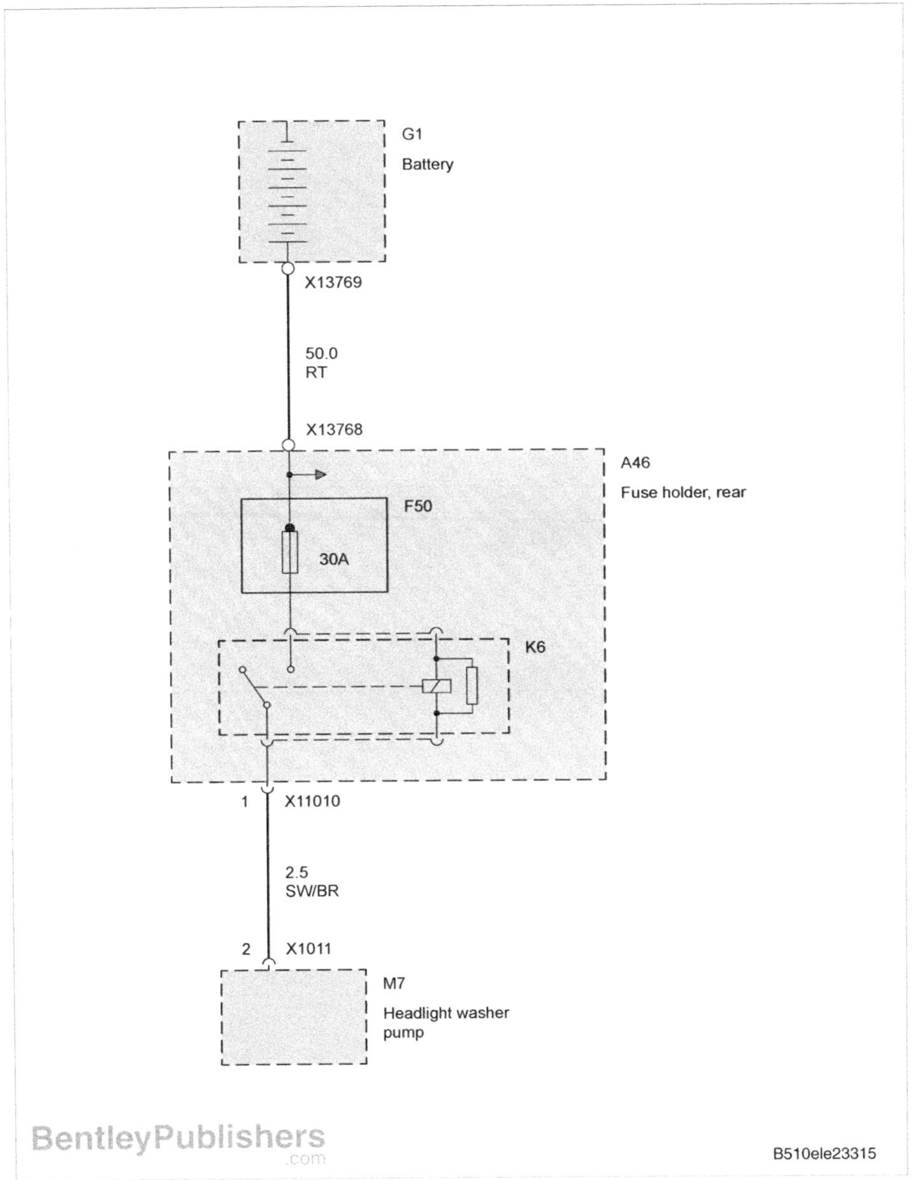

Fuse F51

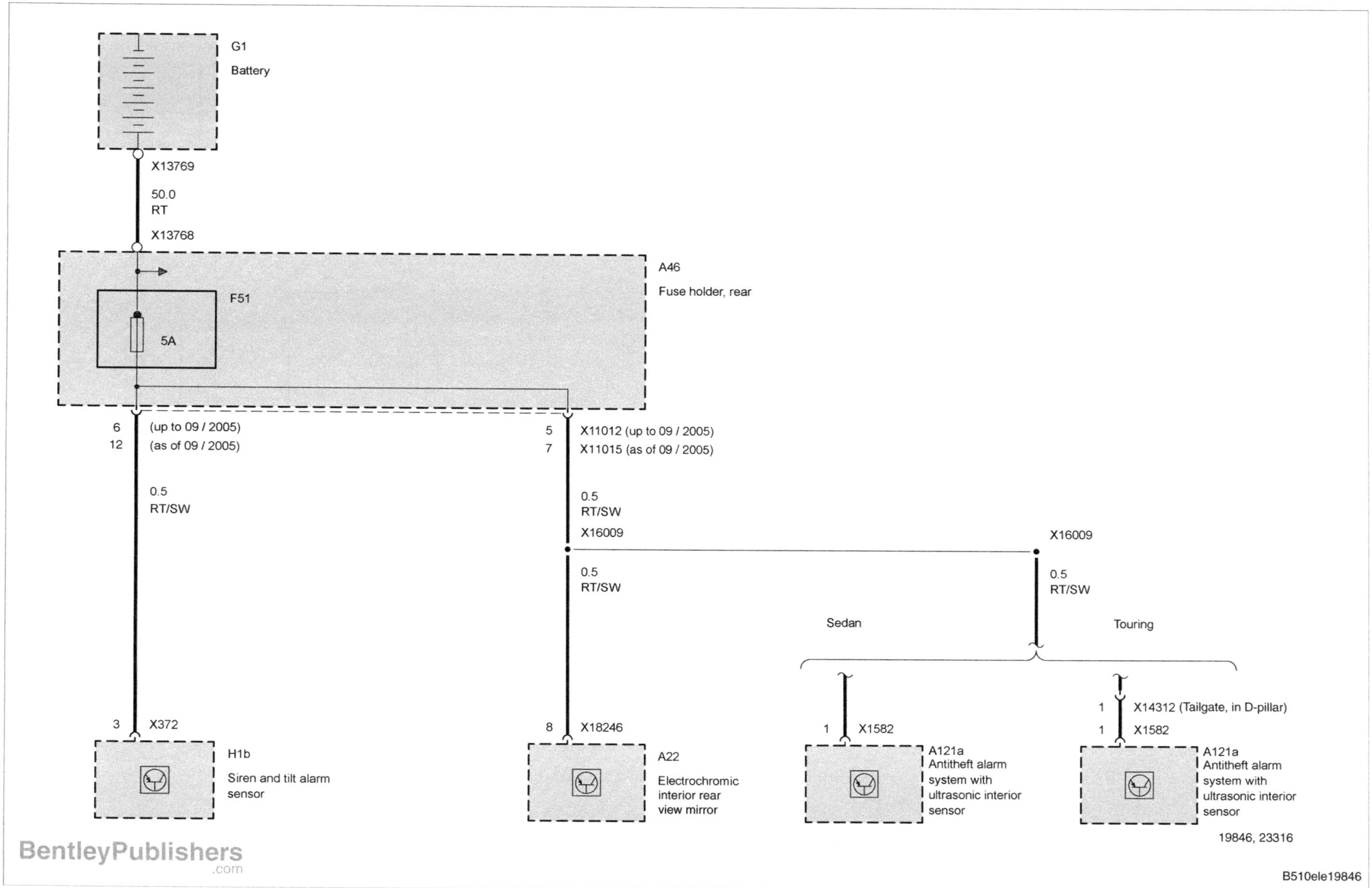

BentleyPublishers.com

B510ele19846

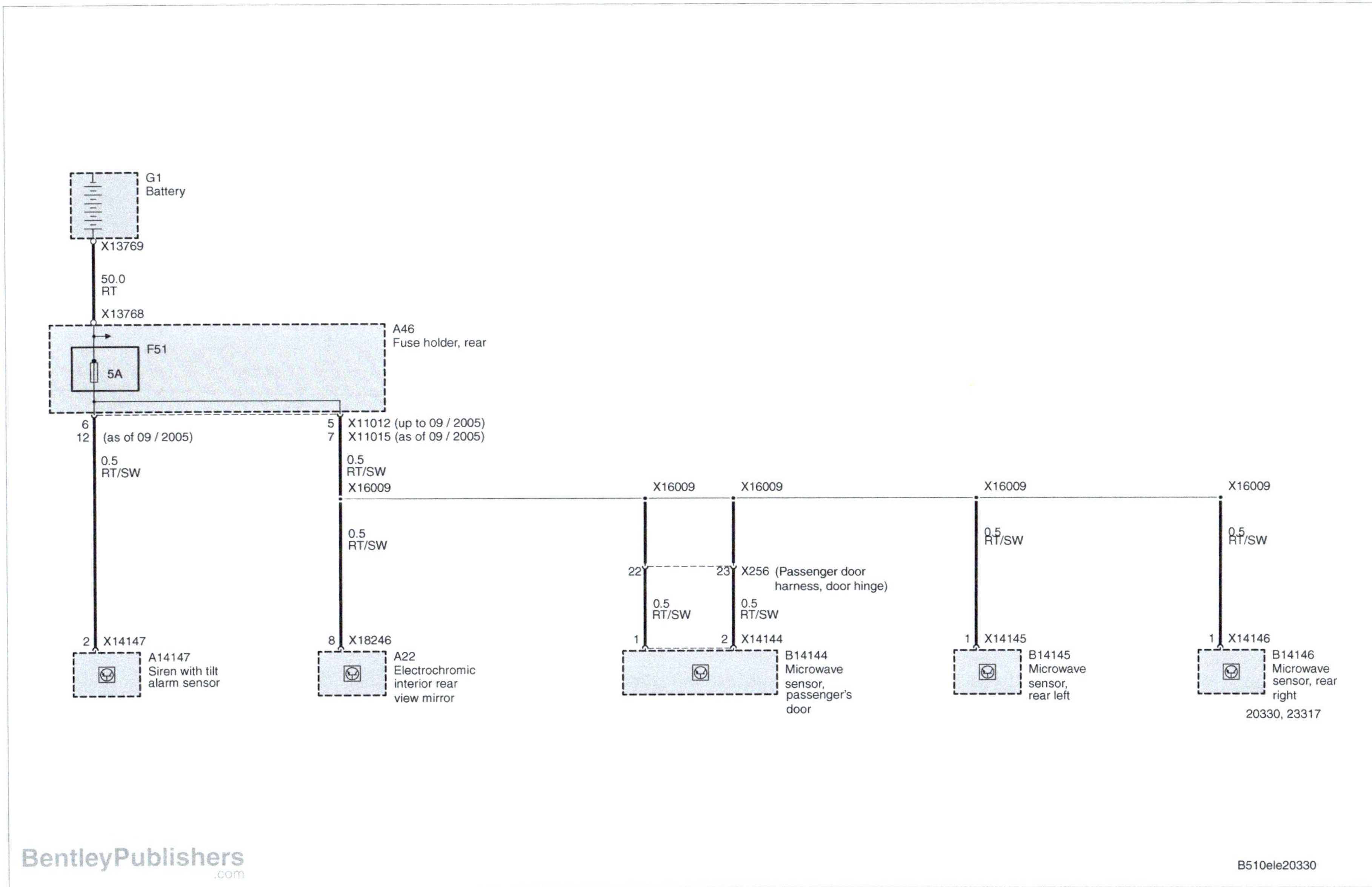

G1
Battery
X13769
50.0
RT
X13768
A46
Fuse holder, rear
F51
5A
6
12
(as of 09 / 2005)
5
7
X11012 (up to 09 / 2005)
X11015 (as of 09 / 2005)
0.5
RT/SW
X16009
X16009
X16009
X16009
X16009
22
23
X256 (Passenger door harness, door hinge)
2
X14147
A14147
Siren with tilt alarm sensor
8
X18246
A22
Electrochromic interior rear view mirror
1
2
X14144
B14144
Microwave sensor, passenger's door
1
X14145
B14145
Microwave sensor, rear left
1
X14146
B14146
Microwave sensor, rear right
20330, 23317
BentleyPublishers.com
B510ele20330

Fuse F52
up to 09 / 2005

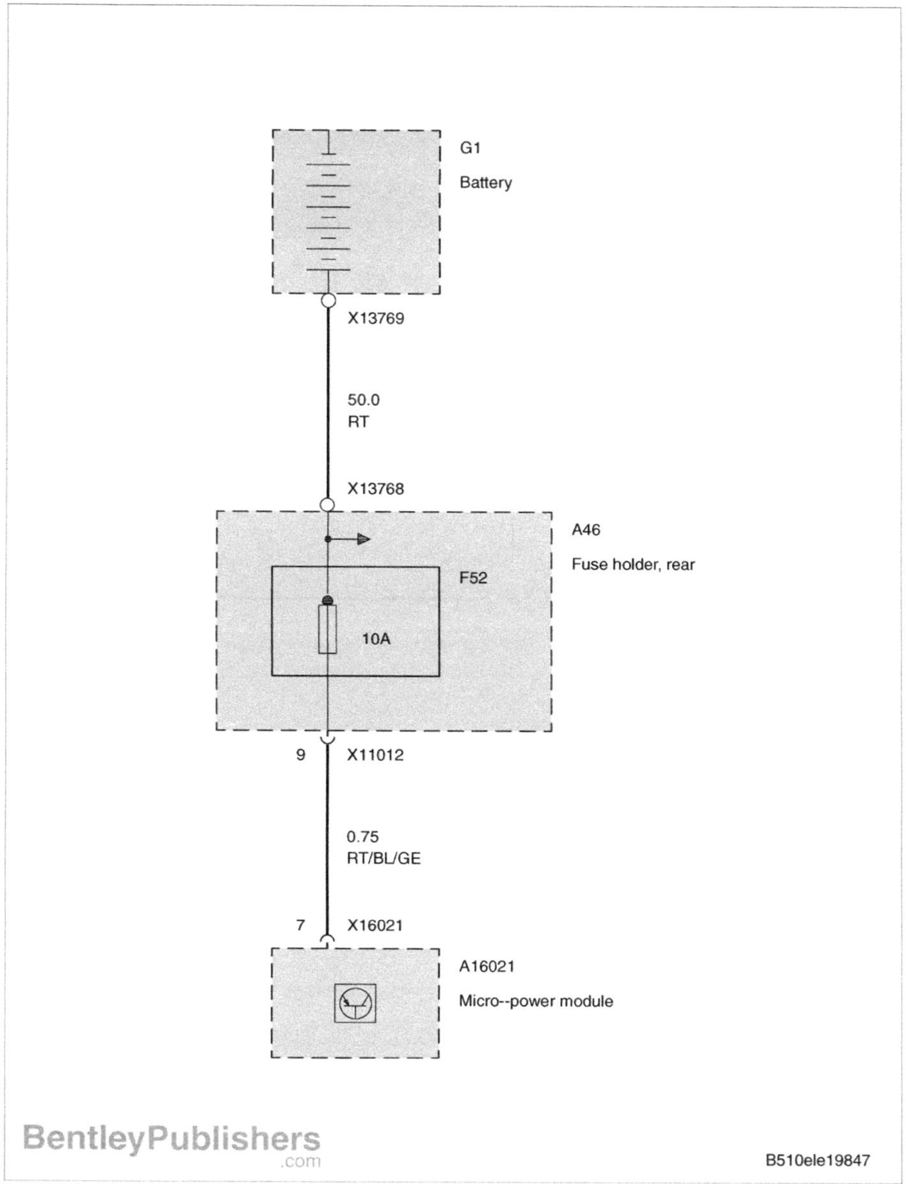

Fuse F52
as of 09 / 2005

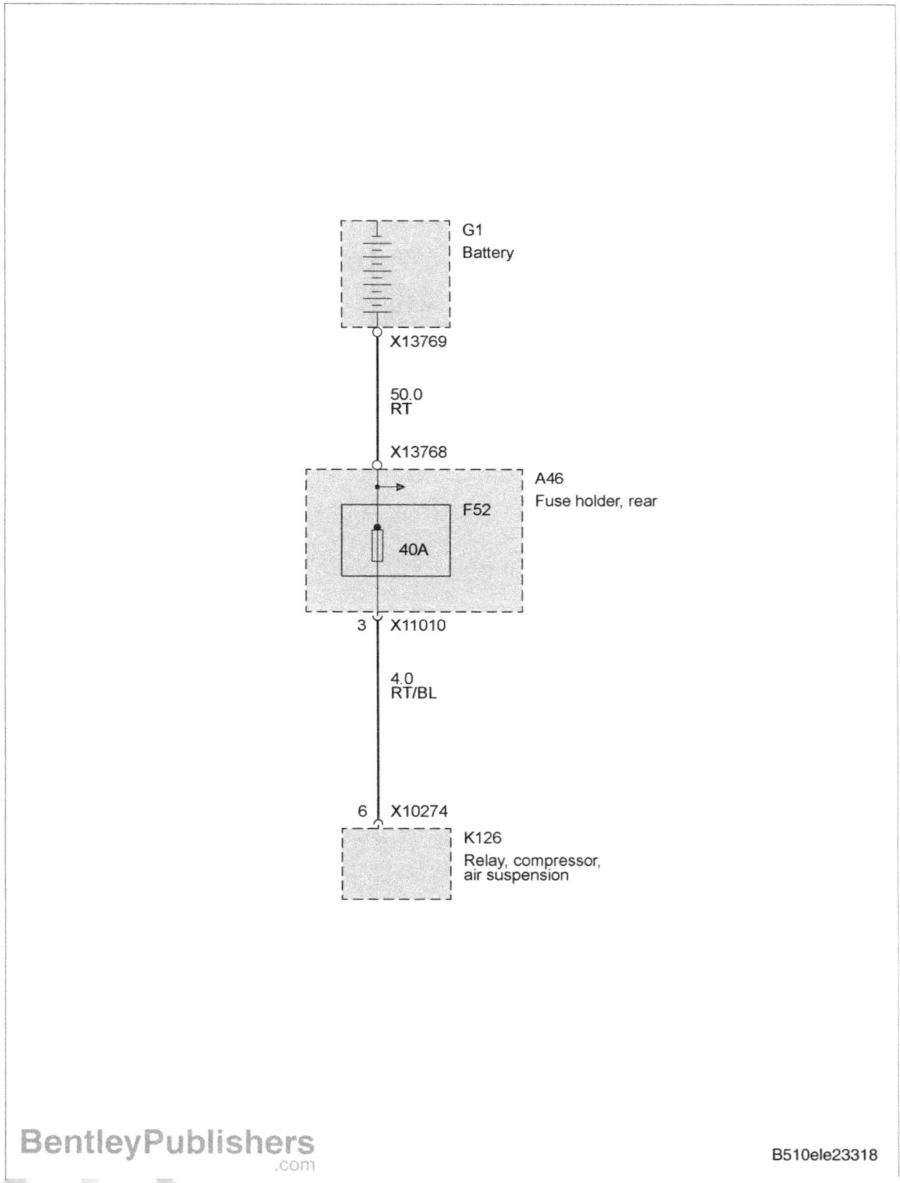

Fuse F53
up to 09 / 2004

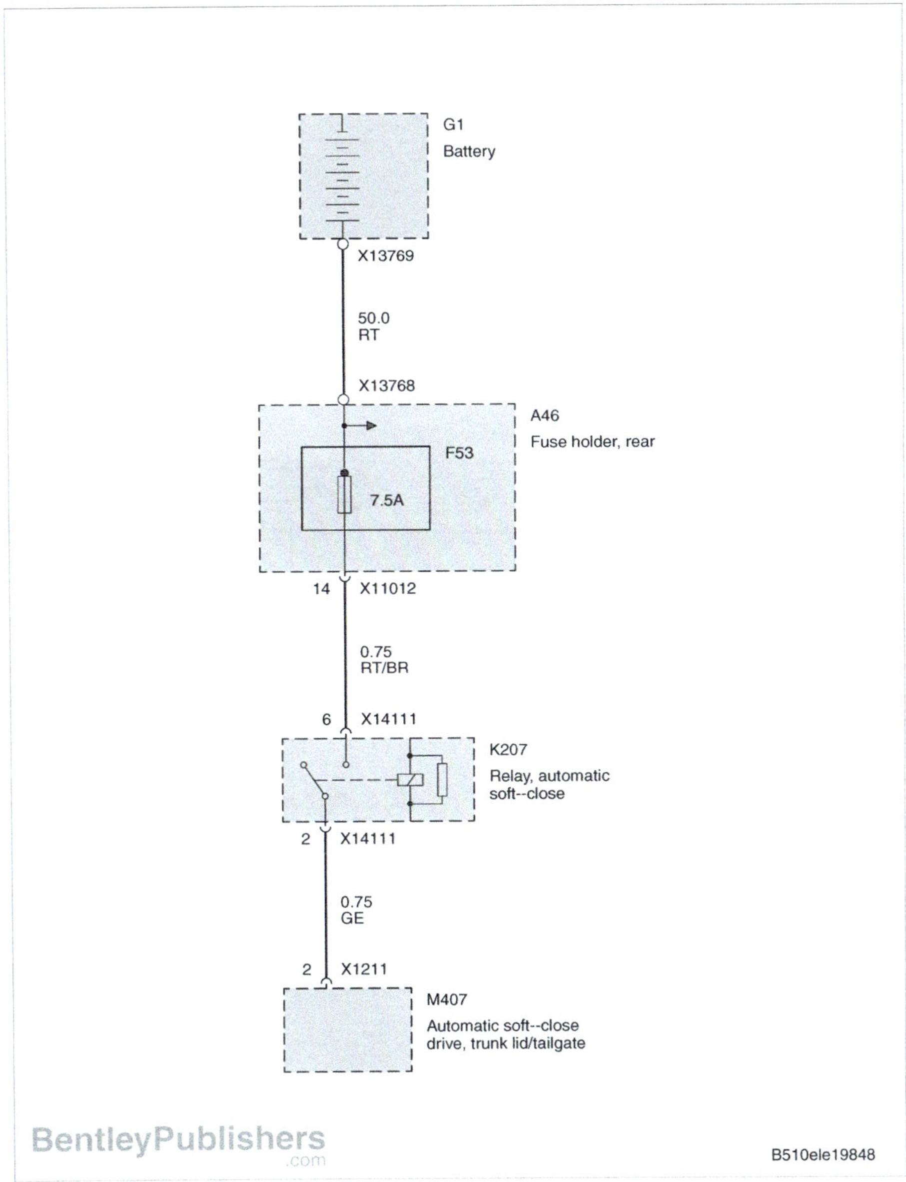

Fuse F53
as of 09 / 2004 up to 09 / 2005

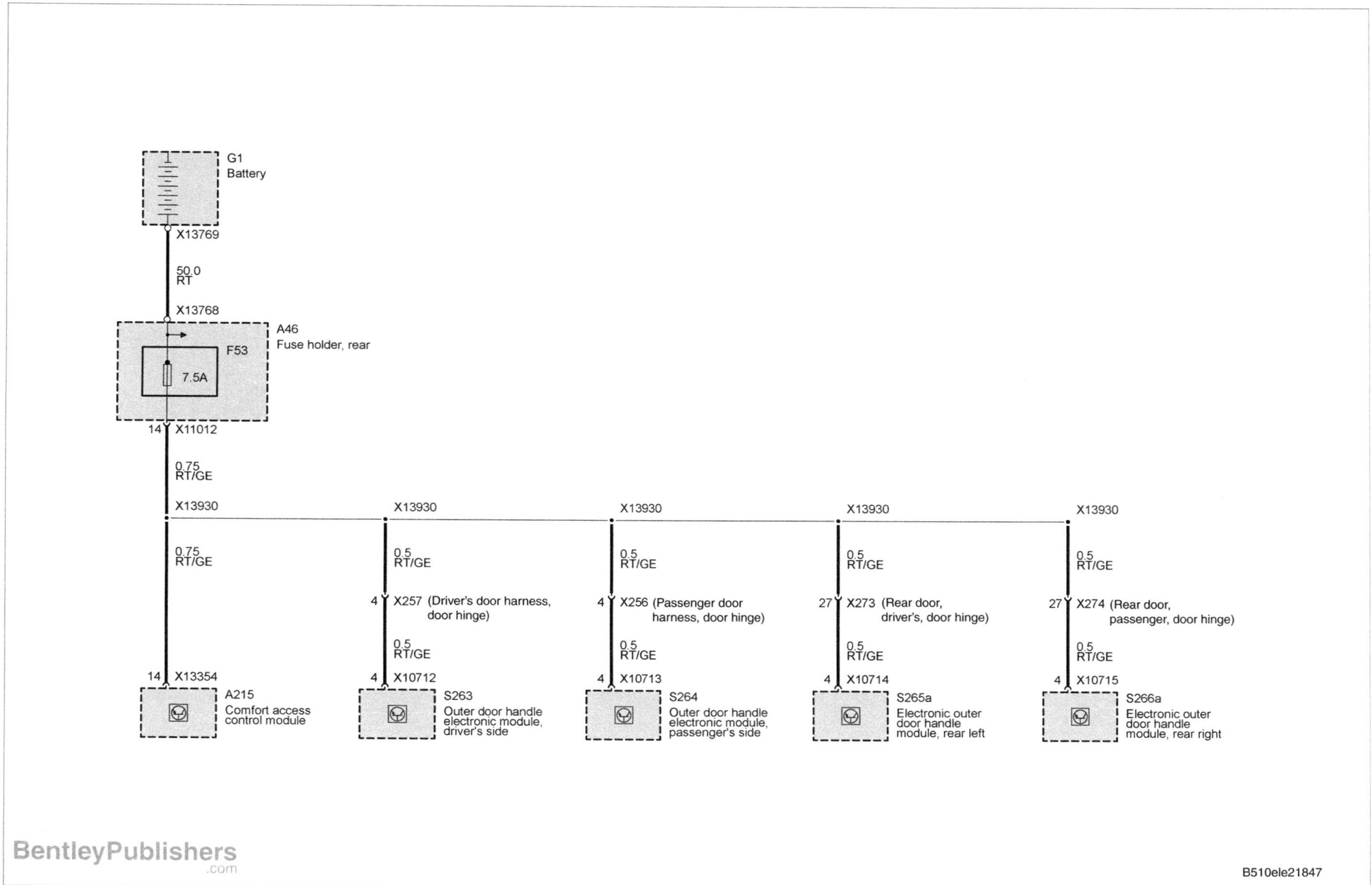

BentleyPublishers
.com

Fuse F53
as of 09 / 2005

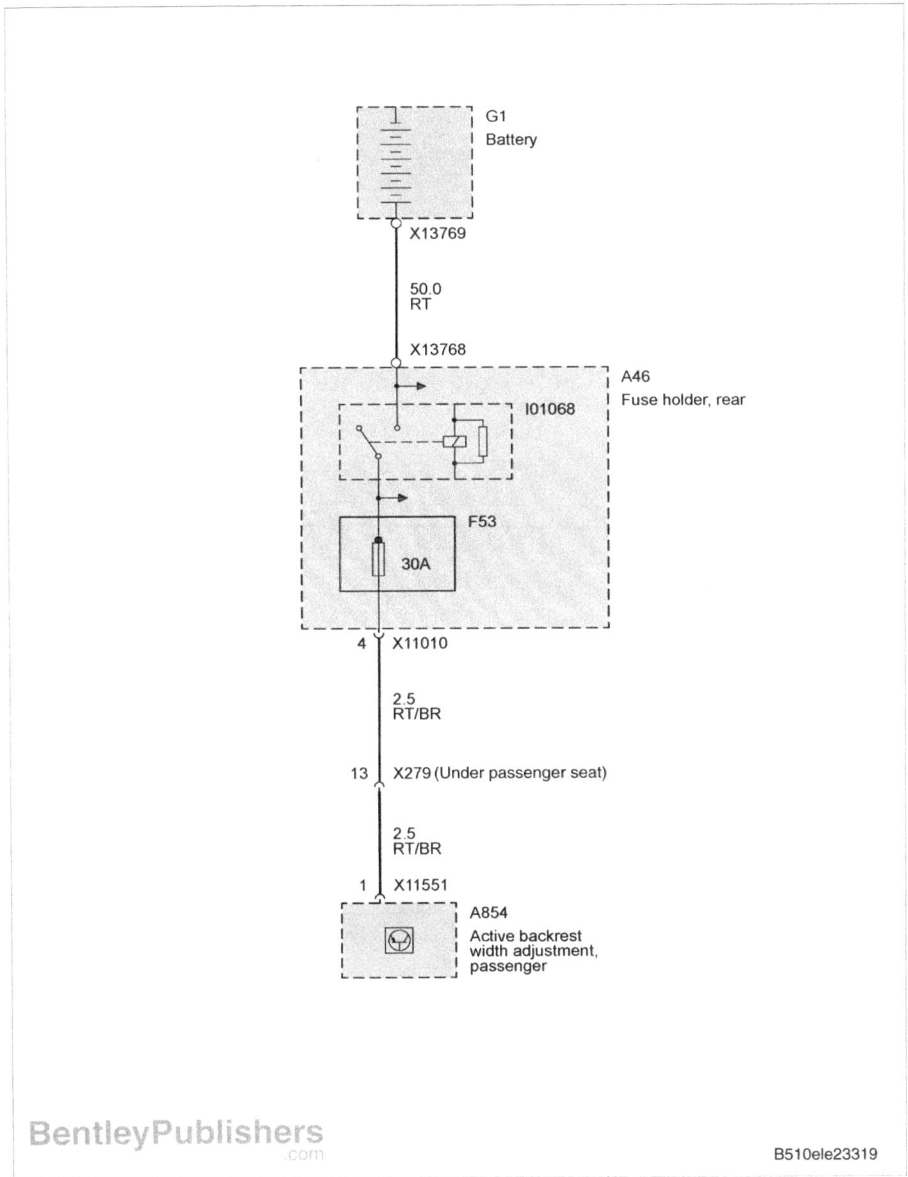

Fuse F54
up to 09 / 2005

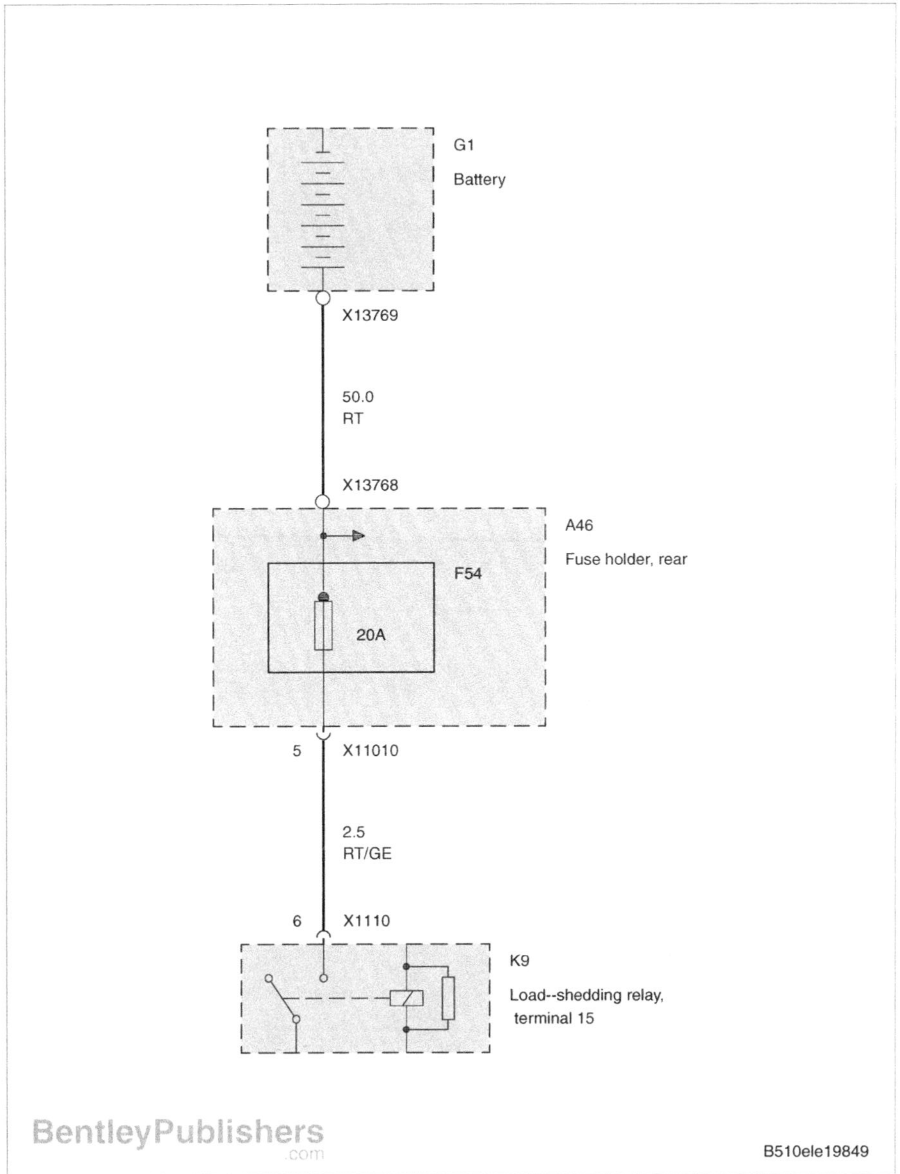

Fuse F54
as of 09 / 2005

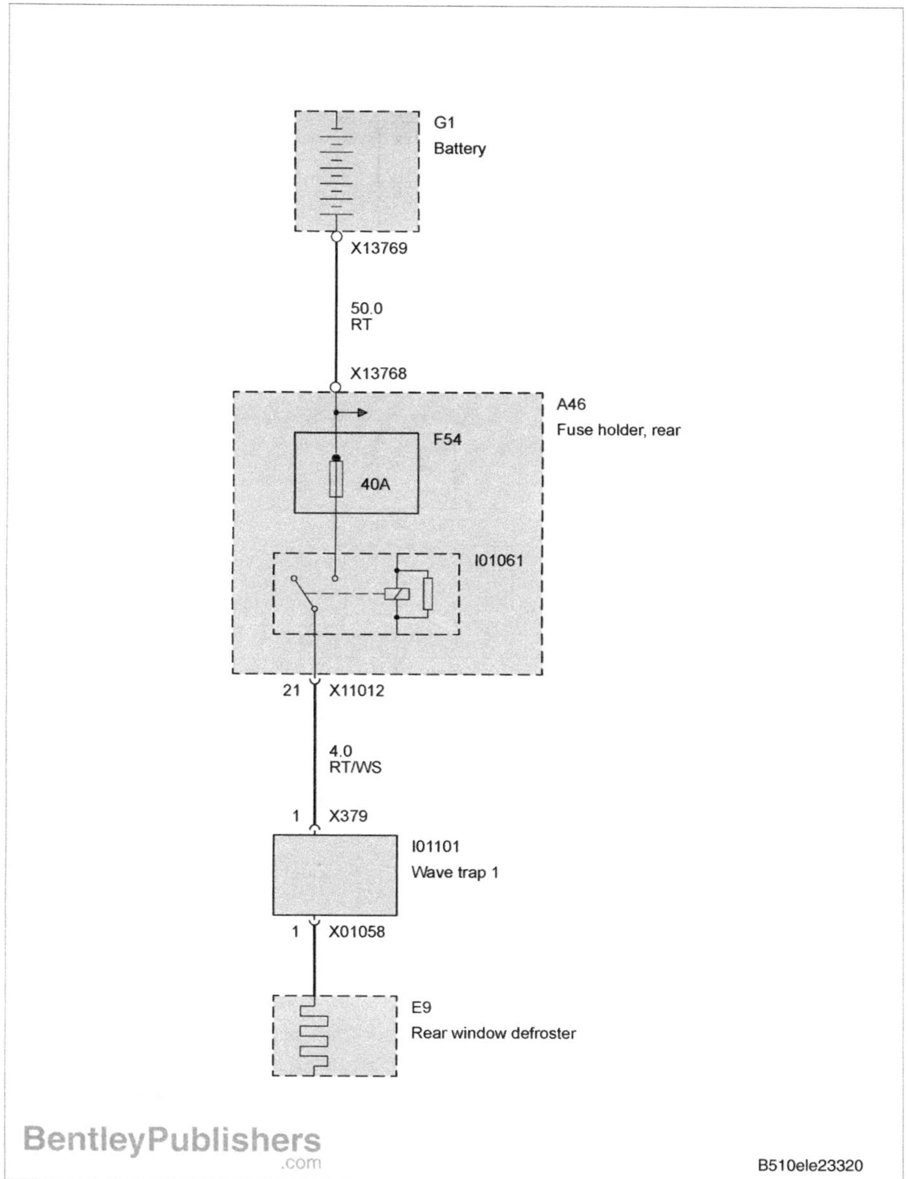

Fuse F55
up to 09 / 2005

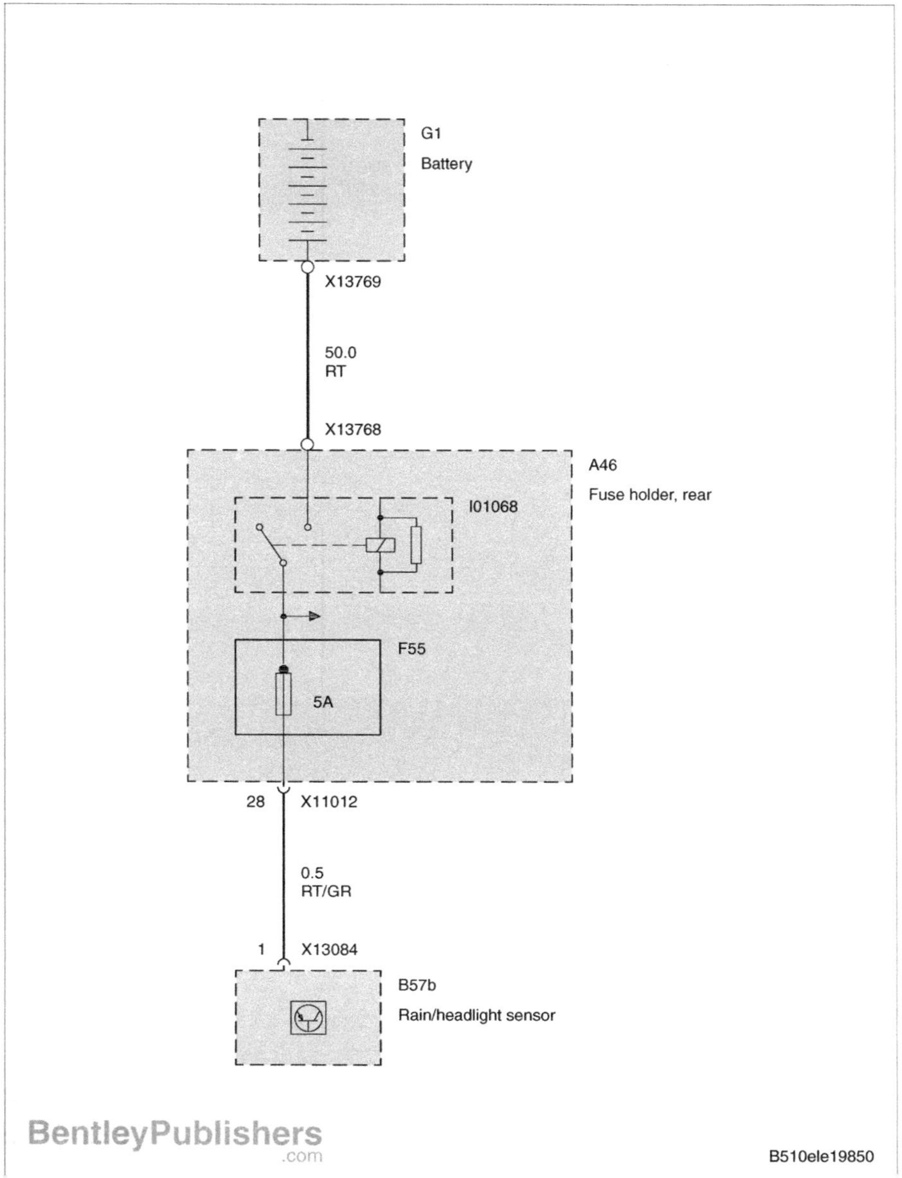

Fuse F55
as of 09 / 2005

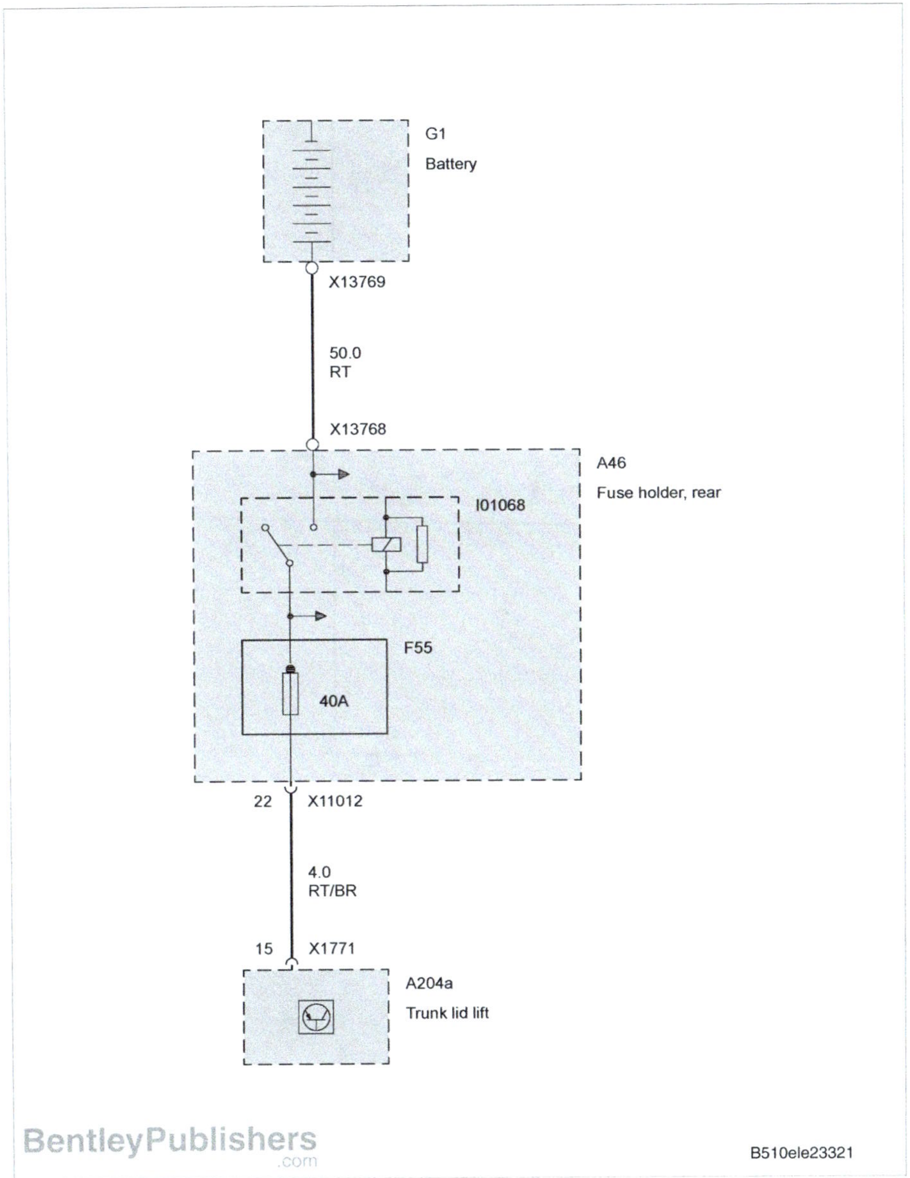

Fuse F56
as of 09 / 2005

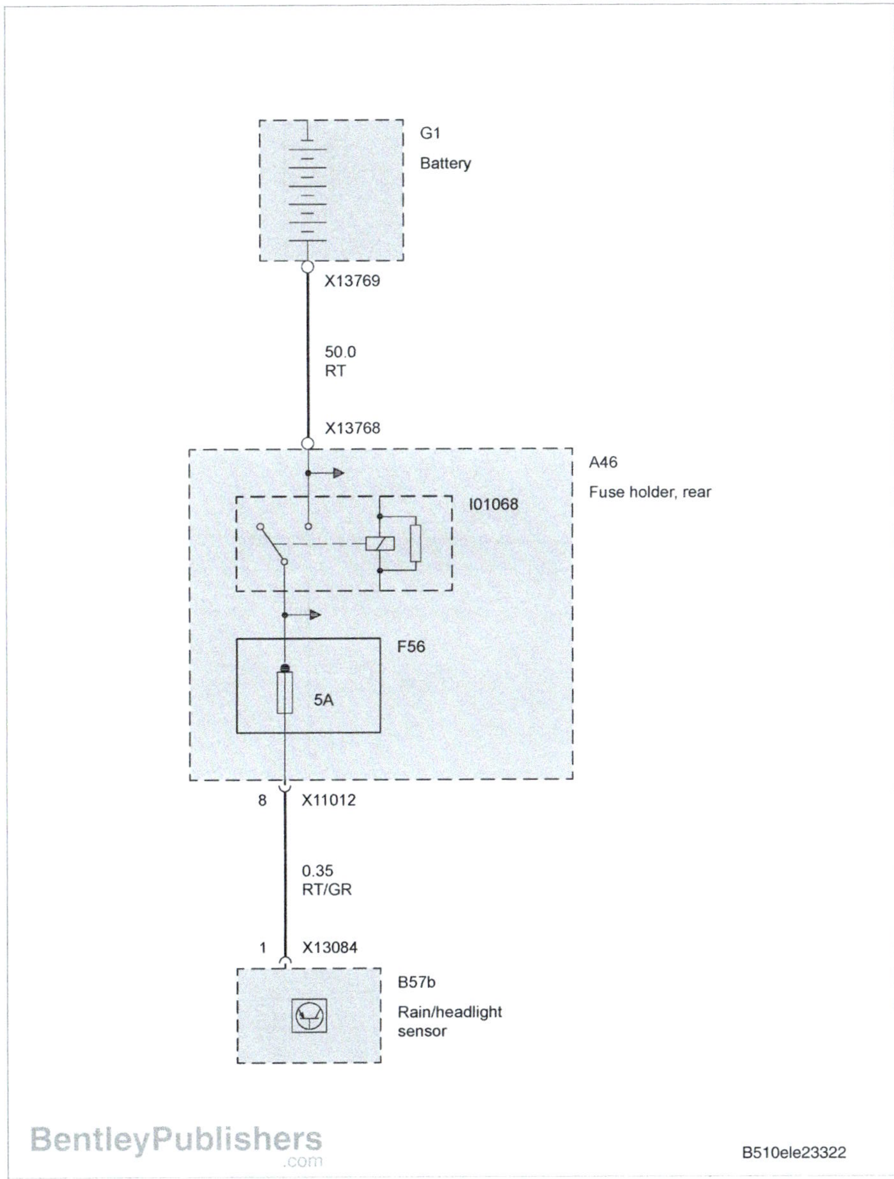

Fuse F57
up to 09 / 2005

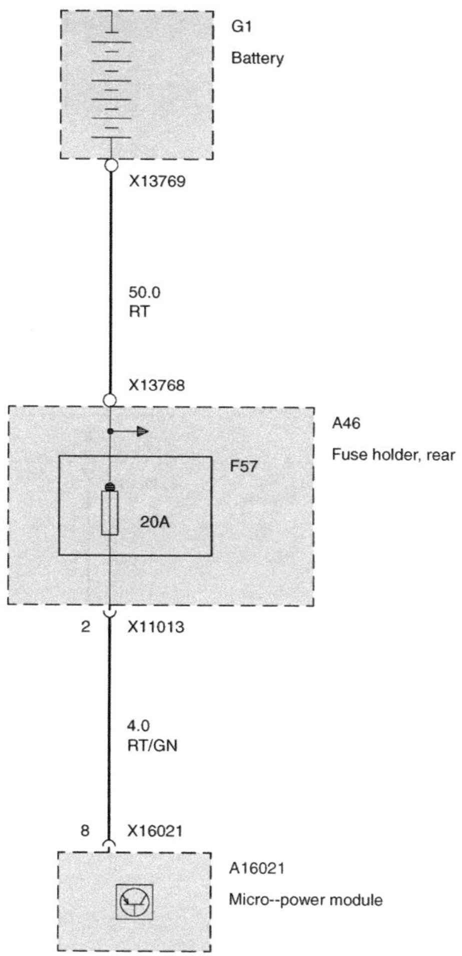

BentleyPublishers.com

B510ele19851

Fuse F57
IHKA Basic as of 09 / 2005

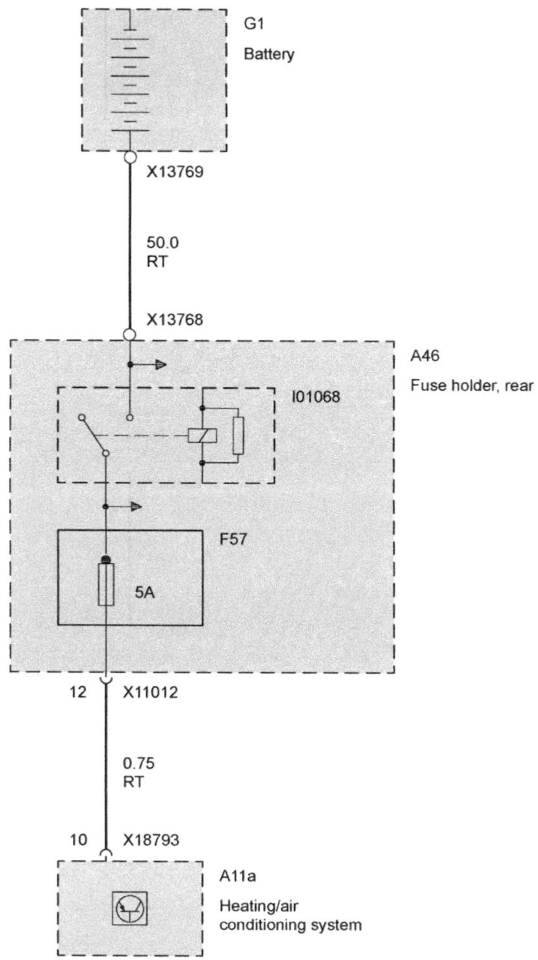

BentleyPublishers.com

B510ele23323

Fuse F58 up to 09 / 2005

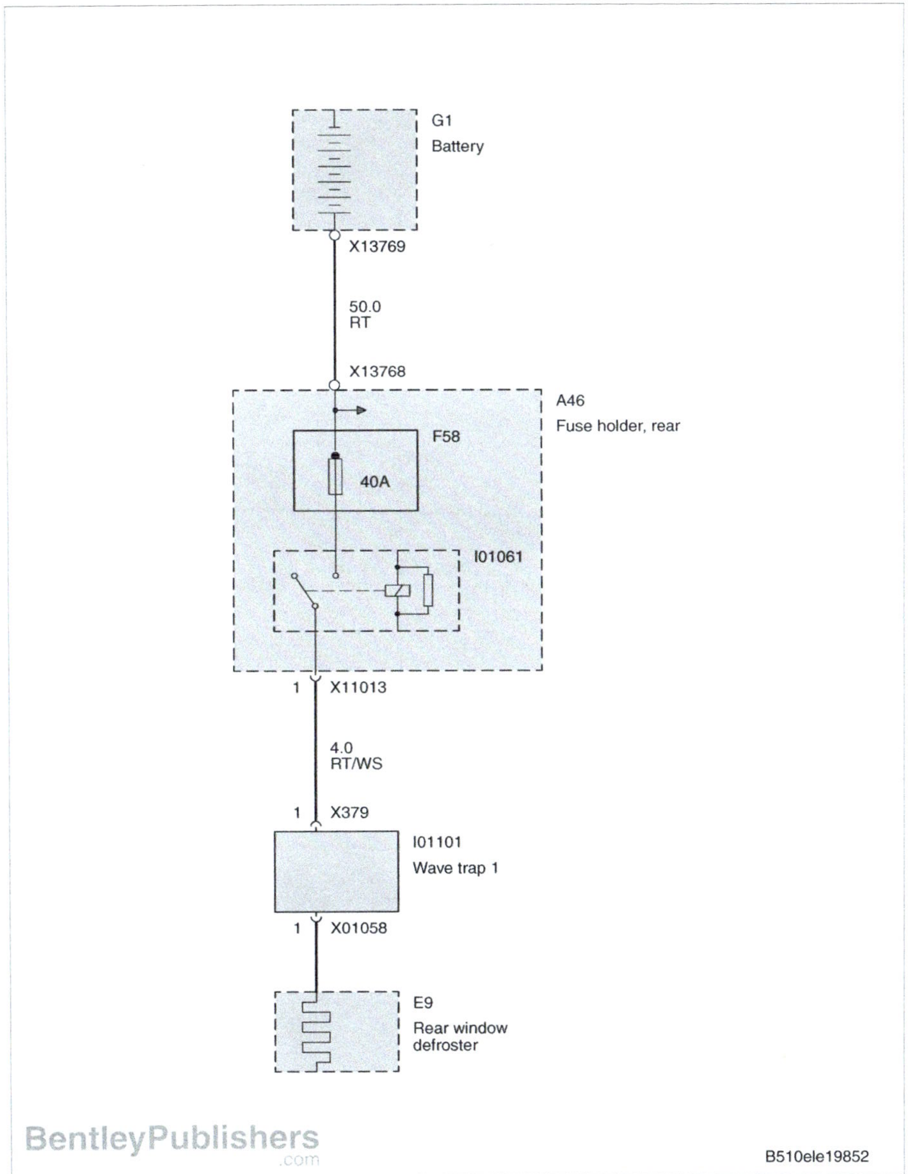

Fuse F58 as of 09 / 2005

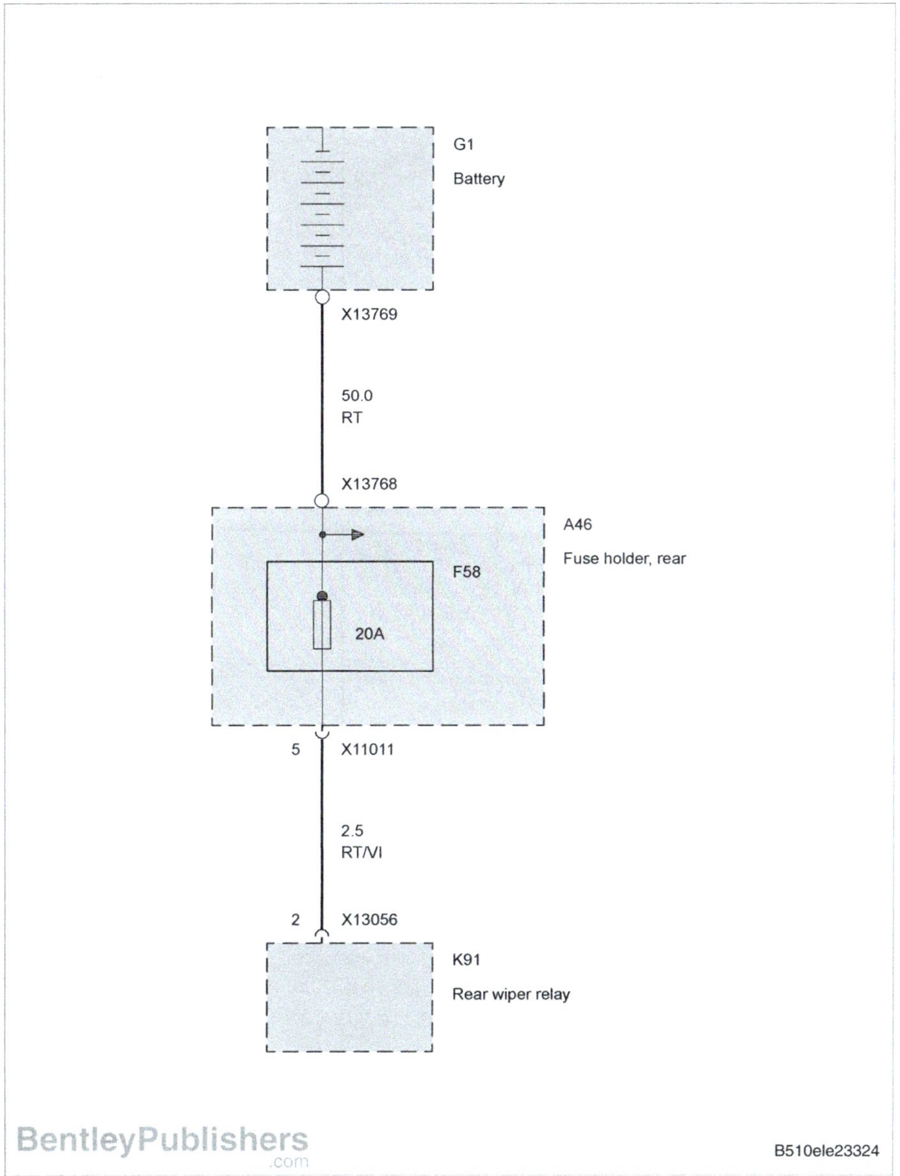

Fuse F59
as of 09 / 2005

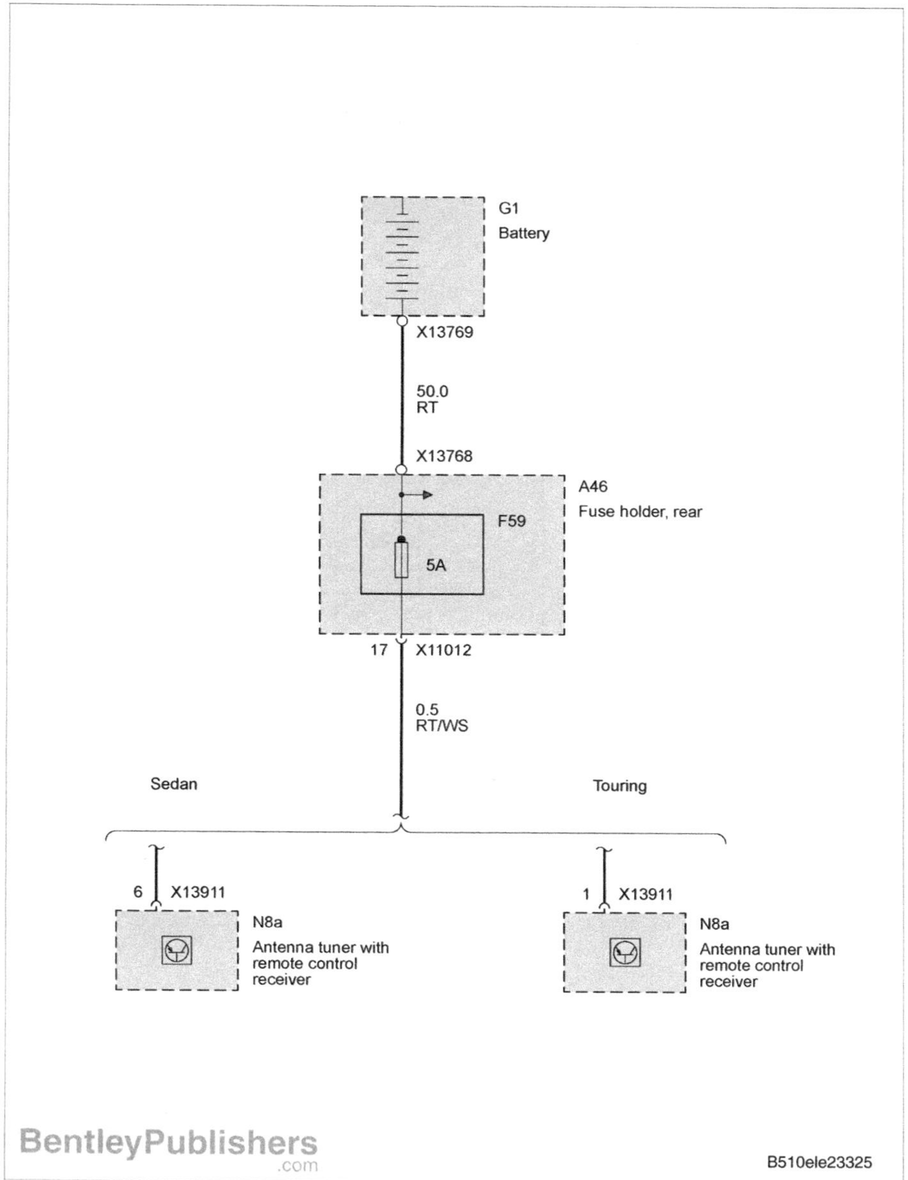

Fuse F59

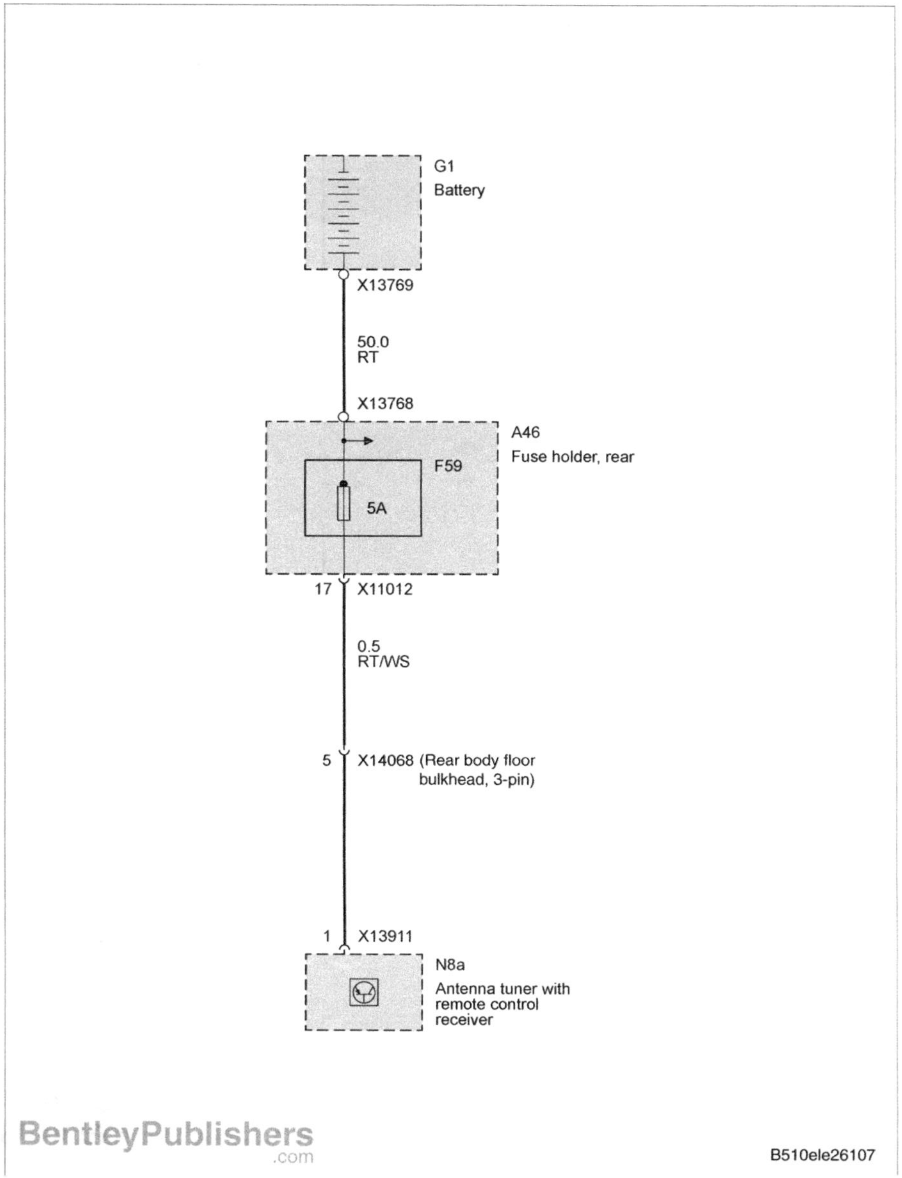

Fuse F60
as of 09 / 2005

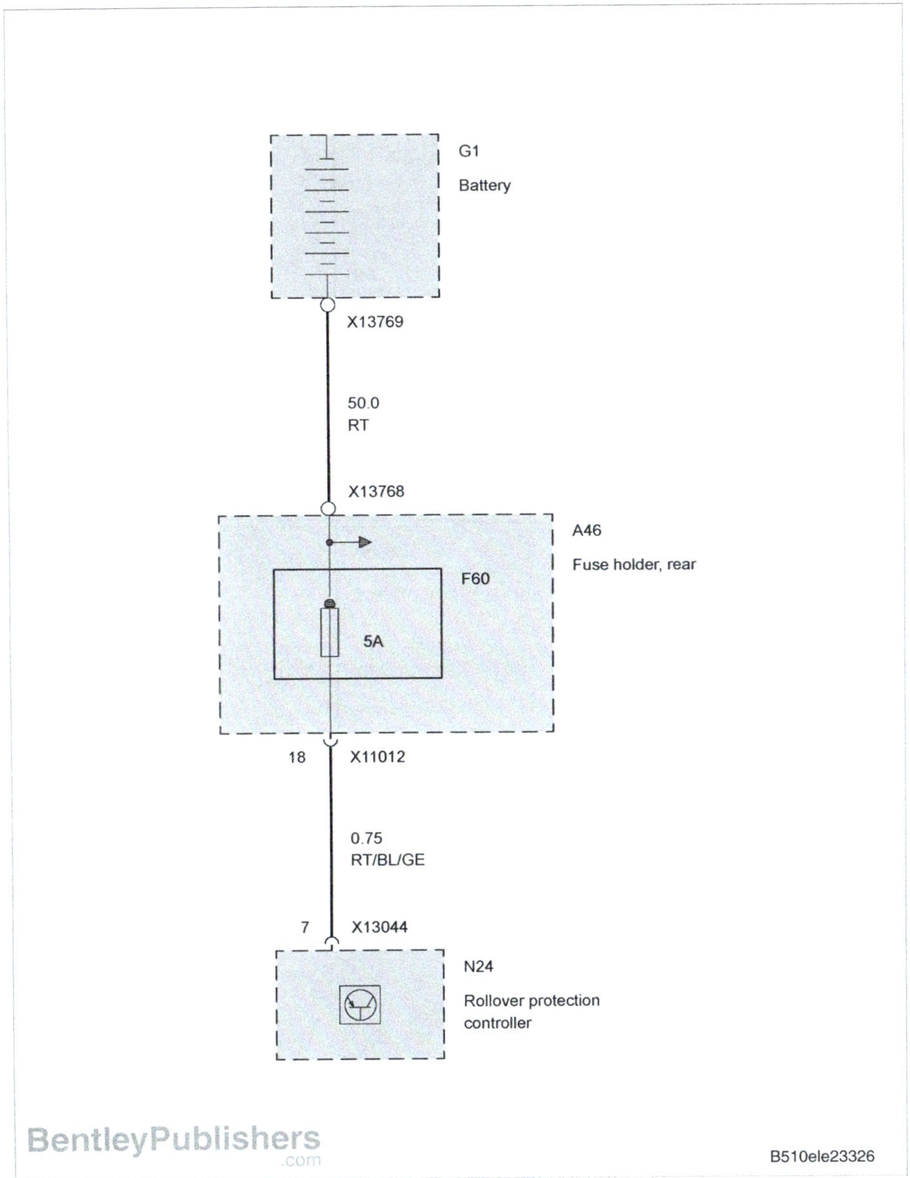

Fuse F61
up to 09 / 2005

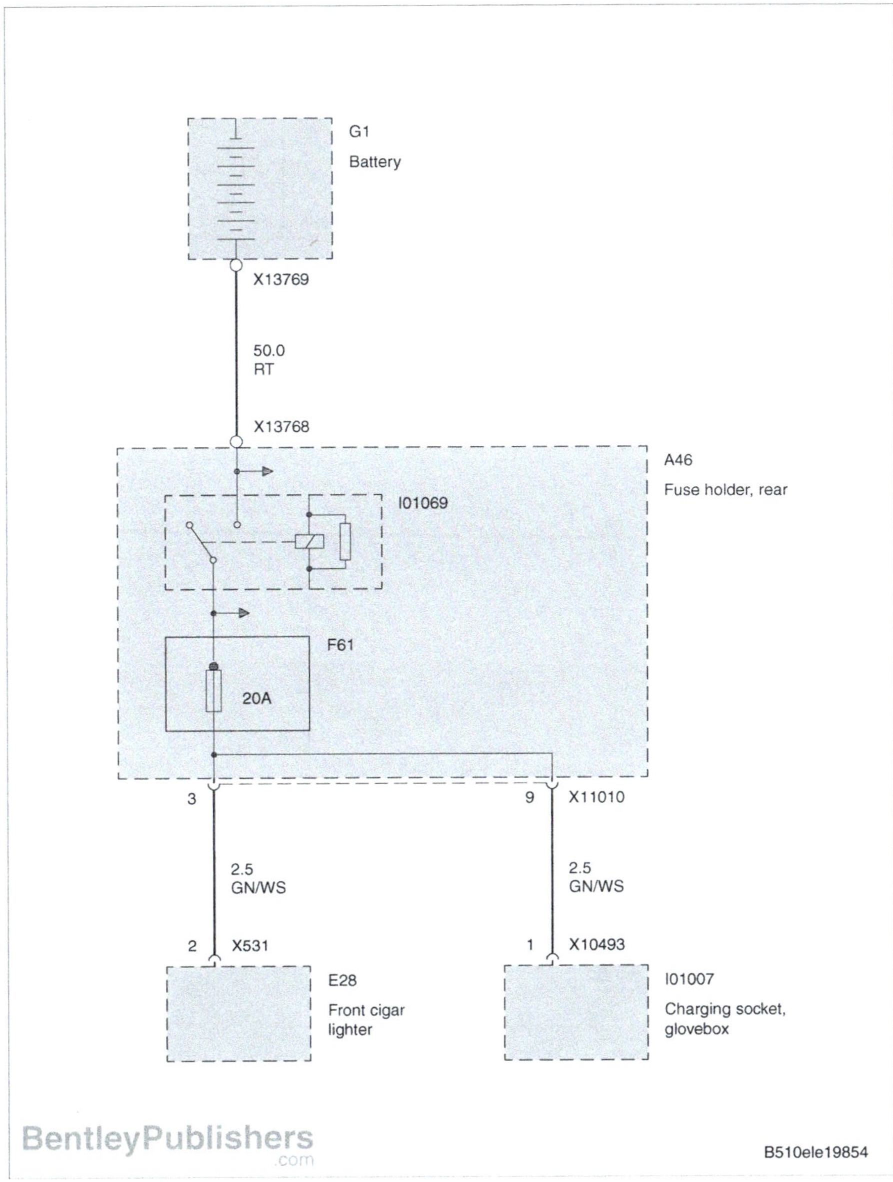

Fuse F61
as of 09 / 2005

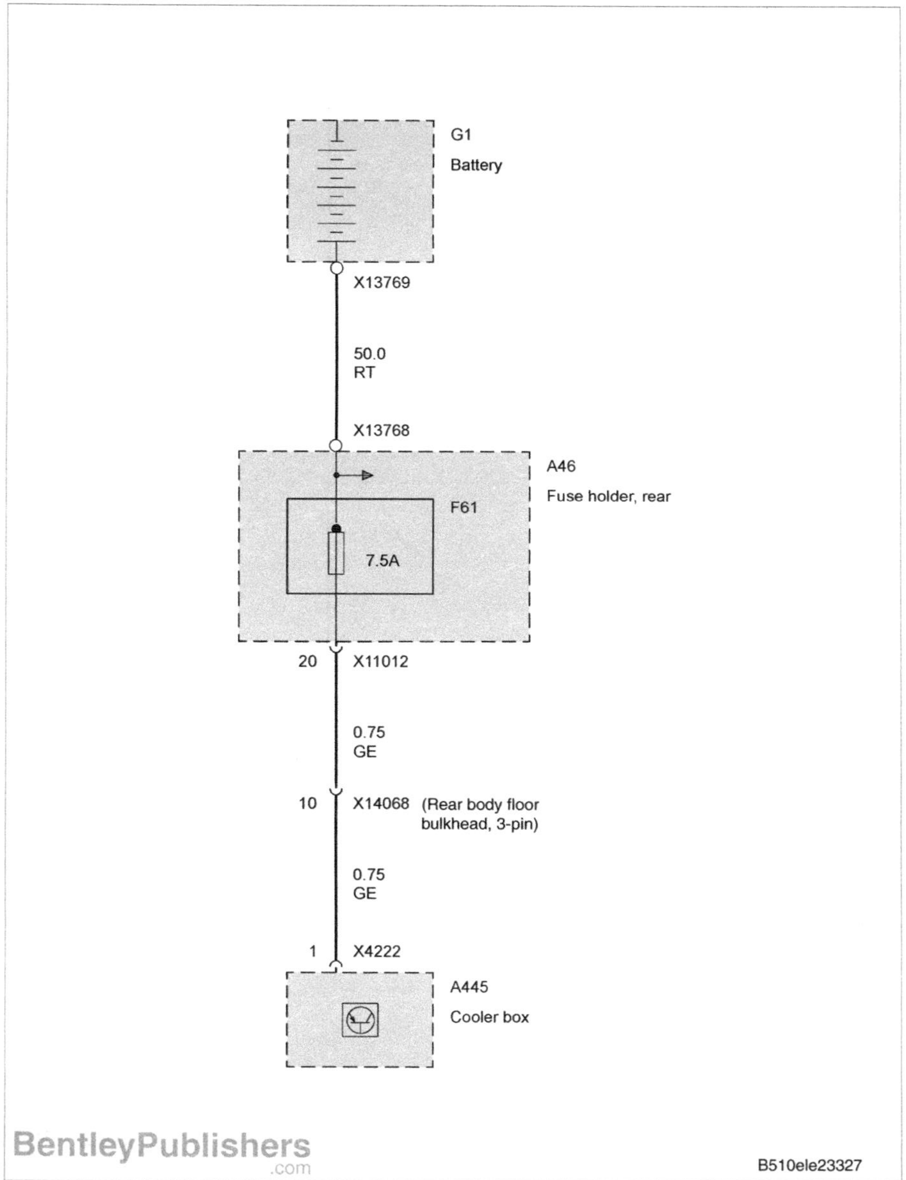

Fuse F62
up to 09 / 2005

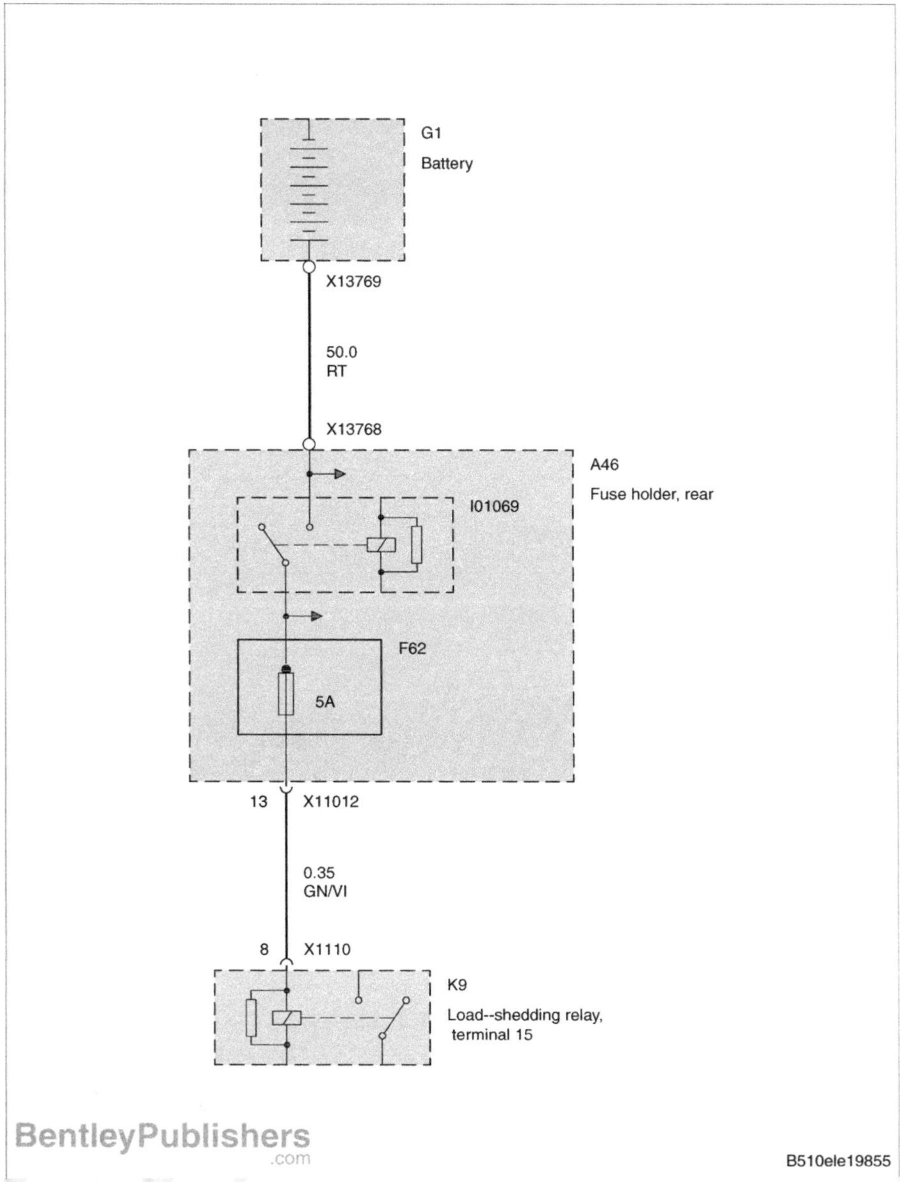

Fuse F62 as of 09 / 2005

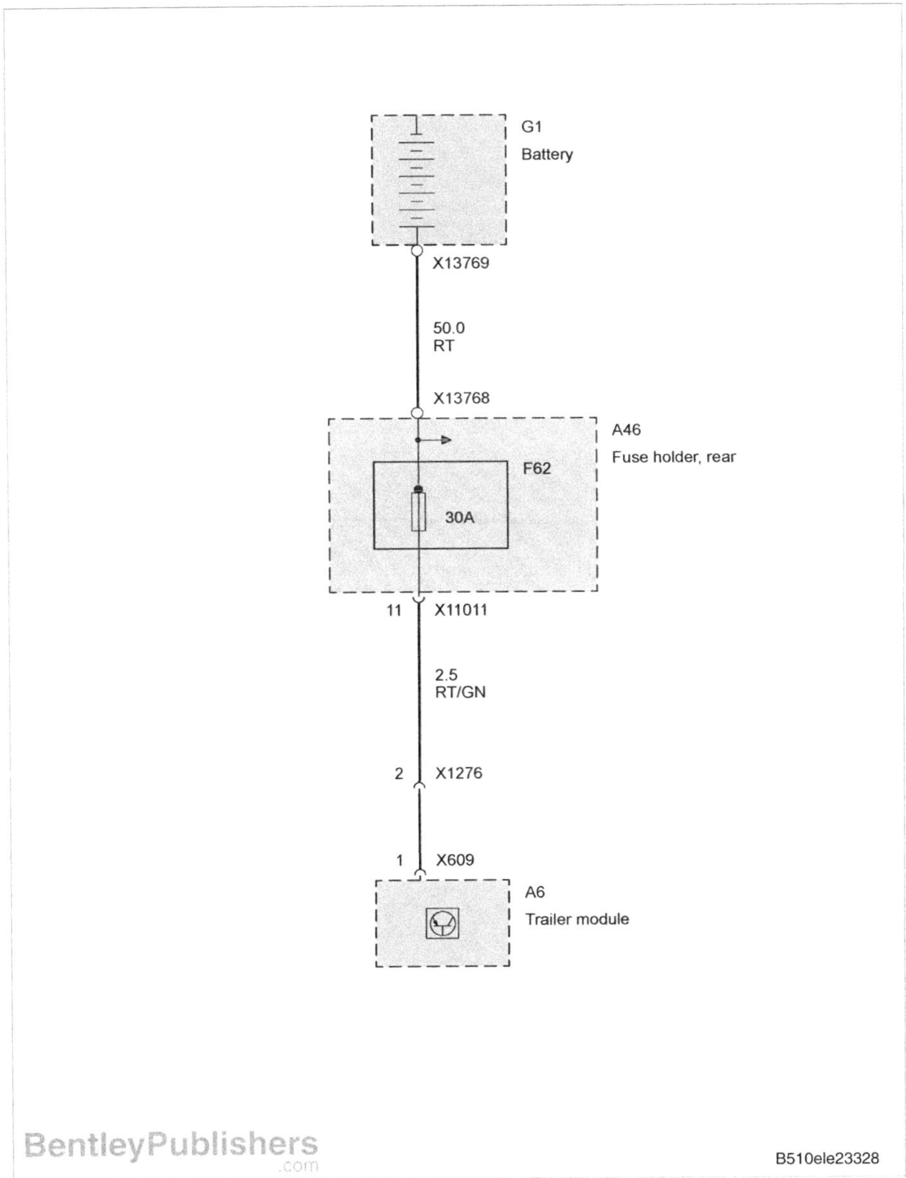

Fuse F63 up to 09 / 2005

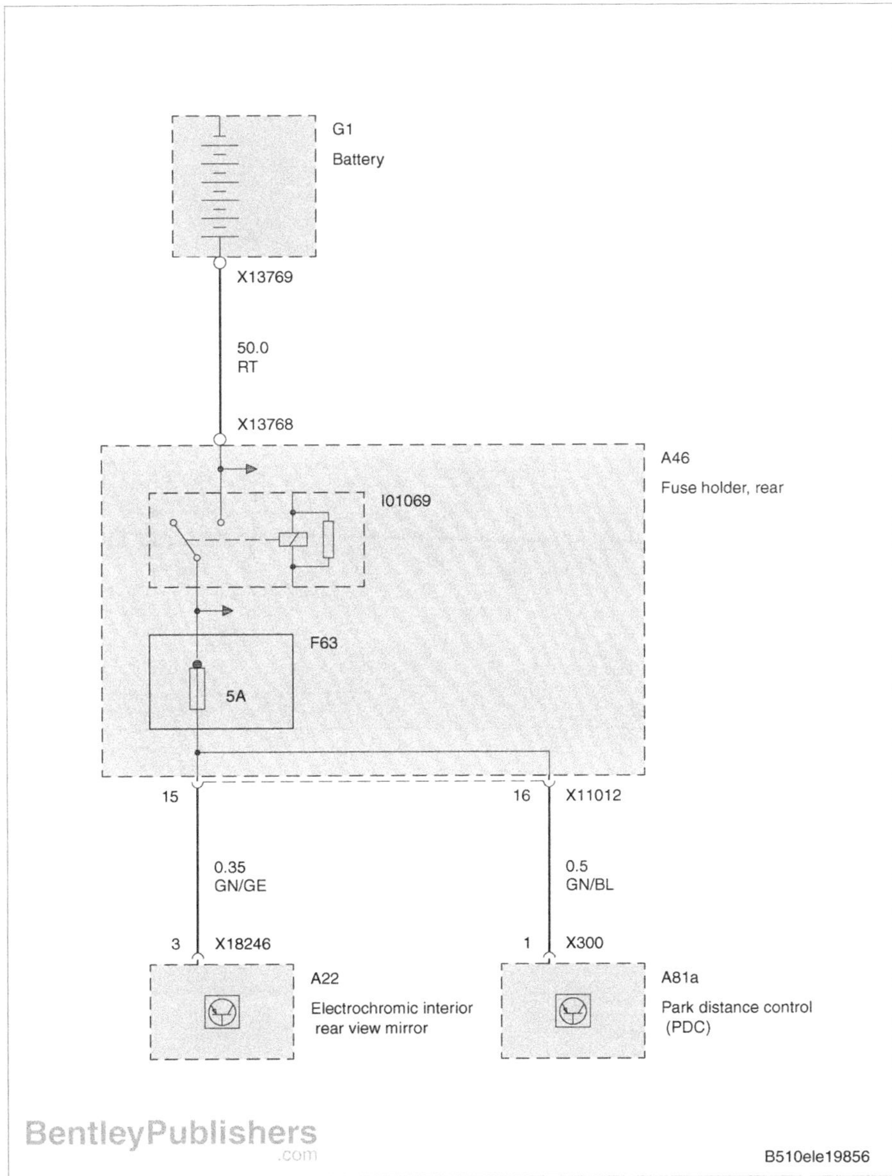

Fuse F63
as of 09 / 2005

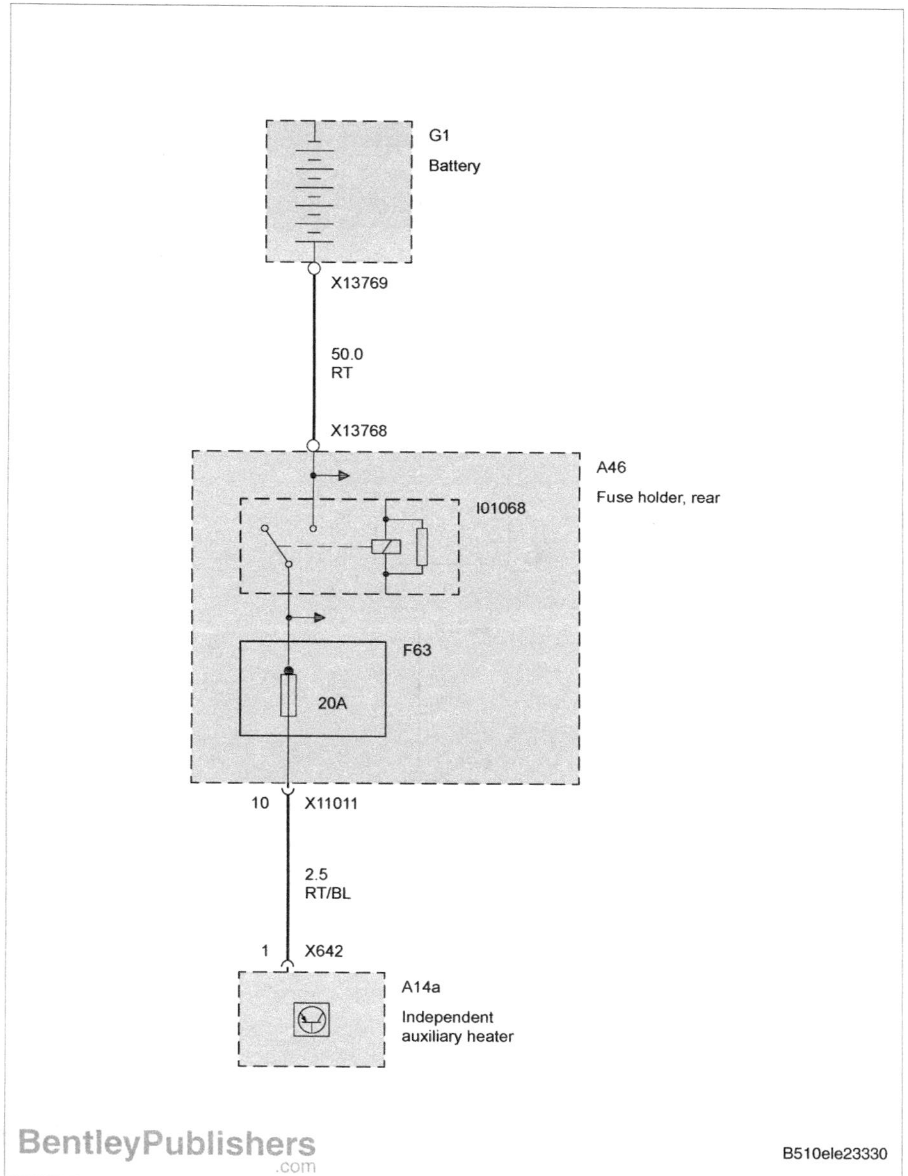

Fuse F64
up to 09 / 2005

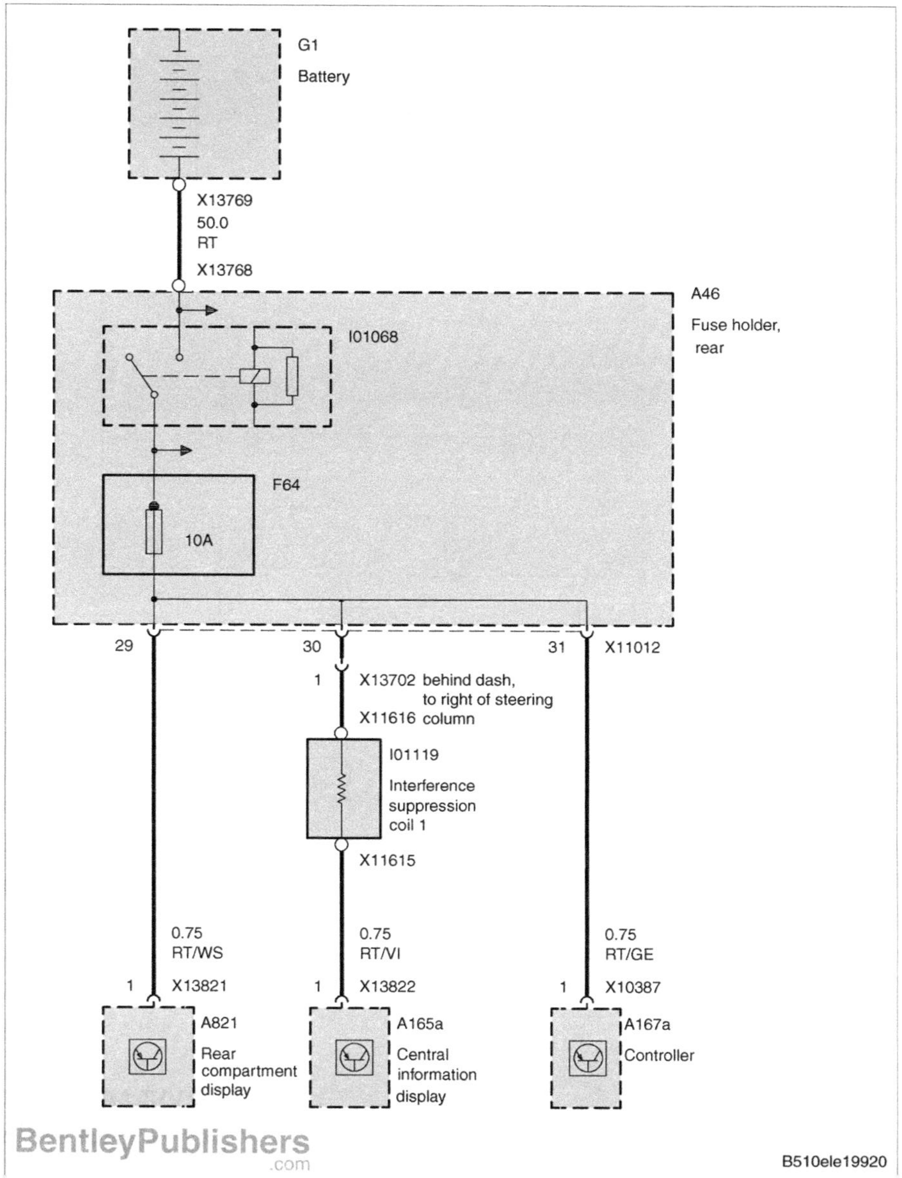

Fuse F64
IHKA High as of 09 / 2005

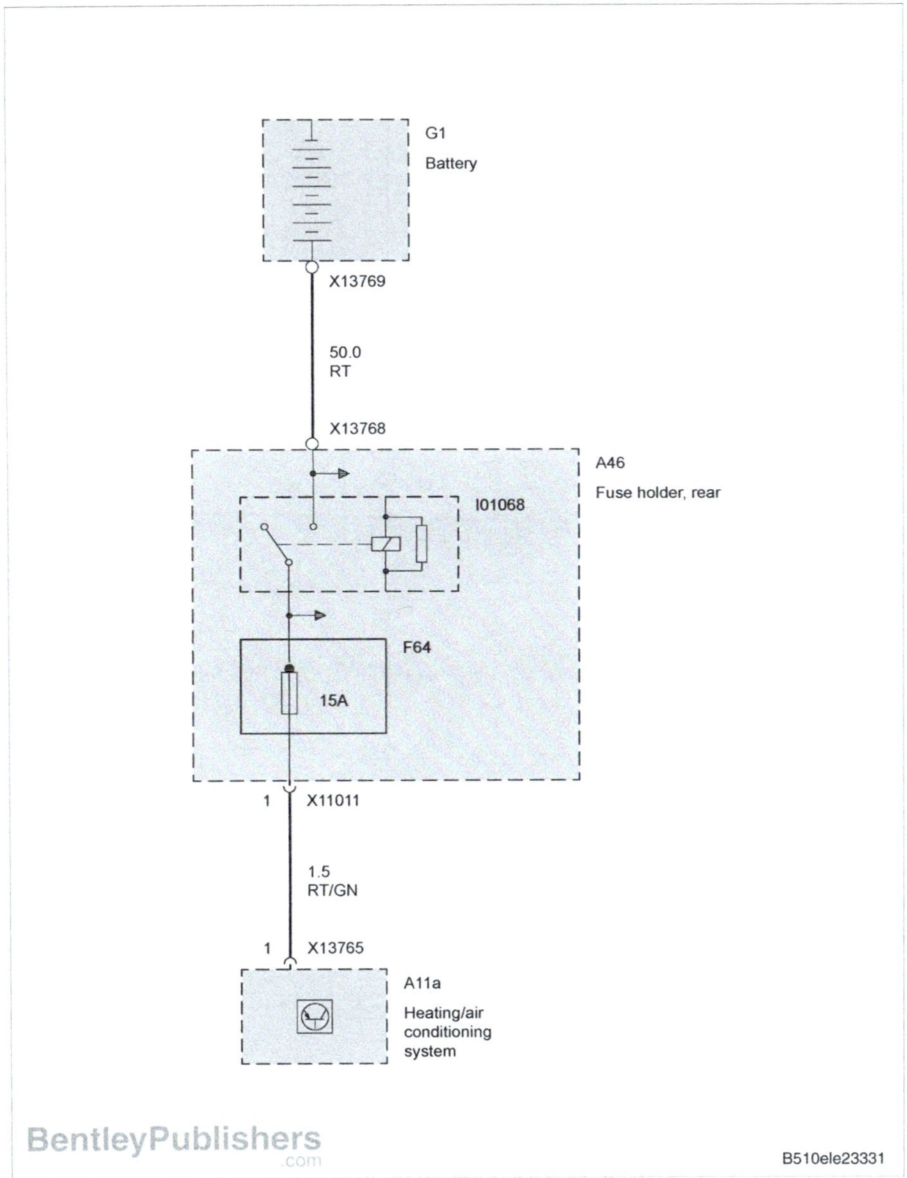

Fuse F65
up to 09 / 2005

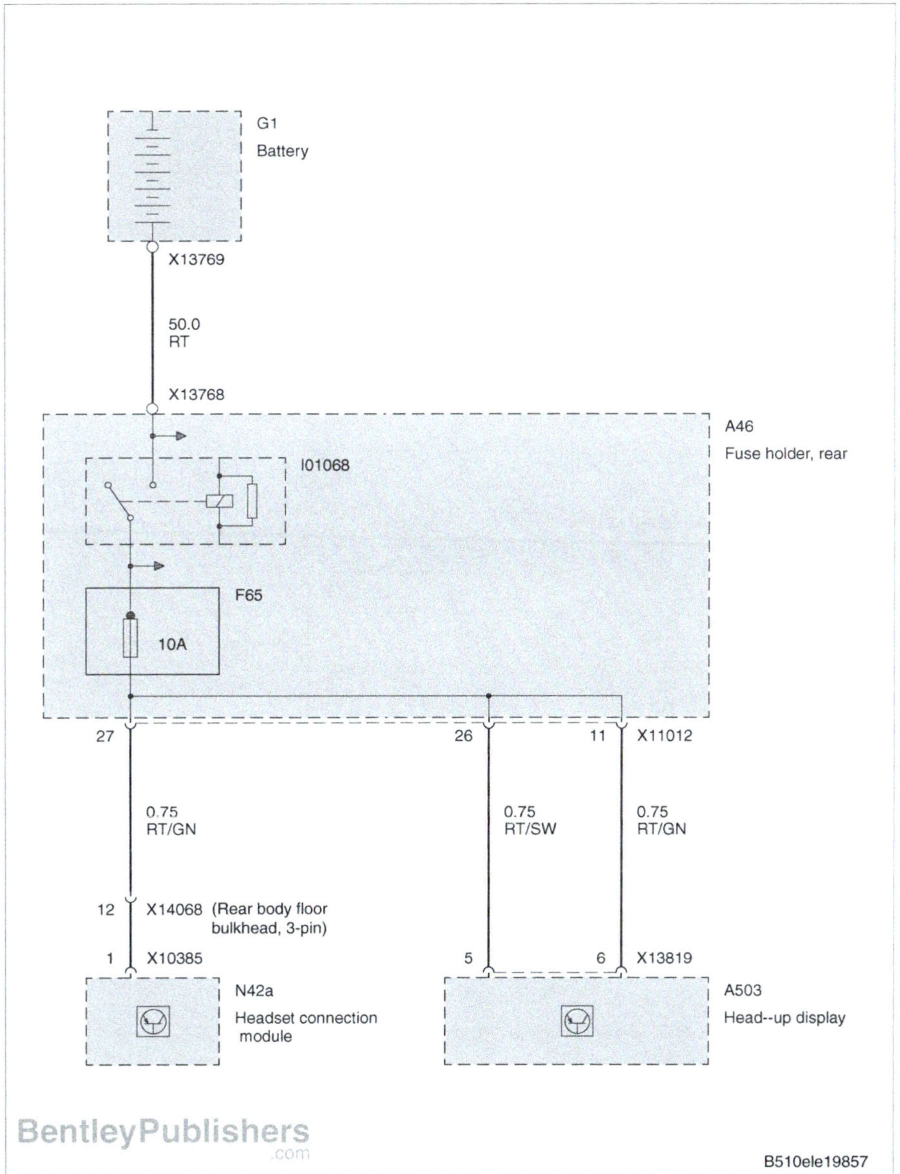

Fuse F65
as of 09 / 2005

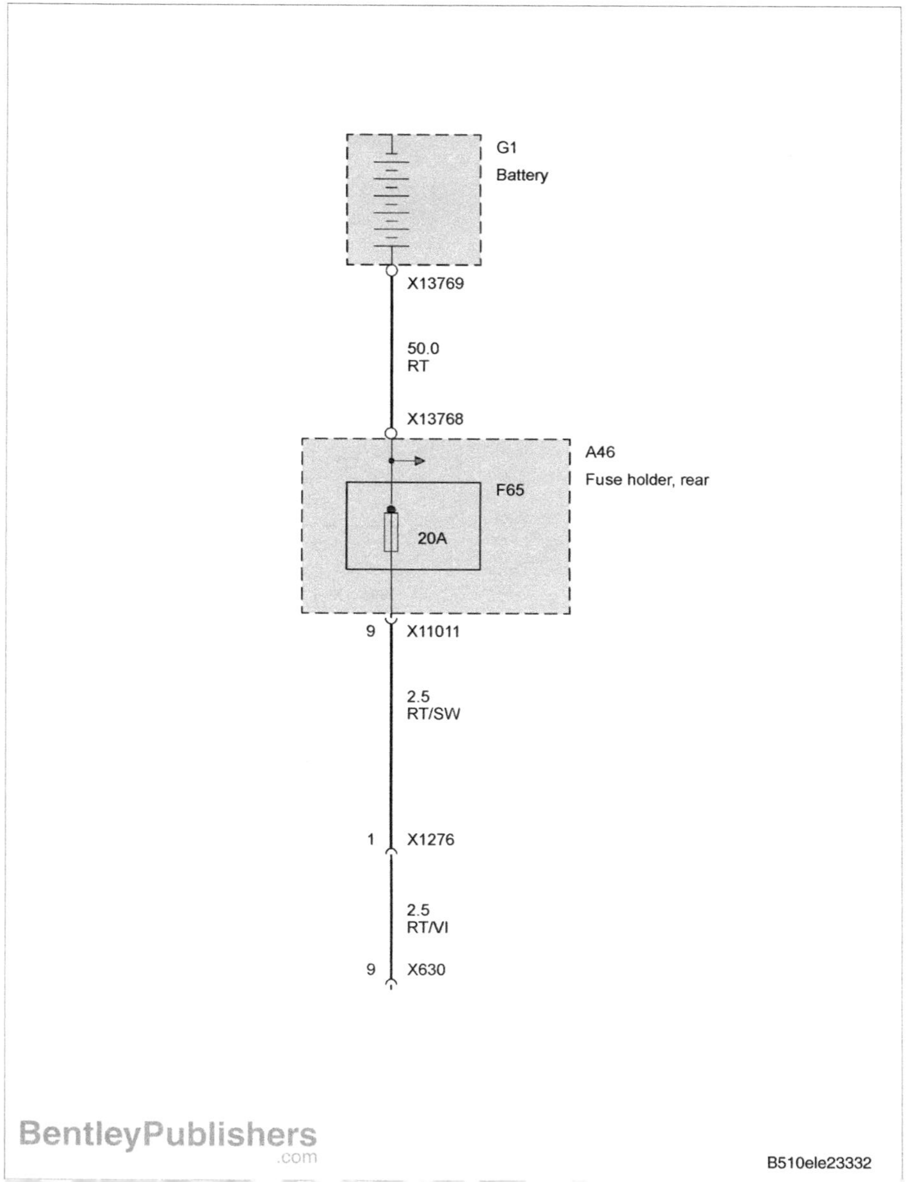

Fuse F66
without VTG Transfer case up to 09 / 2005

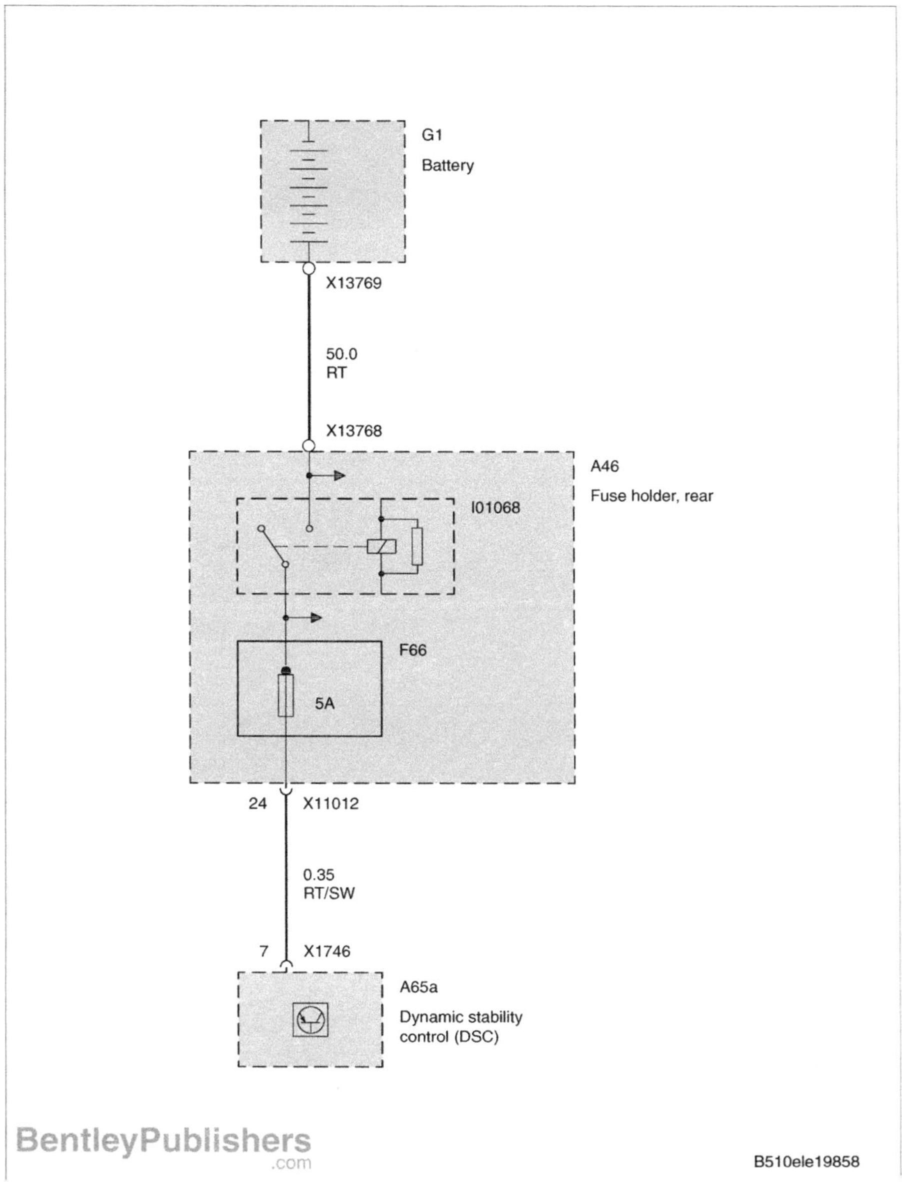

Fuse F66
VTG Transfer case up to 09 / 2005

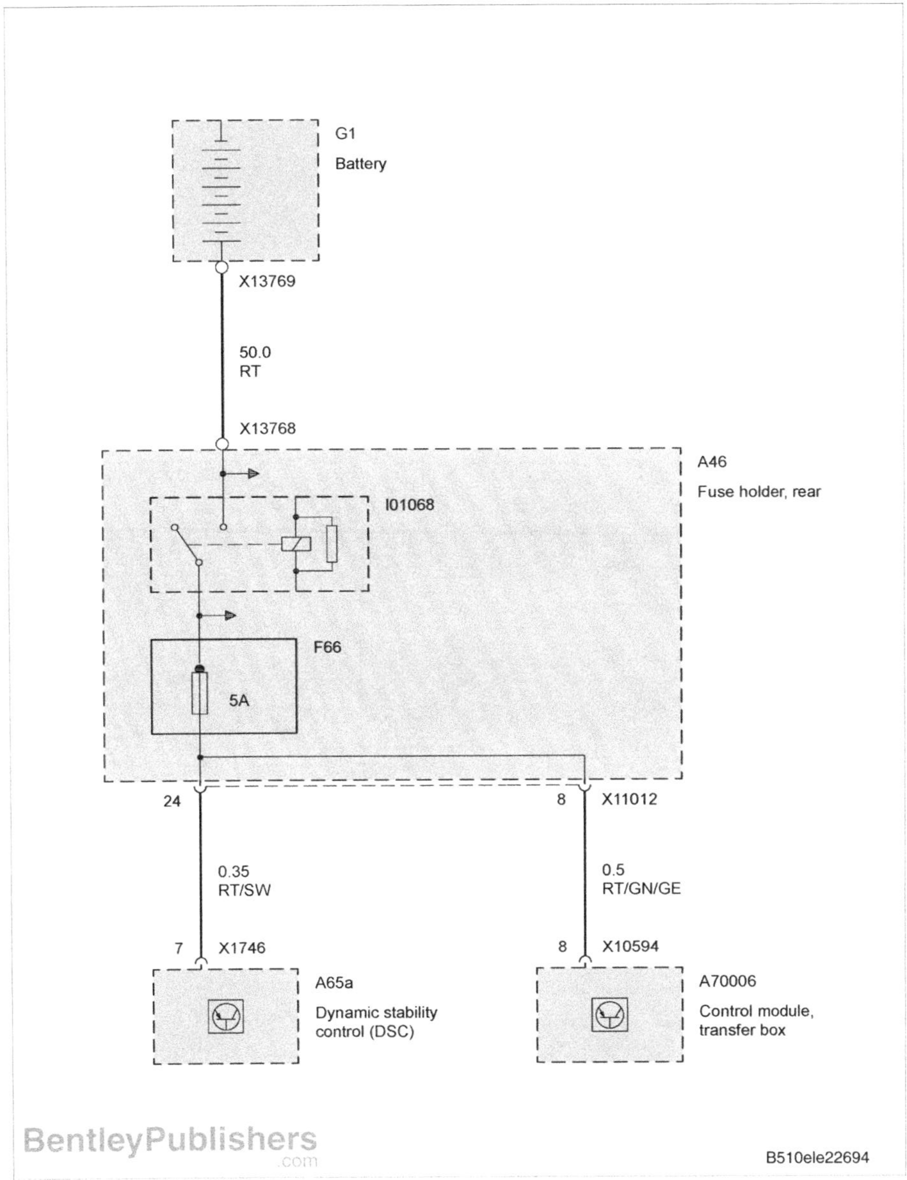

Fuse F66
Sedan as of 09 / 2005

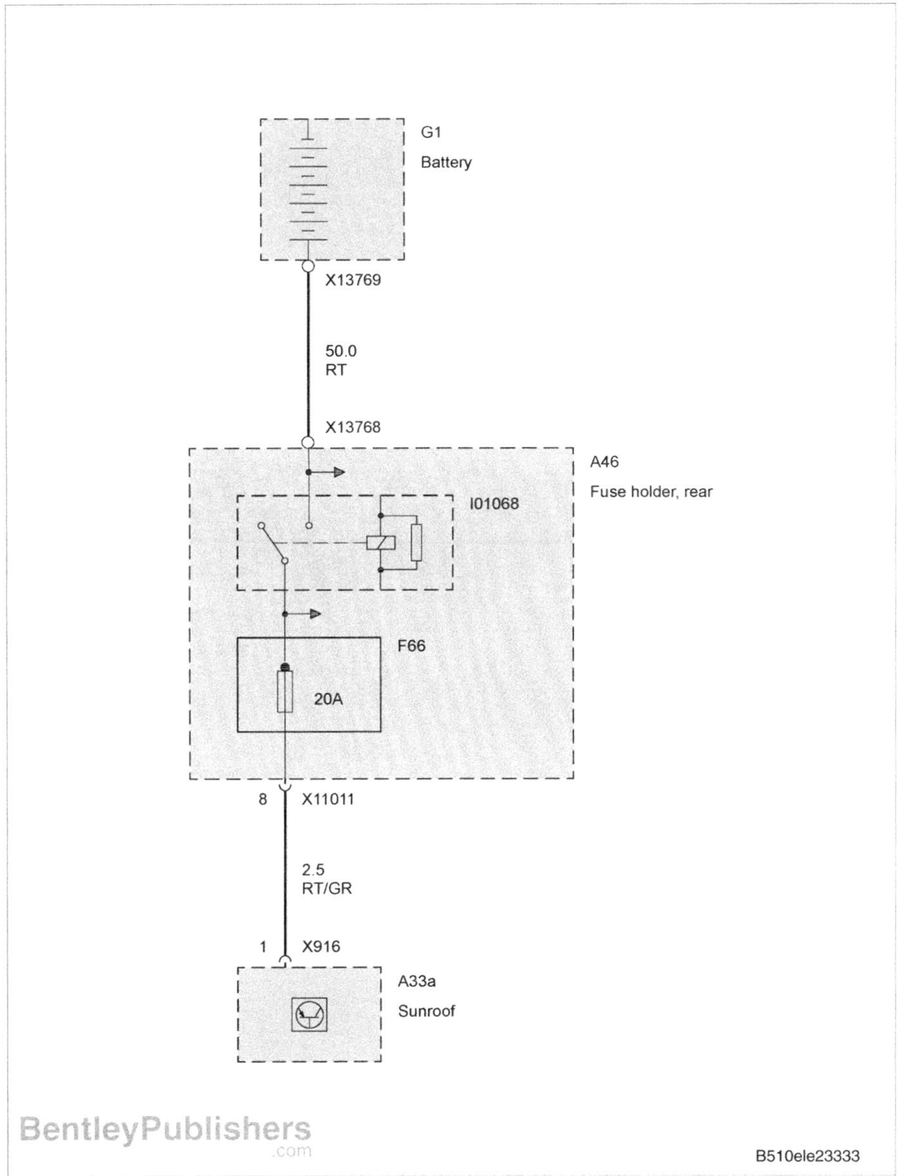

Fuse F66
except Sedan as of 09 / 2005

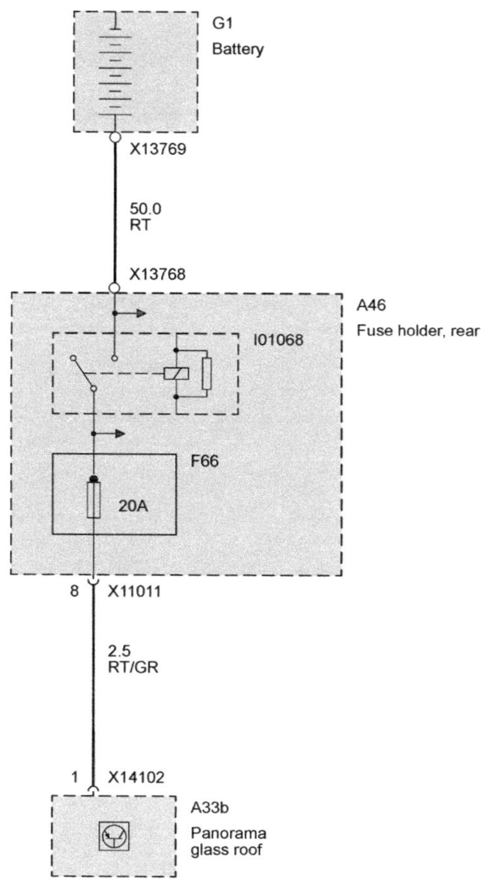

Fuse F67
up to 03 / 2006

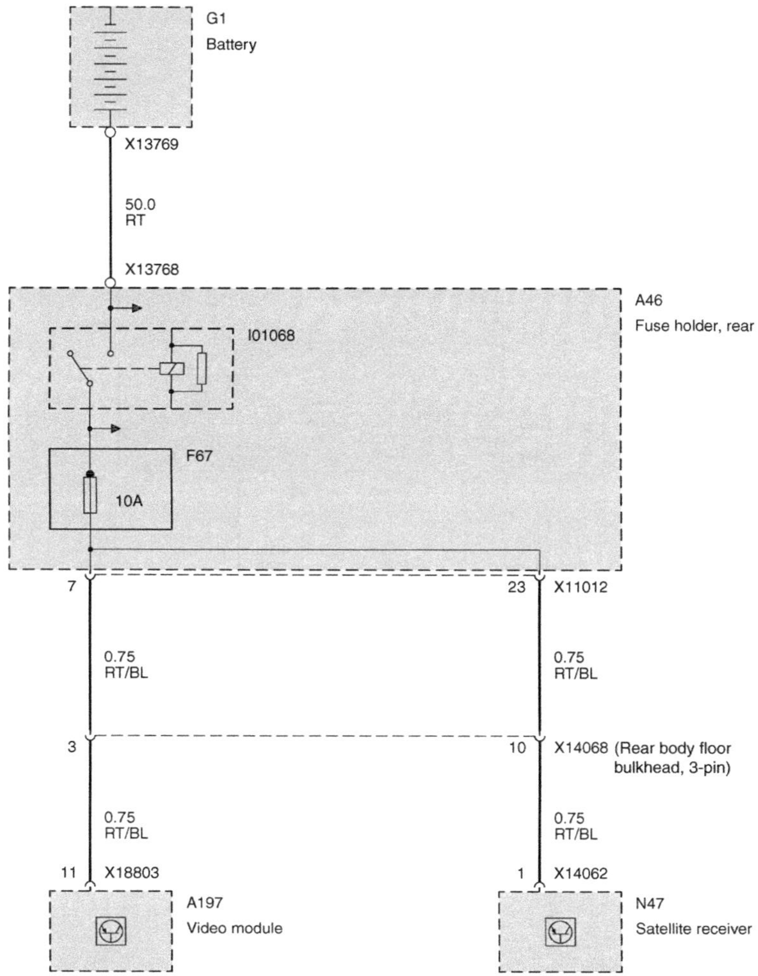

BentleyPublishers.com
BentleyPublishers.com

Fuse F67
as of 03 / 2006

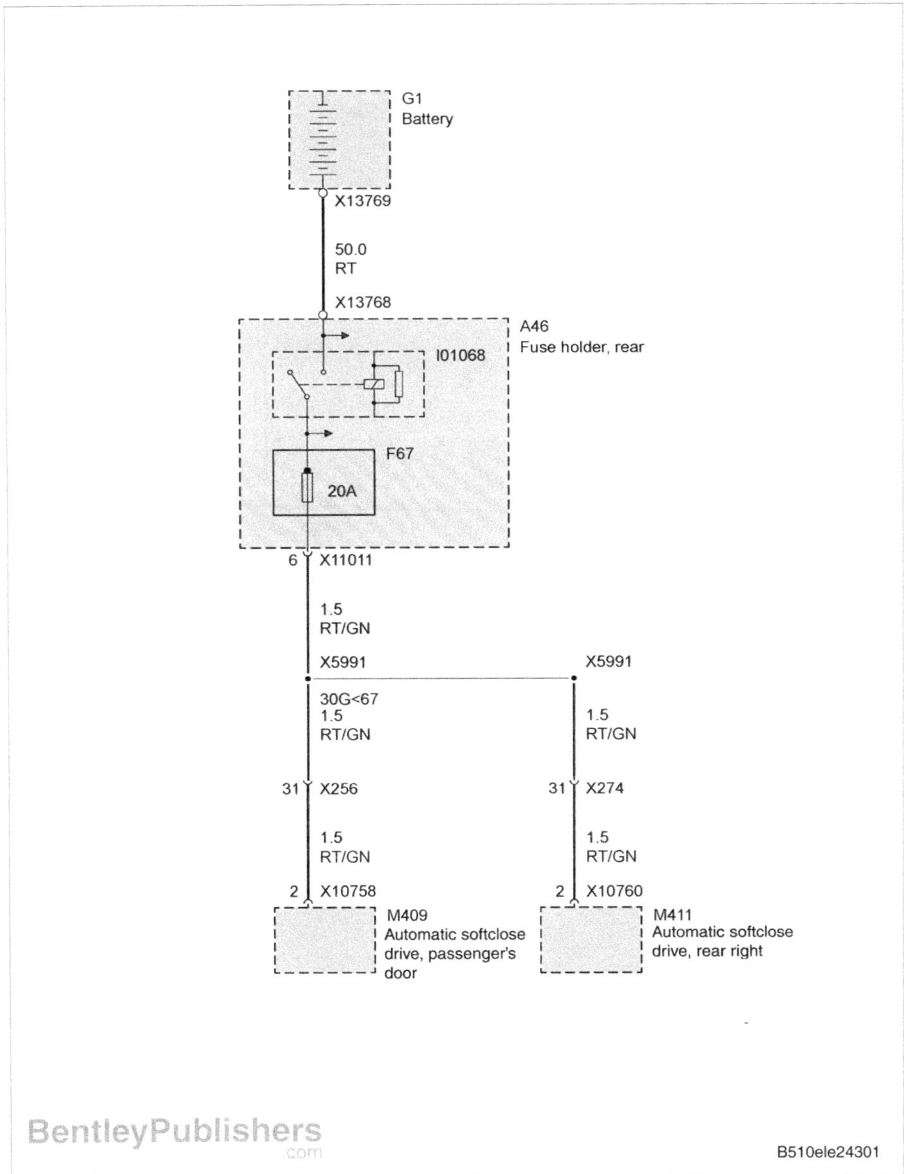

Fuse F68
up to 03 / 2006

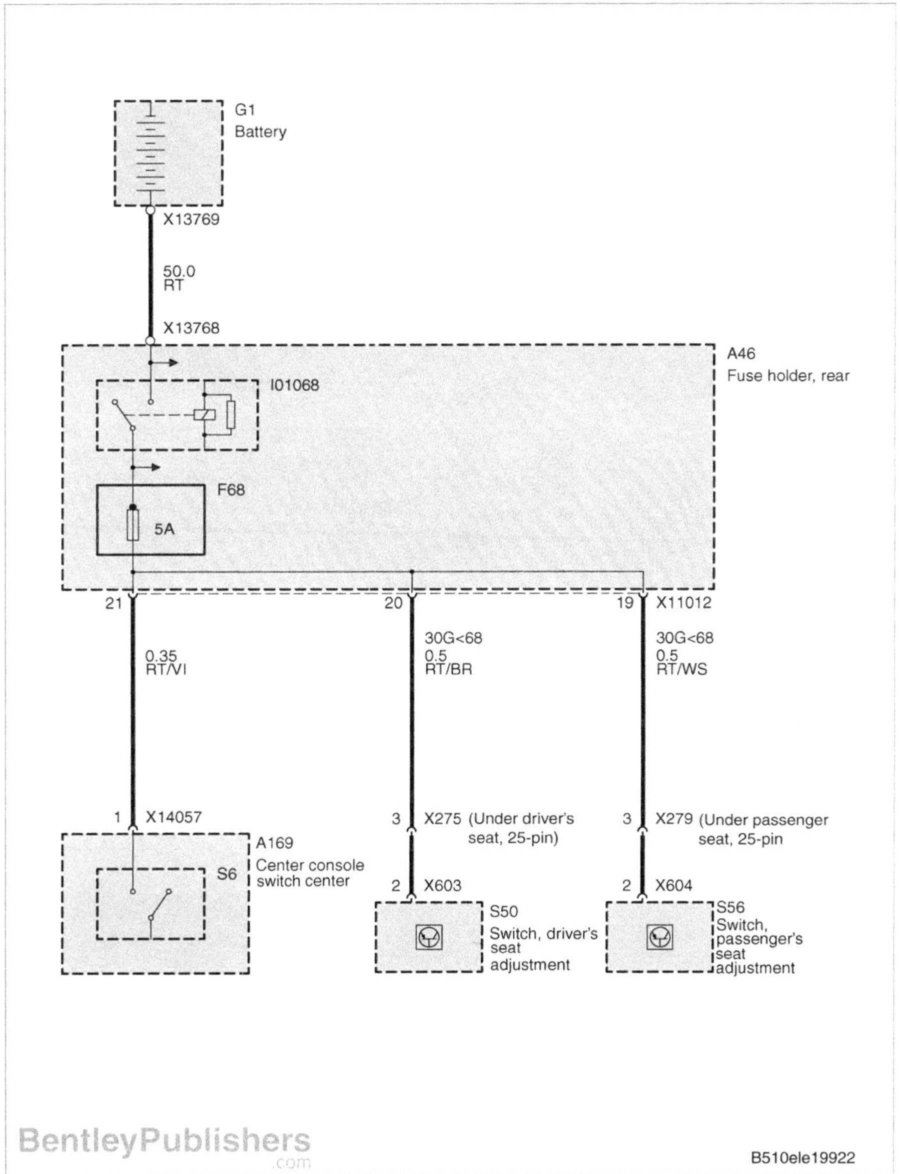

Fuse F68
as of 03 / 2006

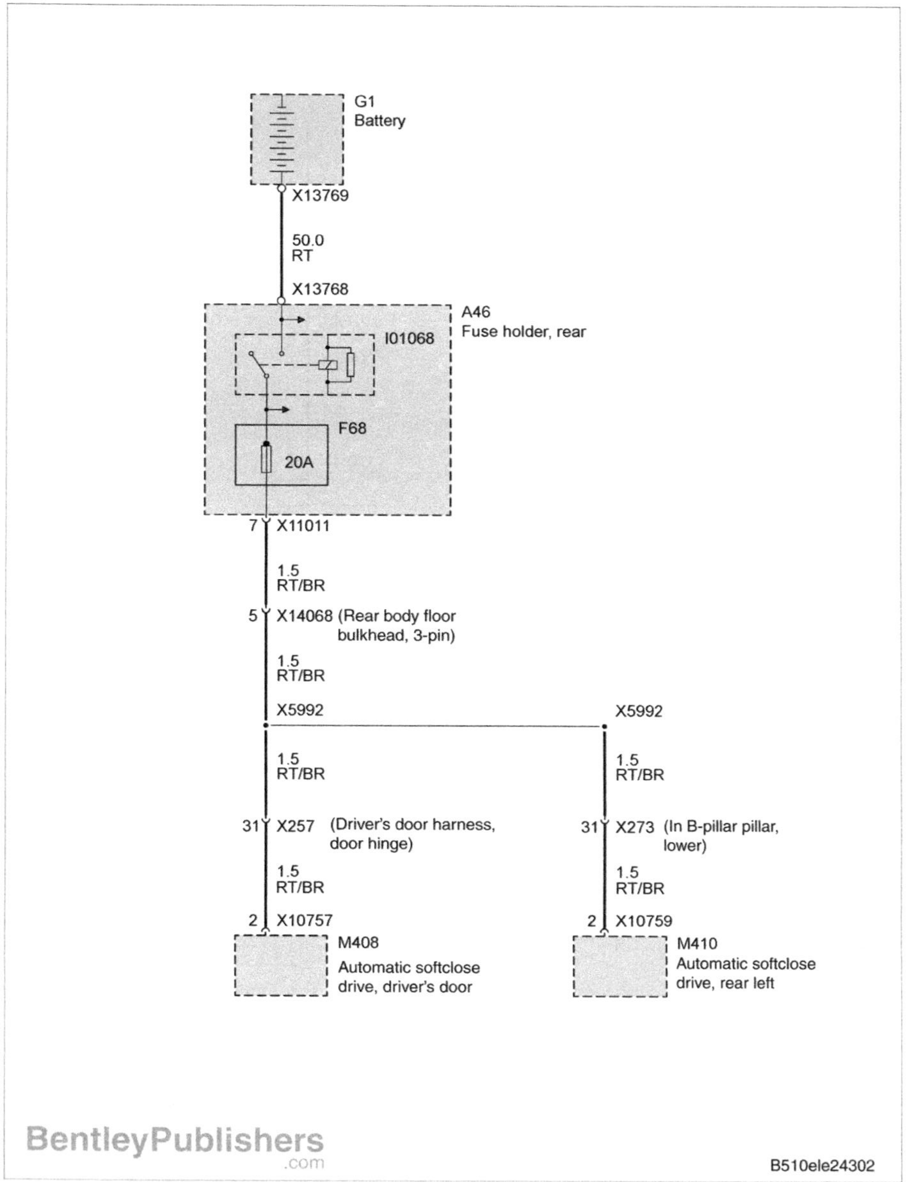

Fuse F69
up to 09 / 2005

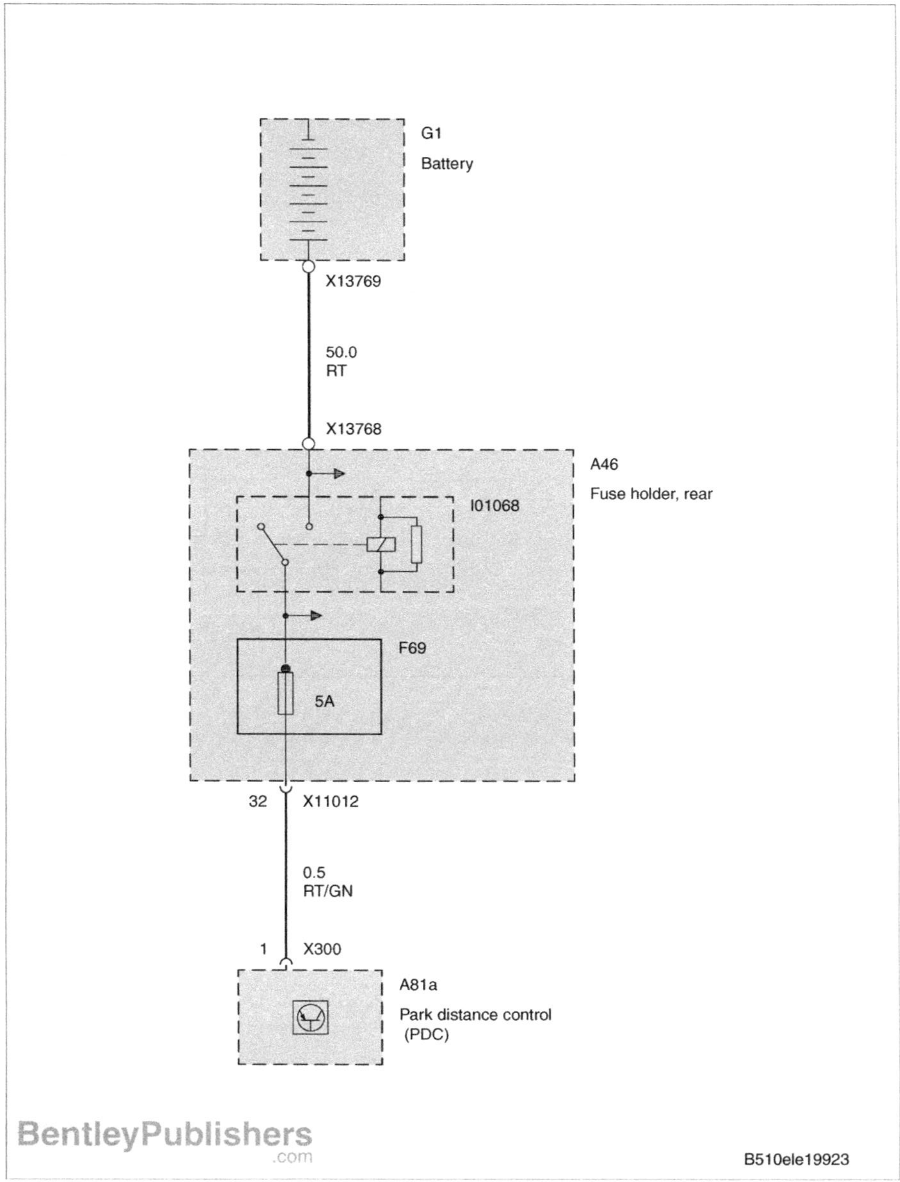

Fuse F69
as of 09 / 2005

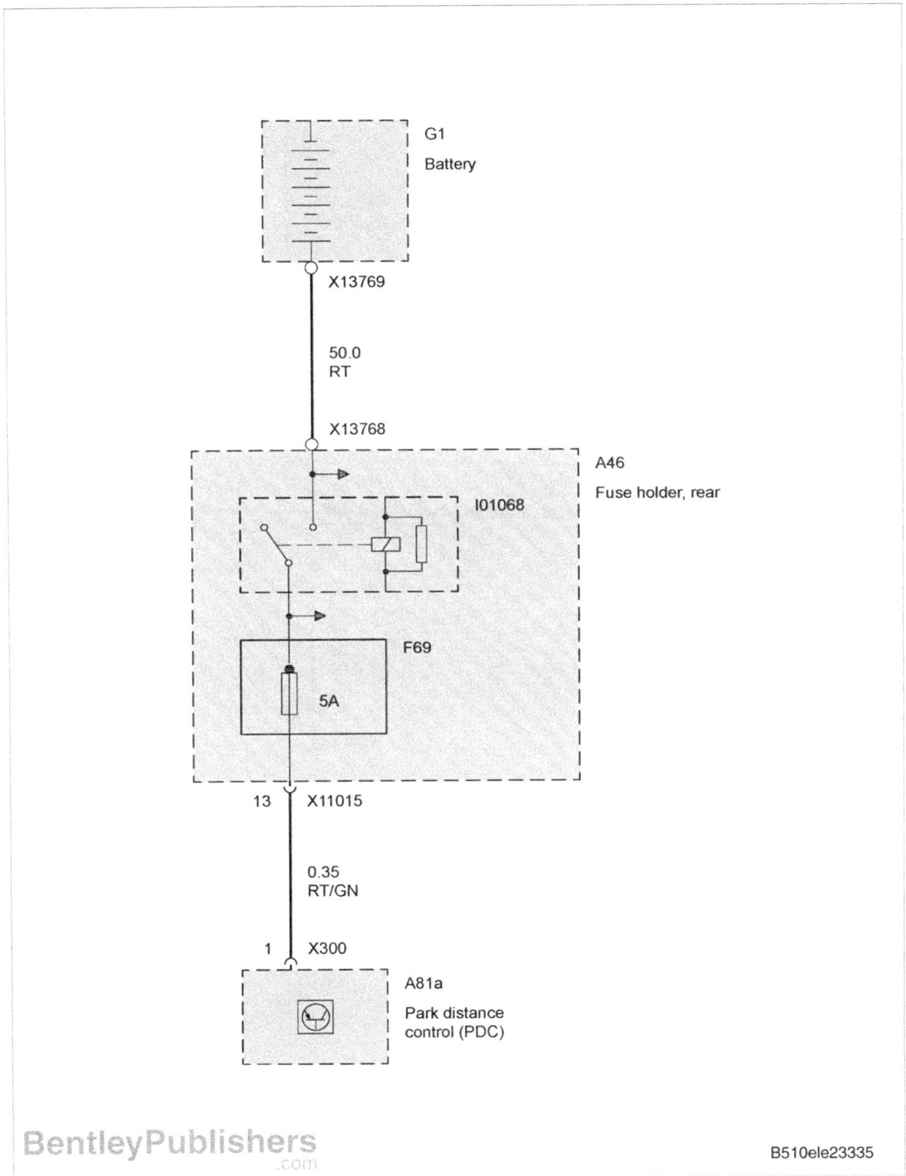

Fuse F70
up to 09 / 2005

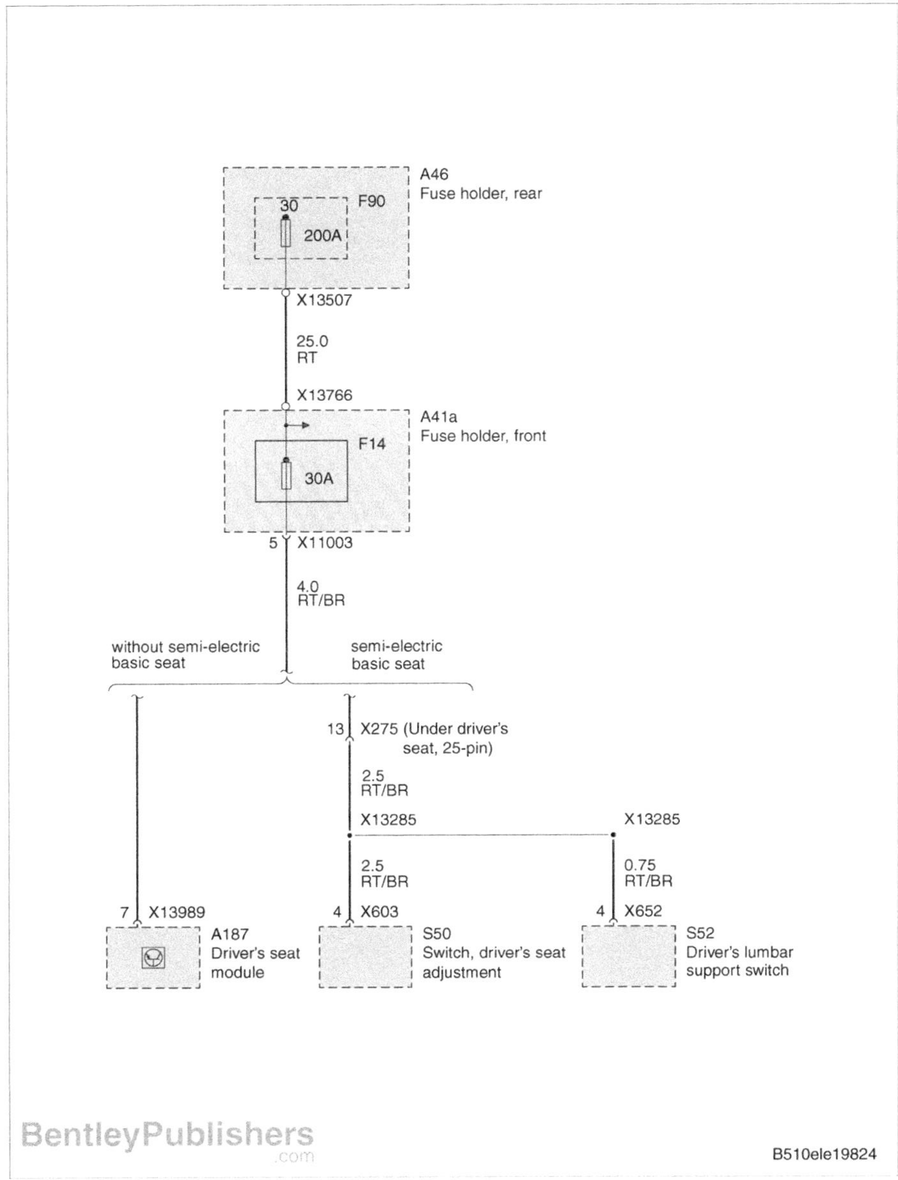

Fuse F70
as of 09 / 2005 up to 03 / 2007

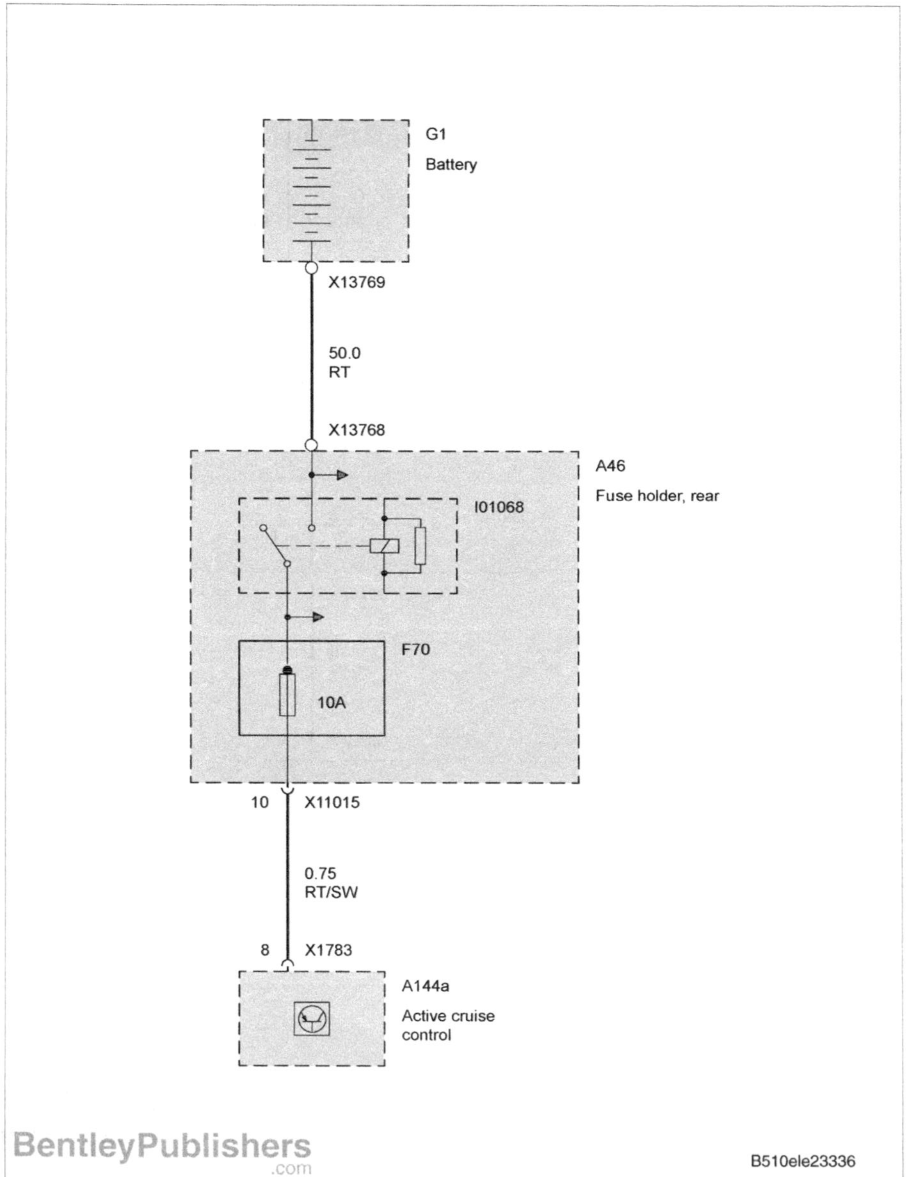

Fuse F70
as of 03 / 2007

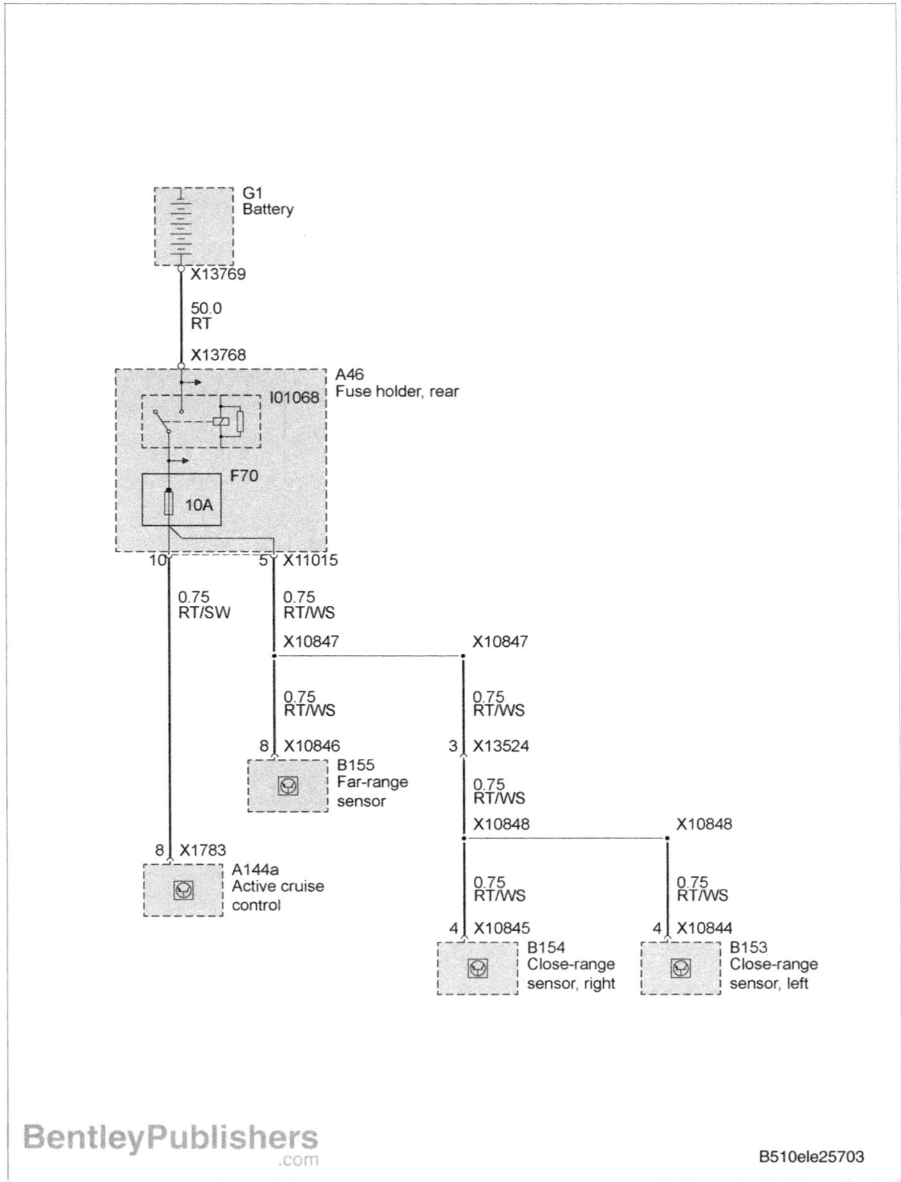

Fuse F71
up to 09 / 2005

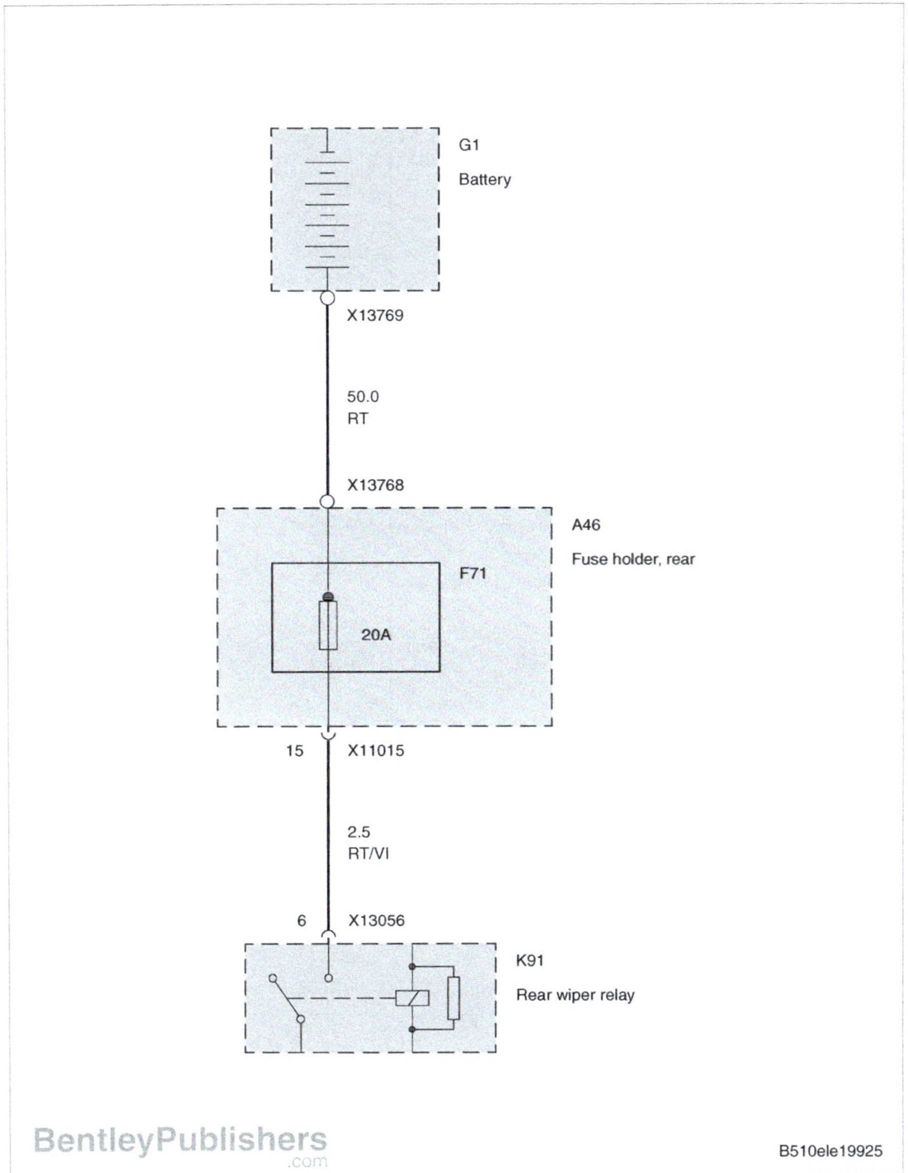

Fuse F71
as of 09 / 2005

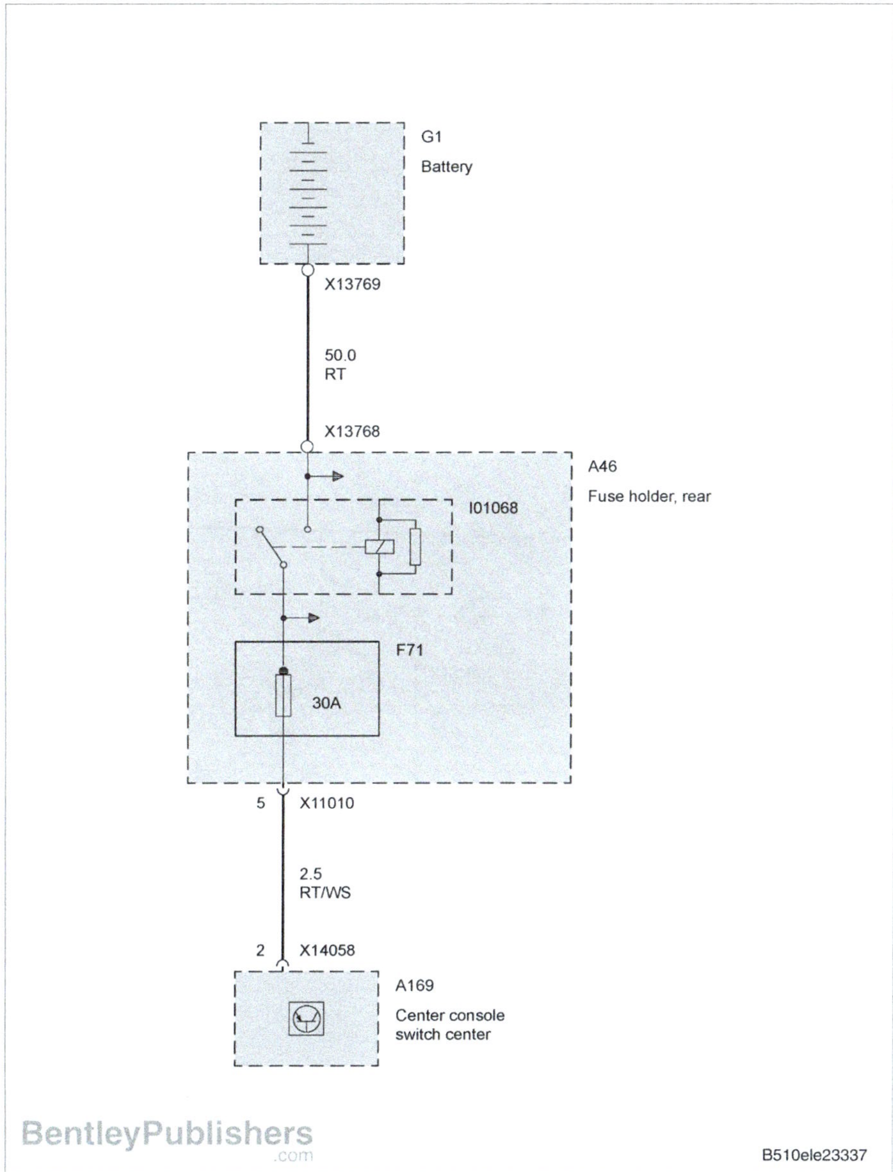

Fuse F72
up to 09 / 2005 (Wagon)

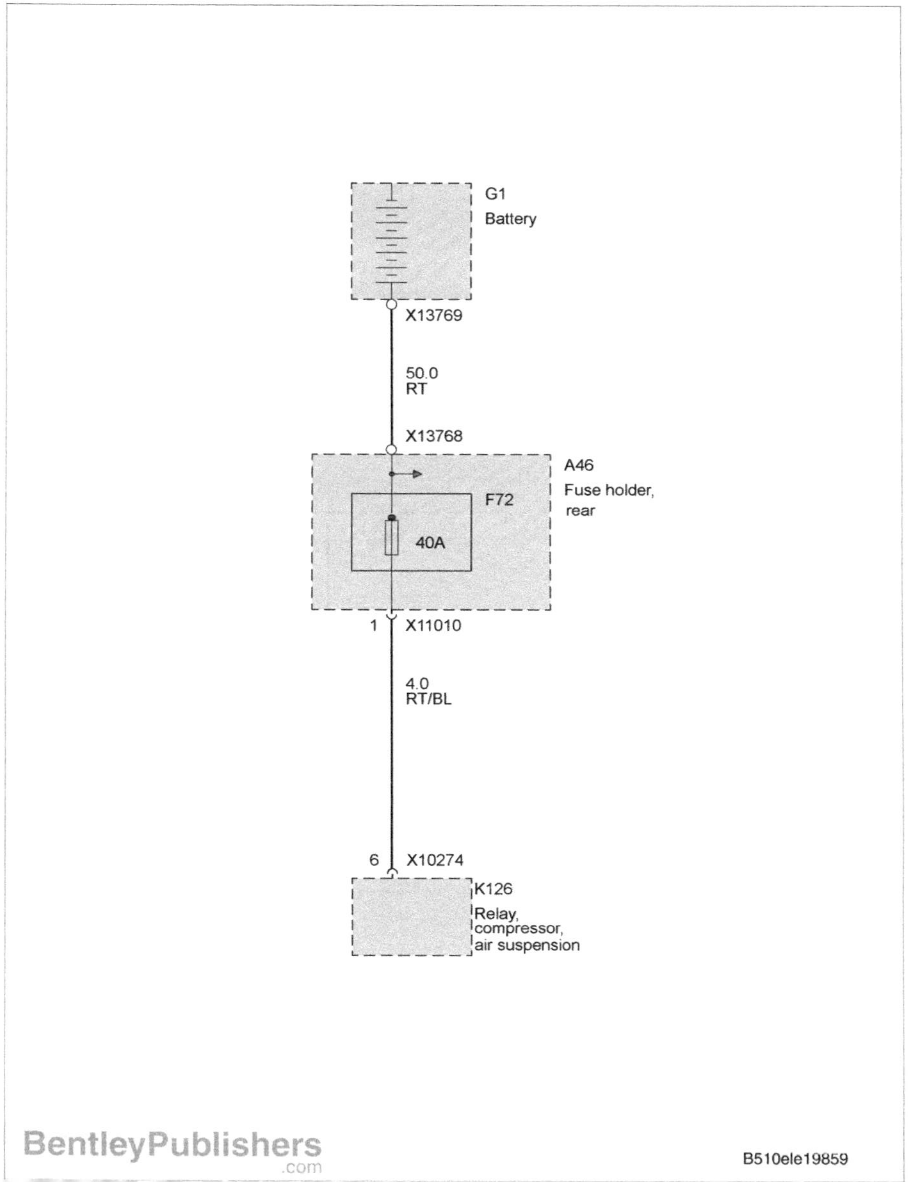

Fuse F72
as of 09 / 2005

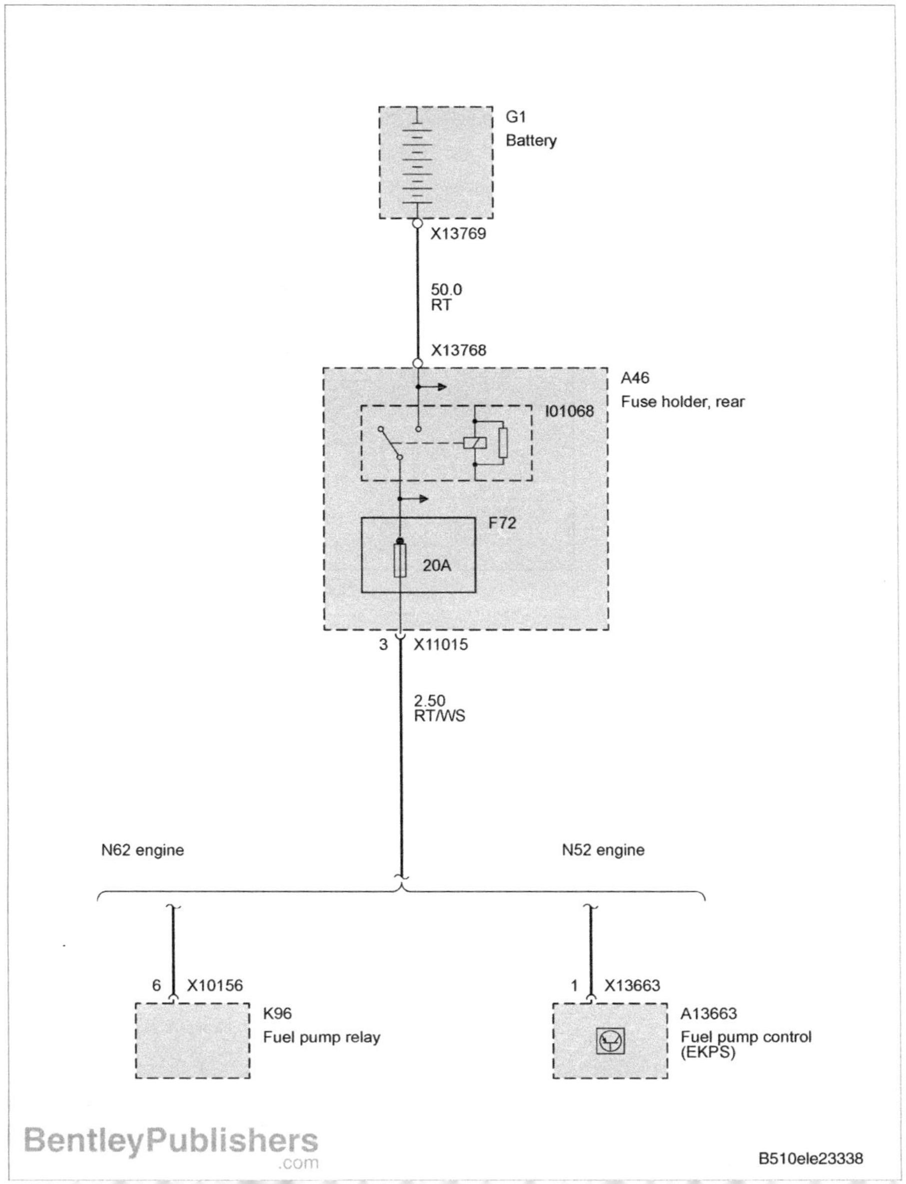

Fuse F73
except Convertible up to 09 / 2005

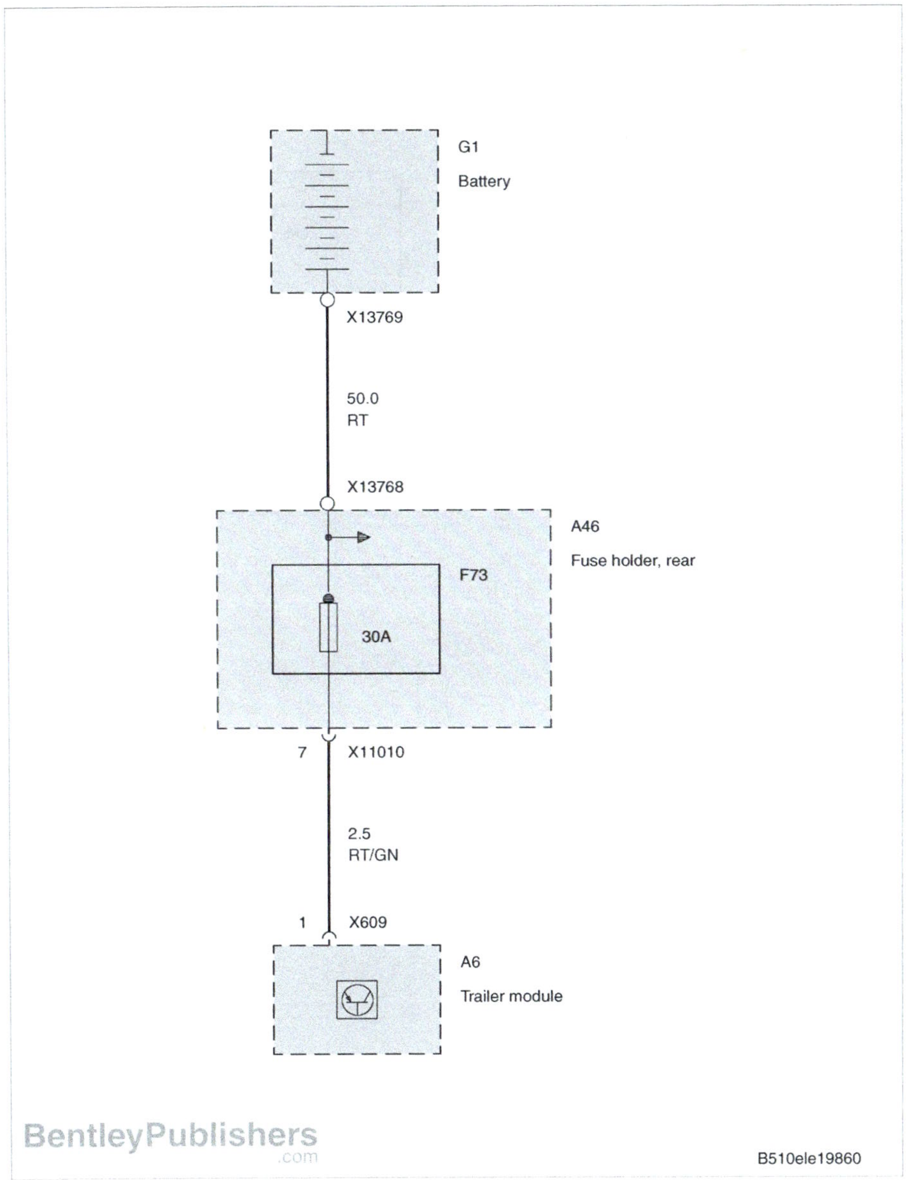

Fuse F73
Convertible up to 09 / 2005

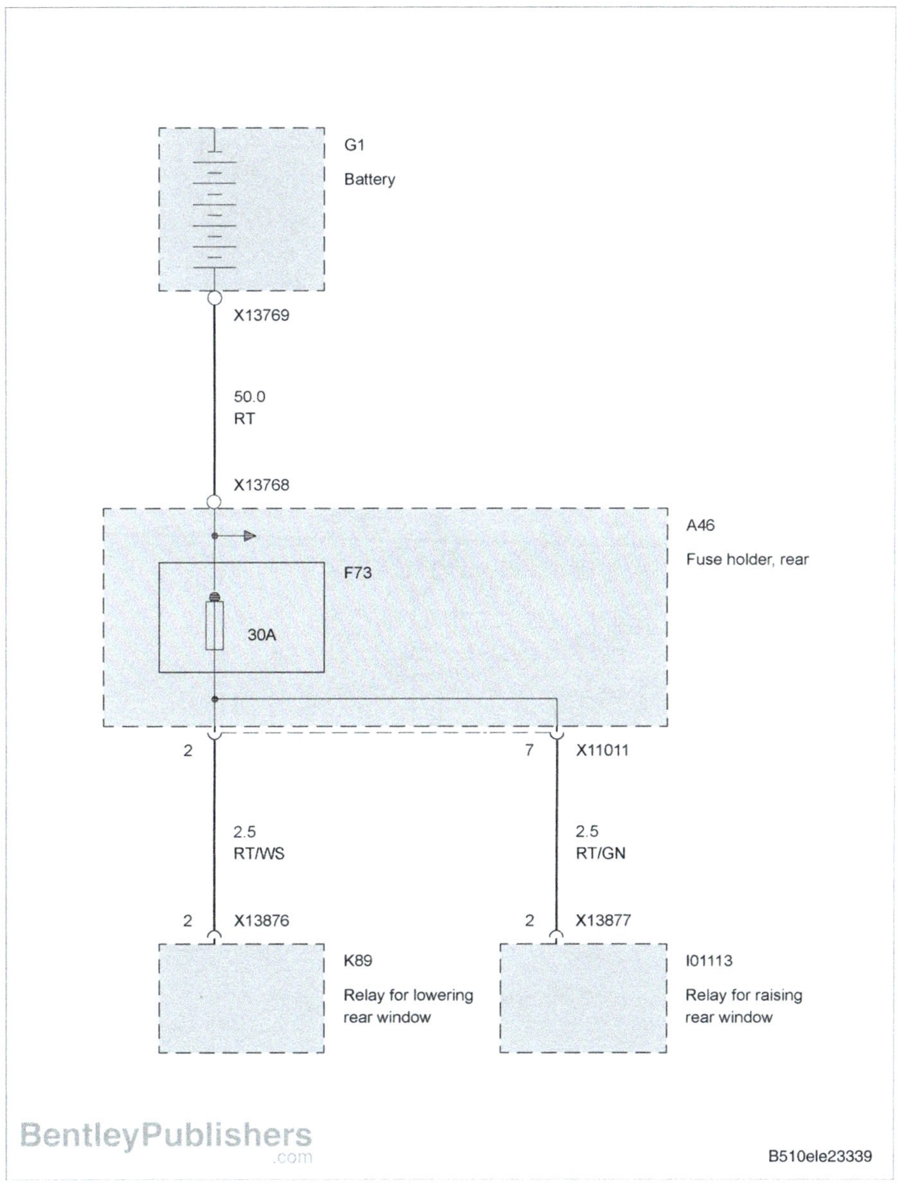

Fuse F73
as of 09 / 2005 up to 09 / 2007

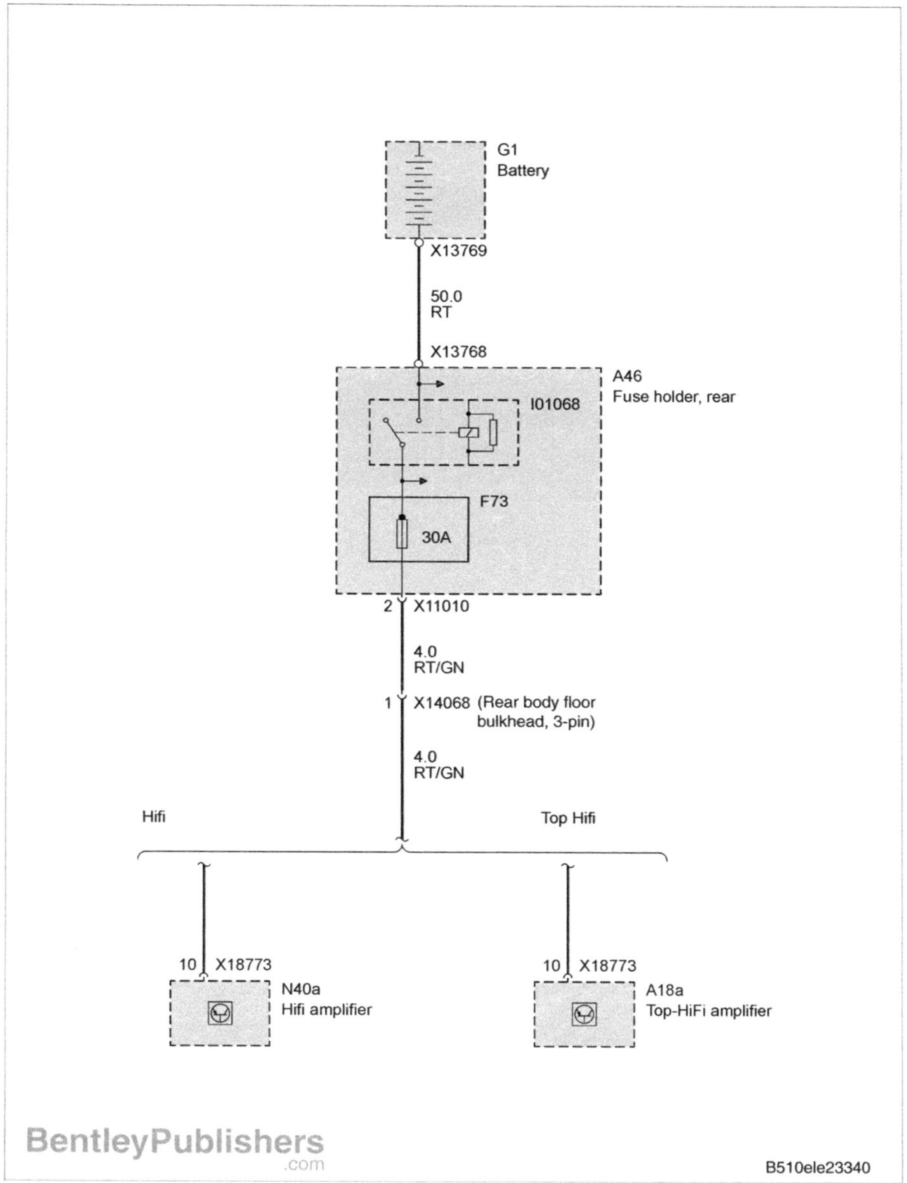

Fuse F73
as of 09 / 2007

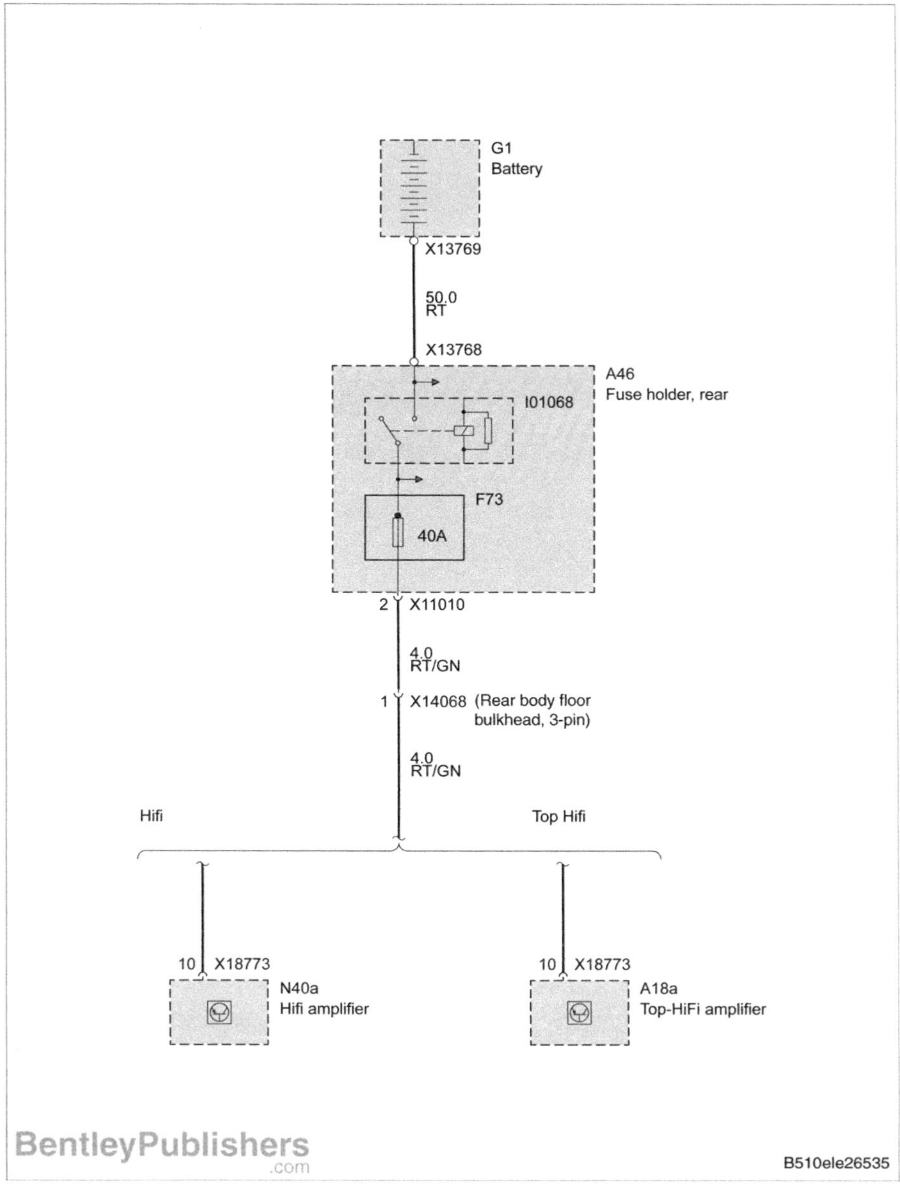

Fuse F74
up to 09 / 2005

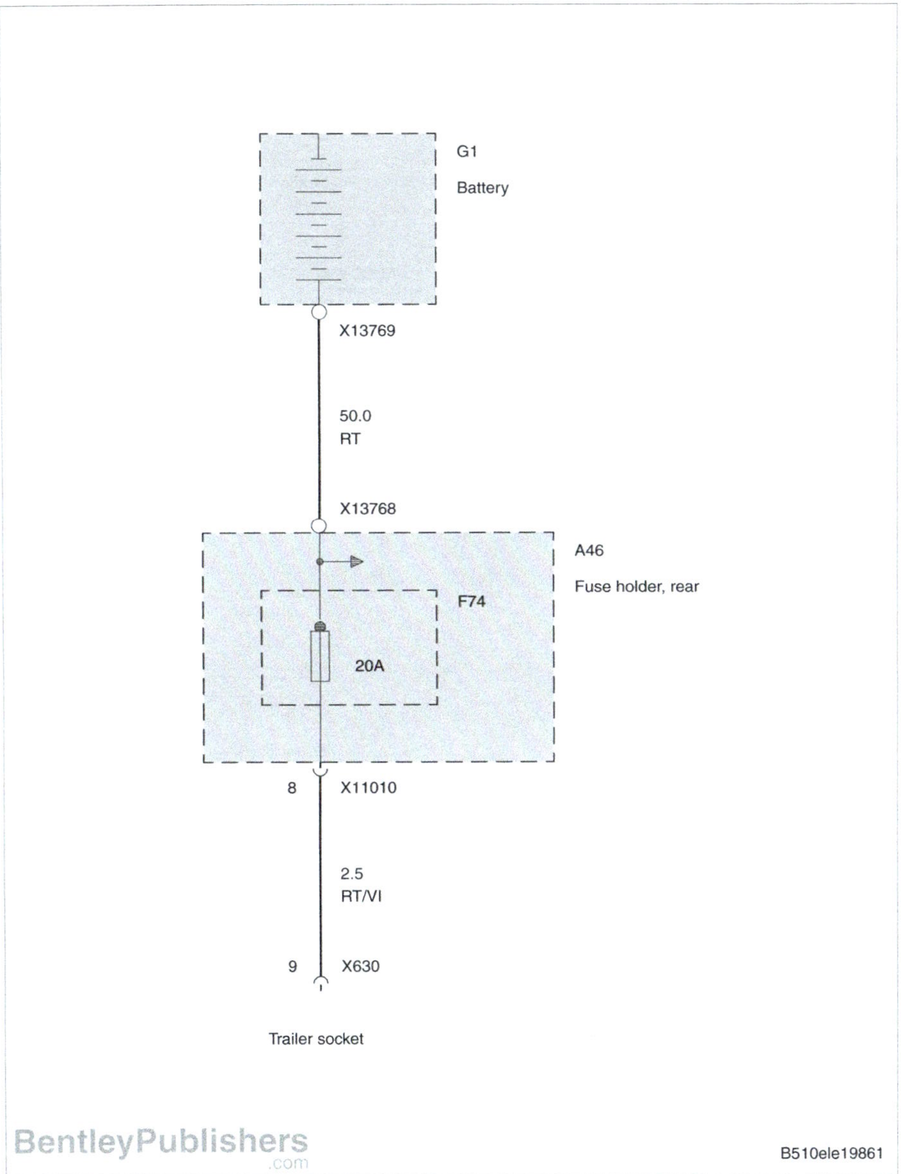

Fuse F74
as of 09 / 2005

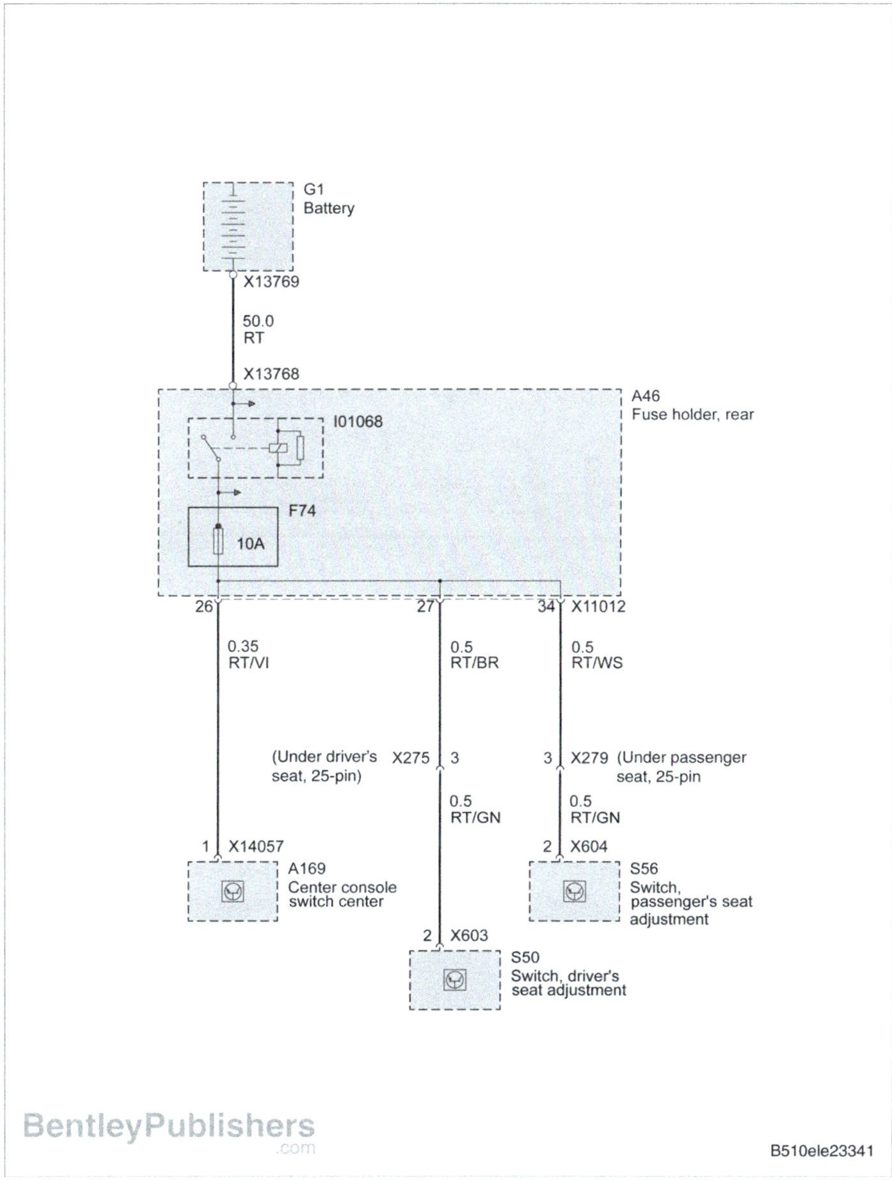

Fuse F74
as of 09 / 2007

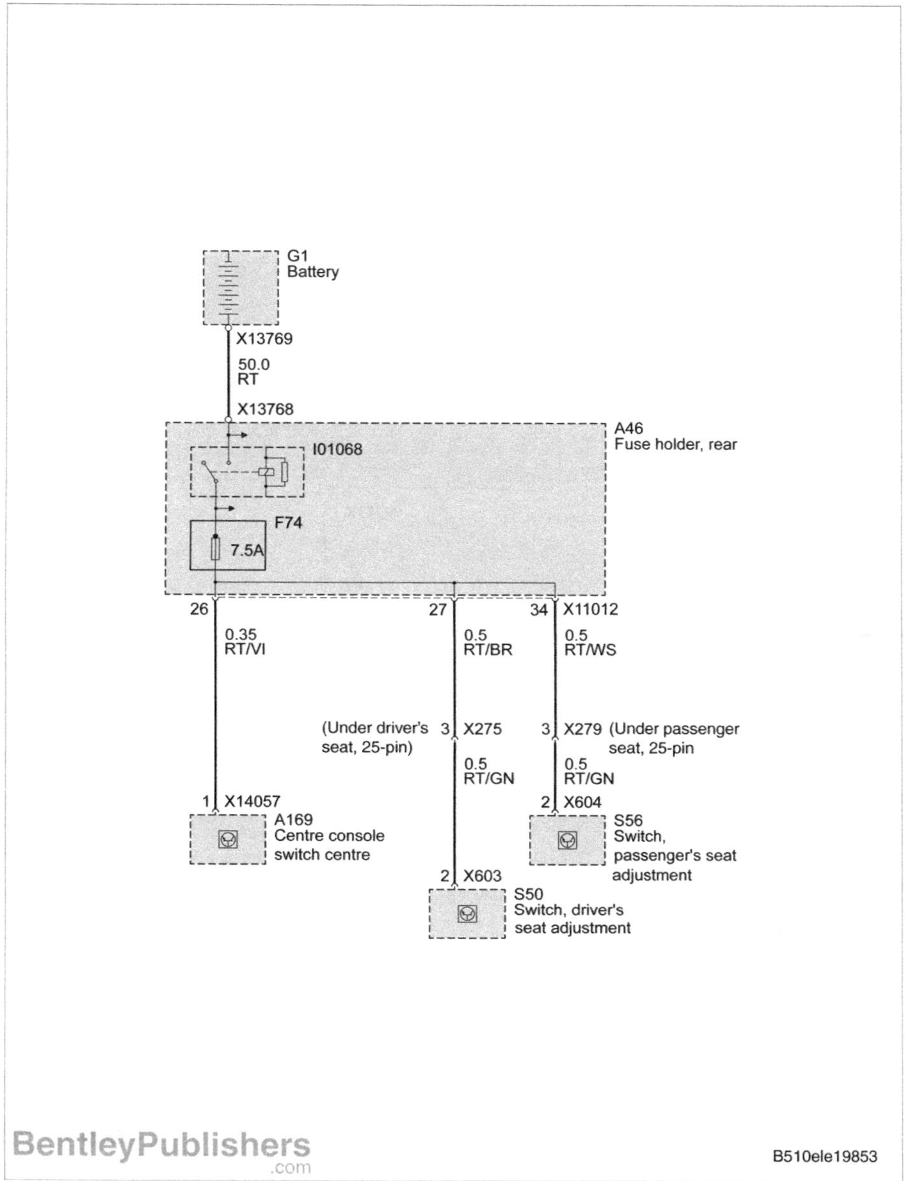

Fuse F75
up to 09 / 2005

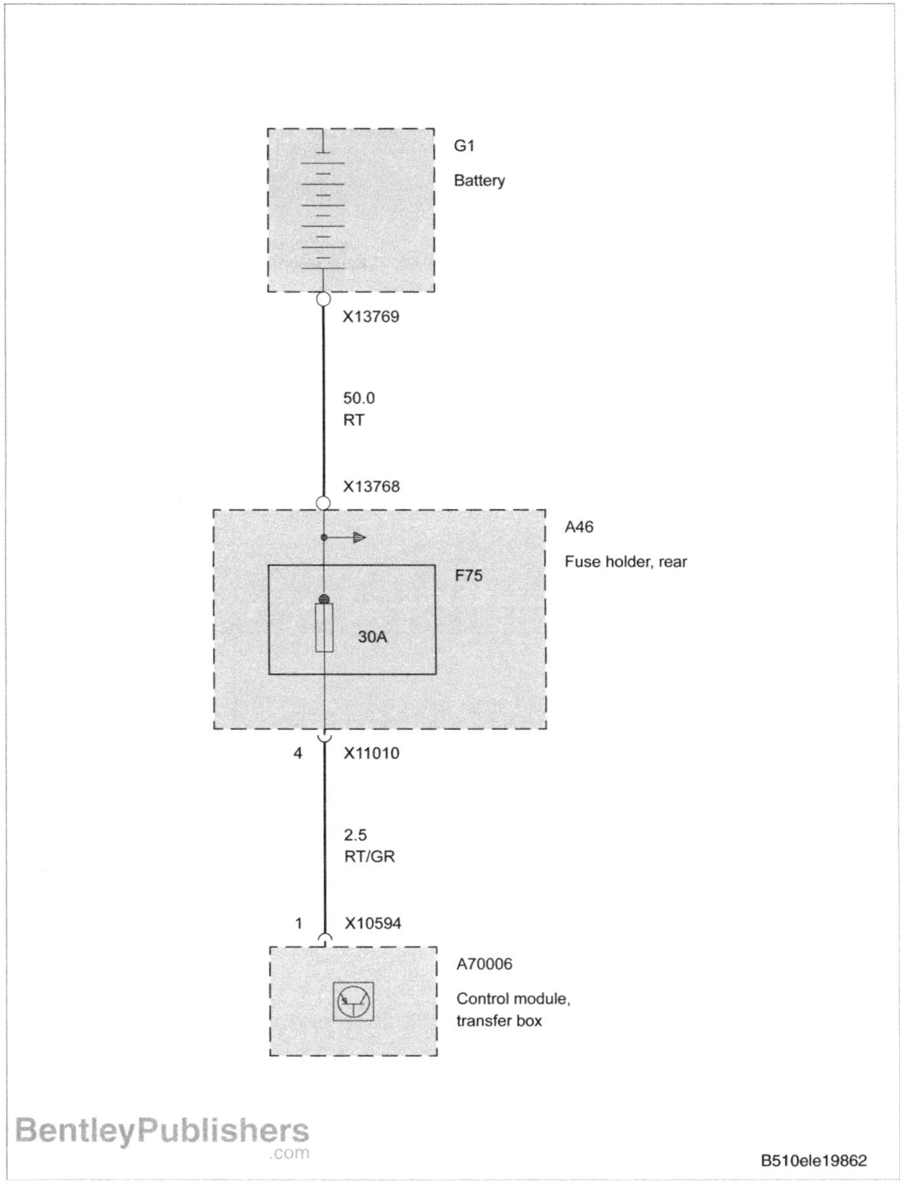

Fuse F75
as of 09 / 2005

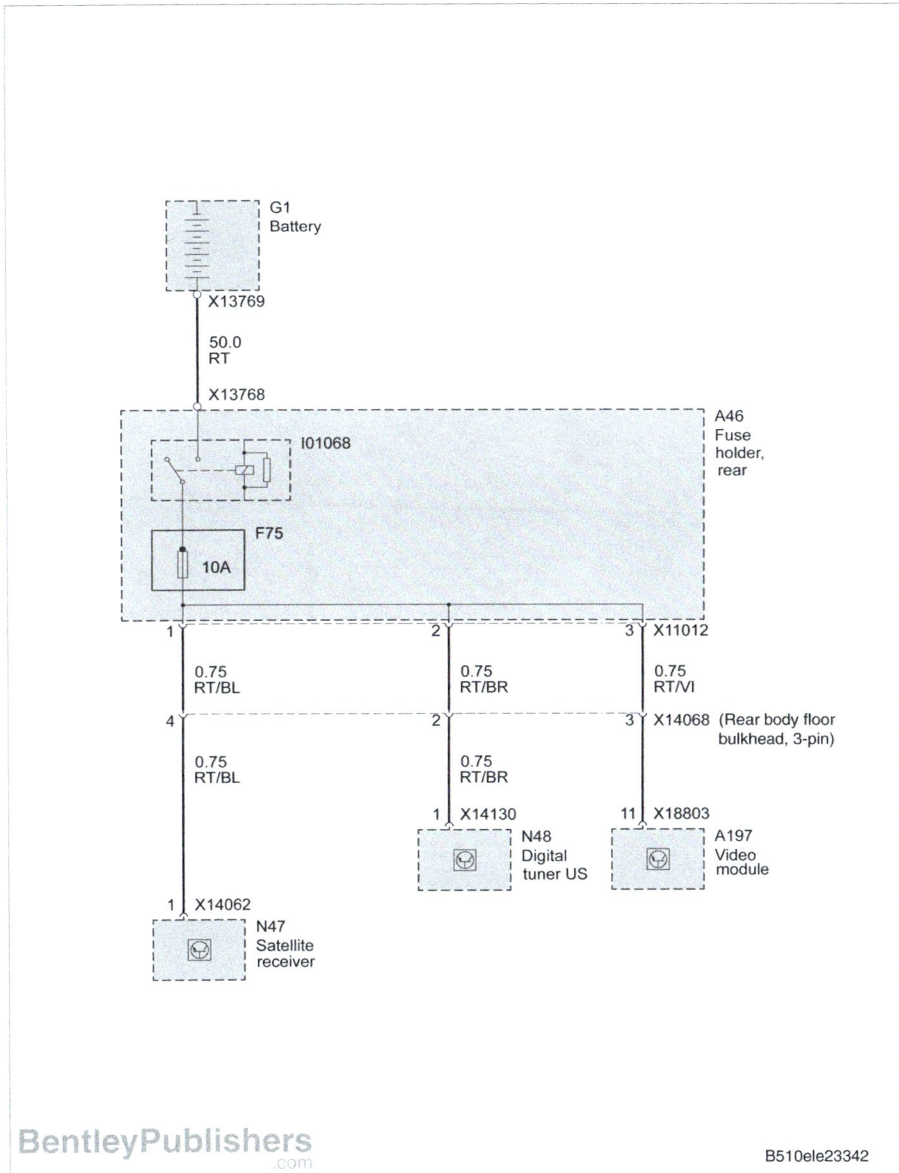

Fuse F76
up to 09 / 2005

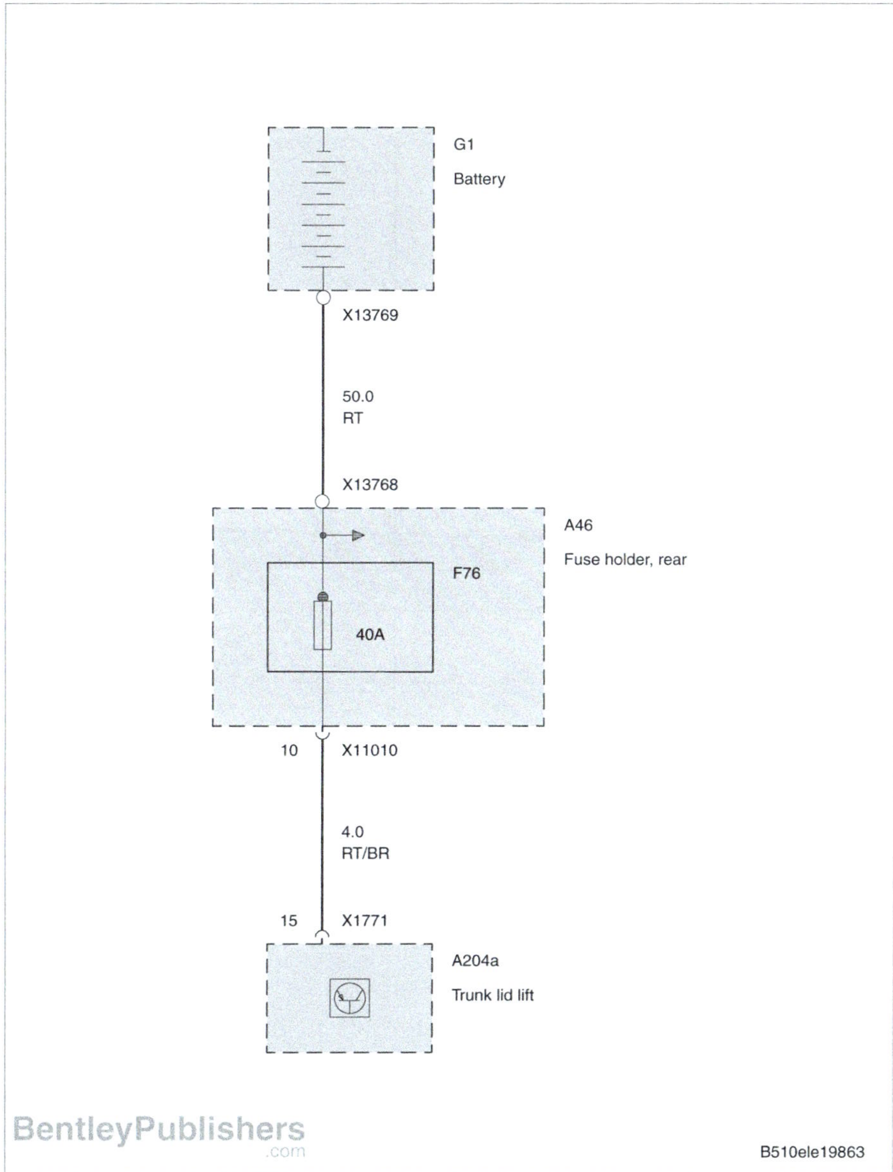

Fuse F76
as of 09 / 2005

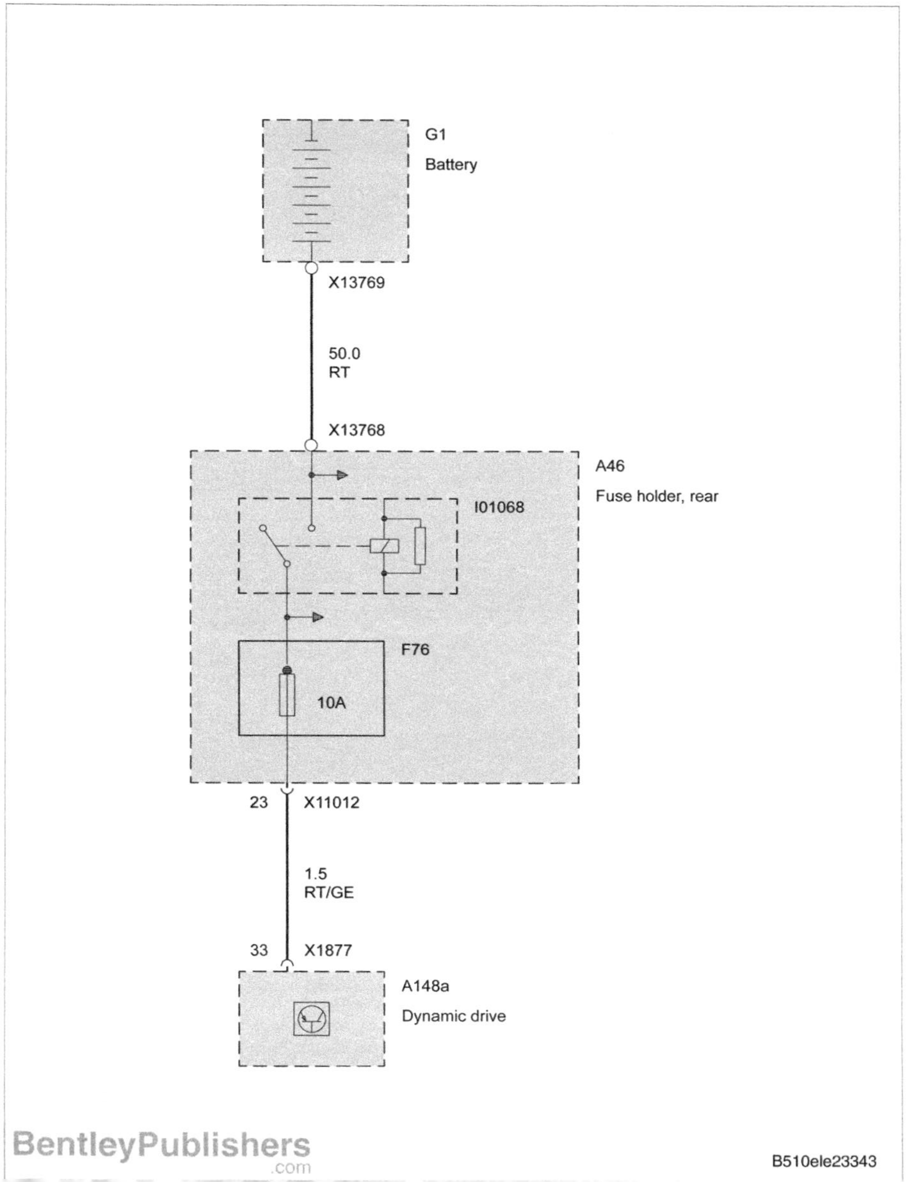

Fuse F77
up to 09 / 2005

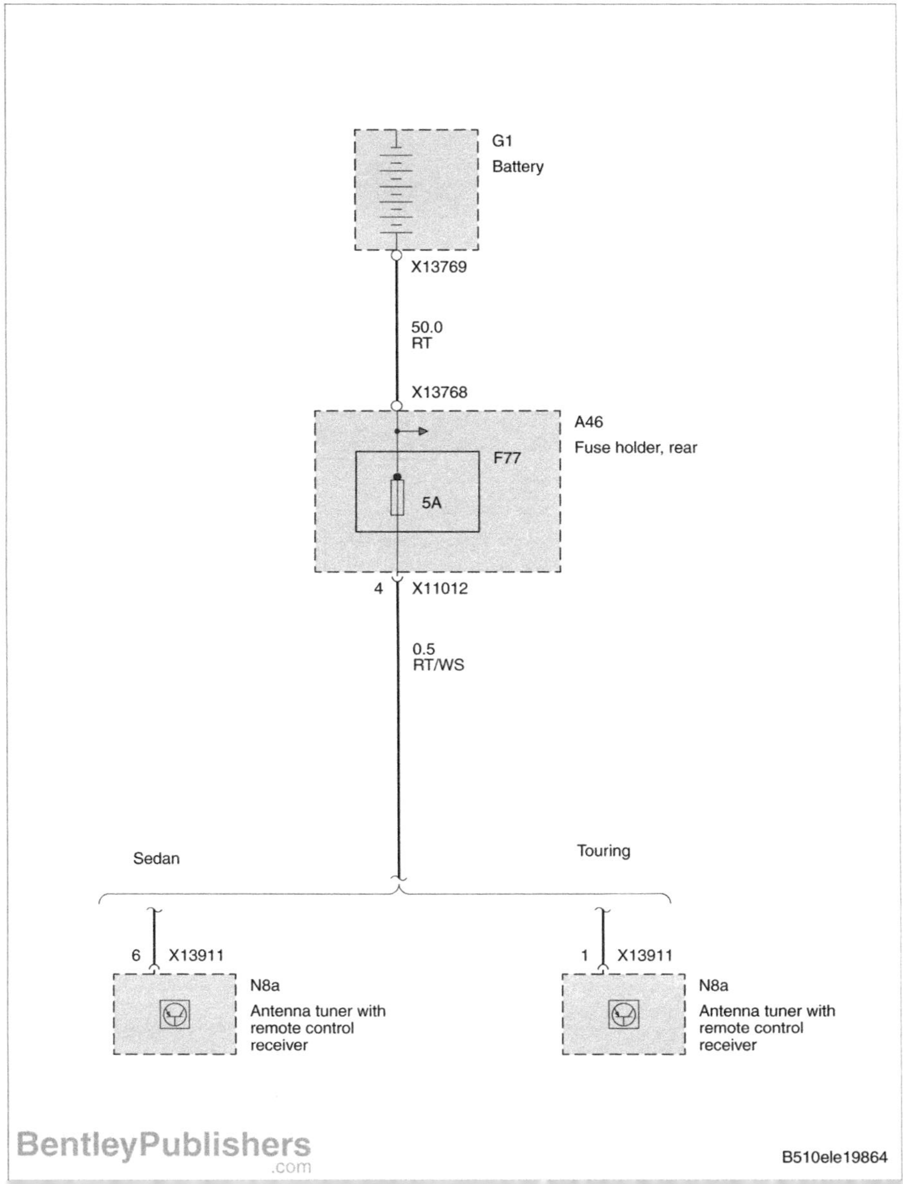

Fuse F77
as of 09 / 2005 up to 03 / 2007

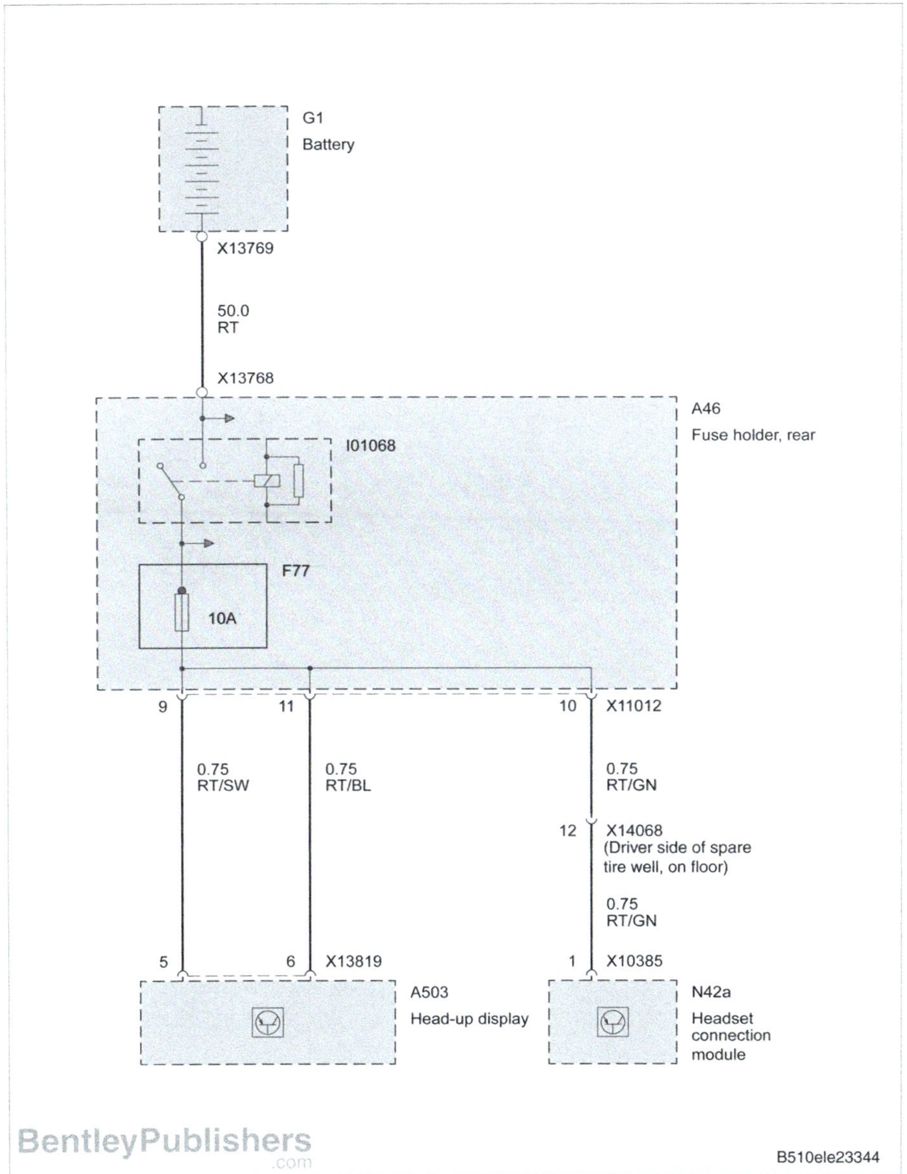

Fuse F77
as of 03 / 2007

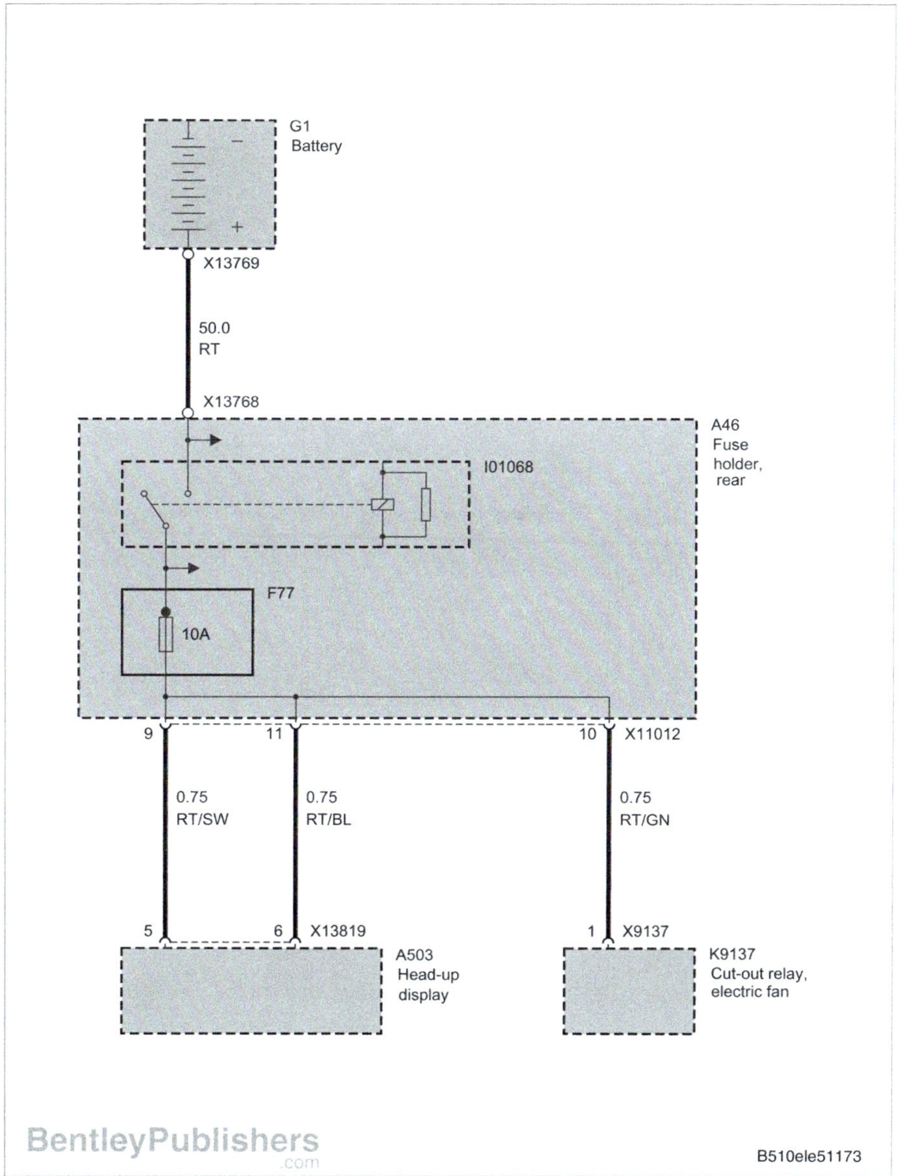

Fuse F78
as of 09 / 2005

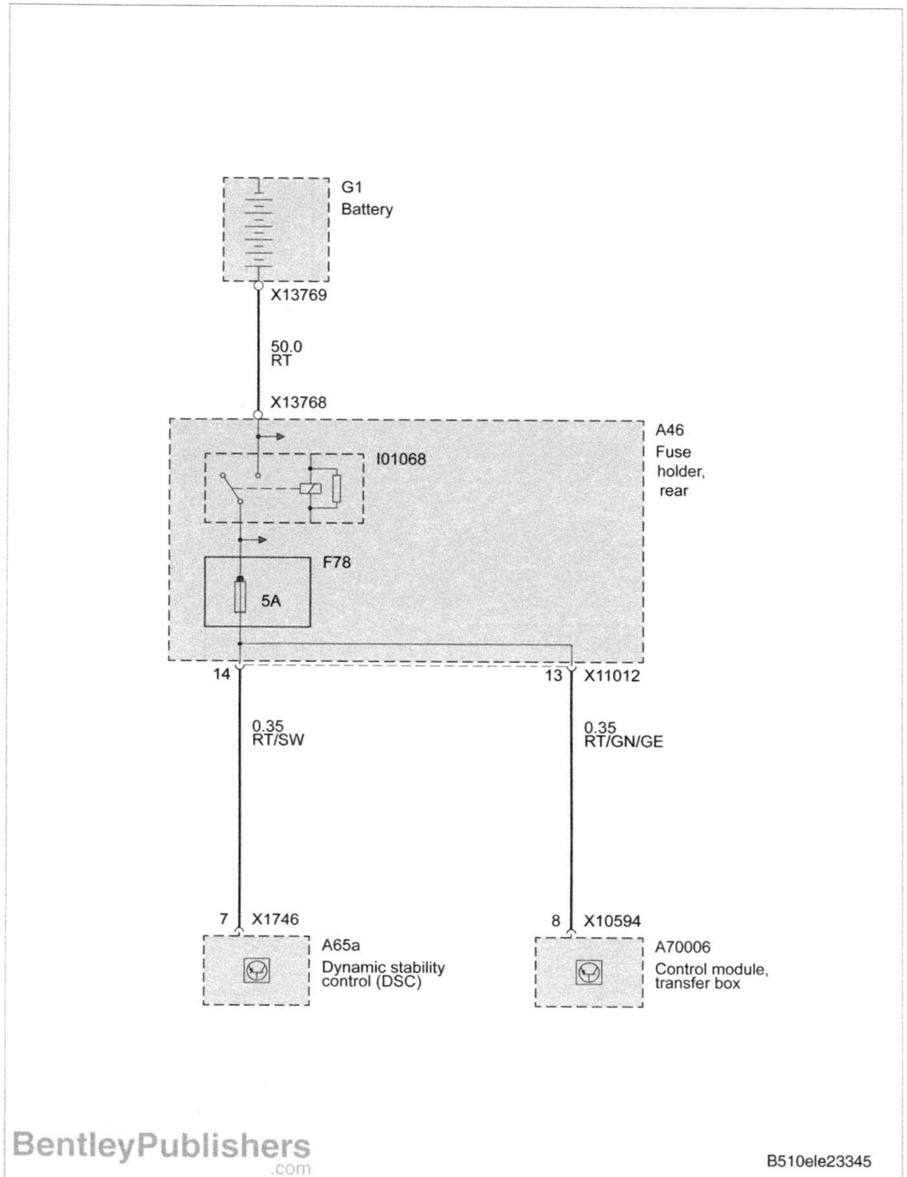

Fuse F79
up to 09 / 2005

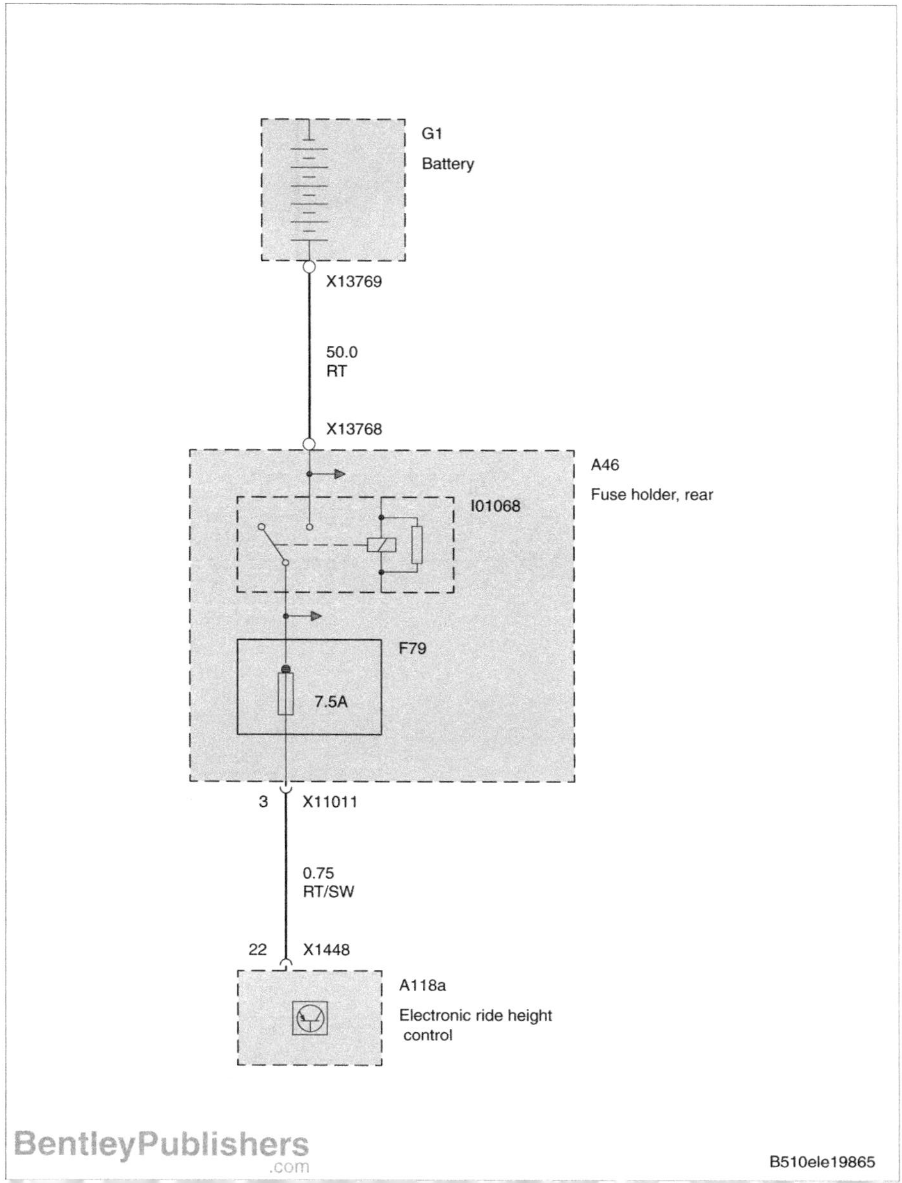

BentleyPublishers.com
BentleyPublishers.com

Fuse F79
as of 09 / 2005

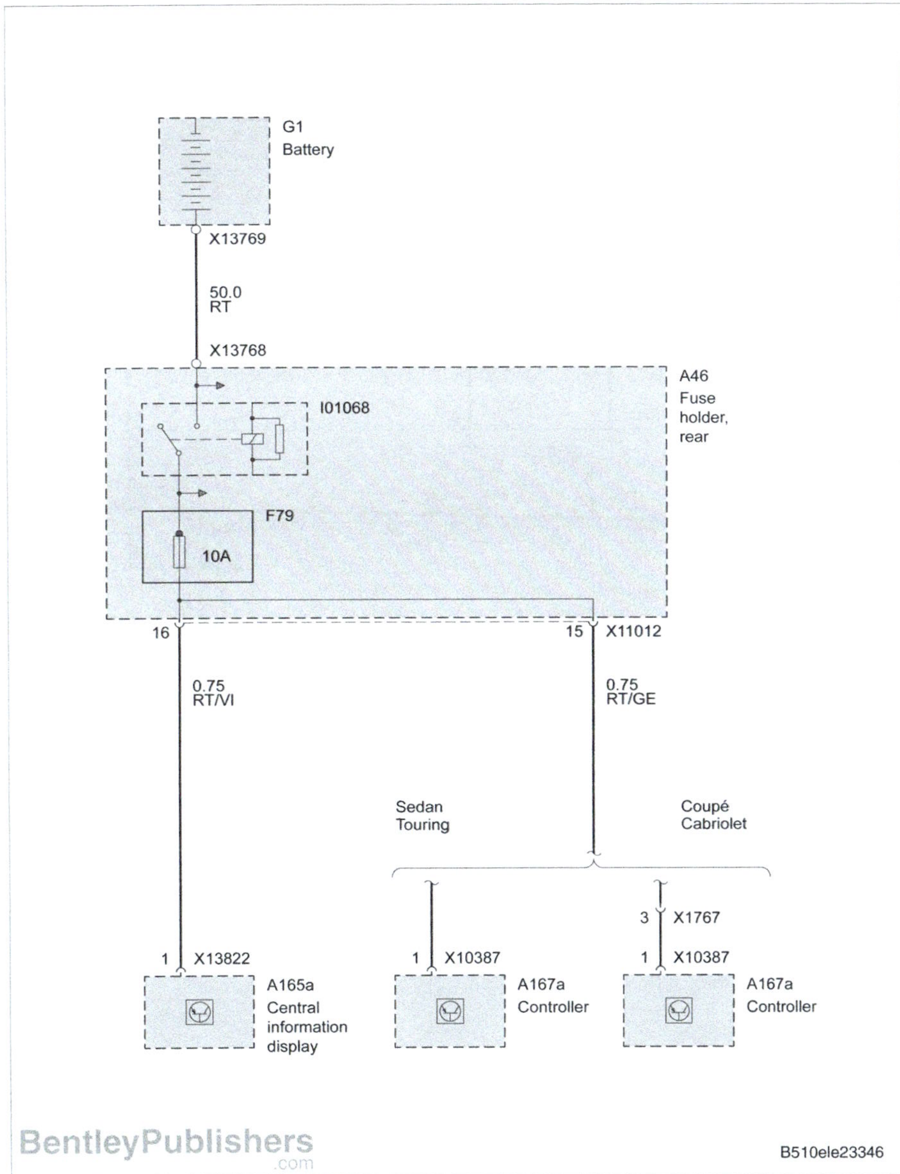

Fuse F80
up to 09 / 2005

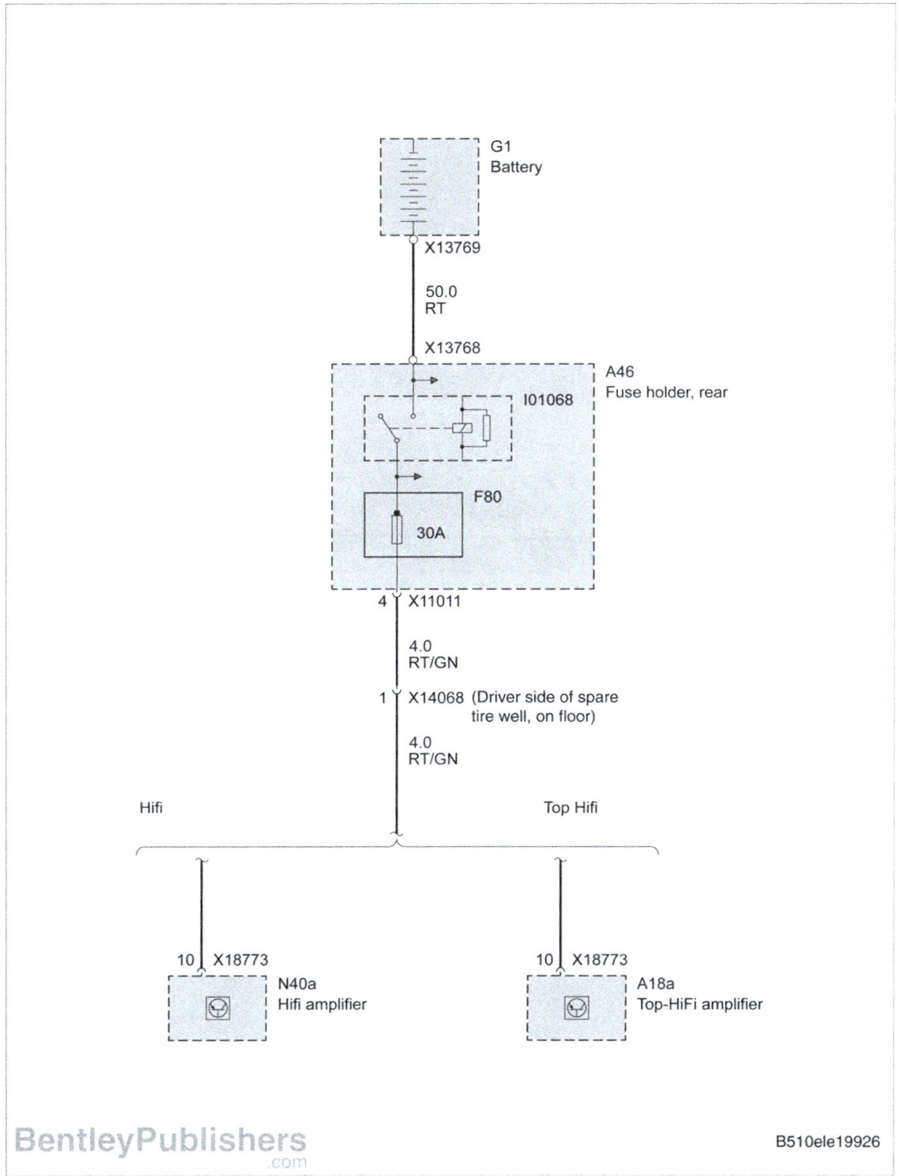

Fuse F80
as of 09 / 2005

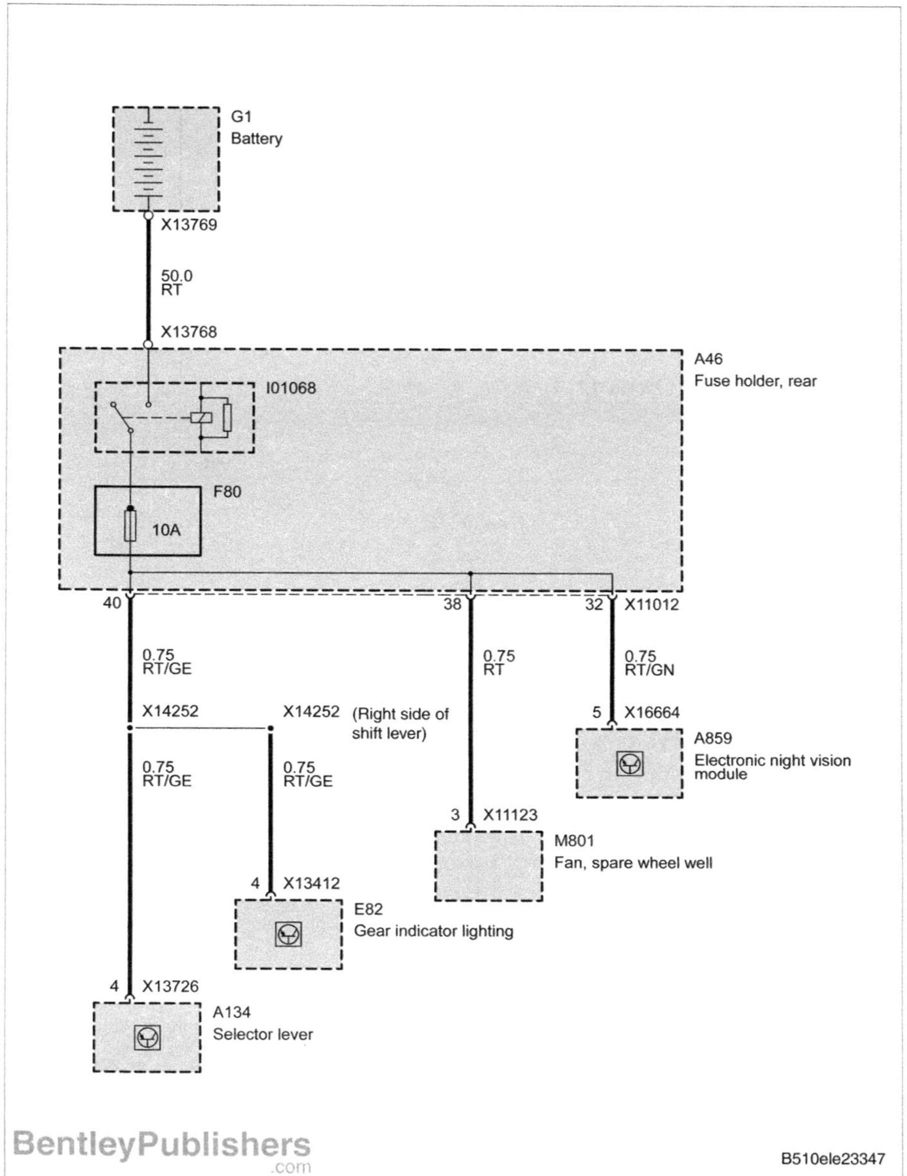

Fuse F81
as of 09 / 2005

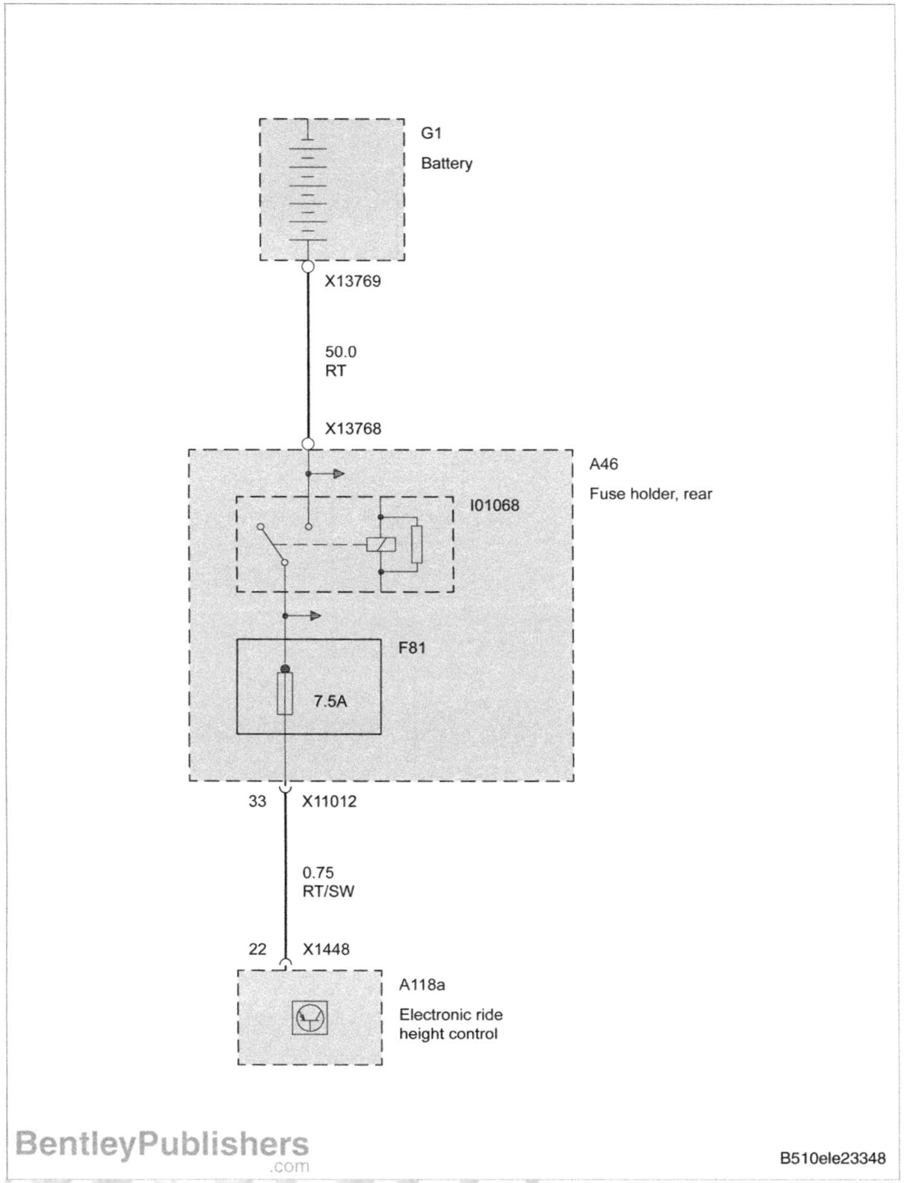

Fuse F82
up to 09 / 2005

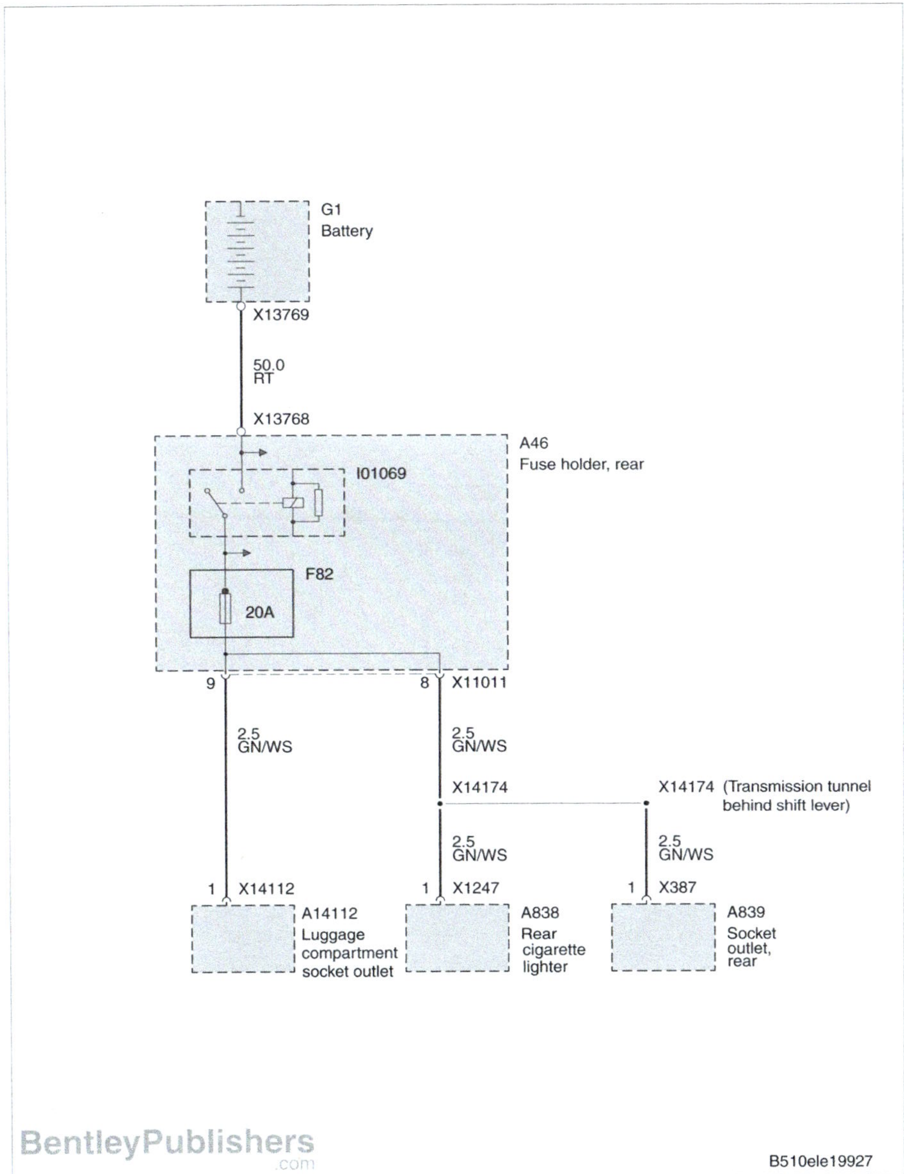

Fuse F82
as of 09 / 2005

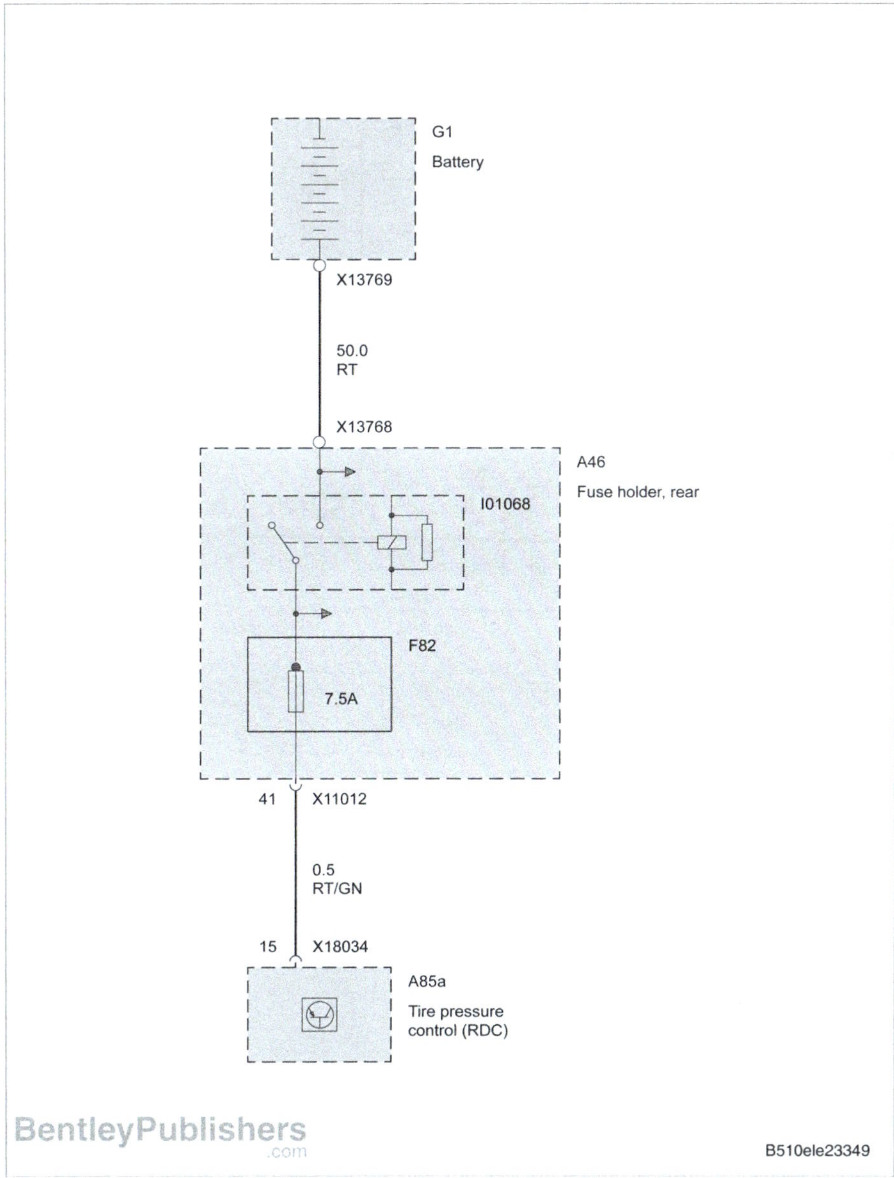

Fuse F83
up to 09 / 2005

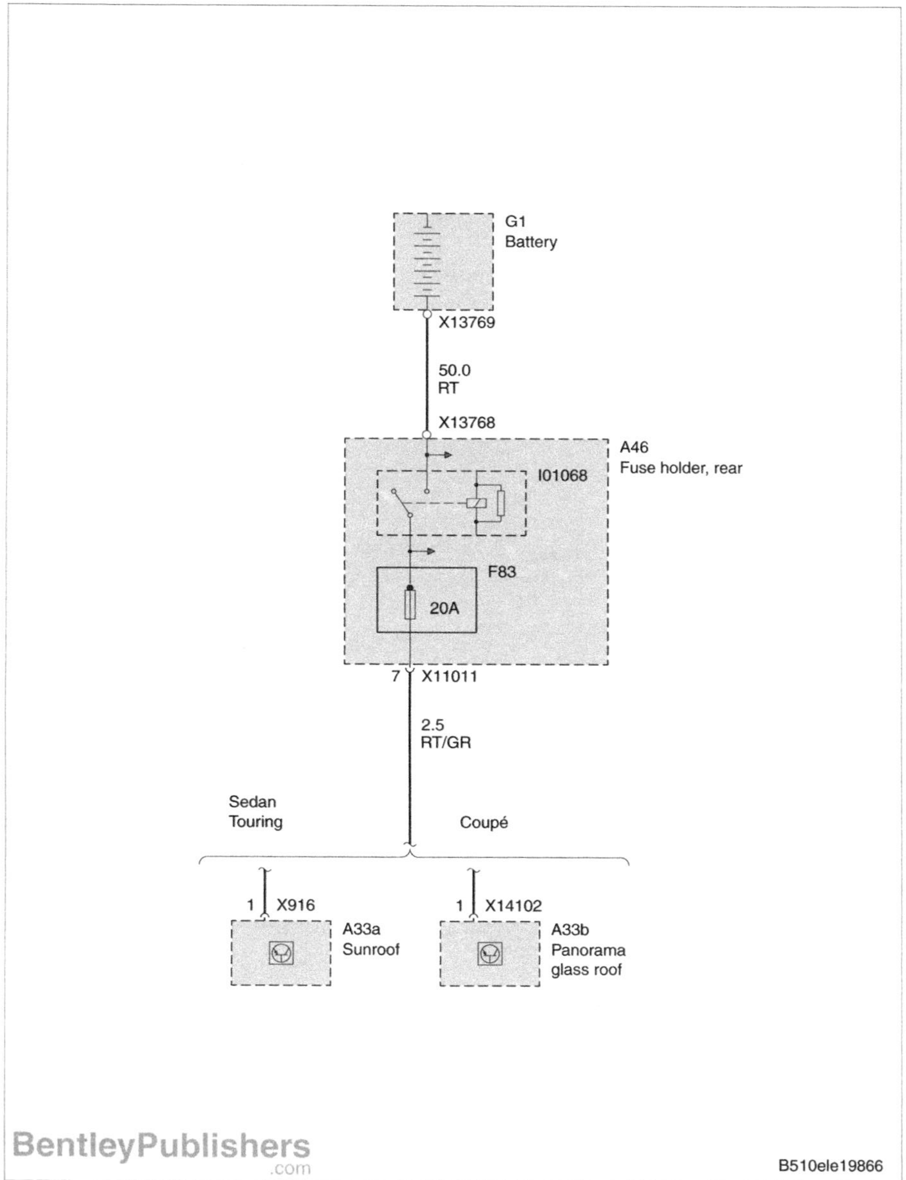

Fuse F83
as of 09 / 2005

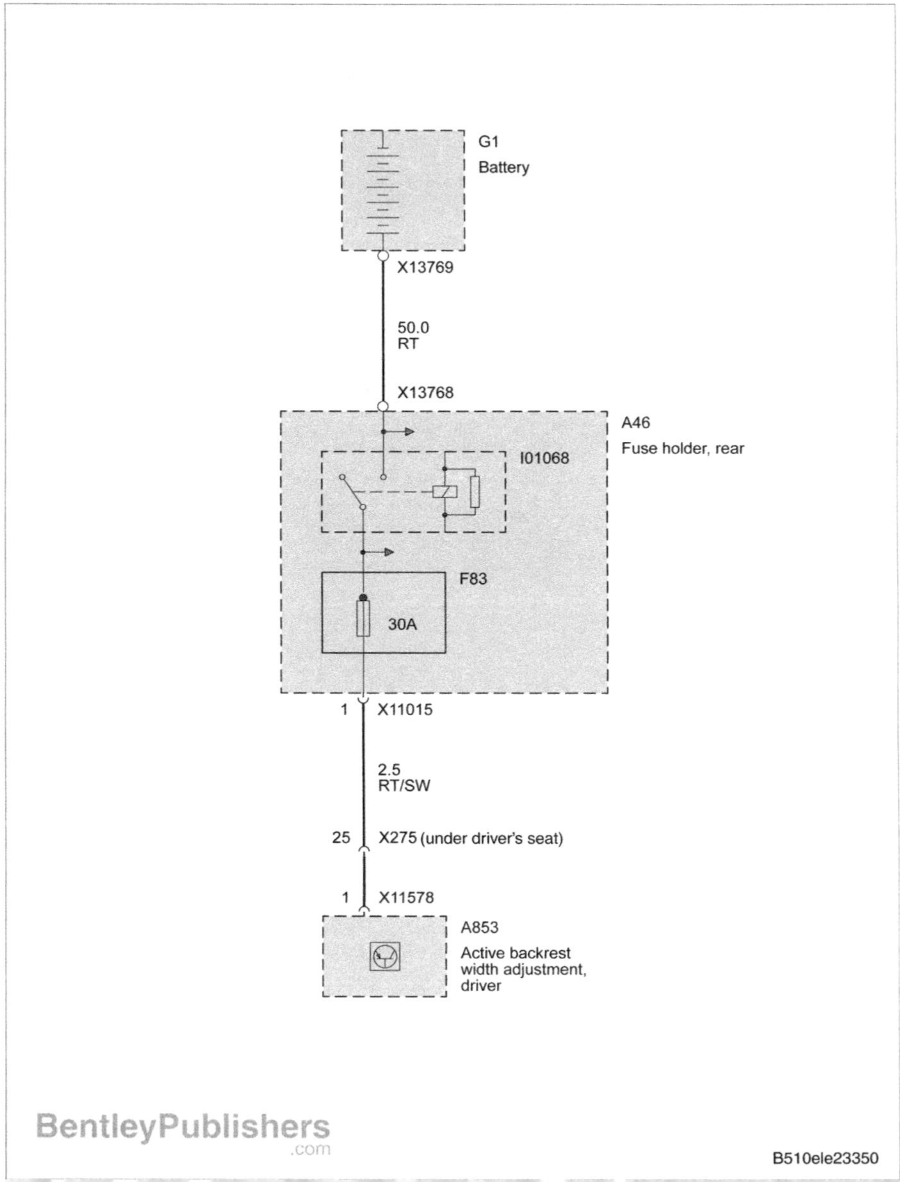

Fuse F84
up to 09 / 2005

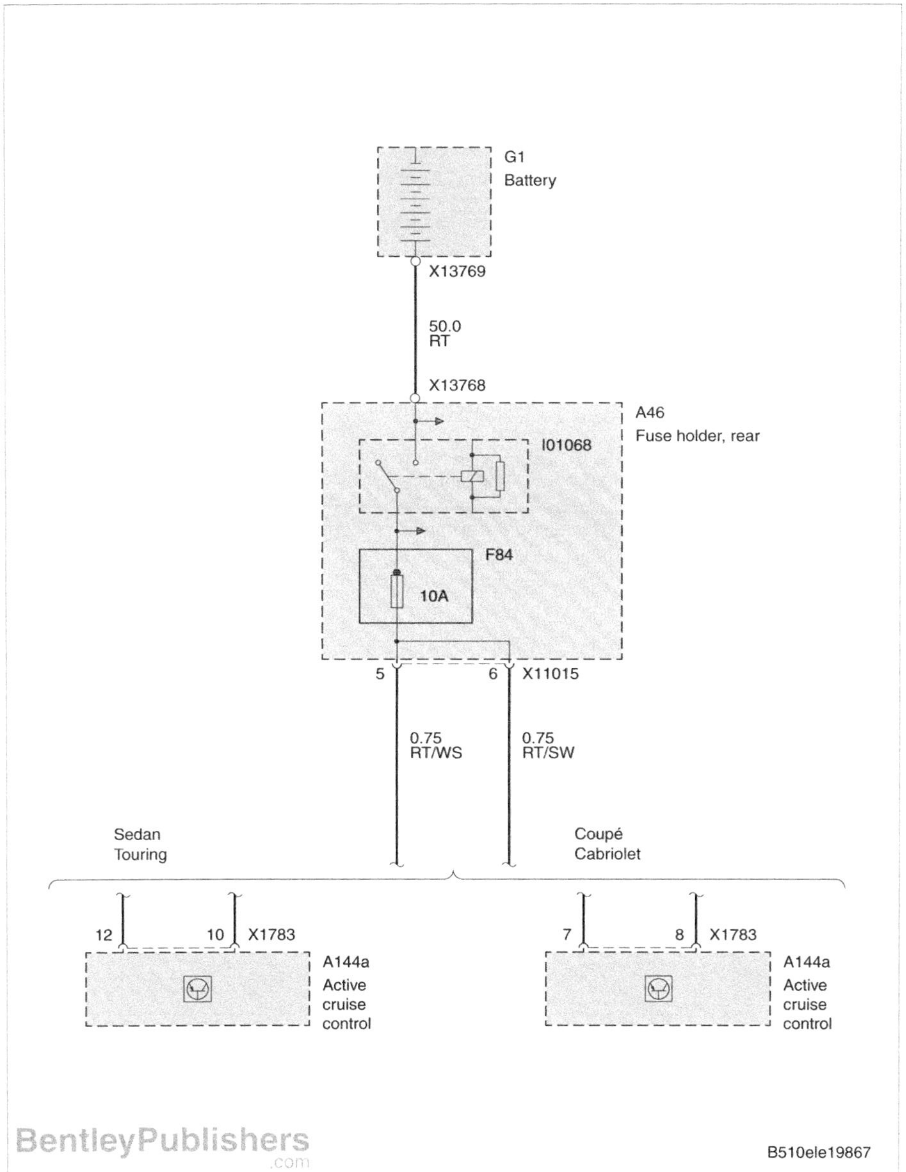

Fuse F84
N62 TU engine
as of 09 / 2005

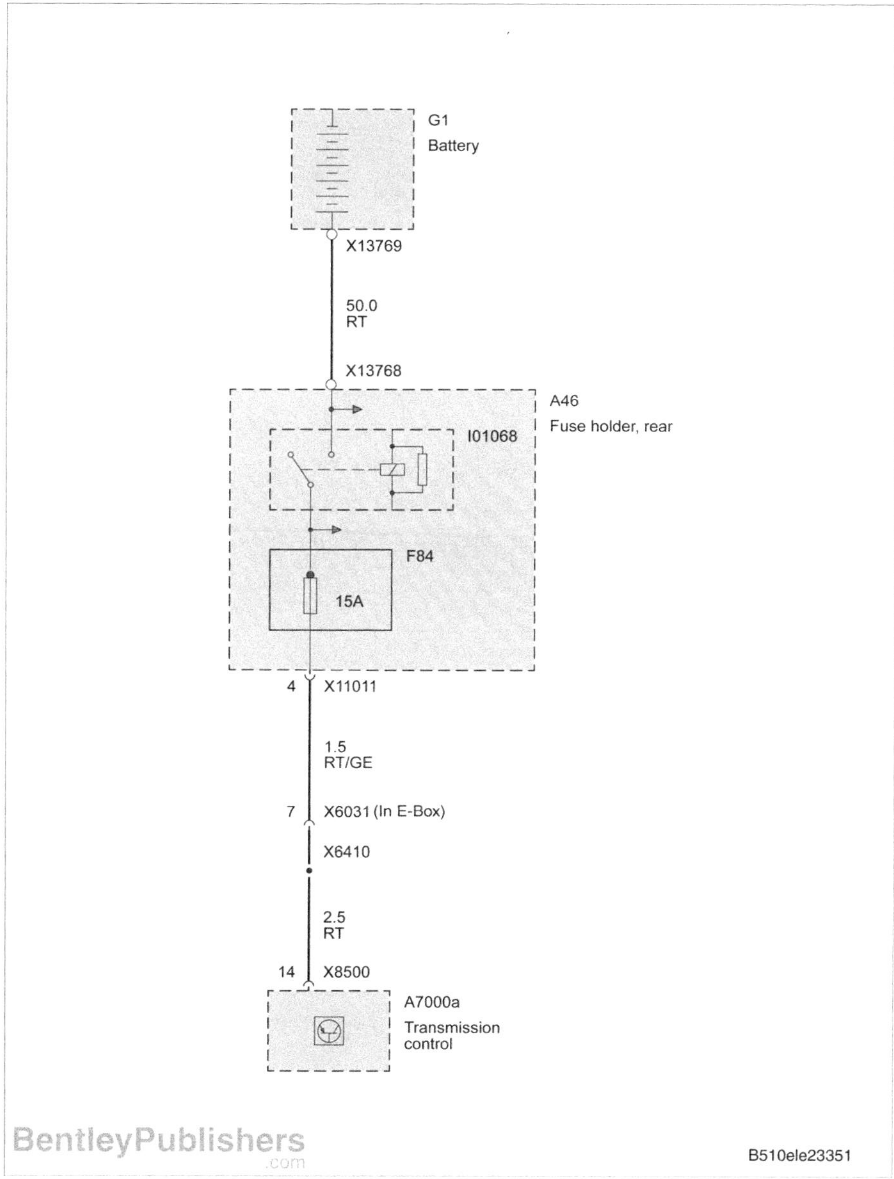

Fuse F84
without N62 TU engine
as of 09 / 2005

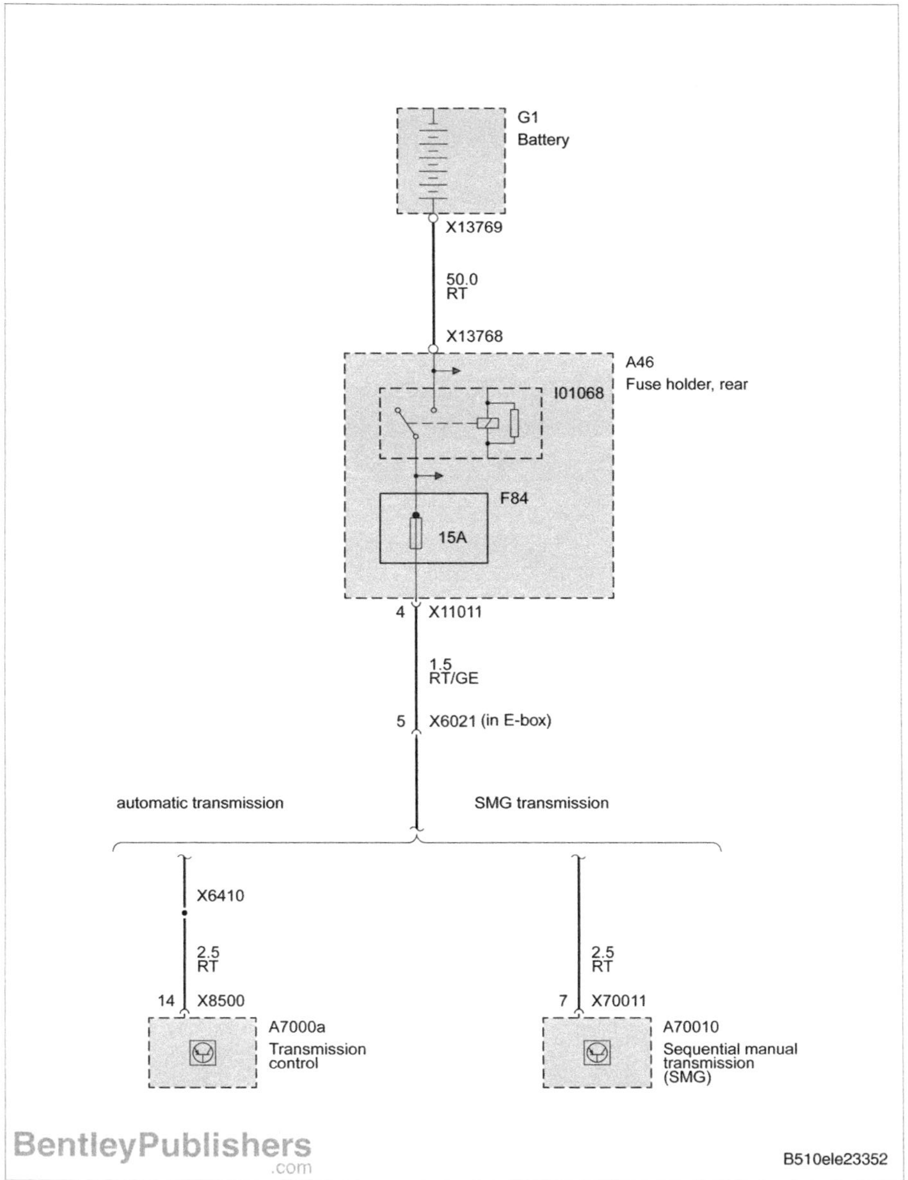

Fuse F85
as of 09 / 2005

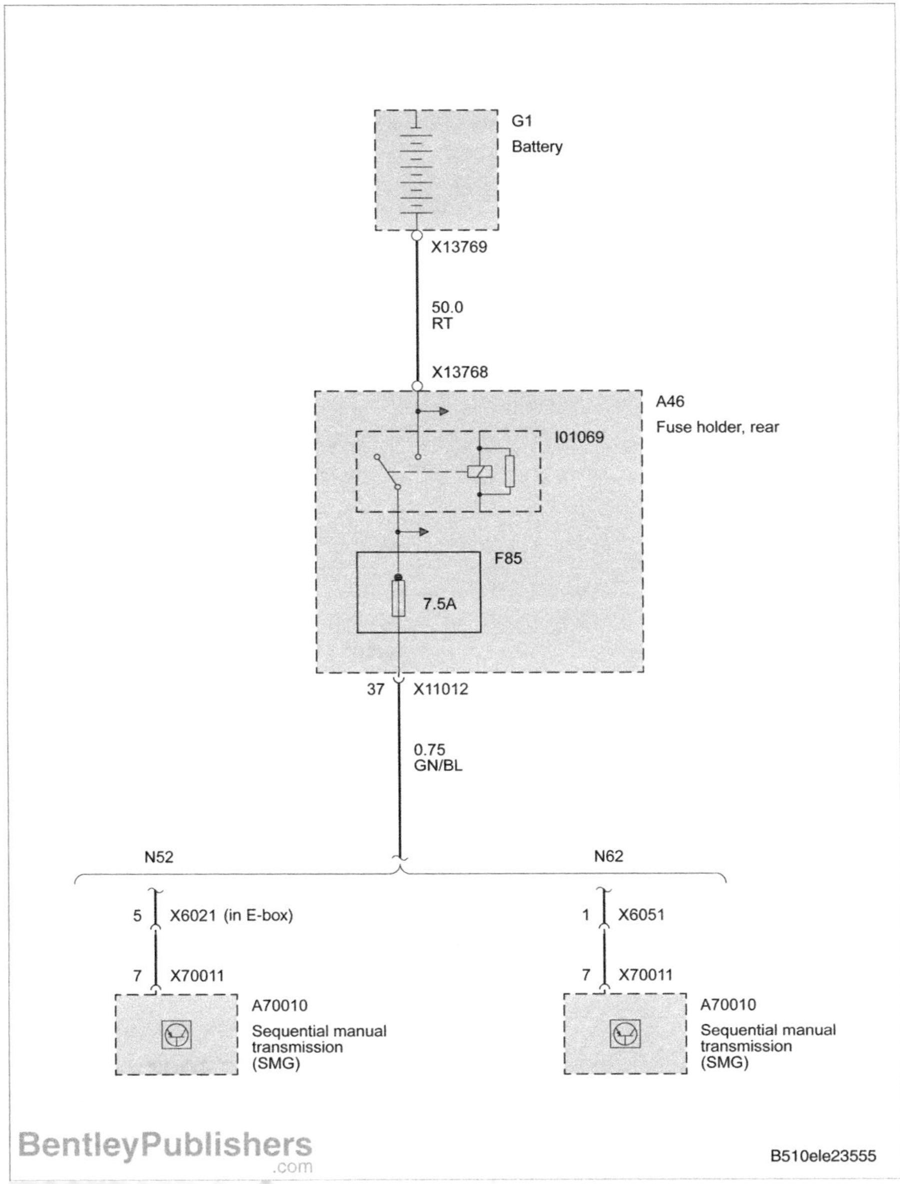

Fuse F86
without N62 TU engine
up to 09 / 2005

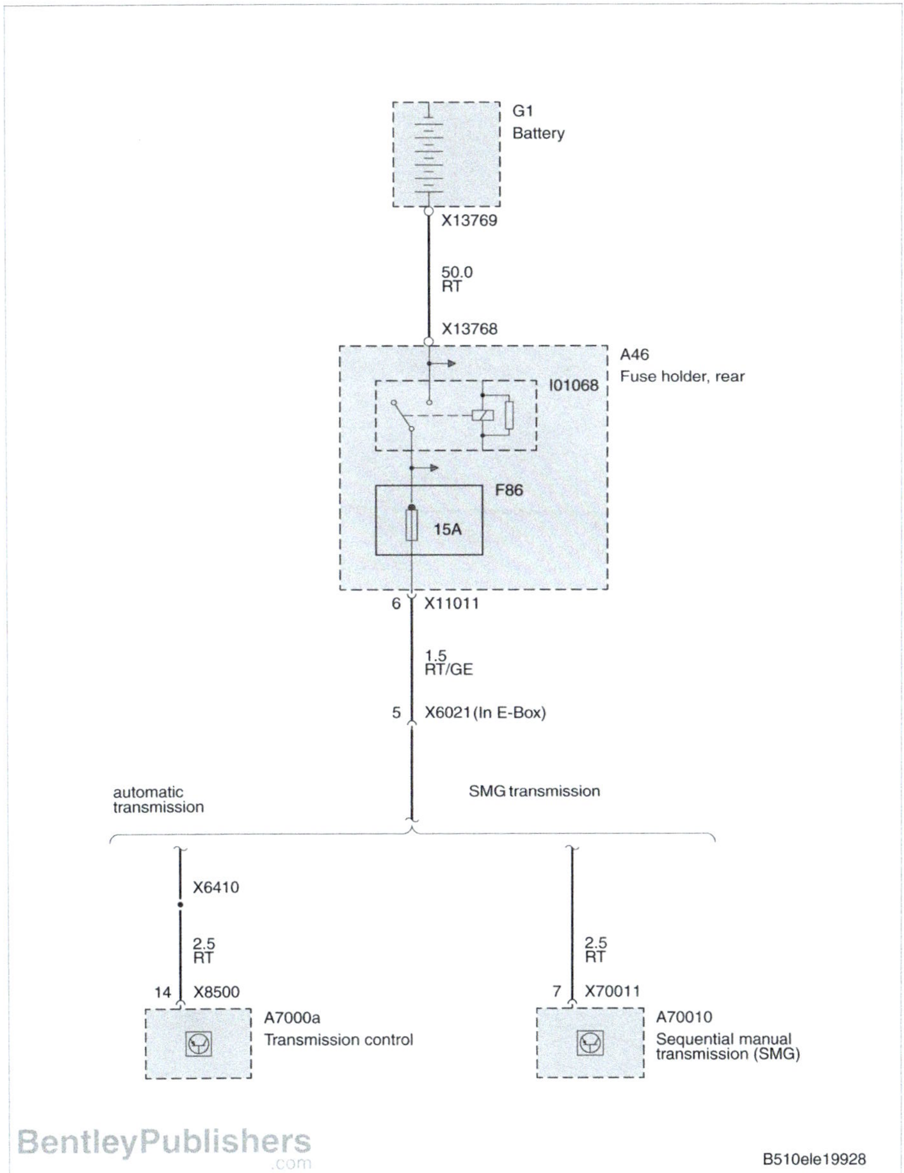

Fuse F86
N62 engine
up to 09 / 2005

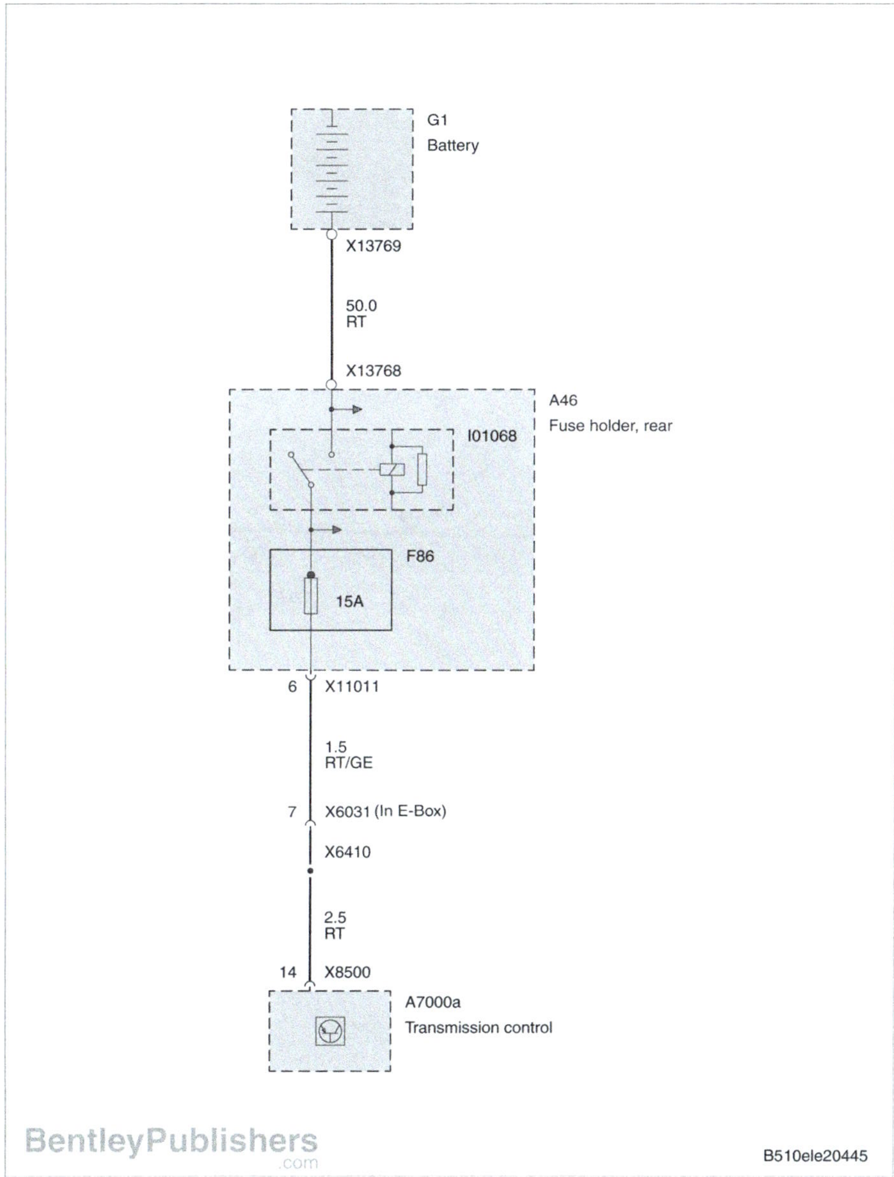

Fuse F86
as of 09 / 2005

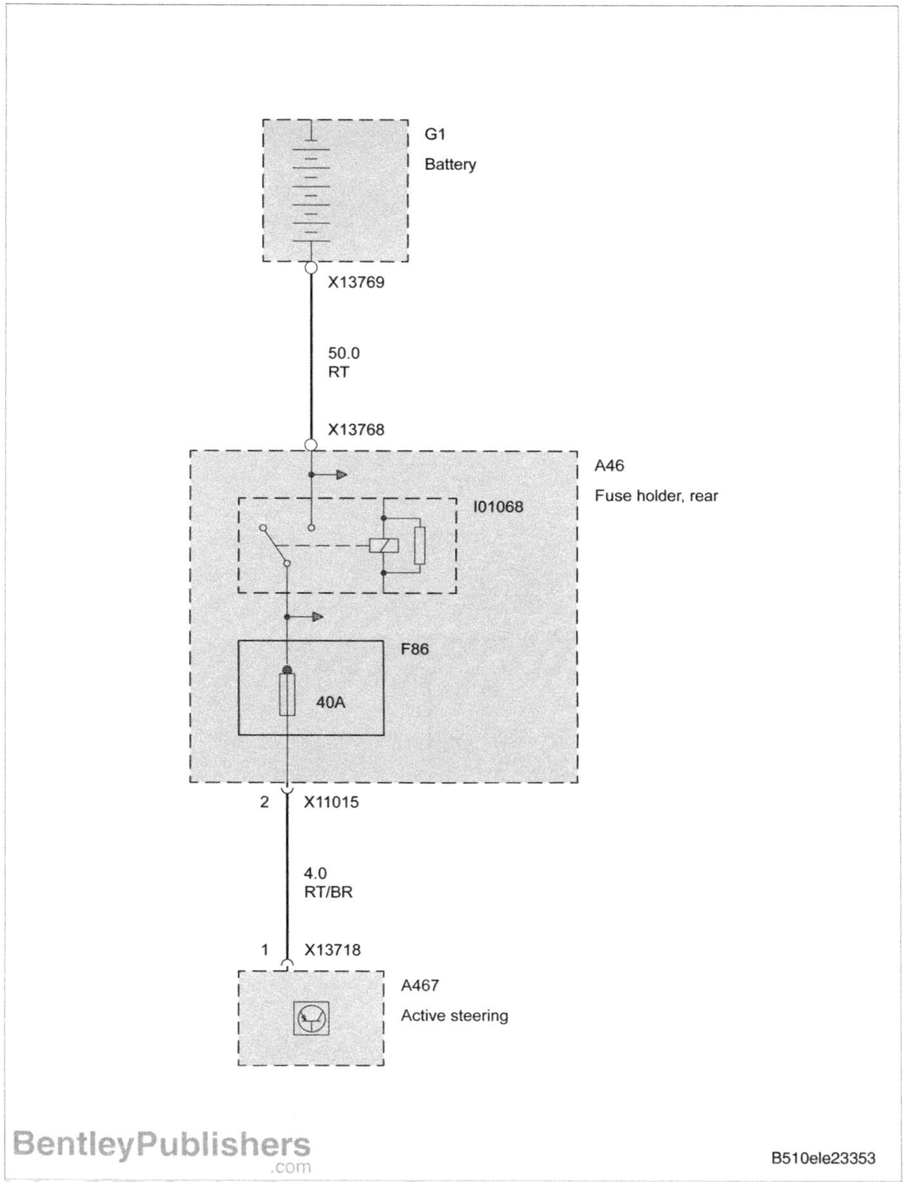

Fuse F87
as of 09 / 2005

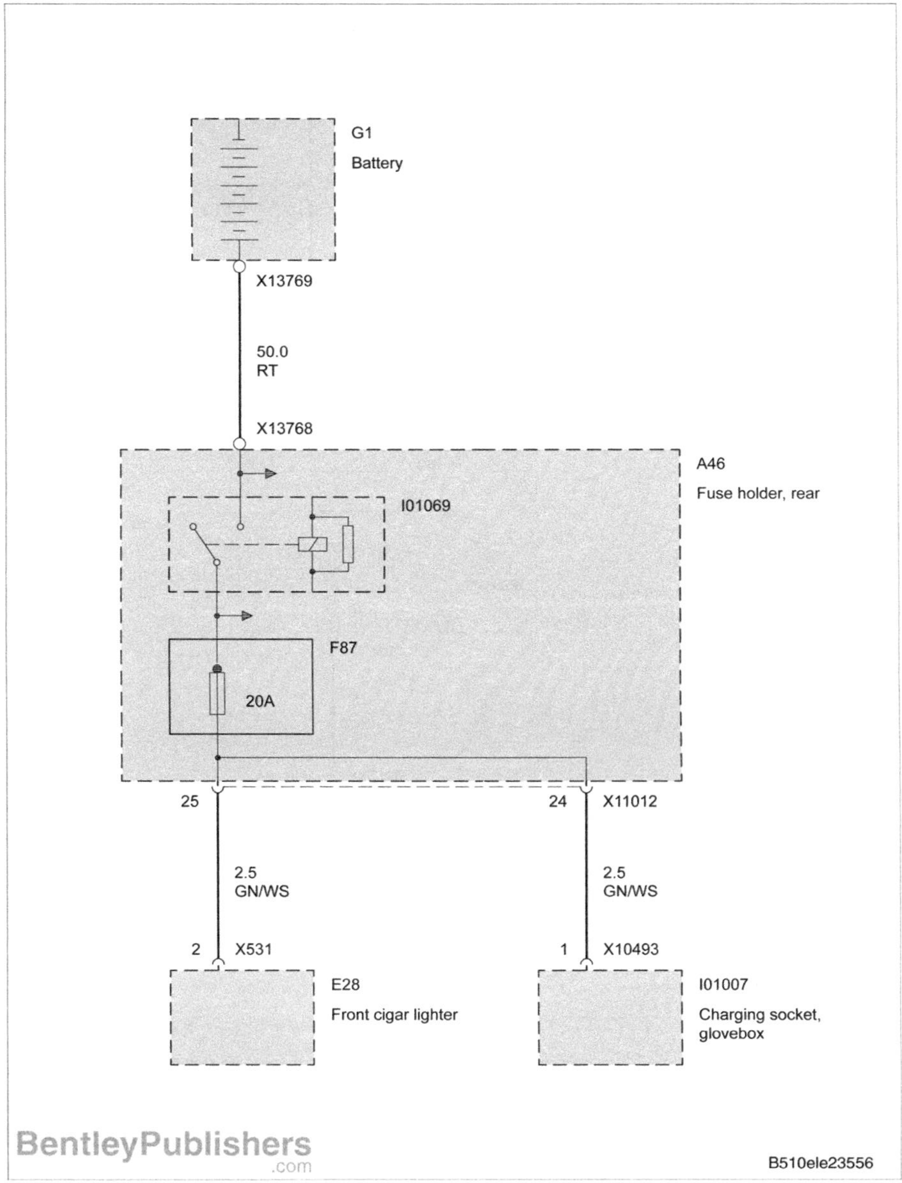

Fuse F88 up to 09 / 2005

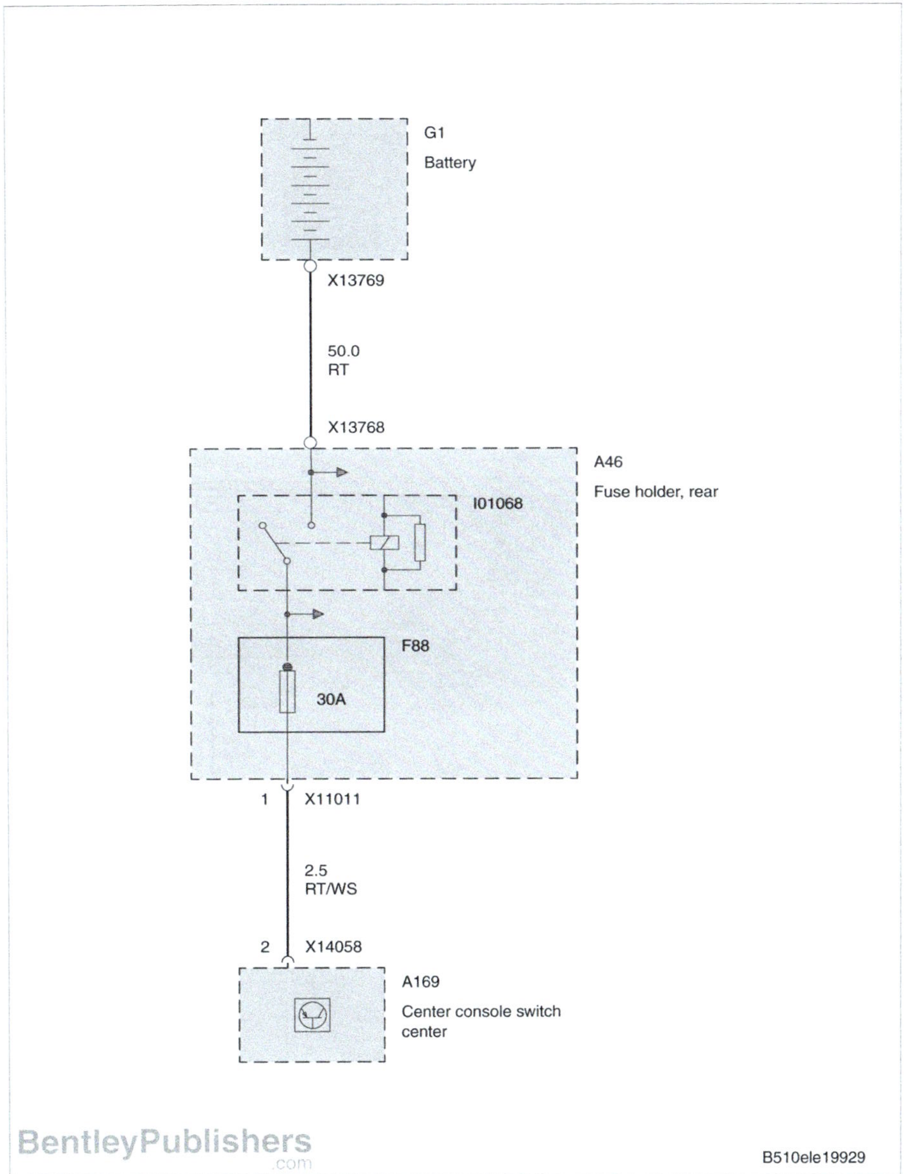

Fuse F88 as of 09 / 2005

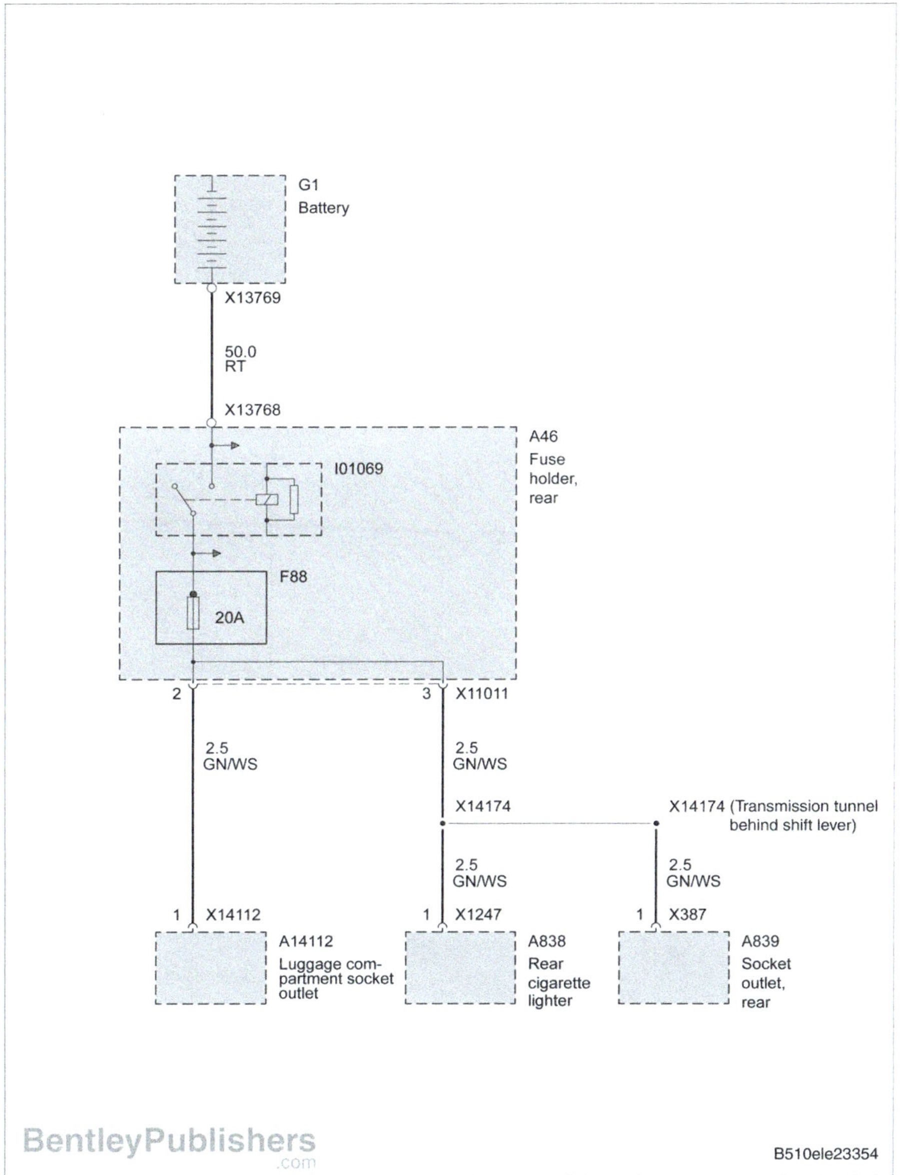

Fuse F89
up to 09 / 2005

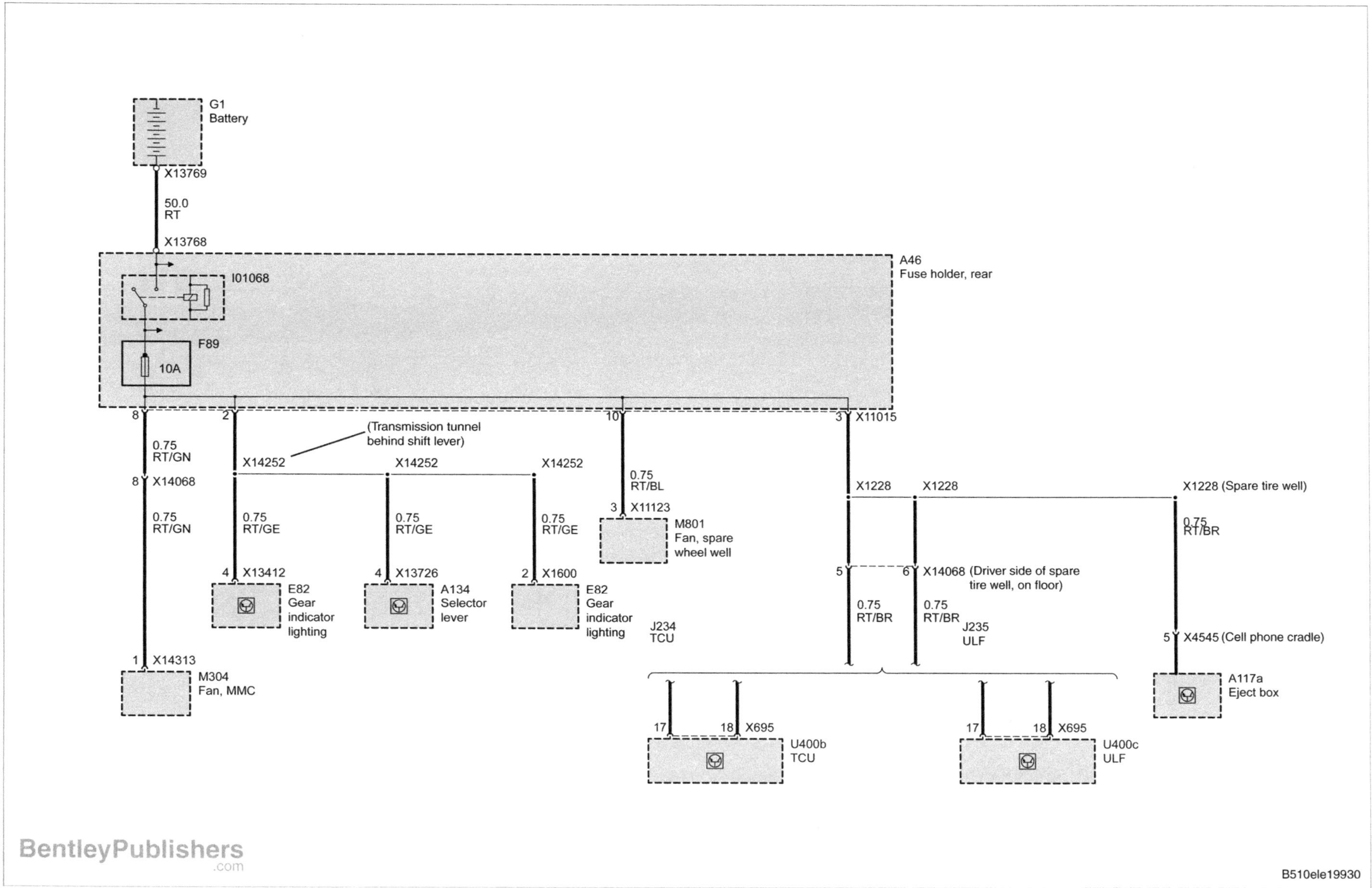

BentleyPublishers
.com

B510ele19930

Fuse F89
as of 09 / 2005

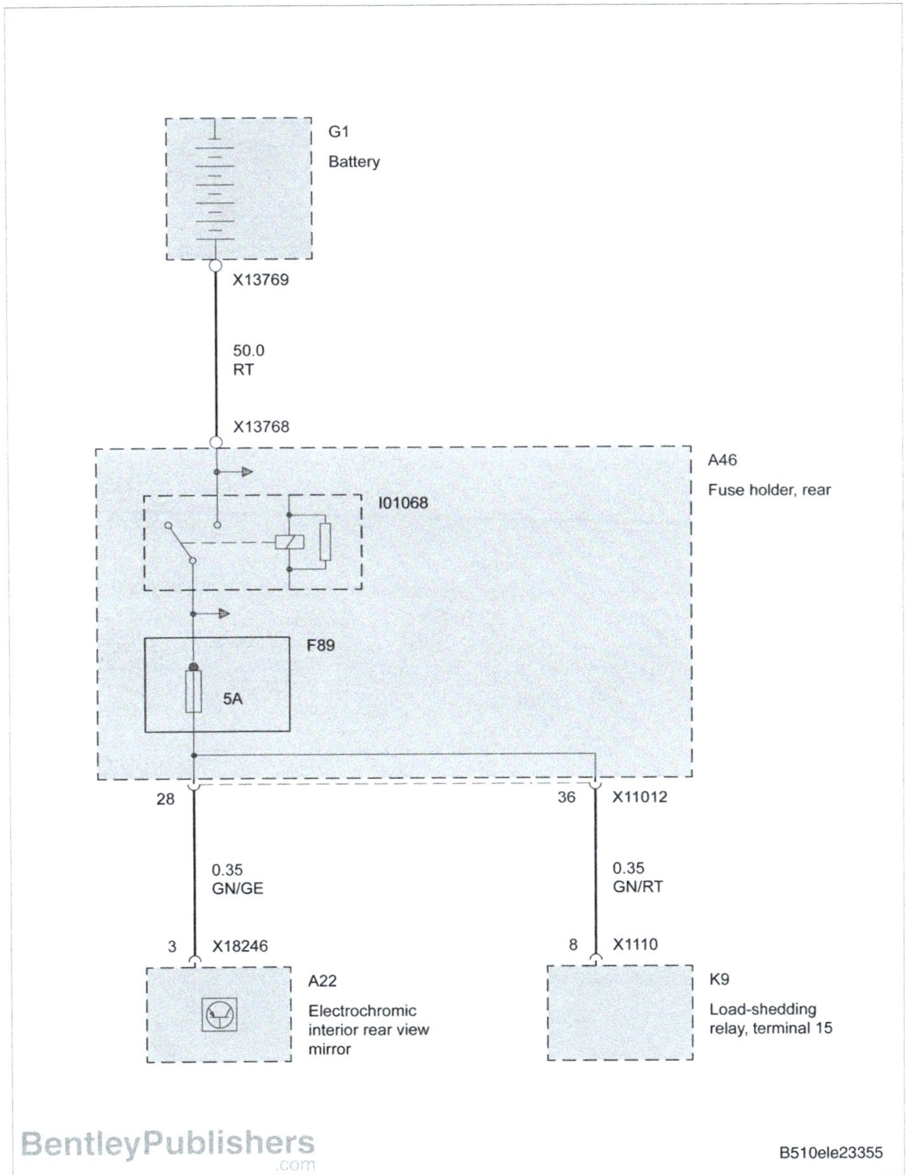

Fuse F90
up to 09 / 2005

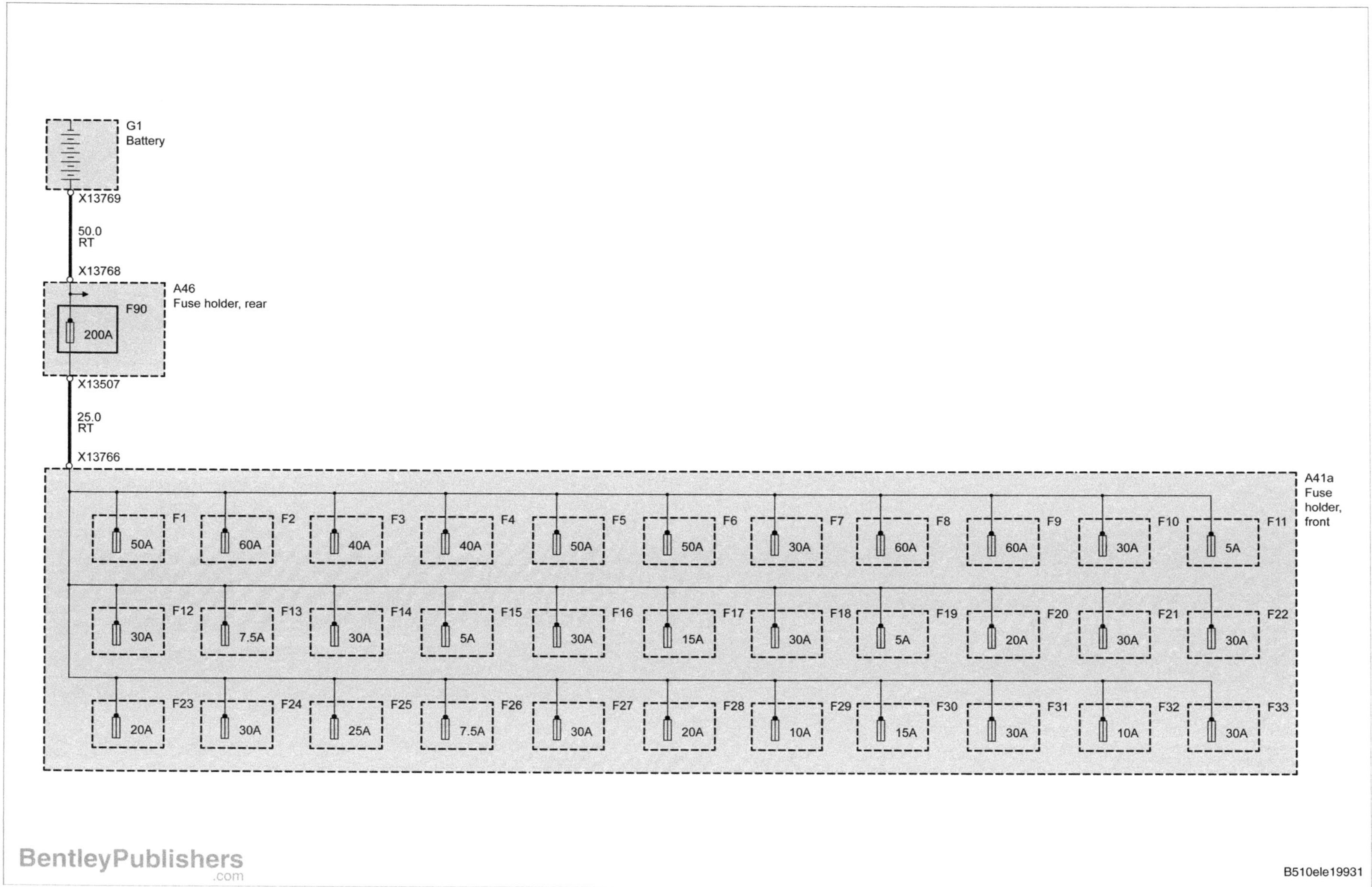

BentleyPublishers.com

Fuse F90
as of 09 / 2005

BentleyPublishers.com

Fuse F91
N62 engine and N62 TU engine

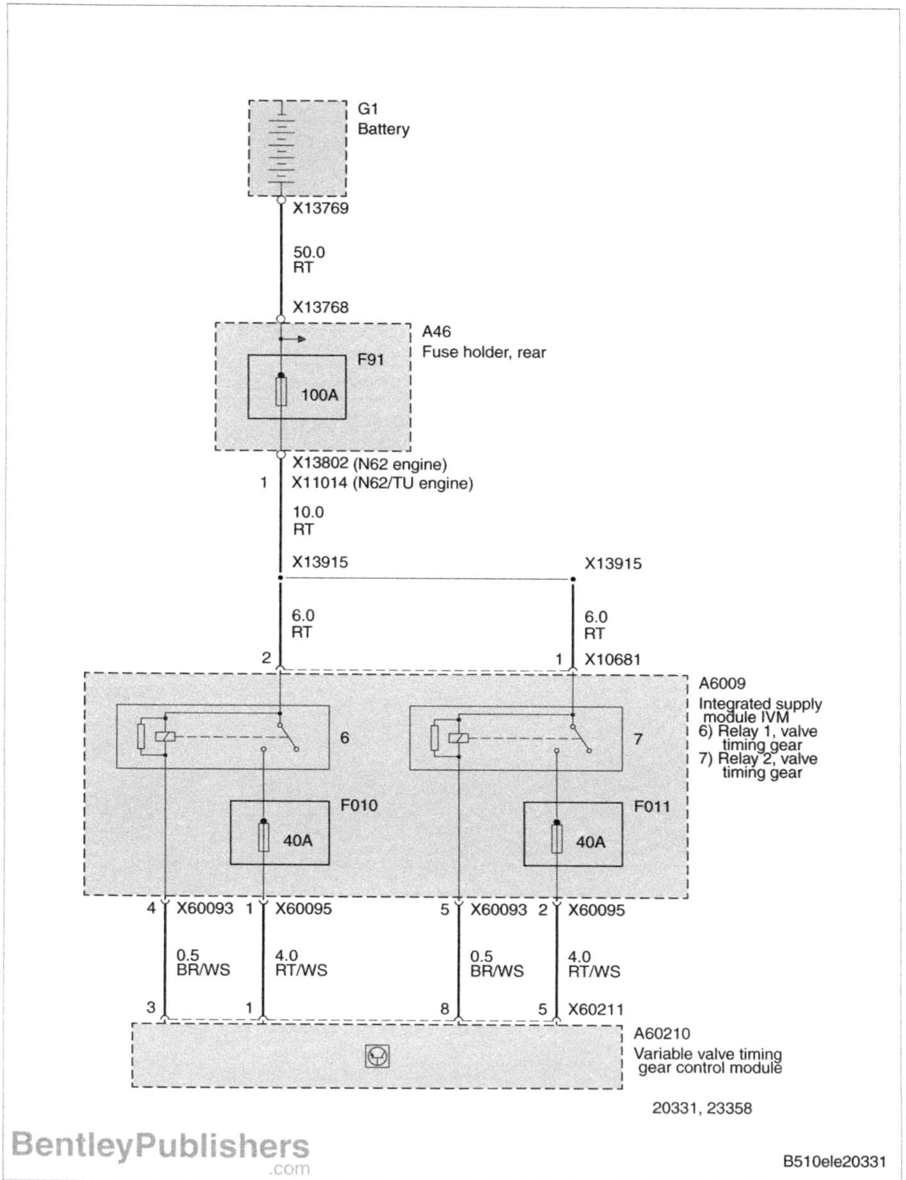

Fuse F92

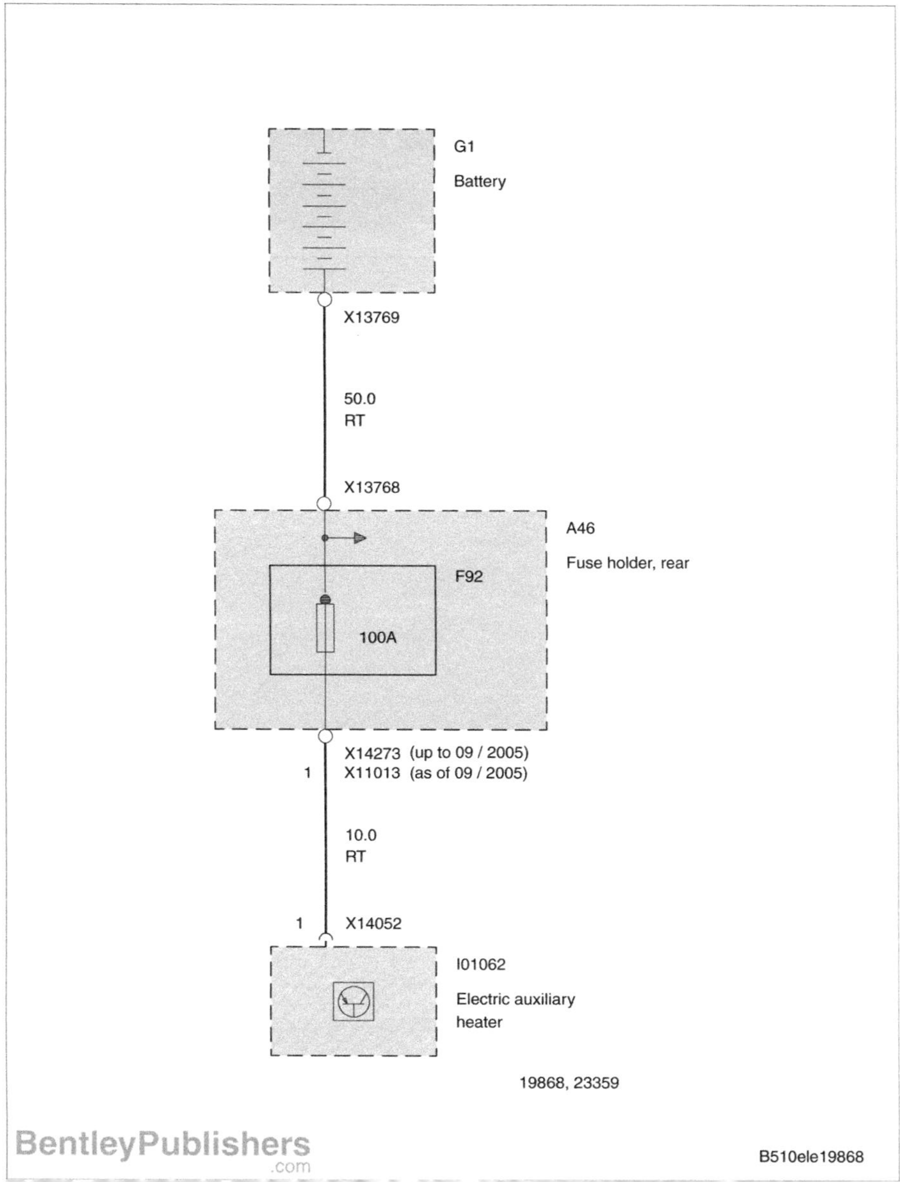

Integrated supply module (IVM)
N62 engine or N62 TU engine up to 09 / 2005
(page 1 of 3)

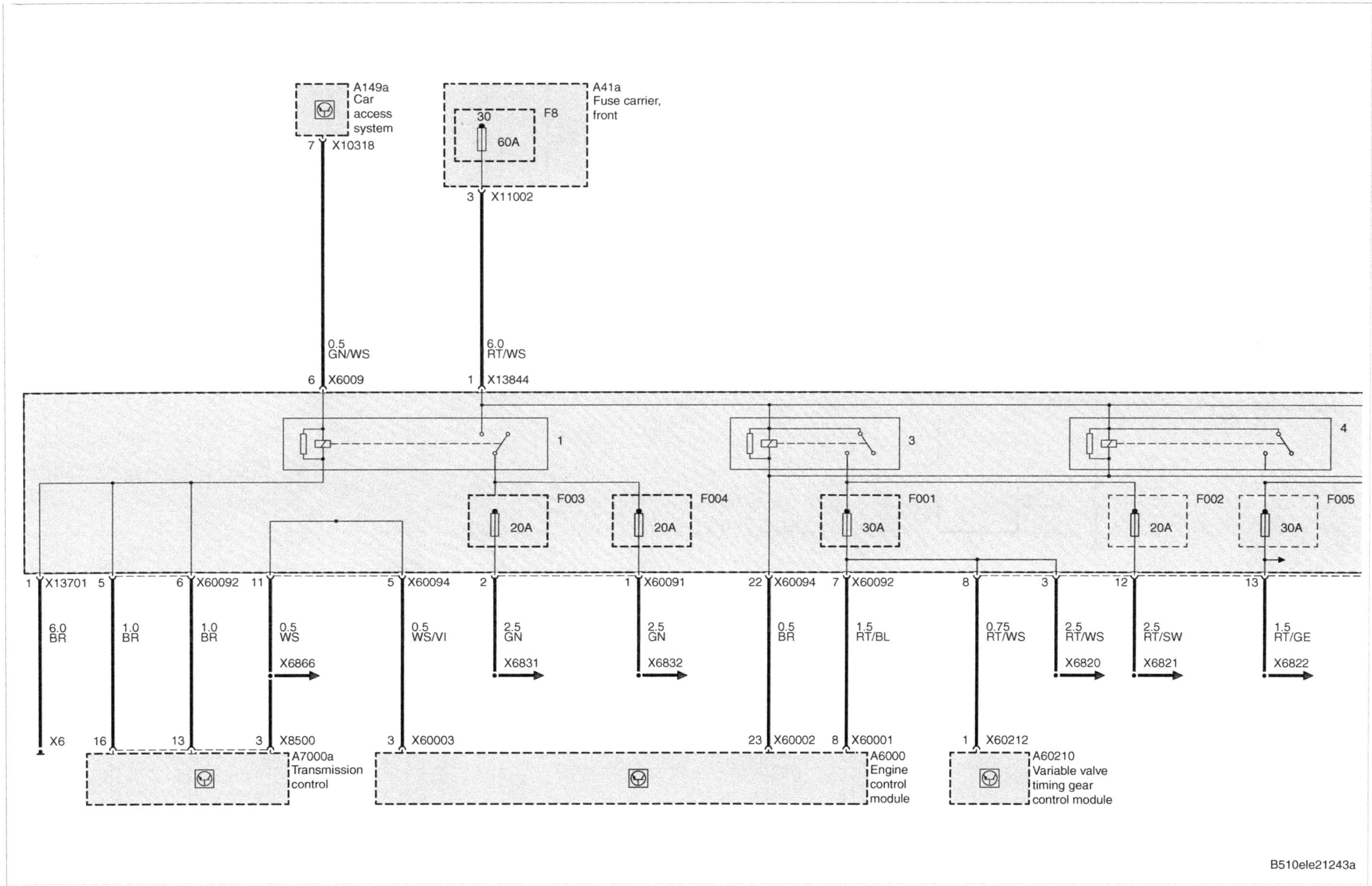

Integrated supply module (IVM)
N62 engine or N62 TU engine up to 09 / 2005
(page 2 of 3)

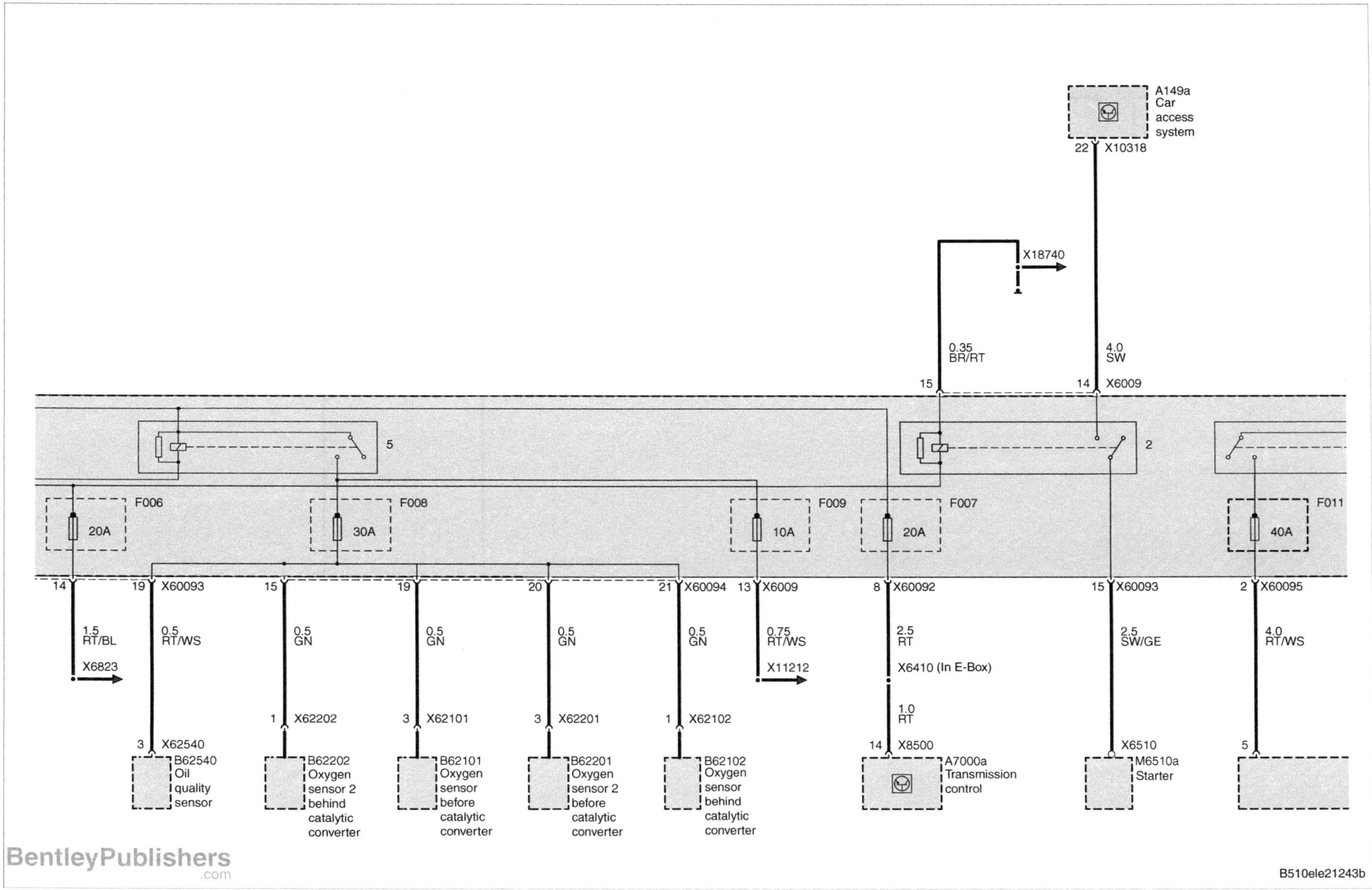

BentleyPublishers.com

Integrated supply module (IVM)
N62 engine or N62 TU engine up to 09 / 2005
(page 3 of 3)

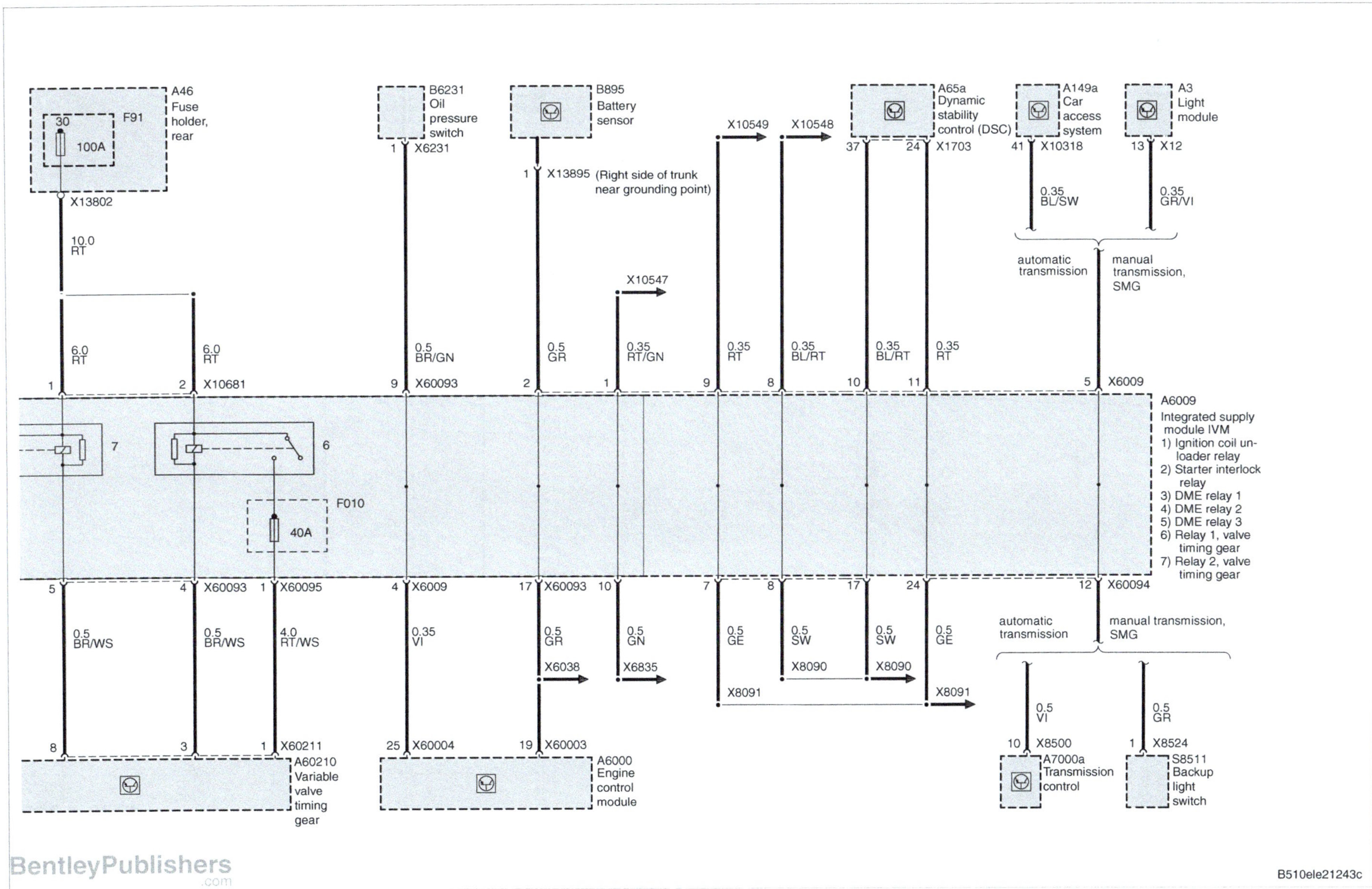

BentleyPublishers.com

Integrated supply module (IVM)
N62 engine or N62 TU engine as of 09 / 2005
(page 1 of 3)

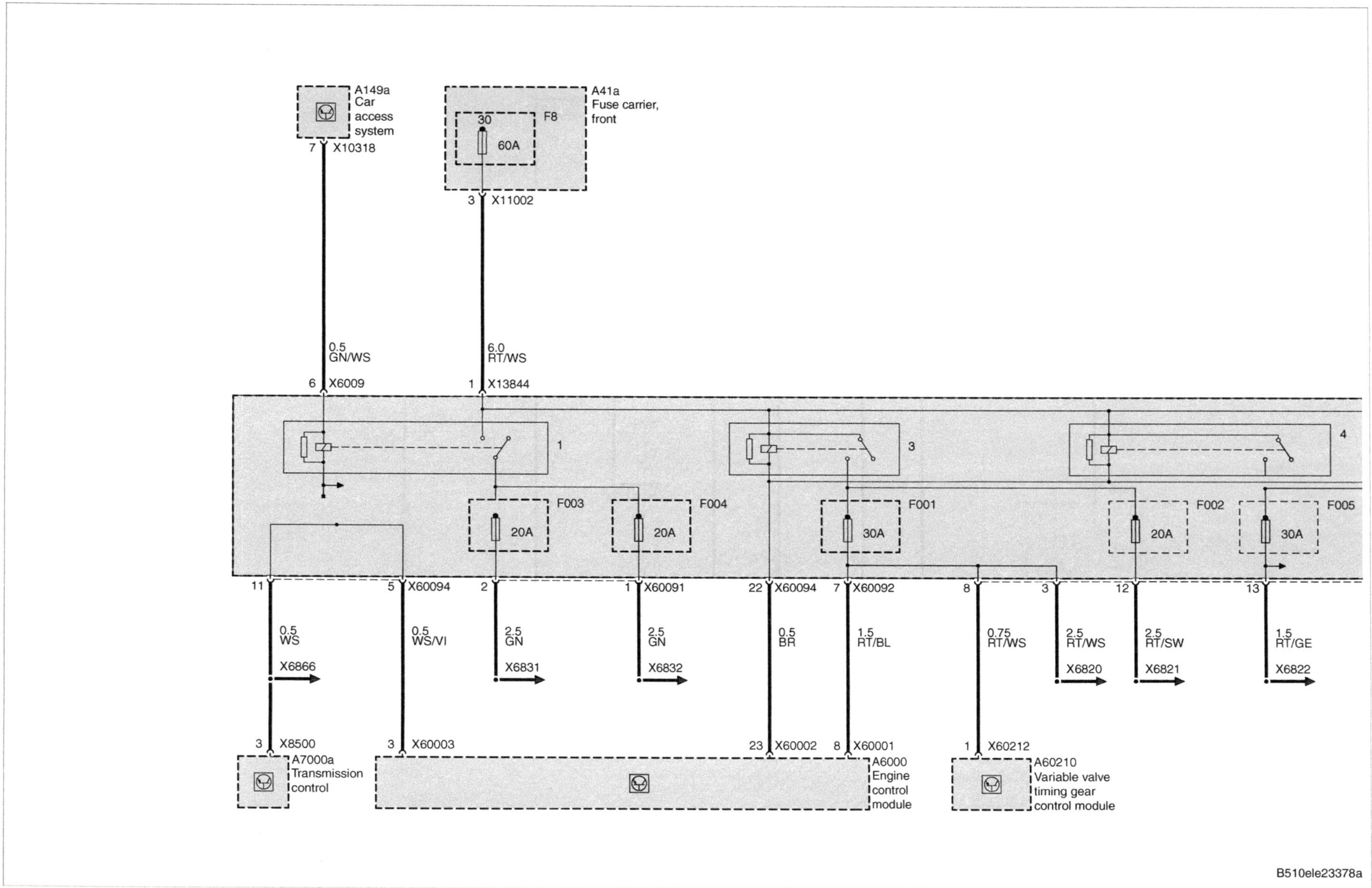

B510ele23378a

Integrated supply module (IVM)
N62 engine or N62 TU engine as of 09 / 2005
(page 2 of 3)

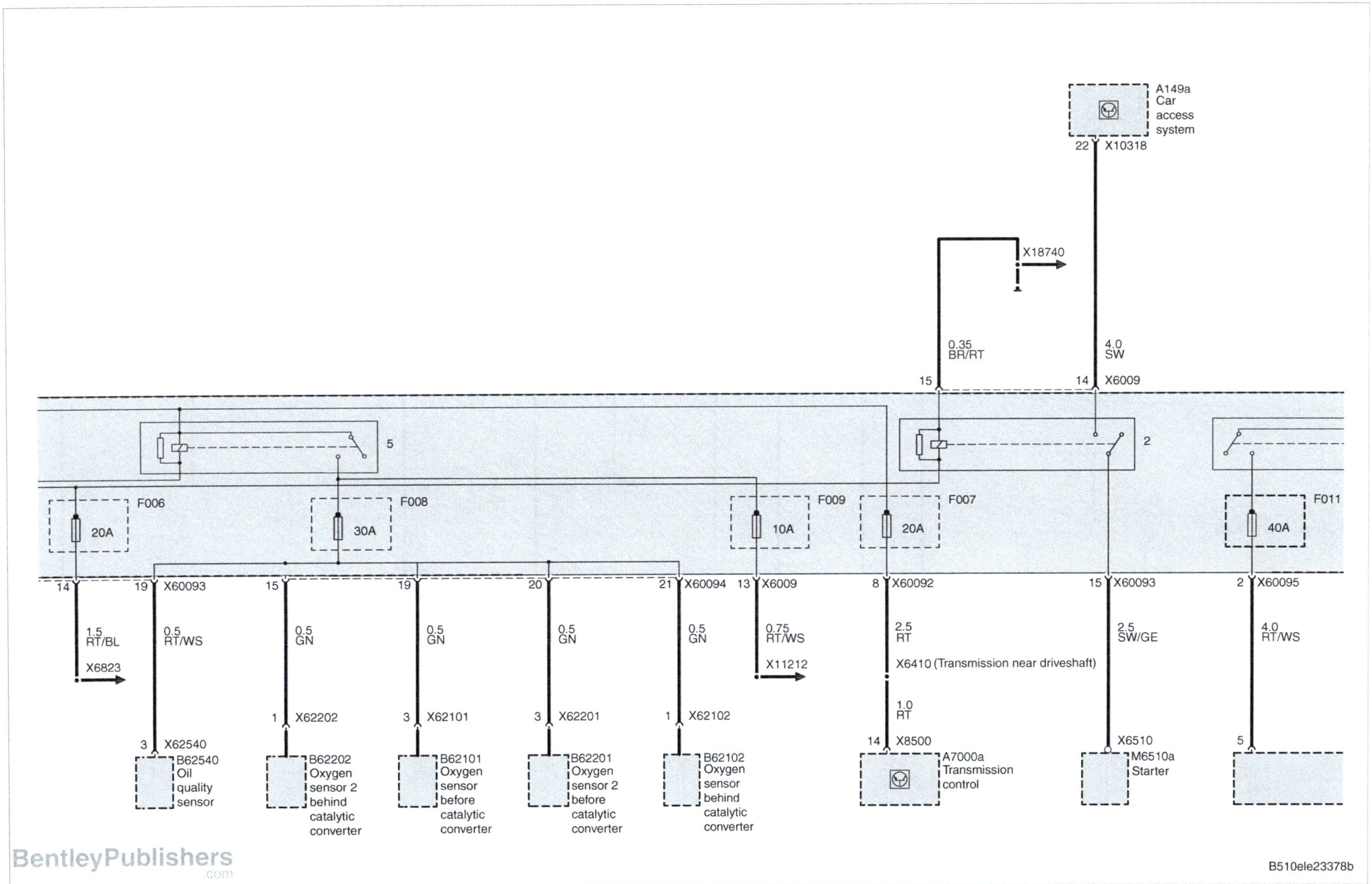

BentleyPublishers.com

Integrated supply module (IVM)
N62 engine or N62 TU engine as of 09 / 2005
(page 3 of 3)

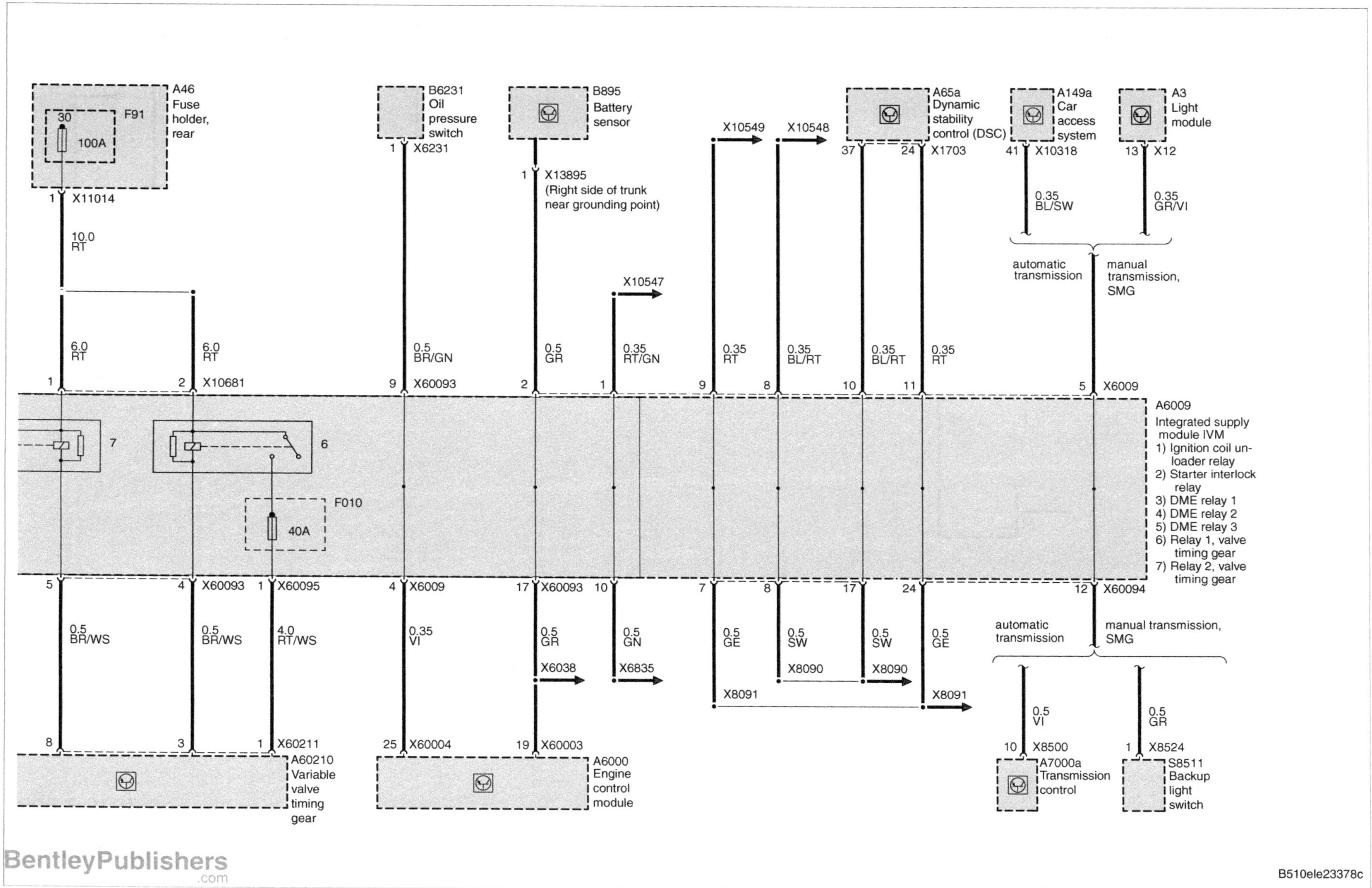

BentleyPublishers.com

B510ele23378c

OBDII socket
M54 engine or N62 engine

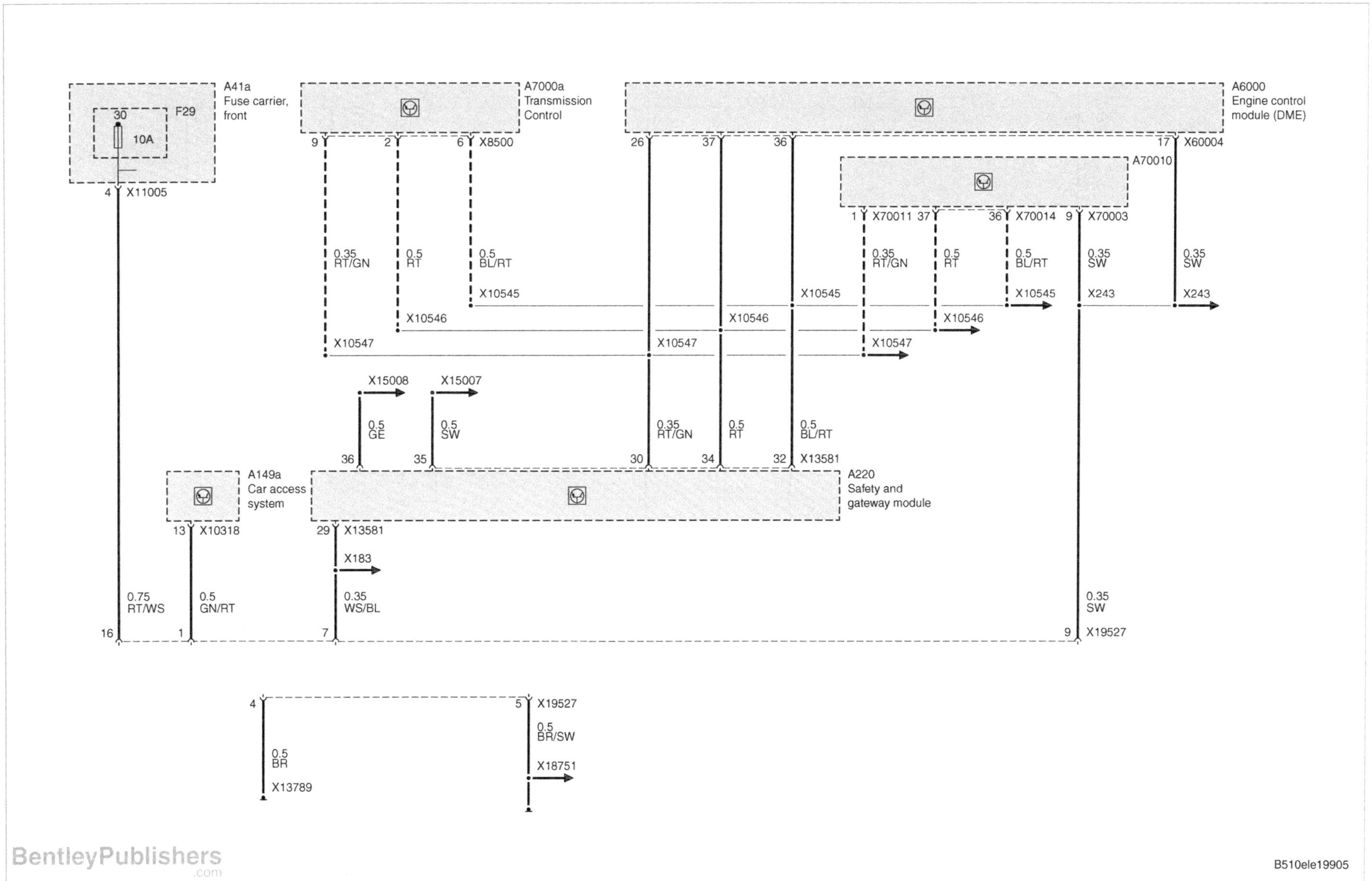

OBDII socket
N52 engine up to 09 / 2005

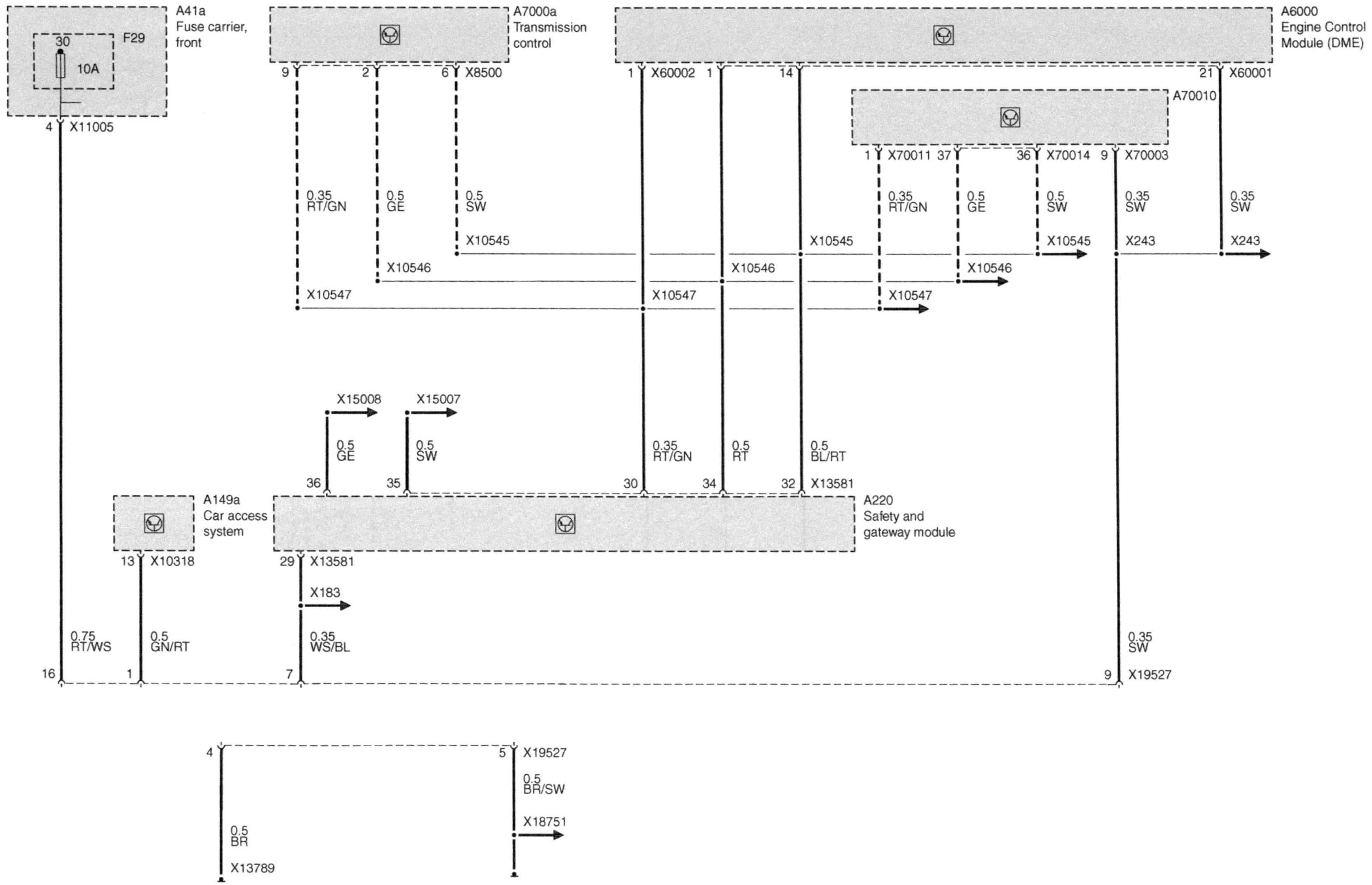

BentleyPublishers.com

B510ele22244

OBDII socket
N52 engine as of 09 / 2005 up to 03 / 2007

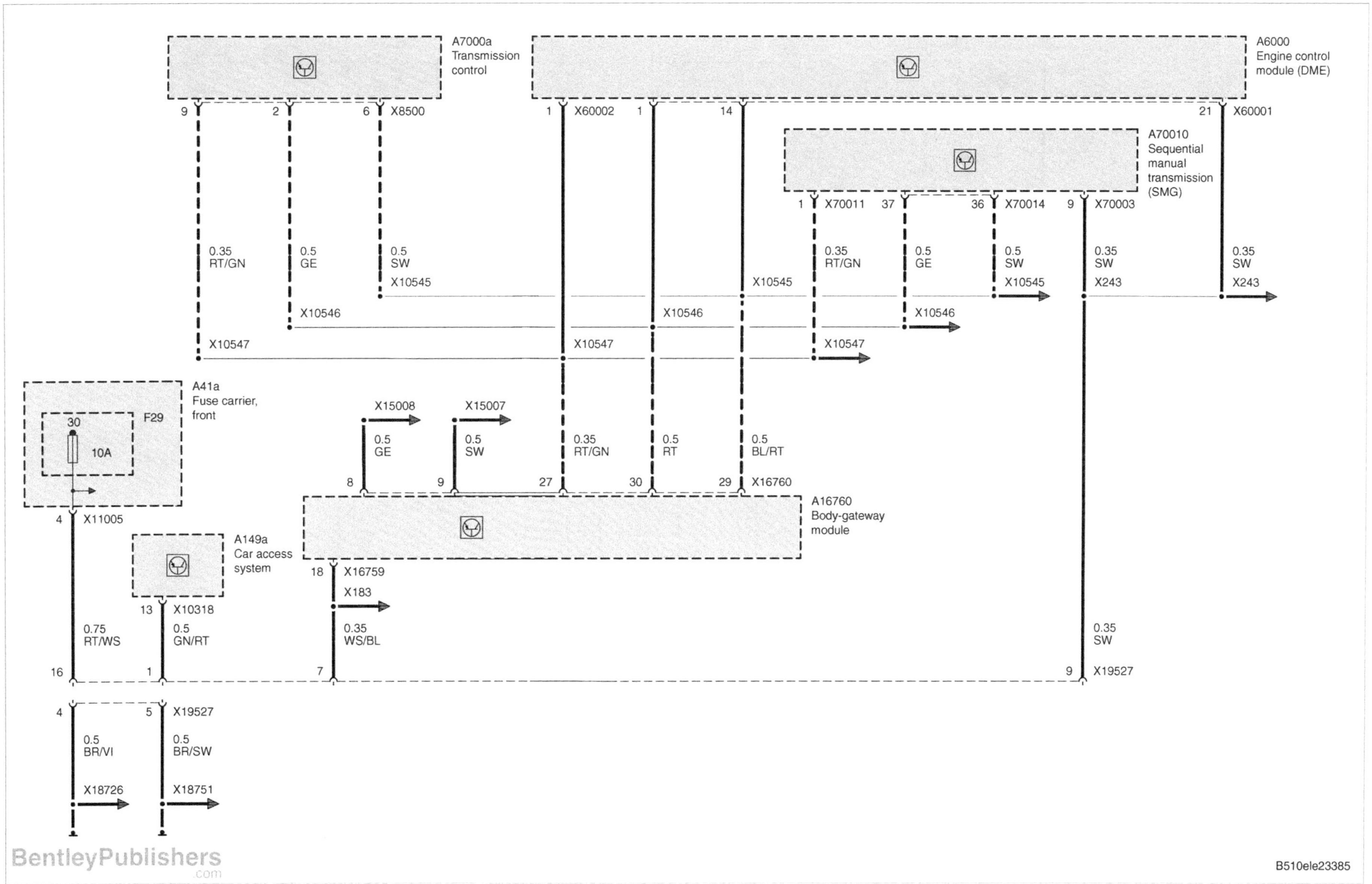

BentleyPublishers.com

OBDII socket
N52 engine or N54 engine as of 03 / 2007 up to 09 / 2008

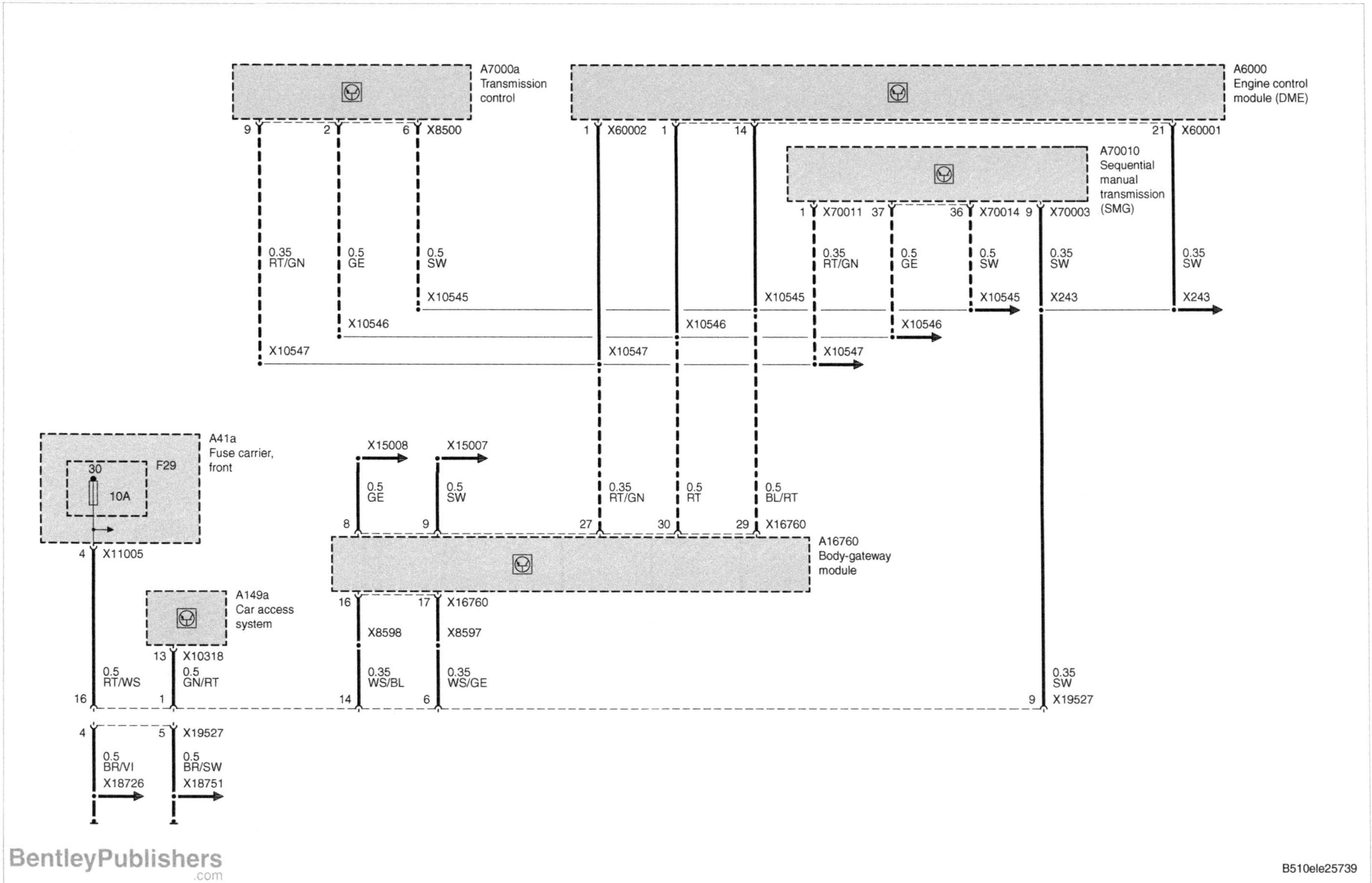

BentleyPublishers.com

OBDII socket
N52 engine or N54 engine as of 09 / 2008

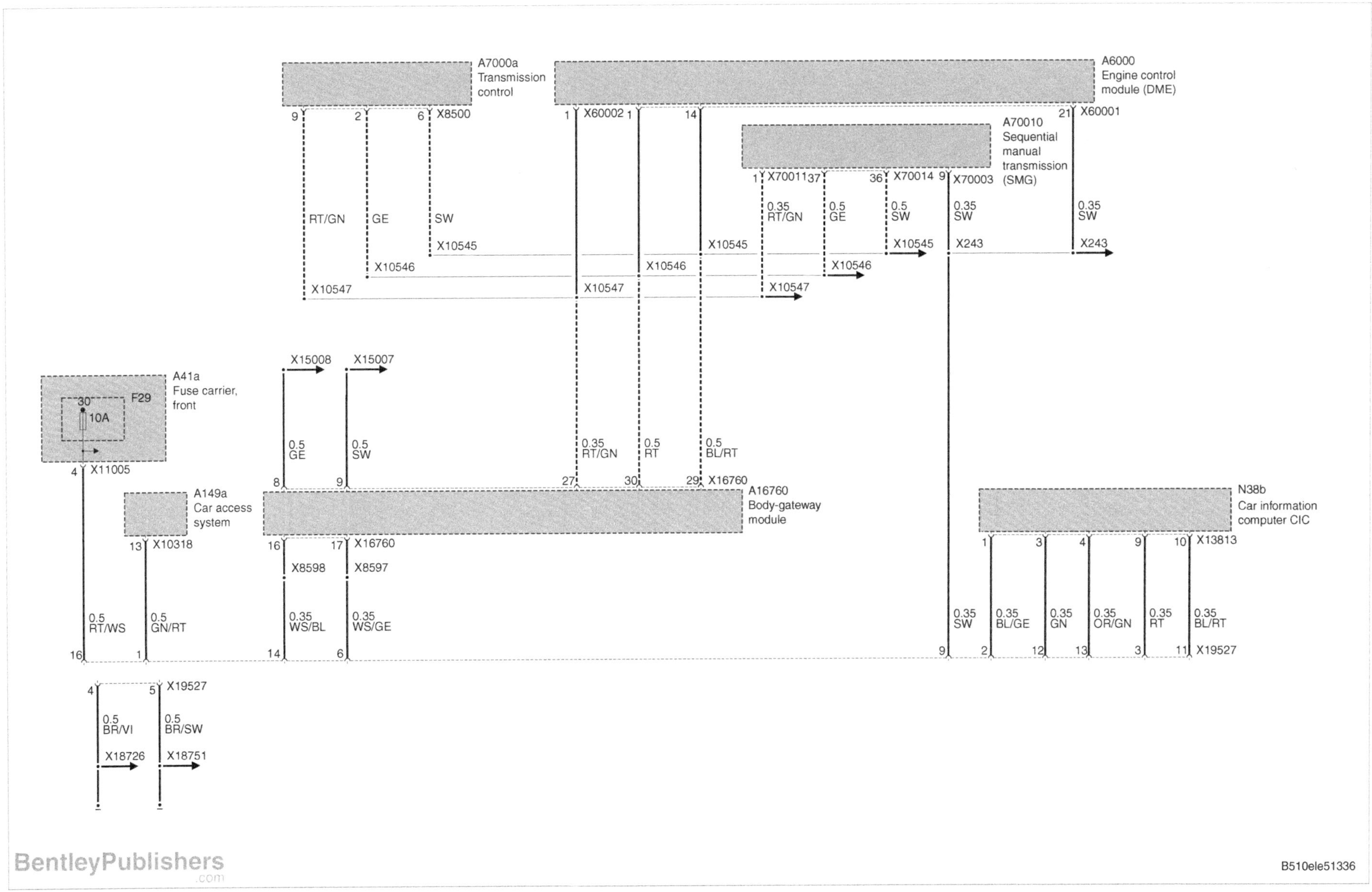

OBDII socket
N62 TU engine up to 03 / 2007

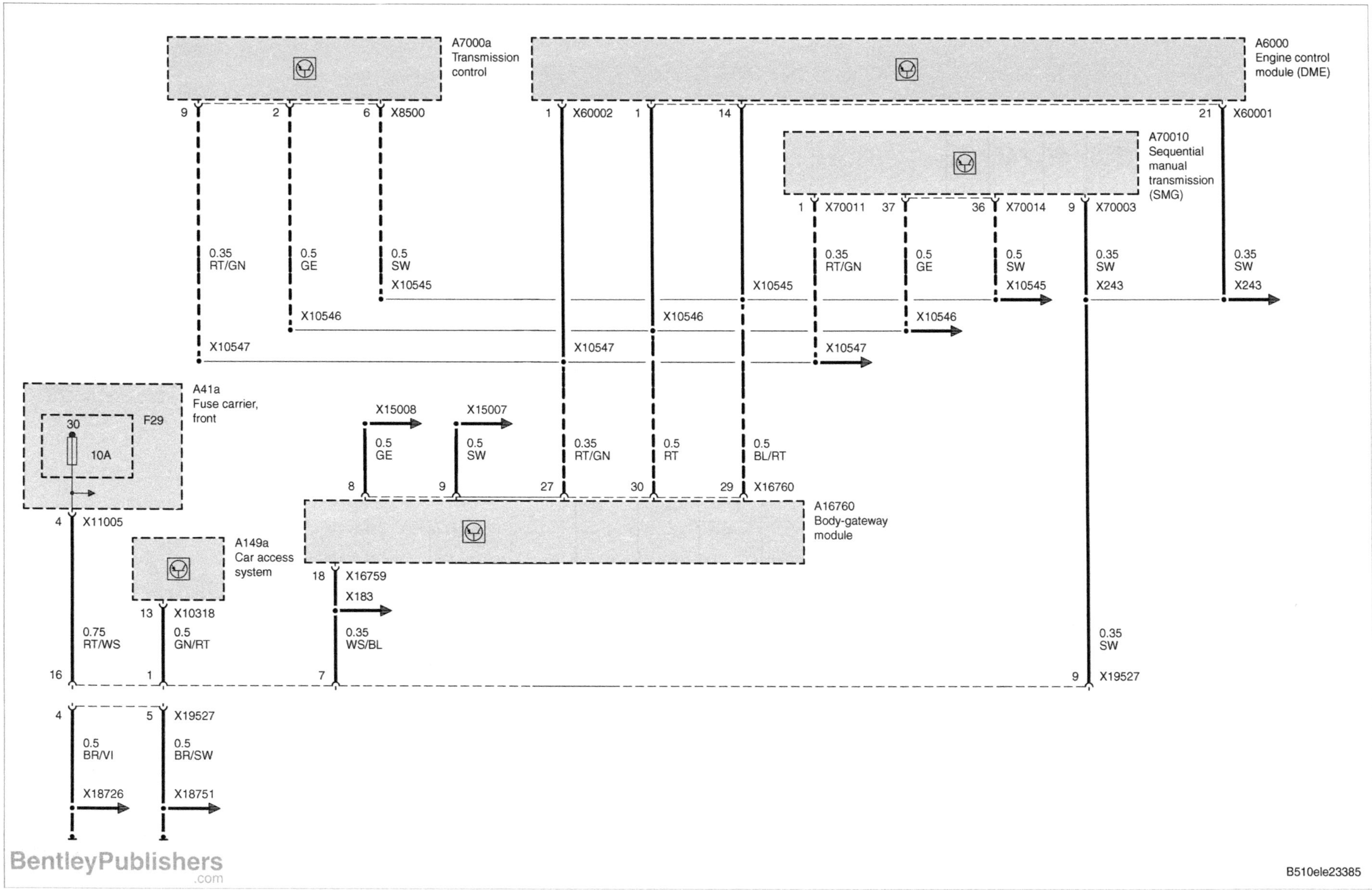

BentleyPublishers.com

OBDII socket
N62 TU engine as of 03 / 2007 up to 09 / 2008

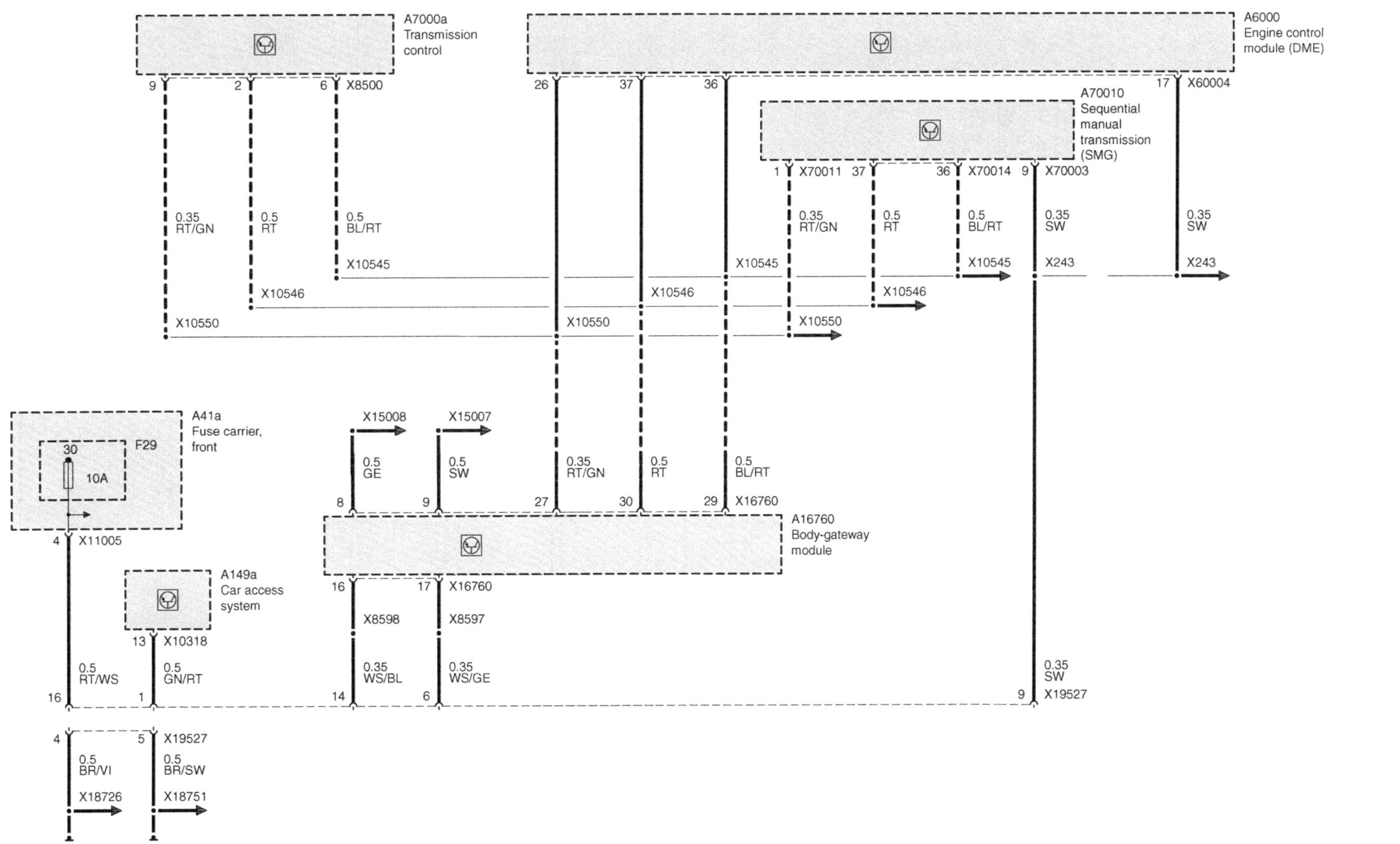

BentleyPublishers.com

OBDII socket
N62 TU engine as of 09 / 2008

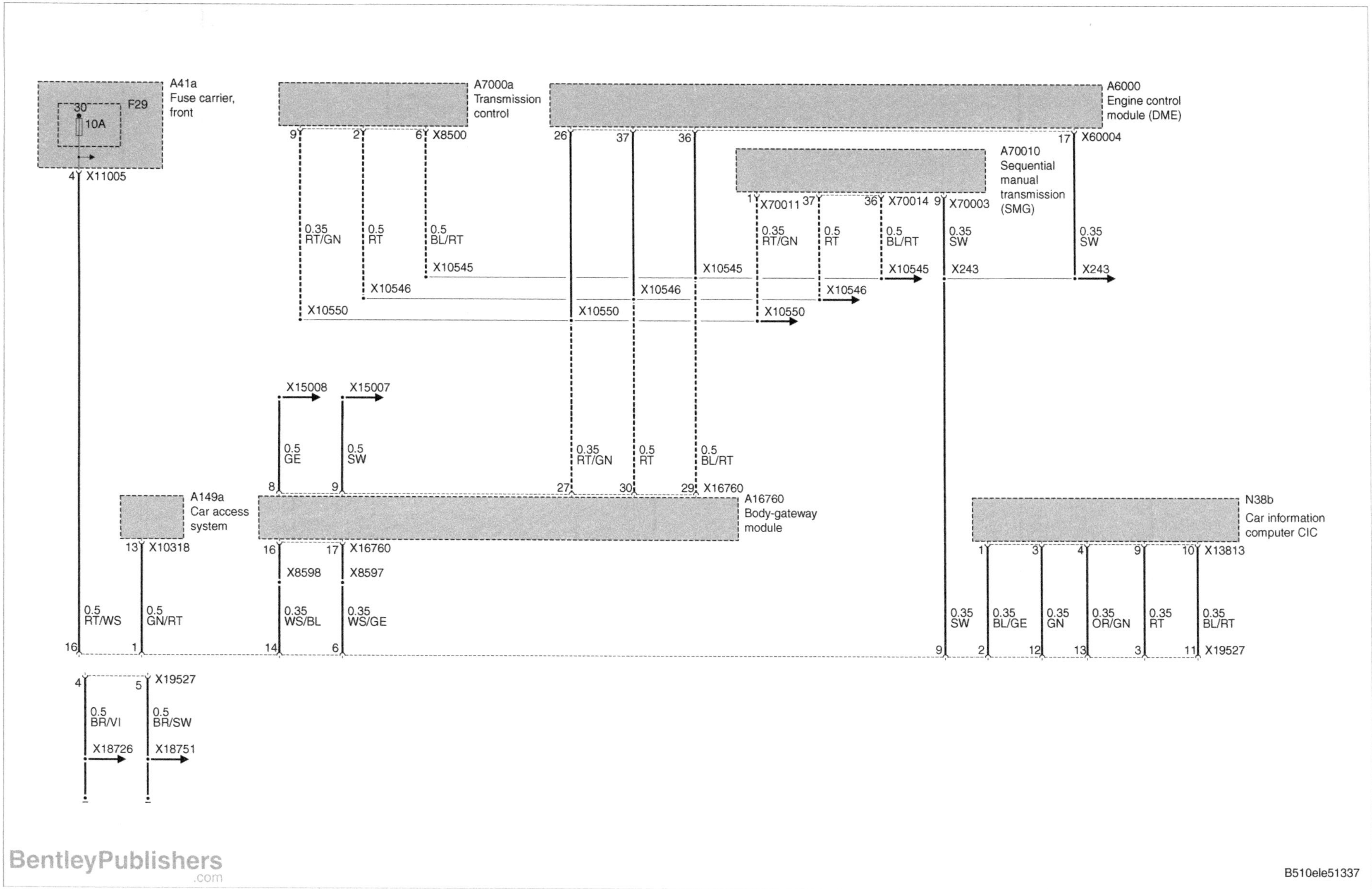

BentleyPublishers.com

Oxygen sensors before catalytic converter
M54 engine

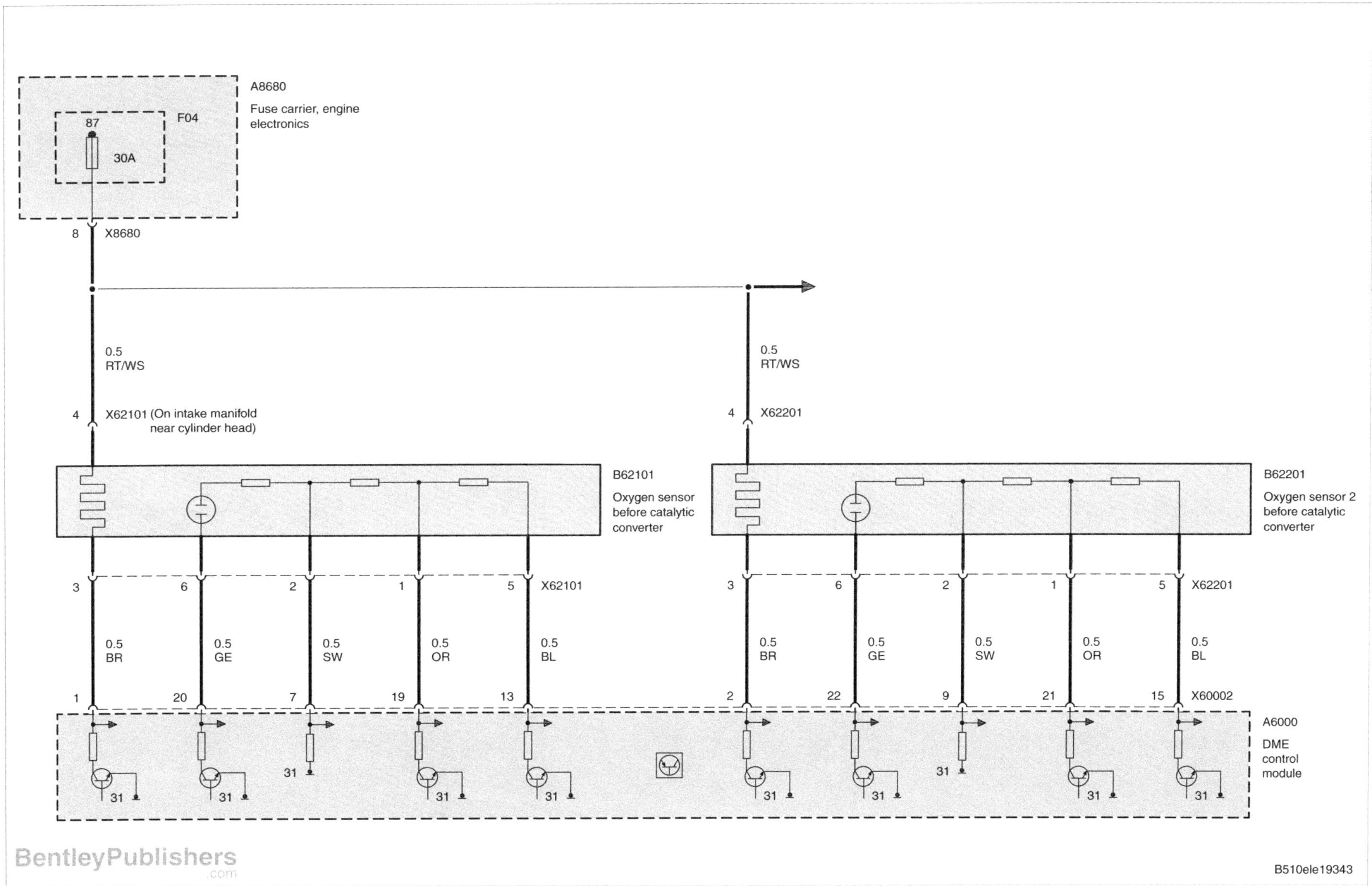

E-box fan
M54 engine

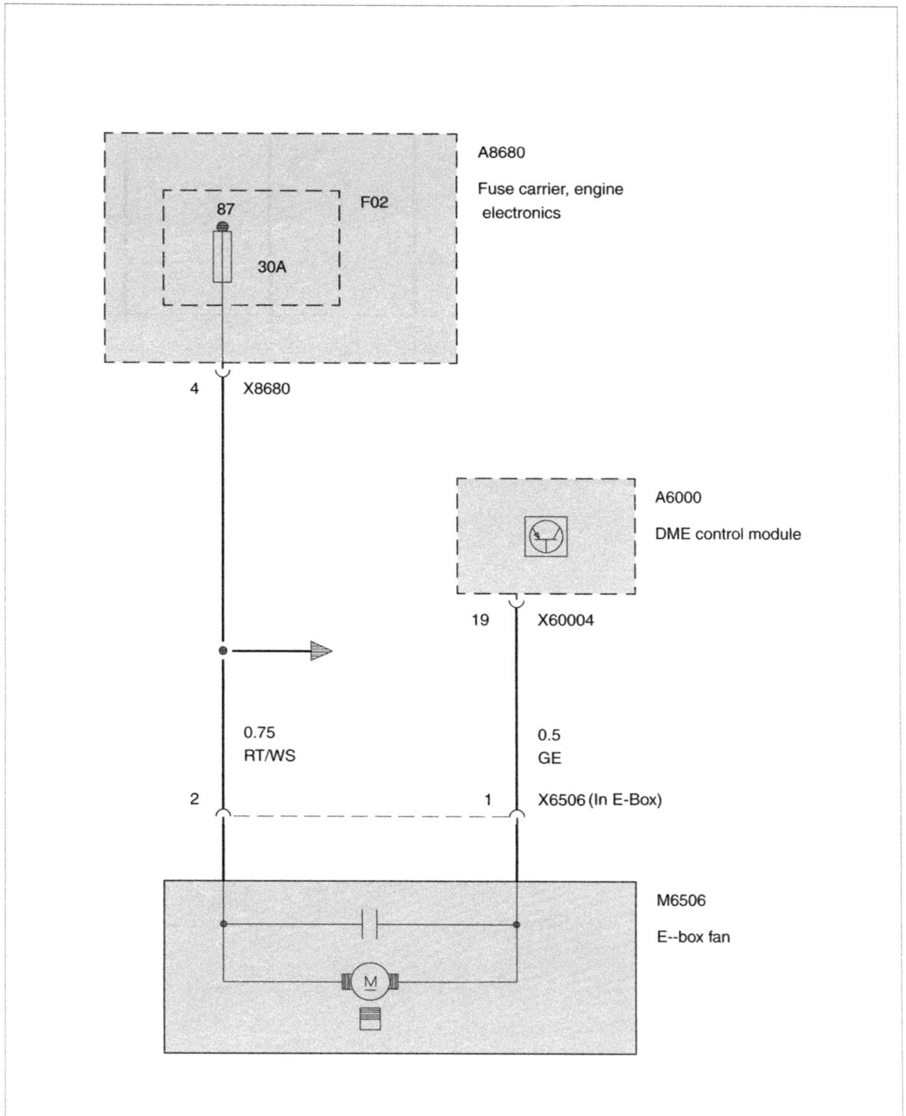

BentleyPublishers.com

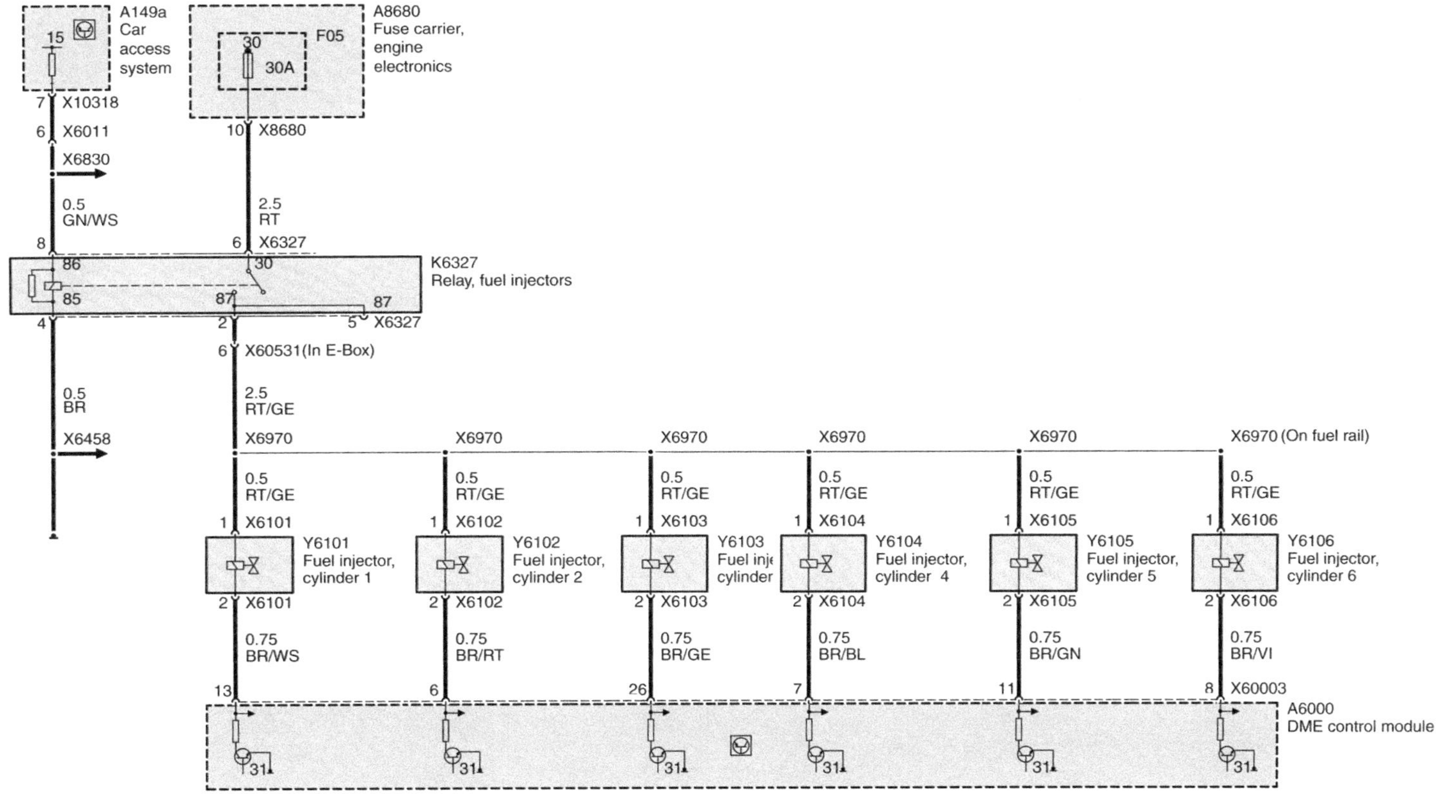

BentleyPublishers.com

B510ele19340

Fuel pump
M54 engine

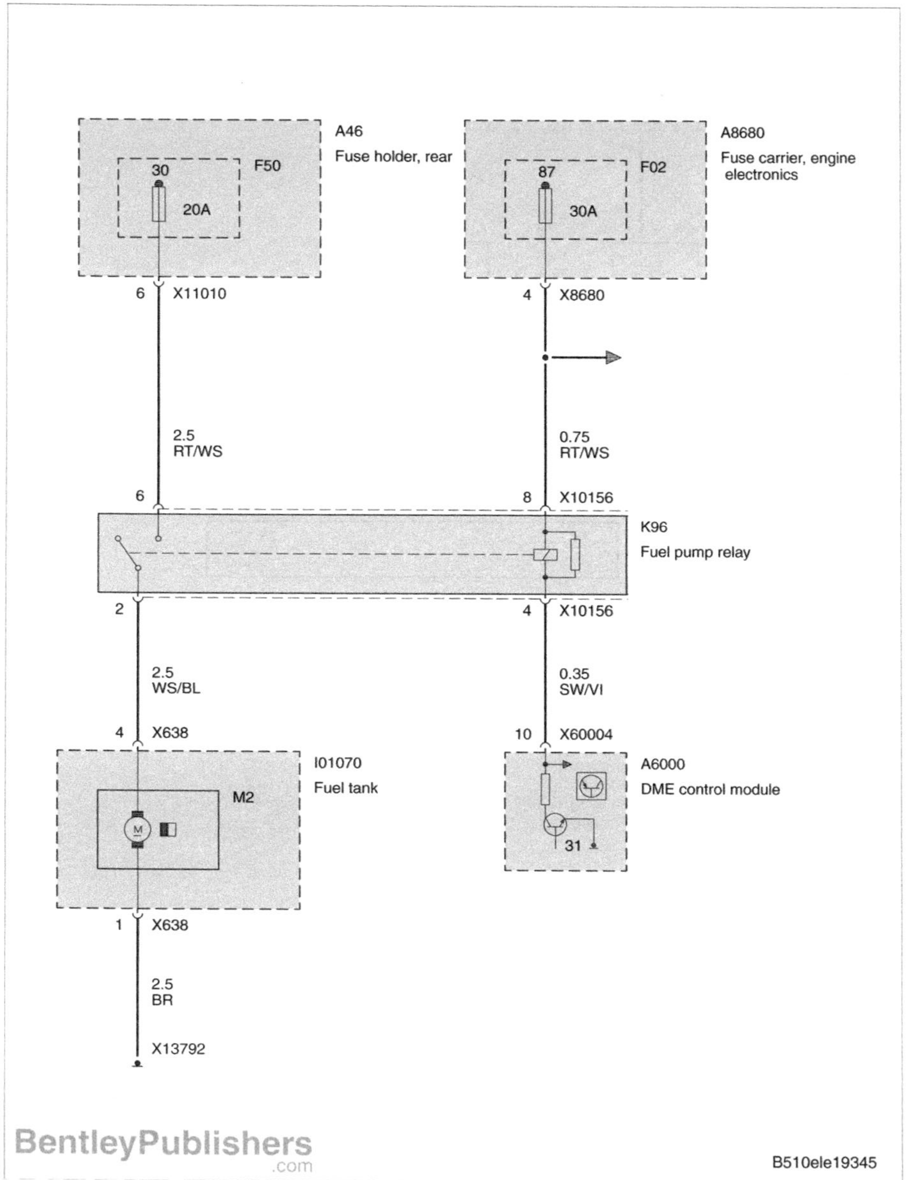

Vehicle speed control
M54 engine

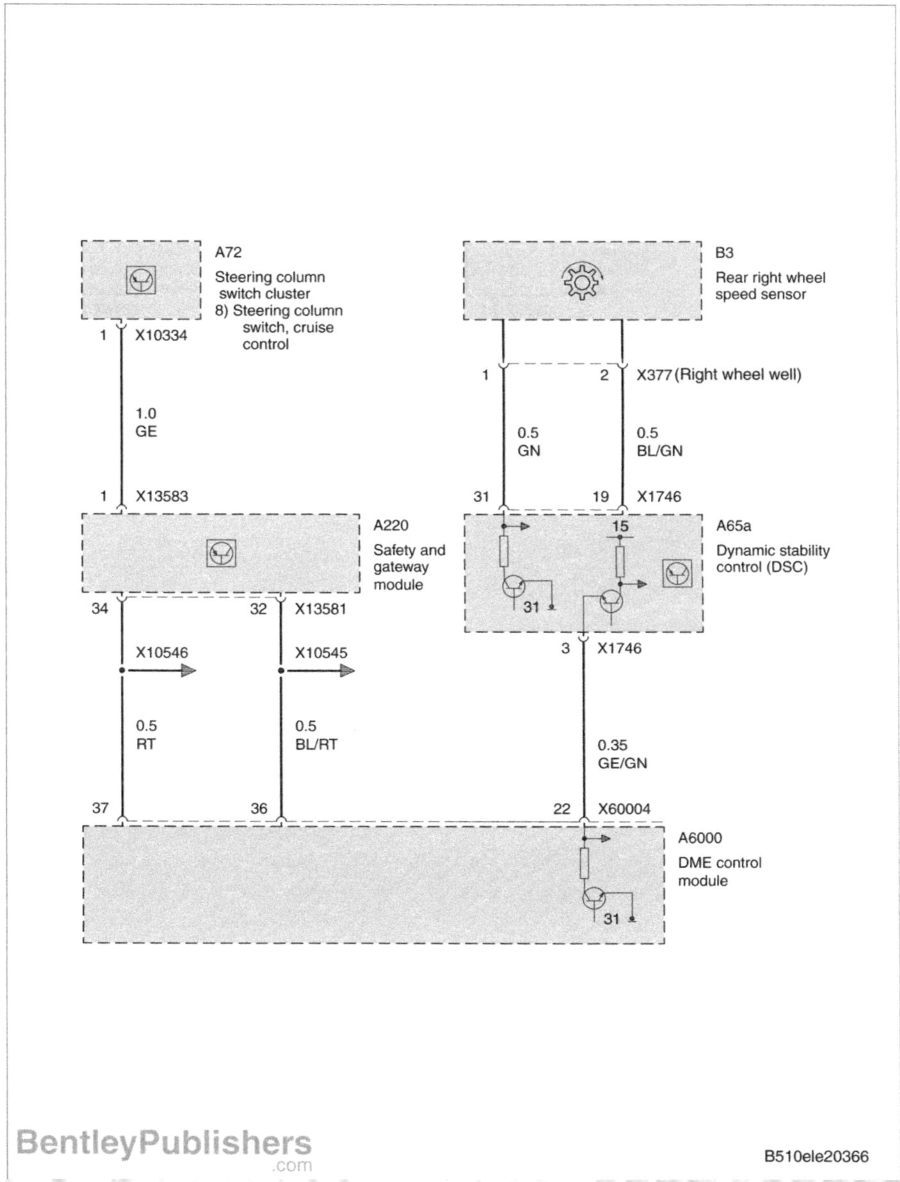

Brake light switch

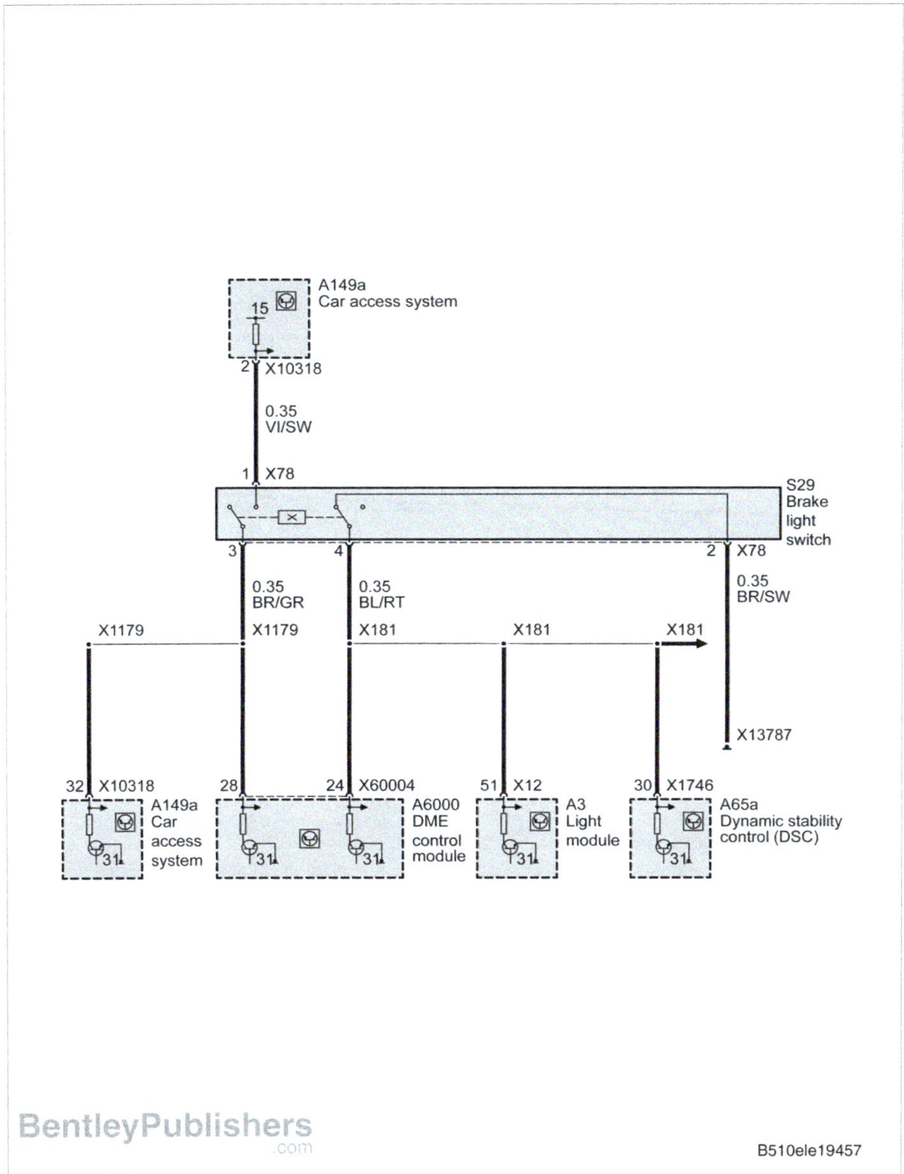

Accelerator pedal module
M54 engine

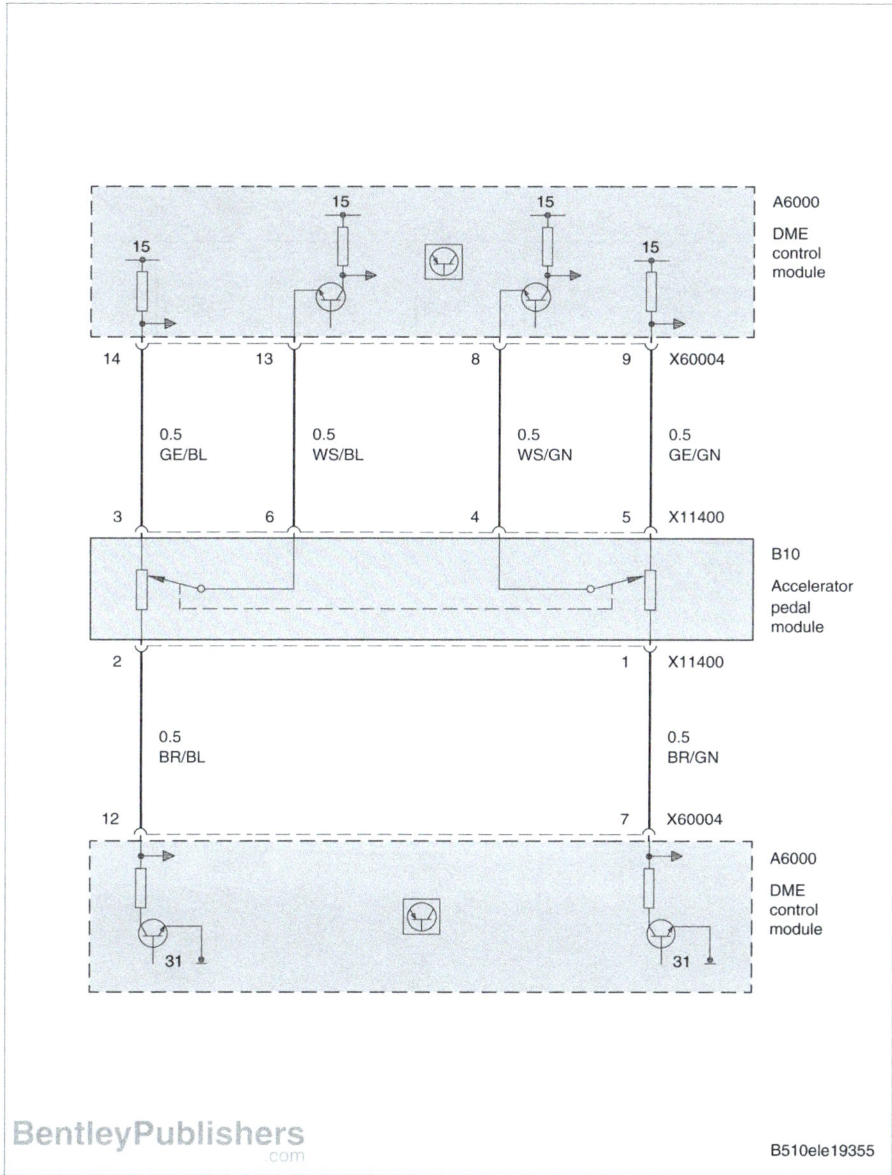

Knock sensors
M54 engine

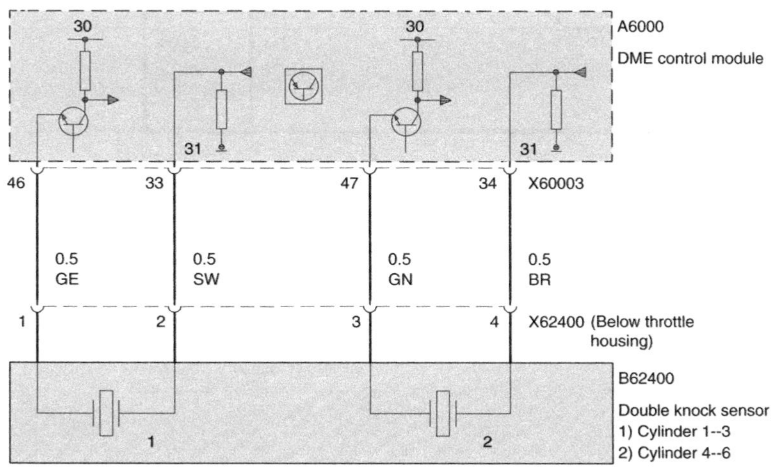

BentleyPublishers.com

B510ele19242

Throttle valve
M54 engine

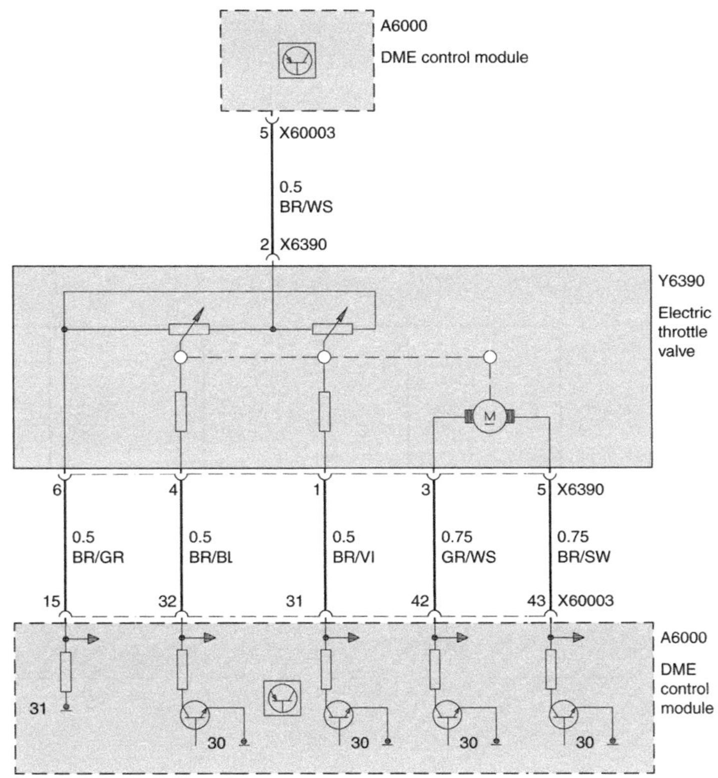

BentleyPublishers.com

B510ele19246

DISA changeover valve
M54 engine

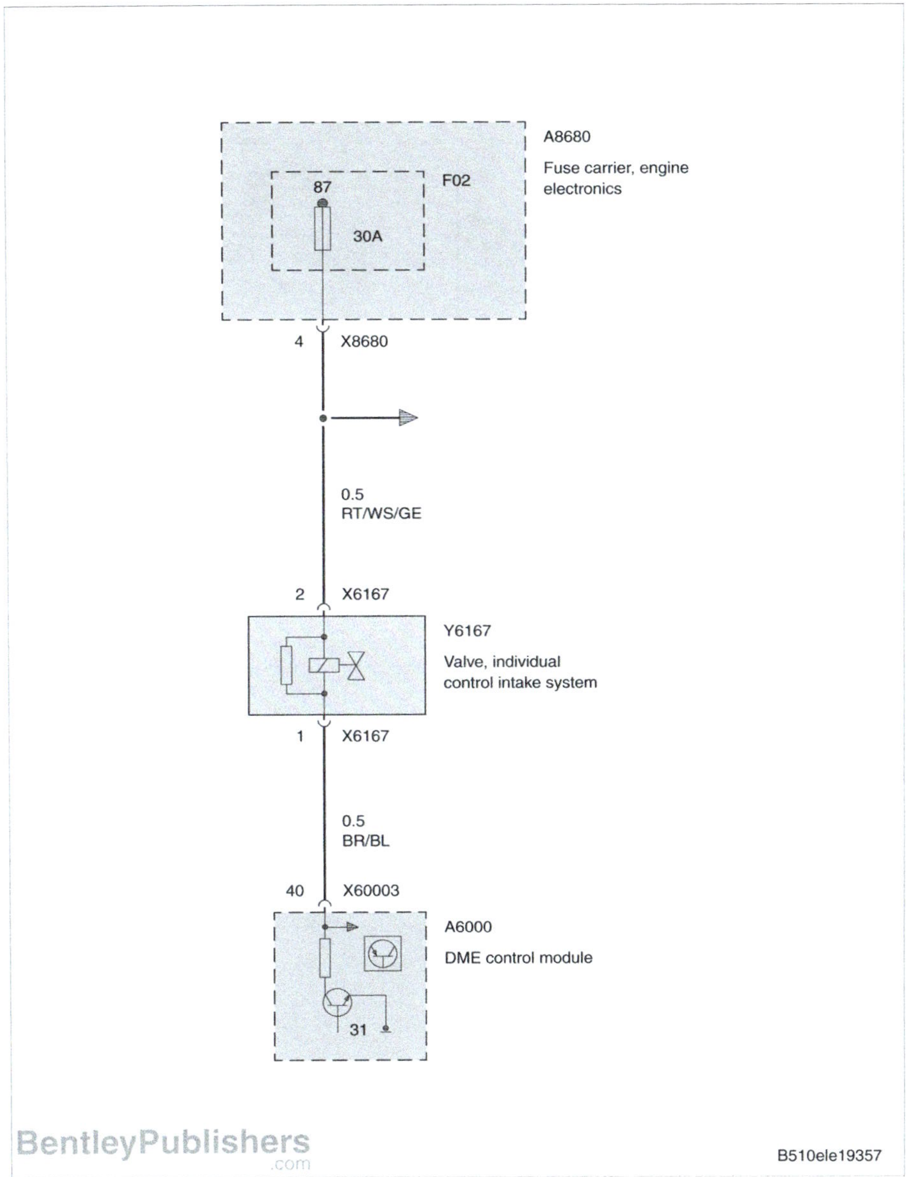

Idle speed control valve
M54 engine

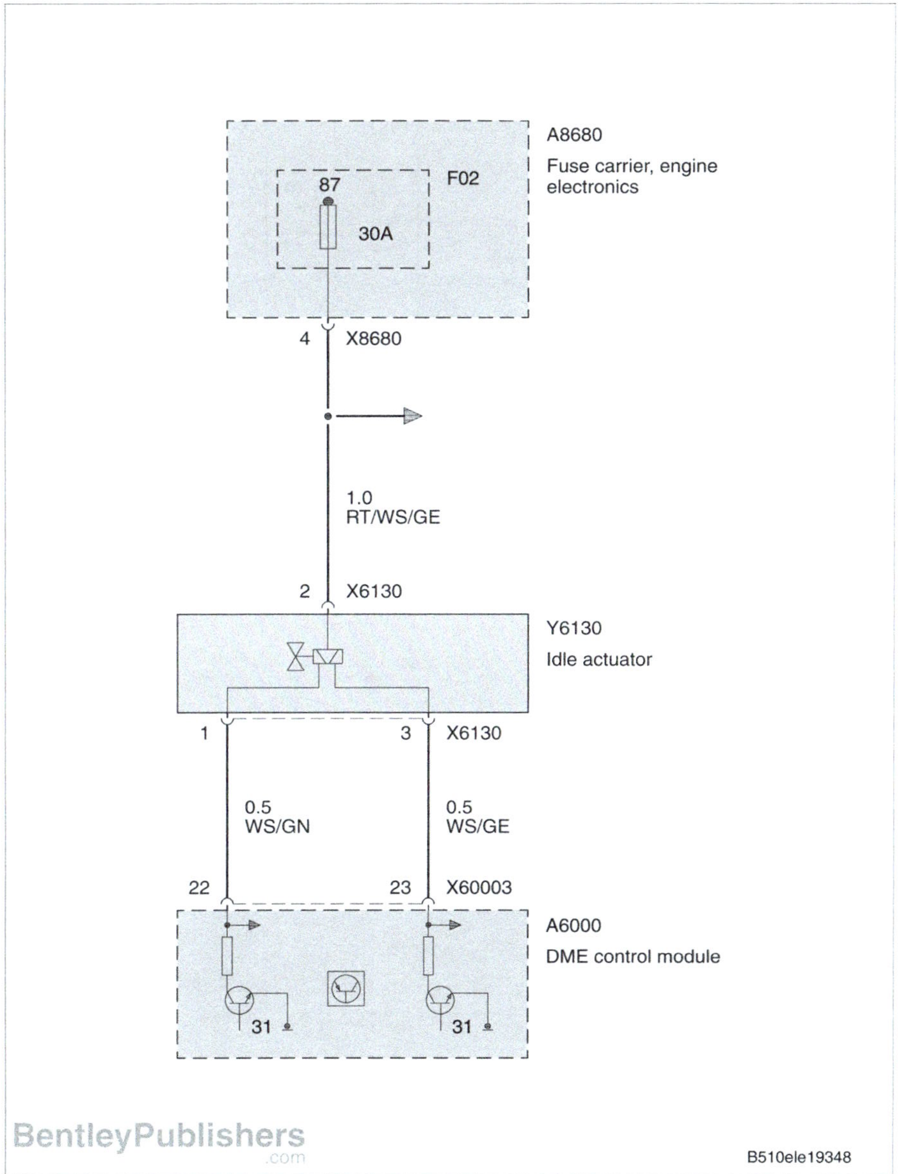

Hot-film mass air flow sensor
M54 engine

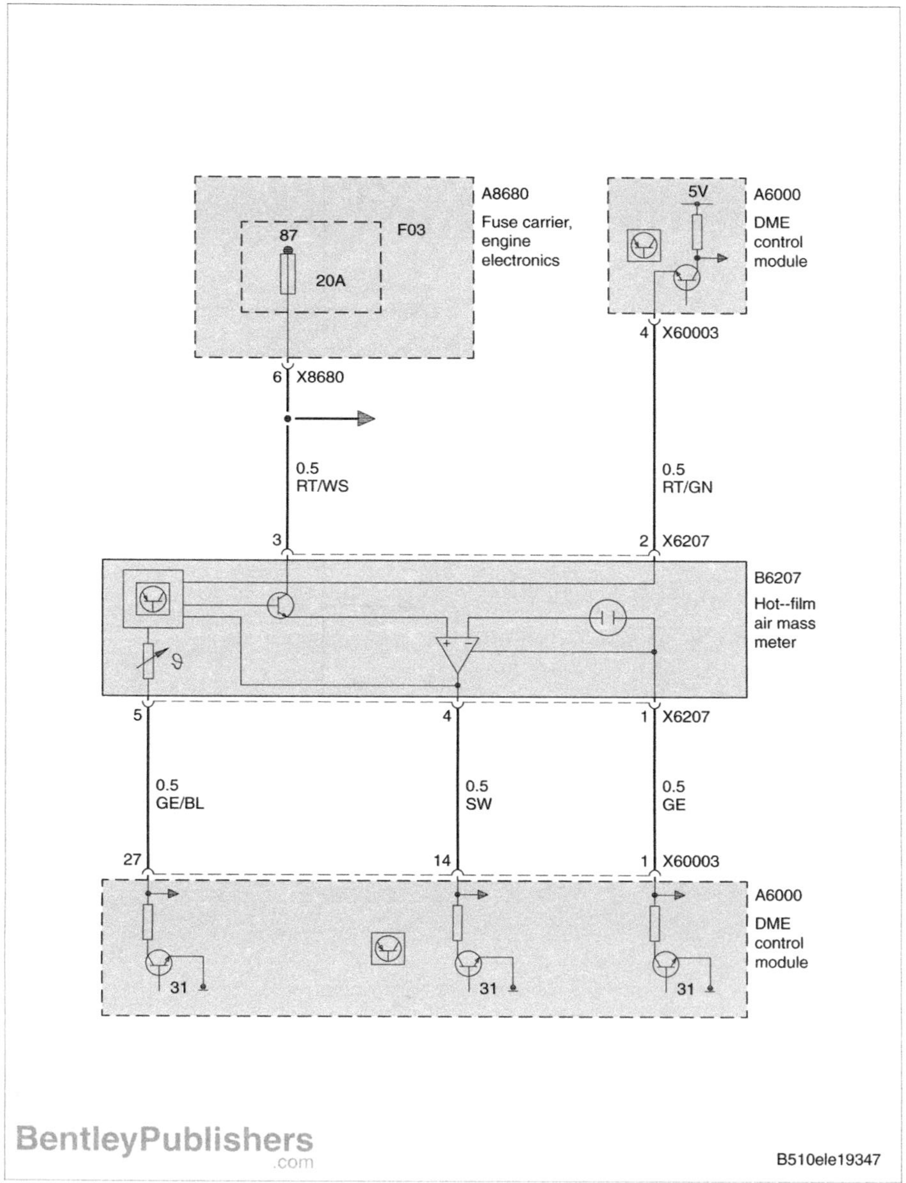

Crankshaft position sensor
M54 engine

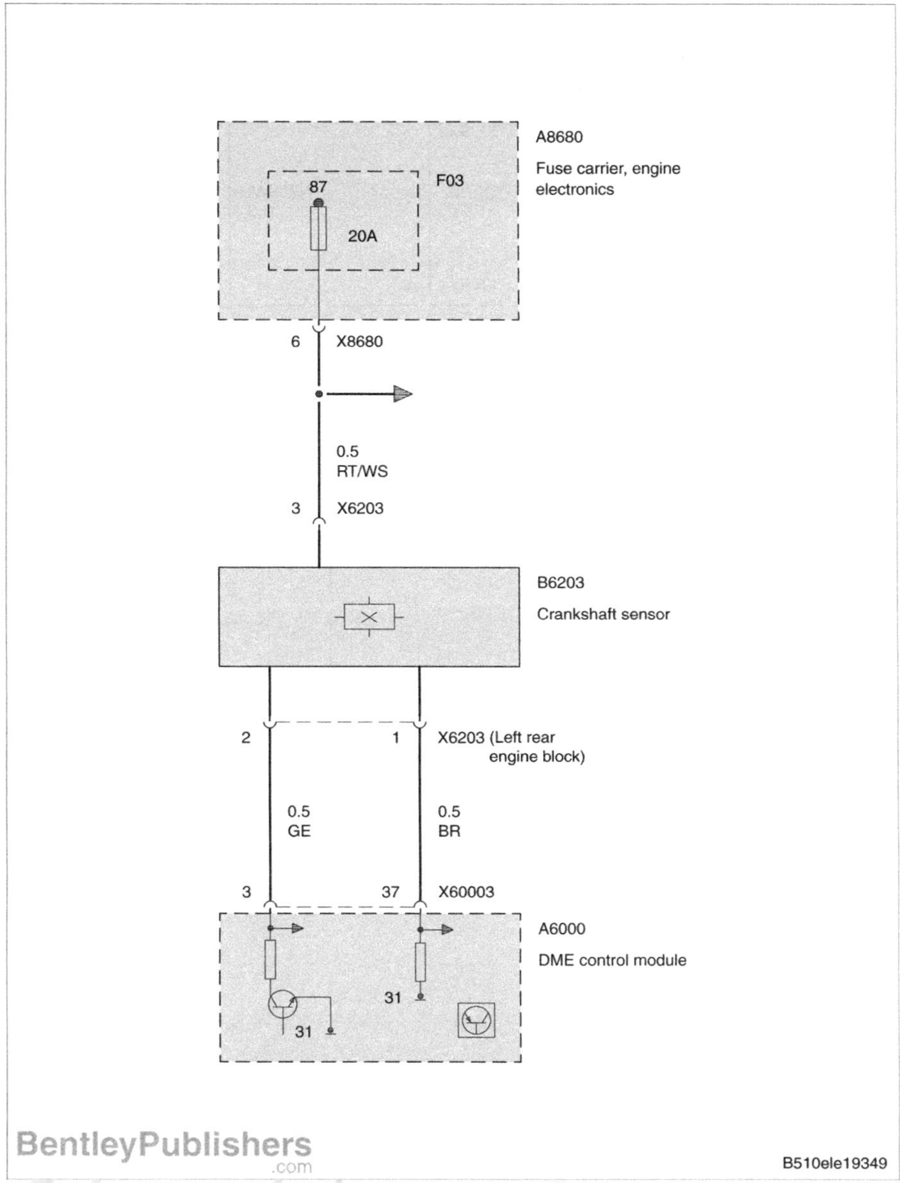

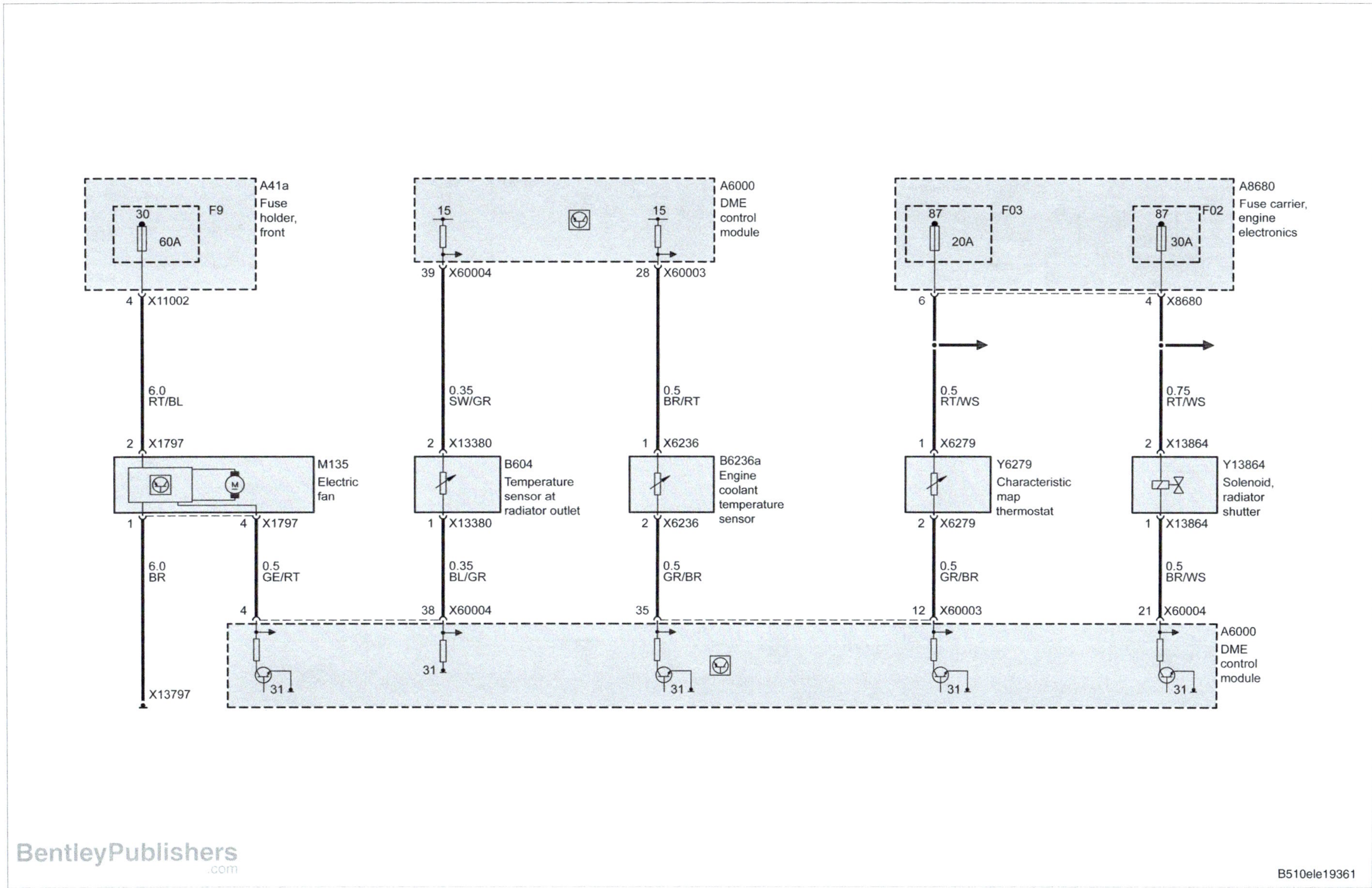
A41a
Fuse holder, front
F9
30
60A
4 X11002
6.0 RT/BL
2 X1797
M135
Electric fan
1
4 X1797
6.0 BR
0.5 GE/RT
X13797
A6000
DME control module
15
39 X60004
0.35 SW/GR
2 X13380
B604
Temperature sensor at radiator outlet
1 X13380
0.35 BL/GR
38 X60004
28 X60003
0.5 BR/RT
1 X6236
B6236a
Engine coolant temperature sensor
2 X6236
0.5 GR/BR
35
A8680
Fuse carrier, engine electronics
F03
87
20A
F02
30A
6
4 X8680
0.5 RT/WS
0.75 RT/WS
1 X6279
Y6279
Characteristic map thermostat
2 X6279
2 X13864
Y13864
Solenoid, radiator shutter
1 X13864
0.5 BR/WS
12 X60003
21 X60004
31
BentleyPublishers.com
B510ele19361

Oil supply
M54 engine

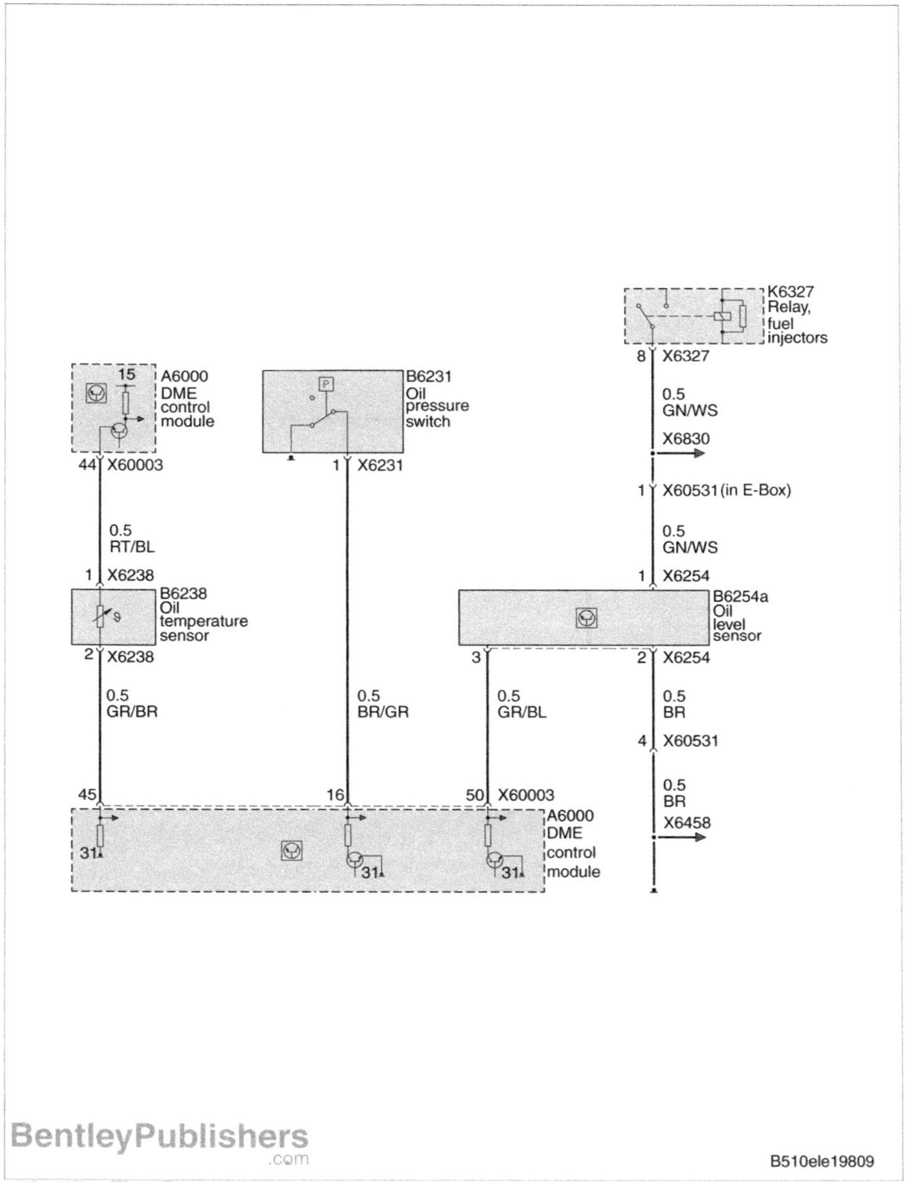

Camshaft position
M54 engine

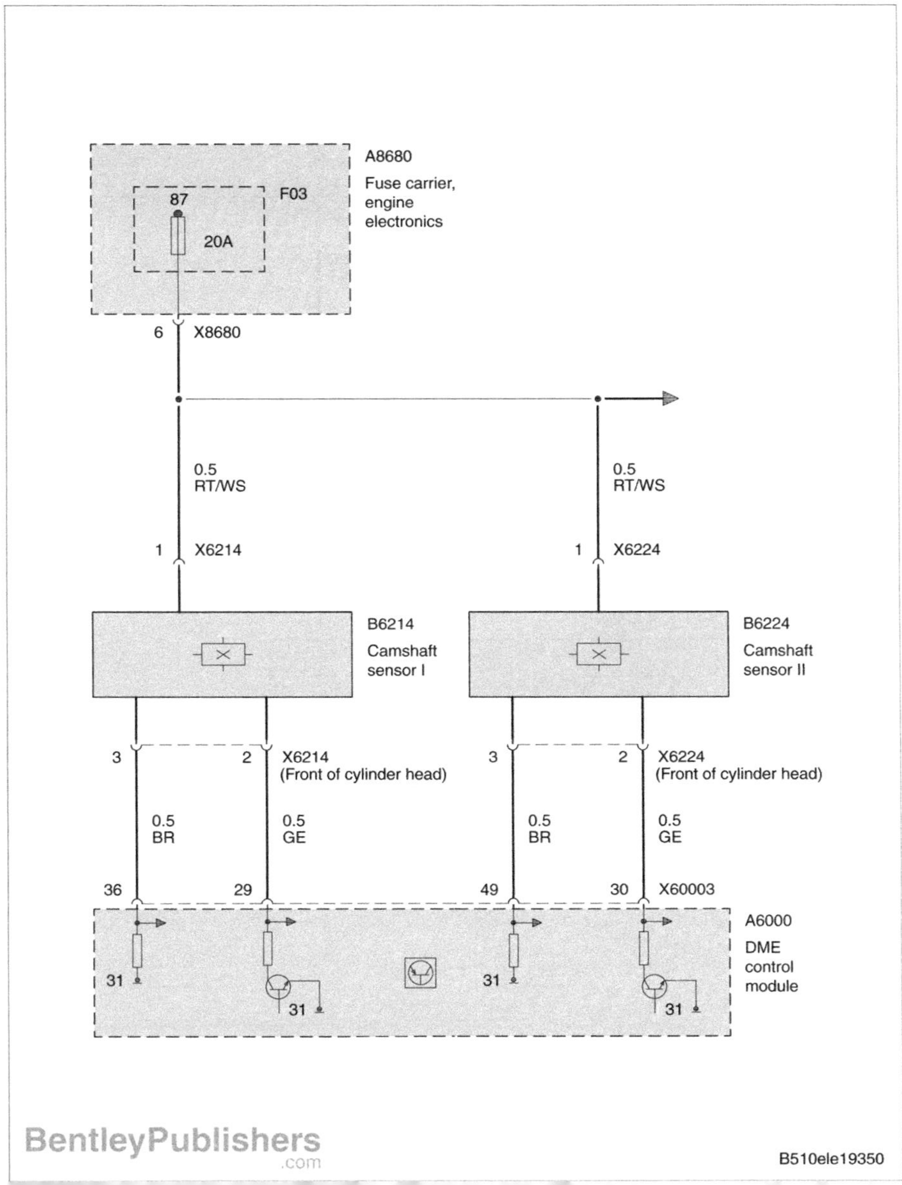

VANOS valves
M54 engine

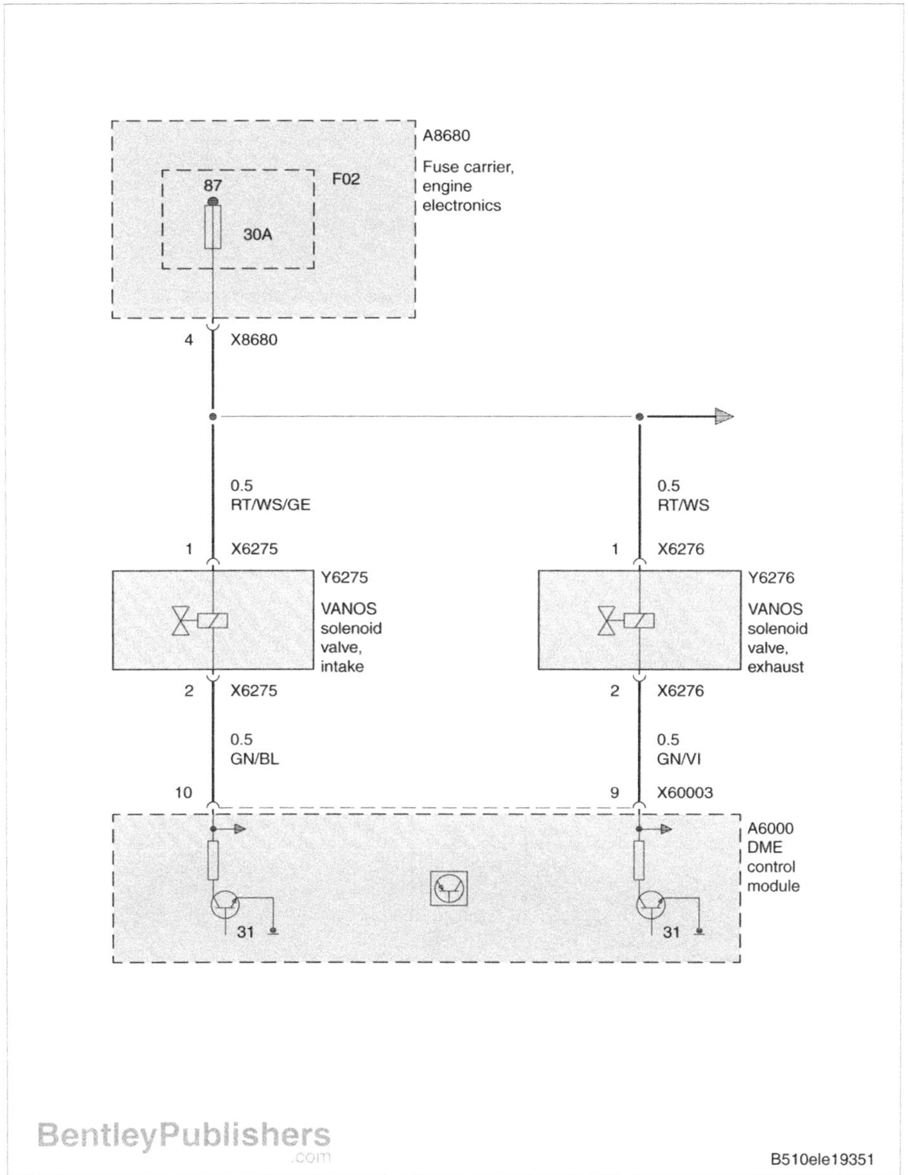

Vehicle speed signal

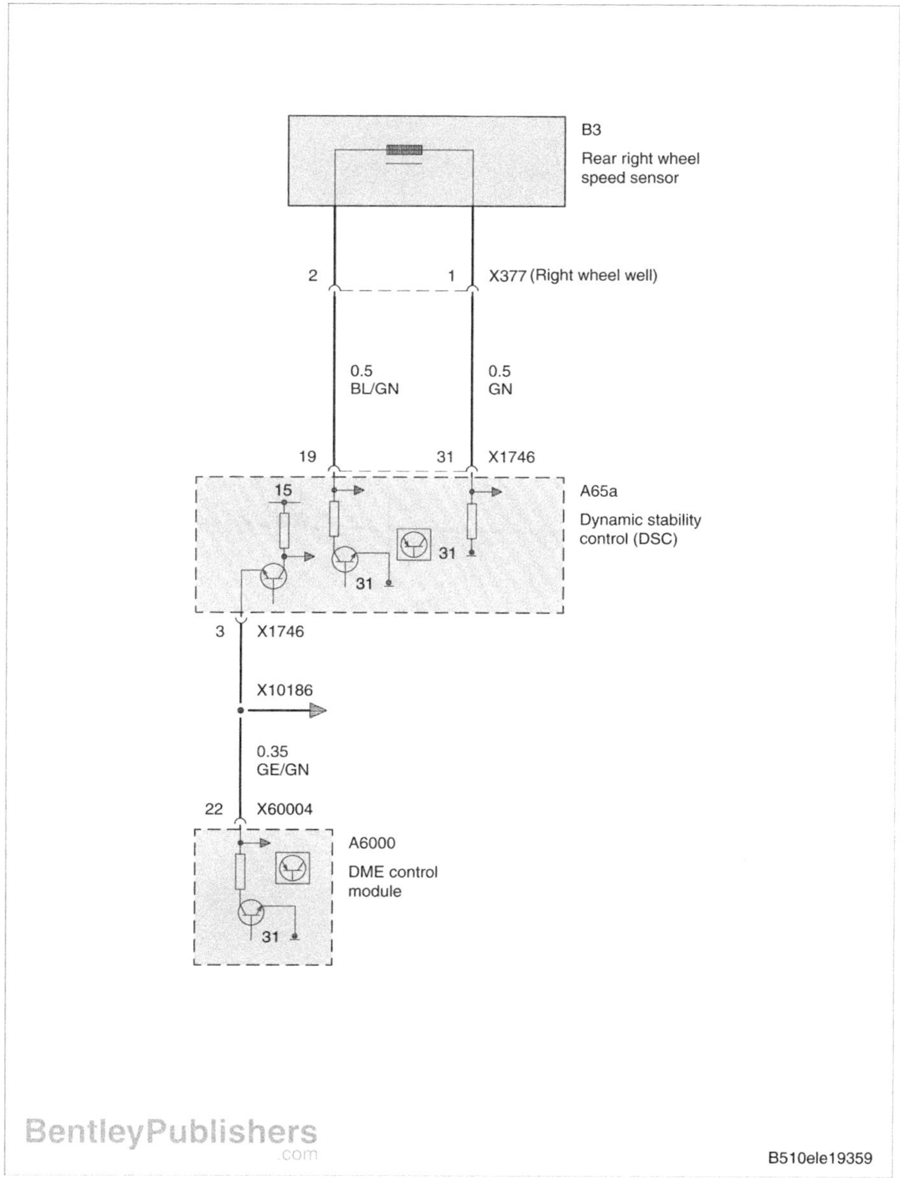

Power train CAN-bus
M54 engine

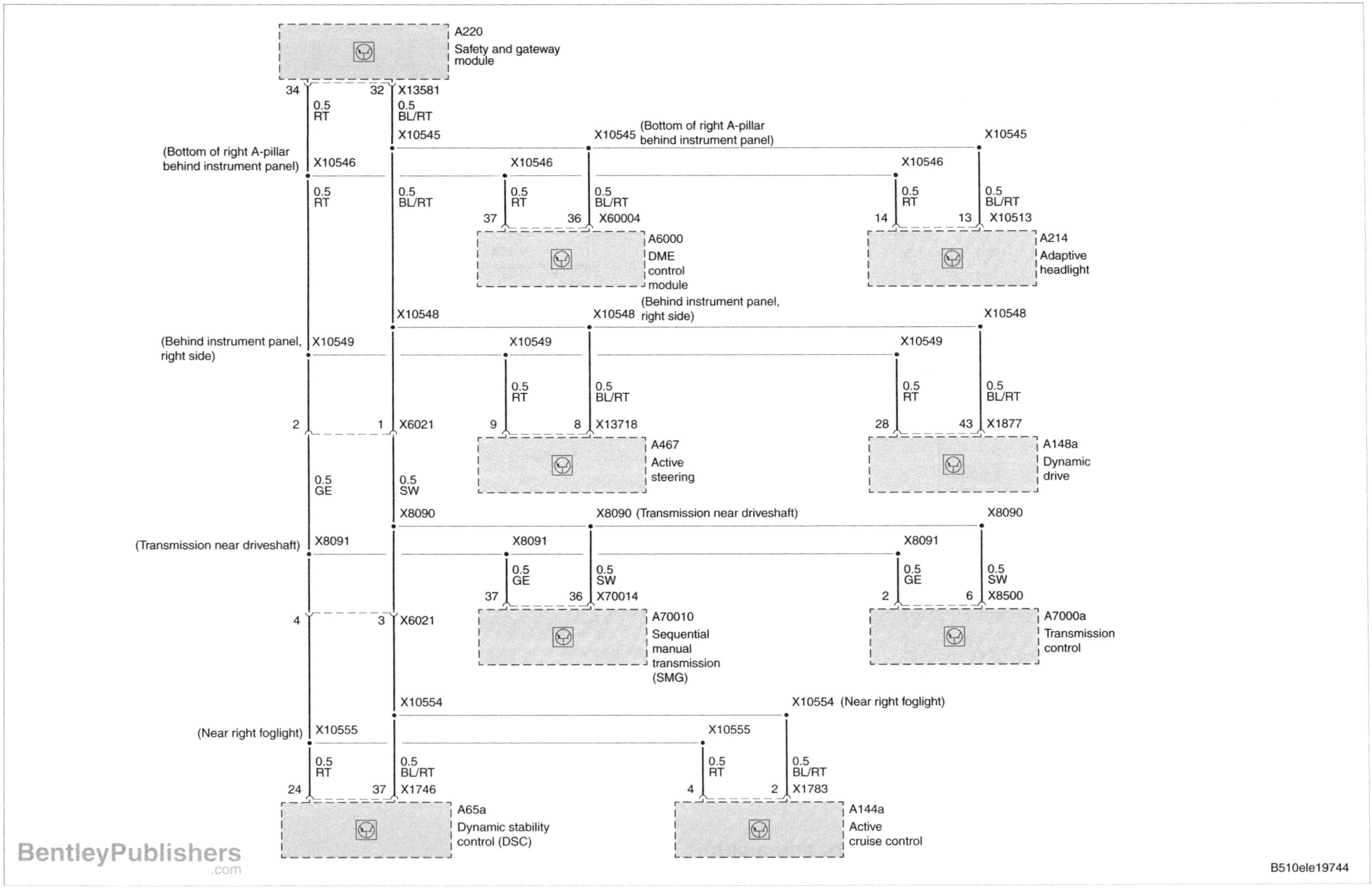

BentleyPublishers.com

Alternator, battery, intelligent battery sensor (IBS) N62 engine or N62 TU engine up to 03 / 2007

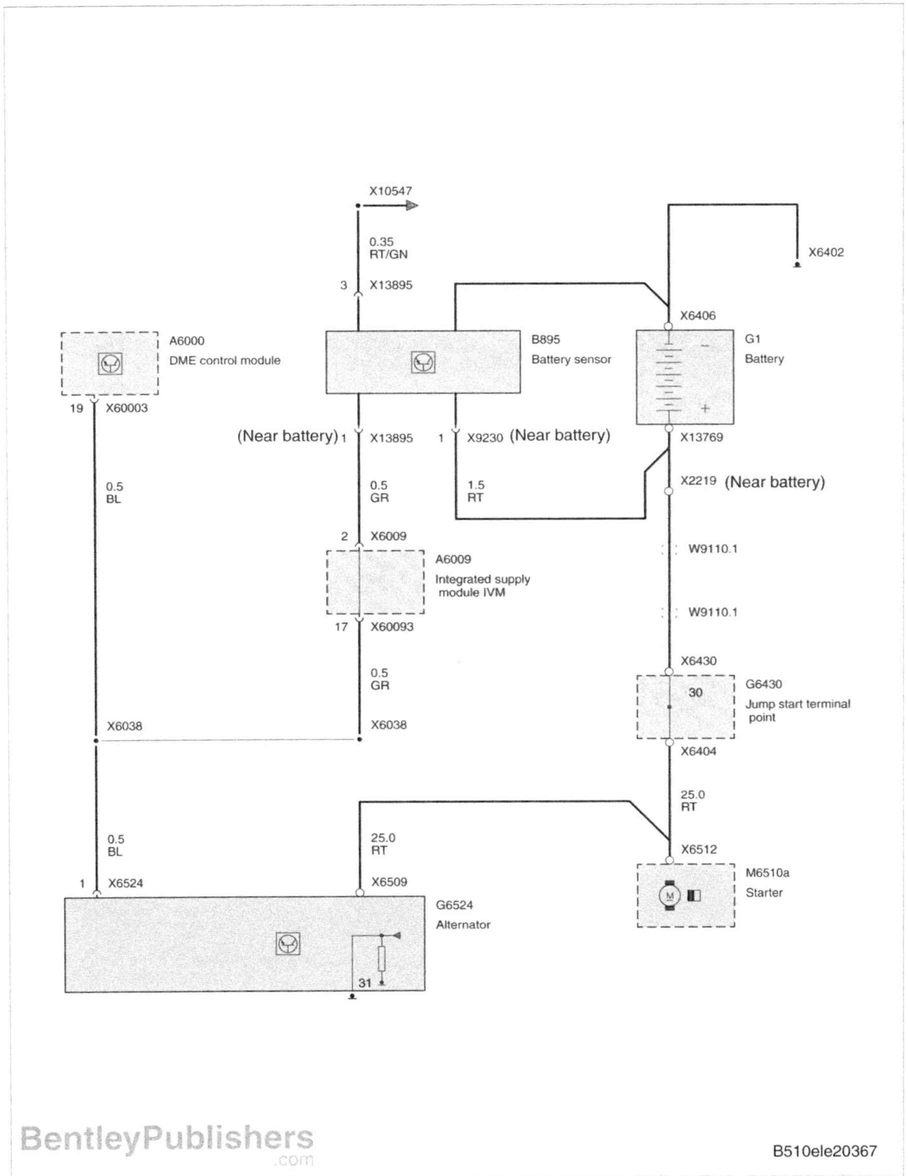

Alternator, battery, intelligent battery sensor (IBS) N62 TU engine as of 03 / 2007

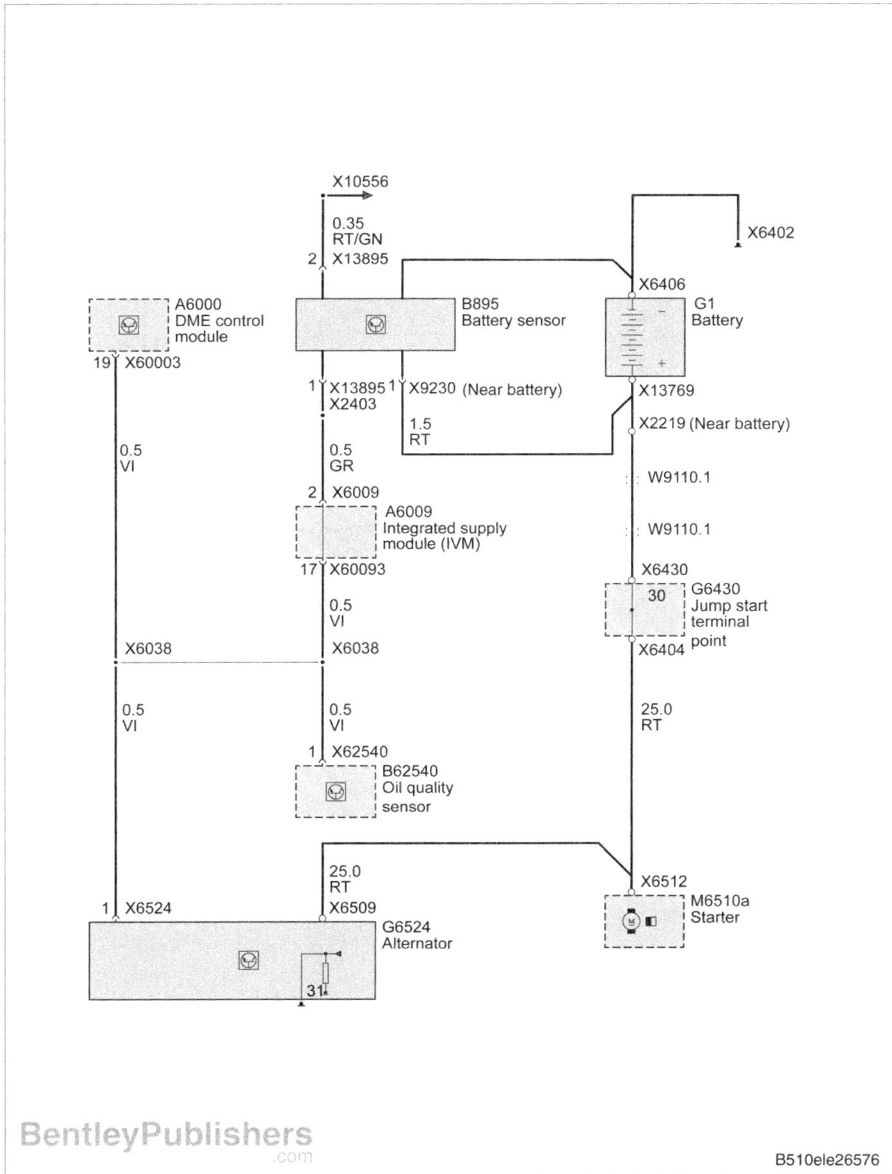

Alternator, battery, intelligent battery sensor (IBS)
M54 engine

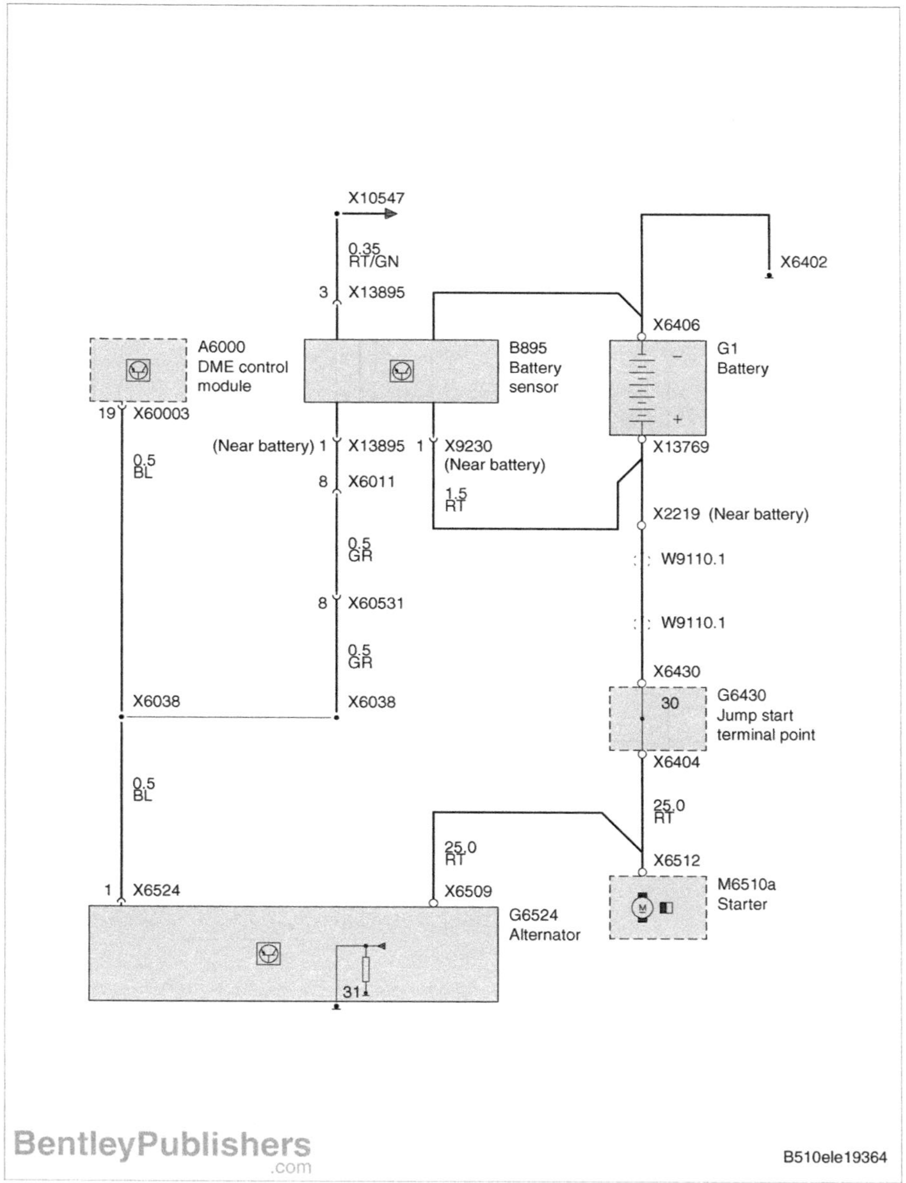

Alternator, battery, intelligent battery sensor (IBS)
N52 engine
up to 03 / 2007

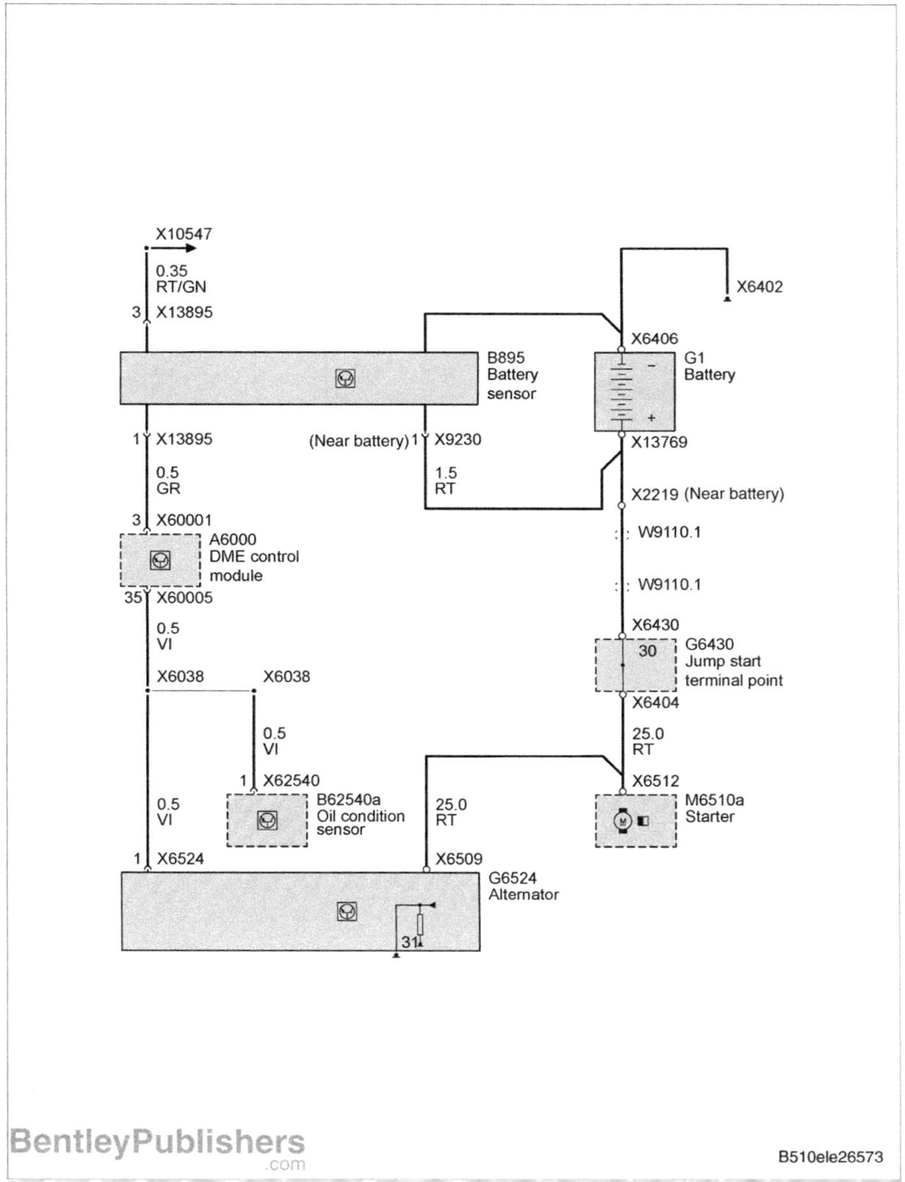

Alternator, battery, intelligent battery sensor (IBS) N52 engine or N54 engine as of 03 / 2007

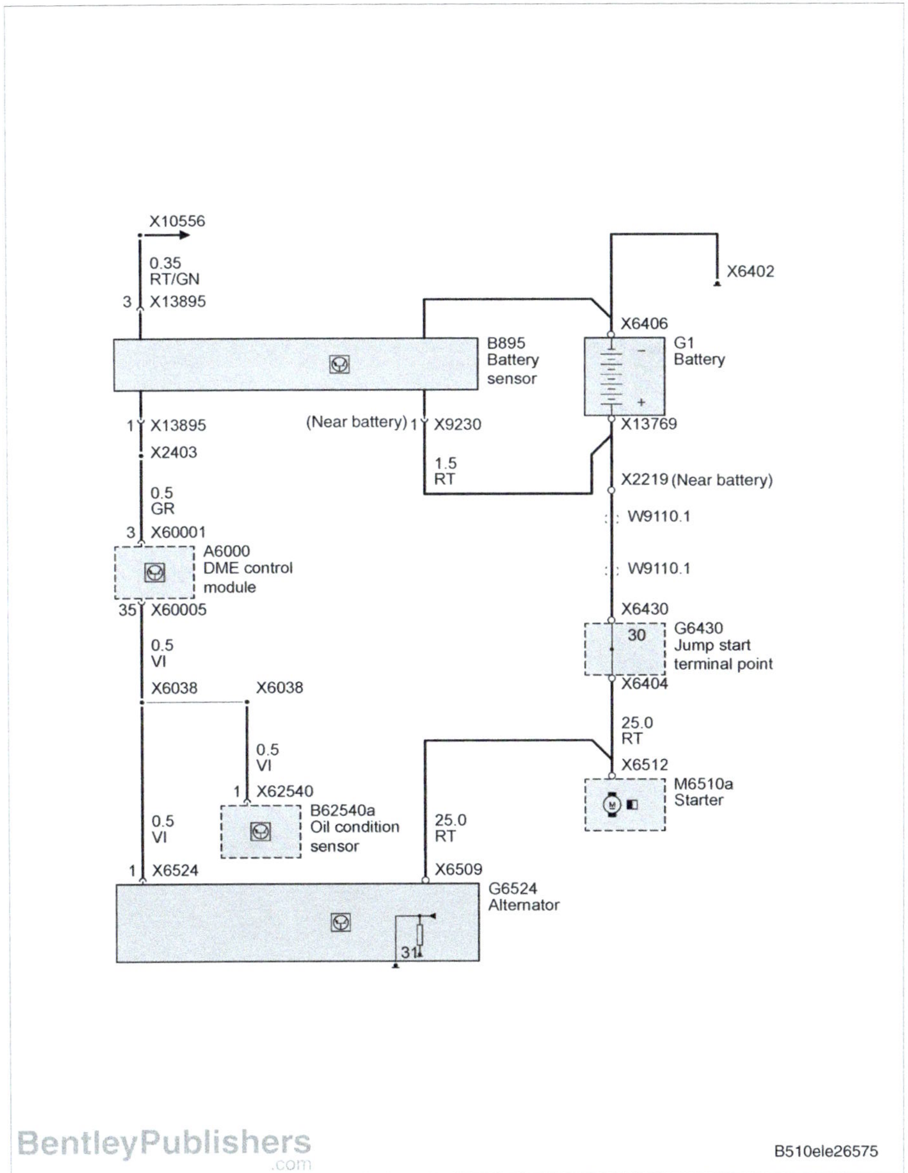

Secondary air system M54 engine

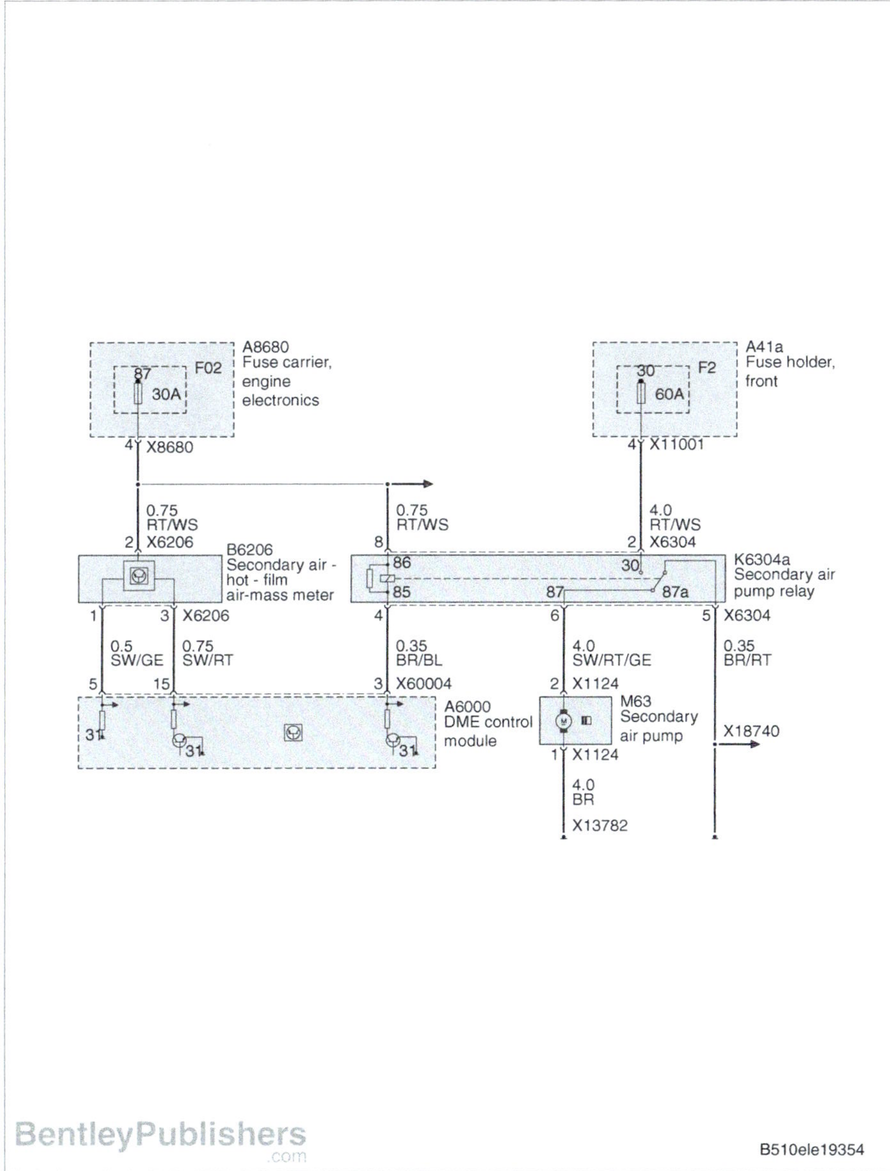

DME main relay
M54 engine

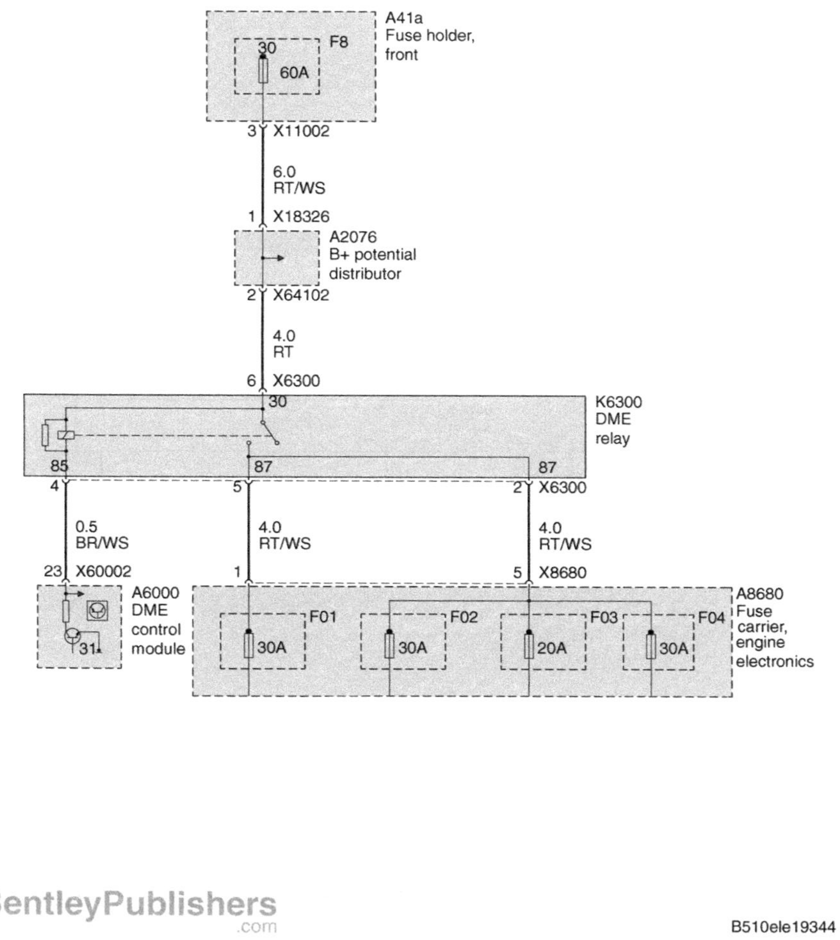

BentleyPublishers.com

DME control module power supply M54 engine

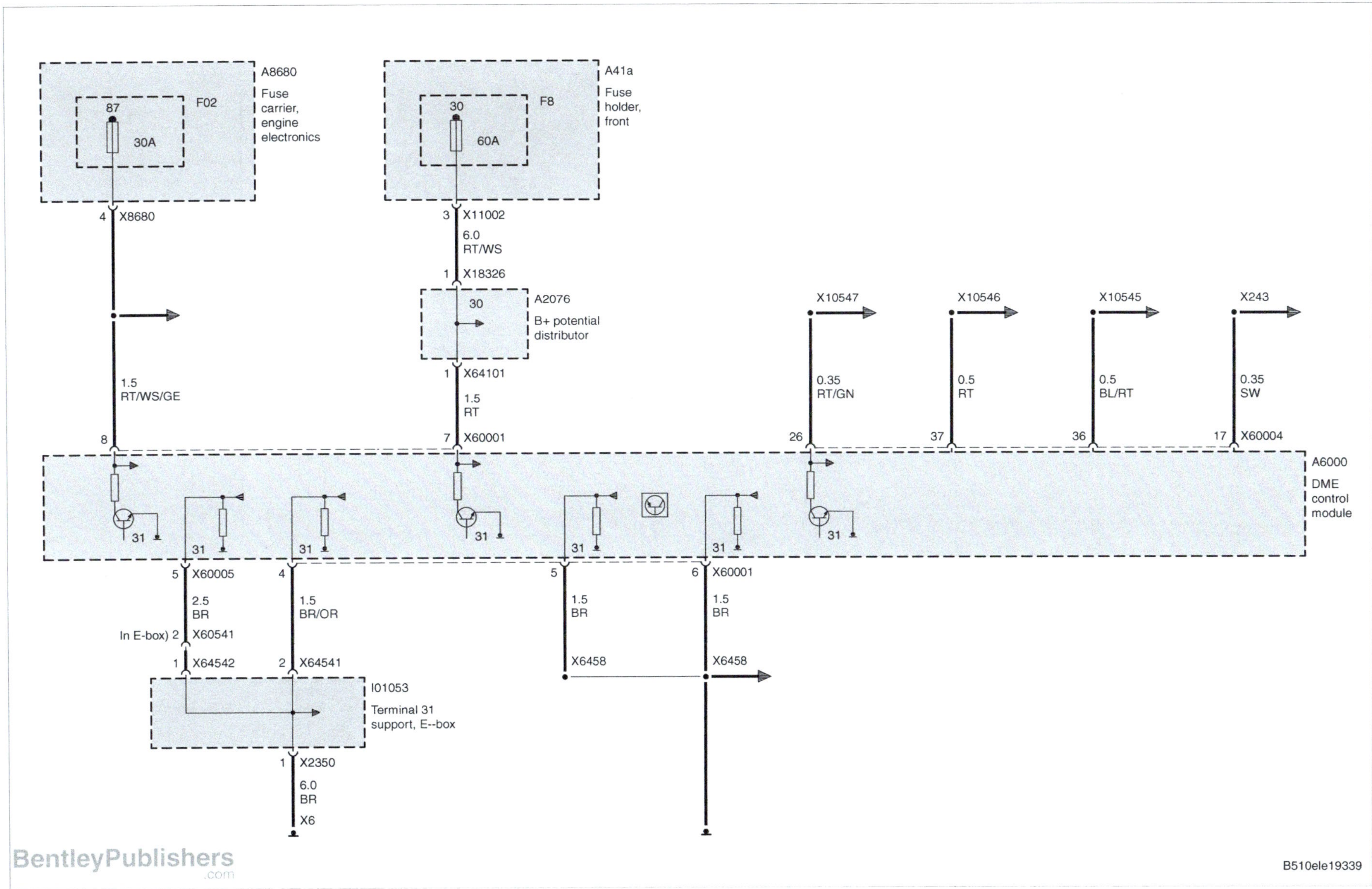

Fuel tank vent valve
M54 engine

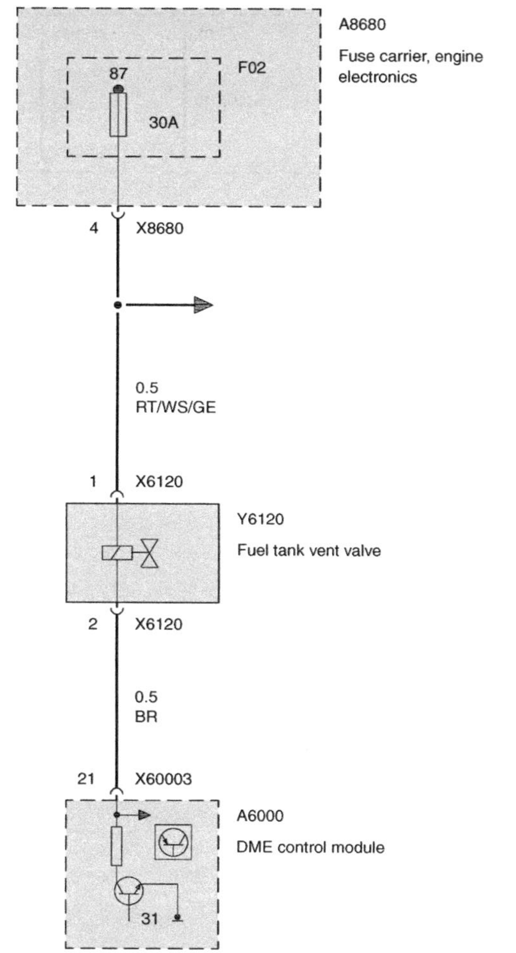

BentleyPublishers.com

B510ele19360

Diagnostic module, fuel tank leakage
M54 engine

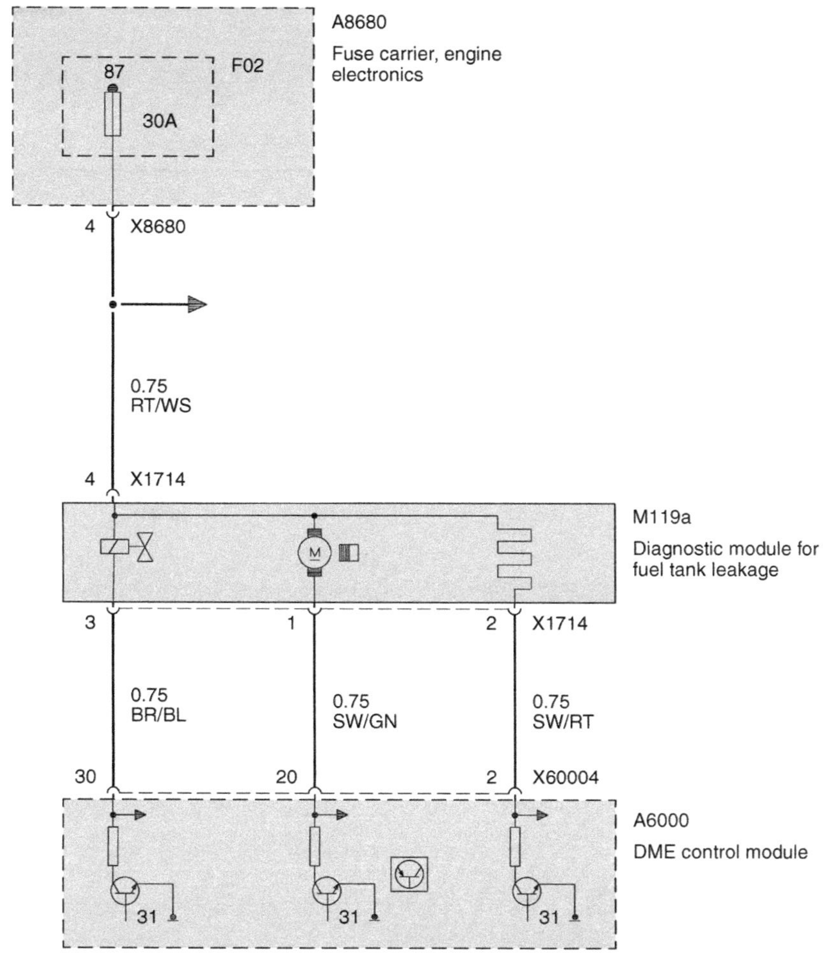

BentleyPublishers.com

B510ele19358

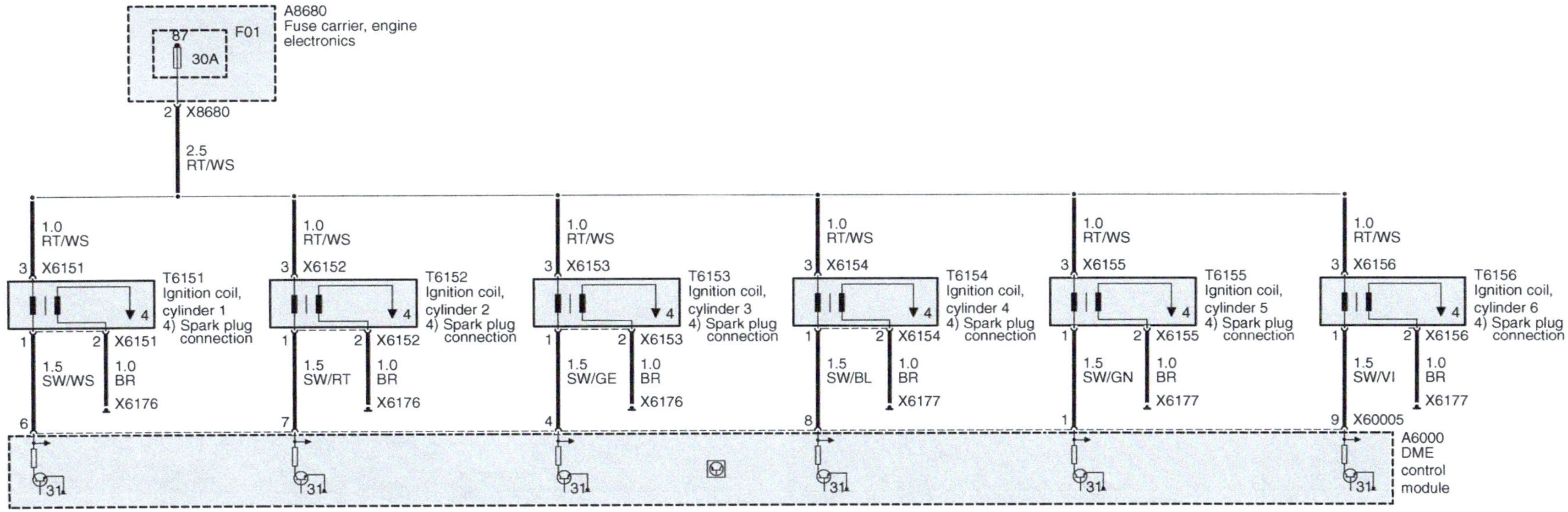

BentleyPublishers.com

Exhaust flap
N62 engine or N62 TU engine

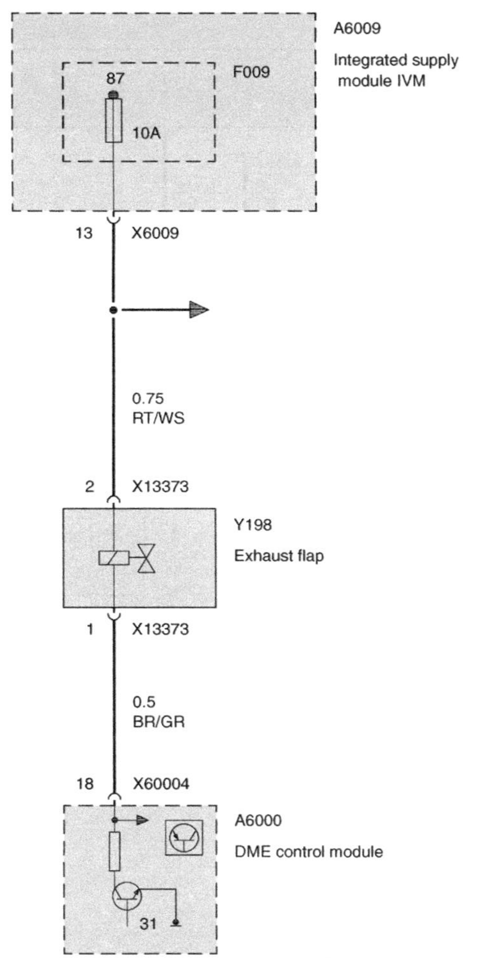

BentleyPublishers.com

B510ele20353

Crankshaft position sensor
N62 engine or N62 TU engine

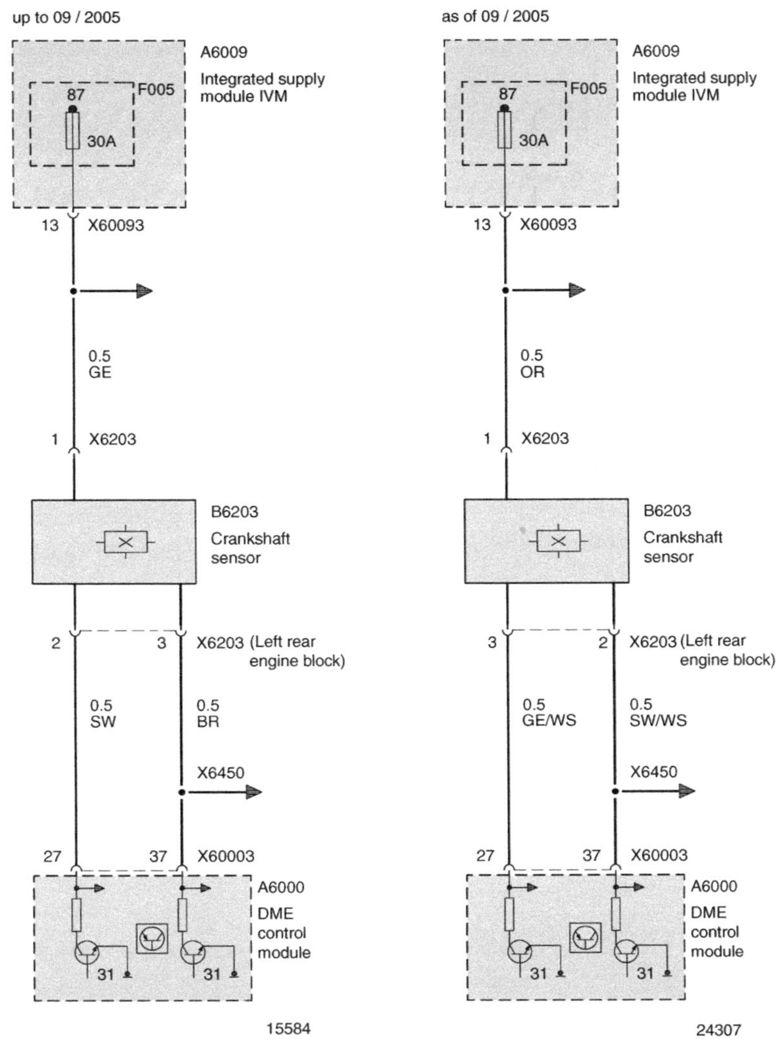

BentleyPublishers.com

B510ele15584

E-box fan
N62 engine or N62 TU engine

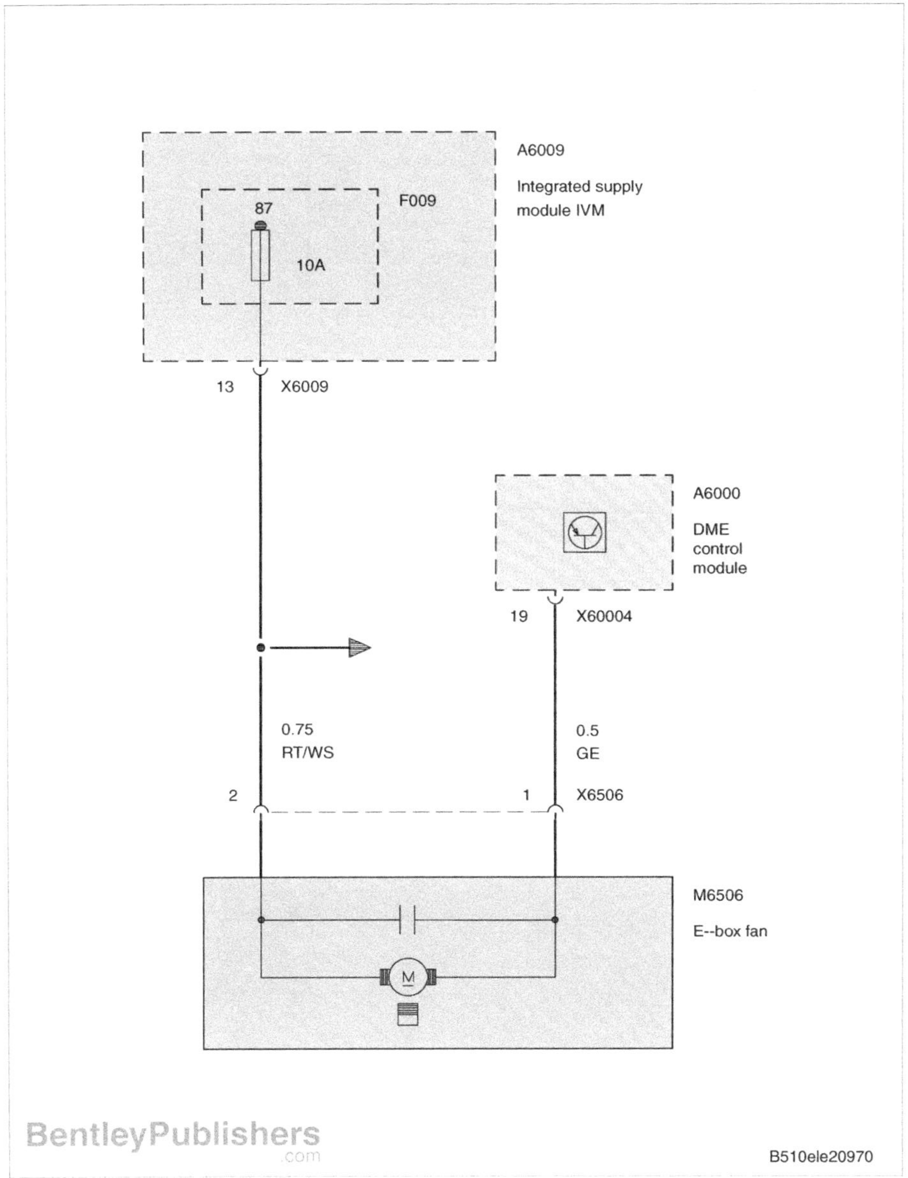

Fuel injectors, cylinders 1 through 4
N62 engine or N62 TU engine

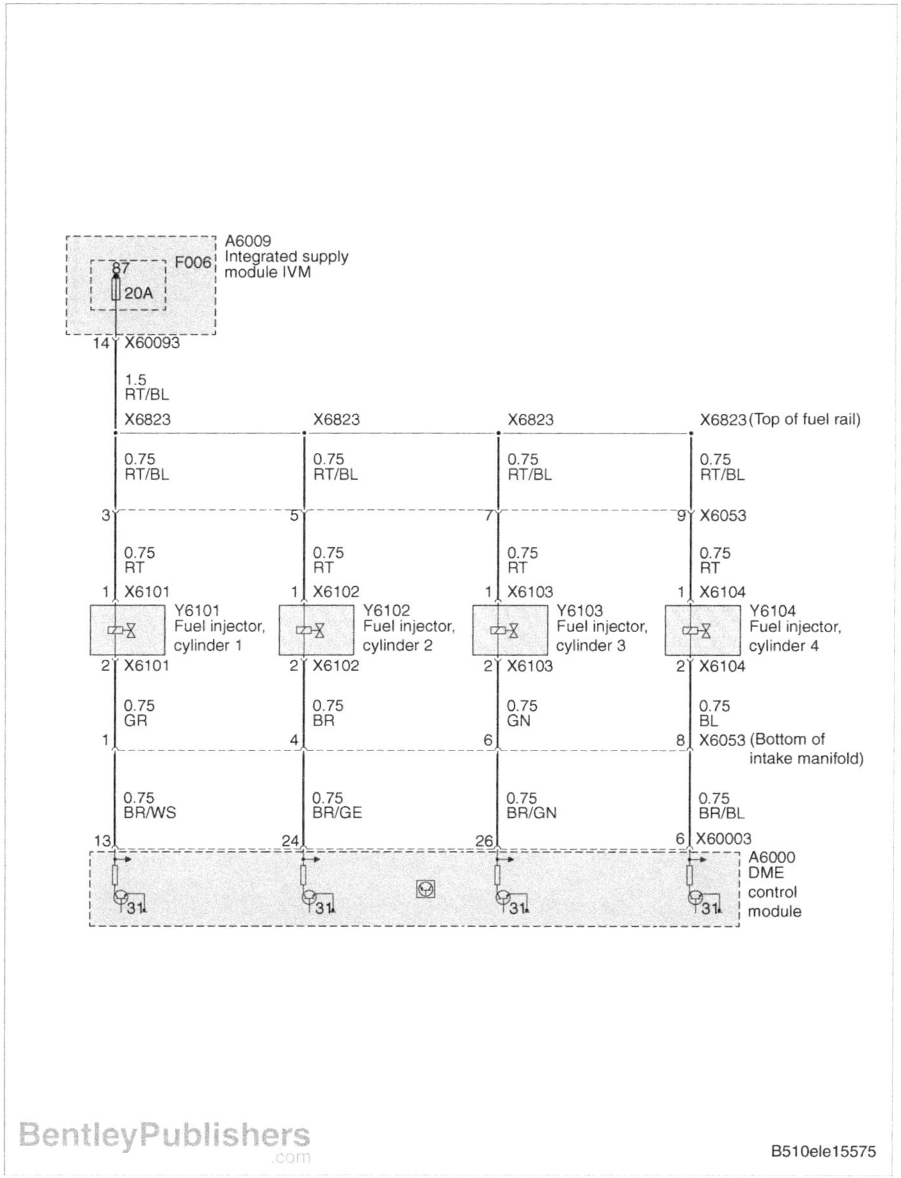

Fuel injectors, cylinders 5 through 8
N62 engine or N62 TU engine

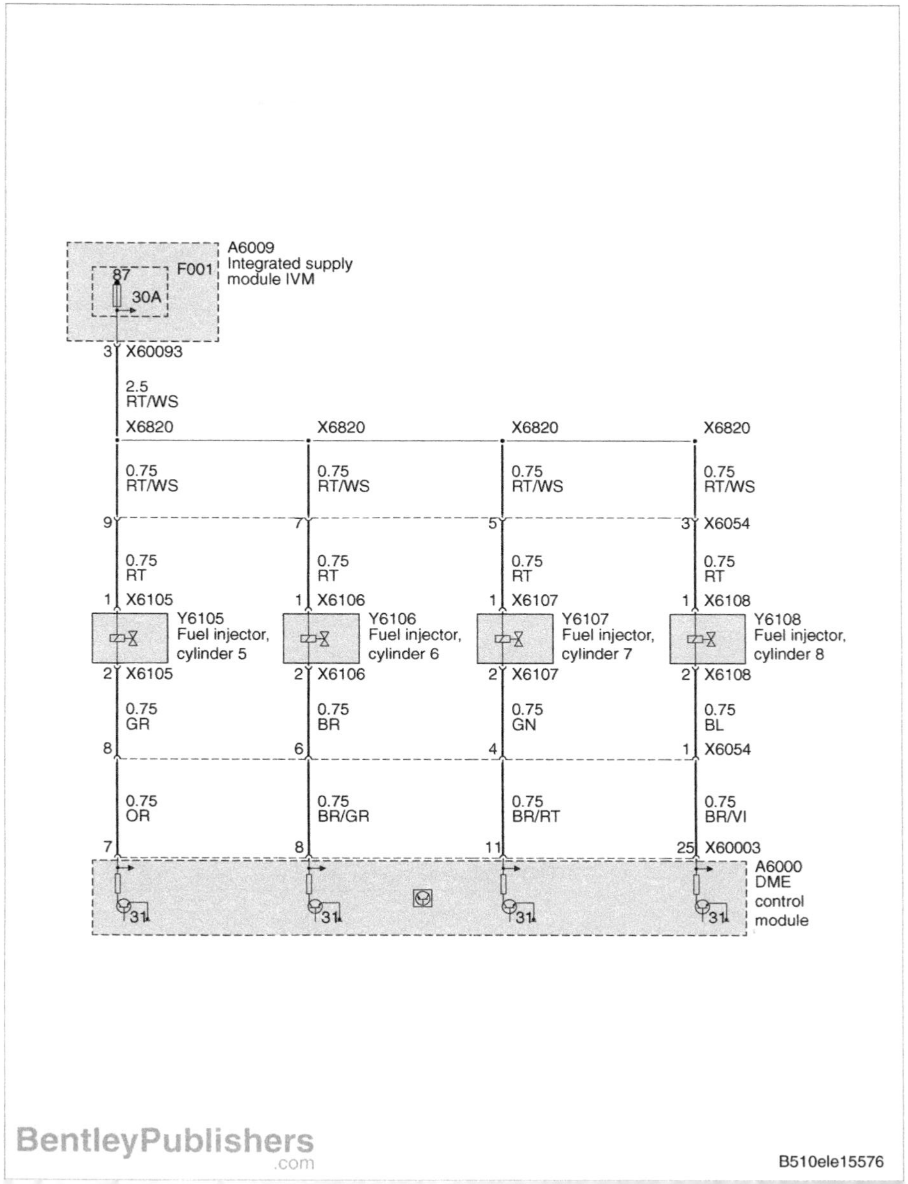

Brake light switch, DSC Dynamic Stability Control Premium
N62 engine or N62 TU engine

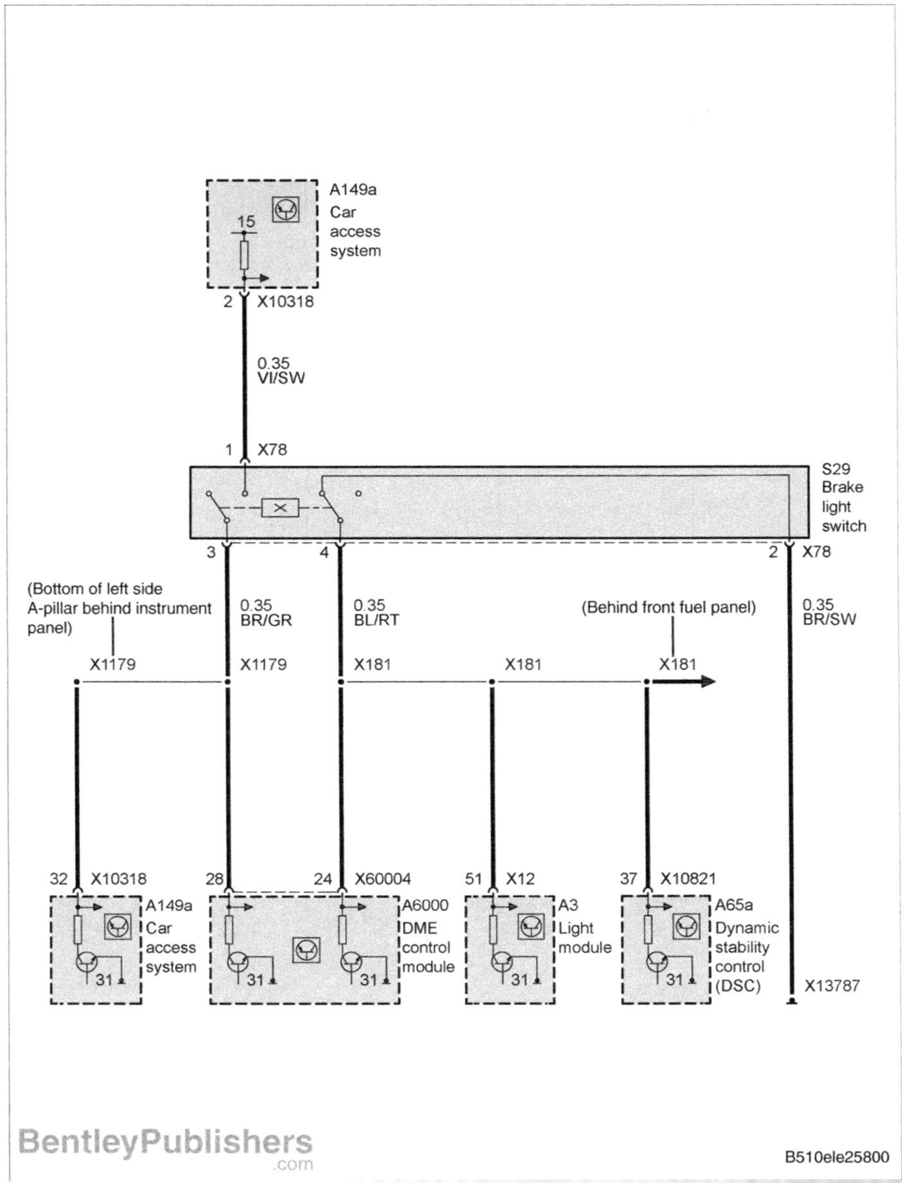

Accelerator pedal module
N62 engine or N62 TU engine

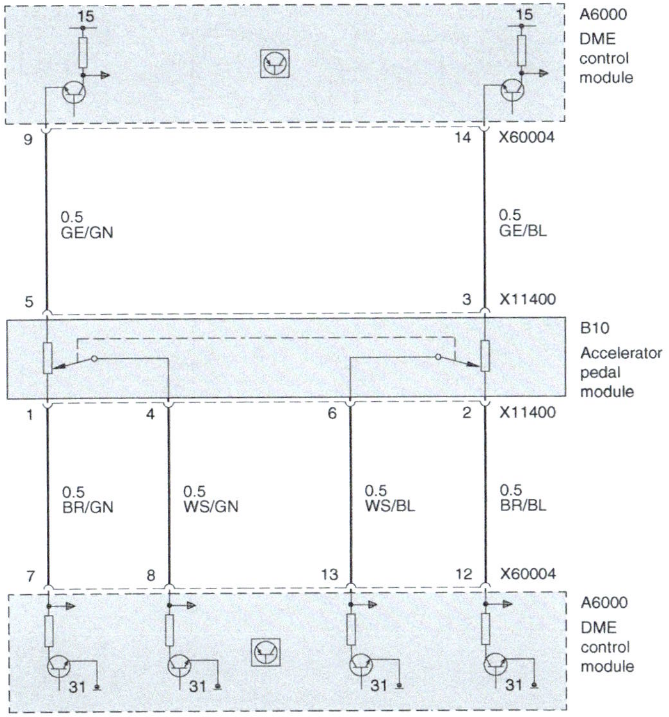

BentleyPublishers.com

B510ele15577

Knock control
N62 engine or N62 TU engine

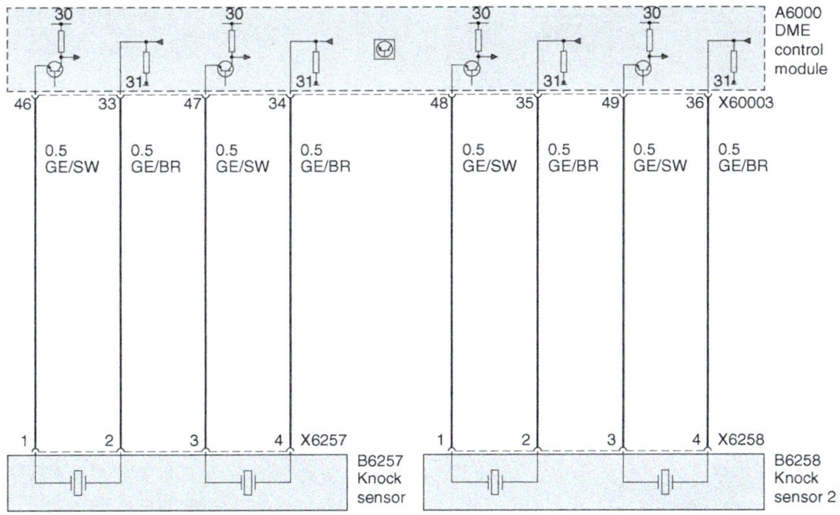

BentleyPublishers.com

B510ele15578

Knock control
N62 engine or N62 TU engine

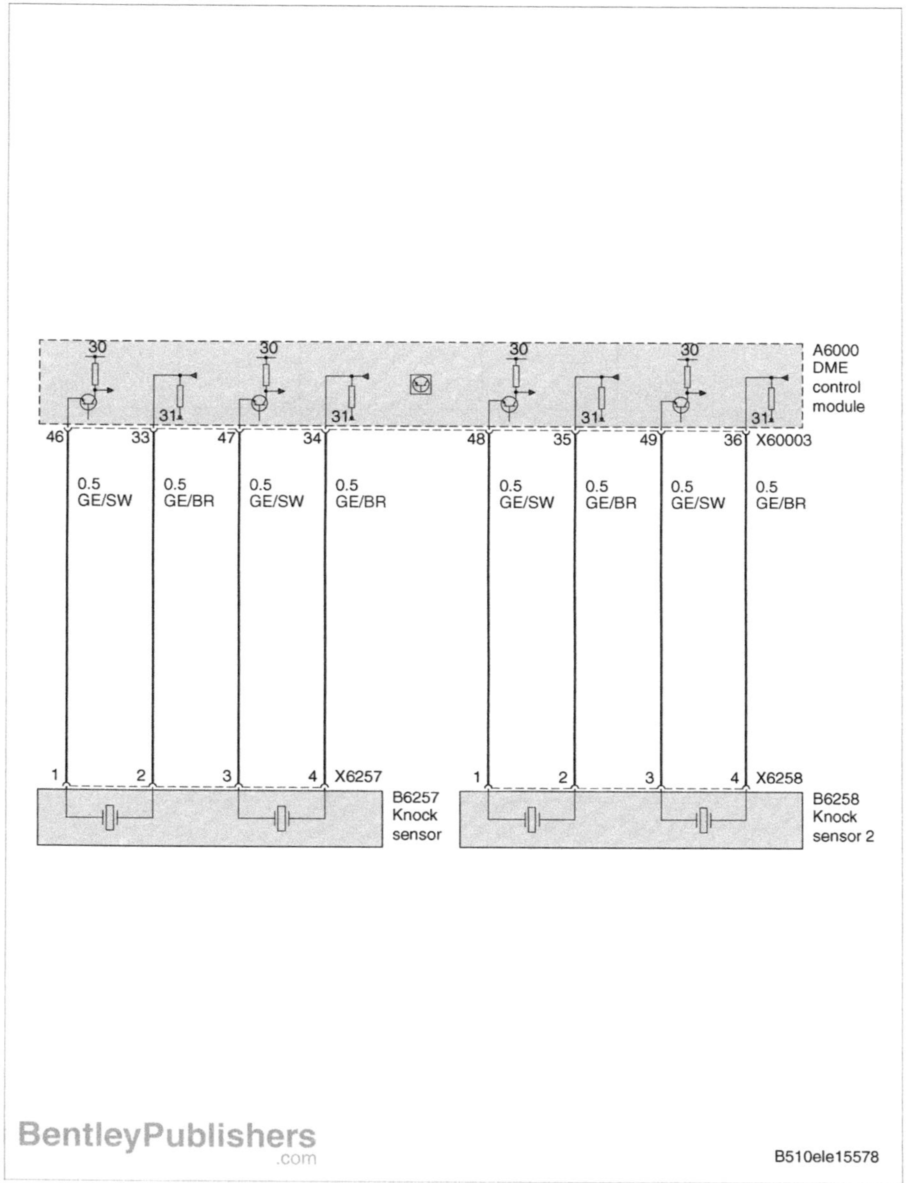

Convenient start (up to 09 / 2005)
N62 engine or N62 TU engine
page 1 of 2

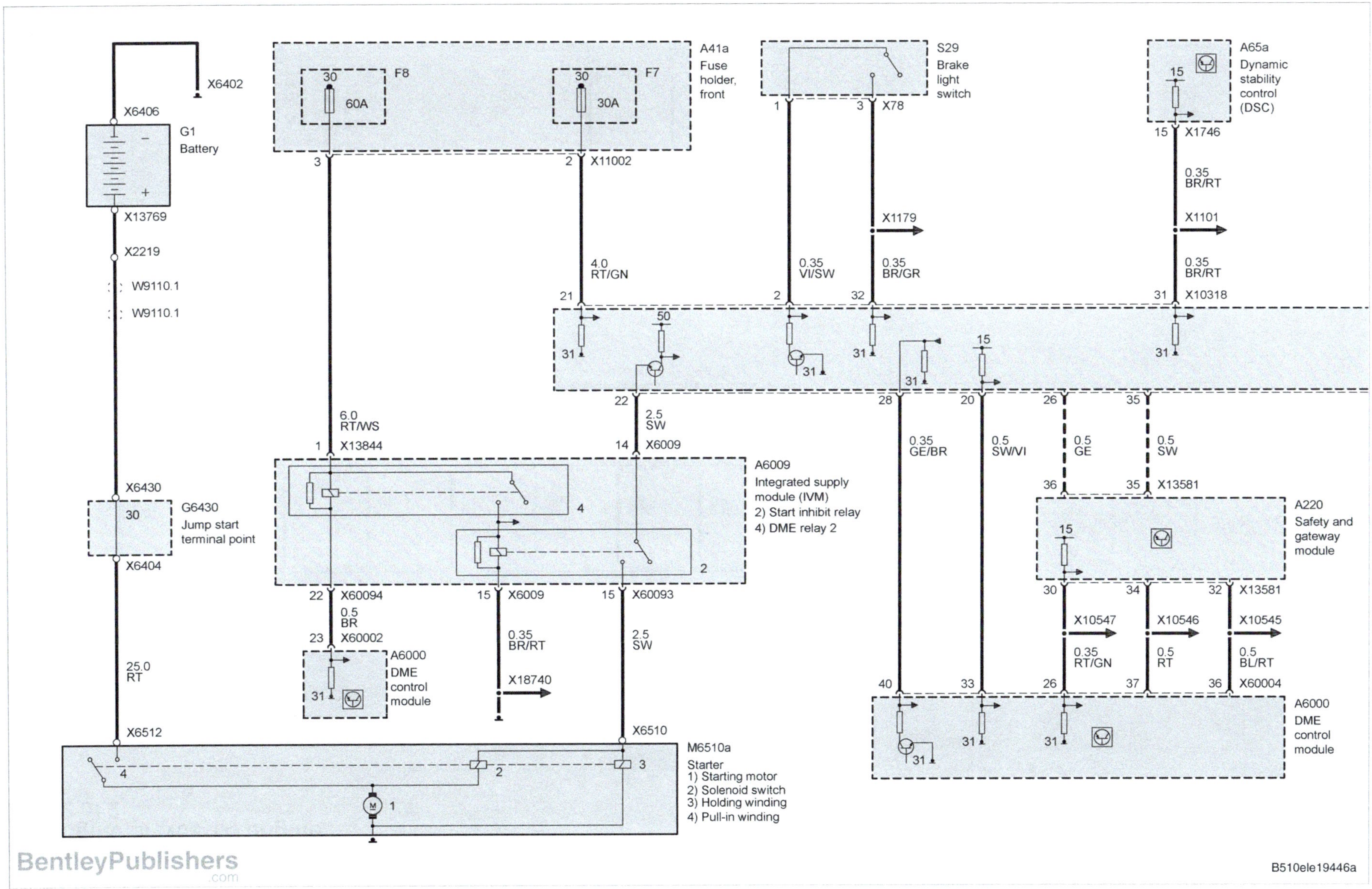

BentleyPublishers.com

Convenient start (up to 09 / 2005)
N62 engine or N62 TU engine
page 2 of 2

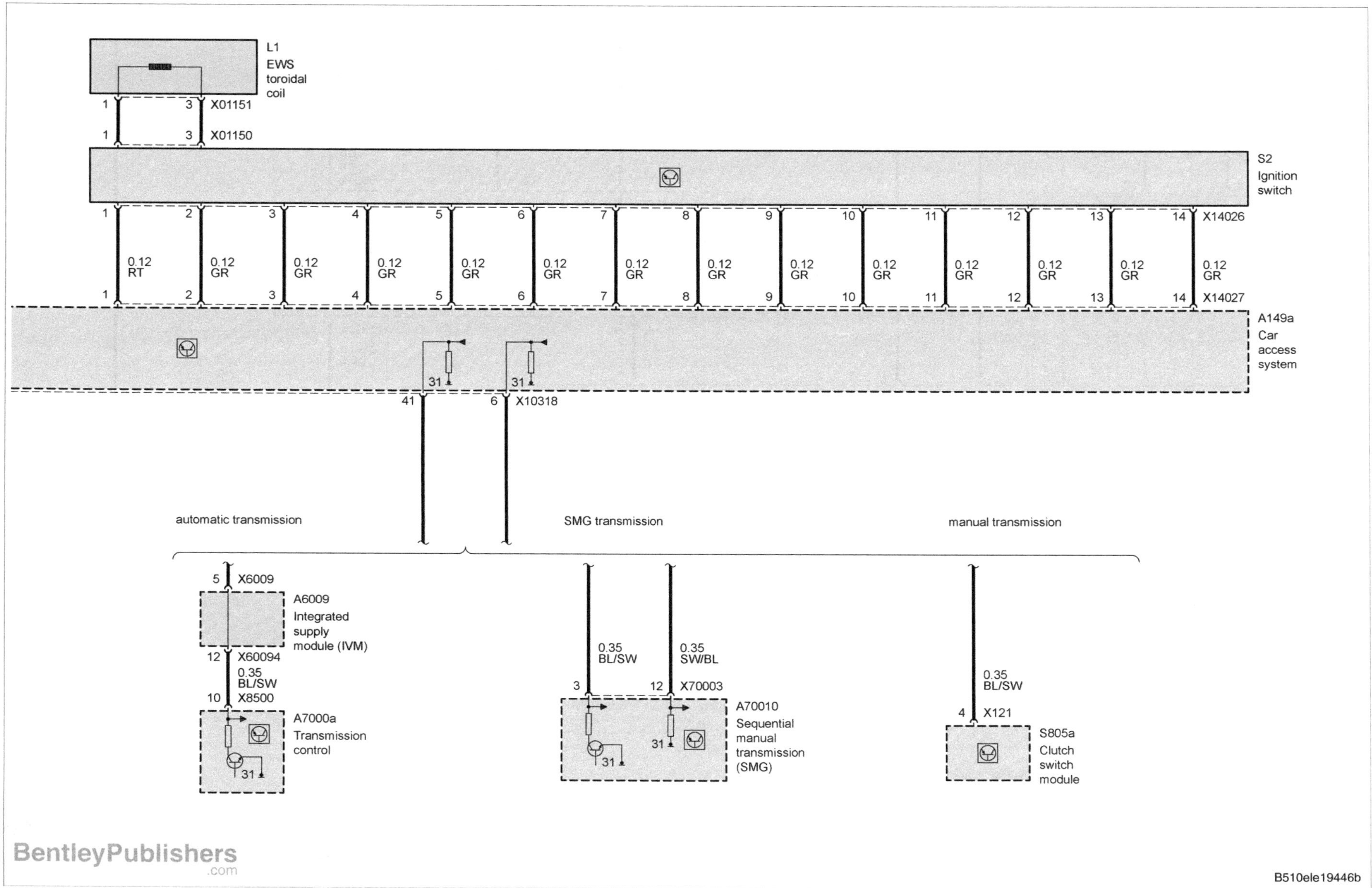

Convenient start (as of 09 / 2005)
N62 engine or N62 TU engine
page 1 of 2

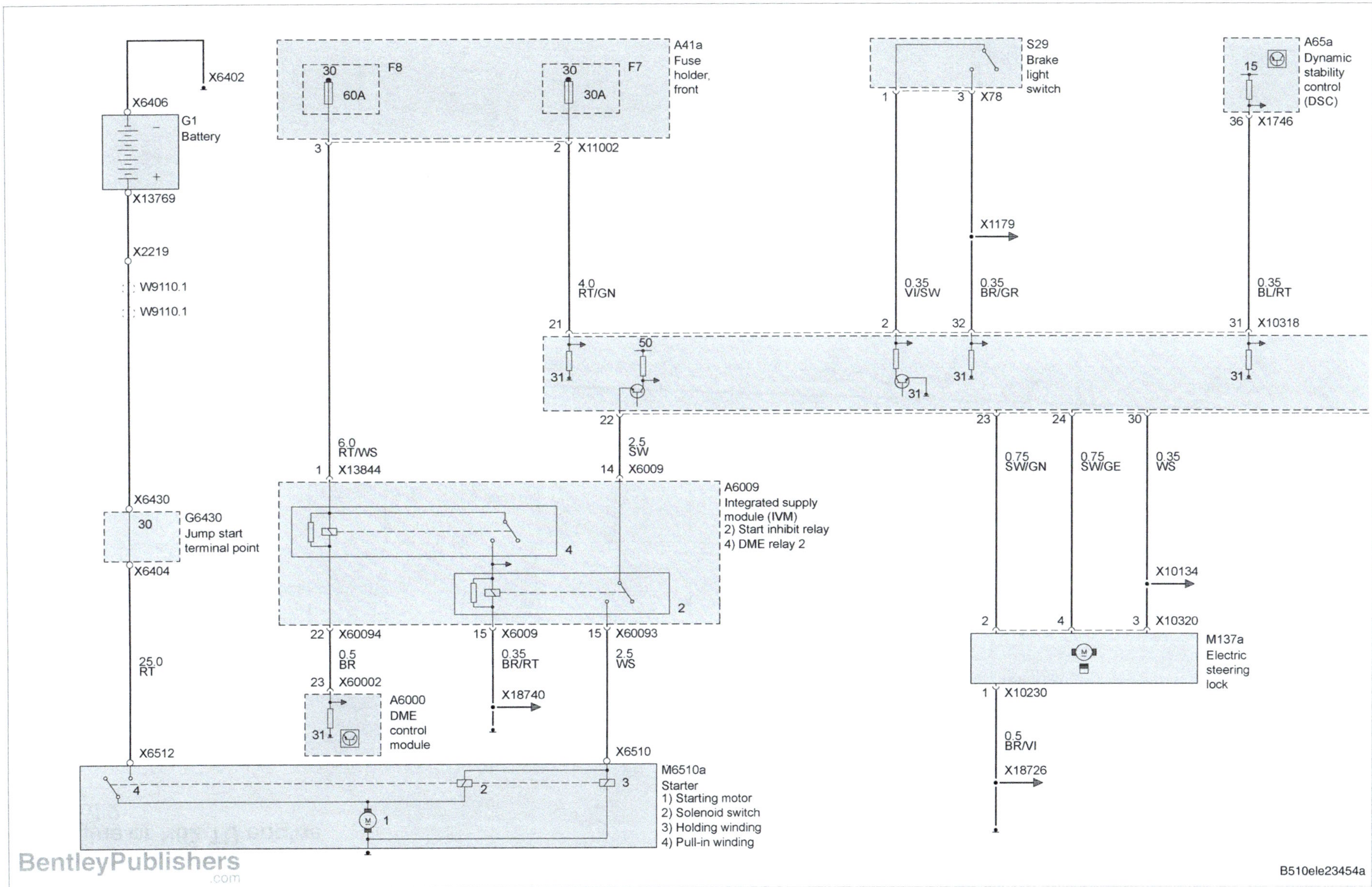

BentleyPublishers.com

Convenient start (as of 09 / 2005)
N62 engine or N62 TU engine
page 2 of 2

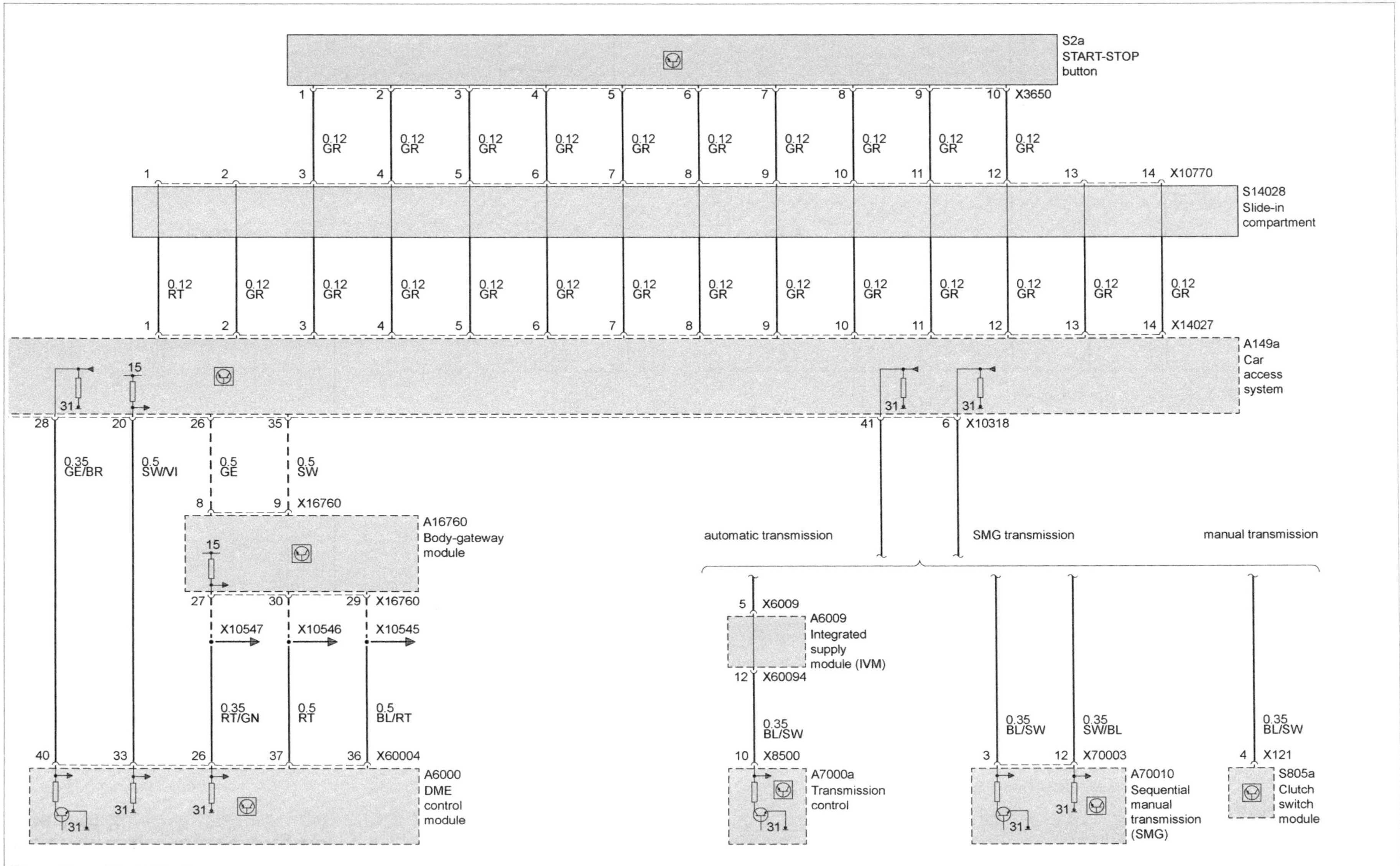

BentleyPublishers.com

B510ele23454b

Fuel pump
up to 09 / 2005
N62 engine or N62 TU engine

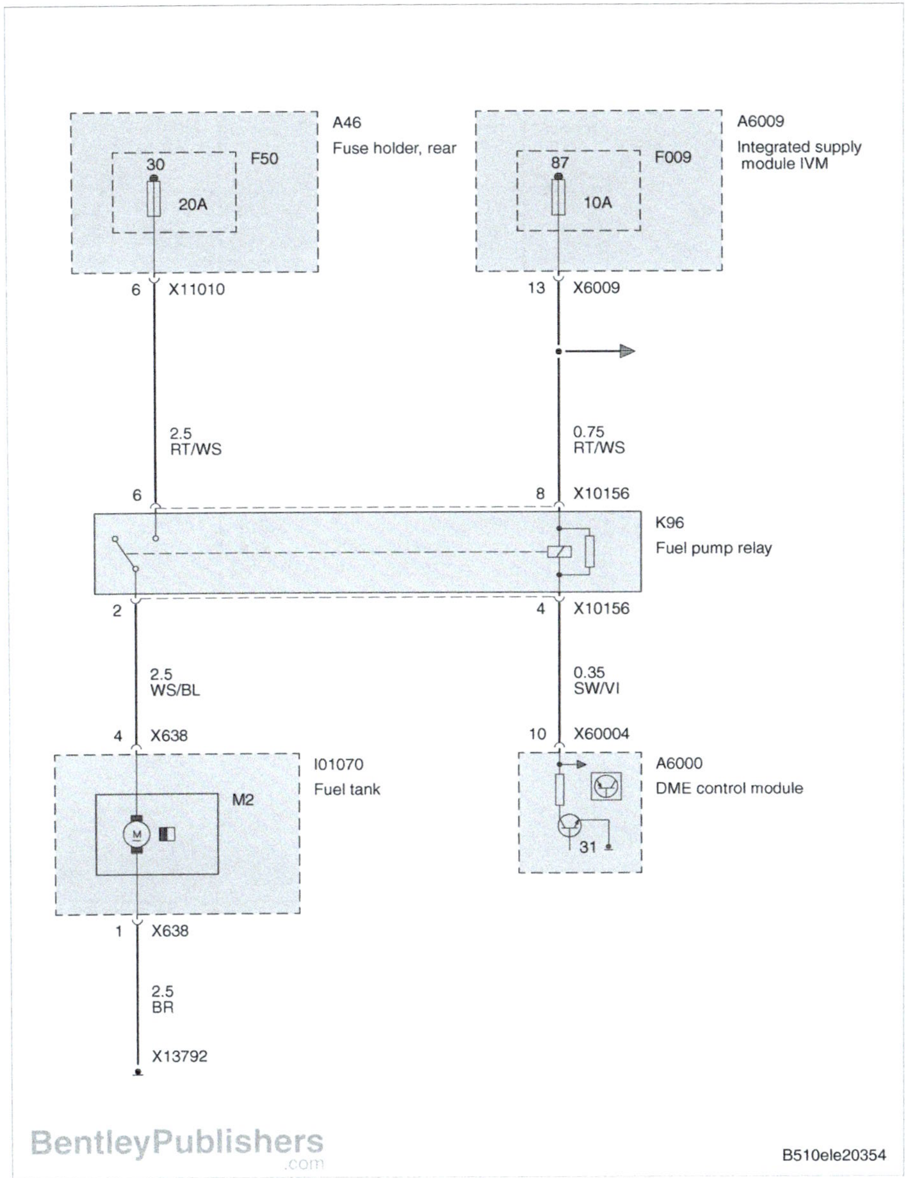

Fuel pump
as of 09 / 2005
N62 engine or N62 TU engine

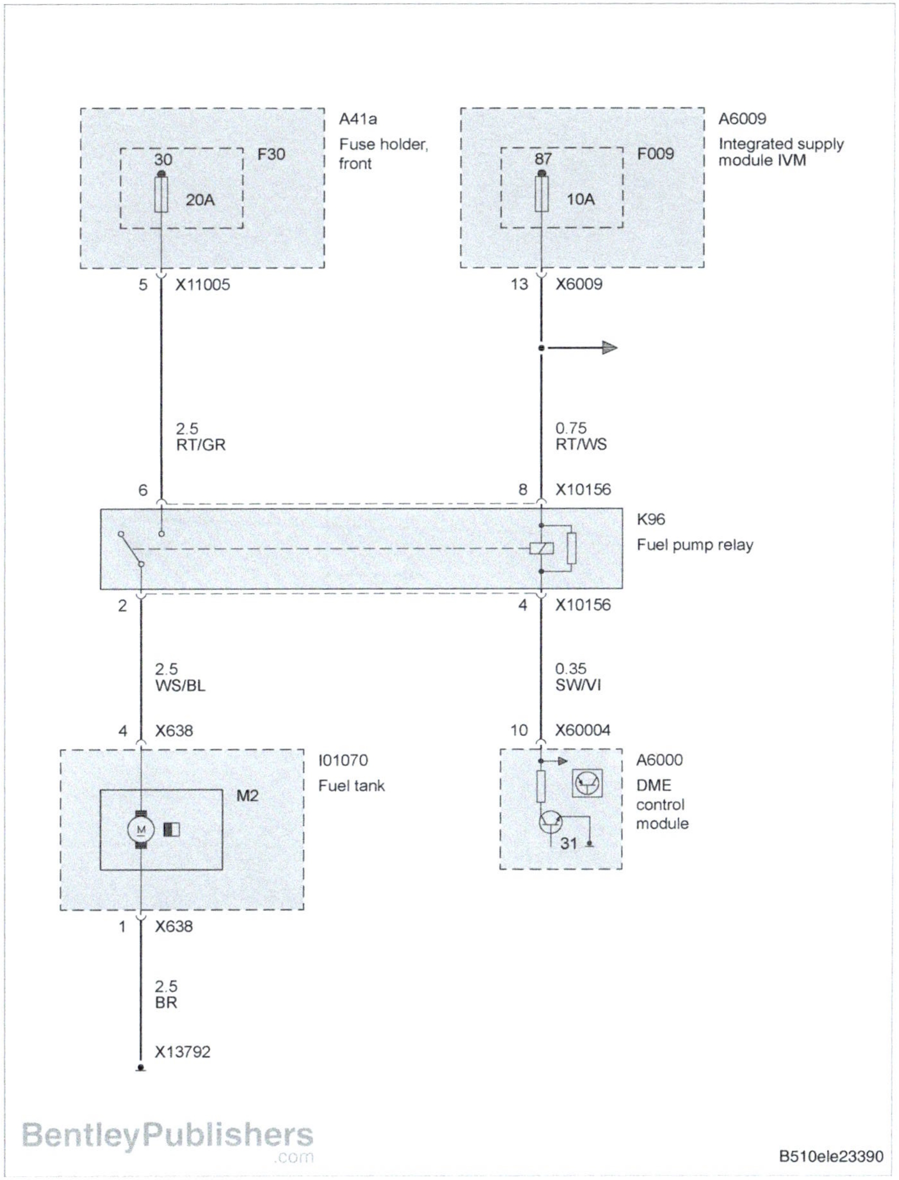

Diagnostic module, fuel tank leakage N62 engine or N62 TU engine

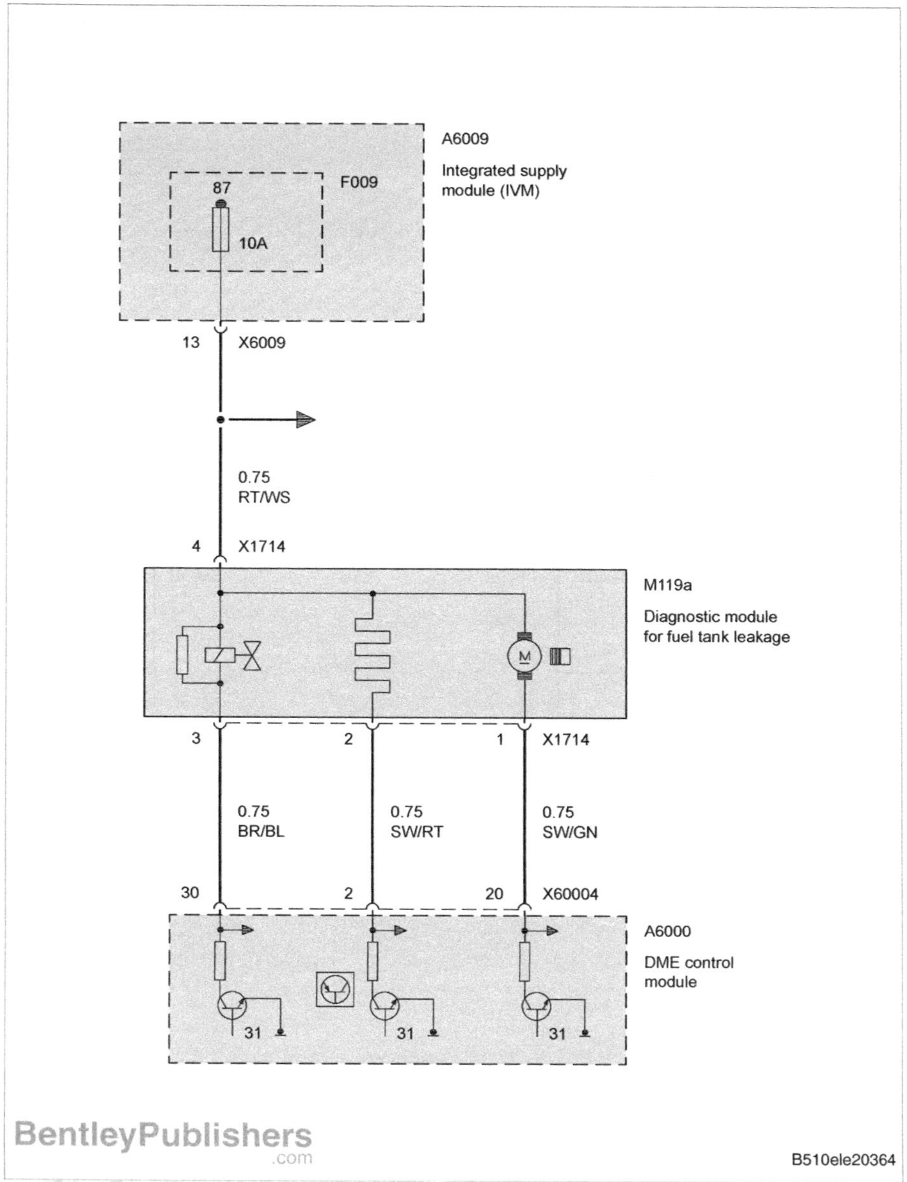

Fuel tank vent valve up to 09 / 2005 N62 engine or N62 TU engine

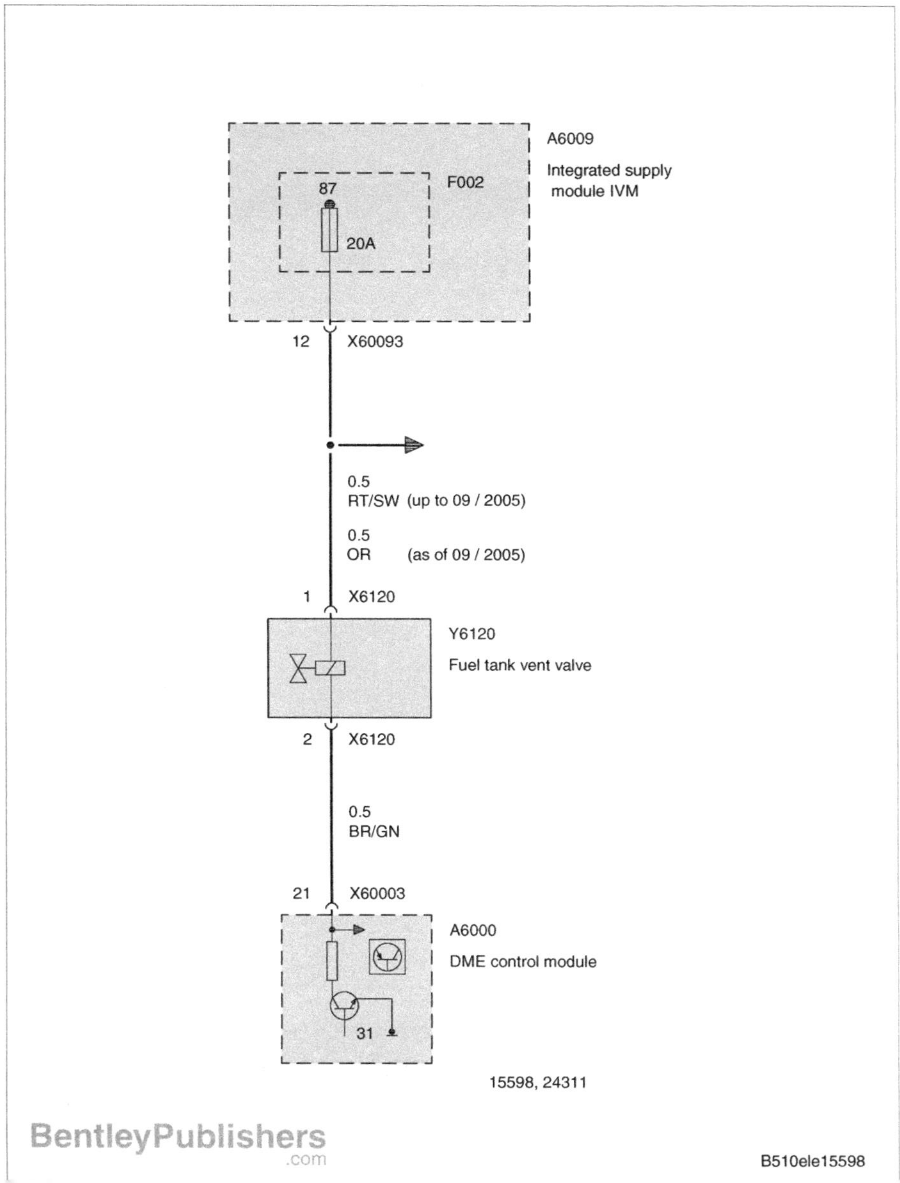

BentleyPublishers.com

BentleyPublishers.com

Oxygen sensors before catalytic converter
N62 engine

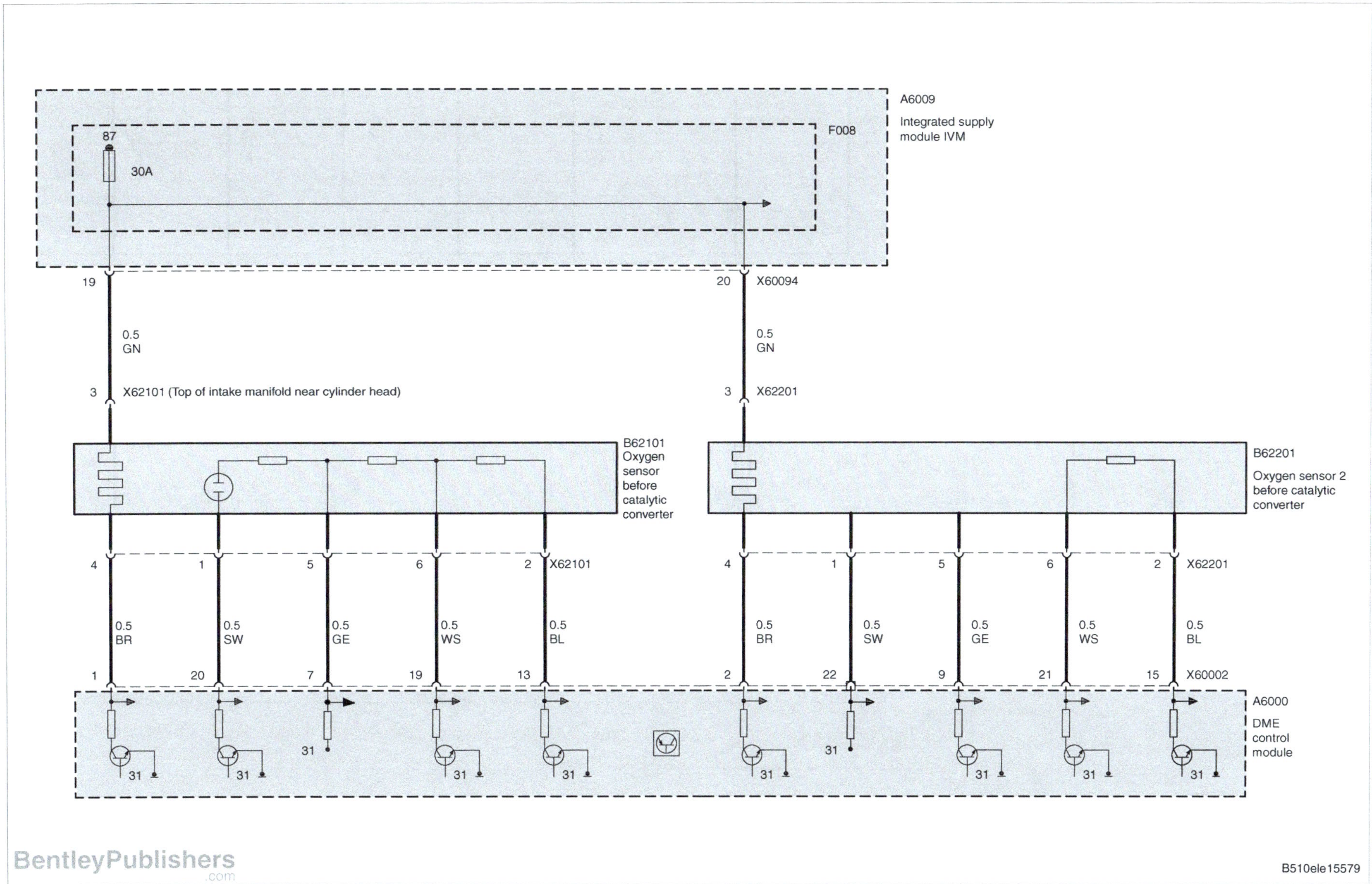

Oxygen sensors before catalytic converter
N62 TU engine

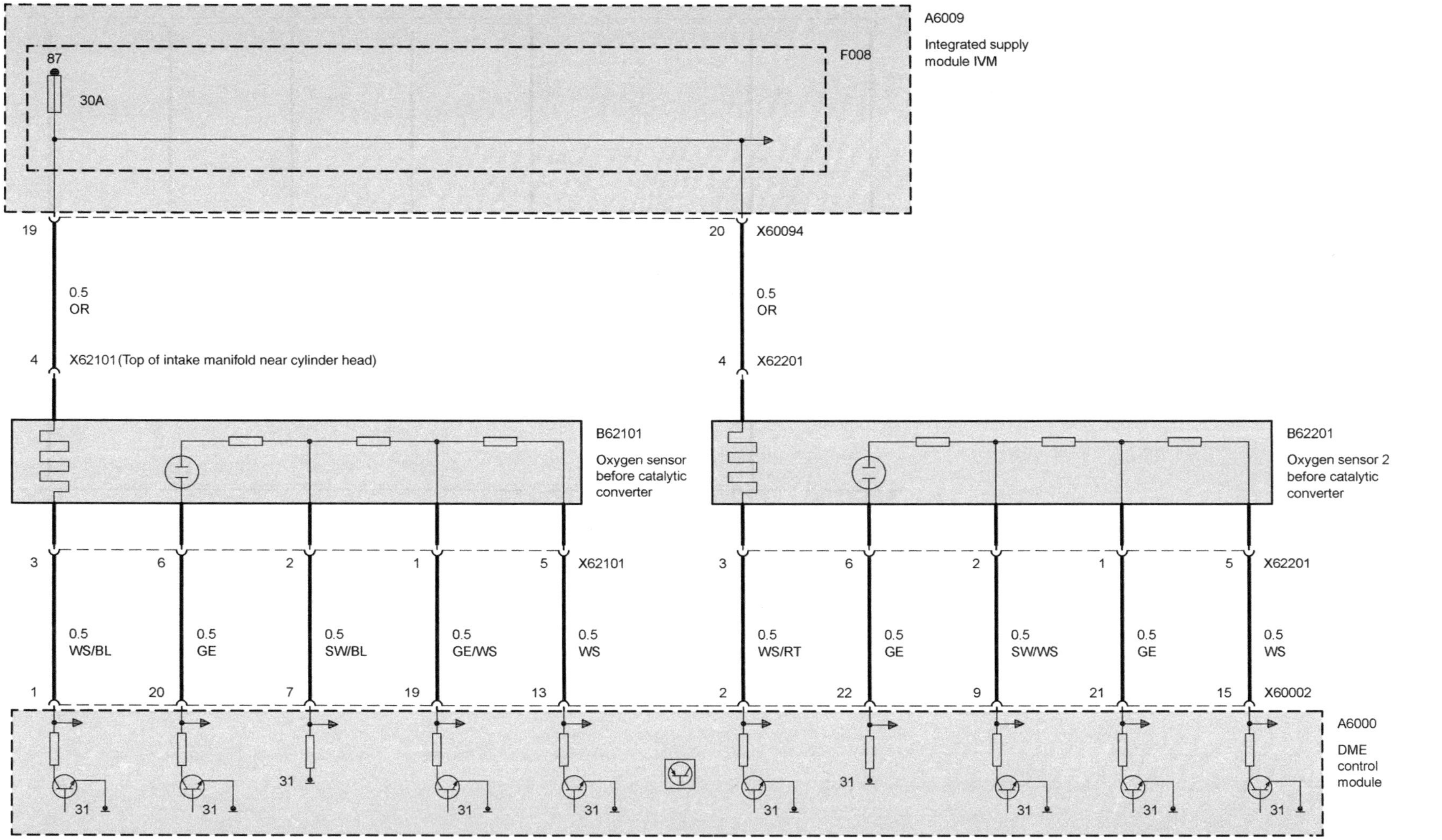

BentleyPublishers.com

B510ele22523

Oxygen sensors after catalytic converter N62 engine

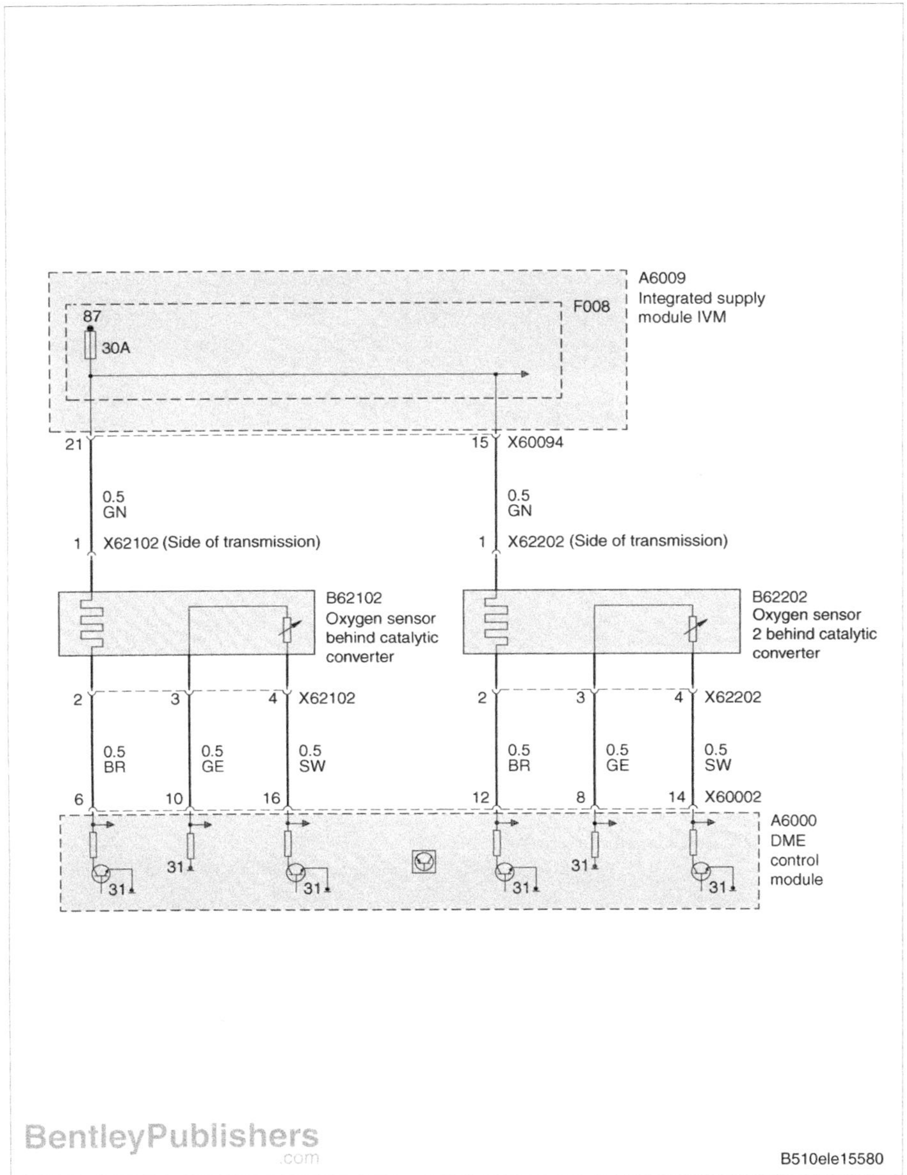

Oxygen sensors after catalytic converter N62 TU engine

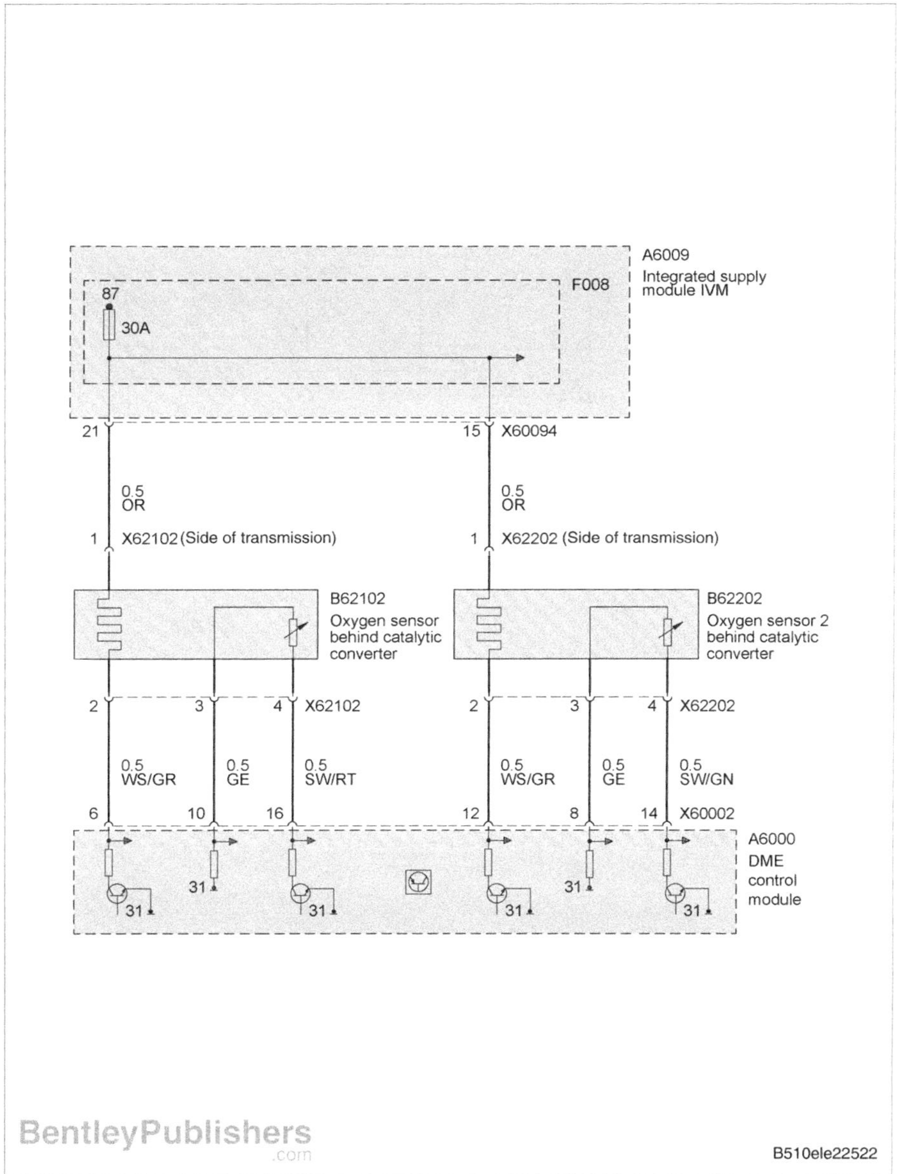

Hot film air mass meter, intake pipe pressure sensor up to 09 / 2005 N62 engine or N62 TU engine

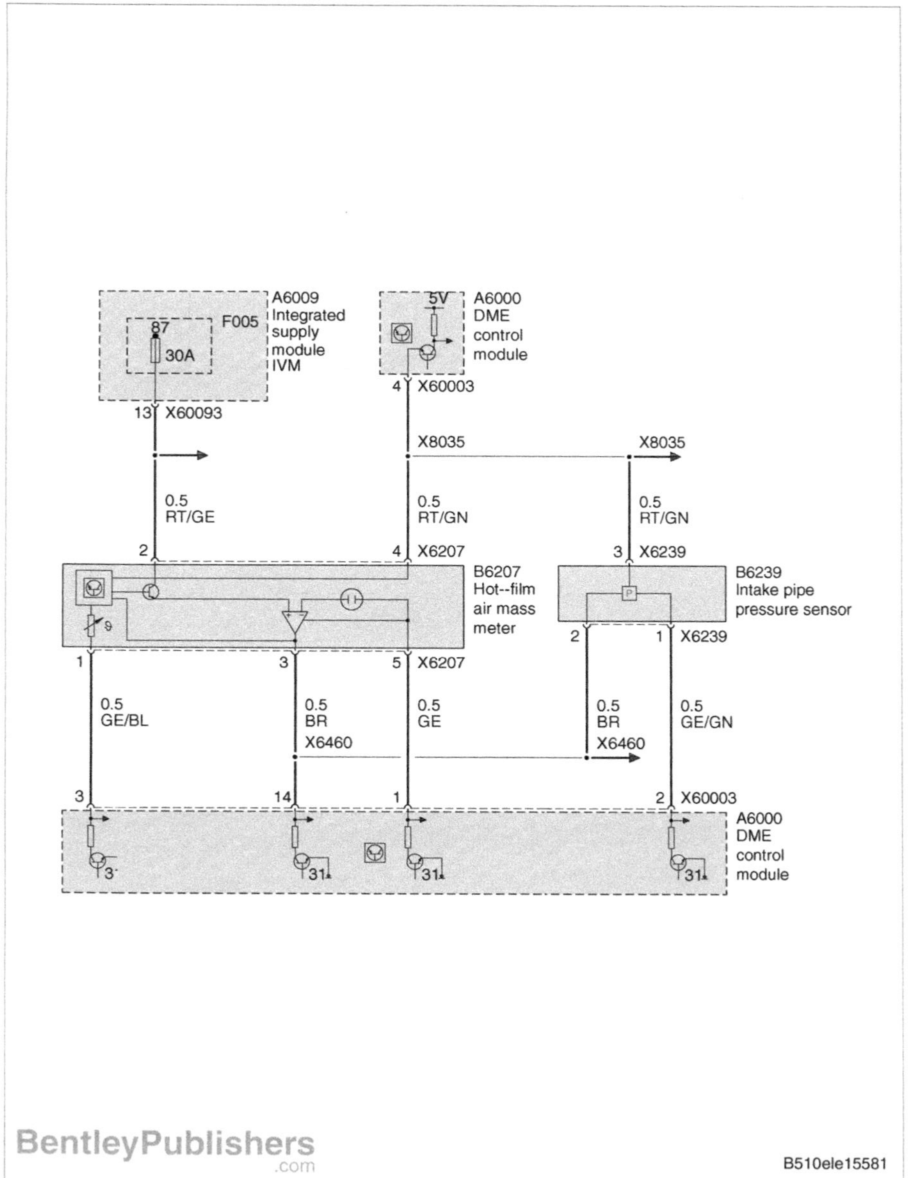

Hot film air mass meter, intake pipe pressure sensor as of 09 / 2005 N62 engine or N62 TU engine

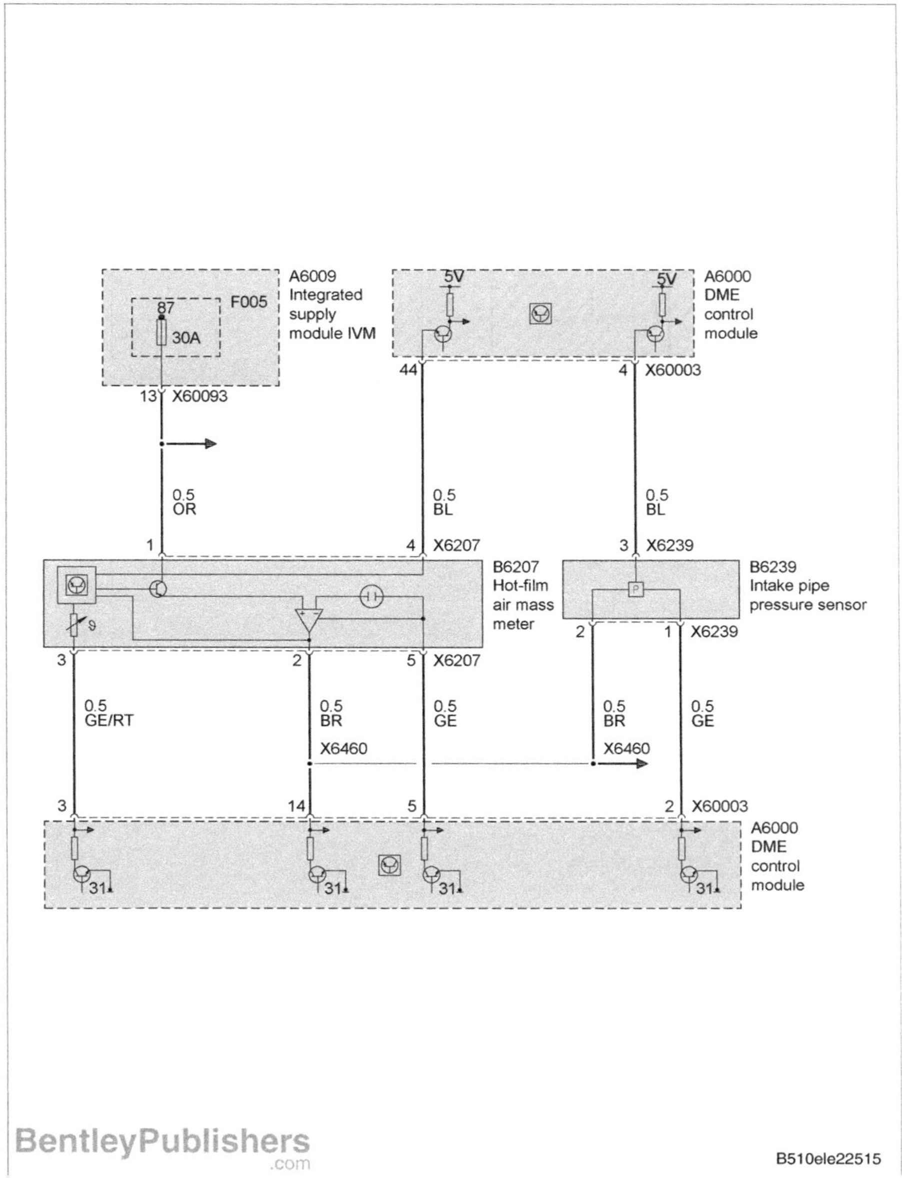

Throttle valve
N62 engine or N62 TU engine

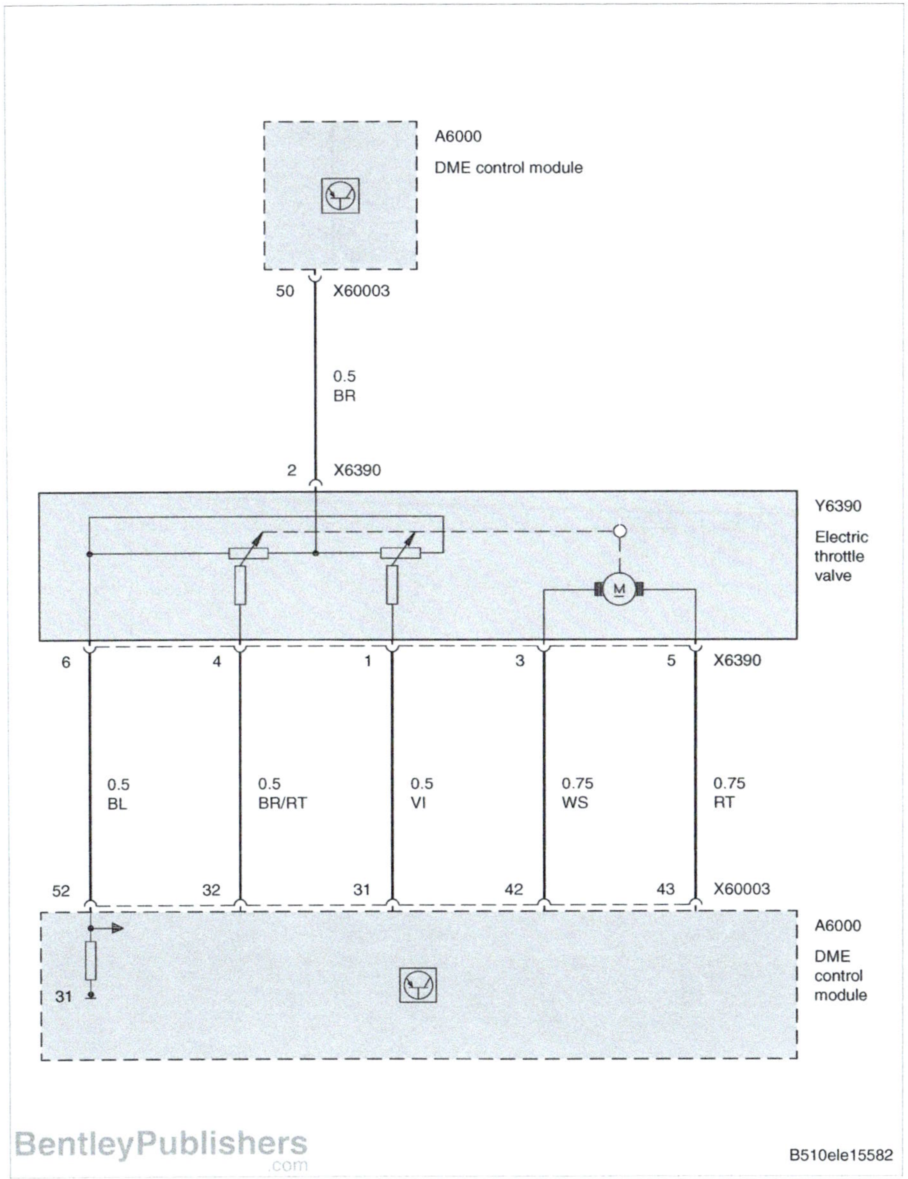

Variable intake system
N62 engine

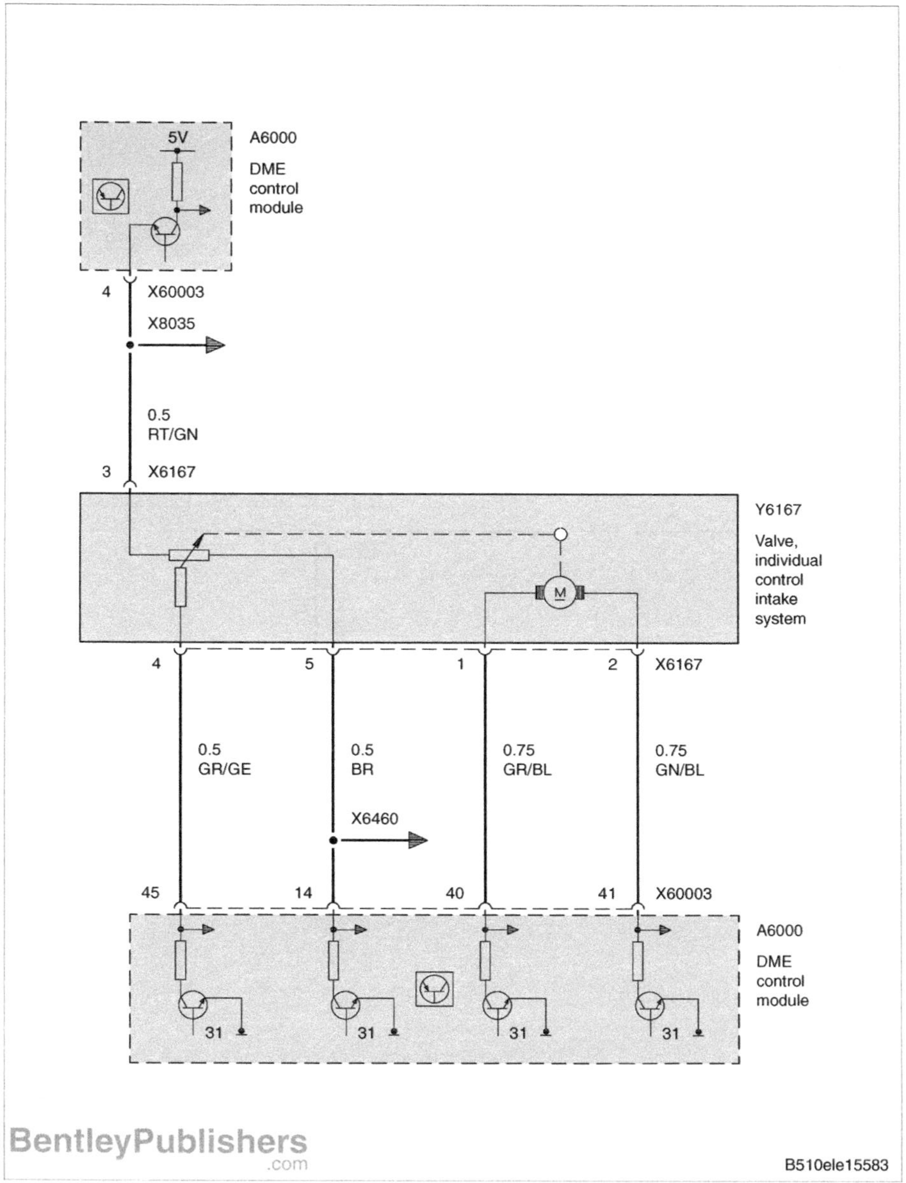

Variable intake system
N62 TU engine

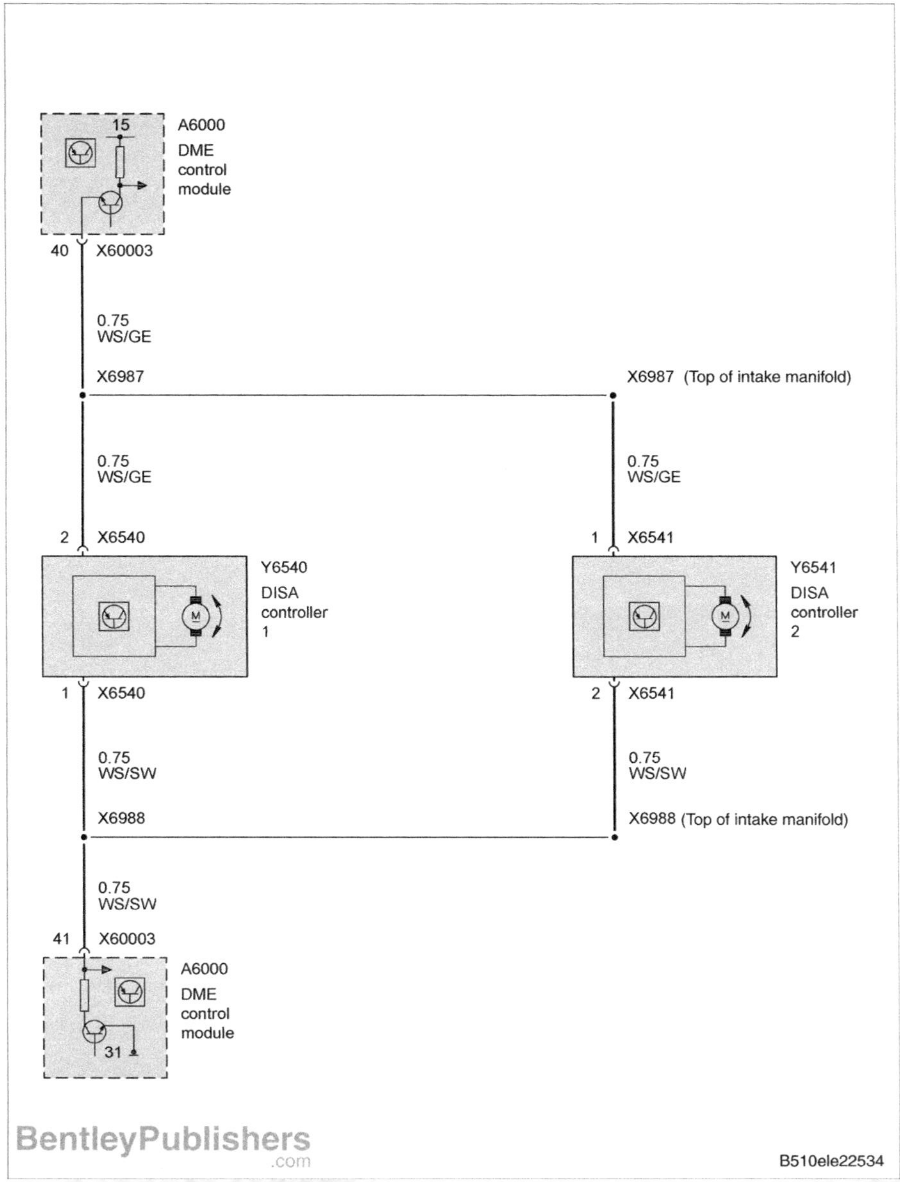

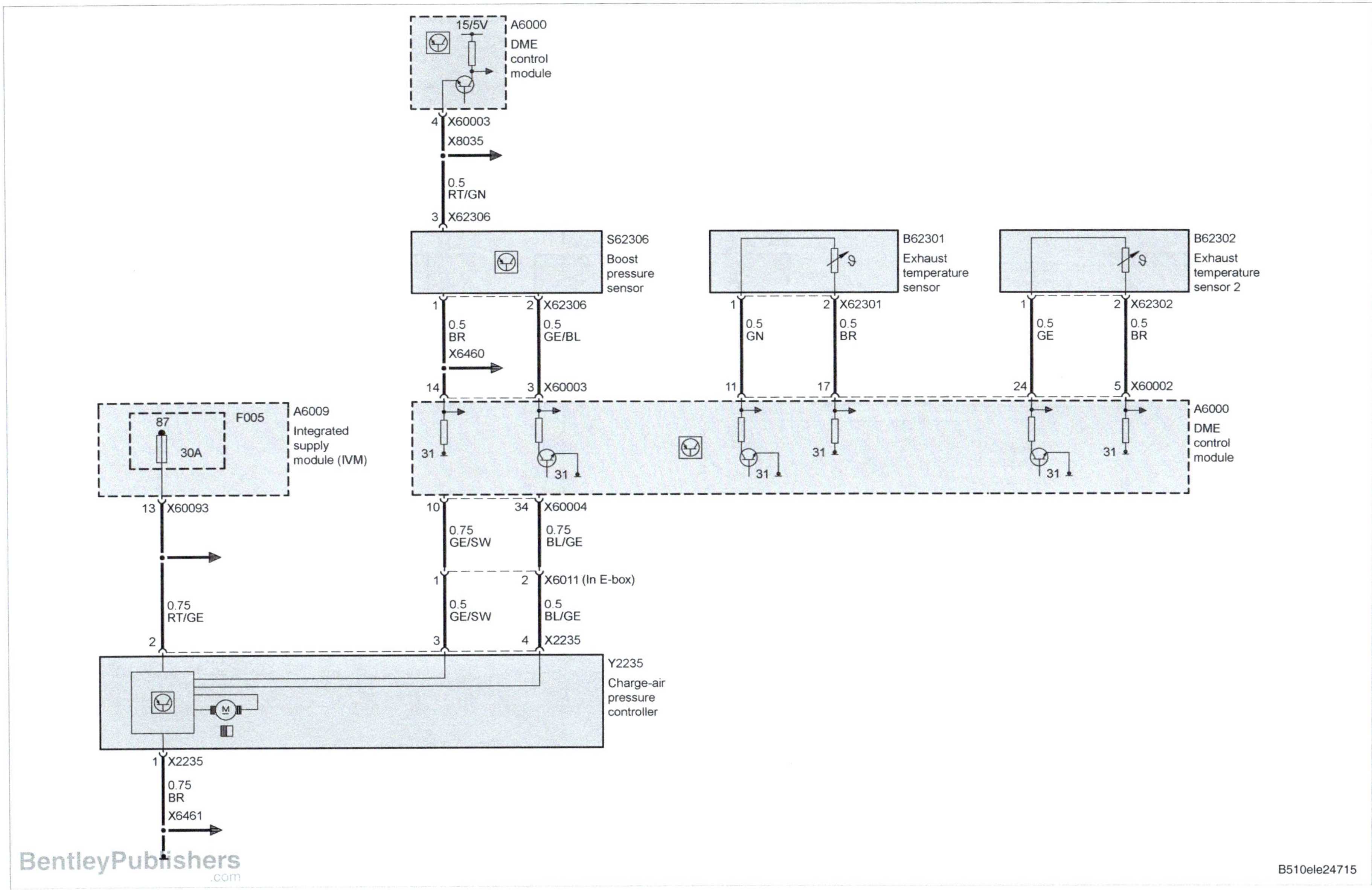

BentleyPublishers
.com

Engine cooling system
up to 03 / 2007
N62 engine or N62 TU engine

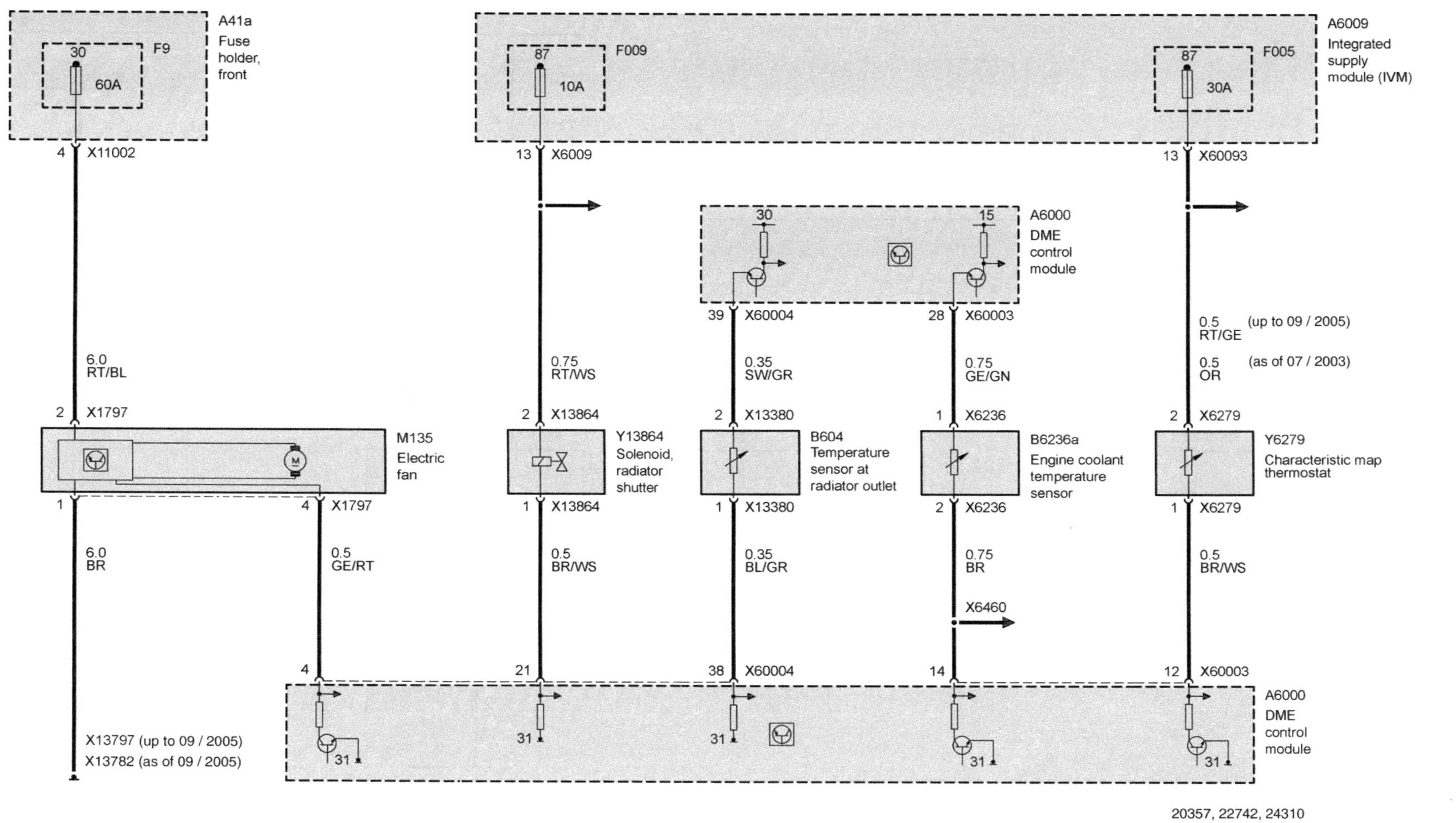

BentleyPublishers.com

B510ele20357

Engine cooling system
as of 03 / 2007
N62 engine or N62 TU engine

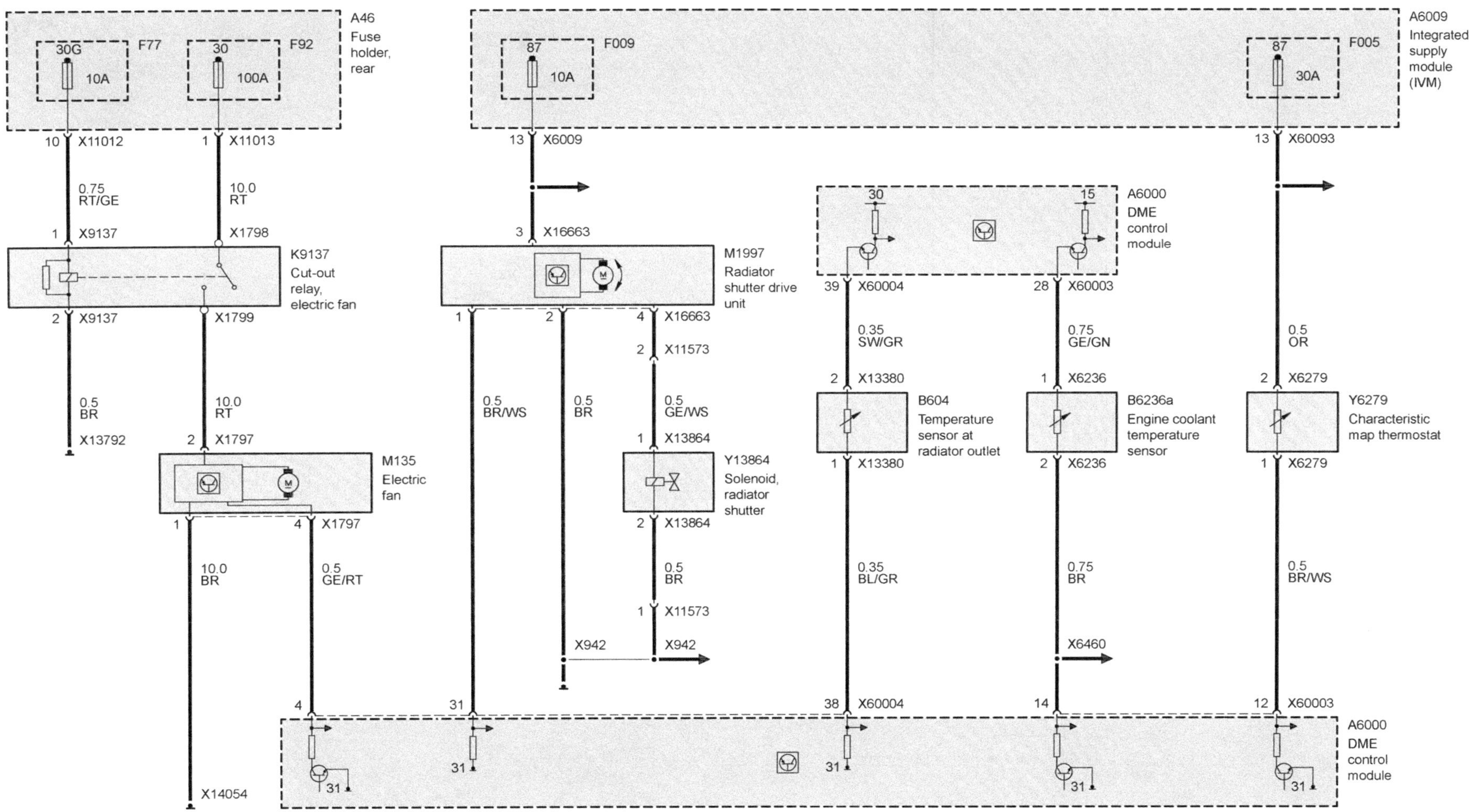

BentleyPublishers.com

B510ele25745

Instrument gauges / indicators
N62 engine or N62 TU engine

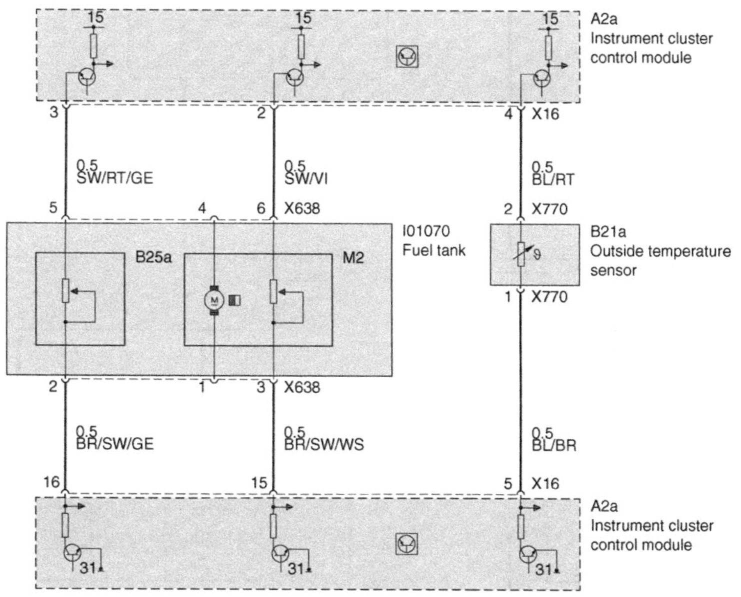

BentleyPublishers.com

B510ele19454

Local CAN-bus
N62 engine or N62 TU engine

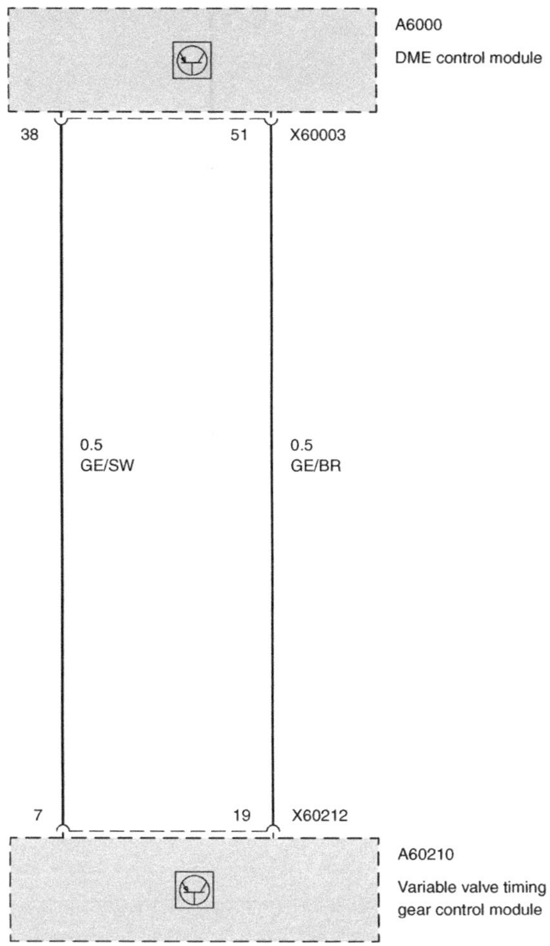

BentleyPublishers.com

B510ele15594

Secondary air injection pump
N62 engine or N62 TU engine

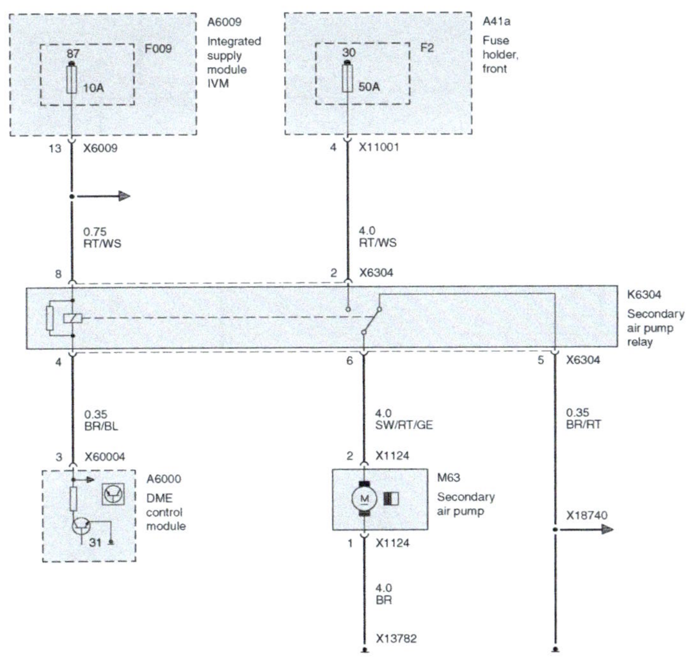

DME control module
power supply
N62 engine or N62 TU engine

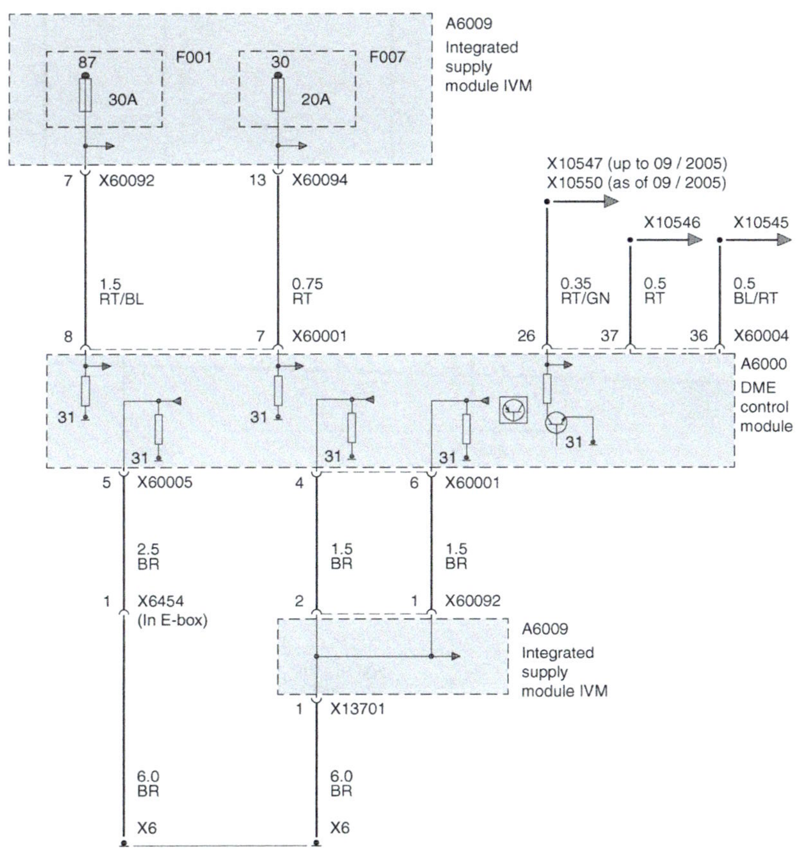

Camshaft position
up to 09 / 2005
N62 engine or N62 TU engine

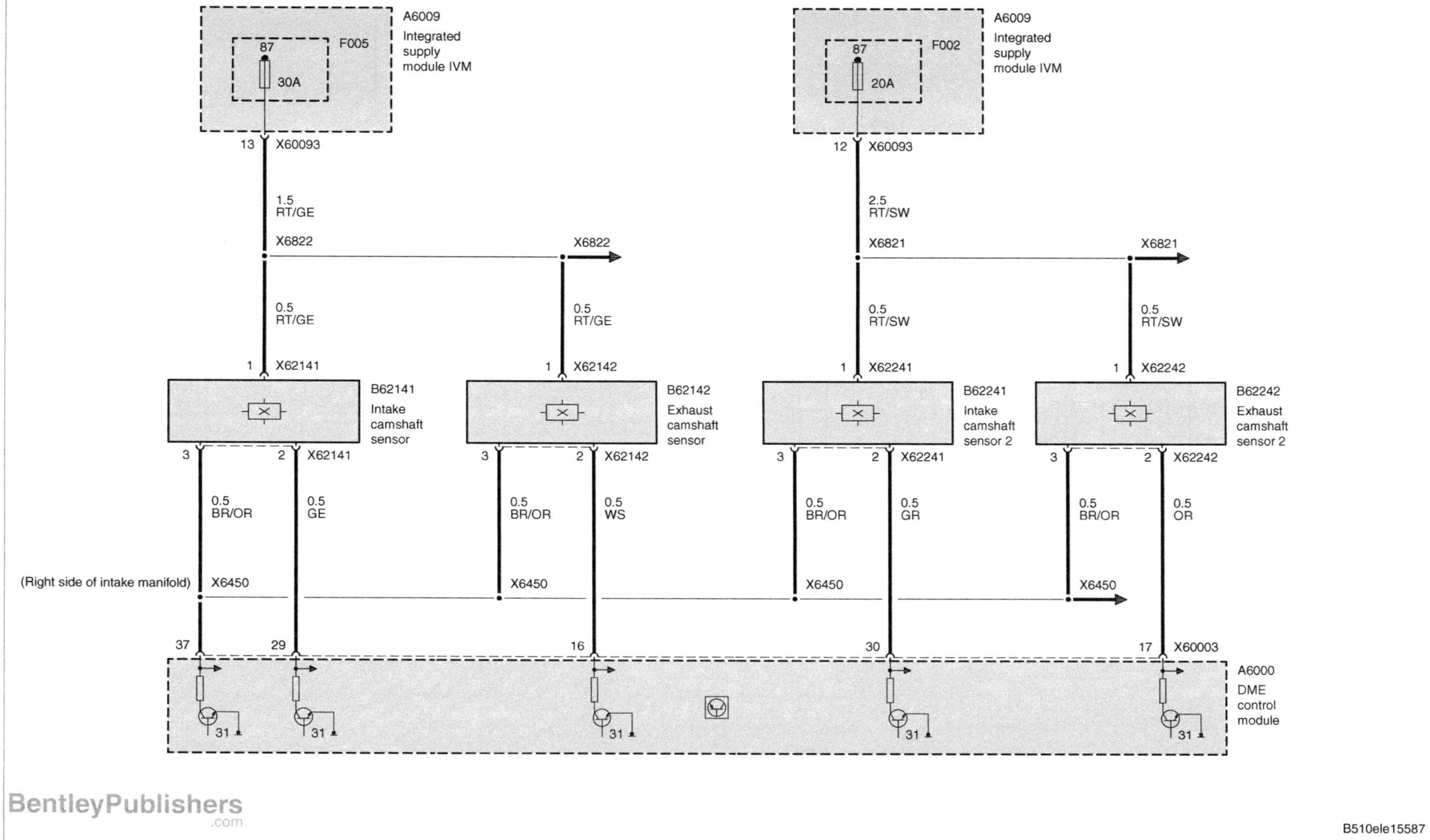

BentleyPublishers.com

Camshaft position
as of 09 / 2005
N62 engine or N62 TU engine

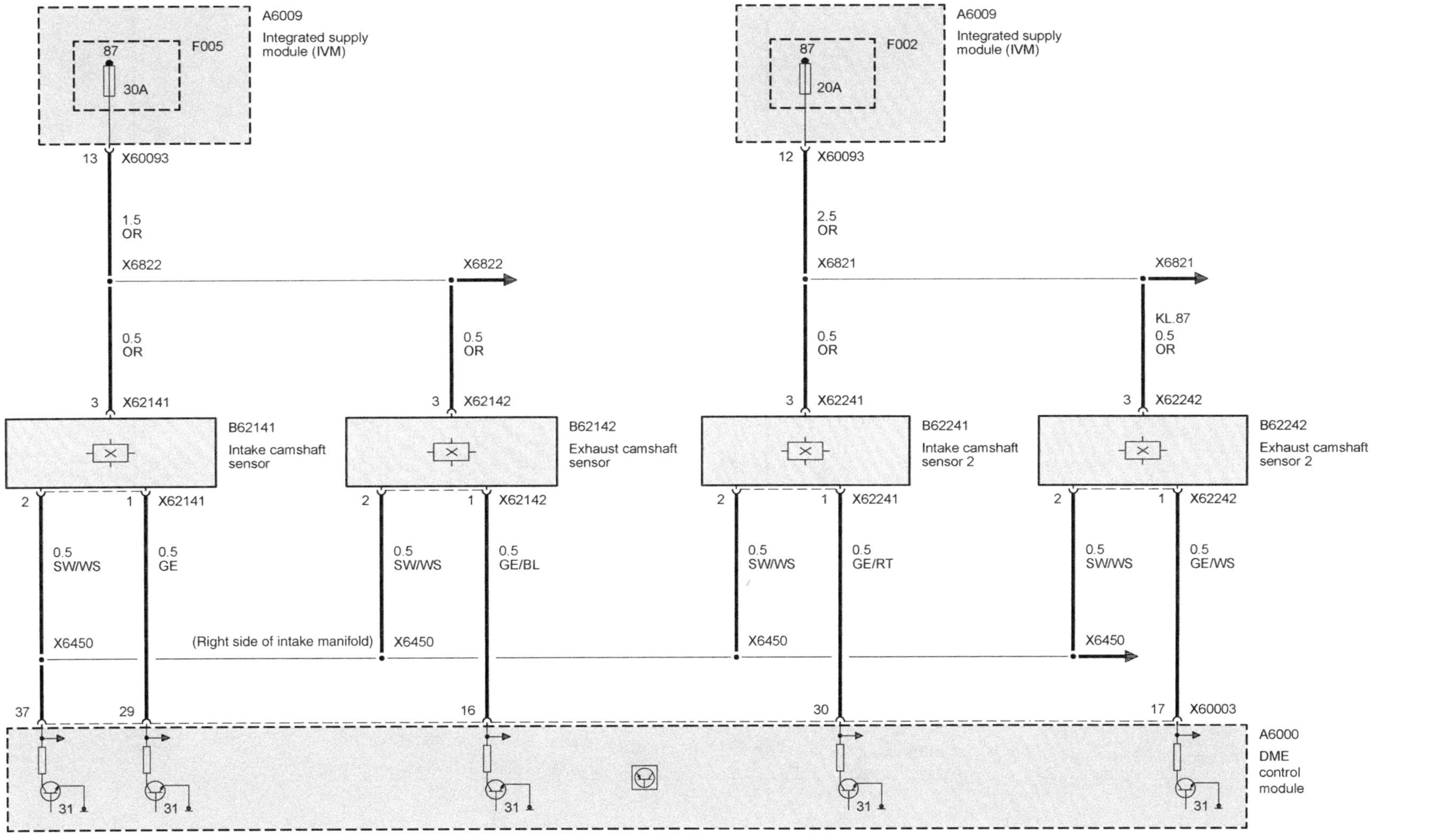

BentleyPublishers.com

VANOS solenoid valves 1 and 2
N62 engine or N62 TU engine

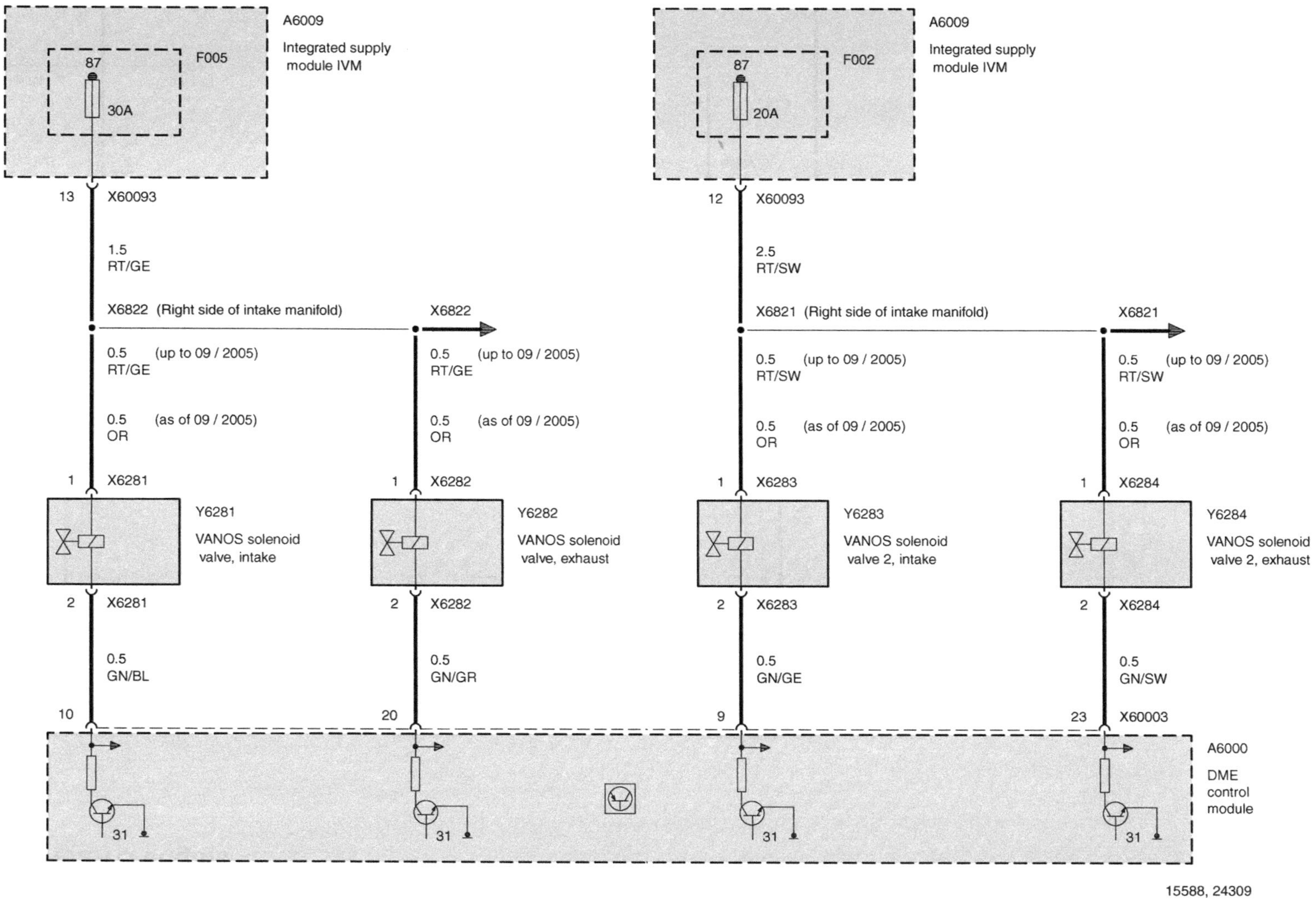

BentleyPublishers.com

B510ele15588

Variable valve timing gear 1
N62 engine or N62 TU engine

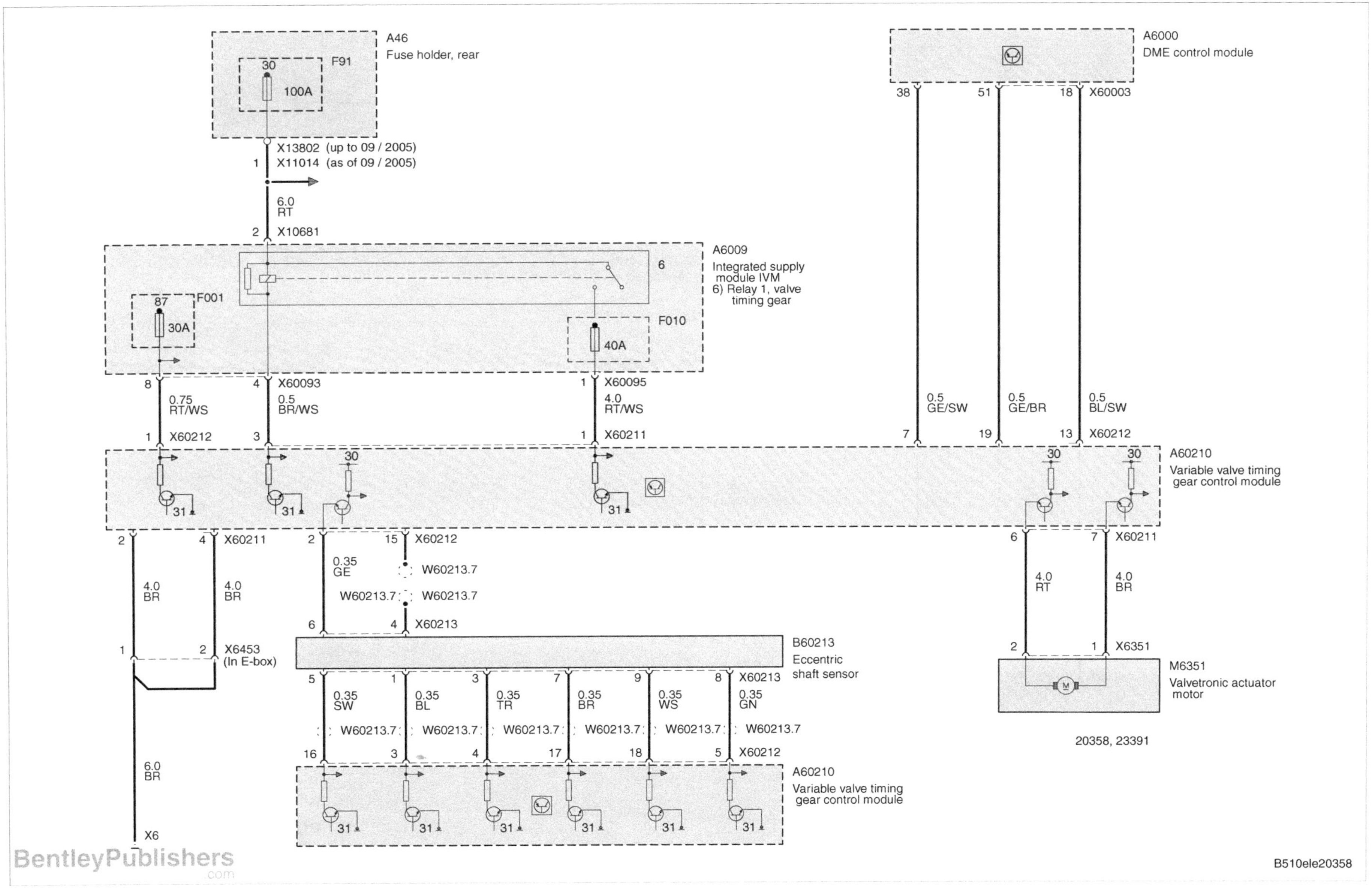

BentleyPublishers.com

Variable valve timing gear 2
N62 engine or N62 TU engine

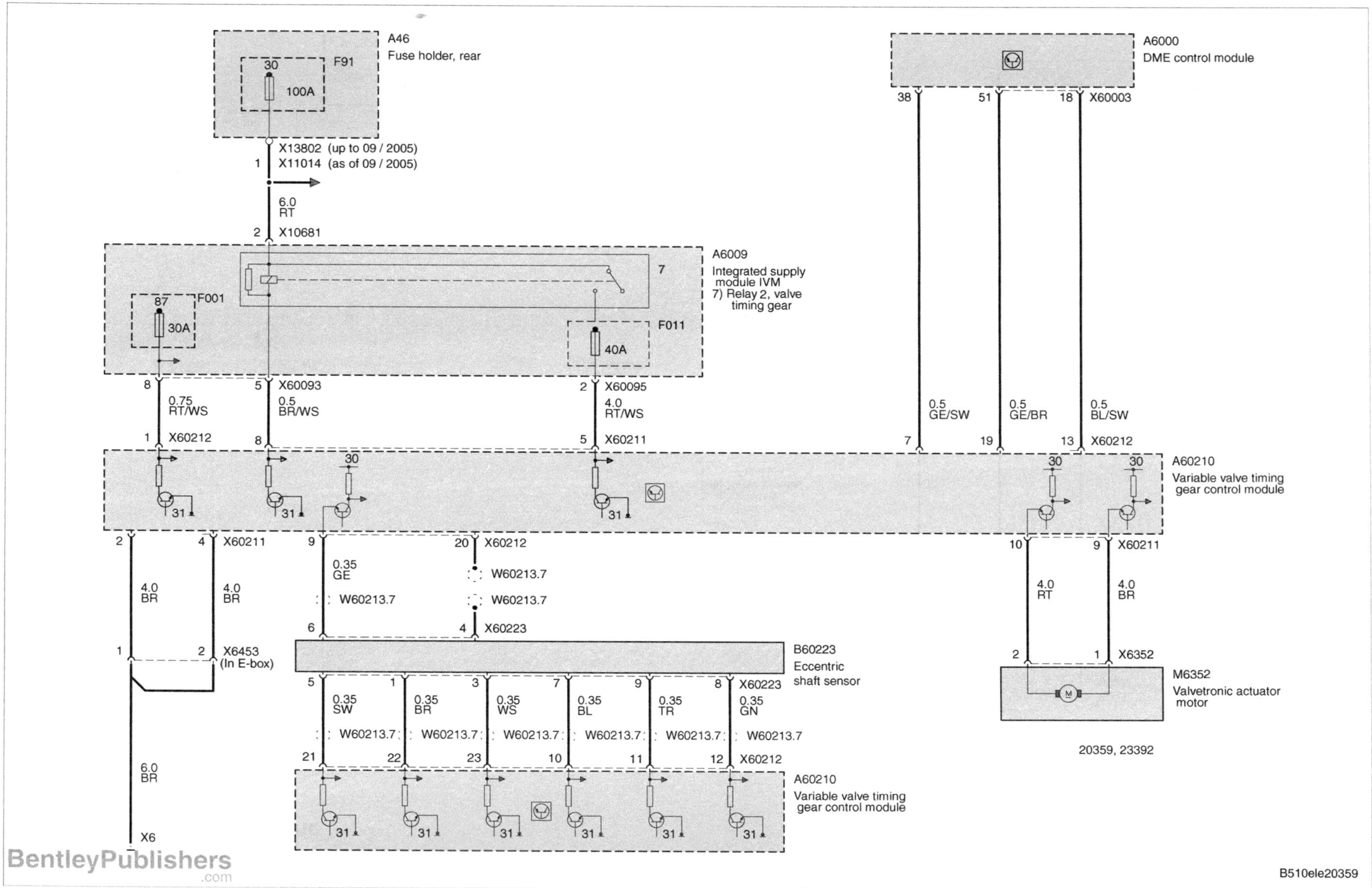

BentleyPublishers.com

Ignition coils 1
N62 engine or N62 TU engine

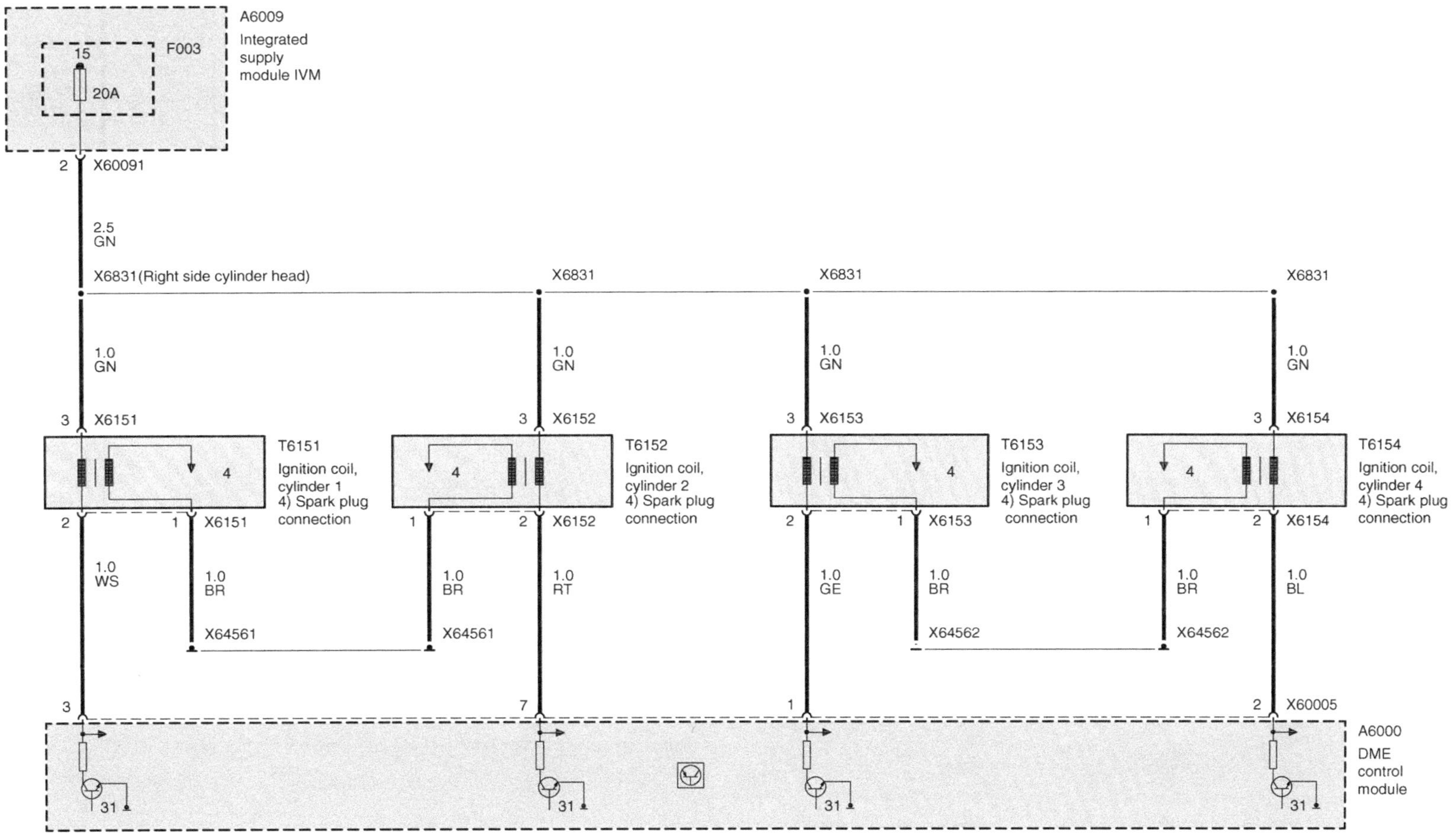

BentleyPublishers.com

B510ele15600

Ignition coils 2
N62 engine or N62 TU engine

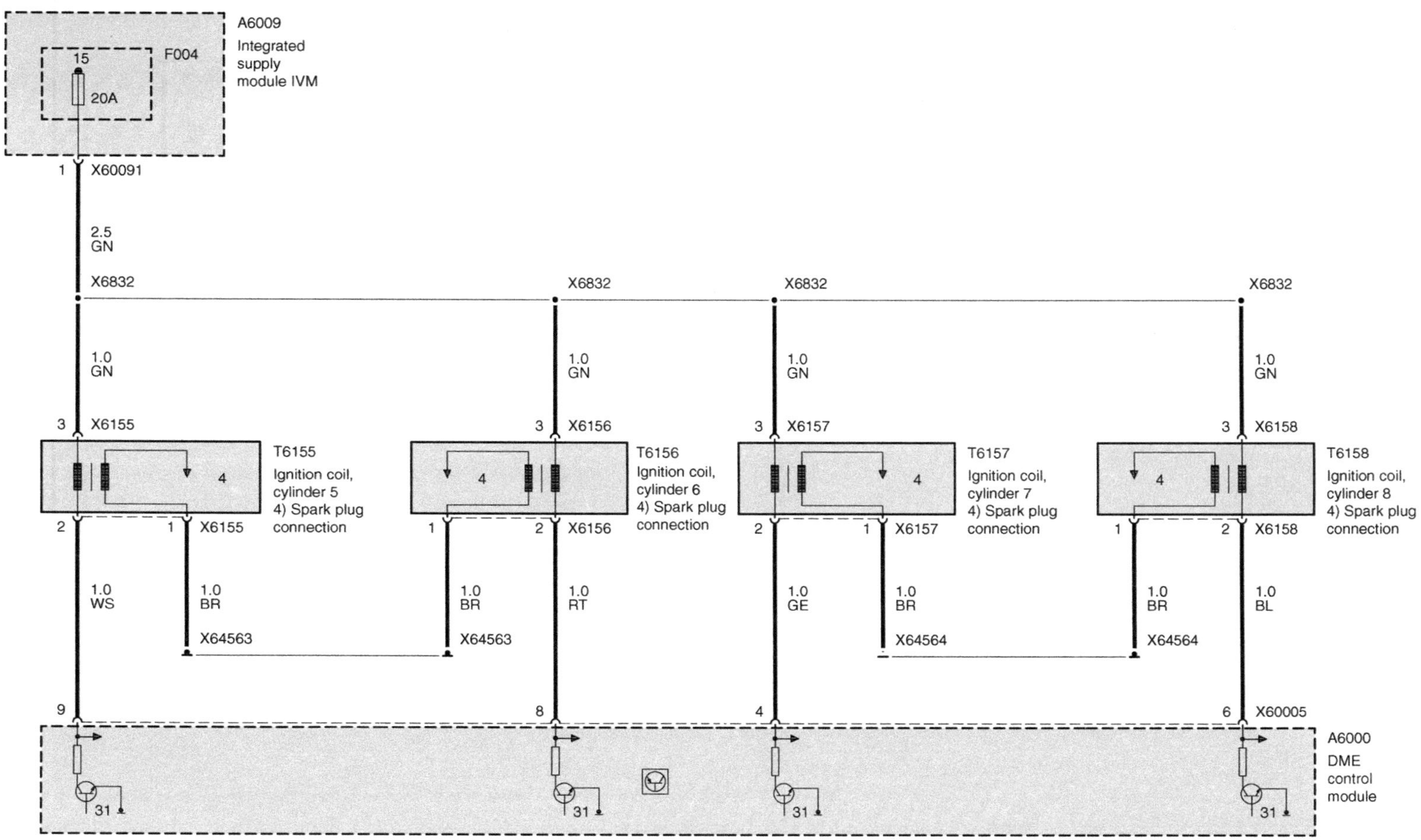

BentleyPublishers.com

B510ele15601

Oil supply
up to 09 / 2005
N62 engine or N62 TU engine

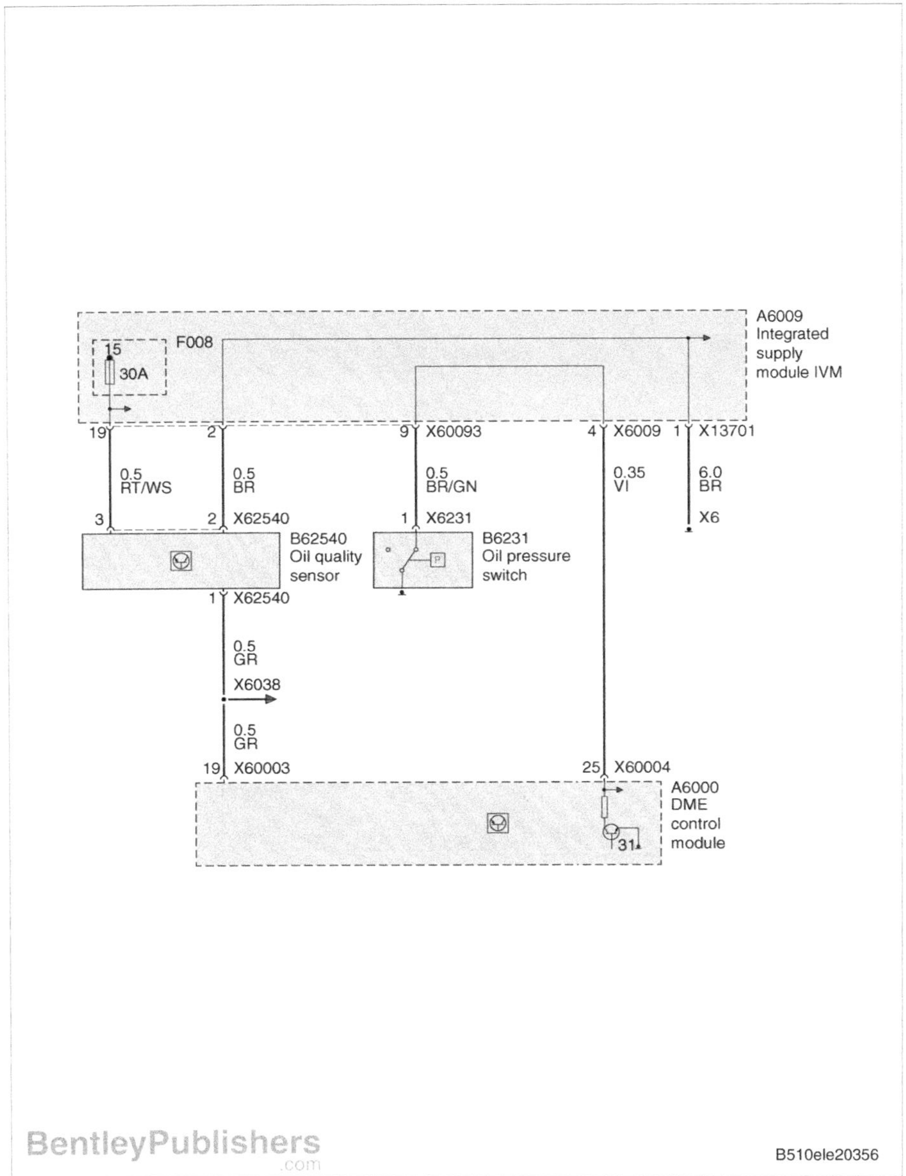

Oil supply
as of 09 / 2005
N62 engine or N62 TU engine

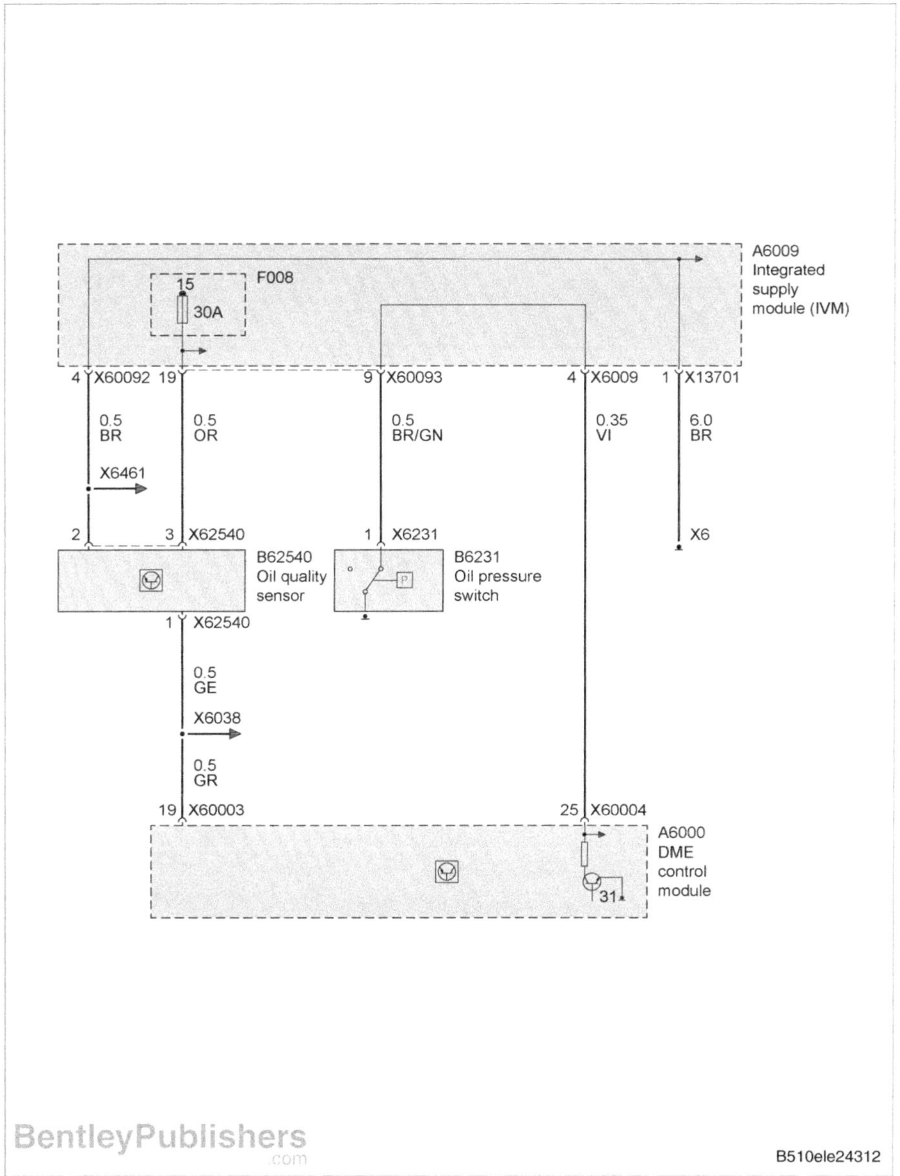

Exhaust flap
N52 and N54 engines

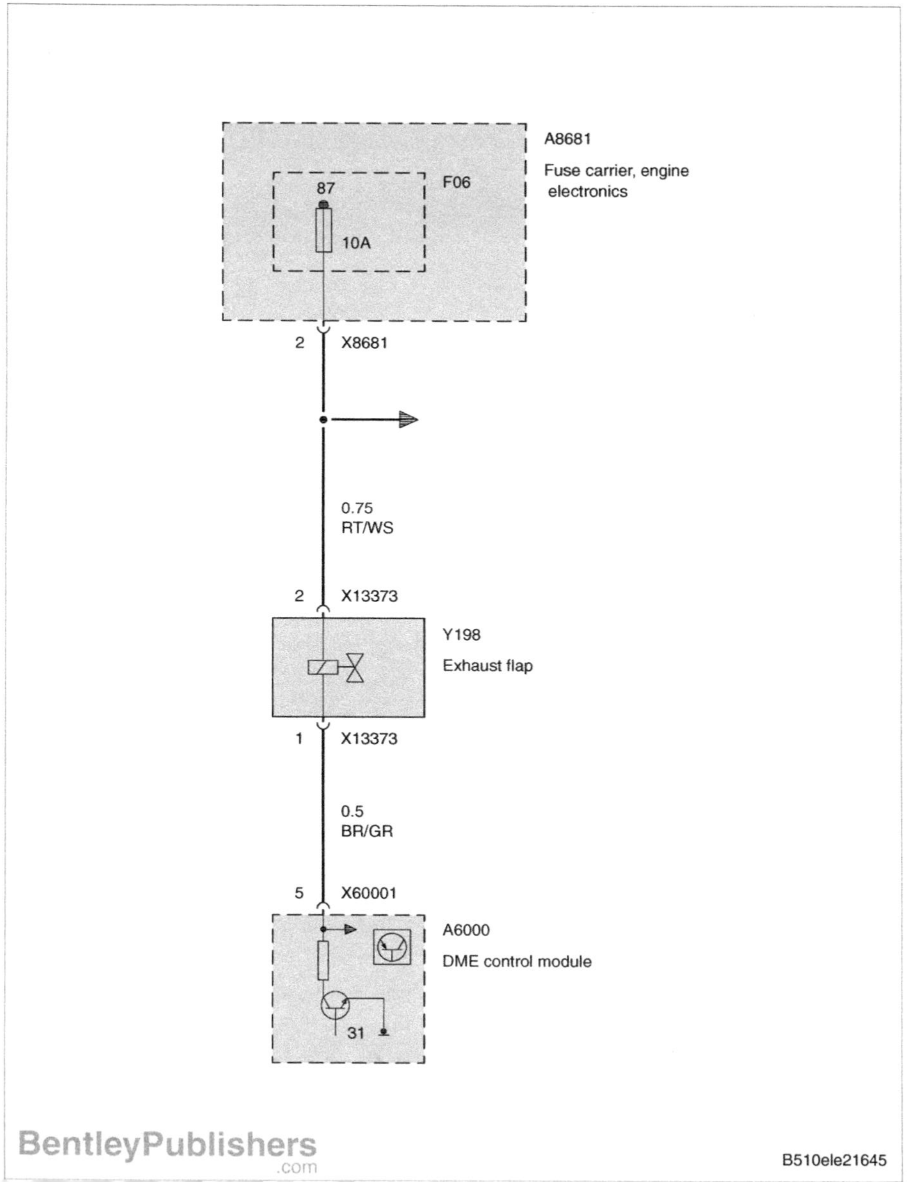

BentleyPublishers.com

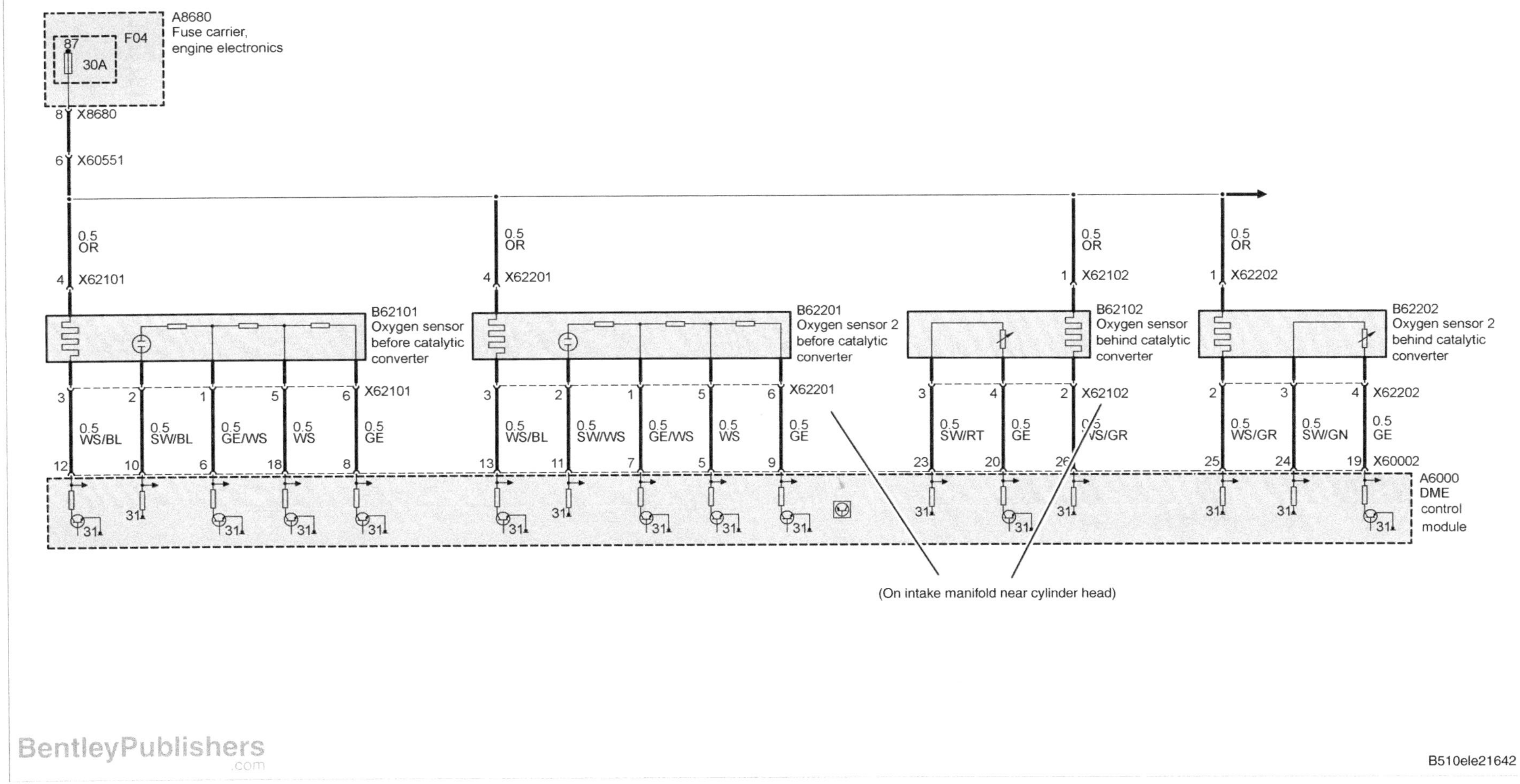

BentleyPublishers.com

Valve gear
N52 engine
(page 1 of 2)

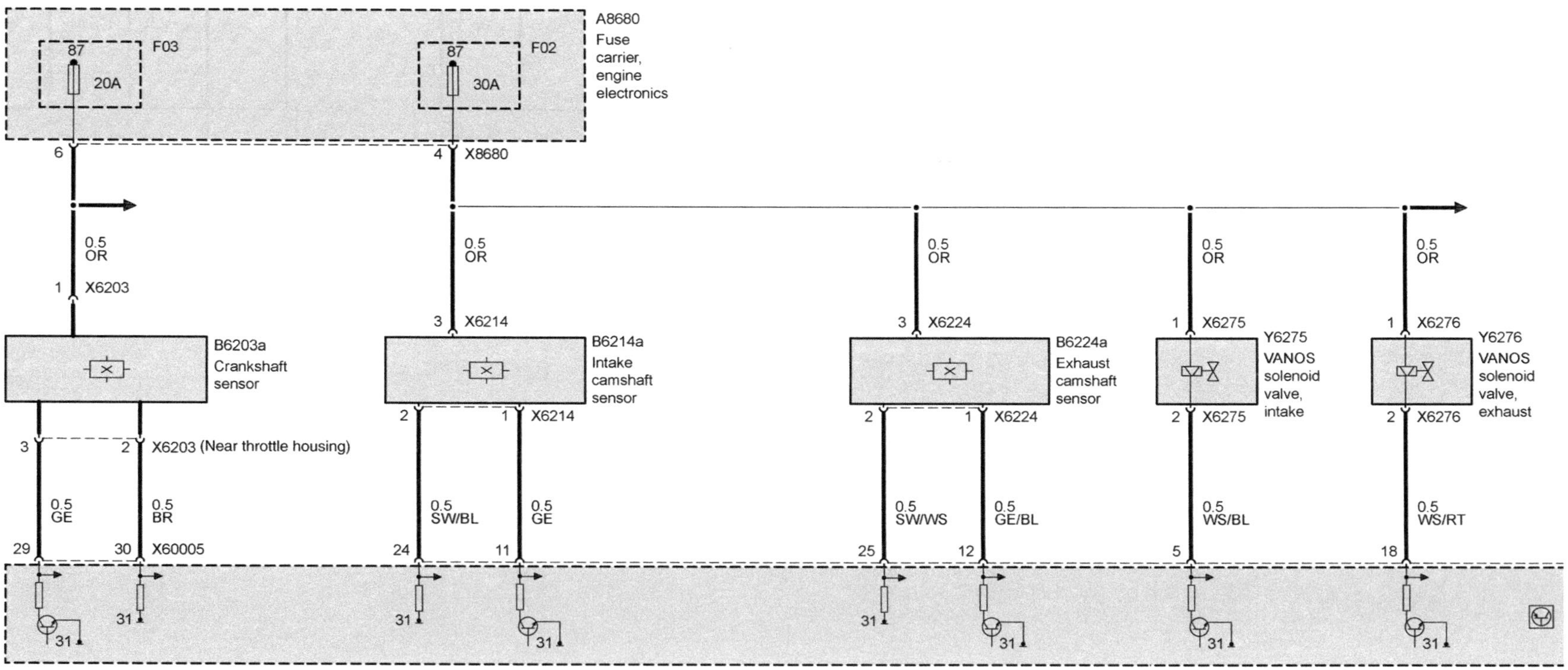

BentleyPublishers.com

B51021650a

Valve gear
N52 engine
(page 2 of 2)

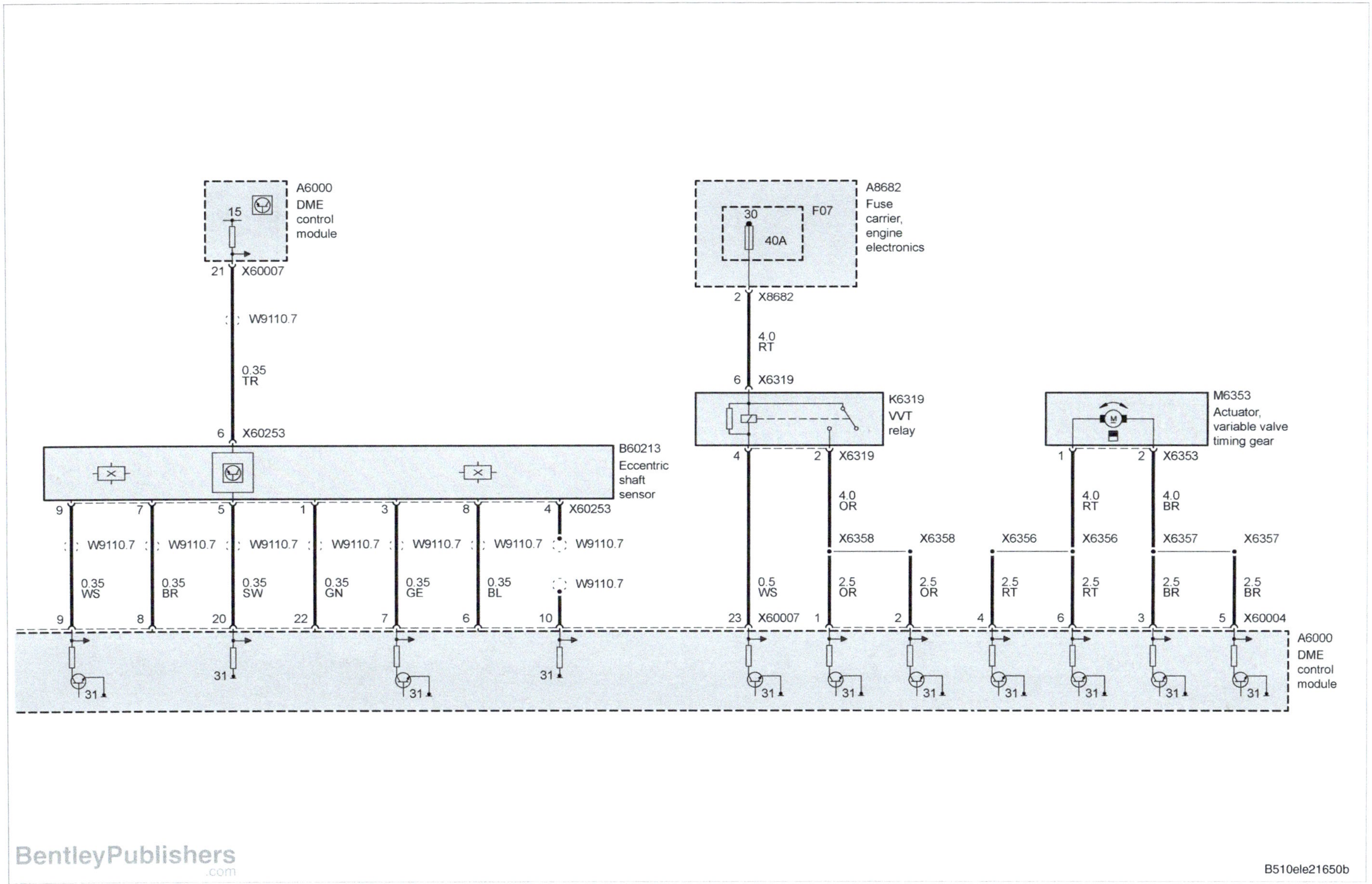

Crankshaft sensor
N54 engine

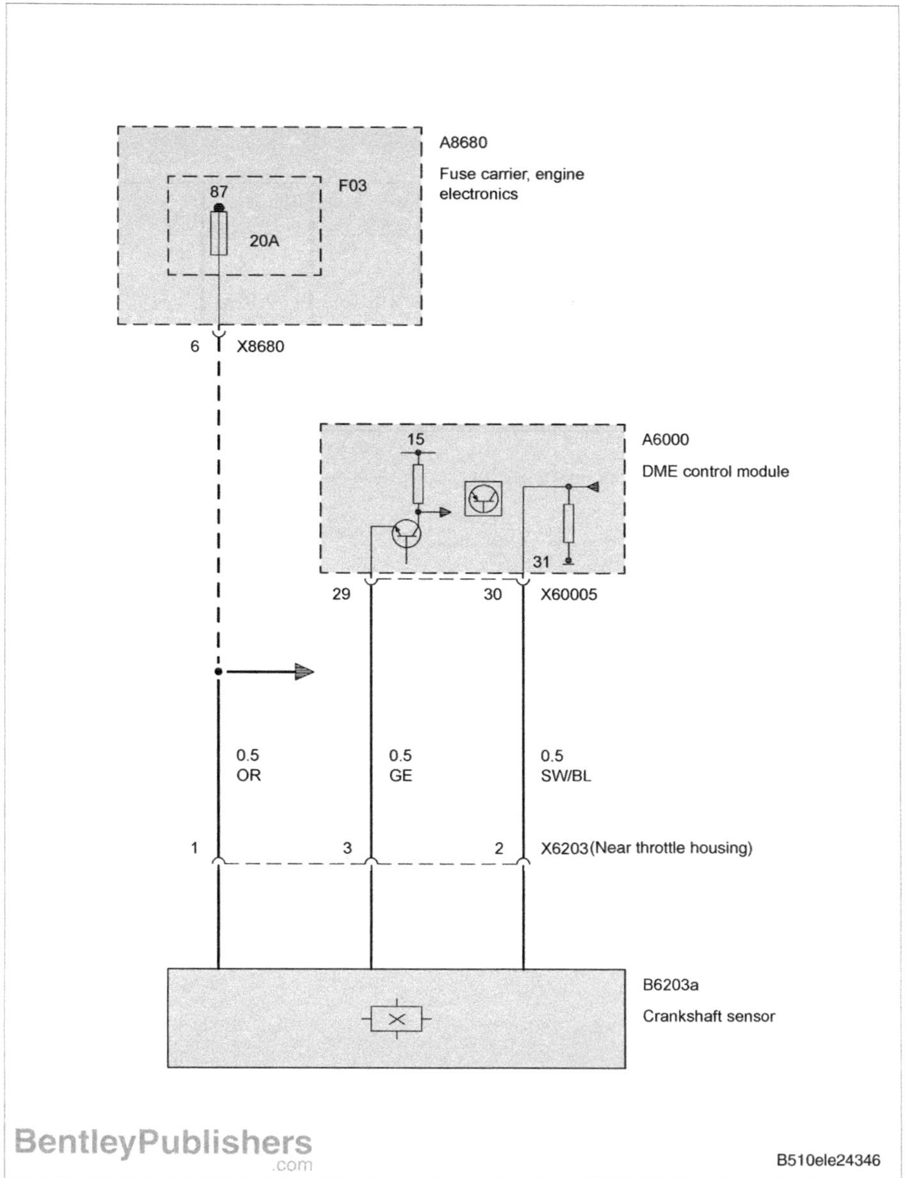

E-box fan
N52 and N54 engines

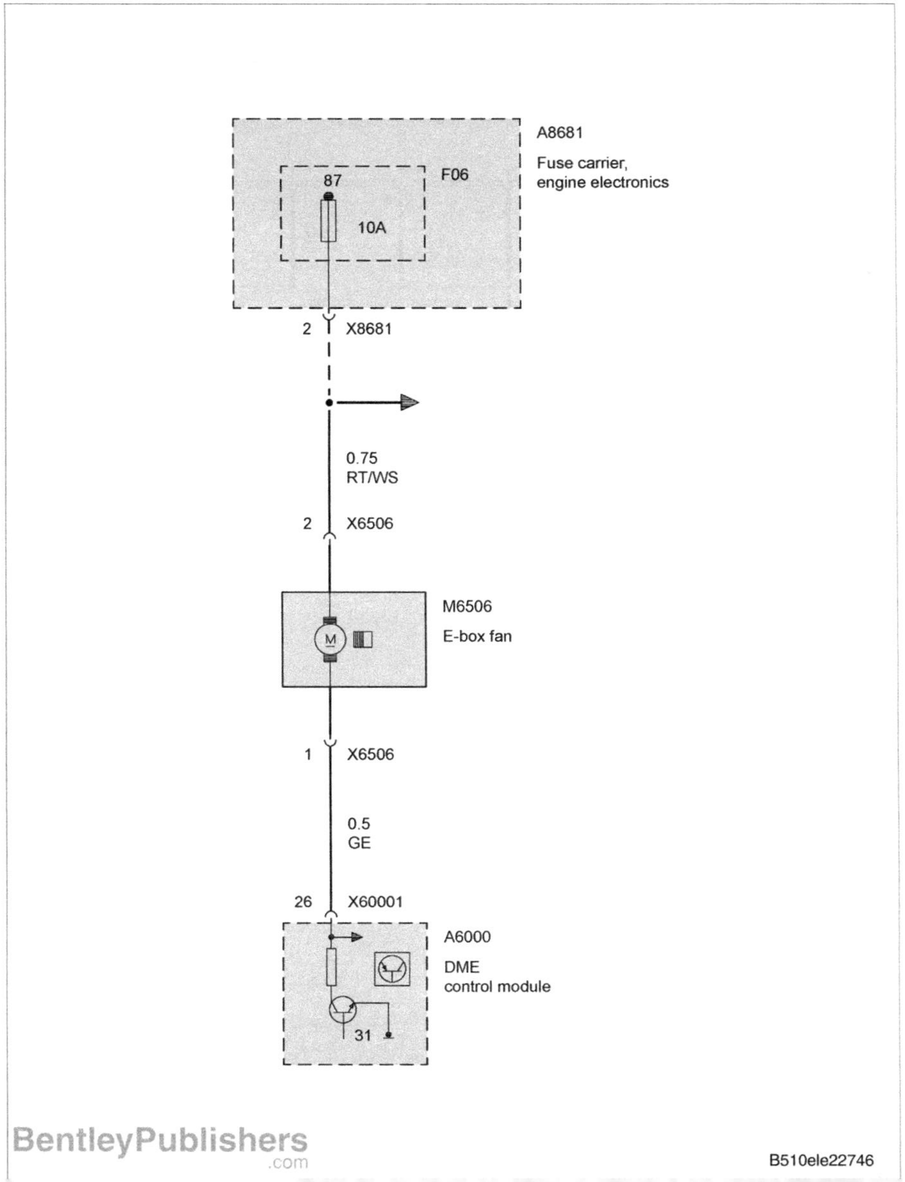

BentleyPublishers.com

Accelerator pedal module, clutch switch module, brake light switch N54 engine

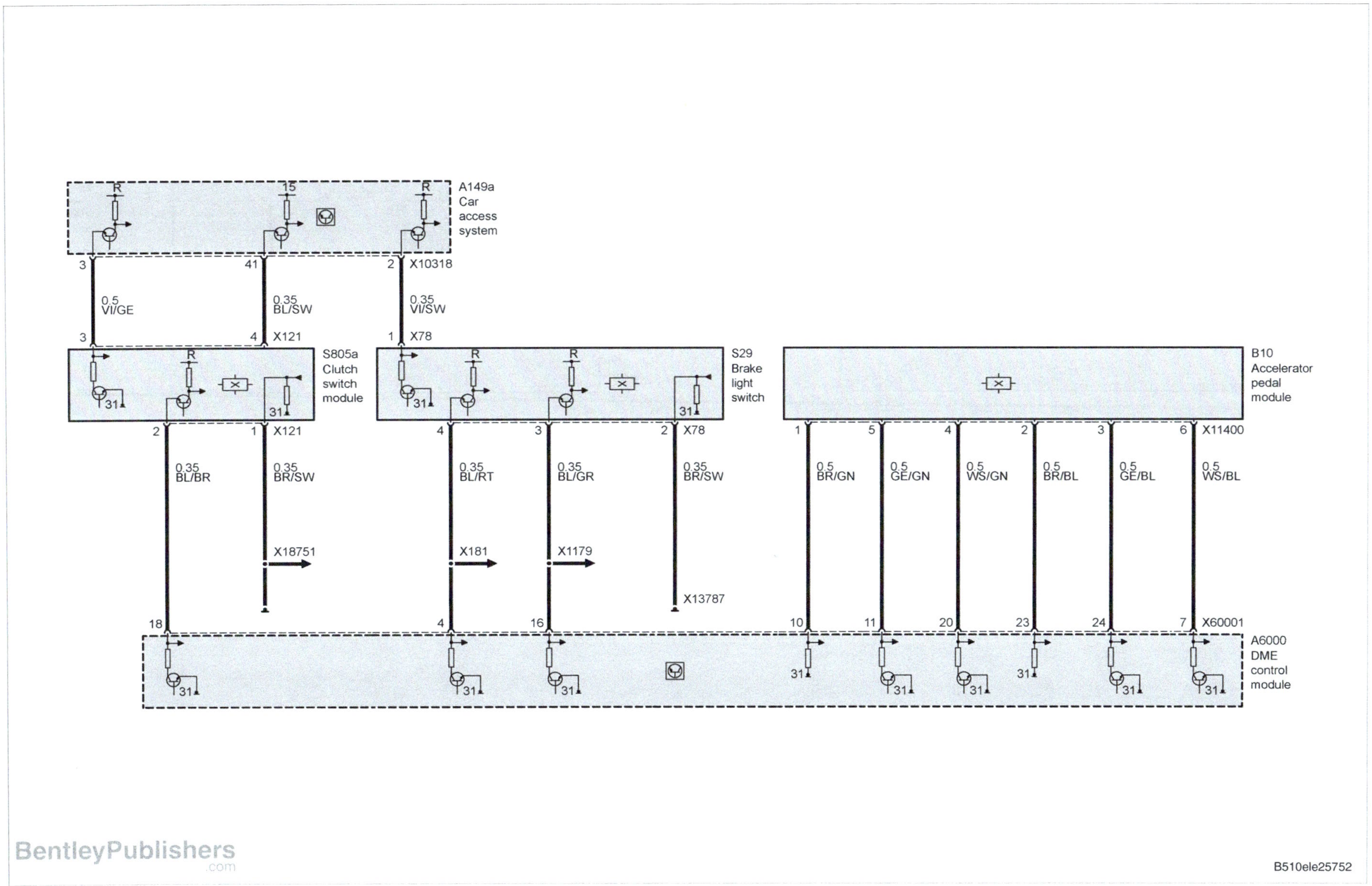

Brake light switch
N52 engine with DSC Dynamic Stability Control PLUS

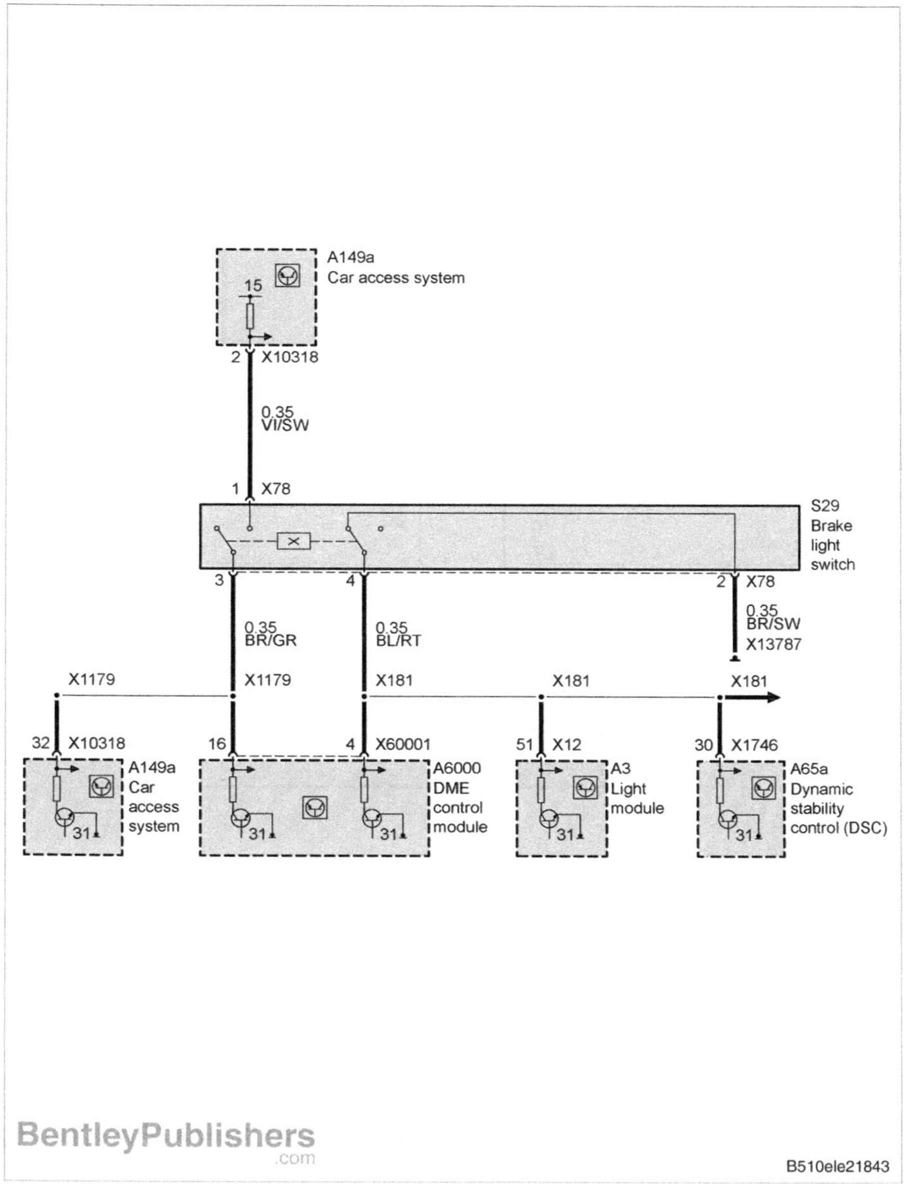

Brake light switch
N52 engine with DSC Dynamic Stability Control Premium

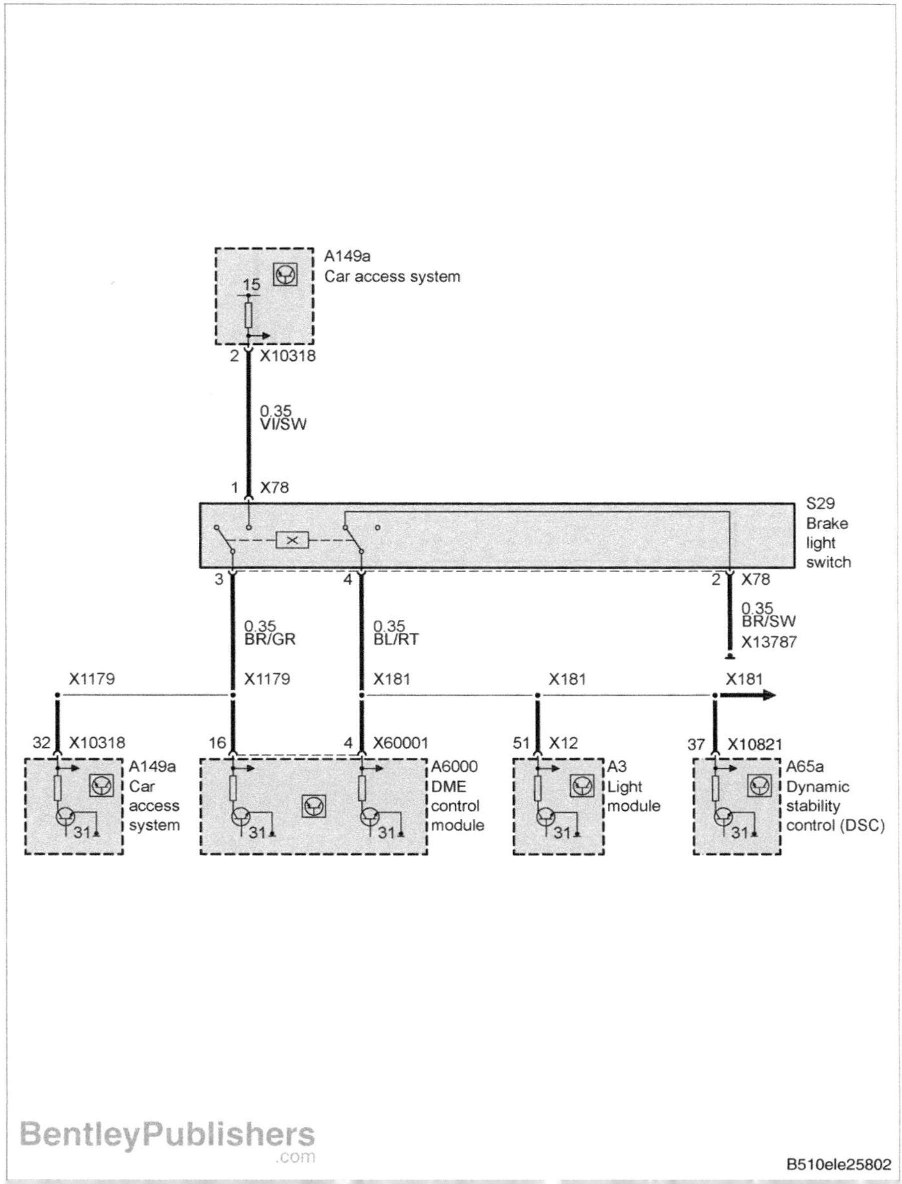

BentleyPublishers.com
BentleyPublishers.com

Clutch switch module except N54 engine

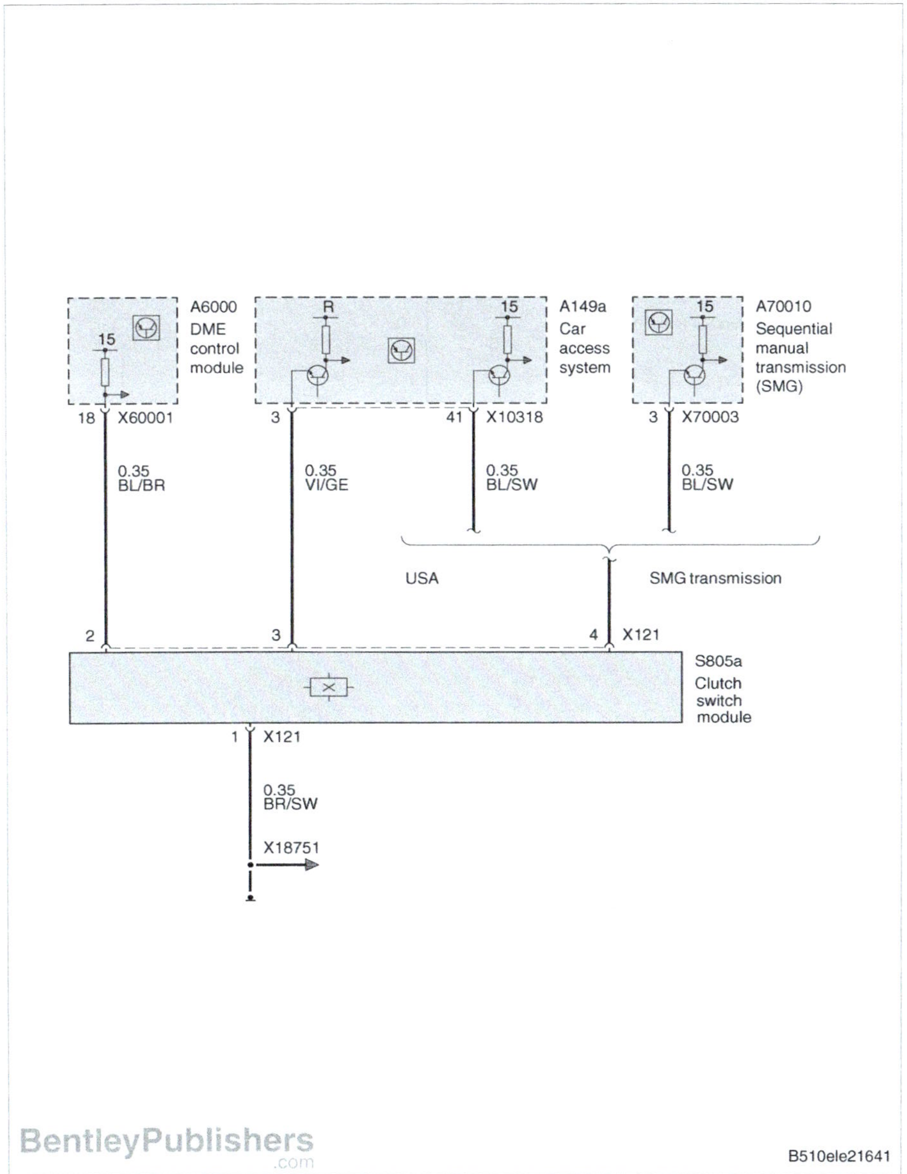

Accelerator pedal module N52 engine

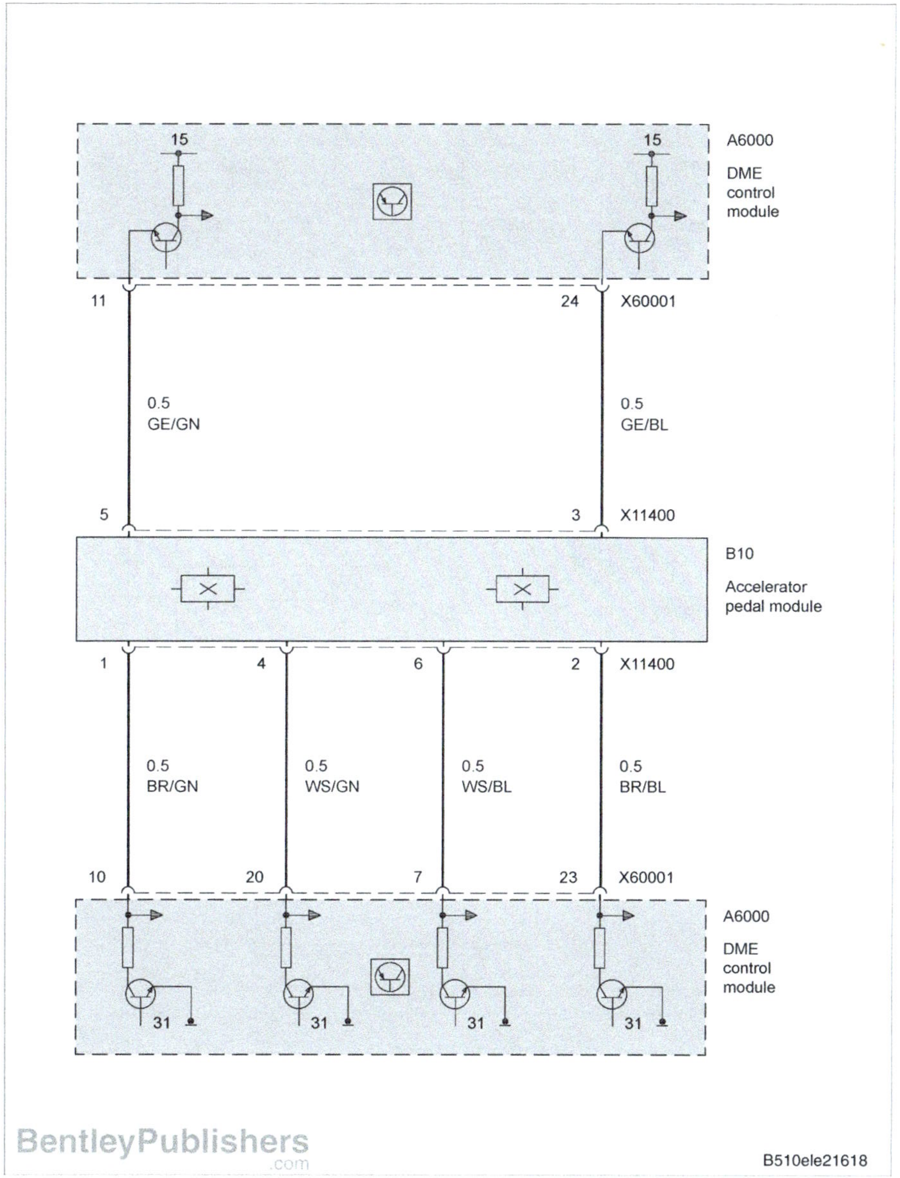

Knock sensors
N52 and N54 engines

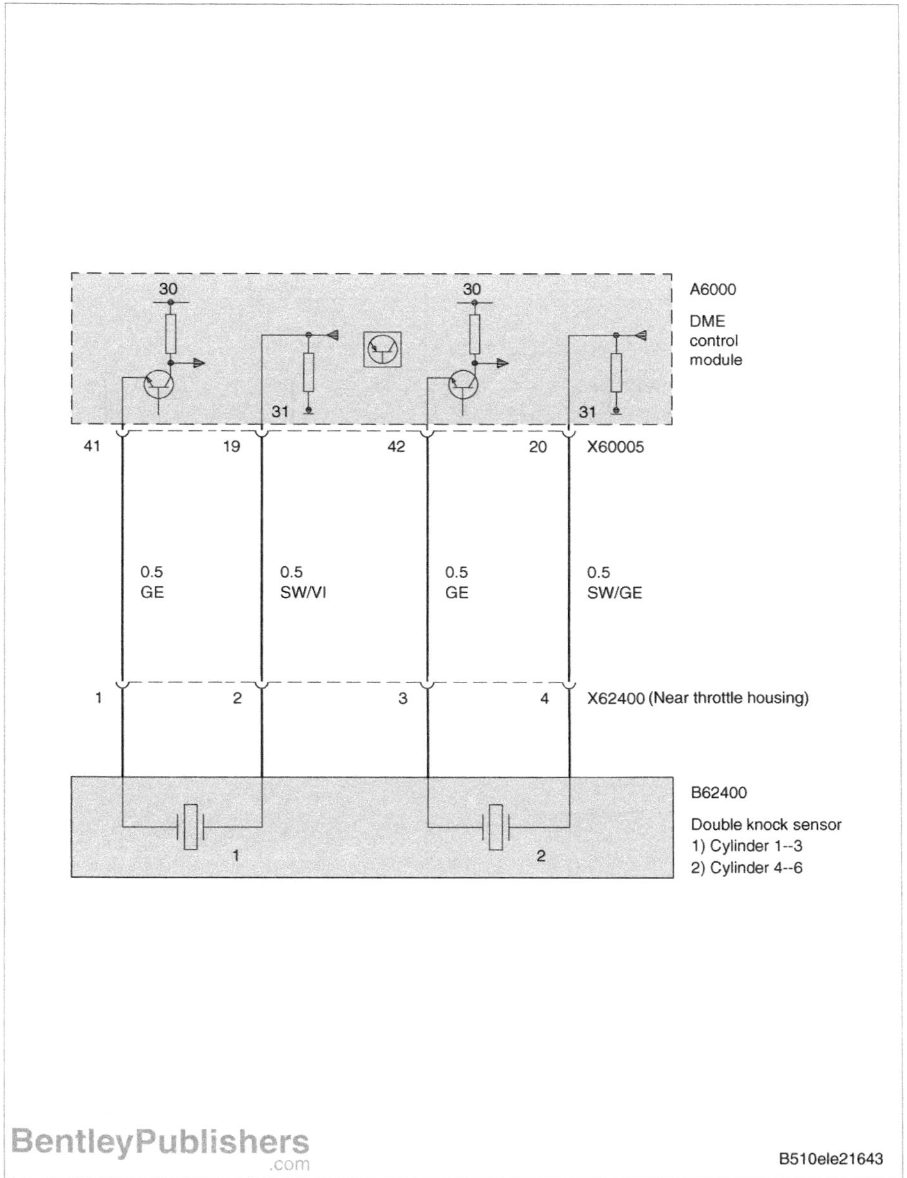

High pressure fuel system
N52 and N54 engines

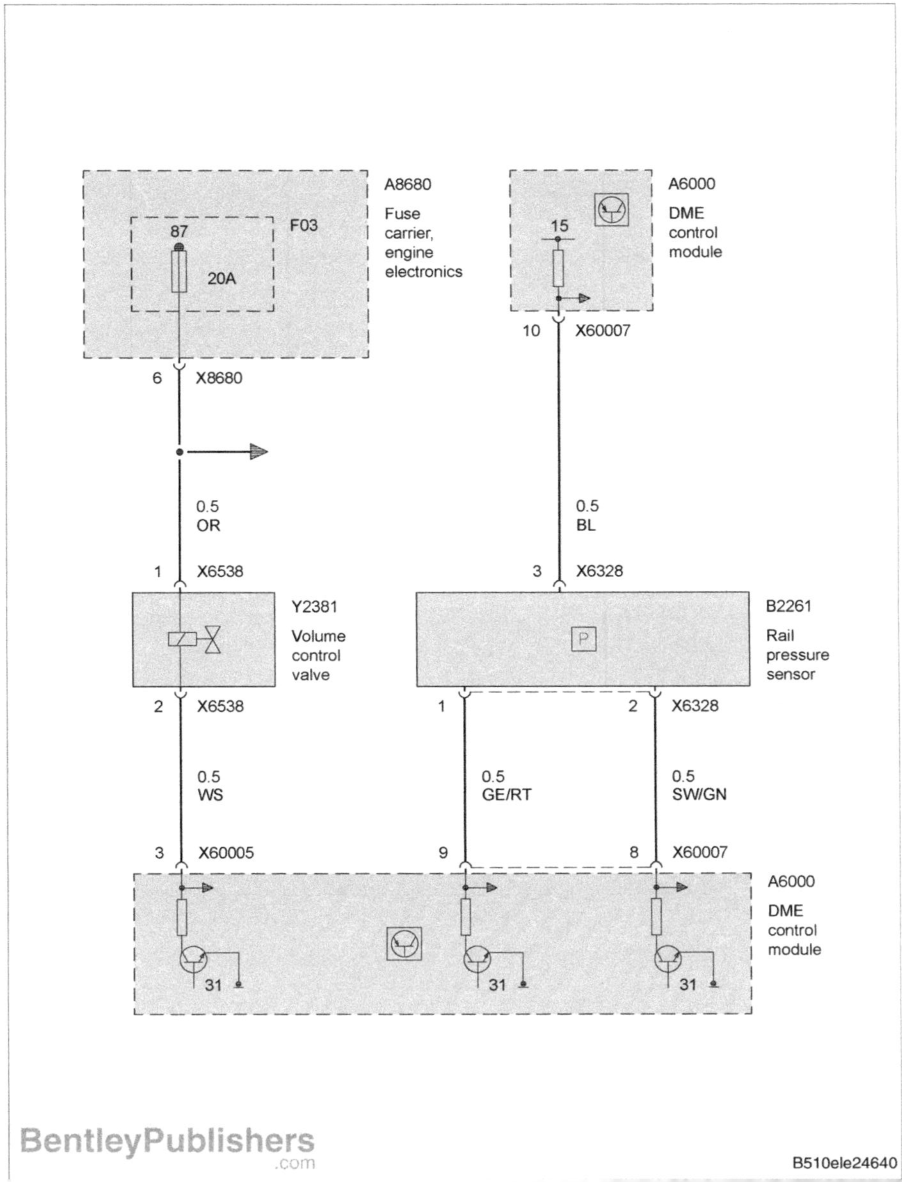

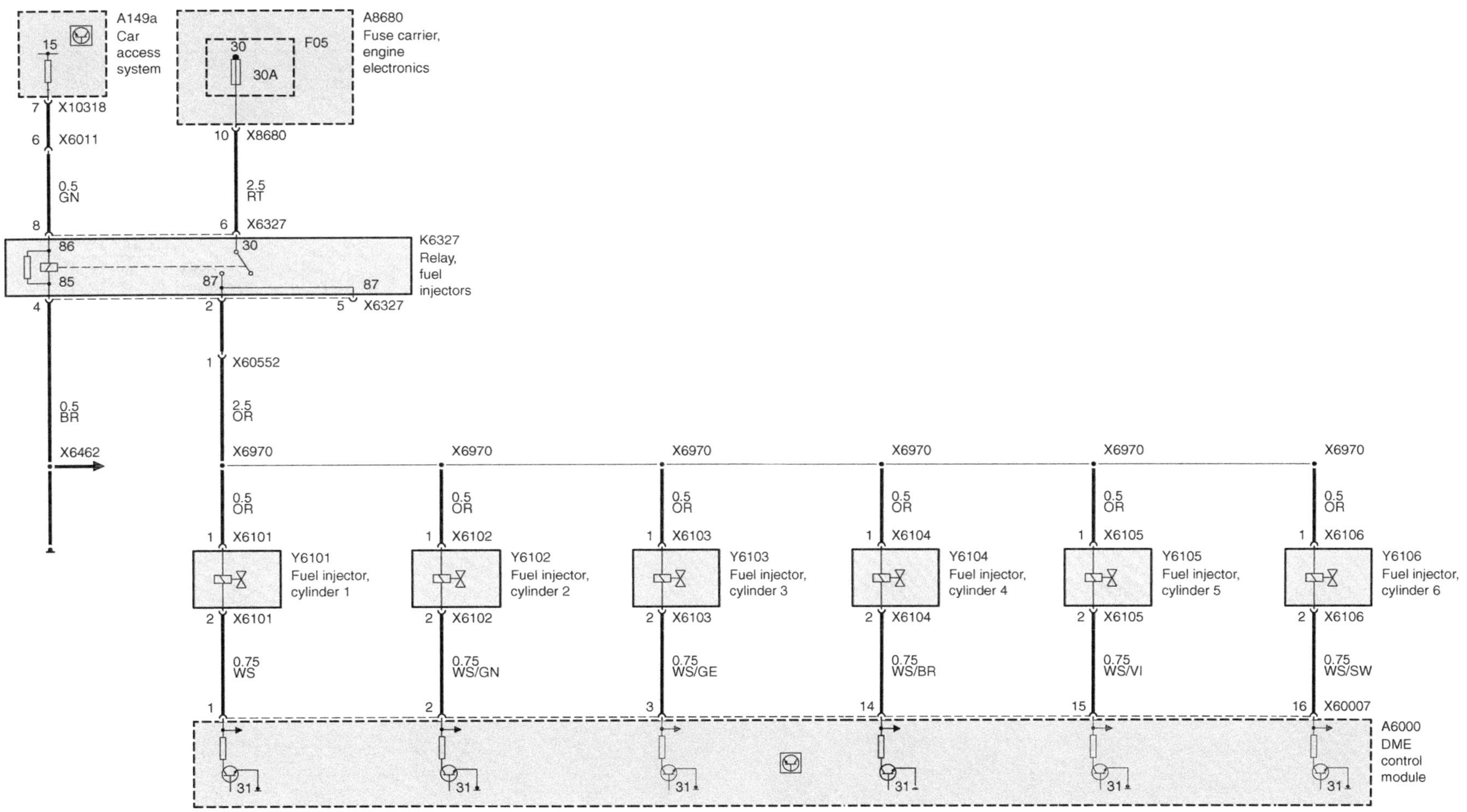

BentleyPublishers
.com

B510ele21640

Fuel injectors
N54 engine

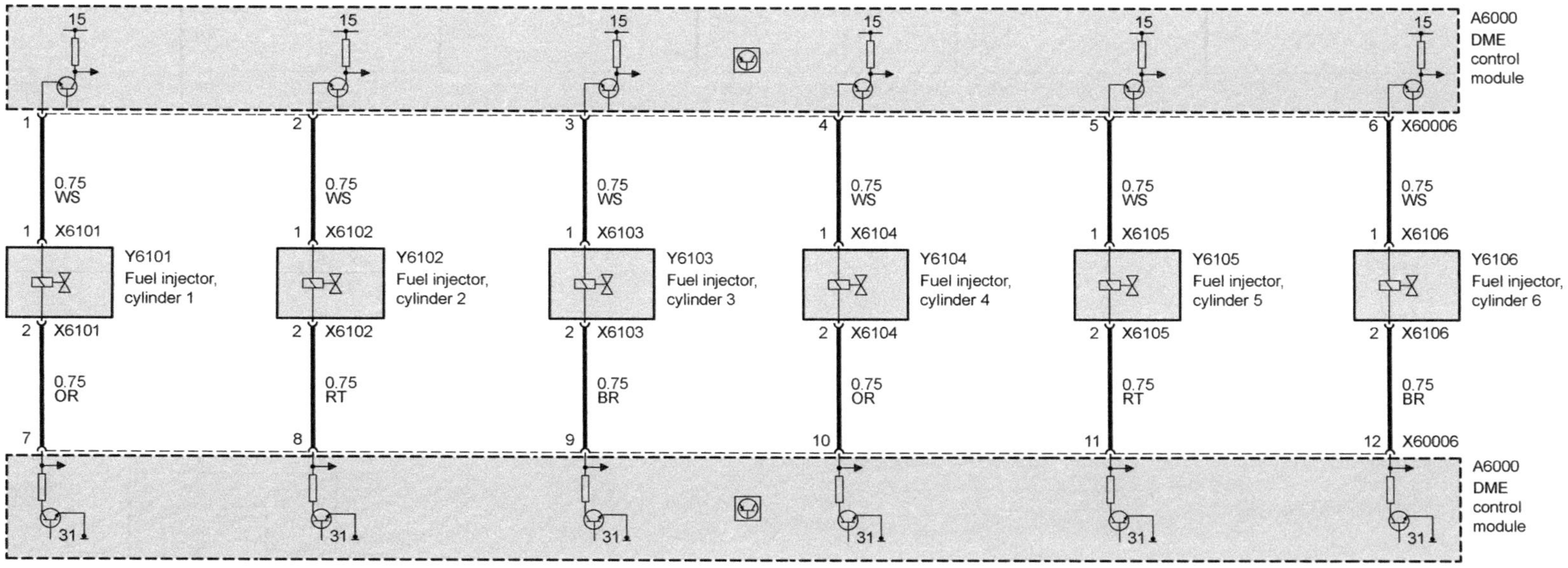

BentleyPublishers
.com

B510ele24342

Fuel pump
N52 engine as of 03 / 2004 up to 09 / 2005

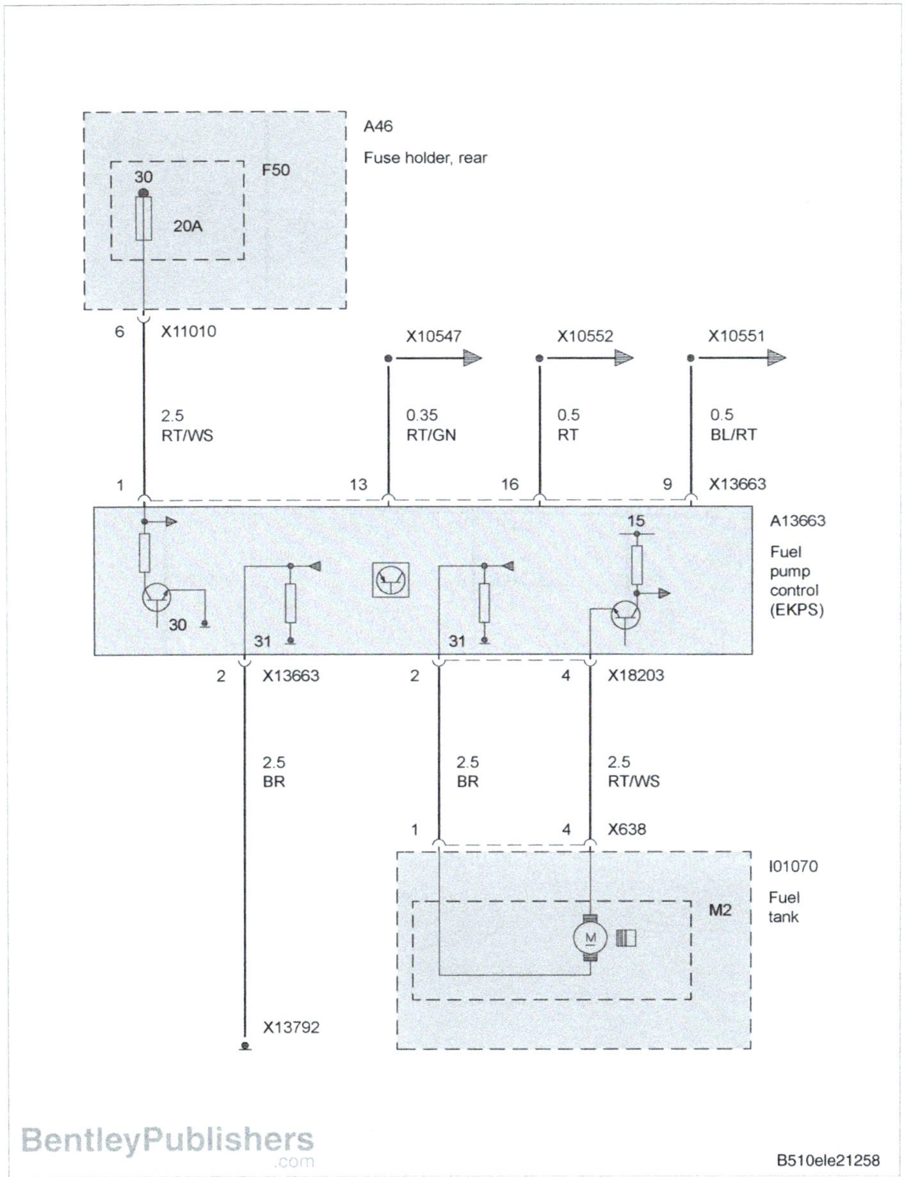

Throttle valve
N52 engine as of 09 / 2005

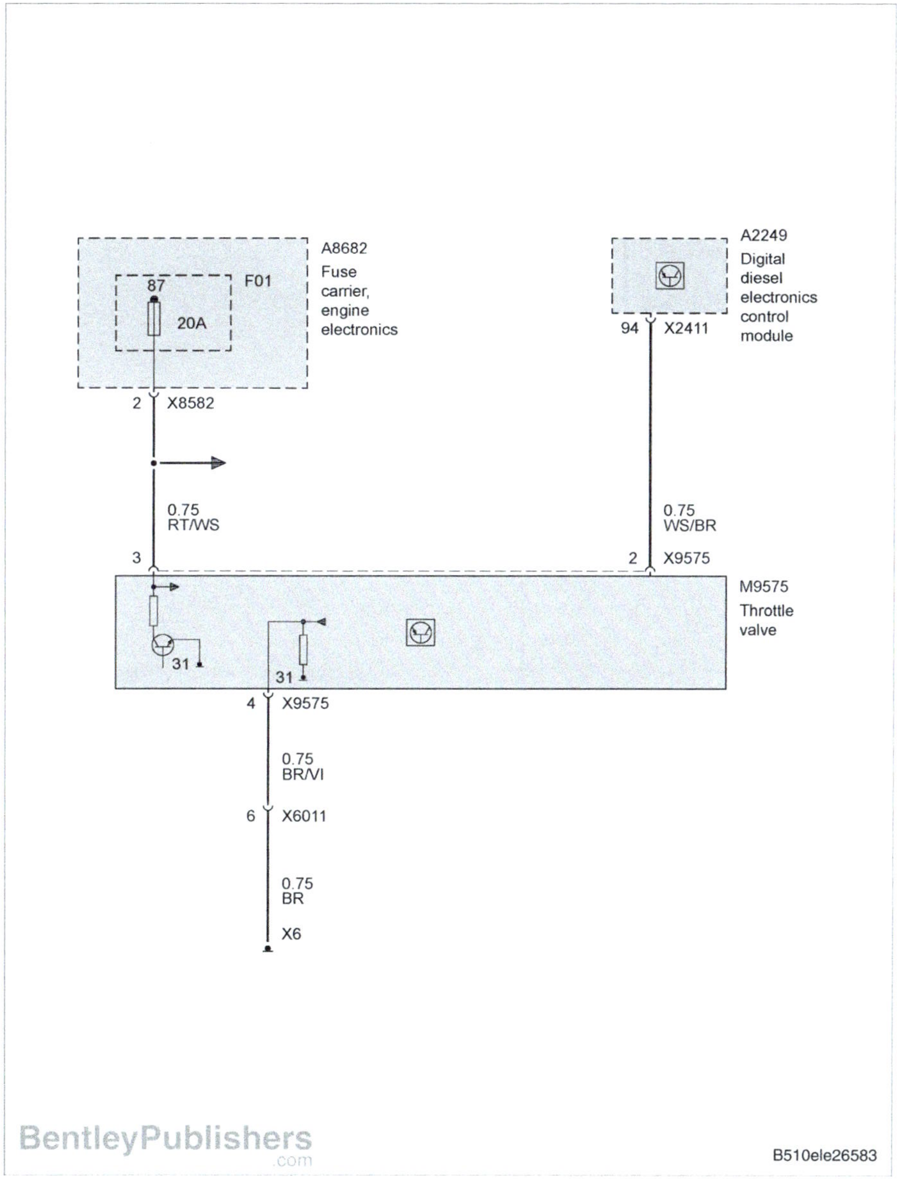

Low pressure fuel system
N54 engine

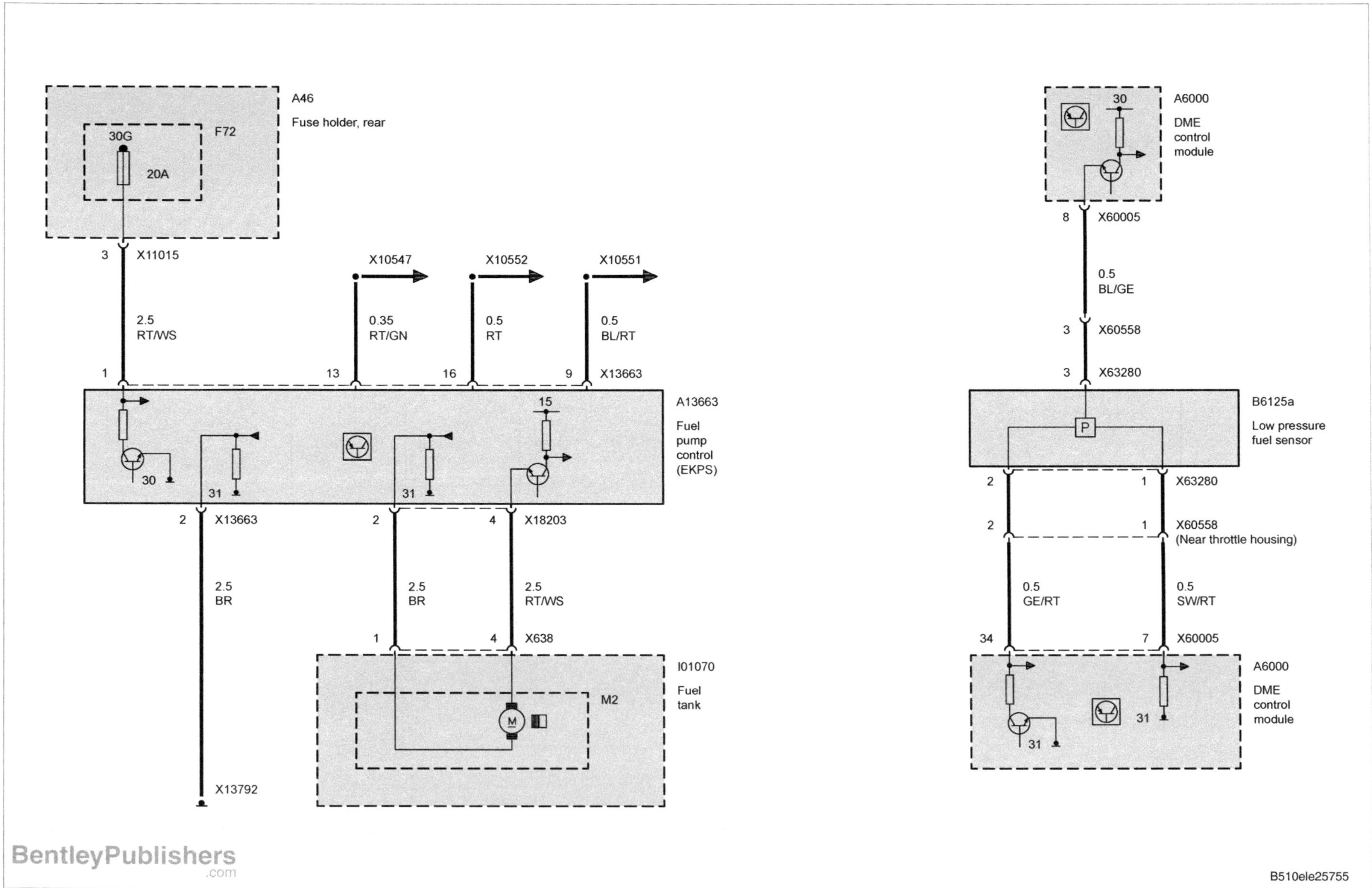

BentleyPublishers
.com

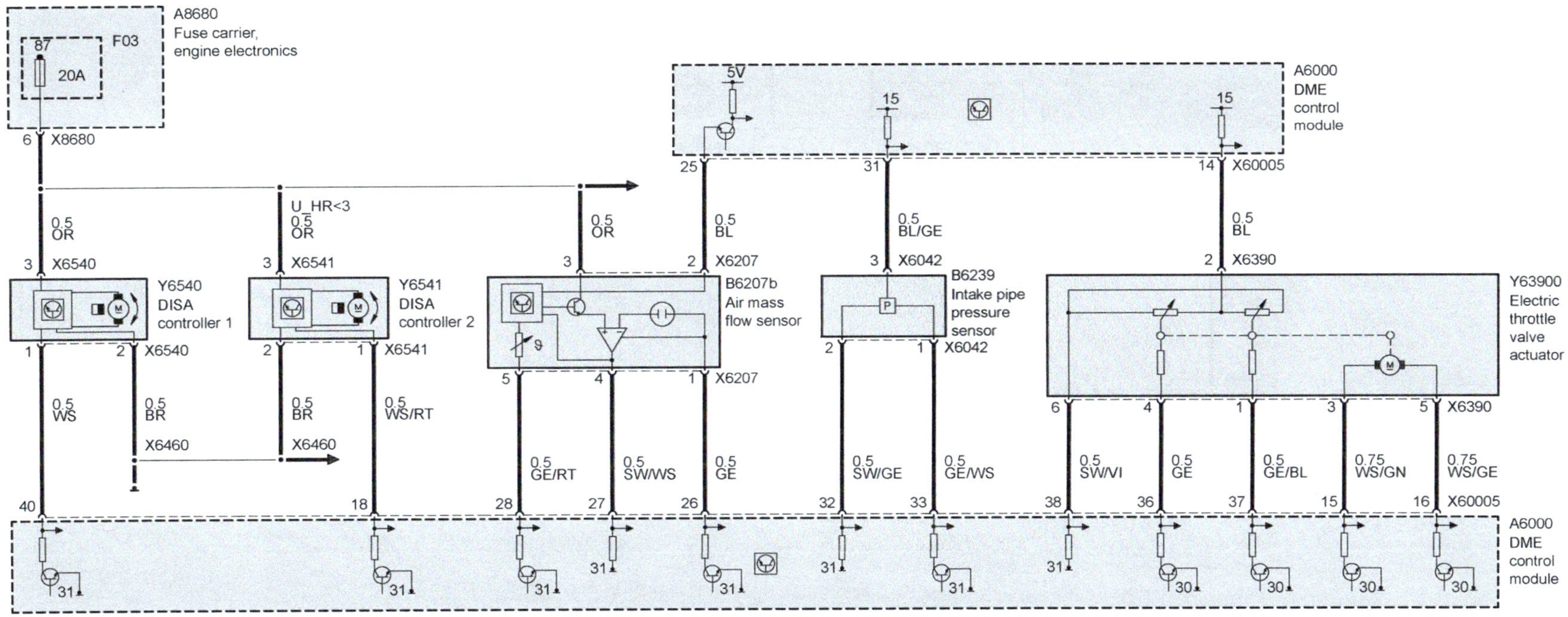

BentleyPublishers
.com

B510ele21644

Air supply
N54 engine

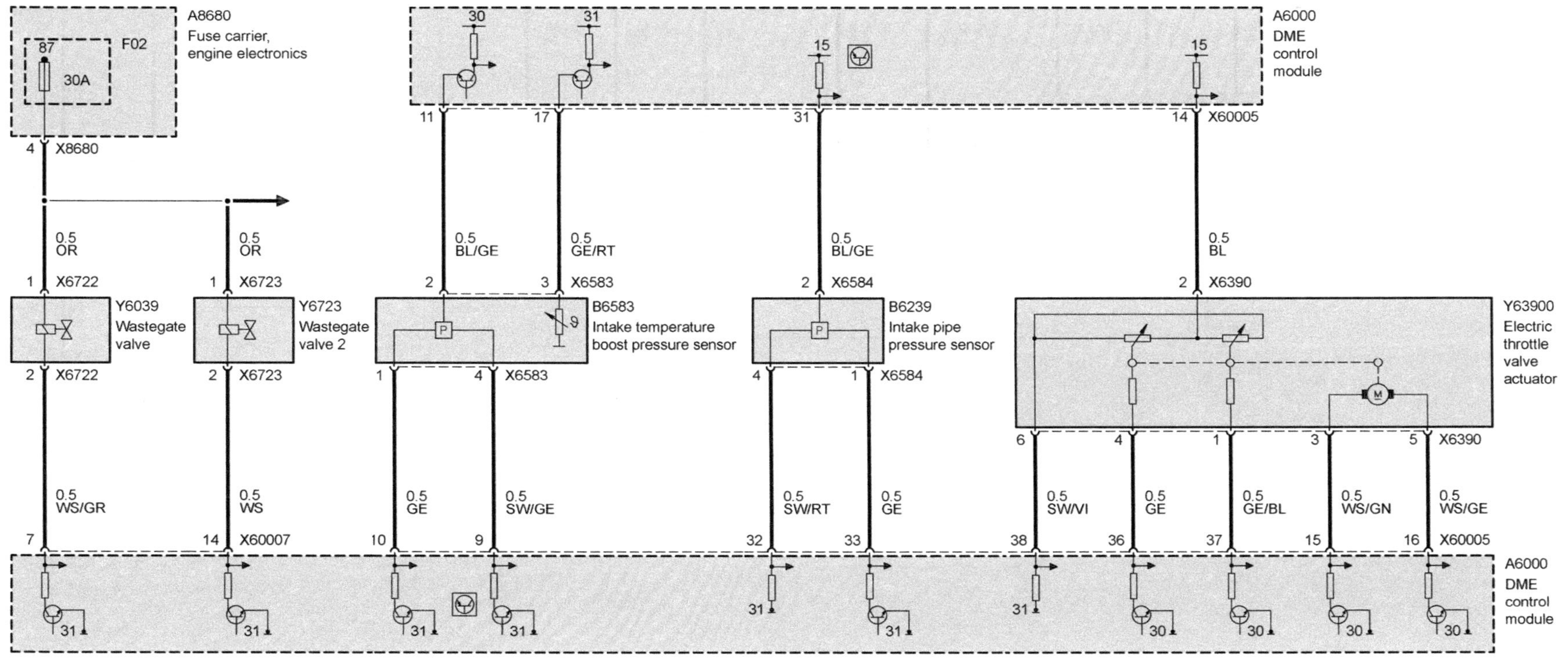

BentleyPublishers.com

B510ele24345

Engine cooling system
N52 engine up to 03 / 2007

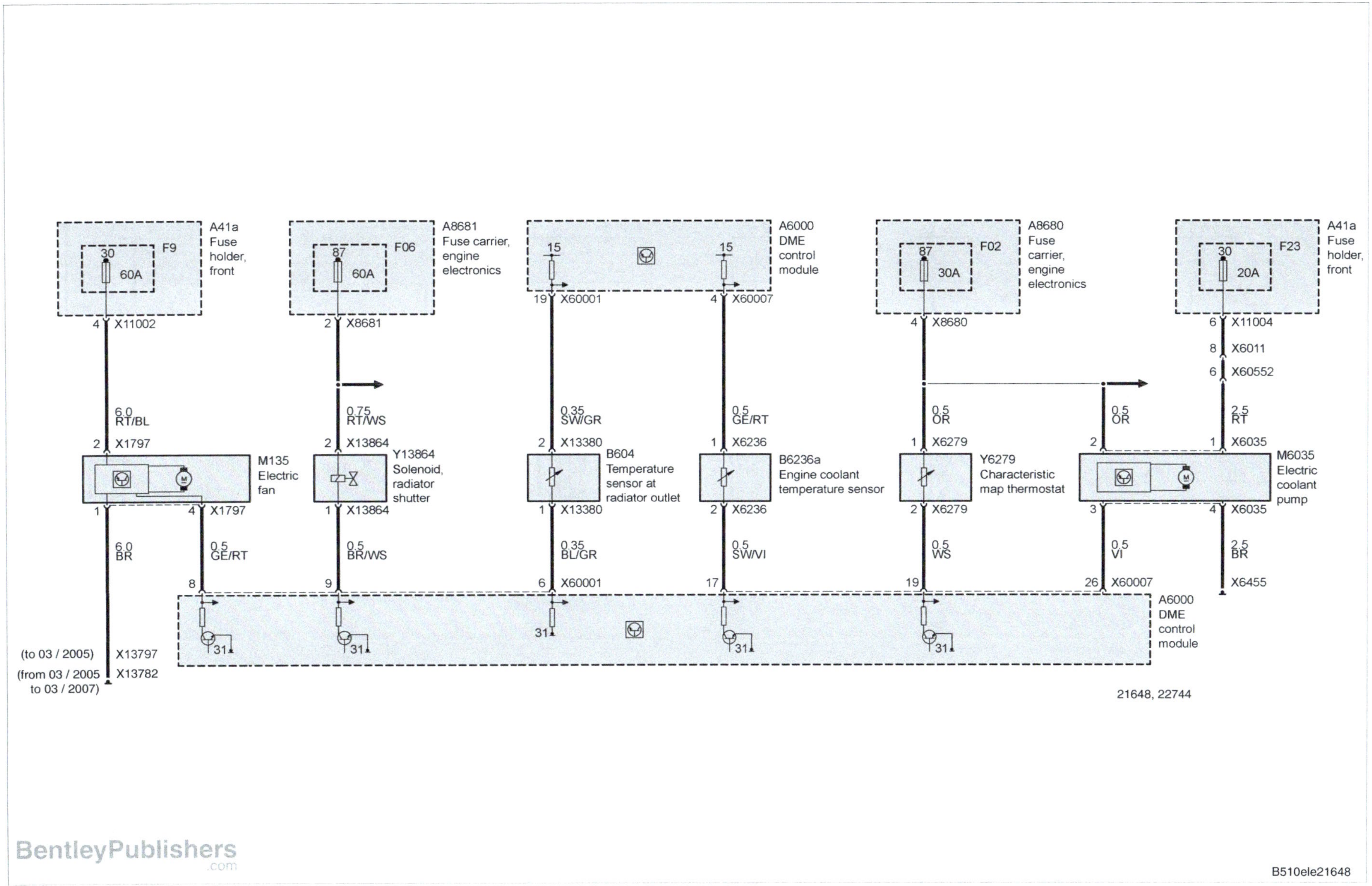

BentleyPublishers.com

Engine cooling system
N52 engine as of 03 / 2007

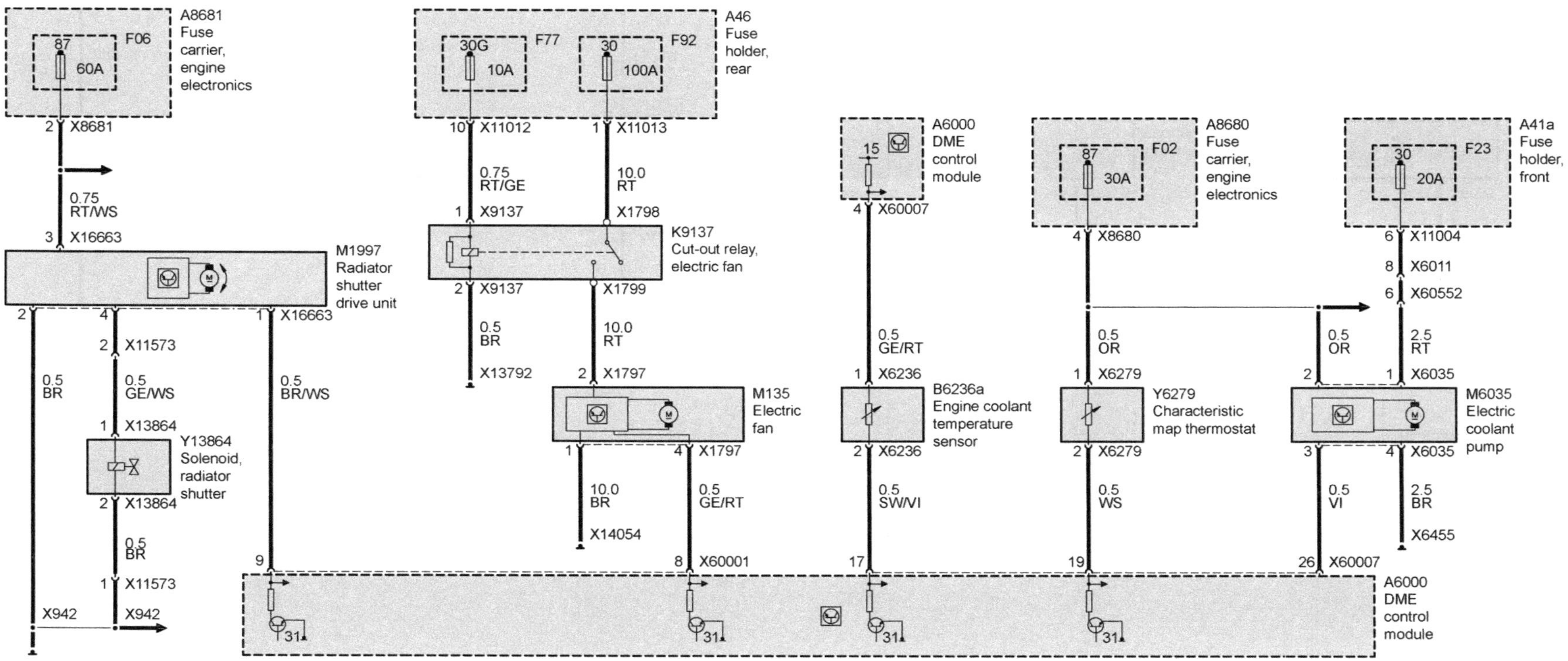

B510ele25749

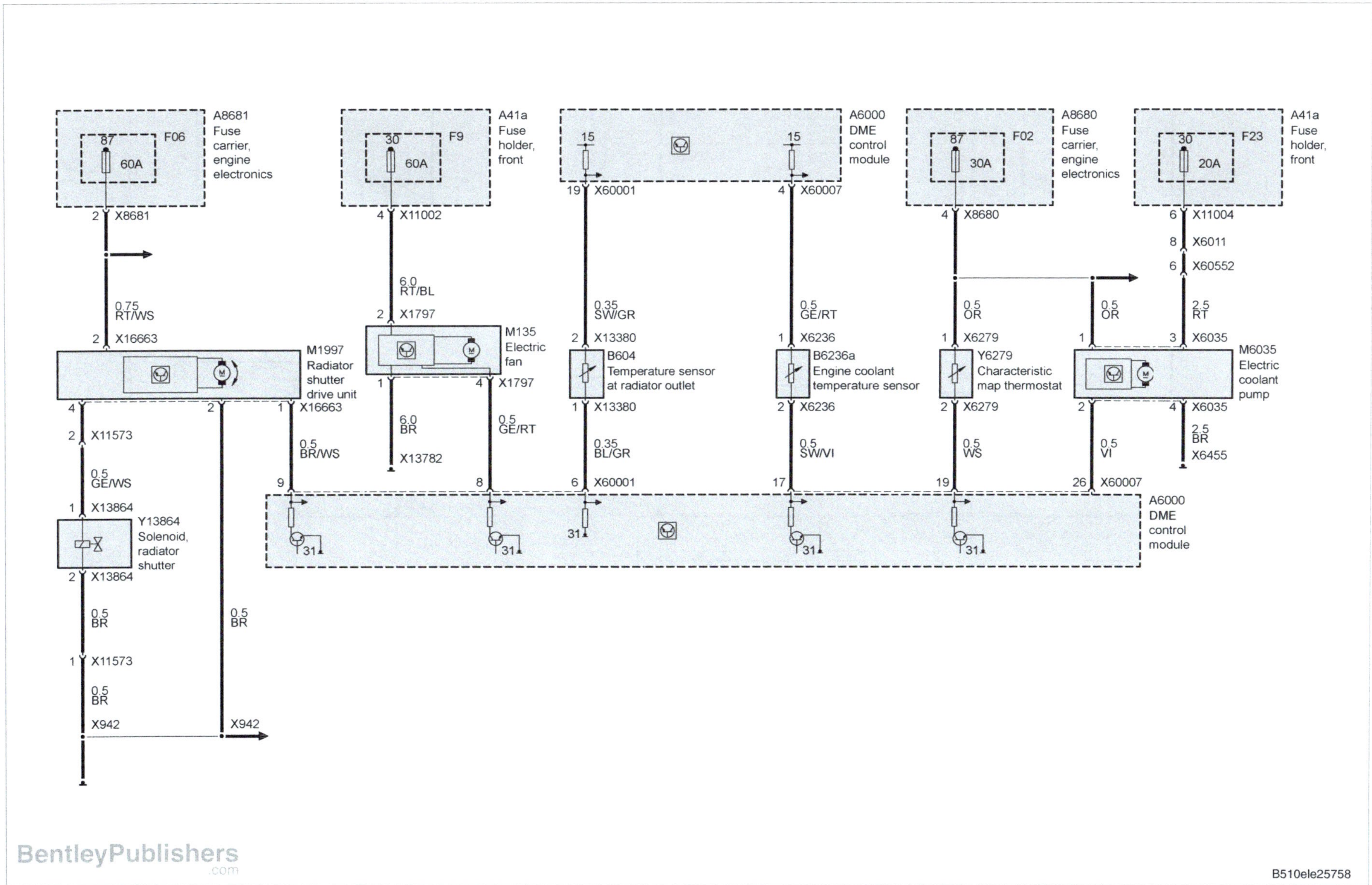

BentleyPublishers.com

Oil supply
N52 engine

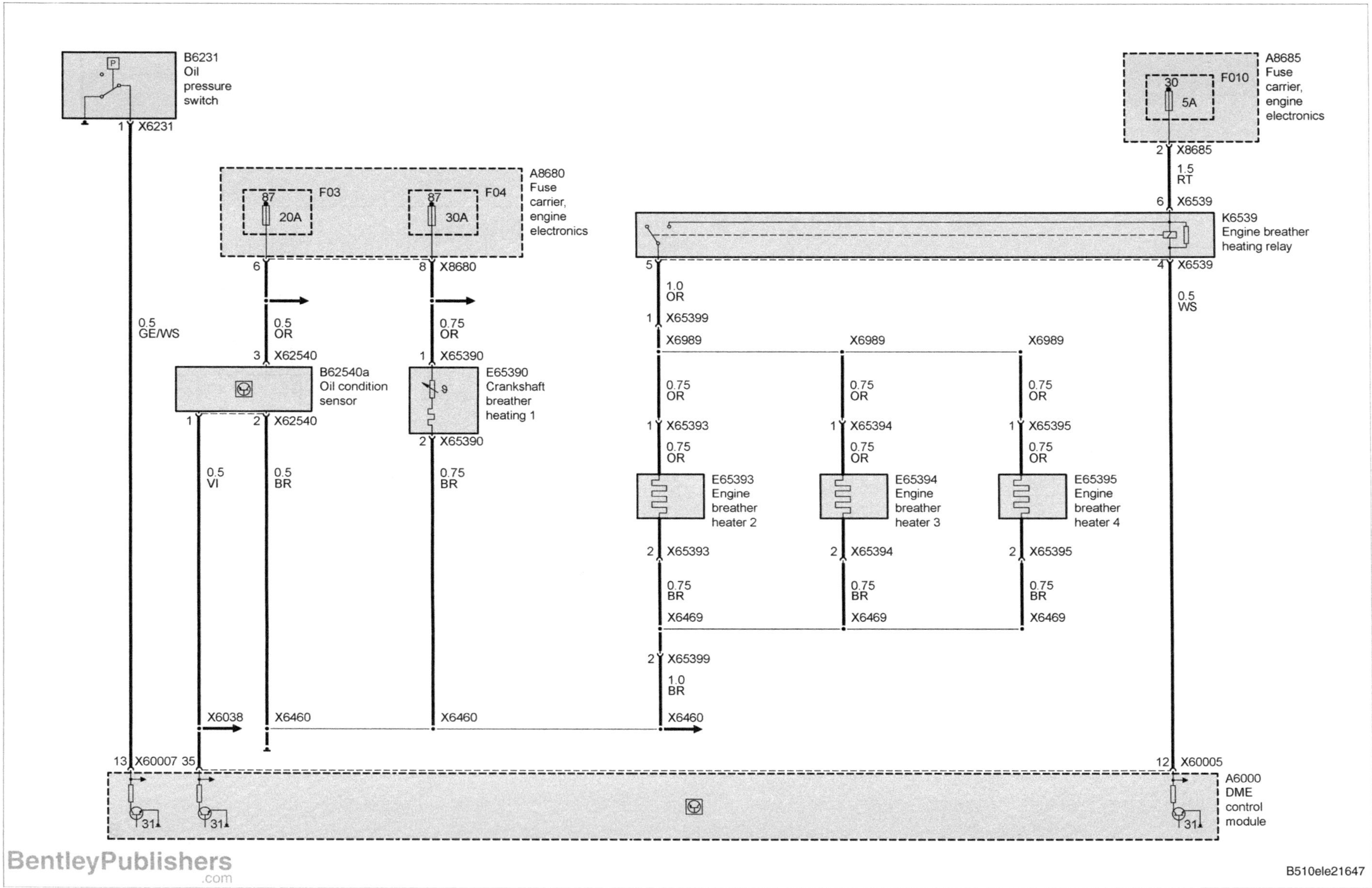

BentleyPublishers
.com

B510ele21647

Oil supply
N54 engine

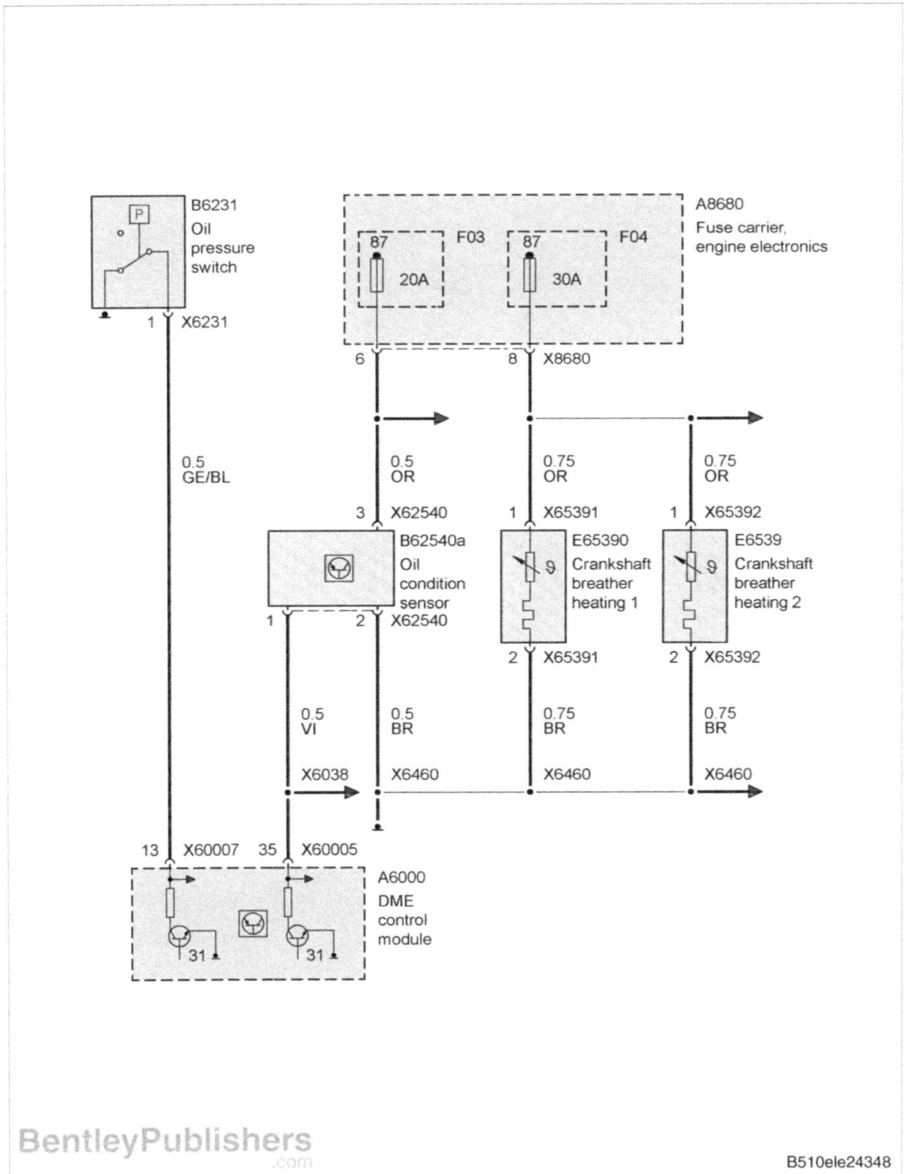

DME inteface signals
N54 engine

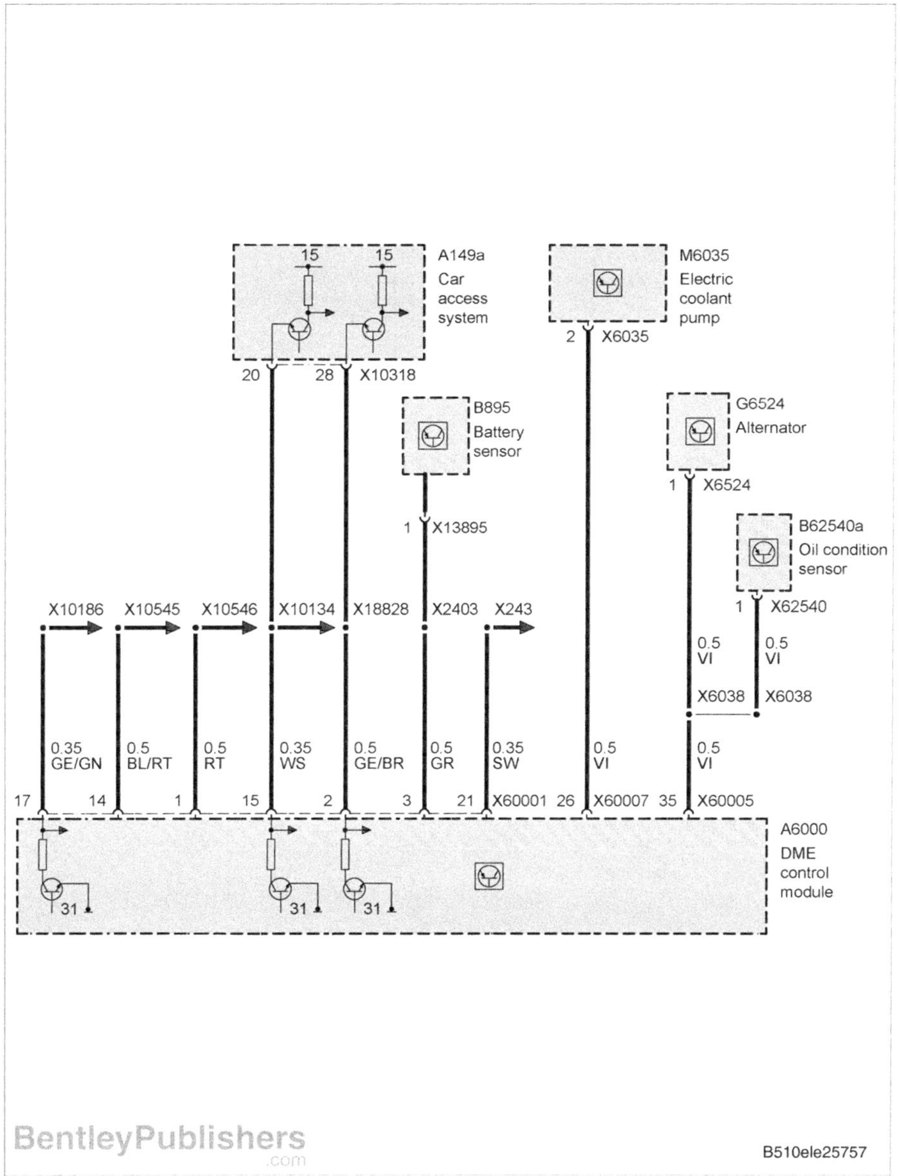

Bit serial data interface
N52 engine up to 03 / 2007

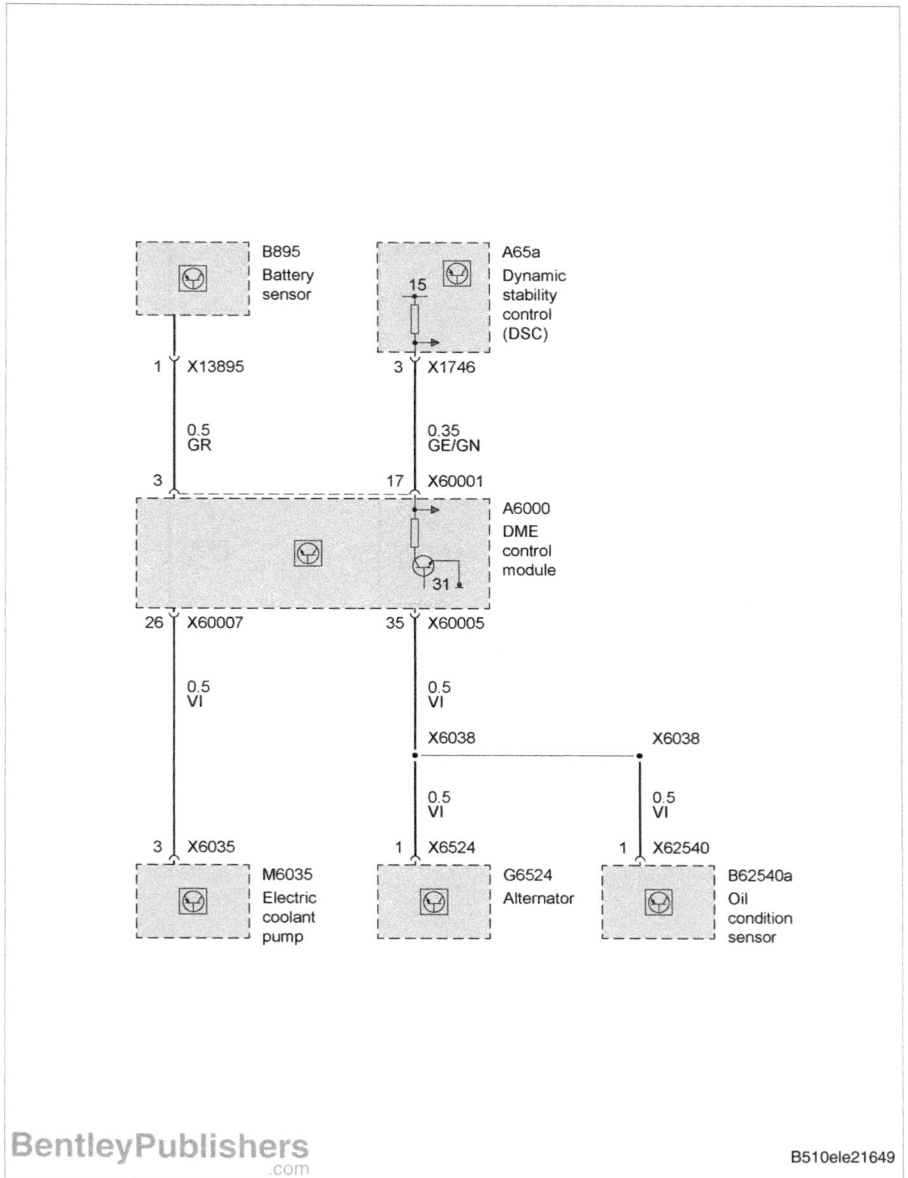

Bit serial data interface
N52 engine as of 03 / 2007

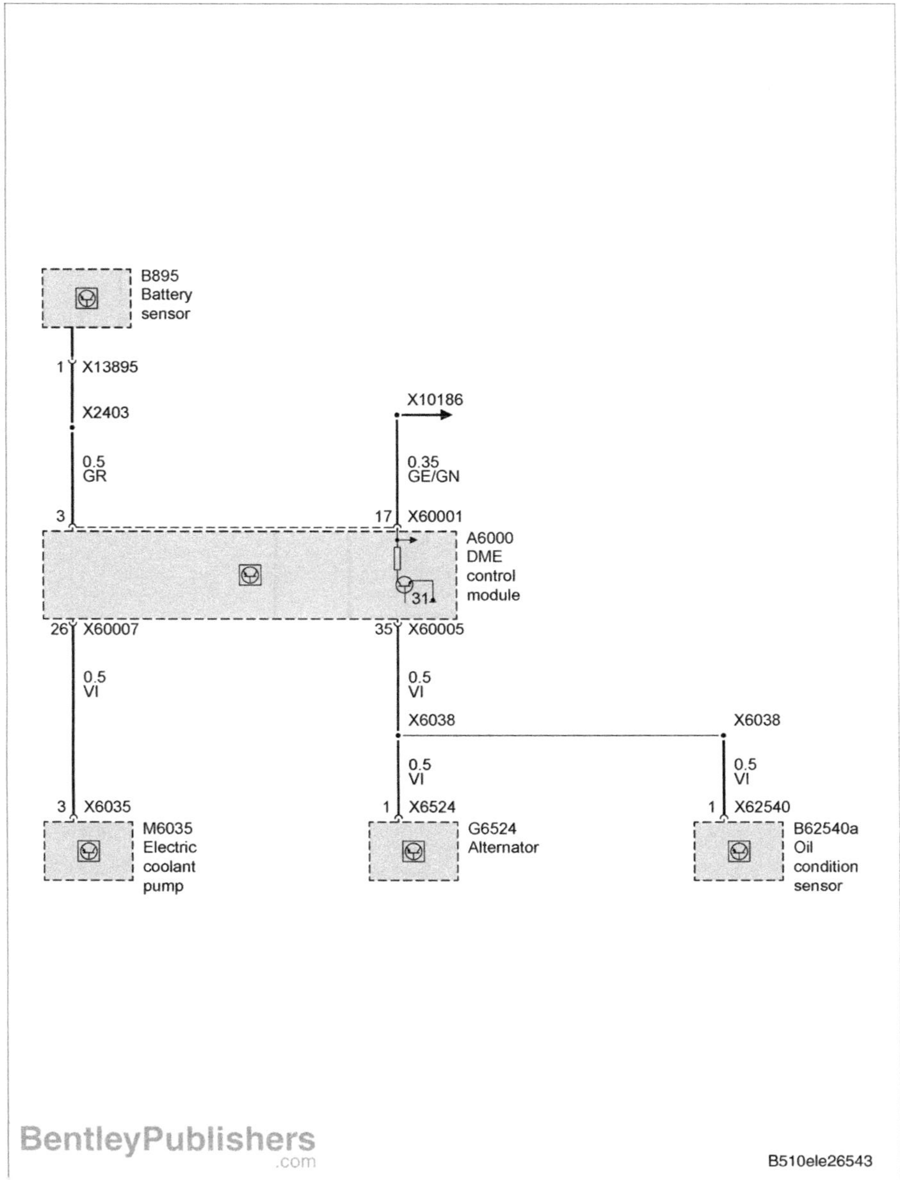

Convenient start
N52 engine up to 09 / 2005
(page 1 of 2)

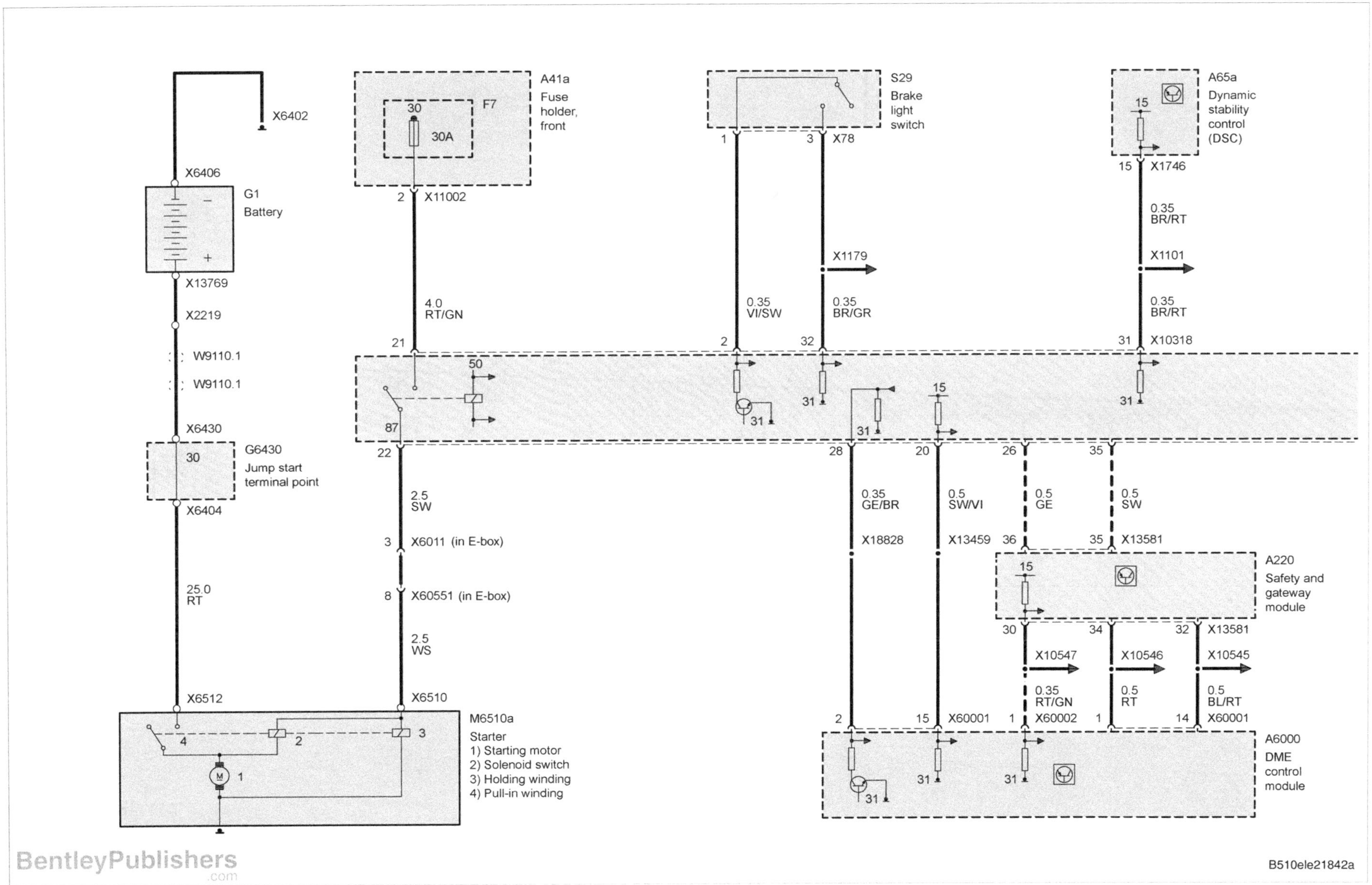

BentleyPublishers.com

Convenient start
N52 engine up to 09 / 2005
(page 2 of 2)

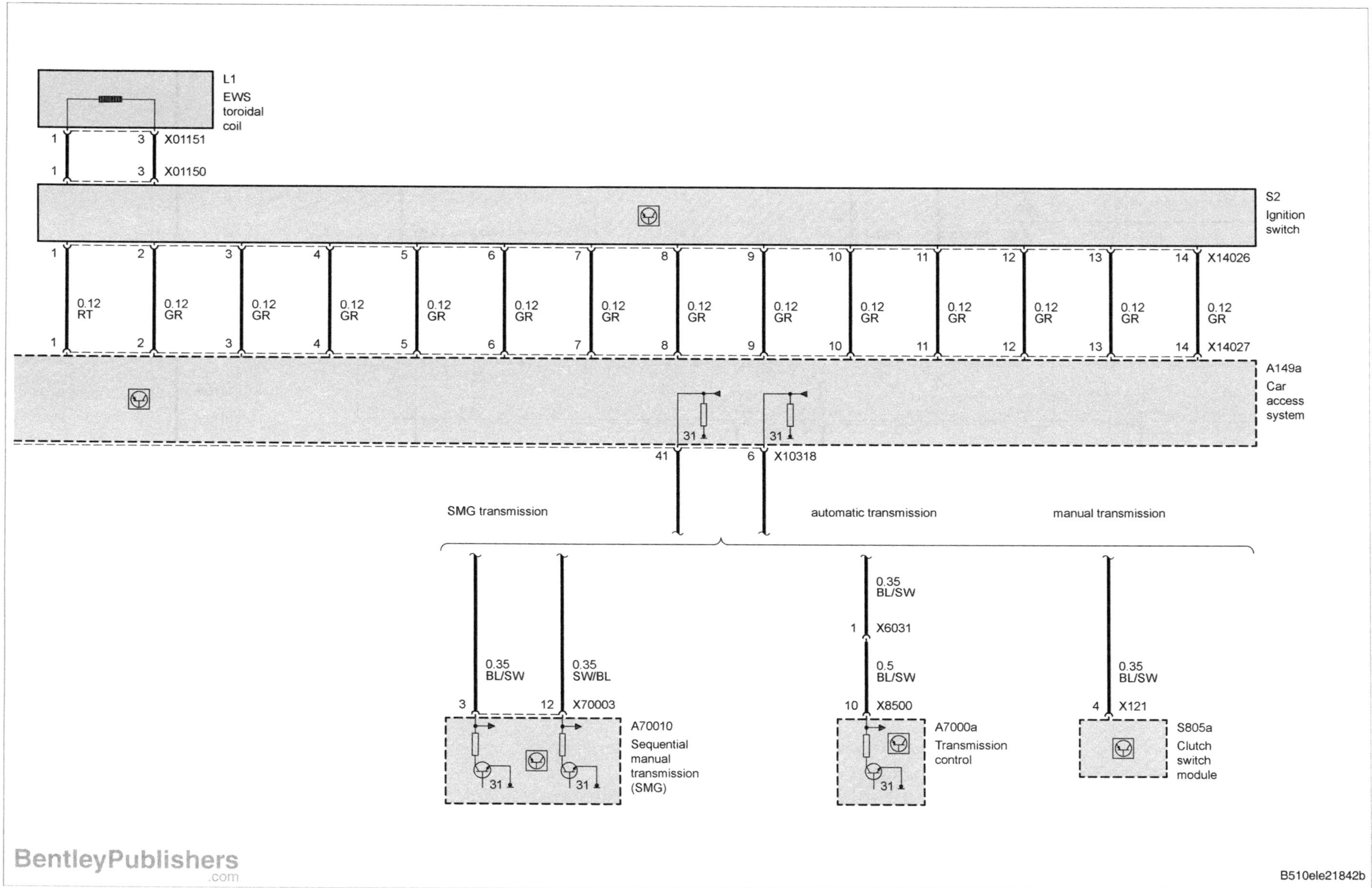

BentleyPublishers.com

B510ele21842b

Vehicle speed control
N52 engine up to 09 / 2005

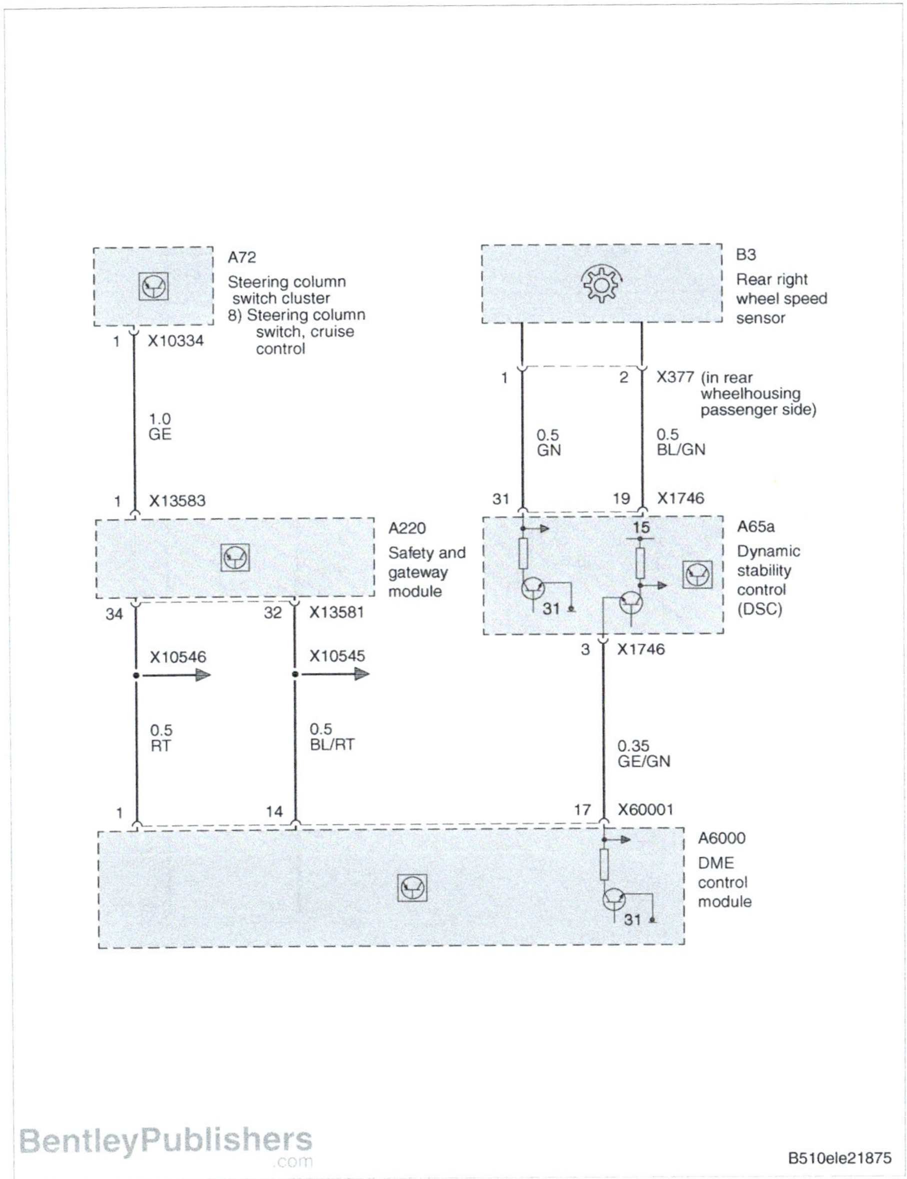

Vehicle speed control
N52 engine as of 09 / 2005

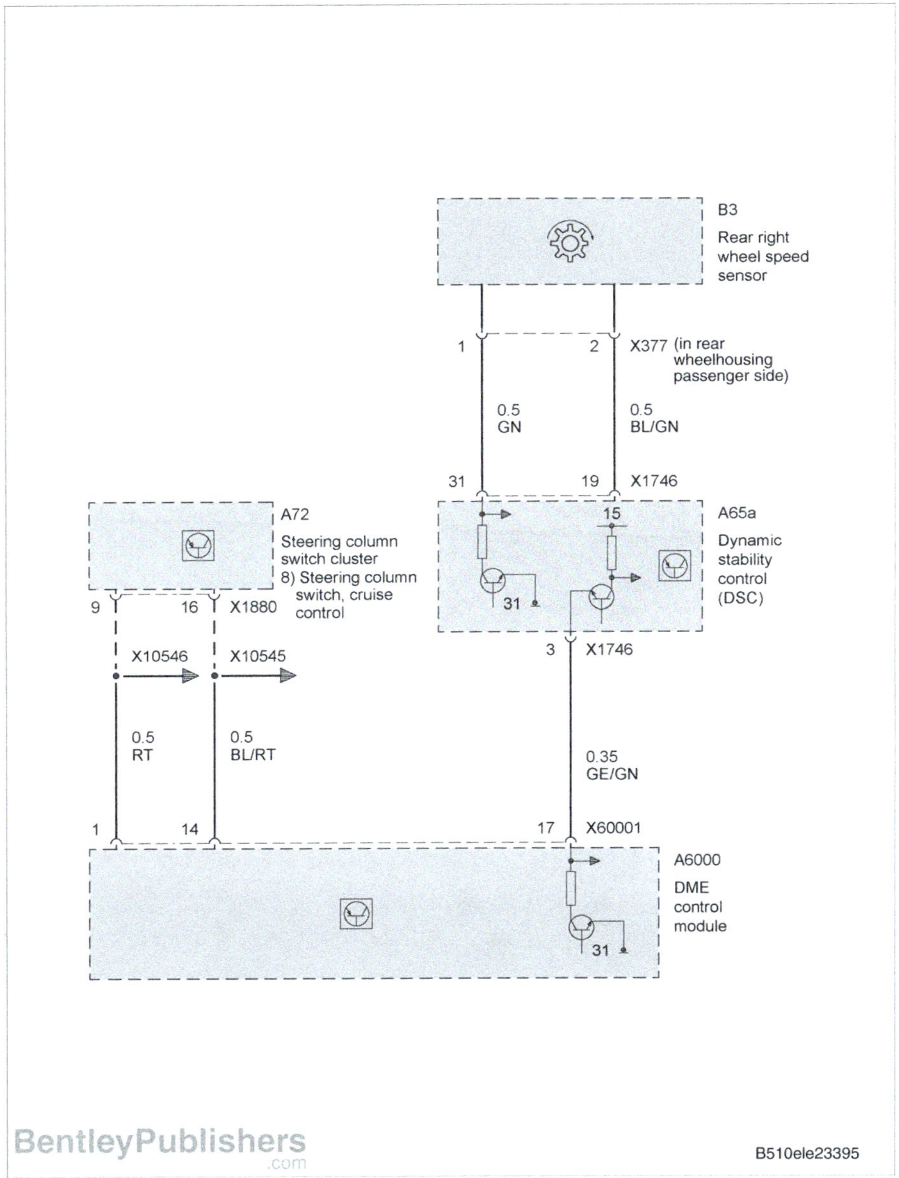

Power train CAN-bus
N52 and N54 engines
up to 03 / 2005

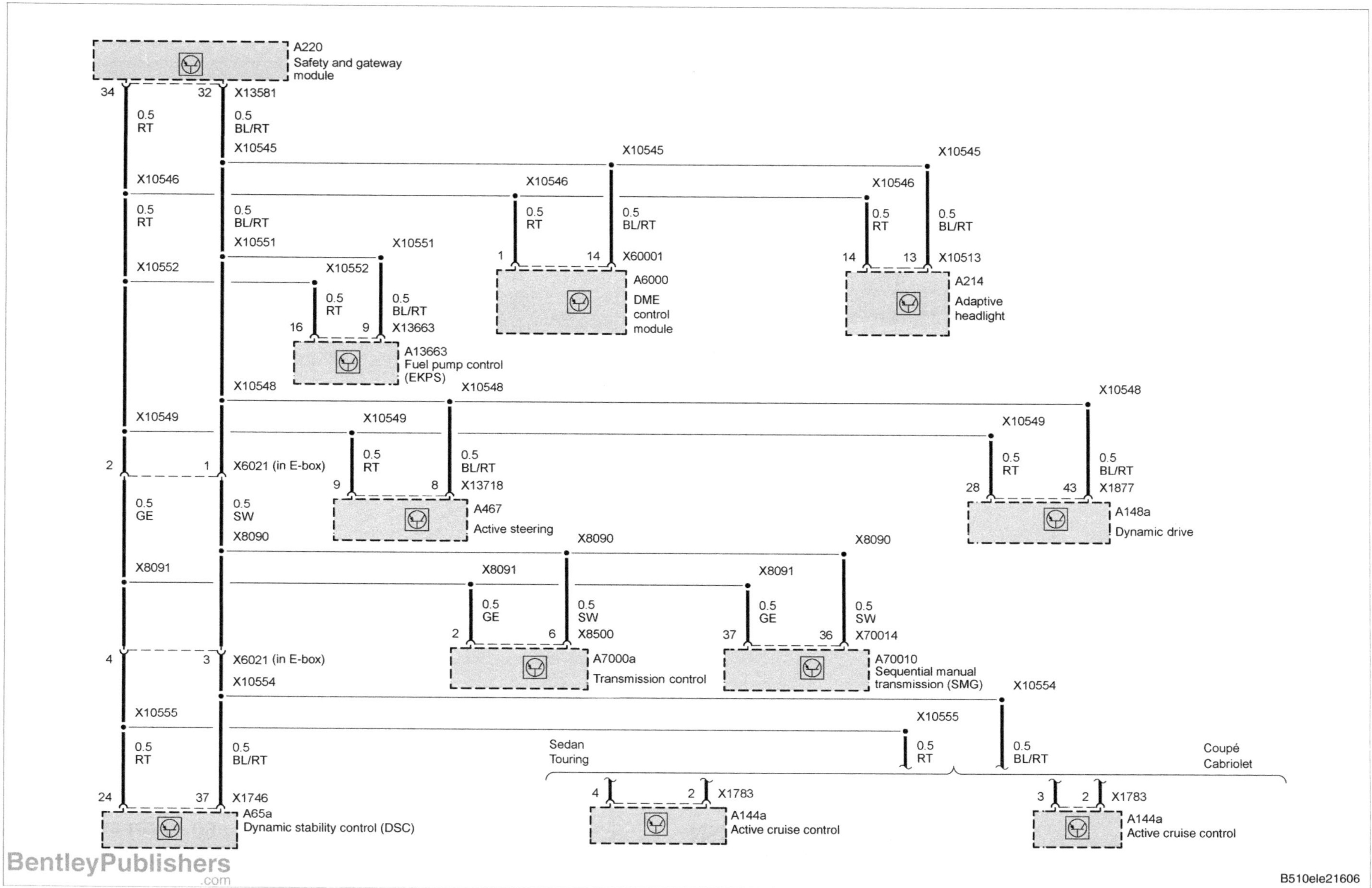

BentleyPublishers.com

Power train CAN-bus
N52 and N54 engines
as of 03 / 2005 with DSC Dynamic Stability Control PLUS

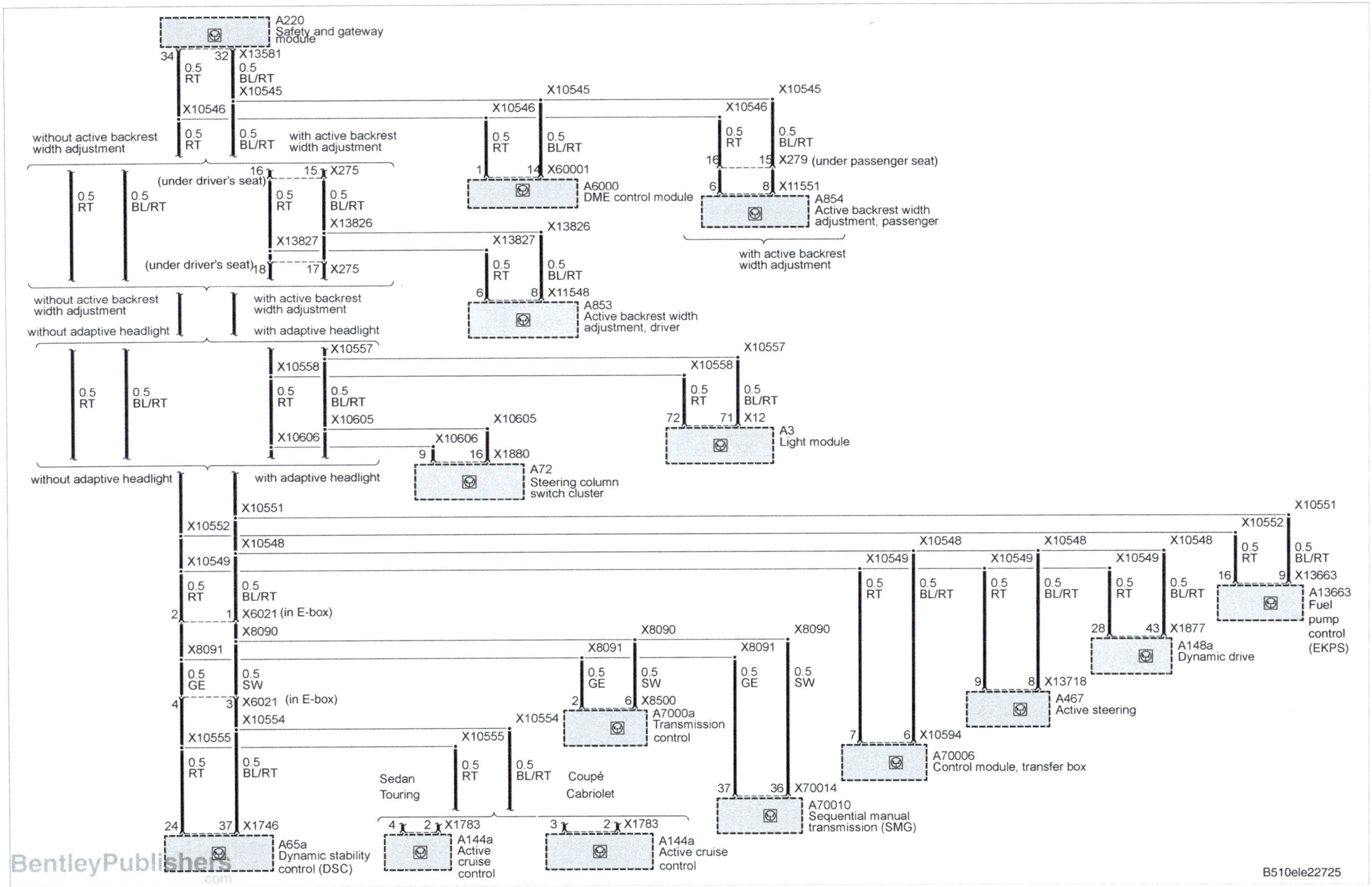

BentleyPublishers.com

Power train CAN-bus
N52 and N54 engines
with DSC Dynamic Stability Control Premium

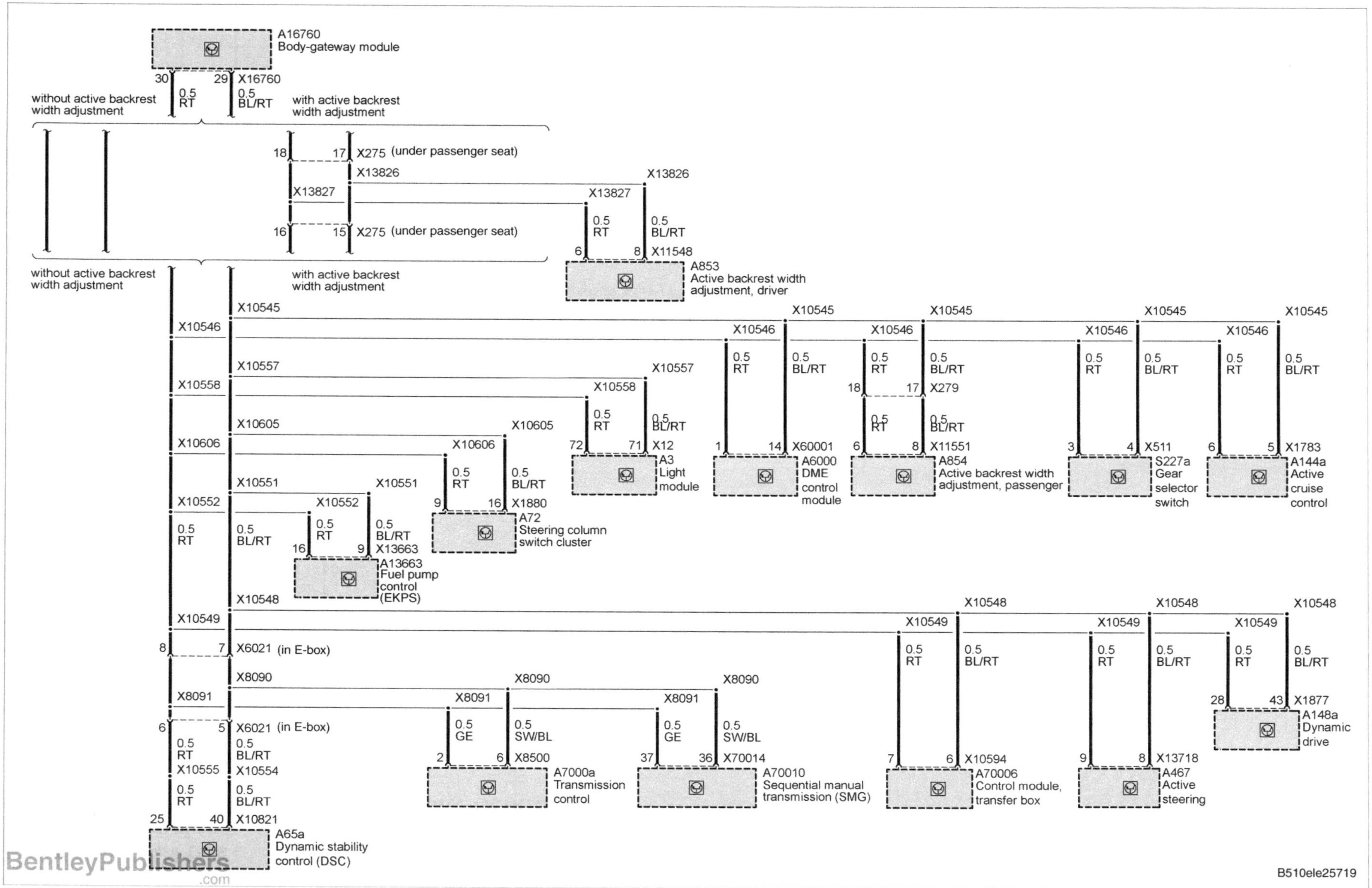

BentleyPublishers.com

Secondary air system
N52 and N54 engines

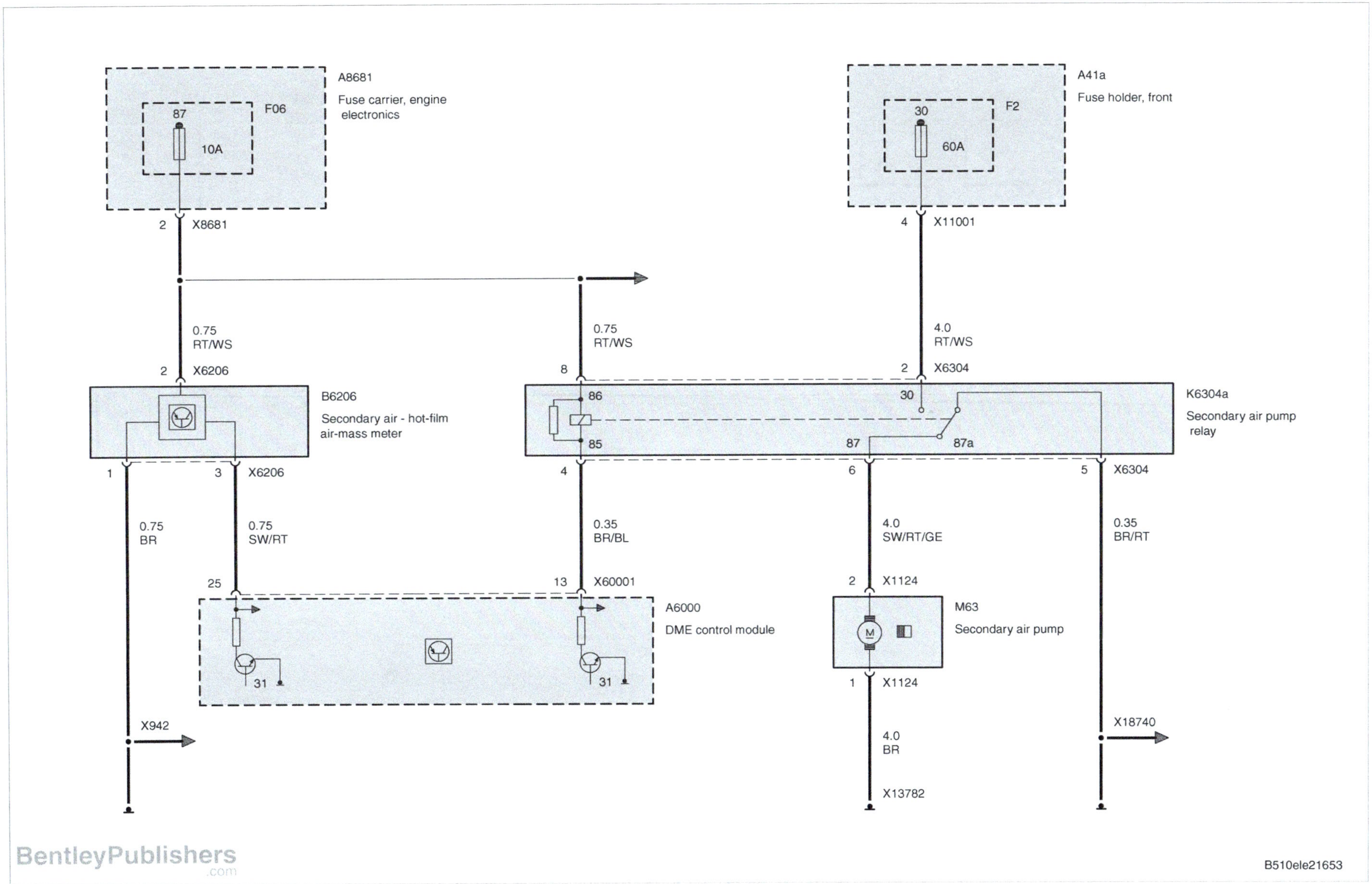

DME control module power supply N52 engine

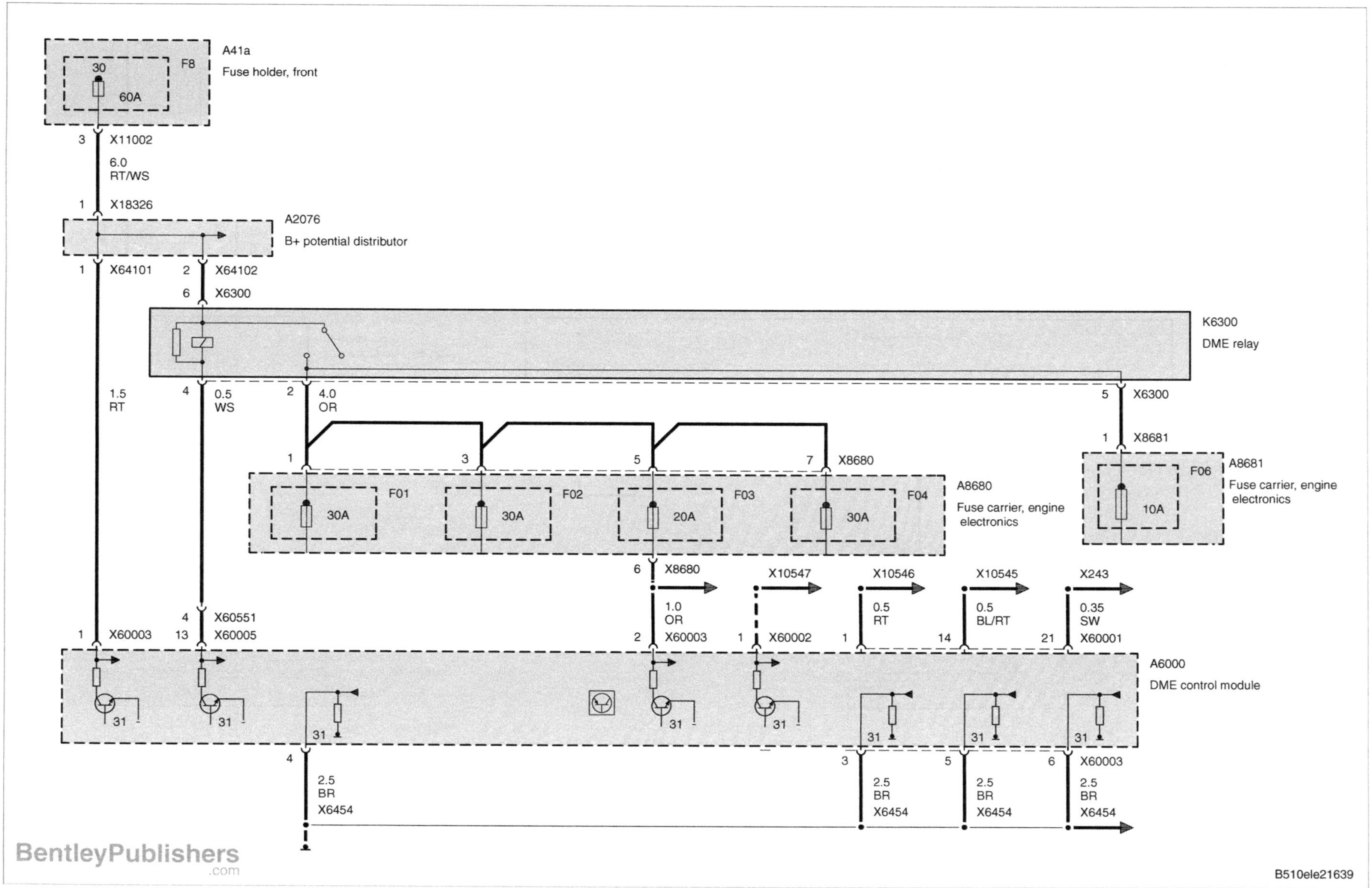

BentleyPublishers.com

B510ele21639

DME control module power supply N54 engine

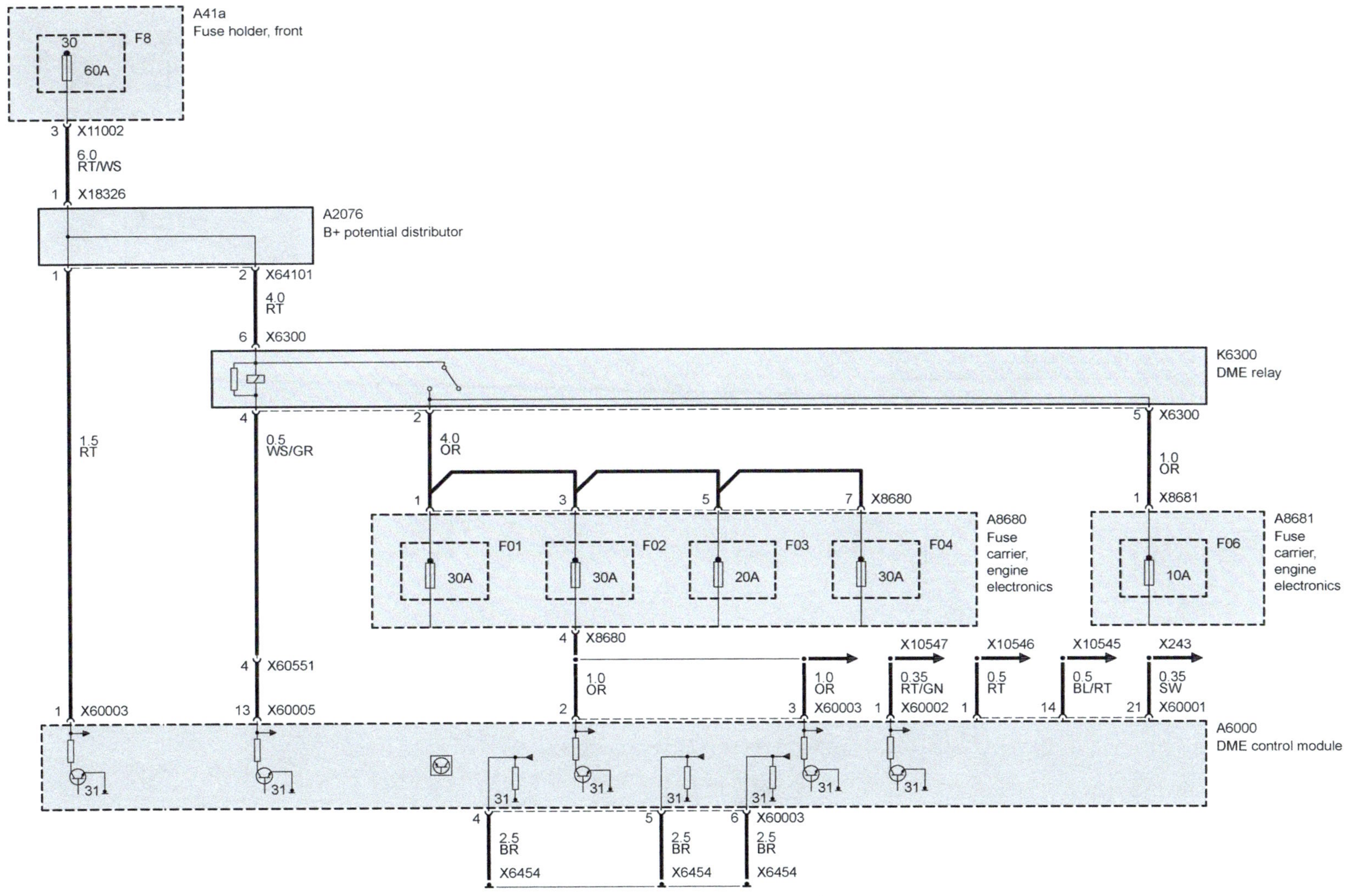

BentleyPublishers.com

Fuel tank vent valve
N52 and N54 engines

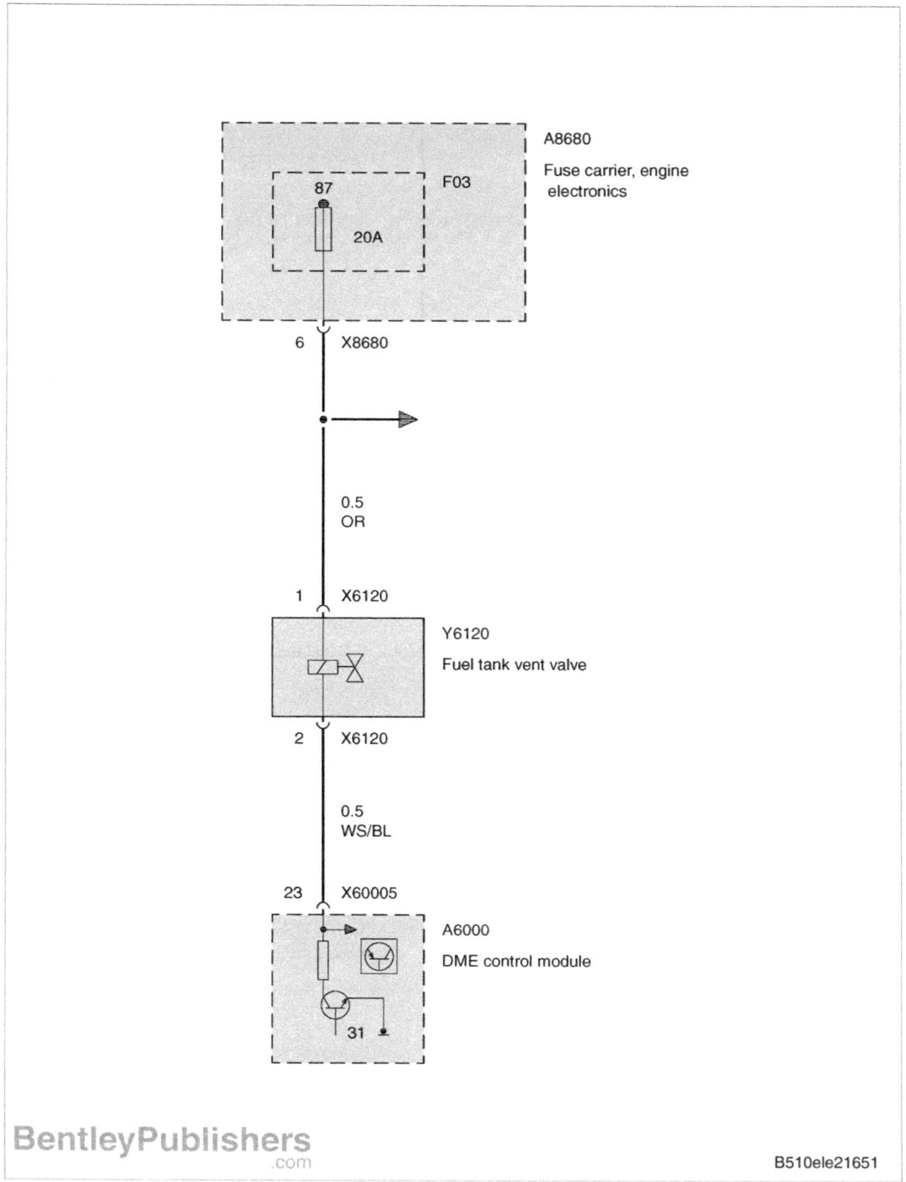

Diagnostic module for fuel tank leakage
N52 and N54 engines

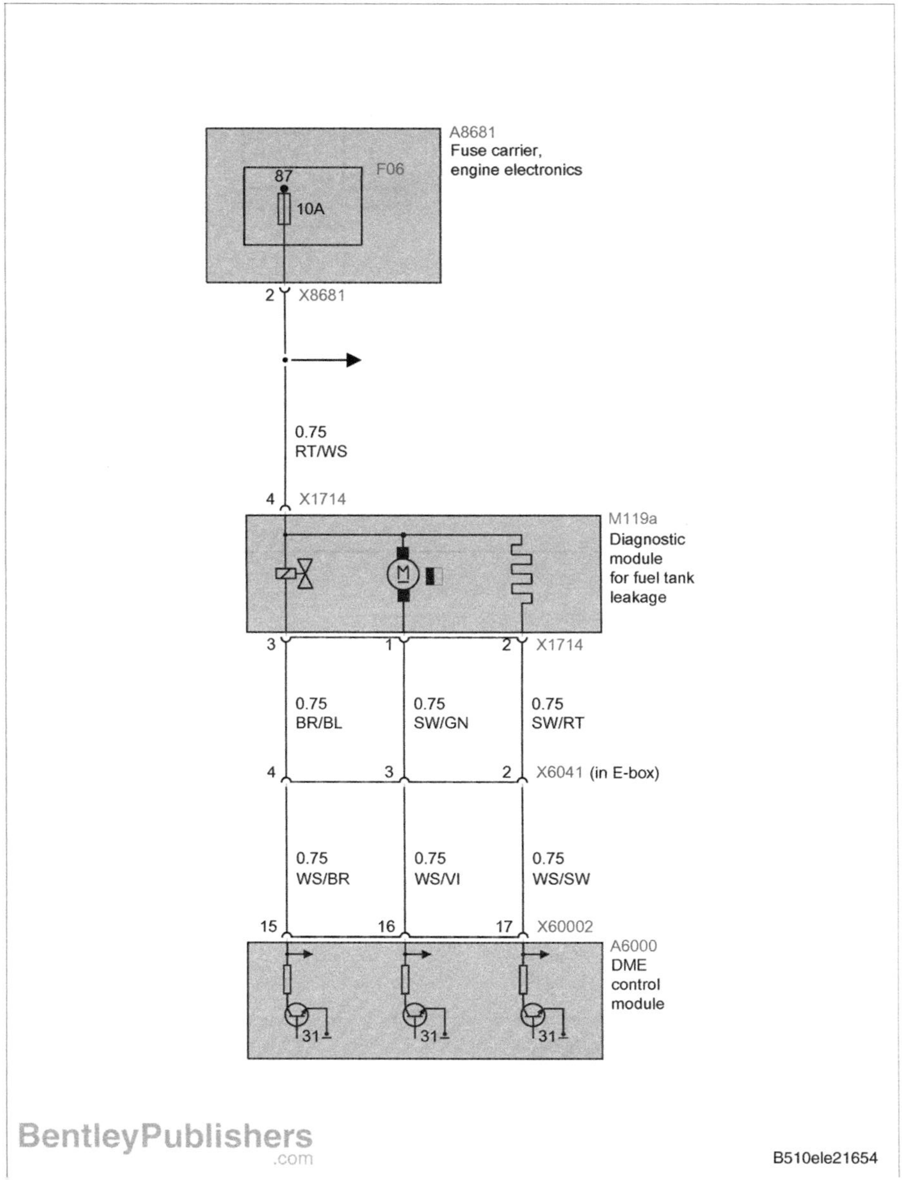

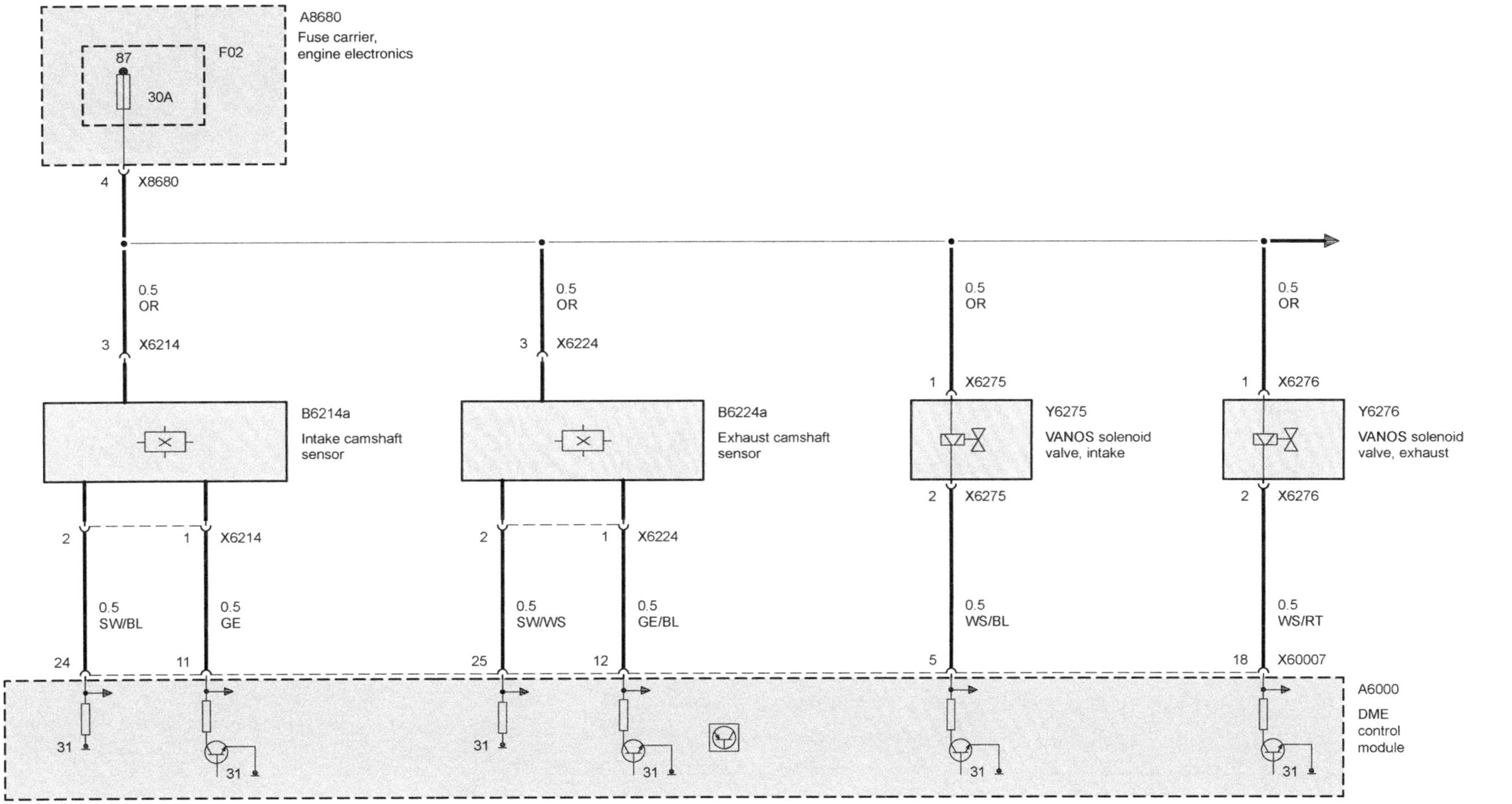

BentleyPublishers.com

Ignition coils
N52 engine
(page 1 of 2)

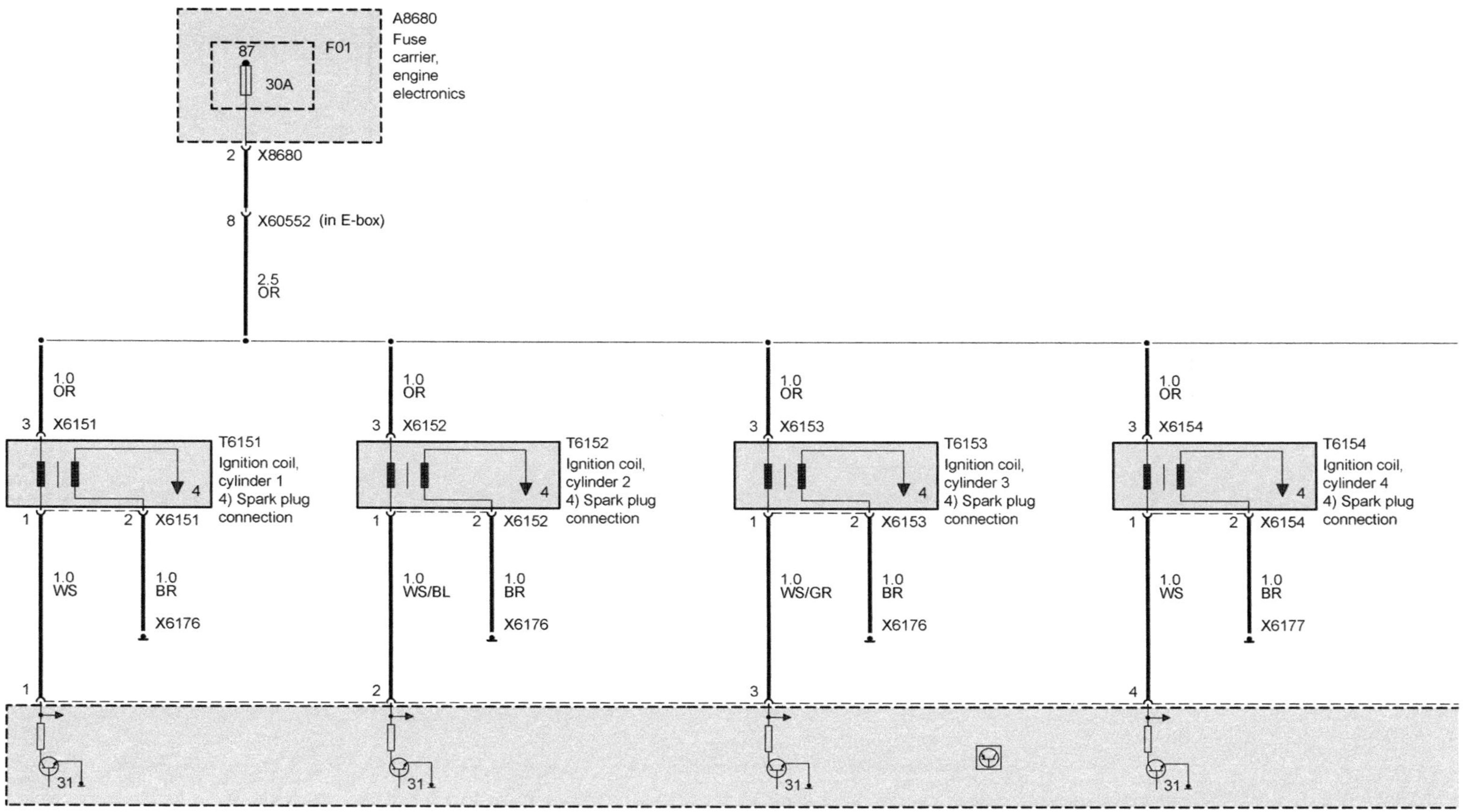

BentleyPublishers.com

B510ele21646a

Ignition coils
N52 engine
(page 2 of 2)

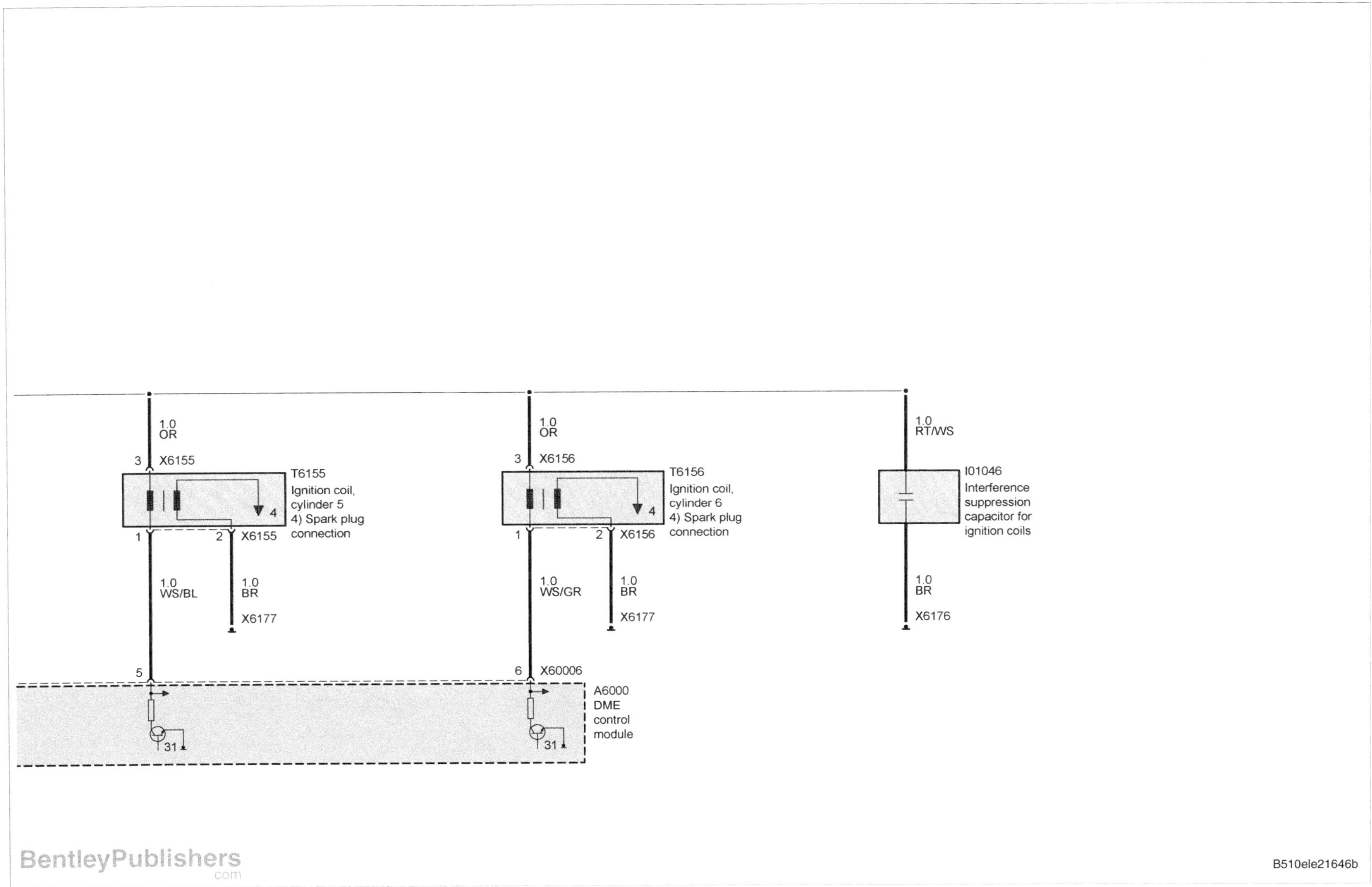

Ignition coils
N54 engine
(page 1 of 2)

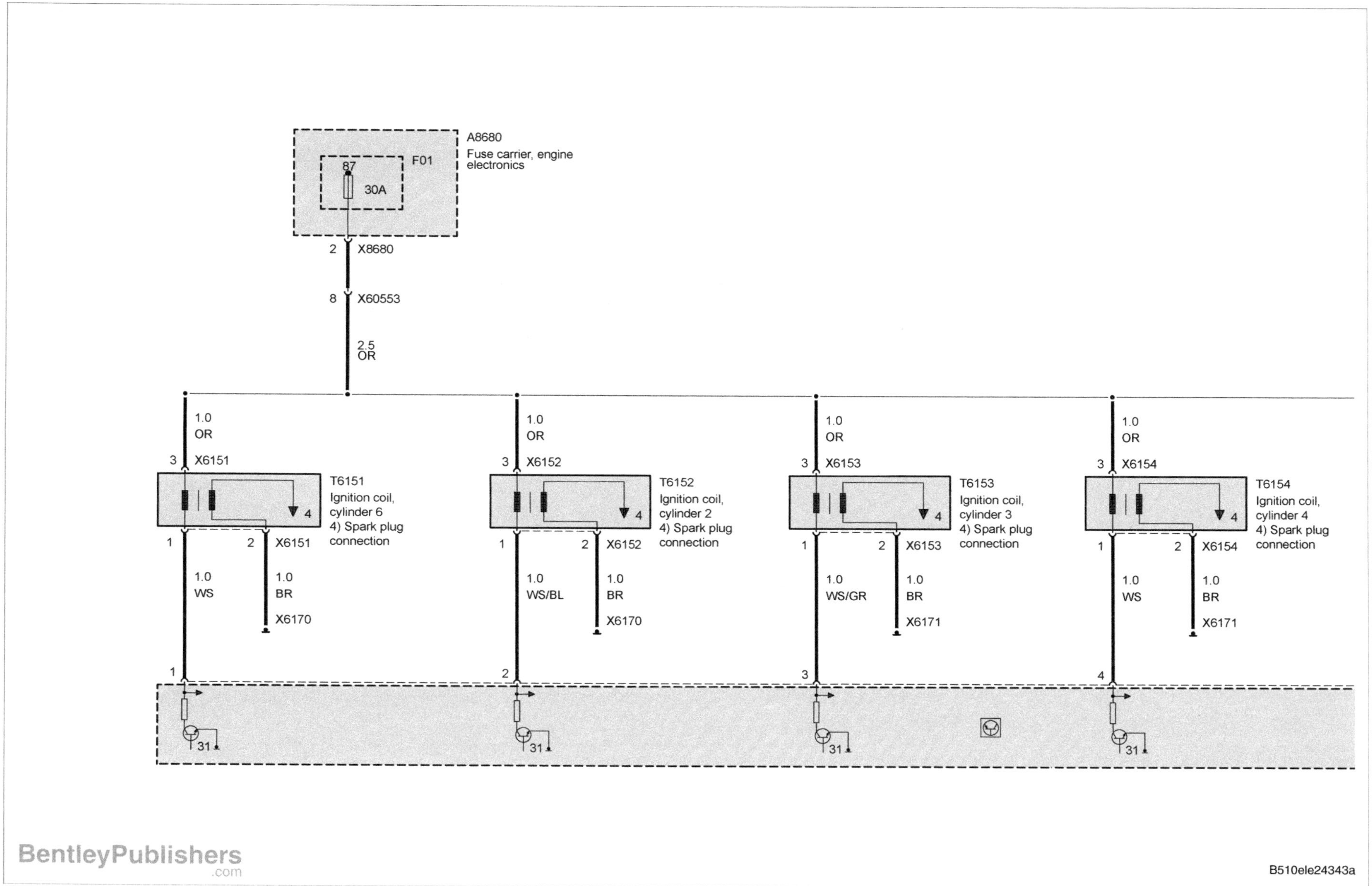

BentleyPublishers.com

Ignition coils
N54 engine
(page 2 of 2)

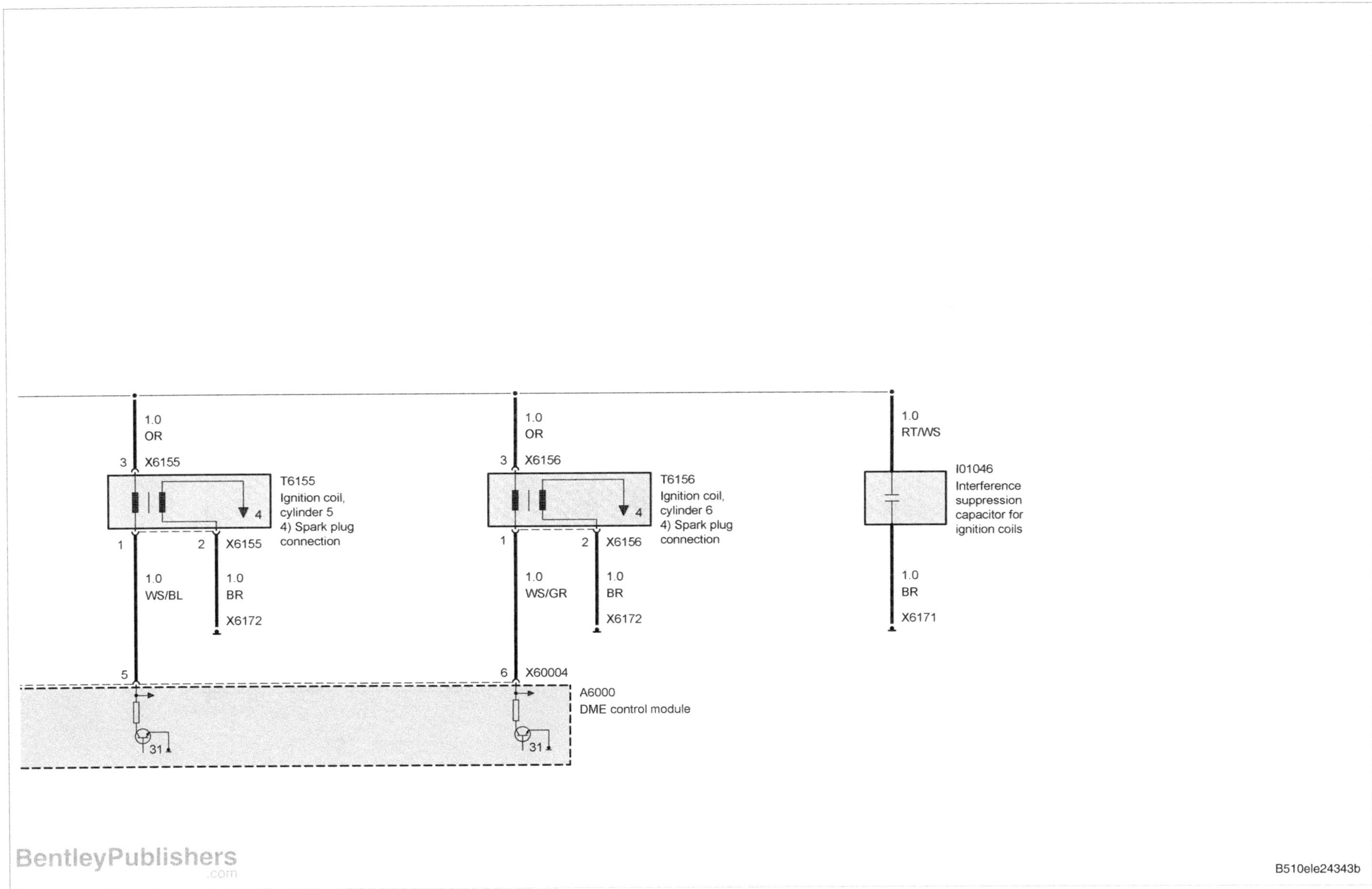

EGS transmission control
Steptronic switch
up to 03 / 2005

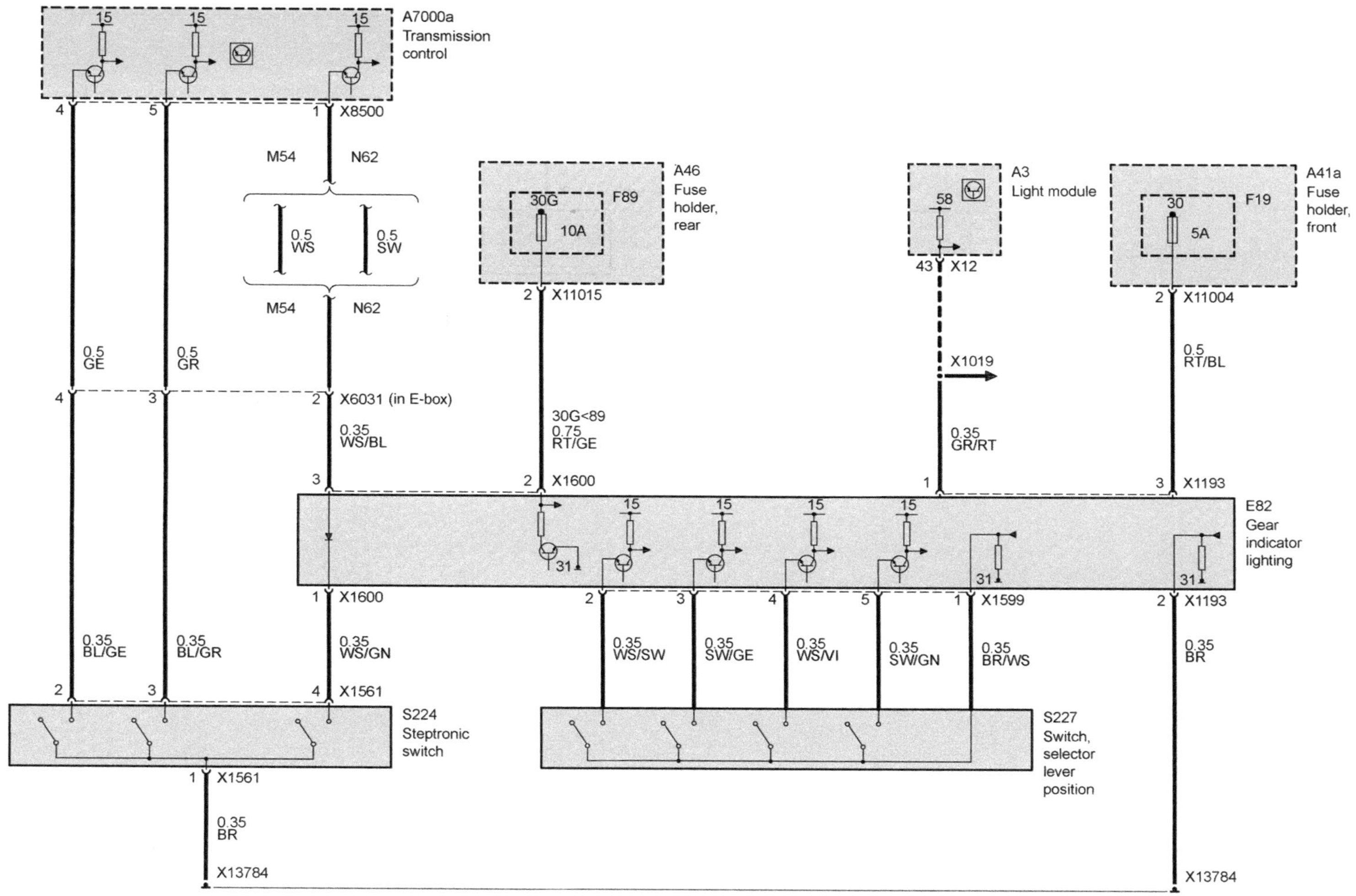

BentleyPublishers.com

B510ele19377

EGS transmission control
Steptronic switch
as of 03 / 2005 up to 09 / 2005 without N62 or N62 TU engine

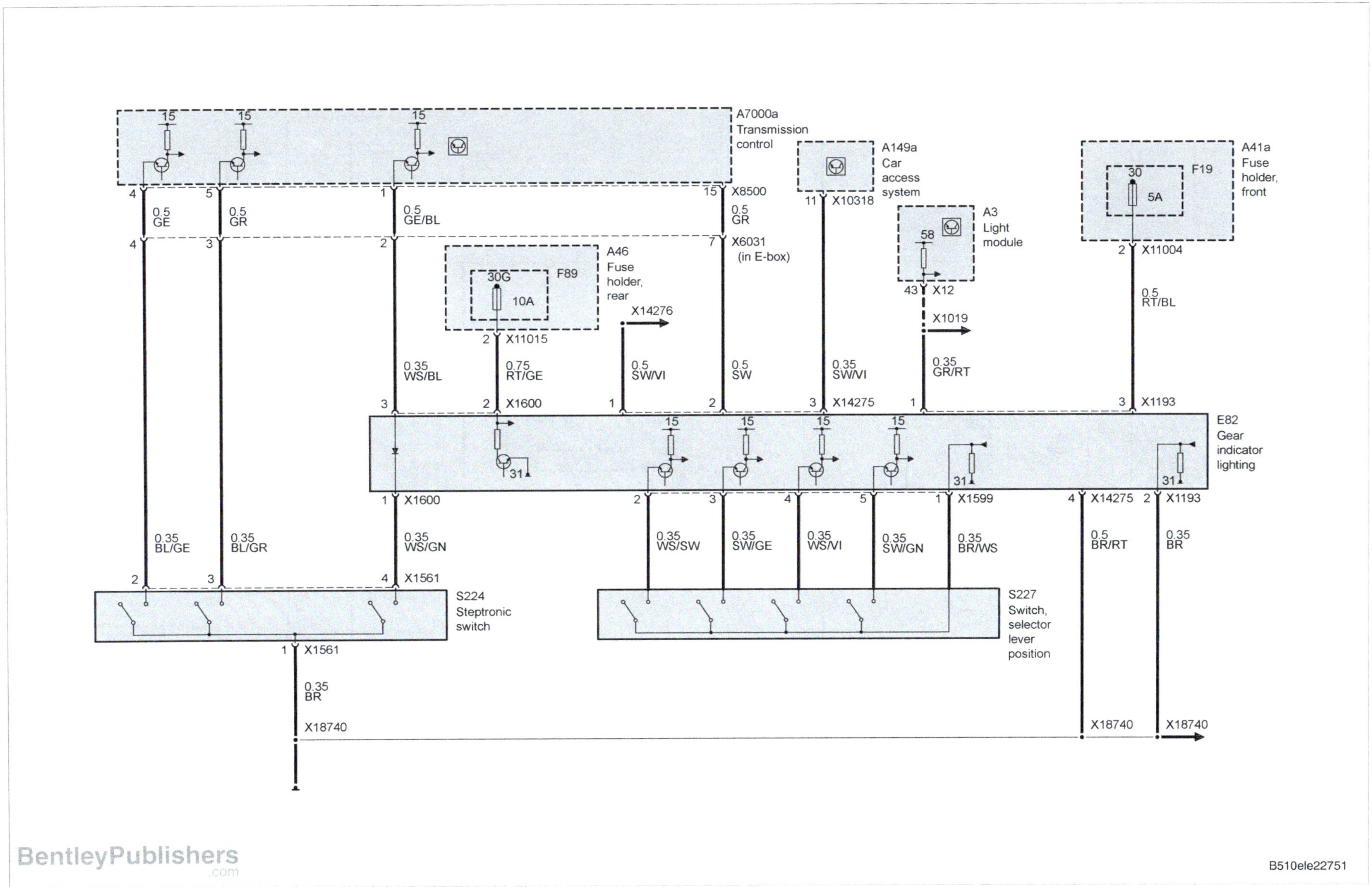

EGS transmission control
Steptronic switch
as of 09 / 2005 without N62 TU engine

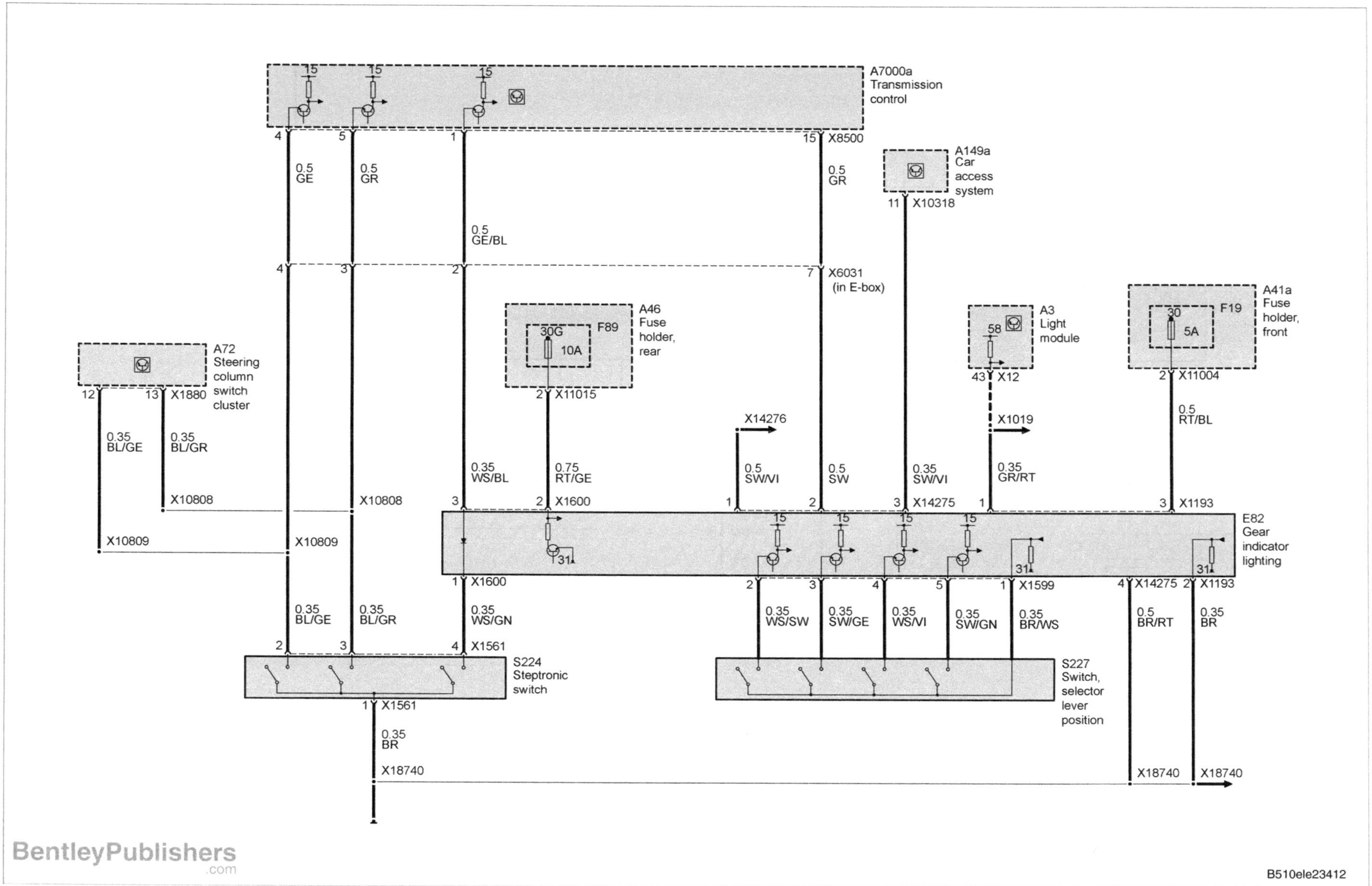

BentleyPublishers.com

B510ele23412

EGS transmission control
Steptronic switch
as of 03 / 2005 up to 09 / 2005 with N62 engine

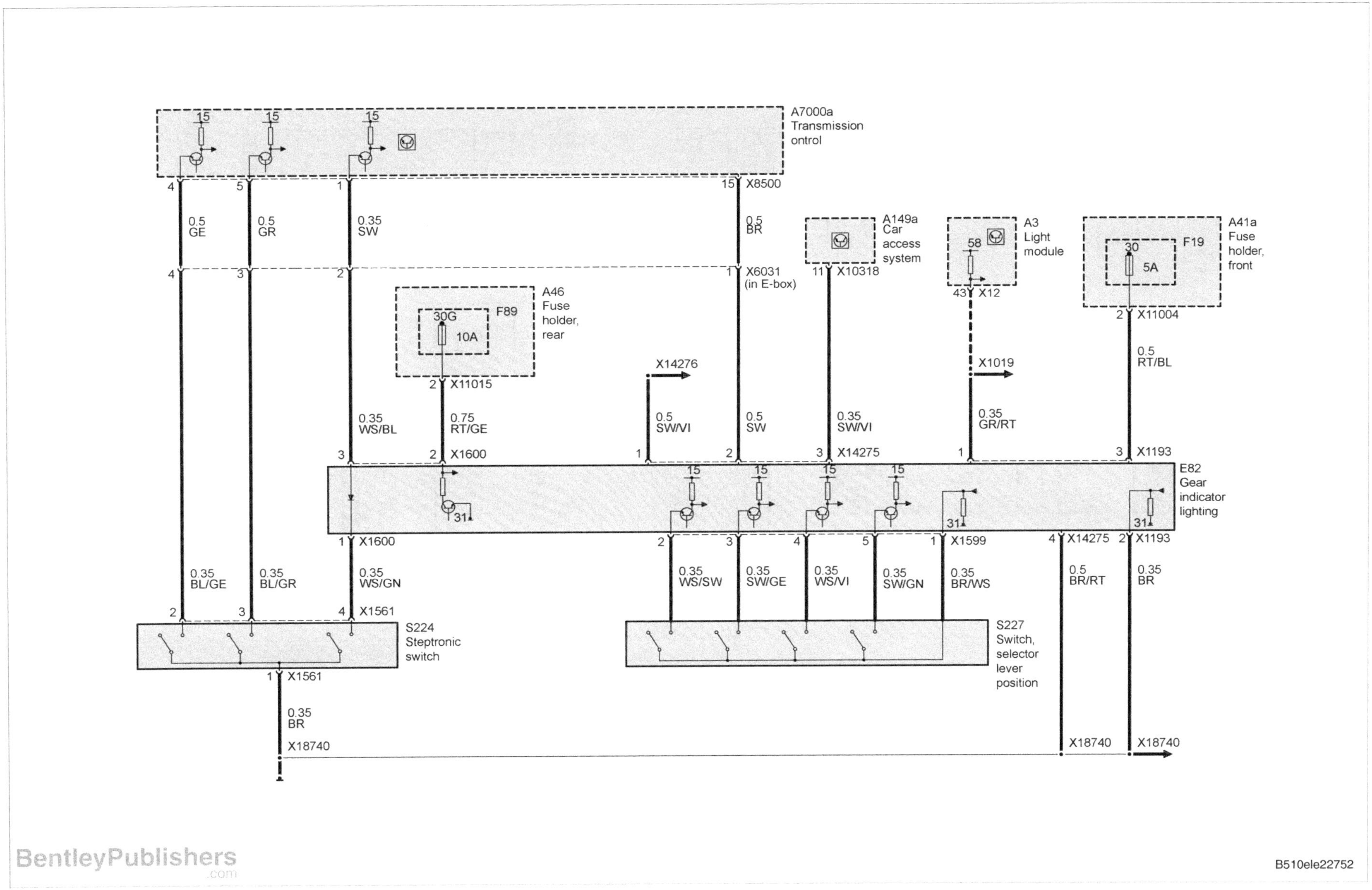

BentleyPublishers.com

EGS transmission control
Steptronic switch
N62 TU engine

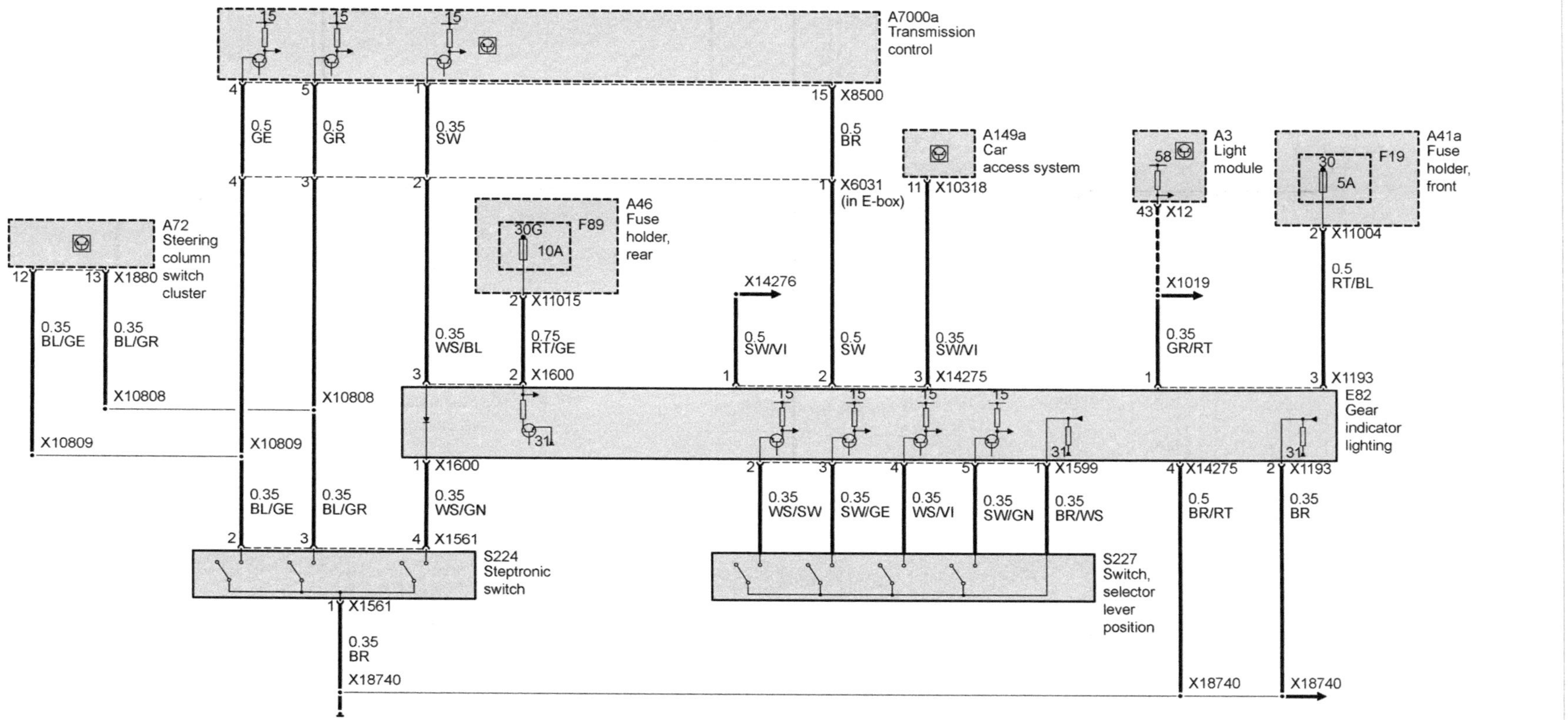

BentleyPublishers.com

B510ele23413

EGS transmission control
Shift lock

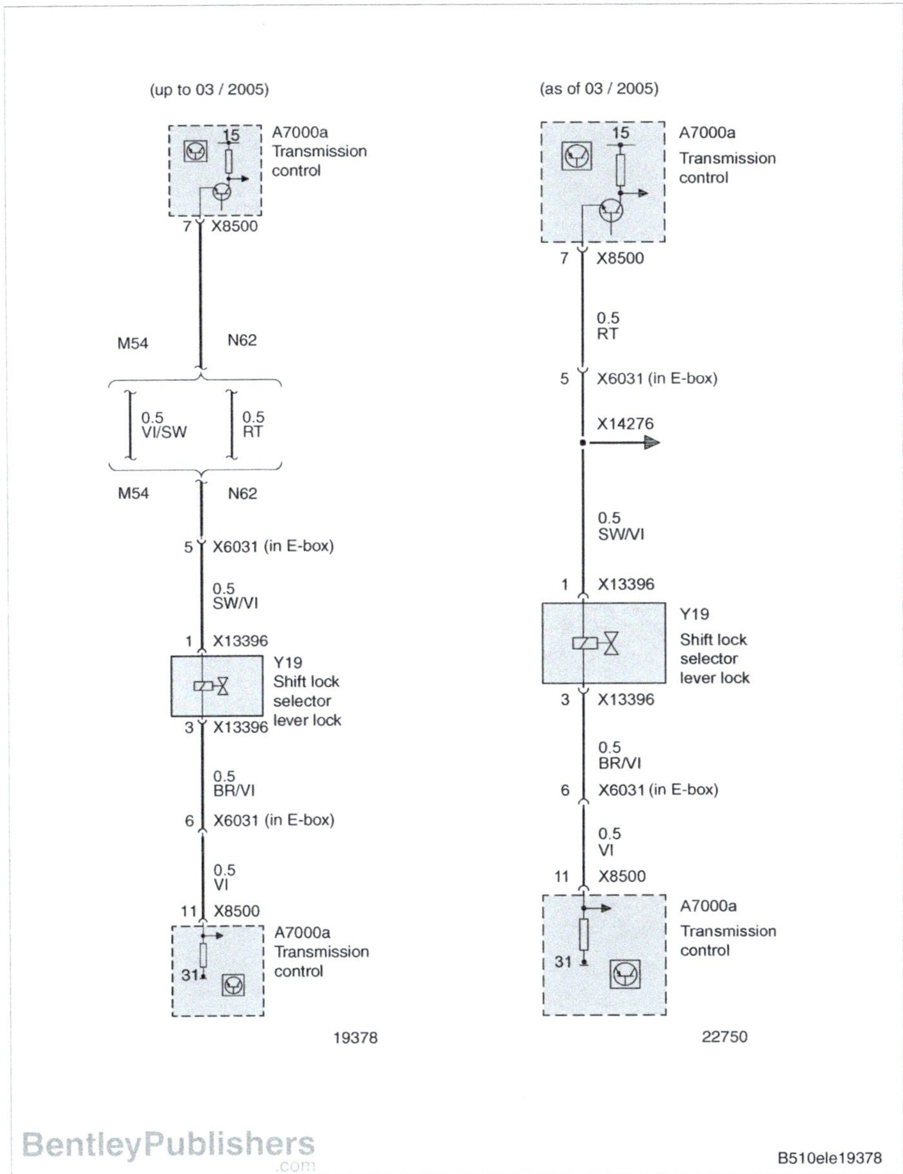

EGS transmission control
Mechatronics, power supply
M54 engine

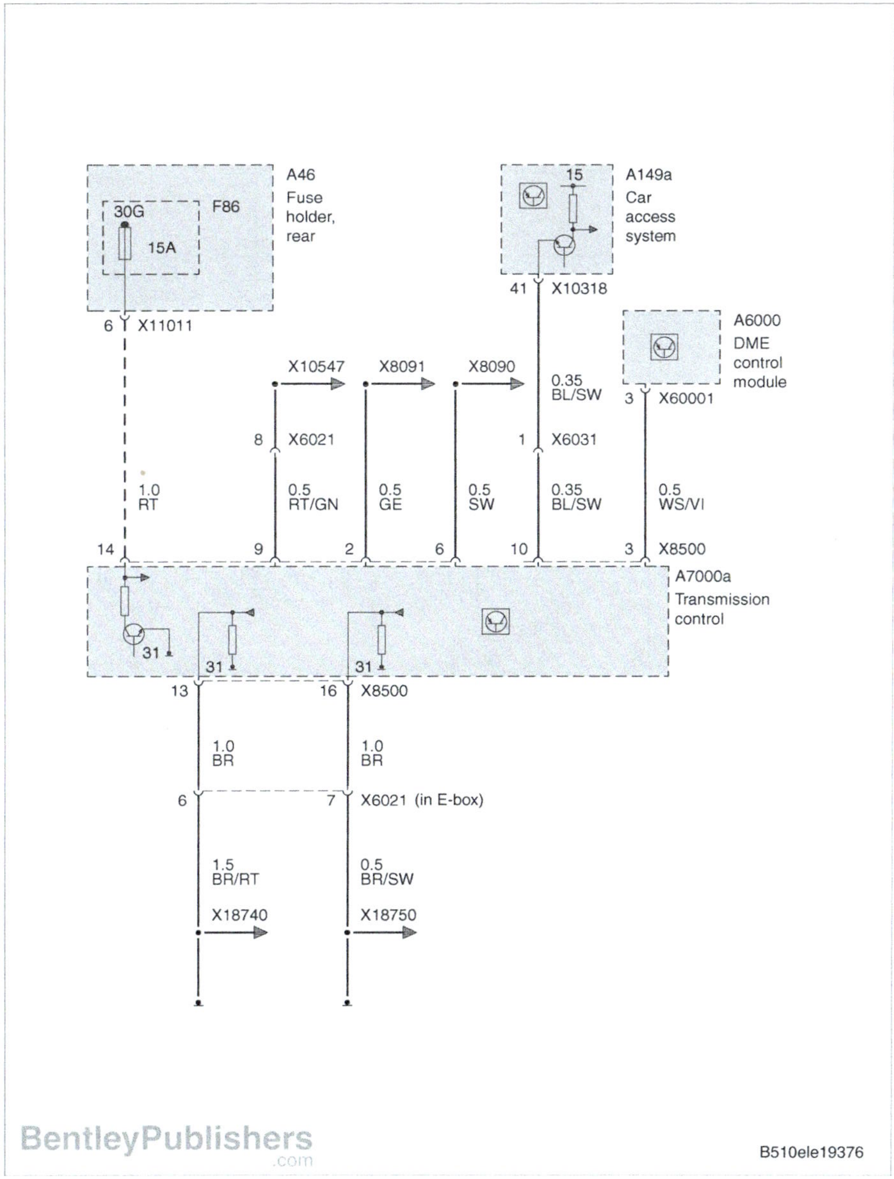

EGS transmission control
Transmission control, power supply
N62 engine

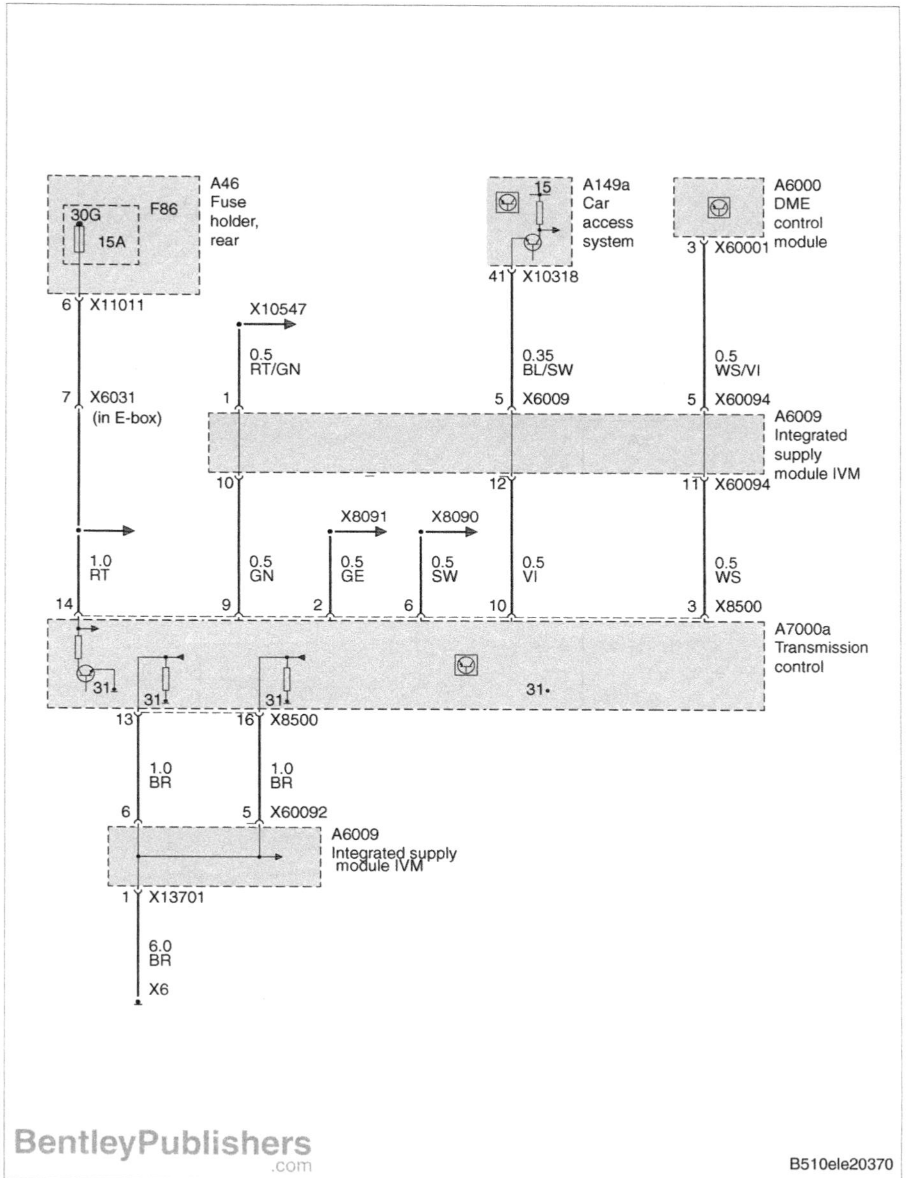

EGS transmission control
Transmission control, power supply
N62 TU engine

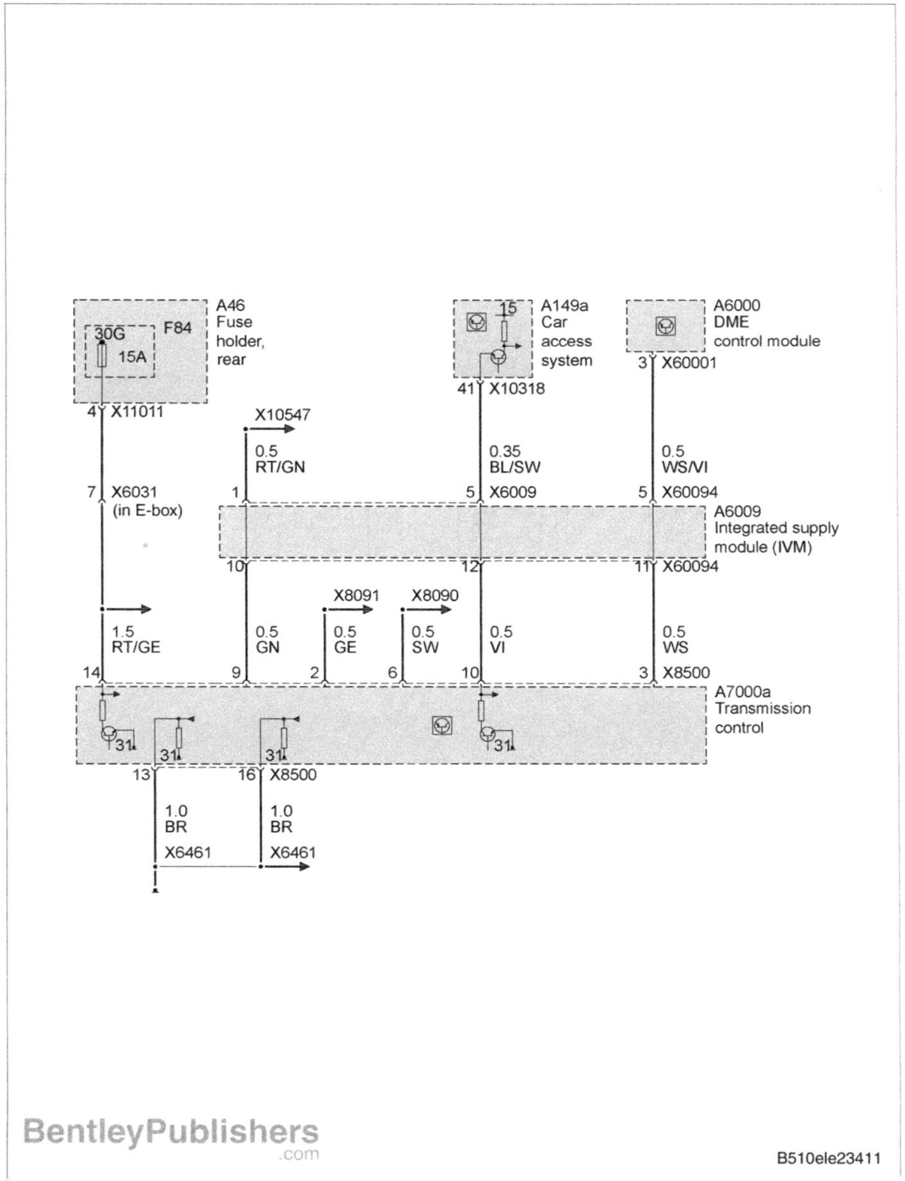

EGS transmission control
Transmission control, power supply
N52 engine up to 09 / 2005

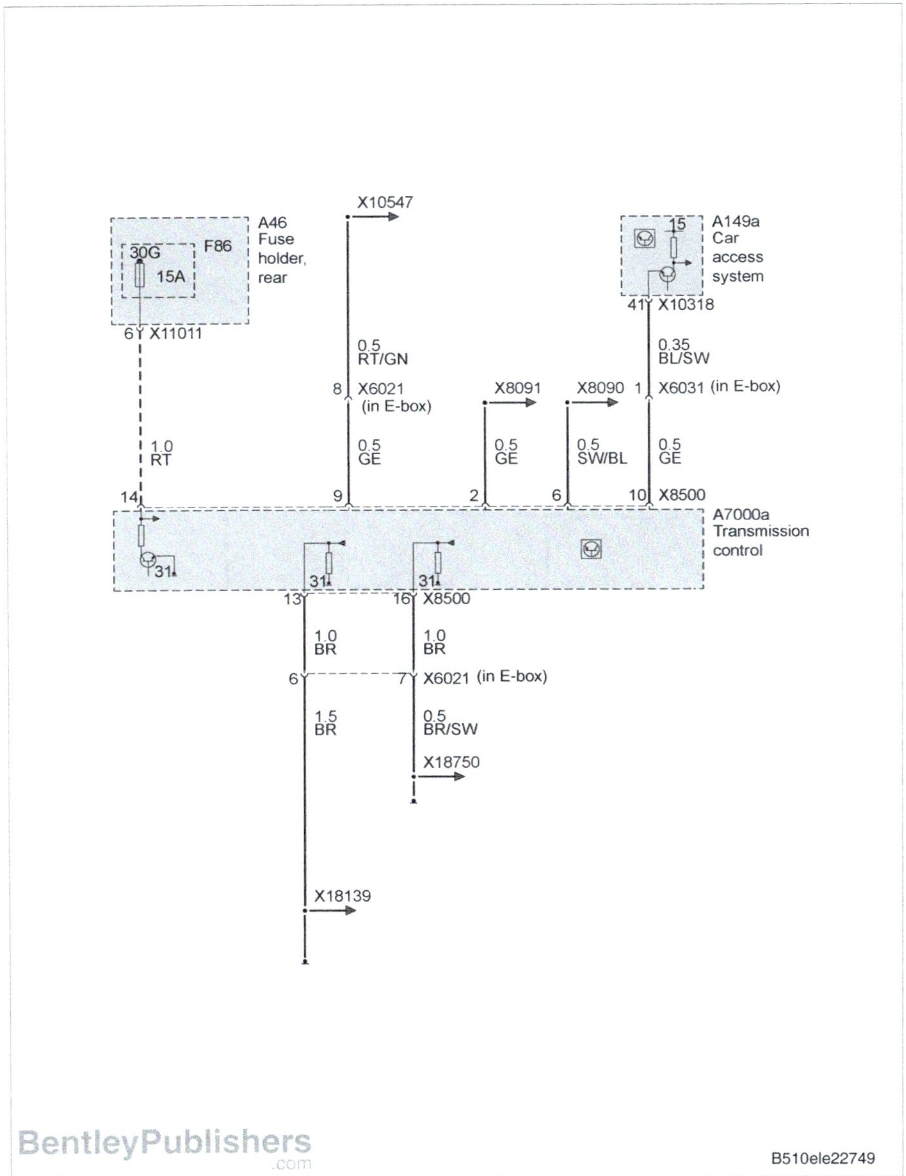

EGS transmission control
Transmission control, power supply
N52 engine as of 09 / 2005

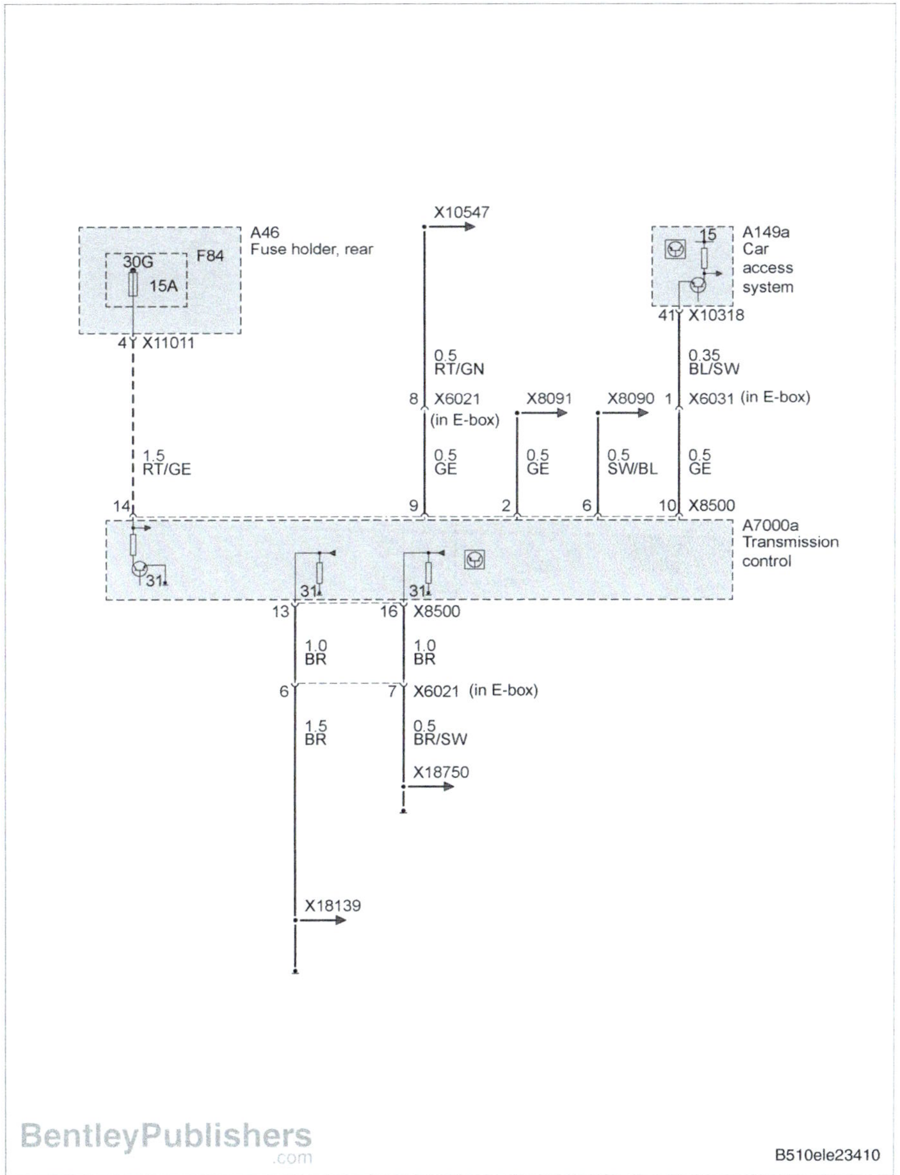

SMG control module, power supply
M54 engine

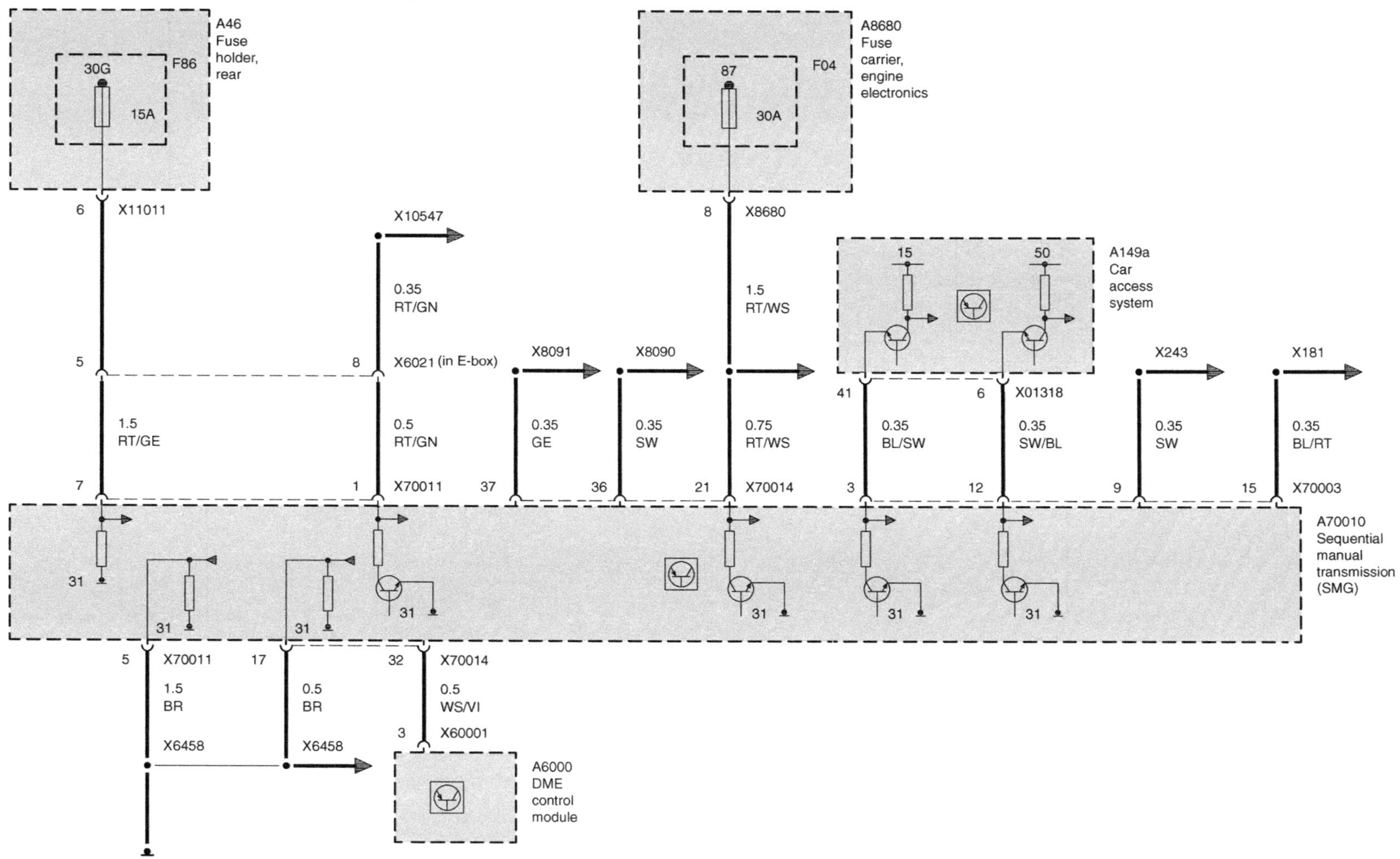

BentleyPublishers.com

B510ele19367

SMG control module, power supply
N62 engine

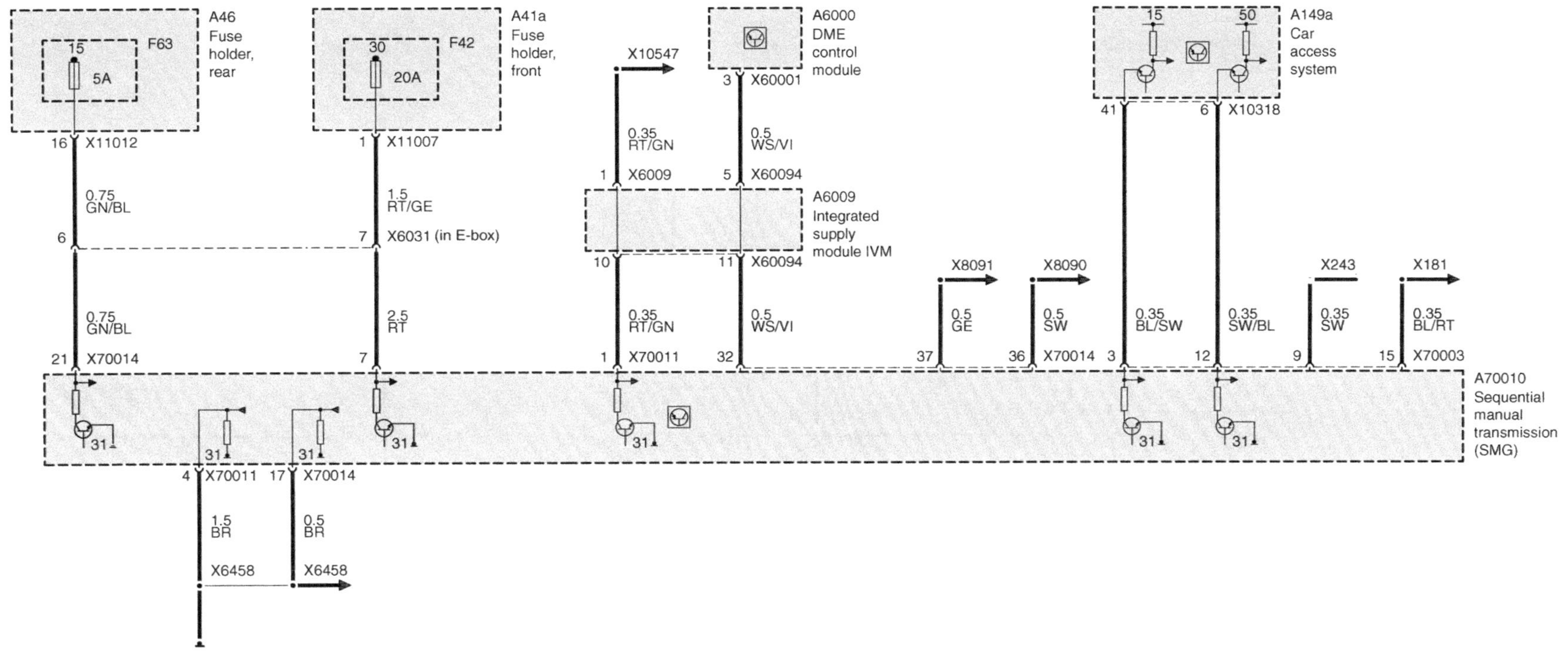

BentleyPublishers
.com

SMG control module, power supply
N52 engine up to 09 / 2005

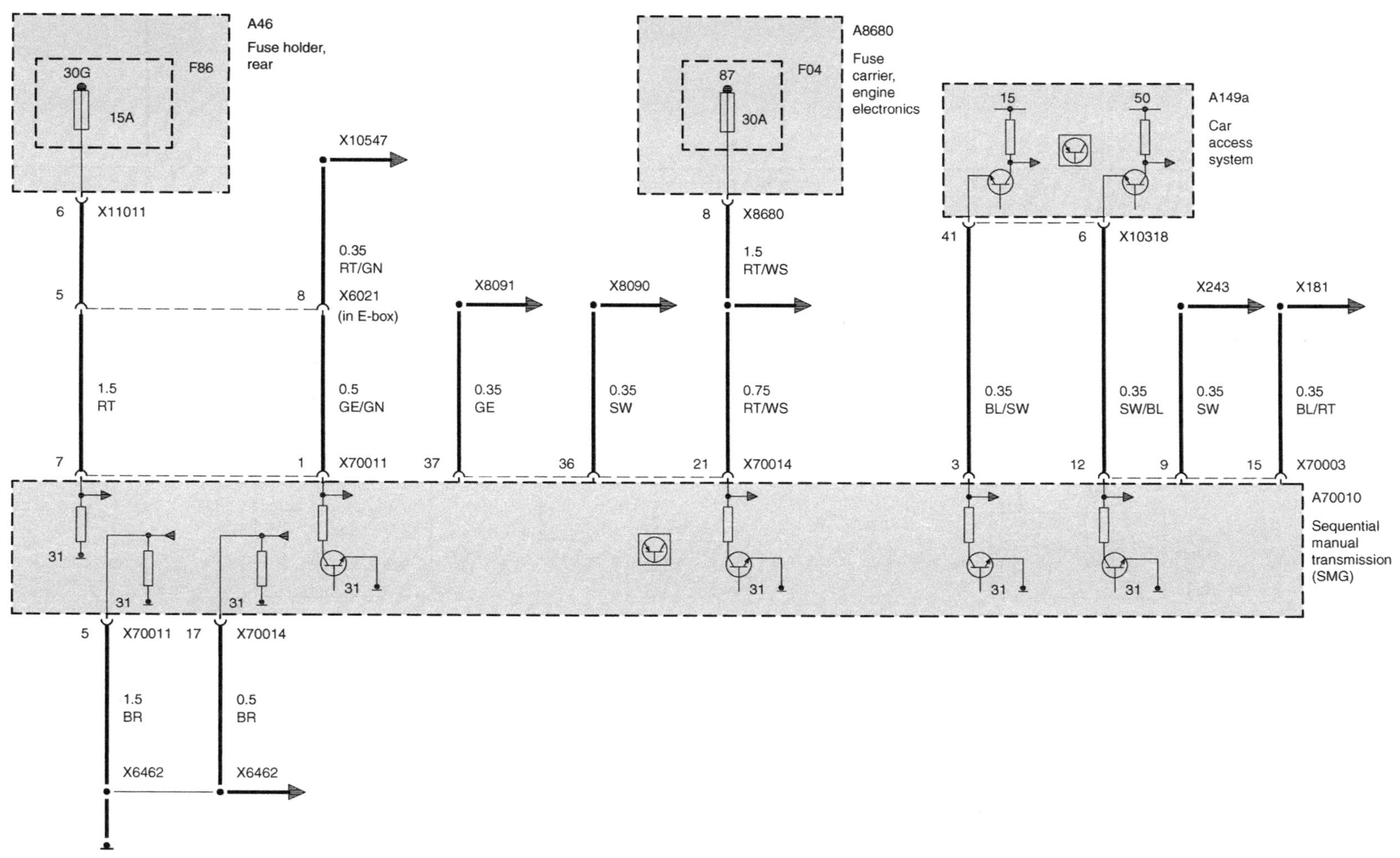

BentleyPublishers.com

B510ele19368

SMG control module, power supply
N52 engine as of 09 / 2005

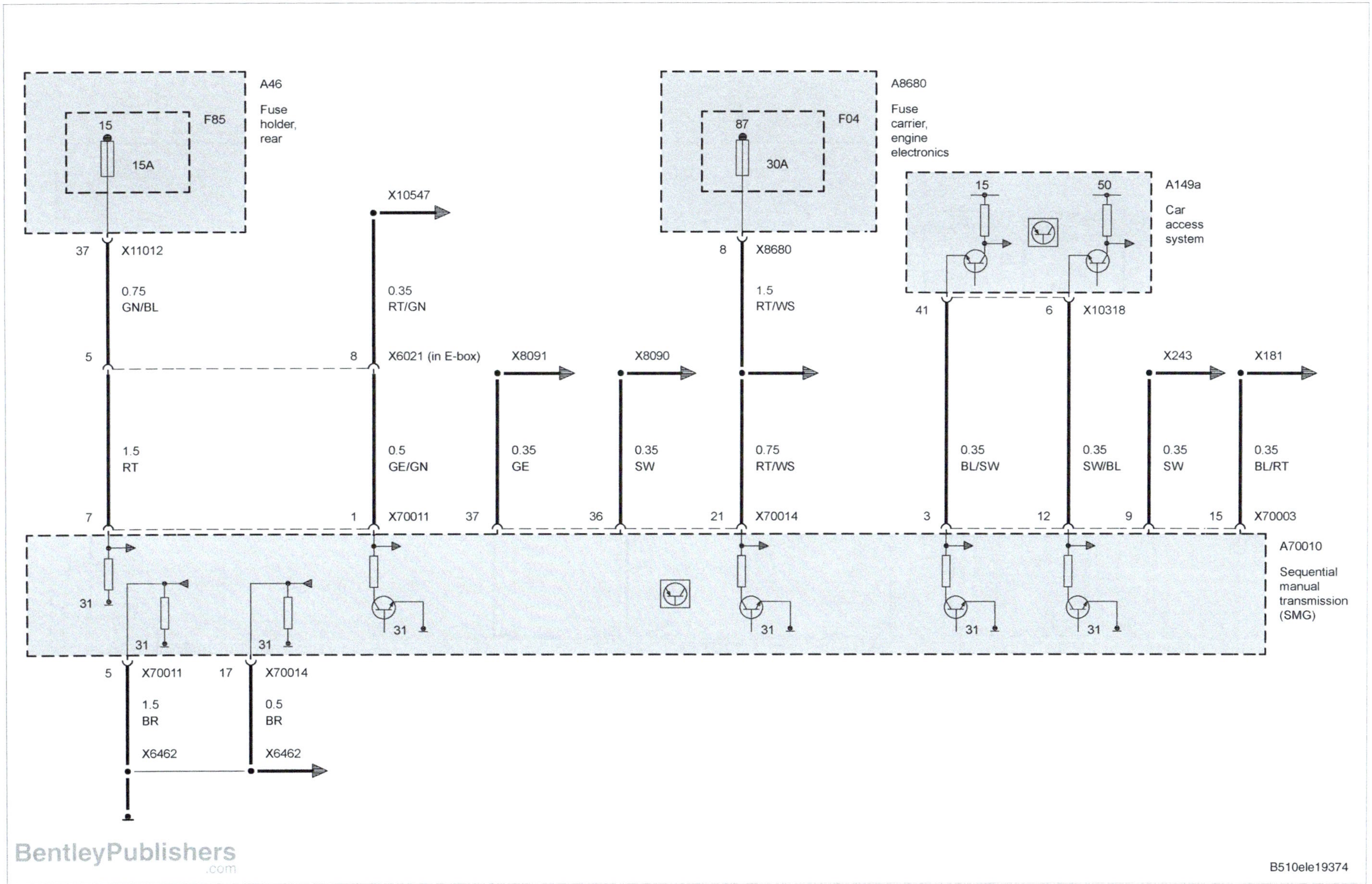

SMG control module, power supply
N62 TU engine as of 09 / 2005

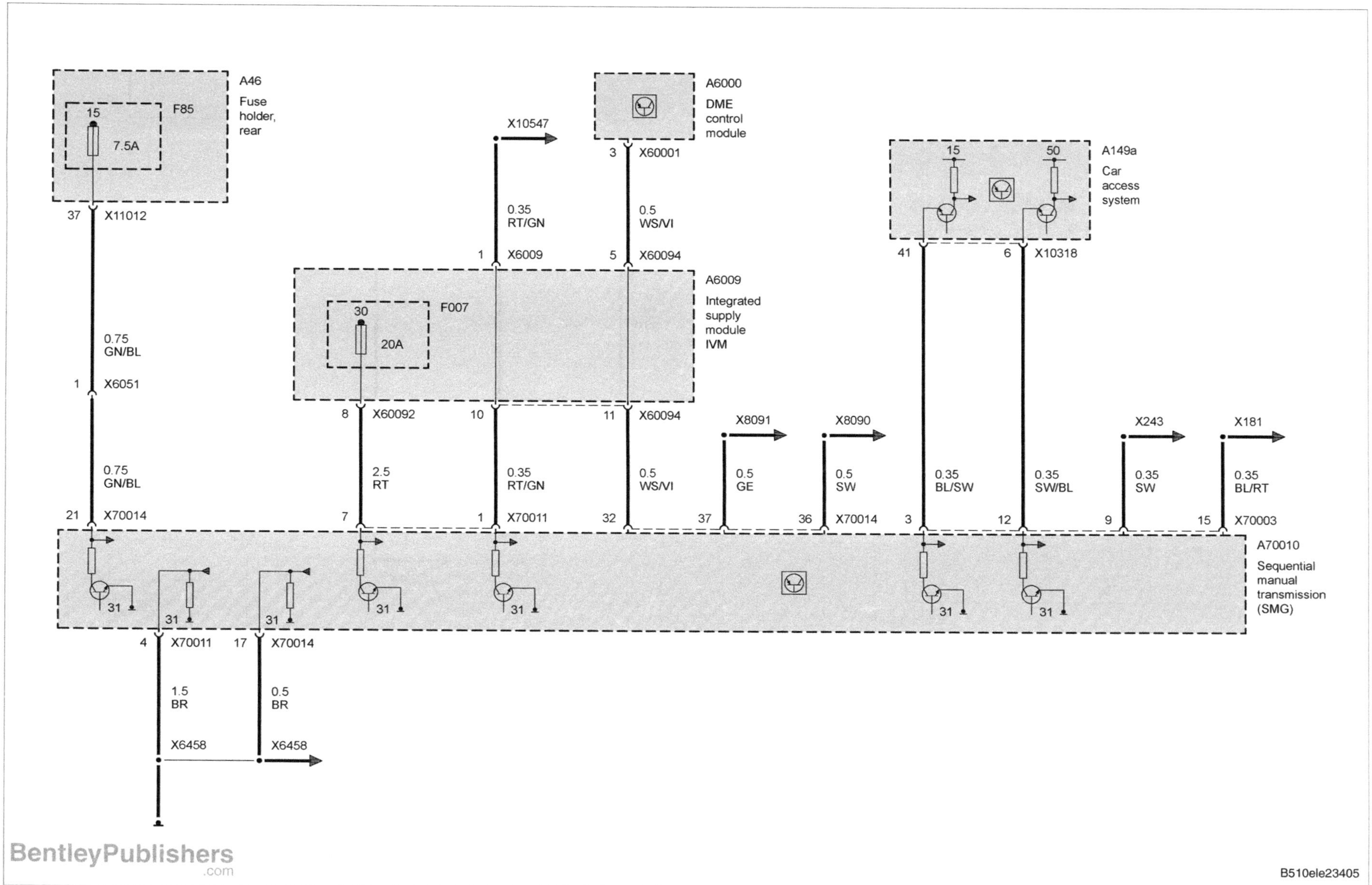

BentleyPublishers.com

SMG transmission control
Light module
power supply, up to 03 / 2005

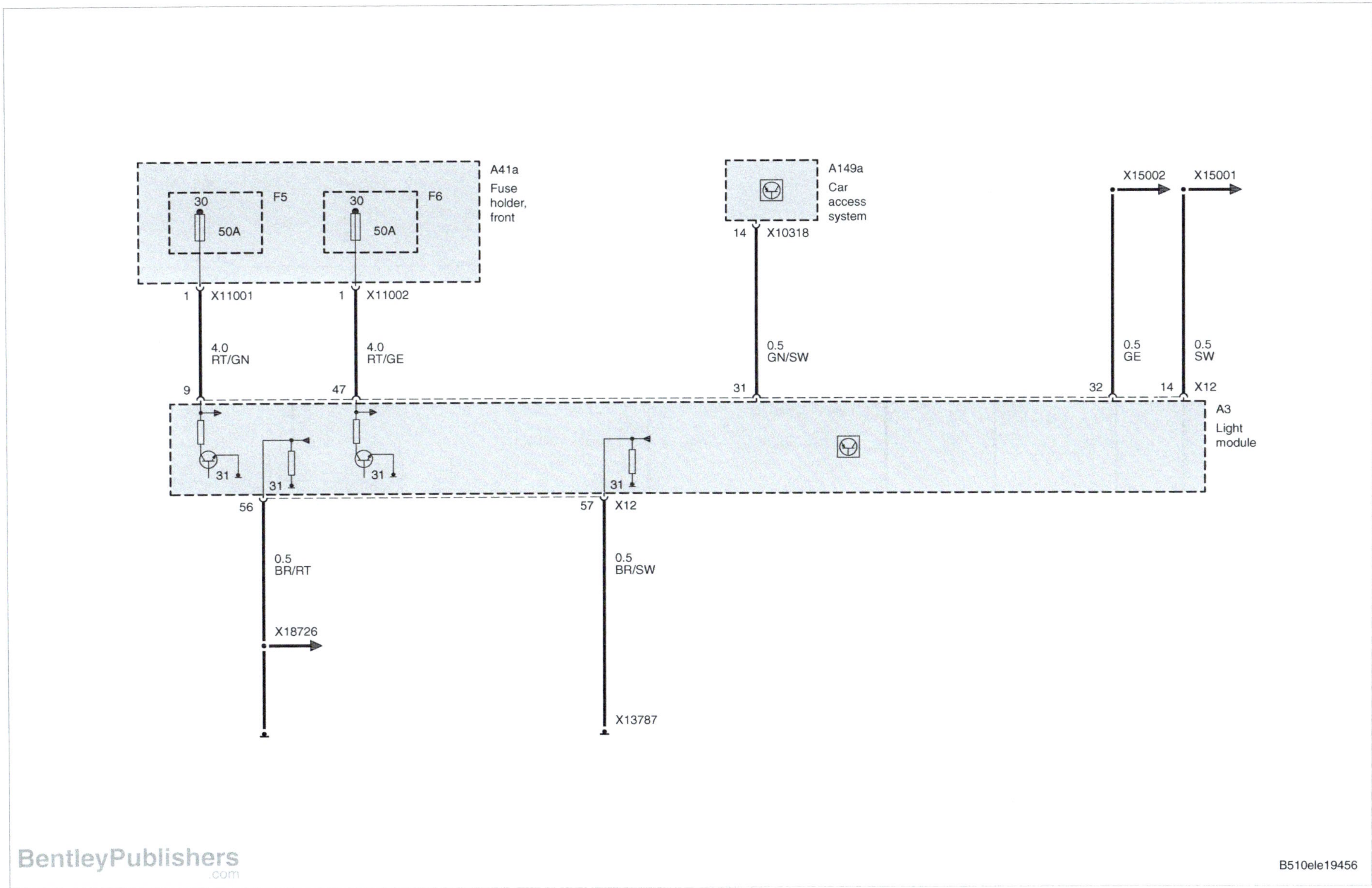

SMG transmission control
Light module
power supply, as of 03 / 2005

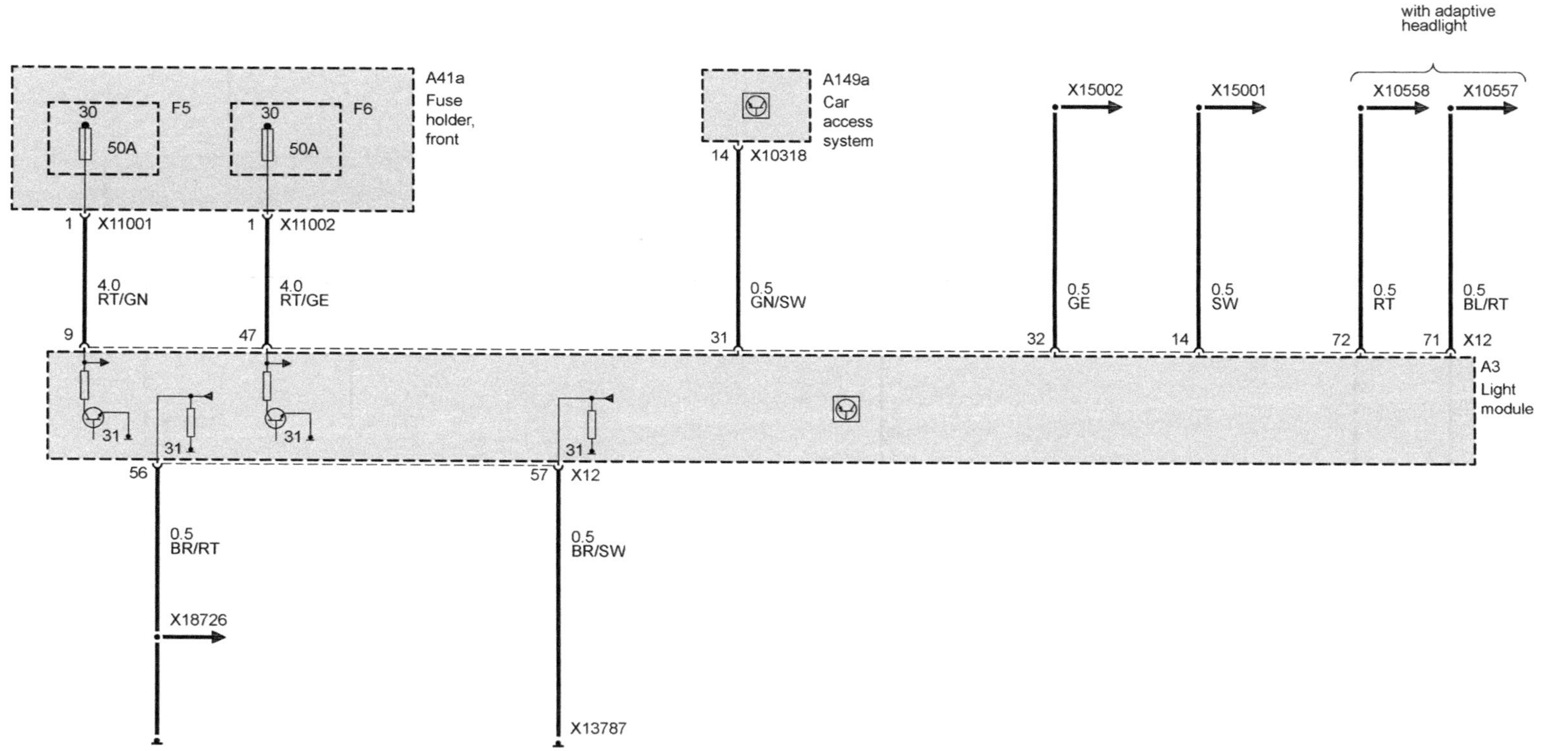

BentleyPublishers
.com

B510ele22930

SMG transmission control
Sensor system
M54, N62, N52 engines

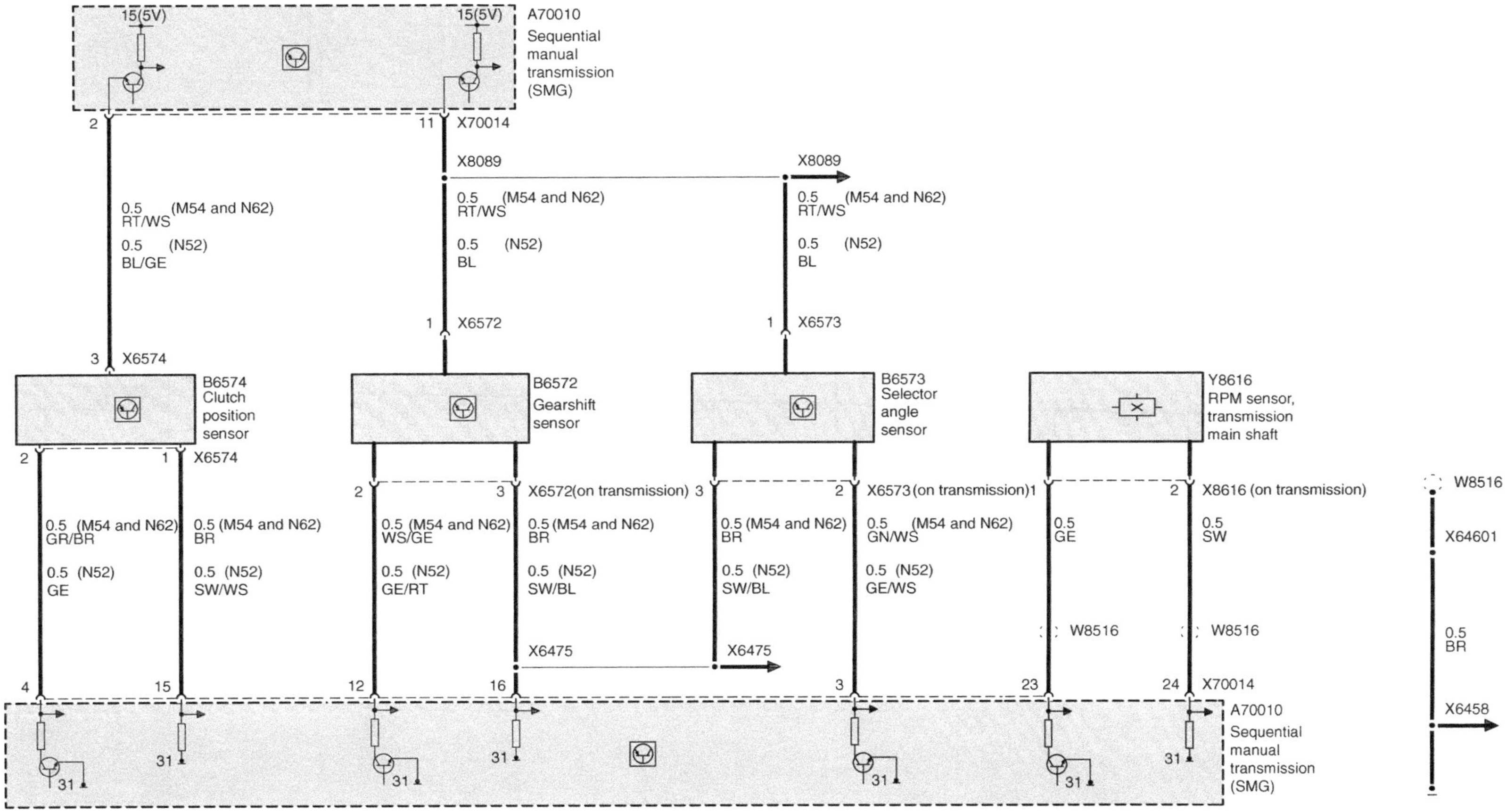

19372, 19370

BentleyPublishers.com

B510ele19372

SMG transmission control
Hydraulic system
M54 engine

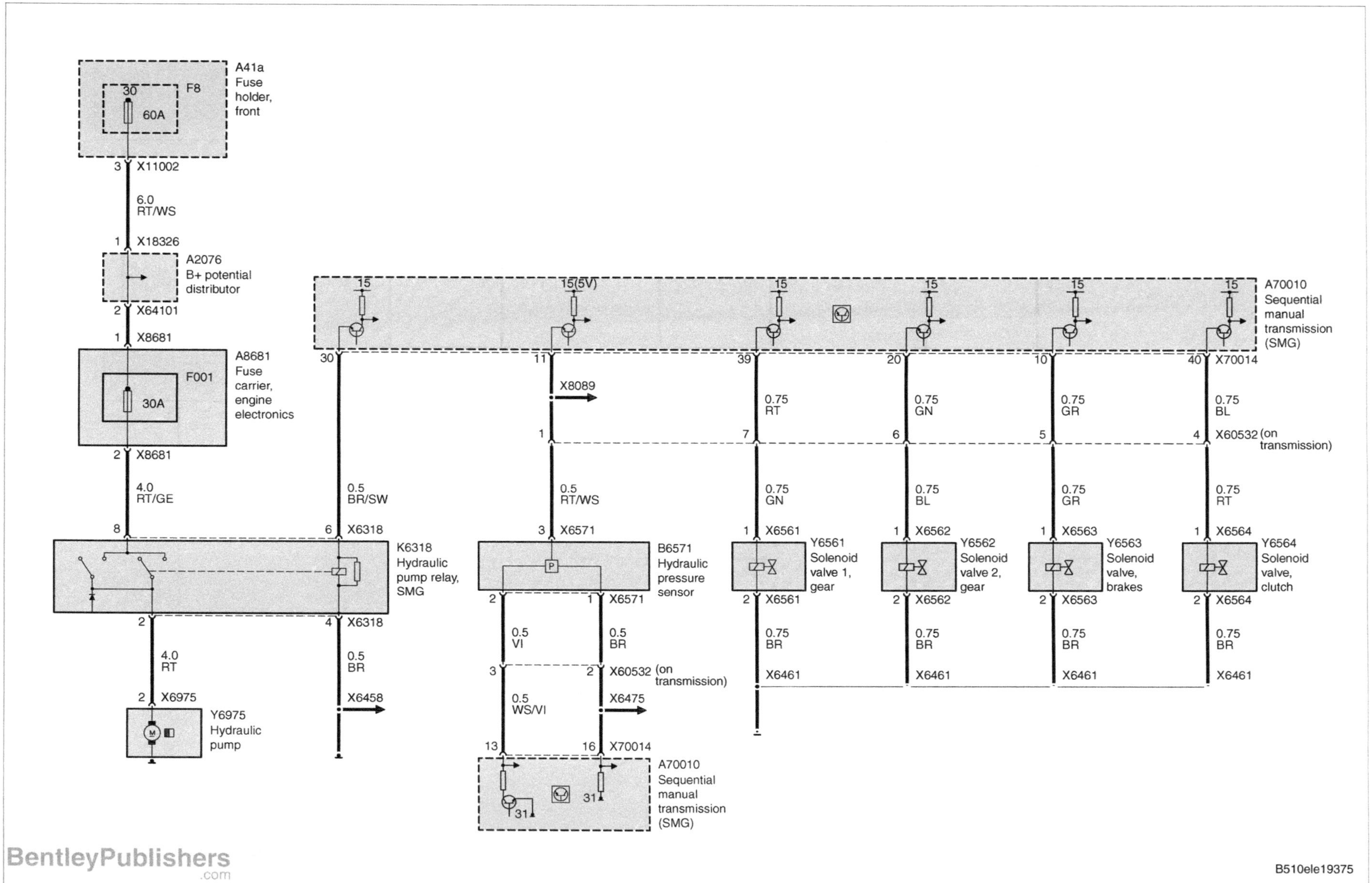

BentleyPublishers
.com

SMG transmission control
Hydraulic system
N62 engine

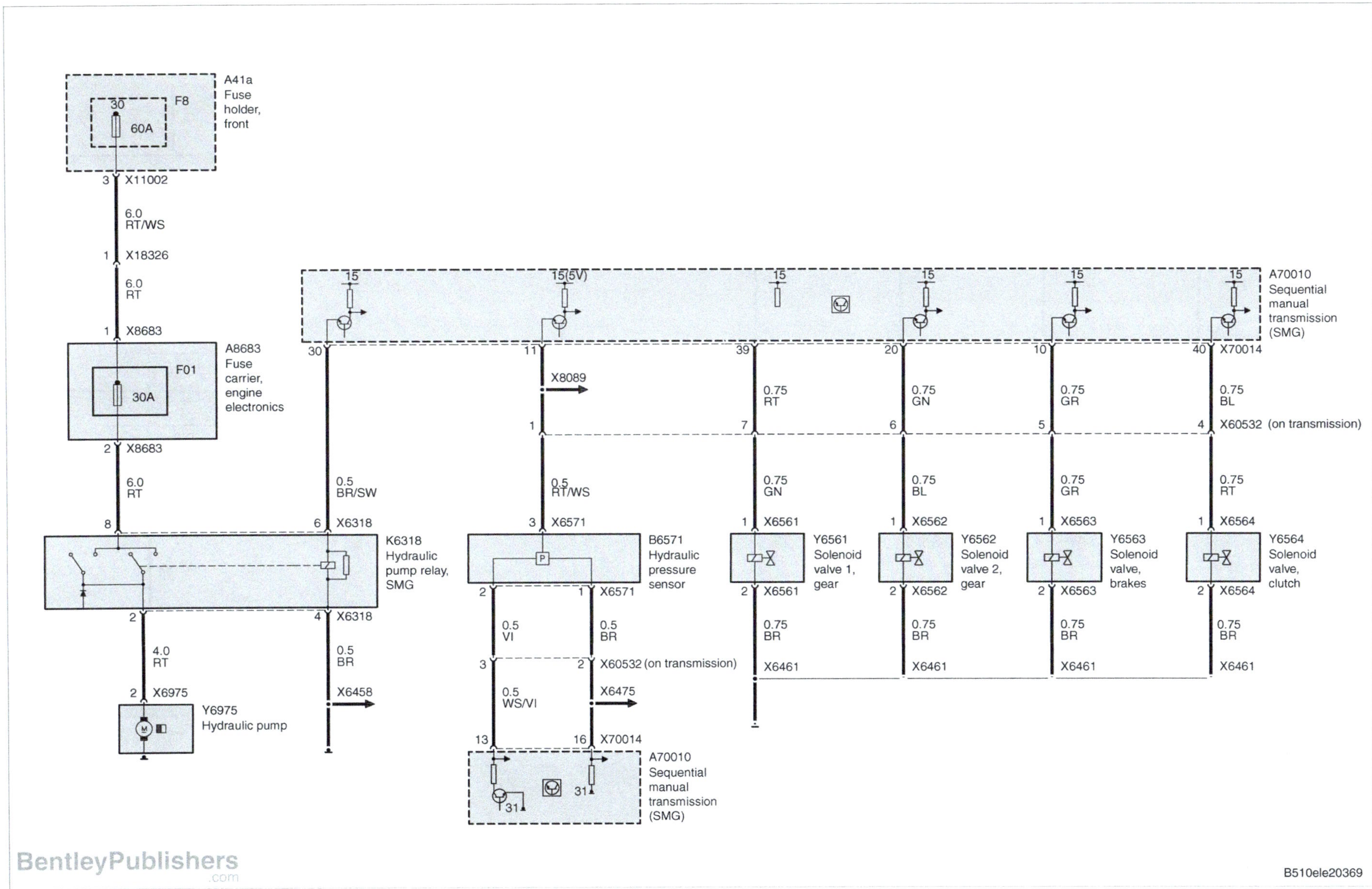

SMG transmission control
Hydraulic system
N52 engine

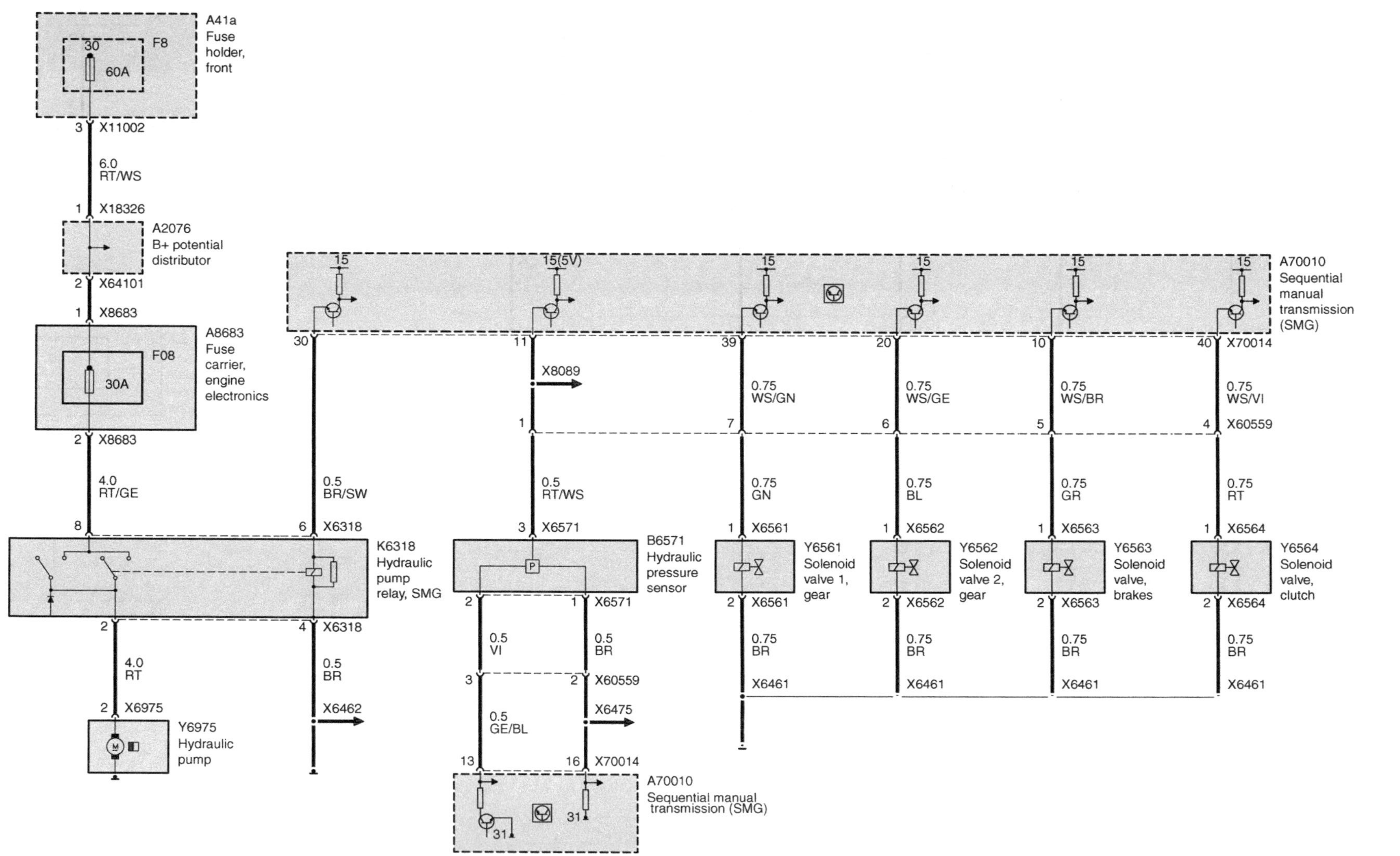

BentleyPublishers.com

B510ele19373

SMG transmission control
Selector lever and gear indicator lighting

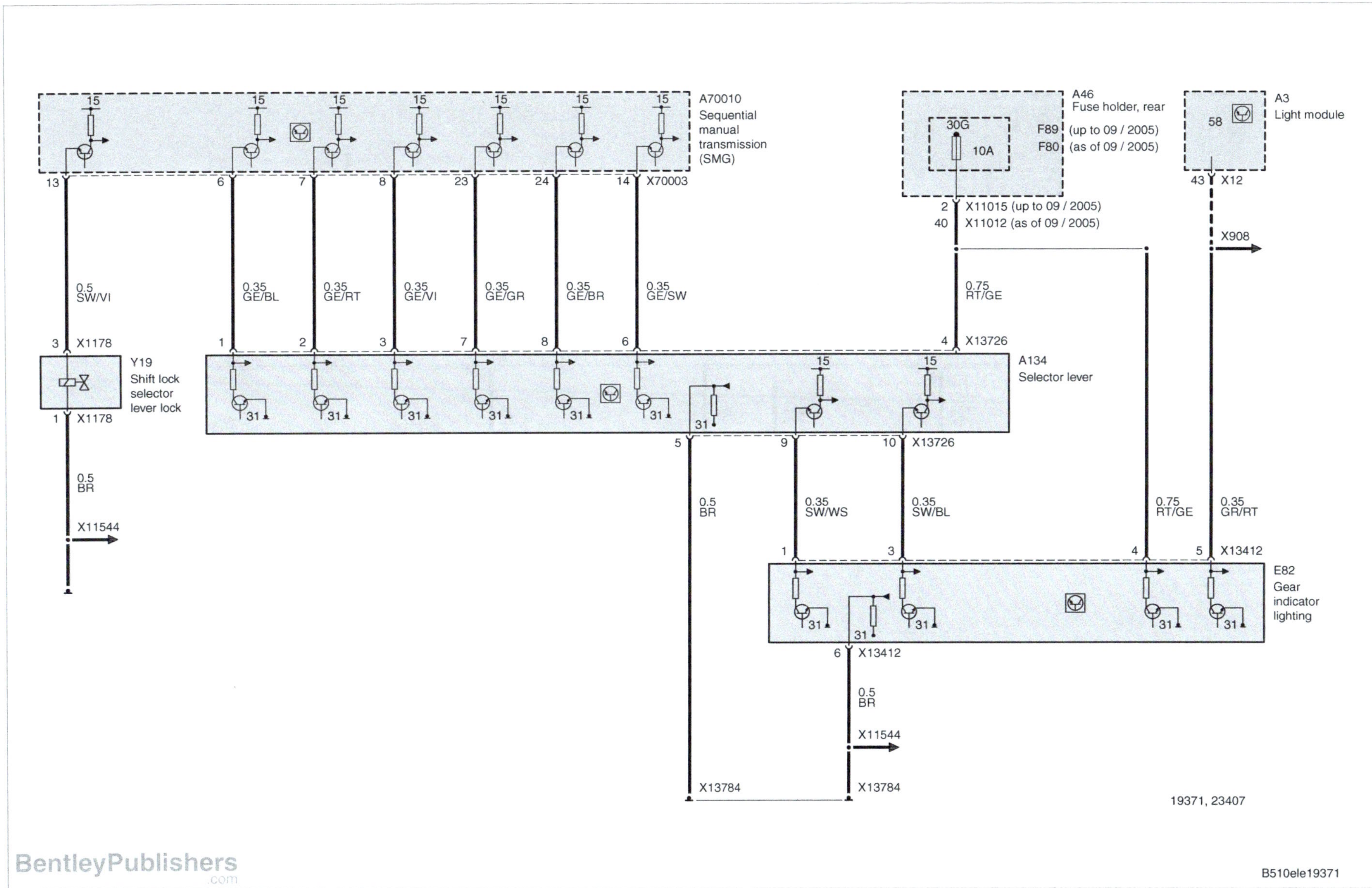

BentleyPublishers.com

Steering column switch center power supply (up to 03 / 2005) (page 1 of 2)

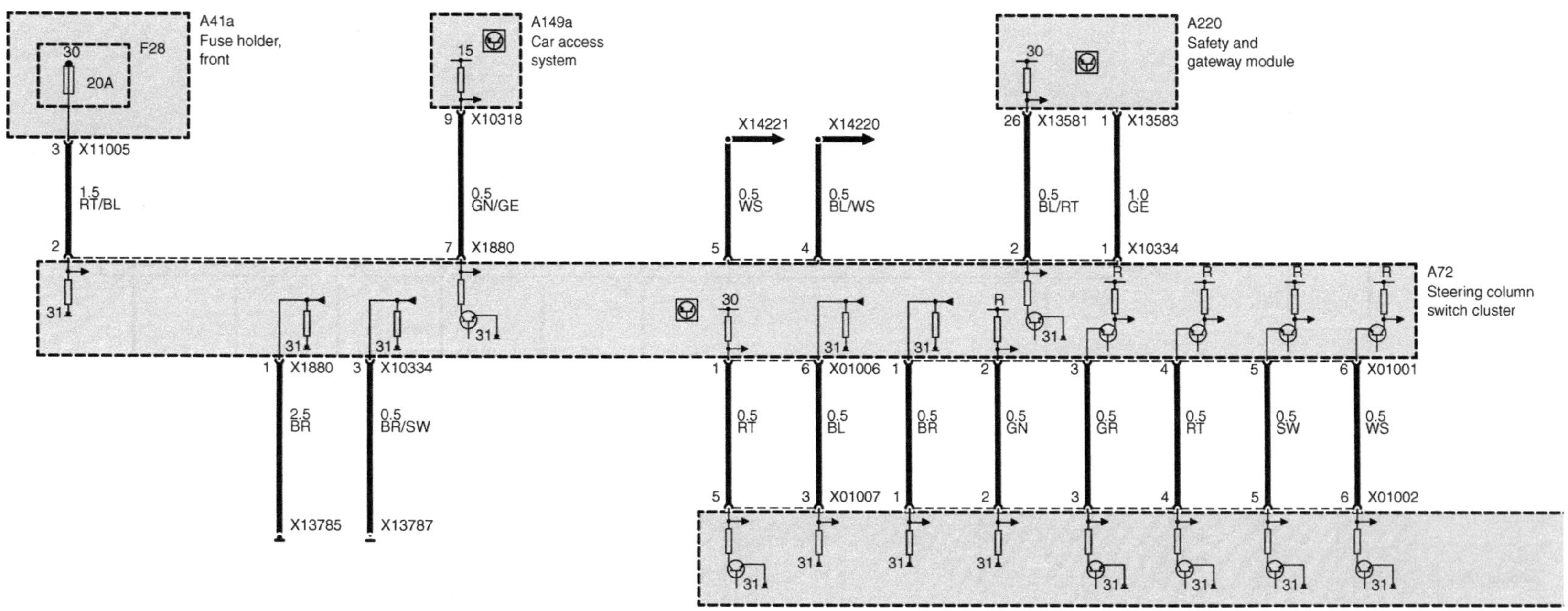

B510ele19914a

BentleyPublishers.com

Steering column switch center power supply (up to 03 / 2005) (page 2 of 2)

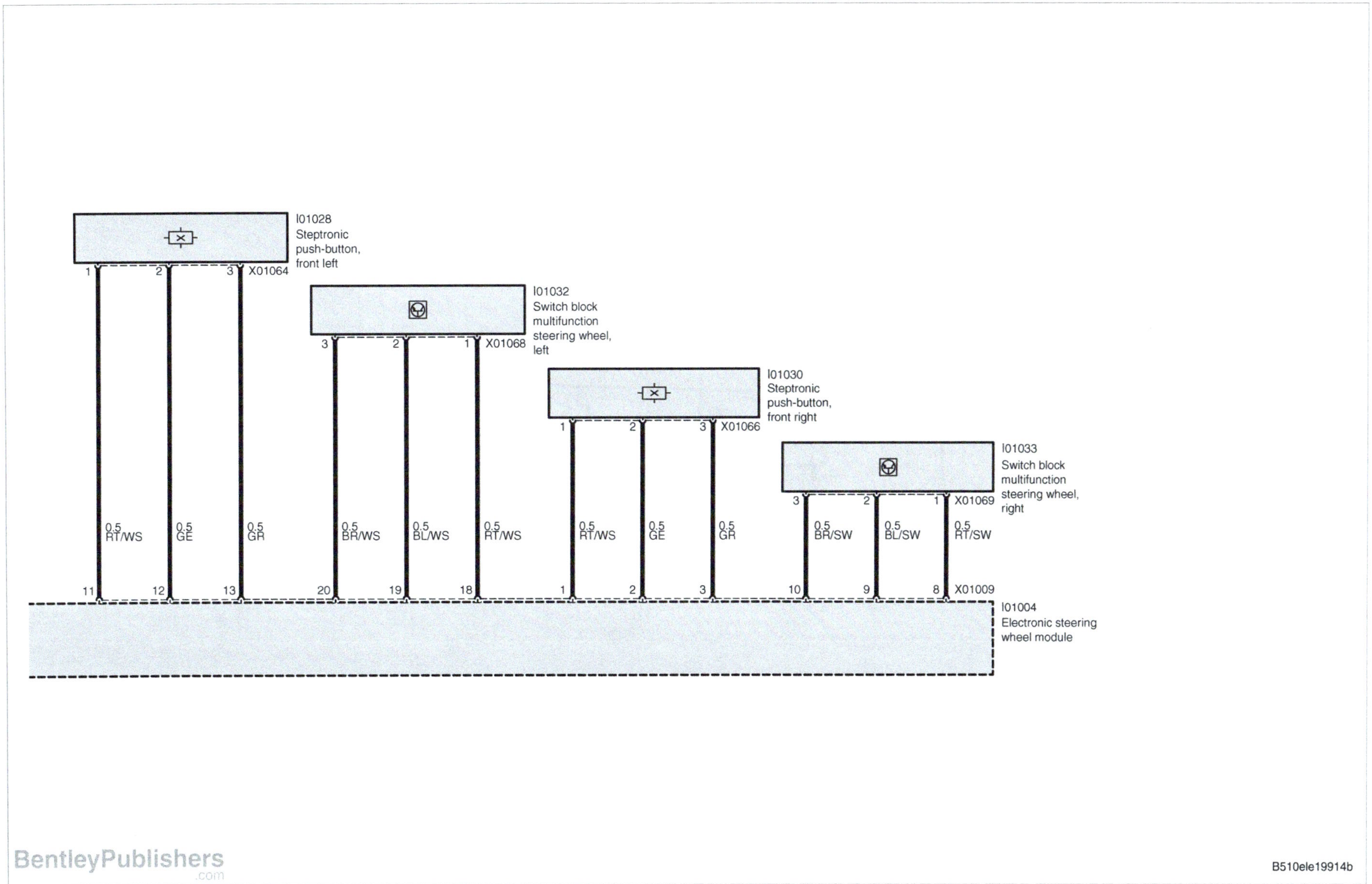

Steering column switch center power supply (as of 03 / 2005 up to 09 / 2005) (page 1 of 2)

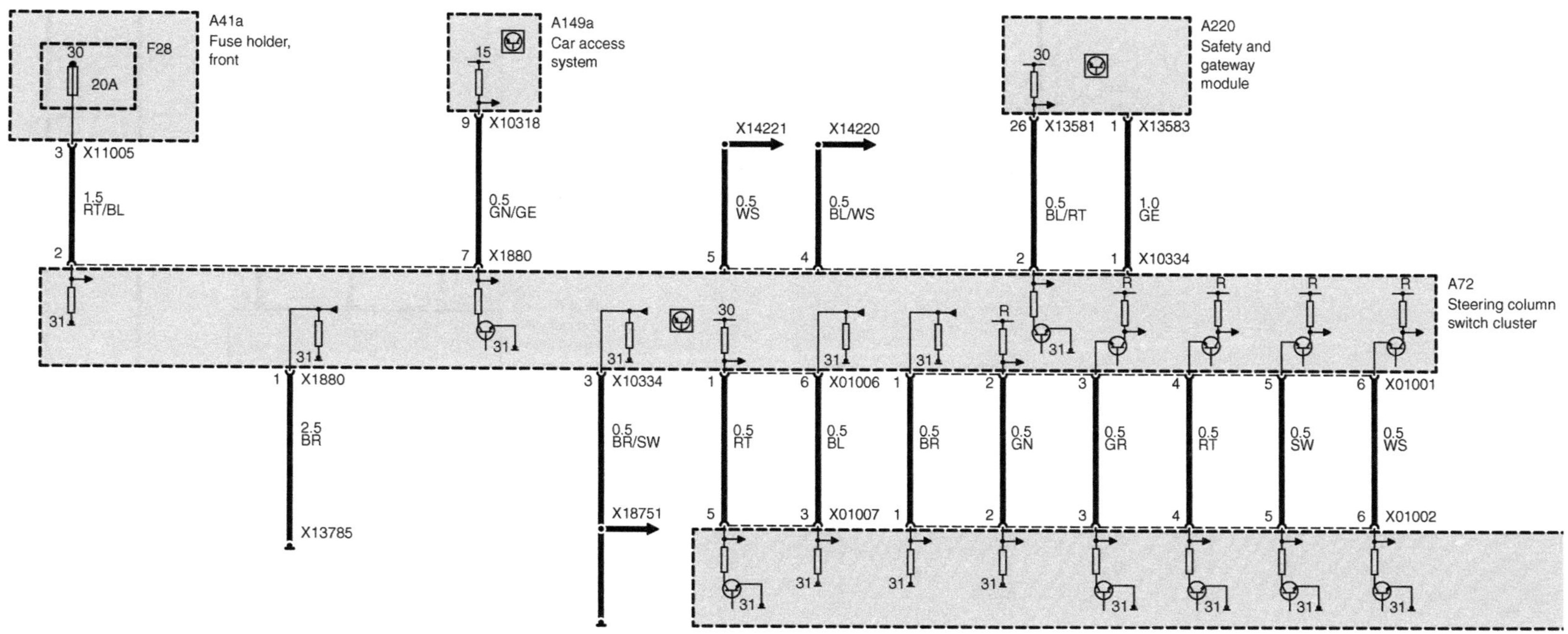

B510ele22755a

Steering column switch center power supply (as of 03 / 2005 up to 09 / 2005) (page 2 of 2)

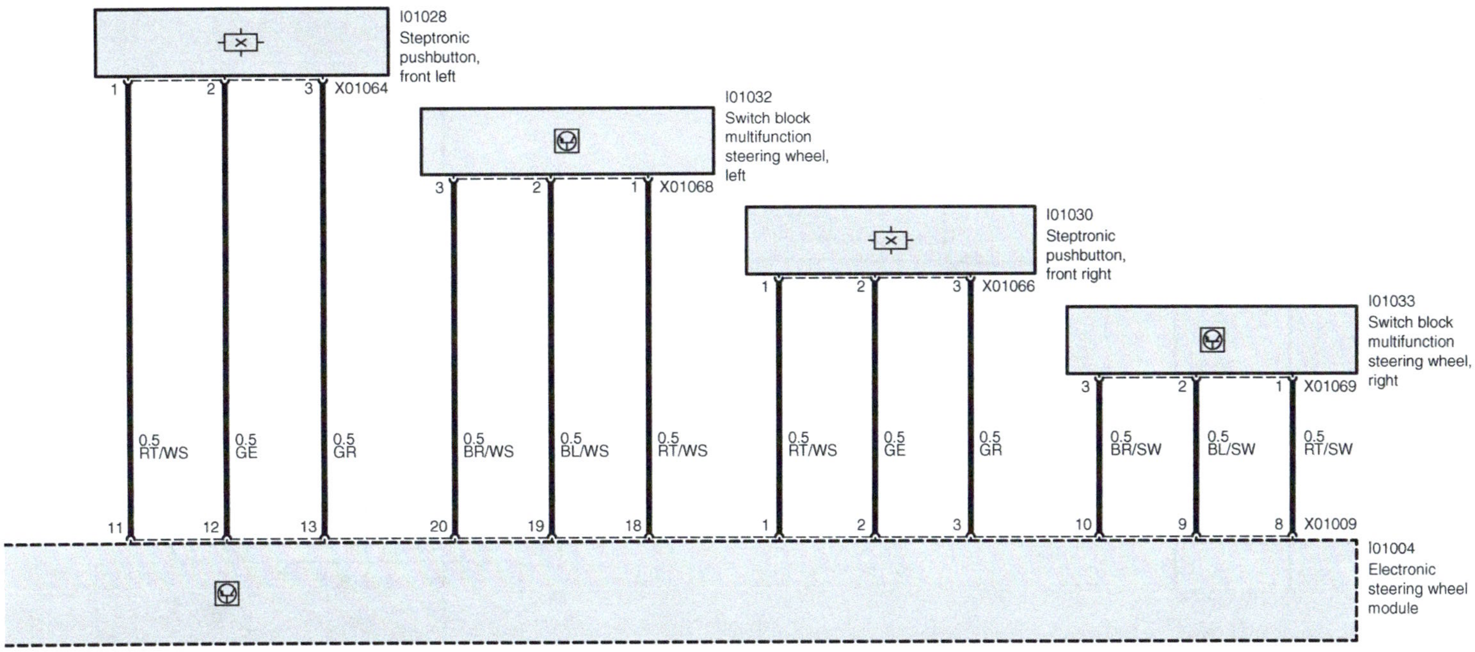

Steering column switch center power supply (as of 09 / 2005 up to 03 / 2007) (page 1 of 2)

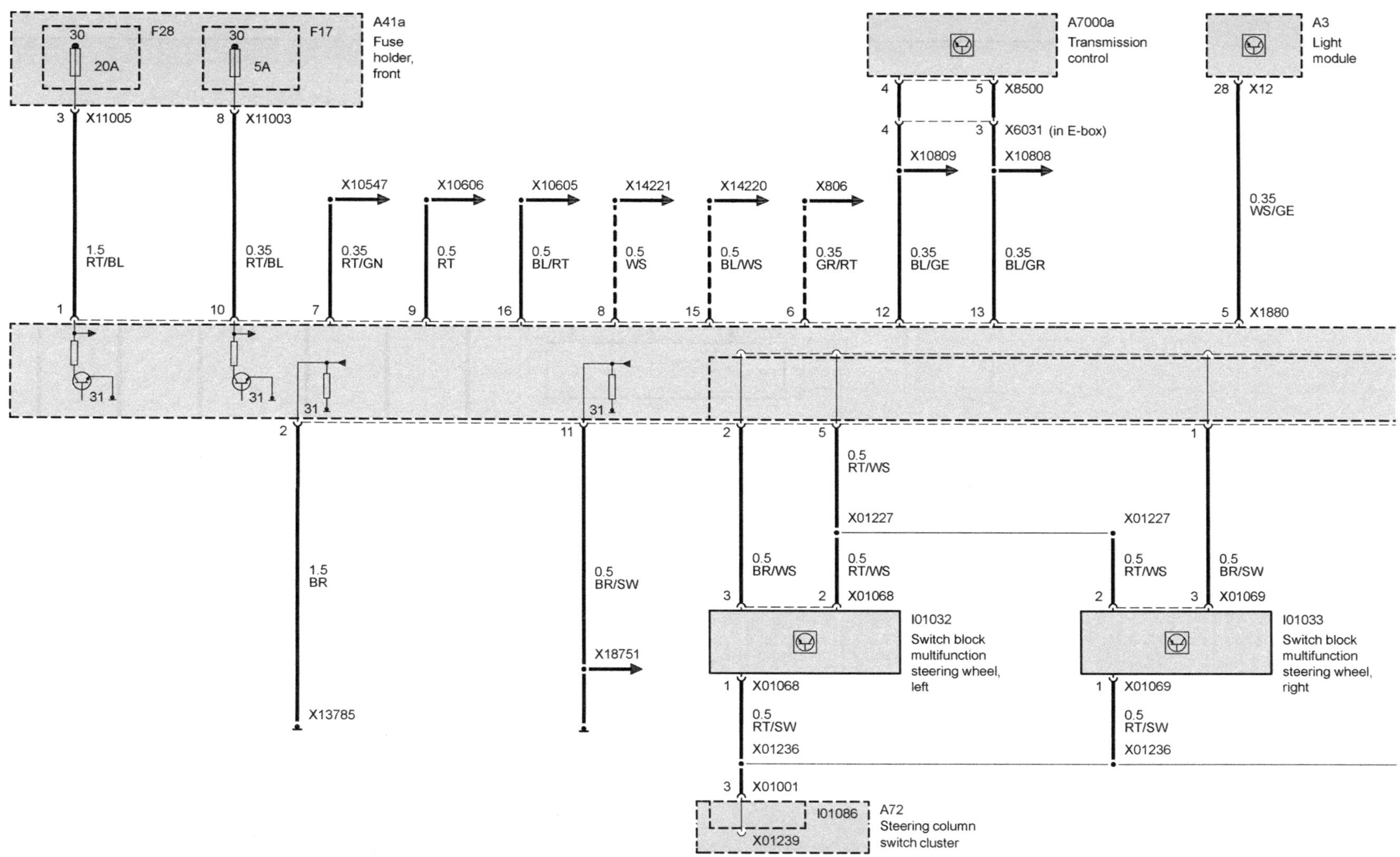

BentleyPublishers.com

Steering column switch center power supply (as of 09 / 2005 up to 03 / 2007) (page 2 of 2)

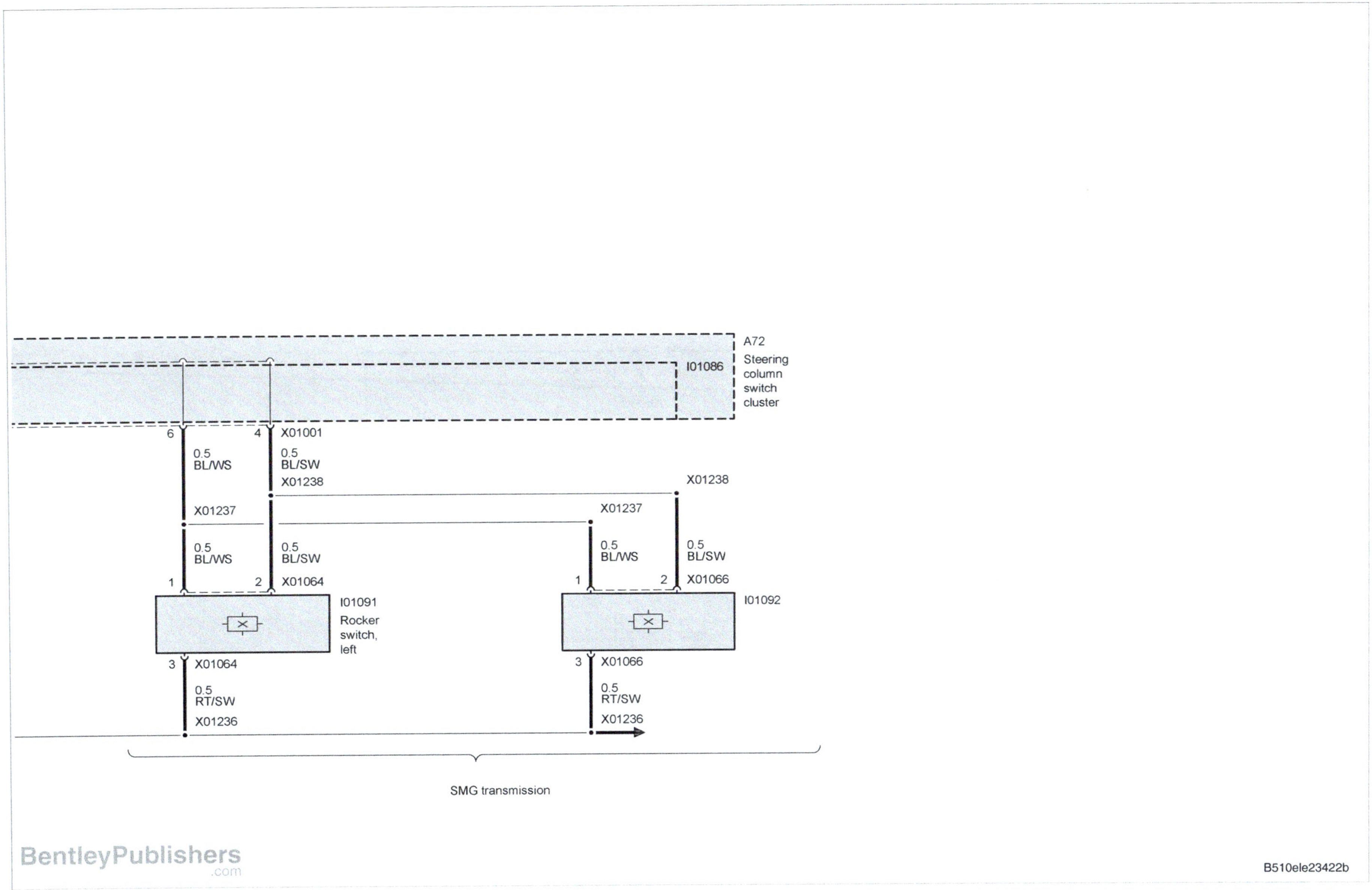

Steering column switch center power supply (as of 03 / 2007 up to 09 / 2007) (page 1 of 2)

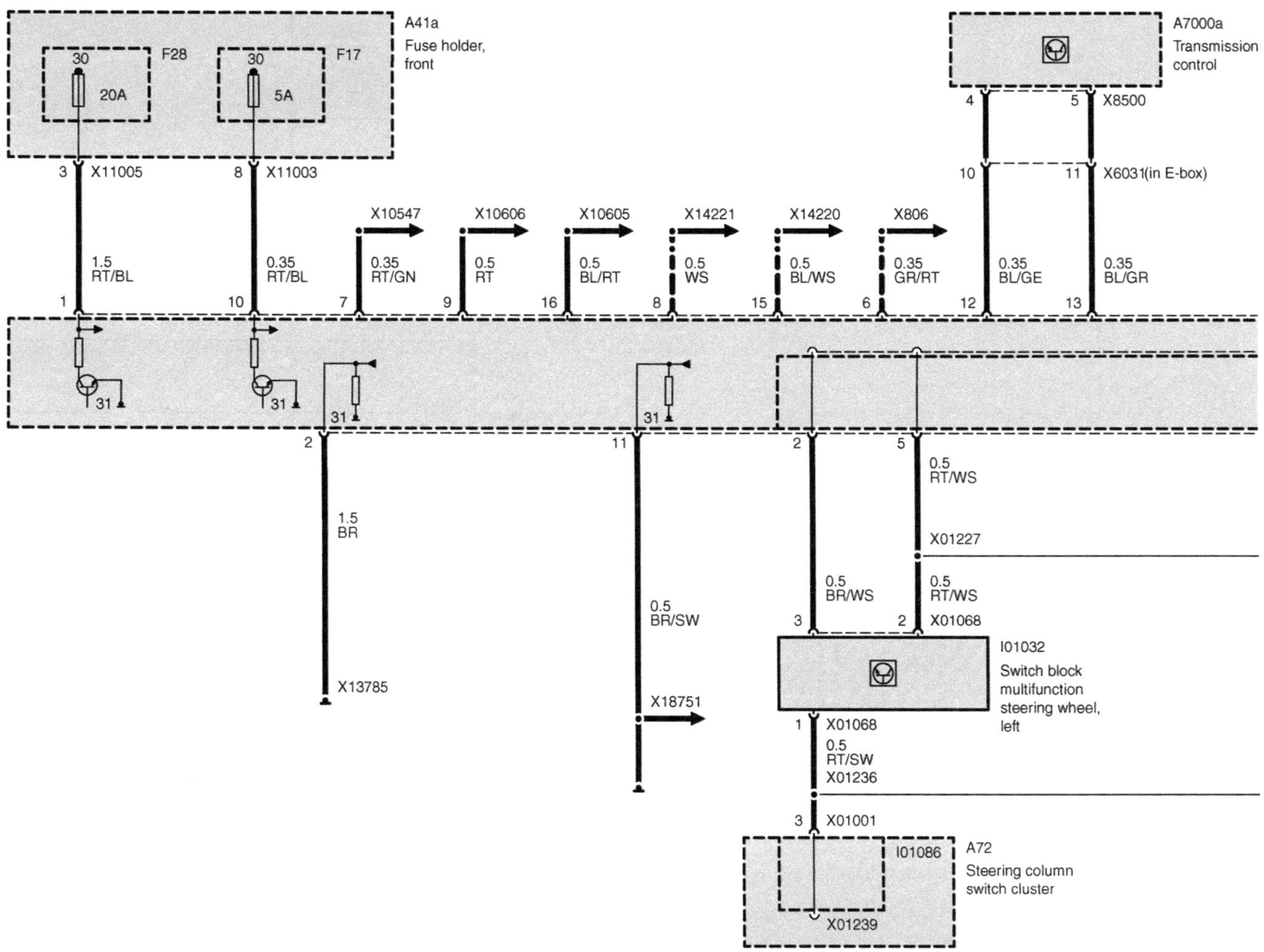

BentleyPublishers.com

Steering column switch center power supply (as of 09 / 2005 up to 03 / 2007) (page 2 of 2)

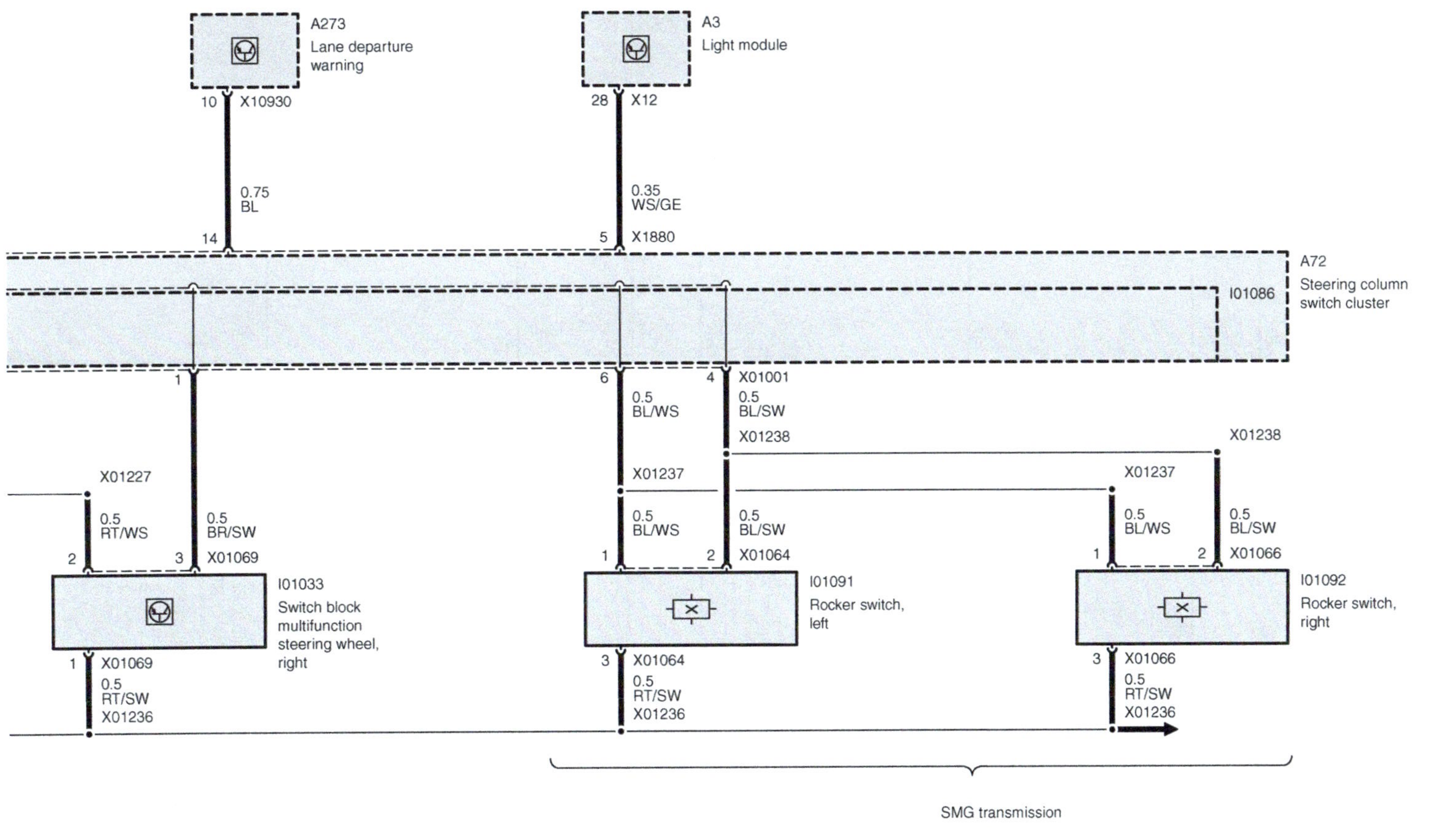

BentleyPublishers.com

Steering column switch center power supply (as of 09 / 2007)

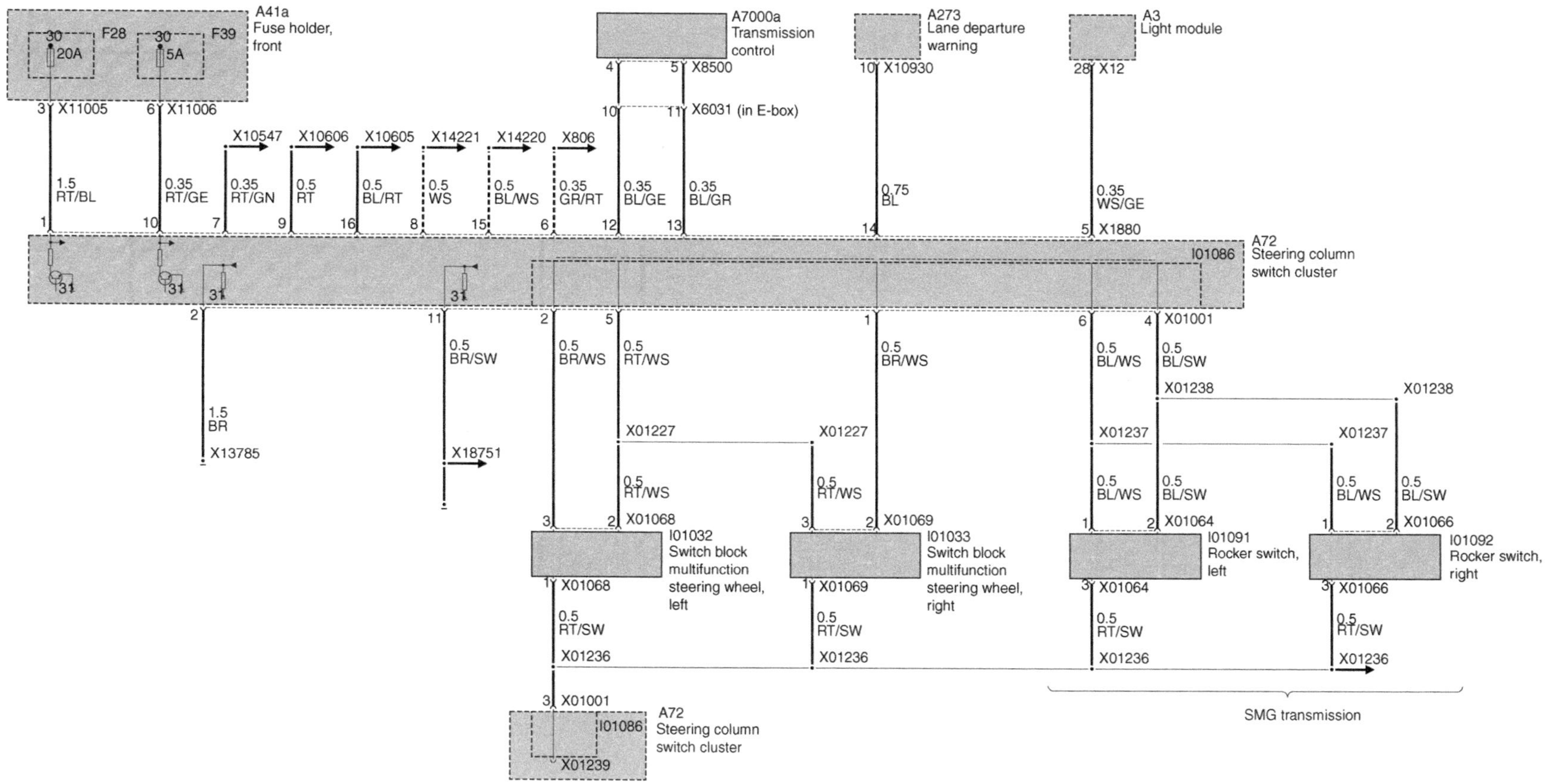

BentleyPublishers.com

B510ele51175

VTG transmission control
up to 09 / 2005

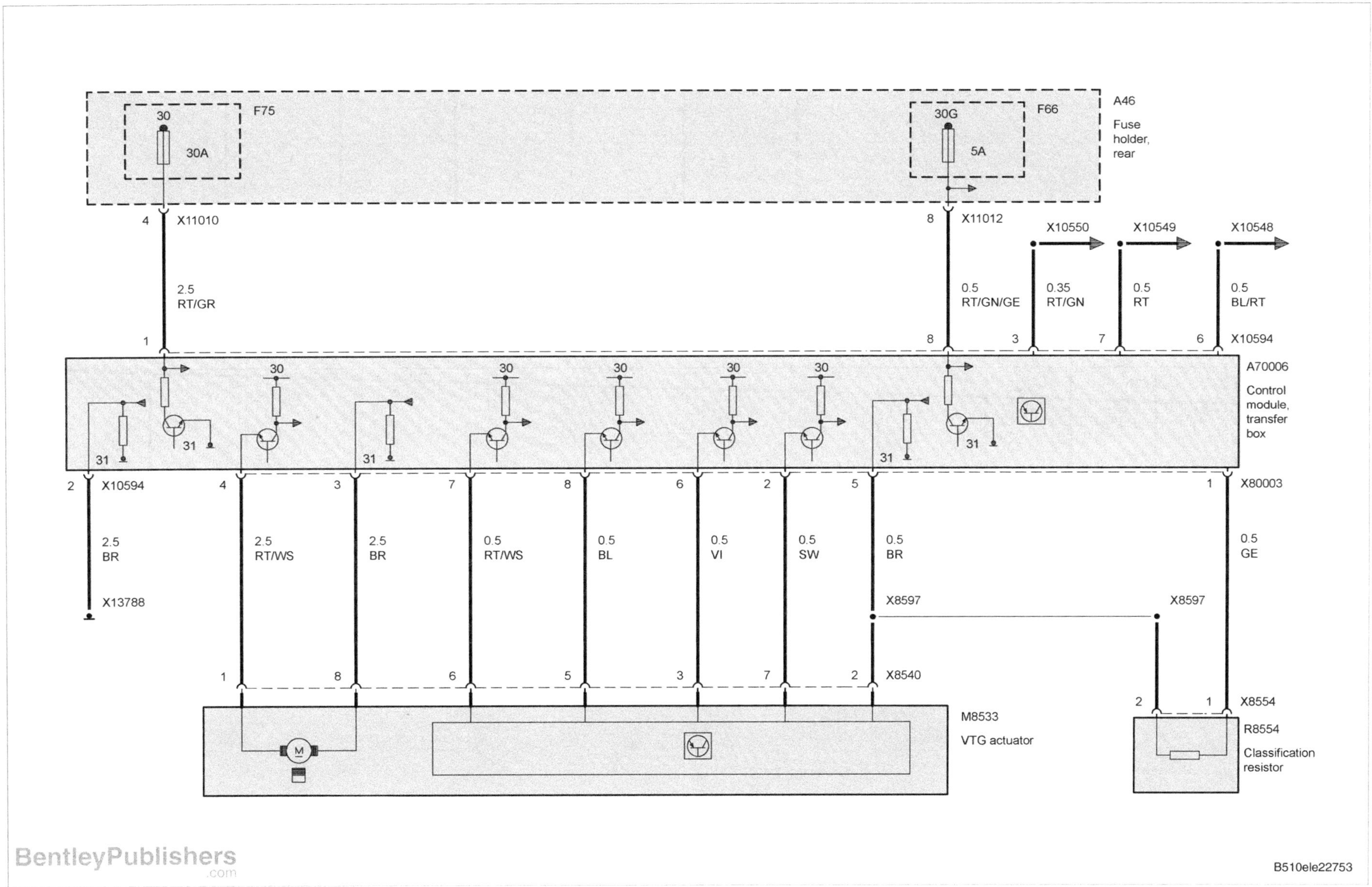

VTG transmission control as of 09 / 2005

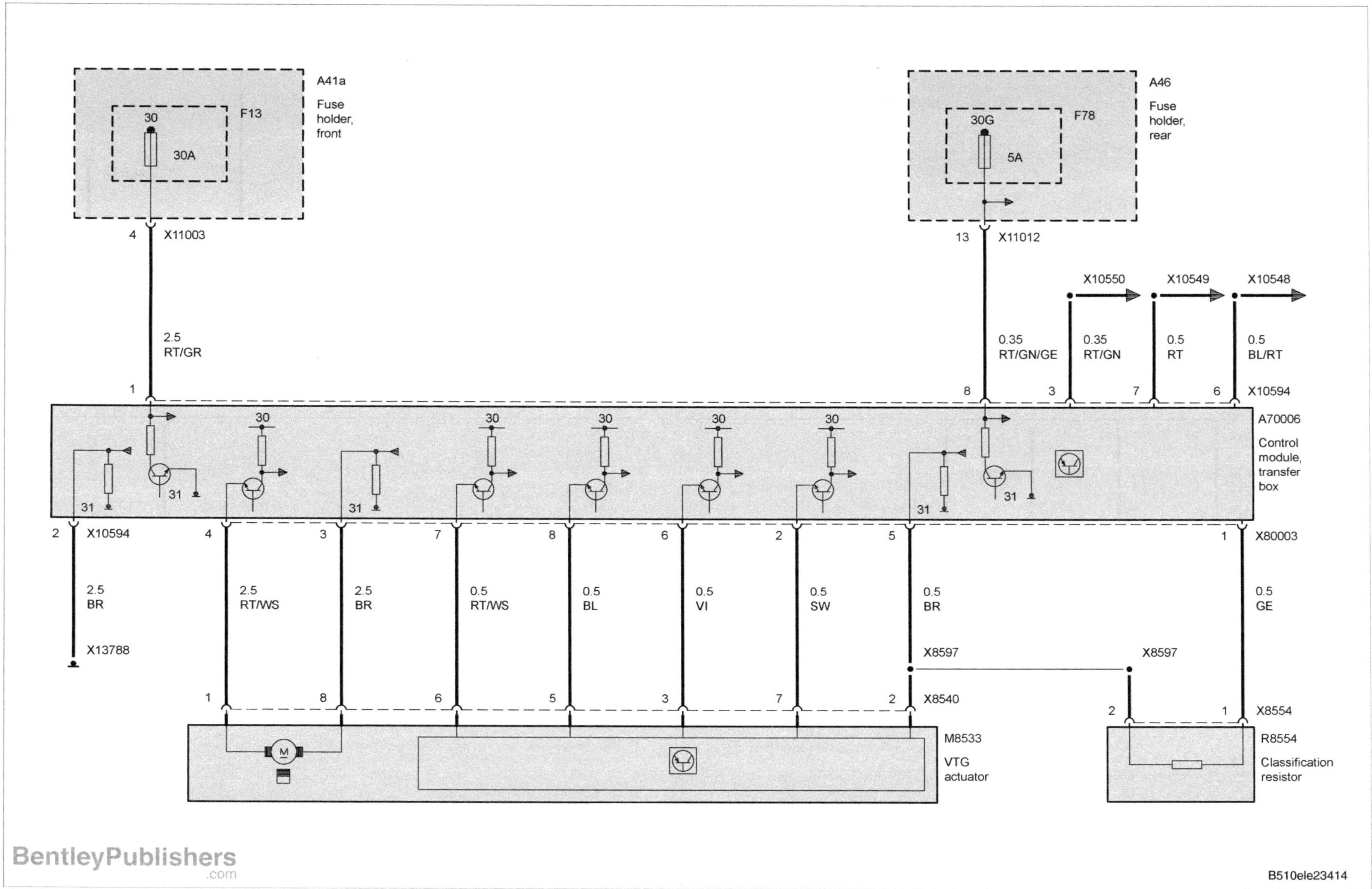

BentleyPublishers.com

EGS transmission control 6HPTU
Gear selector switch

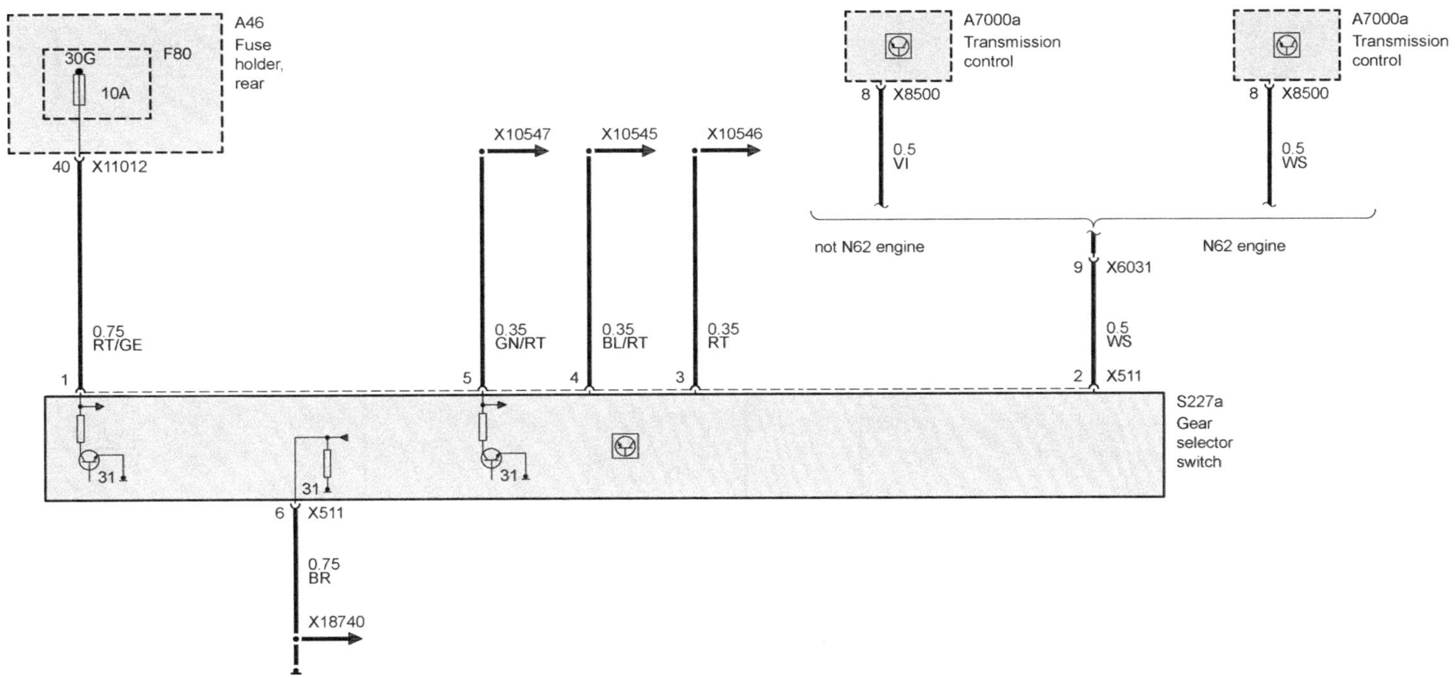

BentleyPublishers.com

B510ele25759

Convenient start M54 engine (page 1 of 2)

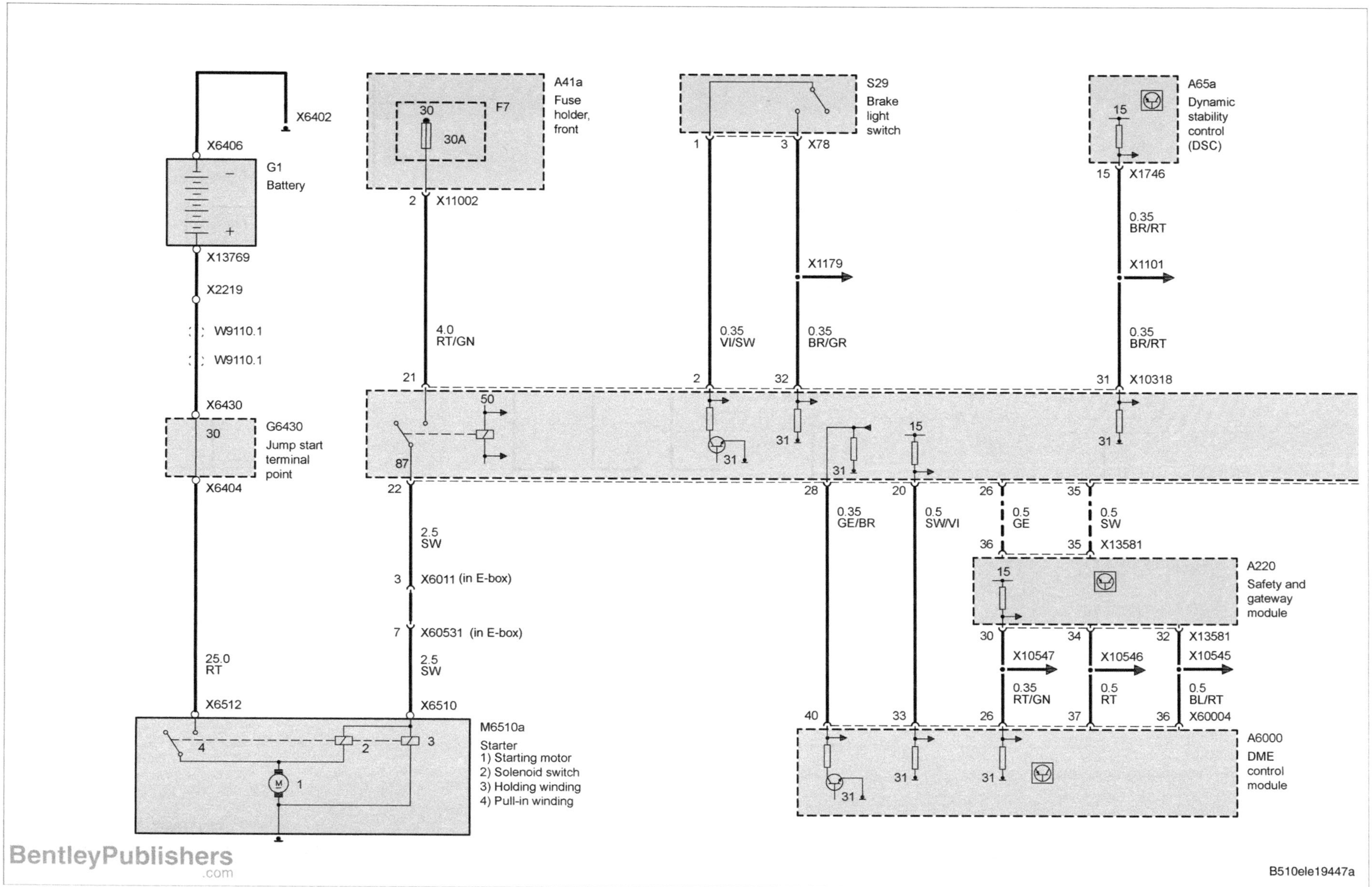

BentleyPublishers.com

B510ele19447a

Convenient start
M54 engine
(page 2 of 2)

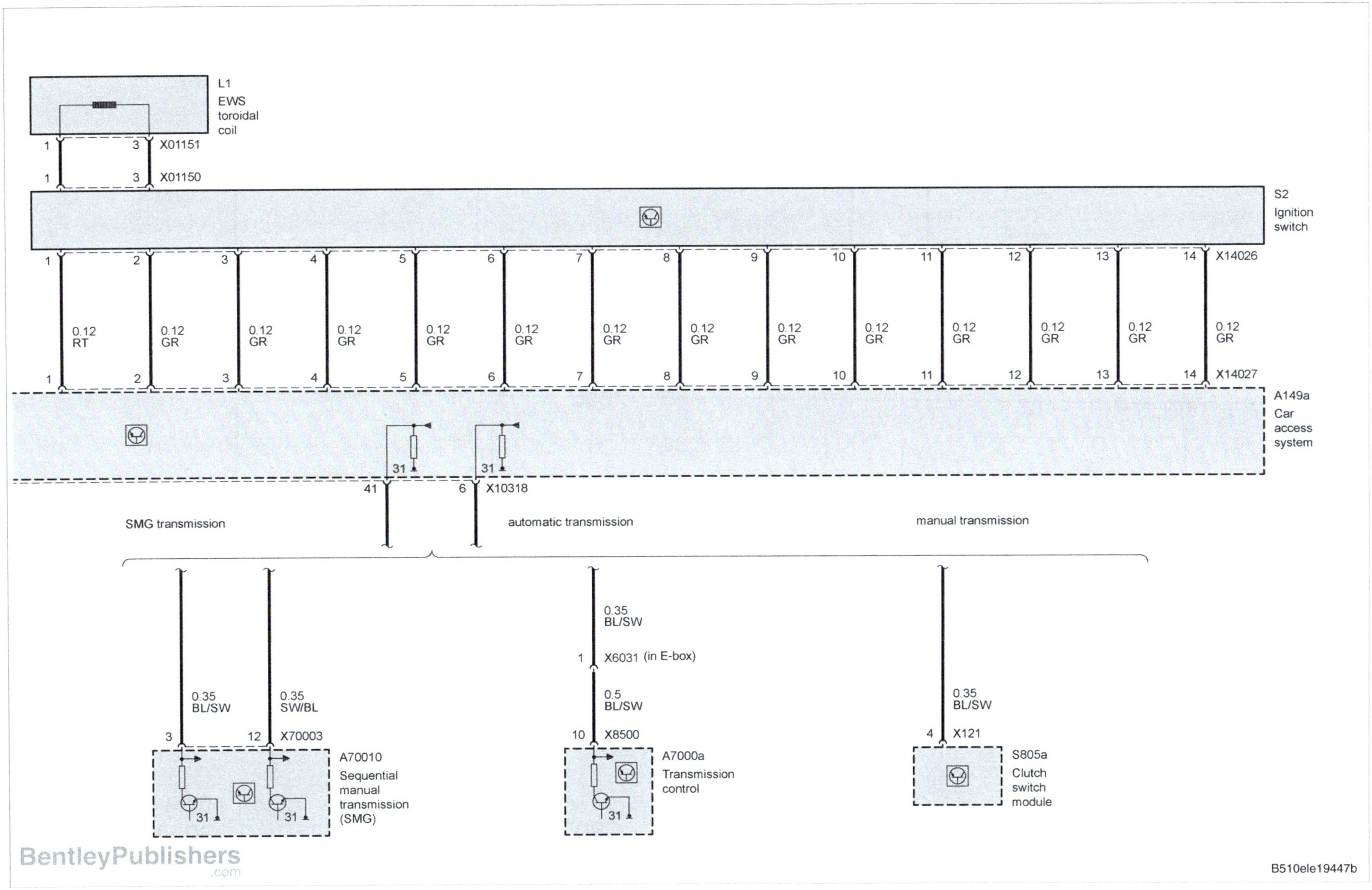

Convenient start
N52 engine with DSC Dynamnic Stability Control PLUS
as of 09 / 2005 (page 1 of 2)

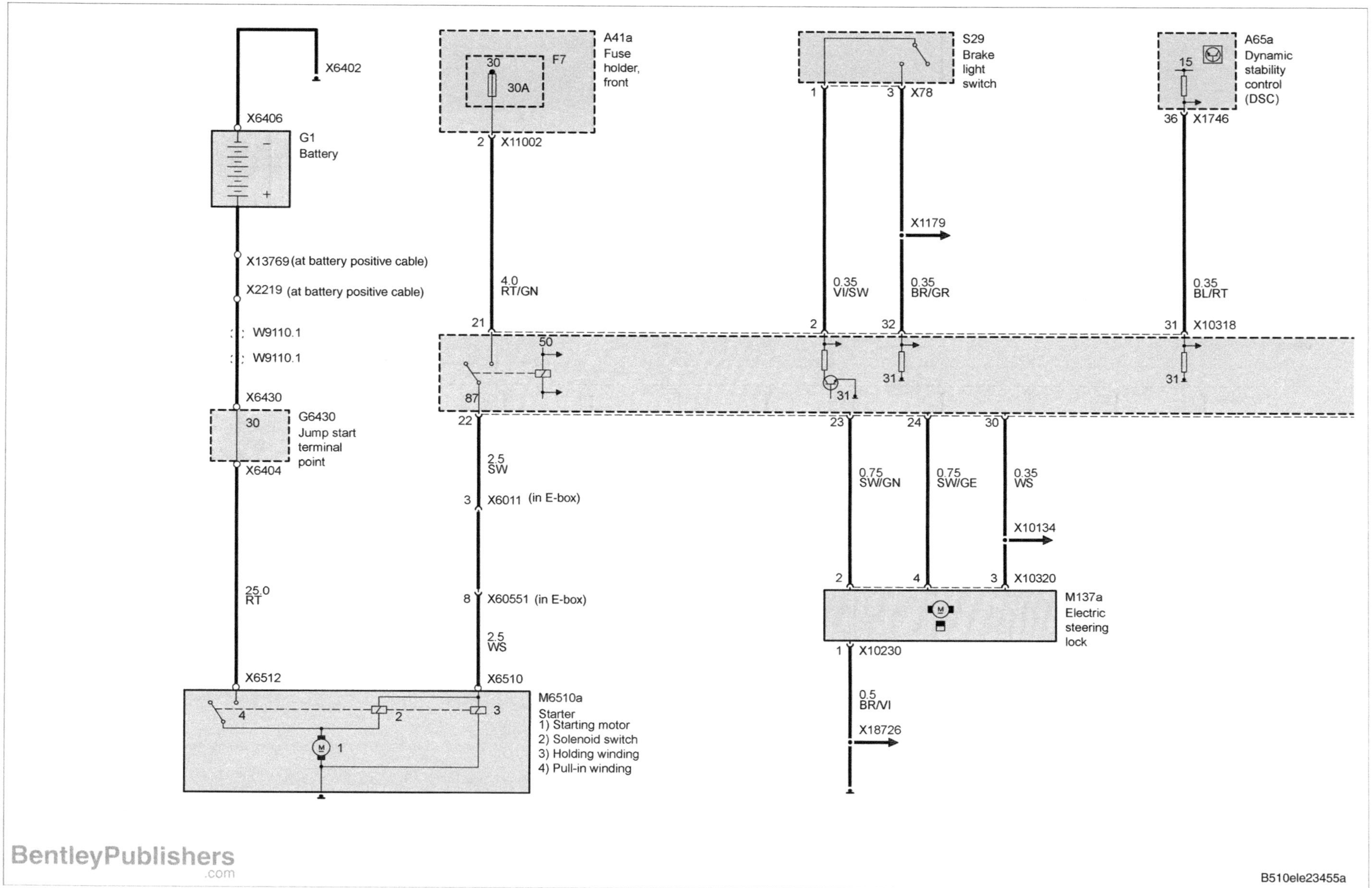

BentleyPublishers.com

B510ele23455a

Convenient start
N52 engine with DSC Dynamnic Stability Control PLUS
as of 09 / 2005 (page 2 of 2)

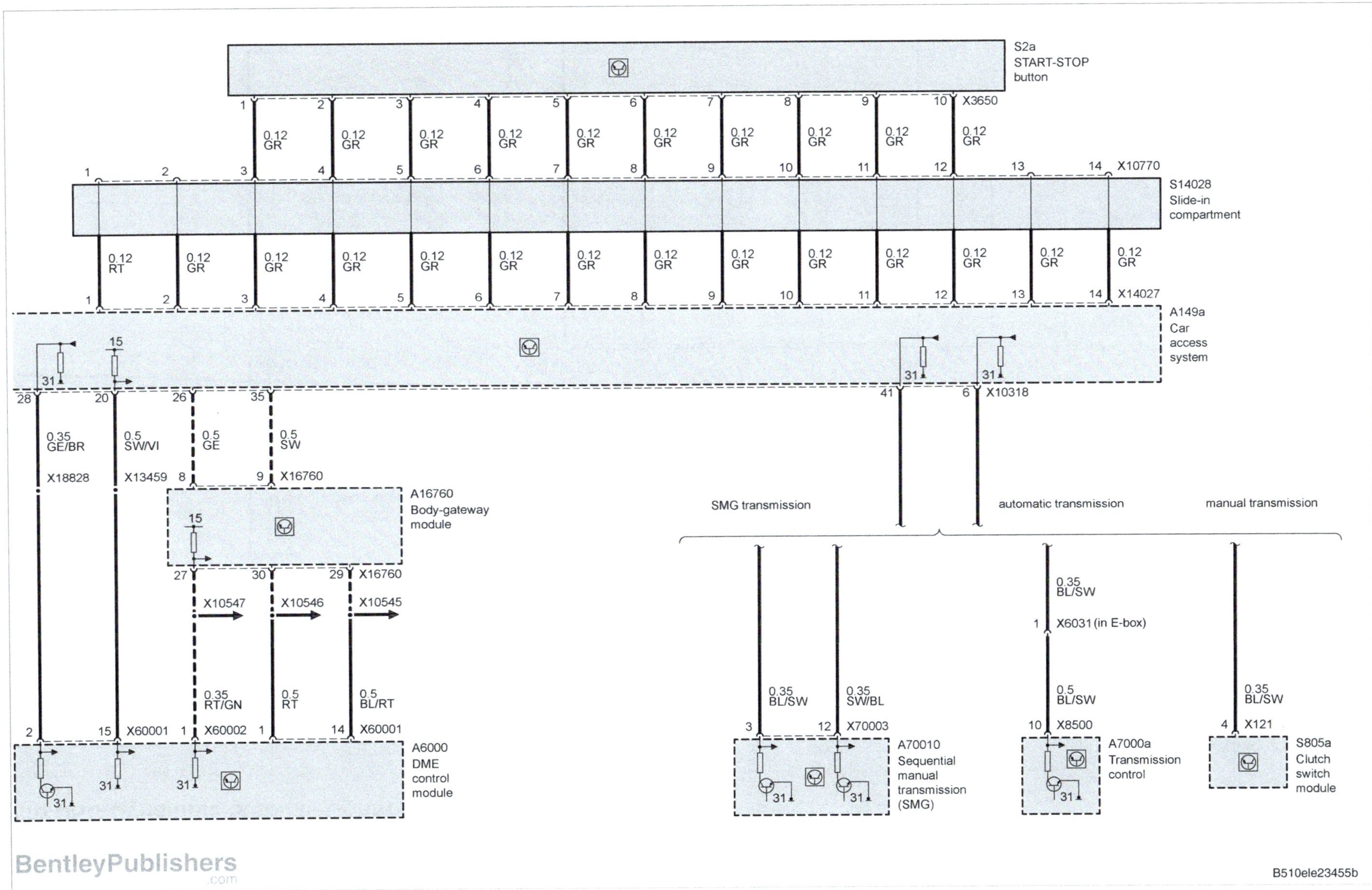

Convenient start
N52 engine or N54 engine
with DSC Dynamnic Stability Control Premium (page 1 of 2)

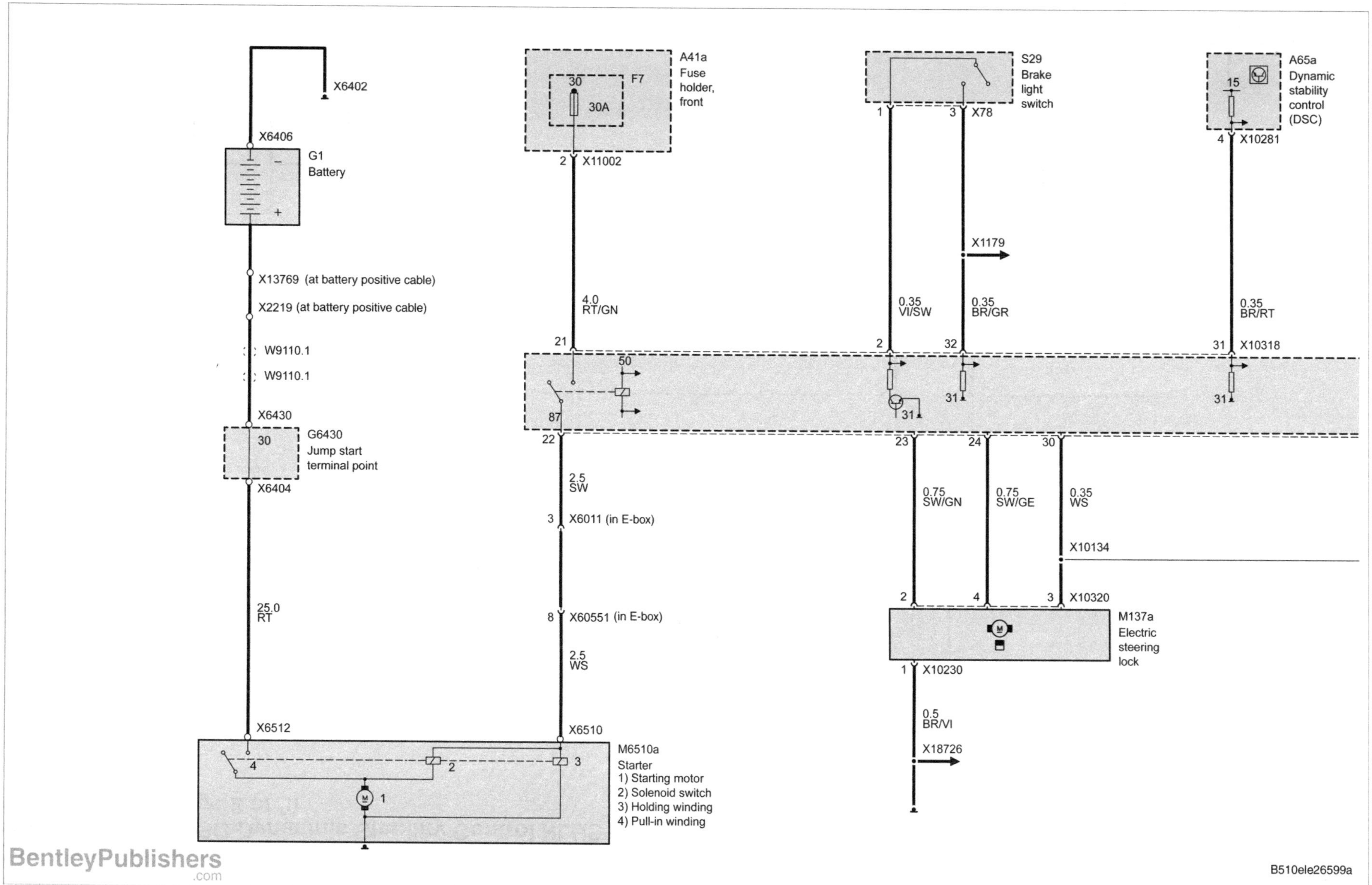

BentleyPublishers.com

Convenient start
N52 engine or N54 engine
with DSC Dynamnic Stability Control Premium (page 2 of 2)

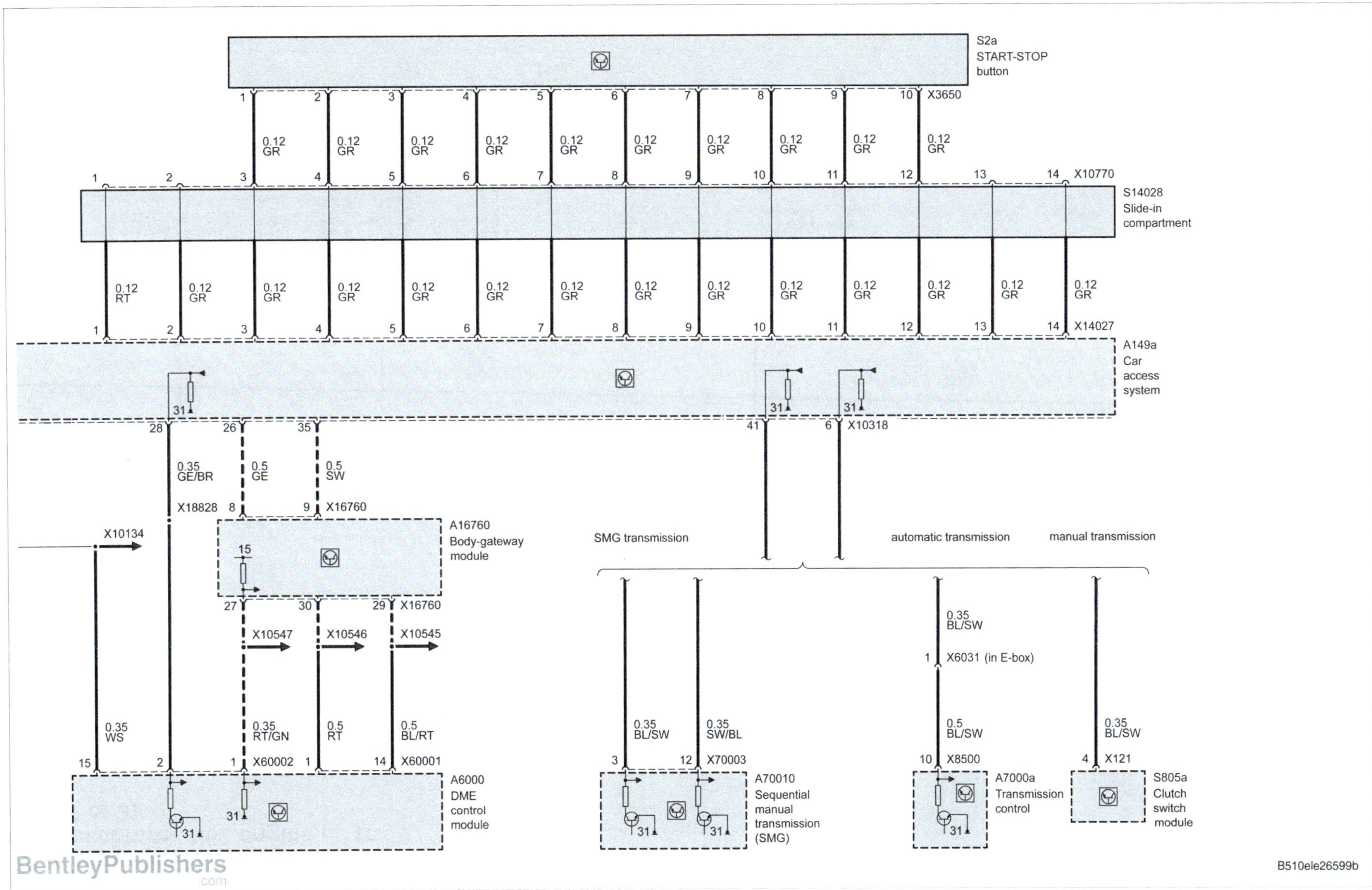

Car access system terminal control, N62 engine up to 09 / 2005 (page 1 of 2)

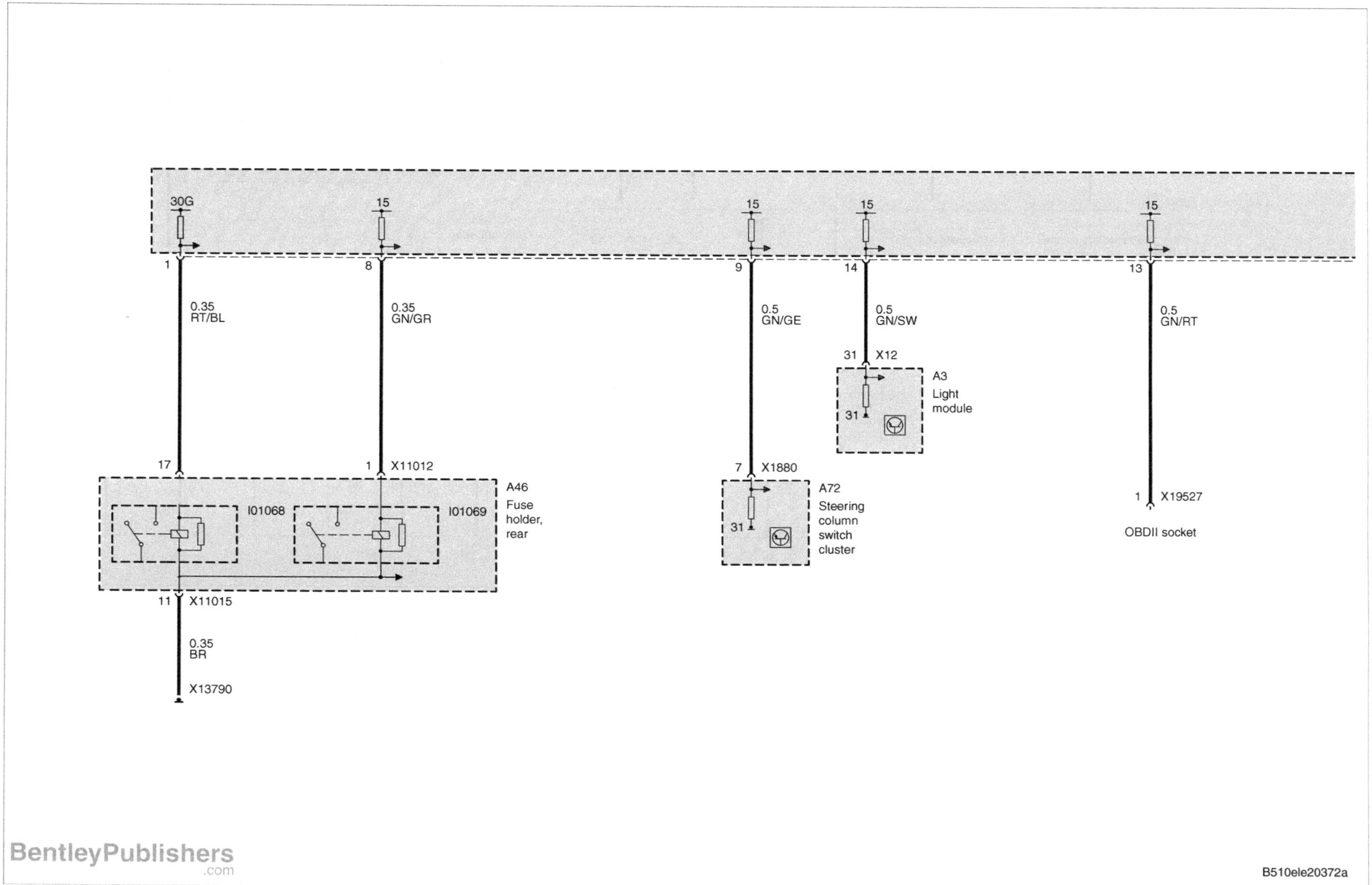

BentleyPublishers.com

Car access system terminal control, N62 engine up to 09 / 2005 (page 2 of 2)

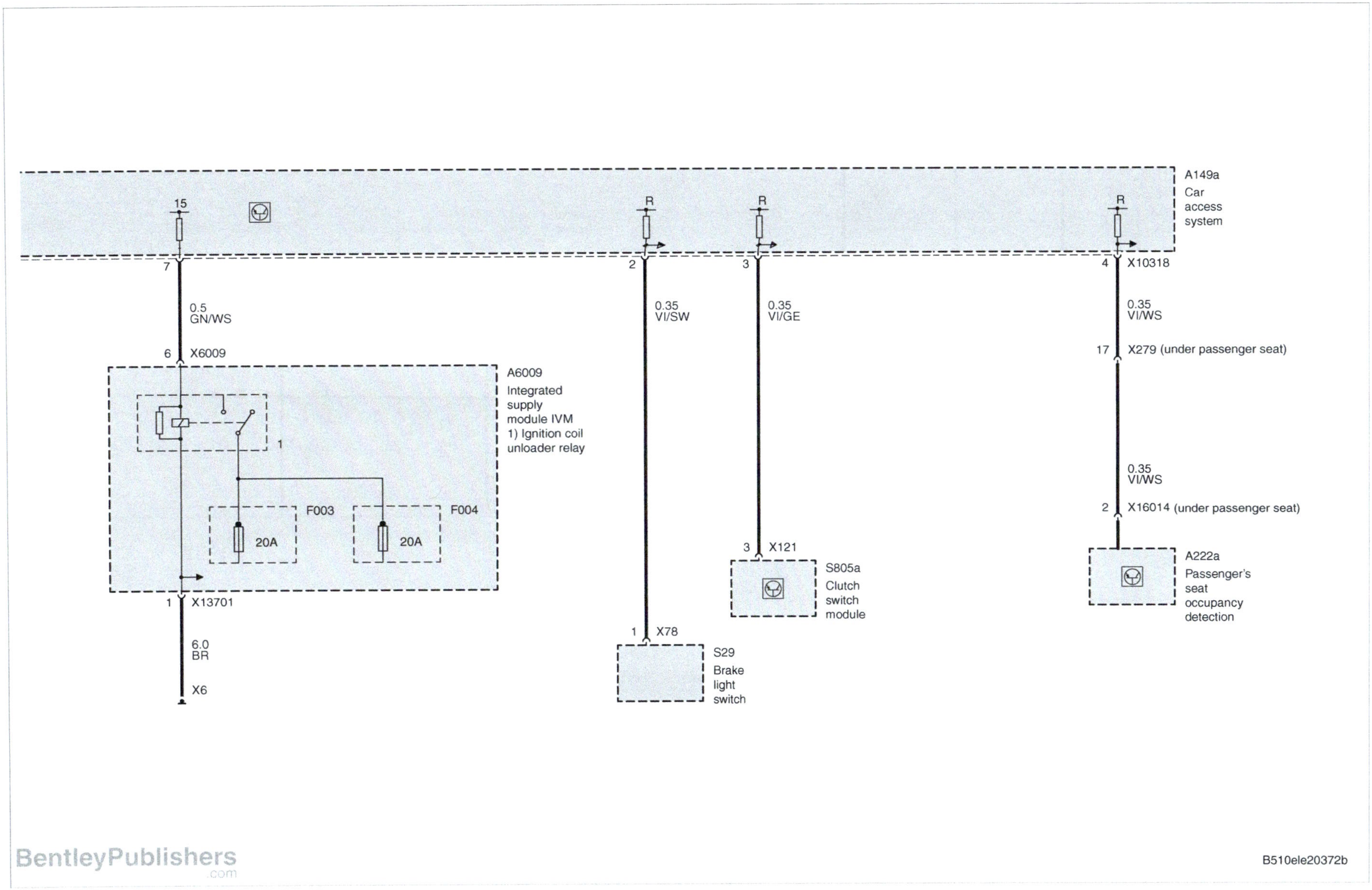

BentleyPublishers.com

Car access system terminal control without N62 engine up to 09 / 2005

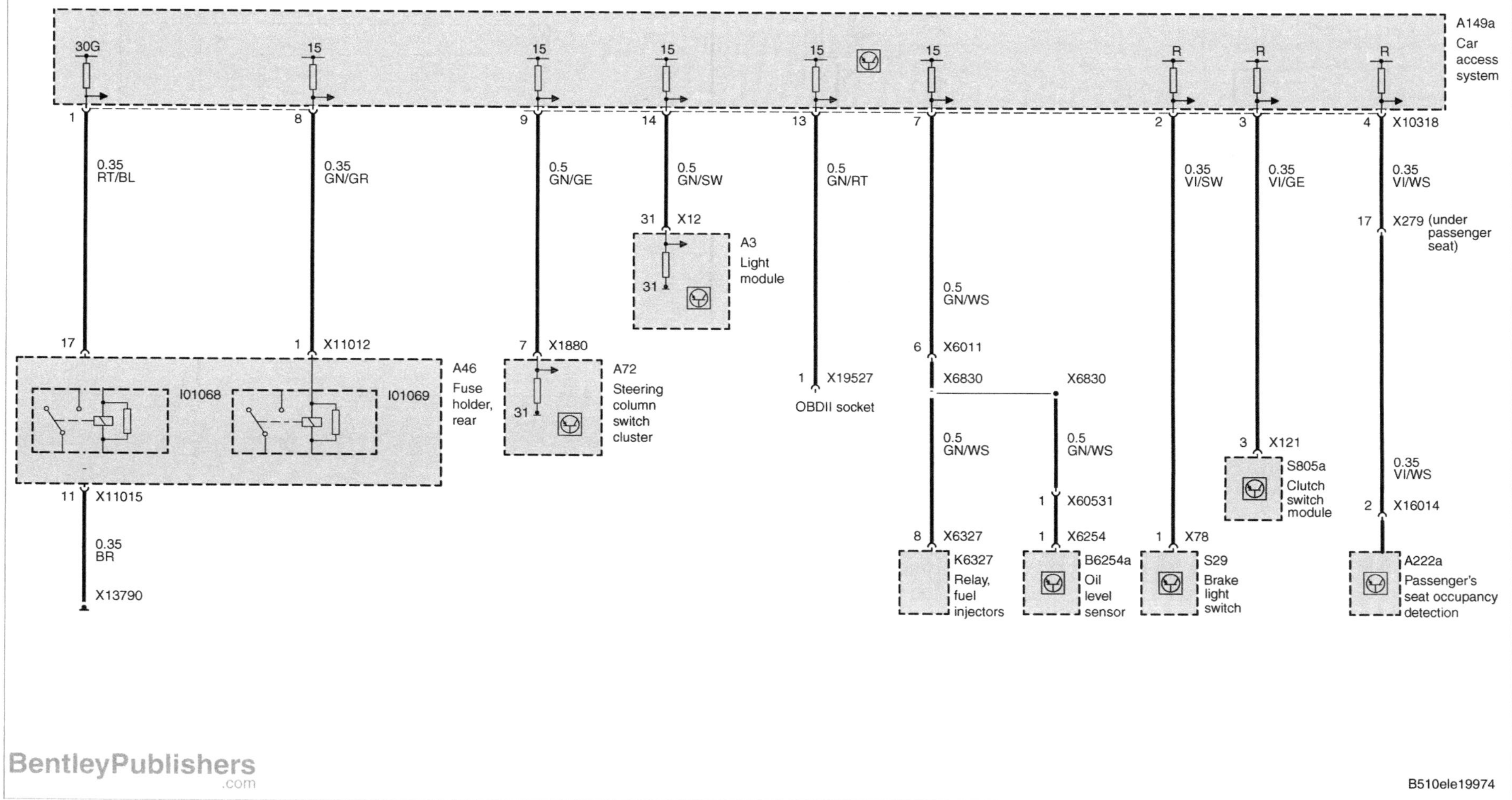

BentleyPublishers.com

Car access system terminal control N62 TU engine as of 09 / 2005

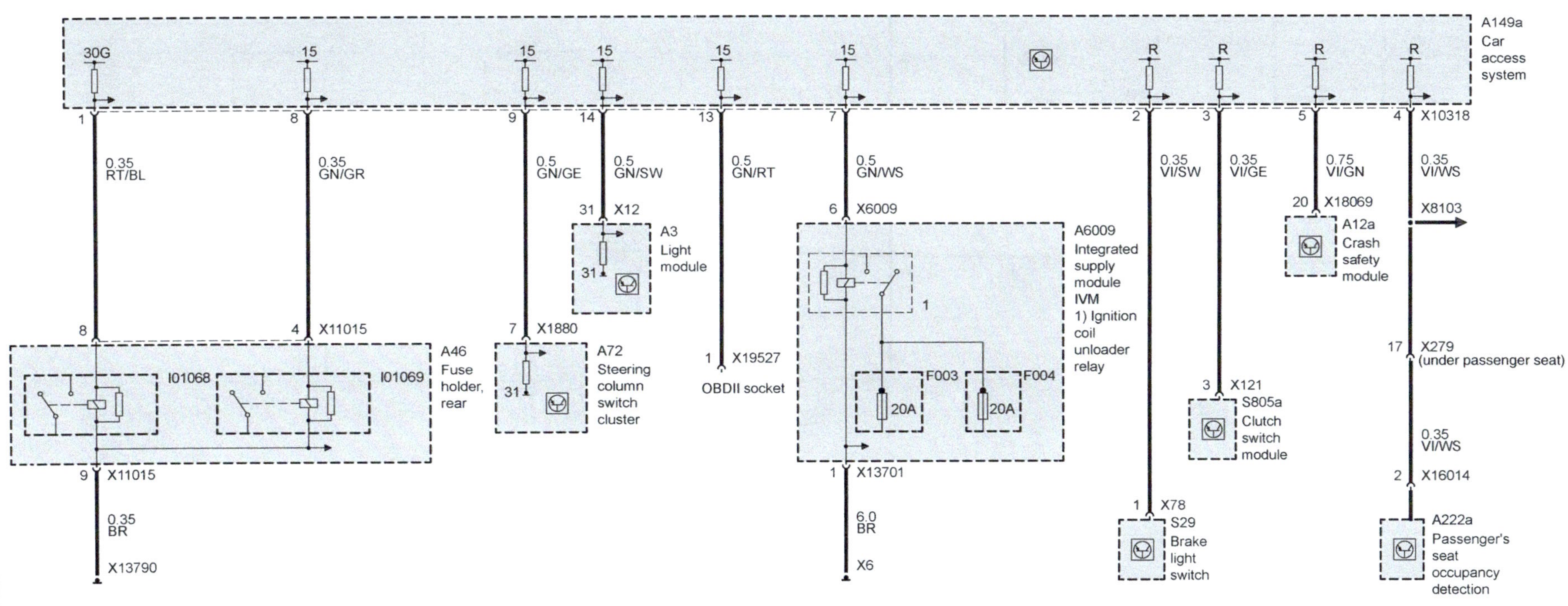

BentleyPublishers.com

Car access system terminal control without N62 TU engine as of 09 / 2005

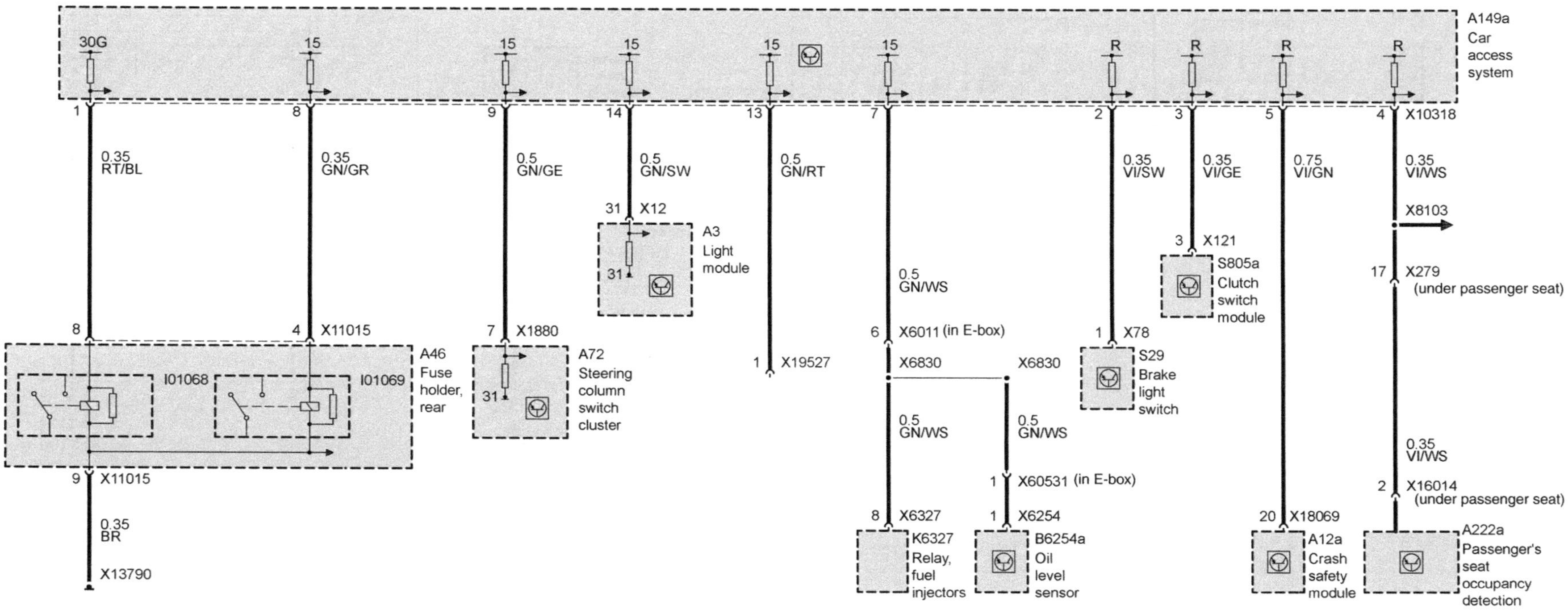

BentleyPublishers
.com

B510ele19775

Car access system power supply up to 03 / 2005

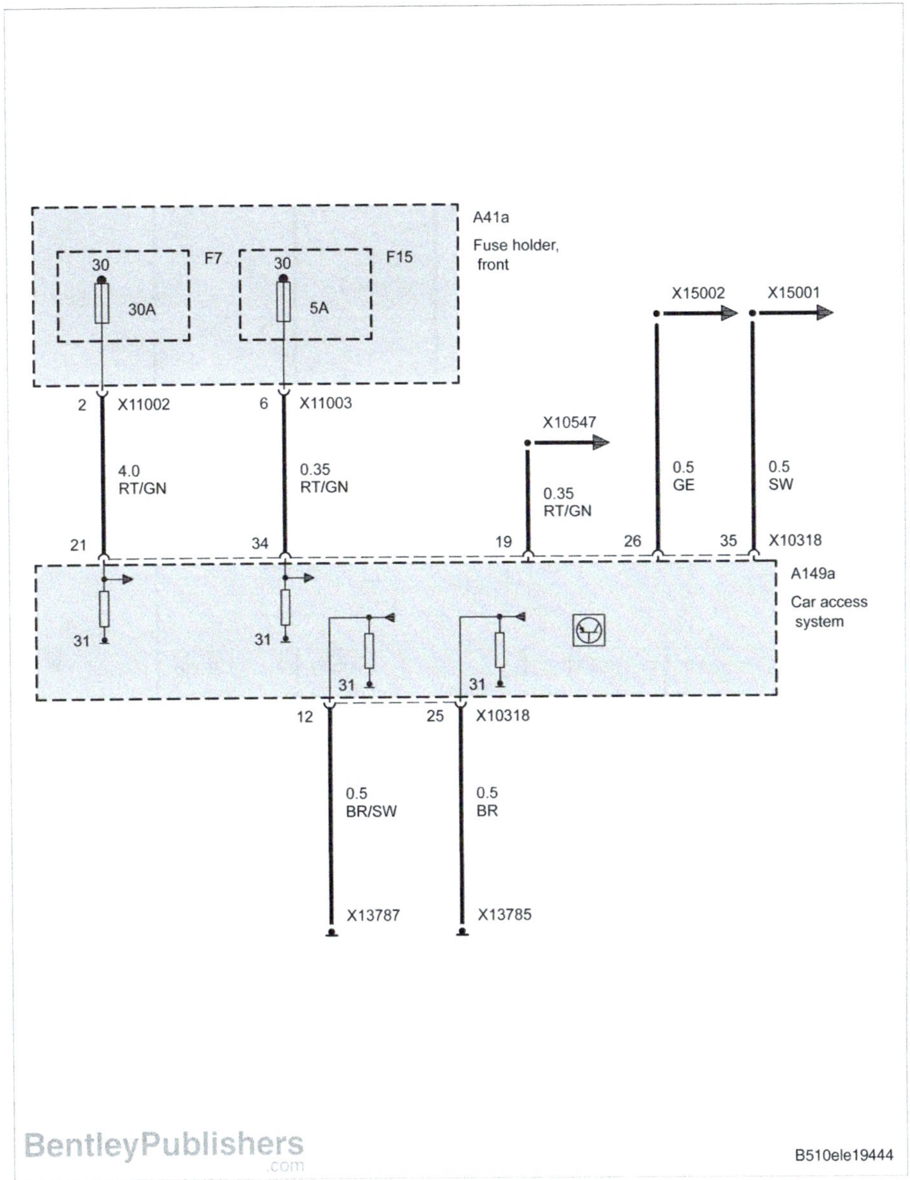

Car access system power supply as of 03 / 2005

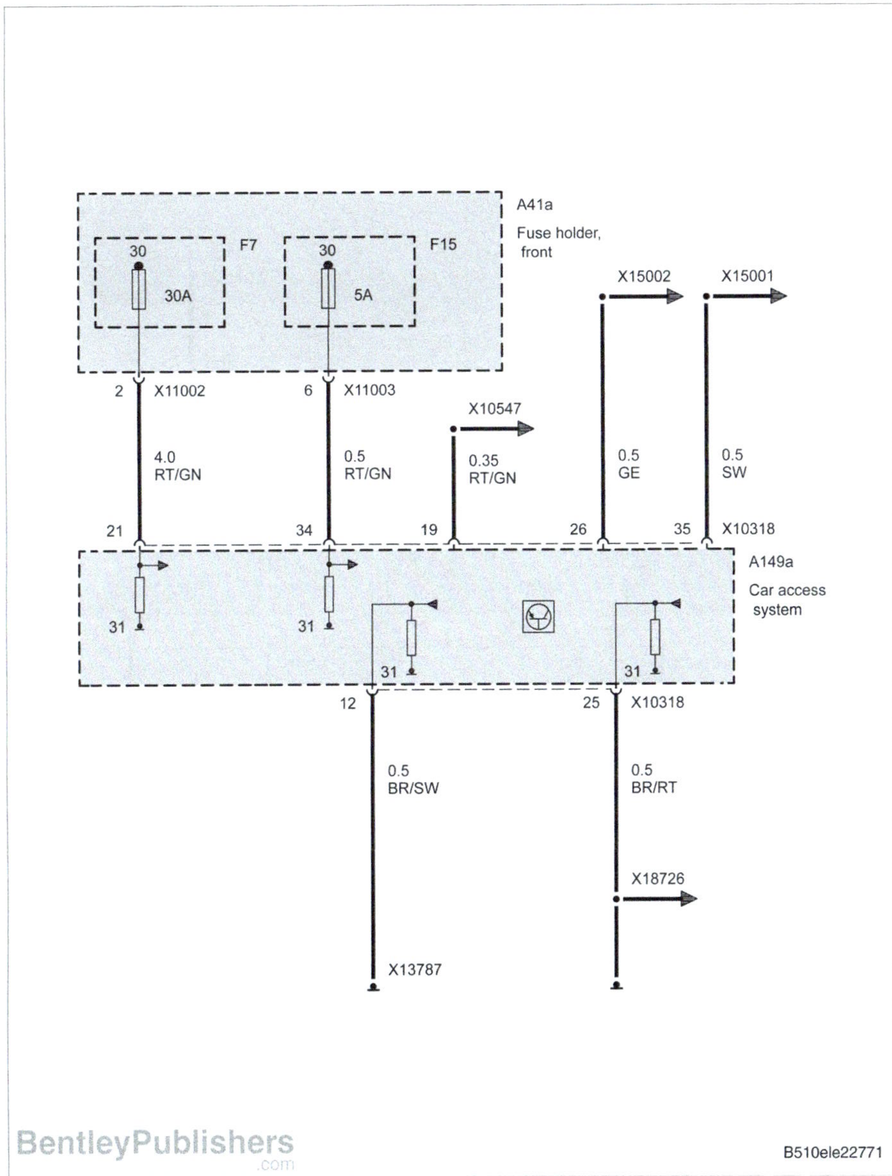

Ground connection X6
M54 engine

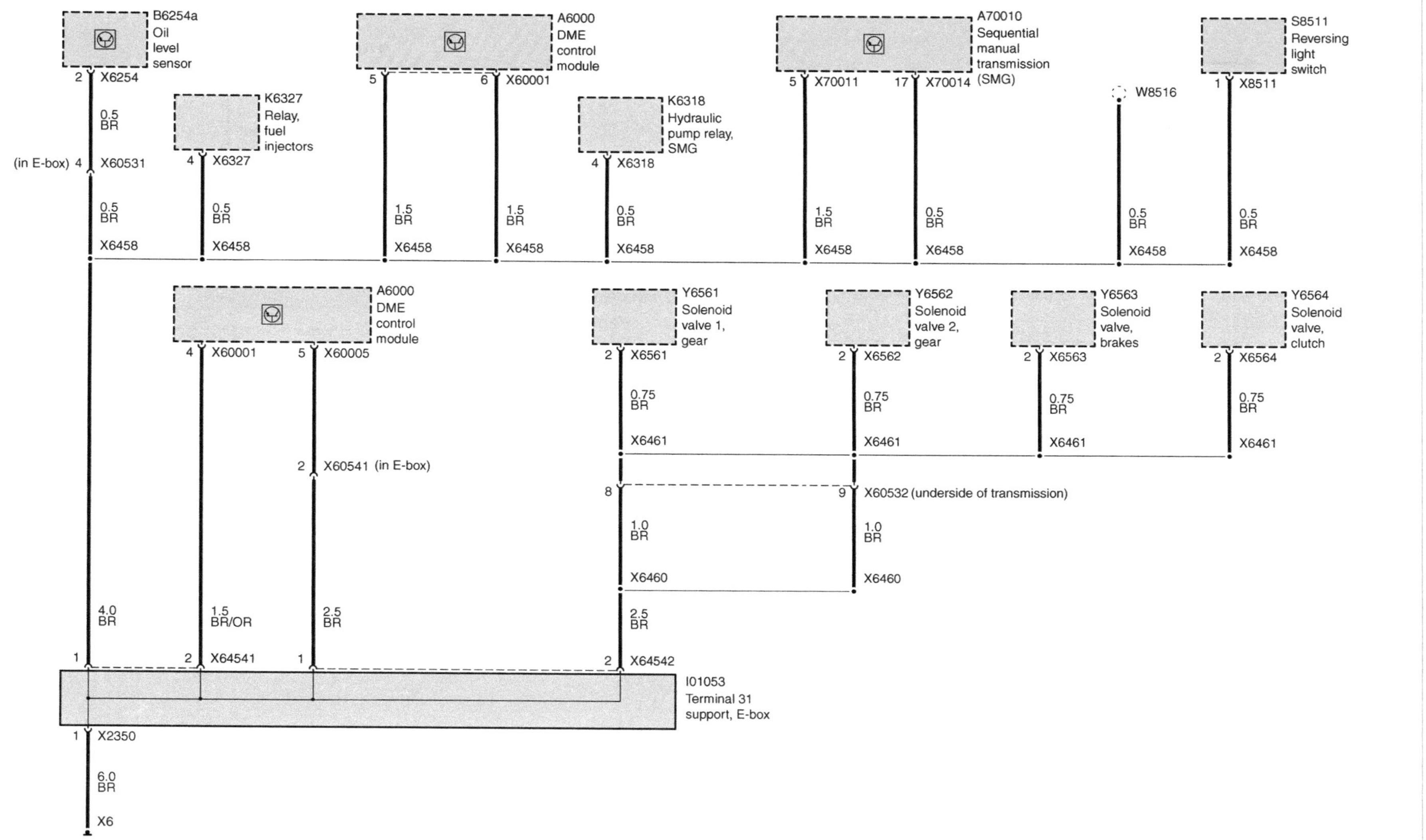

BentleyPublishers
.com

Ground connection X6
N52 engine

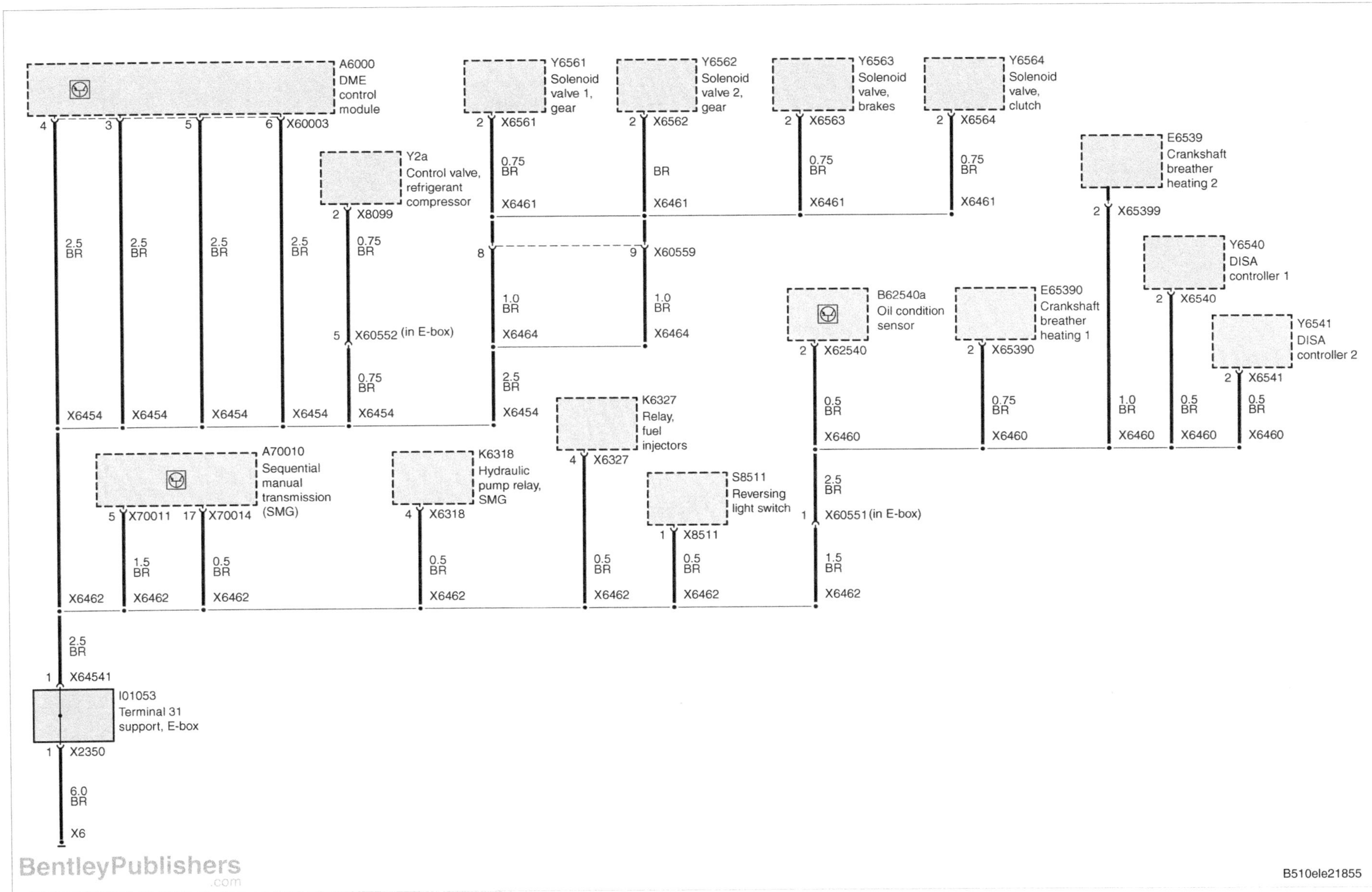

BentleyPublishers.com

Ground connection X7517
N62 engine

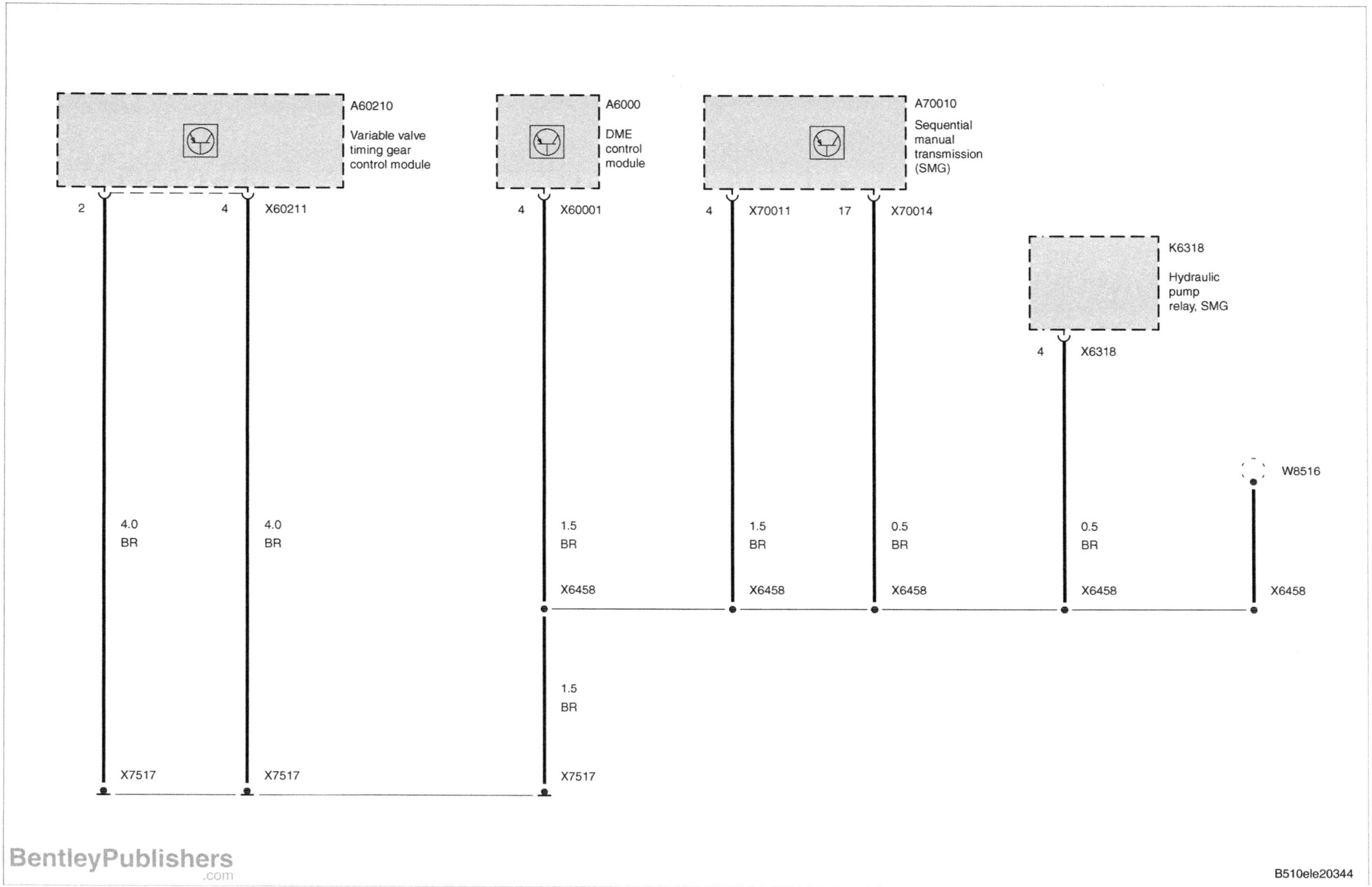

BentleyPublishers.com

DSC control module, power supply
DSC Dynamic stability control
up to 03 / 2007

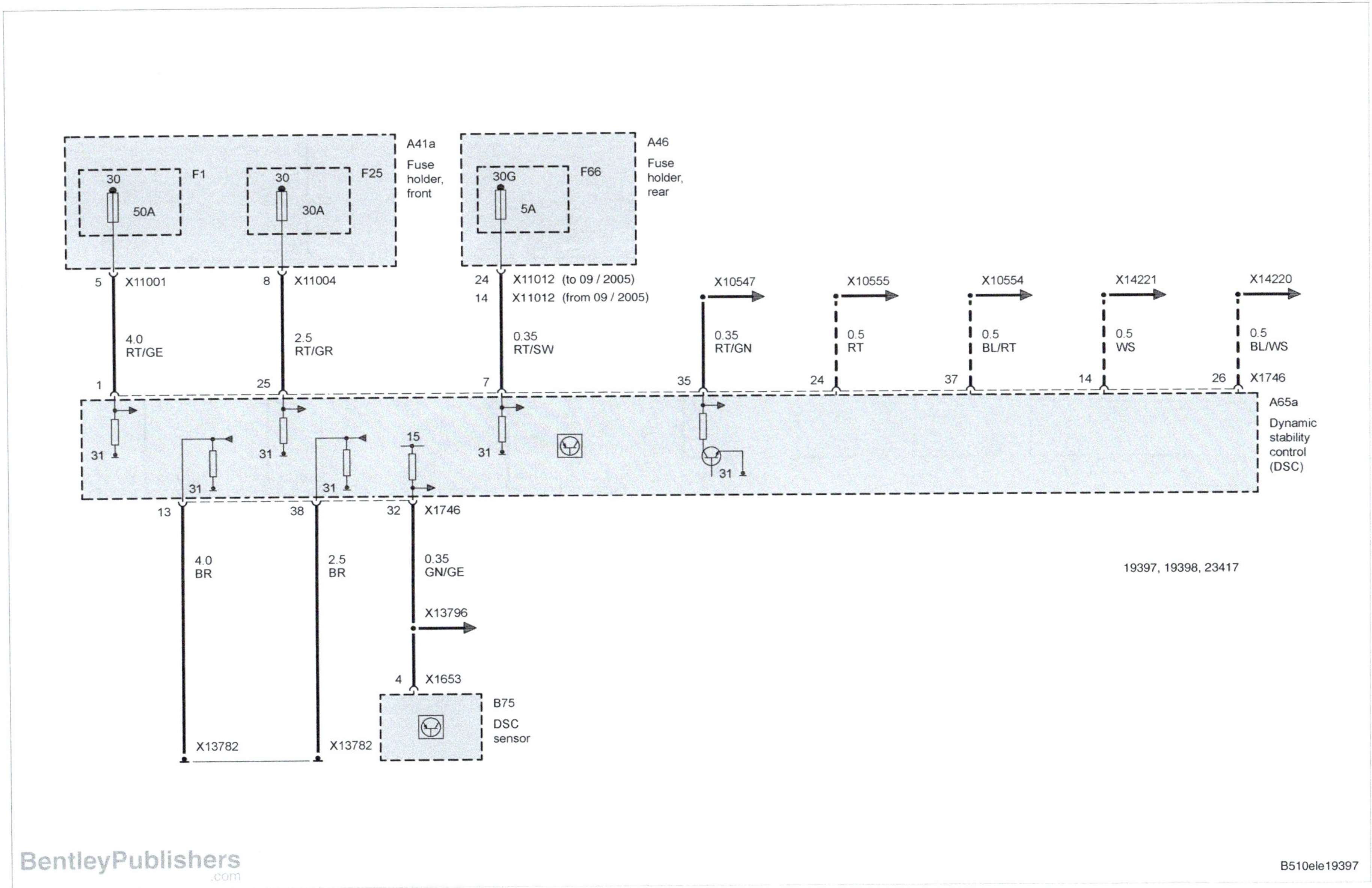

BentleyPublishers.com

DSC control module, power supply
DSC Dynamic stability control
as of 03 / 2007

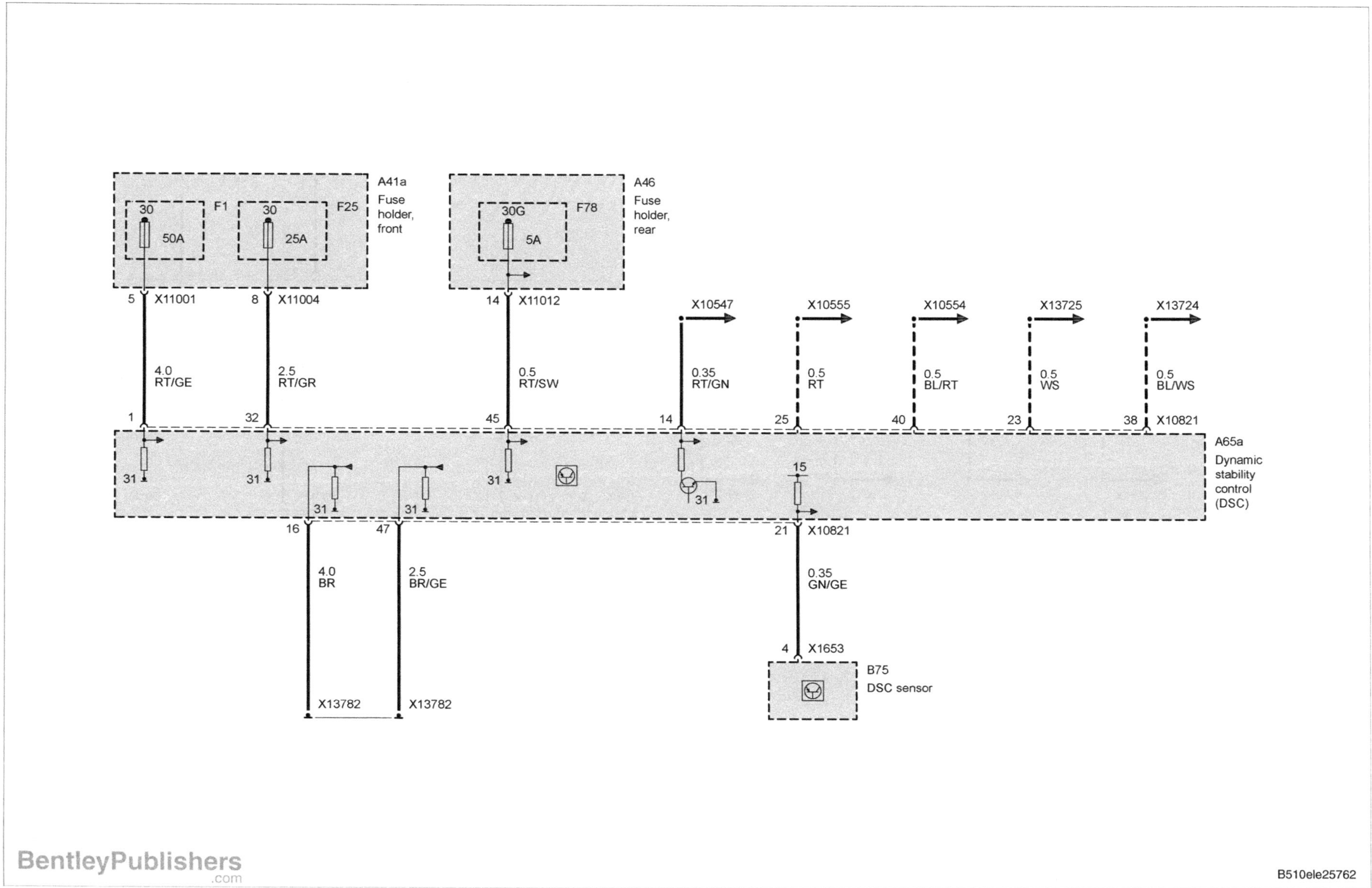

DSC switch functions
DSC Dynamic stability control
up to 03 / 2007

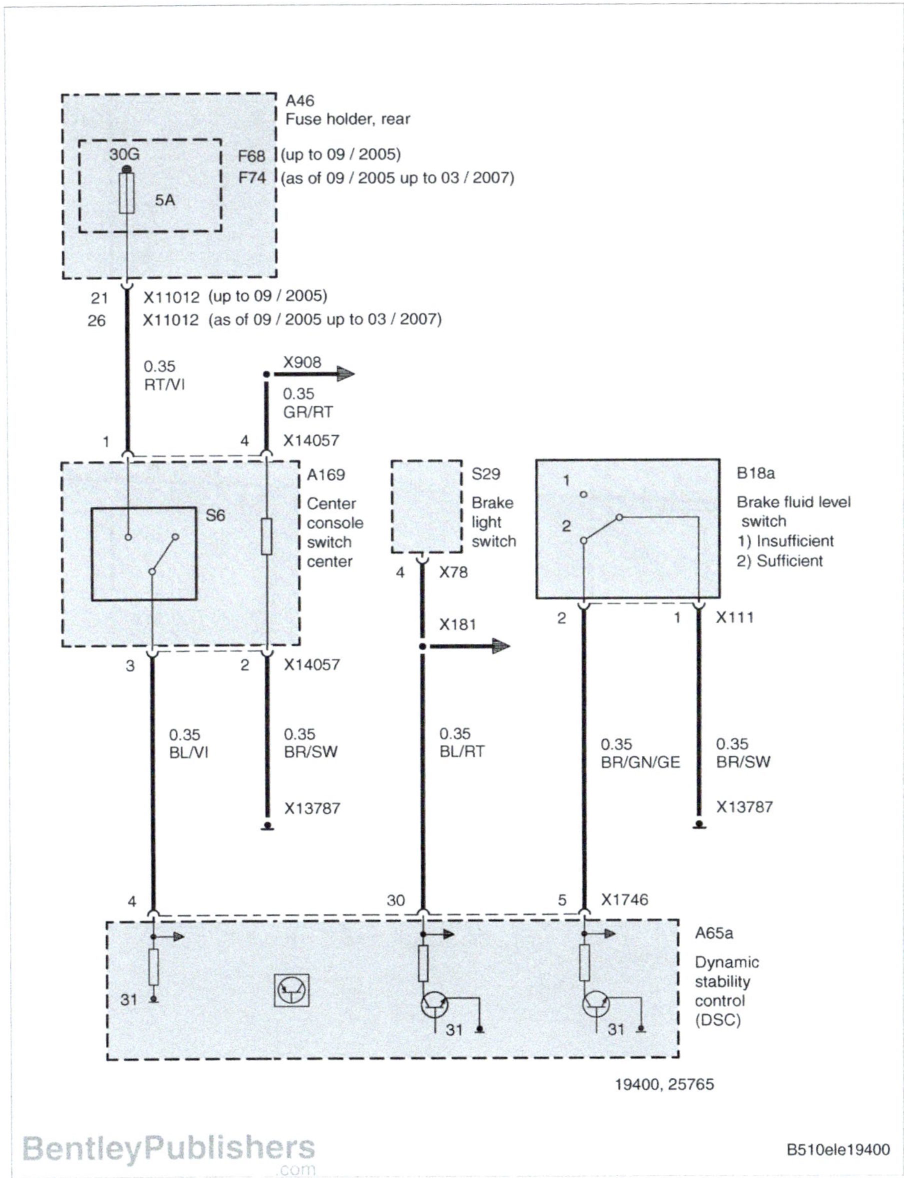

DSC switch functions
DSC Dynamic stability control
as of 03 / 2007

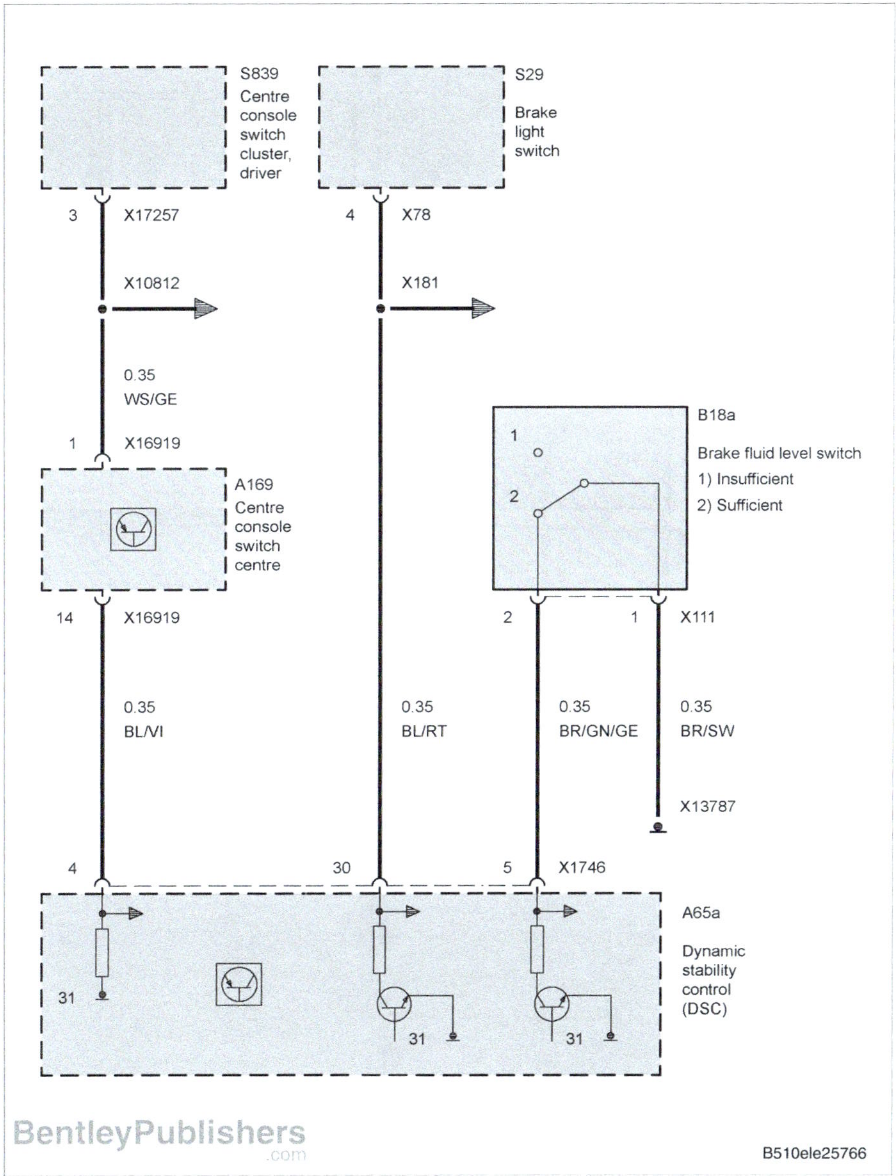

DSC switch functions
DSC Dynamnic Stability Control Premium

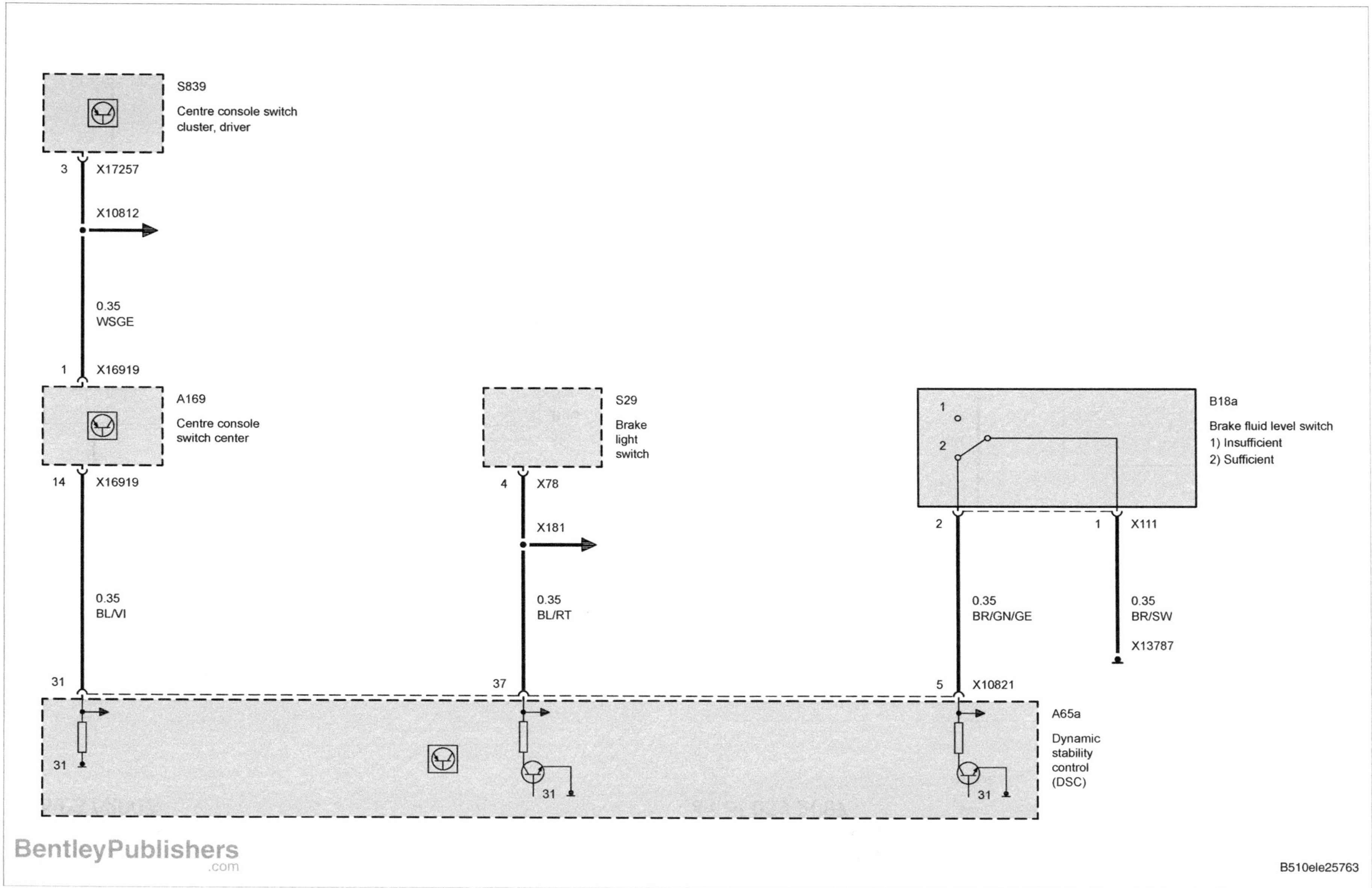

DSC, sensors to 09 / 2005
DSC Dynamic stability control
(page 1 of 2)

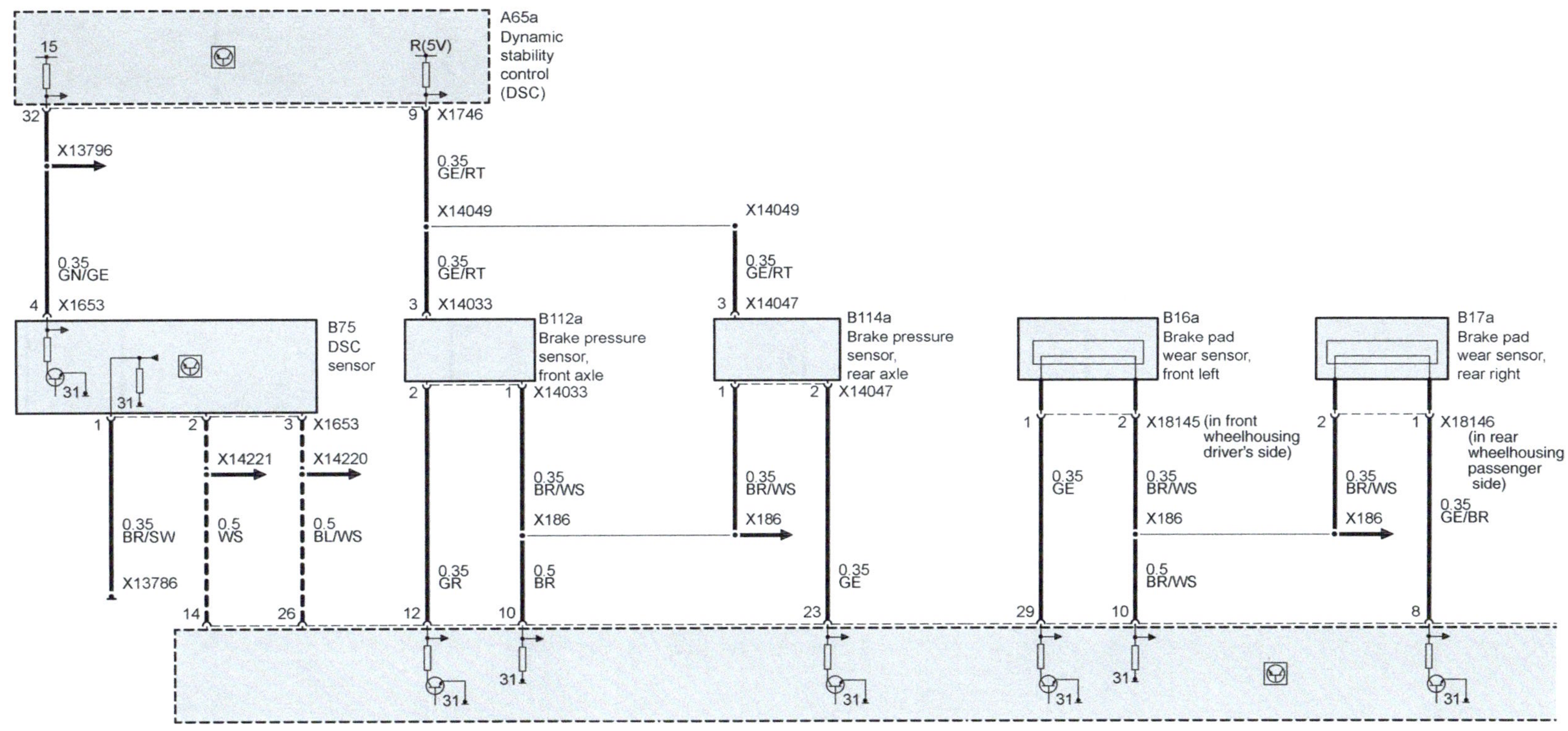

BentleyPublishers
.com

DSC, sensors to 09 / 2005
DSC Dynamic stability control
(page 2 of 2)

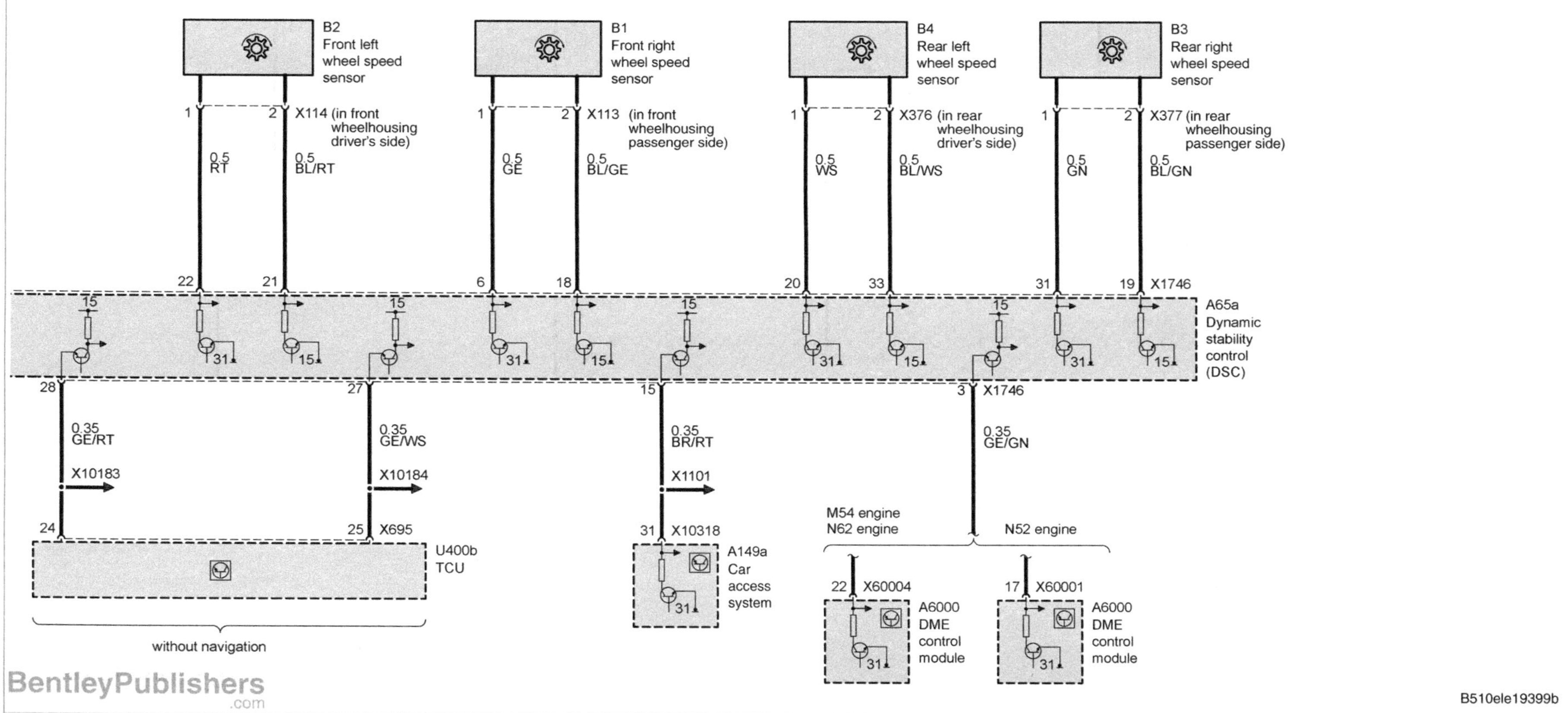

BentleyPublishers.com

DSC, sensors from 09 / 2005
DSC Dynamic stability control
(page 1 of 2)

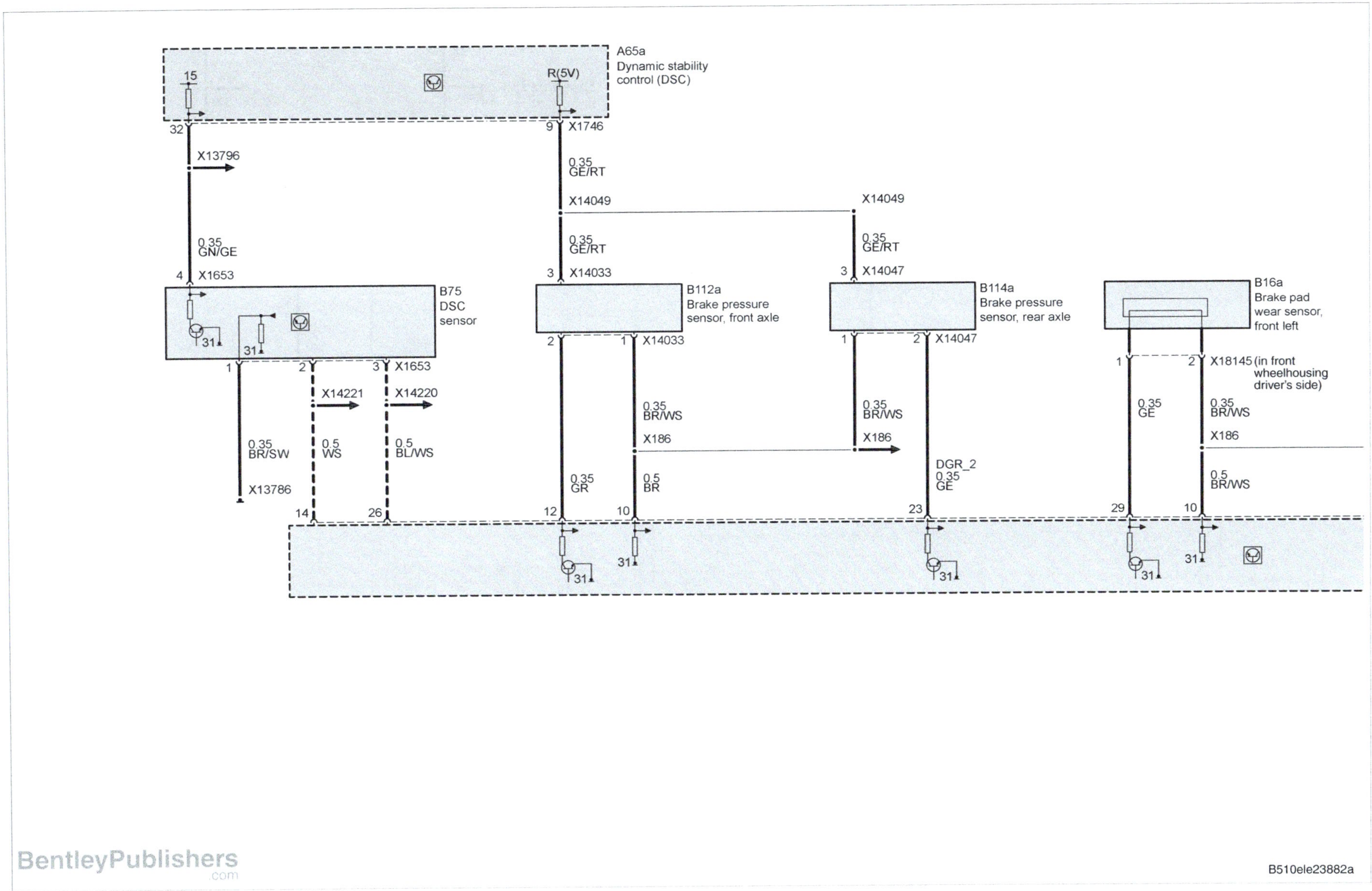

DSC, sensors from 09 / 2005
DSC Dynamic stability control
(page 2 of 2)

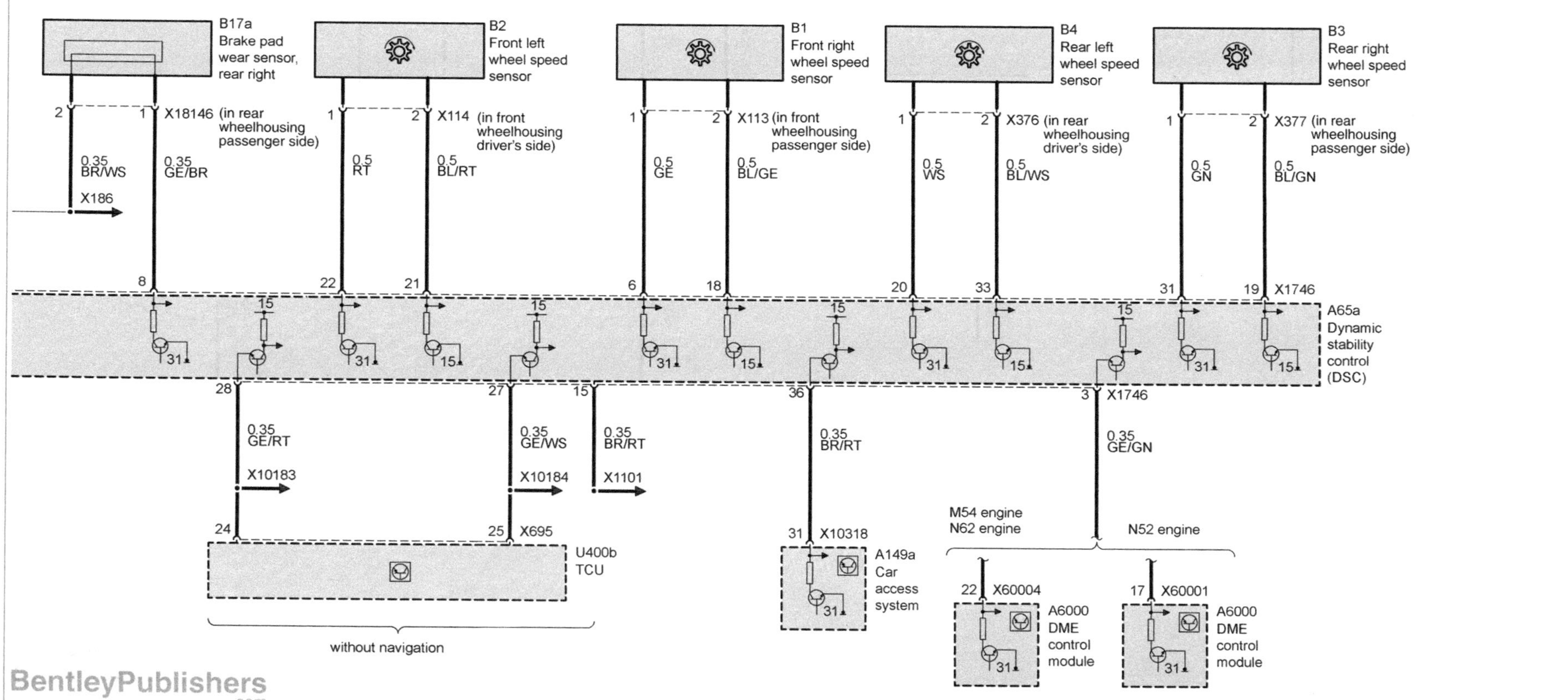

BentleyPublishers.com

B510ele23882b

DSC Dynamnic Stability Control Premium
(page 1 of 2)

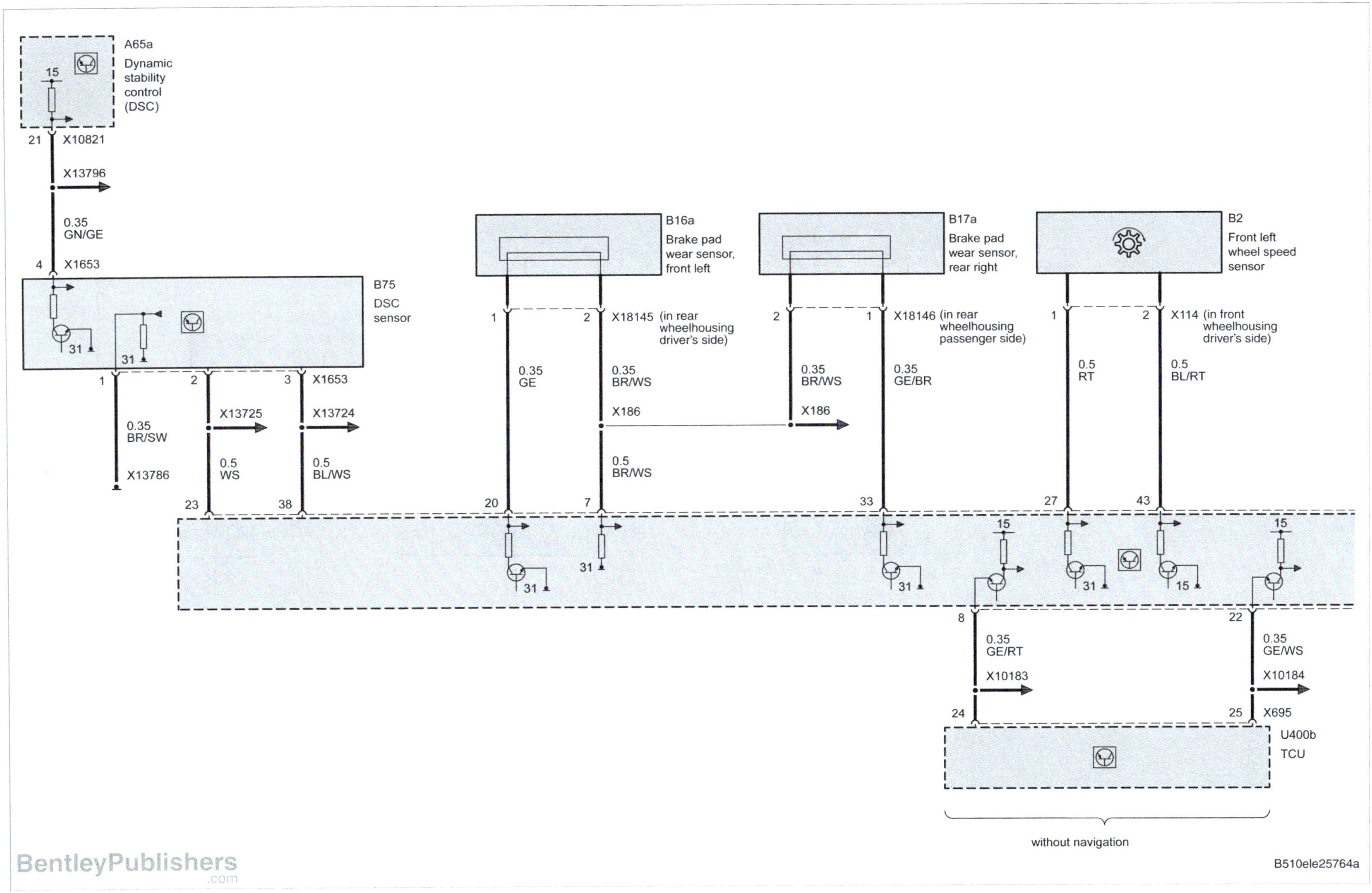

BentleyPublishers.com

DSC, sensors
DSC Dynamic stability control premium
(page 2 of 2)

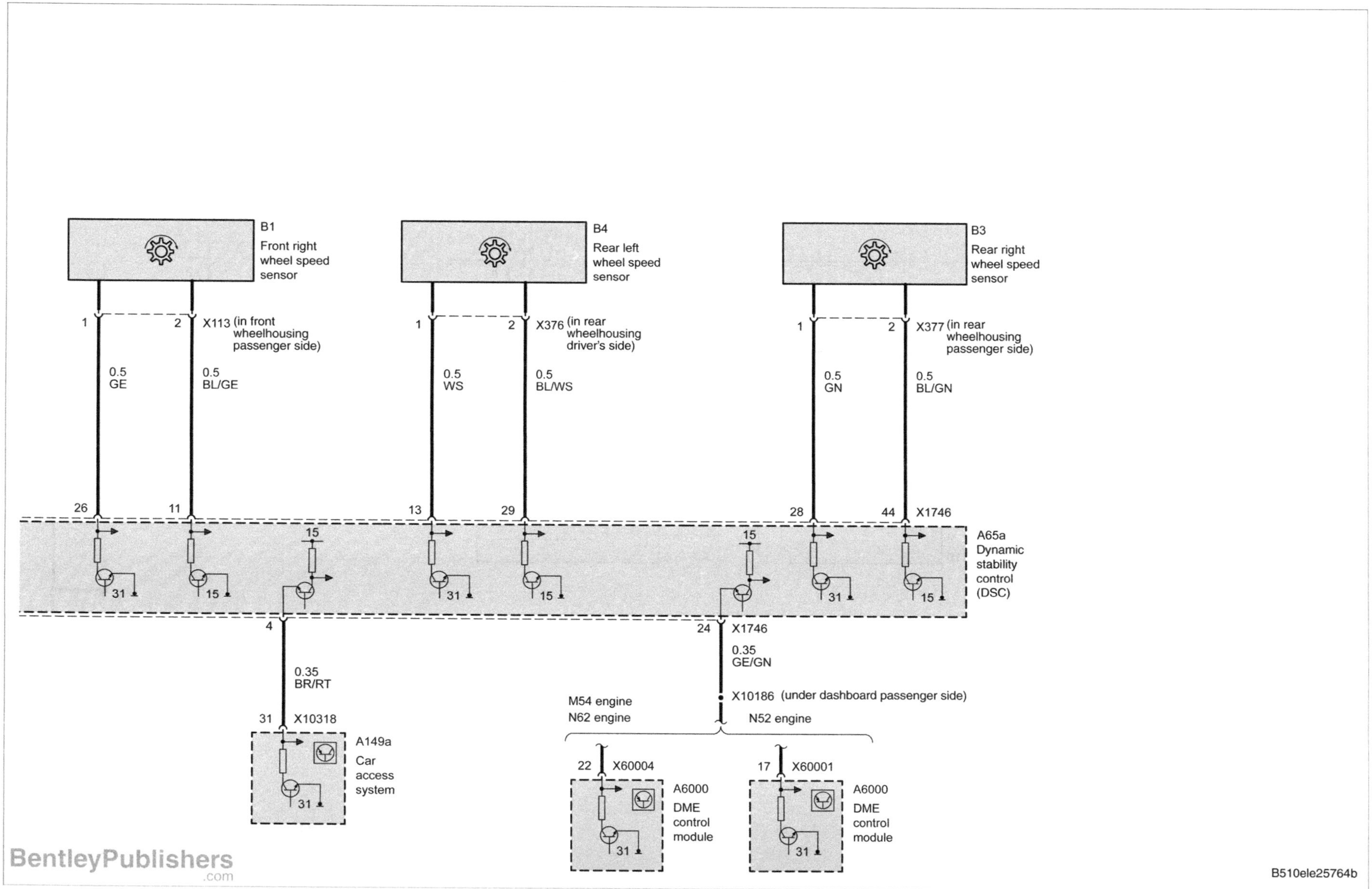

B510ele25764b

Brake vent flaps
N52 engine or N54 engine

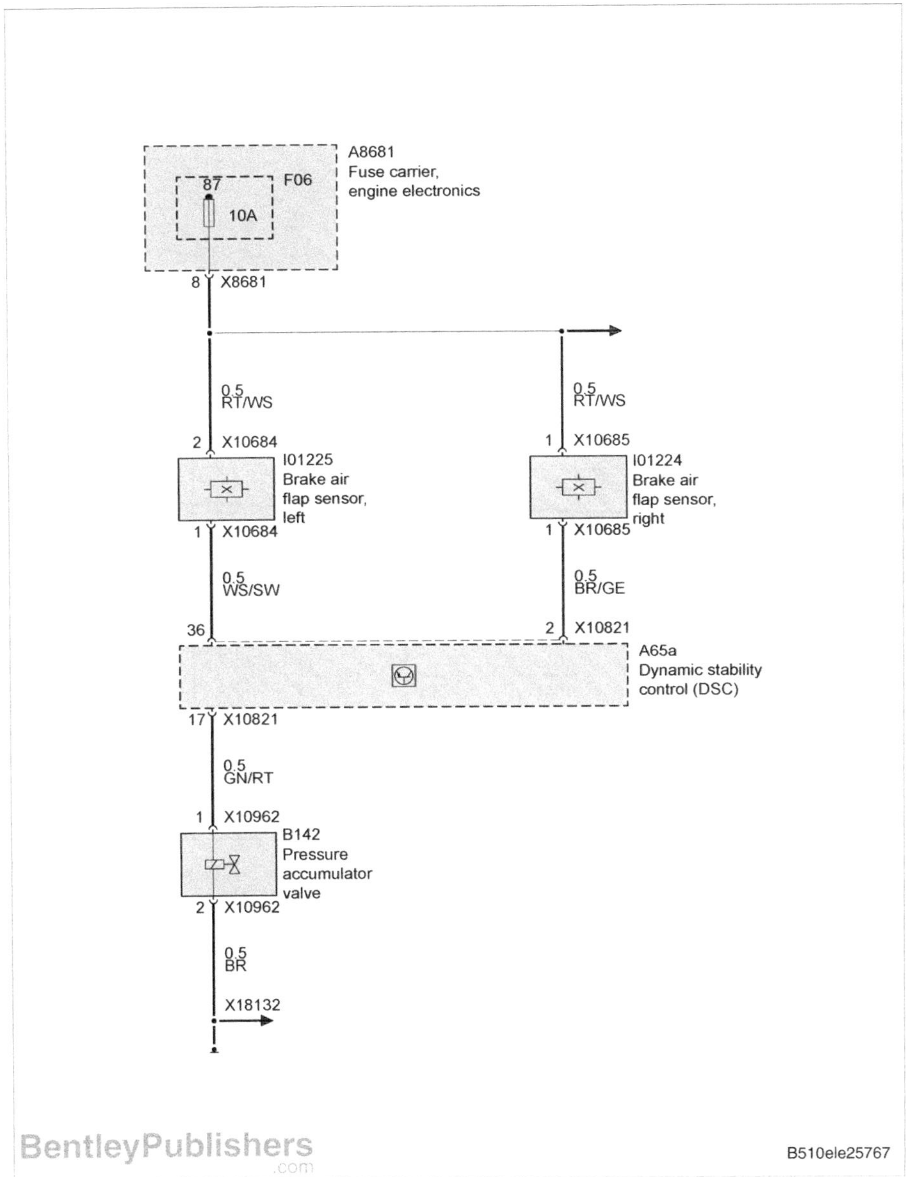

Brake vent flaps
N62 TU engine

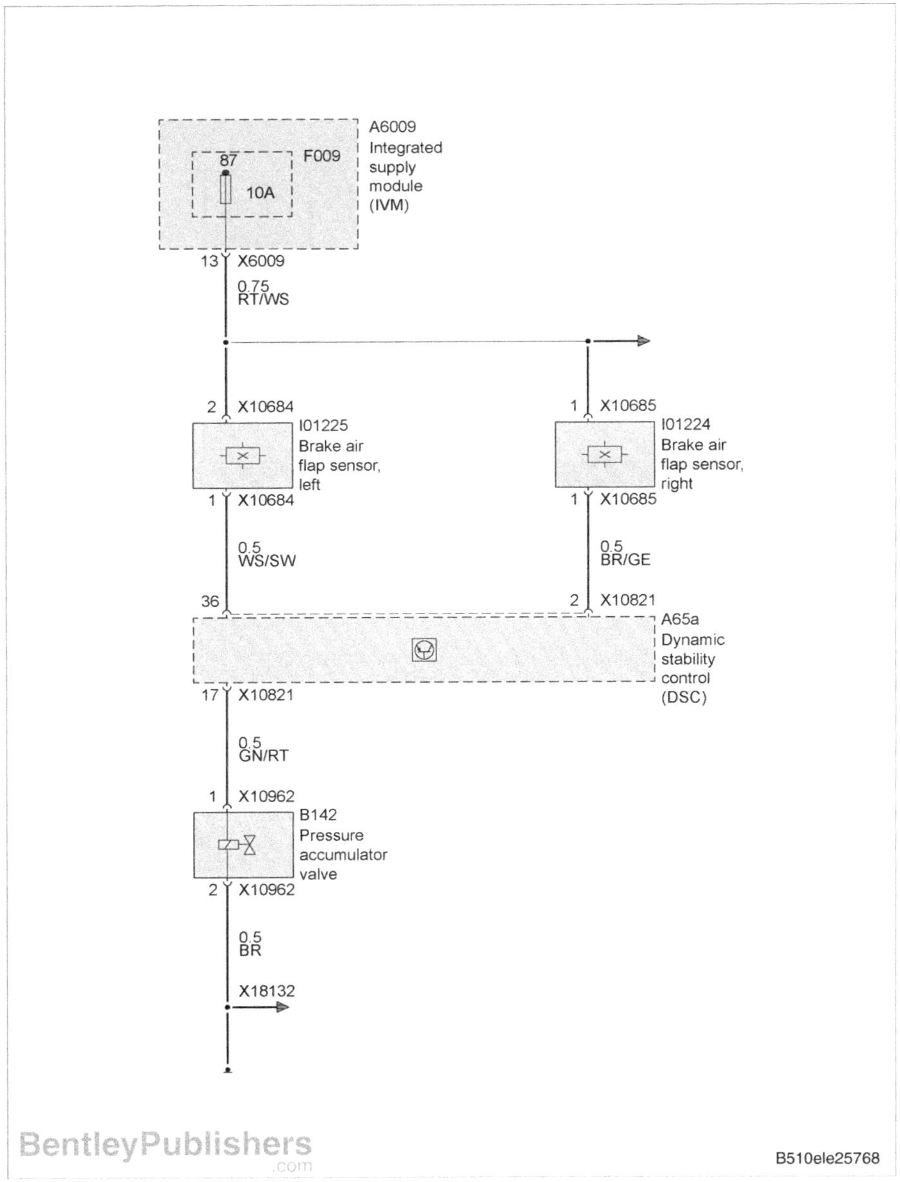

Active steering up to 09 / 2005

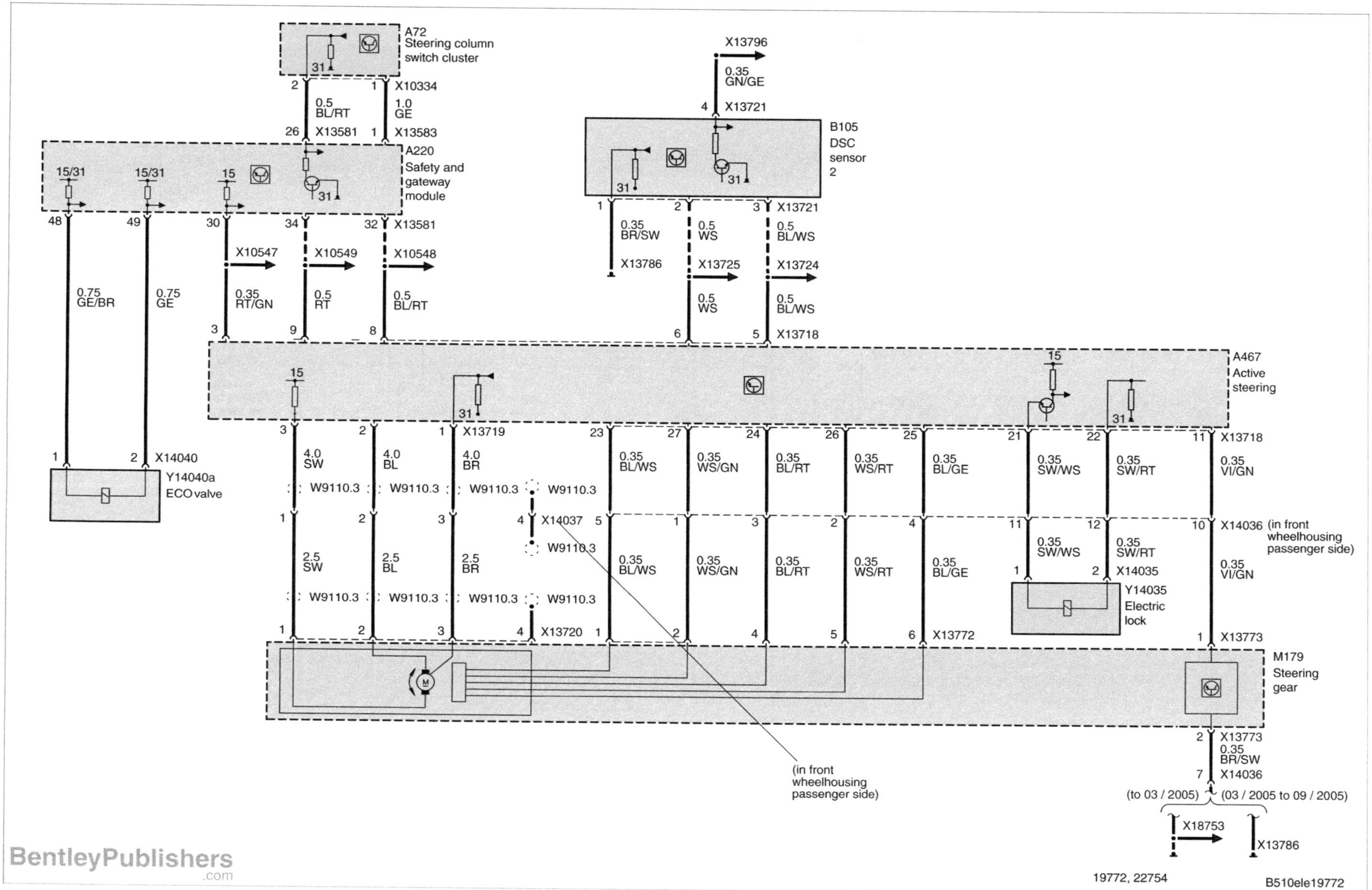

BentleyPublishers.com

Active steering
as of 09 / 2005

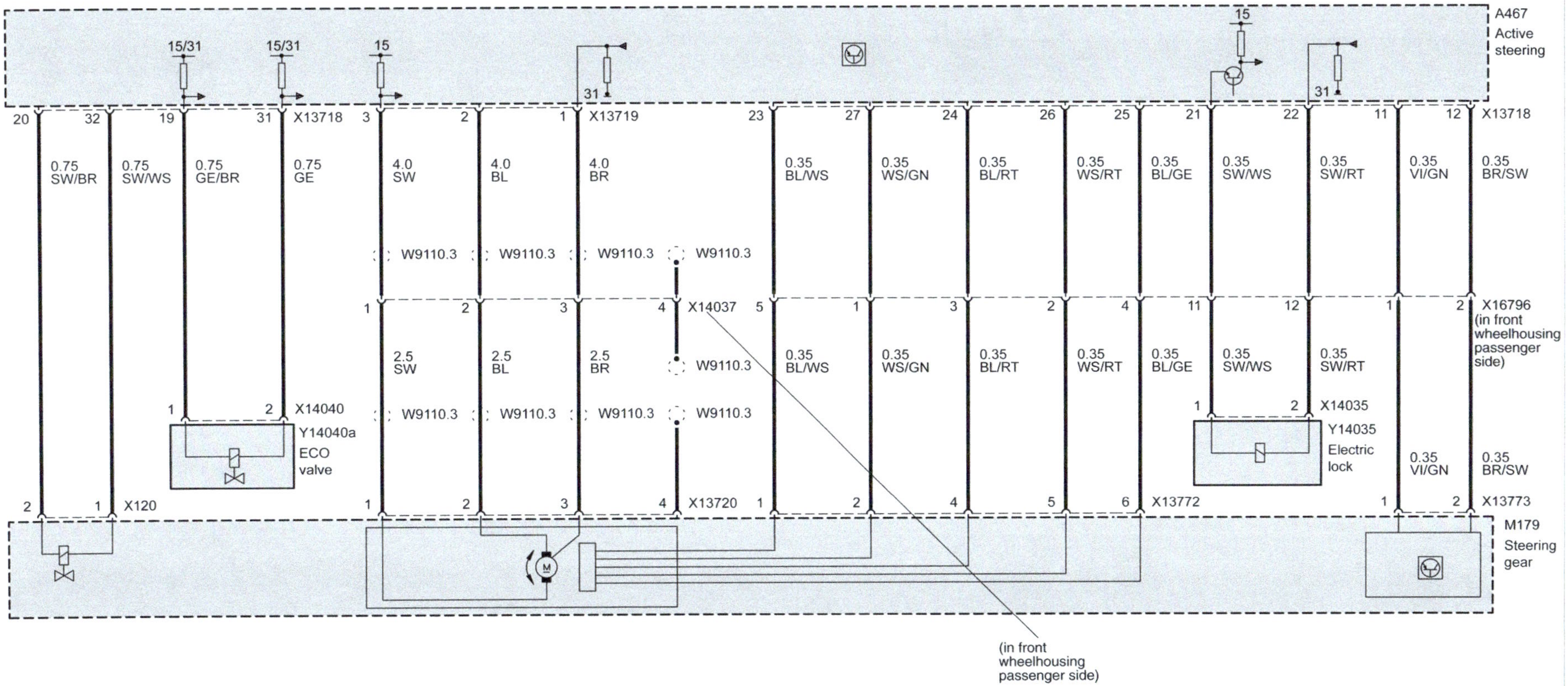

BentleyPublishers.com

B510ele23416

Active steering, power supply AL active steering up to 09 / 2005

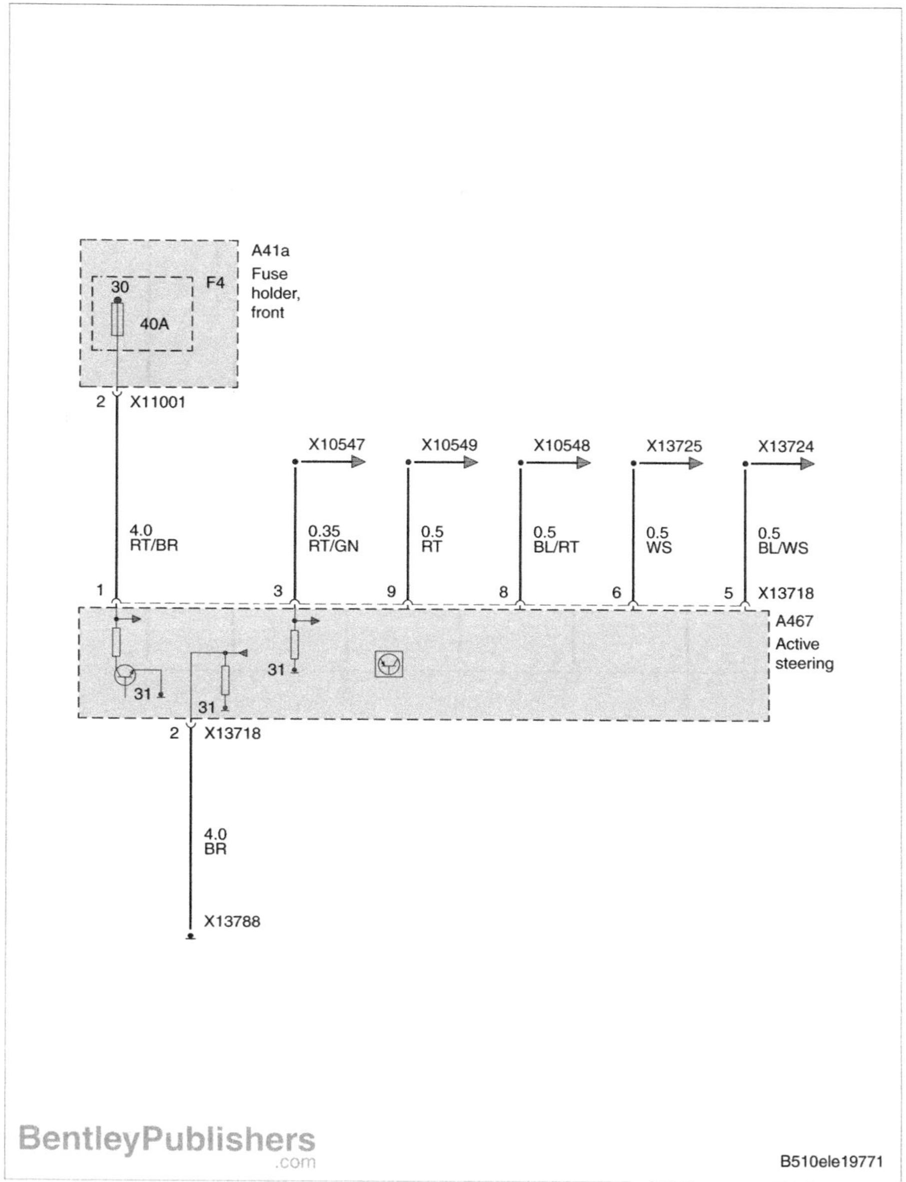

Active steering, power supply AL active steering as of 09 / 2005

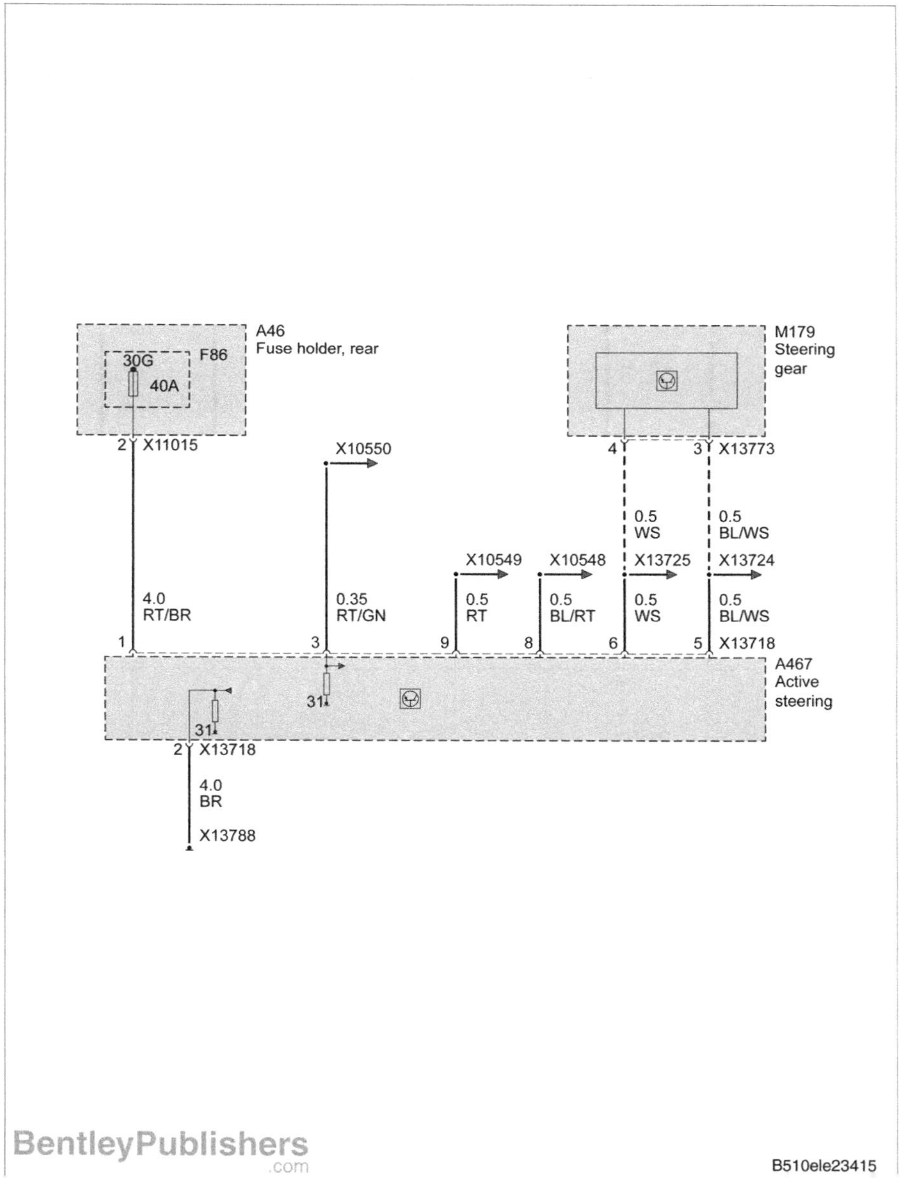

Safety and gateway module power supply

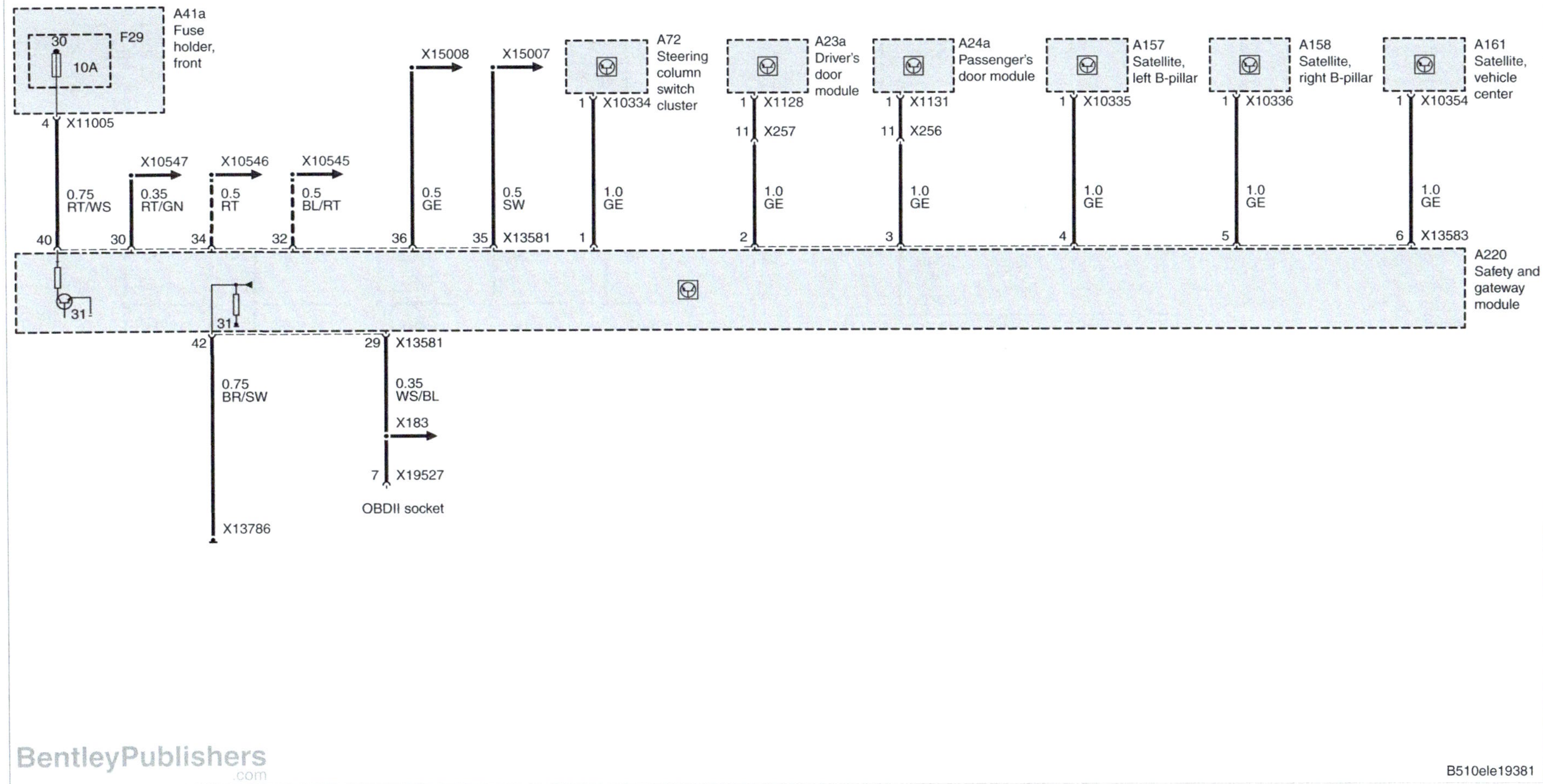

Servotronic up to 09 / 2005

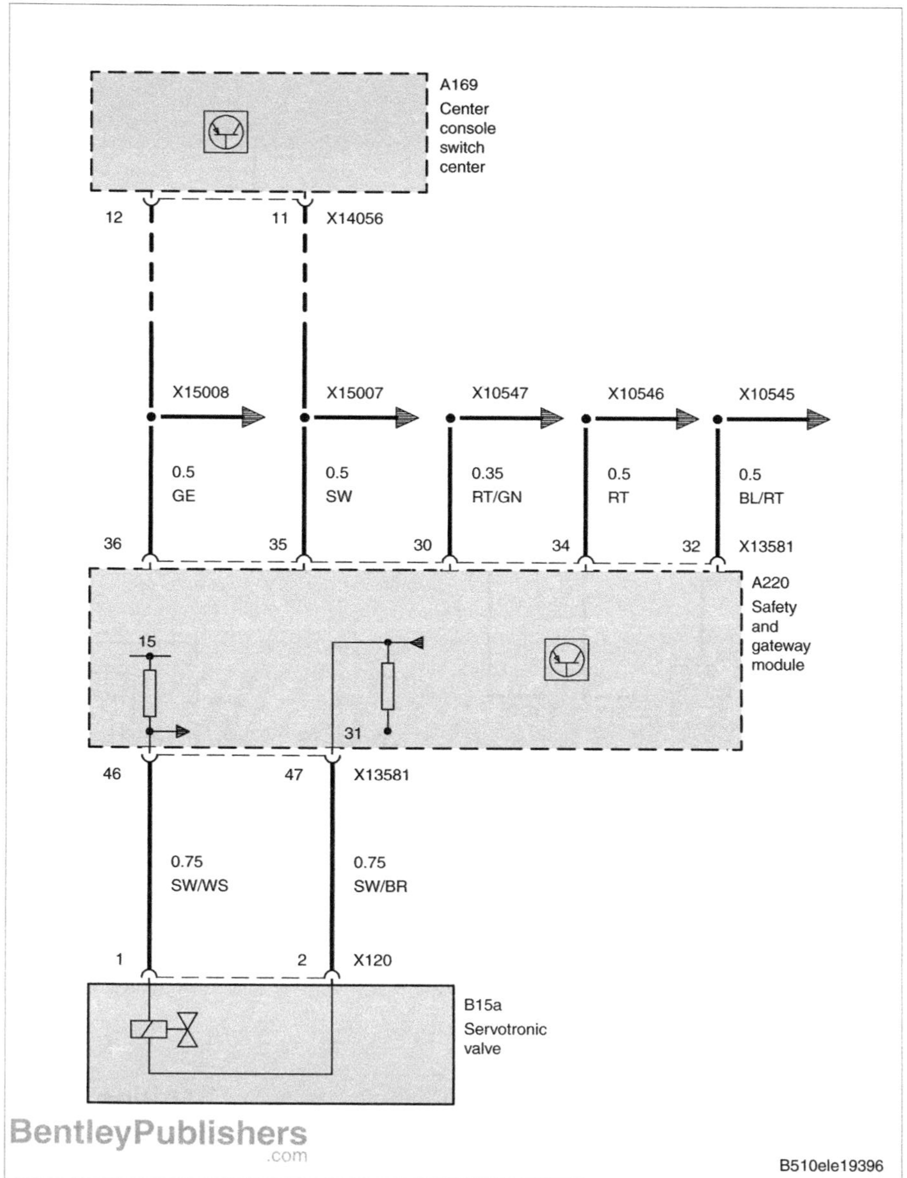

Servotronic as of 09 / 2005

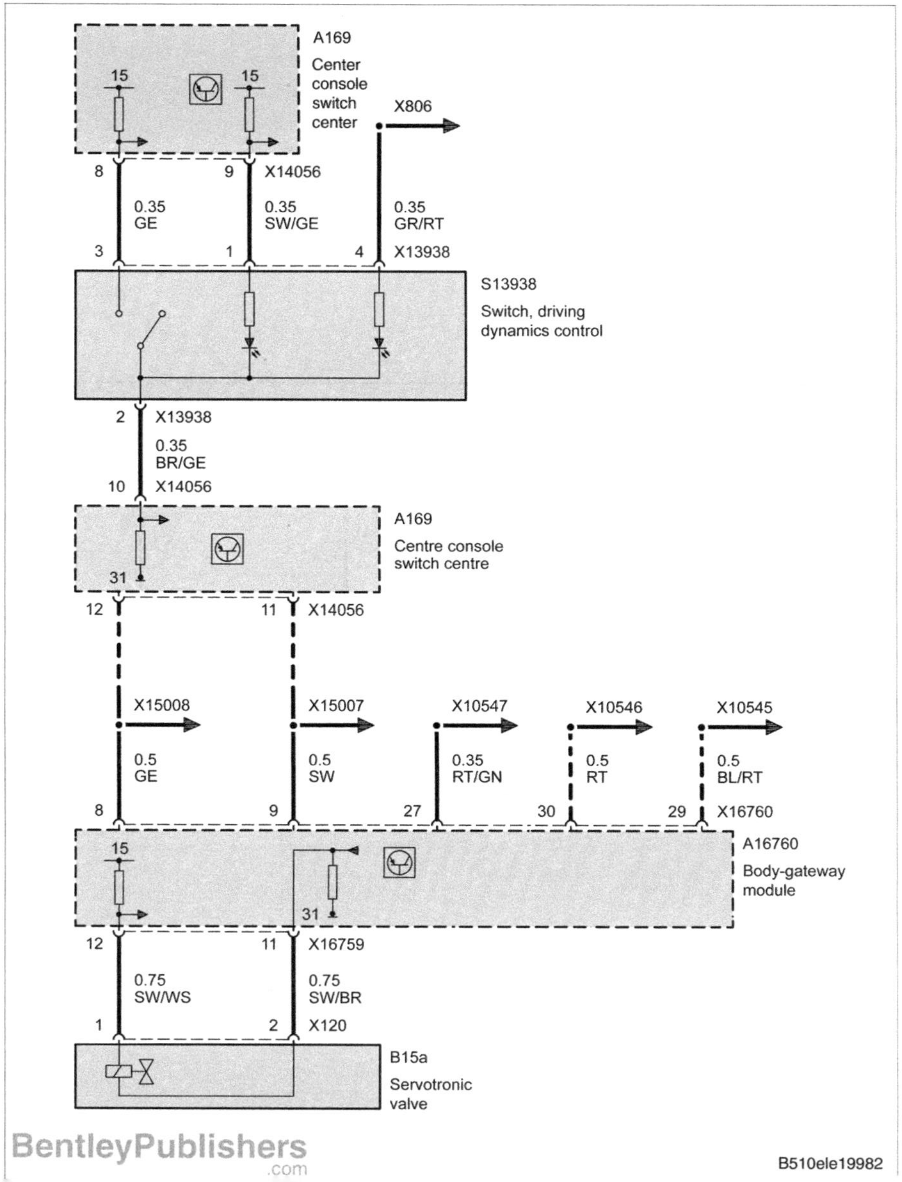

Tire pressure control (RDS)

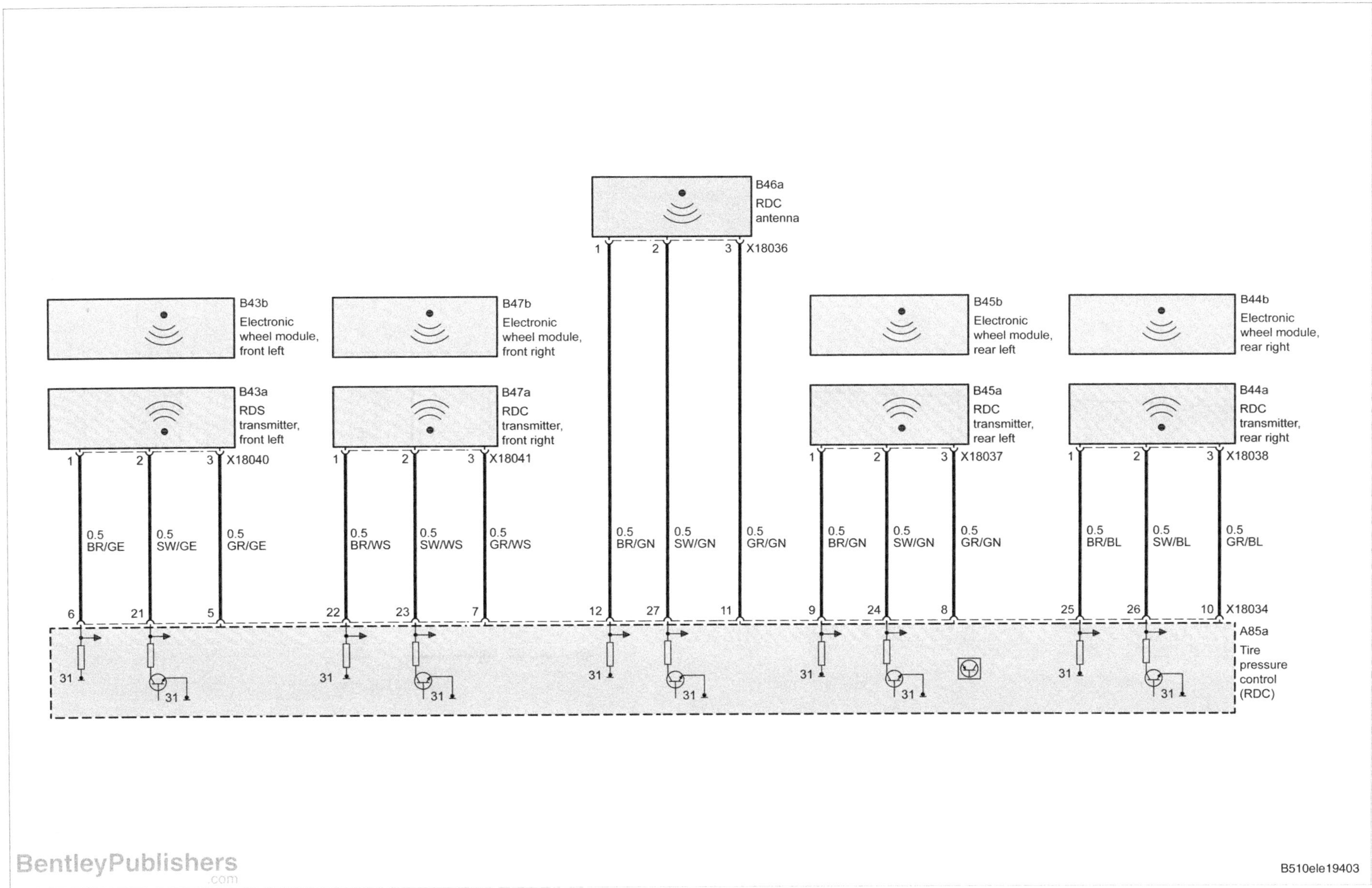

Tire pressure control system power supply

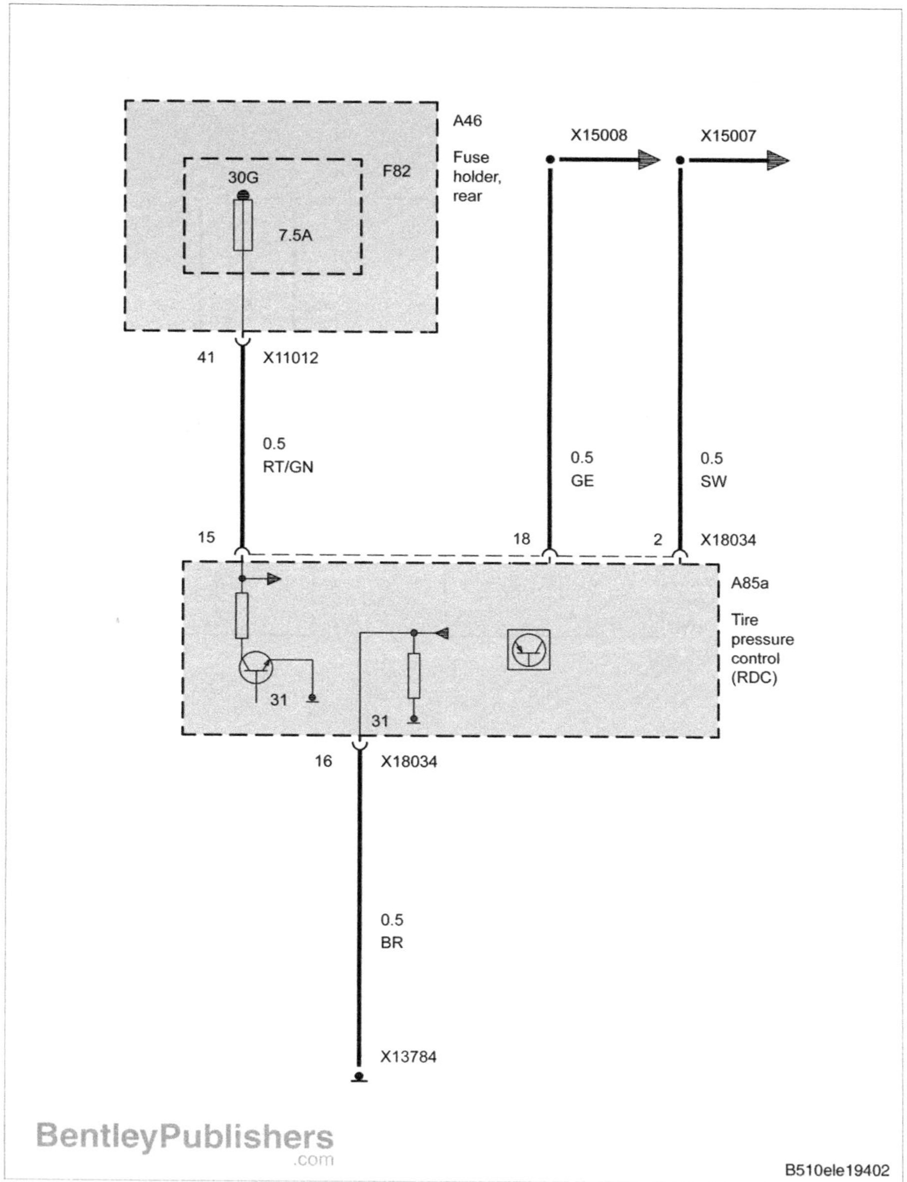

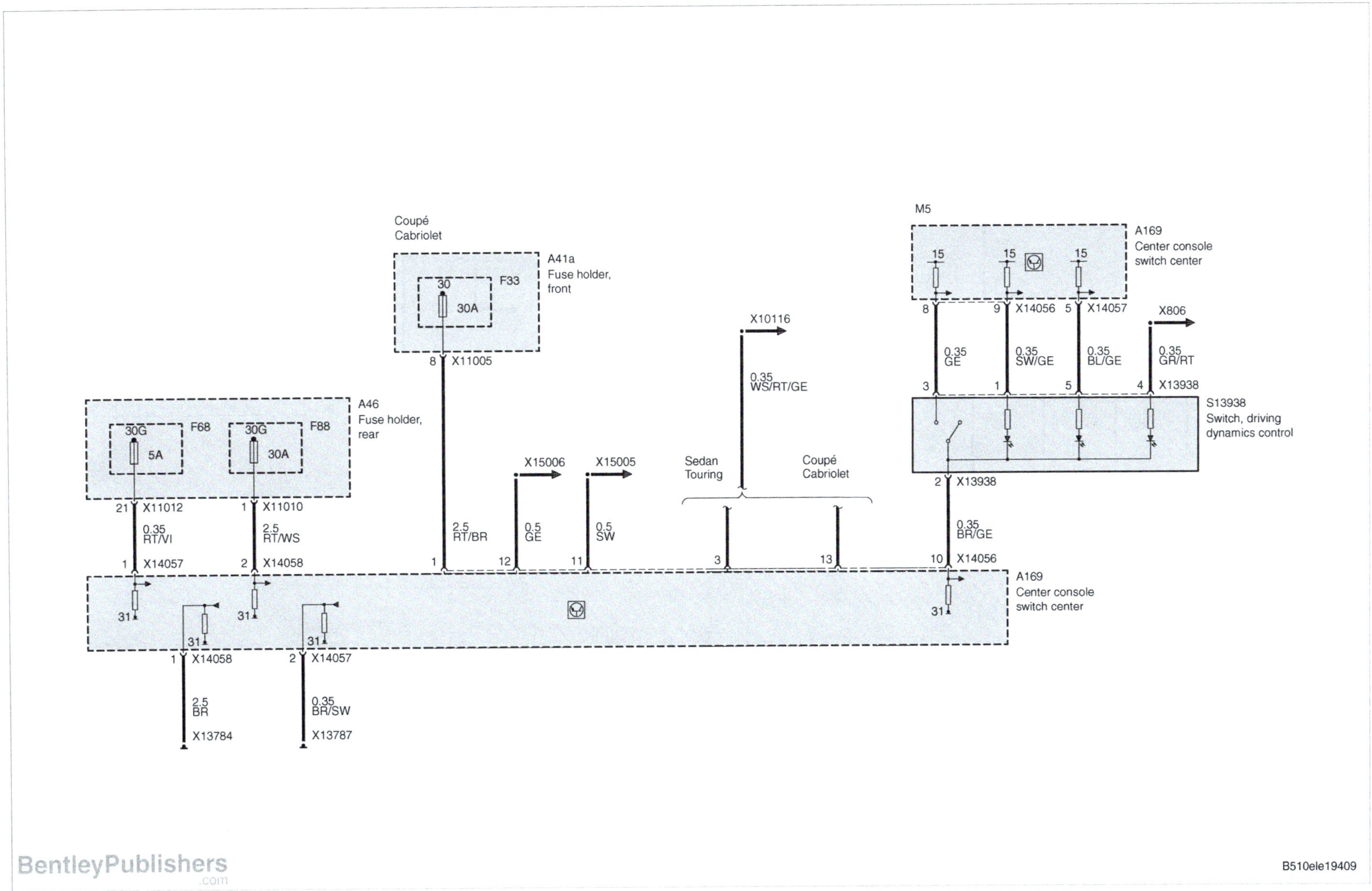
Coupé
Cabriolet
A41a
Fuse holder,
front
F33
30
30A
8 X11005
A46
Fuse holder,
rear
30G
F68
5A
30G
F88
30A
21 X11012
1 X11010
0.35
RT/VI
2.5
RT/WS
1 X14057
2 X14058
2.5
RT/BR
X15006
0.5
GE
X15005
0.5
SW
1
12
11
X10116
0.35
WS/RT/GE
Sedan
Touring
Coupé
Cabriolet
3
13
M5
A169
Center console
switch center
15
15
15
8
9 X14056
5 X14057
X806
0.35
GE
0.35
SW/GE
0.35
BL/GE
0.35
GR/RT
3
1
5
4 X13938
S13938
Switch, driving
dynamics control
2 X13938
0.35
BR/GE
10 X14056
A169
Center console
switch center
31
31
31
31
31
1 X14058
2 X14057
2.5
BR
0.35
BR/SW
X13784
X13787
BentleyPublishers
.com
B510ele19409

Center console switch center, power supply as of 09 / 2005 up to 03 / 2007

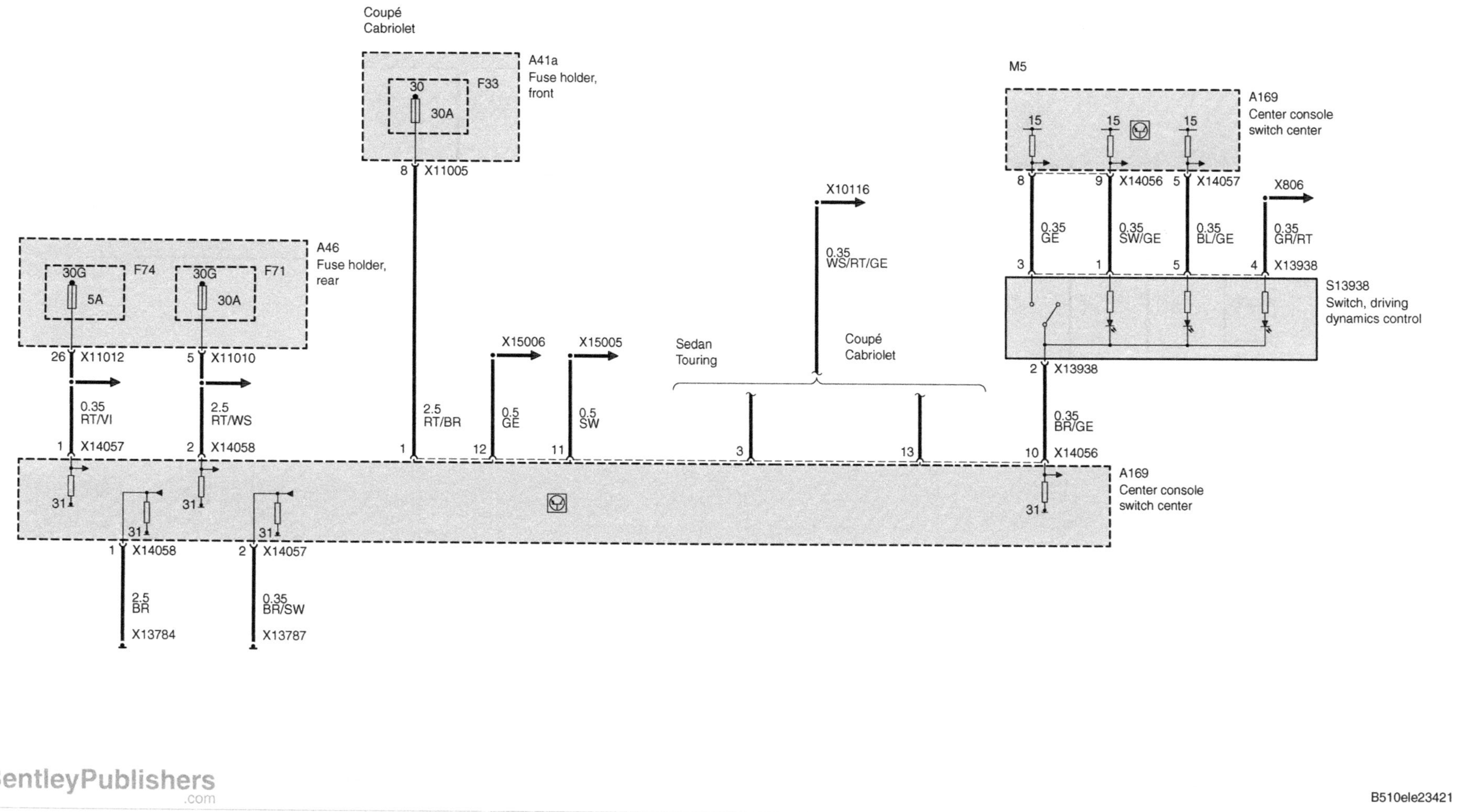

BentleyPublishers.com

Center console switch center, power supply
as of 03 / 2007

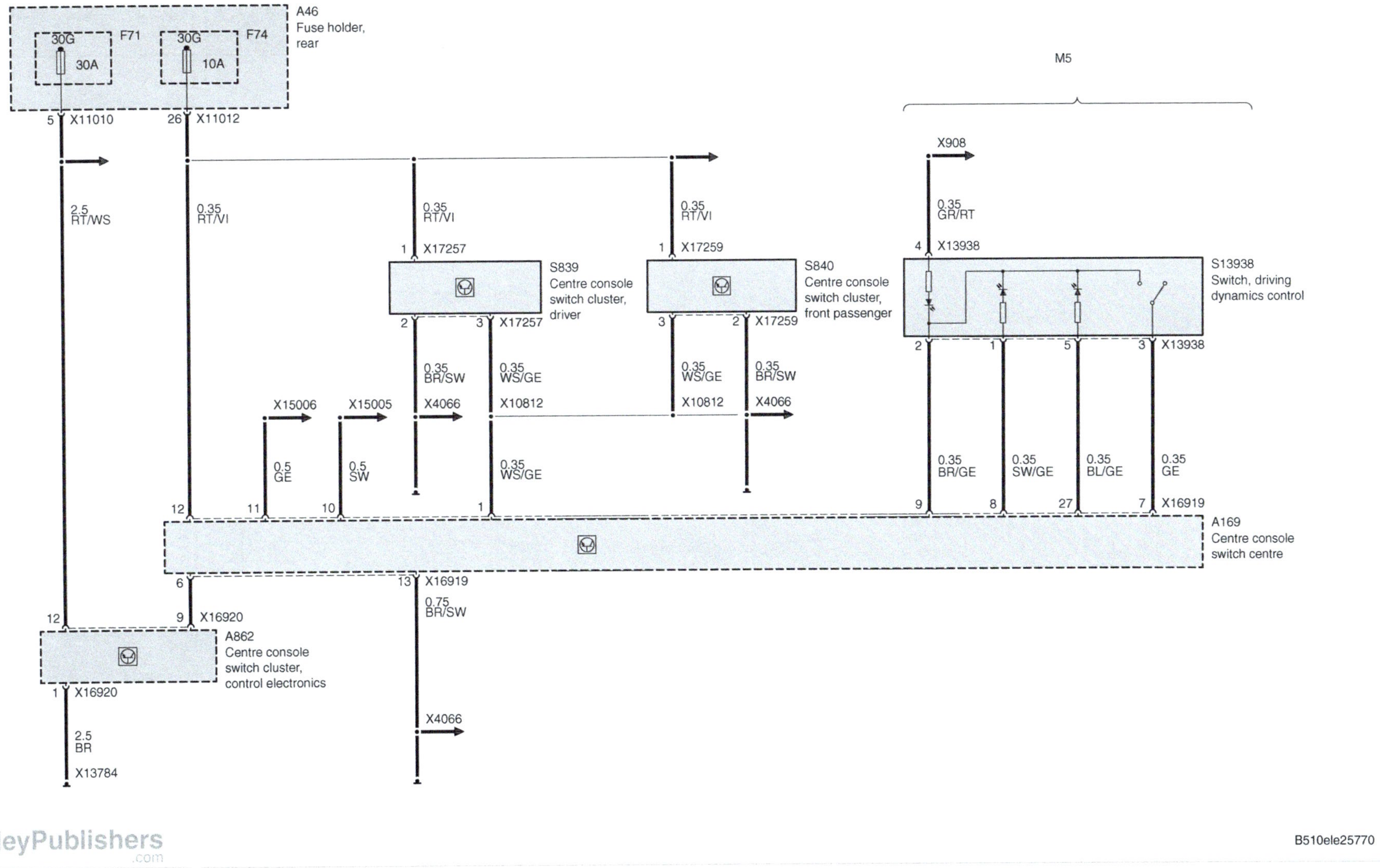

BentleyPublishers.com

Closed circuit current infringement shutdown
KGM body gateway module

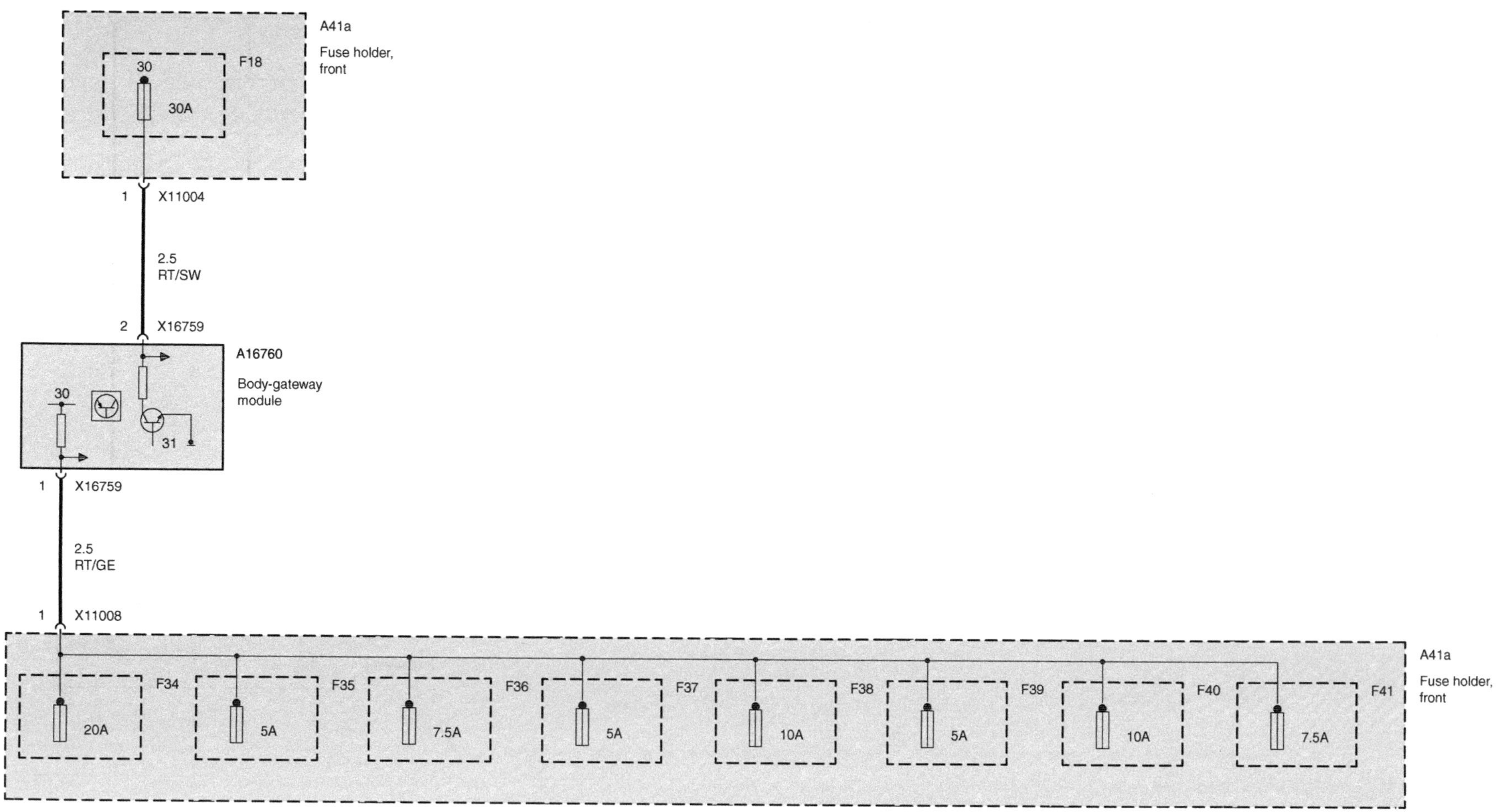

BentleyPublishers.com

B510ele23424

Micro-power module
vehicles without KGM body gateway module

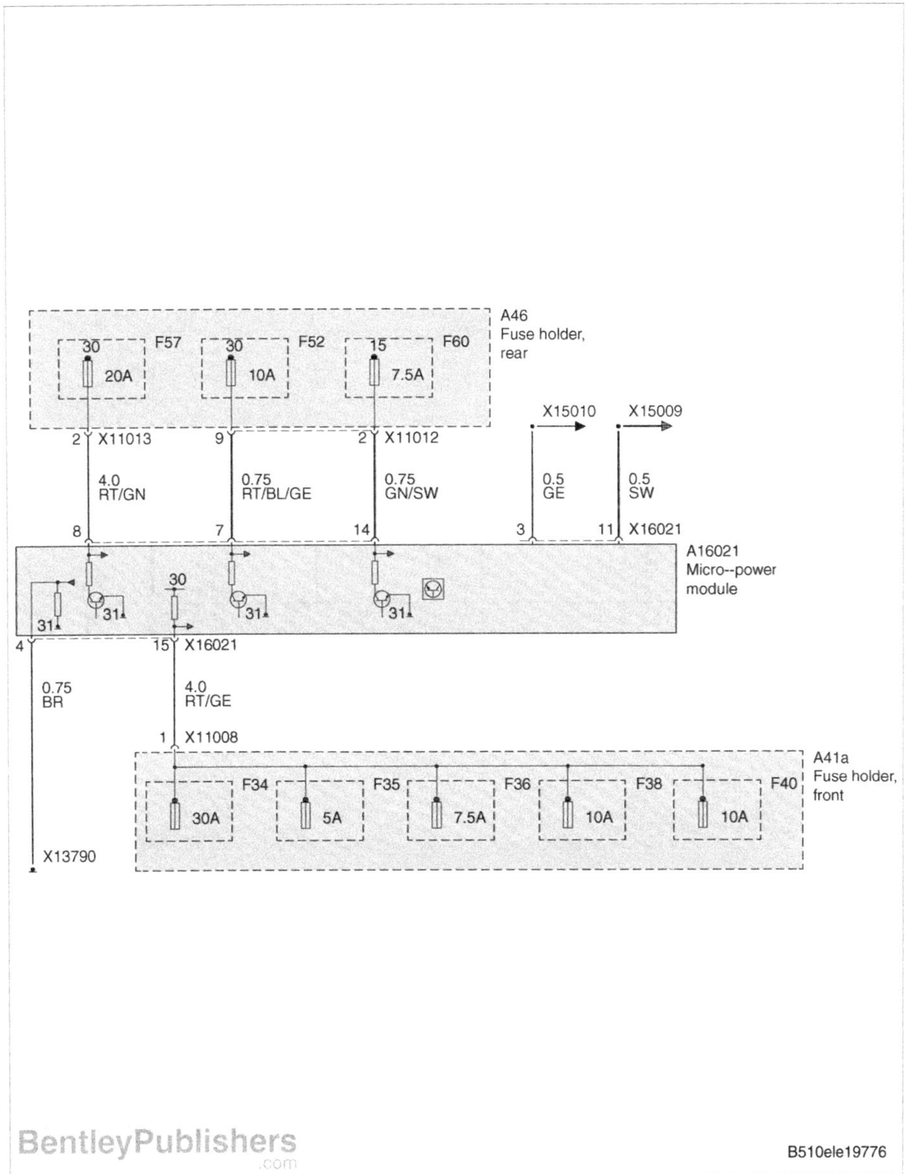

Controller

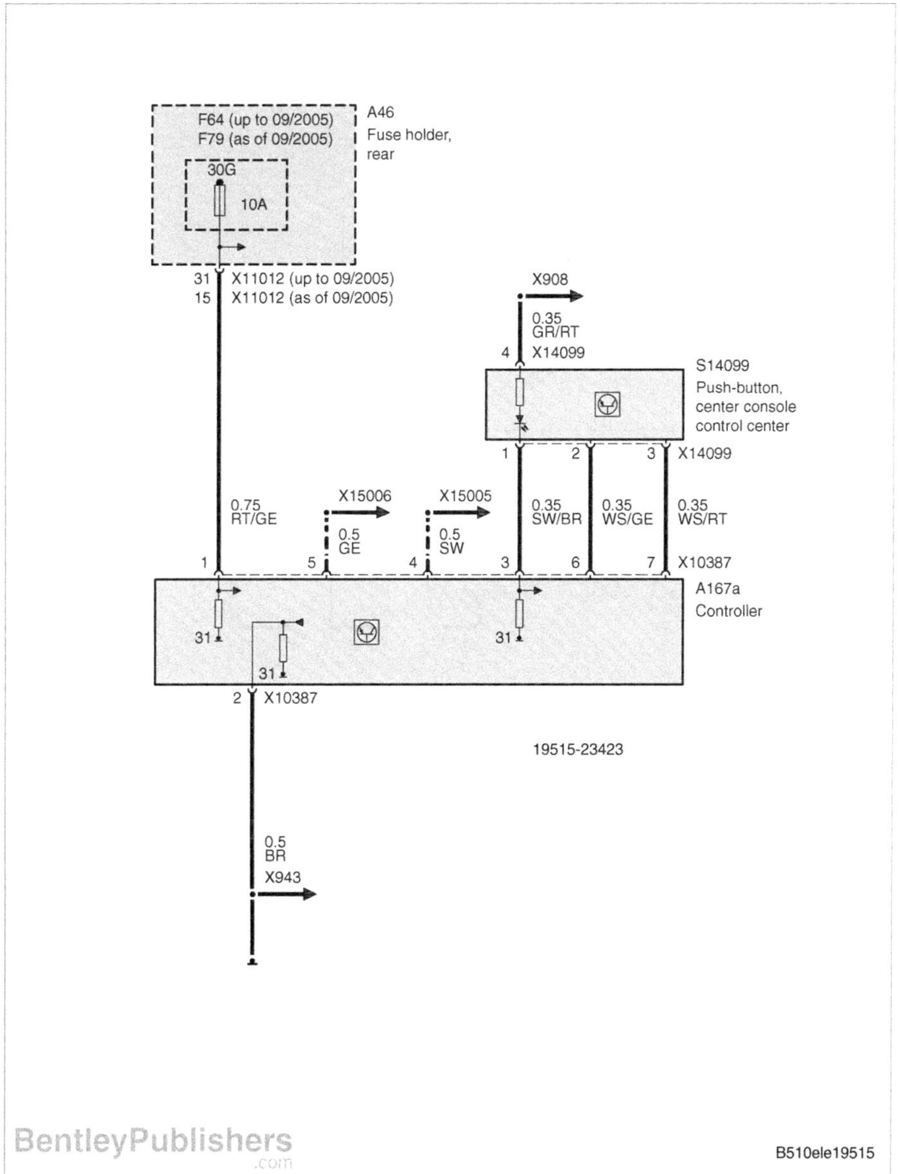

Body gateway module power supply

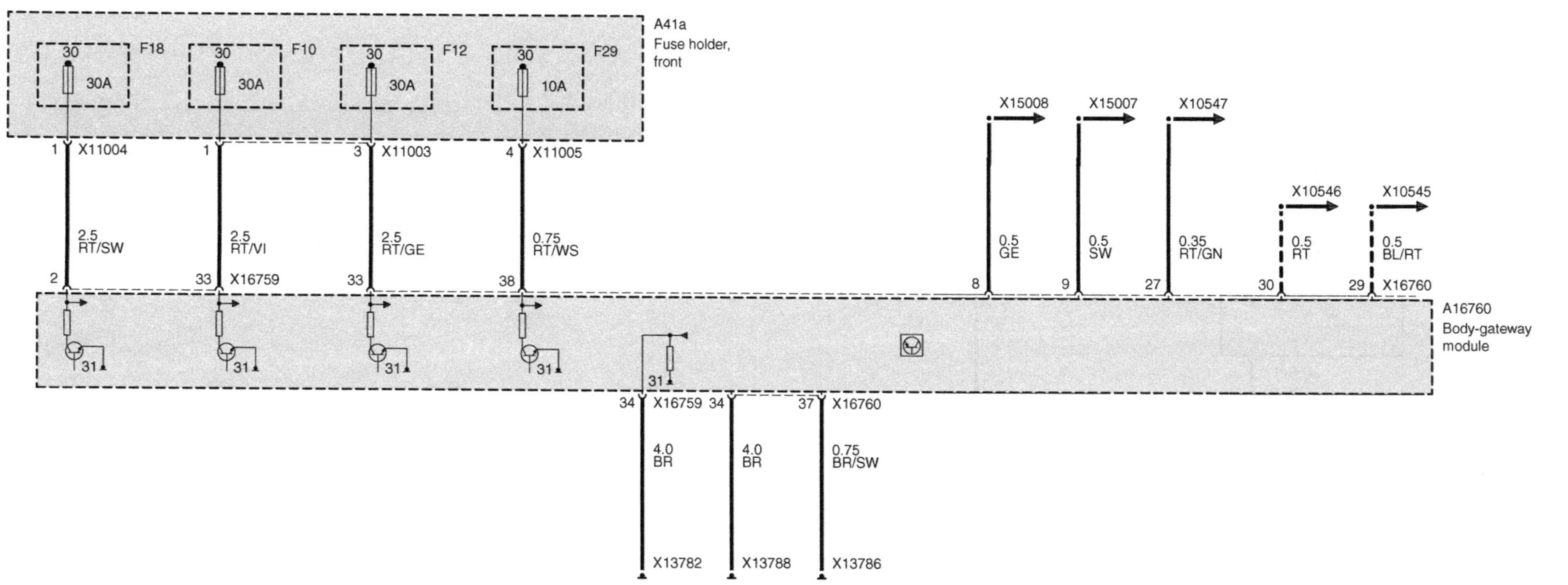

BentleyPublishers.com

Instrument cluster power supply

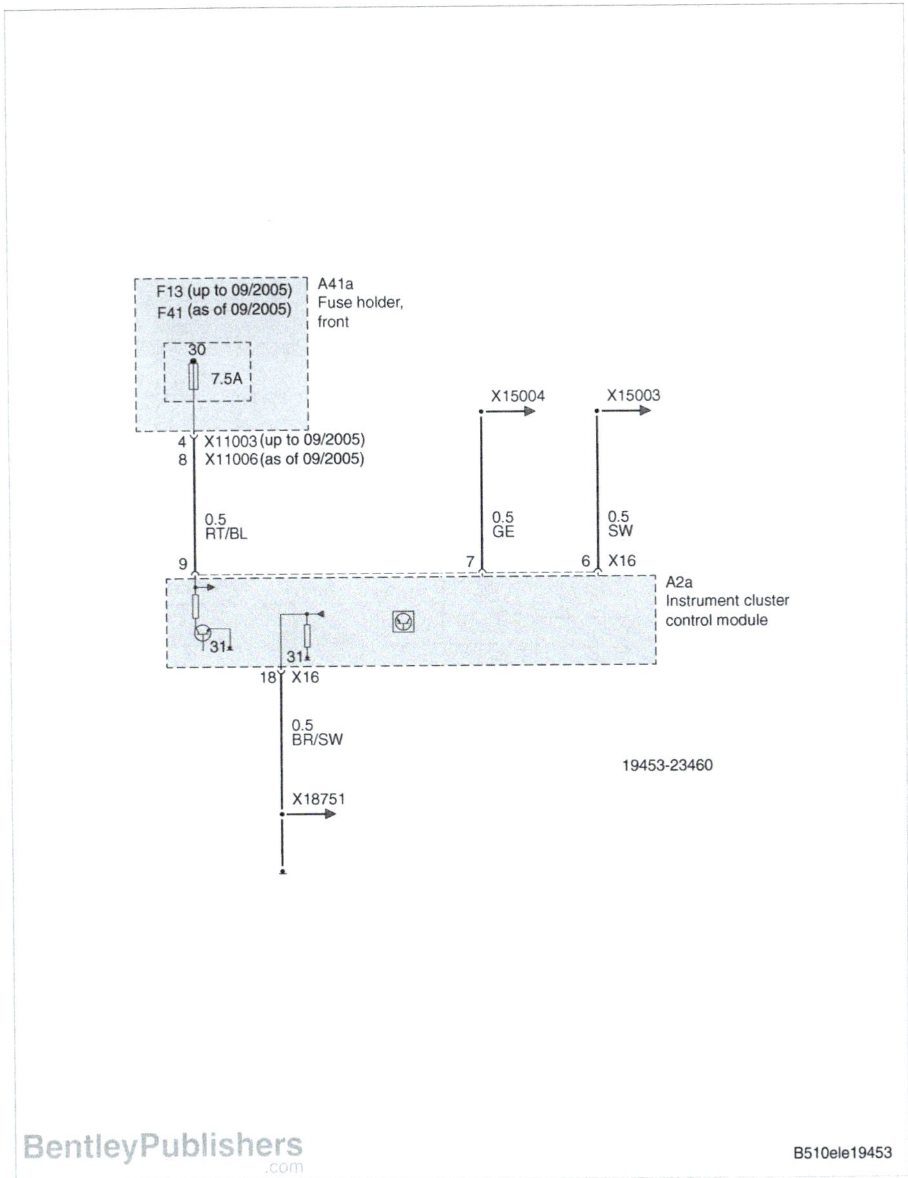

Warning and indicator lights

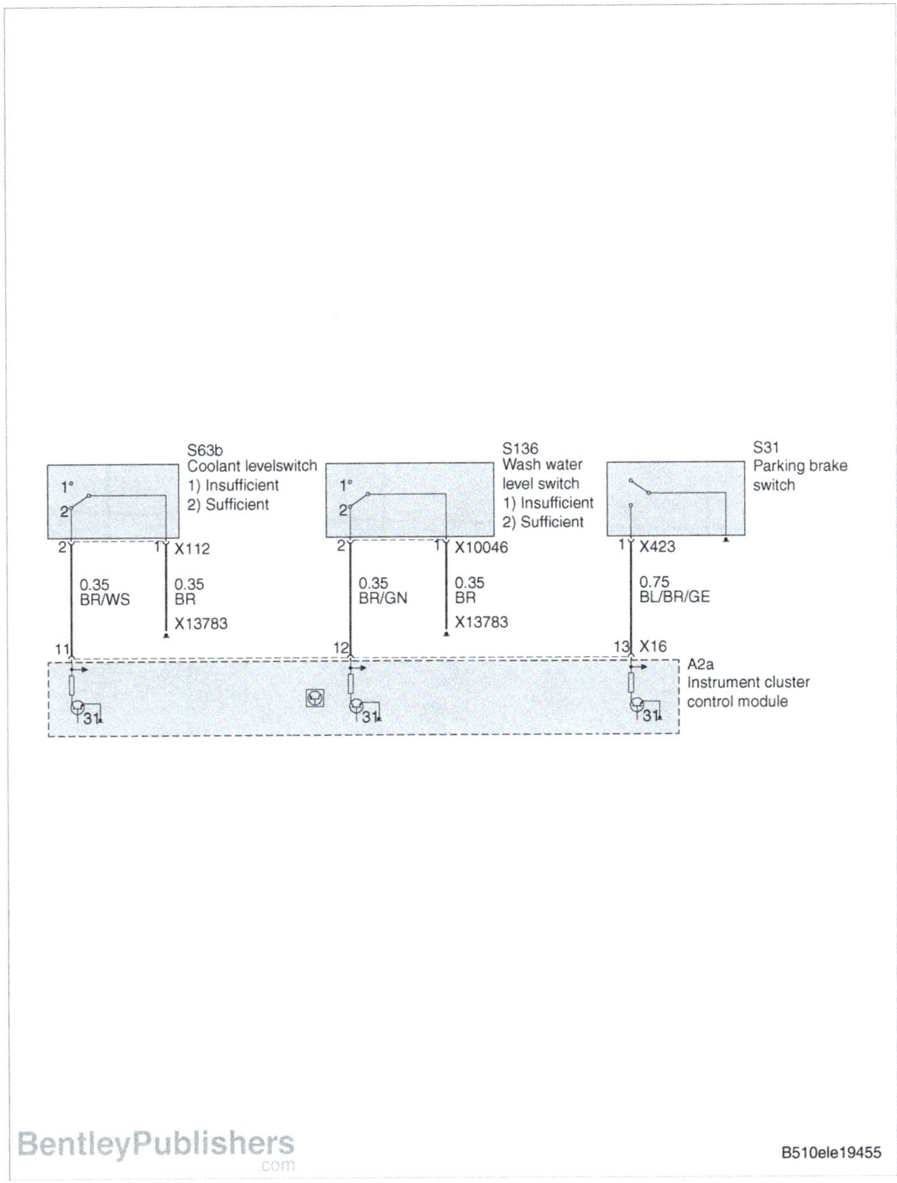

Audio system controller power supply

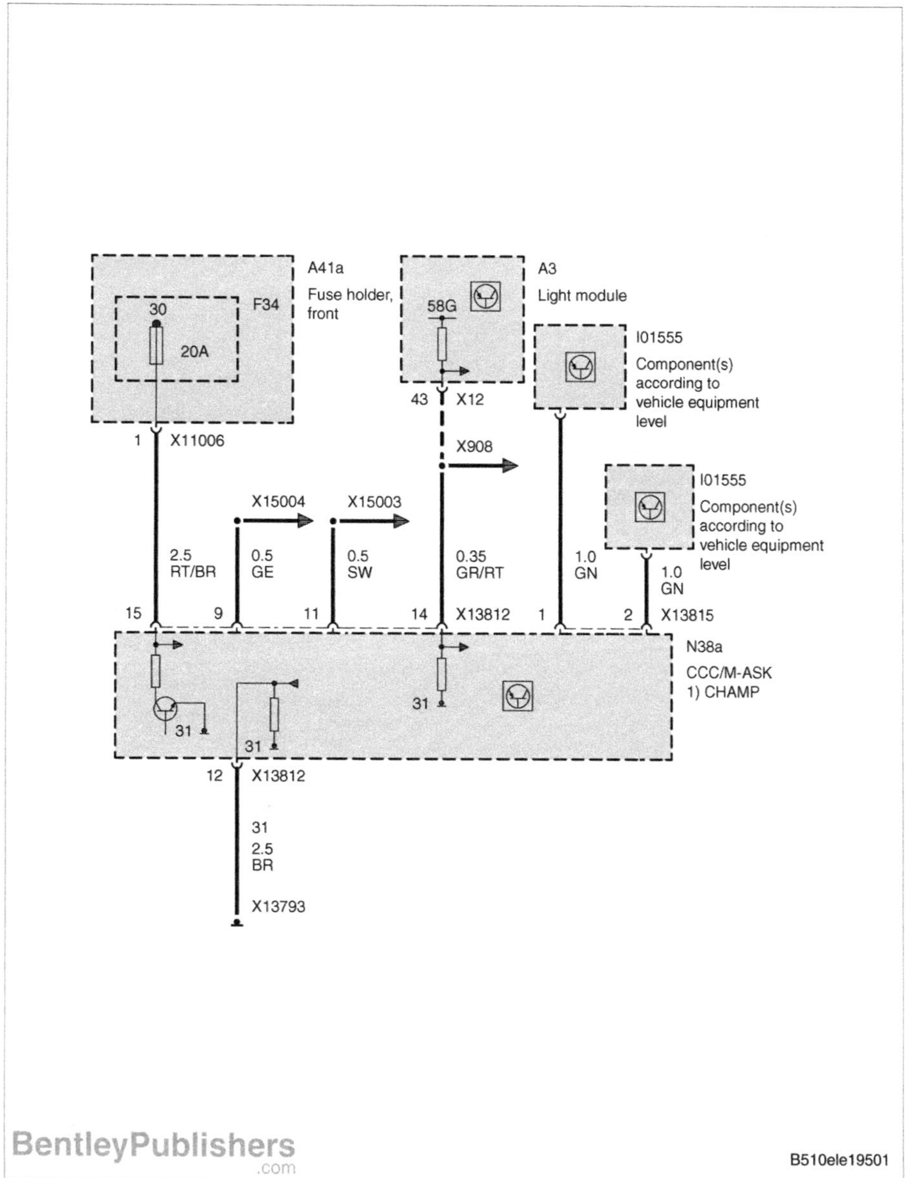

CIC Car information computer power supply

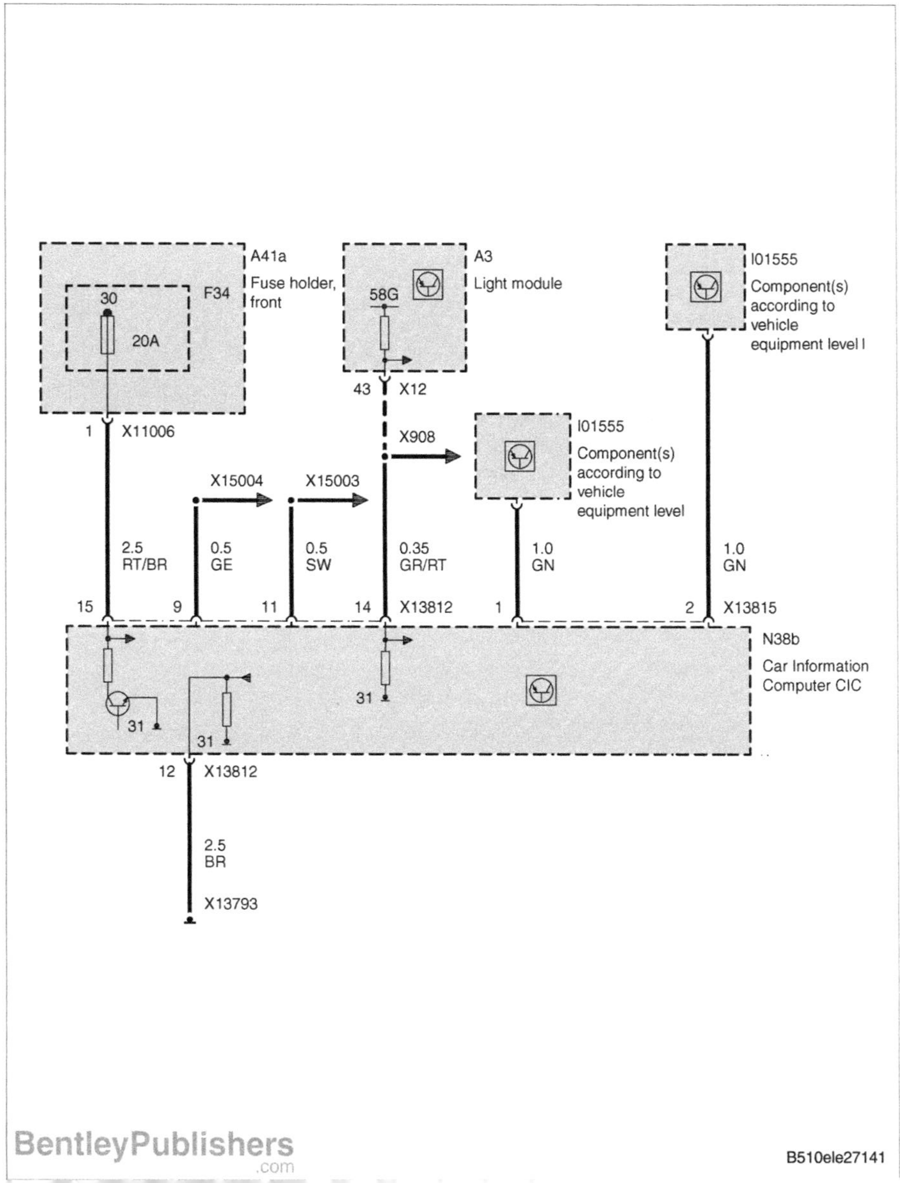

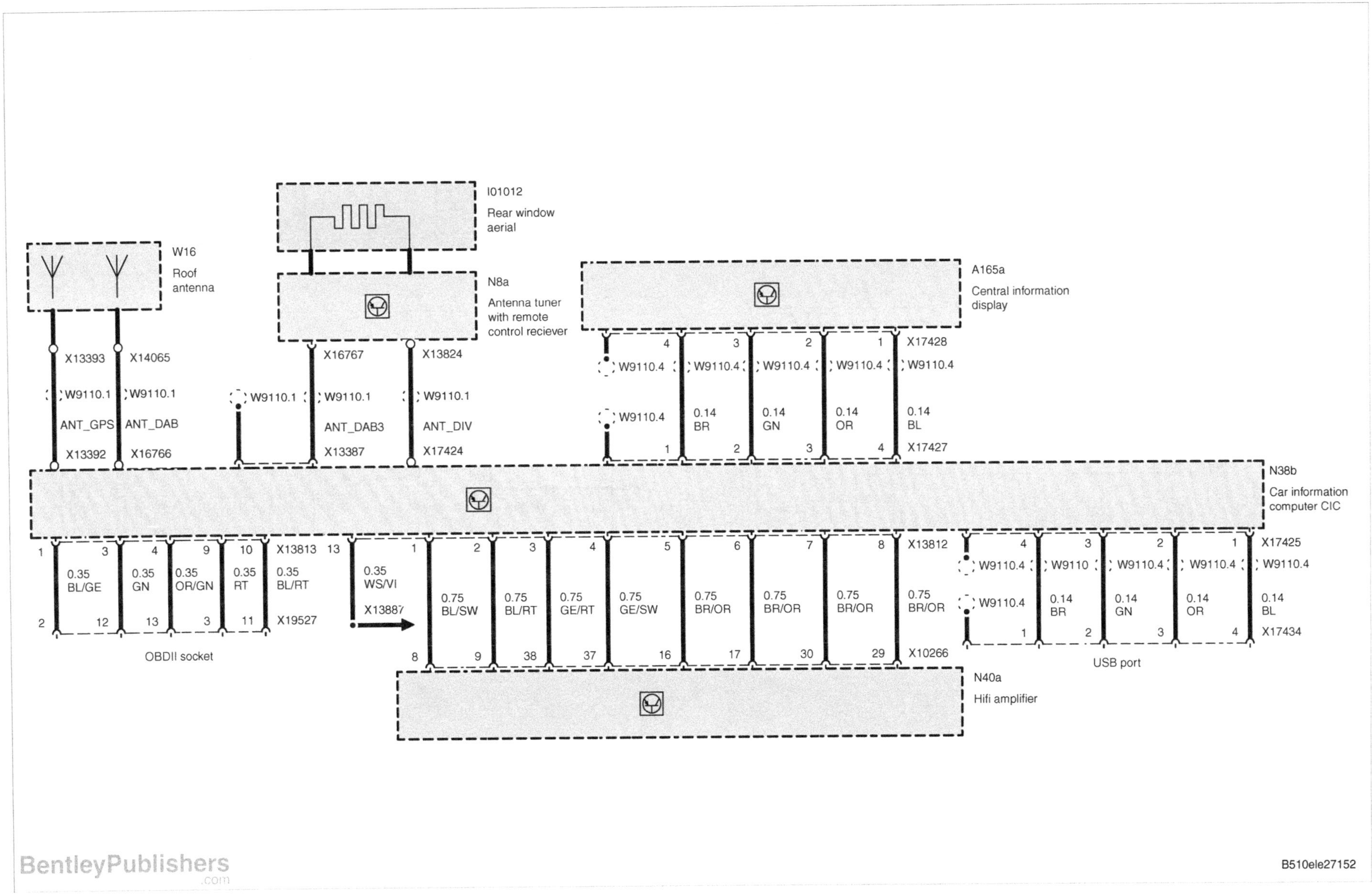

W16
Roof antenna
I01012
Rear window aerial
N8a
Antenna tuner with remote control reciever
A165a
Central information display
N38b
Car information computer CIC
N40a
Hifi amplifier
OBDII socket
USB port
X13393
X14065
W9110.1
ANT_GPS
ANT_DAB
X13392
X16766
X16767
X13824
ANT_DAB3
ANT_DIV
X13387
X17424
X17428
W9110.4
W9110
X17427
0.14 BR
0.14 GN
0.14 OR
0.14 BL
X13813
X19527
0.35 BL/GE
0.35 GN
0.35 OR/GN
0.35 RT
0.35 BL/RT
0.35 WS/VI
X13887
0.75 BL/SW
0.75 BL/RT
0.75 GE/RT
0.75 GE/SW
0.75 BR/OR
X13812
X10266
X17425
X17434
BentleyPublishers.com
B510ele27152

Light module
Sedan with Xenon lights up to 03 / 2005
(page 1 of 3)

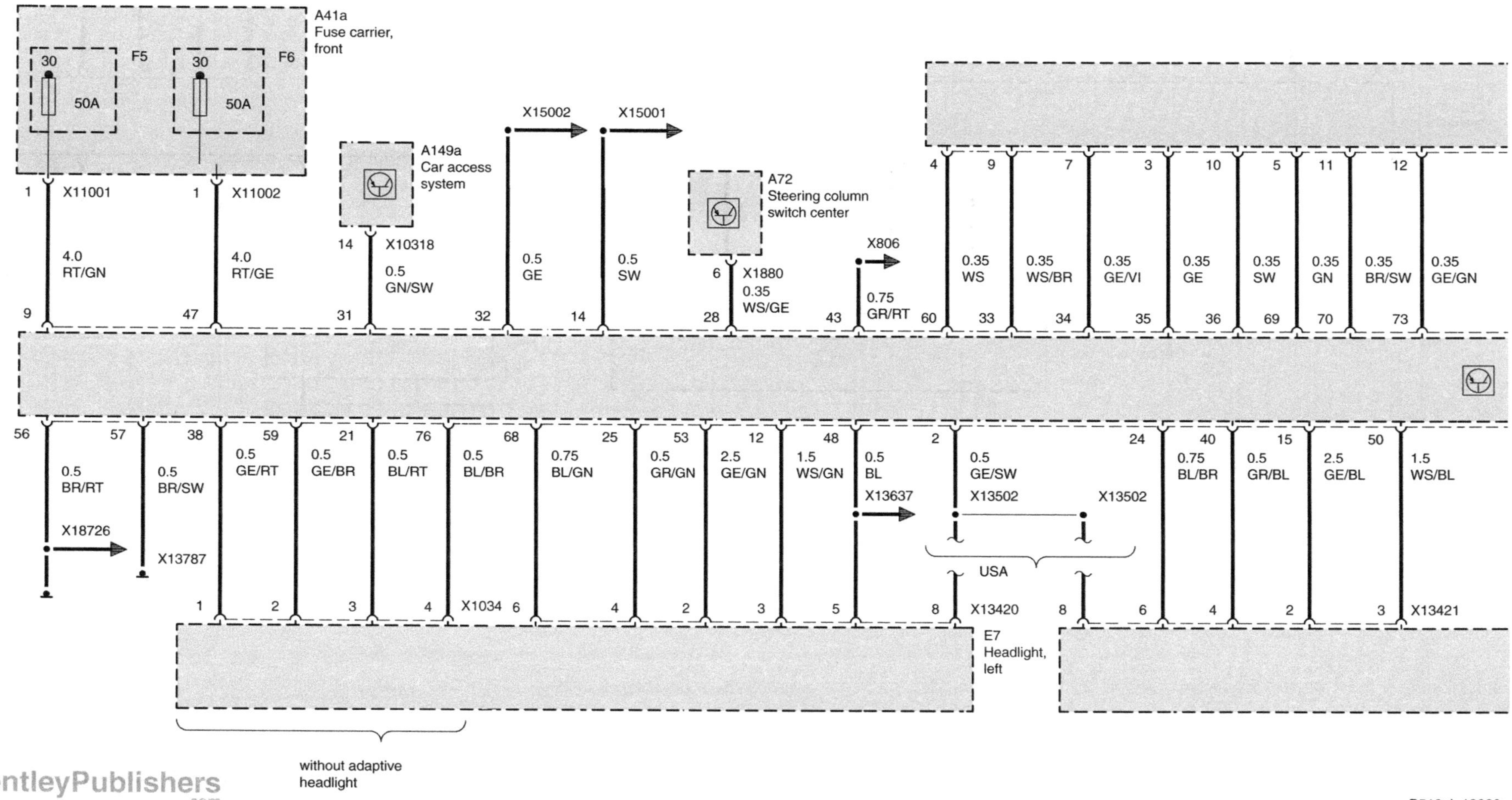

BentleyPublishers.com

B510ele19966a

Light module
Sedan with Xenon lights up to 03 / 2005
(page 2 of 3)

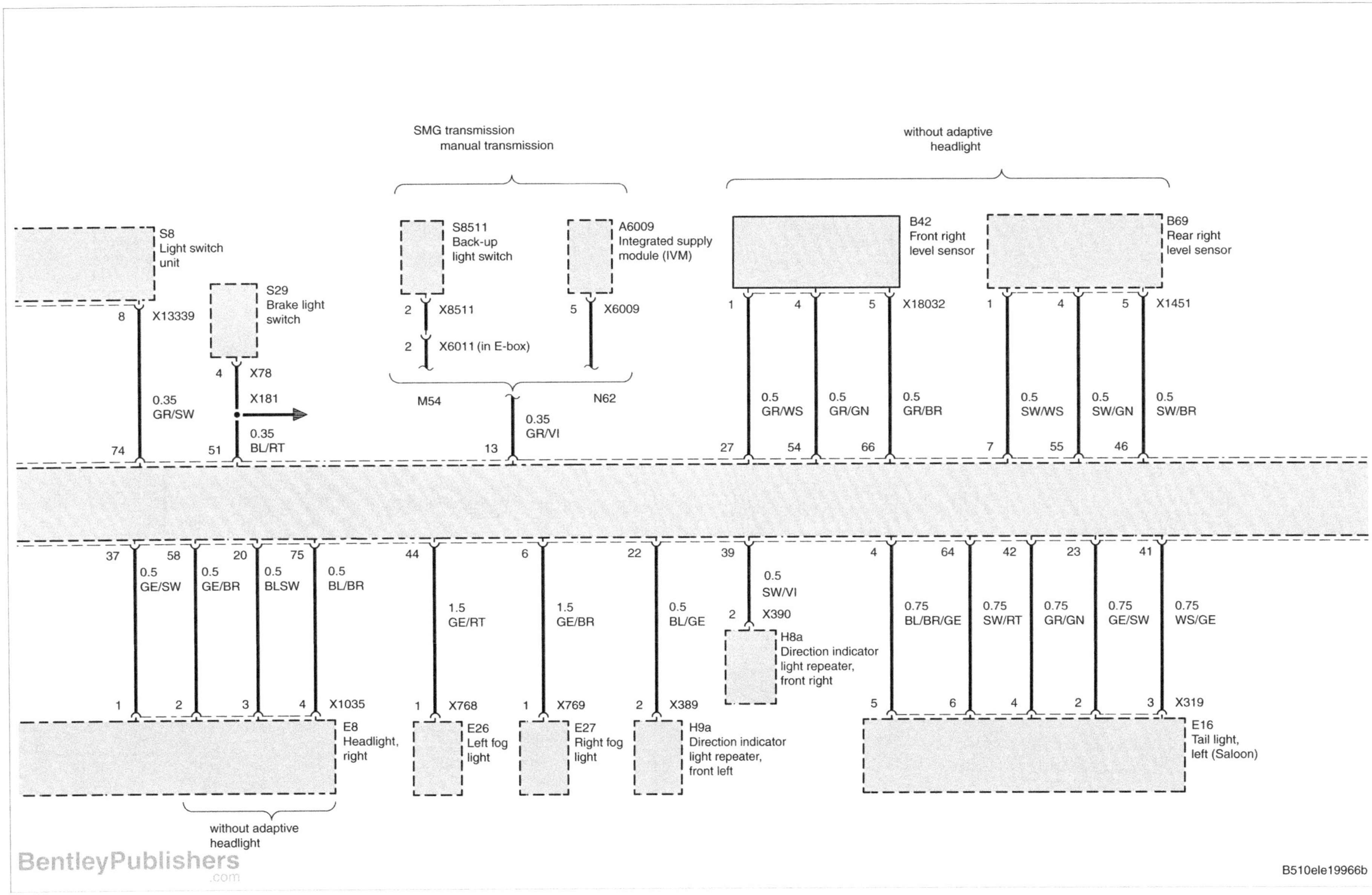

Light module
Sedan with Xenon lights up to 03 / 2005
(page 3 of 3)

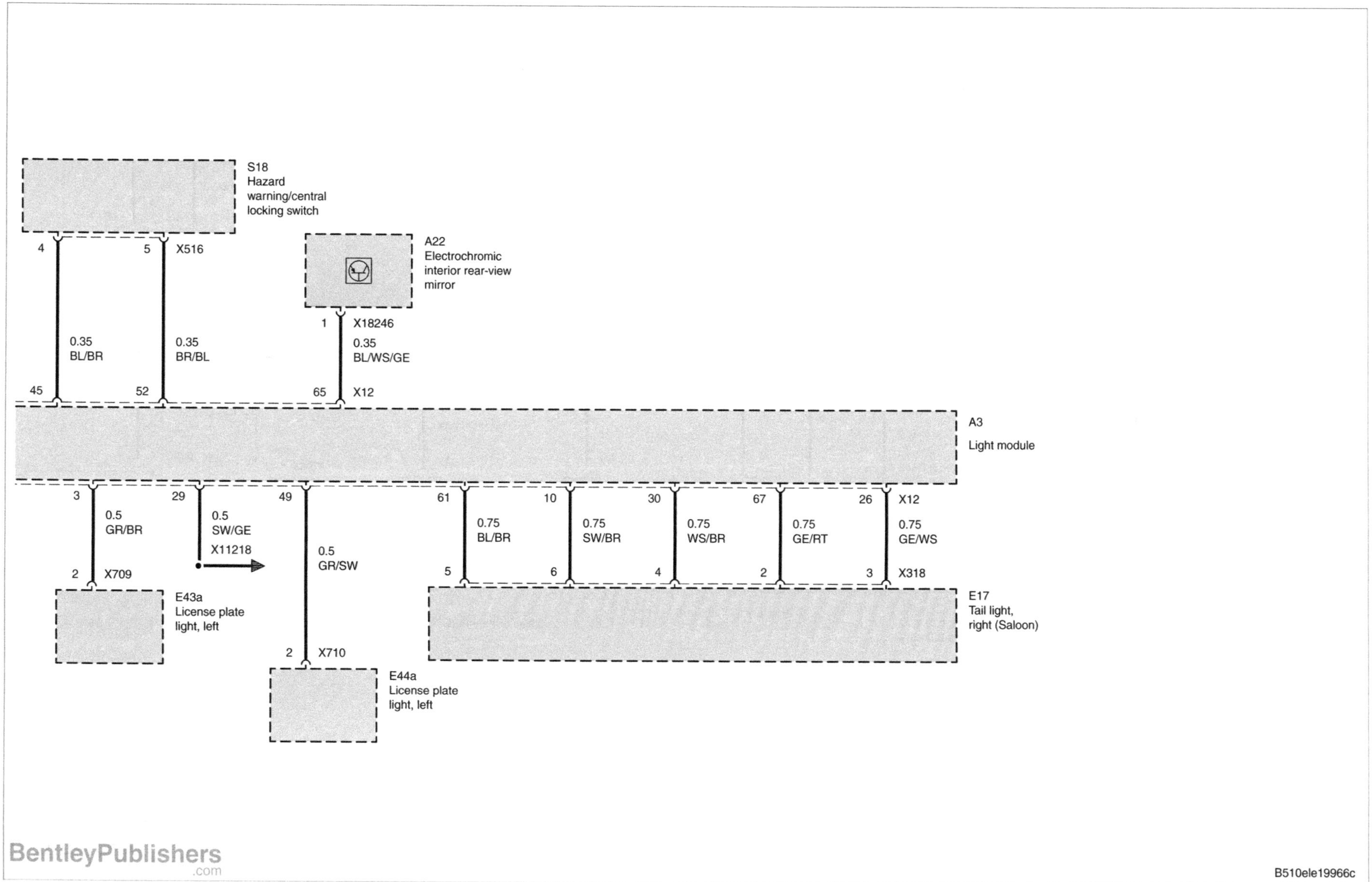

Light module
Sedan without Xenon lights up to 03 / 2005
(page 1 of 3)

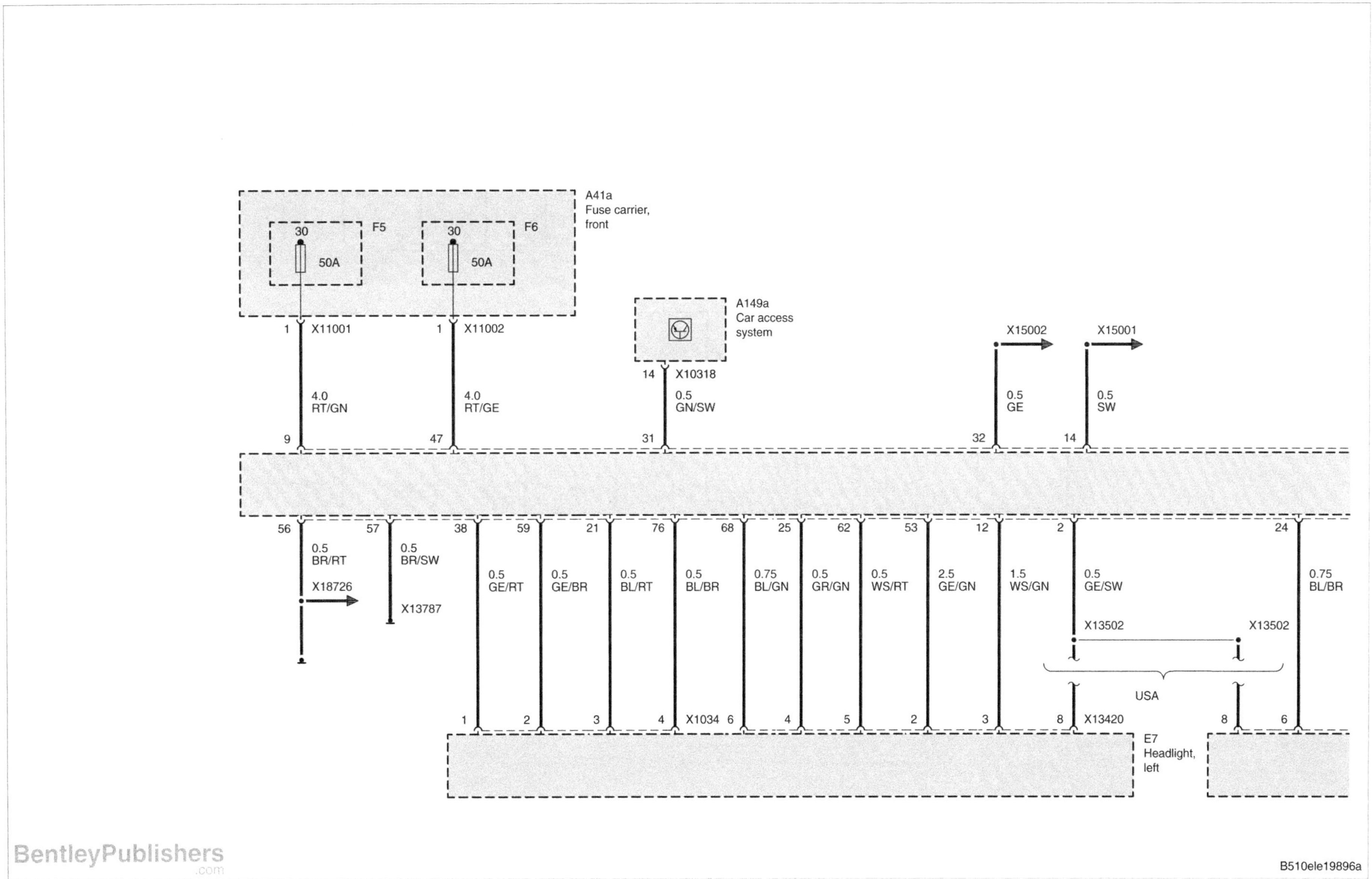

BentleyPublishers.com

Light module
Sedan without Xenon lights up to 03 / 2005
(page 2 of 3)

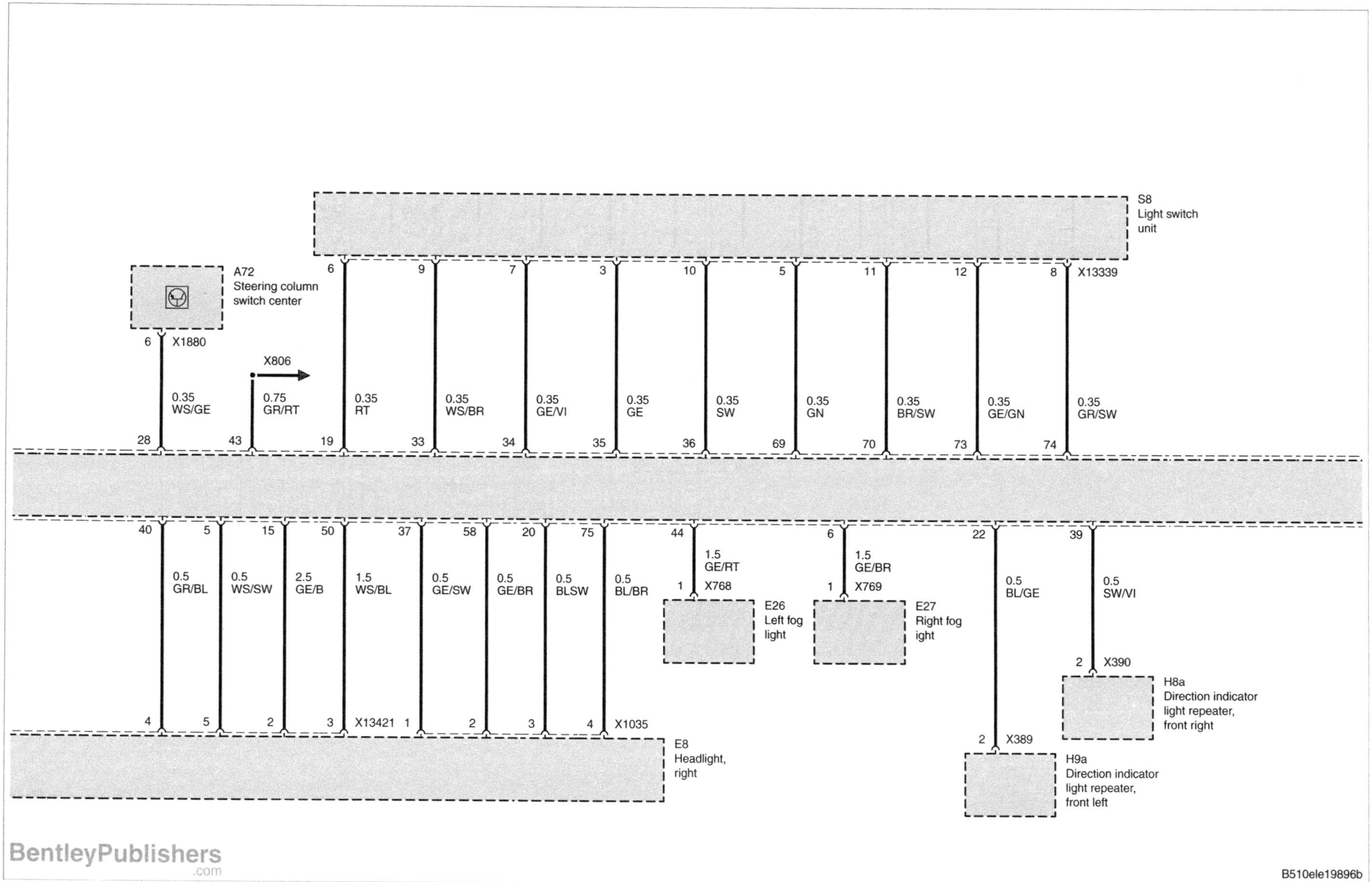

BentleyPublishers
.com

B510ele19896b

Light module
Sedan without Xenon lights up to 03 / 2005
(page 3 of 3)

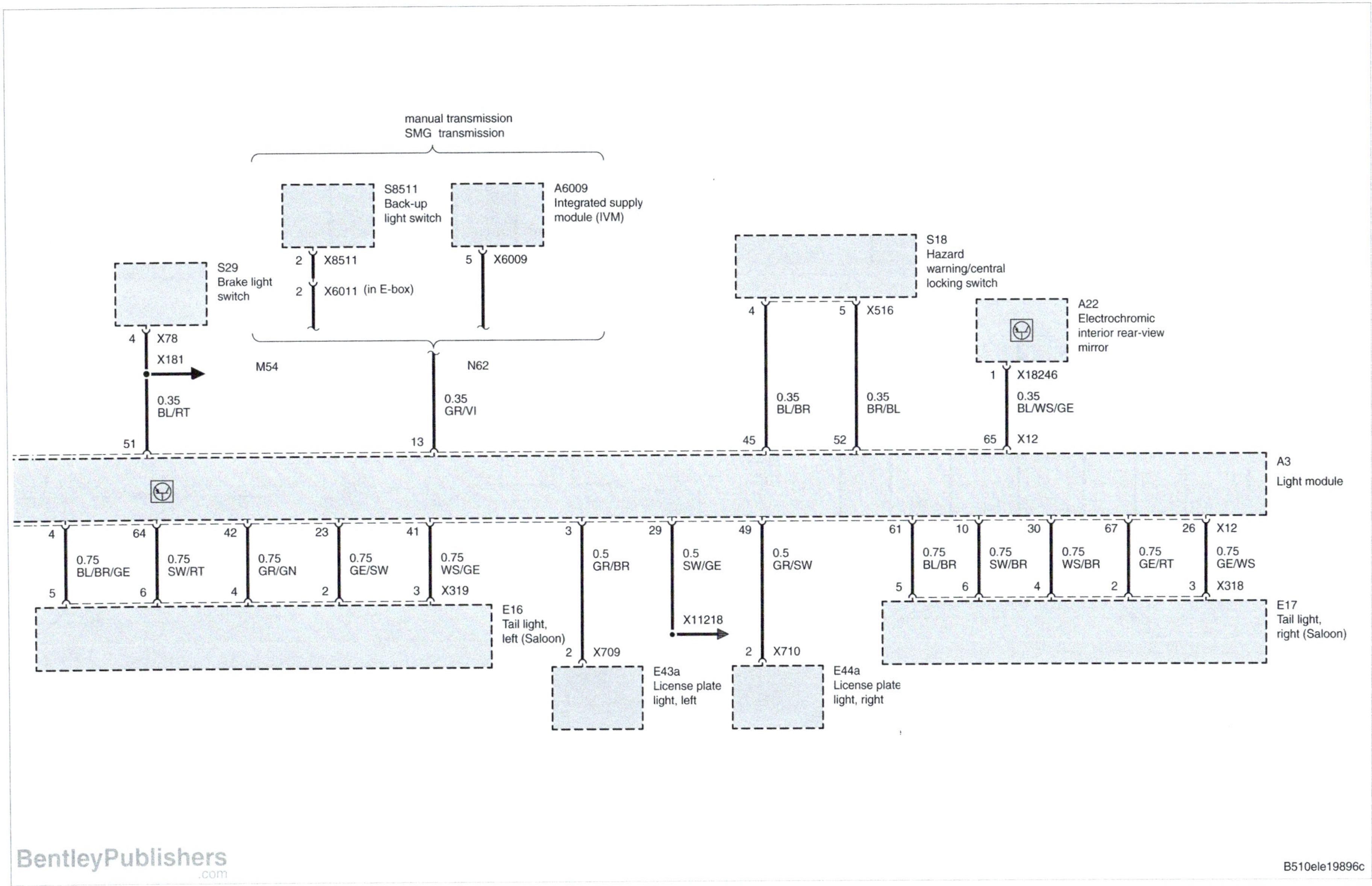

Light module
Wagon with Xenon lights up to 03 / 2005
(page 1 of 2)

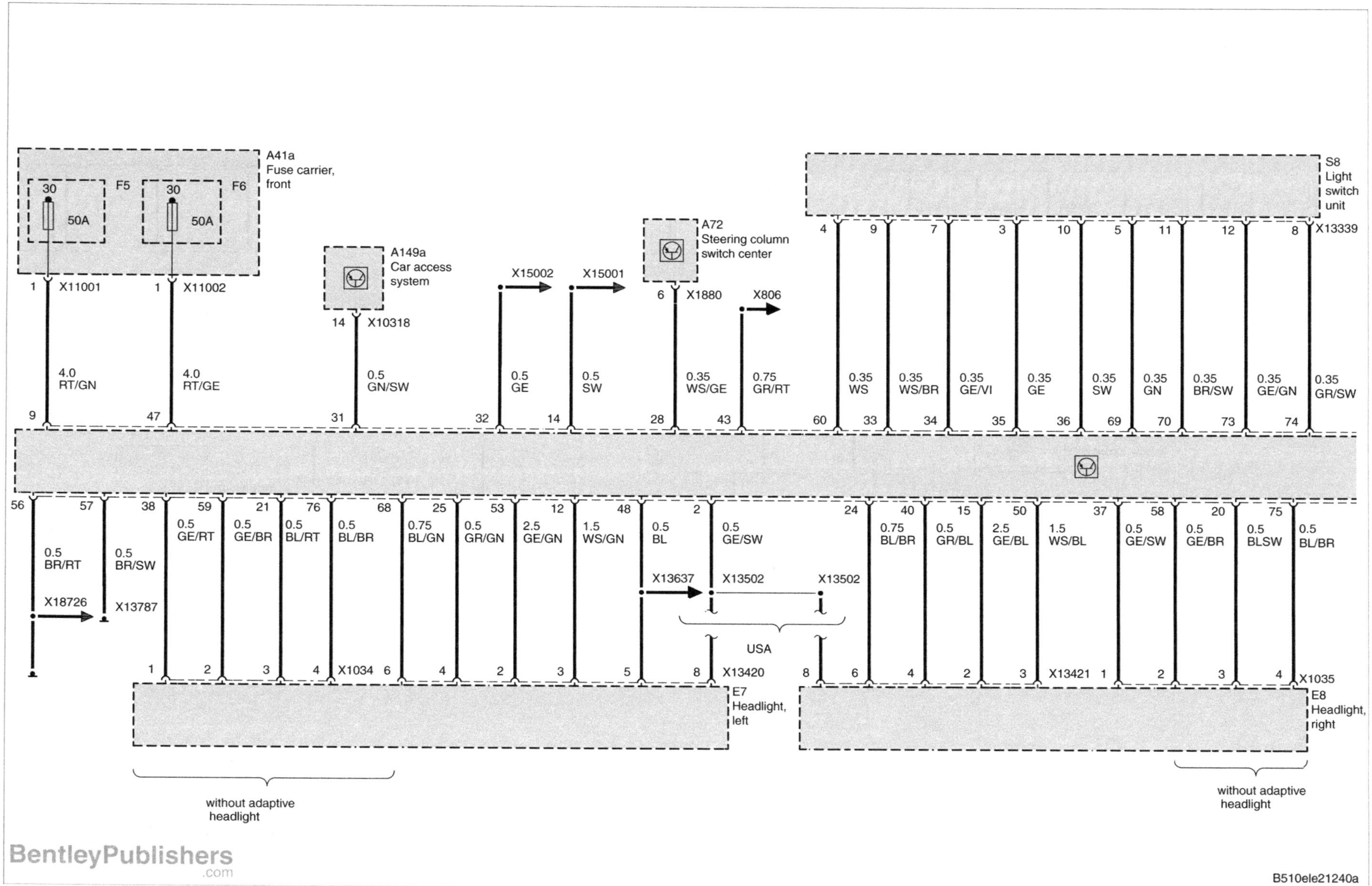

BentleyPublishers
.com

B510ele21240a

Light module
Wagon with Xenon lights up to 03 / 2005
(page 2 of 2)

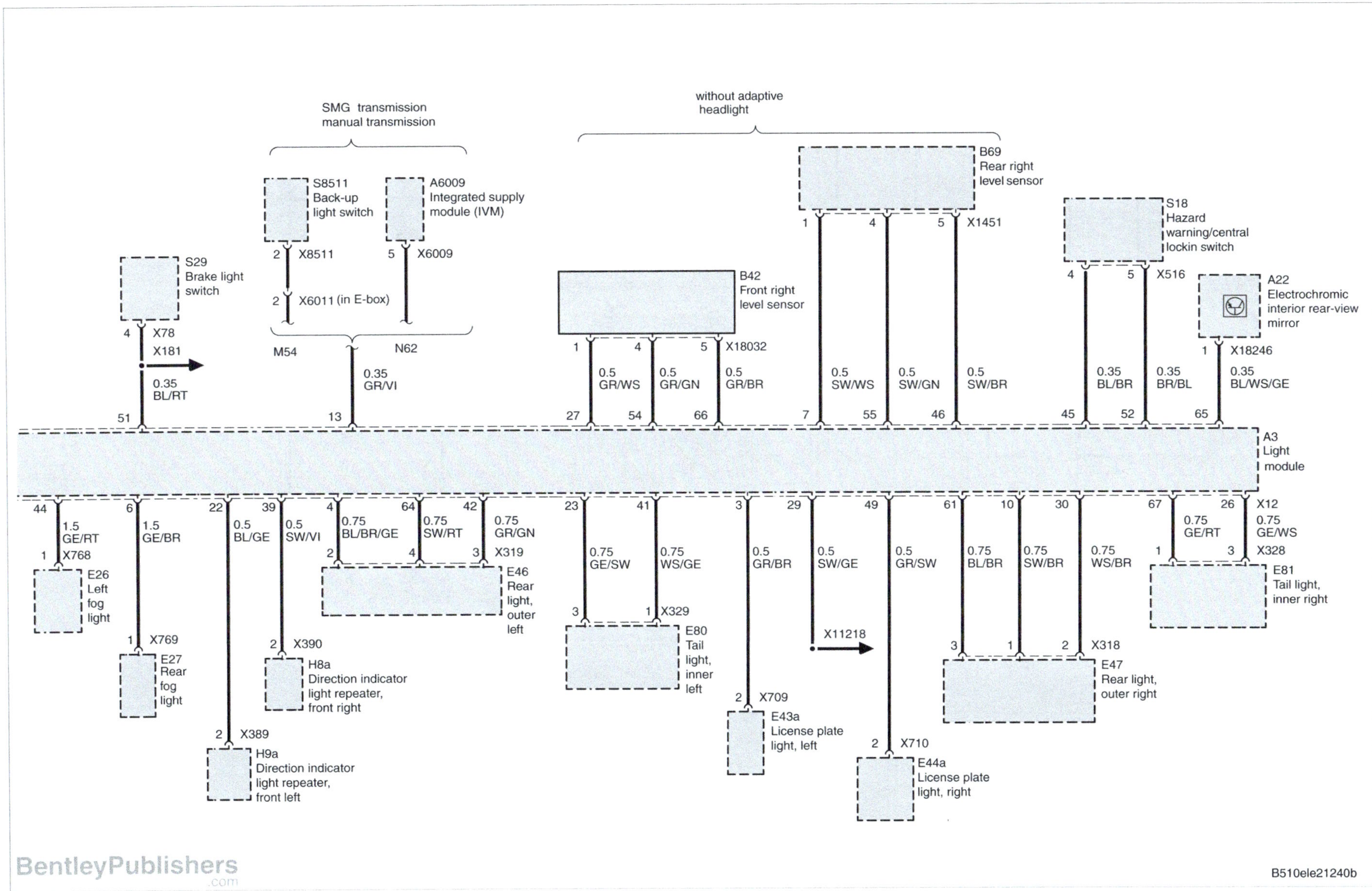

Light module
Wagon without Xenon lights up to 03 / 2005
(page 1 of 3)

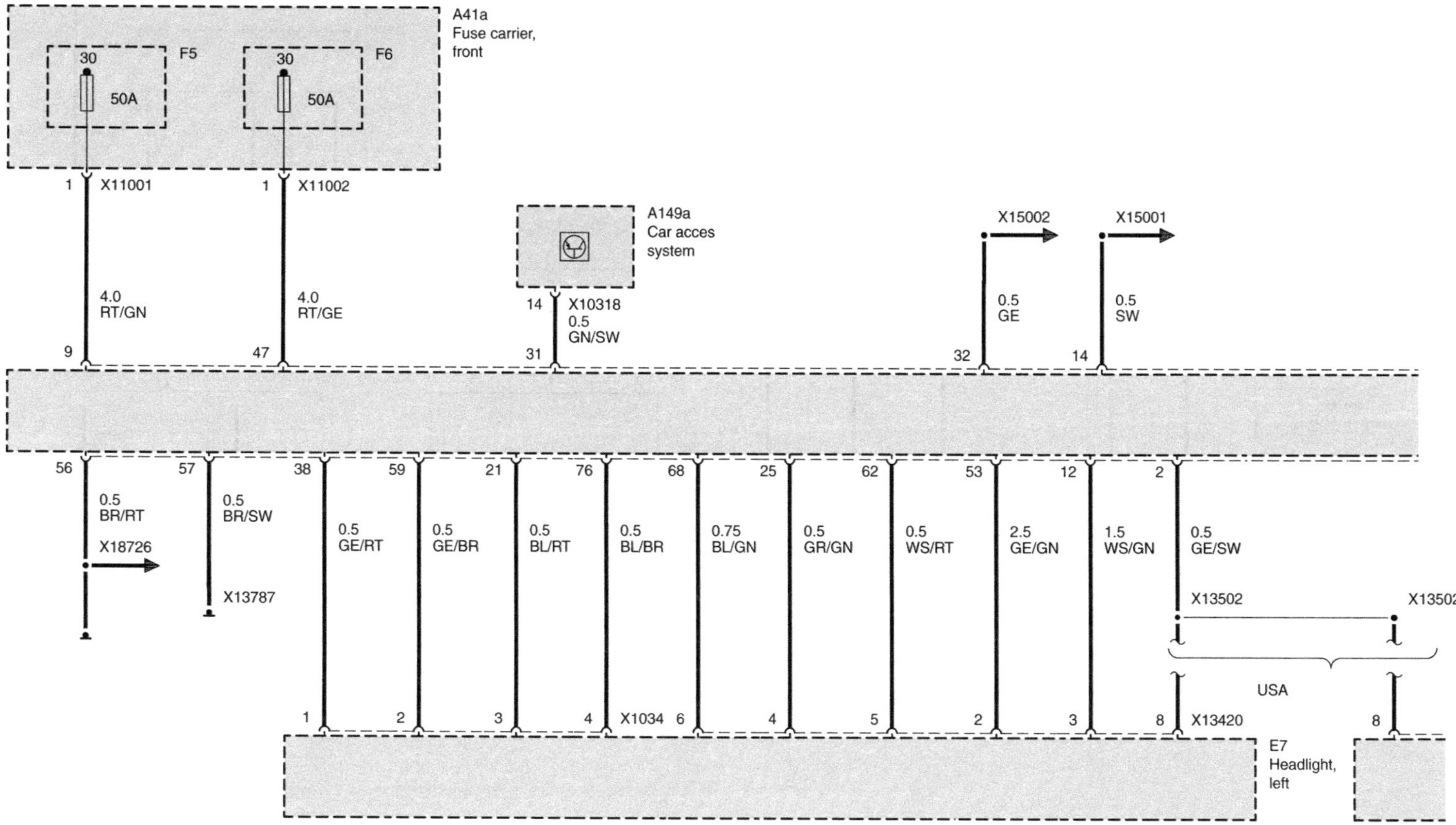

BentleyPublishers.com

B510ele21239a

Light module
Wagon without Xenon lights up to 03 / 2005
(page 2 of 3)

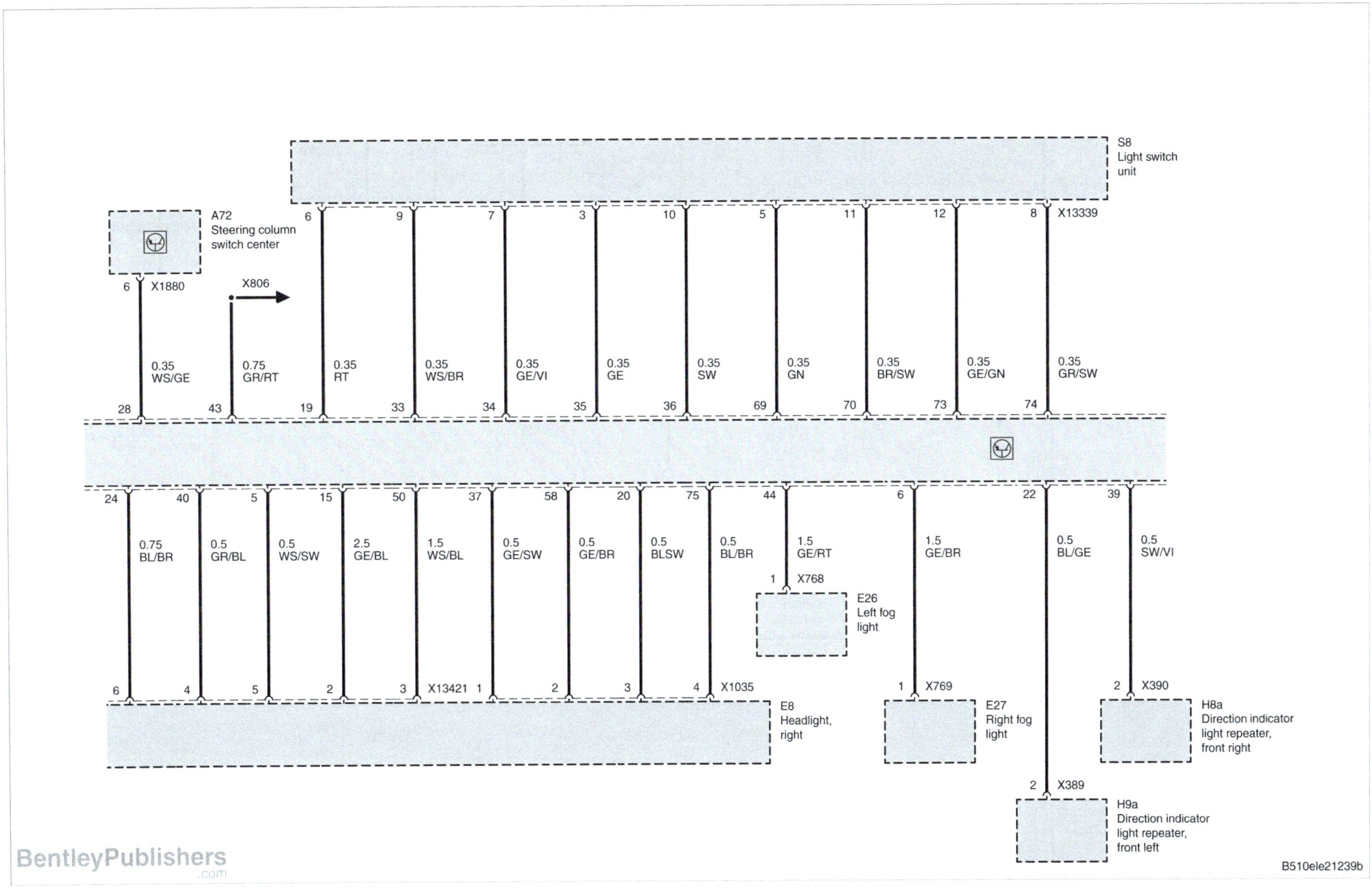

Light module
Wagon without Xenon lights up to 03 / 2005
(page 3 of 3)

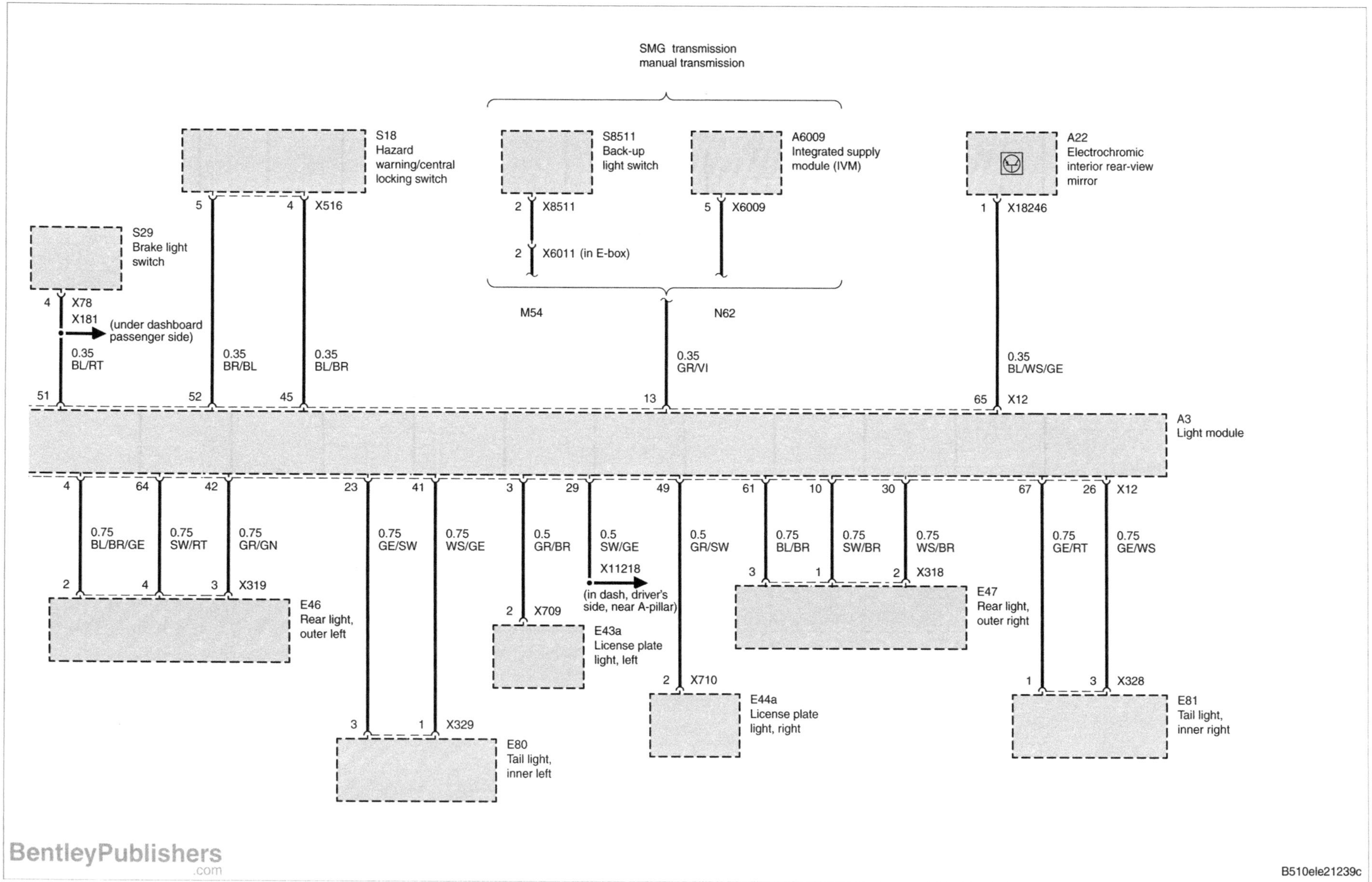

BentleyPublishers.com

B510ele21239c

Light module
Sedan with Xenon lights as of 03 / 2005 up to 03 / 2007
(page 1 of 3)

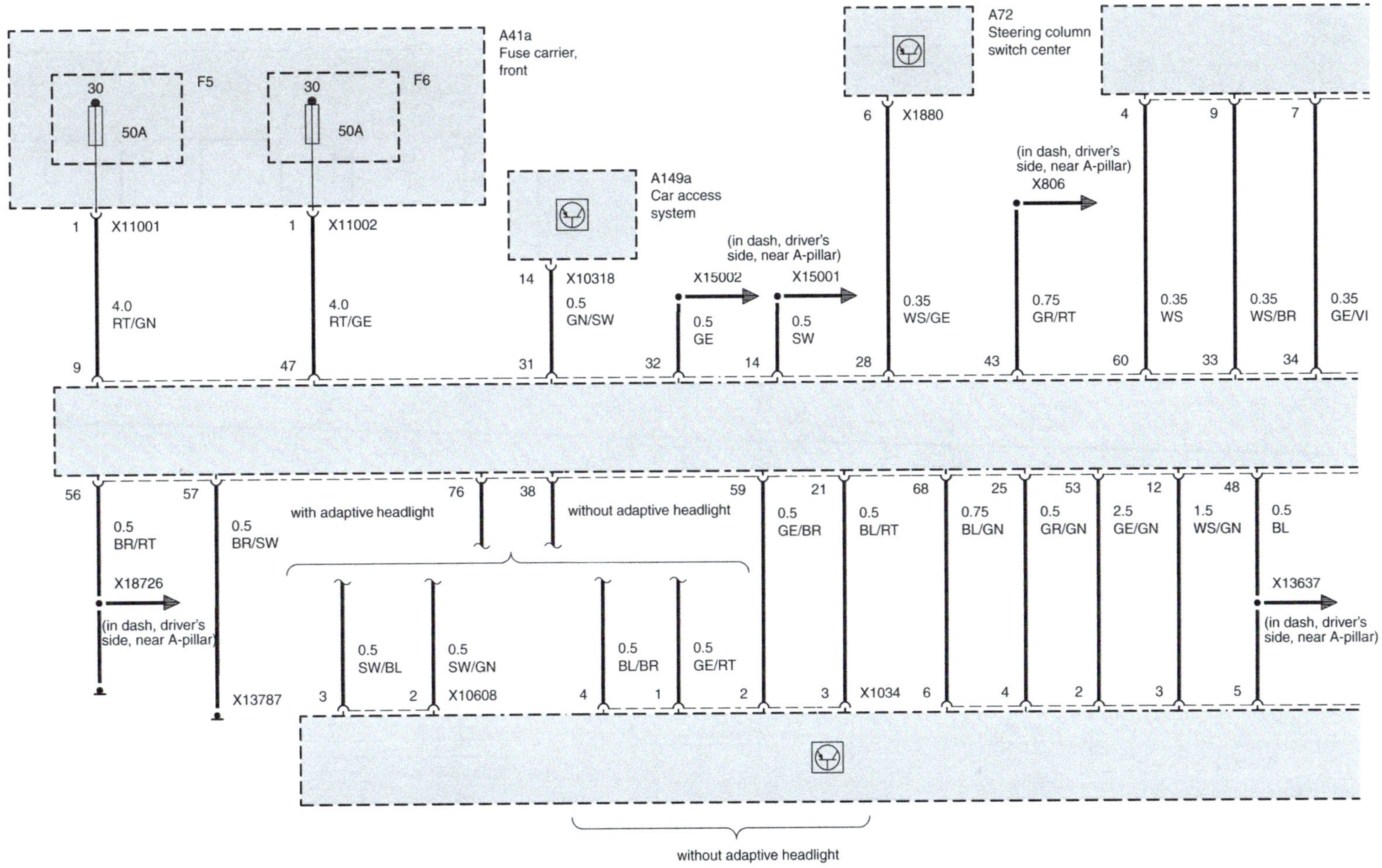

BentleyPublishers.com

B510ele22733a

Light module
Sedan with Xenon lights as of 03 / 2005 up to 03 / 2007
(page 2 of 3)

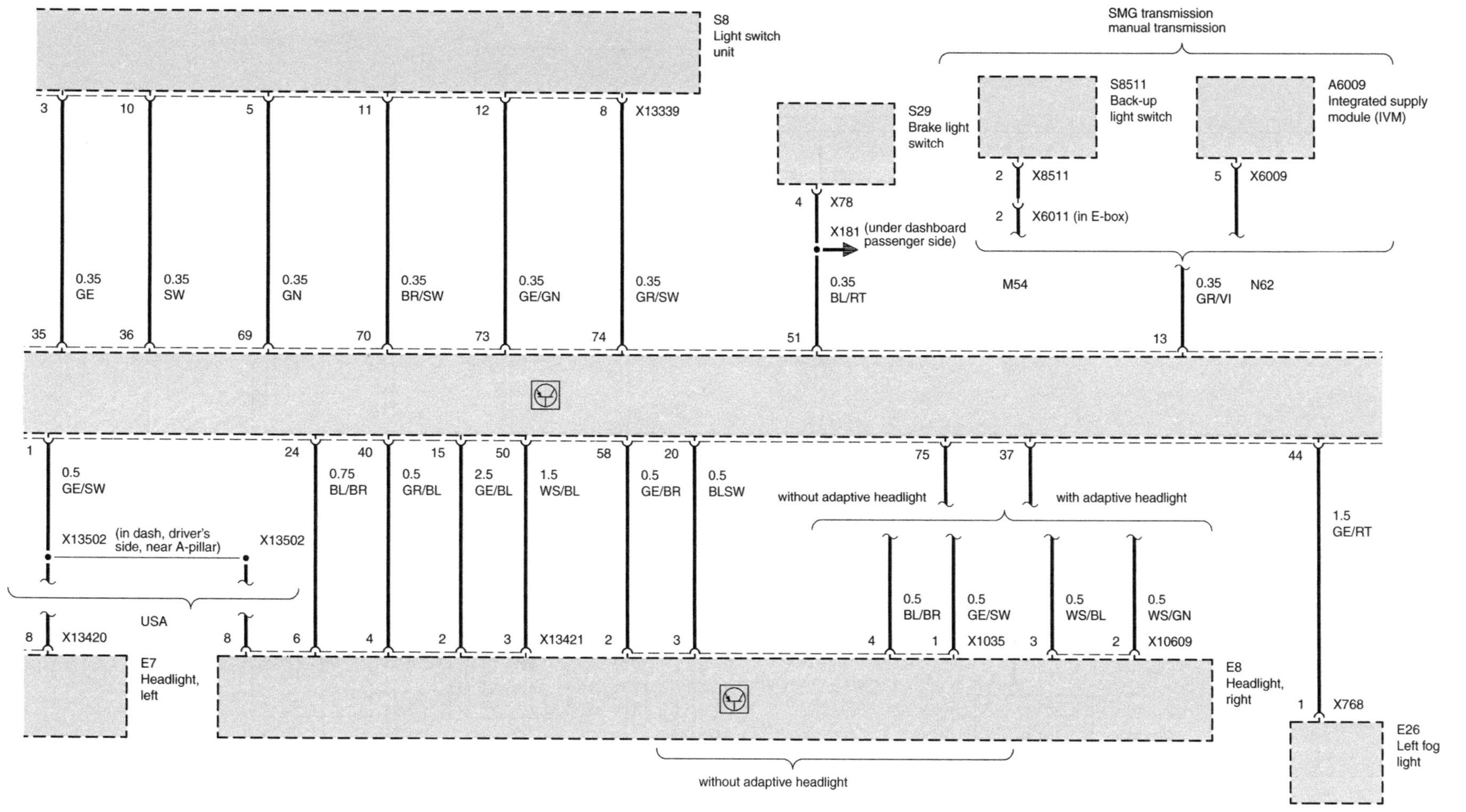

BentleyPublishers.com

B510ele22733b

Light module
Sedan with Xenon lights as of 03 / 2005 up to 03 / 2007
(page 3 of 3)

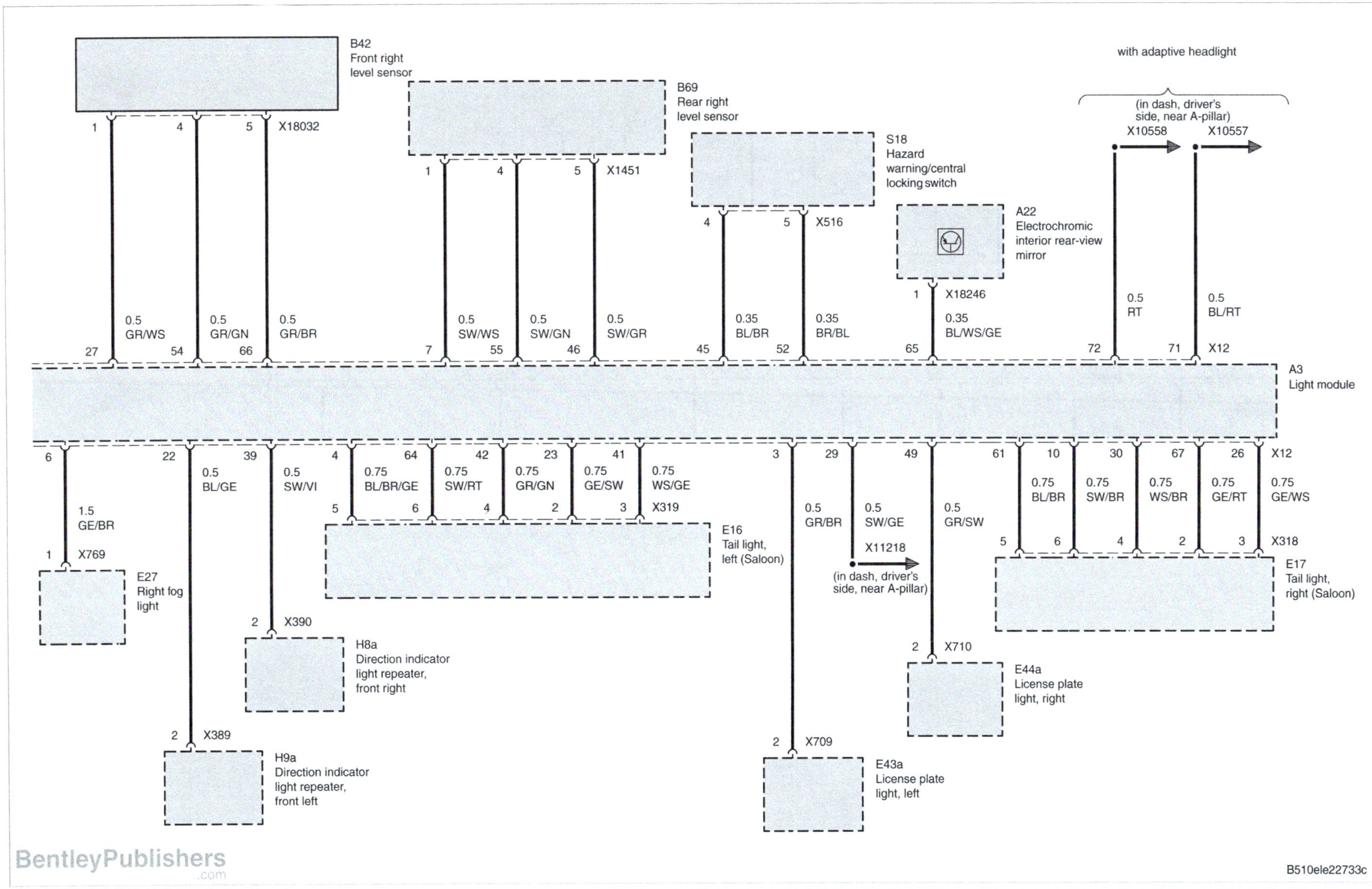

BentleyPublishers.com

Light module
Sedan without Xenon lights as of 03 / 2005 up to 03 / 2007
(page 1 of 3)

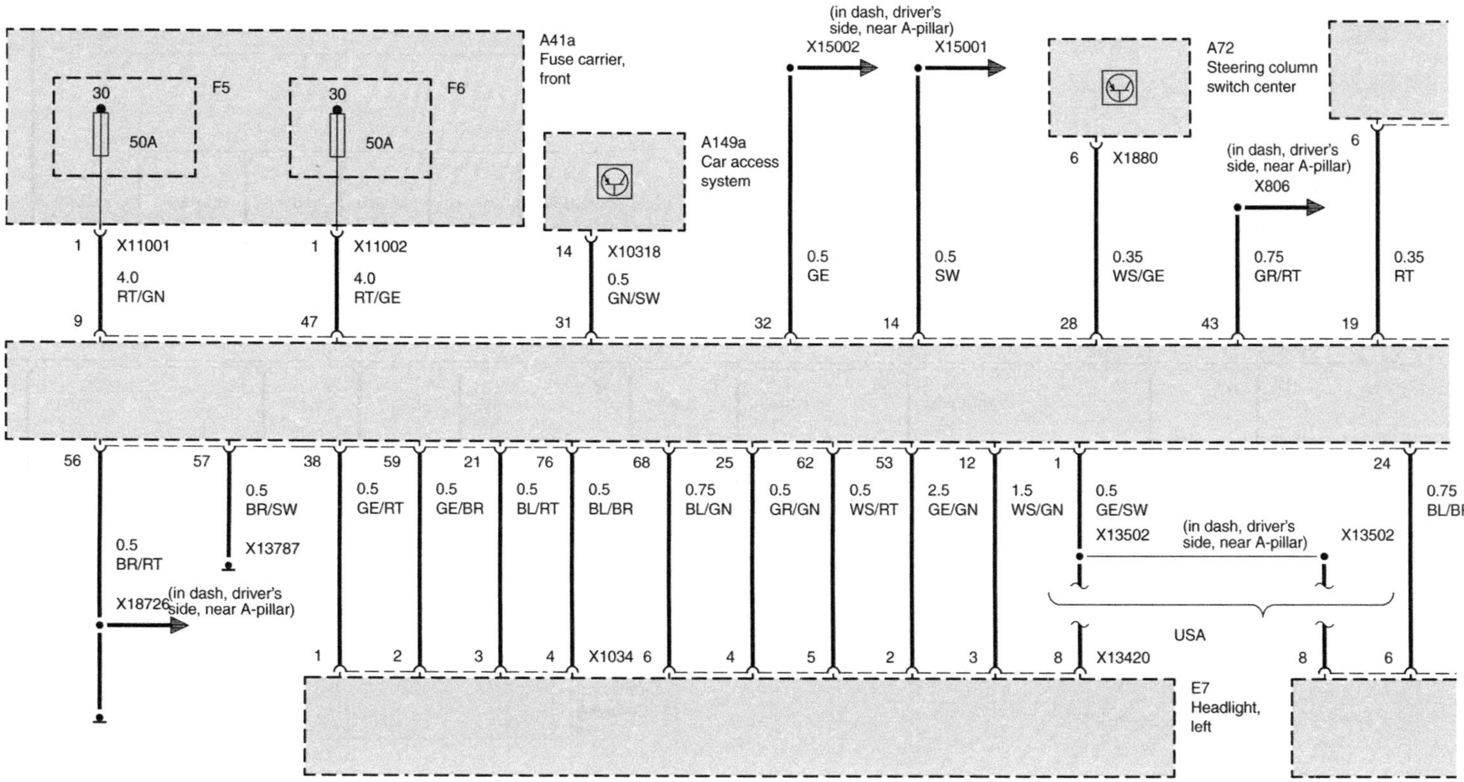

BentleyPublishers.com

B510ele22732a

Light module
Sedan without Xenon lights as of 03 / 2005 up to 03 / 2007
(page 2 of 3)

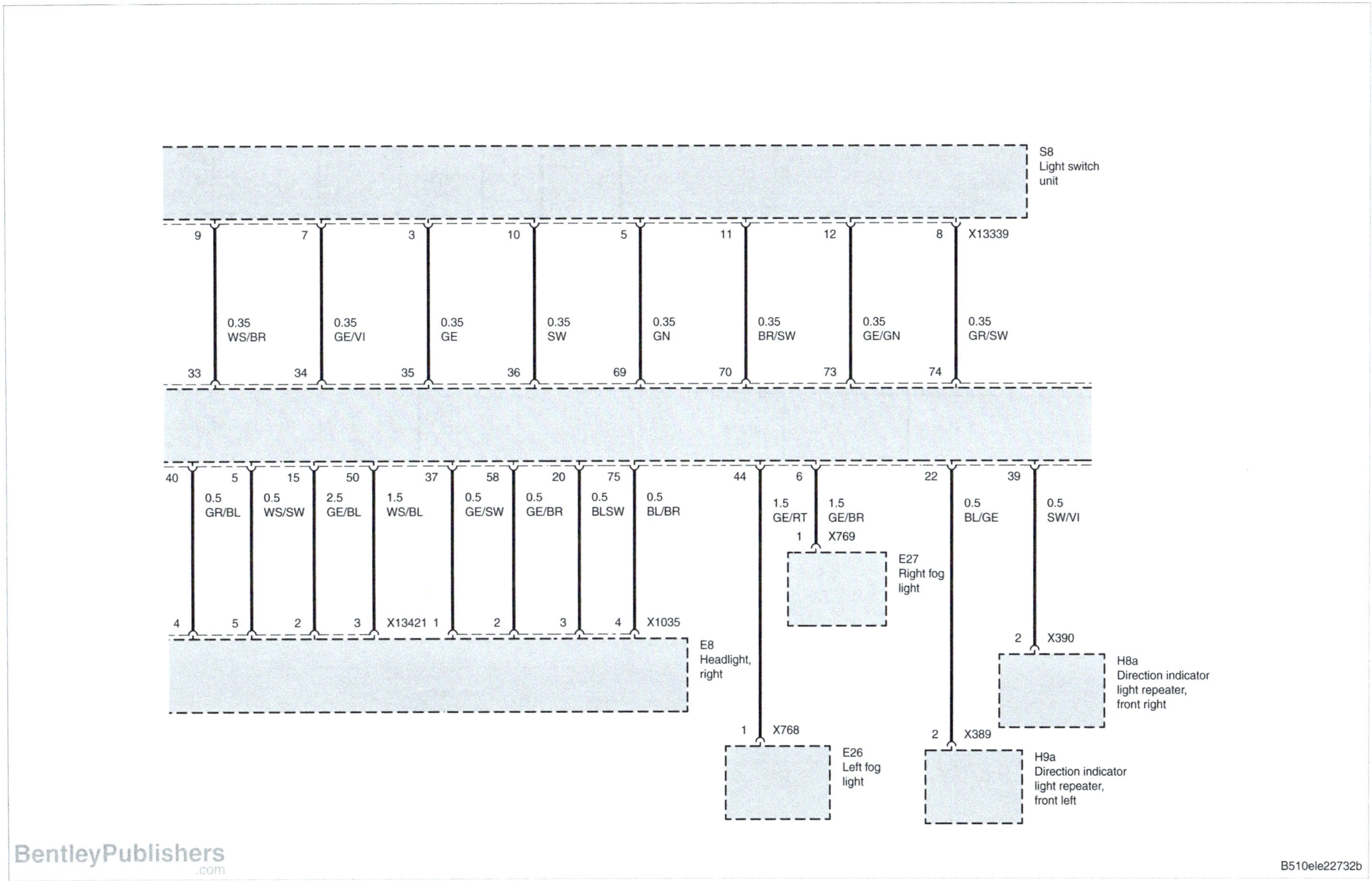

Light module
Sedan without Xenon lights as of 03 / 2005 up to 03 / 2007
(page 3 of 3)

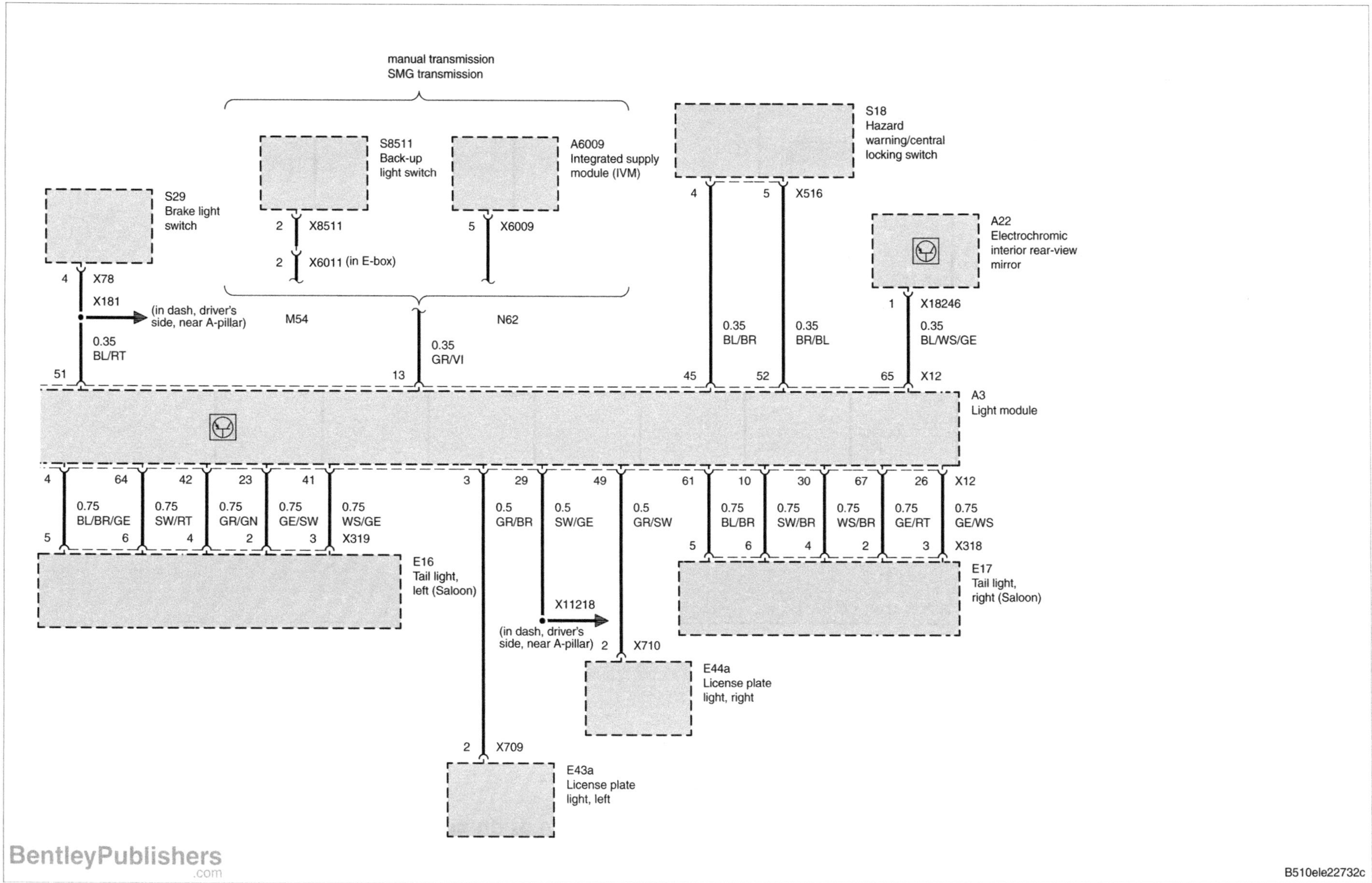

BentleyPublishers.com

Light module
Wagon with Xenon lights as of 03 / 2005
(page 1 of 3)

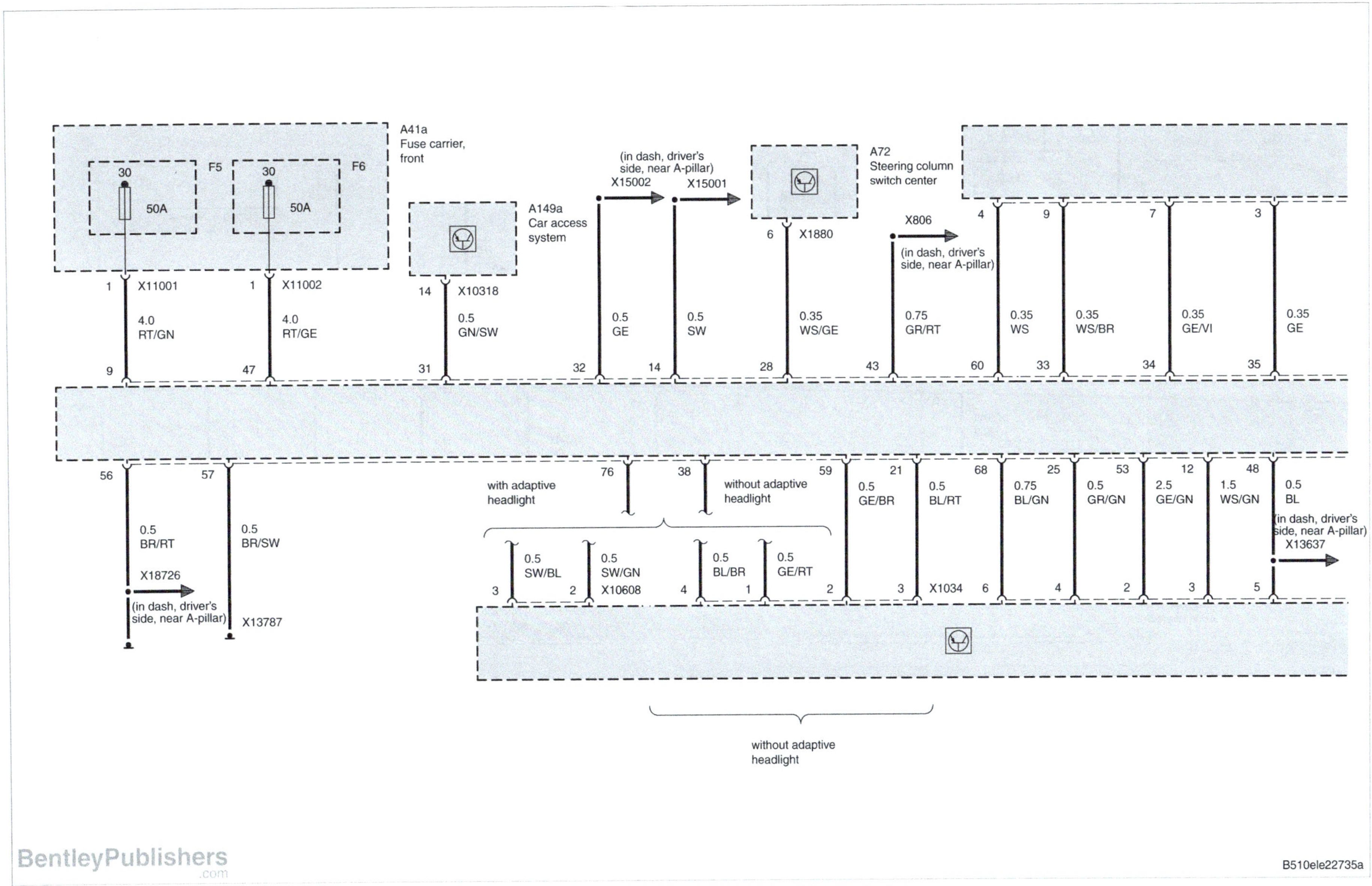

Light module
Wagon with Xenon lights as of 03 / 2005
(page 2 of 3)

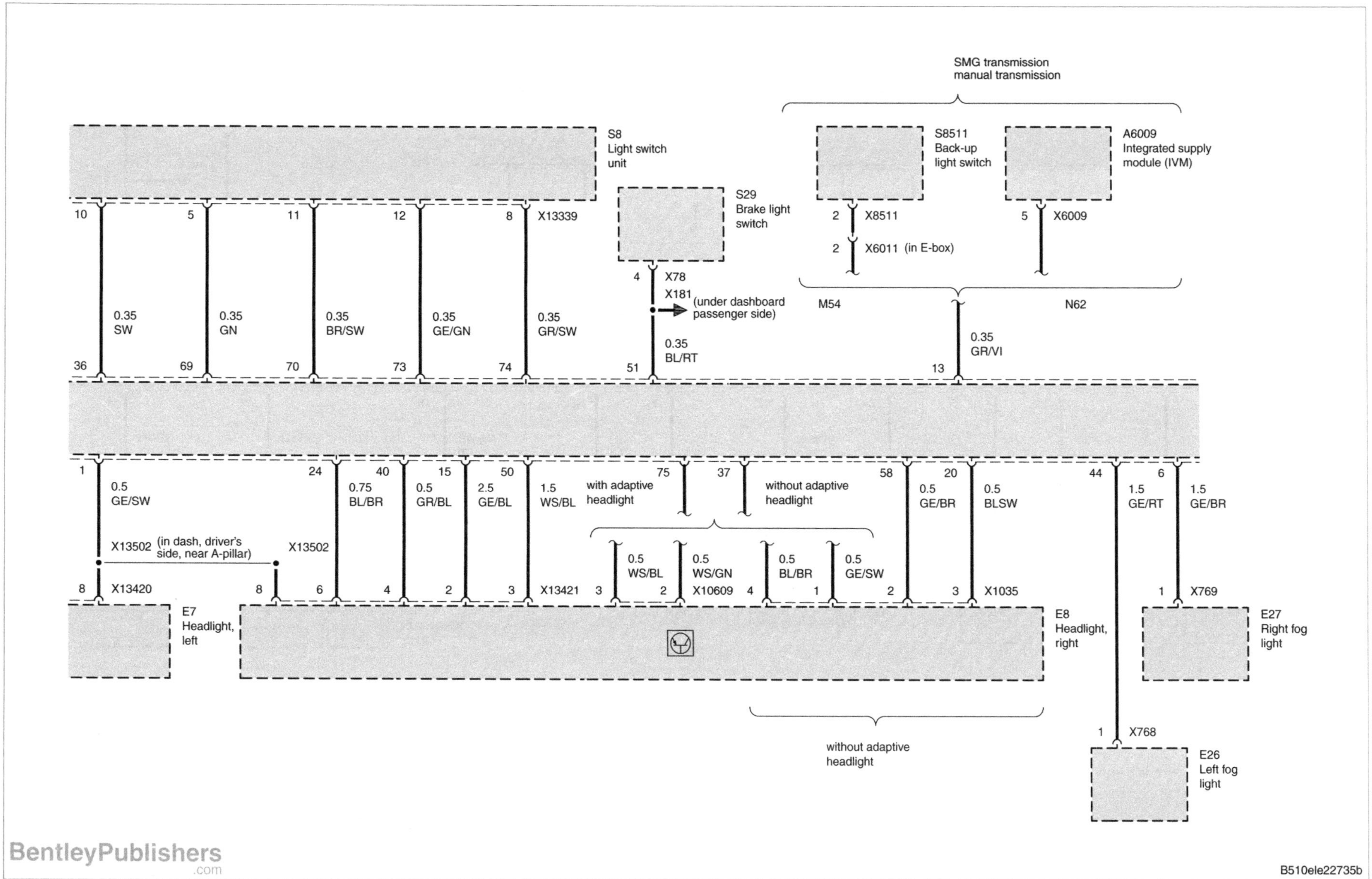

BentleyPublishers.com

B510ele22735b

Light module
Wagon with Xenon lights as of 03 / 2005
(page 3 of 3)

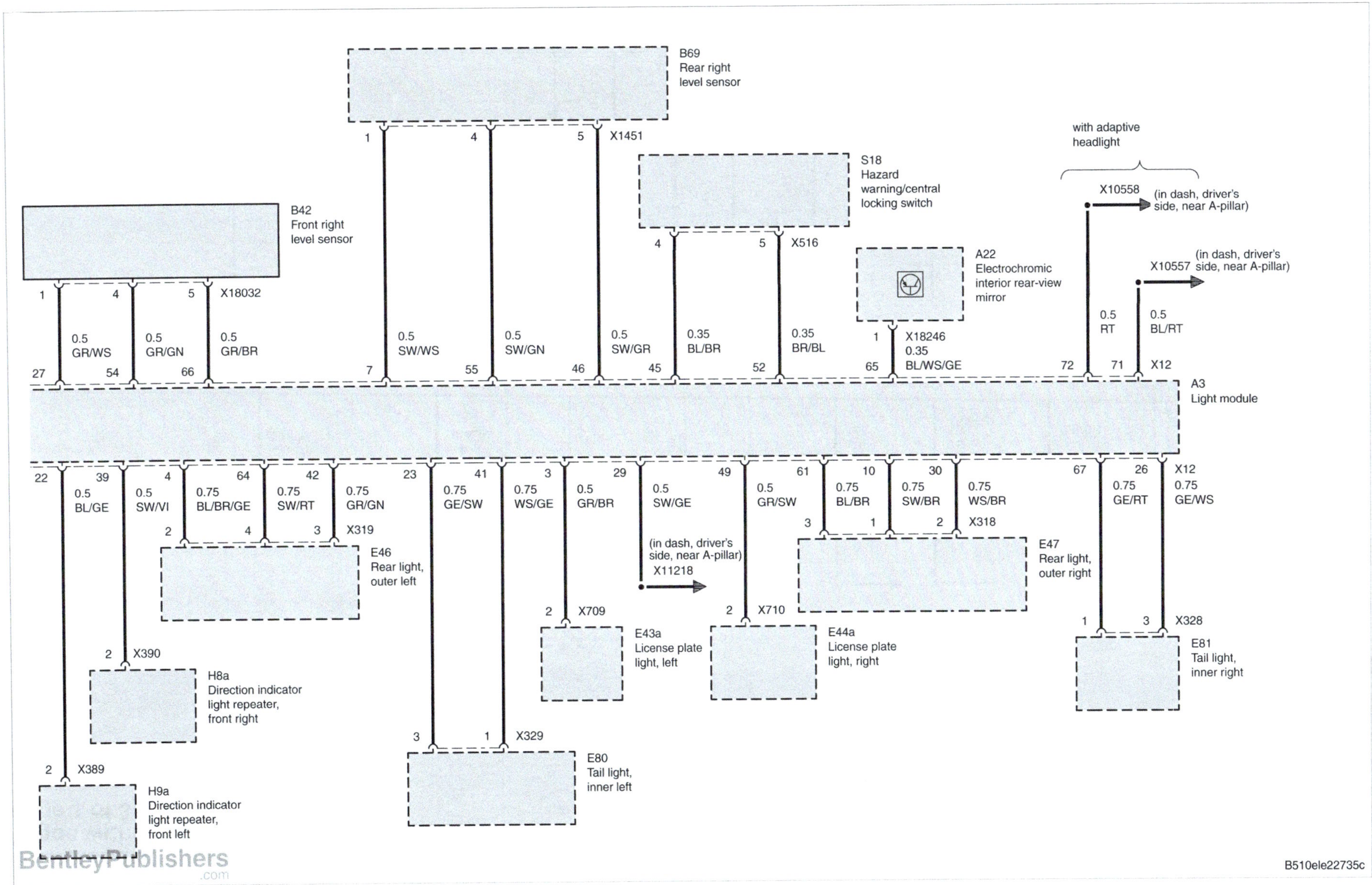

Light module
Wagon without Xenon lights as of 03 / 2005
(page 1 of 3)

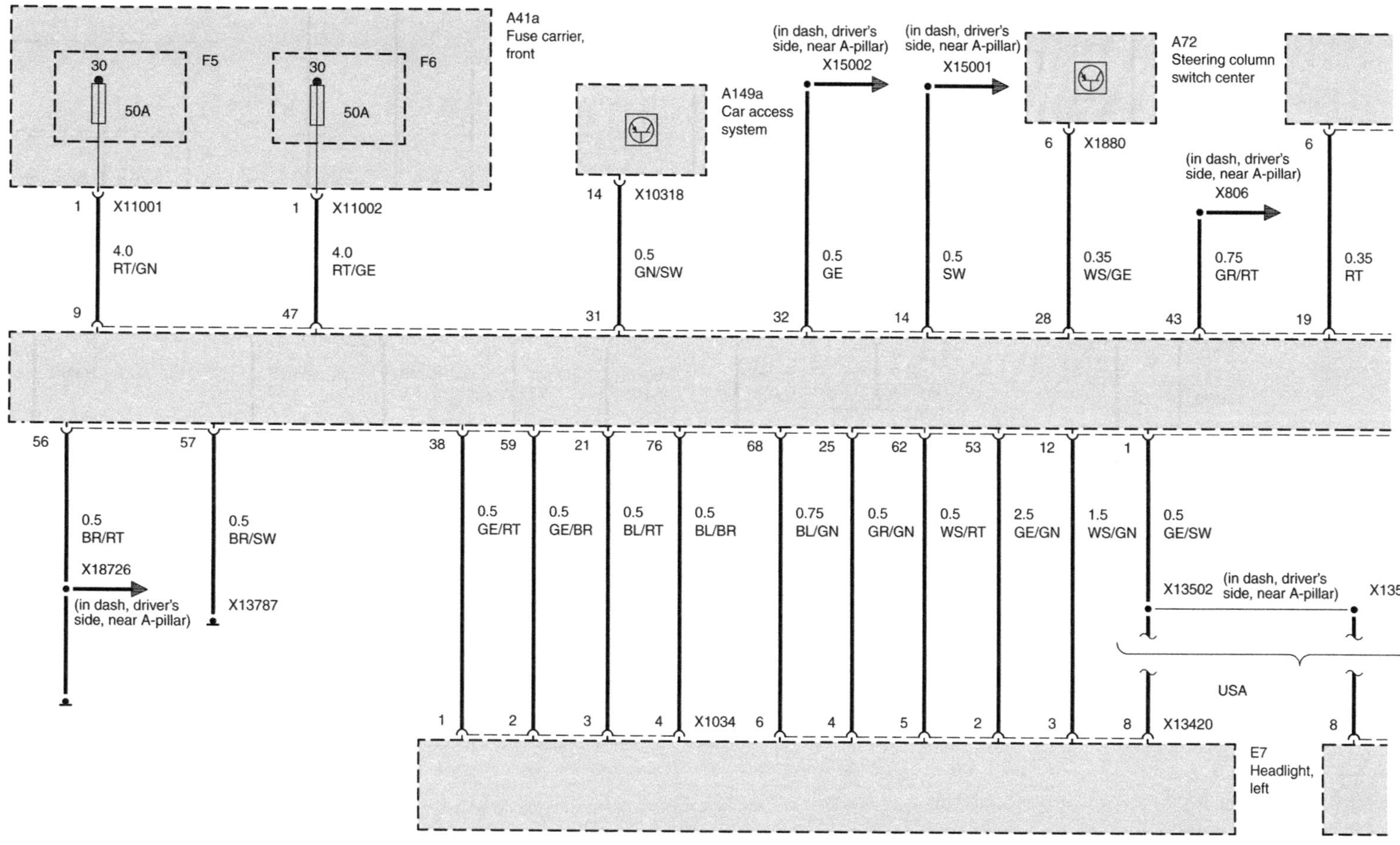

BentleyPublishers.com

B510ele22734a

Light module
Wagon without Xenon lights as of 03 / 2005
(page 2 of 3)

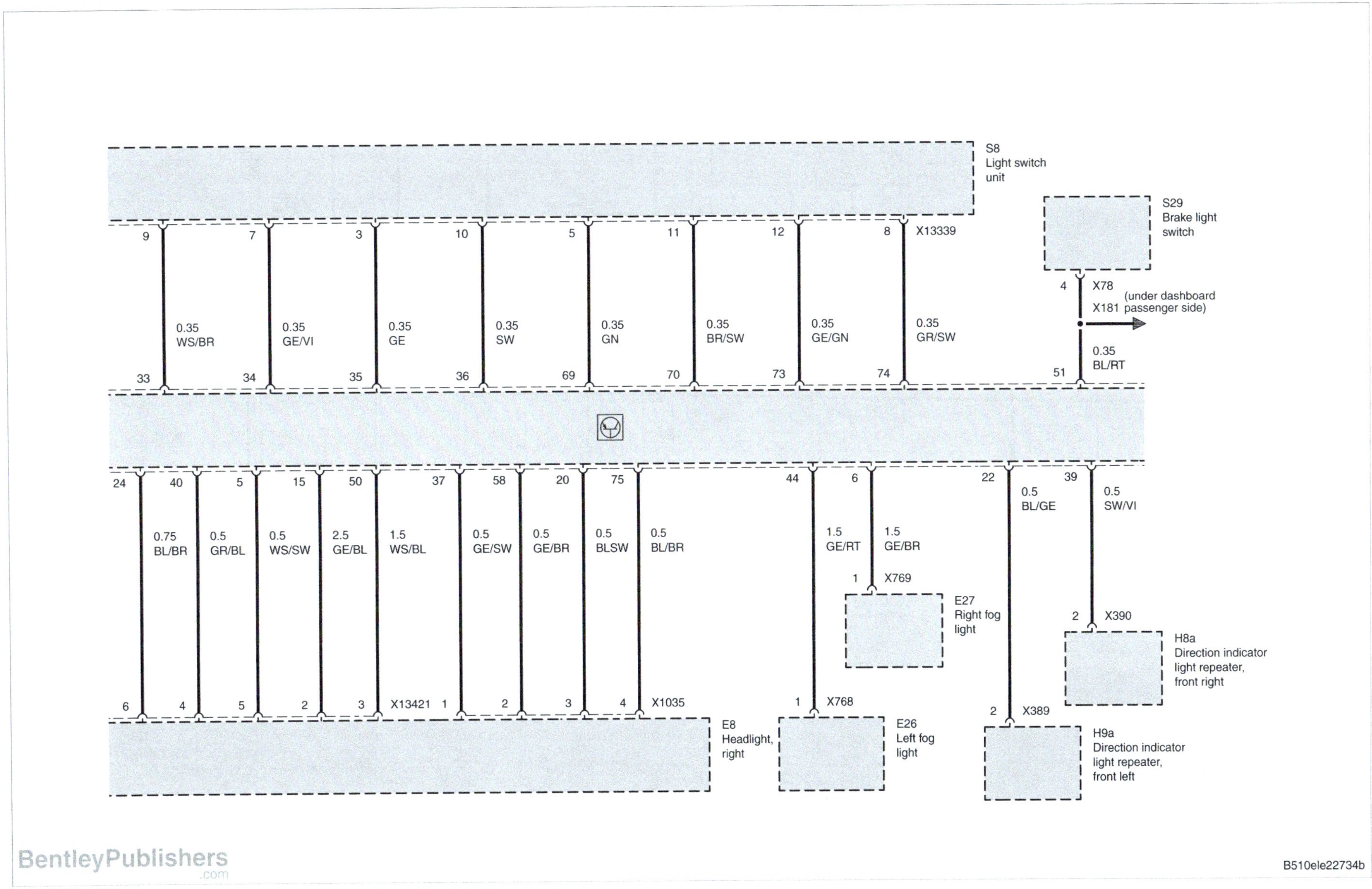

Light module
Wagon without Xenon lights as of 03 / 2005
(page 3 of 3)

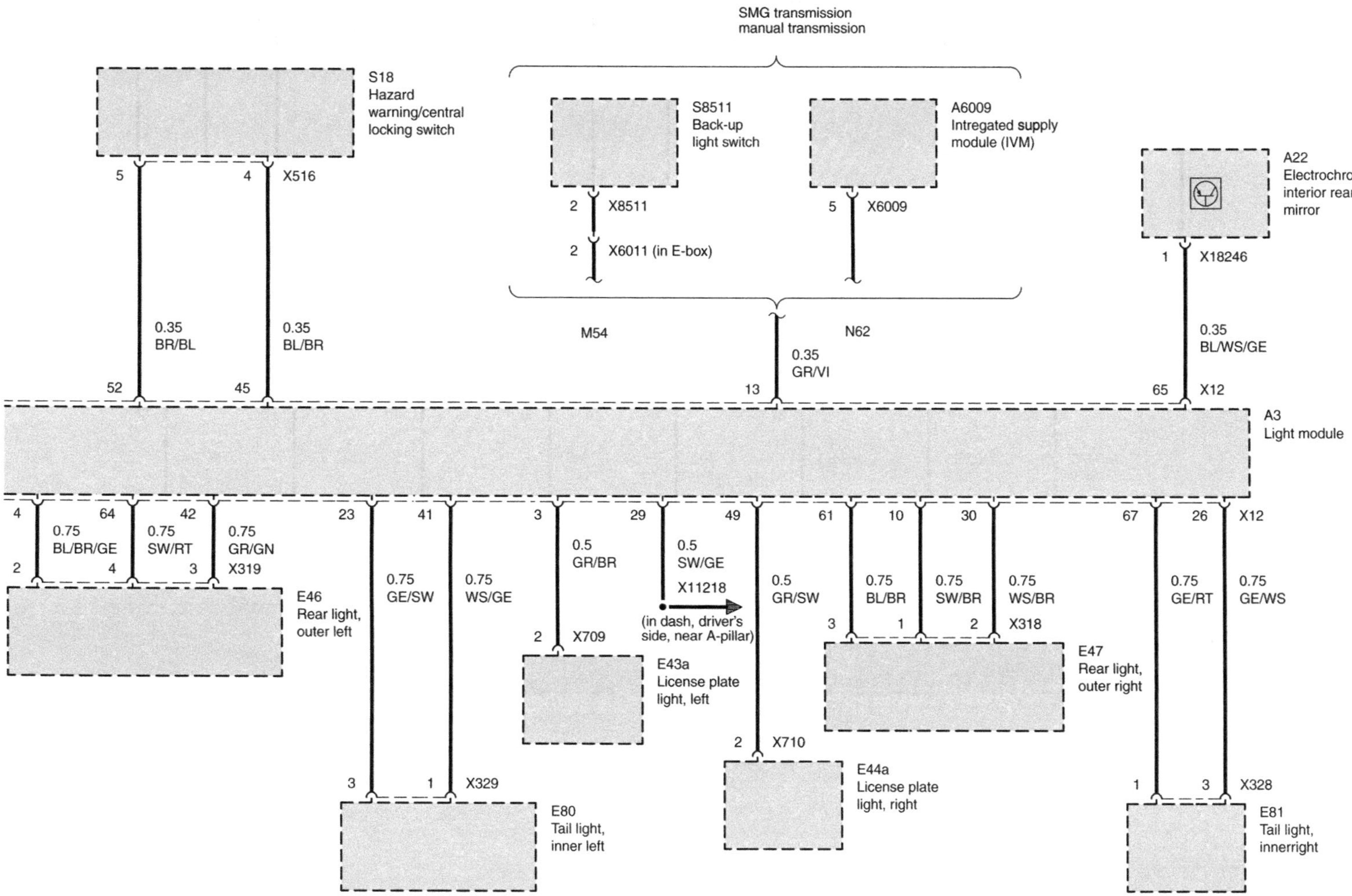

Light module
Sedan with Xenon lights as of 03 / 2007
(page 1 of 3)

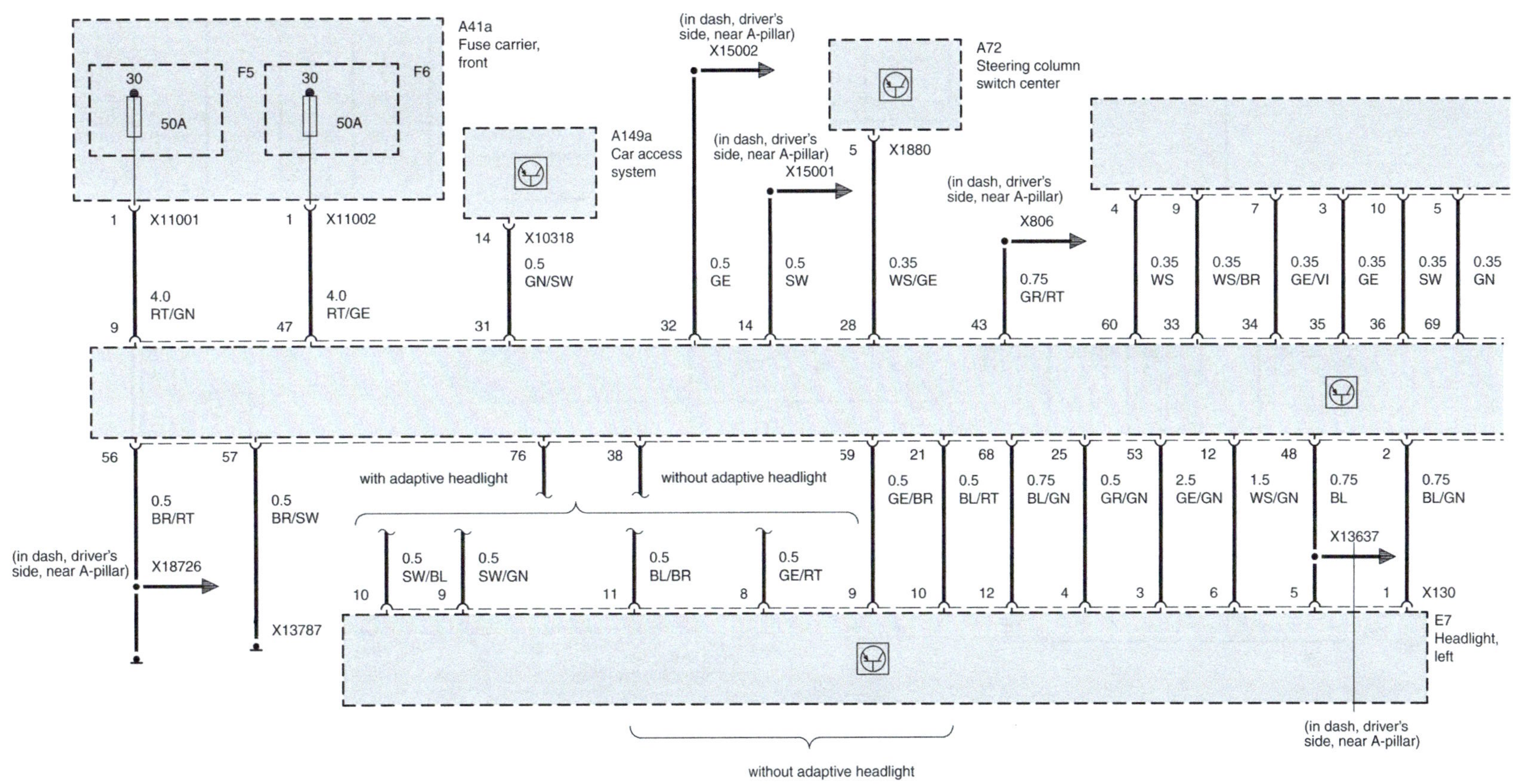

BentleyPublishers.com

Light module
Sedan with Xenon lights as of 03 / 2007
(page 2 of 3)

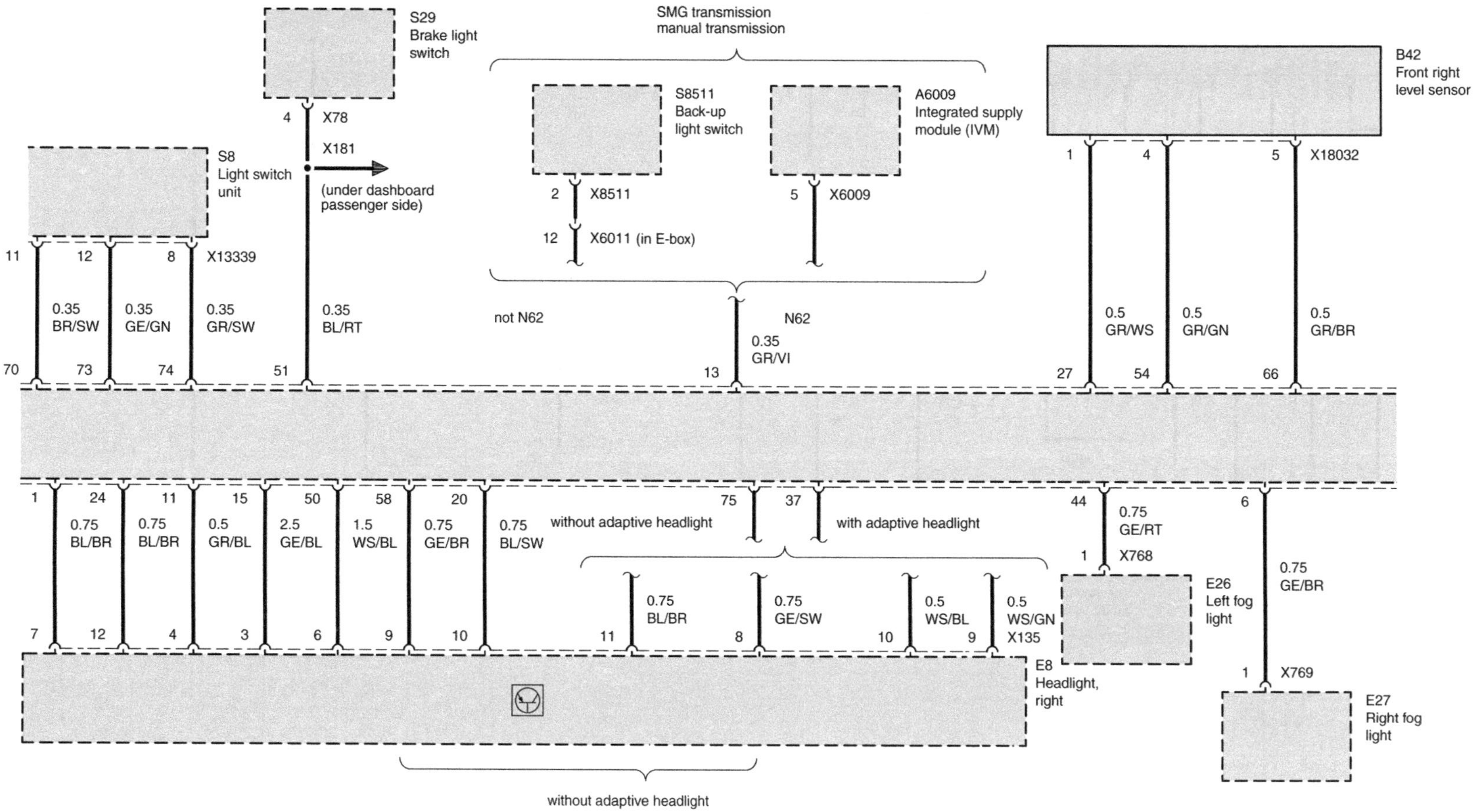

B510ele26117b

Light module
Sedan with Xenon lights as of 03 / 2007
(page 3 of 3)

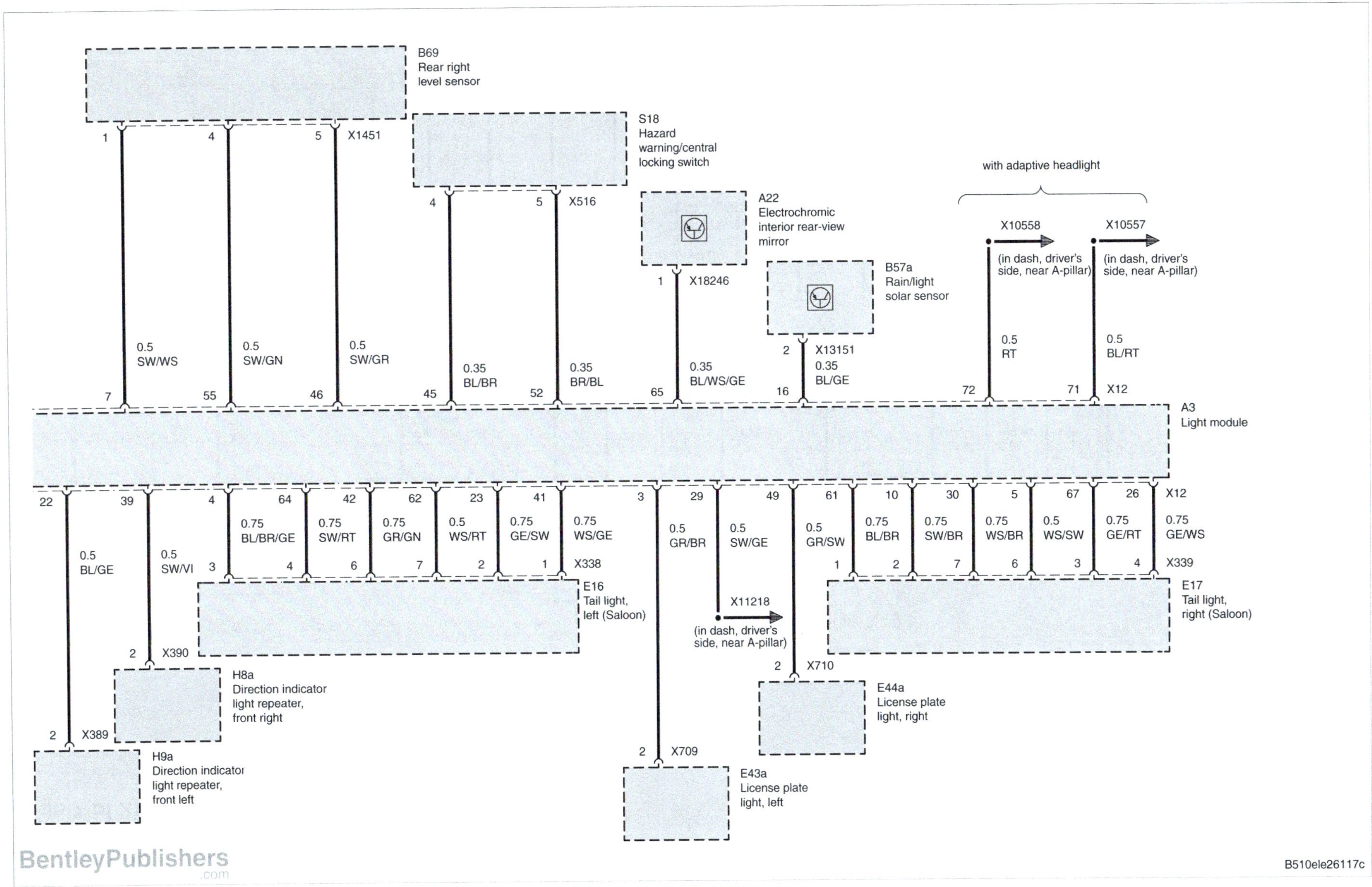

Light module
Sedan without Xenon lights as of 03 / 2007
(page 1 of 2)

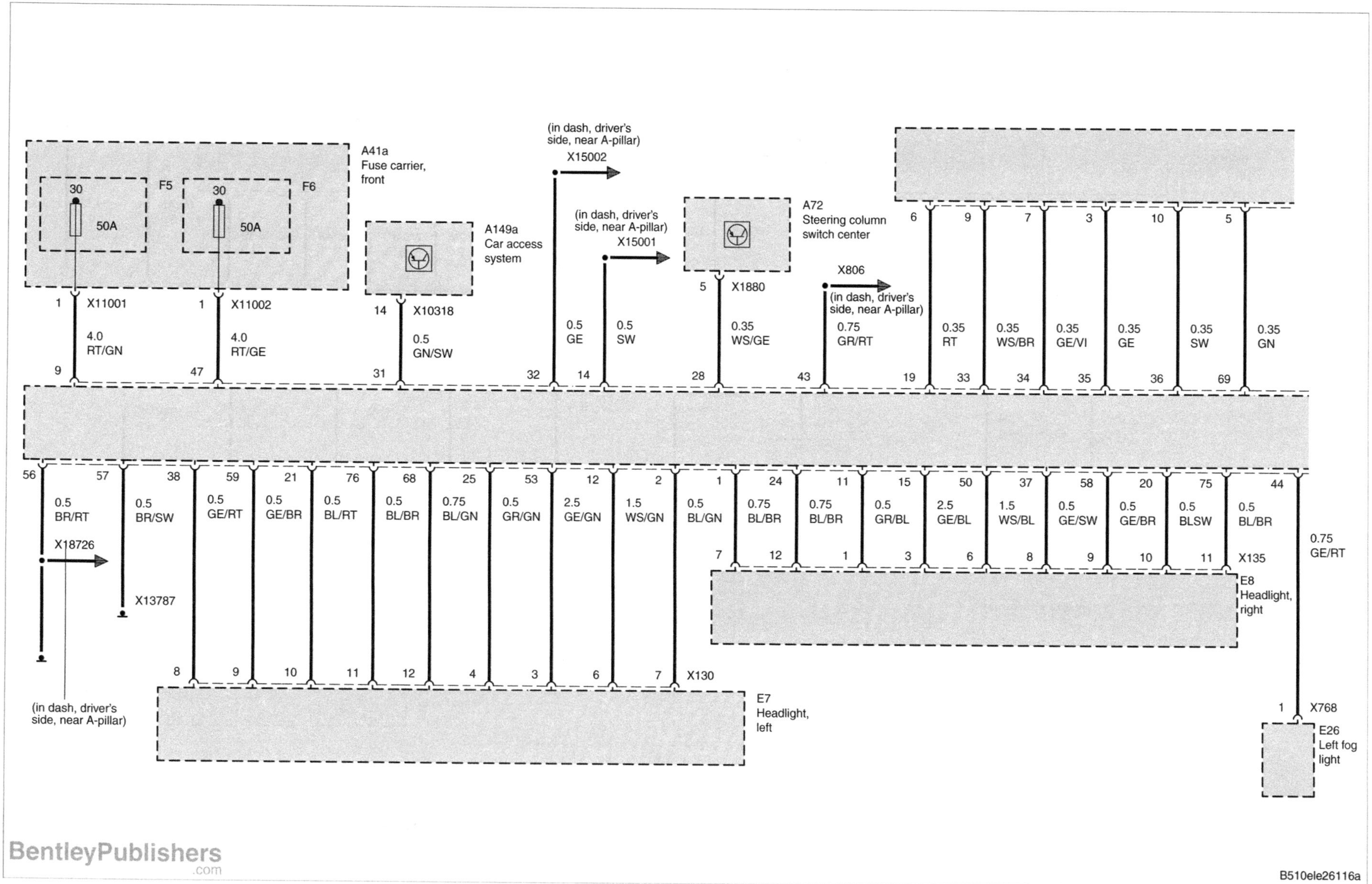

B510ele26116a

BentleyPublishers.com

Light module
Sedan without Xenon lights as of 03 / 2007
(page 2 of 2)

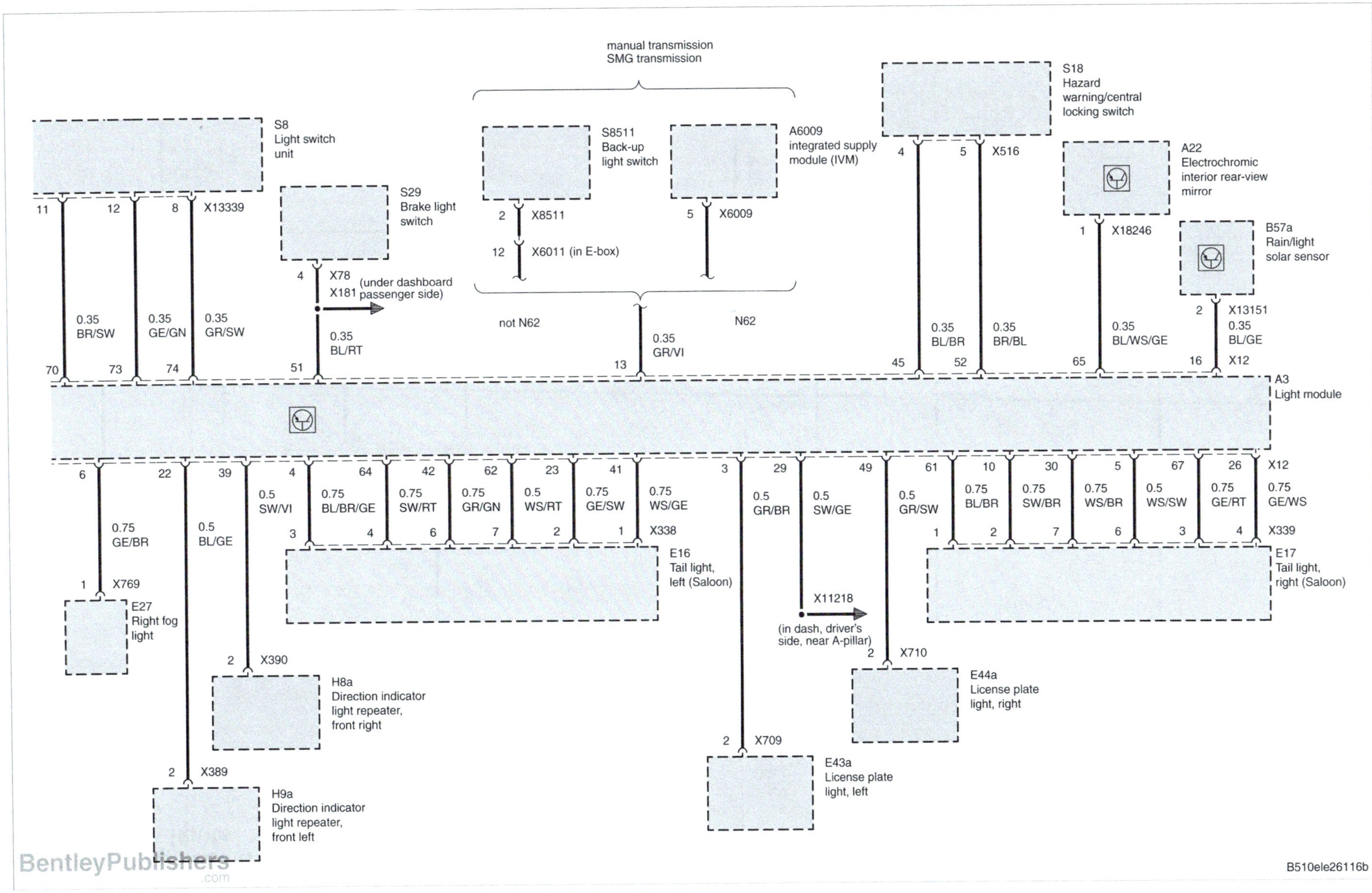

BentleyPublishers.com

Low beam / high beam without Xenon lights up to 03 / 2005

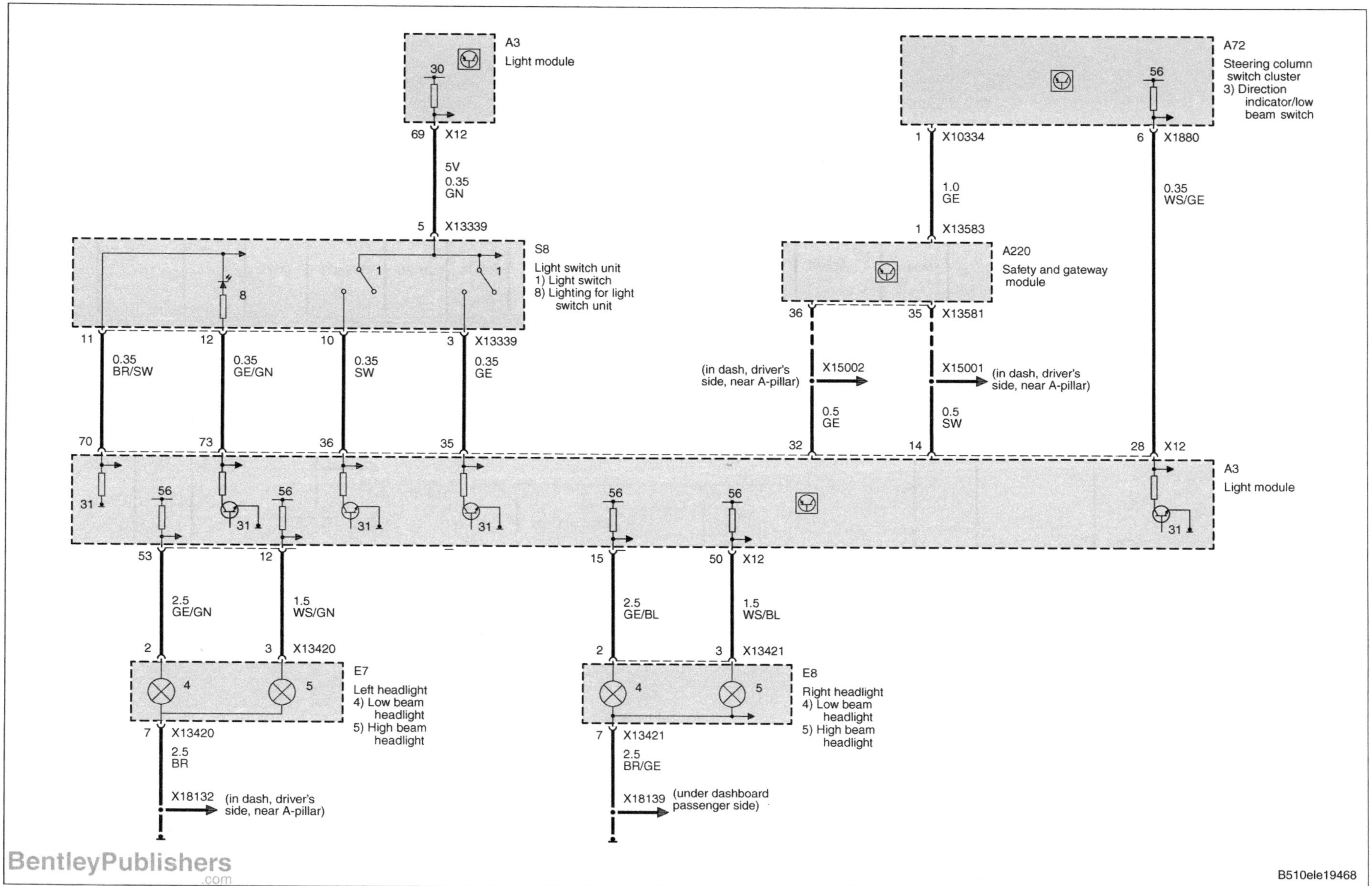

BentleyPublishers.com

Low beam / high beam without Xenon lights as of 03 / 2005 up to 09 / 2005

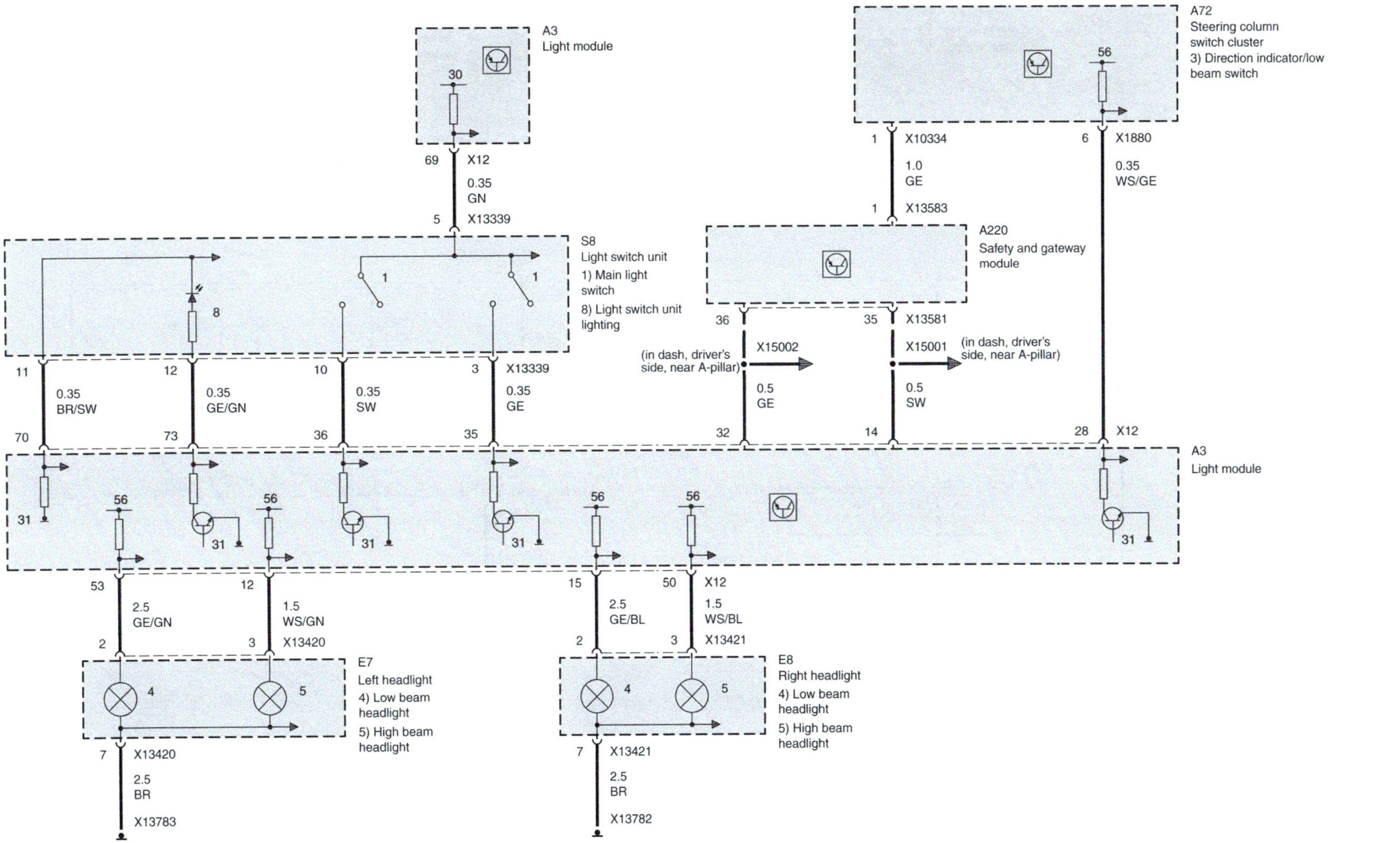

BentleyPublishers.com

Low beam / high beam without Xenon lights as of 09 / 2005 up to 03 / 2007

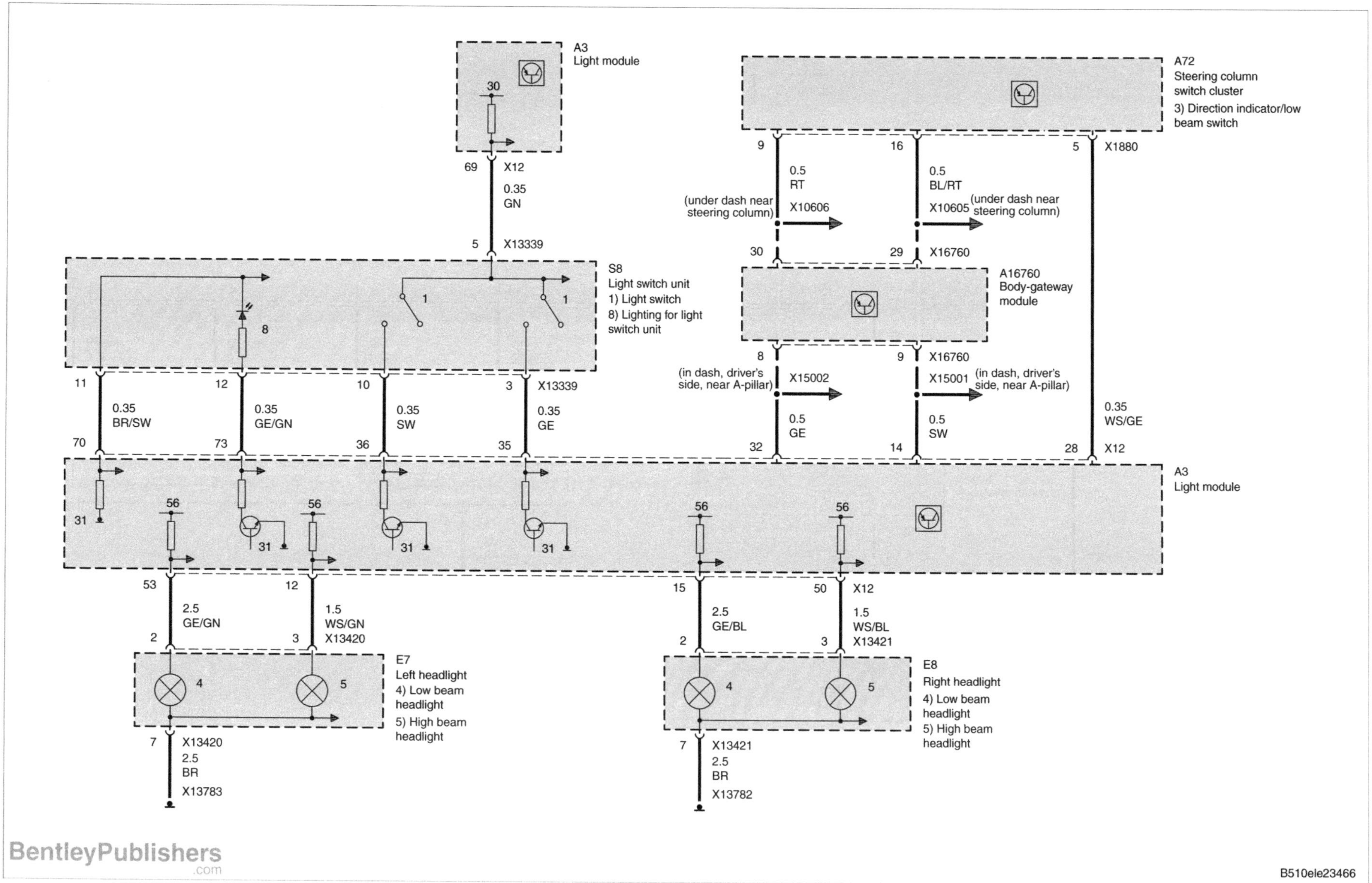

BentleyPublishers.com

Low beam / high beam without Xenon lights as of 03 / 2007

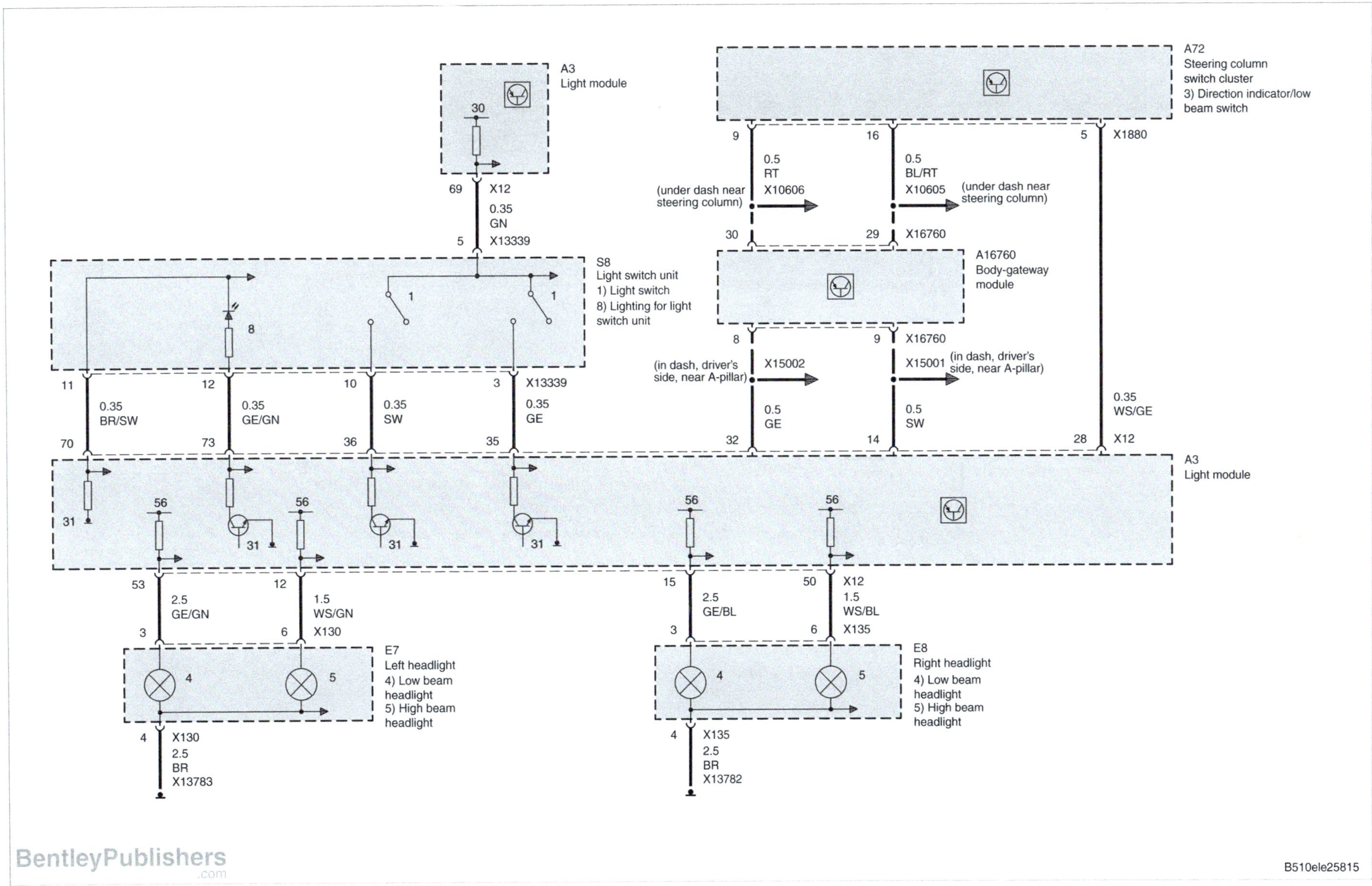

Low beam / high beam with Xenon lights up to 03 / 2005

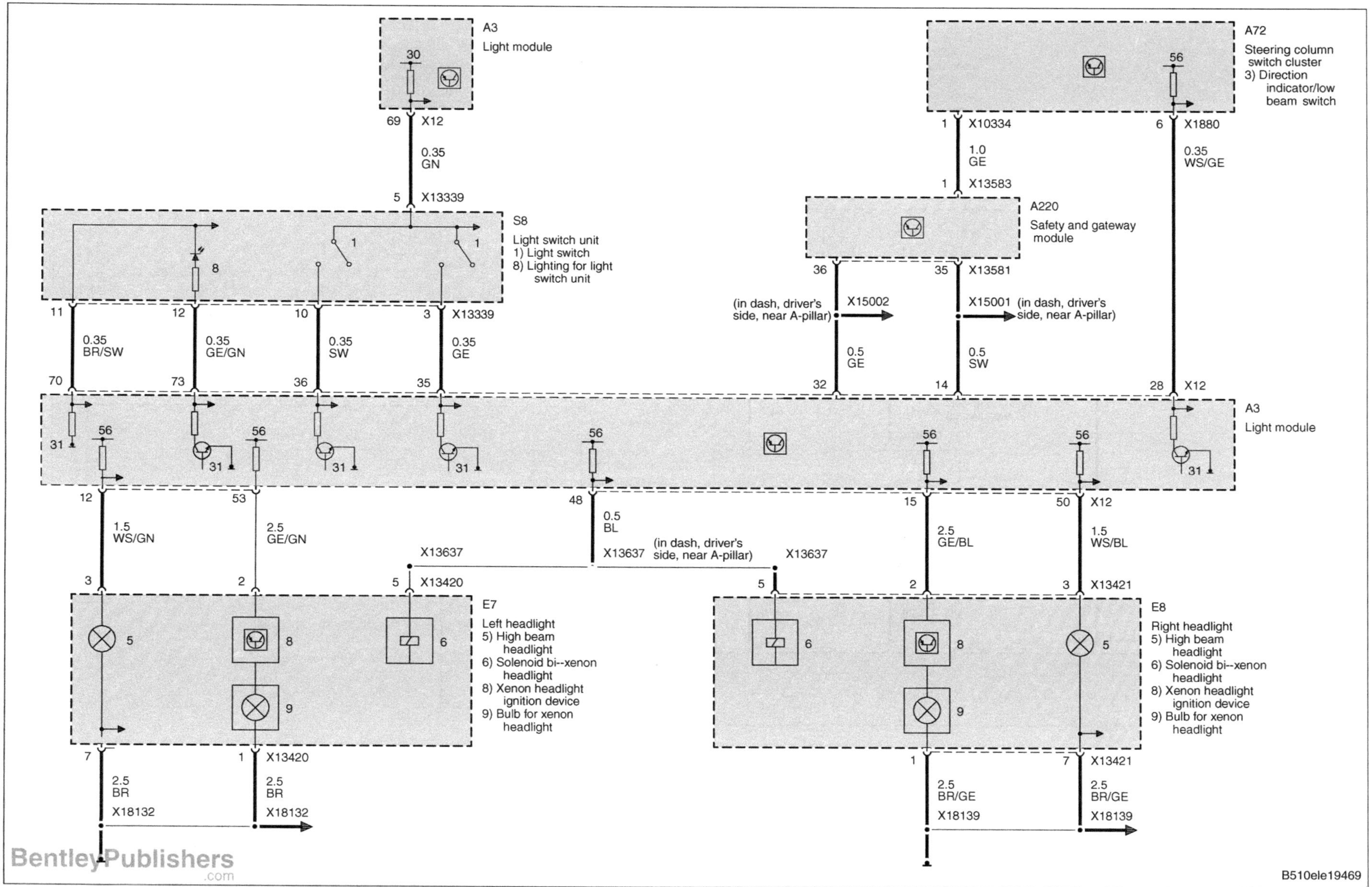

BentleyPublishers.com

B510ele19469

Low beam / high beam with Xenon lights as of 03 / 2005 up to 09 / 2005

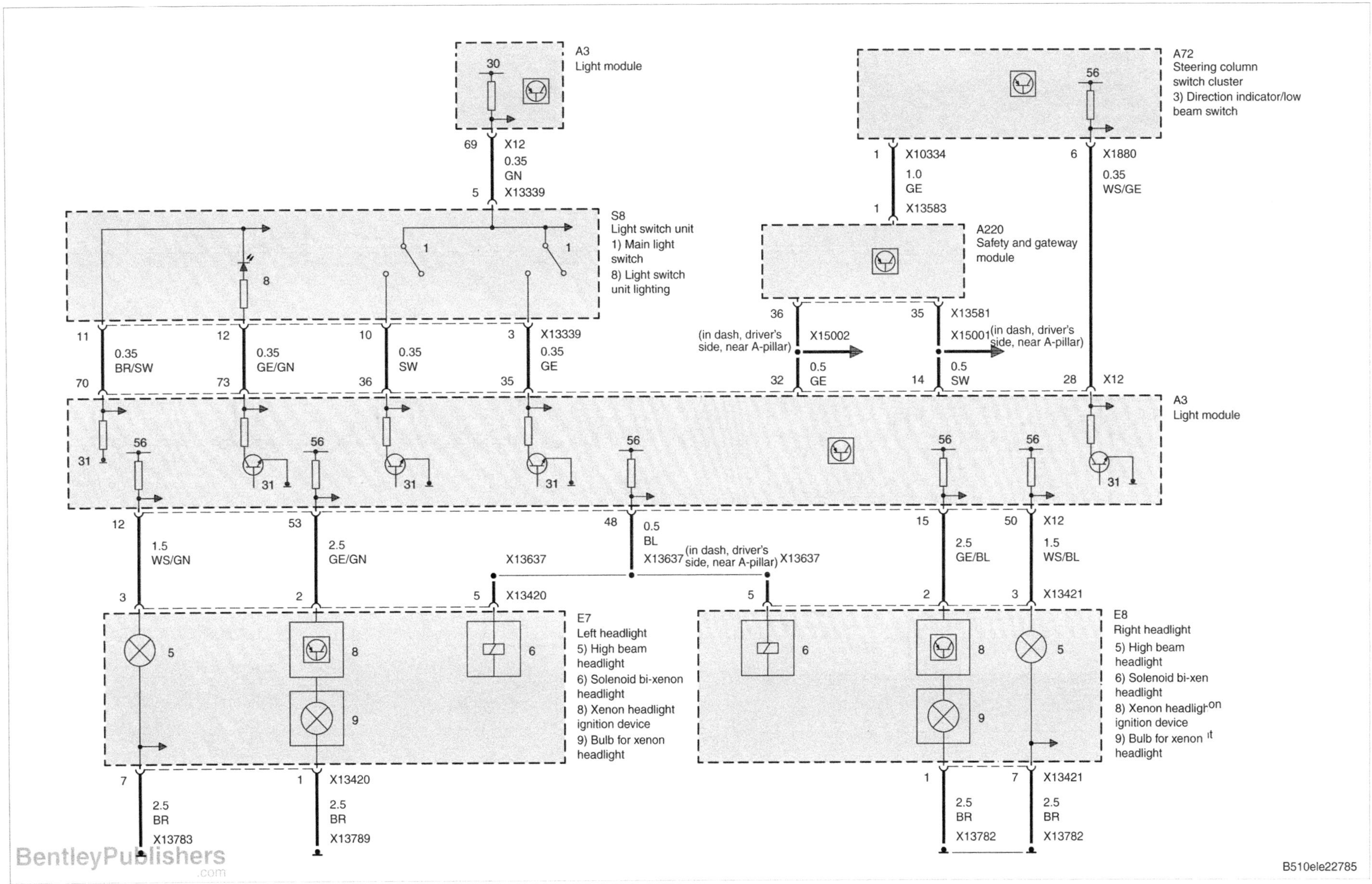

Low beam / high beam with Xenon lights as of 09 / 2005 up to 03 / 2007

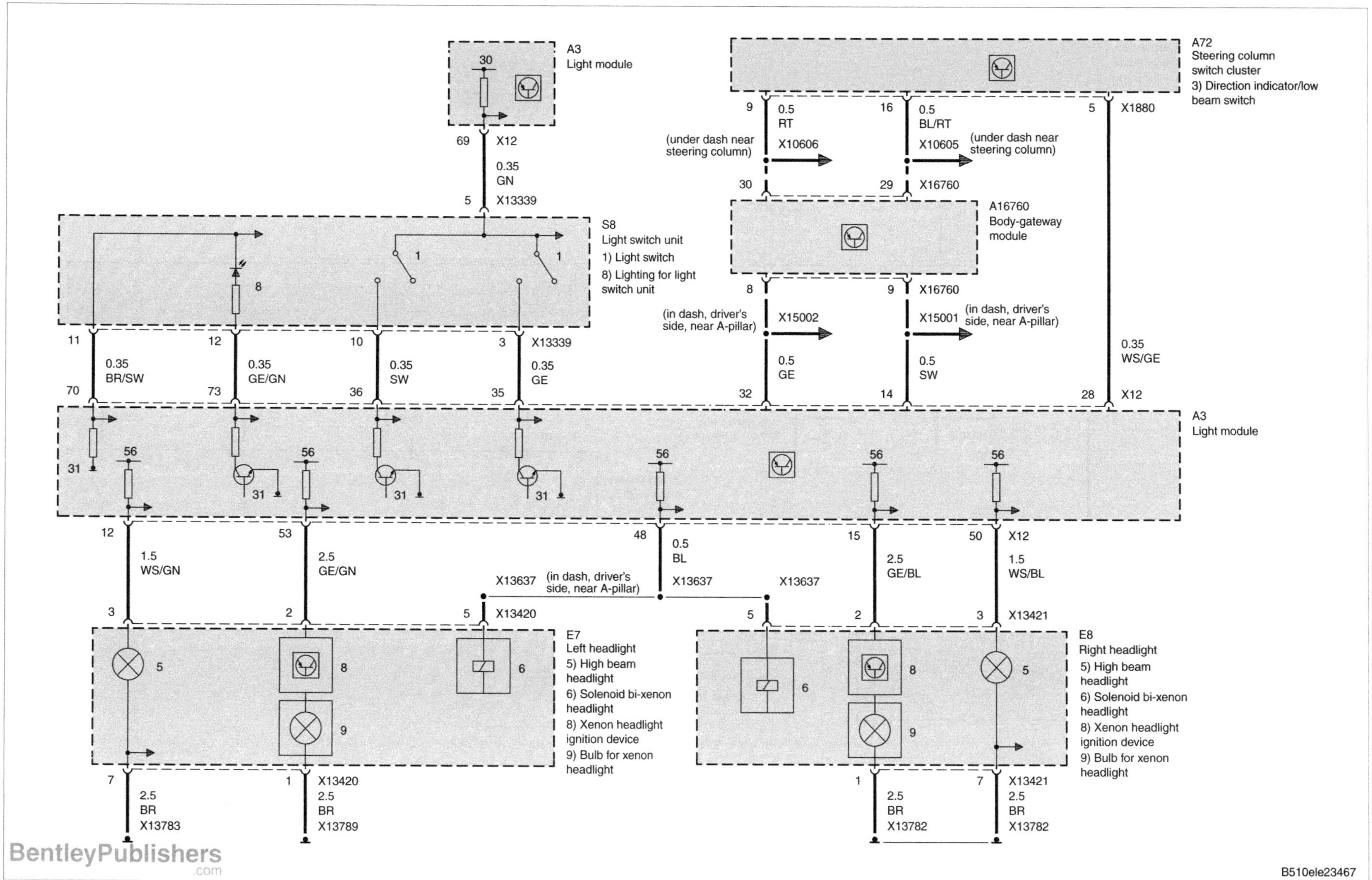

BentleyPublishers.com

B510ele23467

Low beam / high beam with Xenon lights as of 03 / 2007

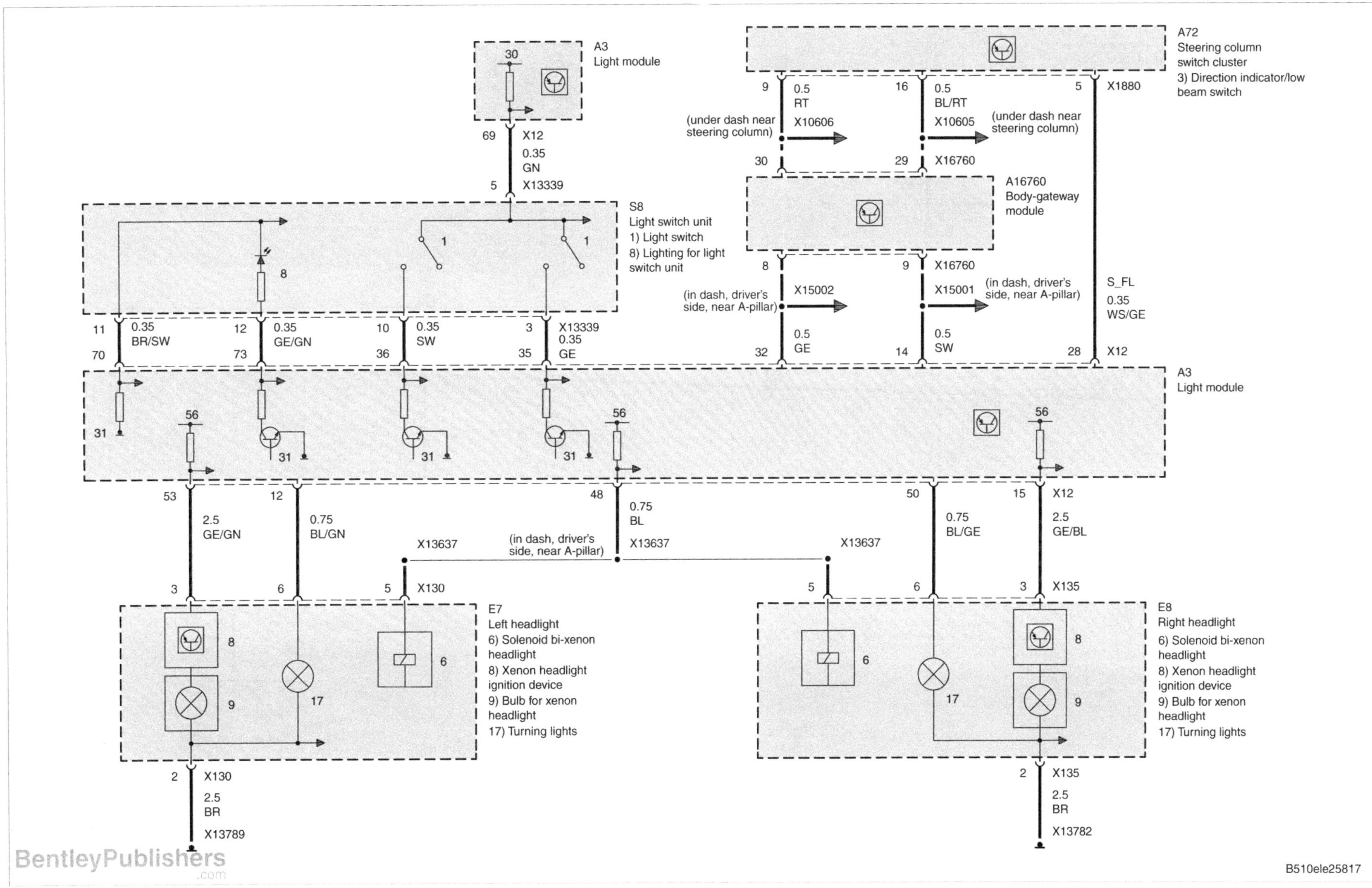

BentleyPublishers.com

Side light, parking light, brake light
Sedan without Xenon lights up to 03 / 2005
(page 1 of 2)

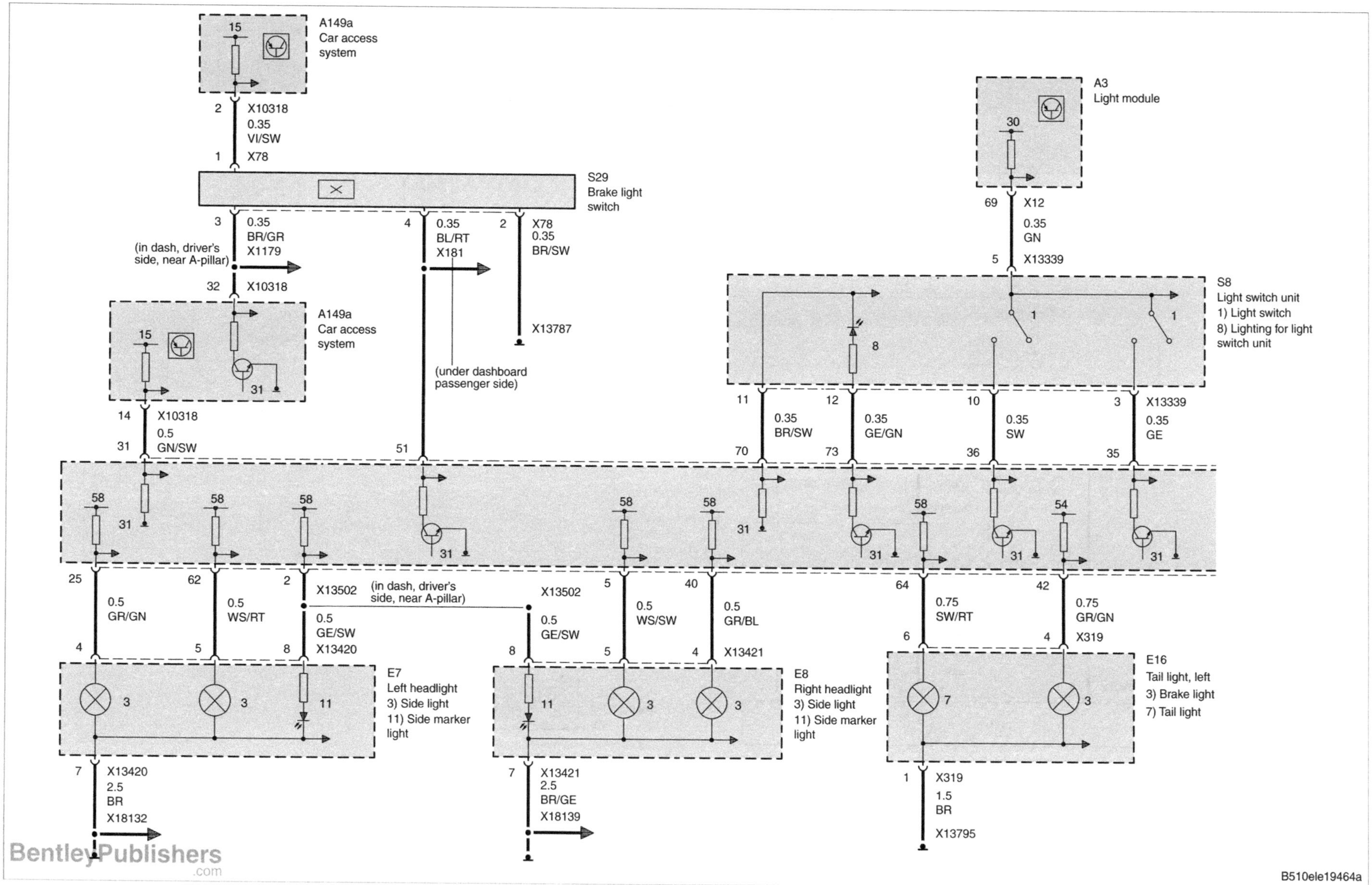

BentleyPublishers.com

Side light, parking light, brake light
Sedan without Xenon lights up to 03 / 2005
(page 2 of 2)

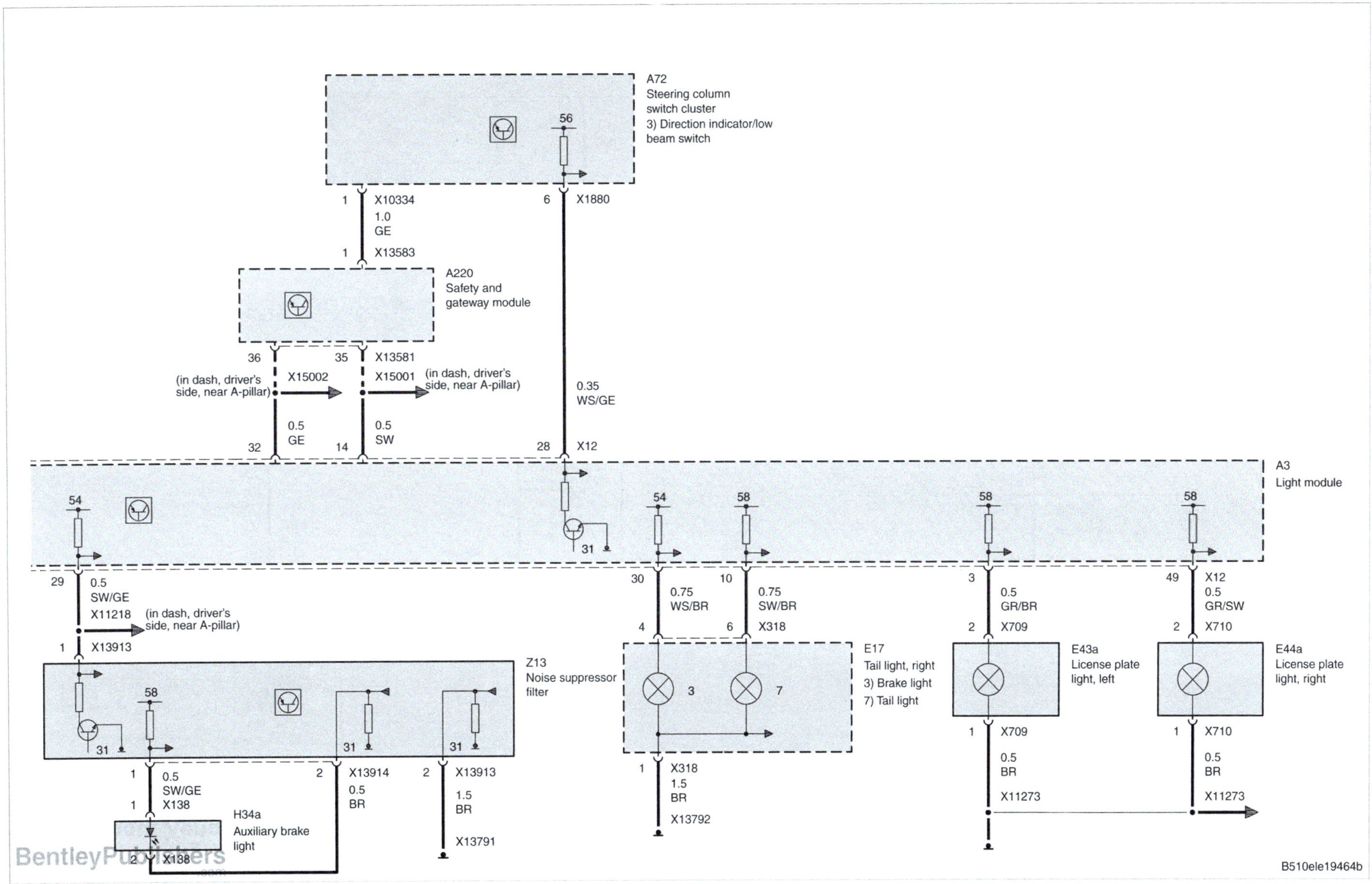

BentleyPublishers.com

Side light, parking light, brake light
Sedan without Xenon lights as of 03 / 2005 up to 09 / 2005
(page 1 of 2)

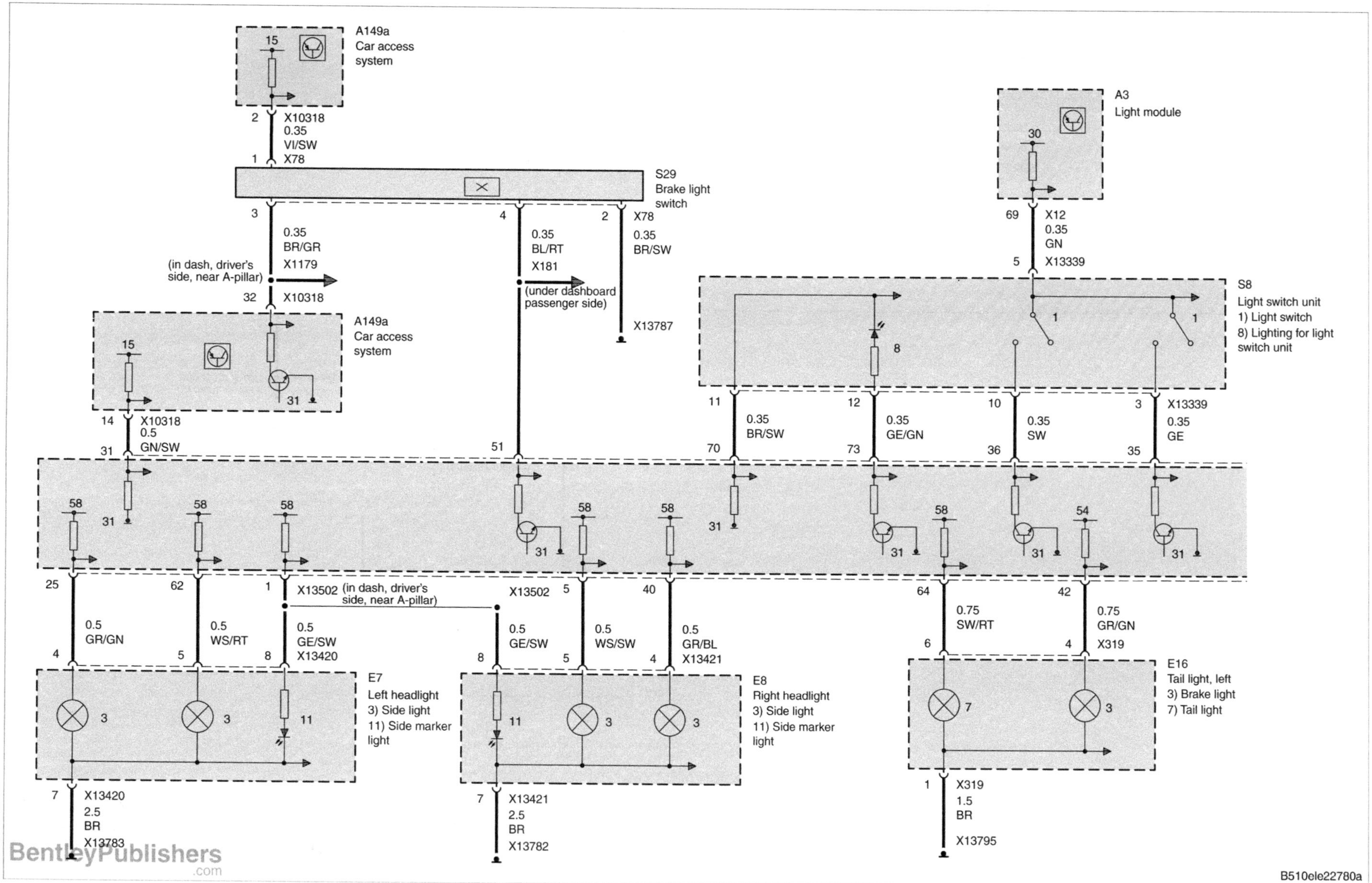

BentleyPublishers.com

Side light, parking light, brake light
Sedan without Xenon lights as of 03 / 2005 up to 09 / 2005
(page 2 of 2)

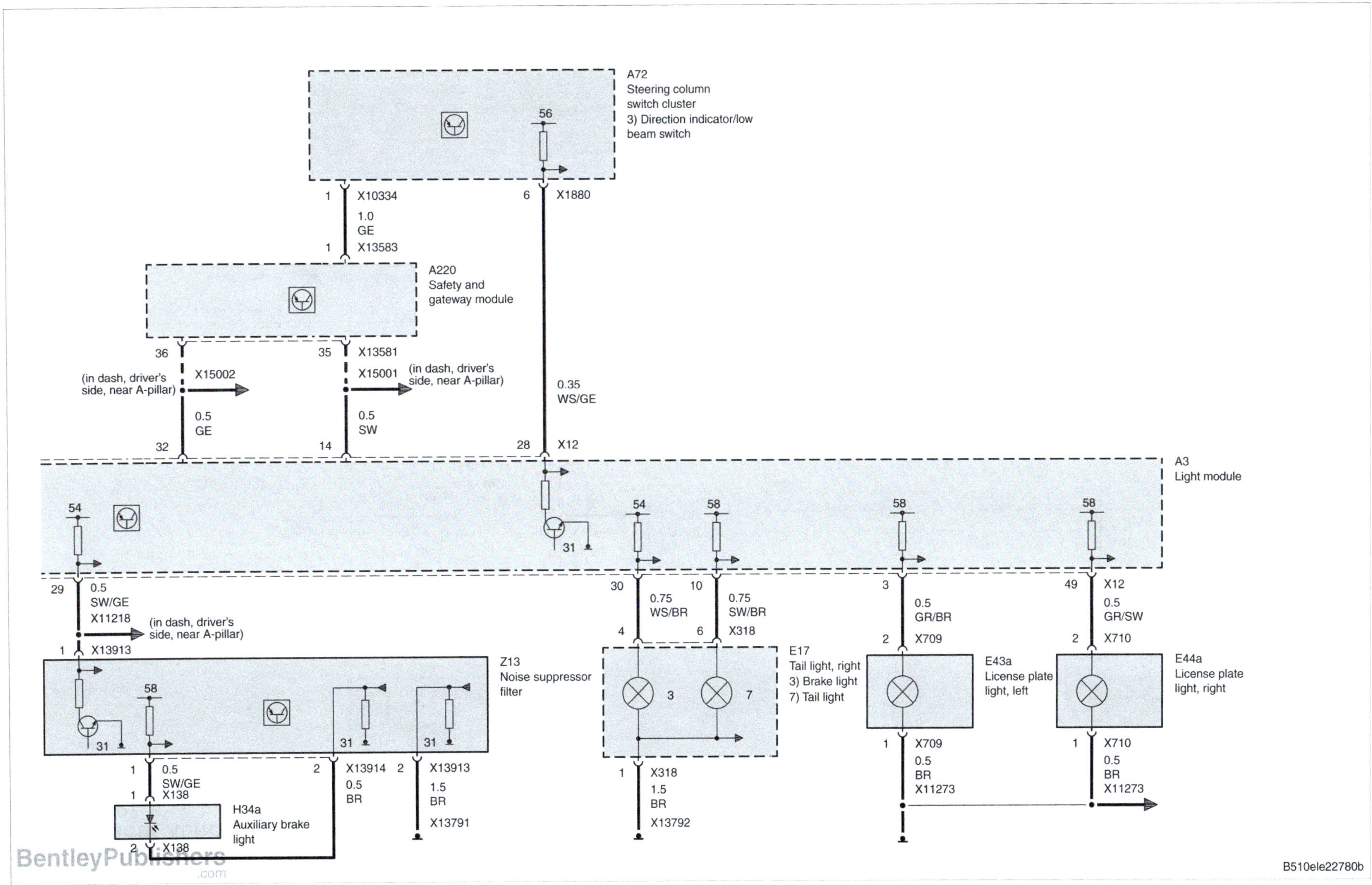

BentleyPublishers.com

Side light, parking light, brake light
Sedan without Xenon lights as of 09 / 2005 up to 03 / 2007
(page 1 of 2)

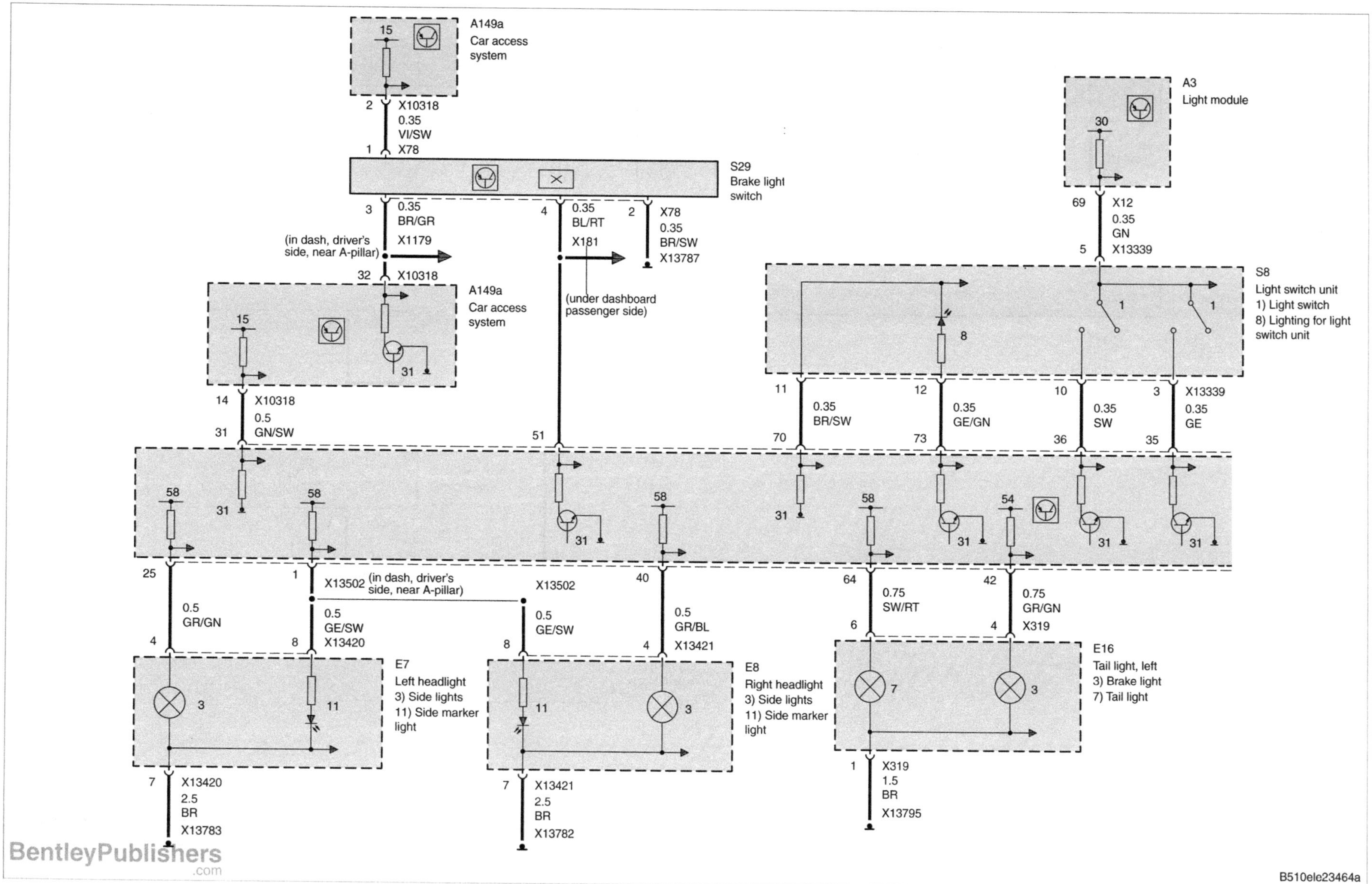

BentleyPublishers.com

Side light, parking light, brake light
Sedan without Xenon lights as of 09 / 2005 up to 03 / 2007
(page 2 of 2)

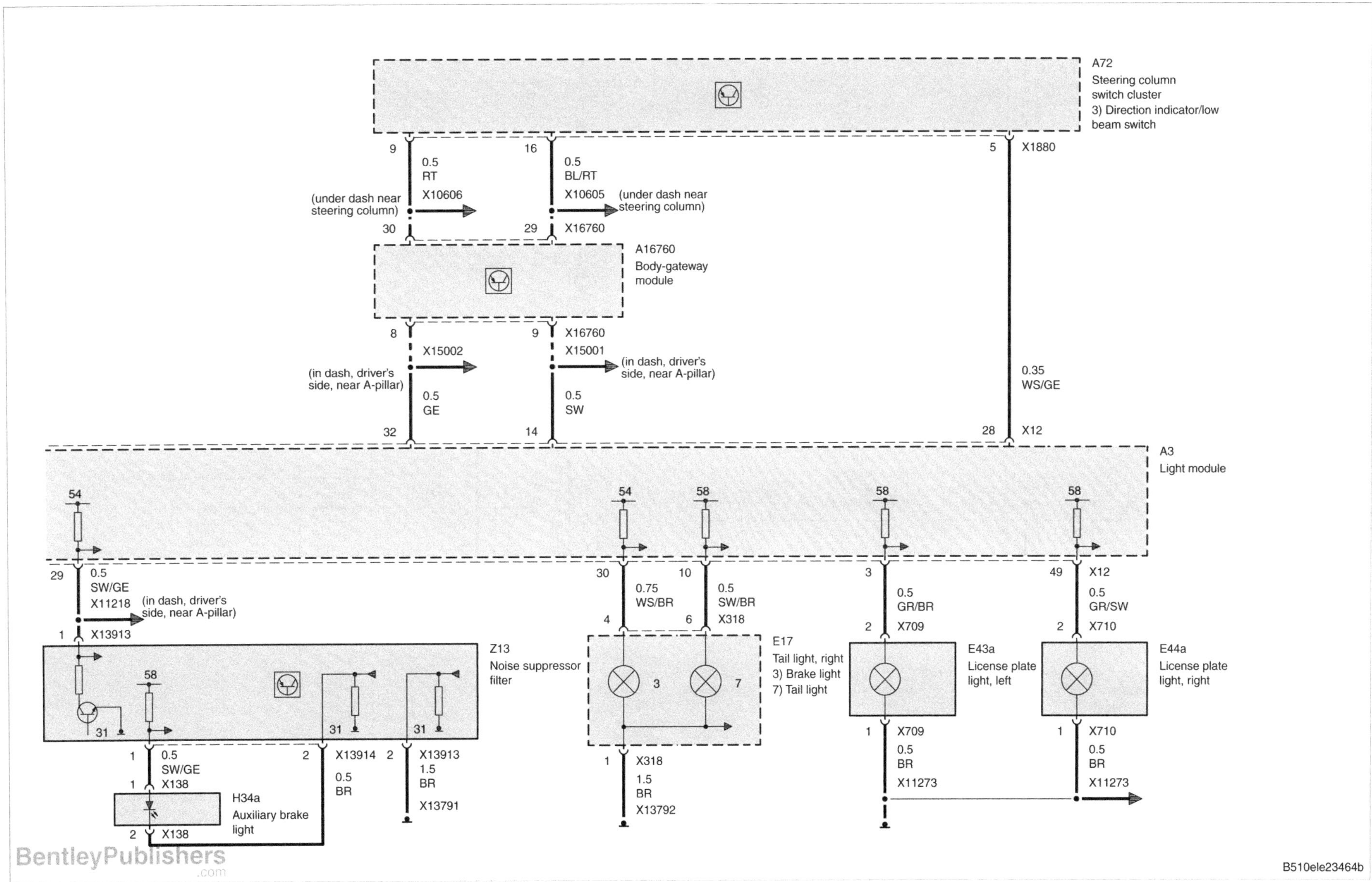

Side light, parking light, brake light
Sedan with Xenon lights up to 03 / 2005
(page 1 of 2)

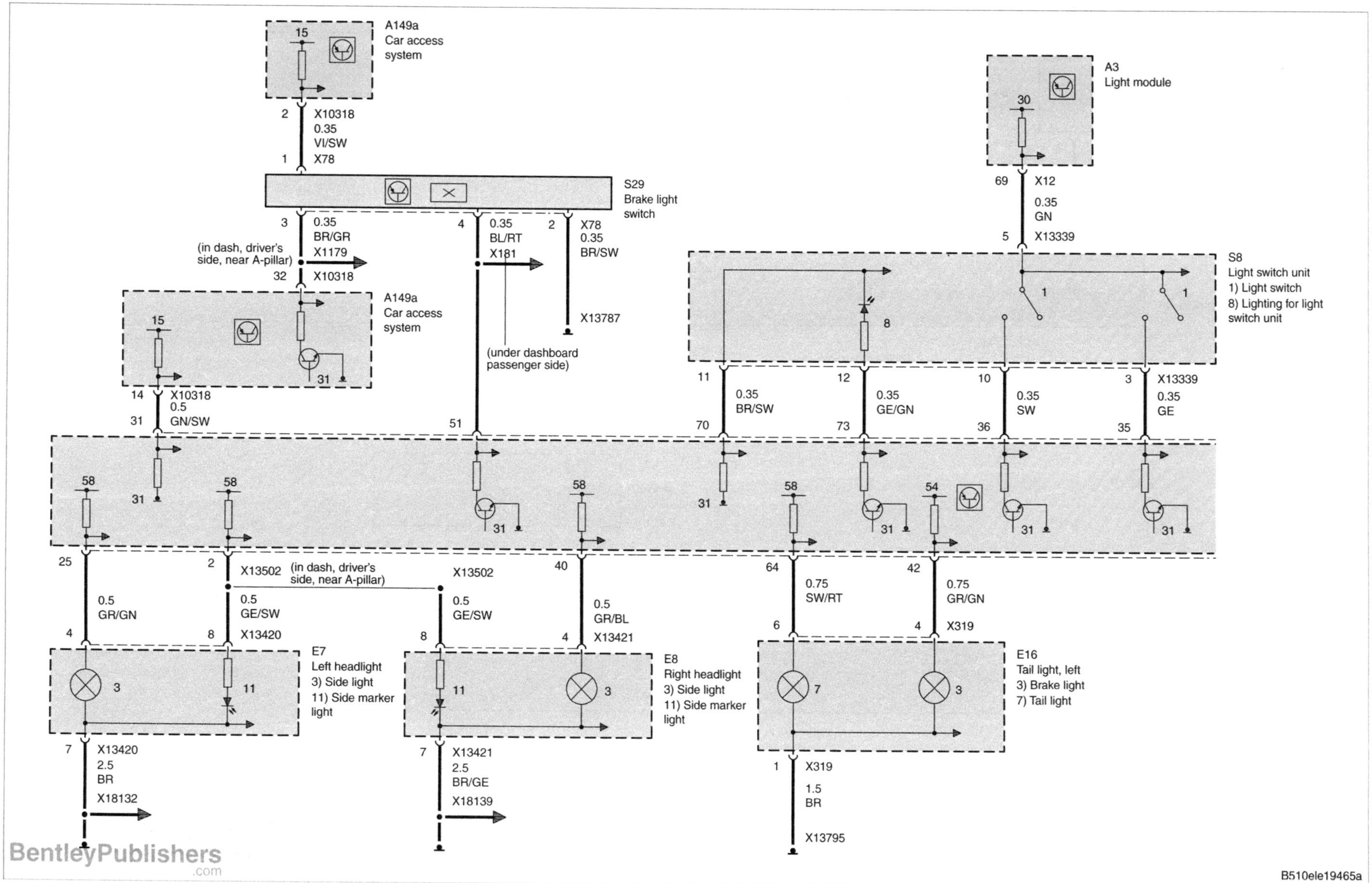

BentleyPublishers.com

B510ele19465a

Side light, parking light, brake light
Sedan with Xenon lights up to 03 / 2005
(page 2 of 2)

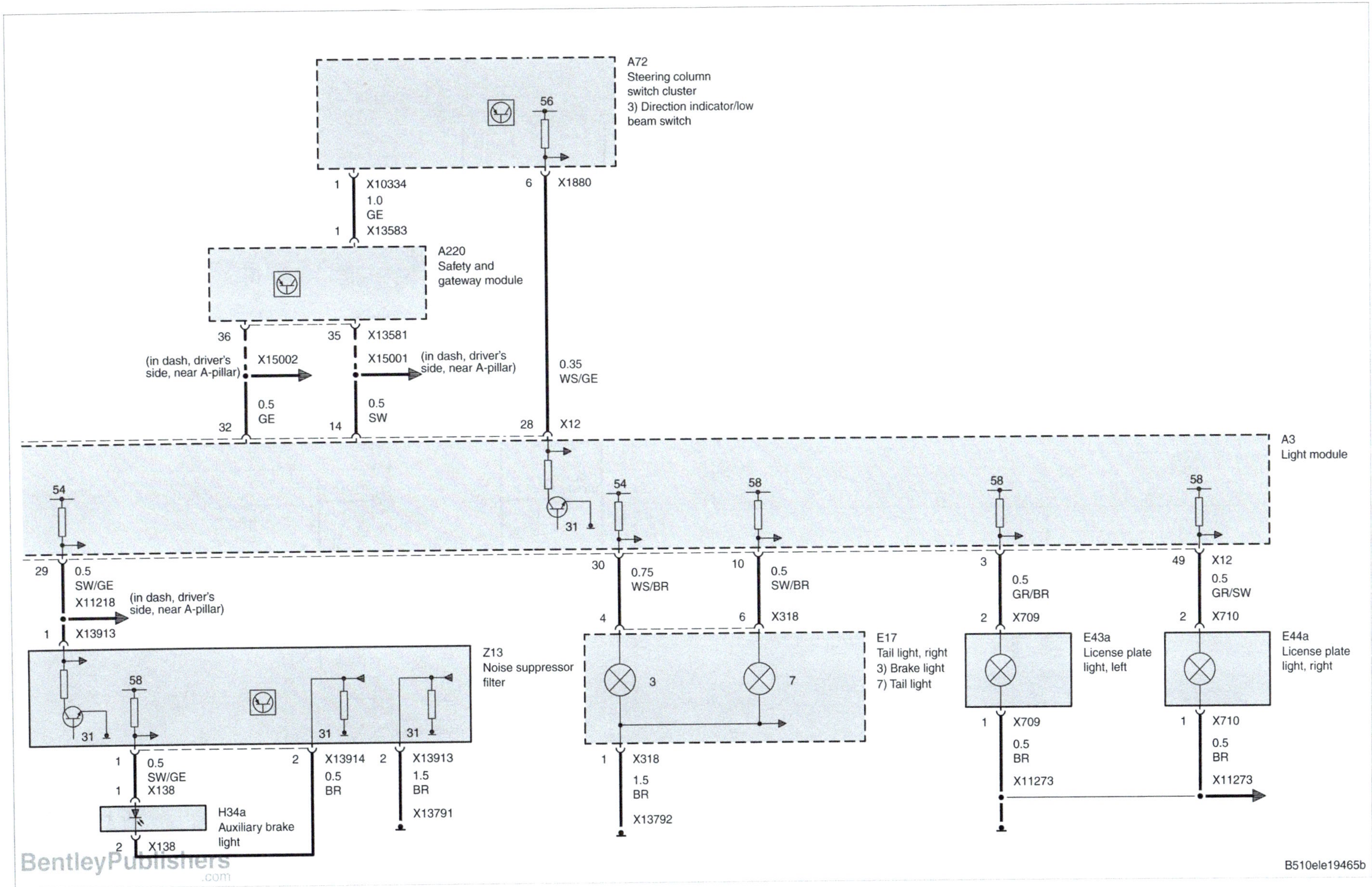

BentleyPublishers.com

Side light, parking light, brake light
Sedan with Xenon lights as of 03 / 2005 up to 09 / 2005
(page 1 of 2)

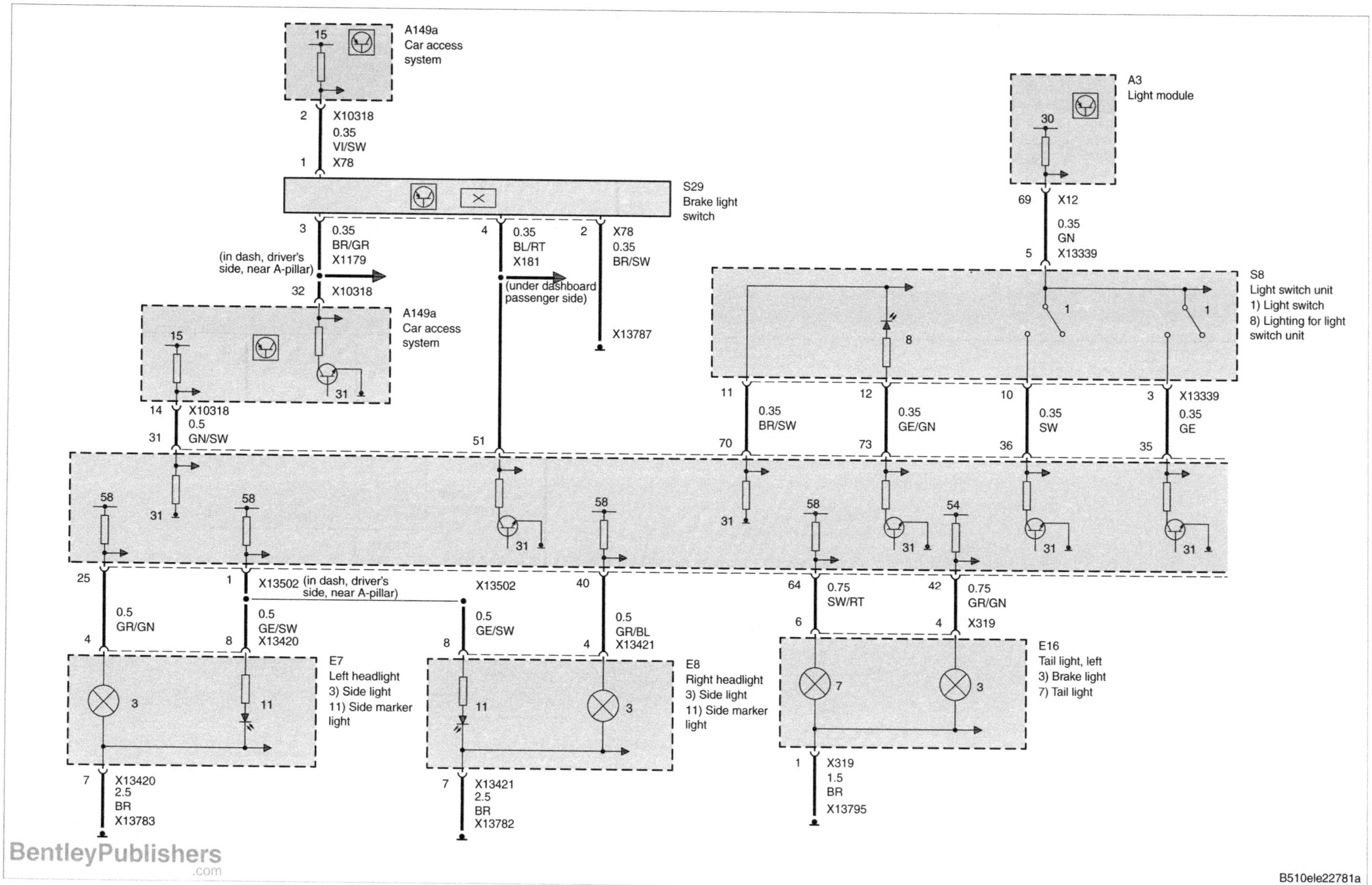

BentleyPublishers.com

B510ele22781a

Side light, parking light, brake light
Sedan with Xenon lights as of 03 / 2005 up to 09 / 2005
(page 2 of 2)

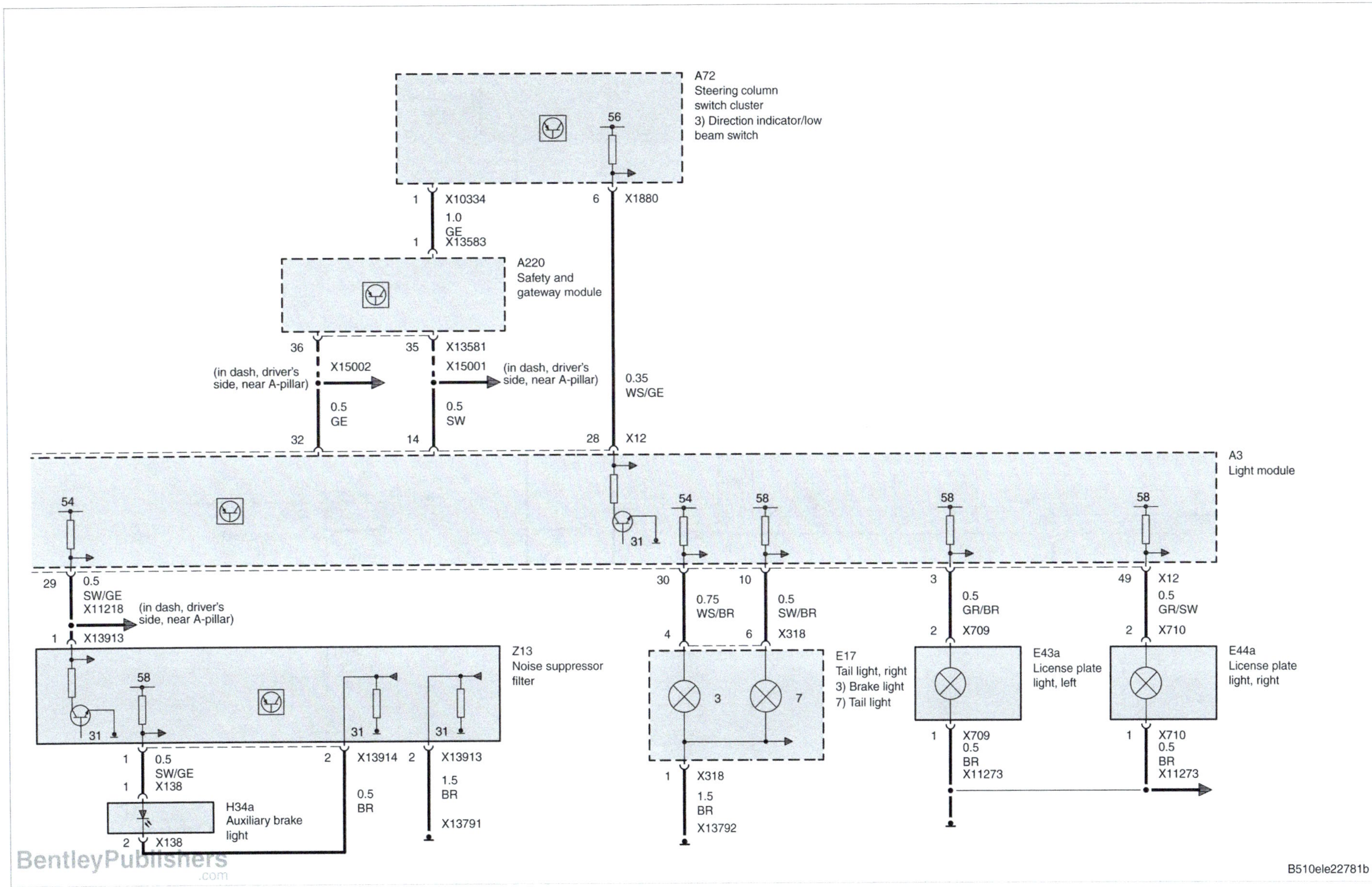

BentleyPublishers.com

Side light, parking light, brake light
Sedan with Xenon lights as of 09 / 2005 up to 03 / 2007
(page 1 of 2)

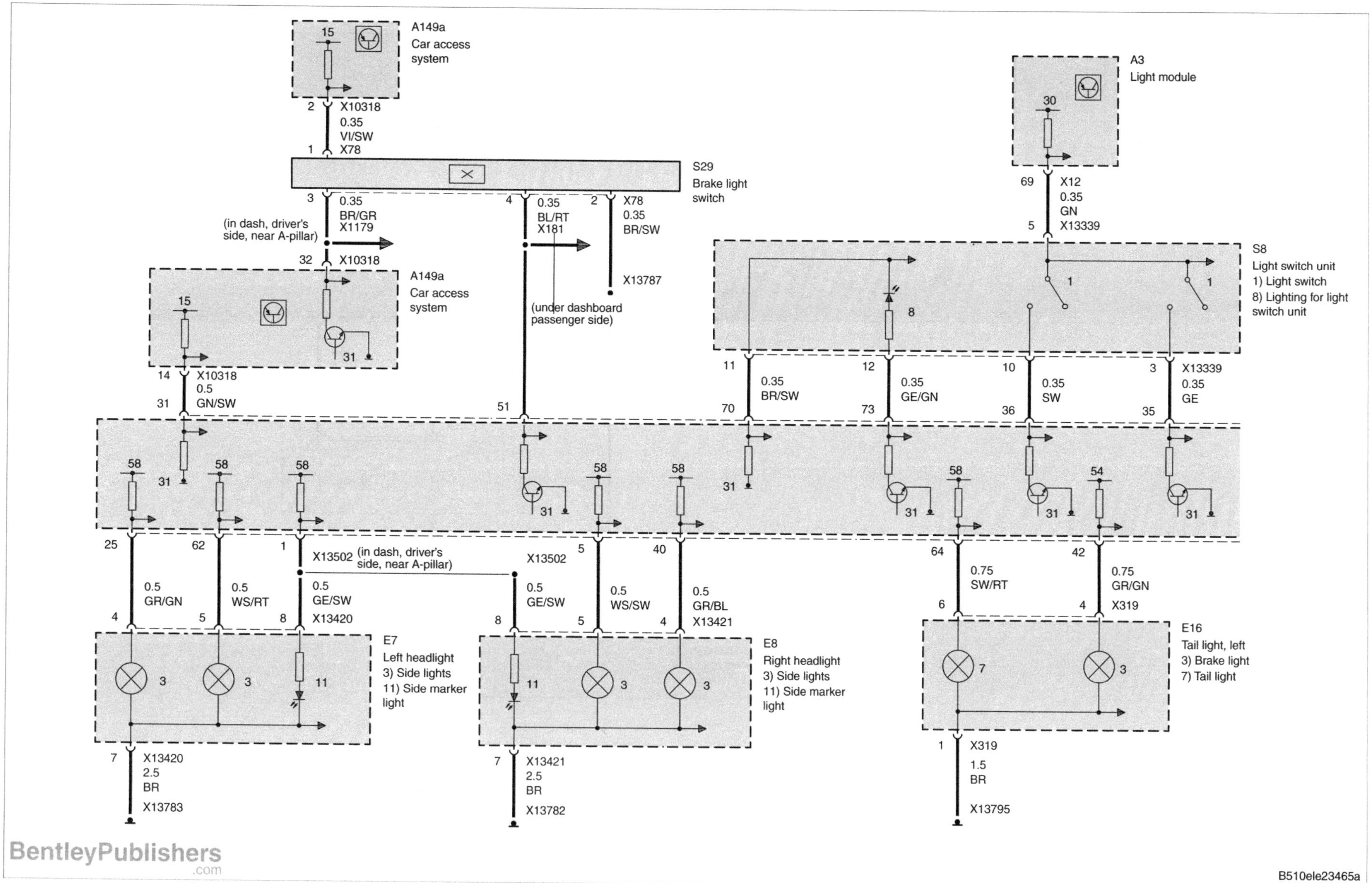

BentleyPublishers.com

B510ele23465a

Side light, parking light, brake light
Sedan with Xenon lights as of 09 / 2005 up to 03 / 2007
(page 2 of 2)

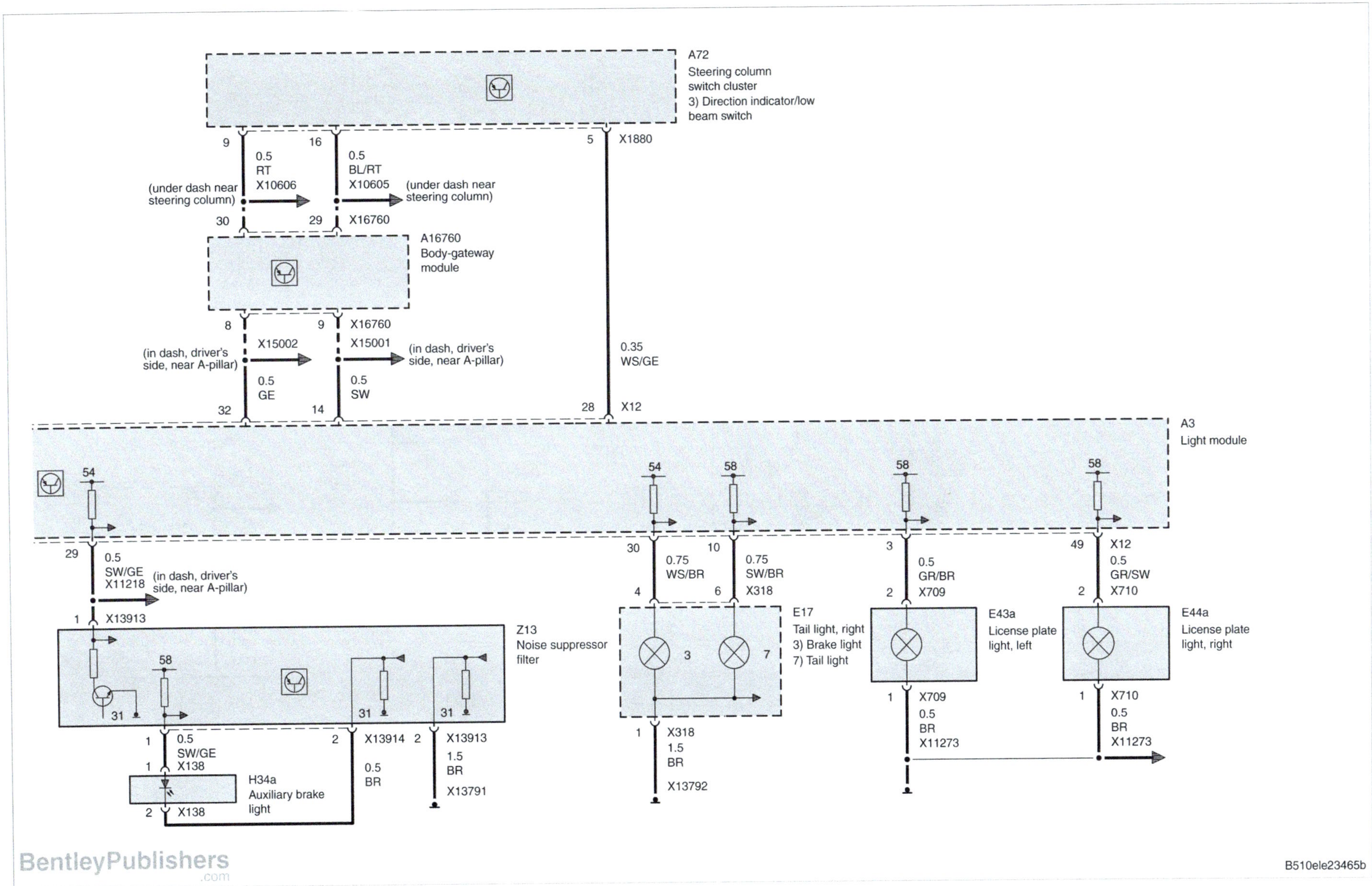

Side light, parking light, brake light
Wagon without Xenon lights up to 03 / 2005
(page 1 of 2)

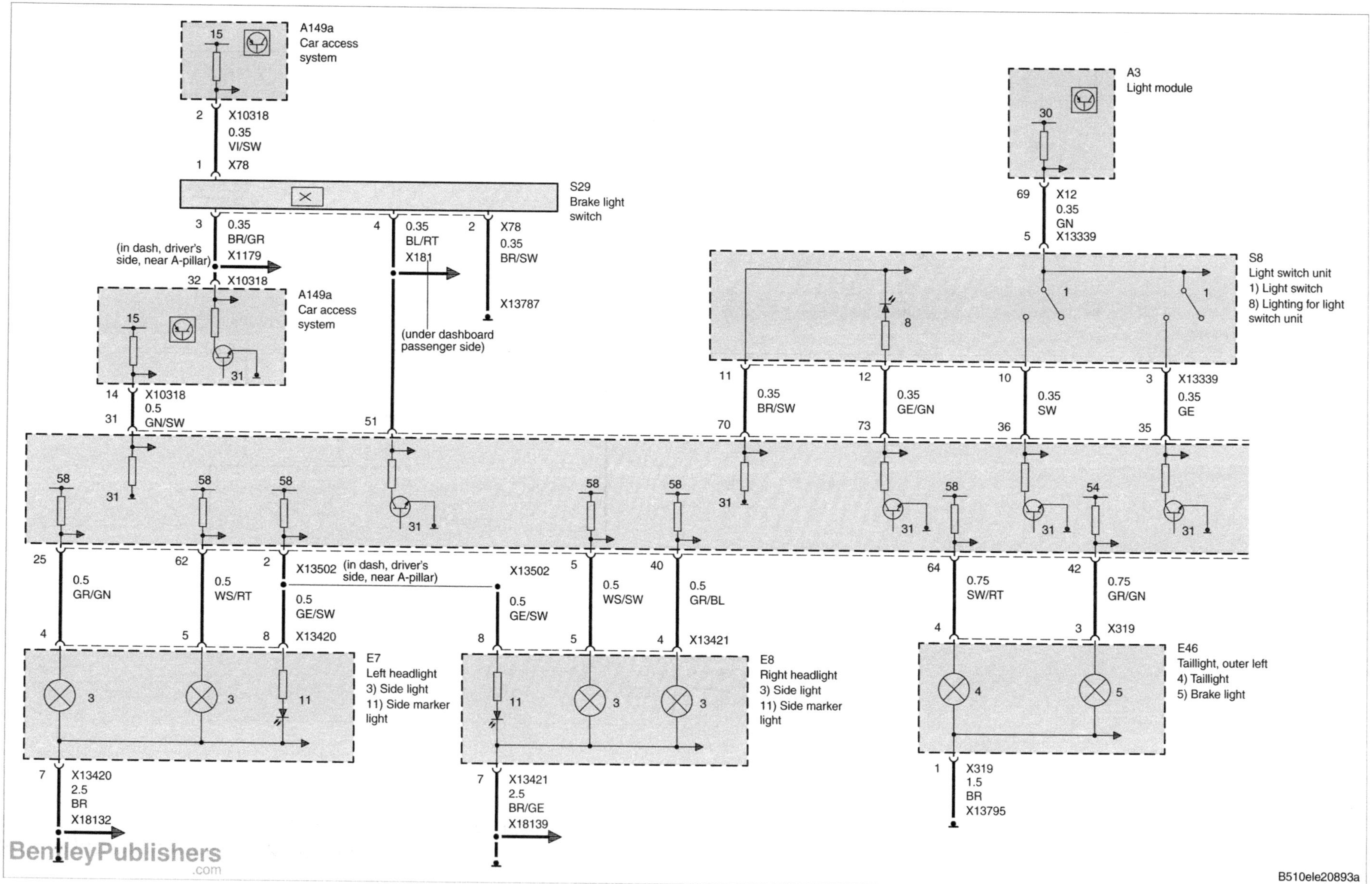

BentleyPublishers.com

Side light, parking light, brake light
Wagon without Xenon lights up to 03 / 2005
(page 2 of 2)

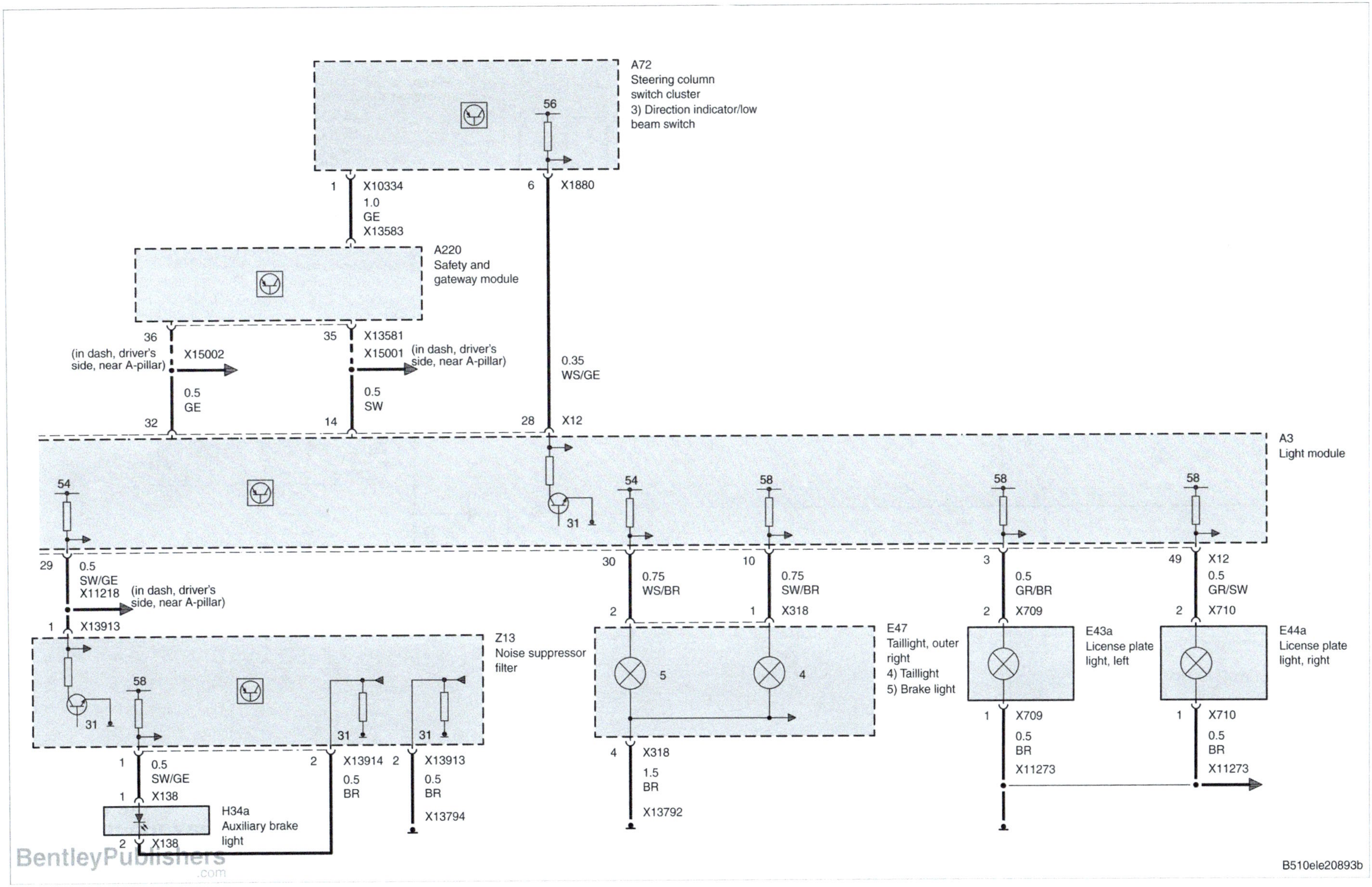

Side light, parking light, brake light
Wagon without Xenon lights as of 03 / 2005 up to 09 / 2005
(page 1 of 2)

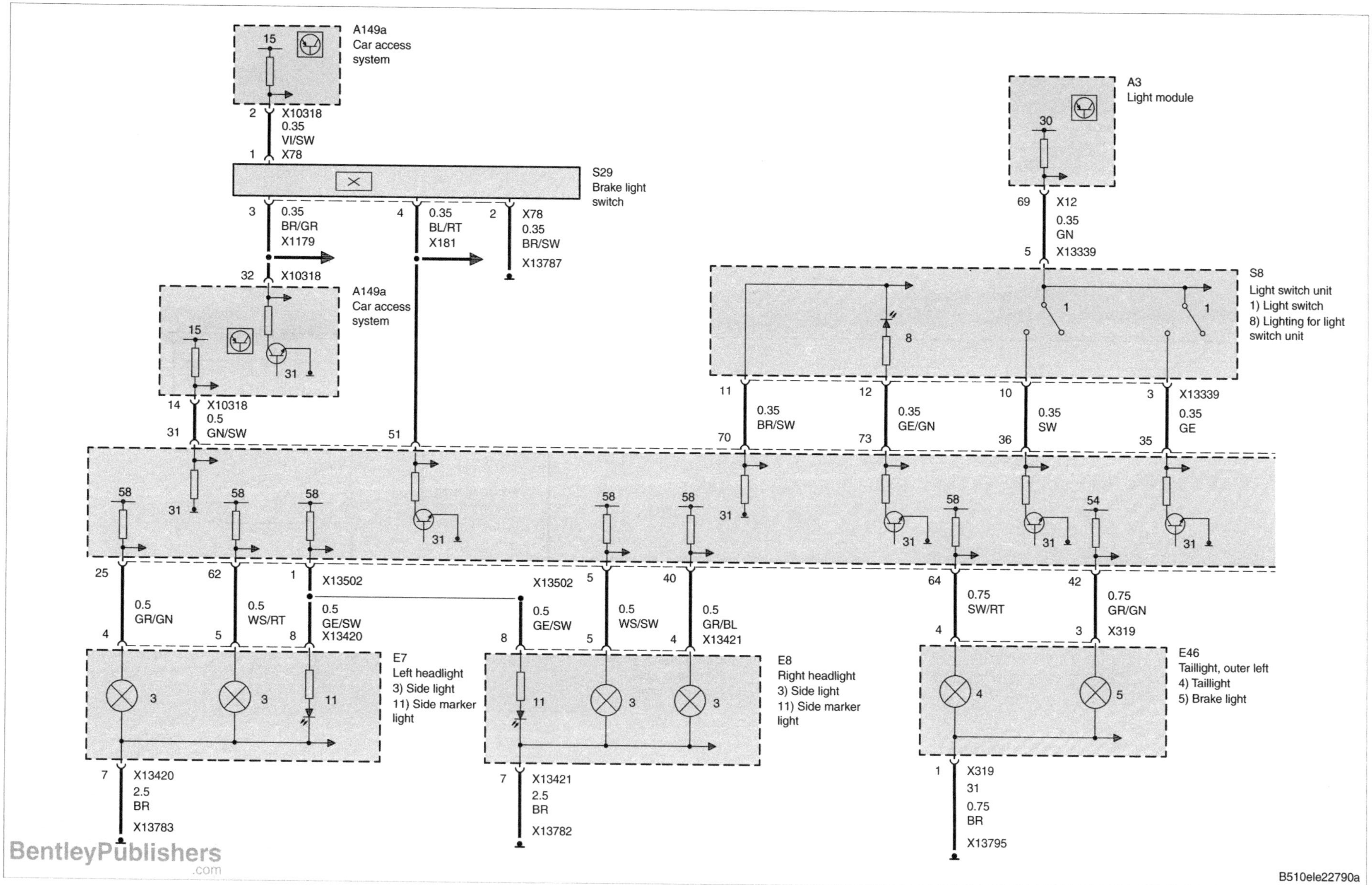

BentleyPublishers.com

Side light, parking light, brake light
Wagon without Xenon lights as of 03 / 2005 up to 09 / 2005
(page 2 of 2)

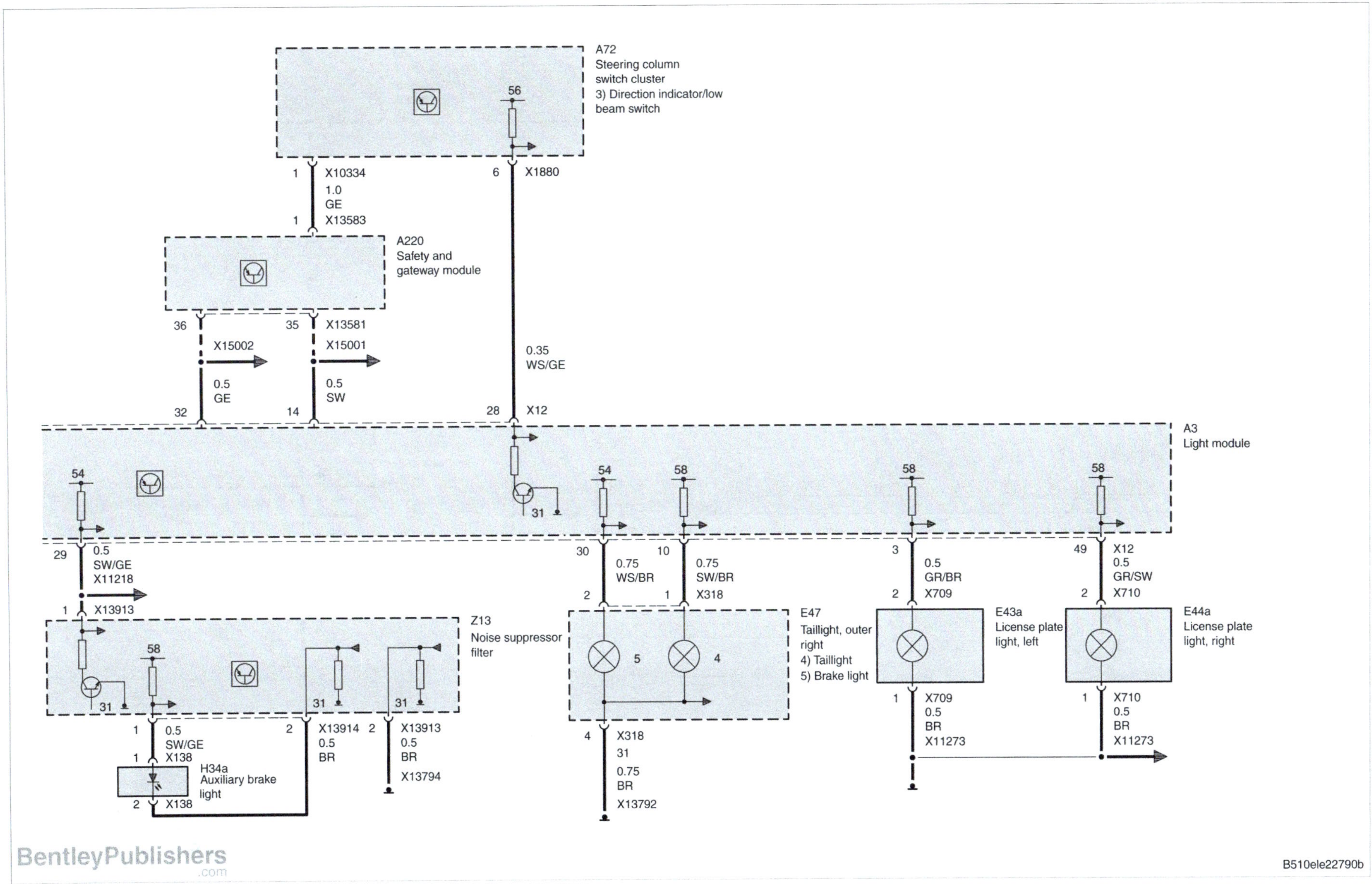

Side light, parking light, brake light
Wagon without Xenon lights as of 09 / 2005 up to 03 / 2007
(page 1 of 2)

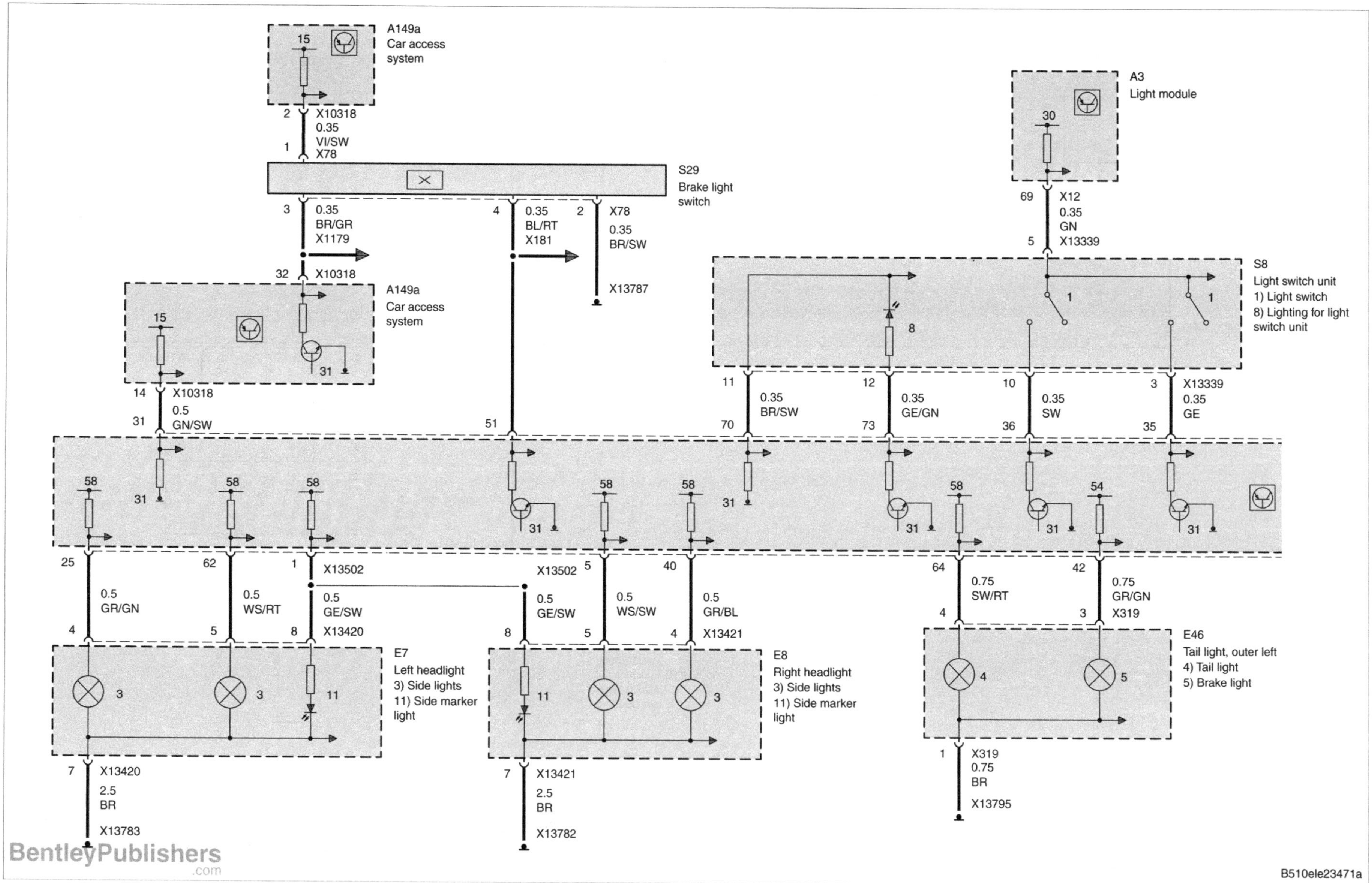

BentleyPublishers
.com

B510ele23471a

Side light, parking light, brake light
Wagon without Xenon lights as of 09 / 2005 up to 03 / 2007
(page 2 of 2)

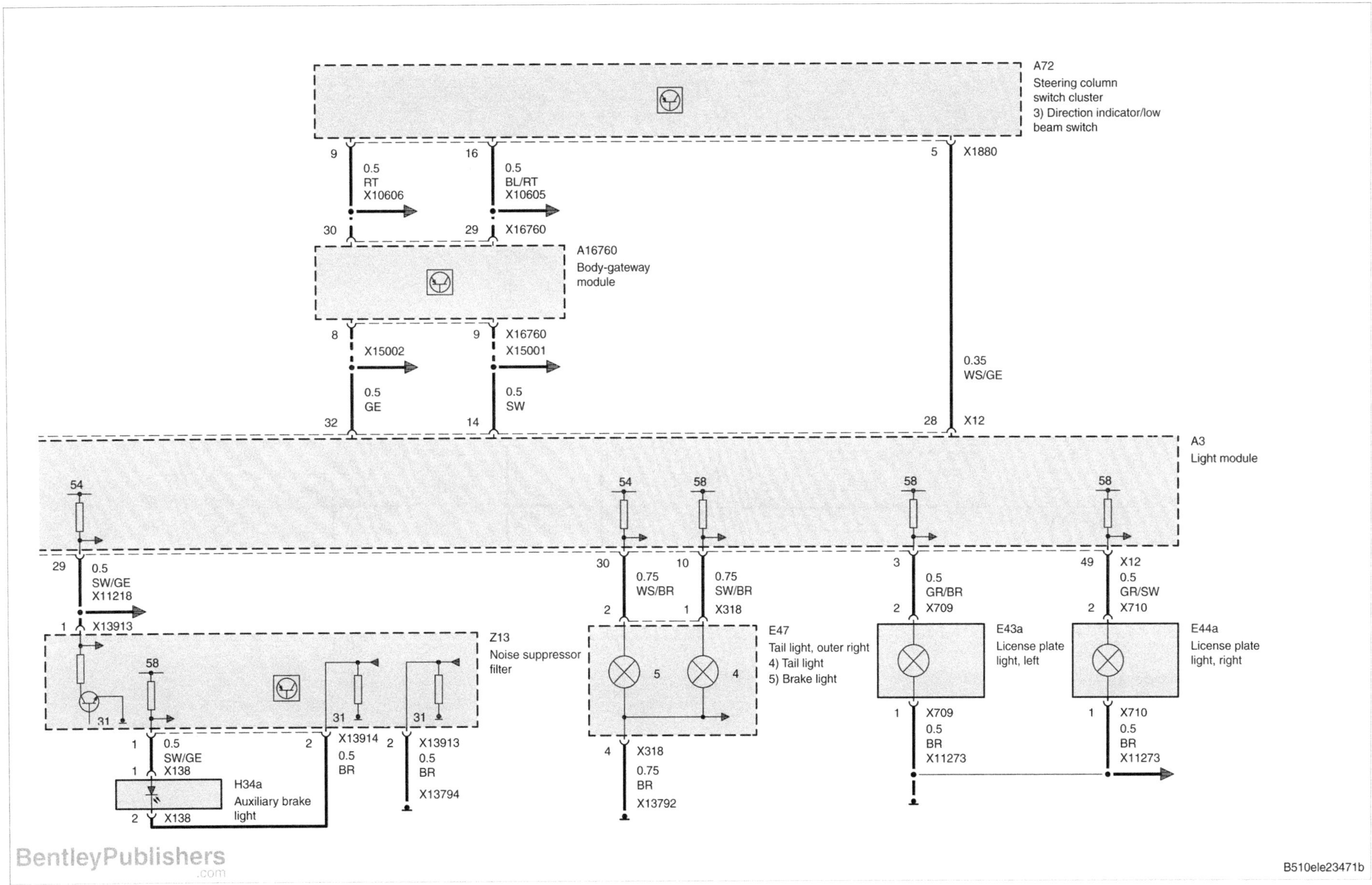

BentleyPublishers.com

Side light, parking light, brake light
Wagon with Xenon lights up to 03 / 2005
(page 1 of 2)

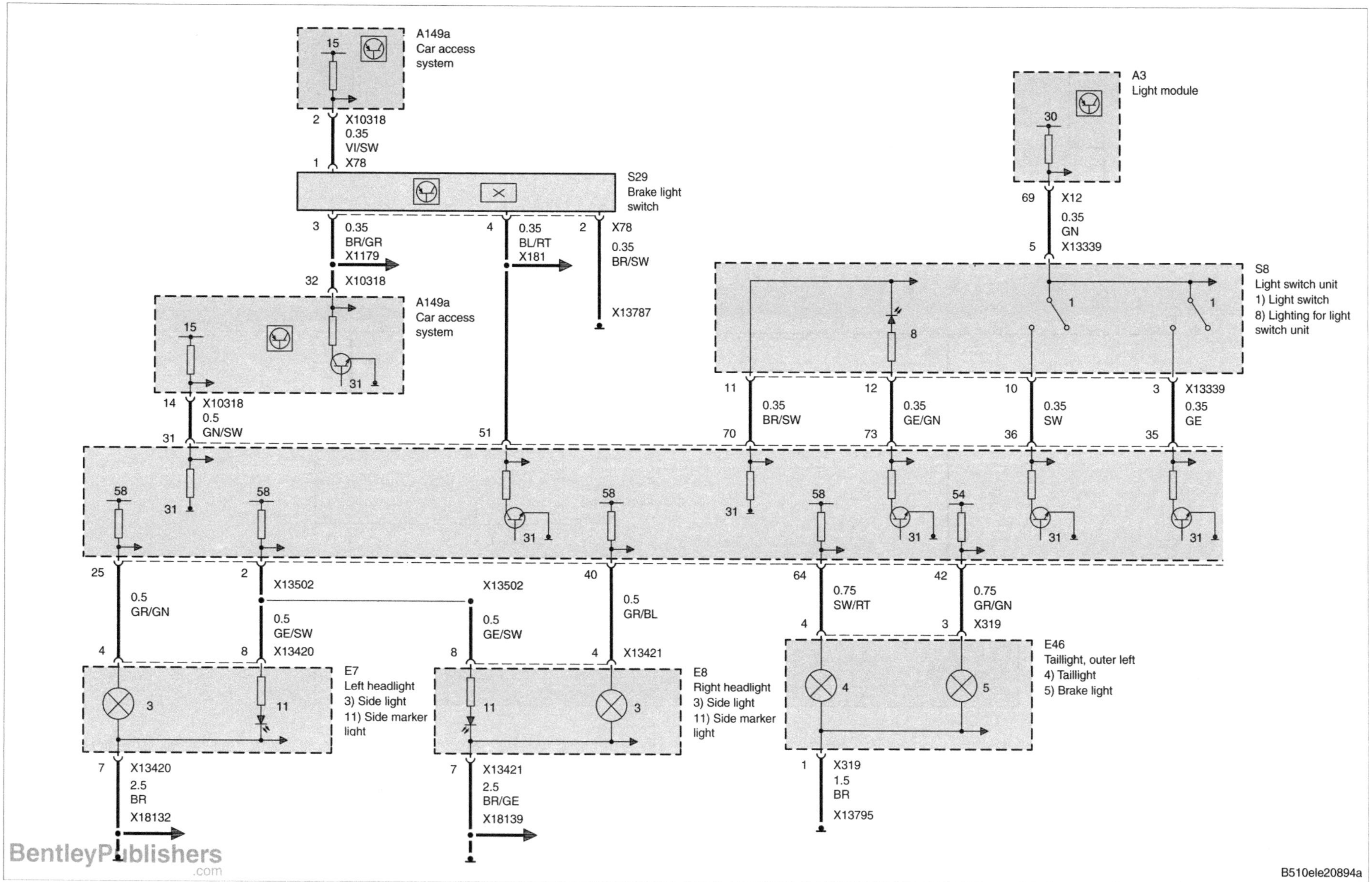

BentleyPublishers.com

B510ele20894a

Side light, parking light, brake light
Wagon with Xenon lights up to 03 / 2005
(page 2 of 2)

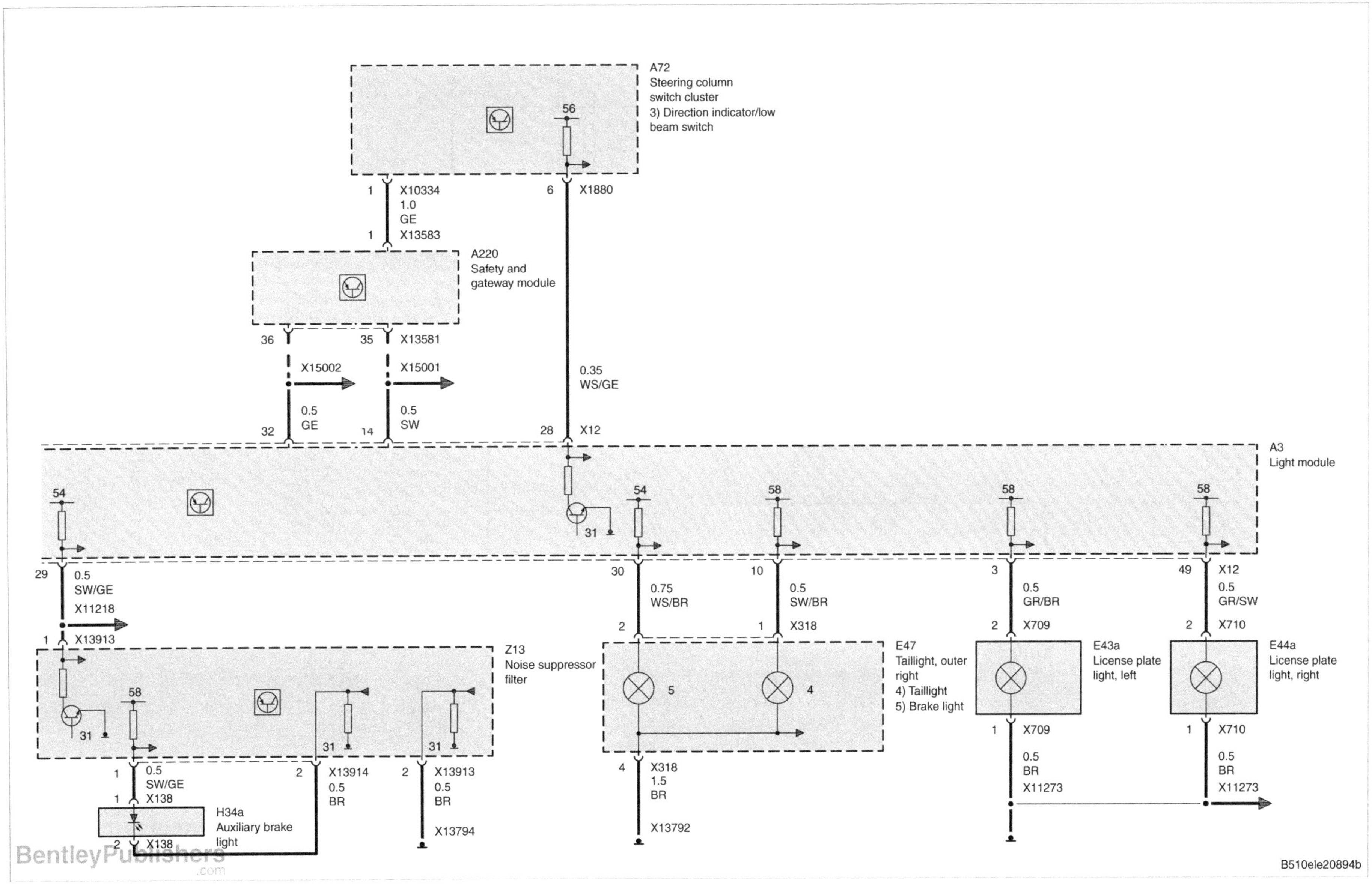

BentleyPublishers
.com

Side light, parking light, brake light
Wagon with Xenon lights as of 03 / 2005 up to 09 / 2005
(page 1 of 2)

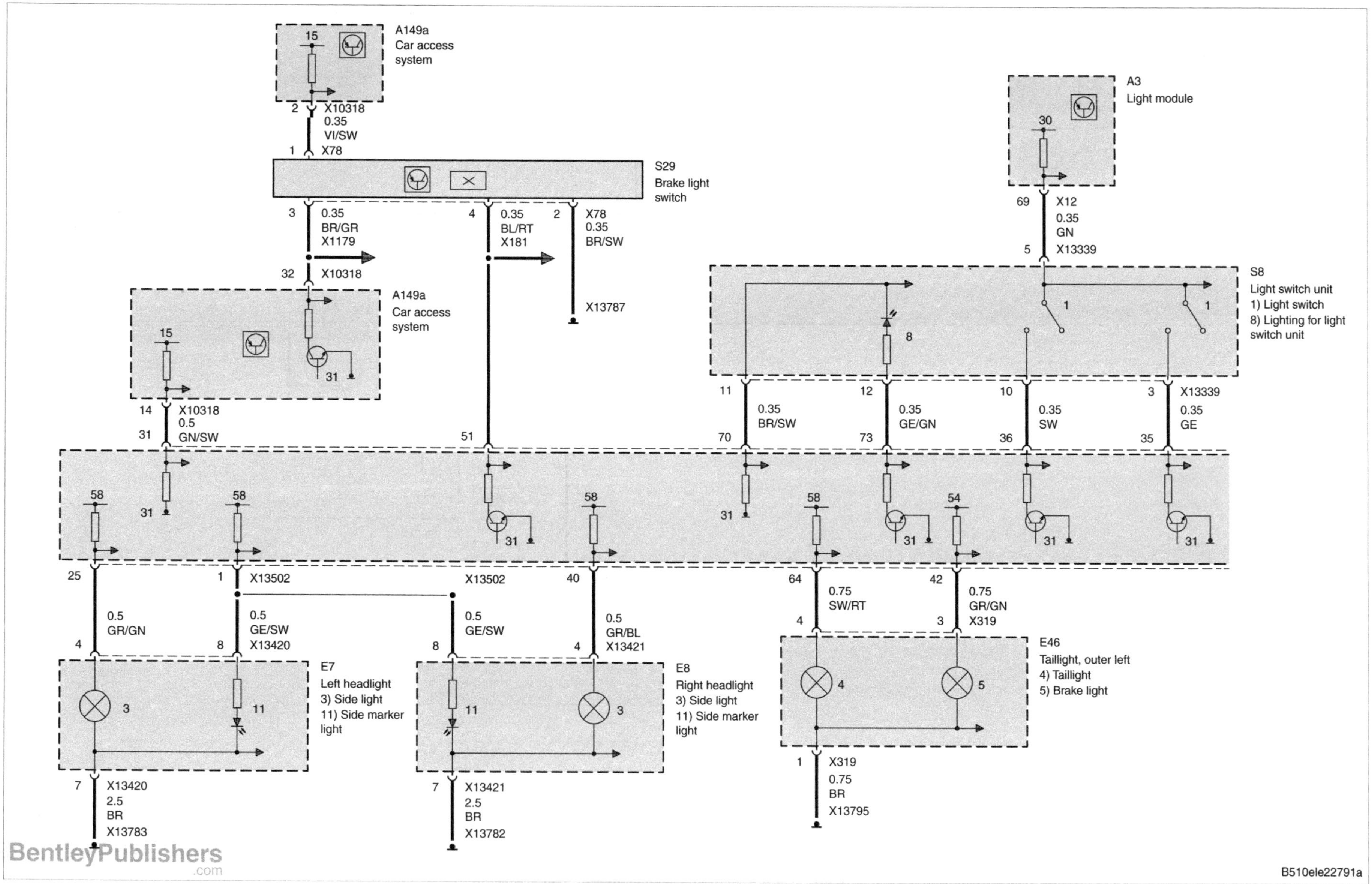

BentleyPublishers.com

Side light, parking light, brake light
Wagon with Xenon lights as of 03 / 2005 up to 09 / 2005
(page 2 of 2)

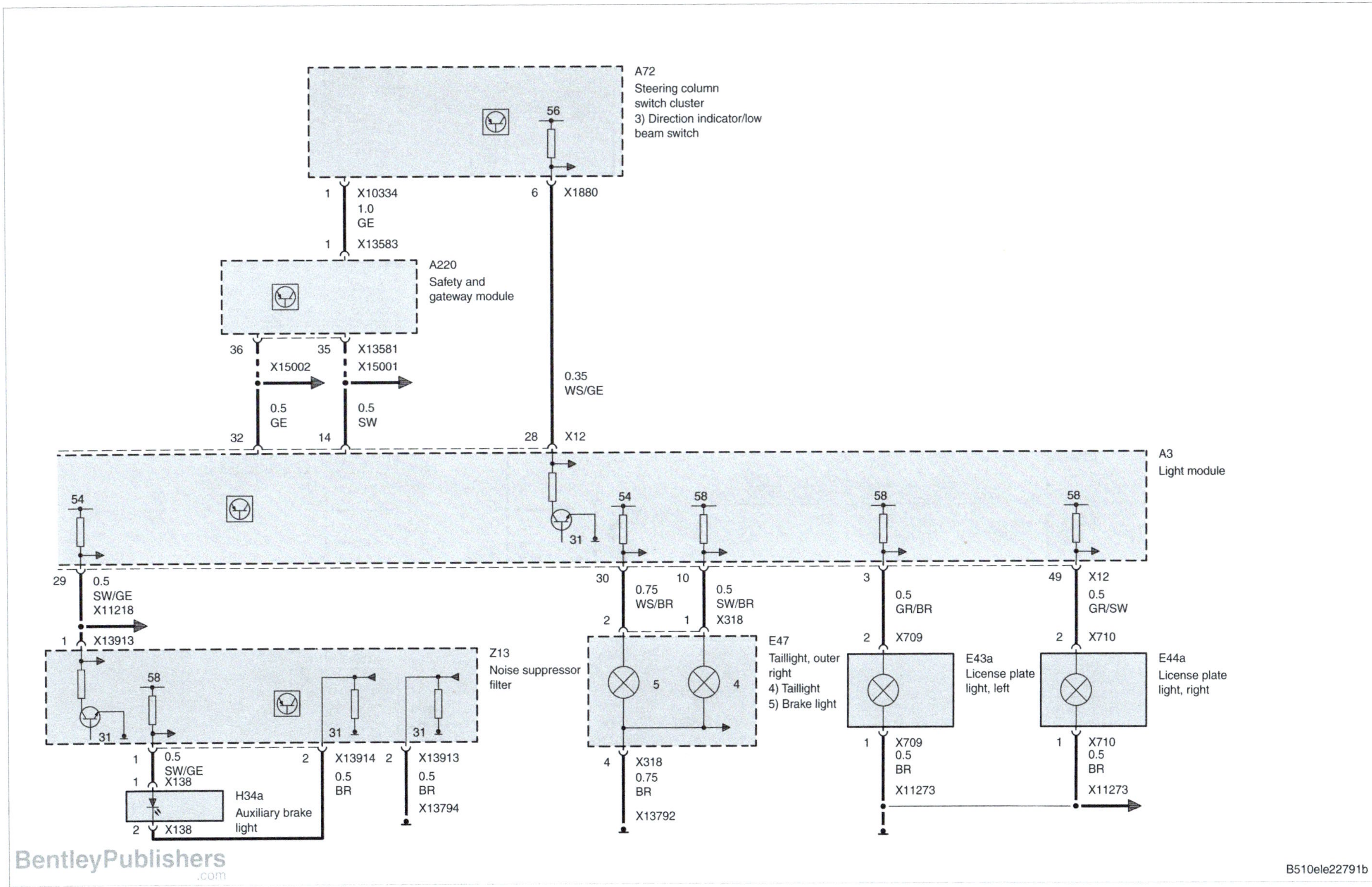

Side light, parking light, brake light
Wagon with Xenon lights as of 09 / 2005 up to 03 / 2007
(page 1 of 2)

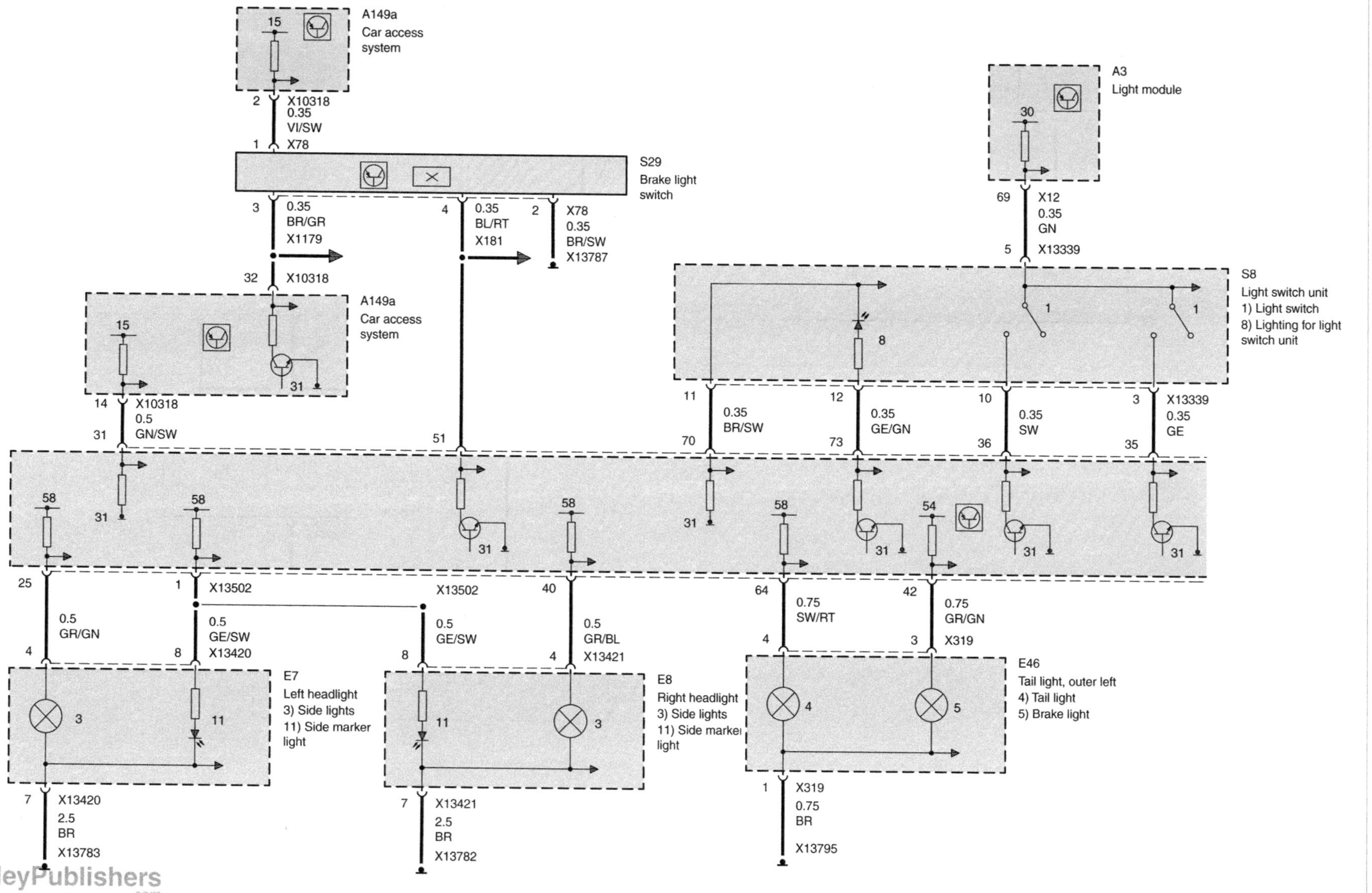

BentleyPublishers
.com

B510ele23472a

Side light, parking light, brake light
Wagon with Xenon lights as of 09 / 2005 up to 03 / 2007
(page 2 of 2)

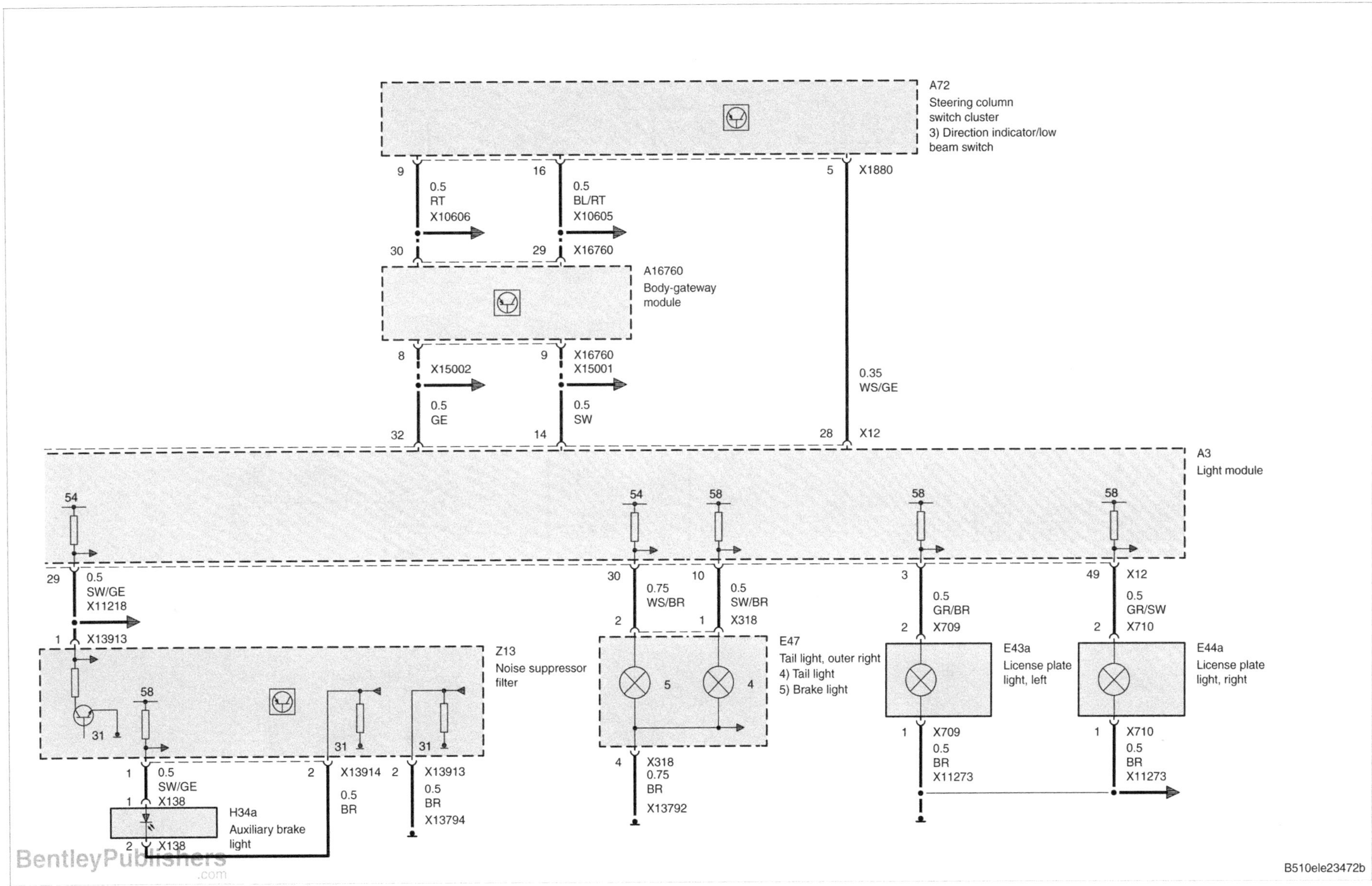

BentleyPublishers.com

Side light, parking light, brake light up to 03 / 2005 (page 1 of 2)

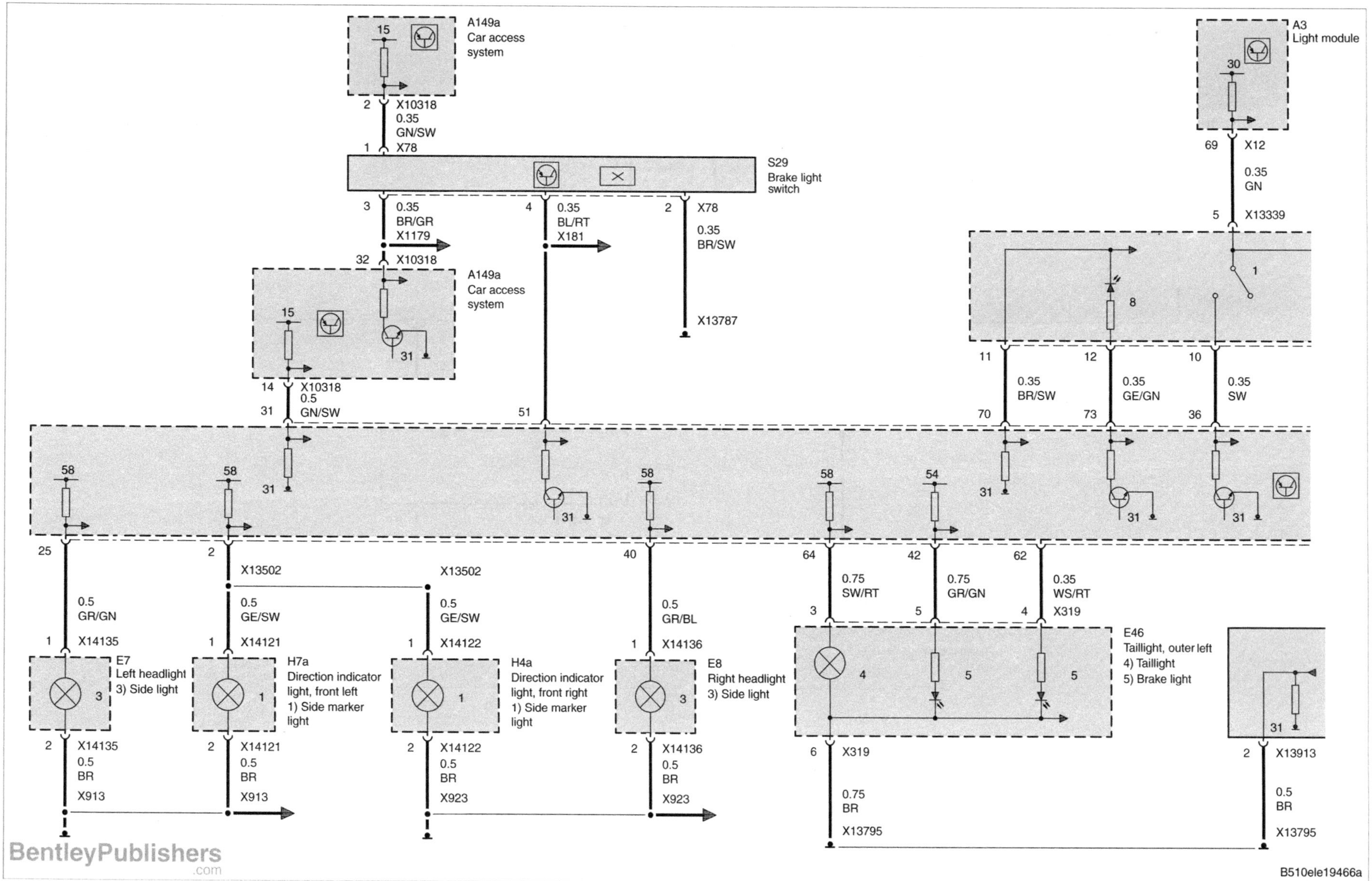

BentleyPublishers
.com

Side light, parking light, brake light up to 03 / 2005 (page 2 of 2)

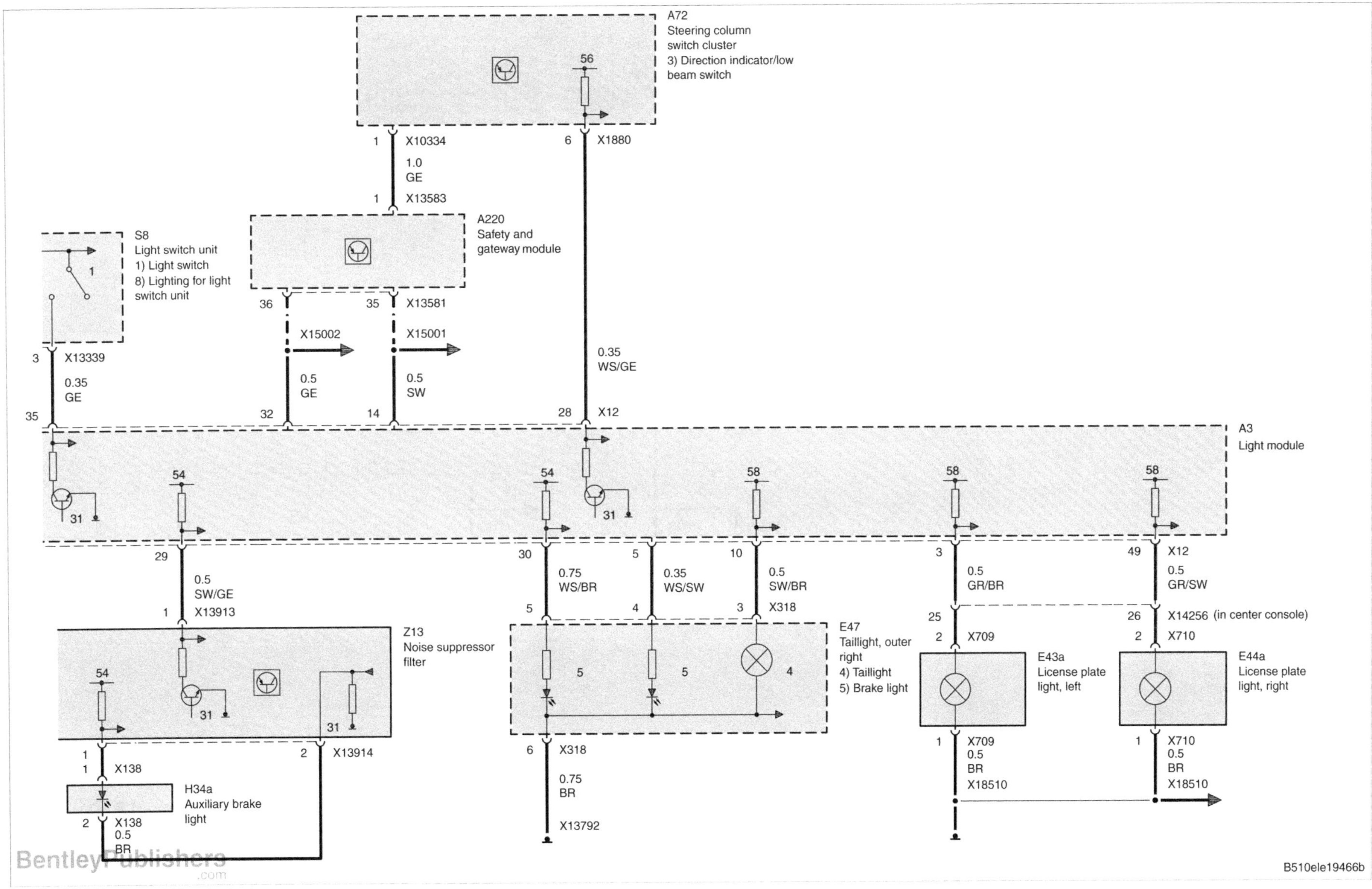

BentleyPublishers.com

Side light, parking light, brake light from 03 / 2005 up to 09 / 2005 (page 1 of 2)

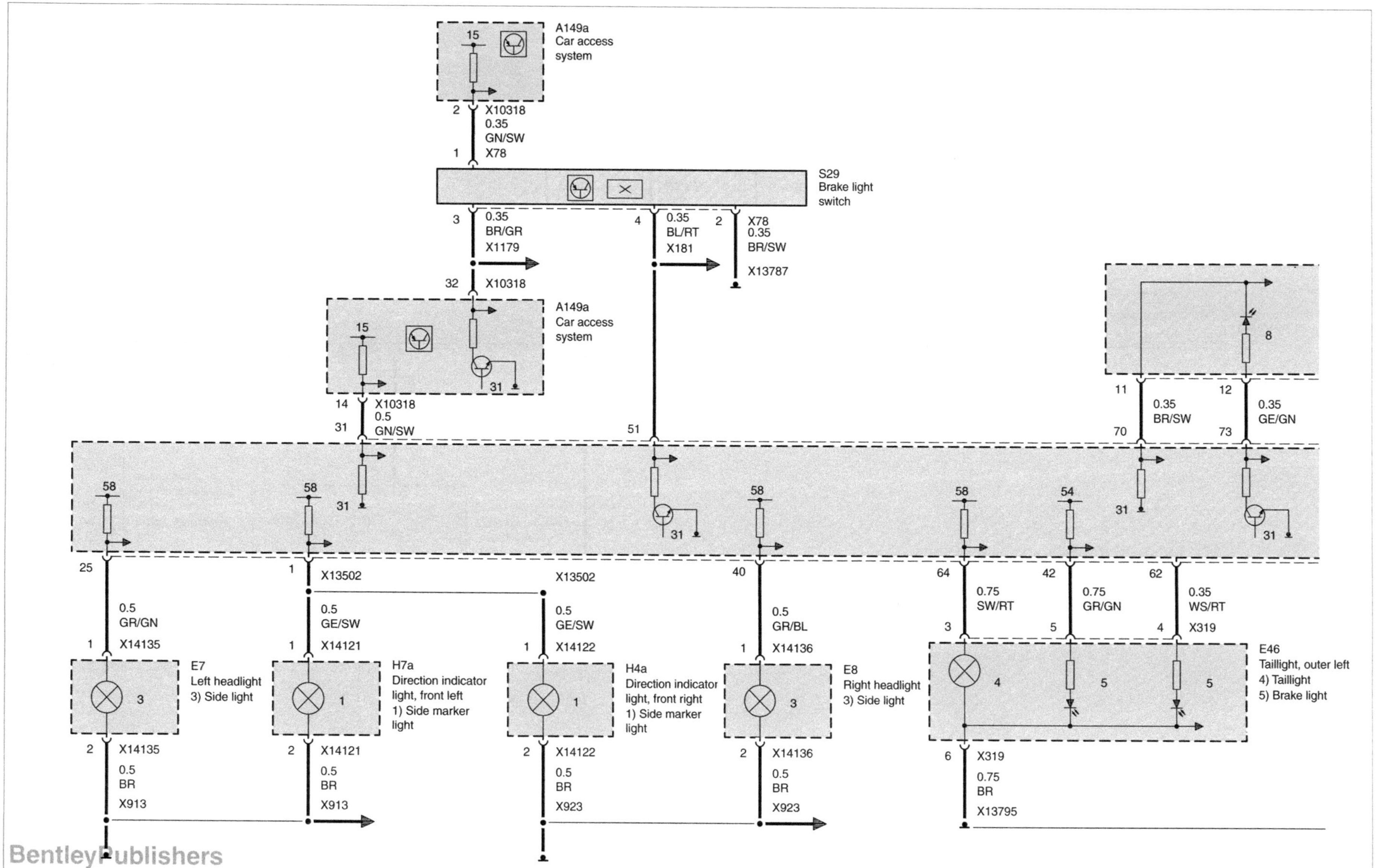

BentleyPublishers.com

B510ele22782a

Side light, parking light, brake light from 03 / 2005 up to 09 / 2005 (page 2 of 2)

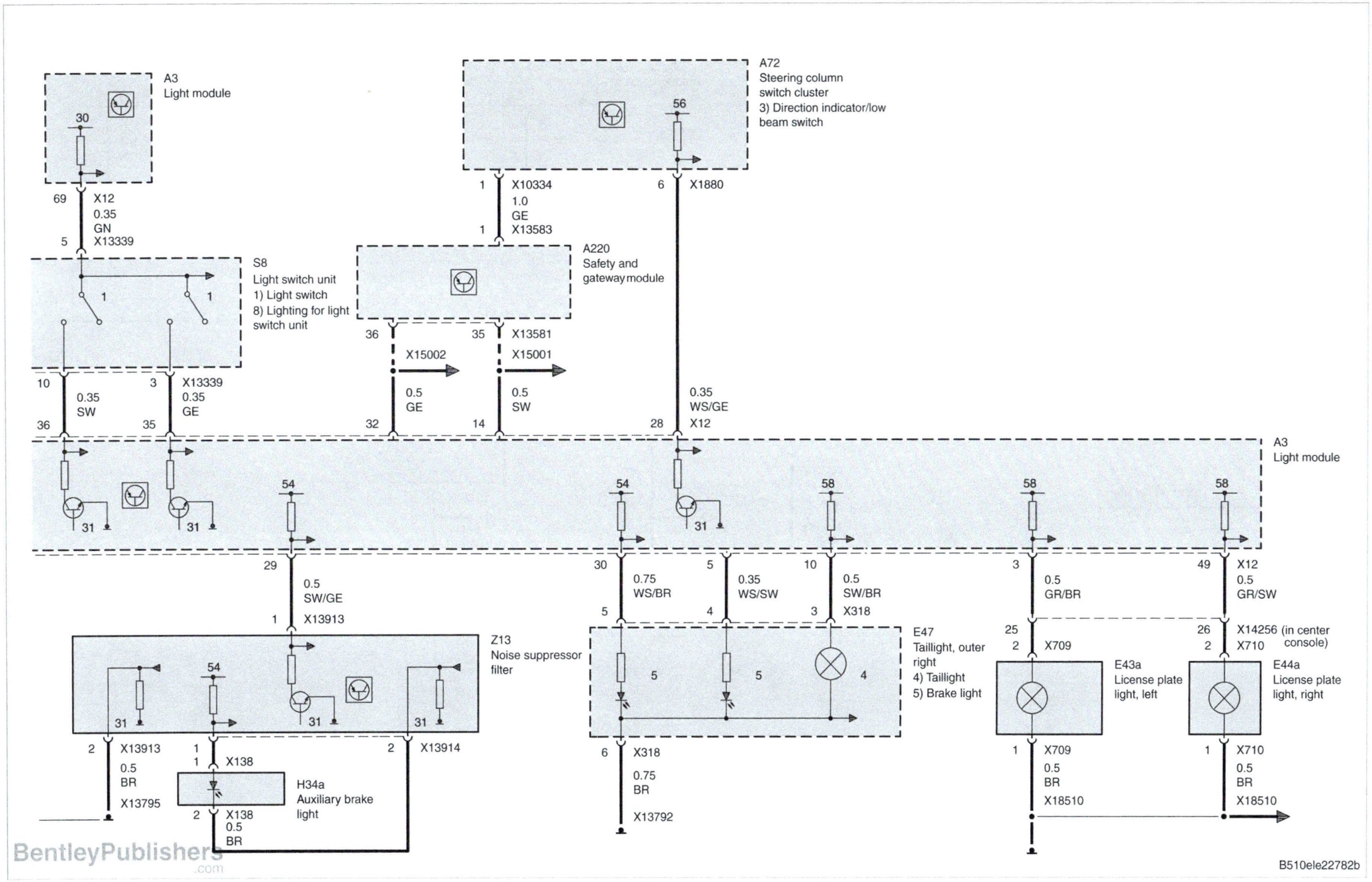

BentleyPublishers
.com

Side light, parking light, brake light from 09 / 2005 up to 03 / 2007 (page 1 of 2)

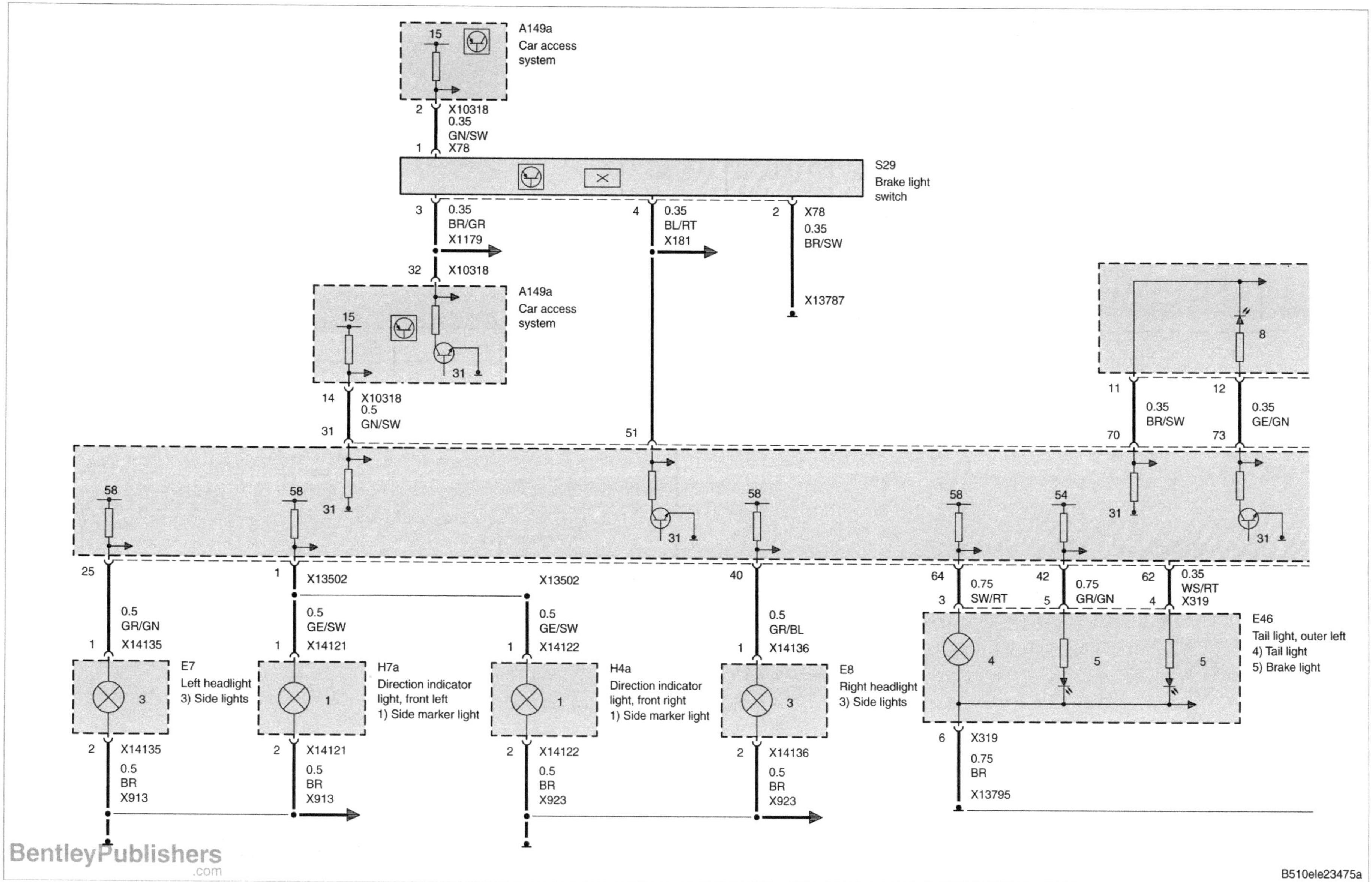

BentleyPublishers.com

B510ele23475a

Side light, parking light, brake light from 09 / 2005 up to 03 / 2007 (page 2 of 2)

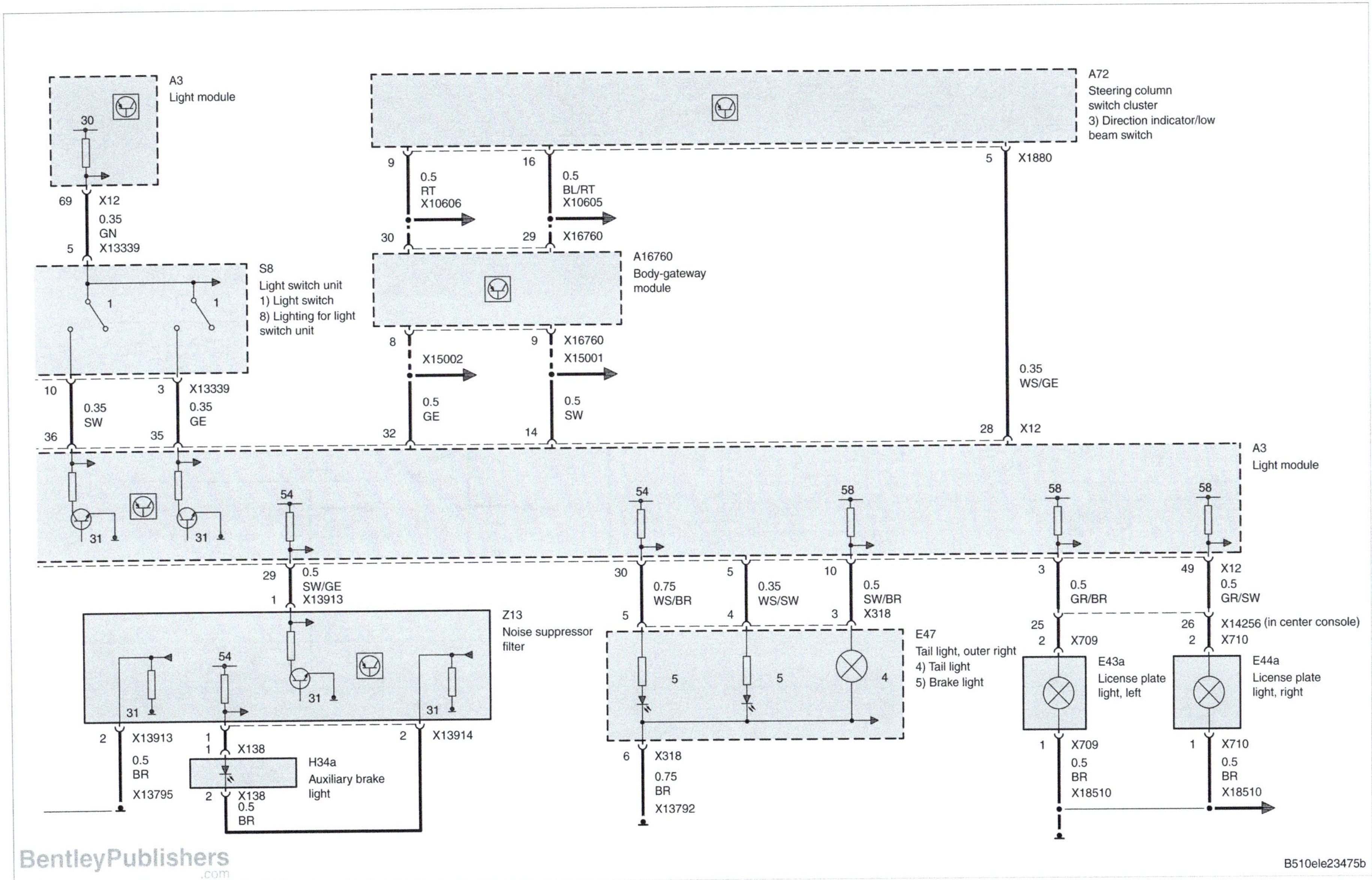

Side light, parking light, brake light
Sedan as of 03 / 2007
(page 1 of 2)

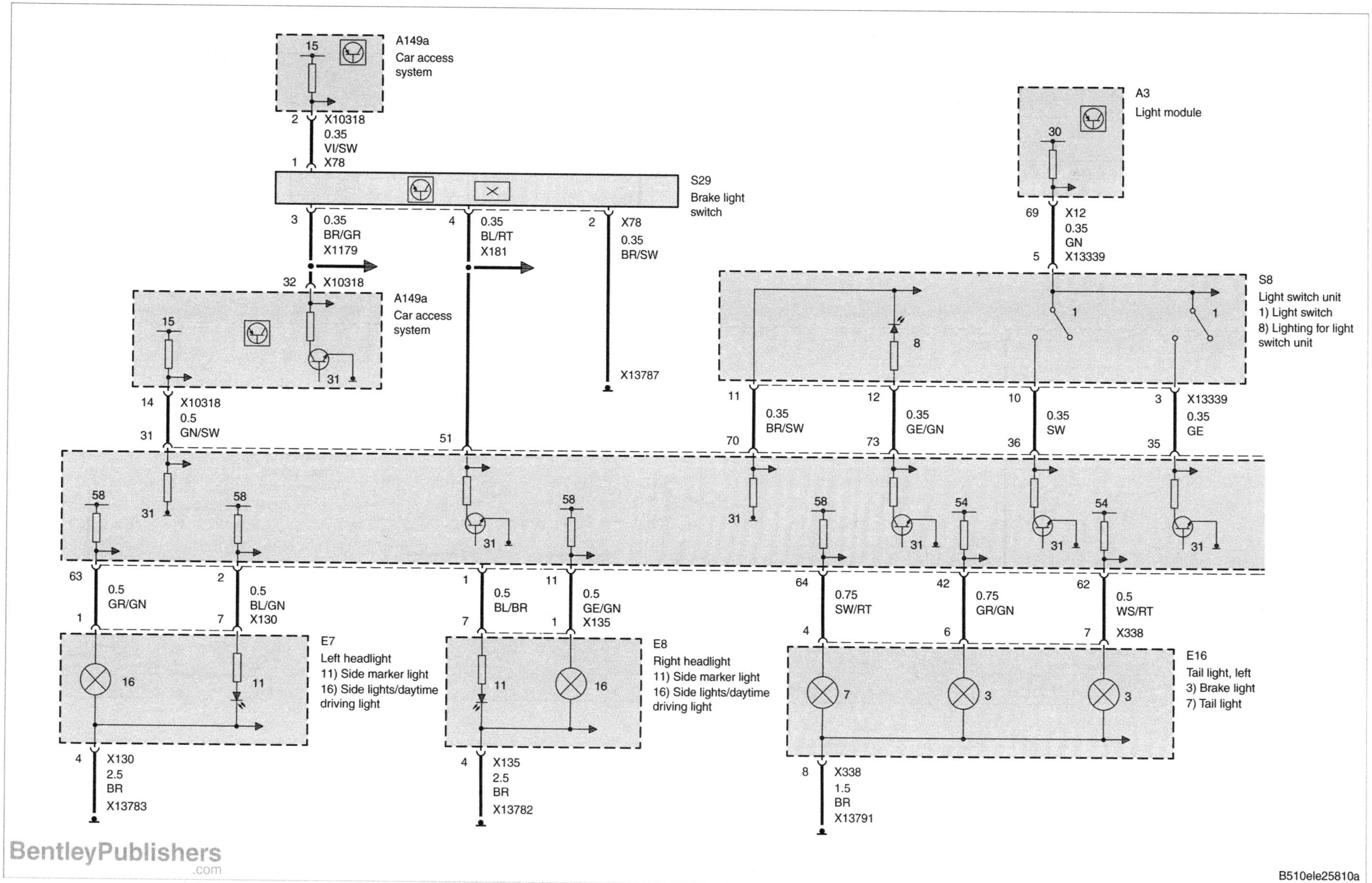

BentleyPublishers.com

B510ele25810a

Side light, parking light, brake light
Sedan as of 03 / 2007
(page 2 of 2)

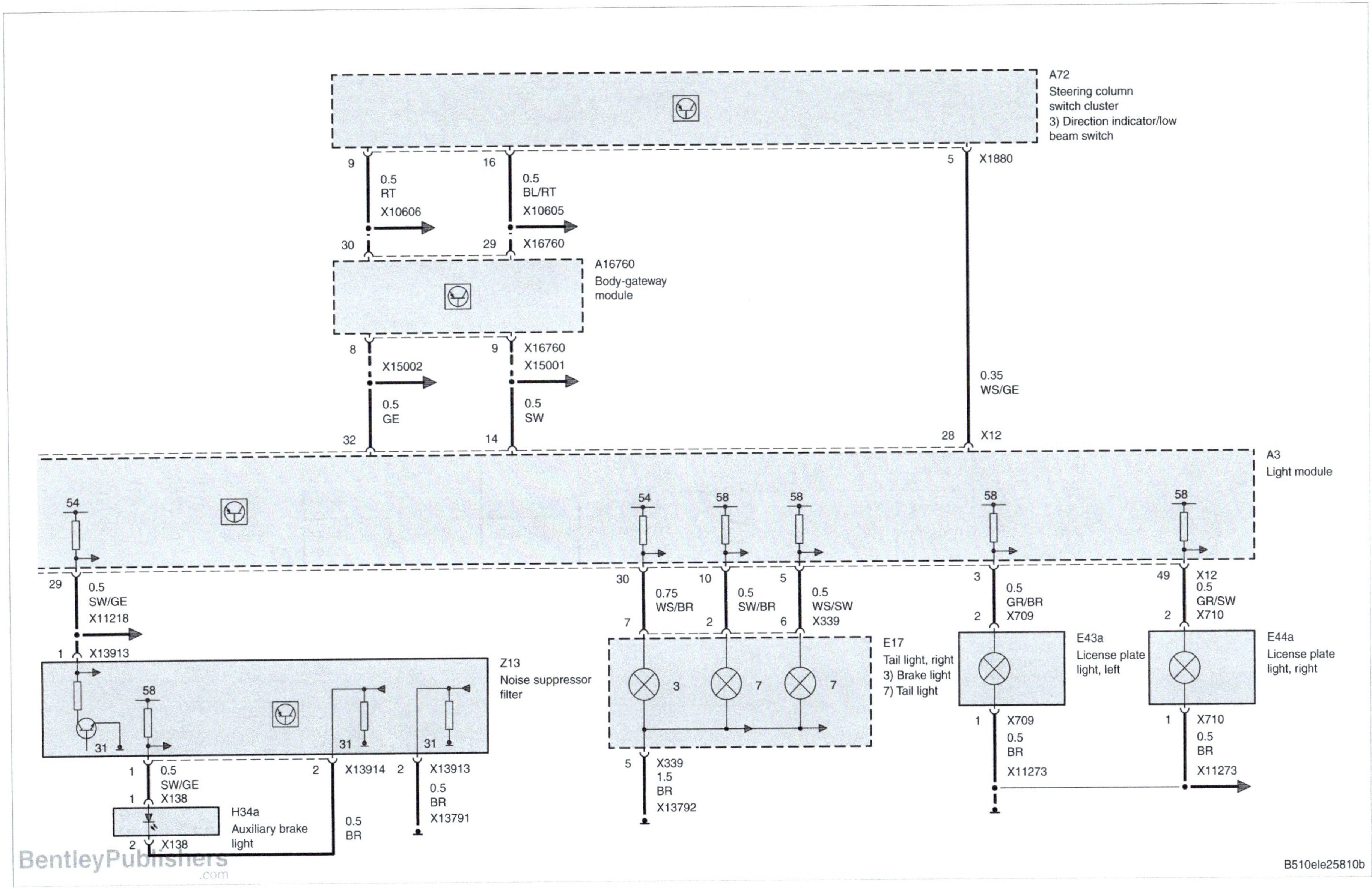

Side light, parking light, brake light
Wagon as of 03 / 2007
(page 1 of 2)

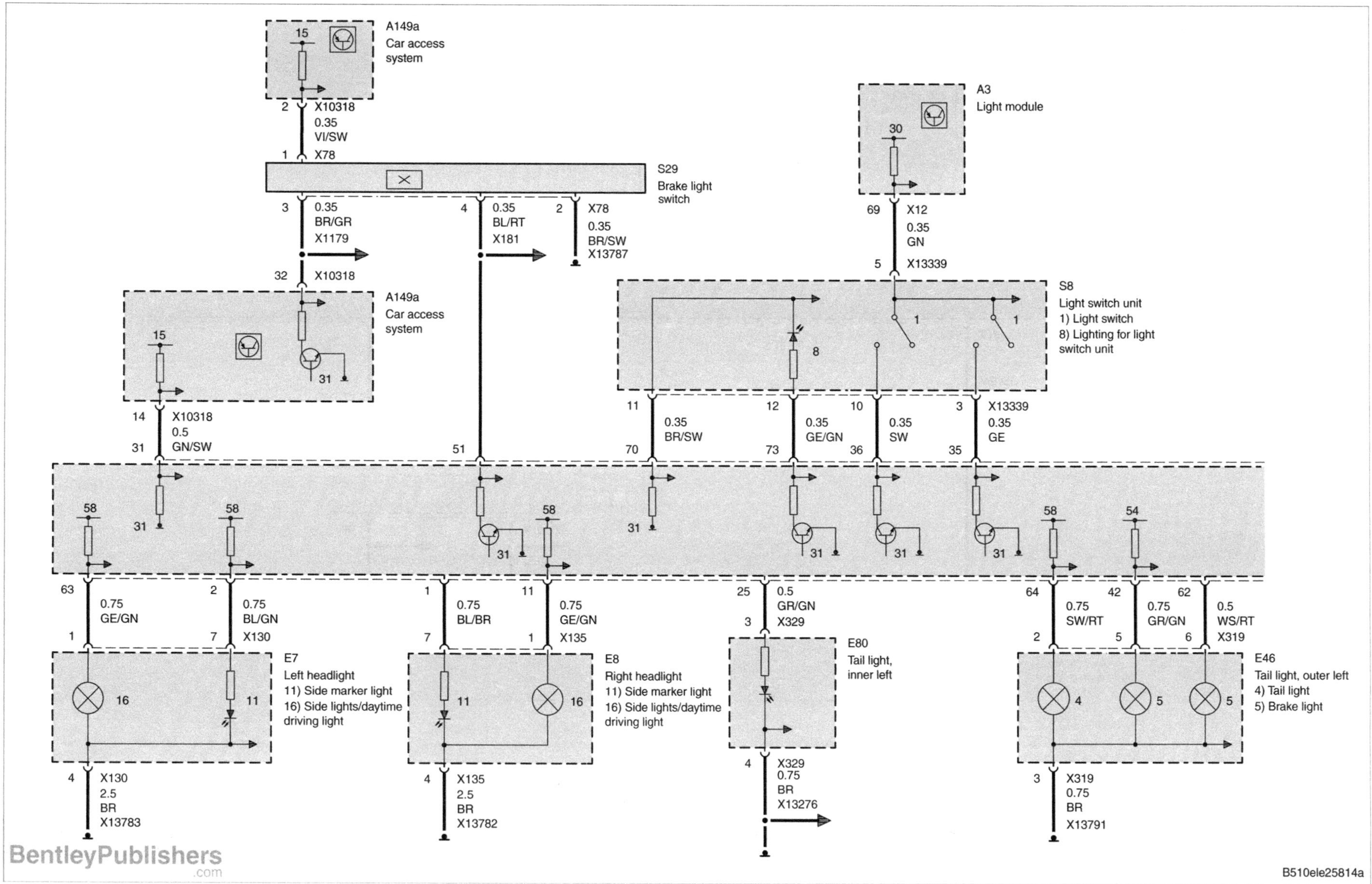

Side light, parking light, brake light
Wagon as of 03 / 2007
(page 2 of 2)

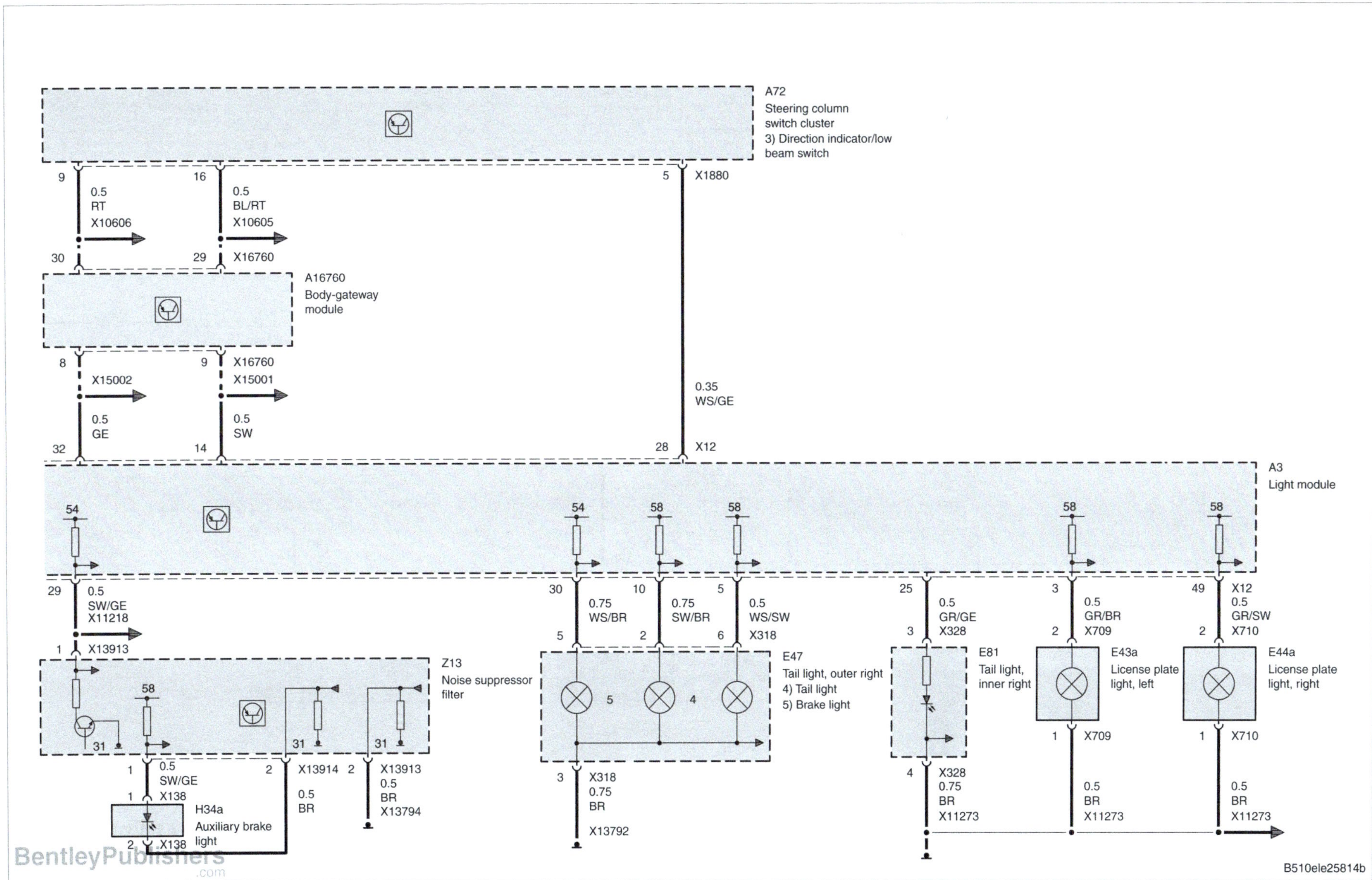

Direction indicator lights
Sedan up to 03 / 2005

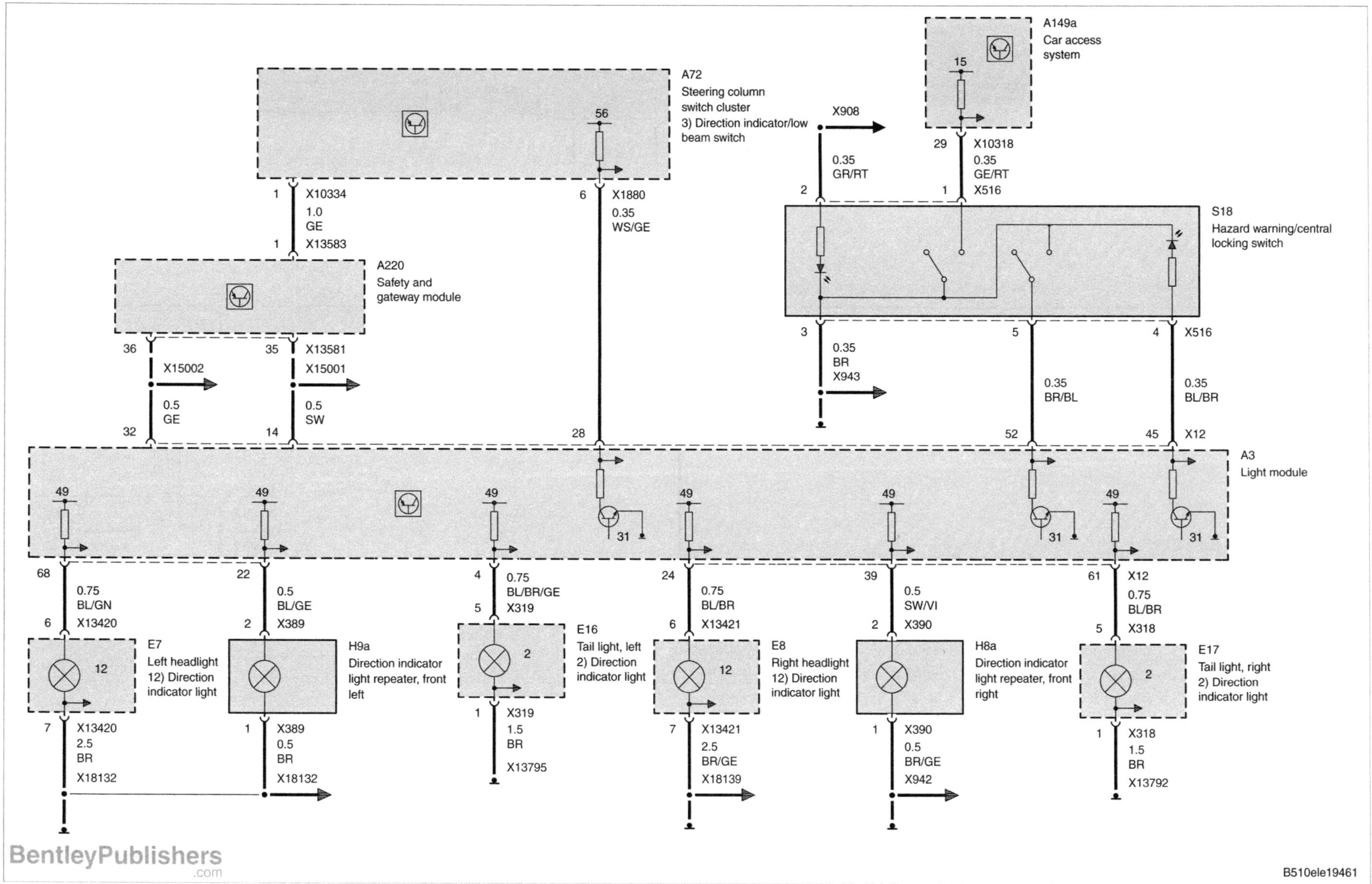

BentleyPublishers.com

Direction indicator lights
Sedan as of 03 / 2005 up to 09 / 2005

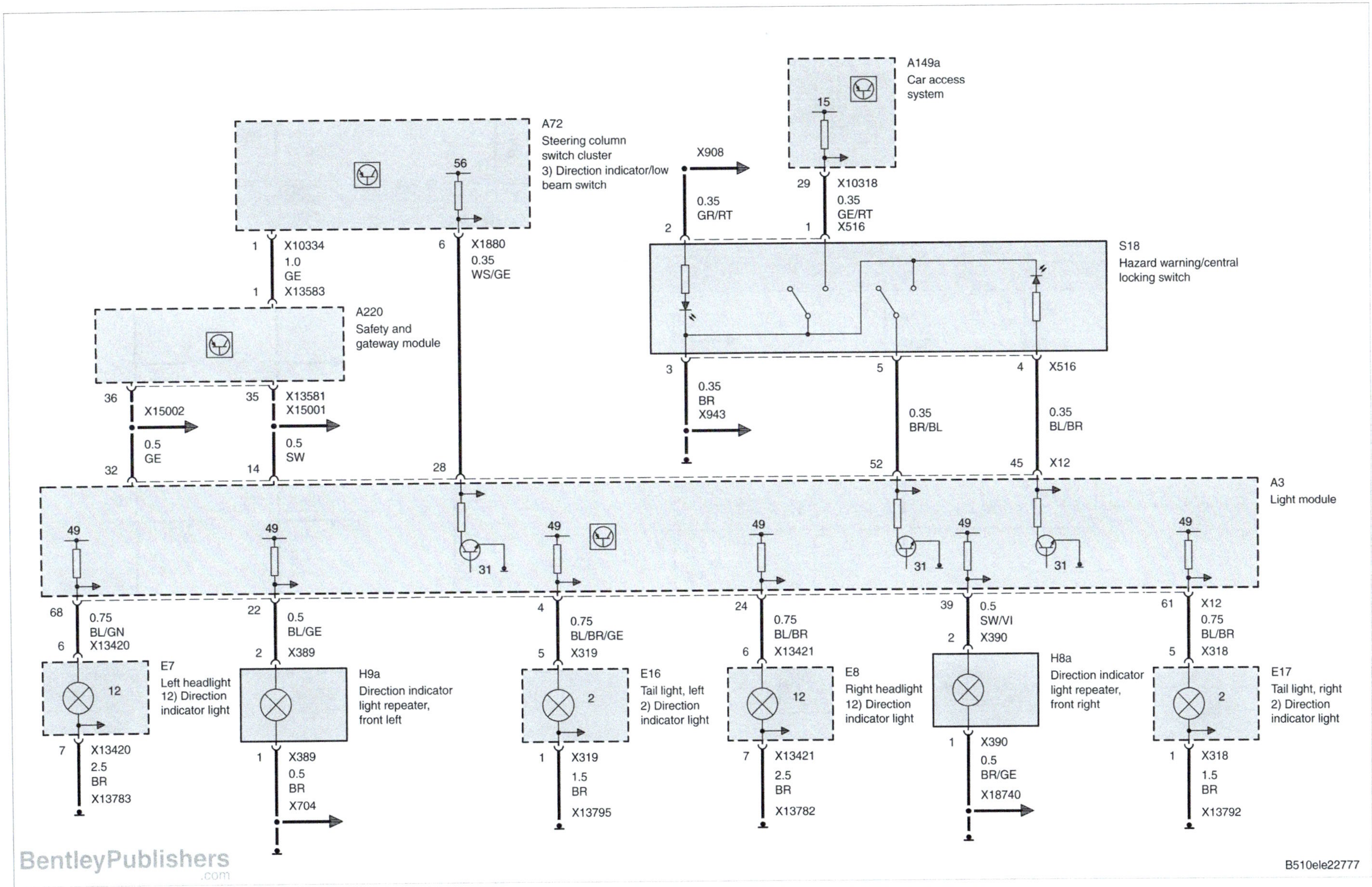

Direction indicator lights
Sedan as of 09 / 2005 up to 03 / 2007

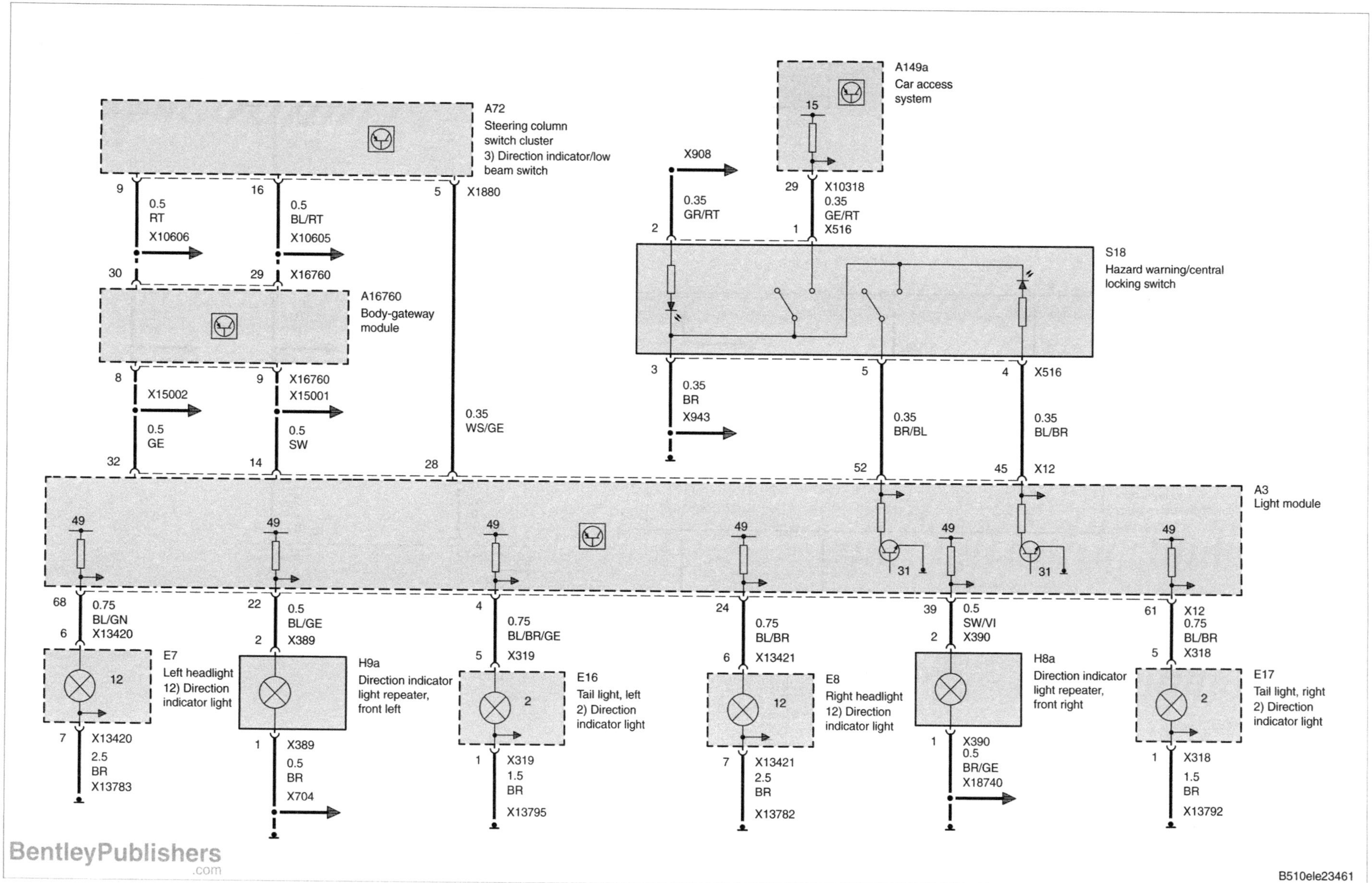

BentleyPublishers.com

Direction indicator lights
Sedan as of 03 / 2007

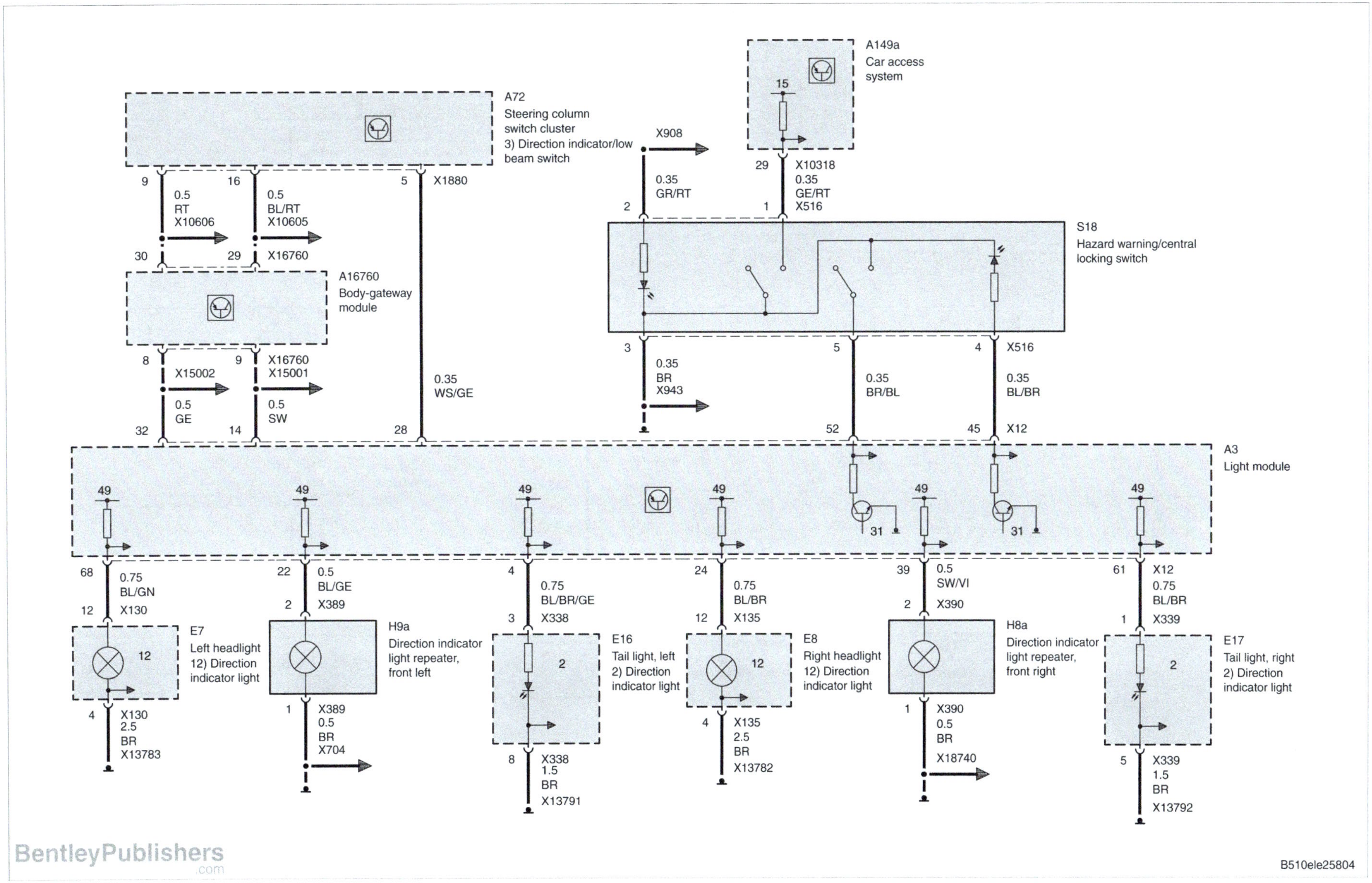

BentleyPublishers.com

Direction indicator lights
Wagon up to 03 / 2005

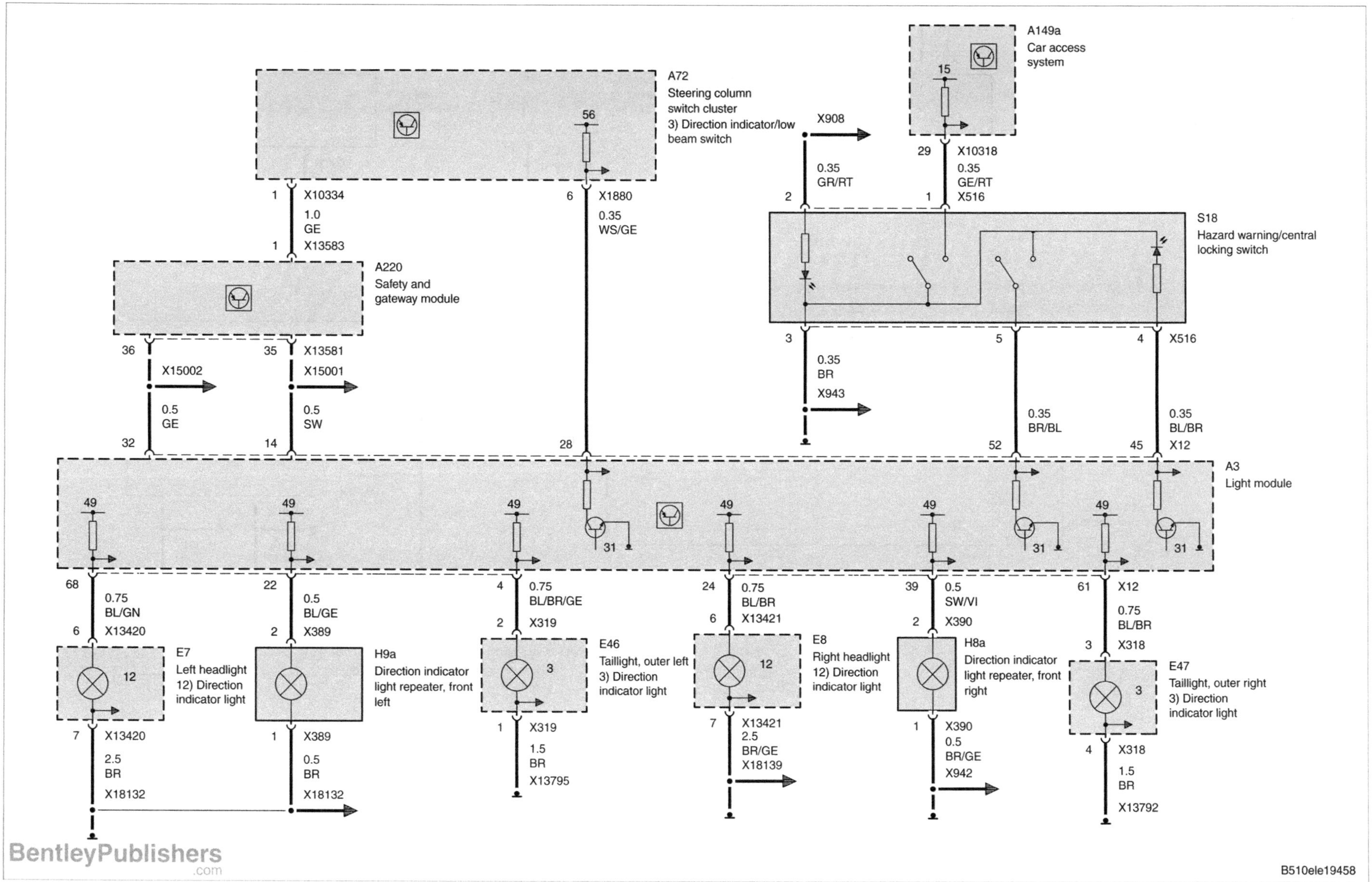

BentleyPublishers.com

Direction indicator lights
Wagon as of 03 / 2005 up to 09 / 2005

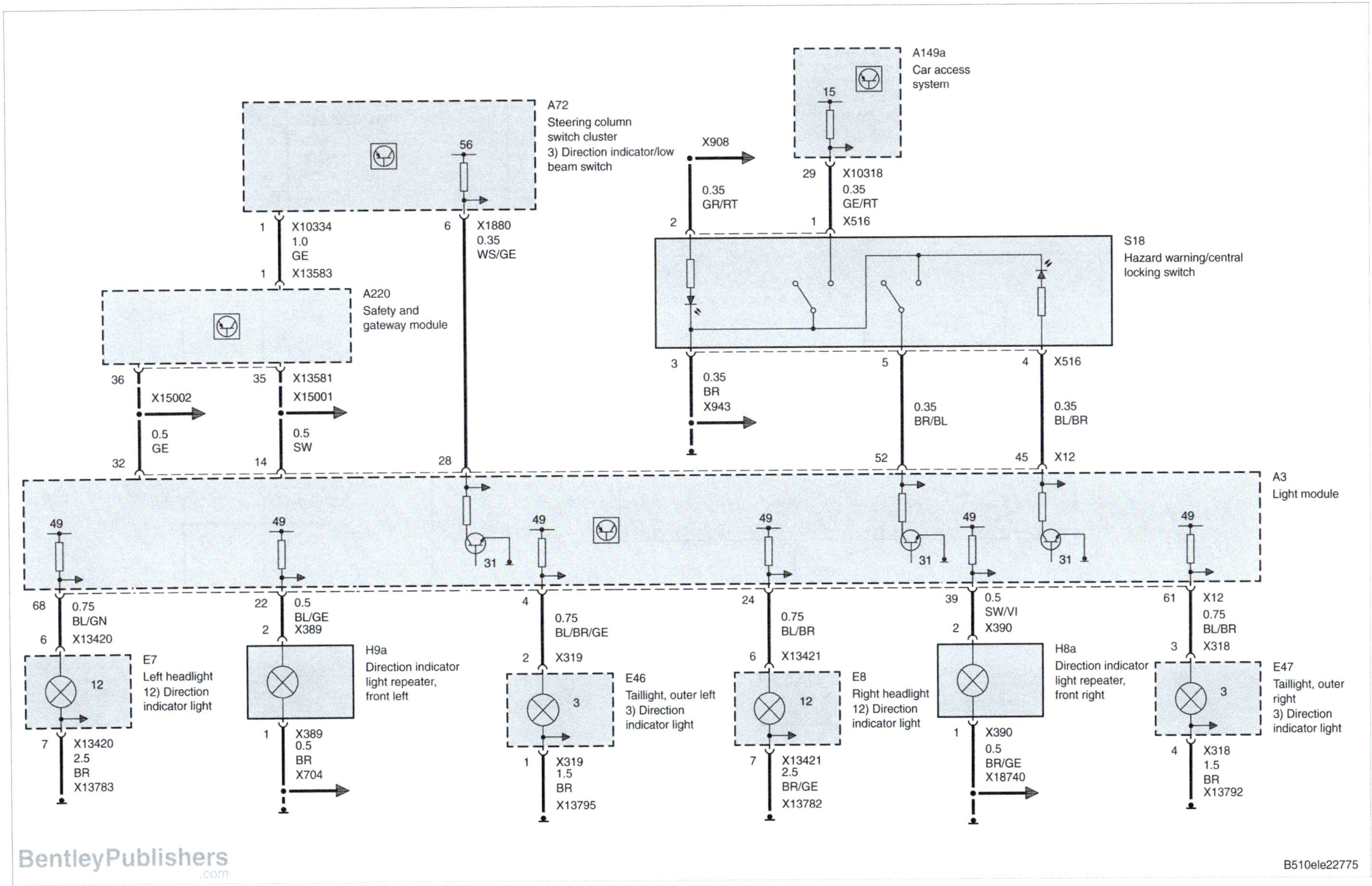

Direction indicator lights
Wagon as of 09 / 2005 up to 03 / 2007

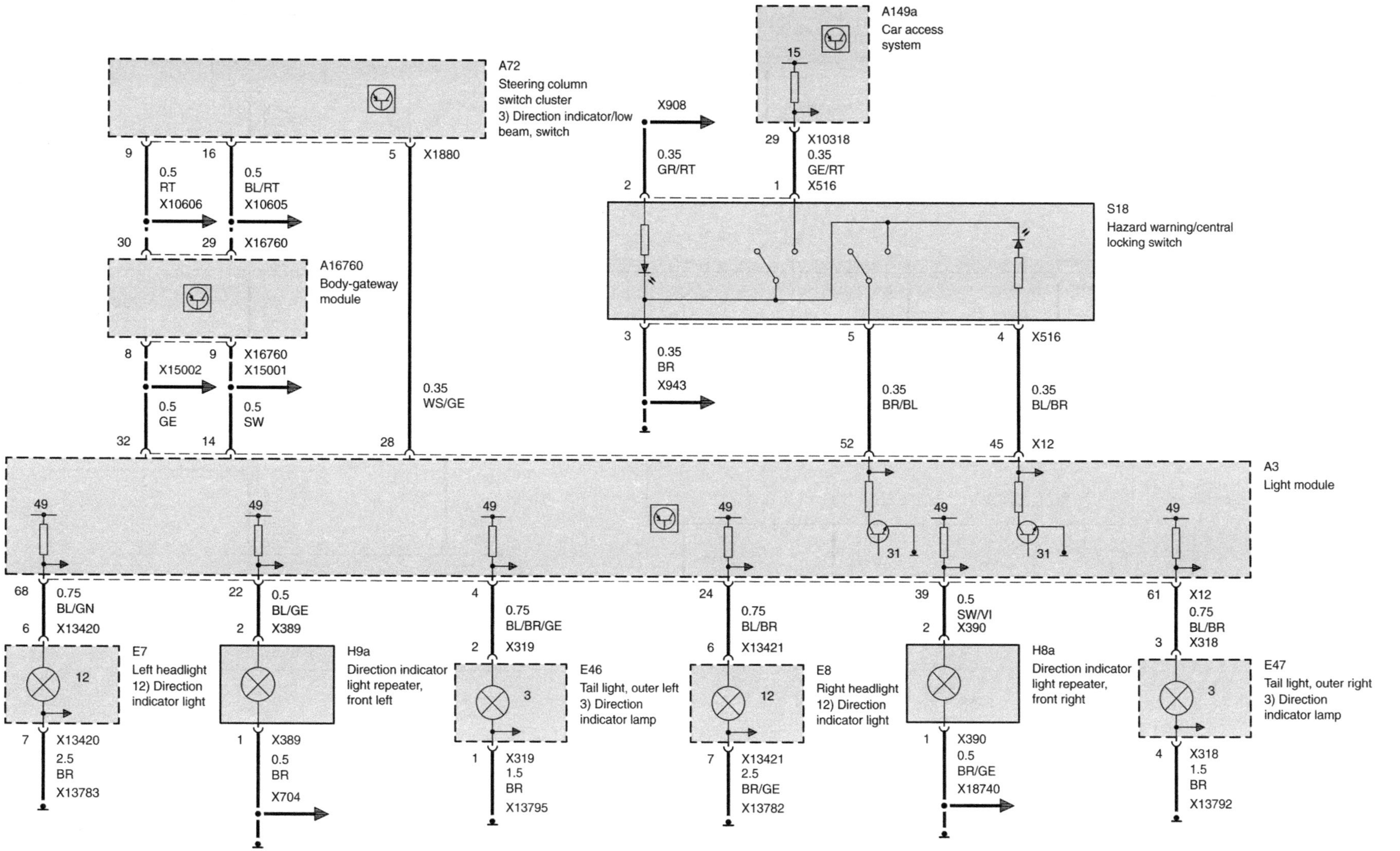

BentleyPublishers.com

B510ele23468

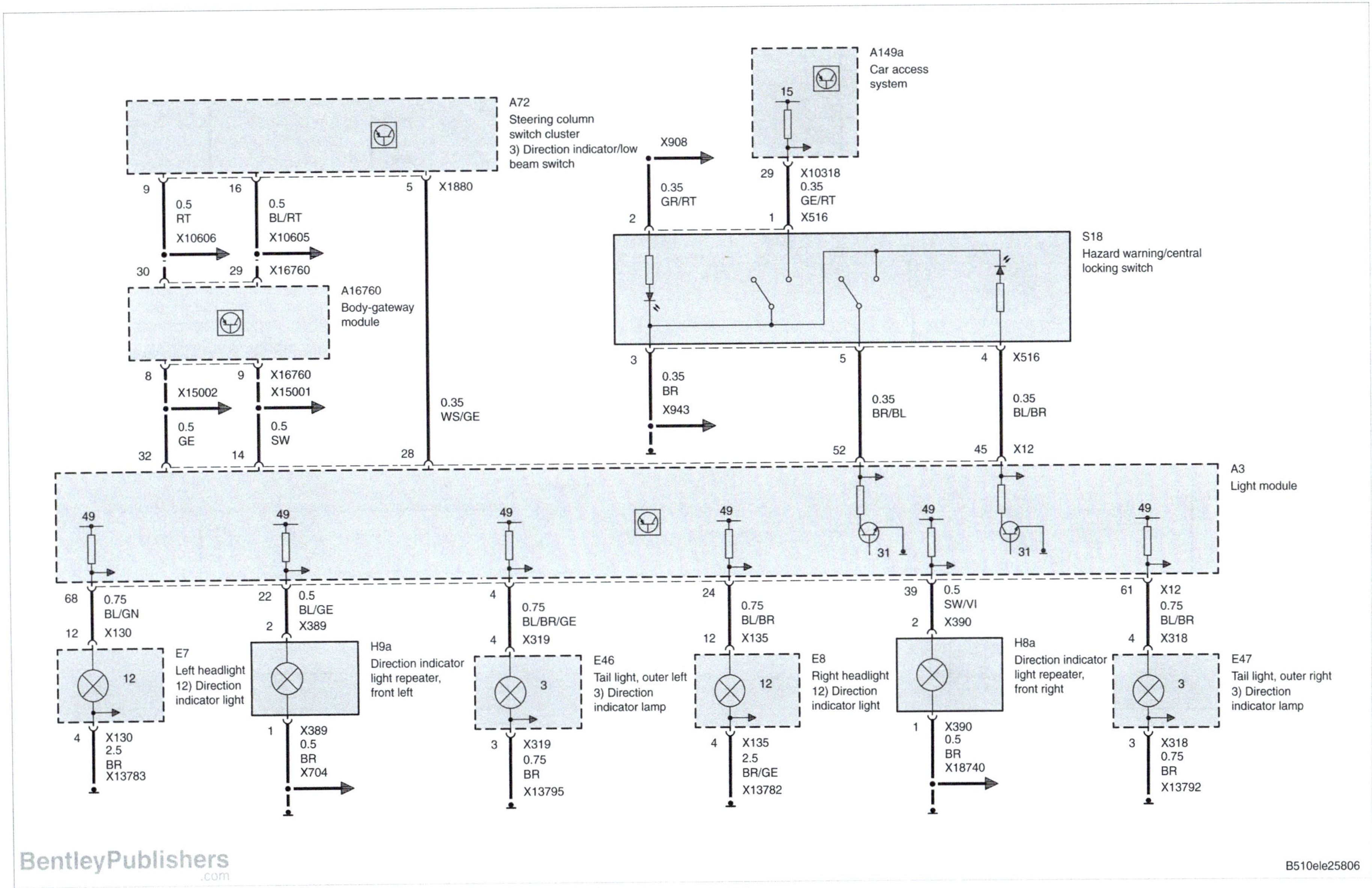

BentleyPublishers.com

B510ele25806

Backup lights
M54 engine with manual transmission

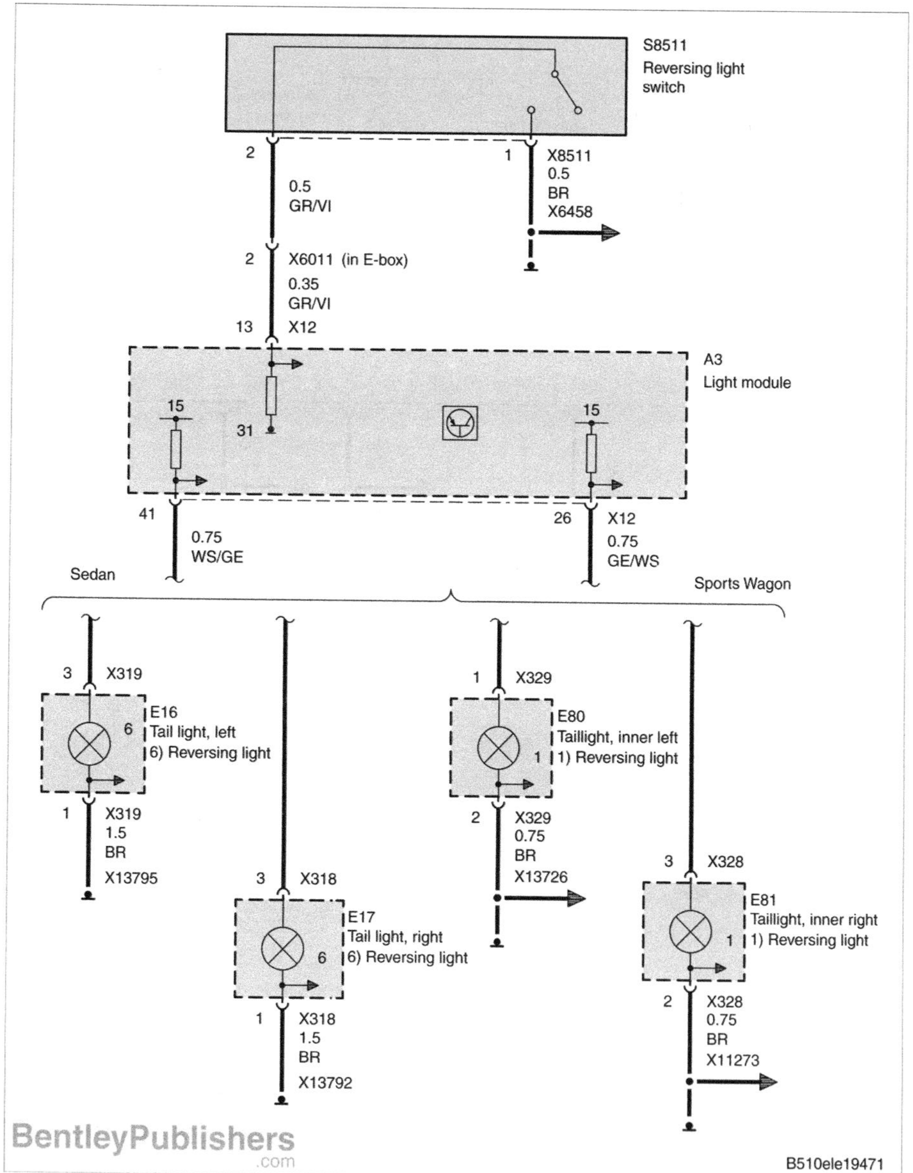

B510ele19471

Backup lights
Sedan with N62 engine with manual transmission

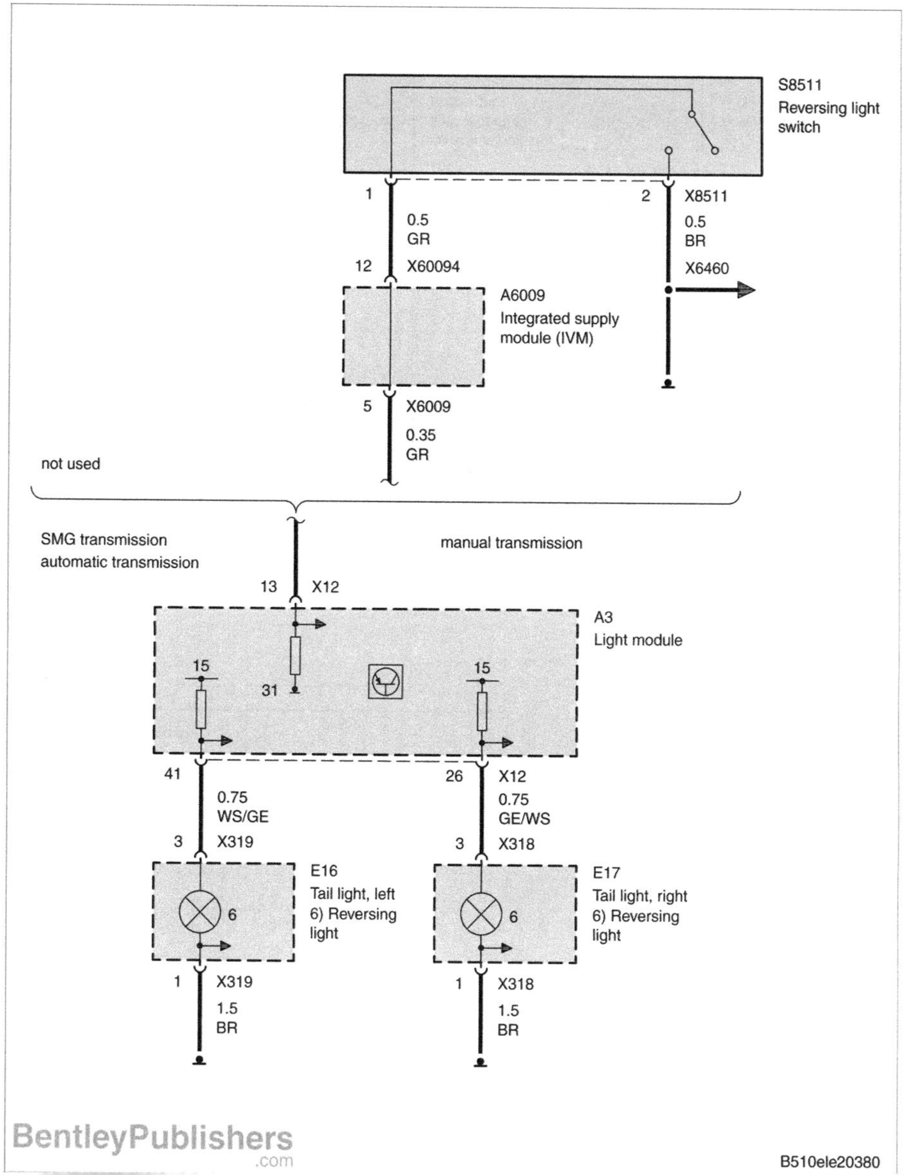

B510ele20380

Backup lights
Wagon with N62 engine with manual transmission

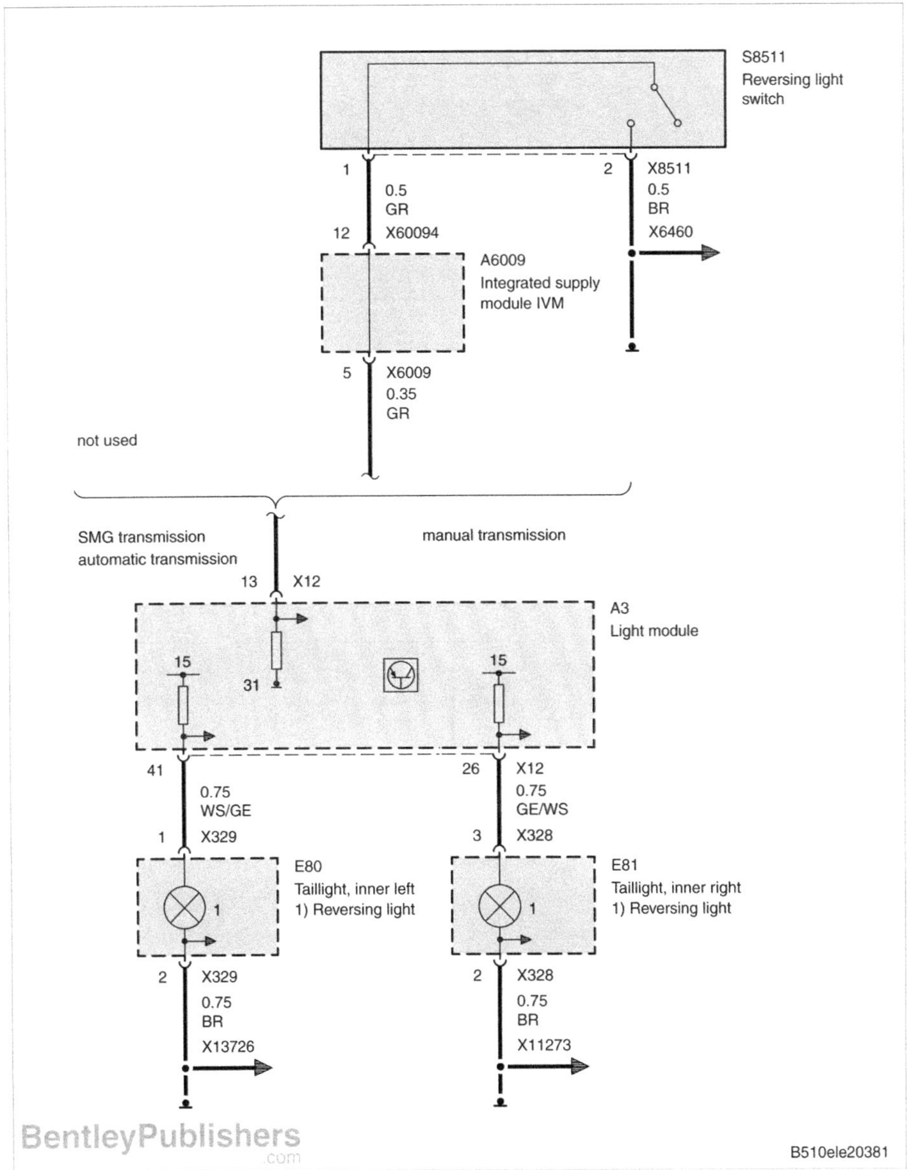

Backup lights
N52 engine up to 09 / 2007

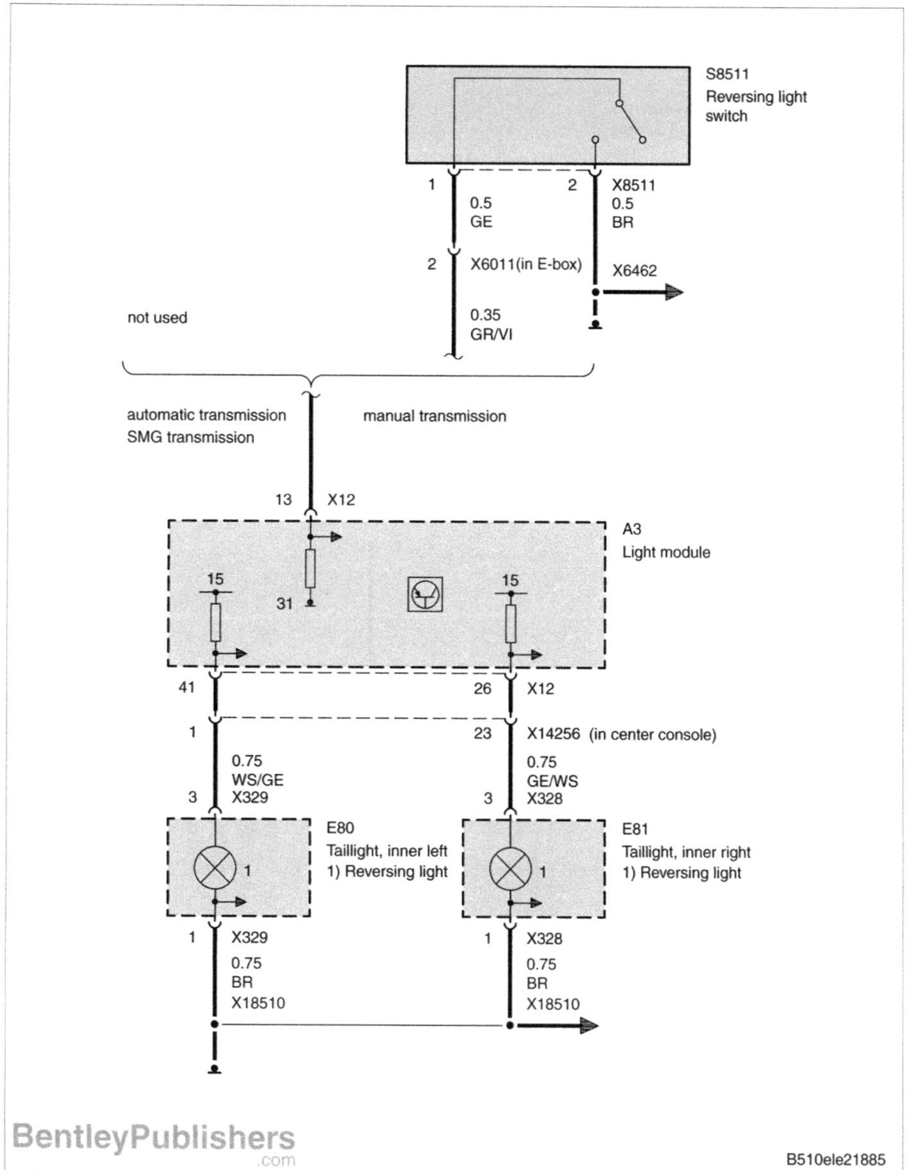

Backup lights
N52 engine as of 09 / 2007

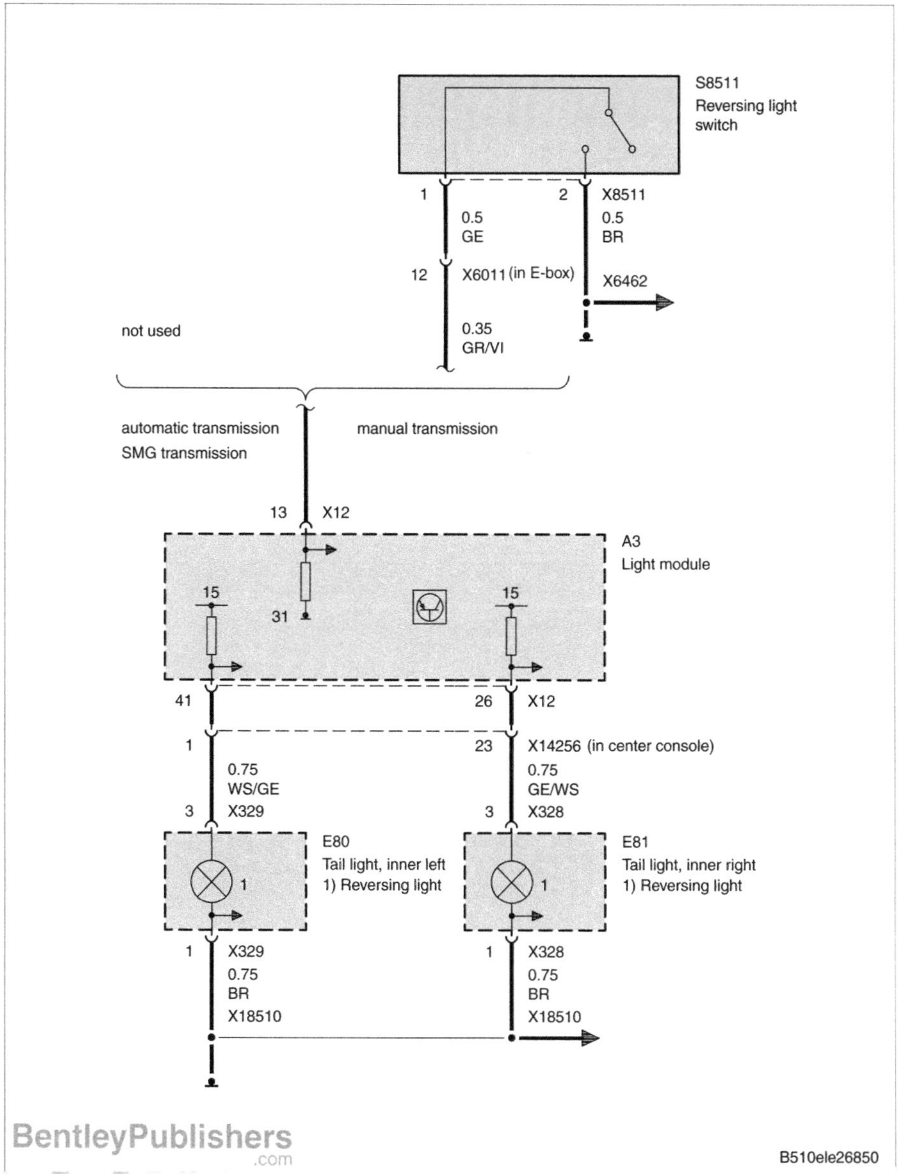

Backup lights automatic transmission

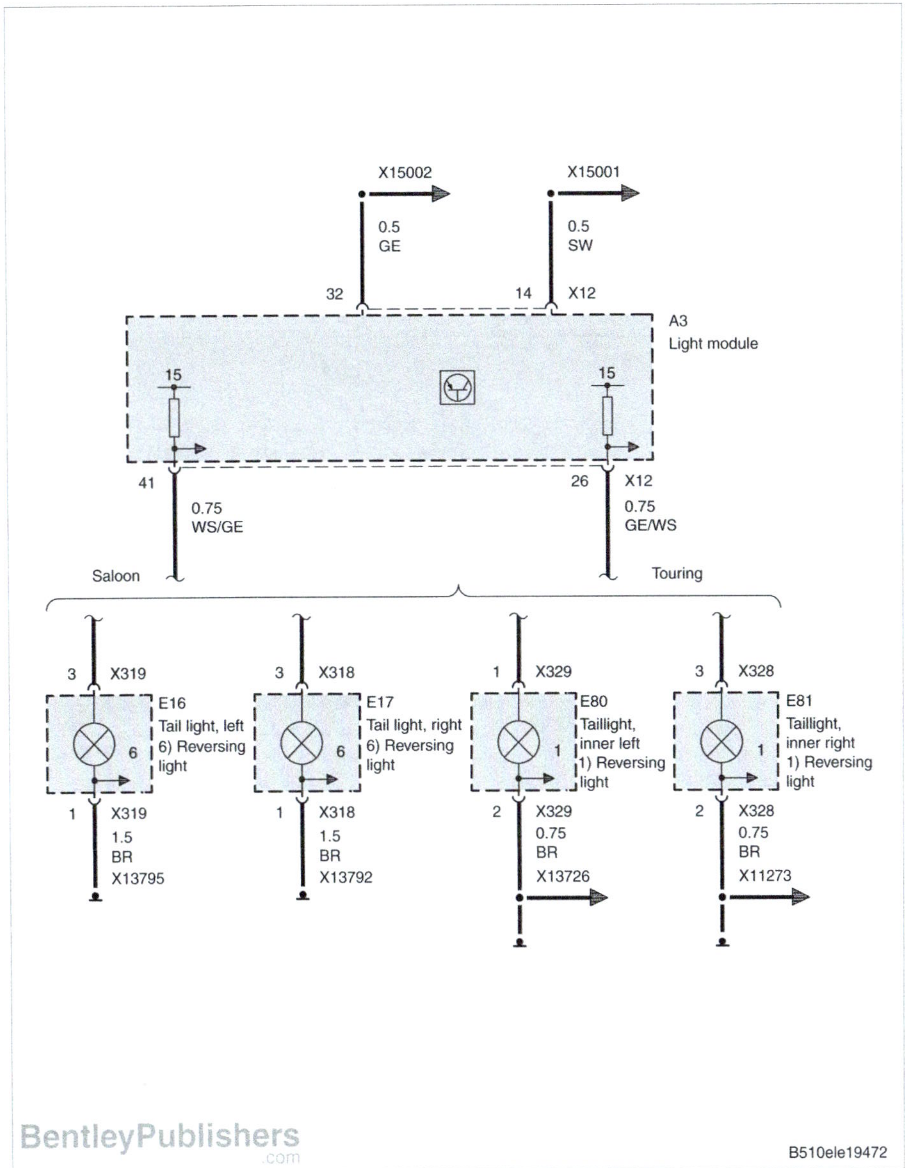

Backup lights automatic transmission as of 03 / 2007

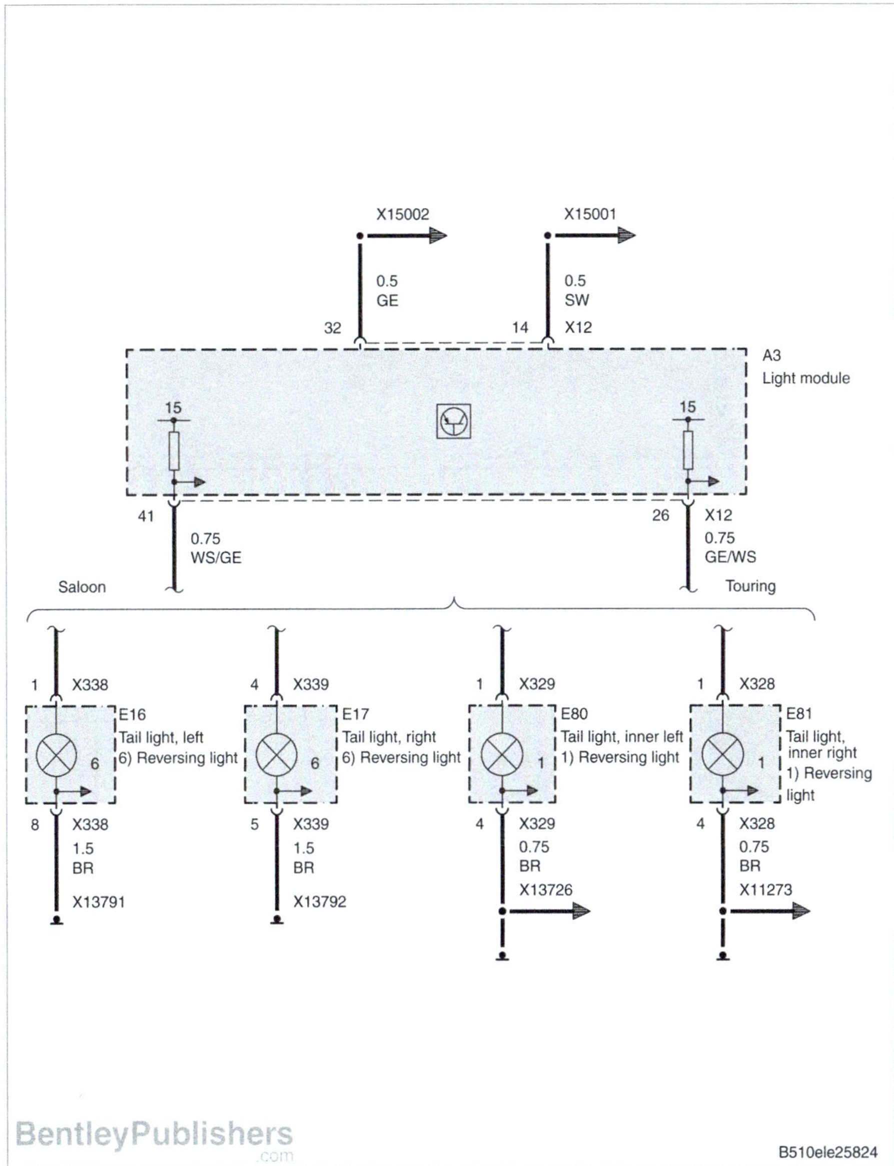

Fog lights
up to 03 / 2005

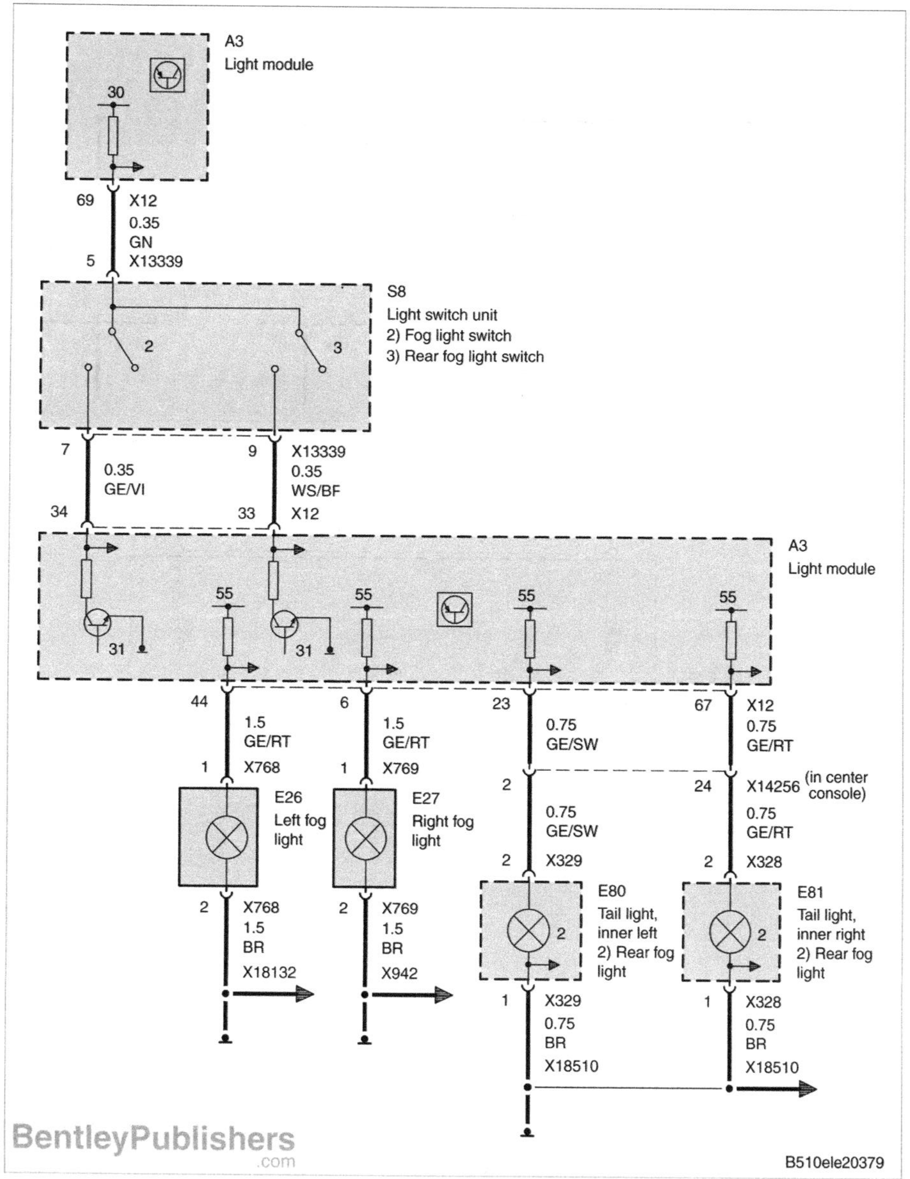

Fog lights
from 03 / 2005 up to 09 / 2007

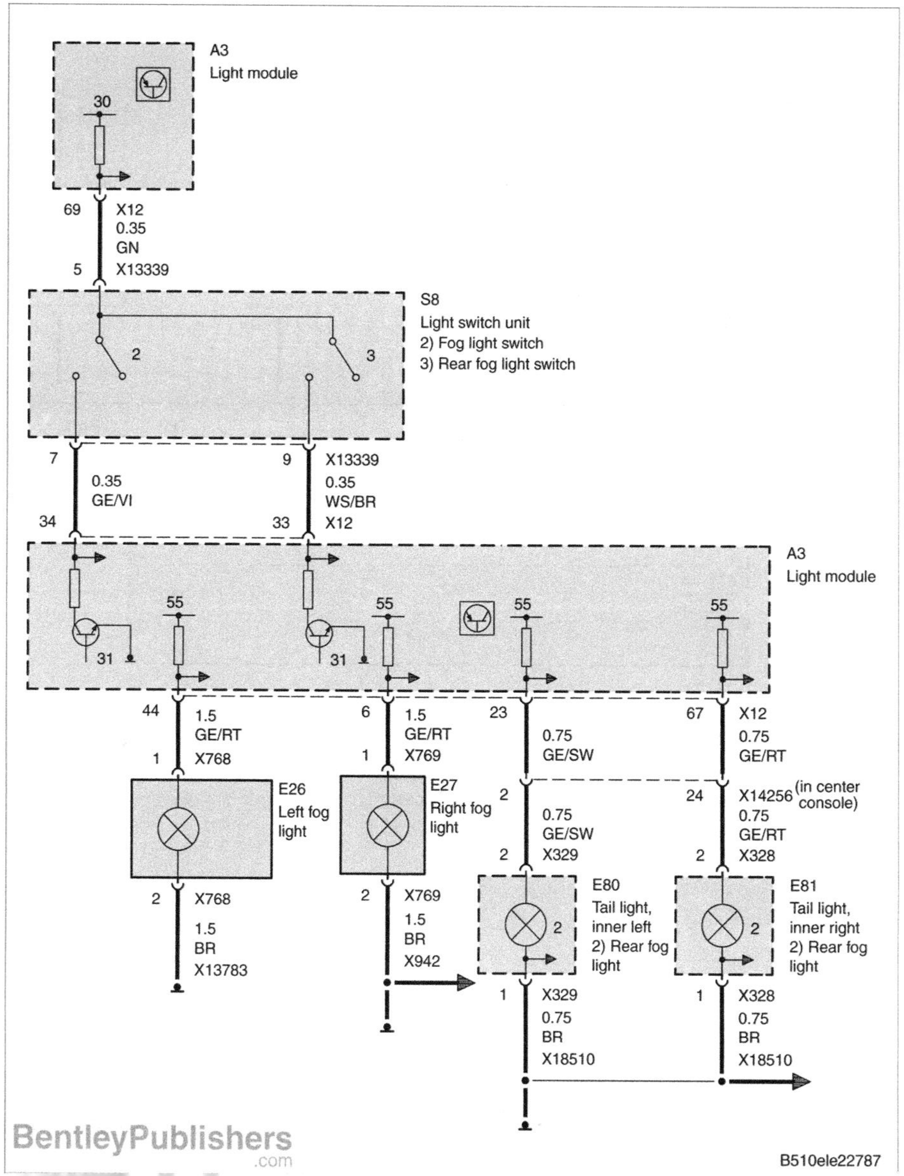

Fog lights as of 09 / 2007

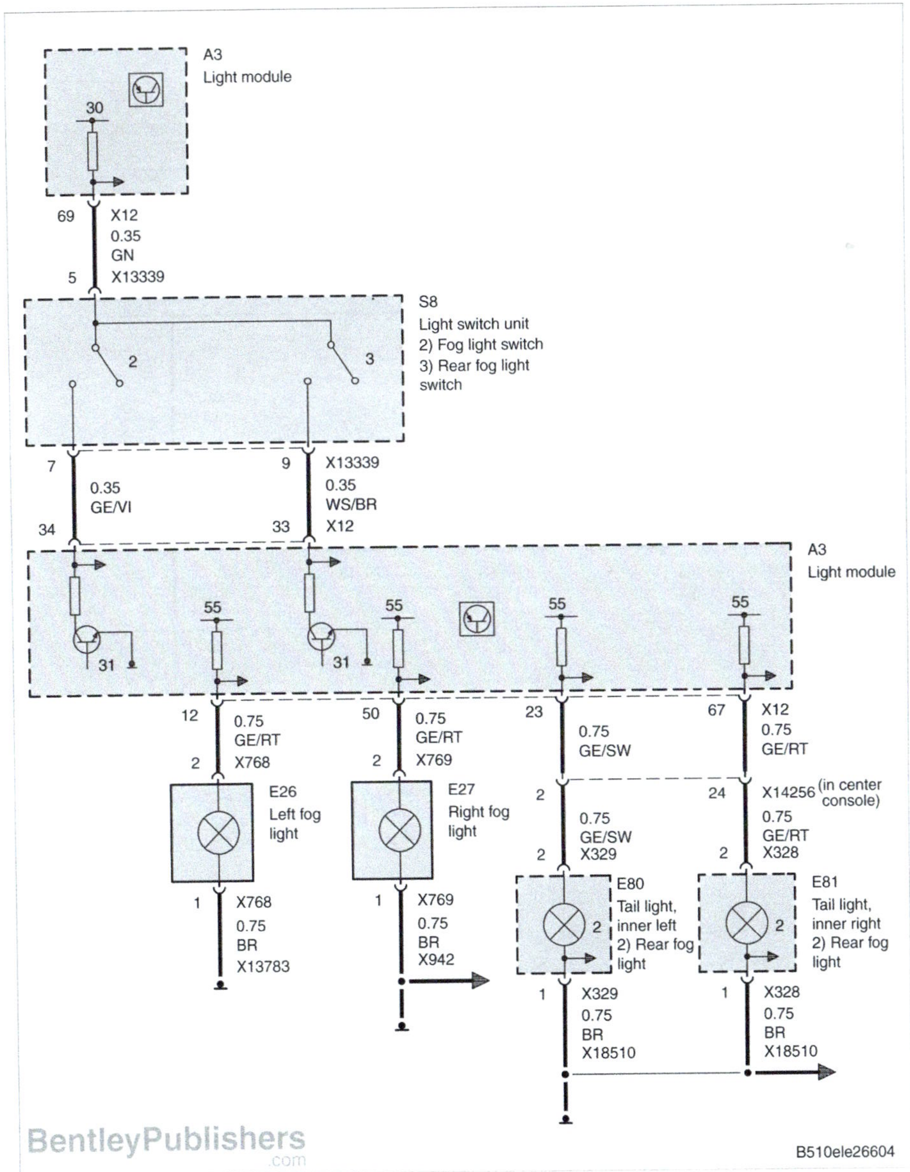

S8 Light switch unit

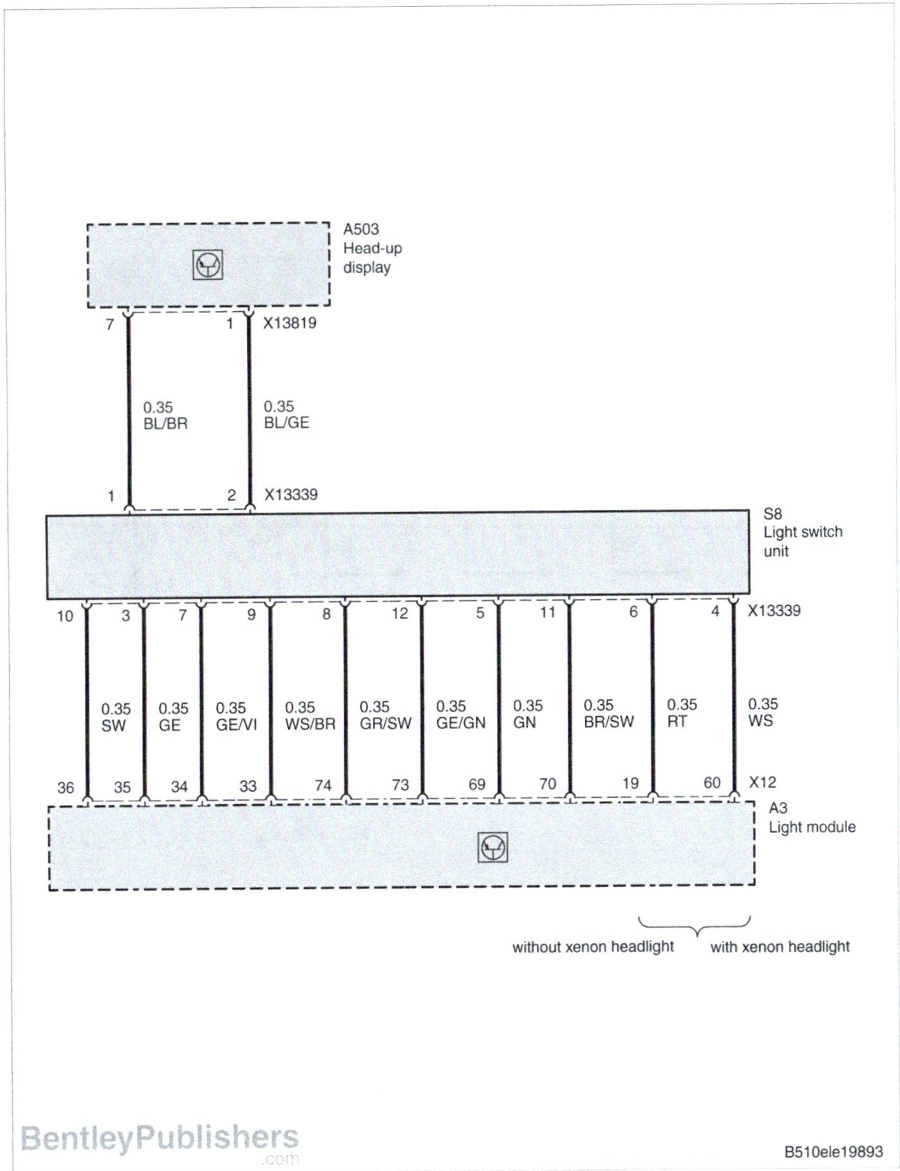

Low beam / rain sensor up to 03 / 2005 RLS Rain / light sensor

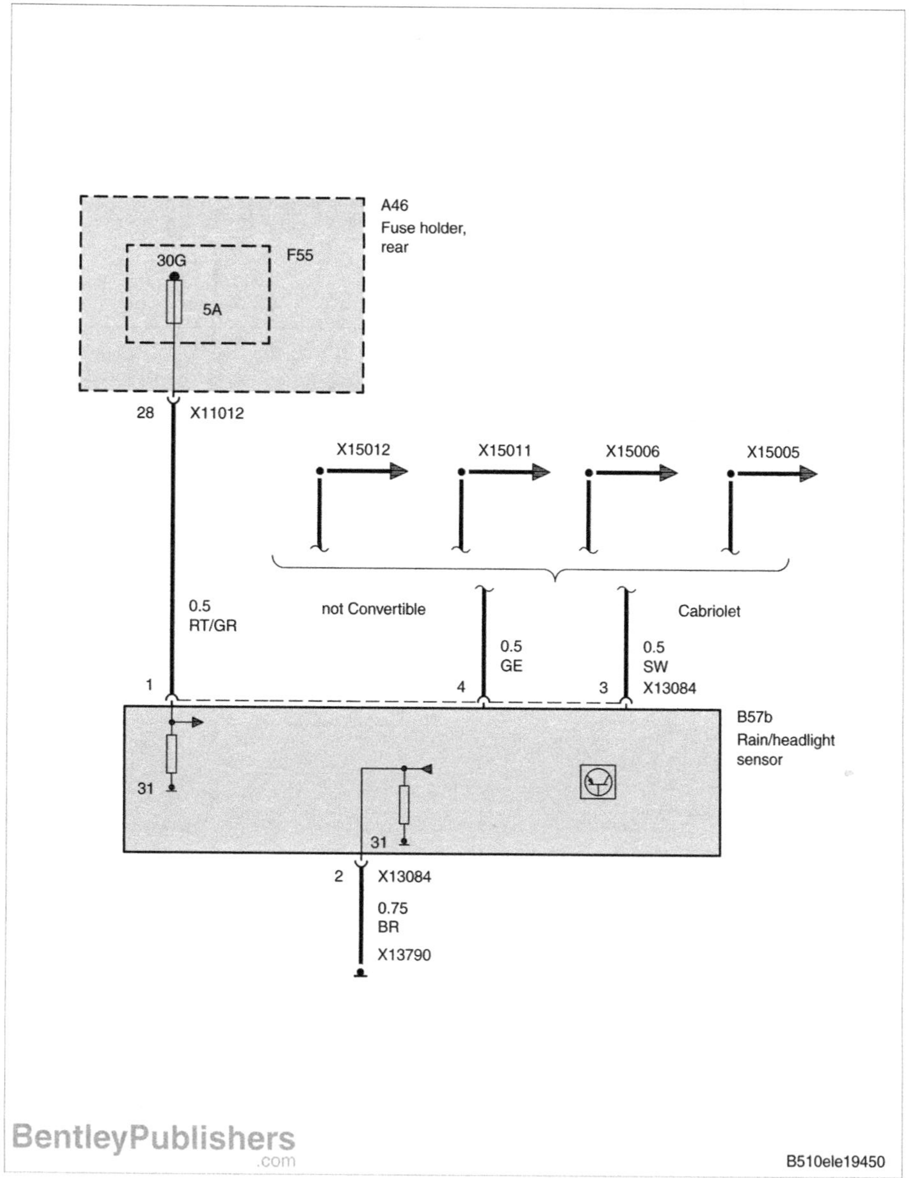

Low beam / rain sensor from 03 / 2005 to 09 / 2005 RLS Rain / light sensor

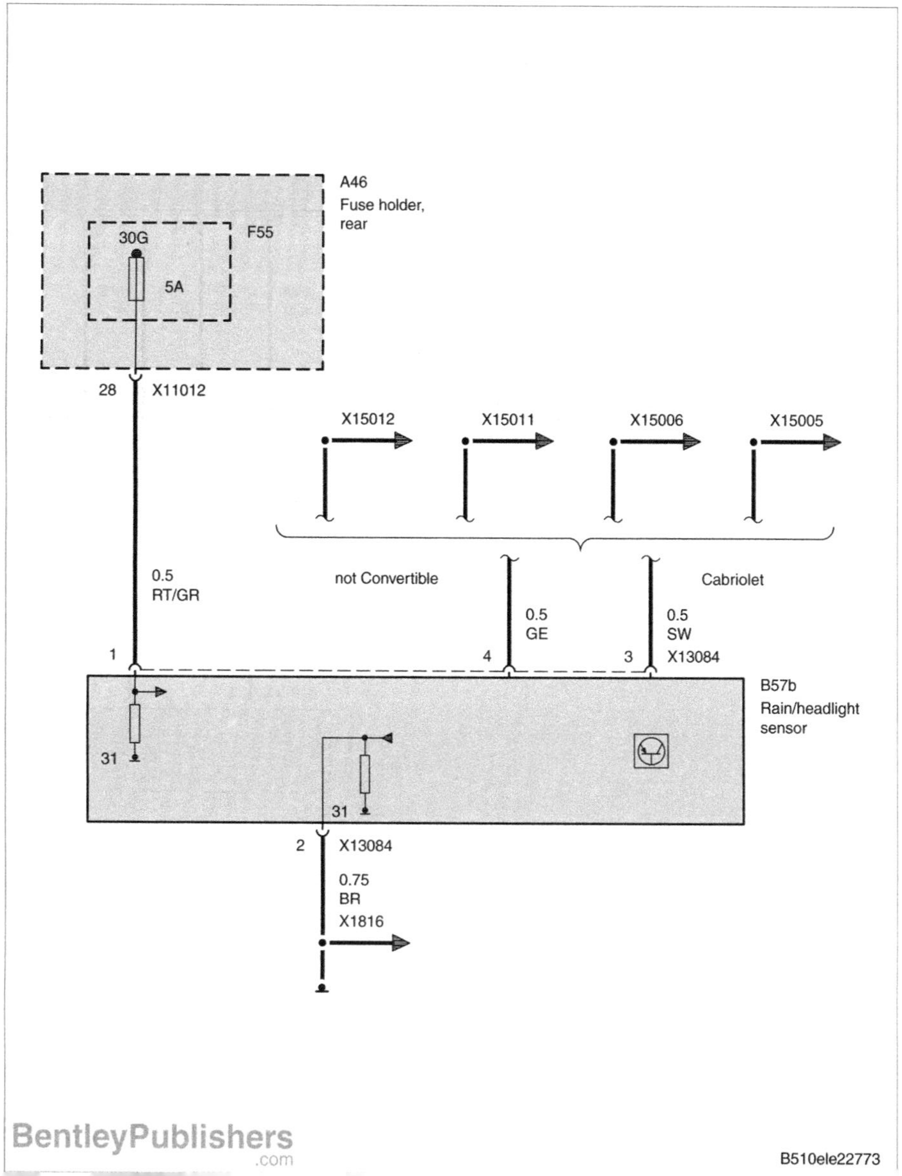

Low beam / rain sensor from 09 / 2005 RLS Rain / light sensor

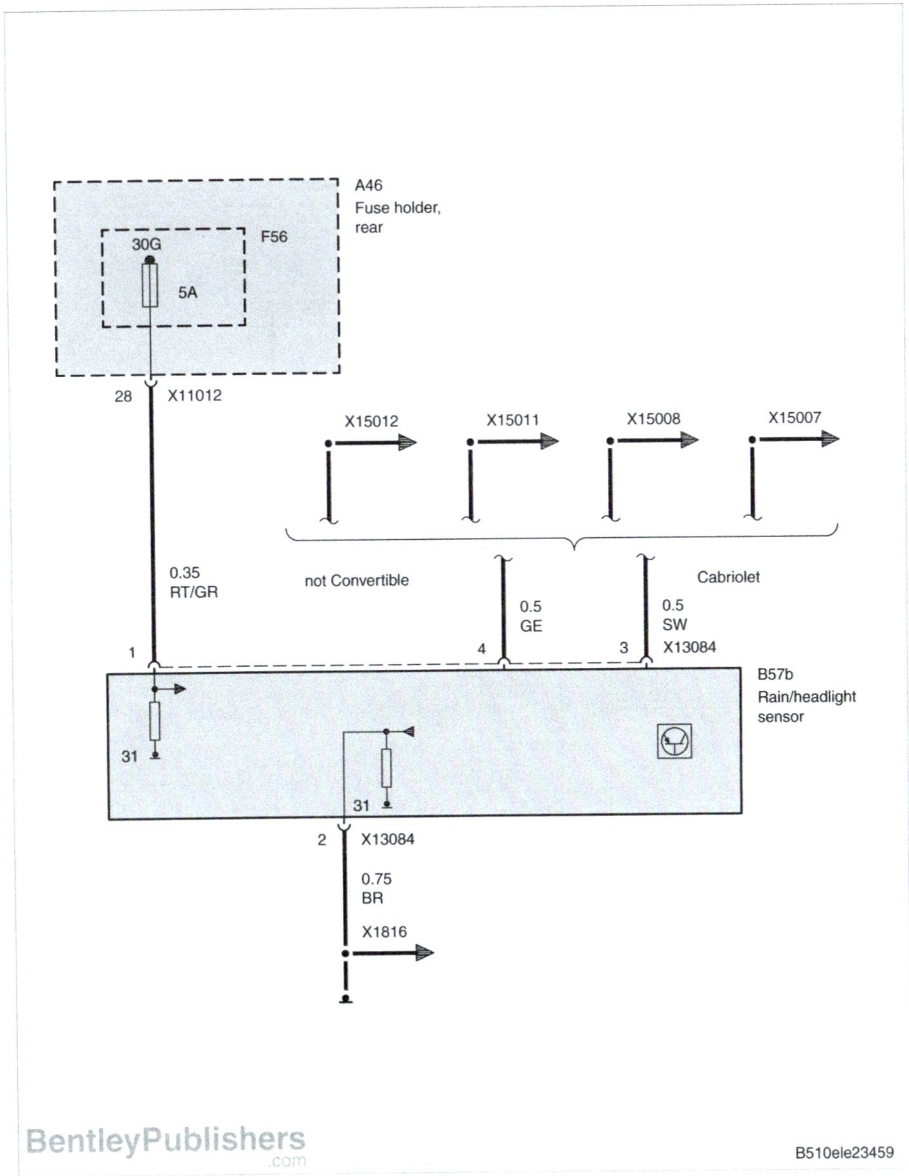

Low beam / rain sensor RLS Rain / light solar sensor

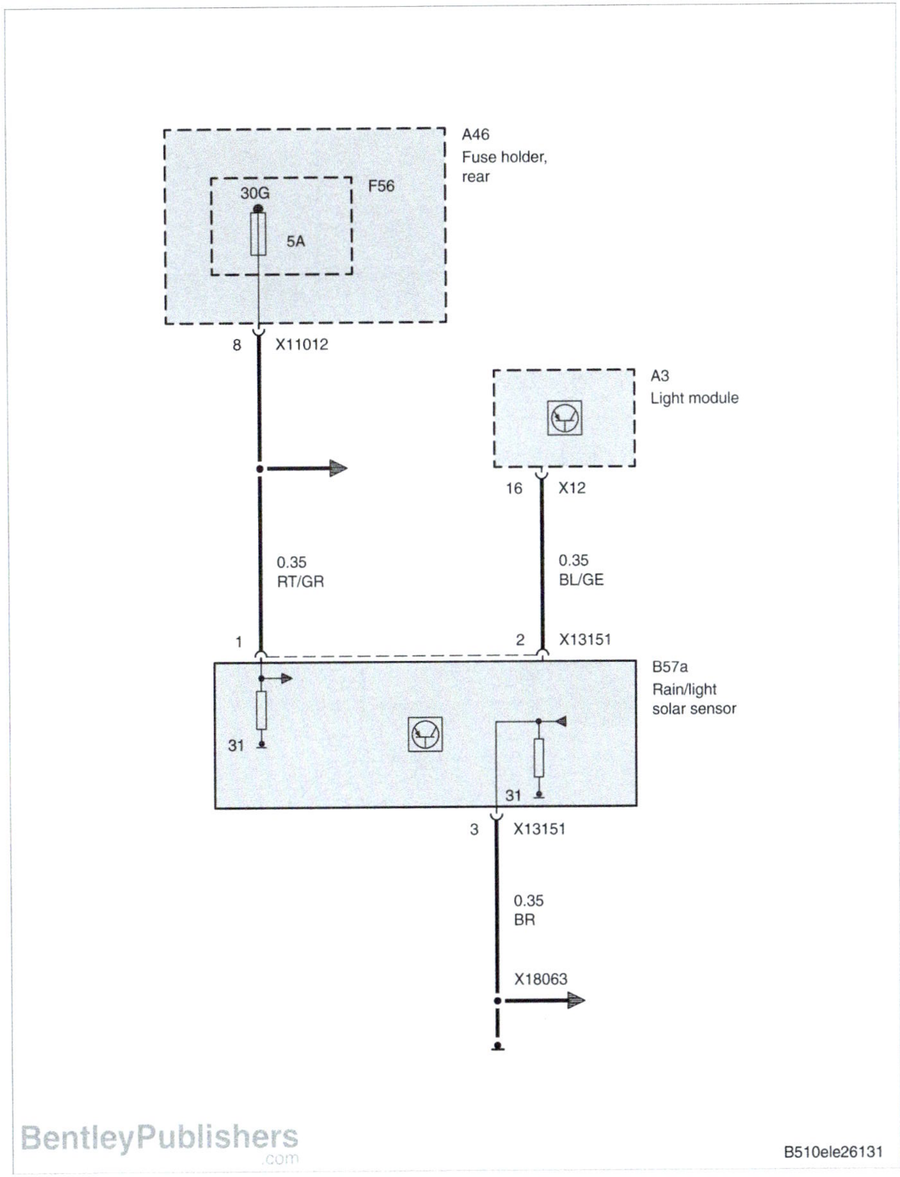

Electrochromic mirror
Sedan and Wagon

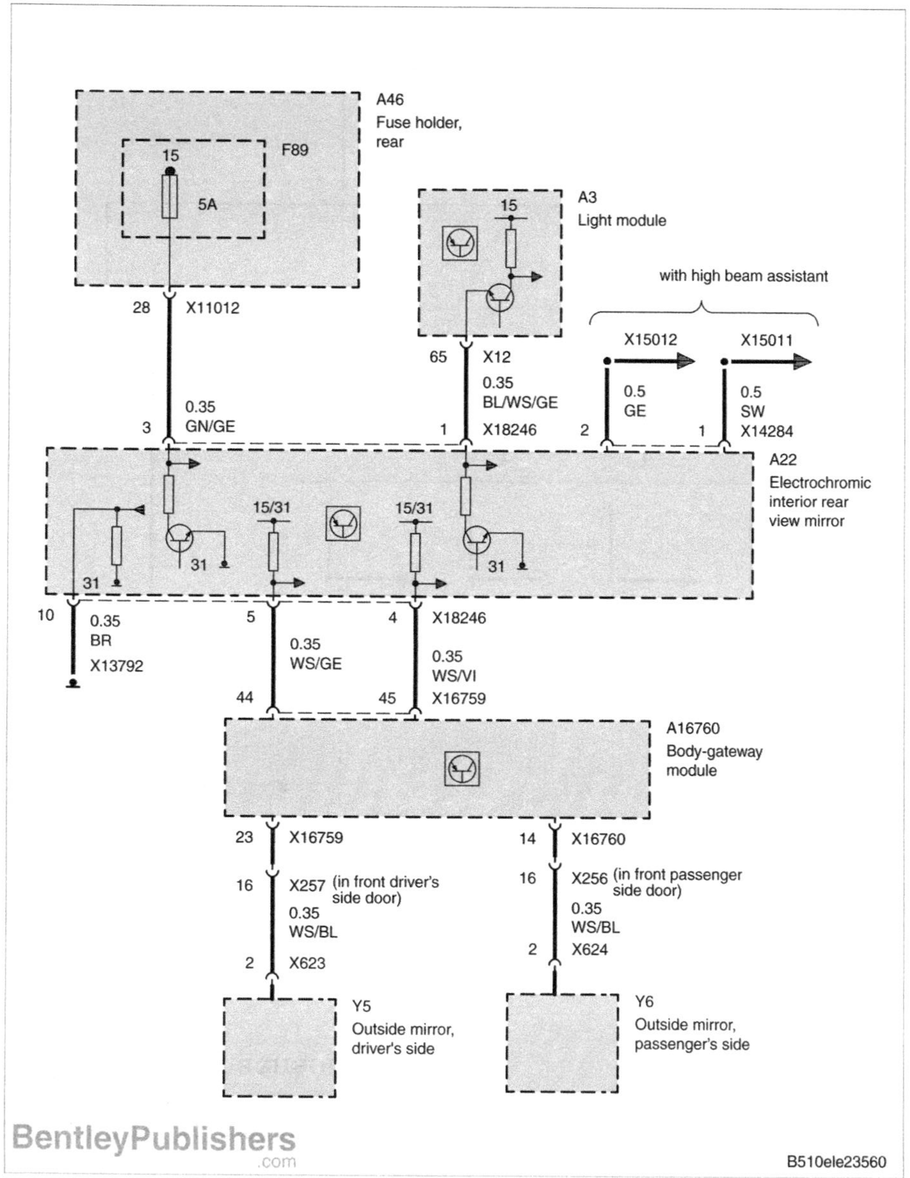

Electrochromic mirror
Coupe and Convertible

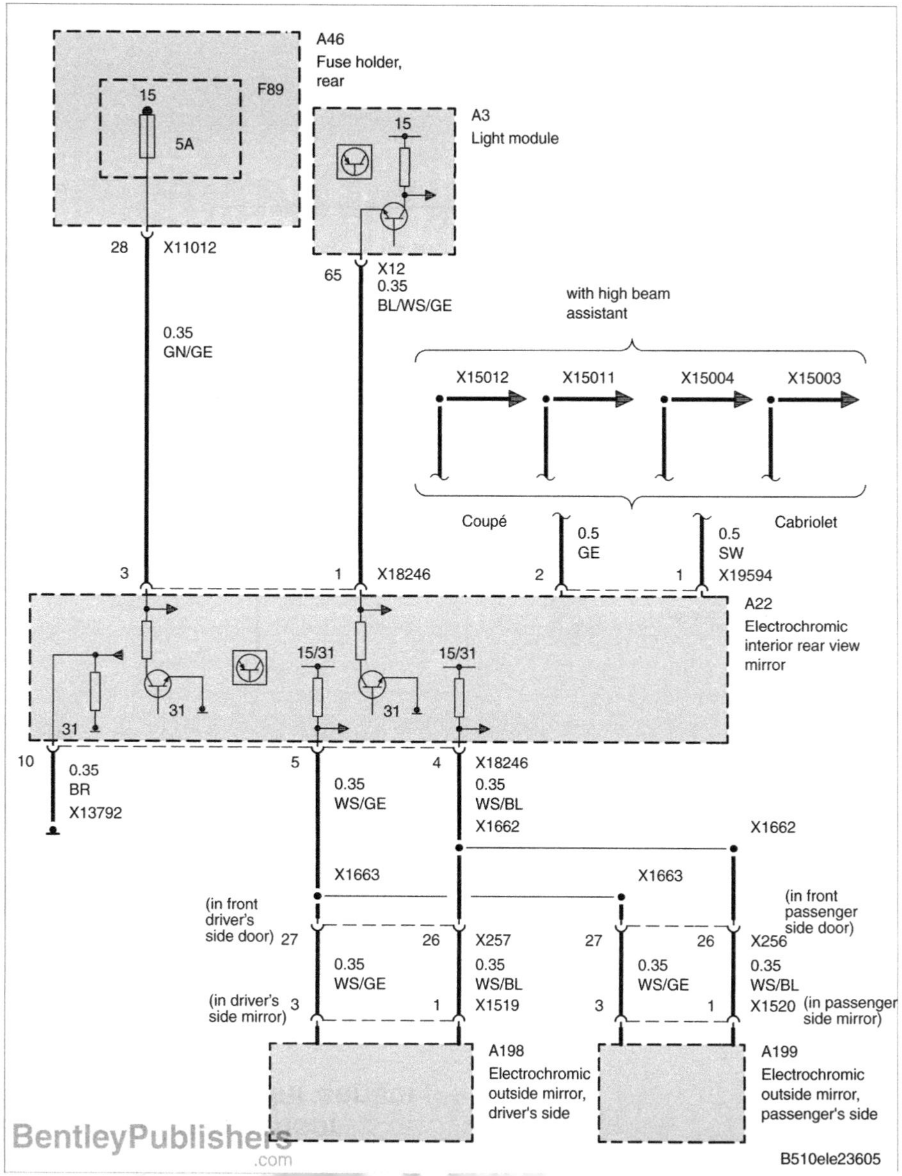

BentleyPublishers.com

Electric outside mirror, driver's side

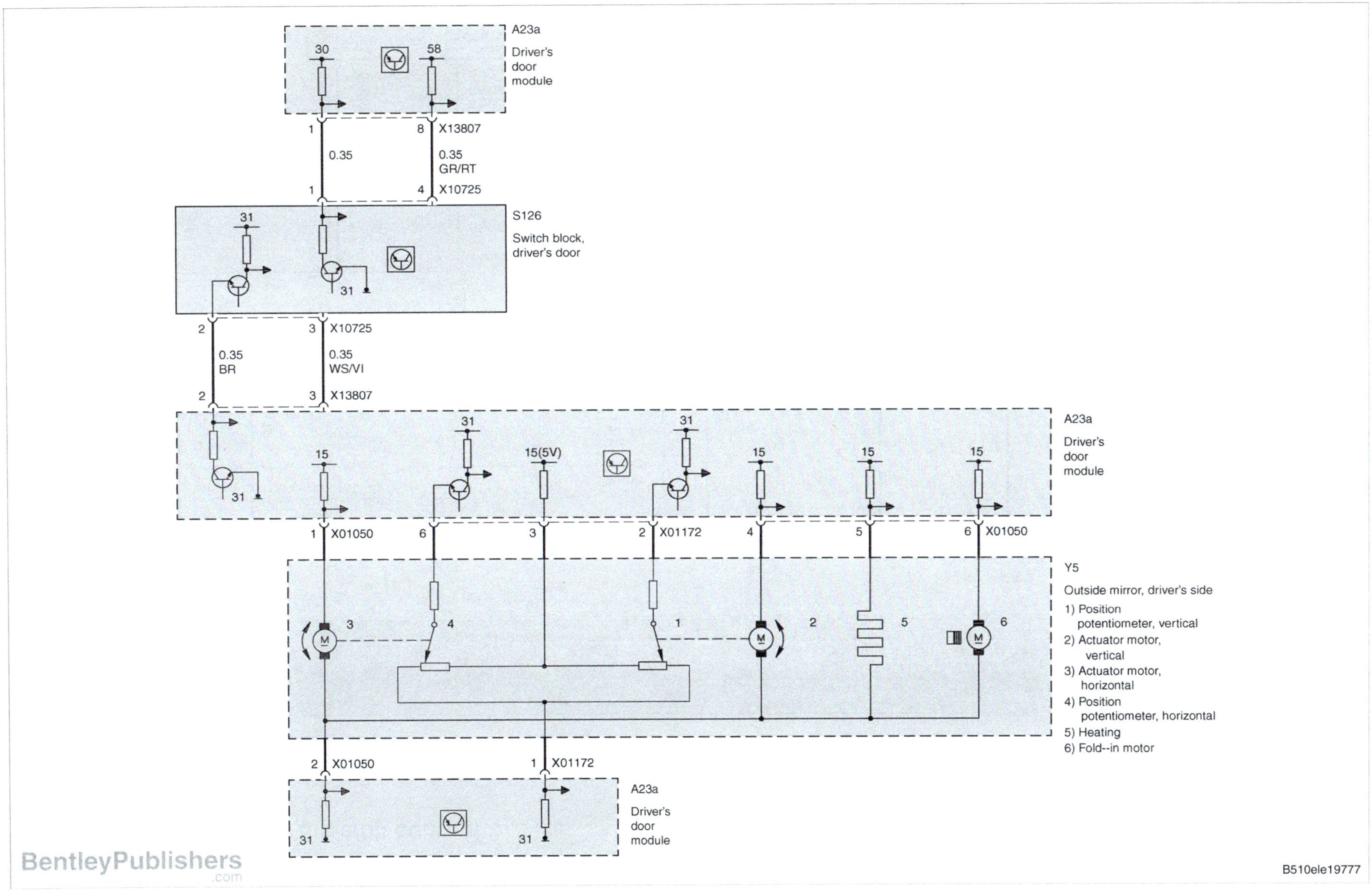

Interior lights
Sedan or Wagon with SHD Slide/tilt sunroof 2

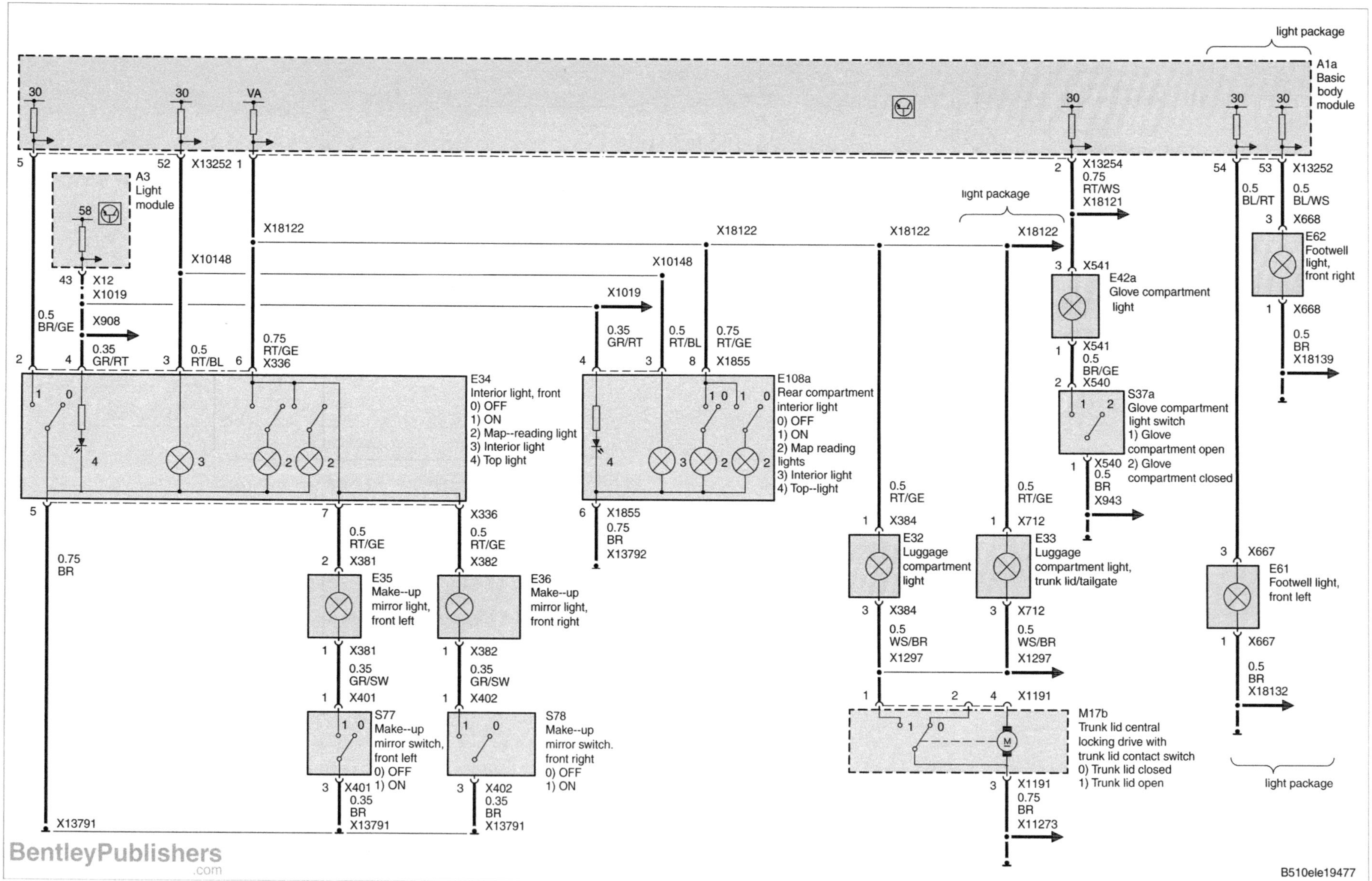

BentleyPublishers.com

Interior lights
Wagon with SHD Slide/tilt sunroof 2

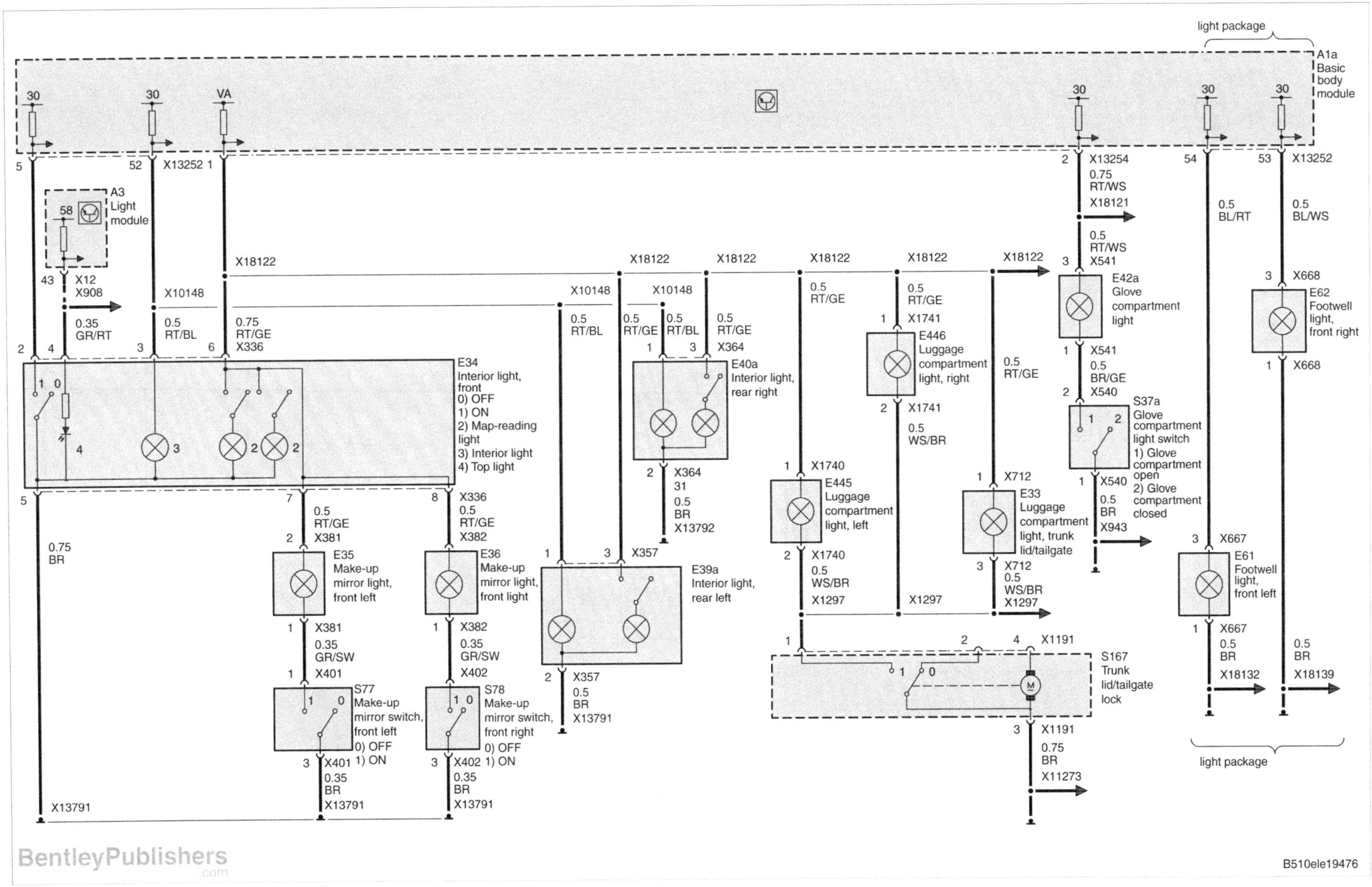

BentleyPublishers.com

Central locking
Driver's door

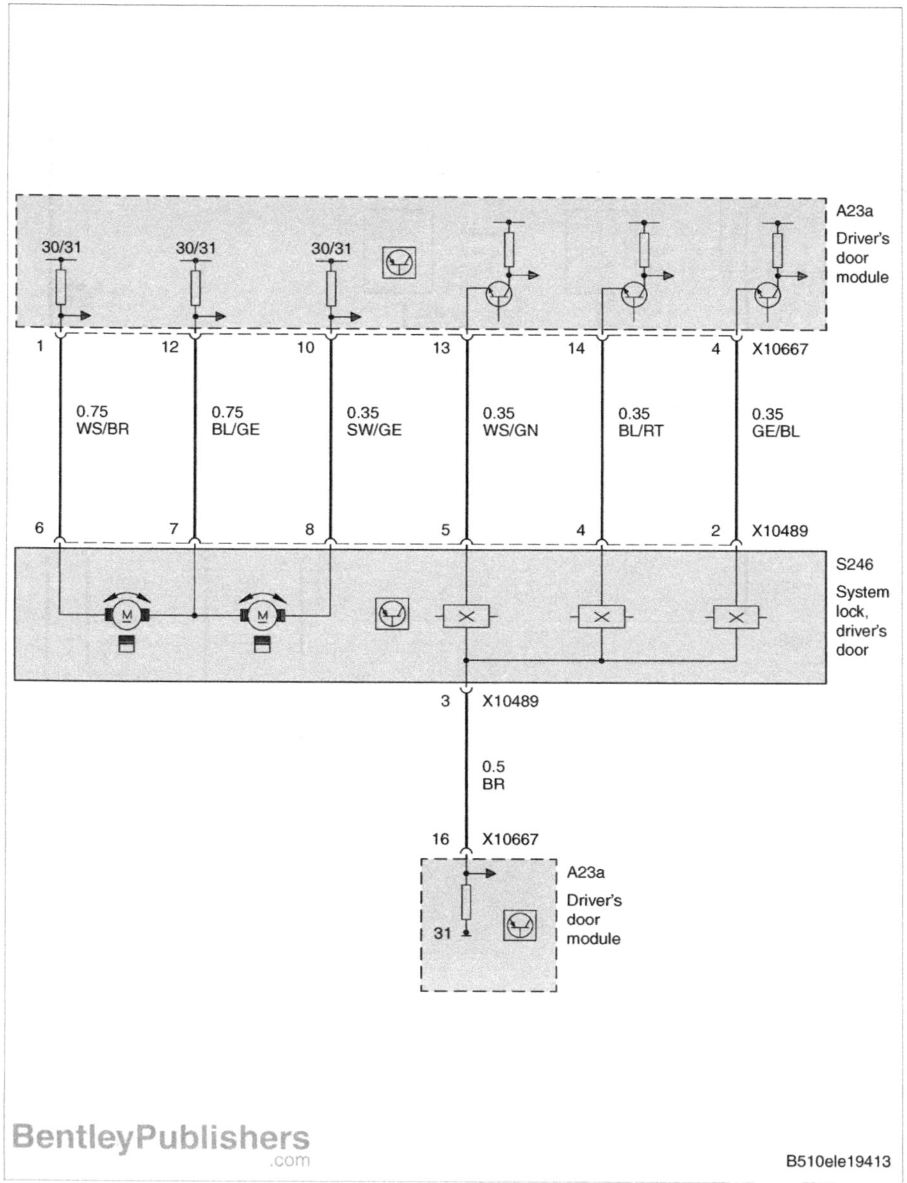

Central locking
Passenger's door

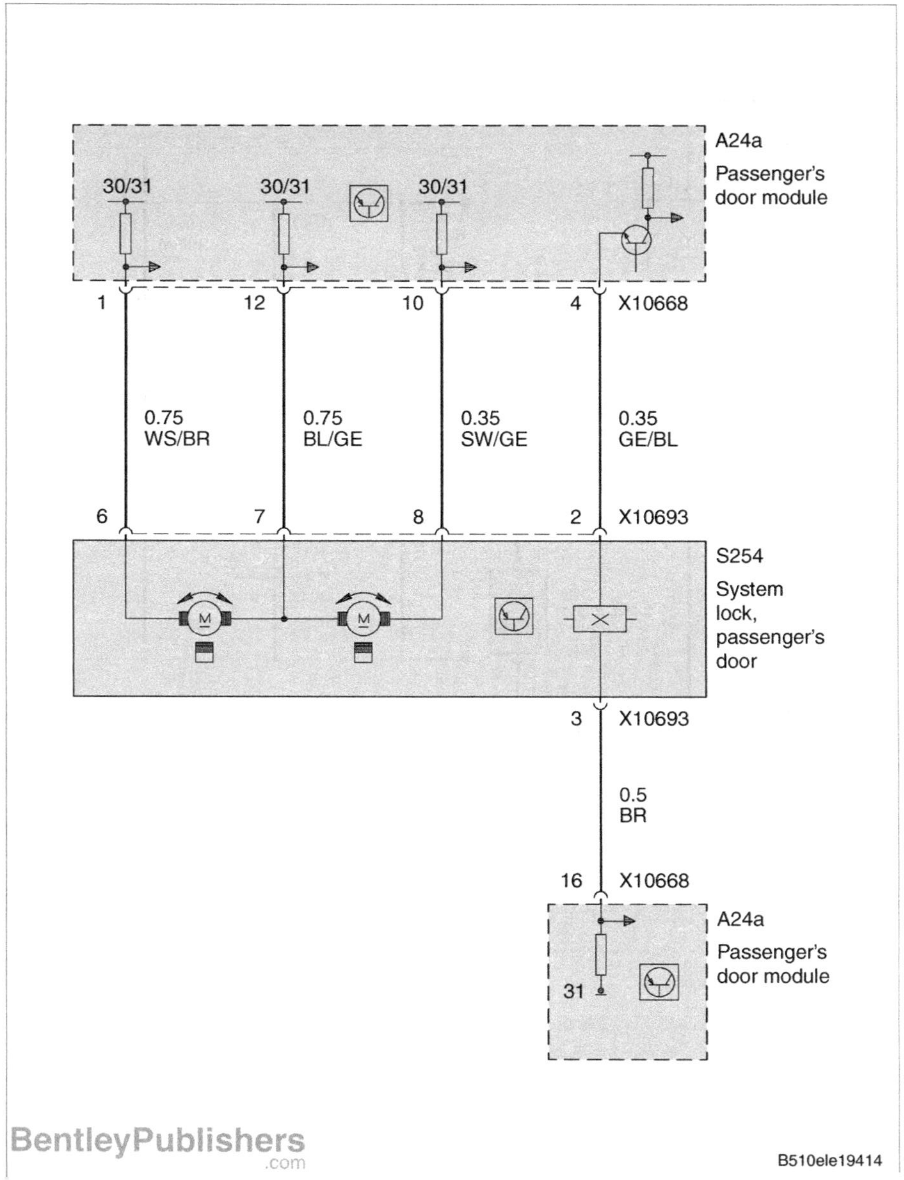

Central locking
Rear doors and fuel filler flap

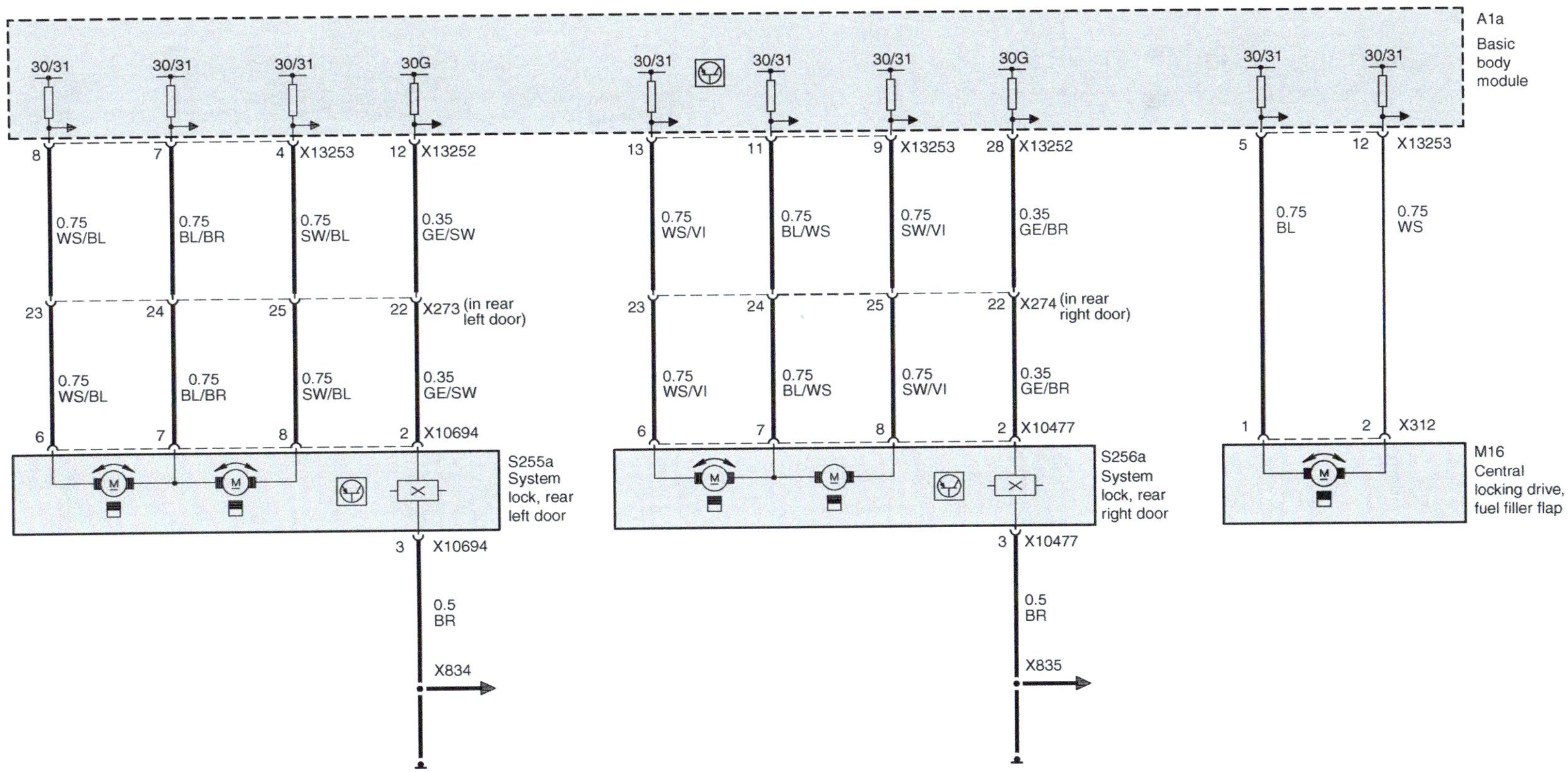

BentleyPublishers.com

B510ele19415

Driver's door module power supply

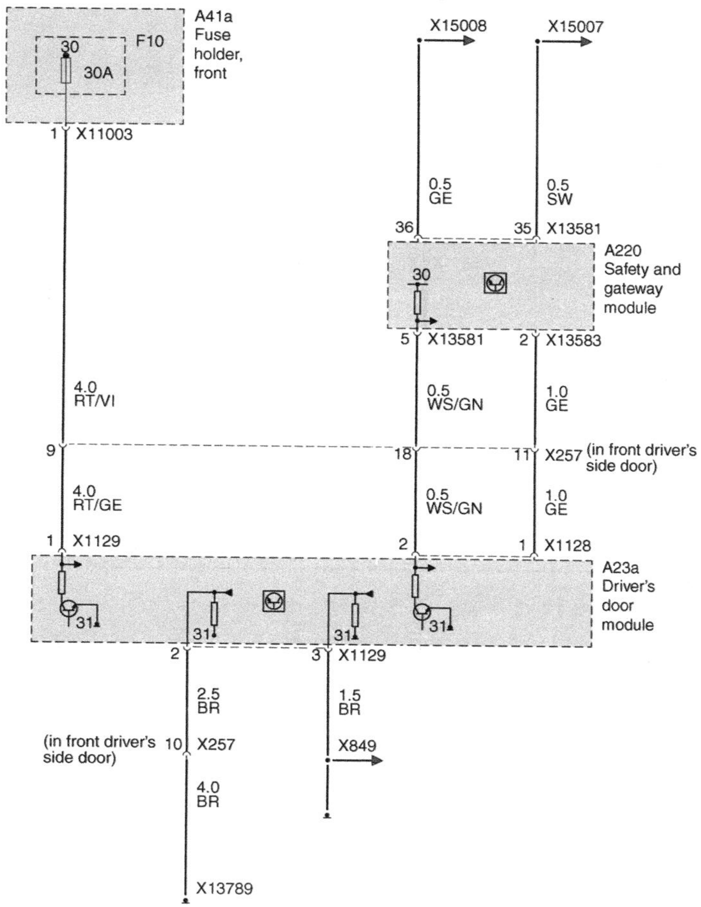

B510ele19411

Passenger's door module power supply

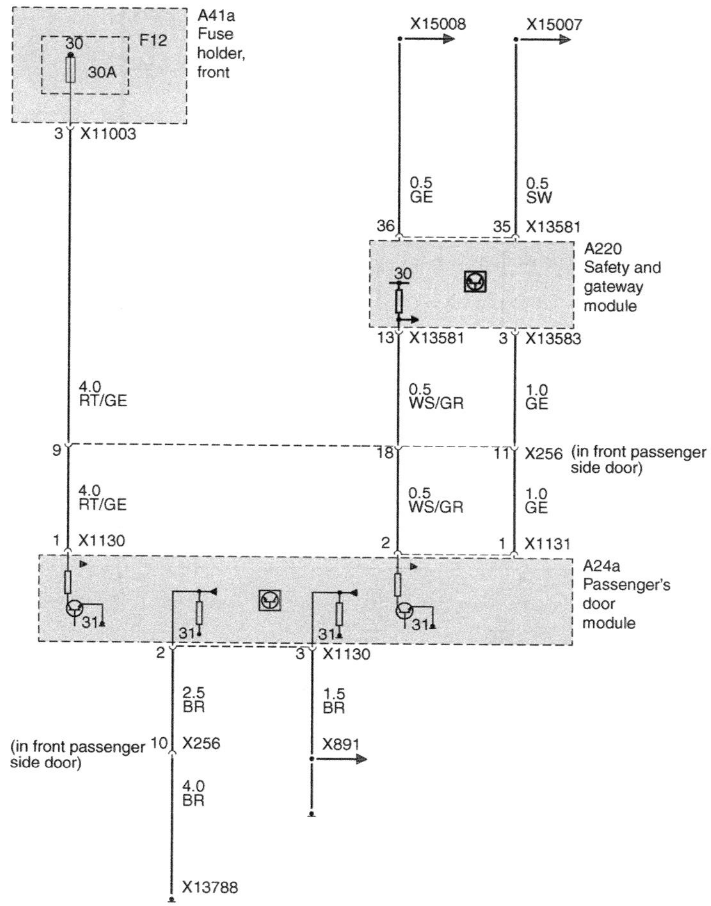

B510ele19412

BentleyPublishers.com
BentleyPublishers.com

Basic body module power supply

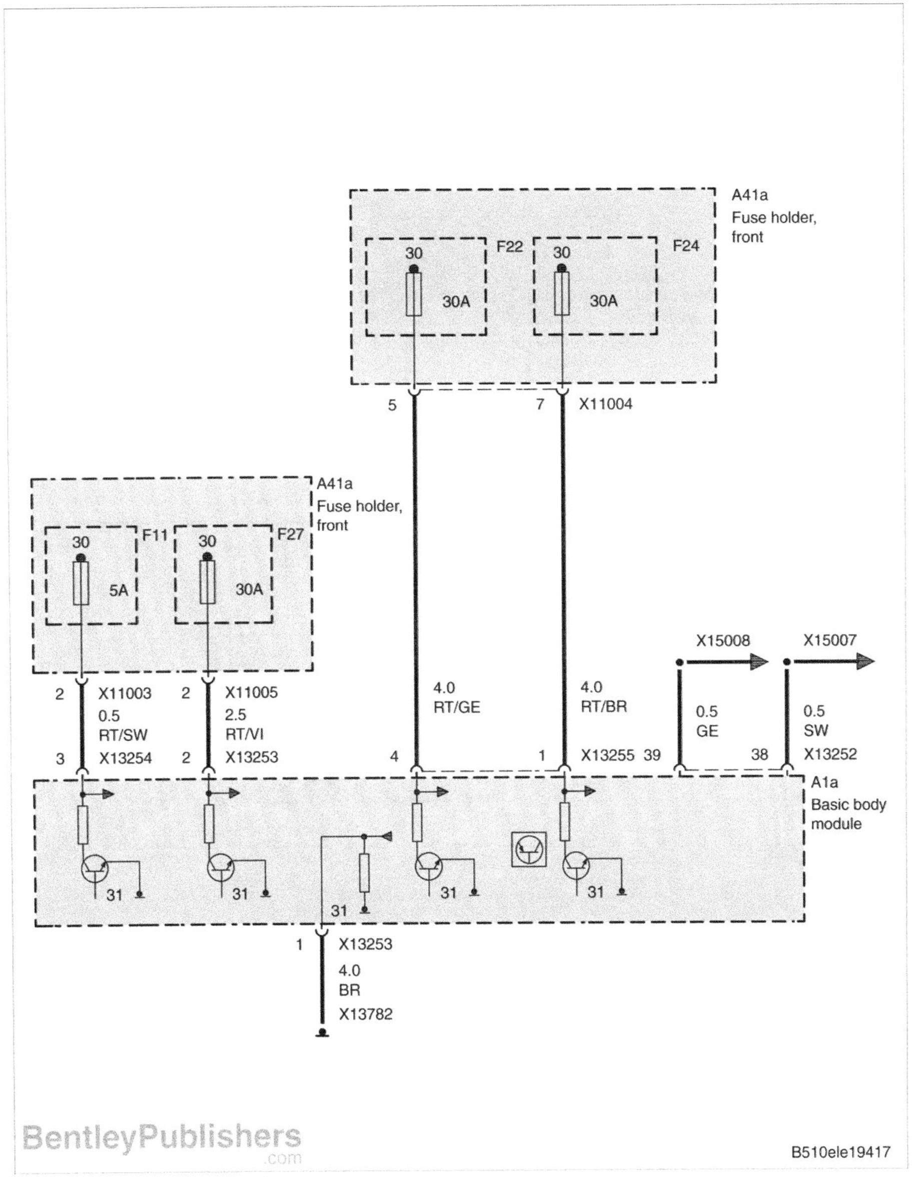

Blower / independent ventilation

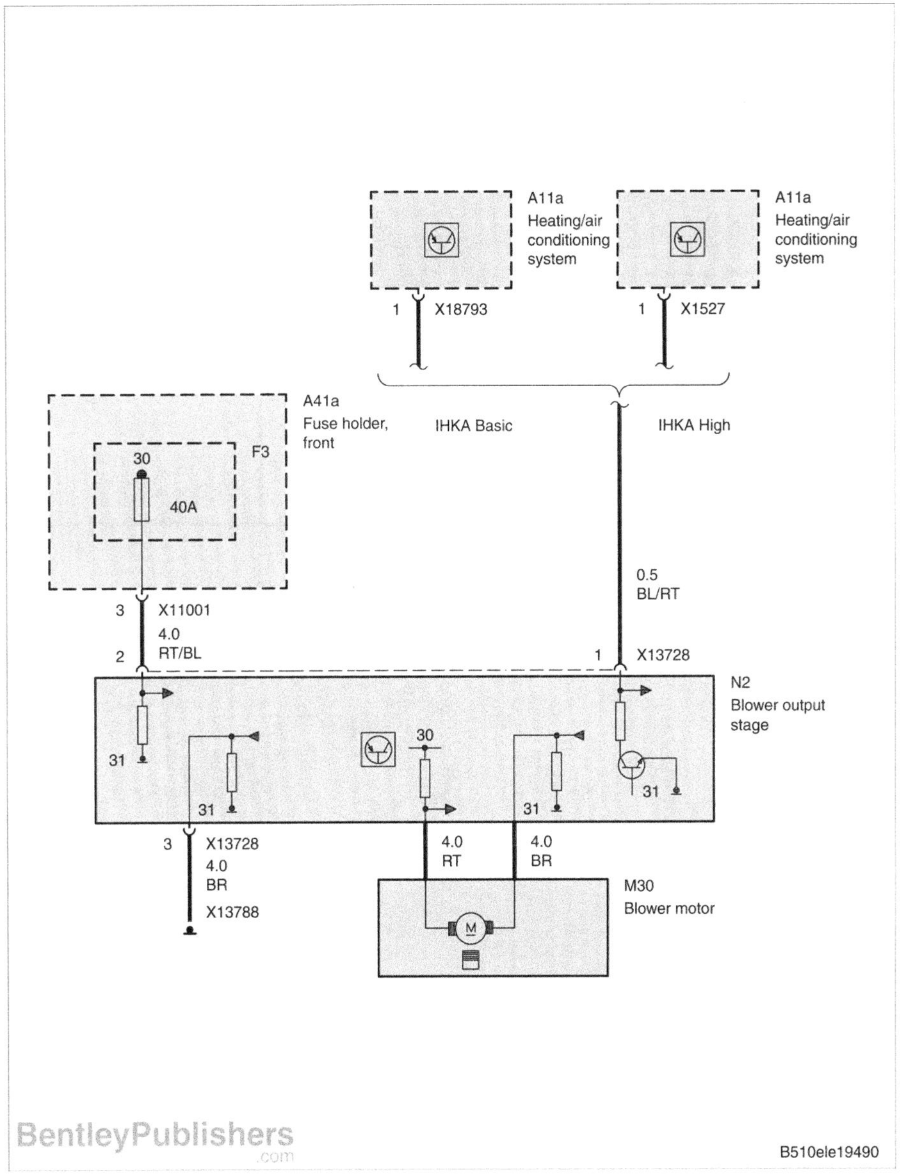

Heating and air conditioning control unit power supply IHKA Basic up to 09 / 2005

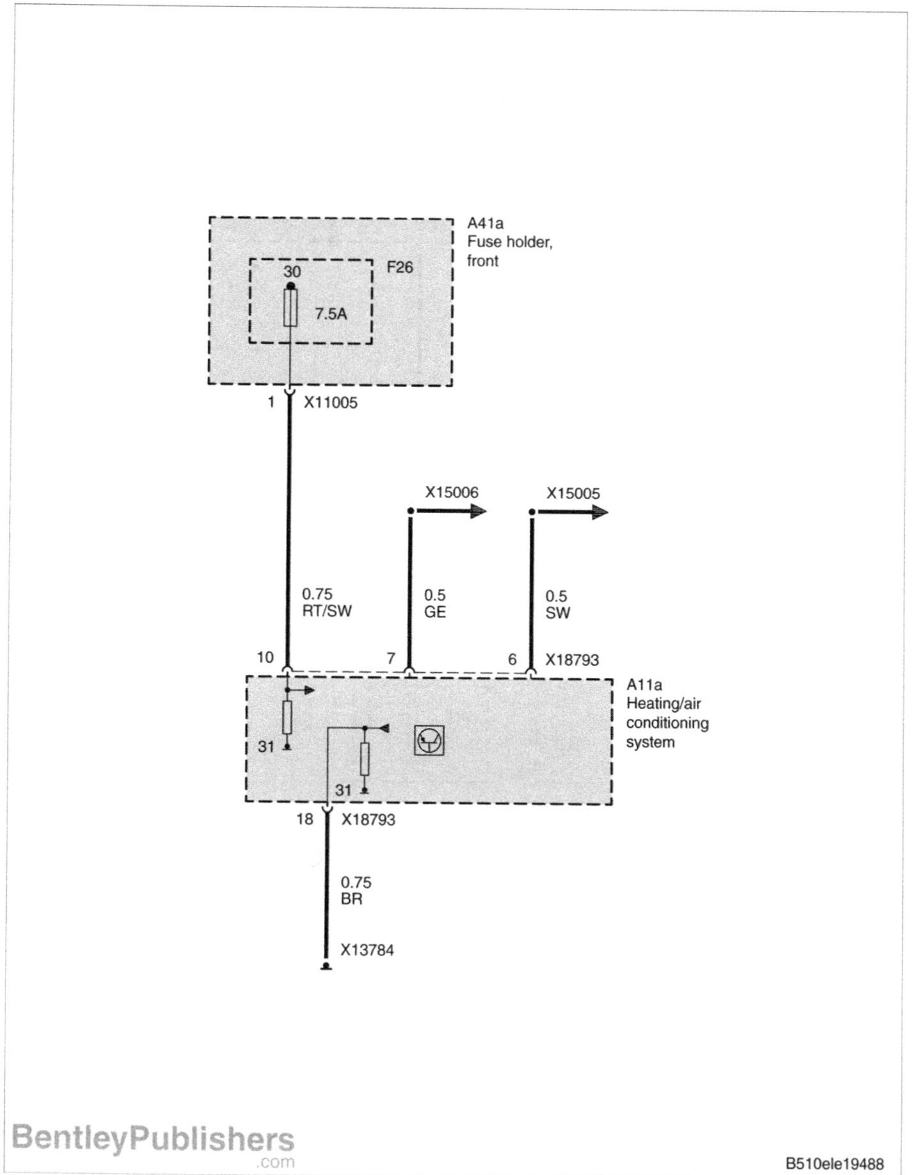

Heating and air conditioning control unit power supply IHKA Basic as of 09 / 2005

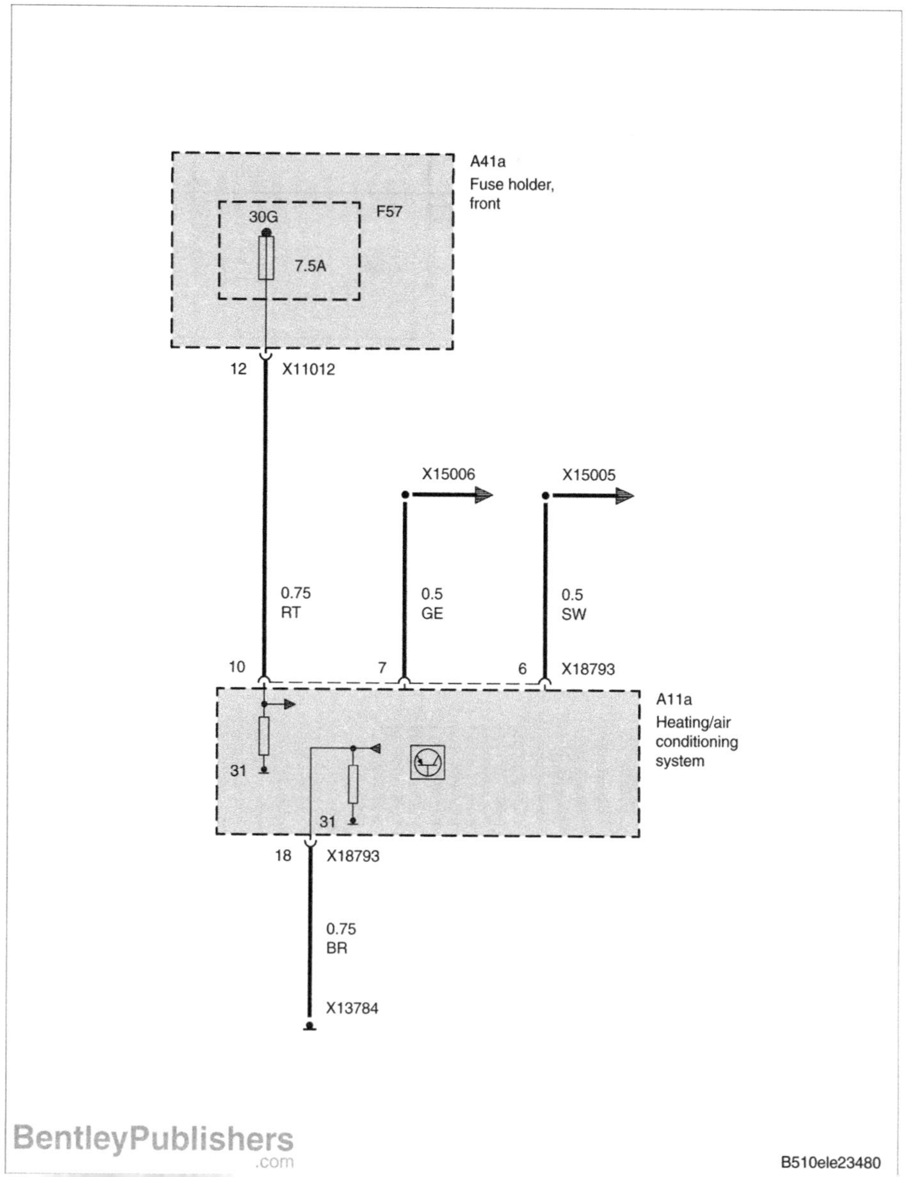

Heating and air conditioning control unit power supply IHKA High up to 03 / 2005

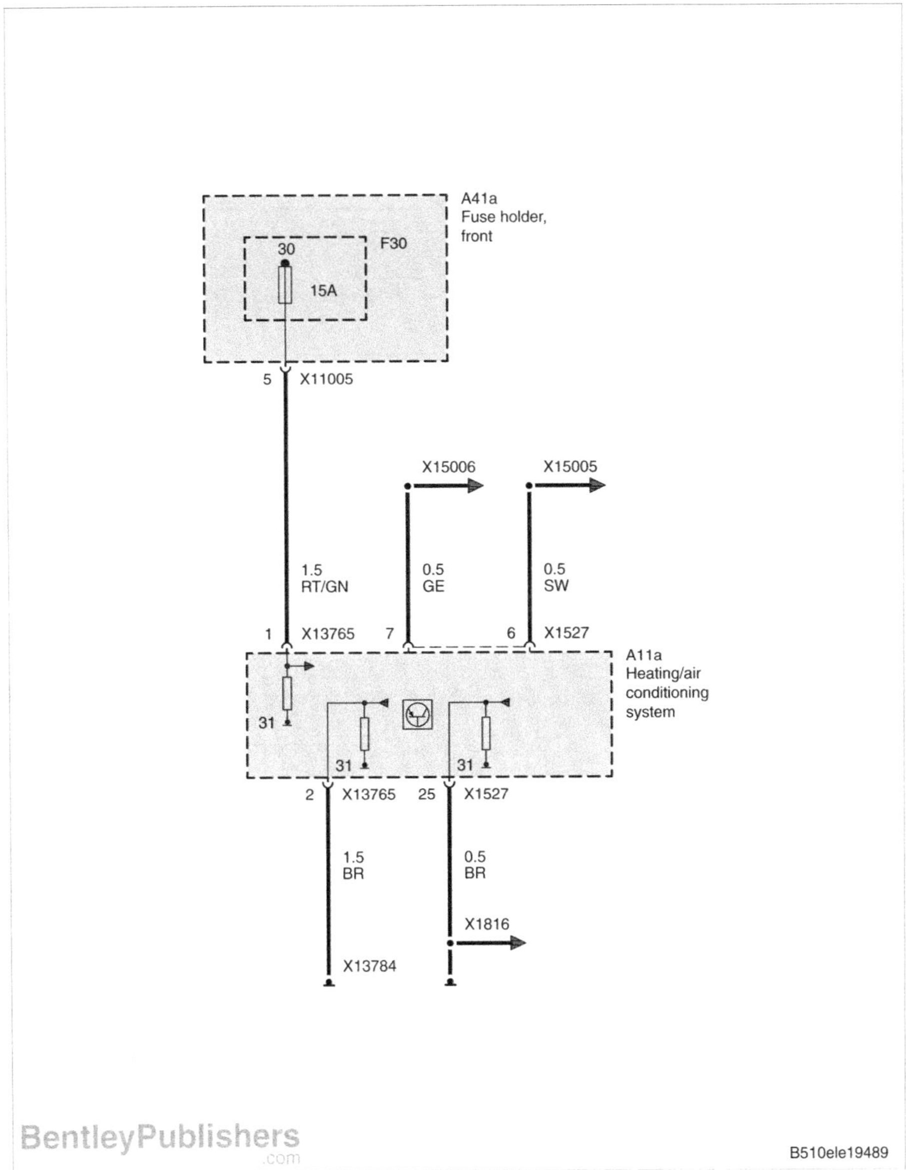

Heating and air conditioning control unit power supply IHKA High as of 03 / 2005 up to 09 / 2005

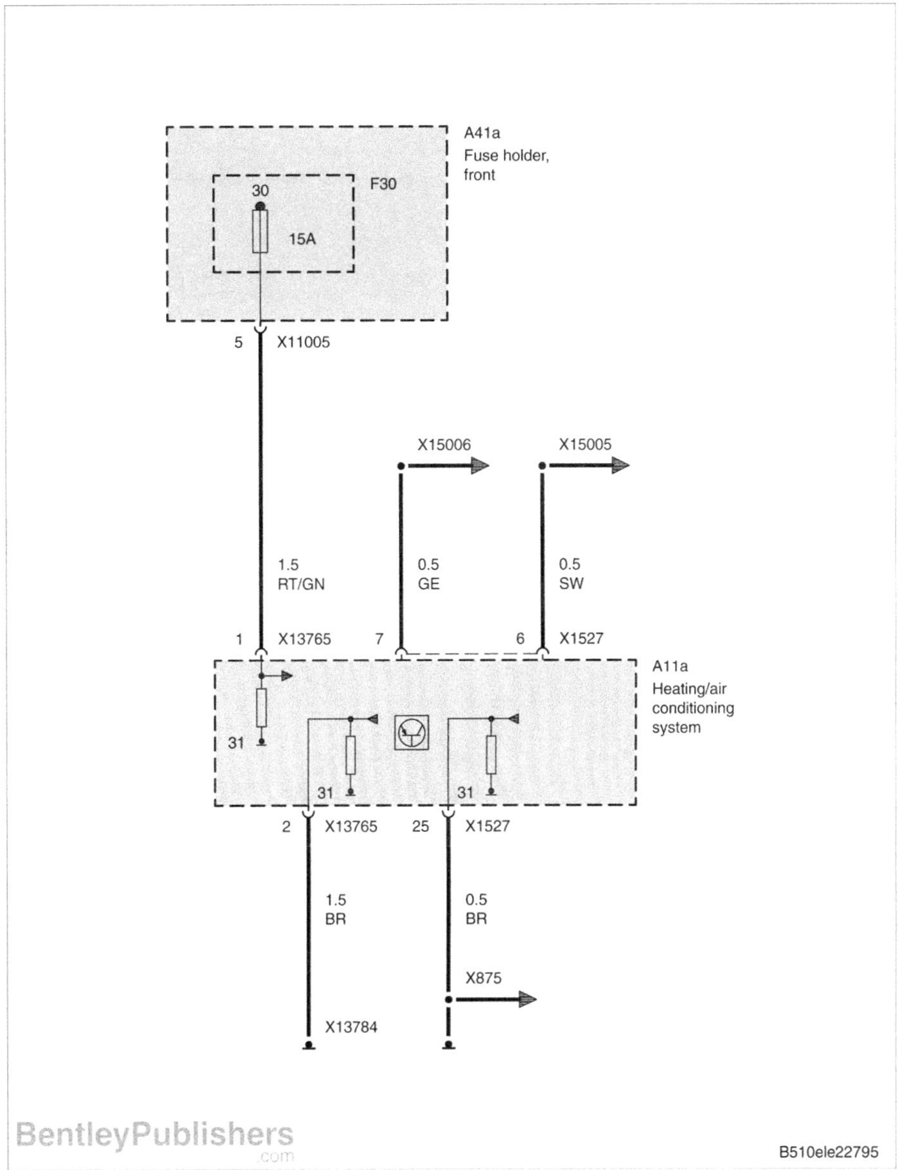

Heating and air conditioning control unit power supply IHKA High as of 09 / 2005

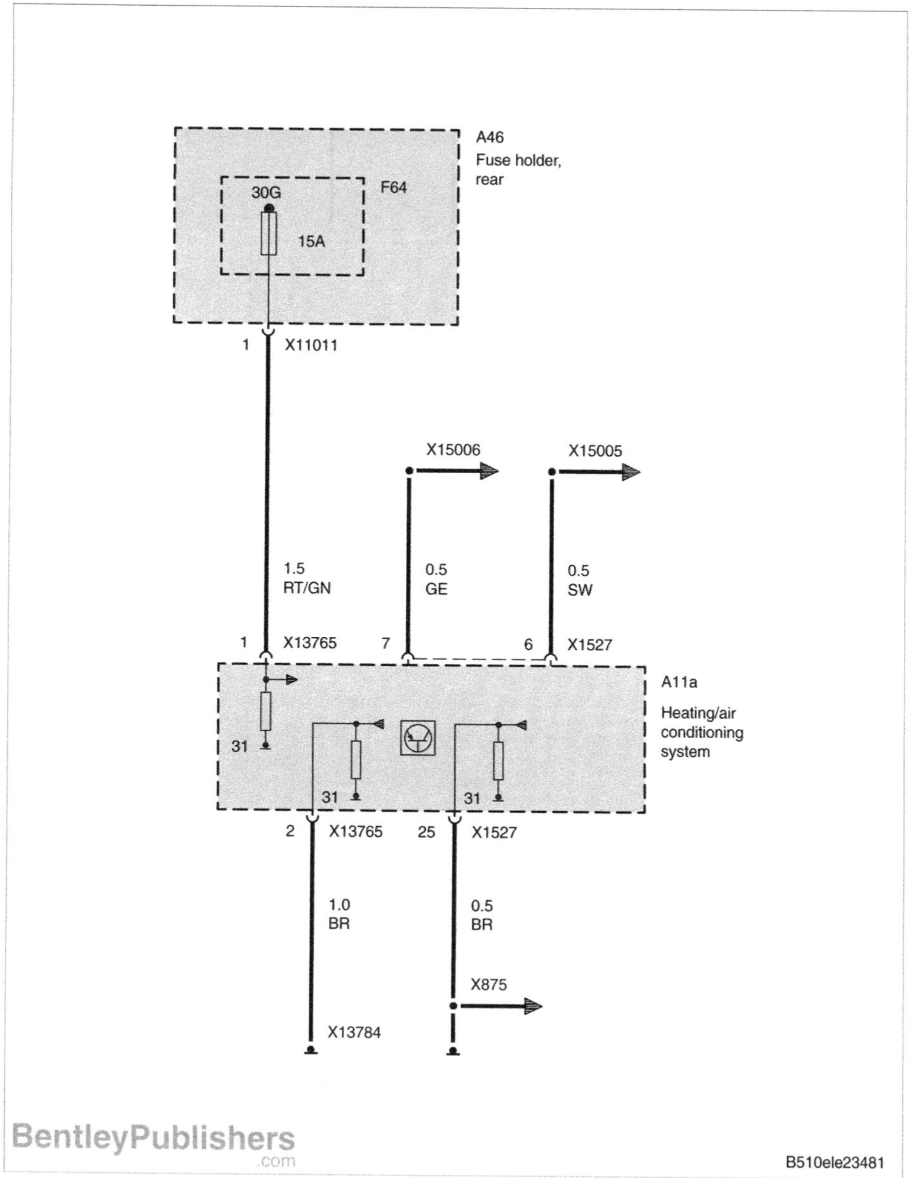

Heated windshield washer jets

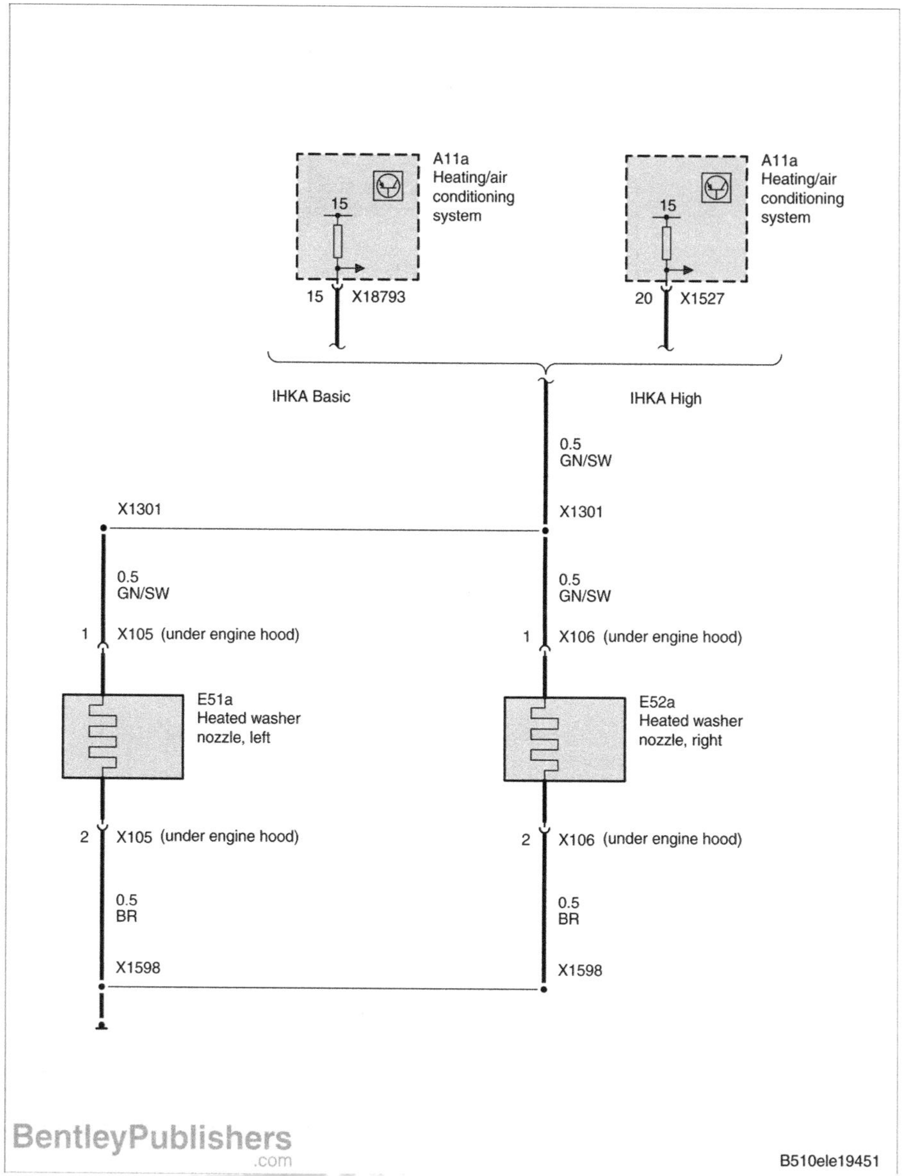

Rear window defogger up to 09 / 2005

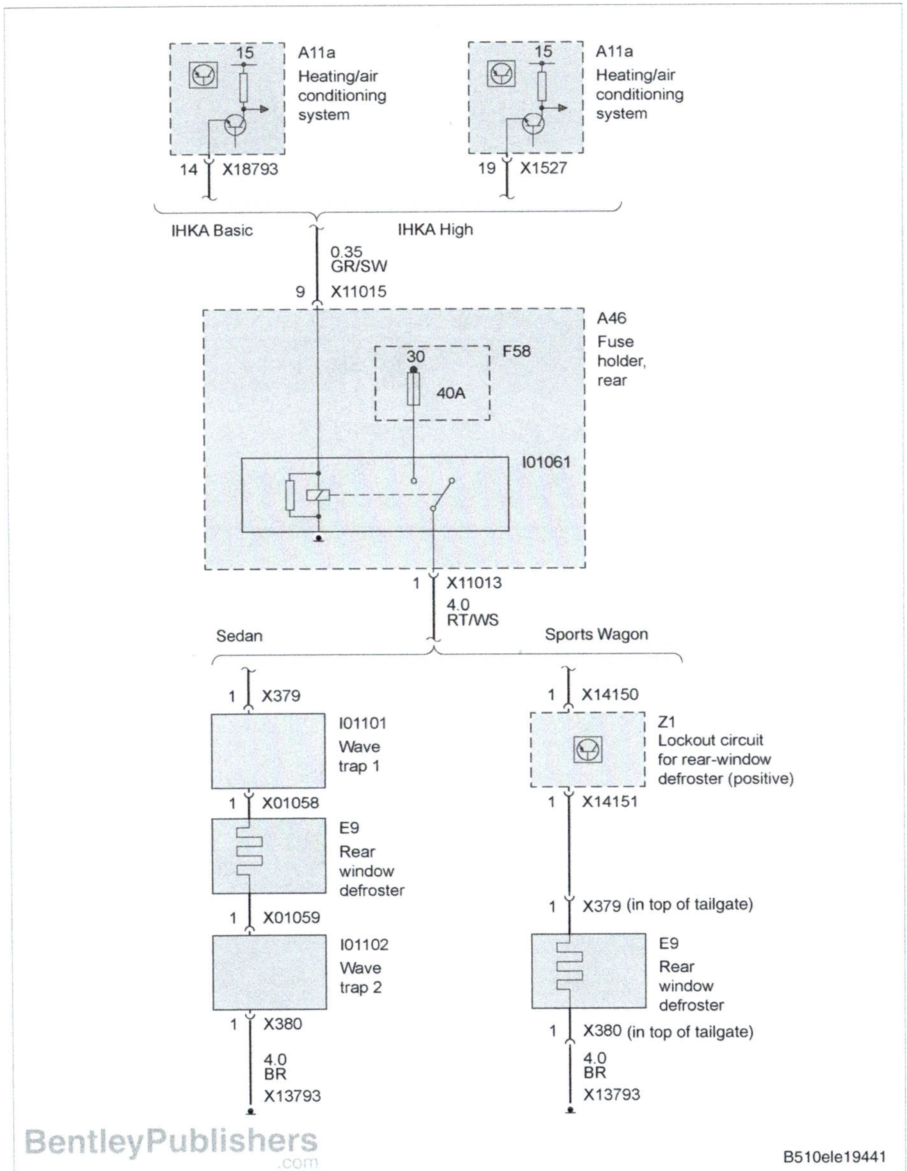

Rear window defogger as of 09 / 2005

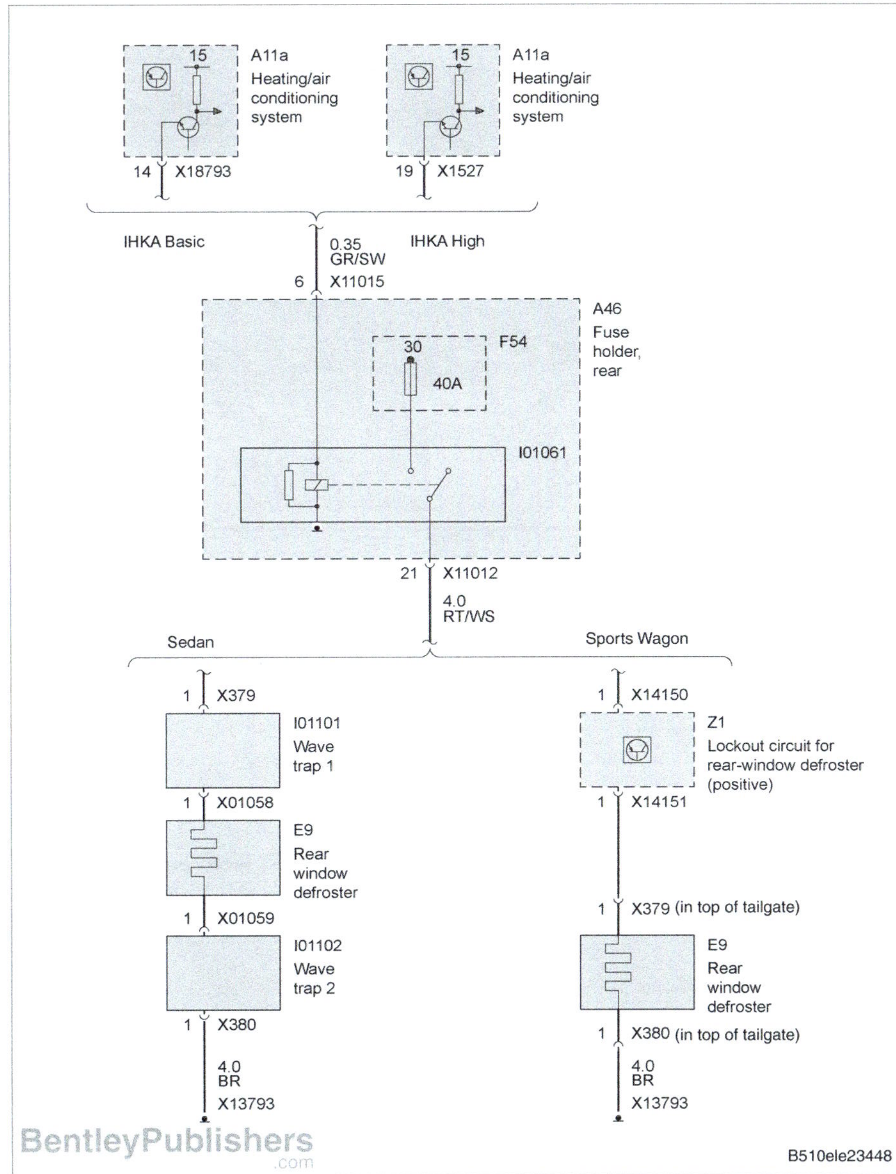

Power window without KGM body gateway module Driver's door

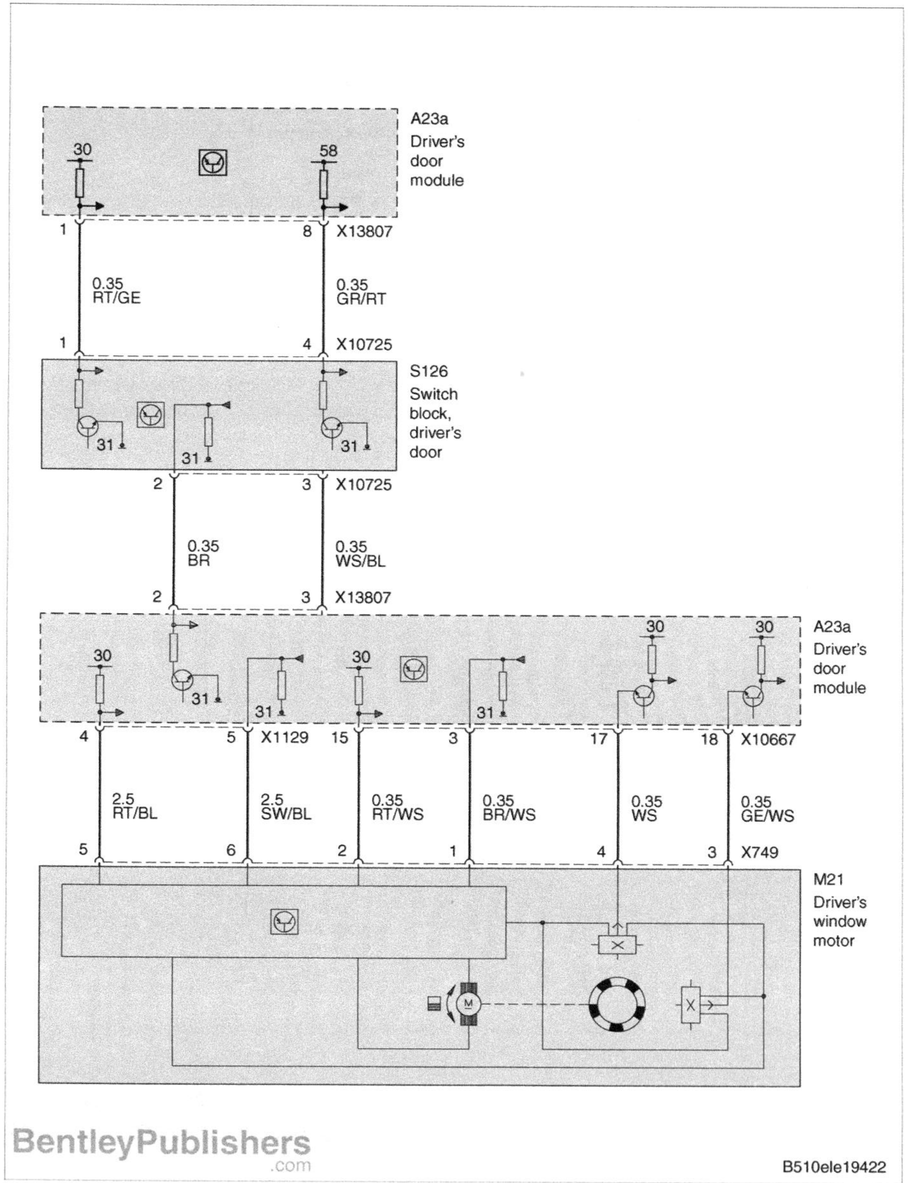

Power window without KGM body gateway module Passenger's door

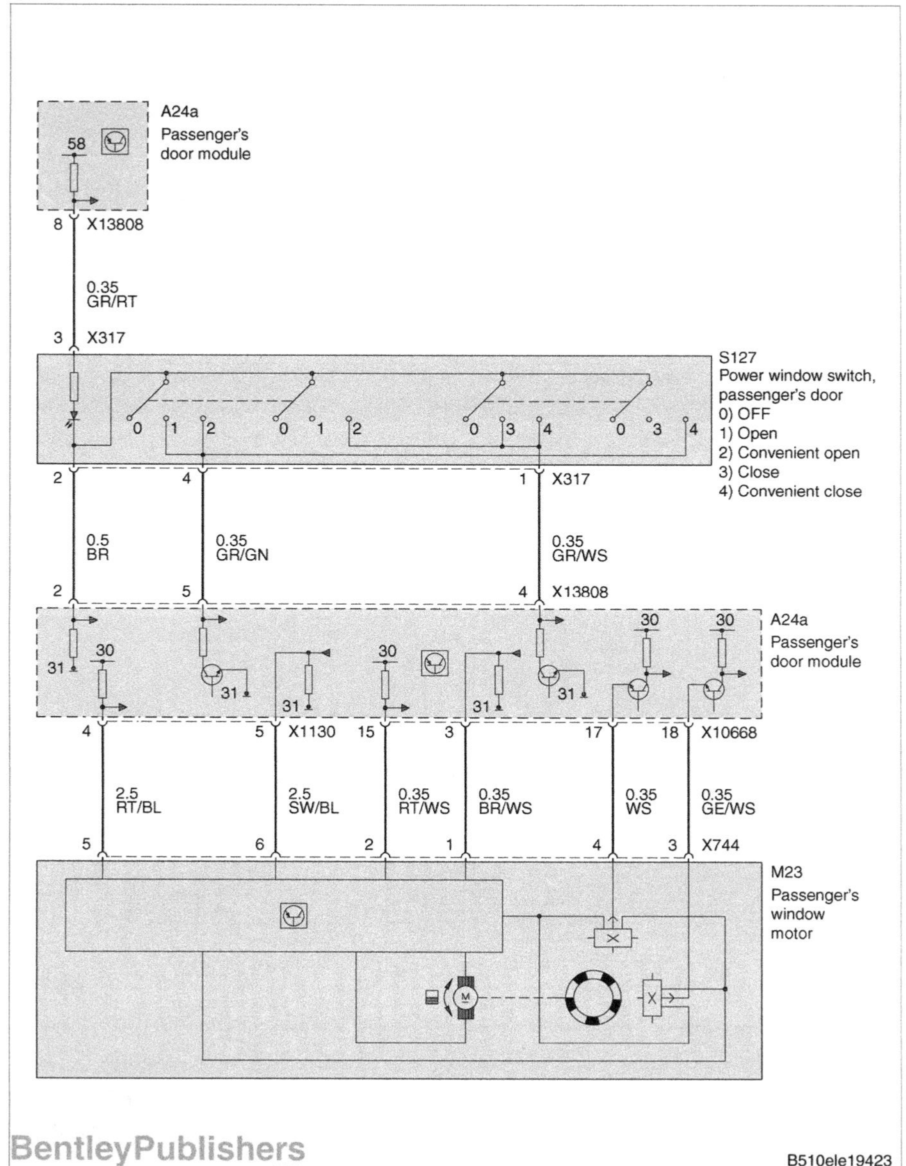

Power window with KGM body gateway module Driver's door

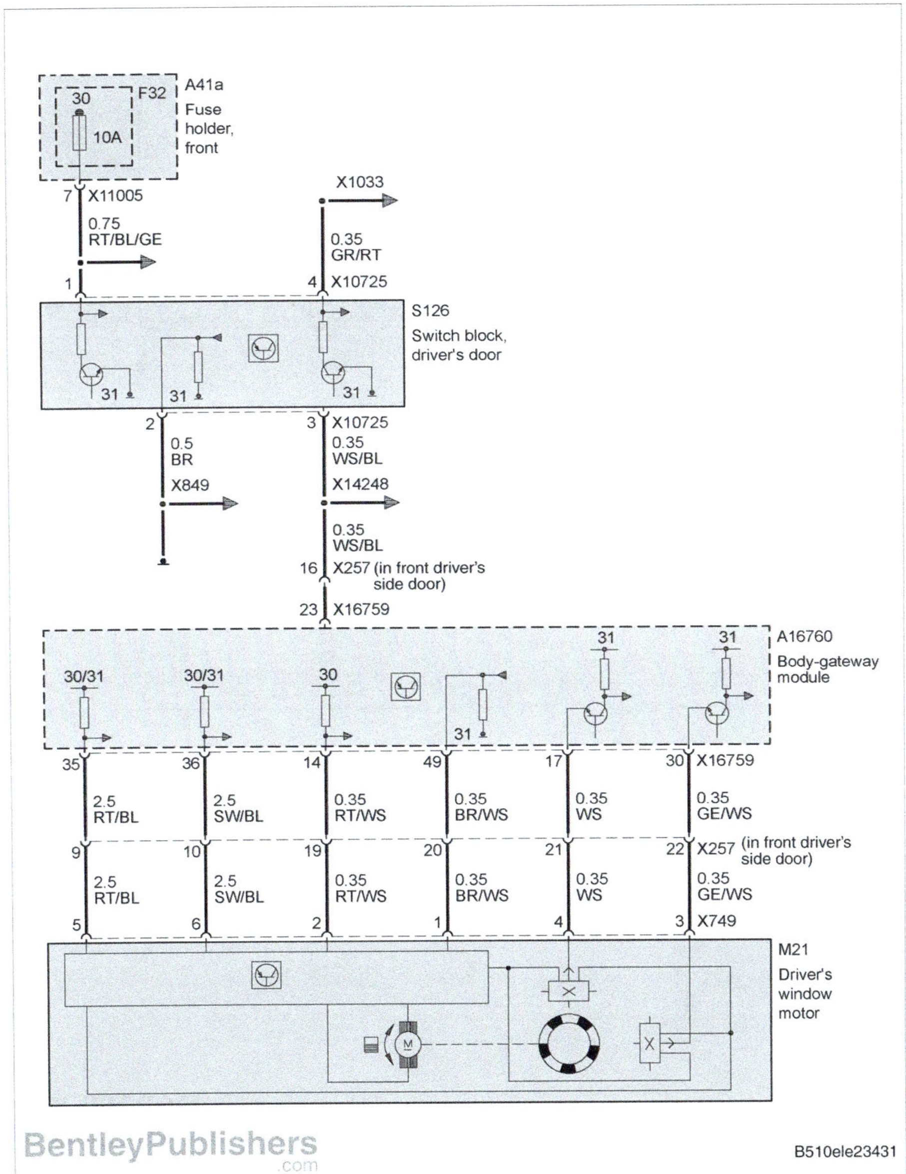

Power window with KGM body gateway module Passenger's door

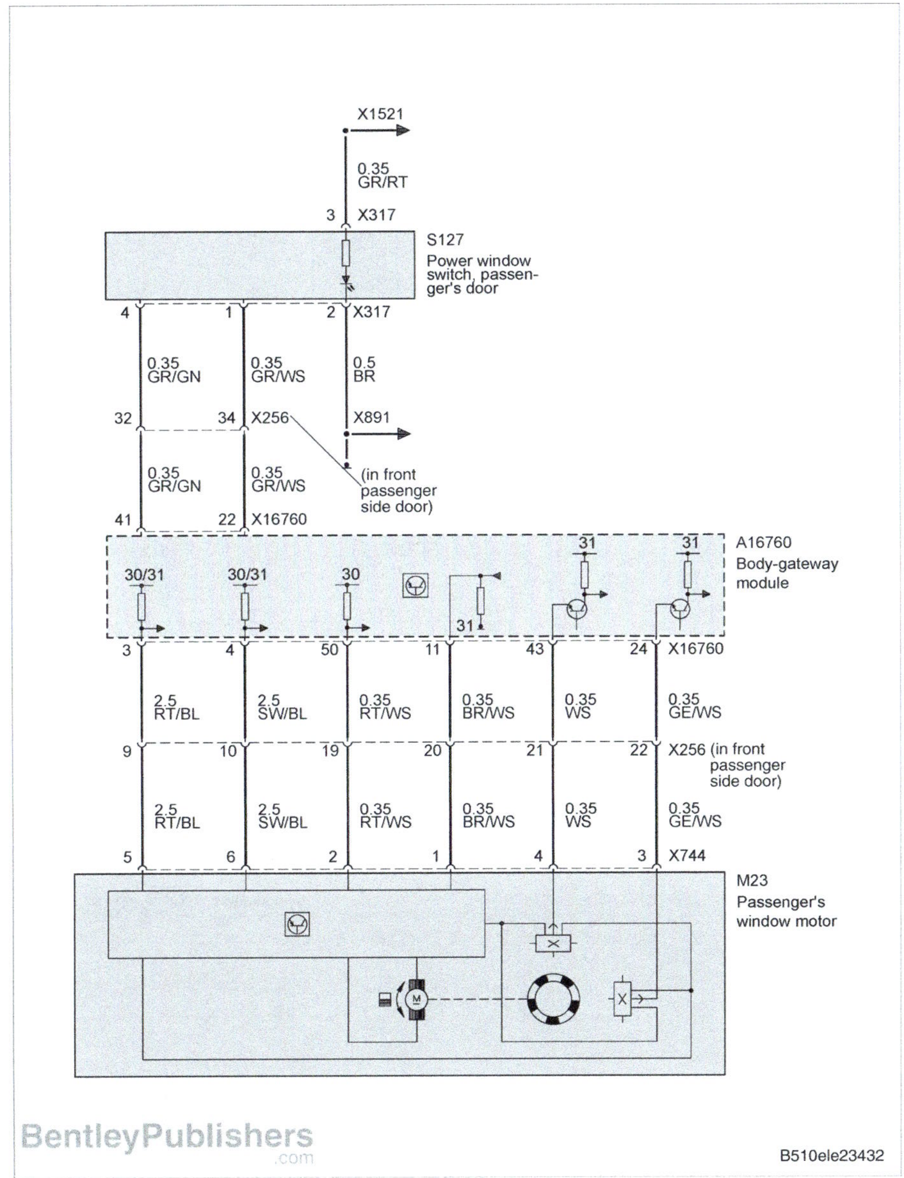

Power window
Rear left door

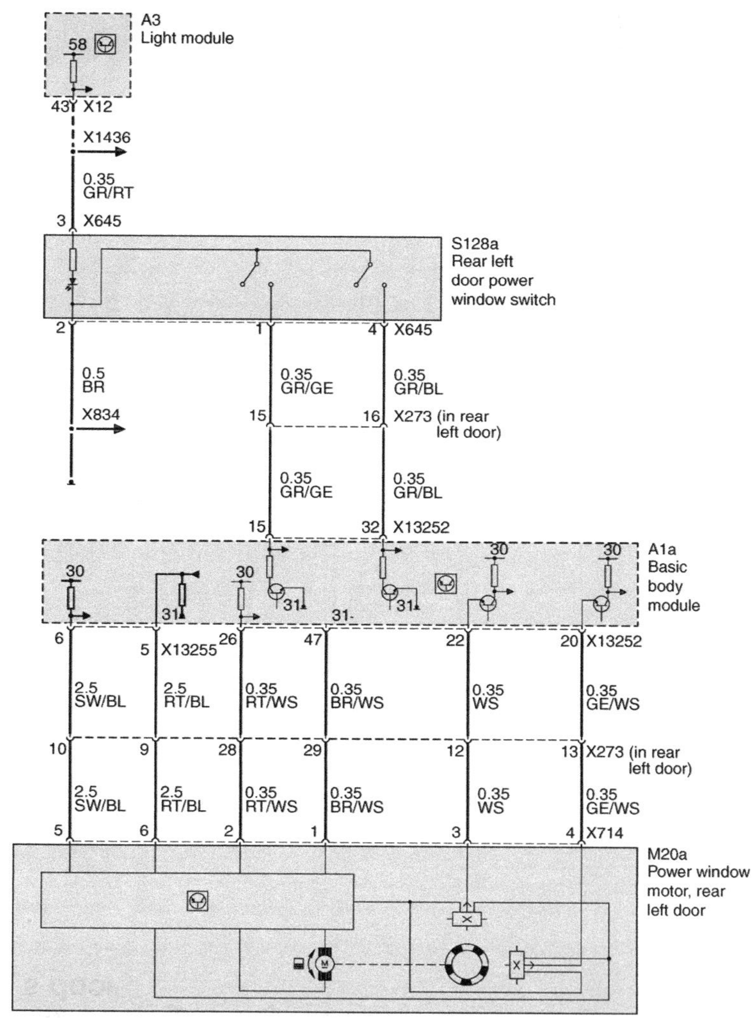

Power window
Rear right door

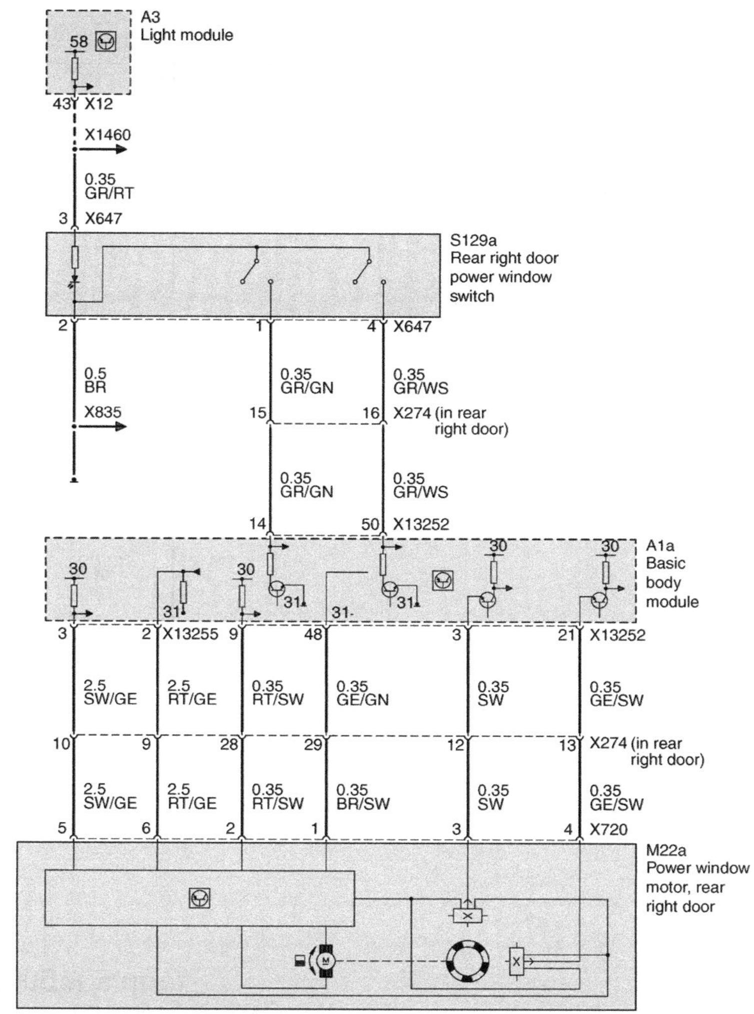

Comfort access module power supply

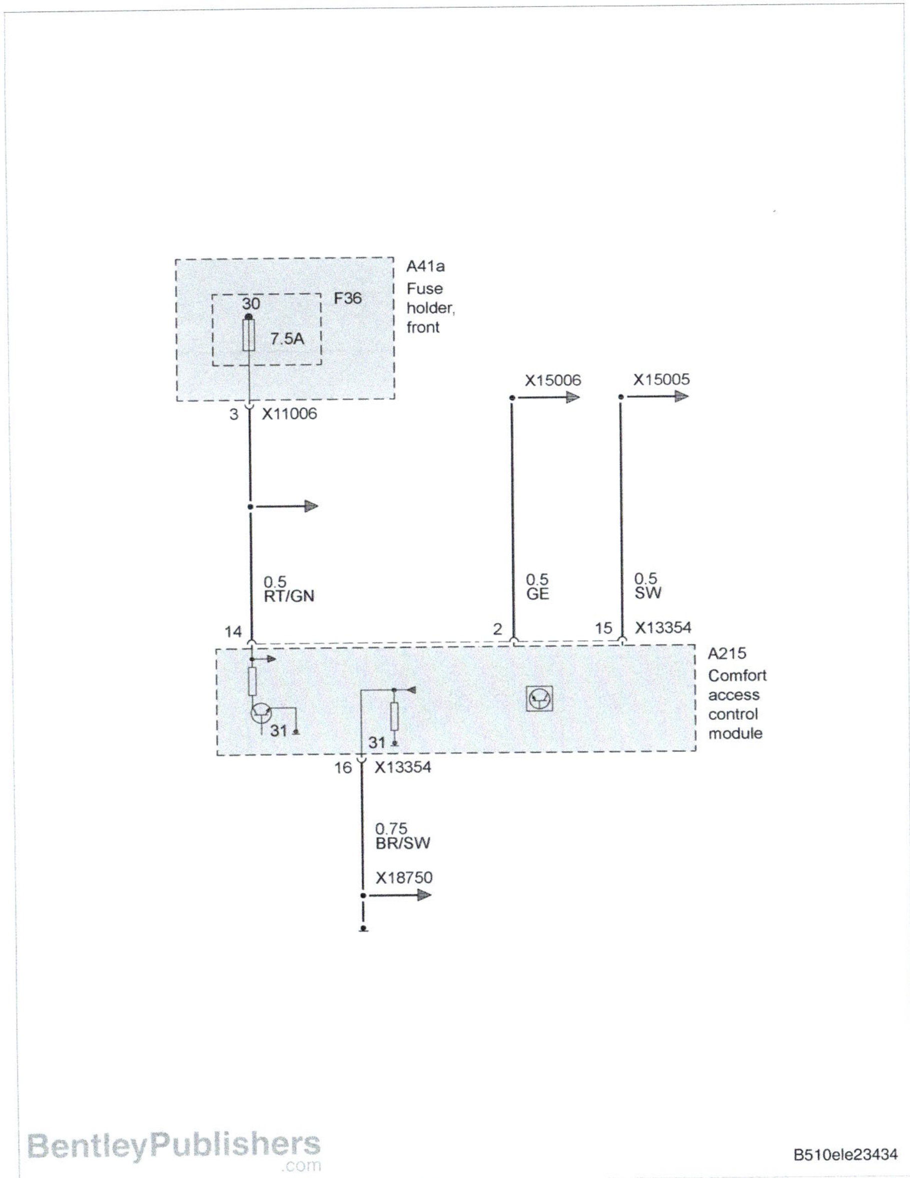

Central locking, front

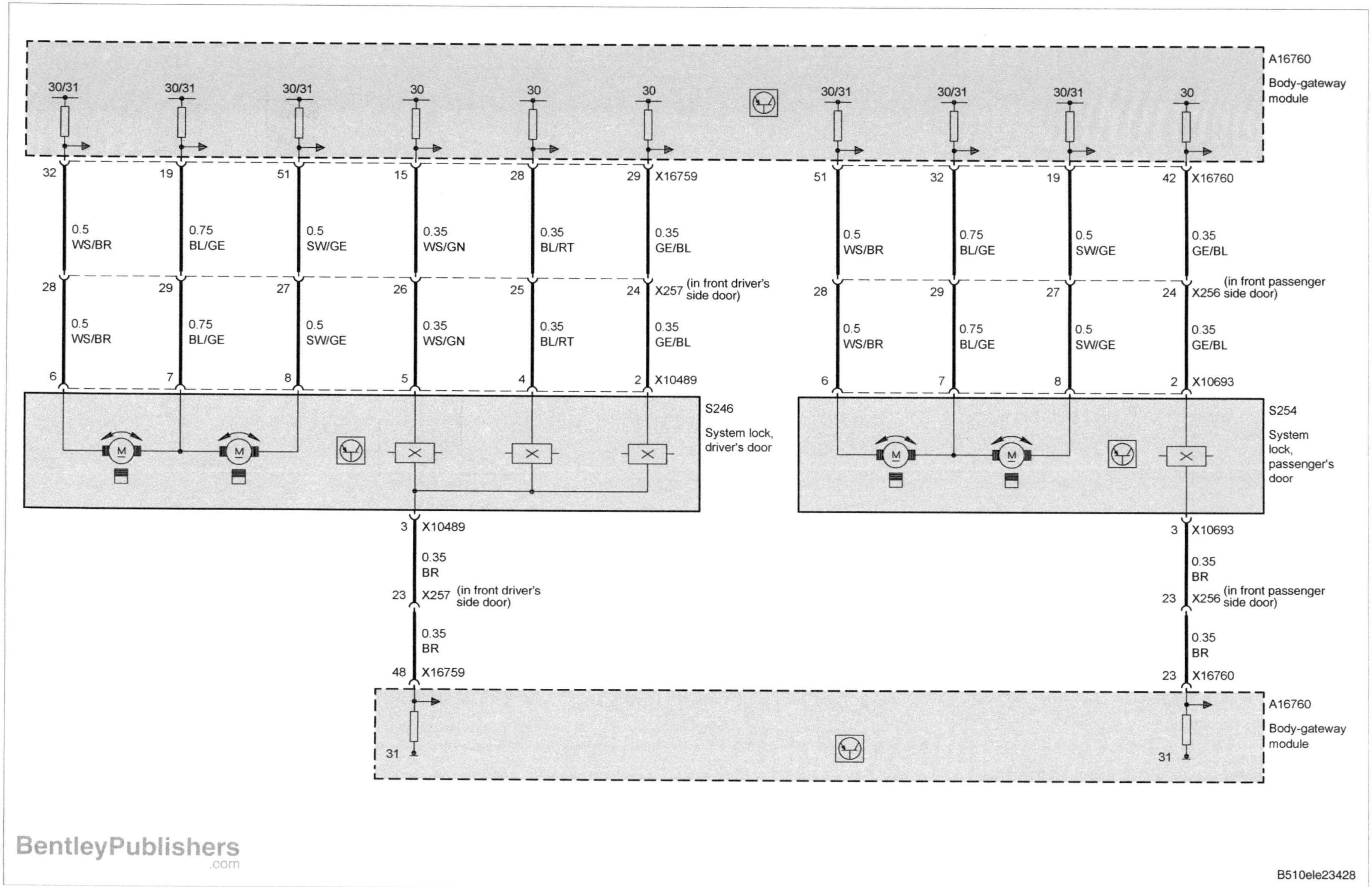

Central locking, trunk lid
Sedan

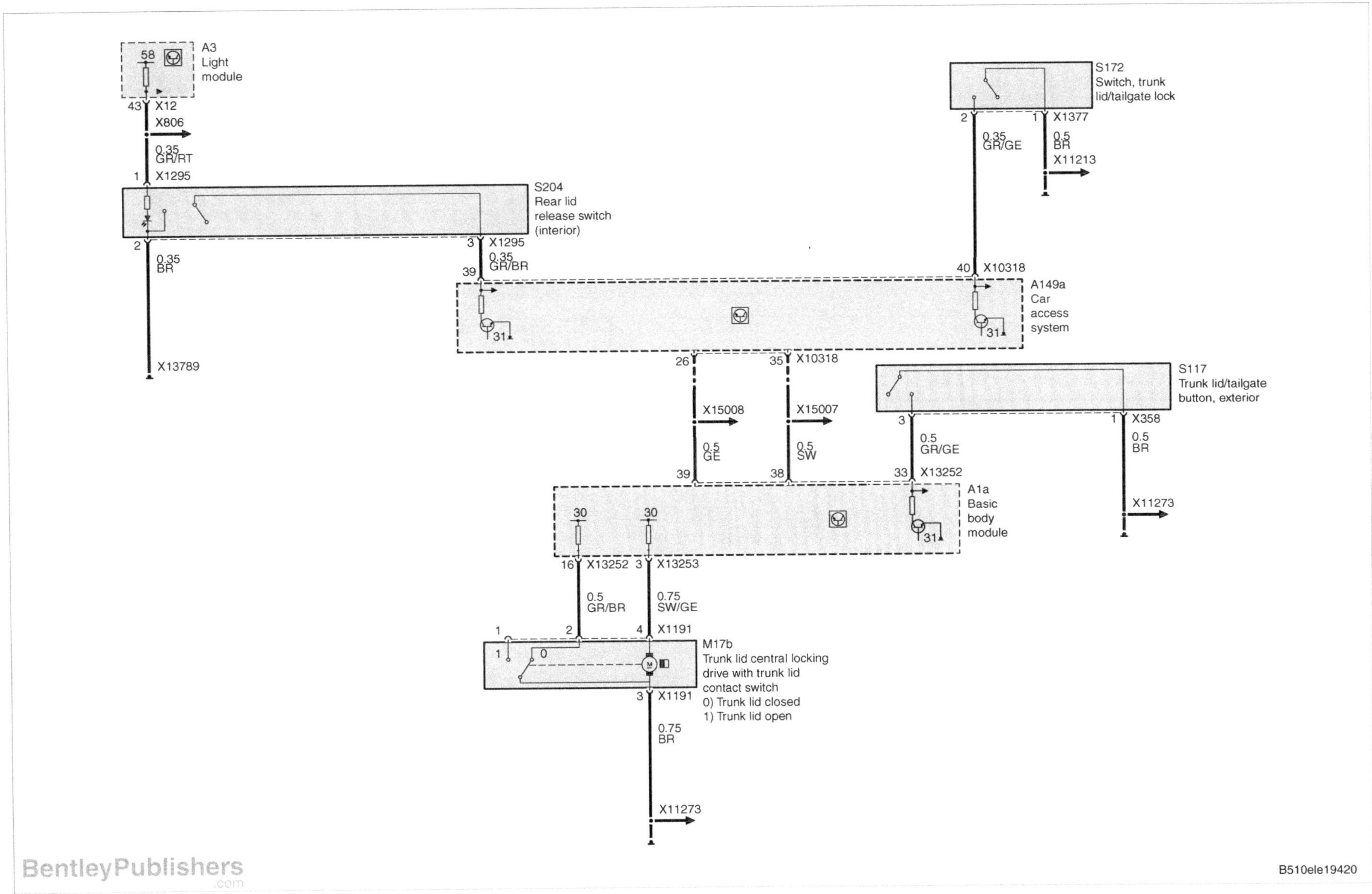

Central locking, tailgate
Wagon to 03 / 2005

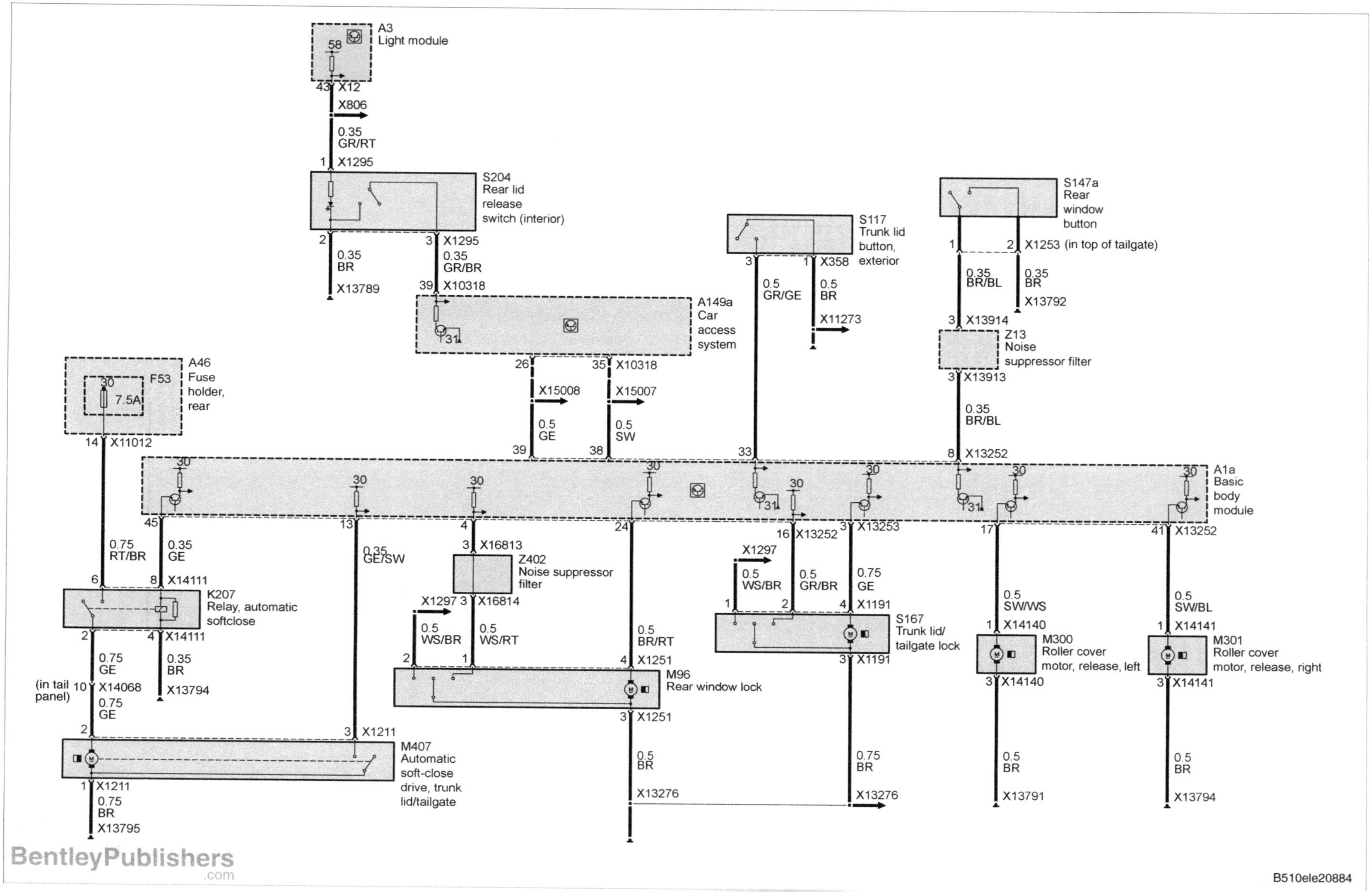

BentleyPublishers
.com

Central locking, tailgate
Wagon from 03 / 2005 to 09 / 2005

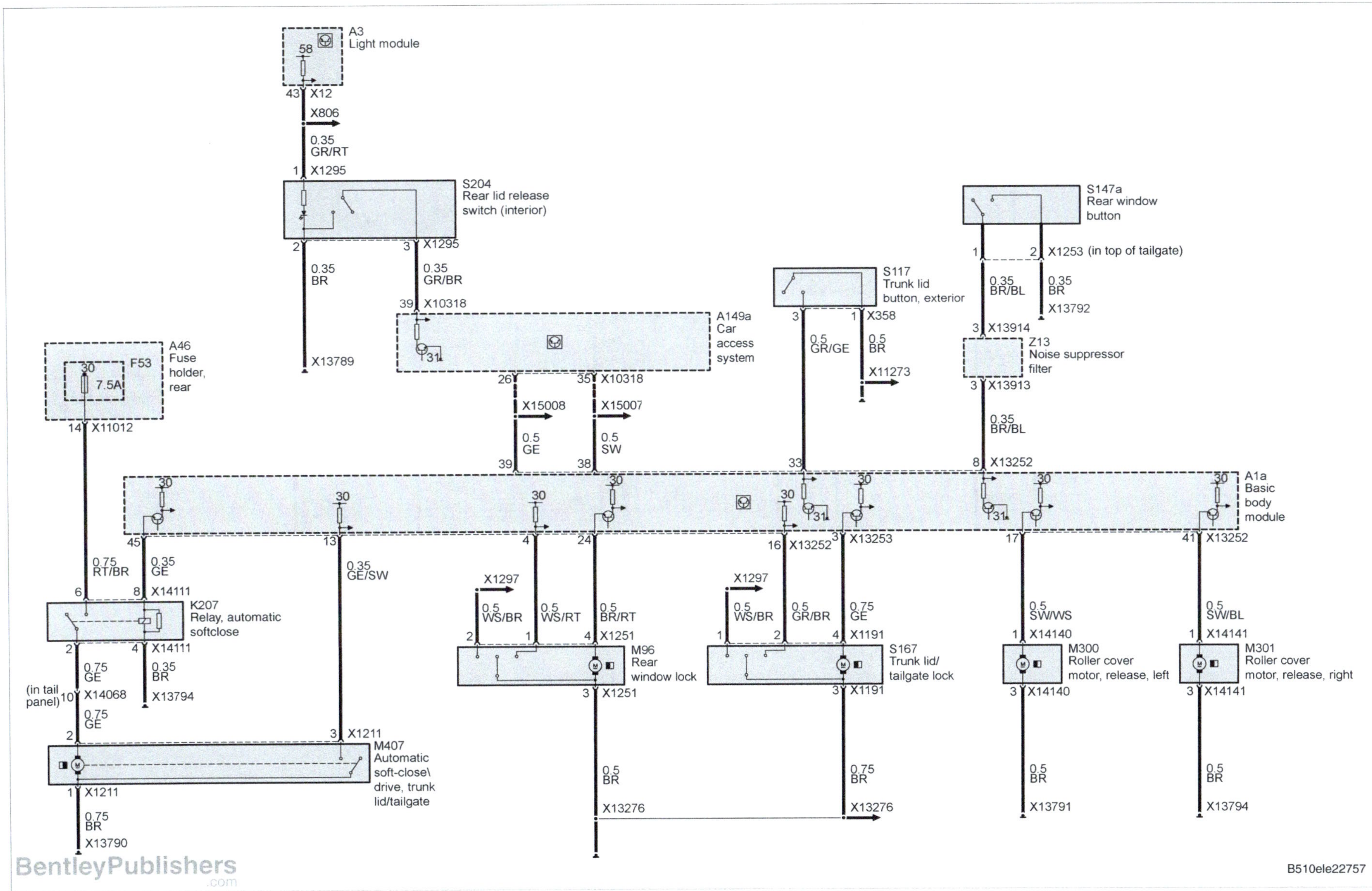

Central locking, tailgate
Wagon from 09 / 2005

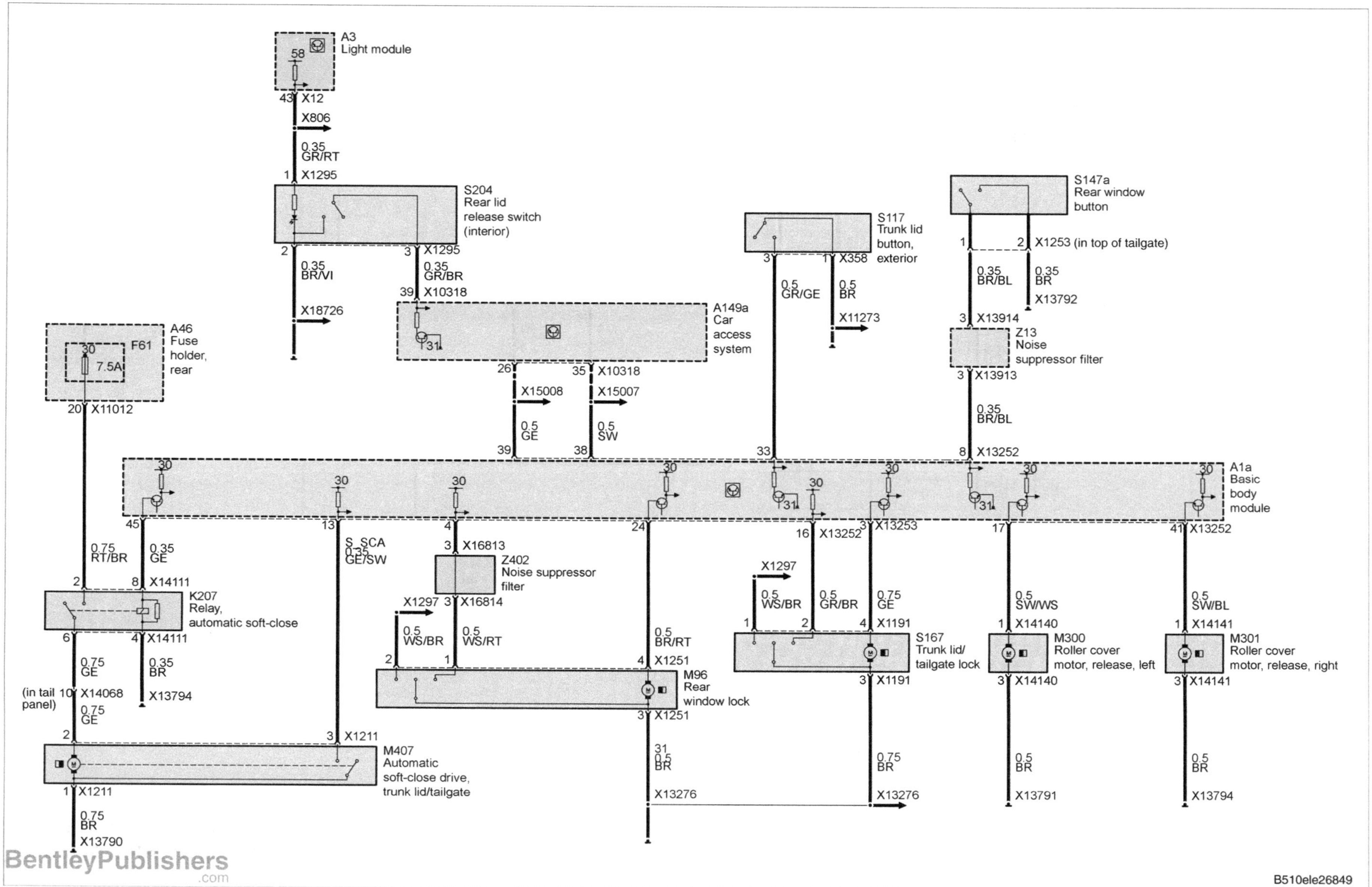

BentleyPublishers.com

B510ele26849

Electric sunroof
Sedan to 09 / 2005

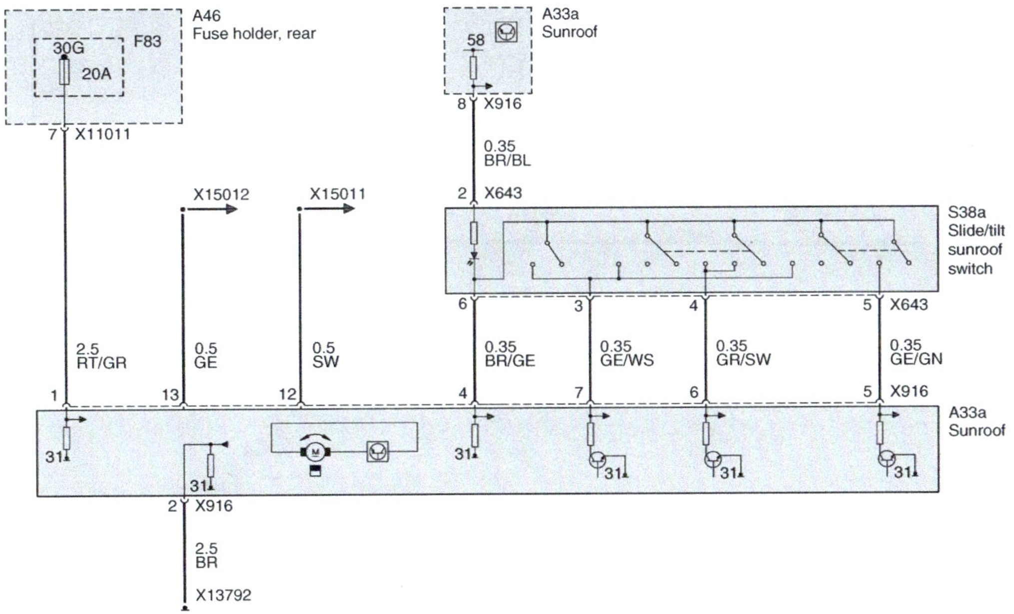

BentleyPublishers.com

B510ele19439

Electric sunroof
Sedan from 09 / 2005

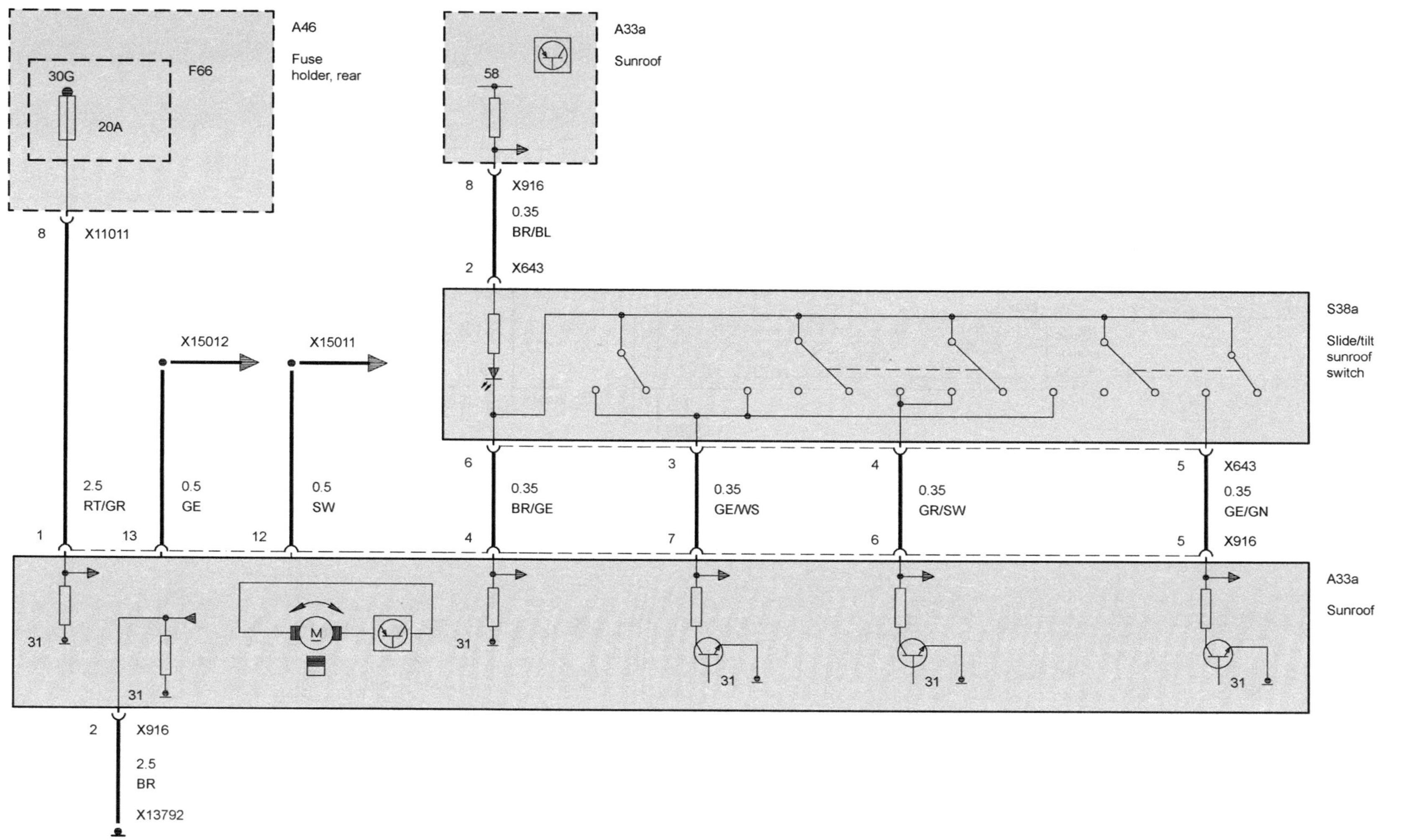

BentleyPublishers.com

B510ele23441

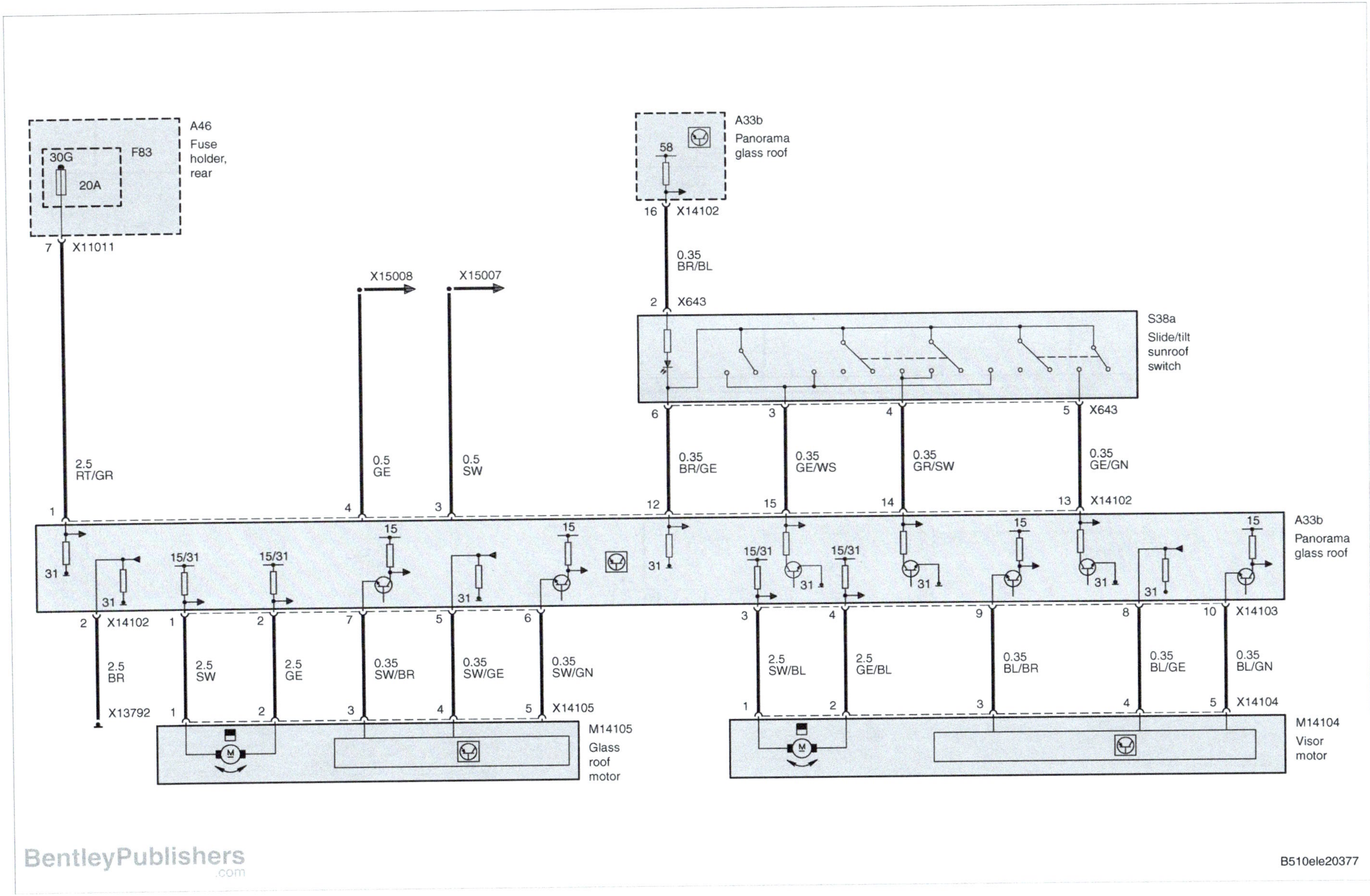
A46
Fuse holder, rear
F83
30G
20A
7
X11011
2.5
RT/GR
X15008
0.5
GE
X15007
0.5
SW
A33b
Panorama glass roof
58
16
X14102
0.35
BR/BL
2
X643
S38a
Slide/tilt sunroof switch
6
3
4
5
X643
0.35
BR/GE
0.35
GE/WS
0.35
GR/SW
0.35
GE/GN
1
4
3
12
15
14
13
X14102
A33b
Panorama glass roof
31
15/31
15
2
X14102
1
2
7
5
6
3
4
9
8
10
X14103
2.5
BR
X13792
2.5
SW
2.5
GE
0.35
SW/BR
0.35
SW/GE
0.35
SW/GN
X14105
M14105
Glass roof motor
2.5
SW/BL
2.5
GE/BL
0.35
BL/BR
0.35
BL/GE
0.35
BL/GN
X14104
M14104
Visor motor
M
BentleyPublishers.com
B510ele20377

Electric sunroof
Wagon from 09 / 2005

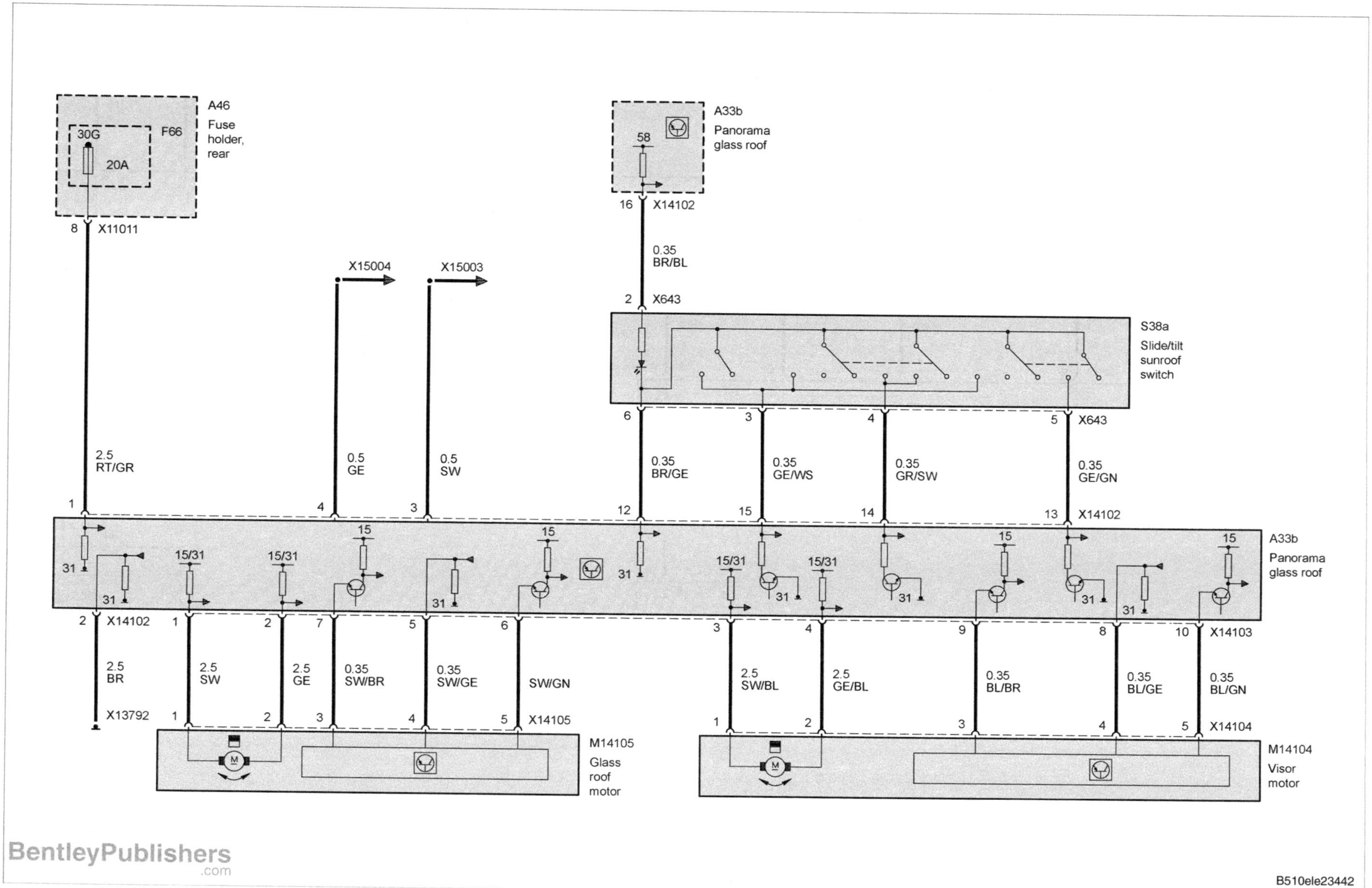

BentleyPublishers.com

B510ele23442

Horn
up to 09 / 2005

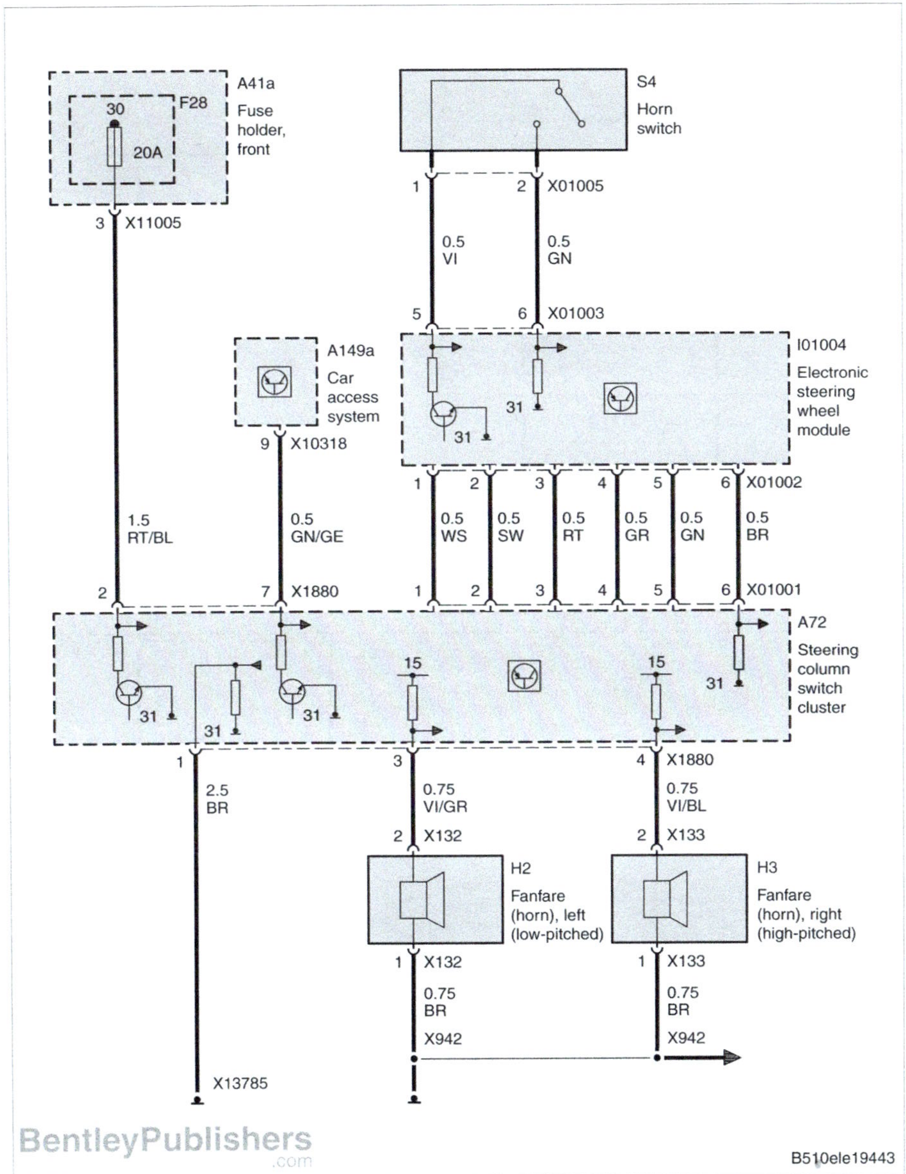

Horn
as of 09 / 2005

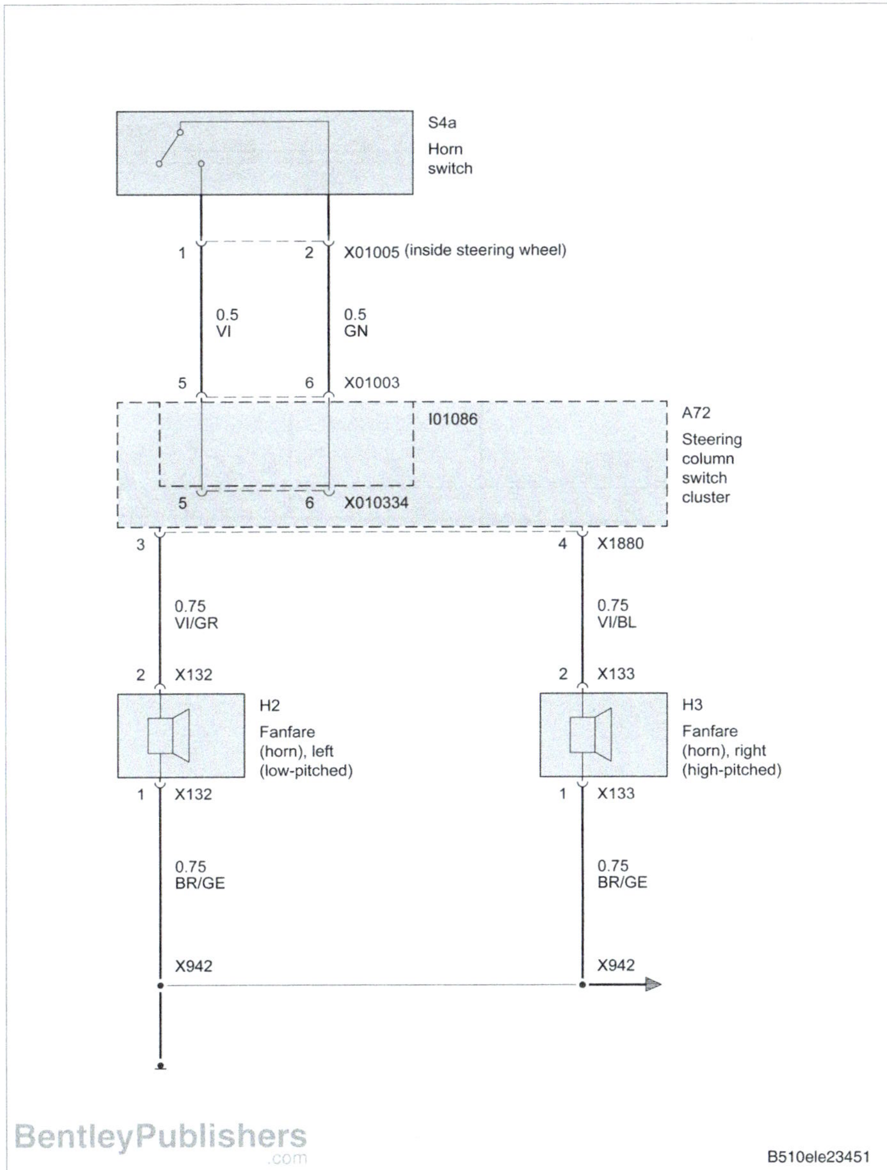

Driver's seat heating with SMFA Seat module, driver up to 03 / 2007

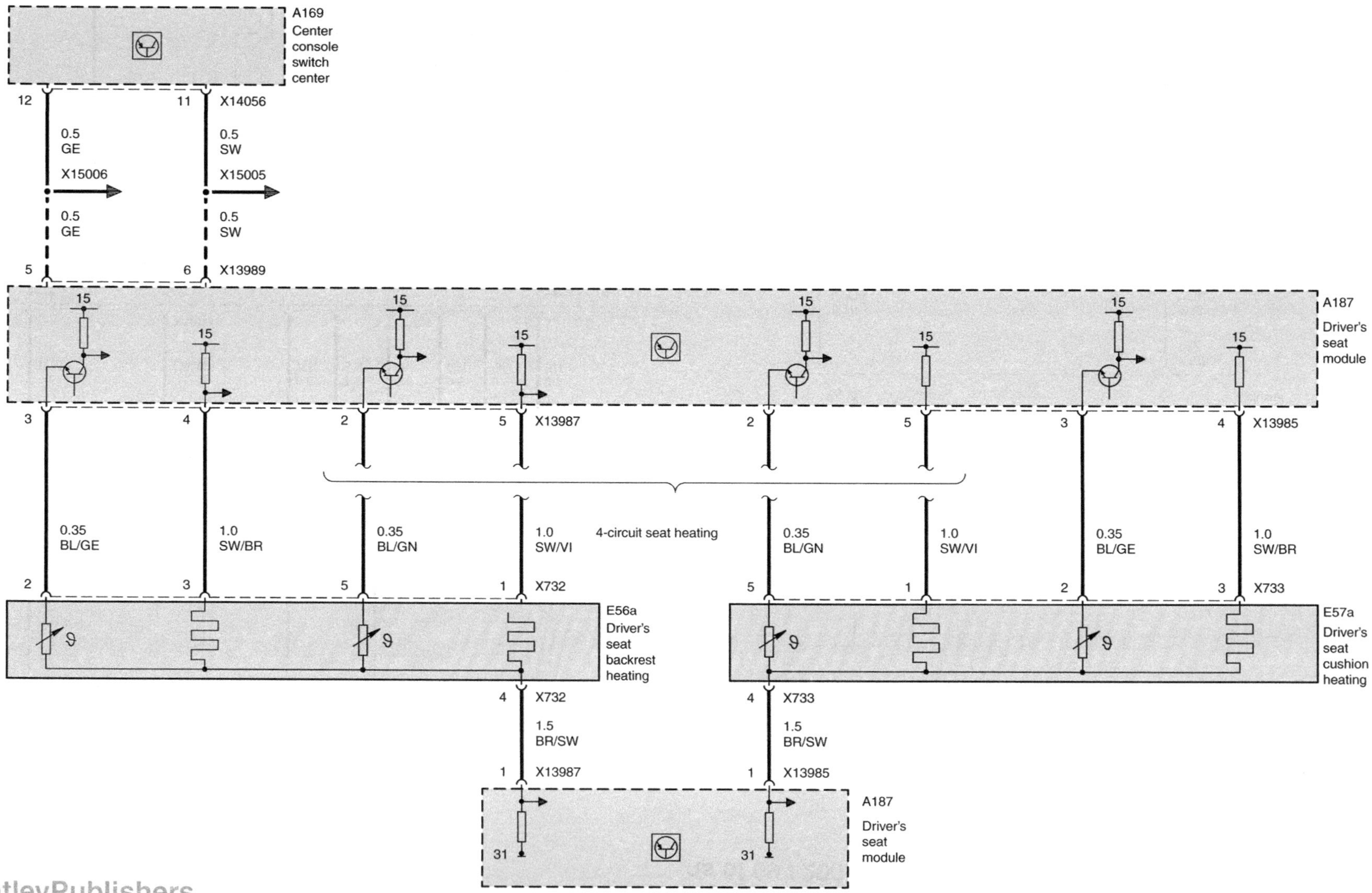

BentleyPublishers
.com

B510ele19437

Driver's seat heating with SMFA Seat module, driver as of 03 / 2007

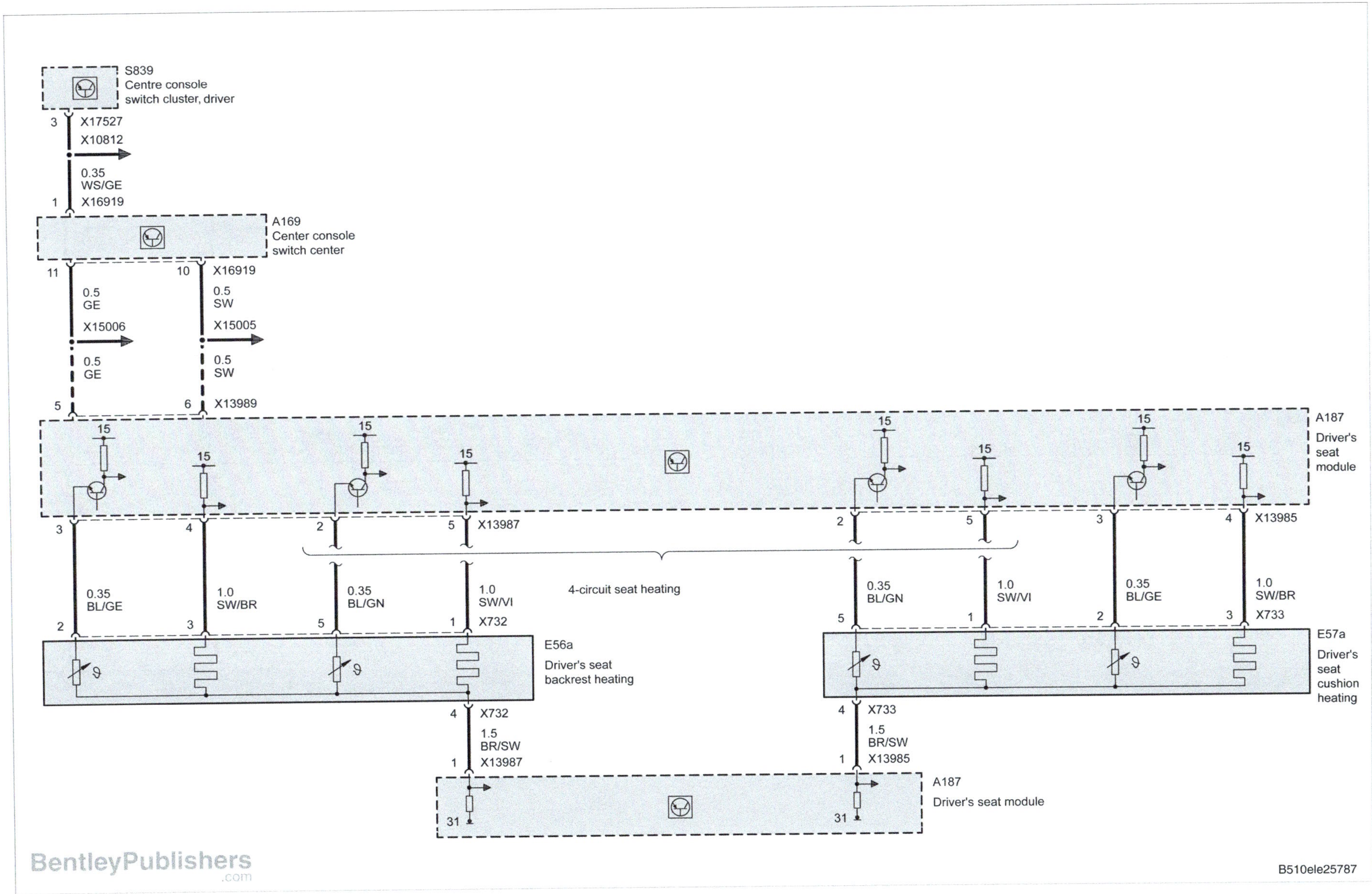

Passenger's seat heating with SMBF Seat module, passenger up to 03 / 2007

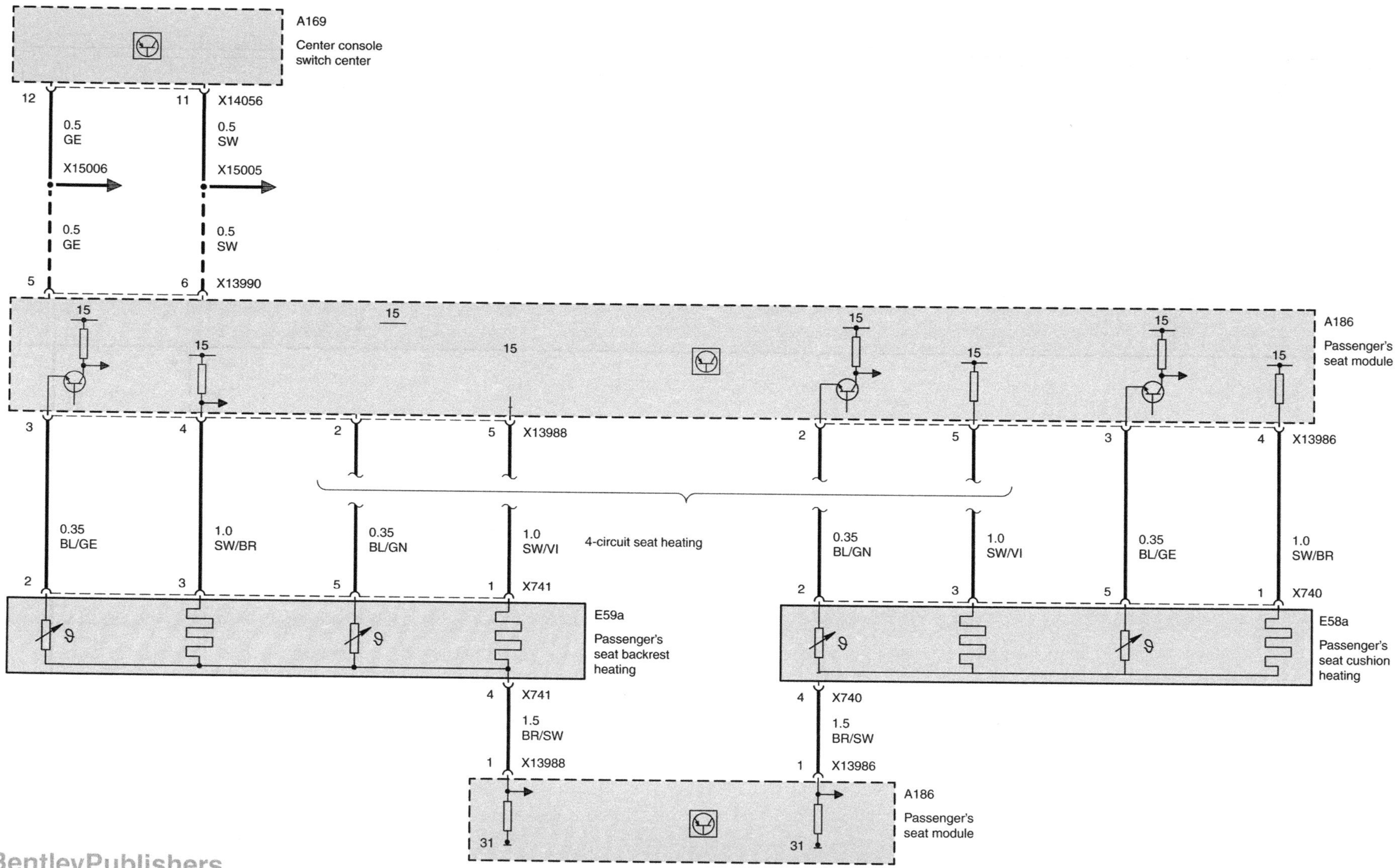

BentleyPublishers.com

B510ele19438

Passenger's seat heating with SMBF Seat module, passenger as of 03 / 2007

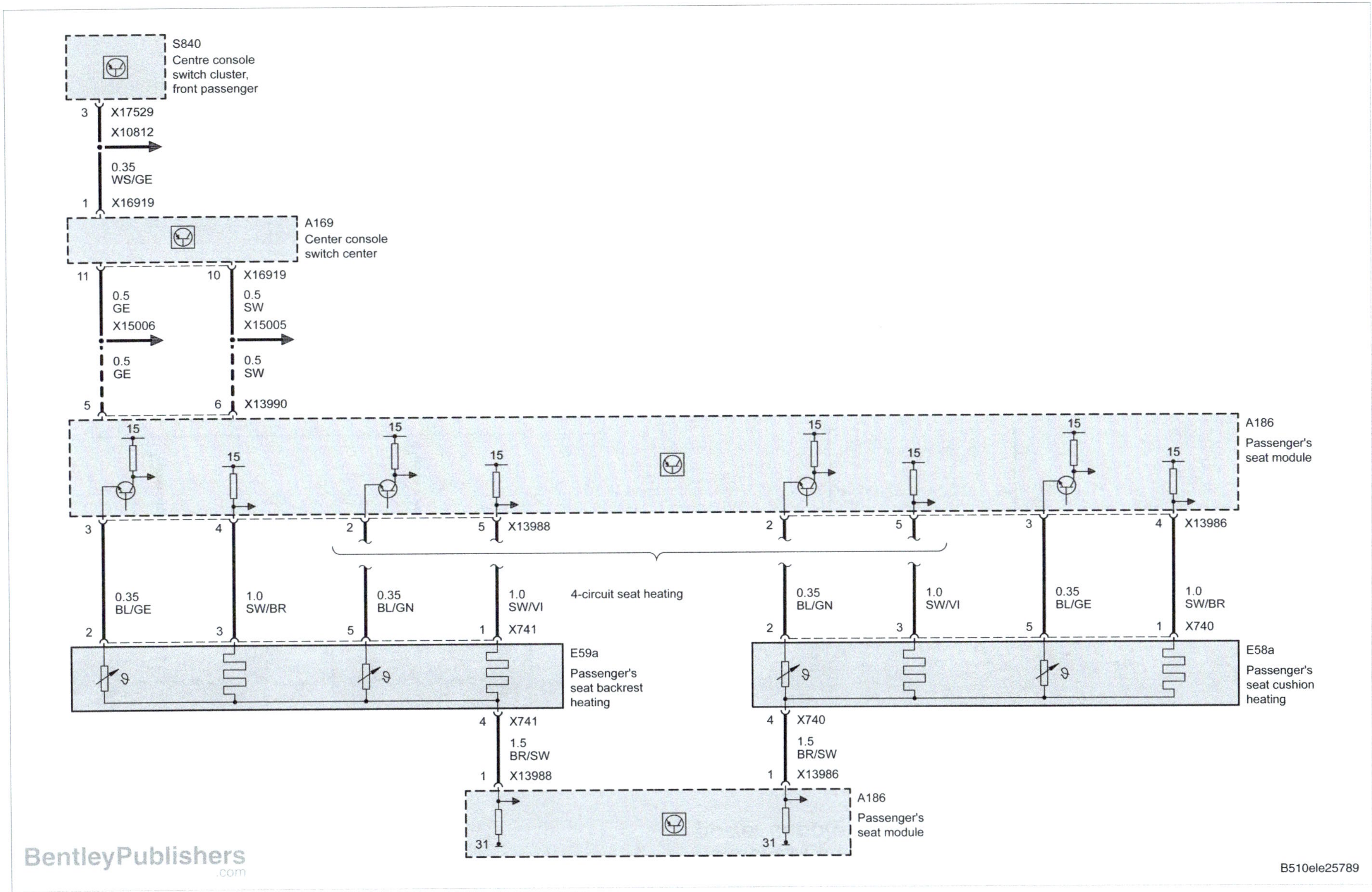

Driver's seat module, power supply

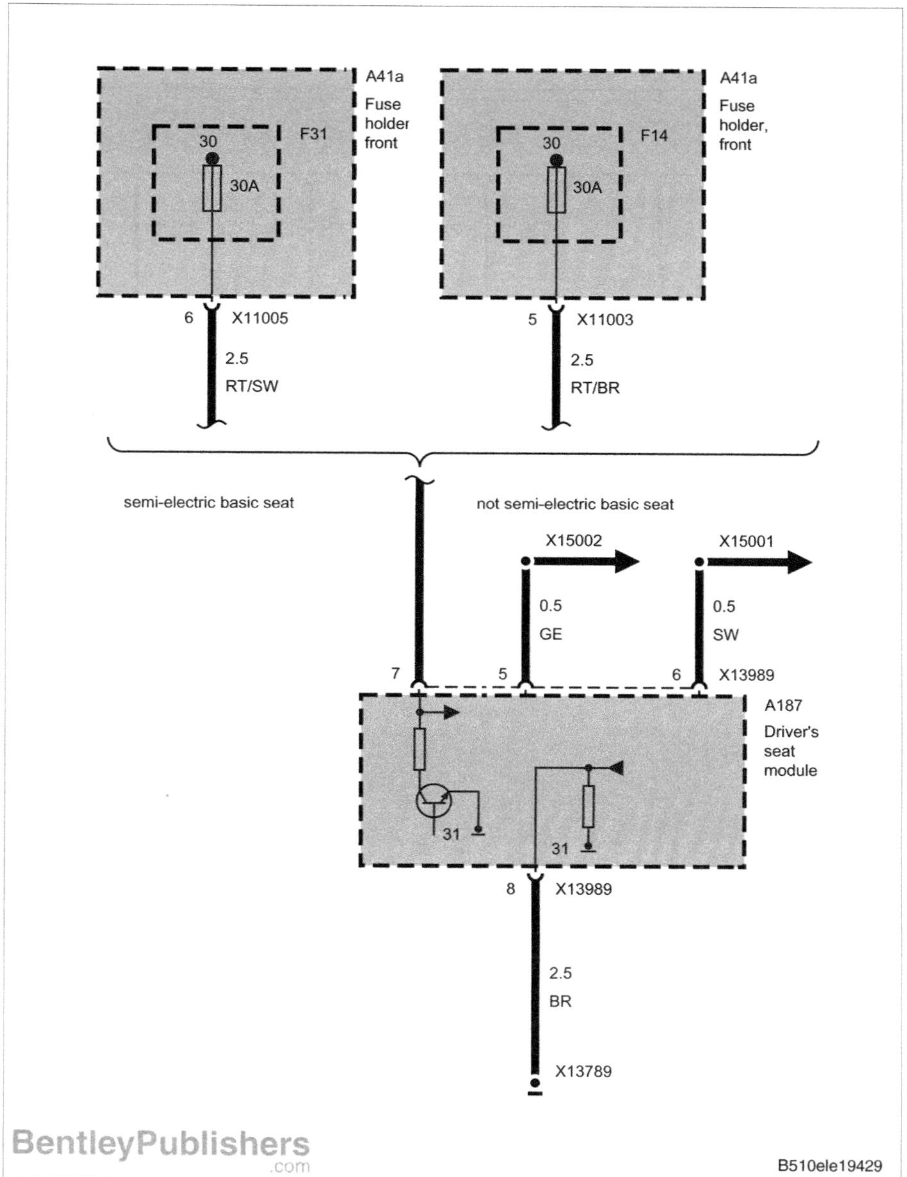

Passenger's seat module power supply

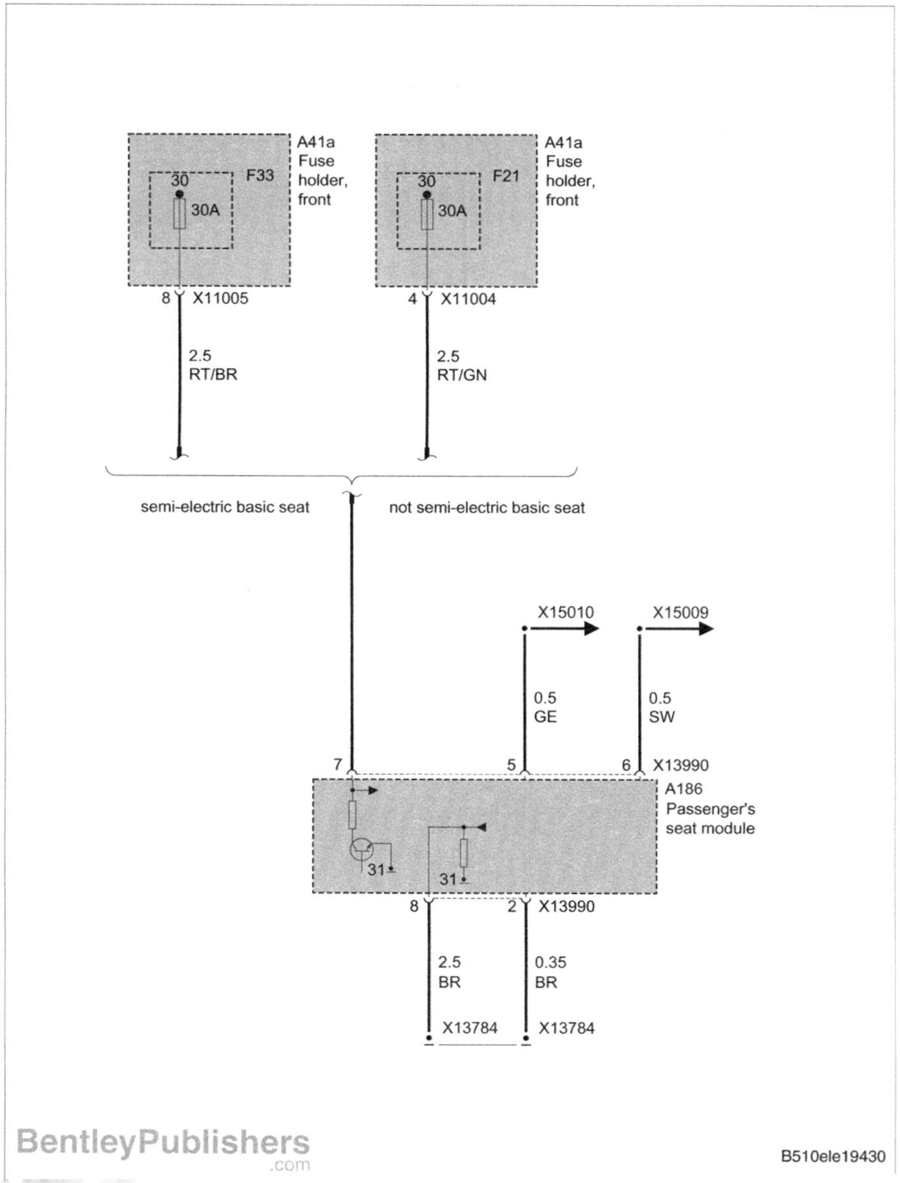

BentleyPublishers.com
BentleyPublishers.com

MOST-bus network
up to 09 / 2006

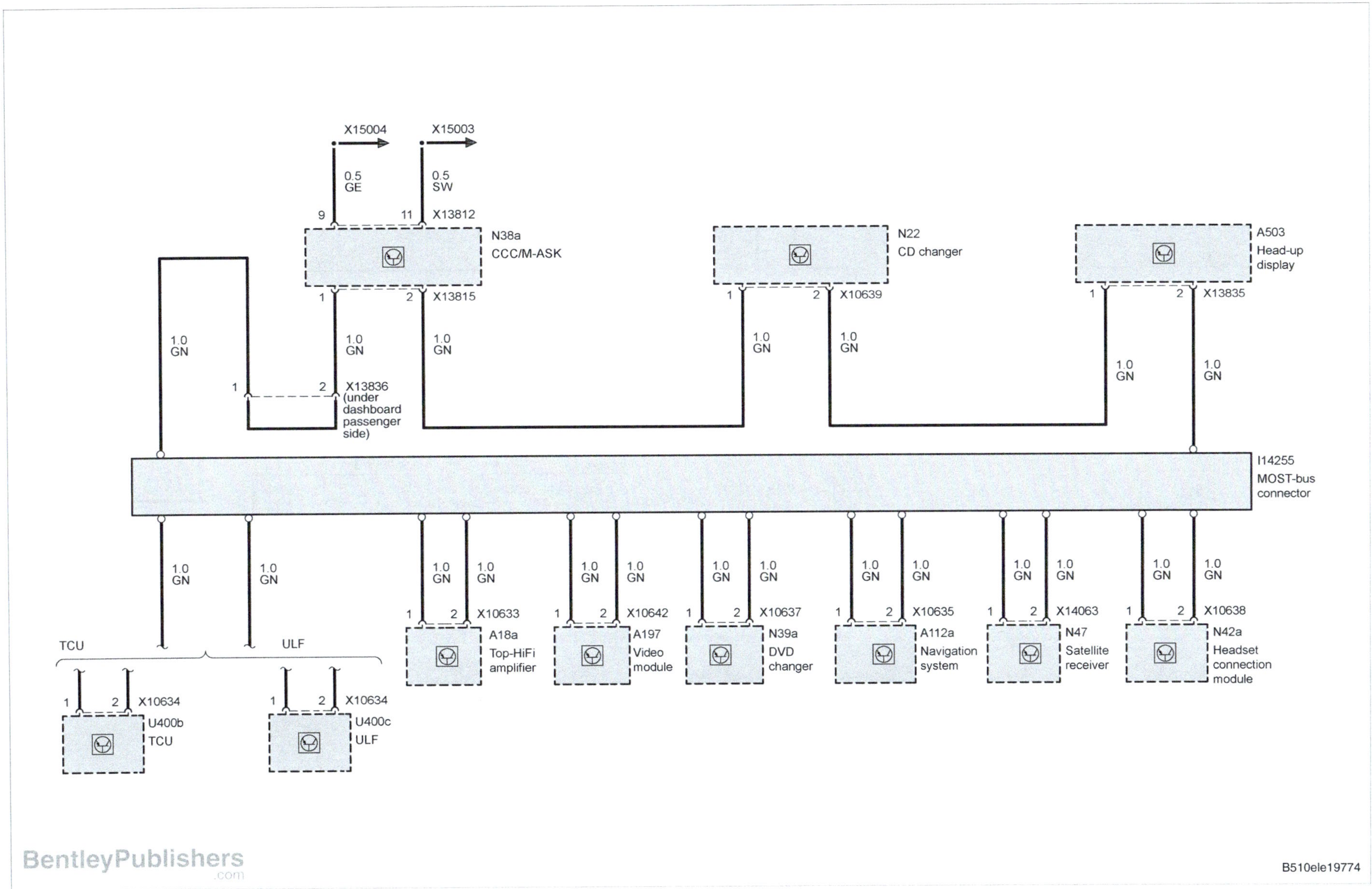

MOST-bus network
as of 09 / 2006 without CIC Car Information Computer

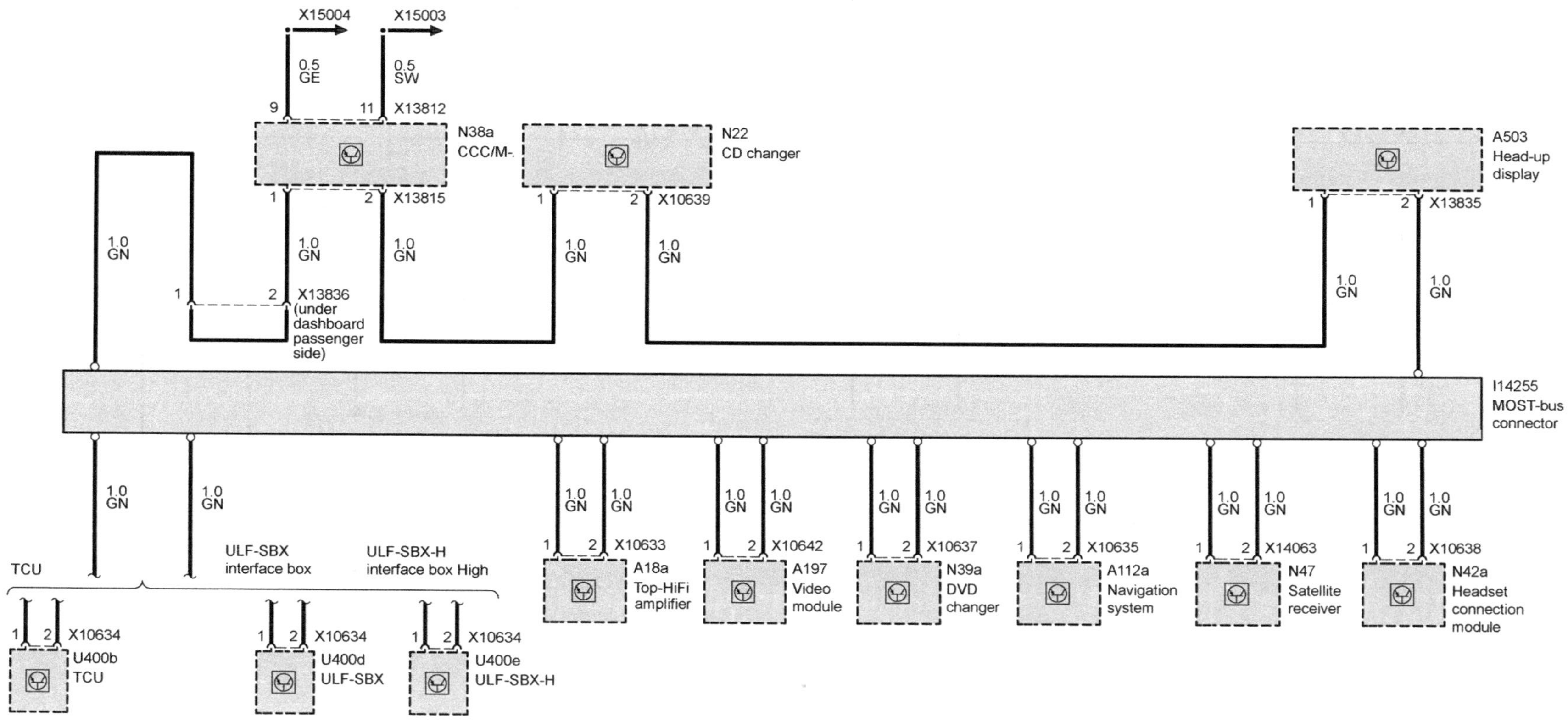

BentleyPublishers.com

B510ele26129

MOST-bus network with CIC Car Information Computer

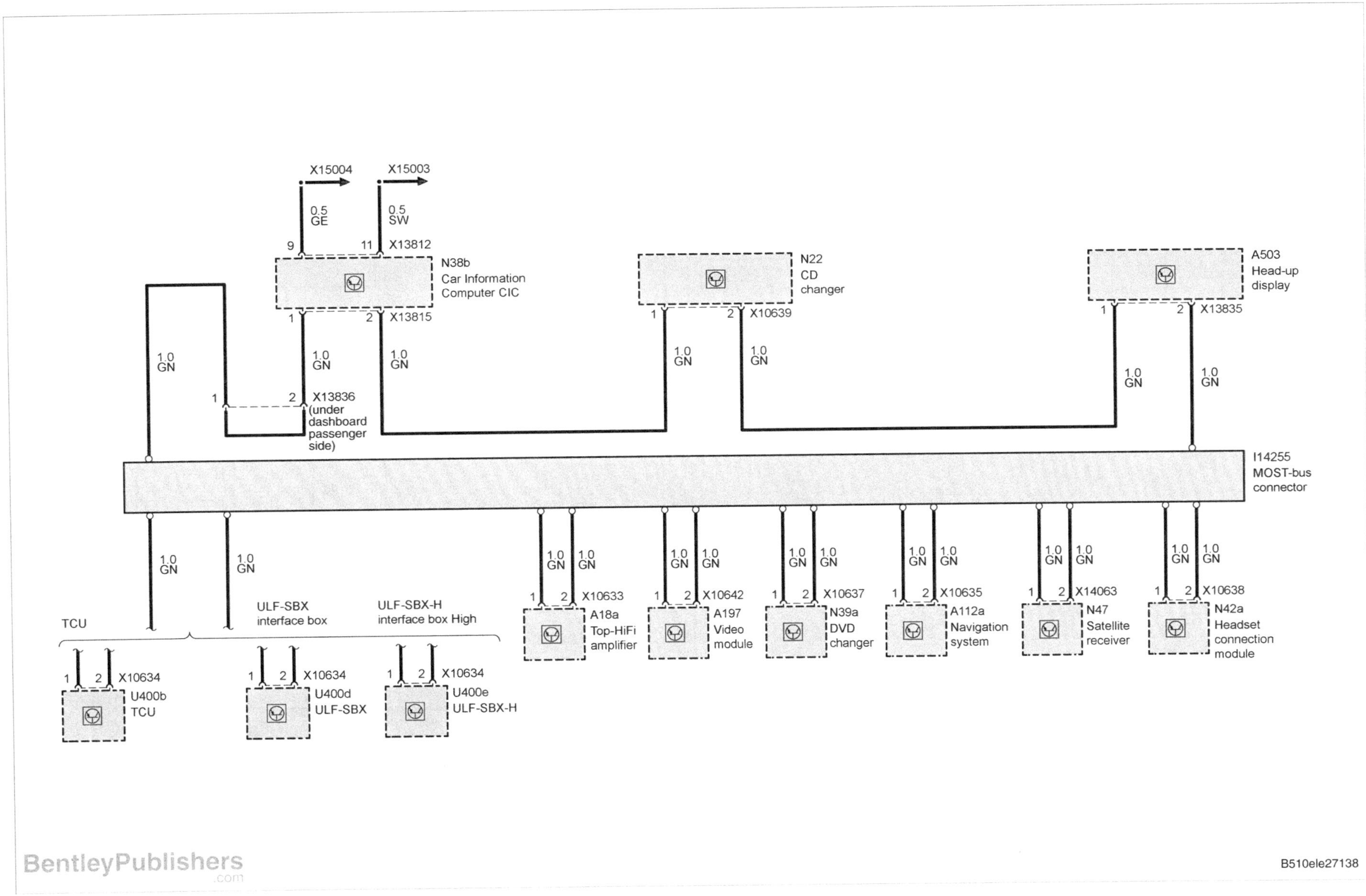

Wiper / washer system up to 03 / 2005

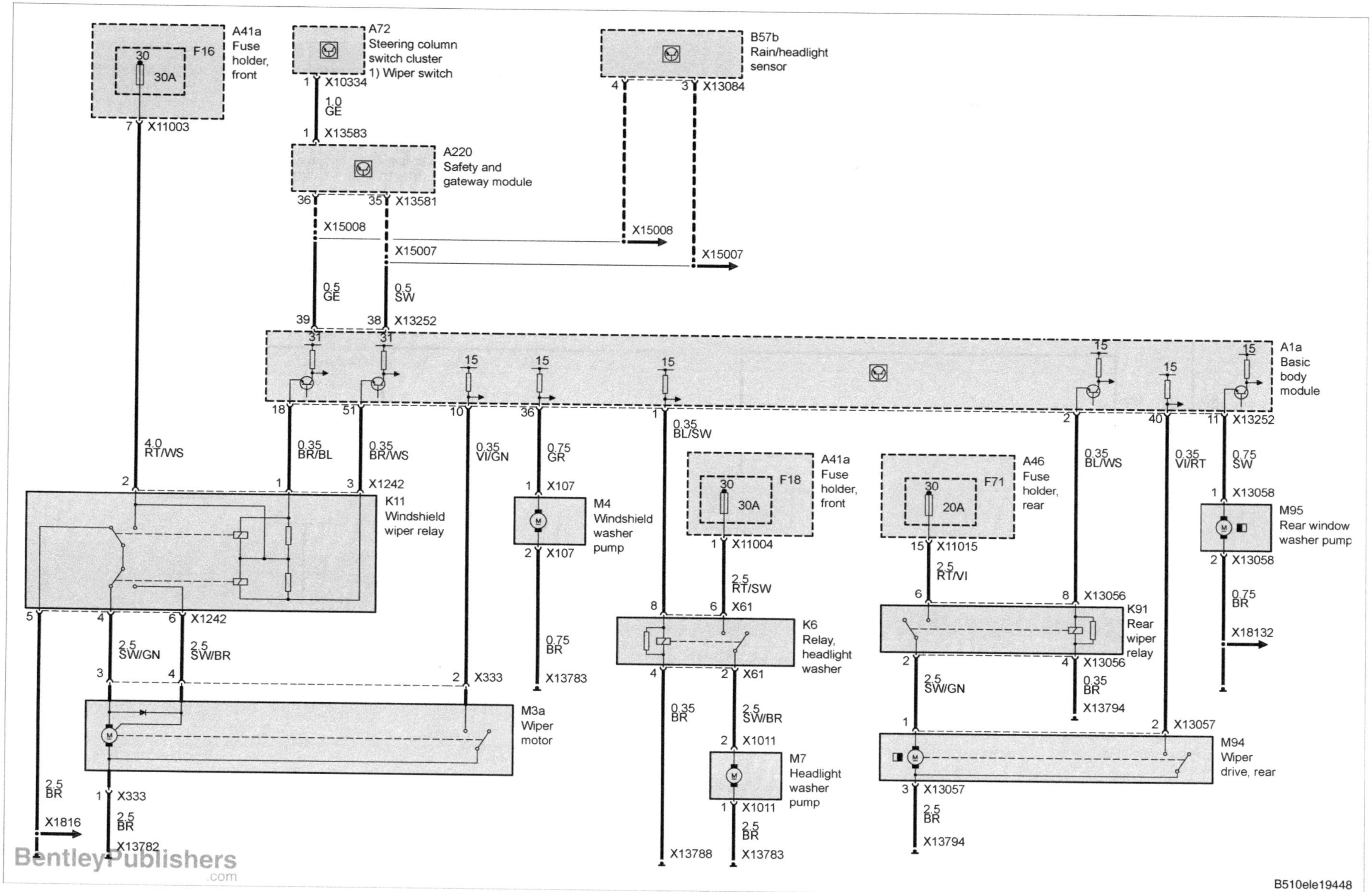

BentleyPublishers.com

B510ele19448

Wiper / washer system
as of 03 / 2005 up to 09 / 2005

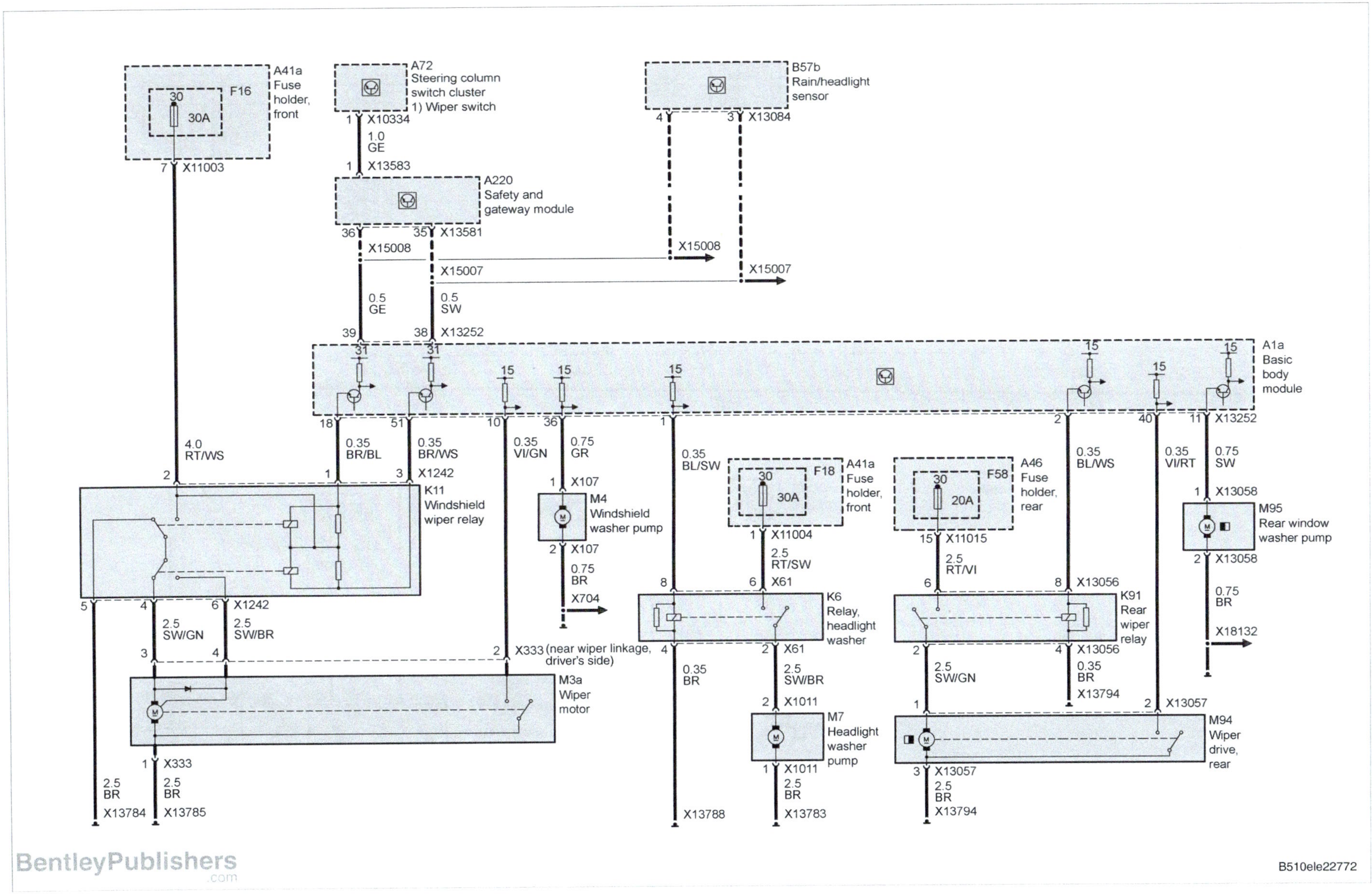

BentleyPublishers.com

Wiper / washer system as of 09 / 2005

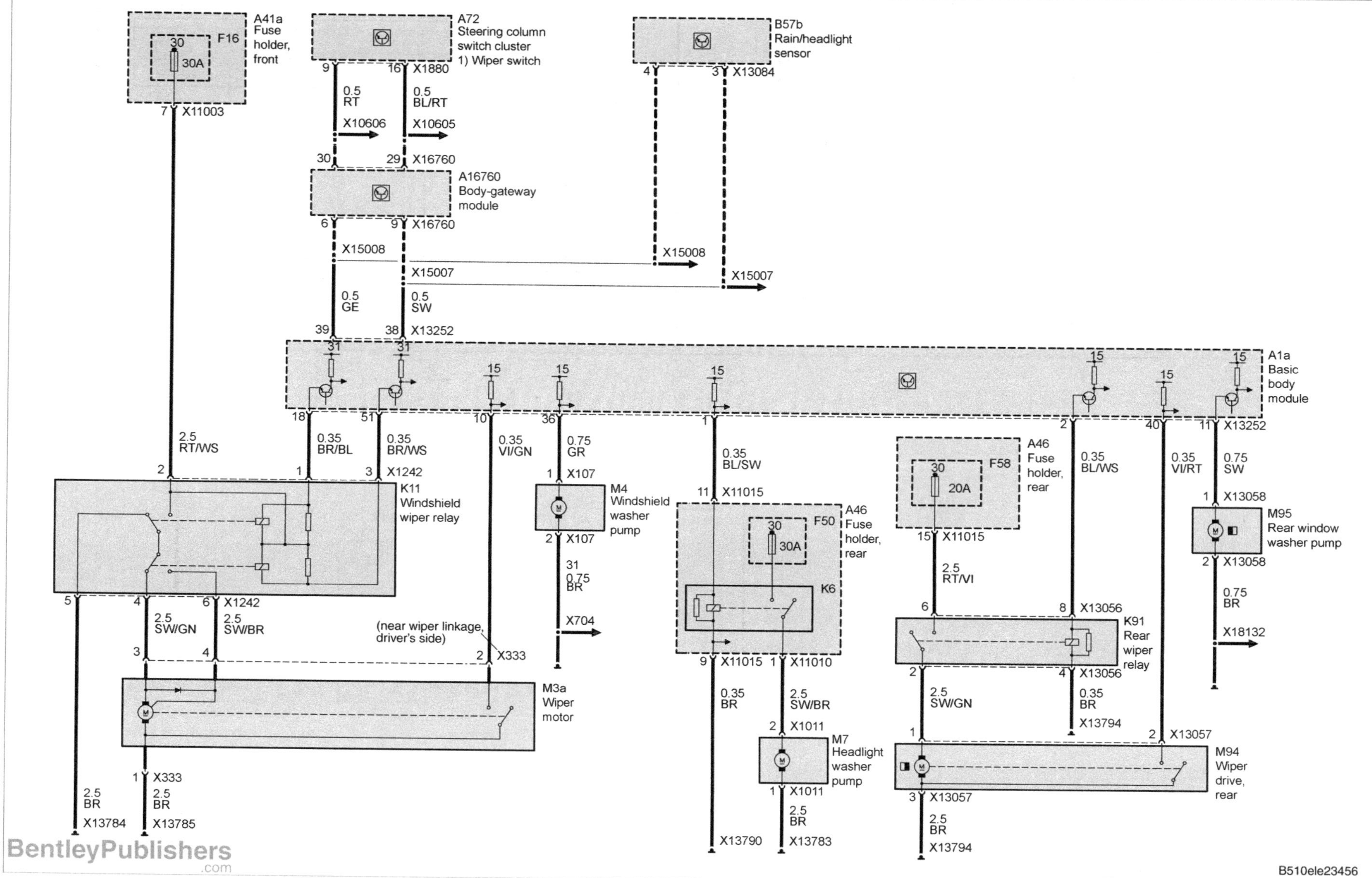

BentleyPublishers.com

OBD On-Board Diagnostics

TABLE

General

This chapter outlines the fundamentals and equipment requirements of On-Board Diagnostics II (OBD II) standards as they apply to BMW vehicles. Also covered here is a listing of BMW and OBD II diagnostic trouble codes (DTCs).

On-Board Diagnostics (OBD II)

OBD II standards were developed by the SAE (Society of Automotive Engineers) and CARB (California Air Resources Board). OBD II is the second generation of on-board self-diagnostic equipment requirements. These standards were originally mandated for California vehicles. Since 1996 they have been applied to all passenger vehicles sold in the United States.

On-board diagnostic capabilities are incorporated into the hardware and software of the engine control module (ECM) to monitor virtually every component that can affect vehicle emissions. The OBD II system works to ensure that emissions remain as clean as possible over the life of the vehicle.

Each emission-influencing component is checked by a diagnostic routine (called a monitor) to verify that it is functioning properly. If a problem or malfunction is detected, the **diagnostic executive** built into the OBD II system illuminates a malfunction indicator light (MIL) on the instrument panel.

The OBD II system also stores diagnostic trouble codes (DTCs) about the detected malfunction in the ECM so that a repair technician can accurately find and fix the problem. Specialized OBD II scan tool equipment is needed to access the fault memory and OBD II data.

BentleyPublishers.com
BentleyPublishers.com – All Rights Reserved

The extra hardware needed to operate the OBD II system consists mainly of the following:

- Additional oxygen sensors downstream of the catalytic converters
- Fuel tank pressure sensor and device to pressurize fuel storage system
- Several engine and performance monitoring devices
- Standardized 16-pin OBD II connector under the dashboard
- Upgraded components for the federally required reliability mandate

Malfunction indicator light (MIL)

OBD II software cause the malfunction indicator light (MIL) to illuminate when emission levels exceed 1.5 times Federal standards.

For 5 Series vehicles covered by this manual, two MIL symbols are used.

MIL illuminates under the following conditions:

- Engine management system fault detected for **two** consecutive OBD II drive cycles. See **Drive cycle** in this repair group.
- Catalyst damaging fault detected.
- Component malfunction causes emissions to exceed 1.5 times OBD II standards.
- Manufacturer-defined specifications exceeded.
- Implausible input signal.
- Misfire faults.
- Leak in evaporative system.
- Oxygen sensors observe no purge flow from purge valve / evaporative system.
- ECM fails to enter closed-loop operation within specified time.
- ECM or automatic transmission control module (TCM) in “limp home” mode.
- Ignition key ON before cranking (bulb check function).

OBD II fault memory (including the MIL) can only be reset using a special scan tool. Removing the connector from the ECM or disconnecting the battery does not erase the fault memory.

Additional MIL information:

- A fault code is stored within the ECM upon the first occurrence of a fault in the system being checked.
- Two **complete** consecutive drive cycles with the fault present illuminate the MIL. The exception to the two-fault requirement is a catalyst-damaging fault, which illuminates the MIL immediately.
- If the second drive cycle was not complete and the fault was not checked, the ECM counts the third drive cycle as the next consecutive drive cycle. The MIL illuminates if the system is checked and the fault is still present.
- Once the MIL is illuminated, it remains illuminated until the vehicle completes three consecutive drive cycles without detecting a fault.
- An existing fault code is cleared from memory automatically when the vehicle completes **40** consecutive drive cycles without the fault being detected.

BentleyPublishers.com
BentleyPublishers.com—All Rights Reserved

In order to automatically clear a catalyst-damaging fault from memory, the condition under which the fault occurred must be evaluated for **80** consecutive drive cycles without the fault reoccurring.

A generic scan tool connected to the OBD II plug can display diagnostic trouble codes (DTCs), along with the conditions associated with the illumination of the MIL. Using a more advanced or BMW-dedicated scan tool, additional proprietary information is normally available.

Scan tool and scan tool display

The complexity of the OBD II system requires that all diagnostics begin by connecting a scan tool to the 16-pin OBD II plug inside the vehicle.

OBD II standards require that the 16-pin OBD II plug be located within three (3) feet of the driver and not require any tools to access.

Professional diagnostic scan tools available at the time of this printing include the BMW factory tools (GT1, ISTA) and a small number of aftermarket BMW-specific tools. See **020 Maintenance**.

In addition to the professional line of scan tools, inexpensive generic OBD II scan tool software programs and handheld units are readily available. Though limited, they are nonetheless powerful diagnostic tools. These tools read live data streams and freeze frame data as well as a host of other valuable diagnostic data.

Diagnostic monitors

Diagnostic monitors run tests and checks on specific emission control systems, components, and functions.

A complete drive cycle is required for the tests to be valid. See **Drive cycle** in this repair group. The diagnostic monitor signals the ECM of the loss or impairment of the signal or component and determines if a signal or sensor is faulty based on 3 conditions:

- Signal or component shorted to ground
- Signal or component shorted to B+
- Signal or component missing (open circuit)

The OBD II system monitors all emission control systems that are installed. Emission control systems vary by vehicle model and year. For example, a vehicle may not be equipped with secondary air injection, so no secondary air readiness code would be present.

OBD II software monitors the following:

- Oxygen sensors
- Catalysts
- Engine misfire
- Fuel tank evaporative control system
- Secondary air injection
- Fuel system

BentleyPublishers.com
BentleyPublishers.com–All Rights Reserved

Oxygen sensor monitoring. When driving conditions allow, response rate and switching time of each oxygen sensor is monitored. The oxygen sensor heater function is also monitored. The OBD II system differentiates between precatalyst and post-catalyst oxygen sensors and reads each one individually. In order for the oxygen sensor to be effectively monitored, the system must be in closed loop operation.

Catalyst monitoring. This strategy monitors the two heated oxygen sensors per bank of cylinders. It compares the oxygen content going into the catalytic converter to the oxygen leaving the converter.

The diagnostic executive knows that most of the oxygen should be used up during the oxidation phase. If it sees higher than programmed values, a fault is set and the MIL illuminates.

Misfire detection. This strategy monitors crankshaft speed fluctuations and determines if an engine misfire occurs by monitoring variations in speed between each crankshaft sensor trigger point. This strategy is so finely tuned that it can determine the severity of the misfire.

The system determines if a misfire is occurring, as well as other pertinent misfire information such as:

- Specific cylinder(s)
- Severity of the misfire event
- Emissions relevant or catalyst damaging

Misfire detection is an on-going monitoring process that is only disabled under certain limited conditions.

Secondary air injection monitoring. Secondary air injection is used to reduce HC and CO emissions during engine warm up. Immediately following a cold engine start (-10° to 40°C), fresh air (and therefore oxygen) is pumped directly into the exhaust manifold. By injecting additional oxygen into the exhaust manifold, catalyst warm-up time is reduced.

Secondary air system components are:

- Electric air injection pump
- Electric pump relay
- Non-return valve
- Vacuum / vent valve
- Mini air mass meter

The secondary air system is monitored via the use of the pre-catalyst oxygen sensors and a mini air mass meter. The oxygen sensors monitor exhaust gas for a lean condition once the air pump is active and air is injected into the system.If the oxygen sensor signal does not change, a fault is set and the faulty bank(s) identified. If after completing the next cold start a fault is again present, the MIL illuminates. The mini air mass meter monitors the air supplied by the secondary air pump when commanded on. If the mini air mass meter detects no air mass or insufficient air mass, a fault is stored in the ECM and the malfunction indicator light (MIL) is activated.

BentleyPublishers.com
BentleyPublishers.com—All Rights Reserved

Fuel system monitoring. This monitor looks at the fuel delivery needed (long / short term fuel trim) for proper engine operation based on programmed data. If too much or not enough fuel is delivered over a predetermined time, a DTC is set and the MIL is turned on.

Fuel trim refers to adjustments to base fuel schedule. Long-term fuel trim refers to gradual adjustments to the fuel calibration adjustment as compared to short term fuel trim. Long term fuel trim adjustments compensate for gradual changes that occur over time.

Fuel system monitoring monitors the calculated injection time (ti) in relation to engine speed, load, and precatalytic converter oxygen sensor(s) signals.

Using this data, the system optimizes fuel delivery for all engine operating conditions.

Evaporative system monitoring. This monitor checks the fuel storage system and related fuel lines for leaks. It can detect very small leaks anywhere in the system.

A leak detection unit (DMTL) is used to pressurize the EVAP system on a continuous basis (as the drive cycle allows) and to check system integrity.

Drive cycle

The OBD II drive cycle is an important concept in understanding OBD II requirements. The purpose of the drive cycle is to run all of the emission-related on-board diagnostics over a broad range of driving conditions.

A drive cycle is considered complete when all of the diagnostic monitors have run their tests without interruption. For a drive cycle to be initiated, the vehicle must be started cold and brought up to 160°F and at least 40°F above its original starting temperature.

Readiness codes

Inspection/maintenance (I/M) readiness codes are mandated as part of OBD II. The readiness code is stored after complete diagnostic monitoring of specified components and systems is carried out. The readiness code function was designed to prevent manipulating an I/M emission test procedure by clearing faults codes or disconnecting the ECM or battery.

Readiness codes indicate whether the OBD II system is actually ready to monitor the various emission control systems on the vehicle. The vehicle must complete a drive cycle to set readiness codes. The code is binary:

- 0 for ready
- 1 for not ready

BentleyPublishers.com
BentleyPublishers.com—All Rights Reserved

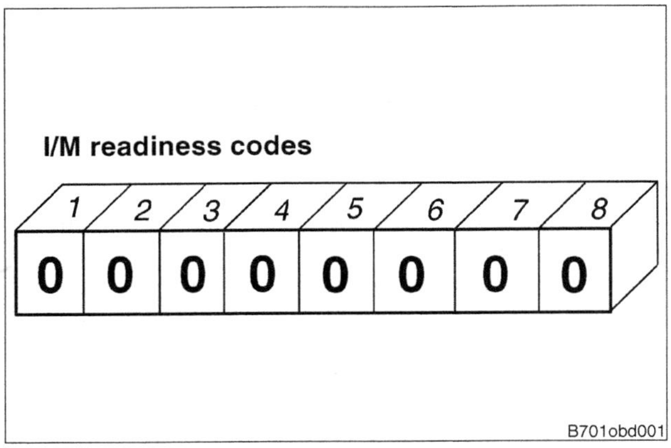

The parameters which are monitored for readiness are:

1. Catalyst efficiency
2. Catalyst heating
3. Fuel tank evaporative control
4. Secondary air injection
5. A/C refrigerant
6. Oxygen sensors
7. Oxygen sensor heaters
8. Exhaust gas recirculation

When all zeros are displayed, the system has established readiness. Readiness codes can be displayed using BMW and aftermarket scan tools.

Readiness codes are set to 1 (not ready) in the following cases:

- The battery or ECM is disconnected.
- DTCs are erased after completion of repairs and a drive cycle is not completed.

BentleyPublishers.com
BentleyPublishers.com—All Rights Reserved

Diagnostic trouble codes (DTCs)

SAE standard J2012 mandates a 5-digit diagnostic trouble code (DTC) standard. Each digit represents a specific value. Emission related DTCs start with the letter P for power train. When the MIL illuminates it indicates that a DTC has been stored:

- DTCs are stored as soon as they occur, whether or not the MIL illuminates.
- DTCs store and display a time stamp.
- DTCs record the current fault status: Present, not currently present, or intermittent.

DTC digit interpretation	
1st digit	
P	powertrain
B	body
C	chassis
2nd digit	
0	SAE
1	BMW
3rd digit	
0	total system
1	air / fuel induction
2	fuel injection
3	ignition system or misfire
4	auxiliary emission control
5	vehicle speed & idle control
6	ECM inputs / outputs
7	transmission
4th - 5th digits	individual circuits or components

DTC example: P 0 3 0 6

- **P:** A powertrain problem
- **0**: SAE sanctioned ('generic')
- **3:** Related to an ignition system / misfire
- **06** Misfire detected at cylinder #6

DTCs provide a freeze frame or snap-shot of a vehicle performance or emissions fault at the moment that the fault first occurs. This information is accessible through generic scan tools.

Freeze frame data contains, but isn't limited to, the following information:

- Engine load (calculated)
- Engine rpm
- Short and long-term fuel trim
- Vehicle speed
- Coolant temperature
- Intake manifold pressure
- Open / closed loop operation
- Fuel pressure (if available)

BentleyPublishers.com
BentleyPublishers.com–All Rights Reserved

Diagnostic trouble codes (DTCs)

Table a. Diagnostic trouble codes (DTCs)

P-code	BMW FC	Definition
P00B2	2EEC	Radiator Coolant Temperature Sensor Circuit Range / Performance
P00B3	2EEA	Radiator Coolant Temperature Sensor Circuit Low
P00B4	2EEA	Radiator Coolant Temperature Sensor Circuit High
P0010	2A80	'A' Camshaft Position Actuator Circuit Open (Bank 1)
P0011		'A' Camshaft Position Timing Over-Advanced or System Performance (Bank 1)
P0012	2A82	'A' Camshaft Position Timing Over-Retarded (Bank 1)
P0013	2A85	'B' Camshaft Position Actuator Circuit Open (Bank 1)
P0014		'B' Camshaft Position Timing Over-Advanced or System Performance (Bank 1)
P0015	2A87	'B' Camshaft Position Timing Over-Retarded (Bank 1)
P0016	2AA4	Camshaft Position Sensor Correlation (Bank 1 Sensor 'A')
P0017	2AA5	Camshaft Position Sensor Correlation (Bank 1 Sensor 'B')
P0018	2A8F	Camshaft / Crankshaft Position Sensor Correlation (Bank 2 Sensor 'A')
P0019	2A91	Camshaft / Crankshaft Position Sensor Correlation (Bank 2 Sensor 'B')
P0020		VANOS Camshaft Timing Inlet Rationality
P0021		VANOS Inlet Camshaft Range Rationality Check
P0022		VANOS Difference Between Actual and Target
P0023		VANOS Outlet Circuit
P0024		VANOS Outlet Rationality Check
P0025		VANOS Outlet Rationality Check
P0030	2C9C	HO2S Heater Control Circuit (Bank 1 Sensor 1)
P0031	2C9C	HO2S Heater Control Circuit Low (Bank 1 Sensor 1)
P0032	2C9C	HO2S Heater Control Circuit High (Bank 1 Sensor 1)
P0036	2C9E	HO2S Heater Control Circuit (Bank 1 Sensor 2)
P0037	2C9E	HO2S Heater Control Circuit Low (Bank 1 Sensor 2)
P0038	2C9E	HO2S Heater Control Circuit High (Bank 1 Sensor 2)
P0040	2C24	HO2S Signals Swapped Bank 1 Sensor 1 / Bank 2 Sensor 1
P0041	2C64	HO2S Signals Swapped Bank 1 Sensor 2/ Bank 2 Sensor 2
P0050	2C9D	HO2S Heater Control Circuit (Bank 2 Sensor 1)
P0051	2C9D	HO2S Heater Control Circuit Low (Bank 2 Sensor 1)
P0052	2C9D	HO2S Heater Control Circuit High (Bank 2 Sensor 1)
P0053	2CA6	HO2S Heater Resistance (Bank 1 Sensor 1)
P0056	2C9F	HO2S Heater Control Circuit (Bank 2 Sensor 2)
P0057	2C9F	HO2S Heater Control Circuit Low (Bank 2 Sensor 2)
P0058	2C9F	HO2S Heater Control Circuit High (Bank 2 Sensor 2)
P0059	2CA7	HO2S Heater Resistance (Bank 2 Sensor 1)
P0070	2F9A	Ambient Air Temperature Sensor Circuit

BentleyPublishers.com
BentleyPublishers.com–All Rights Reserved

Table a. Diagnostic trouble codes (DTCs) (continued)

P-code	BMW FC	Definition
P0071	2F99	Ambient Air Temperature Sensor Circuit Range / Performance
P0072	2F9A	Ambient Air Temperature Sensor Circuit Low
P0073	2F9A	Ambient Air Temperature Sensor Circuit High
P0090	2FBC	Fuel Pressure Regulator 1 Control Circuit Open
p0091	2FBC	Fuel Pressure Regulator 1 Control Circuit Low
P0092	2FBC	Fuel Pressure Regulator 1 Control Circuit High
P0A14	2FAB	Engine Mount ‘A’ Control Circuit Open
P0A15	2FAB	Engine Mount ‘A’ Control Circuit Low
P0A16	2FAB	Engine Mount ‘A’ Control Circuit High
P0A3B	2E97	Generator Over Temperature
P0A3B	2ECF	Generator Over Temperature
P0100	2D16	Mass Air Flow or Volume ‘A’ Circuit
P0101	2D15	Mass or Volume Air Flow Circuit Range/Performance
P0102		Mass or Volume Air Flow Circuit Low Input
P0103		Mass or Volume Air Flow Circuit High Input
P0107		Manifold Absolute Pressure/Barometric Pressure Circuit Low Input
P0108		Manifold Absolute Pressure/Barometric Pressure Circuit High Input
P0111	2F09	Intake Temperature Sensor Bank 1 Temperature Range / Performance
P0112	2F08	Intake Air Temperature Sensor 1 Circuit Low
P0113	2F08	Intake Air Temperature Sensor 1 Circuit High
P0116		Coolant Temperature Sensor rationality Check
P0117	2EE0	Engine Coolant Temperature Circuit Low
P0118	2EE0	Engine Coolant Temperature Circuit High
P0120		General Code Used In Conjuction With Other TPS Fault Code
P0121	2CF6	Throttle/Pedal Position Sensor/Switch 'A' Circuit Range/Performance
P0122	2CF9	Throttle/Pedal Position Sensor/Switch 'A' Circuit Low
P0123	2CF9	Throttle/Pedal Position Sensor/Switch 'A' Circuit High
P0125		Insufficient Coolant Temperature for Closed Loop Fuel Control
P0128	2EF4	Coolant Thermostat (Coolant Temperature Below Thermostat Regulating Temperature)
P0131	2C3F	O2 Sensor Circuit Low Voltage (Bank 1 Sensor 1)
P0132	2C3F	O2 Sensor Circuit High Voltage (Bank 1 Sensor 1)
P0133	2C39	O2 Sensor Circuit Slow Response (Bank 1 Sensor 1)
P0134		O2 Sensor Circuit No Activity Detected (Bank 1 Sensor 1)
P0135	2CA6	O2 Sensor Heater Circuit (Bank 1 Sensor 1)
P0137	2C75	O2 Sensor Circuit Low Voltage (Bank 1 Sensor 2)
P0138	2C73	O2 Sensor Circuit High Voltage (Bank 1 Sensor 2)
P0139	2C7B	O2 Sensor Circuit Slow Response (Bank 1 Sensor 2)

BentleyPublishers.com
BentleyPublishers.com—All Rights Reserved

Table a. Diagnostic trouble codes (DTCs) (continued)

P-code	BMW FC	Definition
P0139		O2 Sensor Circuit Slow Response (Bank 1 Sensor 2)
P0140	2C77	O2 Sensor Circuit No Activity Detected (Bank 1 Sensor 2)
P0141	2CA8	O2 Sensor Heater Circuit (Bank 1 Sensor 2)
P0151	2C40	O2 Sensor Circuit Low Voltage (Bank 2 Sensor 1)
P0152	2C40	O2 Sensor Circuit High Voltage (Bank 2 Sensor 1)
P0153	2C3A	O2 Sensor Circuit Slow Response (Bank 2 Sensor 1)
P0154		O2 Sensor Circuit No Activity Detected (Bank 2 Sensor 1)
P0155	2CA7	O2 Sensor Heater Circuit (Bank 2 Sensor 1)
P0156	2CA7	O2 Sensor Heater Function (Bank 2 Sensor 1)
P0157	2C76	O2 Sensor Circuit Low Voltage (Bank 2 Sensor 2)
P0158	2C74	O2 Sensor Circuit High Voltage (Bank 2 Sensor 2)
P0159	2C7E	O2 Sensor Circuit Slow Response (Bank 2 Sensor 2)
P0159	2C7C	O2 Sensor Circuit Slow Response (Bank 2 Sensor 2)
P0160	2C78	O2 Sensor Circuit No Activity Detected (Bank 2 Sensor 2)
P0161	2CA9	O2 Sensor Heater Circuit (Bank 2 Sensor 2)
P0171	29E0	System Too Lean (Bank 1)
P0171	2A2B	System Too Lean (Bank 1)
P0172	29E0	System Too Rich (Bank 1)
P0172	2A2B	System Too Rich (Bank 1)
P0174	29E1	System Too Lean (Bank 2)
P0174	2A2C	System Too Lean (Bank 2)
P0175	29E1	System Too Rich (Bank 2)
P0175	2A2C	System Too Rich (Bank 2)
P0190	29E2	Fuel Rail Pressure Sensor 'A' Circuit
P0192	29E2	Fuel Rail Pressure Sensor 'A' Circuit Low
P0197		Engine Oil Temperature Sensor Low
P0198		Engine Oil Temperature Sensor High
P02AA	3074	Cylinder 5 Fuel Trim at Maximum Limit
P02AB	3074	Cylinder 5 Fuel Trim at Minimum Limit
P02AE	3075	Cylinder 6 Fuel Trim at Maximum Limit
P02AF	3075	Cylinder 6 Fuel Trim at Minimum Limit
P02A2	3072	Cylinder 3 Fuel Trim at Maximum Limit
P02A3	3072	Cylinder 3 Fuel Trim at Minimum Limit
P02A6	3073	Cylinder 4 Fuel Trim at Maximum Limit
P02A7	3073	Cylinder 4 Fuel Trim at Minimum Limit
P0201	2E30	Injector Circuit/Open - Cylinder 1
P0202	2E31	Injector Circuit/Open - Cylinder 2

BentleyPublishers.com
BentleyPublishers.com—All Rights Reserved

Table a. Diagnostic trouble codes (DTCs) (continued)		
P-code	**BMW FC**	**Definition**
P0203	2E32	Injector Circuit/Open - Cylinder 3
P0204	2E33	Injector Circuit/Open - Cylinder 4
P0205	2E34	Injector Circuit/Open - Cylinder 5
P0206	2E35	Injector Circuit/Open - Cylinder 6
P0207	2E36	Injector Circuit/Open - Cylinder 7
P0208	2E37	Injector Circuit/Open - Cylinder 8
P0221	2CF7	Throttle/Pedal Position Sensor/Switch 'B' Circuit Range/Performance
P0222	2CFA	Throttle/Pedal Position Sensor/Switch 'B' Circuit Low
P0223	2CFA	Throttle/Pedal Position Sensor/Switch 'B' Circuit High
P0234	30FE	Turbocharger ‘A’ Overboost Condition
P0236	2ABD	Turbocharger Boost Sensor ‘A’ Circuit Range / Performance
P0237	2ABC	Turbocharger Boost Sensor ‘A’ Circuit Low
P0238	2ABD	Turbocharger Boost Sensor ‘A’ Circuit High
P0243	30CF	Turbocharger Wastegate Solenoid ‘A’
P0245	30CF	Turbocharger Wastegate Solenoid ‘A’ Low
P0246	30CF	Turbocharger Wastegate Solenoid ‘A’ High
P0247	30D0	Turbocharger Wastegate Solenoid ‘B’
P0249	30D0	Turbocharger Wastegate Solenoid ‘B’ Low
P0250	30D0	Turbocharger Wastegate Solenoid ‘B’ High
P0261	2E30	Cylinder 1 Injector Circuit Low
P0262	2E30	Cylinder 1 Injector Circuit High
P0264	2E31	Cylinder 2 Injector Circuit Low
P0265	2E31	Cylinder 2 Injector Circuit High
P0266	2E32	Cylinder 3 Injector Circuit High
P0267	2E32	Cylinder 3 Injector Circuit Low
P0268	2E32	Cylinder 3 Injector Circuit High
P0269		Injector Circuit Low / High
P0270	2E33	Cylinder 4 Injector Circuit Low
P0271	2E33	Cylinder 4 Injector Circuit High
P0273	2E34	Cylinder 5 Injector Circuit Low
P0274	2E34	Cylinder 5 Injector Circuit High
P0276	2E35	Cylinder 6 Injector Circuit Low
P0277	2E35	Cylinder 6 Injector Circuit High
P0278		Cylinder 7 Injector Circuit Low
P0279		Cylinder 7 Injector Circuit High
P0280		Cylinder 8 Injector Circuit Low
P0281		Cylinder 8 Injector Circuit High

BentleyPublishers.com
BentleyPublishers.com—All Rights Reserved

Table a. Diagnostic trouble codes (DTCs) (continued)

P-code	BMW FC	Definition
P0282		Injector Circuit Low / High
P0283		Injector Circuit Low / High
P029A	3070	Cylinder 1 Fuel Trim at Maximum Limit
P029B	3070	Cylinder 1 Fuel Trim at Minimum Limit
P029E	3071	Cylinder 2 Fuel Trim at Maximum Limit
P029F	3071	Cylinder 2 Fuel Trim at Minimum Limit
P0298	12055	Engine Oil Over Temperature
P0299	30FF	Turbocharger ‘B’ Overboost Condition
P0300	29CC	Cylinder Misfire, Several Cylinders
P0301	29CD	Cylinder 1 Misfire Detected
P0302	29CE	Cylinder 2 Misfire Detected
P0303	29CF	Cylinder 3 Misfire Detected
P0304	29D0	Cylinder 4 Misfire Detected
P0305	29D1	Cylinder 5 Misfire Detected
P0306	29D2	Cylinder 6 Misfire Detected
P0307	29D3	Cylinder 7 Misfire Detected
P0308	29D4	Cylinder 8 Misfire Detected
P0313	29D9	Misfire Detected with Low Fuel
P0326	2E68	Knock Sensor 1 Circuit Range / Performance (Bank 1 or Single Sensor)
P0326	2CA6	Knock Sensor 1 Circuit Range / Performance (Bank 1 or Single Sensor)
P0327	2E68	Knock Sensor 1 Circuit Low (Bank 1 or Single Sensor)
P0328	2E68	Knock Sensor 1 Circuit High (Bank 1 or Single Sensor)
P0332		Knock Sensor 2 Circuit Low (Bank 2)
P0335	2A94	Crankshaft Position Sensor 'A' Circuit
P0336	2B5C	Crankshaft Position Sensor ‘A’ Range
P0339		Crankshaft Position Sensor 'A' Circuit Intermittent
P0340	2AA0	Camshaft Position Sensor 'A' Circuit (Bank 1 or Single Sensor)
P0341	2A9A	Camshaft Position Sensor 'A' Performance (Bank 1)
P0342	2B62	Camshaft Position Sensor 'A' Circuit Low (Bank 1)
P0343	2B62	Camshaft Position Sensor 'A' Circuit High (Bank 1)
P0344	2A9E	Camshaft Position Sensor 'A' Circuit Intermittent (Bank 1 or Single Sensor)
P0345		Camshaft Position Sensor Rationality Check Inlet
P0346		Camshaft Position Sensor Performance Inlet
P0347		Camshaft Position Sensor Short to Ground Inlet
P0348		Camshaft Position Sensor Short to Battery Inlet
P0351	2E24	Ignition Coil ‘A’ Primary / Secondary Circuit
P0352	2E25	Ignition Coil ‘B’ Primary / Secondary Circuit

BentleyPublishers.com
BentleyPublishers.com—All Rights Reserved

Table a. Diagnostic trouble codes (DTCs) (continued)

P-code	BMW FC	Definition
P0353	2E26	Ignition Coil ‘C’ Primary / Secondary Circuit
P0354	2E27	Ignition Coil ‘D’ Primary / Secondary Circuit
P0355	2E28	Ignition Coil ‘E’ Primary / Secondary Circuit
P0356	2E29	Ignition Coil ‘F’ Primary / Secondary Circuit
P0357		Ignition Coil ‘G’ Primary / Secondary Circuit
P0358		Ignition Coil ‘H’ Primary / Secondary Circuit
P0365	2AA1	Camshaft Position Sensor 'B' Circuit (Bank 1)
P0366	2A9B	Camshaft Position Sensor 'B' Performance (Bank 1)
P0367		Camshaft Position Sensor Short to Ground Outlet
P0367	2B63	Camshaft Position Sensor 'B' Circuit Low (Bank 1)
P0368	2B63	Camshaft Position Sensor 'B' Circuit High (Bank 1)
P0369	2A9F	Camshaft Position Sensor 'B' Circuit Intermittent (Bank 1)
P0370	29DB	Smooth Running Segment Timing
P0370	2A96	Timing Reference High Signal ‘A’ Resolution
P0370	2A97	Timing Reference High Signal ‘A’ Resolution
P0372	2B5D	Timing Reference High Signal ‘A’ Resolution Erratic Too Few Pulses
P0373	2A95	Timing Reference High Signal ‘A’ Resolution Erratic
P0390		Camshaft Position Sensor Inlet Outlet Rationality Check
P0391		Camshaft Position Sensor Signal Outlet
P0392		Camshaft Position Sensor Short to Ground Outlet
P0393		Camshaft Position Sensor Short to battery Outlet
P0411		Secondary Air Injection System Low Flow
P0412		Secondary Air Injection System Valve Circuit (Continuity)
P0413		Secondary Air Injection System Switching Valve A Circuit Open
P0414		Secondary Air Injection System Switching Valve A Circuit Shorted
P0418	10755	Secondary Air Injection System Control A Circuit
P0420	29F4	Catalyst System Efficiency Below Threshold (Bank 1)
P0420	29F6	Catalyst System Efficiency Below Threshold (Bank 1)
P0430	29F5	Catalyst System Efficiency Below Threshold (Bank 2)
P0440	2A1A	Evaporative Emission System Incorrect Flow
P0441	2A1A	Evaporative Emission System Incorrect Purge Flow
P0442	2A15	Evaporative Emission System Leak Detected (small leak)
P0443		Evaporative Emission System Purge Control Valve Circuit
P0444	2A19	Evaporative Emission System Purge Control Valve Circuit Open
	2A1A	Fuel Tank Venting Function Fault
P0445		Evaporative Emission System Purge Control Valve Circuit Shorted
P0455		Evaporative Emission System Leak Detected (large leak)

BentleyPublishers.com
BentleyPublishers.com–All Rights Reserved

Table a. Diagnostic trouble codes (DTCs) (continued)

P-code	BMW FC	Definition
P0456	2A16	Evaporative Emission System Leak Detected (very small leak)
P0457	2A1B	Evaporative Emission System Leak Detected (Fuel Filler Cap Loose)
P0458	2A19	Evaporative Emission System Purge Control Valve Circuit Low
P0459	2A19	Evaporative Emission System Purge Control Valve Circuit High
P0460	2A1E	Fuel Level Sensor ‘A’ Circuit
P0461	2A1C	Fuel Level Sensor A Performance
P0462	2DE2	Fuel Level Sensor ‘A’ Circuit Low
P0463	2DE2	Fuel Level Sensor ‘A’ Circuit High
P0475	2F6C	Exhaust Pressure Control Valve
P0476	2F6C	Exhaust Pressure Control Valve Low
P0477	2F6C	Exhaust Pressure Control Valve Low
P0478	2F6C	Exhaust Pressure Control Valve High
P0479		Exhaust Pressure Control Valve Intermittent
P0480	2EFE	Fan 1 Control Circuit
P0491	2A00	Secondary Air Injection System Insufficient Flow (Bank 1)
P0492	2A00	Secondary Air Injection System Insufficient Flow (Bank 2)
P0500	2F4E	Vehicle Speed Sensor 'A'
P0503	2F4F	Vehicle Speed Sensor 'A' Intermittent / Erratic
P0505		Idle Air Control System
P0506	2ADF	Idle Air Control System Lower Than Expected
P0507	2ADF	Idle Air Control System Higher Than Expected
P0512	2F58	Starter Request Circuit
P0520	2F7B	Engine Oil Pressure Switch Circuit
P0521	30C6	Engine Oil Pressure Switch Range / Performance
P0522	30C3	Engine Oil Pressure Switch Open
P0523	30C3	Engine Oil Pressure Switch High
P0524	30C5	Engine Oil Pressure Too Low
P053A	2AE4	Positive Crankcase Ventilation Heater Circuit Open
P053B	2AE4	Positive Crankcase Ventilation Heater Circuit Low
P053C	2AE$	Positive Crankcase Ventilation Heater Circuit High
P0530	2F12	A/C Pressure Sensor ‘A’ Circuit
P0532	2F12	A/C Pressure Sensor ‘A’ Circuit Low
P0533	2F12	A/C Pressure Sensor ‘A’ Circuit High
P0545	2C87	Exhaust Gas Temperature Sensor Circuit Low (Bank1 Sensor 1)
P0546	2C87	Exhaust Gas Temperature Sensor Circuit High (Bank1 Sensor 1)
P0560		System Voltage
P0561		System Voltage Unstable

BentleyPublishers.com
BentleyPublishers.com—All Rights Reserved

Table a. Diagnostic trouble codes (DTCs) (continued)

P-code	BMW FC	Definition
P0562		System Voltage Low
P0563		System Voltage High
P0571	2F63	Brake Switch 'A' Circuit
P0597	2EF5	Thermostat Heater Control Circuit / Open
P0598	2EF5	Thermostat Heater Control Circuit Low
P0599	2EF5	Thermostat Heater Control Circuit High
P060C	2D67	Internal Control Module Internal Processor fault
P0600		Serial Communication Link
P0601		ECM Self Check
P0604	2AB2	Internal Control Module Random Access Memory (RAM) Error
P0605	2D67	Internal Control Module ROM
P0606	2D67	ECM Processor
P061F	2CEF	Internal Control Module Throttle Actuator Performance
P062F	2FA3	Internal Control Module EEPROM Error
P0620	2ECD	Generator Control Circuit
P0620	2E97	Generator Control Circuit
P0645	2F12	A/C Clutch Relay Control Circuit
P0646	2F12	A/C Clutch Relay Control Circuit Low
P0647	2F12	A/C Clutch Relay Control Circuit High
P0668	2F85	ECM / TCM Internal Temperature Sensor Circuit Low
P0669	2F85	ECM / TCM Internal Temperature Sensor Circuit High
P0686	2ACB	ECM Power Relay Control Circuit Low
P0687	2ACB	ECM Power Relay Control Circuit High
P0691	2EFE	Fan 1 Control Circuit Low
P0692	2EFE	Fan 1 Control Circuit High
P0700	2AD0	Transmission Control System MIL Request
P0703	2F64	Brake Switch 'B' Circuit
P0705		Automatic Transmission GM5 Range Switch Rationality
P0706		Position Switch Range
P0709		Automatic Transmission GM5 Range Switch Rationality
P0709		Automatic Transmission 5HP19 Range Switch Rationality
P0710		Automatic Transmission Oil Temperature Sensor Plausibility
P0711		Automatic Transmission Oil Temperature Sensor General
P0712		Automatic Transmission Oil Temperature Sensor Lower Threshold
P0713		Automatic Transmission Oil Temperature Sensor Upper Threshold
P0714		Automatic Transmission Oil Temperature Sensor No Signal
P0715		Automatic Transmission GM5 Input Speed

OBD

BentleyPublishers.com
BentleyPublishers.com—All Rights Reserved

Table a. Diagnostic trouble codes (DTCs) (continued)

P-code	BMW FC	Definition
P0715		Automatic Transmission 5HP19 Input Speed
P0716		Automatic Transmission 5HP19 Input Speed Rationality
P0717		Automatic Transmission GM5 Input Speed Rationality
P0720		Automatic Transmission GM5 Output Speed Performance
P0720		Automatic Transmission 5HP19 Output Speed Rationality
P0721		Automatic Transmission 5HP19 Output Speed Rationality
P0722		Automatic Transmission GM5 Output Speed Rationality
P0727		TCM Rationality
P0729		Gear Ratio Monitoring (6) Plausibility
P0731		Automatic Transmission 1st Gear Ratio Plausibility
P0731		Automatic Transmission 5HP19 Gear Ratio Rationality
P0731		Automatic Transmission GM5 Gear Ratio Rationality
p0732		Automatic Transmission 2nd Gear Ratio Plausibility
P0732		Automatic Transmission 5HP19 Gear Ratio Rationality
P0732		Automatic Transmission GM5 Gear Ratio Rationality
P0733		Automatic Transmission 3rd Gear Ratio Plausibility
P0733		Automatic Transmission 5HP19 Gear Ratio Rationality
P0734		Automatic Transmission 4th Gear Ratio Plausibility
P0734		Automatic Transmission 5HP19 Gear Ratio Rationality
P0734		Automatic Transmission GM5 Gear Ratio Rationality
P0735		Automatic Transmission 5th Gear Ratio Plausibility
P0735		Automatic Transmission 5HP19 Gear Ratio Rationality
P0735		Automatic Transmission GM5 Gear Ratio Rationality
P0736		Gear Ratio Monitoring Plausibility
P0741		Torque Converter Clutch Functional Check
P0741		Torque Converter Clutch Solenoid Stuck ON (MS45)
P0742		Torque Converter Clutch Solenoid Stuck ON
P0745		Pressure Control Valve 1 Rationality
P0748		Pressure Control Valve 1 No Signal
P0750		Shift Solenoid A Rationality
P0751		Shift Solenoid A Short to Power
P0752		Shift Solenoid A Short to Ground
P0753		Shift Solenoid A Circuit Disconnect
P0755		Shift Solenoid B Rationality
P0756		Shift Solenoid B Short to Power
P0757		Shift Solenoid B Short to Ground
P0758		Shift Solenoid B Circuit Disconnect

BentleyPublishers.com
BentleyPublishers.com—All Rights Reserved

Table a. Diagnostic trouble codes (DTCs) (continued)		
P-code	**BMW FC**	**Definition**
P0760		Shift Solenoid C Rationality
P0761		Shift Solenoid C Circuit Disconnect
P0762		Shift Solenoid C Short to Ground
P0763		Shift Solenoid C Short to Power
P0775		Pressure Control Valve 2 Rationality
P0778		Pressure Control Valve 2 No Signal
P0781		Range Shift Monitoring A 1-2 General
P0782		2-3 Shift Rationality No Change
P0783		3-4 Shift Rationality No Change
P0784		4-5 Shift General
P0795		Pressure Control Valve 3 Plausibility
P0798		Pressure Control Valve 3 No Signal
P0829		5-6 Shift General
P0831	2F67	Clutch Pedal Switch 'A' Circuit Low
P0832	2F67	Clutch Pedal Switch 'A' Circuit High
P0932		SMG Hydraulic Pressure Sensor Plausibility
P0934		SMG Hydraulic Pressure Sensor Lower Threshold
P0935		SMG Hydraulic Pressure Sensor Upper Threshold
P0962		Pressure Control Valve 1 Short to Ground
P0963		Pressure Control Valve 1 Short to Power
P0966		Pressure Control Valve 2 Lower Threshold
P0967		Pressure Control Valve 2 Short to Ground
P0968		Pressure Control Valve 2 Short to Power
P0970		Pressure Control Valve 3 Lower Threshold
P0971		Pressure Control Valve 3 Upper Thresholds
P0974		Solenoid Valve 1 Short to Power
P0973		Shift Solenoid A Short to Ground
P0976		Shift Solenoid B Short to Ground
P0979		Shift Solenoid C Short to Ground
P0982		Shiftlock Lower Threshold
P101A	2A39	VVT Self Learning Function Stops Not Learned
P101A	2A46	VVT Self Learning Function Stops Not Learned
P102C	2A34	VVT Eccentric Shaft Sensor Diagnostic Error
P103A	2A45	VVT System Temperature Too High
P104E	2F0A	Turbocharger Intake Air Temperature Sensor 1 Circuit High or Open
P104F	2F0A	Turbocharger Intake Air Temperature Sensor 1 Circuit Input Low
P105A	2A77	Internal Control Module Fault VVT Current Too High

BentleyPublishers.com
BentleyPublishers.com—All Rights Reserved

Table a. Diagnostic trouble codes (DTCs) (continued)

P-code	BMW FC	Definition
P105B	2A77	Internal Control Module Fault VVT Current Too Low
P107A	2A43	VVT Overload Protection Current Too High (Bank 1)
P107B	2A43	VVT Overload Protection (Bank 1)
P107C	2A43	VVT Overload Protection Current Too High (Bank 1)
	2A44	Valvetronic Power Limitation
P1000	2A6F	VVT System Minimum Stroke Adaptation Number of Stops Exceeded
P1001	2A58	VVT Limp Home Request High Input
P1002	2A58	VVT Limp Home Request Low Input
P1003	2A58	VVT Limp Home Request Open Circuit
P1004	2A35	VVT Guiding Sensor Solenoid Loss (Bank 1)
P1005	2A59	VVT Guiding Sensor Reset Error (Bank 1)
P1006	2A31	VVT Eccentric Shaft Sensor Parity Error (Bank 1)
P1007	2A59	VVT Eccentric Shaft Sensor Gradient Error (Bank 1)
P1008	2A5A	VVT Eccentric Shaft Sensor Solenoid Loss (Bank 1)
P1009	2A5A	VVT Guiding Sensor Reset Error (Bank 2)
P1010	2A5A	VVT Eccentric Shaft Sensor Parity Error (Bank 2)
P1011	2A5A	VVT Eccentric Shaft Sensor Gradient Error (Bank 2)
P1012	2A36	VVT Guiding Sensor Solenoid Loss (Bank 1)
P1013	2A5B	VVT Reference Sensor Reset Error (Bank 1)
P1014	2A32	VVT Eccentric Shaft Sensor Parity Error (Bank 1)
P1015	2A5B	VVT Eccentric Shaft Sensor Gradient Error (Bank 1)
P1017	2A37	VVT Guiding Sensor Plausibility (Bank 1)
P1018	2A5E	VVT Sensor Data Plausibility (Bank 2)
P1019	2A30	VVT Eccentric Shaft Sensor Circuit Low
P102B	2A33	VVT Guiding Sensor Diagnostic Error (Bank 1)
P102C	2A34	VVT Reference Sensor Diagnostic Error (Bank 1)
P1020	2A5F	VVT Sensor Supply Voltage Low Input (Bank 1)
P1020	2A30	VVT Eccentric Shaft Sensor Circuit High
P1021	2A60	VVT Sensor Supply Voltage High Input (Bank 2)
P1022	2A60	VVT Sensor Supply Voltage Low Input (Bank 2)
P1023	2A39	VVT Self Learning Function Faulty Adjustment Range (Bank 1)
P1024	2A39	VVT Self Learning Function Faulty Lower Learning Range (Bank 1)
P1025	2A61	VVT Self Learning Function No Positions Stored (Bank 1)
P1026	2A62	VVT Self Learning Function Faulty Adjustment Range (Bank 2)
P1027	2A62	VVT Self Learning Function Faulty Lower Learning Range (Bank 2)
P1028	2A62	VVT Self Learning Function No Positions Stored (Bank 2)
P1030	2A38	VVT Control Motor Position Control Deviation (Bank1)

BentleyPublishers.com
BentleyPublishers.com—All Rights Reserved

Table a. Diagnostic trouble codes (DTCs) (continued)

P-code	BMW FC	Definition
P1031	2A63	VVT Actuator Monitoring Recognition of Direction of Rotation Plausibility (Bank 1)
P1033	2A64	VVT Actuator Monitoring Position Control, Control Deviation (Bank 2)
P1034	2A64	VVT Actuator Monitoring Recognition of Direction of Rotation Plausibility (Bank 2)
P1035	2DDF	VVT CAN Message Faulty (Bank 1)
P1036	2DDF	VVT CAN Message Faulty (Bank 1)
P1037	2DDF	VVT CAN Message Faulty (Bank 1)
P1038	2DDF	VVT CAN Message Faulty (Bank 2)
P1039	2DDF	VVT CAN Message Faulty (Bank 2)
P1040	2DDF	VVT CAN Message Faulty (Bank 2)
P1041	2A3A	Internal VVT Module EEPROM Error
P1042		Valvetronic Module Self Check
P1043		Valvetronic Module Self Check
P1044		Valvetronic Module Self Check
P1045		Valvetronic Module Self Check
P1046		Valvetronic Module Self Check
P1047	2A3D	VVT Control Circuit High (Bank 1)
P1048	2A3D	VVT Control Circuit Low (Bank 1)
P1049	2A3D	VVT Control Circuit Short (Bank 1)
P1050		Valvetronic Module Power Stage Generic
P1051		Valvetronic Module Power Stage Short to Power
P1052		Valvetronic Module Power Stage Short to Ground
P1053		Valvetronic Module Power Stage Short Together
P1054		Valvetronic Module Power Stage General
P1055	2A3F	VVT Control Circuit High (Bank 1)
P1056	2A3F	VVT Control Circuit High (Bank 1)
P1057	2A3C	VVT Motor Supply Voltage (Bank 1)
P1058		Valvetronic Power Supply Range Check Maximum
P1059		Valvetronic Power Supply Range Check Minimum
P1060		Valvetronic Power Supply Range Check Rationality
P1061		Valvetronic fail Safe Mode Not Securely Reached
P1062	2A44	VVT Limp Home Request Full Stroke Position Reached (Bank 1)
P1063	2A6B	VVT Limp Home Request Air Mass Plausibility (Bank 1)
P1064	2A42	VVT Value Comparison Starting / Parking Position Plausibility
P1065		Valvetronic Module BUS Check
P1066		Valvetronic Module BUS Check
P1067	2A5C	VVT Reference Sensor Solenoid Loss (Bank 2)
P1068	2A5C	VVT Reference Sensor Reset Error (Bank 2)

BentleyPublishers.com
BentleyPublishers.com—All Rights Reserved

Table a. Diagnostic trouble codes (DTCs) (continued)

P-code	BMW FC	Definition
p1069	2A5C	VVT Reference Sensor Parity Error (Bank 2)
P1070	2A5C	VVT Reference Sensor Gradient Error (Bank 2)
P1071		Valvetronic Module Watchdog Temperature
P1072		Valvetronic Module Watchdog Temperature
P1073	2A5D	VVT Sensor Data Conformity (Bank 1)
P1074	2A5E	VVT Sensor Data Conformity (Bank 2)
P1075	2A41	VVT Overload Protection (Bank 1)
P1076	2A40	VVT Overload Protection ECM Temperature (Bank 1)
P1077	2A6D	VVT Overload Protection Control Motor Input Temperature (Bank 1)
P1078	2A3E	VVT Overload Protection Current Too High (Bank 1)
P1079	2A6E	VVT Overload Protection (Bank 2)
P1080	2A6e	VVT Overload Protection ECU Temperature (Bank 2)
P1081	2A6E	VVT Overload Protection Control Motor Input Temperature (Bank 2)
p1082		VVT Overload Protection Control Motor Current High (Bank 2)
P1083		Fuel Control Limit Mixture Too Lean (Bank 1 Sensor 1)
P1084		Fuel Control Limit Mixture Too Rich (Bank 1 Sensor 1)
P1085		Fuel Control Limit Mixture Too Lean (Bank 2 Sensor 1)
P1086		Fuel Control Limit Mixture Too Rich (Bank 2 Sensor 1)
P1087		O2 Sensor Circuit Slow Response in Lean Control Range (Bank 1 Sensor 1)
P1088		O2 Sensor Circuit Slow Response in Rich Control Range (Bank 1 Sensor 1)
P1089		O2 Sensor Circuit Slow Response in Lean Control Range (Bank 1 Sensor 2)
P1090		Pre Catalyst Fuel Trim System Too Lean (Bank 1)
P1091		Pre Catalyst Fuel Trim System Too Lean (Bank 2)
P1092		Pre Catalyst Fuel Trim System Too Rich (Bank 1)
P1093		Pre Catalyst Fuel Trim System Too Rich (Bank 2)
P1094		O2 Sensor Circuit Slow Response in Rich Control Range (Bank 2 Sensor 1)
P110D	2D07	Throttle Position Sensor ‘A’ and ‘B’ Range / Performance
P110D	2D61	Throttle Position Sensor ‘A’ and ‘B’ Range / Performance
p110F	2BC1	Ambient Temperature Sensor Faulty CAN Signal
P1104	2D29	Differential Pressure Sensor Intake Manifold Pressure Too Low Bank 1
P1105	2D29	Differential Pressure Sensor Intake Manifold Pressure Too High Bank 1
P111E	2F09	Intake Temperature Sensor Bank 1 Maximum Temperature Implausible
P111F	2F09	Intake Temperature Sensor Bank 1 Temperature Implausible
P1111		Engine Coolant Temperature Sensor Radiator Outlet Low Input
P1112		Engine Coolant Temperature Sensor Radiator Outlet High Input
P112C	2C3D	O2 Sensor Negative Current or Positive Current Control Circuit Open (Bank 1 Sensor 1)
P112D	2C3E	O2 Sensor Negative Current or Positive Current Control Circuit Open (Bank 2 Sensor 1)

Bentley Publishers.com
BentleyPublishers.com—All Rights Reserved

Table a. Diagnostic trouble codes (DTCs) (continued)

P-code	BMW FC	Definition
P112F	2D2E	Manifold Absolute Pressure to Throttle Angle - Too High
P1120		Pedal Position Sensor Circuit
P1121		Pedal Position Sensor 1 Range/Performance Problem
P1122		Pedal Position Sensor 1 Low Input
P1123		Pedal Position Sensor 1 High Input
P1124	2D2A	Differential Pressure Sensor Intake Manifold Pressure Offset Bank 1
P1127		Engine Oil Level Sensor Signal Plausibility
P1128		Engine Oil Level Sensor No Signal
P1129		Engine Oil Level Sensor Oil Level Too Low
P113A	2D14	Mass or Volume Air Flow 1 Correction Signal Plausibility Too Long
p113F	2EE4	Engine Coolant Temperature Sensor 1 Shunt
P113B	2D14	Mass or Volume Air Flow 1 Correction Signal Plausibility Too Short
P1130	2C6D	O2 Sensor Circuit Dynamic Test (Bank 1 Sensor 2)
P1131	2C6E	O2 Sensor Circuit Dynamic Test (Bank 2 Sensor 2)
P1134		O2 Sensor Heater Circuit Signal Intermittent (Bank 1 Sensor 1)
P1135		O2 Sensor Heater Circuit Low Voltage (Bank 1 Sensor 1)
P1136		O2 Sensor Heater Circuit High Voltage (Bank 1 Sensor 1)
P1137		O2 Sensor Heater Circuit Signal Intermittent (Bank 1 Sensor 2)
P1138		O2 Sensor Heater Circuit Low Voltage (Bank 1 Sensor 2)
P1139		O2 Sensor Heater Circuit High Voltage (Bank 1 Sensor 2)
P114A	2C7E	Post Catalyst Fuel Trim Too Rich Bank 1
P114B	2C7E	Post Catalyst Fuel Trim Too Lean Bank 1
P114C	2C7F	Post Catalyst Fuel Trim Too Rich Bank 2
P114D	2C7F	Post Catalyst Fuel Trim Too Lean Bank 2
P114F	2D16	Air Mass Flow Sensor Defective
P1143		O2 Sensor Activity Check Signal Too High (Bank 1 Sensor 2)
P1144		O2 Sensor Activity Check Signal Too Low (Bank 1 Sensor 2)
P1149		O2 Sensor Activity Check Signal Too High (Bank 2 Sensor 2)
P115A	2D15	Mass or Volume Air Flow 'A' Maximum Exceeded
P115B	2D13	Mass Air Volume Flow 'A' Minimum Fallen Below
P115C	2D13	Mass Air Volume Flow 'A' Too Low
P115D	2D13	Mass Air Volume Flow 'A' Too High
P115E	2F0C	Turbocharger Intake Air Temperature Sensor 1 Gradient Implausible
P1150		O2 Sensor Activity Check Signal Too Low (Bank 2 Sensor 2)
P1151		O2 Sensor Heater Circuit Signal Intermittent (Bank 2 Sensor 1)
P1152		O2 Sensor Heater Circuit Low Voltage (Bank 2 Sensor 1)
P1153		O2 Sensor Heater Circuit High Voltage (Bank 2 Sensor 1)

BentleyPublishers.com
BentleyPublishers.com—All Rights Reserved

Table a. Diagnostic trouble codes (DTCs) (continued)

P-code	BMW FC	Definition
P1155		O2 Sensor Heater Circuit Signal Intermittent (Bank 2 Sensor 2)
P1156		O2 Sensor Heater Circuit Low Voltage (Bank 2 Sensor 2)
P1157		O2 Sensor Heater Circuit High Voltage (Bank 2 Sensor 2)
P116C	2D0F	Air Mass Flow Sensor Signal Range
P116E	2D0F	Air Mass Flow Sensor Signal Electrical
P1171		Ambient Pressure Sensor Variant Recognition Value in Boot Range Implausible
P1172		Ambient Pressure Sensor Variant Recognition Error Value Stored in Boot Range
P1173		Ambient Pressure Sensor Variant Recognition Learning Failed
P119A	2D33	Manifold Absolute Pressure Sensor High (Bank 1)
P119B	2D33	Manifold Absolute Pressure Sensor Low (Bank 1)
P119D	2E74	Fuel Trim, Injector Aging Long Term Adaptation Too High (Bank 1)
P119E	2E75	Fuel Trim, Injector Aging Long Term Adaptation Too High (Bank 2)
P1190		Pre Catalyst Fuel Trim System (Bank 1)
P1191		Pre Catalyst Fuel Trim System (Bank 2)
P1192		Post Catalyst Fuel Trim System (Bank 1)
P1193		Post Catalyst Fuel Trim System (Bank 2)
P1197	2D28	Differential Pressure Sensor Intake Manifold High Input Bank 1
P1198	2D28	Differential Pressure Sensor Intake Manifold Low Input Bank 1
P1199	2D29	Differential Pressure Sensor Intake Manifold Low Input Bank 1
P121C	2AF0	NOx Sensor Heater Control Circuit Shorted (Bank 1)
P121E	2AF2	NOx Sensor Heater Control Circuit Open (Bank 1)
P121F	2AF2	NOx Sensor Circuit Shorted (Bank 1)
P1214	2AAE	Fuel Pump Speed Too High
P1215	2AAE	Fuel Pump Speed Too Low
P1216	2AAE	Fuel Pump Emergency Operation
P1217	2AAE	Fuel Pump Over temperature Condition
P122E	2AF6	NOx Sensor Binary Oxygen Sensor Signal Control Circuit Open (Bank 1)
P122F	2AF6	NOx Sensor Binary Oxygen Sensor Signal Control Circuit Shorted (Bank 1)
P1222		Pedal Position Sensor 2 Low Input
P1223		Pedal Position Sensor 2 High Input
P1230	2F94	Fuel Pump Relay Circuit
P1234	2F94	Fuel Pump Relay Circuit Low
P1236	2F94	Fuel Pump Relay Circuit High
P1244	2AAD	Fuel Pump Emergency Cut Off
P126F	30E2	NOx Sensor Linear Oxygen Sensor Signal Too Rich During Deceleration Test (Bank 1)
P127A	30E2	NOx Sensor Signal Too Low During Deceleration Test (Bank 1)
P127B	30E2	NOx Sensor Signal Too High During Deceleration Test (Bank 1)

BentleyPublishers.com
BentleyPublishers.com–All Rights Reserved

Table a. Diagnostic trouble codes (DTCs) (continued)		
P-code	**BMW FC**	**Definition**
P129B	2D2B	Manifold Absolute Pressure Sensor Diagnosis Performance
P1298	2EAF	Serial Communication NOx Sensor (Bank 1)
P1299	2EAF	Serial Communication NOx Sensor (Bank 2)
P130A	2AA3	Camshaft Position Sensor 'B' Segment Timing Error (Bank 1)
P1300	2AA2	Camshaft Position Sensor 'A' Segment Timing Error (Bank 1)
P1301	2E18	Ignition Monitoring Cylinder 1 Spark Duration Too Short
P1302	2E19	Ignition Monitoring Cylinder 2 Spark Duration Too Short
P1303	2E1A	Ignition Monitoring Cylinder 3 Spark Duration Too Short
P1304	2E1B	Ignition Monitoring Cylinder 3 Spark Duration Too Short
P1305	2E1C	Ignition Monitoring Cylinder 5 Spark Duration Too Short
P1306	2E1D	Ignition Monitoring Cylinder 6 Spark Duration Too Short
P1314		Fuel Mixture Deviation Detected with Low Fuel
P1315		Camshaft Position Sensor Rationality Check (Inlet - Outlet)
P1316		Camshaft Position Sensor Rationality Check (Inlet - Outlet)
P1318		Camshaft Position Sensor Rationality Check (Inlet - Outlet)
P1319		Camshaft Position Sensor Rationality Check (Inlet - Outlet)
P1327	2E69	Knock Sensor 2 Circuit Low (Bank 1 or Single Sensor)
P1328	2E69	Knock Sensor 2 Circuit Range / Performance (Bank 1 or Single Sensor)
P1329		Knock Sensor Range Check Minimum
P1329		Knock Sensor 3 Circuit Low
P1330		Knock Sensor 3 Circuit High
P1330		Knock Sensor Range Minimum
P1332		Knock Sensor Range Minimum
P1333		Knock Sensor Range Maximum
P1338		Camshaft Position Sensor Rationality Check (Inlet - Outlet)
P1339		Camshaft Position Sensor Rationality Check (Inlet - Outlet)
P1342		Misfire During Start Cylinder 1
P1343		Misfire Cylinder 1 with Fuel Cut-Off
P1344		Misfire During Start Cylinder 2
P1345		Misfire Cylinder 2 with Fuel Cut-Off
P1346		Misfire During Start Cylinder 3
P1347		Misfire Cylinder 3 with Fuel Cut-Off
P1348		Misfire During Start Cylinder 4
P1349		Misfire Cylinder 4 with Fuel Cut-Off
P135B	2E69	Knock Sensor 2 Circuit Range / Performance (Bank 1 or Single Sensor)
P1350		Misfire during Start Cylinder 5
P1351		Misfire Cylinder 5 with Fuel Cut-Off

BentleyPublishers.com
BentleyPublishers.com—All Rights Reserved

Table a. Diagnostic trouble codes (DTCs) (continued)

P-code	BMW FC	Definition
P1352		Misfire during Start Cylinder 6
P1353		Misfire Cylinder 6 with Fuel Cut-Off
P1377	2B66	Camshaft Position Sensor Master Camshaft Not Defined
P1378		Knock Sensor Module System Error
P1379		Knock Sensor Module System Error
P1380		Knock Sensor Module System Error
P1381		Knock Sensor Module System Error
P1382		Knock Sensor Module System Error
P1383	2E77	Ignition Monitoring Malfunction
P1384		Knock Sensor 3 Circuit
P1386		Knock Sensor Module System Error
P1396	29DA	Crankshaft Position Sensor Segment Timing Plausibility
P140A	2A00	Secondary Air Injection System Flow Bank 1 and Bank 2
P140B		Secondary Air Injection System Low Flow
P140C		Secondary Air Injection System Low Flow
P140E	29DC	Cylinder Injection Cut Off, Fuel Level Too Low
P14C0	2EFF	Fan Mechanical Hardware Defect
P14C1	2F0F	Radiator Shutter Mechanical hardware Defect
P14C2	2AAB	DISA (Differentiated Intake Manifold) Actuator 1 Fault
P14C3	2AAC	DISA (Differentiated Intake Manifold) Actuator 2 Fault
P14C4	2F11	Upper Radiator Shutter Mechanical fault
P14C5	2F11	Upper Radiator Shutter Electrical fault
P14C6	2F10	Lower Radiator Shutter Electrical fault
P140E	29DC	Cylinder Injection Cut Off, Fuel Level Too Low
P1407	2DE2	Fuel Level Signal 1
P1408	2DE1	Fuel Level Signal 2
P1411		Secondary Air Injection System Pump Not Active
P1412		Secondary Air Injection System Max Limit of Leakage Between Air Pump and Valve
P1413	2A03	Secondary Air Injection Pump Relay Control Circuit Signal Low
P1414	2A03	Secondary Air Injection Pump Relay Control Circuit Signal High
P1415	2D06	Mass or Volume Air Flow Too Low
P1417	2D06	Throttle Control Incorrect Air Supply
P1417	2D09	Throttle Control Incorrect Air Supply
P1418		Secondary Air Injection System Valve or Tube Blocked
P142A	2AE4	Crankcase Ventilation Heater Relay Circuit Input High
P142B	2AE4	Crankcase Ventilation Heater Relay Circuit Input Low
P142C	2AE4	Crankcase Ventilation Heater Relay Circuit Input Open

BentleyPublishers.com
BentleyPublishers.com—All Rights Reserved

Table a. Diagnostic trouble codes (DTCs) (continued)		
P-code	**BMW FC**	**Definition**
P142E	29DC	Cylinder Injection Cut Off, Pressure Too Low (High Pressure Fuel System)
P142F	29DC	Cylinder Injection Cut Off, Pressure Too Low (Low Pressure Fuel System)
P1424	2D06	Mass or Volume Air Flow Too High
P1429	2A18	DMTL Heater
P143B	2AD9	Direct Ozone Reduction Catalyst Temperature Sensor Wrong Code
P143C	2AD8	Direct Ozone Reduction Catalyst Temperature / Radiator Temperature Correlation
P1429		DMTL Heater Circuit Open
P143E	2AD8	Direct Ozone Reduction Catalyst Temperature Sensor Gradient Too Low
P1429	2A18	DMTL Heater Activation
P1430	2A18	DMTL Heater Circuit Low
P1431	2A18	DMTL Heater Circuit High
P1434	2A17	DMTL Tank Leakage
P1444		Diagnostic Module Tank Leakage (DMTL) Pump Control Open Circuit
P1445		Diagnostic Module Tank Leakage (DMTL) Pump Control Circuit Signal Low
P1446		Diagnostic Module Tank Leakage (DMTL) Pump Control Circuit Signal High
P1447	2A17	Diagnostic Module Tank Leakage (DMTL) Pump Current Too High during Switching Solenoid Test
P1448	2A17	Diagnostic Module Tank Leakage (DMTL) Pump Current Too Low
P1449	2A17	Diagnostic Module Tank Leakage (DMTL) Pump Current Too High
P1451		Diagnostic Module Tank Leakage (DMTL) Switching Solenoid Control Circuit Signal Low
P1452		Diagnostic Module Tank Leakage (DMTL) Switching Solenoid Control Circuit Signal High
P1453	2A03	Secondary Air Injection Pump Relay Control Circuit Fault
P15AA	30FC	Turbocharger Leak in System
P15A1	30C4	Engine Oil Pressure Mechanical Valve Stuck in De-energized Position
P15A2	30C4	Engine Oil Pressure Mechanical Valve Stuck in Fully Energized Position
P15A3	30C5	Engine Oil Pressure Too High
P15A6	30C6	Engine Oil Pressure Too High Before Start
P15A7	30C6	Engine Oil Pressure Too Low Before Start
P15A9	2FC7	Energy Saving Mode - Transportation Mode
P15B0	2DC3	Terminal 15 Sense Circuit Input High
P15B1	2DC3	Terminal 15 Sense Circuit Input Low
P15B2	2DC3	Terminal 15 Sense Circuit CAS Error
P15B3	2DC3	Terminal 15 Sense Circuit Range / Performance
P150A	2E8B	Battery Sensor Extended Communication Circuit
P150B	2E8B	Battery Sensor Serial Data Interface
P150C	2E8B	Battery Sensor Firmware Implausible
P150D	2E8B	Battery Sensor Temperature Error
P150E	2E8C	Battery Sensor Voltage Error

Bentley Publishers
BentleyPublishers.com—All Rights Reserved

Table a. Diagnostic trouble codes (DTCs) (continued)

P-code	BMW FC	Definition
P150F	28EC	Battery Sensor Current Error
P1500		Idle-Speed Control Valve Stuck Open
P1501		Idle-Speed Control Valve Stuck Closed
P1502		Idle-Speed Control Valve Closing Solenoid Control Circuit Signal High
P1503		Idle-Speed Control Valve Closing Solenoid Control Circuit Signal Low
P1504		Idle-Speed Control Valve Closing Solenoid Control Open Circuit
P1506		Idle-Speed Control Valve Opening Solenoid Control Circuit Signal High
P1507		Idle-Speed Control Valve Opening Solenoid Control Circuit Signal Low
P1508		Idle-Speed Control Valve Opening Solenoid Control Open Circuit
P151A	2E8D	Battery Sensor Terminal 15 / 30 Wakeup Circuit
P151B	2E8D	Battery Sensor Wakeup Circuit
P151C	2E8D	Battery Sensor System Error
P1511	2AA8	DISA (Differentiated Intake Manifold) Control Circuit
P1511	2AA9	DISA (Differentiated Intake Manifold) Control Circuit
	2AAA	Variable Intake System Plausibility
P1512	2AA8	DISA (Differentiated Intake Manifold) Control Circuit Signal Low
P1512	2AA9	DISA (Differentiated Intake Manifold) Control Circuit Signal Low
P1513	2AA8	DISA (Differentiated Intake Manifold) Control Circuit Signal High
P1513	2AA9	DISA (Differentiated Intake Manifold) Control Circuit Signal High
P1515	2F80	Engine OFF Timer Plausibility
P1517	29DD	Poor Road Surface - No Wheel Speed Signal
P1518	29DD	Poor Road Surface - Wheel Speed Too High
P1519	2EA0	Engine Oil Quality Sensor Temperature
P152A	2F4F	Vehicle Speed Sensor Too Low Under Load Compared to Reference Under Load
P152B	2F4F	Vehicle Speed Sensor Too Low During Coast Down Compared to Reference Under Load
P1520	2EA0	Engine Oil Quality Sensor Temperature
P1521	2EA0	Engine Oil Quality Sensor Temperature Communication Error
P1521	2E9F	Engine Oil Quality Sensor Temperature Communication Error
P1522	2EA0	Engine Oil Quality Sensor Temperature
P1523		'A' Camshaft Position Actuator Signal Low (Bank 1)
P1524		'A' Camshaft Position Actuator Control Circuit Signal High (Bank 1)
P1525		'A' Camshaft Position Actuator Control Open Circuit (Bank 1)
P1529		'B' Camshaft Position Actuator Control Circuit Signal Low (Bank 1)
P1530		'B' Camshaft Position Actuator Control Circuit Signal High (Bank 1)
P1531		'B' Camshaft Position Actuator Control Open Circuit (Bank 1)
P1535	2B71	DISA Coil Temperature Limit Exceeded
P1536	2B71	DISA Control Deviation

BentleyPublishers.com
BentleyPublishers.com—All Rights Reserved

Table a. Diagnostic trouble codes (DTCs) (continued)

P-code	BMW FC	Definition
P1537	2B71	DISA Potentiometer Volatge in Lower Diagnosis Range
P1538	2B71	DISA Potentiometer Volatge in Upper Diagnosis Range
P1539	2B72	DISA Coil Temperature Threshold Exceeded
P1540	2AC6	Driving Dynamics Switch Input High
P1541	2AC6	Driving Dynamics Switch Input Low
P155A	2DB7	Multifunction Steering Wheel Toggle-Bit Fault
P1551	2F80	Engine OFF Timer Time-out
P1553	2A99	Engine Position System 'B' Performance (Bank 1)
P1554	2A98	Engine Position System 'A' Performance (Bank 1)
P1561	2AE0	Cold Start Idle RPM Lower Than Expected
P1562	2AE0	Cold Start Idle RPM Higher Than Expected
P1563	2DB6	Multifunction Steering Wheel Rocker Switch Defective
P1565	2DB5	Multifunction Steering Wheel Rocker Switch + / - Pressed Simultaneously
P1567	2DB7	Multifunction Steering Wheel Toggle Bit Error
	2DBE	Adaptive Cruise Control Blocked for Driving Cycle
	2DC0	No Message From LDM
	2DC3	Monitoring Terminal 15
P1576	2DB5	Multifunction Steering Wheel Interface Error
P1577	2F50	Speed Indication Instrument Pack
P1578	2F50	Speed Indication Instrument Pack / ASC Signal Correlation
P1582	30C2	Oil Pump Circuit High
P1583	30C2	Oil Pump Circuit Low
P1584	30C2	Oil Pump Circuit Open
P1586	2E9F	Engine Oil Quality Sensor Temperature Measurement
P1587	2E9F	Engine Oil Quality Sensor Temperature Level Measurement
P1588	2E9F	Engine Oil Quality Sensor Temperature Permeability Measurement
P16A0	2AB3	Internal Control Module Checksum Fault
P16A1	2AB3	Internal Control Module Application Software Checksum Fault
P16A2	2AB3	Internal Control Module Checksum Fault in Data
P16A3	2AB4	Internal Control Module Checksum Fault in Non-Volatile Memory
P16A4	2AB5	Time-out Control Module Knock Sensor SPI-Bus
P16A5	2AB6	Time-out Control Module Multiple Output Stage SPI-Bus
P16A6	2D50	Control Module Self test / Cruise Control Monitoring
P16A7	2D51	Control Module Self Test Hot Film Air Mass Meter Monitoring
P16A8	2D51	Control Module Self Test Throttle Position Monitoring
P16A9	2D52	Control Module Self Test Speed Monitoring Reset
P16B0	2D55	ECM Internal Fault Accelerator Pedal Module

BentleyPublishers.com
BentleyPublishers.com–All Rights Reserved

Table a. Diagnostic trouble codes (DTCs) (continued)

P-code	BMW FC	Definition
P16B1	2D56	Control Module Self Test Idle Air Control System Plausibility
P16B2	2D56	Control Module Self Test Idle Air Control System Component Plausibility
P16B3	2D57	Control Module Self Test Engine Drag Torque
P16B4	2D50	Control Module Self Test Dynamic Cruise Control Monitoring
P16B5	2D57	Control Module Self Test Automatic Manual Transmission
P16B6	2D57	Control Module Self Test ETC Monitoring
P16B7	2D58	Control Module Self Test Clutch Torque Monitoring Maximum value Plausibility
P16B8	2D58	Control Module Self Test Clutch Torque Monitoring
P16B9	2D58	Control Module Self Test Torque Loss Monitoring
P16C0	2D58	Control Module Self Test Driving Dynamics Control Switch Monitoring
	2D5A	Monitoring Engine Torque Limiting
P16C1	2D59	Control Module Self Test Torque Monitoring Current Indicated value Plausibility
P16C2	2D53	Control Module Self Test Speed Limitation Monitoring
P16C3	2D54	Control Module Self Test Speed Limitation Reset
P16C5	2AAC	DME Main Relay Switching Delay
P16C6	2E7C	CAN Time-out Bit Serial Data Interface
P16C8	2DE0	Serial Communication Link EKP (Electronic Fuel Pump)
P164C	2D1D	Pedal Position Sensor Potentiometer Supply Channel 1 Electrical
P16B0	2D55	Control Module Self Test Pedal Position Monitoring
P160A	2DEC	Powermanagement Exhaustive Discharge
P160B	2DEC	Powermanagement Defective
P160C	2DEB	Powermanagement Overvoltage
P160D	2DEB	Powermanagement Undervolatge
P160E	2DEB	Powermanagement Operation Without Battery
P160F	2DED	Powermanagement No Load Current Error
P165F	2CA6	Internal Control Module Measurement Error O2 Sensor Heating (Bank 1 Sensor 1)
P1602		Control Module Self-Test, Control Module Defective
P1603		Control Module Self-Test, Torque Monitoring
P1604		Control Module Self-Test, Speed Monitoring
P1611	2DC8	Serial Communication Link TCM
P1612	2DD0	Serial Communication Link Instrument Panel
P1613	2DCC	Serial Communication Link ASC
P1613	2DCD	Serial Communication Link ASC
P1613	2DCE	Serial Communication Link ASC
P1614	11711	Serial Communication ACC
P1618	2D5C	Control Module Self Test AD-Converter Monitoring
P1619		Map Cooling Thermostat Control Circuit Signal Low

BentleyPublishers.com
BentleyPublishers.com—All Rights Reserved

Table a. Diagnostic trouble codes (DTCs) (continued)		
P-code	**BMW FC**	**Definition**
P1620		Map Cooling Thermostat Control Circuit Signal High
P1624	2D1D	Pedal Position Sensor Potentiometer Supply Channel 1 Electrical (M52: Coolant Thermostat (Coolant Temperature Below Thermostat Regulating Temperature))
P1625	2D1E	Pedal Position Sensor Potentiometer Supply Channel 2 Electrical
P1628		Throttle Valve Spring Does Not Open
P1629		Throttle Valve Spring Does Not Open
P163C	2B9D	Voltage Monitoring Control Module Overload
P163D	2B9D	Voltage Monitoring Control Module Cut-OFF Detected
P163E	2B9D	Voltage Monitoring Control Module Communication Error
P1631		Throttle Valve Actuator Spring Test
P1632	2CFB	Throttle Valve Adaptation Conditions Not Met
P1633	2CFB	Throttle Valve Adaptation Limp-Home Position Unknown
P1634		Throttle Valve Adaptation Spring Test Failed
P1634	2CFC	Throttle Valve Adaptation Spring Test Failed
P1635	2CFE	Throttle Valve Adaptation Lower Mechanical Stop not Adapted
P1636	2CEF	Throttle Valve Control Circuit
P1637	2CEE	Throttle Valve Position Control, Control Deviation
P1638	2CEC	Throttle Valve Position Control Throttle Stuck Temporarily
P1639	2CED	Throttle Valve Position Control Throttle Stuck Permanently
P164E	2D0C	Throttle Deicing Stuck in Closed Direction
P164F	2D0C	Throttle Deicing Stuck in Open Direction
p1641	2D03	Throttle Valve Adaptation Stop Due to Environmental Conditions
P1642	2D03	Throttle Valve Adaptation Stop Due to Environmental Values
P1643	2CFF	Throttle Actuator Start Test Amplifier Balancing Faulty
P1644	2CFB	Throttle Valve Adaptation Relearning Lower Mechanical Stop
P1645	2B99	Internal Control MOdule Fault - RAM Reading Error
P1649	2B99	Internal Control MOdule Fault - RAM Writing Error
P1650		Automatic Start While Engine Running
P166A	2D50	Control Module Self Test, Hot Film Air Mass Meter Monitoring
P166B	2DC0	Longitudinal Dynamics Module Torque Request Inspite of Brake Signal
P166C	2DC0	Longitudinal Dynamics Module Request Implausible
P166F	2CA7	Internal Control Module Measurement Error O2 Sensor Heating (Bank 2 Sensor 1)
P166F	2CAB	Internal Control Module Measurement Error O2 Sensor Heating (Bank 2 Sensor 1)
P1660	2F4A	EWS Telegram Error
P1661	2F4A	Time-out EWS Telegram
P165A	2F4A	EWS Interface to ECM Error
P165B	2F4A	EWS Interface to ECM Checksum Error

BentleyPublishers.com
BentleyPublishers.com–All Rights Reserved

Table a. Diagnostic trouble codes (DTCs) (continued)

P-code	BMW FC	Definition
P165C	2F4B	EWS Data, No Available Storage Possible
P165D	2F4B	EWS Data, Faulty Release Code
P165E	2F4B	EWS Data Checksum Error
P1655	2F46	EWS Rolling COde Faulty Storage in EEPROM
P1660	2F45	EWS Telegram Error
P1661	2F45	EWS Telegram Time-out
P1662	2F45	EWS Telegram Parity Error
P1663	2F46	EWS Rolling Code Faulty Storage in EEPROM
P1664	2F46	EWS Rolling Code Faulty Writing in EEPROM
P1665	2F46	EWS Rolling Code Tampering
P1666		EWS Tampering via KWP2000 / Start VAlue Not Accepted
P1667	2F49	EWS Start Value not yet Programmed
P1668	2F4B	EWS Start Value Destroyed
P167C	2D08	Throttle Heating Relay Circuit Input Low
P167D	2D08	Throttle Heating Relay Circuit
P1675	2CFD	Throttle Valve Actuator Start Test Re-Adaptation Required
P1676	2DBF	ACC Signal Plausibility
P1677	2DBF	ACC No Activity Detected
P1678	2DBF	ACC No Response To Cut-OFF
P1680	2D70	Throttle Valve Control Monitor Level 2/3 ADC Processor fault
P1681	2D75	Throttle Valve Control Monitor Level 2/3 Engine Speed Error
P1686	2D76	Throttle Valve Control Monitor Level 2/3 Pedal Position Sensor Error
P1694	2CFB	Throttle Valve Adaptation Value
P169A	2CFC	Throttle Valve Actuator Start Test Failed Limp Home Position
P17EA		Ratio Upshift 5 - 6
P17EB		Ratio Downshift 6 - 5
P17EC		Ratio Downshift 5 - 4
P17ED		Ratio Downshift 4 - 3
P17EE		Ratio Downshift 3 - 2
P17EF		Ratio Downshift 2 - 1
P17E1		Ratio in Gear 1 - Clutch A
P17E2		Ratio in Gear 3 Clutch B
P17E3		Ratio in Gear 2 Clutch A
P17E4		Ratio in Gear 1 - Clutch D
P17E5		Ratio in Gear 4 Clutch E
P17E6		Ratio Upshift 1 - 2
P17E7		Ratio Upshift 2 - 3

BentleyPublishers.com
BentleyPublishers.com—All Rights Reserved

Table a. Diagnostic trouble codes (DTCs) (continued)		
P-code	**BMW FC**	**Definition**
P17E8		Ratio Upshift 3 - 4
P17E9		Ratio Upshift 4 - 5
P17F0		Ratio in Gear 1 - Clutch A or B
P17F1		Ratio in Gear 2 - Clutch A or B
P17F2		Ratio in Gear 3 - Clutch A or B
P17F3		Ratio in Gear 4 - Clutch A or E
P17F4		Ratio in Gear 5 - Clutch B or E
P17F5		Ratio in Gear 6 - Clutch C or E
P17F6		Ratio in Gear R Clutch B or D
P1694	2CFB	Throttle Valve Actuator Start Test Spring Test and Limp-Home Position Failed
P1700		Automatic Transmission 5HP19 Output Speed, Wheel Speed, Double Fault
P1701		Range Position Switch Plausibility
P171A		Watchdog Check
P171B		Watchdog Check
P171C		Watchdog Check
P171D		Watchdog Check
P171E		Watchdog Check
P1702		Colliding Function or Not Allowed Actuating of valves
P1704		ATF Temperature Sensor Range
P1711		ATF Temperature Sensor Rationality
P1712		Level 2 Calculation Check
P1713		Level 2 Calculation Check
P1714		Level 2 Calculation Check
P1718		ATF Temperature Sensor Rationality
P1719		TCM BUS Check
P1720		TCM BUS Check
P1721		CAN Time-out ASC / DSC
P1722		MC Components
P1723		MC Components
P1724		MC Components
P1725		MC Components
P1727		Engine Speed Signal Plausibility
P1728		Engine Overspeed Plausibility
P1729		CAN Version Plausibility
P1730		Impossible Combination of Substitute Functions or Not Allowed Actuating Valves
P1731		Automatic Transmission 5HP19 Gear Ratio Rationality
P1732		Automatic Transmission 5HP19 4th Gear Ratio Rationality

BentleyPublishers.com
BentleyPublishers.com—All Rights Reserved

Table a. Diagnostic trouble codes (DTCs) (continued)

P-code	BMW FC	Definition
P1736		Automatic Transmission 6th Gear Ratio Plausibility
P1747		TCM BUS Check
P1760		Desired Engine Torque Message on CAN
P1761		Shift Lock Rationality
P1762		Shift Lock Short to Power
P1763		Shift Lock Short to Ground
P1764		Shift Lock Circuit Disconnection
P1765		TCM Rationality
P1766		Engine Speed Plausibility
P1767		CAN Speed Rear Wheels Plausibility
P1771		CAN Generated Engine Torque Plausibility
P1780		TCM Rationality
P1782		CAN Brake Signal Plausibility
P1790		TCM Rationality
P1791		TCM Self Check
P1792		TCM Rationality Micro controller
P1793		TCM Hot Disabling
P1794	2DD3	TCM Checksum Error
P1796		Power Latch Relays Plausibility
P1798		EPROM Plausibility
P1810		Turbine Input Speed Sensor (NTU) Upper Threshold
P1811		Turbine Input Speed Sensor (NTU) Lower Threshold
P1812		Output Speed Sensor Upper Threshold
P1813		Output Speed Sensor Lower Threshold
P1814		Output Speed Sensor Plausibility
P1817		Gear Lever Analog Lines 0 Upper Threshold
P1818		Gear Lever Analog Lines 0 Lower Threshold
P1819		Gear Lever Analog Lines 0 Plausibility
P1820		Gear Lever Analog Lines 1 Upper Threshold
P1821		Gear Lever Analog Lines 1 Lower Threshold
P1822		Gear Lever Analog Lines 1 Plausibility
P1823		Gear Lever Digital Line No Signal
P1850		Shift Position Sensor Upper Threshold
P1851		Shift Position Sensor Lower Threshold
P1852		Shift Position Sensor Plausibility
P1859		Gear Lever Halls Sensors No SIgnal
P1860		Clutch Speed Plausibility

BentleyPublishers.com
BentleyPublishers.com—All Rights Reserved

Table a. Diagnostic trouble codes (DTCs) (continued)		
P-code	**BMW FC**	**Definition**
P1861		2-1 Shift General
P1862		3-2 Shift General
P1863		4-3 Shift General
P1864		5-4 Shift General
P1865		6-5 Shift General
P1866		Gear Lever Analog Lines 2 Upper Threshold
P1867		Gear Lever Analog Lines 2 Lower Threshold
P1868		Gear Lever Analog Lines 2 Upper Plausibility
P1869		Gear Lever Analog Lines 3 Upper Threshold
P1870		Gear Lever Analog Lines 3 Lower Threshold
P1876		Gear Lever Analog Lines 3 Plausibility
P1881		Range Shift Monitoring A 1-2 Upper Threshold
P1882		2-3 Shift Rationality, No Change
P1883		3-4 Shift Rationality Too Big
P1884		4-5 Shift Upper Threshold
P1885		5-6 Shift Upper Threshold
P1887		Hydraulic Pump Failure
P1889		System Voltage Rationality
P1890		Power Supply General
P1891		Power Supply Rationality
P1892		System Voltage Low Input
P1893		Control Solenoid / Shift Solenoid Short to Power
P1894		Control Solenoid / Shift Solenoid Short to Ground
P1895		Control Solenoid / Shift Solenoid Circuit Disconnection
P1896		Valve Voltage Supply Plausibility
P1897		Sensor Voltage Supply Upper Threshold
P1898		Sensor Voltage Supply Lower Threshold
P2065		Fuel Level Sensor 'B' Circuit
P2067	2DE1	Fuel Level Sensor 'B' Circuit Low
P2068	2DE1	Fuel Level Sensor 'B' Circuit High
P2088	2A80	'A' Camshaft Position Actuator Control Circuit Low (Bank 1)
P2089	2A80	'A' Camshaft Position Actuator Control Circuit High (Bank 1)
P2090	2A85	'B' Camshaft Position Actuator Control Circuit Low (Bank 1)
P2091	2A85	'B' Camshaft Position Actuator Control Circuit High (Bank 1)
P2092		VANOS Inlet Camshaft Range Minimum
P2093		VANOS Inlet Camshaft Range Maximum
P2094		VANOS Range Limit Minimum

BentleyPublishers
BentleyPublishers.com–All Rights Reserved

Table a. Diagnostic trouble codes (DTCs) (continued)

P-code	BMW FC	Definition
P2095		VANOS Range Limit Maximum
P2096	2C31	Post Catalyst Fuel Trim System Too Lean (Bank 1)
P2097	2C31	Post Catalyst Fuel Trim System Too Rich (Bank 1)
P2098	2C32	Post Catalyst Fuel Trim System Too Lean (Bank 2)
P2099	2C32	Post Catalyst Fuel Trim System Too Rich (Bank 2)
P2100	2CEF	Throttle Actuator 'A' Control Motor Circuit Open
P2102	2CEF	Throttle Actuator 'A' Control Motor Circuit Low
P2103	2CEF	Throttle Actuator 'A' Control Motor Circuit High
P213F	2AAD	Fuel Pump System Fault Engine Shut Down
P2120	2D1F	Throttle / Pedal Position Sensor 'D' Circuit
P2122	2D1B	Throttle Pedal Position Sensor 'D' Circuit Low
P2123	2D1B	Throttle Pedal Position Sensor 'D' Circuit High
P2127	2D1C	Throttle Pedal Position Sensor 'E' Circuit Low
P2128	2D1C	Throttle Pedal Position Sensor 'E' Circuit High
P213F	2AAD	Fuel Pump Emergency Cut Off
P2138	2D20	Throttle / Pedal Position Sensor 'D' 'E' Voltage Correlation
P2177		Fuel Trim Limit Exceeded System Too Lean
P2178		Fuel Trim Limit Exceeded System Too Rich
P2179		Fuel Trim Limit Exceeded System Too Lean
P2180		Fuel Trim Limit Exceeded System Too Rich
P2183	2EEC	Engine Coolant Temperature Sensor 2 Circuit Range / Performance
P2184	2EEA	Engine Coolant Temperate Sensor 2 Circuit Low
P2185	2EEA	Engine Coolant Temperate Sensor 2 Circuit High
P2187	29E0	System Too Lean at Idle (Bank 1)
P2188	29E0	System Too Rich at Idle (Bank 1)
P2189	29E1	System Too Lean at Idle (Bank 2)
P2190	29E1	System Too Rich at Idle (Bank 2)
P2191	29E5	System Too Lean, Higher Load (Bank 1)
P2192	29E5	System Too Rich, Higher Load (Bank 1)
P2193	2E96	System Too Lean, Higher Load (Bank 2)
P2194	2E96	System Too Rich, Higher Load (Bank 2)
P2195	2C27	O2 Sensor Signal Stuck Lean (Bank 1 Sensor 1)
P2196	2C27	O2 Sensor Signal Stuck Rich (Bank 1 Sensor 1)
P2196	2C2B	O2 Sensor Signal Stuck Rich (Bank 1 Sensor 1)
P2196	2C2C	O2 Sensor Signal Stuck Rich (Bank 1 Sensor 1)
P2197	2C28	O2 Sensor Signal Stuck Lean (Bank 2 Sensor 1)
P2198	2C2C	O2 Sensor Signal Stuck Rich (Bank 2 Sensor 1)

BentleyPublishers.com
BentleyPublishers.com—All Rights Reserved

Table a. Diagnostic trouble codes (DTCs) (continued)

P-code	BMW FC	Definition
P2200	2AF4	NOx Sensor Circuit (Bank 1)
P2205	2AF0	NOx Sensor Heater Control Circuit Open (Bank 1)
P2227		ECM Integrated Barometric Pressure Sensor Rationality Check
P2228	2F76	Barometric Pressure Circuit Low
P2228		ECM Emigrated Barometric Pressure Sensor Circuit
P2229	2F76	Barometric Pressure Circuit High
P2229		ECM Emigrated Barometric Pressure Sensor Circuit
P2231		O2 Sensor Signal Shorted To Heater (Bank 1 Sensor 1)
P2234		O2 Sensor Signal Shorted To Heater (Bank 2 Sensor 1)
P2237		O2 Sensor Reference Circuit Continuity (Bank 1 Sensor 1)
P2240		O2 Sensor Reference Circuit Continuity (Bank 2 Sensor 1)
P2243	2C3D	O2 Sensor Reference Voltage Circuit Open (Bank 1 Sensor 1)
P2247	2C3E	O2 Sensor Reference Voltage Circuit Open (Bank 2 Sensor 1)
P2251		O2 Sensor Reference Circuit Continuity (Bank 1 Sensor 1)
P2254		O2 Sensor Reference Circuit Continuity (Bank 2 Sensor 1)
P2270	2C6B	O2 Sensor Signal Stuck Lean (Bank 1 Sensor 2)
P2271	2C6B	O2 Sensor Signal Stuck Lean (Bank 1 Sensor 2)
P2272	2C6C	O2 Sensor Signal Stuck Lean (Bank 2 Sensor 2)
P2273	2C6C	O2 Sensor Signal Stuck Lean (Bank 2 Sensor 2)
P2297	2C2D	O2 Sensor Signal Out Of Range During Decel (Bank 1 Sensor 1)
P2298	2C2E	O2 Sensor Signal Out Of Range During Decel (Bank 2 Sensor 1)
P2299	2F8F	Brake Pedal Position / Accelerator Pedal Position Incompatible
P2301	30A0	Ignition Coil Primary ‘A’ Circuit High
P2304	30A1	Ignition Coil Primary ‘B’ Circuit High
P2307	30A2	Ignition Coil Primary ‘C’ Circuit High
P2310	30A3	Ignition Coil Primary ‘D’ Circuit High
P2313	30A4	Ignition Coil Primary ‘E’ Circuit High
P2316	30A5	Ignition Coil Primary ‘F’ Circuit High
P2332	2ED1	Generator Mechanical
P232D	2D78	Air Mass Balancing Control Range
P240A	2A18	DMTL Pump Heater Circuit Open
P240B	2A18	DMTL Pump Heater Circuit Low
P240C	2A18	DMTL Pump Heater Circuit High
P2400	2A13	DMTL Activation Control Circuit Open
P2401	2A13	DMTL Activation Control Circuit Low
P2402	2A13	DMTL Activation Control Circuit High
P2414	2C3B	O2 Sensor Oxygen Sample Error (Bank 1 Sensor 1)

BentleyPublishers.com
BentleyPublishers.com—All Rights Reserved

Table a. Diagnostic trouble codes (DTCs) (continued)		
P-code	**BMW FC**	**Definition**
P2415	2C3C	O2 Sensor Oxygen Sample Error (Bank 2 Sensor 1)
P2418	2A12	DMTL Solenoid Control Circuit Open
P2419	2A12	DMTL Solenoid Control Circuit Low
P2420	2A12	Evaporative Emission System Switching Valve Control Circuit High
P2430		Secondary Air Injection System Mass Air Flow Sensor Rationality (Disconnected or Stuck)
P2440		Secondary Air Injection System Air Valve Jammed Open
P2450		Evaporative Emission System Switching Valve Short to Power
P250A	2F9E	Engine Oil Level Sensor Circuit
P250B	2F9E	Engine Oil Level Sensor Range / Performance
P250F	2F9E	Engine Oil Level Too Low
P252A	2EA1	Engine Oil Quality Sensor Circuit
P2420	2A12	DMTL Solenoid Control Circuit High
P2541	29F3	Low Pressure Fuel System Sensor Circuit Low
P2542	29F3	Low Pressure Fuel System Sensor Circuit High
P2568	2ADA	Direct Ozone Reduction Catalyst Temperature Sensor Performance
P2569	2ADA	Direct Ozone Reduction Catalyst Temperature Sensor Circuit Low
P2570	2ADA	Direct Ozone Reduction Catalyst Temperature Sensor Circuit High
P2610		Engine OFF Timer Plausibility Comparison With CAN Timer
P2626	2C3D	O2 Sensor Pumping Current Circuit Open (Bank 1 Sensor 1)
P2629	2C3E	O2 Sensor Pumping Current Circuit Open (Bank 2 Sensor 1)
P2713		Pressure Control Valve 4 Plausibility
P2716		Pressure Control Valve 4 No Signal
P2720		Pressure Regulator Valve 4 Lower Threshold
P2721		Pressure Regulator Valve 4 Upper Threshold
P2722		Pressure Regulator Valve 5 Plausibility
P2725		Pressure Regulator Valve 5 No Signal
P2729		Pressure Control Valve 5 Lower Threshold
P2730		Pressure Control Valve 5 Upper Threshold
P2759		Pressure Control Valve Circuit
P2761		Pressure Control Valve 4 Rationality
P2763		Pressure Control Valve 4 Short to Power
P2764		Pressure Control Valve 4 Short to Ground
P300A	2F0D	Controlled Air Guiding Circuit High
P300B	2F0D	Controlled Air Guiding Circuit Low
P300C	2F0D	Controlled Air Guiding Circuit
P3003	29F2	Fuel Rail Pressure Flow Rate Controlled, Pressure Too High
P3004	29F2	Fuel Rail Pressure Flow Rate Controlled, Maximum Pressure Exceeded

BentleyPublishers.com
BentleyPublishers.com—All Rights Reserved

Table a. Diagnostic trouble codes (DTCs) (continued)

P-code	BMW FC	Definition
P3008		O2 Sensor Low Engine Cold (Bank 1 Sensor 1)
P3009		O2 Sensor Low Engine Cold (Bank 1 Sensor 1)
P3010		O2 Sensor Range Check Cold Sensor (Bank 1 Sensor 2)
P3011		O2 Sensor Range Check Cold Sensor (Bank 2 Sensor 2)
P3012		Adaptation value Too High (Bank 1 Sensor 1)
P3013		Adaptation value Too High (Bank 2 Sensor 1)
P3014		O2 Sensor IC Supply Voltage Too Low (Bank 1 Sensor 1)
P3015		O2 Sensor IC Supply Voltage Too Low (Bank 2 Sensor 1)
P3016		O2 Sensor Ceramic Internal Resistance Adjusted Value Plausibility (Bank 1 Sensor 1)
P3017		O2 Sensor Ceramic Internal Resistance Adjusted Value Plausibility (Bank 2 Sensor 1)
P3018		O2 Sensor Lambda Controller Value Above Threshold Due to Open Pumping Circuit (Bank 1 Sensor 1)
P3019		O2 Sensor Lambda Controller Value Above Threshold Due to Open Pumping Circuit (Bank 2 Sensor 1)
P3020		O2 Sensor Signal Too Low During Coast Down Fuel Cut-OFF (Bank 1 Sensor 1)
P3021		O2 Sensor Signal Too Low During Coast Down Fuel Cut-OFF (Bank 2 Sensor 1)
P3022	2C41	O2 Sensor Disturbed SPI WRAF-IC (Bank 1 Sensor 1)
P3023	2C42	O2 Sensor Disturbed SPI WRAF-IC (Bank 2 Sensor 1)
P3024	2C41	O2 Sensor Initialization Error WRAF-IC (Bank 1 Sensor 1)
P3025	2C42	O2 Sensor Initialization Error WRAF-IC (Bank 1 Sensor 1)
P3026	2CAA	O2 Sensor Operating Temperature Not Reached (Bank 1 Sensor 1)
P3027	2CAA	O2 Sensor Operating Temperature Not Reached (Bank 2 Sensor 1)
P3028	2CA0	O2 Sensor Heater Control No Activity Detected (Bank 1 Sensor1)
P3029	2CA1	O2 Sensor Heater Control No Activity Detected (Bank 2 Sensor1)
P3030		O2 Sensor Heater Internal Resistance (Bank 1 Sensor 1)
P3031		O2 Sensor Heater Internal Resistance (Bank 2 Sensor 1)
P3041		O2 Sensor Lean and Rich Voltage Thresholds not Reached (Bank 2 Sensor 2)
P3090	29F2	Fuel Rail Pressure Flow Rate Controlled, Fallen Below Minimum Pressure
P3094	2A2D	Fuel Low Pressure System Pressure Too High
P3095	2A2D	Fuel Low Pressure System Pressure Maximum Pressure Exceeded
P3096	2A2D	Fuel Low Pressure System Pressure Minimum Pressure Fallen Below
P310B	2E30	Cylinder 1 High Pressure Injector Low Side / High Side Electrical
P310E	2E31	Cylinder 2 High Pressure Injector Low Side / High Side Electrical
P3101	30AC	Cylinder 1 High Pressure Injector Low Side Circuit Low
P3102	30AC	Cylinder 1 High Pressure Injector Low Side Circuit High
P3105	30AD	Cylinder 2 High Pressure Injector Low Side Circuit Low
P3106	30AD	Cylinder 2 High Pressure Injector Low Side Circuit high
P3109	30AE	Cylinder 3 High Pressure Injector Low Side Circuit Low
P311B	2E32	Cylinder 3 High Pressure Injector Low Side / High Side Electrical

Bentley Publishers.com
BentleyPublishers.com—All Rights Reserved

Table a. Diagnostic trouble codes (DTCs) (continued)

P-code	BMW FC	Definition
P311E	2E33	Cylinder 4 High Pressure Injector Low Side / High Side Electrical
P3110	30AE	Cylinder 3 High Pressure Injector Low Side Circuit High
P3113	30AF	Cylinder 4 High Pressure Injector Low Side Circuit Low
P3114	30AF	Cylinder 4 High Pressure Injector Low Side Circuit High
P3117	30B0	Cylinder 5 High Pressure Injector Low Side Circuit Low
P3118	30B0	Cylinder 5 High Pressure Injector Low Side Circuit High
P312B	2E34	Cylinder 5 High Pressure Injector Low Side / High Side Electrical
P312E	2E35	Cylinder 6 High Pressure Injector Low Side / High Side Electrical
P3121	30B1	Cylinder 6 High Pressure Injector Low Side Circuit Low
P3122	30B1	Cylinder 6 High Pressure Injector Low Side Circuit High
P3149	30AC	Cylinder 1 High Pressure Injector High Side Circuit Low
P3150	30AC	Cylinder 1 High Pressure Injector High Side Circuit High
P3152	30AC	Cylinder 2 High Pressure Injector High Side Circuit Low
P3153	30AD	Cylinder 2 High Pressure Injector High Side Circuit High
P3155	30AD	Cylinder 3 High Pressure Injector High Side Circuit Low
P3156	30AE	Cylinder 3 High Pressure Injector High Side Circuit High
P3158	30AF	Cylinder 4 High Pressure Injector High Side Circuit Low
P3159	30AF	Cylinder 4 High Pressure Injector High Side Circuit High
P316A	2EE6	Engine Coolant Temperature Signal Stuck High
P3161	30B0	Cylinder 5 High Pressure Injector High Side Circuit Low
P3162	30B0	Cylinder 5 High Pressure Injector High Side Circuit High
P3164	30B1	Cylinder 6 High Pressure Injector High Side Circuit Low
P3165	30B1	Cylinder 6 High Pressure Injector High Side Circuit High
	2FBF	Fuel Supply Pressure Low During Release of Injection
	2FC0	Fuel Rail Pressure Management Range
P319A	2FBD	Fuel Pressure Control Minimum Characteristic Curve Operation 1 Out of Range
P3194	2FBD	Fuel Pressure Control Basic Characteristic Curve Operation 1 Out of Range
P3196	2EEC	Radiator Coolant Temperature Sensor High
P3197	2EEB	Radiator Coolant Temperature Gradient Too High
P3198	2EE3	Engine Coolant Temperature Gradient Too High
P230D		ECM BUS Check
P3199	2EE2	Engine Coolant Temperature Signal Stuck
P3200		CAN Internal Powertrain Chip
P3201		BUS Check CAN Powertrain Chip
P3202	CD87	Powertrain CAN Chip Cut-Off
P3203		BUS Check Local CAN
P3204		BUS Check Local CAN

BentleyPublishers.com
BentleyPublishers.com—All Rights Reserved

Table a. Diagnostic trouble codes (DTCs) (continued)

P-code	BMW FC	Definition
P3205	CD8B	Local CAN Chip Cut-Off
P3206	11737	CAN Time-out ARS
P3207	11737	CAN Message Monitoring ARS, No Signal
P3208	2DD(	CAN Message ARS
P3209	11735	CAN Message ASC / DSC
P321E	2F77	Ambient Pressure Sensor Maximum Pressure Implausible
P321F	2F77	Ambient Pressure Sensor Minimum Pressure Implausible
P3210	11735	CAN Monitoring ASC / DSC
P3211	2DDA	CAN Message CAS
P3212	2DDA	CAN Message CAS
P3213	2DD3	CAN Message Monitoring ETC Alive Check
P3214		ECM BIS Check
P3215	2DDC	CAN Message SZL
P3216	11727	CAN Time-out Instrument Pack
P3217	11727	CAN Message Instrument Pack
P3219	2DDC	CAN Message SZL Alive Check
P322A		Ambient Pressure Sensor Continuity
P3220	2DDC	CAN Message SZL No Signal
P3221	2DDC	CAN Message SZL
P3223	2E97	Generator Mechanical
P3225	11927	Generator Communication Error
P3231	2D70	Control Module Monitoring Error Response
P3233	2D70	Control Module Monitoring Relative Charge Plausibility
P3226	27F1	E-box Control Fan High Input
P3227	27F1	E-box Control Fan Low Input
P3228	27F1	E-box Control Fan Open Circuit
P323F	2D60	Control Module Monitoring Fuel Volume, Air Mass Injected Fuel Volume Correlation
P3231		Control Module Internal Fault, Throttle Actuator Control MOnitor 2 / 3 ADC Processor Fault
P3232		Control Module Internal Fault, Monitoring of Engine Functions
P3234	2D71	Control Module Monitoring Throttle Actuator Stop Test
P3235	2FA4	Control Module Monitoring Version Coding Plausibility
P3236	2D71	Control Module Monitoring Injection Time Relative Fuel Quantity Plausibility
P3237	2D71	Control Module Monitoring Fuel Correction Error
P3238		Control Module Monitoring TPU Chip Defective
P3239	2FA3	Control Module - No Coding
P324A	2E97	Generator Type Implausible
P324A	2ED3	Generator Type Implausible

OBD

BentleyPublishers.com
BentleyPublishers.com–All Rights Reserved

Table a. Diagnostic trouble codes (DTCs) (continued)

P-code	BMW FC	Definition
P324C	2ED0	Generator Over Temperature Calculated
P324E	2ED2	Generator Regulator Type Implausible
P3247	2B98	Control Module Fault - RAM BAckup
P325A	2ECE	Generator Electrical Error Calculated
P3255	2E96	Generator Voltage In Starting Above Threshold
P3256	2DD2	CAN Time-out Steering Angle Sensor
P3258		Control Module Internal Fault, Hardware
P3259	2D60	Control Module Monitoring Fuel Volume, Lambda Implausible to Operating Mode
P3283	29F1	Fuel Pressure Control Adaptive Fuel Volume Out of Range (Bank 1)
P3284	29F1	Fuel Pressure Control Adaptive Fuel Volume Implausible (Bank 1)
P3300	2E24	Ignition Coil Cylinder 1 High Input
P3301	2E24	Ignition Coil Cylinder 1 High Resistance
P3302	2E24	Ignition Coil Cylinder 1 Cut OFF Due to High Temperature
P3303	2E28	Ignition Coil Cylinder 5 High Input
P3304	2E28	Ignition Coil Cylinder 5 High Resistance
P3305	2E28	Ignition Coil Cylinder 5 Cut OFF Due to High Temperature
P3306	2E27	Ignition Coil Cylinder 4 High Input
P3307	2E27	Ignition Coil Cylinder 4 High Resistance
P3308	2E27	Ignition Coil Cylinder 4 Cut OFF Due to High Temperature
P3309	2E2B	Ignition Coil Cylinder 8 High Input
P3310	2E2B	Ignition Coil Cylinder 8 High Resistance
P3311	2E2B	Ignition Coil Cylinder 8 Cut OFF Due to High Temperature
P3312	2E29	Ignition Coil Cylinder 6 High Input
P3313	2E29	Ignition Coil Cylinder 6 High Resistance
P3314	2E29	Ignition Coil Cylinder 6 Cut OFF Due to High Temperature
P3315		Ignition Coil Cylinder 3 High Input
P3316		Ignition Coil Cylinder 3 High Resistance
P3317	11814	Ignition Coil Cylinder 3 Cut OFF Due to High Temperature
P3318	2E2A	Ignition Coil Cylinder 7 High Input
P3319	2E2A	Ignition Coil Cylinder 7 High Resistance
P3320	2E2A	Ignition Coil Cylinder 7 Cut OFF Due to High Temperature
P3321	11813	Ignition Coil Cylinder 2 High Input
P3322	11813	Ignition Coil Cylinder 2 High Resistance
P3323	11813	Ignition Coil Cylinder 2 Cut OFF Due to High Temperature
P3325	2E26	Ignition Coil Cylinder 3 High Input
P3326	2E26	Ignition Coil Cylinder 3 High Resistance
P3327	2E26	Ignition Coil Cylinder 3 Cut OFF Due to High Temperature

BentleyPublishers.com
BentleyPublishers.com—All Rights Reserved

Table a. Diagnostic trouble codes (DTCs) (continued)		
P-code	**BMW FC**	**Definition**
P3321	2E25	Ignition Coil Cylinder 2 High Input
P3322	2E25	Ignition Coil Cylinder 2 High Resistance
P3323	2E25	Ignition Coil Cylinder 2 Cut OFF Due to High Temperature
P3400		Cylinder Deactivation System (Bank 1)
P3497		Cylinder Deactivation System (Bank 20

BentleyPublishers.com
BentleyPublishers.com–All Rights Reserved

BentleyPublishers.com
BentleyPublishers.com—All Rights Reserved

WARNING

Your common sense, good judgement and general alertness are crucial to safe and successful service work. Before attempting any work on your BMW, be sure to read **001 Warnings and Cautions** *and the copyright page at the front of the manual. Review these warnings and cautions each time you prepare to work on your car. Please also read any warnings and cautions that accompany the procedures in the manual.*

A

BentleyPublishers.com
BentleyPublishers.com–All Rights Reserved

BentleyPublishers.com
BentleyPublishers.com—All Rights Reserved

WARNING

Your common sense, good judgement and general alertness are crucial to safe and successful service work. Before attempting any work on your BMW, be sure to read **001 Warnings and Cautions** *and the copyright page at the front of the manual. Review these warnings and cautions each time you prepare to work on your car. Please also read any warnings and cautions that accompany the procedures in the manual.*

BentleyPublishers.com
BentleyPublishers.com—All Rights Reserved

BentleyPublishers.com
BentleyPublishers.com—All Rights Reserved

WARNING

Your common sense, good judgement and general alertness are crucial to safe and successful service work. Before attempting any work on your BMW, be sure to read **001 Warnings and Cautions** *and the copyright page at the front of the manual. Review these warnings and cautions each time you prepare to work on your car. Please also read any warnings and cautions that accompany the procedures in the manual.*

BentleyPublishers.com
BentleyPublishers.com–All Rights Reserved

BentleyPublishers.com
BentleyPublishers.com–All Rights Reserved

WARNING

Your common sense, good judgement and general alertness are crucial to safe and successful service work. Before attempting any work on your BMW, be sure to read **001 Warnings and Cautions** *and the copyright page at the front of the manual. Review these warnings and cautions each time you prepare to work on your car. Please also read any warnings and cautions that accompany the procedures in the manual.*

BentleyPublishers.com
BentleyPublishers.com–All Rights Reserved

BentleyPublishers.com
BentleyPublishers.com—All Rights Reserved

WARNING

Your common sense, good judgement and general alertness are crucial to safe and successful service work. Before attempting any work on your BMW, be sure to read **001 Warnings and Cautions** *and the copyright page at the front of the manual. Review these warnings and cautions each time you prepare to work on your car. Please also read any warnings and cautions that accompany the procedures in the manual.*

BentleyPublishers.com
BentleyPublishers.com—All Rights Reserved

BentleyPublishers.com
BentleyPublishers.com—All Rights Reserved

WARNING

Your common sense, good judgement and general alertness are crucial to safe and successful service work. Before attempting any work on your BMW, be sure to read **001 Warnings and Cautions** *and the copyright page at the front of the manual. Review these warnings and cautions each time you prepare to work on your car. Please also read any warnings and cautions that accompany the procedures in the manual.*

BentleyPublishers.com
BentleyPublishers.com—All Rights Reserved

BentleyPublishers.com
BentleyPublishers.com—All Rights Reserved

Selected Books and Repair Information From Bentley Publishers

Motorsports

Alex Zanardi: My Sweetest Victory
Alex Zanardi and Gianluca Gasparini
ISBN 978-0-8376-1249-2

The Unfair Advantage *Mark Donohue and Paul van Valkenburgh*
ISBN 978-0-8376-0069-7

Driving Forces: The Grand Prix Racing World Caught in the Malestrom of the Third Reich
Peter Stevenson
ISBN 978-0-8376-0217-2

Equations of Motion – Adventure, Risk and Innovation
William F. Milliken
ISBN 978-0-8376-1570-7

The Racing Driver – The Theory and Practice of Fast Driving
Denis Jenkinson
ISBN 978-0-8376-0201-1

Engineering

Physics for Gearheads *Randy Beikmann*
ISBN 978-0-8376-1615-5

Bosch Automotive Handbook
Updated 9th Edition *Robert Bosch GmbH*
ISBN 978-0-8376-1732-9

Bosch Fuel Injection and Engine Management *Charles O. Probst, SAE*
ISBN 978-0-8376-0300-1

Maximum Boost: Designing, Testing, and Installing Turbocharger Systems
Corky Bell ISBN 978-0-8376-0160-1

Supercharged! Design, Testing and Installation of Supercharger Systems
Corky Bell ISBN 978-0-8376-0168-7

Race Car Aerodynamics *Joseph Katz*
ISBN 978-0-8376-0142-7

Scientific Design of Exhaust and Intake Systems *Phillip H. Smith & John C. Morrison* ISBN 978-0-8376-0309-4

Audi Repair Manuals

Audi A4 Service Manual: 2002-2008, 1.8L Turbo, 2.0L Turbo, 3.0L, 3.2L
Bentley Publishers
ISBN 978-0-8376-1574-5

Audi A4 Service Manual: 1996-2001, 1.8L Turbo, 2.8L *Bentley Publishers*
ISBN 978-0-8376-1675-9

Audi TT Service Manual: 2000-2006, 1.8L turbo, 3.2 L *Bentley Publishers*
ISBN 978-0-8376-1625-4

Audi A6 (C5 platform) Service Manual: 1998-2004 *Bentley Publishers*
ISBN 978-0-8376-1670-4

BMW

BMW 3 Series (F30, F31, F34) Service Manual: 2012-2015 *Bentley Publishers*
ISBN 978-0-8376-1752-7

BMW 3 Series (E36) Service Manual: 1992-1998 *Bentley Publishers*
ISBN 978-0-8376-1709-1

BMW 3 Series (E30) Service Manual: 1984-1990 *Bentley Publishers*
ISBN 978-0-8376-1647-6

BMW X5 (E53) Service Manual: 2000-2006 *Bentley Publishers*
ISBN 978-0-8376-1643-8

BMW Z3 Service Manual: 1996-2002
Bentley Publishers
ISBN 978-0-8376-1617-9

BMW 7 Series (E32) Service Manual: 1988-1994 *Bentley Publishers*
ISBN 978-0-8376-1619-3

BMW 7 Series (E38) Service Manual: 1995-2001 *Bentley Publishers*
ISBN 978-0-8376-1618-6

Memoirs of a Hack Mechanic
Rob Siegel ISBN 978-0-8376-1720-6

Porsche

Porsche 911 (Type 996) Service Manual: 1999-2005 *Bentley Publishers*
ISBN 978-0-8376-1710-7

Porsche 911 Carrera Service Manual: 1984-1989 *Bentley Publishers*
ISBN 978-0-8376-1696-4

Porsche 911 SC Service Manual: 1978-1983 *Bentley Publishers*
ISBN 978-0-8376-1705-3

Porsche 911 (964): Enthusiast's Companion *Bentley Publishers*
ISBN 978-0-8376-0293-6

Porsche Boxster. Boxster S (986) Service Manual: 1997-2004
Bentley Publishers
ISBN 978-0-8376-1645-2

Porsche: Excellence Was Expected
Karl Ludvigsen ISBN 978-0-8376-0235-6

Ferdinand Porsche – Genesis of Genius
Karl Ludvigsen ISBN 978-0-8376-1557-8

Porsche – Origin of the Species
Karl Ludvigsen ISBN 978-0-8376-1331-4

Volkswagen

Volkswagen Jetta, Golf, GTI Service Manual: 1999-2005 *Bentley Publishers*
ISBN 978-0-8376-1678-0

Volkswagen Jetta Service Manual: 2005-2010 *Bentley Publishers*
ISBN 978-0-8376-1616-2

Volkswagen Passat Service Manual: 1998-2005 *Bentley Publishers*
ISBN 978-0-8376-1669-8

Volkswagen Jetta, Golf, GTI: 1993-1999, Cabrio: 1995-2002 Service Manual *Bentley Publishers*
ISBN 978-0-8376-1660-5

Volkswagen GTI, Golf, Jetta Service Manual: 1985-1992 *Bentley Publishers*
ISBN 978-0-8376-1637-7

MINI Repair Manuals

MINI Cooper Service Manual: 2002-2006 *Bentley Publishers*
ISBN 978-0-8376-1639-1

MINI Cooper Service Manual: 2007-2013 *Bentley Publishers*
ISBN 978-0-8376-1730-5

Featured Repair Manuals

BMW X3 (E83) Service Manual: 2004-2010
Bentley Publishers
ISBN 978-0-8376-1731-2

BMW 3 Series (E90, E91, E92, E93) Service Manual: 2006-2011
Bentley Publishers
ISBN 978-0-8376-1723-7

BMW 3 Series (E46) Service Manual: 1999-2005
Bentley Publishers
ISBN 978-0-8376-1657-5

BMW 5 Series (E39) Service Manual: 1997-2003
Bentley Publishers
ISBN 978-0-8376-1672-8

BentleyPublishers®
Automotive Reference

Bentley Publishers has published service manuals and automobile books since 1950. For more information, please contact Bentley Publishers at 1734 Massachusetts Avenue, Cambridge, MA 02138 USA, or visit our web site at
`BentleyPublishers.com`

Notes

Notes

BMW 5 Series
(E60, E61)

Service Manual
525i, 525xi, 528i, 528xi
530i, 530xi, 535i, 535xi
545i, 550i
2004, 2005, 2006, 2007, 2008, 2009, 2010

[B] BentleyPublishers®
.com

[B]® BENTLEY PUBLISHERS® | Automotive Reference™

Bentley Publishers, a division of Robert Bentley, Inc.
1734 Massachusetts Avenue
Cambridge, MA 02138 USA
800-423-4595 / 617-547-4170

Information that makes the difference®

BentleyPublishers®.com

Technical contact information

We welcome your feedback. Please submit corrections and additions to our BMW technical discussion forum at:

`http://www.BentleyPublishers.com`

Updates and corrections

We will evaluate submissions and post appropriate editorial changes online as updates or tech discussion. Appropriate updates will be incorporated with the manual text in future printings. Check for updates and corrections for this book before beginning work on your vehicle. See the following web address for additional information:

`http://www.BentleyPublishers.com/updates/`

WARNING—Important Safety Notice

Do not use this manual for repairs unless you are familiar with automotive repair procedures and safe workshop practices. This manual illustrates the workshop procedures for some maintenance and service work. It is not a substitute for full and up-to-date information from the vehicle manufacturer or for proper training as an automotive technician. Note that it is not possible for us to anticipate all of the ways or conditions under which vehicles may be serviced or to provide cautions as to all of the possible hazards that may result.

We have endeavored to ensure the accuracy of the information in this manual. Please note, however, that considering the vast quantity and the complexity of the service information involved, we cannot warrant the accuracy or completeness of the information contained in this manual.

FOR THESE REASONS, NEITHER THE PUBLISHER NOR THE AUTHOR MAKES ANY WARRANTIES, EXPRESS OR IMPLIED, THAT THE INFORMATION IN THIS MANUAL IS FREE OF ERRORS OR OMISSIONS, AND WE EXPRESSLY DISCLAIM THE IMPLIED WARRANTIES OF MERCHANTABILITY AND OF FITNESS FOR A PARTICULAR PURPOSE, EVEN IF THE PUBLISHER OR AUTHOR HAVE BEEN ADVISED OF A PARTICULAR PURPOSE, AND EVEN IF A PARTICULAR PURPOSE IS INDICATED IN THE MANUAL. THE PUBLISHER AND AUTHOR ALSO DISCLAIM ALL LIABILITY FOR DIRECT, INDIRECT, INCIDENTAL OR CONSEQUENTIAL DAMAGES THAT RESULT FROM ANY USE OF THE EXAMPLES, INSTRUCTIONS OR OTHER INFORMATION IN THIS MANUAL. IN NO EVENT SHALL OUR LIABILITY, WHETHER IN TORT, CONTRACT OR OTHERWISE, EXCEED THE COST OF THIS MANUAL.

Before attempting any work on your BMW, read **001 Warnings and Cautions** and any **WARNING** or **CAUTION** that accompanies a procedure in the manual. Review the **WARNINGS** and **CAUTIONS** each time you prepare to work on your BMW.

Your common sense and good judgment are crucial to safe and successful service work. Read procedures through before starting them. Think about whether the condition of your car, your level of mechanical skill or your level of reading comprehension might result in or contribute in some way to an occurrence which might cause you injury, damage your car or result in an unsafe repair. If you have doubts for these or other reasons about your ability to perform safe repair work on your car, have the work done at an authorized BMW dealer or other qualified shop.

Part numbers listed in this manual are for identification purposes only, not for ordering. Always check with your authorized BMW dealer to verify part numbers and availability before beginning service work that may require new parts.

Special tools required to perform certain service operations are identified in the manual and are recommended for use. Use of improper tools may be detrimental to the car's safe operation as well as the safety of the person servicing the car.

The vehicle manufacturer will continue to issue service information updates and parts retrofits after the editorial closing of this manual. Some of these updates and retrofits will apply to procedures and specifications in this manual. We regret that we cannot supply updates to purchasers of this manual.

This manual is prepared, published and distributed by Bentley Publishers, 1734 Massachusetts Avenue, Cambridge, Massachusetts 02138 USA. All information contained in this manual is based on the information available to the publisher at the time of editorial closing. BMW has not reviewed and does not vouch for the accuracy or completeness of the technical specifications and work procedures described and given in this manual.

© 2010 Robert Bentley, Inc. Bentley Publishers is a trademark of Robert Bentley, Inc.

[B]® is a registered trademark of Bentley Publishers and Robert Bentley, Inc.

All rights reserved. The right is reserved to make changes at any time without notice. No part of this publication may be reproduced, stored in a retrieval system, or transmitted in any form or by any means, electronic, mechanical, photocopying, recording, or otherwise, without the prior written consent of the publisher. This includes text, figures, and tables. All rights reserved under Berne and Pan-American Copyright conventions.

ISBN 978-0-8376-1689-6 (2-volume set)
ISBN 978-0-8376-1690-2 (volume 1)
ISBN 978-0-8376-1691-9 (volume 2) **Job Code: B510-05** **Editorial closing 04 / 2010**

Library of Congress Cataloging-in-Publication Data for the 2010 edition
BMW 5 series, (E60, E61) : service manual : 525i, 525xi, 528i, 528xi, 530i, 530xi, 535i, 535xi, 545i, 550i, 2004, 2005, 2006, 2007, 2008, 2009, 2010
p. cm.
Includes bibliographical references and index.
ISBN 978-0-8376-1621-6 (alk. paper)
1. BMW 5 series automobiles--Maintenance and repair--Handbooks, manuals, etc. I. Robert Bentley, inc.

TL215.B25B57 2010
629.28'722--dc22

2010012318

Cover design and use of blue band on spine and back cover are trade dress and trademarks of Bentley Publishers. All rights reserved.
The paper used in this publication is acid free and meets the requirements of the National Standard for Information Sciences–Permanence of Paper for Printed Library Materials. ∞

Manufactured in the United States of America

202411R001

You are now in **Volume 1**

BentleyPublishers.com
BentleyPublishers.com—All Rights Reserved

Foreword

For the BMW owner with basic mechanical skills and for independent auto service professionals, this manual includes many of the specifications and procedures that were available in an authorized BMW dealer service department as this manual went to press. The BMW owner with no intention of working on his or her car will find that owning and referring to this manual will make it possible to be better informed and to more knowledgeably discuss repairs with a professional automotive technician.

For those intending to do maintenance and repair on their BMW, it is essential that safety equipment be used and safety precautions observed when working on the vehicle. A minimum safety equipment list includes hand protection, eye protection and a fire extinguisher. A selection of good quality hand tools is also needed. This includes a torque wrench to ensure that fasteners are tightened in accordance with specifications. In some cases, the text refers to special tools that are recommended or required to accomplish adjustments or repairs. These tools are often identified by their BMW special tool number and illustrated.

Disclaimer

We have endeavored to ensure the accuracy of the information in this manual. When the vast array of data presented in the manual is taken into account, however, no claim to infallibility can be made. We therefore cannot be responsible for the result of any errors that may have crept into the text. Please also read the **WARNING—Important Safety Notice** on the copyright page at the beginning of this book.

Prior to starting a repair procedure, read the procedure, **001 Warnings and Cautions** and the warnings and cautions that accompany the procedure. Reading a procedure before beginning work helps you determine in advance the need for specific skills, identify hazards, prepare for appropriate capture and handling of hazardous materials, and the need for particular tools and replacement parts such as gaskets.

Bentley Publishers encourages comments from the readers of this manual with regard to errors, and suggestions for improvement of our product. These communications have been and will be carefully considered in the preparation of this and other manuals. If you identify inconsistencies in the manual, you may have found an error. Please contact the publisher and we will endeavor to post applicable corrections on our website. Posted corrections (errata) should be reviewed before beginning work. Please see the following web address:

`http://www.BentleyPublishers.com/updates/`

BMW will continue to issue service information and parts retrofits after the editorial closing of this manual. Some of this updated information may apply to procedures and specifications in this manual. For the latest information, please see the following web address:

`http://www.bmwtechinfo.com/`

BMW offers extensive warranties, especially on components of the fuel delivery and emission control systems. Therefore, before deciding to repair a BMW that may be covered wholly or in part by any warranties issued by BMW of North America, LLC, consult your authorized BMW dealer. You may find that the dealer can make the repair either free or at minimum cost. Regardless of its age, or whether it is under warranty, your BMW is both an easy car to service and an easy car to get serviced. So if at any time a repair is needed that you feel is too difficult to do yourself, a trained BMW technician is ready to do the job for you.

Bentley Publishers

001 Warnings and Cautions

PLEASE READ THESE WARNINGS AND CAUTIONS BEFORE PROCEEDING WITH MAINTENANCE AND REPAIR WORK.

WARNINGS—
See also CAUTIONS

- Read the important safety notice on the copyright page at the beginning of the book.
- Some repairs may be beyond your capability. If you lack the skills, tools and equipment or a suitable workplace for any procedure described in this manual, we suggest you leave such repairs to an authorized BMW dealer service department or other qualified shop.
- Thoroughly read each procedure and the **WARNINGS** and **CAUTIONS** that accompany the procedure. Also review posted corrections at `www.BentleyPublishers.com/updates/` before beginning work.
- If any procedure, tightening torque, wear limit, specification or data presented in this manual does not appear to be appropriate for a specific application, contact the publisher or the vehicle manufacturer for clarification before using the information in question.
- Do not reuse any fasteners that are worn or deformed. Many fasteners are designed to be used only once and become unreliable and may fail when used a second time. This includes, but is not limited to, nuts, bolts, washers, self-locking nuts or bolts, circlips and cotter pins. Replace these fasteners with new parts.
- Do not work under a lifted car unless it is solidly supported on stands designed for the purpose. Do not support a car on cinder blocks, hollow tiles or other props that may crumble under continuous load. Do not work under a car that is supported solely by a jack. Do not work under the car while the engine is running.
- If you are going to work under a car on the ground, make sure that the ground is level. Block the wheels to keep the car from rolling. Disconnect the battery negative (–) terminal to prevent others from starting the car while you are under it.
- Do not run the engine unless the work area is well ventilated. Carbon monoxide kills.
- Remove rings, bracelets and other jewelry so that they cannot cause electrical shorts, get caught in running machinery, or be crushed by heavy parts.
- Tie long hair behind your head. Do not wear a necktie, a scarf, loose clothing, or a necklace when you work near machine tools or running engines. If your hair, clothing, or jewelry were to get caught in the machinery, severe injury could result.
- Do not attempt to work on your car if you do not feel well. You increase the danger of injury to yourself and others if you are tired, upset or have taken medication or any other substance that may keep you from being fully alert.
- Illuminate your work area adequately but safely. Use a portable safety light for working inside or under the car. Make sure the bulb is enclosed by a wire cage. The hot filament of an accidentally broken bulb can ignite spilled fuel, vapors or oil.
- Catch draining fuel, oil, or brake fluid in suitable containers. Do not use food or beverage containers that might mislead someone into drinking from them. Store flammable fluids away from fire hazards. Wipe up spills at once, but do not store the oily rags, which can ignite and burn spontaneously.
- Observe good workshop practices. Wear goggles when you operate machine tools or work with battery acid. Wear gloves or other protective clothing whenever the job requires working with harmful substances.
- Greases, lubricants and other automotive chemicals contain toxic substances, many of which are absorbed directly through the skin. Read the manufacturer's instructions and warnings carefully. Use hand and eye protection. Avoid direct skin contact.
- Disconnect the battery negative (–) terminal whenever you work on the fuel system or the electrical system. Do not smoke or work near heaters or other fire hazards. Keep an approved fire extinguisher handy.
- Friction materials (such as brake pads and shoes or clutch discs) contain asbestos fibers or other friction materials. Do not create dust by grinding, sanding, or by cleaning with compressed air. Avoid breathing dust. Breathing any friction material dust can lead to serious diseases and may result in death.
- Batteries give off explosive hydrogen gas during charging. Keep sparks, lighted matches and open flame away from the top of the battery. If hydrogen gas escaping from the cap vents is ignited, it may ignite gas trapped in the cells and cause the battery to explode.
- The air-conditioning system is filled with chemical refrigerant, which is hazardous. Make sure the system is serviced only by a trained technician using approved refrigerant recovery/recycling equipment, trained in related safety precautions, and familiar with regulations governing the discharge and disposal of automotive chemical refrigerants.

Continued on next page

BentleyPublishers
BentleyPublishers.com – All Rights Reserved

WARNINGS (continued)

- Do not expose any part of the A/C system to high temperatures such as open flame. Excessive heat increases system pressure and may cause the system to burst.
- Some aerosol tire inflators are highly flammable. Be extremely cautious when repairing a tire that may have been inflated using an aerosol tire inflator. Keep sparks, open flame or other sources of ignition away from the tire repair area. Inflate and deflate the tire at least four times before breaking the bead from the rim. Completely remove the tire from the rim before attempting any repair.
- Cars covered by this manual are equipped with a supplemental restraint system (SRS) that automatically deploys airbags and pyrotechnic seat belt tensioners in case of a frontal or side impact. These are explosive devices. Handled improperly or without adequate safeguards, they can be accidently activated and cause serious injury.
- The ignition system produces high voltages that can be fatal. Avoid contact with exposed terminals and use extreme care when working on a car with the engine running or the ignition switched ON.
- Place jack stands only at locations specified by the manufacturer. The vehicle lifting jack supplied with the vehicle is intended for tire changes only. Use a heavy duty floor jack to lift vehicle before installing jack stands. See **020 Maintenance**.
- Battery acid (electrolyte) can cause severe burns. Flush contact area with water, then seek medical attention.
- Aerosol cleaners and solvents may contain hazardous or deadly vapors and are highly flammable. Use only in a well ventilated area. Do not use on hot surfaces (engines, brakes, etc.).
- Due to risk of personal injury, be sure the engine is cold before beginning work on the cooling system.

CAUTIONS—
See also WARNINGS

- If you lack the skills, tools and equipment, or a suitable workshop for any procedure described in this manual, leave such repairs to an authorized BMW dealer or other qualified shop.
- BMW is constantly improving its cars and sometimes these changes, both in parts and specifications, are made applicable to earlier models. Any part numbers listed in this manual are for reference only. Check with your authorized BMW dealer parts department for the latest information.
- Before starting a job, make certain that you have the necessary tools and parts on hand. Read all the instructions thoroughly, and do not attempt shortcuts. Use tools appropriate to the work and use only replacement parts meeting BMW specifications.
- Use pneumatic and electric tools only to loosen threaded parts and fasteners. Do not use these tools to tighten fasteners, especially on light alloy parts. Use a torque wrench to tighten fasteners to the tightening torque specification listed.
- Be mindful of the environment and ecology. Before you drain the crankcase, find out the proper way to dispose of the oil. Do not pour oil onto the ground, down a drain, or into a stream, pond or lake. Dispose of waste in accordance with federal, state and local laws.
- The control module for the anti-lock brake system (ABS) cannot withstand temperatures from a paint-drying booth or a heat lamp in excess of 203°F (95°C). Do not subject to temperatures in excess of 185°F (85°C) for more than two hours.
- Before doing any electrical welding on cars equipped with ABS, disconnect the battery negative (–) terminal (ground strap) and the ABS control module connector.
- Make sure ignition is switched OFF before disconnecting battery.
- Label battery cables before disconnecting. On some models, battery cables are not color coded.
- Disconnecting the battery may erase fault code(s) stored in control module memory. Using special BMW diagnostic equipment, check for fault codes prior to disconnecting the battery cables. If the malfunction indicator light (MIL) is illuminated, see **OBD On-Board Diagnostics** at the back of this manual. (This light may be identified as the Check Engine light or the Service Engine Soon light.) If any other system faults are detected (indicated by an illuminated warning light), see an authorized BMW dealer.
- If a normal or rapid charger is used to charge battery, disconnect the battery remove it from the vehicle in order to avoid damaging the vehicle.
- Do not quick-charge the battery (for boost starting) for longer than one minute. Wait at least one minute before boosting the battery a second time.
- Connect and disconnect a battery charger only with the battery charger switched OFF.
- Sealed or "maintenance free" batteries should be slow-charged only, at an amperage rate that is approximately 10% of the battery's ampere-hour (Ah) rating.
- Do not allow battery charging voltage to exceed 16.5 volts. If the battery begins producing gas or boiling violently, reduce the charging rate. Boosting a sulfated battery at a high charging rate can cause an explosion.
- Do not use steel fasteners on engine components made of aluminum-magnesium alloy. Use aluminum fasteners only. Test fasteners for aluminum composition with magnet.
- Replace aluminum bolts each time they are loosened. Follow torque instructions, including angle of rotation specifications, when installing aluminum fasteners.

Bentley Publishers.com
BentleyPublishers.com—All Rights Reserved

002 Vehicle Identification and VIN Decoder

Some of the information in this manual applies only to cars of a particular model year or range of years. For example, 2004 refers to the 2004 model year but does not necessarily match the calendar year in which the car was manufactured or sold. To be sure of the model year of a particular car, check the vehicle identification number (VIN) on the car.

The VIN is a unique sequence of 17 characters assigned by BMW to identify each individual car. When decoded, the VIN tells the country and year of manufacture; make, model and serial number; assembly plant and even some equipment specifications.

The BMW VIN is on a plate mounted on the top of the dashboard, on the driver's side where the number can be seen through the windshield. The 10th character is the model year code. Example: 4 for 2004. For 2010, A is used. The table below explains some of the codes in the VIN for 2004 through 2010 BMW E60 and E61 5 Series cars covered by this manual.

Sample VIN: WBA NA53 5 0 4 B - - - - - -

position 1 2 3 4 5 6 7 8 9 10 11 12-17

VIN position	Description	Decoding information	
1 - 3	Country of manufacture	WBA WBS	BMW, AG. Munich, Germany BMW Motorsports GmbH, Munich. Germany
4 - 7	Vehcle code	NA53 NE53 NU53 NA73 NE73 NW13 NB33 NB53 NW53 NF33 NV13 NF73 NV93 NN73 PT73 NB93	525i 2004 - 2005 (2003/09/01 to 2005/02/28) 525i 2006 - 2007 (2005/03/01 to 2007/02/28) 528i 2008 - 2010 (2007/03/01 to 2009/12/31) 530i 2004 - 2005 (2003/01/01 to 2005/08/27) 530i 2006 - 2007 (2005/03/01 to 2007/02/28) 535i 2008 - 2010 (2007/03/01 to 2009/12/31) 545i 2004 - 2005 (2003/09/01 to 2005/08/31) 550i 2006 - 2007 (2005/09/01 to 2007/02/28) 550i 2008 - 2010 (2007/03/01 to 2009/12/31) 525xi 2006 - 2007 (2005/04/01 to 2007/02/28) 528xi 2008 - 2010 (2007/03/01 to 2009/12/31) 530xi 2006 - 2007 (2005/04/01 to 2007/02/28) 535xi 2008 - 2010 (2007/03/01 to 2009/12/31) 530xi Sport Wagon 2006 - 2007 (2005/04/01 to 2007/02/28) 535xi Sport Wagon 2008 - 2010 M5
8	Restraint system	5	Multiple restraint system
9	Check digit		0 - 9 or X, calculated by NHTSA
10	Model year	4 5 6 7 8 9 A	2004 2005 2006 2007 2008 2009 2010
11	Assembly plant	B, C	Munich, Germany
12-17	Serial number		Sequential production number for specific vehicle

BentleyPublishers.com - All Rights Reserved

BentleyPublishers.com
BentleyPublishers.com–All Rights Reserved

003 Symptoms and Field Observations

General

This repair group includes a list of symptoms, field observations, and suggested fixes for common problems and issues found on BMW 5 Series (E60, E61) vehicles.

The information is organized by sections that correspond to the layout of the repair manual. Additional information (when available) can be found in the repair group associated with the specific issue:

- **0** General Data and Maintenance
- **1** Engine
- **2** Transmission
- **3** Suspension, Steering and Brakes
- **4** Body
- **5** Body Equipment
- **6** Electrical System
- **7** Equipment and Accessories

This information was developed from varied sources, ranging from professional BMW technician feedback to manufacturer-issued technical service bulletins (TSBs). The content is intended to complement the repair information found in this repair manual.

CAUTION—
- *The information contained in this section is inherently dated material. It was applicable and relevant at the time this manual went to press. Always check BMW factory repair information at www.bmwtechinfo.com or the publisher's website at www.bentleypublishers.com for information that may supersede any information included in this section.*

BentleyPublishers.com
BentleyPublishers.com – All Rights Reserved

0 General Data and Maintenance

	Topic	Notes	Additional information (when available)
	Exiting a vehicle that is double-locked from the outside 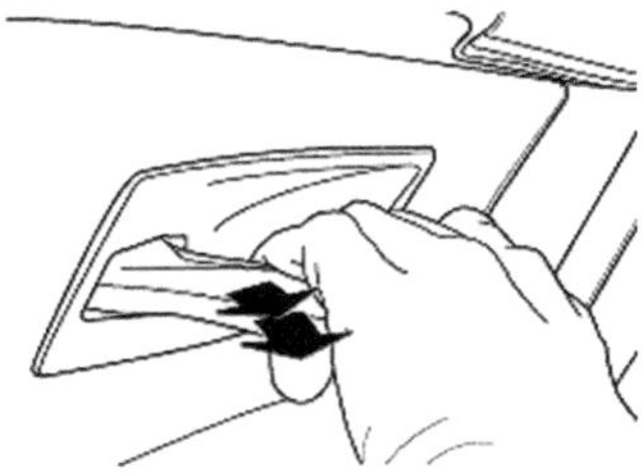	Never lock a vehicle from outside the car with an someone inside the car! In the unlikely that someone gets double-locked inside the car, the person inside the car can exit the vehicle using the following 2-step procedure: • Press the central unlock button to disable “double-lock”. • Pull the door handle twice to open door. ***NOTE—*** • *For vehicles with the factory alarm, opening the door will activate the alarm.*	
N52, N54 engines	**Engine oil, topping up** 	Oil residue may collect on the cylinder head cover near the oil filler cap a few days after topping up or refilling the engine oil. When oil is added, the filler cap is laid partially upside down. In this position, engine oil seeps into the area of the oil cap that is not sealed by the gasket. When the cap is reinstalled, the oil leaks from the unsealed area onto the cylinder head cover surface. If investigating an oil leak in this area, confirm that it is not oil from the cap.	**020**
N52 engine	**Replacing spark plugs, service tip** 	When servicing spark plugs, a standard socket may fit too tightly into the spark plug bore in the cylinder head. If this is the case, remove the heat shield inside plug hole (**arrow**) before servicing spark plugs.	**020**
N54 engine	**Special tool for spark plug removal)** 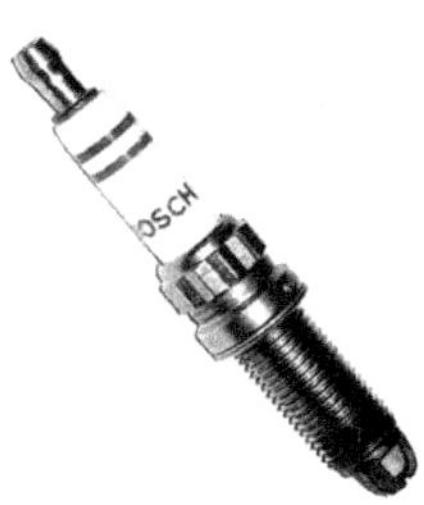	The N54 engine uses a new design spark plug from Bosch that consists of a 12mm thread (compared to the 14mm design on the N52) and a 12-point hex. The plug requires a special tool (BMW special tool part no. 83 30 0 495 560). The special socket is a thin-wall design to facilitate access in the confined area of the N54 cylinder head.	**020**

BentleyPublishers.com
BentleyPublishers.com—All Rights Reserved

0 General Data and Maintenance (continued)

	Topic	Notes	Additional information (when available)
Models from 2006	**Engine oil level, checking after oil change - dynamic check (engine running)** 	Always perform the dynamic oil level measurement (approx. 5 minutes driving time) after an oil change. The oil level could be misinterpreted as the oil level last stored is initially displayed after an oil change. ***NOTE—*** • *There is no oil level initially stored after replacing or reprogramming the engine control unit ("Oil level below min" is displayed). The correct oil level is indicated after running the engine for approx. 5 minutes.* - Start engine. - Select on-board computer function - **OIL** using rocker switch. Press BC switch in once. Oil level check begins. A clock symbol (arrow) may appear while the level measurement is running. ***NOTE—*** • *The clock symbol appears for up to 50 seconds after starting the engine when there is no measured value or the last stored value is not within the tolerance range of the currently measured oil level.*	**020**
V8 engine	**V8 engine A/C belt removing and installing.** **A/C belt installation tool**  B510003002	On the V8 engines, there is no tensioner on the A/C belt. To install a new belt, a special tool (BMW part # 90 88 6 641 040) is required to guide the belt onto the pulley.	**020**
	Battery, replacing 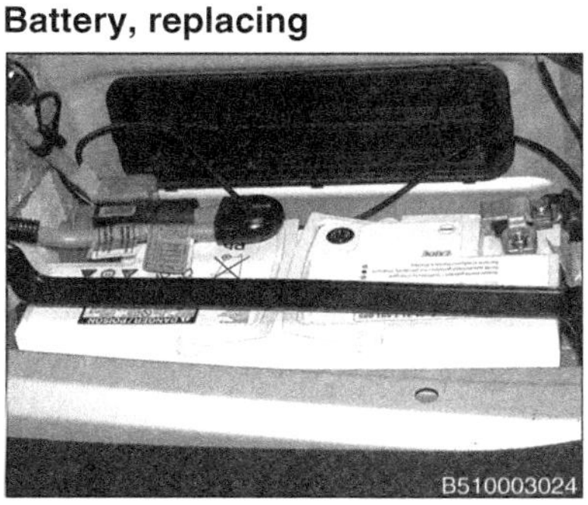 B510003024	When installing a new battery, it must be "registered" to the vehicle to notify the vehicle power management software (engine electronics and intelligent battery sensor) that a new battery was fitted. This operation requires a BMW scan tool. Without new battery registration, various electrical problems as well odd electrical system behavior can occur.	**121**

BentleyPublishers.com
BentleyPublishers.com—All Rights Reserved

0 General Data and Maintenance (continued)

Topic	Notes	Additional information (when available)
No communication with BMW scan tool on DCAN. 	When performing diagnosis on a vehicle with the BMW scan tool, the message "Connect diagnostic connector and switch ON ignition (Terminal 15)!" is displayed. • D-CAN failure in vehicle. Before diagnosing D-CAN in the vehicle, try communicating with another vehicle. If communication cannot be established on other vehicle, try a different scan tool. If communication resumes on the other vehicle, use the following steps to diagnose DCAN. – Disconnect vehicle power supply / charger (if connected). – Disconnect battery. – Check resistance of D-CAN circuit across OBD socket Pin 6 (D_CAN_H) and Pin 14 (D_CAN_L). A good D-CAN measures 60 ohms. The D-CAN circuit is fitted with two 120 ohm terminal resistors. One is fitted in wiring near OBD socket. The second is fitted in gateway module of vehicle (KGM or SGM). – If resistance reading is 0 ohms, the D-CAN wires are shorted together. Check circuit for shorts as per wiring diagram. – If resistance reading is Mega ohms, the D-CAN wiring has an open circuit. Check circuit for opens as per wiring diagram. – If resistance reading is 120 ohms, one of terminal resistors is faulty. Check resistance of both resistors for an open circuit and replace as necessary.	**020, 600**
Intelligent battery sensor (IBS), service precautions 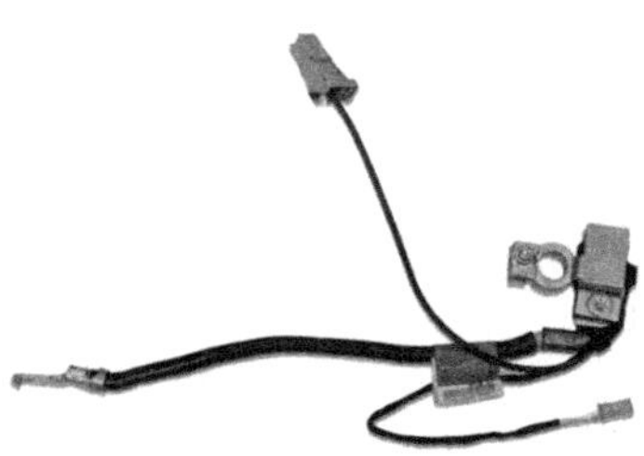	The Intelligent Battery Sensor (IBS) is an electronic device connected directly to the negative battery terminal.The IBS is very sensitive to mechanical stress and strain. Particular attention should be paid to the following points: • Do not make any additional connections at the negative terminal of the battery. • Do not modify the battery ground cable. • Do not make any connections between the IBS and the sensor screw. • Do not use force when disconnecting the ground terminal from the battery. • Never pull on the ground cable. • Do not use the IBS as a pivot point to lever off the ground terminal. • Do not use the connections of the IBS as a lever. • Use only a torque wrench as described in the repair manual during installation. • Do not release or tighten the sensor screw. ***NOTE—*** • A replacement IBS includes the ground cable. The ground cable also serves as a heat dissipater for the IBS.	**121**

BentleyPublishers.com
BentleyPublishers.com—All Rights Reserved

1 Engine

Topic	Notes	Additional information (when available)
Engine knock fault codes 	Engine knock fault codes indicate that the engine is knocking. Detonation or knock is an erratic form of combustion that can cause head gasket failure as well as other engine damage. Prolonged or heavy detonation can be very damaging. The hammer-like shock waves created by detonation subject the head gasket, piston, rings, spark plug and rod bearings to severe overloading. As a first remedy, try higher octane high detergent fuel or switch brands. If the faults codes continue to return, contact your authorized BMW dealer.	120
Diagnosing electric cooling fan 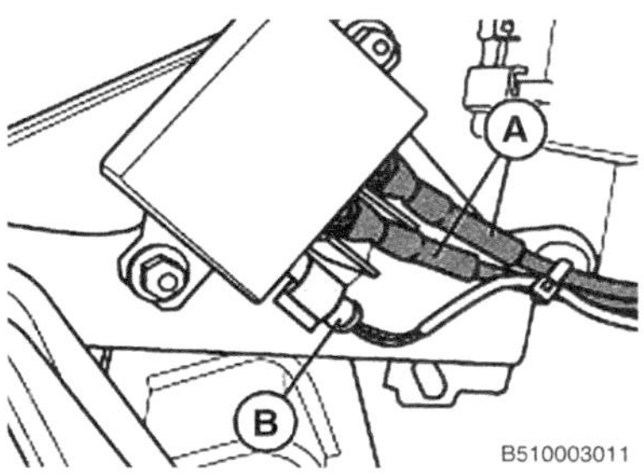	**From 03 / 2007 production vehicle**s: A revised power supply relay for the electric coolant fan power supply was used on these vehicles Coolant fan power supply relay K9137 is located in the back of the rear fuse box. Connector (**B**) provides control signals for relay activation (KL30g from the rear fuse box/power distribution box and local ground KL31). On cars with inoperative electric coolant (with symptoms of A/C malfunction, engine overheating, or fault code 2EFE "Cooling fan activation" stored in the DME), check the power supply relay operation: Check that both B+ connections (**A**) are threaded on tight and free of corrosion. Check KL30g (B+) and KL31 (ground) signals (**B**) for correct operation.	170, ELE
E-box water ingress 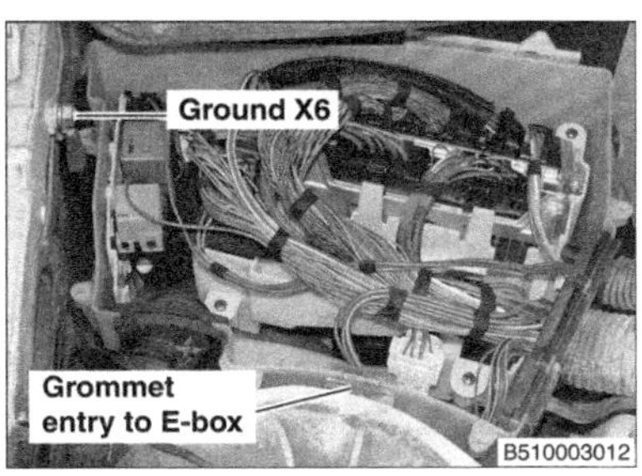	Water leaks into the E-box may cause various drivability complaints, loss of power or the "Service Engine Soon" lamp to illuminate while driving. Water can enter through the grommet for the ground connection X6 on the front of the E-box. If water is present, check that all drains and water channels are clear.	
No crank, no start 	Engine does not crank. Only the SRS (Supplemental Restraint System) and red brake lights show in the instrument cluster with ignition on. FC A111 (ELV voltage supply fault) is stored in CAS (Car Access System). ***NOTE—*** • *As a result of a change in government requirements, the ELV was deleted on most 2007 and later models. Anti-theft protection continues to be provided by drive-away protection in EWS.* Disconnect the IBS connector (3-pin connector with 2 wires) at the negative battery terminal. Follow the white/blue and gray wires from the IBS to locate the connector. Now try starting the vehicle. If the vehicle starts, replace faulty IBS.	121

BentleyPublishers.com—All Rights Reserved

1 ENGINE (CONTINUED)

Topic	Notes	Additional information (when available)
Ignition coil fuel/ oil contamination 	When working on the fuel system, it is important to ensure that the ignition coils are not fouled by fuel or oil (especially critical for N54 engines). The resistance of the silicone material is significantly reduced by fuel contact. This can cause spark over at the spark plug and misfires. Before servicing the fuel system, remove the ignition coils and cover the spark plug hole with a cloth. Ignition coils heavily fouled by fuel must be replaced.	**120**
Valvetrain service note 	When diagnosing engine noise, do not rule out worn timing chain components. This is especially applicable to high mileage engines and engines with extended oil change intervals.	**117**
Valve cover gasket leaks 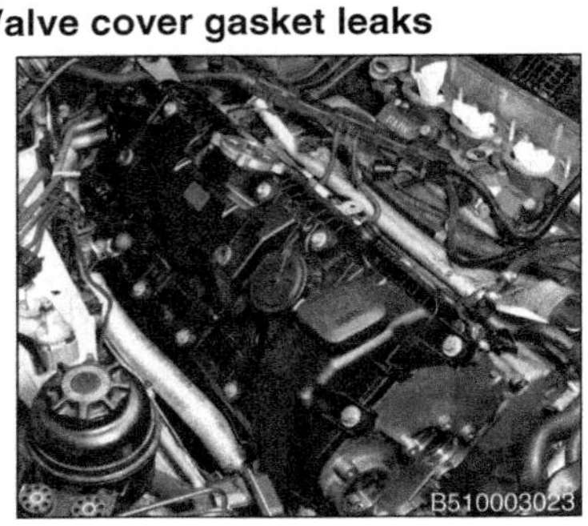	When replacing valve cover gasket, take extra caution when tightening fasteners and sealing at metal joining points. For example front timing cover to cylinder head. There is a small area where silicone should be applied to seal where the gasket may not. Applies to all engines.	**113**
VANOS and silicone sealant 	When working around VANOS units keep area clean and free of silicone sealant. Even a small amount of silicone introduced into VANOS actuator or solenoid can cause a fault.	**117**

BentleyPublishers.com
BentleyPublishers.com—All Rights Reserved

1 Engine (continued)

	Topic	Notes	Additional information (when available)
	Sticky VANOS solenoid sets cam position (CMP) sensor codes and VANOS solenoid codes 	A sticky / jammed VANOS solenoid can set a camshaft position (CMP) sensor fault (tooth error). Models with N52, N52KP, N54 and V8 engines may be difficult to start, run rough and have the Service Engine Soon lamp illuminated with one or more of the following stored faults: • Cam sensor, inlet signal - signal invalid for synchronization • Crankshaft-inlet camshaft, correlation • VANOS intake - stiff, jammed mechanically • Cam sensor, exhaust signal - signal invalid for synchronization • Crankshaft - exhaust camshaft, correlation • VANOS exhaust - stiff, jammed mechanically Remove VANOS solenoid, intake or exhaust, depending on fault codes stored and, using shop air, gently blow out any visible particles and reinstall. If fault returns, try swapping solenoids between banks to see if fault follows solenoid. Replace solenoid if fault cannot be corrected. Possible fault codes: P0011, P0012, P0014, P0015, P0020, P0021, P0022, P0024, P0025, P0340, P0345, P0365, P0390	117
V8 engine	**Do not overfill cooling system**  B5100003016	Be careful not to overfill cooling system during service or when topping up. V8 engine coolant caps have a tendency to not release when pressure exceeds maximum. An overfilled cooling system can amplify this problem.	170
V8 engine	**Coolant pipe leak** 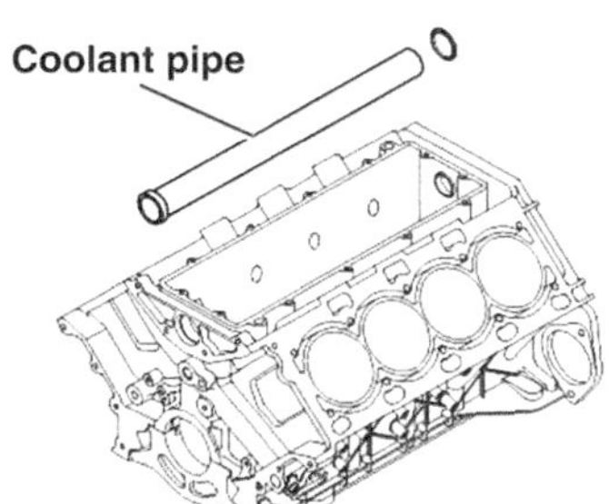 B510003013	The coolant pipe located in valley of V8 engine can develop leaks at the end O-ring seals. To replace the pipe with the OEM pipe, the front of the engine must be fully disassembled, requiring engine removal under most circumstances. Aftermarket options are available (threaded expanding pipe, is one example) that significantly reduce the time and cost required to do the job.	
V8 engine	**Oil leak - alternator bracket profile gasket** 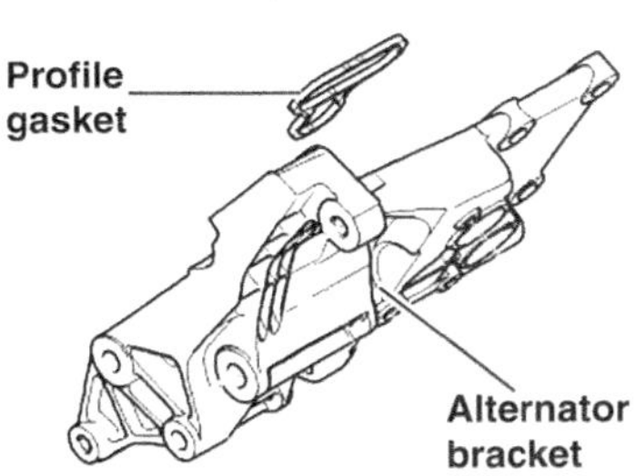 B510003014	An oil leak from the alternator area is usually be caused by a faulty alternator bracket profile gasket. To replace the gasket, remove alternator and bracket.	121

BentleyPublishers.com—All Rights Reserved

1 Engine (continued)

	Topic	Notes	Additional information (when available)
V8 engine	**Vacuum pump diaphragm fails causing oil leak and brake booster failure** 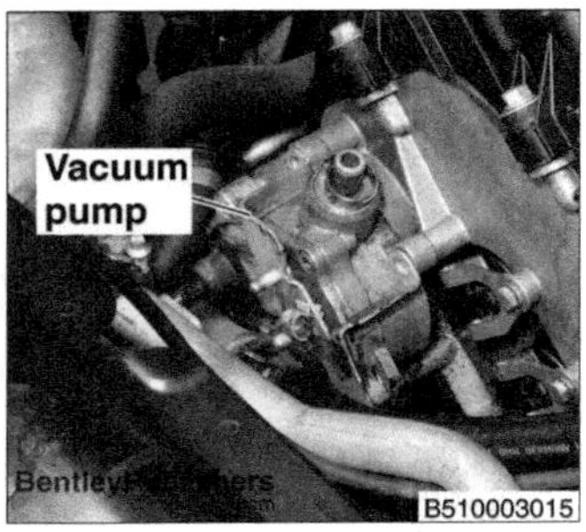	A ruptured vacuum pump diaphragm results in oil being sucked into the brake booster. Remove brake booster vacuum hose to inspect. If oil is present in hose, replace vacuum pump. If large amounts of oil are found inspect brake booster, replacement may be needed.	133
V8 engine	**Throttle housing (EDK) fault codes** 	Fault codes: • 2CF8 – Throttle valve potentiometer, signal • 2CF9 - Throttle valve potentiometer 1, signal • 2CFA - Throttle valve potentiometer 2, signal In most cases, DSC fault 5F77 (DME invalid data) will also be stored. These faults are sporadic and usually not reproducible. The cause is an intermittent interruption of the EDK signal to the DME due to a resistance fluctuation in the EDK harness connector. Replace EDK harness with updated part #12 51 7 602 973.	133
N52 engine	**Intermittent lifter noise** 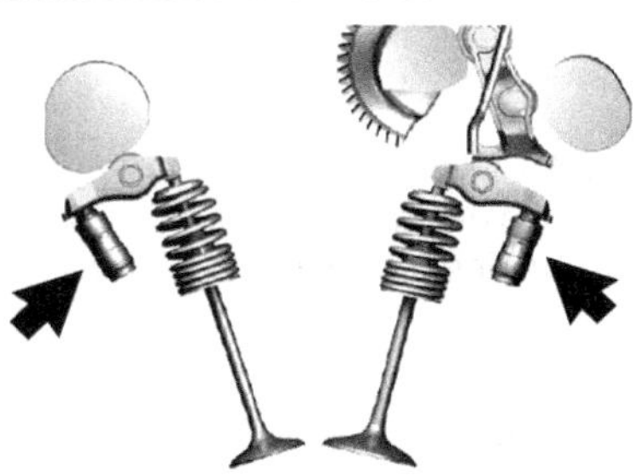	An occasional ticking or rattling noise from hydraulic valve adjusters (HVA) elements may occur during cold engine starts or frequent short-distance driving.This condition will not cause any damage to the engine, and usually remedies itself with a longer driving distance or operating times at full operating temperature. In this situation, the lifters can bleed down or have trapped air causing ticking/rattling noise and may require bleeding, as described below. The bleed time may take anywhere from two minutes to a maximum of 30 minutes. Procedure preconditions: • Engine running at operating temperature • Vehicle stationary on a level surface in a properly ventilated area • Engine oil level correct – not underfilled or overfilled 1. Bring engine up to an operating speed (no load) of 2000-3000 rpm and maintain this speed for three minutes. 2. Let engine idle for 15-30 seconds. Listen for whether the noise is still present: - If engine is quiet, procedure is finished. - If engine noise still present, repeat procedure; perform steps 1 and 2 up to a maximum of 5 times. If the noise remains after performing the bleeding procedure 5 times: proceed by performing the procedure for a final time, also at an engine operating speed of 2000-3000 rpm, but for a total time duration of 15 minutes. If noise returns after multiple bleedings, the cylinder head may be faulty and require replacement.	113

BentleyPublishers.com
BentleyPublishers.com–All Rights Reserved

1 Engine (continued)

	Topic	Notes	Additional information (when available)
N52 engine	**Faulty valvetronic motor linked to DME failure** 	If diagnosing a faulty DME, be sure to check that the valvetronic motor was not the root cause of the damage to DME control unit. If a new control unit is installed and the valvetronic motor is bad (excessive current draw), the new DME will be damaged.	130
N54 engine	**Cylinder-assigned high pressure injectors** 	On N54 engines, mark the cylinder locations on the fuel injectors when removing them. As part of injector manufacturing, tolerance ranges for injection quantity are determined and specified in a six-digit number combination. These values are used by the DME to compensate for variations in the individual injectors. ***NOTE—*** • *When replacing an injector, it is essential to carry out an injector adjustment using the BMW scan tool.*	130
N54 engine	**Servicing high pressure fuel injection system - safety precaution** 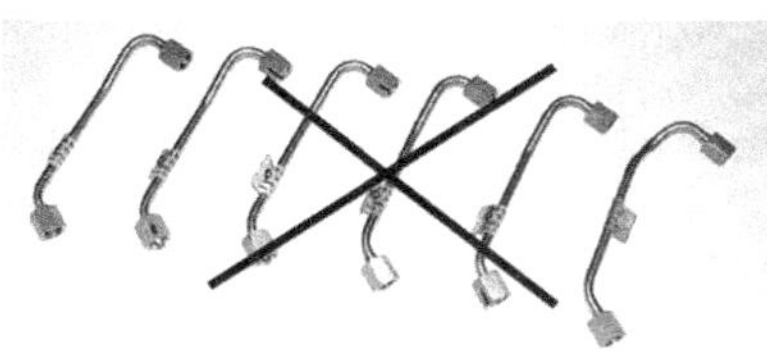	Working on the HPI system must only be carried out after the engine has cooled down. Do not attempt to open any fuel lines or connections on the HPI injection system until the coolant temperature is below 40° C (approx. 104°F) If the engine is not cool, there is danger of residual fuel pressure spray in the high-pressure system. When working on the high-pressure fuel system, take particular care to ensure conditions of absolute cleanliness and follow the work sequences described in the repair instructions. Even the smallest contaminants and/or damage to the threads on the high-pressure lines can cause leaks.	130
N54 engine	**High pressure fuel system faults** 	Service Engine Soon light illuminated, engine enters fail-safe mode with reduced power. Scan tool data shows fault codes 10737 (P3283, P3284), and/or 10738 (P3003, P3090) – "High fuel pressure system, plausibility" stored in MSD80 (DME) control module. These fault codes all relate to the high pressure fuel system. The possible causes are: incorrect adaptation values of the high pressure pump control valve or failure of the high-pressure fuel pump. Begin by resetting adaptations for engine management. If fault codes return immediately after adaptation, the fuel pump may be faulty.	160

BentleyPublishers.com
BentleyPublishers.com—All Rights Reserved

1 Engine (continued)

	Topic	Notes	Additional information (when available)
N54 engine	**High pressure fuel injector, replacing replace teflon seal** 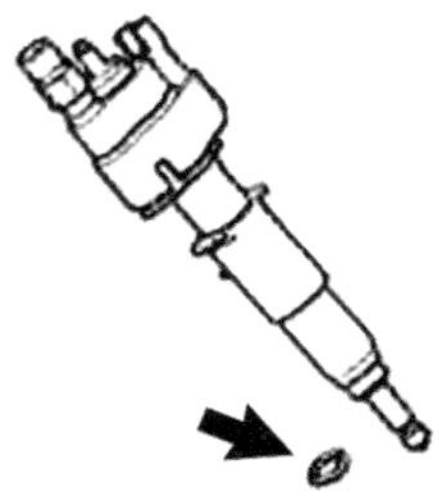	Always replace the teflon O-ring (arrow) when fitting and removing the fuel injector. This also applies to an injector that has just been fitted and has to be removed again after an engine start. A new injector with a new teflon O-ring should be fitted as quickly as possible as the Teflon sealing ring could swell up. Also, be sure that the injector is correctly seated when installing. The injector hold-down must rest on both injector tabs, otherwise the necessary force is not applied. Do not clean the nozzle-needle tip of the piezo-injector.	130
N54 engine	**High pressure fuel pump extended warranty coverage** 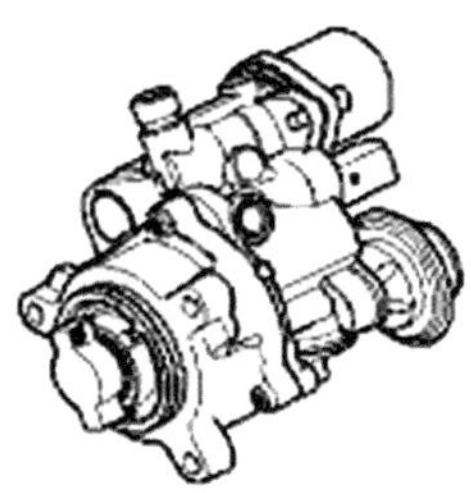	On 2007 - 2009 models with N54 engines, BMW emissions warranty of the high pressure fuel pump (HDP) has been extended from 4 years or 50,000 miles to 10 years or 120,000 miles, whichever comes first. Vehicles with a faulty fuel pump may have the Service Engine Soon lamp illuminated with various low fuel pressure-related faults stored in the engine control module (DME). Affected vehicles may experience long cranking times or reduced engine performance (engine in failsafe mode) when the HDP malfunctions. If the high pressure fuel pump of a MY 2007- 2009 exhibits the symptoms listed above, it should be replaced with the improved part.	160
N54 engine	**Fuel injector faults** 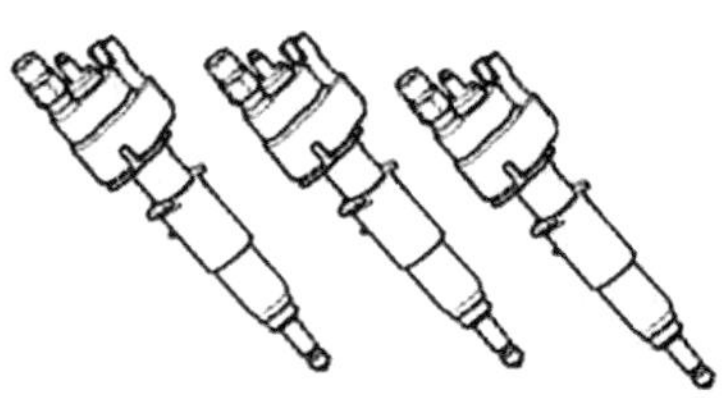	On cars with faulty fuel injectors, various fault codes related to fuel injectors may be stored in the DME control module (Service Engine Soon light is illuminated). The engine may lack power, idle erratically, or have poor throttle response. Faults codes such as injector, cylinder 1-6, short circuit or line break, DME internal fault; Injector driver circuit failure bank 1-3; bank 4-6 or Misfiring, several cylinders may indicate a bank of 3 faulty injectors. Consult your authorized dealer regarding injector replacement as it applies to BMW Service Bulletin SIB 12 02 08 (dated October 2008).	130
N54 engine	**EML light illuminated, limp home mode, no boost pressure control**	In the event of a failure, malfunction or an implausible signal from any of the sensors listed below, activation of the wastegate valves is shut down and the valve flaps are fully opened. Turbocharging ceases at this point. • High-pressure fuel system • Inlet or exhaust VANOS • Crankshaft position sensor • Camshaft position sensor • Boost-pressure sensor • Knock sensors • Intake-air temperature sensor If the ELML light is illuminated, interrogate fault memory and correct any problems found.	

BentleyPublishers.com
BentleyPublishers.com—All Rights Reserved

1 Engine (continued)

	Topic	Notes	Additional information (when available)
N54 engine	**Rattle noise from turbocharger area** 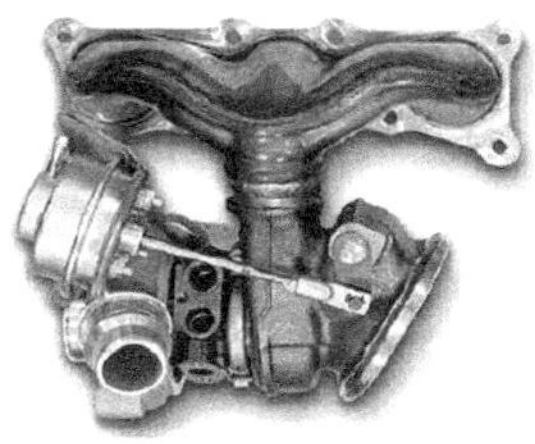	A metallic noise from the exhaust system near the engine or turbocharger area can occur during deceleration from approx. 3,500 RPM or during a high engine speed without a drive gear being engaged. In this situation, the wastegate valve does not completely close due to mechanical tolerances of the wastegate actuator and results in the wastegate valve rattling against its seating surface. Do not replace the turbocharger or adjust the existing wastegate actuator control rod. Instead reprogram the DME. If after programming, the noise is no longer present, take no further action. If the noise is still present, replace both of the turbocharger wastegate actuators.	180
N52, N54 engines	**Coolant contact with magnesium components** 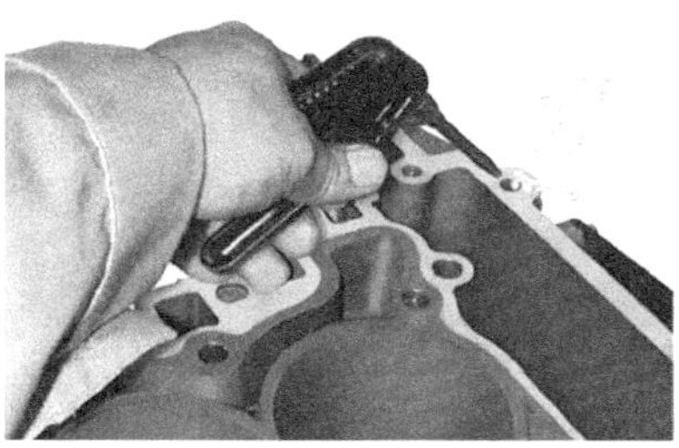	During service procedures which involve coolant, such as cylinder head replacement, the bolt holes must be blown dry immediately. To prevent corrosion, do not allow any traces of coolant to remain in holes.	113
N52, N54 engines	**Damaged connector at throttle module**	When connecting throttle module connector, it is possible to twist the connector before plugging it in, causing damage to the harness and the connector. Be sure to install connector carefully to avoid damage.	
N52, N54 engines	**Electric coolant pump precautions** 	Particular care must be taken to ensure that the coolant pump does not run dry. When the pump is removed, it should be stored filled with coolant. The pump bearings could seize if the pump is not filled with coolant. A inoperative coolant pump could cause serious engine damage. If the pump should ever run dry, the pump impeller should be turned by hand to ensure the pump is not seized before connecting the coolant hoses. The system should then be immediately filled with coolant.	170

BentleyPublishers.com
BentleyPublishers.com—All Rights Reserved

2 Transmission

Topic	Notes	Additional information (when available)
SMG Transmission requires specific socket to remove output flange. 41 mm deep socket  B510003001	An extra deep socket is used to remove the 41mm nut on the SMG output flange. BMW part number 83 30 0 493 421.	
ATF service recommendation 	The automatic transmission fluid in the both the ZF and GM units should be changed at service intervals no greater than 100,000 miles. 60 to 80k miles should be considered a more acceptable maximum interval. When checking the fluid level, it is critical that the temperature of the fluid be maintained at approximately 30°C (86°F). (The factory specification is 30 - 50°C.) Checking the temperature can be done using an infrared temp gun or a more correctly, with a BMW scan tool.	**240**
Vibration or noise in drive train 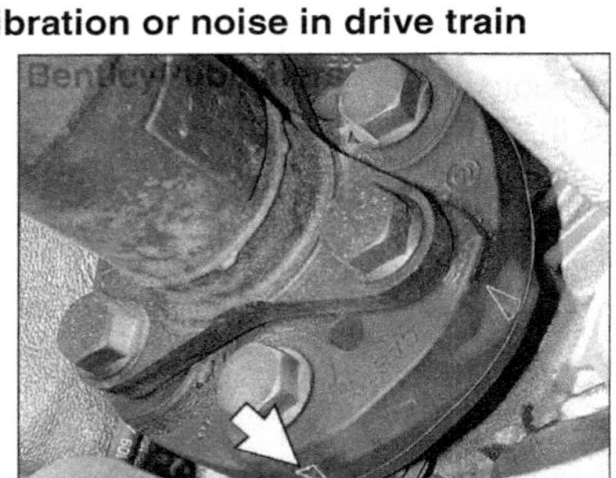 B510003021	One or more of the following may be heard while driving slowly, coming to a stop or at highway speeds: • A clunk can be heard while rolling slowly and downshifting from 2nd to 1st gear coming to a stop. • Load reversal causes a slight clunk noise. • A clunk noise can be heard while shifting into forward or reverse gear from park or neutral. Check for a deterioration or separation of the rubber in the universal joint located between the transmission and drive shaft assembly. To check if the noise is caused by a defective universal joint the driveshaft must first be removed. Once removed, check if any of the metal collars are loose and/or rubber cracking is found. Replace the universal joint if any fault are found.	**240**
Mechatronic sealing sleeve leaking 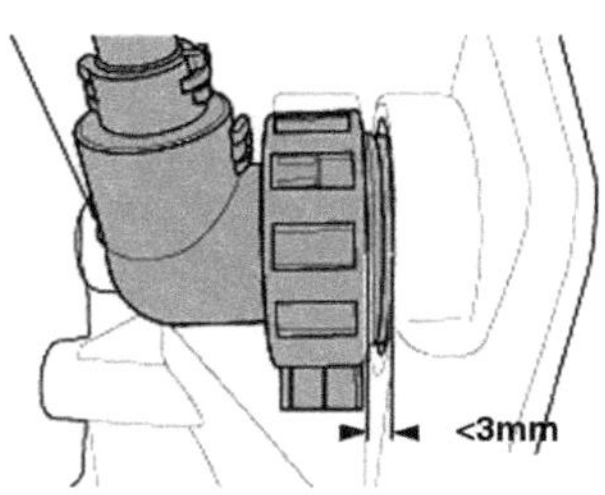	Transmission fluid can be seen leaking from transmission oil pan, or more specifically from the Mechatronic sealing sleeve. This repair requires replacing the sealing sleeve and the transmission oil pan gasket. Refer to BMW Service Bulletin SIB 24 08 06 (dated February 2009) for parts and repair information, including proper sleeve installation instructions. ***NOTE—*** • *After installation of the new sleeve, the correct distance between the transmission harness connector and the housing surface should not exceed 3 mm.*	**240**

BentleyPublishers.com
BentleyPublishers.com—All Rights Reserved

2 TRANSMISSION (CONTINUED)

Topic	Notes	Additional information (when available)
Transmission/transmission control complaints 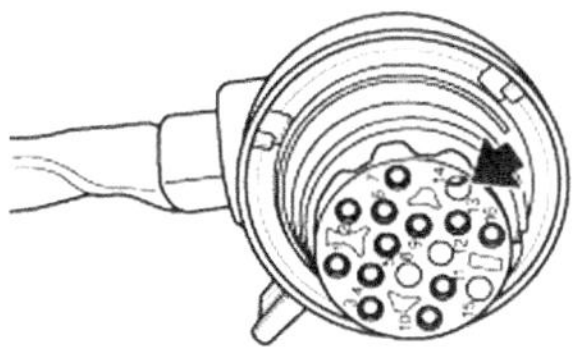	Various complaints associated with inoperative transmission controls. At times the vehicle will not start, gear indicator display not present, transmission failsafe light is on, or there is a transmission failsafe message in the check control. During diagnosis with scan tool, EGS cannot be identified. The transmission harness connector may have a loose or pushed back pin(s). This condition can cause various complaints and many different faults to be stored. Disconnect the battery and check the transmission connector X8500 for loose or pushed back pins. It may be possible when the connector is removed for the pin to move forward, so check that all the pins are tight and locked in place by using a blunt tool like a pick. Look carefully - a pushed in pin may appear as an open slot. ***NOTE—*** • *It may be necessary to remove the exhaust and lower the back of the transmission to properly inspect plug. If this step is necessary be careful to not bend shifter cable bracket otherwise shifter cable operation will be compromised.*	**240**
Transmission in failsafe mode with stored fault (CF33) (from 02/2006 to 08/2007 production with 6HP19TU auto. trans.)*	This is caused by highly-sensitive EGS software. The fault normally sets during a 5-4 downshift in vehicles with a replacement transmission, or after a reset of the adaptation values. Do not replace any parts. Using the factory scan tool (version 49.0 and higher) to run a short test and perform the following test module: Reset learning functions - available using the following diagnostic path: Service Functions / Transmission Control / Calibrations / Reset leaning functions Afterwards, turn ignition off / lock vehicle / let vehicle enter sleep mode. It should then be possible to clear the fault. **This problem is planned to be corrected with August 2007 and later production vehicles equipped with these ZF transmissions.*	**240**

BentleyPublishers.com—All Rights Reserved

3 Suspension, Steering and Brakes

Topic	Notes	Additional information (when available)
Rear coil spring repair procedures 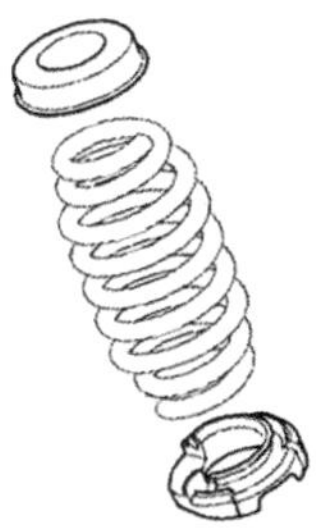	For repair procedures that require removal of the rear coil springs, the spring seats must be properly cleaned prior to reinstalling the springs. Dirt, pebbles, and debris left on the spring pads can become trapped between the spring pads and the lower coil of the springs. This can damage the protective coating of the springs and rust may form on the spring. Over time, rust formation on the spring combined with vehicle operating conditions may cause the spring to break.	**330**
xDrive front pinion seal leak 	Front pinion seal leaks (**arrow**) should be addressed promptly. The front differential has a low fluid capacity and any size leak can pose an immediate problem.	**311**
Brake pad wear sensor, use care when installing / removing 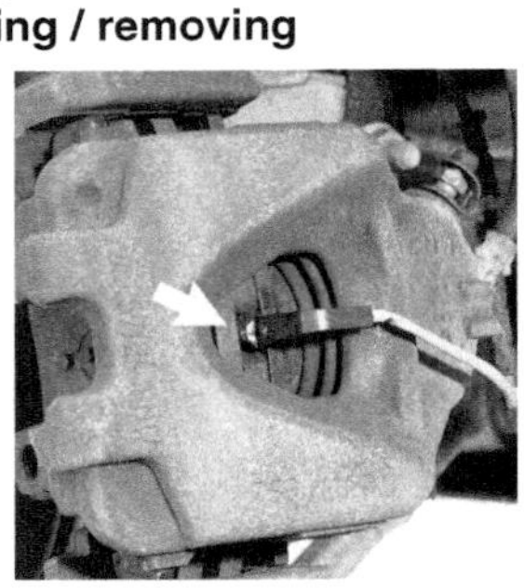	The brake pad wear sensor is easily damage during installation owing to the 90° design. Be careful not to break the sensor when during installation and removal.	**340**
Brake squeal from pad wear sensor	Under light / moderate brake applications, brake squeal can be caused by contact between the brake pad wear sensor and the brake rotor (front or rear brakes). The geometry of the pad wear sensor, combined with individual driving habits, can result in a wear pattern that causes a squeaking and/or squealing noise when the brakes are applied with light to moderate force. If the noise is coming from the brake system, inspect the brake pad wear sensor of the affected axle. If the sensor is in contact with the rotor/disc, remove the sensor and drive the vehicle. If the noise is eliminated, replace the brake pads and wear sensor of the affected axle. If the noise is not eliminated, further diagnosis is required to determine the cause of the noise.	

BentleyPublishers.com
BentleyPublishers.com—All Rights Reserved

3 Suspension, Steering and Brakes (continued)

Topic	Notes	Additional information (when available)
Wheel alignment (track), resetting (cars with active steering) 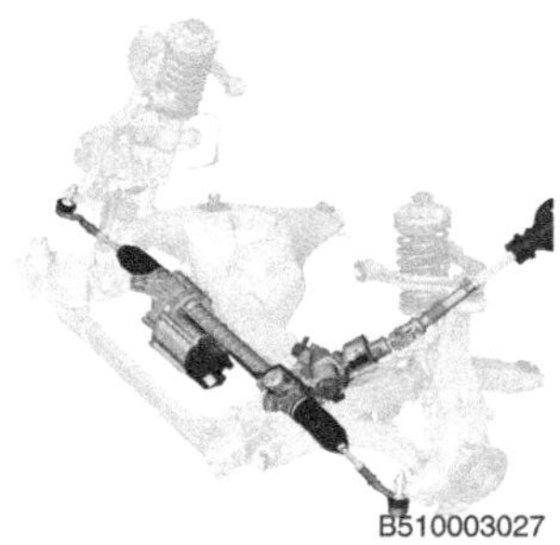 B510003027	On cars with the optional active steering (option SA 214), the front track has to be reset after work on the steering. The procedure for adjusting track on a vehicle with active front steering (total steering angle sensor) is not the same as that for a vehicle with a conventional power-assisted steering system.	320
ELV (Electric Steering Lock) warning shown in the Instrument Cluster 	ELV light on with intermittent no-crank, no-start situation or ELV solenoid in locked position. This fault is usually caused by low voltage during vehicle cranking / starting. Troubleshooting should be carried out using BMW scan tool. ***NOTE—*** • *If terminal KL15 is not switched on, scan tool communication will be inhibited. In this case, start diagnostics by selecting the vehicle manually. Then follow the diagnostics path: Drive / Engine start / Comfort start / button to get into the scan tool test plan.*	320
Front wheel bearing wear 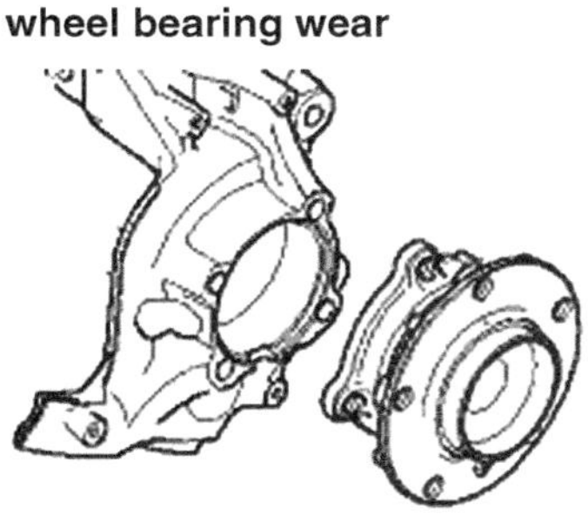	Worn wheel bearings can often produce a knocking sound while driving. Note that special press tools are needed to replace the bearings. • Check the front wheel bearings for wear with the vehicle raised. The bearing will feel tight when spinning the wheel. • Put the vehicle on the ground, pull and push on top of wheel / tire to check for looseness in wheel bearing. Initially it may feel tight. Rock steering wheel back and forth to situate the bearing. Bearing will then feel loose when pushing / pulling on top of tire.	310
Active steering, reinitializing after battery disconnect	Active steering must be reinitialized after the battery has been disconnected. Reinitialize as follows: • Start engine (the "Active Steering Inactive" Check-Control message appears) • Turn steering wheel as far as it will go counter clockwise and then as far as it will go clockwise.	

BentleyPublishers.com
BentleyPublishers.com–All Rights Reserved

3 Suspension, Steering and Brakes (continued)

Topic	Notes	Additional information (when available)
Clicking or cracking noise from rear differential 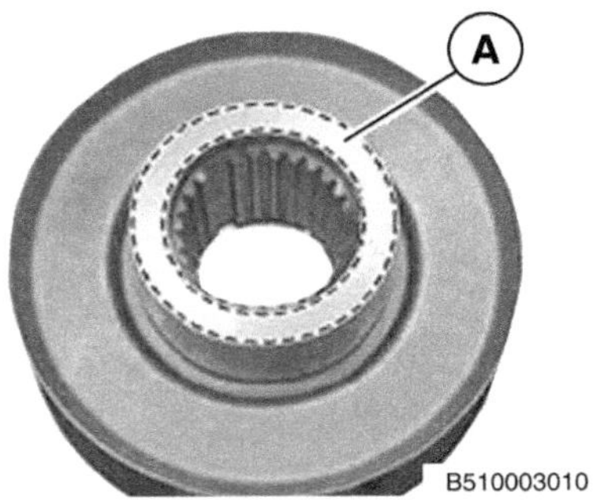 	The differential input flange mating surface with the input drive pinion is insufficiently greased. This is not a failure of the differential assembly. Do not replace the complete differential. Remove the differential drive flange. Apply lubricant P/N 83 23 0 443 864 on the input flange mating surface (**A**), as described in the illustration. Before complete reassembly, replace the input flange seal and collar nut retaining plate	
Tire pressure monitor, damaged wheel electronics (2007-2010 vehicles) 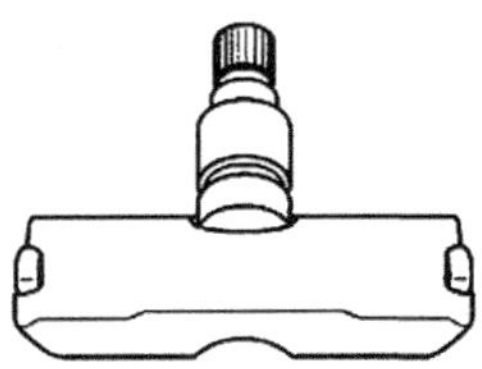	The tire pressure monitor (TPM) system used on 2007 and later cars measures the tire pressures at each of the four wheels. The system utilizes a warning lamp, mounted in the instrument cluster to alert the driver of a low tire pressure condition. The TPM system uses separate wheel electronics mounted at the base of each tire valve inside the wheel. Extra care must be taken when changing tires to avoid damaging the wheel electronics. A damaged wheel electronic can cause the tire warning indicator to illuminate. When diagnosing an illuminated TPM warning lamp, the tires and wheels should be inspected for damage and/or prior replacement. ***NOTE—*** • *The last four digits of the tire's DOT number indicate the calendar week, followed by the year in which the tire was produced. This information can be used to determine if one or more tires have been recently replaced.*	
Flat tire monitoring / tire pressure monitoring, illuminated warning lamp  	The warning lamp is illuminated when the tire pressure decreases by approximately 25%. Temperature fluctuations can cause changes in tire pressure. For every 10° F change in ambient temperature, the tire pressure can change by 1 psi. Therefore, the tire warning indicator may be illuminated by a large change in temperature combined with a low pressure condition. Inspect tires regularly for physical damage and check tire pressures when the tire warning is illuminated. BMW recommends checking tire inflation pressures every two weeks and before long trips. ***CAUTION—*** • *Driving on a significantly under-inflated tire causes the tire to overheat and can lead to tire failure.* • *Under-inflation reduces fuel efficiency and tire tread life, and may affect the vehicle's handling and stopping ability.*	**020**

BentleyPublishers.com
BentleyPublishers.com—All Rights Reserved

4 Body

Topic	Notes	Additional information (when available)
Multiple electrical faults caused by water leak into trunk 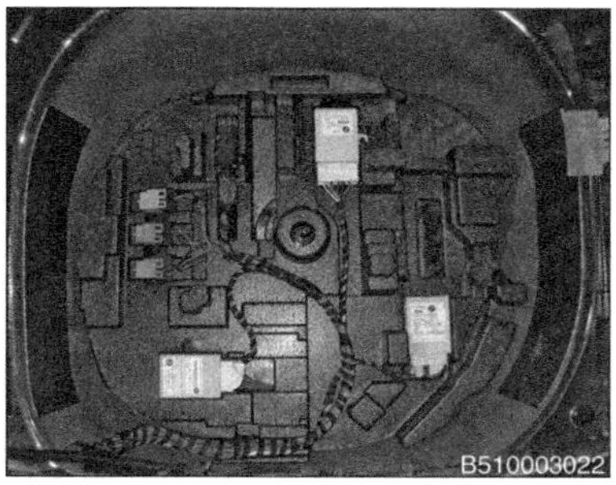 	Electrical components mounted in spare tire well may become damaged by water leaking into trunk. Inspect trunk rubber seal and in particular, the large rubber washers at the bumper fasteners (as viewed from inside trunk).	
Noisy wiper blades, blades chatter	Wiper blades in good condition may make noise during operation, or may not clean the windshield properly. This may be caused by contaminants on the glass or blades, or by the blades or arms being out of adjustment. The first step is to make sure the glass and the blades are clean and free of contaminants. Common contaminants that are not removed by conventional glass cleaners can be removed with BON AMI®, a household cleanser. The cleanser should be applied on a wet cloth or sponge and the windshield must be thoroughly and evenly scrubbed. Wipe off the residue with a clean, dry cloth before the cleanser dries on the glass. If there is evidence of heavy overspray of waxes or silicones on the glass, a wax and silicone remover can be used to remove them. Clean the wiper blades with a clean cloth soaked in rubbing alcohol Once the blades and glass have been cleaned, check the adjustment of the wiper arms. ***NOTE—*** • *To properly adjust the wiper arms, special BMW adjustment gauges and tools are required. See* **611 Wipers and Washers**.	**611**

BentleyPublishers
BentleyPublishers.com–All Rights Reserved

5 Body Equipment

Topic	Notes	Additional information (when available)
Squeaking noise from B-pillar 	A squeaking or grating noise may be heard from the top of the B-pillar when driving over rough roads at very cold ambient temperatures. This noise typically occurs only below approximately 20°F.(For noises from this area at high temperatures, refer to BMW Service Bulletin SI B51 02 05). This is caused by a bad bond between the B-pillar trim and the rear door seal. Replace the door seal with an updated part, which uses a softer rubber compound.	
Sunroof moves in small steps or interior lights not working	Slide/tilt sunroof requires initialization. The motor of the slide/tilt sunroof uses 2 Hall sensors. They are located on the motor shaft and offset by 90 degrees with respect to each other. When the motor is running, this results in two offset Hall signals that are used to register the direction of rotation and for the anti-trapping protection function. ***NOTE—*** • *The control button must remain pressed during the entire initialization procedure. The initialization procedure must be repeated if the button is released.* • *If the power supply is interrupted during initialization the procedure must be repeated.* Press and hold sunroof button in the sunroof tilt direction. After approx. 15 seconds, the initialization begins by storing the tilt-position end stop, then the full open end stop position, etc. Initialization takes about 120 seconds. Initialization is erased under the following conditions: • Removal of the voltage supply to sunroof control module • Hall sensor fault detected • Sunroof position implausible • Modified/changed vehicle coding or coding data faulty	
Tailgate opens slow at low ambient temperatures (Sports Wagon)	Tailgate opens slowly or does not fully open under its own power. In low ambient temperatures, the gas pressure in the tailgate struts is reduced. The result is a less powerful strut. Conversely, at very high ambient temperatures, the gas pressure is higher, thus allowing the tailgate to open quicker. This is a normal condition and the struts are not faulty. However, to ensure that excessive friction is not present in the hinges due to lack of lubrication, be sure to lubricate tailgate and gas strut hinges with grease.	

BentleyPublishers.com
BentleyPublishers.com—All Rights Reserved

5 Body Equipment (continued)

Topic	Notes	Additional information (when available)
Front and rear doors do not unlock at same time	Intermittently, when the vehicle is unlocked using either the remote key or the central locking button, the rear doors do not unlock at the same time as the front doors. This situation is unaffected by whether or not selective locking is active. For the rear doors to unlock, the unlock procedure must be performed an additional time. A High Rate of Battery Discharge check control message may also be displayed. This condition occurs when an excessive closed-circuit current draw is detected. When an excessive draw is detected, the 30gf relay is switched, disconnecting the basic body module (KBM) from the power supply.	

6 Electrical System

Topic	Notes	Additional information (when available)
Changing / setting date format (mm/dd or dd/mm) in the Personal Profile settings **(vehicles without Navigation)**	To change Personal Profile settings: • Press rocker switch in direction indicator stalk, up or down until "Personal Profile" function (key with a check mark) appears, accompanied by the word SET in the instrument cluster. • Press Board Computer (BC) button on steering column stalk. • Press rocker switch, up or down, until "tt/mm" or "mm/tt" is shown. • Press the BC button. • Use the rocker switch to make the selection/change. • Press the BC button to confirm and accept the selection/change. ***NOTE—*** *• This will also change the date format in the radio display.* The following display formats and units of measure can be selected/changed in the same way. • Fuel consumption (l/100km, mpg, km/l) • Distance (km, mls) • Time format (12h/24h) • Temperature (°C, °F) • Reset of display formats / units of measure (factory/default setting)	

BentleyPublishers
BentleyPublishers.com–All Rights Reserved

6 Electrical System (continued)

Topic	Notes	Additional information (when available)
Closed-circuit current (parasitic draw) specification	Increased closed-circuit currents may occur permanently or intermittently and cause the battery to discharge. The increase in closed-circuit current may be caused by a faulty control unit, or by the installation of a non-approved accessory. **E60, E61: 40 mA after 60-70 minutes with TCU (telematics control unit) (30 minutes without TCU)** In general, closed-circuit current consistently over 50 mA must be investigated.	121
Stabilant 22A electrical contact enhancer 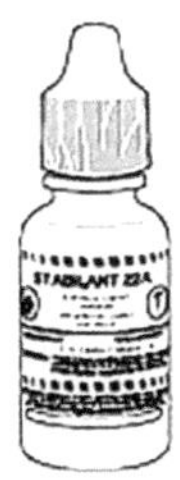	Connectors carrying low current are particularly susceptible to the formation of deposits. These deposits affect the resistance of the circuit and, depending on the sensitivity of the particular circuit, can cause system malfunctions (very often intermittent), and the activation of warning lamps and check control indicators. In difficult to solve electrical resistance situations, Stabilant 22A may help to enhance electrical contact. Stabilant 22A evaporates and leaves a thin polymer film which is conductive between the mating surfaces, while staying non-conductive between adjacent pins. At the same time it prevents the formation of any further harmful deposits. ***NOTE—*** • *Read and understand any Stabilant 22A safety precautions provided by the manufacturer.* • *Do not use Stabilant 22A on Oxygen sensor or SRS connectors.*	
Check Control message: Vehicle Battery! Increased Battery Discharge	If the off-load current exceeds 80 mA after 68 minutes of the key being off, the engine management system outputs the CC message "Vehicle Battery! Increased Battery Discharge" the next time Terminal 15 is switched on. This happens regardless of the battery state of charge. In the event of the above situation, check the vehicle for non-approved electrical equipment installations. Retrofitted after-market equipment (e.g., hands-free systems, navigation systems, GPS or entertainment systems) may have been incorrectly connected to Terminal 30, Terminal 30g-f or directly to the battery. ***NOTE—*** • *The intelligent battery sensor (IBS) monitors the battery's discharge current while the vehicle is off. The measures values are transmitted to the DME through the bit-serial data (BSD) interface. Therefore, equipment or accessories must only be connected to Terminal 15 or Terminal 30g, unless otherwise specified in the installation instructions of the BMW retrofit kit.*	121

BentleyPublishers.com
BentleyPublishers.com—All Rights Reserved

6 Electrical System (continued)

Topic	Notes	Additional information (when available)
Windshield washer, low or no volume 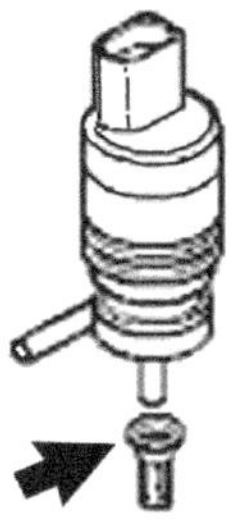	Windshield washers spray a very low volume of fluid, or no fluid at all. The strainer (**arrow**) on the windshield washer fluid pump may become clogged by debris or an algae-like substance. The algae-like substance is typically caused by using non-approved washer fluid or straight water in the washer fluid reservoir. To replace the strainer: • Remove the washer fluid pump, located in lower part right front wheel well. • Drain existing washer fluid in the reservoir and properly discard. • Replace the strainer, which is installed on the end of the pump. An updated strainer has a larger mesh to avoid future occurrences. • Refill the washer fluid reservoir with BMW washer fluid at the specified concentration.	611
Gong warning without visual check control message	A gong warning sounds approximately 10-15 seconds after the engine is started. There is no check control warning in the instrument cluster and no faults are stored. This can be caused by a center door lock button (permanently closed contacts). Test the functionality of the center door lock button and replace as necessary.	515
Remote key not working	The radio remote control function is inoperative: • The vehicle cannot be unlocked or locked with the remote key. • The check control message "Replace Remote Control Battery" is displayed. • Fault Code 9308B0 – ID transmitter: Battery undervoltage is stored. Test remote key using BMW scan tool.	
Instrument cluster mileage display switches to kilometers when key Is switched	The instrument cluster mileage display switches to kilometers after switching keys. This situation occurs only up to 16 minutes after the vehicle is switched off (vehicle not yet asleep). Do not replace any parts or perform any software updates to correct this situation. The problem can be corrected by letting the vehicle go to sleep (longer than 16 minutes after the key is OFF).	
Consumption is Set to mpg UK instead of mpg US 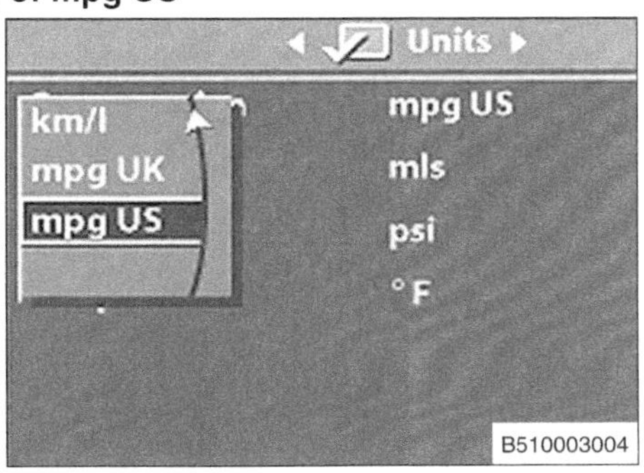	Using iDrive menu change setting under Display Settings.	

BentleyPublishers.com
BentleyPublishers.com–All Rights Reserved

6 Electrical System (continued)

Topic	Notes	Additional information (when available)
A/C blows warm after an extended drive	The air conditioning stops cooling and starts to blow warm air. This usually happens when the weather is very humid or there is a high level of air pollution. Normal operation is restored after the system is turned off for approximately 15 minutes. The air conditioning system temperature regulation is calibrated between 33.4°F (0.8°C) and 34.5°F (1.4°C). Under certain environmental conditions, ice may form on the evaporator, causing loss of air conditioning performance. A software update is required to remedy this problem. Program using BMW scan tool.	
Blower motor stays on with key OFF.	Occasionally after a long drive, the blower motor stays on at a reduced speed from the last setting on the control panel, for up to 16 minutes after the vehicle was turned off. DME fault code 2DED Battery Power Management may also be stored. This complaint will not cause the battery to be discharged. This is caused by a blower motor LIN-Bus communication error (blower in emergency mode). A software update is required to remedy this problem. Program using BMW scan tool.	
A/C compressor does not switch ON 	IHKA (Integrated Heating and Air Conditioning) fault 9C6A – Compressor Coupling is stored. Check for a loose ground connection at compressor. If the ground connection is not loose, continue troubleshooting fault using BMW scan tool test plan	
Testing automatic recirculation (AUC) sensor. 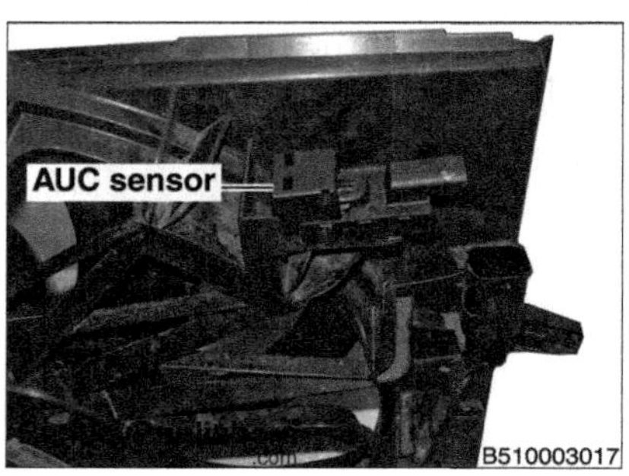	Use the following steps to diagnose AUC-related fault codes. – Remove AUC sensor from mounting point. Electrical connector remains connected. – Turn ignition key on, make sure IHKA system is on. – Blow into AUC sensor air intake. Fresh air flap should slam shut. – If flap does not close, check voltage supply to sensor. If voltage supply is good, replace AUC sensor.	640

BentleyPublishers.com
BentleyPublishers.com–All Rights Reserved

6 Electrical System (continued)

Topic	Notes	Additional information (when available)
Eyebrow lights inoperative after programming 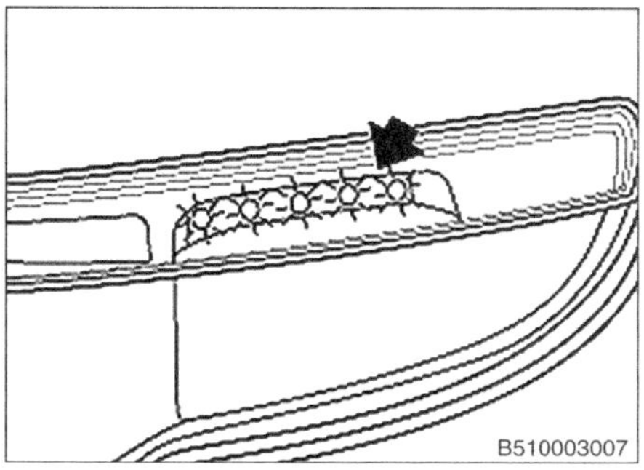 	If eyebrow lights (**arrow**) fail to work after vehicle programming, reprogram complete vehicle using BMW scan tool with latest software.	
Xenon lamp bulbs differ in color after replacement	Replacement xenon bulbs have a yellow-white appearance when compared with the bluish-white of the original bulb. New Xenon bulbs require a burn-in period, which causes a gradual color change from yellow to blue. The greatest color occurs in the first 5 minutes of use, and stabilizes after approximately 100 hours of use.	
LED flashing on headlight switch, no faults stored  	With the light switch switched to the automatic position, the LED on the headlamp switch is flashing and AHL (Active Headlights) is deactivated. Using the BMW scan tool, carry out a short test to identify any faults related to AHL system. If no faults are stored and LED is still flashing, problem may be caused by stuck stepper motor in the headlamp. AHL is integrated into BUS network and relies on communication with other modules to function. Select: "Control Unit Functions/AHL/Component Activation. Activate each of the AHL functions in turn until the problem is identified and replace faulty components.	
Automatic headlights, adjusting sensitivity	Headlights switch on too early or too late, even in daylight, with headlight switch set to automatic position. The factory default for the rain/light sensor is set to "sensitive". With this setting, the rain/light sensor is set for optimum operation. There are three setting options: • Normal (reduced sensitivity) • Sensitive (factory default) • Highly sensitive (most sensitive). If the preference is to have sensor setting either more or less sensitive, the setting can be changed using BMW scan tool.	

BentleyPublishers.com
BentleyPublishers.com–All Rights Reserved

6 Electrical System (continued)

Topic	Notes	Additional information (when available)
Key falls out of, or gets stuck in key holder 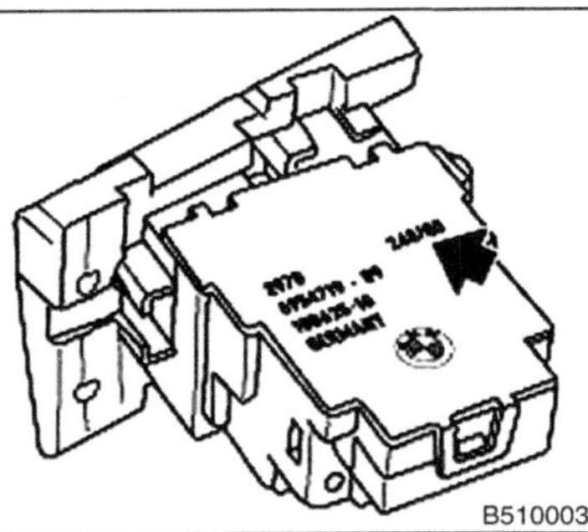 B510003008	Remote key is not adequately retained in the key slot and sometimes falls out. There are also situations where the remote key gets stuck in the key slot and cannot easily be removed. Under these situations, the key slot should be replaced with an updated part (improved mechanism for gripping the remote key). Updated part were introduced in 9 / 2005. The date code on the part is in day/year format. Improved parts have date code 248/05 or later. Illustration shows E90 key holder, E60 is similar.	

7 Equipment and Accessories

Topic	Notes	Additional information (when available)
OC3 (occupancy) mat installation tips. 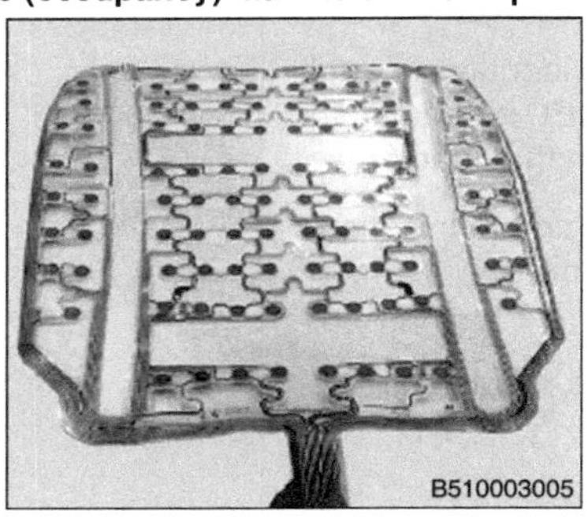 B510003005	When a new OC3 seat mat has been installed use the following tips to avoid activation problems: • With the new seat mat installed in the vehicle, eliminate any tension in the seat mat caused by the installation process by having an adult sit in the seat. • Remove all weight from the seat. • Using BMW scan tool, enable the seat mat. ***NOTE—*** • *OC3 mat warranty has been extended to 10 years / unlimited mileage on vehicles produced to 11 / 2005.*	
Both rear belt tensioners deploy in accident 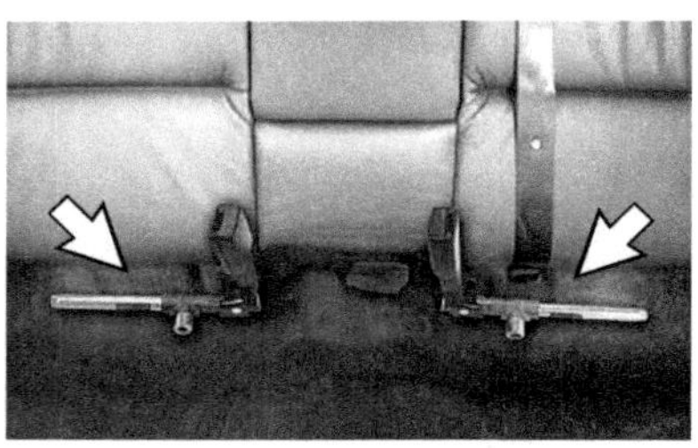	In the case of an accident that would trigger the rear belt tensioners, both tensioners will deploy, regardless of whether the rear seats are occupied, or not. This a normal condition as the rear belt buckles have no switch. ***WARNING—*** • *Always replace deployed airbag components and confirm MRS operational status using BMW scan tool.*	**721**

BentleyPublishers.com
BentleyPublishers.com—All Rights Reserved

010 Product Familiarization

GENERAL

The information in this section is based on introductory material for 2004 through 2010 BMW 5 Series vehicles sold in the USA and Canada. The reference material provided is intended to serve as a product familiarization guide and is subject to change.

> ***WARNING —***
> - *Check BMW factory repair information at www.bmwtechinfo.com or the publisher's website at www.bentleypublishers.com for information that may supersede information in this section.*

B510010002a

PRODUCT OVERVIEW

The 5th generation 5 Series (E60) was introduced in the US in model year 2004 as a replacement for the 1997 - 2003 5 Series (E39) and was produced through model year 2010.

The rear-wheel drive Sedan was introduced first. Sports Wagon (E61) and all wheel xDrive (Sedan and Sports Wagon) versions were introduced in 2006.

Bentley Publishers
BentleyPublishers.com–All Rights Reserved

2004 model introduction	
E60 4-door Sedan	525i (184 hp, 6 cylinder) 530i (225 hp, 6 cylinder) 545i (325 hp, V8)

The E60 5 Series is a larger car than the corresponding E39 chassis it replaced, adding both interior passenger room and wheelbase length.

Body dimensions were increased to improve roominess in the rear compartment (additional 1.97 in knee room) and to increase the luggage compartment capacity (at 520 L the largest capacity of all the BMW vehicles, including the E65).

Compared to the outgoing model, more generous exterior dimensions ensure even greater spaciousness and comfort for all passengers. At 178.0 inches, the length increased by 2.5 inches. The width at 72.7 inches increased by 1.8. Vehicle height is 57.8 inches, an increase of 1.3 inches. The wheelbase is 2.3 inches longer at 113.7 inches.

5 Series sedan with xDrive

BentleyPublishers.com
BentleyPublishers.com—All Rights Reserved

Technical data

BMW 5 Series specifications (2004 Sedan)

	525i	530i	545i
Exterior			
Doors	4	4	4
Height (**A**)	58.0 in (1473 mm)	58.0 in (1473 mm)	58.0 in (1473 mm)
Front track (**B**)	61.3 in (1558 mm)	61.3 in (1558 mm)	61.3 in (1558 mm)
Rear track (**C**)	62.3 in (1582 mm)	62.3 in (1582 mm)	62.3 in (1582 mm)
Width (**D**)	72.7 in (1846 mm)	72.7 in (1846 mm)	72.7 in (1846 mm)
Wheelbase (**E**)	113.7 in (2888 mm)	113.7 in (2888 mm)	113.7 in (2888 mm)
Length (**F**)	190.6 in (4841 mm)	190.6 in (4841 mm)	190.6 in (4841 mm)
Interior			
Front headroom	39.1 in (993 mm)	39.1 in (993 mm)	39.1 in (993 mm)
Rear headroom	38.1 in (967 mm)	38.1 in (967 mm)	38.1 in (967 mm)
Front legroom	41.5 in (1054 mm)	41.5 in (1054 mm)	41.5 in (1054 mm)
Rear legroom	36.0 in (915 mm)	39.1 in (993 mm)	36.0 in (915 mm)
EPA passenger compartment vol.	99.1 cu ft (2806 liters)	38.1 in (967 mm)	99.1 cu ft (2806 liters)
Cargo (trunk) capacity	14.0 cu ft (396 liters)	41.5 in (1054 mm)	14.0 cu ft (396 liters)
Engine			
Engine configuration, cylinders / valves per cyl.	2.5 liter in-line 6 / 4	3.0 liter in-line 6 / 4	4.4 liter V8 / 4
Displacement	2494 cc (152 cu in)	2979 (182)	4398 (268)
Stroke / bore (mm)	84.0 / 75.0	88.0 / 85.0	92.0 / 82.7
Rated power	184 @ 6000 rpm	255 @ 6600 rpm	325 @ 6100
Max. torque (lb-ft)	175 @ 3500 rpm	220 @ 2750rpm	330 @ 3600
Compression	10.5 : 1	10.2 : 1	10.5 : 1
Fuel requirement	Premium unleaded	Premium unleaded	Premium unleaded
General			
Final drive ratio	3.15 : 1 (MT); 3.73 : 1 (AT)	2.93 : 1 (MT); 3.46 :1 (AT)	2.93 : 1 (MT); 3.38 :1 (AT)
Fuel tank capacity	18.5 US gal	18.5 US gal	18.5 US gal
Unladen weight (manual)	3428 lb	3472 lb	3803
Gross vehicle weight (GWR)	4530 lb	4575 lb	4905 lb
Coefficient of drag (C^d)	0.29	0.29	0.29
Max. roof load	100 lbs.	100 lbs.	100 lbs.
Acceleration (0 - 60 mph)	7.8 sec (est)	6.6 sec (est)	5.7 sec (est)

BentleyPublishers.com
BentleyPublishers.com – All Rights Reserved

Vehicle dimensions

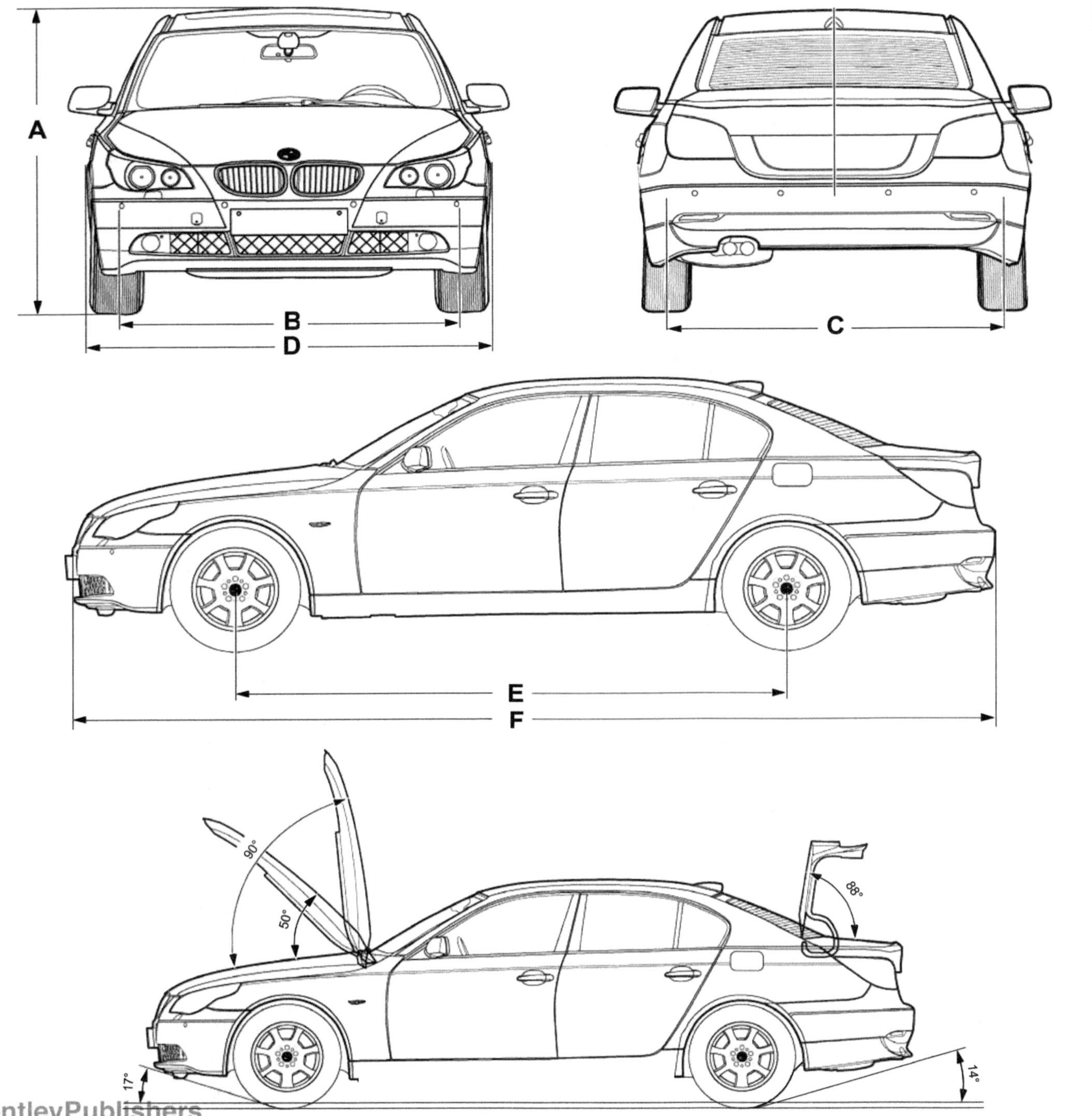

BentleyPublishers.com
BentleyPublishers.com—All Rights Reserved

Model Year Highlights

Below are the 5 Series year-to-year highlights through model year 2010. Major changes are detailed later in this repair group.

Model year 2004

The E60 sedan was introduced with many new features (some optional. Most notable were the following:

- Innovative lightweight construction with an aluminum front end and an all-aluminum suspension. iDrive control with available head-up display that projects important information in the form of a virtual image directly into the driver's field of vision.
- Directional headlights that adapt continuously to the steering angle to illuminate around the bend in the road.

BentleyPublishers.com
BentleyPublishers.com - All Rights Reserved

6-speed Steptronic transmission

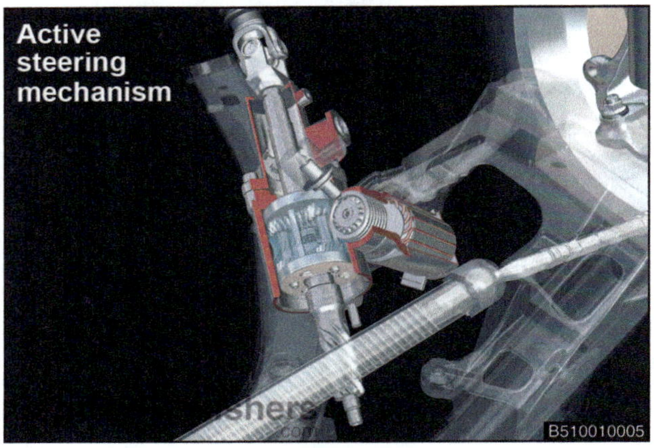

Additional features for 2004:

- Choice of a 2.5- and 3.0-liter inline 6-cylinder engines or a 4.4-liter Valvetronic V8 engine.
- Choice of 6-speed transmissions: manual, STEPTRONIC automatic and Sequential Manual Gearbox (SMG).
- Revised chassis features (suspension, steering, brakes).
- Automatic headlight control and rain-sensing windshield wipers standard on all models.
- iDrive control system for many features and functions
- Active cruise control (ACC).
- Voice command system (included with optional navigation system)
- Flat tire monitor (FTM) monitors tire pressures while vehicle is driven.
- Run-flat tires with all sport packages.
- Adaptive brake lights, enlarged brake light area under heavy braking.
- Sport package with active front steering, active roll stabilization and run-on flat performance tires (standard on 545i 6-speed manual, otherwise optional).
- German built; production of 530i began July 1, 2003, other models September 1, 2003.

Model year 2005

New for 2005 was the launch of the 4th generation E60 M5.

The 5.0 liter 500 horsepower V10 engine was mated to a seven-speed SMG transmission. With this new powertrain, the M5 offered the best power-to-weight ratio in its class. In the tradition of BMW M, the chassis was engineered to keep up with the power while the exterior treatment was distinctively subtle.

M5 (S65) engine specifications	
Configuration	90° V10 with high-pressure double VANOS, steplessly variable valve timing, 10 individual electronically controlled throttles
Construction	Aluminum block and heads
Displacement (liter / cid)	4.999 / 305.1
Bore x stroke (mm)	92.0 x 75.2
Power output (bhp / kW) @ rpm	500 / 373 @ 7750 rpm
Torque (Nm / ft-lb) @rpm	520 / 383 @ 6100
Maximum engine speed	8250
Compression ratio	12.0 : 1
Valves / cyl	4
Engine management	MS S65
0 - 60 mph	4.6 s

BentleyPublishers.com
BentleyPublishers.com—All Rights Reserved

Model year 2006

2006 was a significant update year for the 5 Series: Valvetronic 6-cylinder engines, a more powerful 4.8 liter V8 engine, the introduction of all-wheel drive, and a Sports Wagon body style are the most notable updates.

6-cylinder all-wheel drive models with new badge designations 525xi and 530xi.

N52 engine

6-cylinder models were powered by BMW's advanced N52 engines.

- New generation 6-cylinder engines, called N52, both with 3.0-liter displacement:
  - 525i: 215 hp @ 6250 rpm, 185 lb-ft. torque @ 2750 rpm, single-stage induction system.
  - 530i: 255 hp @ 6600 rpm, 220 lb-ft. torque @ 2750 rpm, 3-stage induction system.
  - Magnesium-aluminum composite construction, Valvetronic variable valve lift, electric coolant pump and many other innovations.

530xi Sports Wagon

The new Sports Wagon, an appealing concept combining 5 Series sophistication with enhanced load-carrying versatility, was introduced as an all-wheel drive model for 2006.

550i

550i model with 4.8-liter Valvetronic V8 engine, 360 hp, 360 lb-ft. torque, a 35 hp increase over the previous 545i.

BentleyPublishers.com
BentleyPublishers.com–All Rights Reserved

Additional features for 2006

Start Stop button replaces former key switch; new-type remote control.

- Comfort access system (can be ordered as an option as of September 2005 production) featuring keyless access.
- Numerically higher final drive ratios for the 6-cylinder models.
- Xenon adaptive headlights standard on 530i and 530xi models.
- All models fitted with chrome vertical grille slats (previously 545i only).

Model year 2007

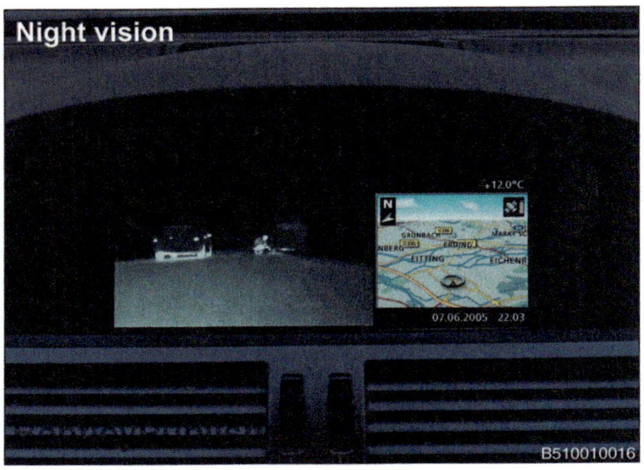

Night vision became an available an option. The system uses a thermal-imaging camera that constantly monitors the road ahead and displays images according to their temperature on the iDrive screen.

- Tire pressure monitor, replacing previous flat tire monitor, constantly monitors tire pressures and displays a warning on the iDrive screen when it detects a significant change in tire pressure.
- High definition (HD) radio available as an option; offers CD quality FM and raises AM reception to the same level as analog FM broadcast.
- Real time traffic information for navigation system, for up-to-the-minute reports on local traffic conditions.
- Auxiliary input for iPod or MP3 player is now standard; located between the two power sockets at rear of center console.
- Revised sport package with 20-way multi-contour seats. Seats include articulated upper backrest, adjustable-width backrest, adjustable thigh support and 4-way lumbar support. Sport package also added 3-spoke steering wheel.

Model year 2008

The biggest visual change for 2008 (as of 03 / 2007 production) was the mid-life update, called Life Cycle Impulse (LCI) by BMW.

LCI exterior updates included:

Reconfigured optical headlight lenses emphasized with chromed tubes.

- New front air dam with reshaped opening and up turned ends.
- Relocated amber DOT lenses; headlight area is completely white.

BentleyPublishers.com
BentleyPublishers.com—All Rights Reserved

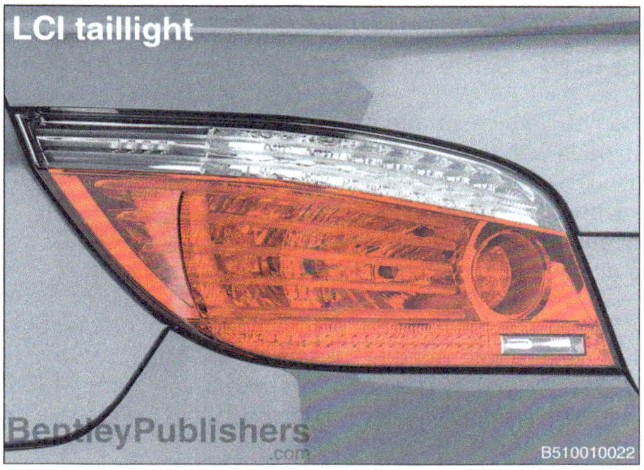

B510010022

Taillights include new white design with fewer reflectors, red lines in optical lenses.

- Reshaped rear bumper, with revised diffuser below bumper.

B51001023

Model year 2008 introduced new models with new engines:

- 528i and 528xi Sedan and Sports Wagon replaced 525i and 525xi. These models received a more powerful 3.0-liter inline six-cylinder engine, generating 230 hp and 200 lb-ft of torque, for an increase of 15 hp and 15 lb-ft over the previous engine.
- 535i and 535xi Sedan and Sports Wagon replaced 530i and 530xi. These models were fitted with the N54 3.0-liter twin-turbocharged inline 6-cylinder engine producing 300 hp and 400 lb-ft of torque, an increase of 45 hp and 80 lb-ft.

Additional features for 2008

- Active cruise control (optional) with stop-and-go braking function
- Bi-xenon headlights
- Optional lane departure warning

B510010024

Freshened interior with additional leather and extended wood trim for a more elegant look.

- Ergonomic changes included more convenient window switches, relocated console ashtray and electronic shifter.
- Technically updated STEPTRONIC automatic transmission with shift-by-wire electronic shifter (a no-cost option in 2008).
- Optional Sport automatic transmission included steering-wheel paddle shifters and rpm-matching during downshifts, with Sport button for quicker, more dynamic shifting (beginning 06 / 2007 production).

Model year 2009

For 2009, the BMW 5 Series had very few changes: new standard wheels on the 528i and 528i xDrive models (17" double-spoke wheels with all-season tires), and on the 535i and 535i xDrive models (17" V-spoke wheels with all-season tires).

Model year 2010

For 2010, the BMW 5 Series experienced two significant changes:

4th-generation iDrive system with new graphics and controls on vehicles equipped with optional on-board navigation system.
High-resolution 8.8 in. control display.
Optimized menus, expanded range of functions.
Greater convenience and more intuitive operation via direct-select buttons at controller, plus more programmable memory keys.

Available M Sport Package included anthracite headliner, exterior aerodynamic package with Shadowline trim, M-specific wheels, door sills, footrest, gearshift lever, and steering wheel. Some M exterior colors were also available on vehicles with M Sport Package.

Engines

There are 4 variants of the 6-cylinder engines and two variants of the V8 engines of the used in the 5 Series. See **100 Engine–General** for a full engine application chart.

M54 engine (2004 - 2005)

2004 and 2005 525i and 530i models are fitted with the M54 engine. This aluminium-block engine features overhead camshafts, four valves per cylinder, hydraulic valve adjustment and double-VANOS system (variable valve timing) to optimize torque, power and emissions.

M54 engine specifications	
Displacement (cm^3) 525i 530i	 2494 2979
Bore x stroke (mm) 525i 530i	 84.0 x 75.0 84.0 x 89.6
Power output (bhp) @ rpm 525i 530i	 184 @ 5900 RPM 225 @ 5800 RPM
Torque (Nm / ft-lb) @rpm 525i 530i	 175 @ 3500 214 @ 3500
Compression ratio 525i 530i	 10.5 : 1 10.2 : 1
Valves / cyl	4
Engine management	Siemens MS 45.1 drive-by-wire throttle

BentleyPublishers.com
BentleyPublishers.com—All Rights Reserved

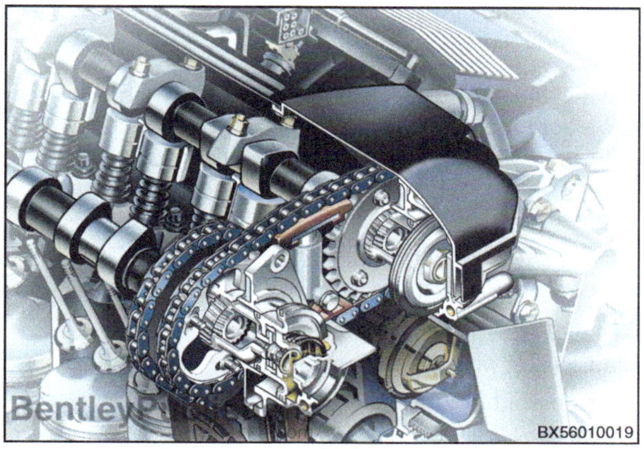

The double-VANOS system is a combined hydraulic and mechanical camshaft control device controlled by the Siemens engine management system.

The VANOS system uses camshaft adjusters to continuously modify intake and exhaust camshaft timing as a function of accelerator pedal position and engine speed.

Double-VANOS significantly enhances emission management, increases output and torque, and offers better idling quality and fuel economy.

N52 engine (2006 - 2007)

N52 3.0 liter Valvetronic engine

In 2006, the new generation N52B30 in-line 6-cylinder engine was introduced. This engine featured many innovations, including a composite crankcase, Valvetronic system and the 3-stage DISA intake manifold. The oil level dipstick was eliminated.

The same basic 3.0 liter engine was shared on the 525i and 530i, with the difference in power ratings being achieved with intake manifold tuning and DME software.

N52 engine specifications	
Displacement (cm^3)	2996
Bore x stroke ore (mm)	85.0 x 88.0
Power output bhp) @ rpm 525i 530i	 215 @ 6250 RPM 255 @ 6600 RPM
Torque (Nm / ft-lb) @rpm 525i 530i	 185 @ 2750 220 @ 2750
Maximum engine speed	7000
Compression ratio	10.7 : 1
Valves / cyl	4
Engine weight (kg)	161 kg (355 lb)
Engine management	Siemens MSV70
Emission compliancy	ULEV 2

Magnesium alloy composite crankcase

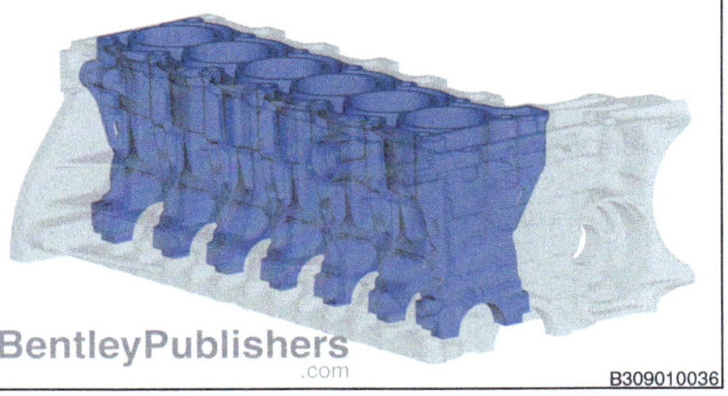

The N52 engine weighed 22 lbs. less than its predecessor M54 engine. The primary weight savings came from the composite magnesium alloy crankcase and the lightweight exhaust manifold.

BentleyPublishers.com
BentleyPublishers.com—All Rights Reserved

N52 engine (2006 - 2007)

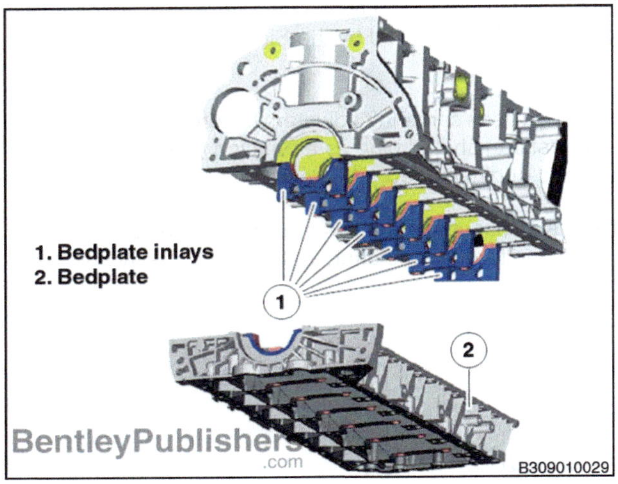

The magnesium bedplate and cylinder head cover also played a role in overall weight reduction.

The traditional belt driven coolant pump from previous engines was replaced by an electric coolant pump. This allowed for the addition of a single belt drive.

Additional N52 engine features:

- Trapezoidal connecting rods (weight optimized).
- Aluminum silicon (AluSil) cylinder head
- Timing case integrated in crankcase and cylinder head
- Cylinder head gasket with silicon sealing lip
- Weight-optimized double VANOS
- Crankcase ventilation with integrated heater
- Volumetric flow controlled oil pump

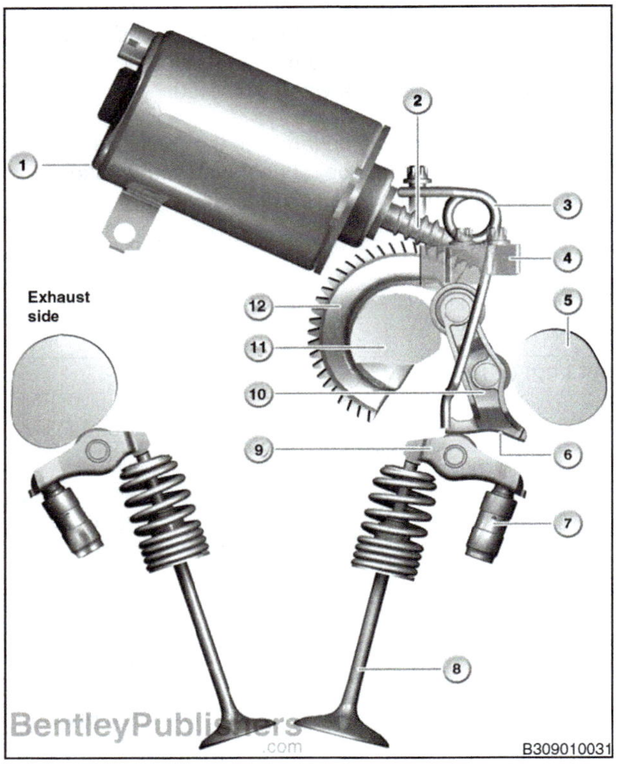

Valvetronic replaces the conventional throttle butterfly with an electromechanical system that controls the amount of lift of the individual intake valves. The Valvetronic engine is able to breathe freely, delivering better performance while using less fuel.

Valvetronic valvetrain uses a stepper motor to control a secondary eccentric shaft fitted with a series of intermediate rocker arms that control the degree of valve lift. The throttle butterfly is no longer needed as a means of controlling the air supply, though for safety reasons it is still fitted as an emergency back-up. Valvetronic improves cold start behavior, lowers exhaust emissions and provides smoother, more immediate power.

1. Actuator
2. Worm shaft
3. Return spring
4. Gate block
5. Intake camshaft
6. Ramp
7. Hydraulic valve adjuster (HVA), intake
8. Intake valve
9. Roller cam follower
10. Intermediate lever
11. Eccentric shaft
12. Worm gear

The compact VANOS units use vane-type construction which is less complex and easier to service than previous VANOS designs. The VANOS units have integrated sprockets and are attached to the camshaft via a central bolt.

NOTE —

- *Due to the different spread ranges for the intake and exhaust camshaft, the VANOS units cannot be interchanged. Doing so would cause engine damage.*

BentleyPublishers.com
BentleyPublishers.com—All Rights Reserved

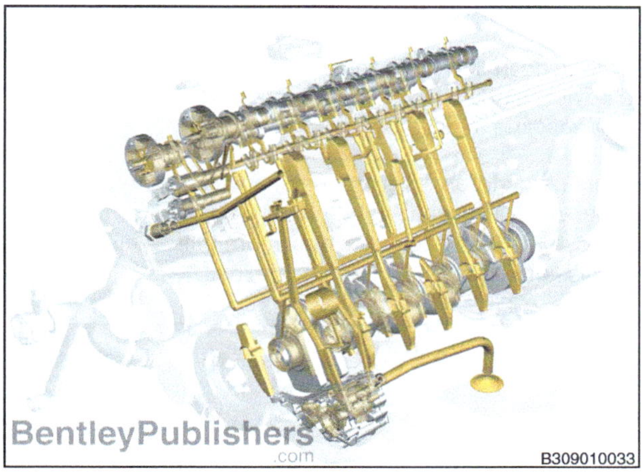

To meet the oiling requirements of the N52, a volumetric-flow controlled oil pump is utilized. Systems such as VANOS require a large volume of oil, particularly at low engine speeds.

This pump only delivers as much oil as each operating range of the engine requires. A conventional oil pump would have to be approximately three times the size of that in the N52 engine. A larger oil pump would also consume excess energy and therefore the new pump design allows:

- Increased power output
- Reduced weight
- Optimized fuel consumption
- Reduced exhaust emissions

N52K engine (2008 - 2010)

For 2008, the N52 base engine from the 525i was replaced by the more powerful N52K engine (230 hp vs. 215 hp). The N52K engine was an improved and cost-optimized version of the N52 and was first available in the 528i and 528xi from March 2007 production.

There are not many changes to the emission systems on this engine. The N52KP continued to use the composite magnesium-aluminum alloy engine block from the existing N52.

The technical highlights of the N52KP include:

- New engine management (MSV80)
- New hot film mass air flow sensor (digital HFM)
- EGAS8 throttle with magnetoresistive position feedback
- Plastic valve cover with integrated crankcase vent valve and oil separation
- Stronger connecting rods
- Exhaust valve stem increased to 6 mm
- Second-generation electric coolant pump
- Lightweight camshafts (hydroformed)

N52K engine specifications	
Displacement (cm^3)	2996
Bore x stroke (mm)	85.0 x 88.0
Power output (bhp) @ rpm	230 @ 6250 RPM
Torque (ft-lb) @rpm	200 @ 3000
Maximum engine speed	7000
Compression ratio	10.7 : 1
Valves / cyl	4
Engine management	Siemens MSV80
Emission compliancy	ULEV 2

BentleyPublishers.com
BentleyPublishers.com - All Rights Reserved

N54 engine (2008 - 2010)

The N54 engine used in the 2008 and later 535i models is the high-performance powerplant utilizing twin-turbo technology and high-pressure direct gasoline injection.

The N54 delivers uncompromising response and high torque output in a unit that is about 150 pounds lighter than a V8 engine. The new technology used in the N54 allows for maximum efficiency and the low exhaust emissions to meet ULEV II guidelines.

N54 3.0 liter turbocharged engine

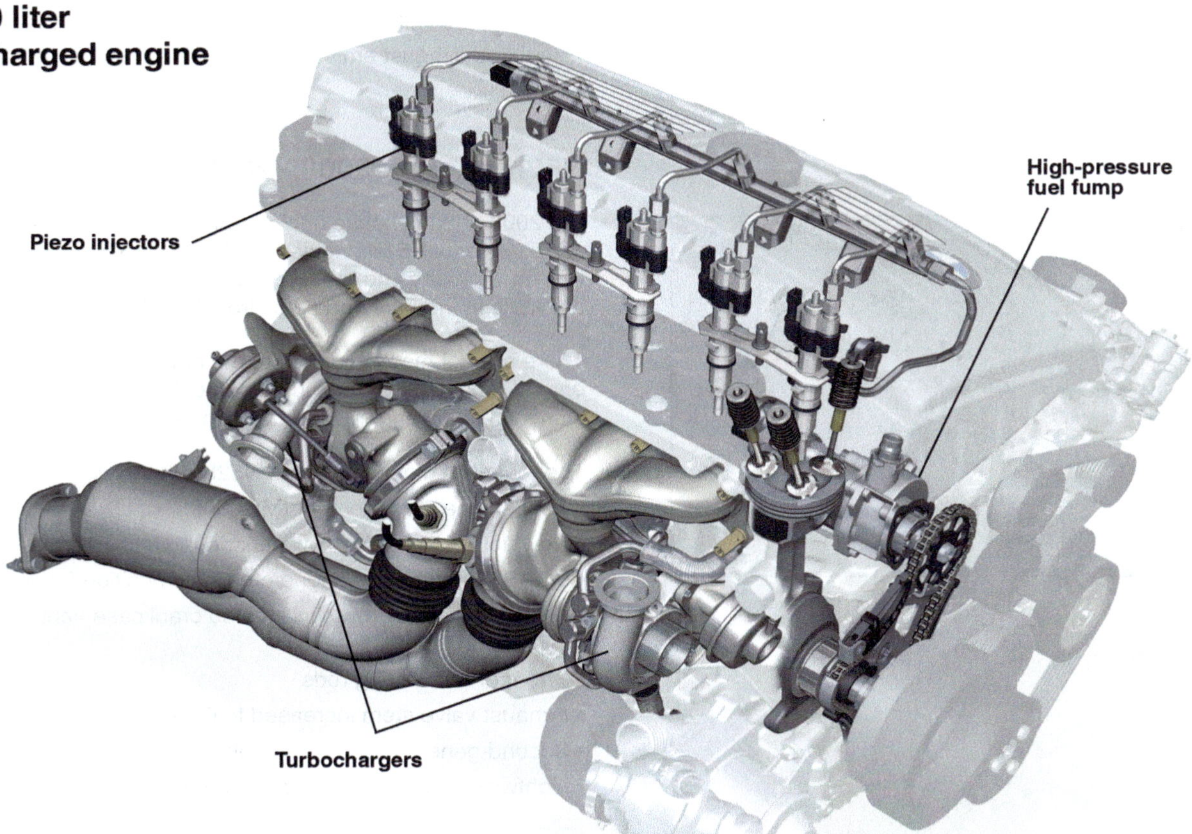

BentleyPublishers.com

B309010040

N54 engine

BentleyPublishers.com

B309010034

The N54 is officially referred to as the N54B30O0. The "O" designates the "upper" output range and the "0' indicates the first generation in this series.

BentleyPublishers.com
BentleyPublishers.com--All Rights Reserved

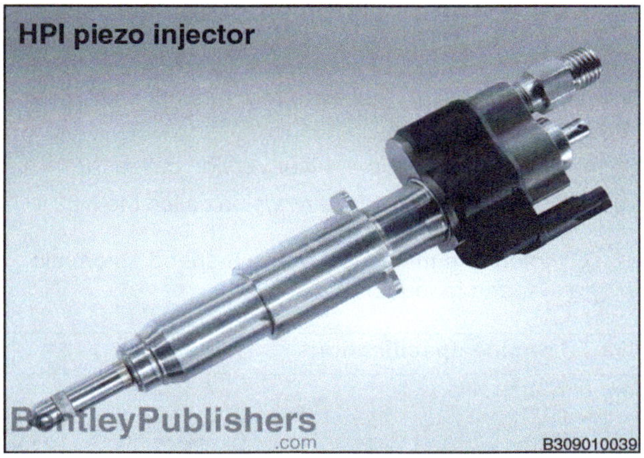

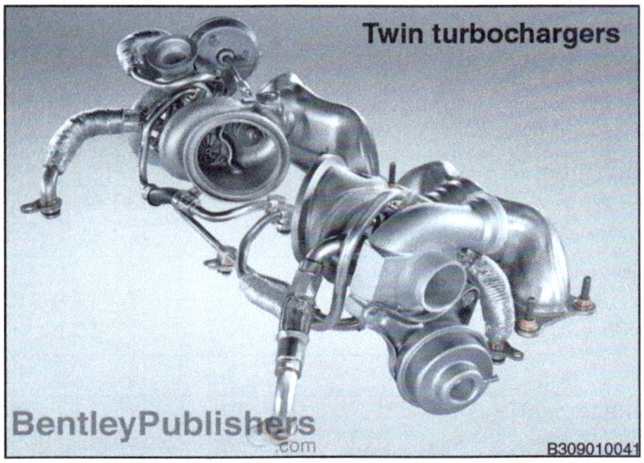

The N54 uses HPI injection system, which is capable of pressures of up to 200 bar (2900 psi). This system features piezo injector technology with outward opening injectors.

Developing 300 hp and peak torque of 400 Nm or 295 lb-ft, this new engine offers the highest standard of spontaneous and superior power and performance.

N54 engine specifications	
Displacement (cm3)	2979
Bore x stroke (mm)	84 x 89.6
Power output (kw / bhp)	225 / 300 @ 5800 RPM
Torque (Nm) @rpm	400 @ 1300 - 5000
Maximum engine speed	7000
Compression ratio	10.2 : 1
Valves / cyl	4
Engine weight	187 kg (412 lb)
Engine management	MSD80, MSD81
Emission compliancy	ULEV 2
Injection system type	HPI (DI 2)

The 3.0 liter N54 engine features the following:

- Exhaust driven twin turbocharger
- Air to air intercooler
- Second generation direct injection (HPI) with piezo injectors
- New engine management (MSD80, MSD81)
- Bi-VANOS
- All aluminum crankcase with iron cylinder liners (similar dimensions to N52)
- External oil cooler
- New high output electric coolant pump (400 W)
- Aluminum cylinder head with plastic valve cover
- Steel crankshaft

Bentley Publishers.com
BentleyPublishers.com–All Rights Reserved

Valvetronic V8 engines (2004 - 2006)

The 2004 through 2010 V8 models were fitted with the new generation Valvetronic engines. There are two variants of the N62 Valvetronic motor:

- 2004 - 2005 models had a 325-hp 4.4 liter version called the N62
- 2006 - 2010 models had 360-hp 4.8 liter version called the N62 TU

N62 and N62 TU motors are mated to either a 6-speed automatic transmission or a 6-speed manual.

N62 and N62 TU engine specifications	
Displacement (cm3) N62 (2004 - 2005) N62TU (2006 - 2010)	 4398 4799
Bore x stroke (mm) N62 (2004 - 2005) N62TU (2006 - 2010)	 92 x 82.7 93 x 88.3
Power output (bhp) @ rpm N62 (2004 - 2005) N62TU (2006 - 2010)	 325 @ 6100 360 @ 6300
Torque (Nm) @rpm N62 (2004 - 2005) N62TU (2006 - 2010)	 330 @ 3600 360 @ 3400
Compression ratio N62 (2004 - 2005) N62TU (2006 - 2010)	 10.0 : 1 10.5 : 1
Valves / cyl	4
Engine management N62 (2004 - 2005) N62TU (2006 - 2010)	 Bosch ME 9.2 Bosch ME 9.2.2
Emission compliancy	ULEV 2

N62 engine (2004 - 2005)

545i model with 4.4-liter V8 engine, 330 hp, 325 lb-ft. torque.

N62 engine features:

- Valvetronic
- Fully variable intake manifold
- Bi-VANOS

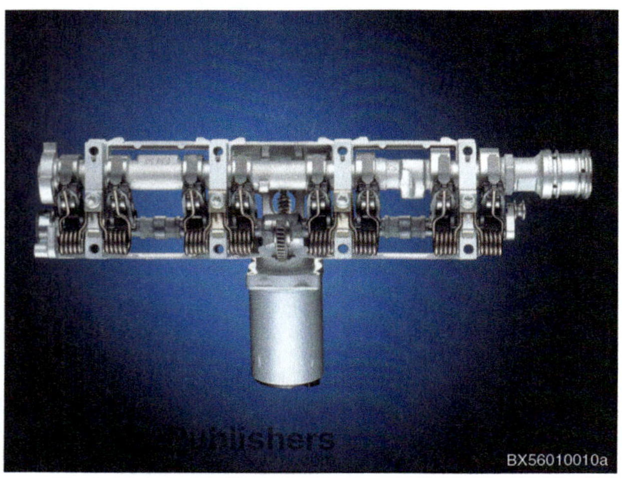

Valvetronic is a patented BMW system that varies valve lift to control engine power. Instead of a conventional throttle, Valvetronic varies valve lift all the way from 0.3 mm (for idling) to 9.7 mm (for full-power operation). By eliminating pumping losses associated with a throttle plate, Valvetronic plays a major role in boosting power on the 4.4 liter engine. It also helps make the engine smoother.

BentleyPublishers.com
BentleyPublishers.com—All Rights Reserved

Valvetronic operation

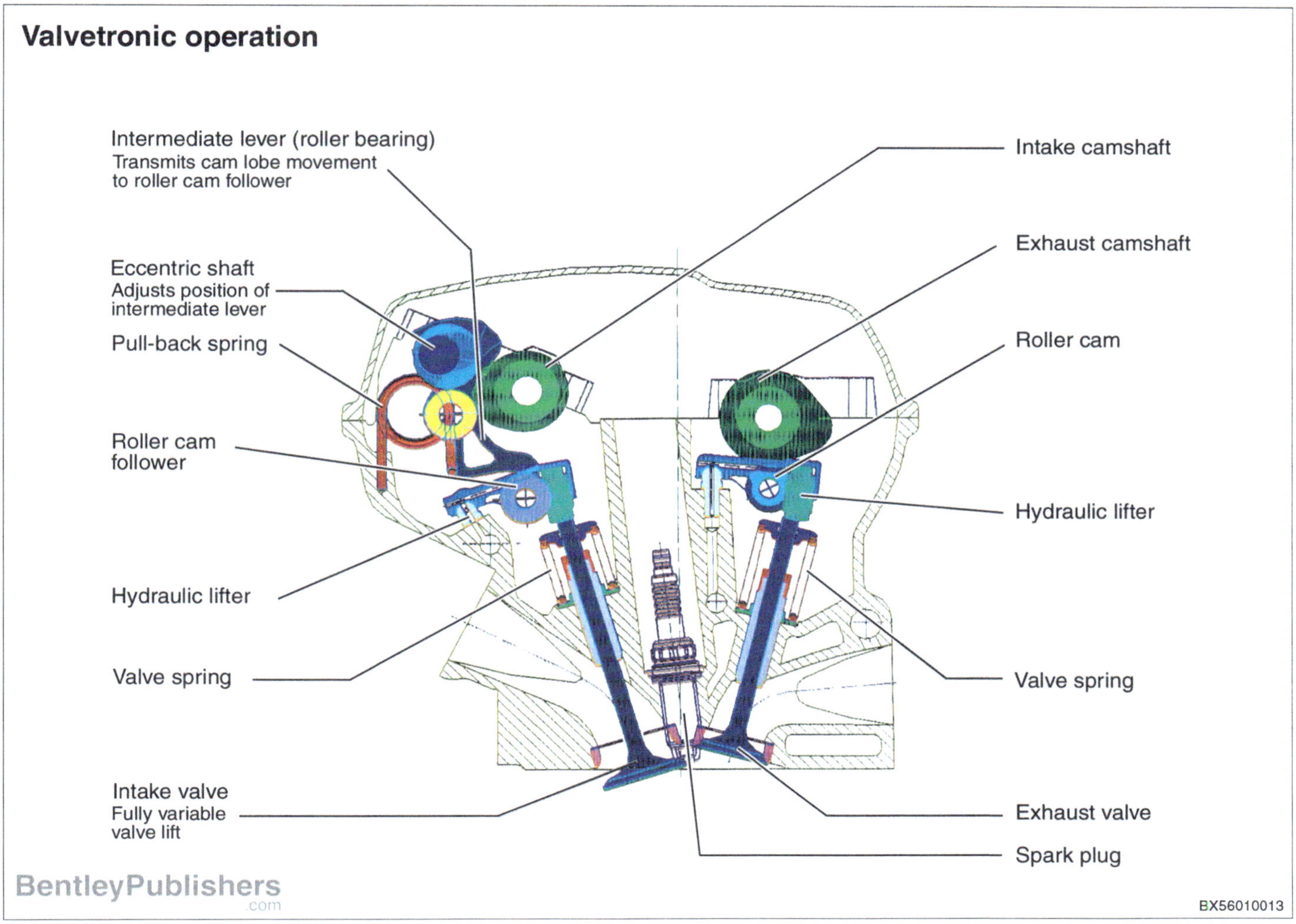

The Valvetronic mechanism sits atop the intake valves on both of the V8 cylinder banks. Each of the engine's 32 valves (4 valves per cylinder) is actuated as the camshaft lobe deflects a finger-type rocker arm with hydraulic clearance adjustment. On the intake side there is an additional element between the cam lobe and rocker arm, called an intermediate lever.

Upon contact by the camshaft lobe, the intermediate lever actuates the rocker arm and, in turn, the valve. The intermediate lever is held in place by an eccentric shaft that is rotated by an electric servo motor. The eccentric shaft position determines the intermediate lever's pivot point, and therefore valve lift. Eccentric shaft position commands come from the Bosch engine management system in response to accelerator pedal movements.

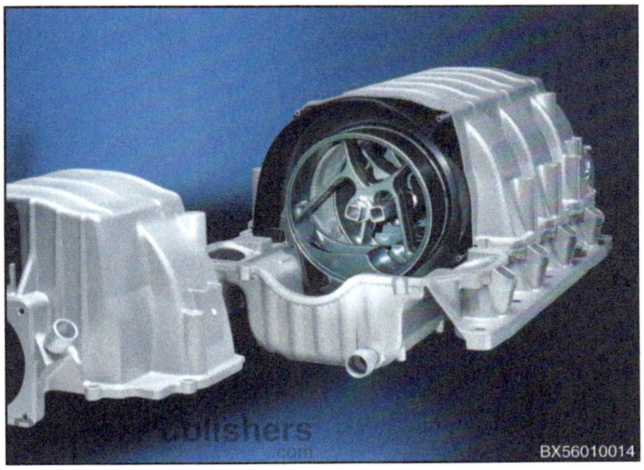

V8 engines employ a steplessly variable intake manifold that optimizes performance across the entire engine speed range.

BentleyPublishers.com
BentleyPublishers.com - All Rights Reserved

In addition to Valvetronic, the N62 engine also uses a double VANOS system similar to that used on the 6-cylinder engine. This arrangement varies valve timing by rotating the intake and exhaust camshafts steplessly to optimize engine power, emissions and fuel efficiency. The combination of VANOS and Valvetronic is exclusive to the new N family of power plants.

N62 TU engine (2006-2010)

The 550i Sedan, replacing the 545i in 2006, is powered by 4.8-liter Valvetronic V8 engine. This engine is very similar to the 4.4 liter Valvetronic engine. The greater displacement and other minor engineering improvements make this unit even more powerful: 360 hp and 360 lb-ft. of torque, up from the previous 4.4-liter engine's 325 hp / 330 lb-ft.

This engine utilizes Bosch ME 9.2.2 and maintains Valvetronic and dual-VANOS controls.

Transmissions

In the E60 5 Series models covered by this manual, all transmission choices are 6-speed.

Manual transmissions

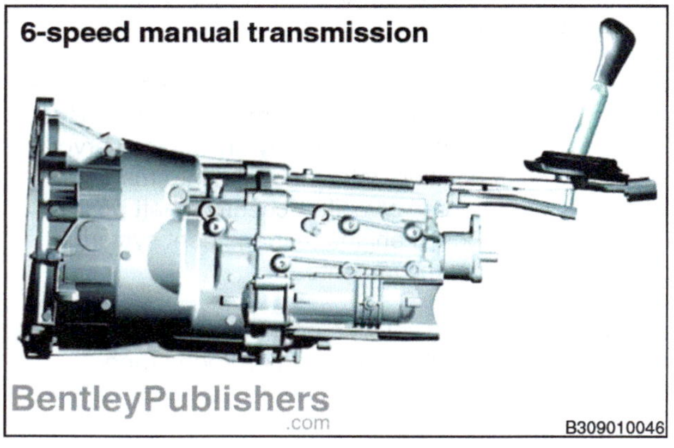

The 6-speed manual gearbox, standard on all models, is engineered to provide precise control and a high level of driving pleasure. The manual transmission has a lifetime oil fill. Both Getrag and ZF manufactured units are installed on the 5 Series cars covered by this manual. See **200 Transmission–General** for application information.

BentleyPublishers.com
BentleyPublishers.com—All Rights Reserved

Sequential Manual Gearbox (SMG)

Sequential Manual Gearbox (SMG)

B510010036

Available as an option, the sequential manual gearbox (SMG) functions like a clutchless manual transmission. Clutch operation is handled via a computer controlled hydraulic unit to engage and disengage the clutch, as well as shift gears. See **200 Transmission–General** for application information.

Automatic transmissions

6-speed automatic

B510010037

A 6-speed automatic transmission is available for 5 Series cars. Compared to the 5-speed unit it replaces, it is 10% lighter, has a more efficient torque converter and operates with fewer internal clutches. See **200 Transmission–General** for application information.

The automatic transmission is controlled by the mechatronics module that is made up of a combination of the valve body and electronic transmission control unit (EGS). The drive range switch (for drive ranges P, R, N and D) is located in the mechatronics module, which is mounted in the ATF sump. It is operated by the selector lever in connection with a bowden cable assembly.

The Steptronic feature offers Normal, Sport and Manual modes. The Sport mode, in which shifts occur at higher engine speeds, is engaged by moving the lever from D; from here, the Manual mode is engaged when the driver manually chooses a gear by tipping the lever forward (for downshifts) or rearward (for upshifts).

Torque converter

B309010045

The torque converter uses an integrated torsion damper which avoids unnecessary slip and power loss. The converter clutch is closed immediately after the vehicle starts off. In this way, it feels similar to a manual transmission, with short response and shift times. With precise control and a low level of converter slip, the 6-speed automatic transmission also improves fuel economy, particularly at highway speeds when 6th gear is in use.

BentleyPublishers.com
BentleyPublishers.com – All Rights Reserved

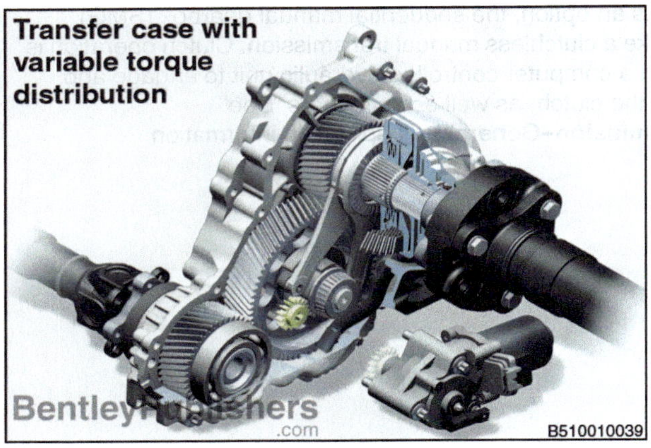

Transfer case with variable torque distribution

BMW xDrive

xDrive is BMW's intelligent all-wheel drive system. It was introduced on the E60 and E61 cars in model year 2006.

xDrive optimizes traction, while enhancing stability on slippery road surfaces. Via the same type of logic that DSC employs to recognize and correct for excessive over- or understeer, xDrive adjusts the front to rear torque split to avoid these tendencies.

NOTE—

- *2009 and later models with all-wheel drive no longer carry xi on the trunk badge. Instead, there is an xDrive badge on the front fender designating the all-wheel drive configuration.*

xDrive operates as follows:

- Driving torque is always transmitted to the front and rear wheels.
- The portion of torque transmitted to the front wheels is controlled by a multi-disc clutch attached to the transfer box. The torque split between front and rear wheels is steplessly variable.
- Engagement pressure on the multi-disc clutch is directed by an electronic control system in response to actual road and driving conditions.

Suspension, Steering and Brakes

E60 and E61 cars feature a highly evolved chassis, including front and rear aluminum suspension systems for low unsprung weight and outstanding handling and ride comfort. The front and rear suspension also include accommodations for active stabilizer bars as part of the dynamic drive system (optional equipment, Sport package).

The steering is an engine-speed-sensitive variable-assist power rack-and-pinion design. Also available as an option was active steering, which electromechanically varies the steering ratio on the basis of vehicle speed and other driving conditions.

Standard equipment includes four-wheel ventilated brake rotors and a new generation dynamic stability control (DSC) system.

Front suspension

The aluminum front suspension utilizes a double pivot axle with MacPherson struts. All models benefit from the reduced sprung and unsprung weight of aluminum for optimal suspension responsiveness.

- Unsprung aluminum components:
  Both lower arms (two per side)
  Steering knuckles
  Strut tubes
- Sprung aluminum components:
  Spring pads
  Subframe
  Front end reinforcement plate
  Steering rack

Other aluminum components further contribute to overall weight savings: engine mounting brackets, transmission crossmember and mounting brackets, and power steering pump housing.

BentleyPublishers.com
BentleyPublishers.com—All Rights Reserved

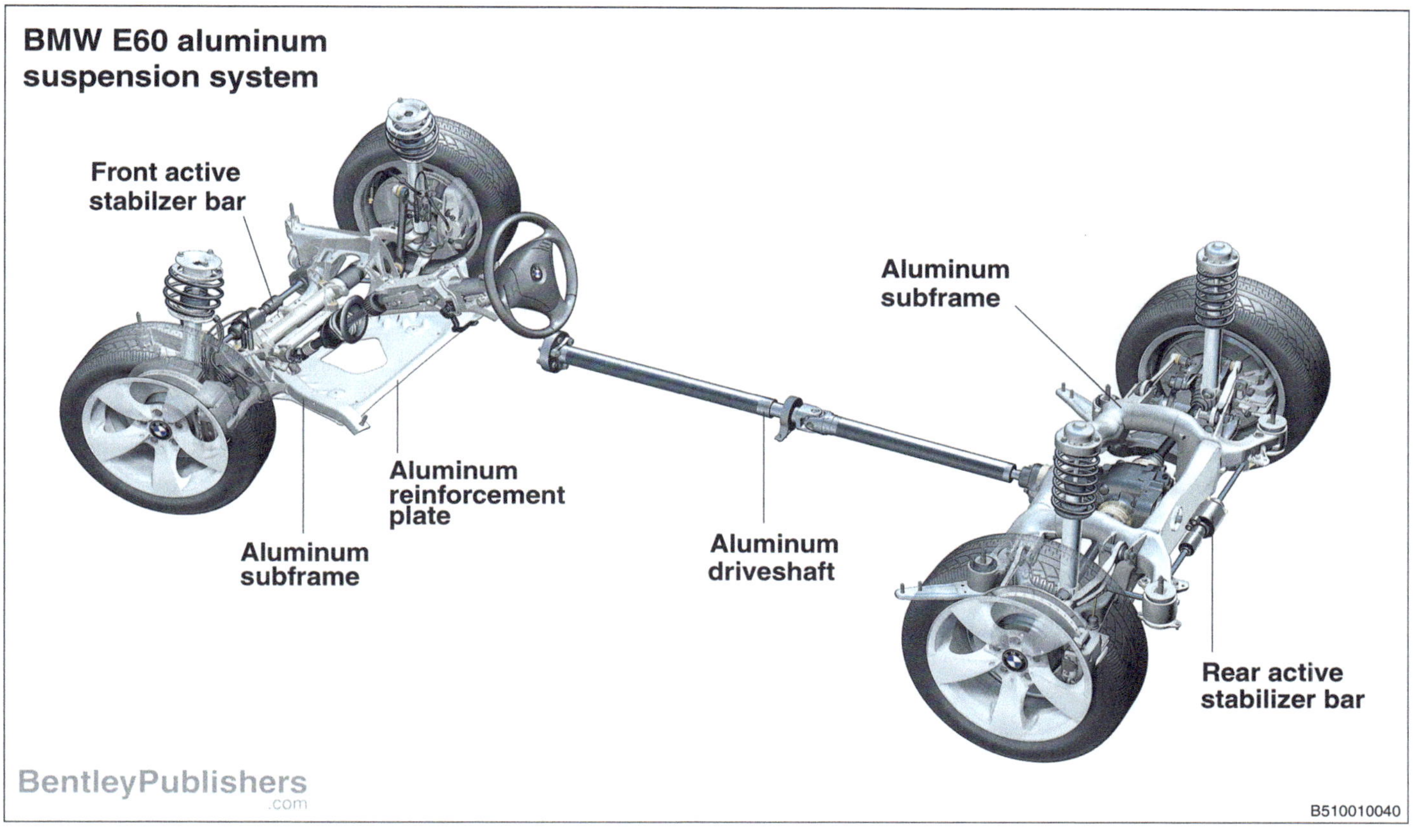

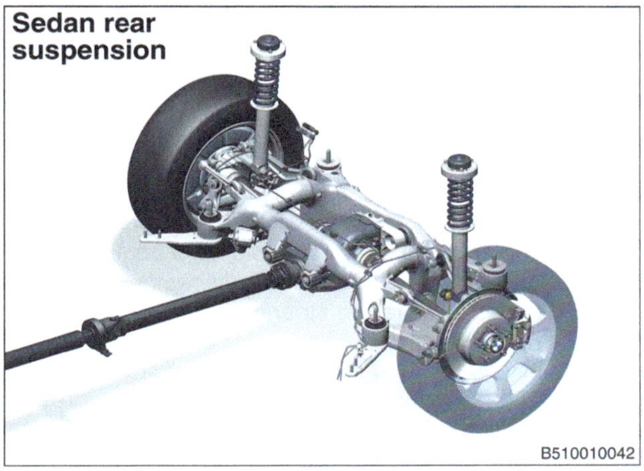

Rear suspension

The rear axle is designed around BMW's highly sophisticated multi-link Integral system.

As in the front suspension, aluminum is extensively used in the rear suspension.

- Sprung aluminum components:
  Subframe
  Spring pads
- Unsprung aluminum components:
  All links of the 4-link Integral system
  Shock absorber tubes

The aluminum rear subframe, which carries the entire rear suspension system and final drive (differential), is mounted to the main structure with four large rubber bushings that help absorb road shocks. The final drive unit is mounted to the subframe through its own rubber bushings. With these two stages of vibration and noise absorption, the differential is acoustically decoupled from the body, minimizing gear and driveline noise.

BentleyPublishers.com
BentleyPublishers.com - All Rights Reserved

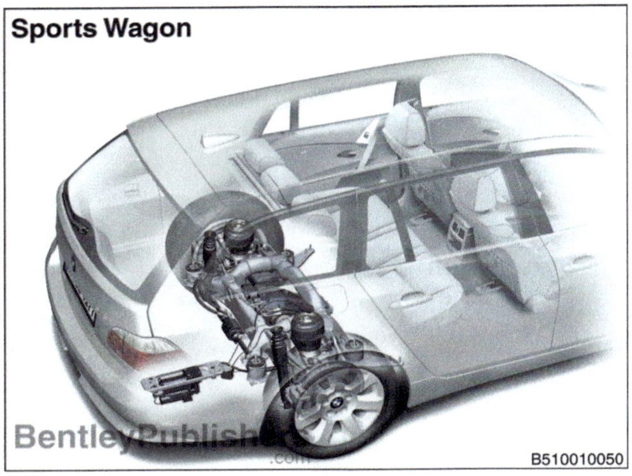

Air suspension (Sports Wagon)

E61 cars are equipped with an automatic self-levelling air suspension (EHC) at the rear axle. This suspension system is a variant of BMW's integral link rear axle and was specifically developed for the Sports Wagon to facilitate a completely flat luggage compartment floor while providing a very large through-load space.

Rear air suspension is particularly advantageous on vehicles with high payload (such as the Sports Wagon) as the load at the rear axle changes when loading and unloading the car.

With conventional steel springs, a change in axle load affects not only the ride height but also vehicle vibration and suspension geometry. There is a change in rear wheel camber at high payloads. This can result in increased load at the inner shoulders of the tires, in turn reducing the load bearing capacity of the tires. EHC eliminates many of these problems by maintaining a constant rear ride height regardless of the load.

Rear air springs and shock absorbers are supported independently on the floor pan of the cargo compartment.

The air supply system is located on the underbody between the spare wheel well and battery box.

Dynamic drive (active stabilizers)

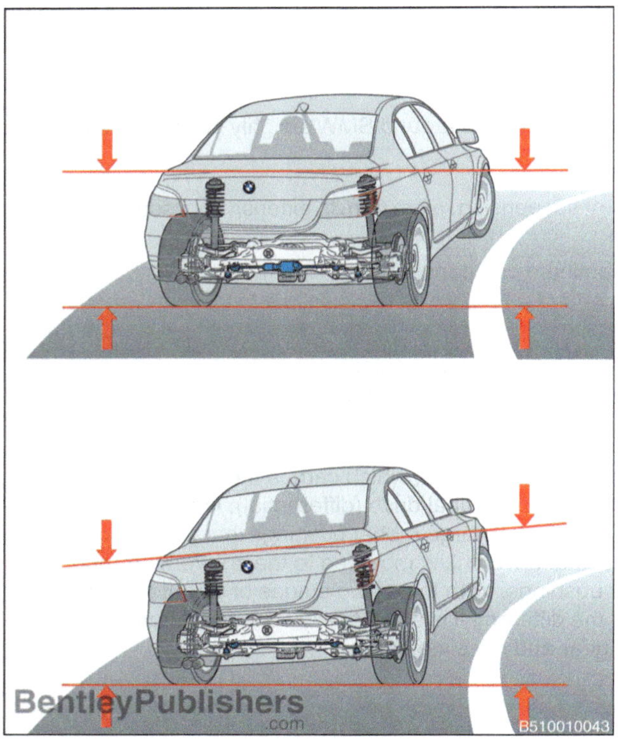

BMW's innovative dynamic drive (also called active stabilizers or ARS) ensures that under- or oversteering is reduced, resulting in enhanced safety, greater comfort and an impressively dynamic driving experience. Dynamic drive minimizes body roll and distributes it between both axles, ensuring a comfortable flat ride.

Dynamic drive is fitted to the 5 Series as part of the Sport Package. In the dynamic drive system, front and rear hydraulically-actuated active stabilizer bars replace conventional solid stabilizer bars.

During cornering, the active bars counteract the vehicle's tendency to roll in a corner. By working together with acceleration and position sensors, an electronic management system and other components, the system ensures a high degree of agility at all speeds, balanced reactions to load changes and precise steering.

The system continually adapts to driving conditions. For example, when driving straight ahead, dynamic drive reduces the pressure on the active stabilizer bars, treating them individually. This results in a highly comfortable ride.

Additional information on dynamic drive, including system troubleshooting, can be found in **300 Suspension, Steering and Brakes–General**.

Steering

The vehicle-speed-sensitive variable-assist power rack-and-pinion steering system was standard on all models. Together with the redesigned suspension and subframe, the steering reaction to inputs is precise and sharp.

BentleyPublishers.com
BentleyPublishers.com—All Rights Reserved

Active front steering

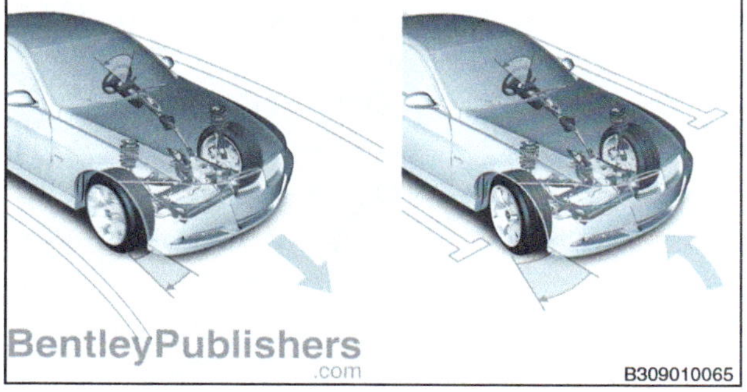

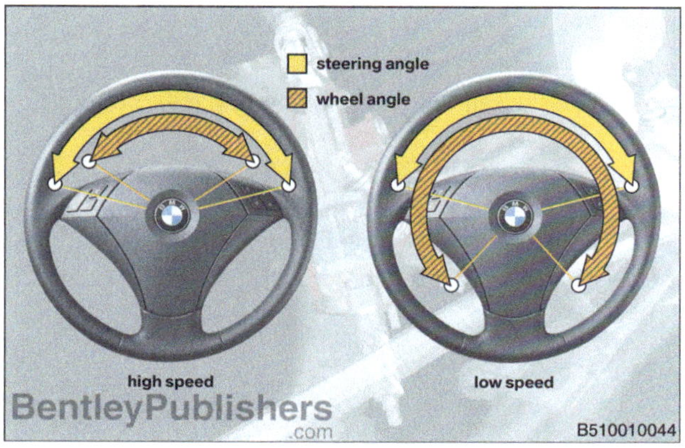

Compound brake rotor

Active steering

Active steering is a revolutionary development from BMW (included in Sport Package option, rear-wheel-drive models). Active steering offers many benefits:

- Wide variable steering ratio; steering-wheel movements for parking maneuvers, U-turns and sharp corners are greatly reduced for amazing agility.
- Vehicle stabilization; in critical situations, can intervene to preserve stability. For example, if the driver applies the brakes while driving on a surface on uneven traction (one side of the road slippery, the other grippy), active steering can steer against the instability.

This sophisticated system electronically varies the degree to which the front wheels turn in relation to steering input from the driver. Under normal driving conditions at low and medium speeds the steering ratio becomes more direct, meaning the wheels turn more in relation to input from the driver, reducing the amount the steering wheel must be turned. Fewer than 2 turns are needed to move the steering wheel from one lock to the other.

At high speeds, on highways for instance, the steering gear ratio becomes increasingly more indirect. The simultaneously increasing steering wheel torque level prevents unintentional steering movements and the driver can feel an improvement in directional stability.

Brakes

The E60 has a hydraulic dual-circuit brake system with front to rear split. The electric precharge pump for the DSC function is no longer used. The brakes use ventilated cast iron rotors all around, and aluminum calipers at the front as a contribution to balanced weight distribution. Braking power is enhanced by large diameter rotors.

A new coating eliminates rusting on portions of the rotors not swept by the brake pads.Separate front and rear pad wear sensors allow the mileage remaining on the front and rear pads to be displayed and appropriate service scheduled. See **020 Maintenance** for accessing the CBS (condition based service) menu.

Some models use lightweight compound brake rotors made of two components: the high-carbon cast iron outer portion, which functions conventionally, and an aluminum hat in the center Advantages include:

- Reduced unsprung weight, complementing the aluminum suspension system. Actual reductions are approximately 2.2 lb. at the front, 1.5 lb. at the rear.
- Reduced rotor deformation under hard braking, meaning less tendency of the brakes to vibrate when very hot.

BentleyPublishers
BentleyPublishers.com – All Rights Reserved

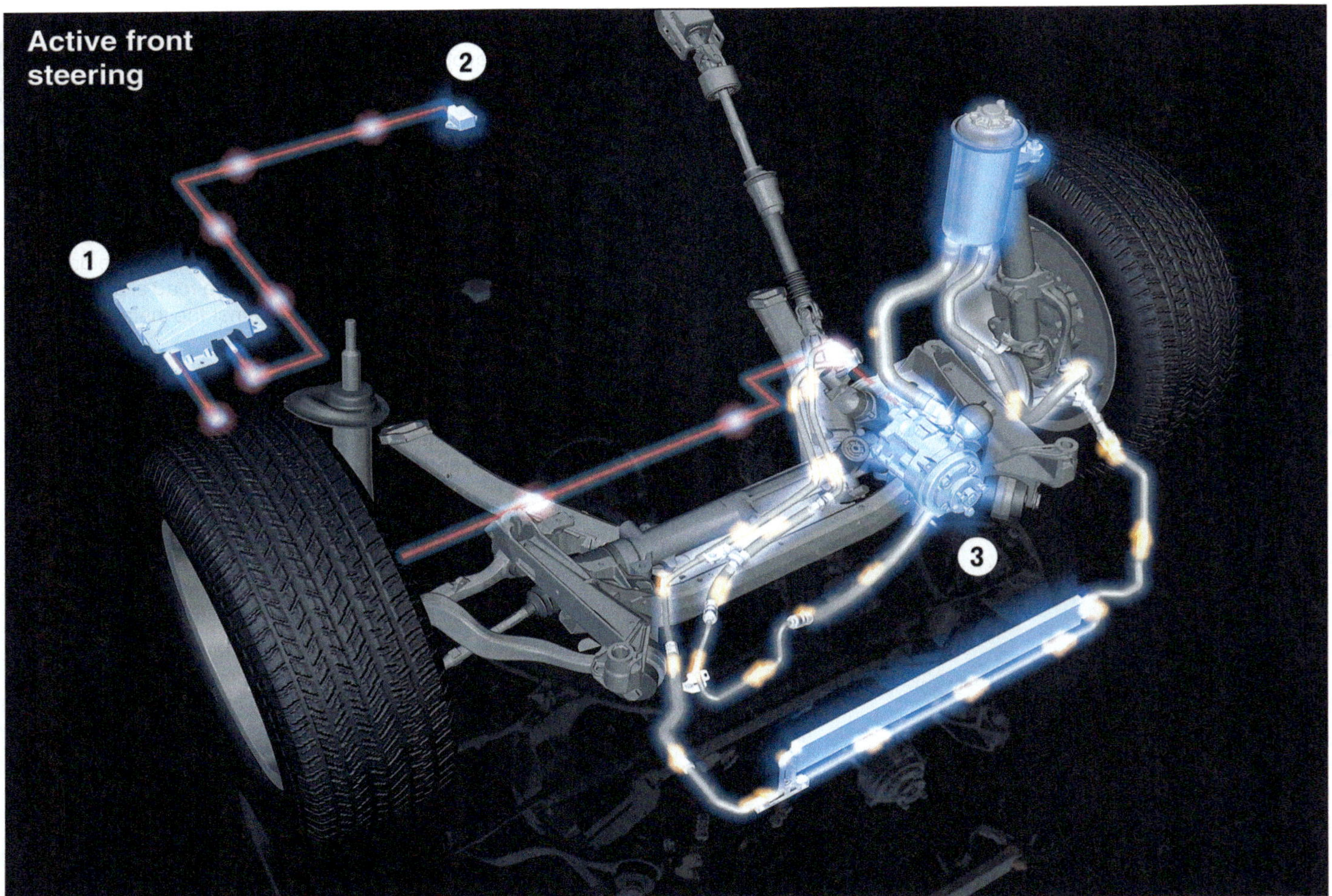

BentleyPublishers
1. Active front steering control module 2. Yaw sensor 3. Active front steering actuating unit B510010045

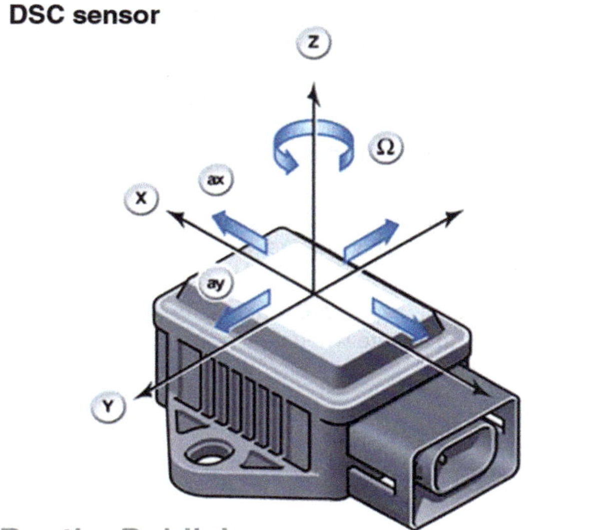

BentleyPublishers
.com
B510010048

Dynamic stability control (DSC)

The 5 Series dynamic stability control (DSC) system is referred to as DSC8 (DSC8+ on later models). The DSC control unit uses vehicle speed, steering angle and lateral acceleration signals to calculate vehicle yaw while cornering. DSC intervention is performed depending on whether the vehicle is oversteering or understeering. The DSC function can be deactivated by means of the DSC button.

The DSC sensor is a key component in the DSC8 system. It is installed under the front passenger's seat next to the transmission tunnel. The DSC sensor contains the yaw rate sensor (Ω), a transverse acceleration sensor (**ay**), a longitudinal acceleration sensor (**ax**) for the drive-off assistant function.

The following advanced functions contribute to increased directional stability, optimized comfort, enhanced system availability and response plus reduced braking distance.

- **ABS** Antilock braking system
- **ASC** Automatic stability control
- **MSR** Engine drag torque control
- **DSC** dynamic stability control

BentleyPublishers.com
BentleyPublishers.com—All Rights Reserved

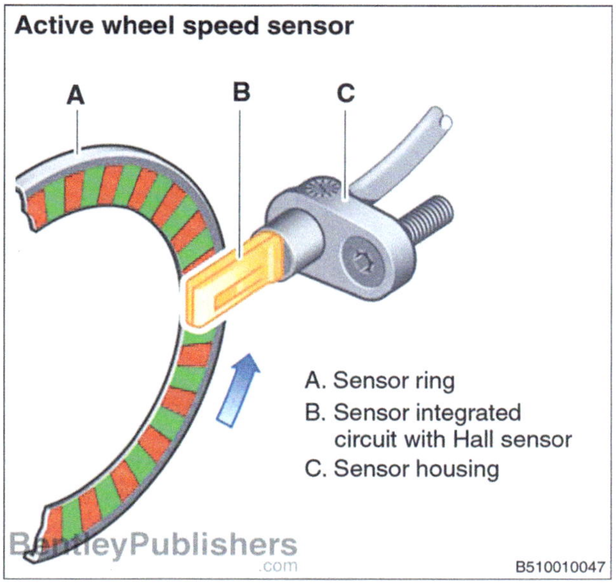

- **DBC** Dynamic brake control
- **CBC** Cornering brake control
- **ECD** Electronically controlled deceleration (active cruise control only)
- **EBV** Electronic brake-force distribution
- **DTC** Dynamic traction control
- **RPA** Tire defect indicator
- BBV Brake-pad wear indication

There are additional DSC8+ functions for 2006 and later cars: Dry braking, brake standby, automatic soft-stop. fading warning and assistance, drive-off assistant, hill descent control

Additional xDrive DSC (also called DCX) functions:

- **TCC** Transfer case control (control of multi-disc clutch in transfer case)
- **ASC-X** Automatic stability control X (special function for all-wheel drive vehicles)
- **ADB-X** Automatic differential brake X (special function for all-wheel drive vehicles)
- **HDC** Hill descent control

Body

A new BMW platform always means exciting new technology and innovative, meaningful features. The most noteworthy feature of the E60 body is the lightweight aluminum chassis technology.

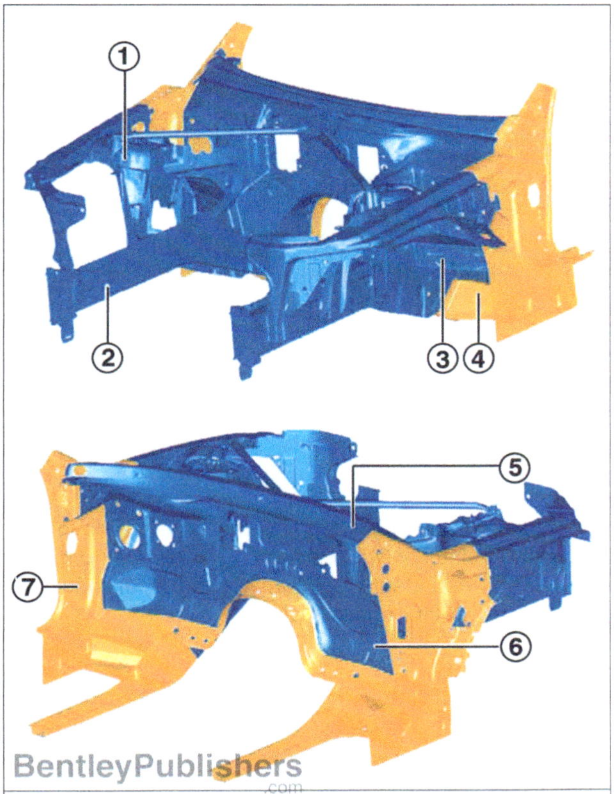

1. Spring Support
2. Engine Support
3. Bulkhead carrier support
4. Outer connection (steel)
5. Bulkhead Cross Member
6. Bulkhead
7. Inner A-pillar (steel)

B510010055

◄ In order to make the E60 significantly lighter, BMW designed a reduced weight aluminum front end (GRAV). The front end consists almost entirely of aluminum (blue shading) while the transition to steel occurs in the front bulkhead area.

The weight saving at the front end not only reduces the total weight of the vehicle but is also a significant factor in the E60 attaining the ideal axle load distribution of 50 / 50.

The transition from the aluminum front engine support rail to the steel rail section joint is produced by bonding and punch riveting. To increase conductivity, ground points are located on the two rails and are connected with a ground strap.

In addition to the underhood strut tower reinforcement brace, the bumper carrier and an aluminum front end reinforcement panel provide additional reinforcement and structural rigidity.

BentleyPublishers.com—All Rights Reserved

GRAV (aluminum front end)

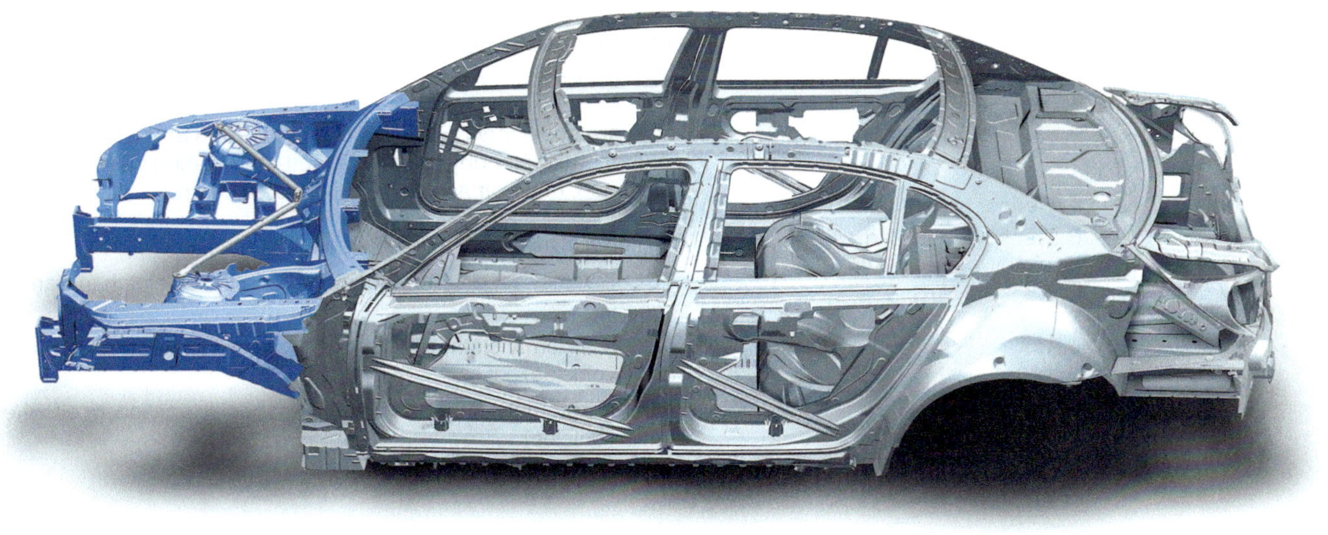

BentleyPublishers.com

B510010049

B510010051

Sports Wagon (E61)

The Sports Wagon offers essentially the same accommodations and features as the Sedan, including the aluminum front end. Front and rear leg room are identical for the two body styles; the Sports Wagon's standard split folding rear seats offer essentially the same functions and benefits as those that are optional in the Sedan.

The big difference is aft of the rear seats where, in terms of load-carrying ability and versatility, the Sports Wagon offers many advantages over the Sedan, such as a high-lift tailgate and separately opening rear window, for convenient loading without opening the tailgate.

The rear hatch is set at a low level in the interests of a low loading sill. The tailgate is comprised of two segments (the rear window can be opened separately). An indentation in the rear bumper reduces the height of the loading sill for easier loading.

BentleyPublishers.com
BentleyPublishers.com—All Rights Reserved

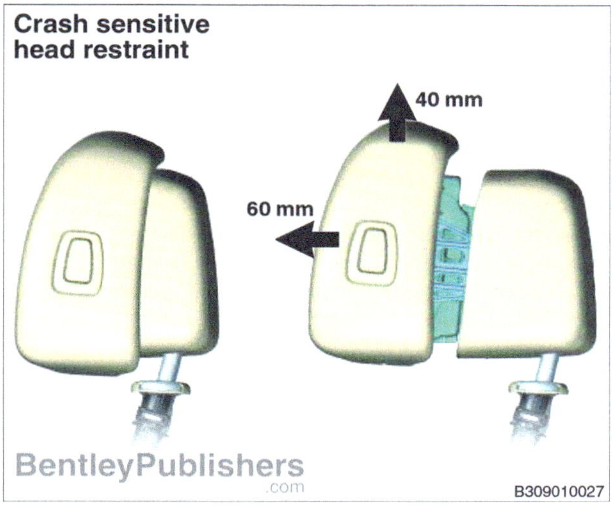

Interior

On the inside, the 5 Series continues the harmonious balance between sportiness and elegance, with an ergonomically designed cockpit for the driver and purposeful materials carefully chosen to please the senses.

Some of the interior highlights include:

- Greater space for passengers and cargo compared to pervious 5 Series: 0.5 in. more shoulder room and 0.3 in. more head room for front-seat occupants. 1.3 in. more shoulder room, 0.7 in. more head room, 1.8 in. more leg room for rear-seat passengers and a 26% increase in trunk volume.
- Evolved iDrive system with optional Navigation: A simplified version of the iDrive system first seen in the 7 Series, with a 6.5-in. color monitor* displaying 256 colors at 400 x 240-pixel resolution. **2010 models with the navigation used a larger screen.*
- Refined automatic climate control, standard rain-sensing windshield wipers.
- Advanced safety and security systems, including front and rear head protection, and optional active (crash sensitive) head restraints.

BentleyPublishers.com
BentleyPublishers.com–All Rights Reserved

Airbag systems

The latest generations of the airbag systems provide enhanced passive protection for vehicle occupants in case of a serious collision. Two system variants were used on the E60 and E61 cars, depending on model year.

- Advanced Safety Electronics (ASE) on 2004 - 2005 models
- Advanced Crash and Safety Management (ACSM) on 2006 - 2010 models.

The differences between the two systems relate mainly to the updated electrical plan on he 2006 and later cars. These sophisticated safety systems differ from previous MRS systems in that the reaction time is decreased and accuracy is increased due to the addition of the door pressure sensors and crash sensors in the engine compartment. See **721 Airbag System (SRS)** for additional information.

ELECTRICAL SYSTEM

Multiple control modules are involved in the operation of the vehicle electrical system. BMW calls this the distributed functions approach. There are several systems (i.e. central locking, power windows and interior lighting) whose functions are distributed across multiple control modules.

The primary control modules are:

- **SGM** Safety gateway module (2004 - 2006)
- **TMFA** Driver's door module (2004 - 2005)
- **TMBF** Passenger door module (2004 - 2005)
- **MPM** Micro power module (2004 - 2005)
- **KGM** Body gateway module (2006 - 2010)
- **CA** Comfort access (2006 - 2010)
- **KBM** Body basic module
- **CAS** Car access system
- **LM** Light module

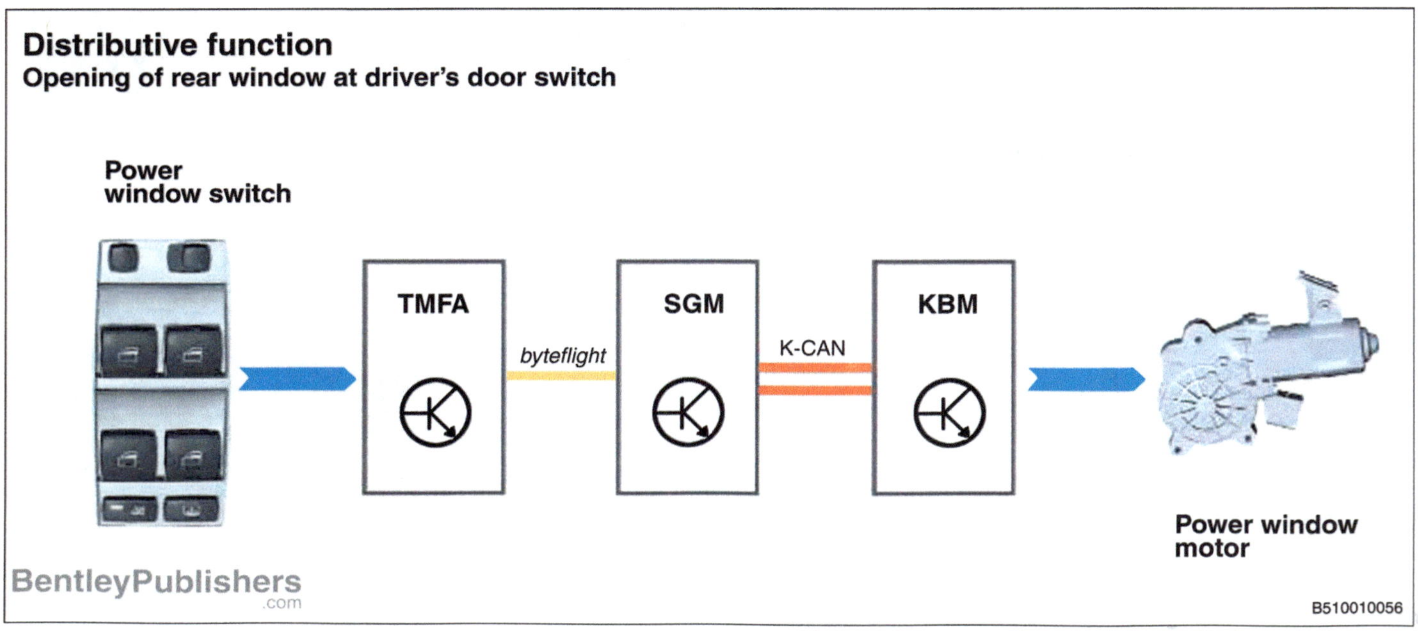

BentleyPublishers.com—All Rights Reserved

Power management and voltage supply

Supply voltage is monitored during both driving and parked conditions to reduce the possibility of unwanted closed current draws and insure that adequate battery power is available for starting. The state of charge (SoC) and the state of health (SoH) of the battery are determined continuously with the E60 power management system.

The power supply system consists of the following components:

- Vehicle ground points: Owing to the aluminum front end construction, additional grounding techniques are employed improves the electromagnetic compatibility (EMC) of the vehicle.
- Battery and battery cable.
- Intelligent battery sensor (IBS) with ground lead: A sensor installed directly at the negative battery terminal. It constantly measures battery terminal voltage, battery charge / discharge current, battery acid temperature and reports this information to the DME.
- Terminal 30g relay prevents increased closed-circuit current consumption by switching off electric loads in a defined manner. This happens approx. 60 minutes after turning the ignition OFF (terminal R off). Terminal 30g relay is actuated by car access system (CAS).
- Micro power module (2004 - 2005 only): In the same way as with terminal 30g, the micro power module (MPM) facilitates defined deactivation of electric loads. The MPM is installed in the spare wheel recess.
- DME: Power management software is contained in the DME. When the vehicle is at rest, the IBS is partially responsible for power management.
- Car access system (CAS): Responsible for ignition and starter functions, remote key, immobilizer (EWS), central locking, power windows, condition based service (CBS), as well as many other distributive electrical functions.

See **600 Electrical System–General** for additional information on the electrical system and buses.

BentleyPublishers
BentleyPublishers.com—All Rights Reserved

Typical voltage supply for 6-cylinder models

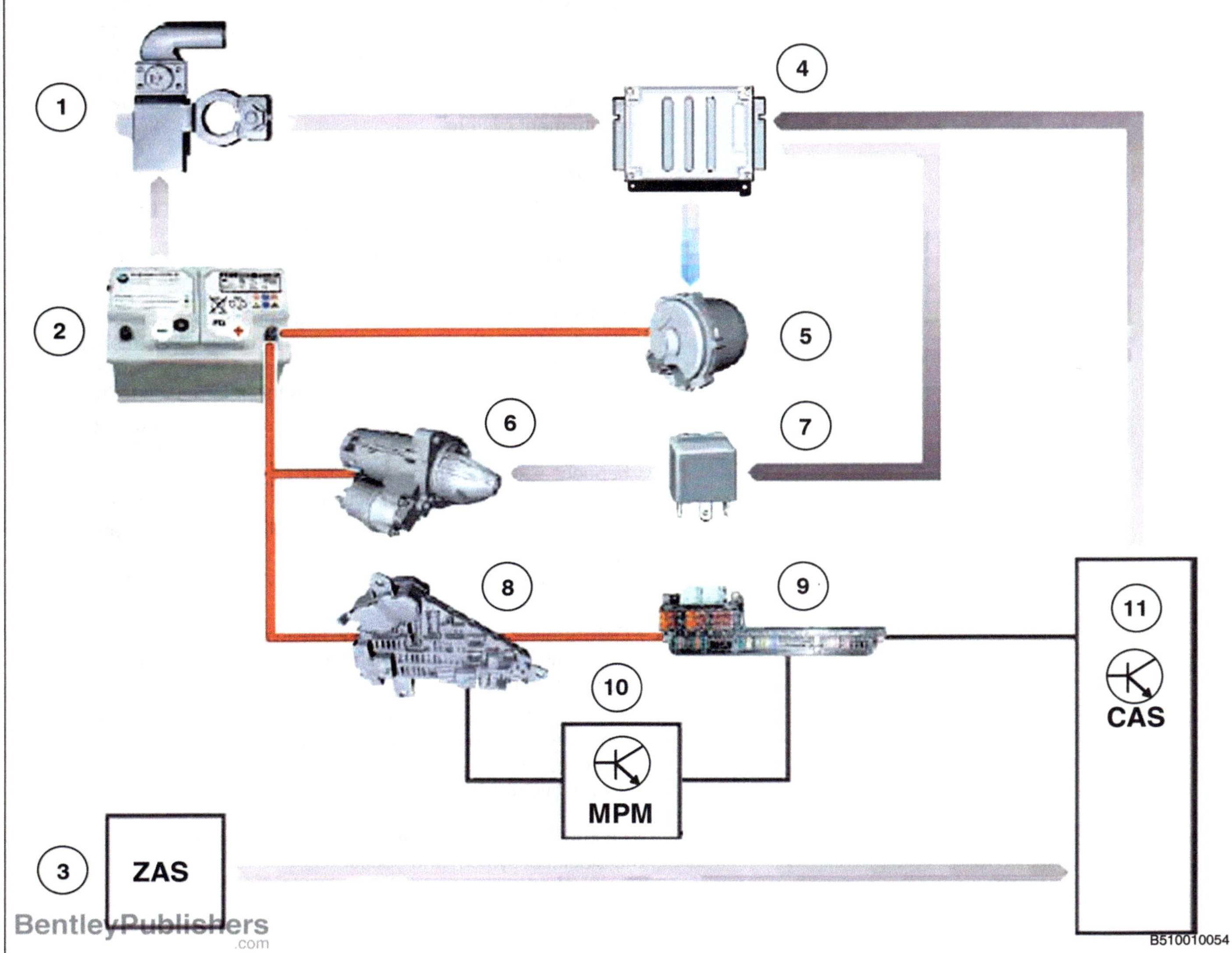

BentleyPublishers.com

1. Intelligent battery sensor (IBS)
2. Battery
3. Ignition and starter switch (ZAS)
4. DME
5. Alternator
6. Starter
7. Starter relay
8. Rear fuse and relay panel with terminal 30g
9. Front fuse and relay panel
10. Car access system (CAS)
11. Micro power module (2004 - 2005)

BentleyPublishers.com
BentleyPublishers.com—All Rights Reserved

Dashboard and controls overview

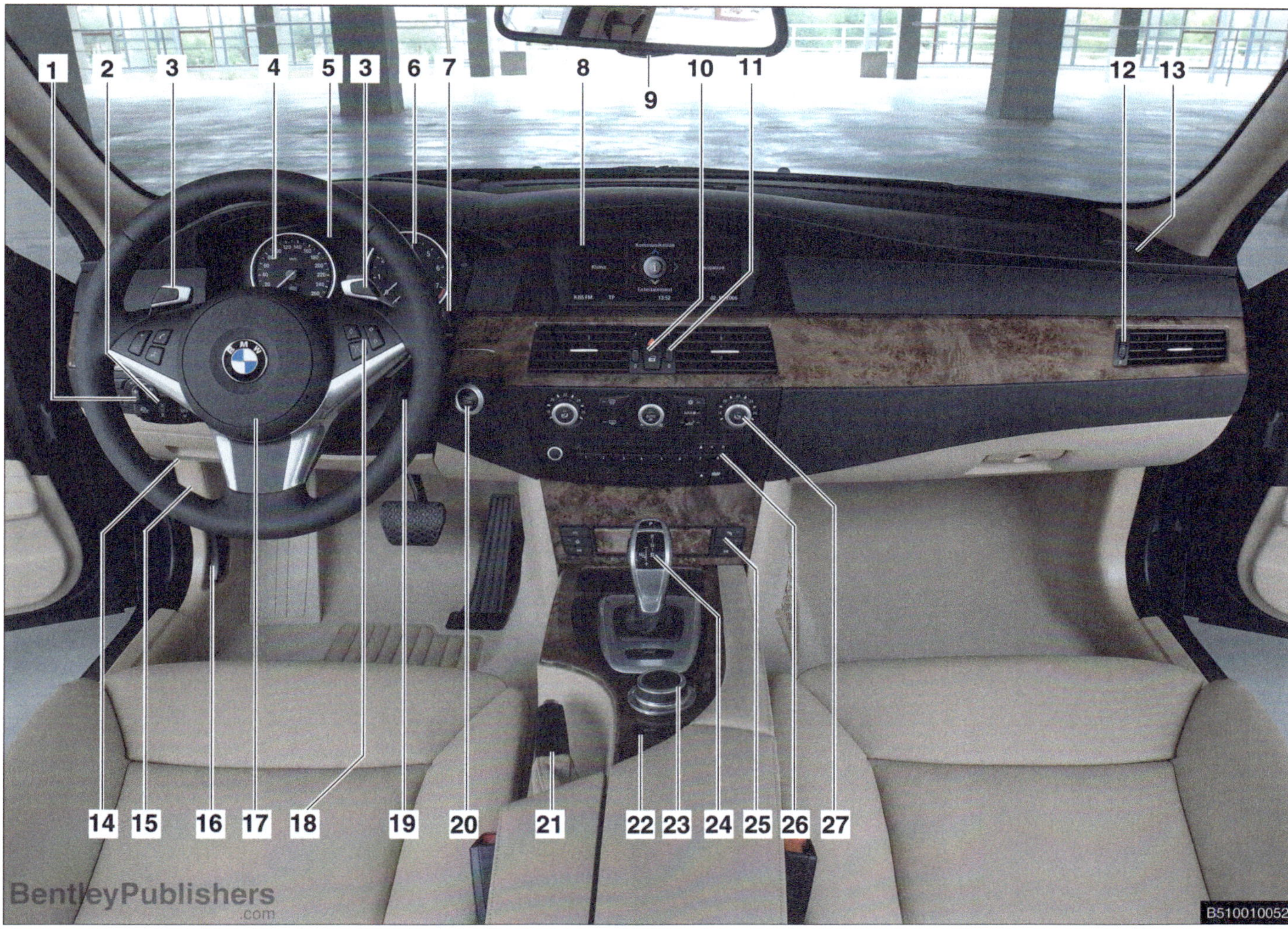

1. Light switch
2. Cruise control stalk switch
3. Paddle shifter
4. Speedometer, fuel gauge
5. Instrument cluster LCD screen
6. Tachometer, engine coolant temperature gauge
7. Wiper and washer stalk switch
8. Navigation and iDrive display screen
9. Central locking and anti-theft alarm indicator
10. Hazard warning and central locking switch
11. Center air outlets adjuster
12. Side air outlet adjuster
13. Side window defroster outlet
14. Trunk lid or tailgate release button
15. OBD II plug cover
16. Engine hood release
17. Horn, airbag
18. Multifunction steering wheel switches
19. Remote key slot
20. START STOP switch
21. Parking brake handle
22. iDrive main menu button
23. iDrive controller
24. Gear shift switch (GWS)
25. Center console switch cluster
26. Radio and navigation control
27. Climate control knobs

NOTE—
- *Features and equipment may vary from the illustration.*

BentleyPublishers
BentleyPublishers.com—All Rights Reserved

B510010057

BentleyPublishers.com
BentleyPublishers.com—All Rights Reserved

020 Maintenance

TABLES

General

This repair group details service and maintenance procedures for the 2004 through 2010 BMW 5 series (E60 and E61).

BMW 5 Series vehicles are equipped with condition based service (CBS) capabilities that monitors many vehicle systems and components and suggests service based on the actual current condition of components. Following the CBS recommended service intervals helps ensure safe and dependable operation of the vehicle.

Aside from keeping your vehicle in the best possible condition, proper maintenance plays a role in maintaining full protection under BMW's new-vehicle warranty coverage. If in doubt about the terms and conditions of your vehicle's warranty, an authorized BMW dealer should be able to explain them.

BMW is constantly updating their recommended maintenance procedures and requirements. The information contained here may not include updates or revisions made by BMW since the publication of the documents supplied with the vehicle. If there is any doubt about what procedures apply to a specific model or model year, or what intervals to follow, consult an authorized BMW dealer.

Warnings, Cautions and Notes

Throughout this manual, there are numerous paragraphs with the headings **WARNING, CAUTION** or **NOTE**. These headings have different meanings.

WARNING—
- *Text under this heading warns of unsafe practices that are very likely to cause injury, either by direct threat to the person(s) doing the work or by increased risk of accident or mechanical failure while driving.*

CAUTION—
- *Text under this heading calls attention to important precautions to be observed during the repair work to help prevent accidentally damaging the vehicle or its parts.*

NOTE—
- *A note contains information, tips, or pointers which help in doing a better job and completing it more easily.*

Read **WARNING, CAUTION** and **NOTE** headings before you begin repair work. See also **001 Warnings and Cautions**.

BentleyPublishers.com
BentleyPublishers.com—All Rights Reserved

CONDITION BASED SERVICE (CBS)

CBS basics

The BMW maintenance system, called condition based service (CBS) notifies the driver of needed service based on the condition of monitored components. In other words, CBS recommends maintenance only for components that are worn.

CBS uses sensors and special algorithms to monitor vehicle driving conditions, then uses this data to calculate current and future service requirements.

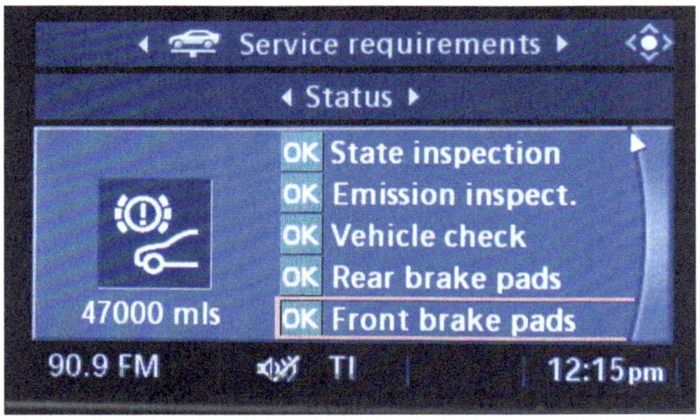

CBS tasks that are monitored are displayed in the instrument cluster or the central control display. The display shows remaining times or distances for the following maintenance tasks.

- Engine oil and filter change
- Front brake pads replacement
- Rear brake pads replacement
- Brake fluid change
- Cabin microfilter replacement*
- Spark plug replacement*
- Vehicle check
- State / emissions inspections.

**The CBS instrument cluster display intervals for spark plugs and cabin microfilter were discontinued as of 2008 models. These items still require periodic replacement. See maintenance tables in this repair group.*

CBS sorts service requirements according to date due and notifies the driver when the vehicle needs a service appointment. Mandatory emissions and safety inspection due dates may also be stored and displayed.

NOTE—

- *Additional maintenance recommended by BMW is listed in* **Table b** *and* **Table d**, *in this repair group.*

Remote with integrated key (2006 and later models)

Service requirement information is stored in the remote key while the vehicle is driven. This information can be accessed by an authorized BMW device center at the time of a service appointment.

For CBS to work correctly, make sure correct local date and time are set in vehicle memory.

BentleyPublishers.com
BentleyPublishers.com—All Rights Reserved

Service recognition

The CBS system recognizes when a service is required using one of the following three methods:

Adaptive. Sensor data and algorithms are used to determine the required maintenance interval for a service based on individual user driving style. Examples:

- Oil change interval is determined by engine oil condition sensor (OZS) monitoring oil condition, oil level, and oil temperature; and from additional algorithms using these parameters: engine load, fuel consumption, time and distance since the last oil change.
- Front and rear brake pad replacement intervals are determined from two-stage brake lining wear sensors on left front and right rear brake pads; and from algorithms using these input parameters: travel distance, wheel speed, braking pressure, braking time, and braking frequency.

Fixed time (months) determines the required maintenance of the following:

- Brake fluid change
- Mandated safety and emissions inspections

Fixed distance (miles) determines the required maintenance of the following:

- Spark plug replacement (dependent on engine)
- Air filter element replacement
- Automatic transmission fluid change
- Microfilter replacement (dependent on vehicle model and CBS version)
- Oxygen sensors replacement
- Vehicle check

CBS items, viewing

The instrument cluster control module communicates service requirements to the central dashboard display where data can be displayed with graphics and text.

NOTE—

- *This procedure is for 2004 - 2009 5 series cars with CID (central information display). Procedure may vary slightly on 2010 and later models fitted the high-definition CIC (car information computer) system and multi-button iDrive controller.*

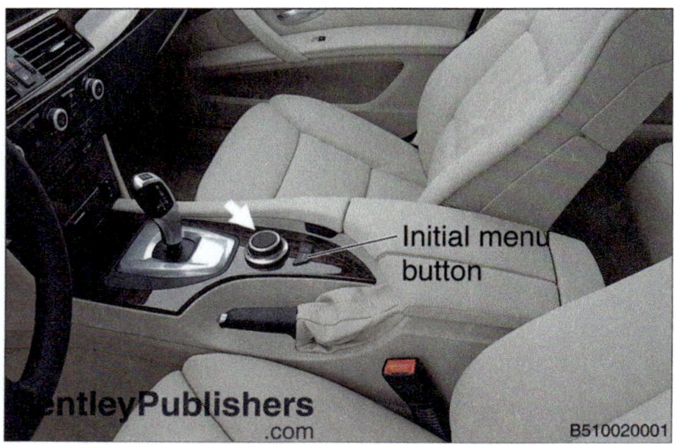

– Switch ignition ON.

◄ Once main screen appears, press down on iDrive controller (**arrow**) to select **Info sources | Settings** menu.

NOTE—

- *Press menu button anytime to return to initial menu.*

BentleyPublishers.com
BentleyPublishers.com—All Rights Reserved

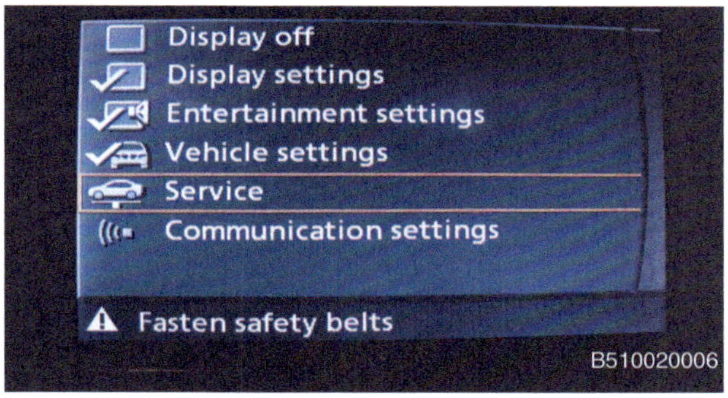

Turn iDrive controller to select **Service**.

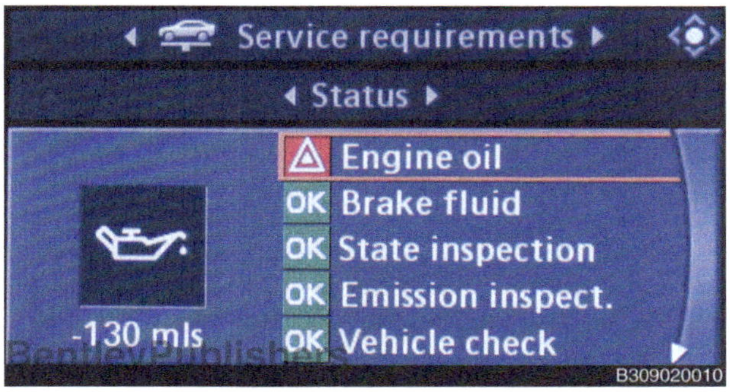

Turn iDrive controller to select maintenance items.

- Service items highlighted in green: No service currently required
- Service items highlighted in yellow: Service deadline is approaching
- Service items highlighted in red: Service deadline has passed

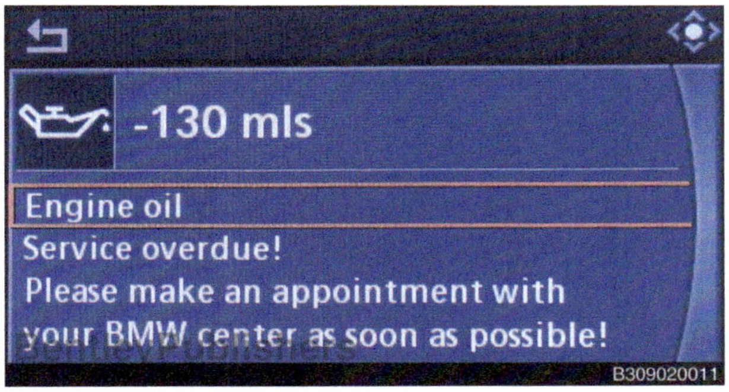

Press down on controller to access additional information for each listed task.

Example: Select **Engine oil** then press down controller to enter **Engine oil** screen. Here you can see engine oil service is overdue by 130 miles.

Push forward (**1**), then down (**2**) on controller to return to **Service Info** menu.

BentleyPublishers.com
BentleyPublishers.com—All Rights Reserved

CBS items, resetting

After carrying out maintenance tasks, e.g. changing engine oil, reset CBS item to its full service interval. This can be done using the BMW factory scan tool, or manually using the instrument cluster set / reset button as described below.

– Enter vehicle and close all doors.

– Switch ignition ON, but do not start engine.
  - Make sure lower displays in cluster shows clock and mileage.

– Press and hold instrument cluster set / reset button for 10 seconds (**arrow**) to enter CBS reset mode.
  - Lower LCD display: Remaining driving distance and date of next scheduled service shown. Service requirements are displayed in order of priority.
  - ⚠ indicates that Check Control messages are stored. See **Check Control** in this repair group.

– Press set / reset button repeatedly to scroll through CBS icons. Hold button until **RESET?** appears. Press button again and hold until distance / time display in the lower window changes to dashes.

– To confirm reset, press BC button while looking at CBS reset icon. Acknowledgment of reset is shown with **OK**.

1. Vehicle check. See **Table d**.
2. Next oil change due. See **Engine oil service** in this repair group.
3. Next safety inspection and road test due.
4. Next emissions inspection due.
5. Next front brake service due. See **Brake Service** in this repair group. See also **340 Brakes**.
6. Next rear brake service due. See **Brake Service** in this repair group. See also **340 Brakes**.
7. Next brake fluid change due. See **340 Brakes**.

NOTE—
- *On 2004 - 2007 cars, a spark plug icon and a cabin microfilter icon are displayed in the CBS menu. On 2008 and later cars, these items are deleted from the display.*

BentleyPublishers.com
BentleyPublishers.com—All Rights Reserved

Car information computer (CIC)

2010 models with navigation option are equipped with a new version of iDrive with multiple menu buttons. The display for this system is known as car information computer (CIC). Functionally, system operations are similar to the earlier version.

RECOMMENDED MAINTENANCE

Table a lists maintenance services prompted by CBS.
Table b lists service and maintenance required at specified mileage intervals (not prompted by CBS).
Table c gives the details of an engine oil service.

> ***CAUTION—***
> - *The publisher of this manual recommends that engine oil and filter be replaced a intervals not exceeding 7500 miles or one year, whichever comes first.*

Table d lists other recommended periodic vehicle inspection tasks.

The number in the "additional repair information" column refers to the repair group in this manual where additional information can be found.

Table a. CBS prompted service	**Additional repair information**
Engine oil service (CBS prompts based on engine oil condition.)	**020**
Front brake pads: Replace. (CBS prompts based on pad thickness.)	**340**
Rear brake pads: Replace. (CBS prompts based on pad thickness.)	**340**
Brake fluid: Change / flush every 2 years. (CBS prompts based on elapsed time since last change)	**340**
Spark plugs: Replace. (CBS prompts based on elapsed mileage.)*	**020**
Cabin microfilter: Replace. (CBS prompts based on elapsed mileage.)*	**020**
State inspection: Emissions (CBS prompts based on elapsed mileage, per state law.)	
State inspection: Safety (CBS prompts based on elapsed mileage, per state law.)	
Vehicle check (CBS prompts after every second oil change.)	
*Items discontinued from CBS menu as of 2008 models. • Spark plugs: Replace every 100,000 miles • Cabin microfilter: Replace at every engine oil service, or approximately every 15,000 miles.	

Bentley Publishers
BentleyPublishers.com - All Rights Reserved

Table b. Additional services	Additional repair information
Every 3rd engine oil service or 45,000 miles	
Air filter element: Replace. Reduce change interval in dusty conditions.	020
Every 100,000 miles	
Oxygen sensors, (N62 engine): Replace.	180
Automatic transmission fluid (ATF): Replace.	240
Every 120,000 miles	
Oxygen sensors (M54, N52, N54, N62TU engines): Replace.	180
Note: BMW states that engine coolant, manual transmission fluid, differential fluid and transfer case fluid are life-time fluids and do not require replacement.	

Table c. Engine oil service	Additional repair information
Engine oil and oil filter, change. See **119 Lubrication System**.	020
Brief diagnostic tests: • Check Control: Check messages. • Indicator and warning lights: Scan for fault codes using BMW scan tool or equivalent. • CBS display: Reset engine oil service interval. • Tires: Check tread, adjust pressures. • Tire pressure monitor: Reset (where applicable, 2006 and later cars).	020
Parking brake: Check operation, adjust as necessary.	340
Washer fluid, top off	

Table d. Vehicle check	Additional repair information
Horn, headlight flasher, hazard warning flasher: Check operation,	
Instrument cluster and control lighting, heater blower: Check operation	
Lighting system (dome lights, glove compartment light, flashlight, trunk or cargo compartment lights, turn signals, hazard warning flasher, back-up lights, brake lights, license plate lights): Check operation.	630
Safety belts: Check condition and function.	720
Wipers and washers: Check.	611
Body: Check for corrosion.	
Tires: Check tread depth, wear pattern, condition, inflation pressure, correct pressure as necessary.	
Run-flat indicator / tire pressure monitor: Initialize	300
Battery: Check state of charge, charge as required.	121
Power steering fluid: Check fluid level.	020, 320

BentleyPublishers.com
BentleyPublishers.com—All Rights Reserved

Table d. Vehicle check (continued)	**Additional repair information**
Airbags: Inspect for torn covers, damage or attached stickers.	**721**
Mirrors: Inspect.	**510**
Engine coolant: Check level and concentration.	**020, 170**
Windshield washer: Intensive cleaning system, check protection level, fluid level, top off if necessary.	
Brake system connections and lines: Check for leaks, damage and correct positioning.	
Underbody (transmission, rear axle, fuel lines, exhaust system): Check for damage, leaks and corrosion.	
Steering components: Check for clearance, leaks, damage and wear.	
Road test: Check for performance, braking, steering, suspension, noises.	

Basic Service Information

Diagnostic trouble codes (DTCs), accessing

Diagnostic trouble codes (DTCs) that are stored in the engine control module (ECM) or other electronic modules may be accessed as follows:

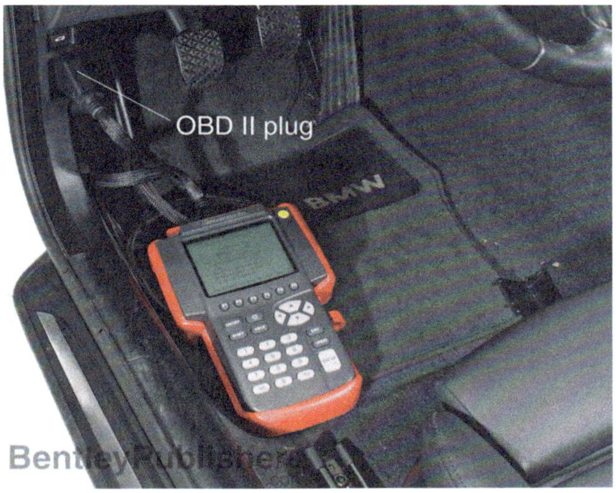

- Place transmission selector lever in PARK or NEUTRAL. Engage parking brake. Make sure ignition is OFF.

- ◄ Connect BMW scan tool to OBD II plug on left door post. Plug is behind plastic trim.

- Turn ignition key on.

- Follow scan tool instructions as they appear on scan tool screen.

- For a listing of scan tool suppliers, see **Tools** in this repair group. For additional information, see **OBD On-Board Diagnostics**.

Non-reusable fasteners

Many fasteners used on the vehicles covered by this manual must be replaced with new ones once they are removed. These include but are not limited to: bolts, nuts (self-locking, nylock, etc.), roll pins, clips and washers. Use genuine BMW replacement parts for this purpose.

Some bolts, known as torque-to-yield fasteners, are designed to stretch during assembly and are permanently altered, rendering them unreliable once removed. Replace these where instructed to do so. Failure to replace these fasteners could cause personal injury or vehicle damage.

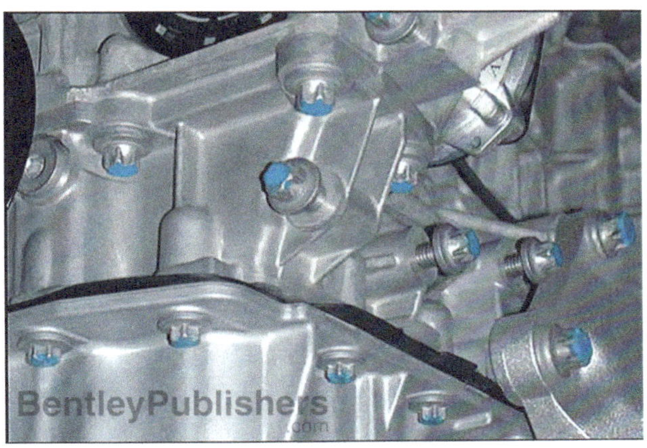

◄ Many components on the later 6-cylinder engines are assembled using aluminum bolts in order to reduce electrochemical corrosion. Most aluminum bolt heads are painted blue. These fasteners also stretch in use; replace with new during reassembly. Always use aluminum bolts where specified.

BentleyPublishers
BentleyPublishers.com—All Rights Reserved

Tightening fasteners

Tighten fasteners on a component gradually and evenly to avoid misalignment or over-stressing any one portion of the component. For components sealed with gaskets, this method helps to ensure that the gasket seals correctly.

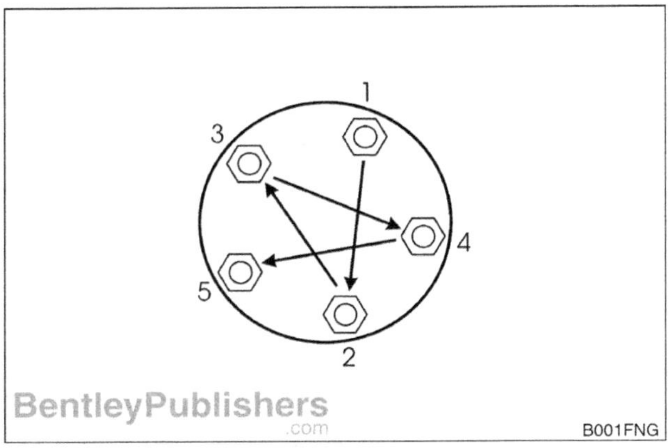

Where there are several fasteners, tighten them in a sequence alternating between opposite sides of the component. Repeat the sequence until all the fasteners are evenly tightened to the proper specification.

For some repairs a specific tightening sequence is necessary, or particular order of assembly is required. Such special conditions are noted in the text, and the necessary sequence is described or illustrated. Where no specific torque is listed, use **Table e** as a general guide for tightening fasteners.

NOTE—
- *Metric bolt classes or grades are marked on the bolt head.*
- *Do not confuse wrench size with bolt diameter.*
- *Values in* **Table e** *are for reference only.*

Table e. General bolt tightening torques in Nm (max. permissible)

	Bolt Class (according to DIN 267)		
Bolt diameter	**8.8**	**10.9**	**12.9**
M5	6	8	10
M6	10	14	17
M8	24	34	40
M10	47	66	79
M12	82	115	140
M14	130	180	220

Buying parts

Many of the maintenance and repair tasks in this manual call for the installation of new parts, or the use of new gaskets and other materials when reinstalling parts. Make sure the parts that are needed are on hand before beginning the job. Read the introductory text and the complete procedure to determine which parts are needed.

For some bigger jobs, partial disassembly and inspection is required to determine a complete parts list. Read the procedure carefully and, if necessary, make other arrangements to get the necessary parts while your vehicle is disassembled.

BentleyPublishers.com
BentleyPublishers.com—All Rights Reserved

Genuine BMW parts

Genuine BMW replacement parts from an authorized BMW dealer are designed and manufactured to the same high standards as the original parts. They are the correct material, manufactured to the same specifications, and guaranteed to fit and work as intended. Most genuine BMW parts carry a limited warranty.

Many independent repair shops make a point of using genuine BMW parts. They know the value of doing the job right with the right parts. Parts from other sources may be as good as BMW parts, particularly if manufactured by one of BMW's original equipment suppliers, but it is often difficult to know.

BMW is constantly updating and improving their vehicles, often making improvements during a given model year. BMW may recommend a newer, improved part as a replacement, and your authorized dealer's parts department will know about it. The BMW parts organization is best equipped to deal with your BMW parts needs.

Non-returnable parts

Some parts cannot be returned. The best example is electrical parts. Buy electrical parts carefully, and be as sure as possible that a replacement is needed, especially for expensive parts such as electronic control modules. It may be wise to let an authorized BMW dealer or other qualified shop confirm your diagnosis before replacing a non-returnable part.

Model and model year

When ordering parts it is important that you know the correct model and engine designation for your vehicle. This manual covers 5 Series sedan and wagon body styles with multiple engine variants in both rear wheel drive and xDrive. For information on engine codes and engine applications, see **100 Engine–General**.

Model year is not necessarily the same as date of manufacture or date of sale. A 2007 model may have been manufactured in late 2006 and perhaps not sold until early 2008. It is still a 2007 model. Model years covered by this manual are 2006 to 2010.

Service

BMW dealers are uniquely qualified to provide service for BMW vehicles. Their authorized relationship with the large BMW service organization means that they have access to special tools and equipment, together with the latest and most accurate repair information.

The BMW dealer's service technicians are highly trained and very capable. Authorized BMW dealers are committed to supporting the BMW product. On the other hand, there are many independent shops that provide quality repair work. Checking with other BMW owners for recommendations on service facilities is a good way to learn of reputable BMW shops in your area.

BentleyPublishers.com
BentleyPublishers.com–All Rights Reserved

Tools

Most maintenance can be accomplished with a small selection of tools. Tools range in quality from inexpensive junk, which may break at first use, to very expensive and well-made tools for the professional. The best tools for most do-it-yourself BMW owners lie somewhere in between.

Many reputable tool manufacturers offer good quality, moderately priced tools with a lifetime guarantee. These are your best buy. They cost a little more, but they are good quality tools that will do what is expected of them. Sears' Craftsman® line is one such source of good quality tools.

Some of the repairs covered in this manual require the use of special tools, such as a custom puller or specialized electrical test equipment. These special tools are called out in the text and can be purchased through an authorized BMW dealer. As an alternative, some specialty and scan tools may be purchased from the following tool manufacturers or distributors:

Specialty tool suppliers

Assenmacher Specialty Tools, Inc.
800-525-2943
www.asttool.com

Samstag Sales
615-735-3388
www.samstagsales.com

Autologic Diagnostics (UK)
44-1865-870050
www.autologic-diagnos.co.uk

Baum Tools Unlimited, Inc.
800-848-6657
www.baumtools.com

Tool Source
888-220-8350
www.toolsource.com

ZDMAK Tools
(877) 938-6657
www.zdmak.com

IDENTIFICATION PLATES AND LABELS

Vehicle identification number (VIN). This is a combination of letters and numbers that identify the particular vehicle. The VIN appears on the state registration document and on the vehicle itself. One location is on the right front strut tower in the engine compartment, another in the lower left corner of the windshield.

For additional information, see **002 Vehicle Identification and VIN Decoder**.

BentleyPublishers.com
BentleyPublishers.com—All Rights Reserved

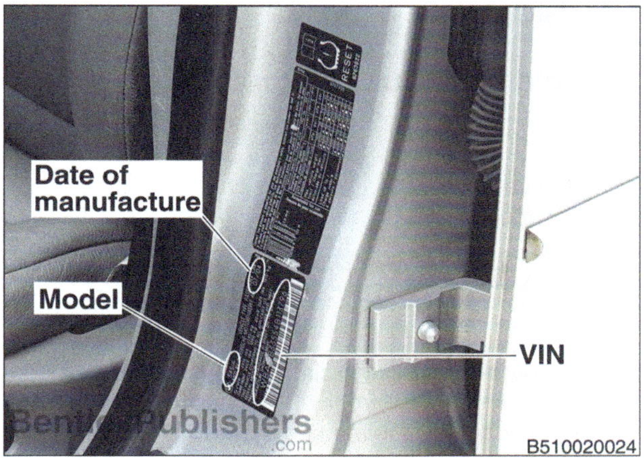

Model and date of manufacture. This information may be necessary when ordering replacement parts or determining if any of the warranty recalls are applicable to your vehicle. The VIN plate on the driver door jamb below the door striker also specifies model, month and year the vehicle was built.

Engine code. 5 Series vehicles covered in this manual are powered by either 6-cylinder or V8 engines. For information on engine codes and engine applications, see **100 Engine–General**.

Transmission code. The transmission type with its identifying code may be important when buying parts such as seals or gaskets. For information on transmission codes and applications, see **200 Transmission–General**.

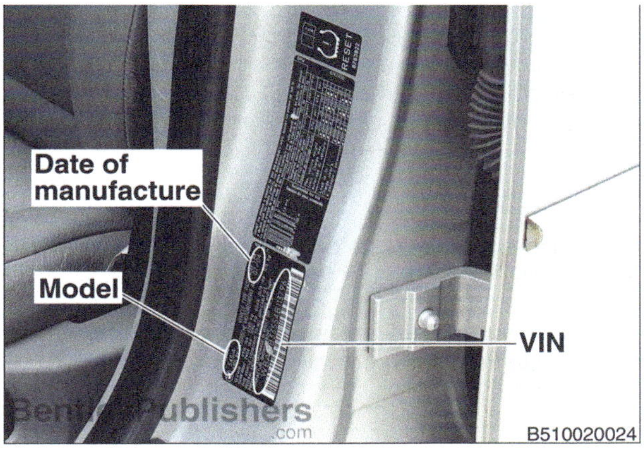

Tire pressures. Recommended tire pressures as well as wheel size, tire size, gross vehicle weight, and carrying capacity are listed on a sticker on the driver door jamb.

NOTE—

- *Observe tire pressure monitor (or flat tire monitor) sticker (**arrow**). On-board system should be reinitialized whenever tire pressure is adjusted. Reinitialize FTM or TPM. See* **300 Suspension, Steering and Brakes–General** *or your owner's manual for instructions.*

Engine serial number is on left side of engine block behind and under intake manifold. 6 cylinder engine shown.

BentleyPublishers.com
BentleyPublishers.com–All Rights Reserved

Raising Vehicle

Raising vehicle safely using floor jack

> ***WARNING —***
> - *When raising the vehicle using a floor jack or hydraulic lift, carefully position the jack pad to prevent damaging the vehicle body.*
> - *Watch the jack closely. Make sure it stays stable and does not shift or tilt.*

B510020027

◄ For safety and to avoid damaging vehicle, use jack only at four jacking pads (**arrows**) just behind front wheels and just in front of rear wheels. Jack pad locations are marked by small molded arrows on rocker panel cover.

> ***CAUTION—***
> - *Check to make sure the hard rubberized jack pads have not fallen off the vehicle. Do not jack the car up if the pad is missing. Damage to the vehicle may result.*

– Park vehicle on flat, level surface.

– If changing a tire, loosen lug bolts before raising vehicle. See **Changing a tire** in this repair group.

– Use wheel chocks to block wheel that is opposite and farthest from jack to prevent vehicle from unexpectedly rolling.

> ***WARNING —***
> - *Do not rely on the transmission or the parking brake to keep the vehicle from rolling. They are not a substitute for positively blocking the opposite wheel.*

B510020029

◄ Place floor jack into position, making sure base of jack is resting on flat, solid surface. Use a board or other support to provide a firm surface for jack, if necessary.

– Raise vehicle slowly while constantly checking position of jack and vehicle.

Raising vehicle using hydraulic lift

◄ Place lifting pads under jacking pads (**arrows**) just behind front wheels and just in front of rear wheels. Raise vehicle carefully, checking to make sure that it is solidly balanced on lift.

BentleyPublishers.com
BentleyPublishers.com—All Rights Reserved

Working under vehicle safely

> *WARNING—*
> - *A jack is a temporary lifting device. Do not use a jack alone to support the vehicle while you are under it.*
> - *Do not work under a lifted vehicle unless it is solidly supported on jack stands that are intended for that purpose.*
> - *Do not use wood, concrete blocks or bricks to support a vehicle. Wood may split. Blocks and bricks, while strong, are not designed for that kind of load and may break or collapse.*
> - *Use care when removing major (heavy) components from one end of the vehicle. The sudden change in weight and balance can cause vehicle to tip off the lift or jack stands.*
> - *Do not support vehicle at engine oil pan, transmission, fuel tank, or on front or rear axle. Serious damage may result.*

– Disconnect negative (-) cable from battery so that vehicle cannot be started. Let others know what you are doing.

> *CAUTION—*
> - *Prior to disconnecting the battery, read the battery disconnection cautions given in* **001 Warnings and Cautions**.

– Raise vehicle slowly. See **Raising vehicle safely using floor jack** in this repair group.

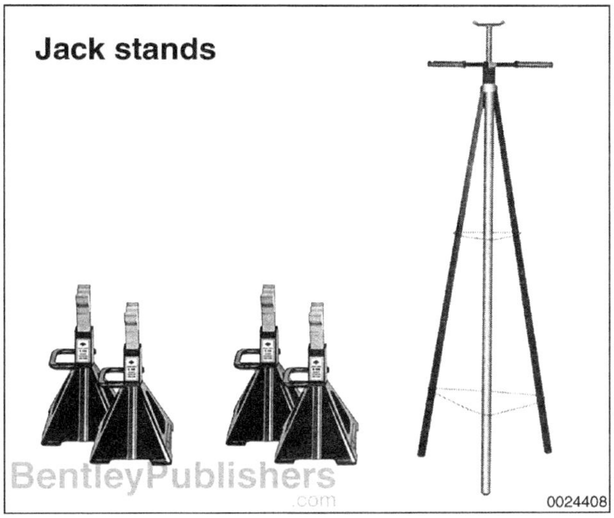

< Use at least two jack stands to support vehicle. Use jack stands designed for the purpose of supporting a vehicle.

– Place jack stands on firm, solid surface. If necessary, use a flat board or similar solid object to provide a firm footing.

– Lower vehicle slowly until its weight is fully supported by jack stands. Watch to make sure that the jack stands do not tip or lean as the vehicle settles on them.

– Observe jacking precautions again when raising vehicle to remove jack stands.

Emergencies

The following is a list of emergency procedures and where they are covered in this manual:

- **Flat tire** warning. See **Flat tire indication** in this repair group.
- **Fuel flap** manual release. See **515 Central Locking and Anti-theft**.
- **Jump start**. See **Jump starting** in this repair group.
- **Sunroof** manual closing. See **540 Sunroof**.
- **Trunk** emergency opening. See **412 Trunk, Tailgate**.
- **Tire** change. See **Tire change** in this repair group.
- **Towing**. See **Towing** in this repair group.

BentleyPublishers.com–All Rights Reserved

Indicator and warning lights

Many vehicle systems are self-monitored for faults while driving. Generally, a red warning light that illuminates while driving should be considered serious. If you cannot immediately determine the seriousness of the warning light, stop the vehicle in a safe place and switch engine OFF as soon as possible. Consult the owner's manual for additional information on the warning light and the recommended action.

If the malfunction indicator light (MIL) (Service Engine Soon warning light) comes on or flashes, it generally indicates an emissions-related fault in the engine management system. Faults such as a bad oxygen sensor or a faulty fuel injector can cause exhaust or evaporative emissions to exceed a specified limit. When these limits are exceeded, the MIL illuminates. The engine can be safely driven with the light on, but the faults should be interrogated with a scan tool as soon as possible. See **OBD On–Board Diagnostics** for more information on the MIL and the on-board diagnostic system.

Flat tire indication

Tire air pressure is automatically checked by either the flat tire monitor (FTM) system or the tire pressure monitoring (TPM) system, depending on model year and equipment installed.

The FTM system uses ABS wheel speeds sensors to calculate tire pressure. TPM system uses electronics in wheels to measure tire pressure.

If a flat tire or substantial loss of tire pressure is detected, warning lights in yellow and red illuminate in the instrument cluster and a tire failure message appears in Check Control and the center dashboard display. In addition, an acoustic signal sounds.

> ***WARNING —***
> - *In case of run-on flat tire failure, drive cautiously and do not exceed a speed of 50 mph or 80 km/h. With tire pressure loss, vehicle handling changes. This includes reduced tracking stability in braking, extended braking distance and altered steering characteristics.*
> - *If unusual vibration or loud noises occur while driving on a damaged tire, this may be an indication that the tire has failed completely. Reduce speed and pull over as soon as possible at a suitable location. Otherwise parts of the tire could come loose, resulting in an accident. Do not continue driving.*

Avoid sudden braking and steering maneuvers. At the next opportunity, check air pressure in all four tires. Tire pressure specifications are given on a sticker on the driver door jamb.

If all four tires are inflated to the correct pressures, reinitialize FTM or TPM. Initialization is covered in **300 Suspension, Steering and Brakes–General**.

If warning lights illuminate in yellow and a tire failure message appears in Check Control, FTM or TPM has a malfunction. Have the system checked as soon as possible.

BentleyPublishers.com
BentleyPublishers.com—All Rights Reserved

Jump starting

Jump start the vehicle with a discharged or dead battery using the good battery from another vehicle. See owner's manual for proper instruction on jump starting.

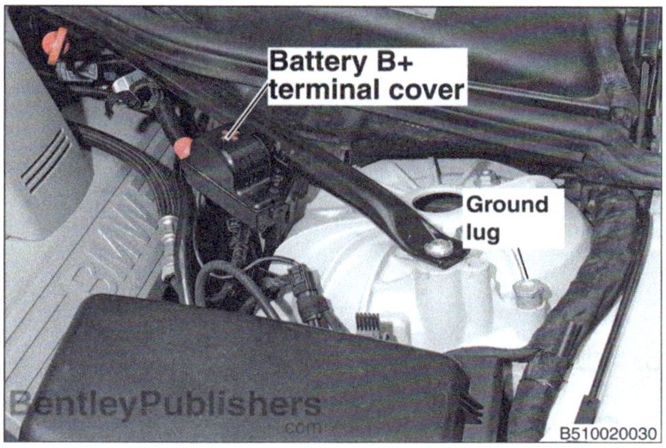

Attach jumper cables to engine compartment B+ terminal (flip cover down to access) and ground lug.

> ***WARNING —***
> • *Do not jump-start the engine if you suspect that the battery is frozen. Trapped gas may explode. Allow the battery to thaw first.*

> ***CAUTION—***
> • *To avoid damage to vehicle electronics, make sure the 12V battery on the car giving assistance is approximately the same capacitance (in Amp-hours) as the dead battery in your 5 Series. This information can be found on the battery.*
> • *Do not quick-charge the battery (for boost starting) for longer than one minute, and do not exceed 16.5 volts at the battery with the boosting cables attached. Wait at least one minute before boosting the battery a second time.*

Tire change

The following details a procedure for changing a tire using floor jack and hand tools. See owner's manual for instructions on using the vehicle-supplied jack.

NOTE —
- *On vehicles equipped with run-on flat tires, there is no need to change a wheel immediately in case of a puncture. See owner's manual for safe driving speed and distance for a run-flat tire with no air pressure.*
- *On cars with run-on flat tires, there is no spare tire or tire changing tools.*

– Chock wheel diagonally opposite to the one being changed.

> ***WARNING —***
> • *Do not rely on the transmission or the parking brake to keep the vehicle from rolling. They are not a substitute for positively blocking the opposite wheel.*

– Loosen lug bolts while vehicle is on ground, but leave them a little snug.

Place jack under jacking point nearest wheel being changed. Use a board to provide a firm footing for jack if ground is soft. Raise vehicle only far enough so that the wheel is fully off ground and then remove lug bolts and wheel.

– Install new wheel. Install lug bolts and tighten hand-tight using lug wrench.

BentleyPublishers.com
BentleyPublishers.com—All Rights Reserved

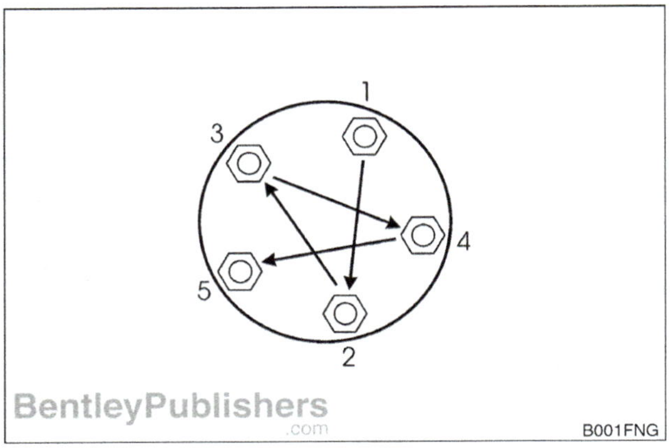

Lower vehicle. With all wheels on ground, fully tighten lug bolts in a crisscross pattern. Check inflation pressure of newly installed tire.

Tightening torque	
Wheel to wheel hub	120 ± 10 Nm (89 ± 7 ft-lb)

Towing

The safest technique for towing vehicles covered by this manual (rear wheel drive or xDrive, manual or automatic transmission) is by using flat bed equipment.

***WARNING*—**
- *Do not tow with rear wheel lifted and front wheels on the ground. There is no steering lock on some vehicles. Vehicle may self-steer and sway dangerously from side to side.*

***CAUTION*—**
- *Rear wheel drive: Do not tow with sling-type equipment. Use a lift bar. Do not use tow hook to tow faster than 30 mph (50 kph) or for a distance greater than 95 miles (150 km).*
- *Rear wheel drive with automatic transmission: Automatic transmission fluid (ATF) does not circulate when the vehicle is on rear wheels. Severe transmission damage may result. Use flat bed equipment.*
- *xDrive: Do not tow with front or rear axle raised individually. Do not use tow hook or cable except for very short distances to winch it to a safe place. Use flat bed equipment.*
- *Winch or tow vehicle with transmission lever in N (neutral).*

To access front threaded towing eye socket, press on lower edge of cover (**arrow**) on front bumper until cover disengages.

To access rear threaded towing eye socket, press on lower edge of cover (**arrow**) on rear bumper until cover disengages.

- Remove towing eye from cargo compartment tool kit and fully screw into threaded hole in bumper until tight.
- Use towing eye for winching vehicle on flat bed.

BentleyPublishers.com
BentleyPublishers.com—All Rights Reserved

Engine Compartment

Engine compartment inspection

Remove upper and lower engine covers as required and check engine compartment for signs of fluid leaks. Fluid leaks attract dust making them easier to spot. Expensive repairs can be avoided by prompt repair of minor fluid leaks.

Working from below, visually inspect for oil and ATF leaks at engine and transmission. Also inspect cooling, fuel, heating and air-conditioning systems for leakage. Visually inspect hoses and hose connections for leaks, worn areas, porosity and brittleness.

Check that fluid levels are between MIN and MAX marks.

The following illustrations show representative engine compartment layouts. There may be minor differences among model years, such as engine cover configuration.

M54 engine compartment

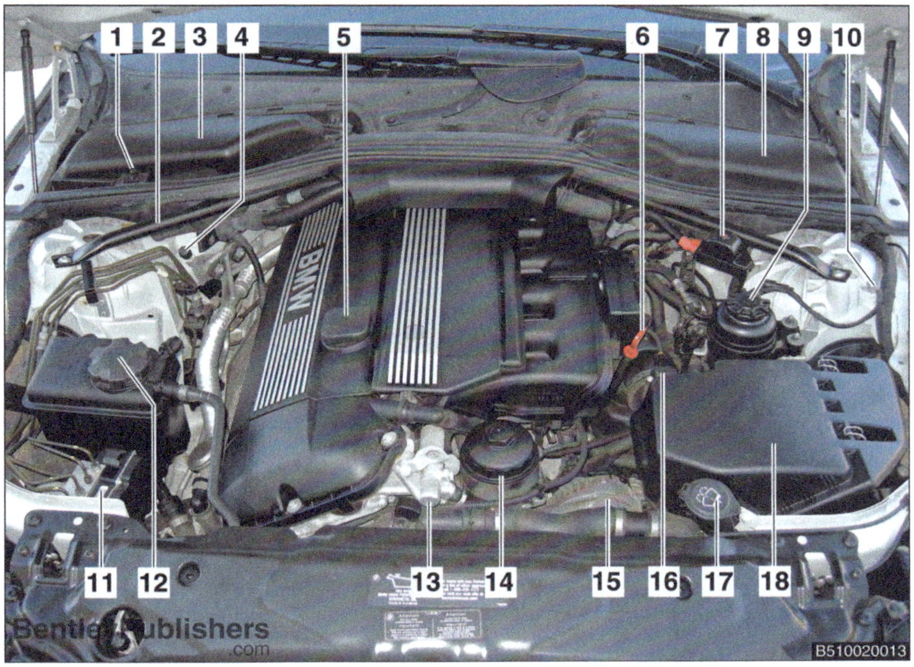

1. Hood alarm switch
2. Strut tower brace
3. Cabin microfilter, right (under cover)
4. A/C refrigerant port
5. Oil filler cap
6. Engine oil dipstick
7. Battery B+ junction
8. Brake fluid reservoir (under cabin microfilter housing)
9. Power steering fluid reservoir
10. Ground lug
11. ABS / DSC control module
12. Coolant reservoir cap
13. Intake VANOS solenoid
14. Oil filter housing
15. Alternator (Generator)
16. Mass air flow sensor
17. Windshield washer reservoir cap
18. Intake air filter housing

BentleyPublishers
BentleyPublishers.com–All Rights Reserved

N52 engine compartment

1. Hood alarm switch
2. Strut tower brace
3. Cabin microfilter, right (under cover)
4. A/C refrigerant port
5. Oil filler cap
6. Battery B+ junction
7. Brake fluid reservoir (under cabin microfilter housing)
8. Power steering fluid reservoir
9. Ground lug
10. ABS / DSC control module
11. Coolant reservoir cap
12. Valvetronic motor (under cover)
13. Oil filter housing
14. Engine oil cooler
15. Alternator (Generator)
16. Mass air flow sensor
17. Windshield washer reservoir cap
18. Intake air filter housing

N62 engine compartment

1. Strut tower brace
2. Hood alarm switch
3. Power steering fluid reservoir
4. A/C refrigerant port
5. Cabin microfilter, right (under cover)
6. Valvetronic motor, right (under cover)
7. Valvetronic motor, left (under cover)
8. Battery B+ junction
9. Brake fluid reservoir (under cabin microfilter housing)
10. Ground lug
11. ABS / DSC control module
12. Coolant reservoir cap
13. Oil filler cap
14. Mass air flow sensor
15. Windshield washer reservoir cap
16. Intake air filter housing

BentleyPublishers.com
BentleyPublishers.com—All Rights Reserved

Engine upper covers (M54 engine), removing

– Open engine hood.

– Remove cabin microfilter housings. See **Cabin microfilter housing, removing** in this repair group.

◁ Unfasten wire harness duct from strut tower brace:

- Unclip front of wiring harness cover and slide up and off.
- Remove rivets (left and right side) holding cable ducting to strut tower brace and unclip ducting from brace.

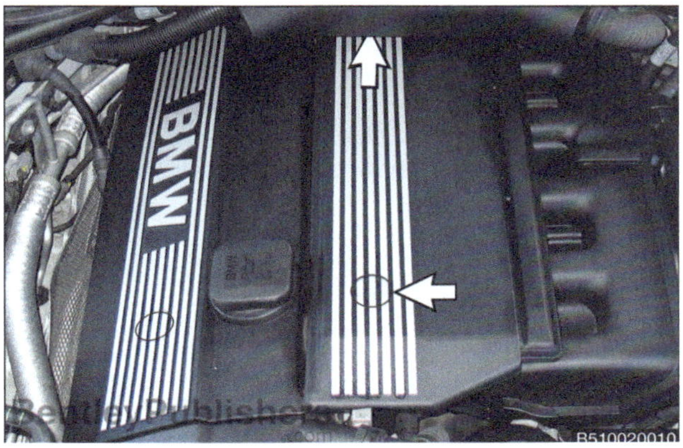

◁ Remove cover above fuel injectors:

- Remove plastic trim covers (**arrows**).
- Remove cover hold down bolts beneath covers and remove cover.

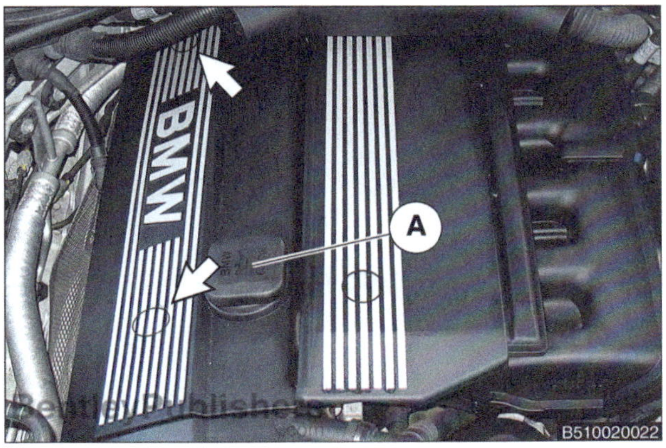

◁ Remove cover above ignition coils:

- Remove oil filter cap (**A**).
- Remove plastic trim covers (**arrows**).
- Remove cover hold down bolts beneath covers and remove cover.
- Reinstall oil filler cap to cylinder head cover.

BentleyPublishers.com
BentleyPublishers.com–All Rights Reserved

Engine upper cover (N52 / N54 engines), removing

– Open engine hood.

– Remove cabin microfilter housings. See **Cabin microfilter housing, removing** in this repair group.

< Unfasten wire harness duct from strut tower brace:

- Unclip front of wiring harness cover and slide up and off (**inset**).
- Remove rivets (**arrows**) holding cable ducting to strut tower brace and unclip ducting from brace.

< Remove cover fasteners (**arrows**) and lift up cover.

- N52 engine with metallic cover shown, N52K and N54 engines similar.

Engine upper covers (N62 / N62TU engines), removing

< N62 engine: Remove cover fasteners (**arrows**) and lift off upper cover.

BentleyPublishers.com
BentleyPublishers.com—All Rights Reserved

N62 TU engine: Remove cover fasteners (**arrows**) and lift off upper cover.

Unclip side cover from grommets on valve cover and remove. When installing, make sure pins on cover fully engage grommets in valve cover (**arrows**). If necessary, use a small amount of lubricant to ease installation.

Engine lower covers (splash shields), removing

Engine splash shield: Unscrew fasteners (**arrows**) and lower rear of cover and pull out of bumper cover.

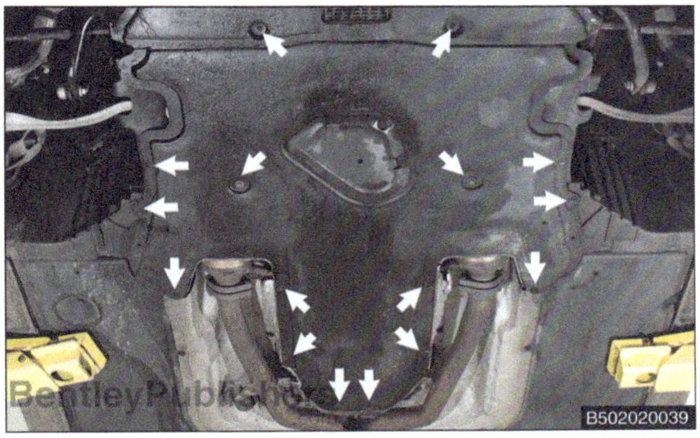

Transmission splash shield: Unscrew fasteners (**arrows**). Rotate cover around lug to remove.

BentleyPublishers.com
BentleyPublishers.com–All Rights Reserved

Air Filter Service

Air filter element, replacing (6-cylinder engines)

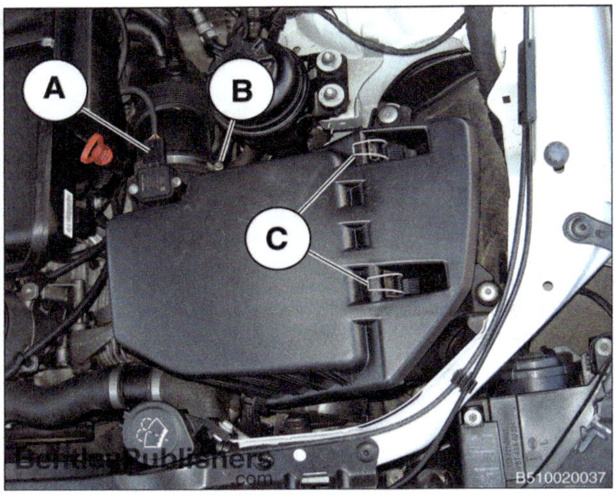

M54 / N52 engines:

- Disconnect connector from mass air flow sensor (**A**).
- Loosen hose clamp (**B**) at mass air flow sensor and disconnect intake pipe from air cleaner housing.
- Unclip spring clamps (**C**).
- Lift off upper section of air filter housing.
- Lift out air filter element and discard.
- Thoroughly clean filter housing and install new element.
- Installation is reverse of removal

N54 engines:

- Unclip spring clamps (**arrows**).
- Lift off upper section of air filter housing.
- Lift out air filter element and discard.
- Thoroughly clean filter housing and install new element.
- Installation is reverse of removal

Air filter element, replacing (V8 engines)

N62 / N62 TU engines:

- Unclip spring clamps (**arrows**).
- Carefully lift up upper section of air filter housing and lift out air filter element and discard.
- Thoroughly clean filter housing and install new element.
- Installation is reverse of removal

BentleyPublishers.com
BentleyPublishers.com—All Rights Reserved

Battery Service

Battery removal is described in **121 Battery, Starter, Alternator**.

Be sure to reinitialize electrical systems and reset vehicle readiness anytime the battery is disconnected. See **121 Battery, Starter, Alternator**.

Access to battery

The battery is in right rear of trunk or cargo compartment.

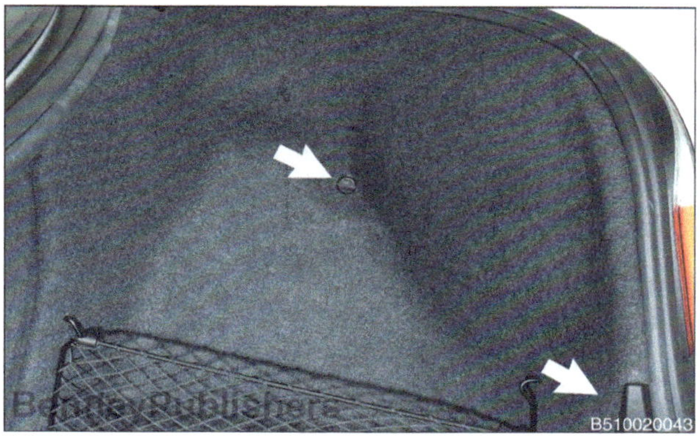

Working in trunk, twist plastic fasteners 90° in right side panel and remove fasteners (**arrows**). Raise floor mat together with cover beneath it. Pull panel out from lower right corner.

Battery access is similar on Sports Wagon models. Remove cargo compartment floor panel. See **513 Interior Trim**. After removing right side panel, remove screw (**arrow**) securing plastic storage shelf (**A**). Remove plastic storage shelf fasteners (**inset**) and remove plastic storage shelf.

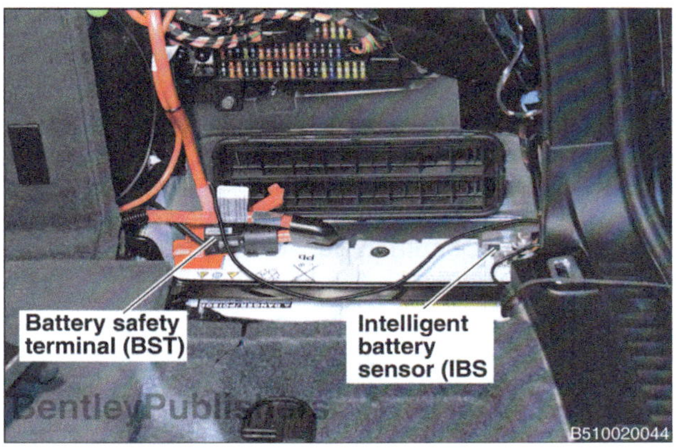

Battery terminals are fitted with battery safety terminal (BST) and intelligent battery sensor (IBS).

WARNING—
- *The BST is and explosive device. Handle with care. Do not pry on the BST to remove the battery terminal.*
- *Batteries give off explosive hydrogen gas during charging. Keep sparks, lighted matches and open flame away from the top of the battery.*
- *Battery acid (electrolyte) can cause severe burns. Flush contact area with water, then seek medical attention.*

CAUTION—
- *The IBS is a delicate electronic device handle gently. Do not pry on the IBS to remove the battery terminal.*
- *The battery is held in place by a hold-down bracket and bolt. A secure battery hold-down is important in order to prevent vibrations and road shock from damaging the battery.*

BentleyPublishers.com
BentleyPublishers.com—All Rights Reserved

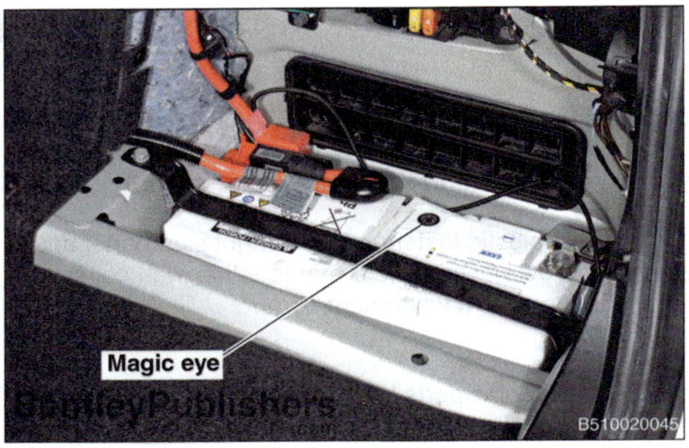

Battery, checking

The original equipment BMW battery is equipped with built-in hydrometer "magic eye".

Battery condition is determined by magic eye color:

- Green: Adequate charge
- Black: Inadequate charge, recharge
- Yellow: Defective battery, replace

See **121 Battery, Starter, Alternator** for additional topics:

- Battery removal and installation
- Replacement battery registration
- Battery reconnection notes

> ***CAUTION—***
> - *Do not charge battery with conventional charger. Use vehicle power supply device for recharging.*

Brake Service

Routine maintenance of the brake system includes checking brake pads for wear, checking parking brake function, and inspecting the system for fluid leaks or other damage:

- Check that brake hoses are correctly routed to avoid chafing or kinking.
- Inspect unions and brake calipers for signs of fluid leaks.
- Inspect rigid lines for corrosion, dents, or other damage.
- Inspect flexible hoses for cracking.
- Replace faulty hoses or lines.

> ***WARNING—***
> - *Incorrect installation or overtightening of hoses, lines and unions may cause chafing or leakage. This can lead to partial or complete brake system failure.*

Note that brake fluid level and brake pad thickness are monitored and replacement intervals recommended by condition based service (CBS) system. See **Condition Based Service (CBS)** in this repair group.

Brake fluid level, checking

The brake fluid reservoir is located on the driver's side near the base of the windshield underneath the cabin microfilter housing.

> ***WARNING—***
> - *Brake fluid is poisonous. Do not ingest. Wash thoroughly with soap and water if it comes into contact with skin.*

– Working in engine bay, remove left (driver's side) cabin microfilter housing to access brake fluid reservoir. See **Cabin air microfilter housing, removing** in this repair group.

BentleyPublishers.com
BentleyPublishers.com—All Rights Reserved

CAUTION—
- *Use only new, previously unopened brake fluid conforming to DOT 4.*
- *Do not let brake fluid come in contact with paint. Wash immediately with soap and water.*
- *Brake fluid absorbs moisture from the air. Store in an airtight container.*
- *Do not use DOT 5 (silicone) brake fluid.*
- *Do not fill brake fluid above* **MAX** *in fluid reservoir.*
- *Do not mix mineral oil products such as gasoline or engine oil with brake fluid. Mineral oil damages rubber seals in the brake system.*
- *Dispose of brake fluid as a hazardous waste.*

Check that fluid level is between **MIN** and **MAX** marks molded on brake fluid reservoir. Fluid level drops slightly as brake pad material wears.

Brake fluid application	
BMW preferred fluid	Low viscosity DOT 4 brake fluid
Low viscosity DOT 4 brake fluid may be used in all E60 vehicles. DOT 4 and low viscosity DOT 4 brake fluid can be mixed.	

Brake fluid absorbs moisture easily. This affects brake performance and reliability. To eliminate moisture, BMW recommends replacing the brake fluid every two years. When replacing or adding brake fluid, use only new fluid from previously unopened containers. Do not use brake fluid that has been bled from the system, even if it is brand new.

See **340 Brakes** for brake fluid flushing and bleeding.

Brake pad and rotor wear, checking

Disc brakes are fitted at all four wheels. Although the brakes are equipped with a brake pad warning system, the system only monitors one wheel per axle. Check brake pad thickness whenever wheels are off or brake work is being done.

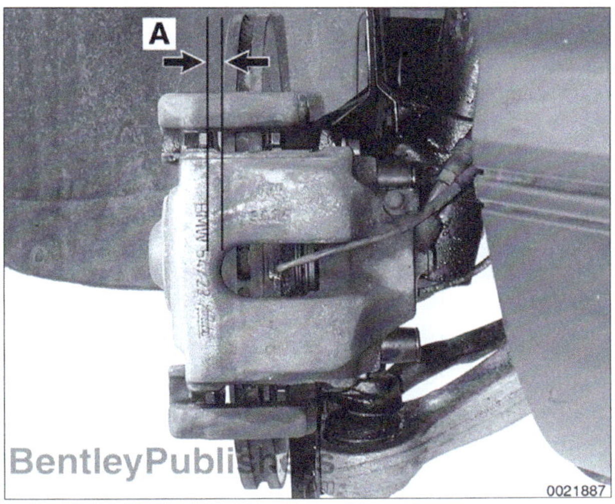

Disc brake pad wear can be checked through opening in caliper:
- Remove road wheel
- Measure distance (**A**) of brake pad "ear" to brake rotor. Compare to specification below.

Brake pad lining minimum thickness	
Dimension **A**	3.0 mm (0.12 in)

Unbolt caliper from caliper carrier to properly inspect. See **340 Brakes** for additional techniques for checking brakes and for brake pad, rotor and caliper replacement procedures.

BentleyPublishers.com
BentleyPublishers.com–All Rights Reserved

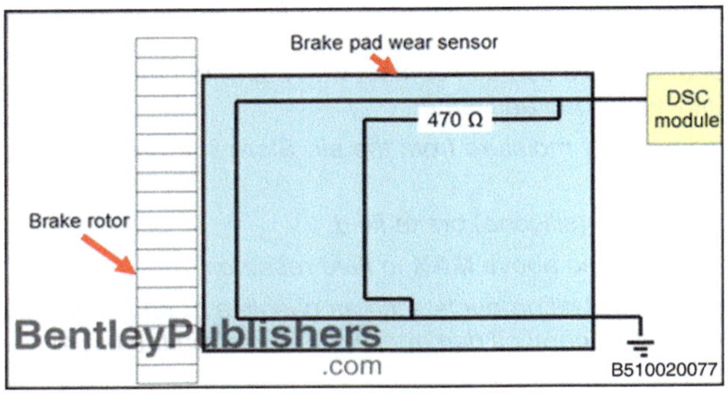

Brake pad wear sensor, testing

BMW E60 models utilize a 2-stage brake pad wear sensor. The wear sensor is used to determine brake pad life and remaining time until service.

If the brake pad wear sensor is partially worn and reinstalled after replacing pads and/or rotors, the service indicator cannot be reset.

Visually inspect sensor. Check that body is intact and not worn.

Sensor that shows wear.

– Disconnect sensor electrical connector.

– Using a DVOM, measure resistance across both terminals
  - Less than 5 ohms indicates a sensor with remaining service life
  - A reading of about 470 ohms indicates a partially worn sensor (stage 1) and requires replacement.
  - A reading of open circuit (infinity) indicates a worn out sensor and requires replacement.

BentleyPublishers.com
BentleyPublishers.com–All Rights Reserved

Brake pad wear indicator, resetting

- Turn ignition to Terminal 15 (ignition on engine off, all warning lights illuminate on instrument panel).
- Close all doors.
- Press and hold instrument cluster set/reset button for 10 seconds.
- Upper display in instrument cluster will be illuminated with a service item (example: An oil can is the designation for oil service). Lower display in instrument cluster will indicate remaining time or mileage left for that service item. Pressing button repeatedly will allow display to scroll through all condition based service items.

- When brake lining indicator appears in message center release trip reset button.
- Reset message will appear, push once more.
- System is now reset.
  - If system will not reset, check all brake pad wear sensors. A slightly worn sensor may cause warning lamp to maintain service needed status.

Parking brake, checking

The parking brake system is independent of the main braking system and may require periodic adjustment depending on use. Adjust the parking brake if the brake lever can be pulled up more than 8 clicks. Check that the cable moves freely. See **340 Brakes** for parking brake adjustment procedure.

The parking brake may lose some of its effectiveness if it is not used frequently. This is due to rust build-up on the parking brake drum. To remove rust, apply the parking brake just until it begins to grip, then pull the lever up one more stop (click). Drive the vehicle approximately 400 meters (1,300 ft.) and release the brake.

CABIN AIR MICROFILTER

The cabin air microfilters filter out dust and pollen from the passenger compartment. There are two filter elements, located at the rear of the engine compartment, left and right sides.

Cabin air microfilters, replacing

- Open engine hood.

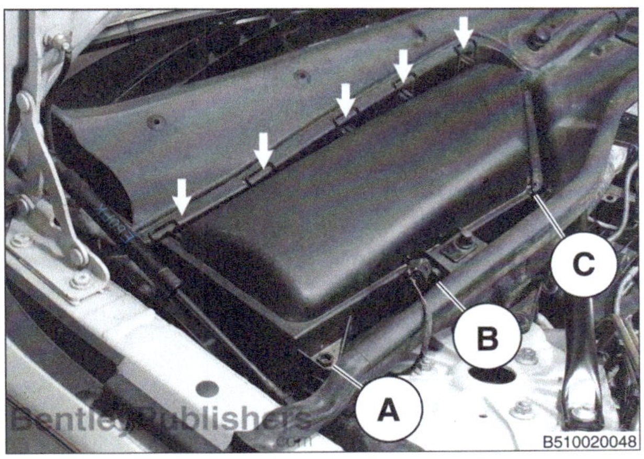

- Remove microfilter cover:
  - Turn quick-release fastener (**A**) approx. 90° counterclockwise.
  - Right side filter: disconnect harness connector at hood alarm contact switch (**B**).
  - Pry filter cover spring clip off cover (**C**).
  - Remove microfilter cover by lifting up and out of rear clips (**arrows**).

BentleyPublishers
BentleyPublishers.com—All Rights Reserved

Unclip filter element from cover and replace.

- Installation is reverse of removal. Make sure filter is correctly seated in cover prior to installation.

Clean filter housing using a disinfectant whenever the filter is replaced. If necessary, remove housing for a thorough cleaning. See **Cabin air microfilter housing, removing**.

Cabin air microfilter housing, removing

- Open engine hood.
- Remove cabin filters together with covers. See **Cabin air filters, replacing**.

Remove center trim from housings. Press down on locking clip (**A**) and slide off in direction of arrow.

Peel out rubber gutter trim pieces from left and right sides.

Unclip left and right ends and peel off bulkhead weather stripping. Remove housing fastener (**arrow**).

BentleyPublishers.com
BentleyPublishers.com–All Rights Reserved

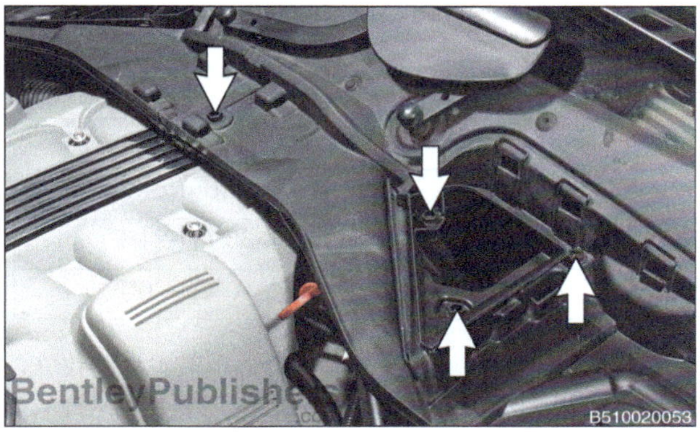
B510020053

Working at left (driver's side) housing, turn quick-release fasteners (**arrows**) approx. 90° counterclockwise to unlock.

- Lift out left cabin air filter housing working from center of engine compartment.
- Remove right housing using instructions for left housing removal.
- Installation is reverse of removal. Install right (passenger side) housing first.

Cooling System Service

Antifreeze raises the boiling point and lowers the freezing point of coolant. BMW recommends engine coolant that is a 50 / 50 mixture of distilled water and ethylene glycol-based antifreeze.

Routine cooling system maintenance consists of maintaining the coolant level and inspecting hoses.

For additional cooling system information and repair procedures, see **170 Radiator and Cooling System**.

Coolant level, checking

> ***WARNING—***
> - *Allow cooling system to cool before opening or draining the cooling system.*

B510170007

With engine fully cooled off, open radiator expansion tank cap. Note that **arrows** on cap and tank line up when cap clicks into fully closed position.

B510170008

Float in expansion tank indicates coolant level when cold.

- When upper mark on float (**MIN**) is level with top of filler neck, coolant is at minimum allowable level.
- When lower mark on float (**MAX**) is level with top of filler neck, coolant is at maximum level.

- Top off coolant if necessary.

> ***CAUTION—***
> - *Use only BMW approved phosphate-free, nitrite-free and amino-free antifreeze when filling the cooling system.*

BentleyPublishers.com—All Rights Reserved

Antifreeze concentration, checking

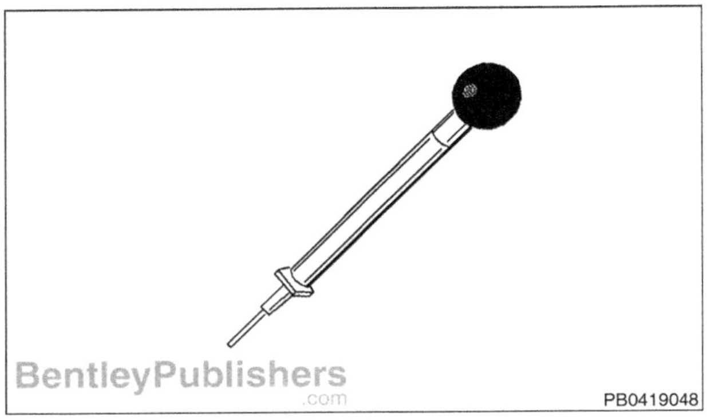

Use a coolant hydrometer to determine antifreeze concentration.

Coolant mixture recommendations	
Concentration	**Cold protection**
50% antifreeze	-35°C (-31°F)
60% antifreeze	-40°C (-40°F)

Do not use a higher concentration of antifreeze than a 60% mixture, as the heat transfer quality of the coolant decreases with higher antifreeze concentrations.

Cooling system hoses, inspecting

Inspect hoses by first checking that all connections are tight and dry. Coolant seepage indicates that either the hose clamp is loose, the hose is damaged, or the connection is dirty or corroded. Dried coolant has a chalky appearance.

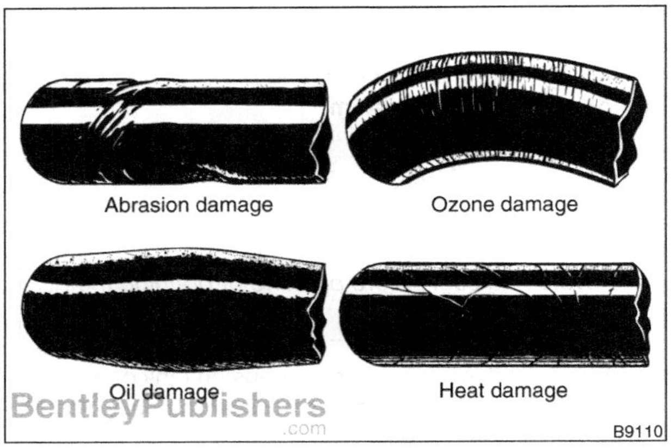

Check hose condition by a visual and tactile inspection, making sure it is firm and springy. Replace hoses that exhibit conditions noted below. (Illustration courtesy of Gates Rubber Company, Inc.)

- Leakage: Dripping, moisture, or seepage near clamps or connectors.

> ***CAUTION—***
> - *When installing or reinstalling coolant hoses, use clamps specified by BMW.*

- Electromechanical degradation: Difficult to see, but detectable by squeezing hose and feeling for cracks, weak areas, and voids.
- Oil damage: Soft and spongy to touch, visible bulges and swelling.
- Abrasion damage: Wear, abrasion, or scuffing, often due to contact with components in engine compartment.
- Heat damage: Internal and external damage, generally due to high under hood temperatures or overheating. Internal heat damage is often indicated by swelling with external damage marked by hardened and cracked areas.
- Ozone damage: Small, parallel cracks in outer layers, but without hardening. Due to exposure to atmospheric conditions.

– As a preventative measure, replace coolant hoses every 4 years.

Electrical Service

See the following repair groups for detailed information on vehicle electrics:

- **121 Battery, Starter, Alternator** for engine electrical components
- **600 Electrical System–General** for basic electrical troubleshooting information
- **ECL Electrical Component Locations** for fuse, relay and control module information
- **ELE Electrical Wiring Diagrams** for detailed electrical schematics.

BentleyPublishers.com
BentleyPublishers.com—All Rights Reserved

Engine Accessory Belt Service

The accessory belt and pulleys transfer power from the engine crankshaft to the alternator, A/C compressor and power steering pump. Note that engine cooling fan is electric. On 2006 and later 6-cylinder engines, the coolant pump is also electric.

Inspect belt with engine OFF. If belt shows signs of wear, cracking, glazing, or missing sections, replace immediately. To reduce the chance of belt failure while driving, replace belt every four years.

When the belt is replaced with a new one, store the old one in the trunk for emergency use.

If accessory belt, power steering pump pulley or front axle carrier are found to be damaged, replace these parts as well as the left engine mount with updated parts.

> ***CAUTION—***
> - *To avoid electrochemical corrosion to engine components made of aluminum-magnesium alloy, do not use steel fasteners. Use aluminum fasteners only.*
> - *For reliable identification, test fasteners for aluminum composition with magnet.*
> - *Replace aluminum bolts each time they are loosened.*
> - *Follow torque instructions, including angle of rotation specifications, when installing aluminum fasteners.*

Accessory belt, replacing (M54, N52)

If planning to reinstall old belt, be sure to mark direction of rotation. If replacing the belt, keep the old belt for emergency use.

– Open engine hood.

– For additional clearance, remove radiator cooling fan cowl with integral fan. See **170 Radiator and Cooling System**.

– **M54 engine**: Remove A/C compressor belt. See **A/C belt, replacing (M54 engine)** in this repair group.

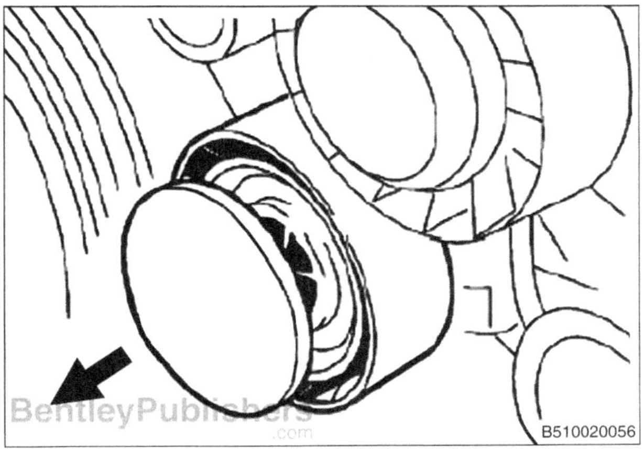

◄ **M54 engine**: pry cover from tensioner.

BentleyPublishers.com
BentleyPublishers.com—All Rights Reserved

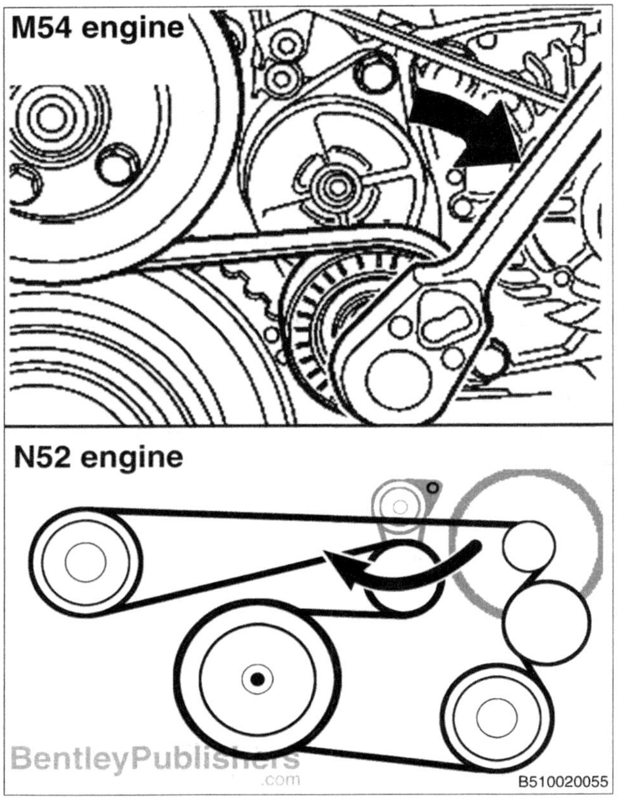

Use Torx socket to rotate belt tensioner clockwise (**arrow**). This releases belt tension.

If applicable, lock tensioner in released position using appropriate diameter pin (BMW special tool 11 3 340).

> ***WARNING —***
> • *Observe care when working with tensioner. Personal injury could result if tensioner springs back into position uncontrolled.*

– Work belt off pulleys and remove.

M54 engine: Install belt using accompanying diagram as guide. If reinstalling used belt, follow direction-of-rotation mark made previously.

BentleyPublishers.com
BentleyPublishers.com–All Rights Reserved

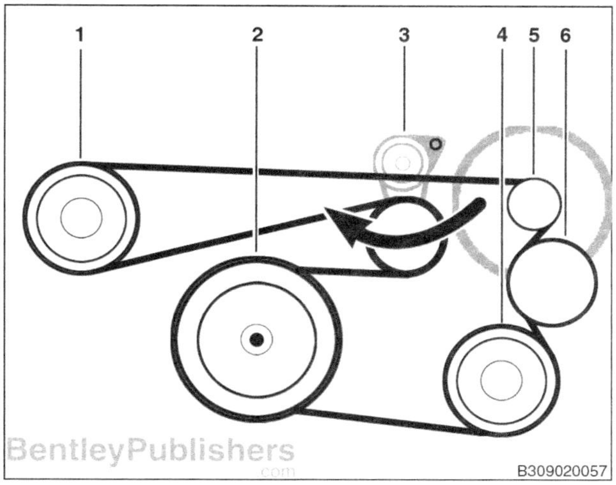

N52 engine: Install belt using accompanying diagram as guide. If reinstalling used belt, follow direction-of-rotation mark made previously.

1. A/C compressor pulley
2. Vibration damper (crankshaft pulley)
3. Tensioner pulley
4. Power steering pump pulley
5. Alternator pulley
6. Idler pulley

– Release tensioner.

– The remainder of installation is reverse of removal. Run engine and check for proper belt installation.

Accessory belt, replacing (N54)

– Open engine hood.

– For additional clearance, remove radiator cooling fan cowl with integral fan. See **170 Radiator and Cooling System**.

– Remove engine splash shield. See **Engine lower covers (splash shields), removing** in this repair group.

– If planning to reinstall old belt, be sure to mark direction of rotation.

Working at left side of engine compartment:

- Loosen hose clamp **A** and detach air duct from air filter housing.
- Rotate quick-release clamp **B** 90° on boost pressure duct. Separate air ducts.

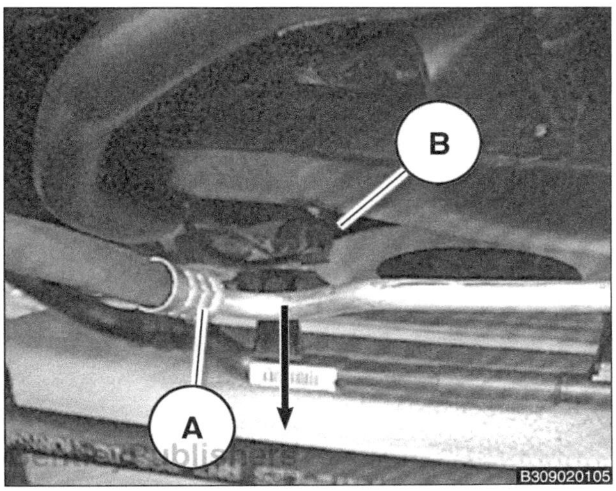

Working underneath radiator, unclip line **A** from clip **B** in direction of **arrow**.

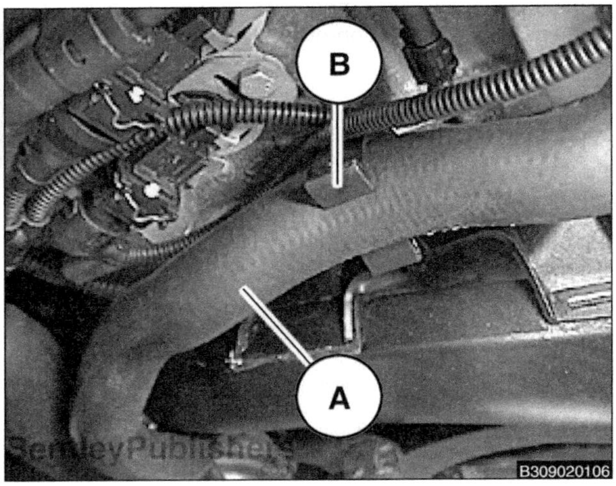

Working at top front of cylinder head, disconnect coolant hose **A** from bracket **B**.

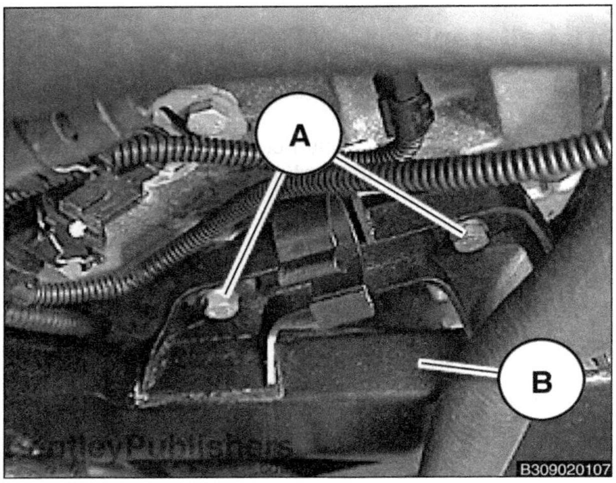

Remove air duct bracket mounting bolts (**A**) at front of engine. Fold air duct (**B**) down without disconnecting or removing.

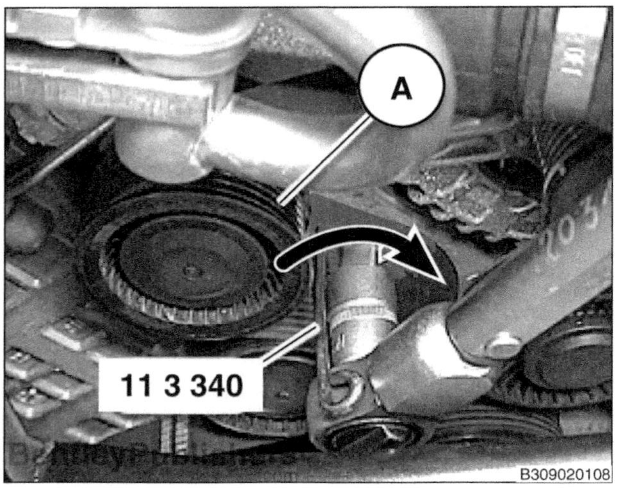

Use Torx socket to rotate belt tensioner (**A**) clockwise (**arrow**). This releases belt tension.

- Lock tensioner in released position using appropriate diameter pin (BMW special tool 11 3 340).

> ***WARNING—***
> *• Observe care when working with tensioner. Personal injury could result if tensioner springs back into position uncontrolled.*

- Work belt off pulleys and remove.

BentleyPublishers.com
BentleyPublishers.com–All Rights Reserved

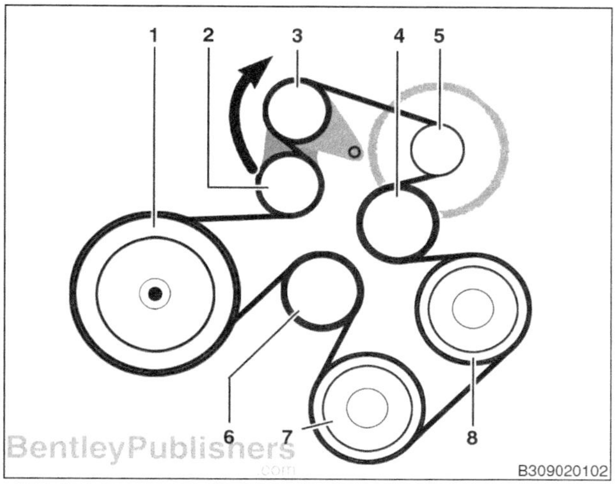

Install belt using accompanying diagram as guide. If reinstalling used belt, follow direction-of-rotation mark made previously.

1. Vibration damper (crankshaft pulley)
2. Tensioner pulley
3. Tensioner-idler pulley
4. Idler pulley
5. Alternator pulley
6. Idler pulley
7. Power steering pump pulley
8. A/C compressor pulley

– Release tensioner.

– The remainder of installation is reverse of removal. Run engine and check for proper belt installation.

Accessory belt, replacing (V8 engines)

If planning to reinstall old belt, be sure to mark direction of rotation. If replacing the belt, keep the old belt for emergency use.

– Open engine hood.

– For additional clearance, remove radiator cooling fan cowl with integral fan. See **170 Radiator and Cooling System**.

Use Torx socket (T60) to rotate belt tensioner fully clockwise (**arrow**). This releases belt tension.

– With tensioner in fully released position, insert lock or appropriate diameter pin (BMW special tool 11 3 340) into bore in bottom of tensioner.

> ***WARNING —***
> *• Observe care when working with tensioner. Personal injury could result if tensioner springs back into position uncontrolled.*

– Work belt off pulleys and remove.

Install belt using accompanying illustration as a guide. If reinstalling used belt, follow direction-of-rotation mark made previously.

– Release tensioner.

– The remainder of installation is reverse of removal. Run engine and check for proper belt installation.

BentleyPublishers.com–All Rights Reserved

A/C belt, replacing (M54 engine)

Two belts are used to drive engine accessories:

- Outer (front) belt: A/C compressor
- Inner (rear) belt: Alternator, coolant pump, steering pump

– If planning to reinstall old belt(s), be sure to mark direction of rotation.

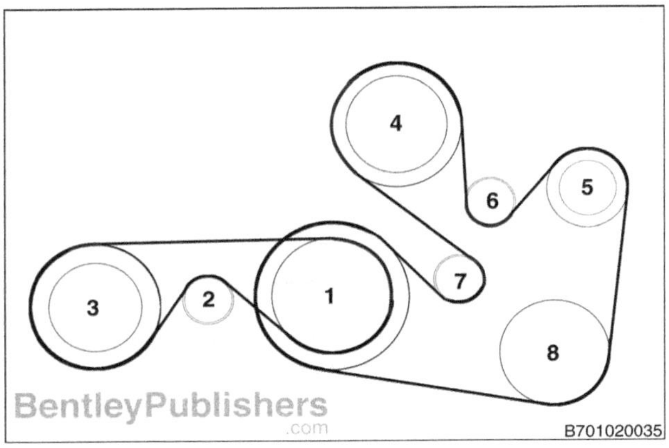

M54 engine accessory belts

1. Crankshaft pulley
2. A/C belt tensioner
3. A/C compressor pulley
4. Coolant pump pulley
5. Alternator pulley
6. Accessory belt tensioner
7. Idler pulley
8. Power steering pump pulley

To remove A/C compressor belt:

- Use hex head (**A**) or Torx socket to rotate A/C belt tensioner in direction of **arrow**.
- Slip belt off pulleys and remove belt.
- Check belt for cracks and coolant or oil residue. Replace if necessary.

– To reinstall A/C compressor belt:

- Rotate A/C belt tensioner in direction of **arrow**.
- Route belt over pulleys and release tensioner.

– When reinstalling belt, check that it is correctly positioned on pulleys. If reinstalling old belt, pay attention to direction of rotation marks made previously.

Tightening torques	
Fan clutch to coolant pump (left- hand thread) • With BMW tool 11 5 040 • Without BMW tool 11 5 040	 30 Nm (22 ft-lb) 40 Nm (29 ft-lb)

BentleyPublishers.com
BentleyPublishers.com—All Rights Reserved

A/C belt, replacing (V8 engine)

Two belts are used to drive engine accessories:

- Outer (rear) belt: Alternator, coolant pump, steering pump
- Inner (front) belt: A/C compressor

– If planning to reinstall old belt, be sure to mark direction of rotation. If replacing the belt, keep the old belt for emergency use.

– Open engine hood.

– Remove accessory belt. See **Accessory belt, replacing (V8 engines)** in this repair group.

– Remove engine splash shield. See **Engine lower covers (splash shields), removing**, in this repair group.

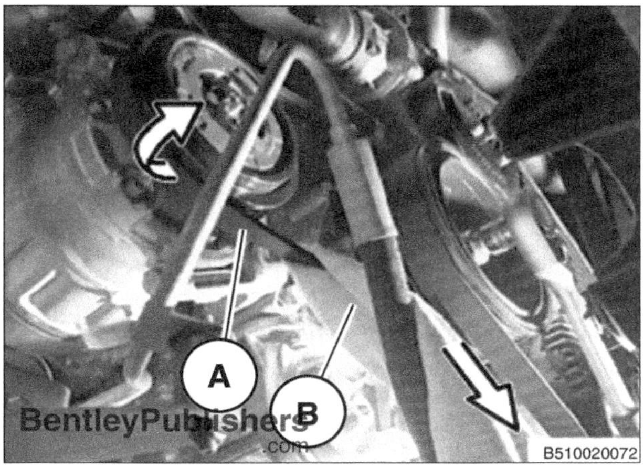

◄ Rotate engine at crankshaft bolt using socket in direction of engine rotation. At the same time, pull off A/C belt (**A**) using clean rag (**B**) in direction of arrow.

◄ Place belt in position and clamp belt to pulley with clamp (BMW special tool **64 1 040**).

◄ Rotate engine at crankshaft bolt using socket in direction of engine rotation until A/C belt is fully seated on pulley. Remove BMW special tool **64 1 040**.

– If reinstalling used belt, follow direction-of-rotation mark made previously.

> ***CAUTION—***
> - *If belt is contaminated with hydraulic fluid, do not reinstall. Install new belt.*

– The remainder of installation is reverse of removal. Run engine and check for proper belt installation.

BentleyPublishers
BentleyPublishers.com–All Rights Reserved

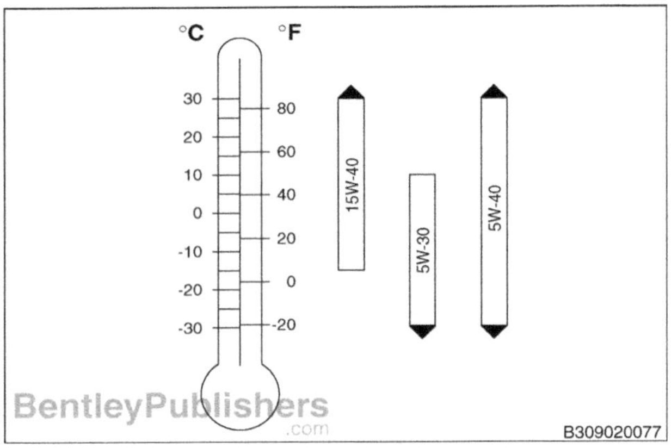

ENGINE OIL SERVICE

> ***CAUTION—***
> *• Use BMW specified oil to top off engine between oil changes.*
> *• Do not use engine oil additives.*

◄ BMW high performance synthetic oil is recommended for use in the cars covered by this manual.

Engine oil specification	
BMW high performance synthetic SAE 5W - 30 or 5W - 40	API rating SM or higher

Engine oil, checking level

◄ **2004 - 2005 models**:

- Check oil level with car on level surface, after engine has been stopped for at least a few minutes.
- Pull out dipstick and wipe it clean. Reinsert dipstick fully and then withdraw it again.
- Oil level is correct if it is between two marks at end of dipstick.
- On 6-cylinder engines, red-handled dipstick is at front left of engine.

2006 - 2010 models: There is no dipstick. Engine oil level and other oil characteristics are monitored electronically by the oil condition sensor (OZS) and displayed via the center dashboard display when called up through the iDrive system. See **119 Engine Lubrication** for detail of OZS operation.

On 2006 and later 6-cylinder engines, the dipstick was eliminated. 2006 and later V8 engines have dipsticks throughout the production range, but oil level can also be checked through iDrive check control as described below.

Monitoring engine oil level electronically ensures that the oil does not reach critically low levels. Overfilling the engine with oil can cause leaks; a corresponding warning is therefore given.

– **2006 - 2009 models**

- Start engine.
- Press down on iDrive **MENU** button to open start menu. Select **Info sources**.

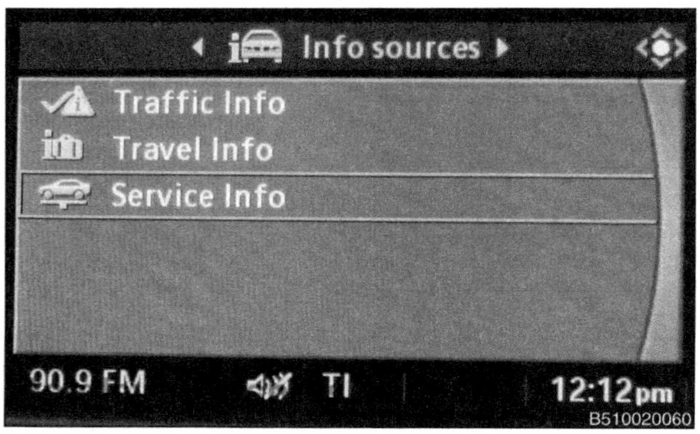

◄ Press down on iDrive controller, then select **Service Info**.

BentleyPublishers.com
BentleyPublishers.com—All Rights Reserved

B510020059

Turn controller until **Service requirements** is selected and press controller. Change to upper field on screen if necessary.

- Turn controller until **Engine oil level** is selected and press controller. Change to second field on screen if necessary. Measurement normally takes about 3 minutes if the vehicle is at a standstill on a level surface, or about 5 minutes if the vehicle is underway.
- One of the following messages is displayed
  Engine oil level O.K.
  No measured value available: Updating engine oil level measurement...
  Engine oil level at the minimum! Refill with 1 quart of engine oil.
  Engine oil level too high. Have this checked immediately.
  Please observe recalculated service interval for engine oil.

— To check oil level on **2010 models**:

- Start engine.

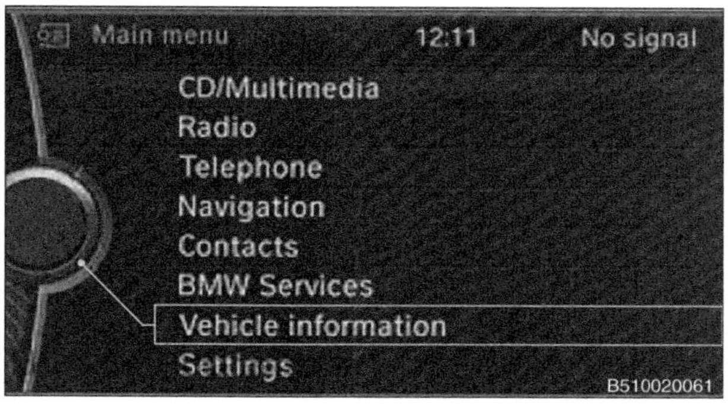

B510020061

Using iDrive controller, select **Vehicle Information** through main menu.

- Select **Vehicle status** and then select **Engine oil level**. Measurement normally takes about 3 minutes if the vehicle is at a standstill on a level surface, or about 5 minutes if the vehicle is underway.
- One of the following messages is displayed
  Engine oil level OK
  Measurement not possible as this time
  Measuring engine oil level...
  Engine oil level below minimum. Add 1 quart!
  Engine oil level too high! Have this checked.
  Measurement inactive. Have this checked

B309020062

If oil level is low, add oil at oil filler (**arrow**). Use correct viscosity and grade oil.

BentleyPublishers.com—All Rights Reserved

Engine oil and filter, changing (6-cylinder)

> ***CAUTION—***
> - *When changing the oil, make sure the engine oil drain plug is started and torqued using hand-tools. Power tools can strip the threads of the plug and the oil pan.*

A complete oil change requires new oil, a new filter insert kit and a new drain plug sealing washer. Run engine for a few minutes to warm engine oil. Switch engine OFF.

◄ Use oil filter wrench to loosen and remove oil filter housing cover (**arrow**). Remove filter cartridge and discard O-ring.

– Raise vehicle and support safely.

> ***WARNING—***
> - *Make sure the vehicle is stable and well supported at all times. Use a professional automotive lift or jack stands designed for the purpose. A floor jack is not adequate support.*

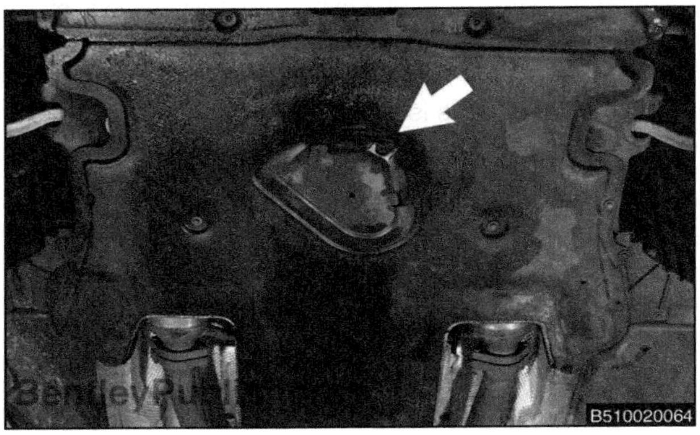

◄ Working underneath engine sump, open service door in engine splash shield (**arrow**) to expose drain plug.

– Place drain pan under oil drain plug. Using a socket or box wrench, loosen drain plug. Remove plug by hand and let oil drain into pan.

> ***WARNING—***
> - *Pull the loose plug away from the hole quickly to avoid being scalded by hot oil. It will run out quickly when the plug is removed. If possible, use gloves to protect your hands.*

– When oil flow has diminished to an occasional drip, reinstall drain plug with a new metal sealing washer and torque plug.

Tightening torque	
Engine oil drain plug to oil pan	25 Nm (18 ft-lb)

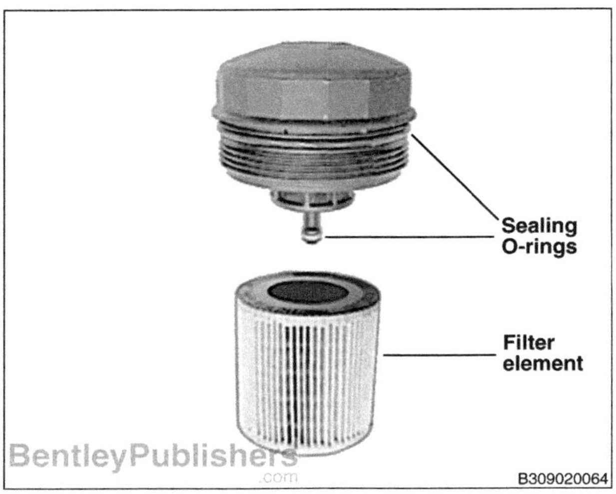

◄ Lubricate and install new oil filter O-ring and new filter cartridge in oil filter housing cover. Install and tighten housing cover.

Tightening torque	
Cover to oil filter housing	25 Nm (18 ft-lb)

– Refill crankcase with oil. Oil capacity is listed in **Table a**.

Table a. Engine oil capacity (6-cylinder)

Engine	**Approx. capacity (incl.oil filter)**
6-cylinder: M54, N52, N54	6.5 liters (6.9 US qt)

– Start engine and check that oil pressure warning light immediately goes out. Check for any leaks.

– Check for correct oil level. See **Engine oil, checking level**, in this repair group.

BentleyPublishers.com
BentleyPublishers.com–All Rights Reserved

Engine oil and filter, changing (V8)

> ***CAUTION—***
> - *When changing the oil, make sure the engine oil drain plug is started and torqued using hand-tools. Power tools can strip the threads of the plug and the oil pan.*

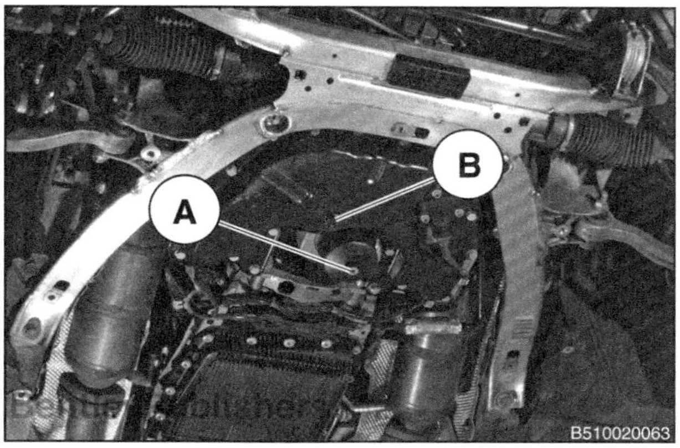

◀ On V8 engines, the oil filter is in the bottom of the engine (**A**) and the engine oil drain plug is in the center of the oil pan (**B**).

A complete oil change requires new oil, a new filter insert kit and a new drain plug O-ring. Run engine for a few minutes to warm engine oil. Switch engine OFF.

– Open engine hood and remove oil filler cap. Store in safe place.

– Raise vehicle and support safely.

> ***WARNING—***
> - *Make sure the vehicle is stable and well supported at all times. Use a professional automotive lift or jack stands designed for the purpose. A floor jack is not adequate support.*

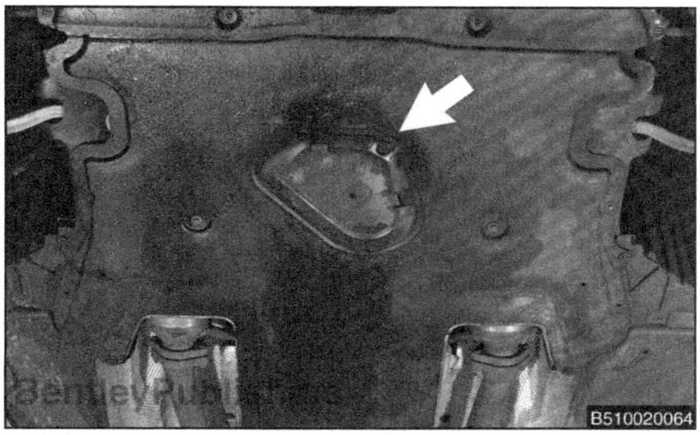

◀ Working underneath engine sump, open service door (**arrow**) in engine splash shield to expose filter cover.

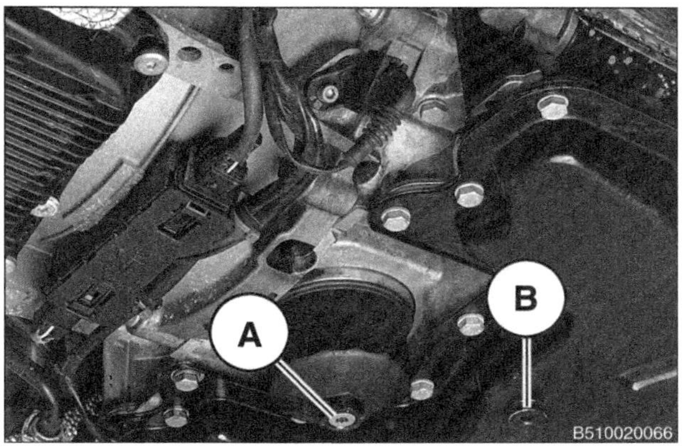

◀ Place drain pan under oil drain plug. Use 6mm allen wrench to loosen drain plug (**A**) from oil filter cover. Remove plug by hand and let oil drain into pan. Remove oil pan drain plug (**B**) and let oil drain into pan.

> ***WARNING—***
> - *Pull the loose plug away from the hole quickly to avoid being scalded by hot oil. It will run out quickly when the plug is removed. Use gloves to protect your hands.*

– When oil flow has diminished to an occasional drip, use an oil filter wrench to loosen and remove the oil filter cover.

– Remove filter cartridge and pick off and discard large O-ring from cover.

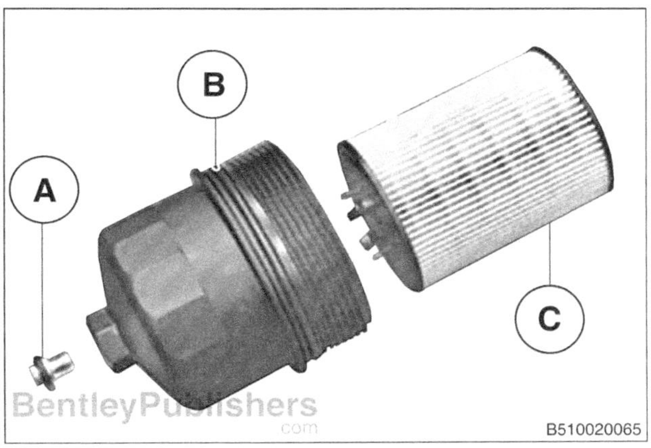

◀ Lubricate and install new oil filter O-ring (**B**) and new filter cartridge (**C**) in oil filter housing cover. Install and tighten cover.

Tightening torque	
Oil filter cover to oil sump	25 Nm (18 ft-lb)

– Reinstall drain plug (**A**) with a new O-ring and torque plug.

Tightening torque	
Engine oil drain plug to oil pan	10 Nm (89 in-lb)

BentleyPublishers.com
BentleyPublishers.com–All Rights Reserved

– Refill crankcase with oil. Oil capacity is listed in **Table b**.

Table b. Engine oil capacity (V8)	
Engine	**Approx. capacity (incl. oil filter)**
V8: M62, M62 TU	8.0 liters (8.5 US qt)

– Start engine and check that oil pressure warning light immediately goes out. Check for any leaks.

– Check for correct oil level. See **Engine oil, checking level**, given earlier.

FUEL FILTER

Models covered by this manual are equipped with a life-time fuel filter with no specified replacement interval.

The filter is located in left lobe of the fuel tank, integral with the fuel level sender and fuel pressure regulator. It is available as a replacement part from BMW. See **160 Fuel Tank and Fuel Pump**.

POWER STEERING FLUID SERVICE

Power steering fluid, checking level

The power steering system is permanently filled with either CHF or ATF hydraulic fluid. Routinely adding fluid is not required unless the system is leaking.

> ***CAUTION—***
> • *Power steering reservoir cap is marked with the type of fluid being used— ATF or CHF. Do not mix.*

To check power steering fluid level in fluid reservoir:

- Park vehicle on level ground with engine OFF.
- Level is correct if it is between **MIN** and **MAX** marks on dipstick.
- If level is below **MIN** mark, add fluid to reservoir to bring level up.

– Hand-tighten reservoir cap.

See **320 Steering and Wheel Alignment** for power steering system bleeding.

BentleyPublishers.com
BentleyPublishers.com–All Rights Reserved

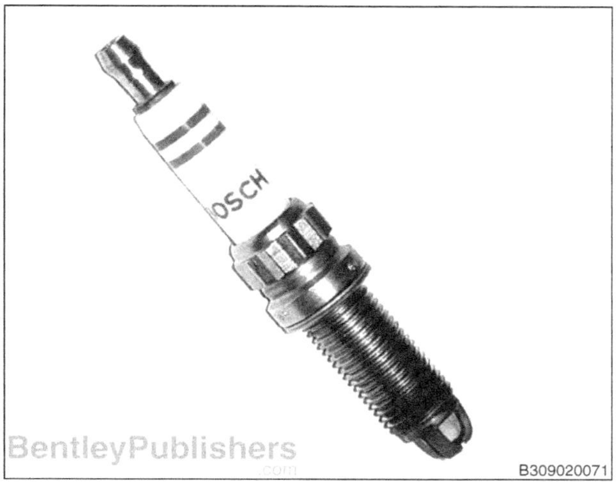

BentleyPublishers.com
B309020071

Spark Plug Service

All E60 5 Series engines use a coil-over-spark plug configuration, with one ignition coil above each spark plug.

Turbocharged (N54) engine spark plugs are particular to that engine. A special thin-walled 12-point spark plug socket, BMW part no. 83 30 0 495 560, is needed for servicing these spark plugs.

Table c. Spark plugs (100k mile change interval)		
Manufacturer	**Bosch**	**NGK**
M54	FGR7DQP	BKR6EQUP
N52 engine	FR7NPP332	ILZFR6D11
N54 engine (turbo)	ZGR6STE2	not available
N62 / N62TU	FGR7DQP	BKR6EQUP
Check for correct applications with authorized BMW dealer parts department.		

Spark plugs, replacing

> ***WARNING—***
> *• To avoid personal injury, be sure the engine is cold before beginning work in the engine compartment.*

– Remove ignition coil covers. See **Engine upper covers, removing** in this repair group.

– Remove ignition coil. See **120 Ignition System**.

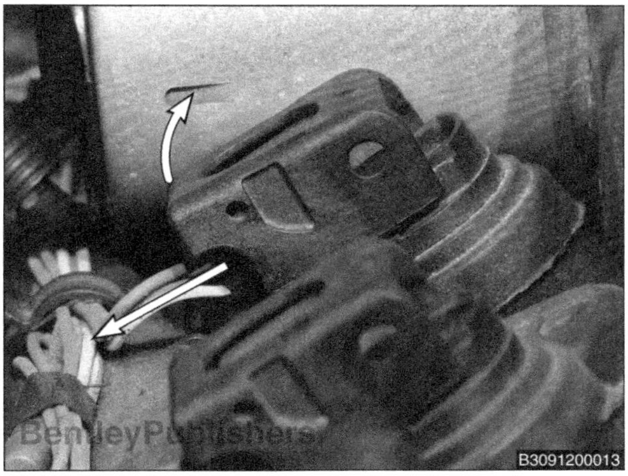
BentleyPublishers
B3091200013

Unlock (**curved arrow**) ignition coil connector. Pull (**straight arrow**) connector off coil.

B3091200014

Slide rod-shaped ignition coil straight out of spark plug hole.

> ***CAUTION—***
> *• Do not allow fuel or oil to contact the ignition coils. Fuel or oil residue can cause a breakdown in the electrical resistance of silicone used in production, resulting in ignition coil failure.*

BentleyPublishers.com—All Rights Reserved

Remove spark plugs:

- **N52 engine**: Use thin-walled spark plug socket.
- **N54 (turbo) engine**: Use special 12-point spark plug socket.

N52 engine: if thin-walled spark plug socket is not available, carefully remove slotted ignition coil sleeves from cylinder head cover holes, then remove spark plugs.

Inspect spark plug electrodes. Light grey indicates normal combustion.

– Installation is reverse of removal, bearing in mind the following:

- Lightly lubricate new spark plug threads with copper-based anti-seize compound.
- Thread plugs into cylinder head by hand to prevent cross-threading.
- **N52 engine**: reinsert or replace slotted ignition coil sleeves, if removed.
- Carefully insert ignition coils in spark plug wells and reattach electrical connectors. Be sure to reattach all removed grounds

Tightening torque	
Spark plug to cylinder head	23 ± 3 Nm (18 ± 2 ft-lb)

BentleyPublishers.com
BentleyPublishers.com—All Rights Reserved

Other Mechanical Maintenance

Differential oil level, checking

The front (if applicable) and rear differentials are filled with lifetime oil that ordinarily does not need to be changed. BMW recommends using only a specially formulated synthetic gear oil (SAF-XO) that is available through an authorized BMW dealer parts department. For additional information on this lubricant and any other lubricants that may be compatible, contact an authorized BMW dealer service department.

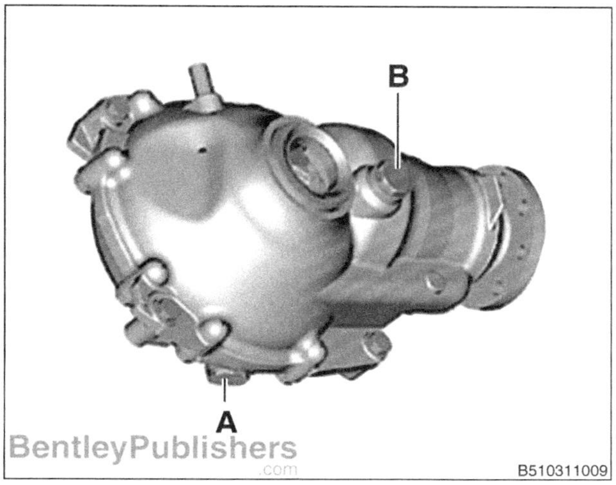

◄ Front differential:

- Remove engine splash shield. See **Engine lower covers (splash shields), removing** in this repair group.
- Remove front reinforcement plate. See **310 Front Suspension**.
- Check oil at fill plug (**arrow**).

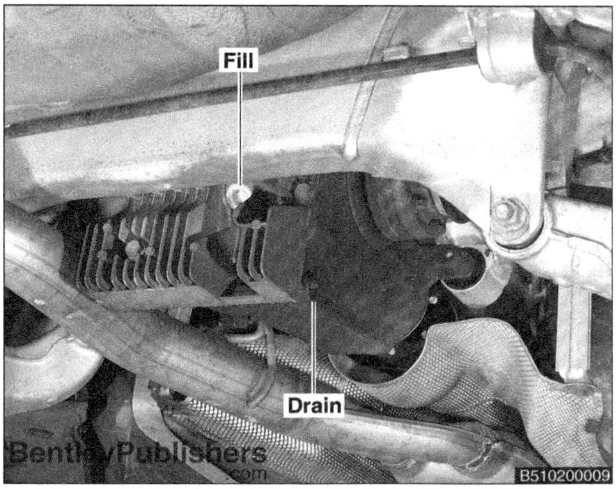

◄ Rear differential: Check oil level at fill plug (**arrow**).

– All: Check lubricant level with vehicle level:

- Use 14 mm or 17 mm Allen socket to remove filler plug.
- Level is correct when fluid just reaches edge of filler hole.
- If necessary, top up fluid.
- Replace filler plug sealing ring.
- Install and tighten oil filler plug when oil level is correct.

Tightening torque	
Fill / drain plug to rear differential housing: • Plug with sealing O-ring (replace O-ring) • Plug without sealing O-ring (replace plug)	 65 Nm (48 ft-lb) 60 Nm (44 ft-lb)

See also **311 Front Axle Differential** or **331 Rear Axle Differential**.

BentleyPublishers.com
BentleyPublishers.com–All Rights Reserved

Drive axle joint (CV joint) boots, inspecting

◀ Inspect CV joint protective boot (**arrow**) for cracks and damage. If the rubber boot fails, water and dirt that enter the joint quickly damage it.

Areas where leaks are most likely to occur are around drive shaft and drive axle mounting flanges.

CV boot replacement is covered in **331 Rear Axle Differential**.

Exhaust system, inspecting

Exhaust system life varies widely according to driving habits and environmental conditions. If short-distance driving predominates, moisture and condensation in the system does not fully dry out. This leads to early corrosion damage and more frequent replacement.

Scheduled maintenance of the exhaust system is limited to inspection.

◀ Check to see that hangers (**arrow**) are in place and properly supporting the system and that parts of the exhaust system do not contact the body.

– Check for restrictions due to dents or kinks.

– Check for weakness or perforation due to rust.

See **180 Exhaust System**.

Fuel tank and fuel lines, inspecting

– Inspect fuel tank and fuel lines for damage or leaks.

– Check for fuel leaks in engine compartment or fuel odors in passenger compartment.

– Check for evaporative emissions hoses that may have become disconnected. Check carefully at charcoal canister and leakage diagnosis pump. See **130 Fuel Injection** and **160 Fuel Tank and Fuel Pump** for component locations and additional information.

Idle speed

Idle speed is electronically adaptive and not adjustable.

BentleyPublishers.com
BentleyPublishers.com–All Rights Reserved

Oxygen sensors

Replacement of oxygen sensors at specified intervals ensures that the engine and emission control system continue to operate as designed. Extending the replacement interval may void the emission control warranty coverage. Oxygen sensor replacement is covered in **180 Exhaust System**.

Suspension and steering, inspecting

Check suspension and steering moving parts for wear and excessive play. Inspect ball joint and tie-rod rubber seals and boots for cracks or tears that allow the entry of dirt and water.

See also:

- **310 Front Suspension**.
- **311 Front Axle Differential**
- **320 Steering and Wheel Alignment**
- **330 Rear Suspension**
- **331 Rear Axle Differential**

Tires, checking inflation pressure

Correct tire pressures are important to handling and stability, fuel economy, and tire wear. Tire pressure changes with temperature. Check tire pressures often during seasonal temperature changes.

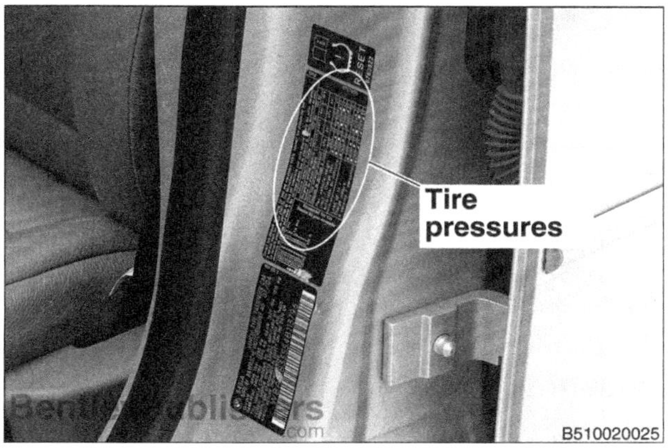

B510020025

◄ Correct inflation pressures can be found on driver door pillar. Note that tire pressure specifications are higher when the vehicle is more heavily loaded.

> ***WARNING —***
> - *Do not inflate any tire to a pressure higher than maximum inflation pressure listed on the sidewall. Use care when adding air to warm tires. Warm tire pressures can increase as much as 4 psi (0.3 bar) over their cold pressures.*

Tires, rotating

Due to the vehicle's suspension design, BMW does not recommend tire rotation. Rotating used or worn tires may adversely affect road handling and tire grip.

Transmission service

Automatic and manual transmissions are filled with lifetime lubrication. No oil change is required for the entire service life of these transmissions. If repairs have to be made to the transmission or transmission oil cooler, use only the approved lifetime lube oil.

Transmission fluid, automatic

The automatic transmission is not equipped with a dipstick. Checking ATF level includes measuring and maintaining a specified ATF temperature during the checking procedure.

For ATF service information, including checking ATF level and ATF filter replacement procedures, see **240 Automatic Transmission**.

BentleyPublishers.com—All Rights Reserved

Transmission fluid, manual

The manual transmissions is filled with a special lifetime fluid. For manual transmission fluid checking and replacement procedures, see **230 Manual Transmission**.

Wheels, aligning

BMW recommends checking the front and rear alignment once a year and whenever new tires are installed. See **320 Steering and Wheel Alignment**.

Body and Interior Maintenance

Body, hinges, locks, lubricating

- Lubricate door locks and lock cylinders with an oil that contains graphite.
- Lubricate body and door hinges, hood latch and door check rods with SAE 30 or SAE 40 engine oil.
- Lubricate seat runners with multipurpose grease.
- Lubricate door weather-strips with silicone spray or talcum powder.
- Lubricate engine hood release and hood release cable with spray grease.

CAUTION—
- *Do not apply oil to rubber parts.*
- *Use an absolute minimum of winter lock deicer spray. Alcohol in deicer washes grease out of lock assemblies and may cause locks to corrode internally or to become difficult to operate.*

Exterior washing

The longer dirt is left on the paint, the greater the risk of damaging the glossy finish, either by scratching or by the chemical effect dirt particles may have on the painted surface.

- Wash with a mixture of lukewarm water and vehicle wash product.
- Rinse using plenty of clear water.
- Wipe body dry with a soft cloth towel or chamois to prevent water-spotting.

CAUTION—
- *Do not wash the vehicle in direct sunlight.*
- *If the engine hood is warm, allow it to cool.*
- *Beads of water not only leave spots when dried rapidly by the sun or heat from the engine, but also can act as small magnifying glasses and burn the finish.*

BentleyPublishers.com
BentleyPublishers.com—All Rights Reserved

Interior care

- Remove dirt spots on upholstery with lukewarm soapy water or dry foam cleaner.

- Use spot remover for grease and oil spots. Do not pour cleaning liquid directly on carpet or fabric. Dampen clean cloth and rub carefully, starting at edge of the spot and working inward.

> ***WARNING —***
> *• Do not use gasoline, naphtha, or other flammable substances.*

Leather upholstery and trim

- Clean leather upholstery and trim periodically.

- Use slightly damp cotton or wool cloth to get rid of dirt in creases and pores that can cause brittleness and premature aging.

- On heavily soiled areas, use a mild detergent (such as Woolite®) or a specially formulated leather cleaner.

- Dry trim and upholstery completely using a soft cloth. Regular use of a good quality leather conditioner reduces drying and cracking of leather.

Seat belts

Dirt and other abrasive particles damage seat belt webbing.

- To clean seat belts, use a mild soap solution.

> ***WARNING —***
> *• Do not clean the seat belt webbing using dry cleaning chemicals, bleach or other strong cleaning agents.*
> *• Allow wet belts to dry before allowing them to retract.*

- Inspect condition of belt webbing and function of retractor mechanisms.

Special cleaning

- Remove tar and insect spots with a bug and tar remover. Do not use gasoline, kerosene, nail polish remover, or other unsuitable solvents.

A bit of baking soda dissolved in the wash water facilitates removal. This method can also be used to remove tree sap spots.

BentleyPublishers.com
BentleyPublishers.com–All Rights Reserved

Underbody visual inspection

– Inspect the following for leaks or damage:
  - Engine
  - Transmission
  - Fuel system
  - Cooling and heating systems
  - Brake system
  - Exhaust system

A small amount of dampness is considered normal in some cases, especially around axle and pulley seals since the leaking fluid helps the seal work properly. On the other hand, expensive repairs can be avoided by prompt repair of minor fluid leaks. Judgement and experience are required to distinguish among the different kinds of fluid leaks.

– Inspect underside of vehicle for damage caused by normal wear and tear or by driving over road debris. Whenever vehicle is raised on a lift, inspect underbody, wheel wells and sill or rocker panels for damage to underbody sealants and coatings. Also inspect after major repairs to vehicle systems.

– Repair damage or defects found. Only use wax-based or tar-based anti-corrosion compounds as specified. Do not use oil-based anti-corrosion sprays due to possible incompatibility with factory applied protection.

Washing chassis

– Periodically wash underside of vehicle, especially in winter, to help prevent accumulation of road salt and rust.

The best time to wash the underside is just after the vehicle has been driven in wet conditions. Spray the chassis with a powerful jet of water. Commercial or self-service vehicle washes may not be best for this, as they may recycle the salt-contaminated water.

Waxing

– For a long-lasting, protective and glossy finish, apply a hard wax after the vehicle has been washed and dried. Use carnauba or synthetic based products.

Waxing is not needed after every washing. You can tell when waxing is required by looking at the finish when it is wet. If the water coats the paint in smooth sheets instead of forming beads that roll off, a new coat of wax is needed. Do not apply wax to black trim pieces, rubber, or other plastic parts.

Windshield wiper blade maintenance

Common problems with windshield wipers include streaking or sheeting, water drops after wiping, and blade chatter. Streaking is usually caused when wiper blades are coated with road film or vehicle wash wax. Clean the blades using soapy water. If cleaning the blades does not cure the problem, replace them. BMW recommends replacing the wiper blades twice a year, before and after the cold season.

BentleyPublishers.com
BentleyPublishers.com–All Rights Reserved

– Check tension spring that forces wiper against glass. Replace wiper arm if spring tension is weak.

– Drops that remain behind after wiping are caused by oil, road film, or diesel exhaust coating the windshield. Use an alcohol or ammonia solution or a nonabrasive cleanser to clean the windshield.

Wiper blade chatter may be caused by dirty or worn blades, by a dirty windshield, or by bent or twisted wiper arms. Clean the blades and windshield as described above. Adjust the wiper arm so that there is even pressure along the blade, and so that the blade is perpendicular to the windshield at rest. Lubricate the wiper linkage with a light oil. The linkage is located under the hood on the driver's side. If the problem persists, the blades are excessively aged or worn and should be replaced.

Wiper blade replacement is covered in **611 Wipers and Washers**.

CHECK CONTROL

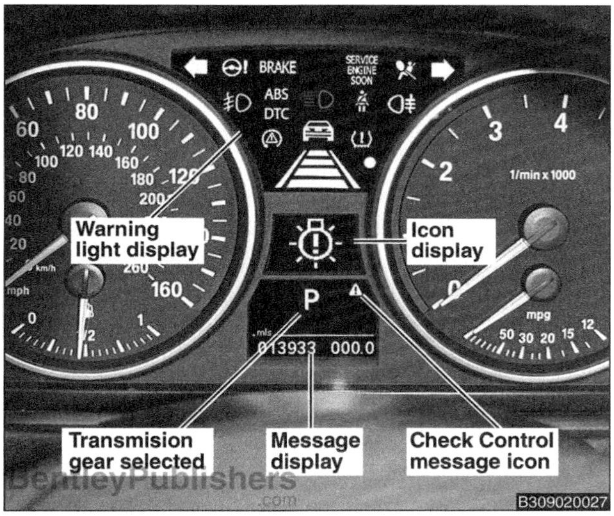

Check Control monitors vehicle functions and alerts the driver to malfunctions using indicator and warning lights in a variety of colors in the instrument cluster. In some circumstances, acoustic signals as well as text messages are given.

⚠ indicates that Check Control messages are stored.

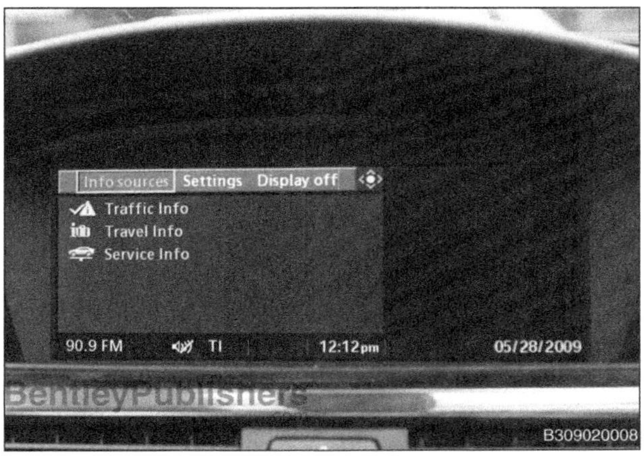

To view Check Control messages in central dashboard display: Press down iDrive controller to brings up **Info sources | Settings** menu.

NOTE —

• *Central dashboard display menus vary slightly on 2010 models.*

BentleyPublishers.com—All Rights Reserved

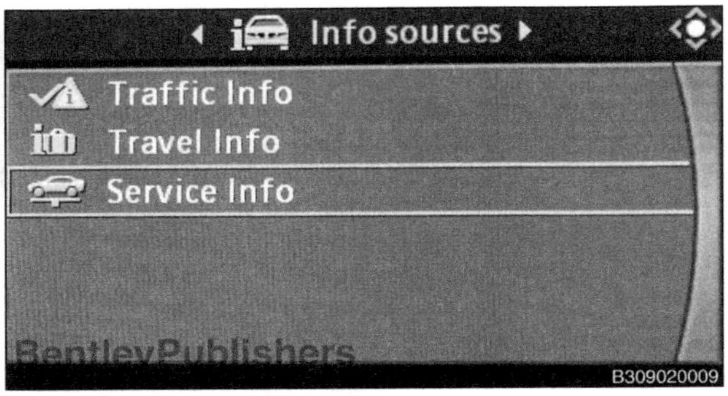

Pull back iDrive controller to enter middle screen, then rotate controller to **Service Info**.

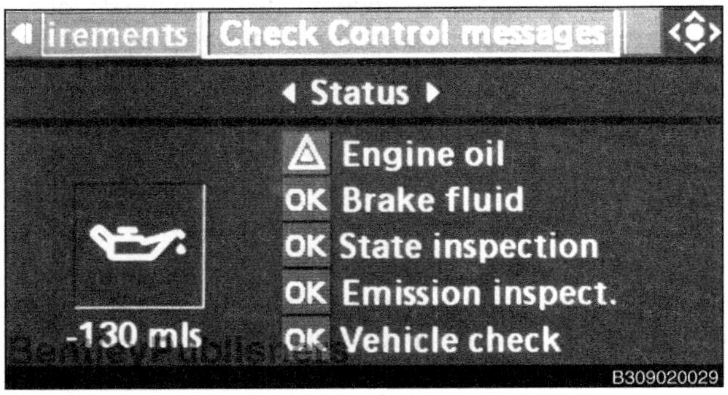

Press down controller to enter **Service Info** (or **Service requirements**) submenu.

– Rotate controller to **Check Control messages**.

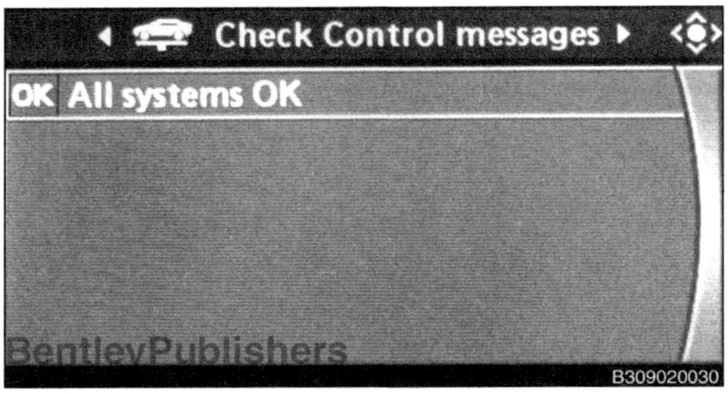

Press down controller to enter **Check Control messages**.

BentleyPublishers.com
BentleyPublishers.com—All Rights Reserved

100 Engine–General

TABLE

General

This section covers general information on engines and engine components. See **Table a** for engine application information and specifications.

For specific repair procedures, refer to:

- **020 Maintenance** for engine compartment views and accessory belt service, oil change and air filter service
- **110 Engine Removal and Installation**
- **113 Cylinder Head Removal and Installation**
- **117 Camshaft Timing Chain**
- **119 Lubrication System** for lubrication system description and crankcase seal service
- **120 Ignition System** for ignition coil, camshaft sensor, crankshaft sensor and knock sensor service
- **130, 131, 132, 133 Fuel Injection** for engine management system description and component replacement (based on engine type)
- **170 Radiator and Cooling System** for radiator, coolant pump and coolant service
- **180 Exhaust System** for exhaust manifold, oxygen sensor and turbocharger service

Bentley Publishers
BentleyPublishers.com–All Rights Reserved

Engine ID tag

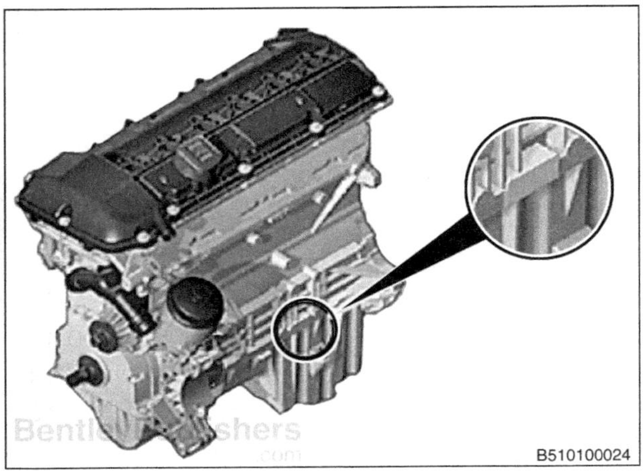

M54 engine ID tag with engine serial number is on left side of engine block, above engine oil pan.

N52 and N54 engine ID tag with engine serial number is on left side of engine block, underneath intake manifold.

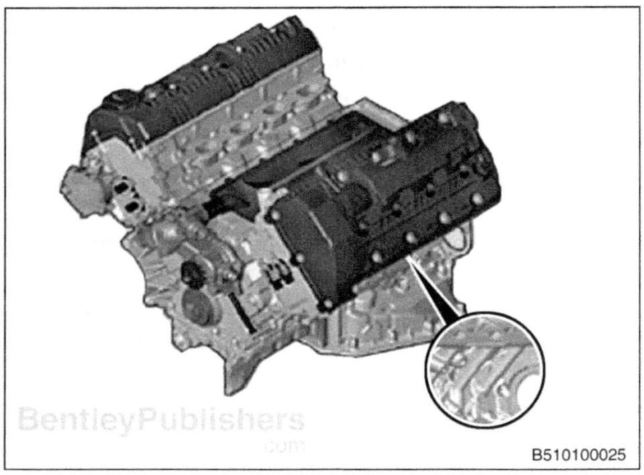

N62 engine ID tag with engine serial number is on left side of engine block, below cylinder head.

BentleyPublishers.com
BentleyPublishers.com–All Rights Reserved

Engine applications

Table a. 5 Series engine applications and specifications							
Year, model	**Engine code, type**	**Displacement cc (cid)**	**Bore x stroke mm (in)**	**Comp. ratio**	**Torque lb-ft @ rpm**	**Horsepower hp @ rpm**	**Engine management**
2.5 liter 6-cylinder							
2004 - 2005 525i	M54B25, double VANOS	2494 (152)	84 x 75 (3.31 x 2.95)	10.5 : 1	175 @ 3500	184 @ 5900	Siemens DME MS 45.1
3.0 liter 6-cylinder (normally aspirated)							
2004 - 2005 530i	M54B30, double VANOS	2979 (182)	84 x 89.6 (3.31 x 3.53)	10.2 : 1	214 @ 3500	225 @ 5800	Siemens DME MS 45.1
2006 - 2007 525i, 525xi Sports Wagon	N52B30, VANOS, Valvetronic	2996 (183)	85.0 x 88.0 (3.35 x 3.46)	10.7 : 1	185 @ 2750	215 @ 6250	Siemens MSV70
2006 - 2007 530i, 530xi, Sports Wagon	N52B30, VANOS, Valvetronic	2996 (183)	85.0 x 88.0 (3.35 x 3.46)	10.7 : 1	220 @ 2750	255 @ 6600	Siemens MSV70
2008 - 2010 528i, 528xi Sports Wagon	N52KB30, VANOS, Valvetronic	2996 (183)	85.0 x 88.0 (3.35 x 3.46)	10.7 : 1	200 @ 2750	230 @ 6500	Siemens MSV80
3.0 liter 6-cylinder (turbocharged)							
2008 - 2010 535i, 535xi 535xi Sports Wagon	N54B30 Turbo, VANOS	2979 (182)	84.0 x 89.6 (3.31 x 3.53)	10.2 : 1	300 @ 3500	300 @ 5800	Siemens MSD80 (up to 03/2008) MSD81 (from 03/2008)
4.4 liter V8							
2004 - 2005 545i	N62B44, VANOS, Valvetronic	4398 (268)	92 x 82.7 (3.62 x 3.26)	10.0 : 1	330 @ 3600	325 @ 6100	Bosch DME ME 9.2, with valvetronic control unit
4.8 liter V8							
2006 - 2010 550i	N62B48TU, Valvetronic, VANOS	4799 (293)	93 x 88.3 (3.66 x 3.48)	10.5 : 1	360 @ 3400	360 @ 6300	Bosch DME ME 9.2.2

BentleyPublishers.com
BentleyPublishers.com–All Rights Reserved

M54 Engine Overview

Engine introduction

525i and 530i E60 and E61 (Sedan and Sports Wagon) models were introduced in 2004 in the North American market equipped with the 3-liter in-line 6-cylinder normally-aspirated M54 engine. Two levels of performance were available:

- Low output engine in 525i models
- High output engine in 530i models

Cylinder block and crankshaft

The cylinder block is cast aluminum alloy ($AlSi_9Cu_3$) with cast iron cylinder liners. The cylinders are exposed on all sides to circulating coolant.

Connecting rods and pistons

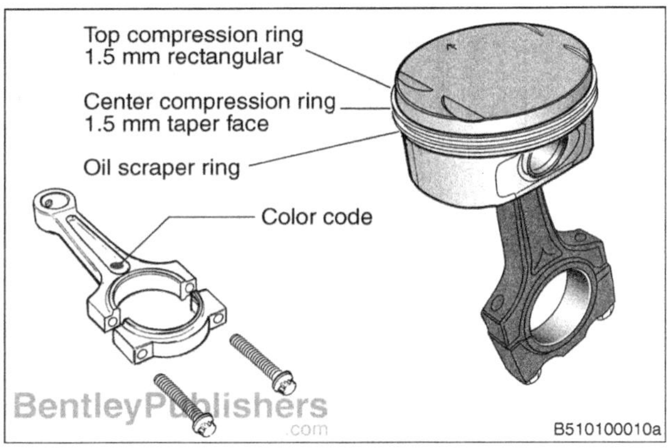

The forged steel connecting rods use replaceable split-shell bearings at the crankshaft end and solid bushings at the piston pin end. Connecting rods weight classes are indicated by a color code system.

The pistons are equipped with three rings with two upper compression rings and a lower one-piece oil scraper ring. Full floating piston pins are retained with circlips.

Graphite coating is used on piston skirts to reduce friction and noise.

Cylinder head and valvetrain

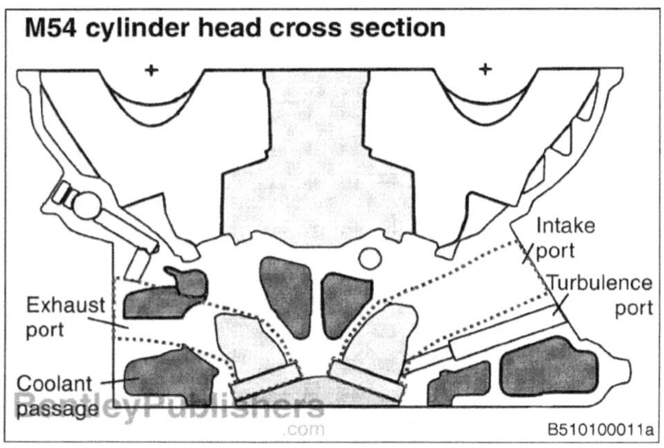

Cooling passages in the M54 cylinder head are designed for optimum coolant circulation, allowing the head to operate at lower temperatures than the cylinder block.

The aluminum cylinder head uses chain-driven double overhead camshafts and four valves per cylinder. The cylinder head employs a crossflow design for greater power and efficiency. Intake air enters the combustion chamber from one side while exhaust gasses exit from the other. The spark plugs are centrally located down the middle of the cylinder head.

BentleyPublishers.com
BentleyPublishers.com–All Rights Reserved

Double VANOS

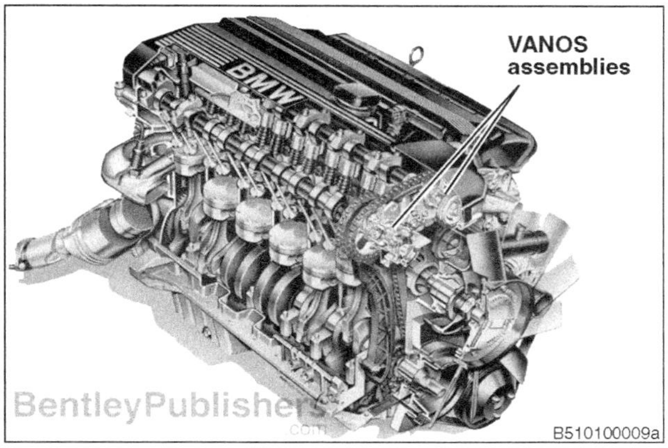

M54 engines are equipped with compact, infinitely variable vane-type VANOS units for both intake and exhaust camshafts. The VANOS system (from the German words *VAriable NOckenwellen Steuerung*) electrohydraulically adjusts valve timing for enhanced mid-range performance and improved emissions. VANOS is controlled by the engine control module (ECM), using engine speed, engine load and engine coolant temperature as the primary inputs. Engine oil pressure is used for powering camshaft adjustment.

Double VANOS components

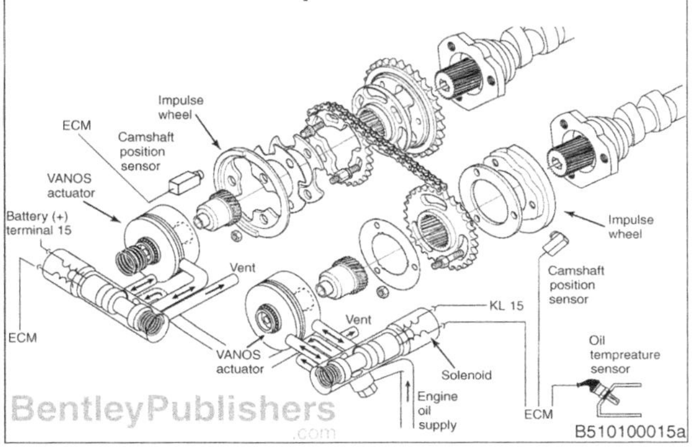

Engine oil pressure is used to position the VANOS actuators. Oil pressure is fed from the engine oil pump up to the VANOS solenoids; oil drains back to the sump as the camshafts are adjusted during engine operation.

With the engine off, the base setting of the camshafts is as follows:

- Intake camshaft: Retarded
- Exhaust camshaft: Advanced

This is also the fail-safe position in case of an electronic control failure. Both camshafts are held in these positions by oil pressure from the engine oil pump. The exhaust camshaft is held additionally by a spring in the VANOS actuator.

When the engine is started, the camshafts remain in these positions until the ECM detects the positions of the camshafts from the camshaft sensors (approximately 50 revolutions or 2 - 5 seconds). The engine is equipped with static Hall-effect camshaft sensors so that camshaft positions are recognized as soon as the ignition is switched ON, before the engine is started.

Once camshaft positions are recognized, the ECM makes an initial camshaft timing adjustment based on oil temperature, oil pressure and engine rpm. Following this initial setting, engine rpm, throttle position signal, intake air temperature and engine coolant temperature are used to adjust camshaft timing.

When the ECM detects that the camshafts are in the desired position, the solenoids are modulated at 100 - 220 Hz, maintaining oil pressure on both sides of the actuators to maintain camshaft timing.

In addition to increasing engine power output, double VANOS offers the following advantages:

- Increased torque at lower and medium rpm ranges
- More efficient combustion and improved idle quality
- Internal EGR in part-load range for lower NO_x emissions
- Quicker warm-up cycle for catalytic converter and faster reduction in emissions
- Overall improved fuel economy

See **117 Camshaft timing chain** for Vanos system repairs.

BentleyPublishers.com
BentleyPublishers.com—All Rights Reserved

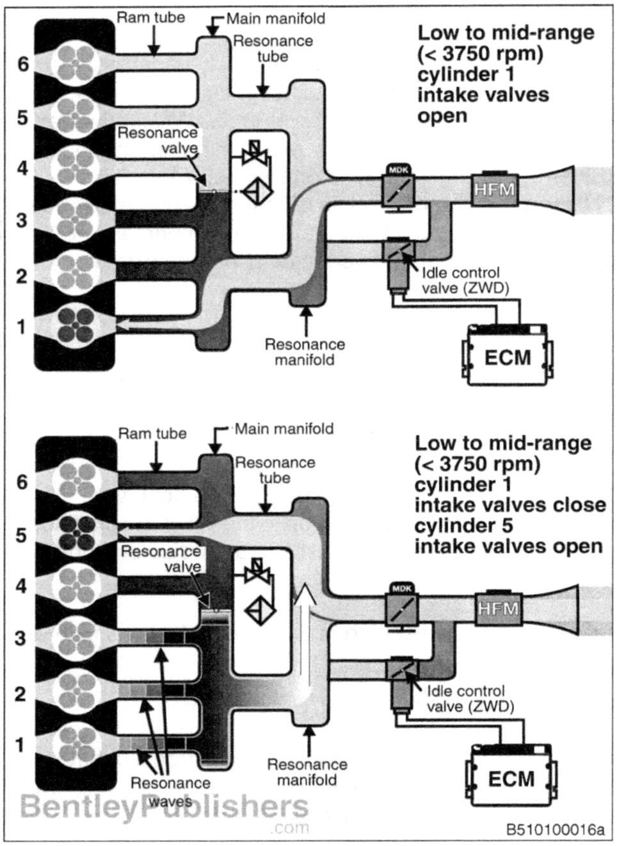

Intake system

The intake manifold, made of molded plastic, is configured as two sets of three runners. This design enhances low end torque by changing the intake air flow configuration for varying engine speeds. This helps achieve optimum torque throughout the entire rpm range.

During engine operation, a closed resonance valve gives the intake air charge the dynamic effect of long intake runners at low to mid-range engine speeds (up to 3750 rpm). This helps increase torque.

During mid-range to high speed operation (above 4100 rpm), the solenoid is de-energized and the resonance valve is sprung open. This allows intake air to be drawn through both resonance tubes, providing the air volume necessary for additional power at the upper rpm range.

The rpm for resonance valve activation varies slightly depending on temperature.

In addition, when the valve is closed, a dynamic effect is produced. For example, as intake air is flowing into cylinder 1, intake valves close. This blocks onrushing air. Cylinder 1 air flow stops and expands backward (resonance back pulse) to fill cylinder 5. The resonance wave along with the intake velocity enhances cylinder filling.

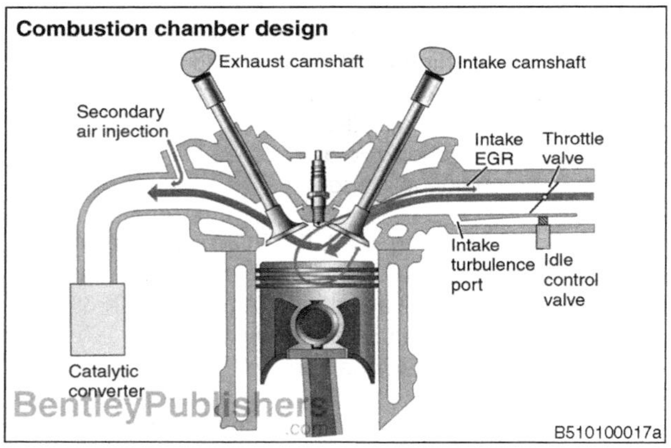

The intake manifold includes intake turbulence ports. The 5.5 mm (0.217 in.) turbulence ports channel idle and low speed air directly from the idle speed control valve to one intake valve of each cylinder.

Routing intake air to one intake valve per cylinder causes the air charge to swirl in the cylinder. Together with the high flow rate of intake air across the small (5.5 mm) port, intake fluctuations are reduced for more stable combustion.

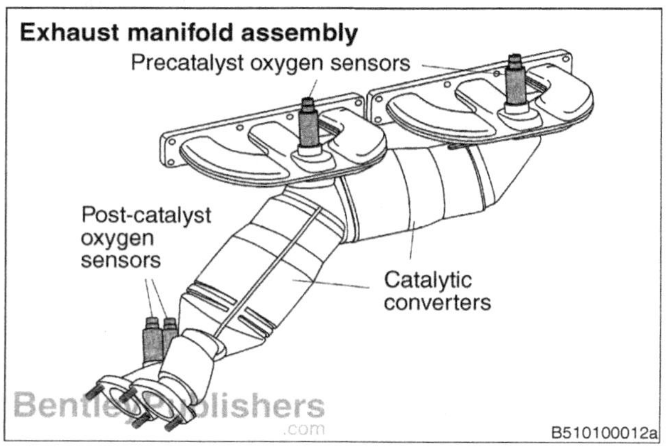

Exhaust manifolds

Each exhaust manifold assembly incorporates a catalytic converter. The converters are mounted close to the engine for faster heat up and light off. Pre- and post-catalyst oxygen sensors are a mounted in each exhaust manifold.

BentleyPublishers.com
BentleyPublishers.com–All Rights Reserved

Cooling system

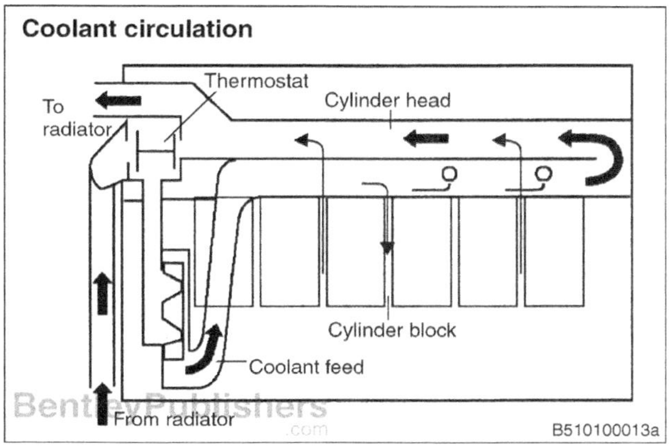

The cooling system is designed so that coolant flows directly from the coolant pump to the cylinder head. The coolant is fed from the coolant pump through a cast coolant feed passage to the rear of the cylinder head. From there it flows forward to the thermostat housing, radiator and heater valve.

525i and 530i models are equipped with a DME-controlled electric cooling fan. See **170 Radiator and Cooling System** for additional information

Lubrication system

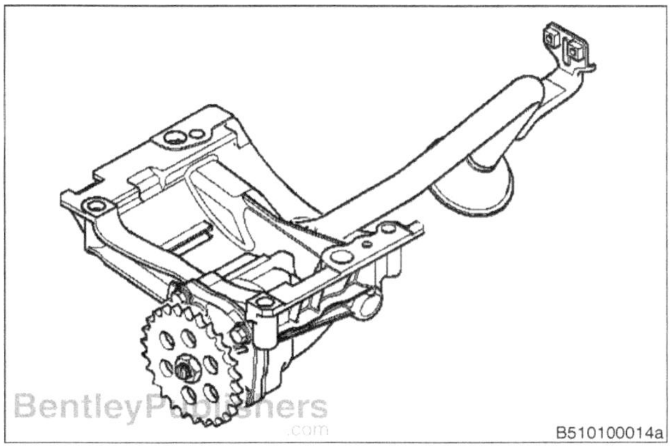

The lubrication system is pressurized whenever the engine is running. The oil pump draws oil through a pickup in the bottom of the oil pan, then forces it through a replaceable oil filter and into the engine oil passages.

Oilways in the cylinder head provide lubrication for the camshafts and valvetrain.

The chain-driven oil pump is bolted to the bottom of the cylinder block inside the oil pan. A pressure relief valve limits the maximum system pressure. A bypass valve prevents the oil filter from bursting and insures engine lubrication should the filter become plugged.

See **119 Lubrication System** for additional information.

BentleyPublishers.com–All Rights Reserved

N52 ENGINE OVERVIEW

Engine introduction

Beginning with 2006, 5 Series models were equipped with 3-liter in-line 6-cylinder normally-aspirated N52 engine. 2-wheel drive and 4-wheel drive models were offered:

- Low output engine in 525i or 525xi models
- High output engine in 530i or 530xi models

Beginning with 2008, 5 Series models were fitted with the N52KP engine. 2-wheel drive and 4-wheel drive (xDrive) models were offered:

- 528i or 528i xDrive models
- 530i or 530i xDrive models

The N52 engine family features the following:

- 4-valves-per-cylinder friction-optimized components
- Two-piece crankcase
- Composite magnesium-aluminum engine block structure
- Trapezoidal weight-optimized connecting rods
- Aluminum silicon (AluSil) cylinder head
- Timing case integrated in crankcase and cylinder head
- Cylinder head gasket with silicon sealing lip
- Valvetronic system
- Weight-optimized double VANOS system
- Volumetric flow-controlled oil pump
- Electrically controlled coolant pump
- Crankcase ventilation with integrated heater
- 3-stage variable intake manifold or DISA

The weight reduction in these systems and components, as compared to the M54 engine, led to increased power-to-weight ratio, reduced emissions and decreased fuel consumption.

Crankcase components

Crankcase

The upper section of the crankcase consists of an aluminum / silicon (AluSil) insert cast in a magnesium alloy. The timing cover is cast as an integral part of the engine block.

BentleyPublishers.com
BentleyPublishers.com—All Rights Reserved

The AluSil cylinder bores are not equipped with iron cylinder liners. Therefore, the cylinder bores cannot be machined, but the deck surface may be planed.

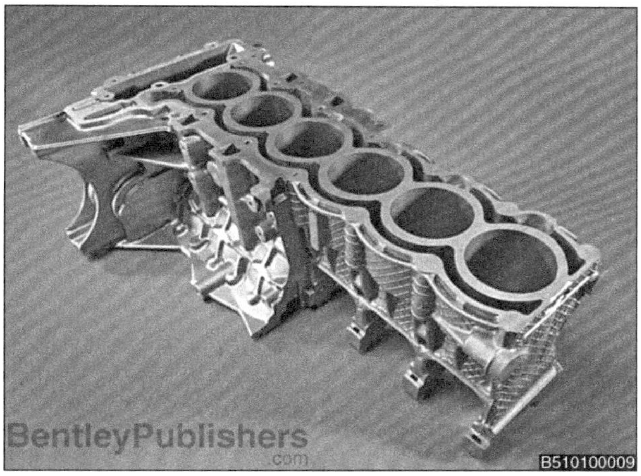

The AluSil insert has threaded bores for transmission, cylinder head and crankshaft main bearing bolts.

The insert provides coolant passages as well. This is to prevent coolant contact with the magnesium portion of the engine block.

The lower section of the crankcase is a bedplate structure made of magnesium with sintered steel inlays (**arrow**) for main bearing support. The magnesium structure is used to increase crankcase rigidity and the steel inlays take up forces which are not suitable for magnesium alone.

A liquid sealant is injected under high pressure into a machined groove between upper and lower crankcase sections. This detail is critical to understand in service applications. As an example, when the crankcase main seals are replaced, a special sealant is injected into the seal grooves and crankcase seams.

BentleyPublishers.com
BentleyPublishers.com–All Rights Reserved

B510100010

Crankshaft

The crankshaft is cast iron with 7 main bearing journals. The trigger wheel for the crankshaft sensor is between cylinders 5 and 6. Due to the design of the timing chain housing (one piece with the crankcase), the crankshaft snout is modified to facilitate timing chain module removal and installation during service.

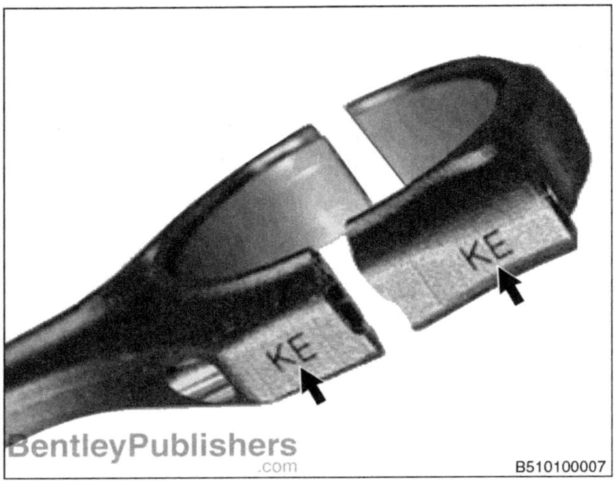

B510100007

Connecting rods

The big end of each connecting rod is cracked to allow centering of the bearing cap without the use of dowel pins. This contributes to the overall weight reduction.

Pairing codes (**arrows**) are stamped on each connecting rod to match the correct rod cap to the connecting rod.

The small end of the rod is tapered, reducing weight without affecting strength.

The connecting rods are separated into weight categories and can only be replaced as a set.

Pistons

The pistons are manufactured from aluminum and have 4 valve reliefs. The undersides of the pistons are cooled with oil spray jets.

Cylinder head and valvetrain

Cylinder head

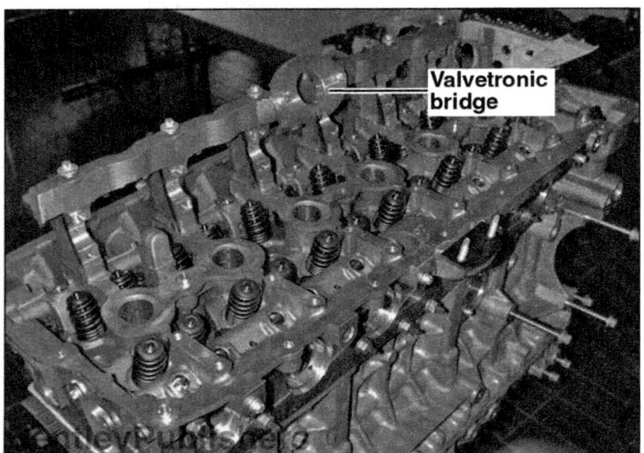

B510100011

The N52 engine uses an AluSil cylinder head with a cast bridge to mount the Valvetronic actuator motor.

BentleyPublishers.com
BentleyPublishers.com–All Rights Reserved

Variable camshaft timing (VANOS)

N52 engines are equipped with compact, infinitely variable vane-type VANOS units for both intake and exhaust camshafts. The VANOS system (from the German words *VAriable NOckenwellen Steuerung*) electrohydraulically adjusts valve timing for enhanced mid-range performance and improved emissions. VANOS is controlled by the engine control module (ECM), using engine speed, engine load and engine coolant temperature as the primary inputs. Engine oil pressure is used for powering camshaft adjustment.

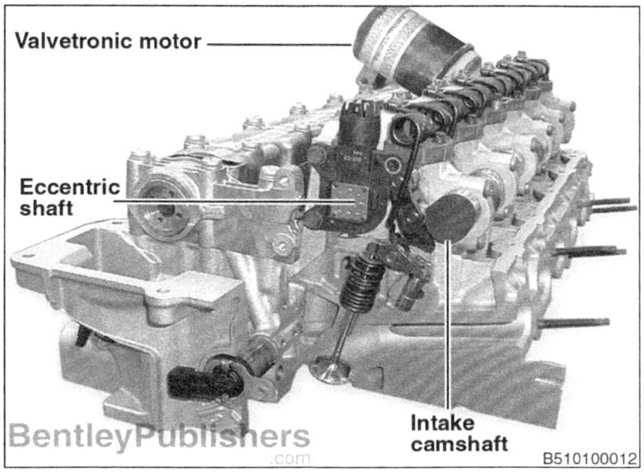

Valvetronic

N52 engine load is controlled via the valve timing gear. This system, called Valvetronic, varies valve lift rather than throttle valve opening to control engine power and torque. It offers the following:

- Increased engine efficiency.
- Improved emission values.
- Top engine speed to 7,000 rpm.
- Power output of 85 hp / liter displacement.
- Engine torque of 100 Nm / liter displacement over a broad engine speed range.
- Reduced CO_2 emissions.

Intake manifold

Variable intake manifold (DISA)

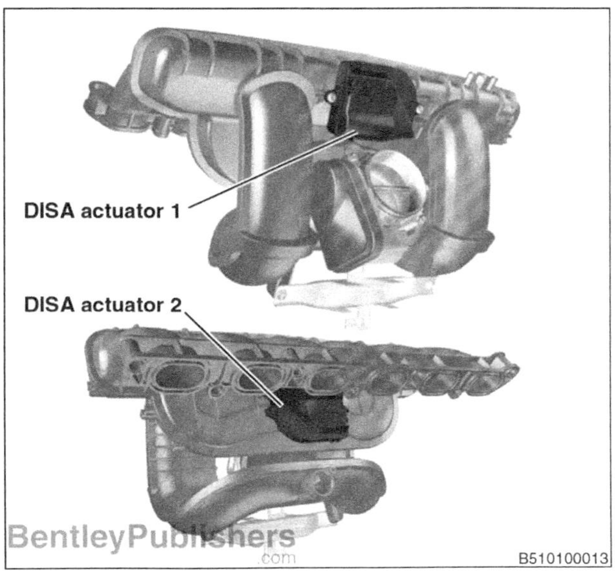

5 Series models with N52 engines are equipped with 3-stage variable intake runners in a system known as DISA.

Valves in the intake manifold are driven by DISA electric motors and gear mechanisms. DISA actuators are driven by pulse width-modulated signals from the engine control module (ECM). Each valve has only two possible positions: Open or closed.

BentleyPublishers
BentleyPublishers.com–All Rights Reserved

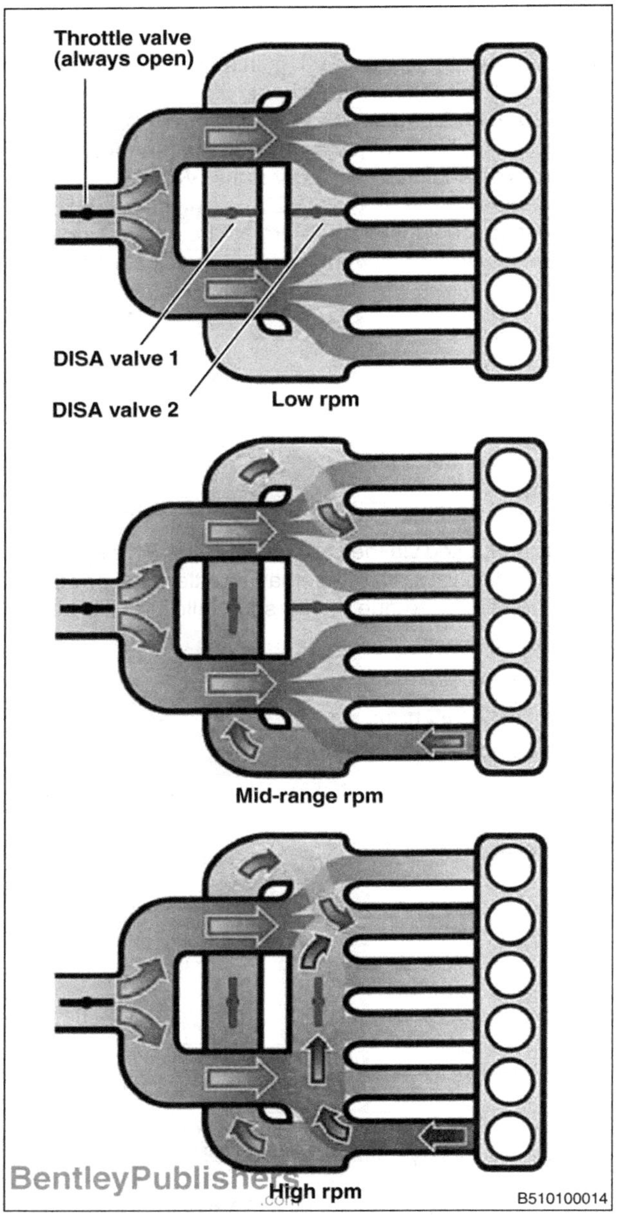

Switched-length intake manifold runners help achieve a high torque over the entire engine speed range using two DISA valves and an overflow pipe in the intake.

- **Stage 1**: Idle and low engine speed range. Both DISA valves closed. Intake air flows past throttle valve into resonance pipe. Air flow is split in resonance pipe and is further routed via collector runner and intake resonating runners into individual cylinders. In this way, a relatively large air mass is made available to cylinders.
- **Stage 2**: Medium engine speed range. DISA valve 2 open. (In this example, cylinder 1 intake valves are just closing.) Air flow produces pressure peaks at closing intake valves. This resonates to next cylinder in firing order, thus improving charge of next cylinder.
- **Stage 3**: High engine speed range. Both DISA valves open. (Again, In this example, cylinder 1 intake valves are just closing.) Pressure peaks ahead of closing intake valves are also utilized in this case. Intake air mass is now routed via resonance, overflow and manifold runners.

BentleyPublishers.com
BentleyPublishers.com—All Rights Reserved

Lubrication, vacuum and cooling systems

Lubrication and vacuum systems

VANOS requires a large volume of oil, particularly at low engine speeds. To meet the oiling requirements of the engine, a high volume volumetric-flow controlled oil pump is utilized. The pump is designed to provide the high volume when needed, but drop back its flow at higher engine speeds when it is not needed. See **119 Engine Lubrication**.

Valvetronic engine load control requires a wide open throttle under most engine operating conditions, resulting in minimal intake manifold vacuum. In order to operate the brake vacuum booster, an auxiliary vacuum pump is fitted to the engine.

A chain assembly is used to drive the oil pump and brake booster auxiliary vacuum pump sprockets.

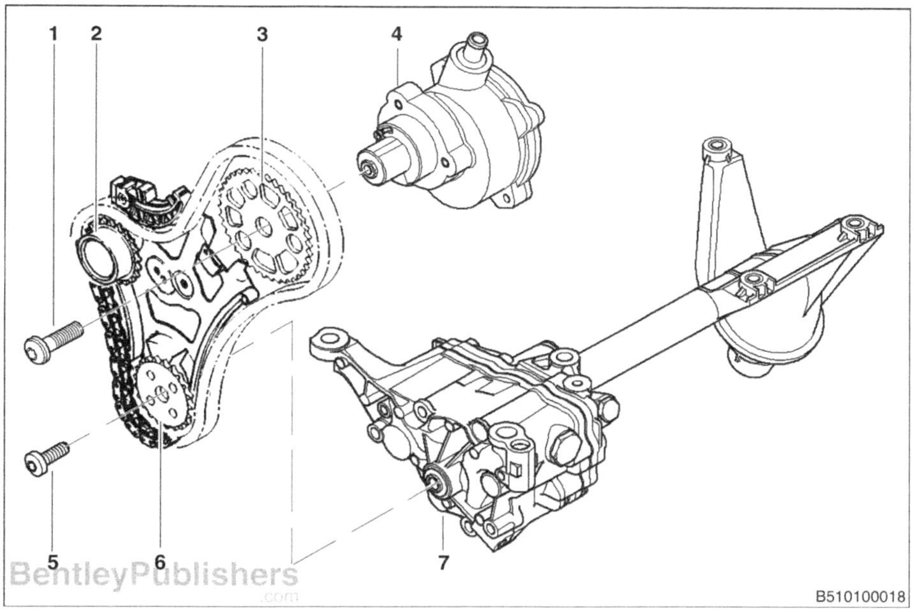

1. Auxiliary vacuum pump sprocket bolt
2. Crankshaft sprocket
3. Auxiliary vacuum pump drive sprocket
4. Auxiliary vacuum pump
5. Oil pump sprocket bolt
6. Oil pump drive sprocket
7. Oil pump

Cooling system

The cooling system utilizes a DME-controlled electric coolant pump which minimizes engine power losses and allows for increased fuel economy by a more efficient method of engine heat management

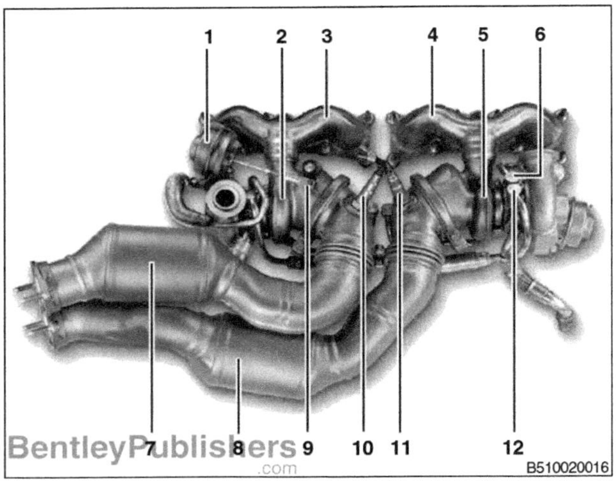

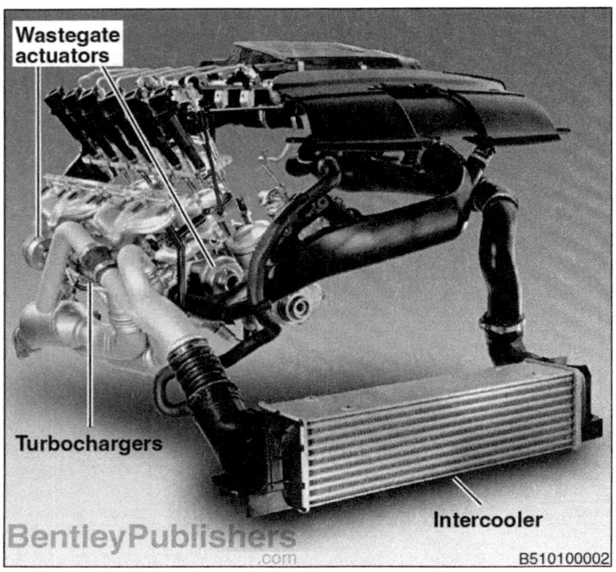

N54 Engine Overview

Engine introduction

Beginning in 2008 the 535i Sedan and Sports Wagon models were equipped with the turbocharged in-line 6-cylinder N54 engine. Available as a 3.0 liter engine, it shares the following features with the N52 engine:

- Double VANOS
- Aluminum cylinder head with plastic cylinder head cover
- Steel crankshaft

The following features are specific to the N54 engine:

- Exhaust driven twin turbochargers (bi-turbo)
- Air to air intercooler
- High output (400 W) electric coolant pump
- Direct injection (high pressure injection or HPI) with piezo injectors
- Bosch MSD81 engine management
- External oil cooler
- All aluminum crankcase with iron cylinder liners

Turbochargers

1. Wastegate actuator, bank 2
2. Turbocharger, bank 2
3. Exhaust manifold, bank 2
4. Exhaust manifold, bank 1
5. Turbocharger, bank 1
6. Coolant return
7. Catalytic converter, bank 2
8. Catalytic converter, bank 1
9. Wastegate actuating lever
10. Planar broad-band oxygen sensor, bank 2
11. Planar broad-band oxygen sensor, bank 1
12. Coolant supply

Intercooler

The charge air is greatly heated when it is compressed in the turbochargers, making it necessary for the air to be cooled again in an intercooler.

BentleyPublishers.com
BentleyPublishers.com–All Rights Reserved

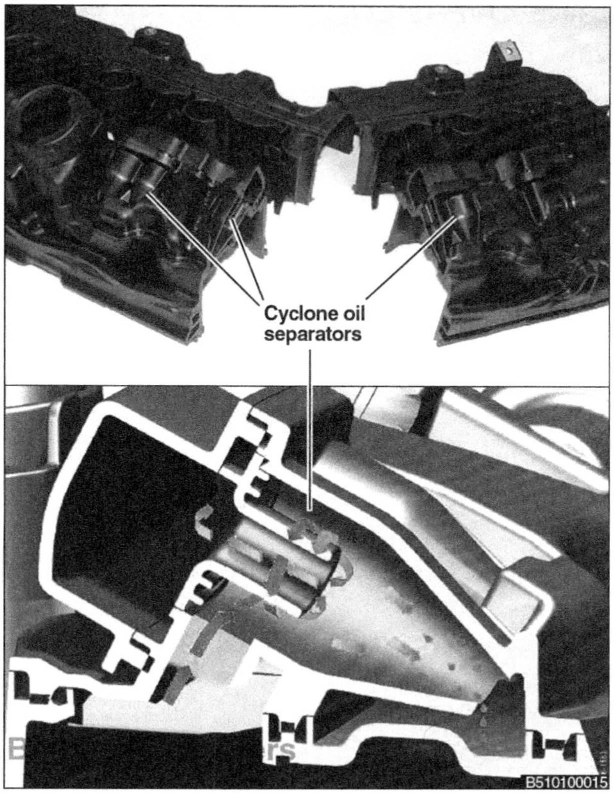

Crankcase breathing

The turbocharged engine crankcase breather valve system is designed so that when the engine is in turbocharged mode, increased manifold pressure does not adversely affect crankcase venting. Instead of a crankcase ventilation valve, the engine breathing system consists of small cyclone separators integrated into the plastic cylinder head cover. The flow of crankcase gases is metered through a series of restrictions which control crankcase pressure.

Engine oil, vacuum and fuel pump drive

As with the N52 engine, the turbo engine oil pump and auxiliary vacuum pump are driven by a single chain powered by a sprocket at the front of the crankshaft. In addition, the high pressure fuel pump is tandem mounted to the rear of the vacuum pump.

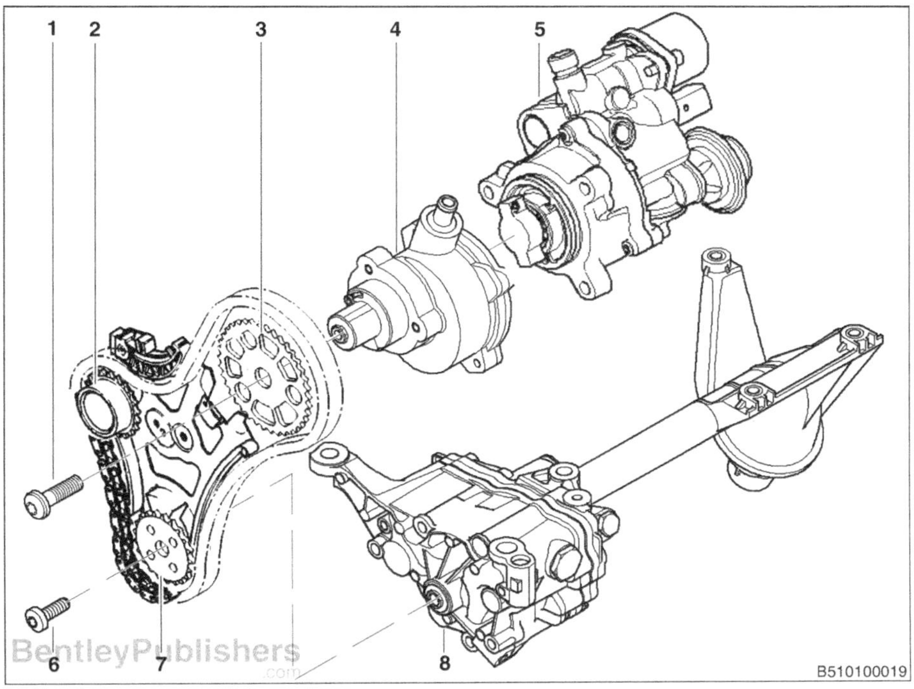

1. Auxiliary vacuum pump sprocket bolt
2. Crankshaft sprocket
3. Auxiliary vacuum pump-high pressure fuel pump drive sprocket
4. Auxiliary vacuum pump
5. High pressure fuel pump
6. Oil pump sprocket bolt
7. Oil pump drive sprocket
8. Oil pump

BentleyPublishers.com–All Rights Reserved

N62 Engine Overview

Engine introduction

545i and 550i models are equipped with the V8 N62 engine with Valvetronic.

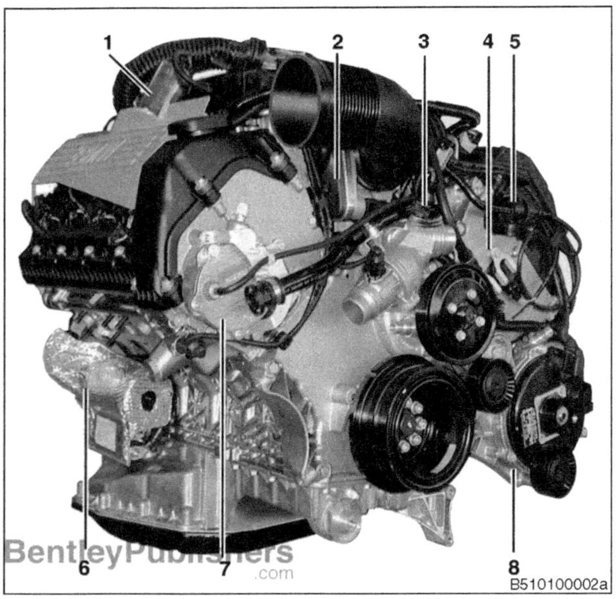

N62 engine from the front:

1. Valvetronic actuator
2. Throttle unit
3. Thermostat housing
4. VANOS solenoid
5. Evaporative emissions valve
6. Starter with heat shield
7. Brake booster vacuum pump
8. Water-cooled alternator

BentleyPublishers.com
BentleyPublishers.com–All Rights Reserved

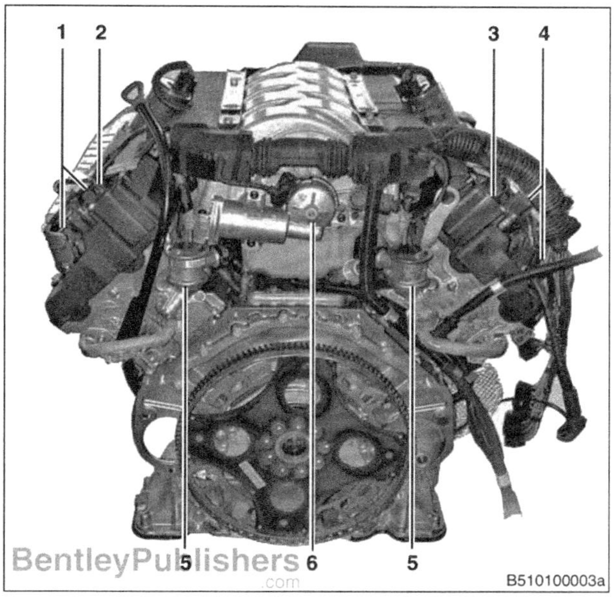

N62 engine from the rear:

1. Camshaft sensors, left bank
2. Valvetronic eccentric shaft position sensor, left bank
3. Valvetronic eccentric shaft position sensor, right bank
4. Camshaft sensors, right bank
5. Secondary air injection valves
6. Variable intake manifold servomotor

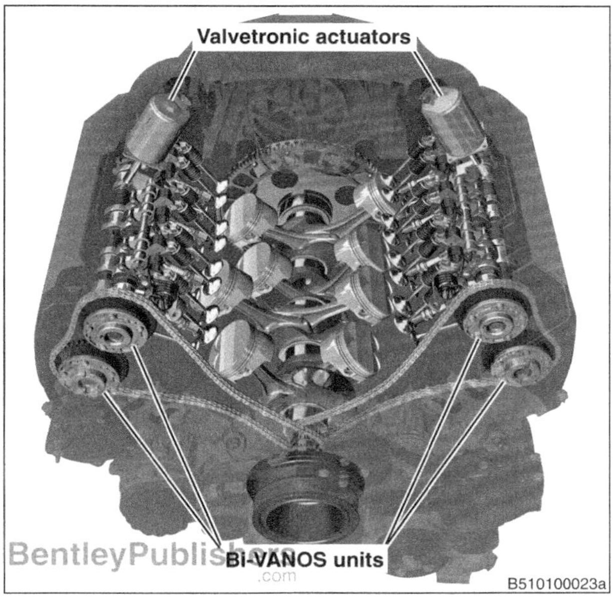

In addition to Valvetronic technology, the N62 engine is equipped with double VANOS.

Double VANOS allows for variable timing on both intake and exhaust camshafts. The VANOS unit is designed as an integral component of the chain drive and is secured to the respective camshaft with a central bolt.

BentleyPublishers.com
BentleyPublishers.com–All Rights Reserved

Valvetronic principles

The Valvetronic system uses hardware and software to eliminate the conventional throttle mechanism. Instead of a throttle valve, intake air is regulated by adjusting valve lift. BMW claims that removing the throttle valve as an obstruction results in 10% fuel savings by minimizing pumping losses.

In order to diagnose and repair the Valvetronic system, be sure to have access to BMW scan tool.

Valvetronic components

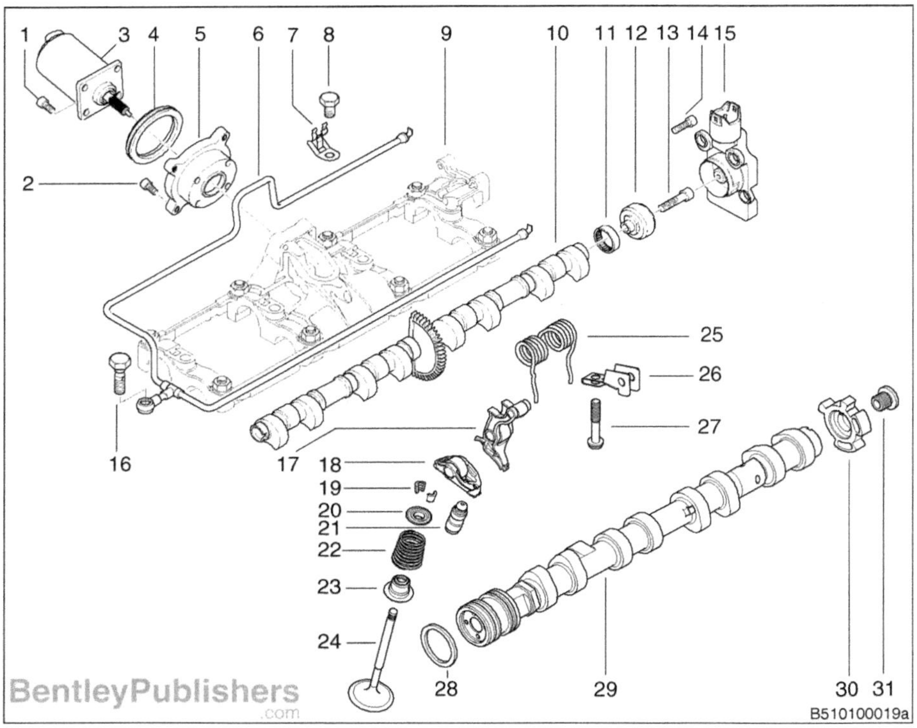

1. Bolt M5 x 12 mm
2. Bolt M5 x 30 mm
3. Valvetronic actuator
4. O-ring
5. Spacer
6. Oil spray tube
7. Locking clip
8. Bolt M7 x 8 mm
9. Bridge support
10. Eccentric shaft
11. Needle bearing
12. Eccentric shaft magnetic pulse generator
13. Bolt M6 x 20 mm
14. Bolt M5 x 16 mm
15. Eccentric shaft position sensor
16. Bolt M8 x 50 mm
17. Intermediate lever
18. Valve rocker (with roller)
19. Valve keeper
20. Upper valve spring seat
21. Hydraulic valve adjuster (HVA)
22. Valve spring
23. Valve stem seal
24. Intake valve
25. Retainer spring
26. Spring support
27. Torx bolt M7 x 44 mm
28. Sealing O-ring
29. Intake camshaft
30. Camshaft position pulse generator wheel
31. Allen bolt M14 x 19 mm

BentleyPublishers.com
BentleyPublishers.com–All Rights Reserved

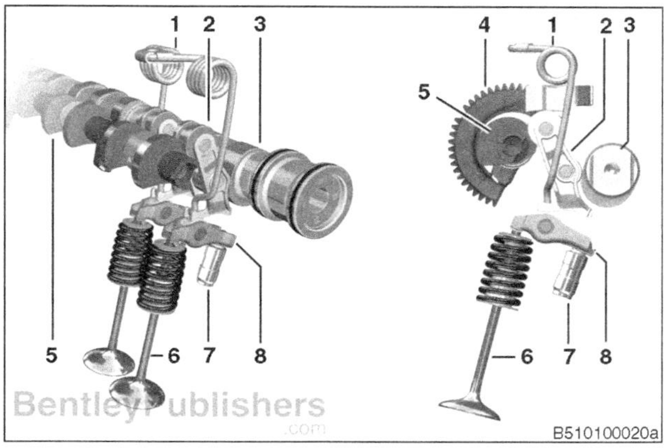

Valvetronic operation

1. Retainer spring
2. Intermediate lever
3. Intake camshaft
4. Cam gear operated by Valvetronic actuator
5. Eccentric shaft
6. Intake valve
7. Hydraulic valve adjuster (HVA)
8. Valve rocker (with roller)

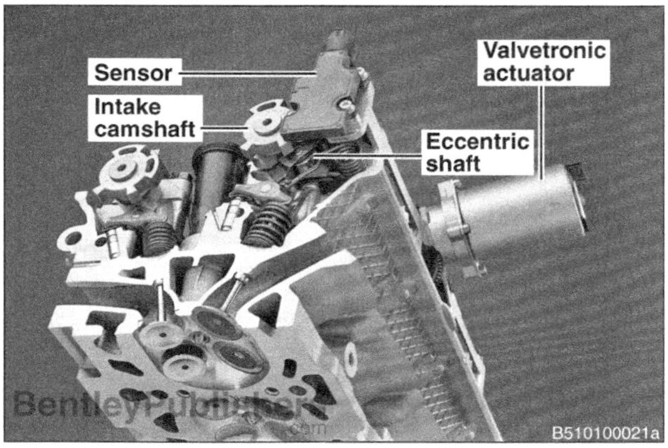

The Valvetronic actuator in each cylinder bank operates an eccentric shaft. Lobes on shaft press on intermediate levers between intake camshaft lobes and each cylinder's pair of intake valve rockers.

Depending on position of eccentric shaft lobes, the size of intake valve opening ranges from about 0.3 to 9.85 mm.

Valvetronic electronic control

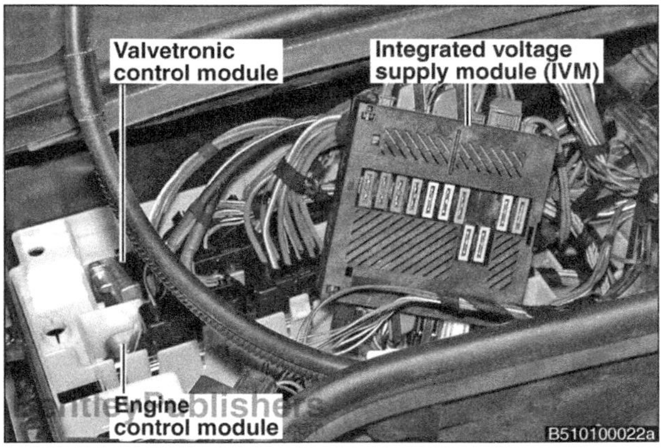

Electrical and electronic controls for the Valvetronic system consist of:

- Valvetronic control module in E-box
- Engine control module (ECM) in E-box
- DME and valvetronic relays in integrated voltage supply module (IVM) in E-box
- Valvetronic actuators
- Eccentric shafts position sensors
- Eccentric shafts magnetic pulse generators

The Valvetronic DC motors (actuators) which adjust the eccentric shafts are operated at 16 kHz for exact control of valve lift. Once the target valve opening is attained, as determined by the ECM, polarity is briefly reversed to generate braking force and to position motors exactly. Adjustment time required to move motors from minimum to maximum valve lift is approximately 300 ms. Motor draw can peak at 100 amperes.

When the engine is stopped, self-stopping motors and worm gear drives prevents position changes. When restarted, if automatic valve opening adjustment is not detected, a fault is recorded and motors are moved back to the target positions.

BentleyPublishers.com
BentleyPublishers.com–All Rights Reserved

BentleyPublishers.com
BentleyPublishers.com—All Rights Reserved

110 Engine Removal and Installation

GENERAL

This repair group includes an overview of M54, N52, N54 (6-cylinder) and N62 (V8) engine removal and installation.

Transmission removal is required before the engine can be removed.

See also:

- **020 Maintenance** for engine covers removal
- **100 Engine–General** for engine codes and applications
- **121 Battery, Starter, Alternator**
- **170 Radiator and Cooling System**
- **180 Exhaust System**
- **230 Manual Transmission**
- **231 Sequential Manual Gearbox (SMG)**
- **240 Automatic Transmission**
- **260 Driveshafts**
- **310 Front Suspension**
- **410 Fenders, Engine Hood**

BentleyPublishers
BentleyPublishers.com—All Rights Reserved

Warnings and Cautions

WARNING—
- *The fuel system is designed to retain pressure even when the ignition is OFF. When working with the fuel system, loosen fuel lines slowly to allow residual fuel pressure to dissipate. Avoid spraying fuel.*
- *Before beginning any work on the fuel system, place a fire extinguisher in the vicinity of the work area.*
- *Fuel is highly flammable. When working around fuel, do not disconnect any wires that could cause electrical sparks. Do not smoke or work near heaters or other fire hazards.*
- *Loosen the fuel tank cap to release pressure in the tank before working on the tank or lines.*
- *When disconnecting a fuel hose, wrap a shop towel around the end of the hose to prevent fuel spray.*
- *Do not use a work light with an incandescent bulb near any fuel. Fuel may spray on the hot bulb causing a fire.*
- *Make sure the work area is properly ventilated.*

CAUTION—
- *If the MIL (malfunction indicator light, also called Check Engine or Service Engine Soon light) is illuminated, see* **OBD On-Board Diagnostics** *for DME fault code information.*
- *If other system faults are indicated by an illuminated ABS, SRS or DSC warning light, see the appropriate repair group in this manual or an authorized BMW dealer for more information on fault codes.*
- *Magnesium crankcase requires aluminum fasteners.*
- *Steel fasteners may not be used in place of aluminum fasteners due to the threat of corrosion.*
- *Aluminum fasteners may be marked with blue paint. Use a magnet to identify.*
- *Replace aluminum fasteners each time they are loosened.*

BentleyPublishers.com
BentleyPublishers.com—All Rights Reserved

ENGINE REMOVAL AND INSTALLATION (M54 ENGINE)

Be sure to cover painted surfaces before beginning the removal procedure. As an aid to installation, label components, wires and hoses before removing them. Do not reuse gaskets, O-rings or seals during reassembly.

> ***WARNING —***
> - *Due to risk of personal injury, be sure the engine is cold before beginning the removal procedure.*

M54 engine, removing and installing

- Place hood in service position. See **410 Fenders, Engine Hood**.
- Remove exhaust system. See **180 Exhaust System**.
- Secure engine with BMW special tool 00 6 000 engine support or equivalent.
- Disconnect negative (–) cable from battery.

> ***CAUTION—***
> - *Prior to disconnecting the battery, read the battery disconnection cautions in* **001 Warnings and Cautions**.

- Remove transmission. See **230 Manual Transmission**, **231 Sequential Manual Gearbox (SMG)** or **240 Automatic Transmission**.
- Drain engine oil.
- Drain engine coolant.
- Remove air cleaner housing and fresh air ducts. See **130 Fuel Injection**.
- Remove fan cowl with electric cooling fan. See **170 Radiator and Cooling System**.
- Remove radiator. See **170 Radiator and Cooling System**.
- Remove coolant expansion tank. See **170 Radiator and Cooling System**.
- Remove intake manifold. See **130 Fuel Injection**.

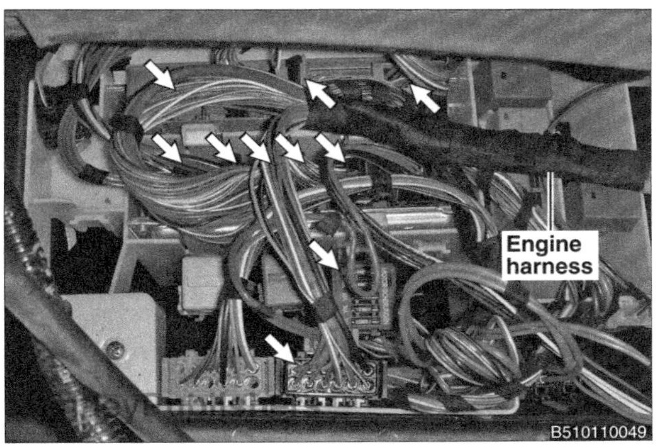

◄ Working in E-box, detach engine wiring harness connectors from ECM, transmission control module, engine electronics fuse carrier and other connectors (**arrows**).

- Disconnect electrical connectors and harness from ignition coils and lay harness aside.

- Disconnect positive battery cable (**arrow**) at intake manifold.
- Remove vacuum line from brake booster.
- Disconnect vapor line from fuel tank vent valve.
- Remove pre-catalyst and post-catalyst oxygen sensors. See **180 Exhaust System**.
- Working at right side engine mount bracket, remove ground cable.
- Remove engine accessory belt. See **020 Maintenance**.
- Remove power steering pump with bracket. See **320 Steering and Wheel Alignment**.

< Remove A/C compressor mounting bolts (**arrows**) and lift A/C compressor off its mounting bracket without disconnecting, distorting, or deforming refrigerant lines. Suspend from body using stiff wire.

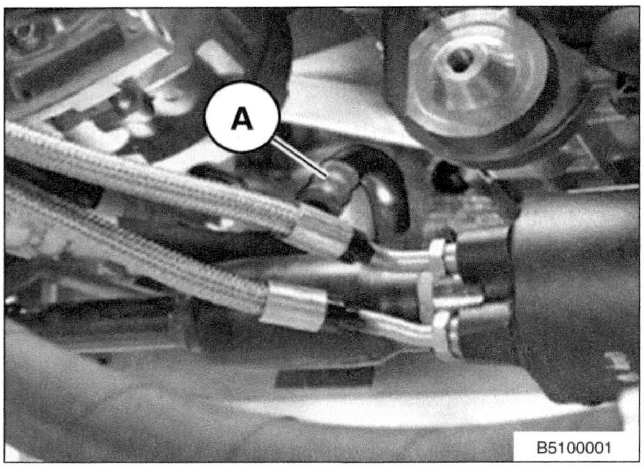

< Vehicle with dynamic drive (active stabilizer bar): Working at front stabilizer bar, remove hydraulic line (**A**).

- Working above engine, remove engine mount fasteners.

BentleyPublishers.com
BentleyPublishers.com–All Rights Reserved

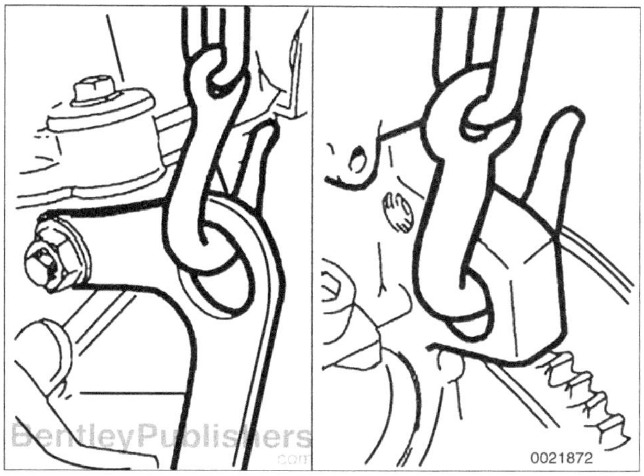

Install engine lifting device (BMW 11 0 000 or equivalent) to the front and rear engine supports and raise engine until its weight is supported.

– Carefully raise engine out of engine compartment, checking for any wiring, fuel lines, or mechanical parts that might become snagged as engine is removed.

– Installation is reverse of removal, noting the following:
  - Replace gaskets, O-rings and seals.
  - Use new fuel injector seals.
  - Check that engine drive belts properly engage pulley grooves.
  - Inspect O-ring seal between mass air flow sensor and air filter housing. To facilitate reassembly, coat seal with acid-free grease.
  - Change engine oil and filter and check all other fluid levels. See **020 Maintenance**.
  - Refill and bleed cooling system. See **170 Radiator and Cooling System**.

> ***CAUTION—***
> - *Use new fasteners for reinstalling front end reinforcement.*
> - *When reattaching throttle assembly harness connector, connector is fully tightened when arrows on connector and plug line up.*

BentleyPublishers.com
BentleyPublishers.com—All Rights Reserved

Engine Removal and Installation (N52, N54 Engines)

Be sure to cover painted surfaces before beginning the removal procedure. As an aid to installation, label components, wires and hoses before removing them. Do not reuse gaskets, O-rings or seals during reassembly.

> ***WARNING —***
> - *Due to risk of personal injury, be sure the engine is cold before beginning the removal procedure.*

N52, N54 engine, removing and installing

- Place hood in service position. See **410 Fenders, Engine Hood**.
- Remove exhaust system. See **180 Exhaust System**.
- Remove transmission. See **230 Manual Transmission** or **240 Automatic Transmission**.
- Drain engine oil.
- Drain engine coolant.
- Disconnect negative (–) cable from battery.

> ***CAUTION—***
> - *Prior to disconnecting the battery, read the battery disconnection cautions in* **001 Warnings and Cautions**.

- Remove air cleaner housing and fresh air ducts. See **131** or **132 Fuel Injection**.
- Remove fan cowl with electric cooling fan. See **170 Radiator and Cooling System**.
- Remove radiator. See **170 Radiator and Cooling System**.
- Remove thermostat. See **170 Radiator and Cooling System**.
- Remove all cooling system hoses from vehicle.
- Remove intake manifold. See **131** or **132 Fuel Injection**.
- Remove vacuum line from brake booster.
- Disconnect electrical connectors and harness from ignition coils and lay harness aside.
- Disconnect engine electrical harness and lay aside.
- Turbo engine: Remove fuel supply line to high pressure fuel pump. See **160 Fuel Tank and Fuel Pump** for high pressure fuel pump removal and installation.

- Remove A/C compressor (**A**) from engine and set on front axle carrier and secure with stiff wire. Do not disconnect A/C lines or discharge system.

BentleyPublishers.com
BentleyPublishers.com—All Rights Reserved

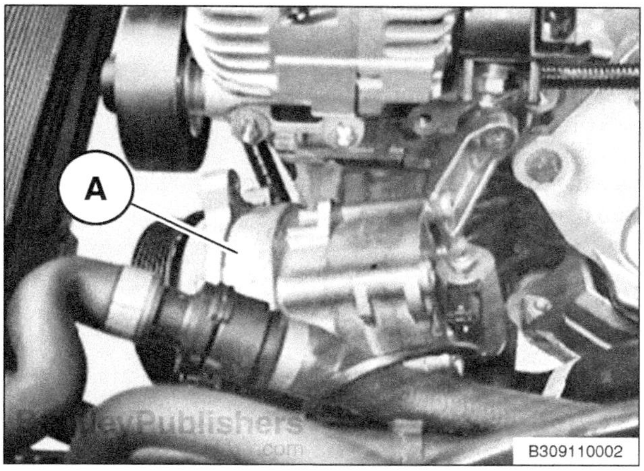

◄ Remove power steering pump (**A**), set on front axle carrier and secure with stiff wire. Do not disconnect hydraulic lines.

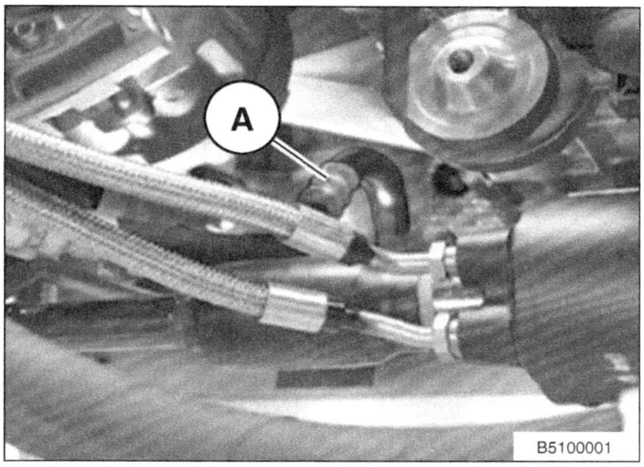

◄ Vehicles with dynamic drive (active stabilizer bar): Working at front stabilizer bar, remove hydraulic line (**A**).

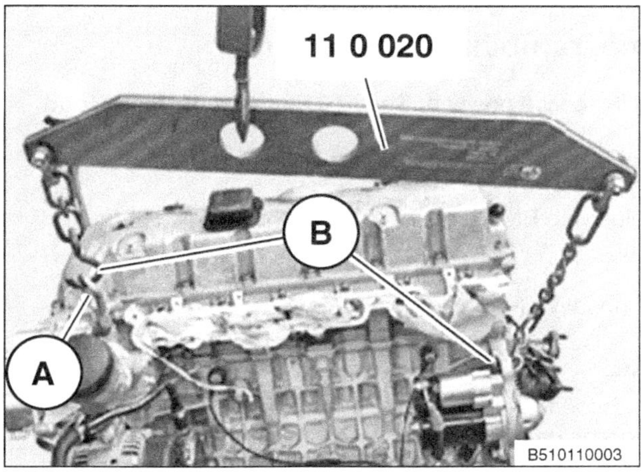

◄ Install engine tow hook (from tool kit in trunk) to front of engine (**A**). Attach BMW special tool 11 0 020 to engine using only specified mounting places (**B**).

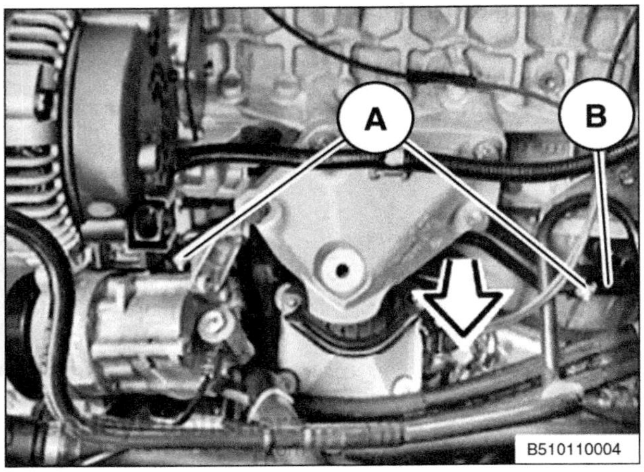

- If equipped with automatic transmission: Raise engine 10 cm (4 in) and remove fasteners (**A**) for ATF cooler lines (**B**).
- Remove engine from vehicle using engine hoist attached to BMW special hook 11 0 020.
- Raise engine slowly while watching carefully to make sure no lines, wires or hoses become snagged.
- Installation is reverse of removal. Remember to:
  - Replace aluminum fasteners.
  - Check for fuel leaks.
  - Check and clear any fault codes using BMW scan tool.
  - Refill and bleed cooling system. See **170 Radiator and Cooling System**.

> ***CAUTION—***
> - *Use new fasteners for reinstalling front end reinforcement.*

ENGINE REMOVAL AND INSTALLATION (V8 ENGINE)

Be sure to cover painted surfaces before beginning the removal procedure. As an aid to installation, label components, wires and hoses before removing them. Do not reuse gaskets, O-rings or seals during reassembly.

> ***WARNING—***
> - *Due to risk of personal injury, be sure the engine is cold before beginning the removal procedure.*

V8 engine, removing and installing

- Place hood in service position. See **410 Fenders, Engine Hood**.
- Remove exhaust system. See **180 Exhaust System**.
- Secure engine with BMW special tool 00 6 000 engine support or equivalent.
- Disconnect negative (–) cable from battery.

> ***CAUTION—***
> - *Prior to disconnecting the battery, read the battery disconnection cautions in* **001 Warnings and Cautions**.

- Remove transmission. See **230 Manual Transmission**, **231 Sequential Manual Gearbox (SMG)** or **240 Automatic Transmission**.
- Drain engine oil.
- Drain engine coolant.
- Remove air cleaner housing and fresh air ducts. See **133 Fuel Injection**.
- Remove radiator. See **170 Radiator and Cooling System**.

BentleyPublishers.com
BentleyPublishers.com—All Rights Reserved

– Remove coolant expansion tank. See **170 Radiator and Cooling System**.

– Remove intake manifold. See **133 Fuel Injection**.

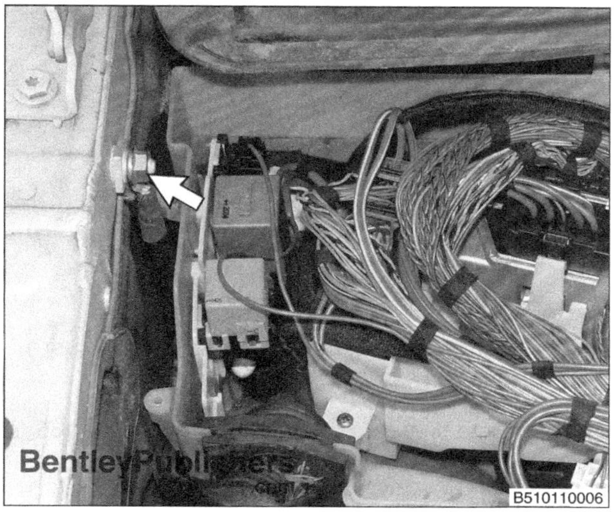
B510110006

◄ Working near right front strut tower, remove ground cable (**arrow**).

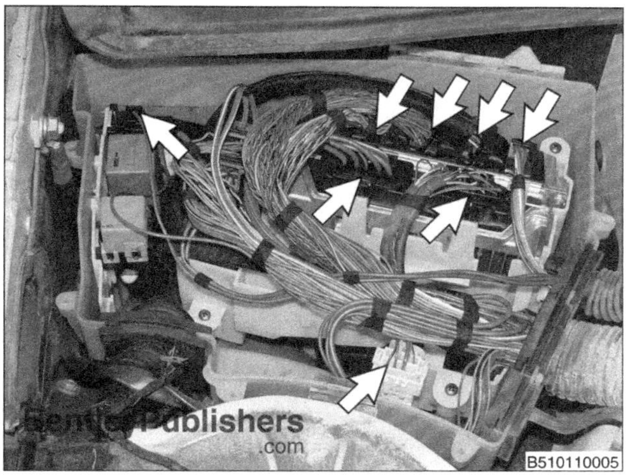
B510110005

◄ Working in E-box, detach engine wiring harness connectors from ECM, transmission control module, engine electronics fuse carrier and other connectors (**arrows**).

– Disconnect electrical connectors and harness from ignition coils and lay harness aside.

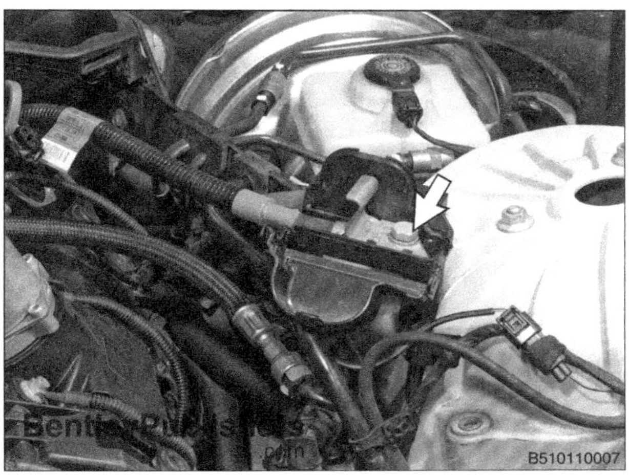
B510110007

◄ Disconnect jump start terminal cable (**arrow**).

BentleyPublishers.com
BentleyPublishers.com—All Rights Reserved

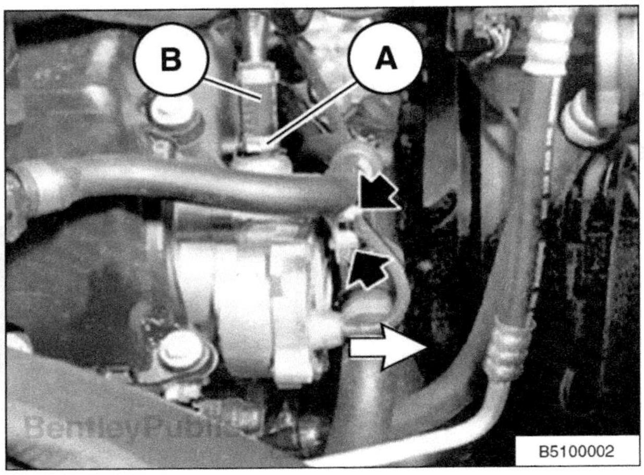

B5100002

◀ Remove clamp (**A**) and discard. Remove line (**B**). Remove hose clamps (**black arrows**) and hoses. Remove vacuum line (**white arrow**).

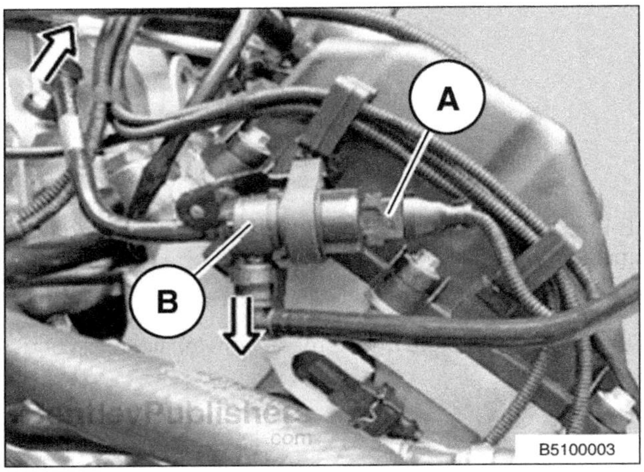

B5100003

◀ Working at fuel tank vent valve (**B**), remove electrical connector (**A**) and emission lines (**arrows**).

– Remove pre-catalyst and post-catalyst oxygen sensors. See **180 Exhaust System**.

– Working at right side engine mount bracket, remove engine ground cable.

– Remove engine accessory belt. See **020 Maintenance**.

– Remove vibration damper (crankshaft pulley). See **117 Camshaft Timing Chain**.

– Remove power steering pump with bracket. See **320 Steering and Wheel Alignment**.

– Remove A/C compressor idler pulley.

B510110008

◀ Remove A/C compressor mounting bolts (**arrows**) and lift A/C compressor off its mounting bracket without disconnecting, distorting, or deforming refrigerant lines. Suspend from body using stiff wire.

BentleyPublishers.com
BentleyPublishers.com–All Rights Reserved

< Remove lower steering shaft universal joint (**arrow**).

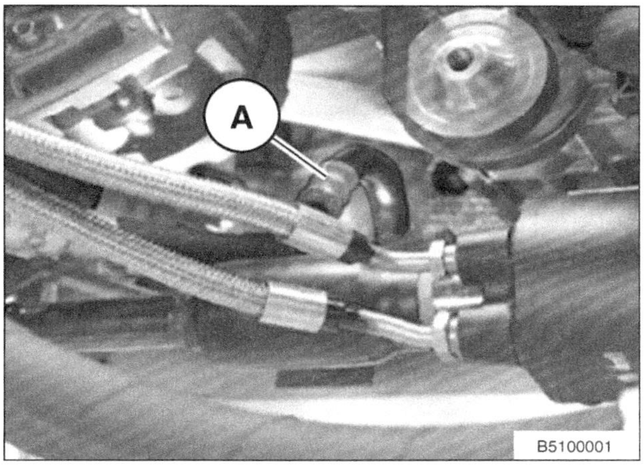

< Vehicles with dynamic drive (active stabilizer bar): Working at front stabilizer bar, remove hydraulic line (**A**).

– Working above engine, remove engine mount fasteners.

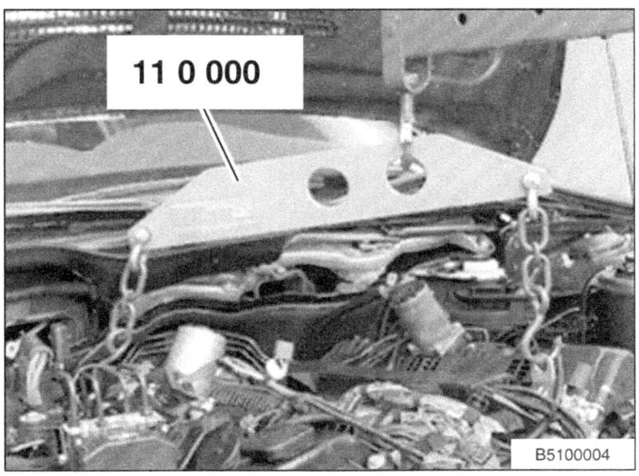

< Install engine lifting device (BMW 11 0 000 or equivalent) to the front and rear engine supports and raise engine until its weight is supported.

– Carefully raise engine out of engine compartment, checking for any wiring, fuel lines, or mechanical parts that might become snagged as engine is removed.

BentleyPublishers.com
BentleyPublishers.com–All Rights Reserved

– Installation is reverse of removal, noting the following:
  - Replace gaskets, O-rings and seals.
  - Check that engine drive belts properly engage pulley grooves.
  - Change engine oil and filter and check all other fluid levels. See **020 Maintenance**.
  - Check and clear any fault codes using BMW scan tool.
  - Refill and bleed cooling system. See **170 Radiator and Cooling System**.

> ***CAUTION—***
> - *Use new fasteners for reinstalling front end reinforcement.*
> - *When reattaching throttle assembly harness connector, connector is fully tightened when arrows on connector and plug line up.*

BentleyPublishers.com
BentleyPublishers.com—All Rights Reserved

113 Cylinder Head Removal and Installation

BentleyPublishers.com
BentleyPublishers.com–All Rights Reserved

Cylinder head:
N52 6-cylinder engine

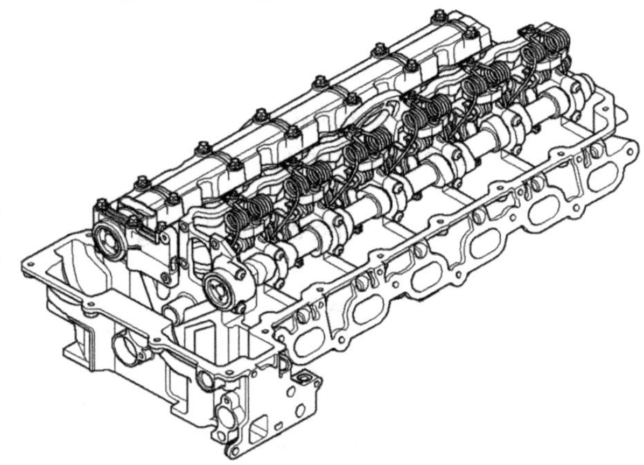

General

This repair group provides removal and installation procedures for the cylinder head and associated components.

The cylinder head is equipped with variable camshaft timing (sometimes referred to as VANOS, from the German words *VAriable NOckenwellen Steuerung*). Special tools and procedures are required to remove and install camshaft adjustment units and to time camshafts. Be sure to read each procedure through before starting work.

See also:

- **020 Maintenance**
- **100 Engine–General** for engine code and application information.
- **117 Camshaft Timing Chain** for camshaft timing chain and camshaft adjuster (VANOS) service
- **130**, **131**, **132**, **133 Fuel Injection** for air filter housing and intake manifold removal
- **170 Radiator and Cooling System**

Engine identification

Engine applications and specifications are detailed in **100 Engine–General**.

Table a. Engine identification

Engine designation	Year, model	Identifying feature
M54	2004 - 2005 525i, 530i	2 upper engine covers
N52	2006 - 2007 525i, 525xi 530i, 530xi 530xi Sports Wagon	Aluminum-magnesium cylinder head cover
N52KP	2008 - 2010 528i, 528xi 530i, 530xi 528i xDrive 530i xDrive	Black plastic cylinder head cover
N54	2008 - 2010 535i, 535xi 535xi Sports Wagon 535i xDrive 535i Sports Wagon 535i xDrive Sports Wagon	Twin-turbo chargers
N62	2004 - 2005 545i 2006 - 2010 550i	V-style engine with two valvetronic motors

Fasteners and fastener torques differ among the versions. Read torque specifications carefully.

BentleyPublishers.com
BentleyPublishers.com– All Rights Reserved

113

Warnings and Cautions

WARNING—
- *To avoid personal injury, be sure the engine is cold before beginning work on engine components.*
- *Use extreme caution when draining and disposing of engine coolant. Coolant is poisonous and lethal to humans and pets. Pets are attracted to coolant because of its sweet smell and taste. Seek medical attention immediately if coolant is ingested.*
- *The fuel system is designed to retain pressure even when the ignition is OFF. When working with the fuel system, loosen fuel lines slowly to allow residual fuel pressure to dissipate. Avoid spraying fuel. Use shop towels to capture leaking fuel.*
- *Before beginning work on the fuel system, place a fire extinguisher in the vicinity of the work area.*
- *Fuel is highly flammable. When working around fuel, do not disconnect wires that could cause electrical sparks. Do not smoke or work near heaters or other fire hazards.*
- *Wear eye protection and protective clothing to avoid injuries from contact with fuel.*
- *Unscrew the fuel tank cap to release pressure in the tank before working on fuel lines.*
- *Do not use a work light with an incandescent bulb near fuel. Fuel may spray on the hot bulb causing a fire.*
- *Make sure the work area is properly ventilated.*

CAUTION—
- *To avoid electrochemical corrosion to engine components made of aluminum-magnesium alloy, do not use steel fasteners. Use aluminum fasteners only.*
- *The end faces of aluminum fasteners are usually painted blue. For reliable identification, test fasteners for aluminum composition with magnet.*
- *Replace aluminum bolts each time they are loosened.*
- *Follow torque instructions, including angle of rotation specifications, when installing aluminum fasteners.*

BentleyPublishers.com
BentleyPublishers.com—All Rights Reserved

Cylinder Head (M54 Engine)

Cylinder head cover, removing and installing (M54 engine)

- Make sure ignition is OFF. Remove ignition key.
- Remove air intake hood above radiator.
- Remove upper engine covers. See **020 Maintenance**.

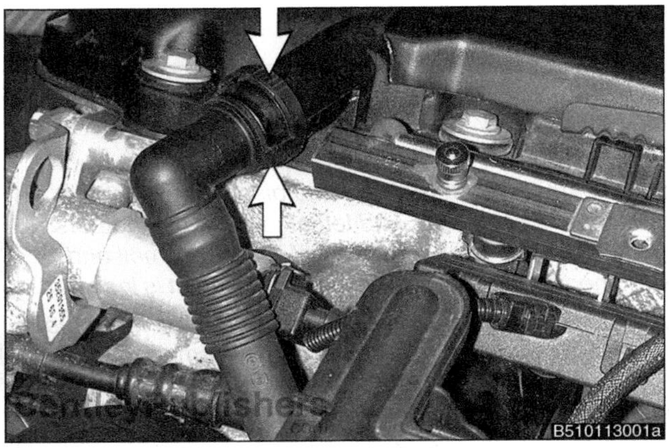

◀ Working at left front of cylinder head cover, pinch crankcase breather hose clip (**arrows**). Detach hose from cover.

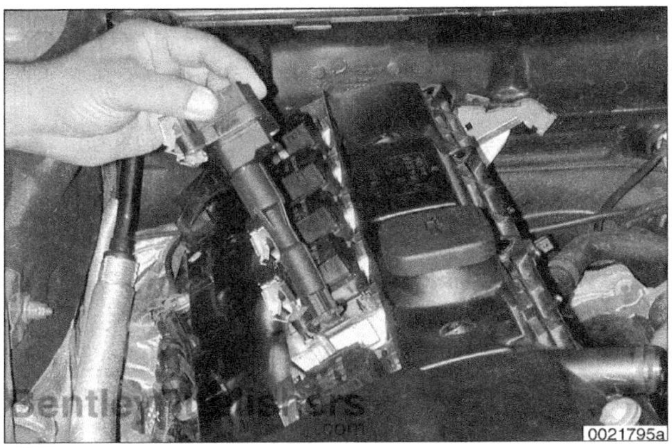

◀ Remove ignition coils:
- Disconnect ignition coil harness connectors.
- Remove coil mounting fasteners.
- Remove coils.
- Remove ground straps.
- Set coil harness to side of engine compartment.

> ***CAUTION—***
> - *Note location of all ground wires. Failure to reinstall grounds can result in permanent damage to engine control module or ignition system components.*

- Remove cylinder head cover mounting fasteners and remove cylinder head cover.

NOTE —
- *Make note of arrangement of fasteners during removal so that they can be reinstalled in their original locations.*

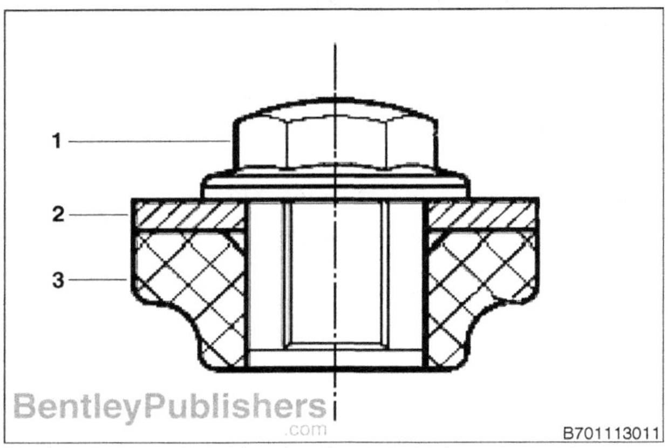

◀ Inspect cylinder head cover fasteners. Replace as necessary.
1. Cap nut
2. Washer
3. Rubber seal

- Clean gasket residue from cylinder head sealing surface. Apply thin beads of 3-Bond 1209® or equivalent sealer at timing chain cover seams.
- Remove inner and outer gaskets from cylinder head cover and clean gasket residue from grooves.
- Coat cylinder head cover grooves with thin layer of anti-friction compound such as glycerine. Press inner and outer gaskets in place, making sure gaskets are not stretched at any point.

◀ Use a small amount of 3-Bond 1209® or equivalent sealer at corners of half-moon seals.

BentleyPublishers.com
BentleyPublishers.com—All Rights Reserved

– Reinstall cylinder head cover.
  - Coat cylinder head and gasket contact surfaces with thin layer of anti-friction compound such as glycerine.
  - Check for correct seating of half-moon seals in back of cylinder head cover.
  - Install cylinder head cover fasteners and tighten gradually in crisscross pattern, starting with inside fasteners.

– Remainder of assembly is reverse of removal.

CYLINDER HEAD REMOVAL AND INSTALLATION (M54 ENGINE)

The 6-cylinder engine cylinder head is equipped with double VANOS. To remove the cylinder head, special tools and procedures are required for removing the VANOS unit. Read the entire procedure before beginning the repair.

> ***CAUTION—***
> - *Disassembly, removal and assembly of camshafts or cylinder head without special tools poses the risk of damage or breakage. The valves may be bent by contact with the piston crown; camshafts may break if stressed incorrectly during removal or installation.*

NOTE—
- *VANOS or variable camshaft timing is from the German words* VAriable NOckenwellen Steuerung.
- *VANOS service is covered in* **117 Camshaft Timing Chain**.

Cylinder head removal and installation procedures are described separately. Remove the intake manifold before removing the cylinder head. See **130 Fuel Injection**.

If you remove a cylinder head and determine that it requires significant reconditioning work, a remanufactured cylinder head is available from an authorized BMW dealer.

Cylinder head, removing (M54 engine)

> ***WARNING—***
> - *Due to risk of personal injury, be sure the engine is cold before beginning the removal procedure.*

– Disconnect negative (–) cable from battery.

> ***CAUTION—***
> - *Prior to disconnecting the battery, read the battery disconnection cautions in* **001 Warnings and Cautions**.

– Remove air intake hood above radiator.

– Remove upper engine covers. See **020 Maintenance**.

– Detach air hose and vacuum line from secondary air valve on right side of cylinder head.

BentleyPublishers
BentleyPublishers.com—All Rights Reserved

Cylinder head, removing (M54 engine)

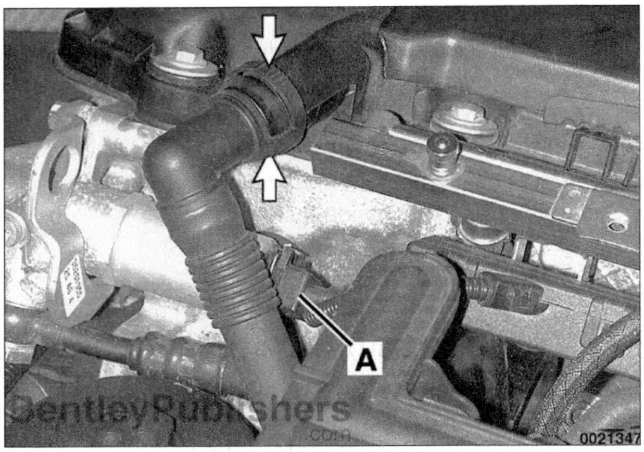

Working above engine, disconnect the following:

- Vent line from cylinder head cover. To remove, pinch clip (**arrows**).
- Electrical harness connector from VANOS solenoid valve (**A**).

– Remove both exhaust manifolds. See **180 Exhaust System**. After removing exhaust manifolds, reinstall engine support arm and engine mount.

– Remove air filter housing and ducts. See **130 Fuel Injection**.

– Remove cylinder head cover. See **Cylinder head cover, removing and installing (M54 engine)** in this repair group.

– Remove spark plugs. Plug spark plug bores to prevent debris from falling into engine.

– Remove intake manifold. See **130 Fuel Injection**. Plug cylinder head intake ports to prevent debris from falling into engine.

WARNING —
- *Unscrew the fuel tank cap to release pressure in the tank before working on the fuel line.*
- *Wrap shop towels around fuel line fitting before disconnecting.*
- *Plug open fuel lines and fittings.*

– Raise car and support safely.

WARNING —
- *Make sure the car is stable and well supported at all times. Use a professional automotive lift or jack stands designed for the purpose. A floor jack is not adequate support.*

Drain engine coolant:

- Remove expansion tank cap on radiator.
- Place a 3 gallon pail underneath engine to capture coolant.
- Remove coolant drain plug (**arrow**) located on exhaust side of cylinder 2 on engine block.

Drain radiator into a 3 gallon pail by removing plastic drain plug (**arrow**) completely.

– Remove engine cooling fan and radiator. See **170 Radiator and Cooling System**.

BentleyPublishers.com
BentleyPublishers.com—All Rights Reserved

Remove thermostat housing:

- Disconnect electrical harness connector at thermostat housing.
- Remove hoses from thermostat housing by releasing locks (**arrows**).
- Unbolt (4 bolts) and remove thermostat housing.

113

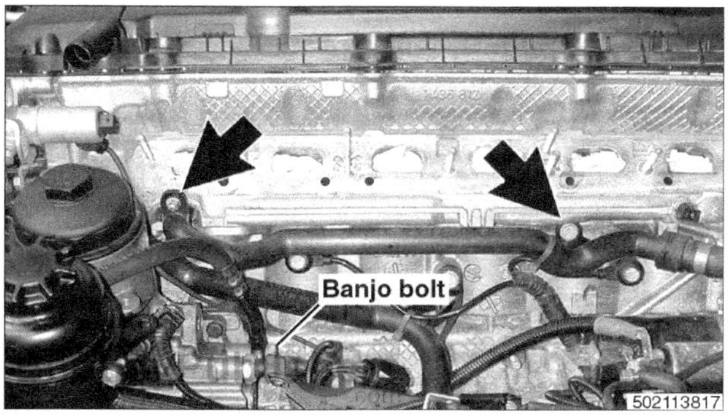

Remove fasteners (**arrows**) from heater bypass tube. Pull tube out of cylinder head and set to side, leaving heater hose connected.

- If necessary, loosen oil pressure line banjo bolt and carefully push oil line aside.

Remove oil baffle cover from above intake camshaft.

Remove banjo bolt from VANOS unit oil pressure line. Use banjo bolt to attach BMW special tool 11 3 450 (compressed air fitting) to VANOS unit.

- Be prepared to catch dripping oil. Do not allow oil to contaminate accessory belts.

Cylinder head, removing (M54 engine)

Cover oil bore (**arrow**) in VANOS unit with shop towel to capture oil which sprays when compressed air is applied.

– Connect compressed air line to air fitting. Apply air pressure set to 2 – 8 bar (30 – 115 psi).

With compressed air line connected, use crankshaft vibration damper bolt to rotate engine in direction of rotation (clockwise) at least twice, until cylinder 1 intake and exhaust camshaft lobes face each other (**arrows**) in top dead center (TDC) position for cylinder 1.

> ***CAUTION—***
> - *Do not rotate engine counterclockwise to reach TDC. If engine rotates past TDC, complete another two complete rotations.*

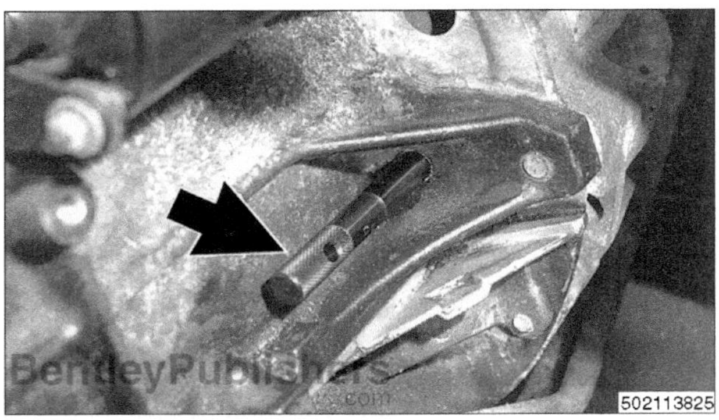

Remove sealing plug from special tool bore on lower left side of engine block near flywheel. Secure crankshaft in TDC position with BMW special tool 11 2 300 (**arrow**).

Unscrew and remove two cylinder head cover studs (**arrows**) at rear of cylinder head.

BentleyPublishers.com
BentleyPublishers.com—All Rights Reserved

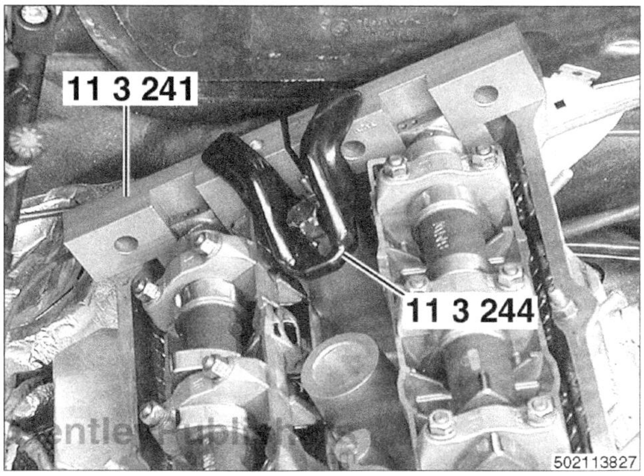

- Secure camshafts in TDC position using BMW special tool set 11 3 240 (includes 11 3 241 and 11 3 244).
- Remove VANOS unit from front of cylinder head. See **117 Camshaft Timing Chain**.
- Remove crankshaft locking tool from transmission bell housing.

- Lift primary chain and hold under tension. Use crankshaft vibration damper bolt to rotate engine in direction opposite to normal rotation (counterclockwise) approximately 30°.

> ***CAUTION—***
> - *By turning crankshaft 30° counterclockwise, pistons are moved away from TDC. This prevents piston-to-valve interference during camshaft servicing.*

- Remove camshaft locking tools (BMW special tool set 11 3 240) from rear of cylinder head.
- Remove camshafts and camshaft bearing carriers. See **117 Camshaft Timing Chain**.

- Remove intake camshaft sensor (**arrow**) from side of cylinder head.

- Remove secondary chain lower guide bolts (**arrow**). Remove chain guide.

BentleyPublishers.com–All Rights Reserved

Remove bolts (**arrows**) holding timing chain cover to cylinder head. (Photo shows camshafts and camshaft bearing caps in place).

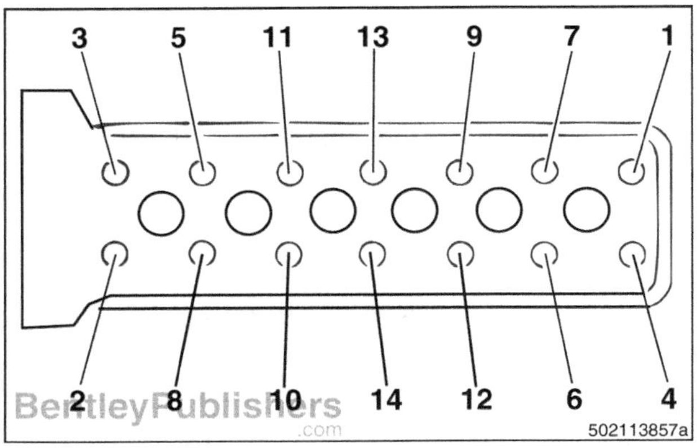

Using BMW special tool 11 2 250 or equivalent, loosen cylinder head bolts in several stages, in sequence shown. Discard head bolts.

NOTE—
- *BMW special tool 11 2 250 is a thin-walled Torx E12 socket with an extended reach. The cylinder head bolts are recessed into the head below the camshaft carriers with little working space.*

– Lift off cylinder head.

Cylinder head, installing (M54 engine)

– Clean cylinder head and gasket surfaces of cylinder block and timing chain covers.

CAUTION—
- *Remove foreign matter and liquid from bolt holes. Trapped oil or coolant in bolt holes can cause damage to the engine block.*
- *Do not use a metal scraper or wire brush to clean the aluminum cylinder head or pistons. If necessary, use a hard wooden or plastic scraper. Also available are abrasive discs to be used in conjunction with an electric drill. Be sure to use the correct disc for the type of metal being cleaned.*

– If the cylinder head is machined, use a special cylinder head gasket available from an authorized BMW dealer. The gasket is 0.3 mm (0.011 in) thicker than standard and is marked accordingly.

– Lubricate camshafts, camshaft carriers, bearing caps, hydraulic lifters and friction washers with assembly lubricant before installation.

Check that two cylinder head aligning sleeves (**arrows**) are correctly positioned in block and are not damaged.

– Apply elastic sealing compound 3-Bond 1209® to corner joints and seam on cylinder block to timing chain cover.

– Place new cylinder head gasket on cylinder block.

BentleyPublishers.com
BentleyPublishers.com—All Rights Reserved

NOTE—

• The word "OBEN" or "TOP", printed on the gasket, faces up. The cylinder head gasket fits correctly in only one orientation.

– Set cylinder head in position, guiding primary chain through cylinder head opening.

– Prior to installing cylinder head bolts, keep in mind the following:
  - Do not reuse cylinder head bolts. They are stretch-type bolts and are replaced whenever loosened.
  - Cylinder head bolt length for aluminum engine block = 110 mm.
  - Check that head bolt washers are in place before installing the bolts. Some washers are staked to the cylinder head.

– Lightly lubricate **new** cylinder head bolts and washers with oil. Install cylinder head bolts and bolts for cylinder head to lower timing chain cover finger tight.

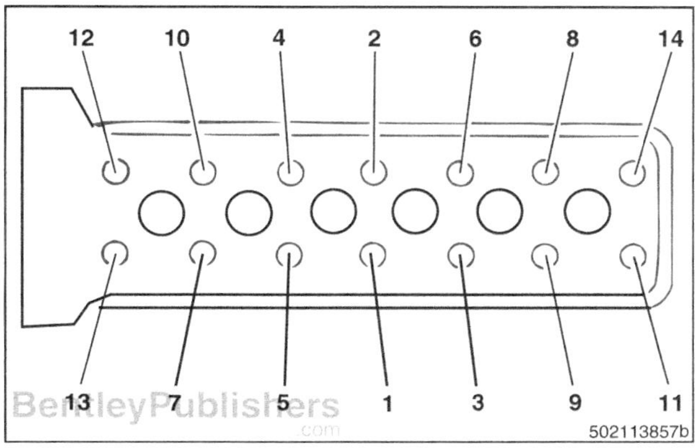

◄ Tighten cylinder head bolts in correct sequence (1 – 14).

CAUTION—

• Tighten the bolts in three stages, as listed below. The final stages require the use of a BMW special tool 11 2 110 or a suitable protractor to tighten the bolts to a specified torque angle.

Tightening torque	
Cylinder head to engine block • Stage 1 • Stage 2 • Stage 3	 40 Nm (30 ft-lb) +90° +90°

◄ Install intake camshaft sensor (**arrow**).

BentleyPublishers.com—All Rights Reserved

Install lower secondary chain guide. Tighten torx bolts (**arrows**) to specifications. Bolt on intake camshaft side is long and extends into engine block.

Tightening torque	
Secondary chain guide to cylinder head (M6)	10 Nm (89 in-lb)

Tighten bolts (**arrows**) for cylinder head to lower timing chain cover. (Photo shows camshafts in place).

Tightening torque	
Cylinder head to lower timing chain cover (M6)	10 Nm (89 in-lb)

– Reinstall camshafts. See **117 Camshaft Timing Chain**.

CAUTION—

- *Make sure the crankshaft, which was rotated approximately 30° opposite the direction of engine rotation from TDC, is still in that position before reinstalling the camshafts. This ensures that pistons are out of TDC position and prevents piston-valve interference when the camshafts are installed.*
- *After the camshafts are installed, a minimum waiting time is required for the hydraulic lifters to bleed down before bringing the crankshaft and pistons back to TDC. When the camshafts are removed, the hydraulic lifters may expand. This expansion can cause increased valve lift when the camshafts are bolted down, possibly resulting in piston-valve interference.*

– Reinstall double VANOS unit. See **117 Camshaft Timing Chain**.

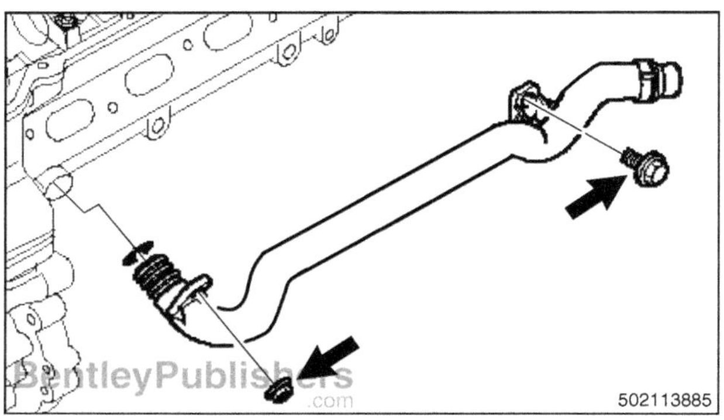

Install coolant pipe fasteners at base of cylinder head and tighten fasteners (**arrows**).

NOTE—

- *Use new sealing O-ring on coolant pipe.*

– Attach oil line to VANOS unit using banjo bolt with new seals.

Tightening torque	
Oil line to VANOS unit (banjo bolt)	32 Nm (24 ft-lb)

BentleyPublishers.com
BentleyPublishers.com—All Rights Reserved

— Install intake camshaft cover and cylinder head cover. See **Cylinder head cover, removing and installing (M54 engine)** in this repair group.

Tightening torque	
Cylinder head cover to cylinder head (M6)	10 Nm (89 in-lb)

— Install exhaust manifolds using new gaskets and nuts. Coat manifold studs with copper paste prior to installing nuts. See **180 Exhaust System**.

Tightening torque	
Exhaust manifold to cylinder head (M7)	20 Nm (15 ft-lb)

— Install electrical harness connectors for oil pressure sender and coolant temperature sensor before installing intake manifold.

— Reinstall intake manifold. See **130 Fuel Injection**.

> ***CAUTION—***
> - *When reattaching throttle assembly harness connector, connector is fully tightened when arrows on connector and plug line up.*

— Installation of the remaining parts is reverse or removal, noting the following:

- Refill and bleed cooling system. See **170 Radiator and Cooling System**.
- Change engine oil and filter. See **020 Maintenance**.
- Reconnect battery.

> ***CAUTION—***
> - *To prevent damaging engine electronic systems, install ground wires previously removed, including the ground wires for the ignition coils.*
> - *Prior to starting engine, be sure to remove crankshaft lock tool.*

Tightening torque	
Coolant drain plug to cylinder block	25 Nm (18 ft-lb)
Radiator cooling fan to coolant pump	40 Nm (30 ft-lb)
Radiator drain screw to radiator	2.5 Nm (22 in-lb)
Spark plug to cylinder head	25 Nm (18 ft-lb)

BentleyPublishers
BentleyPublishers.com—All Rights Reserved

CYLINDER HEAD (N52 ENGINE)

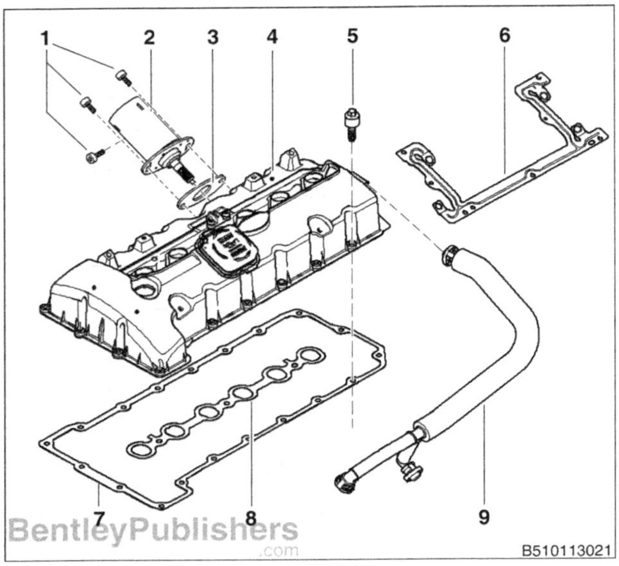

Cylinder head cover components (N52 engine)

Cylinder head cover for 2006 N52 engine is made of aluminum-magnesium alloy. Aluminum stretch bolts and fasteners are used. Replace them whenever the cover is removed.

Cylinder head cover for 2007 and later N52 engine is made of plastic. Steel fasteners may be used and they can be reused.

1. Valvetronic motor mounting bolt (M6 x 16 mm)
   - Tighten to 10 Nm (7 ft-lb)
2. Valvetronic motor
3. Gasket
4. Cylinder head cover
5. Cylinder head cover fastener
6. Bracket (N52 engine only)
7. Cylinder head cover gasket–outer
8. Cylinder head cover gasket–inner (spark plug holes)
9. Crankcase breather hose

Cylinder head cover, removing and installing (N52 engine)

– Use scan tool to read out and record ECM fault memory.

– Switch ignition OFF and remove key.

– Remove cabin microfilter upper and lower housings. See **640 Heating and Air-conditioning**.

– Remove right strut tower brace.

– Remove ignition coil cover (upper engine cover). See **020 Maintenance**.

– Detach ignition coil connectors. Unclip ignition coil harness from cylinder head cover and lift aside.

– Remove ignition coils. See **120 Ignition System**.

Working at cylinder head cover:

- Detach crankcase breather hose.
- If applicable: Pull off metal bracket.
- Remove valvetronic motor mounting bolts (**A**). To disengage valvetronic motor, see **131 Fuel Injection**.

BentleyPublishers.com
BentleyPublishers.com—All Rights Reserved

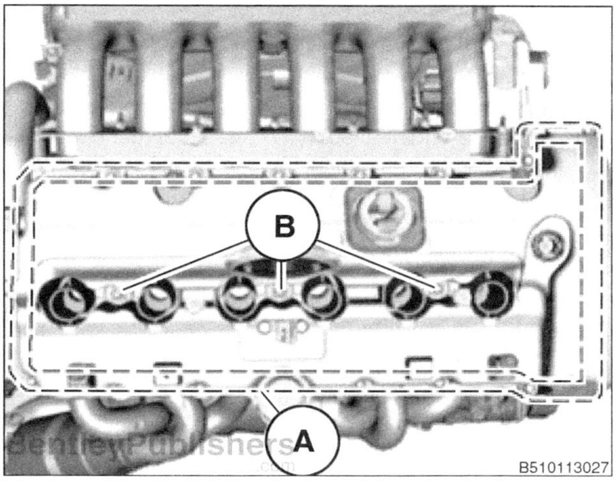

◀ Remove cylinder head cover fasteners (**A**, **B**). Discard if aluminum (blue head).

- Measure and record lengths and positions of bolts and studs. Different fastener lengths are determined by shapes of bushings.

NOTE—

- *The end faces of aluminum fasteners are usually painted blue. For reliable identification, test fasteners for aluminum composition with magnet.*

– Lift off cylinder head cover and discard old gaskets.

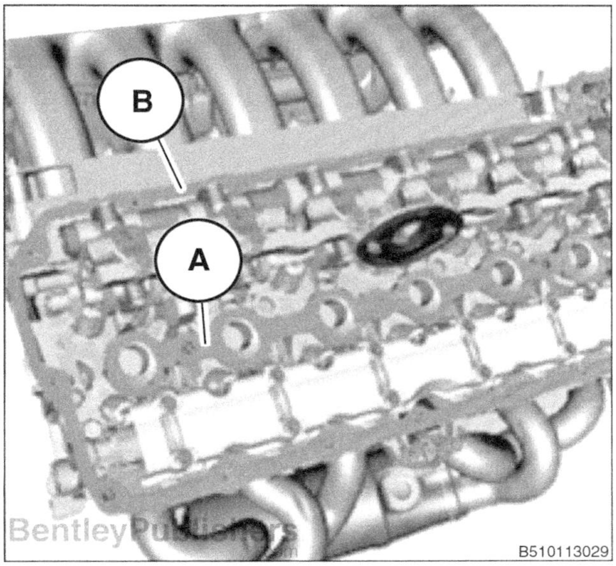

◀ Clean gasket residue from cylinder head and cylinder head cover sealing surfaces (**A**, **B**).

CAUTION—
- *To avoid gouging the cover, do not use metal scraping tool.*

◀ Remove and replace slotted ignition coil sleeves in cylinder head cover.

– Coat cylinder head cover grooves with thin layer of anti-friction compound such as glycerine. Press gaskets in place, making sure they are not stretched at any point.

BentleyPublishers.com—All Rights Reserved

– Reinstall cylinder head cover.
  - Coat cylinder head contact surfaces with thin layer of anti-friction compound such as glycerine.
  - Place small quantity of silicone sealant at metal-to-metal seams (such as timing cover to engine block) and half moon gasket area.
  - Replace aluminum fasteners. Install in correct locations as noted previously.
  - Tighten gradually in crisscross pattern, starting with inside fasteners.

Tightening torques	
N52 engine cylinder head cover to cylinder head (aluminum-magnesium alloy cover, use new aluminum M7 fasteners) • Stage 1 • Stage 2	 7 Nm (63 in-lb) additional 90°
N52KP or N51 engine cylinder head cover to cylinder head (plastic cover, steel M7 fasteners)	9 Nm (7 ft-lb)

– Reinstall valvetronic motor.

Tightening torque	
Valvetronic motor to cylinder head cover (M6 x 16 mm bolts)	10 Nm (7 ft-lb)

– Remainder of assembly is reverse of removal. Make sure cabin microfilter housing seals correctly.

Tightening torques	
Ignition coil cover to cylinder head cover (M6)	4 Nm (35 in-lb)

BentleyPublishers.com
BentleyPublishers.com—All Rights Reserved

Cylinder head and timing chain (N52 engine)

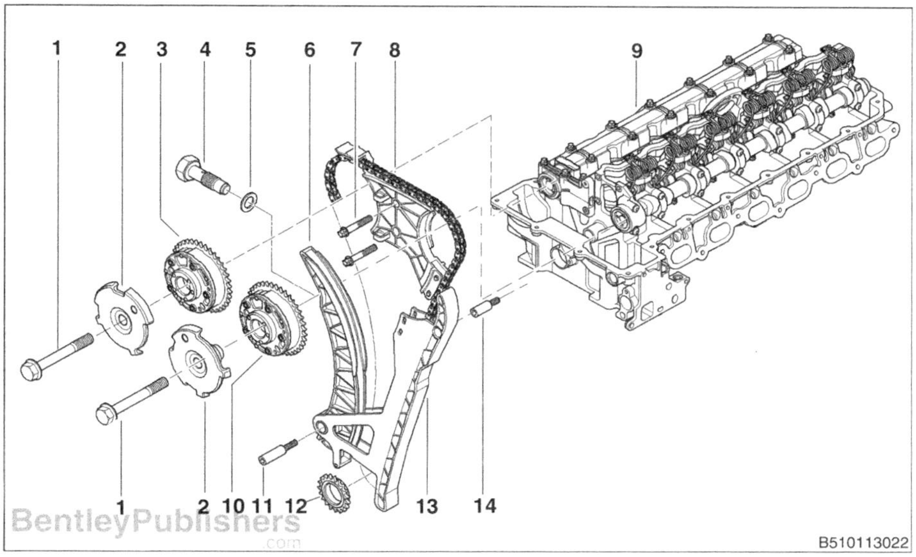

1. Camshaft adjustment unit mounting bolt (M10 x 94 mm)
   - Replace with new
   - Tighten in 2 stages
     Stage 1: 20 Nm (15 ft-lb)
     Stage 2: Additional 180°
2. Camshaft impulse wheel
3. Exhaust camshaft adjustment unit
4. Chain tensioner (M22 x 1.5)
   - Tighten to 55 Nm (41 ft-lb)
5. Sealing O-ring
   - Install new
6. Chain tensioner rail
7. Chain module mounting bolts (M6 x 35 mm Torx)
   - Tighten to 8.5 Nm (6 ft-lb)
8. Timing chain
9. Cylinder head with camshafts
10. Intake camshaft adjustment unit
11. Chain tensioner rail mounting bolt (M8)
    - Tighten to 20 Nm (15 ft-lb)
12. Crankshaft sprocket
13. Chain guide
14. Chain guide mounting bolt (M7)
    - Tighten to 14 Nm (10 ft-lb)

113

Cylinder head, removing and installing (N52 engine)

When removing and installing the cylinder head, special tools and procedures are required to remove and install camshaft adjustment units and valvetronic eccentric shaft and to time camshafts. Read the entire procedure before beginning repairs.

> ***CAUTION—***
> - *Disassembly, removal and assembly of camshafts, camshaft adjustment units or cylinder head without special tools poses the risk of damage or breakage: Valves may be bent by contact with the piston crowns.*

NOTE—
- *Camshaft timing is covered in* **117 Camshaft Timing Chain**.

Cylinder head removal and installation procedures are described separately.

If you remove a cylinder head and determine that it requires significant reconditioning work, a remanufactured cylinder head is available from an authorized BMW dealer.

BentleyPublishers.com
BentleyPublishers.com—All Rights Reserved

Cylinder head, removing (N52 engine)

The cylinder head stretch bolts are aluminum. Replace them any time they are removed. Use angle protractor to torque.

> ***WARNING —***
> • *To avoid personal injury, be sure the engine is cold before beginning the removal procedure.*

– Disconnect negative (–) cable from battery.

> ***CAUTION—***
> • *Prior to disconnecting the battery, read the battery disconnection cautions in* **001 Warnings and Cautions**.

– Raise vehicle and support safely.

> ***WARNING —***
> • *Make sure the vehicle is stable and well supported at all times. Use a professional automotive lift or jack stands designed for the purpose. A floor jack is not adequate support.*

– Remove engine compartment splash shield. See **020 Maintenance**.

– With exhaust system fully cooled off, remove exhaust system and both exhaust manifolds. See **180 Exhaust System**.

– Drain engine coolant. See **170 Radiator and Cooling System**.

– Drain engine oil. See **020 Maintenance**.

– Remove ignition coil cover (upper engine cover). See **020 Maintenance**.

– Remove air filter housing, engine air intake ducts and intake manifold. See **131 Fuel Injection**.

> ***WARNING —***
> • *Unscrew the fuel tank cap to release pressure in the tank before working on fuel lines.*

– Remove cylinder head cover. See **Cylinder head cover, removing and installing (N52 engine)** in this repair group.

– Detach coolant hoses from cylinder head.

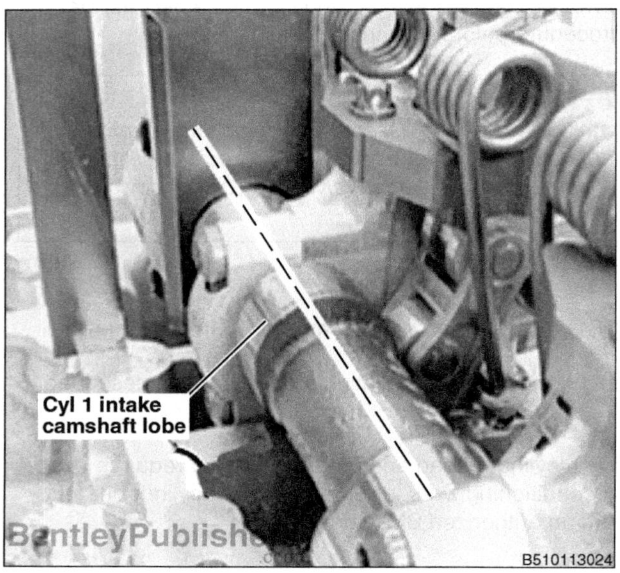

◄ Using vibration damper bolt, rotate crankshaft to place cylinder 1 in TDC firing position. In this position, cylinder 1 intake camshaft lobe points upward at an angle.

BentleyPublishers.com
BentleyPublishers.com—All Rights Reserved

Make sure intake and exhaust camshaft part numbers (**A**) point up.

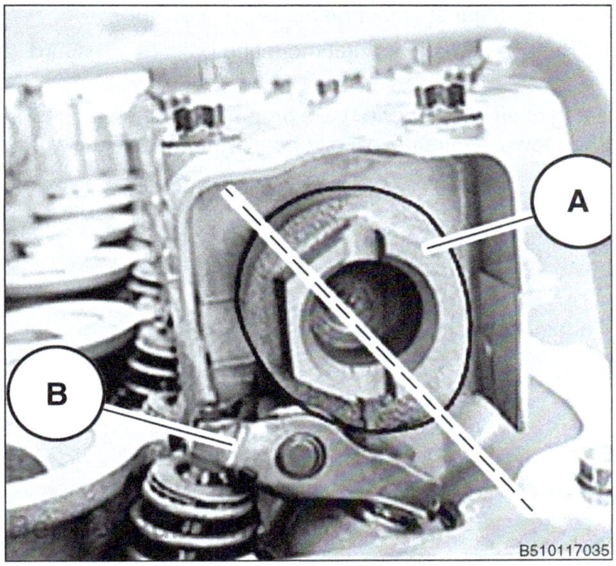

Cylinder 1 in TDC firing position: Cylinder 6 exhaust camshaft lobe (**A**) points downward at an angle (**dashed line**). Cam follower (**B**) is not actuated.

NOTE—
- *Use a mirror to check exhaust camshaft lobe position.*

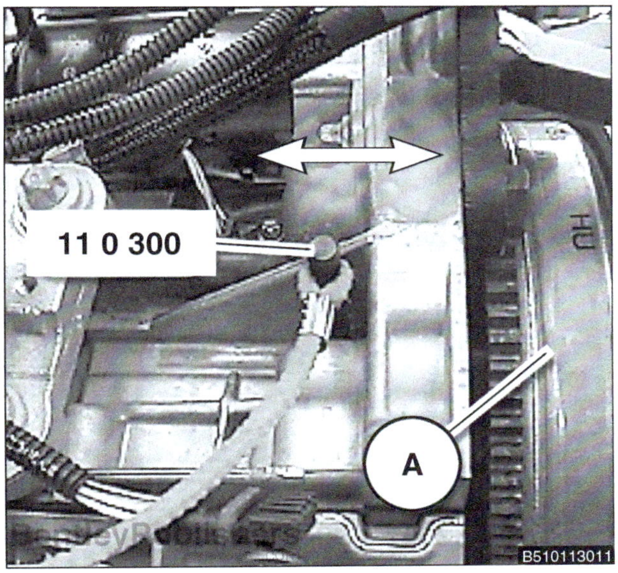

Working underneath engine, slide out protective plug at lower left of engine bell housing flange. Install BMW special tool 11 0 300 (crankshaft locking tool) through bell housing flange port into flywheel (**A**) hole. If necessary, rock flywheel slightly back and forth to line up holes with tool. This locks crankshaft at TDC.

CAUTION—
- *Automatic transmission model: A short distance before the flywheel special tool bore for TDC position, there is a large bore which can be confused with the special tool bore. If the flywheel is locked using the correct bore, the engine can no longer be rotated at the vibration damper bolt.*

BentleyPublishers.com
BentleyPublishers.com—All Rights Reserved

Cylinder head, removing (N52 engine)

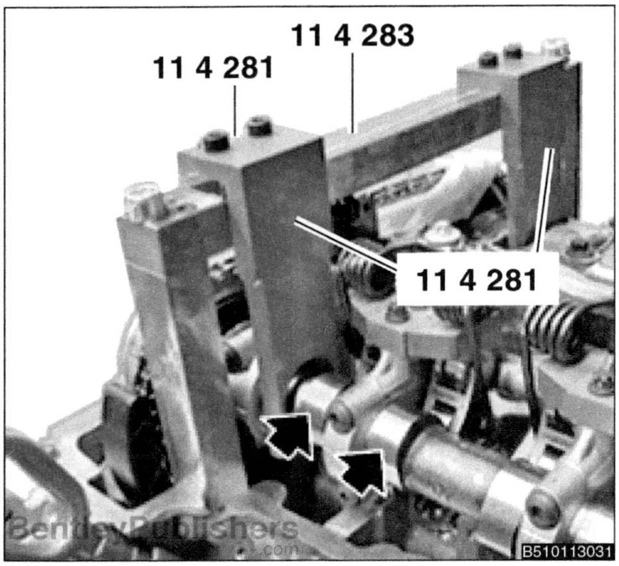

Using BMW special tool set 11 4 280, lock down camshafts.

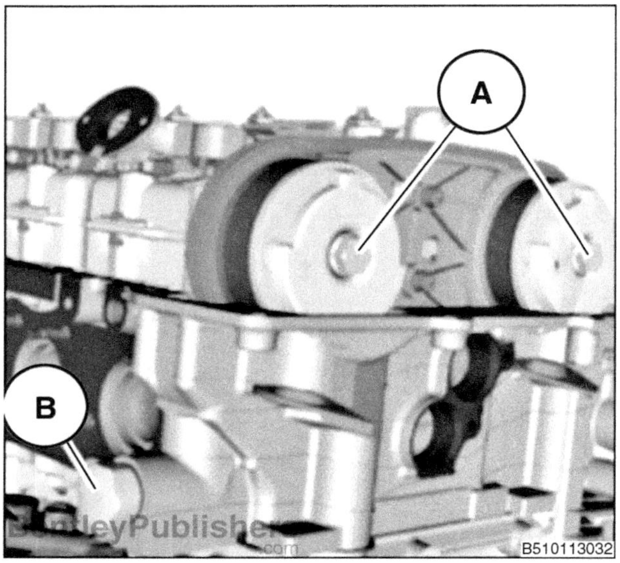

Remove intake and exhaust camshaft adjustment units:

- Loosen and remove adjustment unit mounting bolts (**A**). Discard bolts.
- Remove timing chain tensioner (**B**). Be prepared to catch dripping oil with a shop towel. Do not allow oil to contaminate accessory belt.
- Disengage timing chain from camshaft adjustment unit sprockets. Remove adjustment units and set aside.

NOTE—

- *Illustration does not show special tool set 11 4 280 in place.*

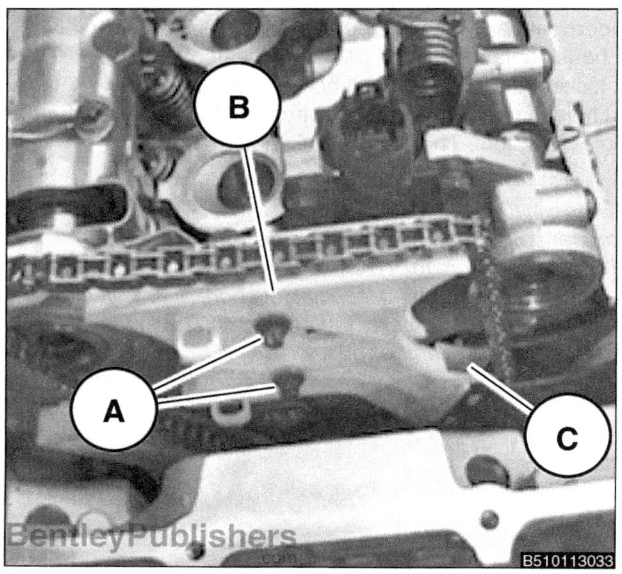

Working at front of cylinder head:

- Remove timing chain module mounting bolts (**A**).
- Unclip chain module (**B**) at junction (**C**) and lift out.
- Place timing chain in cylinder head opening.

CAUTION—

- *Do not remove crankshaft locking tool. Rotating crankshaft may cause the timing chain to jam or jump teeth.*

BentleyPublishers.com
BentleyPublishers.com—All Rights Reserved

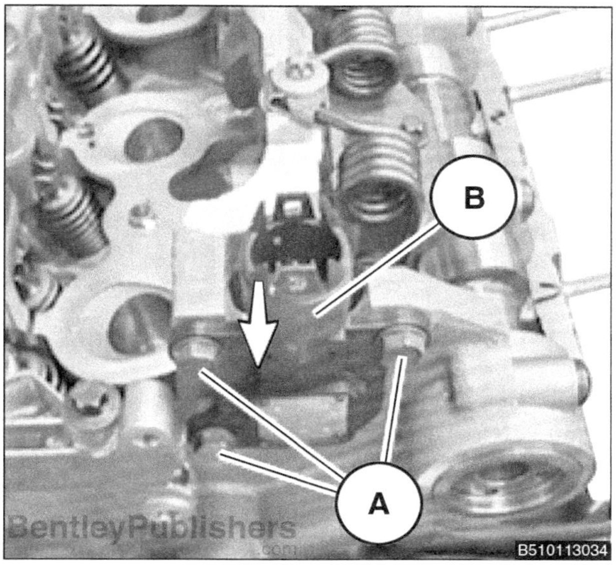

Unbolt and remove valvetronic eccentric shaft sensor mounting bolts (**A**). Slide sensor (**B**) forward (**arrow**) to remove.

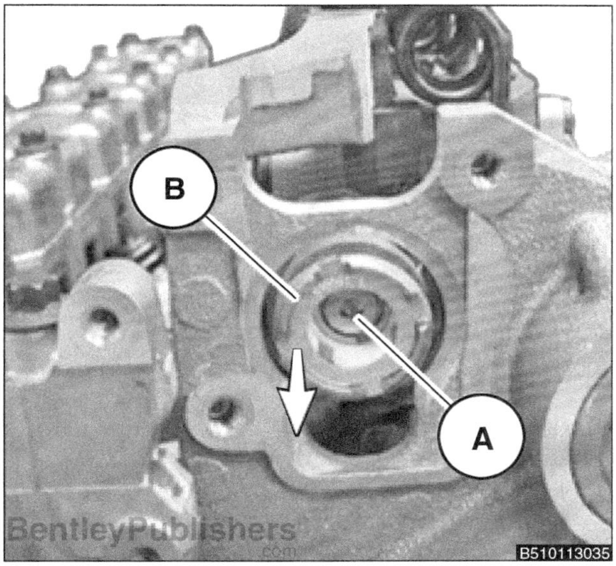

Remove eccentric shaft magnet wheel mounting bolt (**A**). Slide magnet wheel (**B**) forward (**arrow**) to remove.

> ***CAUTION—***
> - *Magnet wheel mounting bolt is non-magnetic. Do not allow it to fall into engine.*
> - *Magnet wheel is highly magnetic. Protect against metal filings by storing in sealed plastic bag.*

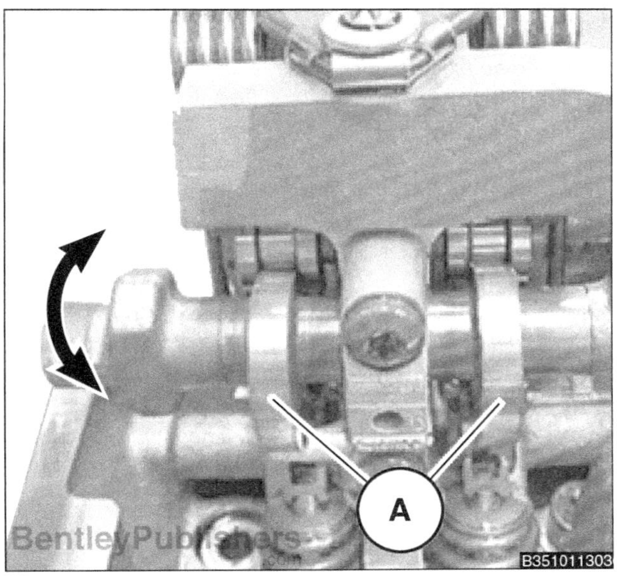

Pretension eccentric shaft (**arrow**). Remove stop screw between cylinders 1 and 2 (**A**).

BentleyPublishers.com—All Rights Reserved

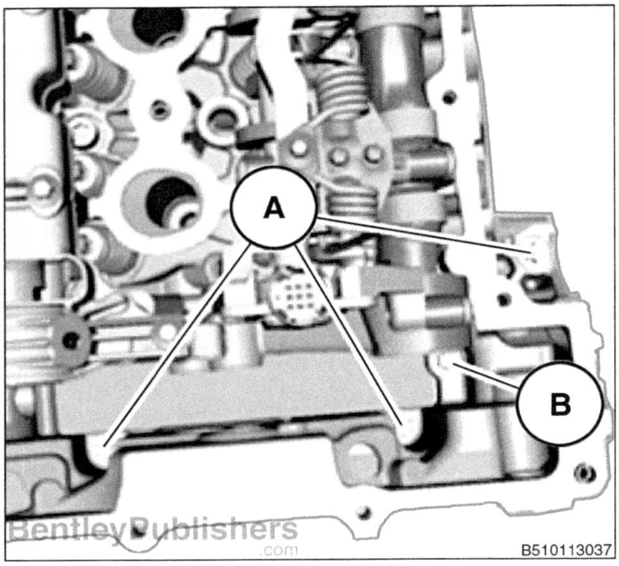

- Remove cylinder head bolts (**A**, **B**) to timing chain housing. Press timing chain rail forward slightly to remove bolt **B**. Discard aluminum bolts.

> ***CAUTION—***
> - *Use mechanical gripper to prevent lower bolt(s) from falling down inside crankcase.*
> - *Measure and note down different bolt lengths for correct reinstallation.*

- Use BMW special tools 11 4 420 (Torx T50) and 11 8 580 (Torx T60) to loosen and remove M9 and M10 cylinder head bolts. (Intake and exhaust camshafts removed for purpose of illustration.)
  - Loosen M9 bolts first.
  - Loosen M10 bolts in crisscross pattern, starting with outside bolts.
  - Note different bolt lengths.
  - Discard aluminum bolts.

- Bolt lifting handles to cylinder head. Lift off head with help of assistant.

> ***CAUTION—***
> - *Cylinder head weighs approx. 40 kg (88 lb).*
> - *Place cylinder head on work bench on its side. Putting it down on sealing surface risks damage to the valves.*
> - *Be sure to clean antifreeze out of cylinder head bolt holes in crankcase immediately.*

Cylinder head, installing (N52 engine)

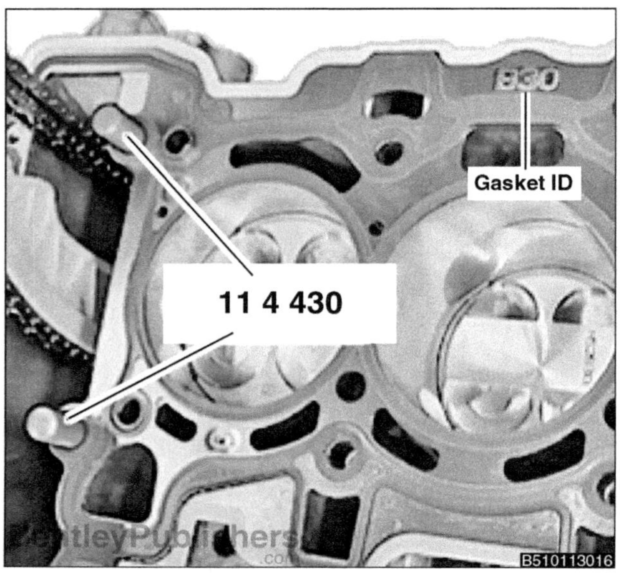

- Before cleaning engine block sealing surface, insert plugs (BMW special tools 11 4 430 or equivalent) into oil passages.
  - Note cylinder head gasket ID at right front edge of gasket.

- Clean cylinder head and gasket surfaces of engine block and timing chain housing.

> ***CAUTION—***
> - *Remove foreign matter and liquid from bolt holes. Trapped oil or coolant in bolt holes can cause damage to the engine block.*
> - *Do not use a metal scraper or wire brush to clean sealing surfaces. If necessary, use a hard wooden or plastic scraper. Also available are abrasive discs to be used in conjunction with an electric drill. Be sure to use the correct disc for the type of metal being cleaned.*

- Inspect cylinder head for visible cracks or other defects.

BentleyPublishers.com
BentleyPublishers.com—All Rights Reserved

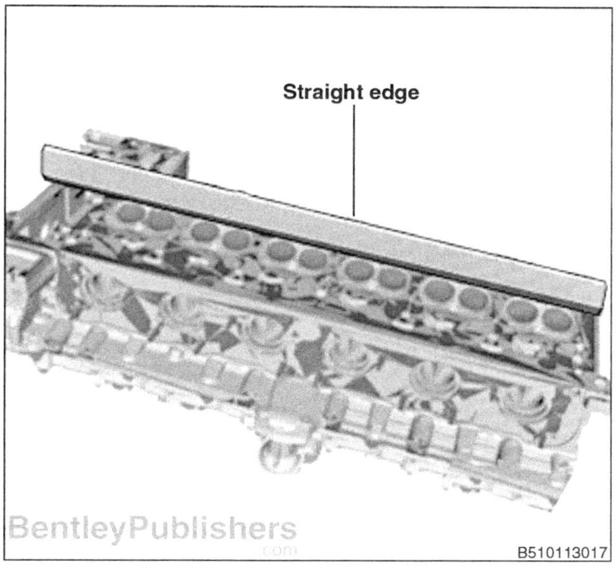

– To check cylinder head for warpage:
  - Remove intake and exhaust camshafts and intermediate intake levers.

> ***CAUTION—***
> - *Keep components in separate compartments and mark them for reassembly in their original positions.*

< Use straight edge to check evenness of cylinder head sealing surface in longitudinal direction.

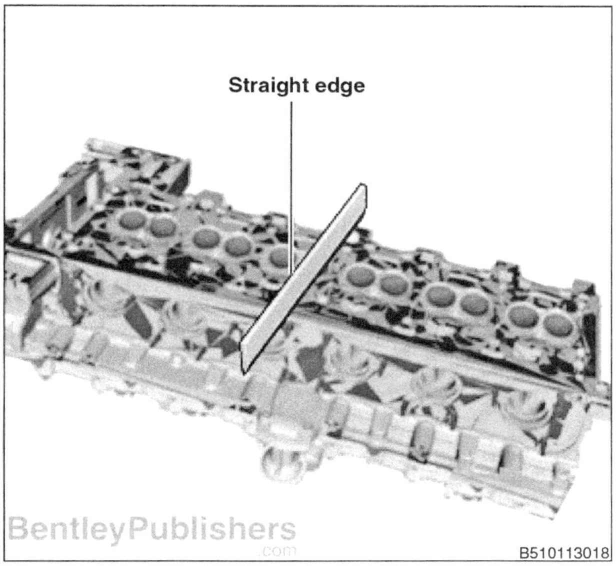

< Use straight edge to check evenness of cylinder head sealing surface in transverse direction.

Cylinder head warpage specifications	
Maximum warpage allowed • Longitudinal • Transverse	 0.10 mm (0.0039 in) 0.05 mm (0.0020 in)

– Have cylinder head tested for coolant leaks and cracks.

– If cylinder head is warped but otherwise sound, machine a maximum of 0.3 mm (0.011 in) off sealing surface.

– Reassemble cylinder head, or use reconditioned BMW cylinder head. See **117 Camshaft Timing Chain**.

Tightening torques	
Oil spray nozzle to valvetronic guide block (M6 x 30 mm bolt)	10 Nm (7 ft-lb)
Valvetronic guide block to timing chain housing (M6 x 23 mm bolt)	10 Nm (7 ft-lb)
Valvetronic torsion spring to cylinder head (M6 x 20 bolt)	10 Nm (7 ft-lb)
Camshaft bearing strip to camshaft (M7 x 70 mm 10.9 steel bolt) • Stage 1 • Stage 2	 8 Nm (71 in-lb) additional 60°

> ***CAUTION—***
> - *Once assembled with valves and camshafts, place cylinder head down on work bench on its side. Putting it down on sealing surface risks damage to the valves.*

BentleyPublishers.com
BentleyPublishers.com–All Rights Reserved

Cylinder head, installing (N52 engine)

- Check that two cylinder head aligning sleeves (**arrows**) are correctly positioned in engine block and are not damaged.
- Replace cylinder head gasket.
- If cylinder head is machined, use a special cylinder head gasket available from an authorized BMW dealer. The gasket is 0.3 mm (0.011 in) thicker than standard and is marked accordingly.
- Place cylinder head on engine block and fit to aligning sleeves.

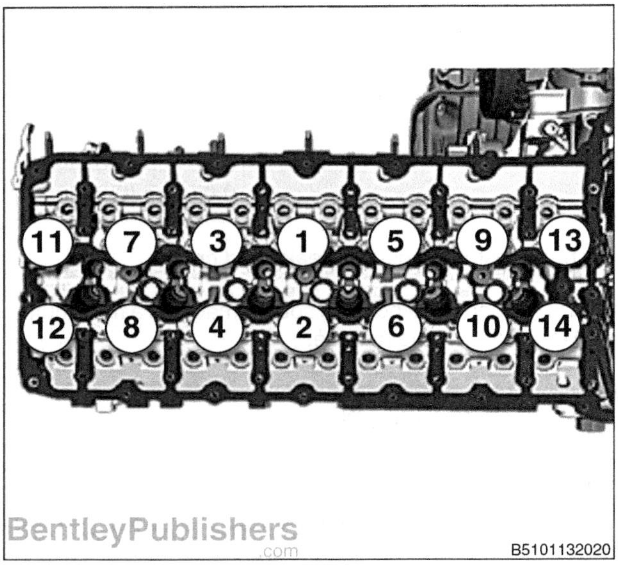

- Insert new aluminum cylinder head bolts:
  - Bolts 1 - 10: M10 x 125 mm
  - Bolts 11 - 14: M9 x 95 mm and M9 x 125 mm. Insert correct length bolts, as marked previously.

NOTE—
• Graphic shows intake and exhaust camshafts removed.

- Use BMW special tool 11 8 850 (Torx T60) to tighten bolts 1 - 10 in sequence shown below.
- Use BMW special tool 11 4 420 (Torx T50) to tighten bolts 11 - 14 in sequence shown below.

Tightening torques and sequence	
Cylinder head to engine block: • Stage 1: All bolts, 1 - 14 • Stage 2: All bolts, 1 - 14 • Stage 3: Only bolts 1 - 10 • Stage 4: All bolts 1 - 14	 30 Nm (22 ft-lb) additional 90° additional 90° additional 45°

CAUTION—
• Stage 3 applies to M10 bolts (1 - 10) only.

BentleyPublishers.com
BentleyPublishers.com—All Rights Reserved

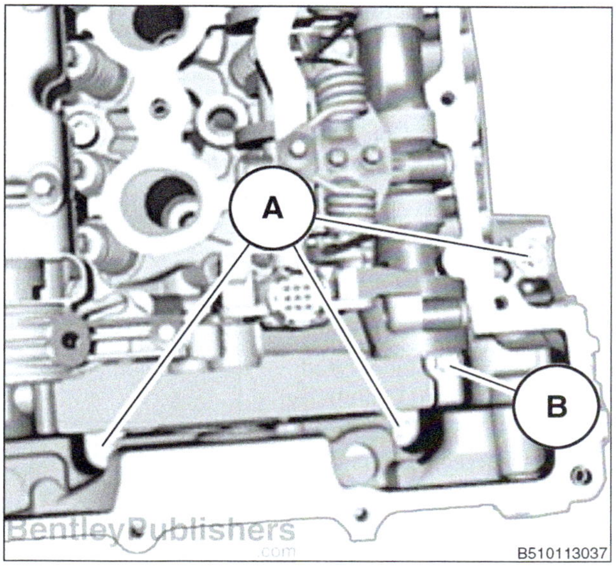

◀ Use new aluminum bolts (**A**, **B**) to attach cylinder head to timing chain housing. Press timing chain rail forward slightly to install lower bolt(s).

> ***CAUTION—***
> *• Use mechanical gripper to prevent lower bolt(s) from falling down inside crankcase.*
> *• Place different bolt lengths in correct bores.*

Tightening torques	
N52 engine cylinder head to timing chain housing (M9 x 30 mm or M9 x 70 mm): • Stage 1 • Stage 2	 15 Nm (11 ft-lb) additional 90°
N52K engine cylinder head to timing chain housing (M9 x 30 mm): • Stage 1 • Stage 2	 10 Nm (7 ft-lb) additional 90°
N52K engine cylinder head to timing chain housing (M9 x 70 mm): • Stage 1 • Stage 2	 10 Nm (7 ft-lb) additional 135°

– Reinstall eccentric shaft stop screw.

Tightening torque	
Eccentric shaft stop screw to cylinder head (M6)	10 Nm (7 ft-lb)

– Replace eccentric shaft magnet wheel.

– Reinstall eccentric shaft sensor.

– Use wire hook to retrieve timing chain from inside timing chain housing. Place chain over chain module and clip module to lower chain rail. Reinstall chain module mounting bolts.

Tightening torque	
Chain module to cylinder head (M6 x 35 mm Torx)	8.5 Nm (6 ft-lb)

◀ Use new bolts to reattach camshaft adjustment (VANOS) units. Note that exhaust and intake units are different and so marked. Camshaft sensor impulse wheel for intake and exhaust are the same.

> ***CAUTION—***
> *• Use special tools shown in* **117 Camshaft Timing Chain** *to lock down camshafts.*

Tightening torque	
Camshaft adjustment (VANOS) unit to camshaft (use new M10 x 94 mm bolt): • Stage 1 • Stage 2	 20 Nm (15 ft-lb) additional 180°

BentleyPublishers.com
BentleyPublishers.com—All Rights Reserved

– Install chain tensioner and set camshaft timing. See **117 Camshaft Timing Chain**.

> ***CAUTION—***
> - *If chain tensioner is reused, drain its oil chamber. Place tensioner on level working surface and compress slowly. Repeat twice.*
> - *No sealing ring is fitted to chain tensioner at the factory. When reassembling engine, be sure to use new sealing O-ring.*

Tightening torque	
Timing chain tensioner to cylinder head (M22 x 1.5) (use new sealing O-ring)	55 Nm (41 ft-lb)

– Remainder of installation is reverse of removal. Remember to:
- Remove camshaft and crankshaft locking tools.
- Assemble cylinder head cover with new gasket.
- Reattach and tighten any ground cables removed.
- Reassemble fuel rail using new O-ring seals.
- Change engine oil and filter. See **020 Maintenance**.
- Fill and bleed cooling system. See **170 Radiator and Cooling System**.

BentleyPublishers.com
BentleyPublishers.com—All Rights Reserved

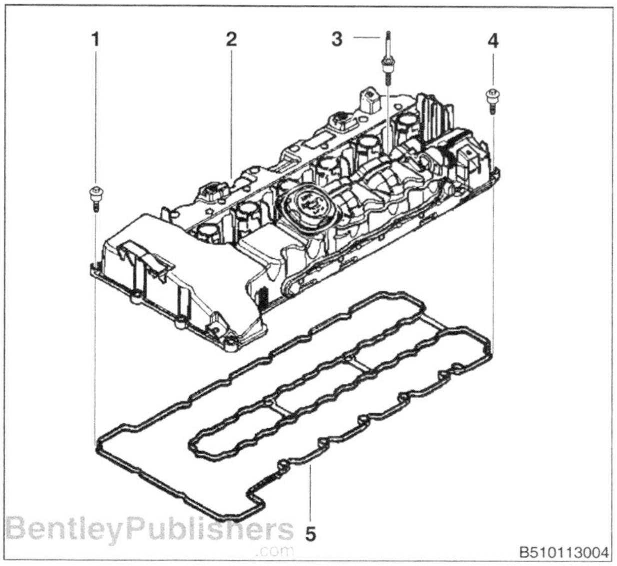

Cylinder Head (N54 Engine)

Cylinder head cover components (N54 engine)

1. Cylinder head cover bolt (M6 x 38 mm)
   - Quantity: 2
   - Tighten to 8.5 Nm (75 in-lb)
2. Cylinder head cover
3. Threaded stud
   - Quantity: 3
   - Remove using BMW special tool 11 8 620 or equivalent deep 10 mm socket wrench.
4. Cylinder head cover bolt (M6 x 32.5 mm)
   - Quantity: 26
   - Tighten to 8.5 Nm (75 in-lb)
5. Cylinder head cover gasket

Cylinder head cover, removing and installing (N54 engine)

– Disconnect negative (-) battery cable and cover battery terminal to keep cable from accidentally contacting terminal.

> ***CAUTION—***
> - *Prior to disconnecting the battery, read the battery disconnection cautions given in* **001 Warnings and Cautions**.

– Use scan tool to read out and record ECM fault memory.

– Switch ignition OFF and remove key.

– Remove upper and lower cabin microfilter housings. See **640 Heating and Air-conditioning**.

– Remove right strut tower brace.

– Remove ignition coil cover (upper engine cover). See **020 Maintenance**.

– Remove ignition coils. See **120 Ignition System**.

– Unclip fuel injector and ignition coil wiring harness and set aside.

BentleyPublishers.com
BentleyPublishers.com—All Rights Reserved

Cylinder head cover, removing and installing (N54 engine)

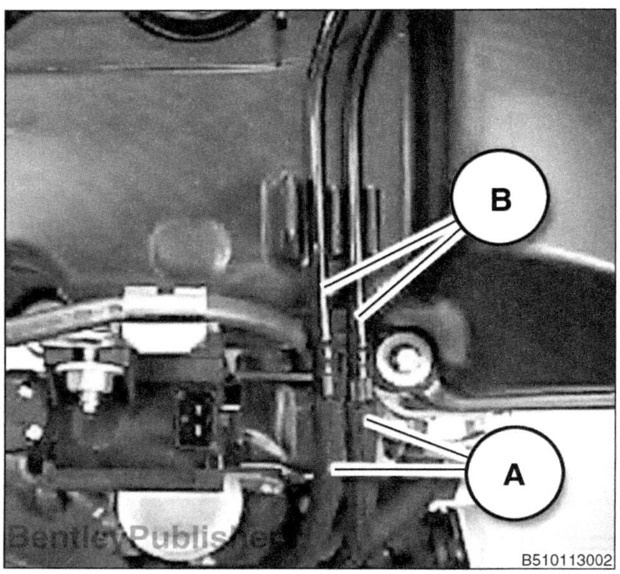

◀ Loosen high pressure fuel line junctions (**arrows**).

> ***WARNING—***
> - *Fuel in the fuel line is under pressure (approx. 200 bar or 2900 psi) and may be expelled forcibly. Do not smoke or work near heaters or other fire hazards. Keep a fire extinguisher handy.*
> - *Work only on fuel system when engine temperature is below 40°C (104°F).*
> - *Unscrew the fuel tank cap to release pressure in the tank before working on the fuel line.*
> - *Plug open fuel lines and fittings.*
> - *Wrap a clean shop towel around fitting before disconnecting. Residual fuel pressure is present in the fuel line.*
> - *To reduce fuel pressure, place towel over injector and carefully tap with a wrench.*

> ***CAUTION—***
> - *Maintain a clean work area when servicing high pressure fuel system. Contaminants can cause a system malfunction.*
> - *Fuel residue can break down the electrical resistance qualities of silicone used in ignition coil production, resulting in coil failure.*
> - *Replace high pressure fuel lines if tightened 10 times.*

– Remove injector lines and mark them in order.

– Remove injectors and mark them in order. Plug fuel injector bores to prevent debris from entering engine. See **132 Fuel Injection**.

◀ Working at left front of cylinder head:
- Detach vacuum hoses (**A**) from vacuum lines.
- Unclip vacuum lines (**B**) from cylinder head cover.

◀ Working at right front of cylinder head:
- Detach vacuum hoses (**A**) from vacuum lines.
- Unclip vacuum lines (**B**) and set aside.

BentleyPublishers.com
BentleyPublishers.com – All Rights Reserved

◄ Use BMW special tool 11 8 620 or equivalent deep 10 mm socket wrench to loosen and remove threaded studs (**arrow**) in center of cylinder head cover valley.

- Quantity: 3 studs

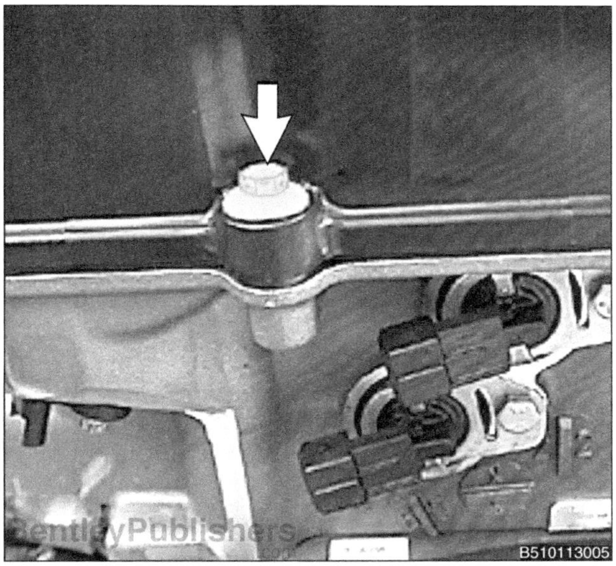

◄ Remove cylinder head cover bolts (**arrow**).

- M6 x 38 mm bolts, quantity: 2
- M6 x 32.5 mm bolts, quantity: 26

NOTE —

- *Make note of arrangement of fasteners during removal so that they can be reinstalled in their original locations.*

– Remove cylinder head cover and discard gasket.

– Inspect cylinder head cover fasteners. Replace as necessary.

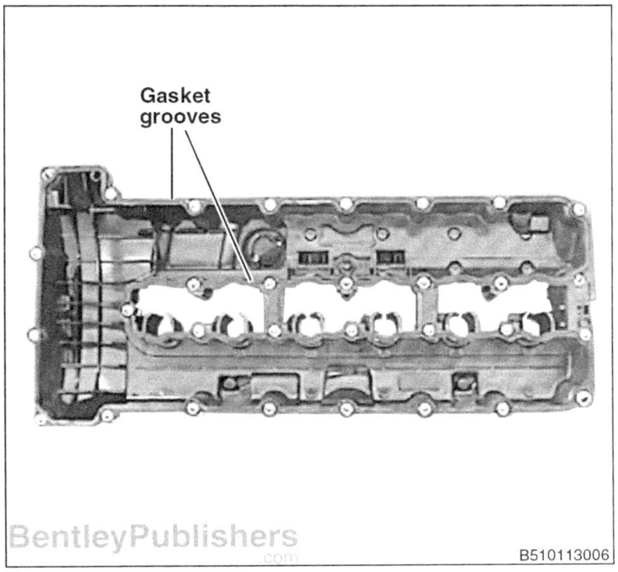

◄ Clean gasket residue from cylinder head sealing surface and from cylinder head cover grooves.

CAUTION—
- *To avoid gouging the cover, do not use metal scraping tool.*

– Coat cylinder head cover grooves with thin layer of anti-friction compound such as glycerine. Press gasket in place, making sure it is not stretched at any point.

BentleyPublishers.com
BentleyPublishers.com–All Rights Reserved

Cylinder head cover, removing and installing (N54 engine)

– Reinstall cylinder head cover.
- Coat cylinder head and gasket contact surfaces with thin layer of anti-friction compound such as glycerine.
- Place small quantity of silicone sealant at metal-to-metal seams (such as timing cover to engine block).
- Install cylinder head cover fasteners in correct locations as noted previously.
- Tighten gradually in crisscross pattern, starting with inside fasteners.
- Replace fuel injector PTFE sealing rings. See **132 Fuel Injection**.

Tightening torque	
Cylinder head cover to cylinder head (M6)	8.5 Nm (75 in-lb)

Reinstall high pressure fuel lines and injectors in correct locations, as marked previously. See **132 Fuel Injection**.
- Coat nut threads with gear oil.
- Connect high pressure line (**A**) finger tight.
- High pressure line may only be retightened 10 times.

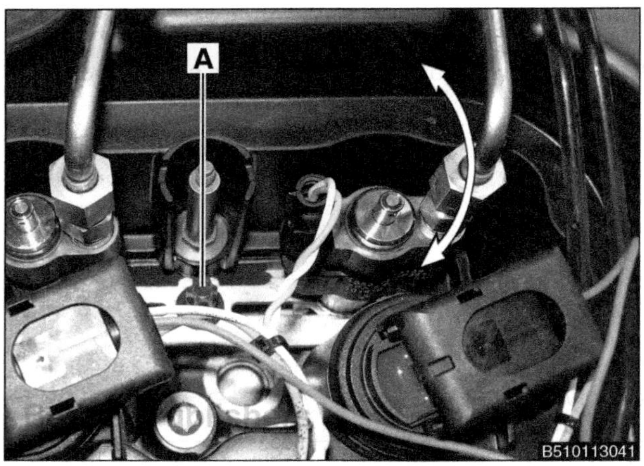

If fuel lines cannot be easily installed, rock injector (**arrows**) to allow fuel line nipple threads to be started by hand.
- Tighten hold-down bracket fastener (**A**).

Tightening torque	
Hold-down bracket to cylinder head	13 Nm (9.5 ft-lb)

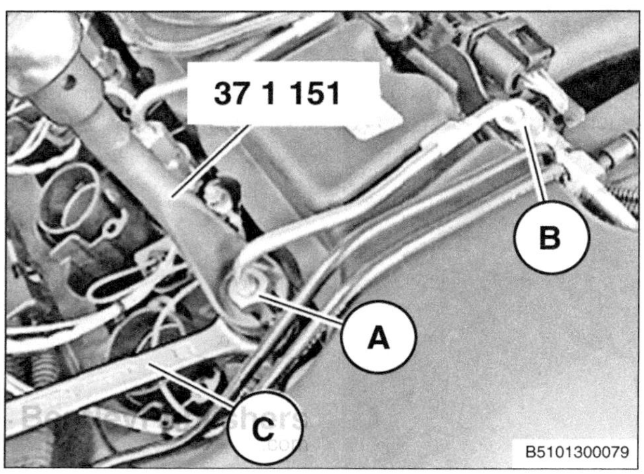

To tighten injector lines:
- Tighten nut (**A**) using special tool 37 1 151 while holding injector with wrench (**C**).
- Tighten nut (**B**) using special tool 37 1 151.

Tightening torque	
High pressure line to injector	23 Nm (17 ft-lb)
High pressure line to fuel rail	23 Nm (17 ft-lb)

BentleyPublishers.com
BentleyPublishers.com—All Rights Reserved

– Clean all fuel from ignition coil hole.

– Remainder of assembly is reverse of removal. Keep in mind the following:
  - Make sure cabin microfilter housing seals correctly.
  - Check fuel lines for leaks once engine starts.

Tightening torque	
Ignition coil cover to cylinder head cover (M6)	4 Nm (35 in-lb)
Strut tower brace to bulkhead (replace M12 bolt) • Stage 1 • Stage 2	 100 Nm (74 ft-lb) additional 90°
Strut tower brace to strut tower (replace M10 bolt) • Stage 1 • Stage 2	 40 Nm (30 ft-lb) additional 60°

Cylinder head and timing chain (N54 engine)

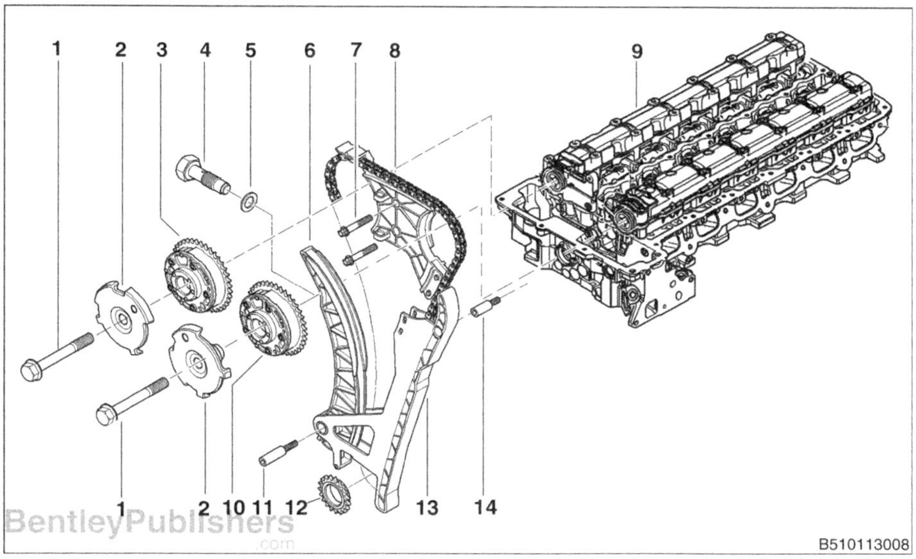

BentleyPublishers.com

1. Camshaft adjustment unit mounting bolt (M10 x 94 mm)
   - Replace with new
   - Tighten in 2 stages
     Stage 1: 20 Nm (15 ft-lb)
     Stage 2: Additional 180°
2. Camshaft impulse wheel
3. Exhaust camshaft adjustment unit
4. Chain tensioner (M22 x 1.5)
   - Tighten to 55 Nm (41 ft-lb)
5. Sealing O-ring
   - Install new
6. Chain tensioner rail
7. Chain module mounting bolts (M6 x 35 mm Torx)
   - Tighten to 8.5 Nm (6 ft-lb)
8. Timing chain
9. Cylinder head with camshafts
10. Intake camshaft adjustment unit
11. Chain guide mounting bolt (M8)
    - Tighten to 20 Nm (15 ft-lb)
12. Crankshaft sprocket
13. Chain guide
14. Chain guide mounting bolt (M7)
    - Tighten to 14 Nm (10 ft-lb)

BentleyPublishers.com
BentleyPublishers.com–All Rights Reserved

Cylinder head, removing and installing (N54 engine)

To remove the turbo engine cylinder head, first remove the engine. Special tools and procedures are required to remove and install camshaft adjustment units and to time camshafts. Read the entire procedure before beginning repairs.

> ***CAUTION—***
> - *Disassembly, removal and assembly of camshafts, camshaft adjustment units or cylinder head without special tools poses the risk of damage or breakage:*
>   *-Valves may be bent by contact with the piston crowns.*
>   *-Camshaft may snap in two if bent by valve springs.*

NOTE—
- *Camshaft timing is covered in* **117 Camshaft Timing Chain**.

Cylinder head removal and installation procedures are described separately.

If you remove a cylinder head and determine that it requires significant reconditioning work, a remanufactured cylinder head is available from an authorized BMW dealer.

Cylinder head, removing (N54 engine)

The cylinder head stretch bolts are aluminum. Replace them any time they are removed. Use angle protractor to torque.

> ***WARNING—***
> - *To avoid personal injury, be sure the engine is cold before beginning the removal procedure.*

– Disconnect negative (–) cable from battery.

> ***CAUTION—***
> - *Prior to disconnecting the battery, read the battery disconnection cautions in* **001 Warnings and Cautions**.

– Remove ignition coil cover (upper engine cover). See **020 Maintenance**.

– With engine fully cooled off, drain engine coolant. See **170 Radiator and Cooling System**.

– Remove engine. See **110 Engine Removal and Installation**.

– Remove cylinder head cover. See **Cylinder head cover, removing and installing (N54 engine)** in this repair group.

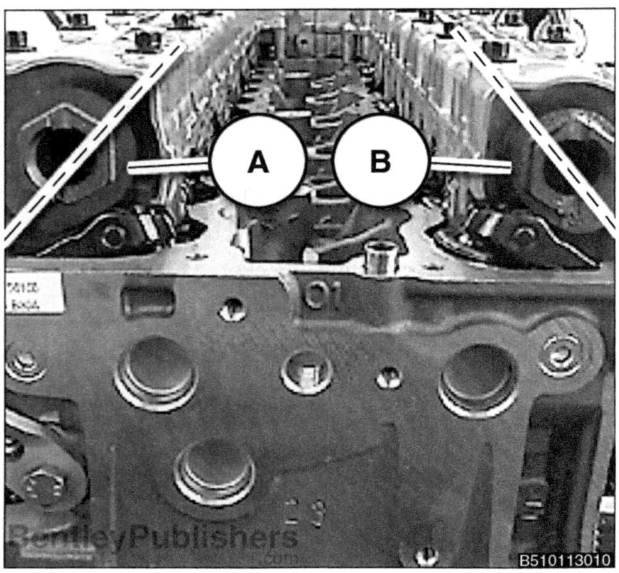

◄ Using vibration damper bolt to rotate crankshaft, place cylinder 1 in TDC firing position. In this position, lobes for cylinder 6 (rear of engine) intake and exhaust camshafts (**A**, **B**) point away from each other and down. Cylinder 6 camshaft lobes do not contact roller cam followers.

BentleyPublishers.com
BentleyPublishers.com—All Rights Reserved

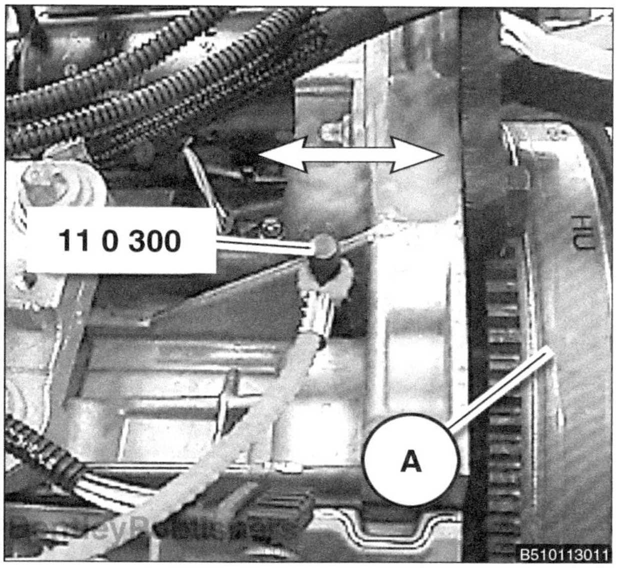

NOTE —
- *Use a mirror to check cylinder 6 cam lobe positions.*

Working underneath engine, slide out protective plug at lower left of engine bell housing flange. Install BMW special tool 11 0 300 (crankshaft locking tool) through bell housing flange port into flywheel (**A**) hole. If necessary, rock flywheel slightly back and forth to line up holes with tool. This locks crankshaft at TDC.

CAUTION—
- *Automatic transmission model: A short distance before the flywheel special tool bore for TDC position, there is a large bore which can be confused with the special tool bore. If the flywheel is locked using the correct bore, the engine can no longer be rotated at the vibration damper bolt.*

— Remove intake and exhaust camshaft adjustment units. See **117 Camshaft Timing Chain**.

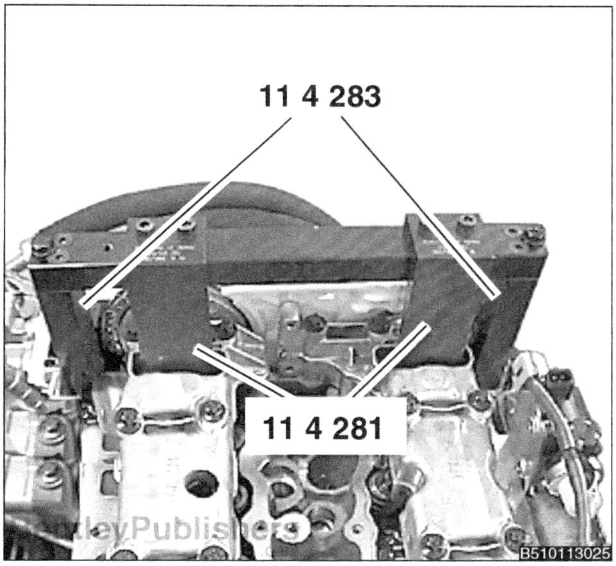

Use BMW special tool set 11 4 280 to lock down camshafts.

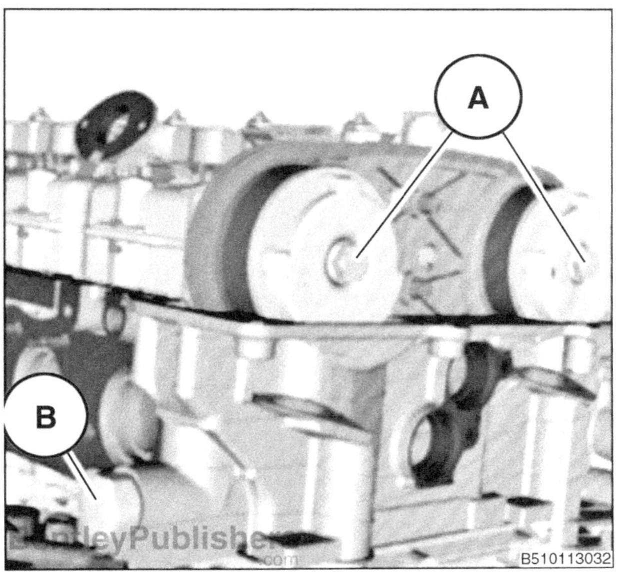

Remove intake and exhaust camshaft adjustment units:
- Loosen and remove adjustment unit mounting bolts (**1**). Discard bolts.
- Remove timing chain tensioner (**2**). Be prepared to catch dripping oil with a shop towel. Do not allow oil to contaminate accessory belt.
- Disengage timing chain from camshaft adjustment unit sprockets. Remove adjustment units and set aside.

NOTE —
- *Illustration does not show special tool set 11 4 280 in place.*

113

BentleyPublishers.com—All Rights Reserved

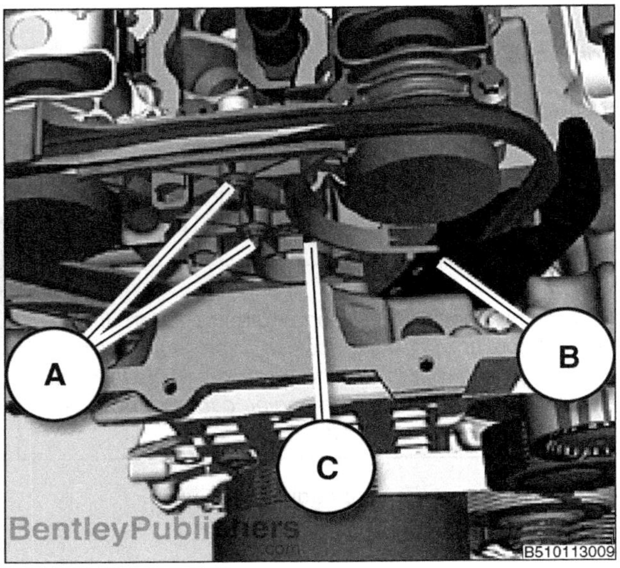

Working at front of cylinder head:

- Remove timing chain module mounting bolts (**A**).
- Unclip chain module (**C**) at junction (**B**) and lift out.
- Place timing chain in cylinder head opening.

> ***CAUTION—***
> - *Do not remove crankshaft locking tool. Rotating crankshaft may cause the timing chain to jam or jump teeth.*

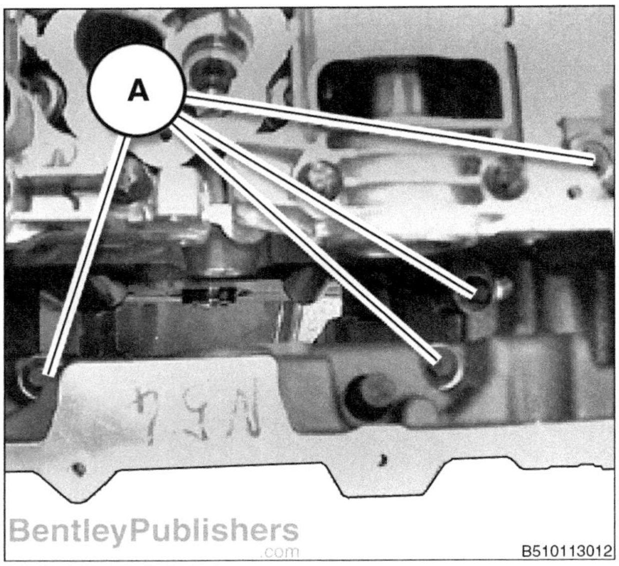

Remove cylinder head bolts (**A**) to timing chain housing. Press timing chain rail forward slightly to remove lower bolt(s).

> ***CAUTION—***
> - *Use mechanical gripper to prevent lower bolt(s) from falling down inside crankcase.*
> - *Measure and note down different bolt lengths for correct reinstallation.*

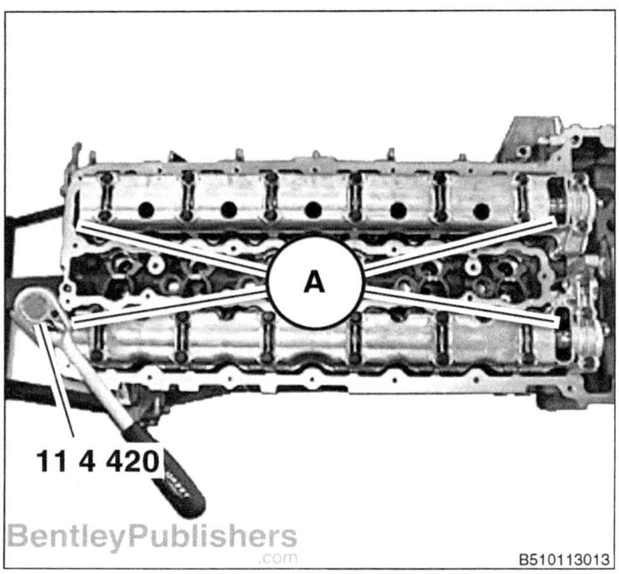

Use BMW special tool 11 4 420 (Torx T50) to loosen and remove M9 cylinder head bolts.

- Note different bolt lengths.
- Discard all aluminum bolts.

BentleyPublishers.com
BentleyPublishers.com—All Rights Reserved

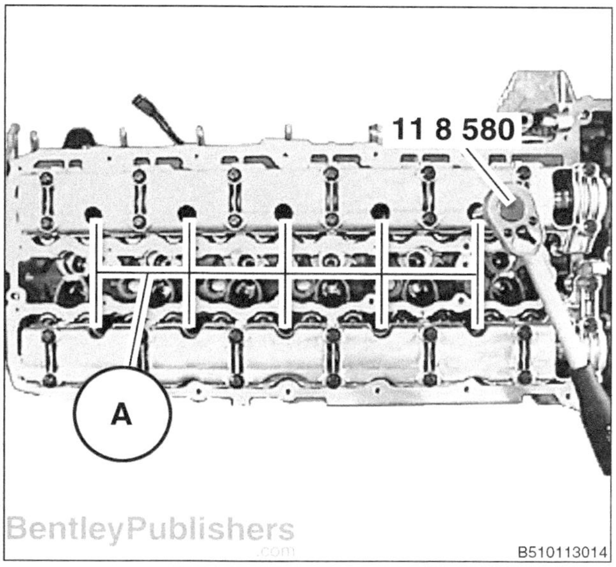

Use BMW special tool 11 8 850 (Torx T60) to loosen and remove M11 cylinder head bolts (**A**).

- Start at outer corners and loosen bolts in crisscross pattern.
- Discard all aluminum bolts.

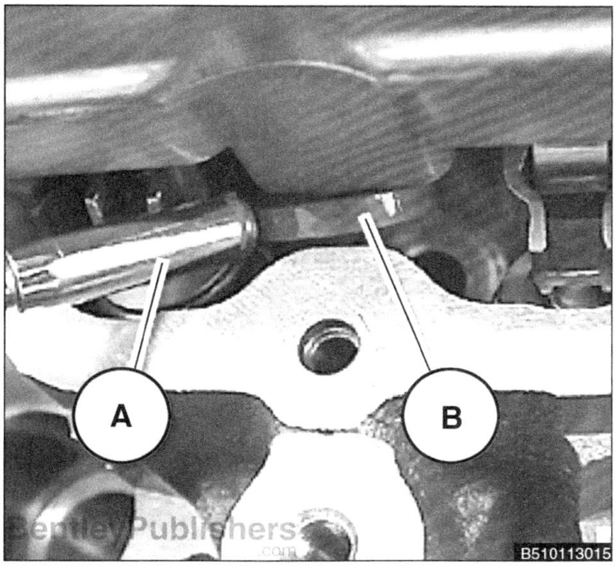

Use magnet (**A**) to lift out cylinder head bolt washers (**B**) between surface of cylinder head and camshaft bearing strip.

– Bolt lifting handles to cylinder head. Lift off head with help of assistant.

> ***CAUTION—***
> - *Cylinder head weighs approx. 40 kg (88 lb).*
> - *Place cylinder head down on work bench on its side. Putting it down on sealing surface risks damage to the valves.*
> - *Be sure to clean antifreeze out of cylinder head bolt holes in crankcase immediately.*

Cylinder head, installing (N54 engine)

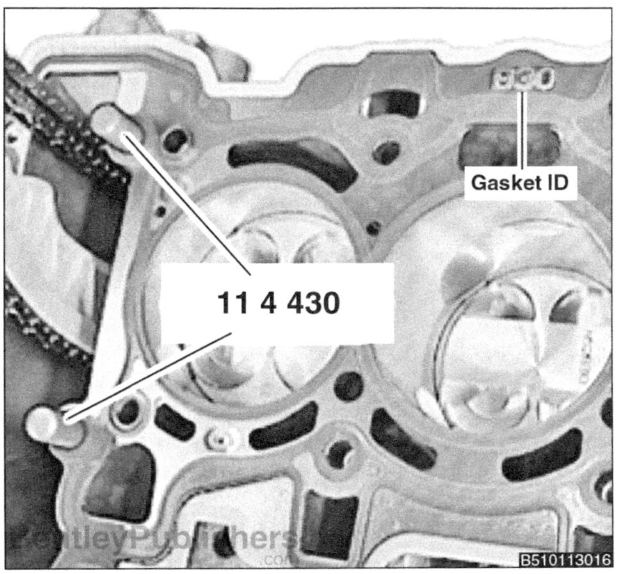

Before cleaning engine block sealing surface, insert plugs (BMW special tools 11 4 430 or equivalent) into oil passages.

– Clean cylinder head and gasket surfaces of engine block and timing chain housing.

> ***CAUTION—***
> - *Remove foreign matter and liquid from bolt holes. Trapped oil or coolant in bolt holes can cause damage to the engine block.*
> - *Do not use a metal scraper or wire brush to clean the aluminum cylinder head or pistons. If necessary, use a hard wooden or plastic scraper. Also available are abrasive discs to be used in conjunction with an electric drill. Be sure to use the correct disc for the type of metal being cleaned.*

– Inspect cylinder head for visible cracks or other defects.

BentleyPublishers.com
BentleyPublishers.com–All Rights Reserved

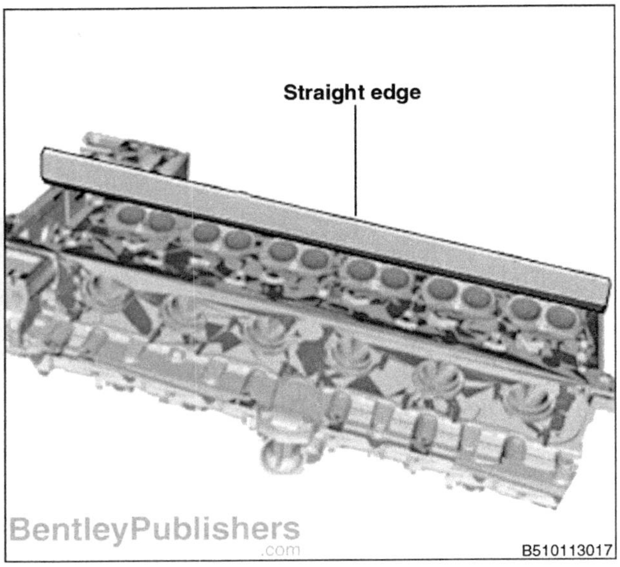

- To check cylinder head for warpage:
  - Remove intake and exhaust camshafts and roller cam followers.
  - Remove valve lifters and valves.

◀ Use straight edge to check evenness of cylinder head sealing surface in longitudinal direction.

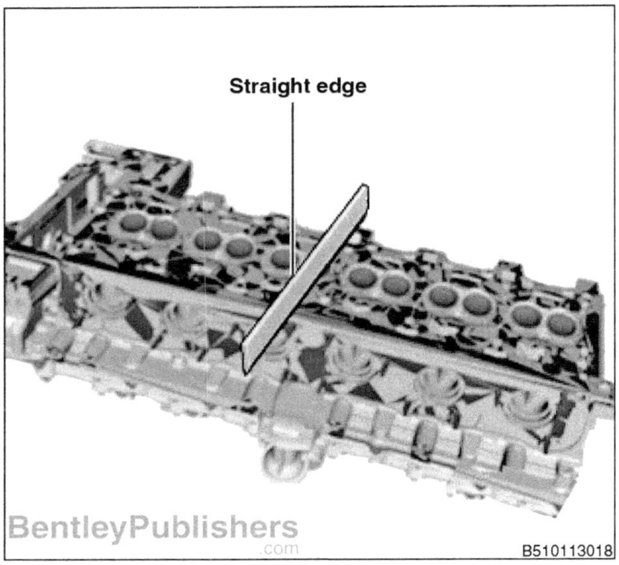

◀ Use straight edge to check evenness of cylinder head sealing surface in transverse direction.

Cylinder head warpage specifications	
Maximum warpage allowed • Longitudinal • Transverse	 0.10 mm (0.0039 in) 0.05 mm (0.0020 in)

- Have cylinder head tested for coolant leaks and cracks.
- If cylinder head is warped but otherwise sound, machine a maximum of 0.3 mm (0.011 in) off sealing surface.
- Reassemble cylinder head, or use reconditioned BMW cylinder head.

Tightening torque	
Camshaft bearing strip to camshaft (M7 x 70 mm 10.9 steel bolt) • Stage 1 • Stage 2	 8 Nm (71 in-lb) additional 60°

> ***CAUTION—***
> - *Once assembled with valves and camshafts, place cylinder head down on work bench on its side. Putting it down on sealing surface risks damage to the valves.*

◀ Check that two cylinder head aligning sleeves (**arrows**) are correctly positioned in engine block and are not damaged.

- Replace cylinder head gasket.
- If cylinder head is machined, use a special cylinder head gasket available from an authorized BMW dealer. The gasket is 0.3 mm (0.011 in) thicker than standard and is marked accordingly.
- Place cylinder head on engine block and fit to aligning sleeves.

BentleyPublishers.com
BentleyPublishers.com—All Rights Reserved

Insert new aluminum cylinder head bolts:

- Bolts 1 - 10: M11 x 135 mm
- Bolts 11 - 14: M9 x 95 mm and M9 x 125 mm. Insert correct length bolts, as marked previously.
- Where necessary, use magnet to fit bolt washer.

NOTE —

- *Graphic shows intake and exhaust camshafts removed.*

– Use BMW special tool 11 8 850 (Torx T60) to tighten bolts 1 - 10 in sequence shown below.

– Use BMW special tool 11 4 420 (Torx T50) to tighten bolts 11 - 14 in sequence shown below.

Tightening torques and sequence	
Cylinder head to engine block: • Stage 1: All bolts, 1 - 14 • Stage 2: All bolts, 1 - 14 • Stage 3: Only bolts 1 - 10 • Stage 4: All bolts 1 - 14	 30 Nm (22 ft-lb) additional 90° additional 90° additional 45°

CAUTION—

- *Stage 3 applies to M11 bolts (1 - 10) only.*

Install cylinder head bolts to timing chain housing (**A**). Press timing chain rail forward slightly to install lower bolt(s).

CAUTION—

- *Use mechanical gripper to prevent lower bolt(s) from falling down inside crankcase.*
- *Place different bolt lengths in correct bores.*

Tightening torque	
Cylinder head to timing chain housing (M9 x 30 mm or M9 x 70 mm)	22 Nm (16 ft-lb)

– Use wire hook to retrieve timing chain from inside timing chain housing. Place chain over chain module and clip module to lower chain rail. Reinstall chain module mounting bolts.

Tightening torque	
Chain module to cylinder head (M6 x 35 mm Torx)	8.5 Nm (6 ft-lb)

BentleyPublishers.com
BentleyPublishers.com—All Rights Reserved

Cylinder head, installing (N54 engine)

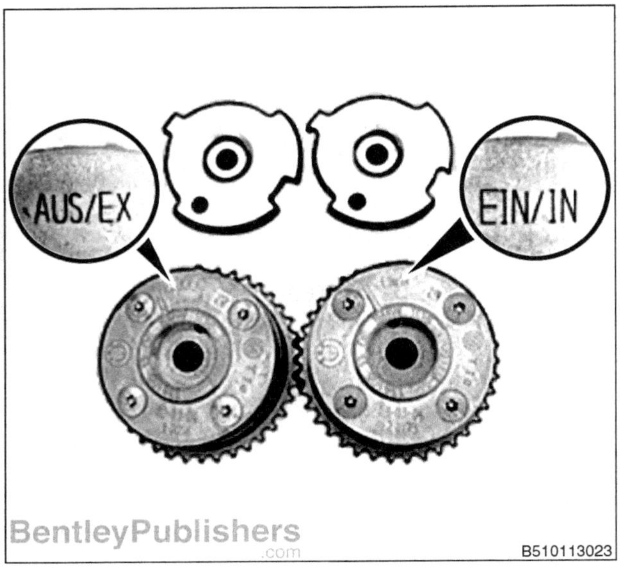

Use new bolts to reattach camshaft adjustment units. Note that exhaust and intake units are different and so marked. Camshaft sensor impulse wheel for intake and exhaust are the same.

> ***CAUTION—***
> - *Use special tools shown in* **117 Camshaft Timing Chain** *to lock down camshafts.*

Tightening torque	
Camshaft adjustment unit to camshaft (use new M10 x 94 mm bolt): • Stage 1 • Stage 2	 20 Nm (15 ft-lb) additional 180°

– Install chain tensioner and set camshaft timing. See **117 Camshaft Timing Chain**.

> ***CAUTION—***
> - *If chain tensioner is reused, drain its oil chamber. Place tensioner on level working surface and compress slowly. Repeat twice.*
> - *No sealing ring is fitted to chain tensioner at the factory. When reassembling engine, be sure to use new sealing O-ring.*

Tightening torque	
Timing chain tensioner to cylinder head (M22 x 1.5) (use new sealing O-ring)	55 Nm (41 ft-lb)

– Remainder of installation is reverse of removal. Remember to:
- Remove camshaft and crankshaft locking tools.
- Assemble cylinder head cover with new gasket.
- Reattach and tighten any ground cables removed.
- Reassemble fuel rail using new O-ring seals.
- Change engine oil and filter. See **020 Maintenance**.
- Fill and bleed cooling system. See **170 Radiator and Cooling System**.

BentleyPublishers.com
BentleyPublishers.com—All Rights Reserved

CYLINDER HEAD (V8 ENGINE)

Cylinder head cover components (V8 engine)

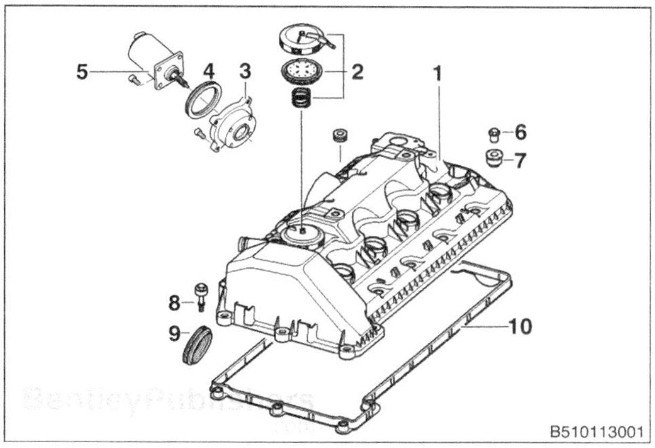

Cylinder head cover for V8 engines is made of plastic. Replace gaskets, seals and O-rings whenever the cover is removed.

1. Cylinder head cover
2. Pressure regulating valve
3. Valvetronic motor spacer
4. O-ring
5. Valvetronic motor
6. Cap nut
7. Rubber seal
8. Bolt
9. Gasket ring
10. Cylinder head cover gasket

Cylinder head cover, removing and installing (V8 engine)

- Use scan tool to read out and record ECM fault memory.
- Switch ignition OFF and remove key.
- Remove upper engine cover. See **020 Maintenance**.
- Remove ignition coil cover. See **020 Maintenance**.

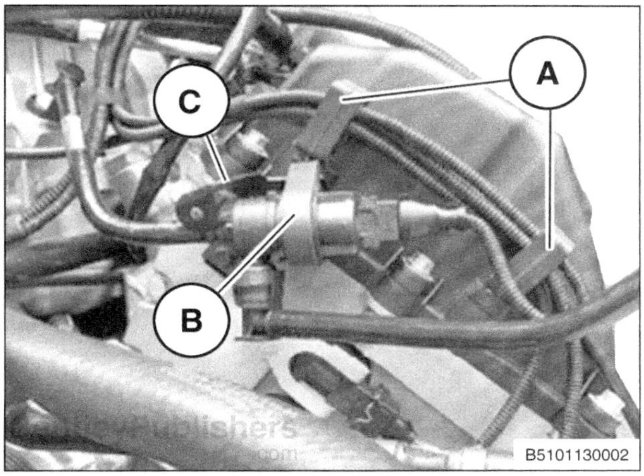

◀ Unclip wiring harness and holder from (**A**) from cylinder head cover. Remove fuel tank vent valve (**B**) off mounting bracket (**C**).

BentleyPublishers.com–All Rights Reserved

Cylinder head cover, removing and installing (V8 engine)

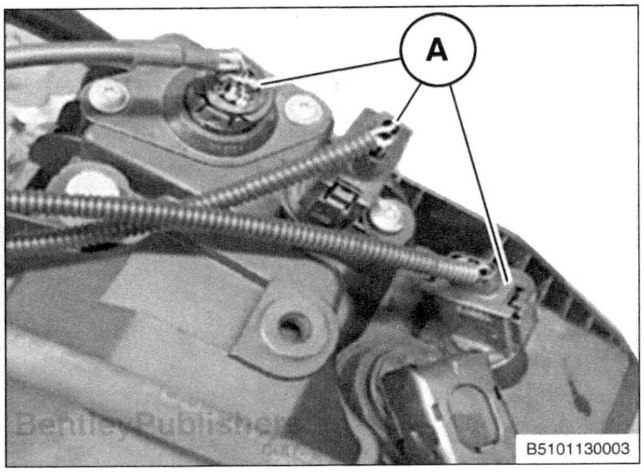

Disconnect electrical connectors (**A**).

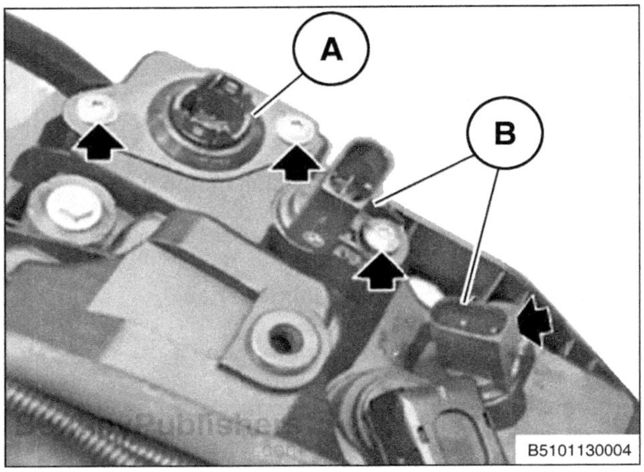

Remove fasteners (**arrows**). Remove wiring harness bulkhead seal (**A**) and discard, remove sensors (**B**).

– Remove ignition coils. See **120 Ignition System**.

Remove ground straps (**arrows**). Unclip and remove wiring harness duct (**A**).

– Remove valvetronic motor. See **133 Fuel Injection**.

BentleyPublishers.com
BentleyPublishers.com–All Rights Reserved

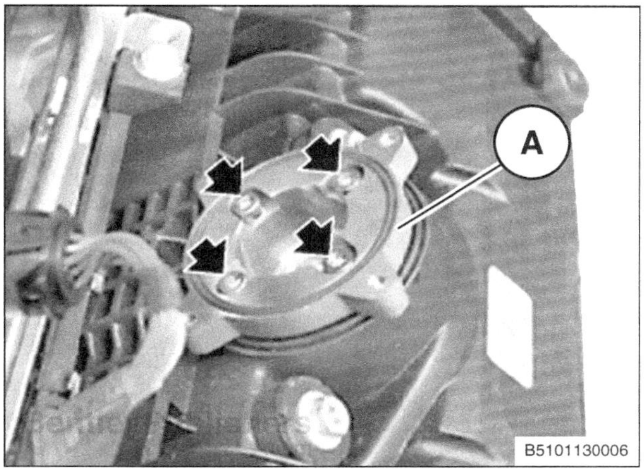

◀ Remove valvetronic motor spacer fasteners (**arrows**) and spacer (**A**). Discard valvetronic motor spacer seal upon removal.

– Remove fuel injector rail. See **133 Fuel Injection**.

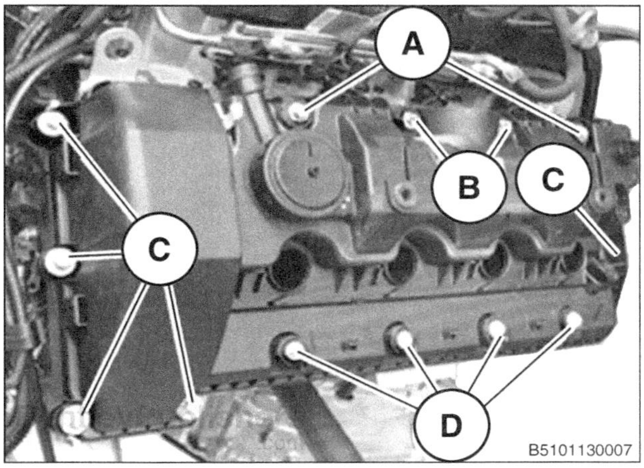

◀ Remove fasteners (**A**, **B**, **C** and **D**). If necessary remove dipstick. Lift and remove cylinder head cover.

– To reinstall cylinder head cover.
  - Coat cylinder head contact surfaces with thin layer of anti-friction compound such as glycerine.
  - Place small quantity of silicone sealant at metal-to-metal seams (such as timing cover to engine block) and half moon gasket areas.
  - Install spark plug domes in cylinder head cover before installation.
  - Replace cylinder head cover gaskets, seals and O-rings.
  - Reinstall valvetronic motor.
  - Tighten cylinder head cover bolts gradually in crisscross pattern, starting with inside fasteners.

Tightening torques	
Cylinder head cover to cylinder head • M6 bolts • M7 bolts • M8 bolts	 10 Nm (7 ft-lb) 15 Nm (11 ft-lb) 17 Nm (12.5 ft-lb)

– Remainder of assembly is reverse of removal.

BentleyPublishers.com–All Rights Reserved

Cylinder Head (V8 Engine)

Cylinder head and timing chain (V8 engine)

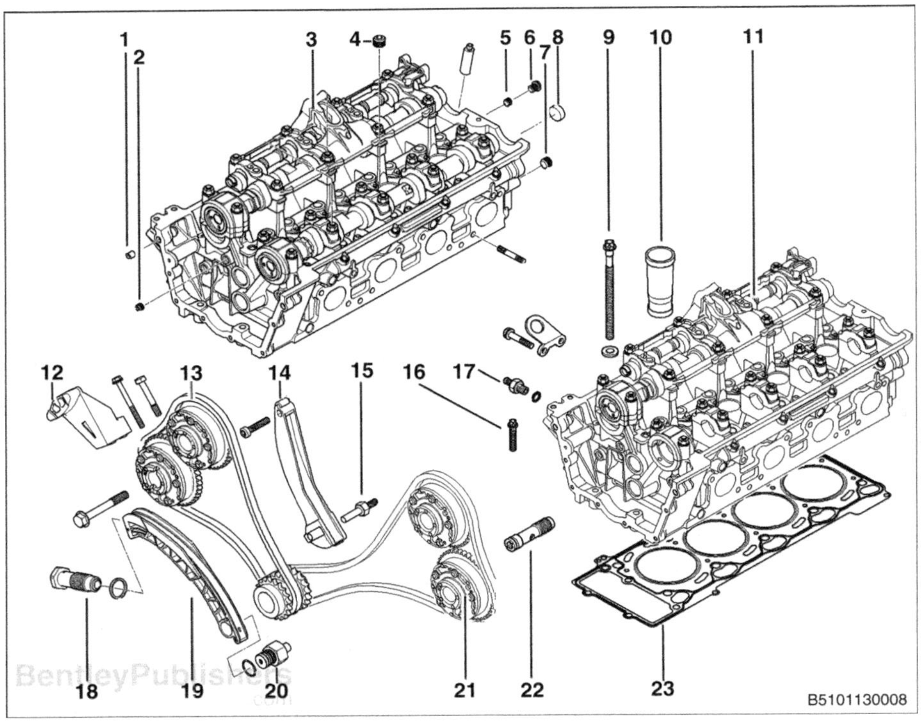

1. Dowel
2. Screw plug
3. Valvetronic motor flange
4. Screw plug
5. O-ring
6. Screw plug
7. Screw plug
8. Cover
9. Cylinder head bolt
10. Spark plug dome
11. Cylinder head assembly
12. Timing chain rail
13. Timing chain
14. Timing chain guide
15. Bearing bolt
16. Cylinder head bolt
17. Connector
18. Timing chain tensioner
19. Timing chain guide
20. Timing chain guide bearing
21. VANOS adjuster
22. Check valve
23. Cylinder head gasket

Cylinder head removal and installation (V8 engine)

V8 engine cylinder heads are equipped with VANOS. To remove the cylinder heads, special tools and procedures are required for removing the VANOS units. Read the entire procedure before beginning the repair.

> ***CAUTION—***
> - *Disassembly, removal and assembly of camshafts or cylinder head without special tools poses the risk of damage or breakage. The valves may be bent by contact with the piston crown; camshafts may break if stressed incorrectly during removal or installation.*

NOTE—
- *VANOS or variable camshaft timing is from the German words VAriable NOckenwellen Steuerung.*
- *VANOS removal is covered in* **117 Camshaft Timing Chain**.

BentleyPublishers.com
BentleyPublishers.com—All Rights Reserved

Right and left cylinder head removal and installation procedures are given separately. Remove the intake manifold before removing either cylinder head. See **133 Fuel Injection**.

If you remove a cylinder head and determine that it requires significant reconditioning work, a remanufactured cylinder head is available from an authorized BMW dealer.

Right cylinder head, removing (V8 engine)

> ***WARNING —***
> - *Due to risk of personal injury, be sure the engine is cold before beginning the removal procedure.*

– Disconnect negative (-) battery cable.

> ***CAUTION—***
> - *Prior to disconnecting the battery, read the battery disconnection cautions in* **001 Warnings and Cautions**.

– Raise car and support safely.

> ***WARNING —***
> - *Make sure the car is stable and well supported at all times. Use a professional automotive lift or jack stands designed for the purpose. A floor jack is not adequate support.*

– Remove right exhaust manifold. See **180 Exhaust System**.

NOTE —
- *To remove exhaust manifolds, secure engine in the installation position with special tool and lower front axle bracket. Once manifolds have been removed Reinstall front axle bracket and remove special tool for securing engine in installation position.*

– Open engine coolant drain plug at right side of engine block. Be prepared to catch draining coolant. Drain remaining engine coolant from radiator. See **170 Radiator and Cooling System**.

– Remove ignition coil cover. See **020 Maintenance**.

– Remove ignition coils. See **120 Ignition System**.

– Remove valvetronic motor. See **133 Fuel Injection**.

– Remove right cylinder head cover. See **Cylinder head cover, removing and installing (V8 engine)** in this repair group.

– Remove intake manifold. Plug intake manifold ports to prevent debris from entering engine See **133 Fuel Injection**.

– Remove upper right timing cover. See **117 Camshaft Timing Chain**.

– Disconnect electrical connectors and lay engine wiring harness aside.

– Remove spark plugs. Plug spark plug bores to prevent debris from entering engine. See **020 Maintenance**.

BentleyPublishers.com
BentleyPublishers.com—All Rights Reserved

Right cylinder head, removing (V8 engine)

– Remove right cylinder head vent hose.

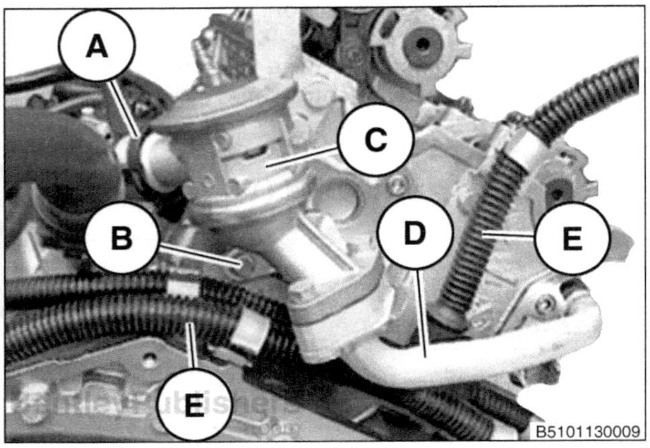

Working at rear of right side cylinder head, remove secondary air hose (**A**). Remove check valve fasteners (**B**) and remove check valve (**C**) with tube (**D**). Unclip wiring harness (**E**) and lay aside.

– Remove intake and exhaust camshaft adjustment (VANOS) units. See **117 Camshaft Timing Chain**.

Loosen timing chain tensioner (**A**) by 1 turn (360°).

Rotate eccentric shaft at dihedral head to reduce tension on torsion spring (**A**). Using a strap (**B**) carefully pull torsion spring (**A**) off roller (**C**) in direction of **arrow**.

BentleyPublishers.com
BentleyPublishers.com—All Rights Reserved

Rotate eccentric shaft at dihedral head in direction of **arrow** to minimum stroke postion.

Remove spring mount fasteners (**arrows**) and feed spring mount (**A**) upwards out of locating pin.

Remove eccentric shaft sensor fasteners (**arrows**). Remove eccentric shaft sensor (**A**) from cylinder head.

BentleyPublishers
BentleyPublishers.com—All Rights Reserved

Right cylinder head, removing (V8 engine)

Remove timing chain guide fastener (**arrow**).

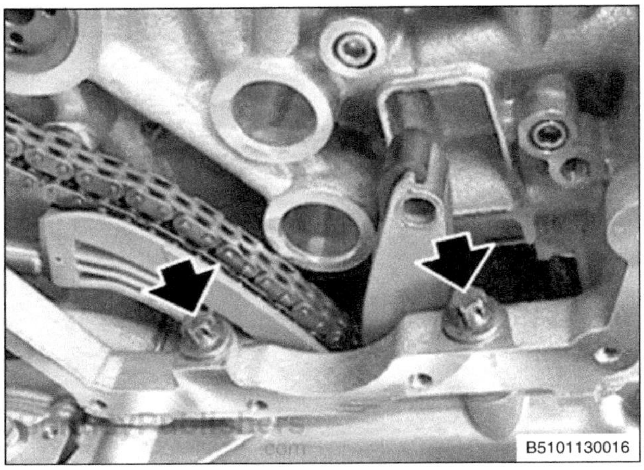

Remove cylinder head to timing cover fasteners (**arrows**).

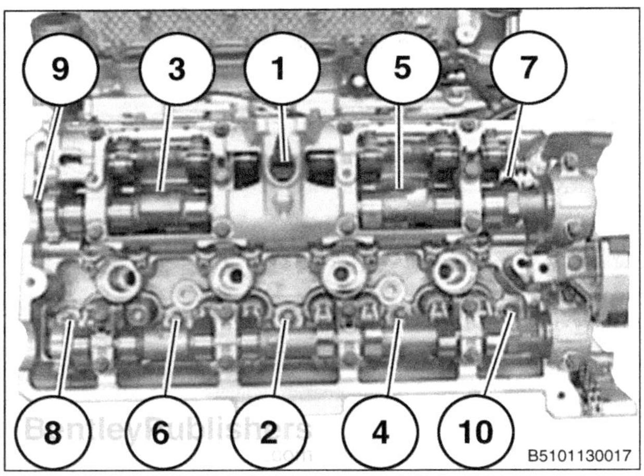

Loosen cylinder head fasteners 1 turn (360°) in sequence 10 - 1.

BentleyPublishers.com
BentleyPublishers.com–All Rights Reserved

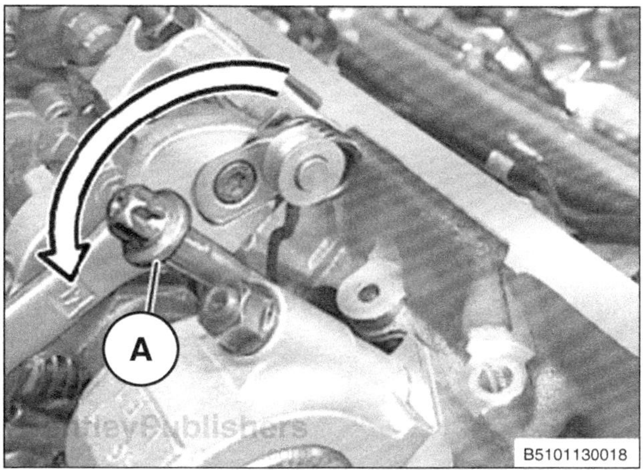

◄ Rotate eccentric shaft at dihedral head in direction of **arrow** to remove cylinder head bolt 7 (**A**) and discard.

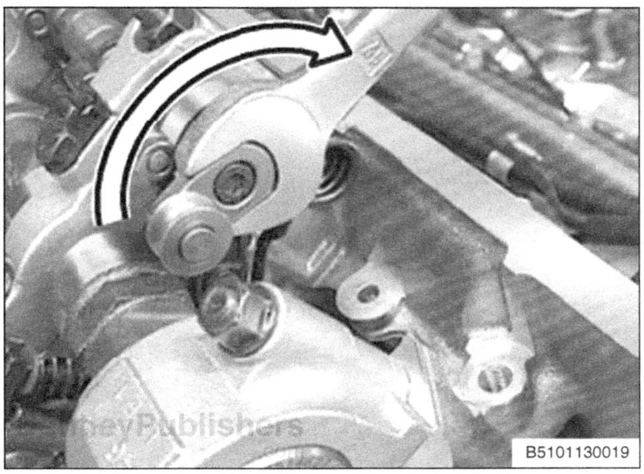

◄ Rotate eccentric shaft at dihedral head in direction of **arrow** to minimum stroke postion.

– Remove remaining cylinder head fasteners and discard.

– Lift off cylinder head with help of assistant.

> ***CAUTION—***
> - *Cylinder head weighs approx. 40 kg (88 lb).*
> - *Place cylinder head on work bench on its side. Putting it down on sealing surface risks damage to the valves.*
> - *Be sure to clean antifreeze out of cylinder head bolt holes in crankcase immediately.*

– Remove and discard cylinder head gasket.

Right cylinder head, installing (V8 engine)

– Clean cylinder head and gasket surfaces of engine block and timing chain housing.

> ***CAUTION—***
> - *Remove foreign matter and liquid from bolt holes. Trapped oil or coolant in bolt holes can cause damage to the engine block.*
> - *Do not use a metal scraper or wire brush to clean sealing surfaces. If necessary, use a hard wooden or plastic scraper. Also available are abrasive discs to be used in conjunction with an electric drill. Be sure to use the correct disc for the type of metal being cleaned.*

– Inspect cylinder head for visible cracks or other defects.

BentleyPublishers.com
BentleyPublishers.com—All Rights Reserved

Right cylinder head, installing (V8 engine)

Remove timing chain tensioner (**A**) and discard tensioner seal.

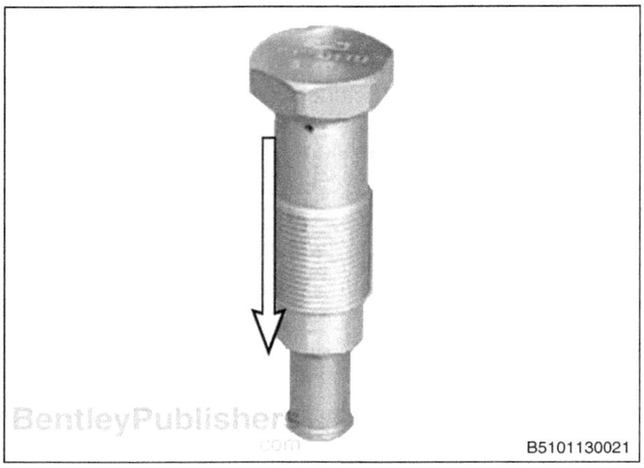

Slowly compress timing chain tensioner piston on a level surface. Repeat procedure two times.

Install timing chain tensioner (**A**) with new seal finger tight.

BentleyPublishers.com
BentleyPublishers.com—All Rights Reserved

Coat joint (**arrows**) between engine block and timing cover with 3-Bond 1209®.

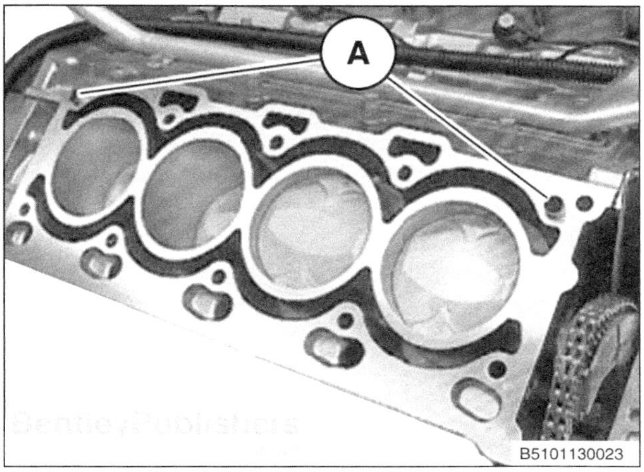

Inspect dowel sleeves (**A**) for damage, replace if necessary.

- Install new cylinder head gasket.
- Place cylinder head onto engine block, being careful not to damage cylinder head gasket.

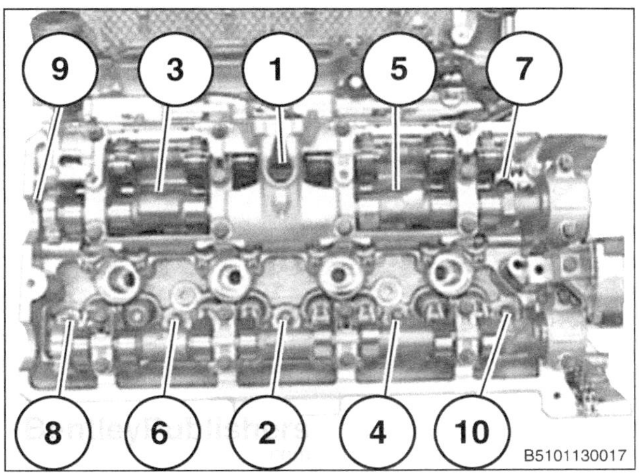

Install new cylinder head fasteners finger tight. Do not wash coating on new cylinder head fasteners.

NOTE —

- *Cylinder head fastener 7 will be installed in next step.*

BentleyPublishers
BentleyPublishers.com—All Rights Reserved

Right cylinder head, installing (V8 engine)

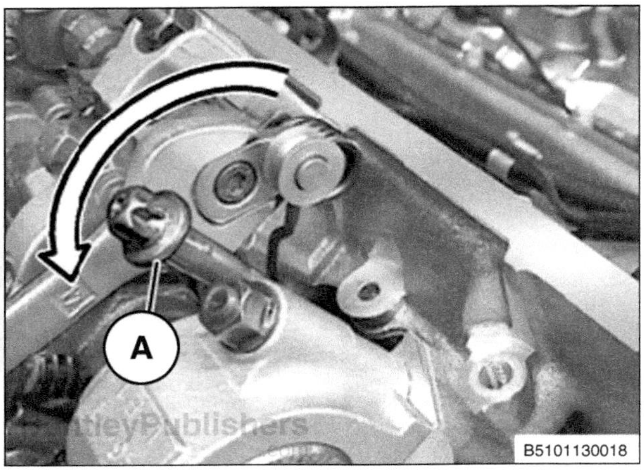

Rotate eccentric shaft at dihedral head in direction of **arrow** to install cylinder head bolt 7 (**A**).

Rotate eccentric shaft at dihedral head in direction of **arrow** to minimum stroke postion.

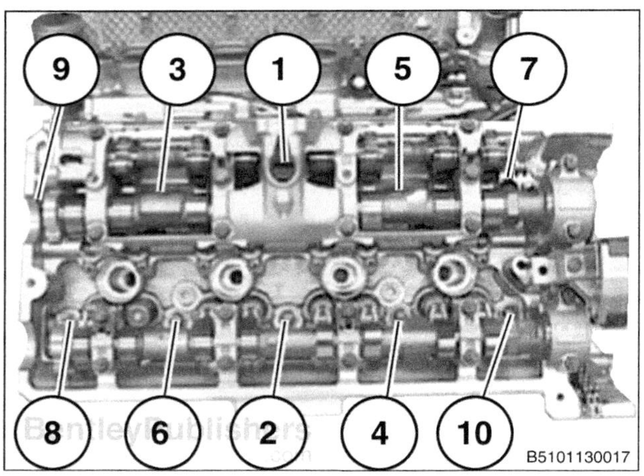

Tighten cylinder head fasteners in sequence 1 - 10.

Tightening torques and sequence	
Cylinder head to engine block: • Stage 1: All bolts, 1 - 10 • Stage 2: All bolts, 1 - 10 • Stage 3: All bolts 1 - 10	 30 Nm (22 ft-lb) additional 90° additional 90°

BentleyPublishers.com
BentleyPublishers.com–All Rights Reserved

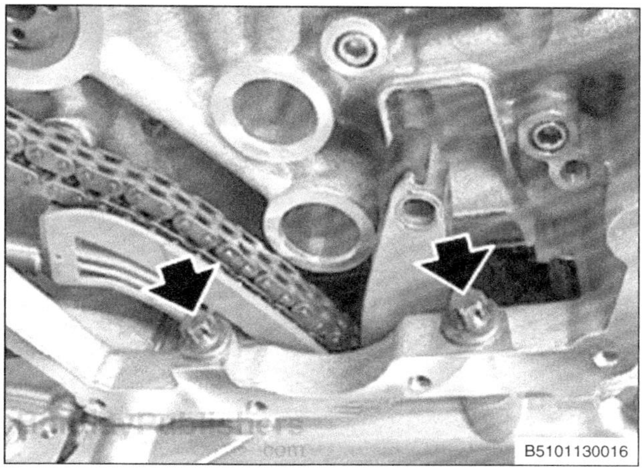

B5101130016

Install and tighten cylinder head to timing cover fasteners (**arrows**).

Tightening torques and sequence	
Cylinder head to timing cover:	10 Nm (7.3 ft-lb)

B5101130015

Install and tighten timing chain guide fastener (**arrow**).

B5101130013

Install eccentric shaft spring mount (**A**). Align locating pin with bore in cylinder head. Install and tighten fasteners (**arrows**).

Right cylinder head, installing (V8 engine)

Rotate eccentric shaft at dihedral head to reduce tension on torsion spring (**A**). Using a strap (**B**) carefully install torsion spring (**A**) onto roller (**C**) in direction of **arrow**.

Rotate eccentric shaft at dihedral head in direction of **arrow** to minimum stroke postion.

Install eccentric shaft sensor (**A**) onto cylinder head and tighten eccentric shaft sensor fasteners (**arrows**).

BentleyPublishers.com
BentleyPublishers.com—All Rights Reserved

Tighten timing chain tensioner (**A**).

Tightening torques and sequence	
Timing chain tensioner to cylinder head	65 Nm (48 ft-lb)

– Install intake and exhaust camshaft adjustment (VANOS) units. See **117 Camshaft Timing Chain**.

– Remainder of installation is reverse of removal. Remember to:
- Remove camshaft and crankshaft locking tools.
- Assemble cylinder head cover with new gasket.
- Reattach and tighten any ground cables removed.
- Reassemble fuel rail using new O-ring seals.
- Change engine oil and filter. See **020 Maintenance**.
- Fill and bleed cooling system. See **170 Radiator and Cooling System**.

Left cylinder head, removing (V8 engine)

> ***WARNING —***
> - *Due to risk of personal injury, be sure the engine is cold before beginning the removal procedure.*

– Disconnect negative (-) battery cable.

> ***CAUTION—***
> - *Prior to disconnecting the battery, read the battery disconnection cautions in* **001 Warnings and Cautions**.

– Raise car and support safely.

> ***WARNING —***
> - *Make sure the car is stable and well supported at all times. Use a professional automotive lift or jack stands designed for the purpose. A floor jack is not adequate support.*

– Remove left exhaust manifold. See **180 Exhaust System**.

NOTE —
- *To remove exhaust manifolds, secure engine in the installation position with special tool and lower front axle bracket. Once manifolds have been removed Reinstall front axle bracket and remove special tool for securing engine in installation position.*

– Open engine coolant drain plug at left side of engine block. Be prepared to catch draining coolant. Drain remaining engine coolant from radiator. See **170 Radiator and Cooling System**.

– Remove ignition coil cover. See **020 Maintenance**.

– Remove ignition coils. Plug spark plug bores to prevent debris from entering engine. See **120 Ignition System**.

– Remove valvetronic motor. See **133 Fuel Injection**.

– Remove left cylinder head cover. See **Cylinder head cover, removing and installing (V8 engine)** in this repair group.

BentleyPublishers.com—All Rights Reserved

113

Left cylinder head, removing (V8 engine)

- Remove intake manifold. Plug intake manifold ports to prevent debris from entering engine. See **133 Fuel Injection**.
- Remove upper left timing cover. See **117 Camshaft Timing Chain**.
- Disconnect electrical connectors and lay engine wiring harness aside.
- Remove spark plugs. See **020 Maintenance**.
- Remove right cylinder head vent hose.

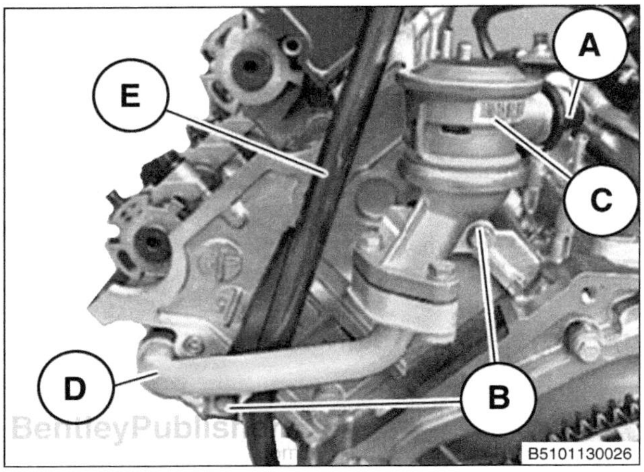

Working at rear of left side cylinder head, remove secondary air hose (**A**). Remove check valve fasteners (**B**) and remove check valve (**C**) with tube (**D**). Do not remove dipstick tube (**E**).

- Remove intake and exhaust camshaft adjustment (VANOS) units. See **117 Camshaft Timing Chain**.

Remove eccentric shaft sensor fasteners (**arrows**). Remove eccentric shaft sensor (**A**) from cylinder head.

Remove timing chain guide fastener (**arrow**).

BentleyPublishers.com
BentleyPublishers.com—All Rights Reserved

Remove cylinder head to timing cover fasteners (**arrows**).

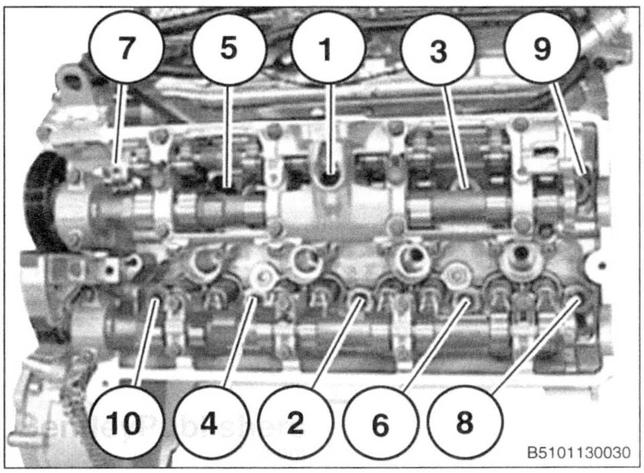

Remove and discard cylinder head fasteners in sequence 10 - 1.

– Lift off cylinder head with help of assistant.

> ***CAUTION—***
> - *Cylinder head weighs approx. 40 kg (88 lb).*
> - *Place cylinder head on work bench on its side. Putting it down on sealing surface risks damage to the valves.*
> - *Be sure to clean antifreeze out of cylinder head bolt holes in crankcase immediately.*

– Remove and discard cylinder head gasket.

Left cylinder head, installing (V8 engine)

– Clean cylinder head and gasket surfaces of engine block and timing chain housing.

> ***CAUTION—***
> - *Remove foreign matter and liquid from bolt holes. Trapped oil or coolant in bolt holes can cause damage to the engine block.*
> - *Do not use a metal scraper or wire brush to clean sealing surfaces. If necessary, use a hard wooden or plastic scraper. Also available are abrasive discs to be used in conjunction with an electric drill. Be sure to use the correct disc for the type of metal being cleaned.*

– Inspect cylinder head for visible cracks or other defects.

Coat joint (**arrows**) between engine block and timing cover with 3-Bond 1209®.

BentleyPublishers.com—All Rights Reserved

Left cylinder head, installing (V8 engine)

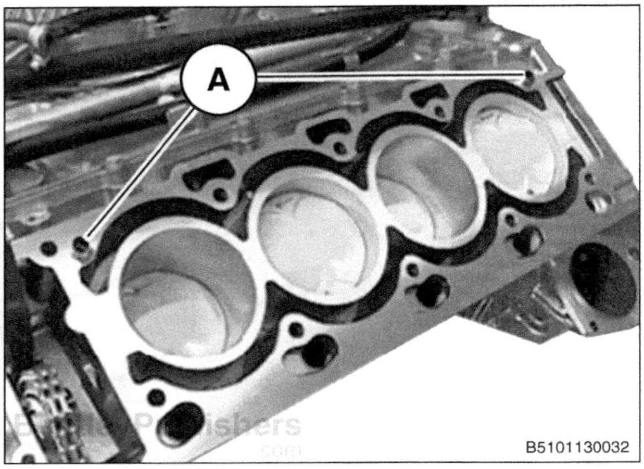

Inspect dowel sleeves (**A**) for damage, replace if necessary.

- Install new cylinder head gasket.
- Place cylinder head onto engine block, being careful not to damage cylinder head gasket.
- Install new cylinder head fasteners finger tight. Do not wash coating on new cylinder head fasteners.

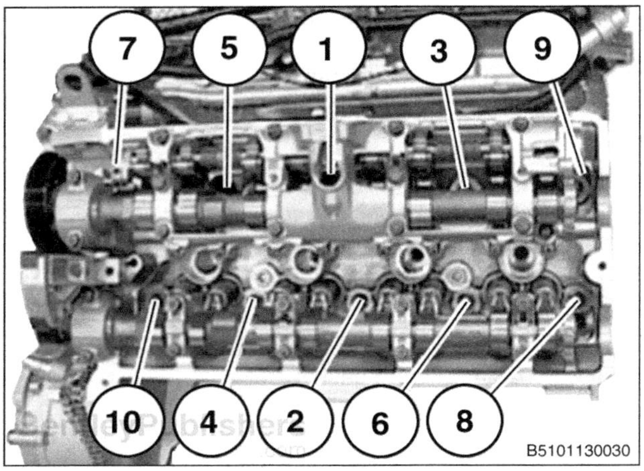

Tighten cylinder head fasteners in sequence 1 - 10.

Tightening torques and sequence	
Cylinder head to engine block: • Stage 1: All bolts, 1 - 10 • Stage 2: All bolts, 1 - 10 • Stage 3: All bolts 1 - 10	 30 Nm (22 ft-lb) additional 90° additional 90°

Install and tighten cylinder head to timing cover fasteners (**arrows**).

Tightening torques and sequence	
Cylinder head to timing cover:	15 Nm (11 ft-lb)

BentleyPublishers.com
BentleyPublishers.com—All Rights Reserved

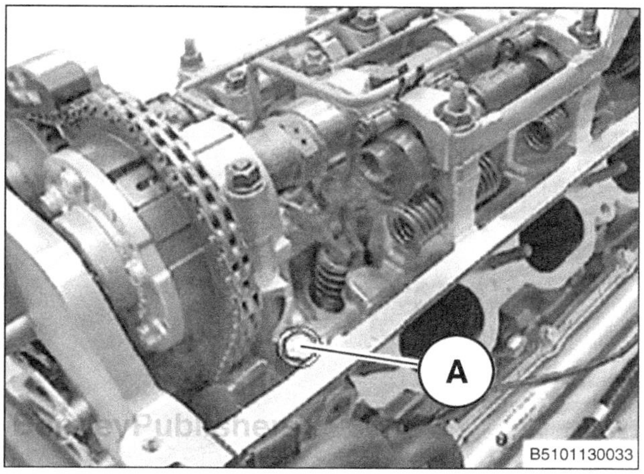

If replacing cylinder head, transfer screw plug (**A**) from old cylinder head to new. Failure to replace screw plug will result in a VANOS malfunction.

Install and tighten timing chain guide fastener (**arrow**).

Install eccentric shaft sensor (**A**) onto cylinder head and tighten eccentric shaft sensor fasteners (**arrows**).

– Install intake and exhaust camshaft adjustment (VANOS) units. See **117 Camshaft Timing Chain**.

– Remainder of installation is reverse of removal. Remember to:
  - Remove camshaft and crankshaft locking tools.
  - Assemble cylinder head cover with new gasket.
  - Reattach and tighten any ground cables removed.
  - Reassemble fuel rail using new O-ring seals.
  - Change engine oil and filter. See **020 Maintenance**.
  - Fill and bleed cooling system. See **170 Radiator and Cooling System**.

BentleyPublishers.com
BentleyPublishers.com—All Rights Reserved

117 Camshaft Timing Chain

GENERAL

This repair group provides removal and repair information for crankshaft vibration damper, engine timing chain and camshaft adjustment (VANOS) units. Camshaft timing procedure is also included.

Repair procedures in this repair group assume that the engine is installed in the engine bay.

The 5 Series engine cylinder head is equipped with variable camshaft timing or VANOS (from the German words *VAriable NOckenwellen Steuerung*). Special tools and procedures are required to remove and install camshaft adjustment units and to time camshafts. Be sure to read each procedure through before starting work.

BentleyPublishers
BentleyPublishers.com—All Rights Reserved

See also:

- **020 Maintenance**
- **100 Engine–General** for engine code and application information.
- **113 Cylinder Head Removal and Installation** for cylinder head cover removal
- **130, 131, 132, 133 Fuel Injection** for air filter housing and intake manifold removal
- **170 Radiator and Cooling System** for radiator drain and refill

Warnings

WARNING —
- *Due to risk of personal injury, be sure the engine is cold before beginning work on engine components.*
- *Use extreme caution when draining and disposing of engine coolant. Coolant is poisonous and lethal to humans and pets. Pets are attracted to coolant because of its sweet smell and taste. Seek medical attention immediately if coolant is ingested.*
- *Fuel is highly flammable. When working around fuel, do not disconnect wires that could cause electrical sparks. Do not smoke or work near heaters or other fire hazards.*

BentleyPublishers.com
BentleyPublishers.com—All Rights Reserved

VIBRATION DAMPER

Vibration damper, removing and installing (M54 engine)

The M54 engine uses a 1-piece crankshaft vibration damper, integrated with the hub.

– Raise front of vehicle and support in a safe manner.

> *CAUTION—*
> - *Make sure the vehicle is stable and well supported at all times. Use a professional automotive lift or jack stands designed for the purpose. A floor jack is not adequate support.*

– Remove splash guard under engine. See **020 Maintenance**.

– Remove electric cooling fan and cooling fan cowl. See **170 Radiator and Cooling System**.

NOTE—
- *Protect radiator from damage with a piece of wood.*

– With engine fully cooled off, mark direction of rotation on engine accessory belts. Remove belts. See **020 Maintenance**.

◀ Bolt BMW special tools 11 8 190 and 11 8 200 together. Insert fingers of special tool 11 8 200 in vibration damper openings. Use this tool setup to counterhold vibration damper hub while loosening vibration damper mounting bolt (**arrow**).

> *CAUTION—*
> - *Make sure an assistant holds special tool.*
> - *If assistant is not available to grip the tool, rest handle of tool on front subframe. Protect subframe from damage with a piece of wood.*
> - *Do not use flywheel locking tool to hold crankshaft stationary to loosen or tighten vibration damper mounting bolt.*

– Remove vibration damper mounting bolt and discard.

– Pull off vibration damper using BMW special tools 11 8 210 and 11 8 200.

> *CAUTION—*
> - *Avoid damaging radiator.*
> - *If using a different puller than the BMW special tools, be careful to avoid damaging crankshaft threads.*

– Inspect front crankshaft seal for leaks. For seal replacement, see **119 Lubrication System**.

BentleyPublishers.com
BentleyPublishers.com—All Rights Reserved

– When reinstalling:
  - Align vibration damper groove to crankshaft woodruff key.
  - Insert new vibration damper mounting bolt and washer.

> ***CAUTION—***
> - *Inspect woodruff key for damage before installing vibration damper. Replace if necessary*

– Use BMW special tools 111 8 190 and 11 8 200 to counterhold vibration damper while tightening mounting bolt.

> ***CAUTION—***
> - *Do not use flywheel locking tool to hold crankshaft stationary to loosen or tighten vibration damper mounting bolt. Use only the recommended special tools, or equivalent tools for counterholding the pulley.*

Tightening torque	
Vibration damper to crankshaft (replace damper bolt)	410 Nm (300 ft-lb)

Vibration damper, removing and installing (N52, N54 engine)

The crankshaft vibration damper is also referred to as the front pulley.

> ***WARNING —***
> - *To avoid personal injury, be sure the engine is cold before beginning the procedure.*

– Raise vehicle and support safely.

> ***WARNING —***
> - *Make sure the vehicle is stable and well supported at all times. Use a professional automotive lift or jack stands designed for the purpose. A floor jack is not adequate support.*

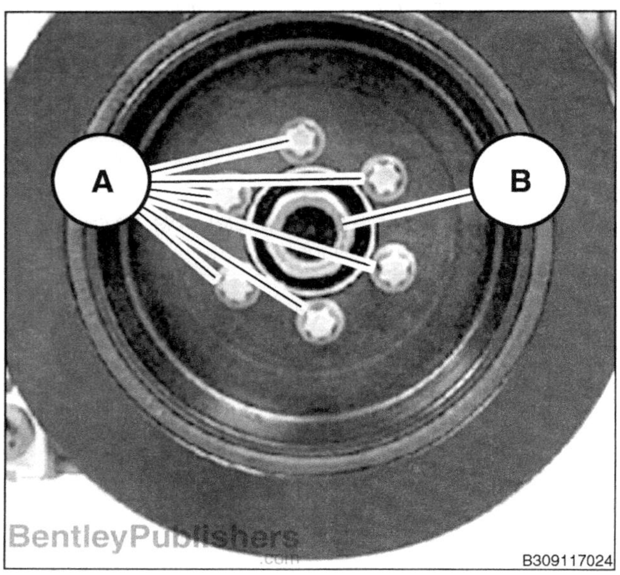

– Remove engine compartment splash shield. See **020 Maintenance**.

– Remove electric cooling fan and cooling fan cowl. See **170 Radiator and Cooling System**.

NOTE —
- *Protect radiator from damage with a piece of wood.*

– With engine fully cooled off, mark direction of rotation on engine accessory belt. Remove belt. See **020 Maintenance**.

< Remove vibration damper mounting Torx bolts (**A**) and discard. Lift off vibration damper.

> ***CAUTION—***
> - *Do not remove vibration damper hub bolt (**B**). If this bolt is removed, the timing chain drive sprocket becomes free to rotate, resulting in valve damage.*

BentleyPublishers.com
BentleyPublishers.com—All Rights Reserved

— Installation is reverse of removal.

Tightening torque	
Vibration damper to hub (replace damper M8 x 16 mm Torx bolts)	35 Nm (26 ft-lb)

Vibration damper, removing and installing (V8 engine)

— Raise front of vehicle and support in a safe manner.

> ***CAUTION—***
> • *Make sure the vehicle is stable and well supported at all times. Use a professional automotive lift or jack stands designed for the purpose. A floor jack is not adequate support.*

— Remove splash guard under engine. See **020 Maintenance**.

— Remove electric cooling fan and cooling fan cowl. See **170 Radiator and Cooling System**.

NOTE—
• *Protect radiator from damage with a piece of wood.*

— With engine fully cooled off, mark direction of rotation on engine accessory belts. Remove belts. See **020 Maintenance**.

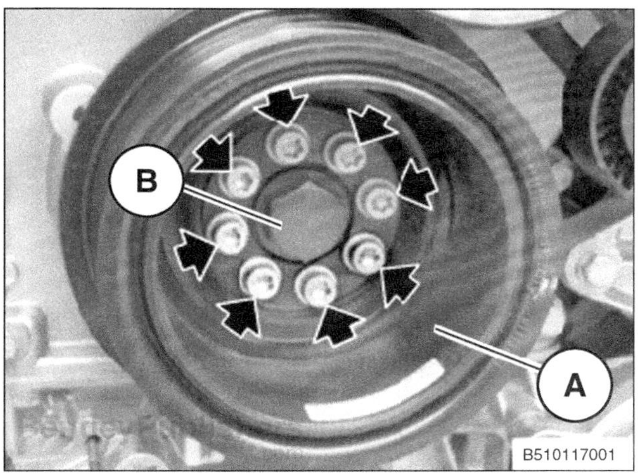

◄ Remove vibration damper mounting bolts (**arrows**) and discard. Lift off vibration damper (**A**).

> ***CAUTION—***
> • *Do not remove vibration damper hub bolt (**B**).*

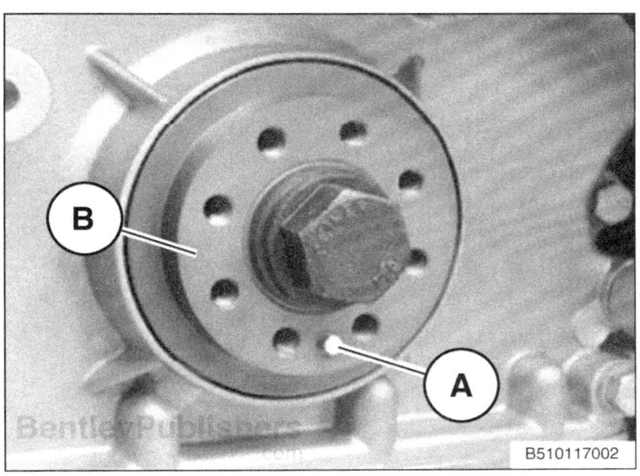

◄ Installation is reverse of removal.

• Align vibration damper with locating pin (**A**) on damper hub (**B**).

Tightening torque	
Vibration damper to hub (replace damper bolts)	22 Nm (16.2 ft-lb)

BentleyPublishers.com—All Rights Reserved

Timing Chain Service (M54 Engine)

The M54 engine uses a pair of chains to drive the camshafts. A primary chain is driven off the crankshaft and drives the primary sprocket on the exhaust camshaft. The secondary chain drives the intake camshaft. Each chain is tensioned by a hydraulic tensioner.

Intake and exhaust camshaft timing is variable and is controlled by the engine control module (ECM). This system is known as double VANOS.

Camshaft timing chain service and repair requires the removal of front crankshaft vibration damper, double VANOS unit, engine oil pan and timing chain housing cover. Be sure to read the procedures before starting work.

Double VANOS unit components (M54 engine)

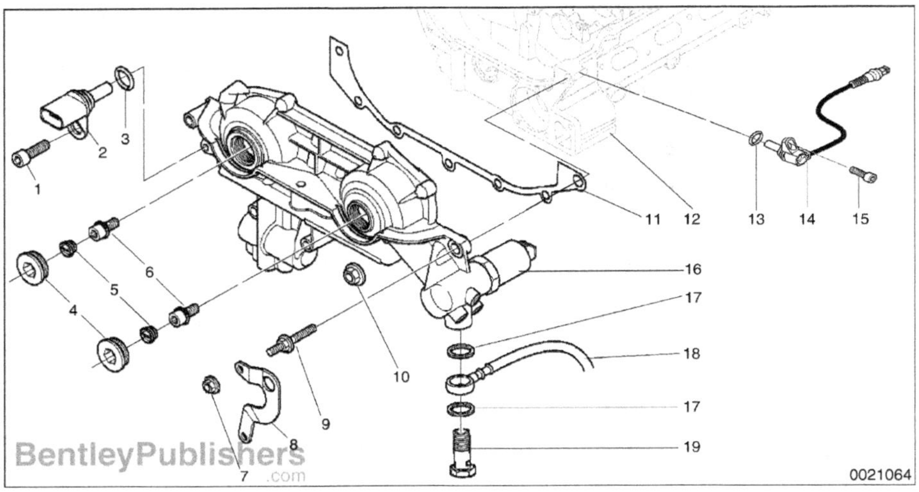

1. Bolt (M6)
2. Exhaust camshaft sensor
3. Sealing ring
4. Camshaft end sealing plug
   - Tighten to 50 Nm (37 ft-lb)
5. Camshaft seal cap
6. VANOS hydraulic piston set screw
   - Left hand thread
     Tighten to 10 Nm (7 ft-lb)
7. Nut (M7)
   - Tighten to 14 Nm (10 ft-lb)
8. Engine lifting hook
9. Stud (M7)
10. Nut (M6)
    - Tighten to 10 Nm (7 ft-lb)
11. Gasket
12. Cylinder head
13. Sealing ring
14. Intake camshaft sensor
15. Bolt (M6)
16. VANOS unit
17. Copper sealing ring
18. VANOS oil feed line
19. Banjo bolt
    - Tighten to 32 Nm (24 ft-lb)

BentleyPublishers.com
BentleyPublishers.com—All Rights Reserved

Double VANOS unit, removing (M54 engine)

> *WARNING—*
> *• Due to risk of personal injury, be sure the engine is cold before beginning the removal procedure.*

> *CAUTION—*
> *• Special BMW service tools are required to remove and install the VANOS unit. Read the procedure through before starting the job.*
> *• If the camshafts are not properly timed, the pistons can contact the valves, causing engine damage.*

– Disconnect negative (–) cable from battery.

> *CAUTION—*
> *• Prior to disconnecting the battery, read the battery disconnection cautions in* **001 Warnings and Cautions**.

– Remove air filter housing with mass air flow sensor. See **130 Fuel Injection**.

– Remove electric cooling fan and cooling fan cowl. See **170 Radiator and Cooling System**.

NOTE—
• Protect radiator from damage with a piece of wood.

– Remove engine covers and cylinder head cover. See **113 Cylinder Head Removal and Installation**.

– Remove spark plugs. Plug spark plug bores to prevent debris from falling into engine.

◄ Remove oil baffle cover from above intake camshaft.

◄ Remove banjo bolt from VANOS unit oil pressure line. Use banjo bolt to attach BMW special tool 11 3 450 (compressed air fitting) to VANOS unit.

- Be prepared to catch spraying oil. Do not allow oil to contaminate accessory belts.

BentleyPublishers.com—All Rights Reserved

Double VANOS unit, removing (M54 engine)

- Cover oil bore (**arrow**) in VANOS unit with shop towel to capture oil which sprays when compressed air is applied.

- Connect compressed air line to air fitting. Apply air pressure set to 2 – 8 bar (30 – 115 psi).

- With compressed air line connected, use crankshaft vibration damper bolt to rotate engine in direction of rotation (clockwise) at least twice, until cylinder 1 intake and exhaust camshaft lobes face each other (**arrows**) in top dead center (TDC) position for cylinder 1.

> ***CAUTION—***
> - *Do not rotate engine counterclockwise to reach TDC. If engine rotates past TDC, continue another two complete rotations.*

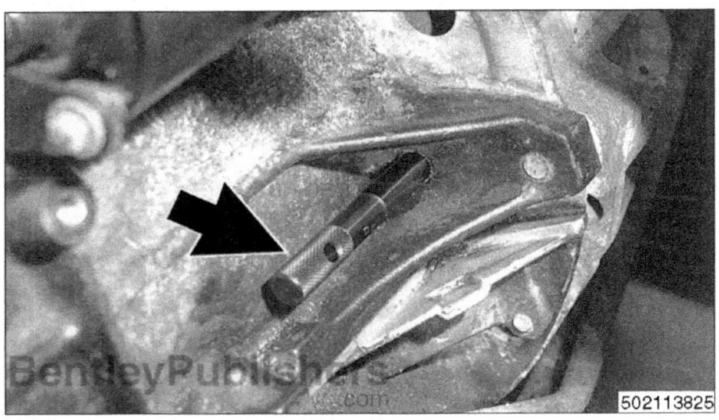

- Remove sealing plug from special tool bore on lower left side of engine block at flywheel bellhousing. Secure crankshaft in TDC position with BMW special tool 11 2 300 (**arrow**).

- Unscrew and remove two cylinder head cover studs (**arrows**) at rear of cylinder head.

BentleyPublishers.com
BentleyPublishers.com—All Rights Reserved

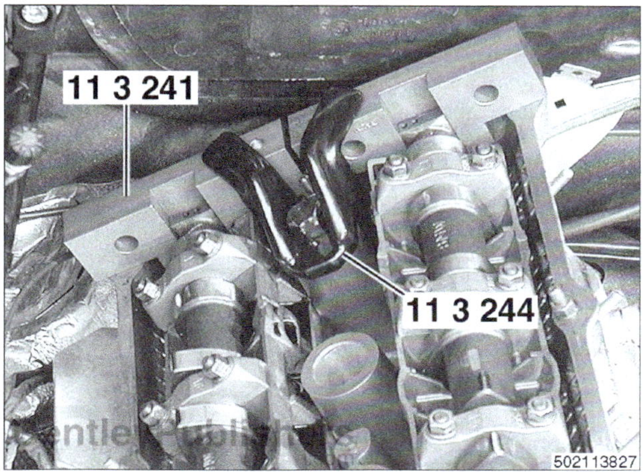

- Secure camshafts in TDC position using BMW special tool set 11 3 240 (includes 11 3 241 and 11 3 244).

– Disconnect compressed air line, leaving compressed air fitting attached to VANOS unit.
  - Be prepared to catch dripping oil. Do not allow oil to contaminate accessory belts.

- Unscrew sealing plugs (**arrows**) from VANOS unit.

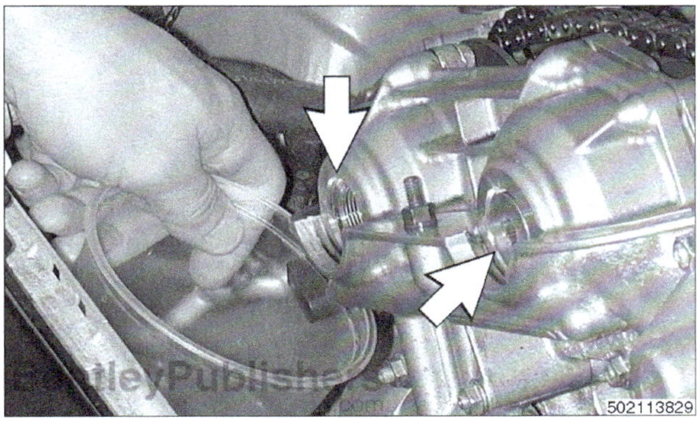

- Be prepared to catch dripping oil. Do not allow oil to contaminate accessory belts.

- Pull sealing caps straight out of VANOS unit with BMW special tool 11 6 170, or with short needle-nosed pliers.
  - Additional oil may drain from VANOS unit.

BentleyPublishers
BentleyPublishers.com—All Rights Reserved

Remove set screws on ends of intake and exhaust camshafts.

> **CAUTION—**
> • *Set screws have left hand threads. Unscrew clockwise.*

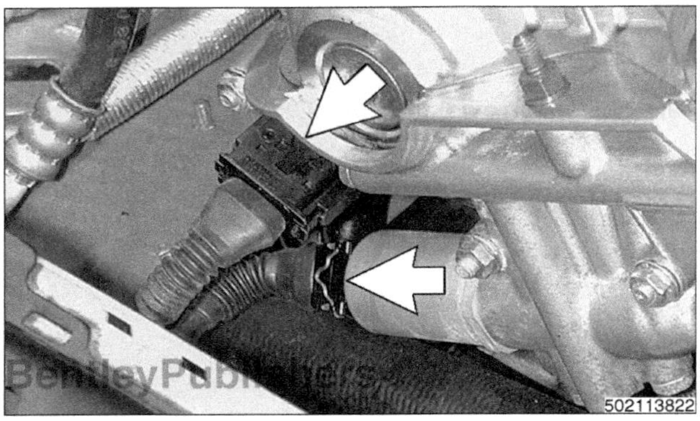

Detach harness connectors at exhaust camshaft position sensor and exhaust camshaft VANOS control valve (**arrows**).

Remove VANOS unit:

- Remove fasteners from engine support eye.
- Remove VANOS mounting nuts (**arrows**) from cylinder head. Slide VANOS unit and metal gasket off.
- The VANOS unit contains residual oil. Place shop towels underneath when removing.

> **CAUTION—**
> • *Do not crank the engine with VANOS unit removed. Damage to engine may result from piston-to-valve interference.*

Double VANOS unit, installing (M54 engine)

Clean contact edges of cylinder head face and VANOS unit and apply a thin coat of permanently elastic 3-Bond 1209® or equivalent to surfaces.

> **CAUTION—**
> • *Check locating dowel (**A**) and dowel sleeve (**B**) at top of cylinder head for damage or incorrect installation.*
> • *Make sure gasket material is removed from face of cylinder head. Clean sealing face and keep free of oil. If foreign material is present on the sealing surface, camshaft timing will be incorrect.*

BentleyPublishers.com
BentleyPublishers.com—All Rights Reserved

– Using new gasket, install VANOS unit to cylinder head.
  • Reinstall engine support eye.

Tightening torques	
VANOS unit to cylinder head • M6 nut • M7 nut	 10 Nm (7 ft-lb) 14 Nm (10 ft-lb)

117

Insert and tighten VANOS hydraulic piston set screws on intake and exhaust camshafts.

Tightening torque	
Hydraulic piston to splined shaft (Left-handed set screw)	10 Nm (7 ft-lb)

CAUTION—
• Set screws have left-hand threads. Tighten counterclockwise.

Replace sealing caps inside VANOS unit with BMW special tool 11 6 170 or short needle-nosed pliers.

Insert and secure VANOS sealing plugs (**arrows**), using new sealing O-rings.

Tightening torque	
Sealing plug to VANOS unit	50 Nm (37 ft-lb)

– Remove compressed air fitting from VANOS unit.

– Fit VANOS oil line banjo bolt with new seals. Attach oil line to VANOS unit.

Tightening torque	
Oil line to VANOS unit (banjo bolt)	32 Nm (24 ft-lb)

– Connect electrical harness connectors to camshaft position sensors and VANOS solenoid valves.

– If double VANOS unit is replaced, or if repair operations are completed that may change camshaft timing, recheck and adjust camshaft timing. See **Camshaft timing, checking and adjusting (M54 engine)** in this repair group.

BentleyPublishers.com
BentleyPublishers.com–All Rights Reserved

Double VANOS unit, installing (M54 engine)

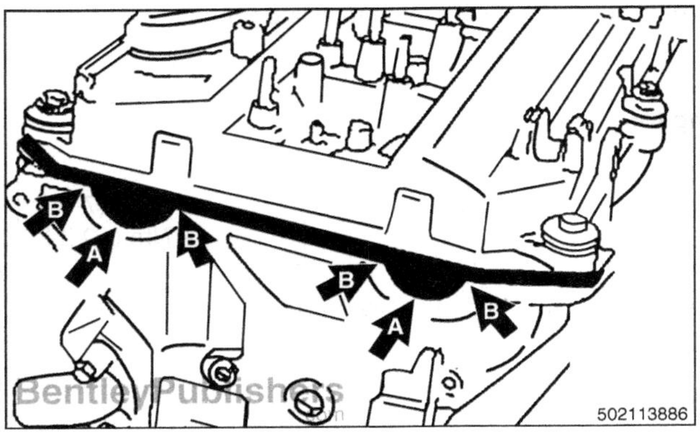

502113886

– Remove BMW special locking tools from rear of cylinder head.

– Remove crankshaft locking tool and replace sealing plug.

◄ Install intake camshaft baffle and cylinder head cover.

- Check for correct seating of half-moon seals (**A**) in back of cylinder head cover.
- Use a small amount of 3-Bond 1209® or equivalent sealant at corners (**B**) of half-moon cutouts.
- Similarly, seat gasket and seal corners in front of cylinder head, at VANOS unit.

Tightening torque	
Cylinder head cover to cylinder head (M6)	10 Nm (7 ft-lb)

– Remainder of installation is reverse of removal.

Tightening torques	
Spark plug to cylinder head	25 Nm (18 ft-lb)

– After completing work, check VANOS operation using BMW scan tool.

BentleyPublishers.com
BentleyPublishers.com—All Rights Reserved

Timing chain components (M54 engine)

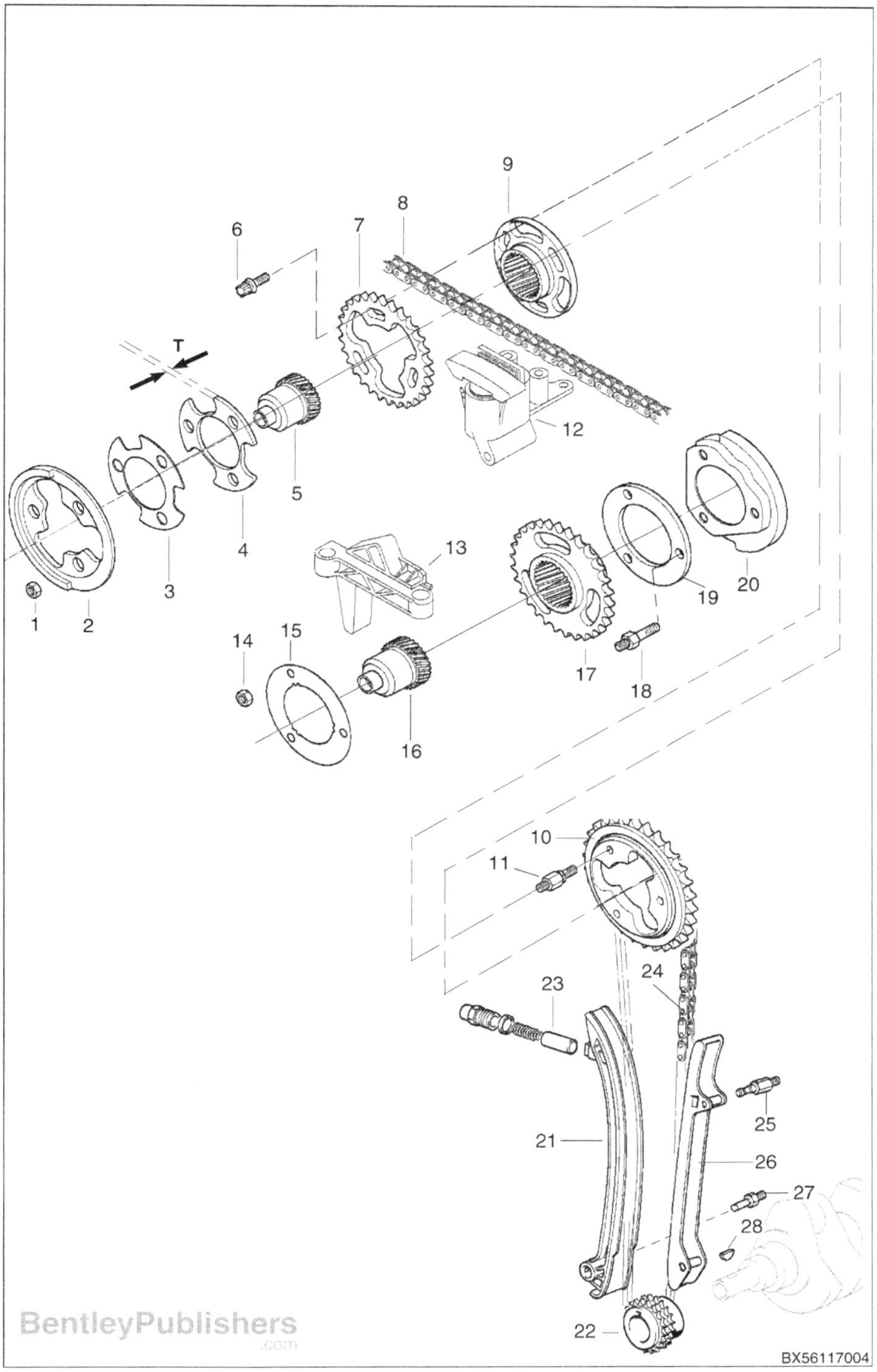

1. Impulse wheel mounting nut
2. Exhaust camshaft impulse wheel
3. Spring plate
4. Thrust spacer
   - Thickness (T) = 3.5 mm
5. Splined shaft
6. Torx screw
7. Exhaust secondary sprocket
8. Secondary timing chain
9. Splined sleeve
10. Primary sprocket
11. Threaded locating stud
12. Secondary chain tensioner
13. Secondary chain lower guide
14. Sprocket mounting nut
15. Spring plate
16. Splined shaft
17. Intake camshaft sprocket
18. Locating stud
19. Thrust spacer
20. Intake camshaft impulse wheel
21. Chain tensioner rail
22. Crankshaft sprocket
23. Primary chain tensioner
24. Primary chain
25. Locating stud
26. Guide rail
27. Locating stud
28. Woodruff key

BentleyPublishers.com
BentleyPublishers.com–All Rights Reserved

Timing chains, removing (M54 engine)

When removing camshaft timing chains, first remove the oil pan. See **119 Lubrication System**. Also required is removal of the VANOS unit and disassembly of the intake and exhaust camshaft sprocket assemblies. See **Double VANOS unit, removing (M54 engine)** in this repair group.

Special BMW service tools are needed for timing chain removal and installation. These tools assure proper timing of the valvetrain. Precise marks to set the timing on the camshafts are not provided for reassembly. Read the procedures through before beginning the job.

> ***WARNING—***
> - *Due to risk of personal injury, be sure the engine is cold before beginning the removal procedure.*

> ***CAUTION—***
> - *Special BMW service tools are required to remove and install the VANOS unit and time the camshafts. Read the procedure through before starting the job.*
> - *If the camshafts are not properly timed, the pistons can contact the valves, causing engine damage.*

– Disconnect negative (–) cable from battery.

> ***CAUTION—***
> - *Prior to disconnecting the battery, read the battery disconnection cautions in* **001 Warnings and Cautions**.

– Remove air intake hood above radiator.

– Remove air filter housing. See **130 Fuel Injection**.

– Remove electric cooling fan and cooling fan cowl. See **170 Radiator and Cooling System**.

– Mark direction of engine rotation on accessory belts. Remove belts. See **020 Maintenance**.

– Remove coolant pump pulley.

– Remove engine covers and cylinder head cover. See **113 Cylinder Head Removal and Installation**.

– Remove spark plugs. Plug spark plug bores to prevent debris from falling into engine.

– Remove VANOS unit. See **Double VANOS unit, removing (M54 engine)** in this repair group.

BentleyPublishers.com
BentleyPublishers.com—All Rights Reserved

– Raise car and support safely.

> ***WARNING—***
> *• Make sure the car is stable and well supported at all times. Use a professional automotive lift or jack stands designed for the purpose. A floor jack is not adequate support.*

– Drain engine cooling system. See **170 Radiator and Cooling System**.

– Drain engine oil and remove oil pan. See **119 Lubrication System**.

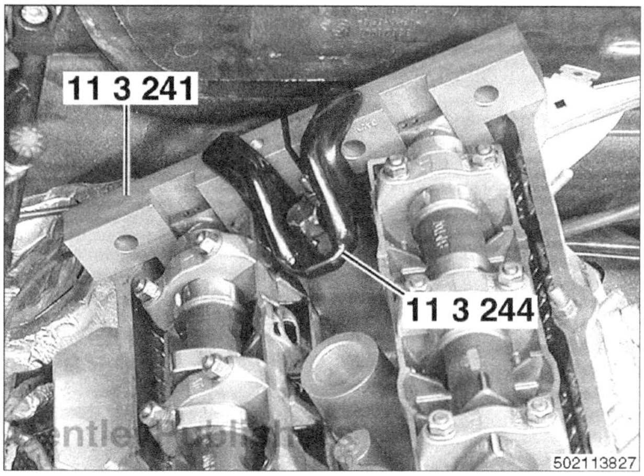

◄ Make sure camshafts are locked in TDC position using BMW special tool set 11 3 240. See **Double VANOS unit, removing (M54 engine)** in this repair group.

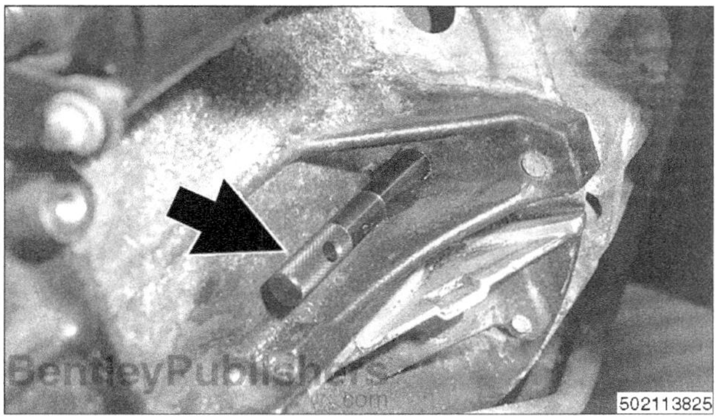

◄ Make sure crankshaft is still locked with BMW special tool 11 2 300. See **Double VANOS unit, removing (M54 engine)** in this repair group.

◄ Remove primary camshaft chain tensioner (**arrow**) from timing chain cover.

> ***CAUTION—***
> *• Chain tensioner piston is under spring pressure.*

BentleyPublishers
BentleyPublishers.com—All Rights Reserved

Press down on secondary chain tensioner and lock into place using BMW special tool 11 3 292 or a similar size pin tool.

Remove exhaust camshaft impulse wheel mounting nuts (**arrows**). Remove impulse wheel (**A**).

Remove spring plate (**A**).

Remove intake camshaft sprocket mounting nuts (**arrows**) and remove spring plate (labeled **FRONT**).

BentleyPublishers.com
BentleyPublishers.com—All Rights Reserved

Remove torx bolts (**arrows**) from exhaust camshaft sprocket.

Lift off exhaust and intake sprockets with secondary chain, thrust spacer (**A**) and splined shaft (**B**).

> ***CAUTION—***
> - *Intake and exhaust camshaft splined shafts share the same part number. Be sure to mark splined shafts so they can be reinstalled in their original positions.*

Remove exhaust camshaft splined sleeve (**A**) and shaft (**B**).

◀ Remove secondary chain tensioner mounting bolts (**arrows**). Remove tensioner while keeping locking pin in place.

◀ Remove primary chain sprocket mounting studs (**arrows**) on exhaust camshaft.

◀ Lift primary chain sprocket off exhaust camshaft. Remove sprocket from chain.

BentleyPublishers.com
BentleyPublishers.com—All Rights Reserved

Hang timing chain on exhaust camshaft end.

Working in cylinder head cavity, remove timing chain cover bolts (**arrows**).

Remove secondary chain lower guide mounting bolts (**arrows**). Remove chain guide.

NOTE—

- *Bolt on intake camshaft side is long and extends into timing chain cover.*

– Remove crankshaft vibration damper. See **Vibration damper, removing and installing (M54 engine)** in this repair group.

> ***CAUTION—***
>
> - *Do not use flywheel locking tool to hold crankshaft stationary to loosen or tighten vibration damper mounting bolt. Use only the special tools specified, or equivalent hub holding tool.*
> - *Do not allow crankshaft to rotate while timing chains are loosened or removed. The pistons can contact the valves, causing engine damage.*

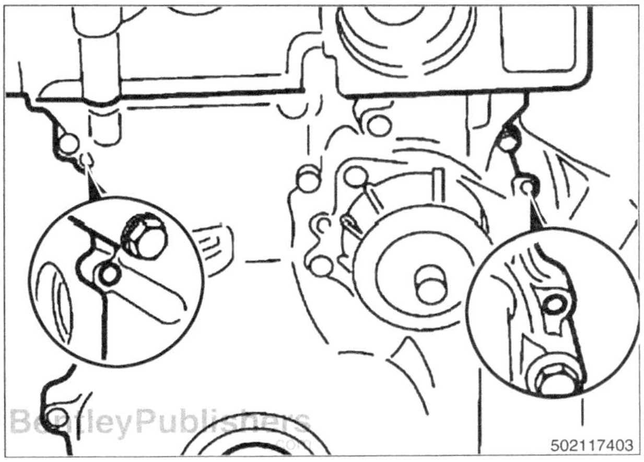

Using a 5 mm or smaller drift, drive two locating dowels (**insets**) in lower timing cover toward rear of car.

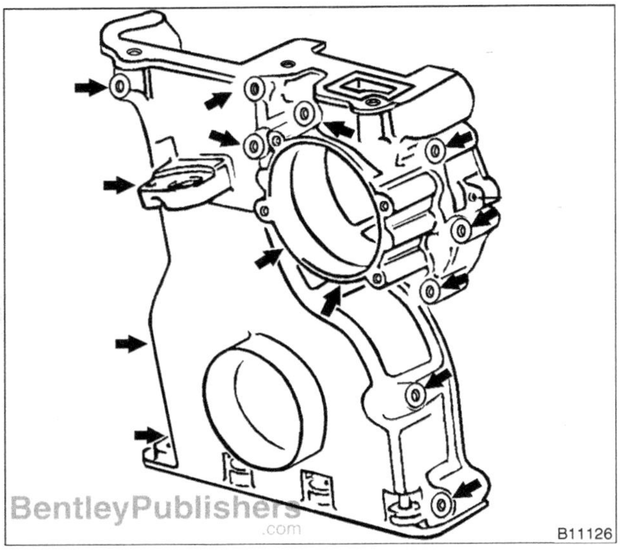

Remove timing cover mounting bolts (**arrows**). Slide cover with coolant pump forward to remove.

> ***CAUTION—***
> - *Use care when removing the cover from the cylinder head gasket. If the cover is stuck, use a sharp knife to separate it from the head gasket.*
> - *If cylinder head gasket is damaged, remove cylinder head and replace head gasket. See* **113 Cylinder Head Removal and Installation**.
> - *Use care when detaching front timing cover from coolant pipe at rear of cover.*

NOTE—
- *For clarity, front timing cover is illustrated with coolant pump removed.*

– Push primary chain tensioner guide rail aside and remove chain.

Timing chains, installing (M54 engine)

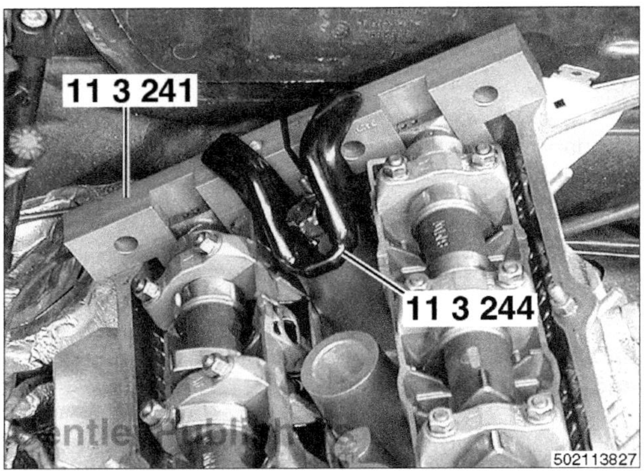

Make sure camshafts are locked in TDC position using BMW special tool set 11 3 240.

– Inspect timing chain sprockets. Inspect guide rail and tensioner rail for damage. Replace any part that is worn or damaged.

NOTE—
- *If any sprockets are replaced due to wear, also replace chain.*

BentleyPublishers.com
BentleyPublishers.com–All Rights Reserved

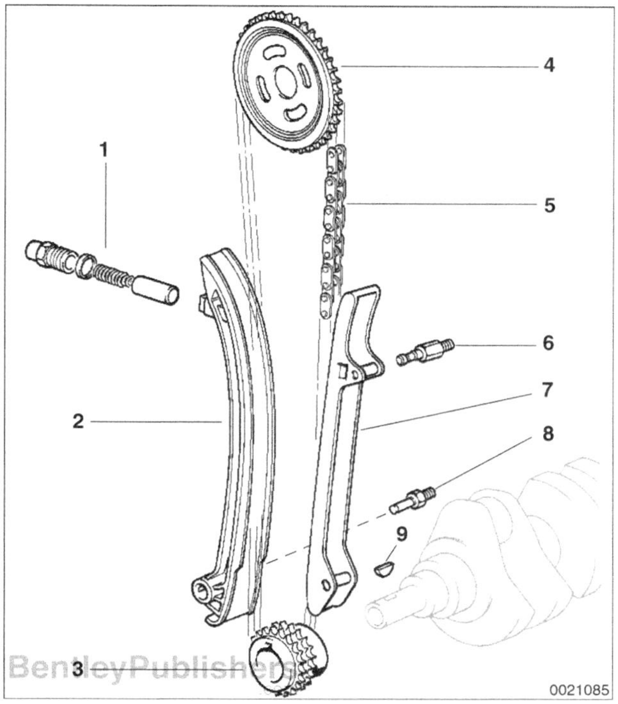

Primary timing chain assembly

1. Primary chain tensioner
2. Chain tensioner rail
3. Crankshaft sprocket
4. Exhaust camshaft sprocket
5. Primary chain
6. Anchor bolt
7. Guide rail
8. Anchor bolt
9. Woodruff key

— Install primary timing chain to crankshaft sprocket and hang upper end from exhaust camshaft.

— If necessary, replace crankshaft front seal in timing chain cover. See **119 Lubrication System**.

— To install lower timing chain cover:

- Clean cover and cylinder block sealing surfaces.
- Use new gaskets and coolant pipe O-ring.
- Remove timing chain cover dowel pins from engine block.
- Drive dowels into cover until they just protrude slightly from sealing surface.
- Apply a small bead of silicon sealer (3-Bond 1209® or equivalent) to corners of cylinder head where timing cover meets cylinder head and engine block.
- Tap cover into position until firmly seated.
- Install all bolts hand tight, including two Torx bolts from above.
- Install secondary chain lower guide.
- Tighten cover mounting bolts alternately and in stages.

Tightening torque	
Lower timing cover to cylinder block (M6)	10 Nm (7 ft-lb)

— Install oil pan with new gasket. See **119 Lubrication System**.

Tightening torques	
Oil pan to engine block (M6) • 8.8 grade • 10.9 grade	 10 Nm (7 ft-lb) 12 Nm (9 ft-lb)

117

BentleyPublishers.com
BentleyPublishers.com—All Rights Reserved

Timing chains, installing (M54 engine)

– Reinstall crankshaft vibration damper. See **Vibration damper, removing and installing (M54 engine)** in this repair group.

> ***CAUTION—***
> - *Do not use BMW special tool 11 2 300 to hold crankshaft stationary to tighten vibration damper mounting bolt. Use only the special tools specified, or equivalent hub holding tool.*
> - *Have a second person hold special tool 11 8 190 while tightening vibration damper mounting bolt.*

Tightening torque	
Vibration damper hub to crankshaft (use new bolt)	410 Nm (300 ft-lb)

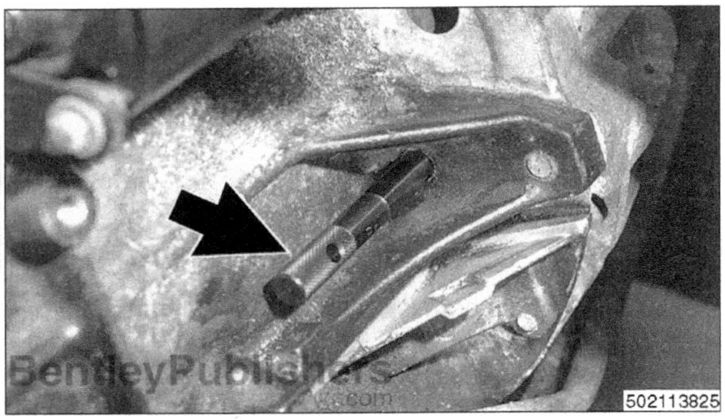

◄ Remove crankshaft locking tool (**arrow**).

◄ Fit top sprocket to primary timing chain and install on exhaust camshaft so that pointer on sprocket (**arrow**) lines up with cylinder head sealing surface (**dashed line**).

– Position sprocket so that chain slack is on tensioner rail side.

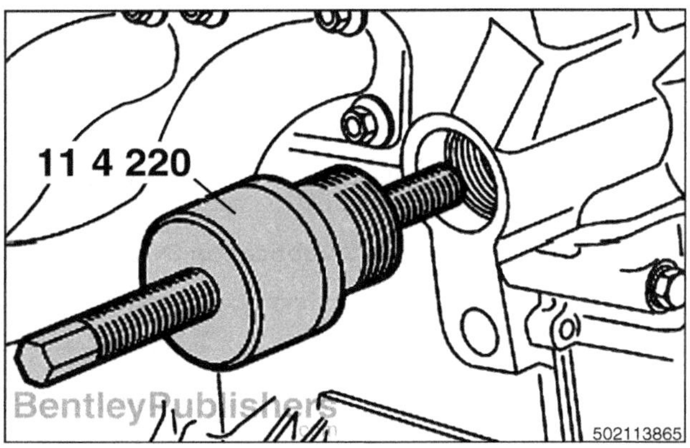

◄ Insert BMW special tool 11 4 220 in cylinder head and screw in adjustment screw by hand just until it contacts tensioning rail and chain slack is removed.

NOTE —
- *BMW special tool 11 4 220 is an adjustable dummy tensioner that simulates the function of the primary chain tensioner.*

– Make sure that arrow on upper primary sprocket still lines up with upper edge of cylinder head. Reposition sprocket if necessary.

◄ Install and tighten down threaded locating studs (**arrows**) in end of exhaust camshaft.

Tightening torque	
Exhaust camshaft locating stud	20 Nm (15 ft-lb)

BentleyPublishers.com
BentleyPublishers.com—All Rights Reserved

502113841

Install secondary chain tensioner on cylinder head (**arrows**). Keep tensioner compressed using BMW special tool 11 3 291 or suitable pin tool.

502113867

Fit exhaust camshaft splined sleeve, aligning gap in sleeve splines with corresponding gap on camshaft splines (**arrows**).

502113868

Slide splined shaft on exhaust camshaft. Be sure that locating tooth of shaft fits into spline gaps of camshaft and splined sleeve (**arrows**).

- Push splined shaft in further to rotate splined sleeve until three threaded holes in primary chain sprocket are centered in splined sleeve slots (**dashed lines**).

BentleyPublishers.com—All Rights Reserved

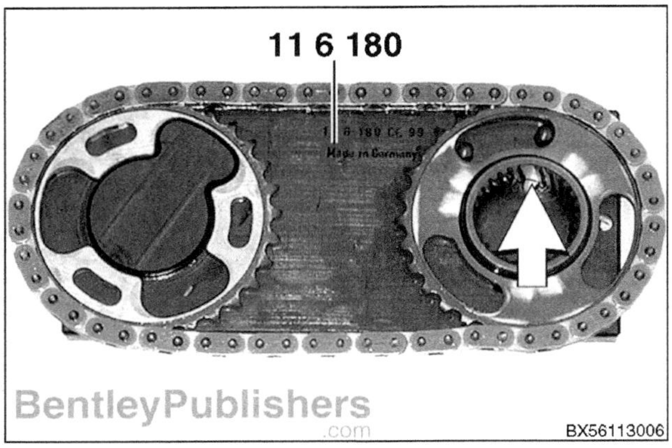

Place intake and exhaust sprockets in BMW special tool 11 6 180. Position spline gap on intake sprocket (**arrow**) as shown and place secondary chain on sprockets.

Remove chain and sprockets from tool and slide sprockets on camshafts. Confirm that gap in intake sprocket splines lines up with corresponding gap in camshaft splines (**arrow**).

> ***CAUTION—***
> - *Do not alter position of sprockets with respect to chain when removing from special tool 11 6 180.*

Align locating tooth of splined shaft with gaps on camshaft and sprocket. Slide splined shaft onto intake camshaft until approx. 1 mm (0.04 in) of splines (**arrows**) are visible.

Install intake camshaft spring plate so that **FRONT** mark is visible. Install mounting nuts (**arrows**) finger tight.

BentleyPublishers.com
BentleyPublishers.com—All Rights Reserved

Insert sprocket mounting bolts (**arrows**) on exhaust side camshaft assembly.

- Initially tighten to approx. 5 Nm (44 in-lb) and then back off by half a turn.

Fit thrust spacer (**A**) on exhaust camshaft.

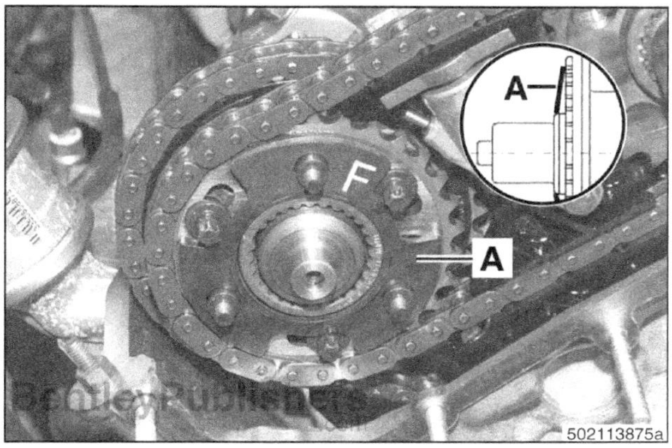

Install spring plate (**A**) to exhaust camshaft. Make sure that **F** mark is visible.

NOTE —

- *If **F** mark is no longer visible, install spring plate as shown in inset: convex side facing forward.*

BentleyPublishers.com
BentleyPublishers.com—All Rights Reserved

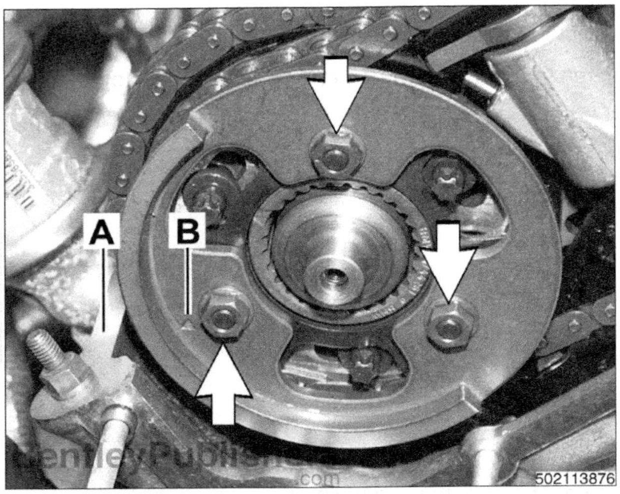

Install exhaust camshaft impulse wheel, aligning pointer (**B**) with top edge of cylinder head (**A**). Install mounting nuts (**arrows**) finger tight.

Pull out exhaust camshaft splined shaft to stop (**arrow**).

– Press down on secondary chain tensioner and remove tensioner lock-down tool.

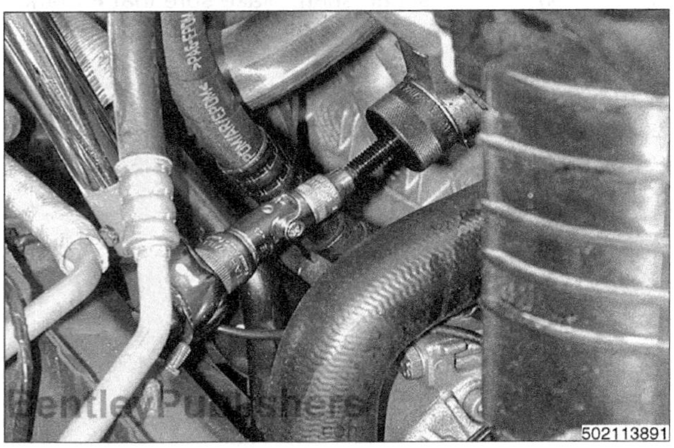

Preload primary chain:

- Tighten adjusting screw on BMW special tool 11 4 220 to specified torque.

Tightening torque	
Primary chain tensioner preload	0.7 Nm (6 in-lb)

Preload exhaust camshaft spring plate by pressing on impulse wheel while tightening mounting nuts (**arrows**) finger tight.

BentleyPublishers.com
BentleyPublishers.com—All Rights Reserved

Install BMW special tool 11 6 150 (VANOS setup bracket) to front of cylinder head timing case. Install nuts (**arrows**) finger tight, then tighten down uniformly until special tool is in full contact with cylinder head.

> ***CAUTION—***
> - *Make sure gasket material is removed from face of cylinder head. Clean sealing face and keep free of oil. If foreign material is present on the sealing surface, camshaft timing will be incorrect.*

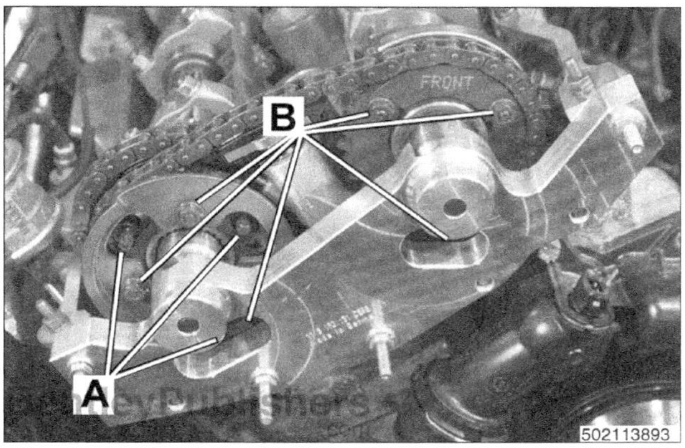

Secure camshaft sprockets and impulse wheels:

- Tighten mounting screws (**A**) on exhaust camshaft impulse wheel to approx. 5 Nm (44 in-lb).
- Tighten mounting nuts (**B**) on exhaust and intake sprocket assemblies to approx. 5 Nm (44 in-lb).
- Torque down mounting screws (**A**) and nuts (**B**) to final specifications.

Tightening torques	
Sprocket assembly wheel to camshaft (initial torque)	5 Nm (44 in-lb)
Sprocket assembly wheel to camshaft • M7 torx screws (**A**) • M6 mounting nuts (**B**)	 20 Nm (15 ft-lb) 10 Nm (8 ft-lb

– Remove crankshaft locking tool so that crankshaft is no longer secured in TDC position.

– Remove camshaft locking tools from cylinder head.

Turn engine over twice in direction of rotation until cylinder 1 intake and exhaust camshaft lobes (**arrows**) face each other.

– Secure crankshaft in TDC position with BMW special tool 11 2 300.

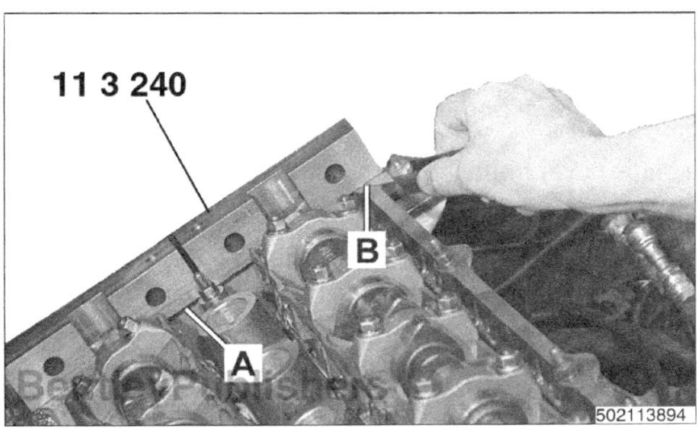

Place BMW special tool set 11 3 240 over camshafts and measure clearances.

- Due to flexible sprocket design, VANOS unit tolerances and play in the VANOS splines, when the camshaft timing is set correctly, the intake side of special tool 11 3 240 (**B**) may be up to 1 mm (0.04 in) above the surface of the cylinder head. This is normal. Reassemble engine.
- Otherwise, reset camshaft timing. See **Camshaft timing, checking and adjusting (M54 engine)** in this repair group.

> ***CAUTION—***
> - *If the exhaust side of the tool (**A**) is not flush with the cylinder head, camshaft timing is incorrect. Reset camshaft timing.*

– Remove BMW special tool 11 6 150 from front of cylinder head.

BentleyPublishers.com—All Rights Reserved

– Reinstall VANOS unit. See **Double VANOS unit, installing (M54 engine)** in this repair group.

– Remove BMW special tool 11 4 220 (adjustable dummy tensioner). Reinstall primary chain tensioner.

Tightening torque	
Primary chain tensioner to timing chain cover (M26 x 1.5)	70 Nm (52 ft-lb)

– Install intake camshaft oil baffle.

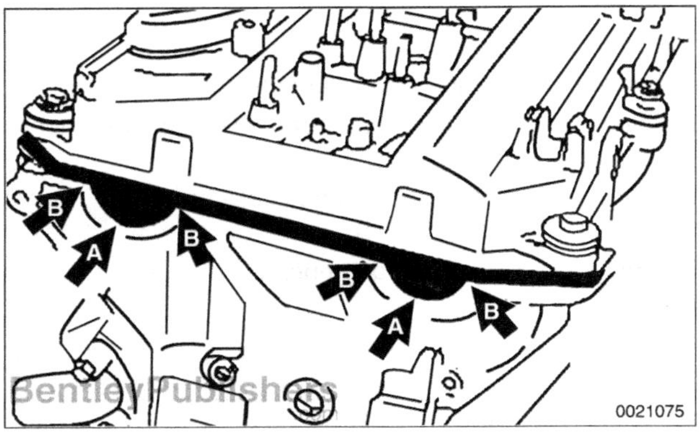

< Install cylinder head cover:

- Check for correct seating of half-moon seals (**A**) in back of cylinder head cover.
- Use a small amount of 3-Bond 1209® or equivalent sealant at corners (**B**) of half-moon cutouts. Seal corners in front of cylinder head at VANOS unit.

Tightening torque	
Cylinder head cover to cylinder head	10 Nm (7 ft-lb)

– Remainder of assembly is reverse of removal, noting the following:

- Secure all coolant hoses, thermostat housing, engine coolant drains.
- Install new engine oil filter and oil. See **119 Lubrication System**.
- Refill and bleed cooling system. See **170 Radiator and Cooling system**.
- Use scan tool to check VANOS operation with engine idling.

Tightening torques	
Coolant drain plug to cylinder block	25 Nm (18 ft-lb)
Radiator drain screw to radiator	2.5 Nm (22 in-lb)
Vibration damper hub to crankshaft (use new bolt)	410 Nm (300 ft-lb)

Camshaft timing, checking and adjusting (M54 engine)

– Remove engine top covers and cylinder head cover. See **113 Cylinder Head Removal and Installation**.

< Remove oil baffle cover from above intake camshaft.

– Remove engine drive belts and cooling fan. See **170 Radiator and Cooling System**.

– Remove spark plugs. Plug spark plug bores to prevent debris from falling into engine.

BentleyPublishers.com
BentleyPublishers.com—All Rights Reserved

Remove primary camshaft chain tensioner cylinder (**arrow**).

> ***CAUTION—***
> *• Chain tensioner piston is under spring pressure.*

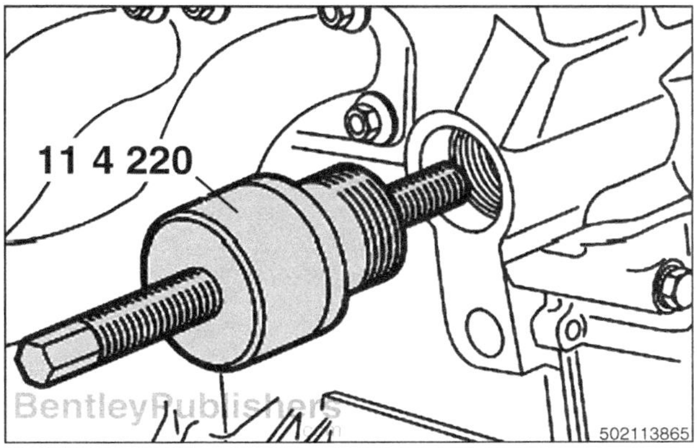

Insert BMW special tool 11 4 220 in cylinder head and screw in adjustment screw by hand just until it contacts tensioning rail and chain slack is removed.

NOTE—

- *BMW special tool 11 4 220 is an adjustable dummy tensioner that simulates the function of the primary chain tensioner.*

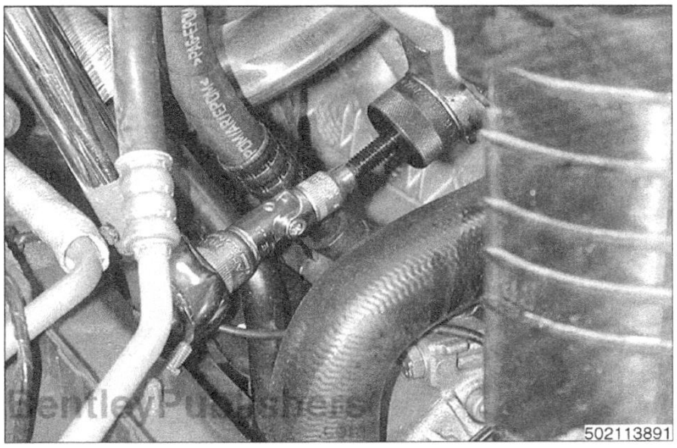

Preload primary chain tensioner rail:

- Tighten adjusting screw on BMW special tool 11 4 220.

Tightening torque	
Primary chain tensioner preload	0.7 Nm (6 in-lb)

Remove banjo bolt from VANOS unit oil pressure line. Use banjo bolt to attach BMW special tool 11 3 450 (compressed air fitting) to VANOS unit.

- Be prepared to catch dripping oil. Do not allow oil to contaminate accessory belts.

BentleyPublishers.com—All Rights Reserved

502113823

Cover oil bore (**arrow**) in VANOS unit with shop towel to capture oil which sprays when compressed air is applied.

– Connect compressed air line to air fitting. Apply air pressure set to 2 – 8 bar (30 – 115 psi).

502113824

With compressed air line connected, use crankshaft vibration damper bolt to rotate engine in direction of rotation (clockwise) at least twice, until cylinder 1 intake and exhaust camshaft lobes face each other (**arrows**) in top dead center (TDC) position for cylinder 1.

> ***CAUTION—***
> - *Do not rotate engine counterclockwise to reach TDC. If engine rotates past TDC, complete another two complete rotations.*

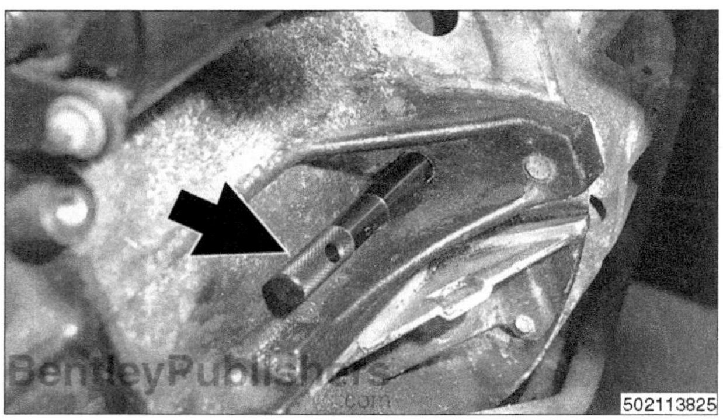

502113825

Remove sealing plug from special tool bore on lower left side of engine block near flywheel. Secure crankshaft in TDC position with BMW special tool 11 2 300 (**arrow**).

502113826

Unscrew and remove two cylinder head cover studs (**arrows**) at rear of cylinder head.

BentleyPublishers.com
BentleyPublishers.com—All Rights Reserved

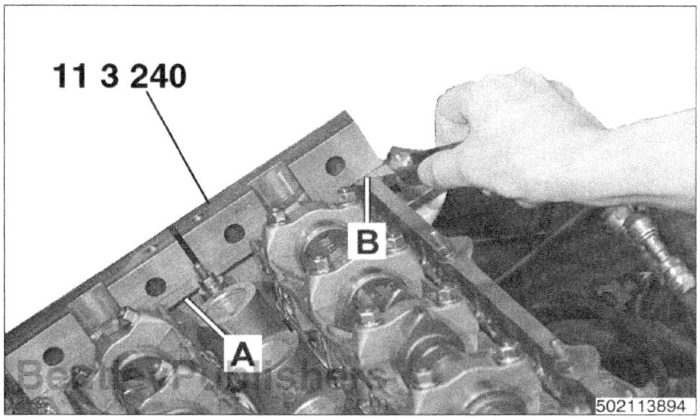

Place BMW special tool set 11 3 240 over camshafts and measure clearances.

- Due to flexible sprocket design, VANOS unit tolerances and play in the VANOS splines, when the camshaft timing is set correctly, the intake side of special tool 11 3 240 (**B**) may be up to 1 mm (0.04 in) above the surface of the cylinder head. This is normal. Reassemble engine.
- Otherwise, continue with camshaft timing adjustment procedure.

> ***CAUTION—***
> - *If the exhaust side of the tool (**A**) is not flush with the cylinder head, camshaft timing is incorrect. Reset camshaft timing.*

— Remove double VANOS unit. See **Double VANOS unit, removing (M54 engine)** in this repair group.

Press down on secondary chain tensioner and lock into place using BMW special tool 11 3 292 or equivalent pin or drift.

Loosen intake camshaft sprocket mounting nuts (**arrows**) by 1 turn.

Loosen exhaust camshaft sprocket mounting fasteners:

- Release bolts **A** ½ turn.
- Release nuts **B** 2 turns.

BentleyPublishers.com
BentleyPublishers.com—All Rights Reserved

Camshaft timing, checking and adjusting (M54 engine)

Pull out intake camshaft splined shaft out so that approx. 1 mm (0.04 in) of splines (**arrows**) can be seen.

Pull out exhaust camshaft splined shaft to stop (**arrow**).

– Make sure camshafts are secured in TDC position using BMW special tools 11 3 240 and 11 3 244.

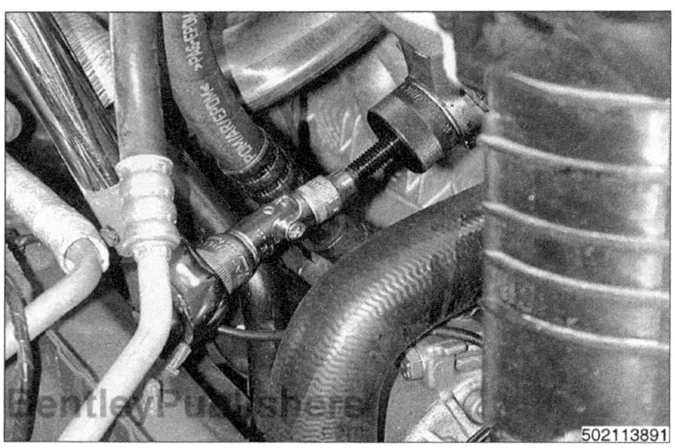

Make sure chain tensioning rail is preloaded.

Tightening torque	
Primary chain tensioner preload	0.7 Nm (6 in-lb)

Preload exhaust camshaft spring plate by pressing on impulse wheel. Tighten mounting nuts (**arrows**) by hand. Do not tighten fully.

BentleyPublishers.com
BentleyPublishers.com—All Rights Reserved

Install BMW special tool 11 6 150 (VANOS setup bracket) to front of cylinder head timing case. Tighten nuts (**arrows**) by hand, and then tighten down uniformly until special tool contacts cylinder head firmly.

> ***CAUTION—***
> • *Make sure gasket material is removed from face of cylinder head. Clean sealing face and keep free of oil. If foreign material is present on the sealing surface, camshaft timing will be incorrect.*

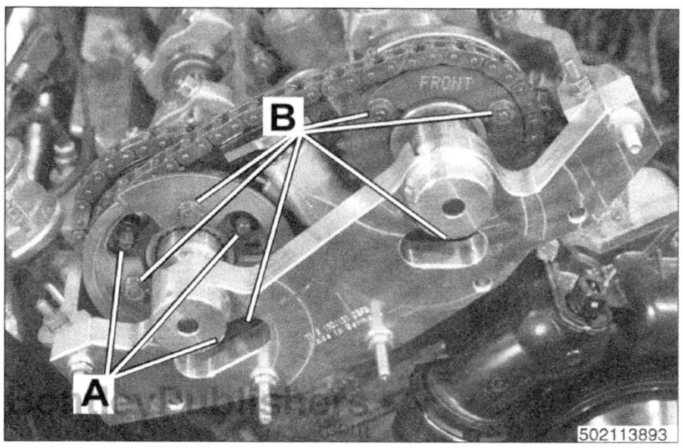

Secure camshaft impulse sprockets and wheels:

- Pretighten Torx screws (**A**) on exhaust camshaft impulse wheel to approx. 5 Nm (44 in-lb).
- Pretighten mounting nuts (**B**) on exhaust and intake sprocket assemblies to approx. 5 Nm (44 in-lb).
- Torque down Torx screws (**A**) and nuts (**B**) to final specifications.

Tightening torques	
Sprocket assembly to camshaft (initial torque)	5 Nm (44 in-lb)
Sprocket assembly to camshaft • M7 Torx screw (**A**) • M6 mounting nut (**B**)	 20 Nm (15 ft-lb) 10 Nm (7 ft-lb)

– Remove flywheel locking tool from transmission bellhousing so that crankshaft is no longer locked.

– Remove camshaft locking tools from rear of cylinder head.

– Use crankshaft vibration damper bolt to crank engine over twice by hand in direction of rotation until cylinder 1 intake and exhaust camshaft lobes face each other again.

– Secure crankshaft with BMW special tool 11 2 300.

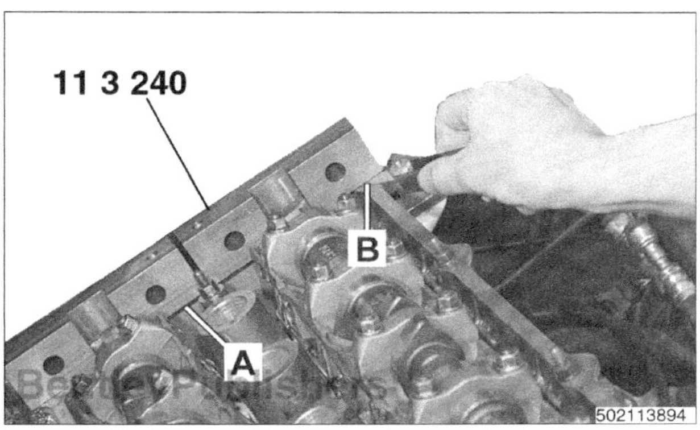

Place BMW special tool set 11 3 240 over camshafts and measure clearances.

- Due to flexible sprocket design, VANOS unit tolerances and play in the VANOS splines, when the camshaft timing is set correctly, the intake side of special tool 11 3 240 (**B**) may be up to 1 mm (0.04 in) above the surface of the cylinder head. This is normal.
- Otherwise, repeat camshaft timing adjustment procedure.

> ***CAUTION—***
> • *If the exhaust side of the tool (**A**) is not flush with the cylinder head, camshaft timing is incorrect. Reset camshaft timing.*

– Remove camshaft locking tools from rear of camshafts.

BentleyPublishers.com
BentleyPublishers.com–All Rights Reserved

- Remove BMW special tool 11 4 220 (adjustable dummy tensioner). Reinstall primary chain tensioner.

Tightening torque	
Primary chain tensioner to timing chain cover (M26 x 1.5)	70 Nm (52 ft-lb)

- Remove crankshaft locking tool from transmission bellhousing. Reinstall sealing plug.
- Remove VANOS setup bracket from front of cylinder head.
- Install VANOS unit. See **Double VANOS unit, installing (M54 engine)** in this repair group.
- Remove compressed air line adapter (BMW special tool 11 3 450) from VANOS unit.
- Fit VANOS oil line banjo bolt with new seals. Attach oil line to VANOS unit.

Tightening torque	
Oil line to VANOS unit (banjo bolt)	32 Nm (24 ft-lb)

- Remainder of installation is reverse of removal. Note the following:
  - Install spark plugs.
  - Install intake camshaft plastic baffle prior to installing cylinder head cover. Install ignition coil packs and connect electrical harness connectors.
  - Install accessory belts, radiator cooling fan and cooling fan shroud.
  - Install upper engine covers.

Tightening torques	
Cylinder head cover to cylinder head	10 Nm (7 ft-lb)
Spark plug to cylinder head	25 Nm (18 ft-lb)

BentleyPublishers.com
BentleyPublishers.com–All Rights Reserved

CAMSHAFT SERVICE (M54 ENGINE)

Camshaft sprockets assembly (M54 engine)

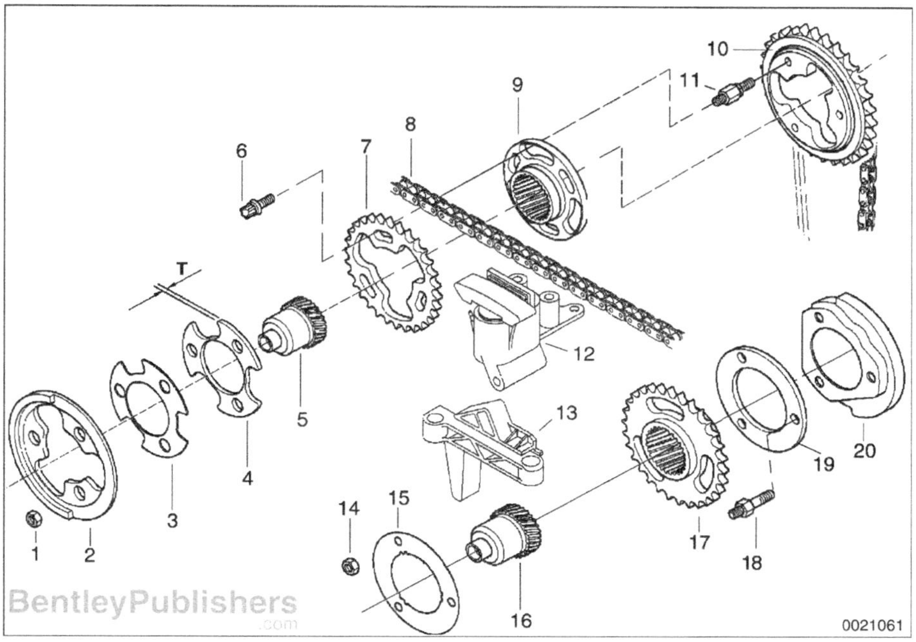

1. Impulse wheel mounting nut
2. Exhaust camshaft impulse wheel
3. **Spring plate, exhaust**
4. Thrust spacer (T = 3.5 mm)
5. Splined shaft
6. Torx screw
7. Exhaust camshaft secondary sprocket
8. Secondary timing chain
9. Splined sleeve
10. Primary sprocket
11. Threaded locating stud
12. Secondary chain tensioner
13. Secondary chain lower guide
14. Sprocket mounting nut
15. Spring plate, intake
16. Splined shaft
17. Intake camshaft sprocket
18. Locating stud
19. Thrust spacer
20. Intake camshaft impulse wheel

Camshafts, removing (M54 engine)

> ***WARNING—***
> - *Due to risk of personal injury, be sure the engine is cold before beginning the removal procedure.*

> ***CAUTION—***
> - *Special BMW service tools are required to remove and install the camshafts and VANOS units. Read the procedure through before starting the job.*
> - *If the camshafts are not properly timed, the pistons can contact the valves, causing engine damage.*

– Disconnect negative (–) cable from battery.

> ***CAUTION—***
> - *Prior to disconnecting the battery, read the battery disconnection cautions in* **001 Warnings and Cautions**.

– Remove air filter housing and ducts. See **130 Fuel Injection**.

BentleyPublishers.com
BentleyPublishers.com—All Rights Reserved

– Remove electric cooling fan and cooling fan shroud. See **170 Radiator and Cooling System**.

– Remove engine covers and cylinder head cover. See **113 Cylinder Head Removal and Installation**.

– Remove spark plugs. Plug spark plug bores to prevent debris from falling into engine.

– Remove VANOS unit. See **Double VANOS unit, removing (M54 engine)** in this repair group.

◄ Remove primary camshaft chain tensioner cylinder (**arrow**).

> ***CAUTION—***
> *• Chain tensioner piston is under spring pressure.*

◄ Press down on secondary chain tensioner and lock into place using BMW special tool 11 3 291, or a thin drift or pin.

◄ Remove exhaust camshaft impulse wheel mounting nuts (**arrows**). Remove impulse wheel (**A**).

BentleyPublishers.com
BentleyPublishers.com—All Rights Reserved

Remove spring plate (**A**).

Remove intake camshaft sprocket mounting nuts (**arrows**) and remove spring plate (labeled **FRONT**).

Remove torx bolts (**arrows**) from exhaust camshaft sprocket.

Lift off exhaust and intake sprockets together with secondary chain, thrust spacer (**A**), and splined shaft (**B**).

> ***CAUTION—***
> - *Intake and exhaust camshaft splined shafts share the same part number. Be sure to mark splined shafts so they can be reinstalled in their original positions.*

BentleyPublishers.com–All Rights Reserved

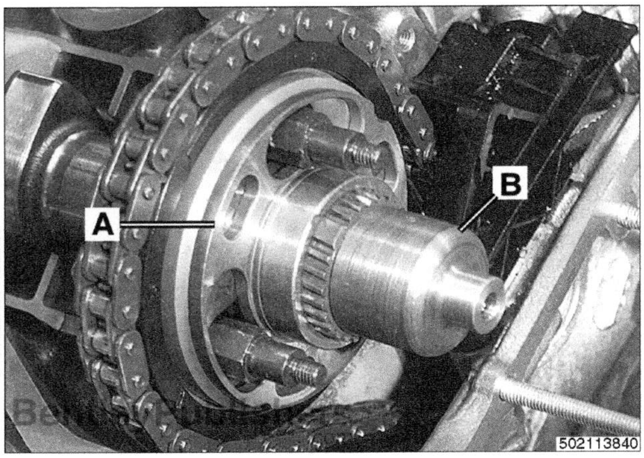

Remove exhaust camshaft splined sleeve (**A**) and shaft (**B**).

Remove secondary chain tensioner mounting bolts (**arrows**) and remove tensioner with retaining pin in place.

Remove sprocket mounting studs (**arrows**) from exhaust camshaft.

BentleyPublishers.com
BentleyPublishers.com—All Rights Reserved

Lift primary chain sprocket off exhaust camshaft. Remove sprocket from chain.

Place timing chain on end of exhaust camshaft.

Remove locating studs (**arrows**) from intake camshaft. Lift off intake camshaft thrust spacer (**A**) and impulse wheel (**B**).

Do not remove Torx® screws (**arrows**) from camshafts.

Bentley Publishers
BentleyPublishers.com – All Rights Reserved

Remove cylinder head cover mounting studs (**arrows**) from center of cylinder head.

– Remove crankshaft locking tool.

Lift primary chain and hold under tension. Use crankshaft vibration damper bolt to rotate engine in direction opposite to normal rotation (counterclockwise) approximately 30°.

CAUTION—
- *By turning crankshaft 30° counterclockwise, pistons are moved away from top dead center. This prevents piston-to-valve interference during camshaft servicing.*

– Remove camshaft locking tools (BMW special tool set 11 3 240) from rear of cylinder head.

Remove intake camshaft bearing cap 1 fasteners (**arrows**). Remove bearing cap.

CAUTION—
- *Do not remove camshaft bearing caps 2 - 6 at this time.*

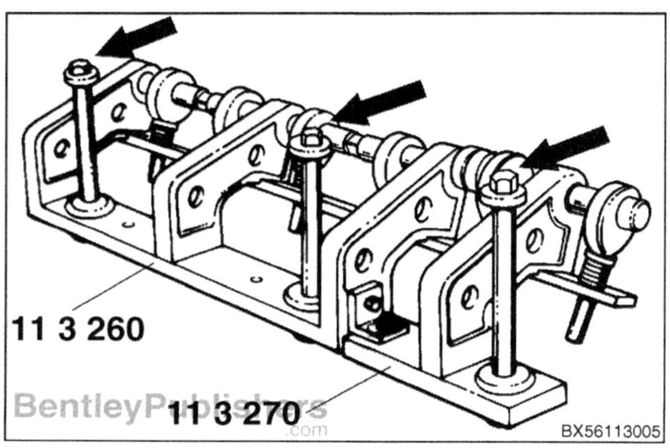

Fit BMW special tools 11 3 260 and 11 3 270 to cylinder head and screw long bolts (**arrows**) into spark plug threads. Align tensioning pins on tool with bearing caps on camshaft.

CAUTION—
- *Do not overtighten bolts in spark plug holes.*

BentleyPublishers.com
BentleyPublishers.com—All Rights Reserved

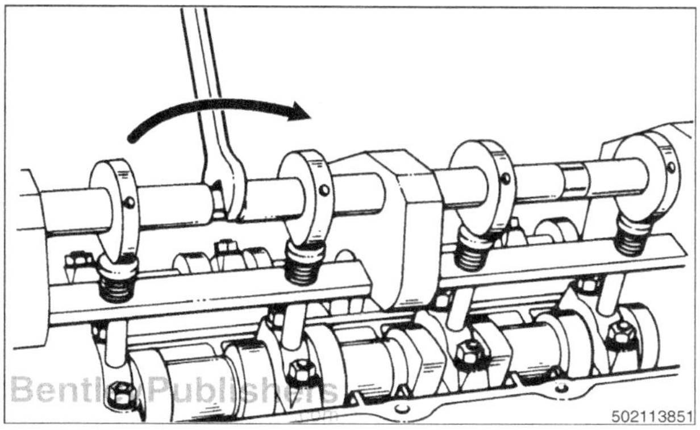

Rotate eccentric shaft of special tool (**arrow**) to pretension intake camshaft bearing caps. Remove remaining intake camshaft bearing cap fasteners.

- Slowly release tension on eccentric shaft and remove BMW special tools. Remove bearing caps and set aside in order. Remove camshaft and store safely.
- Repeat procedure for exhaust camshaft.

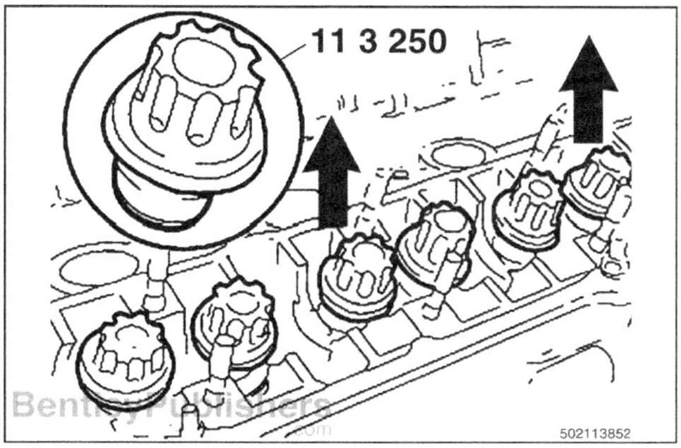

If cylinder head is being removed, secure hydraulic lifters in lifter bores using BMW special tool set 11 3 250, or remove lifters using a magnetic pick-up tool. With lifters secure or removed, lift out camshaft bearing carriers from cylinder head.

> ***CAUTION—***
> - *Do not let hydraulic lifters fall out as camshaft carrier is removed.*
> - *Mark hydraulic lifters so that they can be reinstalled in their original lifter bores.*
> - *Store hydraulic lifters in an upright position.*

Camshafts, installing (M54 engine)

- Lubricate camshafts, camshaft carriers, bearing caps, hydraulic lifters and friction washers with assembly lubricant before installation.

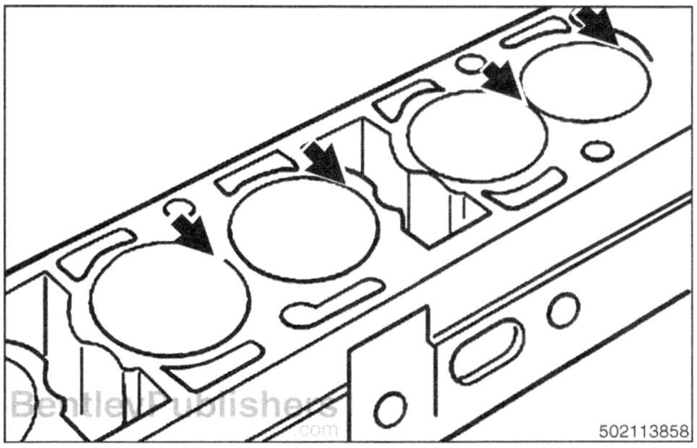

Before installing camshaft carriers, examine hydraulic lifter bore bearing points (**arrows**) on underside of camshaft carriers for signs of wear.

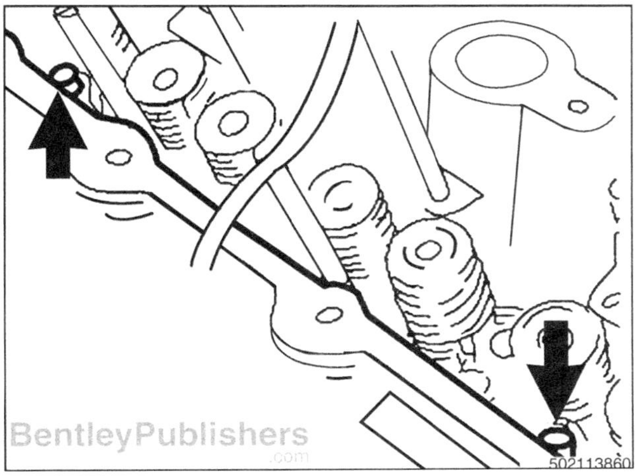

Center camshaft carrier with hydraulic lifters on pins (**arrows**) at bearing positions 2 and 7.

BentleyPublishers.com
BentleyPublishers.com—All Rights Reserved

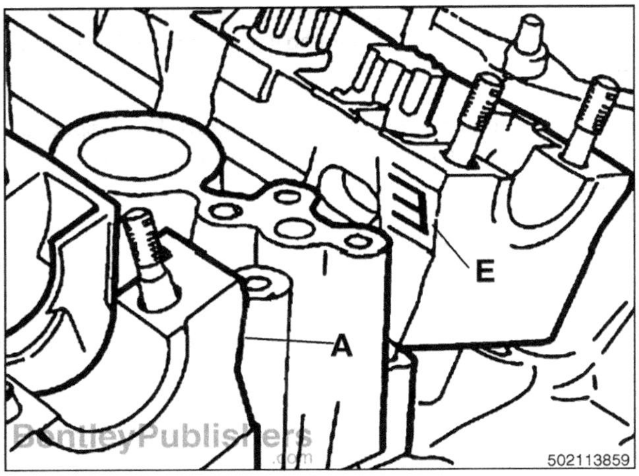

Note marks on camshaft carriers: **E** for intake side and **A** for exhaust side.

– Lift timing chain and place exhaust camshaft on exhaust camshaft carrier. Place intake camshaft on intake camshaft carrier.

Place bearing caps on camshafts and install retaining nuts (**arrows**) finger tight.

- Note that each bearing cap is marked (**inset**):
  E1 to E7 for intake (front to rear)
  A1 to A7 for exhaust (front to rear)

CAUTION—
- *Make sure the crankshaft, which was rotated approximately 30° opposite the direction of engine rotation from TDC, is still in that position before reinstalling the camshafts. This ensures that pistons are out of TDC position and prevents piston-to-valve interference when the camshafts are installed.*

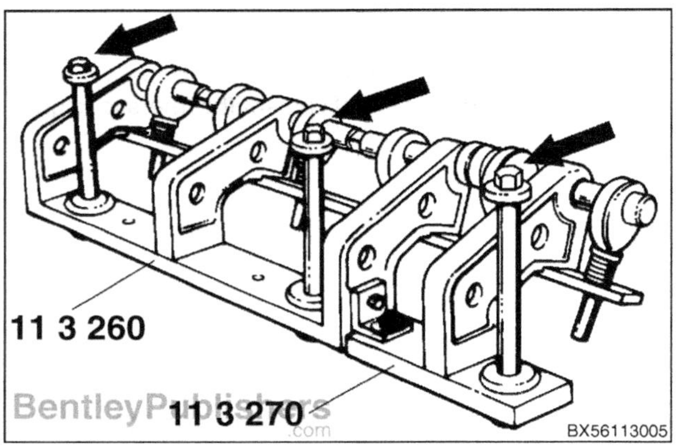

Fit BMW special tools 11 3 260 and 11 3 270 to cylinder head and screw long bolts (**arrows**) into spark plug threads. Align tensioning pins on tool with bearing caps on intake camshaft.

CAUTION—
- *Do not overtighten bolts in spark plug holes.*

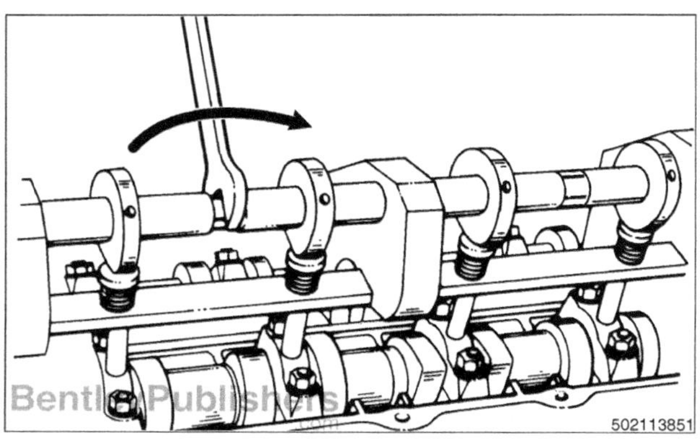

Rotate eccentric shaft of special tool (**arrow**) to pretension bearing caps. Torque bearing cap nuts.

Tightening torque	
Camshaft bearing cap to cylinder head (M7)	14 Nm (10 ft-lb)

– Release tension on eccentric shaft and remove special tools.

BentleyPublishers.com
BentleyPublishers.com—All Rights Reserved

– Repeat procedure for exhaust camshaft.

> ***CAUTION—***
> - *After the camshafts are installed, a minimum waiting time is required for the hydraulic lifters to bleed down before bringing the crankshaft and pistons back to TDC. When the camshafts are removed, the hydraulic lifters may expand. This expansion can cause increased valve lift when camshafts are bolted down, possibly resulting in piston-to-valve interference.*

Hydraulic lifter bleed down waiting times	
68°F (20°C) and higher	4 minutes
50° – 68°F (10° – 20°C)	11 minutes
32° – 50°F (0° – 10°C)	30 minutes

In procedure steps that follow, use 27 mm open-end wrench to rotate camshaft into position. If necessary, grind off outer edges of wrench jaws to prevent damage to cylinder head.

Rotate camshafts until intake and exhaust lobes for cylinder 1 face each other (**arrows**).

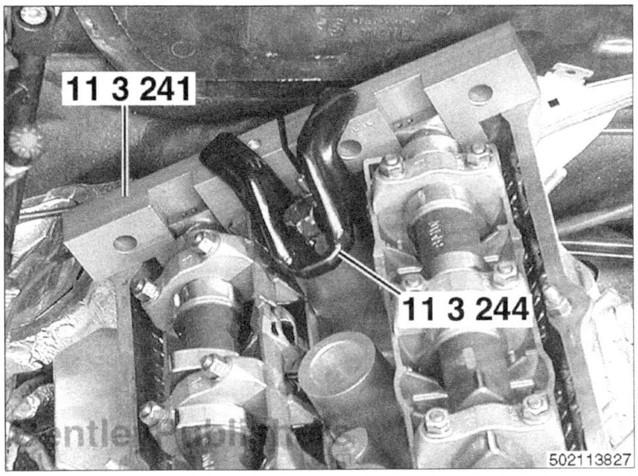

Secure camshafts in TDC position using BMW camshaft TDC tools 11 3 241. Align camshafts so that camshaft TDC tools rest solidly on cylinder head upper surface. Lock down with special tool 11 3 244 bolted to a spark plug bore.

BentleyPublishers.com–All Rights Reserved

Camshafts, installing (M54 engine)

◄ Lift timing chain and hold under tension.

◄ While maintaining tension on timing chain, use crankshaft vibration damper bolt to rotate crankshaft in direction of rotation (clockwise) until cylinder 1 is at TDC position. **O|T** (**arrow**) on front pulley lines up with boss on lower timing chain cover.

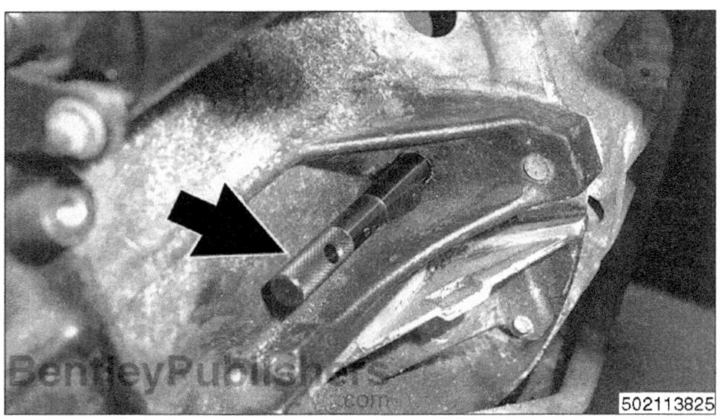

◄ Secure crankshaft in TDC position with BMW special tool 11 2 300 (**arrow**).

◄ Slide impulse wheel on intake camshaft, aligning boss with raised portion on camshaft (**arrow**).

BentleyPublishers.com
BentleyPublishers.com—All Rights Reserved

Fit thrust spacer (**A**) to intake camshaft (**B**) and tighten down with threaded locating studs. Place longer threaded end of studs into camshaft (**arrows**).

Tightening torque	
Impulse wheel studs to intake camshaft	20 Nm (15 ft-lb)

Fit primary sprocket to primary timing chain and install on exhaust camshaft so that pointer on sprocket (**arrow**) lines up with cylinder head sealing surface (**dashed line**).

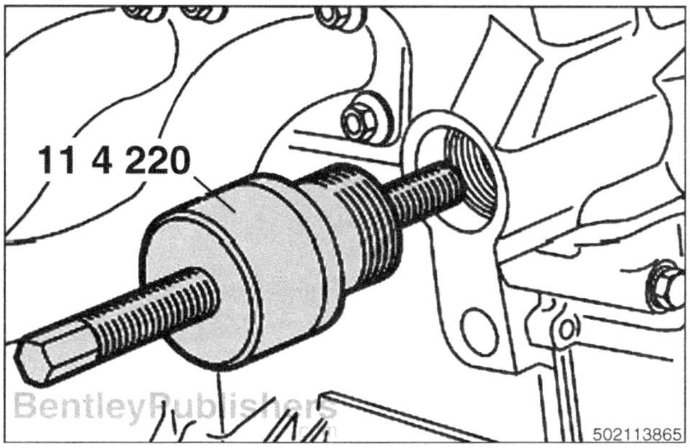

Insert BMW special tool 11 4 220 into timing chain tensioner bore and bring center spindle into contact with tensioning rail. Tighten until slack is removed, but do not pretension timing chain.

NOTE—

- *BMW special tool 11 4 220 is an adjustable dummy tensioner and simulates the function of the primary chain tensioner.*

– Make sure that arrow on upper primary sprocket still lines up with upper edge of cylinder head. Reposition sprocket if necessary.

Insert and tighten down threaded locating studs (**arrows**) in end of exhaust camshaft. Note installed position of studs.

Tightening torque	
Exhaust camshaft locating stud	20 Nm (15 ft-lb)

117

Install secondary chain tensioner on cylinder head (**arrows**). Keep tensioner compressed using BMW special tool 11 3 291 or suitable pin tool.

Install exhaust camshaft splined sleeve. Confirm that gap in sleeve splines aligns with corresponding gap in camshaft splines (**arrows**).

Slide splined shaft on exhaust camshaft. Be sure that locating tooth of shaft fits into spline gaps of camshaft and splined sleeve (**arrows**).

- Push splined shaft in further to rotate splined sleeve until three threaded holes in primary chain sprocket are centered in splined sleeve slots (**dashed lines**).

BentleyPublishers.com
BentleyPublishers.com—All Rights Reserved

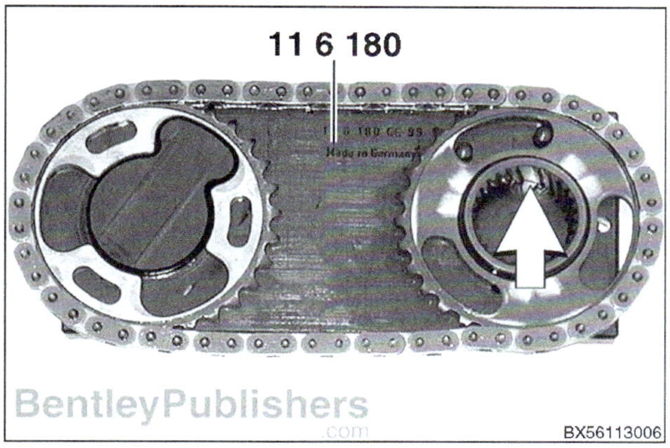

Place intake and exhaust sprockets in BMW special tool 11 6 180. Position spline gap on intake sprocket (**arrow**) as shown and place secondary chain on sprockets.

Remove chain and sprockets from tool and slide sprockets on camshafts. Confirm that gap in intake sprocket splines lines up with corresponding gap in camshaft splines (**arrow**).

CAUTION—
- *Do not alter position of sprockets with respect to chain when removing from special tool 11 6 180.*

Align locating tooth of splined shaft with gaps on camshaft and sprocket. Slide splined shaft onto intake camshaft until approx. 1 mm (0.04 in) of splines (**arrows**) are visible.

Install intake camshaft spring plate so that **FRONT** mark is visible. Install mounting nuts (**arrows**) finger tight.

BentleyPublishers.com
BentleyPublishers.com—All Rights Reserved

Insert sprocket mounting bolts (**arrows**) on exhaust side camshaft assembly.

- Initially tighten to approx. 5 Nm (44 in-lb) and then back off by half a turn.

Fit thrust spacer (**A**) on exhaust camshaft.

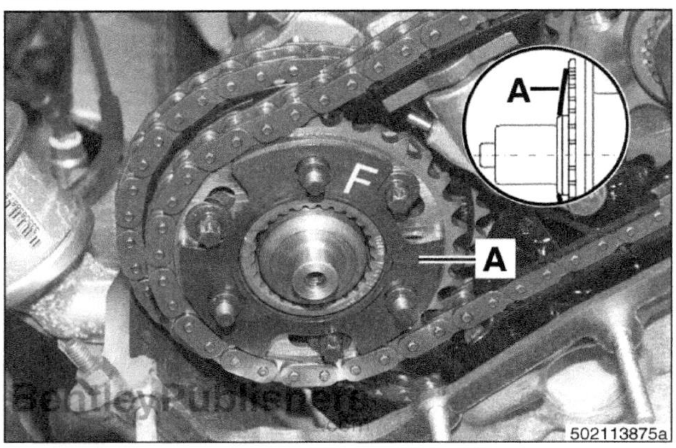

Install spring plate (**A**) to exhaust camshaft. Make sure that **F** mark is visible.

NOTE —

- *If **F** mark is no longer visible, install spring plate as shown in inset: convex side facing forward.*

Install exhaust camshaft impulse wheel, aligning pointer (**B**) with top edge of cylinder head (**A**). Install mounting nuts (**arrows**) finger tight.

BentleyPublishers.com
BentleyPublishers.com—All Rights Reserved

Pull out exhaust camshaft splined shaft to stop (**arrow**).

– Press down on secondary chain tensioner and remove tensioner lock-down tool.

Preload primary chain:

- Tighten adjusting screw on BMW special tool 11 4 220 to specified torque.

Tightening torque	
Primary chain tensioner preload	0.7 Nm (6 in-lb)

Preload exhaust camshaft spring plate by pressing on impulse wheel while tightening mounting nuts (**arrows**) finger tight.

Install BMW special tool 11 6 150 (VANOS setup bracket) to front of cylinder head timing case. Install nuts (**arrows**) finger tight, then tighten down uniformly until special tool is in full contact with cylinder head.

CAUTION—

- *Make sure gasket material is removed from face of cylinder head. Clean sealing face and keep free of oil. If foreign material is present on the sealing surface, camshaft timing will be incorrect.*

Camshafts, installing (M54 engine)

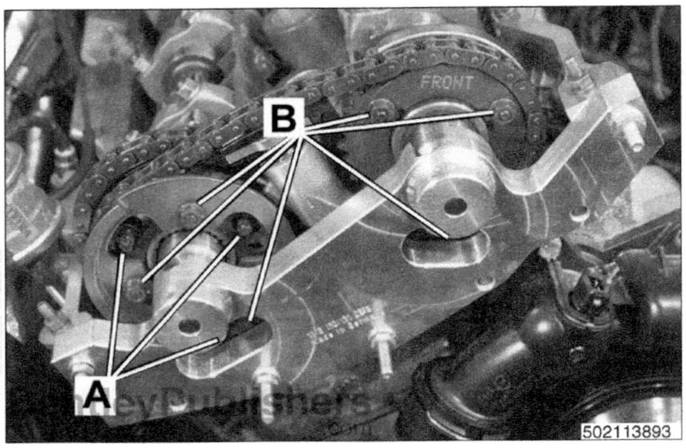

Secure camshaft sprockets and impulse wheels:

- Tighten mounting screws (**A**) on exhaust camshaft impulse wheel to approx. 5 Nm (44 in-lb).
- Tighten mounting nuts (**B**) on exhaust and intake sprocket assemblies to approx. 5 Nm (44 in-lb).
- Torque down mounting screws (**A**) and nuts (**B**) to final specifications.

Tightening torques	
Sprocket assembly wheel to camshaft–(initial torque)	5 Nm (44 in-lb)
Sprocket assembly wheel to camshaft • M7 torx screws (**A**) • M6 mounting nuts (**B**)	 20 Nm (15 ft-lb) 10 Nm (8 ft-lb)

– Remove crankshaft locking tool so that crankshaft is no longer secured in TDC position.

– Remove camshaft locking tools from cylinder head.

Turn engine over twice in direction of rotation until cylinder 1 intake and exhaust camshaft lobes (**arrows**) face each other.

– Secure crankshaft in TDC position with BMW special tool 11 2 300.

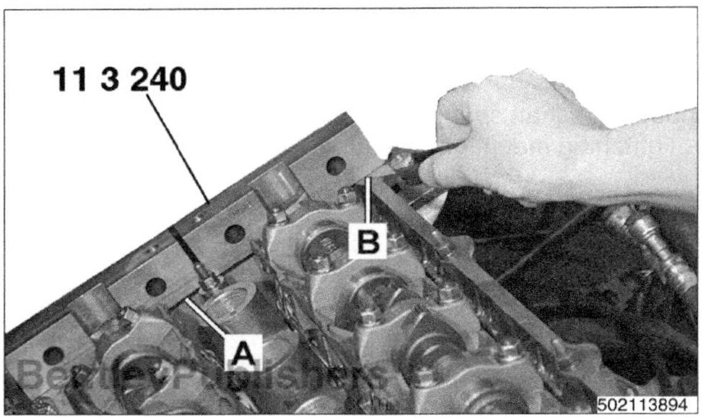

Place BMW special tool set 11 3 240 over camshafts and measure clearances.

- Due to flexible sprocket design, VANOS unit tolerances and play in the VANOS splines, when camshaft timing is set correctly, the intake side of special tool 11 3 240 (**B**) may be up to 1 mm (0.04 in) above the surface of the cylinder head. This is normal.
- Otherwise, reset camshaft timing. See **Camshaft timing, checking and adjusting (M54 engine)** in this repair group.

> ***CAUTION—***
> - *If the exhaust side of the tool (**A**) is not flush with the cylinder head, camshaft timing is incorrect. Reset camshaft timing.*

– Remove BMW special tool 11 6 150 from front of cylinder head.

– Reinstall VANOS unit. See **Double VANOS unit, installing (M54 engine)** in this repair group.

– Remove camshaft and crankshaft locking tools.

– Reassemble engine. Be sure to top off and bleed cooling system

BentleyPublishers.com
BentleyPublishers.com—All Rights Reserved

TIMING CHAIN SERVICE (N52 ENGINE, N54 ENGINE)

Camshaft timing chain components (N52, N54 engine)

Camshaft timing chain, vibration damper (front pulley) and associated components of N52 and N54 engines are similar.

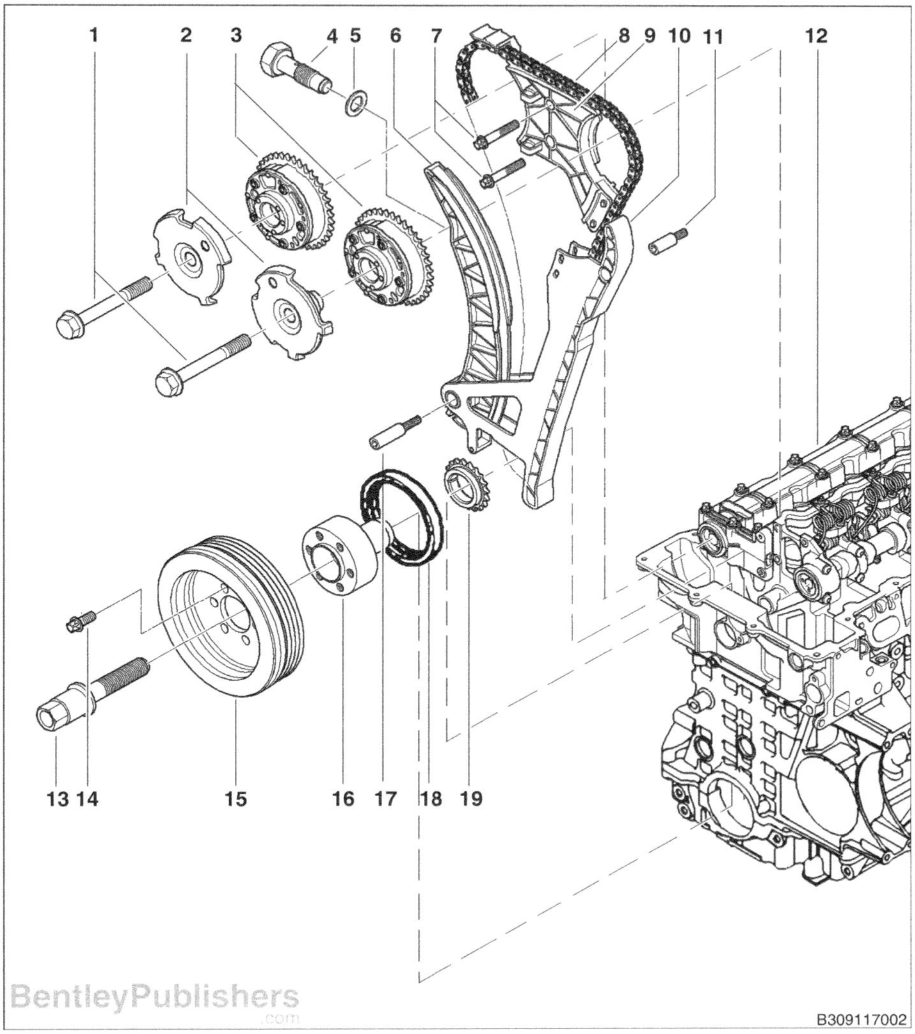

1. VANOS unit mounting bolt (M10 x 94 mm)
   - Replace with new
   - Tighten in 2 stages
     Stage 1: 20 Nm (15 ft-lb)
     Stage 2: Additional 180°
2. Camshaft impulse wheel
3. VANOS unit
   - Different intake and exhaust adjustment units
4. Chain tensioner (M22 x 1.5)
   - Tighten to 55 Nm (41 ft-lb)
5. Sealing O-ring
6. Chain tensioner rail
7. Chain assembly mounting bolts (M6 x 35 mm Torx)
   - Tighten to 8.5 Nm (6 ft-lb)
8. Timing chain
9. Chain assembly
10. Chain guide
11. Chain guide mounting bolt (M7)
    - Tighten to 14 Nm (10 ft-lb)
12. Cylinder head with camshafts
    - N52 cylinder head illustrated
13. Vibration damper hub mounting bolt (M16 x 80 mm)
    - Replace with new
    - Tighten in 2 stages
      Stage 1: 100 Nm (74 ft-lb)
      Stage 2: Additional 360°
14. Vibration damper mounting bolt (M8 x 16 mm Torx)
    - Tighten to 35 Nm (26 ft-lb)
15. Vibration damper (front pulley)
16. Vibration damper hub
17. Chain tensioner rail mounting bolt (M8)
    - Tighten to 20 Nm (15 ft-lb)
18. Crankshaft front seal
19. Crankshaft sprocket

BentleyPublishers.com
BentleyPublishers.com—All Rights Reserved

Camshaft timing chain, removing and installing (N52, N54 engine)

When removing and installing the camshaft timing chain, special tools and procedures are required to remove and install VANOS units and to time camshafts. Read the entire procedure before beginning repairs.

> ***WARNING —***
> - *To avoid personal injury, be sure the engine is cold before beginning the procedure.*

> ***CAUTION—***
> - *Disassembly, removal and assembly of camshafts or VANOS units without special tools poses the risk of damage or breakage: Valves may be bent by contact with the piston crowns.*
> - *Throughout this procedure, unless otherwise specified, crankshaft and camshafts remain locked against rotation using BMW special tools.*

- Use scan tool to read out and record ECM fault memory.
- Switch ignition OFF and remove key.
- Remove upper and lower cabin microfilter housings. See **640 Heating and Air-conditioning**.
- Remove strut tower brace.
- Remove ignition coil cover (upper engine cover). See **020 Maintenance**.
- Detach ignition coil connectors. Lift ignition coil harness off cylinder head cover and set aside.
- Remove ignition coils. See **120 Ignition System**.
- Remove cylinder head cover. See **113 Cylinder Head Removal and Installation**.
- Remove spark plugs.

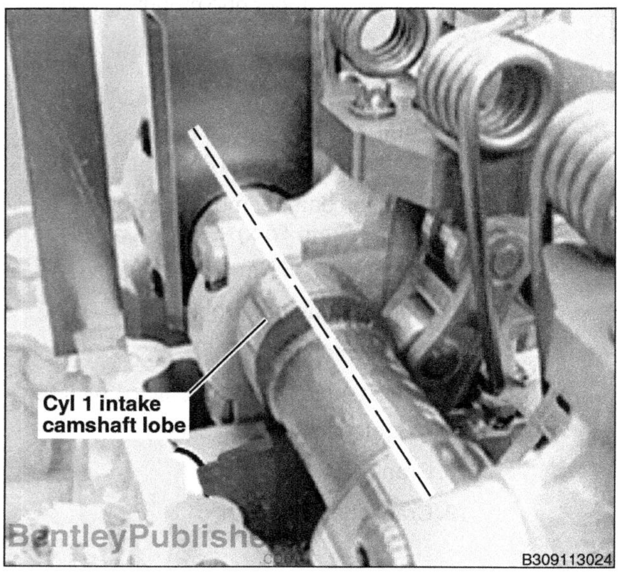

◄ N52 engine: Using vibration damper bolt, rotate crankshaft to place cylinder 1 in TDC firing position. In this position, cylinder 1 intake camshaft lobe points upward at an angle (**dashed line**).

BentleyPublishers.com
BentleyPublishers.com—All Rights Reserved

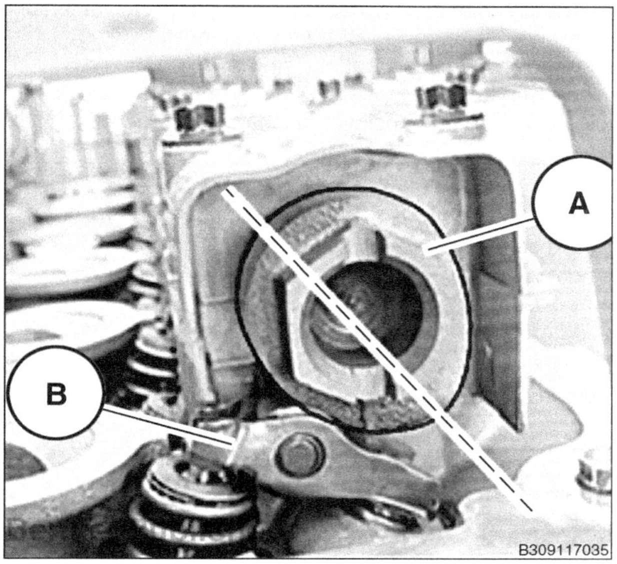

N52 engine: Cylinder 1 in TDC firing position: Cylinder 6 exhaust camshaft lobe (**A**) points downward at an angle (**dashed line**). Cam follower (**B**) is not actuated.

NOTE —

- *Use a mirror to check exhaust camshaft lobe position.*

N54 engine: Using vibration damper bolt to rotate crankshaft, place cylinder 1 in TDC firing position. In this position, lobes for cylinder 6 (rear of engine) intake and exhaust camshafts (**A**, **B**) point away from each other and down. Cylinder 6 camshaft lobes do not contact roller cam followers.

NOTE —

- *Use a mirror to check cylinder 6 cam lobe positions.*

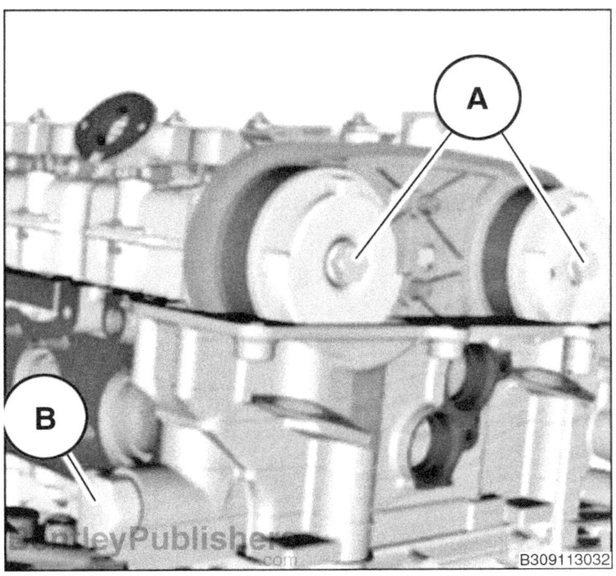

Working at right front of cylinder head, remove timing chain tensioner (**B**). Be prepared to catch dripping oil with a shop towel. Do not allow oil to contaminate accessory belt.

– Raise vehicle and support safely.

WARNING —

- *Make sure the vehicle is stable and well supported at all times. Use a professional automotive lift or jack stands designed for the purpose. A floor jack is not adequate support.*

– Remove engine compartment splash shield. See **020 Maintenance**.

– Remove engine accessory belt. See **020 Maintenance**.

– Remove vibration damper (front pulley). See **Vibration damper, removing and installing (N52, N54 engine)** in this repair group.

– Remove crankshaft front seal. See **119 Lubrication System**.

BentleyPublishers.com
BentleyPublishers.com—All Rights Reserved

Camshaft timing chain, removing and installing (N52, N54 engine)

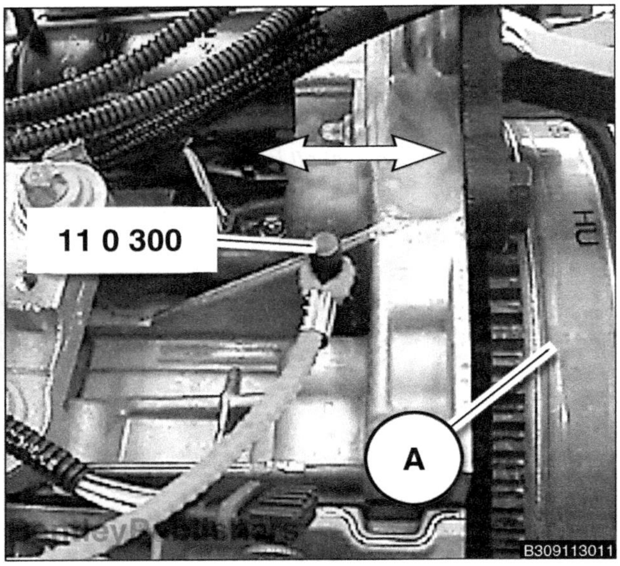

Working underneath engine, slide out protective plug at lower left of engine bell housing flange. Install BMW special tool 11 0 300 (crankshaft locking tool) through bell housing flange port into flywheel (**A**) hole. If necessary, rock flywheel slightly back and forth to line up holes with tool. This locks crankshaft at TDC.

> ***CAUTION—***
> - *Models with automatic transmission: In the flywheel (torque plate), a short distance before the bore for TDC position, there is a large bore which can be confused with the special tool bore. If the flywheel is locked using the correct bore, the engine can no longer be rotated at the vibration damper bolt.*
> - *If special tool 11 0 300 does not slide easily into the bell housing flange bore, sand paint off the tool. Do not enlarge the bore.*

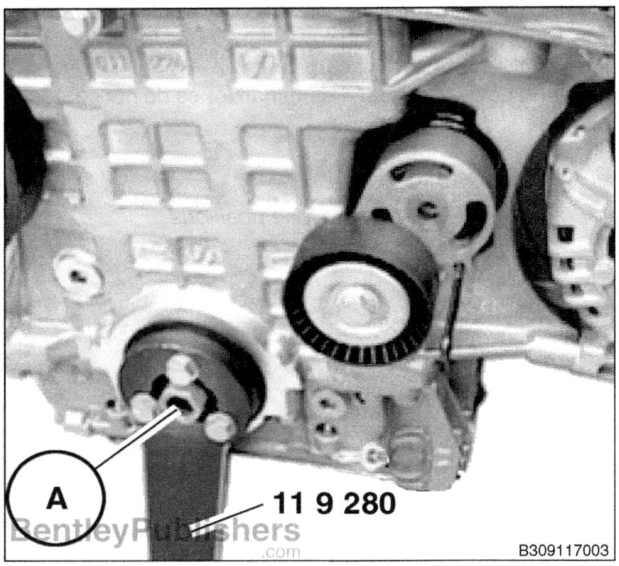

Remove vibration damper hub:

- Bolt BMW special tool 11 9 280 (counter hold tool) to hub.
- With help of assistant counter holding hub, loosen and remove hub bolt (**A**). Discard bolt.
- Pull hub off crankshaft, sliding it out of timing chain lower sprocket.

> ***CAUTION—***
> - *Counter hold hub securely. Do not rely on flywheel lock (special tool 11 0 300) to counter hold against crankshaft rotation.*

Unscrew plug (**arrow**) at upper left of timing chain housing.

BentleyPublishers.com
BentleyPublishers.com—All Rights Reserved

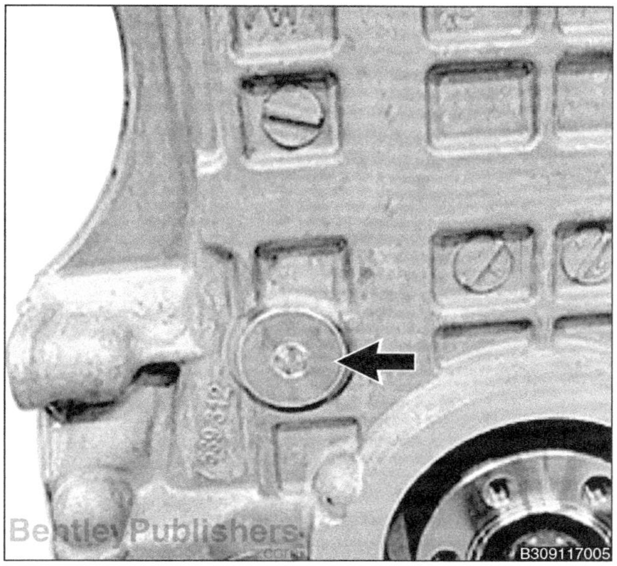

Unscrew plug (**arrow**) at lower right of timing chain housing.

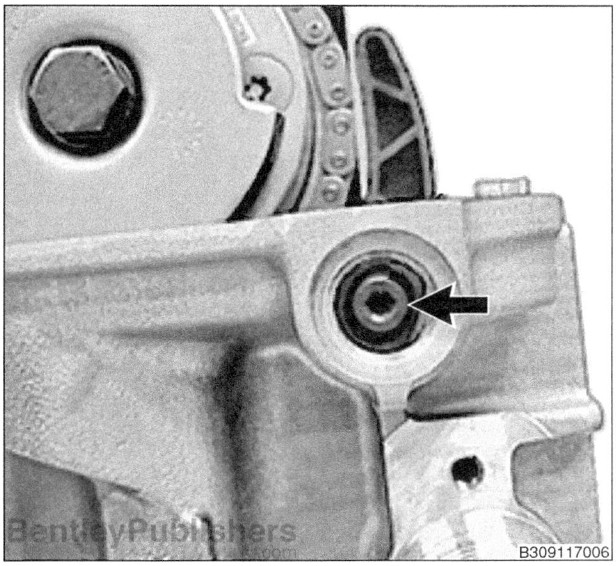

Remove chain guide mounting bolt (**arrow**) at upper left of timing chain housing.

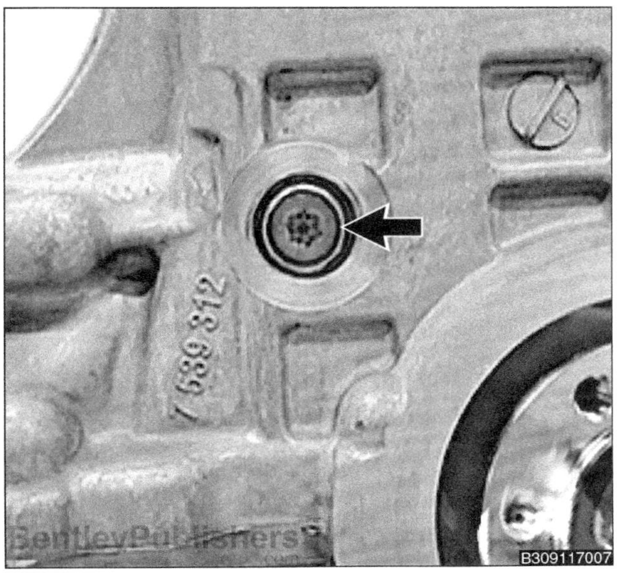

Remove chain tensioning rail mounting bolt (**arrow**) at lower right of timing chain housing.

BentleyPublishers.com
BentleyPublishers.com—All Rights Reserved

Camshaft timing chain, removing and installing (N52, N54 engine)

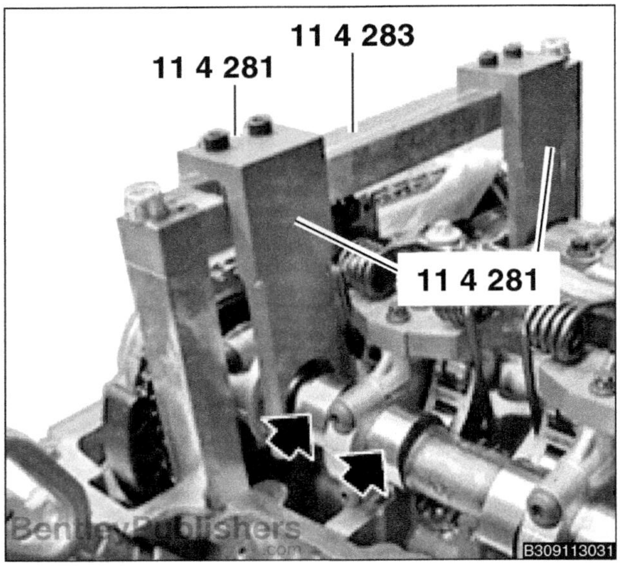

Use BMW special tool set 11 4 280 to lock down camshafts.

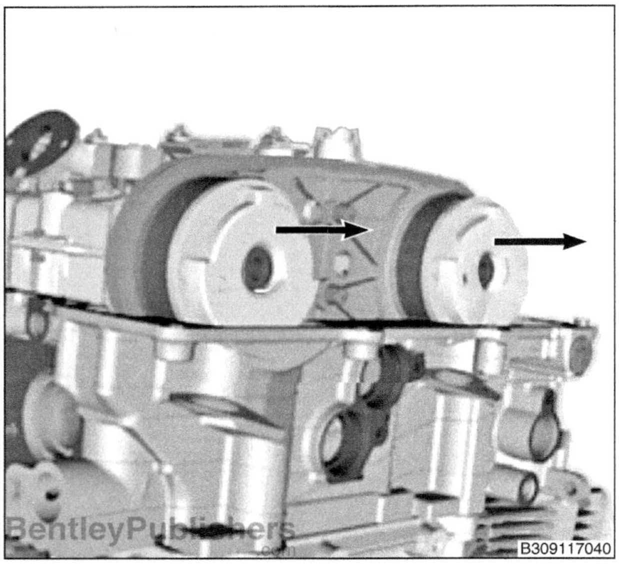

Remove intake and exhaust VANOS units:

- Loosen and remove VANOS unit mounting bolts. Discard bolts.
- Disengage timing chain from VANOS unit sprockets.
- Remove VANOS units (**arrows**) and set aside. See **VANOS units, removing and installing (N52, N54 engine)** in this repair group.

NOTE —

- *Illustration does not show special tool set 11 4 280 in place.*

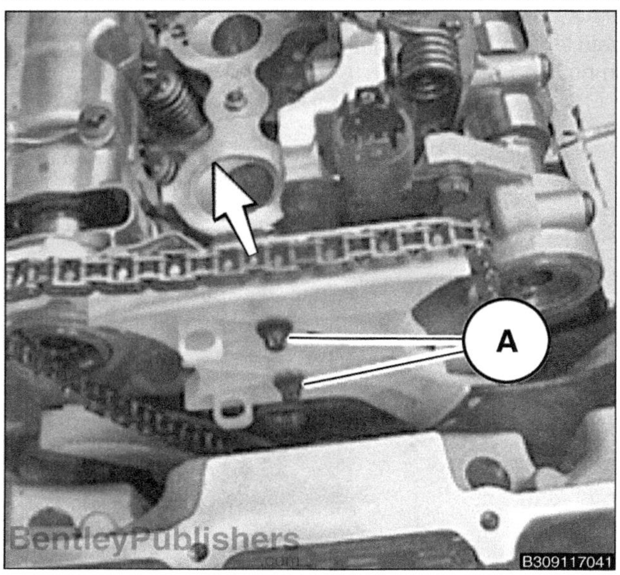

Working at front of cylinder head:

- Remove timing chain assembly mounting bolts (**A**).
- Lift timing chain, timing chain assembly and crankshaft sprocket straight up (**arrow**) out of timing chain housing.

BentleyPublishers.com
BentleyPublishers.com—All Rights Reserved

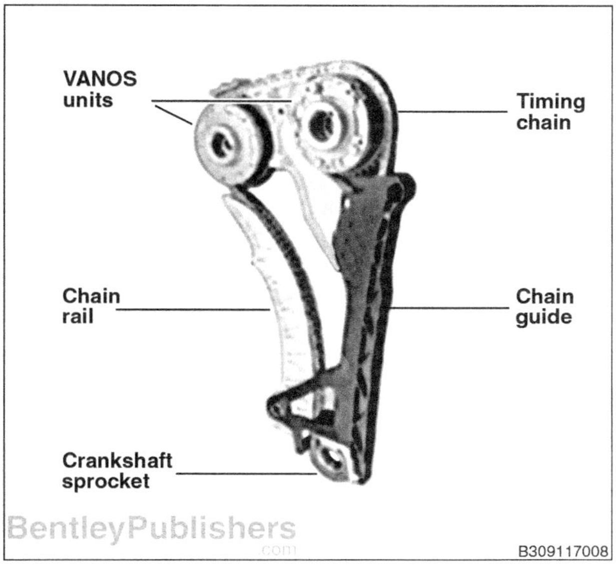

◀ Chain, chain assembly and crankshaft sprocket. VANOS units shown installed on assembly for purposes of illustration.

– Disengage timing chain from guide and rail assembly and crankshaft sprocket and fit new chain.

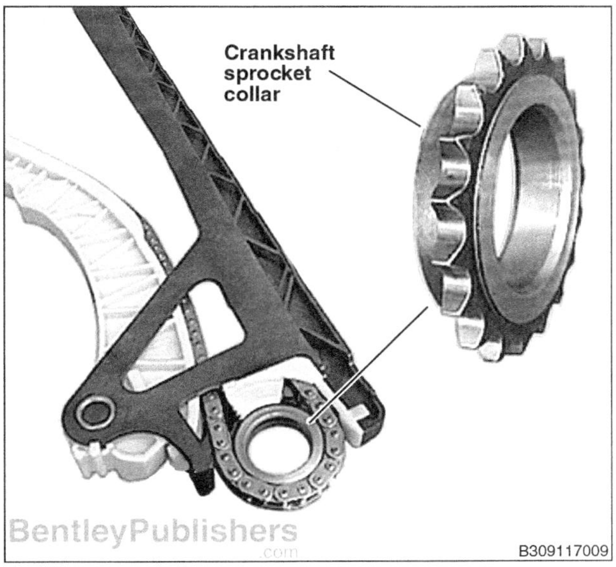

◀ Fit crankshaft sprocket to chain in orientation shown, with collar pointing to engine.

> ***CAUTION—***
> • *Incorrect sprocket assembly results in engine damage.*

◀ N54 engine: Note that friction plates (**arrows**) are fitted between timing chain sprocket and oil pump drive sprocket.

> ***CAUTION—***
> • *If N54 engine is operated without friction plates, engine damage results.*

BentleyPublishers.com
BentleyPublishers.com—All Rights Reserved

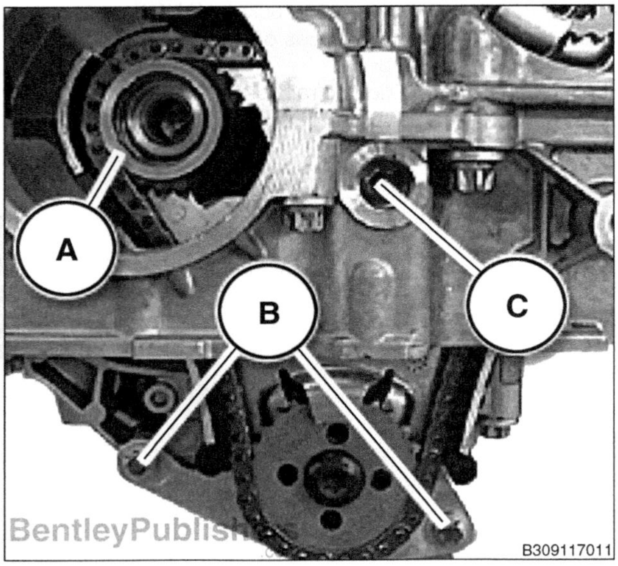

N54 engine: Make sure oil pump drive sprocket friction plate (**A**) is in correct position.

NOTE—

- ***B** and **C** are oil pump and oil pump chain guide mounting bolts.*

– Pull timing chain taut until sprocket engages chain guide. Install timing chain and chain assembly in chain housing in this position, holding chain under tension at all times.

– Install and torque timing chain assembly, rail and guide bolts.

Tightening torques	
Timing chain guide to cylinder head (M7)	14 Nm (10 ft-lb)
Timing chain assembly to cylinder head (M6 x 35 mm Torx)	8.5 Nm (6 ft-lb)
Timing chain rail to crankcase (M8)	20 Nm (14 ft-lb)

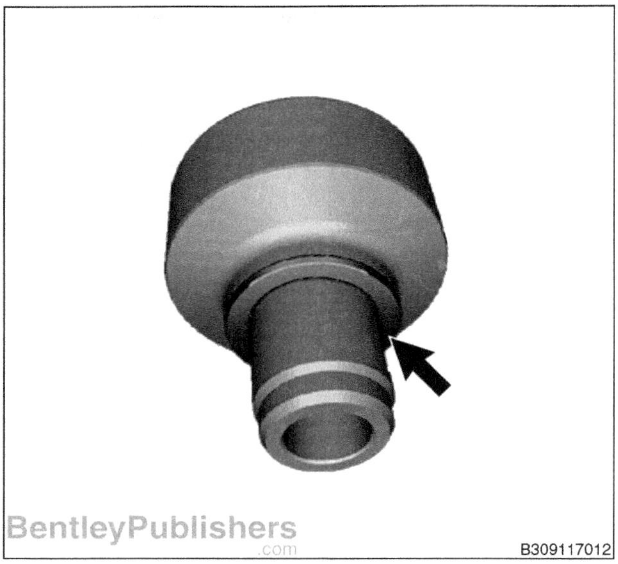

N54 engine: Slide friction plate (**arrow**) on vibration damper hub shaft prior to inserting hub in crankcase bore.

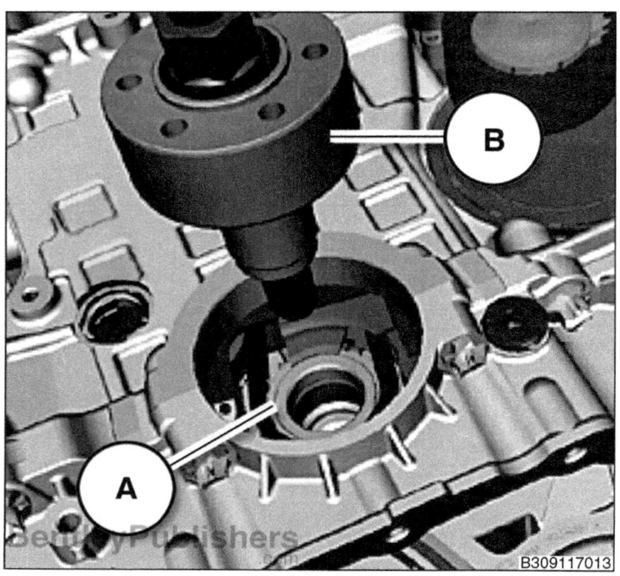

Line up timing chain and oil pump chain sprockets (**A**) inside crankcase bore, then insert vibration damper hub (**B**), sliding it through chain sprockets. Using new vibration damper bolt, attach hub finger tight to crankshaft.

– To achieve final torque on vibration damper hub, follow one of several alternative procedure steps for preventing crankshaft rotation recommended by BMW.

CAUTION—

- *Do not rely on flywheel lock (special tool 11 0 300) to counter hold against crankshaft rotation.*

BentleyPublishers.com
BentleyPublishers.com—All Rights Reserved

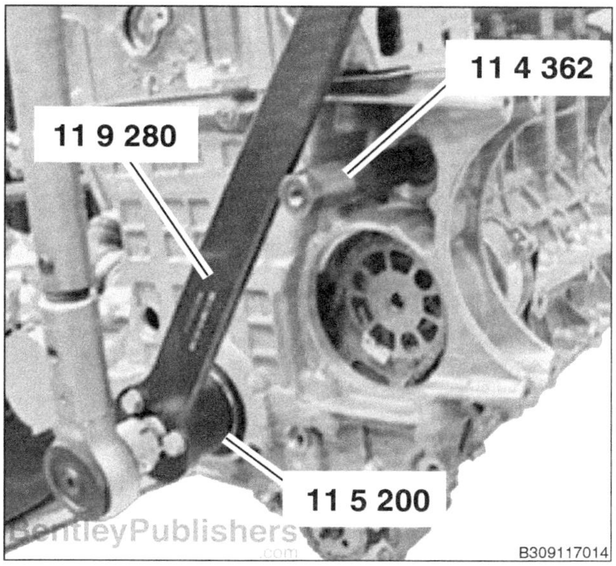

Remove belt tensioner and screw in special tool 11 4 362.

- Bolt special tool 11 5 200 to vibration damper hub.
- Bolt special tool 11 9 280 to tool 11 5 200. Rest tool 11 9 280 against tool 11 4 362.
- Tighten to initial tightening torque.

Tightening torque	
Vibration damper hub to crankshaft • Stage 1	100 Nm (74 ft-lb)

117

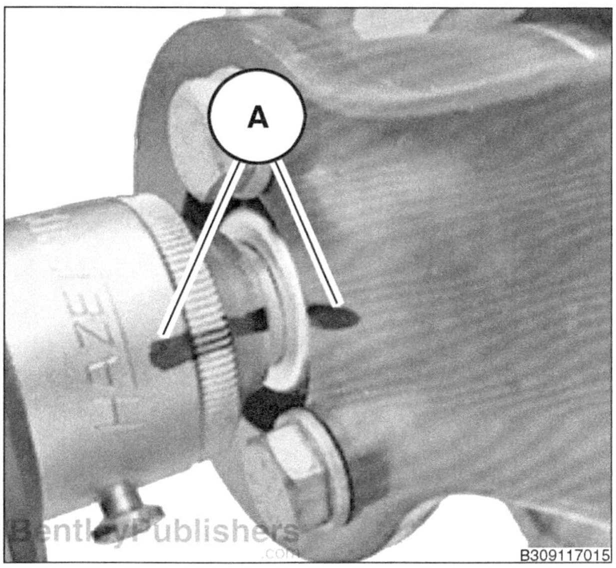

Mark special tool 11 9 280 and socket wrench with paint (**A**).

- Tighten hub bolt to final torque with assistant helping to counter hold tool 11 9 280.

Tightening torque	
Vibration damper hub to crankshaft • Stage 2	Additional 360°

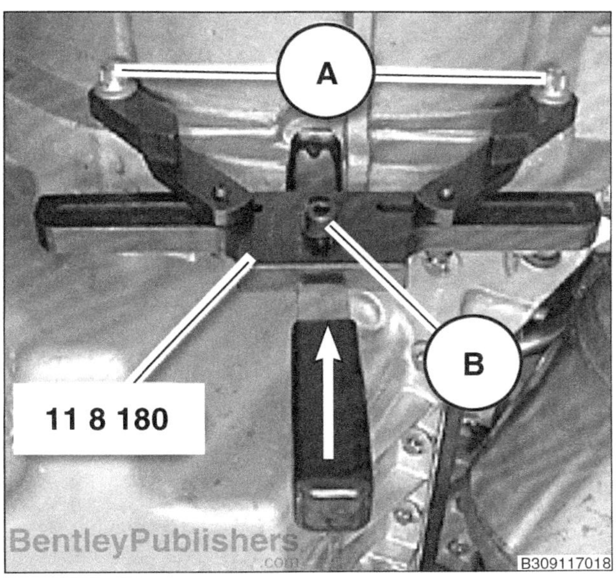

Alternate method for manual transmission vehicle:

- Mount special tool 11 8 180 on transmission using bolts **A**.
- Slide lock handle up (**arrow**) and tighten bolt **B**.

BentleyPublishers
BentleyPublishers.com—All Rights Reserved

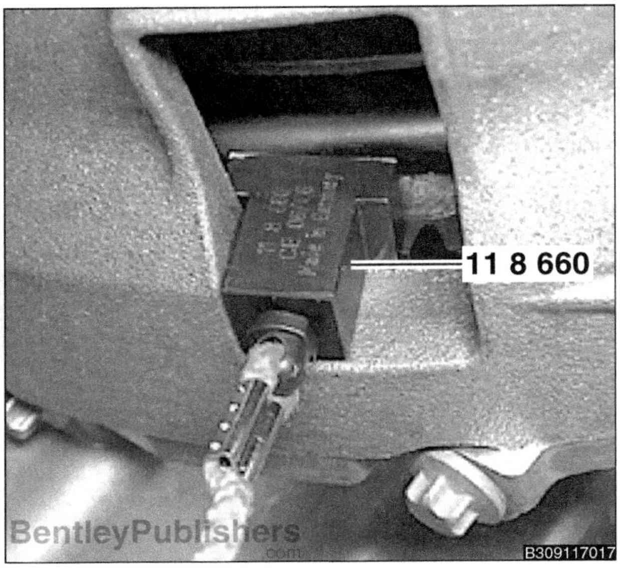

Alternate method for automatic transmission vehicle:

- Secure flywheel with special tool 11 8 660 through transmission bell housing opening.

Alternate method if transmission is removed:

- Secure flywheel with special tool 11 9 260.

Use angle protractor (BMW special tool 00 9 140 or equivalent) to torque hub bolt.

Tightening torque	
Vibration damper hub to crankshaft • Stage 1 • Stage 2	 100 Nm (74 ft-lb) Additional 360°

– Install crankshaft front seal. See **119 Lubrication System**.

BentleyPublishers.com
BentleyPublishers.com—All Rights Reserved

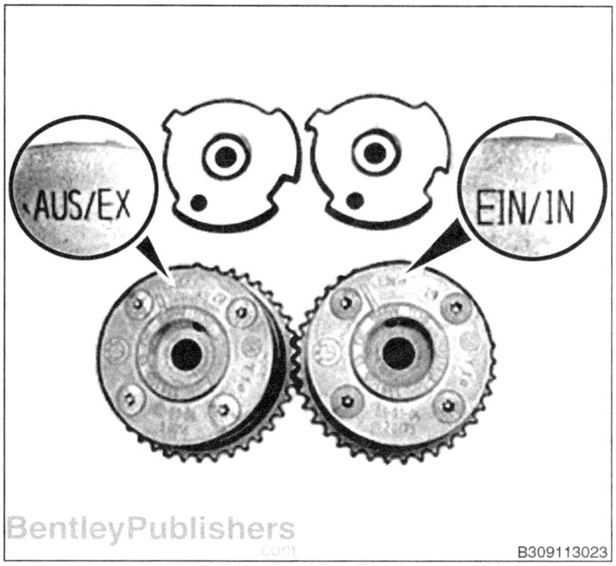

Use new bolts to reattach VANOS units. Note that exhaust and intake units are different and so marked. Camshaft sensor impulse wheel for intake and exhaust are the same. See **VANOS units, removing and installing (N52, N54 engine)** in this repair group.

Tightening torque	
VANOS unit to camshaft (use new M10 x 94 mm bolt): • Stage 1 • Stage 2	20 Nm (15 ft-lb) additional 180°

– Install chain tensioner. See **Timing chain tensioner (N52, N54 engine)** in this repair group.

> ***CAUTION—***
> - *If chain tensioner is reused, drain its oil chamber. Place tensioner on level working surface and compress slowly. Repeat twice.*
> - *No sealing ring is fitted to chain tensioner at the factory. When reassembling engine, be sure to use new sealing O-ring.*

Tightening torque	
Timing chain tensioner to cylinder head (M22 x 1.5) (use new sealing O-ring)	55 Nm (41 ft-lb)

– Remove crankshaft and camshaft locking tools and rotate engine 2 full revolutions. Then check and, if necessary, reset camshaft timing. See **Camshaft timing, adjusting (N52, N54 engine)** in this repair group.

> ***CAUTION—***
> - *Remove crankshaft and camshaft locking tools before rotating or starting engine.*

– Reassemble engine.

Camshaft timing, adjusting (N52, N54 engine)

> ***WARNING—***
> - *To avoid personal injury, be sure the engine is cold before beginning the procedure.*

– Use scan tool to read out and record ECM fault memory.

– Switch ignition OFF and remove key.

– Remove upper and lower cabin microfilter housings. See **640 Heating and Air-conditioning**.

– Remove strut tower brace.

– Remove ignition coil cover (upper engine cover). See **020 Maintenance**.

– Detach ignition coil connectors. Lift ignition coil harness off cylinder head cover and set aside.

BentleyPublishers.com
BentleyPublishers.com—All Rights Reserved

– Remove ignition coils. See **120 Ignition System**.

– Remove cylinder head cover. See **113 Cylinder Head Removal and Installation**.

– Raise vehicle and support safely.

> ***WARNING —***
> - *Make sure the vehicle is stable and well supported at all times. Use a professional automotive lift or jack stands designed for the purpose. A floor jack is not adequate support.*

– Remove engine compartment splash shield. See **020 Maintenance**.

◄ N52 engine: Using vibration damper bolt, rotate crankshaft to place cylinder 1 in TDC firing position. In this position, cylinder 1 intake camshaft lobe points upward at an angle (**dashed line**).

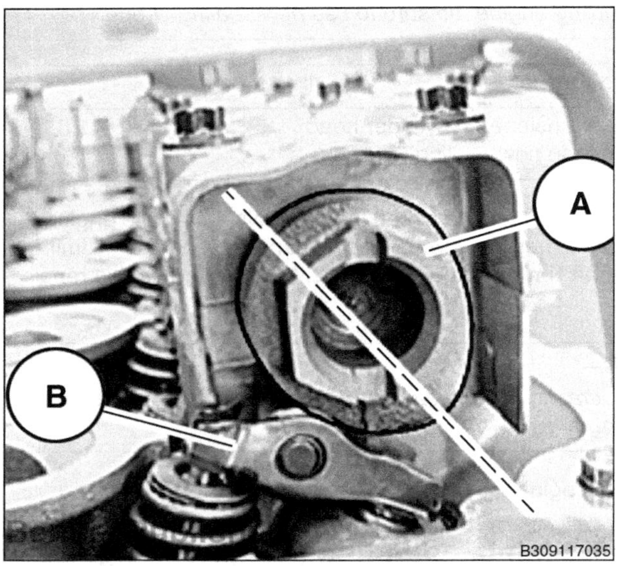

◄ N52 engine: Cylinder 1 in TDC firing position: Cylinder 6 exhaust camshaft lobe (**A**) points downward at an angle (**dashed line**). Cam follower (**B**) is not actuated.

NOTE —
- *Use a mirror to check exhaust camshaft lobe position.*

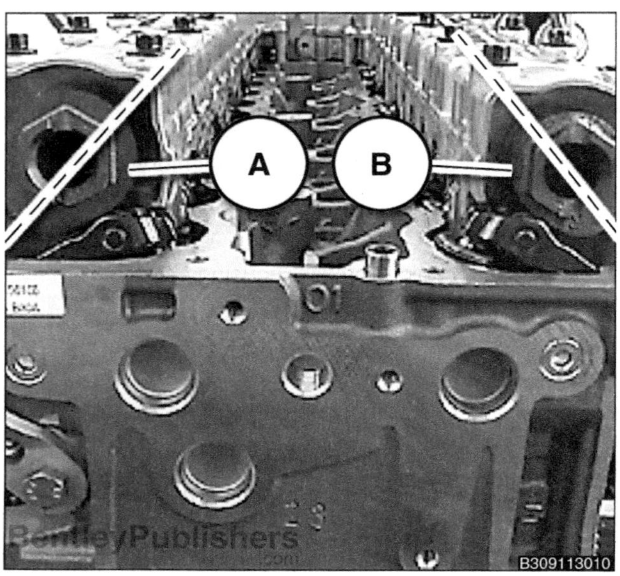

◄ N54 engine: Using vibration damper bolt to rotate crankshaft, place cylinder 1 in TDC firing position. In this position, lobes for cylinder 6 (rear of engine) intake and exhaust camshafts (**A**, **B**) point away from each other and down. Cylinder 6 camshaft lobes do not contact roller cam followers.

NOTE —
- *Use a mirror to check cylinder 6 cam lobe positions.*

BentleyPublishers.com
BentleyPublishers.com—All Rights Reserved

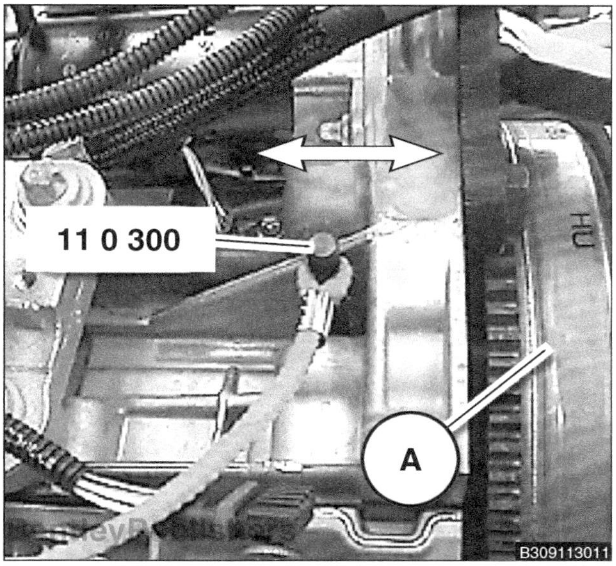

Working underneath engine, slide out protective plug at lower left of engine bell housing flange. Install BMW special tool 11 0 300 (crankshaft locking tool) through bell housing flange port into flywheel (**A**) hole. If necessary, rock flywheel slightly back and forth to line up holes with tool. This locks crankshaft at TDC.

> ***CAUTION—***
> - *Models with automatic transmission: In the flywheel (torque plate), a short distance before the bore for TDC position, there is a large bore which can be confused with the special tool bore. If the flywheel is locked using the correct bore, the engine can no longer be rotated at the vibration damper bolt.*
> - *If special tool 11 0 300 does not slide easily into the bell housing flange bore, sand paint off the tool. Do not enlarge the bore.*

117

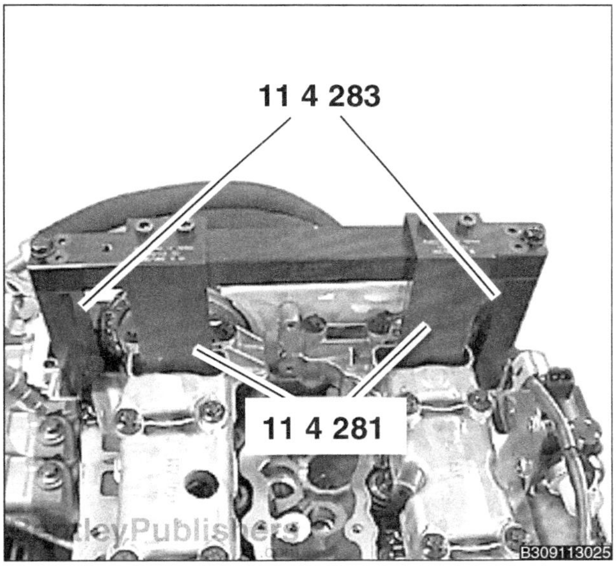

Use BMW special tool set 11 4 280 to lock down camshafts.

– With crankshaft locked, if special tool 11 4 281 fits, camshaft timing is correct. Remove special tools and reassemble engine.

– If special tool 11 4 281 does not fit, use hexagon casting at rear of camshafts to rotate camshafts until tool fits. Proceed with camshaft timing procedure below.

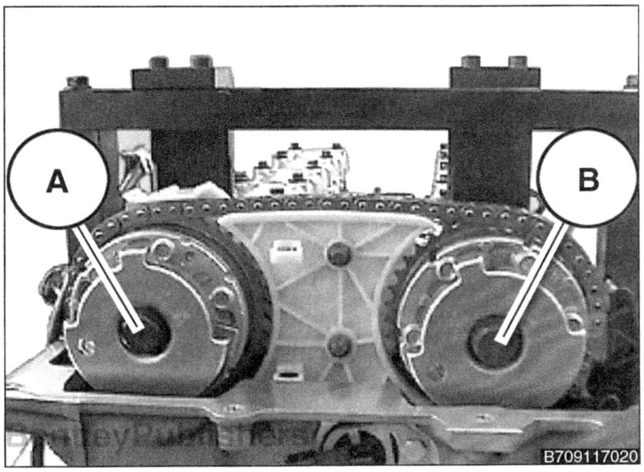

Working at front of camshafts:

- Loosen exhaust and intake camshaft adjustment (VANOS) unit mounting bolts (**A**, **B**).
- Discard bolts and replace with new.
- Tighten bolts finger tight.

BentleyPublishers.com—All Rights Reserved

Install BMW special tool 11 8 520.

- Place tool up against VANOS unit sensor gears.
- Rotate sensor gears (**arrows**) until locating pins on tool line up with bores in sensor gears.
- Push tool firmly against sensor gears and lock down with M6 x 45 mm bolts (**A**).

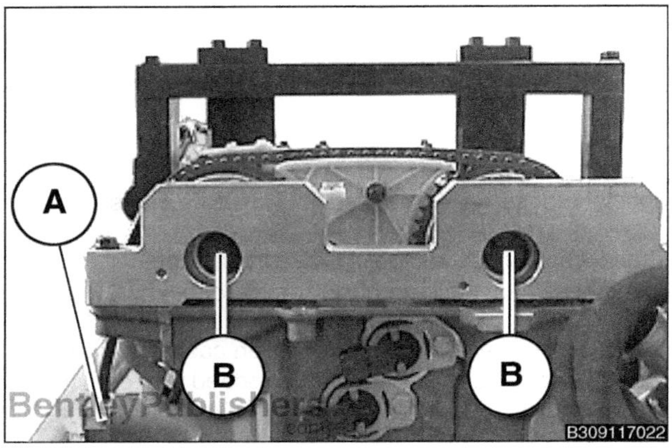

Working at front of cylinder head:

- Remove timing chain tensioner (**A**). Be prepared to catch dripping oil. Do not allow oil to contaminate accessory belt.
- Screw BMW special tool 11 9 340 (dummy chain tensioner) into cylinder head.
- Using BMW special tool 00 9 250 (low torque torque-wrench), rotate stud of special tool 11 9 340 to pretension timing chain to 0.6 Nm (5.3 in-lb).
- Tighten VANOS unit mounting bolts (**B**).

Tightening torque	
VANOS unit to camshaft: • Stage 1 • Stage 2	 20 Nm (15 ft-lb) additional 180°

– Remove special tools and assemble engine. Reinstall chain tensioner using new sealing O-ring. See **Timing chain tensioner (N52, N54 engine)** in this repair group.

Tightening torque	
Timing chain tensioner to cylinder head (M22 x 1.5) (use new sealing O-ring)	55 Nm (41 ft-lb)

Timing chain tensioner (N52, N54 engine)

The timing chain tensioner is screwed into the right front of the cylinder head. When it is removed be sure to catch any dripping oil in a shop towel. Do not allow oil to contaminate accessory belt.

If chain tensioner is reused, drain its oil chamber. Place tensioner vertically on level working surface and compress slowly (**arrow**). Repeat twice.

> ***CAUTION—***
> - *No sealing ring is fitted to chain tensioner at the factory. When reassembling engine, be sure to use new sealing O-ring.*

Tightening torque	
Timing chain tensioner to cylinder head (M22 x 1.5) (use new sealing O-ring)	55 Nm (41 ft-lb)

BentleyPublishers.com
BentleyPublishers.com—All Rights Reserved

VANOS UNITS (N52, N54 ENGINE)

VANOS units, removing and installing (N52, N54 engine)

Special tools and procedures are required to remove and install VANOS units and to time camshafts. Read the entire procedure before beginning repairs.

WARNING—
- *To avoid personal injury, be sure the engine is cold before beginning the procedure.*

CAUTION—
- *Disassembly, removal and assembly of VANOS units without special tools poses the risk of damage or breakage: Valves may be bent by contact with the piston crowns.*
- *Throughout this procedure, unless otherwise specified, crankshaft and camshafts remain locked against rotation using BMW special tools.*
- *Keep VANOS components free of silicone sealant. A small amount of silicone introduced into VANOS actuator or solenoid can set faults.*

— Use scan tool to read out and record ECM fault memory.

— Switch ignition OFF and remove key.

— Remove upper and lower cabin microfilter housings. See **640 Heating and Air-conditioning**.

— Remove strut tower brace.

— Remove ignition coil cover (upper engine cover). See **020 Maintenance**.

— Detach ignition coil connectors. Lift ignition coil harness off cylinder head cover and set aside.

— Remove ignition coils. See **120 Ignition System**.

— Remove cylinder head cover. See **113 Cylinder Head Removal and Installation**.

— Raise vehicle and support safely.

WARNING—
- *Make sure the vehicle is stable and well supported at all times. Use a professional automotive lift or jack stands designed for the purpose. A floor jack is not adequate support.*

— Remove engine compartment splash shield. See **020 Maintenance**.

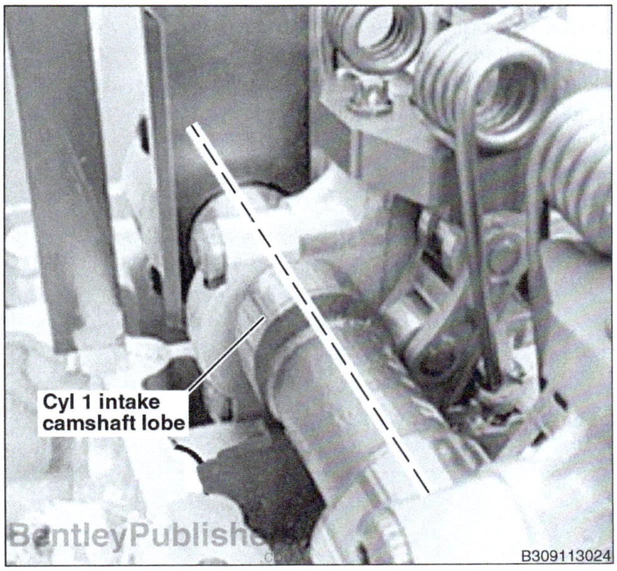

◄ N52 engine: Using vibration damper bolt, rotate crankshaft to place cylinder 1 in TDC firing position. In this position, cylinder 1 intake camshaft lobe points upward at an angle.

BentleyPublishers.com
BentleyPublishers.com—All Rights Reserved

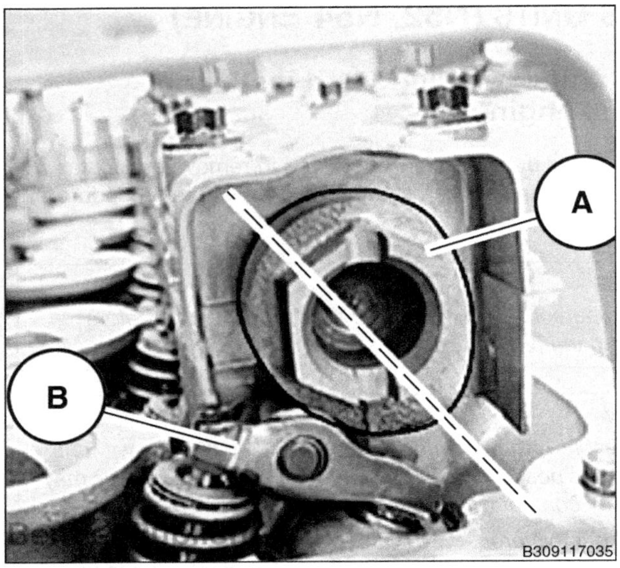

N52 engine: Cylinder 1 in TDC firing position: Cylinder 6 exhaust camshaft lobe (**A**) points downward at an angle (**dashed line**). Cam follower (**B**) is not actuated.

NOTE—
- *Use a mirror to check exhaust camshaft lobe position.*

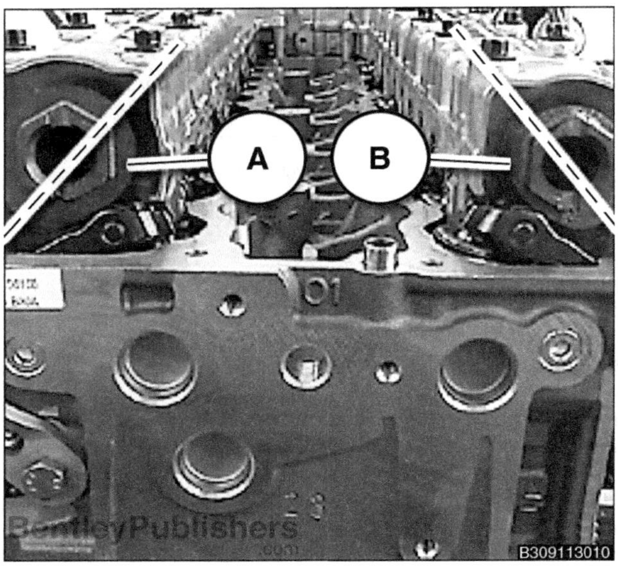

N54 engine: Using vibration damper bolt to rotate crankshaft, place cylinder 1 in TDC firing position. In this position, lobes for cylinder 6 (rear of engine) intake and exhaust camshafts (**A**, **B**) point away from each other and down. Cylinder 6 camshaft lobes do not contact roller cam followers.

NOTE—
- *Use a mirror to check cylinder 6 cam lobe positions.*

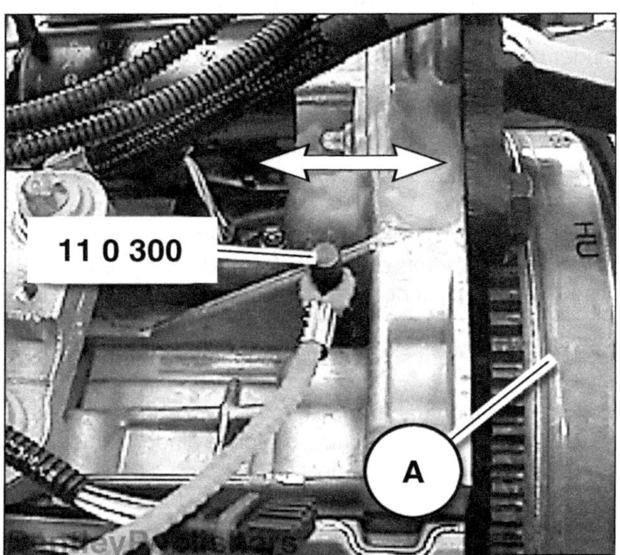

Working underneath engine, slide out protective plug at lower left of engine bell housing flange. Install BMW special tool 11 0 300 (crankshaft locking tool) through bell housing flange port into flywheel (**A**) hole. If necessary, rock flywheel slightly back and forth to line up holes with tool. This locks crankshaft at TDC.

CAUTION—
- *Models with automatic transmission: In the flywheel (torque plate), a short distance before the bore for TDC position, there is a large bore which can be confused with the special tool bore. If the flywheel is locked using the correct bore, the engine can no longer be rotated at the vibration damper bolt.*
- *If special tool 11 0 300 does not slide easily into the bell housing flange bore, sand paint off the tool. Do not enlarge the bore.*

BentleyPublishers.com
BentleyPublishers.com—All Rights Reserved

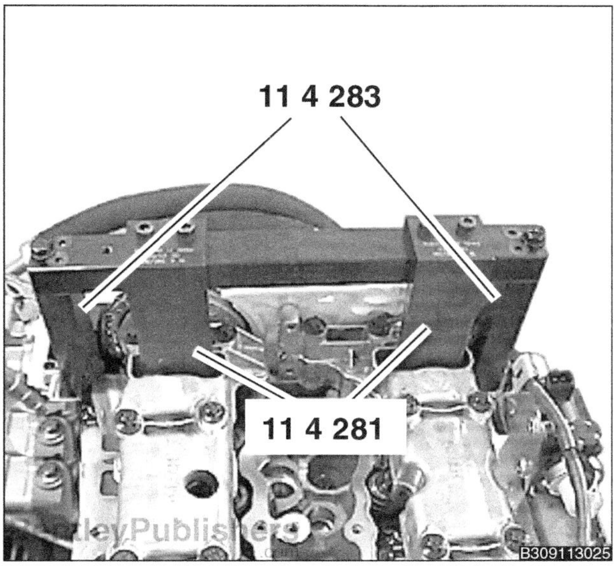

Use BMW special tool set 11 4 280 to lock down camshafts.

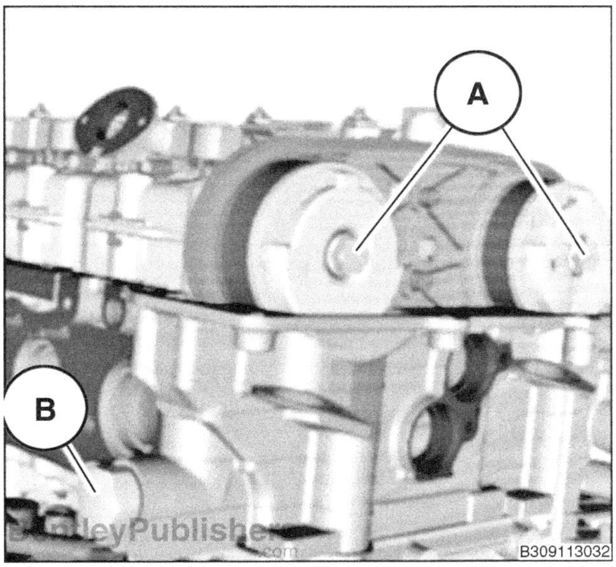

Working at right front of cylinder head, remove timing chain tensioner (**B**). Be prepared to catch dripping oil with a shop towel. Do not allow oil to contaminate accessory belt.

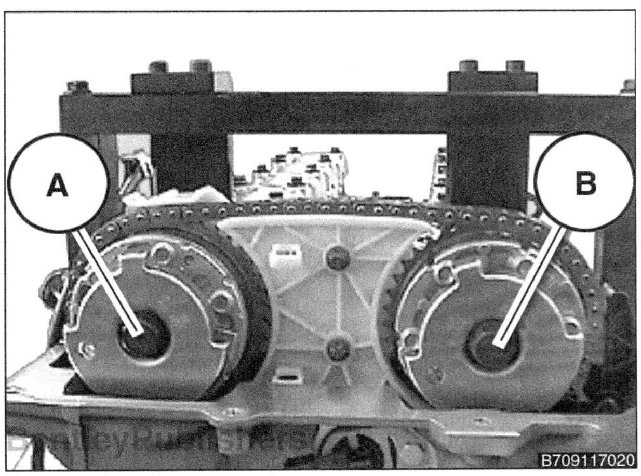

Working at front of camshafts:

- Loosen and remove exhaust and intake VANOS unit mounting bolts (**A**, **B**). Discard bolts.
- Disengage timing chain from camshaft adjustment unit sprockets.
- Tilt VANOS units down to remove.

BentleyPublishers.com
BentleyPublishers.com—All Rights Reserved

BentleyPublishers.com B309113023

Use new bolts to reattach VANOS units.

- Note that exhaust and intake units are different and so marked.
- Camshaft sensor impulse wheel for intake and exhaust are the same.
- VANOS units and sensor wheels may be installed in any position initially. Finger-tighten mounting bolts for now.

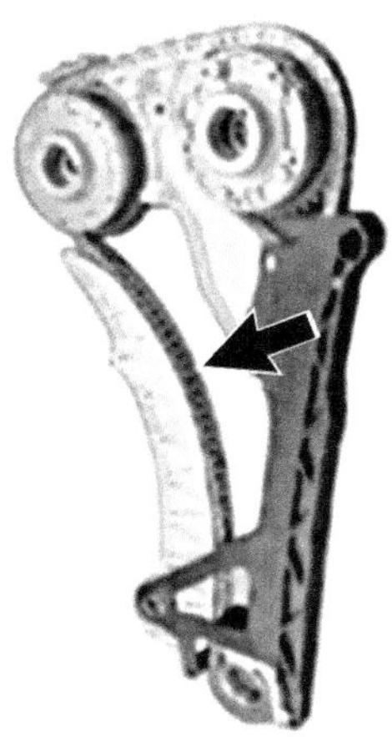

BentleyPublishers.com B309117039

Press chain into chain tensioner rail (**arrow**) by hand to make sure it is routed correctly. (Chain assembly and VANOS units shown removed from engine for purpose of illustration.)

– Adjust camshaft timing. See **Camshaft timing, adjusting (N52, N54 engine)** in this repair group. Then torque VANOS mounting bolts.

Tightening torque	
VANOS unit to camshaft (use new M10 x 94 mm bolt): • Stage 1 • Stage 2	 20 Nm (15 ft-lb) additional 180°

– Install chain tensioner. See **Timing chain tensioner (N52, N54 engine)** in this repair group.

CAUTION—
- *If chain tensioner is reused, drain its oil chamber. Place tensioner on level working surface and compress slowly. Repeat twice.*
- *No sealing ring is fitted to chain tensioner at the factory. When reassembling engine, be sure to use new sealing O-ring.*

Tightening torque	
Timing chain tensioner to cylinder head (M22 x 1.5) (use new sealing O-ring)	55 Nm (41 ft-lb)

– Reassemble engine.

CAUTION—
- *Remove crankshaft and camshaft locking tools before rotating or starting engine.*

BentleyPublishers.com
BentleyPublishers.com—All Rights Reserved

Timing Chain Service (V8 Engine)

Timing chain components (V8 engine)

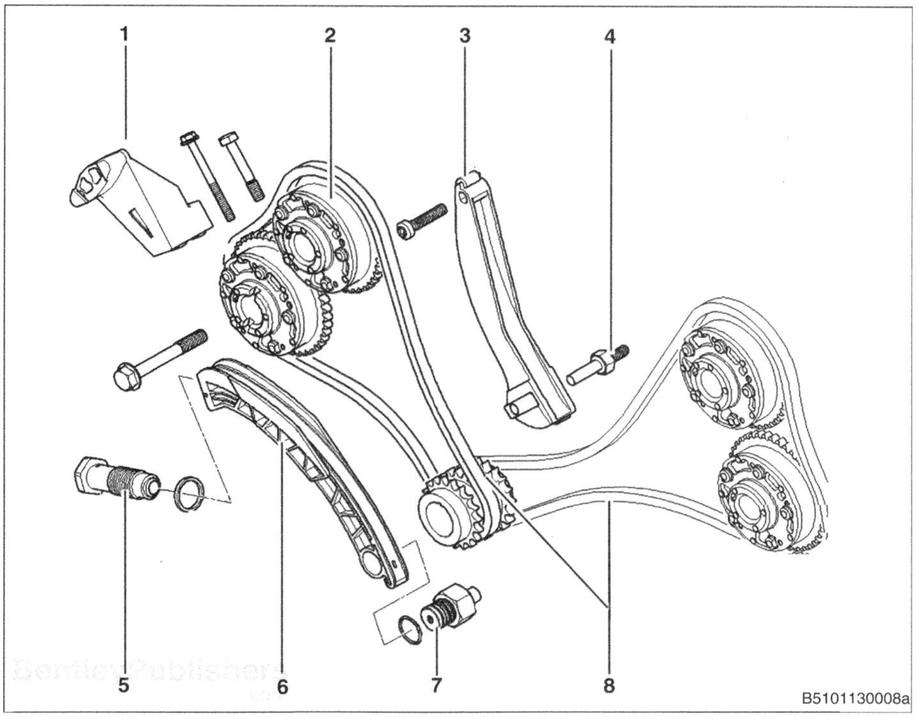

1. Timing chain rail
2. Camshaft adjuster (VANOS) unit
3. Timing chain guide
4. Support stud
5. Timing chain tensioner
6. Timing chain guide
7. Timing chain guide bearing
8. Timing chain

Right camshaft timing, adjusting (V8 engine)

– Remove right side valvetronic motor. **See 133 Fuel Injection**.

– Remove spark plugs. Plug spark plug bores to prevent debris from entering engine.

– Remove right side cylinder head cover. See **113 Cylinder Head Removal and Installation**.

– Remove engine cooling fan and cooling fan shroud. See **170 Radiator and Cooling System**.

– Remove air filter housing and ducts. See **133 Fuel Injection**.

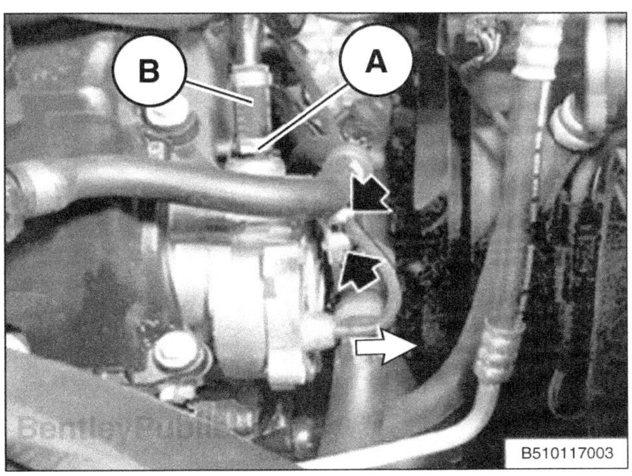

◄ Working at front of right cylinder head:
- Remove clamp (**A**) and remove line (**B**).
- Loosen hose clamps (**black arrows**).
- Detach vaccum line (**white arrow**).

BentleyPublishers.com
BentleyPublishers.com—All Rights Reserved

Right camshaft timing, adjusting (V8 engine)

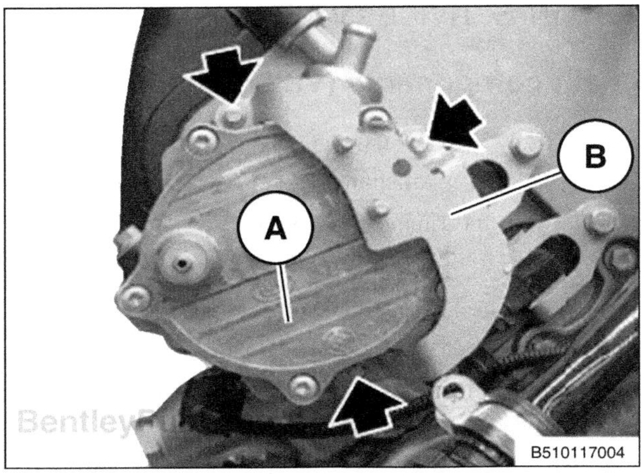

Remove vaccum pump fasteners (**arrows**) and vacuum pump (**A**) with bracket (**B**).

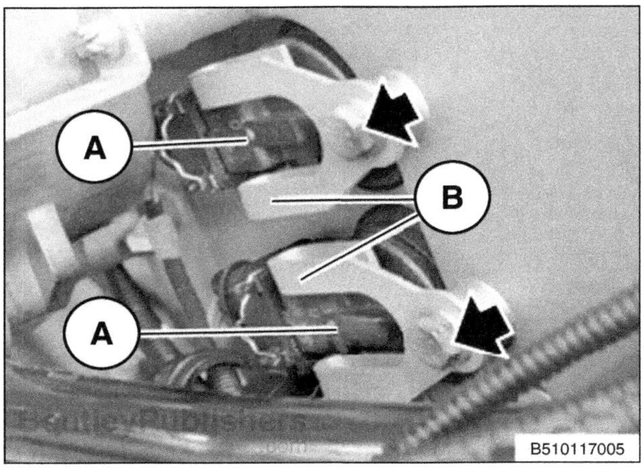

Disconnect electrical connectors from VANOS adjuster solenoids (**A**). Remove fasteners (**B**), brackest (**arrows**) and remove actuator solenoids (**A**). Discard actuator solenoids O-rings.

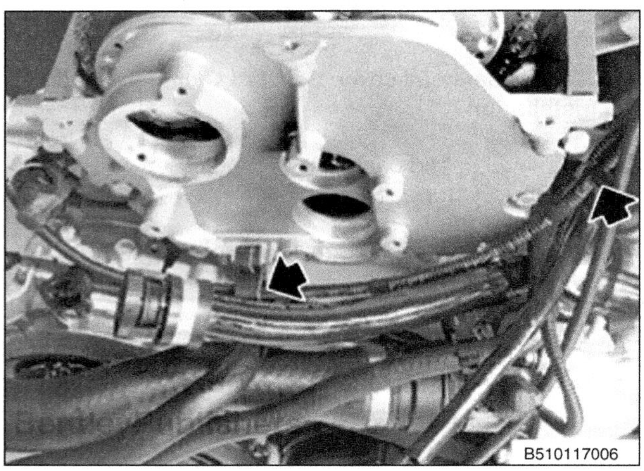

Remove vacuum lines from bracket (**arrow**).

BentleyPublishers.com
BentleyPublishers.com—All Rights Reserved

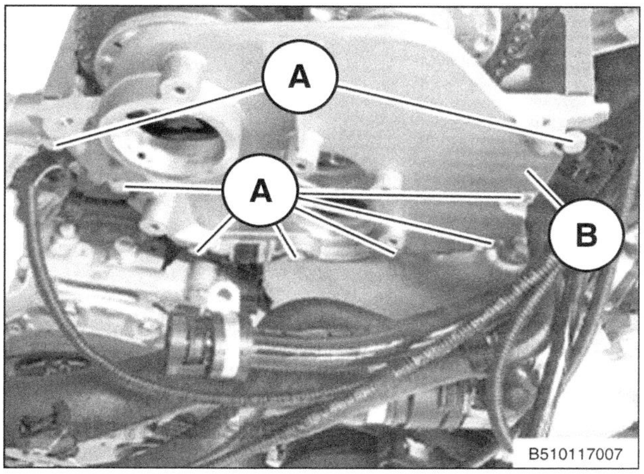

Remove upper timing chain cover fasteners (**A**) and remove upper cover (**B**).

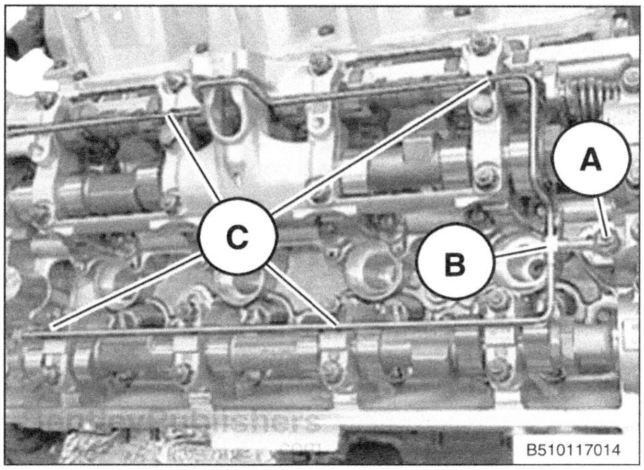

Remove banjo bolt (**A**). Unclip oil line (**B**) from retianers (**C**) and remove.

> ***CAUTION—***
> *• Do not misplace oil line banjo bolt.*

Using vibration damper center bolt, turn engine in direction of rotation to cylinder 1 TDC position.

NOTE—

- *Cylinder 1 exhaust lobe points upwards at an angle. Cylinder 1 intake lobe points downwards at an angle.*

BentleyPublishers.com—All Rights Reserved

Right camshaft timing, adjusting (V8 engine)

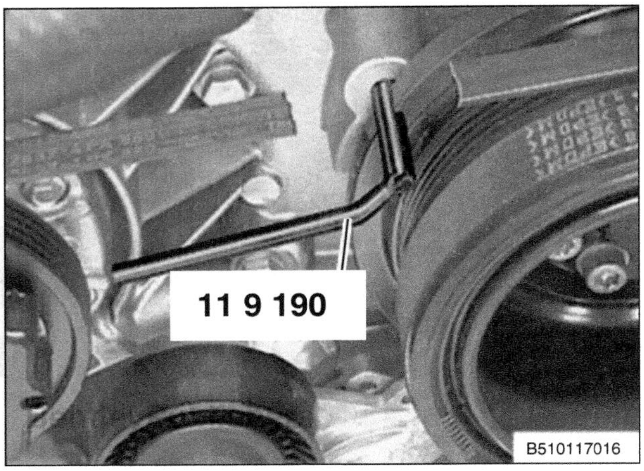

Using BMW special tool 11 9 190, lock engine in cylinder 1 TDC position.

To avoid incorrect timing adjustment, check VANOS adjustment unit locking using the following steps.

Grip hexagon (**black arrow**) of intake camshaft and attempt to rotate intake camshaft carefully in direction opposite (**white arrow**) to engine rotation. If there is no fixed connection between intake camshaft and intake VANOS unit, rotate intake camshaft in direction opposite to engine rotation as far as it will go.

Grip hexagon (**black arrow**) of exhaust camshaft and attempt to rotate exhaust camshaft carefully in the direction of engine rotation (**white arrow**). If there is no fixed connection between exhaust camshaft and exhaust VANOS unit, rotate exhaust camshaft in the direction of engine rotation as far as it will go.

BentleyPublishers.com
BentleyPublishers.com—All Rights Reserved

B510117008

Grip each camshaft at hexagon and loosen VANOS adjuster bolts (**arrows**). Do not remove VANOS adjuster bolts.

B510117009

Install BMW special tool 11 9 461 on intake camshaft so that it rests flush on cylinder head.

B510117010

Install BMW special tool 11 9 463, secure holder (**A**) in thread for oil line hand tight.

BentleyPublishers.com—All Rights Reserved

Right camshaft timing, adjusting (V8 engine)

Grip intake camshaft at hexagon and tighten intake VANOS adjuster bolt (**A**).

Tightening torque	
VANOS adjuster to camshaft (replace bolt)	80 Nm (59 ft-lb)

Remove holder (**A**) and BMW special tools 11 9 461, 11 9 463.

Install BMW special tool 11 9 462 on exhaust camshaft so that it rests flush on cylinder head.

BentleyPublishers.com
BentleyPublishers.com—All Rights Reserved

Install BMW special tool 11 9 463 and secure holder (**A**) in oil line thread hand tight.

Grip exhaust camshaft at hexagon and tighten exhaust VANOS adjuster bolt (**A**).

Tightening torque	
VANOS adjuster to camshaft (replace bolt)	80 Nm (59 ft-lb)

Remove holder (**A**) and BMW special tools 11 9 462, 11 9 463.

Right camshaft timing, adjusting (V8 engine)

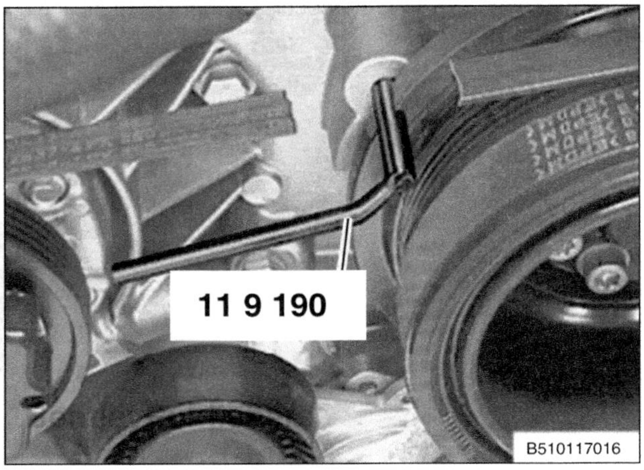

Remove BMW special tool 11 9 190. Using vibration damper center bolt turn engine 2 full rotations. Using BMW special tool 11 9 190 lock engine in cylinder 1 TDC position.

Place BMW special tool 11 9 461 on intake camshaft and check timing. Timing is correctly adjusted when BMW special tool 11 9 461 rests flat on cylinder head or protrudes by up to 0.5 mm to exhaust side.

NOTE—

- *Due to flexible sprocket design, VANOS unit tolerances and play in the VANOS units, when the camshaft timing is set correctly, the exhaust side of special tool 11 9 461 may be up to 0.5 mm (0.02 in) above the surface of the cylinder head. This is normal.*

– Remove special tool 11 9 461 from inlet camshaft.

Place BMW special tool 11 9 462 on exhaust camshaft and check timing. Timing is correctly adjusted when BMW special tool 11 9 461 rests flat on cylinder head or protrudes by up to 0.5 mm to exhaust side.

NOTE—

- *Due to flexible sprocket design, VANOS unit tolerances and play in the VANOS units, when the camshaft timing is set correctly, the exhaust side of special tool 11 9 462 may be up to 0.5 mm (0.02 in) above the surface of the cylinder head. This is normal.*

– Remove special tool 11 9 462 from exhaust camshaft.

– Reassemble engine.

– Remember to replace:
- VANOS unit bolts.
- VANOS solenoid O-rings.

BentleyPublishers.com
BentleyPublishers.com–All Rights Reserved

Left camshaft timing, adjusting (V8 engine)

- Remove left side valvetronic motor. See **133 Fuel Injection**.
- Remove spark plugs. Plug spark plug bores to prevent debris from entering engine.
- Remove left side cylinder head cover. See **113 Cylinder Head Removal and Installation**.
- Remove engine cooling fan and fan shroud. See **170 Radiator and Cooling System**.
- Remove air filter housing and ducts. See **133 Fuel Injection**.
- Disconnect negative (–) cable from battery.

> ***CAUTION***—
> - *Prior to disconnecting the battery, read the battery disconnection cautions in* **001 Warnings and Cautions**.

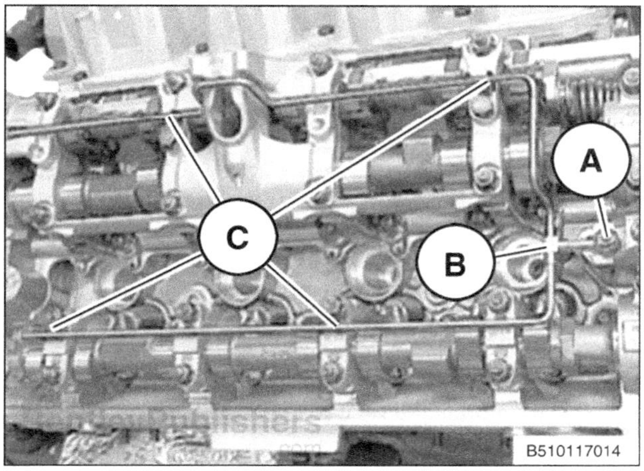

B510117014

Remove banjo bolt (**A**). Unclip oil line (**B**) from retainers (**C**) and remove.

> ***CAUTION***—
> - *Do not misplace oil line banjo bolt.*

B510117019

Using vibration damper center bolt, turn engine in direction of rotation to cylinder 1 TDC position.

NOTE—
- *With cylinder 1 in TDC position, cylinder 5 camshaft intake and exhaust camshaft lobes point upwards at an angle (**arrows**).*

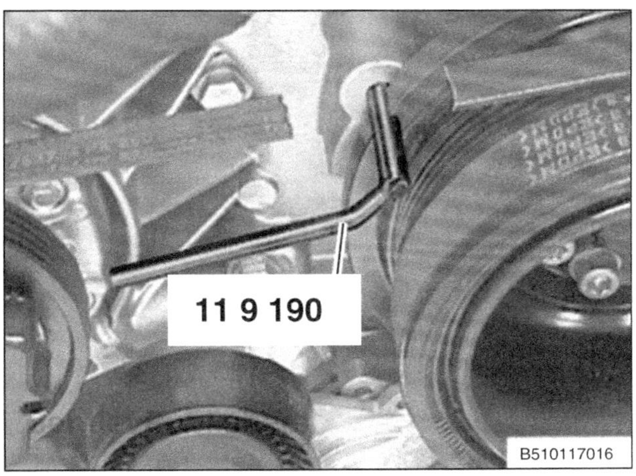

B510117016

Using BMW special tool 11 9 190, lock engine in cylinder 1 TDC position.

117

To avoid incorrect timing adjustment, check VANOS adjustment unit locking using the following steps.

Grip hexagon (**black arrow**) of intake camshaft and attempt to rotate intake camshaft carefully in direction opposite (**white arrow**) to engine rotation. If there is no fixed connection between intake camshaft and intake VANOS unit, rotate intake camshaft in direction opposite to engine rotation as far as it will go.

Grip hexagon (**black arrow**) of exhaust camshaft and attempt to rotate exhaust camshaft carefully in the direction of engine rotation (**white arrow**). If there is no fixed connection between exhaust camshaft and exhaust VANOS unit, rotate exhaust camshaft in the direction of engine rotation as far as it will go.

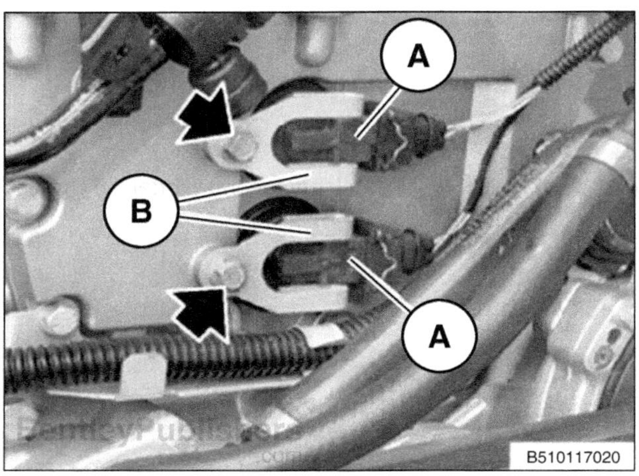

Disconnect electrical connectors from VANOS adjuster solenoids (**A**). Remove fasteners (**arrows**), brackest (**B**) and remove actuator solenoids (**A**). Discard actuator solenoids O-rings.

- Remove alternator belt. See **020 Maintenance**.

BentleyPublishers.com
BentleyPublishers.com—All Rights Reserved

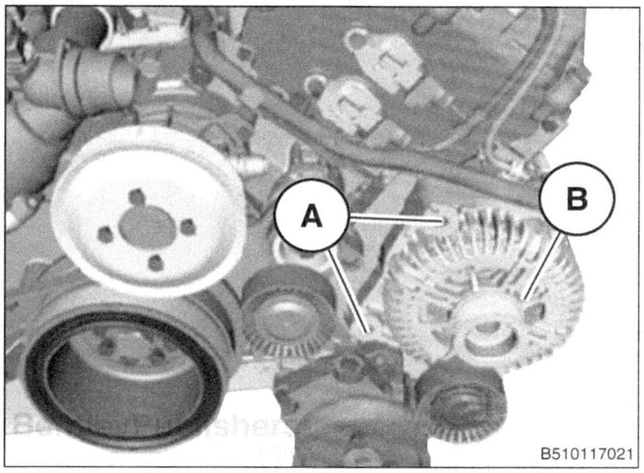

Remove alternator fasteners (**A**) and lay alternator (**B**) aside.

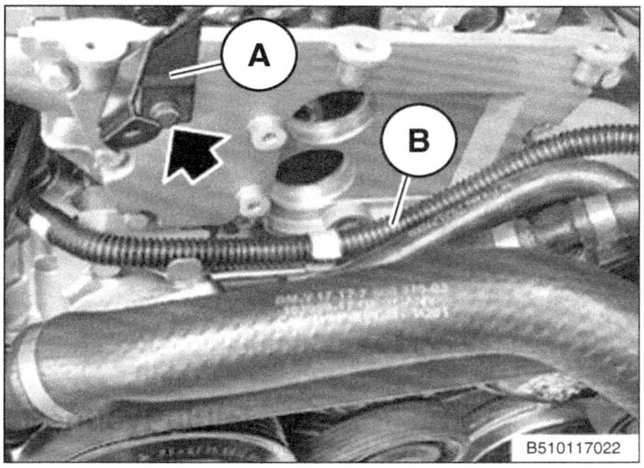

Remove fastener (**arrow**) and bracket (**A**). Unclip wiring harness (**B**) from bracket.

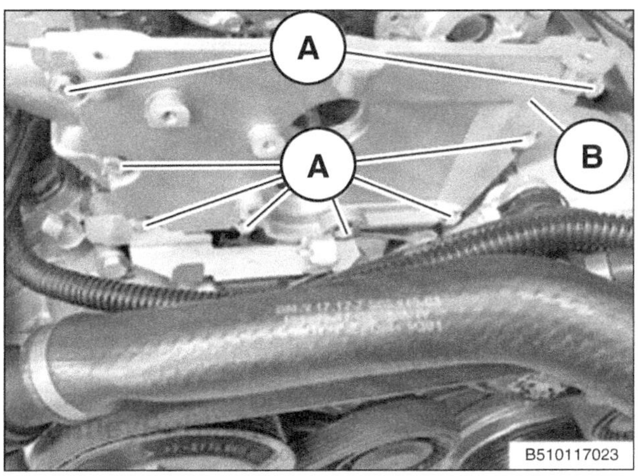

Remove upper timing chain cover fasteners (**A**) and remove cover (**B**).

Left camshaft timing, adjusting (V8 engine)

Grip each camshaft at hexagon and loosen VANOS adjuster bolts (**arrows**). Do not remove VANOS bolts.

Install BMW special tool 11 9 461 on intake camshaft so that it rests flush on cylinder head.

Install BMW special tool 11 9 463 and secure holder (**A**) in oil line thread hand tight.

BentleyPublishers.com
BentleyPublishers.com—All Rights Reserved

Grip intake camshaft at hexagon and tighten intake VANOS adjuster bolt (**A**).

Tightening torque	
VANOS adjuster to camshaft (replace bolt)	80 Nm (59 ft-lb)

Remove holder (**A**) and BMW special tools 11 9 461, 11 9 463.

Install BMW special tool 11 9 462 on exhaust camshaft so that it rests flush on cylinder head.

BentleyPublishers.com
BentleyPublishers.com—All Rights Reserved

Left camshaft timing, adjusting (V8 engine)

Install BMW special tool 11 9 463 and secure holder (**A**) in oil line thread hand tight.

Grip exhaust camshaft at hexagon and tighten intake VANOS adjuster bolt (**A**).

Tightening torque	
VANOS adjuster to camshaft (replace bolt)	80 Nm (59 ft-lb)

Remove holder (**A**) and BMW special tools 11 9 462, 11 9 463.

BentleyPublishers.com
BentleyPublishers.com—All Rights Reserved

◀ Remove BMW special tool 11 9 190. Using vibration damper center bolt turn engine 2 full rotations. Using BMW special tool 11 9 190 lock engine in cylinder 1 TDC position.

◀ Place BMW special tool 11 9 461 on intake camshaft and check timing. Timing is correctly adjusted when BMW special tool 11 9 461 rests flat on cylinder head or protrudes by up to 0.5 mm to exhaust side.

NOTE—

- *Due to flexible sprocket design, VANOS unit tolerances and play in the VANOS units, when the camshaft timing is set correctly, the exhaust side of special tool 11 9 462 may be up to 0.5 mm (0.02 in) above the surface of the cylinder head. This is normal.*

– Remove special tool 11 9 461 from inlet camshaft.

◀ Place BMW special tool 11 9 462 on exhaust camshaft and check timing. Timing is correctly adjusted when BMW special tool 11 9 461 rests flat on cylinder head or protrudes by up to 0.5 mm to exhaust side.

NOTE—

- *Due to flexible sprocket design, VANOS unit tolerances and play in the VANOS units, when the camshaft timing is set correctly, the exhaust side of special tool 11 9 462 may be up to 0.5 mm (0.02 in) above the surface of the cylinder head. This is normal.*

– Remove special tool 11 9 462 from exhaust camshaft.

– Reassemble engine.

– Remember to replace:
  - VANOS unit bolts.
  - VANOS solenoid O-rings.

BentleyPublishers.com—All Rights Reserved

Timing chain tensioner (V8 engine)

A timing chain tensioner is screwed into the left front of each cylinder head. When it is removed be sure to catch any dripping oil in a shop towel. Do not allow oil to contaminate accessory belt.

If chain tensioner is reused, drain its oil chamber. Place tensioner vertically on level working surface and compress slowly (**arrow**). Repeat twice.

> ***CAUTION—***
> - *When reassembling engine, be sure to use new sealing O-ring.*

Tightening torque	
Timing chain tensioner to cylinder head (M22 x 1.5) (use new sealing O-ring)	65 Nm (48 ft-lb)

VANOS Units (V8 Engine)

Right VANOS unit, removing (V8 engine)

- Remove right side valvetronic motor. **See 133 Fuel Injection**.
- Remove spark plugs. Plug spark plug bores to prevent debris from entering engine.
- Remove right side cylinder head cover. See **113 Cylinder Head Removal and Installation**.
- Remove engine cooling fan and cooling fan shroud. See **170 Radiator and Cooling System**.
- Remove air filter housing and ducts. See **133 Fuel Injection**.

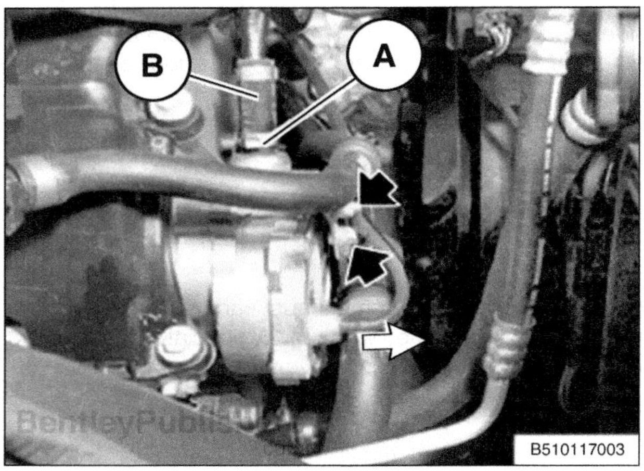

Working at front of right cylinder head:
- Remove clamp (**A**) and remove line (**B**).
- Remove hose clamps (**black arrows**).
- Remove vaccum line (**white arrow**).

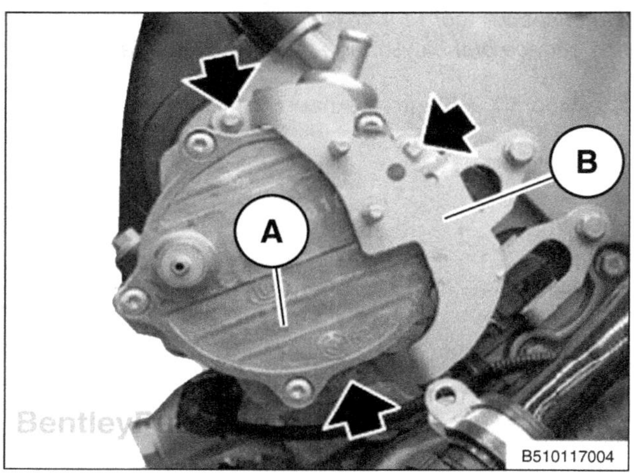

Remove vaccum pump fasteners (**arrows**) and vacuum pump (**A**) with bracket (**B**).

BentleyPublishers.com
BentleyPublishers.com—All Rights Reserved

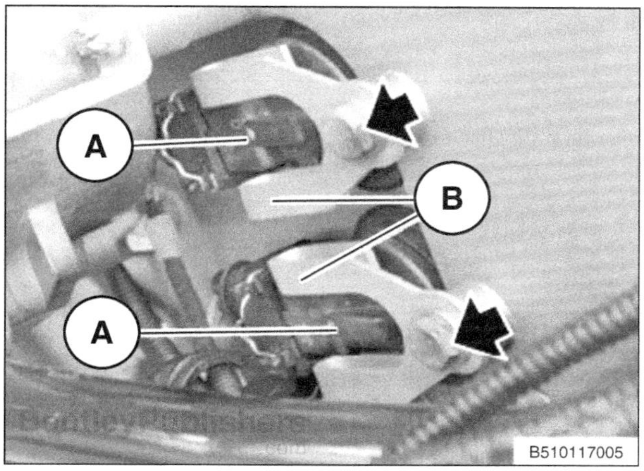

Disconnect electrical connectors from VANOS adjuster solenoids (**A**). Remove fasteners (**B**), brackest (**arrows**) and remove actuator solenoids (**A**). Discard actuator solenoids O-rings.

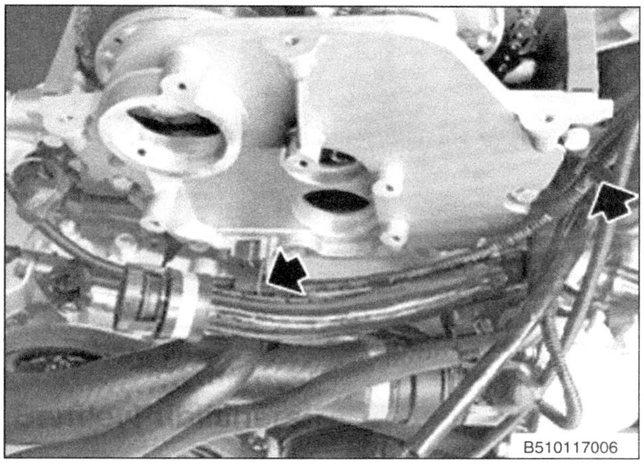

Remove vacuum lines from bracket (**arrow**).

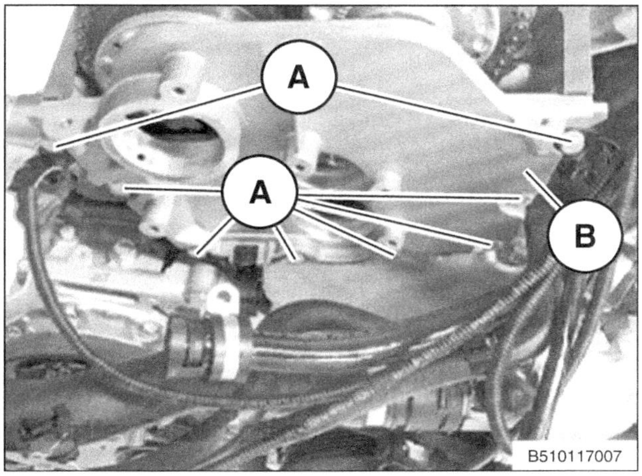

Remove upper timing chain cover fasteners (**A**) and remove upper cover (**B**).

BentleyPublishers.com
BentleyPublishers.com–All Rights Reserved

Right VANOS unit, removing (V8 engine)

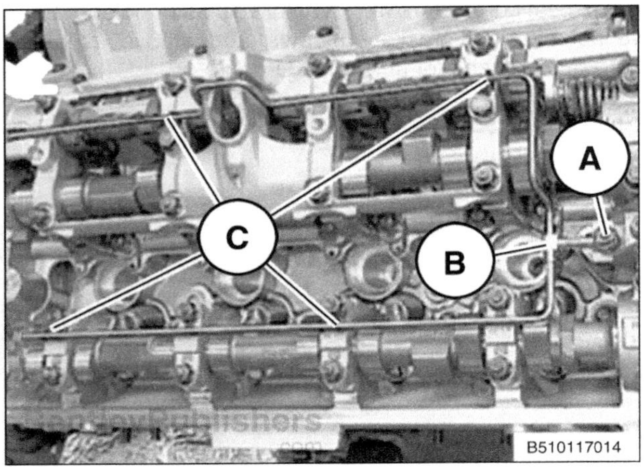

Remove banjo bolt (**A**). Unclip oil line (**B**) from retainers (**C**) and remove.

> ***CAUTION—***
> *• Do not misplace oil line banjo bolt.*

Using vibration damper center bolt, turn engine in direction of rotation to cylinder 1 TDC position.

NOTE —

• Cylinder 1 exhaust lobe points upwards at an angle. Cylinder 1 intake lobe points downwards at an angle.

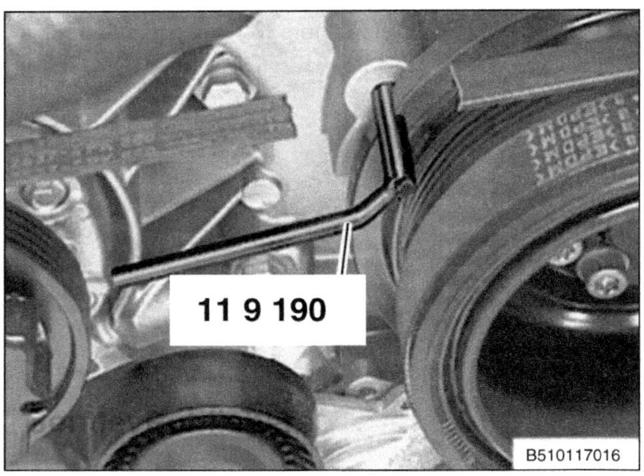

Using BMW special tool 11 9 190, lock engine in cylinder 1 TDC position.

BentleyPublishers.com
BentleyPublishers.com—All Rights Reserved

To avoid incorrect timing adjustment, check VANOS adjustment unit locking using the following steps.

B510117017

Grip hexagon (**black arrow**) of intake camshaft and attempt to rotate intake camshaft carefully in direction opposite (**white arrow**) to engine rotation. If there is no fixed connection between intake camshaft and intake VANOS unit, rotate intake camshaft in direction opposite to engine rotation as far as it will go.

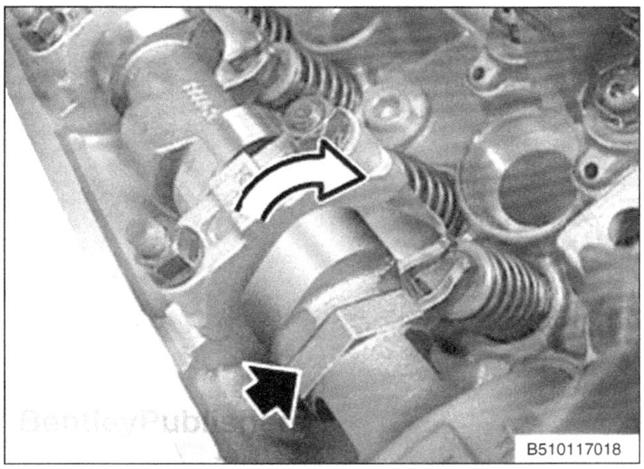
B510117018

Grip hexagon (**black arrow**) of exhaust camshaft and attempt to rotate exhaust camshaft carefully in the direction of engine rotation (**white arrow**). If there is no fixed connection between exhaust camshaft and exhaust VANOS unit, rotate exhaust camshaft in the direction of engine rotation as far as it will go.

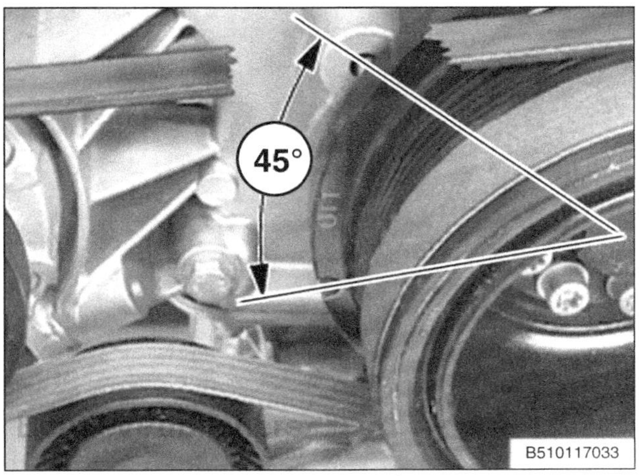

B510117033

Using vibration damper center bolt, turn engine in direction opposite to rotation to 45° before cylinder 1 TDC position.

BentleyPublishers.com
BentleyPublishers.com—All Rights Reserved

Right VANOS unit, removing (V8 engine)

Grip each camshaft at hexagon and loosen VANOS adjuster bolt (**arrows**). For now, do not remove VANOS bolts.

– Remove intake VANOS adjuster bolt and discard.

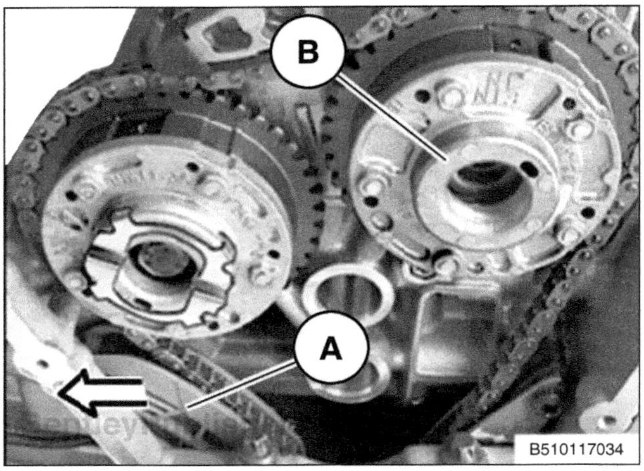

Compress tensioner rail (**A**) several times to remove oil from tensioner. Remove intake VANOS adjuster (**B**).

– Remove exhaust VANOS adjuster bolt and discard.

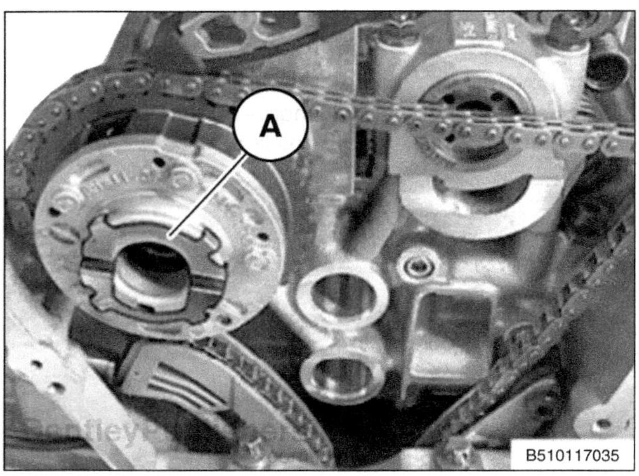

Remove exhaust VANOS adjuster (**A**). Secure timing chain to prevent it from falling into timing cover.

BentleyPublishers.com
BentleyPublishers.com—All Rights Reserved

Right VANOS unit, installing (V8 engine)

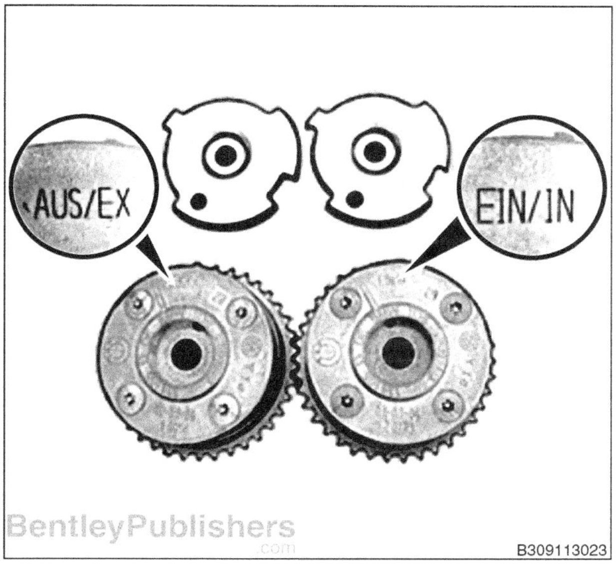

B309113023

Use new bolts to reattach VANOS units.

- Exhaust and intake units are different and so marked.
- VANOS units may be installed in any position initially. Finger-tighten mounting bolts for now.

B510117036

Exhaust VANOS adjuster has a metal clip (**A**) for driving vacuum pump.

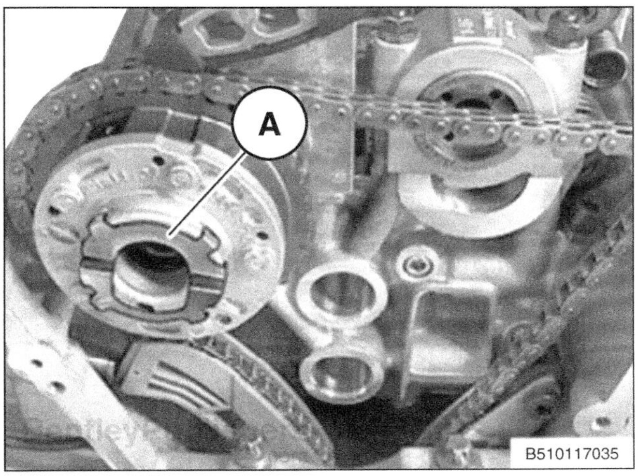

B510117035

Pull timing chain up and place over exhaust VANOS sprockets. Install VANOS adjuster bolt hand tight, then loosen by ½ turn.

BentleyPublishers.com
BentleyPublishers.com—All Rights Reserved

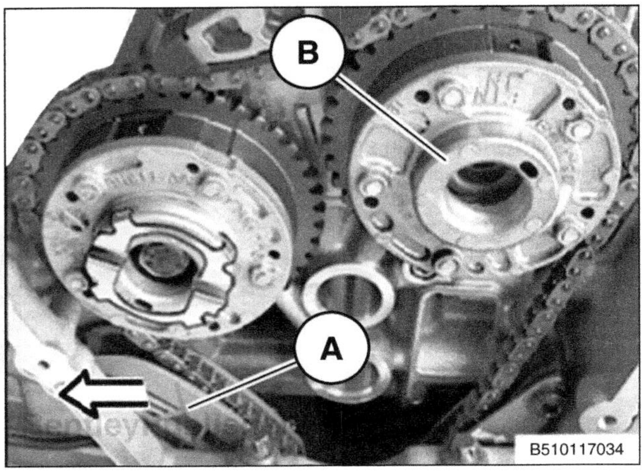

Compress tensioner rail (**A**) and place chain over sprockets (**B**). Install VANOS adjuster bolt hand tight, then loosen by ½ turn

NOTE—
- *If tensioner rail cannot be compressed enough to instal chain loosen timing chain tensioner 3 turns.*

– Check and adjust camshaft timing. See **Right camshaft timing, adjusting (V8 engine)** in this repair group.

– Reassemble engine.

Tightening torque	
VANOS adjuster to camshaft (replace bolt)	80 Nm (59 ft-lb)
Timing chain tensioner to cylinder head	65 Nm (48 ft-lb)

Left VANOS unit, removing (V8 engine)

– Remove left side valvetronic motor. **See 133 Fuel Injection**.

– Remove spark plugs. Plug spark plug bores to prevent debris from entering engine.

– Remove left side cylinder head cover. See **113 Cylinder Head Removal and Installation**.

– Remove engine cooling fan and cooling fan shroud. See **170 Radiator and Cooling System**.

– Remove air filter housing and ducts. See **133 Fuel Injection**.

– Disconnect negative (–) cable from battery.

> ***CAUTION—***
> - *Prior to disconnecting the battery, read the battery disconnection cautions in* **001 Warnings and Cautions**.

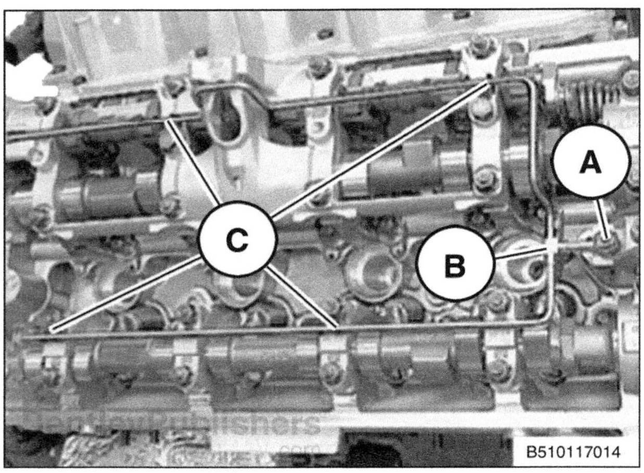

Remove banjo bolt (**A**). Unclip oil line (**B**) from retainers (**C**) and remove.

> ***CAUTION—***
> - *Do not misplace oil line banjo bolt.*

Using vibration damper center bolt, turn engine in direction of rotation to cylinder 1 TDC position.

NOTE—
- *With cylinder 1 in TDC position, lobes of intake and exhaust camshafts at cylinder 5 point upwards at an angle (***arrows***).*

BentleyPublishers.com
BentleyPublishers.com—All Rights Reserved

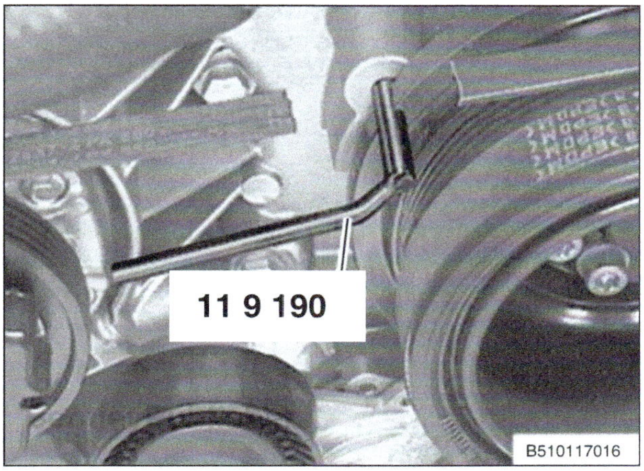

B510117016

Using BMW special tool 11 9 190, lock engine in cylinder 1 TDC position.

To avoid incorrect timing adjustment, check VANOS adjustment unit locking using the following steps.

B510117017

Grip hexagon (**black arrow**) of intake camshaft and attempt to rotate intake camshaft carefully in direction opposite (**white arrow**) to engine rotation. If there is no fixed connection between intake camshaft and intake VANOS unit, rotate intake camshaft in direction opposite to engine rotation as far as it will go.

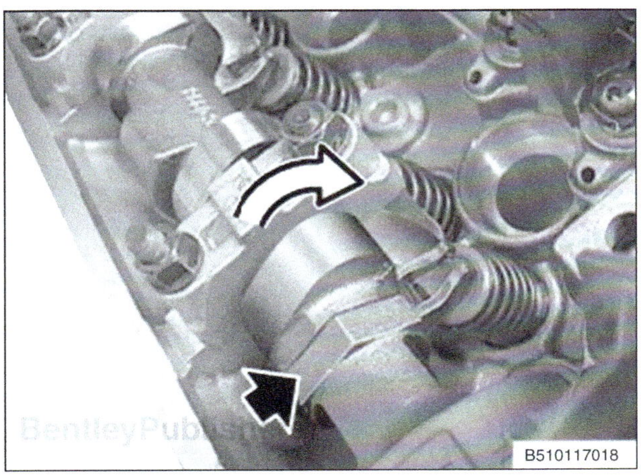

B510117018

Grip hexagon (**black arrow**) of exhaust camshaft and attempt to rotate exhaust camshaft carefully in the direction of engine rotation (**white arrow**). If there is no fixed connection between exhaust camshaft and exhaust VANOS unit, rotate exhaust camshaft in the

BentleyPublishers.com
BentleyPublishers.com—All Rights Reserved

Left VANOS unit, removing (V8 engine)

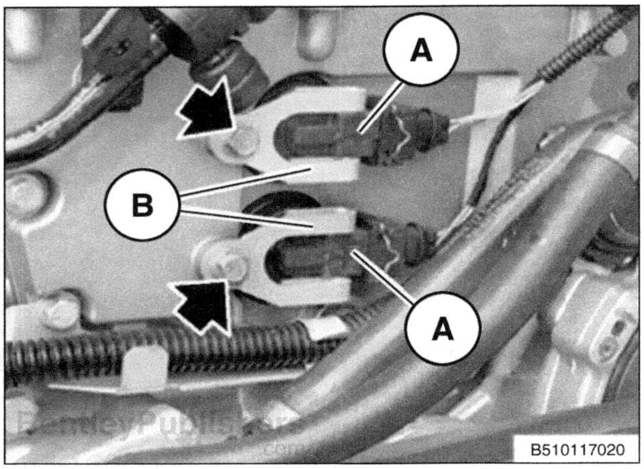

Disconnect electrical connectors from VANOS adjuster solenoids (**A**). Remove fasteners (**arrows**), brackest (**B**) and remove actuator solenoids (**A**). Discard actuator solenoids O-rings.

– Remove alternator belt. See **020 Maintenance**.

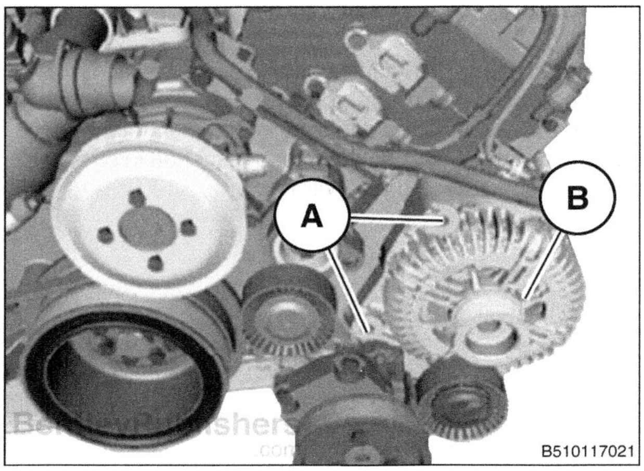

Remove alternator fasteners (**A**) and lay alternator (**B**) aside.

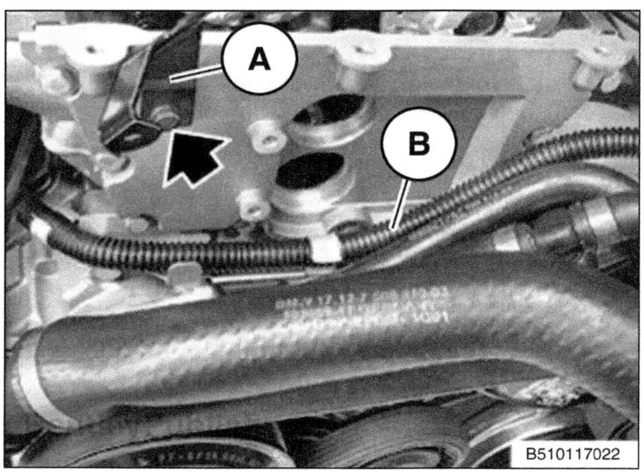

Remove fastener (**arrow**) and bracket (**A**). Unclip wiring harness (**B**) from bracket.

BentleyPublishers.com
BentleyPublishers.com—All Rights Reserved

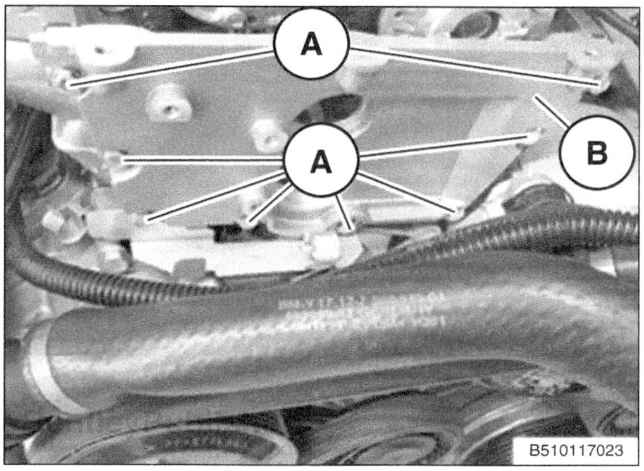

Remove upper timing chain cover fasteners (**A**) and remove cover (**B**).

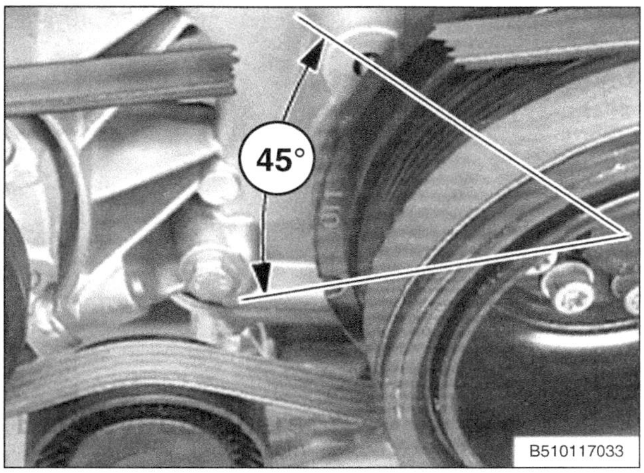

Using vibration damper center bolt, turn engine in direction opposite to rotation to 45° before cylinder 1 TDC position.

Grip each camshaft at hexagon and loosen VANOS adjuster bolt (**arrows**). For now, do not remove VANOS bolts.

- Remove and exhaust VANOS adjuster bolt.

BentleyPublishers.com—All Rights Reserved

Compress tensioner rail (**A**) several times to remove oil from tensioner. Remove exhaust VANOS adjuster (**B**).

– Remove intake VANOS adjuster bolt and discard.

Remove intake VANOS adjuster (**A**). Secure timing chain to prevent it from falling into timing cover.

Left VANOS unit, installing (V8 engine)

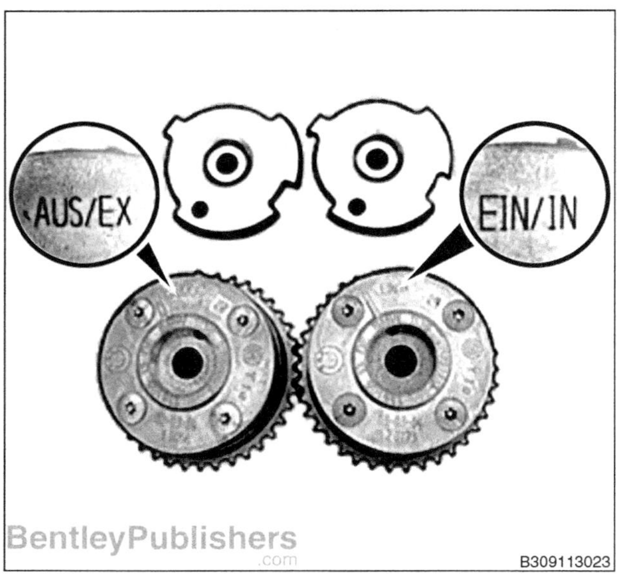

Use new bolts to reattach VANOS units.

- Exhaust and intake VANOS units are different and so marked.
- VANOS units may be installed in any position initially. Finger-tighten mounting bolts for now.

BentleyPublishers.com
BentleyPublishers.com—All Rights Reserved

Pull timing chain up and place chain over sprocket (**A**). Install VANOS adjuster bolt hand tight, then loosen by ½ turn.

Compress tensioner rail (**A**) and place chain over sprocket (**B**). Install VANOS adjuster bolt hand tight, then loosen by ½ turn

NOTE—

- *If tensioner rail cannot be compressed enough to install chain remove timing chain tensioner. Remember to replace sealing O-ring.*

– Check and adjust camshaft timing. See **LLeft camshaft timing, adjusting (V8 engine)** in this repair group.

– Reassemble engine.

Tightening torque	
VANOS adjuster to camshaft (replace bolt)	80 Nm (59 ft-lb)
Timing chain tensioner	65 Nm (48 ft-lb)

BentleyPublishers.com
BentleyPublishers.com—All Rights Reserved

119 Lubrication System

BentleyPublishers
BentleyPublishers.com–All Rights Reserved

GENERAL

This repair group covers lubrication system service and repair, including crankshaft seal replacement.

See also:

- **020 Maintenance** for oil and oil filter change
- **100 Engine–General** for engine identification and application
- **170 Radiator and Cooling System** for N54 and N62TU oil cooler removal.

Warnings and Cautions

WARNING—
- *Due to risk of personal injury, be sure the engine is cold before beginning work on engine components.*

CAUTION—
- *Prior to disconnecting the battery, read the battery disconnection cautions in* **001 Warnings and Cautions**.
- *To avoid electrochemical corrosion to engine components made of aluminum-magnesium alloy, do not use steel fasteners. Use aluminum fasteners only. For reliable identification, test fasteners for aluminum composition with magnet.*
- *Replace aluminum bolts each time they are loosened.*
- *Follow torque instructions, including angle of rotation specifications, when installing aluminum fasteners.*
- *Cover alternator with shop towel to protect from oil drips.*

BentleyPublishers.com
BentleyPublishers.com—All Rights Reserved

ENGINE LUBRICATION

Engine lubricant

Engine oil specification	
M54 engine: BMW high performance synthetic • SAE 5W - 30 or 15W - 40	API rating SJ / CF or higher
N52, N54, V8 (N62) engine: BMW long-life rating (LL-01) synthetic oils for US market: • Castrol Syntec European Formula SAE 0W - 30 • Mobil 1 SAE 0W - 40 • Pennzoil Platinum European Formula Ultra SAE 5W - 30 • Valvoline SynPower SAE 5W - 30	API rating SM or higher

Engine oil capacity	
Engine	**Approximate capacity (incl. oil filter)**
M54	7.5 liters (8.0 US qt)
N52, N54	6.5 liters (6.9 US qt)
V8 (N62)	8 liters (8.45 US qt)

Engine lubrication (M54 engine)

Oil pressure is generated by a gear-type pump bolted to the bottom of the engine block. The oil pump is chain driven off the front of the crankshaft.

The oil pressure warning system consists of an oil pressure switch mounted in the oil circuit and an instrument panel warning light. Other safety features include:

- Filter bypass to provide lubrication should the oil filter become clogged.
- Oil pump pressure relief valve to prevent excessive system pressure.

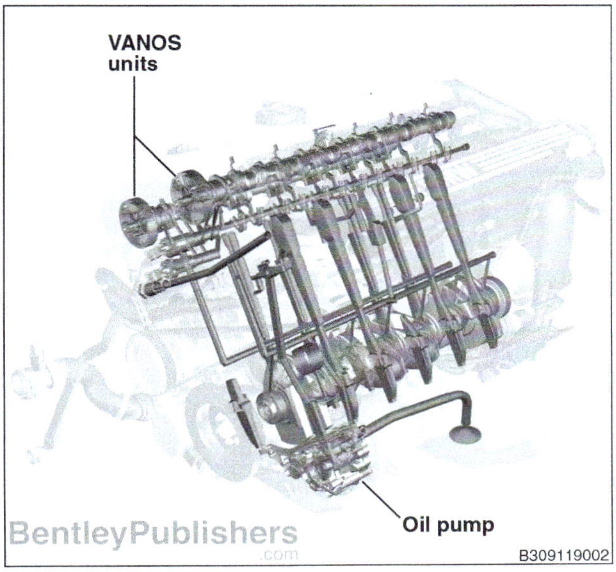

Engine lubrication (N52, N54 engine)

To meet the oiling requirements of N52 and N54 engines, a chain-driven volumetric-flow controlled oil pump is utilized. The VANOS system requires a large volume of oil, particularly at low engine speeds.

The volumetric-flow controlled pump design allows:

- Increased power output
- Reduced weight
- Optimized fuel consumption
- Reduced exhaust emissions

BentleyPublishers.com
BentleyPublishers.com—All Rights Reserved

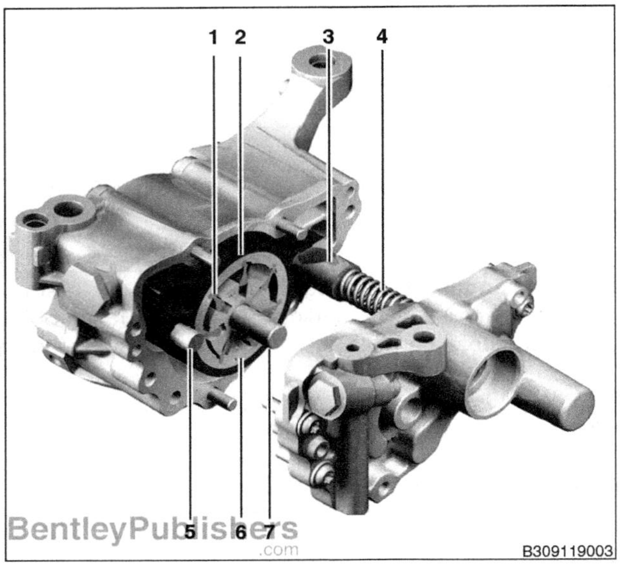

Engine oil pump (N52, N54 engine)

1. Sliding vane valve
2. Pivoting rotor housing
3. Beveled control piston
4. Compression spring
5. Pivot pin
6. Rotor
7. Pump shaft

The volumetric-flow controlled oil pump delivers only as much oil as is necessary. No surplus quantities of oil are delivered in low load operating ranges. This reduces the fuel consumption of the engine and slows down the oil wear rate.

The pump is designed with sliding vane valves. The pump shaft and rotor are positioned off-center in the pivoting rotor housing and the vanes are displaced radially during rotation. The vanes form chambers of differing volume depending on the position of the rotor housing.

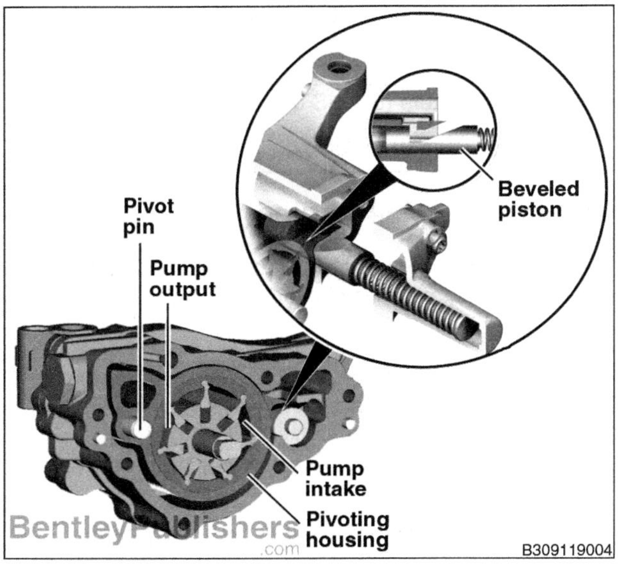

When the pump rotates, oil pressure acts on the control piston against the force of the compression spring. The beveled piston varies the position of the pivoting housing.

When the housing is centered on the rotor, changes in intake and output chamber volumes are small: Delivered oil volume is low. When the rotor housing is off-center in relation to the rotor, changes in input and output chamber volumes are greater: Delivered oil volume is high.

When the oil volume required by the engine increases, for example during VANOS operation, oil pressure in the lubricating system drops, reducing pressure on the beveled piston. In response, the compression spring presses the beveled piston against the pivoting housing control dog, forcing the housing further off-center and increasing oil volume. The opposite occurs as engine oil pressure increases and the need for delivered oil volume drops.

Engine lubrication (V8 engine)

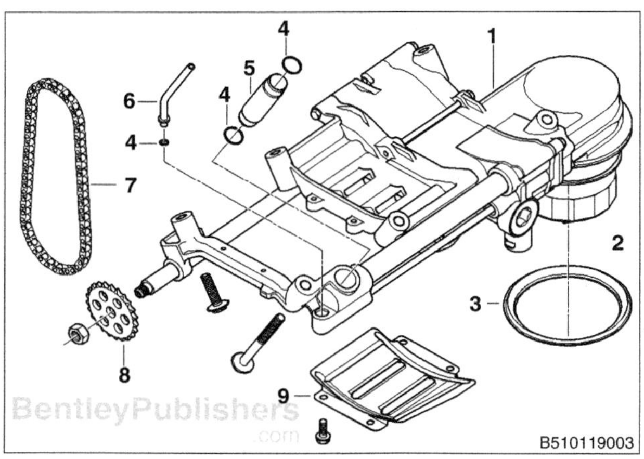

Oil pressure is generated by a two-stage gear-type pump with two parallel switched gear clusters. It is bolted to the bottom of the engine block at an angle. The oil pump is chain driven off the front of the crankshaft.

1. Oil pump
2. Oil filter
3. Profile gasket
4. O-ring
5. Oil pump inlet
6. Oil pipe
7. Oil pump drive chain
8. Oil pump drive gear
9. Oil deflector

BentleyPublishers.com
BentleyPublishers.com—All Rights Reserved

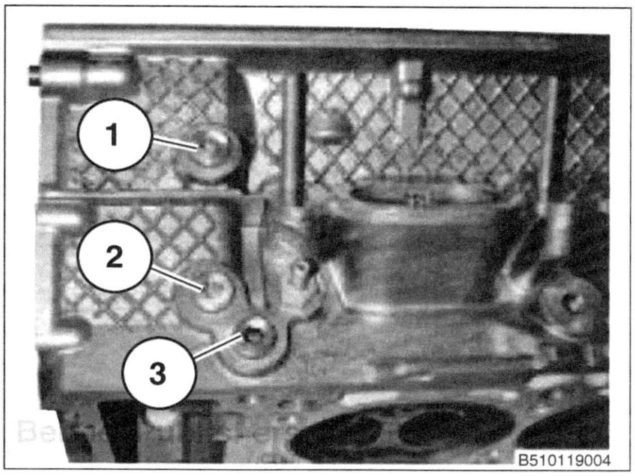

Engine oil is supplied by the oil pump to lubrication points in the engine block and is pumped into the cylinder heads. Three oil check valves are installed in each cylinder head. This prevents engine oil from draining out of the cylinder head and VANOS units. The check valves are serviceable from the outside casting of the cylinder head.

1. Intake VANOS check valve
2. Exhaust VANOS check valve
3. Cylinder head oil supply check valve

OIL PRESSURE WARNING SYSTEM

> ***CAUTION—***
> - *If the red oil pressure warning light comes on or flashes on while driving, assume that the oil pressure is low. Stop the engine immediately and make arrangements to test oil pressure.*

Oil pressure, checking

Test oil pressure by removing oil pressure switch and installing oil pressure gauge in its place.

– 6-cylinder engines: Loosen oil filter cover to allow engine oil to drain back down into oil pan. Tighten cover.

M54 engine: Access to oil pressure warning switch (**arrow**) is restricted (intake manifold removed for purposes of photo).

BentleyPublishers.com—All Rights Reserved

N52, N54 engine: Detach harness connector (**arrow**) from oil pressure switch. If necessary, remove ignition coil cover. See **020 Maintenance**.

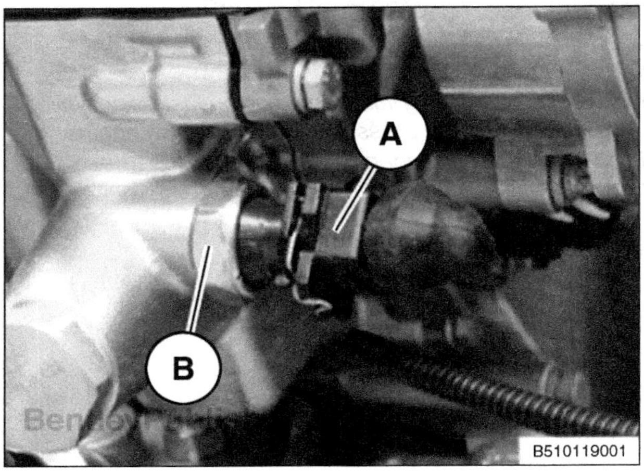

V8 engine: Working at front of right cylinder head, detach electrical connector (**A**) from oil pressure switch (**B**).

– Remove switch. Be prepared to catch leaking oil with a shop towel.

CAUTION—

- *Thoroughly clean around the oil pressure switch before removing it.*

– Install pressure gauge in place of oil pressure switch. If necessary, use BMW special adapter 11 4 050 in place of oil pressure switch.

– With gauge installed, start engine and allow to reach operating temperature. Check oil pressure. See **Table a**.

NOTE—

- *For the most accurate test results, make sure the engine oil and filter are new and the oil of the correct grade.*

Table a. Engine oil pressure	
At idle, engine at operating temperature • M54 engine • N52, N54 engine • V8 engine	 0.5 bar (7 psi) 1.5 bar (22 psi) 1.0 bar (14.5 psi)
Maximum regulated pressure, engine at operating temperature • M54 engine • N52, N54 engine • V8 engine	 4 ± 0.5 bar (58 ± 7 psi) 4.0 - 6.0 bar (58 - 87 psi) 4.0 - 6.0 bar (58 - 87 psi)

BentleyPublishers.com
BentleyPublishers.com—All Rights Reserved

– Remove pressure gauge and reinstall pressure switch with new sealing washer.

Tightening torque	
Oil pressure switch to oil filter housing M54 engine (M12 x 1.5)	27 Nm (20 ft-lb)
Oil pressure switch to oil filter housing N52, N54 engine (M12 x 1.5)	20 Nm (15 ft-lb) additional 16°
Oil pressure switch to cylinder head V8 engine (M12 x 1.5)	27 Nm (20 ft-lb)

If testing shows low oil pressure, one or more of the following conditions may be indicated:

- Worn or faulty oil pump or faulty pump pressure relief valve
- Worn or damaged engine bearings
- Severe engine wear

Any of these conditions indicate the need for major repairs.

Oil pressure warning switch, replacing

M54 engine: Access to oil pressure warning switch (**arrow**) is restricted (intake manifold removed for purposes of photo).

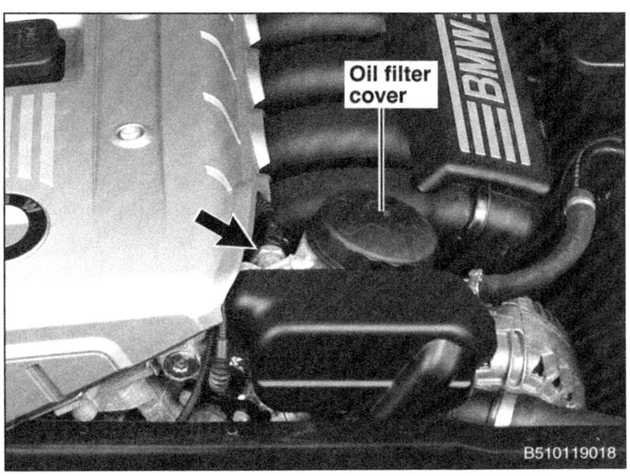

N52, N54 engine: If necessary, remove ignition coil cover. See **020 Maintenance**.

BentleyPublishers.com
BentleyPublishers.com—All Rights Reserved

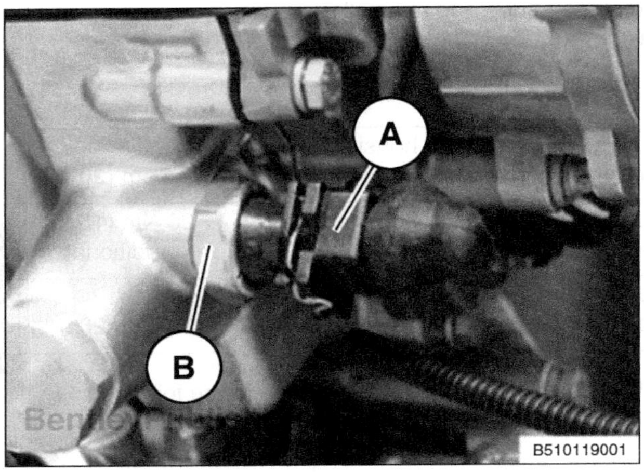

V8 engine: Oil pressure switch (**B**) is located at right front cylinder head.

- 6-cylinder engine: Loosen oil filter cap to allow engine oil to drain back down into oil pan.

- Disconnect harness connector from oil pressure switch and remove switch. Be prepared to catch leaking oil with a shop towel.

> ***CAUTION—***
> - *Thoroughly clean around the oil pressure switch before removing it.*

- Installation is reverse of removal. Tighten oil filter housing cap and top off oil, if necessary.

Tightening torque	
Oil pressure switch to oil filter housing M54 engine (M12 x 1.5)	27 Nm (20 ft-lb)
Oil pressure switch to oil filter housing N52, N54 engine (M12 x 1.5)	20 Nm (15 ft-lb) additional 16°
Oil pressure switch to cylinder head V8 engine (M12 x 1.5)	27 Nm (20 ft-lb)

OIL MONITORING

Oil level and temperature sensor [M54, V8 engine (to 2006)]

Engine oil level and temperature is measured by the oil level and temperature sensor and indicated in the instrument cluster or central information display.

Oil condition information is processed by the engine control module (ECM), then routed via PT-CAN and K-CAN buses to the instrument cluster and central information display. The ECM software determines the optimal time for oil change interval.

Checking oil level is covered in **020 Maintenance**.

Oil condition sensor (OZS) [N52, N54, V8 engine (from 2006)]

Engine oil level is measured by the oil condition sensor (OZS) and indicated in the instrument cluster or central information display. Engine oil temperature and condition are also monitored by the OZS. N52, N54 engines are not equipped with a dipstick.

Oil condition information is processed by the engine control module (ECM), then routed via PT-CAN and K-CAN buses to the instrument cluster and central information display. The ECM uses condition based service (CBS) software to determine optimal oil change interval.

Checking oil level is covered in **020 Maintenance**.

BentleyPublishers.com
BentleyPublishers.com—All Rights Reserved

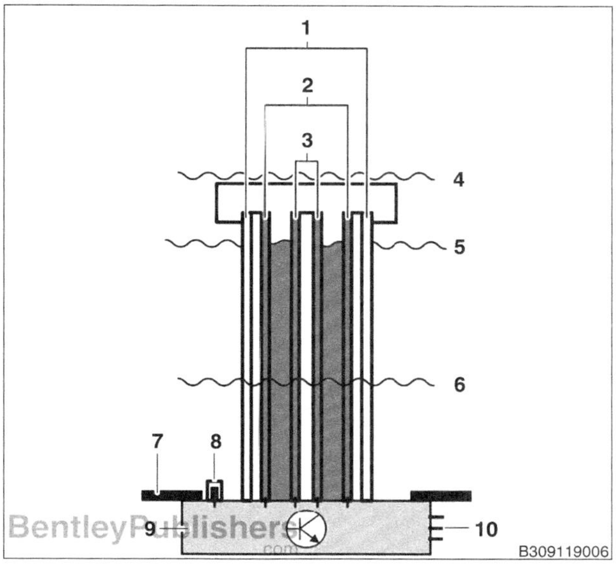

OZS operation, (N52, N54, V8 engine)

1. Housing
2. Outer metal tube
3. Inner metal tube
4. Engine oil (maximum)
5. Engine oil (average)
6. Engine oil (minimum)
7. Oil pan
8. Oil temperature sensor
9. Sensor electronics
10. Sensor connector

OZS consists of two cylindrical capacitors arranged one above the other. Oil condition is determined by the lower, smaller capacitor. Two metal tubes (**1**, **2**), arranged one inside the other, serve as capacitor electrodes. The dielectric is the engine oil between the electrodes. The electrical property of engine oil changes with age and break-down of oil additives, changing the OZS capacitance. This change is processed in the sensor electronics (**9**) and converted to a digital signal. The digital sensor signal is transferred to the ECM, which uses it to calculate the next oil change service due.

Engine oil level is determined by the capacitor of the OZS. As oil level drops, sensor capacitance changes accordingly. This change is processed in the sensor electronics (**9**) and converted to a digital signal. The digital sensor signal is transferred to the ECM and displayed to the driver.

A platinum temperature sensor (**8**) is installed at the base of the oil condition sensor.

Engine oil level, temperature and condition are monitored continuously as long as voltage is applied at terminal 15 (ignition ON). OZS is powered via terminal 87.

OZS fault evaluation

OZS electronic circuitry features a self-diagnosis function. In case of a fault in the sensor, an error message is sent to the ECM. The fault can be viewed using a BMW scan tool.

Oil level and temperature sensor, replacing (M54 engine)

Oil level warning switch is located at bottom of engine oil pan.

– Raise vehicle and support safely.

> ***WARNING—***
> *• Make sure the vehicle is stable and well supported at all times. Use a professional automotive lift or jack stands designed for the purpose. A floor jack is not adequate support.*

– Remove splash shield underneath engine. See **020 Maintenance**.

– Drain engine oil. See **020 Maintenance**.

BentleyPublishers.com
BentleyPublishers.com—All Rights Reserved

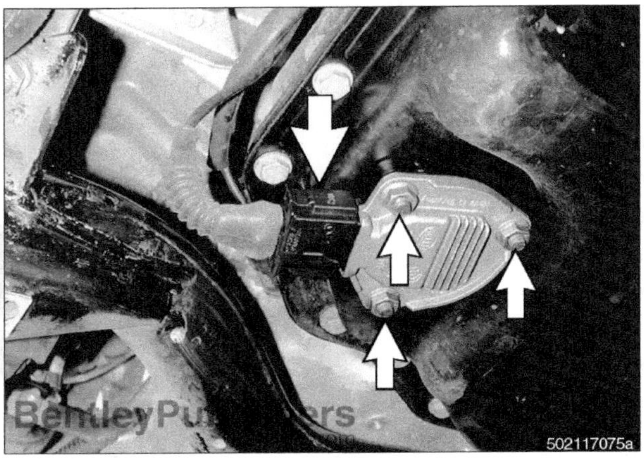

- Disconnect electrical connector and remove mounting nuts (**arrows**). Lower level sensor. be prepared to catch dripping oil.
- When reinstalling, replace sealing O-ring.
- Be sure to refill engine and check for leaks after work is completed.

OZS, replacing (N52, N54, V8 engine)

N52 engine shown. Other engines are similar.

- Raise vehicle and support safely.

> ***WARNING —***
> - *Make sure the vehicle is stable and well supported at all times. Use a professional automotive lift or jack stands designed for the purpose. A floor jack is not adequate support.*

- Remove splash shield underneath engine. See **020 Maintenance**.
- Drain engine oil. See **020 Maintenance**.

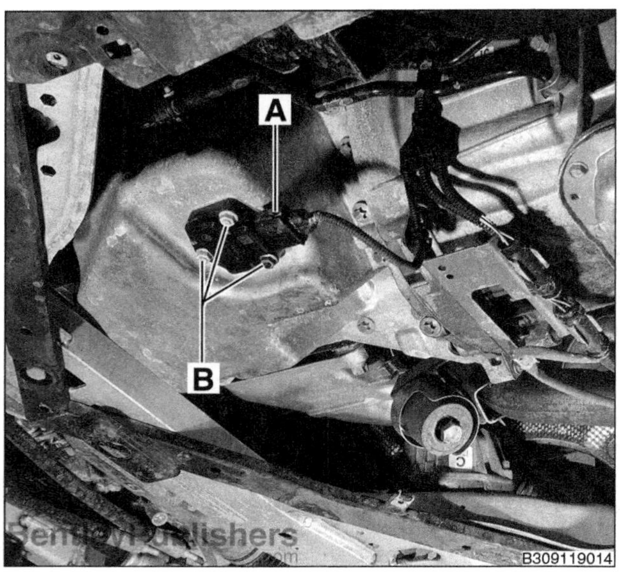

- With engine fully cooled off, disconnect OZS electrical connector (**A**) and remove mounting nuts (**B**). Lower level sensor. Be prepared to catch dripping oil.
- When reinstalling, replace sealing O-ring.

Tightening torque	
OZS to oil pan • Stage 1 • Stage 2	 2 Nm (18 lb-in) 8.5 Nm (75 lb-in)

- Refill engine with oil and check for leaks after work is completed.

BentleyPublishers.com
BentleyPublishers.com—All Rights Reserved

CRANKSHAFT SEALS (M54 ENGINE)

Crankshaft front oil seal, replacing (M54 engine)

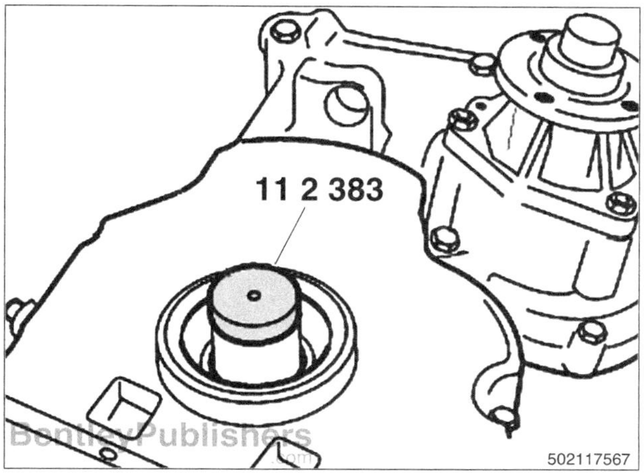

- With engine fully cooled off, remove electric cooling fan, engine accessory belts and crankshaft vibration damper and pulley. Be sure to mark belts with direction of rotation. See **117 Camshaft Timing Chain**.

◀ Install cap (BMW special tool 11 2 383) on end of crankshaft.

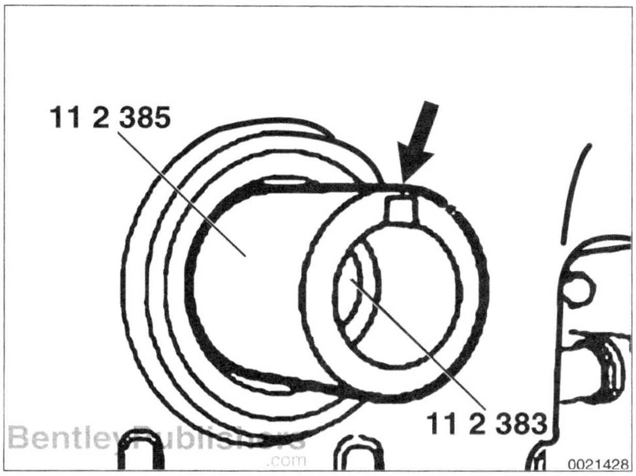

◀ Install seal guide (BMW special tool 11 2 385) over end of crankshaft. Be sure to align groove with keyway on crankshaft (**arrow**).

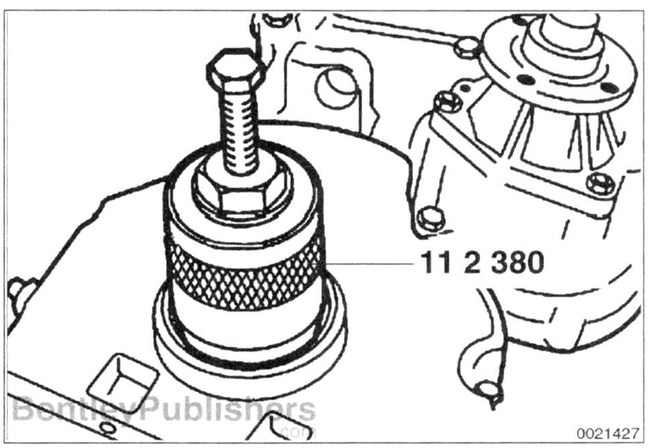

◀ Install oil seal puller (BMW special tool 11 2 380) and screw in body of puller until it contacts seal. Tighten bolt in center of tool to draw seal out of cover.

BentleyPublishers.com
BentleyPublishers.com—All Rights Reserved

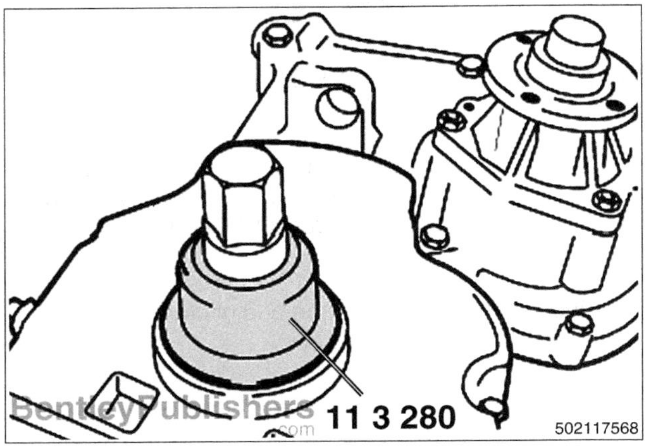

Coat new seal with oil and position in timing chain cover. Use BMW special tool 11 3 280 and crankshaft center bolt to draw seal in flush with timing case cover.

– Reinstall crankshaft vibration damper and pulley.

Tightening torque	
Vibration damper hub to crankshaft (use new bolt)	410 Nm (302 ft-lb)

– Remainder of installation is reverse of removal, noting the following:
  - Install engine accessory belts. See **020 Maintenance**.
  - Install cooling fan. See **170 Radiator and Cooling System**.

Crankshaft rear main seal, replacing (M54 engine)

Crankshaft rear main seal (flywheel seal) replacement requires removal of the transmission and flywheel.

– Remove transmission. See **230 Manual Transmission**, **231 Sequential Manual Gearbox (SMG)** or **240 Automatic Transmission**.

– Manual transmission vehicles: Remove clutch pressure plate and disc. See **210 Clutch**.

– Remove flywheel. See **210 Clutch** or **240 Automatic Transmission**.

– Drain engine oil.

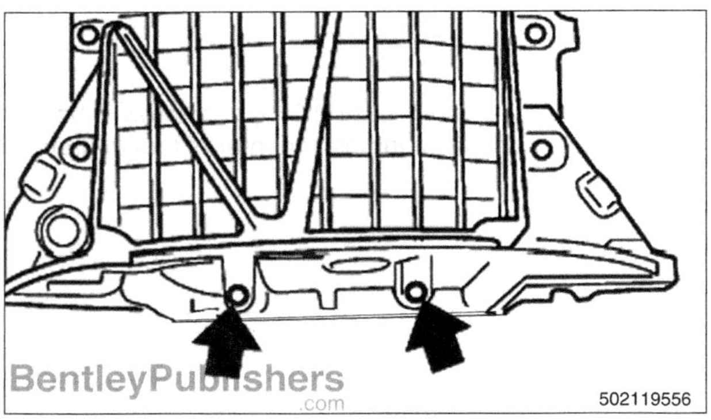

Remove seal carrier bolts from oil pan (**arrows**). Loosen all oil pan bolts.

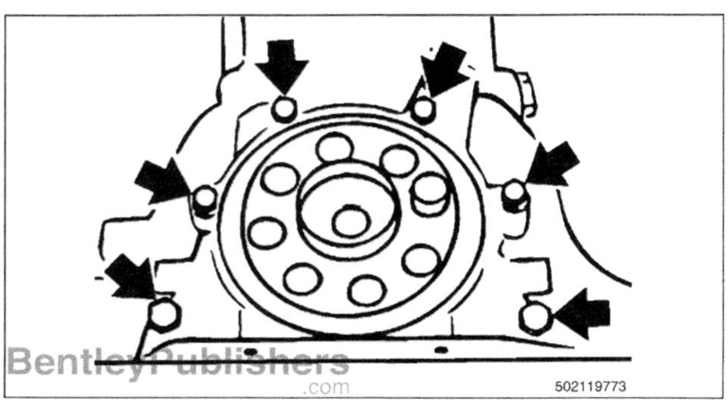

Remove remaining seal carrier bolts (**arrows**) and carefully remove seal carrier without damaging oil pan gasket.

> ***CAUTION—***
> - *After removing seal carrier, check oil pan gasket for damage. If gasket is damaged during removal, remove oil pan and replace gasket. See* **Oil pan, removing and installing (M54 engine)** *in this repair group.*

NOTE—
- *BMW offers the rear main seal carrier and seal in a kit. A plastic installation sleeve, required for sliding the seal over the crankshaft, is included with the kit.*

BentleyPublishers.com
BentleyPublishers.com—All Rights Reserved

– Install new seal carrier to crankcase. Remember to:
  - Check locating sleeves and bores in lower seal housing for damage.
  - Replace gasket between rear main seal carrier and crankcase.
  - Apply thin coat of 3 Bond® 1209 or equivalent sealant to oil pan gasket sealing surfaces.
  - Lubricate crankshaft seal contact surface.

> ***CAUTION—***
> - *Do not kink or damage the sealing lip. Do not touch with fingers.*

Use plastic support bushing (**A**) that comes with seal kit as an installation guide when fitting seal over end of crankshaft. Use care to push seal carrier and installation sleeve on straight, without tilting sideways. Remove installation sleeve.

– Install seal carrier bolts and torque to specifications.

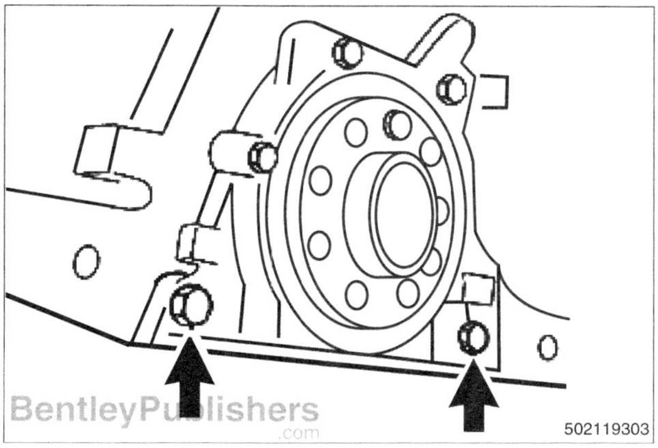

Use new bolts (**arrows**) at base of seal carrier. Coat threads with sealing compound before installing.

– Install and tighten oil pan bolts.

– Remainder of assembly is reverse of disassembly.
  - Install flywheel, clutch or torque converter and transmission.
  - Fill engine with oil.
  - Run engine and check for leaks.

Tightening torques	
Oil drain plug to oil pan (M12)	25 Nm (18 ft-lb)
Oil pan to engine block or seal carrier • M6 (8.8 grade) • M6 (10.9 grade) • M8 (8.8 grade)	 10 Nm (89 in-lb) 12 Nm (9 ft-lb) 22 Nm (16 ft-lb)
Rear main seal carrier to crankcase • M6 • M8	 10 Nm (89 in-lb) 22 Nm (16 ft-lb)

BentleyPublishers.com
BentleyPublishers.com—All Rights Reserved

CRANKSHAFT SEALS (N52, N54 ENGINE)

Crankshaft front seal, removing and installing (N52, N54 engine)

> ***WARNING—***
> *• To avoid personal injury, be sure the engine is cold before beginning the procedure.*

> ***CAUTION—***
> *• When the crankshaft seal is replaced, its grooves and crankcase seams are filled with special Loctite® sealant to avoid oil leaks. Be sure to read the entire procedure before starting work.*

– Raise vehicle and support safely.

> ***WARNING—***
> *• Make sure the vehicle is stable and well supported at all times. Use a professional automotive lift or jack stands designed for the purpose. A floor jack is not adequate support.*

– Remove splash shield underneath engine. See **020 Maintenance**.

– Remove engine accessory belt. See **020 Maintenance**.

◄ Remove vibration damper (front pulley). See **117 Camshaft Timing Chain**.

> ***CAUTION—***
> *• Do not remove vibration damper hub bolt. If this bolt is removed, the timing chain drive sprocket becomes free to rotate, resulting in valve damage.*

◄ Place BMW special tool 11 9 221 against vibration damper hub and tighten down using bolts (special tool 11 9 224). This pushes crankshaft seal inward approx. 1 mm, thus loosening it for subsequent removal.

Tightening torque	
Special tool 11 9 221 to vibration damper hub	20 Nm (15 ft-lb)

– Remove special tools.

BentleyPublishers.com
BentleyPublishers.com—All Rights Reserved

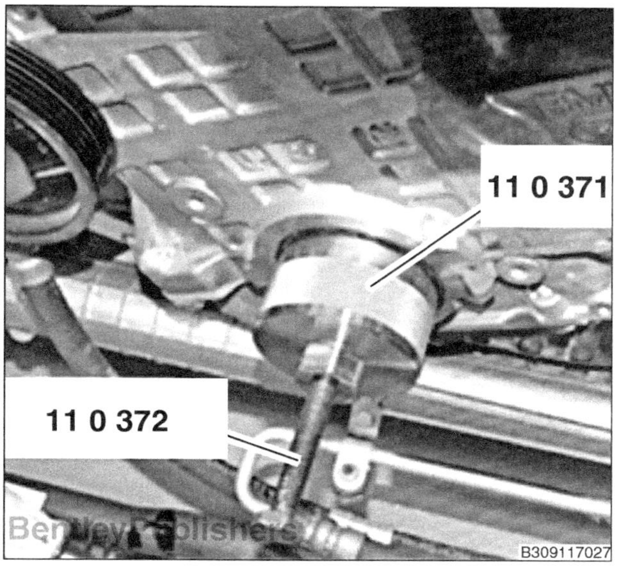

Screw BMW special tool 11 0 371 into crankshaft seal with approx. 80 Nm (59 ft-lb) of torque. Then rotate special tool 11 0 371 clockwise to pull out seal. Repeat process if necessary.

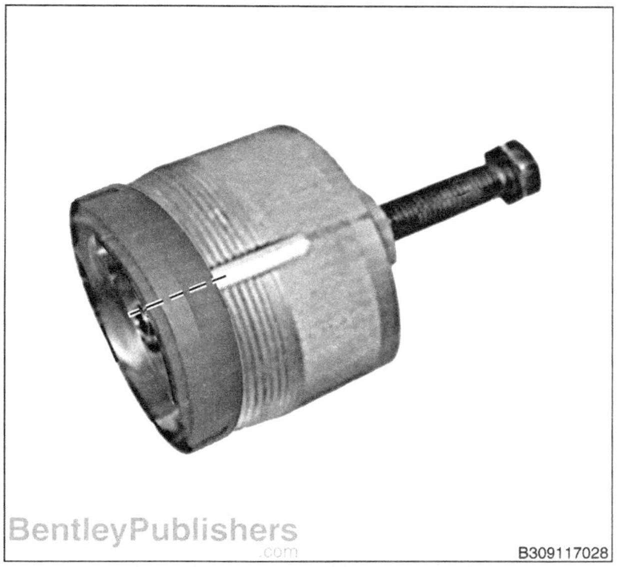

Carefully saw seal at **dashed line** and peel off tool.

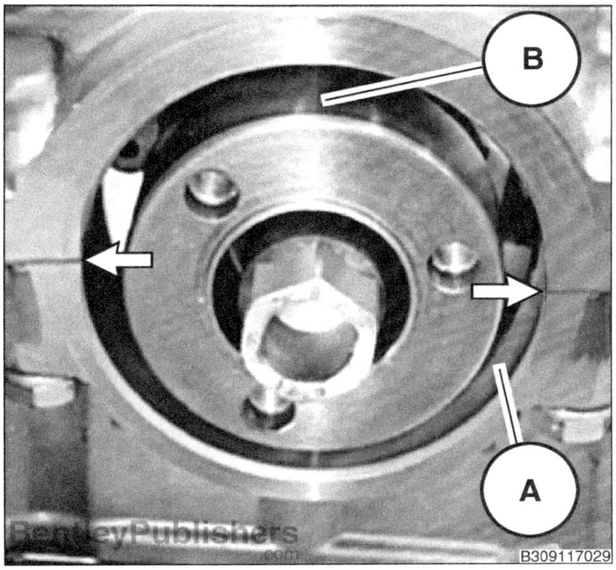

Working at crankcase opening:

- Clean crankshaft seal seating surface (**A**).
- Degrease thoroughly area around crankcase seam (**arrows**).
- Lightly oil sealing surface (**B**) on vibration damper hub.

BentleyPublishers.com
BentleyPublishers.com—All Rights Reserved

B309117030

◄ Attach BMW special tool 11 9 232 to vibration damper hub using bolts (special tool 11 9 234).

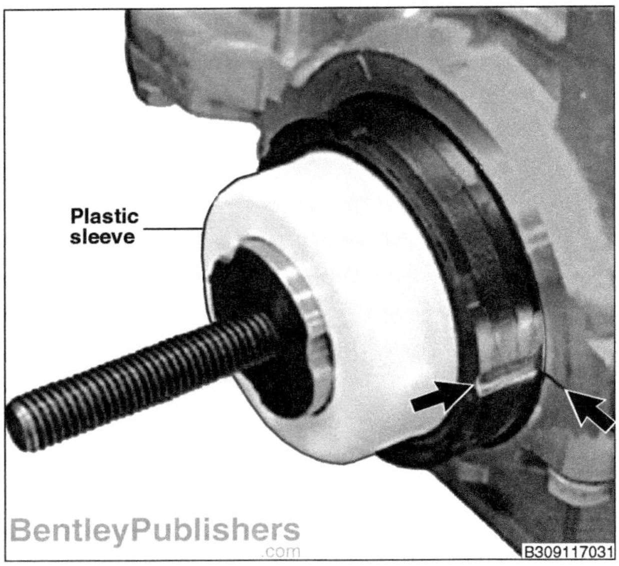

B309117031

◄ Place crankshaft seal and installation sleeve over tool and against crankcase opening.

- Plastic sleeve is supplied with seal and is used as installation guide.
- Make sure seal grooves center on crankcase seam (**arrows**).
- Coat both grooves on seal with Loctite® 171000 (primer). Allow to dry for approx. 1 minute.

B309117032

◄ Use BMW special tools 11 9 231 and 11 9 233 to press in seal.

BentleyPublishers.com
BentleyPublishers.com—All Rights Reserved

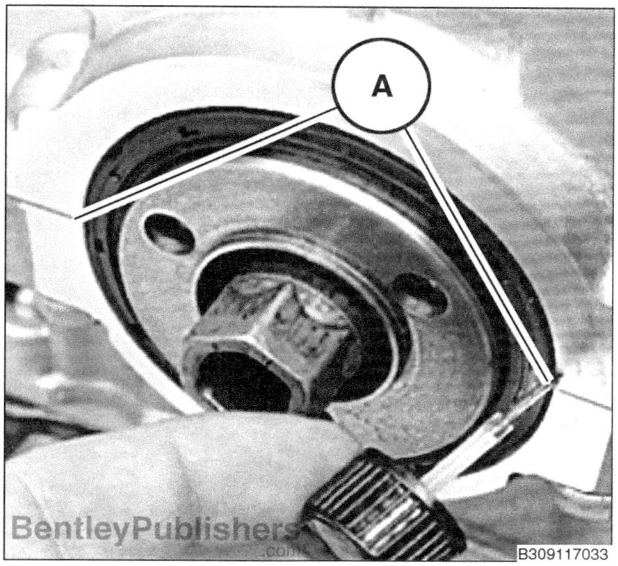

◄ Use brush to coat seal grooves and crankcase seams (**A**) with Loctite® 171000 (primer).

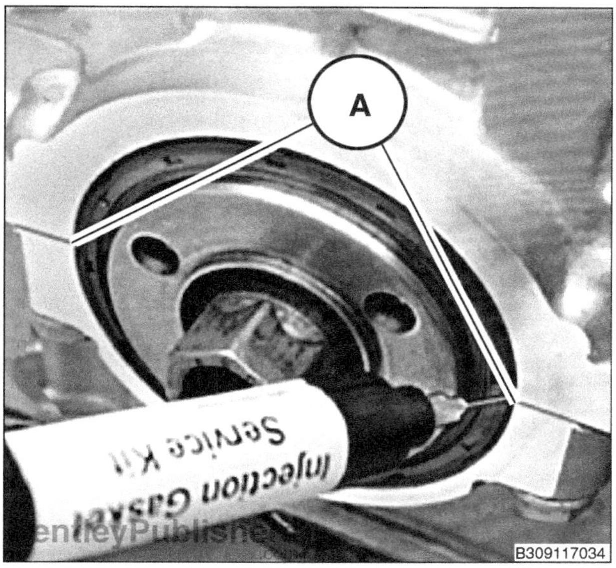

◄ Use injector kit to fill seal grooves and crankcase seams (**A**) with Loctite® 128357 (sealant).

– Coat sealed areas with Loctite® 171000 (primer) to bind sealant.

– Reinstall crankshaft vibration damper.

Tightening torque	
Vibration damper to vibration damper hub (M8 x 16 mm Torx)	35 Nm (26 ft-lb)

– Remainder of assembly is reverse of disassembly.

Crankshaft rear main seal, removing and installing (N52, N54 engine)

Crankshaft rear main seal (flywheel seal) replacement requires removal of the transmission and flywheel.

> ***CAUTION—***
> - *When the crankshaft seal is replaced, its grooves and crankcase seams are filled with special Loctite® sealant to avoid oil leaks. Be sure to read the entire procedure before starting work.*

– Remove transmission. See **230 Manual Transmission**, **231 Sequential Manual Gearbox (SMG)** or **240 Automatic Transmission**.

– Manual transmission vehicles: Remove clutch pressure plate and disc. See **210 Clutch**.

BentleyPublishers
BentleyPublishers.com—All Rights Reserved

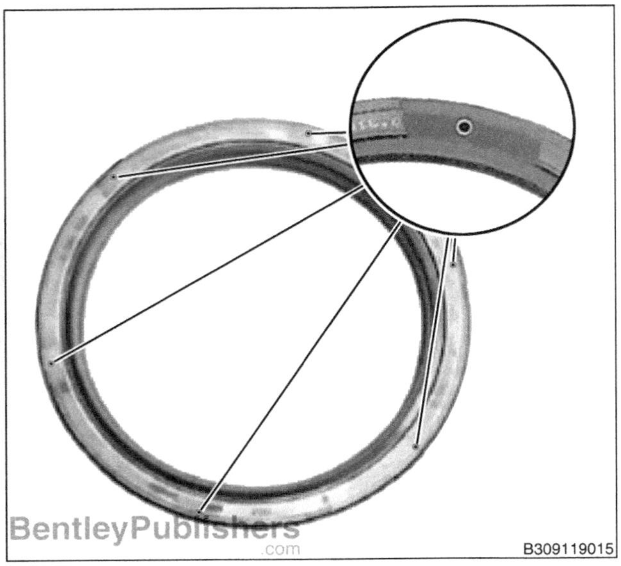

– Remove flywheel. See **210 Clutch** or **240 Automatic Transmission**.

– Drain engine oil.

< Crankshaft rear main seal has six removal openings (**inset**). If necessary, scrape off rubber coating to expose openings.

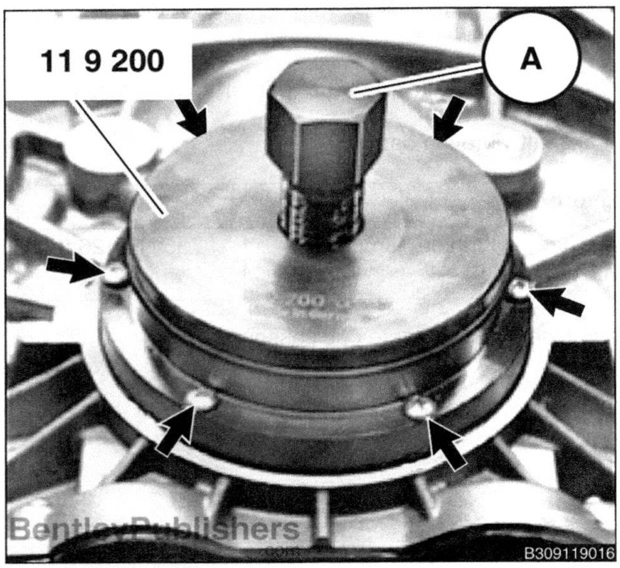

< Fit BMW seal puller, special tool 11 9 200, to seal.

- Insert sheet metal screws (**arrows**) into seal removal openings. Tighten snugly but do not overtighten.
- Screw in seal remover spindle (**A**) slowly to pull out seal.

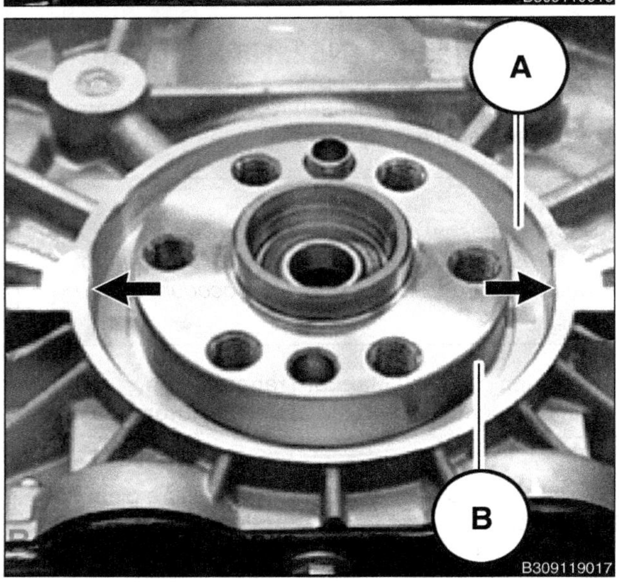

< Before installing new seal:

- Clean seal seating surface (**A**).
- Degrease crankcase seams (**arrows**) thoroughly.
- Apply light coat of engine oil to seal lip running surface (**B**) on crankshaft flange.

BentleyPublishers.com
BentleyPublishers.com—All Rights Reserved

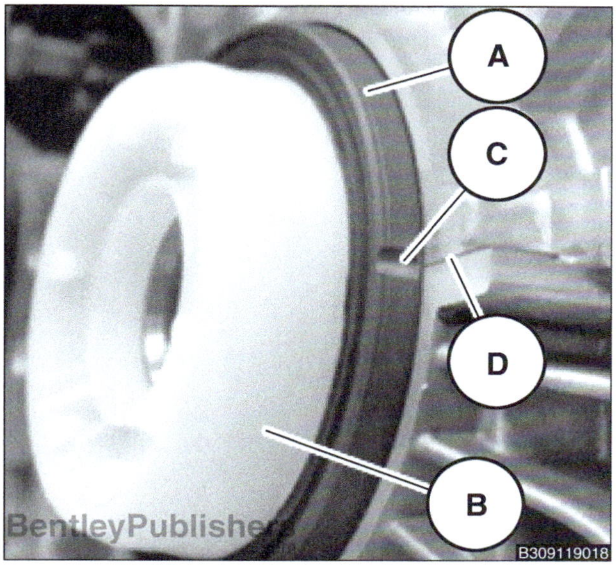

Place crankshaft seal (**A**) and installation sleeve (**B**) against crankcase opening.

> ***CAUTION—***
> *• Do not kink or damage the sealing lip. Do not touch with fingers.*

- Plastic installation sleeve is supplied with seal and is used as installation guide.
- Make sure seal grooves (**C**) center on crankcase seams (**D**).
- Coat both grooves on seal with Loctite® 171000 (primer). Allow to dry for approx. 1 minute.
- Press in seal as far as possible by hand, then remove installation sleeve.

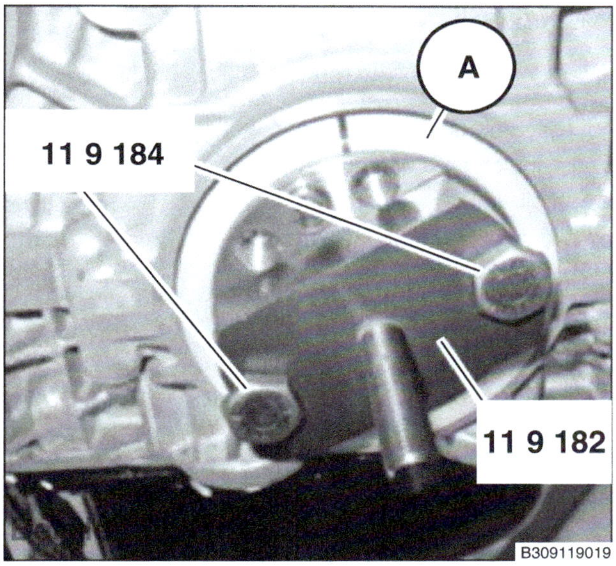

Fit BMW special tools 11 9 182 and 11 9 184 to crankshaft flange.

- Fit spacer ring (**A**) over seal. Spacer ring is supplied with seal.

Use BMW special tools 11 9 181 and 11 9 183 to press in seal.

– Remove special tools and spacer ring.

BentleyPublishers.com
BentleyPublishers.com—All Rights Reserved

◄ Use brush to coat seal grooves and crankcase seams (**A**) with Loctite® 171000 (primer).

◄ Use injector kit to fill seal grooves and crankshaft seams (**A**) with Loctite® 128357 (sealant).

– Coat sealed areas with Loctite® 171000 (primer) to bind sealant.

– Remainder of assembly is reverse of disassembly.
  - Install flywheel, clutch or torque converter and transmission.
  - Fill engine with oil.
  - Run engine and check for leaks.

BentleyPublishers.com
BentleyPublishers.com—All Rights Reserved

Crankshaft Seals (V8 Engine)

Crankshaft front seal, removing and installing (V8 engine)

> ***WARNING —***
> - *To avoid personal injury, be sure the engine is cold before beginning the procedure.*

> ***CAUTION—***
> - *Be sure to read the entire procedure before starting work.*

– Raise vehicle and support safely.

> ***WARNING —***
> - *Make sure the vehicle is stable and well supported at all times. Use a professional automotive lift or jack stands designed for the purpose. A floor jack is not adequate support.*

– Remove splash shield underneath engine. See **020 Maintenance**.

– Remove engine accessory belt. See **020 Maintenance**.

– Remove vibration damper (front pulley). See **117 Camshaft Timing Chain**.

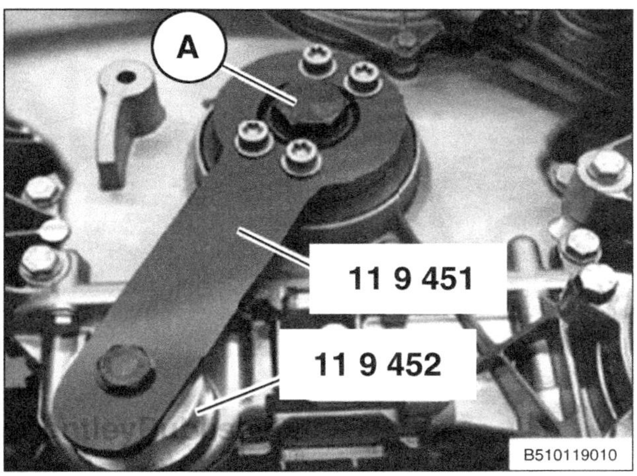

◄ Working at front of engine:

- Install BMW special tool 11 9 451 to vibration damper hub and BMW special tool 11 9 452 to oil sump.
- Remove vibration damper hub bolt (**A**) and discard.
- Remove vibration damper hub.

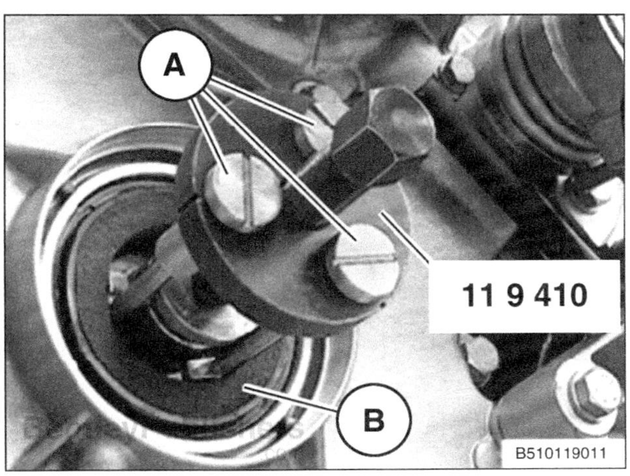

◄ Install BMW special took 11 9 410 onto crankshaft. Rotate puller levers (**A**) to grab behind crankshaft seal (**B**).

– Tighten center bolt of BMW special tool 11 9 410 and remove crankshaft seal (**B**).

BentleyPublishers.com
BentleyPublishers.com—All Rights Reserved

Crankshaft front seal, removing and installing (V8 engine)

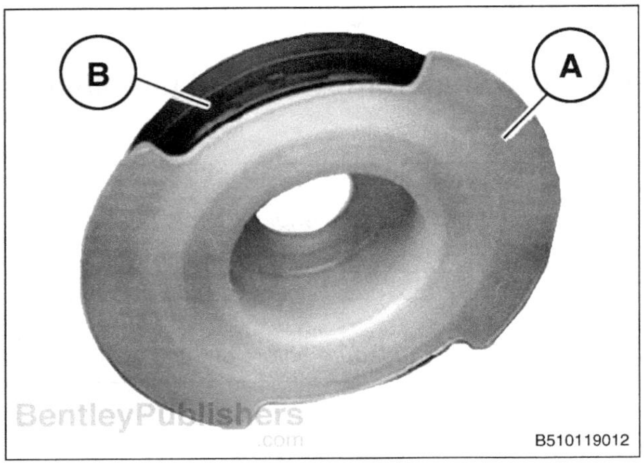

A support sleeve (**A**) is included with a new crankshaft seal. The support sleeve remains in crankshaft seal until installation. Do not remove prior to installation.

> ***CAUTION—***
> - Do not touch the sealing lip with your fingers it is highly sensitive and is easily kinked.

– Remove support sleeve from crankshaft seal.

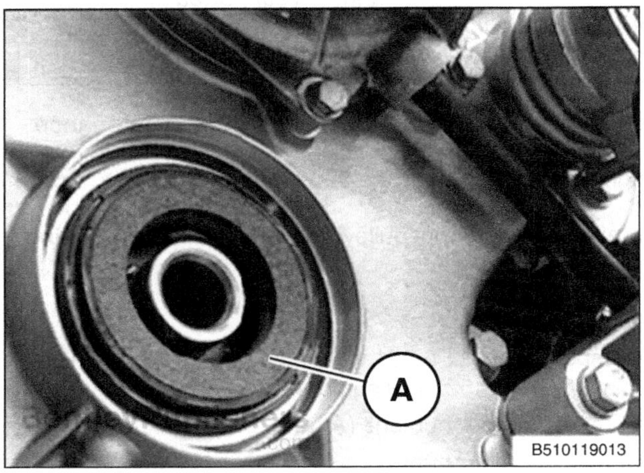

Place crankshaft seal (**A**) into timing cover by hand.

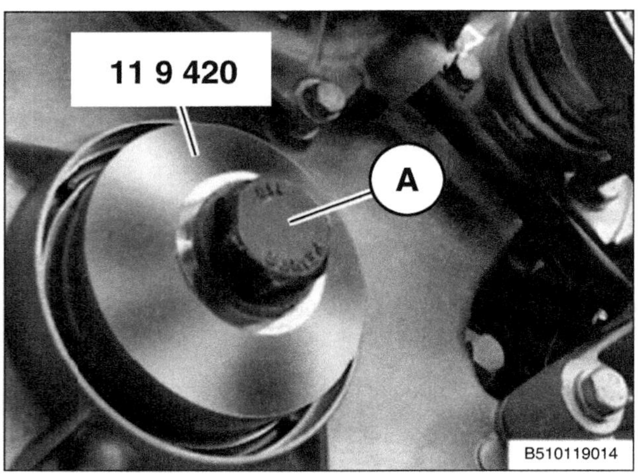

Using BMW special tool 11 9 420, turn hex (**A**) to press crankshaft seal into timing cover until it is flush.

– Install vibration damper hub by hand on crankshaft.

BentleyPublishers.com
BentleyPublishers.com—All Rights Reserved

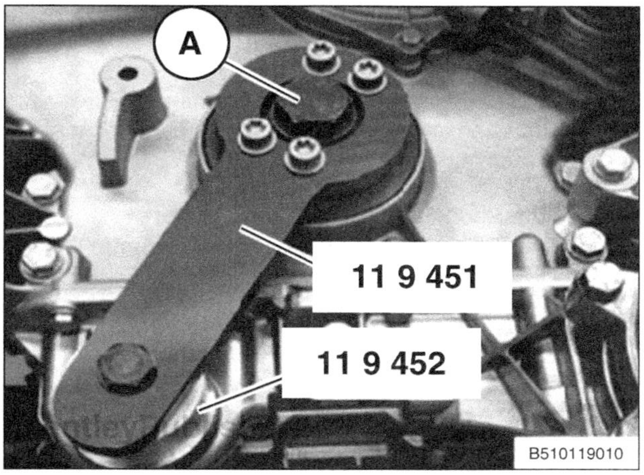

◀ Install BMW special tool 11 9 451 to vibration damper hub and BMW special tool 11 9 452 to oil sump.

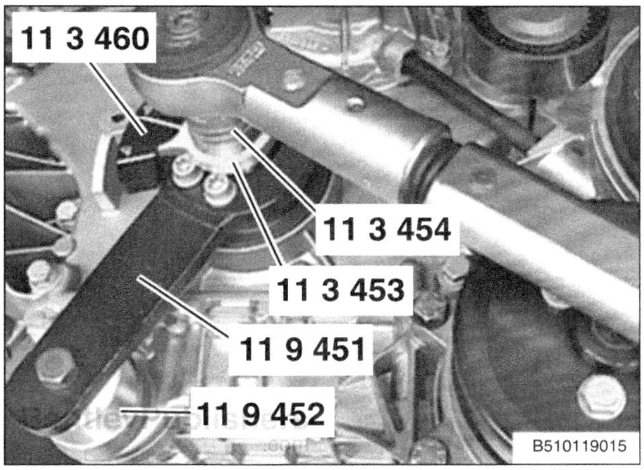

◀ Install BMW special tool 11 3 460 to end of crankshaft.

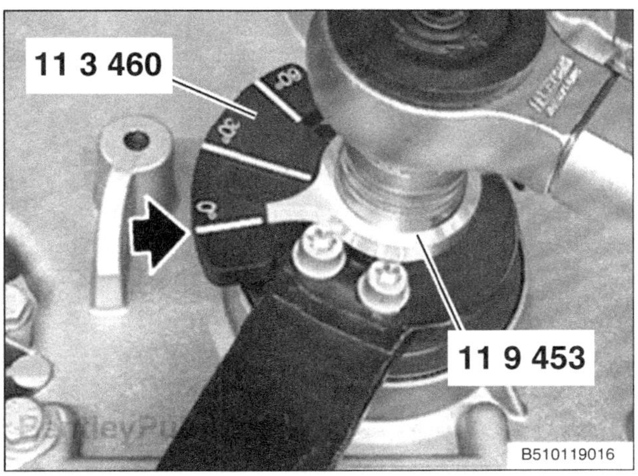

◀ Place pointer on BMW special tool 11 3 454 to 0° (**arrow**).

– Tighten damper bolt to specified torque and angle.

Tightening torque	
Vibration damper hub to crankshaft (replace bolt) (M18 x 1.5) • Stage 1 • Stage 2 • Stage 3 • Stage 4	 100 Nm (73.7 ft-lb) Additional 60° Additional 60° Additional 60°

– Remainder of assembly is reverse of disassembly.
- Install vibration damper.
- Fill engine with oil.
- Run engine and check for leaks.

BentleyPublishers.com
BentleyPublishers.com—All Rights Reserved

Crankshaft rear main seal, removing and installing (V8 engine)

Crankshaft rear main seal (flywheel seal) replacement requires removal of the transmission and flywheel. Crankshaft rear main seal is integral part of rear cover.

- Remove transmission. See **230 Manual Transmission**, **231 Sequential Manual Gearbox (SMG)** or **240 Automatic Transmission**.
- Manual transmission vehicles: Remove clutch pressure plate and disc. See **210 Clutch**.
- Remove flywheel. See **210 Clutch** or **240 Automatic Transmission**.
- Drain engine oil.

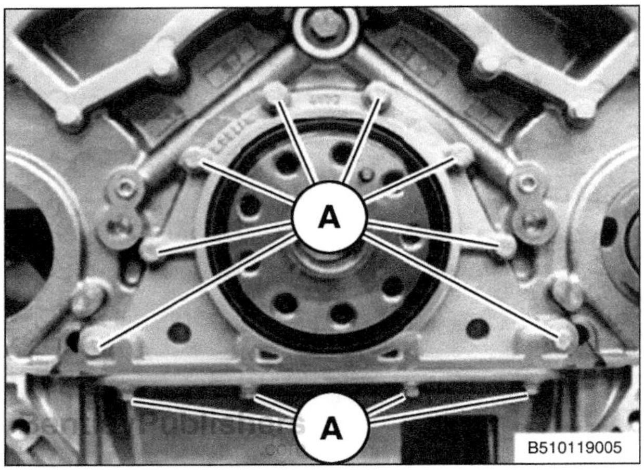

◀ Working at rear of engine block, remove fasteners (**A**).

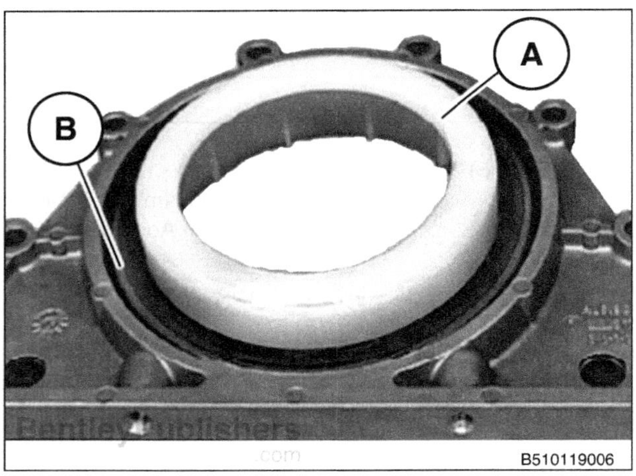

◀ A support sleeve (**A**) is included with new crankshaft rear main seal.The sleeve remains in crankshaft rear main seal during installation.

> ***CAUTION—***
> - *Do not touch the sealing lip with your fingers. It is highly sensitive and is easily kinked.*

BentleyPublishers.com
BentleyPublishers.com—All Rights Reserved

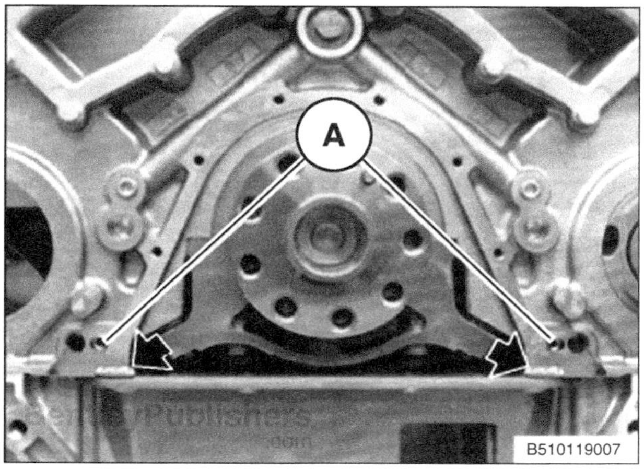

Inspect cover alignment dowels (**A**) for correct installation. Keep sealing area free from oil (**arrows**).

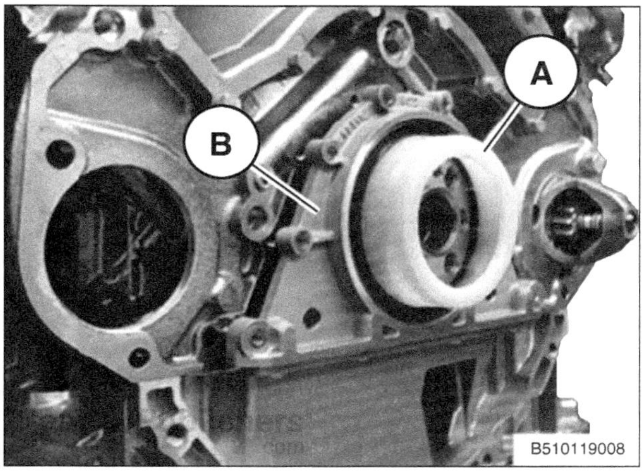

Lightly oil crankshaft. Place cover (**B**) and support sleeve (**A**) over crankshaft and carefully push on.

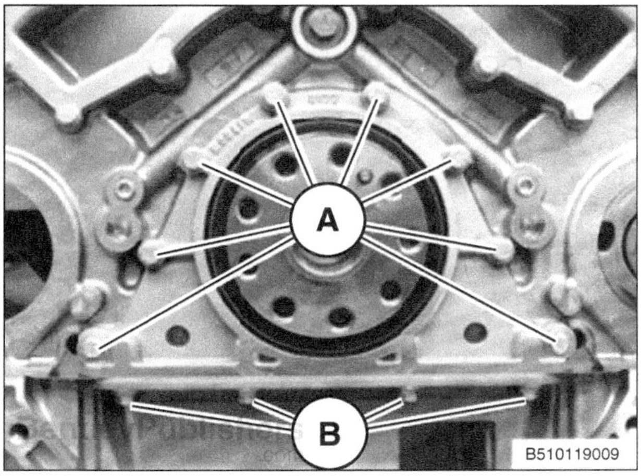

When installing:

- Install fasteners (**A**) finger tight
- Install fasteners (**B**) finger tight
- Tighten fasteners (**A**).
- Tighten fasteners (**B**).

– Remainder of assembly is reverse of disassembly.

- Install flywheel, clutch or torque converter and transmission.
- Fill engine with oil.
- Run engine and check for leaks.

Bentley Publishers
BentleyPublishers.com—All Rights Reserved

Oil Pan and Oil Pump (M54 Engine)

Oil pan, removing and installing (M54 engine)

– Disconnect negative (-) battery cable.

> ***CAUTION—***
> • *Prior to disconnecting the battery, read the battery disconnection cautions in* **001 Warnings and Cautions**.

– Remove air filter housing. See **130 Fuel Injection**.

– Raise engine hood into assembly position. See **410 Fenders, Engine Hood**.

– Raise car and support safely.

> ***WARNING—***
> • *Make sure the car is stable and well supported at all times. Use a professional automotive lift or jack stands designed for the purpose. A floor jack is not adequate support.*

– Drain engine oil. See **020 Maintenance**.

– Remove dipstick guide pipe.

– Remove splash shield and front end reinforcement from under engine.

◄ Install engine support brace across engine bay. Raise engine slightly to release load on engine mounts.

– Working underneath engine, remove right and left engine mount upper nuts (**arrow**).

– Lower front subframe. See **310 Front Suspension**.

BentleyPublishers.com
BentleyPublishers.com—All Rights Reserved

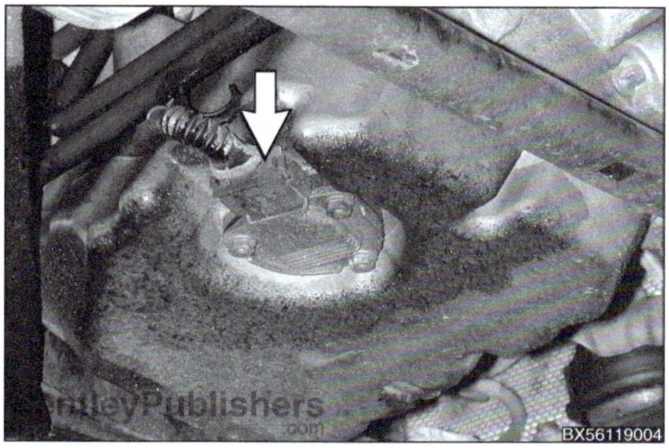

◀ Detach electrical harness connector (**arrow**) at oil level sensor.

– Automatic transmission: Disconnect transmission oil cooler lines from transmission and oil pan.

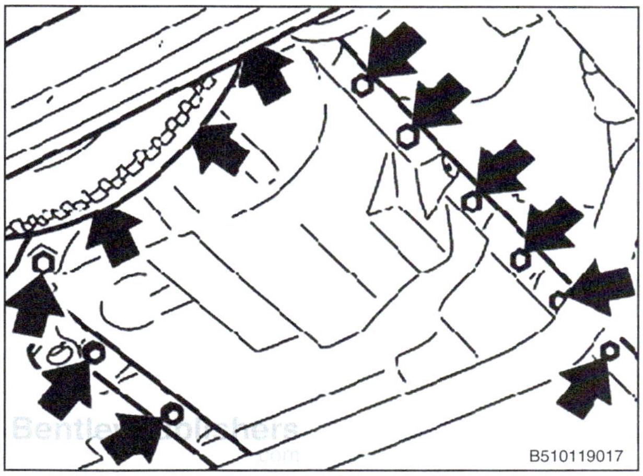

◀ Remove oil pan screws. Slide oil pan backward to remove.

> ***CAUTION—***
> - *If the oil pan does not separate easily from the engine cylinder block, tap lightly with a rubber mallet to break it free. Do not use prying tools.*

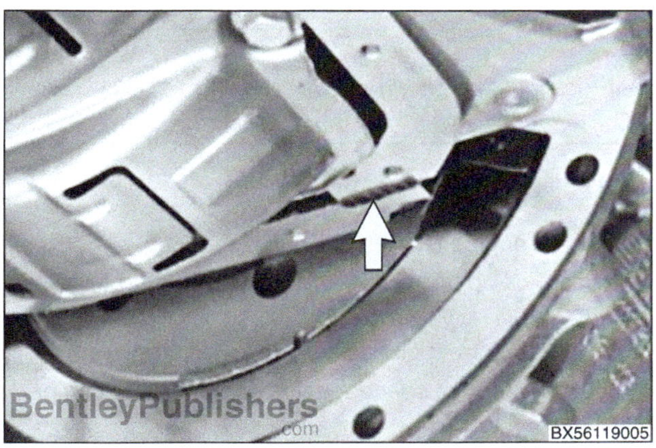

◀ Clean sealing surface of engine block, then apply beads of 3-Bond® 1209 or equivalent approx. 3 mm high and 2 mm wide (**arrow**) to timing cover and rear main seal carrier seams.

– When reinstalling oil pan:
- Thoroughly clean old gasket material from oil pan sealing surface and use a new gasket.
- Insert oil pan bolts.
- Tighten front oil pan bolts first.
- Tighten transmission bellhousing bolts last.

– Remainder of installation is reverse of removal. Fill engine with oil, run engine and check for leaks.

> ***CAUTION—***
> - *Follow correct installation procedure for front end reinforcement in* **310 Front Suspension**.

NOTE—
- *It is not necessary to realign the front end after this procedure.*

BentleyPublishers.com
BentleyPublishers.com—All Rights Reserved

Oil pump, removing and installing (M54 engine)

The M54 engine oil pump is located on the underside of the engine block, inside the oil pan. It is driven by the crankshaft via chain and sprocket assembly. To remove the oil pump, lower the front suspension and remove the oil pan. Read the procedure through before starting work.

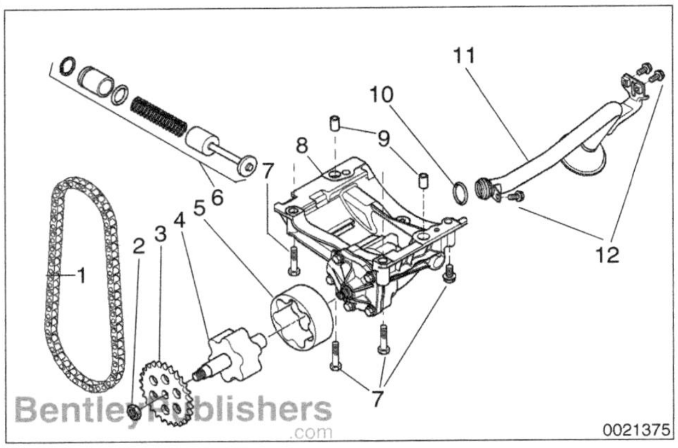

6-cylinder engine oil pump assembly:

1. Drive chain
2. M10 x 1 nut (left hand threads)
   - tighten to 25 Nm (18 ft-lb)
3. Oil pump sprocket
4. Inner pump rotor
5. Outer pump rotor
6. Pressure relief valve
7. M8 bolt
   - tighten to 23 Nm (17 ft-lb)
8. Oil pump housing
9. Locating dowels
10. Pick-up tube sealing O-ring
11. Pick-up tube
12. Pickup tube mounting bolts (M6)
   - Tighten to 10 Nm (7 ft-lb)

– Remove oil pan. See **Oil pan, removing and installing (M54 engine)** in this repair group.

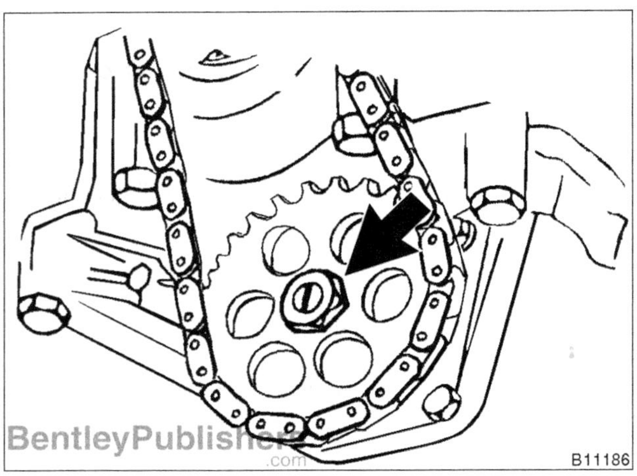

Remove oil pump sprocket mounting nut (**arrow**).

> ***CAUTION—***
> *• Oil pump sprocket nut has left hand threads. Turn clockwise to loosen.*

– Lift sprocket off pump together with drive chain.

– Unbolt oil pump from engine block.
- Remove oil pump pick-up tube bolts and remove pick-up tube before removing pump.
- Remove mounting bolts from oil pump. Withdraw pump.
- Be sure to record location of spacers between pump and engine block, if applicable.

– Spin oil pump shaft and check that gears turn smoothly. Replace pump if gears spin with difficulty.

– Inspect oil pressure relief valve. See **Oil pressure relief valve (M54 engine)** in this repair group.

BentleyPublishers.com
BentleyPublishers.com—All Rights Reserved

– Installation is reverse of removal, noting the following:
  - Inspect pump locating dowels for damage and correct positioning.
  - Replace oil pick-up and return tube sealing O-rings.
  - Align sprocket splines to oil pump shaft splines before tightening sprocket nut.

CAUTION—
- *Oil pump sprocket nut has left hand threads. Turn counterclockwise to tighten.*

Tightening torques	
Oil pump sprocket to oil pump shaft (M10 x 1) (left hand thread)	25 Nm (18 ft-lb)
Oil pump to crankcase (M8)	23 Nm (17 ft-lb)

– Fill engine with oil, run engine and check for leaks.

Oil pressure relief valve (M54 engine)

– Remove oil pump. See **Oil pump, removing and installing (M54 engine)**. The oil pump pressure relief valve is held in the side of the oil pump with a circlip.

Oil pressure relief valve assembly (M54 engine)

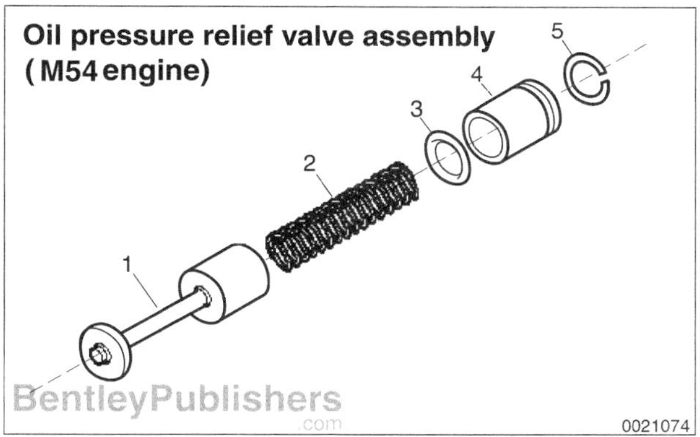

< Inspect oil pump relief valve components and bore for scoring or other damage.

1. Control plunger
2. Spring
3. Sealing O-ring
4. Sleeve
5. Circlip

– Install using a new circlip.

BentleyPublishers.com
BentleyPublishers.com—All Rights Reserved

Oil Cooler

Engine oil cooler, removing and installing (N54, N62TU engines)

External Oil Coooler (535i)

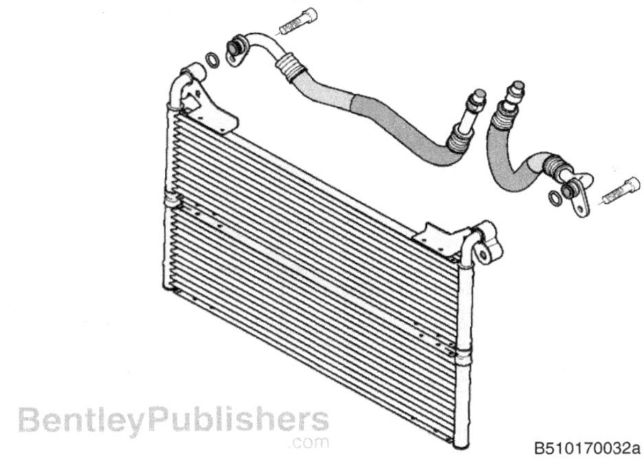

B510170032a

On 535i (N54) and 550i (N62TU) models, an external engine oil cooler is fitted between the radiator and the electric cooling fan. Cars with this external cooler can be identified by the oil lines passing through the radiator cooling fan shroud.

The oil cooler is removed together with the electric cooling fan. Once the fan is removed, the cooler can be unclipped from the fan shroud. For electric cooling fan removal and installation, see **170 Radiator and Cooling System**.

Engine oil cooler, removing and installing (N52 engines)

2006 and later 6 cylinder models (N52 engines) may be equipped with an engine oil cooler mounted to the front of the oil filter housing.

B510119018a

Oil cooler location on N52 engine (**arrow**).

- Drain engine coolant. See **170 Radiator and Cooling System**.
- Loosen oil filter cover to allow oil in filter housing to flow into pan.
- Remove engine upper cover. See **020 Maintenance**.

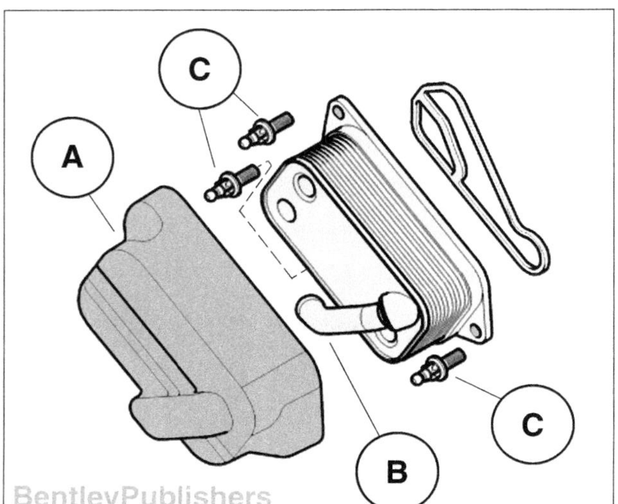

B510119019

- Unclip cover (**A**) from cooler.
- Remove coolant line from hose fitting on cooler (**B**). Use a shop rag to catch coolant.
- Remove bolts (**C**) and remove oil cooler. Use a shop rag to catch oil.
- Installation is reverse of removal, noting the following.
  - Clean mating surfaces and replace seal.
  - Tighten fasteners to specified torque avoid damaging seal.

Tightening torques	
M8	22 Nm (16 ft-lb)

- Refill coolant, see **170 Radiator and Cooling System**.
- Check and top off oil, see **020 Maintenance**.

BentleyPublishers.com
BentleyPublishers.com–All Rights Reserved

OIL FILER HOUSING

Oil filter housing bracket, removing and installing (M54 engine)

- Disconnect battery negative cable.
- Loosen oil filter cover to allow oil in filter housing to flow into oil pan.
- Remove drive belt and unbolt alternator from oil filter housing bracket. See **121 Battery Starter, Alternator**.
- Empty power steering reservoir tank and then unbolt tank and move to one side (lines remain connected).
- Unbolt power steering pump from oil filter housing bracket pump (lines remain connected). See **320 Steering and Wheel Alignment**.
- Unbolt and remove alternator drive belt tensioner.
- Disconnect harness connector from oil pressure switch.
- Disconnect harness connector for oil temperature switch.
- Working at oil filter housing, disconnect oil pressure line for VANOS adjustment unit.

< Remove bolts (**arrows**) and remove complete oil filter housing bracket.

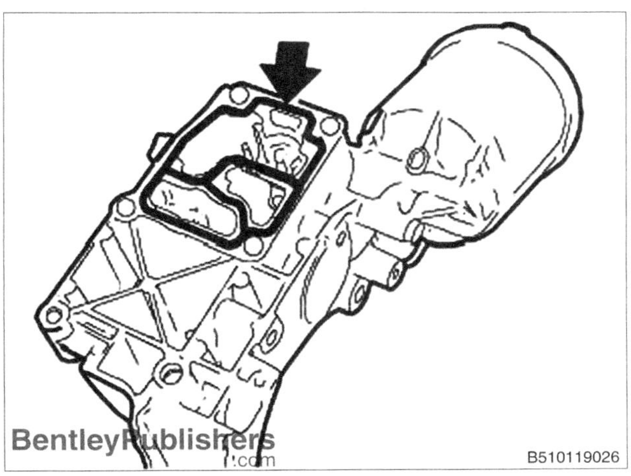

< Remove old profile gasket and install new gasket (**arrow**).

BentleyPublishers.com
BentleyPublishers.com—All Rights Reserved

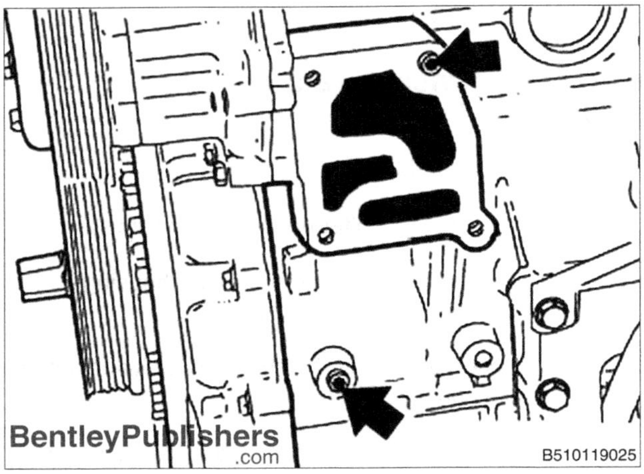

Installation is reverse of removal. Keep in mind the following:

- A unique torque is not given for the oil filter housing bracket bolts. General tightening torque based on bolt grade is given below.
- Be sure to align bracket to locating dowels (**arrows**).
- Reconnect all harness connectors previously removed.
- Check engine oil level and power steering fluid level upon job completion.
- Reconnect battery negative cable.

NOTE—

- *Bolt grade is marked on bolt head.*

Tightening torques	
M8 (8.8 grade) M8 (10.9 grade)	24 Nm (18 ft-lb) 34 Nm 25 ft-lb)

Oil filter housing, removing and installing (N52 engines)

– Drain engine coolant. Remove electric cooling fan. See **170 Radiator and Cooling System**.

– Loosen oil filter cover to allow oil in filter housing to flow into pan.

– Remove intake manifold, see **131 Fuel Injection MSV70, MSV80 (N52 Engine)**.

NOTE—

- *Intake manifold removal is required to access a difficult-to-reach oil filter housing fastener (bolt **D**). This bolt is located at right side of oil filter housing under intake manifold. It may be possible to access this bolt without removing the intake manifold using suitable tools (universal joints / e-torx socket).*

– If necessary, remove oil cooler from oil filter housing. See **Engine oil cooler, removing and installing (N52 engines)** given earlier.

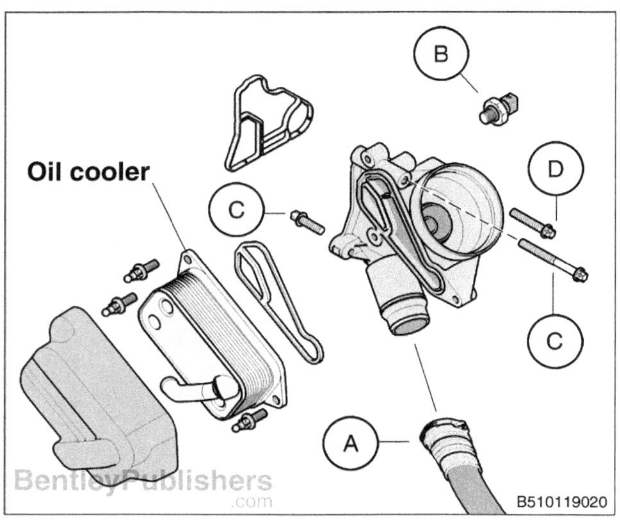

Working at oil filter housing:

- Disconnect coolant line from housing (**A**).
- Remove oil pressure sensor (**B**).
- Remove bolts **C** and **D** and remove oil filter housing from engine.

– Installation is reverse of removal, noting the following:

- Clean mating surfaces and replace o-rings and seals.
- Be sure not to over tighten fasteners to avoid damaging seals.

Tightening torques	
M8	22 Nm (16 ft-lb)

- Refill coolant, see **170 Radiator and Cooling System**.
- Check and top off oil, see **020 Maintenance**.

BentleyPublishers.com
BentleyPublishers.com—All Rights Reserved

Oil filter housing, removing and installing (N54 engine)

– Drain engine coolant. Remove electric cooling fan with oil cooler. See **170 Radiator and Cooling System**.

– Loosen oil filter cover to allow oil in filter housing to flow into pan.

– Remove intake manifold, see **132 Fuel Injection MSD80, MSD81 (N54 Engine)**.

NOTE —

- *Intake manifold removal is required to access a difficult-to-reach oil filter housing fastener (bolt **C**). This bolt is located at right side of oil filter housing under intake manifold. It may be possible to access this bolt without removing the intake manifold using suitable tools (universal joints / e-torx socket).*

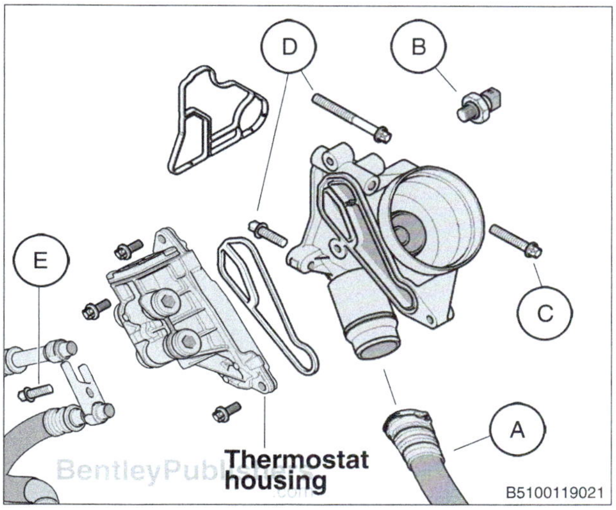

◄ Working at oil filter housing:

- Remove fastener **E** and pull lines from thermostat housing.
- Remove thermostat housing fasteners and remove thermostat housing from oil filter housing.
- Remove coolant line from housing (**A**).
- Remove oil pressure sensor (**B**).
- Remove bolts **C** and **D** and remove oil filter housing from engine.

– Installation is reverse of removal, noting the following.

- Clean mating surfaces and replace o-rings and seals.
- Be sure not to over tighten fasteners to avoid damaging seals.

Tightening torques	
M8	22 Nm (16 ft-lb)

- Refill coolant, see **170 Radiator and Cooling System**.
- Check and top off oil, see **020 Maintenance**.

BentleyPublishers.com
BentleyPublishers.com—All Rights Reserved

BentleyPublishers.com
BentleyPublishers.com–All Rights Reserved

120 Ignition System

TABLE

General

This repair group covers ignition component troubleshooting and replacement.

See also:
- **020 Maintenance** for spark plug replacement
- **100 Engine–General** for engine applications
- **130, 131, 132, 133 Fuel Injection**
- **ECL Electrical Component Locations**
- **ELE Electrical Wiring Diagrams**
- **OBD On-Board Diagnostics**

BentleyPublishers.com–All Rights Reserved

Ignition firing order

Table a. Ignition firing order	
6-cylinder engine	1-5-3-6-2-4
V8 engine	1-5-4-8-6-3-7-2

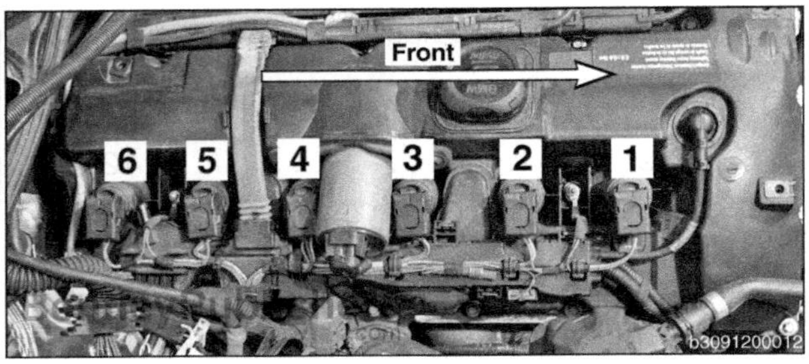

6-cylinder engine: Cylinder 1 is at the front of the engine.

V8 N62 engine:
- Cylinder 1 is at right (passenger side) front.
- Cylinders 1 - 4 are the right bank
- Cylinders 5 - 8 are the left bank.

Disabling ignition system

The ignition system operates in a lethal voltage range. Disable the ignition system any time engine service or repair work is being done that requires the ignition to be switched ON.

One way of disabling the ignition is by removing the DME main relay. The relay is located in the electronics box (E-box) in the right rear of the engine compartment. DME fuse, relay and power supply information is in **130**, **131**, **132**, **133 Fuel Injection** or **ECL Electrical Component Locations**.

> ***CAUTION—***
> - *Relay locations vary. Use care when identifying relays. See* **ECL Electrical Component Locations**.

BentleyPublishers.com
BentleyPublishers.com—All Rights Reserved

Warnings and Cautions

WARNING —

- *Do not touch or disconnect any cables from the ignition coils while the engine is running or being cranked by the starter.*
- *The ignition system produces high voltages that can be fatal. Avoid contact with exposed terminals. Use extreme caution when working on a car with the ignition switched ON or the engine running.*
- *Connect and disconnect the DME system wiring and test equipment leads when the ignition is OFF.*
- *Before operating the starter without starting the engine (for example when testing compression), disable the ignition. See* **Disabling ignition system** *in this repair group.*

CAUTION—

- *Do not attempt to disable the ignition by removing the coils from the spark plugs.*
- *Do not connect any test equipment that delivers a 12-volt power supply to terminal 15 (+) of the ignition coil. The current flow may damage the engine control module (ECM). Connect test equipment only as specified by BMW or the equipment maker.*
- *Connect or disconnect ignition system wires, multiple wire connectors and ignition test equipment leads only while the ignition is OFF. Switch multimeter functions or measurement ranges with test probes disconnected.*
- *Do not disconnect the battery while the engine is running.*
- *Prior to disconnecting the battery cables, read the battery disconnection cautions in* **001 Warnings and Cautions**.
- *Wait at least 1 minute after switching the ignition OFF before removing the ECM connector. If the connector is removed before this time, residual power in the system relay may damage the control module.*
- *Use a digital multimeter for electrical tests. Use an LED test light for quick tests.*
- *To avoid electrochemical corrosion to engine components made of aluminum-magnesium alloy, do not use steel fasteners in place of aluminum. For reliable identification, test fasteners with a magnet for aluminum composition.*
- *Replace aluminum fasteners each time they are loosened.*
- *Follow torque instructions, including angle of rotation specifications, when installing aluminum fasteners.*

BentleyPublishers.com

BentleyPublishers.com—All Rights Reserved

Digital Motor Electronics (DME) Ignition System

BMW 5 Series models are equipped with digital motor electronics (DME), also known as Motronic. In these systems, fuel injection and ignition are controlled by an integrated engine control module (ECM). Application information for DME systems is in **Table b**.

Most DME functions are described in **130**, **131**, **132**, **133 Fuel Injection**. In the topics that follow, functions that are traditionally associated with ignition are given fuller treatment.

Table b. 5 Series engine management applications

Year, model	Engine code	Engine management	Features
2004-2005			
Sedan: 525i, 530i	M54	Siemens DME MS 45.1	Dual VANOS
Sedan: 545i	N62	Bosch DME ME 9.2.1	Valvetronic, VANOS
2006-2007			
Sedan, Sports Wagon: 525i, 525xi, 530i, 530xi	N52	Siemens MSV70 SULEV II	Valvetronic, VANOS
Sedan: 550i	N62TU	Bosch DME ME 9.2.2	Valvetronic, VANOS
2008-2010			
Sedan, Sports Wagon: 528i, 528xi,	N52KP	Siemens MSV80 ULEV II	Valvetronic, VANOS
Sedan, Sports Wagon: 535i, 535x, 535i xDrive	N54	Siemens MSD81	VANOS
Sedan: 550i	N62TU	Bosch DME ME 9.2.2	Valvetronic, VANOS

Ignition coils

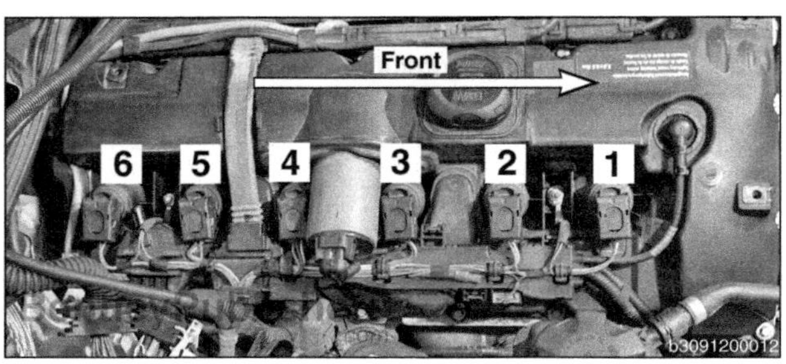

5 Series engines use a distributorless ignition system with individual ignition coils (**numbered**) for each cylinder. There is no distributor cap or ignition rotor. (N52 engine shown in photo.)

BentleyPublishers.com
BentleyPublishers.com—All Rights Reserved

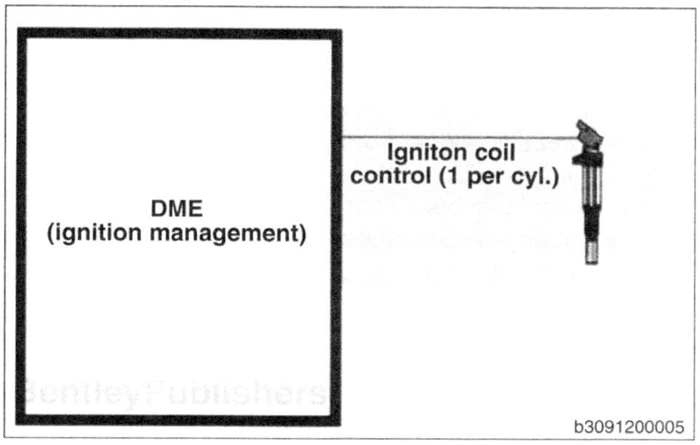

Each coil is separately controlled and monitored by the ECM.

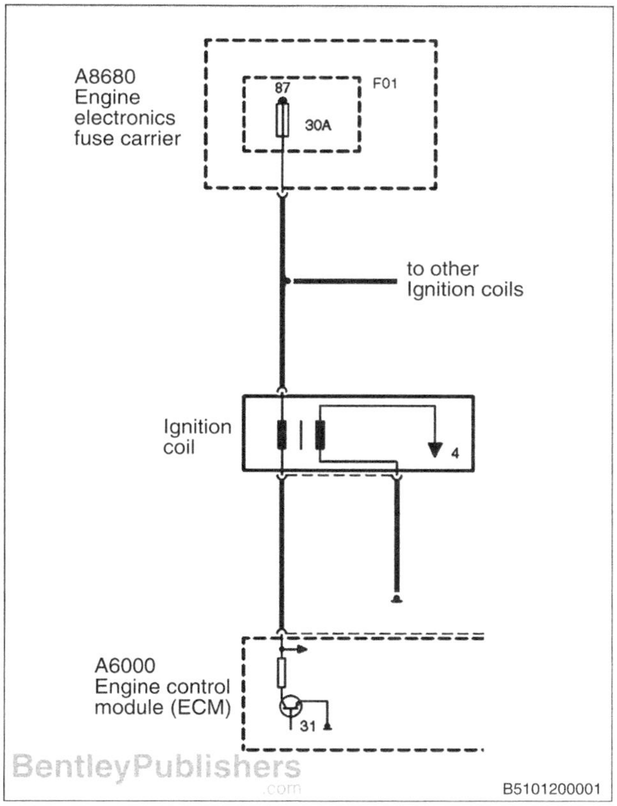

When the ignition is switched ON, the ECM receives wake-up (KL 15w) via the bus network and a hardwire KL15 wake-up signal directly from the car access system (CAS). It then energizes the ECM main relay to supply power to the engine electronics fuse carrier and to the ignition coils.

> ***CAUTION—***
> - *Maintain a high level of cleanliness when servicing ignition coils. Fuel / oil residue can cause a breakdown in the electrical resistance qualities of silicone used in production. This in turn can result in ignition coil failure.*

NOTE—
- *Schematic diagrams shown are for illustrative purposes only. See* **ELE Electrical Wiring Diagrams** *for engine management specific wiring information.*

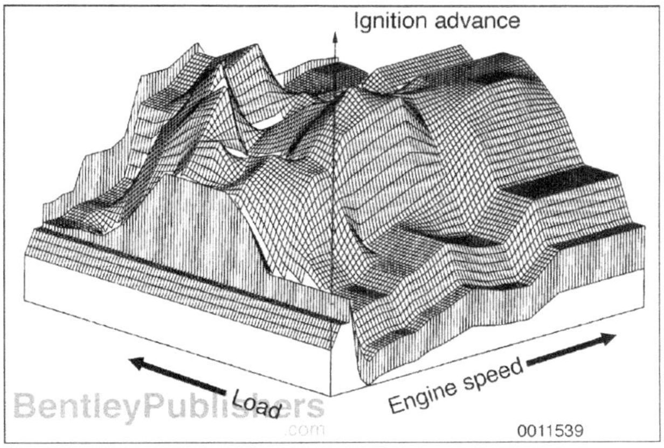

Ignition timing is electronically mapped and not adjustable. The ECM uses engine load, engine speed, coolant temperature, knock detection, and intake air temperature as the basic inputs for timing control.

A three-dimensional map similar to the one shown is digitally stored in the ECM. The initial ignition point is determined by the crankshaft sensor during cranking. Once the engine is running, the ECM refers to the stored map to continually adjust ignition timing based on operating conditions.

Crankshaft sensor

The crankshaft sensor detects crankshaft position and rotation speed via a toothed pulse wheel mounted on flywheel. If the ECM does not receive an impulse signal from the crankshaft sensor during cranking, the engine does not start. If the OBD II system misfire detection protocol detects a catalyst damaging fault due to a malfunction in crankshaft sensor components, the malfunction indicator light (MIL) is illuminated.

The crankshaft position sensor is supplied 12 volts from the engine electronics fuses and ground from the ECM. It produces a 5 volt square wave signal to the ECM.

◄ M54 engine: Crankshaft sensor (**arrow**) is at left rear of cylinder block below starter.

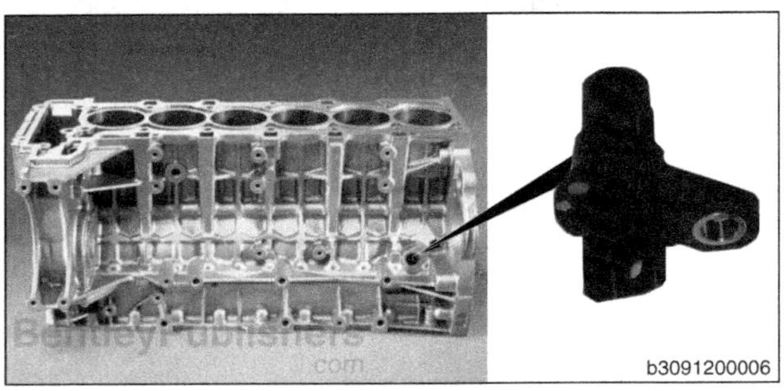

◄ N52, N54 engine: Crankshaft sensor (**arrow**) is at left rear of cylinder block below starter.

◄ V8 engine: Crankshaft sensor (**arrow**) is on right of transmission bell housing.

The crankshaft sensor is also referred to as engine speed / reference sensor.

BentleyPublishers.com
BentleyPublishers.com—All Rights Reserved

Camshaft sensors

Each camshaft sensor responds to an impulse wheel mounted at the end of the camshaft. The signal from camshaft sensors are used by the ECM for cylinder recognition, spark timing, sequential fuel injection, VANOS (camshaft timing) and Valvetronic (variable valve lift) control. If a fault with the camshaft sensor is detected the malfunction indicator light (MIL) is illuminated

The camshaft position sensor is supplied 12 volts from the engine electronics fuses and ground from the ECM. It produces a 5 volt square wave signal to the ECM.

M54 engine intake camshaft sensor (**arrow**): Underneath crankcase breather hose, left side of cylinder head.

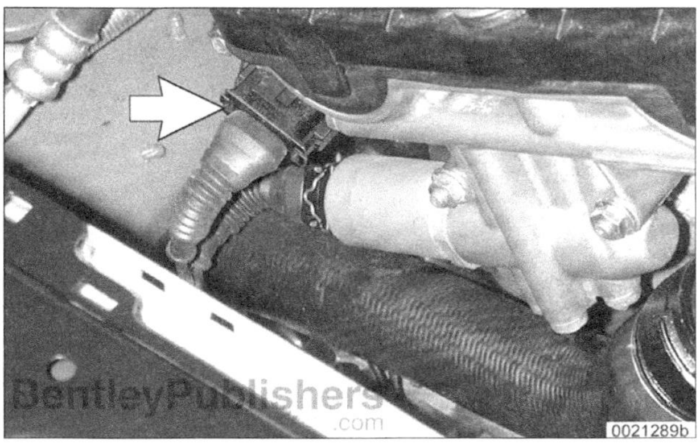

M54 engine exhaust camshaft sensor (**arrow**): Right front of cylinder head.

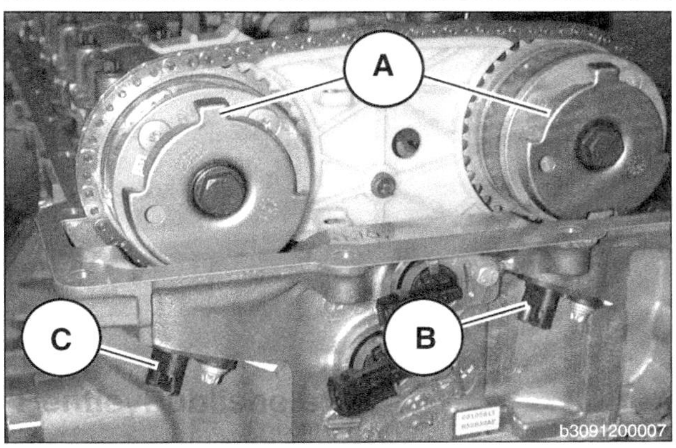

N52, N54 engine:

- Intake camshaft sensor (**B**): Left front of cylinder head.
- Exhaust camshaft sensor (**C**): Right front of cylinder head.

NOTE—

- *N52, N54 engine: A malfunctioning or sticking VANOS solenoid may set camshaft position sensor fault codes. If you suspect this, swap VANOS solenoid from bank with fault code to the other bank. If the camshaft position sensor fault code follows the location of the solenoid, the solenoid is at fault.*

The camshaft sensor is also called the cylinder identification sensor.

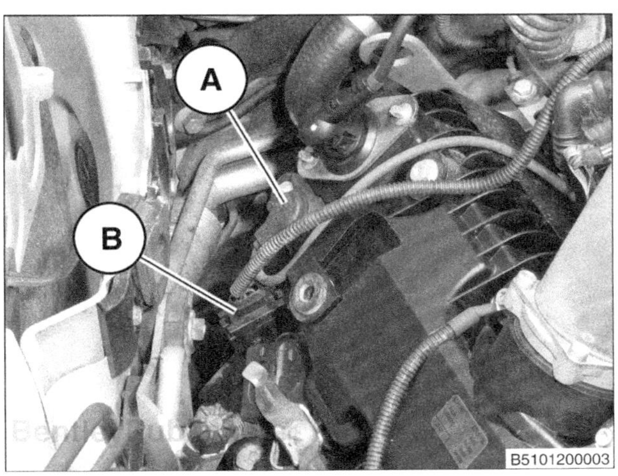

V8 engine: intake (**A**) and exhaust (**B**) camshaft sensors are located on the front top of each cylinder head.

BentleyPublishers.com
BentleyPublishers.com—All Rights Reserved

Knock sensors

Knock sensors monitor the combustion chambers for engine-damaging knock. A knock sensor is a piezoelectric microphone tuned to the frequencies of engine knock or detonation. If engine knock is detected, ignition spark is retarded by the ECM. If a fault with the knock sensor is detected the malfunction indicator light (MIL) is illuminated

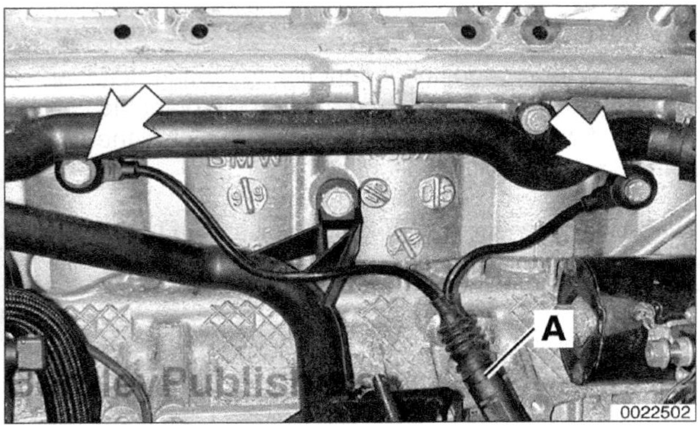

M54 engine: Knock sensors are bolted to left side of cylinder block under intake manifold.

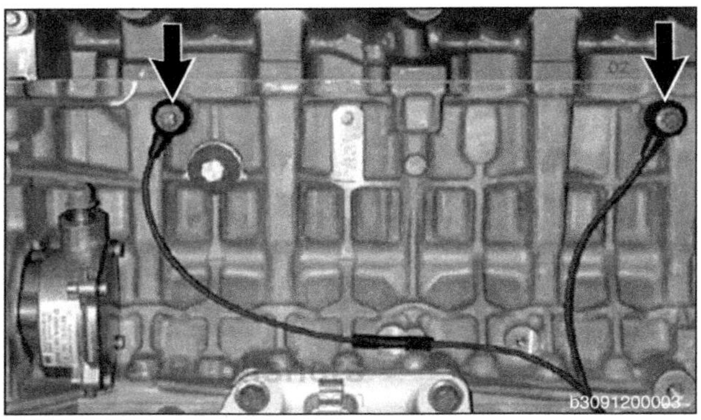

N52, N54 engine: Knock sensors are bolted to left side of cylinder block under intake manifold (**arrows**). They share an electrical connector and have to be replaced as a pair.

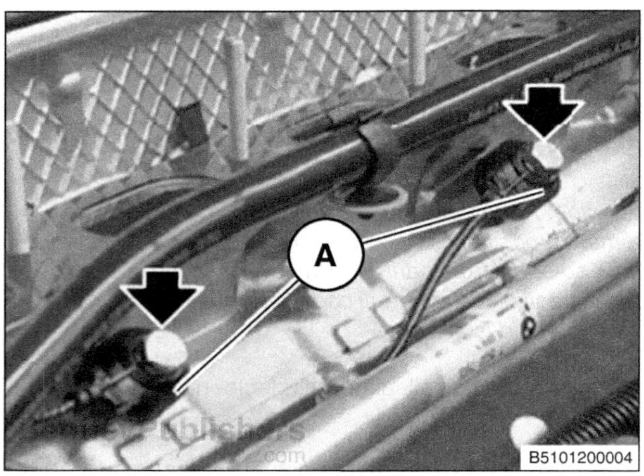

V8 engine: 4 sensors (**A**), 2 per bank of cylinders, are under intake manifold.

BentleyPublishers.com
BentleyPublishers.com—All Rights Reserved

TROUBLESHOOTING

On-board diagnostics

B309120017

If faults arise, or if the malfunction indicator light (MIL) is illuminated, begin troubleshooting by connecting a BMW scan tool to the data link connector (DLC or OBD II plug). For information on how to access diagnostic trouble codes (DTCs), see **020 Maintenance**.

Misfire detection

Engine misfire in one or more cylinders may be caused by malfunctions in various subsystems. The OBD II system incorporated into the engine management systems is designed to detect and warn of misfire faults during engine operation. See **OBD On-Board Diagnostics.**

> ***WARNING—***
> - *Ignition misfires can cause high hydrocarbon exhaust emissions and catalytic converter damage. For this reason, if a severe misfire is detected, the fuel injector to the specific cylinder is switched OFF and the MIL is illuminated. A misfire may also overheat the catalytic converter(s), a fire hazard.*

Oscilloscope diagnostic diagrams

One way to diagnose faulty engine management components or functions is to use an oscilloscope to analyze spark quality with the engine running.

12 7 050

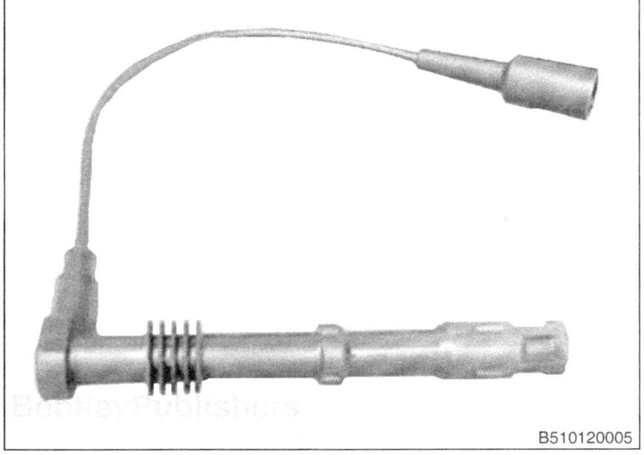

B510120005

BMW engines covered in this manual use a rod type ignition coil. Install BMW special tool 12 7 050 in series with coil to test secondary ignition voltage.

See **Table c** for a list of common ignition coil voltage faults and related causes.

Table c. Ignition secondary voltage diagnostics

	Secondary voltage low	Secondary voltage high
Spark plug electrode gap	Too small	Too big
Spark plug electrode condition		Worn / burnt
Spark plug electrode temperature	Too high	Too low
Engine compression	Too low	Too high
Spark plug wires		Faulty
Fuel air mixture		Too lean

BentleyPublishers
BentleyPublishers.com—All Rights Reserved

Oscilloscope diagnostic diagrams

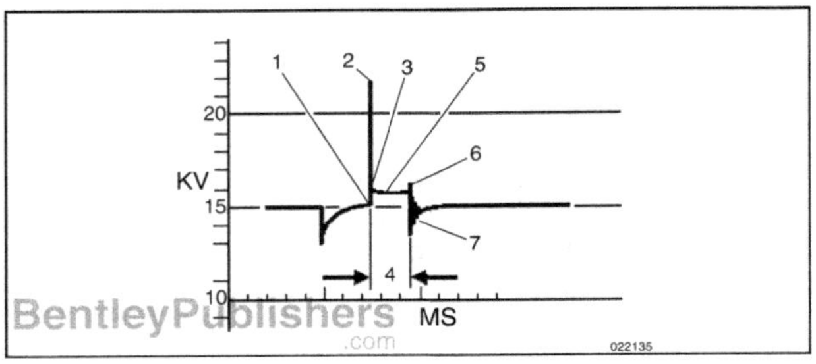

Normal scope trace of spark at idle

1. Start of ignition voltage peak
2. Level of ignition voltage
3. Level of combustion voltage
4. Period of combustion
5. Combustion curve characteristics
6. Start of spark decay
7. Termination oscillations

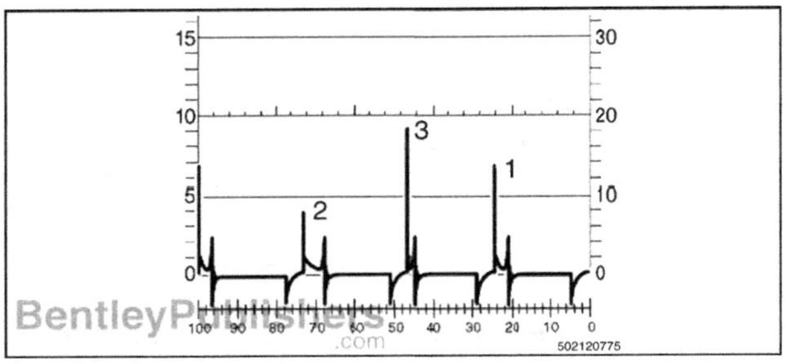

Ignition spark at idle speed

1. Normal ignition voltage peak: Good spark plug
2. Low voltage peak: Closed plug gap
3. High voltage peak: Open plug gap

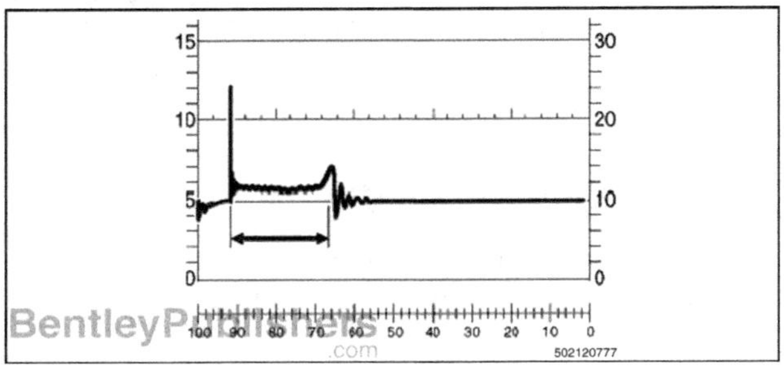

Long combustion period: Small spark plug gap

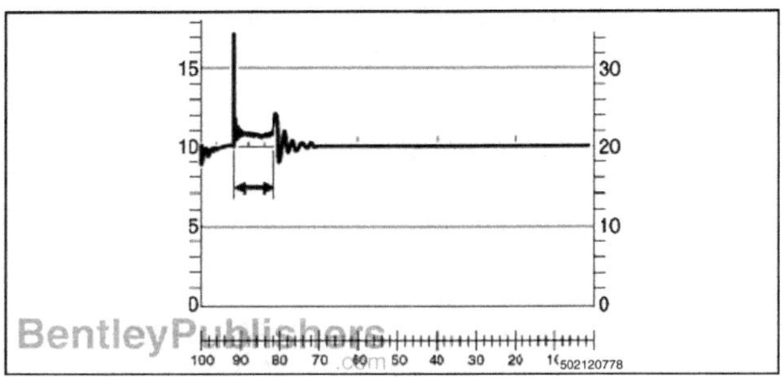

Short combustion period: Large spark plug gap

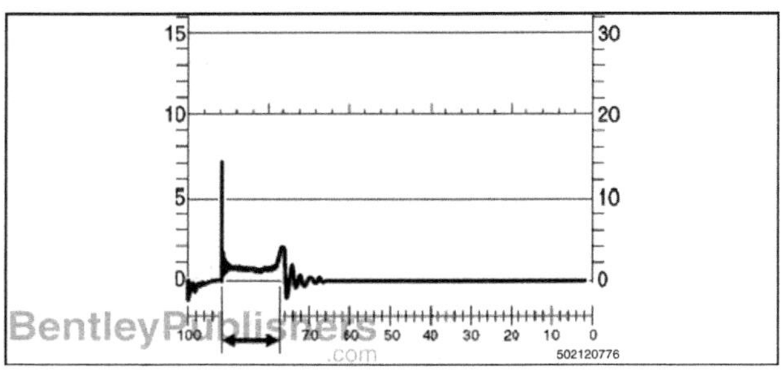

Normal combustion period at idle

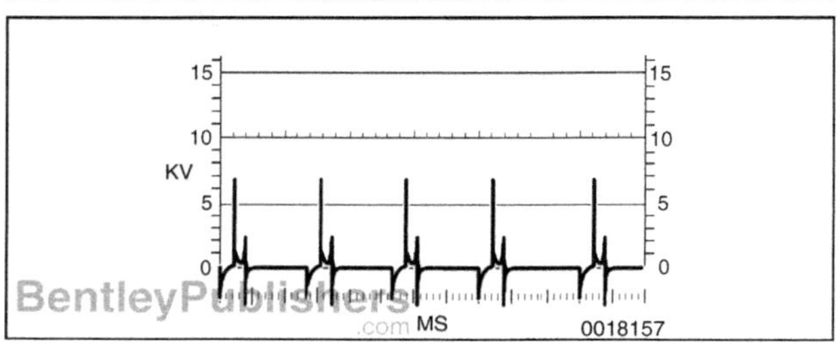

Normal oscilloscope pattern for ignition system at idle

BentleyPublishers.com
BentleyPublishers.com–All Rights Reserved

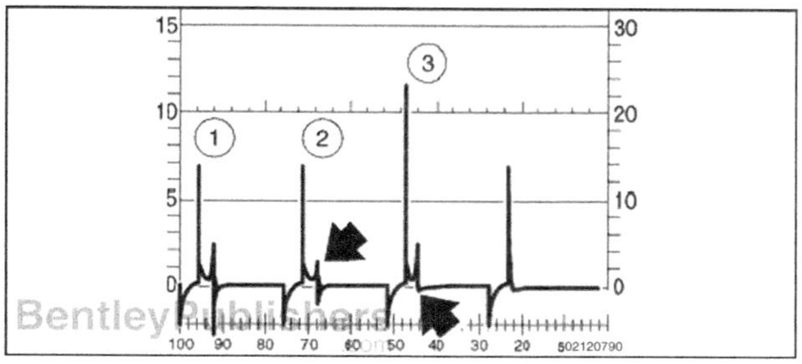

Ignition voltage peaks at idle

1. Normal ignition peaks
2. Downward peak (**arrow**) shortened: Ignition coil defective
3. Downward peak (**arrow**) missing completely: Ignition coil defective

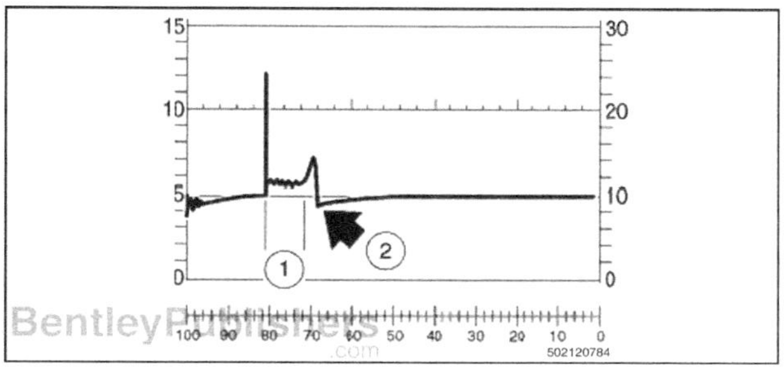

Defective ignition coil

1. Short spark period
2. Spark voltage line (**arrow**) with very slight drop

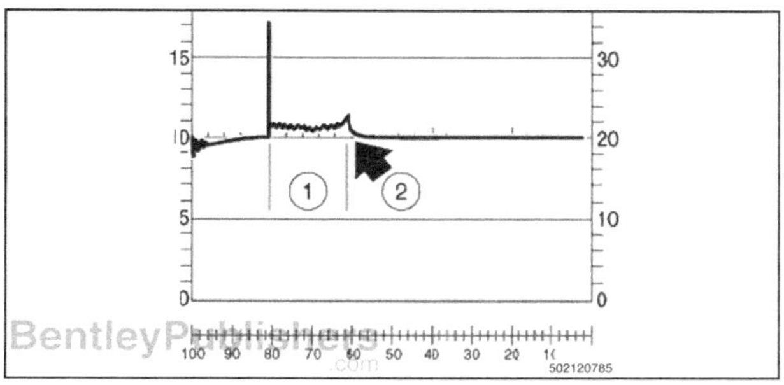

Defective ignition coil

1. Normal combustion period
2. Spark voltage line (**arrow**) absent

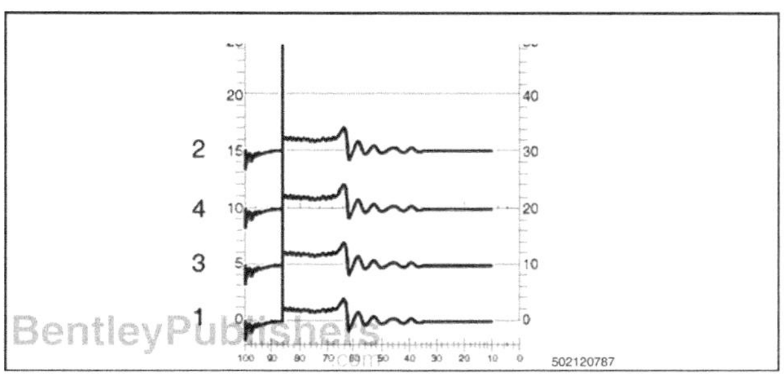

Normal secondary voltage patterns (4–cylinder pattern shown)

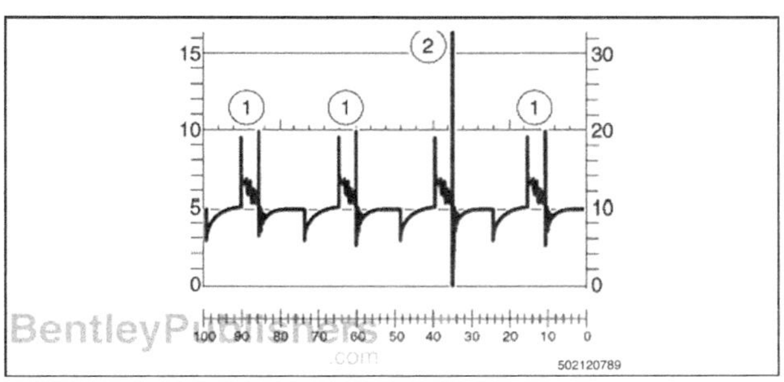

Ignition voltage peaks in response to sudden acceleration load:

1. Normal ignition pattern. Beginning of dying out pattern is not much higher than ignition voltage peak.
2. Beginning of dying out pattern considerably higher than ignition voltage peak. Fault in injection system:
   - Lean fuel mixture
   - Defective fuel injector
   - Low compression in cylinder

BentleyPublishers.com
BentleyPublishers.com—All Rights Reserved

IGNITION COIL SERVICE

There is a separate ignition coil above each spark plug. Spark plug replacement is covered in **020 Maintenance**.

> ***CAUTION—***
> *• Maintain a high level of cleanliness when servicing ignition coils. Fuel / oil residue can cause a breakdown in the electrical resistance qualities of silicone used in production. This in turn can result in ignition coil failure.*

Ignition coil, replacing (M54 engine)

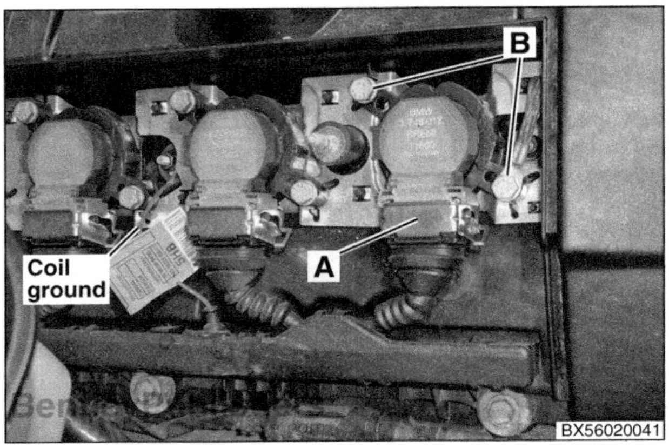

- Make sure ignition is switched OFF.
- Remove ignition coils cover. See **020 Maintenance**.
- ◄ Lift up on lock (**A**) and release ignition coil electrical connector.
- Remove fasteners (**B**). Note position of coil ground wire.

- ◄ Remove ignition coil by pulling straight up and off spark plug.
- Installation is reverse of removal. Be sure to reattach ground cables or wires.
- Check and clear fault codes from ECM memory.

Ignition coil, replacing (N52, N54 engine)

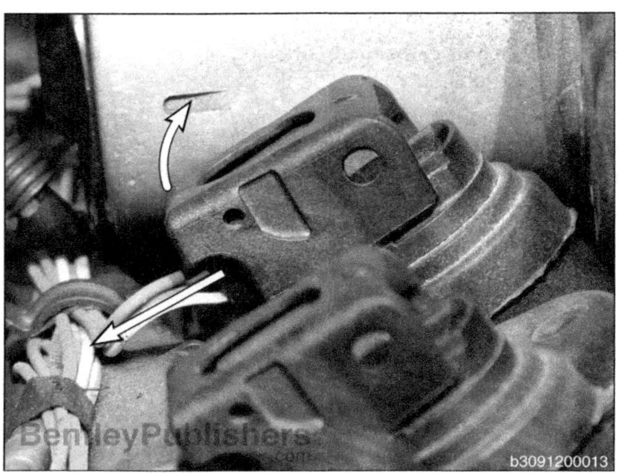

- Make sure ignition is switched OFF.
- Remove ignition coil cover. See **020 Maintenance**.
- ◄ Unlock (**curved arrow**) ignition coil connector. Pull (**straight arrow**) connector off coil.

BentleyPublishers.com
BentleyPublishers.com—All Rights Reserved

- Slide rod-shaped ignition coil straight out of spark plug hole.
- Installation is reverse of removal.
- Check and clear fault codes from ECM memory.

Ignition coil, replacing (V8 engine)

- Make sure ignition is switched OFF.
- Remove ignition coils cover. See **020 Maintenance**.

- Remove ignition coil hold-down bracket fasteners (**A**) and remove hold-down brackets (**arrows**). Left side shown, right is similar.

- Unlock (**curved arrow**) ignition coil connector. Pull (**straight arrow**) connector off coil.

BentleyPublishers
BentleyPublishers.com—All Rights Reserved

Slide rod-shaped ignition coil straight out of spark plug hole.

— Installation is reverse of removal.

— Check and clear fault codes from ECM memory.

CRANKSHAFT SENSOR SERVICE

Crankshaft sensor, replacing (M54 engine)

Crankshaft sensor in M54 engine is on left rear of engine block, underneath intake manifold and starter motor.

— Make sure ignition is OFF.

— Raise car and support in a safe manner.

> ***WARNING —***
> * *Make sure the car is stable and well supported at all times. Use a professional automotive lift or jack stands designed for the purpose. A floor jack is not adequate support.*

— Remove front end reinforcement plate. See **310 Front Suspension**.

Working underneath starter:

- Cut off wire tie securing crankshaft sensor harness connector (**arrow**) to sensor.
- Disconnect harness connector.

Remove sensor mounting fastener (**arrow**) and remove sensor from cylinder block.

— Installation is reverse of removal.

- Use new sealing O-ring when installing sensor.
- Replace self-locking mounting bolt.
- Be sure wiring is rerouted as before.

Tightening torque	
Crankshaft sensor to engine block (M6, replace with new)	10 Nm (7 ft-lb)

— Check and clear fault codes from ECM memory.

BentleyPublishers.com
BentleyPublishers.com—All Rights Reserved

Crankshaft sensor, replacing (N52, N54 engine)

Crankshaft sensor is on left rear of engine block, underneath intake manifold and starter motor.

— Make sure ignition is OFF.

— Raise car and support in a safe manner.

> ***WARNING —***
> - *Make sure the car is stable and well supported at all times. Use a professional automotive lift or jack stands designed for the purpose. A floor jack is not adequate support.*

— Remove front end reinforcement plate. See **310 Front Suspension**.

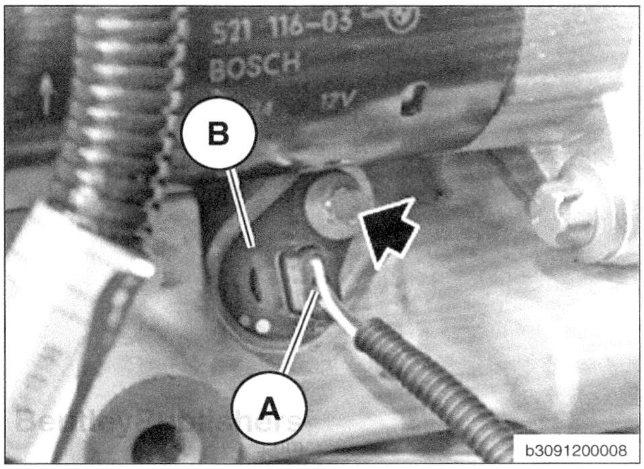

Working underneath starter:
- Disconnect crankshaft sensor harness connector (**A**) from sensor (**B**).

— Remove sensor mounting fastener (**arrow**) and remove sensor from cylinder block. Be prepared to catch dripping oil.

— Installation is reverse of removal. Remember to:
- Use new sealing O-ring when installing sensor.
- Replace mounting bolt.
- Be sure wiring is routed as before.
- Check and clear fault codes from ECM memory.

Tightening torque	
Crankshaft sensor to engine block (M6, replace with new)	8 Nm (6 ft-lb)

Crankshaft sensor, replacing (V8 engine)

Crankshaft sensor is on right bottom of transmission housing.

— Make sure ignition is OFF.

— Raise car and support in a safe manner.

> ***WARNING —***
> - *Make sure the car is stable and well supported at all times. Use a professional automotive lift or jack stands designed for the purpose. A floor jack is not adequate support.*

— Remove front end reinforcement plate. See **310 Front Suspension**.

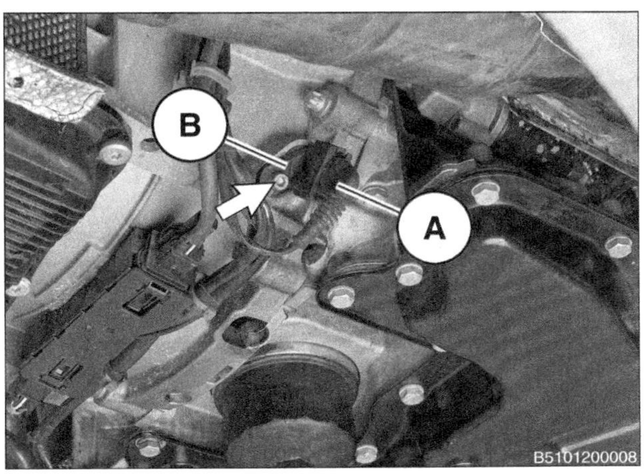

Disconnect crankshaft sensor harness connector (**A**) from sensor (**B**).

— Remove sensor mounting fastener (**arrow**) and remove sensor from cylinder block. Be prepared to catch dripping oil.

BentleyPublishers.com—All Rights Reserved

– Installation is reverse of removal. Remember to:
  - Use new sealing O-ring when installing sensor.
  - Replace mounting bolt.
  - Be sure wiring is routed as before.
  - Check and clear fault codes from ECM memory.

Tightening torque	
Crankshaft sensor to engine block (M6, replace with new)	8 Nm (6 ft-lb)

Camshaft Sensor Service

Intake camshaft sensor, replacing (M54 engine)

Intake camshaft sensor on M54 engine is mounted at left side of cylinder head near front.

– Make sure ignition is OFF.

– Remove upper engine covers. See **020 Maintenance**.

– Remove air filter housing and ducts. See **130 Fuel Injection**.

< Working at left side of cylinder head:
  - Detach engine vent hose from cylinder head cover by pinching spring clips (**arrows**).
  - Detach VANOS solenoid electrical connector (**A**).
  - Unscrew VANOS solenoid.
  - If necessary, unfasten oil supply line on VANOS adjustment unit. Be prepared to catch dripping oil.

< Remove camshaft sensor mounting bolt. Remove sensor (**arrow**) from cylinder head.

– Reach under intake manifold to disconnect camshaft sensor harness.

NOTE —
- *Prior to removal, attach a stiff piece of wire to connector to aid in routing of harness during reinstallation.*

BentleyPublishers.com
BentleyPublishers.com—All Rights Reserved

– Installation is reverse of removal. Remember to:
  - Replace camshaft sensor mounting bolt.
  - Use new sealing O-rings when installing sensor, VANOS solenoid and VANOS oil supply line.
  - Be sure sensor harness is rerouted as before.

Tightening torques	
Intake camshaft sensor to cylinder head (M6, replace bolt)	10 Nm (7 ft-lb)
Oil supply line to VANOS unit (banjo bolt)	32 Nm (24 ft-lb)
VANOS solenoid to VANOS unit	30 Nm (22 ft-lb)

– Check and clear fault codes from ECM memory.

Exhaust camshaft sensor, replacing (M54 engine)

Exhaust camshaft sensor on M54 engine is mounted at right side of cylinder head near front.

– Make sure ignition is OFF.

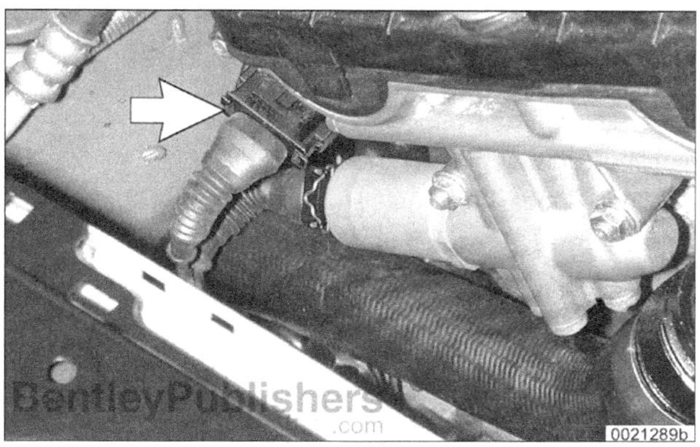

< Working at front of engine on exhaust (right) side, detach exhaust camshaft sensor electrical harness connector (**arrow**).

– Remove sensor mounting bolt and remove sensor. Be prepared to catch dripping oil

– Installation is reverse of removal.
  - Replace camshaft sensor mounting bolt.
  - Use new sealing O-ring when installing sensor.

Tightening torque	
Exhaust camshaft sensor to cylinder head (M6, replace bolt)	10 Nm (7 ft-lb)

– Check and clear fault codes from ECM memory.

Intake camshaft sensor, replacing (N52, N54 engine)

Intake camshaft sensor is mounted to left front of cylinder head.

– Make sure ignition is OFF.

– Remove upper engine covers. See **020 Maintenance**.

– Remove air filter housing and ducts. See **131**, **132 Fuel Injection**.

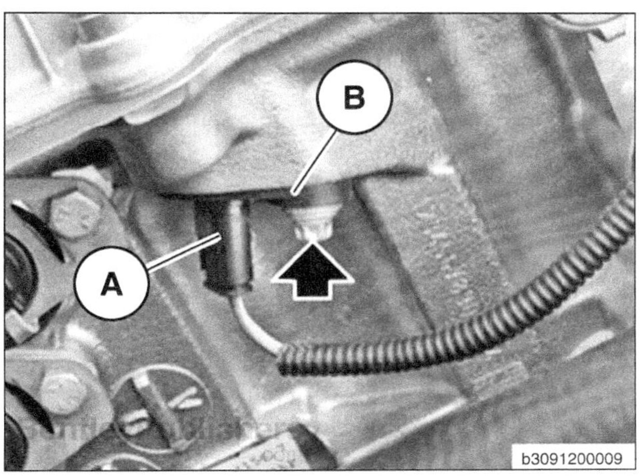

< Detach camshaft sensor electrical connector (**A**).

– Remove camshaft sensor mounting bolt (**arrow**). Remove sensor (**B**) from cylinder head. Be prepared to catch dripping oil.

BentleyPublishers.com—All Rights Reserved

– Installation is reverse of removal. Remember to:
  - Replace camshaft sensor mounting bolt.
  - Use new sealing O-rings when installing sensor.
  - Be sure sensor harness is routed as before.
  - Check and clear fault codes from ECM memory.

Tightening torques	
Intake camshaft sensor to cylinder head (M6, replace with new)	9 Nm (6.5 ft-lb)

Exhaust camshaft sensor, replacing (N52, N54 engine)

Exhaust camshaft sensor is mounted to right front of cylinder head.

– Make sure ignition is OFF.

– Remove upper engine covers. See **020 Maintenance**.

– Remove air filter housing and ducts. See **131**, **132 Fuel Injection**.

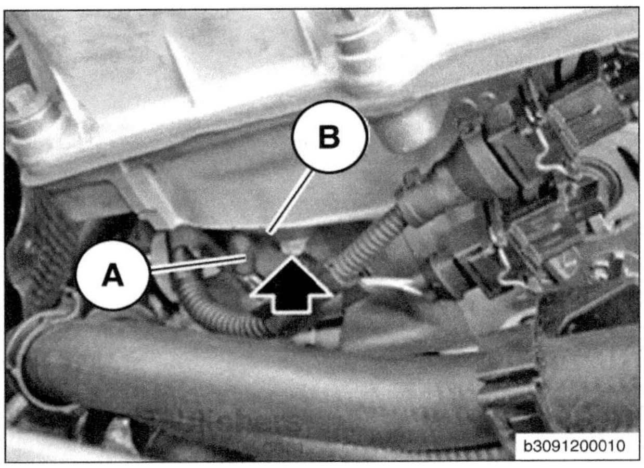

< Detach camshaft sensor electrical connector (**A**).

– Remove camshaft sensor mounting bolt (**arrow**). Remove sensor (**B**) from cylinder head. Be prepared to catch dripping oil.

– Installation is reverse of removal. Remember to:
  - Replace camshaft sensor mounting bolt.
  - Use new sealing O-rings when installing sensor.
  - Be sure sensor harness is routed as before.
  - Check and clear fault codes from ECM memory.

Tightening torques	
Intake camshaft sensor to cylinder head (M6, replace bolt)	9 Nm (6.5 ft-lb)

Camshaft sensor, replacing (V8 engine)

V8 engine: Intake and exhaust camshaft sensors are located on the front top of each cylinder head. Right side intake camshaft sensor replacement is shown. Others are similar.

– Make sure ignition is OFF.

– Remove upper engine covers. See **020 Maintenance**.

Intake camshaft sensor

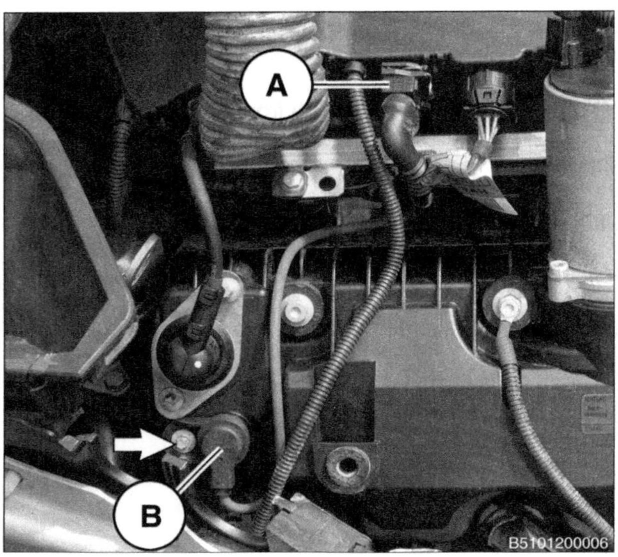

< Detach camshaft sensor electrical connector (**A**).

– Remove camshaft sensor mounting bolt (**arrow**). Remove sensor (**B**) from cylinder head. Be prepared to catch dripping oil.

BentleyPublishers.com
BentleyPublishers.com—All Rights Reserved

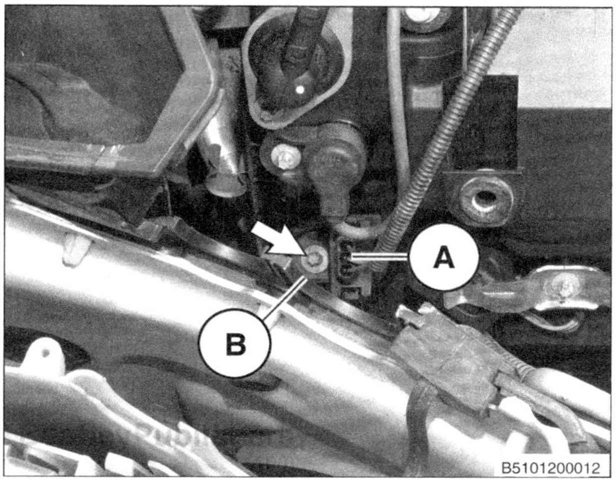

Detach camshaft sensor electrical connector (**A**).

- Remove camshaft sensor mounting bolt (**arrow**). Remove sensor (**B**) from cylinder head. Be prepared to catch dripping oil.
- Installation is reverse of removal. Remember to:
  - Replace camshaft sensor mounting bolt.
  - Use new sealing O-rings when installing sensor.
  - Be sure sensor harness is routed as before.
  - Check and clear fault codes from ECM memory.

Knock Sensor Service

Knock sensors, replacing (M54 engine)

M54 engine knock sensors are under the intake manifold on the left side of the cylinder head.

- Disconnect negative (-) battery cable and cover battery terminal to keep cable from accidentally contacting terminal.

> ***CAUTION—***
> - *Prior to disconnecting the battery, read the battery disconnection cautions given in* **001 Warnings and Cautions**.

- With engine fully cooled, remove intake manifold. See **130 Fuel Injection**.

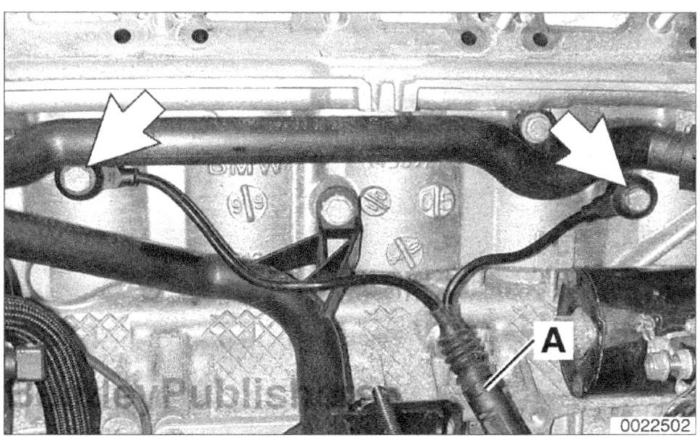

Disconnect knock sensor electrical harness connector (**A**) on left side of engine block.

- Remove knock sensor mounting bolts (**arrows**) on side of cylinder block.
- Remove sensors.

> ***CAUTION—***
> - *Note the installed angle of the knock sensor on the block before removing it. Reinstall the sensor in the same position. Be sure to use a torque wrench when tightening the sensor mounting bolt.*

Clean knock sensor contact surface on engine block and sensor (**arrows**) before installing knock sensor.

> ***CAUTION—***
> - *Do not overtighten knock sensors.*

Tightening torque	
Knock sensor to engine block (M8)	20 Nm (15 ft-lb)

- Check and clear fault codes from ECM memory.

Knock sensors, replacing (N52 engine)

N52 engine knock sensors are under the intake manifold on the left side of the cylinder head.

– Disconnect negative (-) battery cable and cover battery terminal to keep cable from accidentally contacting terminal.

> ***CAUTION—***
> • *Prior to disconnecting the battery, read the battery disconnection cautions given in* **001 Warnings and Cautions**.

– With engine fully cooled, remove intake manifold. See **131 Fuel Injection**.

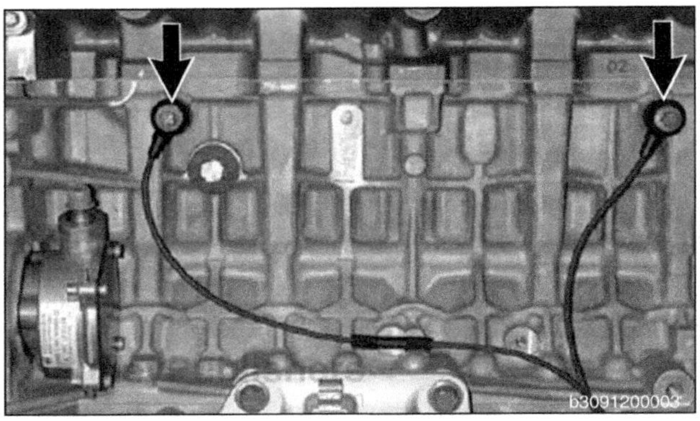

◄ Disconnect knock sensor electrical harness connector.

• Remove knock sensor mounting bolts (**arrows**) on side of cylinder block. Remove sensors.

> ***CAUTION—***
> • *Note the installed angle of the knock sensor on the block before removing it. Reinstall the sensor in the same position. Be sure to use a torque wrench when tightening the sensor mounting bolt.*

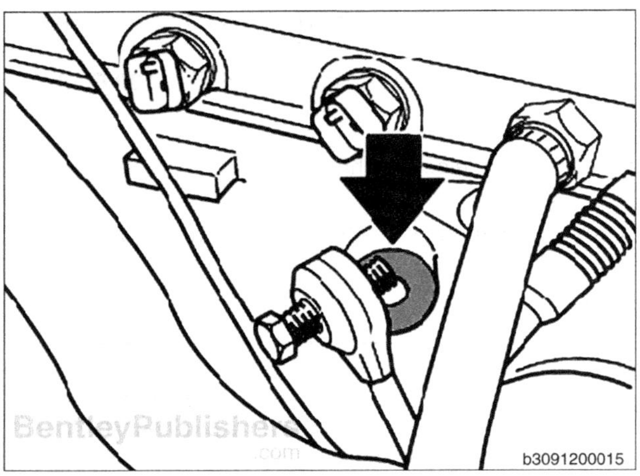

◄ Clean knock sensor contact surface on engine block and sensor (**arrow**) before installing knock sensor.

> ***CAUTION—***
> • *Do not overtighten knock sensors.*

Tightening torques	
Knock sensor to engine block (use new aluminum M8 fasteners) • Stage 1 • Stage 2	 7 Nm (63 in-lb) additional 90°

– Installation is reverse of removal. Remember to:

• Replace knock sensor mounting bolt.

• Be sure sensor harness is routed as before.

– Check and clear fault codes from ECM memory.

BentleyPublishers.com
BentleyPublishers.com—All Rights Reserved

Knock sensors, replacing (N54 engine)

N54 engine knock sensors are under the intake manifold on the left side of the cylinder head.

– Disconnect negative (-) battery cable and cover battery terminal to keep cable from accidentally contacting terminal.

> ***CAUTION—***
> - *Prior to disconnecting the battery, read the battery disconnection cautions given in* **001 Warnings and Cautions**.

– With engine fully cooled, remove intake manifold. See **132 Fuel Injection**.

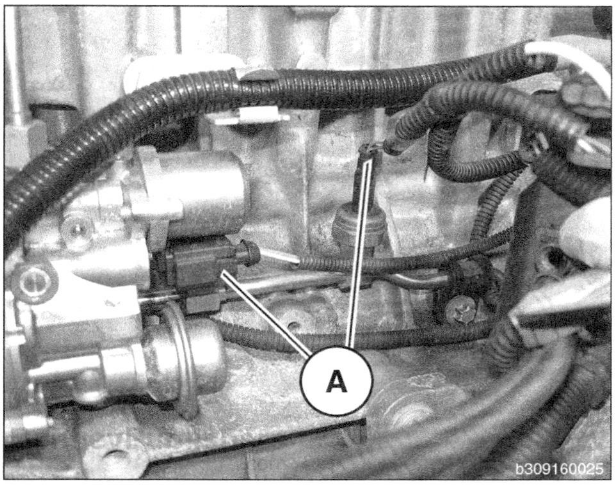

◀ Disconnect electrical connectors (**A**), remove and seal high pressure fuel pump fuel supply line connections. Be prepared to catch dripping fuel.

> ***WARNING —***
> - *The fuel system is designed to retain pressure even when the ignition is OFF. When working with the fuel system, remove fuel lines slowly to allow residual fuel pressure to dissipate. Avoid spraying fuel, Use shop towels to capture leaking fuel*
> - *Electric fuel pump starts automatically when door is opened.*
> - *Carry out installation work on fuel system with coolant temperature below 40°C (104°F).*

> ***CAUTION—***
> - *When working with high pressure fuel lines observe clean working habits. Any dirt or debris introduced into system will cause malfunctions.*

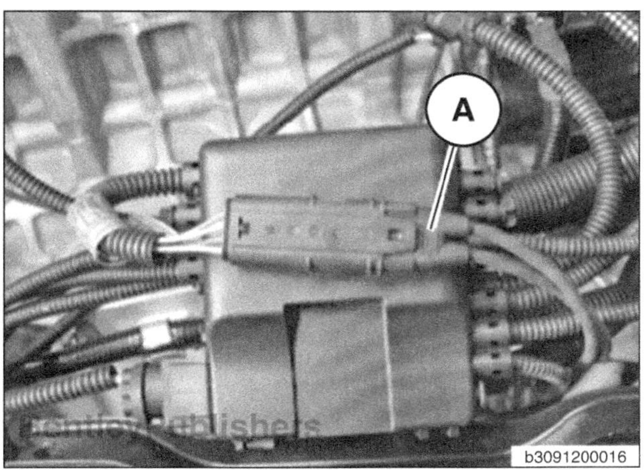

◀ Disconnect knock sensor electrical connector (**A**) on left side of engine block.

BentleyPublishers.com—All Rights Reserved

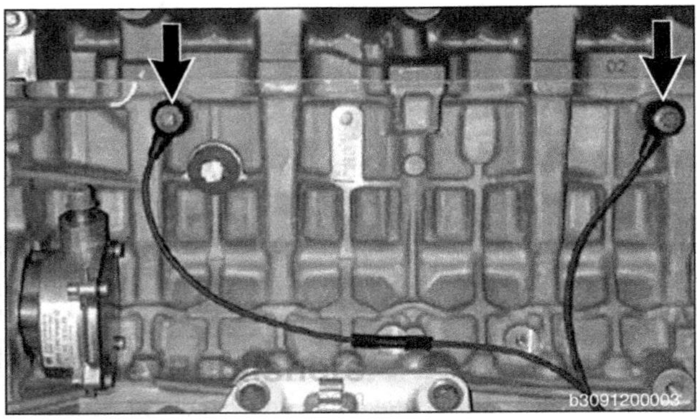

Remove knock sensor mounting bolts (**arrows**) on side of cylinder block. Remove sensors.

> ***CAUTION—***
> • *Note the installed angle of the knock sensor on the block before removing it. Reinstall the sensor in the same position. Be sure to use a torque wrench when tightening the sensor mounting bolt.*

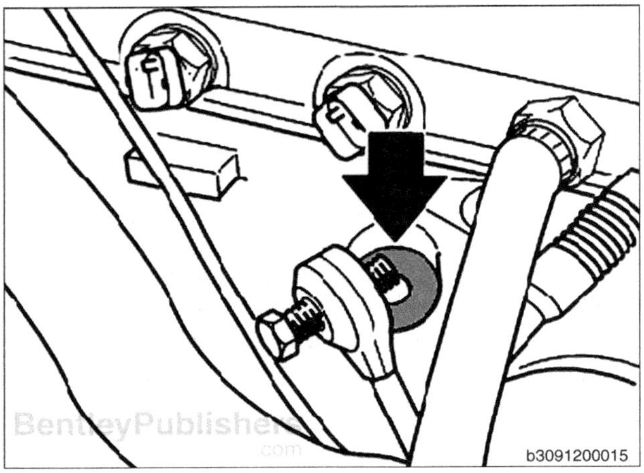

Clean knock sensor contact surface on engine block and sensor (**arrow**) before installing knock sensor.

> ***CAUTION—***
> • *Do not overtighten knock sensors.*

Tightening torques	
Knock sensor to engine block (use new aluminum M8 fasteners) • Stage 1 • Stage 2	 7 Nm (63 in-lb) additional 90°
Fuel line nut to high pressure pump (coat thread with transmission fluid)	30 Nm (22 ft-lb)
Fuel line mounting bracket to engine block	8 Nm (87.5 in-lb)

– Installation is reverse of removal. Remember to:
  - Replace knock sensor mounting bolt.
  - Be sure sensor harness is rerouted as before.
  - Check for fuel system leaks.
  - Check and clear fault codes from ECM memory.

Knock sensor, replacing (V8 engine)

V8 engine is equipped with 4 sensors, 2 per bank of cylinders, installed under intake manifold.

– Disconnect negative (-) battery cable and cover battery terminal to keep cable from accidentally contacting terminal.

> ***CAUTION—***
> • *Prior to disconnecting the battery, read the battery disconnection cautions given in* **001 Warnings and Cautions**.

– With engine fully cooled, remove intake manifold. See **133 Fuel Injection**.

BentleyPublishers.com
BentleyPublishers.com—All Rights Reserved

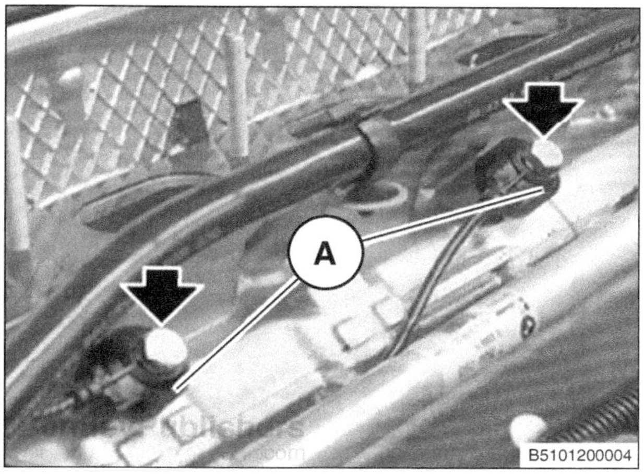

< Disconnect knock sensor electrical harness connector.

- Remove knock sensor mounting bolts (**arrows**) on side of cylinder block. Remove sensors (**A**).

> ***CAUTION—***
> - *Note the installed angle of the knock sensor on the block before removing it. Reinstall the sensor in the same position. Be sure to use a torque wrench when tightening the sensor mounting bolt.*

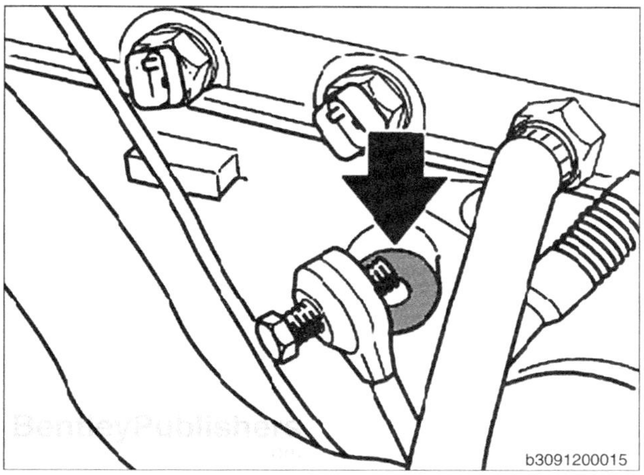

< Clean knock sensor contact surface on engine block and sensor (**arrow**) before installing knock sensor.

> ***CAUTION—***
> - *Do not overtighten knock sensors.*

Tightening torques	
Knock sensor to engine block (M8 replace with new)	21 Nm (14.7 ft-lb)

– Installation is reverse of removal. Remember to:

- Replace knock sensor mounting bolt.
- Be sure sensor harness is routed as before.
- Check and clear fault codes from ECM memory.

BentleyPublishers.com
BentleyPublishers.com—All Rights Reserved

BentleyPublishers.com
BentleyPublishers.com—All Rights Reserved

121 Battery, Starter, Alternator

GENERAL

This repair group covers the battery, alternator, starter and associated components of the electrical system.

See also:

- **020 Maintenance** for accessory belt replacement and routine battery service.
- **600 Electrical System–General** for bus system information, power reset procedure and general electrical troubleshooting
- **721 Airbag System (SRS)** for battery safety terminal (BST)
- **ECL Electrical Component Locations**
- **ELE Electrical Wiring Diagrams**

Engine electrical system

The alternator and starter are wired directly to the battery. To prevent accidental shorts that might blow a fuse or damage wires and electrical components, disconnect the negative (–) battery cable before working on the electrical system.

BentleyPublishers
BentleyPublishers.com–All Rights Reserved

Different versions of alternators, starters and batteries are used in 5 Series models. Replace components according to the original equipment specifications.

Troubleshooting

Basic battery, starter and charging system troubleshooting information is in **Table a**.

Table a. Battery, starter and charging system troubleshooting

Symptom	Probable cause	Corrective action
Engine does not crank.	Fault in car access system (CAS).	Try another ignition key. If problem persists, contact your authorized BMW dealer.
Engine cranks slowly or not at all, solenoid clicks when starter is operated.	Battery cables loose, dirty or corroded.	Clean or replace cables. See **020 Maintenance**.
	Battery discharged.	Charge battery and test. Replace if necessary.
	Battery to body ground cable in trunk loose, dirty or corroded	Inspect ground cable. Clean, tighten or replace if necessary.
	Poor connection at starter motor terminal X6512.	Check connections at starter motor.
	Starter motor or solenoid faulty.	Test starter.
Battery does not stay charged more than a few days.	Short circuit draining battery. (parasitic draw)	Test for excessive current drain with everything electrical off.
	Short driving trips and high electrical drain on charging system does not allow battery to recharge.	Evaluate driving style. Where possible, reduce electrical consumption when making short trips.
	Engine accessory belt loose, worn or damaged.	Inspect or replace accessory belt. See **020 Maintenance**.
	Battery faulty.	Test battery and replace if necessary.
	Battery cables loose, dirty or corroded.	Clean or replace cables. See **020 Maintenance**.
	Alternator faulty.	Test alternator.

Warnings and Cautions

WARNING—
- *Wear goggles, rubber gloves and a rubber apron when working around the battery or battery acid (electrolyte).*
- *Battery acid contains sulfuric acid and can cause skin irritation and burning. If acid is spilled on your skin or clothing, flush the area at once with large quantities of water. If electrolyte gets into your eyes, flush them with large quantities of clean water for several minutes and call a physician.*
- *A battery that is being charged or is fully charged gives off explosive hydrogen gas. Keep sparks and open flames away. Do not smoke.*
- *Battery lead and battery acid are both hazardous materials. When disposing of an old battery, follow federal, state and local regulations on hazardous waste disposal.*

BentleyPublishers.com
BentleyPublishers.com—All Rights Reserved

CAUTION—
- *Use a digital multimeter when testing automotive electrical components.*
- *If a repair procedure specifies disconnecting the battery, follow the instruction for safety reasons.*
- *Prior to disconnecting the battery, read the battery disconnection cautions in* **001 Warnings and Cautions**.
- *Disconnecting the battery cables may erase fault codes stored in ECM memory.*
- *Using a conventional battery charger to jump start the vehicle may damage electronic control modules. Use an electronic vehicle power supply or another 12-volt battery for jump-starting. Jump starting is covered in* **020 Maintenance**.
- *Disconnect the negative (–) battery cable first and reconnect it last. Cover the battery post with an insulating material whenever the cable is removed.*
- *Do not disconnect battery, alternator or starter wires while the engine is running.*
- *Do not reverse the battery cables. Even a momentary wrong connection can damage the alternator or other electrical components.*
- *Do not depend on insulation color to tell battery positive and negative cables apart. Label cables before removing.*
- *Power windows and sunroof may fail to function properly after disconnecting and reconnecting the battery. Reinitialize windows and sunroof. See* **Battery reconnection notes** *in this repair group.*
- *Do not make any additional connections at the negative terminal of the battery.*
- *Do not modify the battery ground cable.*
- *Do not make any connections between the intelligent battery sensor (IBS) and the sensor screw.*
- *Do not use force when disconnecting the ground terminal from the battery.*
- *Do not pull at the ground cable.*
- *Do not use the IBS as a pivot point to lever off the ground terminal.*
- *Do not use the connections of the IBS as a lever.*
- *Use a torque wrench when reinstalling the battery ground cable.*
- *Do not release or tighten the IBS screw (screw with Torx head).*
- *To avoid electrochemical corrosion to engine components made of aluminum-magnesium alloy, do not use steel fasteners. Use aluminum fasteners only. Use a magnet to identify aluminum fasteners.*
- *Replace aluminum fasteners each time they are loosened.*
- *Follow torque instructions, including angle of rotation specifications, when installing aluminum fasteners.*

BentleyPublishers
BentleyPublishers.com—All Rights Reserved

Battery

Battery basics

The wet-cell lead-acid battery is rated by ampere / hours (Ah) and cold cranking amps (CCA) listed on the battery.

Battery location and connections

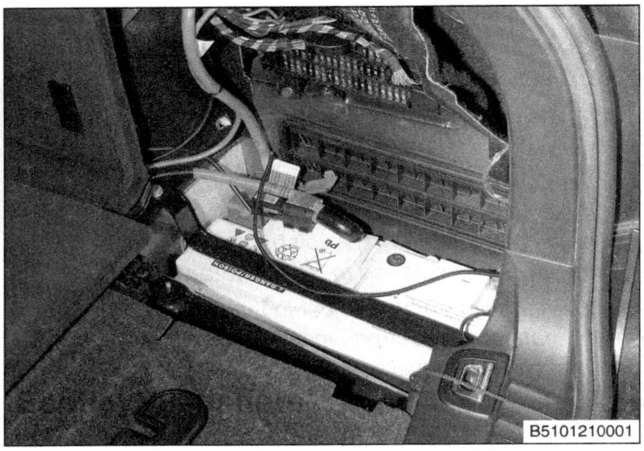
B5101210001

The battery is mounted in the right side of the trunk or cargo compartment, underneath the floor trim.

For rear power distribution panel details, see **ECL Electrical Component Locations**.

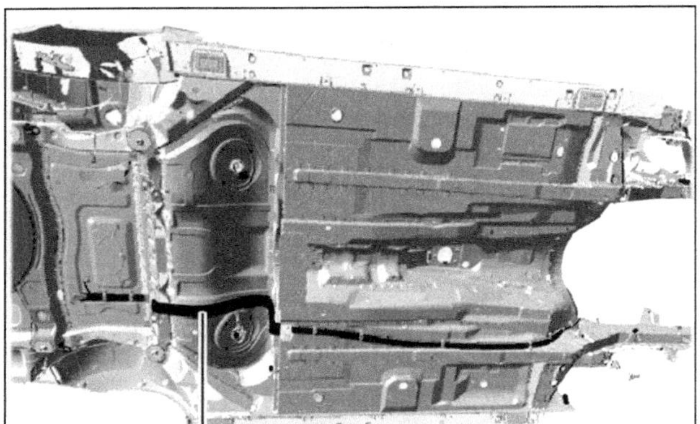

B5101210002

Underneath the vehicle at the battery box are two power transfer points for the two cables that run to the engine compartment.

- One cable goes to the jump start terminal in the engine compartment and then connects to the starter and alternator.
- The second cable is used to supply power to engine electronics (DME and Valvetronic).
- The power cables are routed in a protected area underneath the body of the vehicle.

Battery safety terminal (BST)

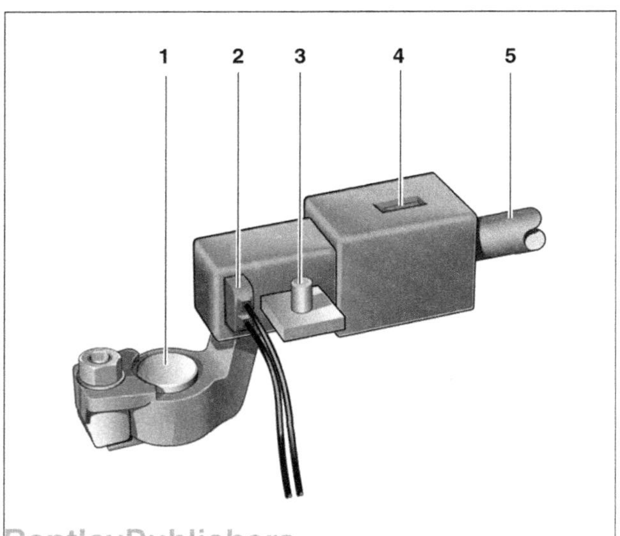

B301121003

1. Positive battery terminal (B+)
2. BST igniter connector
3. B+ to rear power distribution box (A46)
4. BST cover retaining clip
5. B+ to starter and alternator

The battery safety terminal (BST) is controlled by the multiple restraint system (MRS) control module. The system disconnects electrical power to the engine compartment in case of a significant crash. MRS fires an encapsulated pyrotechnic device in the positive (+) battery terminal that disconnects power to the engine compartment but maintains power to the exterior lights and interior of the vehicle.

See also **721 Airbag System (SRS)**.

BentleyPublishers.com
BentleyPublishers.com—All Rights Reserved

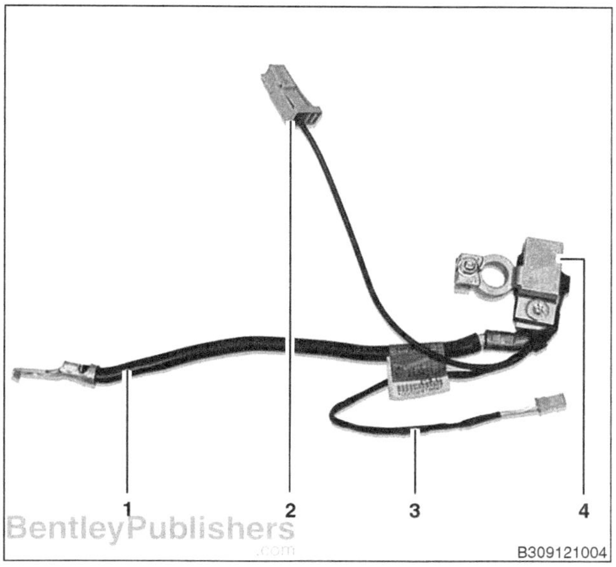

Intelligent battery sensor (IBS)

Intelligent battery sensor (IBS) is connected to the negative battery terminal.

1. Battery ground cable
2. B+ connection to IBS
3. Bit-serial data (BSD) interface
4. Intelligent battery sensor (IBS)

The IBS is sensitive to mechanical stress and strain. See **Warnings and Cautions** in this repair group.

Replace IBS as a complete unit with the battery ground cable. The ground cable serves as a heat sink for the IBS.

If IBS is replaced, allow vehicle to rest unused at least 3 hours. Only then can the new IBS determine battery status.

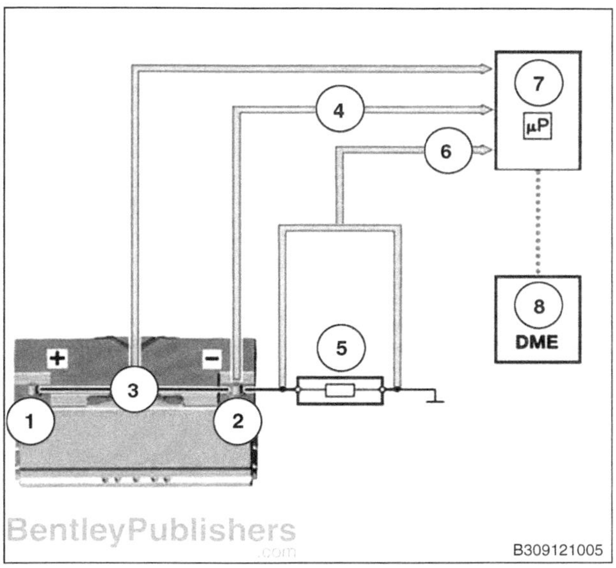

The IBS microprocessor monitors and measures battery conditions such as:

- Terminal voltage via measurement from B+ to ground
- Charge / discharge current via integrated shunt resistor
- Temperature of battery acid via integrated temperature sensor

1. B+ battery positive terminal
2. Battery ground terminal
3. Voltage measurement
4. Temperature measurement
5. Integrated shunt resistor
6. Current measurement
7. IBS microprocessor
8. Engine control module (ECM)

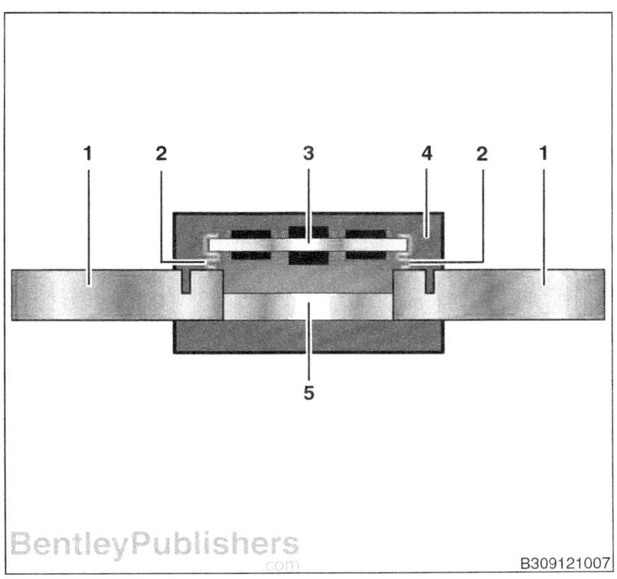

IBS components:

1. Copper plate
2. Gull wing tabs
3. PC board with evaluation electronics
4. Housing
5. Manganin copper alloy resistor maintains constant temperature regardless of current flow, used to measure current flow.

IBS evaluation function

IBS electronics continuously measure the following values under all vehicle operating conditions:

- Voltage (6 v to 16.5 v)
- Starting current (0 A to 1000 A)
- Charging current (20 A to 200 A)
- Temperature (-40°C to 105°C or -40°F to 221°F)
- Parasitic draw or closed circuit current (0 A to 10 A)

BentleyPublishers.com—All Rights Reserved

When the vehicle is stationary, the IBS is programmed to wake up every 40 seconds and makes battery condition measurements within approx. 50 ms in order to save power. The measured values, communicated to the engine control module (ECM) via binary serial data (BSD) interface, are used to calculate the following:

- State of charge (SoC) shows the current charge of the battery. SoC is used during key OFF periods to insure that the battery maintains a sufficient charge to start the engine at least one more time.
- State of health (SoH) tracks the history of the battery in the vehicle. Charge / discharge cycles and times are monitored. SoH helps the ECM determine correct battery charging rates and anticipated battery life. Internal resistance of the battery is calculated by IBS from current and voltage drop during engine start. These values are used by the ECM to calculate the SoH of the battery.

Software contained in the IBS microprocessor utilizes the measured values to calculate battery SoC during vehicle sleep mode and compares this information with that received from the ECM pertaining to the battery SoC / SoH during the period between engine OFF and deactivation of the DME main relay.

Battery data is stored every 2 hours over a 6 hour time frame, providing three snapshots of battery SoC information. The stored information / snapshot data is overwritten every 6 hours. Whenever the ignition is activated, IBS updates the ECM with the current closed circuit histogram / battery status information. The ECM evaluates the new data and, if a parasitic draw (closed-circuit current draw) is identified, a fault is stored in fault memory.

IBS diagnosis

A fault code is stored in the ECM when the IBS is defective. The ECM boosts idle speed in order to sufficiently charge the battery.

The IBS can only be diagnosed through the ECM. The self-diagnosis function checks the voltage, current, temperature, terminal 15 wake up signal as well as system errors in the IBS.

Voltage measurement

If IBS is shorted to ground, ECM fault code displays “Voltage Fault DME ON”. The IBS is unable to wake up the ECM.

If IBS is shorted to B+, ECM fault code displays “Voltage Fault, DME not ON”. The vehicle does NOT enter sleep mode.

Current measurement

Current measurement is a dynamic process. If an implausible current flow value is detected, ECM fault code displays “Current Fault”.

IBS wake-up

The IBS constantly monitors battery SoC, even when the vehicle is asleep. A wake-up signal is sent to the ECM if the SoC falls below a predetermined threshold. Upon wake-up, the ECM switches OFF auxiliary electrical loads in order to preserve battery charge.

IBS wakes up the vehicle once during a key-off cycle. Once awakened, the vehicle returns to sleep mode.

BentleyPublishers.com
BentleyPublishers.com—All Rights Reserved

Wake-up signal faults

The IBS detects wake-up line faults under the following conditions:

- ECM switched ON.
- Ignition switched ON (voltage high at IBS).
- Terminal 15 running via alternator output at BSD.

If voltages at IBS terminal 15 and terminal 15 via the BSD are not equal, a fault is indicated either in the BSD line or the IBS.

Battery, testing

BMW batteries are rated as follows:

- **Ah rating** is determined by the average amount of current the battery can deliver over time without dropping below a specified voltage.
- **CCA rating** is determined by the battery's ability to deliver starting current at 0°F (-18°C) without dropping below a specified voltage.

Battery testing determines battery condition and state of charge. Before testing the battery, check that the battery cables are tight and free of corrosion.

Electronic battery testing

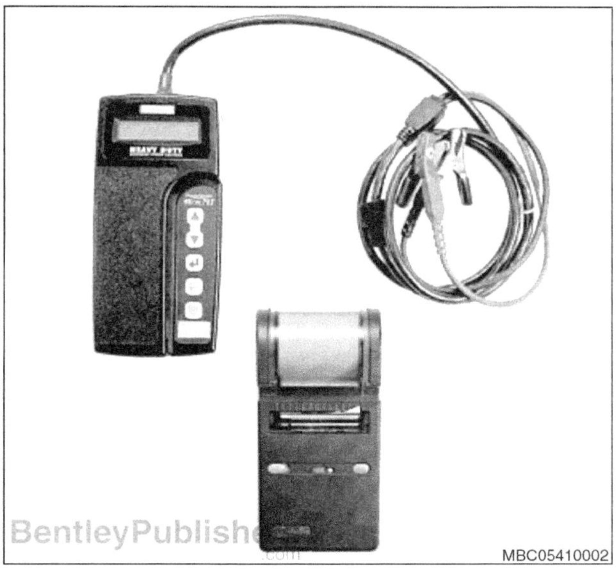

MBC05410002

< Electronic battery testers such as Midtronics®MCR 717 are recommended for testing all battery types.

Hydrometer testing

On conventional or low-maintenance batteries, check the specific gravity of the electrolyte using a hydrometer. The more dense the concentration of sulfuric acid in the electrolyte, the higher the state of charge.

Before hydrometer testing, load the battery with 15 amperes for one minute. If the battery is installed in the vehicle, this can be done by switching the headlights ON for about one minute without the engine running. The state of battery charge based on specific gravity values are in **Table b**.

Note that electrolyte temperature affects hydrometer reading. Check the electrolyte temperature with a thermometer. Add 0.004 to the hydrometer reading for every 6°C (10°F) that the electrolyte is above 27°C (80°F). Subtract 0.004 from the reading for every 6°C (10°F) that the electrolyte is below 27°C (80°F).

Table b. Specific gravity of battery electrolyte at 27°C (80°F)

Specific gravity	State of charge
1.265	Fully charged
1.225	75% charged
1.190	50% charged
1.155	25% charged
1.120	Fully discharged

BentleyPublishers.com—All Rights Reserved

If the specific gravity is at or above 1.225 but the battery lacks power for starting, determine the battery's service condition with a load voltage test. See **Battery load voltage test** in this repair group.

If the average specific gravity of the six cells is below 1.225, charge battery and retest. See **Battery charging notes** in this repair group.

Battery open-circuit voltage test

– Before testing, load battery with 15 amperes for one minute with battery load-tester or switch headlights ON for about one minute without engine running. Connect digital voltmeter across battery terminals. Open-circuit voltage levels are given in **Table c**.

Table c. Open-circuit voltage and battery charge	
Open-circuit voltage	**State of charge**
12.6 or more	Fully charged
12.4	75% charged
12.2	50% charged
12.0	25% charged
11.7 or less	Fully discharged

– If open-circuit voltage is OK but battery still lacks power for starting, perform a load voltage test. See **Battery load voltage test** in this repair group.

– If open-circuit voltage is below 12.4 volts, recharge battery and retest. See **Battery charging notes** in this repair group.

Battery load voltage test

A battery load tester is required for a load voltage test. The test is made by applying a high resistive load to the battery terminals and simultaneously measuring battery voltage. For the most accurate results, make sure the battery is fully charged. Disconnect battery cables before making the test. Replace the battery if the voltage is below that listed in **Table d**.

> ***WARNING —***
> - *Wear protective goggles and clothing when performing a load test.*

Table d. Battery load test–minimum voltage	
Ambient temperature	**Voltage***
27°C (80°F)	9.6
16°C (60°F)	9.5
4°C (40°F)	9.3
-7°C (20°F)	8.9
-18°C (0°F)	8.5
* Measure after applying a 200 amp load for 15 seconds.	

BentleyPublishers.com
BentleyPublishers.com—All Rights Reserved

Battery parasitic draw, testing

If the vehicle battery is discharged for an unknown reason, perform the parasitic draw test designed to detect excessive current flow from the battery when everything in the vehicle is shut down. This is also known as a closed circuit current measurement.

- Check that battery voltage is 12 volts or higher. If lower, recharge battery.

Battery isolation switch

BentleyPublishers.com

B309121029

- Remove negative cable and install battery isolation switch (OTC tool 7645 or equivalent).
- Close switch. Connect digital multimeter in series across switch. Set meter to milliamp (mA) scale.
- Turn trunk lock to locked position using screwdriver or similar (simulates trunk lid being closed).
- To simulate normal parasitic draw conditions: Switch ignition ON and activate all electrical consumers, including any accessories. Switch ignition OFF. Open and close driver's door (simulates somebody getting out). Lock car, arming alarm if installed. With exception of trunk, keep all other doors and lids closed.
- Open battery isolation switch and monitor current draw.
  - Vehicle without telematics (TCU): Wait 30 minutes.
  - Vehicle with telematics (TCU): Wait 60 - 70 minutes.
- Maximum allowable parasitic draw after waiting: Approx. 40 mA.
- If draw exceeds maximum allowable, find consumer(s) at fault by removing individual fuses and noting change in parasitic draw. Repair or replace component(s) found at fault.

Fuse locations and applications are in **ECL Electrical Component Locations**.

Battery charging notes

430-watt Deutronic power supply

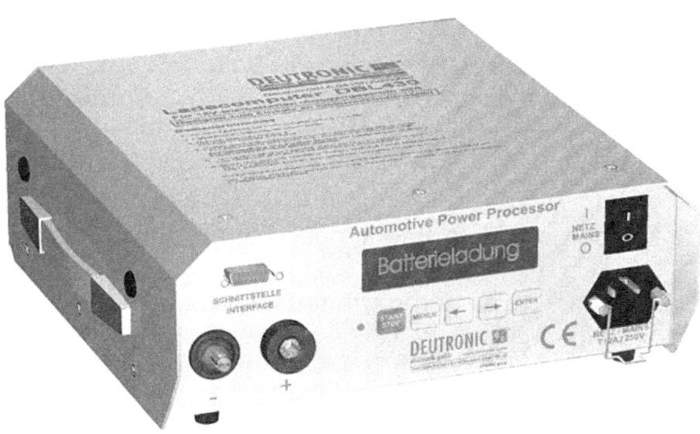

BentleyPublishers.com

B309121013

Consistent, clean voltage is essential when charging the battery or maintaining battery charge while coding or programming control modules. A transformer-based battery charger is inadequate for this task. BMW recommends the use of a vehicle power supply instead of a battery charger.

A power supply is a constant voltage, high-amperage power station with electronically controlled functions. It insures clean DC voltage and eliminates AC. It also provides fast, efficient and safe battery charging. Once proper battery voltage is obtained, the power supply floats the charge to maintain that voltage.

Battery charging voltage	
BMW recommendation	14.8 v

Power supply units from different manufacturers are illustrated on next page.

BentleyPublishers.com
BentleyPublishers.com—All Rights Reserved

70-amp ACCTIVA Professional power supply

If using a power supply, recharge battery without removing it. Otherwise, remove the battery from the trunk or cargo compartment during charging. See **Battery, removing and installing** in this repair group.

Prolonged charging with a conventional charger can cause battery electrolyte evaporation to a level that can damage. It is best to use a low-current charger (6 amperes or less) to prevent battery damage caused by overheating.

> ***WARNING —***
> - *Hydrogen gas given off by the battery during charging is explosive. Do not smoke. Keep open flames away from the top of the battery, and prevent electrical sparks by turning off the battery charger before connecting or disconnecting it.*

> ***CAUTION—***
> - *Battery electrolyte (sulfuric acid) can damage the car. If electrolyte is spilled, clean the area with a solution of baking soda and water.*
> - *Allow a frozen battery to thaw before attempting to recharge it.*
> - *Do not exceed 16.5 charging voltage at the battery.*

Battery, charging

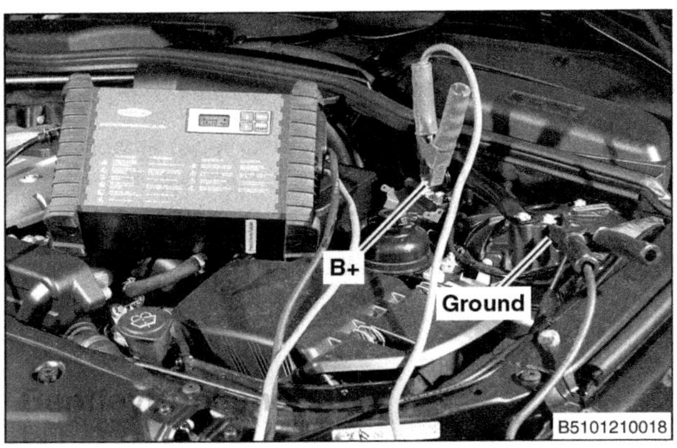

Connect vehicle power supply charger to engine compartment B+ and ground. Follow manufacturer's instructions for setting charging voltage.

Battery, removing and installing

> ***CAUTION—***
> - *Prior to disconnecting the battery, read the battery disconnection cautions given in* **001 Warnings and Cautions**.

- Switch ignition OFF.
- Working in trunk or cargo compartment, open battery compartment.
- Sports Wagon: Remove cargo compartment floor panel. See **513 Interior Trim**.

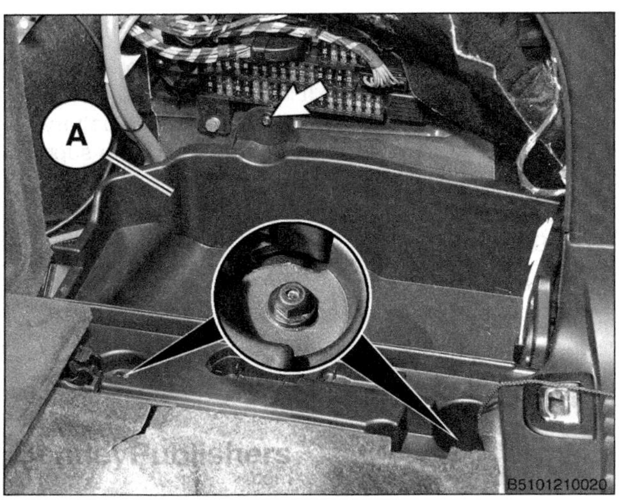

Sports Wagon: Remove screw (**arrow**) securing plastic storage shelf (**A**). Remove plastic storage shelf fasteners (**inset**) and remove plastic storage shelf.

BentleyPublishers.com
BentleyPublishers.com—All Rights Reserved

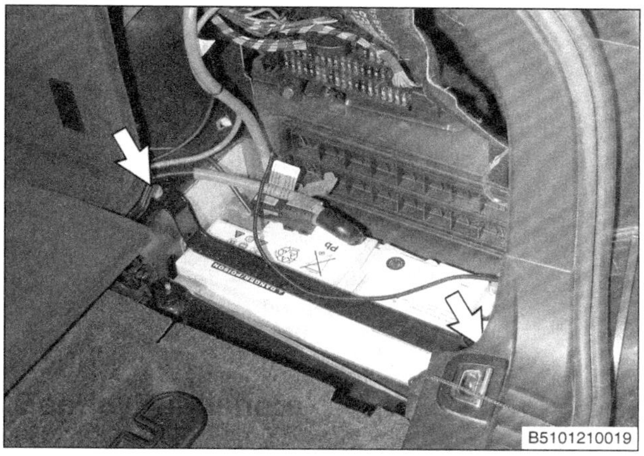

< Working at top of battery, remove protective bracket mounting fasteners (**arrows**) and lift off bracket.

< Working at top of battery:

- Unlock battery safety terminal plastic cover and loosen battery cable fasteners (**inset**).
- Disconnect negative cable first.
- Carefully lift positive cable, negative cable and intelligent battery sensor off battery and lay aside.

> ***CAUTION—***
> - *Do not damage intelligent battery sensor (IBS), which is attached to ground cable. See* **Warnings and Cautions** *in this repair group.*

- Disconnect battery vent hose at side of battery, if applicable.

— Lift battery out of trunk or cargo compartment.

— Clean away corrosion in and around battery tray and on cables ends.

— When reinstalling battery, reconnect negative cable last.

Tightening torques	
Battery post pinch bolt • M6 • M8	 5 Nm (44 in-lb) 15 Nm (11 ft-lb)

Once the battery is reinstalled, the following steps are necessary:

- If higher rating battery is installed, recode CAS with correct battery specifications.
- Register battery. See **Replacement battery registration** in this repair group.
- Follow **Battery reconnection notes** in this repair group.

BentleyPublishers.com
BentleyPublishers.com—All Rights Reserved

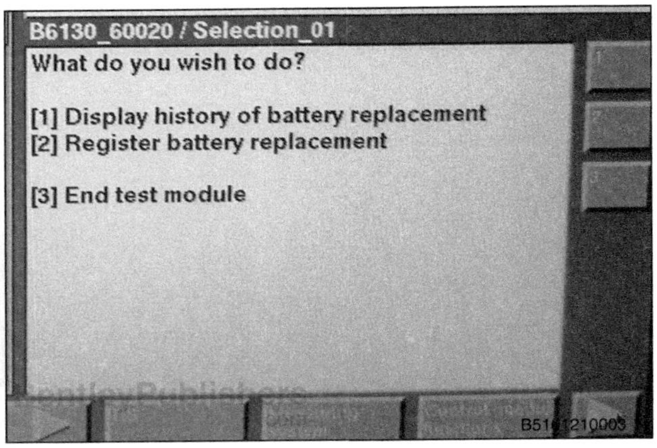

Replacement battery registration

After fitting a new battery, connect BMW scan tool and run service function "Register battery replacement" to notify vehicle power management (software in the engine electronics and intelligent battery sensor) that a new battery is fitted in the vehicle.

Without new battery registration, CAS and vehicle power management does not run properly and this can lead for example: reduction or deactivation of individual consumer functions.

During battery registration, the following operations are performed:

- Current mileage is stored.
- Stored values from old battery (charge status, current, voltage, temperature, etc.) are deleted.
- Power management is initialized.

Battery reconnection notes

In addition to **Warnings and Cautions** in this repair group and in **001 Warnings and Cautions,** observe the following whenever the battery is disconnected or accidentally discharged and then reconnected.

Clock

– Reset.

Radio

– Enter presets.

Steering angle sensor

AFS (active front steering) only:

– Adjust using BMW scan tool.

Active front steering

– Initializing and commissioning, See **320 Steering and Wheel Alignment**.

Active stabilizer

– Commissioning, See **300 Suspension, Steering and Brakes–General**.

Automatic tailgate

– Setting readiness, See **412 Trunk Lid, Tailgate**.

Power window regulator motors

Initialize as follows:

– Switch ignition ON and open window completely.

– Hold switch in Open one-touch operation position and hold for 15 to 25 seconds.

BentleyPublishers.com
BentleyPublishers.com—All Rights Reserved

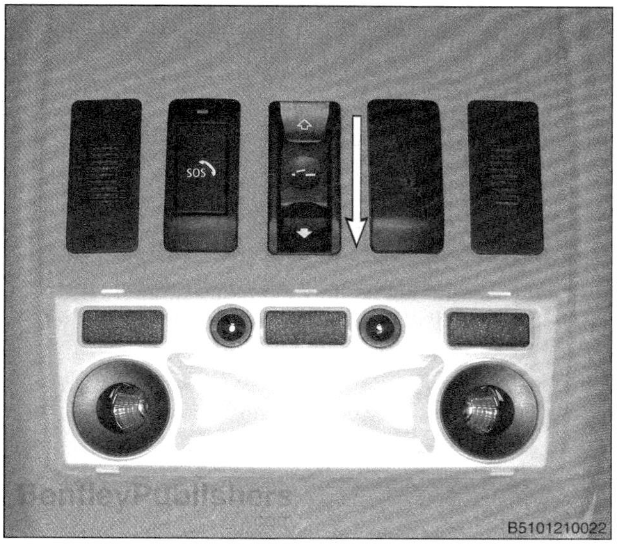

– Release window button, then hold switch in Close direction.

The window fully opens and then closes. Initialization is completed once the window has closed.

– Check to make sure one-touch control, anti-trapping and comfort functions are active.

See **512 Door Windows** for additional information.

Sunroof

Initialize as follows (if applicable):

◄ Switch ignition ON and press and hold sunroof switch in Raised direction (**arrow**).

– After sunroof reaches Raised position, keep switch pressed for approx. 15 seconds longer.

– Initialization is complete when sunroof attempts briefly to lift again.

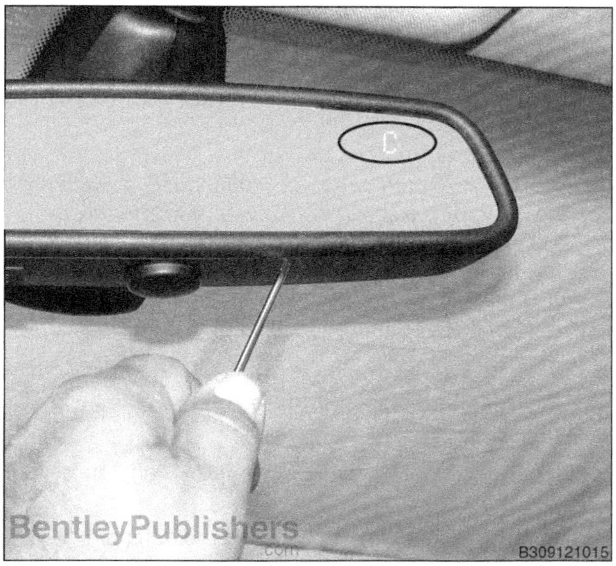

Compass in interior mirror

Calibration (if applicable):

– Switch ignition ON

◄ Using thin bladed screwdriver or equivalent tool, press push button at base of mirror until C appears in compass display (**oval**).

– Drive vehicle 2 - 3 times in a circle at approx. 10 mph.

– Calibration is complete when direction display appears.

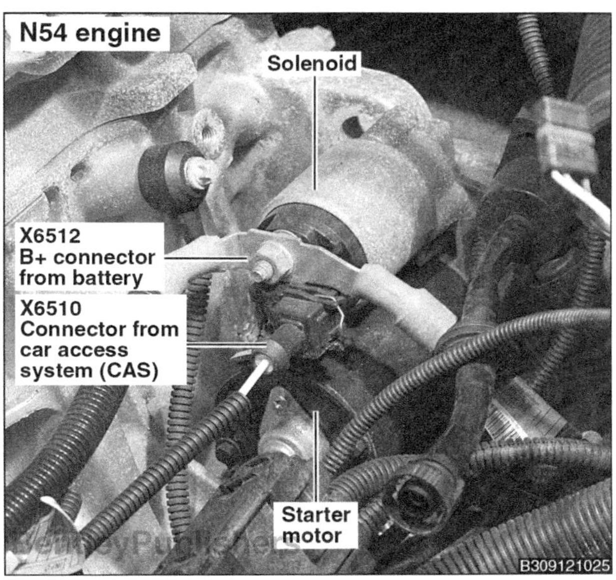

Starter

Starter troubleshooting

◄ M54, N52, N54 engine: Starter is bolted to engine bell housing under intake manifold. Starter wiring connects to solenoid, above starter, as follows:

- Large wire (X6512 or B+ connector) is direct battery voltage.
- Smaller wire (X6510) operates starter solenoid via ignition switch and car access system (CAS).

BentleyPublishers
BentleyPublishers.com—All Rights Reserved

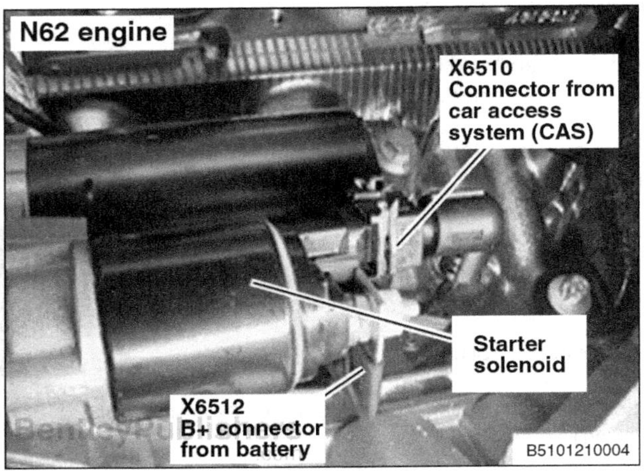

N62 engine: Starter is bolted to engine bell housing under right exhaust manifold. Starter wiring connects to solenoid, above starter, as follows:

- Large wire (X6512 or B+ connector) is direct battery voltage.
- Smaller wire (X6510) operates starter solenoid via ignition switch and car access system (CAS).

– If starter and solenoid click repeatedly without starting vehicle:

- Check battery state of charge.
- Inspect battery and starter cables, terminals and ground connections for good contact. In particular, make sure ground connections between battery, body and engine are completely clean and tight.
- If no faults are found, starter may be faulty and should be replaced.

– If starter fails to operate, check the following:

- Car access system (CAS). Try another ignition key. If no faults can be found, check operation of CAS using a BMW scan tool. CAS replacement is covered in **CAS module, removing and installing** in this repair group.
- Clutch switch module (manual transmission). Make sure clutch is pressed all the way to the floor. See **210 Clutch**.

NOTE —

- *Automatic transmission neutral safety switch function is built into transmission mechatronics electronics. Failure of this function causes a no-start condition.*

– Access to starter is difficult as it is under the intake manifold. If possible, check for battery voltage (12.6 volts) at B+ terminal (X6512).

– Check for battery voltage at terminal X6510 of starter motor with ignition in START position.

- If voltage is not present, check wiring between CAS and starter terminal.
- Check CAS and other inputs that disrupt power to solenoid. See **ELE Electrical Wiring Diagrams**.
- If voltage is present and no other visible wiring faults can be found, problem is most likely in starter motor.

CAS module, removing and installing

CAUTION—

- *CAS module is static sensitive. Static discharge may damage it permanently. Handle using static prevention equipment and techniques. See* **600 Electrical System–General**.

– Remove left side dashboard lower trim (pedal cluster trim). See **513 Interior Trim**.

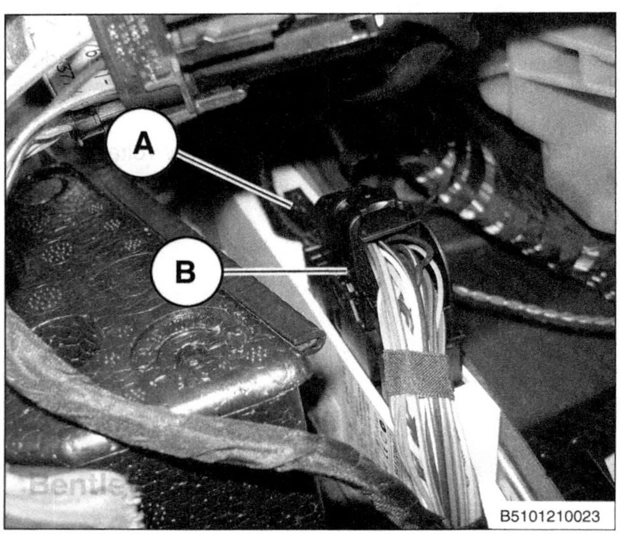

Working to left of steering column:

- Detach ribbon harness (**A**) from CAS module.
- Detach harness connector (**B**) from CAS module.

BentleyPublishers.com
BentleyPublishers.com—All Rights Reserved

B5101210005

Remove module mounting screw (**arrow**) and withdraw module to left and down to remove from dashboard.

- Inspect ribbon cable and replace if damaged. (Ribbon cable connects CAS to ignition key slot and START / STOP button.)
- When installing, make sure CAS module guide fits correctly into mounting slot.
- After reassembly, carry out vehicle coding and programming. See **600 Electrical System–General**.

Starter, removing and installing (M54 engine)

The starter is removed from below. Access to starter bolts is difficult. Be sure to have the proper tools on hand before proceeding.

- Disconnect negative (–) battery cable.

> ***CAUTION—***
> - *Prior to disconnecting the battery, read the battery disconnection cautions in* **001 Warnings and Cautions**.

- Raise vehicle and support safely.

> ***WARNING—***
> - *Make sure the car is stable and well supported at all times. Use a professional automotive lift or jack stands designed for the purpose. A floor jack is not adequate support.*

- Remove front suspension reinforcement plate.

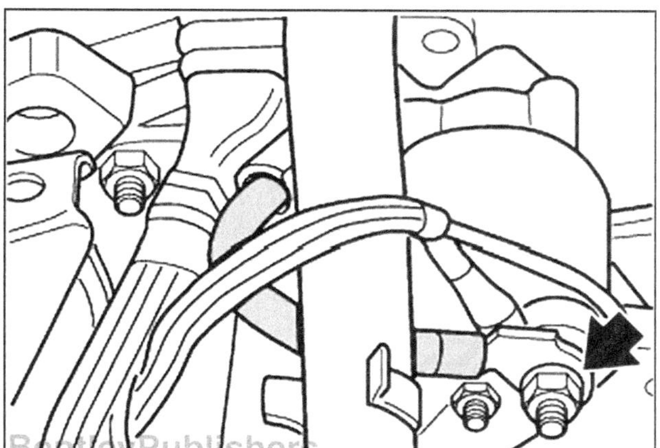

502121034

Working underneath car, disconnect battery positive (+) lead from starter (**arrow**).

NOTE—
- *Note routing for cables. Cable routing can vary. When reinstalling, make sure cables do not chafe.*

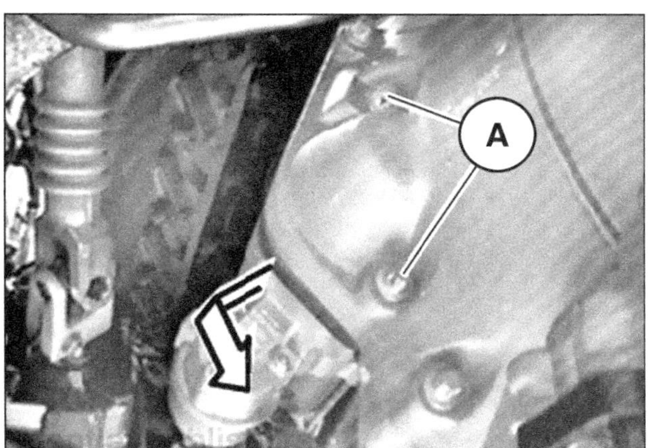

B5101210009

Using a long extension and universal joint from output end of transmission, remove bolts (**A**) fastening starter to transmission bell housing and engine block.

- Pull starter out of transmission bell housing in direction of **arrow** and detach remaining starter wiring.
- Remove starter from below.

BentleyPublishers.com
BentleyPublishers.com–All Rights Reserved

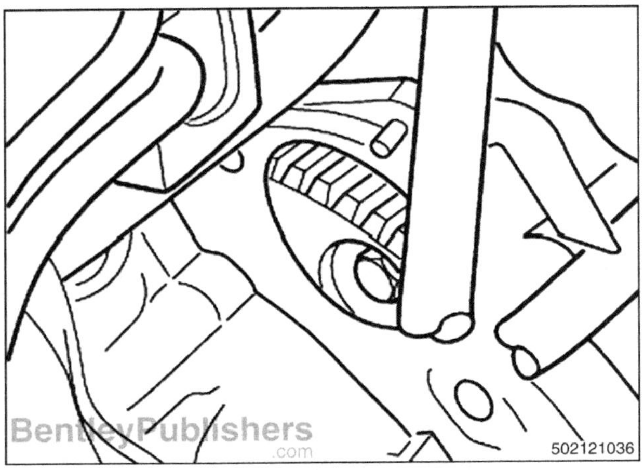

Check starter pinion gear and flywheel teeth for damage.

– Installation is reverse of removal.

Tightening torques	
Starter to transmission bell housing	45 Nm (33 ft-lb)
Wire terminals to starter • M5 • M6 • M8	 5 Nm (44 in-lb) 7 Nm (62 in-lb) 13 Nm (10 ft-lb)

Starter, removing and installing (N52, N54 engine)

– Switch ignition OFF. Disconnect negative (–) battery cable and cover battery terminal to keep cable from accidentally contacting terminal.

> ***CAUTION—***
> • *Prior to disconnecting the battery, read the battery disconnection cautions in* **001 Warnings and Cautions**.

– Remove upper and lower cabin air microfilter housings. See **020 Maintenance**.

– Remove engine air filter housing and ducts. See **131**, **132 Fuel Injection**.

– Remove ignition coil cover. See **020 Maintenance**.

– Remove intake manifold. See **131**, **132 Fuel Injection**.

> ***CAUTION—***
> • *Do not detach fuel lines or injectors.*

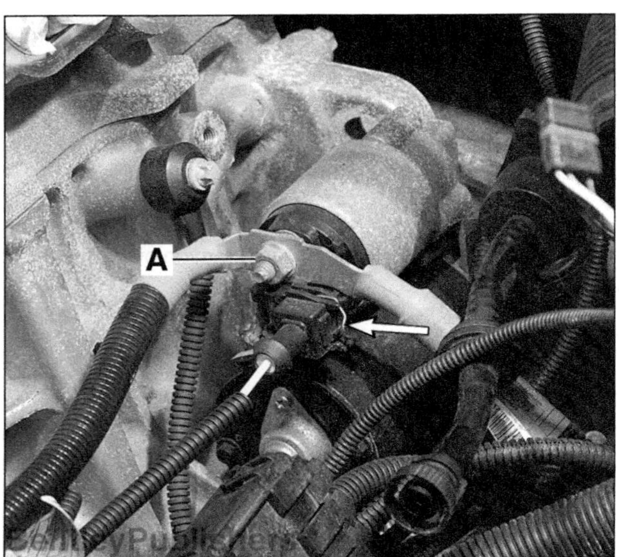

Working at starter solenoid electrical connections:
- Remove fastener (**A**) and detach B+ cable from solenoid.
- Press in wire clip (**arrow**) to release solenoid wire connector.

BentleyPublishers.com
BentleyPublishers.com–All Rights Reserved

- Remove starter mounting bolts (**arrows**). Discard aluminum bolts. (Aluminum bolt heads are not always marked with blue paint. Use a magnet to identify them.)

- Lift out starter motor.

- Before installing new starter, rotate engine and inspect flywheel or torque plate gear teeth through starter hole. Replace if damaged. See **210 Clutch** or **240 Automatic Transmission**.

- Reinstall starter and reattach electrical connectors.

Tightening torques	
B+ cable to starter (M8)	13 Nm (10 ft-lb)
Starter to bell housing (turbo)	38 Nm (28 ft-lb)
Starter to bell housing (non-turbo) (replace M10 aluminum bolts) • Stage 1 • Stage 2	 20 Nm (15 ft-lb) additional 180°

Starter, removing and installing (V8 engine)

- Switch ignition OFF. Disconnect negative (–) battery cable and cover battery terminal to keep cable from accidentally contacting terminal.

> ***CAUTION—***
> *• Prior to disconnecting the battery, read the battery disconnection cautions in* **001 Warnings and Cautions**.

- Remove right side exhaust manifold. See **180 Exhaust System**.

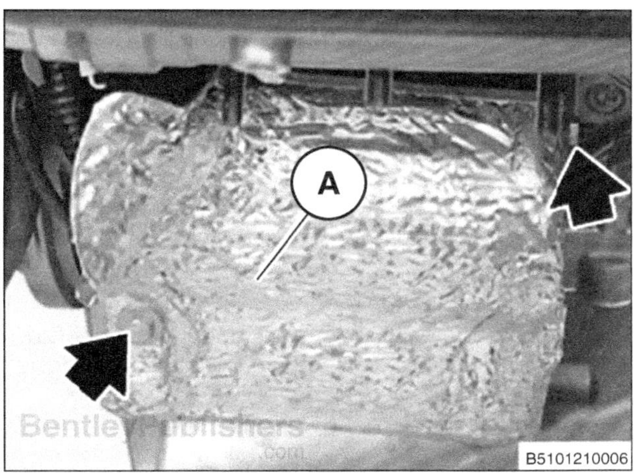

- Remove starter heat shield fasteners (**arrows**) then remove heat shield (**A**).

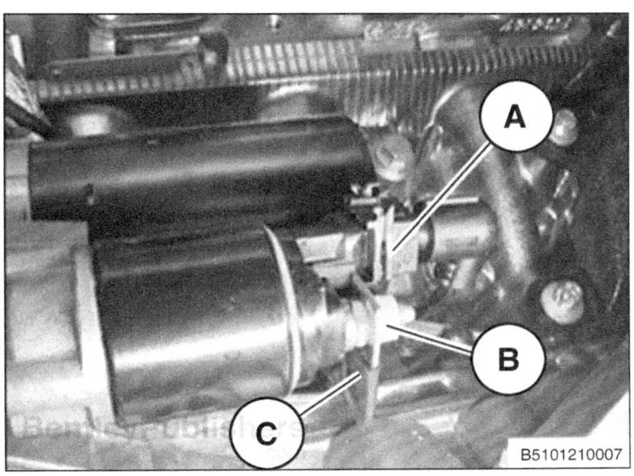

- Disconnect electrical connector (**A**). Remove fastener (**B**) and remove starter cable (**C**).

BentleyPublishers.com
BentleyPublishers.com—All Rights Reserved

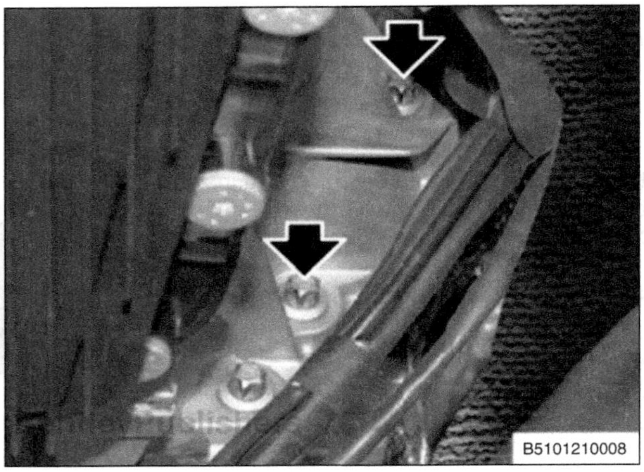

- Remove starter mounting bolts (**arrows**).
- Lift out starter motor.
- Before installing new starter, rotate engine and inspect flywheel or torque plate gear teeth through starter hole. Replace if damaged. See **210 Clutch** or **240 Automatic Transmission**.
- Reinstall starter and reattach electrical connectors.

Tightening torques	
B+ cable to starter (M8)	13 Nm (10 ft-lb)
Starter to bell housing (replace bolts)	45 Nm (33 ft-lb)
Starter heat shield	6 Nm (4.4 ft-lb)

ALTERNATOR AND CHARGING SYSTEM

Alternator basics

The alternator is bolted to the left side of the engine block. It is driven by the engine accessory belt. Accessory belt replacement is covered in **020 Maintenance**.

Several versions of alternators are used in E60 vehicles. Be sure that the replacement alternator is correct for the application.

Alternator interface

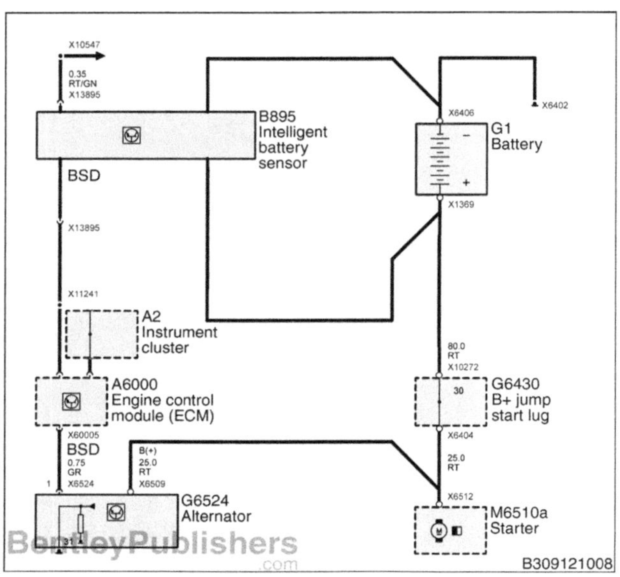

The alternator communicates with the engine control module (ECM) via bit-serial data (BSD) interface. This single wire connection allows the ECM to adapt its calculations and control data to alternator output. The ECM controls the following functions:

- Alternator activation and deactivation.
- Informing the voltage regulator of the nominal voltage value to be set.
- Alternator load response control.
- BSD line diagnosis.
- Alternator fault code storage.
- Charge indicator light activation in instrument cluster.

BentleyPublishers.com
BentleyPublishers.com—All Rights Reserved

Charging system troubleshooting

Some charging system tests require special test equipment. If the test equipment is not available, charging system fault diagnosis can be performed by an authorized BMW dealer or other qualified repair shop. See **Table a** for general electrical component troubleshooting.

Before checking the alternator, make sure the battery is fully charged and capable of holding a charge. Check that the battery terminals are clean and tight and the engine accessory drive belt is properly tensioned and not severely worn. For drive belt service see **020 Maintenance**.

Charging system quick-check

– Use a digital multimeter to measure voltage across the battery terminals with ignition OFF and then again with engine running. Battery voltage should be about 12.6 volts with key OFF and between 13.2 and 14.5 volts with engine running.

– Run engine at about 2000 rpm and switch ON electrical loads (fans, lights, rear window defroster, and wipers). With all accessories on, battery voltage should be above 12.6 volts.

The regulated voltage (engine running, battery charged, accessories and lights OFF) is usually between 13.2 and 14.5, depending on temperature and operating conditions. If the voltage is higher than 14.8, the voltage regulator is most likely faulty.

Charging system, checking

> ***CAUTION—***
> - *Do not disconnect the battery while the engine is running. Damage to the alternator or engine electronic systems may result.*
> - *Only use a digital multimeter when testing charging system components.*

◄ Switch ignition ON. Check that battery warning light comes ON for self test. If warning light does not come ON, check for fault codes.

- Check for battery voltage at alternator terminal 30 (B+):
  - Pull off insulating cover at terminal 30 (**arrow**) at rear of alternator.
  - Connect voltmeter between terminal 30 and ground and check for battery voltage. If voltage is not present, check wiring for faults.

– Connect oscilloscope across battery terminals to check alternator function.

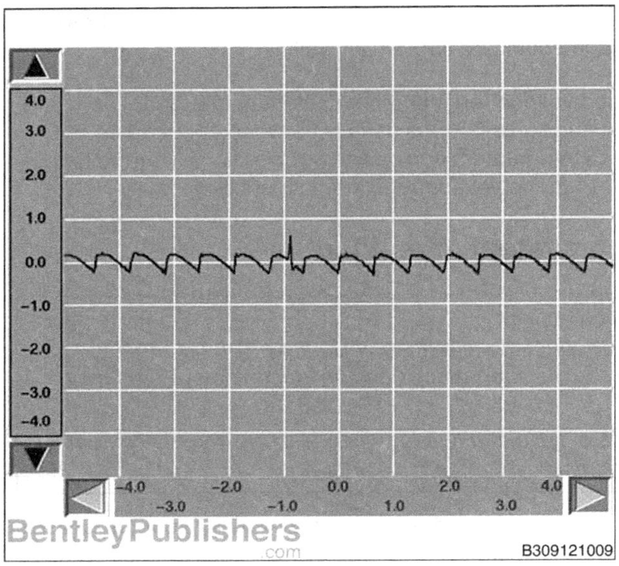

– Connecting oscilloscope:
  - Positive test lead of oscilloscope to positive battery terminal.
  - Negative test lead of oscilloscope to negative battery terminal
  - Set oscilloscope to A/C volts and a time base of 500ms.

- Normal alternator pattern.

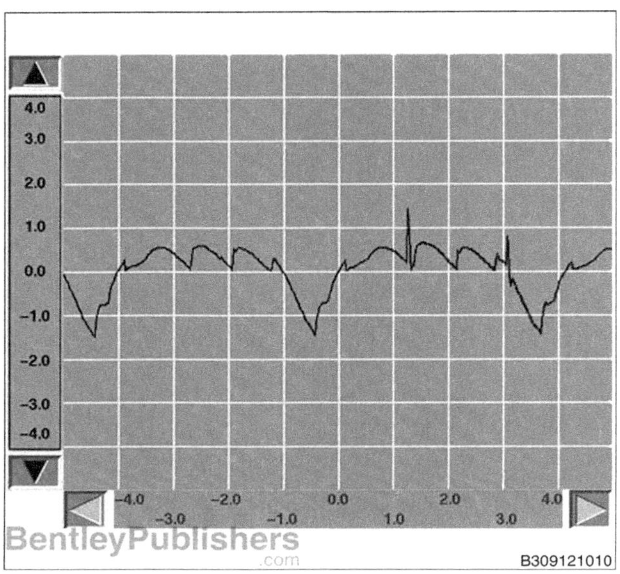

- Alternator with defective diode.

BentleyPublishers.com
BentleyPublishers.com–All Rights Reserved

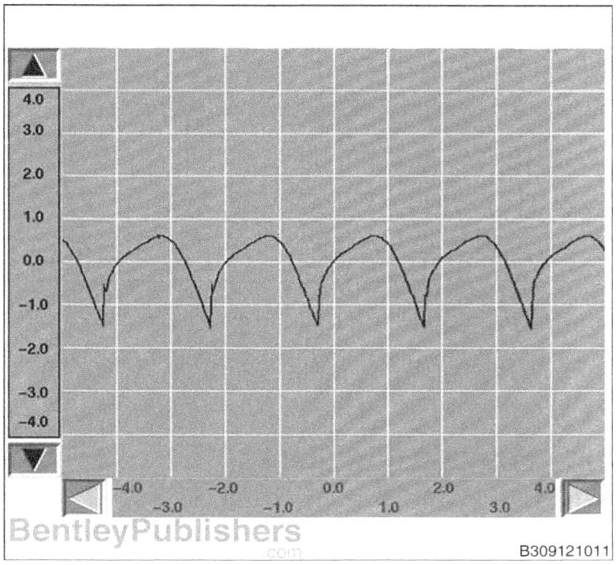

B309121011

Alternator with broken winding.

– If test indicate a defective alternator, replace unit. Repair parts are not available for the alternator.

Alternator, removing and installing (M54 engine)

A replacement alternator should have the same rating as the original. Alternator manufacturer and ampere rating are normally marked on the alternator housing.

– Switch ignition OFF. Disconnect negative (–) battery cable and cover battery terminal to keep cable from accidentally contacting terminal.

> ***CAUTION—***
> *• Prior to disconnecting the battery, read the battery disconnection cautions in* **001 Warnings and Cautions**.

– Remove air intake duct and filter housing. See **130 Fuel Injection**.

– Remove engine accessory belt. See **020 Maintenance**. Mark direction of rotation on belt before removing.

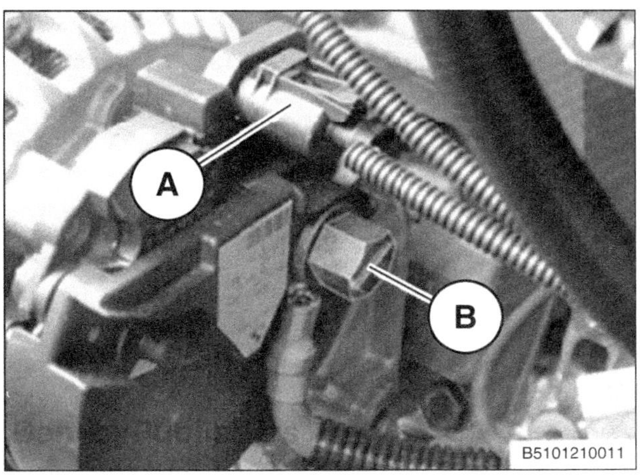

B5101210011

Working at rear of alternator, remove electrical connector (**A**) and battery positive cable fastener (**B**).

B5101210012

Working at front of alternator (**B**), remove cover from idler pulley (**A**) and remove alternator fasteners (**arrows**).

– Rotate alternator forward and then up and out of engine compartment, using care to not damage radiator.

> ***CAUTION—***
> *• Protect radiator with heavy cardboard or sheet metal.*

BentleyPublishers.com
BentleyPublishers.com—All Rights Reserved

– Installation is reverse of removal.Remember to:
  - Reinstall accessory belt using previously made direction-of-rotation marks.
  - Make sure accessory belt grooves engage pulleys correctly.

Tightening torques	
Pulley to alternator	70 Nm (44 ft-lb)
Terminal 30 (B+) wire to alternator (M8)	13 Nm (10 ft-lb)
Alternator to crankcase (M8)	21 Nm (15.4 ft-lb)

– After reconnecting battery, see **Battery reconnection notes** in this repair group.

Alternator, removing and installing (N52 engine)

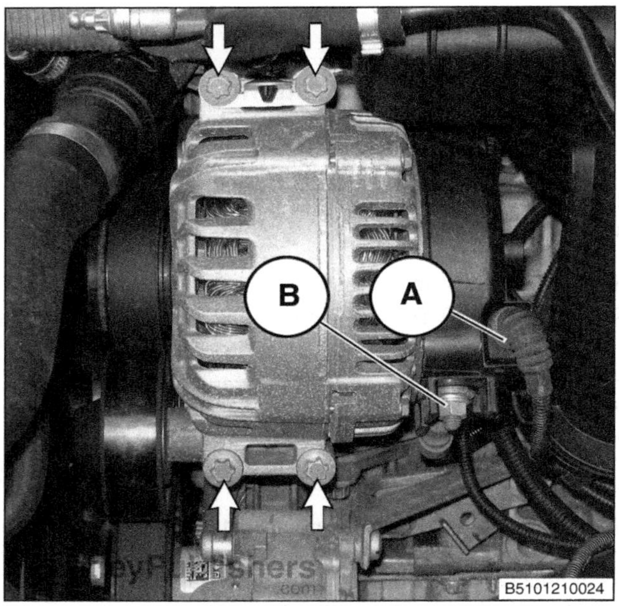

– Switch ignition OFF. Disconnect negative (–) battery cable and cover battery terminal to keep cable from accidentally contacting terminal.

> ***CAUTION—***
> - *Prior to disconnecting the battery, read the battery disconnection cautions in* **001 Warnings and Cautions**.

– Remove engine accessory belt. See **020 Maintenance**. Mark direction of rotation on belt before removing.

– Remove air intake duct and filter housing. See **131 Fuel Injection**.

◀ Working at rear of alternator:
  - Detach harness connector (**A**)
  - Remove nut (**B**) and detach alternator cable.
  - Remove alternator mounting bolts (**arrows**) and discard (aluminum bolts).
  - Lift out alternator.

◀ With alternator on work bench:
  - Remove idler pulley (**C**) fastener plastic cover, if applicable.
  - Remove idler pulley and transfer to new alternator.

Tightening torques	
• Idler pulley to alternator housing	80 Nm (59 ft-lb)

– Reinstall alternator using new aluminium bolts. Reattach electrical harnesses.

Tightening torques	
Alternator cable to alternator (M8)	19 Nm (14 ft-lb)
Alternator to engine block (M8 x 82 mm aluminum bolts, replace with new) • Stage 1 • Stage 2	 10 Nm additional 180°

BentleyPublishers.com
BentleyPublishers.com—All Rights Reserved

– Reassemble front of engine. Remember to:
  - Reinstall accessory belt using previously made direction-of-rotation marks.
  - Make sure accessory belt grooves engage pulleys correctly.
  - After reconnecting battery, see **Battery reconnection notes** in this repair group.

Alternator, removing and installing (N54 engine)

– Switch ignition OFF. Disconnect negative (–) battery cable and cover battery terminal to keep cable from accidentally contacting terminal.

> ***CAUTION—***
> - *Prior to disconnecting the battery, read the battery disconnection cautions in* **001 Warnings and Cautions**.

– Raise vehicle and support safely.

> ***WARNING —***
> - *Make sure the vehicle is stable and well supported at all times. Use a professional automotive lift or jack stands designed for the purpose. A floor jack is not adequate support.*

– Remove splash shield underneath engine compartment. See **020 Maintenance**.

– Remove electric fan cowl and fan. See **170 Radiator and Cooling System**.

– Remove engine accessory belt. See **020 Maintenance**. Mark direction of rotation on belt before removing.

– Remove air intake duct and filter housing. See **132 Fuel Injection**.

– Remove left side charge air duct. See **132 Fuel Injection**.

– Remove A/C compressor from mounting bracket and hang aside. See **640 Heating and Air-conditioning**. Do not detach A/C refrigerant lines.

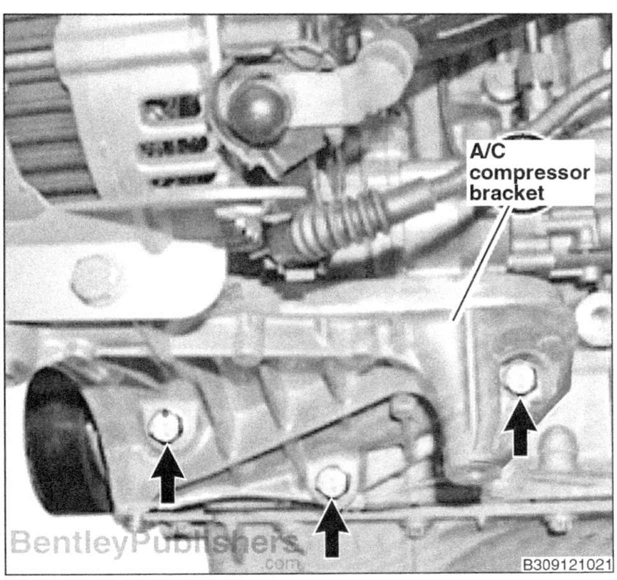

◄ Loosen A/C compressor bracket bolts (**arrows**) a few turns. Do not remove bracket.

BentleyPublishers.com—All Rights Reserved

◄ Working at rear of alternator:

- Detach harness connector (**A**)
- Pull off plastic protective cover and remove nut (**B**) to detach alternator terminal 30 (B+) cable.

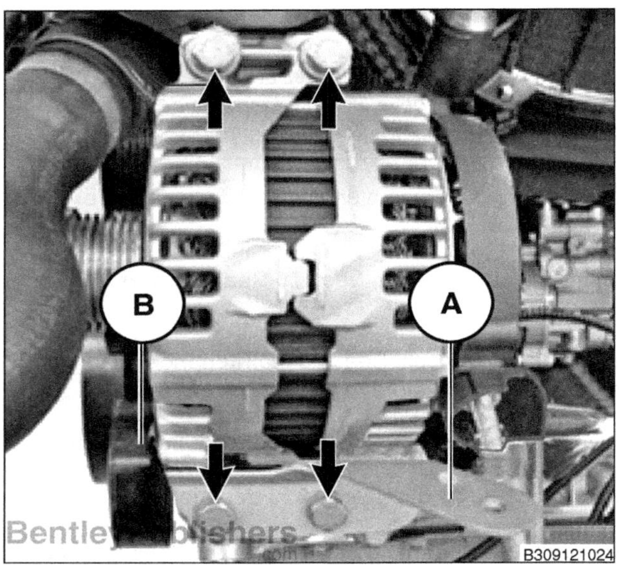

◄ Working at left side of engine:

- Remove alternator mounting bolts (**arrows**).
- Lift out alternator with bracket (**A**).

– With alternator on work bench:

- Remove idler pulley (**B**) fastener plastic cover, if applicable.
- Remove idler pulley and transfer to new alternator.

Tightening torques	
Idler pulley to alternator housing	80 Nm (59 ft-lb)

– Reinstall alternator and reattach electrical harnesses.

Tightening torques	
Alternator cable to alternator (M8)	19 Nm (14 ft-lb)
Alternator to engine block (M8)	19 Nm (14 ft-lb)

– Tighten A/C compressor bracket bolts, then reinstall compressor.

Tightening torque	
A/C compressor bracket to engine block	19 Nm (14 ft-lb)

– Reassemble front of engine. Remember to:

- Reinstall accessory belt using previously made direction-of-rotation marks.
- Make sure accessory belt grooves engage pulleys correctly.
- After reconnecting battery, see **Battery reconnection notes** in this repair group.

BentleyPublishers.com
BentleyPublishers.com—All Rights Reserved

Alternator, removing and installing (V8 engine)

– Switch ignition OFF. Disconnect negative (–) battery cable and cover battery terminal to keep cable from accidentally contacting terminal.

> ***CAUTION—***
> • *Prior to disconnecting the battery, read the battery disconnection cautions in* **001 Warnings and Cautions**.

– Remove air intake duct and filter housing. See **133 Fuel Injection**.

– Remove electric fan cowl and fan. See **170 Radiator and Cooling System**.

– Remove alternator drive belt. See **020 Maintenance**. Mark direction of rotation on belt before removing.

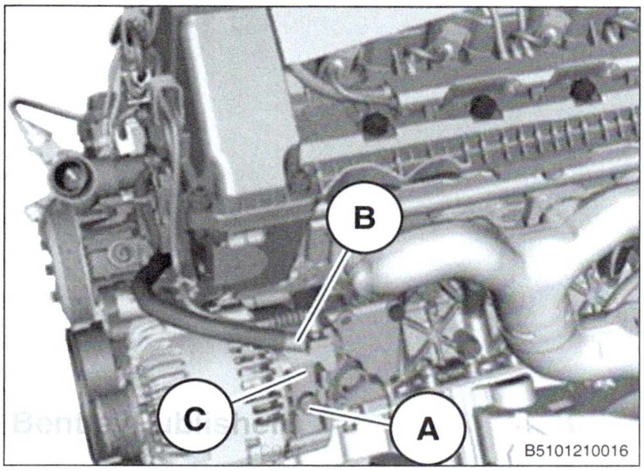

◄ Disconnect alternator electrical connector (**B**). Remove fastener (**A**) and alternator cable (**C**).

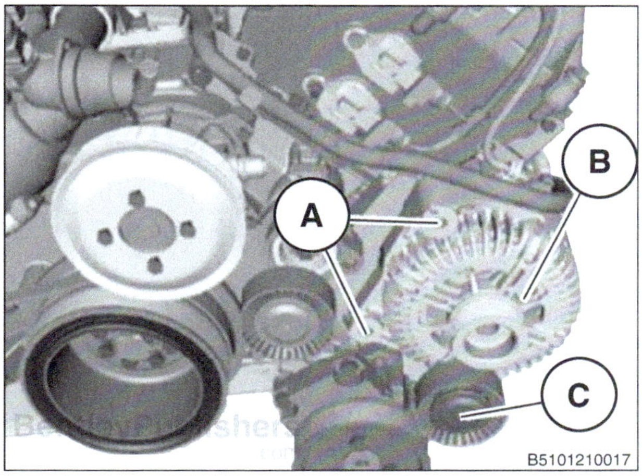

◄ Remove alternator fasteners (**A**) and lift out alternator (**B**).
- Remove idler pulley (**C**) fastener plastic cover, if applicable.
- Remove idler pulley and transfer to new alternator.

– Installation is reverse of removal.

Tightening torques	
Terminal 30 (B+) wire to alternator (M8)	13 Nm (10 ft-lb)
Idler pulley to alternator	45 Nm (33 ft-lb)
Drive pulley to alternator	70 Nm (51 ft-lb)
Alternator to crankcase	21 Nm (15.4 ft-lb)

BentleyPublishers
BentleyPublishers.com–All Rights Reserved

BentleyPublishers.com
BentleyPublishers.com—All Rights Reserved

130 Fuel Injection MS45.1 (M54 Engine)

TABLE

General

This repair group covers service and repair of engine management system for 5 Series models with M54 engine and Siemens DME MS45.1.

See also:

- **020 Maintenance** for air filter, fuel filter and spark plug replacement
- **100 Engine–General** for engine identification
- **120 Ignition System** for ignition coil, camshaft sensor, crankshaft sensor and knock sensor service
- **160 Fuel Tank and Fuel Pump** for fuel pump and fuel pump relay service, fuel pressure tests and fuel pressure regulator replacement
- **180 Exhaust System** for oxygen sensor service
- **ECL Electrical Component Locations**
- **ELE Electrical Wiring Diagrams**
- **OBD On-Board Diagnostics**

BentleyPublishers.com
BentleyPublishers.com–All Rights Reserved

DME applications

BMW 5 Series models are equipped with digital motor electronics (DME), also known as Motronic. In these systems, fuel injection and ignition are controlled by an integrated engine control module (ECM). Application information for DME systems is in **Table a**.

Table a. 5 Series engine management applications

Year, model	Engine code	Engine management	Features
2004-2005			
Sedan: 525i, 530i	M54	Siemens MS 45.1	Double VANOS
Sedan: 545i	N62	Bosch DME ME 9.2 with Valvetronic control module	Valvetronic, VANOS
2006 - 2007			
Sedan, Sports Wagon: 525i, 525xi, 530i, 530xi Sports Wagon	N52	Siemens MSV70	Valvetronic, VANOS
2008 - 2010			
Sedan, Sports Wagon: 528i, 528xi	N52KP	Siemens MSV80	Valvetronic, VANOS
Sedan, Sports Wagon: 535i, 535xi	N54	Siemens MSD80 (to 03 / 2008) MSD81 (from 03 / 2008)	VANOS, turbocharged
2004-2005			
Sedan: 545i	N62	Bosch DME ME 9.2 with valvetronic control unit	Valvetronic, VANOS
2006 - 2010			
Sedan: 550i	N62TU	Bosch DME ME 9.2.2	Valvetronic, VANOS

BentleyPublishers.com
BentleyPublishers.com—All Rights Reserved

Warnings and Cautions

WARNING—

- *The fuel system is designed to retain pressure even when the ignition is OFF. When working with the fuel system, loosen the fuel lines slowly to allow residual fuel pressure to dissipate. Avoid spraying fuel. Use shop towels to capture leaking fuel.*
- *Fuel in fuel lines is under pressure (approx. 3 - 5 bar or 45 - 75 psi) and may be expelled forcibly. Do not smoke or work near heaters or other fire hazards. Keep a fire extinguisher handy.*
- *Unscrew the fuel tank cap to release pressure in the tank before working on fuel lines.*
- *Plug open fuel lines and fittings.*
- *Before beginning work on the fuel system, place a fire extinguisher in the vicinity of the work area.*
- *Work only on the fuel system when engine temperature is below 40°C (104°F).*
- *When disconnecting a fuel line, clamp off the line and wrap a clean shop towel around the fitting before disconnecting. Residual fuel pressure is present in the line.*
- *Fuel is highly flammable. When working around fuel, do not disconnect wires that could cause electrical sparks. Do not smoke or work near heaters or other fire hazards.*
- *Wear eye protection and protective clothing to avoid injuries from contact with fuel.*
- *When working on an open fuel system, wear suitable hand protection, as prolonged contact with fuel can cause illnesses and skin disorders.*
- *Unscrew the fuel tank cap to release pressure in the tank before working on fuel lines.*
- *Do not use a work light with an incandescent bulb near fuel. Fuel may spray on the hot bulb causing a fire.*
- *Make sure the work area is properly ventilated.*
- *Due to risk of personal injury, be sure the engine is cold before beginning work on engine components.*
- *The ignition system produces high voltages that can be fatal. Avoid contact with exposed terminals. Use extreme caution when working on a car with the ignition switched ON or the engine running.*

BentleyPublishers
BentleyPublishers.com—All Rights Reserved

CAUTION—

- *Renew fuel system hoses, clamps and O-rings any time they are removed.*
- *Prior to disconnecting the battery, read the battery disconnection cautions in* **001 Warnings and Cautions**.
- *Connect and disconnect the DME system wiring and test equipment leads only when the ignition is switched OFF.*
- *Wait at least 1 minute after switching the ignition OFF before disconnecting engine control module (ECM) connectors. If the connectors are removed before this time, residual power in the system relay may damage the control module.*
- *Tests or repair procedures in this section may set fault codes (DTCs) in the ECM and illuminate the MIL. After repairs are completed, access and clear DTC memory using a BMW scan tool. See* **On-board diagnostics** *in this repair group.*
- *Fuel system cleaners and other chemical additives other than those specifically recommended by BMW may damage catalytic converters, oxygen sensors or other fuel supply components.*
- *Do not connect any test equipment that delivers a 12-volt power supply to terminal 15 (+) of the ignition coils. The current flow may damage the ECM. Connect test equipment only as specified by BMW or the equipment maker.*
- *Relay positions can vary. Be sure to confirm relay location and function by identifying the wiring in the socket using the wiring diagrams found in* **ELE Electrical Wiring Diagrams**.
- *Use a digital multimeter for electrical tests. Use an LED test light for quick tests.*

BentleyPublishers.com
BentleyPublishers.com—All Rights Reserved

Driveability Troubleshooting

The self-diagnostic DME engine management systems monitor and store diagnostic trouble codes (DTCs). If the malfunction indicator light (MIL) illuminates, it indicates that an emissions-related fault has occurred and that one or more DTCs are stored in the engine control module (ECM).

If faults arise, or if the MIL is illuminated, begin troubleshooting by connecting BMW scan tool to the OBD II plug under the dashboard.

The capabilities of OBD II software has the potential to save hours of diagnostic time and to help avoid incorrect component replacement and possible damage to system components. See **On-board diagnostics** in this repair group.

Basic engine settings

Idle speed, idle mixture (%CO), and ignition timing are not adjustable. The DME system is adaptive and automatically compensates for changes in the engine due to age, minor wear or small problems such as a disconnected vacuum hose. However, the adaptive range is limited. Once the limits are exceeded, driveability problems become noticeable.

Poor initial driveability may be encountered when the battery is disconnected and reconnected. When the battery is disconnected, ECM adaptive memory may be reset. The system readapts after about ten minutes of driving.

System voltage

Digital motor electronics (DME) requires that the system (battery) voltage be maintained within a narrow range of DC voltage. DC voltage levels beyond or below the operating range, or any AC voltage in the electrical system can cause havoc.

- When troubleshooting an illuminated MIL, make sure the battery is fully charged and capable of delivering all its power to the electrical system. An undercharged battery can amplify AC alternator output ripple.

- To make a quick check of battery charge, measure voltage across battery terminals with all cables attached and ignition OFF. A fully charged battery measures 12.6 volts or slightly more, compared to 12.15 volts for a battery with a 25% charge.

The DME system operates at low voltage and current levels, making it sensitive to small increases in resistance. The electrical system is routinely subject to corrosion, vibration and wear, so faults or corrosion in the wiring harness and connectors are not uncommon.

- Check battery terminals for corrosion or loose cable connections. See **121 Battery, Starter, Alternator**.

– If a battery cable connection has no visible faults but is still suspect, measure voltage drop across the connection. A large drop indicates excessive resistance, meaning that the connection is corroded, dirty or damaged. Clean or repair and retest. See **600 Electrical System–General** for voltage drop test procedure.

– Visually inspect wiring, connectors, switches and fuses. Loose or damaged connectors can cause intermittent problems, especially small terminals in ECM connectors. Disconnect wiring harness connectors to check for corrosion, and use electrical cleaning spray to remove contaminants.

Main grounds

Good grounds are critical to proper DME operation. If a ground connection has no visible faults but is still suspect, measure the voltage drop across the connection. A large voltage drop means high resistance. Clean or repair the connection and retest.

For voltage drop testing, see **600 Electrical System-General**.

For ground locations, see **ECL Electrical Component Locations**.

On-board diagnostics

Second generation on-board diagnostics (OBD II) software and hardware is incorporated in the engine management systems. The OBD II system monitors components that influence exhaust and evaporative emissions. If a problem is detected, the OBD II system stores the associated diagnostic trouble code (DTC) and condition.

If vehicle emissions levels exceed 1.5 times Federally mandated criteria, the OBD II system illuminates a malfunction indicator light (MIL) in the instrument cluster.

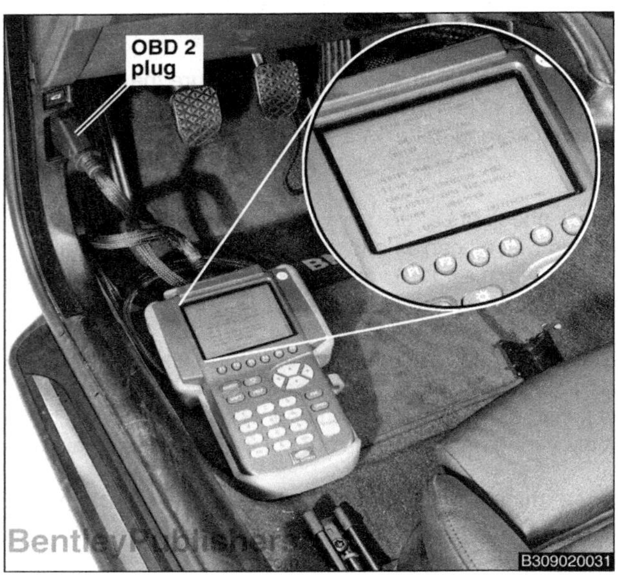

Scan tool connected to OBD II plug on floor post below left dashboard.

Professional diagnostic scan tools available at the time of this printing include the BMW GT1, ISTA and a small number of aftermarket BMW-specific tools.

In addition to professional scan tools, there are inexpensive generic OBD II scan tool software programs and handheld units available. Although these have limited capabilities as compared to the dedicated tools, they are powerful diagnostic tools.

For the DIY owner, a simple aftermarket DTC reader is available through http://www.peakeresearch.com. This tool is capable of checking for DTCs as well as switching the illuminated MIL OFF and resetting service indicator lights.

NOTE —

- *OBD II DTC memory (including an illuminated MIL) can only be reset using a BMW scan tool or equivalent. Removing the connector from the ECM or disconnecting the battery does not erase DTC memory.*

BentleyPublishers.com
BentleyPublishers.com—All Rights Reserved

ENGINE MANAGEMENT FUSES AND POWER SUPPLY COMPONENTS

Power supply fuses

See **ECL Electrical Component Locations** for fuse panel access information.

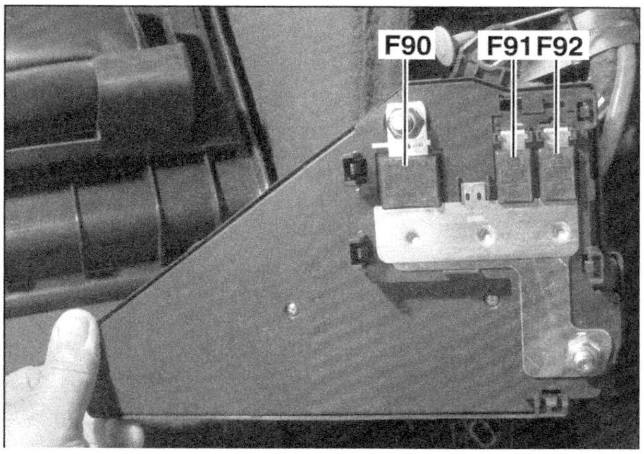

Rear distribution panel (A46) in trunk, installed in right rear fender:

- **F90** 200A: Power supply to fuse panel A41a.

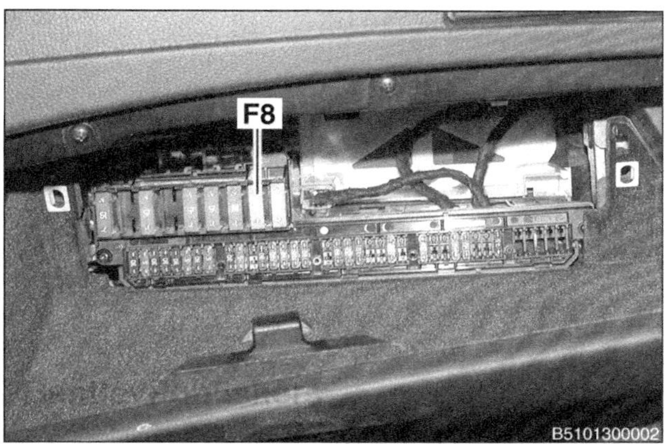

A41a: fuse and relay panel behind glove compartment:

- **F8** 60A:
  Engine control module (ECM) (A6000)
  ECM main relay (K6300)
  Engine electronics fuse carrier (A8680)
  Engine electronics fuse carrier (A8681)

See **ECL Electrical Component Locations** for specific vehicle fuse locations.

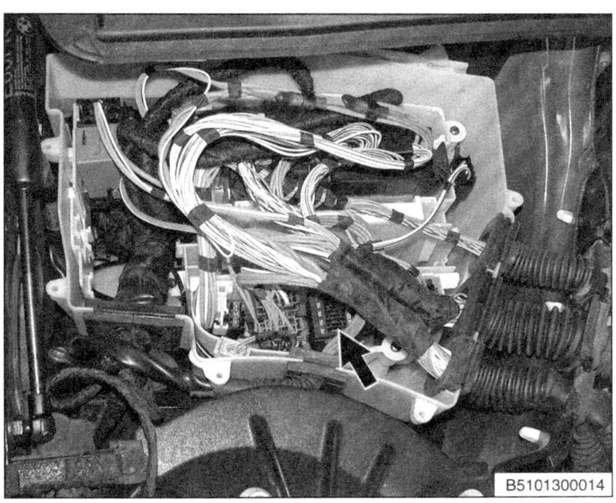

Engine electronics fuse carrier A8680 (**arrow**).

BentleyPublishers.com–All Rights Reserved

SIEMENS DME MS45.1 (M54 ENGINE)

DME MS45.1 system description

DME MS45.1 manages and monitors the following functions:

Air

- Idle speed valve
- Electronic throttle
- Hot film mass air flow sensor
- Resonance-turbulence intake control

Fuel

- Fuel supply (returnless)
- Fuel injection

Ignition

- Direct ignition
- Knock control
- Primary / secondary ignition monitoring

Emissions

- OBD II compliance
- Secondary air injection and monitoring
- Pre- and post-catalyst oxygen sensors
- Electrically heated DME-mapped thermostat
- Misfire detection
- Evaporative emission control and leak detection
- Ambient pressure sensing
- Malfunction indicator light (MIL)

Performance controls

- Double VANOS control
- Electric radiator cooling fan
- E-box cooling fan
- CAN-bus communication
- CAS (car access system) - EWS 3.3
- Cruise control
- Comfort start
- ECM programming
- Alternator interface (BSD, IBS)
- Dynamic stability control (DSC) interface

BentleyPublishers.com
BentleyPublishers.com—All Rights Reserved

Engine control module (ECM)

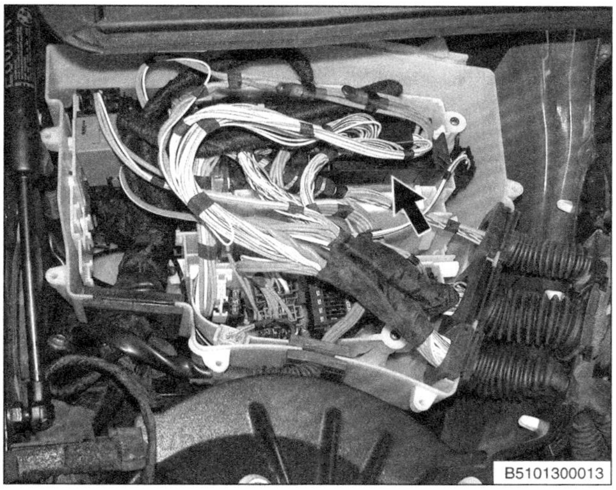

B5101300013

The engine control module (ECM) (**arrow**) is mounted in the electronics box (E-box) at right rear of engine compartment.

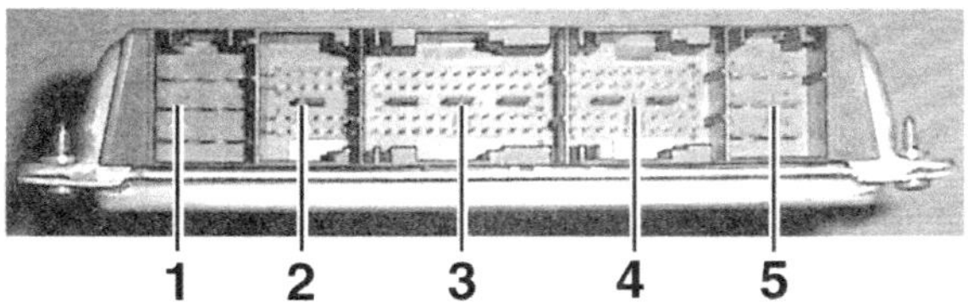

BentleyPublishers.com

B701130045

The MS45.1 ECM is flash-programmable and features 5 electrical harness connectors with a total of 134 pins.

- **Connector 1:** Voltages and grounds
- **Connector 2:** Ancillary signals (oxygen sensors, etc.)
- **Connector 3:** Engine signals
- **Connector 4:** Vehicle signals
- **Connector 5:** Ignition signals

The EEPROM (chip) in the MS45.1 ECM is coded to the vehicle. It can be flashed (reprogrammed) up to 13 times.

Fuel metering

The ECM meters pressurized fuel by changing the opening time (pulse width) of the fuel injectors. The exact amount of fuel injected is determined by the amount of time the injectors are open. To ensure that injector pulse width is the only factor that determines fuel metering, fuel pump pressure is maintained by a pressure regulator. The injectors are mounted to a common fuel rail.

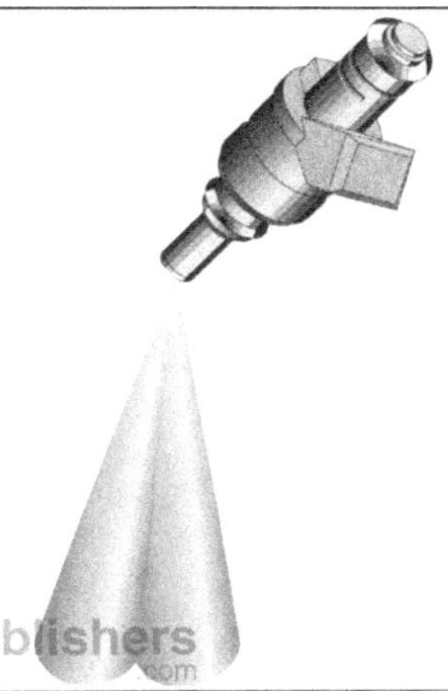

BentleyPublishers.com

0021010

The fuel-injectors inject fuel at an angle in a dual cone spray pattern.

The ECM monitors engine operating conditions to determine injector opening duration. Each injector can be individually controlled for cylinder selective fuel trim.

Air intake

Air entering the engine passes through a pleated paper element in the air filter housing. Intake air mass is then measured by a mass air flow sensor. A reference current is used to heat a thin film in the sensor when the engine is running. The current needed to hold the temperature of the film constant is the basis of the electronically converted voltage measurement corresponding to the mass of the intake air.

BentleyPublishers.com
BentleyPublishers.com—All Rights Reserved

Idle speed control

Idle speed is electronically controlled via the idle speed control valve by bypassing varying amounts of air around the closed throttle valve. Idle speed is not adjustable. The ECM determines idle speed by controlling a dual-winding rotary idle control valve. The basic functions and parameters of idle speed control are as follows:

- Control of cold air intake volume. For example, at air temperatures below 0°C (32°F), electronic throttle control (EDK) is also opened during idle.
- Smooth idle speeds regardless of load and inputs
- Smooth transition from acceleration to deceleration

Idle speed stabilization is active during the following conditions:

- Engine warm up
- Heating or A/C activation
- Drive gear selected (automatic transmission)
- Varying engine cooling fan speeds

If the ECM detects a fault in the idle control valve, it increases or decreases air flow, depending on the nature of the fault:

- If the fault causes decreased air flow (idle control valve closed), the electronic throttle control compensates to maintain idle.
- If the fault causes increased air flow (idle control valve failed open), VANOS and knock control are deactivated. This reduces engine performance noticeably.

Throttle control

In order to integrate the driver's wish with the requirements of the traction control system (DSC) and cruise control, DME systems feature electronic throttle control (EDK). There is no throttle cable between the accelerator pedal and the throttle housing

Knock (detonation) control

Knock sensors monitor and control ignition knock through the ECM. Knock sensors function like microphones and are able to convert mechanical vibration (knock) into electrical signals. The ECM is programmed to react to frequencies that are characteristic of engine knock and adapt the ignition timing point accordingly.

Knock sensor replacement is covered in **120 Ignition System**.

Secondary air injection

The secondary air system pumps ambient air into the exhaust stream after a cold engine start to reduce catalytic converter warm-up time and to reduce HC and CO emissions. The ECM controls and monitors the secondary air injection system.

The electric air pump draws in ambient air and supplies it to the secondary air valve. The air injection valve is opened by a solenoid using engine vaccum and is closed by an internal spring.

The secondary air valve is bolted to the right front of the cylinder head. Cast passageways within the cylinder head duct the secondary air directly into the cylinder head exhaust ports.

BentleyPublishers.com
BentleyPublishers.com—All Rights Reserved

Secondary air injection system components

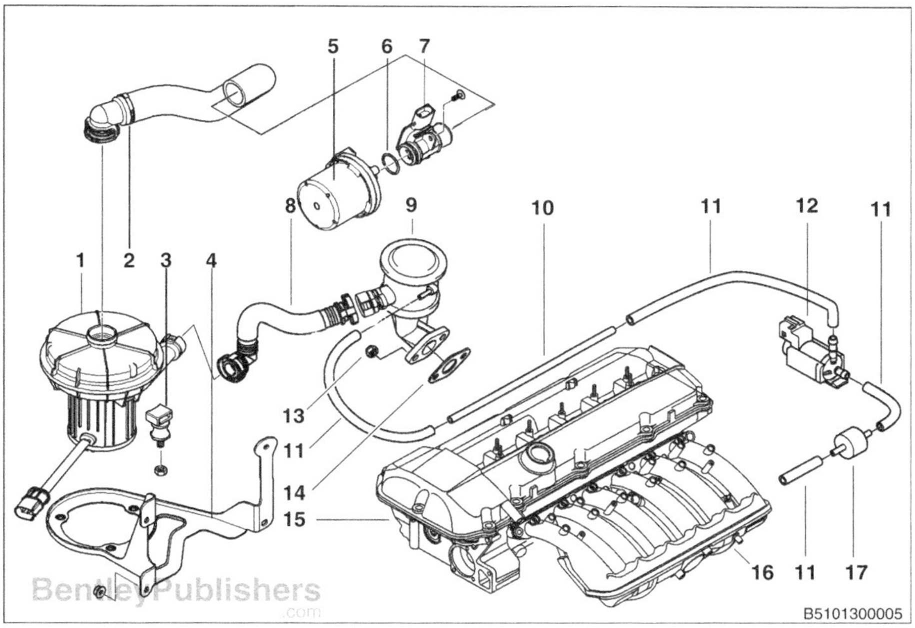

B5101300005

1. **Air pump**
2. **Filter air duct**
3. **Air pump mount**
4. **Air pump bracket**
5. **Air pump filter**
6. **O-ring**
7. **Air pump mini mass air flow sensor**
8. **Pressure hose**
9. **Secondary air valve**
10. **Vacuum pipe**
11. **Vacuum hose (3.3 mm)**
12. **Vacuum solenoid (ECM controlled)**
13. **Mounting nut M6**
14. **Gasket**
15. **Cylinder head**
16. **Intake manifold**
17. **Check valve**

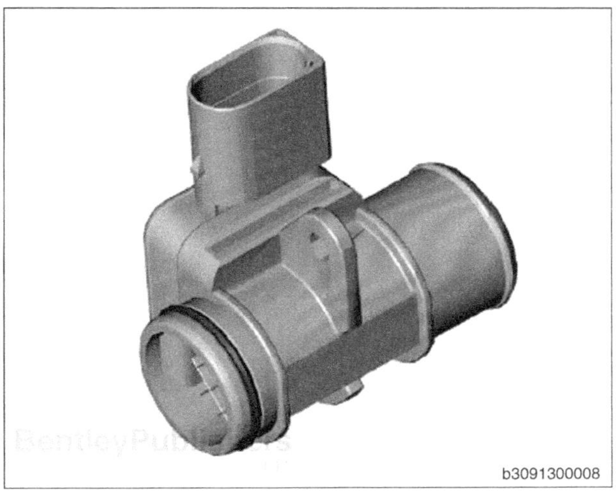
b3091300008

A miniature mass air flow sensor in the secondary air system detects the air mass supplied by the secondary air pump. This function monitors the secondary air system for OBD II compliance.

When the mini mass air flow sensor detects no or insufficient air flow, a fault is stored in the ECM and the malfunction indicator light (MIL) is illuminated.

Secondary air injection illustrations

BX56130009

Secondary air pump is in right front fender behind wheel well liner.

BentleyPublishers
BentleyPublishers.com–All Rights Reserved

Secondary air injection illustrations

B2606 Mini mass air flow sensor.

Secondary air valve is bolted to right front of cylinder head.

BentleyPublishers.com
BentleyPublishers.com—All Rights Reserved

Air filter housing and ducts, removing and installing

For air filter servicing, see **020 Maintenance**.

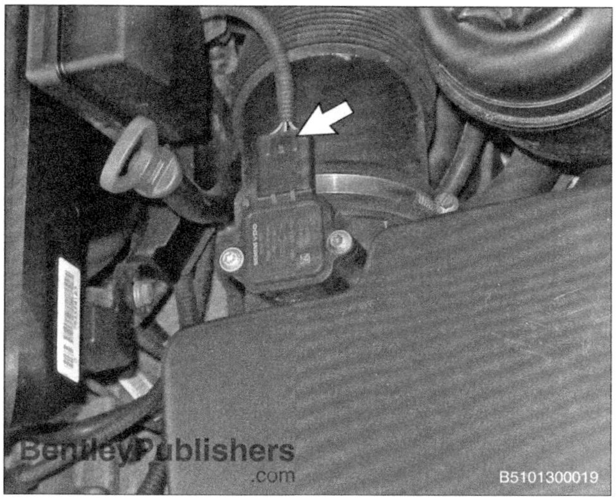

◄ Disconnect mass air flow sensor connector (**arrow**).

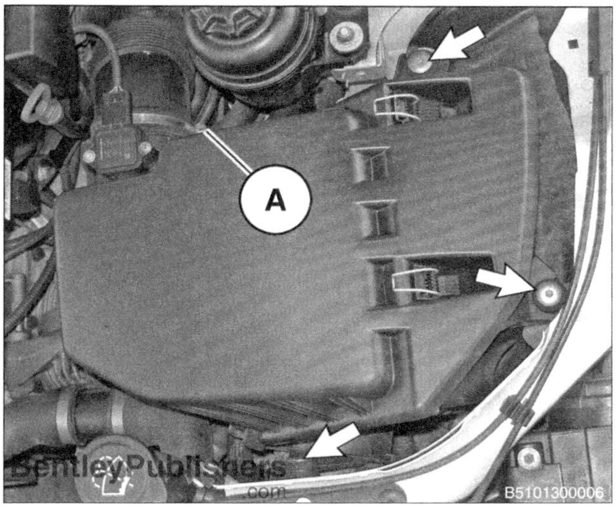

◄ Loosen hose clamp (**A**) and remove air filter housing fasteners (**arrows**).

- Remove air filter housing by lifting up and out in direction of **arrow**.
- Working at throttle body, loosen duct clamp. Remove duct from throttle body.
- Installation is reverse of removal.
- After reinstalling, check and clear fault codes from ECM memory.

BentleyPublishers
BentleyPublishers.com—All Rights Reserved

Mass air flow, measurement

The mass air flow sensor, incorporating an electrically heated film in the air flow stream, sends to the ECM a varying voltage (approx. 0.5 - 4.5 volts) representing the mass of the intake air.

The ECM provides operating voltage and ground to the air flow sensor. As air flows through the sensor, the film is cooled. To maintain the film at a constant temperature, additional current is necessary. It is this additional current that is the basis for the air mass signal.

If there is no output signal from the air flow sensor, the ECM operates the engine using throttle position and engine rpm inputs. A faulty air flow sensor illuminates the MIL.

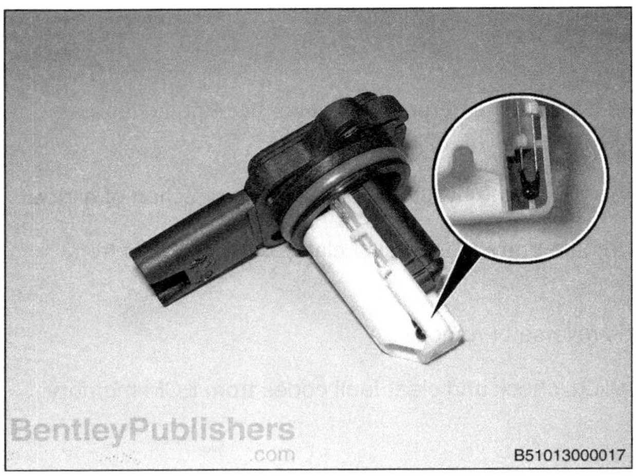

The intake air temperature sensor (**inset**) is integrated into the mass air flow sensor. The sensor is an NTC thermistor which receives a 5 volt reference current and ground from the ECM.

A faulty air flow sensor can produce the following problems:

- Difficult to restart when engine is hot.
- Engine starts then stalls.
- Engine starts and runs only with accelerator pedal depressed.

Mass air flow sensor, replacing

– Switch ignition OFF and remove key.

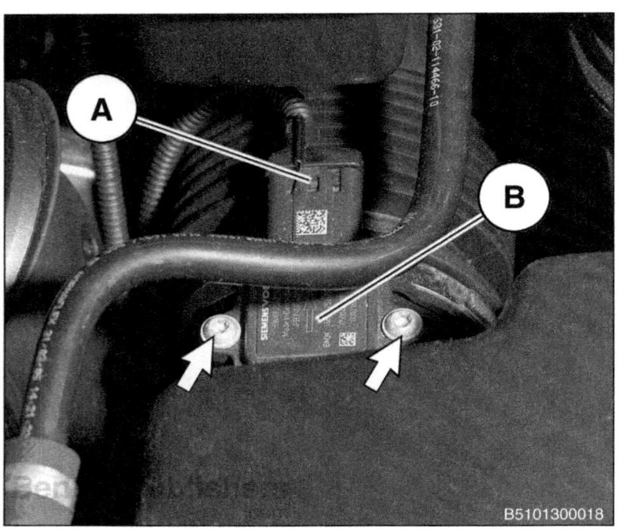

Working at left front of engine compartment at air filter housing:

- Detach mass air flow sensor electrical connector (**A**).
- Remove fasteners (**arrows**).
- Remove mass air flow sensor (**B**).

– Check intake ducts for cracks and vacuum leaks.

BentleyPublishers.com
BentleyPublishers.com—All Rights Reserved

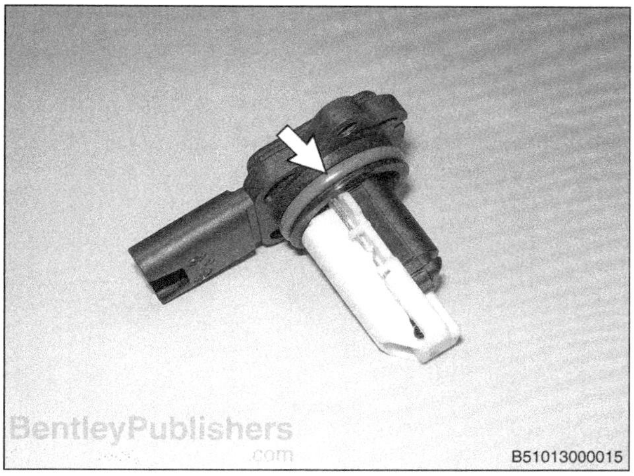

Installation is reverse of removal.

- Inspect O-ring (**arrow**) and replace if necessary. Keep O-ring and duct sealing area free from oil.

– After reinstalling, check and clear fault codes from ECM memory.

Tightening torques	
Mass air flow sensor to housing	3 Nm (26 in-lb)

Throttle housing (EDK), replacing

There is no mechanical (cable) link between the accelerator pedal and the throttle plate.

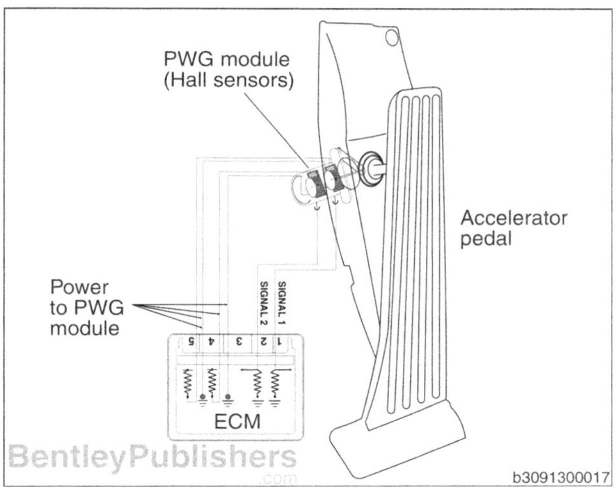

The accelerator pedal module (PWG) at the pedal assembly communicates pedal position directly to the ECM. The module provides two variable voltage signals (via two Hall sensors) to the ECM for pedal position and rate of movement.

The ECM provides an independent voltage and ground supply for each Hall sensor. Each Hall sensor is provided with 5 volts and ground. As the accelerator pedal is moved from rest to full throttle, the sensors produce a variable voltage signal.

The output of the Hall sensors is checked for plausibility. The voltage range of Hall sensor 1 is approximately 0.5 to 4.5 volts. Hall sensor 2 ranges from approximately 0.5 to 2.5 volts.

The throttle housing unit EDK is nonadjustable. If found to be faulty, replace as a complete unit.

After replacing the throttle housing, use BMW scan tool to reset adaptation values.

> ***CAUTION—***
> - *If the adaptation process is not completed correctly, the engine does not start.*

– Switch ignition OFF and remove key.

– Remove air filter housing and air intake ducts. See **Air filter housing and ducts, removing and installing** in this repair group.

BentleyPublishers
BentleyPublishers.com–All Rights Reserved

Throttle housing (EDK), replacing

◄ Working at side of intake manifold, disconnect electrical connectors:

- Intake manifold resonance valve (**A**)
- Idle speed control valve (**B**)

– Disconnect electrical connectors at oil pressure sender and oil temperature sender at base of oil filter housing.

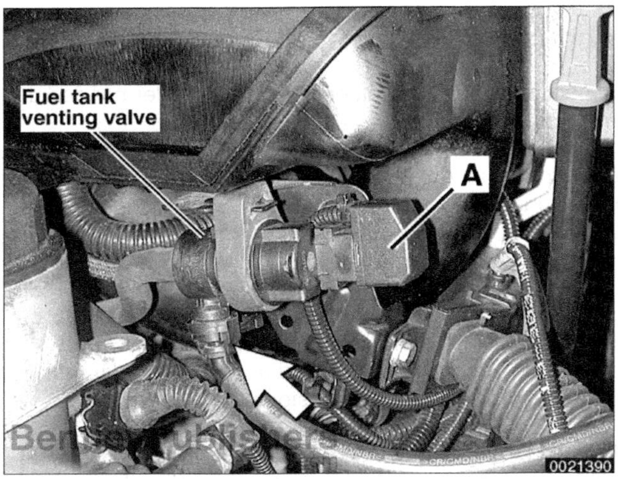

◄ Working at left side and underneath intake manifold:

- Detach electrical connector (**A**) at fuel tank venting valve.
- Disconnect hose at quick disconnect fitting (**arrow**).
- Remove valve from manifold and set aside.

◄ Remove wiring harness housing mounting fasteners (**arrows**) and pull housing away from throttle assembly.

– Remove engine oil dipstick tube.

BentleyPublishers.com
BentleyPublishers.com—All Rights Reserved

Working at throttle housing, unscrew harness plug counterclockwise (**arrow**) to disconnect.

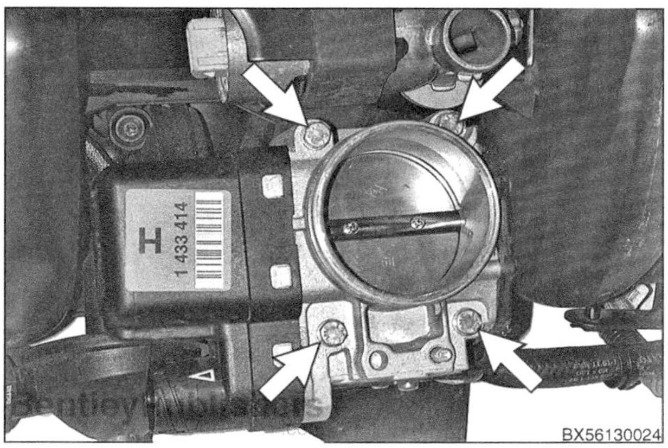

Remove throttle housing mounting fasteners (**arrows**) and pull assembly off manifold.

- Installation is reverse of removal.
  - Replace profile gasket (sealing O-ring) between throttle assembly and intake manifold.
  - Reattach throttle assembly connector with care. Arrow on fully tightened connector must line up with corresponding arrow on throttle assembly housing.
- Using BMW scan tool:
  - Check and clear fault codes from ECM memory.
  - Reset throttle adaptation.

Idle speed control valve, replacing

The idle speed control valve, mounted directly above the throttle housing, regulates idle speed by redirecting air around the throttle valve. The valve, supplied with battery voltage from the DME main relay, incorporates a two-coil rotary actuator. Coil grounds are pulsed simultaneously by the ECM. The duty cycle of each circuit is varied to achieve the required idle speed.

A faulty idle speed control valve sets a DTC. The MIL illuminates if OBD II fault criteria are exceeded.

- Switch ignition OFF and remove key.
- Disconnect battery negative (-) cable in cargo compartment.

> ***CAUTION—***
> - *Prior to disconnecting the battery, read the battery disconnection cautions in* **001 Warnings and Cautions***.*

- Remove air filter housing and air intake ducts. See **Air filter housing and ducts, removing and installing** in this repair group.

BentleyPublishers.com
BentleyPublishers.com—All Rights Reserved

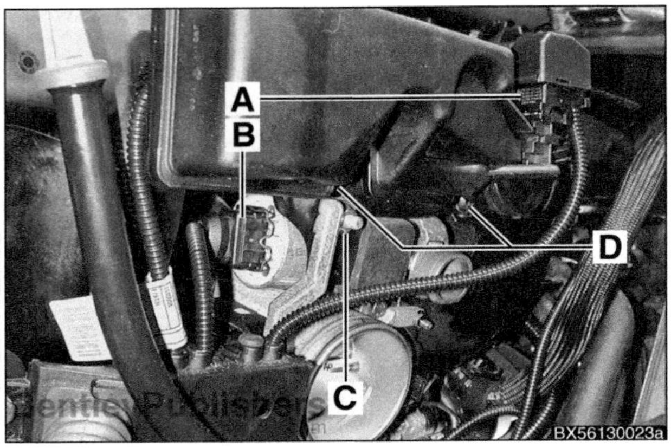

Working at side of intake manifold:

- Disconnect electrical connector to intake manifold resonance valve (**A**).
- Disconnect electrical connector to idle speed control valve (**B**).
- Remove upper wiring harness housing mounting fastener (**C**). Pull harness housing back slightly.
- Remove idle speed valve mounting bracket screws (**D**).
- Pull idle regulator off intake manifold. It is fitted to manifold using a sealing grommet.

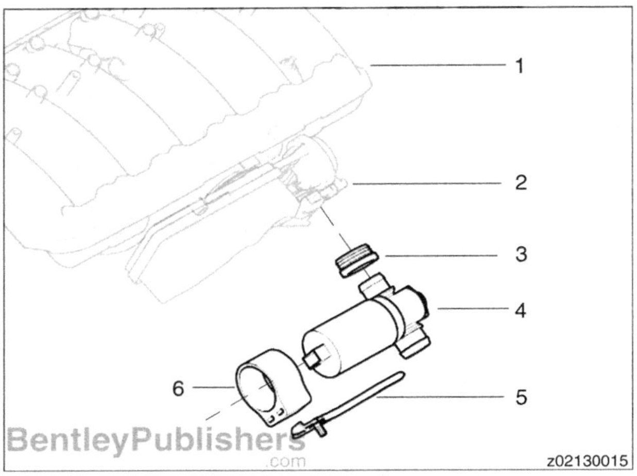

When reinstalling, replace rubber grommet. Coat grommet on inside with assembly lubricant, then install in manifold. Press idle valve into place.

1. Intake manifold
2. Intake manifold resonance valve
3. Sealing grommet
4. Idle speed control valve
5. Mounting bracket
6. Rubber mounting

– Reattach wiring harnesses. Reinstall air filter assembly and ducts.

– Check and clear fault codes from ECM memory.

Fuel rail and injectors, removing and installing

Fuel injectors are electrically controlled solenoid valves that provide precisely metered and atomized fuel into the engine intake ports.

Injectors receive voltage from the DME main relay. The engine control module (ECM) controls injector opening by activating the ground circuit for the injector valve solenoids. The ECM varies the duration (in milliseconds) of injector opening to regulate air / fuel ratio.

– To access injectors, remove engine covers. See **020 Maintenance**.

– For a quick check of injectors, run engine and touch each injector with a screwdriver or stethoscope. You should feel a vibration or hear a buzzing. Switch engine OFF.

– Disconnect battery negative (-) cable.

> ***CAUTION—***
> *• Prior to disconnecting the battery, read the battery disconnection cautions in* **001 Warnings and Cautions**.

– Remove upper engine covers. See **020 Maintenance**.

BentleyPublishers.com
BentleyPublishers.com—All Rights Reserved

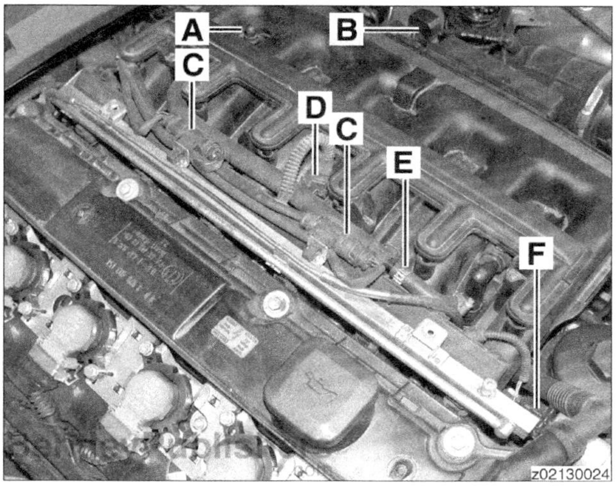

Working above engine, detach the following:

- **B** Resonance valve electrical connector
- **C** Oxygen sensor connectors
- **D** Electrical connector for intake air temperature sensor
- **E** Manifold vacuum line
- **F** VANOS solenoid electrical connector

> ***CAUTION—***
> - *Be sure to mark oxygen sensor connectors so that they can be reassembled as before.*

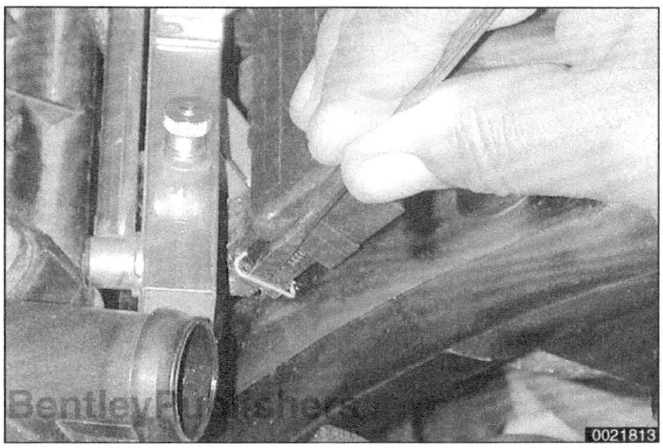

Disconnect fuel injector electrical connectors from injectors:

- Use small screwdriver to pry one corner of wire lock clip on fuel injector 1 connector.
- Repeat for all injectors.
- Lift off connector loom and set aside.

Unscrew Schræder valve cap (**arrow**) from fuel rail.

- Unscrew fuel tank cap to release pressure.
- Using a brief burst of compressed air (maximum of 3 bar or 43.5 psi), blow fuel back into fuel tank.

> ***WARNING —***
> - *Wrap a clean shop towel around fitting before connecting. Residual fuel pressure is present in the fuel rail.*

BentleyPublishers.com–All Rights Reserved

Fuel rail and injectors, removing and installing

Working at rear of manifold, use BMW special tool 16 1 050 or equivalent to press locking clip inside end of fuel line fitting (**arrow**) to disconnect fuel supply line from fuel rail.

> ***WARNING —***
> *• Wrap a clean shop towel around fitting before disconnecting. Residual fuel pressure is present in the fuel line.*

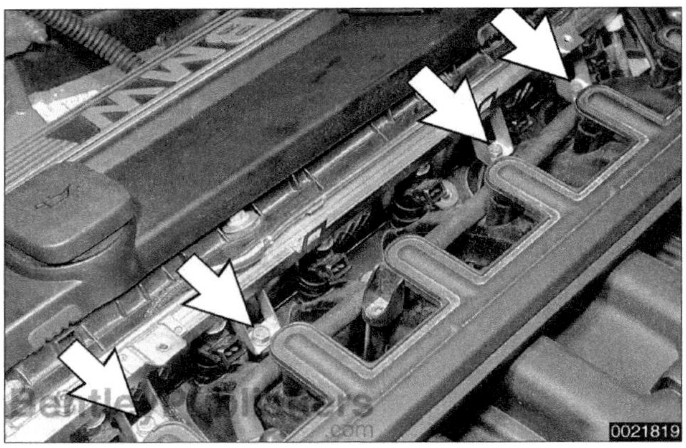

Remove fuel rail mounting bolts (**arrows**).

- Carefully pry fuel rail and injectors off manifold.
- Separate fuel line support bracket at rear of intake manifold.
- Guide fuel line(s) out of rear of engine compartment while lifting fuel rail out.

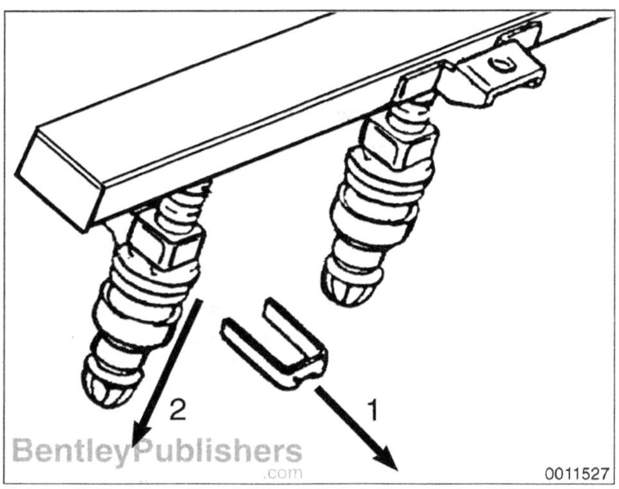

Remove individual injectors:

- Pry retaining clip from injector (**1**).
- Pull injector from rail (**2**).

– Installation is reverse of removal. Remember to:

- Fit new sealing O-rings when installing injectors. For ease of installation, lightly lubricate O-rings with assembly lubricant.
- Check that injector electrical connections are correctly fitted and that injectors are fully seated prior to installing fuel rail mounting bolts.
- Check fuel line rubber O-rings and replace as necessary. For ease of installation, lightly lubricate O-rings with assembly lubricant.
- Replace wire ties.

Tightening torque	
Fuel rail to cylinder head	10 Nm (7 ft-lb)

– Check and clear fault codes from ECM memory.

BentleyPublishers.com
BentleyPublishers.com—All Rights Reserved

Intake manifold, removing and installing

– Switch ignition OFF and remove key.

– Disconnect battery negative (-) cable.

> ***CAUTION—***
> - *Prior to disconnecting the battery, read the battery disconnection cautions in* **001 Warnings and Cautions**.

– Remove upper engine covers. See **020 Maintenance**.

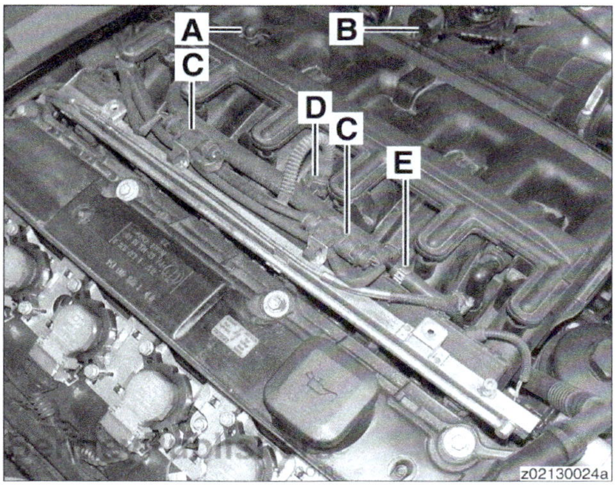

◄ Working above engine, detach the following:

- **A** B+ lead bracket
- **B** Resonance valve electrical connector
- **C** Oxygen sensor connectors
- **D** Electrical connector for intake air temperature sensor
- **E** Manifold vacuum line

> ***CAUTION—***
> - *Be sure to mark oxygen sensor connectors so that they can be reassembled as before.*

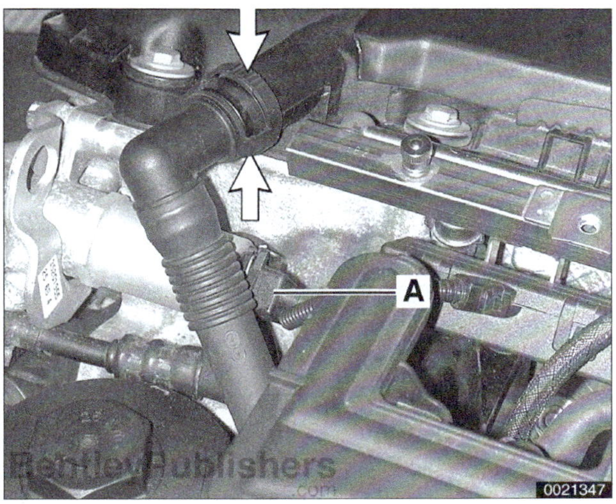

◄ Working at left side of cylinder head:

- Detach engine vent hose from cylinder head cover by pinching spring clips (**arrows**).
- Detach VANOS solenoid electrical connector (**A**).

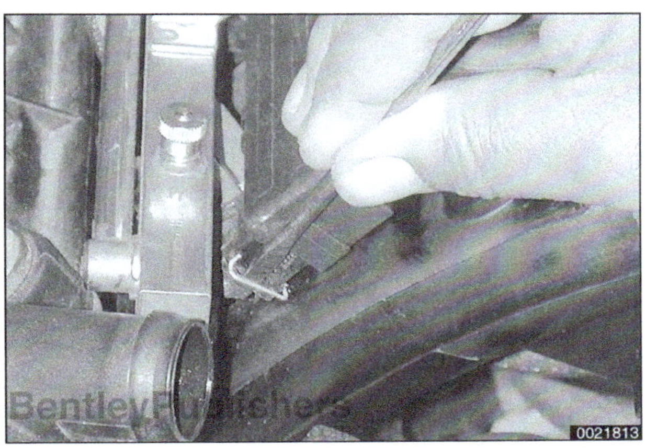

◄ Detach fuel injector electrical connectors from injectors:

- Use small screwdriver to pry one corner of wire lock clip on fuel injector 1 connector.
- Repeat for all injectors.
- Remove two main harness hold-down fasteners. Lift off entire harness and lay aside.

BentleyPublishers
BentleyPublishers.com—All Rights Reserved

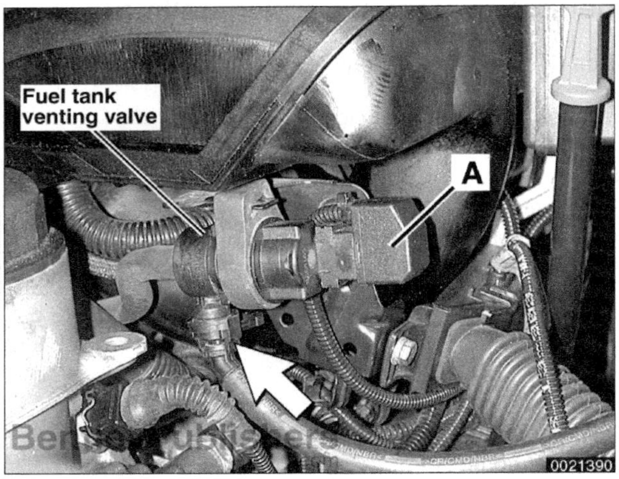

◀ Working at left side and underneath intake manifold:

- Detach electrical connector (**A**) at fuel tank venting valve.
- Disconnect hose at quick disconnect fitting (**arrow**).
- Remove valve from manifold and set aside.

◀ Disconnect electrical connector (**arrow**) at idle speed control valve.

– Remove throttle housing. See **Throttle housing (EDK), replacing** in this repair group.

◀ Unscrew Schræder valve cap (**arrow**) from fuel rail.

- Unscrew fuel tank cap to release pressure.
- Using a brief burst of compressed air (maximum of 3 bar or 43.5 psi), blow fuel back into fuel tank.

> ***WARNING —***
> - *Wrap a clean shop towel around fitting before connecting. Residual fuel pressure is present in the fuel rail.*

BentleyPublishers.com
BentleyPublishers.com—All Rights Reserved

Working at rear of manifold, use BMW special tool 16 1 050 or equivalent to press locking clip inside end of fuel line fitting (**arrow**) to disconnect fuel supply line from fuel rail.

> ***WARNING —***
> - *Wrap a clean shop towel around fitting before disconnecting. Residual fuel pressure is present in the fuel line.*

Working underneath car, remove lower intake manifold support mounting bolt (**arrow**), located adjacent to left engine mount.

– Detach knock sensor connector from bracket at base of manifold.

– Remove manifold lower support nuts.

Remove intake manifold mounting nuts (**arrows**).

– Remove intake manifold from cylinder head while carefully checking for electrical connections or hoses that might become snagged.

> ***CAUTION—***
> - *Plug open intake ports to prevent parts or debris from falling into the engine intake.*

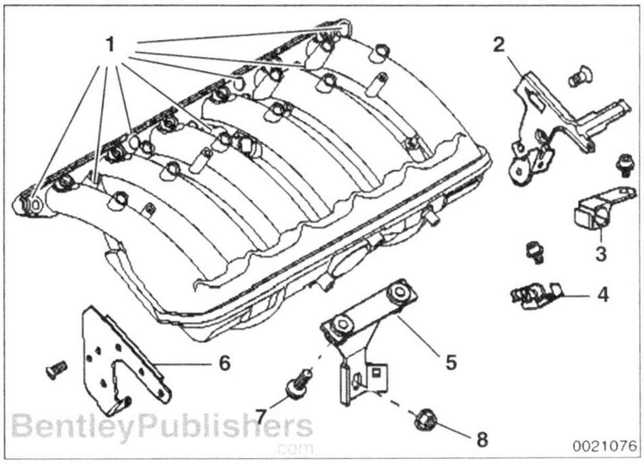

Note intake manifold brackets and mounts:

1. Manifold mounting nuts
   - tighten to 15 Nm (11 ft-lb)
2. Fuel pipe bracket
3. Fuel pipe bracket
4. Vacuum pump bracket
5. Manifold mounting bracket
6. Tank venting valve bracket
7. Mounting bracket to manifold bolt (M6)
   - tighten to 10 Nm (7 ft-lb)
8. Mounting bracket to engine block nut (M10)
   - tighten to 47 Nm (33 ft-lb)

– Intake manifold installation is reverse of removal.
  - Use new fuel injector seals.
  - Carefully check intake manifold gaskets and replace if necessary.
  - Check condition of rubber seals at throttle housing and idle speed control valve. Replace as necessary.
  - Inspect O-ring seal between mass air flow sensor and air filter housing. To facilitate reassembly, coat seal with assembly fluid.

> ***CAUTION—***
> - *When reattaching throttle assembly harness connector, make sure arrows on connector and plug line up when fully tightened.*

Tightening torques	
Intake manifold to cylinder head (M7)	15 Nm (11 ft-lb)
Mounting bracket to engine block (M10)	47 Nm (33 ft-lb)
Mounting bracket to intake manifold (M6)	10 Nm (7 ft-lb)

Crankcase breather valve components

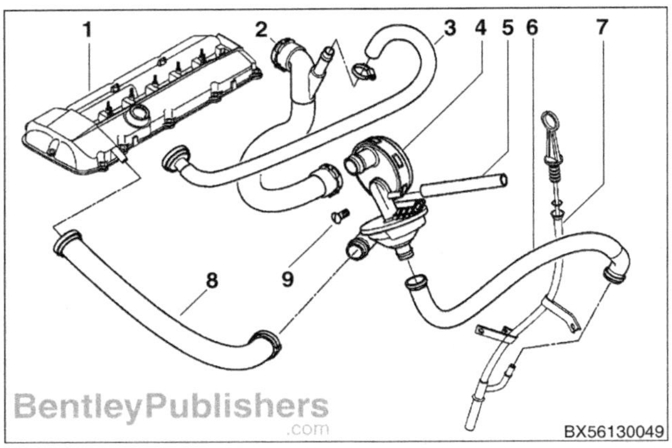

1. Cylinder head cover
2. Connecting hose
3. Oil return hose
4. Breather valve
5. Vacuum hose
6. Dipstick guide tube vent hose
7. Dipstick guide tube
8. Vent pipe
9. Screw

The M54 engine crankcase is ventilated by breather valve under the intake manifold. If the diaphragm valve in the breather housing leaks, full intake vacuum may be applied to the crankcase, resulting in excessive oil consumption, irregular idle, whistling or howling noises or oil smoke in the exhaust.

To gain access to breather valve, remove air filter housing and ducts. See **Throttle housing, removing and installing** in this repair group.

Engine coolant temperature (ECT) sensor, replacing

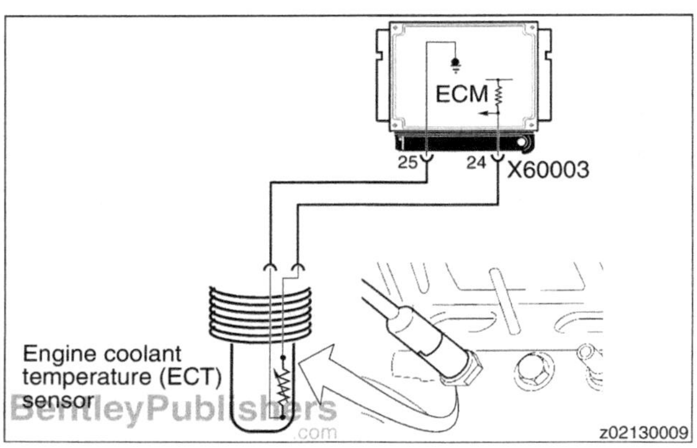

The ECT sensor is an NTC (negative temperature coefficient) sensor. As coolant temperature rises, resistance through the sensor decreases.

The ECM varies ignition timing and air / fuel mixture based on engine coolant temperature. The ECT sensor is supplied a 5 volt reference voltage. The voltage drop across the sensor varies as the coolant temperature (sensor resistance) changes.

BentleyPublishers.com
BentleyPublishers.com—All Rights Reserved

If the ECT sensor input is faulty or not plausible, the MIL is illuminated. The ECM assumes a substitute value (80°C / 176°F) to maintain engine operation. The ignition timing is set to a safely conservative basic setting.

The ECT sensor is located below cylinder 6 intake port at the rear of the engine, underneath the intake manifold.

– With engine fully cooled off, remove intake manifold. See **Intake manifold, removing and installing** in this repair group.

< Disconnect ECT sensor connector (**arrow**).

– Unscrew ECT sensor from cylinder head. Be prepared to catch small amount of coolant.

– Installation is reverse of removal:
- Use new copper sealing washer when installing sensor.
- Replace lost coolant.
- Check and clear fault codes from ECM memory.

Tightening torque	
Temperature sensor to cylinder head	13 Nm (10 ft-lb)

Radiator outlet coolant temperature sensor

< The radiator outlet temperature sensor (**inset**) is an NTC (negative temperature coefficient) sensor. As coolant temperature rises, resistance through the sensor decreases.

The ECM varies cooling fan speed based on radiator outlet coolant temperature. The radiator outlet coolant temperature sensor is supplied a 5 volt reference voltage. The voltage drop across the sensor varies as the coolant temperature (sensor resistance) changes.

BentleyPublishers.com—All Rights Reserved

DME main relay, testing

The DME main relay is energized via the ECM and supplies battery positive (B+) power to the following:

- ECM
- Ignition coils
- Fuel injectors
- Fuel tank leak detection pump
- Idle air control valve
- Camshaft sensors
- Evaporative emissions valve
- Mass air flow sensor
- Fuel pump relay
- Oxygen sensor heaters

If the DME main relay is faulty, the engine does not start.

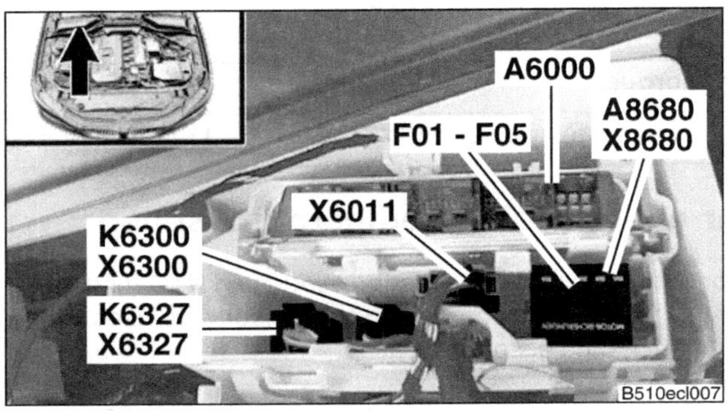

With ignition off, remove DME main relay (K6300) (**arrow**) in electronics box (E-box) at right rear of engine compartment.

NOTE —

- *Relay locations can vary. Confirm relay identification by matching wiring colors and terminal numbers. See* **ELE Electrical Wiring Diagrams**.

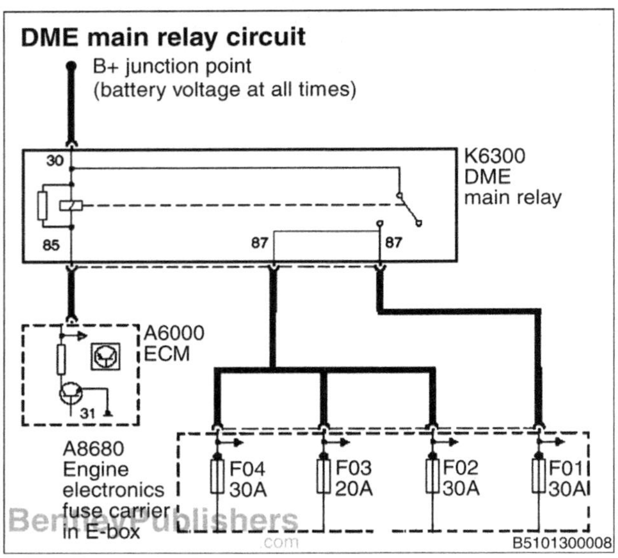

Check for voltage at terminal **6** of relay socket (corresponds to terminal **30** on relay).

– If battery voltage is present, continue testing.

– If battery voltage is not present, check the following:
  - Large red wire in relay socket
  - A2076 (battery positive junction) in E-box, right rear of engine compartment
  - Fuse F8 in fuse and relay panel above glove compartment. See **Power supply fuses** in this repair group.
  - See **ELE Electrical Wiring Diagrams** for more details

– Check for ground at relay socket **4** (corresponds to terminal **85** on relay).

– If ground is present, continue testing.

– If ground is not present, signal from ECM is missing. Check wire between relay and ECM 24-pin connector X60002. See **ECM pin assignments (Siemens DME MS45.1)** in this repair group.

– If no faults are found:
  - Check ECM grounding.
  - ECM may be defective.

BentleyPublishers.com
BentleyPublishers.com—All Rights Reserved

– With ignition ON and relay installed, check for battery voltage at relay sockets **2** and **5** (correspond to terminals **87** on relay).

NOTE—
- *In some models there is only one terminal 87.*

– If battery voltage is present, relay has energized and is functioning correctly.

– If battery voltage is not present and all earlier tests are OK, relay is faulty. Replace.

– If no faults are found during relay testing but power is not reaching ECM or other components, check fuses in engine electronics fuse carrier (A8680) in E-box. See fuses in **ECL Elelctrical Component Locations**.

– When finished testing, check for DTCs and clear and reset ECM memory.

ECM pin assignments (Siemens DME MS45.1)

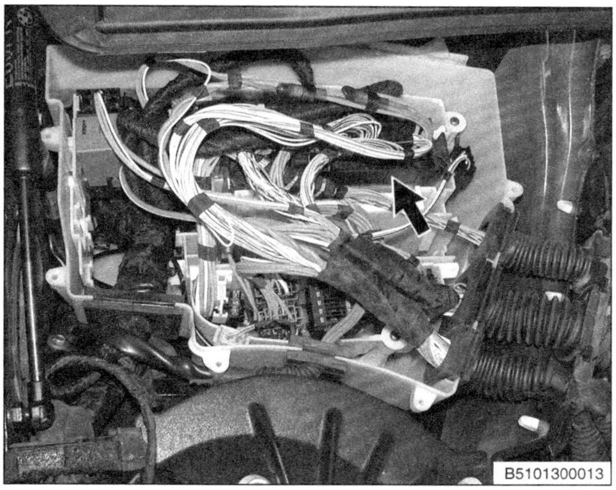

B5101300013

The engine control module (ECM) (**arrow**) is located in the right rear of the engine compartment in the E-box.

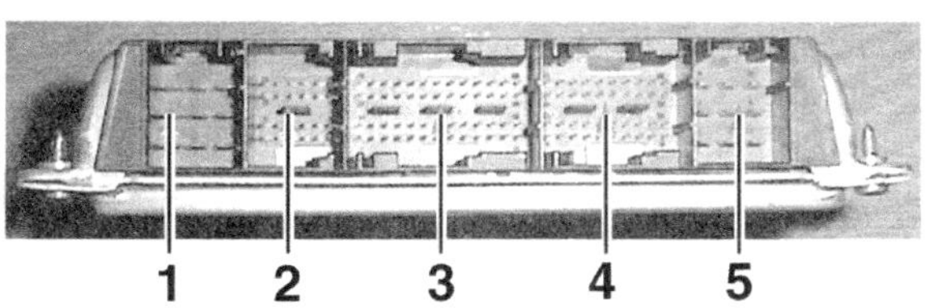

BentleyPublishers.com

B701130045

The ECM has 5 electrical connectors with a total of 134 pins:
- Connector **1** (X60001), 9 pins: Power and grounds
- Connector **2** (X60002), 24 pins: Ancillary signals (oxygen sensors, CAN-bus, etc.)
- Connector **3** (X60003), 52 pins: Engine signals
- Connector **4** (X60004), 40 pins: Vehicle signals
- Connector **5** (X60005), 9 pins: Ignition signals

ECM pin assignments for Siemens DME MS45.1 are given in **Table b**. This information can be helpful when diagnosing faults to or from the ECM.

Generally, absence of voltage or continuity means there is a wiring or connector problem. Test results with incorrect values do not necessarily mean that a component is faulty. Check for loose, broken or corroded connections and wiring before replacing components. For engine management system electrical schematics, see **ELE Electrical Wiring Diagrams**.

When making checks at the ECM, use a break-out box to allow tests to be made with the connector attached to the ECM. This prevents damage to the small terminals in the connector. As an alternative, the connector housing can be separated so that electrical checks can be made from the back of the connector.

CAUTION—
- *Wait at least one minute after switching the ignition OFF before removing the connector from the ECM. If the connector is removed before this time, residual power in the system relay may damage the ECM.*
- *Connect or disconnect the control module connector and meter probes with the ignition OFF.*

BentleyPublishers.com
BentleyPublishers.com–All Rights Reserved

Table b. Siemens DME MS45.1 ECM pin assignments			
Connector X60001 9-pin (power, grounds)			
Pin	**Type**	**Component or function**	**Note**
1	-	Not used	
2	-	Not used	
3	input / output	TXD signal	Solder connector
4	ground	ECM	
5	ground	ECM	
6	ground	ECM	
7	input	Power supply (terminal 30)	Battery voltage (B+) at all times, fuse F105
8	input	Terminal 87, fuse F2	Voltage supply
	-	Not used	

Connector X60002 24-pin (ancillary signals)			
Pin	**Type**	**Component or function**	**Note**
1	output	Oxygen sensor heater (pre-catalyst cyl. 1-3)	Negative activation
2	output	Oxygen sensor heater (pre-catalyst cyl. 4-6)	Negative activation
3	-	Not used	
4	-	Not used	
5	-	Not used	
6	output	Oxygen sensor heater (post-catalyst cyl. 1-3)	Negative activation
7	input	Oxygen sensor (pre-catalyst, cyl. 1-3)	Negative Signal
8	input	Oxygen sensor (post-catalyst, cyl. 4-6)	Negative Signal
9	input	Oxygen sensor (pre-catalyst, cyl. 4-6)	Negative Signal
10	input	Oxygen sensor (post-catalyst, cyl. 1-3)	Negative Signal
11	-	Not used	
12	output	Oxygen sensor heater (post-catalyst cyl. 4-6)	Negative activation
13	input	Oxygen sensor (pre-catalyst, cyl. 1-3)	Signal
14	output	Oxygen sensor (post-catalyst, cyl. 4-6)	Signal
15	output	Oxygen sensor (pre-catalyst, cyl. 4-6)	Signal
16	output	Oxygen sensor (post-catalyst, cyl. -3)	Signal
17	-	Not used	
18	-	Not used	
19	input	Oxygen sensor (pre-catalyst, cyl. 1-3)	Signal
20	input	Oxygen sensor (pre-catalyst, cyl. 1-3)	Signal
21	input	Oxygen sensor (pre-catalyst, cyl. 4-6)	Signal
22	input	Oxygen sensor (post-catalyst, cyl. 4-6)	Signal
23	input	DME system power supply, terminal 87	Battery voltage from DME main relay
	-	Not used	

BentleyPublishers.com
BentleyPublishers.com—All Rights Reserved

Connector X60003 52-pin (engine signals)			
Pin	Type	Component or function	Note
1	input	Mass air flow sensor	Signal
2	-	Not used	
3	input	Crankshaft sensor	Signal
4	output	Mass air flow sensor	Voltage supply
5	output	Throttle valve (EDK)	Voltage supply
6	output	Fuel injector 2	Pulsed ground
7	output	Fuel injector 4	Pulsed ground
8	output	Fuel injector 6	Pulsed ground
9	output	VANOS solenoid exhaust	Signal
10	output	VANOS solenoid intake	Signal
11	output	Fuel injector 5	Pulsed ground
12	output	Thermostat heating	Signal
13	output	Fuel injector 1	Pulsed ground
14	ground	Mass air flow sensor	
15	ground	Throttle valve (EDK)	
16	ground	Oil pressure switch	
17	-	Not used	
18	-	Not used	
19	input / output	Alternator BSD	Interface
20	-	Not used	
21	output	Fuel tank vent valve	Signal
22	output	Idle speed control valve	Signal
23	output	Idle speed control valve	Signal
24	-	Not used	
25	-	Not used	
26	output	Fuel injector 3	Pulsed ground
27	input	Intake air temperature sensor	Signal
28	input	Coolant temperature sensor	Signal
29	input	Camshaft position sensor 1	Signal
30	input	Camshaft position sensor 2	Signal
31	input	Throttle position signal	Throttle valve (EDK)
32	input	Throttle position signal	Throttle valve (EDK)
33	input	Knock sensor	Double knock sensor
34	input	Knock sensor, cyl. 4 - 6	Signal
35	ground	Coolant temperature sensor	
36	ground	Camshaft position sensor 1	
37	ground	Crankshaft position sensor	

BentleyPublishers.com
BentleyPublishers.com—All Rights Reserved

ECM pin assignments (Siemens DME MS45.1)

Connector X60003 52-pin (engine signals) (continued)			
Pin	**Type**	**Component or function**	**Note**
38	-	Not used	
39	output	Secondary air injection vacuum solenoid	Signal
40	output	DISA changeover valve	Signal
41	-	Not used	
42	output	Throttle valve (EDK)	Signal
43	output	Throttle valve (EDK)	Signal
44	input	Oil temperature sensor	Signal
45	ground	Oil temperature sensor	Signal
46	input	Knock sensor	Double knock sensor
47	input	Knock sensor	Double knock sensor
48	-	Not used	
49	ground	Camshaft position sensor 2	
50	input	Oil level sensor	Signal
51	-	Not used	
52	-	Not used	

Connector X60004 40-pin (vehicle signals)			
Pin	**Type**	**Component or function**	**Note**
1	-	Not used	
2	output	Diagnostic module tank leak (DM-TL) detection	Signal
3	output	Secondary air injection pump relay	Signal
4	input	Electric cooling fan	Signal
5	ground	Mass air flow sensor	
6	-	Not used	
7	ground	Pedal position sensor	
8	input	Pedal position sensor 1	Signal
9	output	Pedal position sensor	Voltage supply
10	output	Fuel pump relay	Signal
11	-	Not used	
12	ground	Pedal position sensor	
13	input	Pedal position sensor 2	Signal
14	output	Pedal position sensor	Voltage supply
15	Input	Mass air flow sensor	Signal
16	-	Not used	
17	output	Crankshaft speed signal	OBD II connector or 20-pin DLC
18	-	Not used	
19	output	E-box cooling fan	Signal

BentleyPublishers.com
BentleyPublishers.com—All Rights Reserved

Connector X60004 40-pin (vehicle signals) (continued)			
Pin	**Type**	**Component or function**	**Note**
20	output	Diagnostic module tank leak (DM-TL) detection	Signal
21	output	Radiator shutter solenoid	Signal
22	input	Wheel speed sensor	Signal
23	input	Clutch pedal position switch	Signal
24	input	Brake light switch	Signal
25	-	Not used	
26	input	Terminal 15, fuse F31 (wake up signal)	Switched power
27	-	Not used	
28	input	Brake light switch	Signal
29	-	Not used	
30	input	Fuel tank leakage diagnosis pump (DM-TL)	Signal
31	-	Not used	
32	-	Not used	
33	input	EWS (electronic immobilizer)	Car access system
34	-	Not used	
35	-	Not used	
36	input/output	CAN-bus high	Signal
37	input/output	CAN-bus low	Signal
38	ground	Radiator coolant outlet temperature sensor	
39	input	Radiator coolant outlet temperature sensor	Signal
40	input	Start signal	Car access system

Connector X60005 9-pin (ignition signals)			
Pin	**Type**	**Component or function**	**Note**
1	output	Ignition coil 5	Signal
2	-	Not used	
3	-	Not used	
4	output	Ignition coil 3	Signal
5	ground	ECM ground	
6	ground	Ignition coil 1	
7	output	Ignition coil 2	Signal
8	output	Ignition coil 4	Signal
9	output	Ignition coil 6	Signal

BentleyPublishers.com
BentleyPublishers.com—All Rights Reserved

BentleyPublishers.com
BentleyPublishers.com—All Rights Reserved

131 Fuel Injection MSV70, MSV80 (N52 Engine)

General

This repair group covers service and repair of engine management systems for 5 Series models with N52 engine and Siemens DME MSV70 or MSV80.

See also:

- **020 Maintenance** for air filter, fuel filter and spark plug replacement
- **100 Engine–General** for engine identification and application information
- **120 Ignition System** for ignition coil, camshaft sensor, crankshaft sensor and knock sensor service
- **160 Fuel Tank and Fuel Pump** for fuel pump and fuel pump relay service, fuel pressure tests and fuel pressure regulator replacement
- **180 Exhaust System** for oxygen sensor service
- **ECL Electrical Component Locations**
- **ELE Electrical Wiring Diagrams**
- **OBD On-Board Diagnostics**

BentleyPublishers
BentleyPublishers.com–All Rights Reserved

DME applications

BMW 5 Series models are equipped with digital motor electronics (DME), also known as Motronic. In these systems, fuel injection and ignition are controlled by an integrated engine control module (ECM). Application information for DME systems is in **Table a**.

Table a. 5 Series engine management applications			
Year, model	**Engine code**	**Engine management**	**Features**
2004-2005			
Sedan: 525i, 530i	M54	Siemens MS 45.1	Double VANOS
Sedan: 545i	N62	Bosch DME ME 9.2 with Valvetronic control module	Valvetronic, VANOS
2006 - 2007			
Sedan, Sports Wagon: 525i, 525xi, 530i, 530xi Sports Wagon	N52	Siemens MSV70	Valvetronic, VANOS
2008 - 2010			
Sedan, Sports Wagon: 528i, 528xi	N52KP	Siemens MSV80	Valvetronic, VANOS
Sedan, Sports Wagon: 535i, 535xi	N54	Siemens MSD80 (to 03 / 2008) MSD81 (from 03 / 2008)	VANOS, turbocharged
2004-2005			
Sedan: 545i	N62	Bosch DME ME 9.2 with valvetronic control unit	Valvetronic, VANOS
2006 - 2010			
Sedan: 550i	N62TU	Bosch DME ME 9.2.2	Valvetronic, VANOS

BentleyPublishers.com
BentleyPublishers.com—All Rights Reserved

Warnings and Cautions

WARNING—
- *The fuel system is designed to retain pressure even when the ignition is OFF. When working with the fuel system, loosen the fuel lines slowly to allow residual fuel pressure to dissipate. Avoid spraying fuel. Use shop towels to capture leaking fuel.*
- *Fuel in fuel lines is under pressure (approx. 3 - 5 bar or 45 - 75 psi) and may be expelled forcibly. Do not smoke or work near heaters or other fire hazards. Keep a fire extinguisher handy.*
- *Unscrew the fuel tank cap to release pressure in the tank before working on fuel lines.*
- *Plug open fuel lines and fittings.*
- *Before beginning work on the fuel system, place a fire extinguisher in the vicinity of the work area.*
- *Work on the fuel system only when engine temperature is below 40°C (104°F).*
- *Fuel is highly flammable. When working around fuel, do not disconnect wires that could cause electrical sparks. Do not smoke or work near heaters or other fire hazards.*
- *Wear eye protection and protective clothing to avoid injuries from contact with fuel.*
- *When working on an open fuel system, wear suitable hand protection, as prolonged contact with fuel can cause illnesses and skin disorders.*
- *Unscrew the fuel tank cap to release pressure in the tank before working on fuel lines.*
- *Do not use a work light with an incandescent bulb near fuel. Fuel may spray on the hot bulb causing a fire.*
- *Make sure the work area is properly ventilated.*
- *Due to risk of personal injury, be sure the engine is cold before beginning work on engine components.*
- *The ignition system produces high voltages that can be fatal. Avoid contact with exposed terminals. Use extreme caution when working on a car with the ignition switched ON or the engine running.*

BentleyPublishers.com
BentleyPublishers.com—All Rights Reserved

CAUTION—

- *Renew fuel system hoses, clamps and O-rings any time they are removed.*
- *Prior to disconnecting the battery, read the battery disconnection cautions in* **001 Warnings and Cautions**.
- *Connect and disconnect the DME system wiring and test equipment leads only when the ignition is switched OFF.*
- *Wait at least 1 minute after switching the ignition OFF before disconnecting engine control module (ECM) connectors. If the connectors are removed before this time, residual power in the system relay may damage the control module.*
- *Tests or repair procedures in this section may set fault codes (DTCs) in the ECM and illuminate the MIL. After repairs are completed, access and clear DTC memory using a BMW scan tool. See* **On-board diagnostics** *in this repair group.*
- *Fuel system cleaners and other chemical additives other than those specifically recommended by BMW may damage catalytic converters, oxygen sensors or other fuel supply components.*
- *Do not connect any test equipment that delivers a 12-volt power supply to terminal 15 (+) of the ignition coils. The current flow may damage the ECM. Connect test equipment only as specified by BMW or the equipment maker.*
- *Relay positions can vary. Be sure to confirm relay location and function by identifying the wiring in the socket using the wiring diagrams found in* **ELE Electrical Wiring Diagrams**.
- *Use a digital multimeter for electrical tests. Use an LED test light for quick tests.*

BentleyPublishers.com
BentleyPublishers.com—All Rights Reserved

Driveability Troubleshooting

The self-diagnostic DME engine management systems monitor and store diagnostic trouble codes (DTCs). If the malfunction indicator light (MIL) illuminates, it indicates that an emissions-related fault has occurred and that one or more DTCs are stored in the engine control module (ECM).

If faults arise, or if the MIL is illuminated, begin troubleshooting by connecting BMW scan tool to the OBD II plug under the dashboard.

131

The capabilities of OBD II software has the potential to save hours of diagnostic time and to help avoid incorrect component replacement and possible damage to system components. See **On-board diagnostics** in this repair group.

Basic engine settings

Idle speed, idle mixture (%CO), and ignition timing are not adjustable. The DME system is adaptive and automatically compensates for changes in the engine due to age, minor wear or small problems such as a disconnected vacuum hose. However, the adaptive range is limited. Once the limits are exceeded, driveability problems become noticeable.

Poor initial driveability may be encountered when the battery is disconnected and reconnected. When the battery is disconnected, ECM adaptive memory may be reset. The system readapts after about ten minutes of driving.

System voltage

Digital motor electronics (DME) requires that the system (battery) voltage be maintained within a narrow range of DC voltage. DC voltage levels beyond or below the operating range, or any AC voltage in the electrical system can cause havoc.

- When troubleshooting an illuminated MIL, make sure the battery is fully charged and capable of delivering all its power to the electrical system. An undercharged battery can amplify AC alternator output ripple.

- To make a quick check of battery charge, measure voltage across battery terminals with all cables attached and ignition OFF. A fully charged battery measures 12.6 volts or slightly more, compared to 12.15 volts for a battery with a 25% charge.

The DME system operates at low voltage and current levels, making it sensitive to small increases in resistance. The electrical system is routinely subject to corrosion, vibration and wear, so faults or corrosion in the wiring harness and connectors are not uncommon.

- Check battery terminals for corrosion or loose cable connections. See **121 Battery, Starter, Alternator**.

BentleyPublishers.com
BentleyPublishers.com–All Rights Reserved

- If a battery cable connection has no visible faults but is still suspect, measure voltage drop across the connection. A large drop indicates excessive resistance, meaning that the connection is corroded, dirty or damaged. Clean or repair and retest. See **600 Electrical System–General** for voltage drop test procedure.

- Visually inspect wiring, connectors, switches and fuses. Loose or damaged connectors can cause intermittent problems, especially small terminals in ECM connectors. Disconnect wiring harness connectors to check for corrosion, and use electrical cleaning spray to remove contaminants.

Main grounds

Good grounds are critical to proper DME operation. If a ground connection has no visible faults but is still suspect, measure the voltage drop across the connection. A large voltage drop means high resistance. Clean or repair the connection and retest.

For voltage drop testing, see **600 Electrical System-General**.

For ground locations, see **ECL Electrical Component Locations**.

On-board diagnostics

Second generation on-board diagnostics (OBD II) software and hardware is incorporated in the engine management systems. The OBD II system monitors components that influence exhaust and evaporative emissions. If a problem is detected, the OBD II system stores the associated diagnostic trouble code (DTC) and condition.

If vehicle emissions levels exceed 1.5 times Federally mandated criteria, the OBD II system illuminates a malfunction indicator light (MIL) in the instrument cluster.

◄ Scan tool connected to OBD II plug below left dashboard.

Professional diagnostic scan tools available at the time of this printing include the BMW GT1, ISTA and a small number of aftermarket BMW-specific tools.

In addition to professional scan tools, there are inexpensive generic OBD II scan tool software programs and handheld units available. Although these have limited capabilities as compared to the dedicated tools, they are powerful diagnostic tools.

For the DIY owner, a simple aftermarket DTC reader is available through http://www.peakeresearch.com. This tool is capable of checking for DTCs as well as switching the illuminated MIL OFF and resetting service indicator lights.

NOTE—
- *OBD II DTC memory (including an illuminated MIL) can only be reset using a BMW scan tool or equivalent. Removing the connector from the ECM or disconnecting the battery does not erase DTC memory.*

BentleyPublishers.com
BentleyPublishers.com—All Rights Reserved

Engine Management Fuses and Power Supply Components

Power supply fuses

See **ECL Electrical Component Locations** for fuse panel access information.

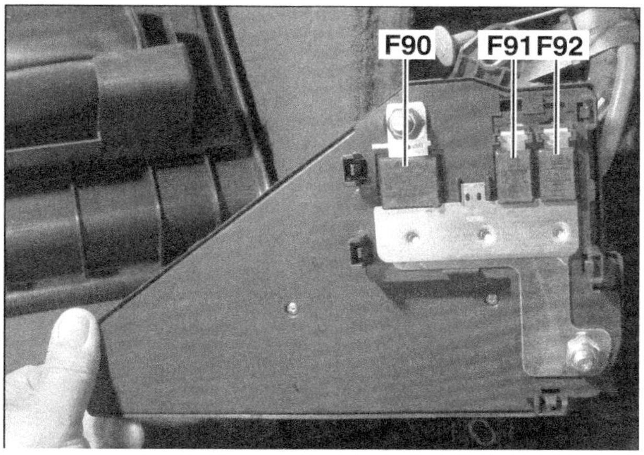

◄ Rear distribution panel (A46) in trunk, installed in right rear fender:

- **F90** 200A: Power supply to fuse panel A41a.

B5101300002

◄ **A41a**: fuse and relay panel behind glove compartment:

- **F8** 60A:
  Engine control module (ECM) (A6000)
  ECM main relay (K6300)
  Engine electronics fuse carrier (A8680)
  Engine electronics fuse carrier (A8681)

See **ECL Electrical Component Locations** for specific vehicle fuse locations.

B5101300014

◄ Engine electronics fuse carrier A8680 (**arrow**).

BentleyPublishers.com
BentleyPublishers.com–All Rights Reserved

SIEMENS DME MSV70, MSV80 (N52 ENGINE)

DME engine management system description

The DME manages and monitors the following functions:

Air

- Electronic throttle
- Mass air flow
- Resonance-turbulence intake control
- Valvetronic II

Fuel

- Fuel supply
- Fuel injection

Ignition

- Direct ignition
- Knock control
- Primary / secondary ignition monitoring

Emissions

- OBD II compliance
- Pre and post-catalyst oxygen sensors
- Electrically heated DME-mapped thermostat
- Misfire detection
- Evaporative emission control and leak detection
- Malfunction indicator light (MIL)

Performance controls

- Dual VANOS control
- Output of injection signal (TI) for fuel economy gauge
- Output of engine rpm (TD) for tachometer
- A/C compressor control
- Electric radiator cooling fan
- CAN-bus communication
- Dynamic stability control (DSC) interface
- CAS (car access system) - EWS 3.3
- Cruise control
- ECM programming

BentleyPublishers.com
BentleyPublishers.com—All Rights Reserved

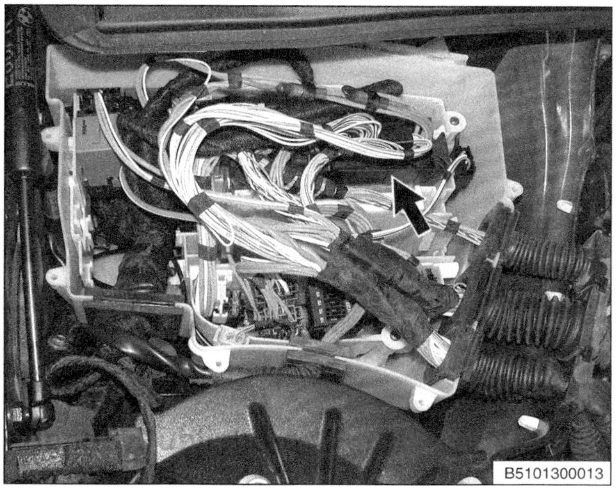

B5101300013

Engine control module (ECM)

The engine control module (ECM) (**arrow**) is mounted in the electronics box (E-box) at the right rear of the engine compartment.

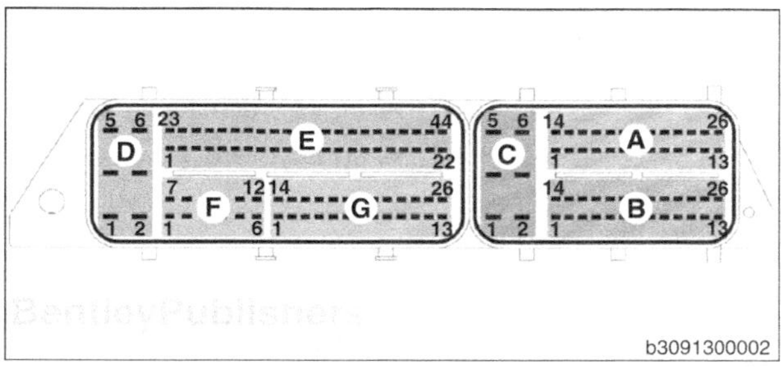

b3091300002

The ECM is flash-programmable and features 2 main electrical harness connectors, one with 4 modular connections and the other with 3 for a total of 7 subconnectors and 147 pins.

- **A:** X60001 (signals)
- **B:** X60002 (signals)
- **C:** X60003 (voltage, ground supply)
- **D:** X60004 (valvetronic)
- **E:** X60005 (signals)
- **F:** X60006 (ignition coils)
- **G:** X60007 (signals)

The EEPROM (chip) in the ECM is coded to the vehicle. The ECM cannot be swapped for testing purposes.

Fuel metering

The ECM meters pressurized fuel by changing the opening time (pulse width) of the fuel injectors. The exact amount of fuel injected is determined by the amount of time the injectors are open. To ensure that injector pulse width is the only factor that determines fuel metering, fuel pump pressure is maintained by a pressure regulator. The injectors are mounted to a common fuel rail.

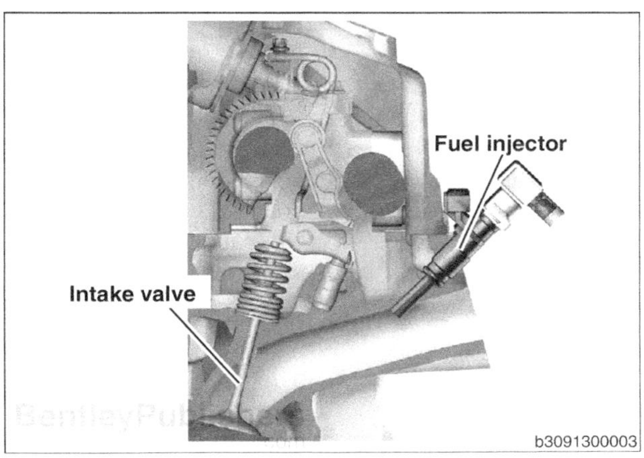

b3091300003

The fuel injectors are mounted into a machined bore in the cylinder head. This design allows the fuel injectors to be closely mounted to the intake valves.

The injectors are a compact design manufactured by Deka with a resistance value of approximately 12 ohms each.

The ECM monitors engine operating conditions to determine injector opening duration. Each injector can be individually controlled for cylinder selective fuel trim.

BentleyPublishers.com–All Rights Reserved

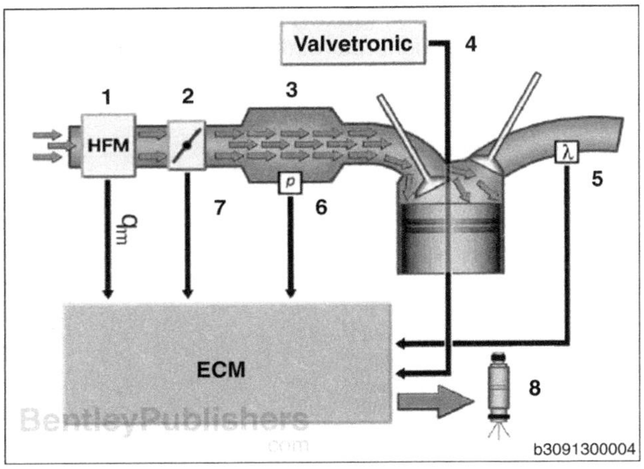

Air intake

Air entering the engine passes through a pleated paper element in the air filter housing. Intake air mass is then measured by a mass air flow sensor. A reference current is used to heat a thin film in the sensor when the engine is running. The current needed to hold the temperature of the film constant is used to calculate the mass of the intake air.

1. Air mass measurement (HFM)
2. Throttle valve
3. Intake manifold
4. Variable intake valve lift (Valvetronic)
5. Residual oxygen measurement in exhaust
6. Intake manifold vacuum
7. Engine speed
8. Injection timing

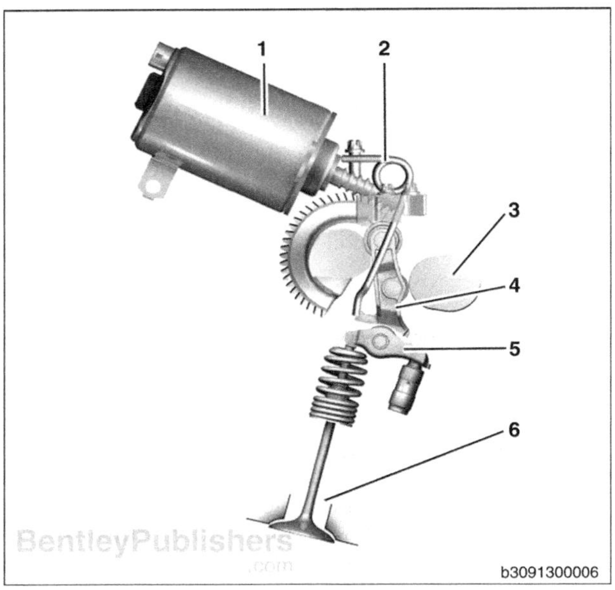

Idle speed control

The ECM controls Idle speed by varying intake valve lift via the Valvetronic system.

1. Valvetronic motor
2. Return spring
3. Intake camshaft
4. Intermediate lever
5. Intake rocker arm
6. Intake valve

Idle speed is not adjustable. The basic functions and parameters of idle speed control are as follows:

- Control of cold air intake volume.
- Smooth idle speeds regardless of load and inputs.
- Smooth transition from acceleration to deceleration.

Idle speed stabilization is active during the following conditions:

- Engine warm up
- Heating or A/C activation
- Drive gear selected (automatic transmission)
- Varying cooling fan speeds

Throttle control

The throttle module (EDK) is not needed for engine load control. Which is carried out by the Valvetronic function of the ECM. However, the throttle may be slightly closed to allow sufficient manifold vacuum for the crankcase ventilation and canister purge systems.

BentleyPublishers.com
BentleyPublishers.com—All Rights Reserved

Knock (detonation) control

Knock sensors monitor and control ignition knock through the ECM. The knock sensors function like microphones and are able to convert mechanical vibration (knock) into electrical signals. The ECM is programmed to react to frequencies that are characteristic of engine knock and adapt the ignition timing point accordingly.

Knock sensor replacement is covered in **120 Ignition System**.

Air filter housing and ducts, removing and installing

For air filter element replacement, see **020 Maintenance**.

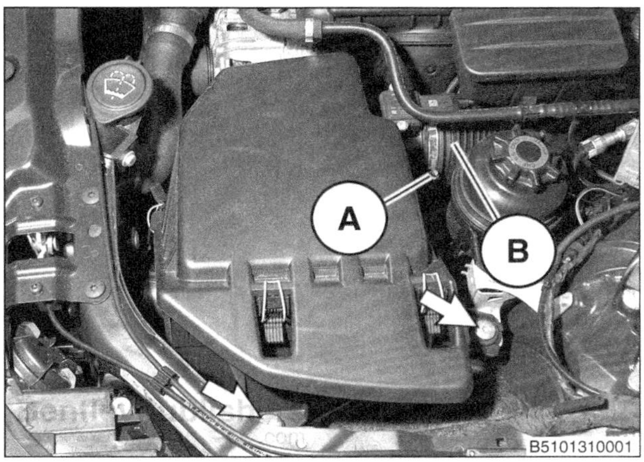

Working at air filter housing:

- Loosen hose clamp (**A**) at mass air flow sensor and detach air intake duct (**B**) from filter housing.
- Remove filter housing mounting bolts (**arrows**) and lift housing out of engine compartment.

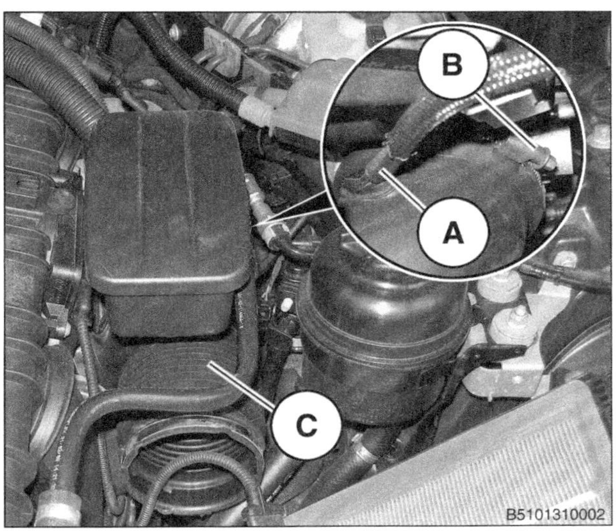

Disconnect hose (**A**) from duct. Loosen duct clamp (**B**). Remove duct (**C**) from throttle body.

– Installation is reverse of removal.

– After reinstalling, check and clear fault codes from ECM memory.

BentleyPublishers.com
BentleyPublishers.com—All Rights Reserved

Mass air flow, measurement

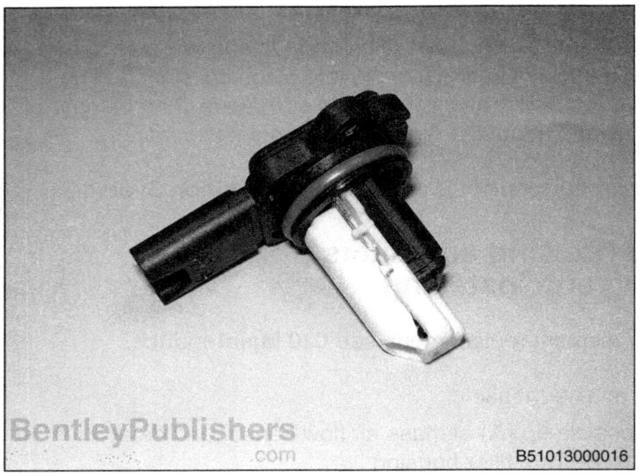

5 Series N52 engines utilize two different air mass measurement systems.

Analog mass air flow sensor is used in models produced to 09 / 2006. The sensor, incorporating an electrically heated film in the air flow stream, sends to the ECM a varying voltage (approx. 0.5 - 4.5 volts) representing the mass of the intake air.

The ECM provides operating voltage and ground to the air flow sensor. As air flows through the sensor, the film is cooled. To maintain the film at a constant temperature, additional current is necessary. It is this additional current that is the basis for the input signal.

If there is no output signal from the air flow sensor, the ECM operates the engine using throttle position and engine rpm inputs. A faulty air flow sensor illuminates the MIL.

Digital mass air flow sensor is used in models from 09 / 2006. In this system, a duty cycle signal corresponds to changes in intake air mass. This eliminates the need for signal conversion in the ECM.

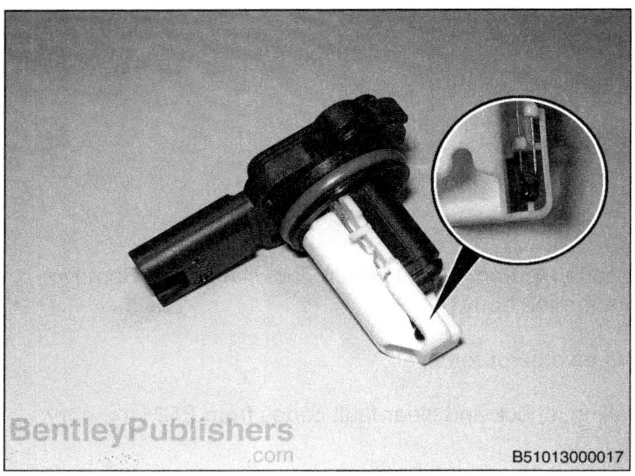

The intake air temperature sensor is integrated into the mass air flow sensor. The sensor is an NTC thermistor which receives a 5 volt reference current and ground from the ECM.

A faulty air flow sensor can produce the following problems:

- Difficult to restart when engine is hot.
- Engine starts then stalls.
- Engine starts and runs only with accelerator pedal depressed.

Mass air flow sensor, replacing

– Switch ignition OFF and remove key.

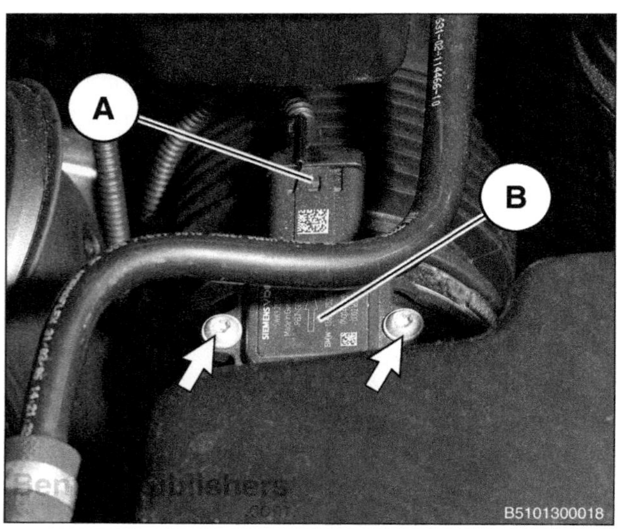

Working at left front of engine compartment at air filter housing:

- Detach mass air flow sensor electrical connector (**A**).
- Remove fasteners (**arrows**).
- Remove mass air flow sensor (**B**).

– Check intake ducts for cracks and vacuum leaks.

BentleyPublishers.com
BentleyPublishers.com—All Rights Reserved

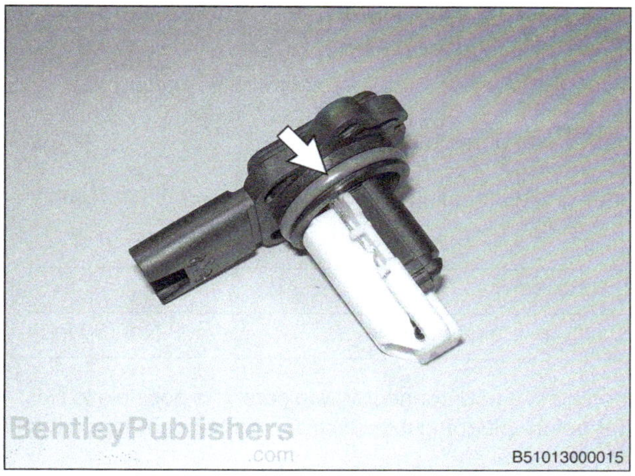

Installation is reverse of removal.

- Inspect O-ring (**arrow**) and replace if necessary.

– After reinstalling, check and clear fault codes from ECM memory.

Tightening torques	
Mass air flow sensor to housing	3 Nm (26 in-lb)

Throttle housing (EDK), replacing

The throttle housing (EDK) is not used for engine load control. Which is carried out by the valvetronic function of the ECM. However, the throttle may be slightly closed to allow sufficient manifold vacuum for the crankcase ventilation and canister purge systems. There is no mechanical (cable) link between the accelerator pedal and the throttle plate.

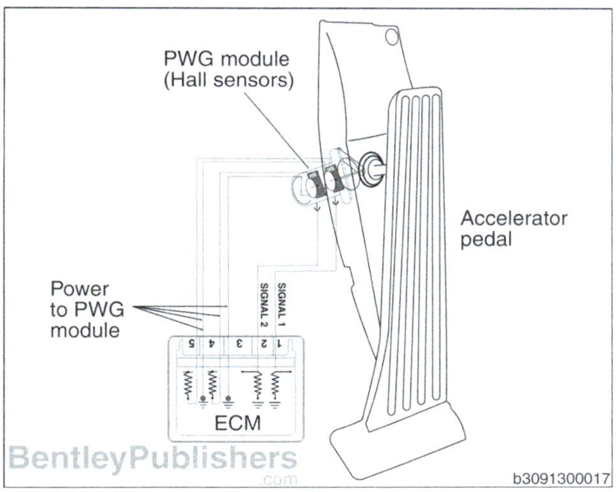

The accelerator pedal module (PWG) at the pedal assembly communicates pedal position directly to the ECM. The module provides two variable voltage signals (via two Hall sensors) to the ECM for pedal position and rate of movement.

The ECM provides an independent voltage and ground supply for each hall sensor. Each Hall sensor is provided with 5 volts and ground. As the accelerator pedal is moved from rest to full throttle, the sensors produce a variable voltage signal.

The output of the Hall sensors is checked for plausibility. The voltage range of Hall sensor 1 is approximately 0.5 to 4.5 volts. Hall sensor 2 ranges from approximately 0.5 to 2.5 volts.

The throttle housing unit is nonadjustable. If found to be faulty, replace as a complete unit.

After replacing the throttle housing, use BMW scan tool or equivalent to reset adaptation values.

> ***CAUTION—***
> - *If the adaptation process is not completed correctly, the engine does not start.*

– Switch ignition OFF and remove key.

– Remove air filter housing and air intake ducts. See **Air filter housing and ducts, removing and installing** in this repair group.

BentleyPublishers
BentleyPublishers.com—All Rights Reserved

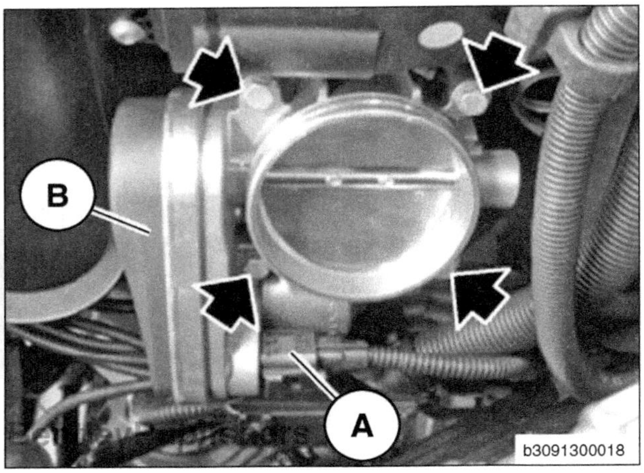

Working at left side of intake manifold:

- Disconnect electrical connector (**A**).
- Remove fasteners (**arrows**) and remove throttle housing (**B**).

– Installation is reverse of removal.

– Replace profile gasket (sealing O-ring) between throttle assembly and intake manifold.

Tightening torques	
Throttle housing to intake manifold	9 Nm (80 in-lb)

– Reattach throttle assembly connector with care. It is possible to twist the connector before plugging it in. This can cause damage to the harness and connector.

– After reinstalling, check and clear fault codes from ECM memory. Reset throttle plate adaptation values following scan tool on-screen instructions.

Valvetronic motor, removing and installing

– Switch ignition OFF and remove key.

– Remove upper and lower cabin microfilter housings. See **020 Maintenance**.

– Remove ignition coil cover. See **020 Maintenance**.

– Disconnect battery negative (-) cable.

> ***CAUTION—***
> - *Prior to disconnecting the battery, read the battery disconnection cautions in* **001 Warnings and Cautions**.

Disconnect ignition coil electrical connectors and remove ignition coil overlay harness (**arrows**).

Gently screw valvetronic motor in clockwise direction (**arrow**) until you feel resistance of the eccentric shaft stop.

BentleyPublishers.com
BentleyPublishers.com—All Rights Reserved

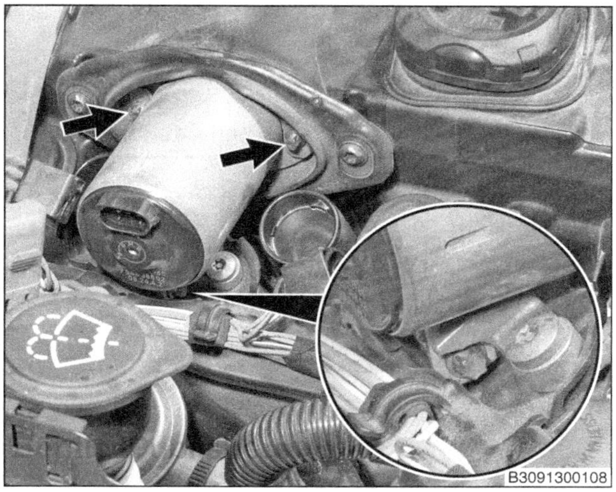

< Remove valvetronic motor fasteners (**arrows**) and Valvetronic motor support bracket fastener (**inset**).

< Gently screw valvetronic motor in counterclockwise direction while sliding motor out of cylinder head cover (**arrow**).

– Installation is reverse of removal. Remember to:
- Replace aluminum fasteners.
- Route ignition harness correctly

Tightening torque	
Valvetronic motor to cylinder head	4 Nm (2.9 ft-lb)

– After reinstalling, check and clear fault codes from ECM memory.

Fuel rail and injectors, removing and installing

Fuel injectors are electrically controlled solenoid valves that provide precisely metered and atomized fuel into the engine intake ports.

Injectors receive voltage from the DME main relay. The engine control module (ECM) controls injector opening by activating the ground circuit for the injector valve solenoids. The ECM varies the duration (in milliseconds) of injector opening to regulate air / fuel ratio.

– Switch ignition OFF and remove key.

– Remove upper and lower cabin microfilter housing. See **020 Maintenance**.

– Remove ignition coil cover. See **020 Maintenance**.

– For a quick check of injectors, run engine and touch each injector with a screwdriver or stethoscope. You should feel a vibration or hear a buzzing. Switch engine OFF.

BentleyPublishers.com—All Rights Reserved

– Disconnect battery negative (-) cable.

> ***CAUTION—***
> - *Prior to disconnecting the battery, read the battery disconnection cautions in* **001 Warnings and Cautions**.

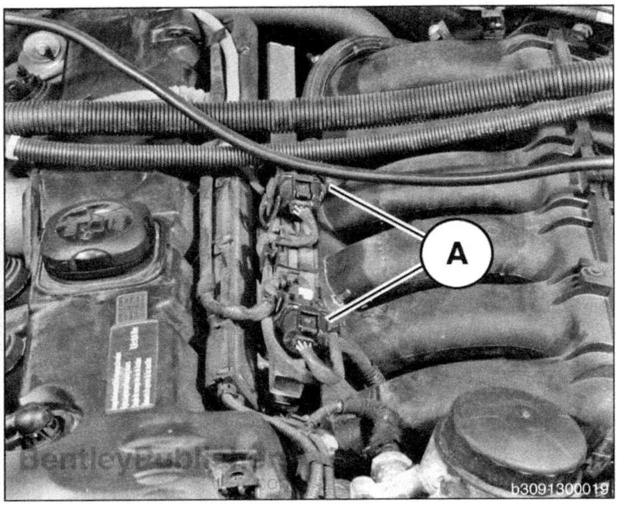

◀ Working above engine, detach oxygen sensor connectors (**A**) and unclip wiring harness from holder.

> ***CAUTION—***
> - *Be sure to mark oxygen sensor connectors so that they can be reassembled as before.*

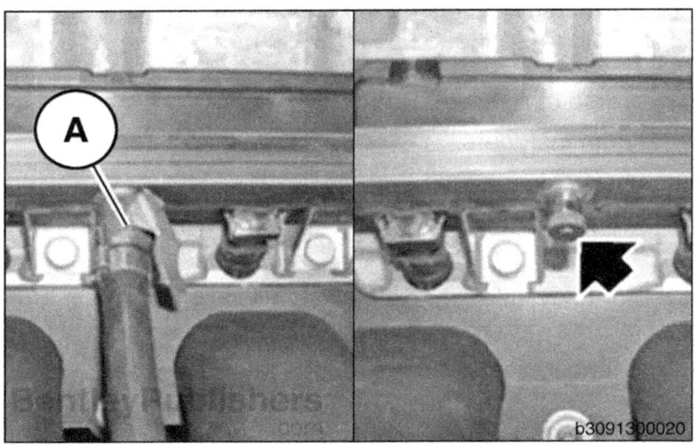

◀ Unscrew schræder valve cap (**arrow**) from fuel rail. Connect air line adapter (**A**) to fitting.

- Unscrew fuel tank cap to release pressure.
- Using a brief burst of compressed air (maximum of 3 bar or 43.5 psi), blow fuel back into fuel tank.

> ***WARNING —***
> - *Fuel in the fuel line is under pressure (approx. 3 - 5 bar or 45 - 75 psi) and may be expelled forcibly. Do not smoke or work near heaters or other fire hazards. Keep a fire extinguisher handy.*
> - *Unscrew the fuel tank cap to release pressure in the tank before working on the fuel line.*
> - *Plug open fuel lines and fittings.*

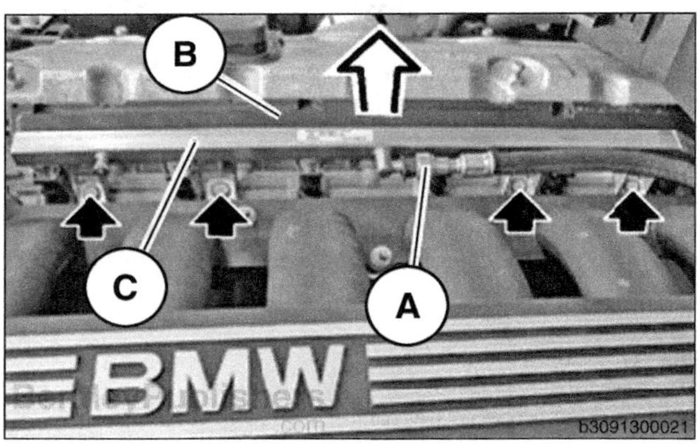

◀ Working above engine:

- Remove and detach fuel line (**A**).
- Disconnect connector strip (**B**) in direction of arrow.
- Remove fasteners (**arrows**).
- Remove fuel rail (**C**).

> ***WARNING —***
> - *Wrap a clean shop towel around fitting before disconnecting. Residual fuel pressure is present in the fuel line.*

BentleyPublishers.com
BentleyPublishers.com—All Rights Reserved

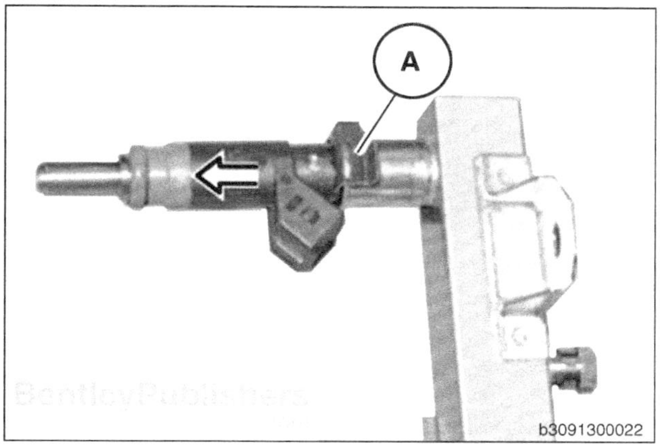

Remove individual injectors:

- Pry retaining clip from injector (**A**).
- Pull injector from rail (**arrow**).

— Installation is reverse of removal. Remember to:

- Fit new sealing O-rings when installing injectors. For ease of installation, lightly lubricate O-rings with assembly lubricant.
- Check that injector electrical connections are correctly fitted and that injectors are fully seated prior to installing fuel rail mounting bolts.
- Replace wire ties.

Tightening torque	
Fuel rail to cylinder head	10 Nm (7 ft-lb)

— After reinstalling, check and clear fault codes from ECM memory.

Intake manifold, removing and installing

— Disconnect battery negative (-) cable in cargo compartment.

> ***CAUTION—***
> - *Prior to disconnecting the battery, read the battery disconnection cautions in* **001 Warnings and Cautions**.

— Remove upper engine covers. See **020 Maintenance**.

— Remove air filter housing and air intake ducts. See **Air filter housing and ducts, removing and installing** in this repair group.

Remove and discard tension strut fasteners (**arrows**). Remove tension strut (**A**).

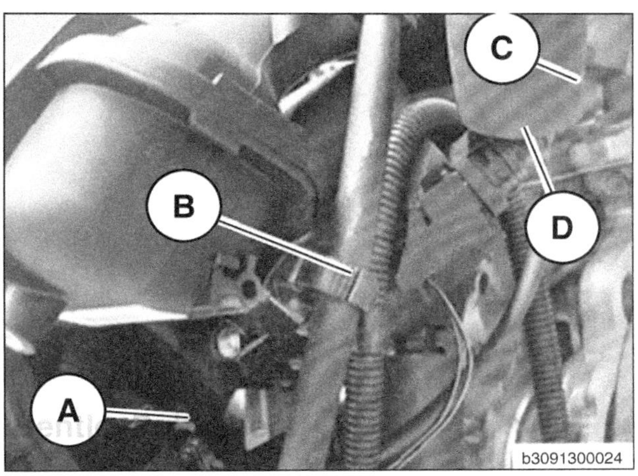

Working at left side and underneath intake manifold:

- Open harness holder (**B**).
- Disconnect electrical connector (**A**).
- Release both crankcase breather connections (**C**,**D**).

BentleyPublishers
BentleyPublishers.com—All Rights Reserved

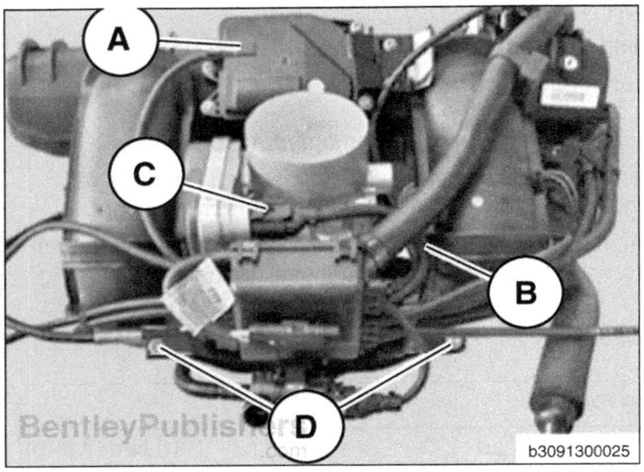

Working at left side and underneath intake manifold:

- Disconnect electrical connectors (**A**,**C**).
- Remove fasteners (**D**).
- Detach engine wiring harness (**B**) from intake manifold and lay aside.

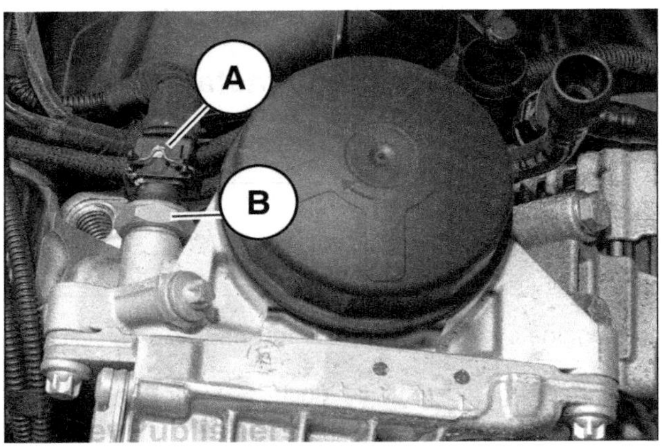

Disconnect electrical connector (**A**) at oil pressure switch (**B**).

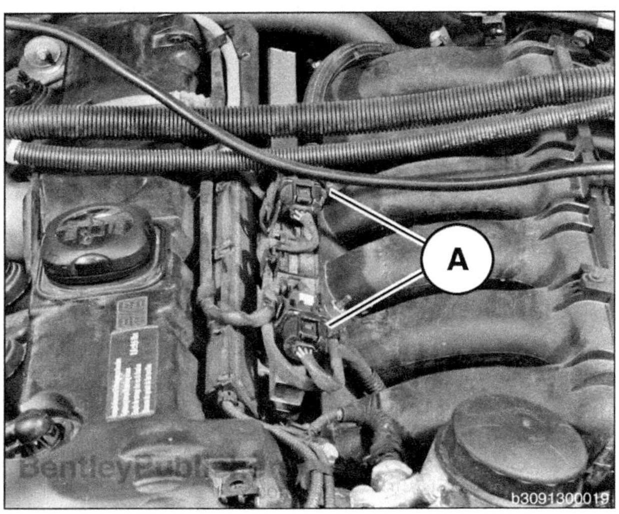

Working above engine, detach oxygen sensor connectors (**A**) and unclip wiring harness from holder.

> ***CAUTION—***
> - *Be sure to mark oxygen sensor connectors so that they can be reassembled as before.*

BentleyPublishers.com
BentleyPublishers.com–All Rights Reserved

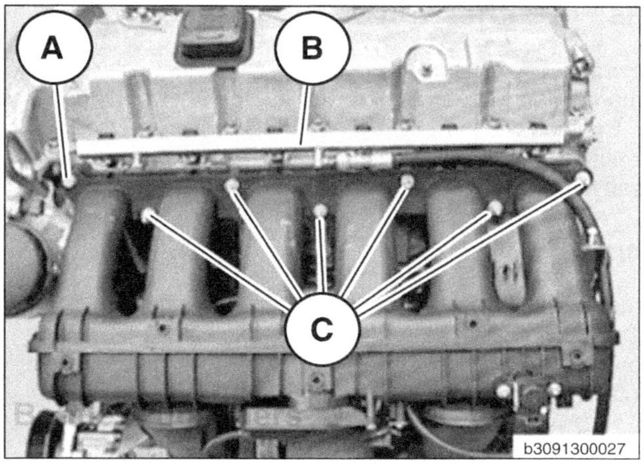

Working above engine:

- Remove and lay fuel rail (**B**) aside. Do not disconnect fuel line.
- Remove fastener (**A**).
- Remove nuts (**C**).

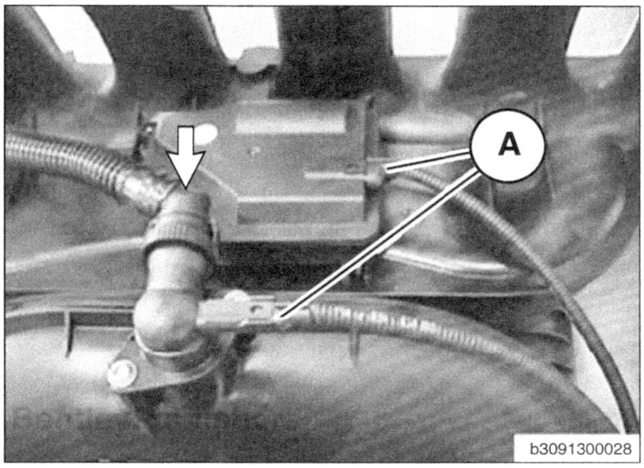

Raise intake manifold approximately 10 cm (4 in). Working underneath:

- Disconnect electrical connector (**A**).
- Release fuel tank vent line (**arrow**) behind throttle valve assembly.
- Raise and remove intake manifold.

> ***CAUTION—***
> *• Plug open intake ports to prevent parts or debris from falling into the engine intake.*

– Installation is reverse of removal. Remember to replace all seals.

Tightening torque	
Manifold to cylinder head	15 Nm (11 ft-lb)
Tension strut to body (M8) replace fasteners	30.5 Nm (22.4 ft-lb)

BentleyPublishers.com
BentleyPublishers.com—All Rights Reserved

Crankcase breather valve

N52 engine crankcase is ventilated by either a breather valve under the intake manifold or a breather valve integrated in the cylinder head cover. Engines with a metal cylinder head cover utilize an external breather valve, engines with a plastic valve cover utilize an internal breather valve.

External crankcase breather

b3091300031

The crankcase ventilation system has an electrically heated crankcase breather valve. The pressure control valve and cyclonic oil separator are combined into one unit. The assembly is located under the intake manifold (**arrow**).

The pressure control valve varies the vacuum applied to the crankcase breather depending on engine load. The valve is balanced between spring pressure and the amount of manifold vacuum.

At idle when the intake manifold vacuum is high, vacuum reduces the valve opening, allowing a small amount of crankcase vapors to be drawn into the intake manifold.

At part to full load conditions when intake manifold vacuum is lower, the spring opens the valve and additional crankcase vapors are drawn into the intake manifold.

Crankcase breather heating

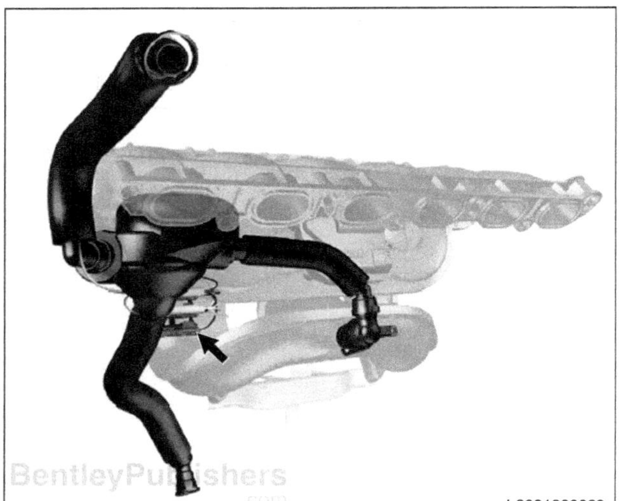
b3091300029

Integrated into the design of the crankcase breather valve is an electric heating system (**arrow**) designed to prevent moisture buildup. Moisture buildup can eventually lead to ice at low ambient temperatures leading to malfunctions of the crankcase breather valve. The crankcase breather valve is insulated by a protective foam covering to provide additional shelter from low ambient temperatures.

The heating elements are integrated into the crankcase ventilation valve and hose assemblies. There is a junction point on the intake manifold which provides a connection point for the individual heating elements.

There is also a heating element located on the centrally located port on the intake manifold. This port is also provided with a separate heating circuit controlled by a PTC thermistor. For wiring schematic **ELE Electrical Wiring Diagrams**.

BentleyPublishers.com
BentleyPublishers.com—All Rights Reserved

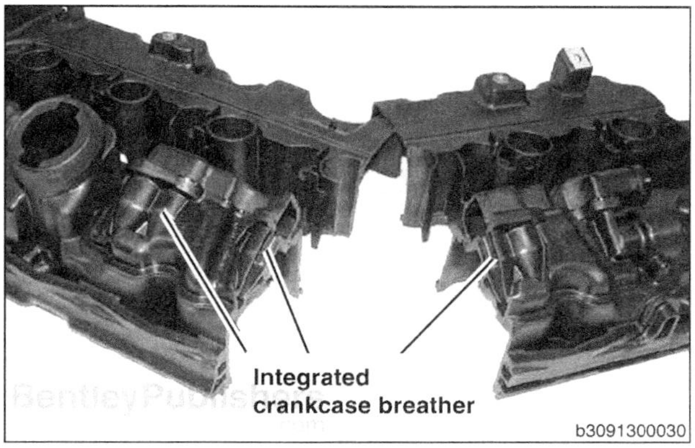

Integrated crankcase breather

The internal crankcase ventilation system is integrated into the plastic cylinder head cover. The crankcase breather valve is part of the cylinder head cover and is not serviceable as a separate component.

Oil separation is carried out via a labyrinth system and two cyclone separators incorporated into the cylinder head cover. By having the system components integrated into the cylinder head cover, the crankcase gases are heated by the engine rather than an electric heater as on the external system. However, there is still one electric heating element at the manifold inlet. Once the liquid oil is separated from the crankcase vapors, the oil is allowed to drain back through check valves back into the engine.

If the diaphragm valve in the breather housing leaks, full intake vacuum may be applied to the crankcase, resulting in excessive oil consumption, irregular idle, whistling or howling noises or oil smoke in the exhaust.

Crankcase pressure specifications for N52 engines

The pressure control valve sets the following pressure range:

- N52 24 to 28 mbar
- N52K 28 to 32 mbar

Deviation to the current ambient pressure: +20 mbar to -60 mbar

A faulty crankcase breather valve causes significant deviations from the specified value:

- Crankcase ventilation clogged: Ambient pressure +100 mbar
- Crankcase ventilation internal leak: ambient pressure -170 mbar

To gain access to external breather valve, remove throttle housing. See **Throttle housing, removing and installing** in this repair group.

To replace internal breather valve. Replace cylinder head cover. See **113 Cylinder Head, Removal and Installation**.

BentleyPublishers.com–All Rights Reserved

Engine coolant temperature (ECT) sensor, replacing

The engine coolant temperature (ECT) sensor is an NTC (negative temperature coefficient) sensor. As coolant temperature rises, resistance through the sensor decreases.

The ECM varies ignition timing and air / fuel mixture based on engine coolant temperature. The ECT sensor is supplied a 5 volt reference voltage. The voltage drop across the sensor varies as the coolant temperature (sensor resistance) changes.

If the ECT sensor input is faulty or not plausible, the MIL is illuminated when OBD II fault criteria are exceeded. The ECM assumes a substitute value (80°C / 176°F) to maintain engine operation. The ignition timing is set to a safely conservative basic setting.

The ECT sensor is located in the left front of the engine near oil filter housing.

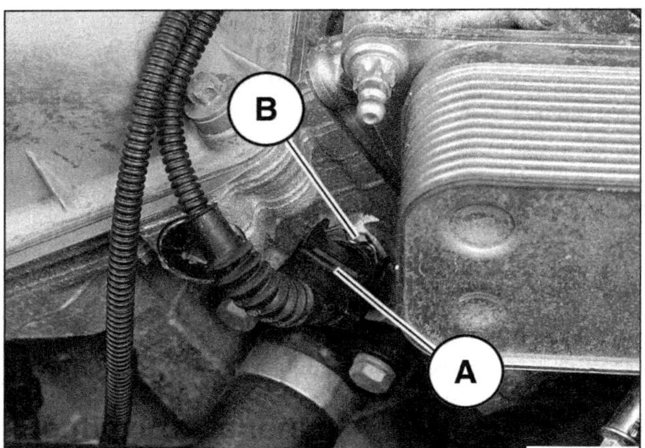

B5101310004

- With engine fully cooled off, disconnect ECT sensor connector (**A**).
- Unscrew ECT sensor (**B**) from cylinder head. Be prepared to catch small amount of coolant.
- Installation is reverse of removal:
  - Use new copper sealing washer when installing sensor.
  - Replace lost coolant.

Tightening torque	
Temperature sensor to cylinder head	18 Nm (13 ft-lb)

- After reinstalling, check and clear fault codes from ECM memory.

Radiator outlet coolant temperature sensor

B51013000010

- N52 engine to 03 / 2007: Radiator outlet temperature sensor (**inset**) is NTC (negative temperature coefficient) sensor. As coolant temperature rises, resistance through sensor decreases.

ECM varies cooling fan speed based on radiator outlet coolant temperature. Radiator outlet coolant temperature sensor is supplied a 5 volt reference voltage. Voltage drop across sensor varies as coolant temperature (sensor resistance) changes.

BentleyPublishers.com
BentleyPublishers.com–All Rights Reserved

DME main relay, testing

The DME main relay is energized via the ECM and supplies battery positive (B+) power to the following:

- ECM
- Ignition coils
- Tailpipe flap
- Mass air flow sensor
- Fuel tank leak detection pump (DMTL)
- Crankshaft sensor
- Camshaft sensors
- Evaporative emissions valve
- Crankcase breather valve heater
- Oil condition sensor
- Variable intake manifold (DISA) controllers
- Electric coolant pump
- Thermostat
- Oxygen sensor heaters

If the DME main relay is faulty, the engine does not start.

◄ With ignition off, remove DME main relay (**K6300**) in electronics box (E-box) at right rear of engine compartment.

NOTE—

- *Relay locations can vary. Confirm relay identification by matching wiring colors and terminal numbers. See* **ELE Electrical Wiring Diagrams**.

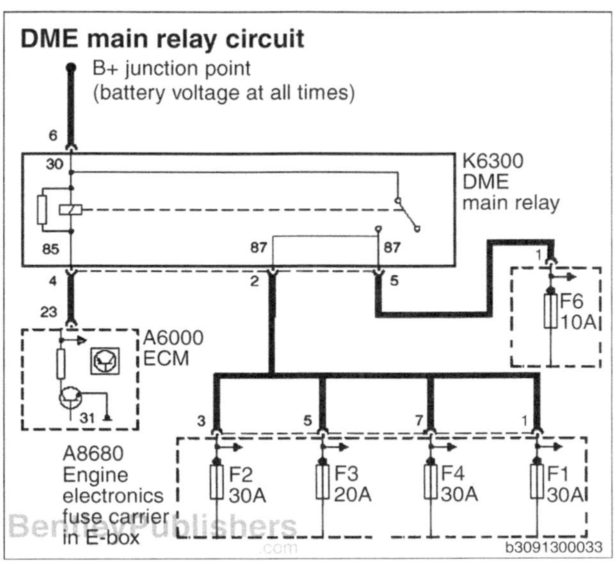

◄ Check for voltage at terminal **6** of relay socket (corresponds to terminal **30** on relay).

– If battery voltage is present, continue testing.

– If battery voltage is not present, check the following:
  - Large red wire in relay socket
  - A2076 (battery positive junction) in E-box, right rear of engine compartment
  - Fuse F8 in fuse and relay panel behind glove compartment. See **Power supply fuses** in this repair group.
  - See **ELE Electrical Wiring Diagrams** for more details

– Check for ground at relay socket **4** (corresponds to terminal **85** on relay).

– If ground is present, continue testing.

– If ground is not present, signal from ECM is missing. Check wire between relay and ECM 44 pin connector X60005. See **ECM pin assignments** in this repair group.

BentleyPublishers.com
BentleyPublishers.com—All Rights Reserved

– If no faults are found:
  - Check ECM grounding.
  - ECM may be defective.

– With ignition ON and relay installed, check for battery voltage at relay sockets **2** and **5** (correspond to terminals **87** on relay).

NOTE —
- *In some models there is only one terminal 87.*

– If battery voltage is present, relay has energized and is functioning correctly.

– If battery voltage is not present and all earlier tests are OK, relay is faulty. Replace.

– If no faults are found during relay testing but power is not reaching ECM or other components, check fuses in engine electronics fuse carrier (A8680) in E-box and Front fuse panel (A41). See **Power supply fuses** in this repair group.

– When finished testing, check and clear fault codes from ECM memory.

BentleyPublishers.com
BentleyPublishers.com—All Rights Reserved

ECM pin assignments

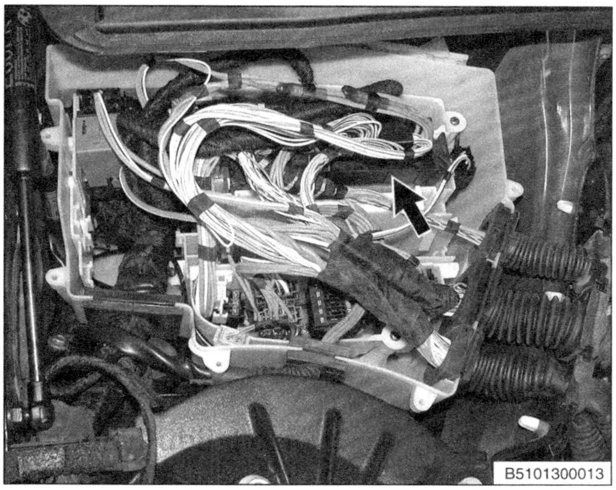

◄ The engine control module (ECM) (**arrow**) is located in the right rear of the engine compartment in the E-box.

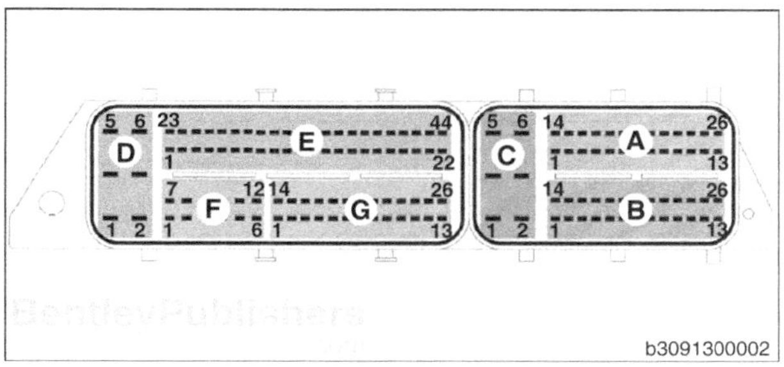

◄ The ECM has 2 main electrical harness connectors, one with 4 modular connections and the other with 3 for a total of 7 sub connectors and 147 pins.

- **A:** X60001 (signals)
- **B:** X60002 (signals)
- **C:** X60003 (voltage, ground supply)
- **D:** X60004 (valvetronic)
- **E:** X60005 (signals)
- **F:** X60006 (ignition coils)
- **G:** X60007 (signals)

ECM pin assignments for non-turbo engines are given in **Table b**. This information can be helpful when diagnosing faults to or from the ECM.

Generally, absence of voltage or continuity means there is a wiring or connector problem. Test results with incorrect values do not necessarily mean that a component is faulty. Check for loose, broken or corroded connections and wiring before replacing components. For engine management system electrical schematics, see **ELE Electrical Wiring Diagrams**.

When making checks at the ECM, use a break-out box to allow tests to be made with the connector attached to the ECM. This prevents damage to the small terminals in the connector. As an alternative, the connector housing can be separated so that electrical checks can be made from the back of the connector.

> ***CAUTION—***
> - *Wait at least one minute after switching the ignition OFF before removing the connector from the ECM. If the connector is removed before this time, residual power in the system relay may damage the ECM.*
> - *Connect or disconnect the control module connector and meter probes with the ignition OFF.*

BentleyPublishers
BentleyPublishers.com—All Rights Reserved

Table b. Siemens DME MSV70 MSV80 ECM pin assignments			
Connector X60001 26-pin (signals)			
Pin	**Type**	**Component or function**	**Note**
1	input/output	Signal PT-CAN low	Powertrain CAN-bus
2	input	Start signal	Car access system
3	input/output	BSD signal	Battery sensor
4	input	Brake light signal	Brake light switch
5	output	Exhaust flap signal	Exhaust flap
6	input	Radiator outlet temperature sensor signal	Temperature sensor at radiator outlet
7	input	Accelerator pedal module signal	Accelerator pedal module
8	output	Electric fan signal	Electric engine cooling fan
9	output	Radiator shutter solenoid	Signal
10	ground	Accelerator pedal module	Accelerator pedal module
11	output	Accelerator pedal module supply voltage	Accelerator pedal module
12	-	Not used	
13	-	Not used	
14	input/output	Signal PT-CAN high	Powertrain CAN-bus
15	input	EWS signal to 03/2007 PA bus signal from 03/2007	Car access system
16	input	Brake light test signal	Brake light switch
17	input	Right rear wheel speed sensor signal	Connector X10186 (right rear wheel speed sensor)
18	input	Clutch switch signal	Clutch switch module
19	input	Radiator outlet temperature sensor signal	Temperature sensor at radiator outlet
20	input	Accelerator pedal module signal	Accelerator pedal module
21	input/output	TD signal	TD signal connector
22	input	Not used to 03/2007 Signal RXIN	 Car access system
23	ground	Accelerator pedal module	Accelerator pedal module
24	output	Accelerator pedal module supply voltage	Accelerator pedal module
25	-	Not used	
26	output	E-box fan signal	E-box fan

Connector X60002 26-pin (signals)			
Pin	**Type**	**Component or function**	**Note**
1	input	Wake up signal, terminal 15	Terminal 15 wake up connector
2	-	Not used	
3	-	Not used	
4	-	Not used	
5	input	Oxygen sensor signal	Bank 2 sensor 1
6	input	Oxygen sensor signal	Bank 1 sensor 1

BentleyPublishers.com
BentleyPublishers.com—All Rights Reserved

Connector X60002 26-pin (signals) (continued)			
Pin	Type	Component or function	Note
7	input	Oxygen sensor signal	Bank 2 sensor 1
8	input	Oxygen sensor signal	Bank 1 sensor 1
9	input	Oxygen sensor signal	Bank 2 sensor 1
10	ground	Oxygen sensor	Bank 1 sensor 1
11	ground	Oxygen sensor	Bank 2 sensor 1
12	input	Oxygen sensor heating	Bank 1 sensor 1
13	input	Oxygen sensor heating	Bank 2 sensor 1
14	-	Not used	
15	input	DMTL valve signal	Diagnosis module for fuel tank leakage
16	input	DMTL pump signal	Diagnosis module for fuel tank leakage
17	input	DMTL heating signal	Diagnosis module for fuel tank leakage
18	input	Oxygen sensor signal	Bank 1 sensor 1
19	input	Oxygen sensor signal	Bank 2 sensor 2
20	input	Oxygen sensor signal	Bank 1 sensor 2
21	-	Not used	
22	-	Not used	
23	ground	Oxygen sensor	Bank 1 sensor 2
24	input	Oxygen sensor	Bank 2 sensor 2
25	input	Oxygen sensor	Bank 2 sensor 2
26	output	Oxygen sensor	Bank 1 sensor 2

Connector X60003 6-pin (voltage, ground supply)			
Pin	Type	Component or function	Note
1	input	Terminal 30	B+ potential distributor
2	input	Terminal 87	Fuse F03
3	ground	Ground	Ground point
4	ground	Ground	Ground point
5	ground	Ground	Ground point
6	ground	Ground	Ground point

Connector X60004 6-pin (valvetronic)			
Pin	Type	Component or function	Note
1	input	Terminal 87	VVT relay
2	input	Terminal 87	VVT relay
3	output	Valvetronic actuator signal	Valvetronic actuator timing actuator
4	output	Valvetronic actuator signal	Valvetronic actuator timing actuator
5	output	Valvetronic actuator signal	Valvetronic actuator timing actuator
6	output	Valvetronic actuator signal	Valvetronic actuator timing actuator

BentleyPublishers.com
BentleyPublishers.com—All Rights Reserved

Connector X60005 44-pin (signals)			
Pin	**Type**	**Component or function**	**Note**
1	-	Not used	
2	-	Not used	
3	-	Not used	
4	-	Not used	
5	-	Not used	
6	-	Not used	
7	-	Not used	
8	-	Not used	
9	-	Not used	
10	-	Not used	
11	-	Not used	
12	output	Crankcase breather valve heating activation	Crankcase breather valve heating relay
13	output	DME main relay activation	DME main relay
14	output	Throttle valve voltage supply	Throttle valve
15	input	Throttle valve signal	Throttle valve
16	input	Throttle valve signal	Throttle valve
17	-	Not used	
18	output	DISA changeover valve signal	DISA controller 2
19	input	Knock sensor signal	Knock sensor
20	input	Knock sensor signal	Knock sensor
21	-	Not used	
22	-	Not used	
23	output	Fuel tank vent valve signal	Fuel tank vent valve
24	-	Not used	
25	output	Mass air flow sensor voltage supply	Mass air flow sensor
26	input	Mass air flow sensor signal	Mass air flow sensor
27	ground	Mass air flow sensor	Mass air flow sensor
28	input	Intake air temperature sensor signal	Intake air temperature sensor
29	input	Crankshaft position sensor signal	Crankshaft position sensor
30	ground	Crankshaft position sensor	Crankshaft position sensor
31	output	Intake manifold pressure sensor voltage supply	Intake manifold pressure sensor
32	ground	Intake manifold pressure sensor	Intake manifold pressure sensor
33	input	Intake manifold pressure sensor signal	Intake manifold pressure sensor
34	-	Not used	
35	input/output	BSD signal	BSD signal connector
36	input	Throttle valve signal	Throttle valve
37	input	Throttle valve signal	Throttle valve

BentleyPublishers.com
BentleyPublishers.com—All Rights Reserved

Connector X60005 44-pin (signals) (continued)			
Pin	**Type**	**Component or function**	**Note**
38	ground	Throttle valve	Throttle valve
39	-	Not used	
40	output	DISA changeover valve signal	DISA controller 1
41	input	Knock sensor signal	Knock sensor
42	input	Knock sensor signal	Knock sensor
43	-	Not used	
44	-	Not used	

Connector X60006 12-pin (ignition coils)			
Pin	**Type**	**Component or function**	**Note**
1	output	Ignition coil signal	Ignition coil signal 1
2	output	Ignition coil signal	Ignition coil signal 2
3	output	Ignition coil signal	Ignition coil signal 3
4	output	Ignition coil signal	Ignition coil signal 4
5	output	Ignition coil signal	Ignition coil signal 5
6	output	Ignition coil signal	Ignition coil signal 6
7	-	Not used	
8	-	Not used	
9	-	Not used	
10	-	Not used	
11	-	Not used	
12	-	Not used	

Connector X60007 26-pin (signals)			
Pin	**Type**	**Component or function**	**Note**
1	output	Fuel injector signal	Fuel injector 1
2	output	Fuel injector signal	Fuel injector 2
3	output	Fuel injector signal	Fuel injector 3
4	input	Engine coolant temperature signal	Engine coolant temperature sensor
5	output	Intake VANOS solenoid signal	Intake VANOS solenoid
6	input	Eccentric shaft sensor signal	Eccentric shaft sensor
7	input	Eccentric shaft sensor signal	Eccentric shaft sensor
8	input	Eccentric shaft sensor signal	Eccentric shaft sensor
9	input	Eccentric shaft sensor signal	Eccentric shaft sensor
10	ground	Wire shielding	Eccentric shaft sensor
11	input	Camshaft position sensor 1 signal	Camshaft position sensor 1
12	input	Camshaft position sensor 2 signal	Camshaft position sensor 2
13	input	Oil pressure switch signal	Oil pressure switch

BentleyPublishers
BentleyPublishers.com—All Rights Reserved

Connector X60007 26-pin (signals) (continued)			
Pin	**Type**	**Component or function**	**Note**
14	output	Fuel injector signal	Fuel injector signal 4
15	output	Fuel injector signal	Fuel injector signal 5
16	output	Fuel injector signal	Fuel injector signal 6
17	ground	Engine coolant temperature sensor	Engine coolant temperature sensor
18	output	Exhaust VANOS solenoid signal	Exhaust VANOS solenoid
19	input	Thermostat signal	Thermostat
20	ground	Eccentric shaft sensor	Eccentric shaft sensor
21	output	Eccentric shaft sensor voltage supply	Eccentric shaft sensor
22	input	Eccentric shaft sensor signal	Eccentric shaft sensor
23	output	Valvetronic actuator relay signal	Valvetronic actuator relay
24	ground	Camshaft position sensor 1	Camshaft position sensor 1
25	ground	Camshaft position sensor 2	Camshaft position sensor 2
26	input/output	BSD signal	Electric coolant pump

BentleyPublishers.com
BentleyPublishers.com—All Rights Reserved

132 Fuel Injection MSD80, MSD81 (N54 Engine)

TABLE

GENERAL

This repair group covers service and repair of engine management systems for turbocharged N54 engines with Siemens DME MSD80 or MSD81.

See also:

- **020 Maintenance** for air filter, fuel filter and spark plug replacement
- **100 Engine–General** for engine identification and application
- **120 Ignition System** for ignition coil, camshaft sensor, crankshaft sensor and knock sensor service
- **160 Fuel Tank and Fuel Pump** for fuel pump and fuel pump relay service, fuel pressure tests and fuel pressure regulator replacement
- **180 Exhaust System** for oxygen sensor service
- **ECL Electrical Component Locations**
- **ELE Electrical Wiring Diagrams**
- **OBD On-Board Diagnostics**

BentleyPublishers.com
BentleyPublishers.com – All Rights Reserved

DME applications

BMW 5 Series models are equipped with digital motor electronics (DME), also known as Motronic. In these systems, fuel injection and ignition are controlled by an integrated engine control module (ECM). Application information for DME systems is in **Table a**.

Table a. 5 Series engine management applications

Year, model	Engine code	Engine management	Features
2004-2005			
Sedan: 525i, 530i	M54	Siemens MS 45.1	Double VANOS
Sedan: 545i	N62	Bosch DME ME 9.2 with Valvetronic control module	Valvetronic, VANOS
2006 - 2007			
Sedan, Sports Wagon: 525i, 525xi, 530i, 530xi Sports Wagon	N52	Siemens MSV70	Valvetronic, VANOS
2008 - 2010			
Sedan, Sports Wagon: 528i, 528xi	N52KP	Siemens MSV80	Valvetronic, VANOS
Sedan, Sports Wagon: 535i, 535xi	N54	Siemens MSD80 (to 03 / 2008) MSD81 (from 03 / 2008)	VANOS, turbocharged
2004-2005			
Sedan: 545i	N62	Bosch DME ME 9.2 with valvetronic control unit	Valvetronic, VANOS
2006 - 2010			
Sedan: 550i	N62TU	Bosch DME ME 9.2.2	Valvetronic, VANOS

BentleyPublishers.com
BentleyPublishers.com—All Rights Reserved

Warnings and Cautions

WARNING —
- *The fuel system is designed to retain pressure even when the ignition is OFF. When working with the fuel system, loosen the fuel lines slowly to allow residual fuel pressure to dissipate. Avoid spraying fuel. Use shop towels to capture leaking fuel.*
- *Fuel in fuel lines is under pressure (approx. 3 - 5 bar or 45 - 75 psi) and may be expelled forcibly. Do not smoke or work near heaters or other fire hazards. Keep a fire extinguisher handy.*
- *Unscrew the fuel tank cap to release pressure in the tank before working on fuel lines.*
- *Plug open fuel lines and fittings.*
- *Before beginning work on the fuel system, place a fire extinguisher in the vicinity of the work area.*
- *Work only on the fuel system when engine temperature is below 40°C (104°F).*
- *Fuel is highly flammable. When working around fuel, do not disconnect wires that could cause electrical sparks. Do not smoke or work near heaters or other fire hazards.*
- *Wear eye protection and protective clothing to avoid injuries from contact with fuel.*
- *When working on an open fuel system, wear suitable hand protection, as prolonged contact with fuel can cause illnesses and skin disorders.*
- *Unscrew the fuel tank cap to release pressure in the tank before working on fuel lines.*
- *Do not use a work light with an incandescent bulb near fuel. Fuel may spray on the hot bulb causing a fire.*
- *Make sure the work area is properly ventilated.*
- *Due to risk of personal injury, be sure the engine is cold before beginning work on engine components.*
- *The ignition system produces high voltages that can be fatal. Avoid contact with exposed terminals. Use extreme caution when working on a car with the ignition switched ON or the engine running.*

BentleyPublishers.com
BentleyPublishers.com—All Rights Reserved

CAUTION—

- *Renew fuel system hoses, clamps and O-rings any time they are removed.*
- *Prior to disconnecting the battery, read the battery disconnection cautions in* **001 Warnings and Cautions**.
- *Connect and disconnect the DME system wiring and test equipment leads only when the ignition is switched OFF.*
- *Wait at least 1 minute after switching the ignition OFF before disconnecting engine control module (ECM) connectors. If the connectors are removed before this time, residual power in the system relay may damage the control module.*
- *Tests or repair procedures in this section may set fault codes (DTCs) in the ECM and illuminate the MIL. After repairs are completed, access and clear DTC memory using a BMW scan tool. See* **On-board diagnostics** *in this repair group.*
- *Fuel system cleaners and other chemical additives other than those specifically recommended by BMW may damage catalytic converters, oxygen sensors or other fuel supply components.*
- *Do not connect any test equipment that delivers a 12-volt power supply to terminal 15 (+) of the ignition coils. The current flow may damage the ECM. Connect test equipment only as specified by BMW or the equipment maker.*
- *Relay positions can vary. Be sure to confirm relay location and function by identifying the wiring in the socket using the wiring diagrams found in* **ELE Electrical Wiring Diagrams**.
- *Use a digital multimeter for electrical tests. Use an LED test light for quick tests.*

BentleyPublishers.com
BentleyPublishers.com—All Rights Reserved

Driveability Troubleshooting

The self-diagnostic DME engine management system monitors and stores diagnostic trouble codes (DTCs). If the malfunction indicator light (MIL) illuminates, it indicates that an emissions-related fault has occurred and that one or more DTCs are stored in the engine control module (ECM).

If faults arise, or if the MIL is illuminated, begin troubleshooting by connecting BMW scan tool to the OBD II plug under the dashboard.

The capabilities of OBD II software have the potential to save hours of diagnostic time and to help avoid incorrect component replacement and possible damage to system components. See **On-board diagnostics** in this repair group.

Basic engine settings

Idle speed, idle mixture (%CO), and ignition timing are not adjustable. The DME system is adaptive and automatically compensates for changes in the engine due to age, minor wear or small problems such as a disconnected vacuum hose. However, the adaptive range is limited. Once the limits are exceeded, driveability problems become noticeable.

Poor initial driveability may be encountered when the battery is disconnected and reconnected. When the battery is disconnected, ECM adaptive memory may be reset. The system readapts after about ten minutes of driving.

System voltage

Digital motor electronics (DME) requires that the system (battery) voltage be maintained within a narrow range of DC voltage. DC voltage levels beyond or below the operating range, or any AC voltage in the electrical system can cause havoc.

- When troubleshooting an illuminated MIL, make sure the battery is fully charged and capable of delivering all its power to the electrical system. An undercharged battery can amplify AC alternator output ripple.

- To make a quick check of battery charge, measure voltage across battery terminals with all cables attached and ignition OFF. A fully charged battery measures 12.6 volts or slightly more, compared to 12.15 volts for a battery with a 25% charge.

The DME system operates at low voltage and current levels, making it sensitive to small increases in resistance. The electrical system is routinely subject to corrosion, vibration and wear, so faults or corrosion in the wiring harness and connectors are not uncommon.

- Check battery terminals for corrosion or loose cable connections. See **121 Battery, Starter, Alternator**.

BentleyPublishers.com–All Rights Reserved

– If a battery cable connection has no visible faults but is still suspect, measure voltage drop across the connection. A large drop indicates excessive resistance, meaning that the connection is corroded, dirty or damaged. Clean or repair and retest. See **600 Electrical System–General** for voltage drop test procedure.

– Visually inspect wiring, connectors, switches and fuses. Loose or damaged connectors can cause intermittent problems, especially small terminals in ECM connectors. Disconnect wiring harness connectors to check for corrosion, and use electrical cleaning spray to remove contaminants.

Main grounds

Good grounds are critical to proper DME operation. If a ground connection has no visible faults but is still suspect, measure the voltage drop across the connection. A large voltage drop means high resistance. Clean or repair the connection and retest.

For voltage drop testing, see **600 Electrical System-General**.

For ground locations, see **ECL Electrical Component Locations**.

On-board diagnostics

Second generation on-board diagnostics (OBD II) software and hardware is incorporated in engine management systems. The OBD II system monitors components that influence exhaust and evaporative emissions. If a problem is detected, the OBD II system stores the associated diagnostic trouble code (DTC) and condition.

If vehicle emissions levels exceed 1.5 times Federally mandated criteria, the OBD II system illuminates a malfunction indicator light (MIL) in the instrument cluster.

Scan tool connected to OBD II plug below left dashboard.

Professional diagnostic scan tools available at the time of this printing include the BMW GT1, ISTA and a small number of aftermarket BMW-specific tools.

In addition to professional scan tools, there are inexpensive generic OBD II scan tool software programs and handheld units available. Although these have limited capabilities as compared to the dedicated tools, they are powerful diagnostic tools.

For the DIY owner, a simple aftermarket DTC reader is available through http://www.peakeresearch.com. This tool is capable of checking for DTCs as well as switching the illuminated MIL OFF and resetting service indicator lights.

NOTE—

- *OBD II DTC memory (including an illuminated MIL) can only be reset using a BMW scan tool or equivalent. Removing the connector from the ECM or disconnecting the battery does not erase DTC memory.*

BentleyPublishers.com
BentleyPublishers.com—All Rights Reserved

Engine Management Fuses and Power Supply Components

Power supply fuses

See **ECL Electrical Component Locations** for fuse panel access information.

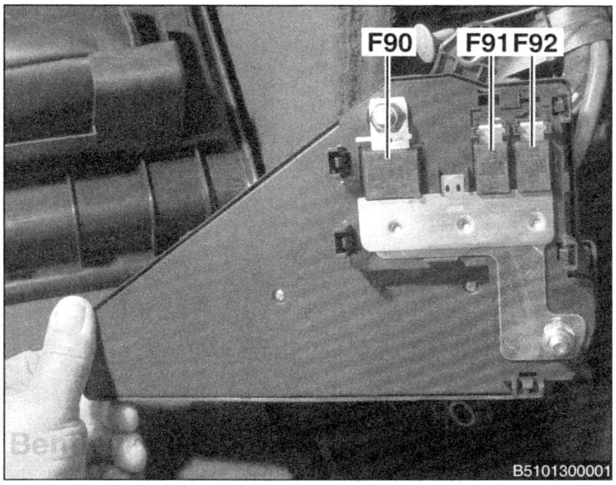

Rear distribution panel (A46) in trunk, installed in right rear fender:

- **F90** 200A: Power supply to fuse panel A41a.

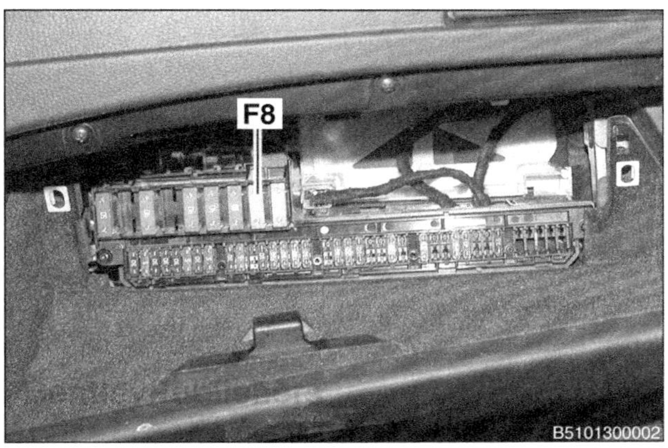

A41a: fuse and relay panel behind glove compartment:

- **F8** 60A:
  Engine control module (ECM) (A6000)
  ECM main relay (K6300)
  Engine electronics fuse carrier (A8680)
  Engine electronics fuse carrier (A8681)

See **ECL Electrical Component Locations** for specific vehicle fuse locations.

Engine electronics fuse carrier A8680 (**arrow**).

BentleyPublishers.com—All Rights Reserved

SIEMENS DME MSD80, MSD81 (N54 ENGINE)

DME engine management system description

The DME manages and monitors the following functions:

Air

- Twin-turbo chargers
- Electronic throttle
- Charge air intercoolers

Fuel

- Fuel supply
- Direct fuel injection

Ignition

- Direct ignition
- Knock control
- Primary / secondary ignition monitoring

Emissions

- OBD II compliance
- Pre- and post-catalyst oxygen sensors
- Electrically heated DME-mapped thermostat
- Misfire detection
- Evaporative emission control and leak detection
- Malfunction indicator light (MIL)

Performance controls

- Double VANOS control
- Output of injection signal (TI) for fuel economy gauge
- Output of engine rpm (TD) for tachometer
- A/C compressor control
- Electric radiator cooling fan
- CAN-bus communication
- Dynamic stability control (DSC) interface
- Car access system (CAS) with EWS 3.3
- Cruise control
- ECM programming

BentleyPublishers.com
BentleyPublishers.com—All Rights Reserved

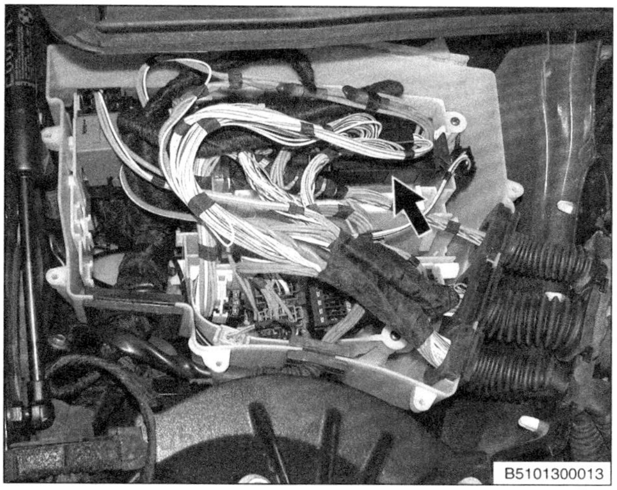

Engine control module (ECM)

The engine control module (ECM) (**arrow**) is mounted in the electronics box (E-box) at the right rear of the engine compartment.

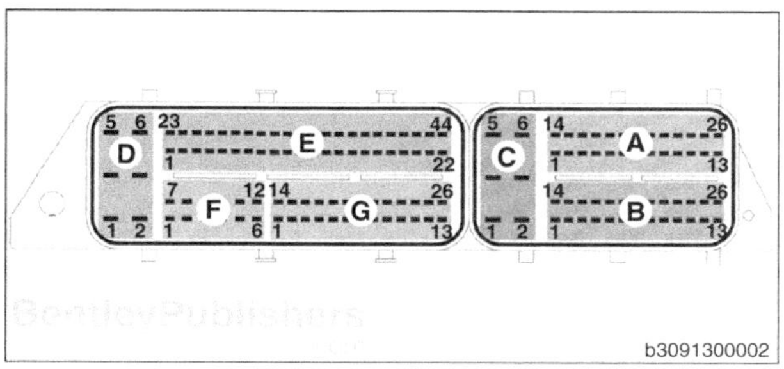

The ECM is flash-programmable and features 2 main electrical harness connectors, one with 4 modular connections and the other with 3 for a total of 7 subconnectors and 147 pins.

- **A:** X60001 (signals)
- **B:** X60002 (signals)
- **C:** X60003 (voltage, ground supply)
- **D:** X60004 (ignition coils)
- **E:** X60005 (signals)
- **F:** X60006 (fuel injectors)
- **G:** X60007(signals)

The EEPROM (chip) in the ECM is coded to the vehicle.

Fuel metering

The ECM meters pressurized fuel by changing the opening time (pulse width) of the fuel injectors. The exact amount of fuel injected is determined by the amount of time the injectors are open.

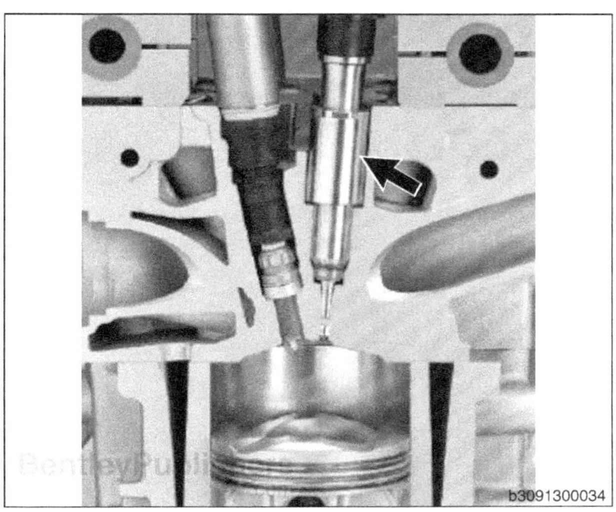

5 Series model with N54 engine utilizes direct fuel injection. Each fuel injector (**arrow**) is mounted into a machined hole in the cylinder head near the spark plug.

A direct injection turbocharged engine achieves a higher compression ratio when compared to ones utilizing manifold injection. At the same time, the exhaust gas temperature is reduced under full load. Another advantage of this injection process is improved efficiency in part-load operation.

The ECM monitors engine operating conditions to determine injector opening duration. Each injector can be individually controlled for cylinder selective fuel trim.

BentleyPublishers
BentleyPublishers.com—All Rights Reserved

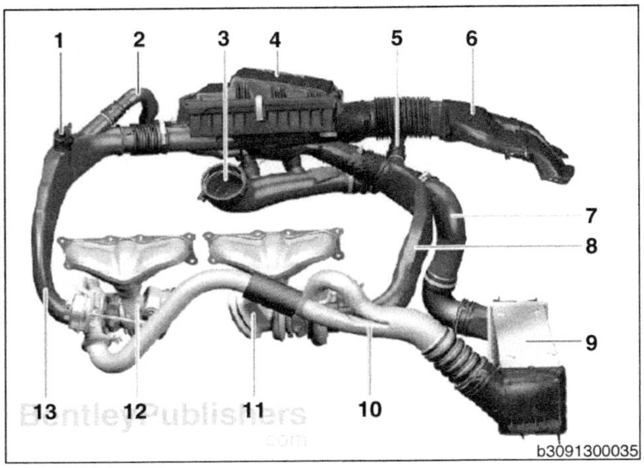

Air intake

Air intake ducting plays a significant role due to the requirements of a turbocharged engine. In principle, the energy of escaping exhaust gases is utilized to precompress the inducted fresh air and thus introduce a greater air mass into the combustion chamber. This is only possible if air intake ducting is leak free and installed properly.

1. Heater for blow by gases (PTC)
2. Recirculated air line, bank 2
3. Connection to throttle valve
4. Air filter housing
5. Recirculated air line, bank 1
6. Air intake snorkel
7. Charge air pressure line
8. Charge air suction line, bank 1
9. Charge air intercooler
10. Charge air manifold
11. Turbocharger, bank 1
12. Turbocharger, bank 2
13. Charge air suction line, bank 2

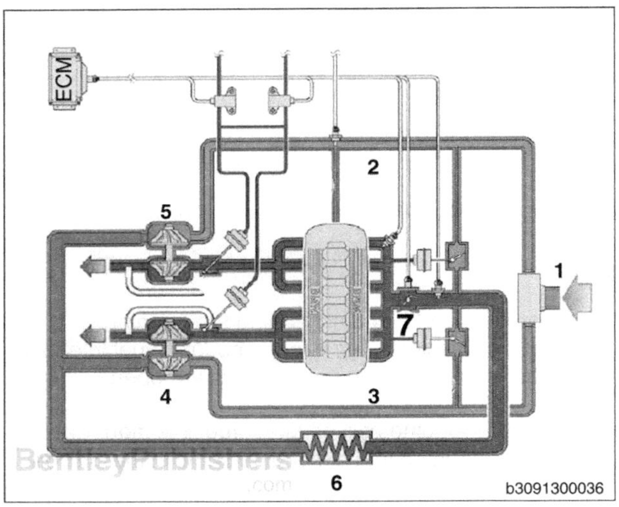

Fresh air is drawn in via the air cleaner (**1**) and charge-air suction lines (**2**, **3**) to the turbochargers (**4**, **5**) and compressed. Because the turbochargers can get very hot during operation, they are connected to the engine coolant and engine oil circuits. The charge air is greatly heated when compressed in the turbocharger, making it necessary for the air to be cooled in an intercooler (**6**).

The compressed and cooled charge air is routed from the intercooler via the throttle valve (**7**) into the intake manifold. The system is equipped with several sensors and actuators in order to ensure that the volume of air is optimally adapted to engine operating conditions.

Idle speed control

The ECM controls idle speed by varying throttle valve actuation. Idle speed is not adjustable. The basic functions and parameters of idle speed control are as follows:

- Control of cold air intake volume.
- Smooth idle speeds regardless of load and inputs.
- Smooth transition from acceleration to deceleration.

Idle speed stabilization is active during the following conditions:

- Engine warm-up
- Heating or A/C activation
- Drive gear selected (automatic transmission)
- Varying cooling fan speeds

BentleyPublishers.com
BentleyPublishers.com—All Rights Reserved

Throttle control

The N54 engine features electronic throttle control (EDK). There is no throttle cable between the accelerator pedal and the throttle housing. EDK integrates driver commands with the requirements of the traction control system (DSC) and cruise control.

Knock (detonation) control

Knock sensors monitor and control ignition knock through the ECM. Knock sensors function like microphones and are able to convert mechanical vibration (knock) into electrical signals. The ECM is programmed to react to frequencies that are characteristic of engine knock and adapt the ignition timing point accordingly.

Knock sensor replacement is covered in **120 Ignition System**.

Air filter housing and ducts, removing and installing

- Working at top of engine, detach air intake duct (**arrow**) and lift out of radiator support.

- Remove upper engine covers. See **020 Maintenance**.

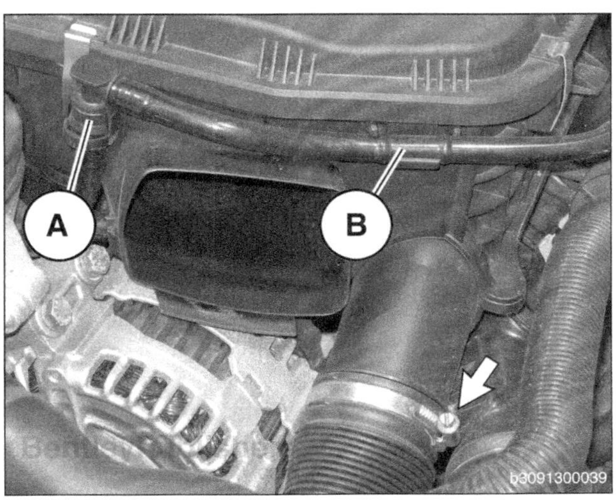

- Remove complete air cleaner housing, disconnecting and labeling connectors, ducting and hoses as necessary:
  - Air duct hose clamp (**arrow**).
  - Disconnect line (**A**) and detach from air intake housing (**B**).

BentleyPublishers.com—All Rights Reserved

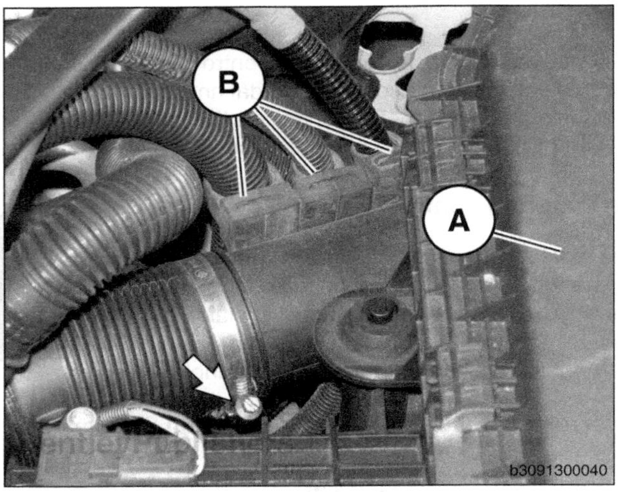

Loosen clamp (**arrow**) and remove air intake hose.

- Pull up wiring harness rubber mounts (**B**) to detach.
- Detach intake filter housing (**A**) from rubber mounts, then lift and remove.

– Installation is reverse of removal.

> ***CAUTION—***
> - *Turbocharger failure may result if charge air ducts are not installed dry and free of grease. Install using antiseize compound only.*

– After reinstalling, check and clear fault codes from ECM memory.

Charge air ducts, left side, removing and installing

– Remove air filter housing and ducts. See **Air filter housing and ducts, removing and installing** in this repair group.

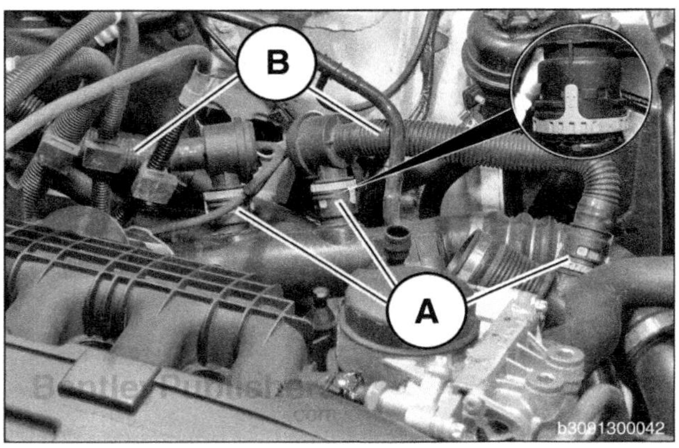

Turn quick release coupling locks (**A**) 90° to disconnect. Detach recirculated air hoses (**B**) and lay aside. Note position of arrows on quick connector for reinstallation.

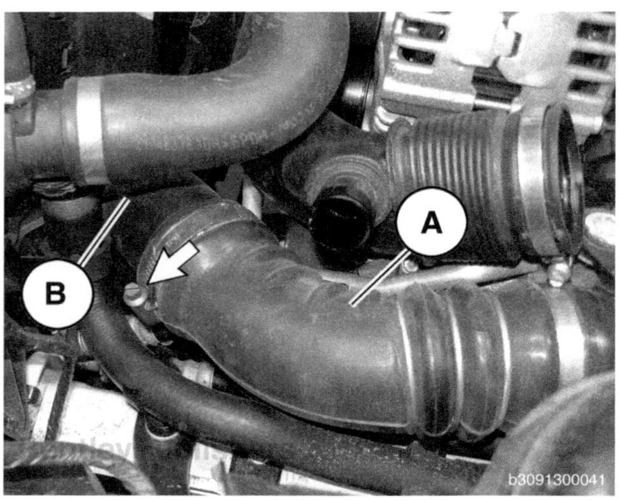

Working at left charge air duct:

- Loosen clamp (**arrow**) and remove charge air hose (**A**) from duct (**B**).

BentleyPublishers.com
BentleyPublishers.com—All Rights Reserved

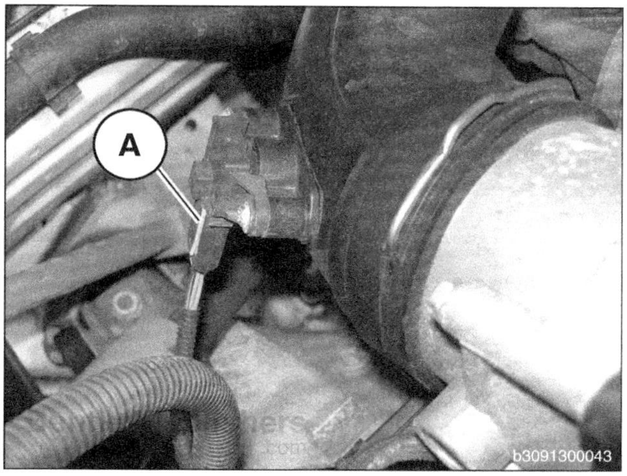

Disconnect electrical connector (**A**) and place harness aside.

Remove fastener (**arrow**).

- Unlock quick connect (**A**) and remove charge air duct (**B**) from throttle valve.

– Raise vehicle and support safely.

> ***WARNING—***
> - *Make sure the vehicle is stable and well supported at all times. Use a professional automotive lift or jack stands designed for the purpose. A floor jack is not adequate support.*

– Remove splash shield underneath engine. See **020 Maintenance**.

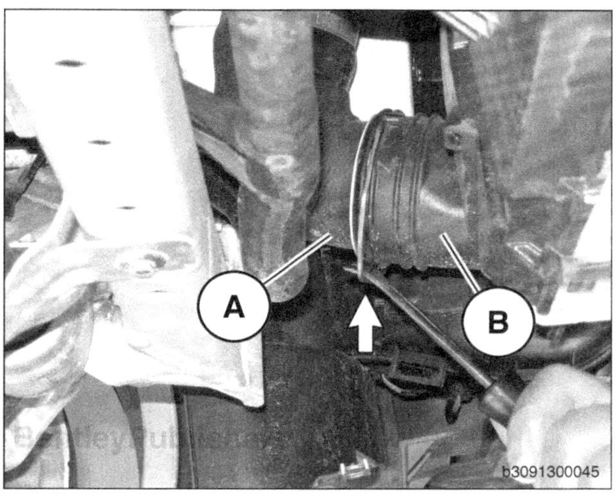

Working under vehicle at right front air charge duct:

- Unlock quick connect (**arrow**) and disconnect charge air duct (**A**) from intercooler (**B**).

BentleyPublishers.com—All Rights Reserved

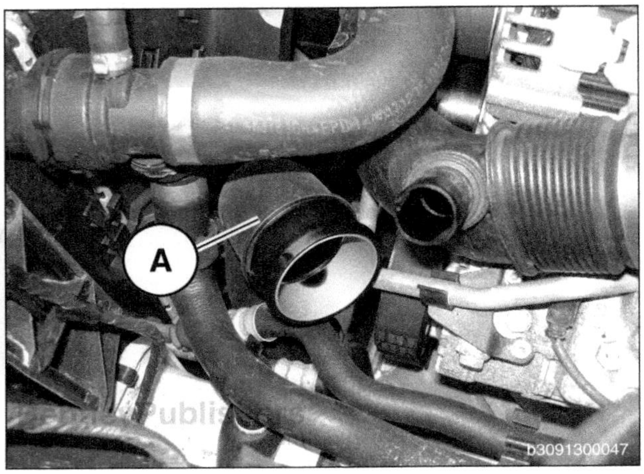

- Feed air charge duct (**A**) through top and remove.

– Installation is reverse of removal.

> ***CAUTION—***
> - *Turbocharger failure may result if charge air ducts are not installed dry and free of grease. Install using antiseize compound only.*

– After reinstalling, check and clear fault codes from ECM memory.

Charge air ducts, right side, removing and installing

– Drain coolant and remove coolant expansion tank. See **170 Radiator and Cooling System**.

– Remove splash shield underneath engine. See **020 Maintenance**.

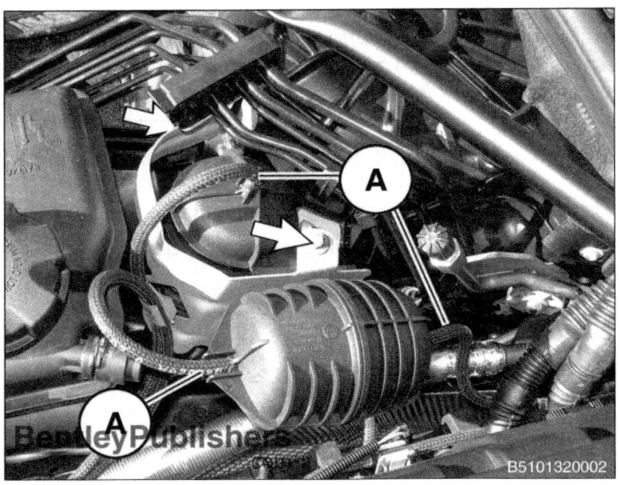

- Working in right front engine compartment:
  - Label and detach vacuum lines (**A**) from canisters.
  - Remove fasteners (**arrows**).

– Raise bracket slightly. Label and detach vacuum hose from bottom of canisters. Remove assembly.

BentleyPublishers.com
BentleyPublishers.com—All Rights Reserved

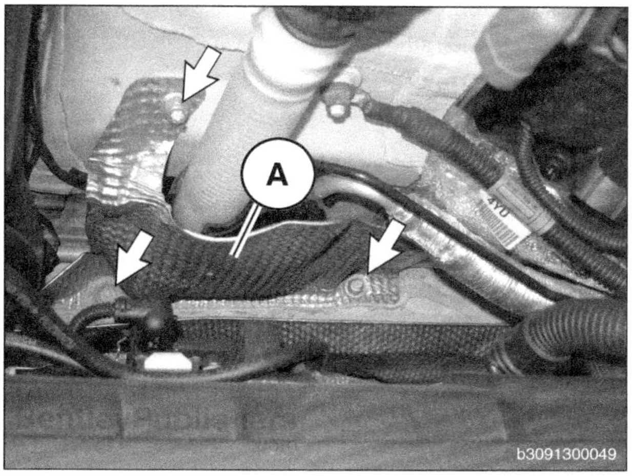

Working at right strut tower, remove nuts (**arrows**) and remove heat shield (**A**) if equipped.

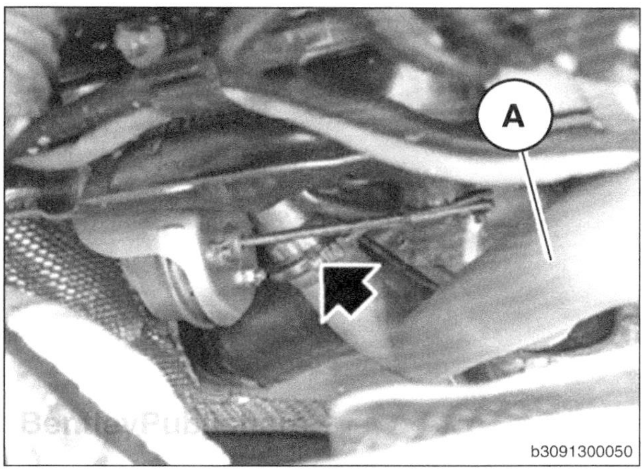

Working above turbocharger, loosen clamp (**arrow**) and remove charge air duct (**A**).

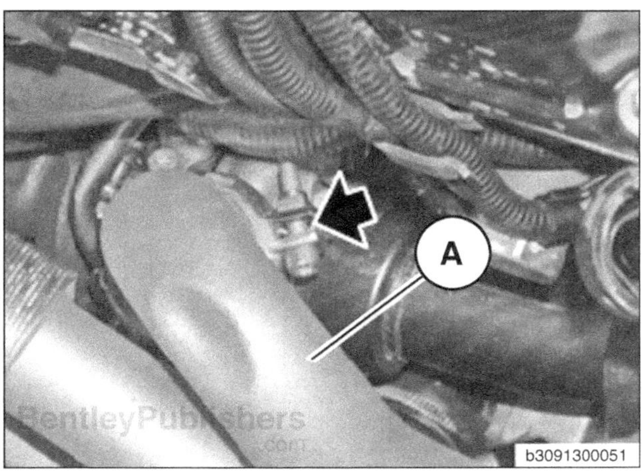

Working above turbocharger, loosen clamp (**arrow**) and remove charge air duct (**A**).

BentleyPublishers.com—All Rights Reserved

Charge air ducts, right side, removing and installing

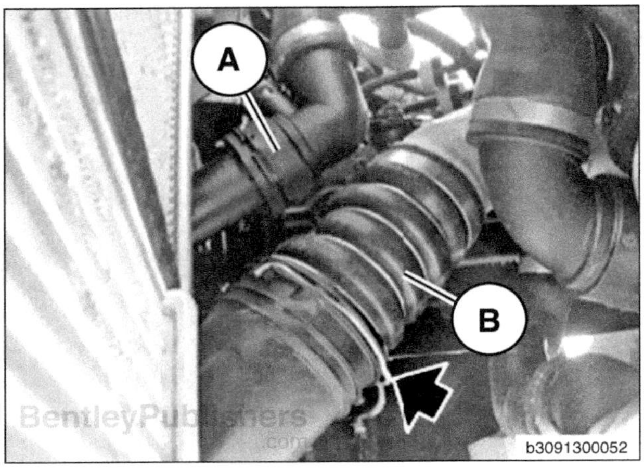

Working beneath right front of vehicle:

- Unlock quick connect and remove radiator hose (**A**).
- Unlock quick connect (**arrow**) and remove charge air duct (**B**).

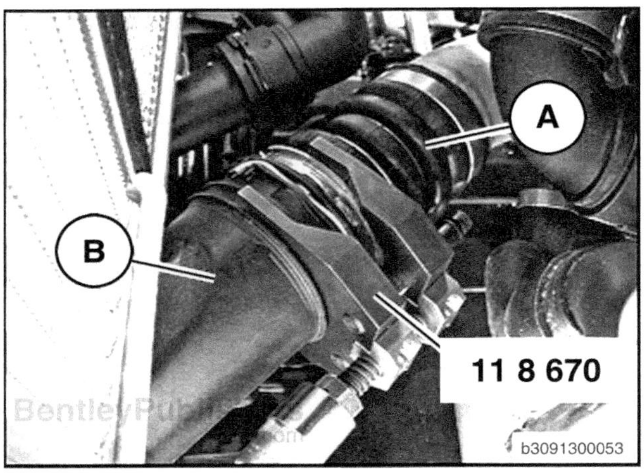

When reinstalling charge air duct to intercooler use BMW special tool 11 8 670.

Coat charge air duct O-ring (**arrow**) with antiseize before reinstalling.

– Remainder of installation is reverse of removal.

> ***CAUTION—***
> - *Turbocharger failure may result if charge air ducts are not installed dry and free of grease. Install using antiseize compound only.*

– After reinstalling, check and clear fault codes from ECM memory.

BentleyPublishers.com
BentleyPublishers.com—All Rights Reserved

Mass air flow, measurement

The N54 engine utilizes a calculated air mass system. This calculation makes use of the following signals:

- VANOS setting (load acquisition)
- Throttle setting
- Intake air temperature (air density correction)
- Engine temperature (air density correction)
- Engine speed (cylinder charge)
- Intake manifold pressure (throttle correction)
- Ambient pressure (air density, altitude correction)

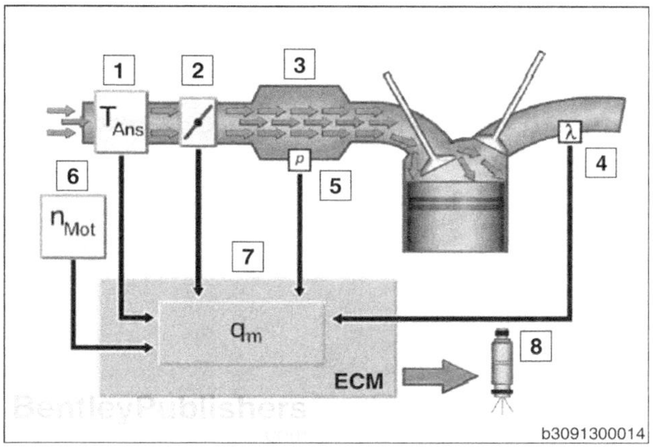

The calculated air mass is adjusted with the oxygen sensor signal (mixture ratio) and injector timing and corrected if necessary.

1. Intake air temperature sensor
2. Throttle valve position
3. Intake manifold
4. Residual oxygen measurement in exhaust
5. Intake manifold vacuum
6. Engine speed
7. ECM with charge calculation model
8. Injection timing

In the event of an oxygen sensor failure, a fault code is stored in the ECM (air mass plausibility). The calculated air mass is not adjusted in this case.

Charge air duct pressure sensor, replacing

– Switch ignition OFF and remove key.

Remove air filter housing and ducts. See **Air filter housing and ducts, removing and installing** in this chapter.

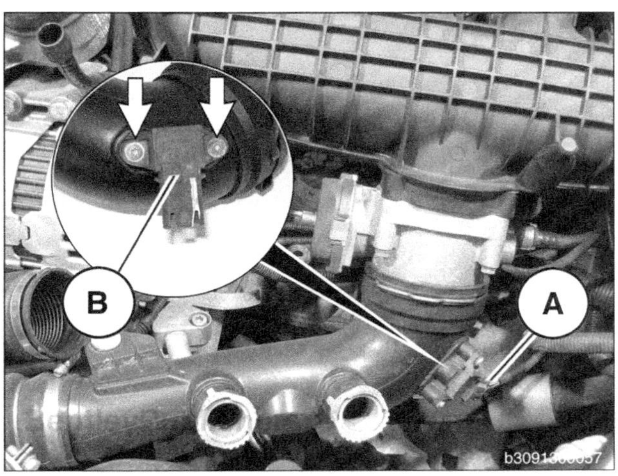

Working at throttle body charge air duct:

- Detach sensor electrical connector (**A**).
- Remove fasteners (**arrows**).
- Remove sensor (**B**).

BentleyPublishers.com - All Rights Reserved

- Installation is reverse of removal.
  - Inspect sealing O-ring (**arrow**) and replace in necessary.
- After reinstalling, check and clear fault codes from ECM memory.

Tightening torques	
Sensor to charge air duct	3.5 Nm (31 in-lb)

Intake manifold charge air pressure sensor, replacing

- Switch ignition OFF and remove key.
- Remove upper engine covers. See **020 Maintenance**.

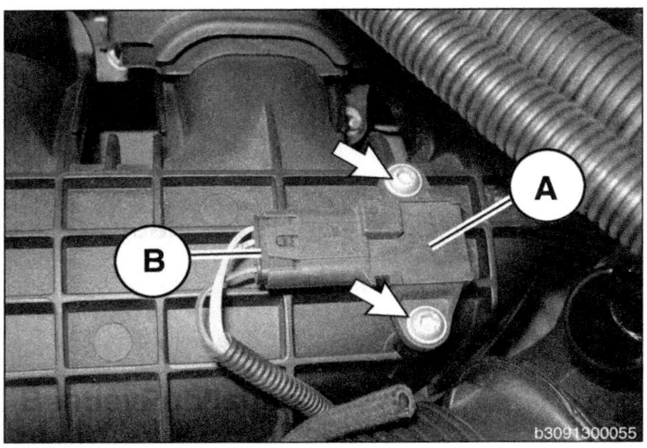

- Working at rear of intake manifold:
  - Detach sensor electrical connector (**B**).
  - Remove fasteners (**arrows**).
  - Remove sensor (**A**).

- Installation is reverse of removal.
  - Inspect sealing O-ring (**arrow**) and replace if necessary.
- After reinstalling, check and clear fault codes from ECM memory.

Tightening torques	
Sensor to intake manifold	3.5 Nm (31 in-lb)

BentleyPublishers.com
BentleyPublishers.com—All Rights Reserved

Throttle housing (EDK), replacing

There is no mechanical (cable) link between the accelerator pedal and the throttle plate.

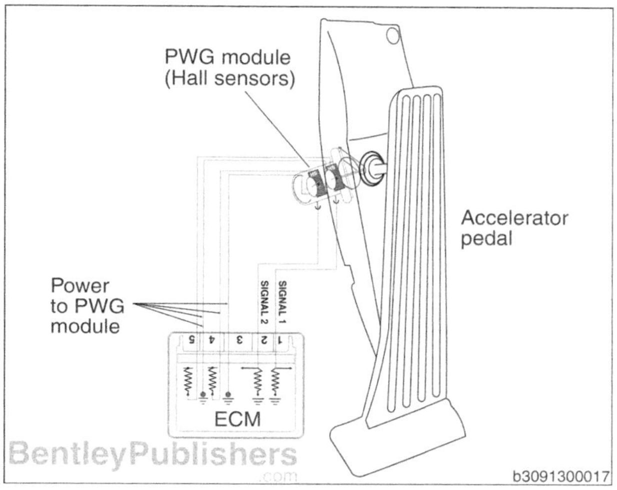

◄ The accelerator pedal module (PWG) at the pedal assembly communicates pedal position directly to the ECM. The module provides two variable voltage signals (via two Hall sensors) to the ECM for pedal position and rate of movement.

The ECM provides an independent 5V and ground supply for each Hall sensor. As the accelerator pedal is moved from rest to full throttle, the sensors produce a variable voltage signal.

The output of the Hall sensors is checked for plausibility. The voltage range of Hall sensor 1 is approximately 0.5 to 4.5 volts. Hall sensor 2 ranges from approximately 0.5 to 2.5 volts.

The throttle housing unit is non-adjustable. If found to be faulty, replace as a complete unit.

After replacing the throttle housing, use BMW scan tool or equivalent to reset adaptation values.

> ***CAUTION—***
> - *If the adaptation process is not completed correctly, the engine will not start.*

– Switch ignition OFF and remove key.

– Remove air filter housing and ducts. See **Air filter housing and ducts, removing and installing** in this chapter.

– Remove charge air ducts. See **Charge air ducts, left side, removing and installing** in this chapter.

◄ Working at side of throttle assembly, disconnect line (**arrow**).

132

BentleyPublishers.com—All Rights Reserved

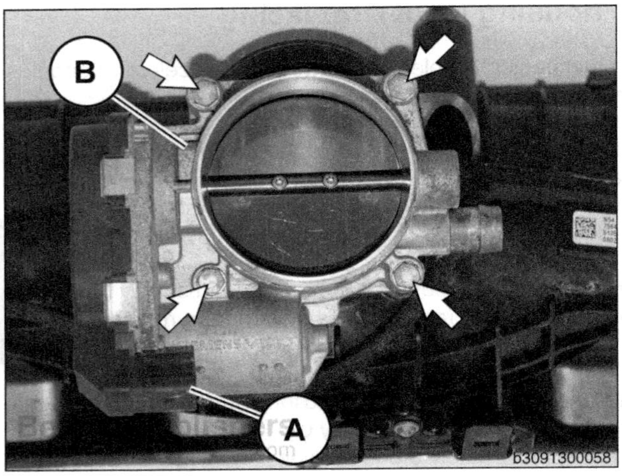

- Working under throttle assembly, disconnect electrical connector (**A**).
  - Remove fasteners (**arrows**) and remove throttle housing (**B**).

- Replace profile gasket (sealing O-ring) between throttle assembly and intake manifold.

Tightening torques	
Throttle housing to intake manifold	8 Nm (70 in-lb)

- Installation is reverse of removal.

- Reattach throttle assembly connector with care.

> ***CAUTION—***
> - *It is possible to twist the connector before plugging it in. This can cause damage to the harness and connector.*

- After reinstalling, check and clear fault codes from ECM memory. Reset throttle plate adaptation values following BMW scan tool on-screen instructions.

Fuel injector service

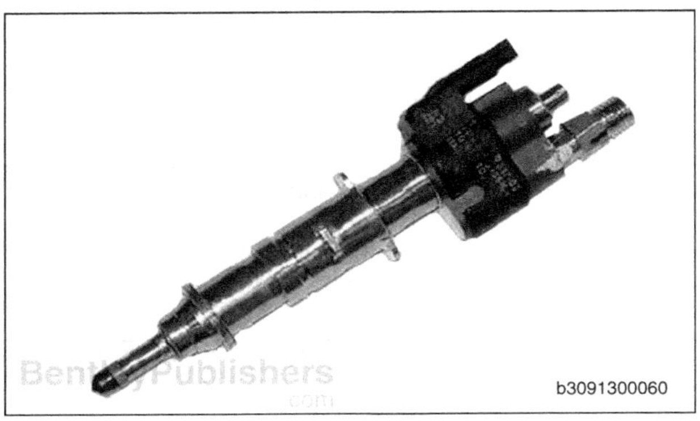

Fuel injectors are electrically controlled solenoid valves that provide precisely metered and atomized fuel into the engine intake ports. The N54 turbocharged engine utilizes a high precision injection (HPI) direct fuel injection system.

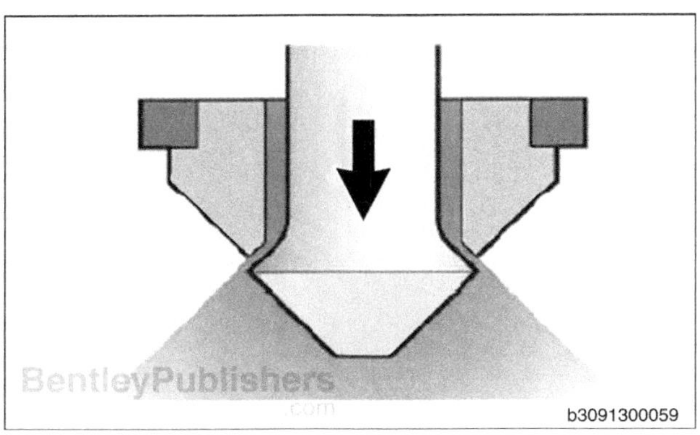

The HPI system uses piezo-electric fuel injectors. These injectors open in an outward direction, which forms a precise tapered spray pattern. With the aid of high system pressure (200 bar), the HPI system is capable of providing a high level of efficiency.

Voltage from the ECM controls injector opening by activating the injector valve piezo element. The ECM varies the duration (in milliseconds) of injector opening to regulate air / fuel ratio.

BentleyPublishers.com
BentleyPublishers.com—All Rights Reserved

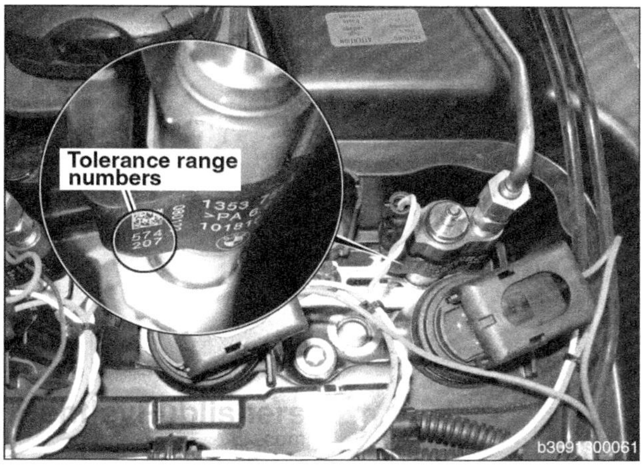

During manufacture, the tolerance ranges for injector-quantity adjustment are determined and specified in a six-digit number combination. Make note of tolerance numbers. They are required for fuel injector adaptation. See **Fuel injector adaptation** in this repair group.

Information on the lift performance of the injector is also added for injector voltage adjustment. Injector adjustment is required because of the individual voltage demand of each piezo actuator. This data is programmed in the ECM. During engine operation, these values are used to compensate for deviations in fuel metering.

When replacing an injector, it is essential to carry out an injector adjustment using BMW factory scan tool or equivalent.

If a fault is diagnosed in the system, such as failure of the high pressure sensor, the fuel-supply control valve is de-energized; fuel then bypasses into the rail. In the event of limp-home mode, turbocharging is deactivated by an opening of the wastegate valves.

Additional causes of limp-home mode can be:

- Implausible high-pressure sensor values
- Failure of the fuel-supply control valve
- Leakage in the high-pressure system
- Failure of the high-pressure pump
- Failure of the high-pressure sensor

Fuel injectors, removing

– Disconnect battery negative (-) cable in cargo compartment.

> ***CAUTION—***
> - *Prior to disconnecting the battery, read the battery disconnection cautions in* **001 Warnings and Cautions**.

– Remove upper engine covers. See **020 Maintenance**.

– Remove ignition coil at corresponding cylinder of fuel injector service. See **120 Ignition System**.

> ***CAUTION—***
> - *Maintain a high level of cleanliness when servicing ignition coils. Fuel, oil residue can break down the electrical resistance qualities of silicone used in ignition coil production, resulting in coil failure.*

BentleyPublishers.com—All Rights Reserved

Fuel injectors, removing

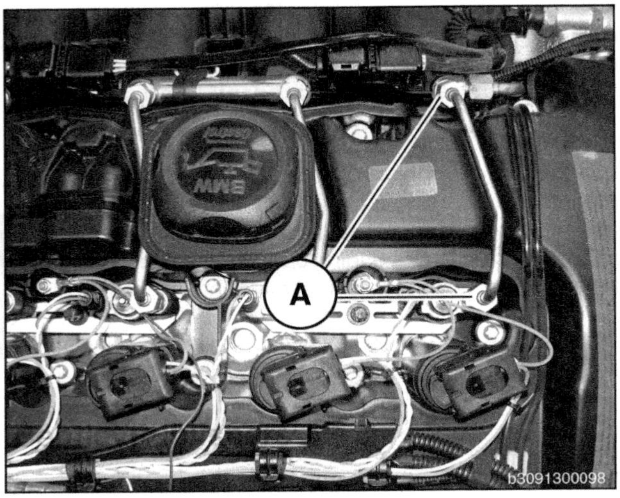

Remove fuel line to corresponding fuel injector (**A**).

> ***WARNING —***
> - *Fuel in the fuel line is under pressure (approx. 200 bar or 2900 psi) and may be expelled forcibly. Do not smoke or work near heaters or other fire hazards. Keep a fire extinguisher handy.*
> - *Work only on fuel system when engine temperature is below 40°C (104°F).*
> - *Unscrew the fuel tank cap to release pressure in the tank before working on the fuel line.*
> - *Plug open fuel lines and fittings.*
> - *Wrap a clean shop towel around fitting before disconnecting. Residual fuel pressure is present in the fuel line.*
> - *To reduce fuel pressure, place towel over injector and carefully tap with a wrench.*

> ***CAUTION—***
> - *Maintain a clean work area when servicing high pressure fuel system. Contaminants can cause a system malfunction.*
> - *Seal open fuel system lines with protective caps.*
> - *Replace high pressure fuel lines if tightened 10 times.*

NOTE —
- *If several injectors are removed, ensure that each injector is reinstalled in its original location (cylinder).*

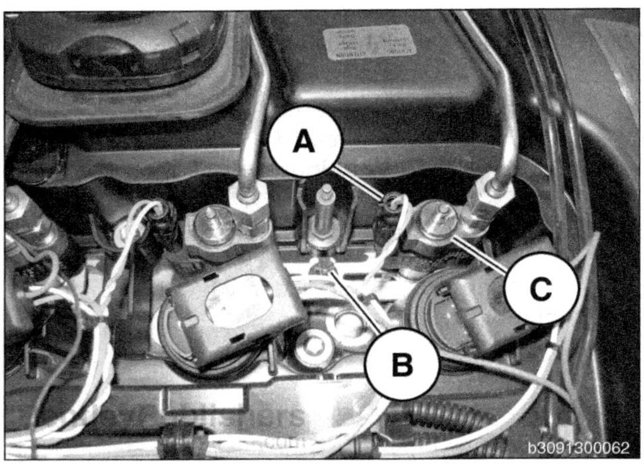

Unlock and remove electrical connector (**A**). Remove fastener and remove hold-down bracket (**B**). Remove injector (**C**).

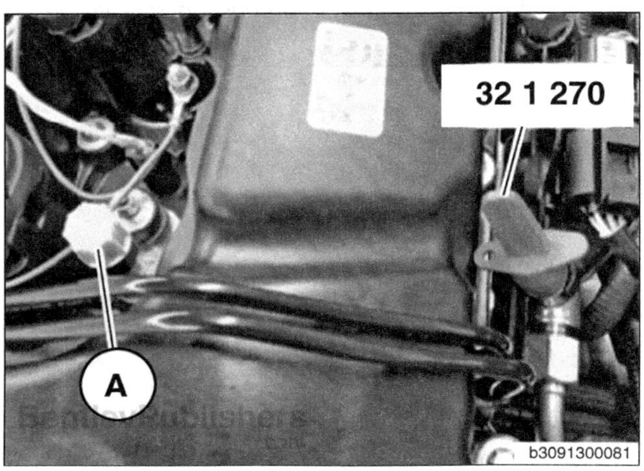

Seal injector (**A**) with cap and fuel rail connections with plug from special tool kit 32 1 270.

BentleyPublishers.com
BentleyPublishers.com—All Rights Reserved

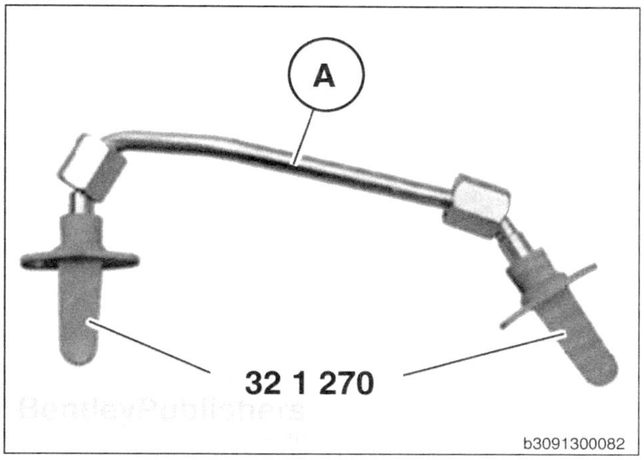

Seal high pressure fuel lines (**A**) with caps from special tool kit 32 1 270.

The following procedure can be used if of an injector is stuck in the cylinder head.

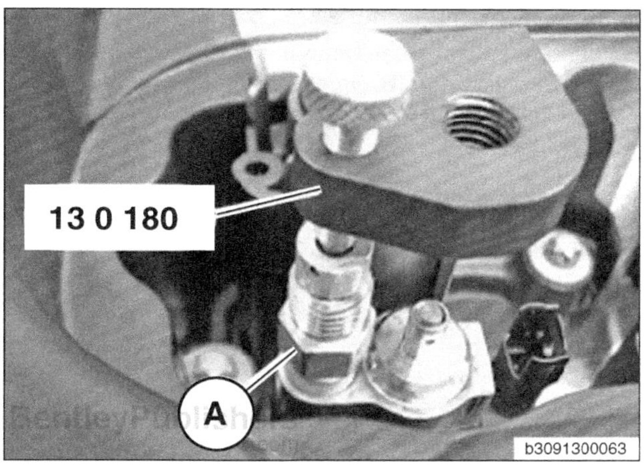

Install special tool 13 0 180 on injector (**A**).

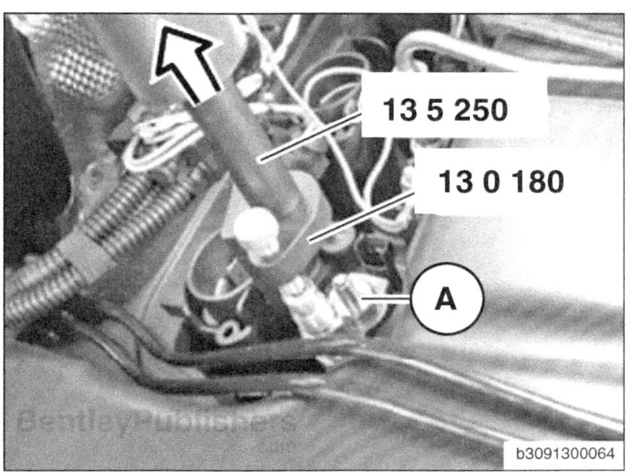

Install special tool 13 5 250 (slide hammer) and carefully knock fuel injector (**A**) out.

BentleyPublishers.com—All Rights Reserved

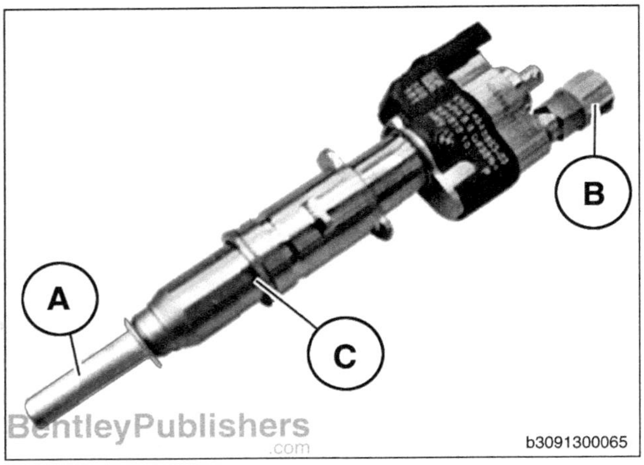

Once injector (**C**) is removed, install sealing caps to fuel line connection (**B**) and injector tip (**A**). Caps are available from BMW dealer parts department.

Injector bore, cleaning

Only use the following procedure if injector bores are dirty.

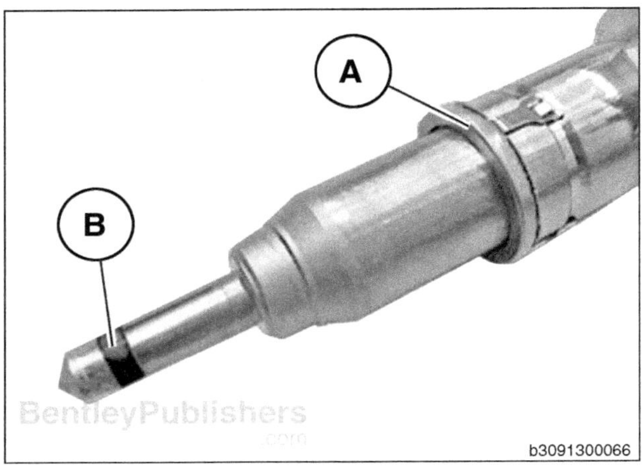

To clean injector bore, slide injector, without uncoupling element (**A**) but with new PTFE sealing ring (**B**), in and out of injector bores several times. Replace PTFE sealing ring after cleaning.

– Clean contact surfaces of uncoupling element and cylinder head. Replace uncoupling element (**A**).

Fuel injector PTFE seal, replacing

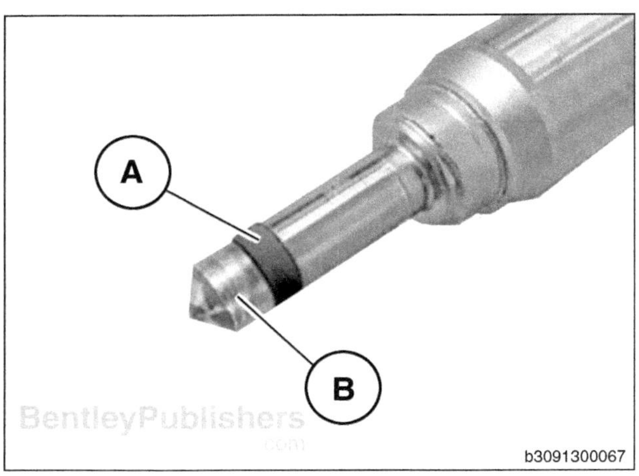

Before replacing PTFE sealing ring (**A**) make sure your hands and work surface are clean and free of oil. Avoid mechanical contact with injector tip (**B**).

BentleyPublishers.com
BentleyPublishers.com—All Rights Reserved

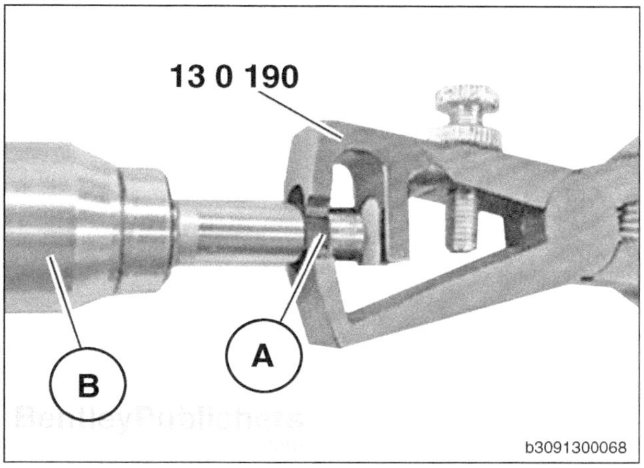

Using special tool 13 0 190, remove PTFE sealing ring (**A**) from injector (**B**). Use a lint-free cloth to remove combustion residue from cylindrical part of injector tip. Do not clean injector tip.

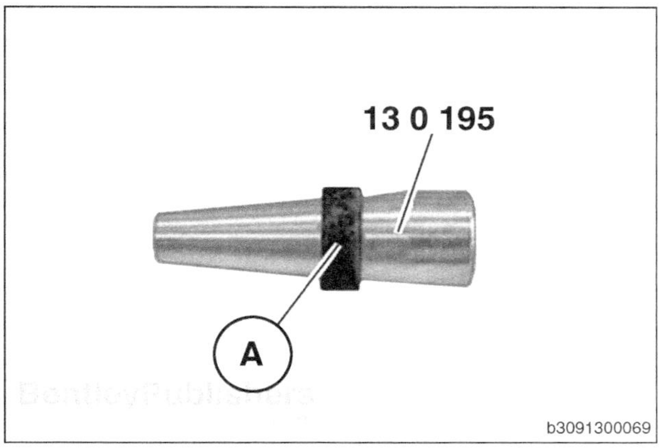

Slide new PTFE sealing ring (**A**) on mounting taper 13 0 195.

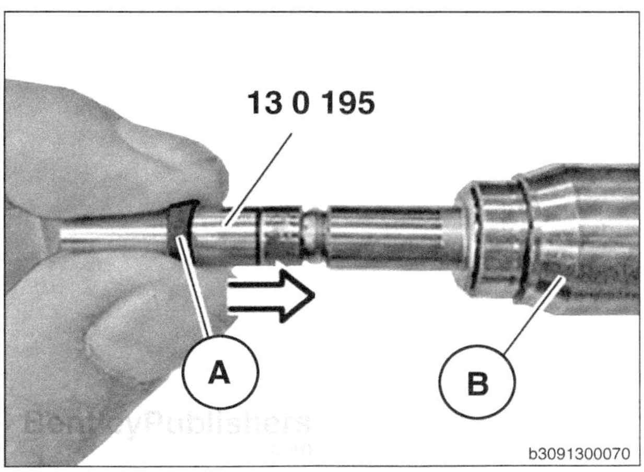

Using fingers and mounting taper (13 0 195) install PTFE sealing ring (**A**) on injector (**B**).

BentleyPublishers
BentleyPublishers.com–All Rights Reserved

Fuel injector PTFE seal, replacing

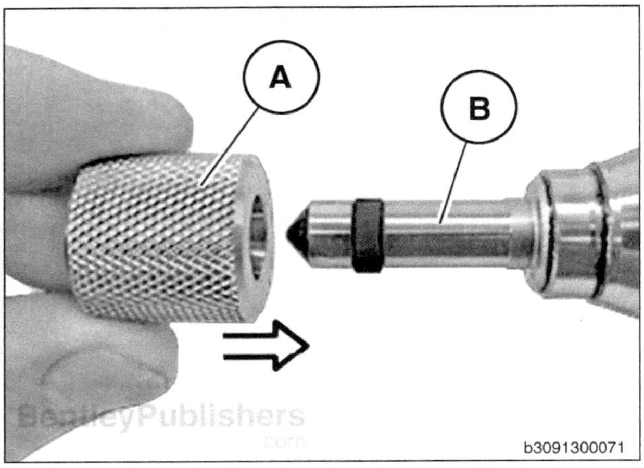

To bring the PTFE sealing ring to its installation dimension, slide three mounting sleeves (**A**) with decreasing diameters on injector (**B**).

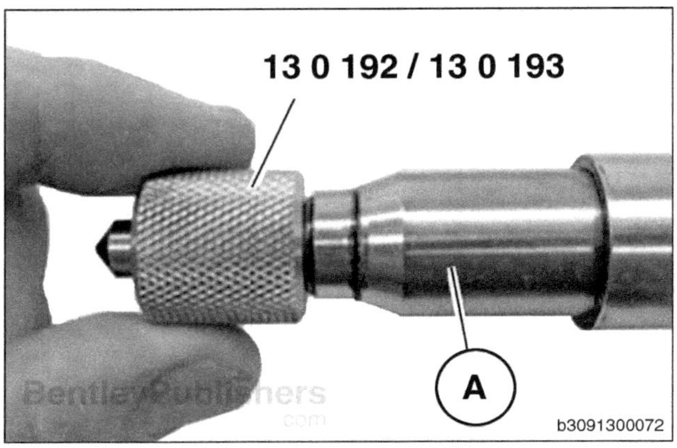

First slide 13 0 192 (large diameter sleeve) on injector (**A**). Then slide mounting sleeve 13 0 193 on injector (**A**).

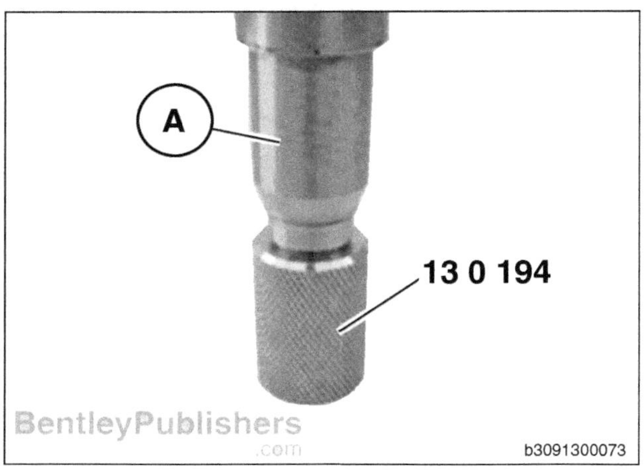

Next press injector (**A**) into mounting sleeve 13 0 194 (small diameter). Install injector within 10 minutes of removing mounting sleeve.

BentleyPublishers.com
BentleyPublishers.com—All Rights Reserved

Fuel injectors, installing

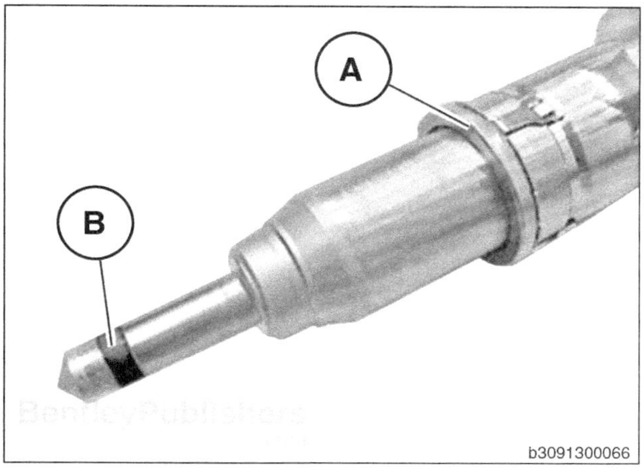

Remember to:

- Replace uncoupling element (**A**).
- Replace PTFE sealing ring (**B**).
- Remove PTFE protective cap a maximum of 10 minutes before installing. Sealing ring swells up once cap is removed.
- If injector adaptation is needed, note injector tolerance range numbers for each corresponding cylinder.

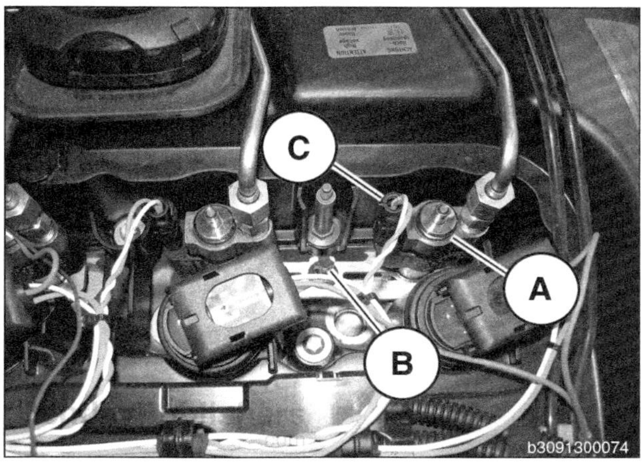

Working at cylinder head:

- Install injector (**A**) into cylinder head.
- Install hold-down bracket (**B**). Fastener (**arrow**) should be finger tight allowing injector (**A**) to be adjusted.
- Install electrical connector (**C**).

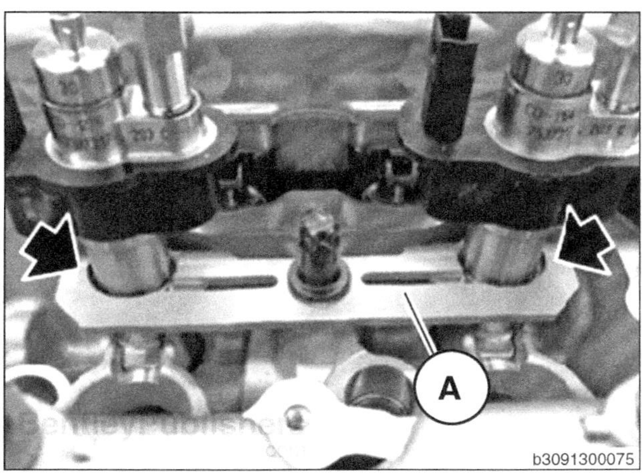

Check that hold-down bracket (**A**) is correctly installed (**arrows**).

BentleyPublishers
BentleyPublishers.com–All Rights Reserved

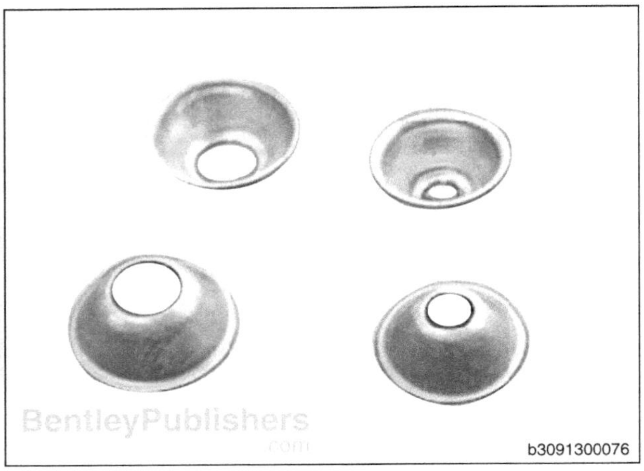

Copper seals on high pressure lines installed at factory are no longer needed. Remove and discard.

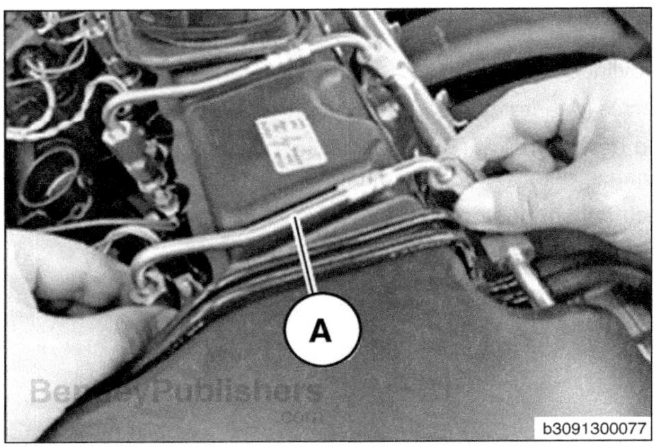

Connect high pressure line (**A**) finger tight.

- If reusing high pressure line, coat nut threads with gear oil.
- High pressure line may only be retightened 10 times.
- If both nuts on line cannot be installed easily by hand, adjust position of injector using following step.

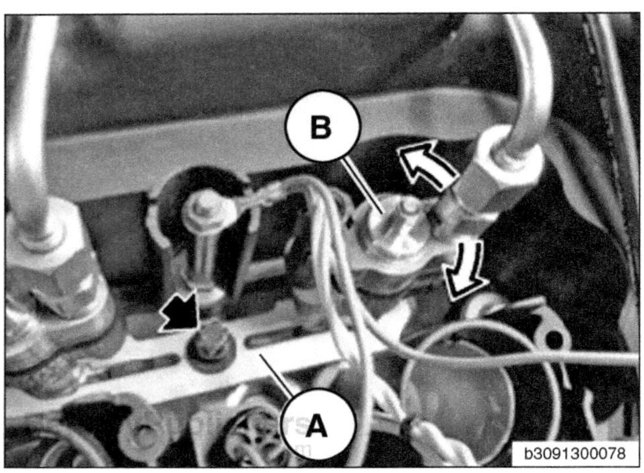

If fuel line union cannot be easily installed by hand, rotate injector in direction of **arrows** until high pressure line nuts can be easily tightened by hand.

– Tighten fastener for hold-down bracket (**arrow**).

Tightening torque	
Hold-down bracket to cylinder head	13 Nm (9.5 ft-lb)

BentleyPublishers.com
BentleyPublishers.com—All Rights Reserved

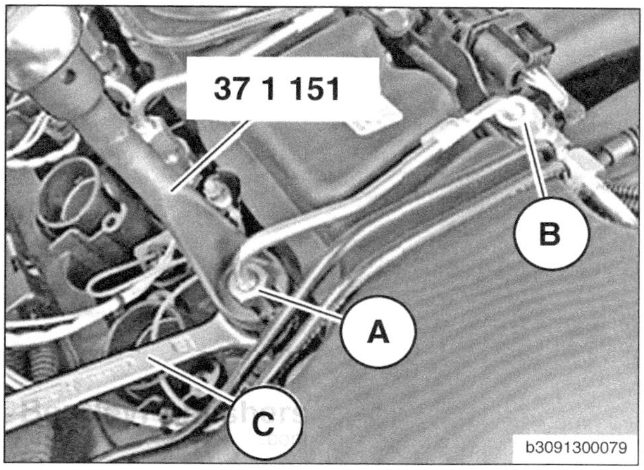

◀ Use following tightening sequence:

- Tighten nut (**A**) using special tool 37 1 151 while holding injector with wrench (**C**).
- Tighten nut (**B**) using special tool 37 1 151.

— Clean all fuel from ignition coil hole.

— Check fuel system for leaks.

— Perform injector adaptation. See **Fuel injector adaptation** in this repair group.

— Check and clear fault codes from ECM memory.

Tightening torque	
High pressure line to injector	23 Nm (17 ft-lb)
High pressure line to fuel rail	23 Nm (17 ft-lb)

Fuel injector adaptation

— Connect BMW scan tool and perform following tasks:

- Identify vehicle.
- Select **Function selection**.
- Select **Service functions**.
- Select **Adjustment programs**.
- Select **Adjust injectors**.
- Change **Test schedule**.

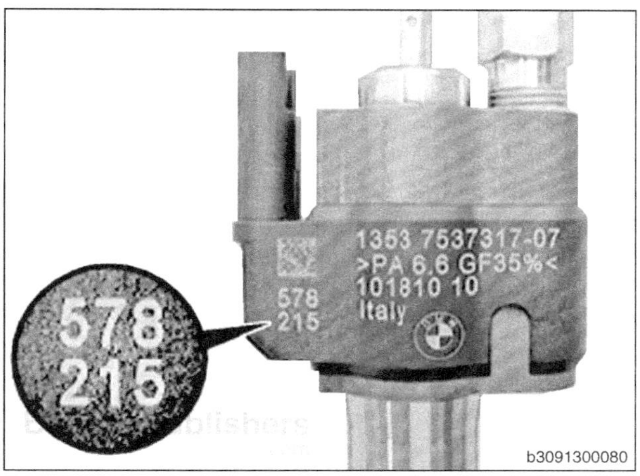

◀ For each replaced injector, enter fuel injector tolerance range (**inset**) to corresponding cylinder.

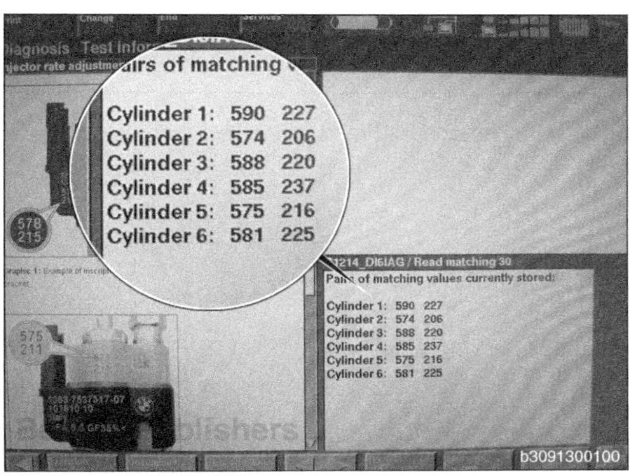

◀ Confirm correct tolerance range (**inset**) has been entered for each injector.

BentleyPublishers
BentleyPublishers.com—All Rights Reserved

Fuel rail, removing

– Disconnect battery negative (-) cable in trunk or cargo compartment.

> ***CAUTION—***
> - *Prior to disconnecting the battery, read the battery disconnection cautions in* **001 Warnings and Cautions**.

– Remove upper engine covers. See **020 Maintenance**.

– Remove ignition coils. See **120 Ignition System**.

> ***CAUTION—***
> - *Maintain a high level of cleanliness when servicing ignition coils. Fuel or oil residue causes a breakdown of the electrical resistance qualities of silicone used in production. Resulting in failure of the ignition coil.*

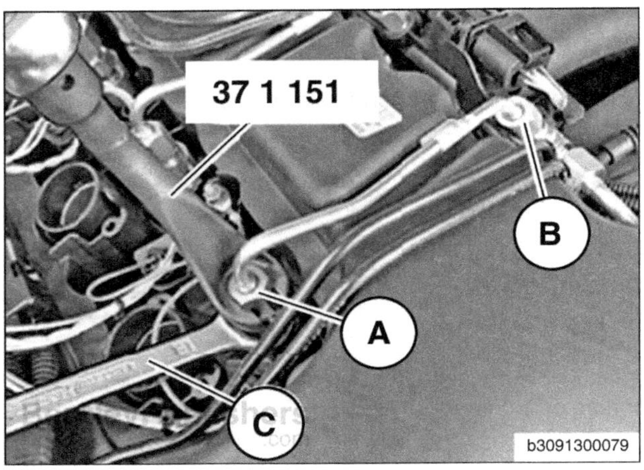

Label position and remove high pressure fuel lines:

- Using special tool 37 1 151 loosen nut (**A**) while holding injector with wrench.
- Using special tool 37 1 151, loosen nut (**B**).

> ***WARNING—***
> - *Fuel in the fuel line is under pressure (approx. 200 bar or 2900 psi) and may be expelled forcibly. Do not smoke or work near heaters or other fire hazards. Keep a fire extinguisher handy.*
> - *Work only on fuel system when engine temperature is below 40°C (104°F).*
> - *Unscrew the fuel tank cap to release pressure in the tank before working on the fuel line.*
> - *Plug open fuel lines and fittings.*
> - *Wrap a clean shop towel around fitting before disconnecting. Residual fuel pressure is present in the fuel line.*
> - *To reduce fuel pressure, place towel over injector and carefully tap with a wrench.*
> - *Protect alternator from dripping fuel and oil.*

> ***CAUTION—***
> - *Maintain a clean work area when servicing high pressure fuel system. Contaminants can cause a system malfunction.*
> - *Seal all fuel system opening with protective caps.*
> - *Replace high pressure fuel lines if tightened 10 times.*

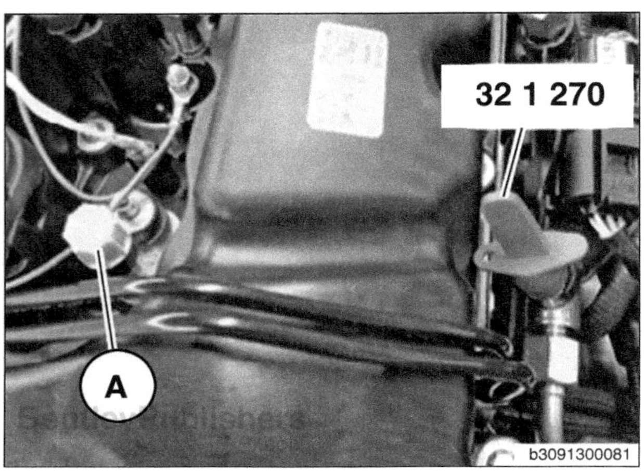

Seal injector (**A**) with cap and fuel rail connections with plug from special tool kit 32 1 270.

BentleyPublishers.com
BentleyPublishers.com—All Rights Reserved

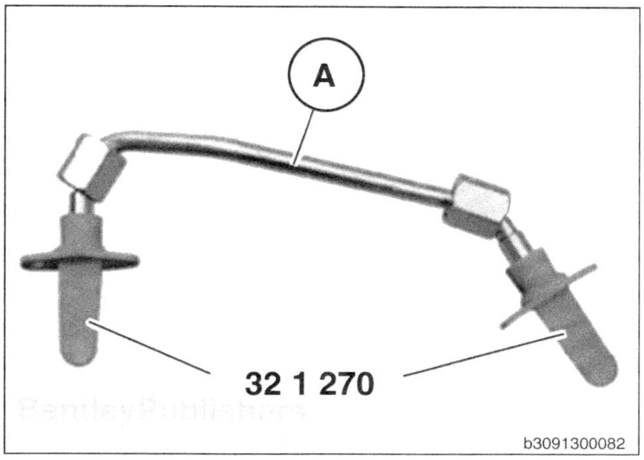

◄ Seal high pressure fuel lines (**A**) with caps from special tool kit 32 1 270.

◄ Unlock and remove electrical connector (**A**) from fuel rail.

– Remove electrical connector brackets from fuel rail.

◄ Detach fuel rail pressure sensor electrical connector (**arrow**).

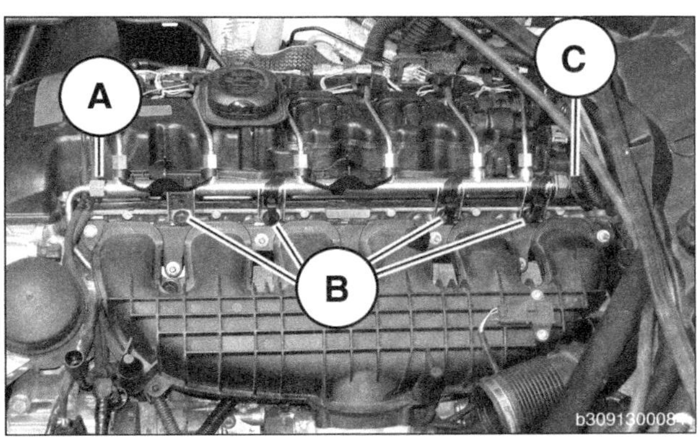

◄ Working at fuel injector rail:

- Loosen fuel line union nut (**A**).
- Plug open fuel lines and fittings.
- Wrap a clean shop towel around fitting before disconnecting. Residual fuel pressure is present in fuel line.
- Remove fasteners (**B**) and remove fuel rail (**C**).
- Seal connections at high pressure line (**A**) and fuel rail (**C**).

BentleyPublishers.com–All Rights Reserved

Fuel rail, installing

Install high pressure rail:

- Connect fuel line union (**A**) finger tight.
- Install fasteners (**B**) finger tight.
- Install electrical connector (**C**).

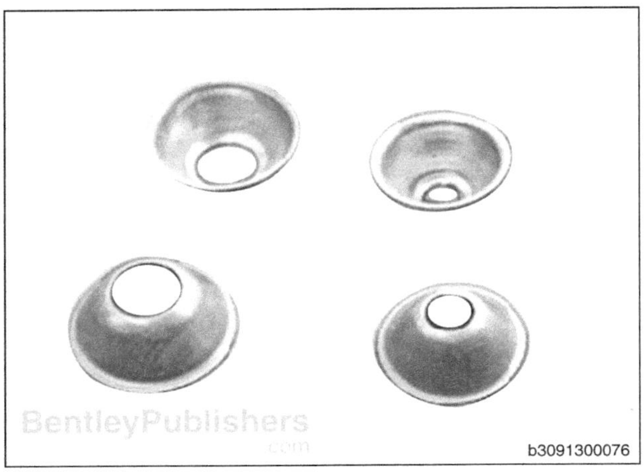

Copper seals on high pressure lines installed at factory are no longer needed. Remove and discard.

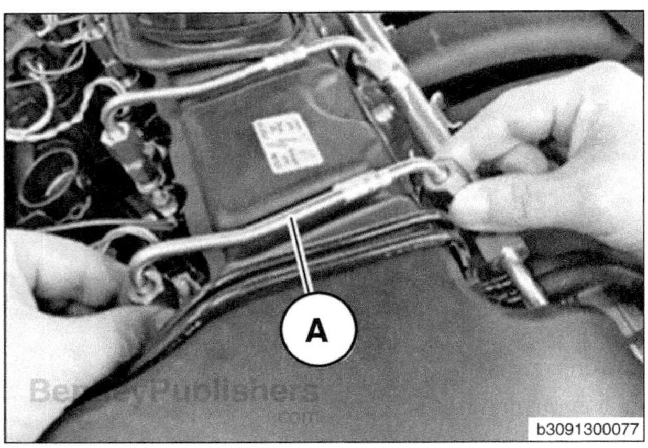

Connect high pressure line (**A**) finger tight.

- If reusing high pressure line, coat nut threads with gear oil.
- High pressure line may only be retightened 10 times.

BentleyPublishers.com
BentleyPublishers.com—All Rights Reserved

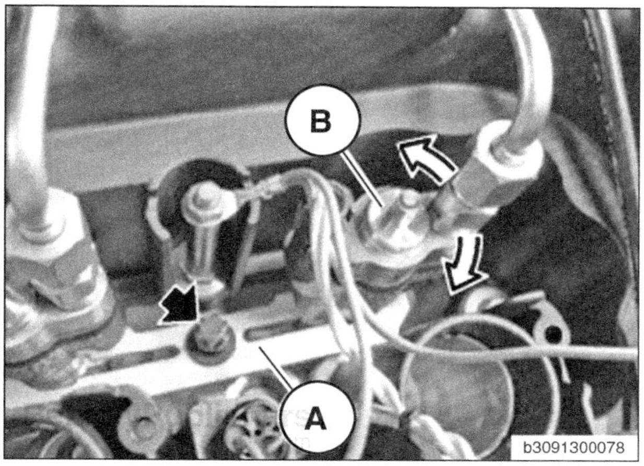

◀ If fuel lines cannot be easily installed by hand:

- Loosen fastener (**arrow**)
- Rotate injector in direction of **arrows** until high pressure line nuts can be easily tightened by hand.

– Tighten fastener for hold-down bracket (**arrow**).

Tightening torque	
Hold-down bracket to cylinder head	13 Nm (9.5 ft-lb)

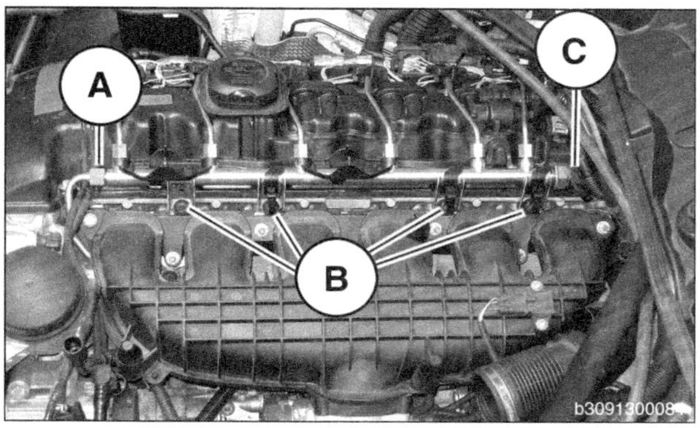

◀ Tighten fuel injector rail (**C**) fasteners (**B**).

– Tighten fuel line nut (**A**).

Tightening torque	
Fuel injector rail to cylinder head	13 Nm (9.5 ft-lb)
High pressure line to fuel rail	30 Nm (22 ft-lb)

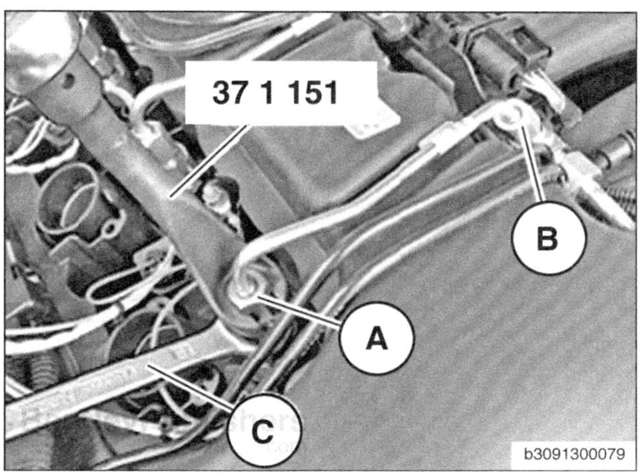

◀ Use following tightening sequence:

- Tighten nut (**A**) using special tool 37 1 151 while counter holding injector with wrench (**C**).
- Tighten nut (**B**) using special tool 37 1 151.

– Clean all fuel from ignition coil hole and install ignition coil.

– Check fuel system for leaks.

– Check and clear fault codes from ECM memory.

Tightening torque	
High pressure line to injector	23 Nm (17 ft-lb)
High pressure line to fuel rail	23 Nm (17 ft-lb)

BentleyPublishers.com—All Rights Reserved

Intake manifold, removing and installing

– Disconnect battery negative (-) cable in trunk or cargo compartment.

> ***CAUTION—***
> *• Prior to disconnecting the battery, read the battery disconnection cautions in* **001 Warnings and Cautions**.

– Remove upper engine covers. See **020 Maintenance**.

– Remove complete air filter housing and air intake ducts. See **Air filter housing and ducts, removing and installing** in this repair group.

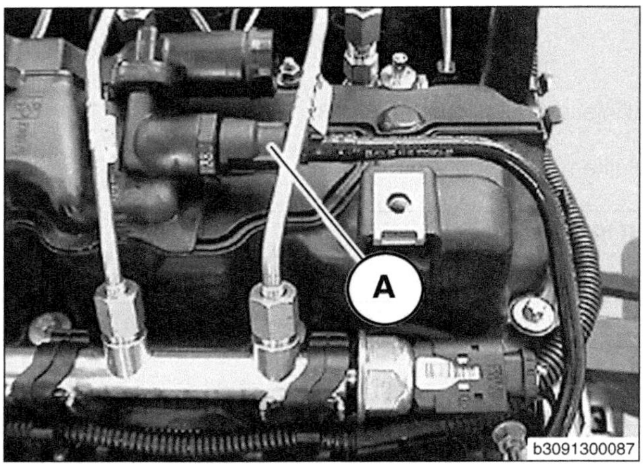

◄ Disconnect crankcase breather line (**A**) at cylinder head cover if equipped.

NOTE—
• Not all models have additional line.

◄ Example of N54 engine without additional crankcase breather line. There is a plug (**arrow**) in its place.

– Remove throttle housing. See **Throttle housing (EDK), replacing** in this repair group.

BentleyPublishers.com
BentleyPublishers.com—All Rights Reserved

Working at right side underneath intake manifold:

- Disconnect vacuum hose (**A**), if equipped.
- Remove fuel tank vent valve (**B**) from mounting bracket.
- Remove vacuum hose (**C**) and lay aside.

Slide wiring harness junction box (**A**) off bracket under intake manifold.

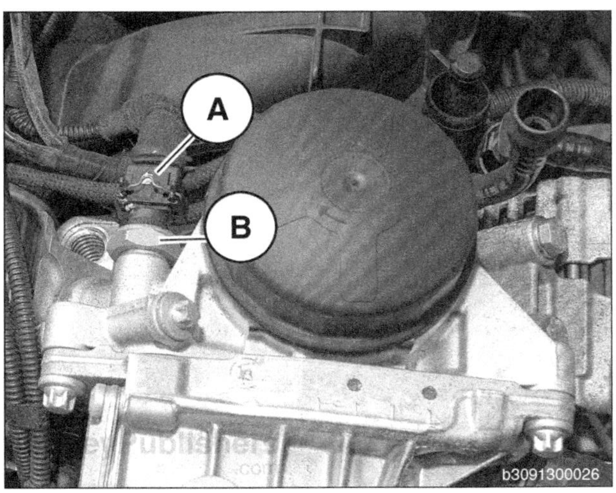

Disconnect electrical connector (**A**) at oil pressure switch (**B**).

BentleyPublishers.com
BentleyPublishers.com—All Rights Reserved

Working at top of intake manifold, unhook electrical harness retainers (**A**) and place harness out of way.

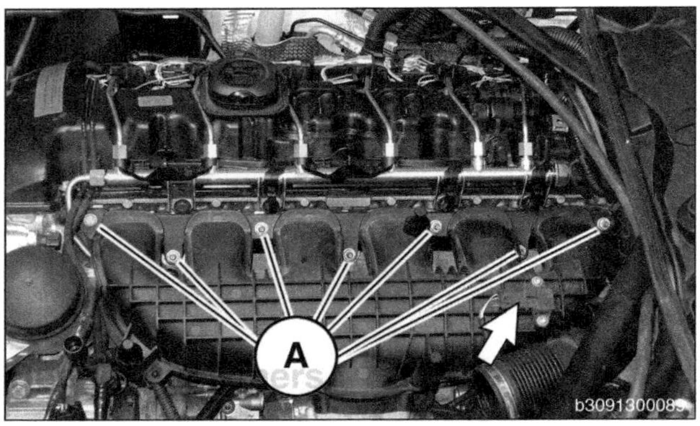

Remove electrical connector (**arrow**) and remove fasteners (**A**).

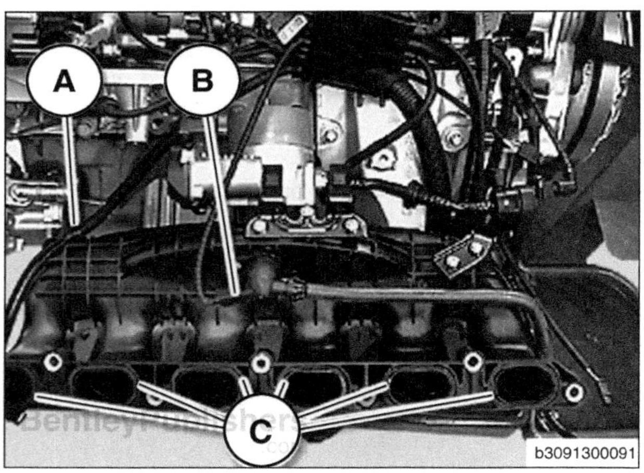

Working at backside of intake manifold:

- Cut cable tie (**A**).
- Disconnect electrical connector (**B**), if equipped

CAUTION—
- *Plug open intake ports to prevent parts or debris from falling into the engine intake.*

– Installation is reverse of removal. Remember to replace intake manifold seals (**C**) if necessary.

Tightening torque	
Manifold to cylinder head	15 Nm (11 ft-lb)

BentleyPublishers.com
BentleyPublishers.com—All Rights Reserved

Crankcase breather valve

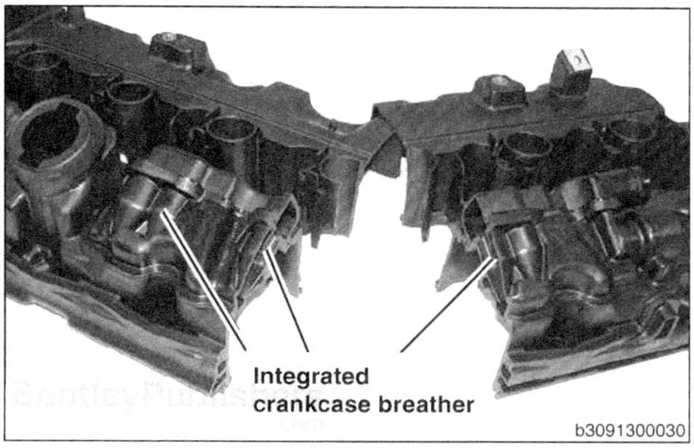

The crankcase breather valve is part of the cylinder head cover and is not serviceable as a separate component.

Oil separation is carried out via a labyrinth system and two cyclone separators incorporated into the cylinder head cover. By having the system components integrated into the cylinder head cover, the crankcase gases are heated by the engine.

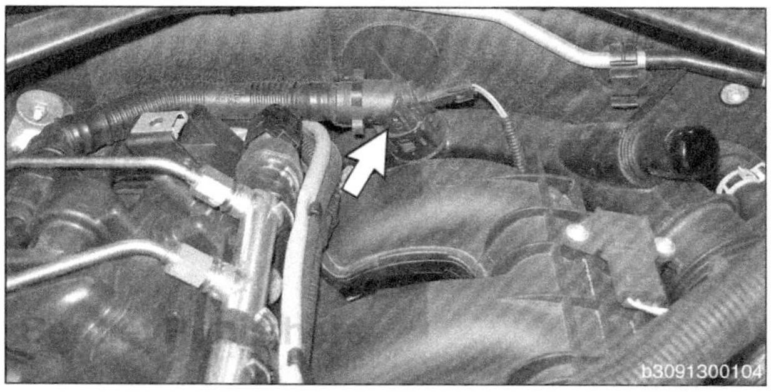

However, there is one electric heating element at the manifold inlet (**arrow**). Once the liquid oil is separated from the crankcase vapors, the oil is allowed to drain back through check valves back into the engine.

If the diaphragm valve in the breather housing leaks, full intake vacuum may be applied to the crankcase, resulting in excessive oil consumption, irregular idle, whistling or howling noises or oil smoke in the exhaust.

Intake vacuum specifications for N54 engines

The breather valve sets a pressure range of 7 to 11 mbar.

A faulty crankcase breather valve causes significant deviations from the specified value:

- Crankcase ventilation clogged: Ambient pressure +100 mbar
- Crankcase ventilation internal leak: Ambient pressure -170 mbar

BentleyPublishers.com—All Rights Reserved

Engine coolant temperature (ECT) sensor, replacing

The engine coolant temperature (ECT) sensor is a negative temperature coefficient (NTC) sensor. As coolant temperature rises, resistance through the sensor decreases.

The ECM varies ignition timing and air / fuel mixture based on engine coolant temperature. The ECT sensor is supplied a 5 volt reference voltage. The voltage drop across the sensor varies as the coolant temperature (sensor resistance) changes.

If the ECT sensor input is faulty or not plausible, the MIL is illuminated when OBD II fault criteria are exceeded. The ECM assumes a substitute value (80°C / 176°F) to maintain engine operation. Ignition timing is set to a safely conservative basic setting.

The ECT sensor is located at front of cylinder head.

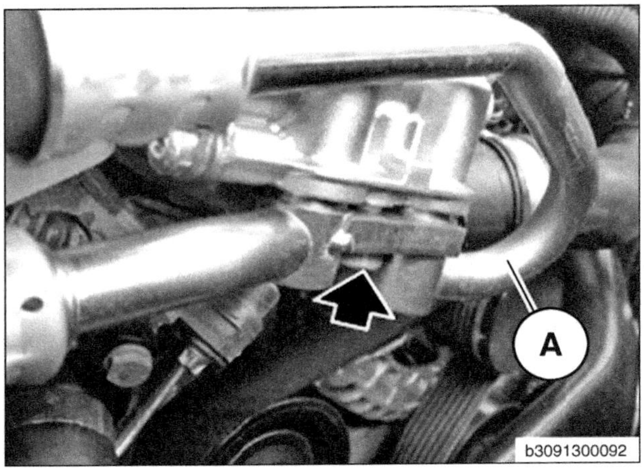

With engine fully cooled off, remove fastener (**arrow**) and line (**A**).

> ***CAUTION—***
> *• Be prepared to catch any dripping coolant*

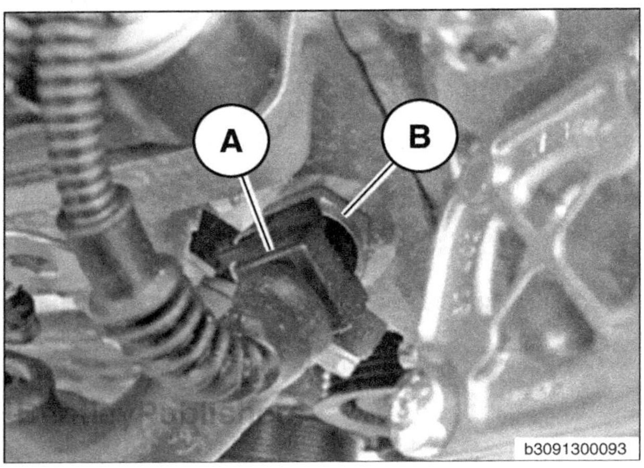

Remove electrical connector (**A**). Unscrew ECT sensor (**B**) from cylinder head. Be prepared to catch small amount of coolant.

– Installation is reverse of removal:
  - Use new copper sealing washer when installing sensor.
  - Replace O-rings on coolant line.
  - Replace lost coolant.

Tightening torque	
Temperature sensor to cylinder head	19 Nm (14 ft-lb)
Coolant pipe to cylinder head	19 Nm (14 ft-lb)

– After reinstalling, check and clear fault codes from ECM memory.

Radiator outlet coolant temperature sensor

The radiator outlet temperature sensor (**inset**) is a negative temperature coefficient (NTC) sensor. As coolant temperature rises, resistance through the sensor decreases.

The ECM varies cooling fan speed based on radiator outlet coolant temperature. The radiator outlet coolant temperature sensor is supplied a 5 volt reference voltage. The voltage drop across the sensor varies as the sensor varies as the coolant temperature (sensor resistance) changes.

BentleyPublishers.com
BentleyPublishers.com—All Rights Reserved

DME main relay, testing

The DME main relay is energized via the ECM and supplies battery positive (B+) power to the following:

- ECM
- Ignition coils
- Tailpipe flap
- E-box cooling fan
- Fuel tank leak detection pump
- Crankshaft sensor
- Camshaft sensors
- Evaporative emissions valve
- Crankcase breather valve heater
- Oil condition sensor
- Electric coolant pump
- Thermostat
- Oxygen sensor heaters

If the DME main relay is faulty, the engine does not start.

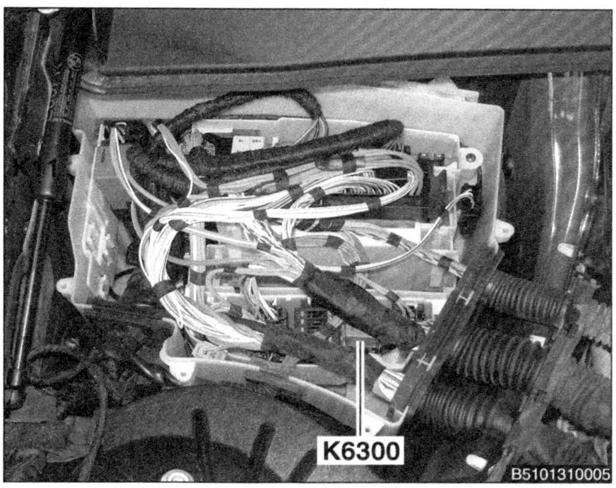

◄ With ignition off, remove DME main relay (**K6300**) in electronics box (E-box) at right rear of engine compartment.

NOTE—

- *Relay locations can vary. Confirm relay identification by matching wiring colors and terminal numbers. See* ***ELE Electrical Wiring Diagrams****.*

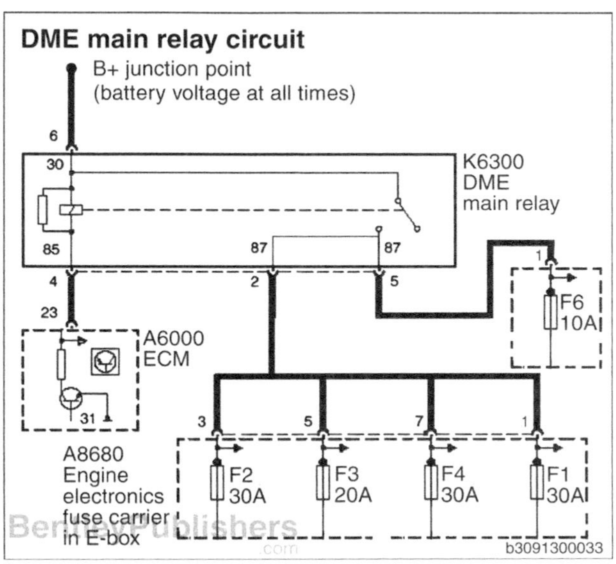

◄ Check for voltage at terminal **6** of relay socket (corresponds to terminal **30** on relay).

– If battery voltage is present, continue testing.

– If battery voltage is not present, check the following:
  - Large red wire in relay socket
  - A2076 (battery positive junction) in E-box, right rear of engine compartment
  - Fuse F8 in fuse and relay panel behind glove compartment. See **Power supply fuses** in this repair group.
  - Wiring can vary. See **ELE Electrical Wiring Diagrams** for more details.

– Check for ground at relay socket **4** (corresponds to terminal **85** on relay).

– If ground is present, continue testing.

– If ground is not present, signal from ECM is missing. Check wire between relay and ECM 44-pin connector X60005. See **ECM pin assignments** in this repair group.

BentleyPublishers.com
BentleyPublishers.com—All Rights Reserved

- If no faults are found:
  - Check ECM grounding.
  - ECM may be defective.

- With ignition ON and relay installed, check for battery voltage at relay sockets **2** and **5** (correspond to terminals **87** on relay).

NOTE *—*
- *In some models there is only one terminal 87.*

- If battery voltage is present, relay has energized and is functioning correctly.

- If battery voltage is not present and all earlier tests are OK, relay is faulty. Replace.

- If no faults are found during relay testing but power is not reaching ECM or other components, check fuses in engine electronics fuse carrier (A8680) in E-box. See **Power supply fuses** in this repair group.

- When finished testing, check and clear fault codes from ECM memory.

BentleyPublishers.com
BentleyPublishers.com—All Rights Reserved

ECM pin assignments

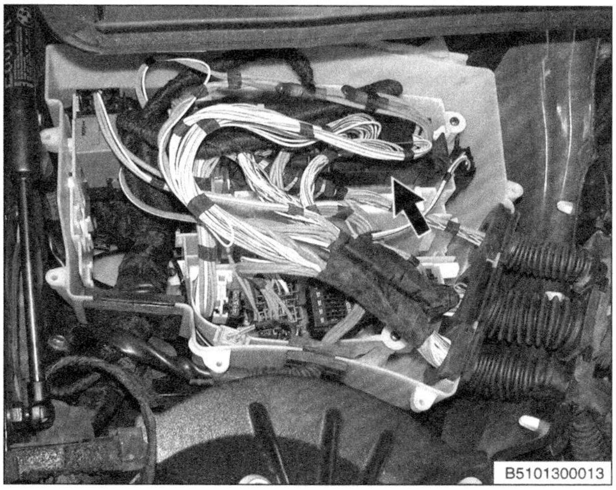

The engine control module (ECM) (**arrow**) is located in the right rear of the engine compartment in the E-box.

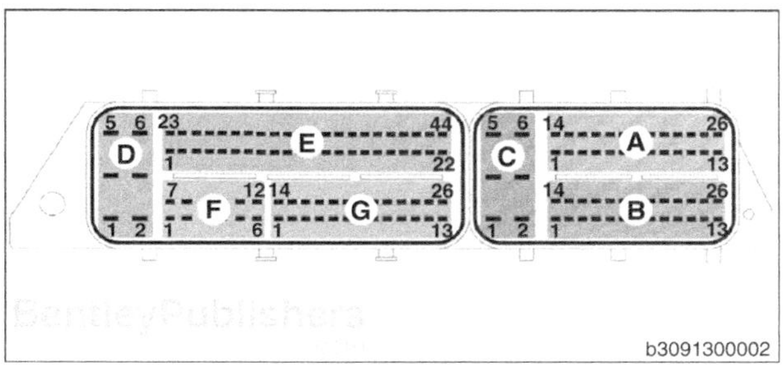

The ECM has 2 main electrical harness connectors, one with 4 modular connections and the other with 3 for a total of 7 sub connectors and 147 pins.

- **A:** X60001 (signals)
- **B:** X60002 (signals)
- **C:** X60003 (voltage, ground supply)
- **D:** X60004 (ignition coils)
- **E:** X60005 (signals)
- **F:** X60006 (fuel injectors)
- **G:** X60007(signals)

ECM pin assignments for turbo engines are given in **Table b**. This information can be helpful when diagnosing faults to or from the ECM.

Generally, absence of voltage or continuity means there is a wiring or connector problem. Test results with incorrect values do not necessarily mean that a component is faulty. Check for loose, broken or corroded connections and wiring before replacing components. For engine management system electrical schematics, see **ELE Electrical Wiring Diagrams**.

When making checks at the ECM, use a break-out box to allow tests to be made with the connector attached to the ECM. This prevents damage to the small terminals in the connector. As an alternative, the connector housing can be separated so that electrical checks can be made from the back of the connector.

CAUTION—

- *Wait at least one minute after switching the ignition OFF before removing the connector from the ECM. If the connector is removed before this time, residual power in the system relay may damage the ECM.*
- *Connect or disconnect the control module connector and meter probes with the ignition OFF.*

BentleyPublishers.com
BentleyPublishers.com—All Rights Reserved

Table b. Siemens DME MSD80 / 81 ECM pin assignments			
Connector X60001 26-pin (signals)			
Pin	**Type**	**Component or function**	**Note**
1	input/output	Signal PT-CAN low	Powertrain CAN-bus
2	input	Start signal	Car access system
3	input/output	BSD signal	Battery sensor
4	input	Brake light signal	Brake light switch
5	-	Not used	
6	input	Radiator outlet temperature sensor signal	Temperature sensor at radiator outlet
7	input	Accelerator pedal module signal	Accelerator pedal module
8	output	Electric fan signal	Electric engine cooling fan
9	-	Not used	
10	ground	Accelerator pedal module	Accelerator pedal module
11	output	Accelerator pedal module supply voltage	Accelerator pedal module
12	-	Not used	
13	-	Not used	
14	input/output	Signal PT-CAN high	Powertrain CAN-bus
15	input	EWS signal	Car access system
16	input	Brake light test signal	Brake light switch
17	input	Right rear wheel speed sensor signal	Connector X10186 (right rear wheel speed sensor)
18	input	Clutch switch signal	Clutch switch module
19	input	Radiator outlet temperature sensor signal	Temperature sensor at radiator outlet
20	input	Accelerator pedal module signal	Accelerator pedal module
21	input/output	TD signal	TD signal connector
22	-	Not used	
23	ground	Accelerator pedal module	Accelerator pedal module
24	output	Accelerator pedal module supply voltage	Accelerator pedal module
25	-	Not used	
26	-	Not used	

Connector X60002 26-pin (signals)			
Pin	**Type**	**Component or function**	**Note**
1	input	Wake up signal, terminal 15	Terminal 15 wake up connector
2	-	Not used	
3	-	Not used	
4	-	Not used	
5	input	Oxygen sensor signal	Bank 2 sensor 1
6	input	Oxygen sensor signal	Bank 1 sensor 1
7	input	Oxygen sensor signal	Bank 2 sensor 1

BentleyPublishers.com
BentleyPublishers.com—All Rights Reserved

Connector X60002 26-pin (signals) (continued)			
Pin	**Type**	**Component or function**	**Note**
8	input	Oxygen sensor signal	Bank 1 sensor 1
9	input	Oxygen sensor signal	Bank 2 sensor 1
10	ground	Oxygen sensor	Bank 1 sensor 1
11	ground	Oxygen sensor	Bank 2 sensor 1
12	input	Oxygen sensor heating	Bank 1 sensor 1
13	input	Oxygen sensor heating	Bank 2 sensor 1
14	-	Not used	
15	-	Not used	
16	input	DMTL pump signal	Diagnosis module for fuel tank leakage
17	input	DMTL heating signal	Diagnosis module for fuel tank leakage
18	input	Oxygen sensor signal	Bank 1 sensor 1
19	input	Oxygen sensor signal	Bank 2 sensor 2
20	input	Oxygen sensor signal	Bank 1 sensor 2
21	-	Not used	
22	-	Not used	
23	ground	Oxygen sensor	Bank 1 sensor 2
24	input	Oxygen sensor	Bank 2 sensor 2
25	input	Oxygen sensor	Bank 2 sensor 2
26	output	Oxygen sensor	Bank 1 sensor 2

Connector X60003 6-pin (voltage, ground supply)			
Pin	**Type**	**Component or function**	**Note**
1	input	Terminal 30	B+ potential distributor
2	input	Terminal 87	Fuse F03
3	ground	Ground	Ground point
4	ground	Ground	Ground point
5	ground	Ground	Ground point
6	ground	Ground	Ground point

Connector X60004 6-pin (ignition coils)			
Pin	**Type**	**Component or function**	**Note**
1	output	Ignition coil signal	Ignition coil signal 1
2	output	Ignition coil signal	Ignition coil signal 2
3	output	Ignition coil signal	Ignition coil signal 3
4	output	Ignition coil signal	Ignition coil signal 4
5	output	Ignition coil signal	Ignition coil signal 5
6	output	Ignition coil signal	Ignition coil signal 6

BentleyPublishers.com—All Rights Reserved

Connector X60005 44-pin (signals)			
Pin	**Type**	**Component or function**	**Note**
1	-	Not used	
2	-	Not used	
3	output	Volume control valve	Control signal
4	-	Not used	
5	-	Not used	
6	-	Not used	
7	output	Low pressure fuel sensor ground	Low pressure fuel sensor
8	output	Low pressure fuel sensor voltage supply	Low pressure fuel sensor
9	input	Boost pressure sensor signal	Boost pressure sensor
10	input	Boost pressure sensor signal	Boost pressure sensor
11	output	Boost pressure sensor voltage supply	Boost pressure sensor
12	-	Not used	
13	output	DME main relay activation	DME main relay
14	output	Throttle valve voltage supply	Throttle valve
15	input	Throttle valve signal	Throttle valve
16	input	Throttle valve signal	Throttle valve
17	input	Intake air temperature signal	Integral to boost pressure sensor
18	-	Not used	
19	input	Knock sensor signal	Knock sensor
20	input	Knock sensor signal	Knock sensor
21	-	Not used	
22	-	Not used	
23	output	Fuel tank vent valve signal	Fuel tank vent valve
24	-	Not used	
25	-	Not used	
26	-	Not used	
27	-	Not used	
28	-	Not used	
29	input	Crankshaft position sensor signal	Crankshaft position sensor
30	ground	Crankshaft position sensor	Crankshaft position sensor
31	output	Intake manifold pressure sensor voltage supply	Intake manifold pressure sensor
32	ground	Intake manifold pressure sensor	Intake manifold pressure sensor
33	input	Intake manifold pressure sensor signal	Intake manifold pressure sensor
34	input	Low pressure fuel sensor signal	Low pressure fuel sensor
35	input/output	BSD signal	BSD signal connector
36	input	Throttle valve signal	Throttle valve
37	input	Throttle valve signal	Throttle valve

BentleyPublishers.com
BentleyPublishers.com—All Rights Reserved

Connector X60005 44-pin (signals) (continued)			
Pin	**Type**	**Component or function**	**Note**
38	ground	Throttle valve	Throttle valve
39	-	Not used	
40	-	Not used	
41	input	Knock sensor signal	Knock sensor
42	input	Knock sensor signal	Knock sensor
43	-	Not used	
44	-	Not used	

Connector X60006 12-pin (ignition coils)			
Pin	**Type**	**Component or function**	**Note**
1	output	Fuel injector voltage supply	Fuel injector 1
2	output	Fuel injector voltage supply	Fuel injector 2
3	output	Fuel injector voltage supply	Fuel injector 3
4	output	Fuel injector voltage supply	Fuel injector 4
5	output	Fuel injector voltage supply	Fuel injector 5
6	output	Fuel injector voltage supply	Fuel injector 6
7	output	Fuel injector trigger	Fuel injector 1
8	output	Fuel injector trigger	Fuel injector 2
9	output	Fuel injector trigger	Fuel injector 3
10	output	Fuel injector trigger	Fuel injector 4
11	output	Fuel injector trigger	Fuel injector 5
12	output	Fuel injector trigger	Fuel injector 6

Connector X60007 26-pin (signals)			
Pin	**Type**	**Component or function**	**Note**
1	-	Not used	
2	-	Not used	
3	-	Not used	
4	input	Engine coolant temperature signal	Engine coolant temperature sensor
5	output	Intake VANOS solenoid signal	Intake VANOS solenoid
6	-	Not used	
7	output	Wastegate control valve signal	Wastegate control valve
8	input	Fuel rail pressure sensor signal	Fuel rail pressure sensor
9	output	Fuel rail pressure sensor signal	Fuel rail pressure sensor
10	output	Fuel rail pressure sensor voltage supply	Fuel rail pressure sensor
11	input	Camshaft position sensor 1 signal	Camshaft position sensor 1
12	input	Camshaft position sensor 2 signal	Camshaft position sensor 2
13	input	Oil pressure switch signal	Oil pressure switch

BentleyPublishers.com
BentleyPublishers.com—All Rights Reserved

Connector X60007 26-pin (signals) (continued)			
Pin	**Type**	**Component or function**	**Note**
14	output	Wastegate control valve signal	Wastegate control valve signal
15	-	Not used	
16	-	Not used	
17	ground	Engine coolant temperature sensor	Engine coolant temperature sensor
18	output	Exhaust VANOS solenoid signal	Exhaust VANOS solenoid
19	input	Thermostat signal	Thermostat
20	-	Not used	
21	-	Not used	
22	-	Not used	
23	-	Not used	
24	ground	Camshaft position sensor 1	Camshaft position sensor 1
25	ground	Camshaft position sensor 2	Camshaft position sensor 2
26	input/output	BSD signal	Electric coolant pump

BentleyPublishers.com
BentleyPublishers.com—All Rights Reserved

133 Fuel Injection ME 9.2, ME 9.2.2 (V8 Engine)

TABLE

GENERAL

This repair group covers service and repair of engine management systems for 5 Series models with V8 engine and Bosch DME ME 9.2 and ME 9.2.2.

See also:

- **020 Maintenance** for air filter, fuel filter and spark plug replacement
- **100 Engine–General** for engine identification and application
- **120 Ignition System** for ignition coil, camshaft sensor, crankshaft sensor and knock sensor service
- **160 Fuel Tank and Fuel Pump** for fuel pump and fuel pump relay service, fuel pressure tests and fuel pressure regulator replacement
- **180 Exhaust System** for oxygen sensor service
- **ECL Electrical Component Locations**
- **ELE Electrical Wiring Diagrams**
- **OBD On-Board Diagnostics**

Bentley Publishers
BentleyPublishers.com–All Rights Reserved

DME applications

BMW 5 Series models are equipped with digital motor electronics (DME), also known as Motronic. In these systems, fuel injection and ignition are controlled by an integrated engine control module (ECM). Application information for DME systems is in **Table a**.

Table a. 5 Series engine management applications

Year, model	Engine code	Engine management	Features
2004-2005			
Sedan: 525i, 530i	M54	Siemens MS 45.1	Double VANOS
Sedan: 545i	N62	Bosch DME ME 9.2 with Valvetronic control module	Valvetronic, VANOS
2006 - 2007			
Sedan, Sports Wagon: 525i, 525xi, 530i, 530xi Sports Wagon	N52	Siemens MSV70	Valvetronic, VANOS
2008 - 2010			
Sedan, Sports Wagon: 528i, 528xi	N52KP	Siemens MSV80	Valvetronic, VANOS
Sedan, Sports Wagon: 535i, 535xi	N54	Siemens MSD80 (to 03 / 2008) MSD81 (from 03 / 2008)	VANOS, turbocharged
2004-2005			
Sedan: 545i	N62	Bosch DME ME 9.2 with Valvetronic control module	Valvetronic, VANOS
2006 - 2010			
Sedan: 550i	N62TU	Bosch DME ME 9.2.2	Valvetronic, VANOS

BentleyPublishers.com
BentleyPublishers.com—All Rights Reserved

Warnings and Cautions

WARNING —

- *The fuel system is designed to retain pressure even when the ignition is OFF. When working with the fuel system, loosen the fuel lines slowly to allow residual fuel pressure to dissipate. Avoid spraying fuel. Use shop towels to capture leaking fuel.*
- *Fuel in fuel lines is under pressure (approx. 3 - 5 bar or 45 - 75 psi) and may be expelled forcibly. Do not smoke or work near heaters or other fire hazards. Keep a fire extinguisher handy.*
- *Unscrew the fuel tank cap to release pressure in the tank before working on fuel lines.*
- *Plug open fuel lines and fittings.*
- *Before beginning work on the fuel system, place a fire extinguisher in the vicinity of the work area.*
- *Work only on the fuel system when engine temperature is below 40°C (104°F).*
- *Fuel is highly flammable. When working around fuel, do not disconnect wires that could cause electrical sparks. Do not smoke or work near heaters or other fire hazards.*
- *Wear eye protection and protective clothing to avoid injuries from contact with fuel.*
- *When working on an open fuel system, wear suitable hand protection, as prolonged contact with fuel can cause illnesses and skin disorders.*
- *Do not use a work light with an incandescent bulb near fuel. Fuel may spray on the hot bulb causing a fire.*
- *Make sure the work area is properly ventilated.*
- *Due to risk of personal injury, be sure the engine is cold before beginning work on engine components.*
- *The ignition system produces high voltages that can be fatal. Avoid contact with exposed terminals. Use extreme caution when working on a car with the ignition switched ON or the engine running.*

BentleyPublishers
BentleyPublishers.com—All Rights Reserved

CAUTION—
- *Renew fuel system hoses, clamps and O-rings any time they are removed.*
- *Prior to disconnecting the battery, read the battery disconnection cautions in* **001 Warnings and Cautions**.
- *Connect and disconnect the DME system wiring and test equipment leads only when the ignition is switched OFF.*
- *Wait at least 1 minute after switching the ignition OFF before disconnecting engine control module (ECM) connectors. If the connectors are removed before this time, residual power in the system relay may damage the control module.*
- *Tests or repair procedures in this section may set fault codes (DTCs) in the ECM and illuminate the MIL. After repairs are completed, access and clear DTC memory using a BMW scan tool. See* **On-board diagnostics** *in this repair group.*
- *Fuel system cleaners and other chemical additives other than those specifically recommended by BMW may damage catalytic converters, oxygen sensors or other fuel supply components.*
- *Do not connect any test equipment that delivers a 12-volt power supply to terminal 15 (+) of the ignition coils. The current flow may damage the ECM. Connect test equipment only as specified by BMW or the equipment maker.*
- *Relay positions can vary. Be sure to confirm relay location and function by identifying the wiring in the socket using the wiring diagrams found in* **ELE Electrical Wiring Diagrams**.
- *Use a digital multimeter for electrical tests. Use an LED test light for quick tests.*

BentleyPublishers.com
BentleyPublishers.com—All Rights Reserved

DRIVEABILITY TROUBLESHOOTING

The self-diagnostic DME engine management systems monitor and store diagnostic trouble codes (DTCs). If the malfunction indicator light (MIL) illuminates, it indicates that an emissions-related fault has occurred and that one or more DTCs are stored in the engine control module (ECM).

B309120017

If faults arise, or if the MIL is illuminated, begin troubleshooting by connecting BMW scan tool to the OBD II plug on left door post.

The capabilities of OBD II software has the potential to save hours of diagnostic time and to help avoid incorrect component replacement and possible damage to system components. See **On-board diagnostics** in this repair group.

Basic engine settings

Idle speed, idle mixture (%CO), and ignition timing are not adjustable. The DME system is adaptive and automatically compensates for changes in the engine due to age, minor wear or small problems such as a disconnected vacuum hose. However, the adaptive range is limited. Once the limits are exceeded, driveability problems become noticeable.

Poor initial driveability may be encountered when the battery is disconnected and reconnected. When the battery is disconnected, ECM adaptive memory may be reset. The system readapts after about ten minutes of driving.

System voltage

Digital motor electronics (DME) requires that the system (battery) voltage be maintained within a narrow range of DC voltage. DC voltage levels beyond or below the operating range, or any AC voltage in the electrical system can cause havoc.

- When troubleshooting an illuminated MIL, make sure the battery is fully charged and capable of delivering all its power to the electrical system. An undercharged battery can amplify AC alternator output ripple.

- To make a quick check of battery charge, measure voltage across battery terminals with all cables attached and ignition OFF. A fully charged battery measures 12.6 volts or slightly more, compared to 12.15 volts for a battery with a 25% charge.

The DME system operates at low voltage and current levels, making it sensitive to small increases in resistance. The electrical system is routinely subject to corrosion, vibration and wear, so faults or corrosion in the wiring harness and connectors are not uncommon.

- Check battery terminals for corrosion or loose cable connections. See **121 Battery, Starter, Alternator**.

BentleyPublishers
BentleyPublishers.com—All Rights Reserved

– If a battery cable connection has no visible faults but is still suspect, measure voltage drop across the connection. A large drop indicates excessive resistance, meaning that the connection is corroded, dirty or damaged. Clean or repair and retest. See **600 Electrical System–General** for voltage drop test procedure.

– Visually inspect wiring, connectors, switches and fuses. Loose or damaged connectors can cause intermittent problems, especially small terminals in ECM connectors. Disconnect wiring harness connectors to check for corrosion, and use electrical cleaning spray to remove contaminants.

Main grounds

Good grounds are critical to proper DME operation. If a ground connection has no visible faults but is still suspect, measure the voltage drop across the connection. A large voltage drop means high resistance. Clean or repair the connection and retest.

For voltage drop testing, see **600 Electrical System-General**.

For ground locations, see **ECL Electrical Component Locations**.

On-board diagnostics

Second generation on-board diagnostics (OBD II) software and hardware is incorporated in the engine management systems. The OBD II system monitors components that influence exhaust and evaporative emissions. If a problem is detected, the OBD II system stores the associated diagnostic trouble code (DTC) and condition.

If vehicle emissions levels exceed 1.5 times Federally mandated criteria, the OBD II system illuminates a malfunction indicator light (MIL) in the instrument cluster.

BentleyPublishers

◄ Scan tool connected to OBD II plug below left dashboard.

Professional diagnostic scan tools available at the time of this printing include the BMW GT1, ISTA and a small number of aftermarket BMW-specific tools.

In addition to professional scan tools, there are inexpensive generic OBD II scan tool software programs and handheld units available. Although these have limited capabilities as compared to the dedicated tools, they are powerful diagnostic tools.

For the DIY owner, a simple aftermarket DTC reader is available through http://www.peakeresearch.com. This tool is capable of checking for DTCs as well as switching the illuminated MIL OFF and resetting service indicator lights.

NOTE—
- *OBD II DTC memory (including an illuminated MIL) can only be reset using a BMW scan tool or. Removing the connector from the ECM or disconnecting the battery does not erase DTC memory.*

BentleyPublishers.com
BentleyPublishers.com—All Rights Reserved

Engine Management Fuses and Power Supply Components

Power supply fuses

See **ECL Electrical Component Locations** for fuse panel access information.

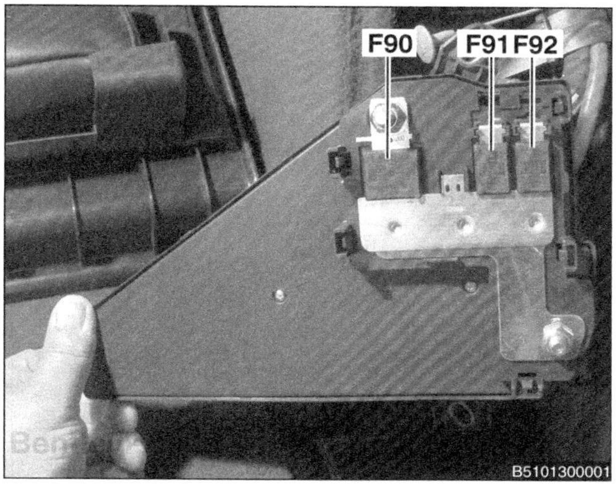

Rear distribution panel (A46) in trunk, installed in right rear fender:

- **F90** 200A: Power supply to fuse and relay panel behind glove compartment (A41a).
- **F91** 100A: Power supply to fuse panel Integrated voltage supply module (IVM) (A6009).

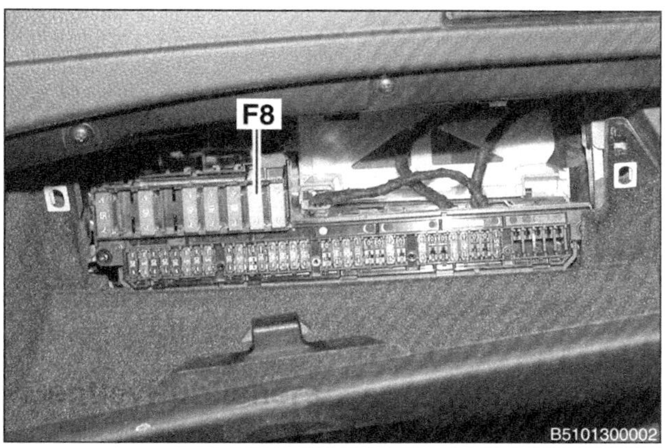

A41a: fuse and relay panel behind glove compartment:

- **F8** 60A: Integrated voltage supply module (IVM) (A6009)

See **ECL Electrical Component Locations** for specific vehicle fuse locations.

BentleyPublishers.com—All Rights Reserved

Integrated voltage supply module (IVM) (A6009)

The integrated voltage supply module (IVM) (A6009) is in the E-box, right rear engine compartment, under the plenum cover. See also:

- **Integrated voltage supply module (IVM) relays** in this repair group for IVM relay information
- **ECL Electrical Component Locations** for E-box access information.

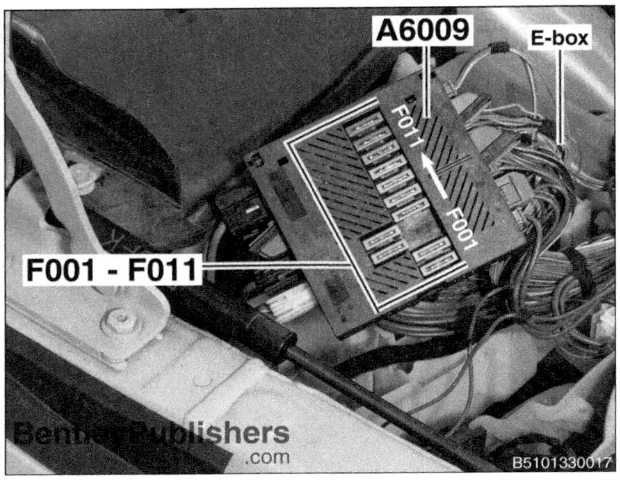

IVM fuse ratings are in **Table b**

Table b. Fuse applications, integrated voltage supply module (IVM) (A6009)		
Fuse	**Rating in amps**	**Protected circuits**
F001	30	Engine control module (ECM) (A6000) Fuel injectors 5 - 8 (Y6105 - Y6108) Valvetronic control module (A60210) Sequential manual gearbox (SMG) (A70010)
F002	20	Camshaft sensors, left bank (B62241, B62242) Ignition coils 5 - 8 (T6155 - T 6158) VANOS solenoids, left bank (Y6283, Y6284) Fuel tank vent solenoid (Y6120)
F003	20	Ignition coils 1 - 4 (T6151 - T6154)
F004	20	Ignition coils 5 - 8 (T6155 - T6158)
F005	30	Camshaft sensors, right bank (B62141, B62142) Crankshaft sensor (B6203) Thermostat, characteristic map (Y6279) VANOS solenoids, right bank (Y6281, Y6282) Mass air flow sensor (B6207)
F006	20	Fuel injectors 1 - 4 (Y6101 - Y6104)
F007	20	Engine control module (ECM) (A6000)) Sequential manual gearbox (SMG) (A70010)
F008	30	Oil quality sensor (B62540) Oxygen sensor heaters
F009	10	Fuel pump relay (K96) Tailpipe flap (Y198) Radiator shutter solenoid (Y13864) E-box fan (M6506) DMTL pump (M119a) Secondary air pump relay (K6304)
F010	40	Valvetronic control module (A60210)
F011	40	Valvetronic control module (A60210)

BentleyPublishers.com
BentleyPublishers.com—All Rights Reserved

Bosch DME ME 9.2, ME 9.2.2

5 series models to 09 / 2005: N62 V8 engine with Valvetronic (4.4 liter) is equipped with Bosch DME ME 9.2 engine management.

5 series models from 09 / 2005: N62TU V8 engine with Valvetronic (4.8 liter) is equipped with Bosch DME ME 9.2.2 engine management.

Both DME engine management systems for N62 and N62TU V8 engine are similar and will be covered together unless otherwise noted.

For a description of the N62 engine, see **100 Engine–General**.

Bosch DME introduction

The Bosch DME engine management system meets stringent emission requirements in addition to reducing fuel consumption and increasing driving performance. A flash EEPROM is used as the storage medium for program data, DTC memory and adaptation values. The engine control module (ECM) works in combination with the Valvetronic control module:

- ECM: Overall engine management
- Valvetronic control module: Variable intake valve lift

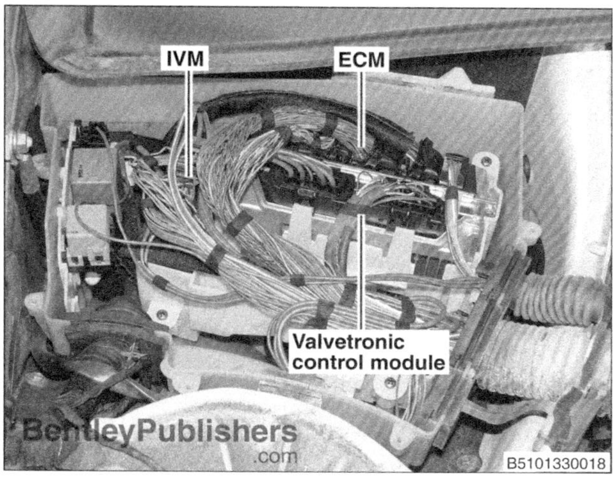

The ECM, Valvetronic control module and integrated voltage supply module (IVM) are in the E-box at the right rear of the engine compartment.

The ECM controls an electric cooling fan (**arrow**) in the base of the E-box to draw in cool air from the passenger compartment.

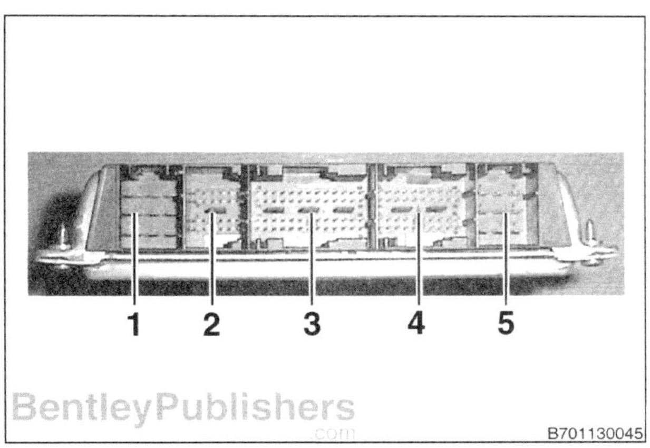

The 134 pin ECM is manufactured by Bosch to BMW specifications. The ECM uses the SKE (standard shell construction) housing and has 5 modular connectors.

BentleyPublishers.com—All Rights Reserved

Integrated voltage supply module (IVM) relays

When ignition is switched ON, the ECM activates a ground path to energize relays 1 - 3 in the IVM.

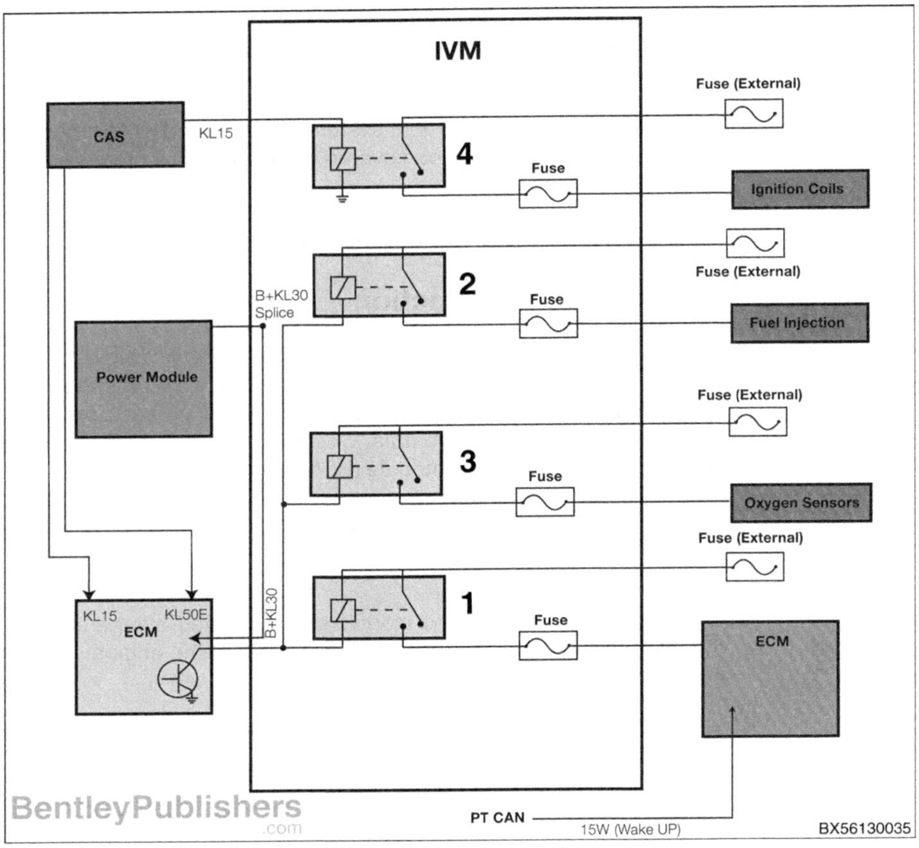

1. Power to ECM
2. Power to fuel injectors

1. Power to oxygen sensor heaters
   - CAS activates relay 4.
2. Power to ignition coils
   - Additional relays in IVM supply power to the Valvetronic system and automatic transmission control module.

See also **Engine Management Fuses and Power Supply Components** in this repair group.

BentleyPublishers.com
BentleyPublishers.com—All Rights Reserved

DME engine management system description

The DME manages and monitors the following functions:

Air

- Electronic throttle
- Mass air flow
- Variable intake manifold
- Valvetronic

Fuel

- Fuel supply
- Fuel injection

Ignition

- Direct ignition
- Knock control
- Primary and secondary ignition monitoring

Emissions

- OBD II compliance
- Pre- and post-catalyst oxygen sensors
- Electrically heated DME-mapped thermostat
- Misfire detection
- Evaporative emission control and leak detection
- Secondary air injection system (to 03 / 2005)
- Malfunction indicator light (MIL)

Performance controls

- Dual VANOS control
- Output of injection signal (TI) for fuel economy gauge
- Output of engine rpm (TD) for tachometer
- A/C compressor control
- Electric radiator cooling fan
- CAN-bus communication
- Dynamic stability control (DSC) interface
- CAS (car access system) with EWS 3.3
- Cruise control
- ECM programming

BentleyPublishers.com
BentleyPublishers.com–All Rights Reserved

Fuel metering

The ECM meters pressurized fuel by changing the opening time (pulse width) of the fuel injectors. The exact amount of fuel injected is determined by the amount of time the injectors are open. To ensure that injector pulse width is the only factor that determines fuel metering, fuel pump pressure is maintained by a pressure regulator.

The fuel injectors are positioned close to the intake valves resulting in large injection spray angles. Greater fuel spray atomization leads to optimum fuel mixing and thereby reduces fuel consumption and exhaust emissions.

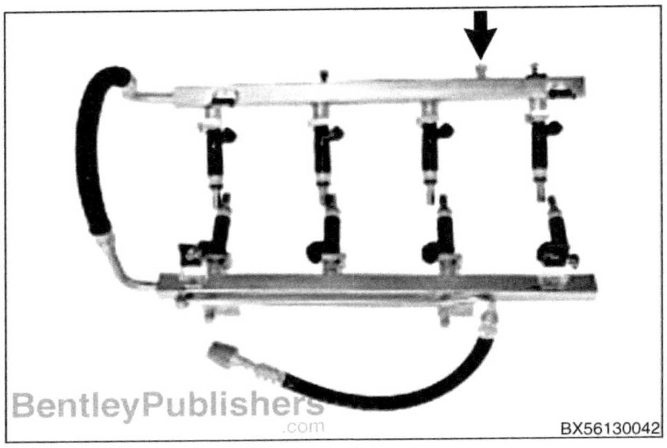

A service port (**arrow**) in the non-return fuel rail is for fuel pressure tests.

The Siemens fuel injectors have a coil resistance of approximately 12 Ω each.

The Bosch DME system uses 8 individual fuel injector circuits so that any one or multiple injectors can be deactivated, if needed. If the ECM detects a catalyst-damaging fault, it deactivates all the injectors.

Variable intake manifold

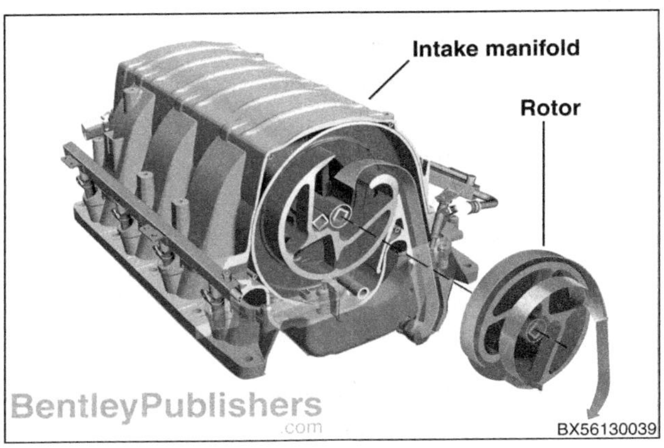

The variable intake manifold is operated by turning a rotor.

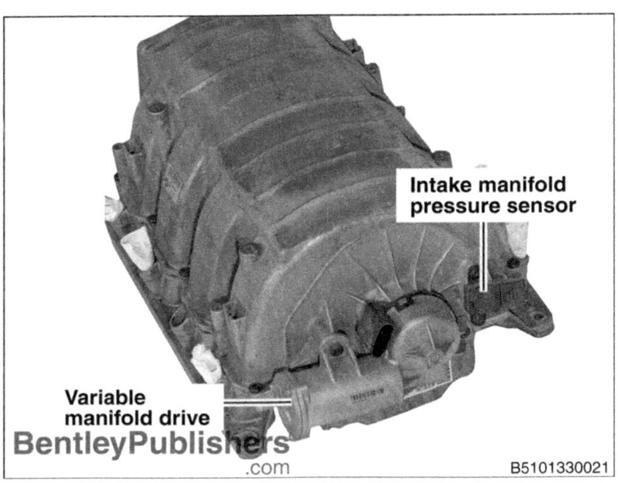

Adjustments to the intake manifold are carried out by the ECM controlling a drive unit. The drive unit is mounted on the rear of the intake manifold.

The drive unit consists of a 12 V electric motor with worm gears and an integral potentiometer for intake manifold position feedback. If the drive unit fails, the system remains in its current position. The driver may notice a loss in power and a fault code may be stored in the ECM.

BentleyPublishers.com
BentleyPublishers.com—All Rights Reserved

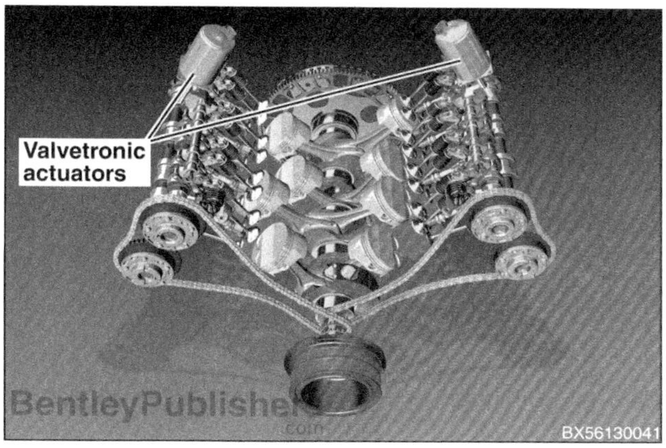

Valvetronic system

The Valvetronic control system varies intake valve opening time and valve lift according to engine speed and load. Extent of valve lift is based on a request from the ECM. Two Valvetronic actuators adjust the two eccentric shafts in the cylinder heads, affecting valve lift. See **100 Engine–General** for additional information.

Faults in the Valvetronic system are detected by the Valvetronic control module and transmitted via CAN-bus to the ECM where they are stored for diagnostics.

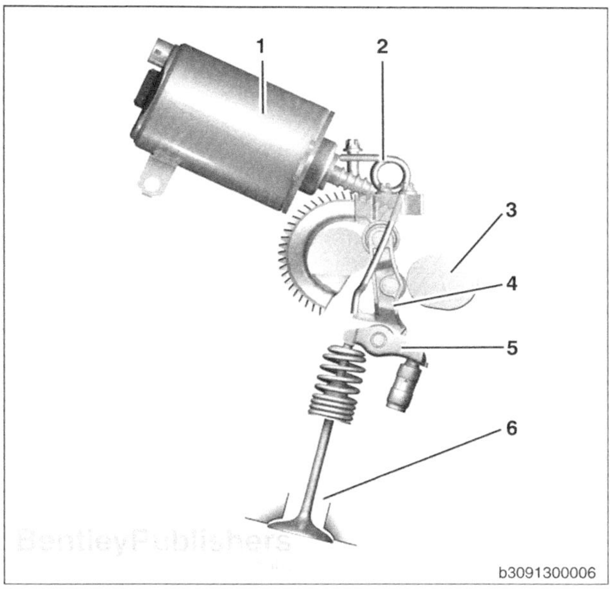

Idle speed control

The ECM controls idle speed by varying intake valve lift via the Valvetronic system.

1. Valvetronic motor
2. Return spring
3. Intake camshaft
4. Intermediate lever
5. Intake rocker arm
6. Intake valve

Idle speed is not adjustable. The basic functions and parameters of idle speed control are as follows:

- Control of cold air intake volume.
- Smooth idle speeds regardless of load and inputs.
- Smooth transition from acceleration to deceleration.

Idle speed stabilization is active during the following conditions:

- Engine warm up
- Heating or A/C activation
- Drive gear selected (automatic transmission)
- Varying cooling fan speeds

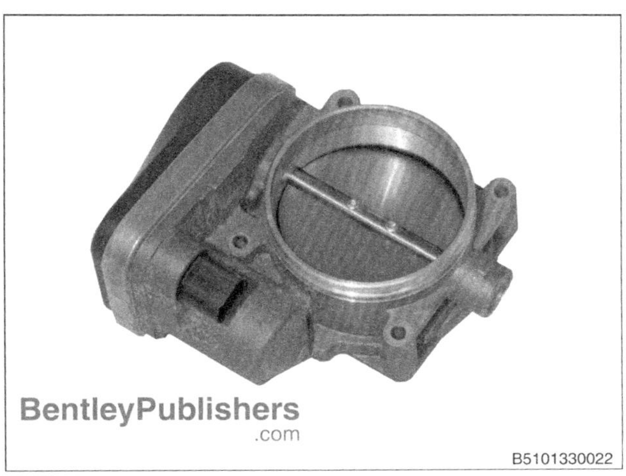

Throttle control

The throttle module (EDK) is not needed for engine load control. Which is carried out by the Valvetronic function of the ECM. However, the throttle may be slightly closed to allow sufficient manifold vacuum for crankcase ventilation and canister purge systems.

The ECM provides operating voltage and ground to the throttle valve for opening and closing the throttle plate.

BentleyPublishers.com
BentleyPublishers.com–All Rights Reserved

Knock (detonation) control

Knock sensors monitor and control ignition knock through the ECM. The knock sensors function like microphones and are able to convert mechanical vibration (knock) into electrical signals. The ECM is programmed to react to frequencies that are characteristic of engine knock and adapt the ignition timing point accordingly.

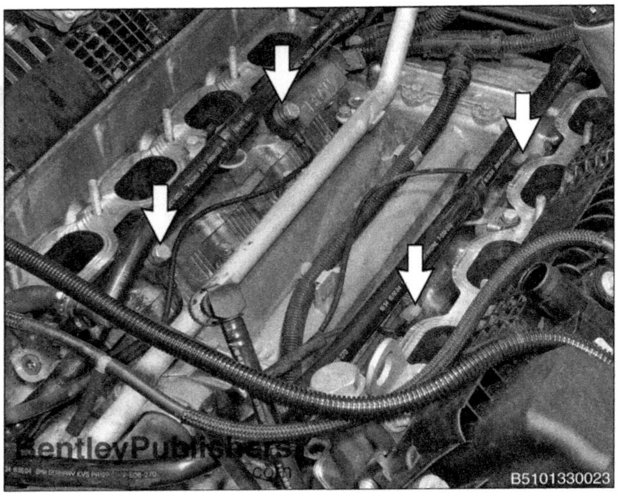

Four knock sensors (**arrows**) are bolted to the cylinder heads between cylinders 1 & 2, 3 & 4, 5 & 6 and 7 & 8. If the signal value exceeds the programmed threshold, the ECM identifies knock and retards the ignition timing for that cylinder.

If a fault is detected with the sensor(s), the ECM deactivates knock control. Then the ignition timing is set to a conservative basic setting based on intake air temperature and a DTC is stored.

Knock sensor replacement is covered in **120 Ignition System**.

Evaporative emissions

Control of evaporative fuel vapors from the fuel tank continues through tighter controls of fuel tank venting and refinements in the leak detection system (DM-TL).

On-board refueling vapor recovery (ORVR)

The ORVR system recovers and stores hydrocarbon (fuel) vapors during refueling. When refueling, the pressure of the fuel entering the tank forces hydrocarbon vapors through the tank refuelling breather hose to the liquid / vapor expansion tank and into the activated charcoal canister. The vapors are stored in the canister and the system can then breathe through the DM-TL and the air filter.

Oxygen sensors

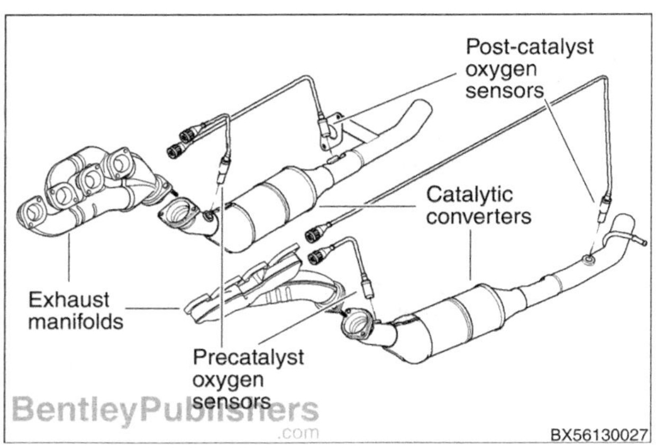

In order to control exhaust emissions, 4 oxygen sensors are utilized, 2 before and 2 after the converters.

The post-catalyst sensors are conventional heated zirconia units which produce a low and relatively stable voltage (approx. 700 - 800 mV) when the system is stable and operating normally.

BentleyPublishers.com
BentleyPublishers.com—All Rights Reserved

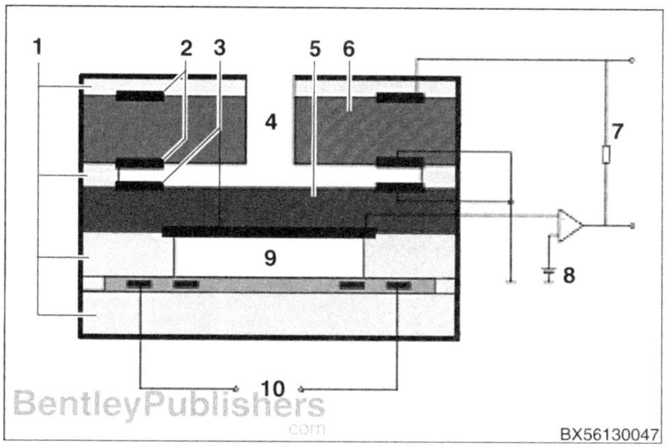

The precatalyst units are Bosch LSU planar wideband oxygen sensors. This compact sensor uses thin layers of zirconium dioxide (ZrO_2) ceramic films. The laminate structure allows the integration of several functions including the heating element which ensures that the minimum operating temperature (750°C / 1382°F) is reached rapidly.

1. Zirconium ceramic film
2. Platinum electrode, pump cell
3. Platinum electrode, reference cell
4. Exhaust gas port
5. Reference cell
6. Pump cell
7. Pump current (I_P)
8. Reference voltage (U_{ref})
9. Reference air channel
10. Heating voltage (U_H)

The wideband oxygen sensor can measure mixture in a wide range, from rich to extremely lean, very rapidly.

To operate effectively, the oxygen sensor requires ambient air as the reference gas inside. This ambient air reaches the inside of the sensor through the harness connector and the electrical harness. In case of oxygen sensor malfunction, check the connector and clean contamination as necessary. Also, disconnect and reconnect the harness connector to remove oxidation from the connector pins.

Secondary air injection

5 series V8 engine models manufactured to 03 / 2005 (545i): The secondary air system pumps ambient air into the exhaust stream after a cold engine start to reduce catalytic converter warm-up time and to reduce HC and CO emissions. The ECM controls and monitors the secondary air injection system.

The electric air pump draws in ambient air and supplies it to the secondary air valve. The air injection valve is opened by a solenoid using engine vacuum and is closed by an internal spring.

The secondary air valve is bolted to the right front of the cylinder head. Cast passageways within the cylinder head duct the secondary air directly into the cylinder head exhaust ports.

BentleyPublishers.com
BentleyPublishers.com–All Rights Reserved

Secondary air injection system components

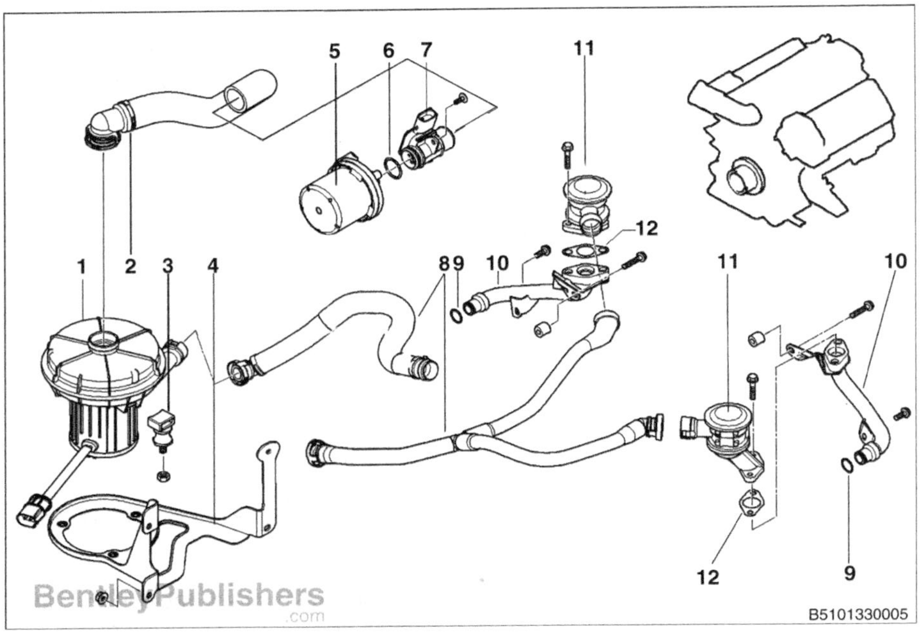

1. Air pump
2. Filter air duct
3. Air pump mount
4. Air pump bracket
5. Air pump filter
6. O-ring
7. Air pump mini mass air flow sensor
8. Pressure hose
9. O-ring
10. Cylinder head air intake tube
11. Secondary air valve
12. Gasket

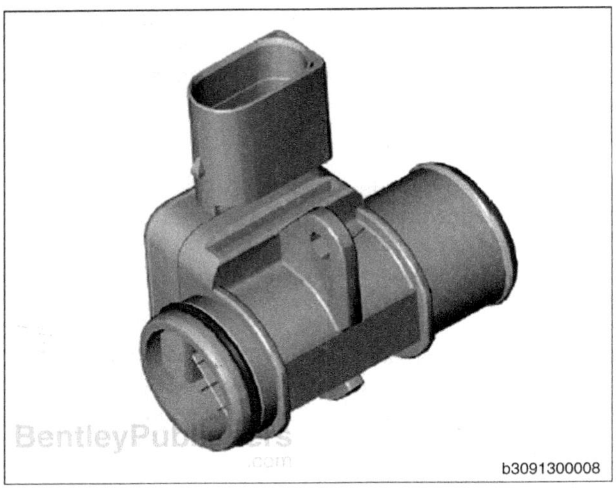

A miniature mass air flow sensor in the secondary air system detects the air mass supplied by the secondary air pump. This function monitors the secondary air system for OBD II compliance.

When the mini mass air flow sensor detects no or insufficient air flow, a fault is stored in the ECM and the malfunction indicator light (MIL) is illuminated.

Secondary air injection illustrations

Secondary air pump is in right front fender behind wheel well liner.

BentleyPublishers.com
BentleyPublishers.com—All Rights Reserved

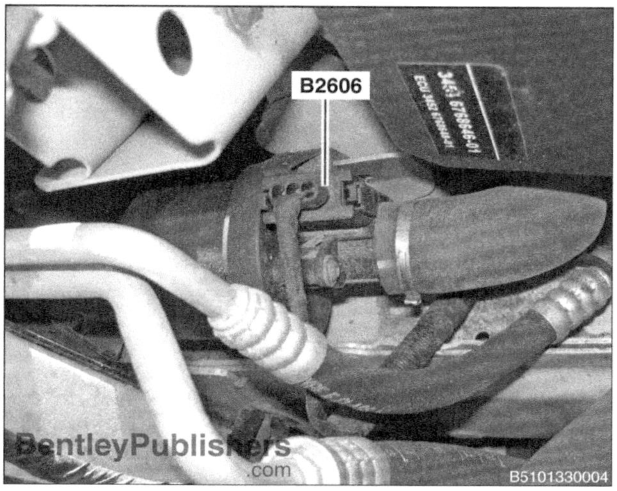

B2606 Mini mass air flow sensor.

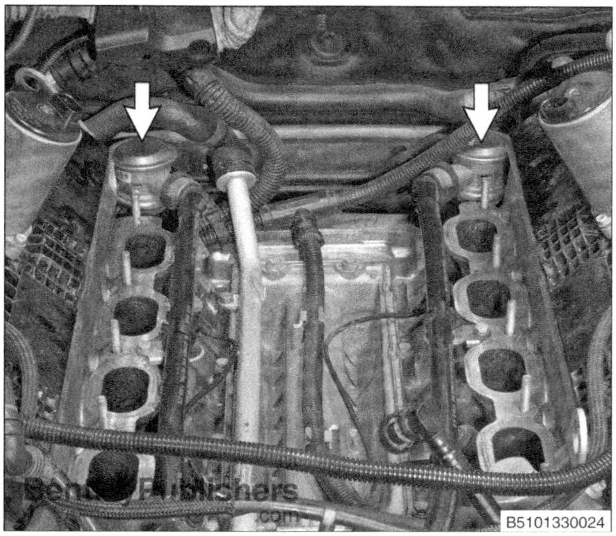

Secondary air valves (**arrows**) are bolted to the rear of each of cylinder head.

BentleyPublishers.com
BentleyPublishers.com–All Rights Reserved

Air filter housing and duct, removing and installing

For air filter element replacement, see **020 Maintenance**.

– Remove upper engine covers. See **020 Maintenance**.

Working at air filter housing:

- Disconnect mass air flow sensor electrical connector (**A**).
- Remove filter housing mounting bolts (**arrows**).

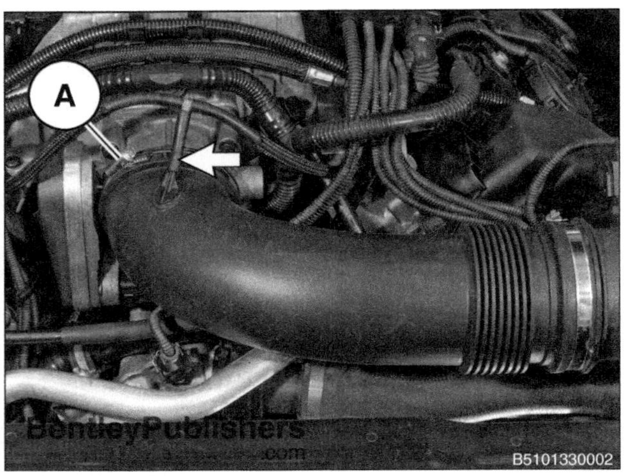

Loosen hose clamp (**A**) at throttle housing and detach vacuum line (**arrow**) from air intake duct.

– Remove air filter housing and duct from vehicle.

– Installation is reverse of removal.

– After reinstalling, check and clear fault codes from ECM memory.

Mass air flow, measurement

The air volume input signal is produced electronically by the hot film mass air flow sensor which uses a heated metal film in the air flow stream to detect the mass of the air flowing past.

The sensor is mounted in the air inlet pipe between the air filter and the throttle valve. As air flows through the sensor, the hot film is cooled, changing the resistance which affects current flow (voltage drop) through the circuit. The ECM monitors this change and regulates the amount of fuel injected.

The mass air flow sensor contains an integral air temperature sensor. The signal from this negative temperature coefficient (NTC) sensor is needed by the ECM to correct the air volume input signal. Changes in intake air temperature (air density) affect the amount of fuel injected, ignition timing and secondary air injection activation.

BentleyPublishers.com
BentleyPublishers.com–All Rights Reserved

Mass air flow sensor, replacing

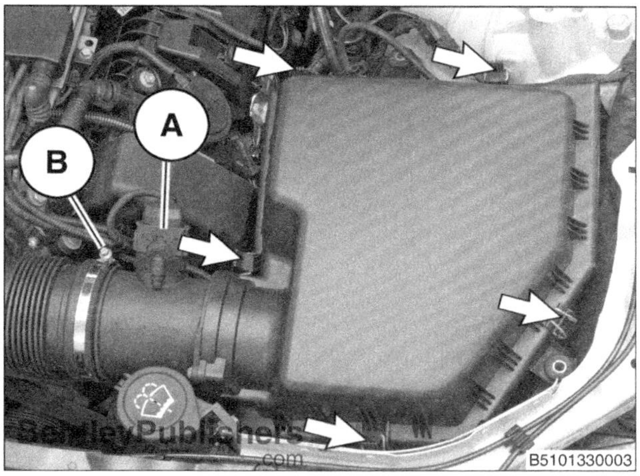

BentleyPublishers.com
B5101330003

– Switch ignition OFF and remove key.

◄ Working at air filter housing:
- Disconnect mass air flow sensor electrical connector (**A**).
- Unclip filter housing lid (**arrows**).
- Loosen air duct clamp (**A**) and remove air filter housing lid from vehicle.

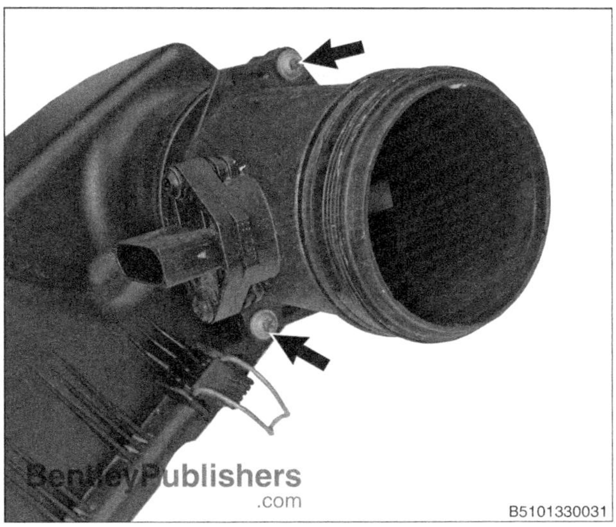
BentleyPublishers.com
B5101330031

◄ Remove fasteners (**arrows**) and remove mass air flow sensor from air filter housing.

– Check intake ducts for cracks and vacuum leaks.

– Installation is reverse of removal.
- Inspect O-ring and replace if necessary.

– After reinstalling, check and clear fault codes from ECM memory.

Throttle housing (EDK), replacing

Accelerator pedal position module (PWG)

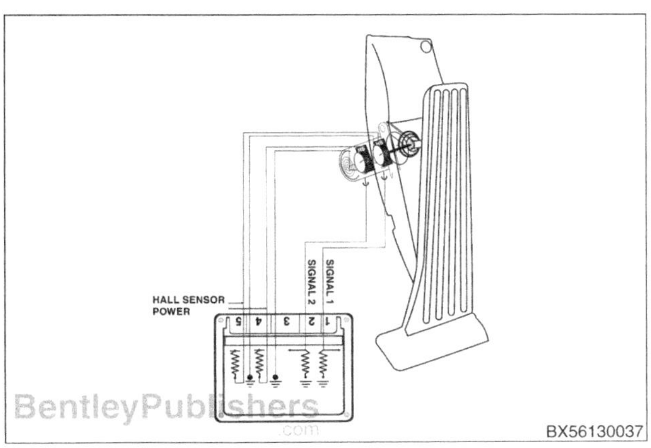

BentleyPublishers.com
BX56130037

◄ The accelerator pedal position module (PWG) provides two variable voltage signals to the ECM that represent accelerator pedal position and rate of movement. The ECM activates the Valvetronic system based on this input.

Dual Hall sensors are used in the PWG. The ECM compares the values from the two for plausibility. If the signals are not within the plausible range, the ECM uses the lower of the two signals as the driver request input. Acceleration response in that case is slower because maximum Valvetronic opening is reduced.

After replacing the throttle housing, use BMW scan tool to reset adaptation values.

> ***CAUTION—***
> - *If the adaptation process is not completed correctly, the engine does not start.*

BentleyPublishers.com
BentleyPublishers.com—All Rights Reserved

– Switch ignition OFF and remove key.

– Remove upper engine covers. See **020 Maintenance**.

◄ Working at air filter housing:
- Disconnect mass air flow sensor electrical connector (**A**).
- Remove filter housing mounting bolts (**arrows**).

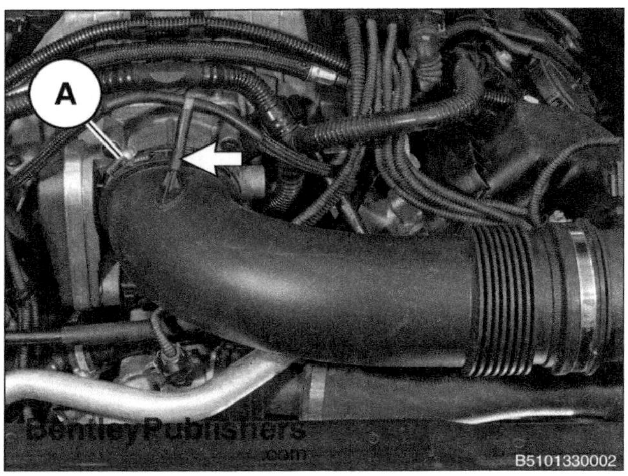

◄ Loosen hose clamp (**A**) at throttle housing and detach vacuum line (**arrow**) from air intake duct.

– Remove air filter housing and duct from vehicle.

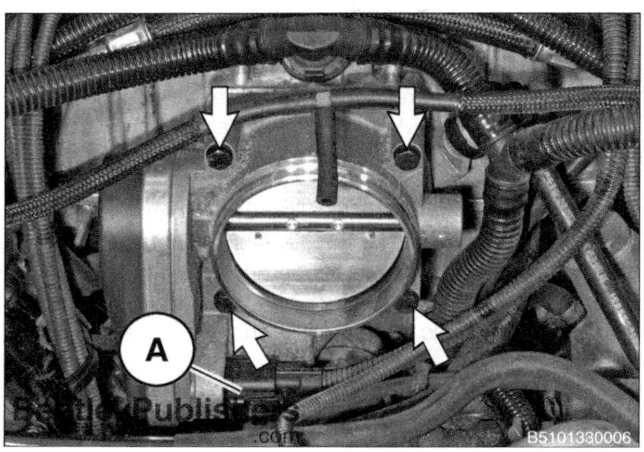

◄ Working at left side of intake manifold:
- Disconnect electrical connector (**A**).
- Remove fasteners (**arrows**) and remove throttle housing.

– Installation is reverse of removal.

– Replace profile gasket (sealing O-ring) between throttle assembly and intake manifold.

– After reinstalling, check and clear fault codes from ECM memory. Reset throttle plate adaptation values following scan tool on-screen instructions.

BentleyPublishers.com
BentleyPublishers.com—All Rights Reserved

Valvetronic motor, removing and installing

Right side valvetronic motor shown. Left side is similar.

- Switch ignition OFF and remove key.
- Remove upper engine covers. See **020 Maintenance**.
- Disconnect battery negative (-) cable.

> ***CAUTION—***
> - *Prior to disconnecting the battery, read the battery disconnection cautions in* **001 Warnings and Cautions**.

◄ Disconnect valvetronic motor electrical connector (**arrow**).

◄ Locate BMW roundel (logo) at back of valvetronic motor. Using a small pick, chisel center area of roundel to expose Allen head (**inset**).

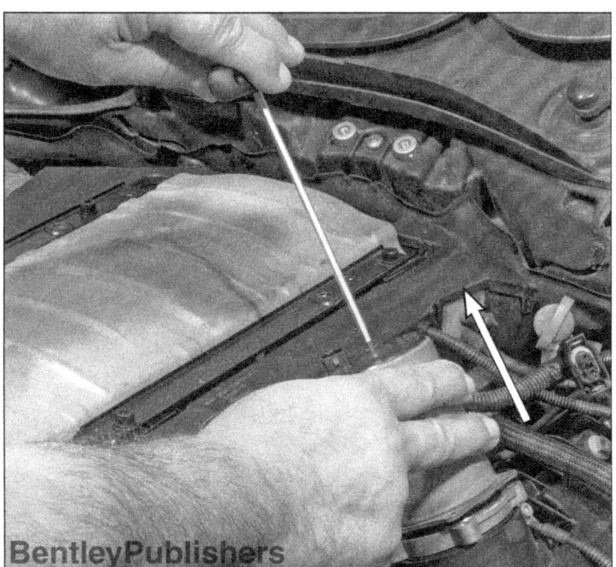

◄ Remove valvetronic motor fasteners (**arrows**).

- Using a 4 mm Allen, gently rotate valvetronic motor shaft counterclockwise until released from eccentric shaft splines.
- Remove valvetronic motor in direction of **arrow**.
- Installation is reverse of removal. Remember to:
  - Route ignition harness correctly
- After reinstalling, check and clear fault codes from ECM memory.

BentleyPublishers.com
BentleyPublishers.com—All Rights Reserved

Fuel rail and injectors, removing and installing

Fuel injectors are electrically controlled solenoid valves that provide precisely metered and atomized fuel into the engine intake ports.

Injectors receive voltage from the DME main relay. The engine control module (ECM) controls injector opening by activating the ground circuit for the injector valve solenoids. The ECM varies the duration (in milliseconds) of injector opening to regulate air / fuel ratio.

– Switch ignition OFF and remove key.

– Remove upper engine covers. See **020 Maintenance**.

– For a quick check of injectors, run engine and touch each injector with a screwdriver or stethoscope. You should feel a vibration or hear a buzzing. Switch engine OFF.

– Disconnect battery negative (-) cable.

> ***CAUTION—***
> *• Prior to disconnecting the battery, read the battery disconnection cautions in* **001 Warnings and Cautions**.

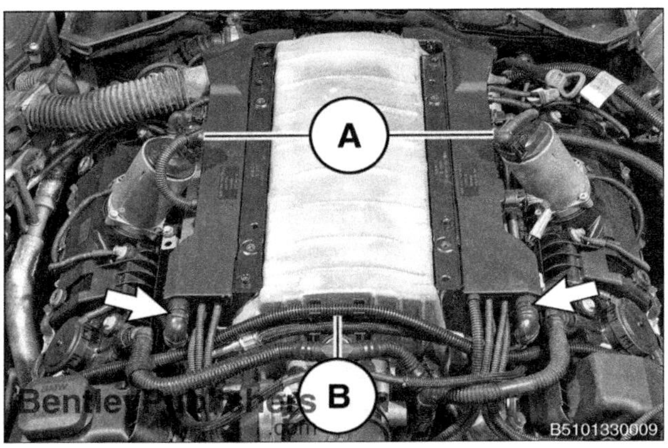

< Working above engine, disconnect valvetronic motor electrical connectors (**A**). Disconnect knock sensor electrical connectors (**arrows**) and unclip wiring harness from holder (**B**).

> ***CAUTION—***
> *• Be sure to mark oxygen sensor connectors so that they can be reassembled as before.*

– Disconnect the following electrical connectors:

1. Fuel injector electrical connectors at fuel rail, left and right side.
2. Throttle housing.
3. Both temperature sensors.
4. All electrical solenoids.
5. Alternator.
6. Manifold pressure sensor.
7. Intake manifold drive unit.
8. Camshaft position sensors, left and right side.
9. Left eccentric shaft sensor.

< Remove wiring harness bracket fasteners (**arrows**). Lift up and lay wiring harness aside.

BentleyPublishers.com
BentleyPublishers.com—All Rights Reserved

Unscrew Schræder valve cap (**arrow**) from fuel rail. Connect air line adapter (**A**) to fitting.

- Unscrew fuel tank cap to release pressure.
- Using a brief burst of compressed air (maximum of 3 bar or 43.5 psi), blow fuel back into fuel tank.

> ***WARNING —***
> - *Fuel in the fuel line is under pressure (approx. 3 - 5 bar or 45 - 75 psi) and may be expelled forcibly. Do not smoke or work near heaters or other fire hazards. Keep a fire extinguisher handy.*
> - *Unscrew the fuel tank cap to release pressure in the tank before working on the fuel line.*
> - *Plug open fuel lines and fittings.*

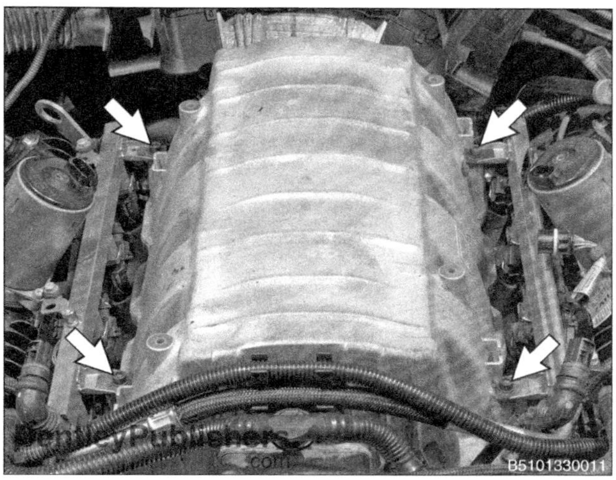

Working above engine:

- Remove fuel rail fasteners (**arrows**).
- Lift left and right side fuel rail up and off engine.

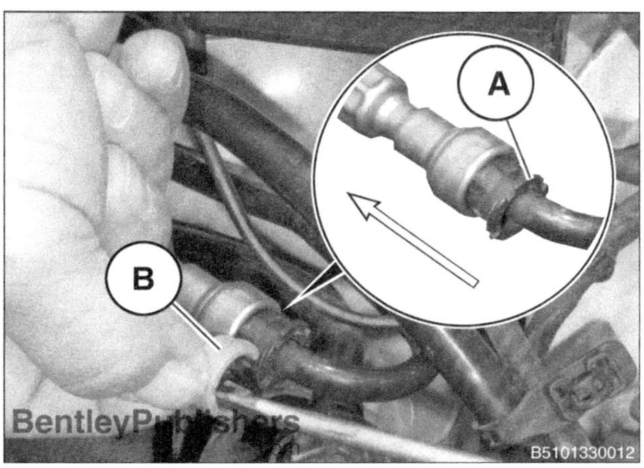

Using a small flat blade screwdriver, remove fuel line locking collar (**B**). Unlock fuel line by pushing coupler (**A**) in direction of **arrow**. Remove fuel line with from fuel rail from vehicle.

> ***WARNING —***
> - *Wrap a clean shop towel around fitting before disconnecting. Residual fuel pressure is present in the fuel line.*

BentleyPublishers
BentleyPublishers.com—All Rights Reserved

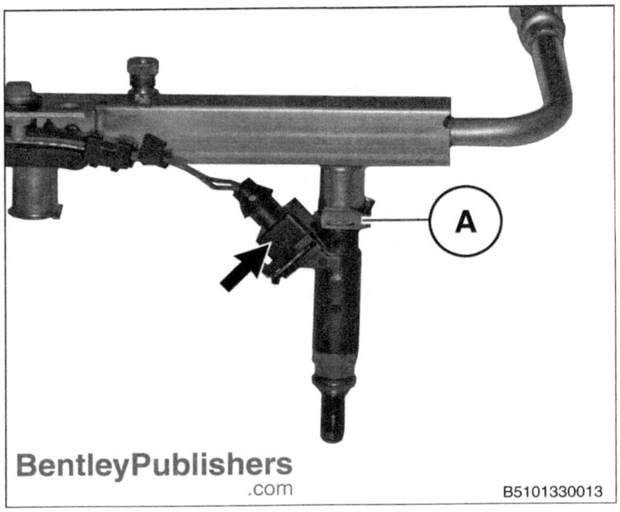

◄ Disconnect fuel injector electrical connectors (**arrow**). Pry retaining clip from injector (**A**).

◄ Remove individual injectors:

- Pull injector from rail (**arrow**).

– Installation is reverse of removal. Remember to:

- Fit new sealing O-rings when installing injectors. For ease of installation, lightly lubricate O-rings with assembly lubricant.
- Check that injector electrical connections are correctly fitted and that injectors are fully seated prior to installing fuel rail mounting bolts.
- Replace wire ties.

Tightening torque	
Fuel rail to cylinder head	10 Nm (7 ft-lb)

– After reinstalling, check and clear fault codes from ECM memory.

Intake manifold, removing and installing

– Disconnect battery negative (-) cable in cargo compartment.

> ***CAUTION—***
> - *Prior to disconnecting the battery, read the battery disconnection cautions in* **001 Warnings and Cautions**.

– Remove upper engine covers. See **020 Maintenance**.

– Remove air filter housing and air intake ducts. See **Air filter housing and ducts, removing and installing** in this repair group.

◄ Remove and discard tension strut fasteners (**arrows**). Remove strut brace (**A**).

– Remove fuel rail. See **Fuel injectors, removing and installing** in this repair group.

BentleyPublishers.com
BentleyPublishers.com—All Rights Reserved

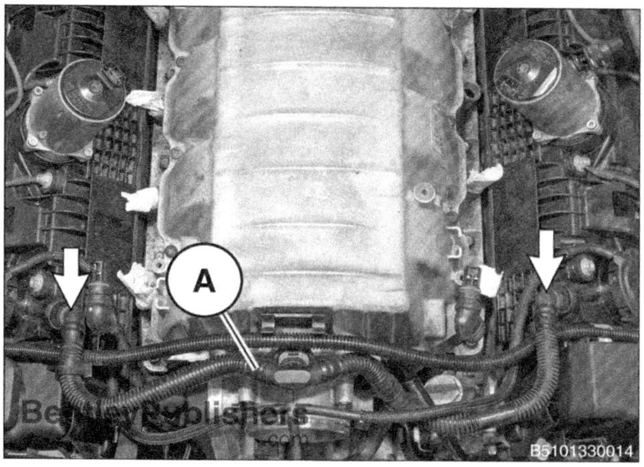

Disconnect crankcase vent hose (**A**) from valve covers (**arrows**). Remove from intake manifold.

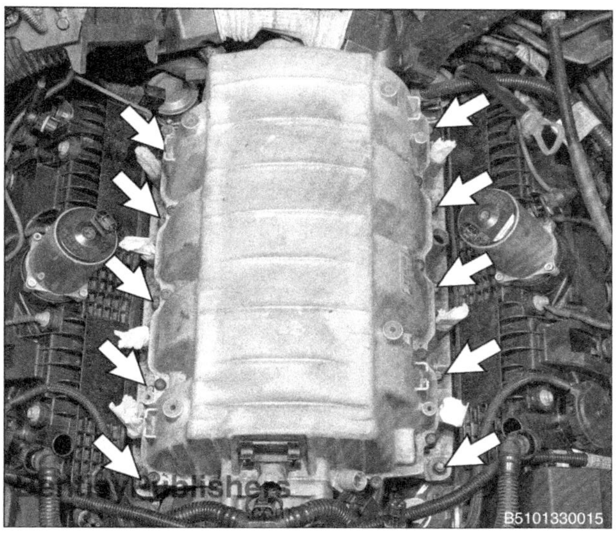

Remove intake manifold fasteners (**arrows**). With help from an assistant, lift intake manifold up and off engine.

– Installation is reverse of removal. Remember to:
  - Install new gaskets to intake manifold before installing.
  - Clean sealing surfaces.
  - Replace all seals.

Tightening torque	
Manifold to cylinder head • M6 • M7 • M8	 10 Nm (7.3 ft-lb) 15 Nm (11 ft-lb) 22 Nm (16.2 ft-lb)
Tension strut to body (replace fasteners)	30.5 Nm (22.4 ft-lb)

Engine coolant temperature (ECT) sensor, replacing

The engine coolant temperature (ECT) sensor is an NTC (negative temperature coefficient) sensor. As coolant temperature rises, resistance through the sensor decreases.

The ECM varies ignition timing and air / fuel mixture based on engine coolant temperature. The ECT sensor is supplied a 5 volt reference voltage. The voltage drop across the sensor varies as the coolant temperature (sensor resistance) changes.

If the ECT sensor input is faulty or not plausible, the MIL is illuminated when OBD II fault criteria are exceeded. The ECM assumes a substitute value (80°C / 176°F) to maintain engine operation. The ignition timing is set to a safely conservative basic setting.

The ECT sensor is located on the front of the engine near thermostat housing.

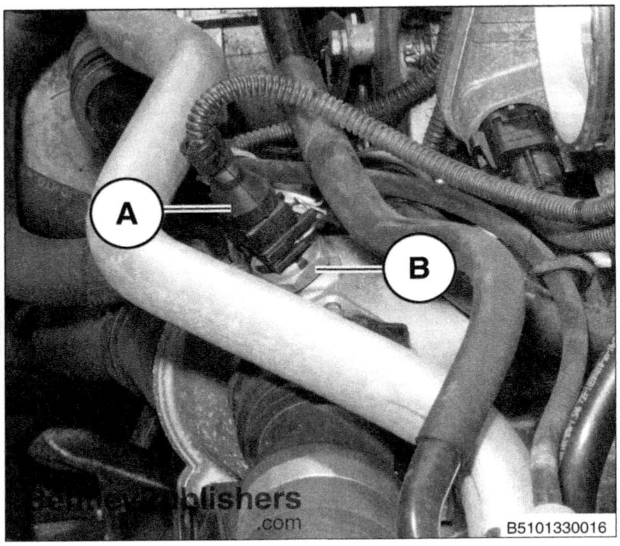

◀ With engine fully cooled off, disconnect ECT sensor connector (**A**).

– Unscrew ECT sensor (**B**) from cylinder head. Be prepared to catch small amount of coolant.

– Installation is reverse of removal:
  - Use new copper sealing washer when installing sensor.
  - Replace lost coolant.

Tightening torque	
Temperature sensor to cylinder head	13 Nm (9.5 ft-lb)

– After reinstalling, check and clear fault codes from ECM memory.

Radiator outlet coolant temperature sensor

◀ The radiator outlet temperature sensor (**inset**) is an NTC (negative temperature coefficient) sensor. As coolant temperature rises, resistance through the sensor decreases.

The ECM varies cooling fan speed based on radiator outlet coolant temperature. The radiator outlet coolant temperature sensor is supplied a 5 volt reference voltage. Voltage drop across the sensor varies as coolant temperature (sensor resistance) changes.

Vacuum pump, removing and installing

– Pump brake pedal 10 times to reduce vacuum pressure in brake booster.

– Remove upper engine covers. See **020 Maintenance**.

– Remove intake air ducts. See **Air filter housing and ducts, removing and installing** in this repair group.

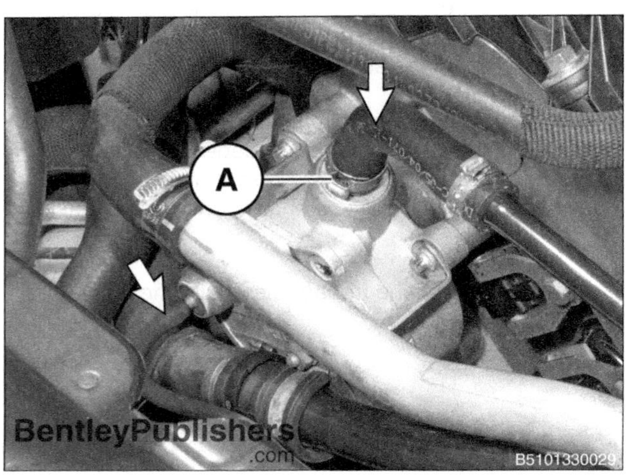

◀ Remove clamp (**A**) and vacuum line from pump (**arrows**).

BentleyPublishers.com
BentleyPublishers.com—All Rights Reserved

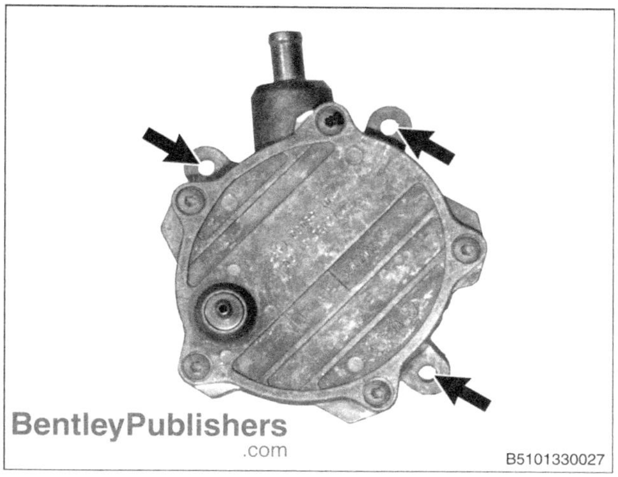

Remove vacuum pump fasteners (**arrows**).

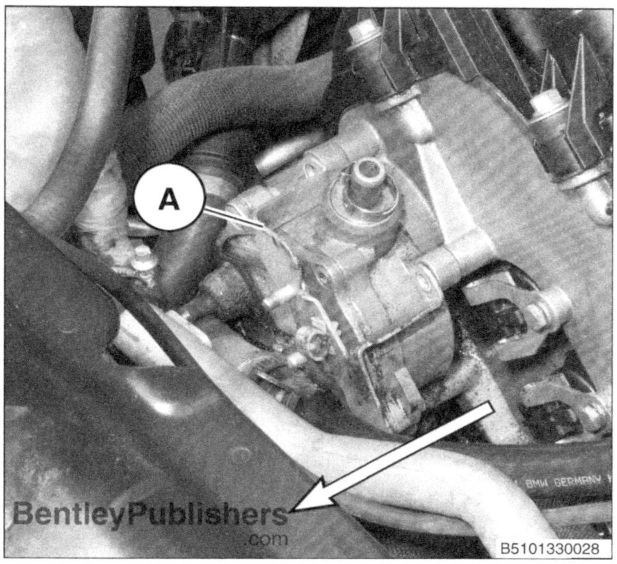

Remove vacuum pump (**A**) from engine.

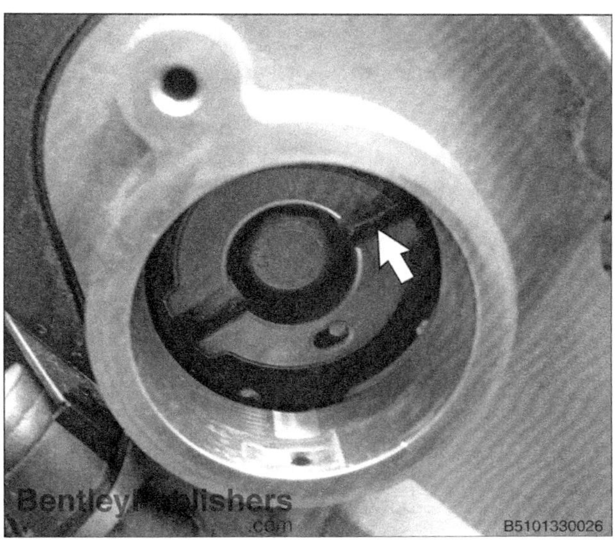

Installation is reverse of removal. Remember to:

- Replace vacuum pump O-ring.
- Properly align pump to camshaft slot (**arrow**).

BentleyPublishers
BentleyPublishers.com–All Rights Reserved

Bosch ME 9.2, ME 9.2.2 ECM pin assignments

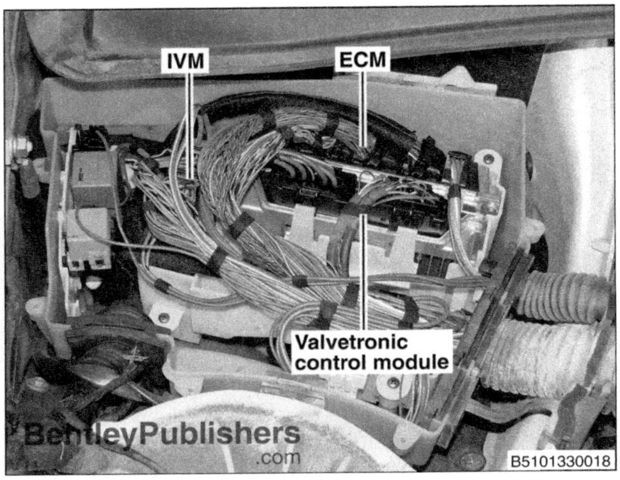

B5101330018

The engine control module (ECM) (**arrow**) is located in the right rear of the engine compartment in the E-box.

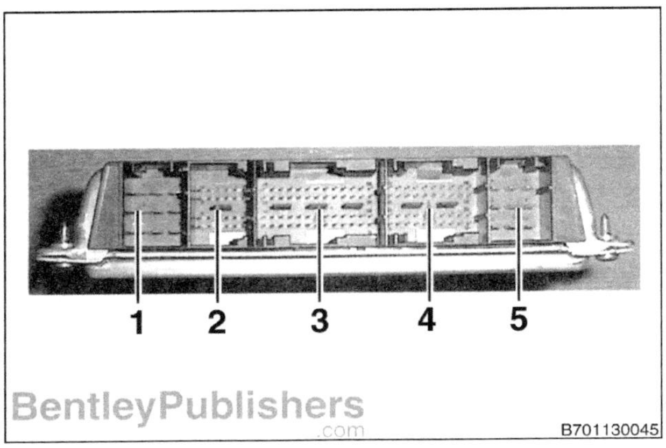

B701130045

The ECM has 5 main electrical harness connectors, for a total 134 pins.

- **1:** X60001 (voltage, ground supply)
- **2:** X60002 (oxygen sensors)
- **3:** X60003 (signals)
- **4:** X60004 (signals)
- **5:** X60005 (ignition coils)

ECM pin assignments for V8 engines are given in **Table c**. This information can be helpful when diagnosing faults to or from the ECM.

Generally, absence of voltage or continuity means there is a wiring or connector problem. Test results with incorrect values do not necessarily mean that a component is faulty. Check for loose, broken or corroded connections and wiring before replacing components. For engine management system electrical schematics, see **ELE Electrical Wiring Diagrams**.

When making checks at the ECM, use a break-out box to allow tests to be made with the connector attached to the ECM. This prevents damage to the small terminals in the connector. As an alternative, the connector housing can be separated so that electrical checks can be made from the back of the connector.

CAUTION—

- *Wait at least one minute after switching the ignition OFF before removing the connector from the ECM. If the connector is removed before this time, residual power in the system relay may damage the ECM.*
- *Connect or disconnect the control module connector and meter probes with the ignition OFF.*

BentleyPublishers.com
BentleyPublishers.com—All Rights Reserved

Table c. Bosch ME 9.2 ME 9.2.2 ECM pin assignments			
Connector X60001 9-pin (signals)			
Pin	**Type**	**Component or function**	**Note**
1	-	Not used	
2	-	Not used	
3	input/output	TXD signal	Integrated voltage supply module (IVM)
4	ground	ECM ground	Integrated voltage supply module (IVM)
5	-	Not used	
6	ground	ECM ground	Integrated voltage supply module (IVM)
7	input	Terminal 30	Integrated voltage supply module (IVM)
8	input	Terminal 87	Integrated voltage supply module (IVM)
9	-	Not used	

Connector X60002 24-pin (signals)			
Pin	**Type**	**Component or function**	**Note**
1	output	Signal, oxygen sensor heating	Bank 1 sensor 1
2	output	Signal, oxygen sensor heating	Bank 2 sensor 1
3	-	Not used	
4	-	Not used	
5	-	Not used	
6	output	Oxygen sensor heating	Bank 1 sensor 2
7	ground	Oxygen sensor	Bank 1 sensor 1
8	ground	Oxygen sensor	Bank 2 sensor 2
9	ground	Oxygen sensor	Bank 2 sensor 1
10	ground	Oxygen sensor	Bank 1 sensor 2
11	-	Not used	
12	output	Oxygen sensor heating	Bank 2 sensor 2
13	input	Oxygen sensor signal	Bank 1 sensor 1
14	input	Oxygen sensor signal	Bank 2 sensor 2
15	input	Oxygen sensor signal	Bank 2 sensor 1
16	input	Oxygen sensor signal	Bank 1 sensor 2
17	-	Not used	
18	-	Not used	
19	input	Oxygen sensor signal	Bank 1 sensor 1
20	input	Oxygen sensor signal	Bank 1 sensor 1
21	input	Oxygen sensor signal	Bank 2 sensor 1
22	input	Oxygen sensor signal	Bank 2 sensor 1
23	input	Terminal 87	Integrated voltage supply module (IVM)
24	-	Not used	

BentleyPublishers.com
BentleyPublishers.com - All Rights Reserved

Connector X60003 52-pin (signals)			
Pin	**Type**	**Component or function**	**Note**
1	input	Mass air flow sensor signal	Mass air flow sensor
2	input	Intake pipe pressure sensor signal	Intake pipe pressure sensor
3	input	Mass air flow sensor signal	Mass air flow sensor
4	output	5 volt reference voltage	Mass air flow sensor
5	-		
6	output	Fuel injector signal	Fuel injector signal 4
7	output	Fuel injector signal	Fuel injector signal 5
8	output	Fuel injector signal	Fuel injector signal 6
9	output	VANOS solenoid signal	VANOS solenoid intake, bank 2
10	output	VANOS solenoid signal	VANOS solenoid intake, bank 1
11	output	Fuel injector signal	Fuel injector signal 7
12	output	Thermostat heating	Characteristic MAP thermostat
13	output	Fuel injector signal	Fuel injector signal 1
14	ground	Ground	Connector X6460
15	-	Not used	
16	input	Camshaft position sensor signal	Camshaft position sensor, exhaust bank 1
17	input	Camshaft position sensor signal	Camshaft position sensor, exhaust bank 2
18	input / output	Valvetronic control module	Valvetronic control module
19	input / output	BSD signal	Connector X6038
20	output	VANOS solenoid signal	VANOS solenoid, bank 1
21	output	Fuel tank vent valve signal	Fuel tank vent valve
22	-	Not used	
23	output	VANOS solenoid signal	VANOS solenoid exhaust, bank 2
24	output	Fuel injector signal	Fuel injector signal 2
25	output	Fuel injector signal	Fuel injector signal 8
26	output	Fuel injector signal	Fuel injector signal 3
27	output	Crankshaft position sensor signal	Crankshaft position sensor
28	input	Engine coolant temperature sensor signal	Engine coolant temperature sensor
29	input	Camshaft position sensor signal	Camshaft position sensor, intake bank 1
30	input	Camshaft position sensor signal	Camshaft position sensor, intake bank 2
31	input	Throttle valve signal	Throttle valve
32	input	Throttle valve signal	Throttle valve
33	input	Knock sensor signal	Knock sensor 1
34	input	Knock sensor signal	Knock sensor 1
35	input	Knock sensor signal	Knock sensor 2
36	input	Knock sensor signal	Knock sensor 2
37	ground	Ground	Connector X6450

BentleyPublishers.com
BentleyPublishers.com—All Rights Reserved

Connector X60003 52-pin (signals) (continued)			
Pin	**Type**	**Component or function**	**Note**
38	input / output	Signal PT-CAN high	Valvetronic control module
39	-	Not used	
40	output	Variable intake manifold changeover valve signal	
41	output	Variable intake manifold changeover valve signal	Knock sensor
42	output	Throttle valve signal	Throttle valve
43	output	Throttle valve signal	Throttle valve
44	-	Not used	
45	input	Variable intake manifold changeover valve signal	
46	input	Knock sensor signal	Knock sensor 1
47	input	Knock sensor signal	Knock sensor 1
48	input	Knock sensor signal	Knock sensor 2
49	input	Knock sensor signal	Knock sensor 2
50	output	Throttle valve signal	Throttle valve
51	input / output	Signal PT-CAN high	Valvetronic control module
52	ground	Throttle valve signal	Throttle valve

Connector X60004 40-pin (signals)			
Pin	**Type**	**Component or function**	**Note**
1	-	Not used	
2	input	DMTL signal	Diagnostic module for fuel tank leakage
3	output	Secondary air pump relay	Secondary air pump
4	output	Engine cooling fan signal	Engine cooling fan
5	input	Secondary air mass air flow signal	Secondary air mass air flow sensor
6	-	Not used	
7	ground	Accelerator pedal module	Accelerator pedal module
8	input	Accelerator pedal module signal	Accelerator pedal module
9	output	Accelerator pedal module	Accelerator pedal module
10	output	Fuel pump relay activation	Fuel pump relay
11	-	Not used	
12	ground	Accelerator pedal module	Accelerator pedal module
13	input	Accelerator pedal module	Accelerator pedal module
14	output	Accelerator pedal module	Accelerator pedal module
15	ground	Secondary air mass air flow sensor	Secondary air mass air flow sensor
16	output	Tailpipe flap signal	Tailpipe flap
17	output	TD signal	Connector X243
18	output	Tailpipe flap signal	Tailpipe flap
19	output	E-box cooling fan signal	E-box cooling fan

BentleyPublishers.com—All Rights Reserved

Connector X60004 40-pin (signals) (continued)			
Pin	**Type**	**Component or function**	**Note**
20	input	DMTL signal	Diagnostic module for fuel tank leakage
21	output	Radiator shutter signal	Radiator shutter
22	input	Processed wheel speed	Right rear wheel
23	input	Clutch switch signal	Clutch switch
24	input	Brake light switch	Connector X181
25	input	Oil pressure switch signal	Oil pressure switch
26	input	Terminal 15	Wake up signal
27	-	Not used	
28	input	Brake light switch signal	Test signal
29	-	Not used	
30	input	DMTL signal	Diagnostic module for fuel tank leakage
31	-	Not used	
32	-	Not used	
33	input	EWS signal	Car access system (CAS)
34	-	Not used	
35	-	Not used	
36	input / output	Signal PT-CAN high	CAN bus connector
37	input / output	Signal PT-CAN high	CAN bus connector
38	input	Engine coolant temperature sensor signal	Engine coolant temperature sensor, radiator outlet
39	input	Engine coolant temperature sensor signal	Engine coolant temperature sensor, radiator outlet
40	input	Start signal	Car access system (CAS)

Connector X60005 9-pin (ignition coils)			
Pin	**Type**	**Component or function**	**Note**
1	output	Ignition coil signal	Ignition coil signal 3
2	output	Ignition coil signal	Ignition coil signal 4
3	output	Ignition coil signal	Ignition coil signal 1
4	output	Ignition coil signal	Ignition coil signal 7
5	ground	Ground	
6	output	Ignition coil signal	Ignition coil signal 8
7	output	Ignition coil signal	Ignition coil signal 2
8	output	Ignition coil signal	Ignition coil signal 6
9	output	Ignition coil signal	Ignition coil signal 5

BentleyPublishers.com
BentleyPublishers.com—All Rights Reserved

160 Fuel Tank and Fuel Pump

TABLE

BentleyPublishers.com–All Rights Reserved

General

This repair group covers service information for the fuel supply system.

See also:

- **100 Engine–General** for model year, engine code, and DME applications
- **130 Fuel Injection (M54 Engine)**
- **131 Fuel Injection (N52 Engine)**
- **132 Fuel Injection (N54 Engine)**
- **133 Fuel Injection (V8 Engine)**
- **ECL Electrical Component Locations** for fuel pump fuse and relay access information

Fuel tank

The saddle-shaped plastic fuel tank is mounted underneath the center of the car (underneath the rear seat). Mounted in the fuel tank are the electric fuel pump, fuel level sending units and suction jet pump. Connecting lines for the evaporative emission control system are also attached to the tank.

Fuel tank capacity for 5 Series models is listed in **Table a**.

Table a. Fuel tank capacity for 5 Series models	
Tank capacity	70 liters (18.5 US gal)
Reserve capacity	10 liters (2.6 US gal)

Fuel delivery systems

5 Series models use two main fuel delivery systems. Naturally aspirated M54, N52, V8 (N62) engines use only an electric fuel pump. Turbocharged N54 engine models use an electric and mechanical fuel pump.

M54 and N62 engine fuel injectors are installed in machine-bored holes in the intake manifold. An in-tank electric fuel pump delivers pressure-regulated fuel to the fuel rail. The fuel pump is controlled by the DME via a fuel pump relay (K96). It operates at a pressure of 3.5 bar (50.7 psi).

N52 engine fuel injectors are installed in machine-bored holes in the cylinder head. An in-tank electric fuel pump delivers pressure-regulated fuel to the fuel rail. The fuel pump is a variable speed pump controlled by the EKP (electric fuel pump control) module and delivers only the amount of fuel needed for engine operation. This eliminates excess fuel delivery and reduces pump wear. It can operate at up to 5 bar (72.5 psi) and 95 liters (25 gallons) per hour.

N54 engine direct fuel injection allows for precise metering of mixture and higher compression. This is made possible by locating a fuel injector for each cylinder centrally between the valves. The electric fuel pump delivers fuel to a high pressure fuel pump. The high pressure fuel pump is driven by a chain and can provide fuel pressure from 5 bar (72.5 psi) up to 200 bar (2900 psi).

BentleyPublishers.com
BentleyPublishers.com—All Rights Reserved

Electric fuel pump

An electric fuel pump is mounted in the fuel tank in tandem with the right side fuel level sender and suction jet pump. A suction jet pump is located in the left side of the tank to transfer fuel to the right side. The combined fuel filter / pressure regulator delivers pressurized fuel to the fuel injection system. The system is a non-return design; there is one fuel line to the fuel injectors and excess fuel flows directly from the fuel filter / pressure regulator back into the tank.

Fuel pump self-diagnostics are built into the DME. A failure in the fuel pump circuit may set a fault code.

160

Fuel system schematic (M54, N62 engines)

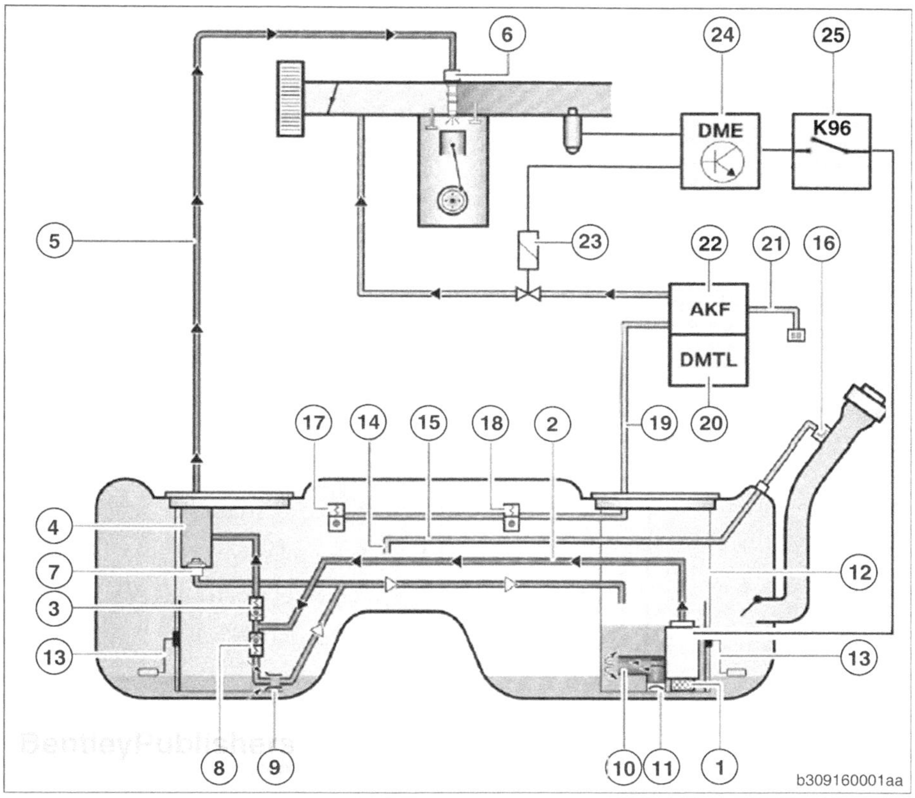

1. Electric fuel pump
2. Fuel supply line
3. Check valve
4. Fuel filter
5. Fuel supply line to engine
6. Fuel injector
7. Pressure regulator
8. Check valve
9. Left suction jet pump
10. Right suction jet pump
11. Pump initial filling valve
12. Fuel baffle
13. Fuel level sender
14. Refueling line connector
15. Refueling ventilation line
16. Refueling ventilation nipple
17. Left operation ventilation valve
18. Right operation ventilation valve
19. Operation ventilation line
20. Diagnosis pump for tank leakage (DMTL)
21. Fresh air vent line
22. Activated carbon canister
23. Fuel tank vent valve
24. Engine control module (ECM)
25. Fuel pump relay (K96)

BentleyPublishers
BentleyPublishers.com—All Rights Reserved

Fuel system schematic (N52 engine)

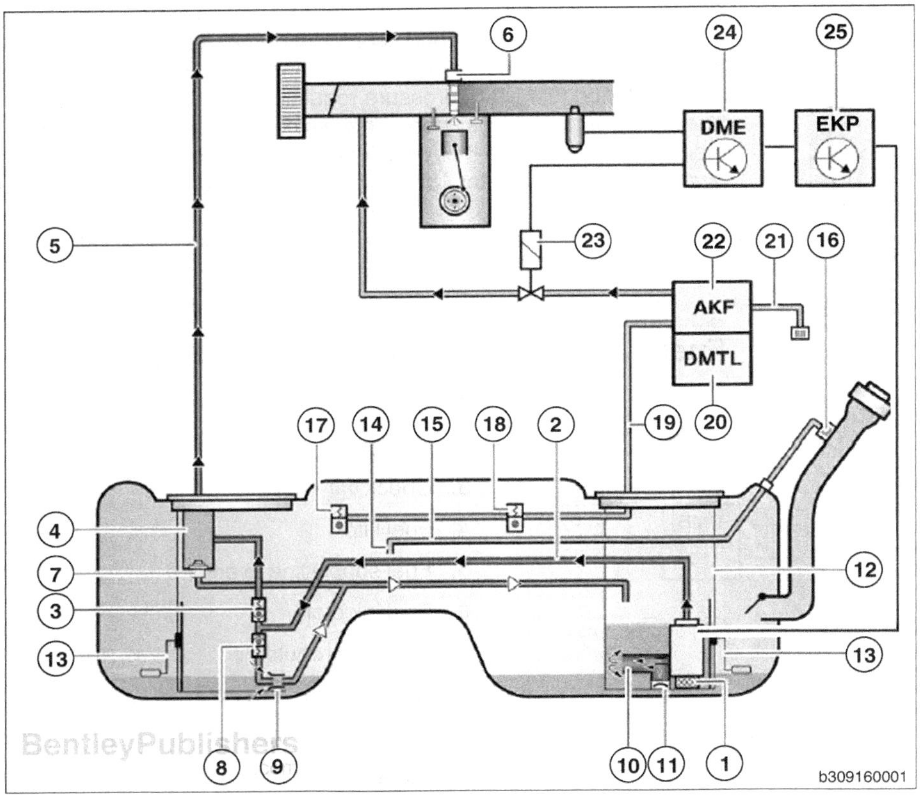

1. Electric fuel pump
2. Fuel supply line
3. Check valve
4. Fuel filter
5. Fuel supply line to engine
6. Fuel injector
7. Pressure regulator
8. Check valve
9. Left suction jet pump
10. Suction jet pump
11. Pump initial filling valve
12. Fuel baffle
13. Fuel level sender
14. Refueling line connector
15. Refueling ventilation line
16. Refueling ventilation nipple
17. Left operation ventilation valve
18. Right operation ventilation valve
19. Operation ventilation line
20. Diagnosis pump for tank leakage (DMTL)
21. Fresh air vent line
22. Activated carbon canister
23. Fuel tank vent valve
24. Engine control module (ECM)
25. Fuel pump module (EKP)

BentleyPublishers.com
BentleyPublishers.com—All Rights Reserved

Fuel system schematic (N54 engine)

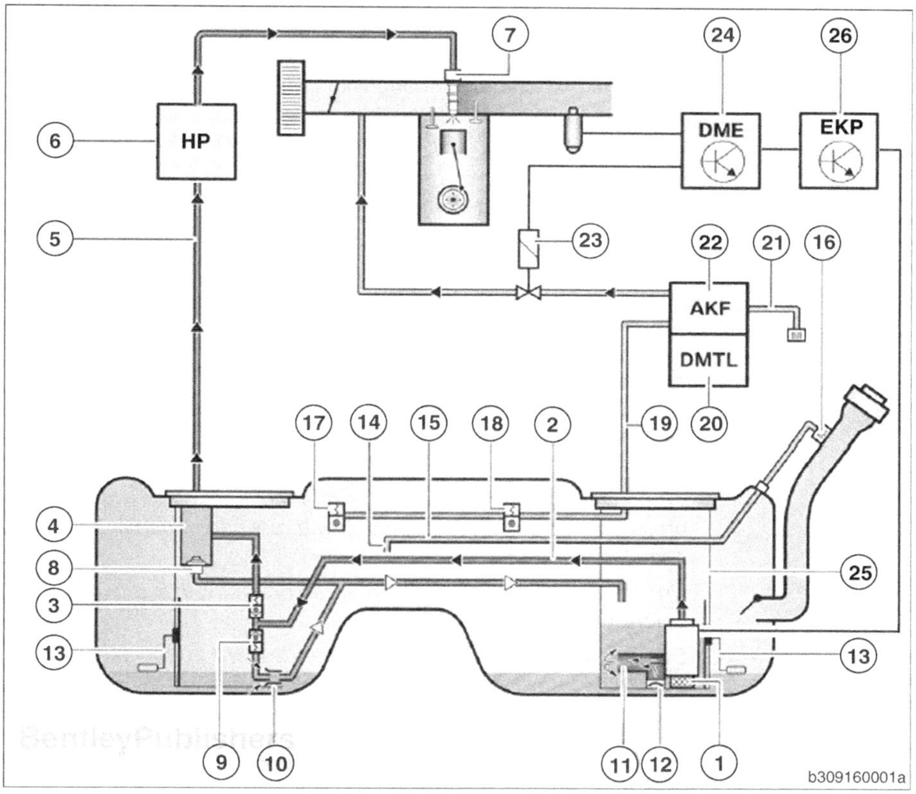

1. Electric fuel pump
2. Fuel supply line
3. Check valve
4. Fuel filter
5. Fuel supply line to engine
6. High pressure fuel pump
7. Fuel injector
8. Pressure regulator
9. Check valve
10. Left suction jet pump
11. Suction jet pump
12. Pump initial filling valve
13. Fuel level sender
14. Refueling line connector
15. Refueling ventilation line
16. Refueling ventilation nipple
17. Left operation ventilation valve
18. Right operation ventilation valve
19. Operation ventilation line
20. Diagnosis pump for tank leakage (DMTL)
21. Fresh air vent line
22. Activated carbon canister
23. Fuel tank vent valve
24. Engine control module (ECM)
25. Fuel baffle
26. Fuel pump module (EKP)

BentleyPublishers.com—All Rights Reserved

Warnings and Cautions

WARNING —
- *The fuel system is designed to retain pressure even when the ignition is OFF. When working with the fuel system, loosen the fuel lines slowly to allow residual fuel pressure to dissipate. Avoid spraying fuel. Use shop towels to capture leaking fuel.*
- *Before beginning work on the fuel system, place a fire extinguisher in the vicinity of the work area.*
- *Work only on the fuel system when engine temperature is below 40°C (104°F).*
- *Electric fuel pump starts each time a door is opened.*
- *Fuel is highly flammable. When working around fuel, do not disconnect wires that could cause electrical sparks. Do not smoke or work near heaters or other fire hazards.*
- *Wear eye protection, face shield and protective clothing to avoid injuries from contact with fuel.*
- *When working on an open fuel system, wear suitable hand protection, as prolonged contact with fuel can cause illnesses and skin disorders.*
- *Unscrew the fuel tank cap to release pressure in the tank before working on fuel lines.*
- *Do not use a work light with an incandescent bulb near fuel. Fuel may spray on the hot bulb causing a fire.*
- *Make sure the work area is properly ventilated.*
- *Due to risk of personal injury, be sure the engine is cold before beginning work on engine components.*

CAUTION—
- *Prior to disconnecting the battery, read the battery disconnection cautions given in* **001 Warnings and Cautions**.
- *Before making any electrical tests with ignition switched ON, disable the ignition system as described in* **120 Ignition System**. *Be sure the battery is disconnected when replacing components.*
- *To prevent damage to the ignition system or other DME components, including the engine control module (ECM), connect and disconnect wires and test equipment with ignition OFF.*
- *Cleanliness is essential when working with the fuel system. Thoroughly clean the fuel line unions before disconnecting any of the lines. Plug open fuel lines and ports.*
- *Use only clean tools. Keep removed parts clean and sealed or covered with a clean, lint-free cloth, especially if completion of the repair is delayed.*
- *Do not move the car while the fuel system is open.*
- *Avoid using high pressure compressed air to blow out lines and components. High pressure can rupture internal seals and gaskets.*
- *Use new seals, O-rings and hose clamps when replacing fuel system components*
- *Replace aluminum fasteners when loosened.*
- *For reliable identification test fastener with magnet for aluminum composition.*

BentleyPublishers.com
BentleyPublishers.com—All Rights Reserved

Fuel Tank Evaporative Control System

Evaporative control, also referred to as running losses control, is designed to prevent fuel system evaporative losses from venting into the atmosphere. The components of this system allow control and monitoring of evaporative losses by on-board diagnostic (OBD II) software incorporated into the engine control module (ECM).

Listed below are the main components of the evaporative control system and their functions:

- Carbon canister stores fuel vapors.
- Plumbing ducts vapors from fuel tank to canister and from canister to intake manifold.
- Carbon canister purge valve is controlled by engine control module (ECM).
- Pressure regulator shunts excess fuel volume directly back to fuel tank before it circulates through fuel lines.
- Leak detection unit (DMTL) pressurizes fuel tank and evaporative system to monitor system leaks.

Evaporative system troubleshooting

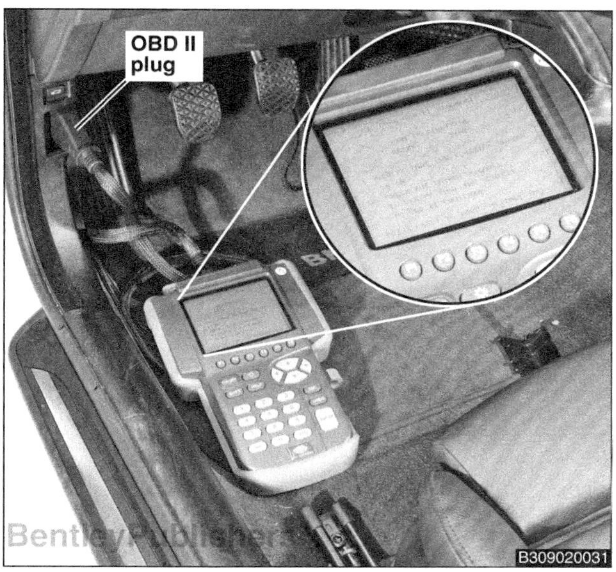

< Start by accessing diagnostic trouble codes (DTCs) using a BMW scan tool.

- For purposes of OBD II emissions compliance, the DME system sets a diagnostic trouble code (DTC) when it detects a leak in the evaporative control system.
- Malfunction indicator light (MIL) is illuminated upon second recurrence of fault. See **OBD On-Board Diagnostics**.

– When leak testing, observe the following conditions to obtain plausible results:

- Fuel tank ¼ to ¾ full.
- Vehicle parked for at least 2 hours to allow fuel to reach ambient temperature. Ideal fuel temperature is 10°- 20°C (50° - 68°F).
- Do not refuel immediately before leak test.

– If a leak is detected, check the following areas:

- Fuel filler cap leaking or off.
- Fuel tank ventilation lines leaking at fuel tank or activated carbon canister.
- Tank ventilation valve leaking (in engine compartment).
- Fuel level sensor and fuel pump assembly cover leaking.

For evaporative system component replacement, see **Fuel Supply Components** in this repair group.

BentleyPublishers
BentleyPublishers.com—All Rights Reserved

Evaporative system leak test

Use the following leak test when diagnosing evaporative system leak fault codes.

– Working underneath left rear bumper: gain access to activated carbon canister. See **Activated carbon canister, removing and installing** in this repair group.

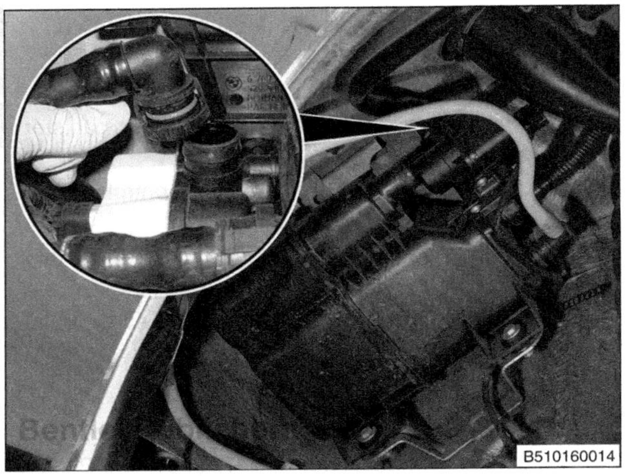

◀ Remove fresh air line (**inset**) from DMTL pump.

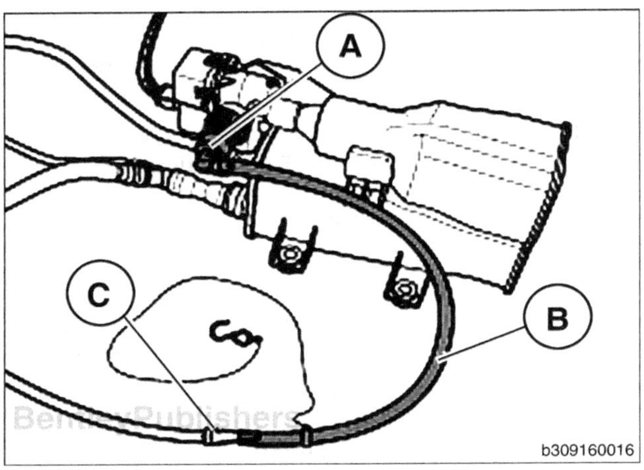

◀ Install BMW test adapter 83 30 0 433 207 (**B**) (or suitable substitute) at DMTL fresh air connection (**A**).

– Connect evaporative system smoke machine (**C**) to test adapter (**B**). BMW recommends VACUTEC 625-522B (nitrogen pressure swing absorption technology).

– Remove fuel filler cap.

– Activate smoke machine and fill fuel system with smoke until it escapes from fuel filler neck.

– Shut off smoke machine.

> ***WARNING —***
> - *Use only smoke machine that utilizes nitrogen to pressurize fuel system.*
> - *The fuel system may release a substantial amount of fuel-rich gasoline vapors during testing. Work in a well ventilated area.*

– Reinstall fuel filler cap.

– Reactivate smoke machine and fill fuel system with smoke.

> ***WARNING —***
> - *Do not exceed a maximum fuel system pressure of 0.3 bar (4 psi) when using smoke machine.*

– Examine fuel system components for signs of escaping smoke. Repair or replace faulty components.

– After completing tests:
- Attach DMTL fresh air line.
- Check and clear fault codes from ECM memory.

BentleyPublishers.com
BentleyPublishers.com—All Rights Reserved

FUEL SYSTEM TROUBLESHOOTING

Fuel pump fuse and relay (M54, N62 engines)

Begin troubleshooting fuel system faults by checking fuel pump fuse and relay. Also, check DME main relay and integrated voltage supply module (IVM). The fuel pump relay is located behind the glove compartment. See **ECL Electrical Component Locations** for access information.

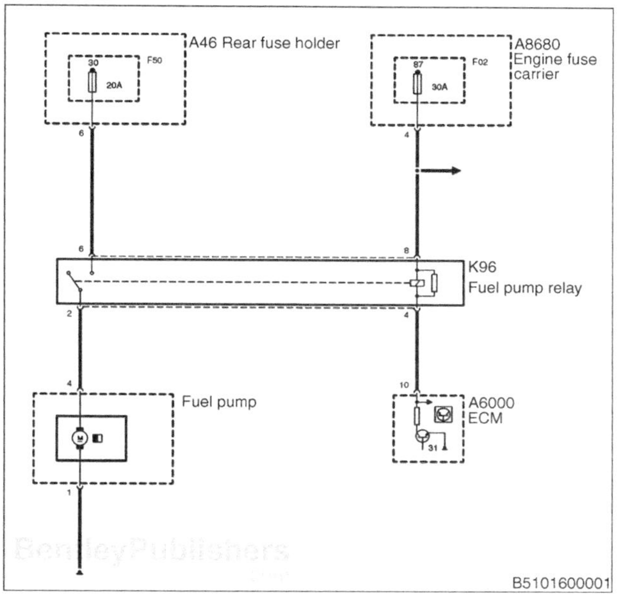

Fuel pump supply circuit is fuse-protected by:

- M54 engine: fuse F50 in the rear fuse holder:
- N62 engine (to 05/2009): fuse F50 in the rear fuse holder
- N62 engine (from 05/2009): fuse F72 in the rear fuse holder

> ***CAUTION—***
> - *Relay and fuse positions and wire colors vary. A good way to verify a relay position or wire colors is to compare the wiring colors at the relay socket to the colors indicated on the wiring diagrams in* **ELE Electrical Wiring Diagrams**.

– The engine control module (ECM) supplies switched ground to relay. During starting, the fuel pump runs as long as ignition switch is in START position and continues to run once engine starts. If electrical system faults interrupt power to fuel pump, the engine does not run.

Fuel pump fuse and module (N52, N54 engines)

Begin troubleshooting fuel system faults by checking fuel pump fuse. The fuel pump fuse is located in the rear fuse holder. See **ECL Electrical Component Locations** for access information.

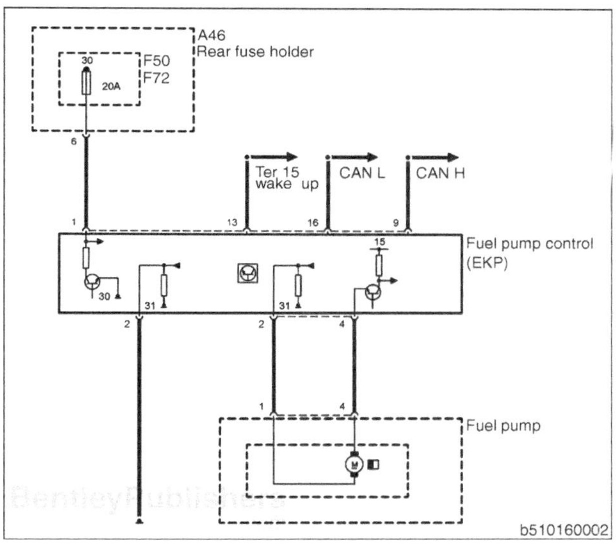

Fuel pump circuit is fuse-protected by:

- N52 engine (to 05/2009): Fuse F50
- N52 engine (from 05/2009): Fuse F72
- N54 engine: Fuse F72

> ***CAUTION—***
> - *Relay and fuse positions and wire colors vary. A good way to verify a relay position or wire colors is to compare the wiring colors at the relay socket to the colors indicated on the wiring diagrams in* **ELE Electrical Wiring Diagrams**.

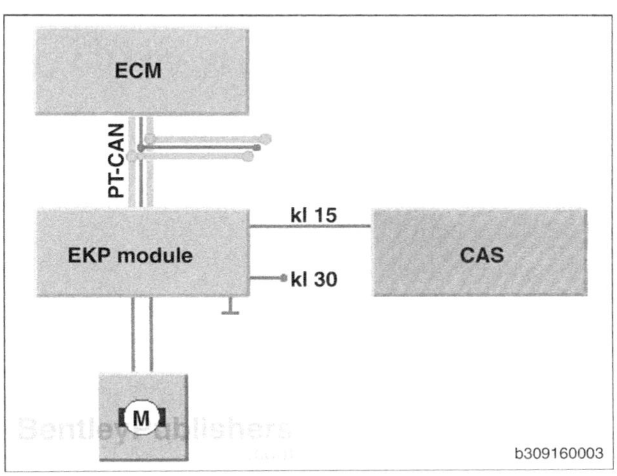

The ECM calculates the amount of fuel required at the given point in time and the electric fuel pump is activated as required. The total volume required is transmitted as a message to the EKP control module via PT-CAN.

The EKP module converts this message into an output voltage which controls the fuel pump. The fuel pump is capable of delivering up to 95 liters (25 gallons) per hour at a pressure of 5 bar (72.5 psi). On turbo engine models, if the ECM detects a fault in the low pressure fuel sensor, the fuel pump is activated to 100% or 5 bar (72.5 psi).

BentleyPublishers.com
BentleyPublishers.com—All Rights Reserved

Operating fuel pump for tests (M54, N62 engines)

This procedure explains how to operate the fuel pump for testing purposes without having to run the engine.

- Remove fuel pump relay (**K96**) located in relay panel behind the glove compartment. See **ECL Electrical Component Locations** for access information.

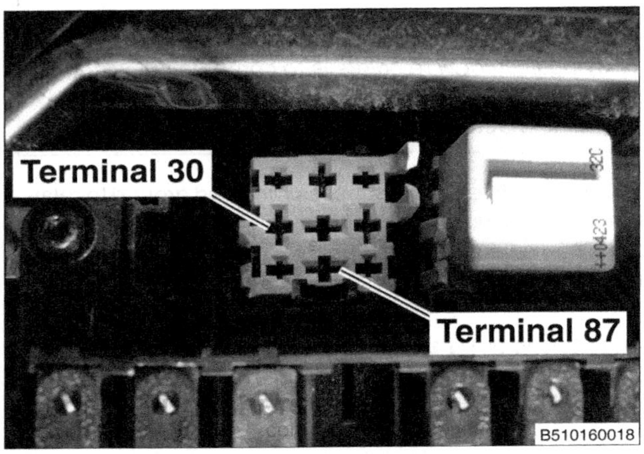

- Connect relay socket terminal 30 to relay socket terminal 87 with fused jumper wire.
  - Use 14 ga wire to make jumper and include an in-line fuse holder with a 15A fuse. To avoid relay socket damage from repeated connecting and disconnecting, also include a toggle switch in jumper harness.
  - Fuel pump operates as soon as jumper wire is attached.

– If pump does not run with jumper installed, fault could be in fuel pump, fuse F50, or wiring to pump. Check pump, fuses and its wiring. See **Fuel pump electrical circuit, testing** in this repair group.

– After completing tests, remove jumper harness and install relay.

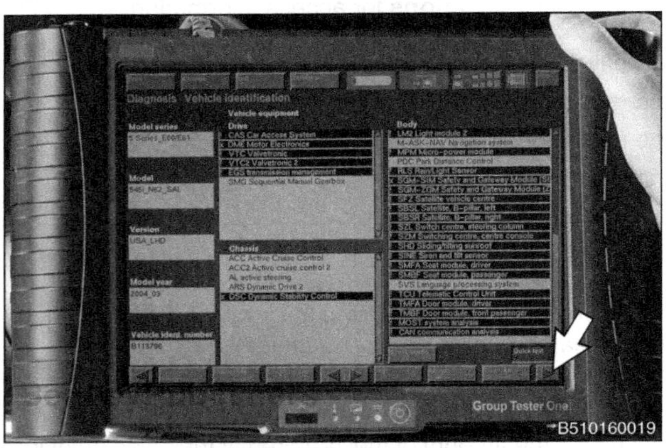

Operating fuel pump for tests (N52, N54 engines)

This procedure explains how to operate the fuel pump for testing purposes without having to run the engine.

- Using BMW scan tool:
  - Identify vehicle
  - Select **Next** (**arrow**)
  - Select **Control Module Functions**
  - Select **Complete Vehicle**
  - Select **Drive**
  - Select **DME (Motor Electronics)**
  - Select **Component Activation**
  - Select **Fuel Pump**

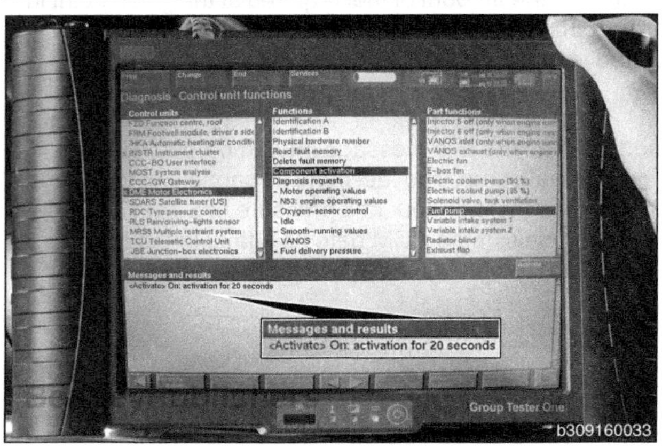

- Fuel pump will run at 100% (5 bar) for 20 seconds.

BentleyPublishers.com
BentleyPublishers.com—All Rights Reserved

Fuel pump electrical circuit, testing

If fuel pump does not operate or there is no fuel pressure, first test fuel pump electrical circuit.

- Remove rear seat to access fuel pump and fuel level sender. See **Accessing fuel pump and fuel level senders** in this repair group.

< Slide electrical connector lock (in direction of **arrow**) and remove connector from fuel pump and fuel level sender.

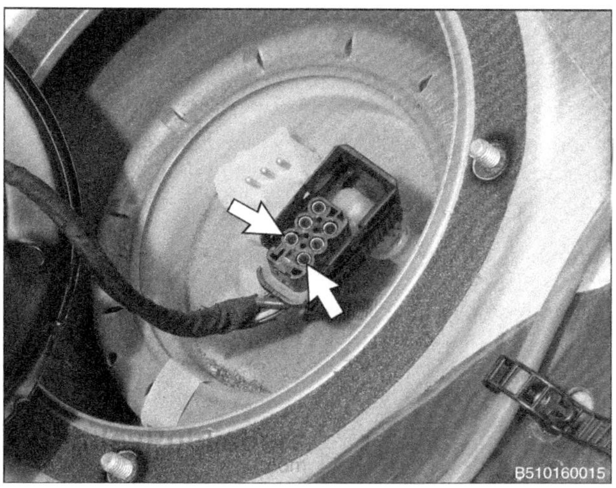

< Connect voltmeter between fuel pump leads (**arrows**) on connector.

- Activate fuel pump. See **Operating fuel pump for tests** in this repair group.
- Use digital multimeter to test for battery voltage at fuel pump connector leads.
- If voltage and ground are present, fuel pump is probably faulty. If there is no voltage, check wiring between fuel pump and fuel pump module.
- After completing tests, reconnect fuel pump harness connector.

BentleyPublishers.com
BentleyPublishers.com—All Rights Reserved

Fuel pump power consumption, testing (M54, N62 engines)

– Remove rear seat to access fuel pump and fuel level sender. See **Accessing fuel pump and fuel level senders** in this repair group.

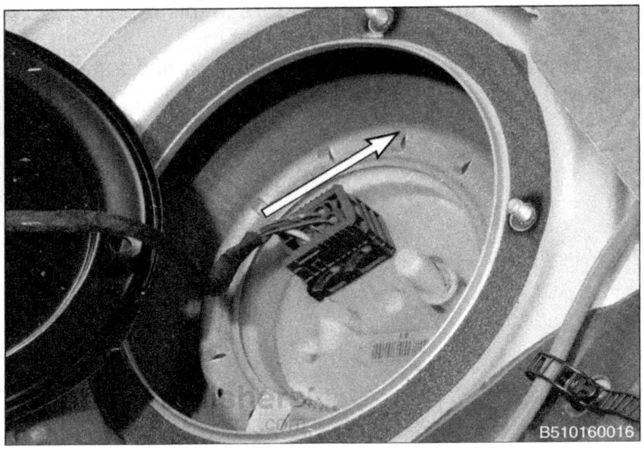

◄ Disconnect harness connector from fuel pump.

– Make sure battery voltage at fuel pump connector is 12.6 volts. Charge battery if necessary.

– Attach digital ammeter between pump lead and harness. Connect jumper wire between pump and harness.

> ***CAUTION—***
> • *Do not allow the test leads to short to ground.*

– Activate fuel pump. See **Operating fuel pump for tests (M54, N62 engines)** in this repair group.

◄ Compare ammeter reading with specification listed in table.

Fuel pump current (M54, N62 engines)	
Current consumption	8 - 13 amps

– Higher than normal power consumption by fuel pump may indicate a worn pump, causing intermittent fuel starvation due to pump overheating and seizure. Replace pump.

– Lower than normal power consumption may indicate blockage in a fuel line. Before replacing fuel pump, be sure to check that return line and pump pickup (inside fuel tank) are not obstructed.

BentleyPublishers.com
BentleyPublishers.com—All Rights Reserved

Fuel pump power consumption, testing (N52, N54 engines)

If fuel delivery is erratic or poor, or if fuel pump makes abnormally loud noises, test pump power consumption.

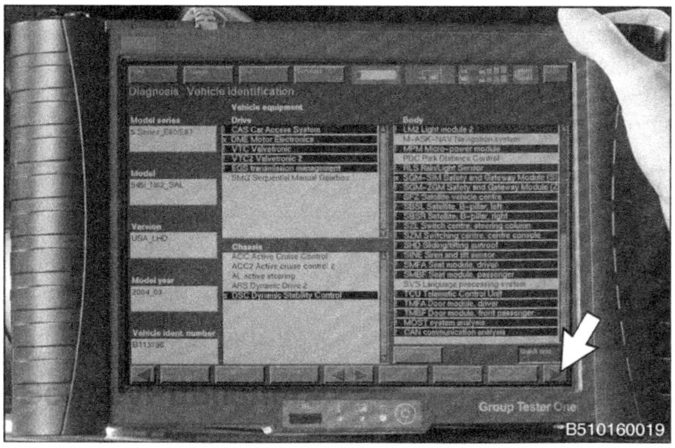

Using BMW factory scan tool or equivalent:

- Identify vehicle
- Select **Next** (**arrow**)
- Select **Function Selection**
- Select **DME (Motor Electronics)**
- Select **Fuel Pump**
- Select **Control Current**
- Select **Next (arrow)**

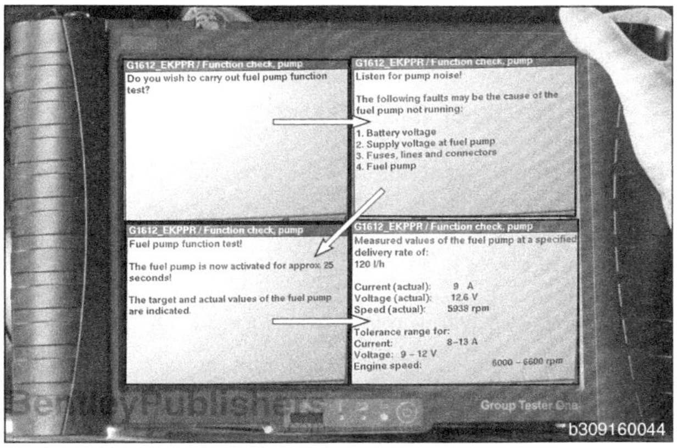

Follow steps on screen to monitor fuel pump current.

Fuel pump current (N52, N54 engines)	
Current consumption (fuel pump activated with BMW scan tool)	8 - 13 amps

An alternate method is to install a digital ammeter in fuel pump circuit using the following steps.

– Remove rear seat to access fuel pump and fuel level sender. See **Accessing fuel pump and fuel level senders** in this repair group.

Slide locking connector in direction of arrow and remove electrical connector from fuel pump.

– Make sure battery voltage at fuel pump connector is 12.6 volts. Charge battery if necessary.

BentleyPublishers.com
BentleyPublishers.com—All Rights Reserved

– Attach digital ammeter between pump lead and harness. Connect jumper wire between pump and harness.

> ***CAUTION—***
> • *Do not allow the test leads to short to ground.*

– Activate fuel pump. See **Operating fuel pump for tests (N52, N54 engines)** in this repair group.

◀ Compare ammeter reading with specification listed in table.

Fuel pump current (N52, N54 engines)	
Current consumption (fuel pump activated with scan tool)	8 - 13 amps

– Higher than normal power consumption by fuel pump may indicate a worn pump, causing intermittent fuel starvation due to pump overheating and seizure. Replace pump.

– Lower than normal power consumption may indicate blockage in a fuel line. Before replacing fuel pump, be sure to check that return line and pump pickup (inside fuel tank) are not obstructed.

Fuel pressure gauge, installing

Working in engine compartment, install fuel pressure gauge at Schræder valve on fuel rail.

– Remove upper engine covers. See **020 Maintenance**.

Use a fuel pressure gauge with a minimum range of 0 to 7 bar (0 to 100 psi).

> ***WARNING—***
> • *The fuel system is designed to retain pressure even when the ignition is OFF. When working with the fuel system, loosen fuel lines slowly to allow residual fuel pressure to dissipate. Avoid spraying fuel. Use shop towels to capture leaking fuel.*
> • *Fuel pump starts each time door is opened.*
> • *Make sure the fuel pressure gauge is securely connected to the fuel rail to prevent it from coming loose under pressure.*

– Remove engine top cover. See **020 Maintenance**.

M54 engine

◀ Remove Schræder valve cap (**arrow**) at fuel rail and connect fuel pressure gauge.

BentleyPublishers.com
BentleyPublishers.com—All Rights Reserved

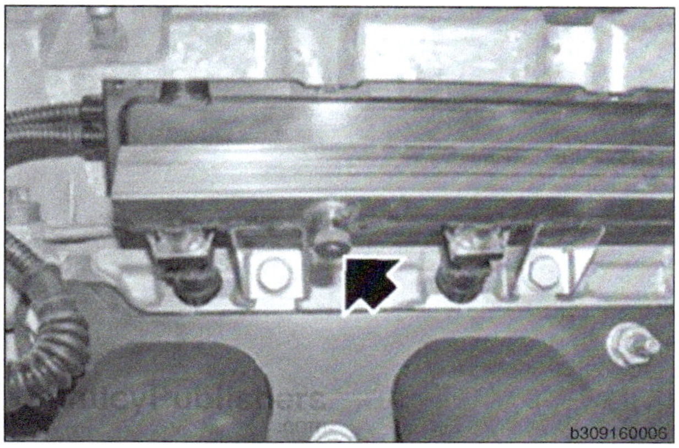

N52 engine

Remove Schræder valve cap (**arrow**) at fuel rail and connect fuel pressure gauge.

N54 engine

Working at left side engine compartment near strut tower, disconnect fuel line (**arrow**).

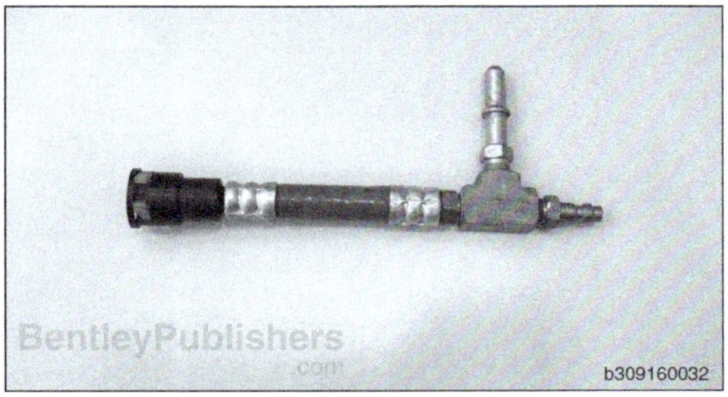

Install compatible T-style test adapter and install fuel pressure gauge.

BentleyPublishers.com
BentleyPublishers.com–All Rights Reserved

V8 engine

◄ Remove Schræder valve cap (**arrow**) at fuel rail and connect fuel pressure gauge.

Fuel pressure regulator

Fuel pressure is created by the fuel pump and maintained by the fuel pressure regulator.

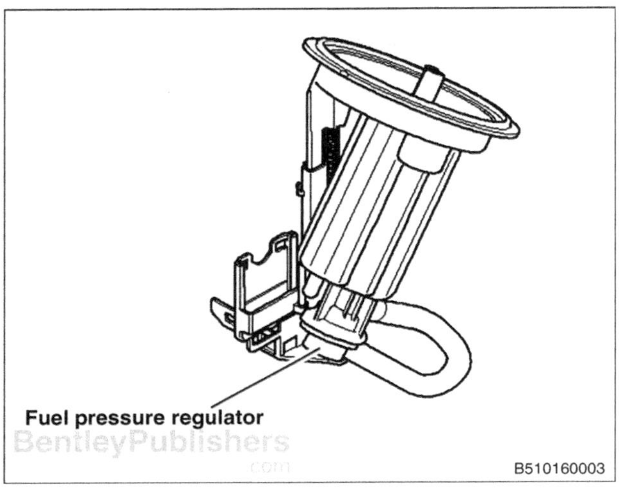

◄ Fuel pressure regulator is integrated into the fuel filter assembly.

BentleyPublishers.com
BentleyPublishers.com–All Rights Reserved

Fuel delivery, testing

Checking fuel delivery is a fundamental part of troubleshooting and diagnosing the DME system. Fuel pressure directly influences fuel delivery. There are two significant fuel delivery values to be measured:

- **System pressure** is created by fuel pump and maintained by pressure regulator. System pressure is adjustable only by the ECM.
- **Residual pressure** is pressure maintained in closed system after engine and fuel pump are shut off.

System pressure

– Attach fuel pressure gauge. See **Fuel pressure gauge, installing** in this repair group.

– Activate fuel pump. See **Operating fuel pump for tests** in this repair group. Compare fuel pressure to specifications in accompanying table.

Fuel system pressure specification	
M54, V8 (N62) engine	3.5 ± 0.2 bar (50.76 ± 2.9 psi)
N52, N54 engine	5 ± 0.2 bar (72.5 ± 2.9 psi)

CAUTION—
- *The fuel pump is capable of developing a higher pressure than that regulated by the pressure regulator. Do not allow pressure to rise above 6.5 bar (94 psi). Damage to the fuel lines or fuel system components could result.*

– After completing tests without running engine, check for fault codes and reset ECM memory.

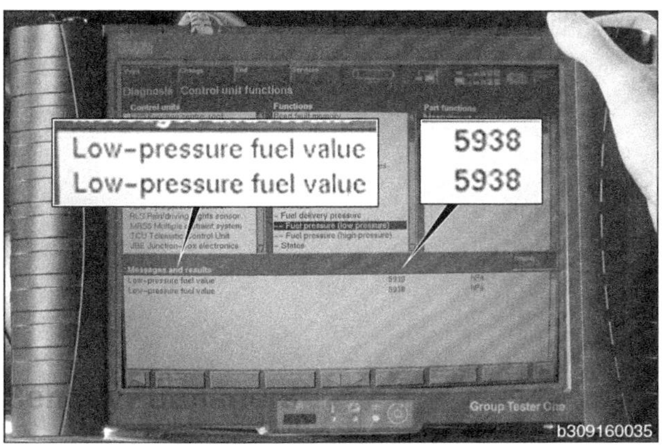

◄ N54 engine: Compare fuel pressure on gauge to data value on BMW scan tool.

– If pressure does not match, replace fuel pressure regulator. See **Fuel pump, removing and installing** in this repair group.

– After completing tests, note the following:
- M54, N52, V8 (N62) engine: replace Schræder valve cap.
- N54 engine: reassemble fuel line.
- Check and clear fault codes from ECM memory.

BentleyPublishers.com—All Rights Reserved

Residual pressure

For quick restarts and to avoid vapor lock when the engine is hot, the fuel injection system retains fuel pressure after the engine is shut off. This residual pressure is primarily maintained by the fuel pressure regulator and a check valve at the fuel pump outlet.

– Attach fuel pressure gauge. See **Fuel pressure gauge, installing** in this repair group.

– Start engine and allow it to run for approximately one minute. Note fuel pressure reading. Shut OFF engine.

– Note fuel pressure after approximately 20 minutes. Make sure pressure does not drop more than 0.5 bar from system pressure.

Fuel system pressure specification	
M54, V8 (N62) engines: System pressure, engine not running	3.5 ± 0.2 bar (50.76 ± 2.9 psi)
N52, N54 engines: System pressure, engine not running	5 ± 0.2 bar (72.5 ± 2.9 psi)

> ***CAUTION—***
> - *The fuel pump is capable of developing a higher pressure than that regulated by the pressure regulator. Do not allow pressure to rise above 6.5 bar (94 psi). Damage to the fuel lines or fuel system components could result.*

– When finished, disconnect pressure gauge and fitting, and replace Schræder valve cap.

– If fuel system does not maintain pressure:
- Visually check for leaks in fuel lines and unions.
- Check for leaking injector(s).
- Check for faulty fuel pump check valve.

BentleyPublishers.com
BentleyPublishers.com—All Rights Reserved

Fuel Pump and Fuel Level Senders

Accessing fuel pump and fuel level senders

NOTE—
- *Right side fuel pump / fuel level sender shown. Accessing left side fuel level sender is similar.*

– Remove rear seat cushion. See **520 Seats**.

– Peel back carpet to expose fuel sender service cover.

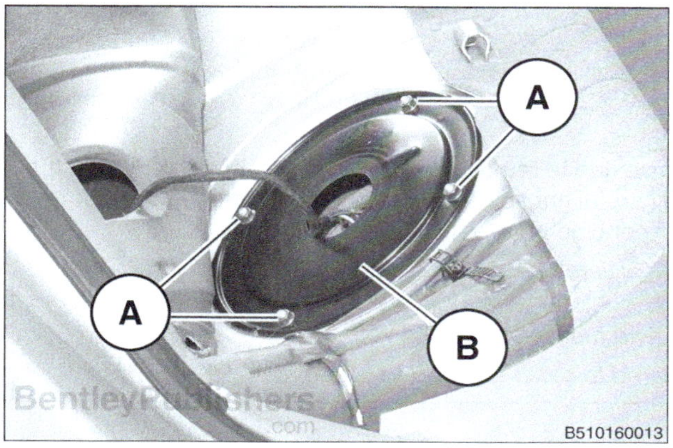

< Working at ride side fuel tank:
- Remove fasteners (**A**).
- Remove fuel pump access cover (**B**).

– Installation is reverse of removal.

Fuel pump and fuel level sender (right side), removing

– Drain fuel tank. See **Fuel tank, draining** in this repair group.

– Disconnect negative cable from battery.

CAUTION—
- *Prior to disconnecting the battery, read the battery disconnection cautions given in* **001 Warnings and Cautions**.

– Remove left side fuel filter and level sender. See **Fuel filter and fuel level sender (left side), removing** in this repair group.

– Remove fuel pump access cover. See **Accessing fuel pump and fuel level senders** in this repair group.

< Slide electrical connector lock in direction of **arrow,** remove connector and fuel lines if equipped.

WARNING —
- *When disconnecting fuel hose, wrap shop towel around end of hose to prevent spray of fuel under pressure.*

BentleyPublishers.com
BentleyPublishers.com—All Rights Reserved

Fuel pump and fuel level sender (right side), removing

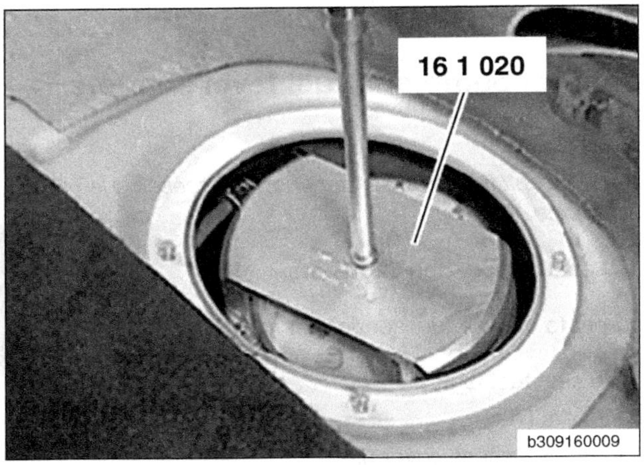

- Using BMW special tool 16 1 020, remove fuel pump service cap.
- Secure a stiff wire to hose pack through left side fuel tank service opening. This will help with installation.

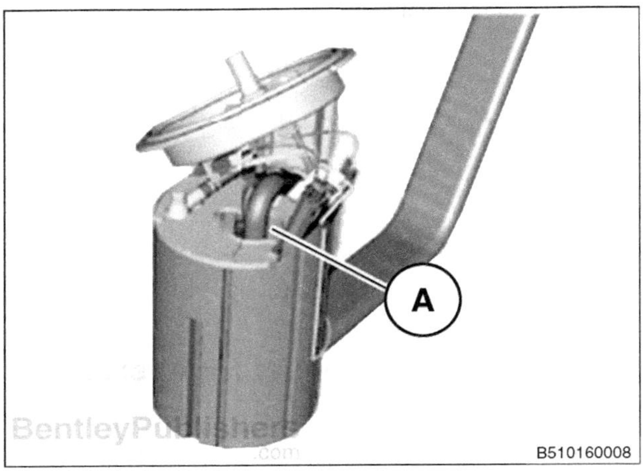

- Carefully raise and to remove sending unit with hose pack from fuel tank, being careful not spill fuel from charge container (**A**). Remove and discard sending unit sealing O-ring.

> ***CAUTION—***
> - *Do not bend fuel sender arm.*
> - *Be prepared to catch any dripping fuel.*

To replace fuel level sender use following step.

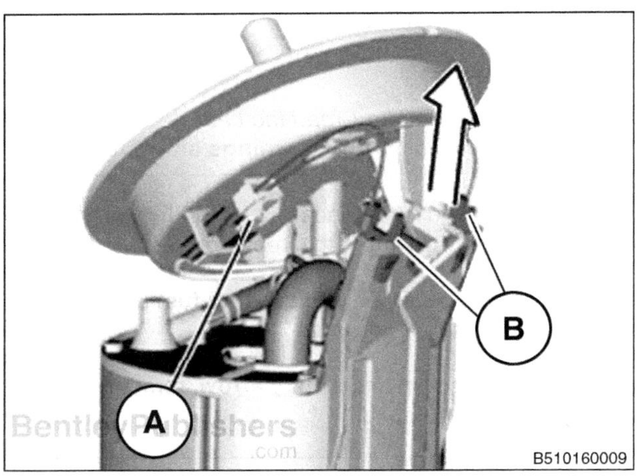

- Remove wire from holders and disconnect electrical connector (**A**). Carefully unlock tabs (**B**) and pull fuel level sensor off in direction of **arrow**.
- Installation of fuel level sender is reverse of removal. **See Fuel pump and fuel level sender (right side), installing** in this repair group for fuel pump installation.

BentleyPublishers.com
BentleyPublishers.com—All Rights Reserved

Fuel pump and fuel level sender (right side), installing

– Working inside fuel tank carefully install fuel pump and fuel level sender. If necessary rotate hose pack and tilt toward fuel level sender.

> **CAUTION—**
> - *To avoid kinking hoses be sure to rotate sensor unit and hoses in the same direction.*
> - *Do not bend fuel sender arm.*

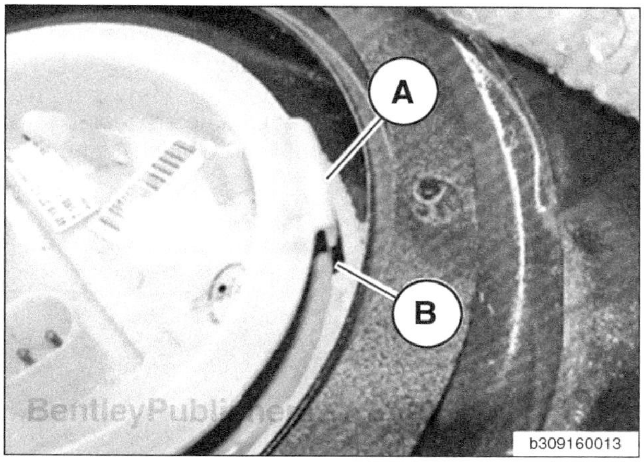

◄ When installing check to see that service cap lug (**A**) engages with corresponding slot (**B**) of fuel tank.

– Install service cap hand tight.

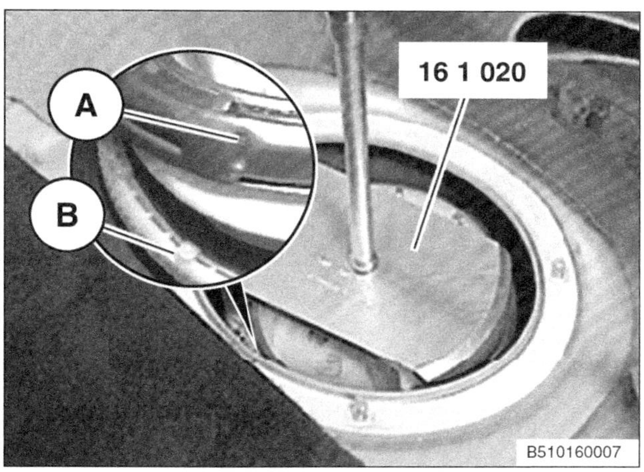

◄ Using BMW special tool 16 1 020 tighten service cap until notch on service cap (**A**) point to notch on fuel tank (**B**).

Tightening torque	
Service cap to fuel tank	45 ± 5 Nm (33 ± 3.6 ft-lb)

◄ Connect electrical connector by sliding lock in direction of **arrow** and connect fuel line if equipped.

– Install fuel tank access cover.

> **CAUTION—**
> - *After finishing repairs but before starting engine, be sure there is at least 5 liters (1.5 gallons) of fuel in the tank. The fuel pump is damaged if run without fuel.*

BentleyPublishers.com–All Rights Reserved

Fuel filter and fuel level sender (left side), removing

– Drain fuel tank. See **Fuel tank, draining** in this repair group.

– Disconnect negative cable from battery.

> ***CAUTION—***
> • *Prior to disconnecting the battery, read the battery disconnection cautions given in* **001 Warnings and Cautions**.

– Remove left side fuel sender access cover. See **Accessing fuel pump and fuel level senders** in this repair group.

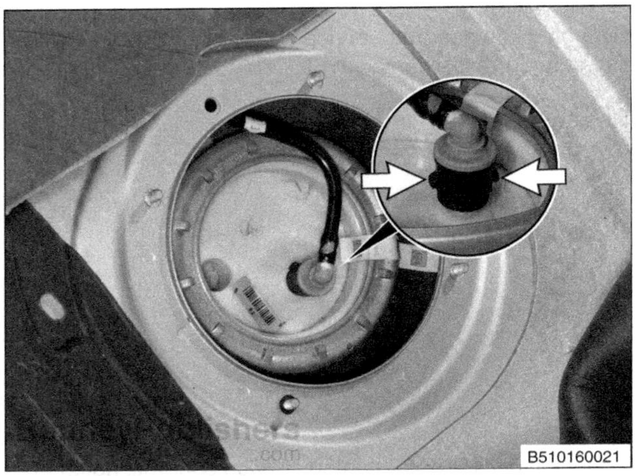

◄ Disconnect electrical connector (if equipped). Using only your fingers press tabs (**inset**) together and pull fuel supply (**B**) line straight up.

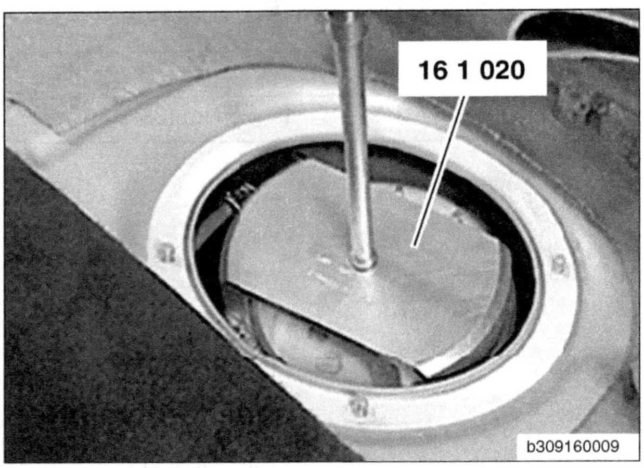

◄ Using special tool 16 1 020, remove fuel pump service cap.

– Carefully lift sender out of tank with hose pack. Remove and discard fuel tank seals (O-rings).

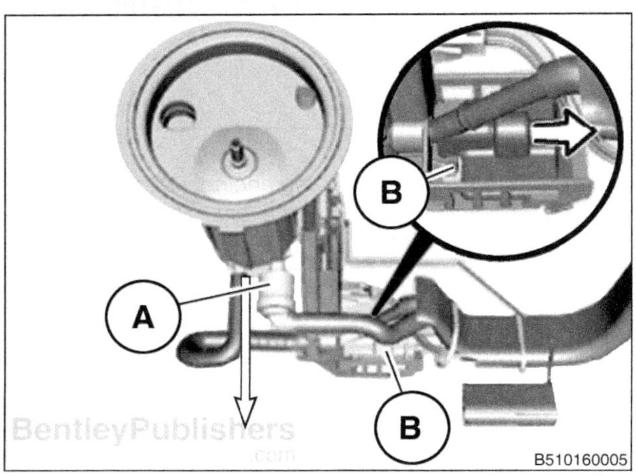

◄ Disconnect line (**A**) at fuel filter in direction of **arrow** and seal connection at fuel filter. Disconnect suction jet pump from fuel filter by pressing catch (**B**) downwards, then detach suction jet pump.

BentleyPublishers.com
BentleyPublishers.com—All Rights Reserved

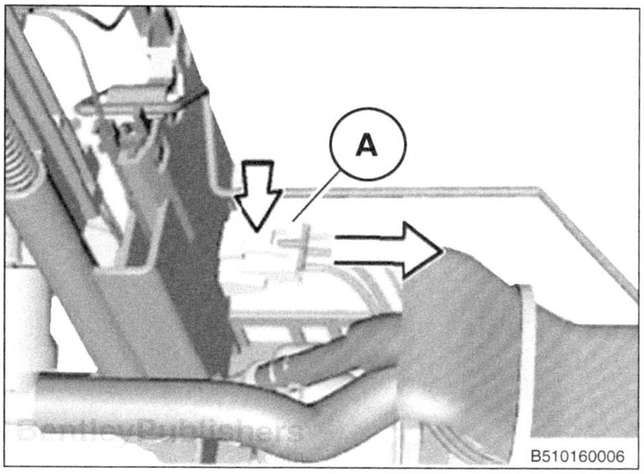

- Unlock and disconnect electrical connector (**A**) on sending unit.
- Remove sending unit from vehicle.

To replace fuel level sender, use following step.

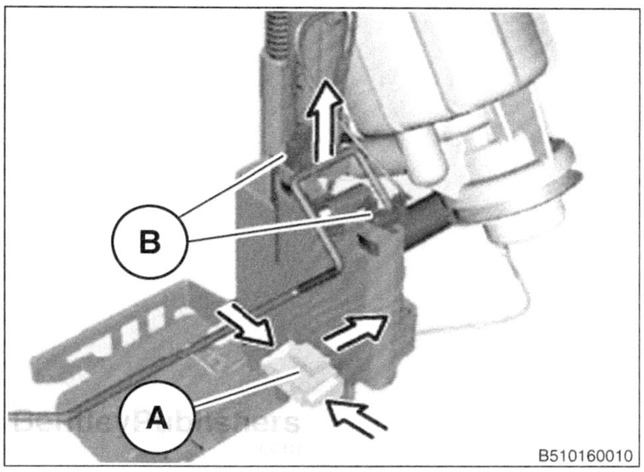

- Squeeze plug (**A**) tabs and remove in direction of **arrow**. Release catches (**B**) and pull fuel level sensor out of sender unit. Disconnect wires from pressure regulator and feed out of wire gland.
- Installation of fuel level sender is reverse of removal. **See Fuel pump and fuel level sender (right side), installing** in this repair group for fuel pump installing.

Fuel filter and fuel level sender (left side), installing

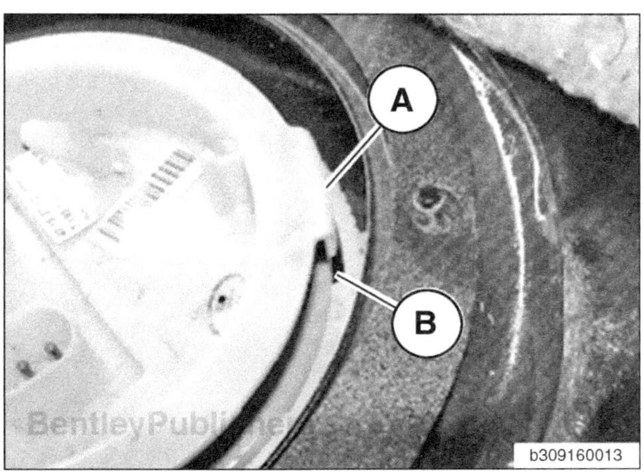

- When installing check to see that service cap lug (**A**) engages with corresponding slot of fuel tank (**B**).
- Install service cap hand tight.

BentleyPublishers.com—All Rights Reserved

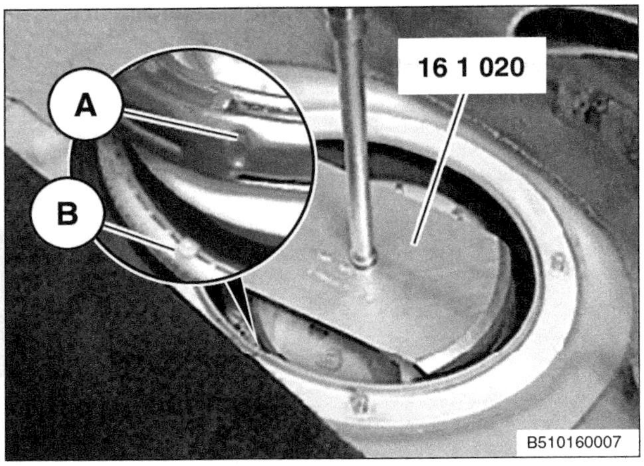

Using BMW special tool 16 1 020 tighten service cap until notch on service cap (**A**) points to notch on fuel tank (**B**).

Tightening torque	
Service cap to fuel tank	45 ± 5 Nm (33 ± 3.6 ft-lb)

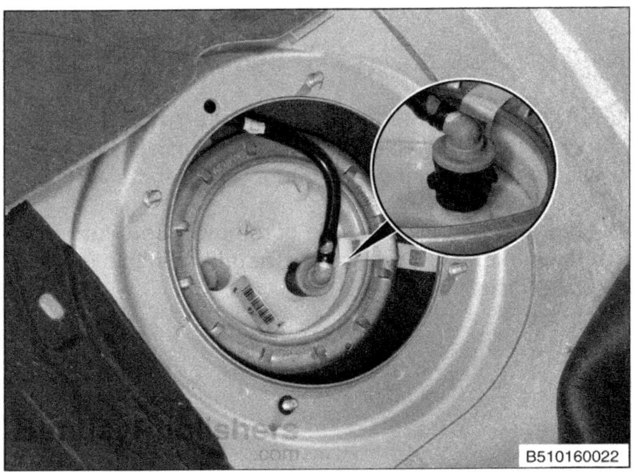

Connect electrical connector (if equipped) and fuel line (**inset**).

– Install fuel tank access cover.

> ***CAUTION—***
> - *After finishing repairs but before starting engine, be sure there is at least 5 liters (1.5 gallons) of fuel in the tank. The fuel pump is damaged if run without fuel.*

High-pressure fuel pump, removing (N54 engine)

– Remove intake manifold. See **132 Fuel Injection (N54 Engine)**.

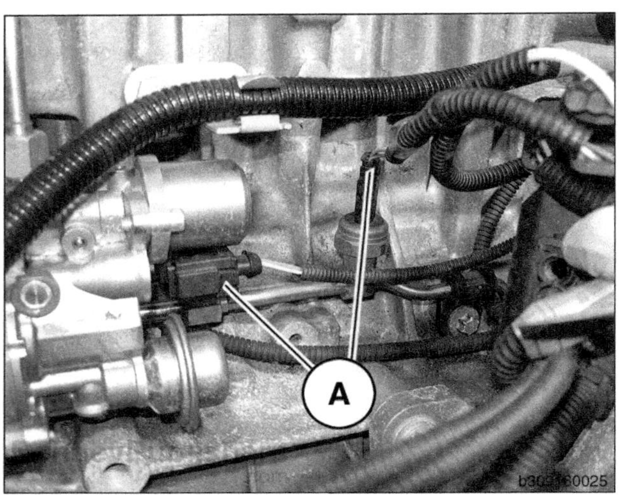

Working at side of engine block, remove electrical connectors (**A**) and lay harness aside.

BentleyPublishers.com
BentleyPublishers.com—All Rights Reserved

High-pressure fuel pump, removing (N54 engine)

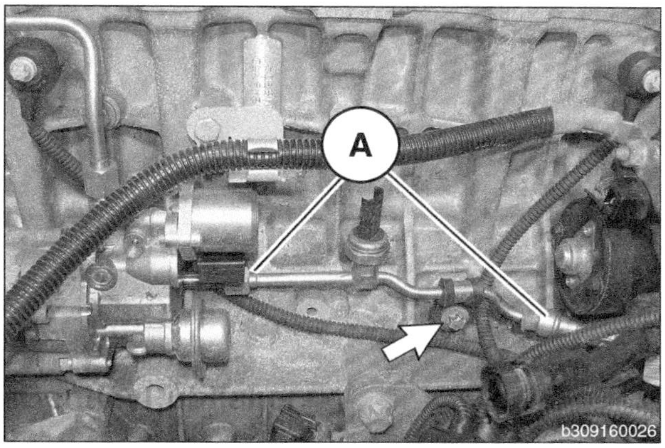

Remove aluminum fastener (**arrow**) and discard. Detach fuel line (**A**). Seal fuel lines with BMW special tools 13 5 281 and 13 5 282. Remember to replace aluminum fastener (**arrow**) with new.

> ***WARNING —***
> - *Wear gloves and face shield when loosening fuel lines.*

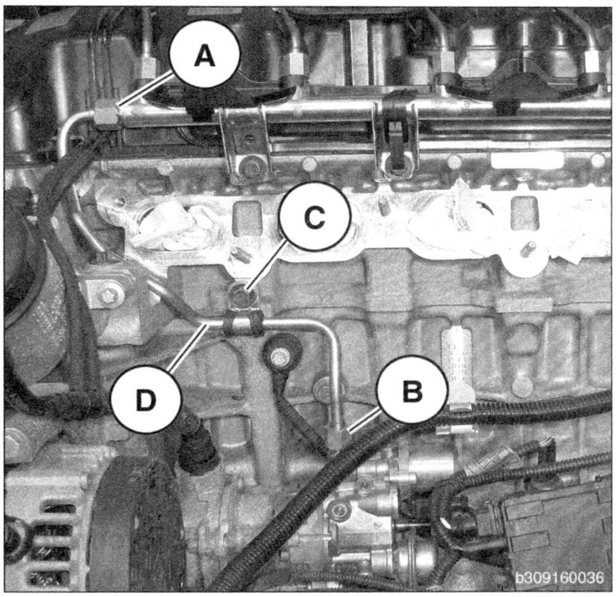

Loosen fuel line union nut (**A**, **B**) in this order. Remove fuel line bracket fastener (**C**) and remove fuel line (**D**). Seal fuel line connections with matching plugs from BMW special tool kit 32 1 270.

> ***WARNING —***
> - *Wear gloves and face shield when loosening fuel lines.*

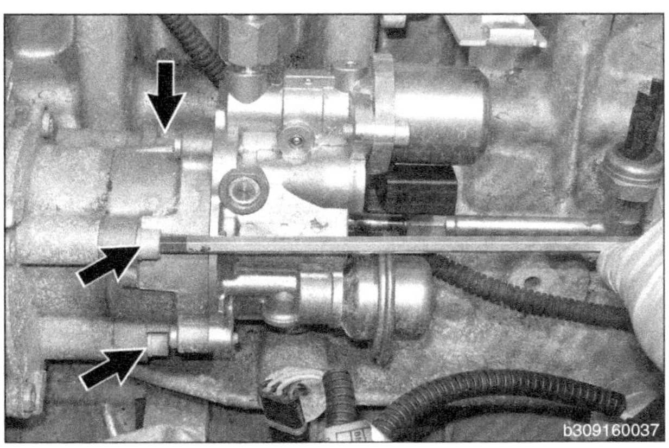

Remove high-pressure fuel pump mounting fasteners (**arrows**) and remove high-pressure fuel pump. Be prepared to catch any dripping oil with shop cloth.

BentleyPublishers.com
BentleyPublishers.com—All Rights Reserved

High-pressure fuel pump, installing (N54 engine)

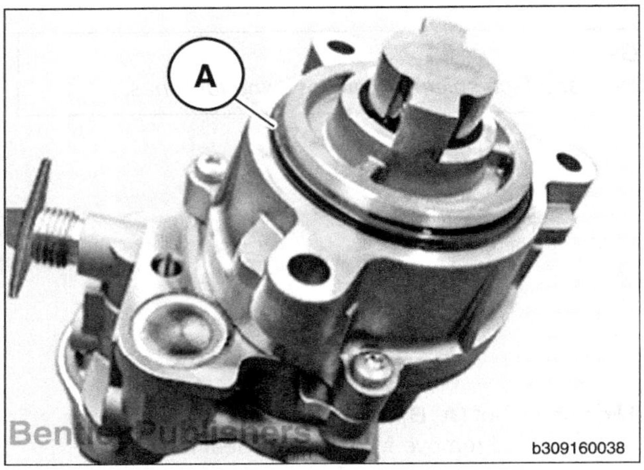

Replace high-pressure fuel pump O-ring (**A**) and clean mounting surfaces.

When installing, rotate pump (**A**) in direction of **arrow** until flush with mounting surface.

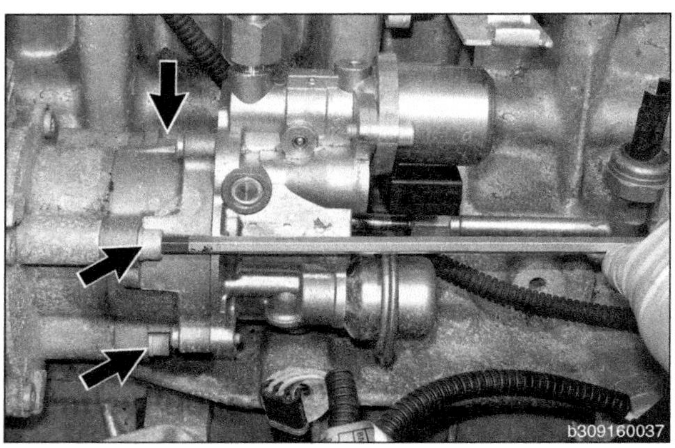

Install high-pressure fuel pump fasteners (**arrows**) finger tight.

BentleyPublishers.com
BentleyPublishers.com—All Rights Reserved

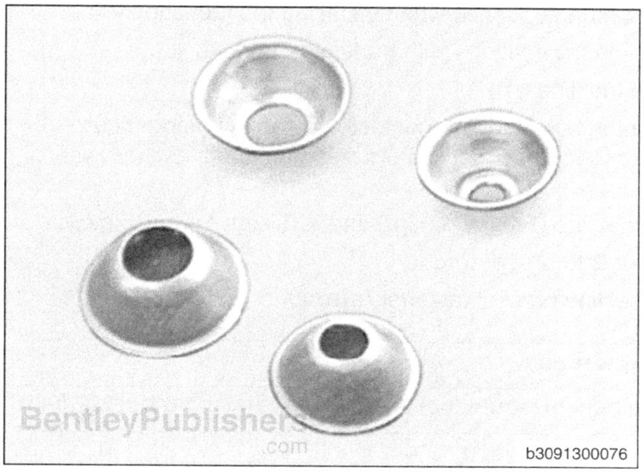

Prior to installing high-pressure fuel lines, dispose of any copper sealing rings installed at factory. They are no longer needed.

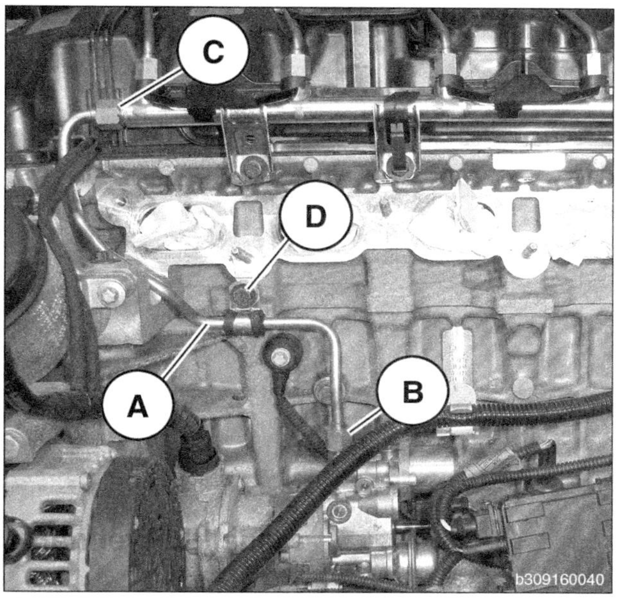

Use the following sequence when installing the fuel line.

- Coat fuel line connections with transmission fluid.
- Preinstall fuel line **A**.
- Tighten union nuts **B**, **C** finger tight.
- Tighten nut **B** to proper torque.
- Tighten union nut **C** to proper torque.
- Tighten fuel line bracket fastener **D**.

Tightening torques	
Fuel line to high-pressure fuel pump	30 Nm (22 ft-lb)
Fuel line to fuel rail	30 Nm (22 ft-lb)
Fuel line mount to cylinder head	13 Nm (9.5 ft-lb)

Before installing high-pressure fuel pump supply line, confirm it has updated elastomer mounting clamp (**arrow**). If it has a soldered hold down clamp, replace fuel supply line with updated part.

BentleyPublishers
BentleyPublishers.com—All Rights Reserved

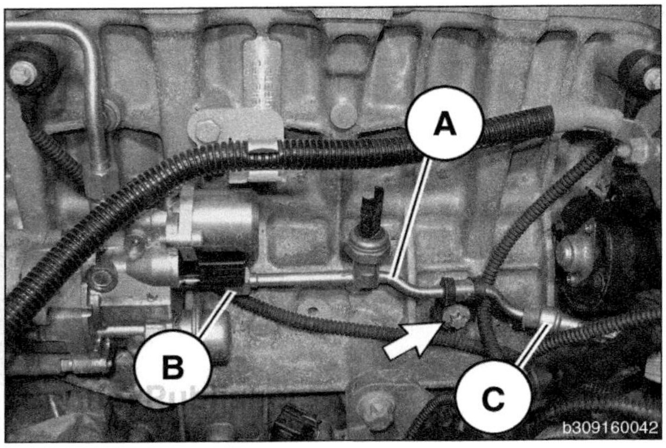

Use the following sequence when installing the fuel supply line.

- Coat fuel line connections with transmission fluid.
- Pre-install fuel line (**A**).
- Tighten union nut **B** and mount fastener (**arrow**) finger tight. Remember to replace fuel line bracket aluminum fastener with new.
- Connect fuel line **C** to fuel supply line **A**. Listen for audible click.
- Tighten nut **B** to proper torque.
- Tighten fuel line bracket fastener (**arrow**).

Tightening torques	
Fuel line to high-pressure fuel pump	30 Nm (22 ft-lb)
Fuel line bracket to engine • Stage 1 • Stage 2	 10 Nm (7.3 ft-lb) Additional 90° ± 15°

– Reassemble engine.

– Check fuel system for leaks.

– Check and clear fault codes from ECM memory.

Fuel Supply Components

Fuel tank, draining

When draining fuel tank, use a safe storage unit and an approved fuel pumping device.

> ***WARNING—***
> - *Before starting to work on tank removal, make sure hot components, such as the exhaust system, are completely cooled down.*
> - *Fuel may be spilled. Do not smoke or work near heaters or other fire hazards.*

– Start engine and allow to run 10 - 15 seconds to fill fuel compensating siphon assembly. This allows both lobes of fuel tank to be drawn off through fuel filler pipe.

– Disconnect negative cable from battery.

> ***CAUTION—***
> - *Prior to disconnecting the battery, read the battery disconnection cautions given in* **001 Warnings and Cautions**.

– Remove fuel tank filler cap.

– Slide suction hose into filler neck about 130 cm (51 in), twisting as necessary. Withdraw fuel into storage unit.

BentleyPublishers.com
BentleyPublishers.com—All Rights Reserved

- Monitor fuel level reduction in both lobes:
  - Remove rear seat cushion and access both fuel tank sender harness connectors. See **Accessing fuel pump and fuel level senders** in this repair group.
  - Use multimeter to measure resistance at both senders. Resistance should drop as fuel level drops.

- If siphoning mechanism is faulty, drain left tank lobe separately by removing sender cover and pumping fuel directly out of left lobe.

- Remove suction hose from tank filler neck carefully to avoid damaging filler neck baffle plate.

> ***CAUTION—***
> - *After finishing repairs but before starting engine, be sure there is at least 5 liters (1.5 gallons) of fuel in tank. The fuel pump is damaged if run without fuel.*

Fuel filter

The fuel filter is a lifetime filter. It is integrated into the left side fuel level sender. See **Fuel level sender (left side), removing** in this repair group.

Fuel pressure regulator

The fuel pressure regulator is integrated into the left side fuel level sender. See **Fuel level sender (left side), removing** in this repair group.

Fuel pump (EKP) module, removing and installing (N52, N54 engine)

- Remove right side luggage compartment trim panel. See **412 Trunk Lid, Tailgate**.

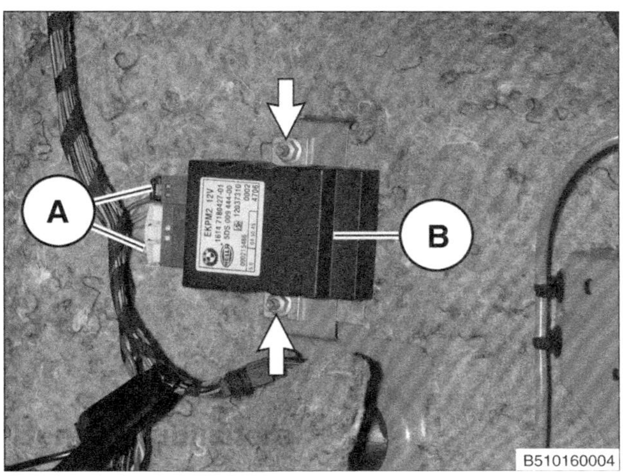

◄ Working at fuel pump (EKP) module:
  - Remove electrical connectors (**A**).
  - Remove fasteners (**arrows**) and remove EKP module (**B**).

- Installation is reverse of removal.
  - Remember to code / program fuel pump (EKP) module if replaced. See **600 Electrical System-General** for information on coding and programming.

Tightening torques	
Fuel pump (EKP) module to body	8 Nm (5.9 ft-lb)

- Check and clear fault codes from ECM memory.

BentleyPublishers
BentleyPublishers.com--All Rights Reserved

Activated carbon canister, removing and installing

The carbon canister is on the right side behind the right rear wheel well liner.

– Raise rear end of car and support safely.

> ***CAUTION—***
> • *Make sure the car is stable and well supported at all times. Use a professional automotive lift or jack stands designed for the purpose. A floor jack is not adequate support.*

– Remove right rear wheel.

– Remove right rear wheel well liner.

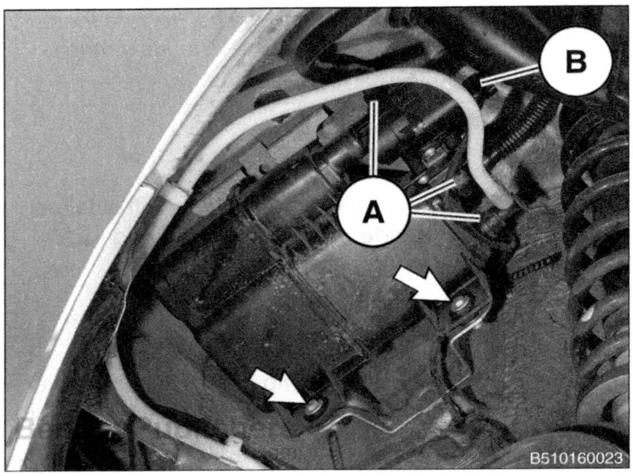

< Remove activated carbon canister:

- Remove line connections at leak detection pump (**A**).
- Disconnect electrical connector (**B**).
- Remove fasteners (**arrows**) and remove carbon canister from vehicle.

– Installation is reverse of removal.

Fuel tank leak detection module (DMTL), removing and installing

The DMTL is mounted to the activated carbon canister.

– Raise car and support safely.

> ***WARNING —***
> • *Make sure the car is stable and well supported at all times. Use a professional automotive lift or jack stands designed for the purpose. A floor jack is not adequate support.*

– Remove activated carbon canister. See **Activated carbon canister, removing and installing** in this repair group.

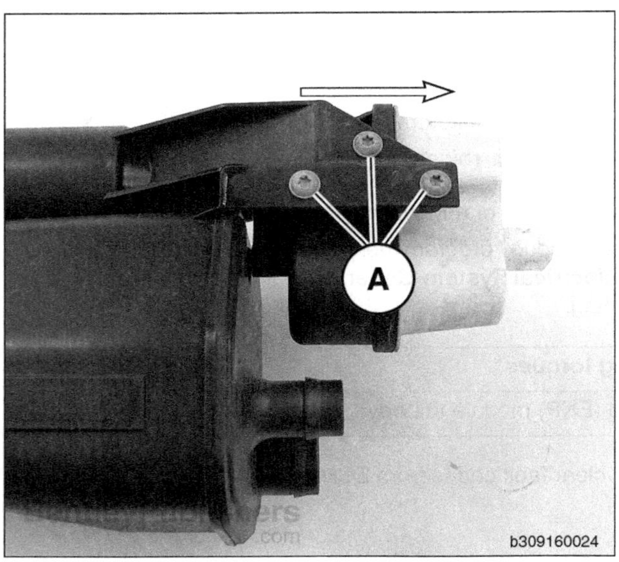

< Remove fasteners (**A**) and pull DMTL unit out of carbon canister in direction of **arrow**.

– Installation is reverse of removal.

BentleyPublishers.com
BentleyPublishers.com—All Rights Reserved

170 Radiator and Cooling System

TABLE

GENERAL

This repair group covers component repair information for the engine cooling system.

See also:

- **020 Maintenance** for coolant hose inspection
- **100 Engine–General** for engine code and application information
- **Fuel Injection (M54 Engine)** for engine coolant temperature (ECT) sensor service
- **131 Fuel Injection (N52 Engine)** for engine coolant temperature (ECT) sensor service
- **132 Fuel Injection (N54 Engine)** for engine coolant temperature (ECT) sensor service
- **133 Fuel Injection (V8 Engine)** for engine coolant temperature (ECT) sensor service

BentleyPublishers.com
BentleyPublishers.com—All Rights Reserved

Coolant pump

M54, V8 (N62) engines

A centrifugal coolant pump is mounted to the front of the engine. The belt-driven pump circulates coolant through the system whenever the engine is running.

N52, N54 engines

B510170001

An externally mounted electric coolant pump is bolted to the right front of the engine block. The electric pump allows for more efficient heating and cooling of the engine. It can operate regardless of whether the engine is ON or OFF.

A low coolant flow rate is used during cold start situations to speed engine warm-up. A high flow rate (including flow with engine OFF) is used for rapid cool-down. In this way the electric coolant pump helps ensure engine efficiency and longevity.

The pump is DME controlled. The DME uses engine load, engine operating range and various temperature sensors to determine coolant pump operation and speed.

The electric coolant pump is self-diagnosable and can identify and store the following faults:

- Impeller speed deviation
- Stiff operation or blocked by foreign objects
- Incorrect mixture ratio coolant / water
- Air in cooling system

Electrically heated thermostat

An electrically heated, DME-controlled thermostat is externally mounted above the coolant pump. The engine control module (ECM) activates the thermostat to maintain engine coolant temperature within a narrow range.

If the electronics fail, the mechanical function of the thermostat acts as a fail-safe.

Radiator and expansion tank

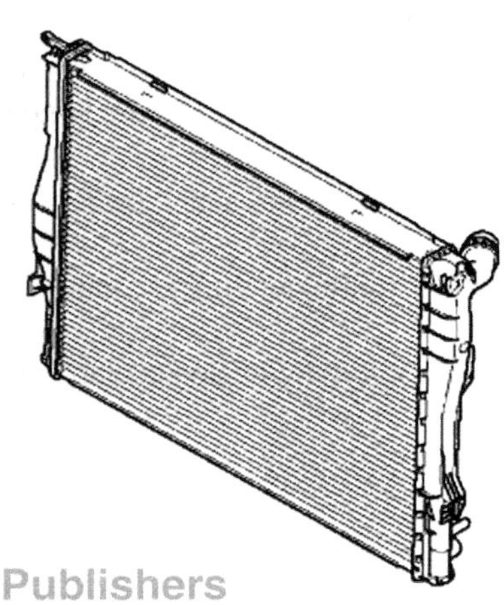

B510170002

The radiator is a crossflow design. An expansion tank provides for coolant expansion at higher temperatures and easy monitoring of the coolant level.

Automatic transmission fluid (ATF) is circulated through an additional heat exchanger (ATF cooler).

BentleyPublishers.com
BentleyPublishers.com—All Rights Reserved

Electric cooling fan

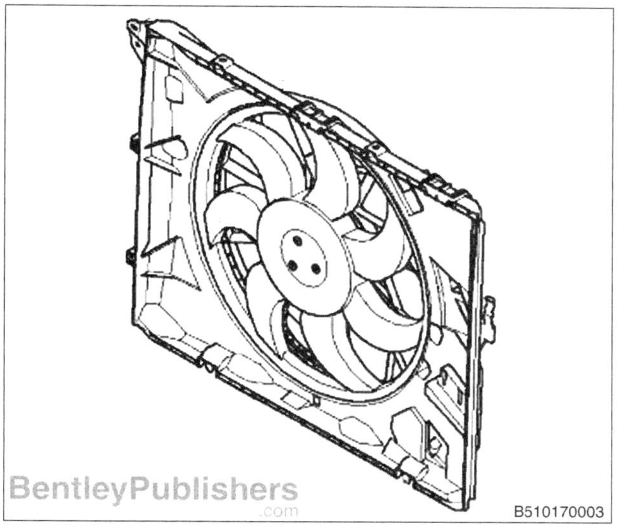

Models covered by this manual use an electric cooling fan mounted on the engine side of the radiator.

The electric cooling fan is controlled by the engine control module (ECM) via the output final stage.

The output final stage is mounted on the fan housing, next to the fan motor. The fan is operated using a pulse width modulated signal. Fan circuit wiring is protected by a high-amp fuse. Electric fan activation is based on the following inputs to the ECM:

- Radiator outlet temperature
- Calculated catalytic converter temperature
- Vehicle speed
- Battery voltage
- Calculated refrigerant pressure

When the vehicle is first started, the ECM activates the electric fan briefly at 20% of its maximum speed, then switches OFF. This is for diagnostic monitoring. The voltage generated by the fan when it slows down (acting as a generator) must match the stored rpm values in the fan output stage to confirm that the fan is operating correctly.

NOTE—

- *If the ECM fault memory indicates a cooling fan fault, check that the fan is not seized and that it spins freely.*
- *When A/C is switched ON, the electric fan is not immediately turned on.*
- *After the engine is switched OFF, the fan may continue to run at varying speeds for up to 10 minutes, based on calculated catalyst temperature.*

Automatic transmission fluid (ATF) cooler

ATF lines connect the transmission to the ATF cooler (heat exchanger) at back of radiator.

On cold engine start-up, the engine coolant is heated more quickly than the transmission fluid. Heat from engine coolant is used to warm up the ATF faster, reducing drag in the transmission and improving fuel mileage.

Once ATF reaches normal operating temperature, it is hotter than the engine coolant. The ATF heat exchanger then acts as a cooling device.

Engine oil cooler

External Oil Coooler (535i)

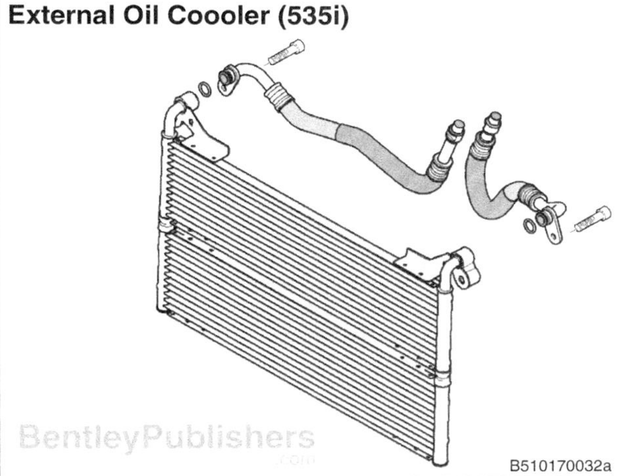

On 535i and 550i models, an additional external engine oil cooler is fitted between the radiator and the electric cooling fan. Cars with this external cooler can be identified by the oil lines passing through the cooling fan shroud.

The oil cooler is removed together with the electric cooling fan. Once the fan is removed, the cooler can be unclipped from the fan shroud. See **Electric cooling fan and cowl, removing and installing**.

170

BentleyPublishers.com
BentleyPublishers.com—All Rights Reserved

Warnings and Cautions

Observe the following warnings and cautions when working on the cooling system.

WARNING—
- *At normal operating temperature the cooling system is pressurized. Allow the system to cool as long as possible before opening (a minimum of one hour), then release the cap slowly to allow safe release of pressure.*
- *Releasing cooling system pressure lowers the coolant boiling point. The coolant may boil suddenly. Use heavy gloves and wear eye and face protection to guard against scalding.*
- *Use extreme care when draining and disposing of engine coolant. Coolant is poisonous and lethal to humans and pets. Pets are attracted to coolant because of its sweet smell and taste. Seek medical attention immediately if coolant is ingested.*

CAUTION—
- *Avoid adding cold water to the coolant while the engine is hot or overheated. If it is necessary to add coolant to a hot system, do so only with the engine running and coolant pump operating.*
- *To avoid excess silicate gel precipitation in the cooling system and loss of cooling capacity, use BMW coolant or equivalent low silicate antifreeze.*
- *If oil enters the cooling system, the radiator, expansion tank and heating circuit must be flushed with cleaning agent. BMW recommends removal of the radiator and expansion tank to flush.*
- *When working on the cooling system, cover the alternator to protect it against coolant drips.*
- *Prior to disconnecting the battery, read the battery disconnection cautions given in* **001 Warnings and Cautions**.
- *Replace aluminum fasteners when loosened.*
- *For reliable identification test fastener with magnet for aluminum composition.*

BentleyPublishers.com
BentleyPublishers.com—All Rights Reserved

TROUBLESHOOTING

Begin the diagnosis of cooling system problems with a thorough visual inspection. If no visual faults are found, check the engine control module (ECM) fault memory for stored diagnostic trouble codes (DTCs) using BMW scan tool.

Common cooling system faults can be grouped into one of 4 categories:

- Cooling system leaks
- Poor coolant circulation
- Radiator cooling fan faults
- Electrical / electronic faults

Cooling system inspection

– Check operation of electric radiator fan:
  - Start engine, turn A/C on, MAX setting. Electric fan should run.

– Check coolant hoses for cracks or softness. Check clamps for looseness. Check coolant level and check for evidence of coolant leaks from engine.

– Check that radiator fins are not blocked with dirt or debris. Clean radiator using low-pressure water or compressed air. Blow outward from engine side out.

– To check mechanical coolant pump:
  - Remove cooling fan. See **Electric cooling fan, removing and installing** in this repair group.
  - Remove accessory belt from coolant pump pulley. See **020 Maintenance**.
  - Firmly grasp opposite sides of pulley and check for play in all directions.
  - Spin pulley and check that shaft runs smoothly without play.

NOTE—
- *The coolant provides lubrication for the pump shaft, so an occasional drop of coolant leaking from the pump is acceptable. If coolant drips steadily from the vent hole, replace the pump.*

– Check operation of electric coolant pump:
  - Turn ignition key on.
  - Adjust heat to max. setting.
  - Press accelerator pedal to floor.
  - Pump should audibly run after about 10 seconds.

– At normal operating temperature, cooling system is pressurized. This raises boiling point of coolant. Leaks may prevent system from becoming pressurized. Pressure test cooling system to help pinpoint hard-to-find leaks. See **Cooling system pressure test**.

BentleyPublishers.com
BentleyPublishers.com—All Rights Reserved

– If cooling system is full of coolant and holds pressure:
  - Use an appropriate scan tool to interrogate engine control module (ECM) for radiator fan or DME control circuit faults.
  - Check for failed thermostat or coolant pump impeller.
  - Check for clogged / plugged radiator or coolant passages.

Cooling system pressure test

A cooling system pressure tester is used to test for coolant leaks, including internal ones. Common sources of internal coolant leaks are a faulty cylinder head gasket, a cracked cylinder head, or a cracked engine block.

> ***WARNING —***
> - *At normal operating temperature the cooling system is pressurized. Allow the system to cool before opening. Remove the cap slowly to allow safe release of pressure.*

– With engine cold, install pressure tester (BMW special tools 17 0 101 / 17 0 113 or equivalent) to expansion tank. Pressurize system to specification listed in **Table a**.
  - Pressure should not drop more than 0.1 bar (1.45 psi) for at least two minutes.
  - If pressure drops rapidly and there is no sign of external leakage, cylinder head gasket may be faulty. Perform compression and leak-down tests.
  - Test expansion tank cap using pressure tester with correct adapter (BMW special tool 17 0 114 or equivalent). Replace faulty cap or cap gasket.

> ***CAUTION—***
> - *Exceeding the specified test pressure could damage the radiator or other system components.*

Table a. Cooling system test pressures

Component	Test pressure
Radiator	1.5 bar (21.75 psi)
Radiator cap	2 bar (29 psi)

Combustion chamber leak test

– If you suspect that combustion chamber pressure is leaking into the cooling system past the cylinder head gasket, use an exhaust gas analyzer to test the vapors rising from the coolant at the expansion tank.

> ***CAUTION—***
> - *Use an extension tube above the reservoir neck to maintain distance between the top of the coolant and the gas analyzer nozzle. The gas analyzer is easily damaged if it is allowed to inhale liquid coolant.*
> - *While running engine to check for causes of overheating, observe coolant temperature carefully in order to avoid engine damage.*

BentleyPublishers.com
BentleyPublishers.com—All Rights Reserved

Thermostat

If the engine overheats or runs too cool and no other cooling system tests indicate trouble, the thermostat may be faulty.

The electrically heated thermostat is monitored by the OBD II diagnostic software. The fault may lie in the DME software or hardware, or it may lie in the wiring to the thermostat. See **OBD On-Board Diagnostics**.

COOLING SYSTEM SERVICE

Coolant, draining and filling

> ***WARNING —***
> - *Allow the cooling system to cool before opening or draining the cooling system.*

– Raise front of car and support safely.

> ***WARNING —***
> - *Make sure the car is stable and well supported at all times. Use a professional automotive lift or jack stands designed for the purpose. A floor jack is not adequate support.*

– Remove splash shield from under engine. See **020 Maintenance**.

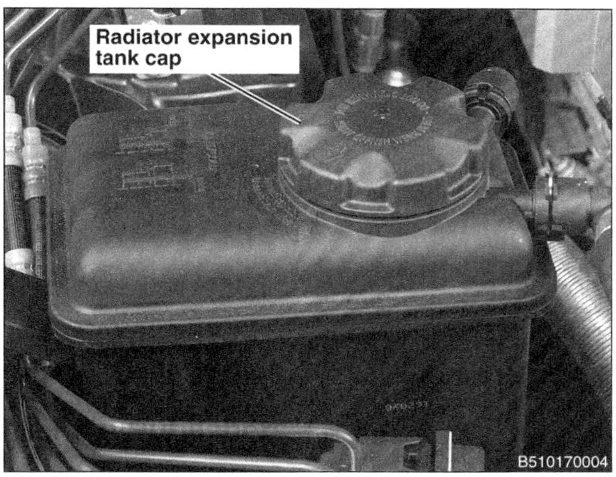

B510170004

◄ Remove cap from radiator expansion tank.

– Place 5-gallon pail underneath radiator.

B510170005

◄ Loosen drain plug (**arrow**) on petcock at bottom of radiator and allow coolant to drain.

NOTE —
- *For radiators without drain plug, remove lower hose to drain radiator.*

> ***WARNING —***
> - *Coolant is poisonous. It is especially lethal to pets. Clean up spills immediately and rinse the area with water. If coolant is ingested, immediately seek medical attention.*

BentleyPublishers.com
BentleyPublishers.com—All Rights Reserved

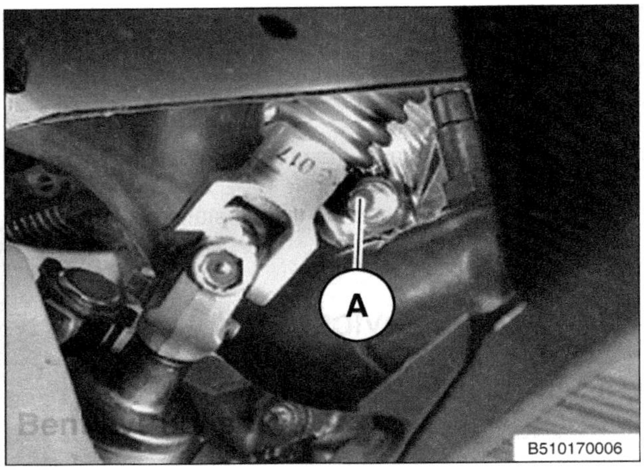

◀ M54 engine block drain plug is located on right side of engine block near cylinder 2. V8 engine block drain plug (**A**) is located on left side of engine block.

– Place 5-gallon pail underneath engine block coolant drain plug.

– Remove engine block coolant drain plug and drain coolant.

– Before refilling radiator:
  - Switch ignition ON. Do not start engine.
  - Set temperature controls to full warm.
  - Set blower control to low.
  - Set seat heat control to maximum.

– Using a coolant mixture of 50% antifreeze and 50% distilled water, fill expansion tank slowly. Continue until coolant emerges from bleed screw. Cooling system capacities are in **Table b**.

NOTE —
- *Tap water may cause corrosion of radiator, engine and coolant hoses.*
- *Coolant can be reused provided it is clean and less than two years old. Do not reuse coolant when replacing damaged engine parts.*

– Bleed cooling system. See **Cooling system, bleeding** in this repair group.

Table b. Cooling system capacities

Engine	Capacity
M54: • Automatic transmission • Manual transmission	 10.6L (11.2 US qt) 10L (10.56 US qt)
N52, N54: • Manual transmission • Automatic transmission	 10.6L (11.2 US qt) 10L (10.56 US qt)
V8 (N62): • Automatic transmission • Manual transmission	 14.2L (15 US qt) 13.8L (14.58 US qt)

Tightening torques	
Block drain plug to engine block: • M14 x 1.5 • M16 x 1.5 • M18 x 1.5	 25 Nm (18 ft-lb) 35 Nm (26 ft-lb) 40 Nm (30 ft-lb)

BentleyPublishers.com
BentleyPublishers.com—All Rights Reserved

Cooling system, bleeding (M54, V8 engines)

Trapped air in the cooling system can prevent proper coolant circulation and cause overheating. Whenever the coolant is drained and filled, bleed the cooling system.

> ***CAUTION—***
> - *Carry out this procedure when replacing cooling system components or refilling the cooling system. Failure to follow this procedure may result in incomplete filling of cooling system, overheating and engine damage.*

– Before refilling radiator:
- Switch ignition ON. (Do not start engine).
- Set temperature controls to full warm.
- Set blower control to low.
- Set seat heat control to maximum.

B510170007

Using a coolant mixture of 50% antifreeze and 50% distilled water, fill expansion tank slowly to MAX.

> ***CAUTION—***
> - *Use genuine BMW coolant or its equivalent to avoid the formation of harmful, clogging deposits in the cooling system.*

B510170008

Start engine and run at idle speed for one minute (cap open). Adjust coolant level to MAX (**A**). Close cap.
- Run engine until it reaches operating temperature.
- After engine has cooled, recheck coolant level.
- Top up so that coolant level indicator is at MAX.
- If trapped air cannot be bled manually, pressure-bleed system using cooling system pressure tester.

Cooling system capacities	
Engine	**Capacity**
M54: • Automatic transmission • Manual transmission	 10.6L (11.2 US qt) 10L (10.56 US qt)
V8 (N62): • Automatic transmission • Manual transmission	 14.2L (15 US qt) 13.8L (14.58 US qt)

BentleyPublishers.com
BentleyPublishers.com—All Rights Reserved

Cooling system, bleeding (N52, N54 engines)

Trapped air in the cooling system can prevent proper coolant circulation and cause overheating. Whenever the coolant is drained and filled, bleed the cooling system.

5 series models equipped with N52 and N54 engines feature an electric coolant pump. The bleeding procedure operates the coolant pump for approximately 12 minutes to remove trapped air and ensure that the cooling system is filled to capacity.

> ***CAUTION—***
> • *Carry out this procedure out when replacing cooling system components or refilling the cooling system. Failure to follow this procedure may result in incomplete filling of cooling system, overheating and engine damage.*

– Before refilling radiator:
- Switch ignition ON. (Do not start engine).
- Set temperature controls to full warm.
- Set blower control to low.
- Set seat heat control to maximum.

◄ Using a coolant mixture of 50% antifreeze and 50% distilled water, fill expansion tank slowly to MAX.

> ***CAUTION—***
> • *Use genuine BMW coolant or its equivalent to avoid the formation of harmful, clogging deposits in the cooling system.*

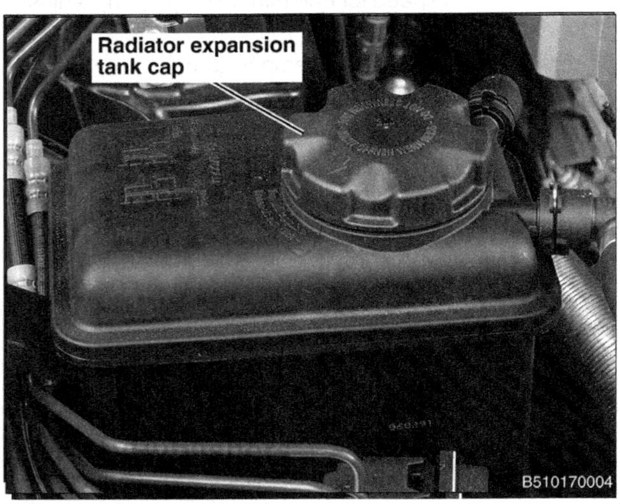

◄ Close cap on coolant expansion tank and perform the following steps:

1. Connect vehicle power supply. See **600 Electrical System–General**.
2. Switch ignition ON.
3. Set heater to MAX temperature and turn fan to lowest speed.
4. Press accelerator pedal to floor for 10 seconds.
5. Bleeding procedure begins when accelerator pedal is pressed and takes approximately 12 minutes. (Electric coolant pump turns ON and OFF automatically after approx. 12 minutes).

> ***CAUTION—***
> • *Do not open expansion tank cap while electric coolant pump is ON.*

6. Fill coolant expansion tank 100 ml above MAX (dimension **A**).
7. Check cooling system for leaks
8. If leaks are found, remove ignition key and wait approximately 3 minutes before repeating cooling system bleeding procedure.

BentleyPublishers.com
BentleyPublishers.com–All Rights Reserved

When bleeding procedure is complete, close coolant expansion tank cap until marks on cap and tank (**arrows**) line up.

Cooling system capacities	
Engine	**Capacity**
N52, N54: • Manual transmission • Automatic transmission	 10.6L (11.2 US qt) 10L (10.56 US qt)

Radiator, removing and installing

– Raise car and support safely

> ***WARNING —***
> *• Make sure the car is stable and well supported at all times. Use a professional automotive lift or jack stands designed for the purpose. A floor jack is not adequate support.*

– Remove splash shield from under engine.

– Drain radiator. See **Coolant draining and filling** in this repair group.

> ***WARNING —***
> *• Allow cooling system to cool before opening or draining system.*

– Remove fan cowl. See **Electric cooling fan and cowl, removing and installing** in this repair group.

– Unlock coolant hoses and detach from radiator.

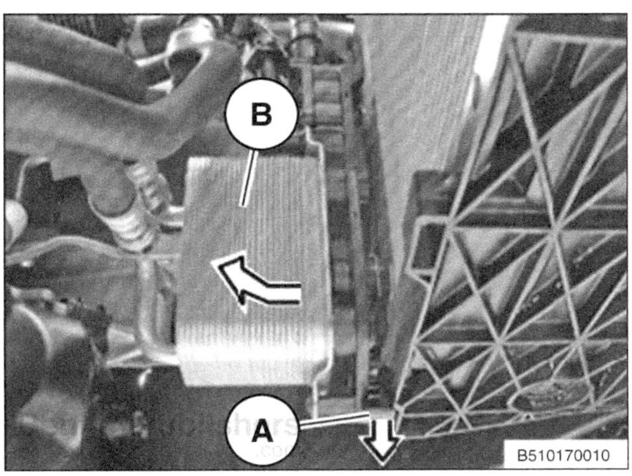

Automatic transmission: Unlock catch (**A**) in direction of **arrow**. Remove ATF heat exchanger (**B**) from radiator in direction of **arrow**.

– Installation is reverse of removal. Remember to:
- Bleed cooling system
- Check cooling system for leaks.

BentleyPublishers.com—All Rights Reserved

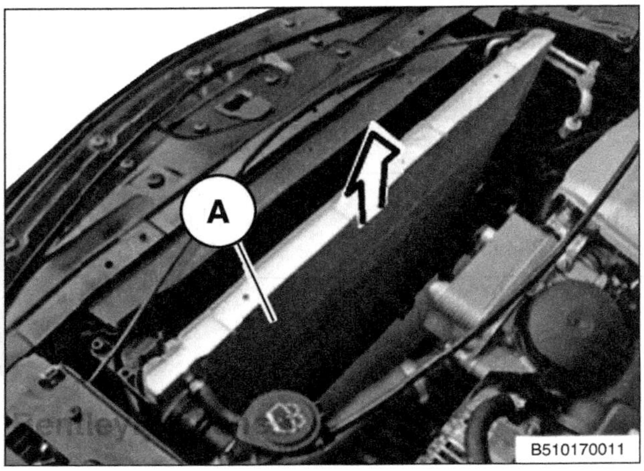

Carefully lift radiator (**A**) up and out of engine compartment. N52 engine shown, other engines similar.

– Installation is reverse of removal. Make sure:
  - Radiator seats firmly in lower mounts.
  - Electric cooling fan cowl clips securely at bottom.
  - Refill and bleed cooling system. See **Cooling system, bleeding** in this repair group.
  - Check cooling system for leaks.

Electric cooling fan and cowl, removing and installing

Working above radiator remove center panel fasteners (**arrows**).

– Lift center panel towards engine. Be careful not to kink hood release bowden cable. Lay center panel to one side.

> ***WARNING —***
> *• Do not place center panel on or near jump start terminals.*

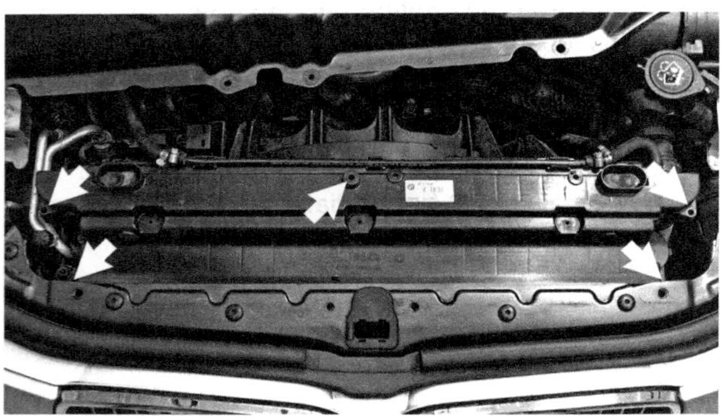

Remove Torx screws (**arrows**) from radiator cover and pull radiator cover up and off.

BentleyPublishers.com
BentleyPublishers.com—All Rights Reserved

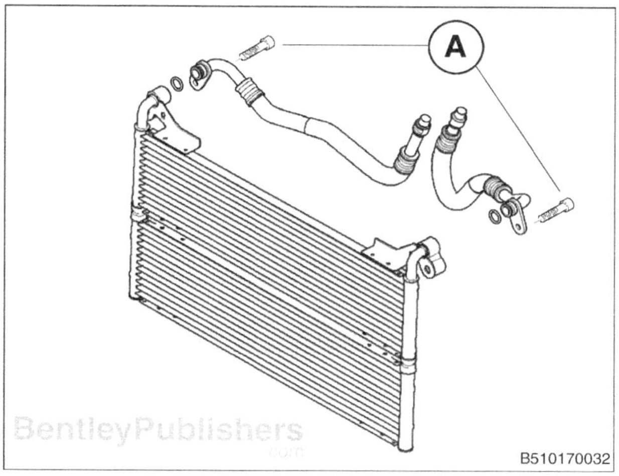

535i models: If equipped with external oil cooler.

- Working at front of fan shroud, remove bolts (**A**) and pull oil lines from oil cooler. Use shop towels to catch any oil that spills from cooler or lines.

170

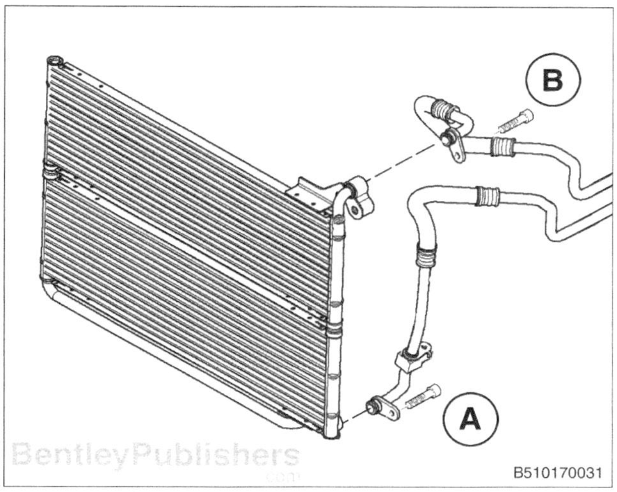

550i models: If equipped with external oil cooler.

- Working beneath vehicle, remove engine lower cover (splash shield). See **020 Maintenance**.
- Place drain pan under vehicle and remove bolt (**A**) and pull oil line from oil cooler. Allow cooler and lines to fully drain.
- Working at top of fan shroud, remove bolt (**B**) and pull oil line from cooler.

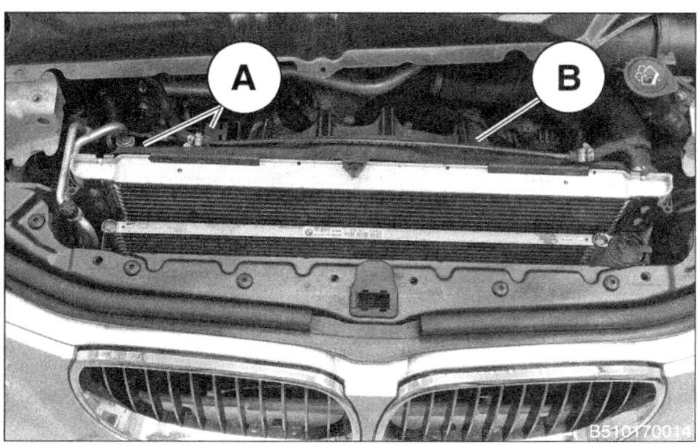

Disconnect electrical connectors from AUC sensor (where applicable) and fan motor (**A**).

- Raise cooling fan cowl (**B**) 3 cm (1.1 in)

BentleyPublishers
BentleyPublishers.com—All Rights Reserved

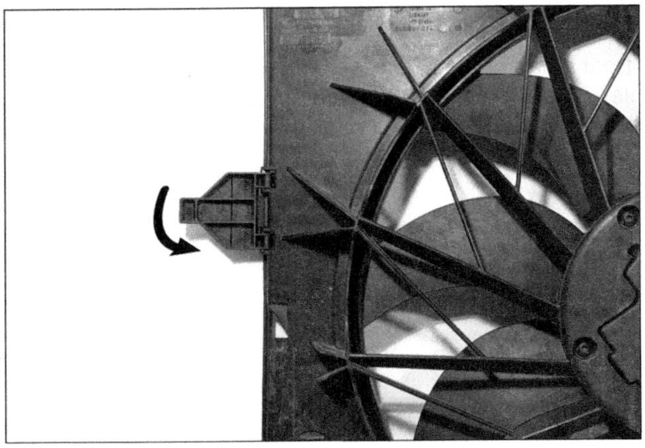

- Fold in retaining tab (**arrow)** on left (driver's) side of fan housing.
- Remove cooling fan cowl by lifting up and out.
  - If equipped with external oil cooler, oil cooler will come out with fan housing.
- Installation is reverse of removal. Remember to:
  - Cars with external engine oil cooler, replace o-rings at oil line fittings.
  - Correctly route all electric harnesses.
  - Check and clear fault codes from ECM memory.
  - Check and top off engine oil if external cooler was removed.

Coolant expansion tank, removing and installing

- Drain cooling system. See **Coolant, draining and filling** in this repair group.

> ***WARNING —***
> *• Allow cooling system to cool before opening or draining system.*

- Release clamps on coolant hoses (**A**).
- Remove expansion tank fasteners (**arrows**).
- Raise tank slightly (if necessary) and release electrical connector for coolant level switch (**inset**) and remove expansion tank.
- Installation is reverse of removal. Remember to:
  - Make sure locating pin at bottom of tank fits securely into rubber grommet.
  - Fill and bleed cooling system. See **Cooling system, bleeding** in this repair group.
  - Check cooling system for leaks.

Tightening torque	
Expansion tank to body	8 Nm (71 in-lb)

BentleyPublishers.com
BentleyPublishers.com—All Rights Reserved

Thermostat, removing and installing (M54 engine)

The electrically-heated thermostat is an integral part of the thermostat housing. The operation of the thermostat is monitored by the ECM. If a faulty thermostat is suspected, interrogate the ECM for stored fault codes using a BMW scan tool.

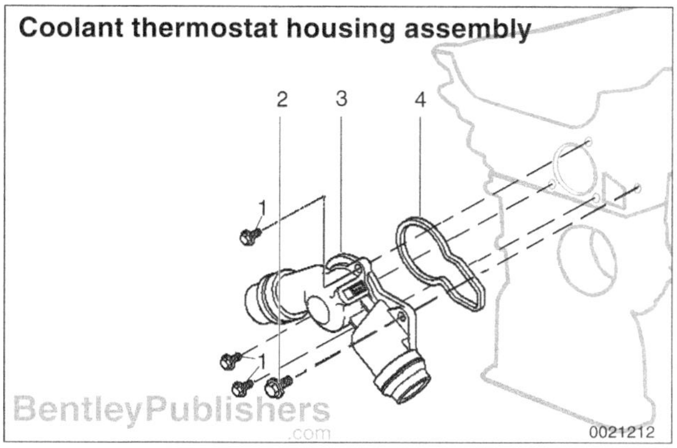

◀ Thermostat housing assembly

1. Bolt M6
2. Bolt M8
3. Housing with heated thermostat
4. Gasket

— Drain radiator and engine block. See **Coolant, draining and filling** in this repair group.

> ***WARNING —***
> • *Allow cooling system to cool before opening system.*

— Remove cooling cowl. See **Electric cooling fan and cowl, removing and installing** in this repair group.

— Disconnect electrical harness connector from thermostat housing.

— Cover alternator to prevent damage from coolant.

◀ Lever out retaining clips (**arrows**) and pull hose fittings off housing.

— Unbolt and remove thermostat housing from front of engine. Loosen nut at top of engine lifting eye to facilitate removal.

— Installation is reverse of removal, noting the following:

- Keep sealing faces free of oil.
- Use new sealing gasket.

— Fill system with coolant and bleed. See **Coolant, draining and filling** in this repair group.

Tightening torques	
Block drain plug to engine block: • M14 x 1.5 • M16 x 1.5 • M18 x 1.5	 25 Nm (18 ft-lb) 35 Nm (26 ft-lb) 40 Nm (30 ft-lb)
Radiator drain plug to radiator	2.5 Nm (22 in-lb)

Cooling system capacities	
Engine	**Capacity**
M54: • Automatic transmission • Manual transmission	 10.6L (11.2 US qt) 10L (10.56 US qt)

BentleyPublishers.com
BentleyPublishers.com—All Rights Reserved

Thermostat, removing and installing (N52, N54 engines)

– Remove splash shield from under engine.

– Drain cooling system. See **Coolant, draining and filling** in this repair group.

> ***WARNING —***
> *• Allow cooling system to cool before opening or draining system.*

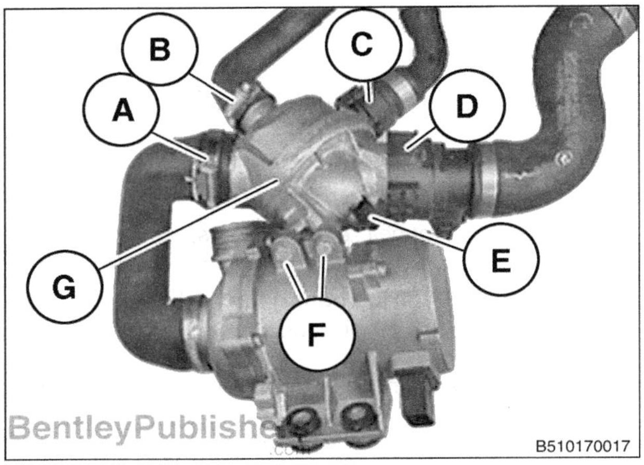

< Working above engine coolant pump, release hose clamps (**A**, **B**) and disconnect hoses.

– Unlock fasteners (**C**, **D**) and remove hoses.

– Disconnect electrical connector (**E**).

– Remove bolts (**F**) and remove thermostat (**G**).

– Installation is reverse of removal. Remember to:
- Refill and bleed cooling system. See **Cooling system, bleeding** in this repair group.
- Check cooling system for leaks.

Tightening torques	
Thermostat to coolant pump: • N52 engine • N54 engine	 9 Nm (80 in-lb) 8 Nm (71 in-lb)

Thermostat, removing and installing (V8 engine)

– Remove splash shield from under engine.

– Drain cooling system. See **Coolant, draining and filling** in this repair group.

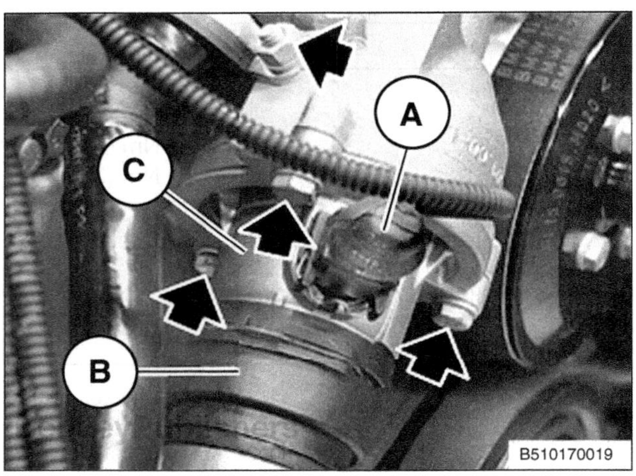

< Working below coolant pump:
- Disconnect electrical connector (**A**).
- Unlock and remove hose (**B**).
- Remove fasteners (**arrows**).
- Remove thermostat (**C**).

– Installation is reverse of removal. Remember to:
- Replace sealing O-ring.
- Refill and bleed cooling system. See **Cooling system, bleeding** in this repair group.
- Check cooling system for leaks.

Tightening torque	
Thermostat to coolant pump	10 Nm (88.5 in-lb)

BentleyPublishers.com
BentleyPublishers.com—All Rights Reserved

Coolant pump, removing and installing (M54 engine)

– Remove cooling cowl. See **Electric cooling fan and cowl, removing and installing** in this repair group.

– Drain cooling system. See **Coolant, draining and filling** in this repair group.

> ***WARNING —***
> *• Allow cooling system to cool before opening or draining system.*

– Remove alternator drive belt. See **020 Maintenance**. Mark direction of rotation on belt.

– Cover alternator to prevent damage from coolant.

– Remove coolant pump pulley bolts and remove pulley from pump.

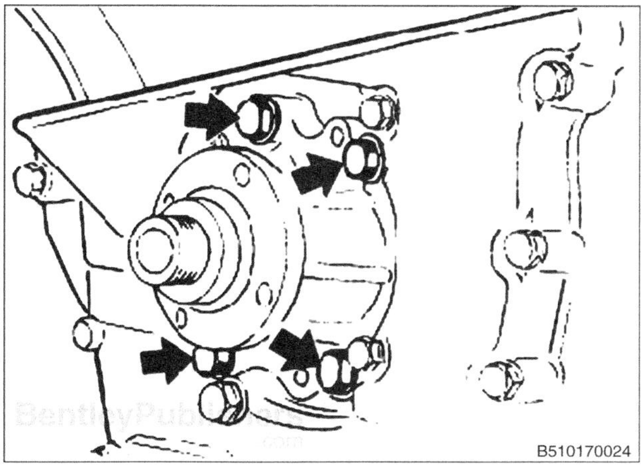

B510170024

< Remove coolant pump fasteners (**arrows**).

B510170025

< Insert two M6 screws (**arrows**) in tapped bores and tighten uniformly until pump is free from timing chain cover.

– Installation is reverse of removal. Remember to:
- Replace sealing O-ring and gaskets.
- Coat O-ring with lubricant during installation.
- Refill and bleed cooling system. See **Cooling system, bleeding (M54, N62 engines)** in this repair group.
- Check cooling system for leaks

Tightening torques	
Coolant pump to timing chain cover	10 Nm (89 in-lb)
Coolant pump pulley to coolant pump	10 Nm (89 in-lb)

BentleyPublishers.com
BentleyPublishers.com—All Rights Reserved

Coolant pump, removing and installing (N52, N54 engines)

– Remove splash shield from under engine.

– Drain cooling system. See **Coolant, draining and filling** in this repair group.

> ***WARNING —***
> *• Allow cooling system to cool before opening or draining system.*

– Cover alternator to prevent damage from coolant.

– Remove coolant thermostat. See **Thermostat, removing and installing** in this repair group.

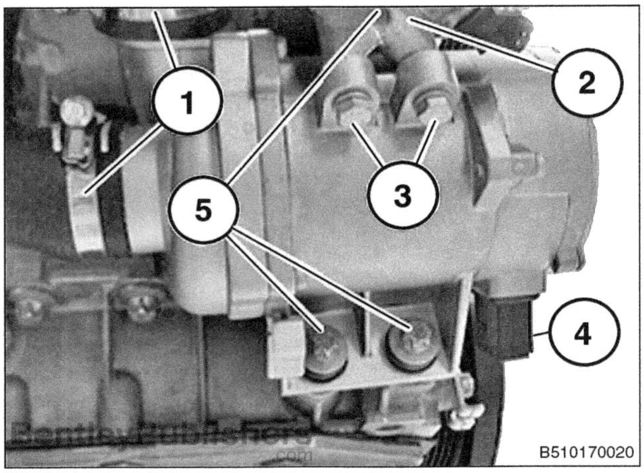

◄ Release hose clamps (**1**) and disconnect coolant hoses from back of pump.

– Disconnect electrical connector (**4**) at front of pump.

– Remove bolts (**5**) and remove pump from side of engine.

> ***CAUTION—***
> *• If coolant pump is to be reused, fill with mixture of 50% antifreeze and 50% distilled water immediately after removing, and plug openings to avoid contamination.*

– Installation is reverse of removal. Remember to:

- Replace aluminum fasteners (fasteners with heads painted blue) with new.

> ***CAUTION—***
> *• Do not use steel fasteners in place of aluminum. Electrochemical corrosion will result.*

- Reinstall thermostat.
- Refill and bleed cooling system. See **Cooling system, bleeding** in this repair group.
- Check cooling system for leaks.

Tightening torques	
Thermostat to coolant pump • N52 engine • N54 engine	 9 Nm (80 in-lb) 8 Nm (71 in-lb)
Coolant pump to engine block (replace aluminum fasteners with new and observe torque angle)	10 Nm (89 in-lb) + 90° turn

BentleyPublishers.com
BentleyPublishers.com—All Rights Reserved

Coolant pump, removing and installing (V8 engine)

— Remove cooling cowl. See **Electric cooling fan and cowl, removing and installing** in this repair group.

— Drain cooling system. See **Coolant, draining and filling** in this repair group.

WARNING —
• Allow cooling system to cool before opening or draining system.

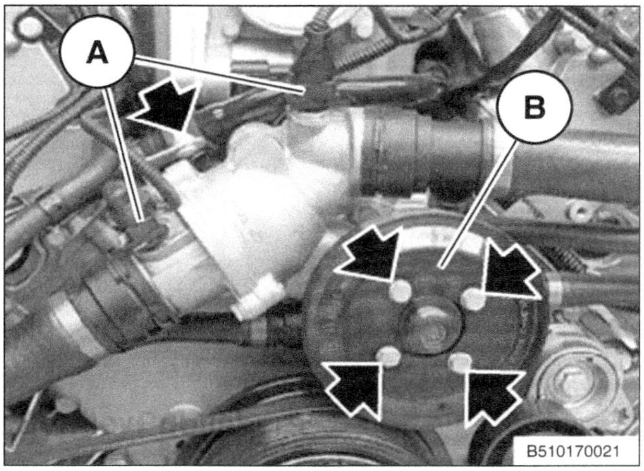

B510170021

— Remove alternator drive belt. See **020 Maintenance**. Mark direction of rotation on belt.

— Cover alternator to prevent damage from coolant.

◄ Working at coolant pump:
- Disconnect electrical connectors (**A**).
- Remove vacuum line holder (**arrow**).
- Unlock and remove coolant hose.
- Remove drive belt pulley (**B**) fasteners (**arrows**).

— Remove crankshaft damper. See **117 Camshaft Timing Chain**.

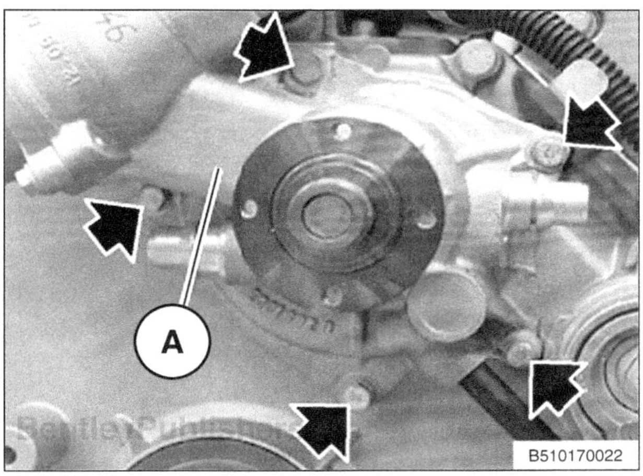

B510170022

◄ Remove coolant pump (**A**) fasteners (**arrows**).

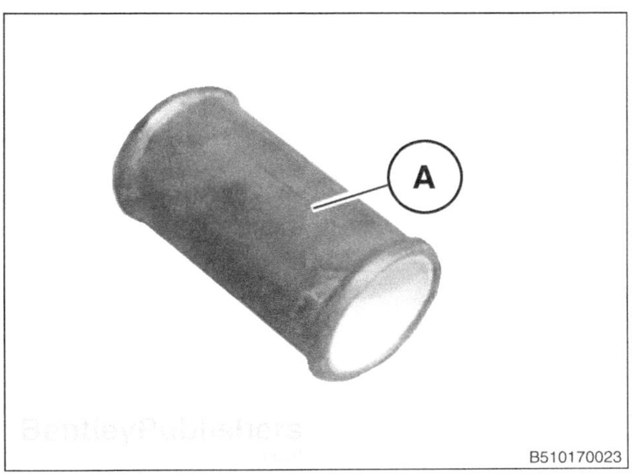

B510170023

◄ Installation is reverse of removal. Remember to:
- Clean all sealing surfaces.
- Replace coolant pipe (**A**) located behind coolant pump.
- Rotate coolant pump impeller one full revolution before installing.
- Refill and bleed cooling system. See **Cooling system, bleeding** in this repair group.
- Check cooling system for leaks.

Tightening torques	
Coolant pump to engine block: • M6 • M7 • M8	 10 Nm (89 in-lb) 15 Nm (11 ft-lb) 22 Nm (16.2 ft-lb)

BentleyPublishers.com
BentleyPublishers.com—All Rights Reserved

Automatic transmission fluid (ATF) cooler, removing and installing

- Remove splash shield from under engine.
- Drain cooling system. See **Coolant, draining and filling** in this repair group.

> ***CAUTION—***
> - *Cover alternator when working on oil, coolant or fuel circuits to protect from damage.*

- Remove coolant hoses from ATF cooler.
- Use BMW special tool 17 0 030 to unlock transmission fluid from hydraulic lines from cooler.

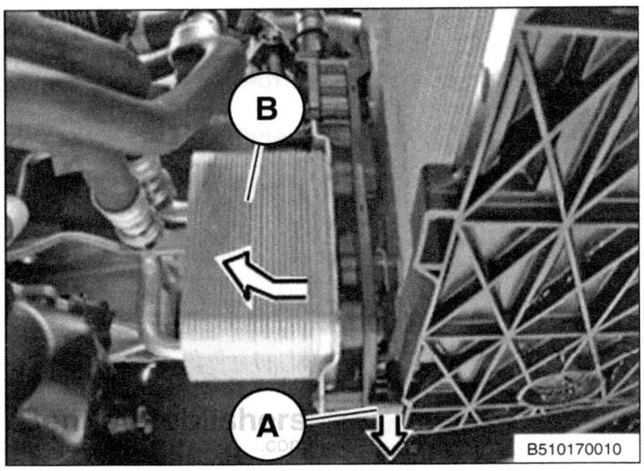

- Unlock catch (**A**) in direction of **arrow**. Remove ATF heat cooler (**B**) from radiator in direction of **arrow**.
- Installation is reverse of removal. Remember to:
  - Check automatic transmission fluid level. See **240 Automatic Transmission**.
  - Refill and bleed cooling system. See **Cooling system, bleeding** in this repair group.
  - Check cooling system for leaks.

Power steering cooling loop, removing and installing

- Raise car and support safely

> ***WARNING—***
> - *Make sure the car is stable and well supported at all times. Use a professional automotive lift or jack stands designed for the purpose. A floor jack is not adequate support.*

- Remove splash shield from under engine.
- Drain radiator. See **Coolant draining and filling** in this repair group.

> ***WARNING—***
> - *Allow cooling system to cool before opening or draining system.*

- Remove fan cowl. See **Electric cooling fan and cowl, removing and installing** in this repair group.
- Remove radiator. See **Radiator, removing and installing** in this repair group.

BentleyPublishers.com
BentleyPublishers.com—All Rights Reserved

- Place a suitable container beneath power steering cooling loop connections to catch dripping fluid. Dispose of properly.

- Using a fluid recovery machine, suction fluid from power steering fluid reservoir.

- Press hydraulic line towards cooling loop. Pull back on locking ring and remove line.

- Secure A/C condenser in place to protect from falling and damage.

> ***CAUTION—***
> *• Do not disconnect A/C refrigerant lines.*

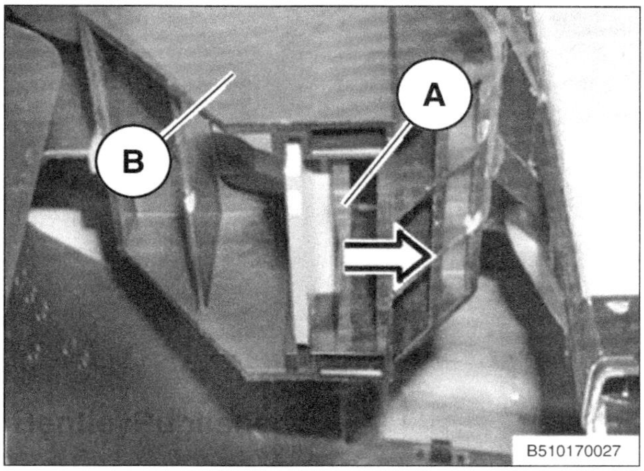

B510170027

◄ Press lock (**A**) and remove bracket (**B**) downwards. Illustration shows left bracket. Same procedure applies to opposite side.

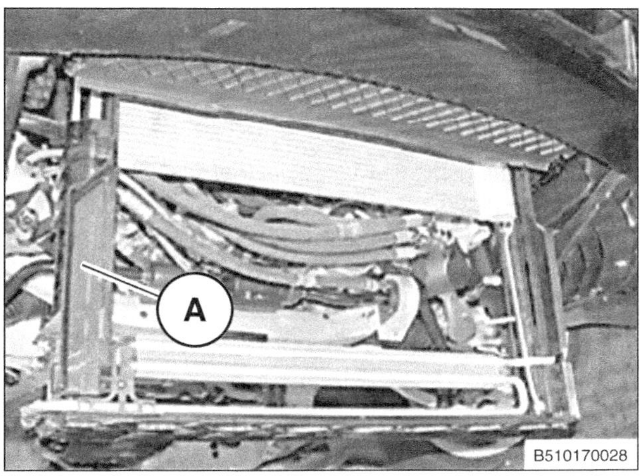

B510170028

◄ Remove cooling module carrier (**A**) towards bottom.

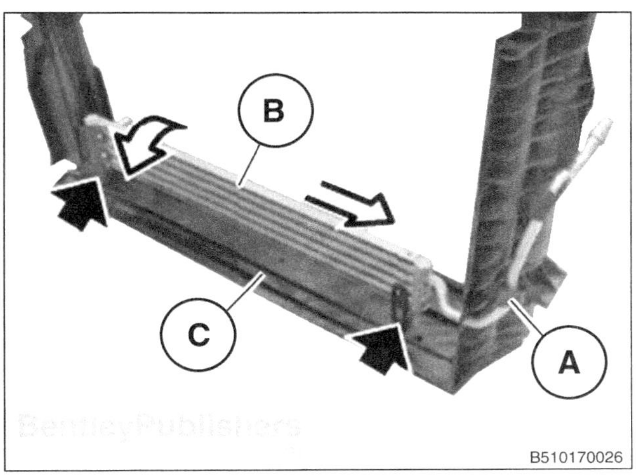

B510170026

◄ Release cooler line mounting catch (**A**). Remove fasteners (**black arrows**).

- Carefully press cooling loop (**B**) on right towards front, then unclip on left and remove from module carrier (**C**).

- Installation is reverse of removal. Remember to:
  - Make sure cooling loop hydraulic lines connect securely.
  - Check that cooling loop lines snap securely into holder (**A**).
  - Refill and bleed cooling system. See **Cooling system, bleeding** in this repair group.
  - Refill and bleed power steering system. See **320 Steering and Wheel Alignment**.
  - Check cooling system for leaks.

Intercooler, removing and installing (N54 engine)

BentleyPublishers.com—All Rights Reserved

– Raise car and support safely

> ***WARNING —***
> - *Make sure the car is stable and well supported at all times. Use a professional automotive lift or jack stands designed for the purpose. A floor jack is not adequate support.*

– Remove splash shield from under engine.

– Disconnect intake air duct from left and right side of intercooler. See **132 Fuel Injection (N54 engine)**.

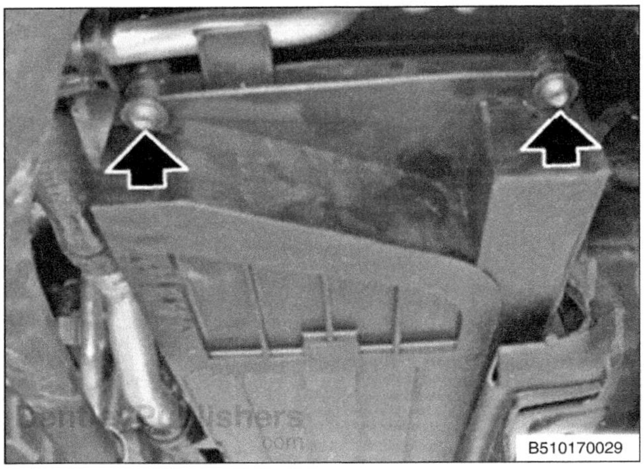

◀ Release screws. (Picture shows right side). Open quick-connect couplings on charge-air hoses.

– Remove intercooler from cooling module carrier.

– When installing, coat sealing rings of intake air ducts with antiseize paste.

NOTE —
- *Fitting air ducts to intercooler without antiseize paste is nearly impossible.*

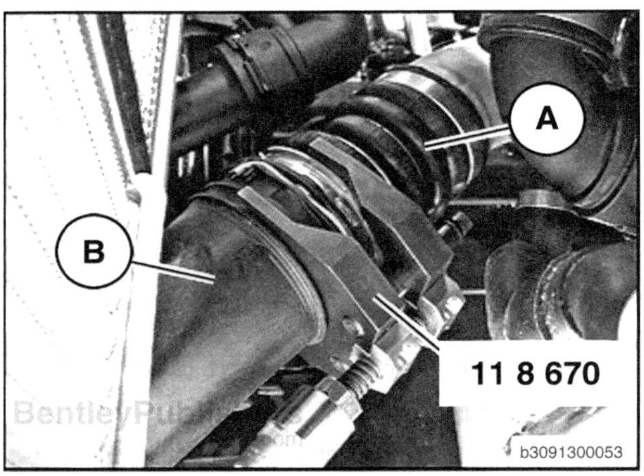

◀ Reattach intake air ducts using BMW special tool 11 8 670. Ducts must snap audibly into place.

> ***CAUTION —***
> - *Special clamping jaws (special tool 11 8 670) are strongly recommended for proper installation and to prevent damage to air ducts.*
> - *Right side shown. Left side is similar.*

BentleyPublishers.com
BentleyPublishers.com—All Rights Reserved

180 Exhaust System

GENERAL

This repair group covers removal and replacement of the exhaust system, including exhaust manifolds, turbochargers and oxygen sensors.

Bentley Publishers
BentleyPublishers.com – All Rights Reserved

Warnings and Cautions

> ***WARNING —***
> - *Exhaust gases are colorless, odorless, and very toxic. Run the engine only in a well-ventilated area. Immediately repair any leaks in the exhaust system or structural damage to the car body that might allow exhaust gases to enter the passenger compartment.*
> - *The exhaust system and catalytic converter operate at very high temperatures. Allow components to cool before servicing. Wear protective clothing to prevent burns. Do not use flammable chemicals near a hot catalytic converter.*
> - *Corroded exhaust system components crumble easily and often have exposed sharp edges. To avoid injury, wear eye protection and heavy gloves when working with old exhaust system parts.*

> ***CAUTION—***
> - *Prior to disconnecting the battery, read the battery disconnection cautions given in* **001 Warnings and Cautions**.
> - *Replace aluminum fasteners when loosened.*
> - *For reliable indentification, test fastener with magnet for aluminum composition.*
> - *Use care not to drag or bang oxygen sensors. Oxygen sensors are easily damaged.*
> - *When detaching and reattaching oxygen sensor connectors, make sure front and rear connectors are not mixed up.*
> - *The exhaust system is heavy. Work with an assistant.*

EXHAUST SYSTEM

The exhaust system is designed to be maintenance free, although regular inspection is warranted due to the harsh operating conditions. Under normal conditions, catalytic converters do not require replacement unless damaged

Use new fasteners, clamps, rubber mounts and gaskets when replacing exhaust components. A liberal application of penetrating oil to the exhaust system nuts and bolts in advance may make removal easier.

Exhaust system, removing and installing (M54, N52 engines)

Remove exhaust system as a complete unit.

Once the complete system is removed from the car, replace individual pipes and mufflers. Rear muffler can be replaced without removing exhaust system.

– Raise vehicle and support safely.

> ***WARNING —***
> - *Make sure the vehicle is stable and well supported at all times. Use a professional automotive lift or jack stands designed for the purpose. A floor jack is not adequate support.*

BentleyPublishers.com
BentleyPublishers.com—All Rights Reserved

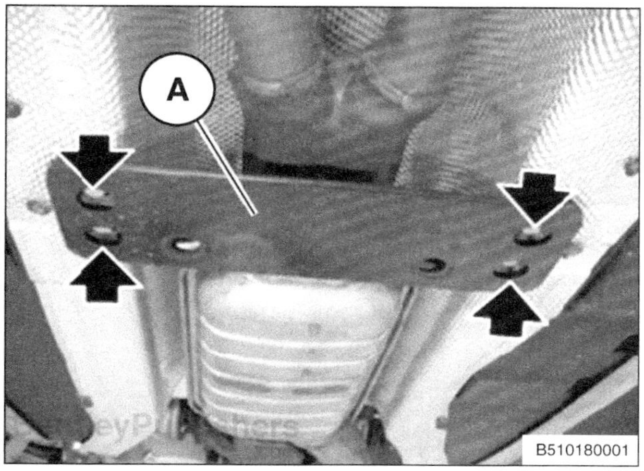

– Support exhaust system using BMW special tool 31 2 220 or equivalent and suitable jack.

NOTE —

- *Exhaust system weighs approximately 40 kg (88 lbs).*

◄ Remove fasteners (**arrows**) and remove front muffler mount (**A**).

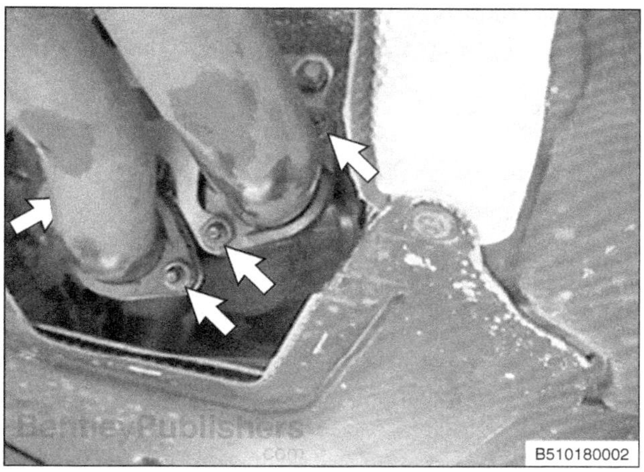

◄ Remove exhaust pipe flange fasteners (**arrows**) and discard.

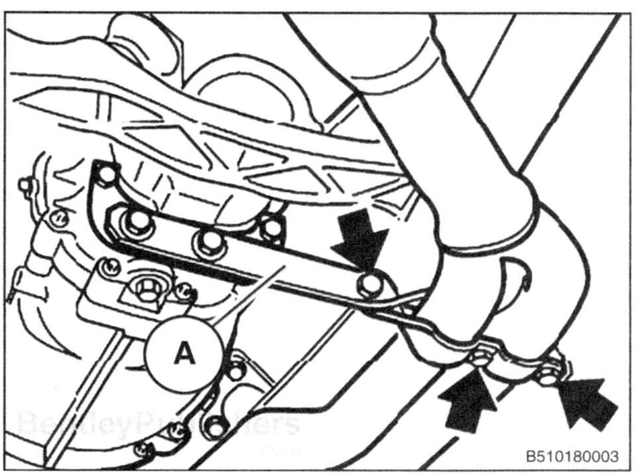

◄ Remove front pipe support fasteners (**arrows**) and remove support (**A**).

BentleyPublishers.com–All Rights Reserved

Exhaust system, removing and installing (M54, N52 engines)

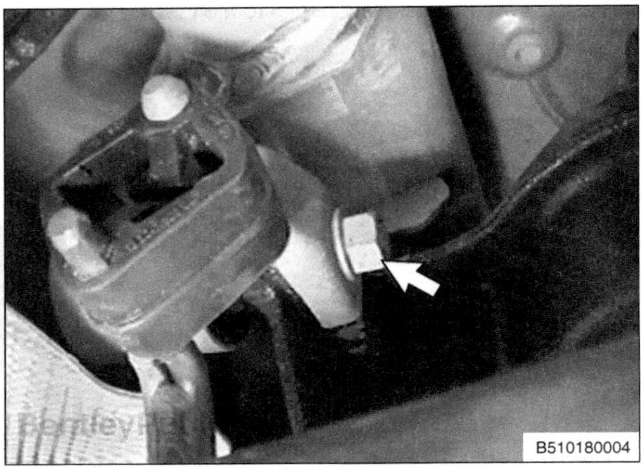

Working next to rear differential, remove exhaust system subframemount fastener (**arrow**).

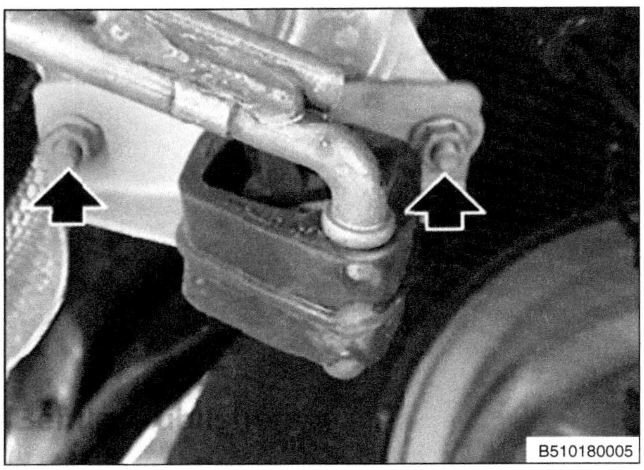

Working at rear subframe, remove exhaust system mount fasteners (**arrows**).

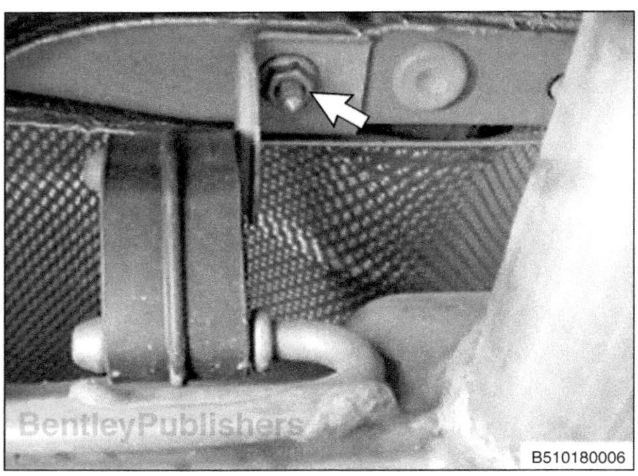

Working above rear muffler, remove exhaust system mount fastener (**arrow**).

– Carefully lower complete exhaust system and remove.

– Installation is reverse of removal. Remember to:
  - Use new gaskets and hardware. Coat hardware threads with anti-seize paste.
  - Make sure there is sufficient clearance between exhaust system and vehicle underbody at every point.
  - Loosely install exhaust system mounting hardware and hangers before tightening fasteners to their final torque.

Tightening torques	
Exhaust pipe to manifold flange	45 Nm (33 ft-lb)
Front exhaust pipe support to transmission	20 Nm (15 ft-lb)
Exhaust system hanger to rear subframe	20 Nm (15 ft-lb)
Exhaust system hanger to rear subframe	20 Nm (15 ft-lb)

– Check exhaust system for leaks.

BentleyPublishers.com
BentleyPublishers.com—All Rights Reserved

Exhaust system, removing and installing (N54 engine)

The removal of the exhaust system as a complete unit is recommended. Once the complete system is removed from the car, individual pipes and mufflers can be more easily replaced. Rear muffler can be replaced without removing exhaust system.

– Raise vehicle and support safely.

> ***WARNING —***
> - *Make sure the vehicle is stable and well supported at all times. Use a professional automotive lift or jack stands designed for the purpose. A floor jack is not adequate support.*

– Support exhaust system using BMW special tool 31 2 220 or equivalent and suitable jack.

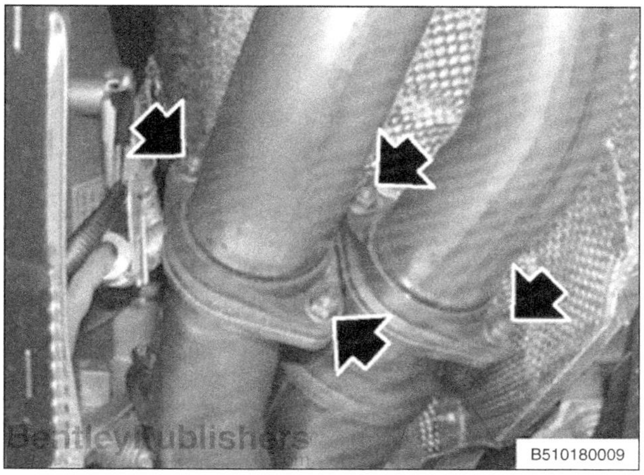

◄ Remove fasteners for front exhaust pipe at exhaust manifold flange (**arrows**).

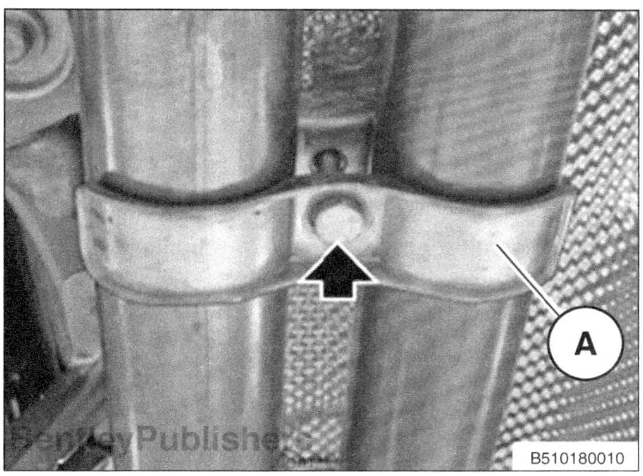

◄ Remove fastener (**arrow**) and front support bracket (**A**).

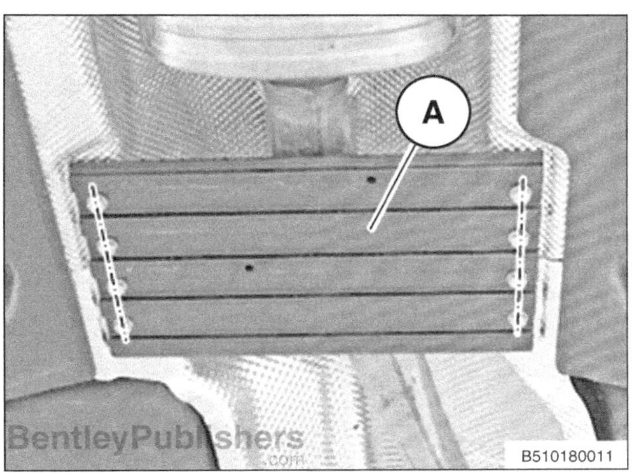

◄ Remove reinforcement plate bolts and remove plate (**A**).

BentleyPublishers
BentleyPublishers.com–All Rights Reserved

Exhaust system, removing and installing (N54 engine)

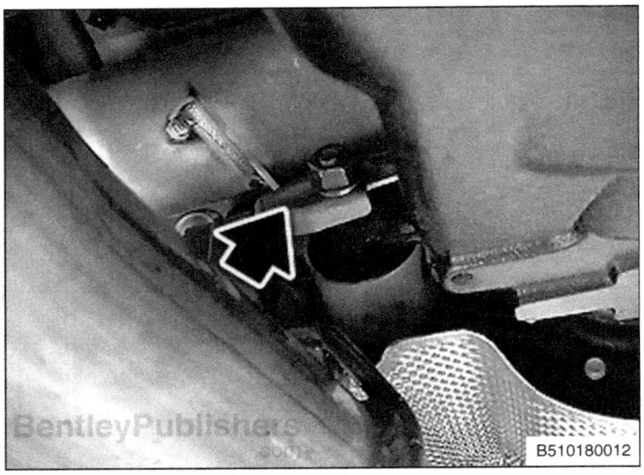

◀ Working at rear differential, remove exhaust system hanger fastener (**arrow**).

◀ Working at rear subframe, remove exhaust system mount fasteners (**arrows**).

> ***CAUTION—***
> *• Exhaust system is heavy. Lower with the help of an assistant.*

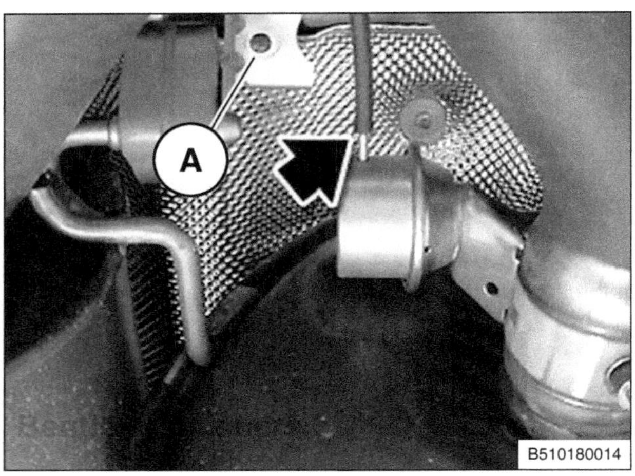

◀ Working at rear muffler, disconnect tailpipe flap vacuum hose (**arrow**) and remove exhaust mount fastener (**A**).

– Carefully lower complete exhaust system and remove.

> ***CAUTION—***
> *• Exhaust system is heavy. Lower with the help of an assistant.*

BentleyPublishers.com
BentleyPublishers.com—All Rights Reserved

- Installation is reverse of removal. Remember to:
  - Use new gaskets and hardware. Coat hardware threads with anti-seize paste.
  - Make sure there is sufficient clearance between exhaust system and vehicle underbody at every point.
  - Loosely install exhaust system mounting hardware and hangers before tightening fasteners to their final torque.
  - Check exhaust system for leaks.

Tightening torques	
Exhaust pipe to manifold flange	21 Nm (15 ft-lb)
Front exhaust pipe clamp to transmission	21 Nm (15 ft-lb)
Exhaust system hanger to rear differential	35 Nm (26 ft-lb)
Exhaust system hanger to wheel carrier	21 Nm (15 ft-lb)
Exhaust system hanger to rear chassis	21 Nm (15 ft-lb)

Exhaust system, removing and installing (V8 engine)

Remove exhaust system as a complete unit.

Once the complete system is removed from the car, replace individual pipes and mufflers. Rear muffler can be replaced without removing exhaust system.

- Raise vehicle and support safely.

> ***WARNING —***
> - *Make sure the vehicle is stable and well supported at all times. Use a professional automotive lift or jack stands designed for the purpose. A floor jack is not adequate support.*

- Support exhaust system using BMW special tool 31 2 220 or equivalent and suitable jack.

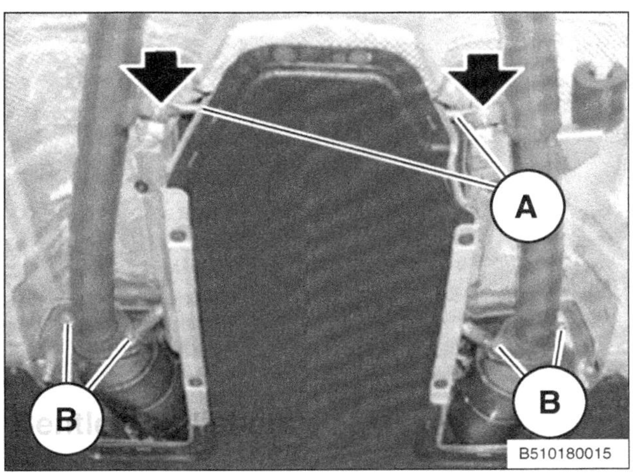

◁ Remove mounting bracket (**A**) fasteners (**arrows**). Remove exhaust pipe flange fasteners (**B**).

BentleyPublishers.com–All Rights Reserved

Exhaust system, removing and installing (V8 engine)

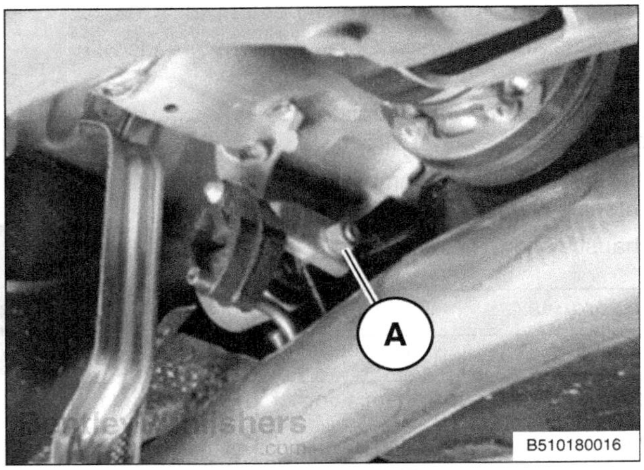

Working at rear differential, remove exhaust system hanger fastener (**A**).

Working at rear subframe, remove exhaust system mount fasteners (**arrows**)

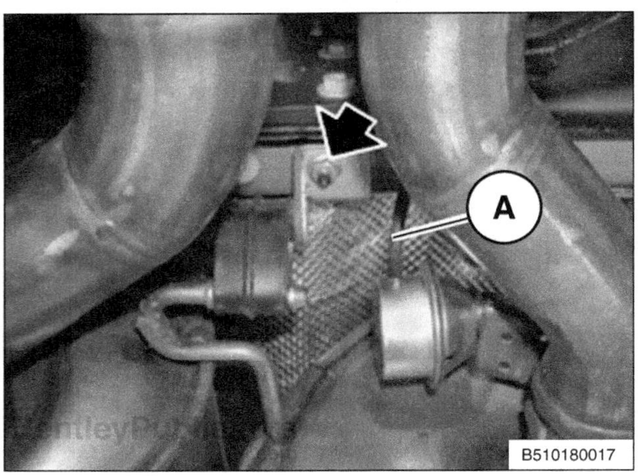

Working at rear muffler, disconnect tailpipe flap vacuum hose (**A**) and remove exhaust mount fastener (**arrow**)

BentleyPublishers.com
BentleyPublishers.com—All Rights Reserved

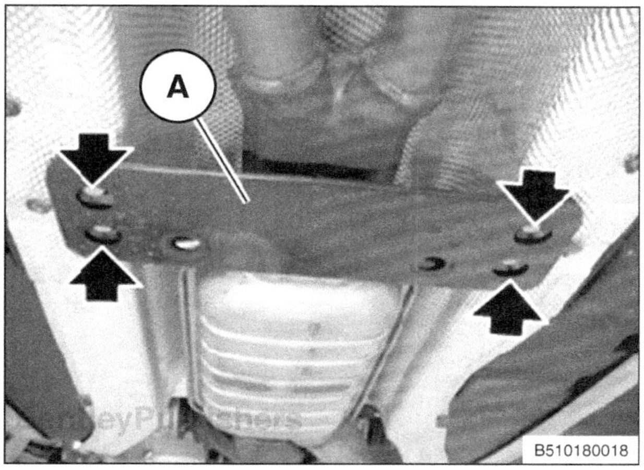

◄ Remove reinforcement plate (**A**) fasteners (**arrows**)

– Carefully lower complete exhaust system and remove.

> ***CAUTION—***
> *• Exhaust system is heavy. Lower with the help of an assistant.*

– Installation is reverse of removal. Remember to:

- Use new gaskets and hardware. Coat hardware threads with anti-seize paste.
- Make sure there is sufficient clearance between exhaust system and vehicle underbody at every point.
- Loosely install exhaust system mounting hardware and hangers before tightening fasteners to their final torque.
- Check exhaust system for leaks.

Tightening torques	
Exhaust pipe to manifold flange	45 Nm (33 ft-lb)
Front exhaust pipe support to transmission	21 Nm (15 ft-lb)
Exhaust system hanger to rear differential	35 Nm (26 ft-lb)
Exhaust system hanger to wheel carrier	21 Nm (15 ft-lb)
Exhaust system hanger to rear chassis	21 Nm (15 ft-lb)

Exhaust system components, removing and installing (M54, N52 engines)

BMW offers individual exhaust components as replacement parts. Originally installed exhaust components are welded together at the factory and must be cut off to be replaced.

Front silencer replacement is best accomplished by first removing the complete exhaust system. Rear muffler can be replaced without removing exhaust system. See **Exhaust system, removing and installing** in this repair group.

– Working at rear muffler, measure old exhaust system parts against replacement parts to locate cutting point.

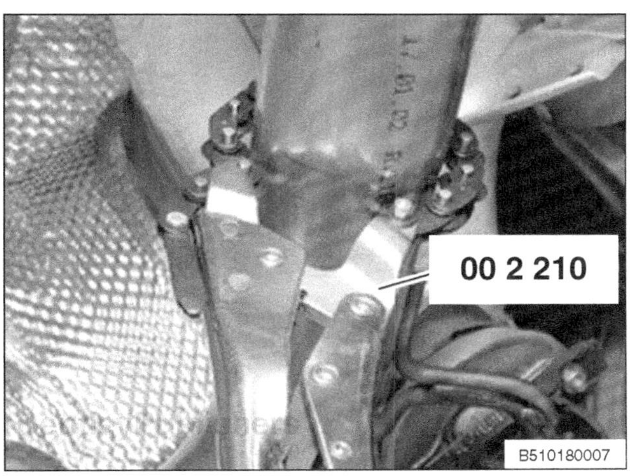

◄ Use exhaust pipe cutter (BMW special tool 00 2 210 or equivalent) to cut pipe(s). Deburr cut pipe end(s).

BentleyPublishers.com—All Rights Reserved

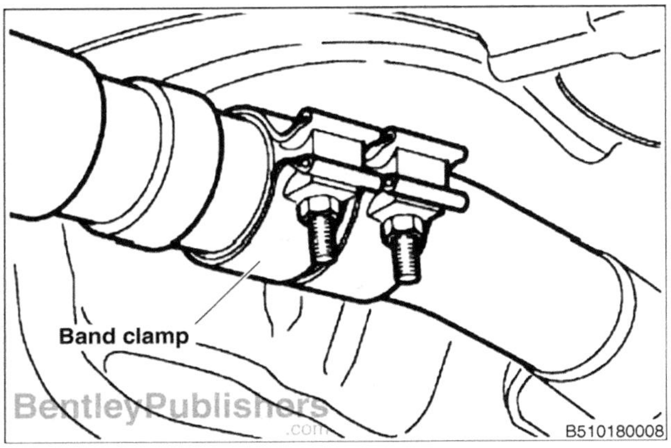

Assemble pipes using band clamps.

– Installation is reverse of removal. Remember to:
  - Make sure there is sufficient clearance between exhaust system and vehicle underbody at every point.
  - Check exhaust system for leaks.

Tightening torque	
Band clamp to exhaust pipes	45 Nm (33 ft-lb)

Exhaust system components, removing and installing (N54 engine)

BMW offers individual exhaust components as replacement parts. Originally installed exhaust components are welded together at the factory and must be cut off to be replaced.

Component replacement is best accomplished by first removing the complete exhaust system. See **Exhaust system, removing and installing** in this repair group.

NOTE—
- *Rear muffler can be replaced without removing exhaust system.*

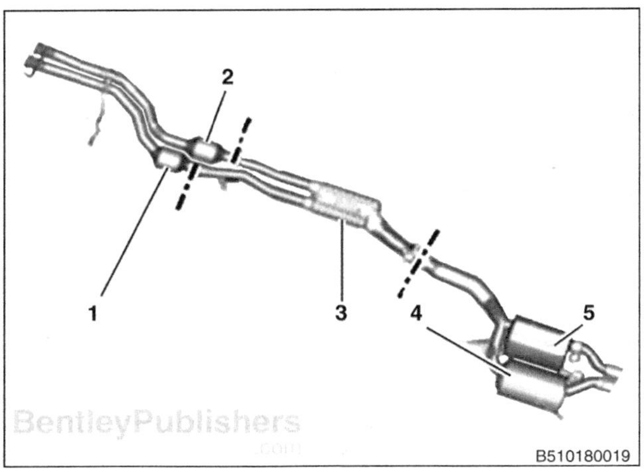

Cut lines for exhaust components are designated by notches in the 535i exhaust system shown.

1. Rear catalytic converter, left
2. Rear catalytic converter, right
3. Center muffler
4. Rear muffler, left
5. Rear muffler, right

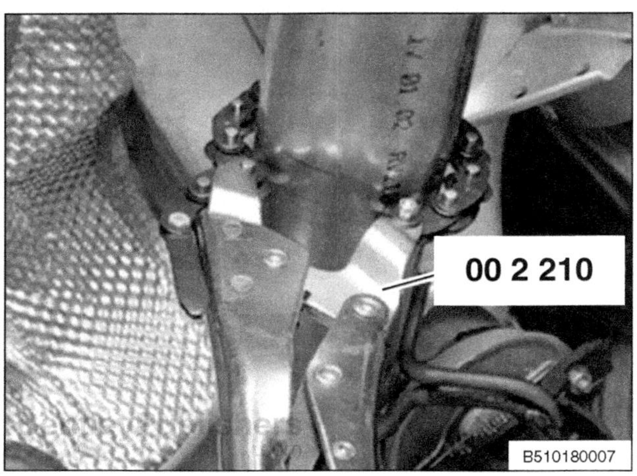

Use exhaust pipe cutter (BMW special tool 00 2 210 or equivalent) to cut pipe(s). Deburr cut pipe end(s).

BentleyPublishers.com
BentleyPublishers.com—All Rights Reserved

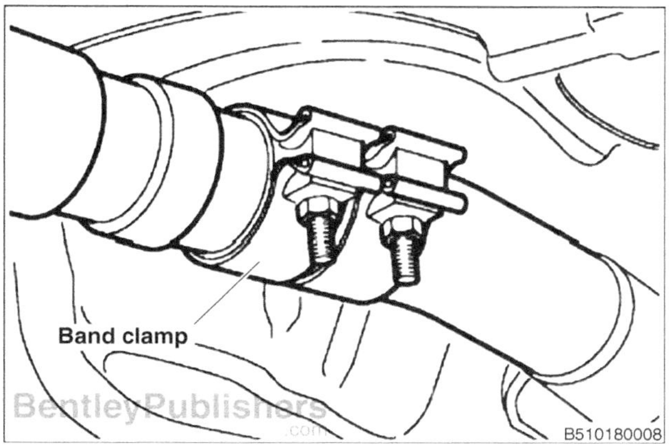

Assemble pipes using band clamps.

– Installation is reverse of removal. Remember to:
- Make sure there is sufficient clearance between exhaust system and vehicle underbody at every point.
- Check exhaust system for leaks.

Tightening torque	
Band clamp to exhaust pipes	45 Nm (33 ft-lb)

Exhaust system components, removing and installing (V8 engine)

BMW offers individual exhaust components as replacement parts. Originally installed exhaust components are welded together at the factory and must be cut off to be replaced.

Component replacement is best accomplished by first removing the complete exhaust system. See **Exhaust system, removing and installing** in this repair group.

Cut lines for exhaust components are designated by notches in the exhaust system. Mark exhaust cutting lines with aid of replacement parts.

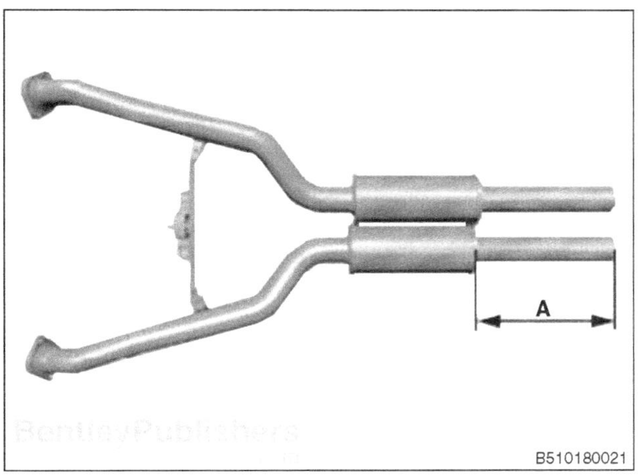

Front muffler: Mark length **A** with aid of replacement muffler.

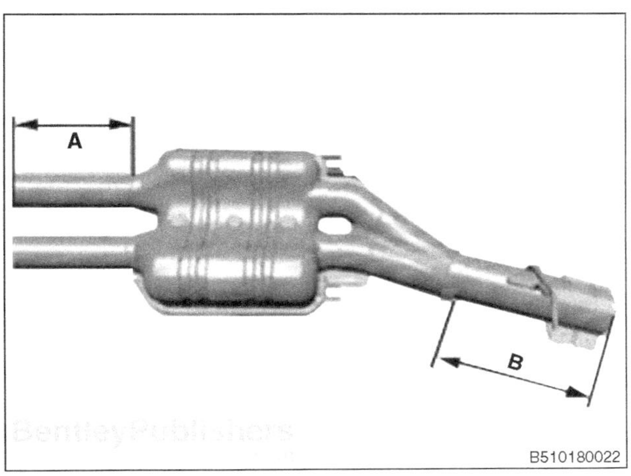

Center muffler. Mark lengths **A** and **B** with aid of replacement muffler.

Exhaust system components, removing and installing (V8 engine)

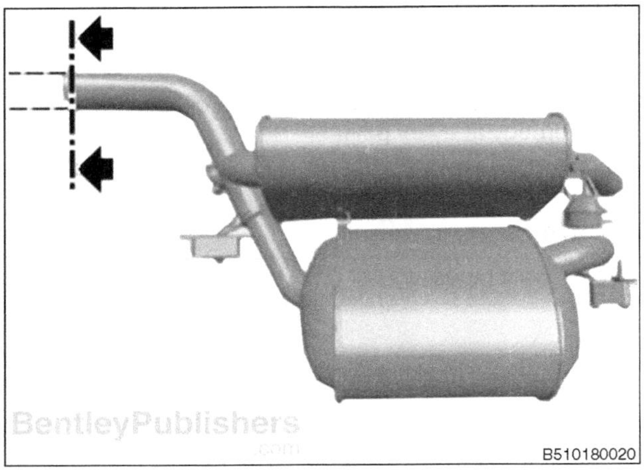

◄ Rear muffler: Mark length of **dotted area** with aid of replacement muffler

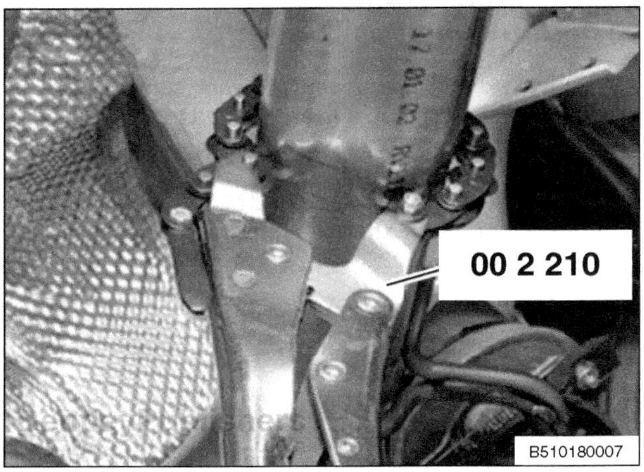

◄ Use exhaust pipe cutter (BMW special tool 00 2 210 or equivalent) to cut pipe(s). Deburr cut pipe end(s).

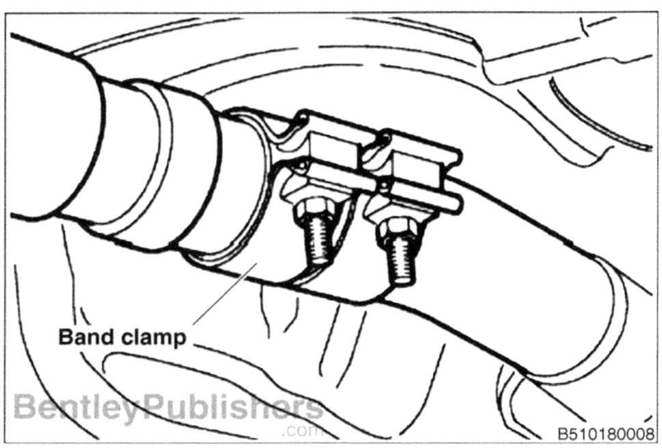

◄ Assemble pipes using band clamps.

– Installation is reverse of removal. Remember to:
  - Make sure there is sufficient clearance between exhaust system and vehicle underbody at every point.
  - Check exhaust system for leaks.

Tightening torque	
Band clamp to exhaust pipes	45 Nm (33 ft-lb)

BentleyPublishers.com
BentleyPublishers.com—All Rights Reserved

Exhaust Manifolds

Exhaust manifolds (cylinders 1 - 6), removing and installing (M54 engine)

The M54 engine uses two exhaust manifolds. Each manifold includes an integral catalytic converter. Should the catalyst need replacing, replace the appropriate exhaust manifold / catalyst unit. See **Exhaust System Components** in this repair group.

– Remove engine cover. See **020 Maintenance**.

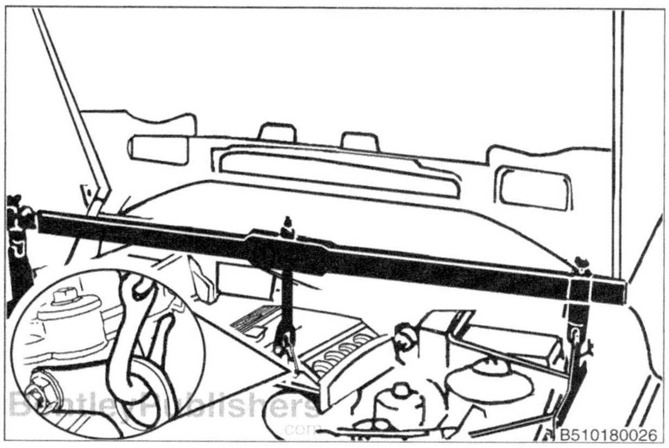

◄ Support engine from above with suitable hoist.

– Remove lower engine covers. See **020 Maintenance**.

– Remove complete exhaust system. See **Exhaust system, removing and installing (M54, N52 engine)** in this repair group.

– Remove oxygen sensors. See **Oxygen sensors, removing and installing (M54, N52 engine)** in this repair group.

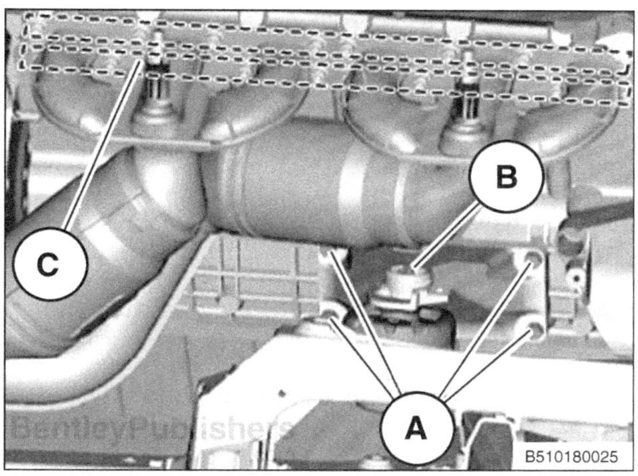

◄ Remove fasteners (**A**) and remove engine mount support (**B**). Remove exhaust manifold fasteners (**C**) and remove exhaust manifolds.

– Installation is reverse of removal. Remember to:
- Clean sealing surfaces and replace gaskets.
- Replace manifold and front pipe mounting nuts.
- Coat fastener threads with anti-seize paste.

Tightening torques	
Exhaust manifold to cylinder head (replace with new M7 nuts)	15 Nm (11 ft-lb)
Front pipe to exhaust manifold (replace with new M10 nuts)	45 Nm (33 ft-lb)
Oxygen sensor to exhaust pipe or catalytic converter (M18)	50 Nm (37 ft-lb)

BentleyPublishers.com–All Rights Reserved

Exhaust manifold (cylinders 1 - 3), removing and installing (N52 engine)

The N52 engine uses two exhaust manifolds. Each manifold includes an integral catalytic converter. Should the catalyst need replacing, replace the appropriate exhaust manifold / catalyst unit. See **Exhaust System Components** in this repair group.

– Remove rear exhaust manifold. See **Exhaust manifold (cylinders 4 - 6), removing and installing (M54, N52 engine)** in this repair group.

– Remove oxygen sensors. See **Oxygen sensors, removing and installing (M54, N52 engine)** in this repair group.

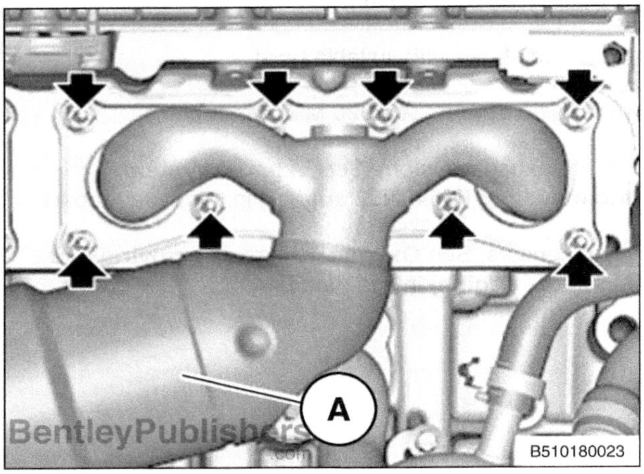

◄ Remove fasteners (**arrows**) from front exhaust manifold (**A**) and remove manifold.

– Installation is reverse of removal. Remember to:
- Clean sealing surfaces of manifold and cylinder head.
- Use new gaskets and hardware.
- Check exhaust system for leaks.

Tightening torque	
Exhaust manifold to cylinder head	20 Nm (15 ft-lb)

Exhaust manifold (cylinders 4 - 6), removing and installing (N52 engine)

– Remove engine cover. See **020 Maintenance**.

– Remove lower engine covers. See **020 Maintenance**.

– Remove complete exhaust system. See **Exhaust system, removing and installing (M54, N52 engine)** in this repair group.

– Remove oxygen sensors. See **Oxygen sensors, removing and installing (M54, N52 engine)** in this repair group.

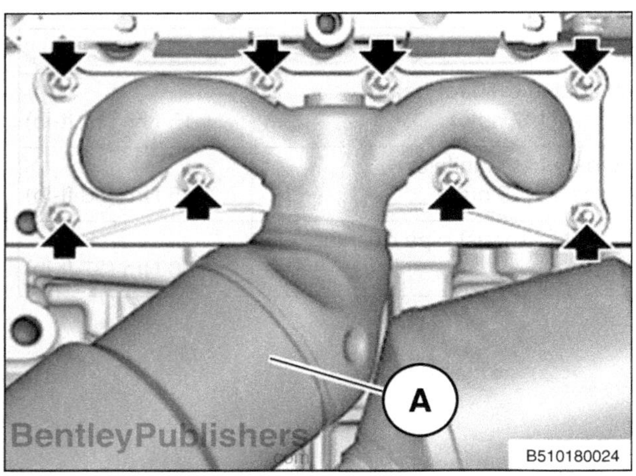

◄ Remove fasteners (**arrows**) from rear exhaust manifold (**A**) and remove manifold.

– Installation is reverse of removal. Remember to:
- Clean sealing surfaces of manifold and cylinder head.
- Use new gaskets and hardware.
- Check exhaust system for leaks.

Tightening torque	
Exhaust manifold to cylinder head	20 Nm (15 ft-lb)

BentleyPublishers.com
BentleyPublishers.com—All Rights Reserved

Exhaust manifold and turbocharger (cylinders 1- 3), removing and installing (N54 engine)

The N54 engine is equipped with two exhaust manifolds each with an integral turbocharger. A catalytic converter is attached downstream of each turbocharger. Should the catalyst need replacing, it is removable and available separately. See **Exhaust System Components** in this repair group.

- Remove both near engine catalytic converters. See **Near engine catalytic converters (N54 engine)** in this repair group.
- Remove intercooler. See **170 Radiator and Cooling System**.
- Drain engine coolant and remove coolant expansion tank. See **170 Radiator and Cooling System**.
- Remove thermostat. See **170 Radiator and Cooling System**.
- Remove coolant pump. See **170 Radiator and Cooling System**.
- Remove turbocharger vacuum reservoirs. See **132 Fuel Injection MSD80, MSD81 (N54 Engine)**.
- Remove air box and air intake duct. See **132 Fuel Injection MSD80, MSD81 (N54 Engine)**.
- Remove right side tie rod from steering rack. See **320 Steering and Wheel Alignment**.

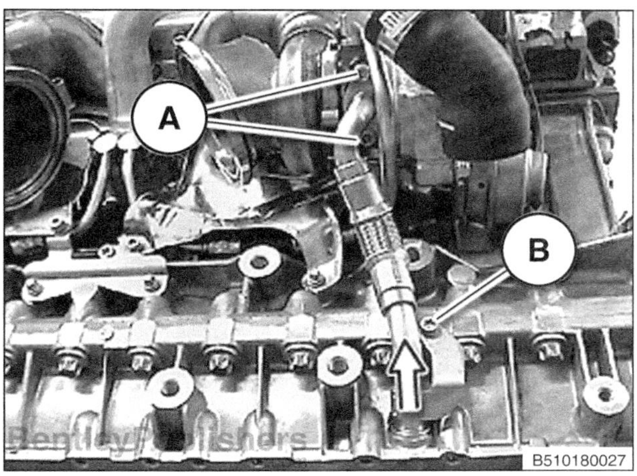

Working below turbo chargers: Remove fasteners on oil return line (**A**) and oil return line bracket (**B**). Gently pull oil return line out of engine block in direction of **arrow**.

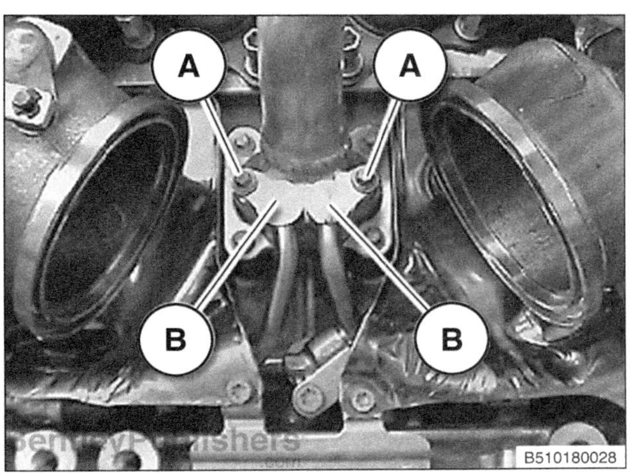

Working between turbochargers, remove fasteners (**A**) and remove coolant supply lines (**B**).

NOTE—

- *If necessary grasp coolant supply line flange with pliers to remove. Do not use pliers on pipes.*

BentleyPublishers.com–All Rights Reserved

Exhaust manifold and turbocharger (cylinders 1- 3), removing and installing (N54 engine)

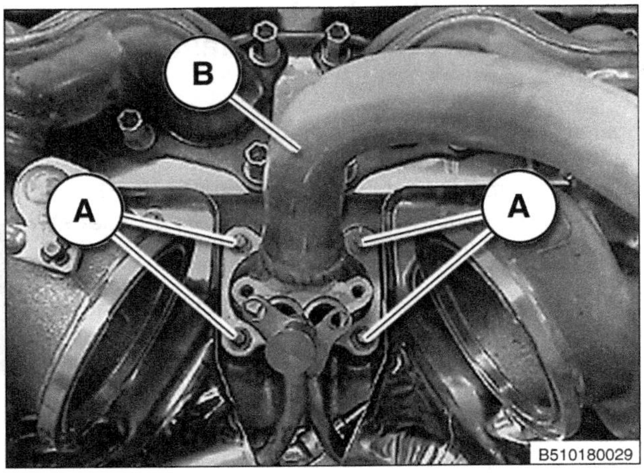

◀ Remove fasteners (**A**) at coolant inlet pipe (**B**) and remove pipe.

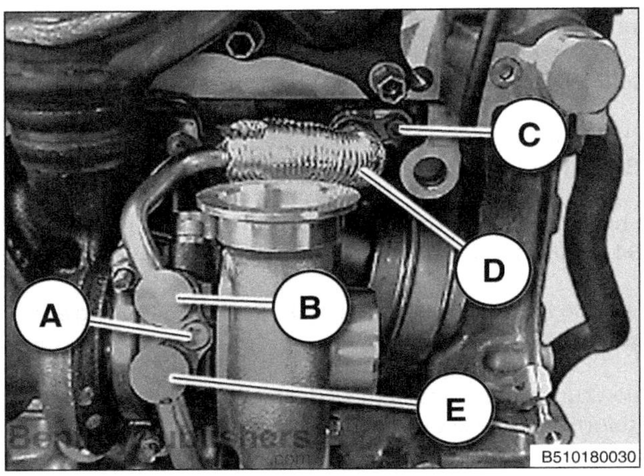

◀ Remove coolant return pipe fasteners (**A**, **C**) and remove return pipe (**D**).

NOTE —

- *If necessary grasp coolant return line flange* (**B**) *with pliers to remove. Do not use pliers on pipes.*

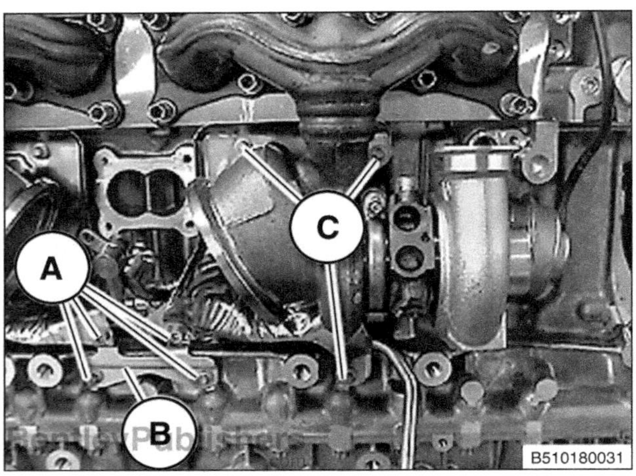

◀ Remove fasteners (**A**) at retaining plate (**B**). Remove heat shield fasteners (**C**).

BentleyPublishers.com
BentleyPublishers.com—All Rights Reserved

Carefuly remove heat shield (**A**) in direction of **arrow** so heat shield and coolant pipes are not damaged.

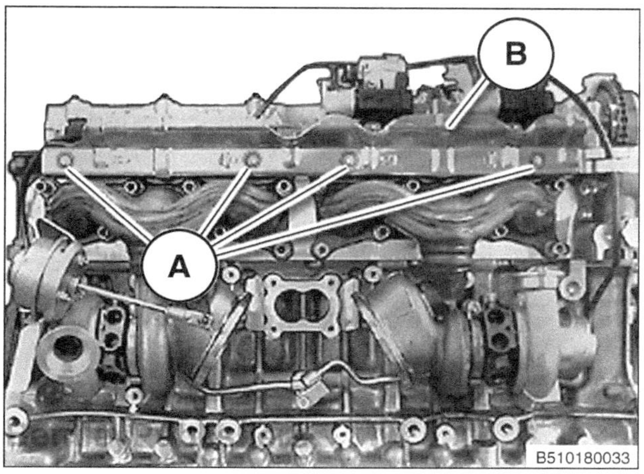

Remove fasteners (**A**) from bracket (**B**). Lay bracket on top of cylinder head.

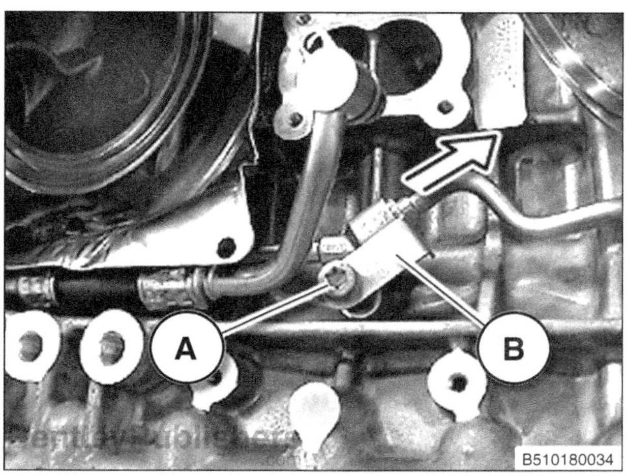

Remove oil supply line bracket (**B**) fastener (**A**). Remove oil supply line in direction of **arrow**.

BentleyPublishers.com–All Rights Reserved

Exhaust manifold and turbocharger (cylinders 1- 3), removing and installing (N54 engine)

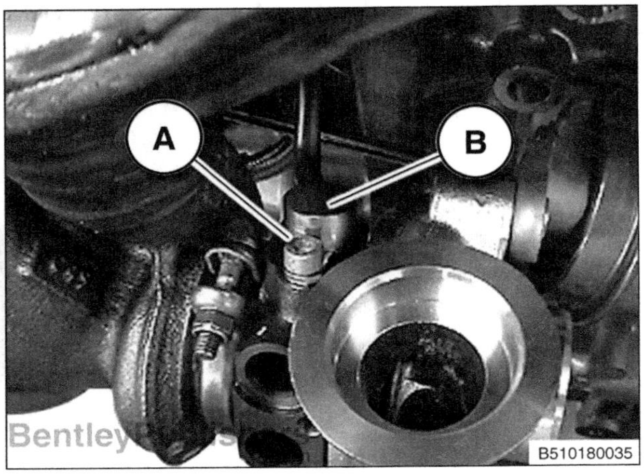

Remove fastener (**A**) and remove oil supply line.

NOTE —
- *If necessary grasp oil supply line flange (**B**) with pliers to remove. Do not use pliers on pipes.*

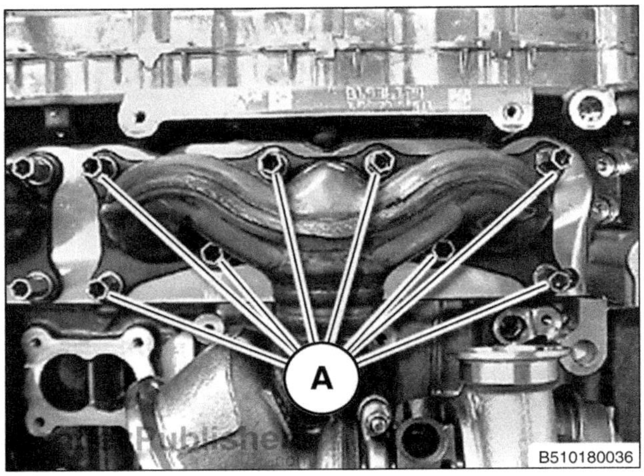

Remove exhaust manifold fasteners (**A**). Detach vacuum hose from wastegate valve and lift out exhaust manifold and turbocharger towards the top.

NOTE —
- *Take care not to damage wastegate valve linkage.*

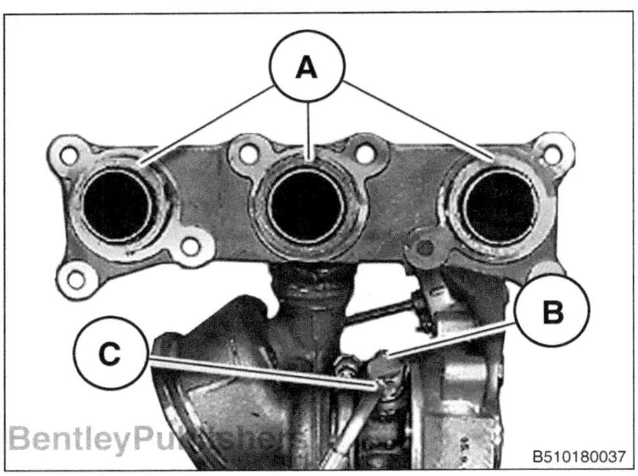

If replacing turbocharger, remove fastener (**B**) for oil supply line (**C**) and install supply line on new unit. Always replace graphite gaskets (**A**).

BentleyPublishers.com
BentleyPublishers.com—All Rights Reserved

– Remainder of installation is reverse of removal. Remember to:
  - Use new O-rings, gaskets and hardware. Coat hardware threads with anti-seize paste.
  - Check and clear fault codes from ECM.
  - Check air intake system for leaks.
  - Check exhaust system for leaks.

Tightening torques	
Bracket to cylinder head	9 Nm (7 ft-lb)
Coolant supply line to crankcase	9 Nm (7 ft-lb)
Coolant return line to crankcase	8 Nm (6 ft-lb)
Coolant return line to turbocharger	8 Nm (6 ft-lb)
Coolant supply line to coolant line	8 Nm (6 ft-lb)
Exhaust manifold heat shield to crankcase	8 Nm (6 ft-lb)
Oil supply line to crankcase	20 Nm (15 ft-lb)
Oil pressure line to turbocharger	8 Nm (6 ft-lb)
Oil return line to crankcase	8 Nm (6 ft-lb)
Oil return line to turbocharger	8 Nm (6 ft-lb)
Turbocharger to exhaust manifold	20 Nm (15 ft-lb)

Exhaust manifold and turbocharger (cylinders 4 - 6), removing and installing (N54 engine)

– Remove both Near engine catalytic converters. See **Near engine catalytic converters (N54 engine)** in this repair group.

– Release right side tie rod from steering rack. See **320 Steering and Wheel Alignment**.

– Remove intercooler. See **170 Radiator and Cooling System**.

– Drain engine coolant and remove coolant expansion tank. See **170 Radiator and Cooling system**.

– Remove thermostat. See **170 Radiator and cooling system**.

– Remove coolant pump. See **170 Radiator and cooling system**.

– Remove turbocharger vacuum reservoirs. See **132 Fuel Injection MSD80, MSD81 (N54 Engine)**. .

– Remove air box and air intake duct. See **132 Fuel Injection MSD80, MSD81 (N54 Engine)**.

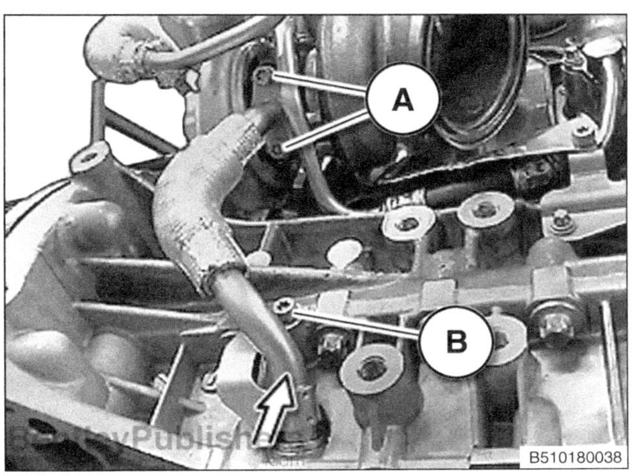

◄ Remove fasteners on oil return line (**A**) and oil return line bracket (**B**). Gently pull oil return line out of engine block in direction of **arrow**.

BentleyPublishers.com–All Rights Reserved

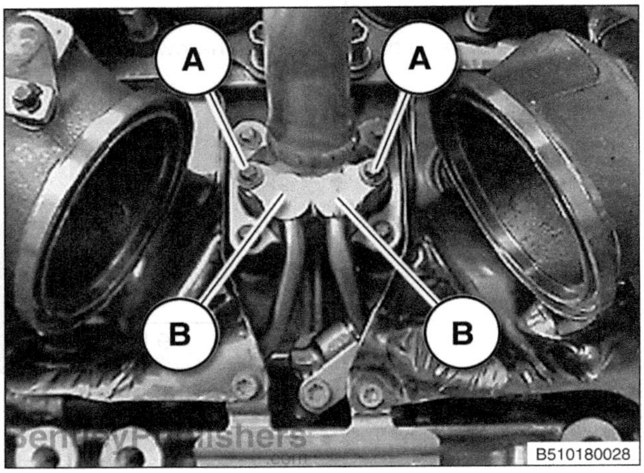

Working between turbochargers, remove fasteners (**A**) and remove coolant supply lines (**B**).

NOTE —
- *If necessary grasp coolant supply line flange with pliers to remove. Do not use pliers on pipes.*

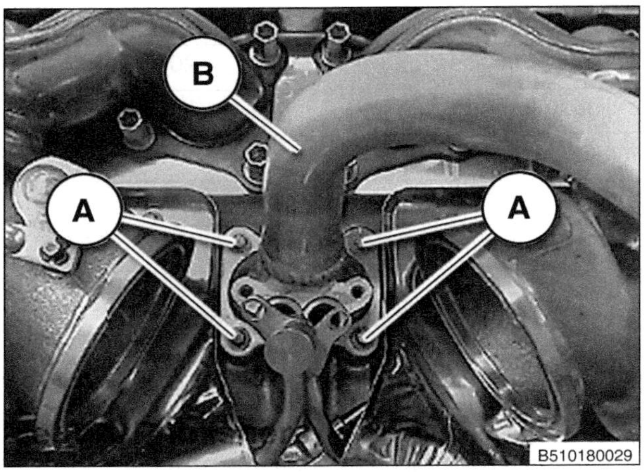

Remove fasteners (**A**) at coolant inlet pipe (**B**) and remove pipe.

Remove fastener (**A**) at coolant return line clamp and remove line.

NOTE —
- *If necessary grasp coolant return line flange with pliers to remove. Do not use pliers on pipes.*

BentleyPublishers.com
BentleyPublishers.com—All Rights Reserved

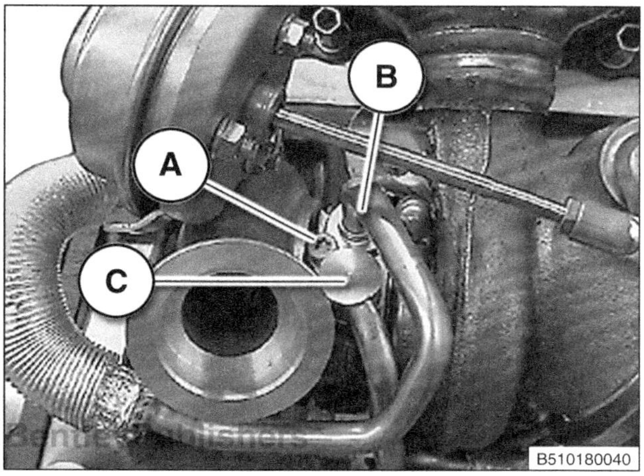

B510180040

Remove fastener (**A**) and remove coolant return line (**B**) and coolant supply line (**C**).

NOTE—
- *If necessary, grasp flange with pliers to remove. Do not use pliers on pipes.*
- *Take care not to damage wastegate valve linkage.*

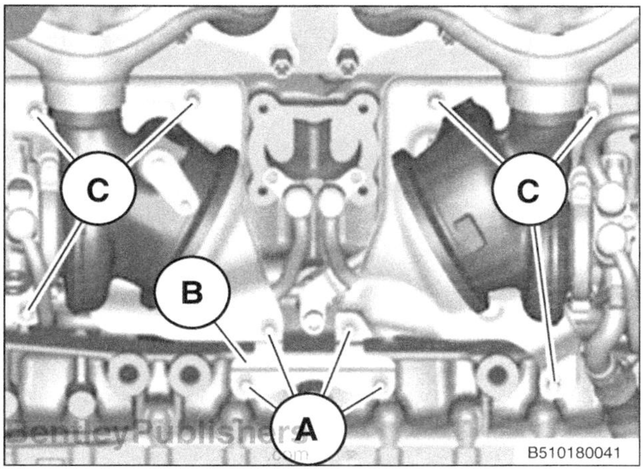

B510180041

Remove fasteners (**A**) at retaining plate (**B**). Remove heat shield fasteners (**C**).

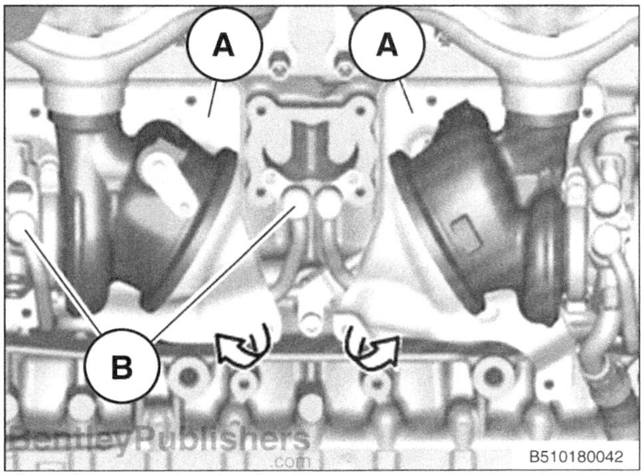

B510180042

Carefully swing out heat shields (**A**) in direction of **arrow**. Coolant supply pipe (**B**) can be removed along with heat shield.

Exhaust manifold and turbocharger (cylinders 4 - 6), removing and installing (N54 engine)

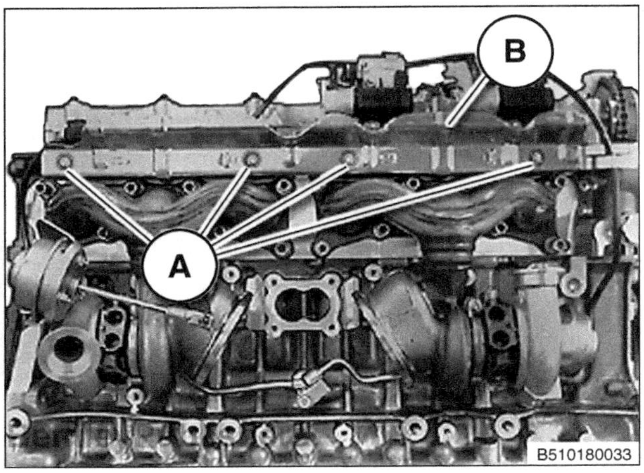

Remove fasteners (**A**) from bracket (**B**). Lay bracket on top of cylinder head.

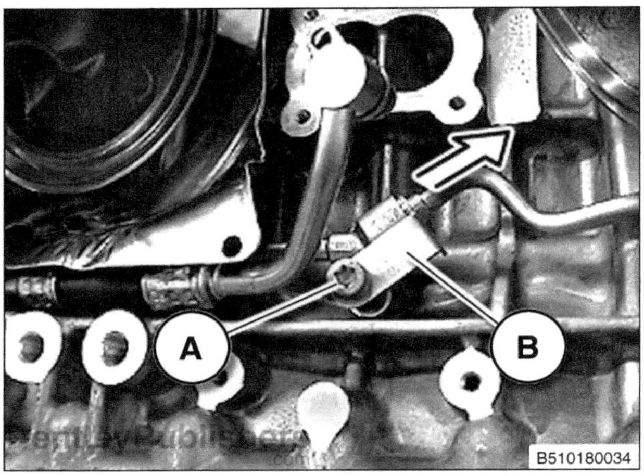

Remove oil supply line bracket (**B**) fastener (**A**). Remove oil supply line in direction of **arrow**.

Remove exhaust manifold fasteners (**A**). Detach vacuum hose from wastegate valve and lift out exhaust manifold and turbocharger towards the top.

NOTE —

- *Take care not to damage wastegate valve linkage.*

BentleyPublishers.com
BentleyPublishers.com—All Rights Reserved

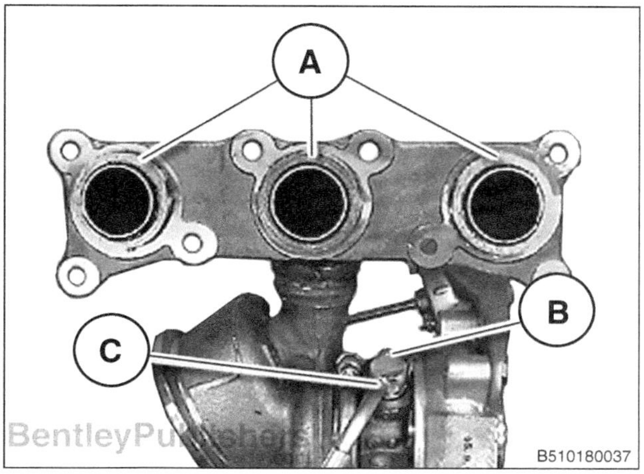

If replacing turbocharger, remove fastener (**B**) for oil supply line (**C**) and install supply line on new unit. Always replace graphite gaskets (**A**).

- Remainder of installation is reverse of removal. Remember to:
  - Use new O-rings, gaskets and hardware. Coat hardware threads with anti-seize paste.
  - Check and clear fault codes from ECM.
  - Check air intake system for leaks.
  - Check exhaust system for leaks.

Tightening torques	
Bracket to cylinder head	9 Nm (7 ft-lb)
Coolant supply line to crankcase	9 Nm (7 ft-lb)
Coolant return line to crankcase	8 Nm (6 ft-lb)
Coolant return line to turbocharger	8 Nm (6 ft-lb)
Coolant supply line to coolant line	8 Nm (6 ft-lb)
Exhaust manifold heat shield to crankcase	8 Nm (6 ft-lb)
Oil supply line to crankcase	20 Nm (15 ft-lb)
Oil pressure line to turbocharger	8 Nm (6 ft-lb)
Oil return line to crankcase	8 Nm (6 ft-lb)
Oil return line to turbocharger	8 Nm (6 ft-lb)
Turbocharger to exhaust manifold	20 Nm (15 ft-lb)

Exhaust manifold (cylinders 1 - 4), removing and installing (V8 engine)

- Remove pre- and post-catalyst oxygen sensors. See **Oxygen sensors, removing and installing (V8 engine)** in this repair group.
- Remove complete exhaust system. See **Exhaust system, removing and installing (V8 engine)** in this repair group.
- Secure engine in place using engine support.
- Lower front subframe. See **310 Front Suspension**.
- Remove right side engine mount and engine mount support.

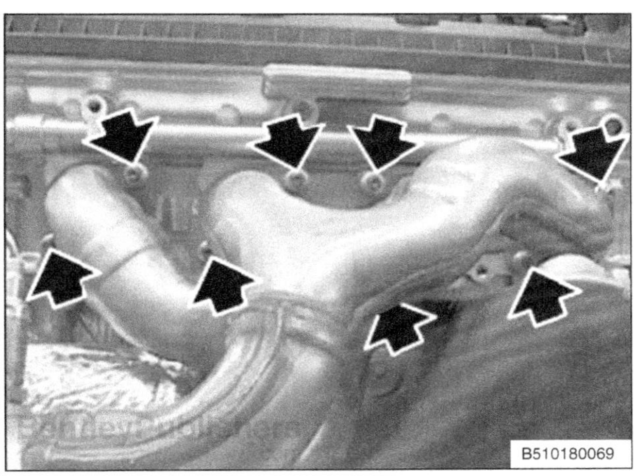

Remove exhaust manifold fasteners (**arrows**). Remove exhaust manifold towards bottom, pull outwards toward rear.

BentleyPublishers.com–All Rights Reserved

- Remainder of installation is reverse of removal. Remember to:
  - Use new gaskets and hardware. Coat hardware threads with anti-seize paste.
  - Check exhaust system for leaks.

Tightening torques	
Exhaust manifold to cylinder head • M6 nuts • M7 nuts • M8 nuts	 10 Nm (7.3 ft-lb) 15 Nm (11 ft-lb) 23 Nm (16.9 ft-lb)

Exhaust manifold (cylinders 5 - 8), removing and installing (V8 engine)

- Remove pre- and post-catalyst oxygen sensors. See **Oxygen sensors, removing and installing (V8 engine)** in this repair group.
- Remove complete exhaust system. See **Exhaust system, removing and installing (V8 engine)** in this repair group.
- Secure engine in place using engine support.
- Lower front subframe. See **310 Front Suspension**.
- Remove steering knuckle. See **Steerng and Wheel Alignment**.

< Remove exhaust manifold fasteners (**arrows**). Remove exhaust manifold towards bottom, pull outwards toward rear.

- Remainder of installation is reverse of removal. Remember to:
  - Use new gaskets and hardware. Coat hardware threads with anti-seize paste.
  - Check exhaust system for leaks.

Tightening torques	
Exhaust manifold to cylinder head • M6 nuts • M7 nuts • M8 nuts	 10 Nm (7.3 ft-lb) 15 Nm (11 ft-lb) 23 Nm (16.9 ft-lb)

BentleyPublishers.com
BentleyPublishers.com—All Rights Reserved

OXYGEN SENSORS

Oxygen sensors, removing and installing (M54, N52 engines)

Oxygen sensors are installed at the front and rear of the catalytic converters. Prior to sensor removal, use BMW scan tool to read and clear ECM fault codes.

> ***CAUTION—***
> - *Use care not to drag or bang oxygen sensors. Oxygen sensors are easily damaged.*
> - *To avoid possible engine damage, do not mix up oxygen sensor connectors. Mark connectors before disassembling.*

– Remove upper engine covers. See **020 Maintenance**.

– Raise vehicle and support safely.

> ***WARNING —***
> - *Make sure the vehicle is stable and well supported at all times. Use a professional automotive lift or jack stands designed for the purpose. A floor jack is not adequate support.*

– Remove front end reinforcement. Discard mounting bolts. See **310 Front Suspension**.

– Label oxygen sensor harness connectors then disconnect.

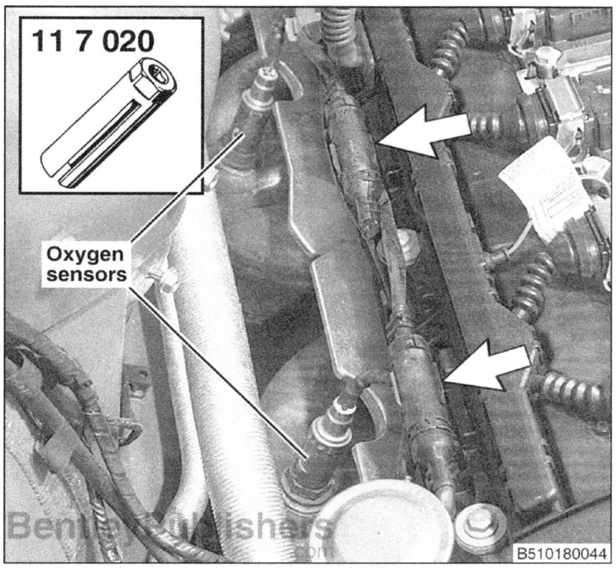

Pre-catalyst oxygen sensors

◄ Working at right side of engine above exhaust manifolds, remove oxygen sensor using slotted 22 mm socket (BMW special tool 11 7 020 or equivalent). M54 engine shown, N52 is similiar.

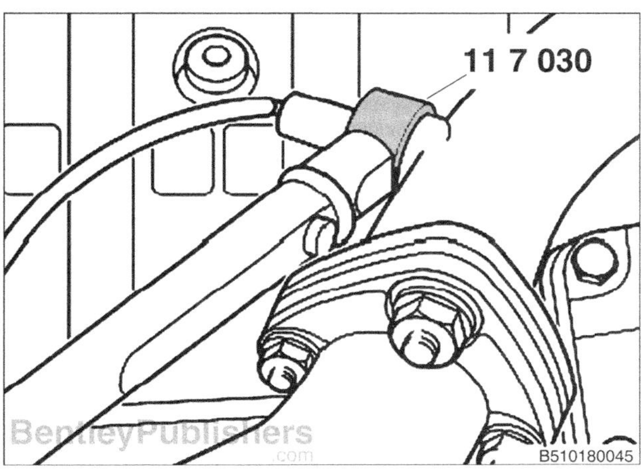

Post-catalyst oxygen sensors

◄ Remove oxygen sensor using crow-foot wrench (BMW special tool 11 7 030 or equivalent).

– Installation is reverse of removal. Remember to:
- Apply a small amount of anti-seize compound to sensor threads.
- Make sure sensor electrical harness connectors are routed as before and connected correctly.

> ***CAUTION—***
> - *Do not overuse anti-seize paste. Do not contaminate tip of sensor with paste or lubricants.*

Tightening torques	
Oxygen sensor to exhaust pipe: • Using BMW special tool 11 7 030 • Using BMW special tool 11 7 020	 47 Nm (35 ft-lb) 50 Nm (37 ft-lb)

BentleyPublishers.com
BentleyPublishers.com—All Rights Reserved

Oxygen sensors, removing and installing (N54 engine)

Oxygen sensors are installed at the front and rear of the catalytic converters. Prior to sensor removal, use BMW scan tool to read and clear ECM fault codes.

> ***CAUTION—***
> - *Use care not to drag or bang oxygen sensors. Oxygen sensors are easily damaged.*
> - *To avoid possible engine damage, do not mix up oxygen sensor connectors. Mark connectors before disassembling.*

– Remove upper engine covers. See **020 Maintenance**.

– Raise vehicle and support safely.

> ***WARNING—***
> - *Make sure the vehicle is stable and well supported at all times. Use a professional automotive lift or jack stands designed for the purpose. A floor jack is not adequate support.*

– Remove front end reinforcement. Discard mounting bolts. See **310 Front Suspension**.

– Label oxygen sensor harness connectors then disconnect.

Pre-catalyst oxygen sensors

– Release right side tie rod from steering rack. See **320 Steering and Wheel Alignment**.

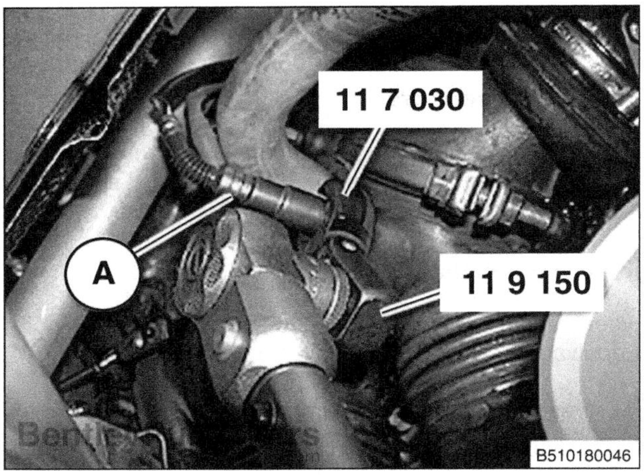

Pre-catalyst oxygen sensor (**A**) for cylinders 1 - 3 shown with BMW special tools 11 7 030 and 11 9 150.

NOTE—
- *Cylinders 1 - 3 harness connector colored black.*

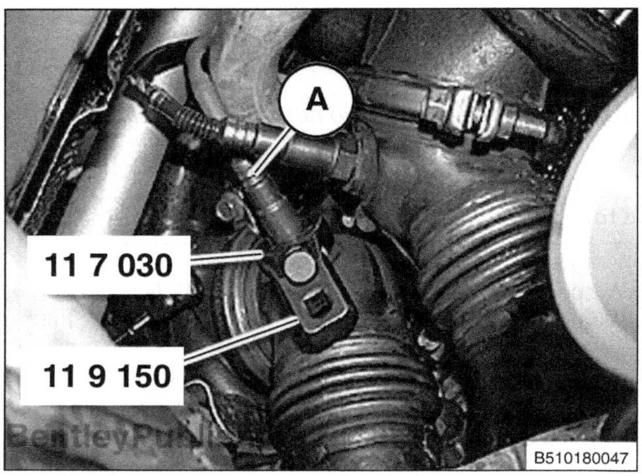

Pre-catalyst oxygen sensor (**A**) for cylinders 4 - 6 shown with BMW special tools 11 7 030 and 11 9 150.

NOTE—
- *Cylinders 4 - 6 harness connector colored grey*

BentleyPublishers.com
BentleyPublishers.com—All Rights Reserved

– Installation is reverse of removal. Remember to:
- Apply a small amount of anti-seize compound to sensor threads.
- Make sure sensor electrical harness connectors are routed as before and connected correctly.

CAUTION—
- *Do not overuse anti-seize paste. Do not contaminate tip of sensor with paste or lubricants.*

Tightening torques	
Oxygen sensor to exhaust pipe: • Using BMW special tool 11 7 030 together with 11 9 150 • Using BMW special tool 11 7 020	 47 Nm (35 ft-lb) 50 Nm (37 ft-lb)

NOTE—
- *When tool 11 7 030 is used together with swivel tool 11 9 150, reduce tightening torque by 3 Nm to 47 Nm (35 ft-lb).*

Post-catalyst oxygen sensors

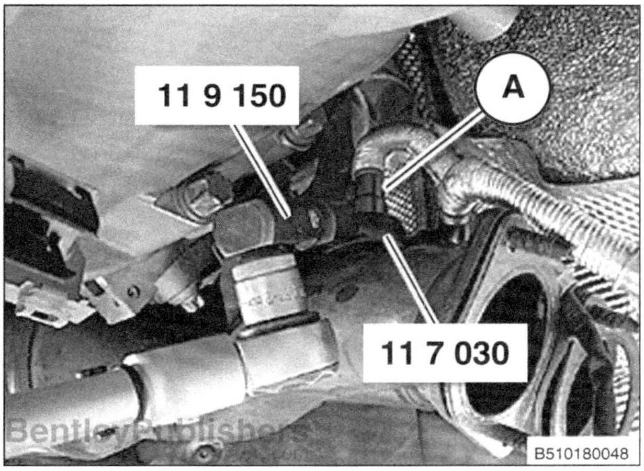

Post-catalyst oxygen sensor (**A**) for cylinders 1 - 3 shown with BMW special tools 11 7 030 and 11 9 150.

NOTE—
- *Cylinders 1 - 3 harness connector colored black.*

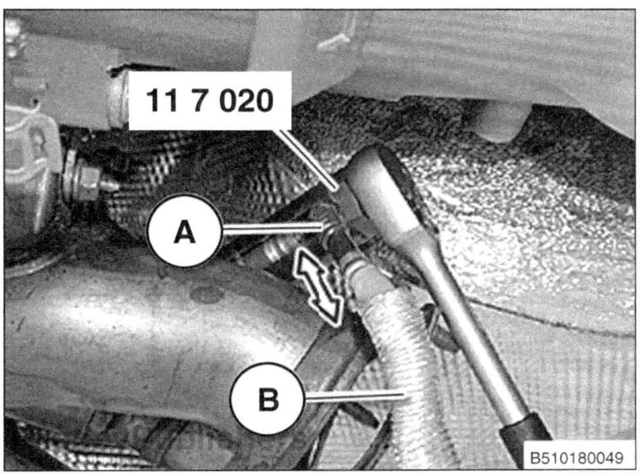

Post-catalyst oxygen sensor (**A**) for cylinders 4 - 6 shown with BMW special tool 11 7 020.

NOTE—
- *Cylinders 4 - 6 harness (**B**) connector colored grey.*

– Installation is reverse of removal. Remember to:
- Apply a small amount of anti-seize compound to sensor threads.
- Make sure sensor electrical harness connectors are routed as before and connected correctly.

CAUTION—
- *Do not overuse anti-seize paste. Do not contaminate tip of sensor with paste or lubricants.*

Tightening torques	
Oxygen sensor to exhaust pipe: • Using BMW special tool 11 7 030 together with 11 9 150 • Using BMW special tool 11 7 020	 47 Nm (35 ft-lb) 50 Nm (37 ft-lb)

NOTE—
- *When tool 11 7 030 is used together with swivel tool 11 9 150, reduce tightening torque by 3 Nm to 47 Nm (35 ft-lb).*

Oxygen sensors, removing and installing (V8 engine)

Oxygen sensors are installed at the front and rear of the catalytic converters. Prior to sensor removal, use BMW scan tool to read and clear ECM fault codes.

> ***CAUTION—***
> - *Use care not to drag or bang oxygen sensors. Oxygen sensors are easily damaged.*
> - *To avoid possible engine damage, do not mix up oxygen sensor connectors. Mark connectors before disassembling.*

– Remove upper engine cover. See **020 Maintenance**.

– Raise vehicle and support safely.

> ***WARNING —***
> - *Make sure the vehicle is stable and well supported at all times. Use a professional automotive lift or jack stands designed for the purpose. A floor jack is not adequate support.*

– Remove front end reinforcement. Discard mounting bolts. See **310 Front Suspension**.

– Label oxygen sensor harness connectors then disconnect.

Pre-catalyst oxygen sensors

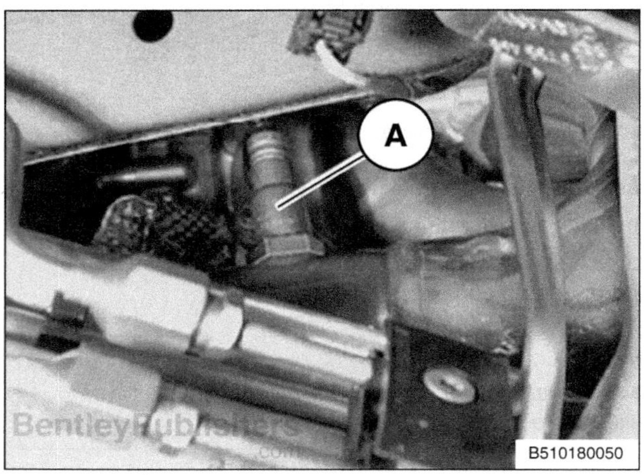

◄ Pre-catalyst oxygen sensor (**A**) for cylinders 1 - 4 shown, cylinders 5 - 8 are similiar. Remove using BMW special tool 11 7 030.

– Installation is reverse of removal. Remember to:
- Apply a small amount of anti-seize compound to sensor threads.
- Make sure sensor electrical harness connectors are routed as before and connected correctly.

> ***CAUTION—***
> - *Do not overuse anti-seize paste. Do not contaminate tip of sensor with paste or lubricants.*

Tightening torques	
Oxygen sensor to exhaust pipe: • Using BMW special tool 11 7 030	50 Nm (37 ft-lb)

BentleyPublishers.com
BentleyPublishers.com—All Rights Reserved

Post-catalyst oxygen sensors

- Remove lower engine cover.

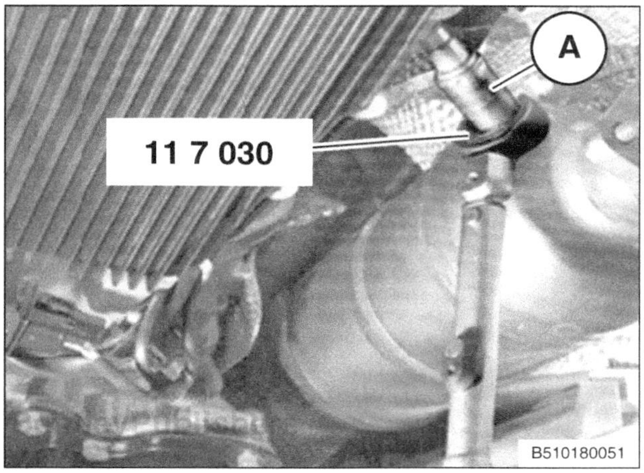

< Post-catalyst oxygen sensor (**A**) for cylinders 1 - 4 shown, cylinders 5 - 8 are similiar.. Remove using BMW special tool 11 7 030.

- Installation is reverse of removal. Remember to:
  - Apply a small amount of anti-seize compound to sensor threads.
  - Make sure sensor electrical harness connectors are routed as before and connected correctly.

> ***CAUTION—***
> - *Do not overuse anti-seize paste. Do not contaminate tip of sensor with paste or lubricants.*

Tightening torques	
Oxygen sensor to exhaust pipe: • Using BMW special tool 11 7 030	50 Nm (37 ft-lb)

Catalytic Converters

Catalytic converters are used in the exhaust system to force oxidation or conversion of unburned hydrocarbons, carbon monoxide and nitrogen oxides.

Near engine catalytic converters (M54, N52, V8 engines)

M54, N52 and N62 engines include an integral catalytic converter in each exhaust manifold. To replace catalyst, see **Exhaust manifold** in this repair group.

Near engine catalytic converters (N54 engine)

N54 engines include a separate catalytic converter clamped to each turbocharger. To replace N54 near engine catalyst, follow the procedures below.

- Remove lower engine cover. See **020 Maintenance**.
- Remove exhaust system. See **Exhaust system, removing and installing (N54 engine)** in this repair group.
- Working at side of transmission: Remove holder for oxygen sensor wiring harness.
- Remove oxygen sensors. See **Oxygen sensors, removing and installing (N54 engine)** in this repair group.
- xDrive models: Remove right side front drive axle bearing block.

BentleyPublishers.com—All Rights Reserved

Near aneign ecatalytic converter, cylinders 1 - 3 (N54 engine)

◄ Remove nuts (**arrows**) from catalytic converter bracket.

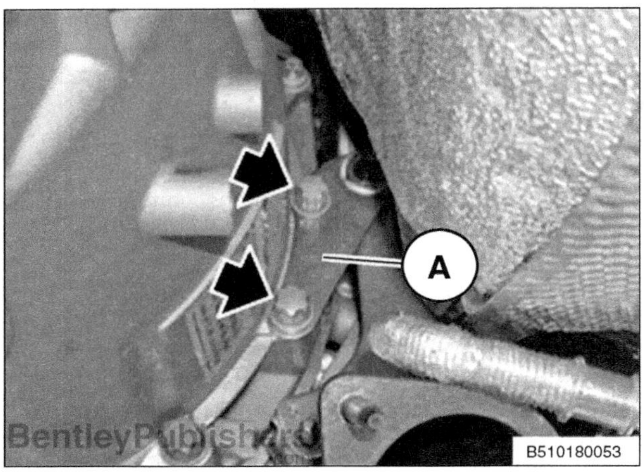

◄ Remove bolts (**arrows**) from catalytic converter bracket (**A**).

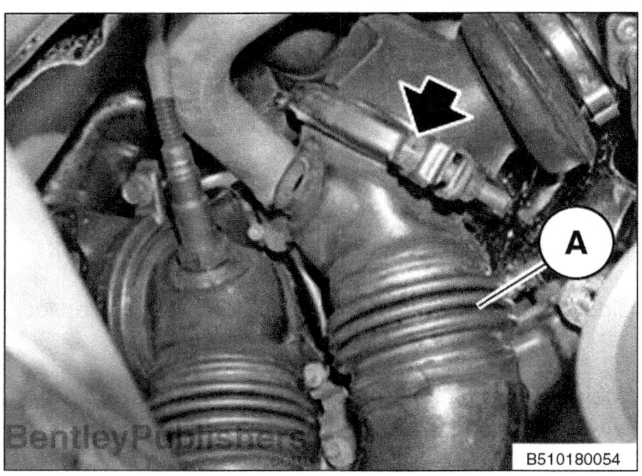

◄ While supporting catalytic converter: Loosen clamp (**arrow**) and lower converter (**A**) to remove.

BentleyPublishers.com
BentleyPublishers.com—All Rights Reserved

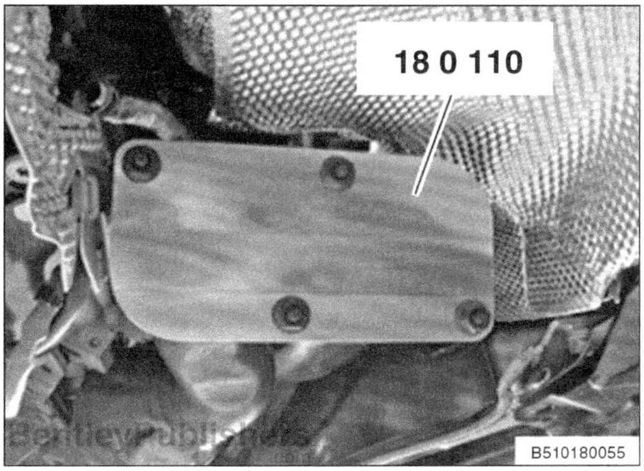

Installation is reverse of removal. Remember to:

- Clean sealing surfaces and replace seal between converter and turbocharger.
- Hold catalytic converter in position using BMW special tool 18 0 110 or equivalent and install clamp.
- Check exhaust system for leaks.

Tightening torque	
Catalytic converter to turbocharger	13 Nm (10 ft-lb)

Near engine catalytic converter, cylinders 4 - 6 (N54 engine)

– Remove near engine catalytic converter for cylinders 1 - 3.

– Remove oxygen sensors. See **Oxygen sensors, removing and installing** in this repair group.

While supporting catalytic converter: Remove clamp (**arrow**) and lower converter (**A**) to remove.

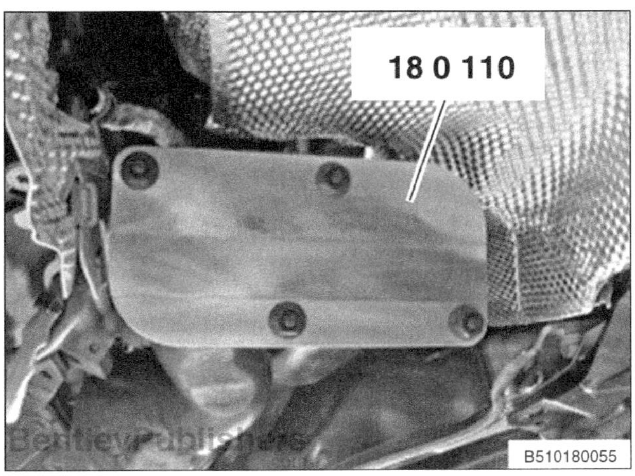

Installation is reverse of removal. Remember to:

- Clean sealing surfaces and replace seal between converter and turbocharger.
- Hold catalytic converter in position using BMW special tool 18 0 110 or equivalent and install clamp.
- Check exhaust system for leaks.

Tightening torque	
Catalytic converter to turbocharger	13 Nm (10 ft-lb)

Downstream catalytic converters

N52 and N54 engines include downstream catalytic converters that are welded into the originally installed exhaust system and must be cut off to be replaced.

To replace N52 or N54 engine downsream catalyst, See **Exhaust system components, removing and installing** in this repair group.

Exhaust System Components

Exhaust system components (M54 engine)

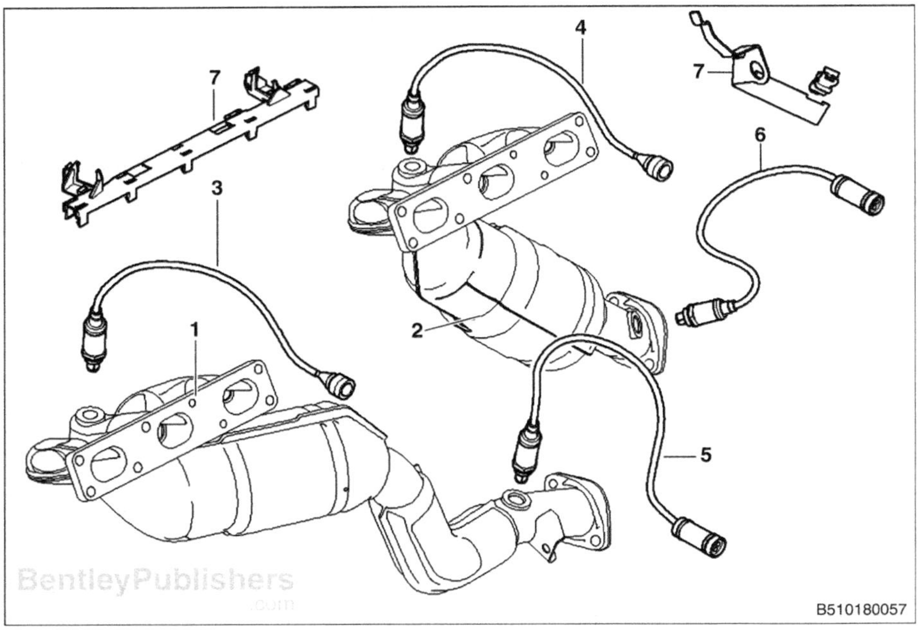

1. Exhaust manifold, cyl 1-3
2. Exhaust manifold, cyl 4-6
3. Pre-catalyst oxygen sensor, cyl 1-3
4. Pre-catalyst oxygen sensor, cyl 4-6
5. Post-catalyst oxygen sensor, cyl 1-3
6. Post-catalyst oxygen sensor, cyl 4-6
7. Bracket

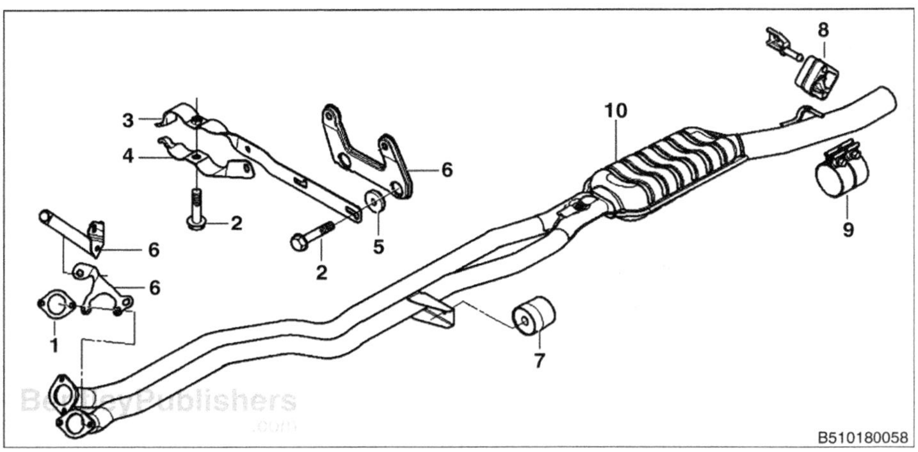

1. Gasket
2. Bolt
3. Exhaust pipe support, lower
4. Exhaust pipe support, upper
5. Washer
6. Bracket
7. Vibration damper
8. Hanger
9. Clamp
10. Center muffler

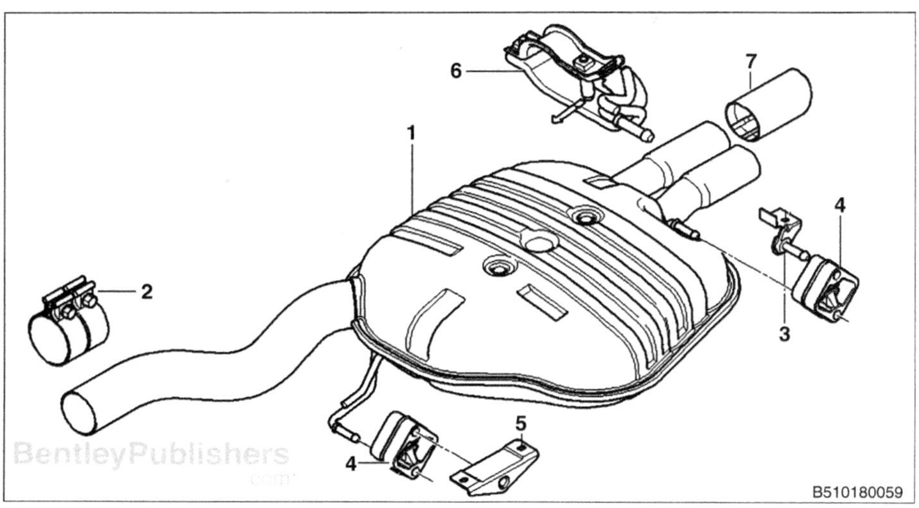

1. Muffler
2. Dual clamp
3. Bracket
4. Bracket hanger
5. Bracket, front
6. Bracket, rear
7. Tailpipe tip

BentleyPublishers.com
BentleyPublishers.com—All Rights Reserved

Exhaust system components (N52 engine)

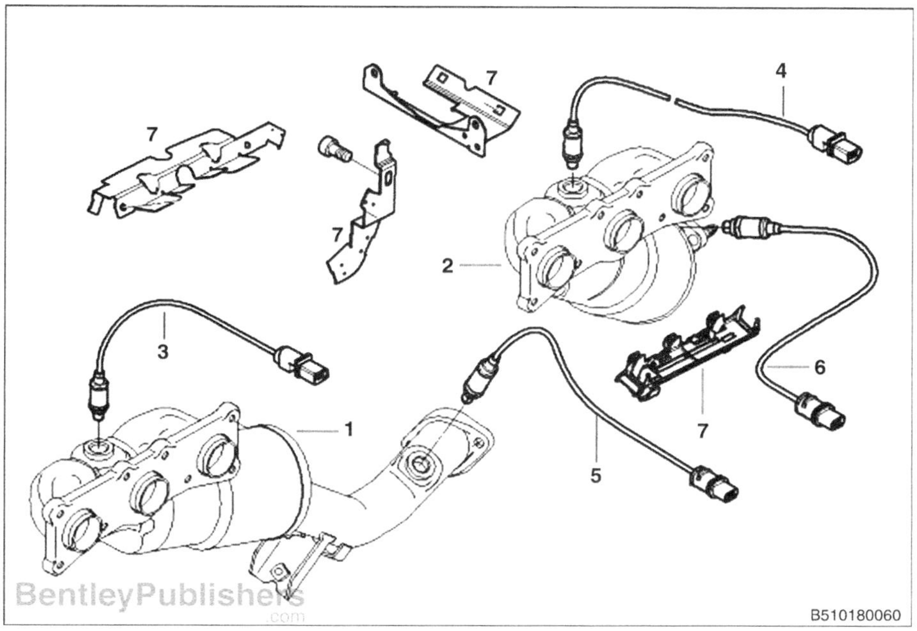

1. Exhaust manifold, cyl 1-3
2. Exhaust manifold, cyl 4-6
3. Pre-catalyst oxygen sensor, cyl 1-3
4. Pre-catalyst oxygen sensor, cyl 4-6
5. Post-catalyst oxygen sensor, cyl 1-3
6. Post-catalyst oxygen sensor, cyl 4-6
7. Bracket

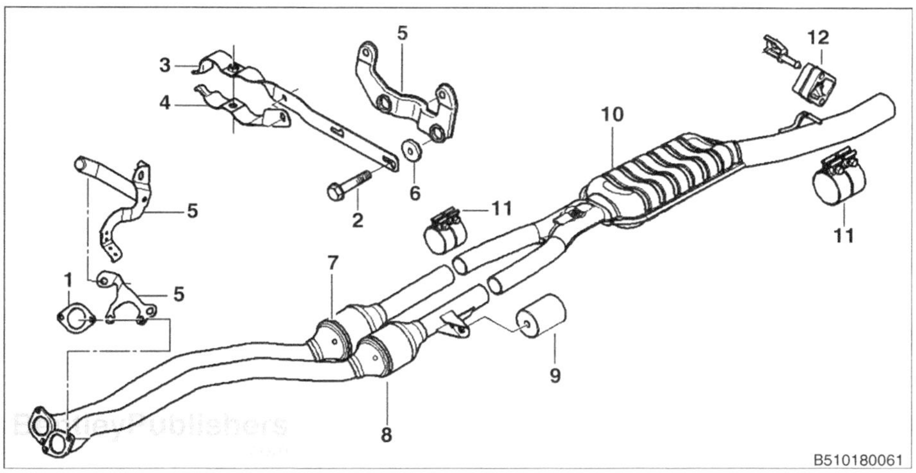

1. Gasket
2. Bolt
3. Exhaust pipe support, lower
4. Exhaust pipe support, upper
5. Bracket
6. Washer
7. Downstream catalytic converter, cyl 1-3
8. Downstream catalytic converter, cyl 4-6
9. Hanger
10. Center muffler
11. Clamp
12. Hanger

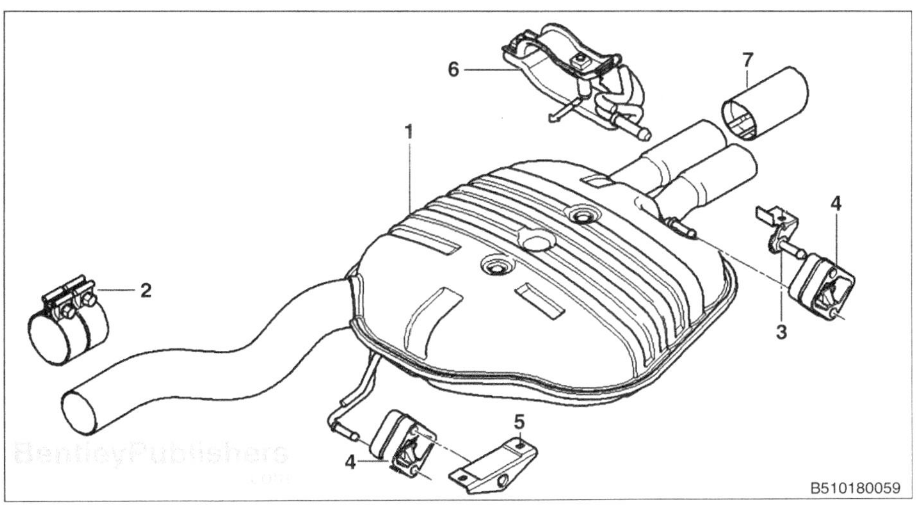

1. Muffler
2. Dual clamp
3. Bracket
4. Bracket hanger
5. Bracket, front
6. Bracket, rear
7. Tailpipe tip

BentleyPublishers.com—All Rights Reserved

Exhaust system components (N54 engine)

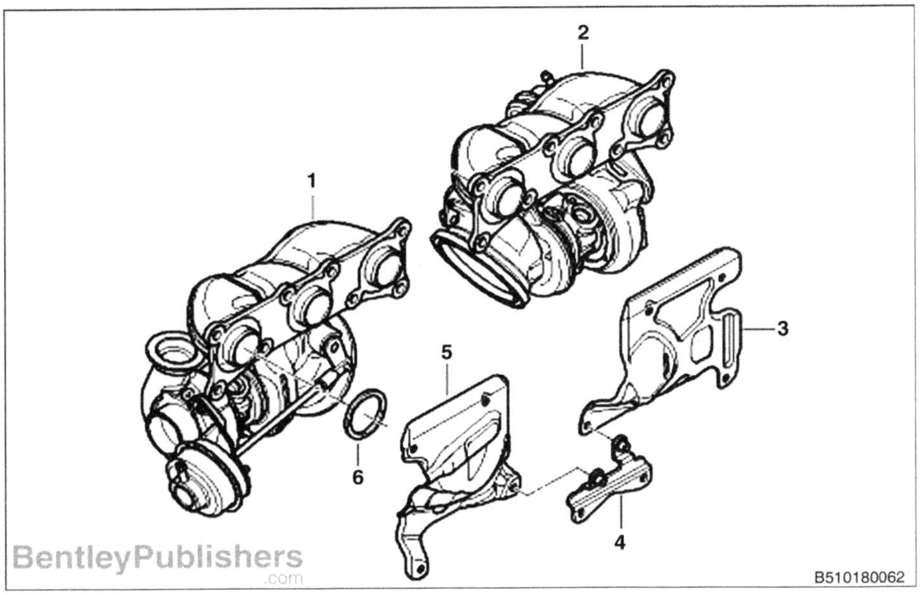

1. Exhaust manifold / turbocharger, cyl 1-3
2. Exhaust manifold / turbocharger, cyl 4-6
3. Heat shield
4. Bracket
5. Heat shield
6. Gasket

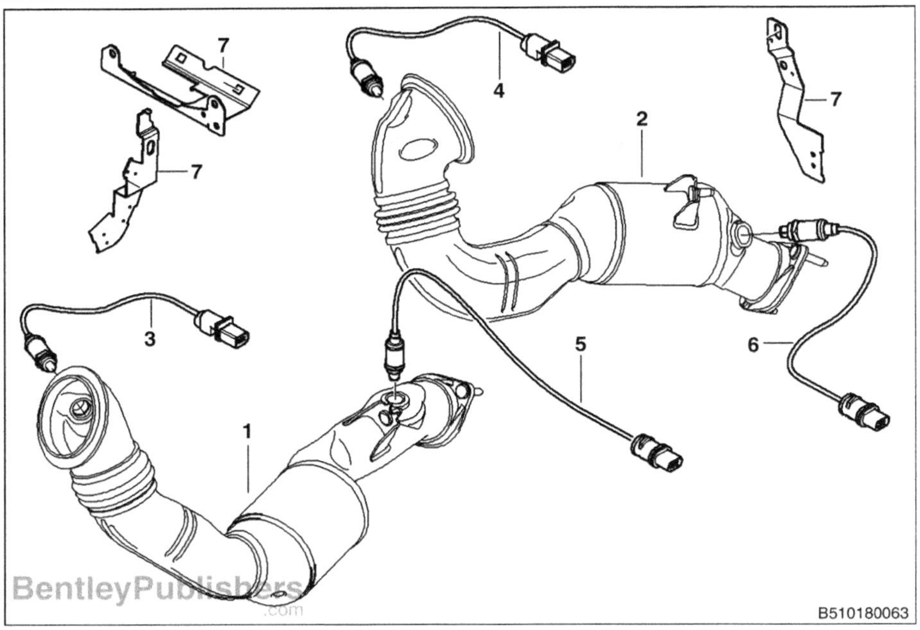

1. Near engine catalytic converter, cyl 1-3
2. Near engine catalytic converter, cyl 4-6
3. Pre-catalyst oxygen sensor, cyl 1-3
4. Pre-catalyst oxygen sensor, cyl 4-6
5. Post-catalyst oxygen sensor, cyl 1-3
6. Post-catalyst oxygen sensor, cyl 4-6
7. Bracket

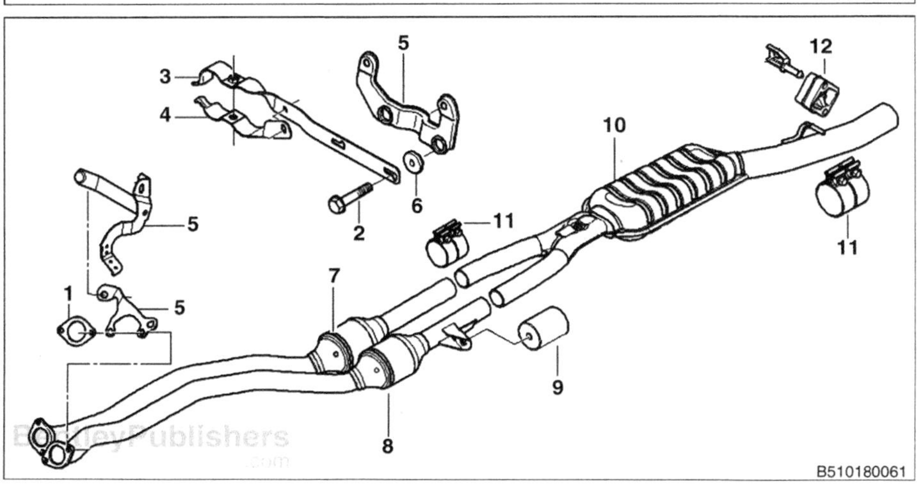

1. Gasket
2. Bolt
3. Exhaust hanger, lower
4. Exhaust hanger, upper
5. Bracket
6. Washer
7. Downstream catalytic converter, cyl 1-3
8. Downstream catalytic converter, cyl 4-6
9. Vibration damper
10. Center muffler
11. Clamp
12. Hanger

BentleyPublishers.com
BentleyPublishers.com—All Rights Reserved

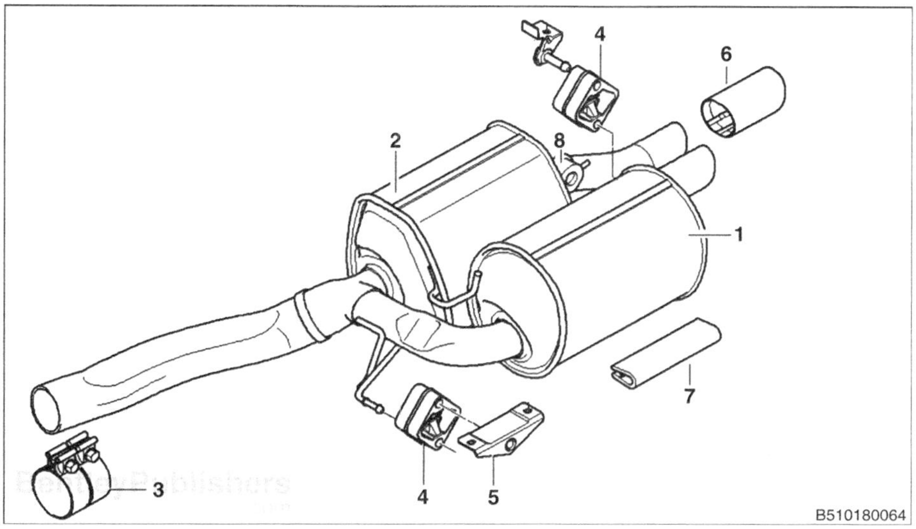

1. Muffler, left
2. Muffler, right
3. Dual clamp
4. Hanger
5. Bracket
6. Tailpipe trim
7. Edge protector
8. Tailpipe flap

Exhaust system components (V8 engine)

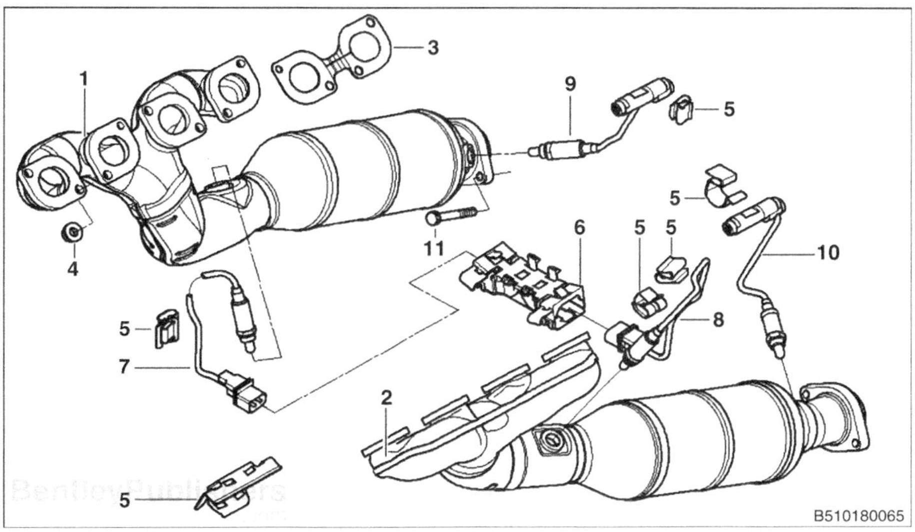

1. Exhaust manifold, cyl 1-4
2. Exhaust manifold, cyl 5-8
3. Gasket
4. Nut
5. Bracket
6. Oxygen sensor bracker connector
7. Pre-catalyst oxygen sensor, cyl 1-4
8. Pre-catalyst oxygen sensor, cyl 5-8
9. Post-catalyst oxygen sensor, cyl 1-4
10. Post-catalyst oxygen sensor, cyl 5-8
11. Bolt

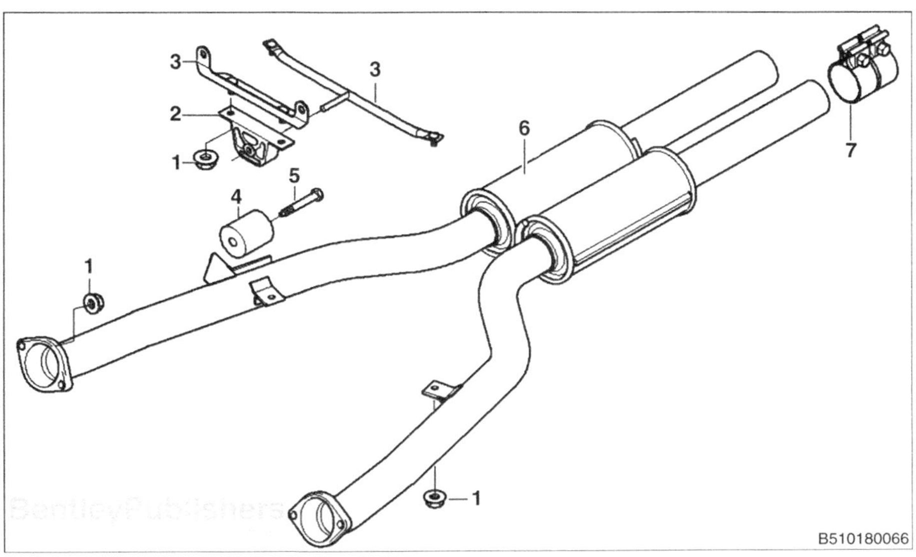

1. Nut
2. Hanger
3. Bracket
4. Vibration damper
5. Bolt
6. Resonator
7. Dual clamp

BentleyPublishers.com–All Rights Reserved

Exhaust system components (V8 engine)

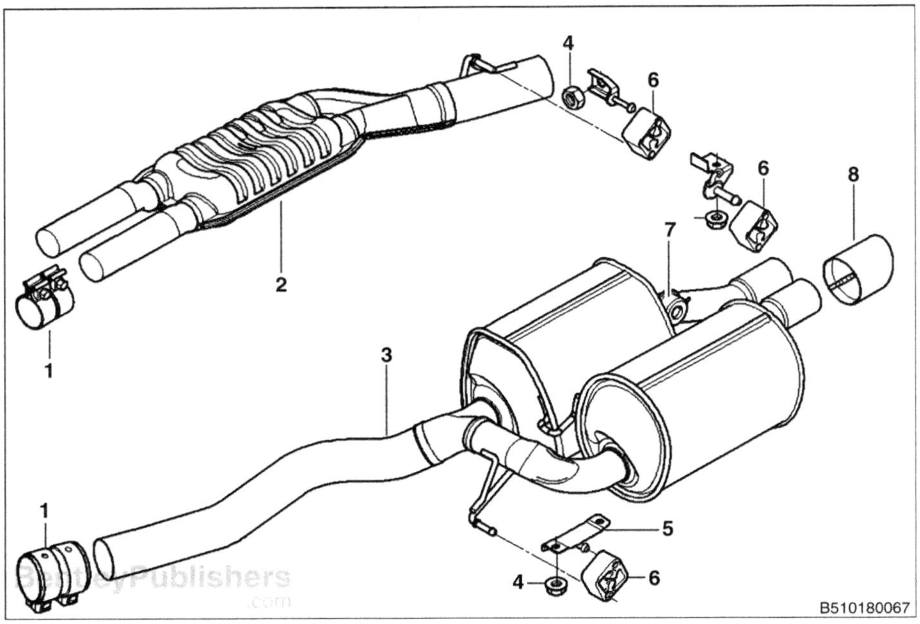

1. Dual clamp
2. Center muffler
3. Rear pipe and mufflers
4. Nut
5. Bracket
6. Hnager
7. Tailpipe flap
8. Tailpipe trim

BentleyPublishers.com
BentleyPublishers.com—All Rights Reserved

200 Transmission–General

TABLES

GENERAL

This section provides general information for drivetrain components such as clutch, transmission, driveshafts and xDrive transfer case.

See also:

- **119 Lubrication System** for rear main seal replacement
- **210 Clutch** for clutch mechanical and hydraulic repairs and flywheel replacement
- **230 Manual Transmission** for manual transmission fluid service, seal replacement and transmission replacement
- **231 Sequential Manual Gearbox (SMG)** for basic SMG service
- **240 Automatic Transmission** for ATF change and transmission replacement
- **260 Driveshafts** for front and rear driveshaft repairs
- **270 Transfer Case** for xDrive transfer case replacement
- **311 Front Axle Differential** for front differential, front drive axles and front CV joints
- **331 Rear Axle Differential** for rear differential, rear drive axles and rear CV joints

BentleyPublishers.com
BentleyPublishers.com–All Rights Reserved

Drivetrain

5 Series models are equipped with a longitudinal drivetrain. The transmission is bolted directly to the rear of the engine.

In rear-wheel drive models, a driveshaft connects the output shaft of the transmission to the rear differential.

On xDrive models, a transfer case is mounted to the rear of the transmission, with drive shafts leading to front and rear differentials. Individual drive axles with integrated constant velocity joints transfer rotational power from the differentials to the drive wheels.

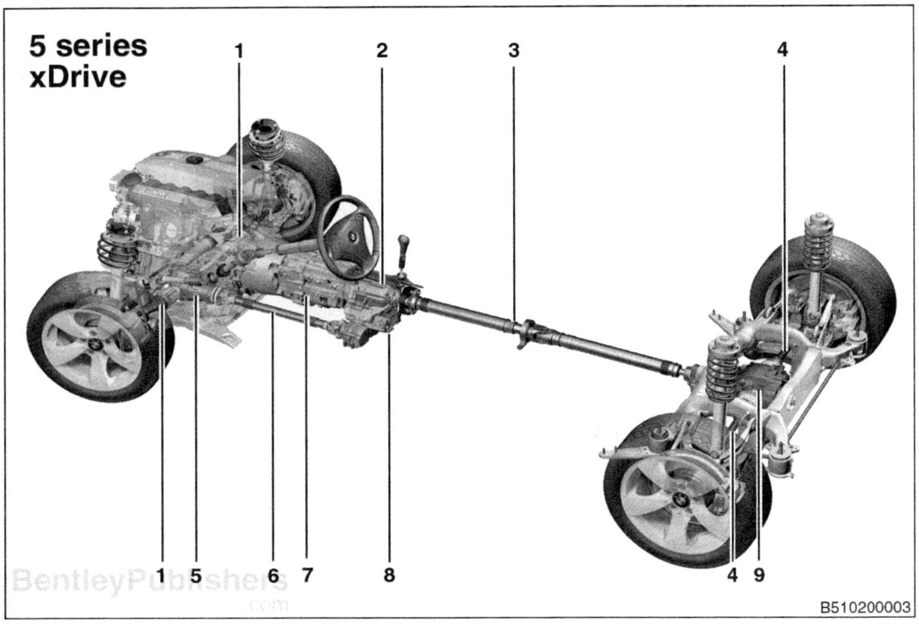

1. Front drive axle
2. xDrive transfer case
3. Rear driveshaft
4. Rear drive axle
5. Front differential
6. Front driveshaft
7. Transmission
8. xDrive clutch (VTG) actuator
9. Rear differential

Manual Transmission

6-speed manual transmission and clutch

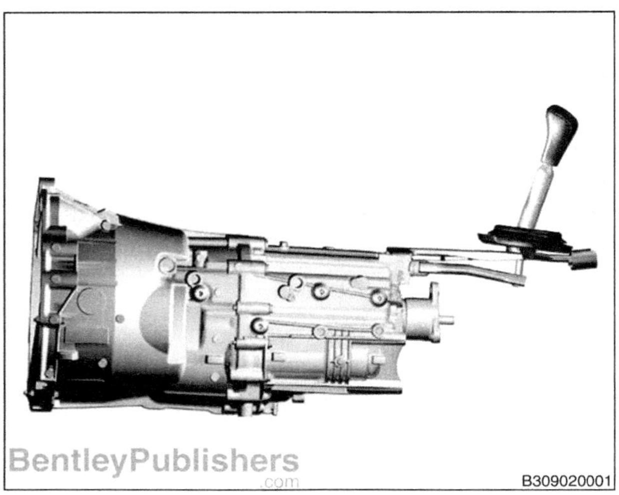

A 6-speed manual transmission is available on all models.

BentleyPublishers.com
BentleyPublishers.com—All Rights Reserved

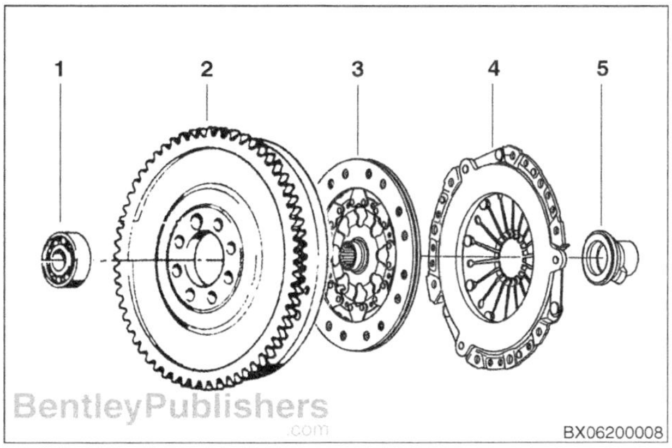

Manual transmission vehicles are equipped with a single-disc clutch and dual-mass flywheel.

1. Pilot bearing
2. Dual-mass flywheel
3. Clutch disc
4. Clutch pressure plate
5. Clutch release bearing

For further information, see **210 Clutch**.

Manual transmission has a metal ID plate mounted on side of transmission. Do not rely on numbers cast on transmission case for identification. For transmission applications, see **Table a**. Transmission gear ratio information is in **230 Manual Transmission**.

Table a. 5 Series manual transmission applications (6-speed)

Model (engine)	Year	BMW designation	Manufacturer designation
525i (M54)	2004 - 2006	GS6-37BZ / DZ	ZF 6HP19 type H
530i (M54, N52)	2004 - 2007	GS6-37BZ / DZ	ZF 6HP19 type H
525i, 528i (N52)	2006 - 2010	GS6-17BG	Getrag type I
525xi, 528xi, 530xi (N52)	2006 - 2010	GS6X37BZ / DZ	ZF 6HP19 type H
535i (N54)	2008 - 2010	GS6-53BZ / DZ	ZF 6HP26 type G
535xi (N54)	2008 - 2010	GS6X53DZ	ZF 6HP26 type G
545i, 550i (N62, N62TU)	2004 - 2010	GS6-53BZ	ZF 6HP26 type G

Sequential Manual Gearbox (SMG)

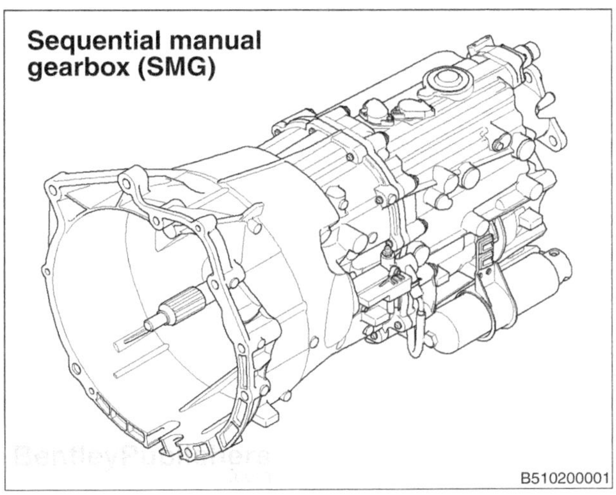

Available as an option, the 6-speed sequential manual gearbox (SMG) functions like a clutchless manual transmission. Clutch operation is handled via a computer controlled hydraulic unit that engages and disengages the clutch automatically, as well as physically shifting the transmission into gear.

SMG application information is in **Table b**.

Table b. 5 Series SMG applications

Model (engine)	Year	Transmission
525i, 530i (M54)	2004 - 2005	GS6S37BZ
530i (N52)	2006 - 2007	GS6S37BZ
545i (N62)	2004 - 2006	GS6S53BZ

BentleyPublishers.com
BentleyPublishers.com–All Rights Reserved

Paddle shift

In addition to the traditional floor-mounted selector lever, SMG models are equipped with paddle shifters above the steering wheel spokes. Pulling or pushing either paddle results in transmission upshift or downshift.

AUTOMATIC TRANSMISSION

6-speed automatic transmission

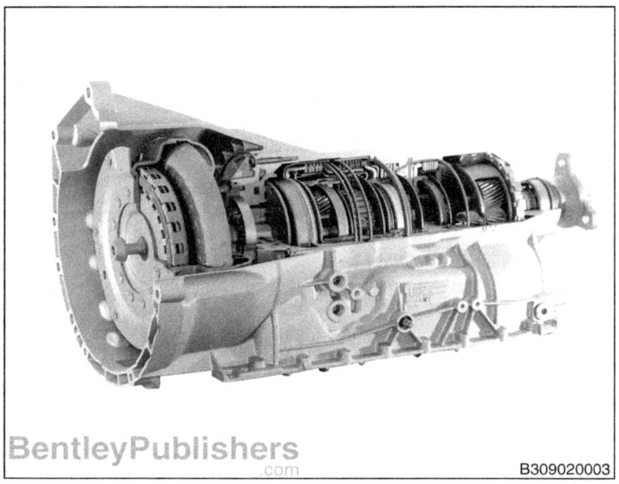

The automatic transmission is a 6-speed Steptronic that offers a choice of Normal, Sport, and Manual mode. Sport mode is engaged by moving the shift lever out of Drive and into the Sport / Manual shift gate. In Sport mode, shifts occur at a higher rpm than normal. Additionally, the driver can tip the lever forward (downshifts) or backward (upshifts) to manually select a gear. Automatic transmission applications are in **Table c**.

Table c. 5 Series automatic transmission applications

Model (engine)	Year	Transmission
525i, 525xi, 530i, 530xi, (M54, N52)	2004 - 2007	GA6HP19Z
545i, 550i (N62, N62TU)	2004 - 2007	GA6HP26Z
528i, 528xi, 535i, 535xi, (N52, N54)	2008 - 2000	GA6HP19ZTU
550i (N62TU)	2008 - 2009	GA6HP26ZTU

BentleyPublishers.com
BentleyPublishers.com—All Rights Reserved

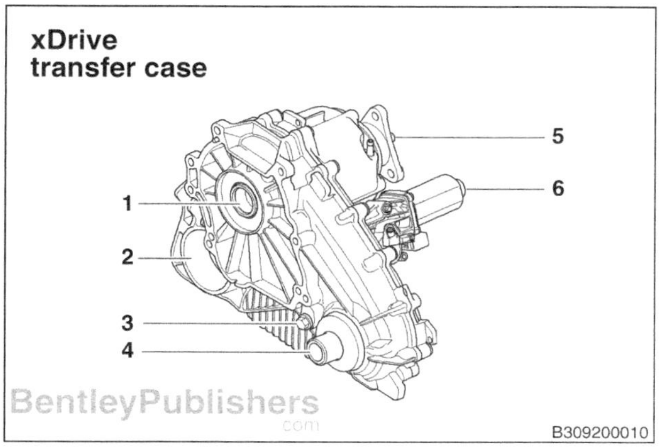

xDRIVE

Transfer case

BMW's advanced all-wheel drive system is known as xDrive. It includes a transfer case with electronically controlled clutch that regulates the front to rear torque split for best traction.

The xDrive transfer case, bolted to the rear of the transmission, directs power from the transmission to both the front and rear differentials via driveshafts. The transfer case is permanently engaged to drive all four wheels all of the time.

1. Transfer case input shaft
2. Transmission and transfer case mount
3. Fill plug
4. Front driveshaft output
5. Rear driveshaft output
6. xDrive clutch actuator

xDrive clutch

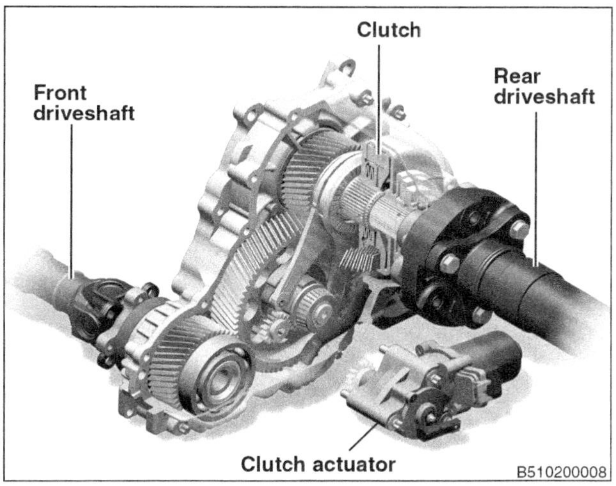

In the xDrive system, an electronically controlled clutch in the transfer case allows for infinitely variable distribution of front to rear torque for best traction.

xDrive functions are integrated with ABS / DSC. Under normal driving conditions, 60% of engine power goes to the rear axle, 40% to the front. If a wheel begins to slip or if the DCS steering angle sensor and acceleration sensors signal excessive yaw (rotational acceleration) in the vehicle, xDrive reacts by transferring power to the wheel or wheels with best traction, usually before a wheel is able to spin. This significantly reduces the risk of under- or oversteer, improving agility and safety on the road.

Torque and engine power are redistributed by xDrive within a few milliseconds. As a result, the driver normally does not notice a change in power distribution.

In addition to controlling power distribution, xDrive and DSC may intervene via braking. If a wheel starts to spin without transmitting power, the brakes are applied at that wheel. In this case, the differential automatically supplies more power to the opposite wheel.

For additional information on xDrive and DSC functions, see **270 Transfer Case** and **340 Brakes**.

Drivetrain Lubricants

Clutch fluids

Manual transmission clutch fluid

Manual transmission hydraulic clutch is operated using brake fluid. Brake fluid level checking and topping off is covered in **020 Maintenance**. Hydraulic clutch bleeding is covered in **210 Clutch**.

SMG clutch fluid

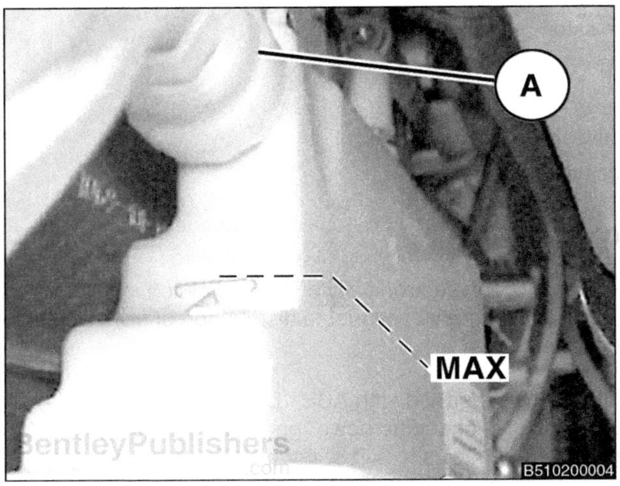

Clutch fluid reservoir (**A**) is at side of transmission. Clutch fluid checking and topping off is covered in **231 Sequential Manual Gearbox (SMG)**.

SMG clutch fluid requirement is in **Table d**.

Table d. SMG clutch fluid	
Hydraulic fluid	Pentosin CHF 11S

Transmission fluids

Manual, SMG and automatic transmissions are filled with lifetime lubrication. No oil change is required for the entire service life of the transmission. If repairs have to be made to the transmission or transmission oil cooler, use only the recommended lifetime fluid to refill. See **Table e** or **Table f**.

Table e. Manual and SMG transmission fluids		
Transmission	**Year range**	**Fluid type**
ZF GS6-37BZ ZF GS6S37BZ (SMG) ZF GS6X37BZ (xDrive) ZF GS6-53BZ ZF GS6S53BZ (SMG) ZF GS6X53BZ (xDrive)	2004 - 2005 from 2006	MTF-LT-2 MTF-LT-3
MTF-LT-2 (Lifetime fluid) BMW part no. 83 22 0 309 031		
MTF-LT-3 (Lifetime fluid) BMW part no. 83 22 0 409 878		

Table f. Automatic transmission fluid		
Type	**BMW part no.**	**Manufacturer part no.**
GA6HP19Z GA6HP19ZTU GA6HP26Z GA6HP26ZTU	83 22 0 142 516	Shell M-1375.4

BentleyPublishers.com
BentleyPublishers.com—All Rights Reserved

A label on manual transmission specifies original lubricant used at the factory.

For manual transmission fluid service, see **230 Manual Transmission**.

For SMG fluid service, see **231 Sequential Manual Gearbox (SMG)**.

ATF is specified on colored label on automatic transmission sump.

For automatic transmission fluid service, including checking fluid level and ATF filter replacement, see **240 Automatic Transmission**.

Transfer case fluid

Transfer case fluid specification for xDrive models is in **Table g**.

Table g. Transfer case fluid specification		
Type	**BMW part no.**	**Manufacturer part no.**
xDrive (all)	83 22 0 397 244	Shell TF0870

For transfer case fluid service, see **270 Transfer Case**.

BentleyPublishers.com—All Rights Reserved

Differential oil

Differentials are filled with lifetime lubrication. No oil change is required for the entire service life of the differential. If repairs have to be made to the differential, use only the recommended lifetime oil to refill. See **Table h**.

Table h. Differential oil (lifetime)	
BMW specification	SAF-XO Synthetic

Front differential in xDrive models is bolted to left side of engine oil pan.

For front differential oil service, see **311 Front Axle Differential**.

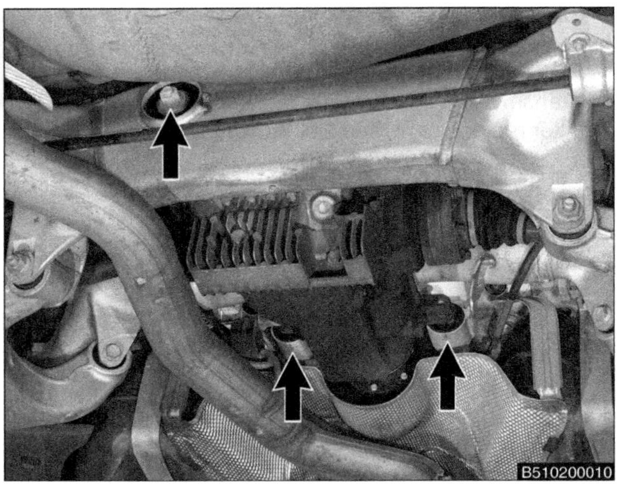

Rear differential is supported at rear suspension subframe using three rubber bushings (**arrows**).

For rear differential oil service, see **331 Rear Axle Differential**.

BentleyPublishers.com
BentleyPublishers.com—All Rights Reserved

210 Clutch

General

This repair group covers replacement of manual transmission clutch mechanical and hydraulic components.

See also:
- **200 Transmission–General** for transmission application information
- **230 Manual Transmission** for transmission removal
- **270 Transfer Case** for xDrive transfer case removal

Clutch system overview

Clutch assembly and hydraulics

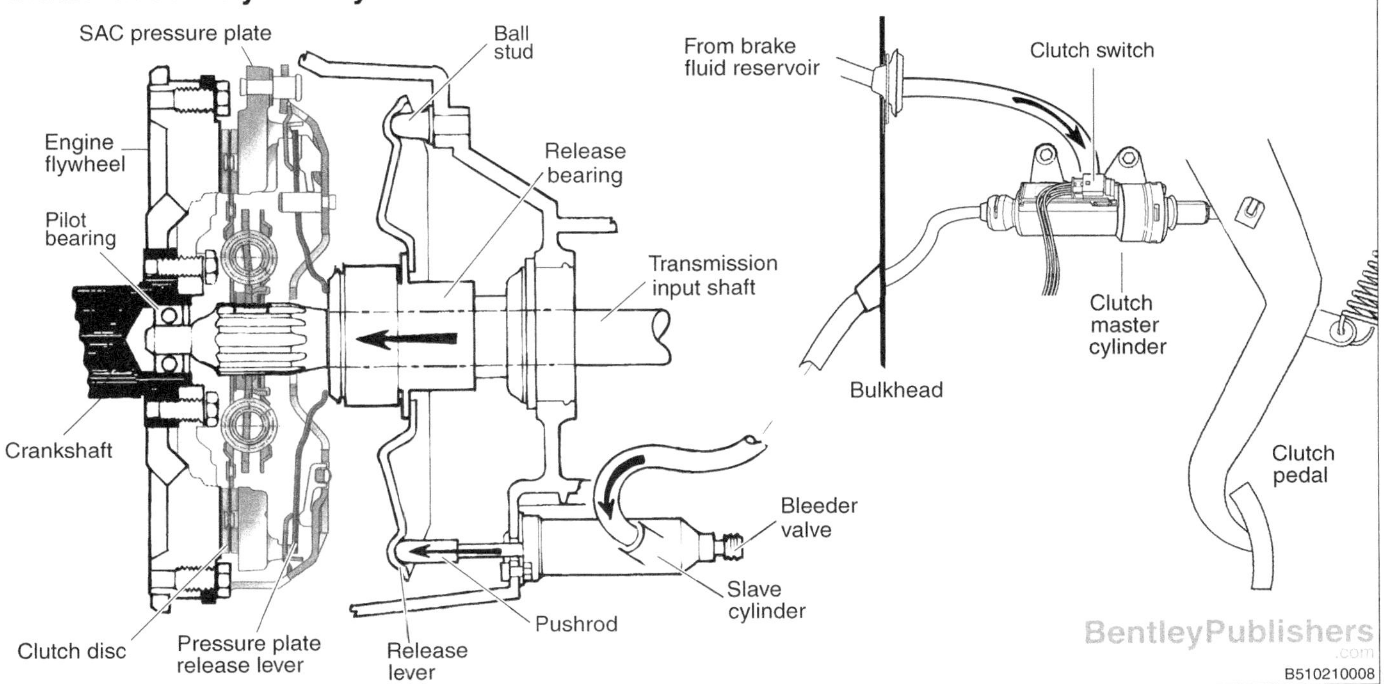

B510210008

BentleyPublishers.com–All Rights Reserved

Cautions

CAUTION—

- *When performing any repair which involves separating the engine and transmission, check that the bell housing dowels are undamaged and properly in place before reassembly. If the alignment of the engine flywheel to the transmission input shaft or torque converter is incorrect any of the following complaints may result: Clutch slipping, shudder, or poor disengagement; noise from transmission input shaft; transmission popping out of gear; difficulty changing gears; or internal damage to the transmission.*
- *When removing interior trim clips, avoid damaging trim pieces by using a plastic prying tool or a screwdriver with the tip wrapped with masking tape.*
- *Brake fluid is poisonous, highly corrosive and dangerous to the environment. Wear safety glasses and rubber gloves when working with brake fluid. Do not siphon brake fluid with your mouth. Immediately clean away any fluid spilled on painted surfaces and wash with water, as brake fluid removes paint.*
- *Use new brake fluid from a fresh, unopened container. Brake fluid absorbs moisture from the air. This can lead to corrosion problems in the hydraulic systems, and also lowers the brake fluid boiling point. Dispose of brake fluid properly.*
- *To avoid electrochemical corrosion to engine components made of aluminum-magnesium alloy, do not use steel fasteners. Use aluminum fasteners only.*
- *The end faces of aluminum fasteners are usually painted blue. For reliable identification, test fasteners for aluminum composition with magnet.*
- *Replace aluminum fasteners each time they are loosened.*
- *Follow torque instructions, including angle of rotation specifications, when installing aluminum fasteners.*

Clutch Hydraulics

The clutch is hydraulically actuated by the master and slave cylinders. Clutch disc wear is automatically taken up through the self-adjusting clutch (SAC) pressure plate springs, making periodic adjustments unnecessary.

A soft or spongy feel to the clutch pedal, long pedal free-play, or grinding noises from the gears while shifting can all indicate problems with clutch hydraulics. In these circumstances it is best to start with a clutch fluid flush, followed, if necessary, by replacement of the hydraulic parts.

The clutch hydraulic system shares the fluid reservoir and fluid with the brake hydraulic system.

BentleyPublishers.com
BentleyPublishers.com—All Rights Reserved

Clutch hydraulic system, bleeding and flushing

If the clutch and brake fluid is murky or muddy, or has not been changed within the last two years, flush the system using a brake system pressure bleeder.

– Raise vehicle and support safely.

> ***WARNING—***
> - *Make sure the vehicle is stable and well supported at all times. Use a professional automotive lift or jack stands designed for the purpose. A floor jack is not adequate support.*

– Remove splash shield under transmission. See **020 Maintenance**.

– Working at left rear of engine compartment, remove left cabin microfilter cover and housing. See **020 Maintenance**.

◄ Remove brake fluid reservoir cap. Using a clean syringe, remove brake fluid from reservoir. Refill reservoir with clean DOT4 low viscosity brake fluid.

– Attach pressure brake bleeder to fluid reservoir and pump bleeder a few times to pressurize hydraulic fluid system.

> ***CAUTION—***
> - *Do not exceed 2 bar (29 psi) pressure at the fluid reservoir when bleeding or flushing the hydraulic system.*
> - *Brake fluid is poisonous, highly corrosive and dangerous to the environment. Wear safety glasses and rubber gloves when working with brake fluid. Do not siphon brake fluid with your mouth. Immediately clean away any fluid spilled on painted surfaces and wash with water, as brake fluid removes paint.*
> - *Use new brake fluid from a fresh, unopened container. Brake fluid absorbs moisture from the air. This can lead to corrosion problems in the hydraulic systems, and also lowers the brake fluid boiling point. Dispose of brake fluid properly.*

Clutch hydraulics, bleeding

◄ Connect a length of hose from clutch slave cylinder bleeder valve (**arrow**) to a container.

– With pressure bleeder attached, open bleeder valve and allow brake fluid to expel until clean fluid comes out free of air bubbles.

– Close bleeder valve and disconnect pressure bleeding equipment from fluid reservoir. Hose on bleeder valve remains connected.

– Slowly operate clutch pedal about 10 times. Fill reservoir with clean fluid as necessary.

BentleyPublishers.com
BentleyPublishers.com—All Rights Reserved

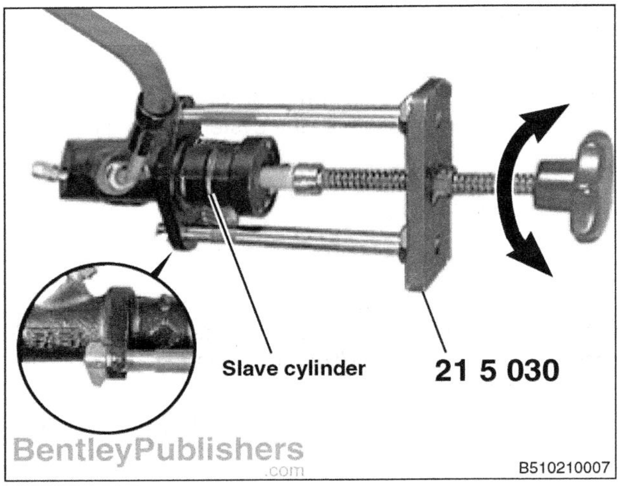

Clutch hydraulics, bleeding using BMW special tool 21 5 030

◄ Working underneath vehicle, unbolt slave cylinder from transmission. Fit BMW special tool 21 5 030 to slave cylinder. Rotate threaded tool spindle to press slave cylinder pushrod completely into slave cylinder.

– With pressure bleeder attached, hold slave cylinder so that bleeder valve is at highest point.
  - Open bleeder valve.
  - If bubble-free brake fluid emerges, use tool spindle to retract slave cylinder push rod slightly, then press in again.
  - If bubbles appear, repeat procedure until fluid runs clear and without bubbles.
  - Once brake fluid appears without air bubbles, close bleeder valve, release pressure bleeder and detach special tool. Disconnect bleeder hose.

> ***CAUTION—***
> - *Do not detach special tool from slave cylinder if fluid is pressurized. The slave cylinder push rod can blow out of the cylinder.*

– Reinstall slave cylinder to transmission. Add clean brake fluid to reservoir as necessary. Check clutch operation.

Tightening torque	
Clutch slave cylinder to transmission	22 Nm (16 ft-lb)

Clutch master cylinder, replacing

The clutch master cylinder is mounted to the pedal assembly directly above the clutch pedal.

– Working at left rear of engine compartment, remove left cabin microfilter cover and housing. See **020 Maintenance**.

◄ Remove brake fluid reservoir cap. Using a clean syringe, remove brake fluid from reservoir.

– Disconnect clutch fluid supply line from brake fluid reservoir. Place a pan under hose to catch dripping fluid.

– Working at pedal cluster, remove lower left trim panel under dashboard (pedal cluster trim). See **513 Interior Trim**.

– Remove light module. See **630 Lights**.

BentleyPublishers.com
BentleyPublishers.com—All Rights Reserved

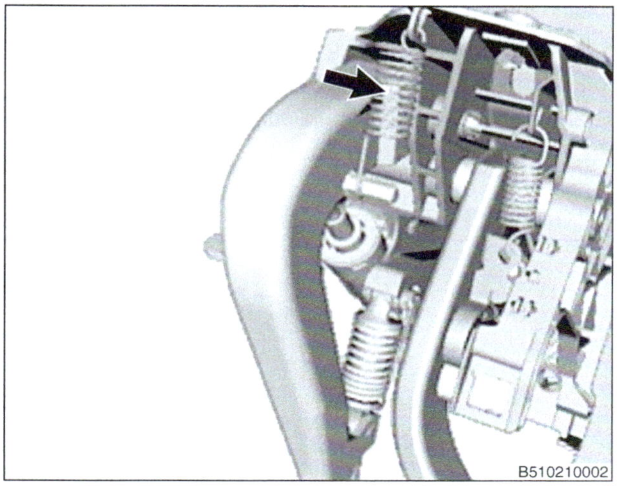

Working underneath left side dashboard at pedal cluster, detach clutch pedal return spring (**arrow**).

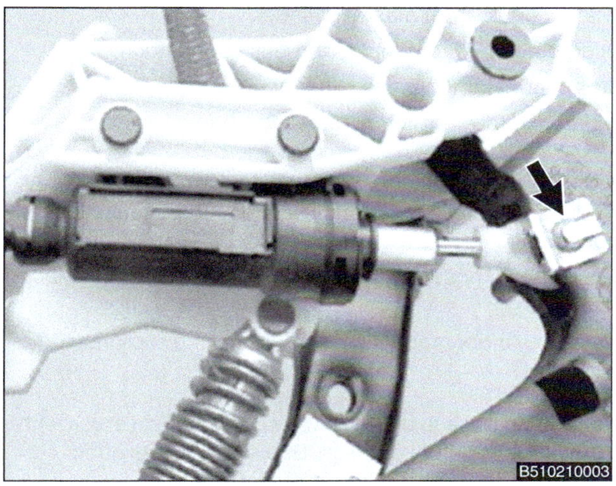

Using screwdriver, remove retaining clip (**arrow**), then push out pin mounting master cylinder push rod to clutch pedal.

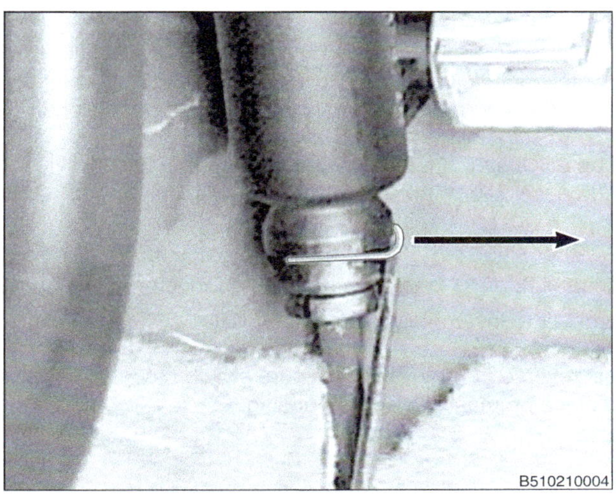

Use thin screwdriver to remove fluid pressure line retaining clip (**arrow**). Detach fluid line from clutch master cylinder.

> ***CAUTION—***
> - *To prevent brake fluid spill, wrap clutch master cylinder with shop towels when removing hydraulic fluid lines from master cylinder.*
> - *Brake fluid damages paint and stains carpets. Clean off any brake fluid on or in the vehicle immediately.*

BentleyPublishers.com
BentleyPublishers.com—All Rights Reserved

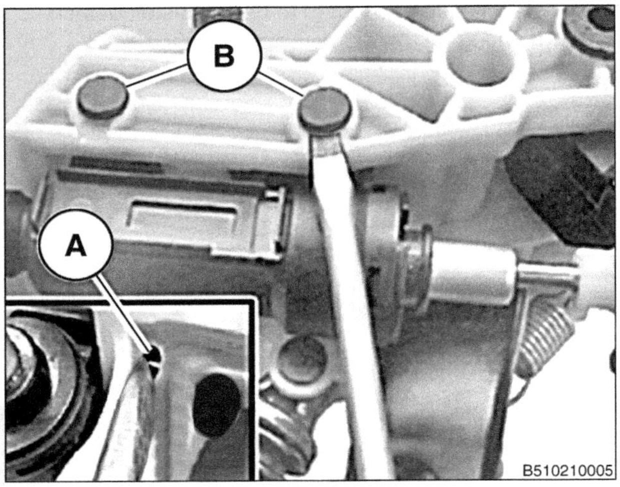

◄ Working at clutch master cylinder mounting pins:

- Use screwdriver to remove locking clip (**A**).
- Pry out mounting pins (**B**).
- Lower master cylinder and detach fluid supply line.

> ***CAUTION—***
> *• Do not pull supply line completely into vehicle interior.*

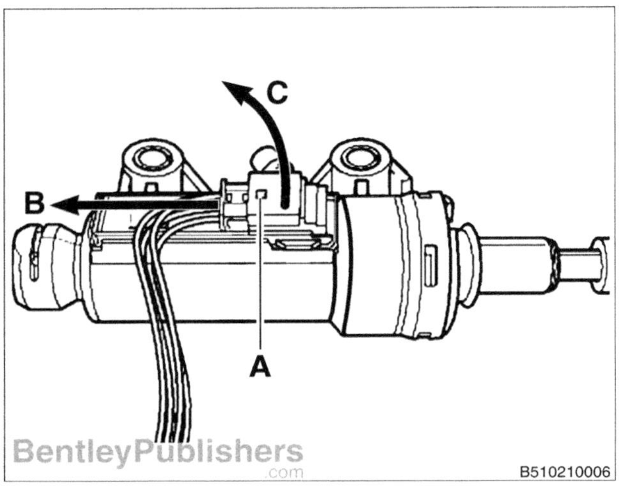

◄ Remove switch module from master cylinder.

- Press locking button (**A**) to release harness connector.
- Pull harness connector (**B**) off switch module.
- Use screwdriver to lever switch module (**C**) off master cylinder.

– Installation is reverse of removal. Keep in mind the following:

- Make sure switch module snaps audibly into place on master cylinder housing.
- Reattach fluid supply hose before placing master cylinder in place on pedal cluster.
- Press in master cylinder mounting pins until locking clips snap into place.

– Reattach fluid supply hose to brake fluid reservoir. Fill reservoir with clean fluid and bleed clutch hydraulics. See **Clutch hydraulic system, bleeding and flushing** in this repair group.

Clutch switch, replacing

The clutch switch module is attached to the clutch master cylinder. It performs both cruise control and starter immobilization functions.

– Working at pedal cluster, remove lower left trim panel under dashboard (pedal cluster trim). See **513 Interior Trim**.

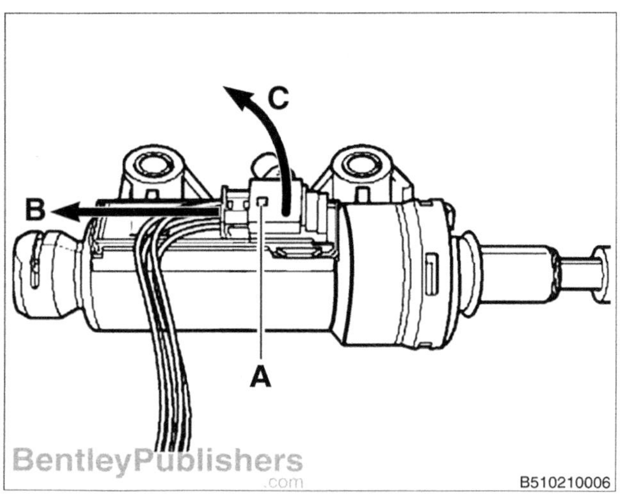

◄ Remove switch module from master cylinder.

- Press locking button (**A**) to release harness connector (**B**).
- Use screwdriver to lever off switch module (**C**).

– Installation is reverse of removal. Make sure switch module snaps audibly into place on master cylinder housing.

BentleyPublishers.com
BentleyPublishers.com—All Rights Reserved

Clutch slave cylinder, replacing

The clutch slave cylinder is bolted to the left side of the transmission housing close to the clutch bell housing.

- Working at left rear of engine compartment, remove left cabin microfilter cover and housing. See **020 Maintenance**.
- Use hose pinch tool 13 3 010 or equivalent to pinch off clutch fluid supply line at brake fluid reservoir.
- Raise vehicle and support safely.

> ***WARNING —***
> - *Make sure the vehicle is stable and well supported at all times. Use a professional automotive lift or jack stands designed for the purpose. A floor jack is not adequate support.*

- Remove lower engine and transmission splash shields as necessary. See **020 Maintenance**.

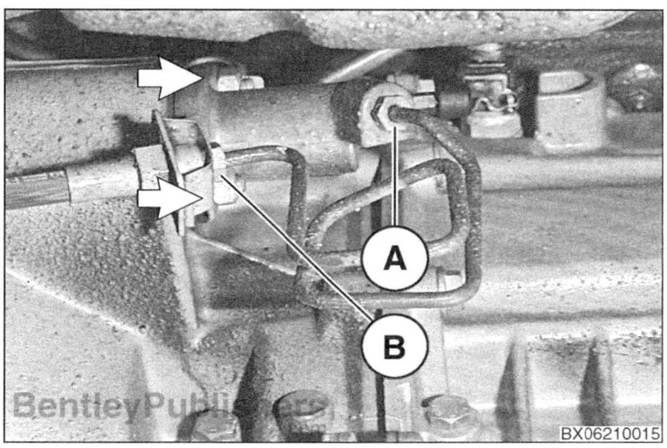

Working underneath transmission:

- Loosen union nut **A** to disconnect fluid line from slave cylinder. Place pan under line to catch dripping fluid.
- Loosen union nut **B** to detach metal line from fluid supply hose.
- Remove slave cylinder mounting nuts (**arrows**) from clutch bell housing and remove cylinder.

- Installation is reverse of removal. Note the following:
  - Check for wear on slave cylinder push rod. Any wear except on tip is caused by misalignment of clutch components.
  - Lightly coat pushrod tip with molybdenum disulfide grease (Molykote® Longterm or equivalent).
  - During installation be sure pushrod tip engages recess in clutch release lever.
  - Fill brake fluid reservoir with clean fluid and bleed clutch hydraulics. See **Clutch hydraulic system, bleeding and flushing** in this repair group.

Tightening torques	
Clutch slave cylinder to transmission	22 Nm (16 ft-lb)
Fluid line to slave cylinder	14.5 Nm (11 ft-lb)

Clutch Mechanical

Clutch replacement requires removal of the transfer case (xDrive models) and transmission.

Clutch disc, pressure plate and release bearing are usually replaced during a clutch overhaul. Check flywheel for wear and scoring. Replace if necessary.

Be sure to check the bottom of the bell housing for oil. If engine oil is found, check for faulty rear crankshaft (rear main) oil seal. Rear main seal replacement is covered in **119 Lubrication System**.

BentleyPublishers
BentleyPublishers.com—All Rights Reserved

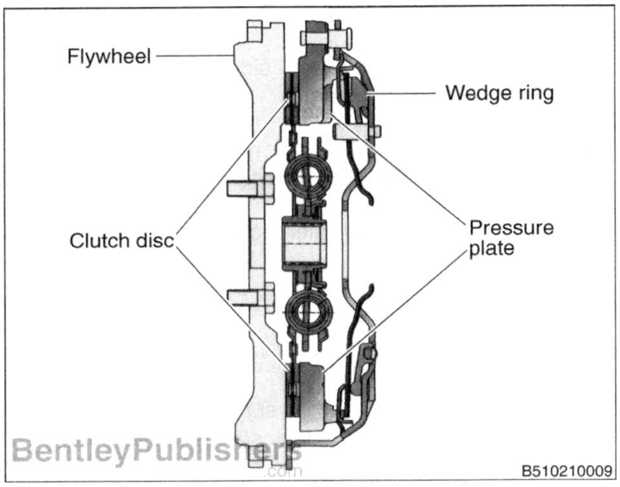

The SAC pressure plate uses a wedge ring which rotates against the diaphragm springs to accommodate for clutch disc wear. The wedge ring adjusts by means of spring tension. Special tools are needed to apply and relieve spring tension as the clutch pressure plate is removed and installed.

Clutch, removing

- Disconnect negative (–) battery cable and cover battery terminal to keep cable from accidentally contacting terminal.

> ***CAUTION—***
> • *Prior to disconnecting the battery, read the battery disconnection cautions in* **001 Warnings and Cautions**.

- Raise vehicle and support safely.

> ***WARNING—***
> • *Make sure the vehicle is stable and well supported at all times. Use a professional automotive lift or jack stands designed for the purpose. A floor jack is not adequate support.*

- Remove splash shield under engine and transmission. See **020 Maintenance**.
- Remove front end reinforcement. Discard mounting bolts.
- Remove complete exhaust system. See **180 Exhaust System**.

> ***WARNING—***
> • *Make sure exhaust system is fully cooled off before starting removal.*

- Remove heat shields under transfer case (xDrive models) and rear driveshaft.
- Detach front (xDrive models) and rear driveshafts and tie to side. See **260 Driveshafts**.
- Remove transmission from engine. See **230 Manual Transmission**.

Lock flywheel in position using BMW special tool 11 9 260 or equivalent.

BentleyPublishers.com
BentleyPublishers.com—All Rights Reserved

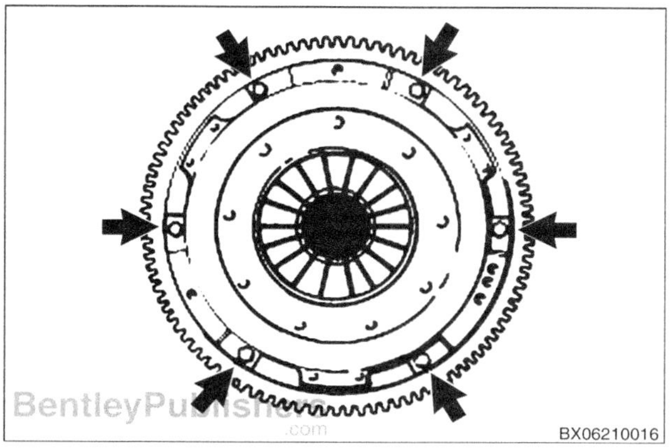

< Loosen 6 clutch mounting bolts (**arrows**) evenly.

– Remove pressure plate and clutch disc.

> ***WARNING —***
> • *Friction material in the clutch disc produces dangerous dust. Do not breathe in dust. Use water to wet down components and collect dripping mixture in a shop towel.*

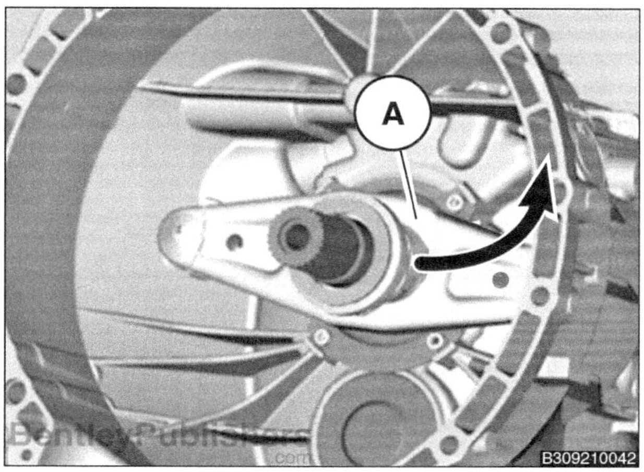

< Remove clutch release bearing and lever module (**A**): Remove module from spring clip and tilt off transmission input shaft in direction of **arrow**.

> ***CAUTION—***
> • *To avoid installing incorrectly, do not remove bearing from release module.*

Clutch components, inspecting

– Inspect clutch disc for wear, cracks, loose rivets, contamination or excessive runout (warping). Replace if necessary.

– Measure depth of clutch lining at lining rivets. If shallowest rivet depth is less than 1 mm (0.04 in), replace clutch disk.

– Inspect flywheel for scoring, hot spots, cracks or loose or worn guide pins. Replace flywheel if any faults are found. See **Flywheel, removing and installing** in this repair group.

> ***CAUTION—***
> • *If flywheel is removed from engine, reinstall using new bolts.*

– Check to make sure transmission pilot bearing rotates smoothly without play. Replace if faulty. See **Transmission pilot bearing, replacing** in this repair group.

– Inspect and clean release bearing module.

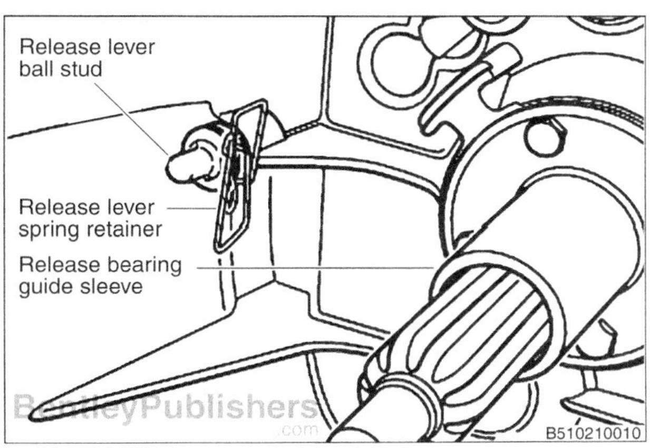

< Check release lever ball stud and spring retainer. Replace if damaged or worn.

– Clean release bearing guide sleeve.

> ***CAUTION—***
> *• Do not grease release bearing module or guide sleeve. If guide sleeve is greased, release bearing may stick to it.*

Clutch, installing (new SAC components)

The SAC pressure plate, when new, comes from the factory with a lock plate at the center which maintains spring tension on the self-adjusting springs. Do not remove this lock plate until the pressure plate is securely installed on the flywheel with clutch disk in place.

> ***CAUTION—***
> *• Avoid contaminating clutch friction surfaces with oil or grease. Do not touch these surfaces.*

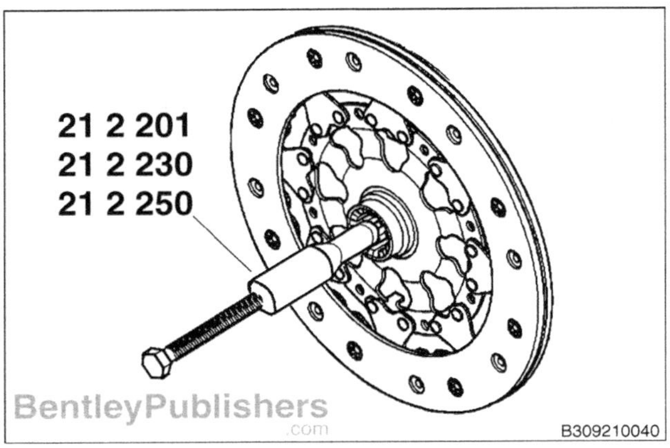

◄ Center clutch disc on flywheel using the appropriate BMW (or equivalent) centering tool.

> ***CAUTION—***
> *• Be sure clutch disc is facing the correct way. The disc is marked "engine side" or "transmission side".*

NOTE—
• The large bolt in the clutch disc centering tool is used to install and remove the tool only. Once the disc is in place on the flywheel, remove the bolt to make room for the SAC pressure plate.

– Lock flywheel in position using BMW special tool 11 9 260 or equivalent.

– Install SAC pressure plate on flywheel dowel pins. Install clutch mounting bolts and tighten each one turn at a time until pressure plate is fully seated. Torque to specification in 2 stages.

Tightening torque	
Clutch to flywheel (use new M8 ZNS bolts) • Stage 1 • Stage 2	 15 Nm (11 ft-lb) additional 90° ± 5°

◄ Use 14 mm Allen wrench to unscrew (clockwise or counterclockwise) lock plate in center of pressure plate.

NOTE—
• The spring lock plate may make snapping noises while being unscrewed.

– Using large bolt, pull out clutch disc centering tool.

– Remove flywheel locking tool.

– Inspect and clean clutch release bearing and lever module.

BentleyPublishers.com
BentleyPublishers.com—All Rights Reserved

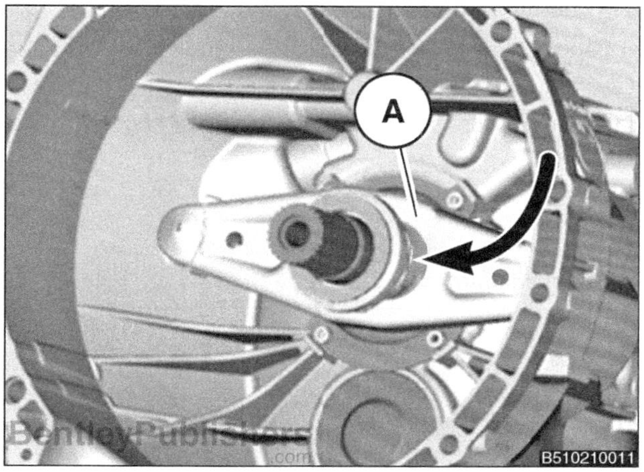

Install clutch release module (**A**) by tilting over transmission input shaft (**arrow**). Be sure it clips securely to retaining spring clip.

– Install transmission. See **230 Manual Transmission**.

> ***CAUTION—***
> - *Replace aluminum bolts each time they are loosened.*
> - *Follow torque instructions, including angle of rotation specifications, when installing aluminum fasteners.*

Tightening torques	
Front end reinforcement to front end (use new fasteners): • Stage 1 • Stage 2	 56 Nm (41 ft-lb) additional 90°
Slave cylinder to transmission	22 Nm (16 ft-lb)
Transmission rear support to body (M8)	19 Nm (14 ft-lb)
Transmission mount to rear support (M8)	19 Nm (14 ft-lb)
Transmission to engine (use new aluminum bolts): • M10 x 30 mm Stage 1 Stage 2 • M10 x 85 mm Stage 1 Stage 2 • M12 Stage 1 Stage 2	 20 Nm additional 90° - 110° 20 Nm additional 180° - 200° 25 Nm additional 130° + 20°
Vibration mount to transmission rear support (M10)	38 Nm (28 ft-lb)

Clutch, installing (used SAC components)

When reinstalling a previously used self-adjusting clutch (SAC), special tools are needed to reset the self-adjusting ring to its original position.

– Place SAC pressure plate on clean work surface.

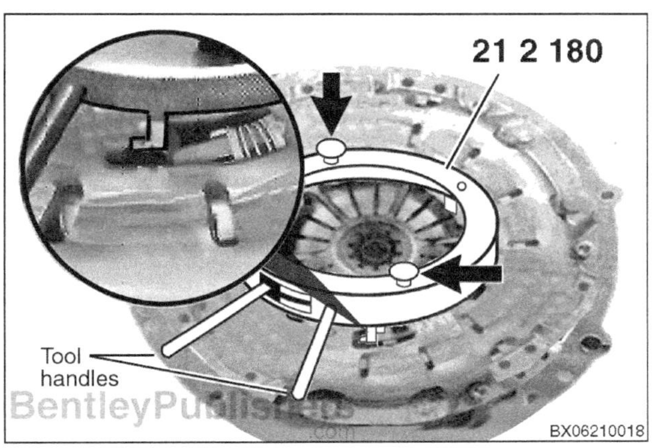

Place BMW special tool 21 2 180 on pressure plate.

- Note that locating hooks of special tool engage pressure plate adjusting ring openings (**inset**).
- Grip tool firmly and squeeze special tool handles together.
- Tighten down special tool knurled screws (**arrows**).
- SAC adjustment ring is now in its original position.

BentleyPublishers
BentleyPublishers.com–All Rights Reserved

Clutch, installing (used SAC components)

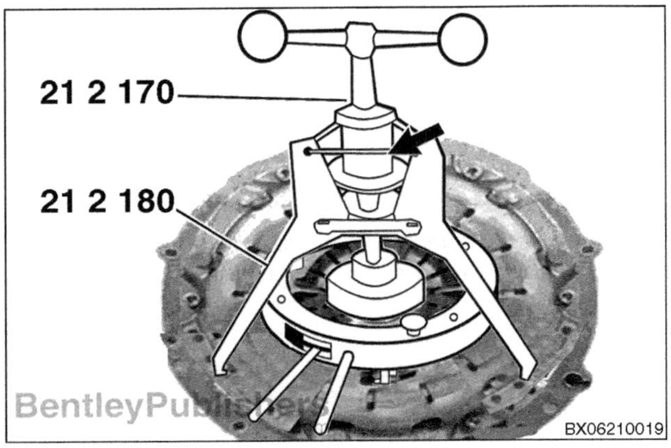

Install BMW special tool 21 2 170 on SAC pressure plate.

- Fit fingers of special tool in slots above dowel pin bores.
- Tighten down knurled nut (**arrow**) finger-tight.
- Screw in T-handle until pressure plate diaphragm spring is pretensioned to stop.

> ***CAUTION—***
> - *Avoid contaminating clutch friction surfaces with oil or grease. Do not touch these surfaces.*

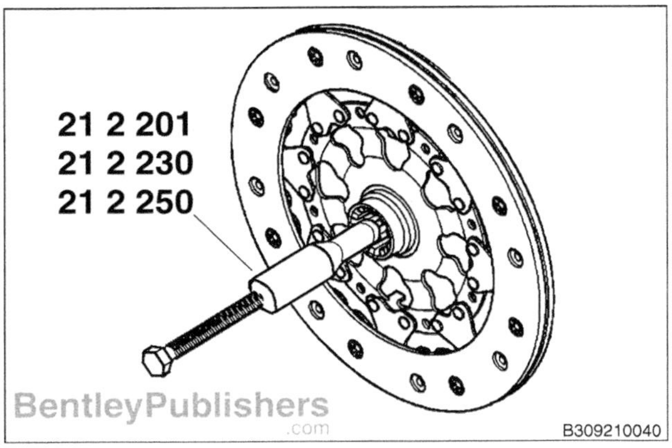

Center clutch disc on flywheel using the appropriate BMW (or equivalent) centering tool.

> ***CAUTION—***
> - *Be sure clutch disc is facing the correct way. The disc is marked "engine side" or "transmission side".*

NOTE —
- *The large bolt in the clutch disc centering tool is used to install and remove the tool only. Once the disc is in place on the flywheel, remove the bolt to make room for the SAC pressure plate.*

– Lock flywheel in position using BMW special tool 11 9 260 or equivalent.

– Install SAC pressure plate on dowel pins at flywheel. Install clutch mounting bolts and tighten each one turn at a time until pressure plate is fully seated. Torque to specification in 2 stages.

Tightening torque	
Clutch to flywheel (use new M8 ZNS bolts) • Stage 1 • Stage 2	 15 Nm (11 ft-lb) additional 90° ± 5°

– Using large bolt, pull out clutch disc centering tool.

– Remove flywheel locking tool.

– Inspect and clean clutch release bearing and lever module.

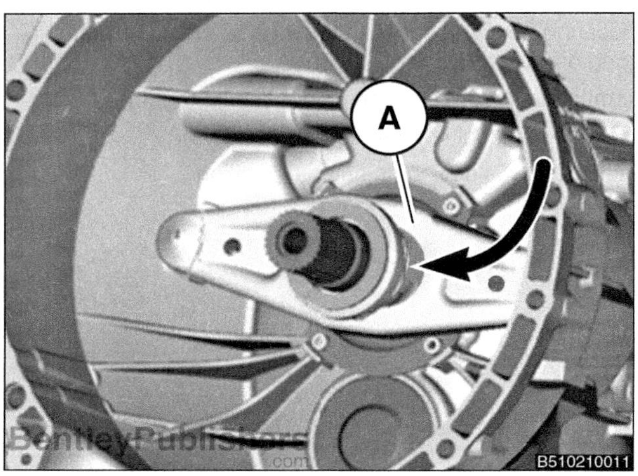

Install clutch release module (**A**) by tilting over transmission input shaft (**arrow**). Be sure it clips securely to retaining spring clip.

BentleyPublishers.com
BentleyPublishers.com—All Rights Reserved

– Install transmission. See **230 Manual Transmission**.

> ***CAUTION—***
> - *Replace aluminum bolts each time they are loosened.*
> - *Follow torque instructions, including angle of rotation specifications, when installing aluminum fasteners.*

Tightening torques	
Front end reinforcement to front end (use new fasteners): • Stage 1 • Stage 2	 56 Nm (41 ft-lb) additional 90°
Slave cylinder to transmission	22 Nm (16 ft-lb)
Transmission rear support to body (M8)	19 Nm (14 ft-lb)
Transmission mount to rear support (M8)	19 Nm (14 ft-lb)
Transmission to engine (use new aluminum bolts): • M10 x 30 mm Stage 1 Stage 2 • M10 x 85 mm Stage 1 Stage 2 • M12 Stage 1 Stage 2	 20 Nm additional 90° - 110° 20 Nm additional 180° - 200° 25 Nm additional 130° + 20°
Vibration damper to transmission rear support (M10)	38 Nm (28 ft-lb)

Self-adjusting clutch (SAC), breaking in

In normal driving, approx. 800 to 1000 gearshifts are needed for a new SAC clutch lining to be fully broken in. To ensure correct operation and long life, break in the SAC gradually, with light to medium loads. Fast sports-style driving and abrupt clutch engagement may destroy a new clutch.The following procedure helps to break in the clutch.

– Drive at 20 mph (30 kph) on level grade. Start, upshift and downshift through the gears at approx. 2000 rpm.

– Increase speed. Upshift and downshift through the gears at approx. 3500 - 4000 rpm.

– On medium grade (approx. 12% and up): Start off at approx. 2500 rpm 3 - 5 times.

BentleyPublishers
BentleyPublishers.com—All Rights Reserved

Flywheel, removing and installing

This procedure is for vehicle with manual transmission. Flywheel and torque plate removal for automatic transmission vehicle is covered in **240 Automatic Transmission**.

- Disconnect negative (–) battery cable and cover battery terminal to keep cable from accidentally contacting terminal.

> ***CAUTION—***
> - *Prior to disconnecting the battery, read the battery disconnection cautions in* **001 Warnings and Cautions**.

- Raise vehicle and support safely.

> ***WARNING —***
> - *Make sure the vehicle is stable and well supported at all times. Use a professional automotive lift or jack stands designed for the purpose. A floor jack is not adequate support.*

- Remove splash shield under engine and transmission. See **020 Maintenance**.
- Remove front end reinforcement. Discard mounting bolts.
- Remove complete exhaust system. See **180 Exhaust System**.

> ***WARNING —***
> - *Make sure exhaust system is fully cooled off before starting removal.*

- Remove heat shields under transfer case (xDrive models) and rear driveshaft.
- Detach front (xDrive models) and rear driveshafts and tie to side. See **260 Driveshafts**.
- Remove transmission from engine. See **230 Manual Transmission**.
- Remove clutch. See **Clutch, removing** in this repair group.
- Inspect flywheel for scoring, hot spots, cracks or loose or worn guide pins. Replace flywheel if any faults are found.

- Working at flywheel:
  - Lock flywheel in position using BMW special tool 11 9 260 or equivalent.
  - Use BMW special tool 11 4 180 or equivalent (Torx T60 socket) to loosen and remove flywheel mounting bolts.
  - Remove flywheel.
- Clean flywheel bolt threads in crankshaft.

BentleyPublishers.com
BentleyPublishers.com—All Rights Reserved

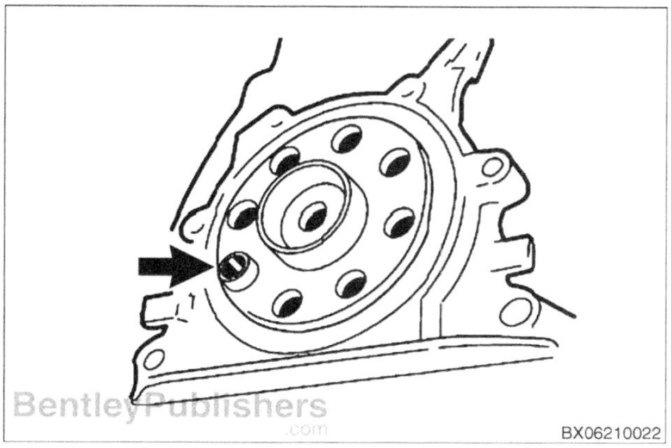

Check flywheel location dowel sleeve (**arrow**) for damage and correct installation.

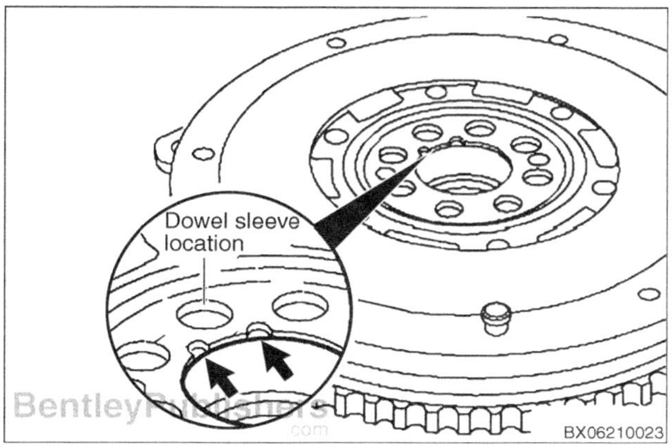

When reinstalling flywheel, position dowel sleeve next to two locating notches (**arrows**).

- Install flywheel using new self-locking bolts.

Tightening torque	
Flywheel to crankshaft (use new self-locking bolts)	120 Nm (89 ft-lb)

> ***CAUTION—***
> • *Use new self-locking bolts to install flywheel. Do not reuse the old stretch-type bolts. Do not install bolts with Loctite® or similar thread-locking compound.*

Transmission pilot bearing, replacing

- Disconnect negative (–) battery cable and cover battery terminal to keep cable from accidentally contacting terminal.

> ***CAUTION—***
> • *Prior to disconnecting the battery, read the battery disconnection cautions in* **001 Warnings and Cautions**.

- Raise vehicle and support safely.

> ***WARNING—***
> • *Make sure the vehicle is stable and well supported at all times. Use a professional automotive lift or jack stands designed for the purpose. A floor jack is not adequate support.*

- Remove splash shield under engine and transmission. See **020 Maintenance**.

- Remove front end reinforcement. Discard mounting bolts.

- Remove complete exhaust system. See **180 Exhaust System**.

> ***WARNING—***
> • *Make sure exhaust system is fully cooled off before starting removal.*

BentleyPublishers.com
BentleyPublishers.com—All Rights Reserved

– Remove heat shields under transfer case (xDrive models) and rear driveshaft.

– Detach front (xDrive models) and rear driveshafts and tie to side. See **260 Driveshafts**.

– Remove transmission from engine. See **230 Manual Transmission**.

– Remove clutch. See **Clutch, removing** in this repair group.

– Remove flywheel. See **Flywheel, removing and installing** in this repair group.

< Use hydraulic press and BMW special tool 21 2 051 to press transmission pilot bearing out of dual-mass flywheel. Press from engine side of flywheel.

> ***CAUTION—***
> *• Do not drive bearing in or out using a chisel or punch.*

– Place new pilot bearing on BMW special tool 21 2 052. Use hydraulic press to press bearing into flywheel as far as it will go. Press from clutch side of flywheel.

BentleyPublishers.com
BentleyPublishers.com—All Rights Reserved

230 Manual Transmission

TABLES

GENERAL

This repair group covers manual transmission external service, including transmission removal and installation. Internal transmission repair is not covered. Special tools and procedures are required to disassemble and service the internal geartrain.

See also:

- **210 Clutch**
- **250 Gearshift Linkage**
- **260 Driveshafts**
- **270 Transfer Case**

6-speed manual transmission (GS6-37BZ)

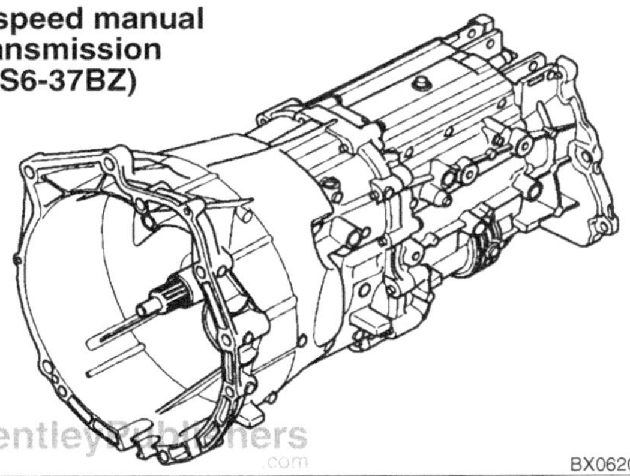

Manual transmission applications and gear ratios

Several manual transmission are used in 5 Series models. See **Table a**. Gear ratios are in **Table b**.

Table a. Manual transmission applications

Model (engine)	Year	BMW designation	Manufacturer designation
525i (M54)	2004 - 2006	GS6-37BZ / DZ	ZF 6HP19 type H
530i (M54, N52)	2004 - 2007	GS6-37BZ / DZ	ZF 6HP19 type H
525i, 528i (N52)	2006 - 2009	GS6-17BG	Getrag type I
525xi, 528xi, 530xi (N52)	2006 - 2009	GS6X37BZ / DZ	ZF 6HP19 type H
535i (N54)	2008 - 2009	GS6-53BZ / DZ	ZF 6HP26 type G
535xi (N54)	2008 - 2009	GS6X53DZ	ZF 6HP26 type G
545i, 550i (N62, N62TU)	2004 - 2009	GS6-53BZ	ZF 6HP26 type G

BentleyPublishers.com—All Rights Reserved

Table b. Manual transmission gear ratios			
Gear	Getrag GS6-17BG	ZF GS6-37BZ / DZ ZF GS6X37BZ / DZ	ZF GS6-53BZ / DZ ZF GS6X53DZ
1st	4.32 : 1	4.35 : 1	4.06 : 1
2nd	2.46 : 1	2.50 : 1	2.40 : 1
3rd	1.66 : 1	1.66 : 1	1.58 : 1
4th	1.23 : 1	1.23 : 1	1.19 : 1
5th	1.00 : 1	1.00 : 1	1.00 : 1
6th	0.85 : 1	0.85 : 1	0.87 : 1
Reverse	3.94 : 1	3.93 : 1	2004 - 2005: 3.77 : 1 2006 - 2009: 3.68 : 1

Cautions

CAUTION—

- *When performing any repair which involves separating the engine and transmission, check that the bell housing dowels are undamaged and properly in place before reassembly. If the alignment of the engine flywheel to the transmission input shaft or torque converter is incorrect any of the following complaints may result: Clutch slipping, shudder, or poor disengagement; noise from transmission input shaft; transmission popping out of gear; difficulty changing gears; or internal damage to the transmission.*
- *When removing interior trim clips, avoid damaging trim pieces by using a plastic prying tool or a screwdriver with the tip wrapped with masking tape.*
- *To avoid electrochemical corrosion to engine components made of aluminum-magnesium alloy, do not use steel fasteners. Use aluminum fasteners only.*
- *The end faces of aluminum fasteners are usually painted blue. For reliable identification, test fasteners for aluminum composition with magnet.*
- *Replace aluminum fasteners each time they are loosened.*
- *Follow torque instructions, including angle of rotation specifications, when installing aluminum fasteners.*
- *Fill manual transmission only with BMW approved synthetic oil. Filling with incorrect lube could cause the following damage*
  - *— Premature synchromesh ring wear.*
  - *— Reduced service life of bearings.*
  - *— Tooth profile damage.*
  - *— Faster wear of selector forks.*
  - *— Damage to radial shaft oil seals.*

TRANSMISSION FLUID SERVICE

Manual transmissions in 5 Series models are provided with lifetime lubrication. No oil change is required for the entire service life of these transmissions. If repairs have to be made to the transmission, use only the recommended lifetime fluid to refill.

BentleyPublishers.com
BentleyPublishers.com—All Rights Reserved

Usually, the transmission fluid sticker near the fill plug identifies the lifetime transmission fluid used at the factory.

Table c. Manual transmission fluids		
Transmission	**Year range**	**Fluid type**
Getrag GS6-17BG	2004 - 2005	MTF-LT-2
ZF GS6-37BZ ZF GS6X37BZ ZF GS6-53BZ ZF GS6X53BZ	from 2006	MTF-LT-3
MTF-LT-2 (Lifetime fluid) BMW part no. 83 22 0 309 031		
MTF-LT-3 (Lifetime fluid) BMW part no. 83 22 0 409 878		

230

Transmission fluid level, checking

– Drive vehicle for a few miles to warm transmission.

– Raise and safely support vehicle to access transmission filler plug. Make sure vehicle is level.

> ***WARNING —***
> - *Make sure the vehicle is stable and well supported at all times. Use a professional automotive lift or jack stands designed for the purpose. A floor jack is not adequate support.*

– Remove plastic splash shield under transmission. See **020 Maintenance**.

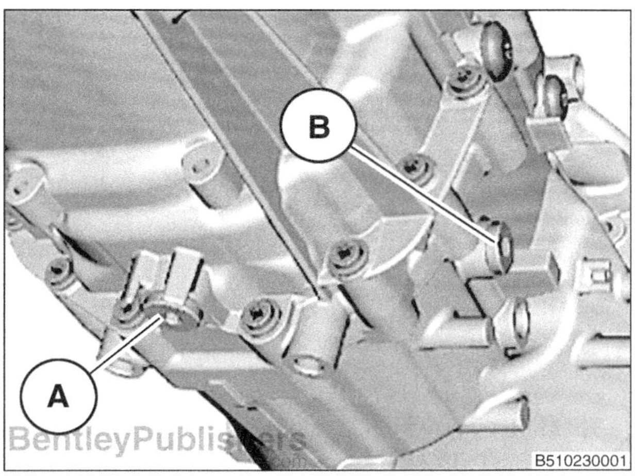

Remove fill plug (**B**). Be prepared to catch dripping fluid.

– Insert gloved finger in fill hole to check fluid level.
- If fluid level is up to bottom of fill hole (finger is wetted by transmission fluid), level is correct.
- If level is low, slowly fill transmission with fluid until fluid overflows fill hole.

– Install and torque fill plug.

Tightening torque	
Drain or fill plug to transmission case: • M12 x 1.5 • M18 x 1.5	 25 Nm (18 ft-lb) 35 Nm (26 ft-lb)

Transmission fluid, replacing

– Drive vehicle for a few miles to warm transmission.

– Raise and safely support vehicle to access transmission drain plug.

> ***WARNING —***
> - *Make sure the vehicle is stable and well supported at all times. Use a professional automotive lift or jack stands designed for the purpose. A floor jack is not adequate support.*

– Remove plastic splash shield under transmission. See **020 Maintenance**.

BentleyPublishers.com
BentleyPublishers.com—All Rights Reserved

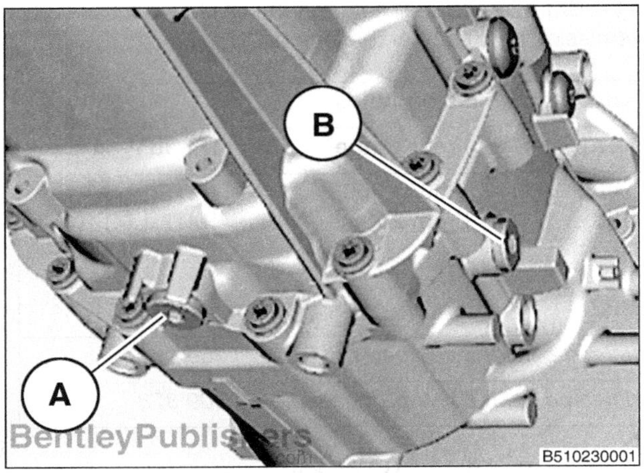

Place drain pan under transmission and remove drain plug (**A**) at bottom of transmission.

- Drain fluid.
- Install and torque drain plug.
- Remove fill plug (**B**) from side of transmission.
- Slowly fill transmission with fluid until fluid overflows fill hole.

Table d. Manual transmission fluid capacities		
Transmission	**Initial fill, new or rebuilt unit**	**Fluid change**
Getrag GS6-17BG	1.4 liter (1.5 US qt)	1.3 liter (1.4 US qt)
ZF GS6-37BZ	1.6 liter (1.7 US qt)	1.5 liter (1.6 US qt)
ZF GS6-53DZ	1.6 liter (1.7 US qt)	1.5 liter (1.6 US qt)

– Install and torque fluid fill plug.

Tightening torque	
Drain or fill plug to transmission case: • M12 x 1.5 • M18 x 1.5	 25 Nm (18 ft-lb) 35 Nm (26 ft-lb)

Transmission External Service

Back-up light switch, replacing

Back-up light switch location on manual transmission varies.

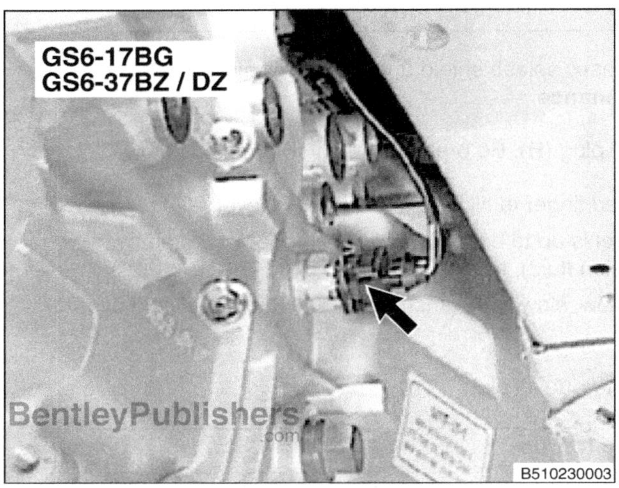

Getrag GS6-17BG or ZF GS6-17BZ / DZ transmission. Back-up light switch (**arrow**) on right side of transmission toward rear.

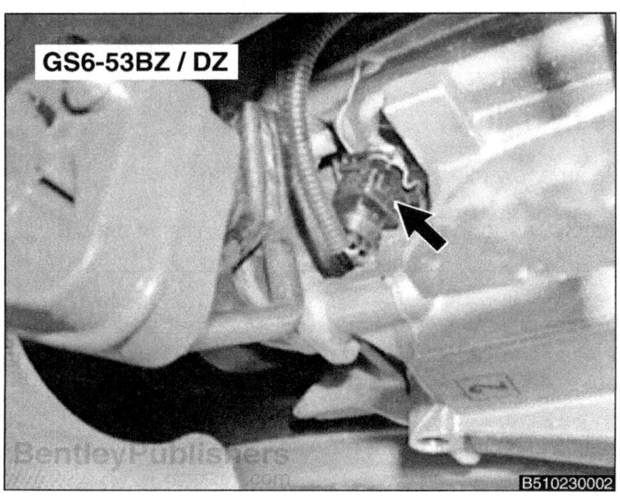

ZF GS6-53BZ / DZ transmission. Back-up light switch (**arrow**) on right side of transmission toward rear.

– Raise and safely support vehicle to access back-up light switch.

> ***WARNING —***
> - *Make sure the vehicle is stable and well supported at all times. Use a professional automotive lift or jack stands designed for the purpose. A floor jack is not adequate support.*

– Remove plastic splash shield under transmission. See **020 Maintenance**.

– Detach back-up light switch harness connector. Unscrew switch from transmission.

BentleyPublishers.com
BentleyPublishers.com—All Rights Reserved

– Install new switch and reconnect harness connector.

Tightening torque	
Back-up light switch to transmission	16 Nm (13 ft-lb)

– Check transmission fluid level before lowering vehicle. Top off as necessary. See **Transmission Fluid Service** in this repair group.

Selector shaft seal, replacing

This repair can be performed with transmission installed in vehicle.

– Place transmission into second gear.

– Raise vehicle and support safely.

> ***WARNING —***
> • *Make sure the vehicle is stable and well supported at all times. Use a professional automotive lift or jack stands designed for the purpose. A floor jack is not adequate support.*

– Remove underbody splash shields.

– Remove exhaust system. See **180 Exhaust System**.

> ***WARNING —***
> • *Make sure exhaust system is fully cooled off before removing.*

– Remove exhaust heat shields.

– xDrive models: Remove front drive shaft. See **260 Driveshafts**.

– All models: Remove rear driveshaft from output flange:

- Remove driveshaft center bearing support fasteners. Support center of driveshaft.
- Detach rear driveshaft flex-disc from transfer case.
- Lower center of driveshaft sufficiently to disengage flex-disc from transfer case flange. Tie driveshaft to side.

> ***CAUTION—***
> • *To prevent damage to rear driveshaft CV joint, do not allow driveshaft to hang down unsupported.*

– Free harnesses for oxygen sensors and back-up light switch from rear transmission support.

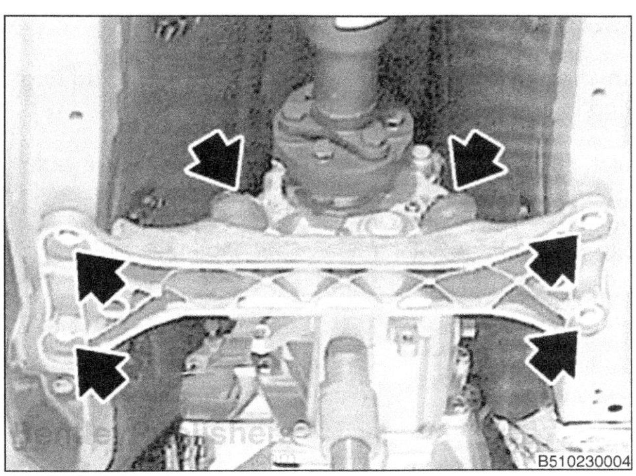

◄ Support end of transmission with suitable jack. Remove mounting fasteners (**arrows**) for rear transmission support.

BentleyPublishers.com
BentleyPublishers.com–All Rights Reserved

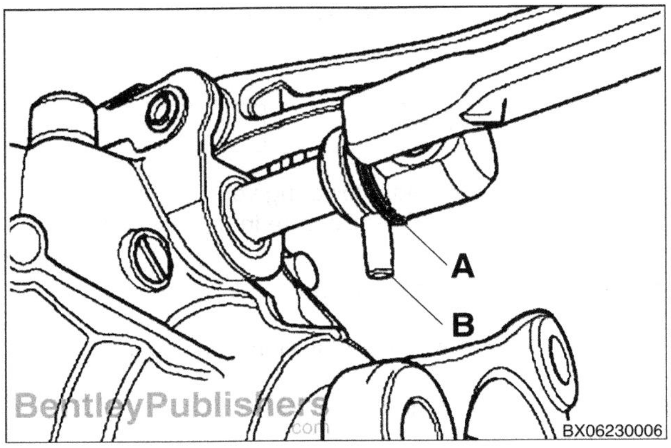

Working at top rear of transmission at shifter console, pry locking spring (**A**) out of groove and press pin (**B**) out of shift coupling.

- Detach shift coupling and shift linkage from selector shaft and tie aside.

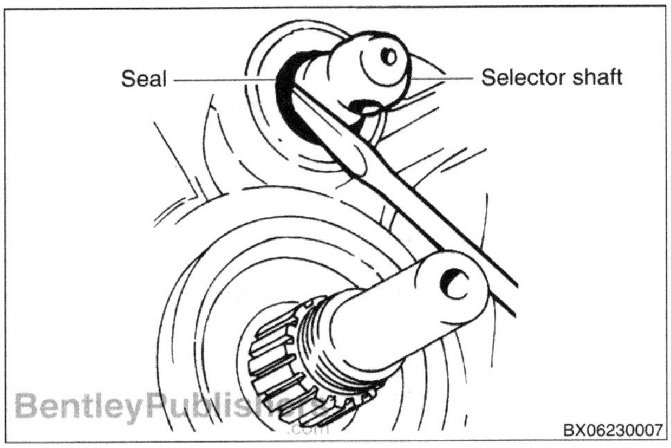

Carefully pry out selector shaft oil seal with BMW special tool 23 0 210, or with a narrow seal remover or small screwdriver.

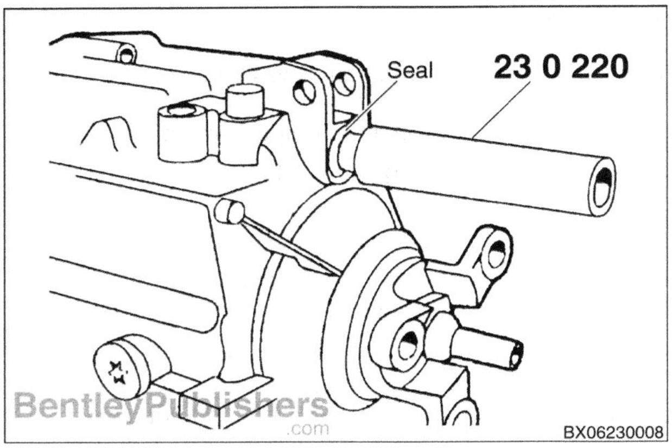

Coat new selector shaft seal with transmission fluid. Drive new seal in flush with housing. Use BMW special seal installation tool 23 0 220 or equivalent and a soft-faced (plastic) hammer.

– Remainder of installation is reverse of removal. Check transmission and transfer case oil level, topping up as necessary.

Tightening torques	
Transfer case to transmission	43 Nm (31 ft-lb)

Transmission input shaft seal, replacing

– Remove transmission from vehicle. See **Transmission, removing and installing** in this repair group.

– Remove clutch release bearing / release lever module from inside bell housing. See **210 Clutch**.

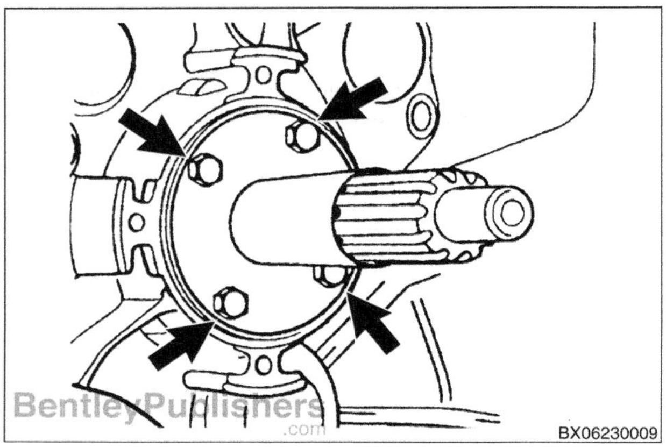

Remove bolts (**arrows**) for clutch release-bearing guide sleeve. Remove guide sleeve.

BentleyPublishers.com
BentleyPublishers.com—All Rights Reserved

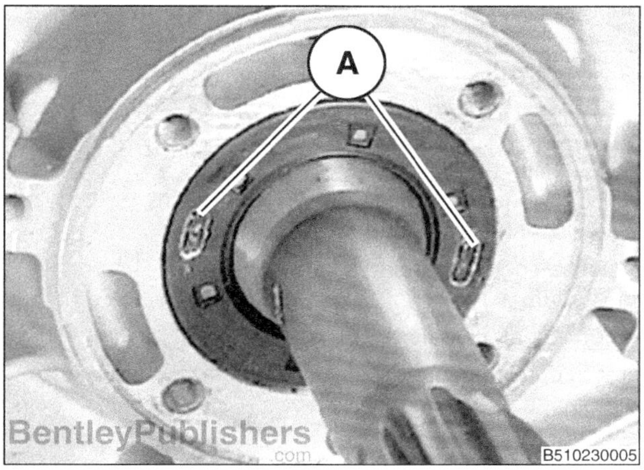

Two covered removal holes (**1**) are present in seal. Thread BMW special tool 23 0 490 or equivalent slide hammer seal puller into one hole and remove seal.

To install new seal:

- Push radial seal on BMW special tool 23 0 320.
- Coat sealing lip of new seal with transmission oil.
- Slide tool onto input shaft.
- Use plastic hammer to drive in seal until fully seated.

– Thoroughly clean clutch release bearing guide sleeve mounting bolts, sealing surfaces, and threads. Apply Loctite®243 or equivalent thread locking compound to guide sleeve bolts. Reinstall guide sleeve and spacer (if equipped).

Tightening torque	
Guide sleeve to transmission housing • M8 X 30 • M8 X 22 • M6	 25 Nm (18 ft-lb) 18 Nm (13 ft-lb) 10 Nm (7 ft-lb)

– Replace clutch release bearing / release lever module. See **210 Clutch**.

– Reinstall transmission.

Transmission output shaft seal, replacing

– Disconnect negative (-) cable from battery.

> ***CAUTION—***
> - *Prior to disconnecting the battery, read the battery disconnection cautions given in* **001 Warnings and Cautions**.

– Raise and safely support vehicle.

> ***WARNING —***
> - *Make sure the vehicle is stable and well supported at all times. Use a professional automotive lift or jack stands designed for the purpose. A floor jack is not adequate support.*

– Remove underbody splash shields.

BentleyPublishers.com
BentleyPublishers.com—All Rights Reserved

– Remove complete exhaust system. See **180 Exhaust System**.

> ***WARNING —***
> *• Make sure exhaust system is fully cooled off before removing.*

– Remove exhaust system heat shields.

– Remove rear driveshaft from output shaft:

- Remove driveshaft center bearing support fasteners. Support center of driveshaft.
- Detach rear driveshaft flex-disc from transfer case.
- Lower center of driveshaft sufficiently to disengage flex-disc from transfer case flange. Tie driveshaft to side.

> ***CAUTION—***
> *• To prevent damage to rear driveshaft CV joint, do not allow driveshaft to hang down unsupported.*

– xDrive models: Remove transfer case. See **270 Transfer Case**.

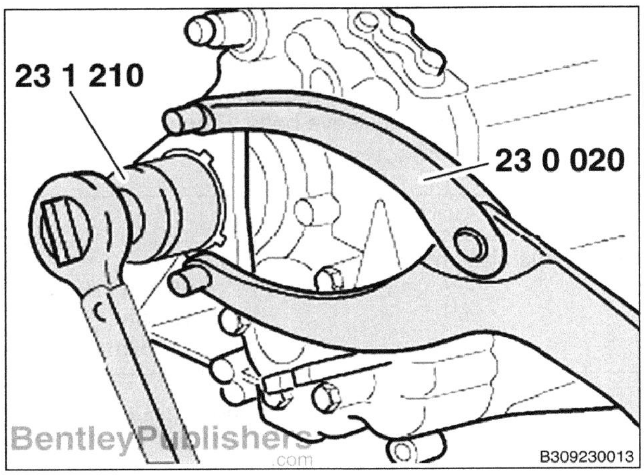

Counterhold output flange with BMW special tool 23 0 020 and release flange collar nut using BMW special tool 23 1 210.

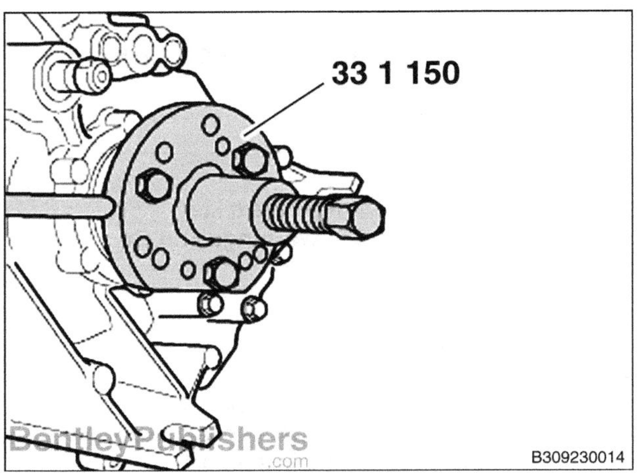

Remove output flange using BMW special tool 33 1 150 or suitable puller.

BentleyPublishers.com
BentleyPublishers.com—All Rights Reserved

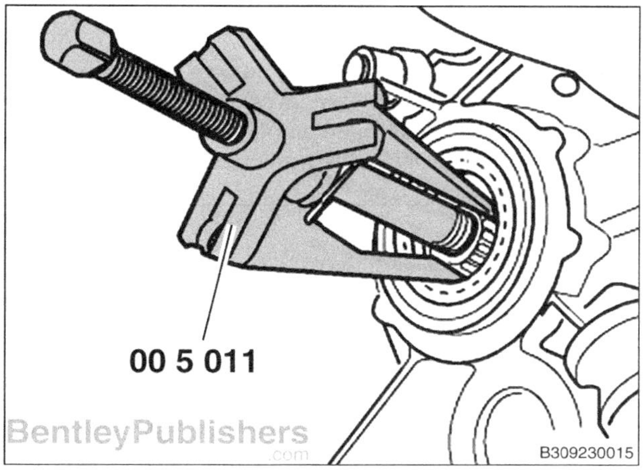

Remove output shaft seal using BMW special tool 00 5 011 or suitable seal puller.

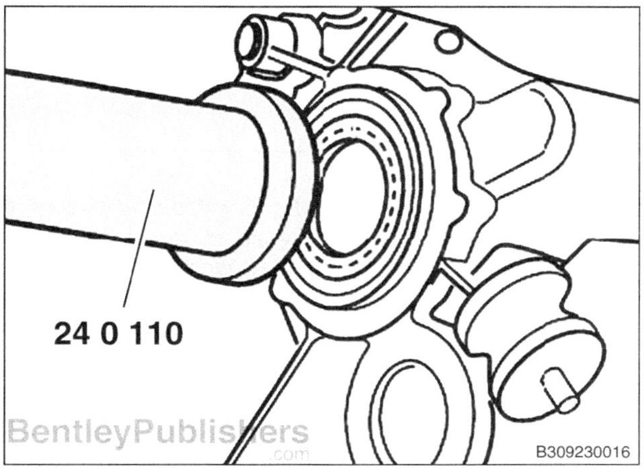

When installing new seal, coat seal lips with transmission fluid. Use BMW special tool 24 0 110 and plastic hammer to drive seal in flush.

– Remainder of installation is reverse of removal. Remember to:
  - Replace collar nut with new.
  - Use Loctite® 431 on output shaft threads when installing collar nut.
  - Follow 3 step tightening procedure (tighten, release, tighten down).
  - Check transmission fluid level and check system for leaks.

Tightening torque	
Transmission output flange to transmission (GS6-17BG and GS6-37BZ transmission) • Pretension • Release • Tighten down	 170 Nm (125 ft-lb) Release tension 120 Nm (89 ft-lb)
Transmission output flange to transmission (GS6-53DZ transmission) • Pretension • Release • Tighten down	 200 Nm (148 ft-lb) Release tension 140 Nm (103 ft-lb)

230

BentleyPublishers.com
BentleyPublishers.com—All Rights Reserved

Transmission Removal and Installation

Removal and installation of the transmission is best accomplished with the vehicle on a lift and using a transmission jack. Support engine using equipment which allows the engine to pivot on its mounts to access the upper bolts at the transmission bell housing.

> ***WARNING—***
> *• The removal of the transmission may upset the balance of the vehicle on a lift.*

Transmission, removing and installing

– Disconnect negative (-) cable from battery.

> ***CAUTION—***
> *• Prior to disconnecting the battery, read the battery disconnection cautions given in* **001 Warnings and Cautions**.

– Raise and safely support vehicle.

> ***WARNING—***
> *• Make sure the vehicle is stable and well supported at all times. Use a professional automotive lift or jack stands designed for the purpose. A floor jack is not adequate support.*

– Remove underbody splash shields. See **020 Maintenance**.

– Remove front end reinforcement plate.

– Remove complete exhaust system. See **180 Exhaust System**.

> ***WARNING—***
> *• Make sure exhaust system is fully cooled off before removing.*

– Remove exhaust system heat shields.

– xDrive models: Remove front drive shaft. See **260 Driveshafts**.

– Disconnect harness connector from back-up light switch on transmission. See **Back-up light switch, replacing** in this repair group.

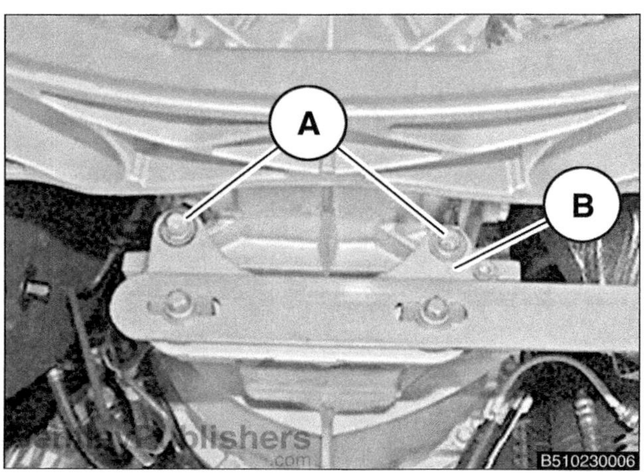

◄ Remove fasteners (**A**) and remove exhaust bracket (**B**).

BentleyPublishers.com
BentleyPublishers.com—All Rights Reserved

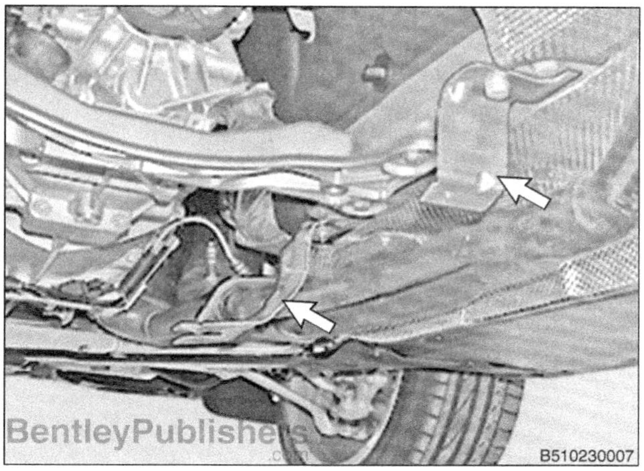

◄ Remove underbody splash shields brackets (**arrows**).

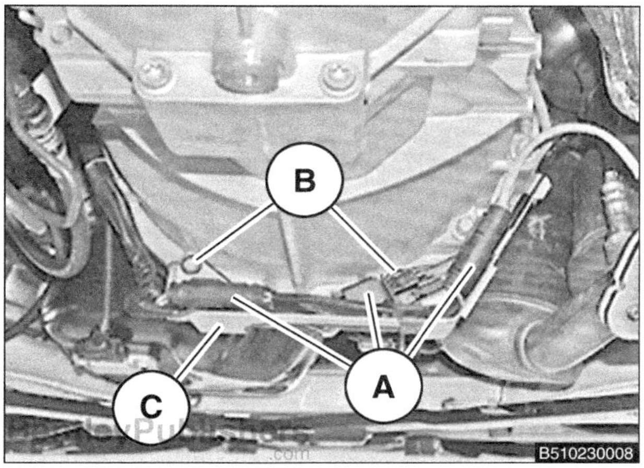

◄ Disconnect oxygen sensor harness connectors (**A**) and remove fasteners (**B**) from connector bracket (**C**).

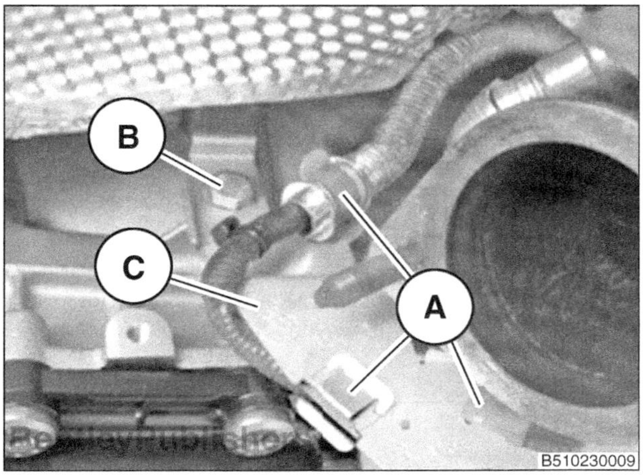

◄ Unclip oxygen sensor harness (**A**) from bracket (**B**). Remove fastener and remove bracket (**C**) from transmission.

BentleyPublishers.com—All Rights Reserved

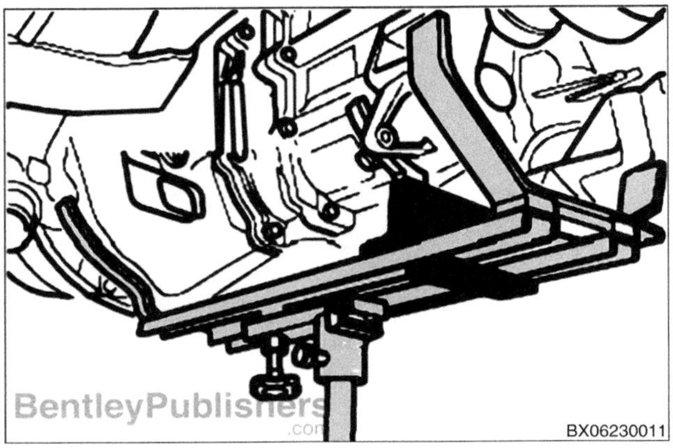

Support transmission on transmission jack. Secure transmission to jack.

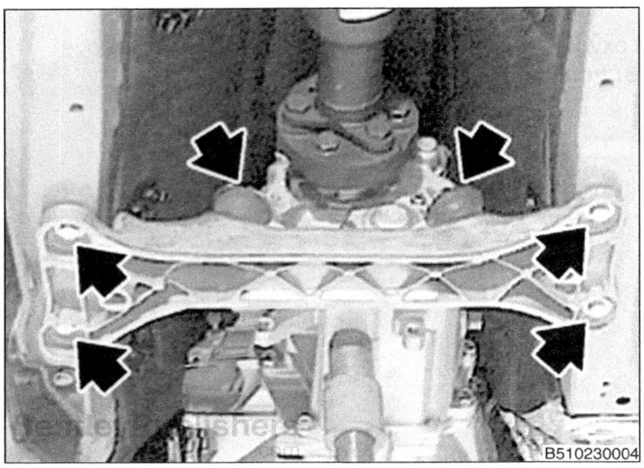

Remove mounting fasteners (**arrows**) for rear transmission support.

- xDrive models: Detach xDrive servomotor harness connectors at transfer case.
- Lower transmission jack slowly, allowing engine and transmission assembly to tilt back and down slightly.

> ***CAUTION—***
> - *Tilting the engine to lower the transmission can lead to damage to engine compartment components due to lack of clearance at rear of engine. Lower transmission carefully while checking engine rear clearance.*

- All models: Remove rear driveshaft from output flange:
  - Remove driveshaft center bearing support fasteners. Support center of driveshaft.
  - Detach rear driveshaft flex-disc from transmission or transfer case output flange.
  - Lower center of driveshaft sufficiently to disengage flex-disc from output flange. Tie driveshaft to side.

> ***CAUTION—***
> - *To prevent damage to rear driveshaft CV joint, do not allow driveshaft to hang down unsupported.*

- Unbolt clutch slave cylinder from side of transmission. Do not disconnect fluid hose. Suspend slave cylinder from chassis using stiff wire.

> ***CAUTION—***
> - *Unbolt clutch slave cylinder slowly to prevent air being drawn into hydraulic system.*
> - *Do not operate clutch pedal with slave cylinder removed from transmission.*

BentleyPublishers.com
BentleyPublishers.com—All Rights Reserved

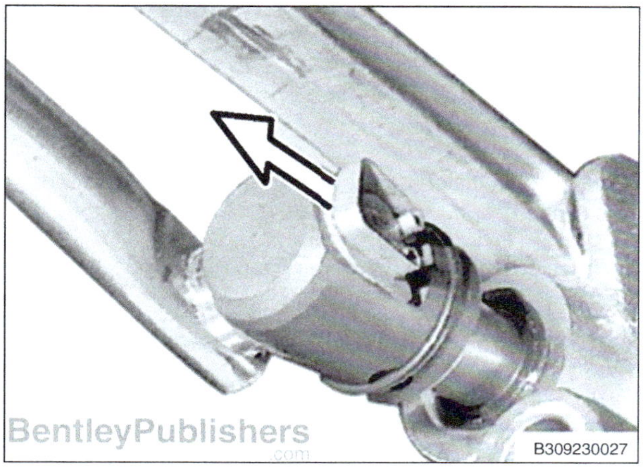

< Remove shift rod clip (**arrow**) and disconnect shift linkage from selector shaft coupling.

NOTE—
- *A new, more secure, shift rod clip is fitted as of 04 / 2008. Use new specification shift rod clip when reassembling.*

– Disconnect shift console from top of transmission. See **250 Gearshift Linkage**.

– Support front of engine with tall jack stand or hydraulic jack.

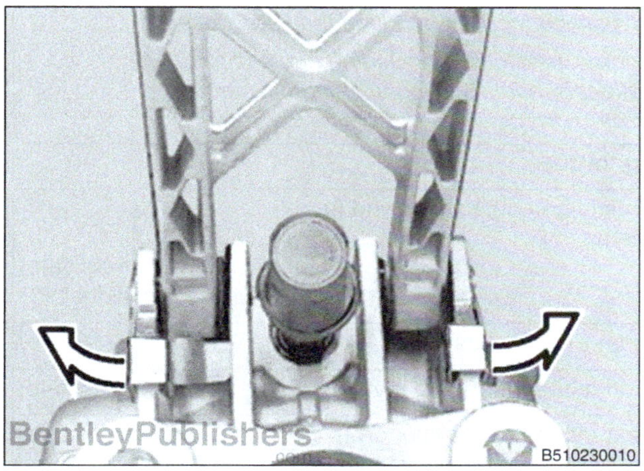

< Release locking clips (**arrows**) and lift out shift arms.

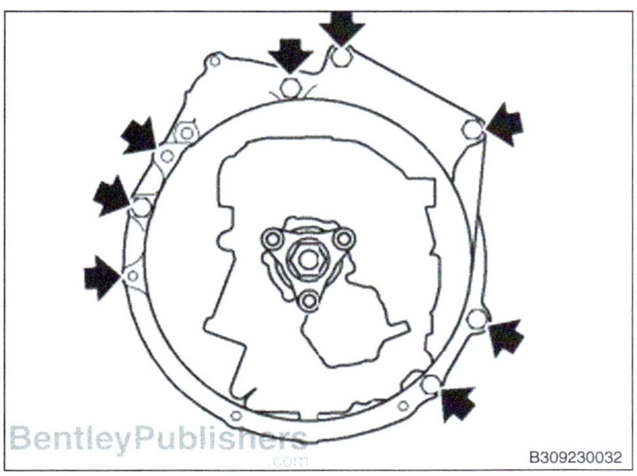

< Remove transmission mounting Torx-head bolts (**arrows**). Note length and location of bolts.

CAUTION—
- *Number, size and location of fasteners varies with transmission installed.*

– Remove transmission by pulling backward until transmission input shaft clears clutch disc splines, then pull downwards. Lower transmission jack to remove transmission.

CAUTION—
- *Do not allow transmission to hang from input shaft.*

BentleyPublishers
BentleyPublishers.com—All Rights Reserved

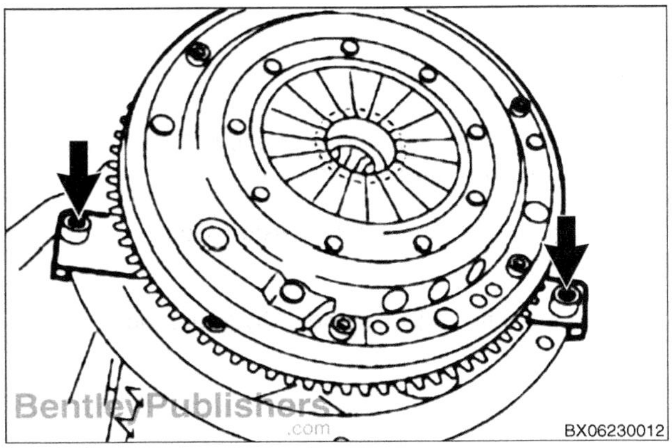

Be sure bell housing alignment dowel sleeves (**arrows**) are located correctly. Replace damaged sleeves.

– Inspect clutch, clutch release bearing and flywheel. See **210 Clutch**. Repair as necessary.

– Check and repair transmission seals. See **Transmission External Service** in this repair group.

– Reinstall transmission and xDrive transfer case (if equipped), keeping in mind the following:
  - If installing a new transmission, be sure to transfer parts from old transmission.
  - Thoroughly clean input shaft and clutch disc splines. Lightly lubricate transmission input shaft splines before installing.

> ***CAUTION—***
> - *Replace aluminum bolts each time they are loosened.*
> - *Follow torque instructions, including angle of rotation specifications, when installing aluminum fasteners.*

Tightening torques	
Front end reinforcement to front end (use new fasteners): • Stage 1 • Stage 2	 56 Nm (41 ft-lb) additional 90°
Slave cylinder to transmission	22 Nm (16 ft-lb)
Transmission rear support to body (M8)	19 Nm (14 ft-lb)
Transmission mount to rear support (M8)	19 Nm (14 ft-lb)
Transmission to engine (use new aluminum bolts): • M10 x 30 mm Stage 1 Stage 2 • M10 x 85 mm Stage 1 Stage 2 • M12 Stage 1 Stage 2	 20 Nm additional 90° - 110° 20 Nm additional 180° - 200° 25 Nm additional 130° + 20°
Vibration damper to transmission rear support (M10)	38 Nm (28 ft-lb)

BentleyPublishers.com
BentleyPublishers.com—All Rights Reserved

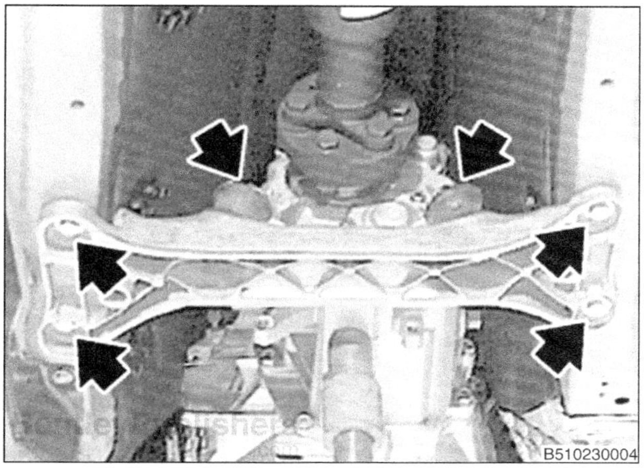

Reinstall transmission support. Center rear of transmission in driveshaft tunnel before tightening support bracket bolts (**arrows**).

Tightening torques	
Transfer case to transmission	43 Nm (31 ft-lb)
Transfer case support • To chassis (M10) • To transfer case rubber mount (M12)	 42 Nm (30 ft-lb) 74 Nm (55 ft-lb)

– Reinstall driveshafts using new self-locking nuts. See **260 Driveshafts**.

> ***CAUTION—***
> - *To avoid damaging flex-disc rubber, be sure to hold bolts stationary while tightening nuts.*

Tightening torques	
Front driveshaft flex-disc to front differential flange (replace fasteners) (M10 10.9)	60 Nm (44 ft-lb)
Rear driveshaft flex-disc to transfer case output flange	100 Nm (74 ft-lb)

– xDrive models: Refill transfer case. See **270 Transfer Case**.

Tightening torque	
Fill plug to transfer case	33 Nm (24 ft-lb)

– Refill transmission before starting or towing vehicle. See **Transmission Fluid Service** in this repair group.

Tightening torque	
Drain or fill plug to transmission case: • M12 x 1.5 • M18 x 1.5	 25 Nm (18 ft-lb) 35 Nm (26 ft-lb)

BentleyPublishers
BentleyPublishers.com—All Rights Reserved

BentleyPublishers.com
BentleyPublishers.com—All Rights Reserved

231 Sequential Manual Gearbox (SMG)

TABLES

GENERAL

This repair group covers sequential manual gearbox (SMG) description and operation, as well as transmission removal and installation. Transmission repair is not covered.

See also:
- **230 Manual Transmission** for transmission seal replacement
- **ECL Electrical Component Locations** for SMG components

SMG applications and gear ratios

SMG application information is in **Table a**. Gear ratios are in **Table b**.

Table a. SMG applications

Model (engine)	Year	Transmission
525i, 530i (M54)	2004 - 2005	GS6S37BZ
530i (N52)	2006 - 2007	GS6S37BZ
545i (N62)	2004 - 2006	GS6S53BZ

Table b. SMG gear ratios

Gear	ZF GS6S37BZ	ZF GS6S53BZ
1st	4.35 : 1	4.06 : 1
2nd	2.50 : 1	2.40 : 1
3rd	1.66 : 1	1.58 : 1
4th	1.23 : 1	1.19 : 1
5th	1.00 : 1	1.00 : 1
6th	0.85 : 1	0.87 : 1
Reverse	3.93 : 1	3.77 : 1

BentleyPublishers
BentleyPublishers.com—All Rights Reserved

SMG overview

Introduction

The sequential manual transmission (SMG) had its origins in racing cars. Modified for use in street vehicles, it was available as an option in selected 2004 - 2007 5 Series rear-wheel drive models.

In contrast to a conventional manual transmission, SMG clutch control is automated. By driver choice, shifting is either manual or automatic. A computerized electrohydraulic system operates the clutch and can take over gearshifts as well.

For SMG, 6-speed manual transmission housings for GS637BZ and GS653BZ transmissions were modified to accommodate the following components:

- Hydraulic unit with actuators for engaging and disengaging the clutch and for switching gears.
- Additional sensors: Shift travel sensor, selector angle sensor, inductive sensor for transmission input speed.

The construction of the transmission and the clutch remain unchanged. The version used in V8 models (GS6S53BZ) is equipped with a more powerful clutch actuator.

SMG maximum torque capabilities	
GS6S37BZ	370 Nm (273 lb-ft)
GS6S53BZ	600 Nm (443 lb-ft)

GS6S37BZ external views

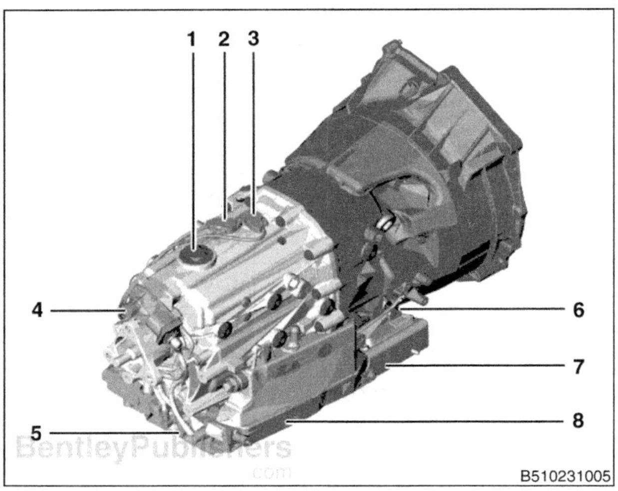

Right side:

1. Assembly port (access to selector shaft, gearshift and selector angle actuator)
2. Shift travel sensor
3. Selector angle sensor
4. Gearshift and selector angle actuator
5. Hydraulic unit
6. Transmission input speed sensor
7. Hydraulic pump
8. Expansion tank

BentleyPublishers.com
BentleyPublishers.com—All Rights Reserved

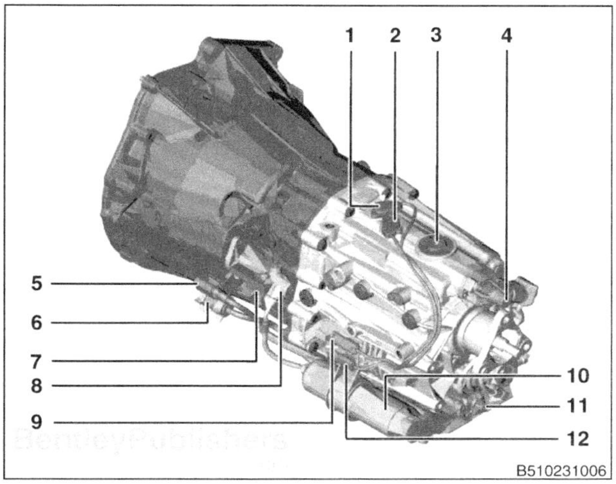

Left side:

1. Selector angle sensor
2. Shift travel sensor
3. Assembly port (access to selector shaft, gearshift and selector angle actuator)
4. Gearshift and selector angle actuator
5. Hydraulic pump power supply connector
6. Hydraulic unit components electrical connector (4 solenoid valves, hydraulic pressure sensor)
7. Clutch travel sensor
8. Clutch actuator
9. Selector angle sensor connector
10. Pressure accumulator
11. Hydraulic unit
12. Shift travel sensor connector

GS6S53BZ external views

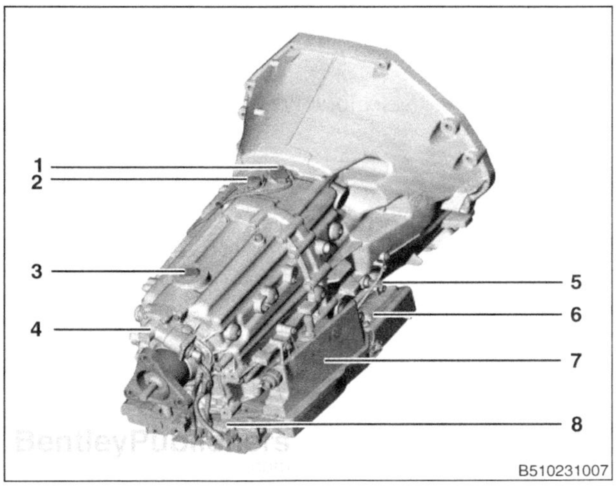

Right side:

1. Selector angle sensor
2. Shift travel sensor
3. Assembly port (access to selector shaft, gearshift and selector angle actuator)
4. Transmission input speed sensor
5. Hydraulic pump
6. Gearshift and selector angle actuator
7. Expansion tank
8. Hydraulic unit

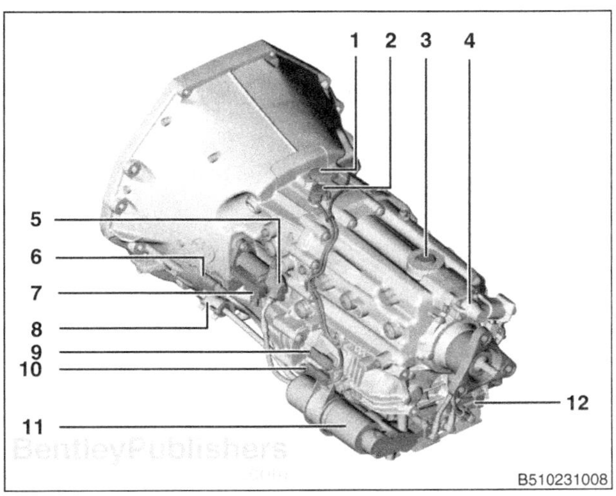

Left side:

1. Selector angle sensor
2. Shift travel sensor
3. Assembly port (access to selector shaft, gearshift and selector angle actuator)
4. Gearshift and selector angle actuator
5. Clutch actuator
6. Hydraulic pump power supply connector
7. Clutch travel sensor
8. Hydraulic unit components electrical connector (4 solenoid valves, hydraulic pressure sensor)
9. Selector angle sensor connector
10. Shift travel sensor connector
11. Pressure accumulator
12. Hydraulic unit

BentleyPublishers.com—All Rights Reserved

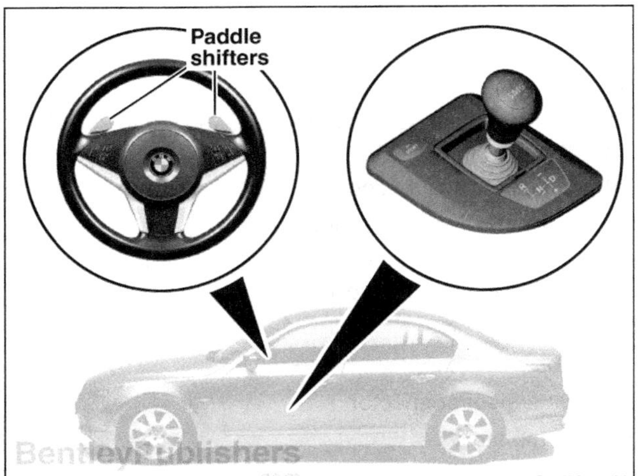

B510231002

Selector lever

SMG can be shifted using either the center-console-mounted selector lever or steering-wheel-mounted shift paddles.

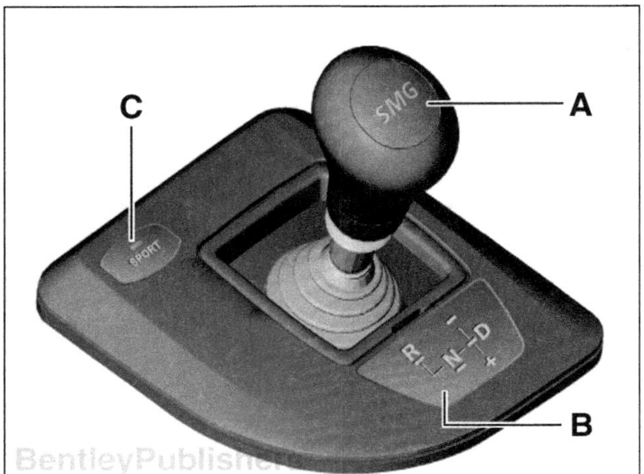

B510231003

SMG operation modes are controlled by the selector lever (**A**). The gear position display (**B**) indicates the modes that can be selected:

- N = Neutral position.
- R = Reverse gear. With foot on brake pedal, press selector down, left and forward.
- + or - = Manual mode. Flick selector forward or backward to change gears.
- D = Automatic mode.

Pressing the SPORT button (**C**) in either manual or automatic mode engages dynamic driving control (FDC). This intervention in engine and transmission controls reduces gear shift times. In automatic mode, up-shifts occur at higher engine speeds.

The SPORT button also switches the steering system Servotronic or active steering setting to sports mode.

The selector lever assembly is equipped with shiftlock function.

Shift paddles

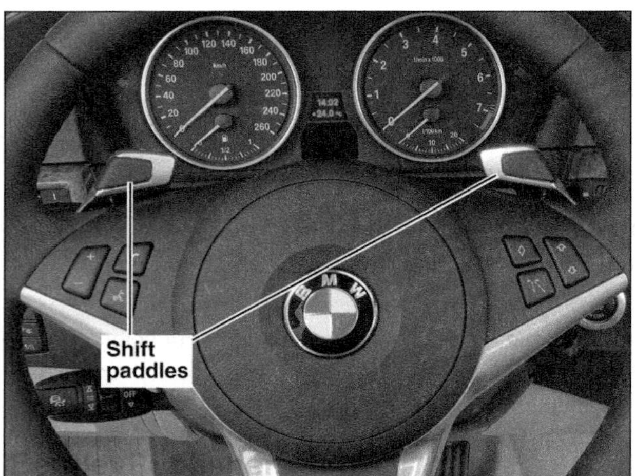

B510231004

The steering wheel with 2 gearshift paddles allows manual gear changes without the driver having to take hands off the steering wheel. Pull one or both of the paddles to shift up a gear. Press one or both of the paddles to shift down a gear.

The two gearshift paddles are identical in construction and are wired in parallel.

Power is supplied to the gearshift paddles from the steering column switch cluster (SZL).

BentleyPublishers.com
BentleyPublishers.com—All Rights Reserved

Manual mode

Gears are selected sequentially by briefly pressing the selector lever or shift paddle(s) in + or - direction. Gears can be skipped by rapidly pressing the shifter repeatedly.

Gears are changed when a shift pulse is transmitted electrically to the SMG control module. With the shift-by-wire gearshift method, there is no mechanical linkage between the gear lever or gearshift paddles and the transmission.

There is no clutch pedal. The clutch is operated automatically.

Automatic mode

When the selector lever is in D, shift points are automatically controlled according to characteristic curves and adapted to the driving situation.

SMG control module

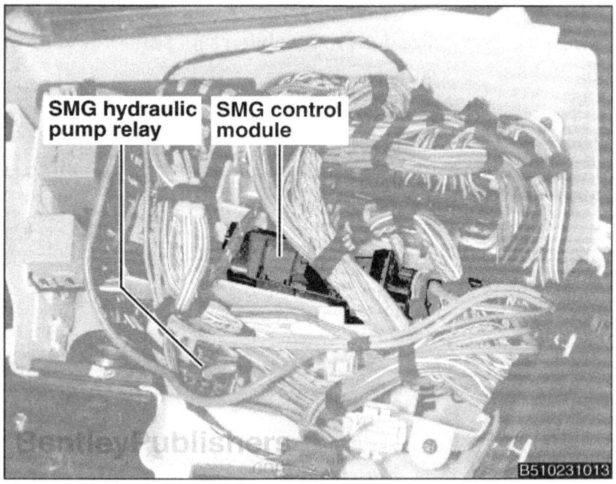

The SMG control module, in E-box at right rear of engine compartment, computes solenoid actuation for clutch operation and gear change. The control module also communicates with other control modules (for example, ECM).

The SMG control module can be programmed and coded.

Hydraulic unit

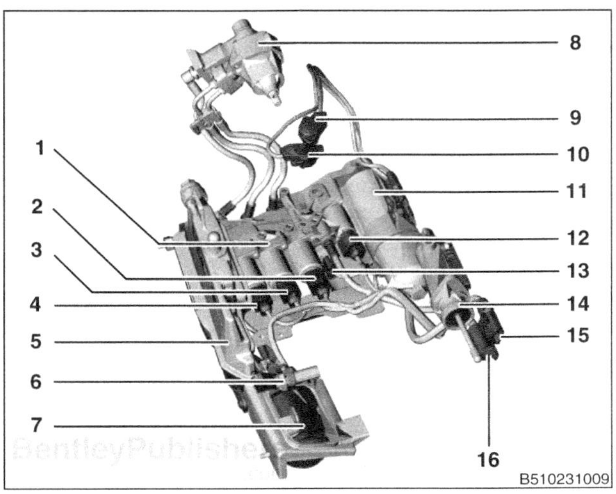

The hydraulic unit assembly, bolted to the bottom of the transmission sump, is precharged and tested. Within this assembly are the actuators for transmission operation.

1. Valve block
2. Solenoid MV2: 2nd, 4th, 6th gear shifts
3. Solenoid MV1: 1st, 3rd, 5th gear shifts
4. Solenoid MV0: Clutch actuator
5. Expansion tank
6. Transmission input speed inductive sensor
7. Hydraulic pump
8. Gearshift and selector angle actuator
9. Shift level sensor
10. Selector angle sensor
11. Pressure accumulator
12. Solenoid MV3: Selector angle brake, pressure relief valve
13. Hydraulic pressure sensor
14. Clutch actuator with clutch travel sensor
15. Hydraulic pump power supply
16. Hydraulic unit components electrical connector (4 solenoid valves, hydraulic pressure sensor)

BentleyPublishers.com–All Rights Reserved

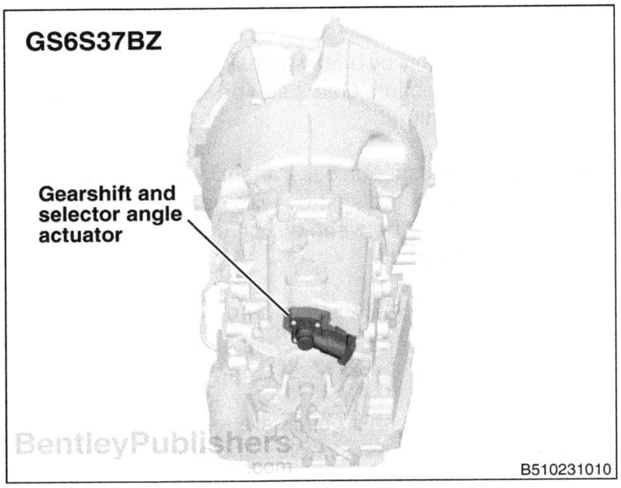

Gearshift and selector angle actuator

The gearshift and selector angle actuator shifts the selector shaft along its longitudinal axis and turns the selector shaft to change shift gate.

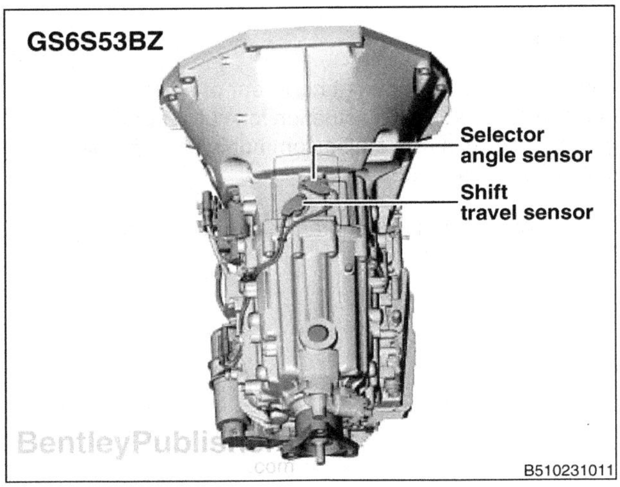

Shift travel sensor and selector angle sensor

The gear engaged is recognized from the shift travel sensor and the selector angle sensor:

- The shift travel sensor detects the forward and backward movements of the selector shaft in the shift gate.
- The selector angle sensor detects how far the selector shaft turns, and so recognizes a change of shift gate.

GS6S53BZ transmission illustrated. GS6S37BZ transmission is similar.

Pressure accumulator

The pressure accumulator, fitted on the lower left of the transmission, ensures that there is always sufficient oil pressure for a gearshift to be executed.

A freely moving piston subdivides the pressure accumulator into 2 chambers. One chamber is permanently charged with nitrogen at approx. 27 bar and -25°C. The other chamber fills with oil from the hydraulic pump. During transmission operation, the oil exerts pressure on one side of the piston, compressing the nitrogen and maintaining oil pressure during shifting.

Hydraulic pressure sensor

The hydraulic pressure sensor is installed directly on the hydraulic unit valve block and reports the current oil pressure to the SMG control module. The hydraulic pump is switched ON and OFF as required.

BentleyPublishers.com
BentleyPublishers.com—All Rights Reserved

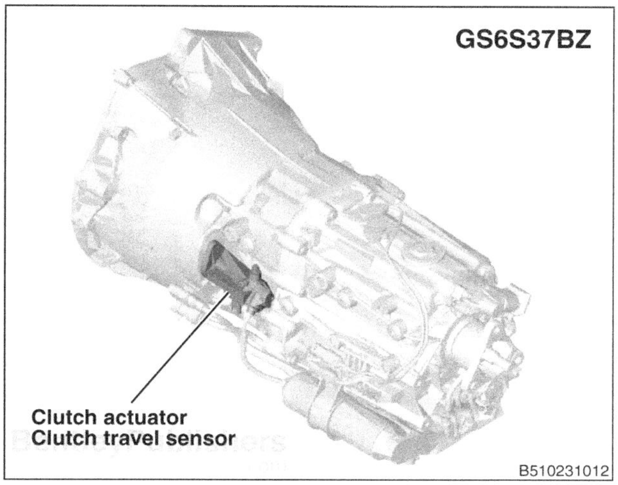

Clutch actuator with clutch travel sensor

The clutch actuator executes clutch operations. The clutch travel sensor registers clutch release travel.

Hydraulic pump with electric motor

The hydraulic pump with electric motor generates 55 bars (798 psi) of system pressure for gearshifts. The pump is actuated by the SMG control module.

Transmission input speed sensor

The SMG control module compares transmission speed, measured by the sensor, with engine speed and computes clutch closing speed (i.e., how fast the clutch can establish the optimal power flow between the engine and the transmission).

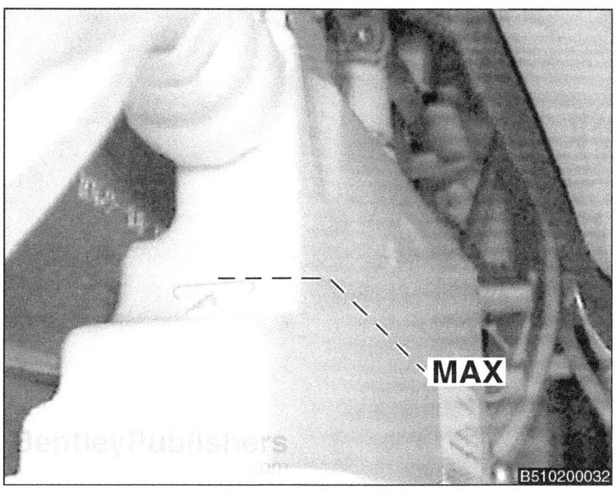

Expansion tank

The expansion tank ensures that there is always sufficient hydraulic fluid available in the system. There are MAX and MIN marks on the expansion tank for adjusting oil level.

Valve block

The valve block contains the solenoid valves, the hydraulic pressure sensor, the pressure accumulator and a check valve.

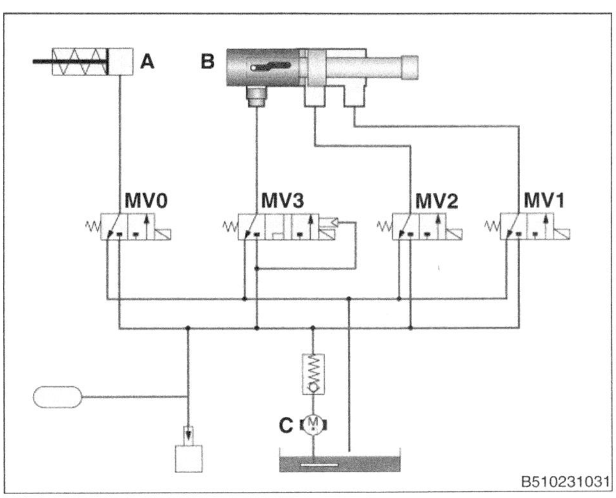

Solenoid valves

Four solenoid valves are installed in the valve block:

- Solenoid MV0 controls clutch actuator (**A**).
- Solenoid MV1 (shift travel valve) controls shifting for 1st, 3rd, 5th and reverse gears.
- Solenoid MV2 controls shifting for 2nd, 4th, and 6th gears.
- Solenoid MV3:
  —Controls selector angle brake for switching shift gate (**B**).
  —Serves as pressure relief valve for hydraulic system.

Pressurized fluid to solenoid valves is supplied by a hydraulic pump (**C**).

BentleyPublishers.com
BentleyPublishers.com—All Rights Reserved

Back-up light function

Back-up lights are switched ON based on messages from the SMG control module to the light module (LM).

- To 09 / 2005: SMG control module -> PT-CAN -> safety and gateway module (SGM) -> K-CAN -> LM
- From 09 / 2005: SMG control module -> PT-CAN -> body-gateway module (KGM) -> K-CAN -> LM

Instrument cluster display

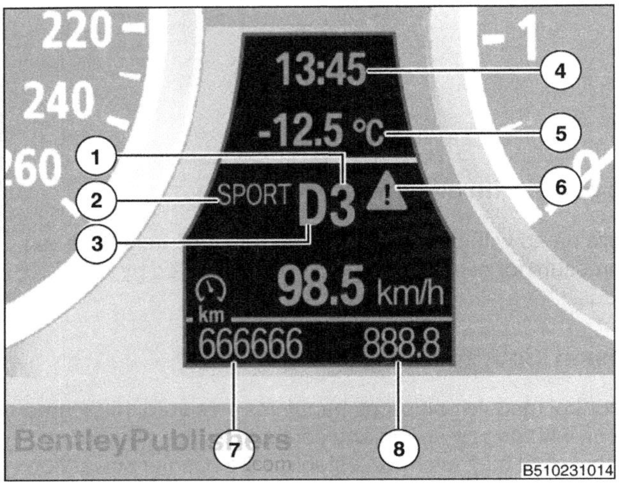

The following information is displayed on the instrument cluster LCD display:

1. Engaged gear
2. SPORT button activated for dynamic drive control (FDC)
3. Engaged drive position
4. Time
5. Ambient temperature
6. Warning triangle indicates Check Control message are present. See **020 Maintenance** for an explanation of Check Control.
7. Total mileage
8. Trip mileage

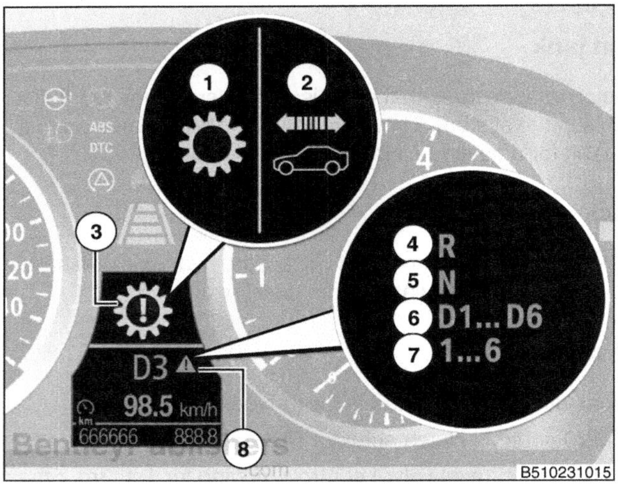

Details of SMG display in LCD:

1. Clutch overheated
   Select N position, then depress brake to start engine
   Select gear again (yellow symbol)
2. Transmission in N: Vehicle can roll (yellow symbol)
3. Emergency program (limp-home mode) activated: Transmission malfunction (yellow or red symbol)
4. Reverse gear
5. Neutral (idle)
6. Automatic mode (D1 - D6)
7. Manual mode: Steptronic (1st - 6th gear)
8. Check Control message(s) present

Fault messages (from SMG control module via PT-CAN, SGM or KGM and K-CAN) illuminate Check Control warning. Check Control messages are displayed in the upper LCD display when the Check Control button on the turn-signal stalk switch is pressed.

Brake light switch

The brake light switch signal is transmitted via PT-CAN by the ECM. For safety reasons, the signal is also transmitted by a direct wire from the brake light switch to the SMG control module. The module uses the signal for the following:

- Shift lever lock (shift-lock) release.
- Brake application detection.
- Shift to N when engine is shut down.

BentleyPublishers.com
BentleyPublishers.com—All Rights Reserved

Starter inhibitor

Car access system (CAS) enables the starter when the SMG control module satisfies the following conditions:

- Gear shift lever in position N (neutral). Instrument cluster LCD shows N, indicating that transmission is also in neutral.
- Clutch is disengaged.

SMG functions

SMG functions include the following:

- Engine start
- Gearshift
- Driving in gear lever position D (automatic mode)
- Launch control (acceleration assist system)
- Rock free (e.g., from snow)
- Vehicle standstill
- Gear lever position N and ignition OFF
- Emergency program (limp-home mode)

Engine start. To start the engine, the gear lever and transmission are in N (neutral). When the engine is started, sufficient system pressure immediately builds up to disengage the clutch. For this reason, the SMG control module is activated by a wake-up signal from CAS when the vehicle is unlocked.

The SMG control module measures system pressure via the hydraulic pressure sensor. If system pressure is too low, the hydraulic pump is actuated and builds pressure. When system pressure is reached, the control module switches the hydraulic pump OFF again.

With the prompt signal, the gear lever position indicator in the center console is activated. If terminal 15 (wake-up wire) is not activated approx. 15 minutes after vehicle unlocking, the gear lever position display is switched OFF again.

Gearshift. If a gearshift is initiated in manual or automatic mode, the SMG control module actuates the corresponding solenoid valve in the valve block. The pressurized hydraulic fluid flows into the clutch actuator and opens or closes the clutch.

At the same time, the SMG control module sends information to the ECM that a gearshift is initiated and that engine torque is to be reduced.

In the next step, the gearshift and selector angle actuator is actuated. At the same time, the gearshift and selector angle actuator executes forward, backward and side-to-side movements of the selector shaft.

During the gearshift, engine speed may be raised as required to allow clutch engagement to be completed more quickly and more smoothly.

When the gear is engaged, the clutch automatically engages. The SMG control module receives precise clutch position from the clutch travel sensor.

BentleyPublishers.com
BentleyPublishers.com—All Rights Reserved

After the gear change, the solenoid valves are actuated without the use of current. The course of the entire gearshift depends on accelerator pedal position, road speed, engine speed and other factors (e.g., engine oil temperature).

Just before the vehicle reaches a standstill, SMG shifts into 1st gear and disengages the clutch. The transmission of power between engine and transmission is broken.

Gearshifts that result in an excessively high or low engine speed are suppressed.

Gear lever position D (automatic mode):

- Constant-speed travel. Continuous changes of accelerator pedal position (opening and closing the throttle) are registered in the SMG control module. The control module suppresses a pendulum shift between adjacent gears (up and down, e.g., on a highway with constantly changing speed).
- Upshift suppression. Based on accelerator pedal position, the SMG control module recognizes when the throttle is closed rapidly, e.g. before applying the brake. The module suppresses the upshift. This ensures that engine braking stays in effect.
- Driving uphill. The SMG control module recognizes when the vehicle is driving uphill from the data from the wheel speed sensors and accelerator pedal position The module prevents unnecessary upshifts and irritating pendulum shifts (e.g., at constantly high engine speed and foot taken off the accelerator pedal).
- Driving downhill. The SMG control module recognizes when the vehicle is driving downhill from throttle valve potentiometer (load), wheel speed sensor and brake light switch signals. To maintain and increase engine braking effect, the control module shifts down a gear as road speed increases.
- Braking deceleration and automatic downshift. The SMG control module recognizes braking deceleration from brake light switch, wheel speed sensor and accelerator pedal position sensor signals. The control module shifts down according to road speed to maintain engine braking effect. When this happens, several gears may be skipped. The correct gear is automatically selected when the vehicle accelerates again.
- Manual intervention in automatic operation. When the gearshift lever or the gearshift paddles are operated in automatic mode, the system automatically changes to manual mode. If automatic mode is required, the driver selects it again.

Launch control (acceleration assist system). Launch control enables optimum vehicle acceleration from standstill. Upshifts are automatic even in manual mode when launch control is active. This prevents the engine from reaching critical speeds.

Preconditions for launch control consist of the following:

- High-grip road surface.
- Dynamic stability control (DSC) deactivated (DSC button pressed for longer than 3 seconds).
- SPORT button pressed.
- Accelerator pedal pressed to floor (kick down position)

BentleyPublishers.com
BentleyPublishers.com—All Rights Reserved

To deactivate launch control:

- Release accelerator pedal (release from kick down position).
- Manual shift request.
- Switch OFF dynamic driving control (FDC) using SPORT button.
- Activate DSC.

> ***CAUTION—***
> - *During the running-in phase of a new vehicle, do not accelerate the vehicle using launch control.*
> - *Frequent use of launch control leads to increased component wear.*

Rocking free. To rock the vehicle free (e.g., from snow) move gearshift lever rapidly from 1st to R and then back. Complete the gearshift within the space of 0.8 seconds. Otherwise the gear lever lock (shift-lock) becomes activated.

Vehicle standstill. The clutch is disengaged and 1st gear selected when the SMG control module detects that the vehicle is stationary. Vehicle standstill is detected by means of the accelerator pedal position, wheel speeds and engine speed. The clutch is disengaged and 1st gear selected irrespective of the drive program (manual or automatic mode).

An audible warning is triggered immediately if the driver door is opened while the engine is running and the transmission is not in position N or P. At this time, N also flashes nine times in the instrument cluster LCD.

If, despite these warnings, neither the pedals, the gearshift paddles nor the gearshift lever are moved, the selected gear is automatically cancelled after approx. 2 seconds.

Gear lever position neutral and ignition OFF. When the gear lever is in N (neutral) and ignition is switched OFF, N flashes in the instrument cluster LCD. In addition, the audio system controller (M-ASK) emits an acoustic warning.

Emergency program. In the emergency program (limp-home mode), the vehicle remains operational, but with limitations. Only gearshift positions R, N and 1st - 3rd gears are available. Depending on the type of transmission fault, the vehicle may no longer be operational after a vehicle standstill.

BentleyPublishers.com
BentleyPublishers.com–All Rights Reserved

Cautions

CAUTION—
- *When performing any repair which involves separating the engine and transmission, check that the bell housing dowels are undamaged and properly in place before reassembly. If the alignment of the engine flywheel to the transmission input shaft is incorrect, any of the following complaints may result:*
  —Clutch slipping, shudder, or poor disengagement.
  —Noise from transmission input shaft.
  —Transmission popping out of gear.
  —Difficulty changing gears.
  —Internal damage to transmission.
- *To avoid electrochemical corrosion to engine components made of aluminum-magnesium alloy, do not use steel fasteners. Use aluminum fasteners only.*
- *The end faces of aluminum fasteners are usually painted blue. For reliable identification, test fasteners for aluminum composition with magnet.*
- *Replace aluminum fasteners each time they are loosened.*
- *Follow torque instructions, including angle of rotation specifications, when installing aluminum fasteners.*
- *Fill manual transmission only with BMW approved synthetic oil. Filling with incorrect lube could cause the following damage:*
  — Premature synchromesh ring wear.
  — Reduced service life of bearings.
  — Tooth profile damage.
  — Faster wear of selector forks.
  — Damage to radial shaft oil seals.
- *When removing interior trim clips, avoid damaging trim pieces by using a plastic prying tool or a screwdriver with the tip wrapped with masking tape.*

TRANSMISSION FLUID SERVICE

SMG transmissions installed in 5 Series models use lifetime manual transmission fluid and do not require periodic fluid changes.

Table c. SMG fluids		
Transmission	**Year range**	**Fluid type**
ZF GS637SBZ ZF GS6S53BZ	2004 - 2005 from 2006	MTF-LT-2 MTF-LT-3
MTF-LT-2 (Lifetime fluid) BMW part no. 83 22 0 309 031		
MTF-LT-3 (Lifetime fluid) BMW part no. 83 22 0 409 878		
Clutch hydraulic fluid	Pentosin CHF 11S	

Transmission fluid level, checking

Checking the transmission fluid level involves removing the side fill plug and inserting a gloved finger into the hole to check the fluid level. If the fluid level is up to the bottom of the fill hole (finger is wetted by transmission fluid), the level is correct.

BentleyPublishers.com
BentleyPublishers.com—All Rights Reserved

Transmission fluid, replacing

- Drive vehicle for a few miles to warm transmission.
- Raise and safely support vehicle to access drain plug.

> ***CAUTION—***
> *• Make sure the car is stable and well supported at all times. Use a professional automotive lift or jack stands designed for the purpose. A floor jack is not adequate support.*

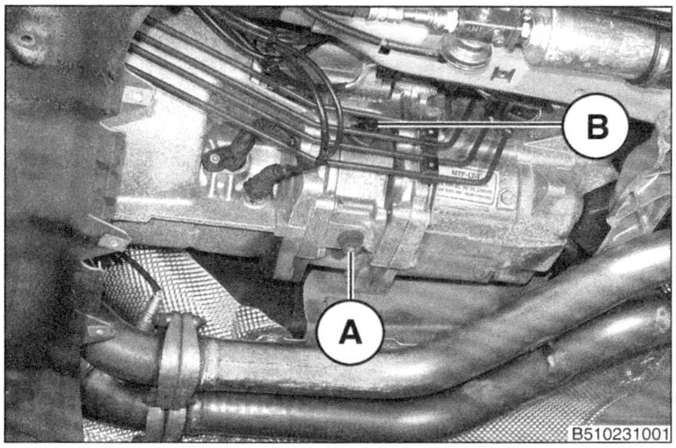

Place a drain pan underneath transmission and remove drain plug (**A**) at bottom of transmission.

- Drain fluid.
- Install and torque drain plug.
- Remove fill plug (**B**) from side of transmission.
- Slowly fill transmission with fluid until fluid overflows fill hole.

231

Table d. SMG fluid capacities		
Transmission	**Initial fill, new or rebuilt unit**	**Fluid change**
ZF GS6S37BZ	1.6 liter (1.7 US qt)	1.5 liter (1.6 US qt)
ZF GS6S53DZ	1.6 liter (1.7 US qt)	1.5 liter (1.6 US qt)

- Install and torque fluid fill plug.

Tightening torque	
• Drain or fill plug to transmission housing	35 Nm (26 ft-lb)

Clutch hydraulic fluid, filling

The clutch in the SMG system is operated via hydraulic fluid.

- Switch ignition ON without starting vehicle.
- Use BMW scan tool to depressurize clutch hydraulic system.

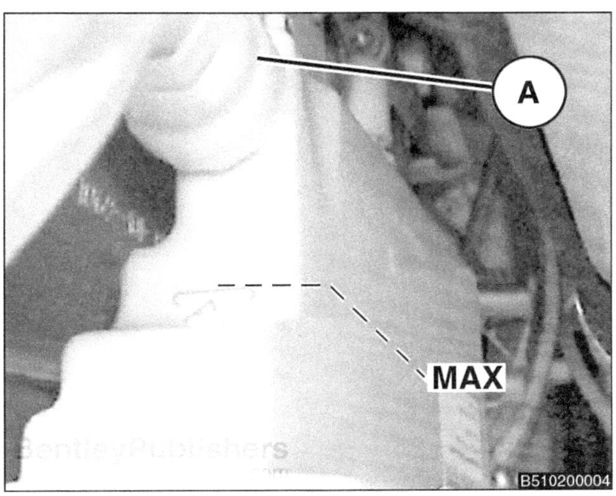

Working at side of transmission, open hydraulic fluid expansion tank (**A**) and fill to MAX level with **Pentosin CHF 11S** hydraulic fluid.

BentleyPublishers.com—All Rights Reserved

Transmission Removal and Installation

Removal and installation of the transmission is best accomplished with the vehicle on a lift and using a transmission jack. Support engine using equipment which allows the engine to pivot on its mounts to access the upper bolts at the transmission bell housing.

> ***WARNING—***
> - *The removal of the transmission may upset the balance of the vehicle on a lift.*

Transmission, removing and installing

- Disconnect negative (-) cable from battery.

> ***CAUTION—***
> - *Prior to disconnecting the battery, read the battery disconnection cautions given in* **001 Warnings and Cautions**.

- Raise and safely support vehicle.

> ***WARNING—***
> - *Make sure the vehicle is stable and well supported at all times. Use a professional automotive lift or jack stands designed for the purpose. A floor jack is not adequate support.*

- Remove underbody splash shields.
- Remove front end reinforcement plate.
- Remove complete exhaust system. See **180 Exhaust System**.

> ***WARNING—***
> - *Make sure exhaust system is fully cooled off before removing.*

- Remove oxygen sensors and store in safe location.

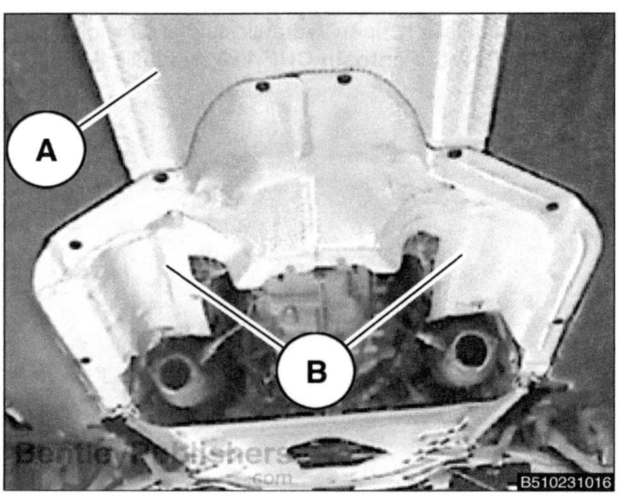

◀ Working underneath vehicle, remove heat shields (**A**, **B**)

BentleyPublishers.com
BentleyPublishers.com—All Rights Reserved

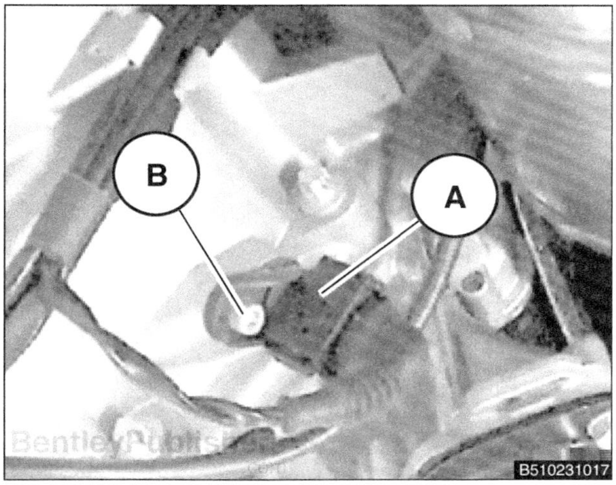

Working at crankshaft sensor:

- Detach electrical connector (**A**).
- Remove mounting bolt (**B**) and remove crankshaft sensor.

Detach electrical connector (**arrow**) from oil level sensor. Unclip harness from harness support clips.

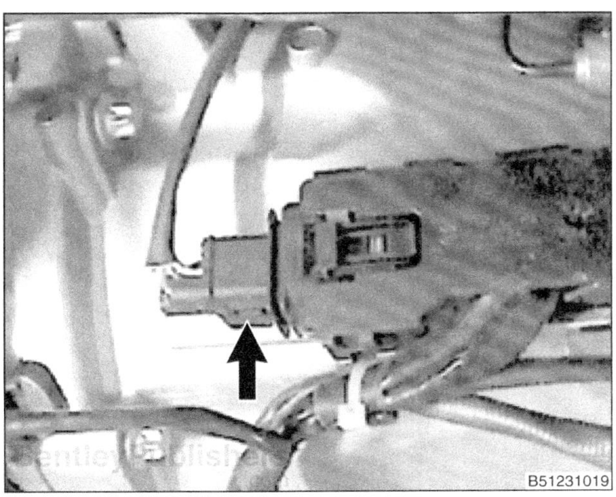

Detach electrical connector (**arrow**) from hydraulic unit components electrical plug.

BentleyPublishers.com–All Rights Reserved

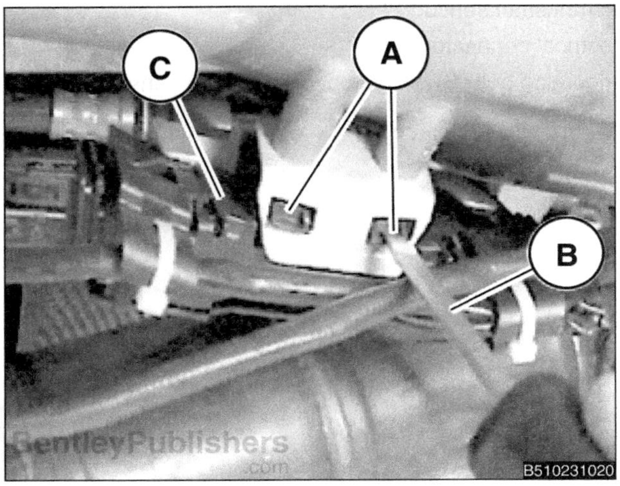

Press in retaining lugs (**A**) with screwdriver (**B**) and remove cable duct (**C**). Tie wire loom to one side.

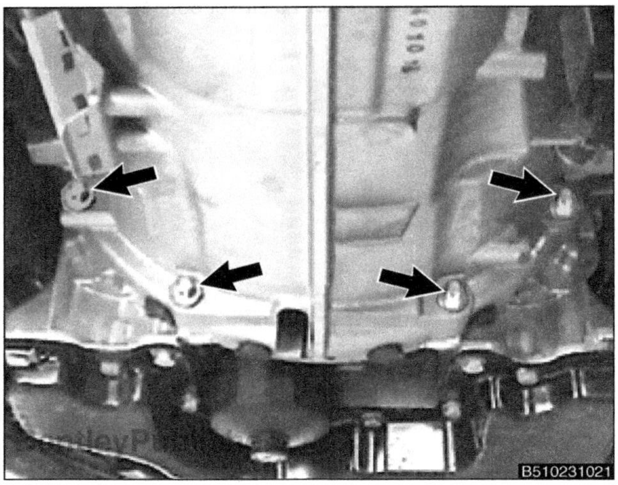

Remove transmission bell housing bolts (**arrows**).

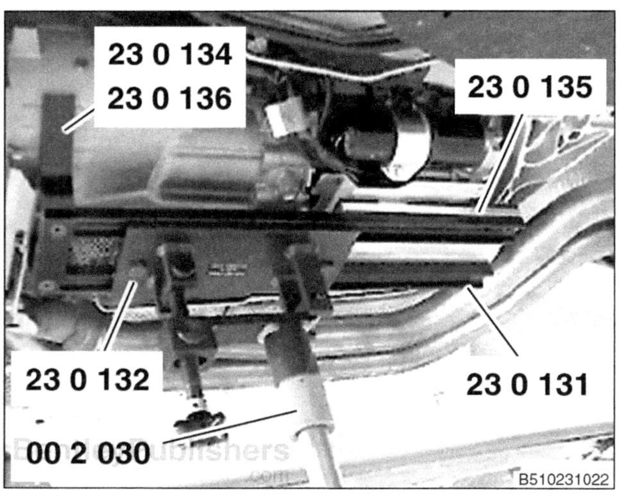

Support transmission with BMW special tools 00 2 030, 23 0 123, 23 0 131, 23 0 134, and 21 0 135.

BentleyPublishers.com
BentleyPublishers.com—All Rights Reserved

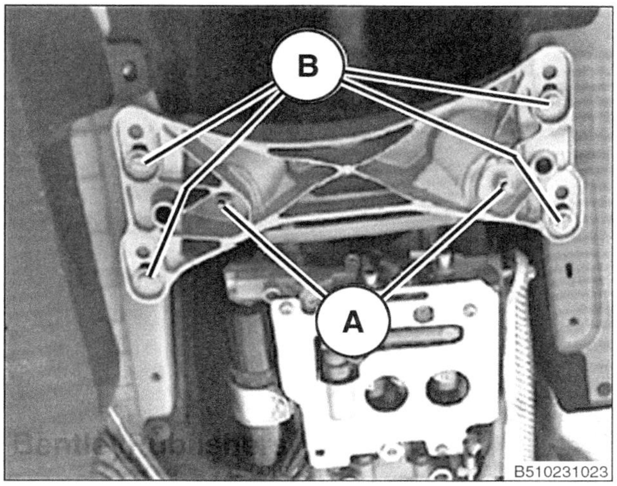

◀ Remove transmission mounting fasteners (**A**). Remove transmission rear support fasteners (**B**). Remove support.

– Remove rear driveshaft from transmission output flange:
  - Remove driveshaft center bearing support fasteners. Support center of driveshaft.
  - Unbolt rear driveshaft flex-disc from transmission flange.
  - Lower center of driveshaft sufficiently to disengage flex-disc from transmission output flange. Tie driveshaft to side.

> ***CAUTION—***
> - *To prevent damage to rear driveshaft CV joint, do not allow driveshaft to hang down unsupported.*

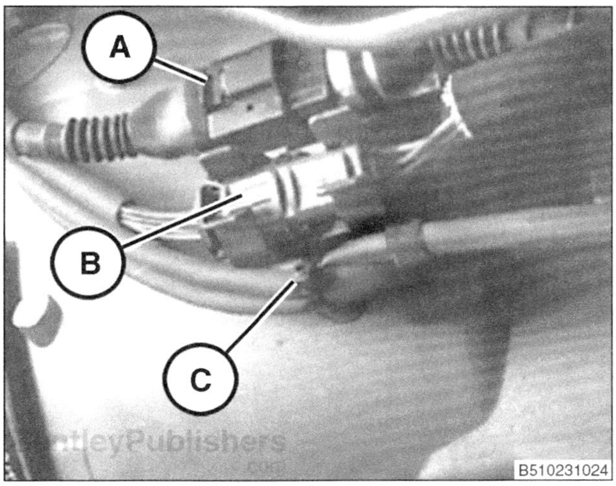

◀ Disconnect electrical harness connectors (**A**) and (**B**). Cut wire tie (**C**).

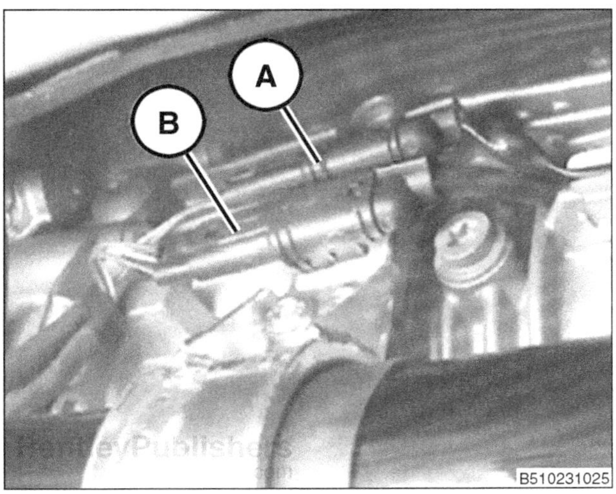

◀ Disconnect electrical harness connectors (**A**, **B**), uncoupling harness from bracket.

BentleyPublishers.com–All Rights Reserved

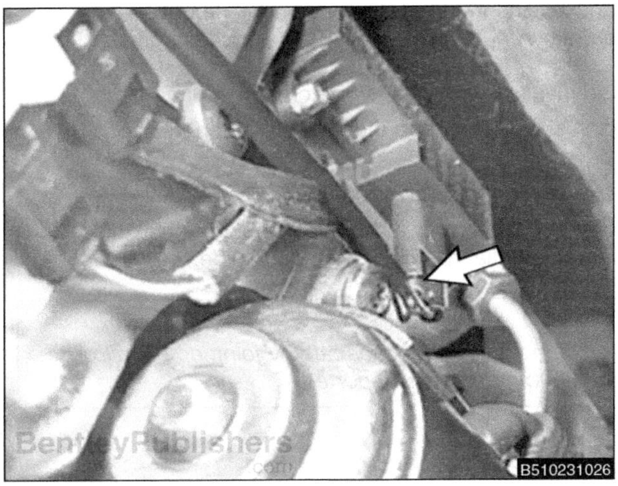
B510231026

Disconnect electrical harness connector (**arrow**) from clutch travel sensor.

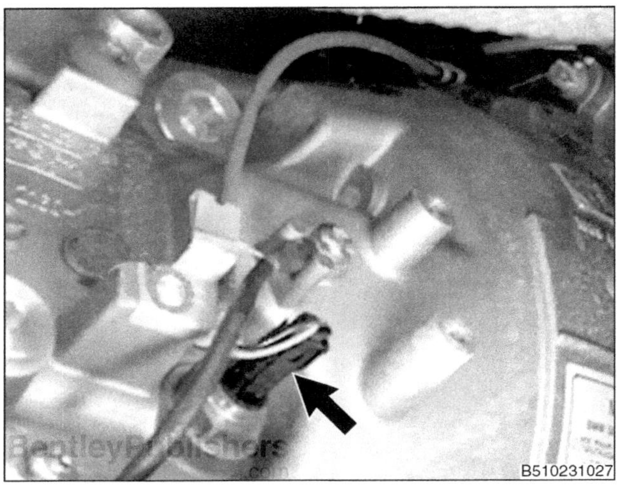
B510231027

Disconnect electrical harness connector (**arrow**) from sensor.

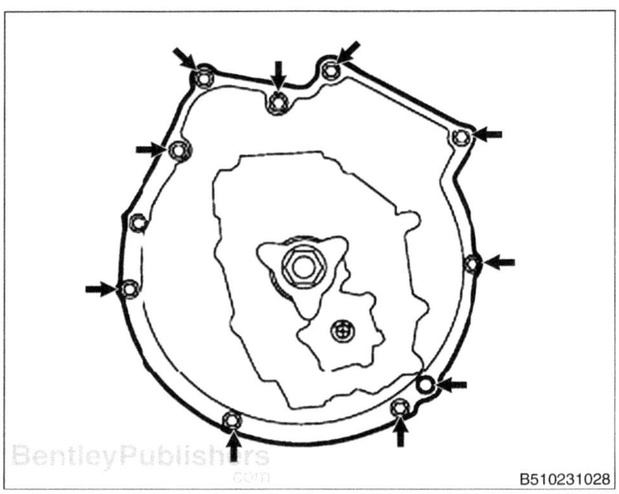
B510231028

Remove bell housing to engine block mounting bolts (**arrows**) and slide transmission off engine.

> ***CAUTION—***
> - *Do not allow transmission to hang on input shaft while removing and installing as the clutch disc can be damaged.*
> - *Tilting the engine to lower the transmission can lead to damage to various components in the engine compartment due to lack of clearance.*

BentleyPublishers.com
BentleyPublishers.com–All Rights Reserved

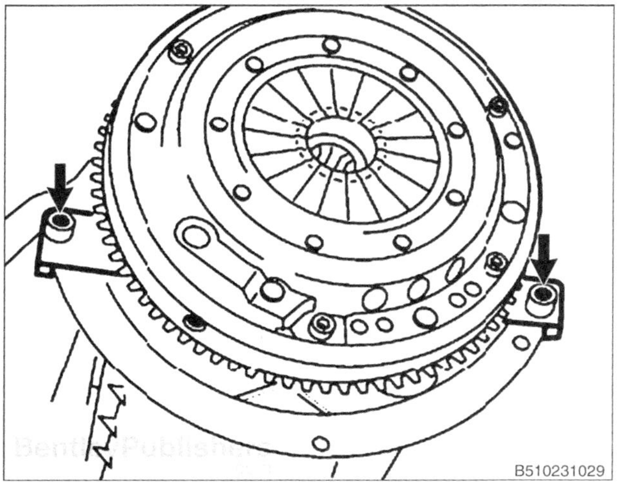

Before installing transmission, make sure bell housing centering dowels (**arrows**) are intact and lined up correctly. Replace if damaged.

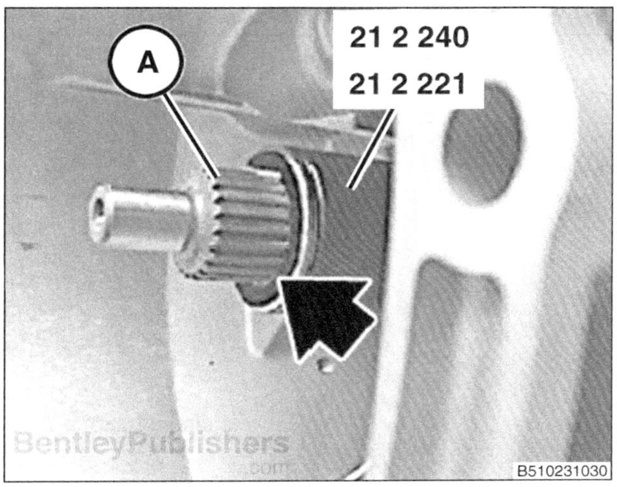

Remove and clean release bearing and release lever.

- Push grease scraper, BMW special tool 21 2 221, on input shaft (**A**) as far as it will go.
- Grease splines of input shaft, using Unirex S2 grease.
- Slide grease scraper off input shaft.

– Installation is reverse of removal, keeping in mind the following:
  - When installing a new transmission, be sure to transfer parts from old transmission, if applicable.
  - Install driveshaft and preload center bearing bracket. Use new nuts when mounting driveshaft to transmission flex disc and final drive. See **260 Driveshafts**.
  - Refill transmission with appropriate lubricant before starting or towing car. See **Transmission Fluid Service** in this repair group.
  - Check and clear fault codes from ECM and SMG control module memory.

Tightening torques	
Front end reinforcement to front end (use new fasteners): • Stage 1 • Stage 2	 56 Nm (41 ft-lb) additional 90°
Slave cylinder to transmission	22 Nm (16 ft-lb)
Transmission rear support to body (M8)	19 Nm (14 ft-lb)
Transmission mount to rear support (M8)	19 Nm (14 ft-lb)
Transmission to engine (use new aluminum fasteners): • M10 x 30 mm Stage 1 Stage 2 • M10 x 85 mm Stage 1 Stage 2 • M12 Stage 1 Stage 2	 20 Nm additional 90° - 110° 20 Nm additional 180° - 200° 25 Nm additional 130° + 20°
Vibration damper to transmission rear support (M10)	38 Nm (28 ft-lb)

BentleyPublishers.com
BentleyPublishers.com—All Rights Reserved

240 Automatic Transmission

GENERAL

This repair group covers maintenance and replacement of the automatic transmission.

Automatic transmission internal repairs are not covered. Transmission repairs require special service equipment and knowledge. If transmission internal service is required, consult an authorized BMW dealer about a factory reconditioned unit or a transmission rebuild.

See also:
- **119 Lubrication System** for crankshaft rear main (flywheel) seal.
- **170 Radiator and Cooling System** for ATF cooler and heat exchanger.
- **200 Transmission–General** for drivetrain information.
- **250 Gearshift Linkage** for manual override of gearshift lock
- **260 Driveshafts** for front (xDrive) and rear driveshaft.
- **270 Transfer Case** for xDrive equipped models.

BentleyPublishers.com
©BentleyPublishers.com–All Rights Reserved

Automatic transmission applications

The 6-speed automatic transmission in 5 Series vehicles was jointly developed by BMW and ZF (*Zahnradfabrik Friedrichshafen*). Application information is in **Table a**.

Table a. Automatic transmission applications

Model (engine)	Year	Transmission	Max. torque
525i, 530i (M54)	2004 - 2005	GA6HP19Z	350 Nm (258 lb-ft)
525i, 525xi, 530i, 530xi, (N52)	2006 - 2007	GA6HP19Z	400 Nm (295 lb-ft)
545i, 550i (N62, N62TU)	2004 - 2007	GA6HP26Z	600 Nm (443 lb-ft)
528i, 528xi, 535i, 535xi, (N52K, N54)	2008* - 2000	GA6HP19ZTU	400 Nm (295 lb-ft)
550i (N62TU)	2008* - 2009	GA6HP26ZTU	650 Nm (479 lb-ft)

* *2008 model year starts with 03 / 2007 production.*

Automatic transmission assembly

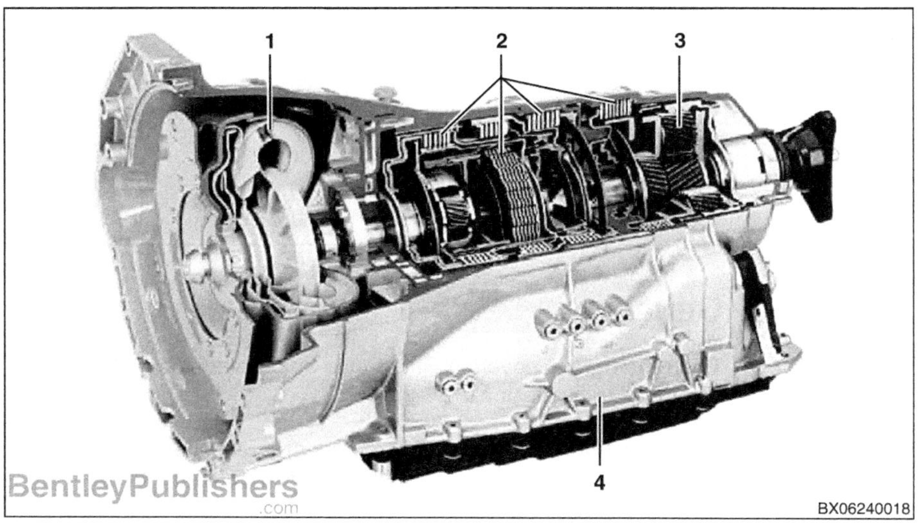

1. Torque converter
   - Capable of torque multiplication
2. Multi-plate clutch packs
3. Planetary gear assembly
4. Mechatronics module
   - Valve body (hydraulic unit)
   - Transmission control module (A7000a)

Automatic transmission features

The 6-speed automatic transmission system features the following components or signals:

- Mechatronics module
- Steptronic shift lever (2004 - 2007)
- Gear shift switch (GSW) (2008 - 2009)
- Gearshift paddles (2008 - 2009)
- Instrument cluster display of gear range
- Brake light switch signal
- ECM signal
- DSC signal
- Accelerator pedal signal
- Bus connections

BentleyPublishers.com
BentleyPublishers.com—All Rights Reserved

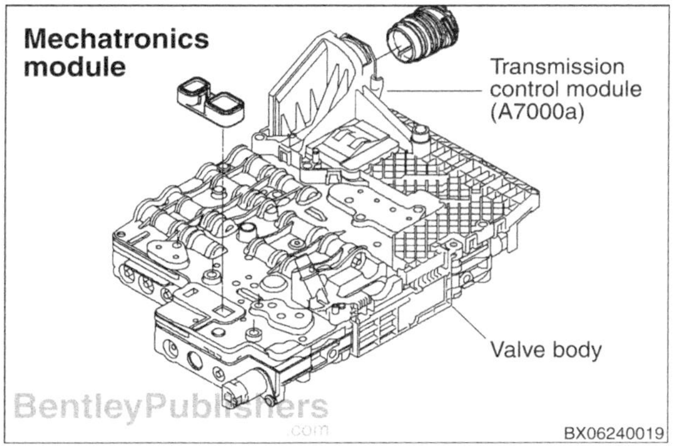

Mechatronics module

Mechatronics module is inside the transmission housing, combining hydraulic unit (valve body) and transmission control module (TCM) (A7000a). TCM operating software is known as electronic transmission control (EGS).

The TCM includes the following components and functions:

- Power supply for solenoid valves and electronic pressure control valves.
- Transmission sensor system:
  —Transmission input speed sensor.
  —Transmission output speed sensor.
  —Transmission oil temperature sensor.
- Mechatronics connector for connection to vehicle wiring harness

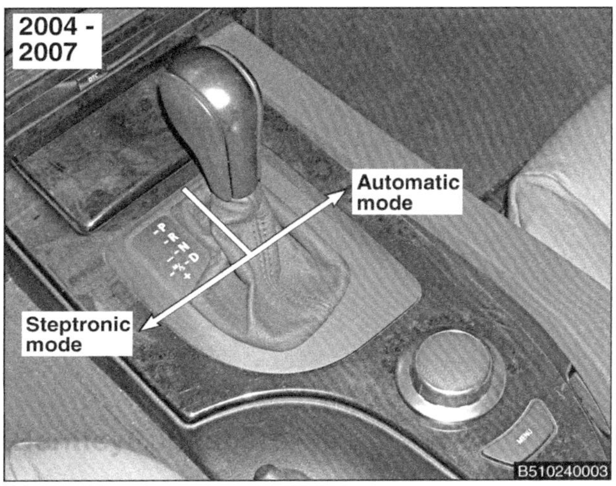

Steptronic shift lever

Steptronic shift lever is equipped with positions P(ARK), R(EVERSE) and N(EUTRAL) as well as D(RIVE) for automatic mode and M/S for manual and Steptronic mode.

The Steptronic function makes it possible to shift the automatic transmission manually. Manual mode is engaged when the shift lever is moved left from automatic gate into manual gate. In manual mode, pressing the shift lever forward or backward closes electrical contacts, resulting in upshift or downshift.

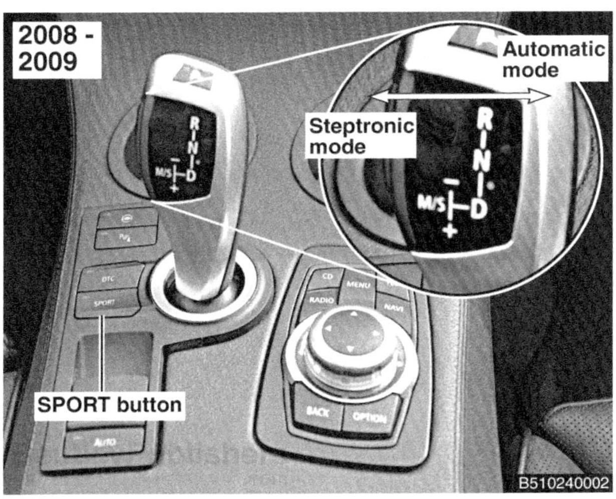

Gear shift switch (GWS)

In models built from 03 / 2007 (2008 model year), the shift lever is configured as a control module referred to as gear shift switch or GWS. Transmission gear range selection in these models is actuated electronically instead of via shift cable.

From 06 / 2007, a Sport automatic transmission is also available, featuring a SPORT button on the shifter console. Dynamic driving control (FDC) is activated with the SPORT button: Gearshift times are shortened and shift characteristics are made more sporty.

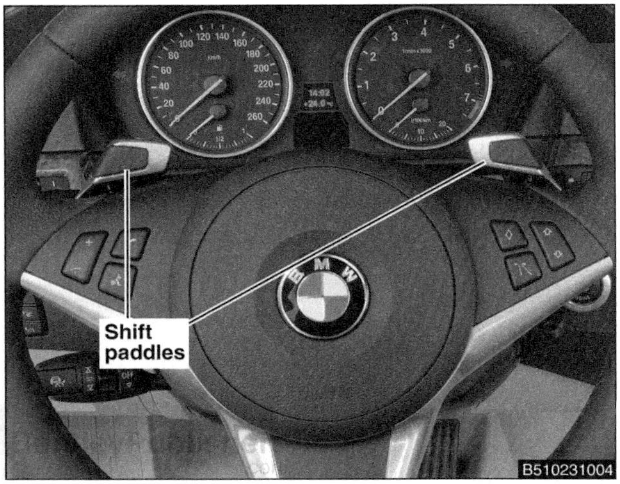

Gear shift paddles

From 06 / 2007 (2008 model year), the Sport automatic transmission also features gearshift paddles. In addition to sequential gear selection with the shift lever, the gears can also be selected using gearshift paddles on the steering wheel.

A shift pulse is transmitted electrically to the TCM. With this shift-by-wire method there is no mechanical linkage between the gearshift paddles and the transmission.

Pull one of the gearshift paddles briefly to change UP a gear. Press one of the gearshift paddles briefly to change DOWN a gear.

Instrument cluster display

Instrument cluster displays transmission gear range as well as Check Control message via K-CAN. Check Control also alerts the driver with acoustic signals. See **020 Maintenance** for a discussion of Check Control.

Brake light switch signal

Brake light switch signal affects function of shiftlock and adaptive transmission control.

Engine control module (ECM) signal

Engine control module (ECM) signals engine speed, torque and other parameters. TCM signals current transmission operating status back to ECM.

Dynamic stability control (DSC) signal

DSC control module signals cornering, acceleration and traction data.

Accelerator pedal signal

Accelerator pedal module signals driver power demand data.

Bus connections

Bus connections are used to handle electrical complexity throughout the 5 Series chassis. See **600 Electrical System–General** for details of bus systems.

BentleyPublishers.com
BentleyPublishers.com—All Rights Reserved

Automatic transmission functions

Automatic transmission functions include the following:
- Adaptive transmission control (AGS)
- Overlap control
- Starter inhibitor
- Parking lock
- Warm-up program
- Engine intervention
- Downshift inhibitor
- Reverse gear inhibitor
- Shift lever lock (shiftlock)
- Standstill decoupling
- Interlock
- Emergency program (limp-home mode)

Adaptive transmission control (AGS)

The transmission offers the choice between a comfort program and a sport program. AGS (a component of EGS) adapts transmission shift characteristics to the driver and the driving situation.

In both programs, AGS can switch from a basic shift map to a performance-orientated map, taking the following conditions into account:
- Comfort program (shift lever position **D**) uses 2 maps: XE (extreme economy) and E (economy).
- Sport program (shift lever position **M/S**) produces dynamic, sporty shift characteristics. There is a basic S (sport) map and a performance-orientated XS (extreme sport) map.

Other AGS functions are as follows:

Adaptation to driver. AGS adapts to different driving styles using the following criteria:
- Kick-fast. If the accelerator pedal is depressed rapidly, the shift program is changed to E in comfort program and XS in sport program.
- Cornering detection reacts to lateral acceleration with an indirect adaptation to the driver type. Lateral acceleration is considered an indirect statement of preferred driving dynamics and does not cause an immediate gearshift response (shifts when cornering could adversely affect driving stability). Lateral acceleration is calculated from wheel-speed, yaw rate and road speed sensor signals.
- Braking is evaluated in a similar way to kick-fast. Braking deceleration is measured and compared with threshold values in the TCM.
- Driver adaptation is restarted each time the vehicle pulls away from a standstill.

Driving uphill or trailer towing function is based on a comparison of vehicle acceleration with a nominal value. From the current engine operating situation, acceleration with normal load on a level surface is calculated. If the actual acceleration is significantly below

BentleyPublishers.com
BentleyPublishers.com—All Rights Reserved

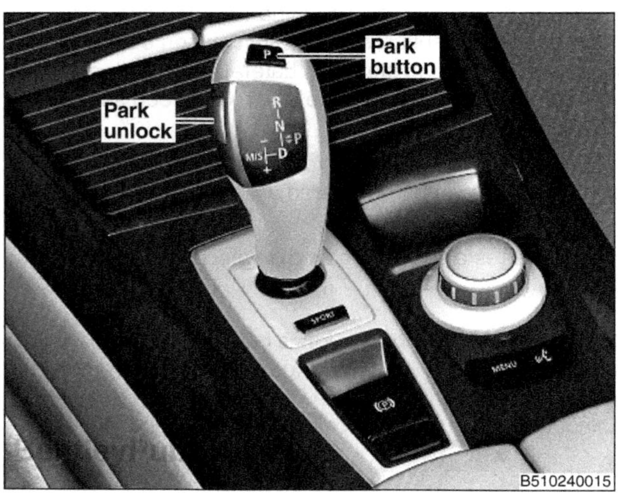

the theoretical value, driving uphill or trailer towing function is activated.

Driving downhill. If the TCM detects downhill driving, it automatically shift down a gear if vehicle speeds up. This enhances the engine braking effect. A downshift only occurs if the engine speed is below the maximum speed of the lower gear.

TCM detects downhill driving from signals from the throttle valve potentiometer (load), wheel speed sensors (road speed) and brake light switch (brake activation).

Braking deceleration and automatic upshift. To decelerate the vehicle, the foot is taken off the gas and the brake depressed as necessary. The gearshift map triggers an upshift when the throttle is closed. These gearshifts are not necessary in conjunction with brake applications as they prevent the engine braking effect from being exploited.

The intention to apply the brakes is anticipated from the accelerator pedal being rapidly released to the zero position. If such an action is detected, transmission upshift is suppressed for as long as the accelerator pedal is in the zero position and the vehicle is in overrun mode.

Winter program (only with shift lever in position D) is automatically activated if the drive wheels start to spin, even at a low rate of acceleration. The winter program ensures better traction by not using 1st gear. Early upshifts reduce the reactions caused by load changes. Winter program is deactivated if the wheels do not spin for several seconds despite high drive torque.

Overlap control

Overlap control provides gentle shift characteristics as follows: During a shift, pressure is reduced in the active multi-plate clutch and at the same time built up in the clutch that is about to be activated. Gradual reduction and build-up of pressure is maintained until the shift is complete. Overlap control is active for all gearshifts from 1st to 6th gear and from 6th to 1st gear.

Starter inhibitor

It is only possible to start the engine when vehicle is in P or N. Car access system (CAS) with electronic immobilizer (EWS) evaluates signals from the TCM via direct wire or CAN-bus signal indicating that shift lever is in P or N.

Parking lock

Parking lock feature locks the transmission output shaft so that the vehicle is prevented from rolling unexpectedly. Parking lock is designed to provide a reliable brake on gradients up to 32%.

2004 - 2007 models: When the vehicle is stationary, parking lock is engaged purely mechanically via Bowden cable from shift lever to mechatronics module.

2008 - 2009 models: The parking lock is signalled electronically via a button on the gear shift switch (shift lever). It is actuated via a mechanical spring and cable system.

BentleyPublishers.com
BentleyPublishers.com—All Rights Reserved

Parking lock is activated under the following conditions:

- Pressure on shift lever button and road speed signal less than 1.2 mph (2 kph).
- Ignition key removed from key slot and road speed signal of 0.
- Engine ON.
  —Shift lever in D or R.
  —Driver door open.
  —Drivers seat belt not fastened in belt buckle.
  —Pedals not depressed [road speed signal less than 2 mph (3 kph)].
- START / STOP button pressed while engine is ON.
  —Shift lever in D or R.
  —Road speed signal 0.

Parking lock is only released when the engine is ON because hydraulic pressure is required to open the parking lock. In case of power failure (such as dead battery) parking lock can be released manually. See **250 Gearshift Linkage**.

Warm-up program

If the engine is started and coolant temperature is lower than approx. 176°F (60°C), the warm-up program is activated. Until the engine warms up, the transmission remains in performance map E or XS, shifting gears at higher speeds. This enables the engine and catalytic converters to reach operating temperature more quickly.

Up to a transmission oil temperature of approximately 95°F (35°C), the torque converter lockup clutch is neither controlled nor closed.

Engine intervention

During a gearshift, the TCM emits signals to retard ignition timing for a few milliseconds. This action briefly reduces the torque, improves shift quality, reduces transmission load and shortens shift time.

Downshift inhibitor

TCM program prevents shifting to a lower gear until engine speed is below maximum speed for the next gear down. Engine speed signal is transmitted by the ECM to the TCM. The downshift inhibitor prevents damage to the engine and transmission.

Reverse gear inhibitor

Shifting into reverse gear is inhibited when the vehicle is moving forward at speeds above 3 mph (5 kph).

Shift lever lock (shiftlock)

The shift lever is locked in positions P or N by an electromagnet actuated by the TCM. Shiftlock engages with shift lever in P or N and ignition ON. Shifting from P or N into a drive gear or reverse is only possible at road speeds less than 3 mph (5 kph) and brake applied at engine speed below 2500 rpm.

BentleyPublishers
BentleyPublishers.com—All Rights Reserved

Standstill decoupling

The torque converter is decoupled when the vehicle is at a standstill. Thus only a minimum load remains on the engine and fuel consumption is reduced. Decoupling is effected by clutch action within the transmission.

Interlock

2004 - 2007 models: The ignition lock is mechanically linked to the shift lever by a Bowden cable. Interlock allows the ignition key to be removed only when the shift lever is in P. Conversely, the shift lever can only be moved from P when the ignition key is in the ignition lock and turned at least to ignition ON.

2008 - 2009 models: The remote key (ignition key) can only be removed when CAS registers a road speed signal below 0.6 mph (1 kph). If the remote control is removed from the ignition slot after the engine is switched OFF, the parking lock automatically applies.

Emergency program

The emergency program (limp home mode) activates if transmission management fails or detects a malfunction that could lead to critical driving conditions. In limp home, the vehicle remains operational but with limitations.

If TCM fails, the following forward gears can be engaged:

- Failure in 1st - 3rd gear: Emergency 3rd gear activated.
- Failure in 4th - 6th gear: Emergency 5th gear activated.
- Engine restart is possible in 3rd gear.

Torque converter

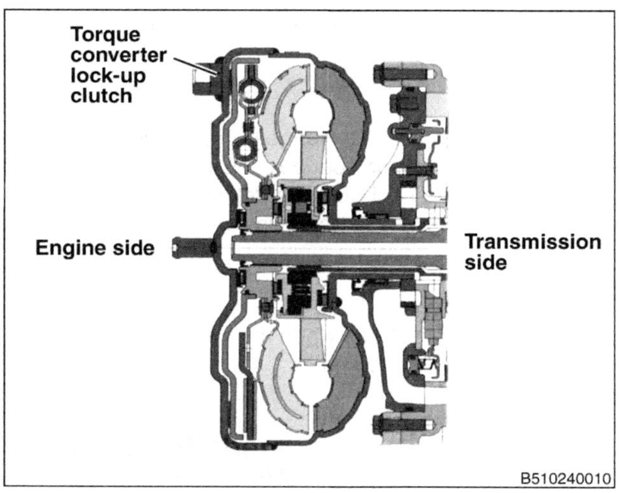

◄ Torque is transmitted from the engine to the transmission by a hydro-dynamic torque converter (fluid coupling). The converter is equipped with slip-controlled lock-up clutch.

BentleyPublishers.com
BentleyPublishers.com—All Rights Reserved

Cautions

> ***CAUTION—***
> - *When performing any repair which involves separating the engine and transmission, check that bell housing dowels are undamaged and in place before reassembly. If the alignment of the engine flywheel to the transmission input shaft or torque converter is incorrect, any of the following complaints may result:*
>   *—Noise from transmission input shaft.*
>   *—Internal damage to the transmission.*
> - *When removing interior trim clips, avoid damaging trim pieces by using a plastic prying tool or a screwdriver with the tip wrapped with masking tape.*
> - *To avoid electrochemical corrosion to engine components made of aluminum-magnesium alloy, do not use steel fasteners. Use aluminum fasteners only.*
> - *The end faces of aluminum fasteners are usually painted blue. For reliable identification, test fasteners for aluminum composition with magnet.*
> - *Replace aluminum fasteners each time they are loosened.*
> - *Follow torque instructions, including angle of rotation specifications, when installing aluminum fasteners.*
> - *Fill transmission only with BMW approved synthetic oil. Filling with incorrect lube could cause the following damage*
>   *— Premature wear.*
>   *— Reduced service life of bearings.*
>   *— Tooth profile damage.*
>   *— Damage to radial shaft oil seals.*

TROUBLESHOOTING

Minor automatic transmission problems may be corrected by changing the automatic transmission fluid (ATF) and filter. See **Automatic Transmission Fluid (ATF) Service** on this repair group.

Begin by checking ATF level and condition. Check to see if the fluid is dirty or has a burned odor indicating overheated fluid. The burned odor may be the results of burned discs in the clutch packs. The friction material from the burned disc can clog valve body passages.

Software in the transmission control module (TCM) monitors transmission operation for faults and alerts the driver by illuminating the transmission fault indicator on the instrument panel. The self-diagnostic software stores diagnostic trouble codes (DTCs) which may be accessed as follows:

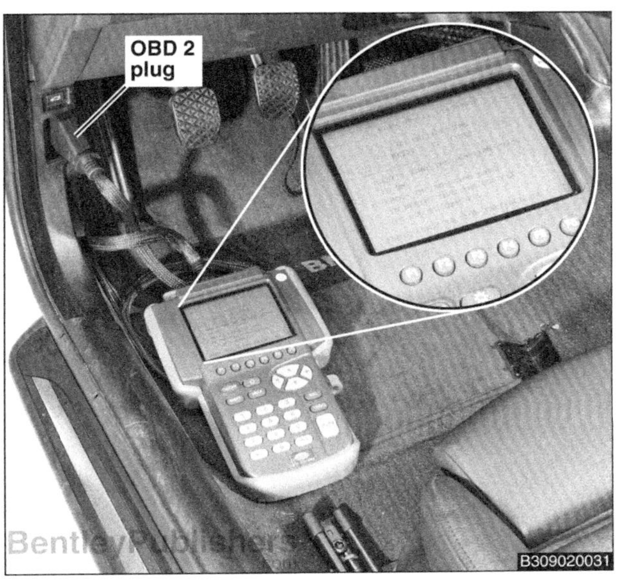

◄ Connect BMW scan tool to OBD 2 plug on left door post. (Plug is behind plastic trim.)

– Place transmission shift lever in P or N. Engage parking brake.

– Switch ignition ON. Follow scan tool on-screen instructions.

– For a listing of scan tool suppliers, see **020 Maintenance**.

BentleyPublishers.com—All Rights Reserved

Mechatronics service notes

The mechatronics electrical connection is delicate. Use care when removing and replacing the multi-pin connector. If pins are damaged, replace the complete mechatronic module.

Leaks around the mechatronics sleeve can be misdiagnosed as coming from the transmission pan gasket. If a leak at the mechatronics sleeve is confirmed, replace the sleeve.

An incorrectly installed mechatronics sleeve can cause damaged connector pins and intermittent transmission communication faults. Example of faults include:

- Serial wire message missing (BMW fault code 5079).
- Serial line signal plausibility (BMW fault code 507A).
- Transmission emergency program light illuminated.
- Transmission entering failsafe mode.

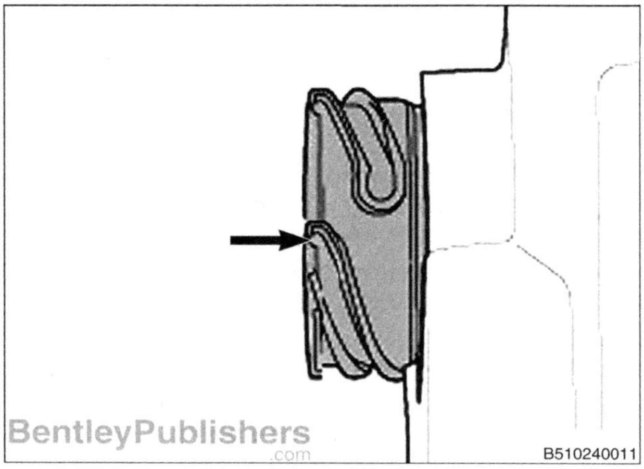

< Push sealing sleeve (**arrow**) into mechatronics housing. Use new O-rings and lubricate sleeve with transmission fluid.

– Reattach mechatronics connector, making sure connector pins are not damaged. Close white locking tab on connector. Do not use excessive force. Excessive force indicates an incorrectly installed sleeve.

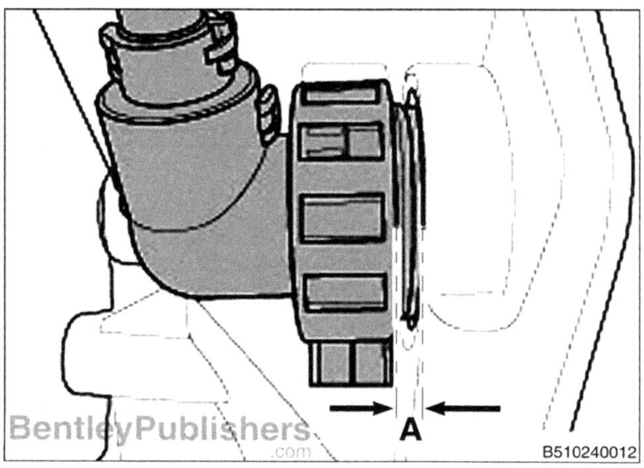

< With harness connector installed and locked, make sure distance between connector and housing surface (**A**) does not exceed 1/8" (3 mm).

– If distance between connector and housing surface (**A**) measures approximately 1/4" - 5/16" (7 - 8 mm), sealing sleeve installation is incorrect and may permanently damage connector pins.

Visible fluid leaks

If visual inspection reveals oil drops or other evidence of oil leaks from transmission, check the following:

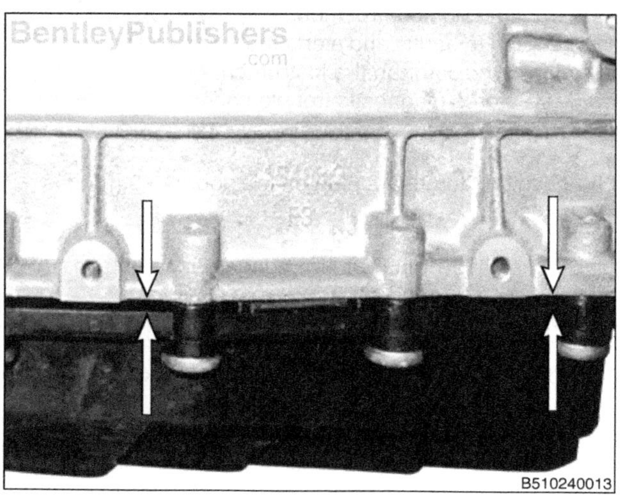

< Check transmission sump gasket for leaks at spots marked with **arrows**.

BentleyPublishers.com
BentleyPublishers.com—All Rights Reserved

< Check transmission output shaft seal (**arrow**) for leaks.

< Check mechatronics connector sleeve (**arrow**) for leaks.

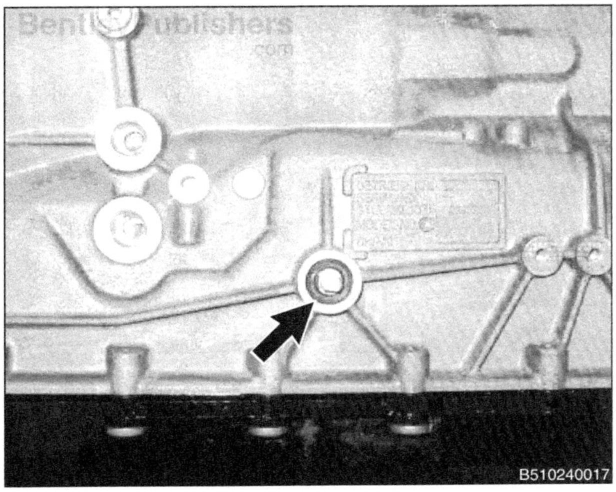

< Check selector shaft bearing seal (**arrow**) for leaks.

Automatic Transmission Fluid (ATF) Service

5 Series automatic transmission fluid (ATF) has a condition-based service (CBS) interval of approximately 100,000 miles (160,000 km). For an explanation of condition-based service, see **020 Maintenance**.

Check ATF level if there is evidence of a leak, a complaint related to fluid level or after transmission repairs. If service or repairs have to be made to the transmission or ATF cooler, use only the approved transmission fluid.

The automatic transmission is not equipped with a dipstick. Check ATF level while monitoring ATF temperature with special equipment. Make sure transmission is at operating temperature and vehicle is level throughout tests.

Be sure necessary equipment, catch bin, transmission fluid and fluid pump are available before starting the fluid level checking procedure.

ATF service interval	
Condition-based: check instrument cluster to monitor service interval. See **020 Maintenance**.	approximately 100,000 miles

ATF level, checking

Transmission fluid expands with temperature. BMW requires that the ATF be checked when fluid temperature is between 30 - 50°C (86 - 122°F).

- Drive vehicle to warm up ATF to operating temperature.

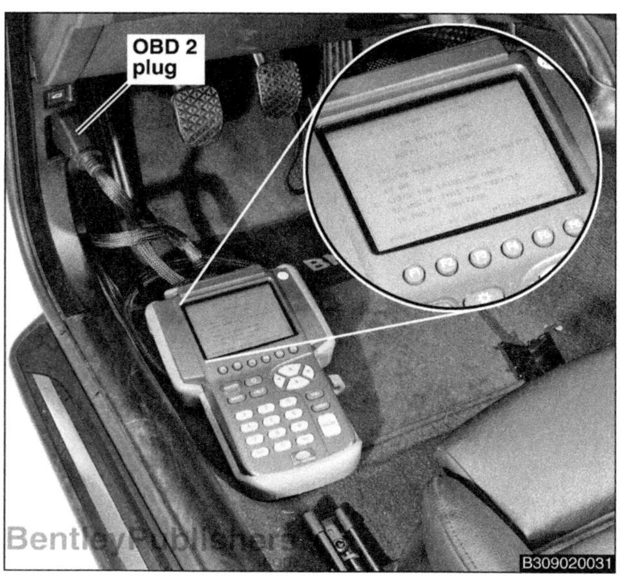

- ◄ Connect BMW scan tool and call up "Service functions (drive)." Set scan tool to measure ATF temperature.
- With engine running, switch on air-conditioning to increase engine idle speed.
- Apply parking brake.
- While applying foot brake firmly move gear shift lever through all gear positions, pausing in each gear briefly.
- Raise and safely support vehicle. Place oil drip pan underneath.

> ***CAUTION—***
> • *Make sure the vehicle is stable and well supported at all times. Use a professional automotive lift or jack stands designed for the purpose. A floor jack is not adequate support.*

- Remove splash shields under transmission. See **020 Maintenance**.

BentleyPublishers.com
BentleyPublishers.com—All Rights Reserved

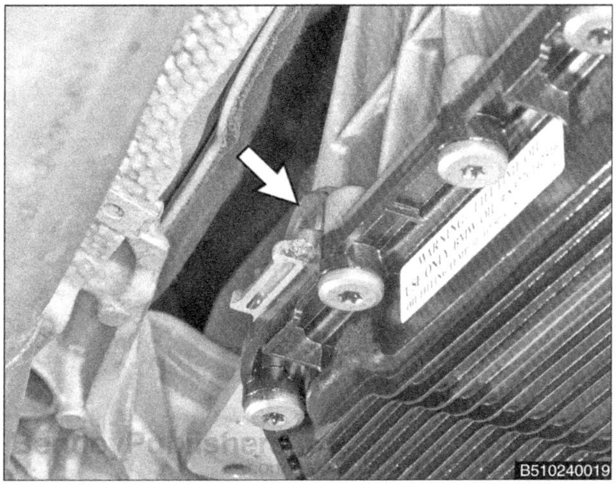

With engine running, shift lever in P and ATF temperature as specified, remove ATF fill plug (**arrow**).

- For best results, check fluid level at 30°C (85°F).
- An accurate level check is not possible if fluid temperature rises above 50°C (120°F).

ATF level checking	
Fluid temperature	30° - 50°C (86° - 122°F)

– Level is correct if small stream of fluid runs out of fill hole.

– If no fluid runs out, add fluid until it starts to overflow.

> ***CAUTION—***
> - *Do not mix BMW transmission oils, and do not replace with another oil.*

Automatic transmission fluid (ATF)	
BMW recommended fluid	BMW part no. 83 22 0 142 516 Shell M-1375.4

– Install fill plug using new sealing ring.

Tightening torque	
Fill plug to transmission housing	35 Nm (26 ft-lb)

– Press on brake pedal. With engine running at idle speed, shift through gears several times. Shift to P.

– Connect BMW scan tool and call up “Service functions (drive).” Follow on-screen instructions to complete ATF level check.

ATF, draining and filling

– Drive vehicle to warm up ATF to operating temperature.

– Raise vehicle and support safely.

> ***WARNING—***
> - *Make sure the vehicle is stable and well supported at all times. Use a professional automotive lift or jack stands designed for the purpose. A floor jack is not adequate support.*

– Remove splash shield underneath transmission. See **020 Maintenance**. If necessary to improve access, remove splash shield and exhaust brackets.

– Place 5-gallon drain pan underneath transmission sump.

Remove ATF drain plug (**arrow**) and allow to drain.

> ***WARNING—***
> - *Use gloves to protect your hands. Hot ATF scalds.*

– Replace drain plug and tighten.

Tightening torque	
Drain plug to plastic sump	8 Nm (6 ft-lb)

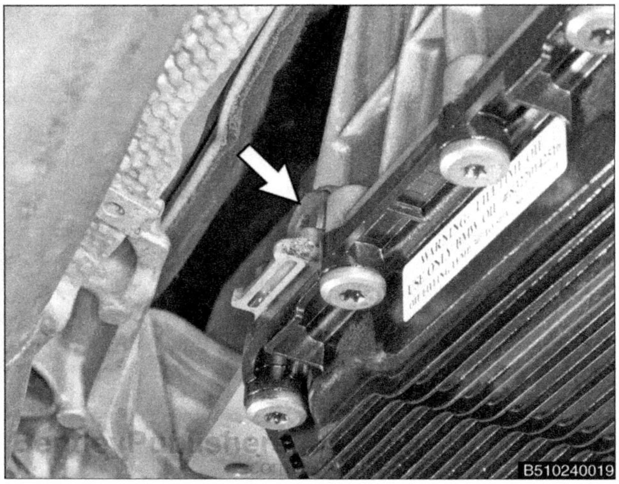

◄ Remove ATF fill plug (**arrow**). Fill transmission until ATF flows out of fill hole.

Table b. Automatic transmission fluid (ATF)	
BMW recommended fluid	BMW part no. 83 22 0 142 516 Shell M-1375.4
Fill capacity	approx. 9 - 10 liters (9.5 - 10.6 US qt)

> ***CAUTION—***
> - *Do not overfill with ATF. Transmission fill capacity given in specifications includes torque converter volume. Torque converter does not drain when transmission is drained using drain plug.*
> - *ATF expands when transmission is warmed up. Final fluid level is set with ATF at a temperature of 30° - 50°C (86° - 122°F).*

– Install fill plug.

– Press on brake pedal. With engine running at idle speed, shift through gears several times. Shift to P.

– Connect BMW scan tool and call up “Service functions (drive).” Follow on-screen instructions to complete ATF level check. See **ATF level, checking** in this repair group.

– Reinstall fill plug using new sealing ring.

Tightening torques	
Fill plug to transmission housing	35 Nm (26 ft-lb)

ATF sump, removing and installing

– Raise and safely support vehicle. Place oil drip pan underneath.

> ***CAUTION—***
> - *Make sure the vehicle is stable and well supported at all times. Use a professional automotive lift or jack stands designed for the purpose. A floor jack is not adequate support.*

– Remove splash shield underneath transmission. See **020 Maintenance**. If necessary to improve access, remove splash shield and exhaust brackets.

BentleyPublishers.com
BentleyPublishers.com—All Rights Reserved

– Remove ATF drain plug and allow to drain. Reinstall and torque drain plug. See **ATF, draining and filling** in this repair group.

> ***WARNING —***
> *• Use gloves to protect your hands. Hot ATF scalds.*

Tightening torque	
Drain plug to plastic sump	8 Nm (6 ft-lb)

Remove ATF sump retaining bolts (**arrows**) and lower sump.

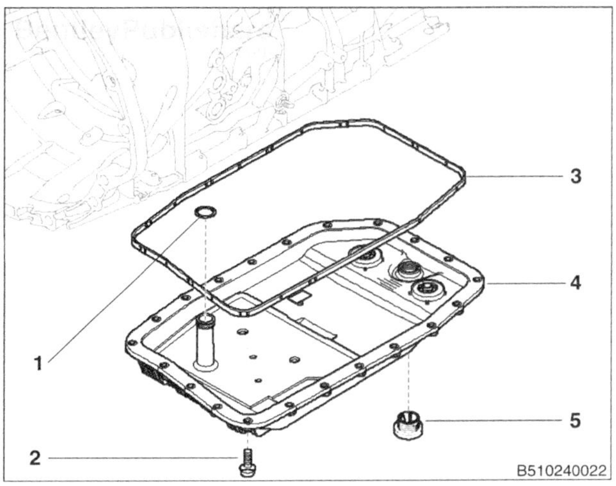

ATF sump components:

1. Oil pickup tube sealing O-ring
2. M6 bolt
   - If necessary, replace with M6 x 28.5 mm Torx T40
   - Tighten to 10 Nm (7 ft-lb)
3. Gasket
4. ATF sump (plastic)
5. Drain plug
   - Tighten to 8 Nm (6 ft-lb)

– Remove and discard old sump gasket and oil pickup tube sealing O-ring. Use shop towels to clean sump interior and seal and gasket grooves.

> ***CAUTION—***
> *• Do not degrease sump with cleaning agent.*

– Install new sealing O-ring.

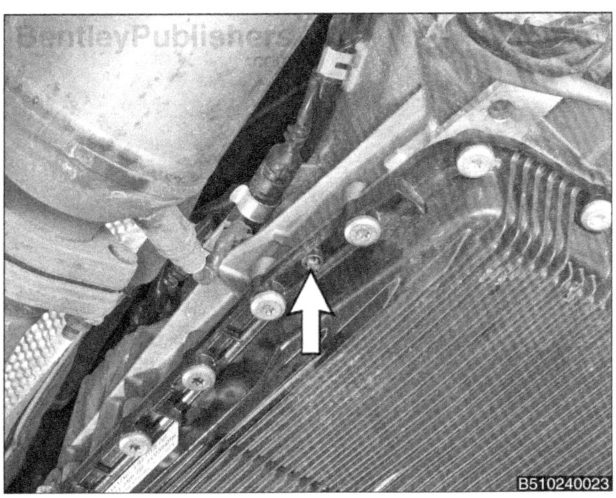

Make sure locating tab (**arrow**) on new gasket mates correctly with bore at edge of sump.

– Reinstall sump. Install retaining bolts hand-tight. Tighten bolts gradually and in crisscross pattern.

Tightening torque	
ATF sump to transmission housing	10 Nm (7 ft-lb)

BentleyPublishers.com
BentleyPublishers.com—All Rights Reserved

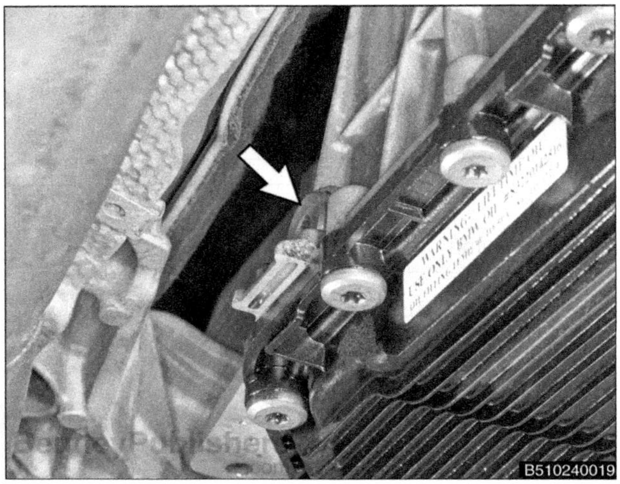

Remove ATF fill plug (**arrow**). Fill transmission until ATF flows out of fill hole.

Automatic transmission fluid (ATF)	
BMW recommended fluid	BMW part no. 83 22 0 142 516 Shell M-1375.4
Fill capacity	approx. 9 - 10 liters (9.5 - 10.6 US qt)

CAUTION—
- *Do not overfill with ATF. Transmission fill capacity given in specifications includes torque converter volume. Torque converter does not drain when transmission is drained using drain plug.*
- *ATF expands when transmission is warmed up. Final fluid level is set with ATF at a temperature of 30° - 50°C (86° - 122°F).*

– Install fill plug.

– Press on brake pedal. With engine running at idle speed, shift through gears several times. Shift to P.

– Connect BMW scan tool and call up "Service functions (drive)." Follow on-screen instructions to complete ATF level check. See **ATF level, checking** in this repair group.

– Reinstall fill plug using new sealing ring.

Tightening torques	
Fill plug to transmission housing	35 Nm (26 ft-lb)

BentleyPublishers.com
BentleyPublishers.com—All Rights Reserved

TRANSMISSION REMOVAL AND INSTALLATION

Removal and installation of the transmission is best accomplished on an automotive lift using a transmission jack. Use caution and safe workshop practices when working underneath vehicle and lowering transmission. Be sure to have appropriate tools on hand before starting the job.

> ***CAUTION—***
> - *Replace aluminium fasteners each time they are released. Do not use steel fasteners in place of aluminium. Electrochemical corrosion will result.*

Transmission, removing and installing (6HP19Z transmission, M54 engine)

> ***WARNING—***
> - *Allow engine and transmission to cool down before starting work on the transmission.*

Transmission, removing (M54 engine)

– Disconnect negative (-) cable from battery.

> ***CAUTION—***
> - *Prior to disconnecting the battery, read the battery disconnection cautions given in* **001 Warnings and Cautions**.

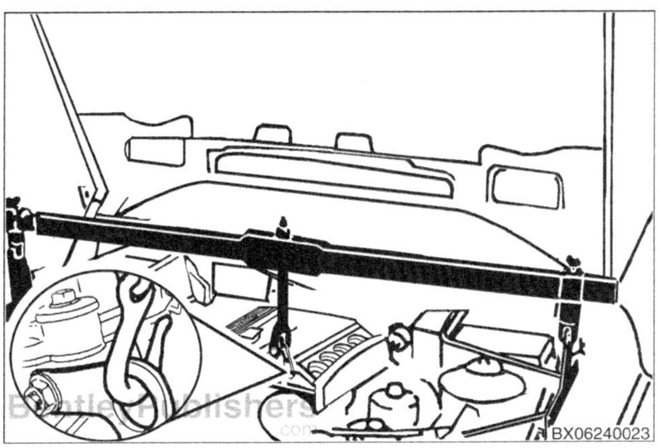

< Support engine from above with suitable hoist.

– Raise and safely support vehicle.

> ***WARNING—***
> - *Make sure the vehicle is stable and well supported at all times. Use a professional automotive lift or jack stands designed for the purpose. A floor jack is not adequate support.*

– Remove splash shields and supporting brackets under engine and transmission. See **020 Maintenance**.

– Remove front end reinforcement. Discard mounting bolts. See **310 Front Suspension**.

– Remove complete exhaust system. See **180 Exhaust System**.

> ***CAUTION—***
> - *To avoid damaging oxygen sensors, transport and store exhaust system carefully.*

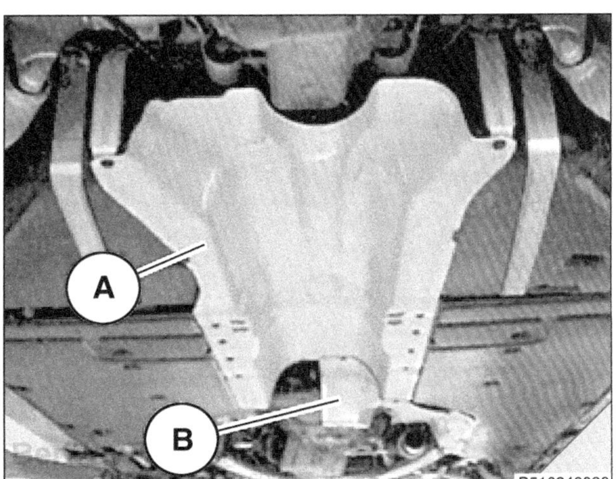

< Remove heat shields (**A**, **B**).

BentleyPublishers.com
BentleyPublishers.com—All Rights Reserved

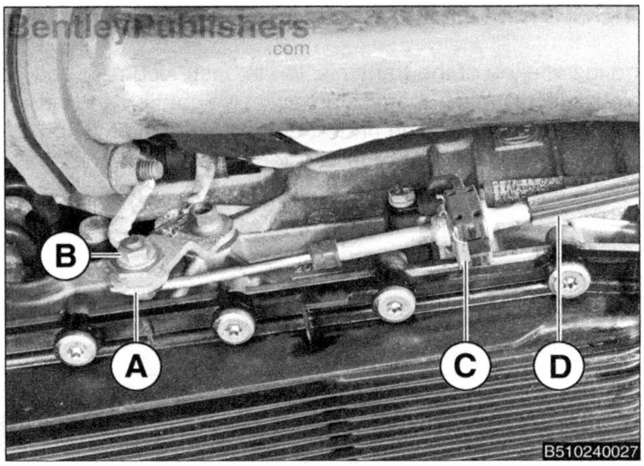

Working at shift cable connection on left side of transmission:

- Counterhold clamping sleeve (**A**) and loosen nut (**B**). Use a screwdriver to pry off retainer clip (**C**) and pull cable (**D**) out of support bracket.

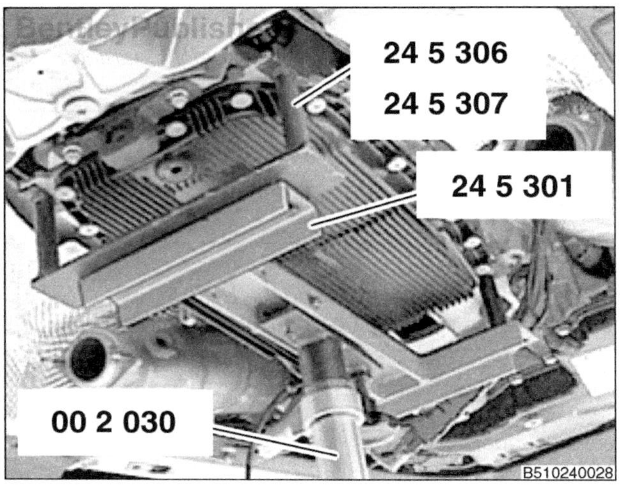

Support transmission with transmission jack. Illustration shows BMW special tools.

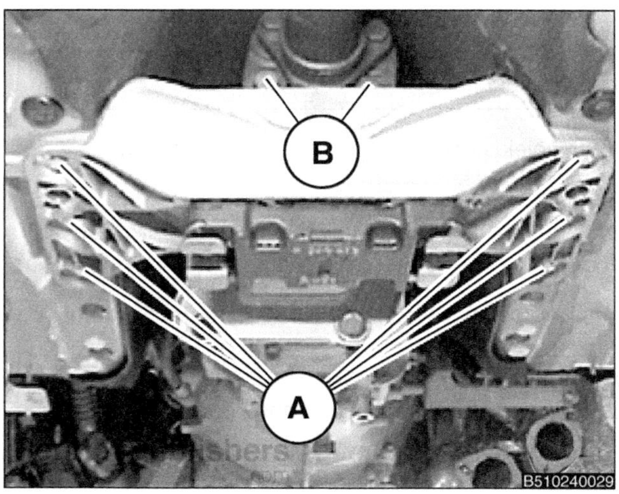

Remove transmission rear support fasteners (**A**, **B**). Remove rear support.

– Detach rear driveshaft from transmission:

- Remove driveshaft center bearing support fasteners. Support center of driveshaft.
- Unbolt rear driveshaft flex-disc from transmission flange.
- Lower center of driveshaft sufficiently to disengage flex-disc from transmission output flange. Tie driveshaft to side with stiff wire. See **260 Driveshafts**.

> ***CAUTION—***
> - *To prevent damage to rear driveshaft CV joint, do not allow driveshaft to hang down unsupported.*

BentleyPublishers.com
BentleyPublishers.com—All Rights Reserved

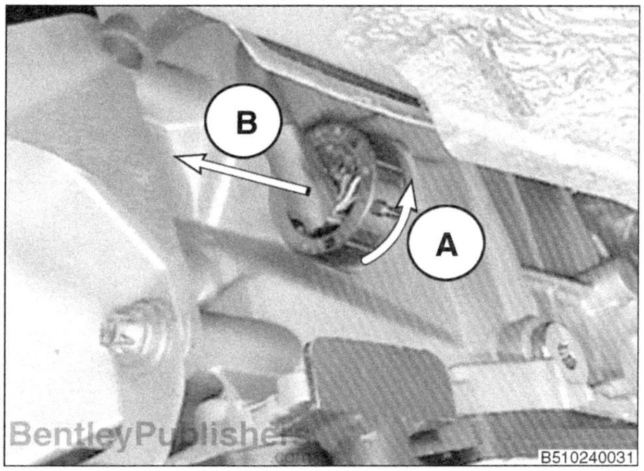

Working at rear of transmission housing:

- Twist mechatronics electrical connector bayonet lock in direction of curved arrow (**A**) to release.
- Carefully pull connector off pins (**B**).
- Insert BMW special tool 24 2 390 in place of mechatronics connector to protect delicate pins.

> ***CAUTION—***
> *• Do not touch pins.*

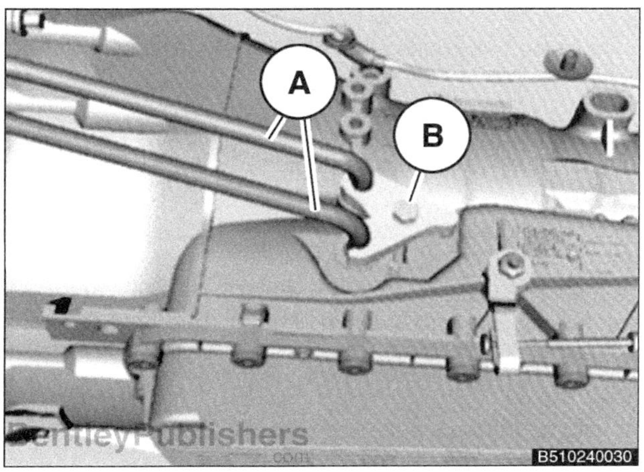

Working at ATF cooler lines (**A**) on left side of transmission:

- Remove fluid line bracket hold-down bolt (**B**)
- Detach fluid lines from transmission housing.

> ***CAUTION—***
> *• Be prepared to catch dripping fluid.*
> *• Remove and discard fluid line O-ring seals. Be sure to use new ones when reassembling.*

– If necessary, remove small cover over torque plate mounting bolts access hole.

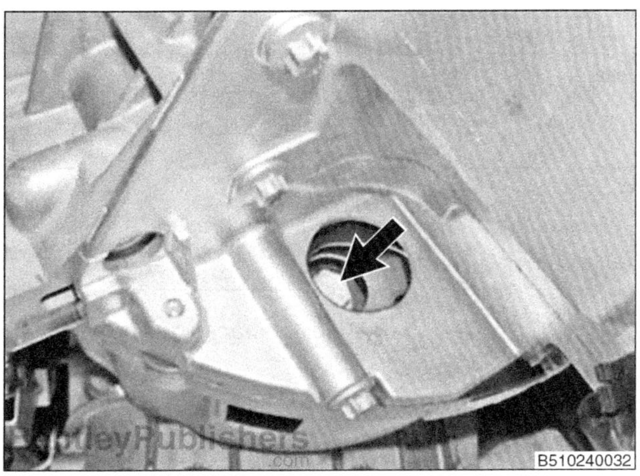

Rotate engine by hand in direction of rotation using bolt on crankshaft vibration damper until torque plate mounting bolt (**arrow**) is visible through access hole in engine bell housing flange.

- Loosen and remove bolt.
- Rotate engine to gain access to and remove remaining torque plate bolts.

BentleyPublishers
BentleyPublishers.com—All Rights Reserved

Transmission, removing and installing (6HP19Z transmission, M54 engine)

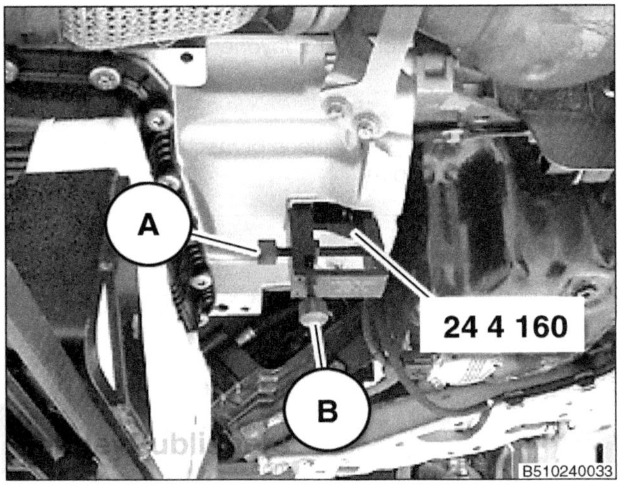

- Use BMW special tool 24 4 160 or equivalent to secure torque converter in bell housing before you separate transmission from engine. This prevents torque converter from pulling off transmission input shaft and damaging torque converter oil seal.
  - Insert tool in bell housing opening and secure in place lightly using knurled knob **A**.
  - Twist in knurled knob **B** and tighten.
  - Tighten down knob **A**.

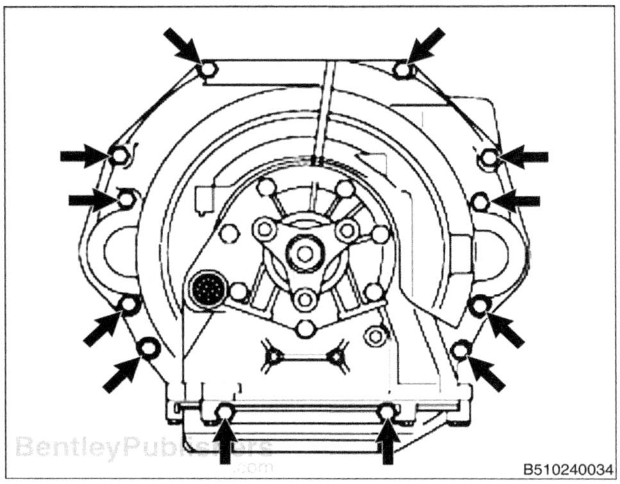

- Remove transmission bell housing bolts (**arrows**). Discard aluminum bolts.

– Remove transmission by pulling back and down. Lower jack slowly while watching carefully to make sure no lines, hoses or wires become snagged.

> ***WARNING —***
> - *Be sure the vehicle is properly supported. The removal of the transmission may upset the balance of the vehicle on the lift.*

> ***CAUTION—***
> - *Tilting the engine to remove the transmission can lead to damage to engine compartment components due to lack of clearance.*
> - *Do not allow the torque converter to fall off the transmission input shaft.*

– Blow out oil cooler lines with low-pressure compressed air and flush cooler with clean ATF twice.

> ***CAUTION—***
> - *Wear safety glasses when working with compressed air.*
> - *Do not reuse ATF used for flushing.*

– Inspect engine torque plate and flywheel for cracks or elongated holes. Replace if necessary.

– If torque converter seal is leaky, or torque converter position on transmission input shaft was disturbed during removal, replace torque converter seal. See **Torque converter oil seal, removing and installing** in this repair group.

BentleyPublishers.com
BentleyPublishers.com—All Rights Reserved

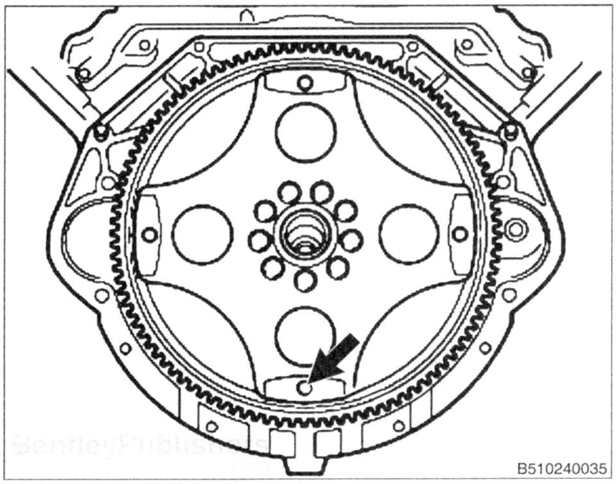

Transmission, installing

< Rotate engine until torque plate bore for mounting bolt (**arrow**) is accessible through opening in engine bell housing.

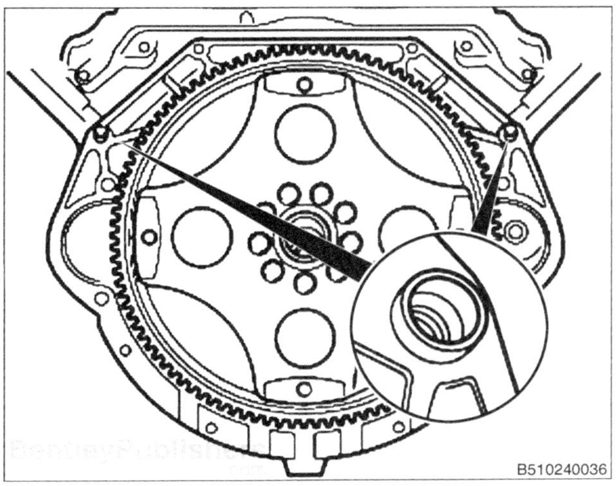

< Check that bell housing dowel sleeves (**arrows**) are in good shape and correctly seated. Replace if needed.

– Check that torque converter is seated correctly in transmission.

– Rotate torque converter until threaded mounting hole in converter is lined up with bore in torque plate.

– Join transmission to engine. Make sure torque converter and torque plate bolt bores are lined up by installing one bolt finger-tight.

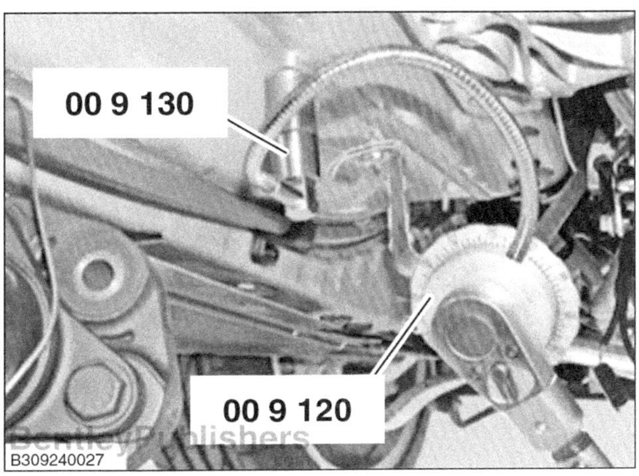

< Using new aluminum bolts, tighten bell housing fasteners to initial torque.

- Use torque angle tool (BMW special tools 00 9 120 / 00 9 130 or equivalent) and tighten fasteners to final torque angle.
- Install and tighten remaining torque converter bolts.

Tightening torques	
Torque converter to torque plate: • M10 x 10.9	56 Nm (41 ft-lb)
Transmission rear support to body (M8)	19 Nm (14 ft-lb)
Transmission rear support to transmission	19 Nm (14 ft-lb)
Transmission to engine (M12 aluminum): • Initial torque • Torque angle	25 Nm (18 ft-lb) + 130°
Transmission to engine (M10 x 30 aluminum): • Initial torque • Torque angle	20 Nm (15 ft-lb) + 90 - 110°
Transmission to engine (M10 x 85 aluminum): • Initial torque • Torque angle	20 Nm (15 ft-lb) + 180 - 200°

BentleyPublishers.com—All Rights Reserved

– Remainder of installation is reverse of removal. Remember to:
  - Install new sealing washers or O-rings on ATF cooler line fittings.
  - Adjust gearshift mechanism. See **250 Gearshift Linkage**.
  - Driveshaft tightening torques are in **260 Driveshafts**. To avoid damaging driveshaft flex-disc, counterhold flex-disc bolt heads while tightening nuts.
  - Fill transmission with clean ATF. See **ATF, draining and filling** in this repair group.
  - Use new bolts to reinstall front end reinforcement plate. See **310 Front Suspension**.

Tightening torques	
Drain plug to plastic sump	8 Nm (6 ft-lb)
Fill plug to transmission housing	35 Nm (26 ft-lb)

Automatic transmission fluid (ATF)	
BMW recommended fluid	BMW part no. 83 22 0 142 516 Shell M-1375.4
Fill capacity	approx. 9 - 10 liters (9.5 - 10.6 US qt)

CAUTION—
- *Do not overfill with ATF. Transmission fill capacity given in specifications includes torque converter volume.*
- *ATF expands when transmission is warmed up. Final fluid level is set with ATF at a temperature of 30° - 50°C (86° - 122°F).*

BentleyPublishers.com
BentleyPublishers.com—All Rights Reserved

Transmission, removing and installing (6HP19Z transmission, N52, N54 engines)

> ***WARNING —***
> • *Allow engine and transmission to cool down before starting work on the transmission.*

Transmission, removing (N52, N54 engines)

– 2008 - 2009 models (starting production date 03 / 2007): Release parking lock manually. See **250 Gearshift Linkage**.

– Disconnect negative (-) cable from battery.

> ***CAUTION—***
> • *Prior to disconnecting the battery, read the battery disconnection cautions given in* **001 Warnings and Cautions**.

– Support engine from above with suitable hoist.

– Raise and safely support vehicle.

> ***WARNING —***
> • *Make sure the vehicle is stable and well supported at all times. Use a professional automotive lift or jack stands designed for the purpose. A floor jack is not adequate support.*

– Remove splash shields and supporting brackets under engine and transmission. See **020 Maintenance**.

– Remove front end reinforcement. Discard mounting bolts. See **310 Front Suspension**.

– Remove complete exhaust system. See **180 Exhaust System**.

> ***CAUTION—***
> • *To avoid damaging oxygen sensors, transport and store exhaust system carefully.*

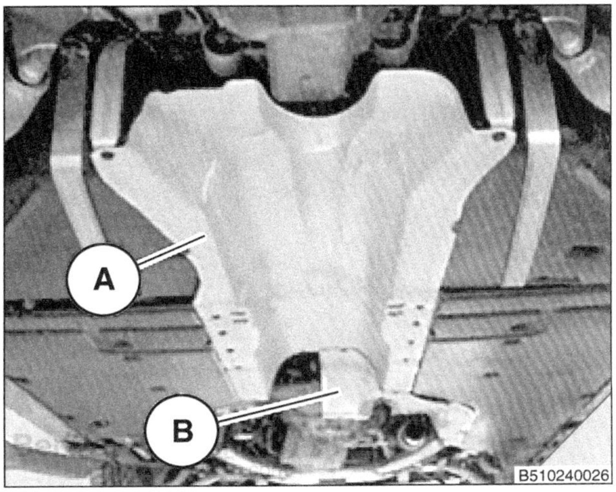

◄ Remove heat shields (**A**, **B**).

BentleyPublishers.com—All Rights Reserved

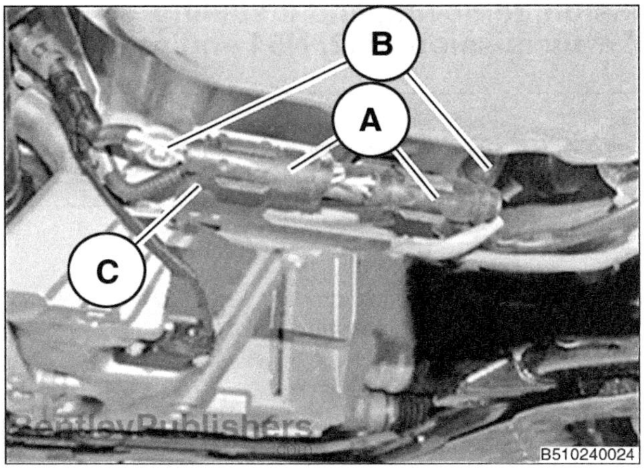

Separate oxygen sensor electrical connectors (**A**). Remove fasteners (**B**) and remove retaining plate (**C**).

Use BMW special tool 00 9 010 (Torx E18 socket) to remove aluminium fastener next to electrical connector brackets. Discard fastener.

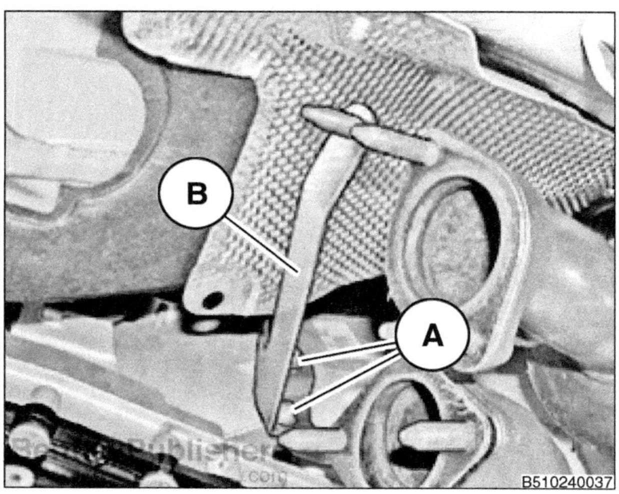

N52 engine: Working at front exhaust pipe flanges, remove heat shield bracket fasteners (**A**) and remove bracket (**B**).

BentleyPublishers.com
BentleyPublishers.com—All Rights Reserved

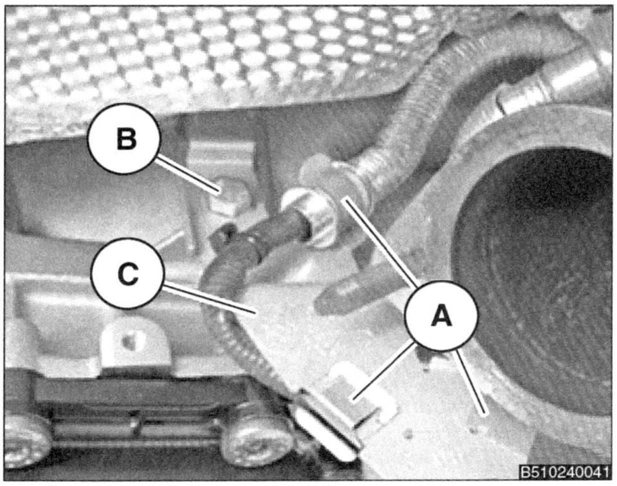

N54 engine: Working at front exhaust pipe flanges:

- Detach electrical harness from bracket (**A**).
- Remove wire bracket mounting fastener (**B**).
- Remove bracket (**C**).

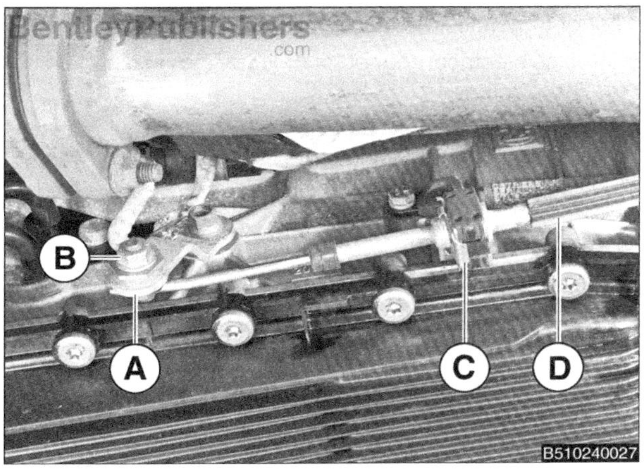

2006 - 2007 models: Working at shift cable connection on left side of transmission:

- Counterhold clamping sleeve (**A**) and loosen nut (**B**). Use a screwdriver to pry off retainer clip (**C**) and pull cable (**D**) out of support bracket.

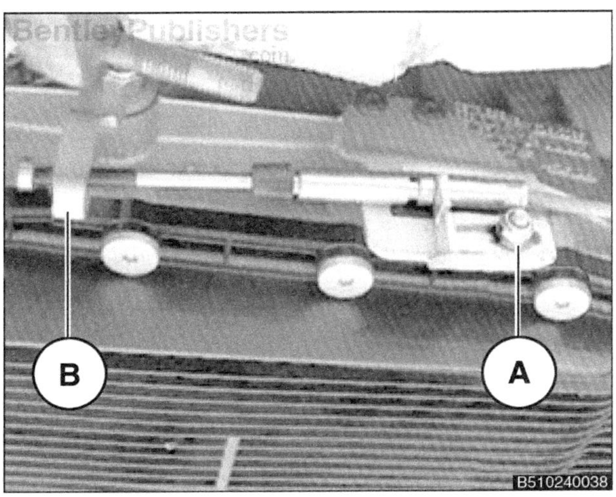

2008 - 2009 models: Working at parking lock cable on left side of transmission:

- Remove cable bracket fastener (**A**).
- Disengage cable end from parking lock lever (**B**).

– Support transmission with transmission jack. Secure transmission to jack with suitable strap.

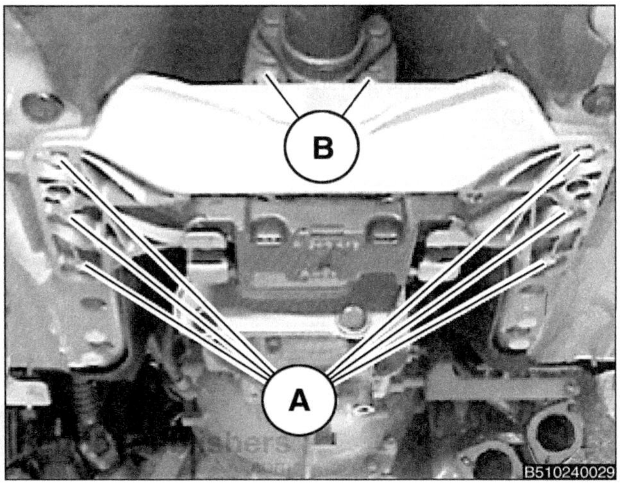

Remove transmission rear support fasteners (**A**, **B**). Remove rear support.

– Detach rear driveshaft from transmission:
  - Remove driveshaft center bearing support fasteners. Support center of driveshaft.
  - Unbolt rear driveshaft flex-disc from transmission flange.
  - Lower center of driveshaft sufficiently to disengage flex-disc from transmission output flange. Tie driveshaft to side using stiff wire. See **260 Driveshafts**.

> ***CAUTION—***
> - *To prevent damage to rear driveshaft CV joint, do not allow driveshaft to hang down unsupported.*

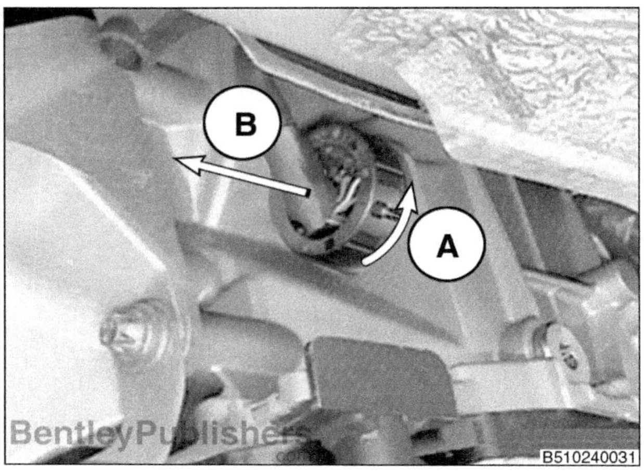

Working at rear of transmission housing:
- Twist mechatronics electrical connector bayonet lock in direction of curved arrow (**A**) to release.
- Carefully pull connector off pins (**B**).
- Insert BMW special tool 24 2 390 in place of mechatronics connector to protect delicate pins.

> ***CAUTION—***
> - *Do not touch pins.*

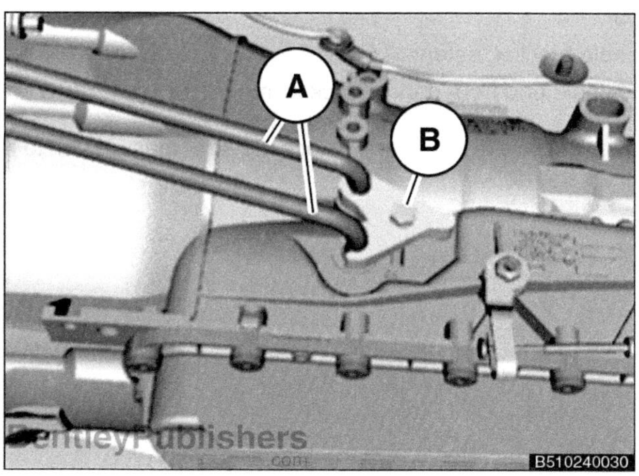

Working at ATF cooler lines (**A**) on left side of transmission:
- Remove fluid line bracket hold-down bolt (**B**)
- Detach fluid lines from transmission housing.

> ***CAUTION—***
> - *Be prepared to catch dripping fluid.*
> - *Remove and discard fluid line O-ring seals. Be sure to use new ones when reassembling.*

BentleyPublishers.com
BentleyPublishers.com—All Rights Reserved

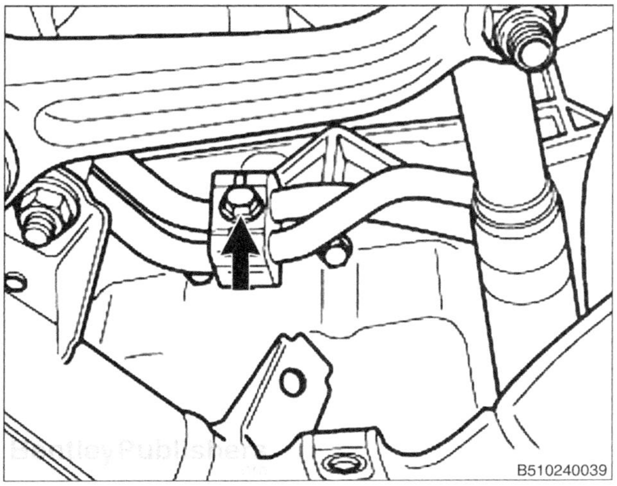

Remove transmission fluid cooler line bracket mounting bolt (**arrow**) at engine oil pan.

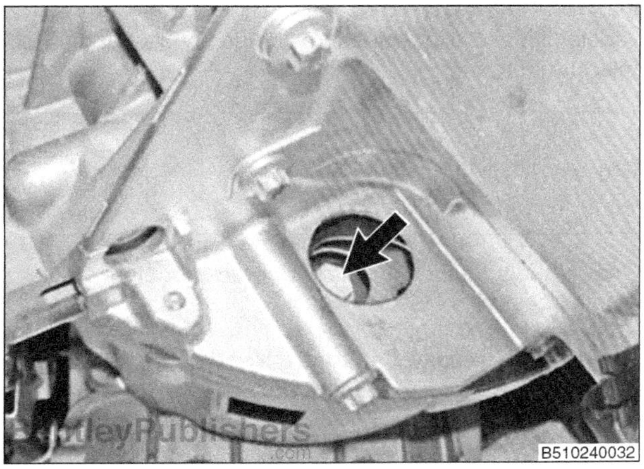

Rotate engine by hand in direction of rotation using bolt on crankshaft vibration damper until torque plate mounting bolt (**arrow**) is visible through access hole in engine bell housing flange.

- Loosen and remove bolt.
- Rotate engine to gain access to and remove remaining torque plate bolts.

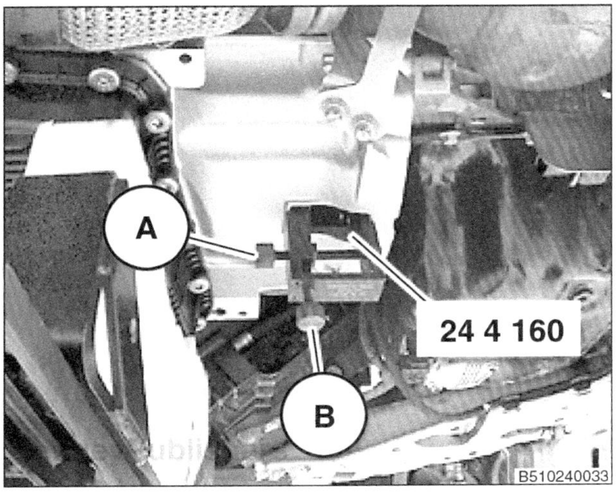

Use BMW special tool 24 4 160 or equivalent to secure torque converter in bell housing before you separate transmission from engine. This prevents torque converter from pulling off transmission input shaft and damaging torque converter oil seal.

- Insert tool in bell housing opening and secure in place lightly using knurled knob **A**.
- Twist in knurled knob **B** and tighten.
- Tighten down knob **A**.

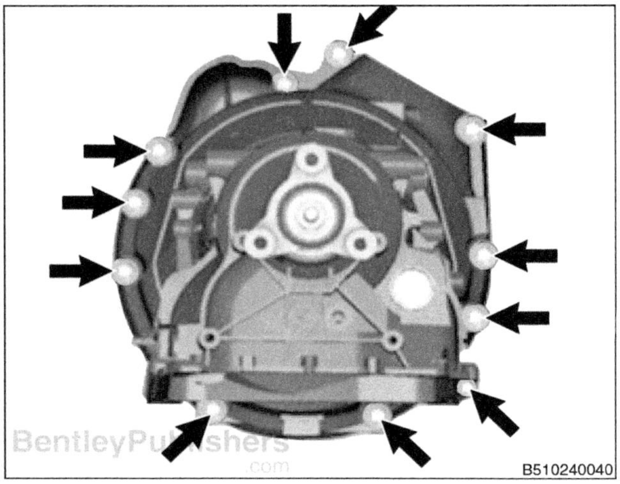

B510240040

◀ Remove transmission bell housing bolts (**arrows**). Discard aluminum bolts.

– Remove transmission by pulling back and down. Lower jack slowly while watching carefully to make sure no lines, hoses or wires become snagged.

> ***WARNING —***
> *• Be sure the vehicle is properly supported. The removal of the transmission may upset the balance of the vehicle on the lift.*

> ***CAUTION—***
> *• Tilting the engine to remove the transmission can lead to damage to engine compartment components due to lack of clearance.*
> *• Do not allow the torque converter to fall off the transmission input shaft.*

– Blow out oil cooler lines with low-pressure compressed air and flush cooler with clean ATF twice.

> ***CAUTION—***
> *• Wear safety glasses when working with compressed air.*
> *• Do not reuse ATF used for flushing.*

– Inspect engine torque plate and flywheel for cracks or elongated holes. Replace if necessary.

– If torque converter seal is leaky, or torque converter position on transmission input shaft was disturbed during removal, replace torque converter seal. See **Torque converter oil seal, removing and installing** in this repair group.

Transmission, installing (N52, N54 engines)

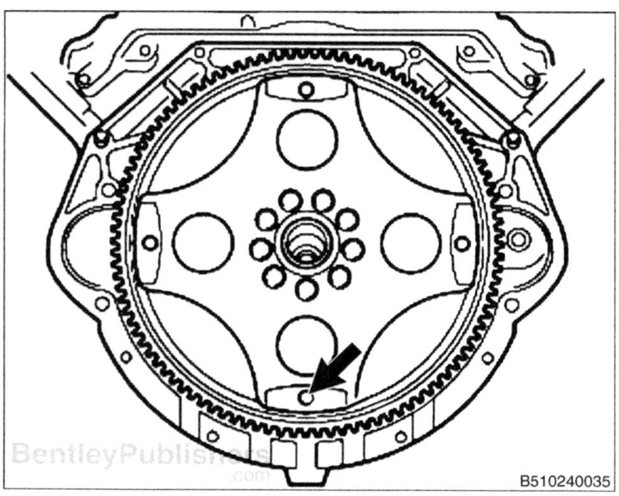

B510240035

◀ Rotate engine until torque plate bore for mounting bolt (**arrow**) is accessible through opening in engine bell housing.

BentleyPublishers.com
BentleyPublishers.com—All Rights Reserved

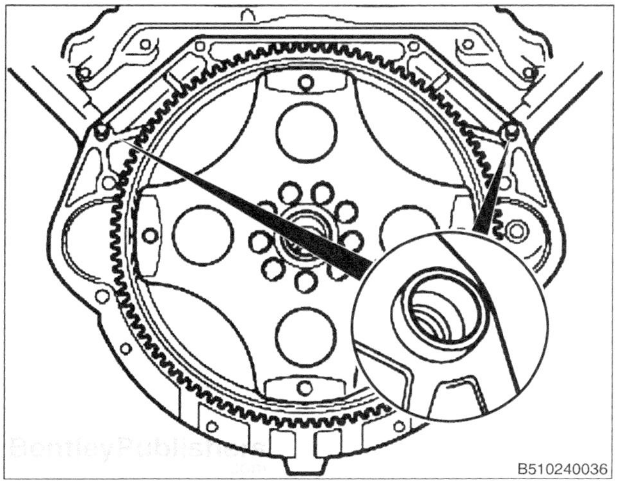

◀ Check that bell housing dowel sleeves (**arrows**) are in good shape and correctly seated. Replace if needed.

– Check that torque converter is seated correctly in transmission.

– Rotate torque converter until threaded mounting hole in converter is lined up with bore in torque plate.

– Join transmission to engine. Make sure torque converter and torque plate bolt bores are lined up by installing one bolt finger-tight.

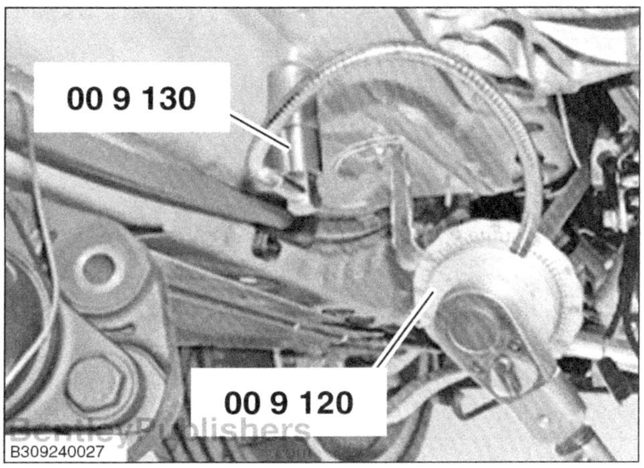

◀ Using new aluminum bolts, tighten bell housing fasteners to initial torque.

- Use torque angle tool (BMW special tools 00 9 120 / 00 9 130 or equivalent) and tighten fasteners to final torque angle.
- Install and tighten remaining torque converter bolts.

Tightening torques	
Torque converter to torque plate: • M10 x 10.9	56 Nm (41 ft-lb)
Transmission rear support to body (M8)	19 Nm (14 ft-lb)
Transmission rear support to transmission	19 Nm (14 ft-lb)
Transmission to engine (M12 aluminum): • Initial torque • Torque angle	25 Nm (18 ft-lb) + 130°
Transmission to engine (M10 x 30 aluminum): • Initial torque • Torque angle	20 Nm (15 ft-lb) + 90 - 110°
Transmission to engine (M10 x 85 aluminum): • Initial torque • Torque angle	20 Nm (15 ft-lb) + 180 - 200°

BentleyPublishers.com—All Rights Reserved

– Remainder of installation is reverse of removal. Remember to:
  - Install new sealing washers or O-rings on ATF cooler line fittings.
  - 2006 - 2007 models: Adjust gearshift mechanism. See **250 Gearshift Linkage**.
  - 2008 - 2009 models: Adjust parking lock cable free play. See **250 Gearshift Linkage**.
  - Driveshaft tightening torques are in **260 Driveshafts**. To avoid damaging driveshaft flex-disc, counterhold flex-disc bolt heads while tightening nuts.
  - Fill transmission with clean ATF. See **ATF, draining and filling** in this repair group.
  - Use new bolts to reinstall front end reinforcement plate. See **310 Front Suspension**.

Tightening torques	
Drain plug to plastic sump	8 Nm (6 ft-lb)
Fill plug to transmission housing	35 Nm (26 ft-lb)

5 Series automatic transmission fluid (ATF)	
BMW recommended fluid	BMW part no. 83 22 0 142 516 Shell M-1375.4
Fill capacity	approx. 9 - 10 liters (9.5 - 10.6 US qt)

CAUTION—
- *Do not overfill with ATF. Transmission fill capacity given in specifications includes torque converter volume.*
- *ATF expands when transmission is warmed up. Final fluid level is set with ATF at a temperature of 30° - 50°C (86° - 122°F).*

BentleyPublishers.com
BentleyPublishers.com—All Rights Reserved

Transmission, removing and installing (6HP19Z xDrive transmission)

> ***WARNING —***
> *• Allow engine and transmission to cool down before starting work on the transmission.*

Transmission, removing (xDrive transmission)

– 2008 - 2009 models (starting production date 03 / 2007): Release parking lock manually. See **250 Gearshift Linkage**.

– Disconnect negative (-) cable from battery.

> ***CAUTION—***
> *• Prior to disconnecting the battery, read the battery disconnection cautions given in* **001 Warnings and Cautions**.

– Support engine from above with suitable hoist.

– Raise and safely support vehicle.

> ***WARNING —***
> *• Make sure the vehicle is stable and well supported at all times. Use a professional automotive lift or jack stands designed for the purpose. A floor jack is not adequate support.*

– Remove splash shields and supporting brackets under engine and transmission. See **020 Maintenance**.

– Remove front end reinforcement. Discard mounting bolts. See **310 Front Suspension**.

– Remove complete exhaust system. See **180 Exhaust System**.

> ***CAUTION—***
> *• To avoid damaging oxygen sensors, transport and store exhaust system carefully.*

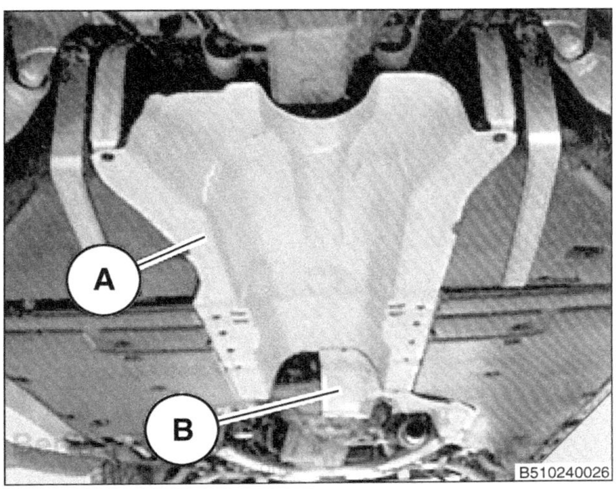

◀ Remove heat shields (**A**, **B**).

BentleyPublishers.com—All Rights Reserved

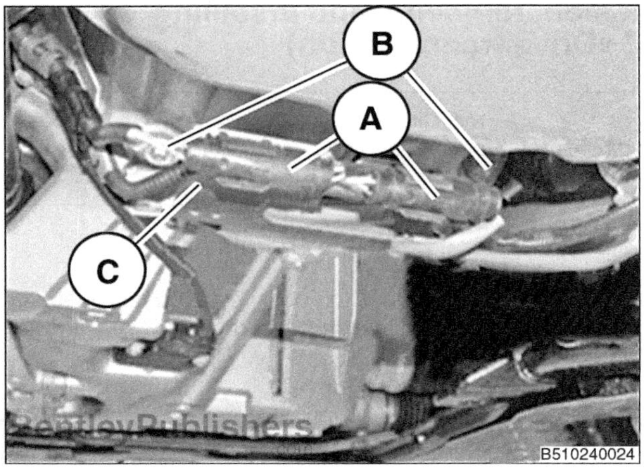

Separate oxygen sensor electrical connectors (**A**). Remove fasteners (**B**) and remove retaining plate (**C**).

Use BMW special tool 00 9 010 (Torx E18 socket) to remove aluminium fastener next to electrical connector brackets. Discard fastener.

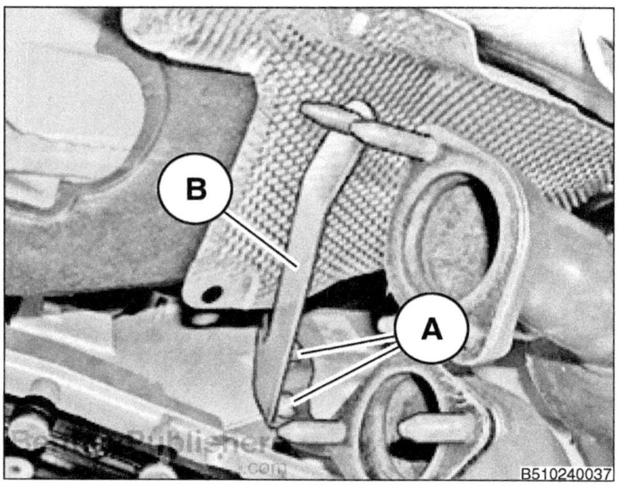

N52 engine: Working at front exhaust pipe flanges, remove heat shield bracket fasteners (**A**) and remove bracket (**B**).

BentleyPublishers.com
BentleyPublishers.com—All Rights Reserved

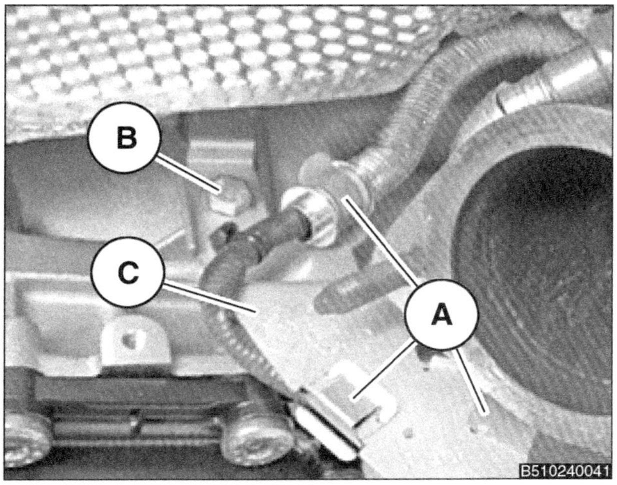

N54 engine: Working at front exhaust pipe flanges:

- Detach electrical harness from bracket (**A**).
- Remove wire bracket mounting fastener (**B**).
- Remove bracket (**C**).

240

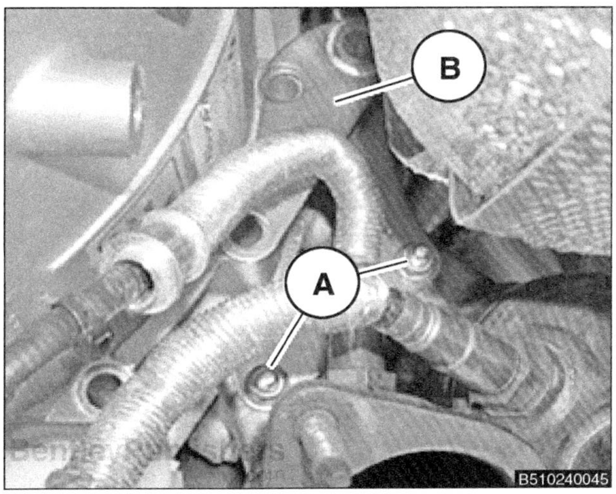

N54 engine: Working at front exhaust pipe flanges:

- Remove wire bracket mounting fastener (**A**).
- Remove bracket (**B**).

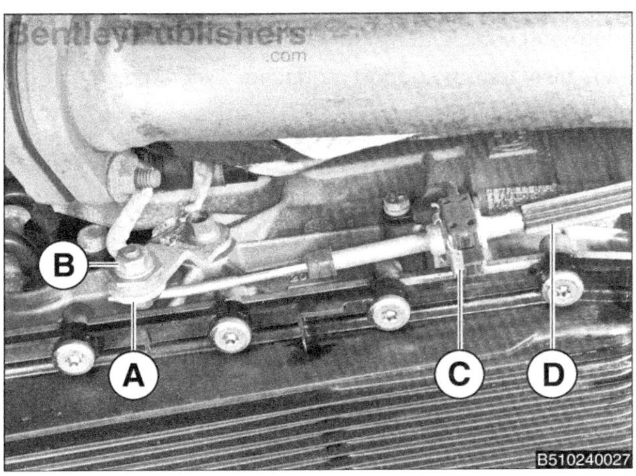

2006 - 2007 models: Working at shift cable connection on left side of transmission:

- Counterhold clamping sleeve (**A**) and loosen nut (**B**). Use a screwdriver to pry off retainer clip (**C**) and pull cable (**D**) out of support bracket.

BentleyPublishers.com—All Rights Reserved

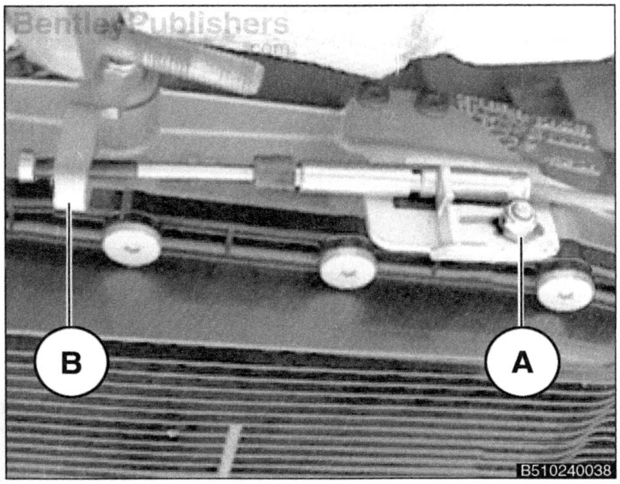

◀ 2008 - 2009 models: Working at parking lock cable on left side of transmission:

- Remove cable bracket fastener (**A**).
- Disengage cable end from parking lock lever (**B**).

– Support transmission with transmission jack. Secure transmission to jack with suitable strap.

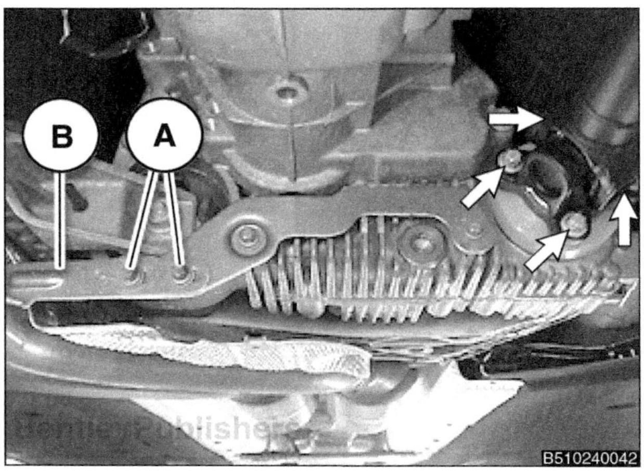

◀ Working underneath transmission at front of transfer case:

- Remove front driveshaft mounting bolts (**arrows**) and tie driveshaft to one side.
- Remove mounting bolts (**A**) and remove exhaust bracket (**B**).

– Detach rear driveshaft from transfer case:

- Remove driveshaft center bearing support fasteners. Support center of driveshaft.
- Unbolt rear driveshaft flex-disc from transfer case flange.
- Lower center of driveshaft sufficiently to disengage flex-disc from transfer case output flange. Tie driveshaft to side. See **260 Driveshafts**.

> ***CAUTION—***
> - *To prevent damage to rear driveshaft CV joint, do not allow driveshaft to hang down unsupported.*

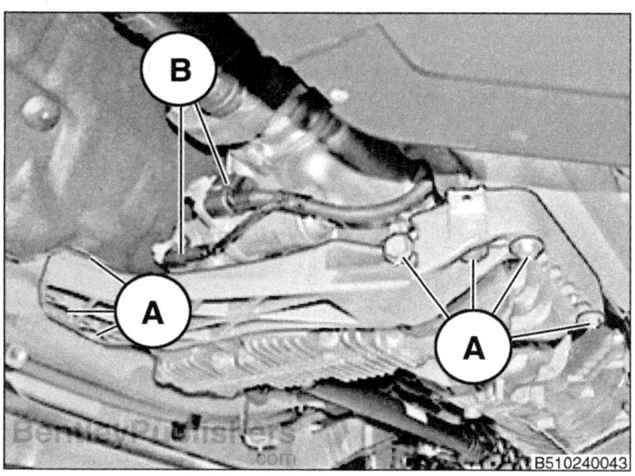

◀ Working underneath vehicle at rear of transfer case:

- Remove transmission rear support mounting bolts (**A**). Remove support.
- Detach transfer case electrical connectors (**B**).

BentleyPublishers.com
BentleyPublishers.com—All Rights Reserved

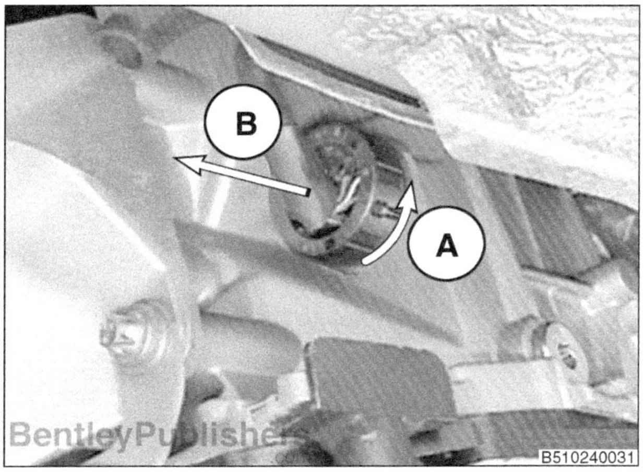

Working at rear of transmission housing:

- Twist mechatronics electrical connector bayonet lock in direction of curved arrow (**A**) to release.
- Carefully pull connector off pins (**B**).
- Insert BMW special tool 24 2 390 in place of mechatronics connector to protect delicate pins.

> ***CAUTION—***
> - *Do not touch pins.*

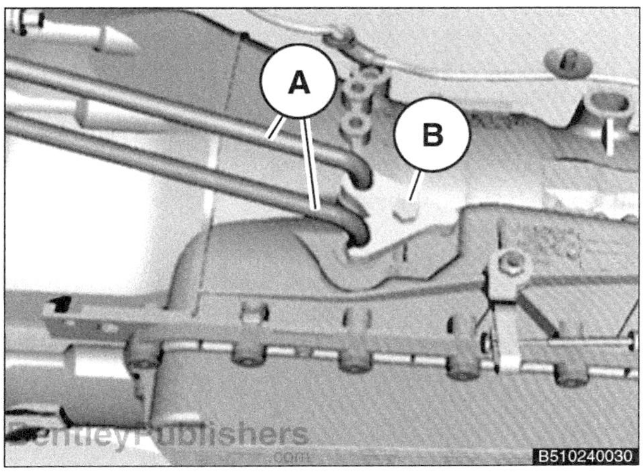

Working at ATF cooler lines (**A**) on left side of transmission:

- Remove fluid line bracket hold-down bolt (**B**)
- Detach fluid lines from transmission housing.

> ***CAUTION—***
> - *Be prepared to catch dripping fluid.*
> - *Remove and discard fluid line O-ring seals. Be sure to use new ones when reassembling.*

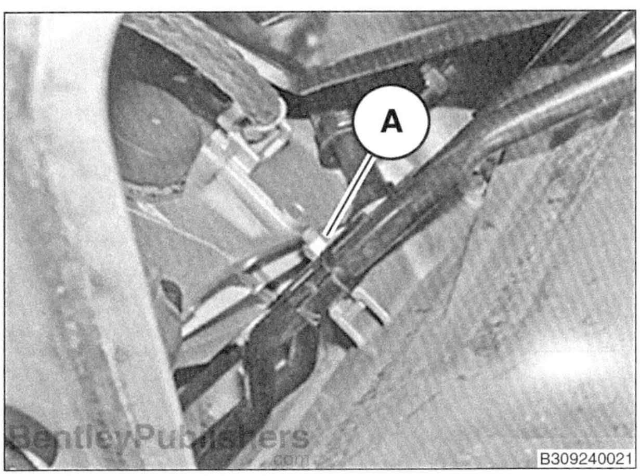

Remove transmission fluid cooler line bracket mounting bolt (**arrow**) at engine oil pan.

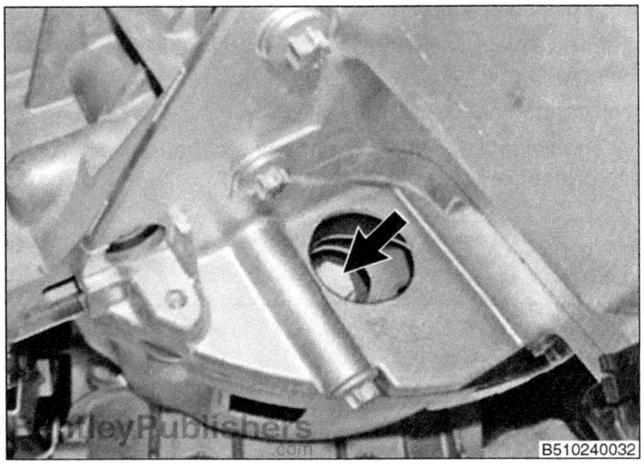

Rotate engine by hand in direction of rotation using bolt on crankshaft vibration damper until torque plate mounting bolt (**arrow**) is visible through access hole in engine bell housing flange.

- Loosen and remove bolt.
- Rotate engine to gain access to and remove remaining torque plate bolts.

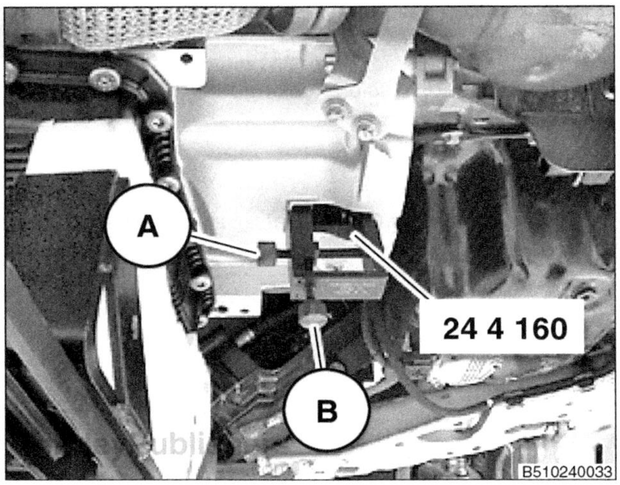

Use BMW special tool 24 4 160 or equivalent to secure torque converter in bell housing before you separate transmission from engine. This prevents torque converter from pulling off transmission input shaft and damaging torque converter oil seal.

- Insert tool in bell housing opening and secure in place lightly using knurled knob **A**.
- Twist in knurled knob **B** and tighten.
- Tighten down knob **A**.

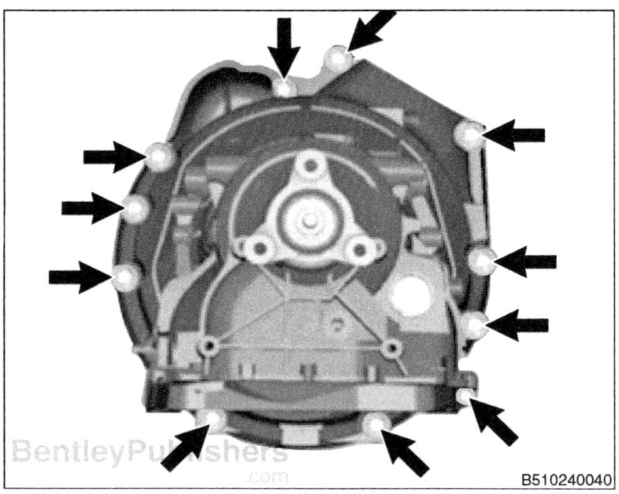

Remove transmission bell housing bolts (**arrows**). Discard aluminum bolts.

– Remove transmission by pulling back and down. Lower jack slowly while watching carefully to make sure no lines, hoses or wires become snagged.

WARNING —
- *Be sure the vehicle is properly supported. The removal of the transmission may upset the balance of the vehicle on the lift.*

CAUTION —
- *Tilting the engine to remove the transmission can lead to damage to engine compartment components due to lack of clearance.*
- *Do not allow the torque converter to fall off the transmission input shaft.*

BentleyPublishers.com
BentleyPublishers.com—All Rights Reserved

— Blow out oil cooler lines with low-pressure compressed air and flush cooler with clean ATF twice.

> ***CAUTION—***
> • *Wear safety glasses when working with compressed air.*
> • *Do not reuse ATF used for flushing.*

— Inspect engine torque plate and flywheel for cracks or elongated holes. Replace if necessary.

— If torque converter seal is leaky, or torque converter position on transmission input shaft was disturbed during removal, replace torque converter seal. See **Torque converter oil seal, removing and installing** in this repair group.

Transmission, installing (xDrive transmission)

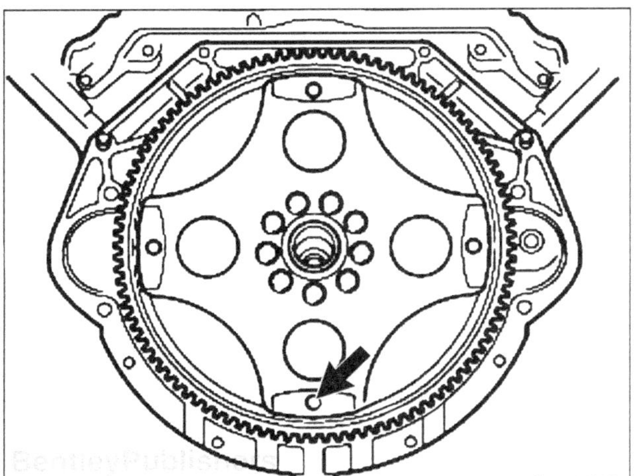
B510240035

◄ Rotate engine until torque plate bore for mounting bolt (**arrow**) is accessible through opening in engine bell housing.

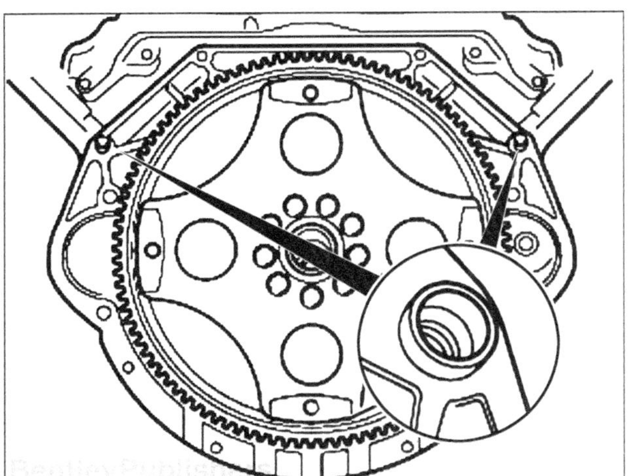
B510240036

◄ Check that bell housing dowel sleeves (**arrows**) are in good shape and correctly seated. Replace if needed.

— Check that torque converter is seated correctly in transmission.

— Rotate torque converter until threaded mounting hole in converter is lined up with bore in torque plate.

— Join transmission to engine. Make sure torque converter and torque plate bolt bores are lined up by installing one bolt finger-tight.

BentleyPublishers.com
BentleyPublishers.com–All Rights Reserved

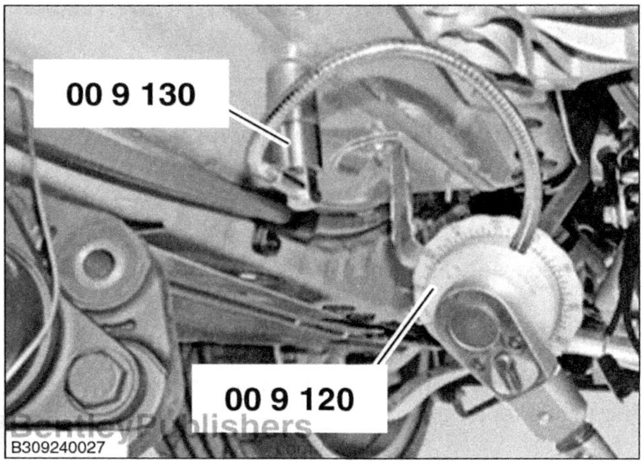

- Using new aluminum bolts, tighten bell housing fasteners to initial torque.
  - Use torque angle tool (BMW special tools 00 9 120 / 00 9 130 or equivalent) and tighten fasteners to final torque angle.
  - Install and tighten remaining torque converter bolts.

Tightening torques	
Torque converter to torque plate: • M10 x 10.9	56 Nm (41 ft-lb)
Transmission rear support to body (M8)	19 Nm (14 ft-lb)
Transmission rear support to transmission	19 Nm (14 ft-lb)
Transmission to engine (M12 aluminum): • Initial torque • Torque angle	25 Nm (18 ft-lb) + 130°
Transmission to engine (M10 x 30 aluminum): • Initial torque • Torque angle	20 Nm (15 ft-lb) + 90 - 110°
Transmission to engine (M10 x 85 aluminum): • Initial torque • Torque angle	20 Nm (15 ft-lb) + 180 - 200°

– Remainder of installation is reverse of removal. Remember to:
- Install new sealing washers or O-rings on ATF cooler line fittings.
- 2006 - 2007 models: Adjust gearshift mechanism. See **250 Gearshift Linkage**.
- 2008 - 2009 models: Adjust parking lock cable free play. See **250 Gearshift Linkage**.
- Driveshaft tightening torques are in **260 Driveshafts**. To avoid damaging driveshaft flex-disc, counterhold flex-disc bolt heads while tightening nuts.
- Fill transmission with clean ATF. See **ATF, draining and filling** in this repair group.
- Use new bolts to reinstall front end reinforcement plate. See **310 Front Suspension**.

Tightening torques	
Drain plug to plastic sump	8 Nm (6 ft-lb)
Fill plug to transmission housing	35 Nm (26 ft-lb)

5 Series automatic transmission fluid (ATF)	
BMW recommended fluid	BMW part no. 83 22 0 142 516 Shell M-1375.4
Fill capacity	approx. 9 - 10 liters (9.5 - 10.6 US qt)

CAUTION—
- *Do not overfill with ATF. Transmission fill capacity given in specifications includes torque converter volume.*
- *ATF expands when transmission is warmed up. Final fluid level is set with ATF at a temperature of 30° - 50°C (86° - 122°F).*

BentleyPublishers.com
BentleyPublishers.com—All Rights Reserved

Transmission, removing and installing (6HP26Z, V8 engine)

Transmission, removing (V8 engine)

- 2008 - 2009 models (starting production date 03 / 2007): Release parking lock manually. See **250 Gearshift Linkage**.
- Disconnect negative (-) cable from battery.

> ***CAUTION—***
> - *Prior to disconnecting the battery, read the battery disconnection cautions given in* **001 Warnings and Cautions**.

- Support engine from above with suitable hoist.
- Raise and safely support vehicle.

> ***WARNING—***
> - *Make sure the vehicle is stable and well supported at all times. Use a professional automotive lift or jack stands designed for the purpose. A floor jack is not adequate support.*

- Remove splash shields and supporting brackets under engine and transmission. See **020 Maintenance**.
- Remove front end reinforcement. Discard mounting bolts. See **310 Front Suspension**.
- Remove exhaust system from catalytic converters back. See **180 Exhaust System**.
- Remove oxygen sensors and store them safely.

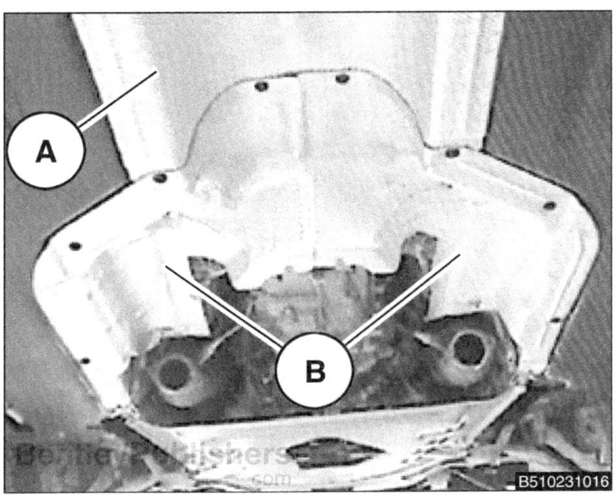

< Remove heat shields (**A**, **B**).

BentleyPublishers.com—All Rights Reserved

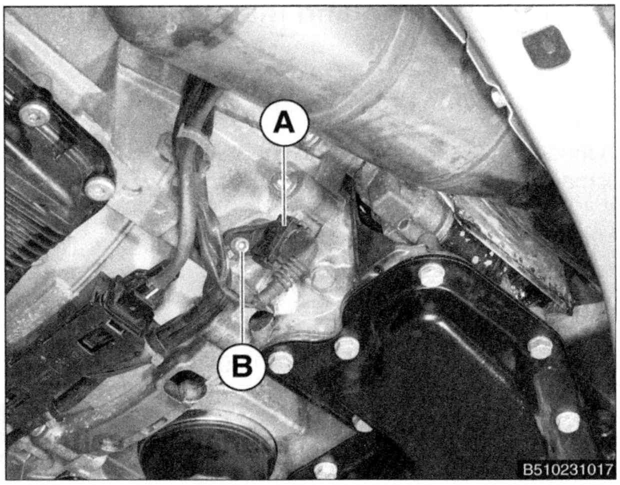

Working underneath right side of engine bellhousing:

- Detach plug (**A**) from crankshaft sensor.
- Remove sensor mounting screw (**B**) and remove speed sensor.

Working underneath engine oil pan:

- Detach electrical connector (**arrow**) from oil level sensor.
- Unclip harness from harness support clips.

Working underneath transmission bell housing, detach harness connector (**arrow**) from cable loom. Unclip loom from transmission housing and tie to one side.

BentleyPublishers.com
BentleyPublishers.com—All Rights Reserved

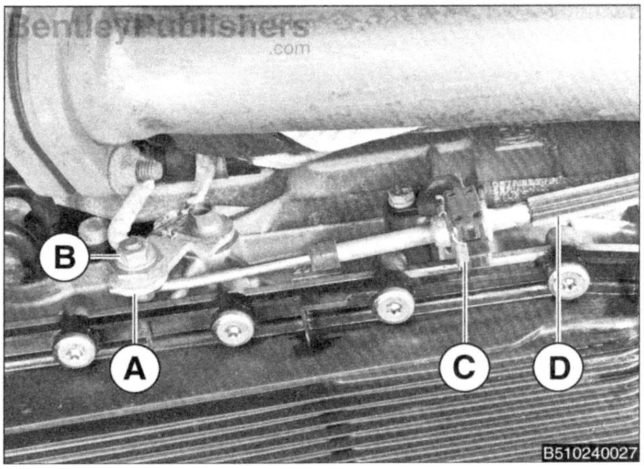

2006 - 2007 models: Working at shift cable connection on left side of transmission:

- Counterhold clamping sleeve (**A**) and loosen nut (**B**). Use a screwdriver to pry off retainer clip (**C**) and pull cable (**D**) out of support bracket.

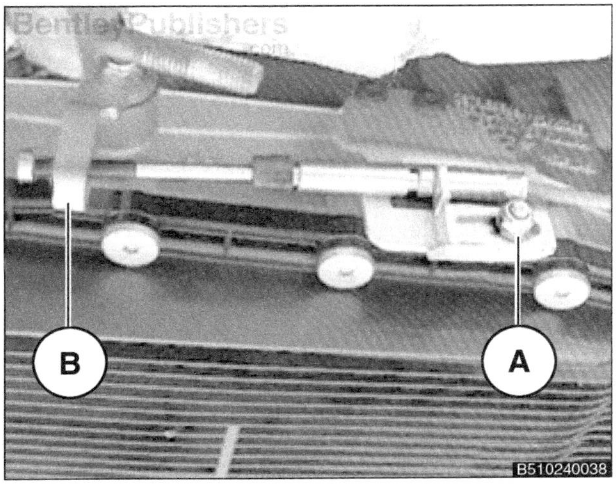

2008 - 2009 models: Working at parking lock cable on left side of transmission:

- Remove cable bracket fastener (**A**).
- Disengage cable end from parking lock lever (**B**).

– Support transmission with transmission jack. Secure transmission to jack with suitable strap.

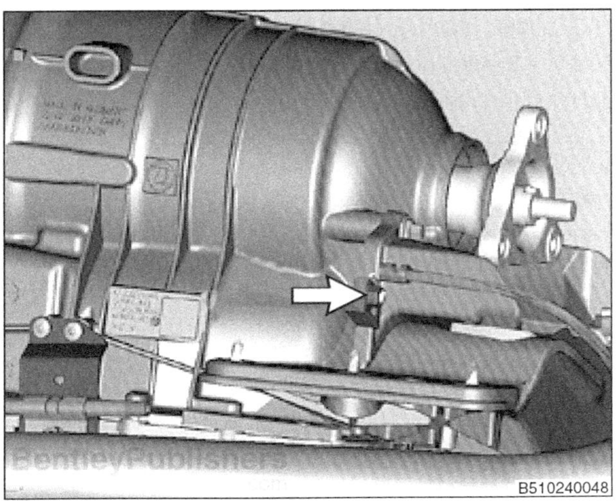

If necessary, remove fastener (**arrow**) and detach transmission ground strap.

BentleyPublishers.com–All Rights Reserved

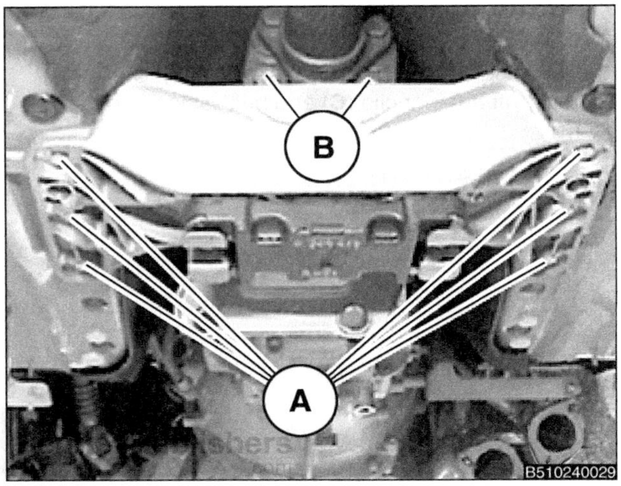

Remove transmission rear support fasteners (**A**, **B**). Remove rear support.

- Detach rear driveshaft from transmission:
  - Remove driveshaft center bearing support fasteners. Support center of driveshaft.
  - Unbolt rear driveshaft flex-disc from transmission flange.
  - Lower center of driveshaft sufficiently to disengage flex-disc from transmission output flange. Tie driveshaft to side using stiff wire. See **260 Driveshafts**.

> ***CAUTION—***
> *• To prevent damage to rear driveshaft CV joint, do not allow driveshaft to hang down unsupported.*

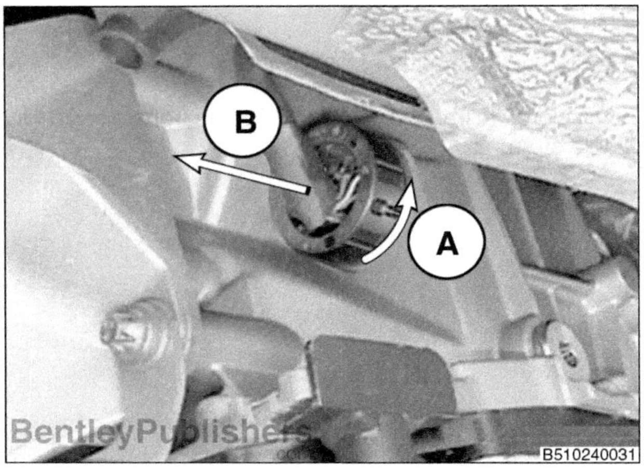

Working at rear of transmission housing:

- Twist mechatronics electrical connector bayonet lock in direction of curved arrow (**A**) to release.
- Carefully pull connector off pins (**B**).
- Insert BMW special tool 24 2 390 in place of mechatronics connector to protect delicate pins.

> ***CAUTION—***
> *• Do not touch pins.*

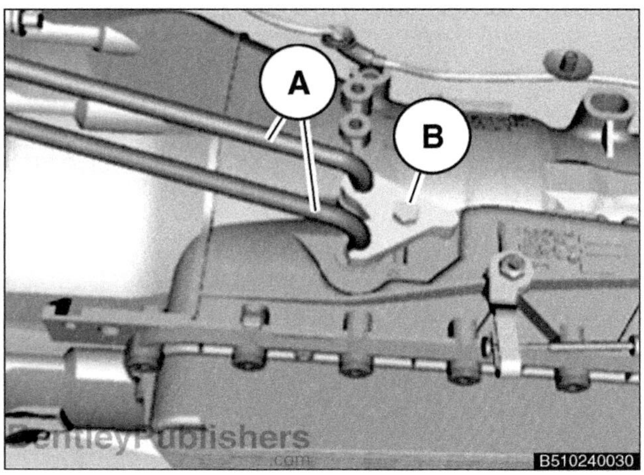

Working at ATF cooler lines (**A**) on left side of transmission:

- Remove fluid line bracket hold-down bolt (**B**)
- Detach fluid lines from transmission housing.

> ***CAUTION—***
> *• Be prepared to catch dripping fluid.*
> *• Remove and discard fluid line O-ring seals. Be sure to use new ones when reassembling.*

BentleyPublishers.com
BentleyPublishers.com—All Rights Reserved

Rotate engine by hand in direction of rotation using bolt on crankshaft vibration damper until torque plate mounting bolt (**arrow**) is visible through access hole in engine bell housing flange.

- Loosen and remove bolt.
- Rotate engine to gain access to and remove remaining torque plate bolts.

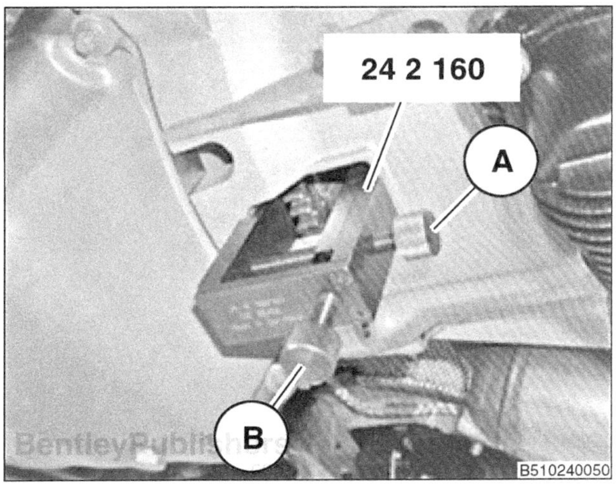

Use BMW special tool 24 4 160 or equivalent to secure torque converter in bell housing before you separate transmission from engine. This prevents torque converter from pulling off transmission input shaft and damaging torque converter oil seal.

- Insert tool in bell housing opening and secure in place lightly using knurled knob **A**.
- Twist in knurled knob **B** and tighten.
- Tighten down knob **A**.

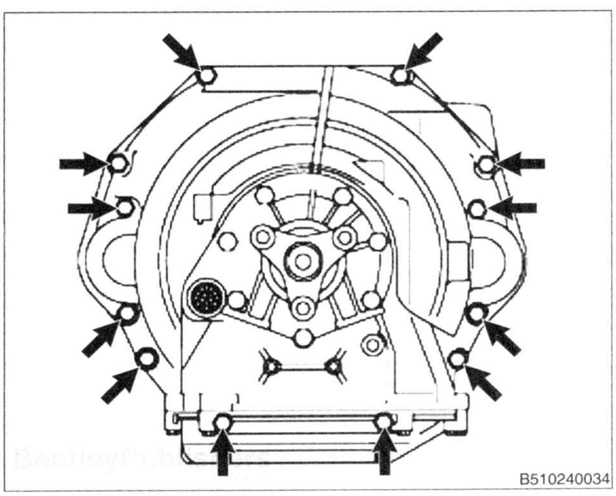

Remove transmission bell housing bolts (**arrows**). Discard aluminum bolts.

– Remove transmission by pulling back and down. Lower jack slowly while watching carefully to make sure no lines, hoses or wires become snagged.

> ***WARNING —***
> - *Be sure the vehicle is properly supported. The removal of the transmission may upset the balance of the vehicle on the lift.*

> ***CAUTION—***
> - *Tilting the engine to remove the transmission can lead to damage to engine compartment components due to lack of clearance.*
> - *Do not allow the torque converter to fall off the transmission input shaft.*

BentleyPublishers
BentleyPublishers.com—All Rights Reserved

– Blow out oil cooler lines with low-pressure compressed air and flush cooler with clean ATF twice.

> ***CAUTION—***
> *• Wear safety glasses when working with compressed air.*
> *• Do not reuse ATF used for flushing.*

– Inspect engine torque plate and flywheel for cracks or elongated holes. Replace if necessary.

– If torque converter seal is leaky, or torque converter position on transmission input shaft was disturbed during removal, replace torque converter seal. See **Torque converter oil seal, removing and installing** in this repair group.

Transmission, installing (V8 engine)

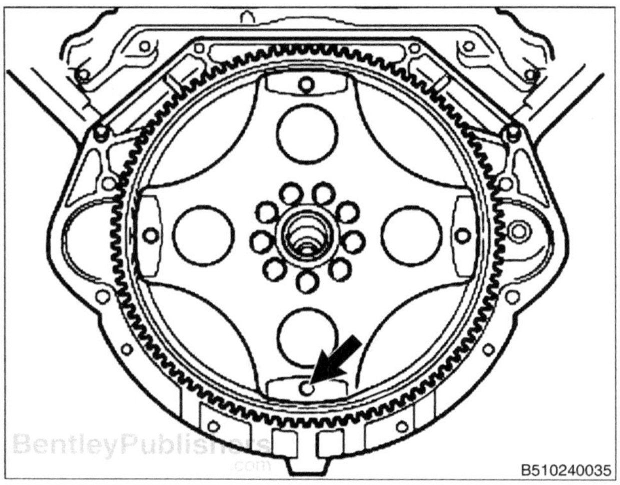

◄ Rotate engine until torque plate bore for mounting bolt (**arrow**) is accessible through opening in engine bell housing.

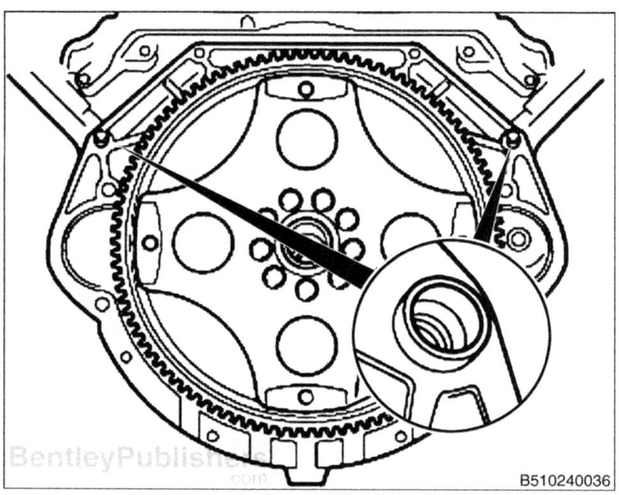

◄ Check that bell housing dowel sleeves (**arrows**) are in good shape and correctly seated. Replace if needed.

– Check that torque converter is seated correctly in transmission.

– Rotate torque converter until threaded mounting hole in converter is lined up with bore in torque plate.

– Join transmission to engine. Make sure torque converter and torque plate bolt bores are lined up by installing one bolt finger-tight.

BentleyPublishers.com
BentleyPublishers.com—All Rights Reserved

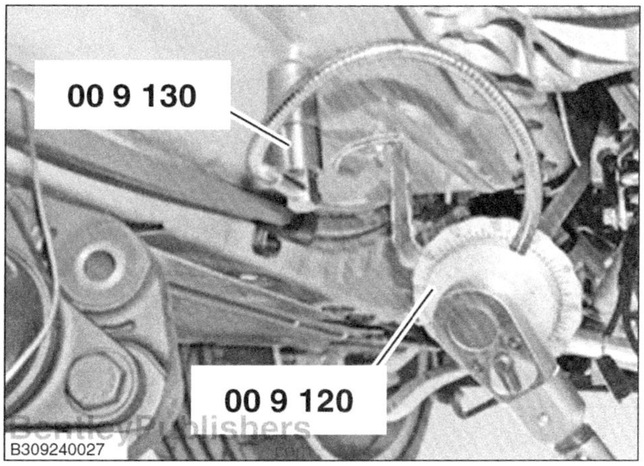

B309240027

Using new aluminum bolts, tighten bell housing fasteners to initial torque.

- Use torque angle tool (BMW special tools 00 9 120 / 00 9 130 or equivalent) and tighten fasteners to final torque angle.
- Install and tighten remaining torque converter bolts.

Tightening torques	
Torque converter to torque plate: • M10 x 10.9	56 Nm (41 ft-lb)
Transmission rear support to body (M8)	19 Nm (14 ft-lb)
Transmission rear support to transmission	19 Nm (14 ft-lb)
Transmission to engine (M12 aluminum): • Initial torque • Torque angle	25 Nm (18 ft-lb) + 130°
Transmission to engine (M10 x 30 aluminum): • Initial torque • Torque angle	20 Nm (15 ft-lb) + 90 - 110°
Transmission to engine (M10 x 85 aluminum): • Initial torque • Torque angle	20 Nm (15 ft-lb) + 180 - 200°

— Remainder of installation is reverse of removal. Remember to:

- Install new sealing washers or O-rings on ATF cooler line fittings.
- 2006 - 2007 models: Adjust gearshift mechanism. See **250 Gearshift Linkage**.
- 2008 - 2009 models: Adjust parking lock cable free play. See **250 Gearshift Linkage**.
- Driveshaft tightening torques are in **260 Driveshafts**. To avoid damaging driveshaft flex-disc, counterhold flex-disc bolt heads while tightening nuts.
- Fill transmission with clean ATF. See **ATF, draining and filling** in this repair group.
- Use new bolts to reinstall front end reinforcement plate. See **310 Front Suspension**.

Tightening torques	
Drain plug to plastic sump	8 Nm (6 ft-lb)
Fill plug to transmission housing	35 Nm (26 ft-lb)

5 Series automatic transmission fluid (ATF)	
BMW recommended fluid	BMW part no. 83 22 0 142 516 Shell M-1375.4
Fill capacity	approx. 9 - 10 liters (9.5 - 10.6 US qt)

CAUTION—
- *Do not overfill with ATF. Transmission fill capacity given in specifications includes torque converter volume.*
- *ATF expands when transmission is warmed up. Final fluid level is set with ATF at a temperature of 30° - 50°C (86° - 122°F).*

BentleyPublishers.com—All Rights Reserved

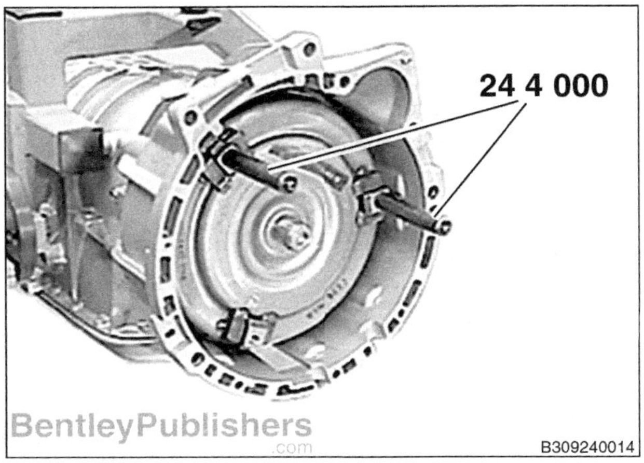

Torque Converter Service

Torque converter, removing and installing

– Remove transmission. See **Transmission Removal and Installation** in this repair group.

◄ Screw BMW special tools 24 4 000 into torque converter. Pull converter straight off transmission input shaft.

> ***CAUTION—***
> - *When torque converter is removed, transmission fluid will drain out. Be prepared to catch dripping fluid.*

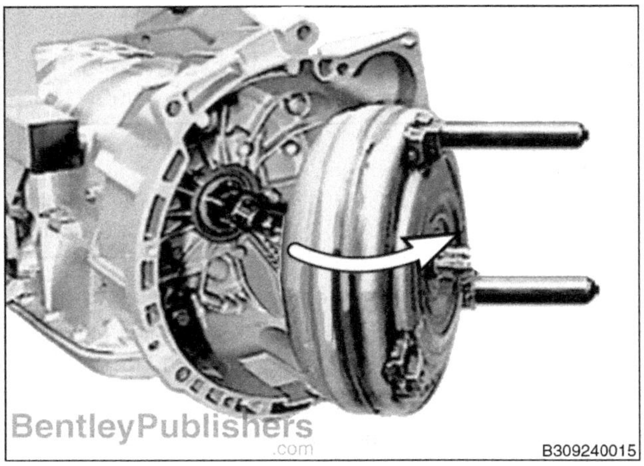

◄ Remove torque converter slowly and set down vertically to avoid spilling additional transmission fluid.

– Replace torque converter seal. See **Torque converter oil seal, removing and installing** in this repair group.

– Reinstall torque converter, taking care to not damage new seal. Lightly oil converter seal surface and rotate converter during installation, applying slight pressure until recesses in converter locate audibly in ATF pump. Then press converter in firmly.

> ***CAUTION—***
> - *If torque converter is not installed correctly, impeller driver in converter is destroyed when transmission is mated to engine.*

Torque converter oil seal, removing and installing

ATF leaking from the torque converter seal usually collects at the bottom of the bell housing and drips out.

Torque converter oil seal leakage is often caused by a worn or scored bushing in the torque converter hub. Check bushing when replacing seal. A damaged bushing rapidly wears the new seal.

– Remove transmission. See **Transmission Removal and Installation** in this repair group.

– Carefully slide torque converter off transmission input shaft. Be prepared to catch ATF as it flows out of torque converter. See **Torque converter, removing and installing** in this repair group.

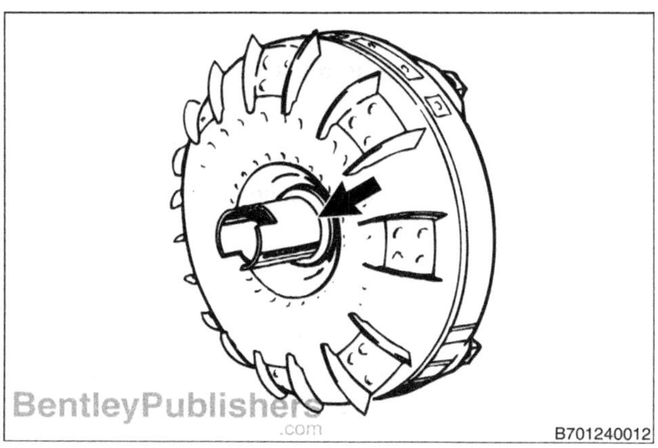

◄ Check converter hub bushing surface (**arrow**) for scoring or wear.
- Remove sharp edges and burrs with fine emery cloth.
- If hub is deeply scored, replace torque converter.

BentleyPublishers.com
BentleyPublishers.com—All Rights Reserved

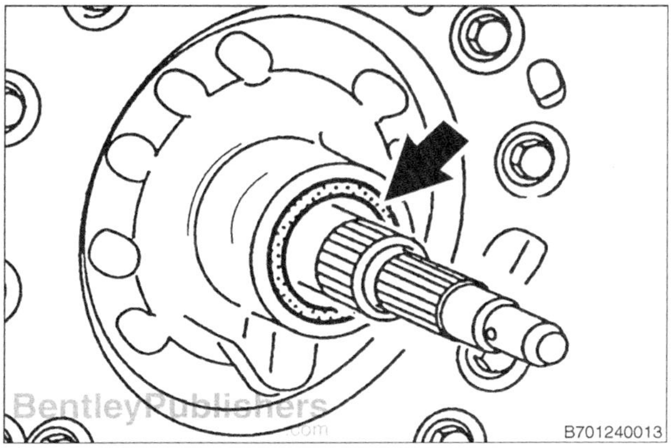

Use small screwdriver to pry out seal retaining circlip (if equipped) (**arrow**) from ATF pump at front of transmission case.

Attach BMW special tool 24 2 353 or 24 2 354 (depending on transmission) to transmission input shaft (**A**).

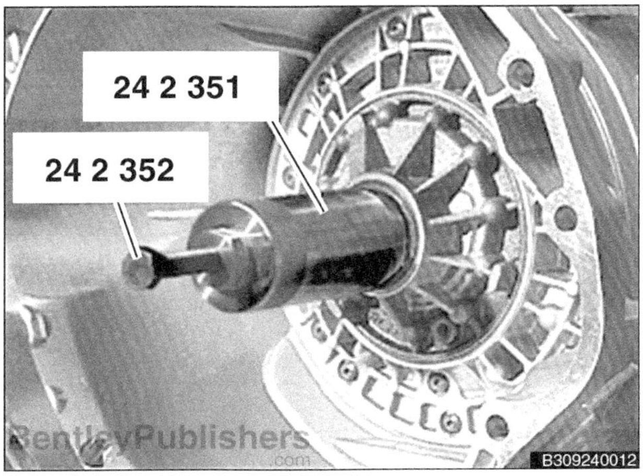

Screw BMW special tool 24 2 351 into seal. Use BMW special tool 24 2 352 to gradually draw out seal.

BentleyPublishers.com—All Rights Reserved

Coat sealing lips of new seal with transmission fluid. Use BMW special tool 24 2 360 or 24 2 400 (depending on transmission) and soft-faced hammer to drive seal (**A**) into transmission as far as it will go.

– Reinstall retaining circlip (if equipped).

– Reinstall torque converter, taking care to not damage new seal. Lightly oil converter seal surface and rotate converter during installation, applying slight pressure until recesses in converter locate audibly in ATF pump. Then press converter in firmly.

> ***CAUTION—***
> - *If torque converter is not installed correctly, impeller driver in converter is destroyed when transmission is mated to engine.*

Torque plate and flywheel, removing and installing

Crankshaft rear main (flywheel) seal replacement is covered in **119 Lubrication System**.

– Remove transmission. See **Transmission, removing and installing** in this repair group.

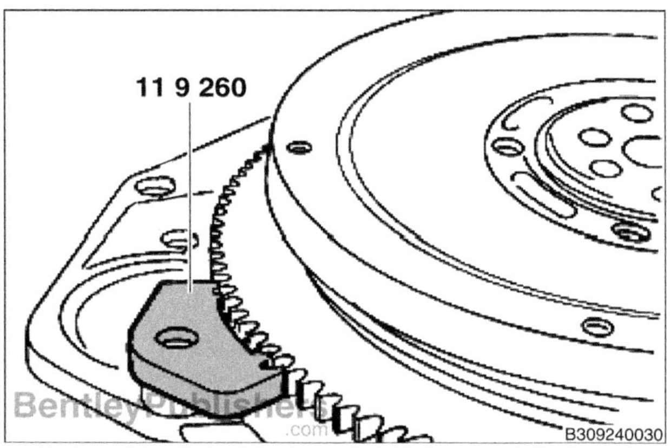

Use BMW special tool 11 9 260 or equivalent to lock flywheel.

Loosen and remove flywheel bolts (**arrows**) and discard.

– Remove torque plate and flywheel.

– Clean bolt threads in crankshaft.

BentleyPublishers.com
BentleyPublishers.com—All Rights Reserved

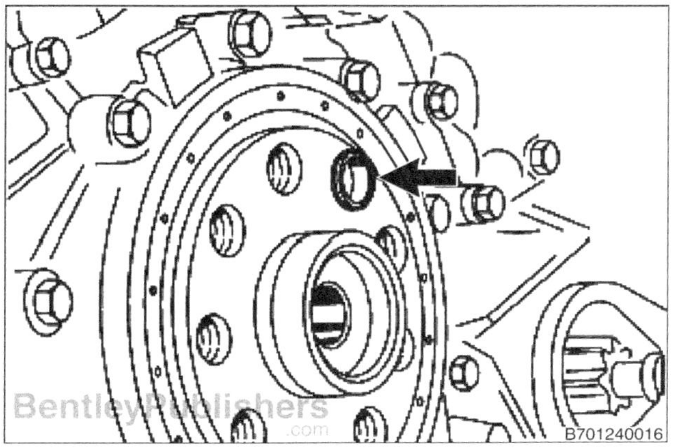

When installing:

- Note that flywheel is located with dowel sleeve (**arrow**).
- Make sure torque converter mounting holes in torque plate and flywheel line up.
- Use new flywheel bolts.

Tightening torque	
Flywheel and torque plate to crankshaft (M12 x 1.5) (use new bolts)	130 Nm (96 ft-lb)

– Reinstall transmission.

BentleyPublishers
Bentley Publishers.com—All Rights Reserved

BentleyPublishers.com
BentleyPublishers.com—All Rights Reserved

250 Gearshift Linkage

GENERAL

This repair group covers transmission gearshift for manual, sequential manual gearbox (SMG) and automatic transmission models.

See also:

- **180 Exhaust System**
- **260 Driveshafts**
- **310 Front Suspension** for front end reinforcement plate

Neutral safety switch function

In models with automatic transmission, starter operation is prevented unless the gear position is (P)ARK or (N)EUTRAL.

BentleyPublishers.com
BentleyPublishers.com—All Rights Reserved

Gearshift knob, removing

This procedure applies to the following:

- Manual transmission models
- Automatic models to 2007

Make sure the engine is OFF when removing gearshift knob.

Pull gearshift knob straight up (**arrow**) to remove. It may require up to 90 lbs. of force to do this.

> ***CAUTION—***
> - *Automatic transmissions for 2008 - 2009 models (starting with 03 / 2007 production) are equipped with gear shift switch (GWS) instead of a shift lever. The procedure given here for gearshift knob removal does not apply.*
> - *Some versions of shift boot and knob are one piece. Release boot from console before removing knob.*
> - *To prevent damage to shift knob locating tabs, do not twist while removing.*

Caution

> ***CAUTION—***
> - *When removing interior trim clips, avoid damaging trim pieces by using a plastic prying tool or a screwdriver with the tip wrapped with masking tape.*

TROUBLESHOOTING

Shift lever blocked

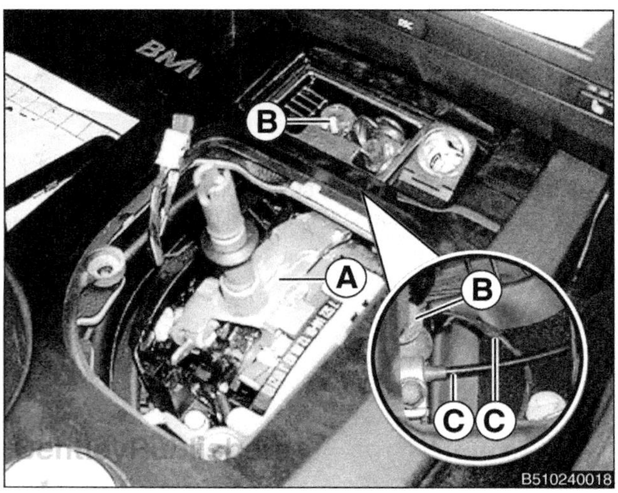

Automatic transmission: If shift lever (**A**) cannot be moved to P or is otherwise blocked from moving, check to see if coins (**B**) from coin holder are lodged underneath shifter Bowden cable (**C**) pivots.

BentleyPublishers.com
BentleyPublishers.com—All Rights Reserved

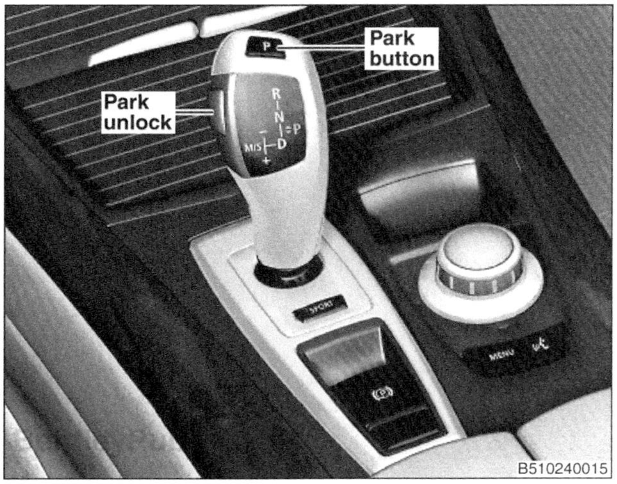

Transmission does not shift to P

2008 - 2009 models (starting production date 03 / 2007) with automatic transmission: Transmission parking lock is engaged using a dedicated cable. Parking lock is triggered when P button is pressed on shift lever.

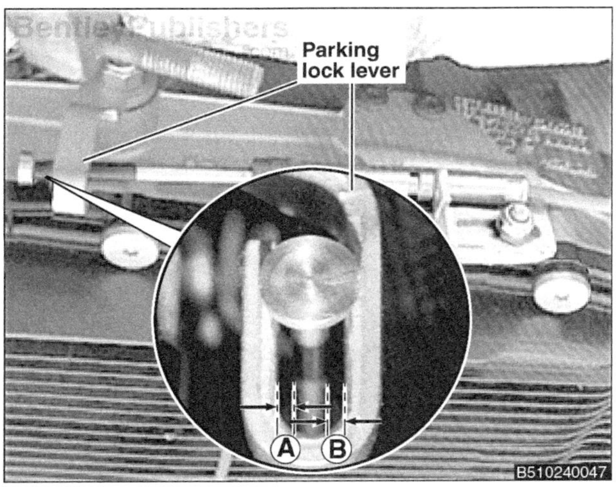

If P button is pressed but transmission does not shift into parking lock (BMW fault code 507B), inspect position of parking lock cable end in parking lock lever (left side of transmission, underneath vehicle):

- If cable end button is centered in lever hook (dimensions **A** and **B** are equal), parking lock lever is OK.
- If cable end button is offset (dimension **A** and **B** are unequal), parking lock lever is bent.

Manual Transmission Gearshift

Manual gearshift linkage

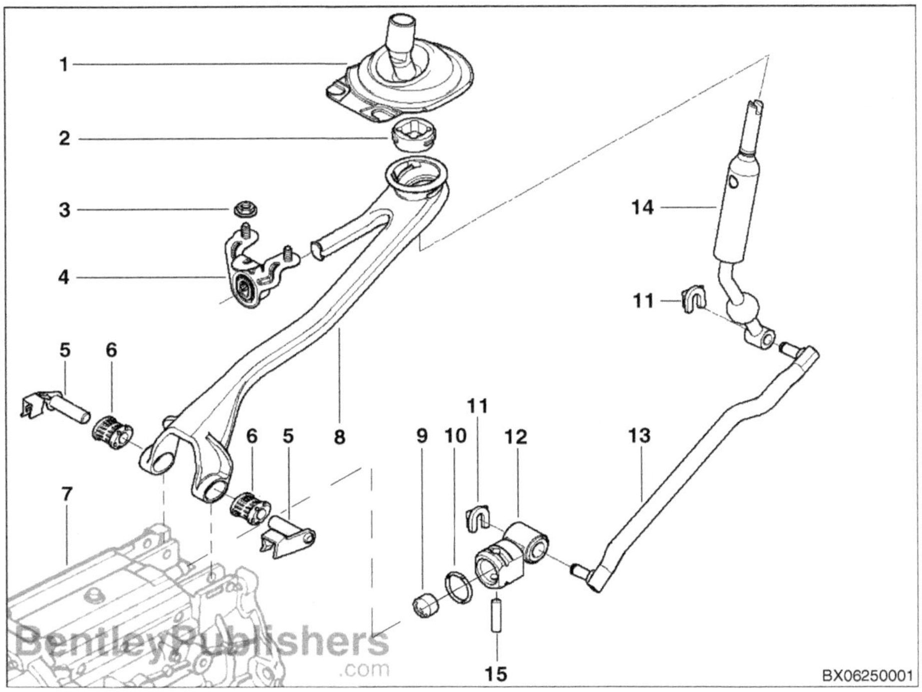

1. Rubber boot
2. Shift lever bearing
3. M8 nut
4. Shift arm support bearing
5. Lock pin
6. Bushing
7. Transmission housing
8. Shift arm
9. Bushing
10. Lock ring
11. Circlip
12. Shifter coupling
13. Selector rod
14. Shift lever
15. Dowel

Gearshift lever, removing and installing (manual transmission)

– Remove shift boot and shift knob. See **Gearshift knob, removing** in this repair group.

> ***CAUTION—***
> - *Some versions of shift boot and knob are one piece. Release boot from console before removing knob.*
> - *To prevent damage to locating tabs on shift knob, do not twist while removing.*

– Remove sound insulation at base of shift lever.

– Raise and safely support vehicle.

> ***WARNING —***
> - *Make sure the vehicle is stable and well supported at all times. Use a professional automotive lift or jack stands designed for the purpose. A floor jack is not adequate support.*

– Remove splash shield under transmission. See **020 Maintenance**.

– Remove complete exhaust system. See **180 Exhaust System**.

> ***CAUTION—***
> - *To avoid damaging oxygen sensors, transport and store exhaust system carefully.*

BentleyPublishers.com
BentleyPublishers.com–All Rights Reserved

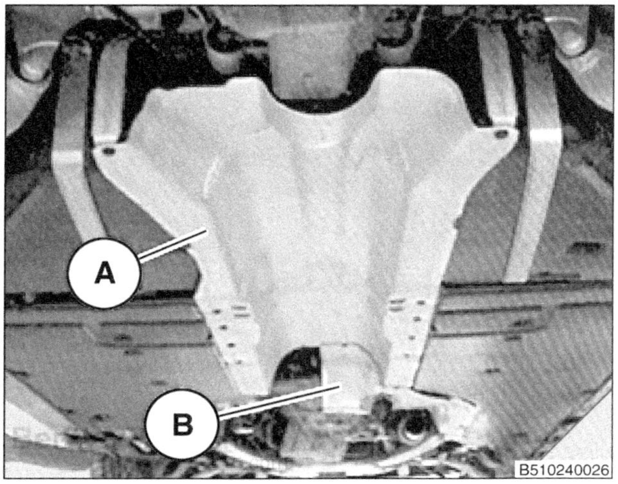

< Remove heat shields (**A**, **B**).

- Support transmission with a suitable jack. Remove transmission rear support and lower transmission slightly to gain access to gearshift lever.

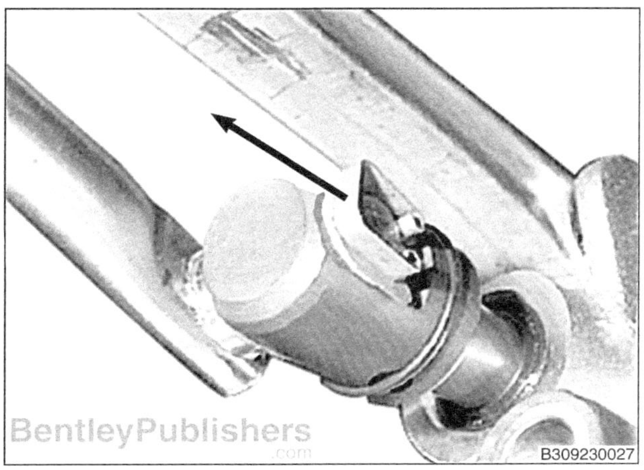

< Working above transmission, disconnect selector rod:
- Pull off selector rod retaining circlip (**arrow**) and discard.
- Disengage selector rod from shifter coupling to.

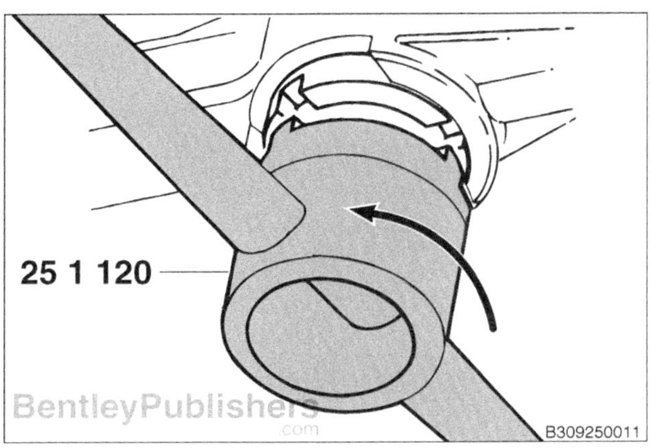

< Release shift lever bearing from below using BMW special spanner 25 1 120. Turn tool 90° (¼ turn) counterclockwise.
- Push shift lever and bearing up to remove from shift arm.

- Lower vehicle. Working from inside passenger compartment, pull up on shift lever to remove it together with inner rubber boot and shift lever bearing.

- Clean old grease from shift lever ball and mounting ring. Also clean shift arm bowl. Lubricate with molybdenum disulfide grease.

- Insert shift lever together with inner boot and bearing into shift arm.

Gearshift lever, removing and installing (manual transmission)

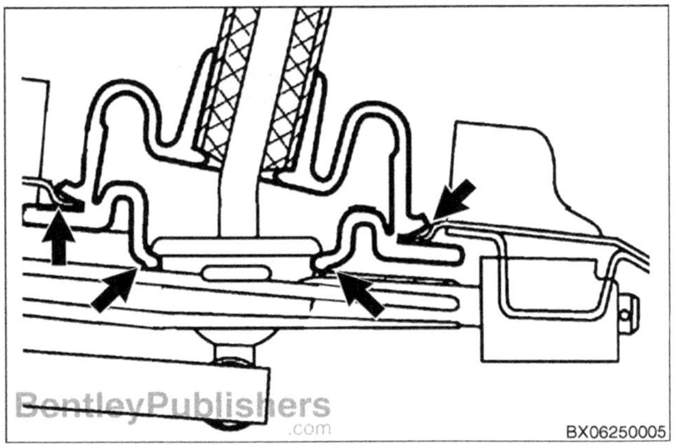

Pull inner rubber boot over shift arm bowl (**arrows**) to seal base of shift lever. Install rubber boot with arrow pointing toward front of vehicle.

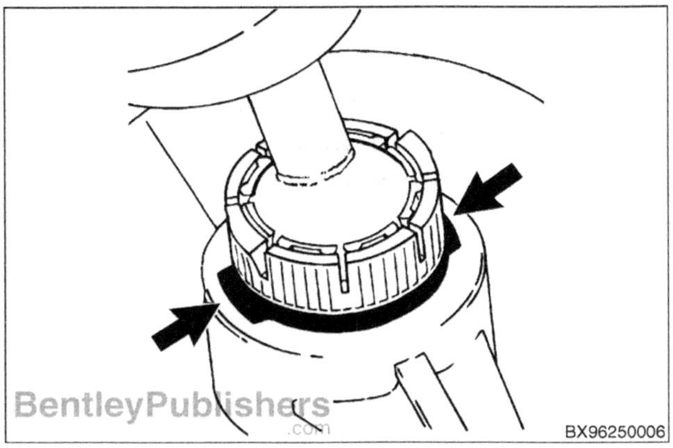

When installing shift lever bearing:

- Align locking tabs with slots (**arrows**) in shift arm bowl.
- Make sure arrow on shift lever bearing points to front of vehicle.
- Press down until bearing snaps into place.

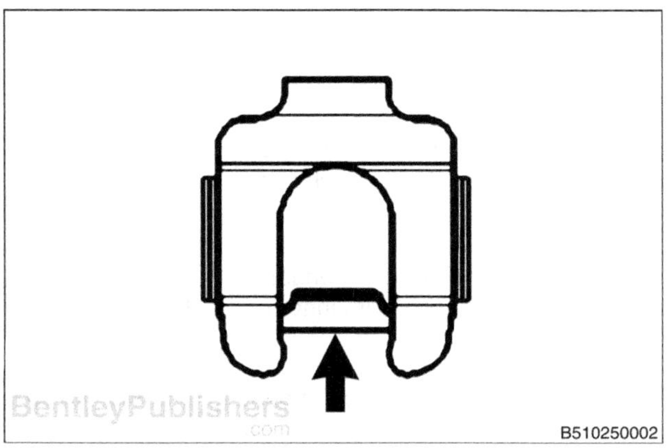

When reinstalling selector rod, be sure to use new shift rod circlip with locking tang (**arrow**).

BentleyPublishers.com
BentleyPublishers.com—All Rights Reserved

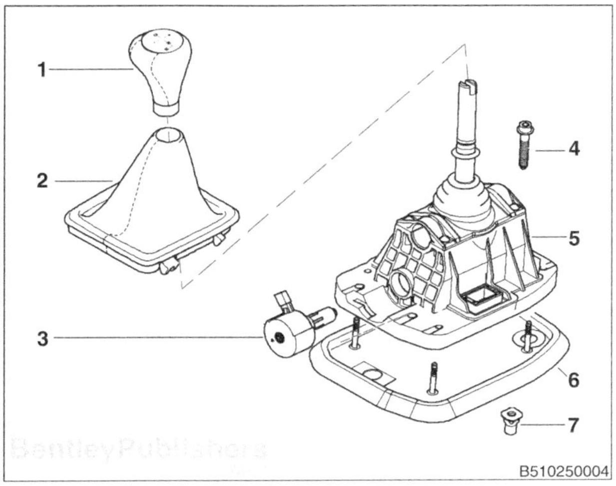

SMG Gearshift

SMG gearshift assembly

1. Shift lever knob
2. Shift lever boot
3. Shiftlock magnetic solenoid
4. M6 x 35 mm bolt
5. Gearshift housing
6. Sealing frame
7. Blind rivet

Automatic Transmission Gearshift

Steptronic gearshift assembly (2004 - 2007)

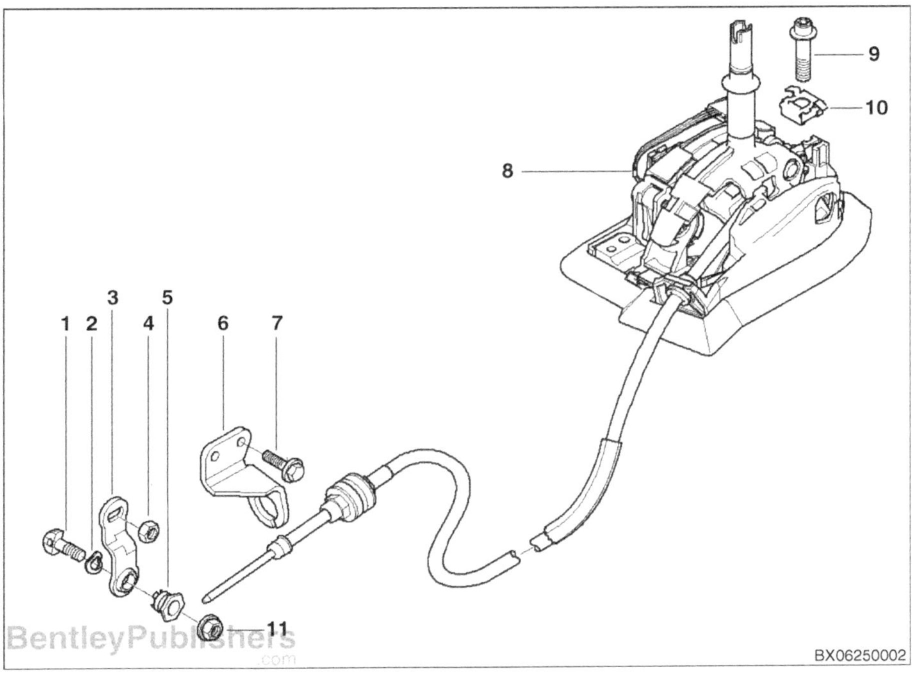

1. Clamping bolt
2. Washer
3. Transmission shift arm on side of transmission
4. M8 self-locking nut
5. Bushing
6. Support bracket
7. M6 bolt
8. Steptronic shift mechanism
9. M6 Torx screw
10. Cage nut
11. M8 self-locking nut

Gearshift mechanism, adjusting (2004 - 2007)

Automatic gearshift mechanism varies with model. Adjustment is similar.

– Position gearshift lever in P(ARK).

– Raise vehicle to gain access to shift linkage.

> ***WARNING —***
> • *Make sure the vehicle is stable and well supported at all times. Use a professional automotive lift or jack stands designed for the purpose.*

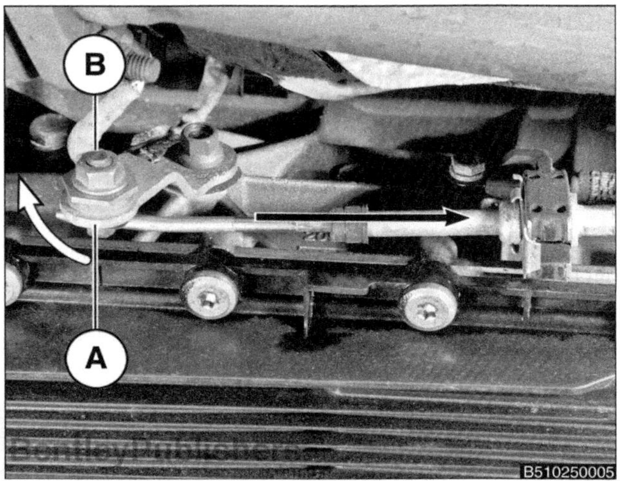

Working at transmission selector lever at side of transmission:

- Counterhold cable clamping bushing (**A**).
- Loosen cable clamping nut (**B**).
- Push transmission shift arm forward toward engine (**curved arrow**) (PARK position).
- Push shifter cable end backward away from engine (**arrow**). Release pressure on cable.
- Tighten clamping nut.

> ***CAUTION—***
> *• Do not overtighten the nut so that it twists the cable.*

Tightening torque	
Shift cable clamping nut	10 Nm (7 ft-lb)

Automatic shiftlock, checking function (2004 - 2007)

The automatic shiftlock uses an electric solenoid to lock the selector lever in P(ARK) or N(EUTRAL). Depressing the foot brake with the ignition ON energizes the solenoid, allowing the lever to be moved into R(EVERSE) or D(RIVE). The solenoid is energized only when engine speed is below 2,500 rpm and vehicle speed is below 5 kph (3 mph). The solenoid is mounted in the right side of the selector lever housing.

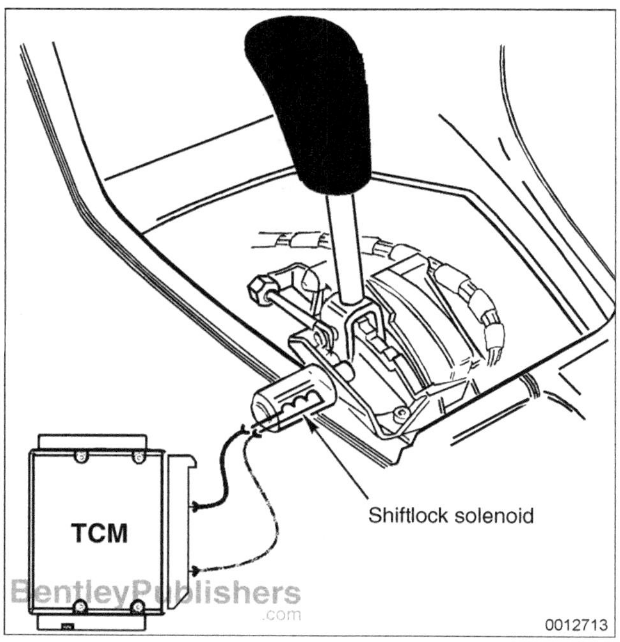

Automatic shiftlock prevents drive gear selection until brake pedal is depressed.

- With engine running and vehicle stopped, place selector lever in P.
- Without depressing brake pedal, check that selector lever is locked in P.
- Depress brake pedal firmly and listen to solenoid clicking audibly.
- Check that selector lever can now be moved out of P.

NOTE—
• Perform the next test in an open area with the parking brake ON and with extreme caution.

– With selector lever in P and brake pedal depressed, raise engine above 2,500 rpm. Check that selector lever cannot be moved out of P.

– If any faults are found:

- Check electrical operation of shiftlock solenoid
- Check for wiring faults to or from transmission control module (TCM). See **ECL Electrical Component Locations** and **ELE Electrical Wiring Diagrams**.

NOTE—
• The solenoid is controlled via the transmission control module (TCM), using brake pedal position, engine speed, and road speed as controlling inputs.

BentleyPublishers.com
BentleyPublishers.com—All Rights Reserved

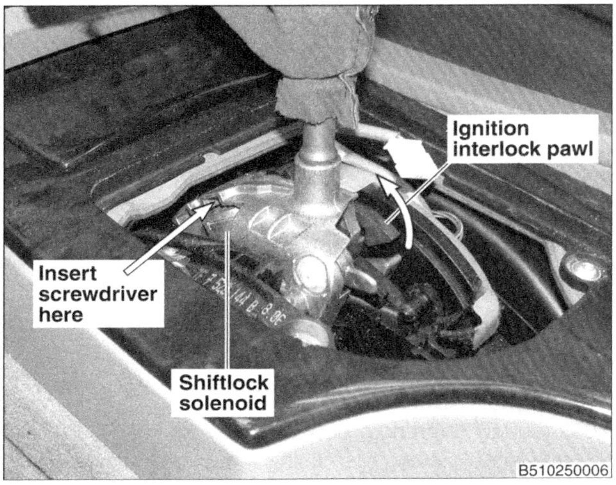

Automatic shiftlock, releasing manually (2004 - 2007)

Normally, the automatic transmission shift lever can only be moved out of P if ignition is ON (or engine running) and brake pedal depressed. In case of power failure (such as dead battery), release move gearshift manually and shift to N before moving vehicle or working.

> *CAUTION—*
> - *Attempting to move vehicle without taking transmission out of P may cause transmission damage.*

- Engage parking brake so that vehicle cannot roll.
- Unclip and pull up shift lever bezel.

◄ Working at shift lever:

- Switch ignition ON to disengage ignition interlock pawl (**curved arrow**).
- Using thin screwdriver, depress shiftlock solenoid plunger and move shifter to N.

250

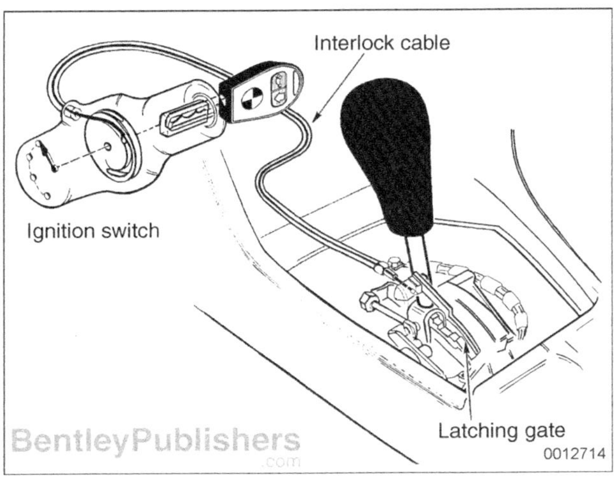

Ignition interlock, checking function (2004 - 2006)

◄ Ignition interlock uses a cable between ignition switch and selector lever to lock lever in P(ARK) position when ignition key is OFF or removed. This feature also prevents key from being removed from ignition lock unless selector lever is in P.

- Shift selector lever to P and switch ignition OFF.
- Remove ignition key. Check that selector lever cannot be shifted out of P.
- Switch ignition ON and depress brake pedal. Check that selector lever moves freely from gear to gear.
- With selector lever in a gear position other than P, attempt to turn key to OFF position and remove. Key should not go into OFF position.
- If any faults are found:
  - Check interlock cable for kinks.
  - Check cable attachment points for damage or faults.

NOTE—
- *Make sure ignition key can only be removed with selector lever in P(ARK).*
- *Do not kink shift interlock cable.*
- *Ignition interlock cable replacement is covered in* **320 Steering and Wheel Alignment**.

BentleyPublishers
BentleyPublishers.com—All Rights Reserved

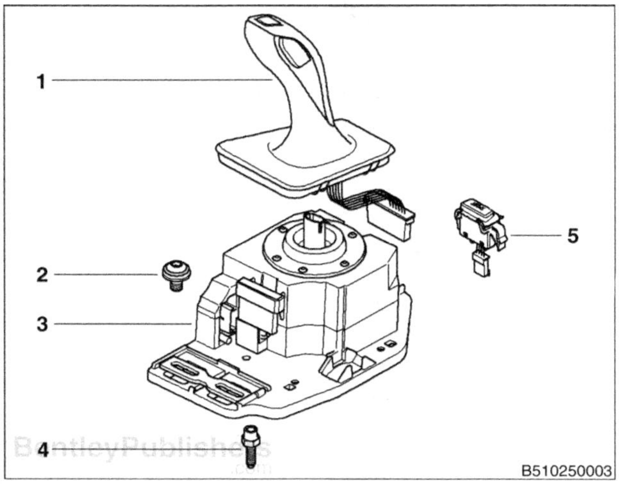

Gear shift switch (GWS) (2008 - 2009)

In models built from 03 / 2007 (2008 model year), the shift lever is configured as a control module (referred to as gear shift switch or GWS). Transmission gear range selection in these models is actuated electronically instead of via shift cable.

1. Shift lever handle
2. M6 x 18 mm screw
3. Gear shift switch (GWS) housing
4. Spacer pin
5. SPORT switch

Parking lock cable bracket, adjusting (2007 - 2008)

On 2008 - 2009 models (starting production date 03 / 2007), transmission parking lock is engaged using a dedicated cable. To adjust cable correctly:

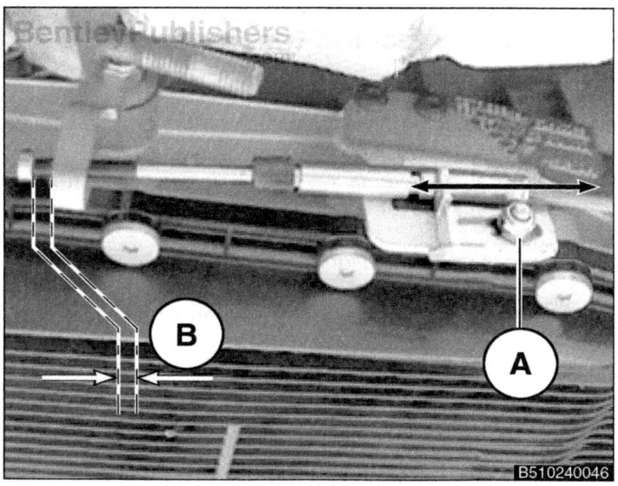

Loosen cable bracket fastener (**A**) and slide bracket to achieve correct free play.

Parking lock cable end free play	
Gap **B**	1 mm (0.04 in)

See also **Transmission does not shift to P** in this repair group.

Parking lock, releasing manually (2008 - 2009)

Normally, the automatic transmission shift lever can only be moved out of P if ignition is ON (or engine running) and brake pedal depressed. In case of power failure (such as dead battery), release transmission parking lock before moving vehicle or working on transmission.

> ***CAUTION—***
> *• Attempting to move vehicle without releasing parking lock may cause transmission damage.*

– Engage parking brake so that vehicle cannot roll.

– Unclip and pull up shift lever bezel.

BentleyPublishers.com
BentleyPublishers.com—All Rights Reserved

Insert screwdriver into opening (**A**). Pull screwdriver toward rear until it is heard to engage (**B**). Leave screwdriver inserted in opening. This releases transmission park lock.

– To relock parking lock, pull out screwdriver.

BentleyPublishers.com—All Rights Reserved

BentleyPublishers.com
BentleyPublishers.com—All Rights Reserved

260 Driveshafts

GENERAL

This repair group covers repair and replacement of front and rear driveshafts and driveshaft components.

See also:

- **270 Transfer case**
- **310 Front Suspension** for front end components
- **311 Front Axle Differential** for front drive axle repairs
- **331 Rear Axle Differential** for rear drive axle repairs

Driveshafts overview

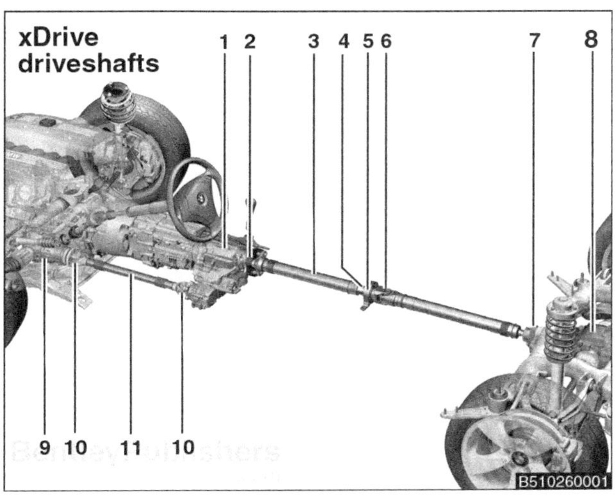

xDrive models are equipped with a transfer case bolted to the rear of the transmission, with front and rear driveshafts transferring power to front and rear differentials. Rear-drive models are equipped with a single driveshaft attached directly to the rear of the transmission.

1. xDrive transfer case
2. Flex-disc
3. Rear driveshaft
4. Crash element (aluminum driveshaft only)
5. Center bearing and mount
6. Center universal joint
7. CV joint
8. Rear differential
9. Front differential
10. Universal joint
11. Front driveshaft

Front driveshaft repairs are limited to replacement of the complete driveshaft. Rear driveshaft repairs consist of replacement of the flex-disc, CV joint, and center mount and bearing. If the rear driveshaft universal joint is defective, replace the complete driveshaft.

BentleyPublishers.com—All Rights Reserved

6-cylinder vehicles are equipped with aluminium or hybrid (aluminum-steel) rear driveshafts, whereas V8 model rear driveshafts are steel. Driveshaft are mounted with ZNS (shiny zinc-coated) fasteners. Be sure to replace ZNS fasteners with same.

Cautions

CAUTION—
- *Be sure the wheels are off the ground before removing the driveshaft. Set the parking brake before removing the driveshaft.*
- *Use only ZNS (zinc-coated) fasteners to mount driveshafts.*
- *ZNS fasteners are designed to be used only once. Replace during reassembly.*
- *Do not move vehicle using engine power once a driveshaft is removed.*
- *Driveshafts are balanced to close tolerances. Whenever a driveshaft is removed or disassembled, mark mounting flanges and driveshaft sections with paint before proceeding with work. This ensures that the driveshaft can be reassembled and installed in exactly the original orientation.*
- *Do not allow rear driveshaft to hang from flex-joint or rear CV joint. The weight of the driveshaft may damage rubber joints or boots.*

TROUBLESHOOTING

Driveshaft problems

Sources of driveline vibrations and noise may be difficult to pinpoint. Engine, transmission, front or rear axle or wheel vibrations may be transmitted through the driveshafts to the car body.

Noises in the interior of vehicle may be caused by:
- Differential problems
- Faulty wheel bearings
- Damaged drive axle(s)
- Worn or improperly inflated tires

Driveshaft noise or vibrations are usually caused by worn or damaged components.

– To inspect or remove rear driveshaft, remove exhaust system and exhaust heat shield. See **180 Exhaust System**.

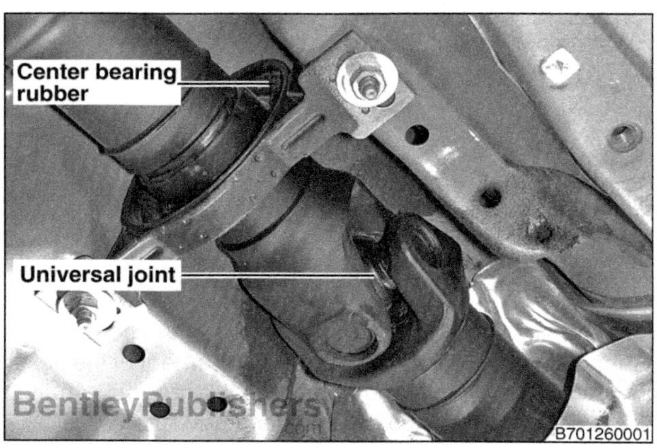

Check driveshaft universal joint for play. Pull and twist driveshaft while watching universal joint. Specified tolerance for play is very small, so almost any noticeable play indicates a problem.

Universal joint play	
Maximum allowable	0.15 mm (0.006 in)

– Check fastener torque at flange connections.

– Check rear driveshaft flex-disc and center bearing rubber for deterioration or tearing.

BentleyPublishers.com
BentleyPublishers.com—All Rights Reserved

– Check driveshaft for broken or missing balance weights. Weights are welded tabs on driveshaft tubes.

– Use paint to matchmark front and rear flanges. Remove driveshaft. See **Rear driveshaft, removing** in this repair group.

– Check rear driveshaft universal joint for wear or binding. If it is difficult to move or binds, replace driveshaft.

– Check rear driveshaft centering guide at output flange for damage or misalignment.

– Check runout at transmission or transfer case output flange and output shaft, and at rear differential input flange.

Driveshaft flange wear limits	
Runout at transfer case output flange (max. allowable): • Axial play • Radial play	 0.10 mm (0.004 in) 0.07 mm (0.003 in)
Rear differential input flange radial play (measured at driveshaft center lip)	0.07 mm (0.003 in)

– Check bolt bores at transmission or transfer case output flange and front and rear differential input flanges for wear and elongation.

– Spin rear driveshaft center bearing and check for smooth operation without play.

If inspection reveals nothing wrong with driveshaft, it may need to be rebalanced. This can be done by a speciality driveshaft repair shop.

Sometimes a minor rear driveshaft vibration can be corrected by disconnecting at front flange and repositioning it 120° or 240° in relation to differential input flange.

Table a lists symptoms of rear driveshaft problems and their probable causes. Most of the repair information is contained within this repair group. There are references to other repair groups, where applicable.

Table a. Rear driveshaft problems and causes

Symptom	**Probable cause**	**Corrective action**
Vibration when starting off (forward or reverse).	Center bearing rubber deteriorated.	Inspect center bearing and rubber. Replace if necessary. See **Center bearing assembly, replacing** in this repair group.
	Flex-disc damaged or worn.	Inspect flex-disc. Replace if necessary. See **Flex-disc, replacing** in this repair group.
	Engine or transmission mounts faulty.	Inspect engine and transmission mounts. Align or replace, if necessary.
	Front centering sleeve worn, or driveshaft mounting flanges out of round.	Check front centering sleeve and replace if necessary. See **Front centering sleeve, replacing** in this repair group. Also check runout of driveshaft flanges.
	Universal joint worn or seized.	Check universal joint play and movement. Replace driveshaft if necessary.

BentleyPublishers.com
BentleyPublishers.com–All Rights Reserved

Table a. Rear driveshaft problems and causes (continued)		
Symptom	**Probable cause**	**Corrective action**
Noise during on / off throttle or when engaging clutch.	Differential components worn or damaged (excessive pinion-to-ring-gear clearance).	Remove differential and repair. See **311 Front Axle Differential** or **331 Rear Axle Differential**.
	Drive axle or CV joint faulty.	Inspect drive axles and CV joints. Repair or replace as necessary. See **311 Front Axle Differential** or **331 Rear Axle Differential**.
Vibration at 25 to 30 mph (40 to 50 kph).	Front centering sleeve worn, or driveshaft mounting flanges out of round or damaged.	Check front centering sleeve and replace if necessary. See **Front centering sleeve, replacing** in this repair group. Also check runout of driveshaft flanges.
	Universal joint worn or seized.	Check universal joint play and movement. Replace driveshaft if necessary.
	Flex-disc damaged or worn.	Inspect flex-disc. Replace if necessary. See **Flex-disc, replacing** in this repair group.
	Center bearing rubber deteriorated.	Inspect center bearing. Replace if necessary. See **Center bearing assembly, replacing** in this repair group.
Vibration, audible rumble over 35 mph (60 kph).	Front centering sleeve worn, or driveshaft mounting flanges out of round or damaged.	Check front centering sleeve and replace if necessary. See **Front centering sleeve, replacing** in this repair group. Also check runout of driveshaft flanges.
	Mounting flange bolts loose or holes worn.	Remove rear driveshaft and check input and output flange bolts and holes. Replace if necessary.
	Rear driveshaft unbalanced.	Check driveshaft for loose or missing balance weights. Have driveshaft rebalanced or replace if necessary.
	Universal joint worn or seized.	Check universal joint play and movement. Replace rear driveshaft if necessary.
	Rear driveshaft center bearing faulty.	Replace center bearing. See **Center bearing assembly, replacing** in this repair group.
	Rear differential rubber bushing(s) faulty.	Inspect differential rubber bushing(s). Replace as necessary. See **33 Rear Axle Differential**.

FRONT DRIVESHAFT SERVICE

If detecting vibration or noise from the front driveshaft, check front universal joints for play. Pull and twist driveshaft while watching universal joint. Specified tolerance for play is very small, so almost any noticeable play indicates a problem.

Universal joint play	
Maximum allowable	0.15 mm (0.006 in)

Front driveshaft, removing and installing

– Raise car and support safely.

> ***WARNING —***
> *• Make sure the car is stable and well supported at all times. Use a professional automotive lift or jack stands designed for the purpose. A floor jack is not adequate support.*

– Remove splash shields underneath engine and transmission.

BentleyPublishers.com
BentleyPublishers.com—All Rights Reserved

– Remove front end reinforcement plate. Discard fasteners.

– If driveshaft is to be reinstalled, match mark mounting flanges with paint to preserve original orientation.

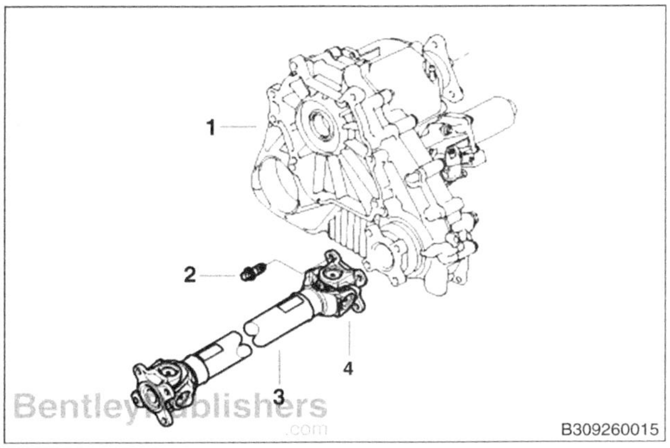

< Front driveshaft components:

1. Transfer case
2. M8 ZNS bolt (replace with new)
   - Tighten to 20 Nm (15 ft-lb) + additional 45°
3. Driveshaft
4. Universal joint

– Working at front differential input flange, remove universal joint bolts.

– Working at transfer case flange, remove universal joint bolts and remove driveshaft.

– Installation is reverse of removal. Remember to:
- Observe marks made previously (if applicable).
- Replace ZNS fasteners with new.

Tightening torque	
Front driveshaft to transfer case or differential (use new M8 ZNS bolts)	20 Nm (15 ft-lb) + additional 45°

Rear Driveshaft Service

Rear driveshaft components

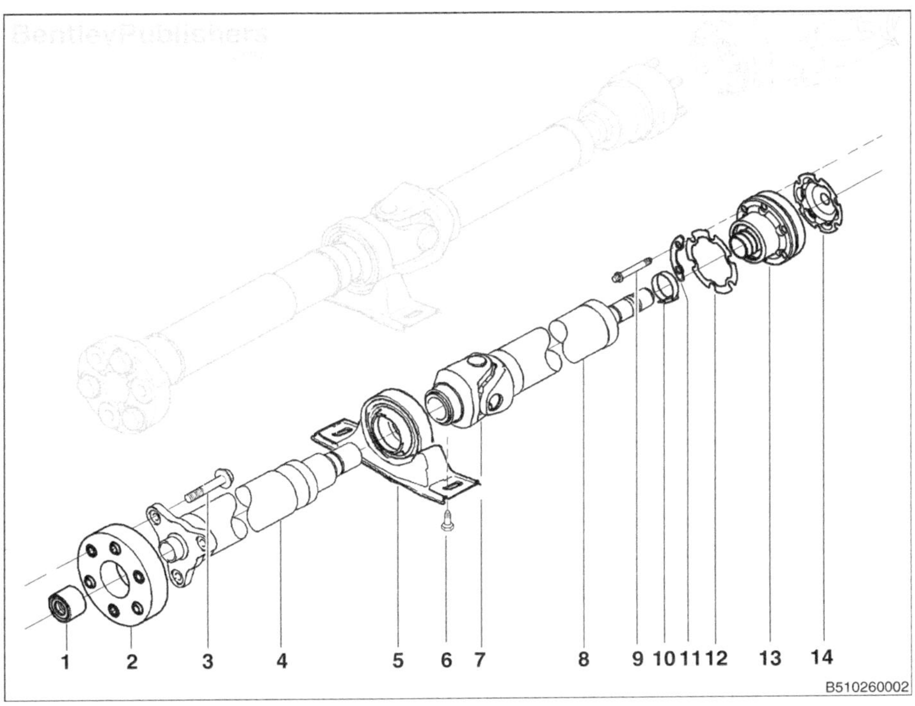

1. Centering sleeve
2. Flex-disc
3. ZNS bolt
   - Replace with new
   - Bolt size and torque values vary. See torque table.
4. Front section
5. Bearing mount
6. M8 center bearing mounting bolt
   - Tighten to 21 Nm (16 ft-lb)
7. Universal joint
8. Rear section
9. M10 x 52 mm ZNS bolt
   - Replace with new
   - Torque to 20 Nm (15 ft-lb) + add'l 90°
10. Hose clamp
11. Reinforcement
12. Metal gasket
13. CV joint
14. End cover

BentleyPublishers.com—All Rights Reserved

Rear driveshaft, removing

– Raise vehicle and support safely.

> ***WARNING —***
> *• Make sure the vehicle is stable and well supported at all times. Use a professional automotive lift or jack stands designed for the purpose. A floor jack is not adequate support.*

– Remove exhaust system and exhaust heat shields. See **180 Exhaust System**.

> ***WARNING —***
> *• Make sure exhaust system is fully cooled off before removing.*

– N54 engine: Support transmission with transmission jack, then remove transmission rear support.

– Match mark driveshaft mounting flanges and sections with paint to preserve original orientation.

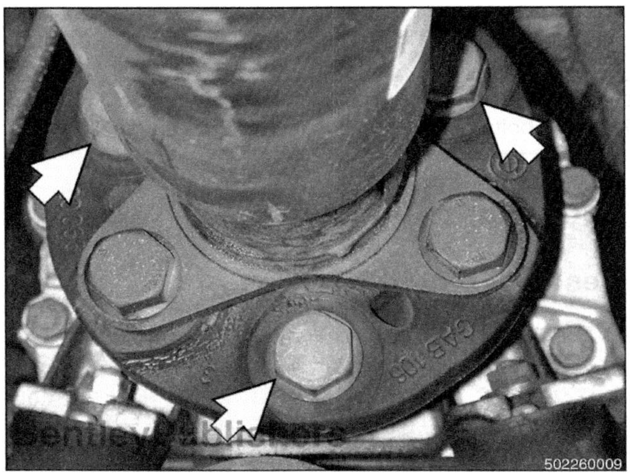

< Working at driveshaft front flex-disc, counterhold bolt heads (**arrows**) and remove nuts at transmission or transfer case output flange. Discard old fasteners.

> ***CAUTION—***
> *• To avoid damaging flex-disc rubber, be sure to hold bolts stationary while loosening nuts.*

– Hang driveshaft front section from body using stiff wire.

< Working at rear differential flange, loosen and remove rear driveshaft CV joint fasteners (**arrows**).

BentleyPublishers.com
BentleyPublishers.com—All Rights Reserved

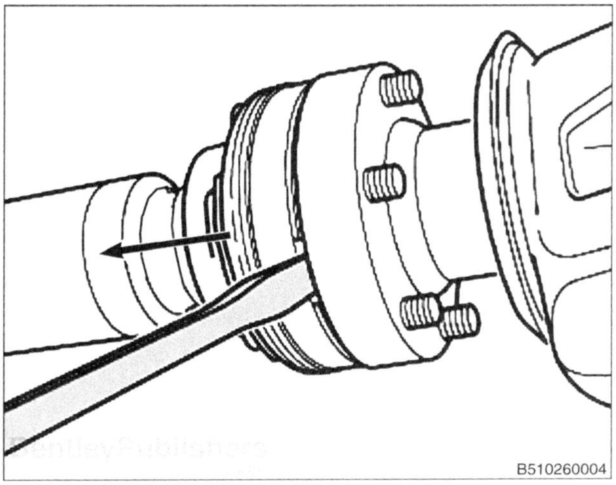

◄ Use screwdriver to pry CV joint off differential flange. Discard old fasteners.

– Hang driveshaft rear section from body using stiff wire.

◄ Remove driveshaft center bearing support mounting nuts (**arrows**).

– Lower driveshaft, releasing front and rear sections from stiff wire.

Rear driveshaft, inspecting

– Check driveshaft universal joint, center bearing, CV joint and flex-disc for defects. See **Troubleshooting** in this repair group. Replace components as needed.

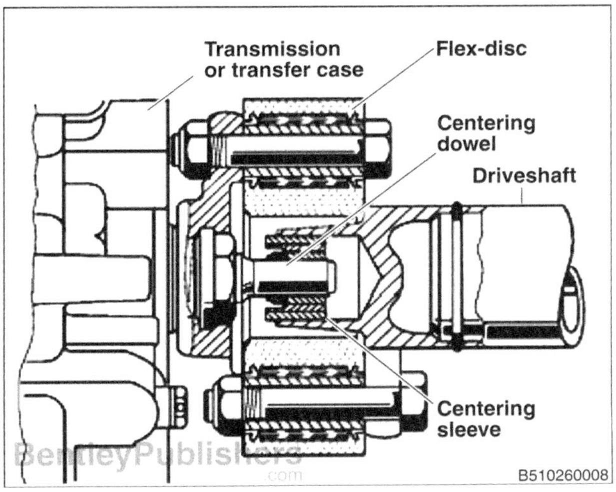

◄ Check driveshaft front centering sleeve for damage. Replace if necessary. See **Front centering sleeve, replacing** in this repair group.

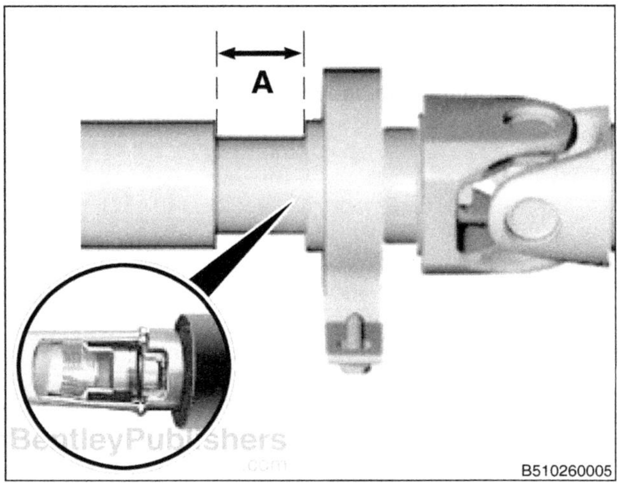

Aluminum driveshaft: Check dimension of crash element ahead of center bearing carrier. Replace driveshaft if crash element is deformed shorter than specification.

Driveshaft crash element specification	
Dimension **A** minimum	approx. 100 mm (4 in)

Rear driveshaft, installing

– Apply thin coat of molybdenum disulfide grease (Molykote Longterm 2 Plus® or equivalent) to centering sleeve.

– Bend driveshaft slightly at center universal joint to install.

– Align matchmarks on driveshaft to matchmarks on front and rear flanges.
  - Rear CV joint: Using new ZNS fasteners, alternately tighten two opposite nuts to draw joint evenly into flange. Once seated, tighten down remaining nuts.
  - Front flex-disc: Attach to transmission or transfer case output flange using new fasteners.

> ***CAUTION—***
> - *To avoid damaging flex-disc rubber, be sure to hold bolts stationary while tightening nuts.*

Tightening torques	
Rear driveshaft CV joint to differential: Replace M10 x 52 mm 10.9 grade ZNS fasteners with new	20 Nm (15 ft-lb) + additional 90°
Rear driveshaft flex-disc to transfer case: Replace M10 x 47 mm 10.9 grade ZNS fasteners with new	20 Nm (15 ft-lb) + additional 90°
Rear driveshaft flex-disc to transmission: Replace M12 10.9 grade ZNS fasteners with new	55 Nm (41 ft-lb) + additional 90°

BentleyPublishers.com
BentleyPublishers.com—All Rights Reserved

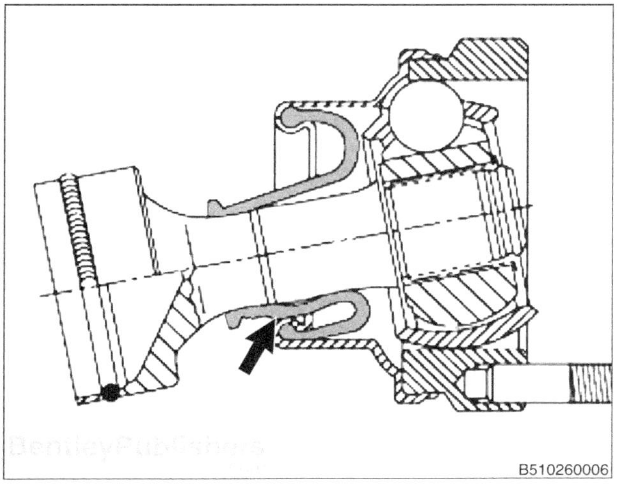

Do not allow rear driveshaft to hang from flex-joint or rear CV joint. Driveshaft weight may damage rubber joints or boots (**arrow**).

- Reinstall center bearing support fasteners.

Tightening torque	
Driveshaft center bearing support to body	21 Nm (15 ft-lb)

- Install heat shields. Install exhaust system. See **180 Exhaust System**.

- Road test vehicle to check for noise or vibration.

Flex-disc, replacing

- Raise vehicle and support safely.

> ***WARNING —***
> - *Make sure the vehicle is stable and well supported at all times. Use a professional automotive lift or jack stands designed for the purpose. A floor jack is not adequate support.*

- Remove exhaust system and exhaust heat shields. See **180 Exhaust System**.

> ***WARNING —***
> - *Make sure exhaust system is fully cooled off before removing.*

- Support transmission with transmission jack, then remove transmission rear support.

- Check flex-disc for cracks, tears, missing pieces or distortion. Check for worn bolt holes in flange.

- Mark driveshaft mounting flanges and sections with paint to preserve original orientation.

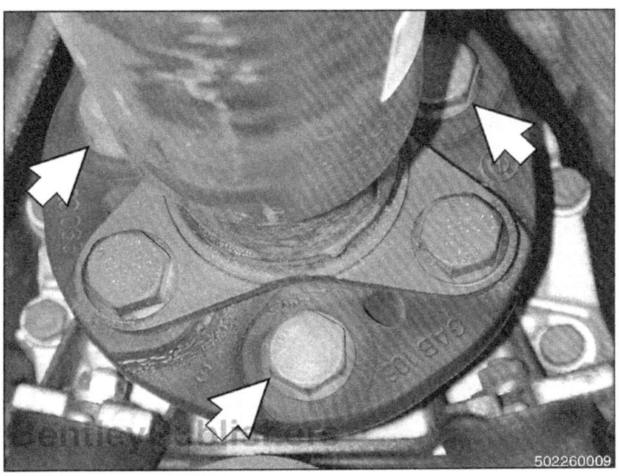

Counterhold flex-disc bolt heads (**arrows**) and remove nuts at transmission or transfer case output flange. Discard old fasteners.

> ***CAUTION—***
> - *To avoid damaging flex-disc rubber, be sure to hold bolts stationary while loosening nuts.*

- Remove driveshaft center bearing support fasteners. Support center of driveshaft.

> ***CAUTION—***
> - *To avoid damaging rear driveshaft CV joint, do not allow driveshaft to hang down unsupported.*

- Lower center of driveshaft sufficiently to disengage flex-disc from transmission or transfer case flange. Tie driveshaft to side.

BentleyPublishers.com—All Rights Reserved

- Remove fasteners and detach flex-disc from driveshaft.
- Before reinstalling flex-disc, check driveshaft front centering sleeve for damage. Replace if necessary. See **Front centering sleeve, replacing** in this repair group.
- Apply thin coat of molybdenum disulfide grease (Molykote Longterm 2 Plus® or equivalent) to centering sleeve.

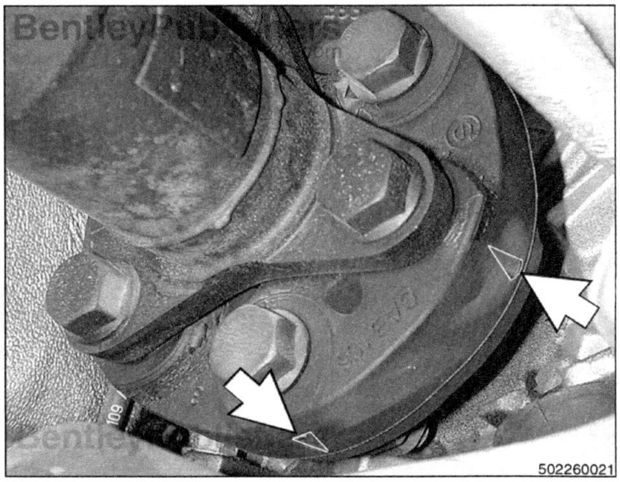

◀ Reinstall flex-disc using new ZNS fasteners.
- Molded arrows (**arrows**) on coupling point toward flange arms.
- Counterhold bolts while tightening nuts.

Tightening torques	
Rear driveshaft flex-disc to transfer case: Replace M10 x 47 mm 10.9 grade ZNS fasteners with new	20 Nm (15 ft-lb) + additional 90°
Rear driveshaft flex-disc to transmission: Replace M12 10.9 grade ZNS fasteners with new	55 Nm (41 ft-lb) + additional 90°

Center bearing assembly, replacing

The rear driveshaft center bearing assembly consists of a grooved ball bearing in a rubber mount. The bearing assembly is pressed on the driveshaft.

- Remove driveshaft. See **Rear driveshaft, removing** in this repair group.
- Use paint to matchmark front and rear driveshaft sections before separating.

> ***CAUTION—***
> - *Do not scratch or punch marks in driveshaft metal.*

◀ Remove bolt (**arrow**) securing front and rear halves of driveshaft and pull driveshaft apart.

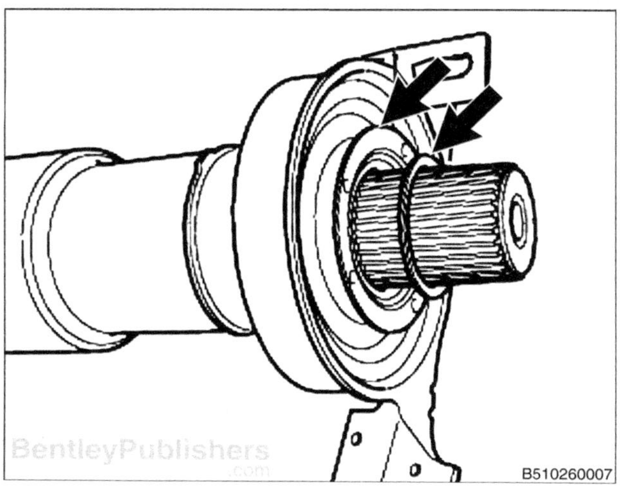

◀ Remove shim and dust shield (**arrows**), where fitted.

- Use puller to remove center bearing assembly from driveshaft.
- On installation, drive new center bearing firmly to stop. Reassemble driveshaft using matchmarks made prior to disassembly.
- Clean threads and install bolt securing driveshaft halves. Use thread locking compound Loctite® or equivalent.

Tightening torque	
Driveshaft front half to rear half (use thread locking compound)	97 Nm (72 ft-lb)

- Reinstall driveshaft. See **Rear driveshaft, installing** in this repair group.

BentleyPublishers.com
BentleyPublishers.com—All Rights Reserved

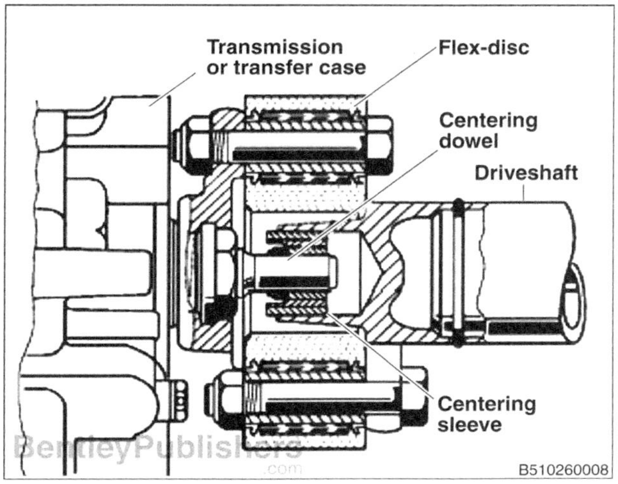

Front centering sleeve, replacing

The front centering sleeve centers the rear driveshaft in relation to transmission or transfer case flange. The guide is press-fit into a cavity in the flex-disc and slides on a dowel on the output flange.

No specifications are given for wear of the guide. Check to make sure that the sleeve fits snugly over the dowel.

- Remove rear driveshaft. See **Rear driveshaft, removing** in this repair group.
- Pack cavity behind centering guide with heavy grease until grease is flush with bottom edge of guide.
- Insert 14 mm (approximately 9/16 in) diameter mandrel (metal rod) into guide. Strike mandrel with hammer to force centering guide out. Make sure mandrel fits snugly in centering guide so that grease cannot escape around sides of mandrel.
- Remove old grease from driveshaft, lubricate new centering guide with molybdenum disulfide grease (Molykote Longterm 2 Plus® or equivalent) and drive through center of flex-disc into driveshaft.

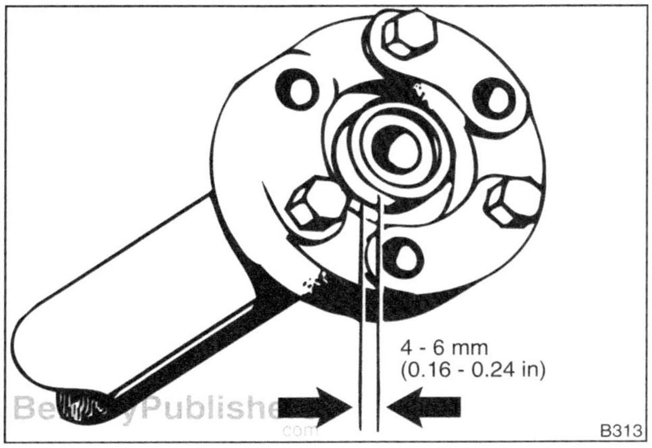

When installing new driveshaft centering guide, be sure sealing lip of guide faces outward. Drive guide into driveshaft to a protrusion depth of 4 - 6 mm (0.16 - 0.24 in).

- Apply thin coat of molybdenum disulfide grease (Molykote Longterm 2 Plus® or equivalent) to centering sleeve.
- Install driveshaft. See **Rear driveshaft, installing** in this repair group.

Constant velocity (CV) joint, replacing

The CV joint at the rear of the rear driveshaft is available from BMW as a separate part.

- Remove rear driveshaft. See **Rear driveshaft, removing** in this repair group.

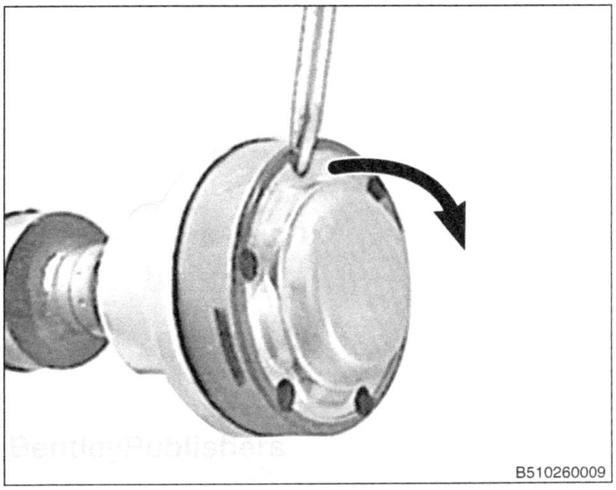

Use screwdriver to pry end cover off CV joint. Discard cover.

260

BentleyPublishers.com—All Rights Reserved

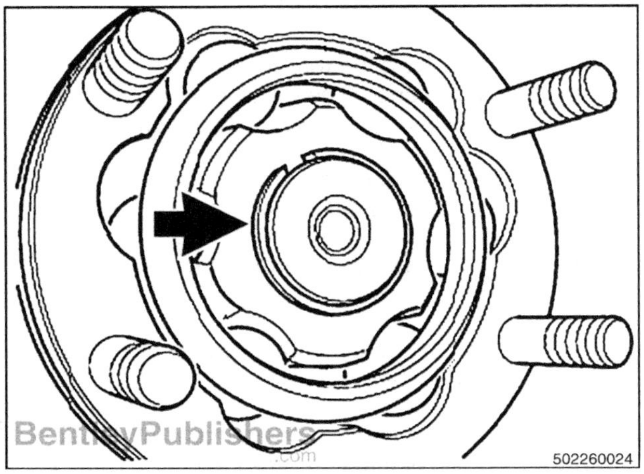

Clean grease off CV joint. Remove retaining circlip (**arrow**) from CV joint end of driveshaft. Discard clip

– Drive out bolts from CV joint.

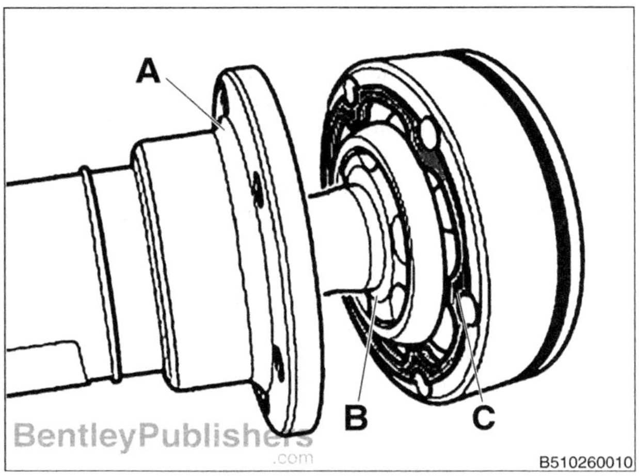

Remove CV joint from driveshaft:
- Pry back protective boot (**A**).
- Using BMW special tool 26 1 070 or 26 1 110 on inner race (**B**), press or pull joint off shaft.

– Check protective boot for damage. Replace if necessary.

– Install new CV joint:
- Working from both sides of joint, inject approximately 80 grams of CV joint grease into joint.
- Install new sealing gasket (**C**) on joint face that mates with protective boot. Seal fits in groove on outer race of joint.
- Clean grease off driveshaft and mating splines on joint.
- Apply Loctite® 648 to driveshaft splines.
- Press new CV joint on shaft as far as it will go.

> ***CAUTION—***
> *• Only press on inner CV joint race.*

– Install new retaining circlip to end of driveshaft. Install new end cover with new sealing gasket.

– Insert new CV joint mounting bolts. Be sure to reuse reinforcing washers.

– Install driveshaft. See **Rear driveshaft, installing** in this repair group.

BentleyPublishers.com
BentleyPublishers.com—All Rights Reserved

270 Transfer Case

General

This repair group covers the all-wheel drive (xDrive) transfer case. Internal transfer case repair is not covered. Special press tools and procedures are required to disassemble and service the internal geartrain.

See also:

- **180 Exhaust System**
- **200 Transmission–General**
- **260 Driveshafts**

xDrive overview

The BMW all-wheel drive (xDrive) control system is integrated with dynamic stability control (DSC).

The xDrive transfer case, bolted to the rear of the transmission, directs power from the transmission to both the front and rear differentials via driveshafts. The transfer case is permanently engaged to drive all four wheels all of the time.

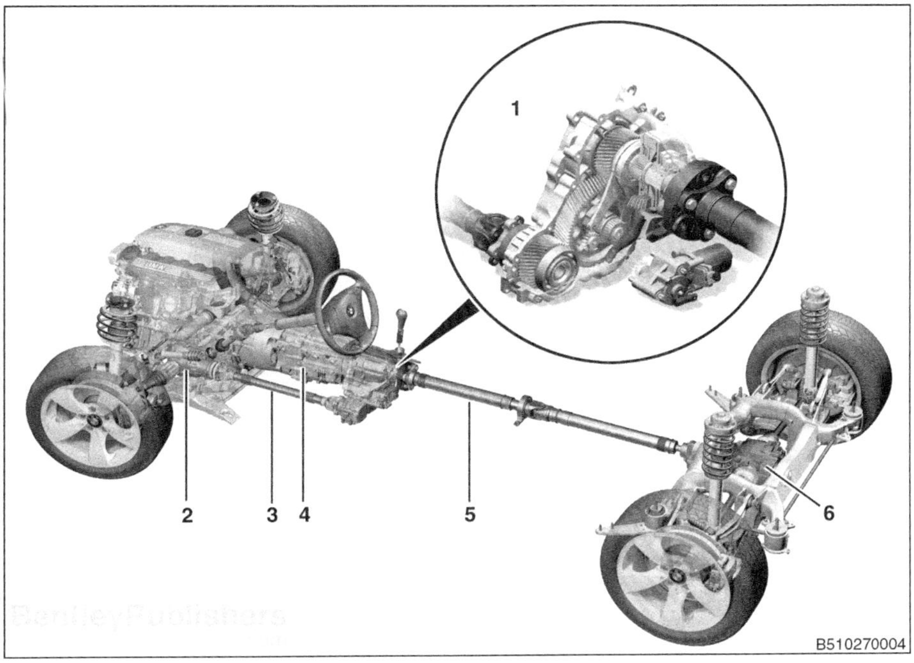

1. xDrive transfer case
   - Active torque control (ATC) 300
2. Front differential
3. Front driveshaft
4. Transmission
5. Rear driveshaft
6. Rear differential

BentleyPublishers
BentleyPublishers.com–All Rights Reserved

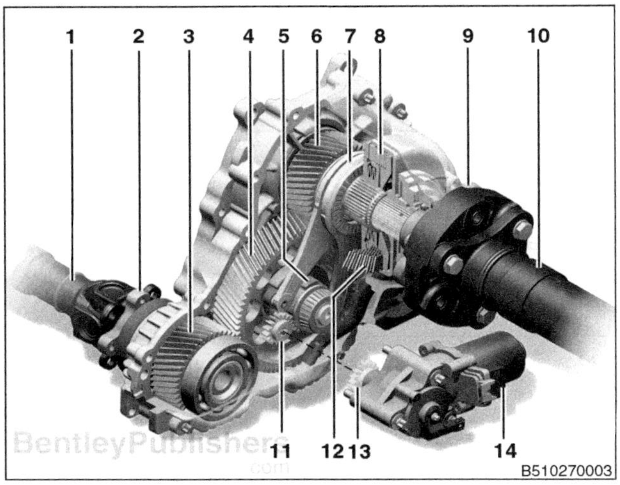

Transfer case (ATC 300)

The active torque control (ATC 300) transfer case uses an L-shaped series of spur gears to transfer power to the front driveshaft.

1. Front driveshaft
2. Front output flange
3. Front output gear
4. Idler gear
5. Clutch control cam
6. Drive gear
7. Control lever
8. Multi-disc clutch pack
9. Rear output flange
10. Rear driveshaft
11. Drive pinion
12. Multi-disc clutch disc pack
13. Clutch actuator drive gear
14. Clutch actuator (VTG) servo motor

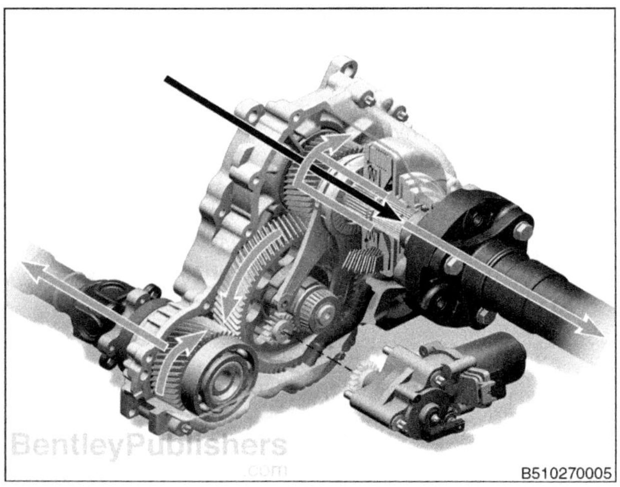

Principles of operation (power flow)

When the multi-disc clutch in the transfer case is disengaged, no driving torque is transmitted to the front axle. All of the driving torque is distributed to the rear axle. This is because the input shaft is splined to the rear driveshaft output, providing a permanent connection to the rear axle.

The multi-disc clutch couples the drive gear to the front driveshaft output flange. Torque to the front axle is increased or decreased by regulating locking pressure of the multi-disc clutch, providing a stepless coupling of the front axle to the drivetrain. This depends on DSC assessment of driving situations and road conditions.

When the multi-disc clutch is fully engaged, the front and rear axles turn at the same speed. Torque distribution front to rear is based on available traction at each axle. For example, when traction is identical on the front and rear axles and the driver accelerates from a stop in first gear at full throttle, the rear axle is capable of transferring greater torque to the road wheels without slipping, as vehicle weight shifts from front to rear. The clutches disengage, transferring most of the power to the rear axle.

Another example: If the front axle is on a high traction surface and the rear axle is on ice, virtually 100% of the available driving torque is transmitted to the front axle. Based on available traction, virtually no driving torque can be transferred to the driving wheels by the rear axle. When more torque is transmitted to the front axle, torque to the rear axle is proportionally reduced due to lack of traction.

The advantages of variable distribution of drive torque at front and rear axles are:

- Optimum utilization of lateral cornering and wheel peripheral forces applied at front and rear axles.
- Brake interventions by DSC reduced, thus increasing overall comfort.
- Torque distribution front to rear wheels more closely matched to differing friction values at the front and rear axles.

BentleyPublishers.com
BentleyPublishers.com—All Rights Reserved

Even when DSC is switched OFF, transfer case control is active to ensure maximum traction and vehicle dynamics.

Permanent all-wheel drive is cancelled to a large extent or completely in three situations:

- When negotiating extremely tight corners with little engine torque, in order to allow speed equalization between the front and rear axle (e.g., when parking).
- At speeds in excess of 180 kph (112 mph).
- In extreme understeer driving situations.

Transfer case control module

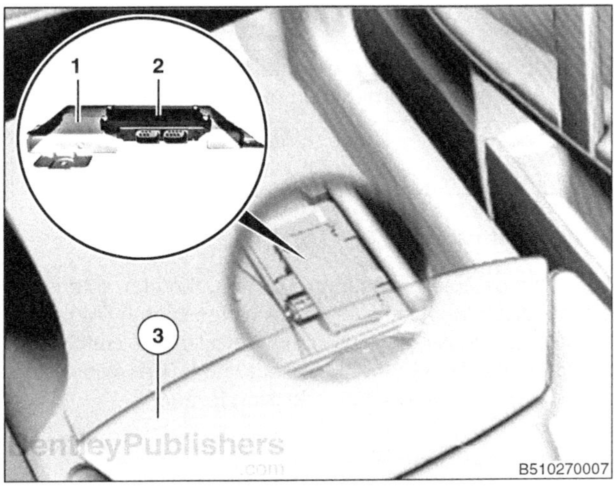

The transfer case (xDrive) control module is installed under the rug forward of the passenger front seat.

1. Kick guard
2. Transfer case control module\
3. Passenger seat

The control module is connected via powertrain CAN-bus (PT-CAN) to the DSC control module, engine control module (ECM), transmission control module (TCM) and body gateway module (KGM).

The control module regulates transfer case multi-disc clutch engagement to control the distribution of torque between front and rear axles. Torque adjustment requests are received from the DSC control module and trigger the power electronics circuitry required for the clutch actuator drive. Control module pin assignments are in **Table a** and **Table b**.

In case of DSC failure, an emergency program (limp-home mode) for locking up transfer case clutches allows all-wheel drive vehicle operation to be continued.

Transfer case control module pin assignments

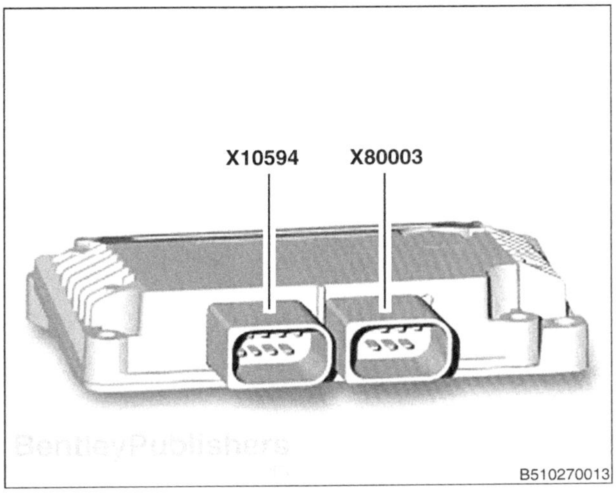

Table a. Pin assignments for connector X10594 to vehicle electrical system via PT-CAN

Pin	Type	Description
1	Power	Terminal 30
2	Ground	Terminal 31
3	Input	Terminal 15 wake-up wire
4		not used
5		not used
6	Input / output	PT-CAN high
7	Input / output	PT-CAN low
8	Power	Terminal 30

BentleyPublishers.com–All Rights Reserved

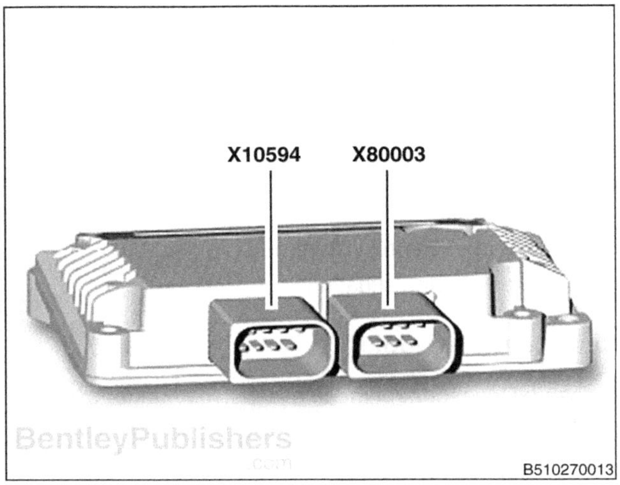

Table b. Pin assignments for connector X8003 to clutch actuator (VTG) motor		
Pin	**Type**	**Description**
1	Input	Classification resistor
2	Input	Temperature sensor in VTG motor
3	Input	Negative wire from VTG motor
4	Output	Positive wire to VTG motor
5	Input	Negative lead from incremental sensor and classification resistor
6	Input	Positive lead for incremental sensor
7	Input	Incremental sensor (indicates direction of VTG motor rotation)
8	Input	Incremental sensor (indicates VTG motor position)

Classification resistor

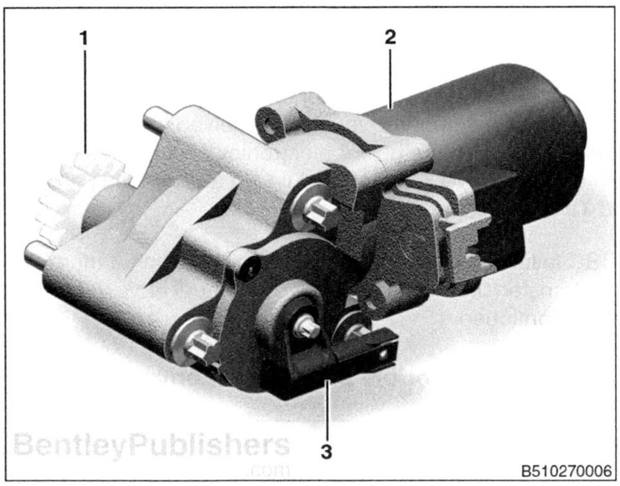

Because of mechanical tolerances in production, the characteristic curve of the multi-disc clutch locking torque varies slightly. At the factory, once the actual locking torque is measured on the clutch test bench, a resistor is attached to the clutch actuator. The resistor value (classification) determines clutch locking torque characteristics.

1. Clutch actuator drive gear
2. Clutch actuator (servo motor)
3. Classification resistor

Each time the engine is started, the transfer case control module measures the resistor value once and selects the optimum program map for transfer case engagement.

Warning lights

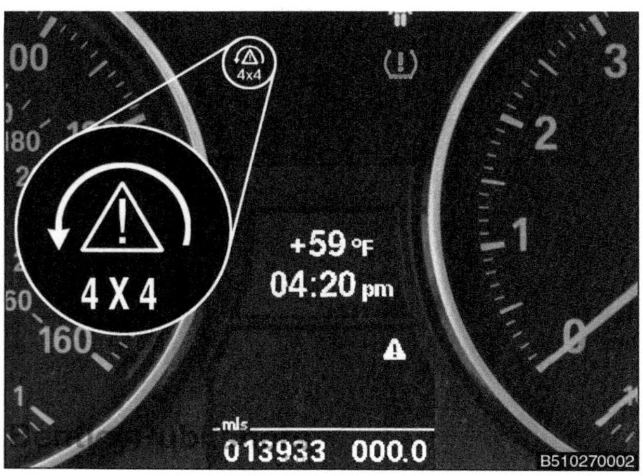

In case of DSC or xDrive system failure, instrument cluster warning lights and a gong warn the driver of malfunction(s). Use BMW scan tool to read out and diagnose faults.

BentleyPublishers.com
BentleyPublishers.com—All Rights Reserved

TRANSFER CASE SERVICE

Transfer case fluid service

xDrive transfer case oil monitoring is performed by the transfer case control module to determine when oil change is due.

– After performing work on transfer case, check fluid level. Be sure to use correct specification fluid.

< Remove drain plug (**arrow**) to drain transfer case.

Tightening torque	
Drain plug to transfer case (M22)	60 Nm (44 ft-lb)

< Fill case to lower edge of filler plug (**arrow**).

Transfer case fluid	
Capacity	0.52 liter (0.55 US qt)
Fluid specification	BMW part no. 83 22 0 397 244

> ***CAUTION—***
> • *In order to prevent serious damage to transfer case, use only the recommended fluid.*

Tightening torque	
Fill plug to transfer case (M22)	60 Nm (44 ft-lb)

BentleyPublishers.com
BentleyPublishers.com—All Rights Reserved

Transfer case, removing and installing

> ***CAUTION—***
> - *If replacing transfer case with new or used unit, be sure to also replace classification resistor corresponding to replacement case.*

– Raise vehicle and support safely.

> ***WARNING —***
> - *Make sure the vehicle is stable and well supported at all times. Use a professional automotive lift or jack stands designed for the purpose. A floor jack is not adequate support.*

– Remove splash shields underneath engine and transmission.

– Remove front end reinforcement plate. Discard fasteners.

– Remove exhaust system and exhaust heat shield. See **180 Exhaust System**.

> ***WARNING —***
> - *Make sure exhaust system is fully cooled off before removing.*

– Remove support brackets for heat and splash shields as needed.

– Use paint to match mark front and rear driveshaft mounting flanges to preserve original orientation.

B510270010

◀ Remove front driveshaft fasteners (**arrows**) at transfer case output flange. Tie driveshaft to side.

– N54 engine: Support transmission with transmission jack, then remove transmission rear support.

– Match mark rear driveshaft mounting flange with paint to preserve original orientation.

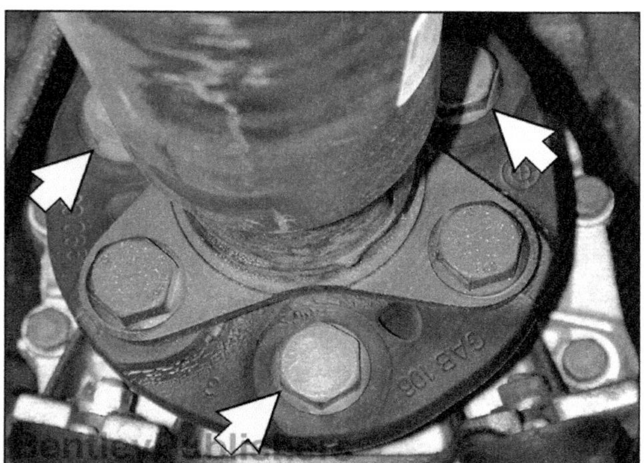

502260009

◀ Working at rear driveshaft front flex-disc, counterhold bolt heads (**arrows**) and remove nuts at transfer case output flange. Discard old fasteners.

> ***CAUTION—***
> - *To avoid damaging flex-disc rubber, be sure to hold bolts stationary while loosening nuts.*

BentleyPublishers.com
BentleyPublishers.com—All Rights Reserved

– Remove driveshaft center bearing support fasteners. Support center of driveshaft.

> ***CAUTION—***
> *• To prevent damage to rear driveshaft CV joint boot, do not allow driveshaft to hang down unsupported.*

– Detach rear driveshaft flex-disc from transfer case. Hang driveshaft front section from body using stiff wire.

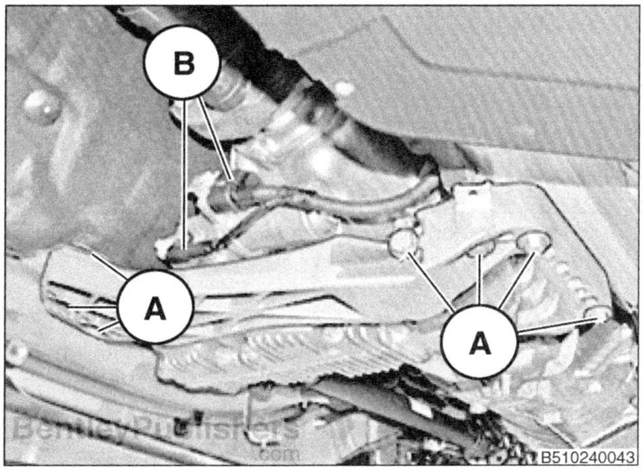

◄ With transmission supported on transmission jack:

- Remove transmission rear support mounting bolts (**A**). Remove support.
- Detach transfer case electrical connectors (**B**).

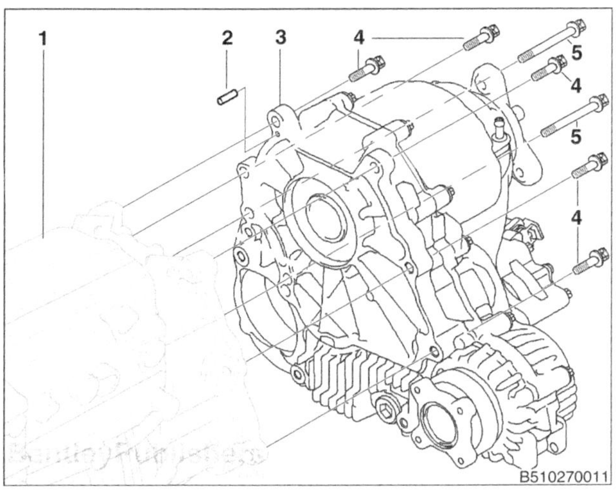

◄ While supporting transfer case, remove transfer case mounting bolts from transmission housing.

1. Transmission housing
2. Dowel pin
3. Transfer case ATC 300
4. M10 x 35 mm ZNS 8.8 grade bolt
   - Tighten to 43 Nm (32 ft-lb)
5. M10 x 80 mm ZNS 8.8 grade bolt
   - Tighten to 43 Nm (32 ft-lb)

– Lower transfer case. Check carefully to make sure no lines, hoses or wires become snagged.

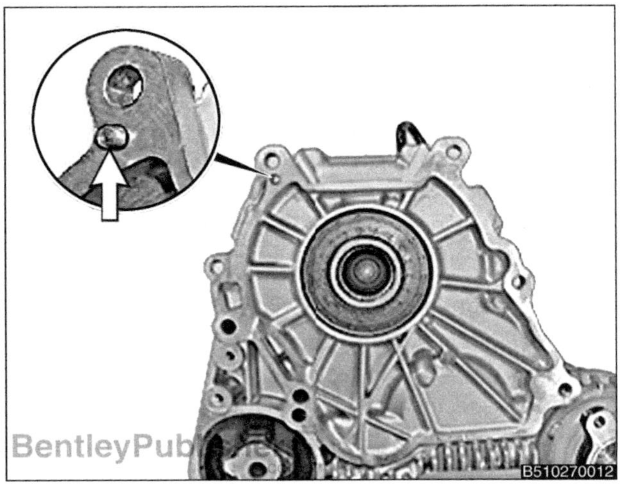

When installing:

- Inspect dowel pin (**arrow**) and replace if damaged.
- Transfer vents hose, covers and protective caps to new transfer case
- Replace transfer case input shaft sealing O-ring, if applicable.
- Apply thin coat of Weicon® antiseize grease to dowel pin and input shaft splines.

Tightening torques	
Front driveshaft to transfer case or differential: Replace new M8 ZNS bolts with new	20 Nm (15 ft-lb) + additional 45°
Rear driveshaft flex-disc to transfer case: Replace M10 x 47 mm 10.9 grade ZNS fasteners with new	20 Nm (15 ft-lb) + additional 90°
Transfer case to transmission (M10 x 35 mm or M10 x 80 mm ZNS 8.8 grade bolt)	43 Nm (32 ft-lb)
Transmission rear support to body (M8)	19 Nm (14 ft-lb)
Transmission rear support to transmission	19 Nm (14 ft-lb)

– After completion of work, check and top off fluid level. See **Transfer case fluid service** in this repair group.

– When installing front end reinforcement plate, use new bolts. See **310 Front Suspension**.

– After filling transfer case, drive vehicle approximately 200 meters (600 ft) and recheck fluid level.

– Use BMW scan tool to carry out "Repair" service function and complete transfer case programing and coding. See **600 Electrical System–General**.

BentleyPublishers.com
BentleyPublishers.com—All Rights Reserved

300 Suspension, Steering and Brakes–General

TABLE

General

This repair group includes overview information for the front and rear suspension, braking and steering systems. Also see:

- **200 Transmission–General** for xDrive transfer case information
- **600 Electrical System–General** for bus information
- **ECL Electrical Component Locations**

Suspension, steering, brakes (xDrive Sedan)

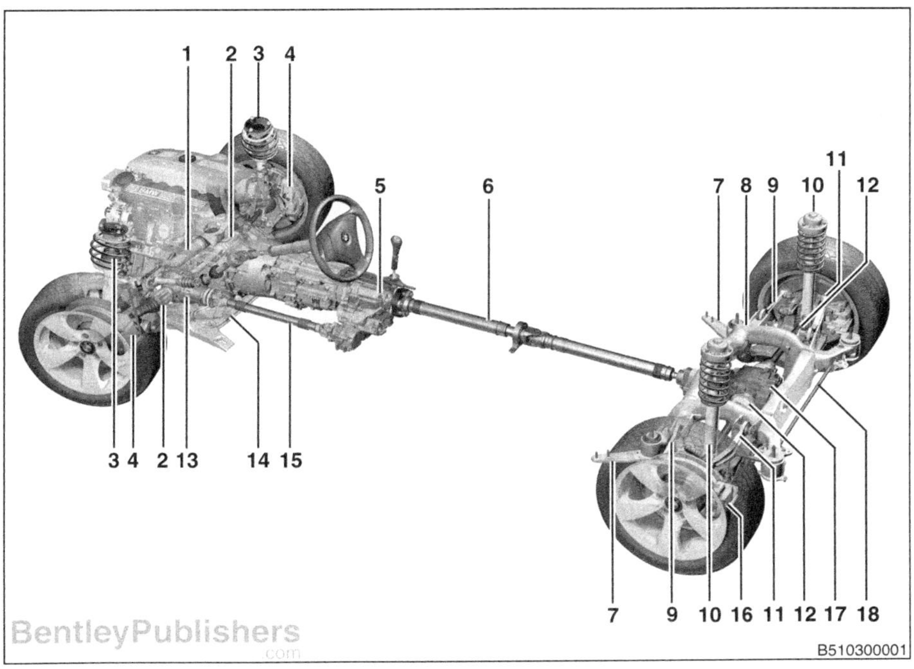

1. Steering rack
2. Front drive axle
3. Front strut and spring assembly
4. Front brake caliper
5. xDrive transfer case
6. Rear driveshaft
7. Thrust rod
8. Rear axle carrier (rear subframe)
9. Traction strut
10. Rear strut and spring assembly
11. Upper control arm
12. Rear drive axle
13. Front differential
14. Front end reinforcement
15. Front driveshaft
16. Rear brake caliper
17. Rear differential
18. Stabilizer bar

BentleyPublishers.com
BentleyPublishers.com–All Rights Reserved

Aluminum suspension components

In order to reduce vehicle weight, aluminum suspension components are used. A test with a magnet easily reveals the metal used. When working with aluminum components, follow the cautions below.

> ***CAUTION—***
> - *Due to the chemical and corrosion characteristics of aluminum, do not bring into contact with battery acid.*
> - *Do not use wire brushes with brass or iron bristles. Only use brushes with stainless steel bristles.*
> - *Do not expose to flying sparks from grinding or cutting operations.*
> - *Do not subject to steel welding splashes.*
> - *Do not expose to temperatures over 80°C (176°F), even for short periods. Temperatures in painting facilities are not a problem.*

FRONT SUSPENSION

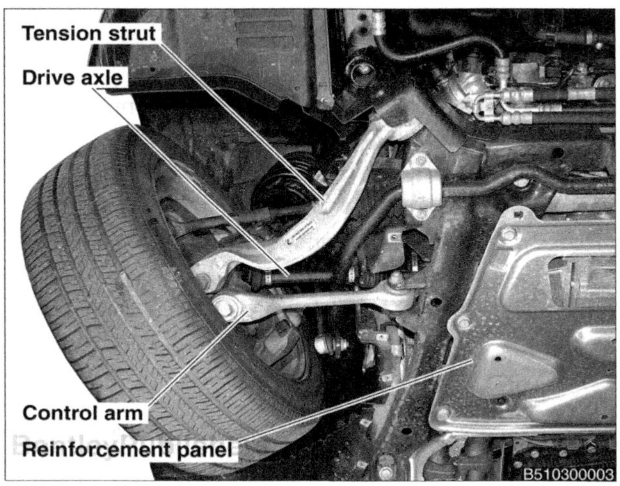

xDrive model right front suspension.

The 5 Series front suspension is a double-pivot, strut-type design with dual lower control arms. Many of its components are manufactured from aluminum alloy. Overall benefits of this front suspension design include:

- Increased vehicle agility.
- Improved comfort due to reduction of unsprung weight.
- Reduction in fuel consumption due to lowering gross vehicle weight and better axle load distribution.

The aluminum axle carrier (front subframe) is made of cast alloy high tensile strength sections welded to extruded sections. It forms a rigid mounting platform for the front suspension arms, stabilizer bar, steering rack, engine mounts, heat shields and underbody splash panels.

The lower control arms (known as control arm and tension strut) are connected to the steering knuckles in a double pivot configuration. The tension struts (or thrust arms) are connected to the front subframe using hydraulic mounts.

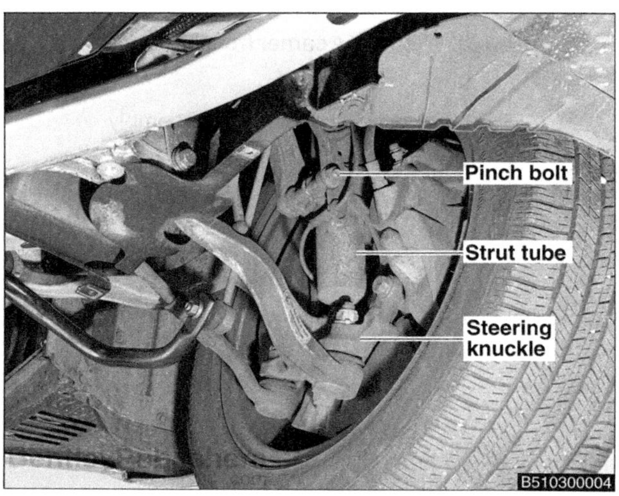

Each front strut assembly includes a tubular strut and a coil spring. The upper strut mount includes a bearing. The lower end of each strut housing is pinch-fitted to the steering knuckle.

On rear drive models, wheel bearings and hub assemblies are bolted to the steering knuckles.

on xDrive models, wheel bearing are pressed into steering knuckles. Use press tools to remove and install.

BentleyPublishers.com
BentleyPublishers.com–All Rights Reserved

The aluminum thrust zone reinforcement panel, bolted to the front subframe to increase transverse rigidity, improves handling, noise level and crash performance. BMW specifies the following:

- Do not drive the vehicle with the front end reinforcement removed.
- When reinstalling the panel after repairs, replace panel mounting fasteners with new.

The front suspension also includes provisions for active stabilizer system (ARS).

Front suspension repairs are covered in **310 Front Suspension**.

The front differential is bolted to the left side of the engine oil pan. Drive axles with constant velocity (CV) joints at each end drive the front wheels.

Front differential and drive axle repairs are covered in **311 Front Axle Differential**.

REAR SUSPENSION

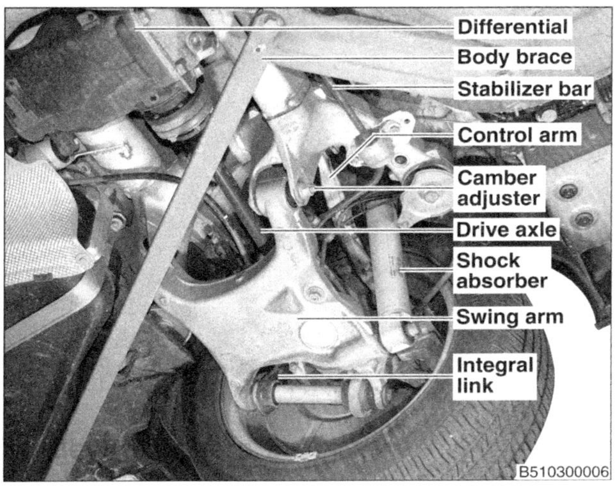

Sports Wagon rear suspension.

The aluminum rear subframe mounts to the body through four large bushings. Rear suspension components attach to the subframe.

The rear suspension is an integral multi-link design. Three control arms on each side locate the rear wheels. The lower control arm, called the swing arm, is attached to the wheel bearing carrier through a pivoting integral link. Aluminum components reduce unsprung weight.

Sedan and Sports Wagon rear axle design, though basically the same, differ as follows:

- Sedan rear suspension uses conventional MacPherson struts and coil springs.
- Sports Wagon is equipped with self-leveling air springs, also known as electronic height control (EHC). By separating the air springs from the shock absorbers and placing the shock absorbers at a lower level and more widely spaced, the Sports Wagon design provides a larger and wider cargo space.

Rear suspension repairs are covered in **330 Rear Suspension**.

The rear differential mounts to the subframe using two rubber bushings in front and one hydromount at rear. Drive axles with constant-velocity (CV) joints at both ends transfer power from the differential to the road wheels.

Rear differential and drive axle repairs are covered in **331 Rear Axle Differential**.

ACTIVE STABILIZERS (DYNAMIC DRIVE)

5 Series vehicles may be equipped with the optional active stabilizer system (acronym ARS). The system, also referred to as dynamic drive, is used to improve handling and suspension comfort.

With this system, front and rear stabilizer bars are split in half, with the halves connected to each other via hydraulic oscillating motors. An electronic control module operates through a valve body to control coupling of the two stabilizer bar halves. The stabilizer bar at each wheel is non-interacting with the stabilizer on the other wheel of the same axle and the suspension system as a whole does not respond if only one wheel is deflected by an irregularity in the road.

Suspension compliance adapts to suit driving conditions, varying from performance-oriented during cornering to smooth and comfort-oriented when the vehicle is proceeding in a straight line. During vehicle operation, the system responds to suspension travel by continuously cycling through closed-loop hydraulic control pulses lasting only fractions of a second.

Front active stabilizer component replacement is covered in **310 Front Suspension**. Rear active stabilizer component replacement is covered in **330 Rear Suspension**.

Active stabilizer components overview

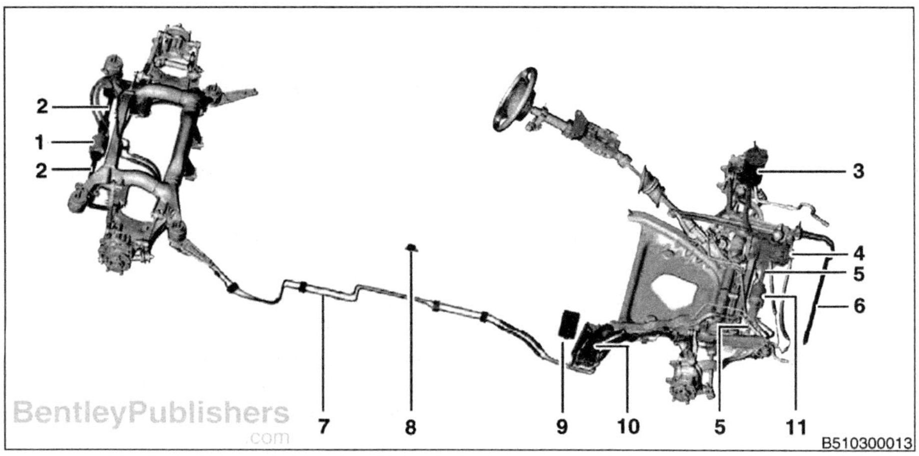

1. Rear oscillating motor
2. Rear stabilizer bar halves
3. Hydraulic fluid reservoir
4. Tandem fluid pump
5. Front stabilizer bar halves
6. Hydraulic fluid cooler
7. Hydraulic lines
8. Lateral acceleration sensor
9. Dynamic drive control module
10. Valve block
11. Front oscillating motor

BentleyPublishers.com
BentleyPublishers.com—All Rights Reserved

Active stabilizer system inputs - outputs

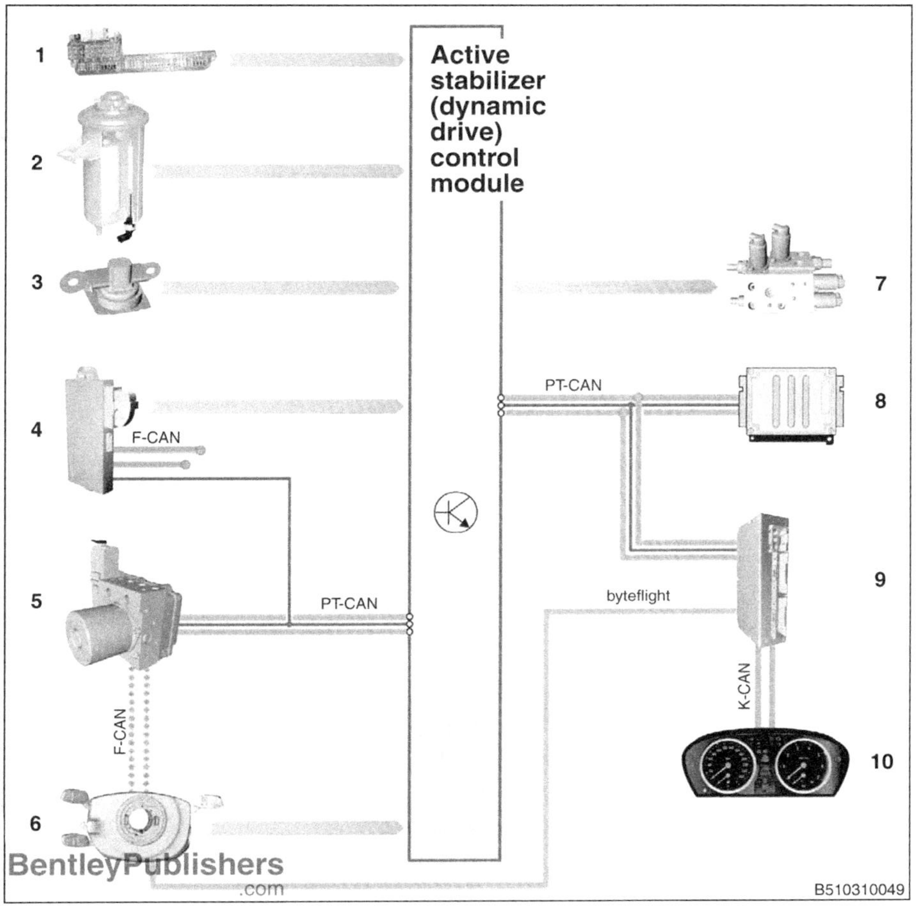

1. Power supply, front fuse panel, behind glove compartment
2. Hydraulic fluid reservoir fluid level sensor
3. Lateral acceleration sensor, underneath passenger seat
4. Car access system (CAS) control module, underneath left side dashboard
5. ABS / DSC module, engine compartment
6. Steering-angle sensor, steering column switch cluster (SZL)
7. Valve block, behind right front wheel housing liner
8. Engine control module (ECM)
9. Safety and gateway module (SGM), behind glove compartment
10. Instrument cluster

Active stabilizer control module

The active stabilizer control module, located behind the passenger kick-panel, is supplied with power via terminal 30 (ignition ON) and is protected by a 10A fuse. The module uses input signals to calculate the operating commands for the valve block. Input signals are also checked for plausibility and used for system monitoring purposes.

When the vehicle is started, the following electronic tests are performed by the control module:

- An authentication process compares the vehicle identification number from CAS with the vehicle identification number encoded in the stabilizer module.
- Control module hardware and software are checked.
- Valve block solenoids are checked for faults. In case of a fault, the system switches to fail-safe mode.
- The control module switches off if there is overvoltage.
- The control module learns the steering angle offset and the lateral acceleration zero-point during start-up and while driving.

If the control module is replaced, use BMW scan tool to code and program the new unit.

BentleyPublishers.com
BentleyPublishers.com–All Rights Reserved

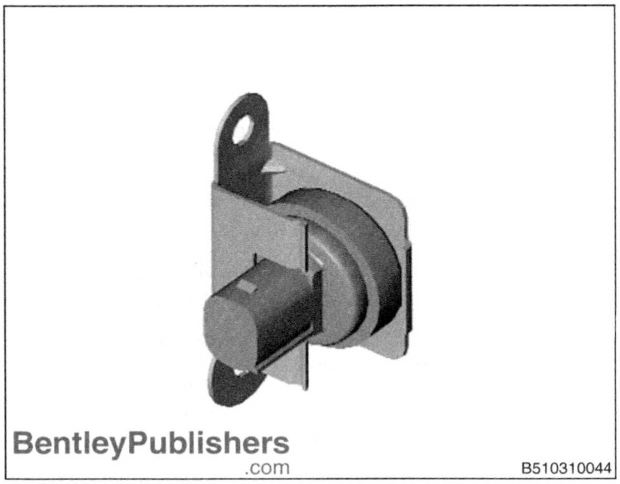

Lateral acceleration sensor

The lateral acceleration sensor, mounted underneath the passenger seat on the floor, supplies the main sensor signal to the control module. When the vehicle corners, the sensor detects lateral acceleration up to a measuring range of approx. 1.1 g.

Tandem pump

The tandem pump, driven by the engine via accessory belt, consists of a radial piston pump for the active stabilizer system and a vane pump for power steering.

Valve block

The valve block is behind the right front wheel housing trim.

1. Rear axle pressure control valve
2. Front axle pressure control valve
3. Line 1, front axle oscillating motor
4. Line 2, front axle oscillating motor
5. Front axle pressure sensor
6. Rear axle pressure sensor
7. Tandem pump connection
8. Selector position recognition sensor
9. Fail-safe valve
10. Directional valve
11. Line 2, rear axle oscillating motor
12. Hydraulic reservoir connection
13. Line 2, rear axle oscillating motor

Pulse-modulated signals from the control module activate the valve block.

Pressure control valves adjust the actuation pressures for the front and rear axle stabilizer bars.

Directional valve specifies the direction of high-pressure fluid flow when vehicle is cornering left or right.

Fail-safe valve shuts off the front axle oscillating motor in case of a power failure or other detected fault. Thus the front active stabilizer bar behaves like a normal mechanical stabilizer.

BentleyPublishers.com
BentleyPublishers.com—All Rights Reserved

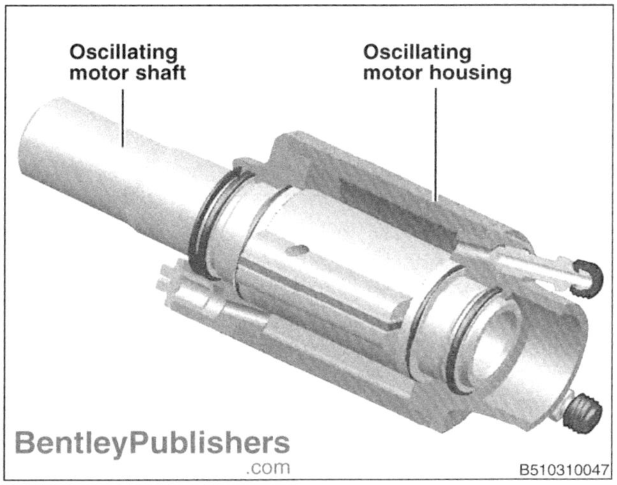

Oscillating motor

The active stabilizer bar consists of the halves of the stabilizer bar mounted to the oscillating motor and the oscillating motor housing.

The oscillating motor has three functions:

- It guides torque into the two halves of the stabilizer bar.
- It decouples the two halves of the stabilizer bar.
- In case of system failure (fail-safe mode), the front oscillating motor creates sufficient damping via hydraulic fluid (hydraulic locking) so that the stabilizer works like a conventional stabilizer bar.

Since one half of the stabilizer bar is connected to the shaft and the other to the housing, the two halves turn in opposite directions, thus damping irregular suspension movements.

If the oscillating motor chambers no longer contain any fluid as a result of a leak, the front axle stabilizer bar can no longer create damping.

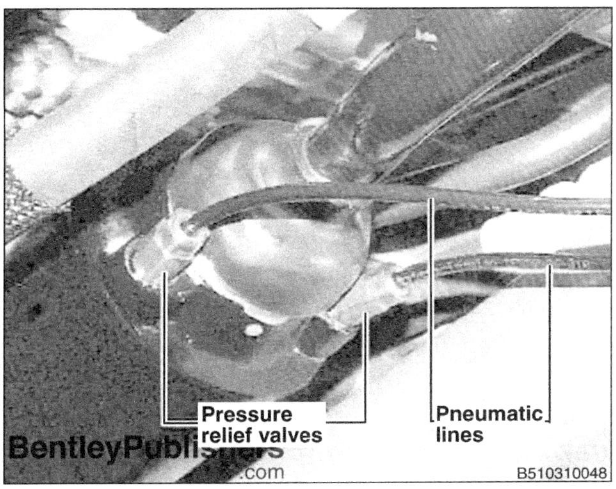

Pressure relief valves

When the vehicle is driven on poor road surfaces, stabilizer bar movements give rise to cavitation in the fluid inside the oscillating motors and this may produce rattling noises. The pressure relief valves on the front oscillating motor eliminate these noises by allowing filtered air to flow into the oscillating motor. This small quantity of air is absorbed by the hydraulic fluid to form an emulsion, which is discharged during the next activations of the oscillating motor. The air separates out in the hydraulic fluid reservoir.

Because noises from the rear axle cannot be heard inside the vehicle, pressure relief valves are not installed on the rear stabilizer bar oscillating motor.

Service notes

- If the active stabilizer system fails, DSC can no longer be deactivated or if it is already deactivated, it does not switch back on automatically.
- Hydraulic connections for each components are different in dimensions and lengths so that they cannot be swapped.
- When installing cables and fluid lines, make sure they are correctly isolated and free of tension. Make sure they are covered by underbody protection.

Steering angle adjustment

The active stabilizer system is dependent on the precise setting of the steering angle sensor. After working on steering or active stabilizer components or replacing steering column switch (SZL) components, reset steering angle sensor. Connect vehicle to BMW scan tool and carry out steering angle sensor adjustment system function.

Active stabilizer, commissioning

> ***WARNING —***
> - *During commissioning, the vehicle performs very strong and rapid rolling motions. For this reason, make sure no persons are near the vehicle, particularly in the proximity of moving vehicle components. Also, make sure that there are no objects underneath the vehicle or in the wheel housings.*
> - *Make sure that commissioning is carried out with the vehicle standing on a firm, flat surface with all four wheels on the ground. Do not perform commissioning with the vehicle on a lift, wheel alignment platform, shock absorber test platform, brake testing platform or dynamometer.*

The purpose of commissioning is to teach the active stabilizer control module the parameters and zero point of system components. Commission the system any time the system is coded and programmed, hydraulics are opened or parts replaced.

Commissioning requires the use of BMW scan tool. The process is divided into 5 phases and takes approx. 25 seconds:
- Check of direction valve.
- Low pressure test.
- Front axle high pressure test.
- Rear axle high pressure test.
- Pressure control valve test.

Before commissioning, make sure of the following:
- Hydraulic fluid temperature of 50° - 70°C (122° - 158°F).
- Vehicle parked level and on all four wheels.
- Vehicle unladen.
- Engine at idle.
- Vehicle in rest status with doors closed and no persons inside.

Active stabilizer troubleshooting

The following may appear as Check Control messages in iDrive display screen:

Vehicle instability icon in **yellow**. Active stabilizer system disabled. Vehicle leans more in curves. To continue driving:
- Check fluid level in power steering and active stabilizer reservoir.
- If there is no significant fluid loss, continue driving, but carefully around corners.

Bring vehicle to qualified repair facility as soon as possible.

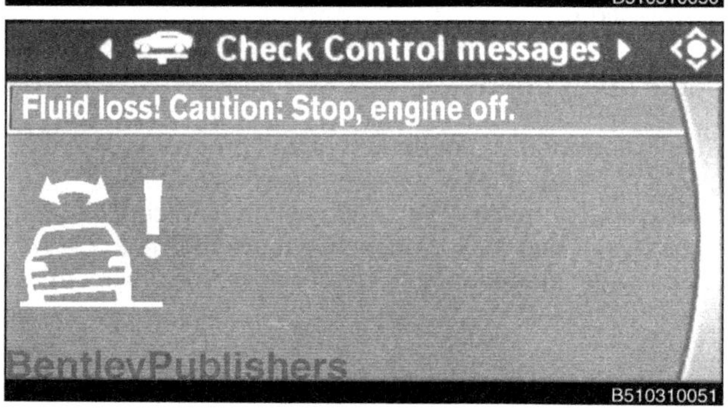

Vehicle instability icon in **red**. Fluid loss in steering and active stabilizer system. Do not continue to drive. Have vehicle towed for diagnosis and repairs.

BentleyPublishers.com
BentleyPublishers.com—All Rights Reserved

Table a lists commonly encountered active stabilizer faults and suggested solutions.

Table a. Active stabilizer troubleshooting

Problem	Possible cause	Suggested solutions
Noises (banging, rattling) during straight-ahead driving and gentle steering through center.	Front or rear suspension bolts loose	Check front and rear suspension fastener torques.
	Stabilizer bar link loose or defective	Check stabilizer link ball joints for damage or play. Tighten or replace stabilizer link as necessary.
	Stabilizer bar mount loose	Tighten stabilizer bar mount bolts.
	Hydraulic lines distorted	Check hydraulic line alignment. Make sure that lines are installed in brackets tension-free, not twisted.
	Hydraulic lines fitted too closely or have insufficient clearance (elastic movement of line during pressure buildup)	Check hydraulic line alignment. Make sure that lines are installed with sufficient clearance.
	Hydraulic line brackets loose, missing or incorrectly fitted	Check hydraulic line bracket fasteners for tightness and fit. Replace if missing or damaged.
	Hydraulic line rubber isolating elements damaged or missing	Replace defective or missing rubber isolating elements.
	Valve block rubber isolating elements damaged or missing	• Replace defective or missing rubber isolating elements. • Replace valve block bracket.
	Valve block defective	If after bleeding and initialization, faults are detected, replace valve block.
Rattling noises on bumps and straight-ahead driving (without altering steering wheel position); noises to a large extent independent of temperature	Front or rear suspension bolts loose	Check front and rear suspension fastener torques.
	Stabilizer bar link loose or defective	Check stabilizer link ball joints for damage or play. Tighten or replace stabilizer link as necessary.
	Stabilizer bar mount loose	Tighten stabilizer bar mount bolts.
	Hydraulic lines distorted	Check hydraulic line alignment. Make sure that lines are installed in brackets tension-free, not twisted.
	Hydraulic lines fitted too closely or have insufficient clearance (elastic movement of line during pressure buildup)	Check hydraulic line alignment. Make sure that lines are installed with sufficient clearance.
	Hydraulic line brackets loose, missing or incorrectly fitted	Check hydraulic line bracket fasteners for tightness and fit. Replace if missing or damaged.
	Hydraulic line rubber isolating elements damaged or missing	Replace defective or missing rubber isolating elements.
	Protective rubber cap (black) not removed from front fluid lines Y-connector	Remove Y-connector rubber cap.
	Vent valve leaking	If black plastic air filter element is filled with fluid, replace vent valve and filter.
	Valve block defective	If after bleeding and initialization, faults are detected, replace valve block

300

Bentley Publishers
BentleyPublishers.com—All Rights Reserved

Table a. Active stabilizer troubleshooting		
Problem	**Possible cause**	**Suggested solutions**
Rattling noises on bumps and straight-ahead driving (without altering steering wheel position); noises only at higher ambient temperature or when vehicle is heated up	Valve block defective	During test drive, check if noises occur during straight-ahead driving or road bumps on one side. If after bleeding and initialization, faults are detected, replace valve block
Hissing noise	Hydraulic lines damaged	Check hydraulic lines to see if they are bent or crushed. Repair or replace as necessary.
	Valve block fluid flow problem	Use BMW scan tool to diagnose system. If fault in valve block is detected, replace valve block.
Whining noises during cornering	Tandem pump belt problem	Check whether noises are from pump or belt. If necessary, install noise-optimized belt pulley,
Pump noise from engine compartment depending on engine speed or only during cornering	Fluid pressure line damaged	Check fluid pressure line from tandem pump. If crushed or bent, replace line.
	Tandem pump defective	Check for pump damage. If other causes are ruled out, replace pump.

STEERING

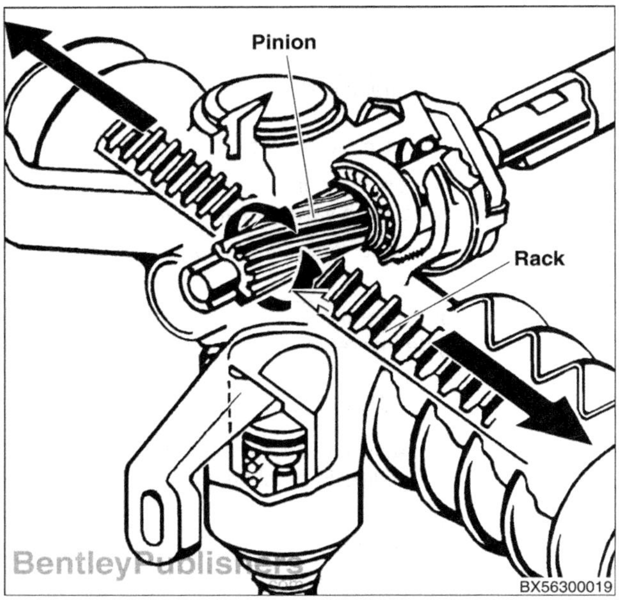

The rack-and-pinion steering system is powered by an hydraulic pump driven by the engine accessory belt.

Power assist to the steering system is regulated in one of two ways:

- BMW Servotronic provides speed-sensitive variable power assist to the steering gear and is regulated by software in the safety and gateway module (SGM).
- Active steering (AS) incorporates Servotronic, but in addition provides variable steering ratio.

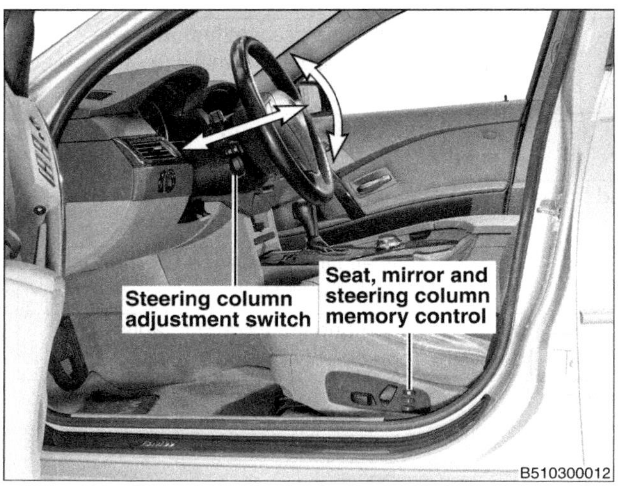

The steering column is electrically adjustable for length and height. Seat and mirror memory control also adjusts steering wheel position.

BentleyPublishers.com
BentleyPublishers.com—All Rights Reserved

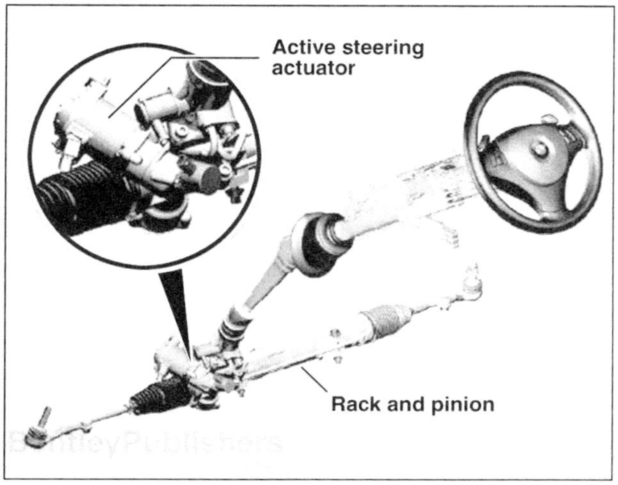

Active steering

Active steering (AFS or AS), an option on 5 Series vehicles, allows variable steering ratio based on road speed and conditions such as yaw and lateral acceleration.

Active steering manipulation of steering ratio from 10 : 1 to 20 : 1 is actuated by a planetary gear set driven by a 3-phase DC motor and controlled by the active steering control module.

BRAKE SYSTEM

5 Series vehicles covered by this manual are equipped with power disc brakes with integral antilock brakes (ABS) and dynamic stability control (DSC). See **340 Brakes** for details.

The parking brake is a dual-drum system integrated with the rear brake rotors.

Lightweight brake rotors are two-piece: The braking surface is made from grey cast iron, while the hubs are made from aluminum. The two parts are riveted together.

Each disc brake uses a caliper with a single hydraulic cylinder. Brake pads in the left front and right rear contain wear sensors. When brake pads need replacement, the sensors illuminate a light on the dashboard.

Power assist is provided by a vacuum booster when the engine is running. The brake pedal pushrod is connected directly to the master cylinder, so failure of the vacuum booster does not normally result in total brake failure.

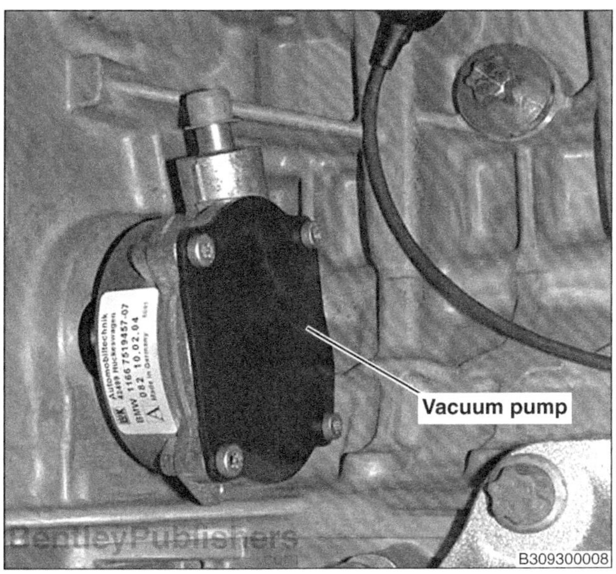

Vacuum booster pump

In order to provide adequate vacuum for the brake booster, an additional vacuum pump is fitted, except in models with M54 engine.

In 6-cylinder models, the pump is driven off the oil pump chain.

N52 or N52KP engine: Vacuum pump is attached to rear of timing chain housing.

BentleyPublishers
BentleyPublishers.com—All Rights Reserved

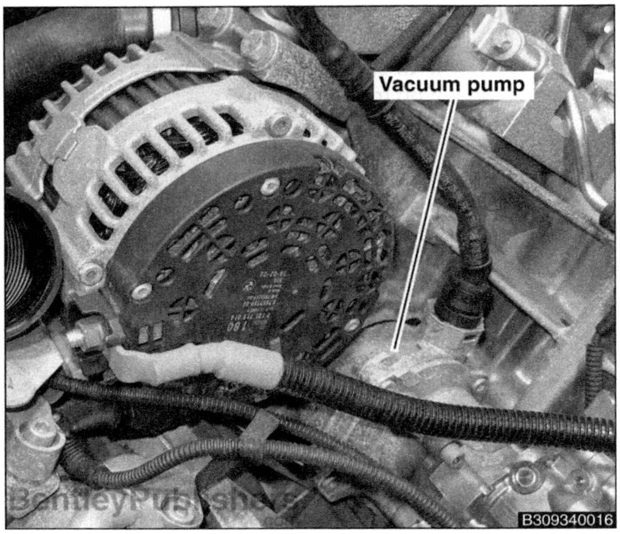

N54 engine: Vacuum pump in tandem with high pressure fuel pump is attached to rear of timing chain housing, below and to right of alternator.

V8 engine: Vacuum pump is attached to front of right timing chain cover and driven off the right exhaust camshaft.

See **100 Engine–General** for additional information.

WHEEL ALIGNMENT

Use electronic alignment equipment to check and set camber and toe for all 4 wheels. Be sure to check and adjust ride height prior to attempting wheel alignment.

For ride height and alignment specifications and procedures, see **320 Steering and Wheel Alignment**.

BentleyPublishers.com
BentleyPublishers.com–All Rights Reserved

Wheels and Tires

Run-on-flat tires are optional on 5 Series cars covered by this manual. Several different styles of wheels in 16, 17, 18 and 19 inch diameters are available from an authorized BMW dealer.

Flat tire monitor (FTM)

Flat tire monitor (FTM) uses input from the ABS wheel speed sensors to keep track of tire inflation pressures as the vehicle is driven. In case of pressure loss in a tire, the rolling radius and therefore the rotation speed of the wheel changes. This change is detected and the system alerts the driver using warning lights in the instrument cluster.

FTM electronic control is integrated in DSC control module.

FTM, initializing using iDrive

In order to assure reliability, use iDrive controller and display to initialize FTM system after tire pressures are set correctly.

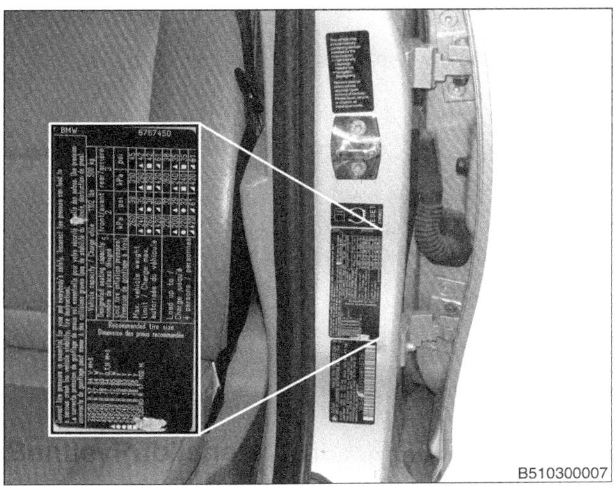

◄ B-pillar door sticker (**inset**) provides factory-recommended tire inflation pressures.

Repeat initialization each time tire inflation pressures are corrected and after every tire or wheel change.

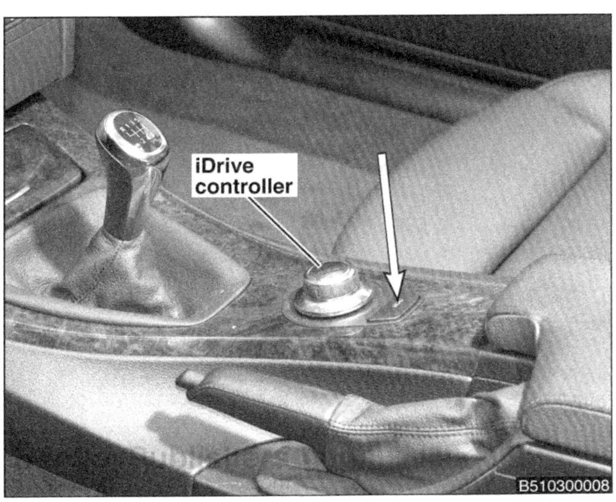

◄ With ignition ON, press down iDrive menu button (**arrow**) to open **Start** menu.

NOTE—

- *Early style iDrive shown. Later version iDrive operates similarly.*

- Press iDrive controller to open **info sources** menu.
- Select **Vehicle settings** and press controller.

BentleyPublishers.com—All Rights Reserved

B510300009

◀ Select **FTM** and press controller.

- Start engine but do not drive.
- Select **Set tire pressure** and press controller.

B510300010

◀ Select **Yes** and press controller.

- Start to drive. The message **Initializing** is displayed.
- Initialization finishes during driving and **Status: FTM active** appears again.

  If driving is interrupted before initialization is complete, initialization continues automatically when driving resumes.

  Do not initialize the system when driving with snow chains or compact spare tire.

B510300011

Flat tire indication

◀ In case of a flat tire or extensive loss of air in a tire, the instrument warning light lights up red and a Check Control message appears. In addition, an acoustic signal sounds.

- Reduce speed. Stop vehicle carefully. Avoid sudden braking and steering maneuvers.
- Identify damaged tire and replace. See **020 Maintenance**.

Run-on-flat tires

In case of a flat tire warning in the instrument cluster:

- Cautiously reduce speed to below 50 mph (80 kph). Avoid sudden braking and steering maneuvers. Do not exceed a speed of 50 mph (80 kph).

> ***WARNING —***
> - *If the vehicle is not equipped with run-on-flat tires, do not continue to drive. Continuing to drive with a flat tire could cause a severe accident.*

- At the next opportunity, check air pressure in all four tires.
- If tire pressure in all four tires is correct, FTM may not have initialized correctly. Initialize system.

BentleyPublishers.com
BentleyPublishers.com—All Rights Reserved

– In case of complete tire pressure loss, use **Table b** to estimate safe distance for continued driving.

Table b. Driving distance with flat run-on flat tire	
Condition	**Approx. safe distance to drive**
1 - 2 persons, no luggage	155 miles (250 km)
2 persons, full luggage 4 persons, no luggage	94 miles (150 km)
Full load: 4 persons, full luggage	30 miles (50 km)

WARNING —
- *In case of a completely flat run-on-flat tire, drive cautiously and do not exceed a speed of 50 mph (80 kph).*
- *Vehicle handling changes with a flat tire. This can mean reduced tracking stability in braking, extended braking distance and altered steering characteristics.*
- *If unusual vibration or loud noises occur while driving on a damaged tire, this may be an indication that the tire has failed completely. Reduce speed and pull over as soon as possible at a suitable location. Otherwise parts of the tire could come loose, resulting in an accident. Do not continue driving.*

System limits

FTM cannot indicate sudden severe tire damage caused by outside factors. It also does not detect an even pressure drop in all four tires. In the following situations, the system could be delayed or malfunction:

- Uninitialized system.
- Driving on snowy or slippery road surface.
- Sporty driving style leading to slip in the drive wheels, high lateral acceleration.
- Driving with snow chains.
- Driving with compact spare.

Tire pressure monitor (TPM)

Beginning with 2006 models, tire pressures are automatically monitored by the tire pressure monitoring system (TPM).

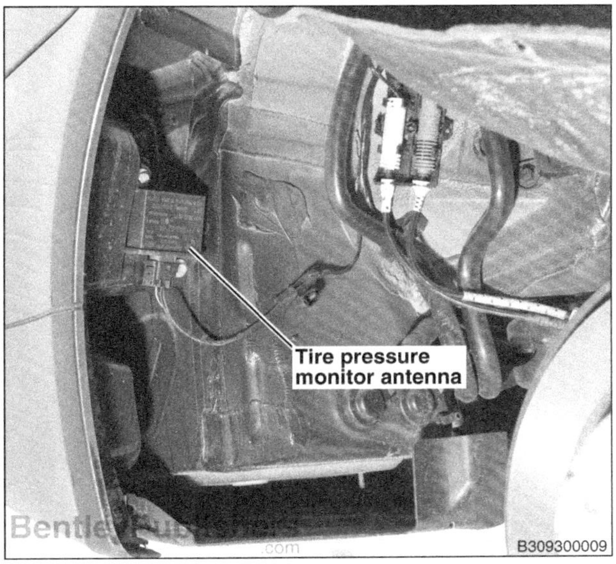

A TPM antenna in each wheel communicates tire pressure data to a pick-up antenna in each wheel housing.

TPM electronic control is integrated in DSC control module.

BentleyPublishers.com
BentleyPublishers.com—All Rights Reserved

Tire pressure data from run-on-flat tires is monitored by the tire pressure monitoring (TPM) system. If a substantial loss of tire pressure is detected, warning lights in red illuminate in the instrument cluster and a tire failure message appears in Check Control display. In addition, an acoustic signal sounds.

> ***WARNING —***
> - *In case of a completely flat run-on-flat tire, drive cautiously and do not exceed a speed of 50 mph (80 kph).*
> - *Vehicle handling changes with a flat tire. This can mean reduced tracking stability in braking, extended braking distance and altered steering characteristics.*
> - *If unusual vibration or loud noises occur while driving on a damaged tire, this may be an indication that the tire has failed completely. Reduce speed and pull over as soon as possible at a suitable location. Otherwise parts of the tire could come loose, resulting in an accident. Do not continue driving.*
> - *TPM does not work correctly if it is not initialized.*
> - *TPM cannot indicate a flat tire if a wheel without TPM electronics is mounted.*

Avoid sudden braking and steering maneuvers. At the next opportunity, check air pressure in all four tires.

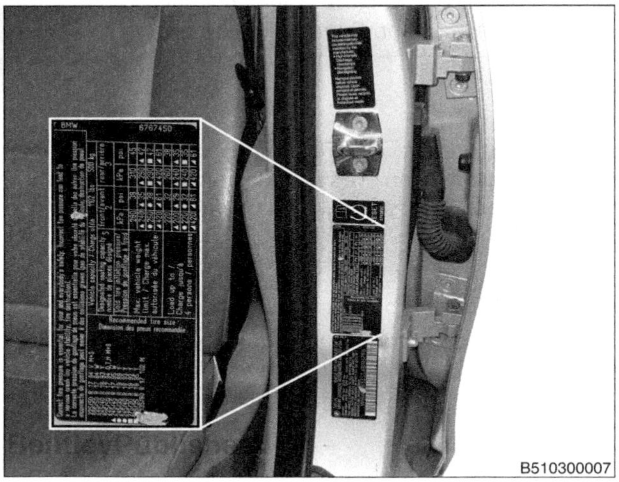

B-pillar door sticker (**inset**) provides factory-recommended tire inflation pressures.

Immediately after correcting tire pressures or mounting a new wheel, reset TPM.

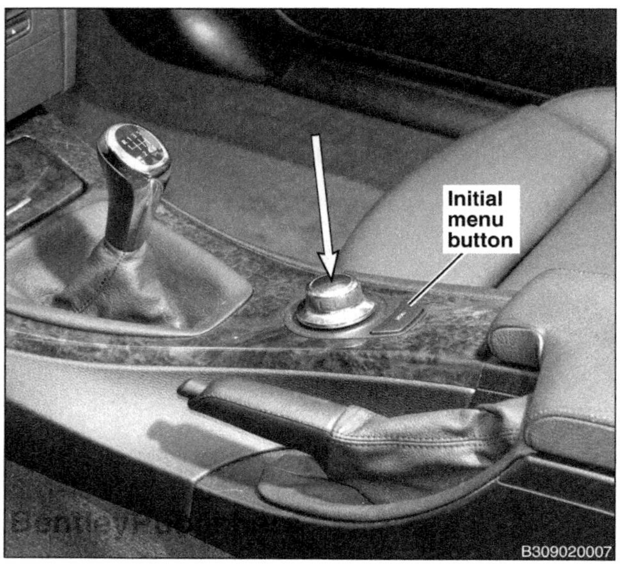

TPM, resetting

– Switch ignition ON.

Press down iDrive controller (**arrow**) to select **Info sources | Settings** menu.

> ***NOTE —***
> - *Early style iDrive shown. Later version iDrive operates similarly.*

BentleyPublishers.com
BentleyPublishers.com—All Rights Reserved

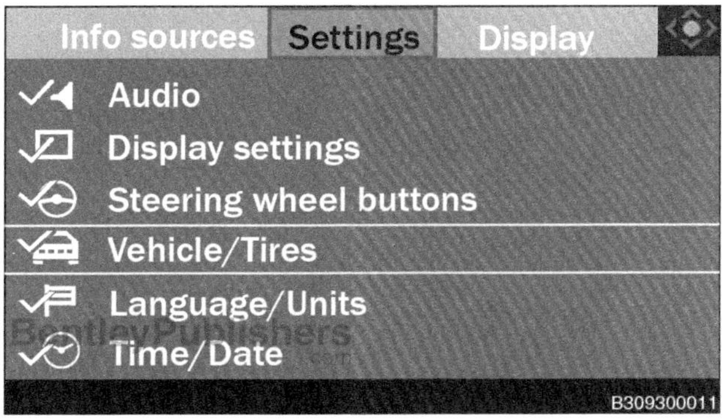

◀ Rotate controller to **Settings**.

- Press down controller to select **Settings**, then rotate to **Vehicle/Tires**.

◀ Press down controller to select **Vehicle/Tires**. In similar manner, navigate to **TPM** screen.

- Start engine, but do not drive off.
- Select **Reset** and press controller.
- Select **Yes** and press controller.
- Start to drive. Initialization completes automatically while vehicle is on the move. Correctly inflated tires show up as green on iDrive display.

300

BentleyPublishers
BentleyPublishers.com—All Rights Reserved

BentleyPublishers.com
BentleyPublishers.com–All Rights Reserved

310 Front Suspension

GENERAL

This repair group covers the repair and replacement of front suspension components.

See also:

- **300 Suspension, Steering and Brakes–General** for a general description of the suspension system
- **311 Front Axle Differential** for front drive components
- **320 Steering and Wheel Alignment** for ride height specifications

Front axle description

Most front axle components are made of aluminum. Weight and space are saved and front axle rigidity increased by the use of an aluminum reinforcement panel bolted to the front subframe. BMW specifies the following:

- Do not drive the vehicle with the front end reinforcement removed.
- When reinstalling the reinforcement panel after repairs, replace mounting fasteners with new.

BentleyPublishers.com
BentleyPublishers.com—All Rights Reserved

The suspension features double-pivot ball-joints at the base of each steering knuckle and a MacPherson strut at the top.

All models are equipped with rack-and-pinion steering. During wheel alignment, adjust toe at the tie rods. Correct camber by removing a locating pin on the strut top bearing and rotating the strut top inside the strut tower. See **320 Steering and Wheel Alignment**.

Front suspension components (rear-wheel drive)

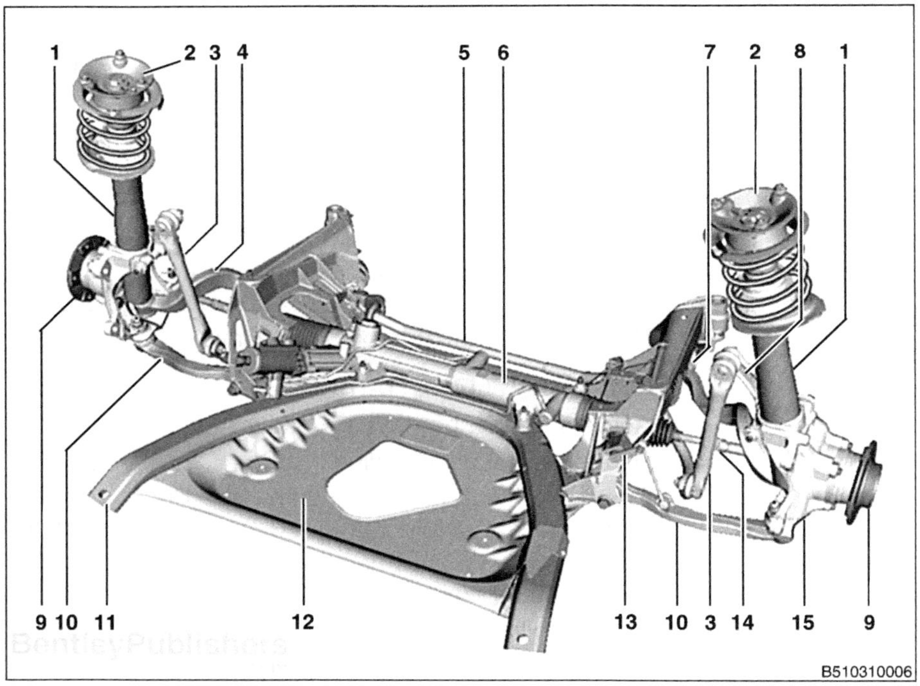

1. Strut assembly
2. Strut top bearing
3. Stabilizer bar link
4. Tension strut
5. Stabilizer bar
6. Steering rack
7. Hydromount
8. Stabilizer link bracket
9. Wheel hub
10. Control arm
11. Front subframe (axle carrier)
12. Front end reinforcement
13. Ride level sensor
14. Steering tie rod
15. Steering knuckle (swivel bearing) and wheel bearing carrier

BentleyPublishers.com
BentleyPublishers.com—All Rights Reserved

Front suspension components (xDrive)

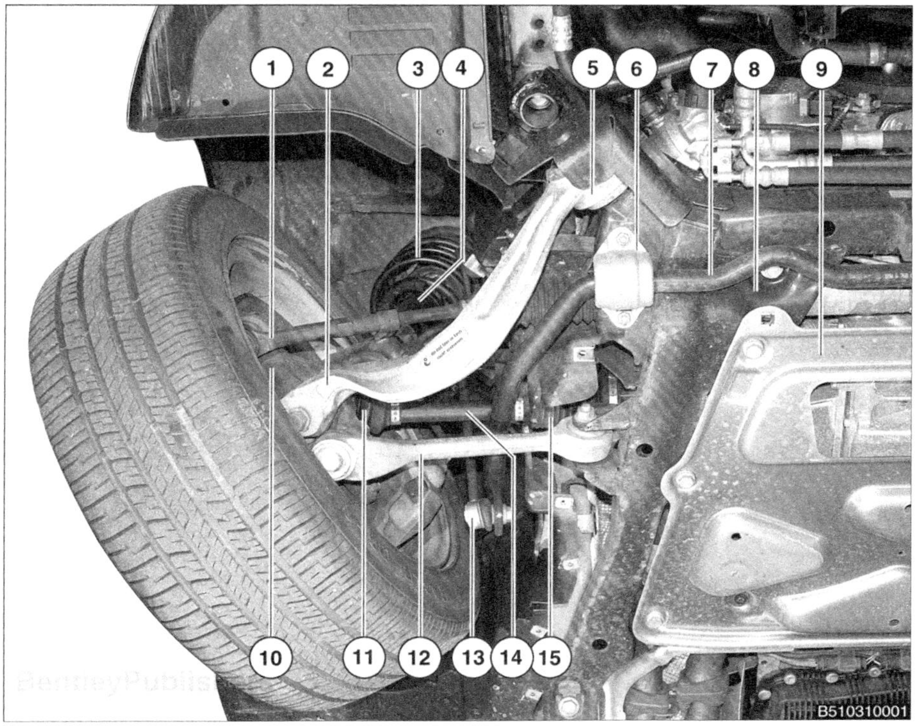

1. Steering tie-rod
2. Tension strut
3. Coil spring
4. Strut tube
5. Hydromount
6. Stabilizer bar mount
7. Stabilizer bar
8. Front subframe
9. Front end reinforcement
10. Steering knuckle (swivel bearing) and wheel bearing carrier
11. Drive axle outer CV joint
12. Control arm
13. Stabilizer arm link
14. Drive axle
15. Drive axle inner CV joint

Front end reinforcement panel, removing and installing

the aluminum front end reinforcement panel provides additional rigidity to the subframe (front axle carrier).

> ***CAUTION—***
> - *Do not drive the vehicle with the front end reinforcement panel removed.*

– Raise vehicle and support safely.

> ***WARNING —***
> - *Make sure the vehicle is stable and well supported at all times. Use a professional automotive lift or jack stands designed for the purpose. A floor jack is not adequate support.*

– Remove underbody splash shields. See **020 Maintenance**.

BentleyPublishers.com—All Rights Reserved

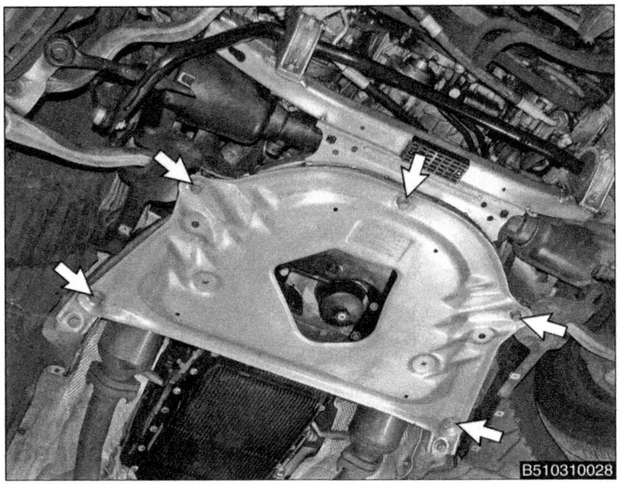

- Remove fasteners (**arrows**) and remove reinforcement panel.

- Installation is reverse of removal. Be sure to replace fasteners with new.

Tightening torque	
Front end reinforcement panel to subframe (replace M10 x 35 mm 10.9 grade ZNS bolts with new): • Stage 1 • Stage 2	 56 Nm (41 ft-lb) + additional 90°

Strut tower brace, removing and installing

A strut tower brace in the engine compartment provide additional rigidity to the front suspension.

> ***CAUTION—***
> • *Do not drive the vehicle with strut tower brace removed.*

- Working in engine compartment, detach wiring harness and remove upper and lower cabin microfilter housings on right and left. See **020 Maintenance**.

- If necessary, remove plastic strut tower covers. Remove plastic shield over strut tower brace at bulkhead.

- Remove strut tower brace fasteners (**arrows**) and lift off brace.

- When installing, replace bolts with new.

Tightening torque	
Brace to strut tower or bulkhead (use new M8 Torx bolts)	30.5 Nm (22 ft-lb)

BentleyPublishers.com
BentleyPublishers.com—All Rights Reserved

Warnings and Cautions

WARNING—
- *Do not reuse self-locking fasteners. They are designed to be used only once and may fail if reused. Replace with new.*
- *Pentosin fluid (in active stabilizer hydraulic system) is injurious to skin and eyes. Wear protective gloves and eyewear.*

CAUTION—
- *Do not drive the vehicle with the front end reinforcement panel removed. See* **Front end reinforcement panel, removing and installing** *in this repair group.*
- *When reinstalling the panel after repairs, replace panel mounting fasteners with new.*
- *Due to the chemical and corrosion characteristics of aluminum, follow these precautions when working with aluminum components:*
  —Do not bring into contact with battery acid.
  —Do not use wire brushes with brass or iron bristles. Only use brushes with stainless steel bristles.
  —Do not expose to flying sparks from grinding or cutting operations.
  —Do not subject to steel welding splashes.
  —Do not expose to temperatures over 80°C (176°F), even for short periods. Temperatures in painting facilities are not a problem.
- *When filling or bleeding hydraulic system, do not allow contamination of fluid or reservoir.*

FRONT STRUT ASSEMBLY

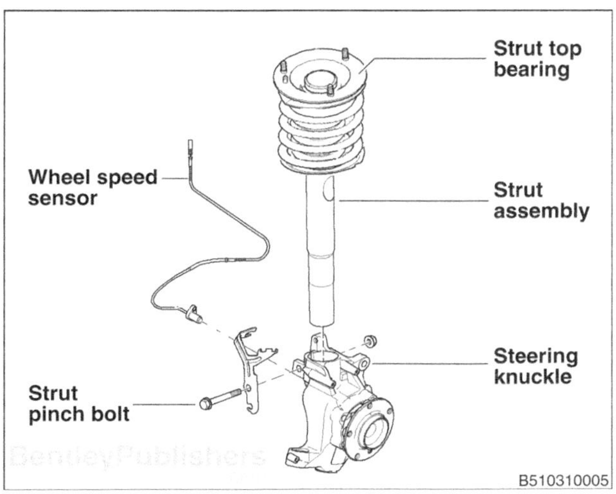

The front suspension shock absorbers are MacPherson struts. The strut is a major component of the front suspension and supports the spring. Most strut assembly components are available as replacement parts.

If a shock absorber or strut is defective on one side of the vehicle, replace both shock absorbers on that axle. If a coil spring is broken due to corrosion, replace both coil springs on the axle.

Front strut, strut top bearing mount or spring replacement is a two-step procedure:

- Removal of strut assembly from vehicle.
- Disassembly and replacement of components on work bench.

Procedures for removal of the front strut assembly differ between rear-wheel drive and all-wheel drive (xDrive) models. Be sure to read the complete procedure before starting work.

BentleyPublishers.com—All Rights Reserved

Front strut assembly, removing and installing (rear-wheel drive)

– Raise car and remove front wheel.

> ***WARNING —***
> *• Make sure the vehicle is stable and well supported at all times. Use a professional automotive lift or jack stands designed for the purpose. A floor jack is not adequate support.*

– Detach front brake caliper from steering knuckle and suspend aside using stiff wire. Do not detach brake fluid hose. See **340 Brakes**.

– Remove brake rotor. See **340 Brakes**.

– Remove ABS wheel speed sensor from steering knuckle. See **340 Brakes**.

– Detach control arm and tension strut from steering knuckle. See **Front Suspension Arms** in this repair group.

– Detach tie rod end from steering knuckle. See **320 Steering and Wheel Alignment**.

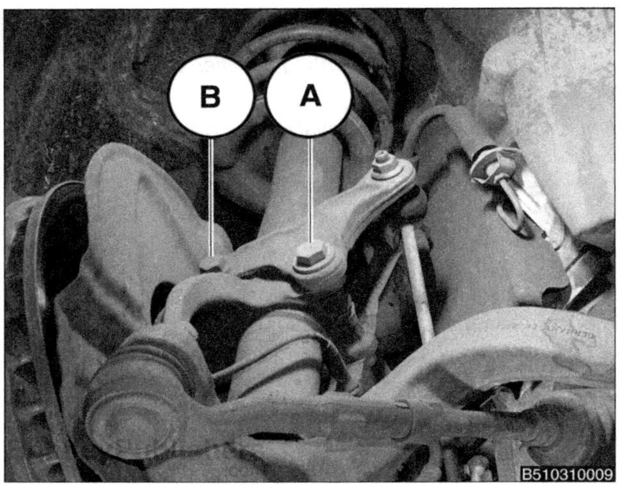

Working at top of steering knuckle:

- Remove strut pinch bolt (**A**) from upper steering knuckle.
- Remove stabilizer link bracket bolt (**B**) and detach bracket from steering knuckle.

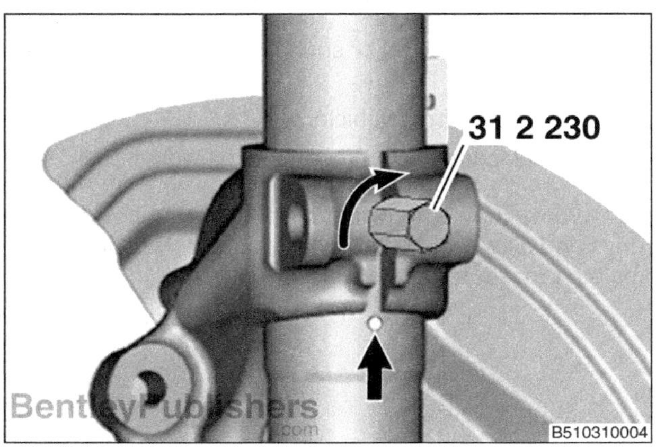

If necessary, use BMW special tool 31 2 230 to spread (**curved arrow**) steering knuckle pinch collar. Slide steering knuckle downward to disengage from strut tube.

- Note locating boss (**arrow**) for strut. During reassembly, boss slides in steering knuckle pinch collar gap.

BentleyPublishers.com
BentleyPublishers.com—All Rights Reserved

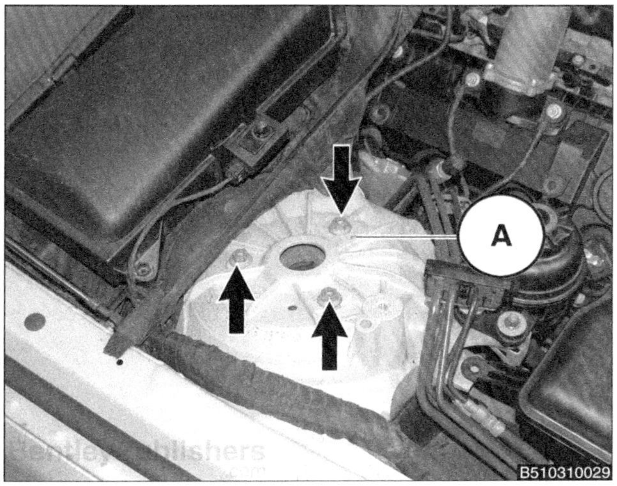

< Working in engine compartment at strut top bearing:

- Note position of strut centering pin (**A**). If centering pin is missing, mark position of strut top bearing studs on strut tower to maintain original camber.
- Secure strut against falling out.
- Remove strut top bearing fasteners (**arrows**).

– Lower strut out of wheel arch.

– If replacing strut, coil spring or strut top bearing, disassemble and reassemble strut assembly on bench. See **Front strut assembly, disassembling and assembling** in this repair group.

– Remainder of installation is reverse of removal. Note the following:

- Letter on strut tube designates mounting location. Struts are marked **R** for right and **L** for left.
- Replace self-locking fasteners.
- Strut top bearing without centering pin: Mount strut top to strut tower using marks made previously.
- Make sure locating boss on strut tube slides in steering knuckle pinch collar gap.
- Face strut pinch bolt head in direction of travel.
- Make sure steering knuckle contacts stop on strut before tightening pinch bolt.
- Clean wheel speed sensor bore hole and lubricate with Staburags® NBU 12/K or equivalent grease.
- Align wheels when job is complete.

Tightening torques	
Brake caliper to steering knuckle replace M12 x 1.5 x 43 mm ZNS bolts with new)	110 Nm (81 ft-lb)
Brake rotor to wheel hub (replace M8 x 14 mm Allen bolt with new)	16 Nm (12 ft-lb)
Control arm to steering knuckle (replace M14 x 1.5 self-locking flange nut with new)	165 Nm (122 ft-lb)
Road wheel to hub	120 ± 10 Nm (89 ± 7 ft-lb)
Strut assembly pinch bolt (replace M12 x 100 mm 10.9 grade bolt with new)	81 Nm (60 ft-lb)
Strut top bearing to strut tower (replace M8 flanged nut with new)	34 Nm (25 ft-lb)
Tension strut to steering knuckle (replace M14 x 1.5 self-locking flange nut with new)	165 Nm (122 ft-lb)
Tie rod end to steering knuckle (replace M14 x 1.5 ZNS3 nut with new)	165 Nm (122 ft-lb)
Wheel speed sensor to steering knuckle	8 Nm (6 ft-lb)

BentleyPublishers.com
BentleyPublishers.com—All Rights Reserved

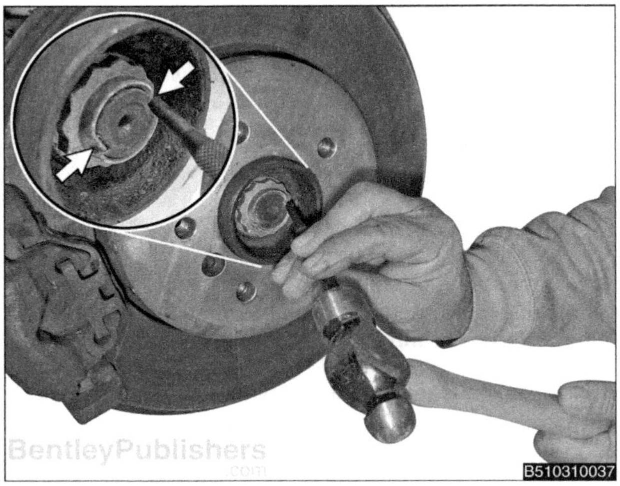

Front strut assembly, removing and installing (xDrive)

– Raise car and remove front wheel.

> ***WARNING —***
> *• Make sure the vehicle is stable and well supported at all times. Use a professional automotive lift or jack stands designed for the purpose. A floor jack is not adequate support.*

◄ Using a suitable punch, straighten staked sides of wheel hub collar nut (**arrows**).

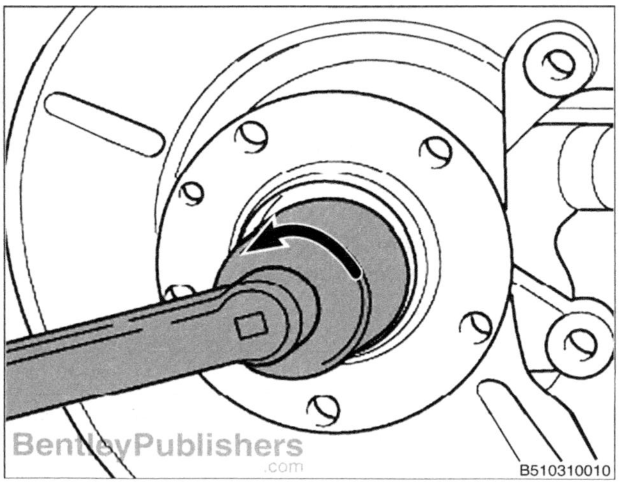

◄ With an assistant holding down brake pedal, remove collar nut and discard.

> ***CAUTION—***
> *• If necessary, reinstall road wheel and place vehicle on floor before breaking collar nut free.*

– Detach front brake caliper from steering knuckle and suspend aside using stiff wire. Do not detach brake fluid hose. See **340 Brakes**.

– Remove brake rotor. See **340 Brakes**.

– Remove wheel speed sensor from steering knuckle. See **340 Brakes**. Detach sensor electrical harness from bracket on strut.

– Detach control arm and tension strut from steering knuckle. See **Front Suspension Arms** in this repair group.

– Detach tie rod end from steering knuckle. See **320 Steering and Wheel Alignment**.

– Detach stabilizer link from strut tube. Use Torx tool (T30) to counterhold shaft of stabilizer bar link ball joint while removing nut.

– Rotate steering knuckle to one side and use BMW special tools 32 2 200 or equivalent to press drive axle stub out of wheel bearing hub. Tie up drive axle end using stiff wire.

> ***CAUTION—***
> *• In order to avoid damaging drive axle threads, do not drive out stub axle using impact tool.*

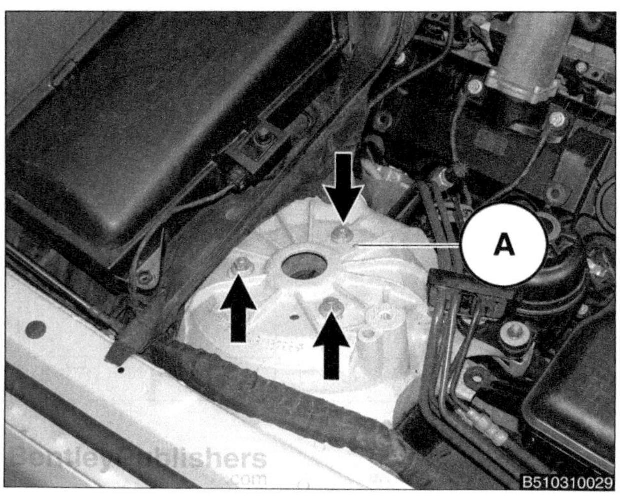

◄ Working at strut top bearing:

- Note position of strut centering pin (**A**). If centering pin is missing, mark position of strut top bearing studs on strut tower to maintain original camber.
- Support steering knuckle with floor jack to secure assembly against falling out.
- Remove strut top bearing fasteners (**arrows**).

– Lower strut and steering knuckle out of wheel arch.

BentleyPublishers.com
BentleyPublishers.com—All Rights Reserved

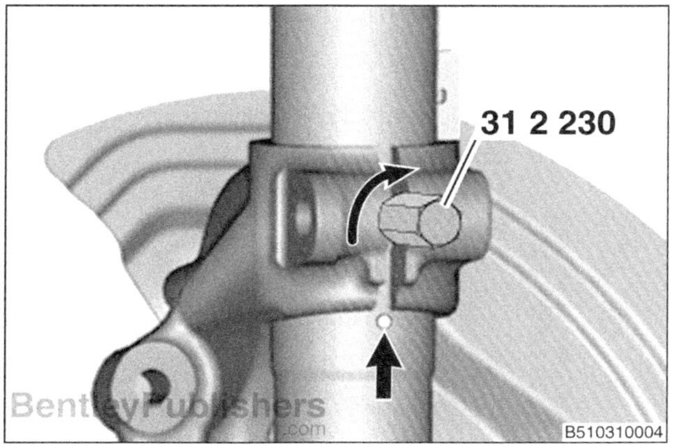

– Working at top of steering knuckle, remove strut pinch bolt.

◄ If necessary, use BMW special tool 31 2 230 to spread (**curved arrow**) steering knuckle pinch collar. Slide strut tube out of steering knuckle.

- Note locating boss (**arrow**) for strut. During reassembly, boss slides in steering knuckle pinch collar gap.

– If replacing strut, disassemble and reassemble strut assembly on work bench. See **Front strut assembly, disassembling and assembling** in this repair group.

– Installation is reverse of removal. Note the following:

- Letter on strut tube designates mounting location. Struts are marked **R** for right and **L** for left.
- Make sure locating boss on strut tube slides in steering knuckle pinch collar gap.
- Make sure steering knuckle contacts stop on strut before tightening pinch bolt. Face strut pinch bolt head in direction of travel.
- Replace self-locking fasteners.
- Strut top bearing without centering pin: Mount strut top to strut tower using marks made previously.
- Use Torx tool (T30) to counterhold shaft of stabilizer bar link ball joint while tightening nut.
- Clean ABS wheel speed sensor bore hole and lubricate with Staburags® NBU 12/K or equivalent grease.
- Oil collar nut and wheel hub contact surfaces lightly. Do not allow oil to contaminate drive axle threads. Final tightening with wheel on ground. Stake collar nut after final tightening.
- Align wheels when job is complete.

Tightening torques	
Brake caliper to steering knuckle (replace M12 x 1.5 x 43 mm ZNS bolts with new)	110 Nm (81 ft-lb)
Brake rotor to wheel hub (replace M8 x 14 mm Allen bolt with new)	16 Nm (12 ft-lb)
Control arm to steering knuckle (replace M14 x 1.5 self-locking flange nut with new)	165 Nm (122 ft-lb)
Drive axle to wheel hub (replace M27 x 1.5 collar nut with new)	420 Nm (310 ft-lb)
Road wheel to hub	120 ± 10 Nm (89 ± 7 ft-lb)
Stabilizer link to strut tube (replace M10 ZNS3 nut with new)	65 Nm (48 ft-lb)
Strut assembly pinch bolt (replace M12 x 1.5 x 80 mm ZNS3 bolt with new)	100 Nm (74 ft-lb)
Strut top bearing to strut tower (replace M8 flanged nut with new)	34 Nm (25 ft-lb)
Tension strut to steering knuckle ball joint (replace M14 x 1.5 self-locking flange nut with new)	165 Nm (122 ft-lb)
Tie rod end to steering knuckle (replace M14 x 1.5 ZNS3 nut with new)	165 Nm (122 ft-lb)
Wheel speed sensor to steering knuckle	8 Nm (6 ft-lb)

BentleyPublishers.com
BentleyPublishers.com—All Rights Reserved

Front strut assembly components

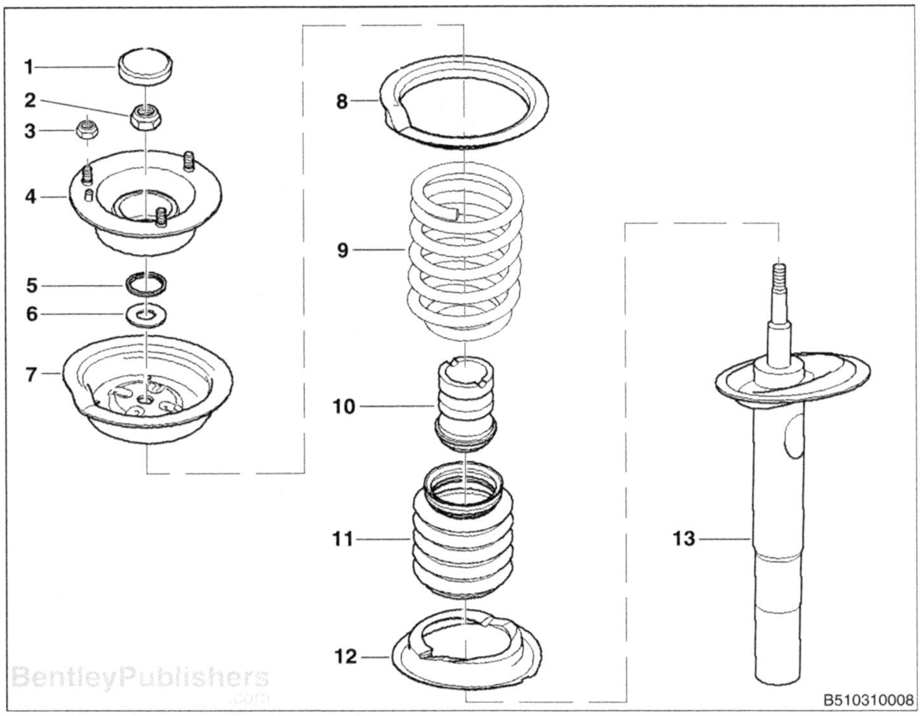

1. Cap
2. Upper strut self-locking nut
   - Use new nut
   - Tighten to 64 Nm (47 ft-lb)
3. Strut top bearing mounting nut
   - Use new nut
   - Tighten to 34 Nm (25 ft-lb)
4. Strut top bearing
5. O-ring seal
6. Washer
7. Upper spring seat
8. Upper spring pad
9. Coil spring
10. Rubber bump stop
11. Dust boot
12. Lower spring pad
13. Strut with lower spring seat

Front strut assembly, disassembling and assembling

If replacing strut, upper strut mount or coil spring, first remove the strut assembly from the vehicle. See **Front strut assembly, removing and installing (rear-wheel drive)** or **Front strut assembly, removing and installing (xDrive)** in this repair group.

BMW recommends the use of BMW spring compressor (special tool 31 3 340). A number of different spring compressor tools are available. Photos in this section illustrate the function of one such tool in use.

> ***WARNING —***
> - *Do not attempt to disassemble the strut assembly without a spring compressor designed specifically for this job.*
> - *Prior to each use, check special tool for functionality.*
> - *Do not use a damaged tool.*
> - *Do not make any modifications to tool.*
> - *Use correct size spring retainers when compressing coil spring.*
> - *When assembling BMW spring compressor (special tool 31 3 340), make sure spring retainer plates are felt and heard snapping into place. Check seating of spring retainers carefully.*

Clamp spring compressor in shop vise.

BentleyPublishers.com
BentleyPublishers.com—All Rights Reserved

Position coil spring between spring retainers so that 3 coils lie between retainers (**arrows**). Compress spring.

> ***WARNING —***
> - *Make sure that when tensioned, the spring coil rests completely in the spring holder recess.*
> - *Do not tighten or loosen spring compressor with an impact tool.*
> - *Tighten down the coil spring until stress on the strut top bearing is relieved.*
> - *Only loosen upper strut nut if spring coils are completely inserted in the spring holder grooves. If necessary, loosen compressor, reposition and recompress.*

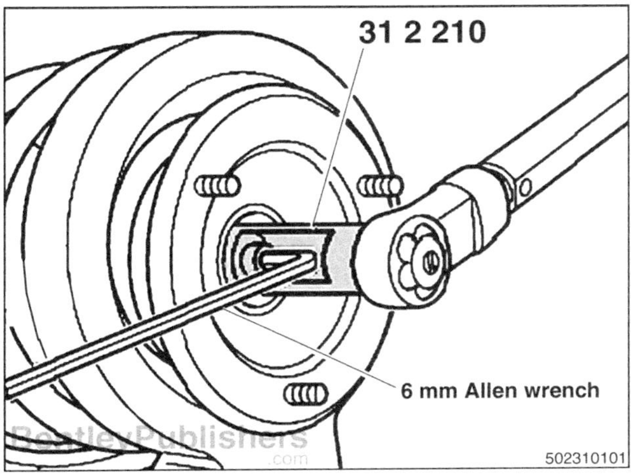

Use BMW special tool 31 2 210 or equivalent to remove upper strut nut. Counterhold strut shaft using 6 mm Allen wrench.

> ***CAUTION—***
> - *Do not remove strut nut with impact tool.*

– Remove upper strut bearing and related components.

– If a new coil spring is being installed, relieve tension on spring compressor and remove coil spring.

– Check strut dust boot, rubber stop and spring pads. Replace as necessary.

– Replace strut, upper strut mount or spring, as needed. If one spring is corroded or damaged, replace both springs as a matched pair.

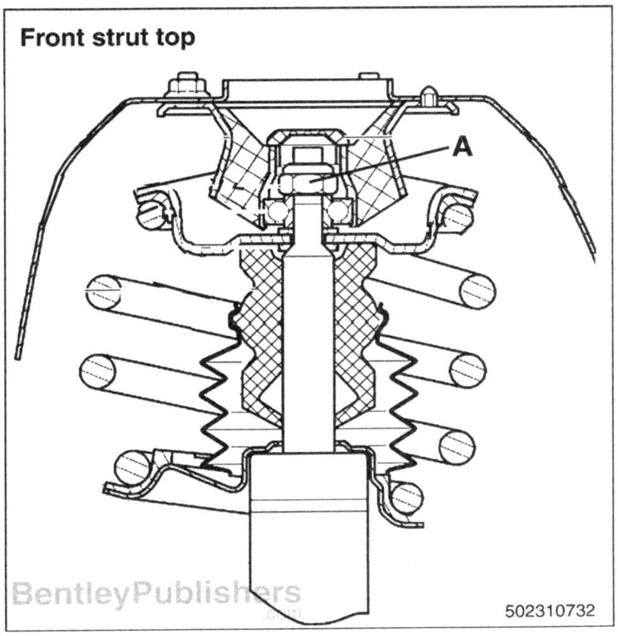

Reassemble components with spring held compressed.

- Use a new upper strut self-locking nut (**A**). Tighten nut fully before releasing spring compressor.
- Illustration shows a typical BMW strut assembly. 5 Series models are similar but not identical.

> ***CAUTION—***
> - *Do not tighten strut nut with impact tool.*

Tightening torques	
Strut shaft to strut top bearing (replace self-locking nut with new)	64 Nm (47 ft-lb)

– Be sure coil spring ends and upper and lower spring pads are correctly installed to spring seats before releasing spring compressor. See following illustrations.

310

BentleyPublishers.com
BentleyPublishers.com—All Rights Reserved

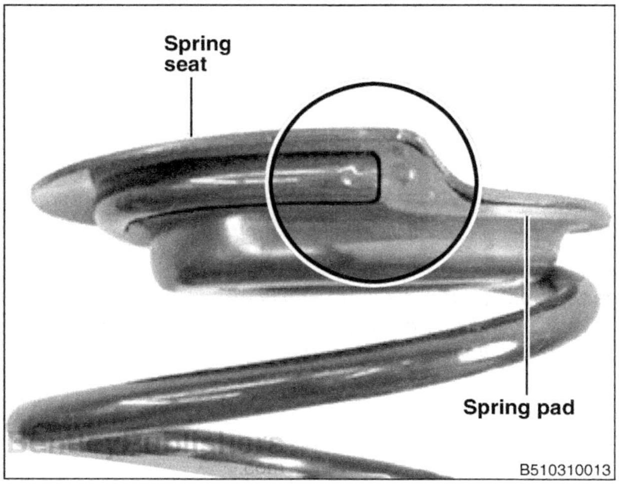

◀ Upper coil spring end:

- Make sure spring pad aligns correctly with indentation in upper spring seat.
- Place coil spring end in spring pad stop (**inset**).

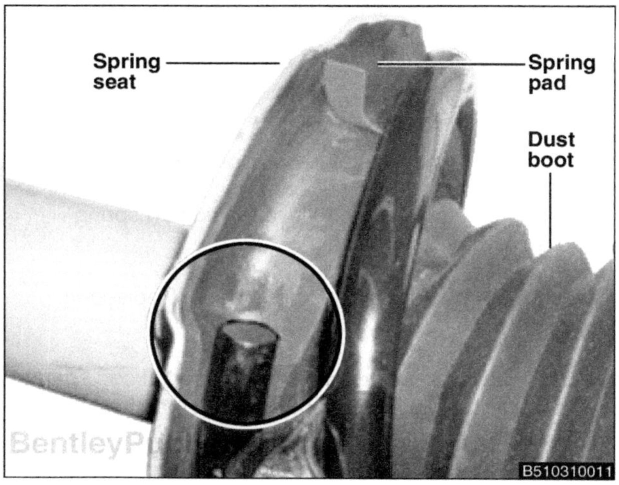

◀ Rear-wheel drive lower coil spring end:

- Align lower end of coil spring (**inset**) to spring pad as illustrated.
- Check installation position of dust boot and correct folds if necessary.

Note that end of coil spring does not rest against shoulder of spring pad. Recess in pad serves to reinforce and correctly seat lower spring pad and does not serve as a stop for the spring end.

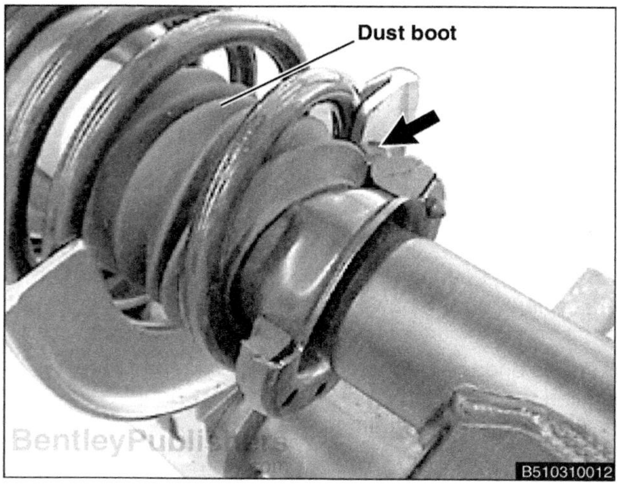

◀ xDrive lower coil spring end:

- Make sure spring end (**arrow**) rests on spring pad stop.
- Check installation position of dust boot and correct folds if necessary.

– Release spring compressor carefully and evenly, allowing spring to expand slowly.

– Reinstall strut assembly in vehicle.

– Have vehicle professionally aligned when job is complete.

BentleyPublishers.com
BentleyPublishers.com—All Rights Reserved

Front Suspension Arms

Control arms and tension struts are collectively known as suspension arms. Tension struts are sometimes called thrust arms or thrust rods.

Control arms and tension struts are attached to the front subframe through rubber bushings and to the steering knuckles by ball joints.

Note that steering wheel vibration during braking (usually at road speeds of 50 to 60 mph) may be caused by faulty suspension arm bushings, not out-of-true brake rotors.

Inspect ball joints for wear and looseness. Inspect bushings for wear or fluid leaks. On xDrive vehicles, the tension strut ball joint is available as a separate part. Otherwise, ball joints can only be replaced as part of an entire suspension arm assembly. Tension strut hydromount bushings are available as replacement parts. Replace bushings in pairs.

Special tools may be required to remove suspension arms and to replace bushings. Read the procedure through before beginning the job.

Control arm, removing and installing

– Raise car and remove front wheel.

> ***WARNING —***
> - *Make sure the vehicle is stable and well supported at all times. Use a professional automotive lift or jack stands designed for the purpose. A floor jack is not adequate support.*

– Remove splash shield underneath vehicle front. See **020 Maintenance**.

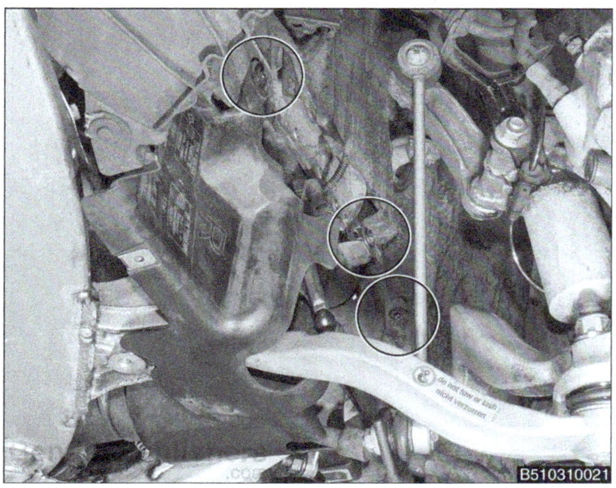

◄ Remove suspension side splash shield screws (**insets**). Remove side splash shield.

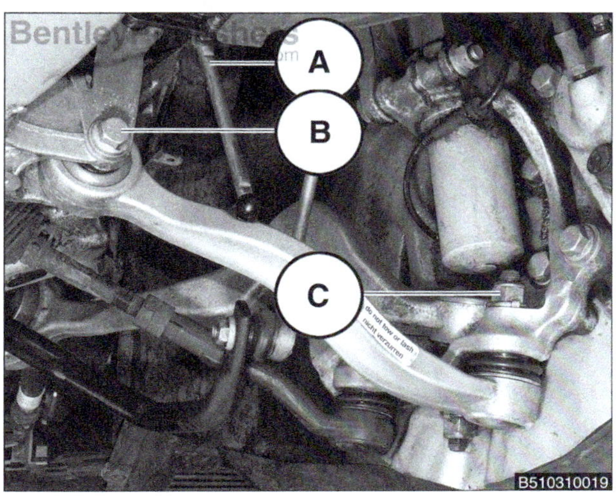

◄ Working at control arm:

- If equipped, detach ride height sensor link rod (**A**).
- Remove control arm fastener (**B**) at subframe. Lightly tap mounting bolt out of subframe.
- If necessary, counterhold ball joint shaft with Torx T40 socket while loosening ball joint nut (**C**).
- xDrive: For space reasons, back off ball joint nut 3 - 4 threads before using special tool to break loose ball joint press fit.

BentleyPublishers.com
BentleyPublishers.com—All Rights Reserved

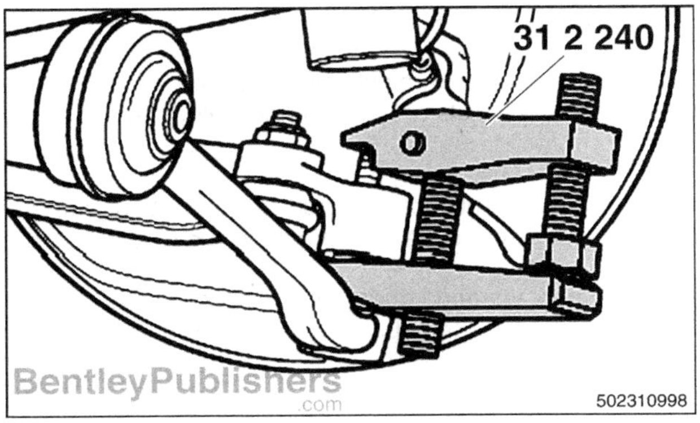

Use BMW special tool 31 2 240 or equivalent to separate ball joint from steering knuckle.

– Inspect control arm ball joint and bushing. Replace both control arms if defects are found.

– Installation is reverse of removal. Remember to:
- Make sure thread bores, bolts, nuts and mating surfaces are clean.
- Install bushing end of control arm to subframe using new fasteners. Bolt head points to front of vehicle. Do not fully tighten nut at this time.
- Install control arm ball joint to steering knuckle using new self-locking nut. Tighten fully.
- If equipped, reattach ride height sensor link rod to control arm.
- Install wheel and lower car.
- Tighten bushing end of control arm to subframe with car on ground and loaded. Bounce suspension a few times before final tightening.
- Use BMW scan tool to carry out steering angle sensor alignment.
- Align wheels when job is complete.

Tightening torques	
Control arm to steering knuckle (replace M14 x 1.5 self-locking flange nut with new)	165 Nm (122 ft-lb)
Control arm to subframe (replace M12 x 1.5 x 100 mm 10.9 grade bolt with new) • Stage 1 • Stage 2	 100 Nm (74 ft-lb) additional 90° ± 15°
Road wheel to hub	120 ± 10 Nm (89 ± 7 ft-lb)

Tension strut, removing and installing (rear-wheel drive)

– Raise car and remove wheel.

> ***WARNING —***
> - *Make sure the vehicle is stable and well supported at all times. Use a professional automotive lift or jack stands designed for the purpose. A floor jack is not adequate support.*

– Remove splash shield underneath vehicle front. See **020 Maintenance**.

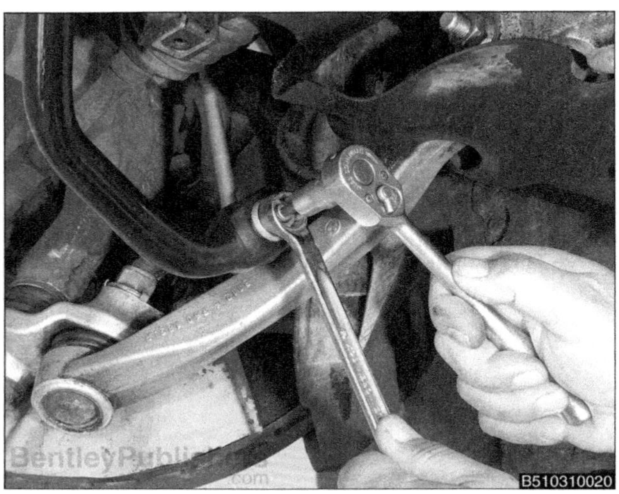

Detach stabilizer link from stabilizer bar. Use Torx tool (T30) to counterhold shaft of stabilizer bar link ball joint while loosening nut.

– Detach control arm from steering knuckle. See **Control arm, removing and installing** in this repair group.

BentleyPublishers.com
BentleyPublishers.com—All Rights Reserved

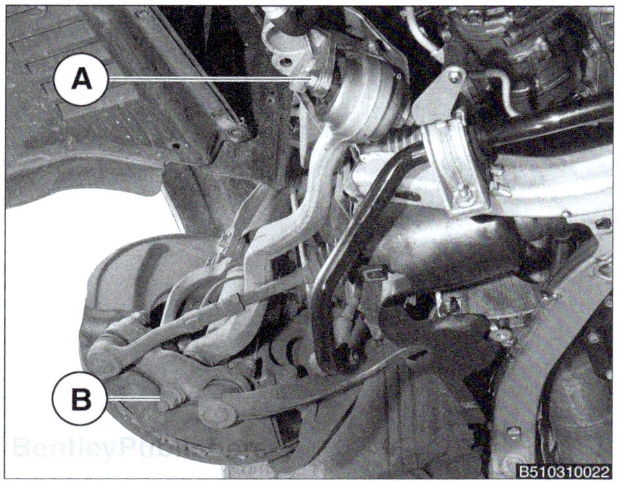

Working at tension strut:

- Remove tension strut bushing fastener (**A**). Lightly tap tension strut bushing bolt out of front subframe.
- Remove tension strut ball joint nut (**B**).

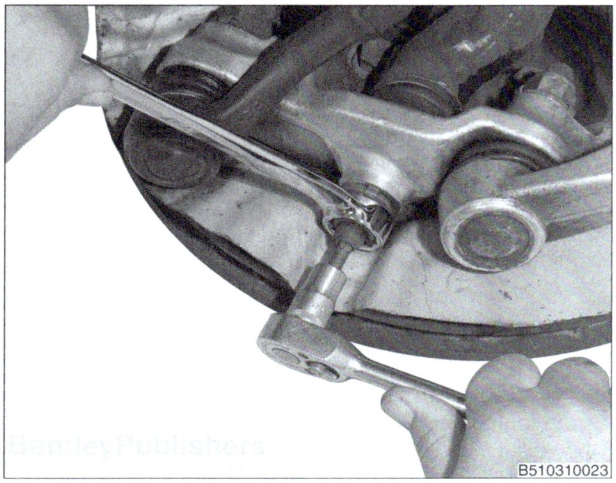

If necessary, counterhold ball joint shaft with Torx tool (T40) while loosening ball joint nut.

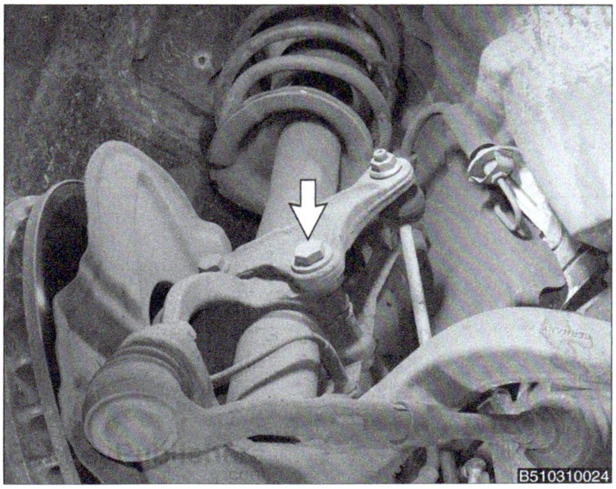

Support steering knuckle using floor jack. Working at top of steering knuckle, remove strut pinch bolt (**arrow**).

BentleyPublishers
BentleyPublishers.com—All Rights Reserved

Tension strut, removing and installing (rear-wheel drive)

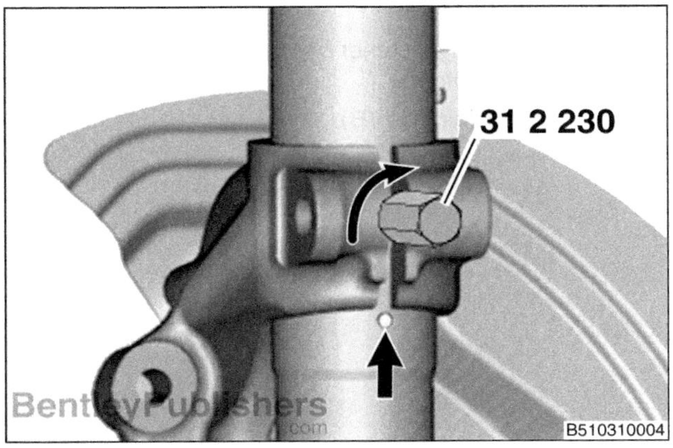

If necessary, use BMW special tool 31 2 230 to spread (**curved arrow**) steering knuckle pinch collar. Slide strut tube out of steering knuckle approx. 1 inch (2.5 cm).

- Note locating boss (**arrow**) for strut. During reassembly, boss slides in steering knuckle pinch collar gap.

– Detach tension strut ball joint from steering knuckle and remove tension strut.

– Inspect torsion strut ball joint and bushing.

- If ball joint is defective, replace both torsion struts.
- If bushing is defective. replace both bushings, but this may only be done once per torsion strut. See **Tension strut bushing, replacing** in this repair group.

– Installation is reverse of removal

- Make sure thread bores, bolts, nuts and mating surfaces are clean.
- Install bushing end of tension strut to subframe using new fasteners. Do not fully tighten nut at this time.
- Install tension strut ball joint to steering knuckle using new self-locking nut. Tighten fully.
- Slide steering knuckle up in strut tube and tighten pinch bolt. Make sure steering knuckle contacts stop on strut before tightening pinch bolt. Face strut pinch bolt head in direction of travel.
- Use Torx tool (T30) to counterhold shaft of stabilizer bar link ball joint while tightening nut.
- Install wheel and lower car.
- Tighten bushing end of tension strut to subframe with car on ground and loaded. Bounce suspension a few times before final tightening.
- Use BMW scan tool to carry out steering angle sensor alignment.
- Align wheels when job is complete.

Tightening torques	
Control arm to steering knuckle (replace M14 x 1.5 self-locking flange nut with new)	165 Nm (122 ft-lb)
Road wheel to hub	120 ± 10 Nm (89 ± 7 ft-lb)
Stabilizer link to stabilizer bar (replace M10 ZNS3 nut with new)	65 Nm (48 ft-lb)
Strut assembly pinch bolt (replace M12 x 100 mm 10.9 grade bolt with new)	81 Nm (ft-lb)
Tension strut to steering knuckle (replace M14 x 1.5 self-locking flange nut with new)	165 Nm (122 ft-lb)
Tension strut to subframe (replace M12 x 1.5 x 100 mm 10.9 grade bolt with new) • Stage 1 • Stage 2	 100 Nm (74 ft-lb) additional 90°

BentleyPublishers.com
BentleyPublishers.com—All Rights Reserved

Tension strut, removing and installing (xDrive)

— Raise car and remove wheel.

> ***WARNING —***
> *• Make sure the vehicle is stable and well supported at all times. Use a professional automotive lift or jack stands designed for the purpose. A floor jack is not adequate support.*

— Remove splash shield underneath vehicle front. See **020 Maintenance**.

— If equipped, detach ride height sensor link rod from tension strut.

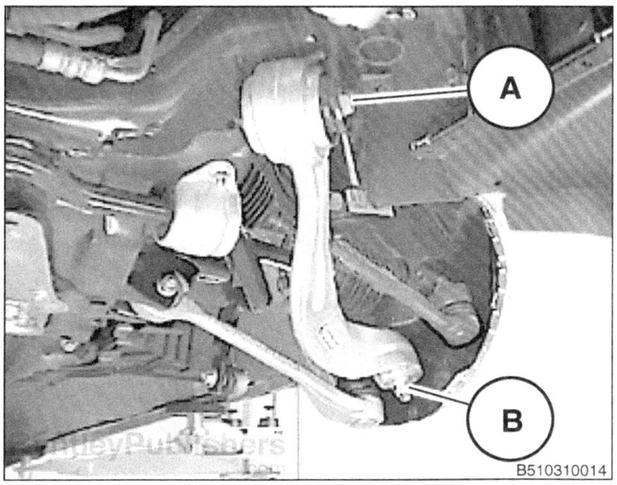

◄ Working at tension strut:

- Remove torsion strut bushing fastener (**A**).
- Remove ball joint nut. If necessary, counterhold ball joint shaft with Torx tool while loosening nut.
- Use ball joint press tool BMW special tool 31 2 240 or equivalent to separate tension strut from steering knuckle.

— Inspect torsion strut ball joint and bushing.

- If ball joint is defective, replace ball joint. See **Tension strut ball joint, replacing (xDrive)** in this repair group.
- If tension strut bushing is defective. replace both bushings, but this may only be done once per torsion strut. See **Tension strut bushing, replacing** in this repair group.

— Installation is reverse of removal

- Make sure thread bores, bolts, nuts and mating surfaces are clean.
- Install bushing end of tension strut to subframe using new fasteners. Do not fully tighten nut at this time.
- Install tension strut to steering knuckle ball joint using new self-locking nut. Tighten fully.
- Install wheel and lower car.
- Tighten bushing end of tension strut to subframe with car on ground and loaded. Bounce suspension a few times before final tightening.
- Use BMW scan tool to carry out steering angle sensor alignment.
- Align wheels when job is complete.

Tightening torques	
Road wheel to hub	120 ± 10 Nm (89 ± 7 ft-lb)
Strut assembly pinch bolt (replace M12 x 1.5 x 80 mm ZNS3 bolt with new)	100 Nm (74 ft-lb)
Tension strut to steering knuckle ball joint (replace M14 x 1.5 self-locking flange nut with new)	165 Nm (122 ft-lb)
Tension strut to subframe (replace M12 x 1.5 x 100 mm 10.9 grade bolt with new) • Stage 1 • Stage 2	 100 Nm (74 ft-lb) additional 90°

310

BentleyPublishers.com—All Rights Reserved

Tension strut bushing, replacing

Replace tension strut bushings in pairs. Bushing may only be replaced once on a tension strut.

– Raise car and remove wheel.

> ***WARNING —***
> - *Make sure that the car is firmly supported on jack stands designed for the purpose. Place jack stands underneath structural chassis points. Do not place jack stands under suspension parts.*

– Rear-wheel drive: Test tension strut ball joint. If defective, replace both tension struts. If ball joint is ok, proceed with bushing replacement procedure.

– Remove tension strut. See **Tension strut, removing and installing (rear-wheel drive)** or **Tension strut, removing and installing (xDrive)** in this repair group.

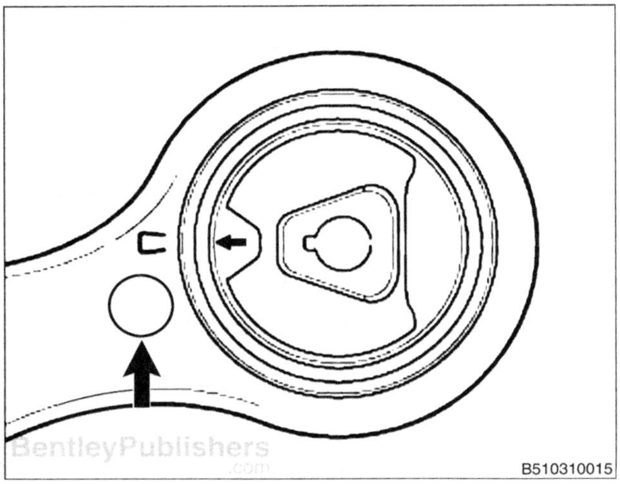

◄ Check circled area (**arrow**) of tension strut for punch mark:

- If strut was previously punch-marked, replace both tension struts.
- If not, punch mark strut to indicate bushing replacement.

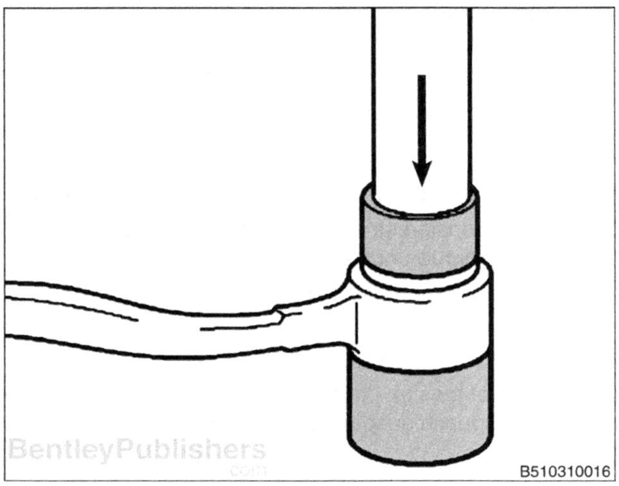

◄ Using a service press and appropriate press tools, press bushing out of tension strut.

> ***CAUTION—***
> - *Make sure press tools are exactly flush with walls of bushing bore.*

– Make sure bushing bore and new bushing are clean and grease-free.

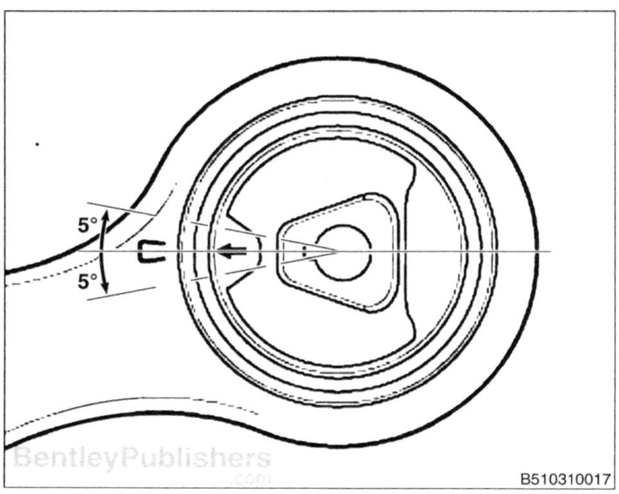

◄ Press in new bushing. Make sure mark on bushing (**small arrow**) is aligned with mark on tension strut. Maximum deviation allowed = ± 5°.

> ***CAUTION—***
> - *To avoid damaging the new bushing, press only on the outer steel sleeve during installation.*

BentleyPublishers.com
BentleyPublishers.com–All Rights Reserved

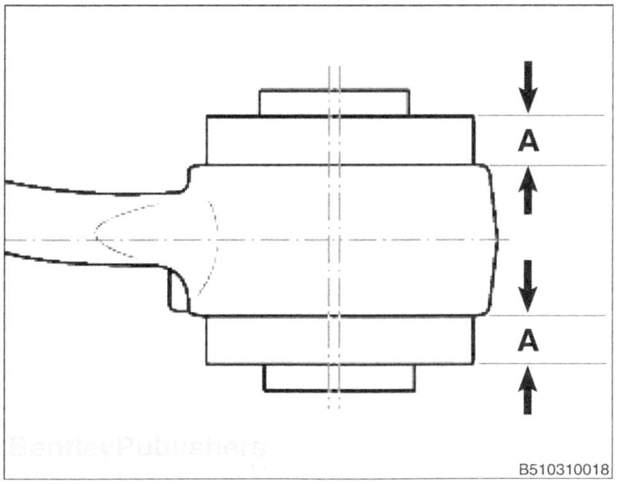

Press bushing in so it protrudes equally (dimension **A**) from both sides of tension strut.

— Install tension strut. See **Tension strut, removing and installing (rear-wheel drive)** or **Tension strut, removing and installing (xDrive)** in this repair group.

> ***CAUTION—***
> *• Tighten the tension strut bushing through-bolt to its final torque only with the vehicle on the ground and the suspension loaded.*

Tension strut ball joint, replacing (xDrive)

— Raise car and remove wheel.

> ***WARNING —***
> *• Make sure that the car is firmly supported on jack stands designed for the purpose. Place jack stands underneath structural chassis points. Do not place jack stands under suspension parts.*

— Detach tension strut from ball joint. See **Tension strut, removing and installing (xDrive)** in this repair group.

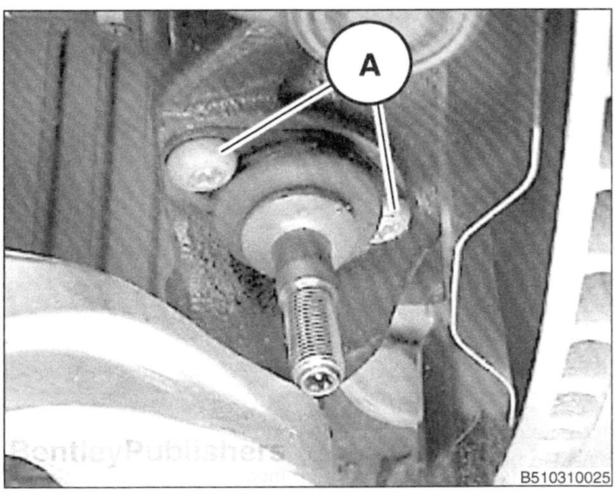

Remove ball joint mounting bolts (**A**).

— Remove ball joint from steering knuckle. If necessary, use slide hammer to drive ball joint out of seat.

> ***CAUTION—***
> *• Do not reuse ball joint once it has been forcibly removed from steering knuckle.*

— Install new ball joint using new self-locking bolts.

Tightening torque	
Ball joint to steering knuckle (replace M10 x 20 mm 10.9 grade ZNS self-locking bolts with new)	60 Nm (44 ft-lb)

— Attach tension strut to ball joint using new self-locking nut.

Tightening torque	
Tension strut to steering knuckle ball joint (replace M14 x 1.5 self-locking flange nut with new)	165 Nm (122 ft-lb)

BentleyPublishers.com—All Rights Reserved

Steering Knuckle and Front Wheel Bearing

The steering knuckle (also called swivel bearing) serves as the outboard attachment point for the suspension arms and as the wheel hub / wheel bearing carrier. The wheel bearings are permanently sealed and require no maintenance.

Steering knuckle components (rear-wheel drive)

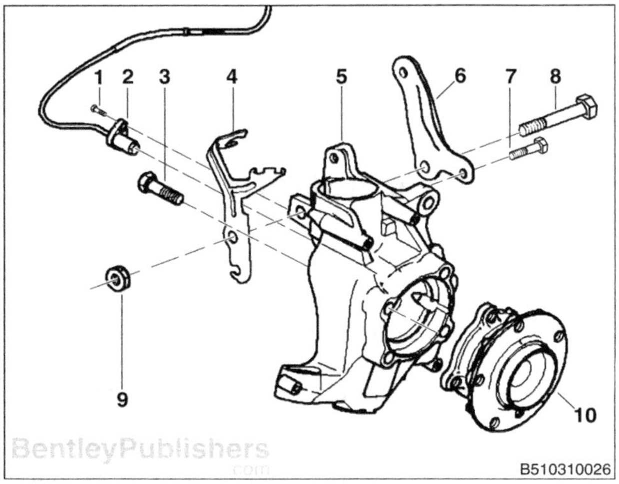

The wheel bearing and hub assembly is bolted to the steering knuckle.

1. M6 x 16 mm 8.8 grade ZNS3 bolt
   - Tighten to 8 Nm (6 ft-lb)
2. ABS wheel speed sensor
   - Lubricate with Staburags® NBU 12/K or equivalent grease before installing.
3. M12 x 1.5 x 55 mm Z1 bearing hub bolt
   - Replace with new
   - Tighten to 110 Nm (81 ft-lb)
4. Electrical harness bracket
5. Steering knuckle (swivel bearing)
6. Stabilizer bar link bracket
7. M10 x 20 mm 10.9 grade bolt
8. M12 x 100 mm 10.9 grade strut pinch bolt
   - Replace with new
   - Tighten to 81 Nm (60 ft-lb)
9. M12 x 1.5 x ZNS3 self-locking collar nut
10. Wheel bearing hub

Steering knuckle, removing and installing (rear-wheel drive)

– Raise car and remove front wheel.

> ***WARNING —***
> - *Make sure the vehicle is stable and well supported at all times. Use a professional automotive lift or jack stands designed for the purpose. A floor jack is not adequate support.*

– Detach front brake caliper from steering knuckle and suspend aside using stiff wire. Do not detach brake fluid hose. See **340 Brakes**.

– Remove brake rotor. See **340 Brakes**.

– Remove wheel speed sensor from steering knuckle. See **340 Brakes**.

– Detach control arm and tension strut from steering knuckle. See **Front Suspension Arms** in this repair group.

– Detach tie rod end from steering knuckle. See **320 Steering and Wheel Alignment**.

BentleyPublishers.com
BentleyPublishers.com—All Rights Reserved

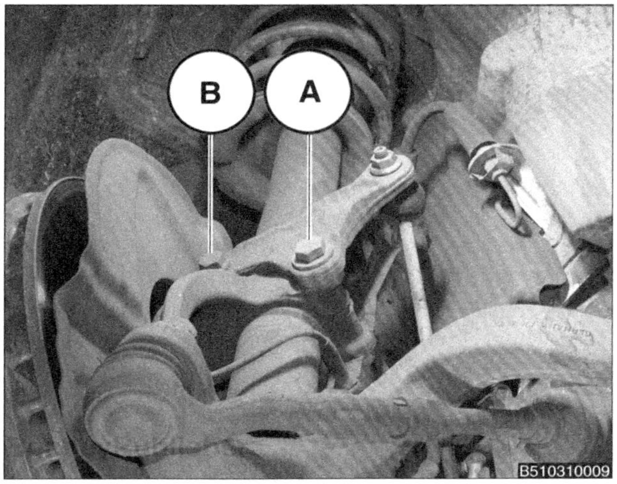

Working at top of steering knuckle:

- Remove strut pinch bolt (**A**) from upper steering knuckle.
- Remove stabilizer link bracket bolt (**B**) and detach bracket from steering knuckle.

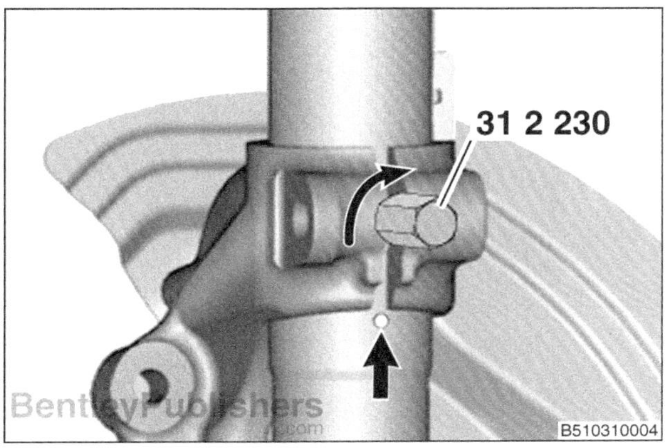

If necessary, use BMW special tool 31 2 230 to spread (**curved arrow**) steering knuckle pinch collar. Slide steering knuckle downward to disengage from strut tube.

- Note locating boss (**arrow**) for strut. During reassembly, boss slides in steering knuckle pinch collar gap.

– Replace wheel bearing if necessary. See **Front wheel bearing, replacing (rear-wheel drive)** in this repair group.

– Installation is reverse of removal. Note the following:

- Replace self-locking fasteners.
- Make sure locating boss on strut tube slides in steering knuckle pinch collar gap.
- Face strut pinch bolt head in direction of travel.
- Make sure steering knuckle contacts stop on strut before tightening pinch bolt.
- Clean wheel speed sensor bore hole and lubricate with Staburags® NBU 12/K or equivalent grease.
- Align wheels when job is complete.

Tightening torques	
Brake caliper to steering knuckle (replace M12 x 1.5 x 43 mm ZNS bolts with new)	110 Nm (81 ft-lb)
Brake rotor to wheel hub (replace M8 x 14 mm Allen bolt with new)	16 Nm (12 ft-lb)
Control arm to steering knuckle (replace M14 x 1.5 self-locking flange nut with new)	165 Nm (122 ft-lb)
Road wheel to hub	120 ± 10 Nm (89 ± 7 ft-lb)
Strut assembly pinch bolt (replace M12 x 100 mm 10.9 grade bolt with new)	81 Nm (60 ft-lb)
Tension strut to steering knuckle (replace M14 x 1.5 self-locking flange nut with new)	165 Nm (122 ft-lb)
Tie rod end to steering knuckle (replace M14 x 1.5 ZNS3 nut with new)	165 Nm (122 ft-lb)
Wheel speed sensor to steering knuckle	8 Nm (6 ft-lb)

BentleyPublishers.com–All Rights Reserved

Front wheel bearing, replacing (rear-wheel drive)

– Raise car and remove front wheel.

> ***WARNING —***
> *• Make sure the vehicle is stable and well supported at all times. Use a professional automotive lift or jack stands designed for the purpose. A floor jack is not adequate support.*

– Remove steering knuckle. See **Steering knuckle, removing and installing (rear-wheel drive)** in this repair group.

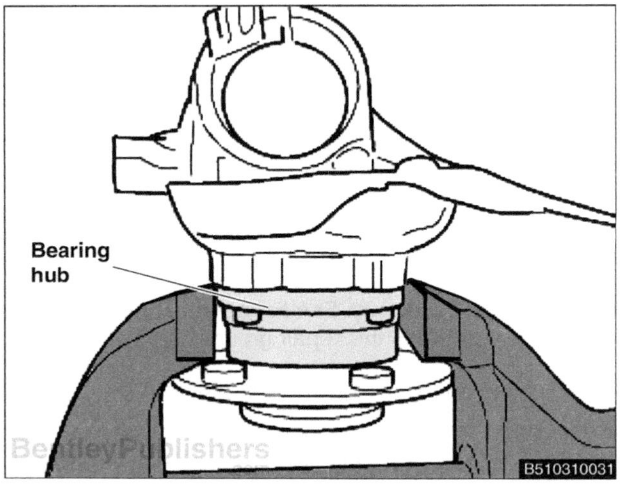

< Clamp wheel bearing hub in shop vise as illustrated.

> ***CAUTION—***
> *• Make sure unit is only clamped at bearing hub.*

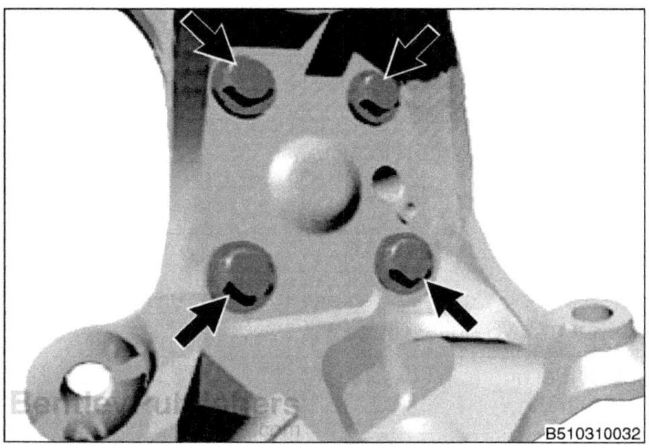

< Remove bearing hub mounting bolts (**arrows**). Lift off steering knuckle.

BentleyPublishers.com
BentleyPublishers.com—All Rights Reserved

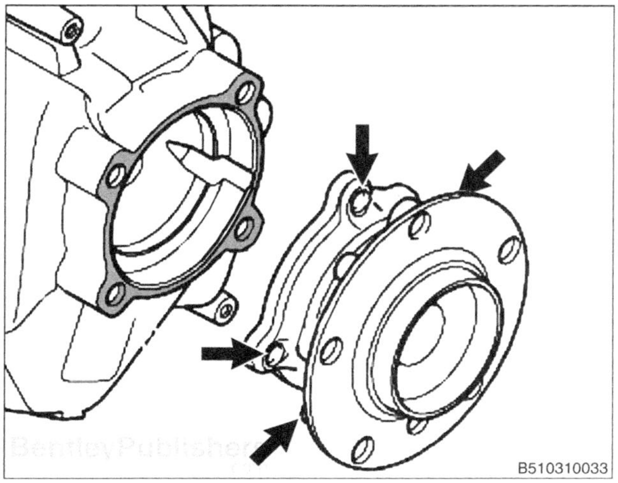

When installing:

- Keep bearing hub contact area (**grey**) clean and grease-free.
- If reusing bearing hub, tap mounting bolt threaded holes (**arrows**).
- Replace mounting bolts with new.

Tightening torque	
Bearing hub to steering knuckle (replace M12 x 1.5 x 55 mm Z1 bolt with new)	110 Nm (81 ft-lb)

Steering knuckle components (xDrive)

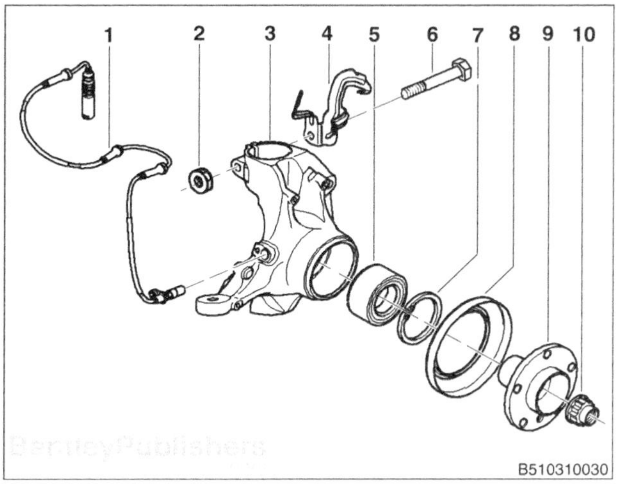

The wheel bearing is pressed on the steering knuckle and requires special tools to remove. The wheel bearing is destroyed when the wheel hub is removed.

1. ABS wheel speed sensor
   - Lubricate with Staburags® NBU 12/K or equivalent grease before installing.
2. M12 x 1.5 x ZNS3 self-locking collar nut
3. Steering knuckle (swivel bearing)
4. Brake hose bracket
5. Wheel bearing
6. M12 x 1.5 x 80 mm ZNS strut pinch bolt
   - Replace with new
   - Tighten to 100 Nm (74 ft-lb)
7. Circlip
8. Dust shield
9. Hub
10. M27 x 1.5 collar nut
    - Replace with new
    - Tighten (with wheel on ground) to 420 Nm (310 ft-lb)

310

BentleyPublishers.com
BentleyPublishers.com—All Rights Reserved

Steering knuckle, removing and installing (xDrive)

– Raise car and remove front wheel.

> ***WARNING —***
> - *Make sure the vehicle is stable and well supported at all times. Use a professional automotive lift or jack stands designed for the purpose. A floor jack is not adequate support.*

◄ Using a suitable punch, straighten staked sides of wheel hub collar nut (**arrows**).

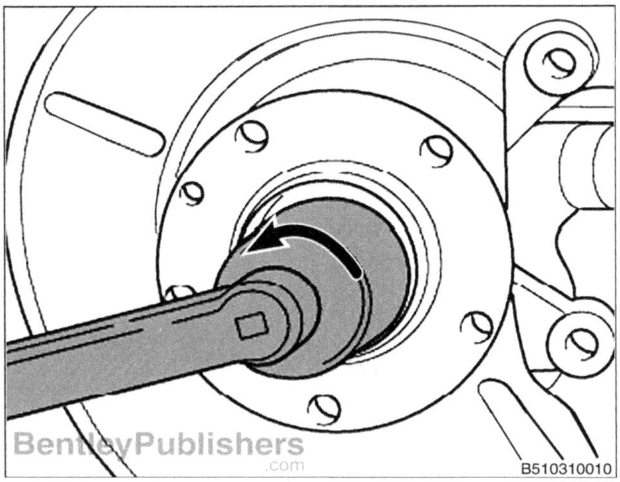

◄ With an assistant holding down brake pedal, remove collar nut and discard.

> ***CAUTION—***
> - *The drive axle collar nut is tightened to a high torque. If necessary, reinstall road wheel and place vehicle on floor before breaking collar nut free.*

– Detach front brake caliper from steering knuckle and suspend aside using stiff wire. Do not detach brake fluid hose. See **340 Brakes**.

– Remove brake rotor. See **340 Brakes**.

– Remove wheel speed sensor from steering knuckle. See **340 Brakes**. Detach sensor electrical harness from bracket on strut.

– Detach control arm and tension strut from steering knuckle. See **Front Suspension Arms** in this repair group.

– Detach tie rod end from steering knuckle. See **320 Steering and Wheel Alignment**.

– Detach stabilizer link from strut tube. Use Torx tool (T30) to counterhold shaft of stabilizer bar link ball joint while removing nut.

– Rotate steering knuckle to one side and press drive axle stub out of wheel bearing hub. Tie up drive axle end using stiff wire.

> ***CAUTION—***
> - *In order to avoid damaging drive axle threads, do not drive out stub axle using impact tool.*

– Support steering knuckle with floor jack. Working at top of steering knuckle, remove strut pinch bolt.

BentleyPublishers.com
BentleyPublishers.com—All Rights Reserved

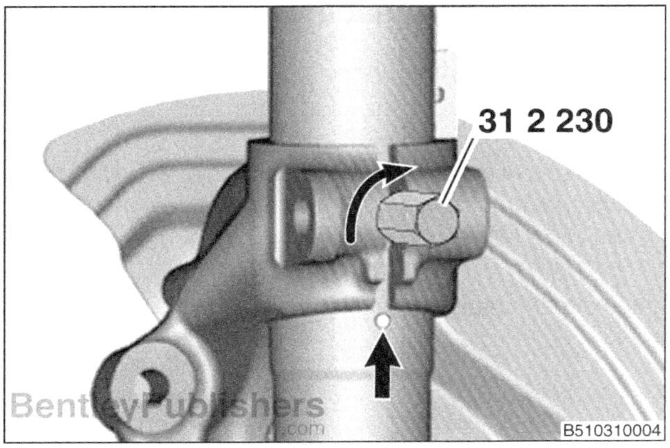

- If necessary, use BMW special tool 31 2 230 to spread (**curved arrow**) steering knuckle pinch collar. Slide steering knuckle down and off strut tube.
  - Note locating boss (**arrow**) for strut. During reassembly, boss slides in steering knuckle pinch collar gap.

– If necessary, replace torsion strut ball joint. See **Tension strut ball joint, replacing (xDrive)**.

– If necessary, replace wheel bearing. See **Front wheel bearing, replacing (xDrive)** in this repair group.

– Installation is reverse of removal. Note the following:
  - Make sure locating boss on strut tube slides in steering knuckle pinch collar gap.
  - Make sure steering knuckle contacts stop on strut before tightening pinch bolt. Face strut pinch bolt head in direction of travel.
  - Replace self-locking fasteners.
  - Use Torx tool (T30) to counterhold shaft of stabilizer bar link ball joint while tightening nut.
  - Clean ABS wheel speed sensor bore hole and lubricate with Staburags® NBU 12/K or equivalent grease.
  - Oil and wheel hub contact surfaces lightly. Do not allow oil to contaminate drive axle threads. Final tightening with wheel on ground. Stake collar nut after final tightening.
  - Align wheels when job is complete.

Tightening torques	
Brake caliper to steering knuckle (replace M12 x 1.5 x 43 mm ZNS bolts with new)	110 Nm (81 ft-lb)
Brake rotor to wheel hub (replace M8 x 14 mm Allen bolt with new)	16 Nm (12 ft-lb)
Control arm to steering knuckle (replace M14 x 1.5 self-locking flange nut with new)	165 Nm (122 ft-lb)
Drive axle to wheel hub (replace M27 x 1.5 collar nut with new)	420 Nm (310 ft-lb)
Road wheel to hub	120 ± 10 Nm (89 ± 7 ft-lb)
Stabilizer link to strut tube (replace M10 ZNS3 nut with new)	65 Nm (48 ft-lb)
Strut assembly pinch bolt (replace M12 x 1.5 x 80 mm ZNS3 bolt with new)	100 Nm (74 ft-lb)
Tension strut to steering knuckle ball joint (replace M14 x 1.5 self-locking flange nut with new)	165 Nm (122 ft-lb)
Tie rod end to steering knuckle (replace M14 x 1.5 ZNS3 nut with new)	165 Nm (122 ft-lb)
Wheel speed sensor to steering knuckle	8 Nm (6 ft-lb)

BentleyPublishers.com
BentleyPublishers.com—All Rights Reserved

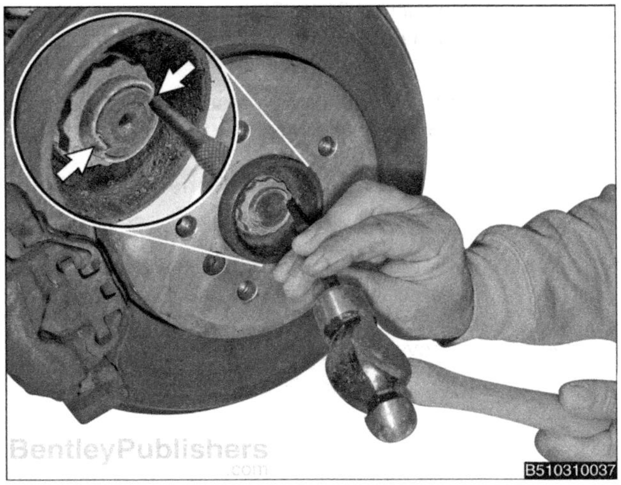

Front wheel bearing, replacing (xDrive)

– Raise car and remove front wheel.

> ***WARNING —***
> *• Make sure the vehicle is stable and well supported at all times. Use a professional automotive lift or jack stands designed for the purpose. A floor jack is not adequate support.*

◀ Using a suitable punch, straighten staked sides of wheel hub collar nut (**arrows**).

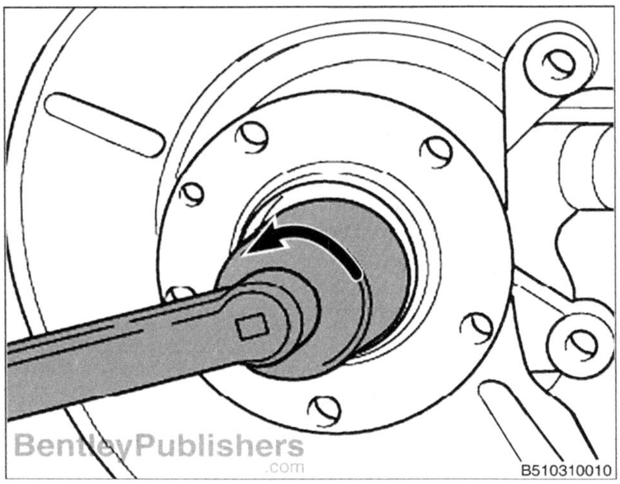

◀ With an assistant holding down brake pedal, remove collar nut and discard.

> ***CAUTION—***
> *• The drive axle collar nut is tightened to a high torque. If necessary, reinstall road wheel and place vehicle on floor before breaking collar nut free.*

– Detach front brake caliper from steering knuckle and suspend aside using stiff wire. Do not detach brake fluid hose. See **340 Brakes**.

– Remove brake rotor. See **340 Brakes**.

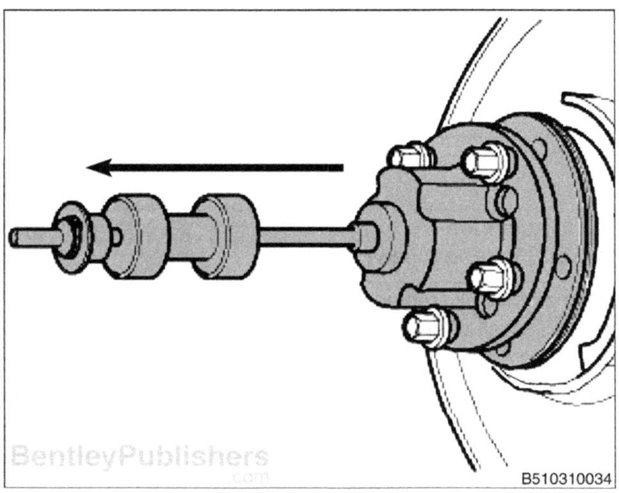

◀ Attach slide hammer hub puller to bearing hub flange using 5 wheel bolts. Drive hub off wheel bearing.

– Note that wheel bearing inner race is removed with flange, thus destroying bearing. Separate race from flange and discard race. Do not damage flange.

– Remove steering knuckle. See **Steering knuckle, removing and installing (xDrive)** in this repair group. Clamp in shop vice with aluminum jaws.

BentleyPublishers.com
BentleyPublishers.com—All Rights Reserved

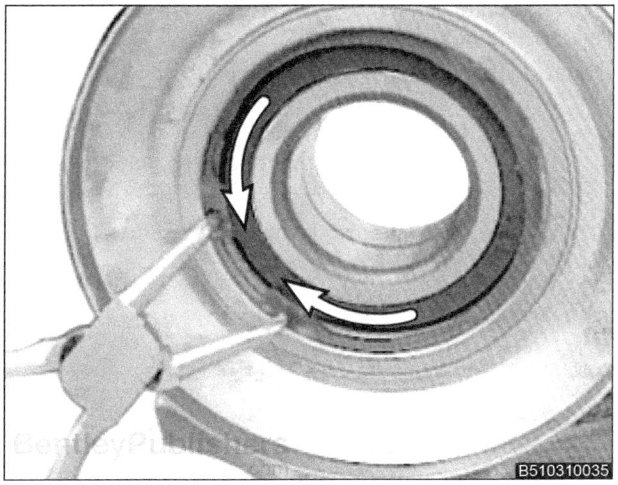

Squeeze wheel bearing retaining circlip using needle-nosed pliers and lift out of hub.

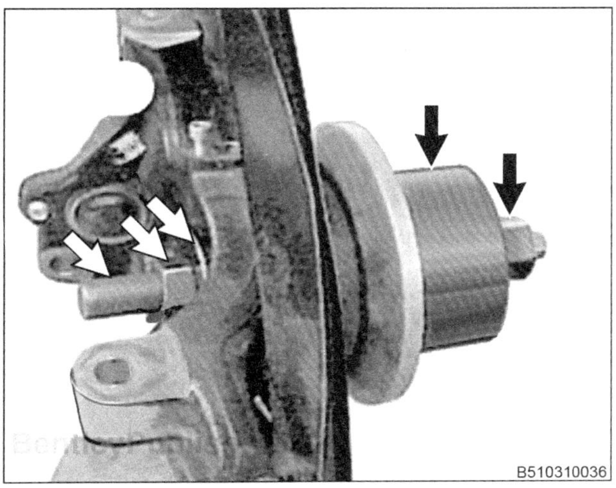

Use bearing removal tools (**arrows**) or shop press to remove bearing from steering knuckle.

- Before installing new bearing:
  - Clean press-fit surface in steering knuckle and keep free of oil and grease.
  - Coat 50% of press-fit surface in steering knuckle with Loctite® 638.

Use bearing installation tools or shop press to press in new wheel bearing. Make sure wide champfer on bearing (**arrow**) points to steering knuckle.

> ***CAUTION—***
> - *Press only on outer race when installing bearing in steering knuckle.*

- Reinstall bearing retaining circlip using needle-nosed pliers. Make sure circlip is seated firmly in steering knuckle groove.

BentleyPublishers.com
BentleyPublishers.com—All Rights Reserved

Use bearing installation tools (**arrows**) or shop press to press wheel hub flange into new bearing.

- Make sure dust shield is intact. Replace if necessary.

> ***CAUTION—***
> *• Press only on inner race when installing hub in wheel bearing.*

- Install steering knuckle. See **Steering knuckle, removing and installing** in this repair group.
- Draw front axle into steering knuckle.
- Install front brake rotor and caliper. See **340 Brakes**.
- Apply light coating of oil to contact surfaces. With an assistant holding down brake, tighten axle collar nut to specified torque.

Tightening torques	
Drive axle to wheel hub (replace M27 x 1.5 collar nut with new)	420 Nm (310 ft-lb)
Road wheel to hub	120 ± 10 Nm (89 ± 7 ft-lb)

- Using a suitable drift, stake side of axle collar nut to stub axle.
- Align wheels when job is complete.

BentleyPublishers.com
BentleyPublishers.com—All Rights Reserved

Front Stabilizer Bar

Active stabilizer bar (dynamic drive)

5 Series vehicles may be equipped with the optional active stabilizer system (acronyms ARS). The system is also referred to as dynamic drive.

With this system, front and rear stabilizer bars are split in half, with the halves connected to each other via hydraulic oscillating motors. An electronic control module operates through a valve body to control coupling of the two stabilizer bar halves.

For additional details, including active stabilizer troubleshooting and commissioning, see **300 Suspension, Steering and Brakes–General**.

Active stabilizer bar components, front

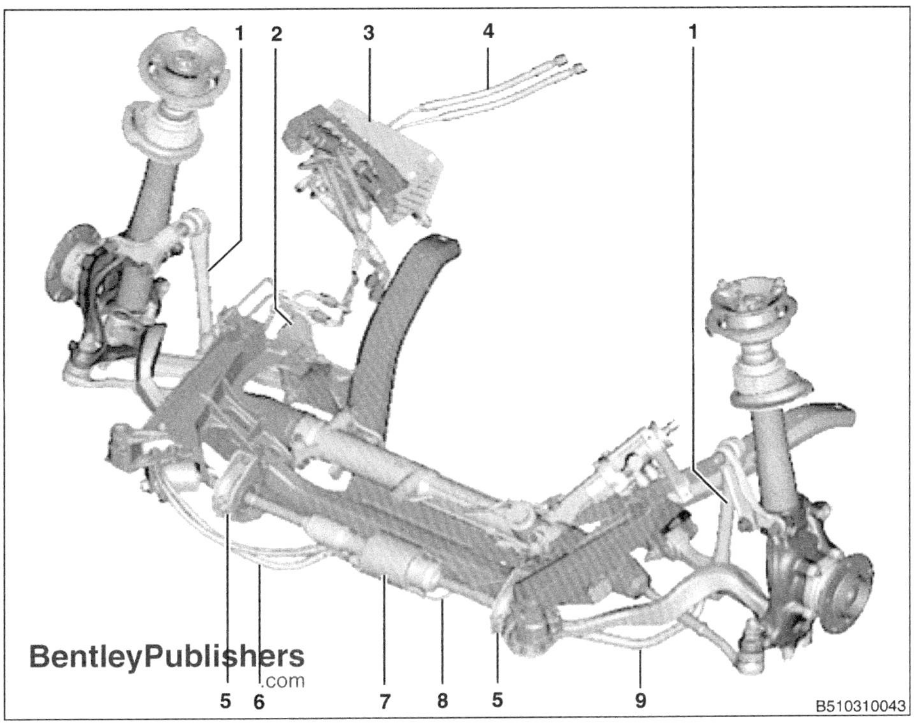

1. Stabilizer bar link
2. Front ride-height sensor
3. Active stabilizer valve block
4. Rear oscillating motor pressure lines
5. Stabilizer bar mount
6. Front oscillating motor pressure lines
7. Front oscillating motor
8. Vent hose
9. Stabilizer bar

BentleyPublishers.com
BentleyPublishers.com—All Rights Reserved

Hydraulic system, bleeding

> ***CAUTION—***
> - *When filling or bleeding hydraulic system, do not allow contamination of fluid or reservoir.*
> - *Using contaminated equipment to add fluid may introduce dirt particles into the fluid reservoir and significantly reduce the service life of system components.*

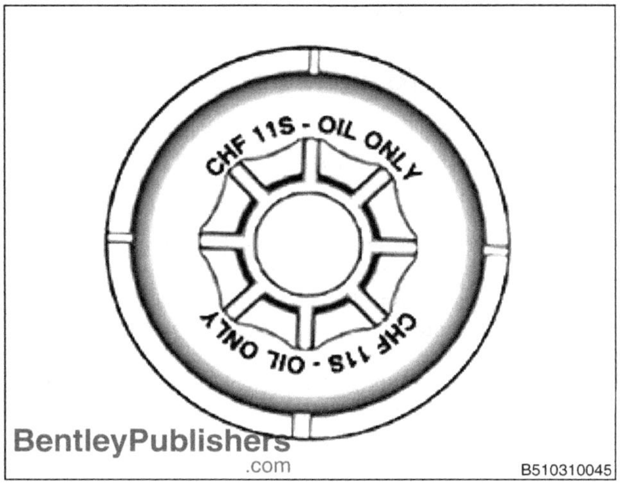

Active stabilizer system and power steering share the fluid reservoir and fluid cooler. The reservoir cap indicates the fluid to be used.

Active stabilizer system fluid	
Hydraulic fluid for tandem pump system	Pentosin CHF 11S

– Thoroughly clean fluid reservoir and its immediate surroundings before opening reservoir cap. Check and top off fluid level with fluid temperature at approx. 20°C (68°F) and the engine stopped.

Make sure reservoir cap is fully screwed in before checking fluid level. Check that fluid level is at MAX mark. Top off with correct fluid if necessary.

> ***CAUTION—***
> - *With fluid at operating temperature (50°-60°C or 122°-140°F), full fluid level is approximately 15 mm (5/8 in) above MAX marking. Do not siphon out fluid in that situation.*

– Start engine.

– Turn steering wheel left and right twice in each direction to full lock. if necessary, top off hydraulic fluid again.

– Move steering wheel to straight-ahead position and switch engine OFF. Check and correct fluid level.

– To make sure all air is out of system, drive vehicle on a short trip, then perform bleed procedure again.

– Check hydraulic system for leaks. If extreme air or fluid leaks are detected, or rattling noises continue in the front end, replace the pressure relief valves and fluid lines, then bleed system again.

Front stabilizer bar, removing and installing (mechanical stabilizer)

The front stabilizer is attached to the front subframe.

– Raise front of vehicle and support safely.

> ***WARNING —***
> - *Make sure the vehicle is stable and well supported at all times. Use a professional automotive lift or jack stands designed for the purpose. A floor jack is not adequate support.*

– Remove engine splash shield and front end reinforcement panel.

BentleyPublishers.com
BentleyPublishers.com—All Rights Reserved

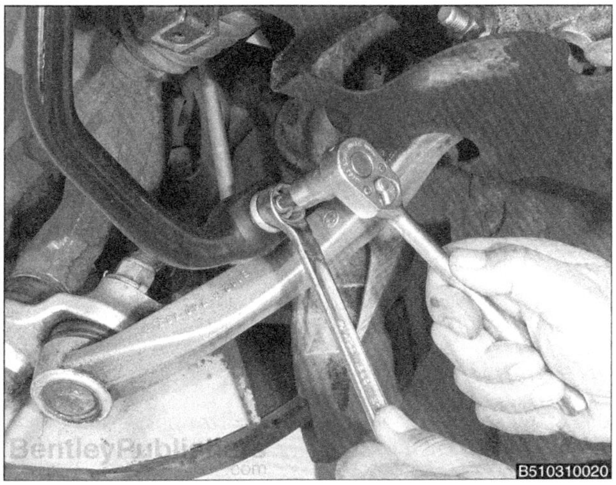
B510310020

< Detach stabilizer links on both sides from stabilizer bar. Use Torx tool (T30) to counterhold shaft of stabilizer bar link ball joint while loosening nut.

B510310040

< Rear-wheel drive right side: Remove power steering line bracket mounting bolt (**arrow**) from stabilizer bar mount.

B510310041

< Remove stabilizer bar mount fasteners (**arrows**) and lower bar.

– Check rubber mount mounts for damage. Replace if necessary.

– Installation is reverse of removal. Note the following.

- Replace self-locking fasteners.
- Use Torx tool (T30) to counterhold shaft of stabilizer bar link ball joint while tightening nut.
- Install front end reinforcement using new fasteners. See **Front end reinforcement panel, removing and installing** in this repair group.

Tightening torques	
Stabilizer bar mount to front subframe (replace M8 fasteners with new)	30.5 Nm (22 ft-lb)
Stabilizer link to stabilizer bar (replace M10 ZNS3 nut with new)	65 Nm (48 ft-lb)

BentleyPublishers.com
BentleyPublishers.com—All Rights Reserved

Front stabilizer bar, removing and installing (active stabilizer)

The front stabilizer is attached to the front subframe.

– Raise front of vehicle and support safely.

> ***WARNING —***
> - *Make sure the vehicle is stable and well supported at all times. Use a professional automotive lift or jack stands designed for the purpose. A floor jack is not adequate support.*

– Remove engine splash shield and front end reinforcement panel.

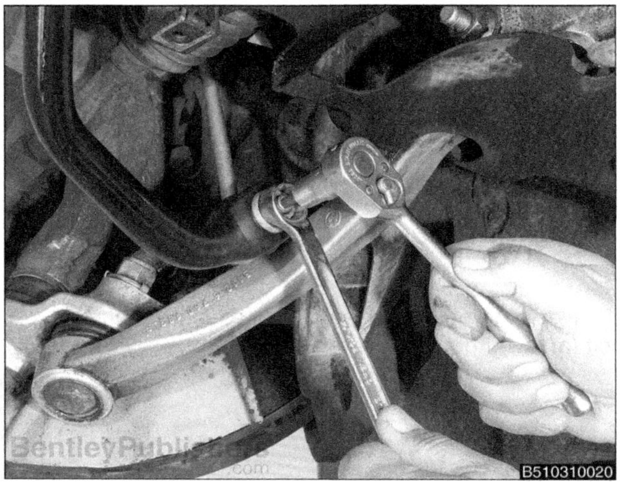

◄ Detach stabilizer links on both sides from stabilizer bar. Use Torx tool (T30) to counterhold shaft of stabilizer bar link ball joint while loosening nut.

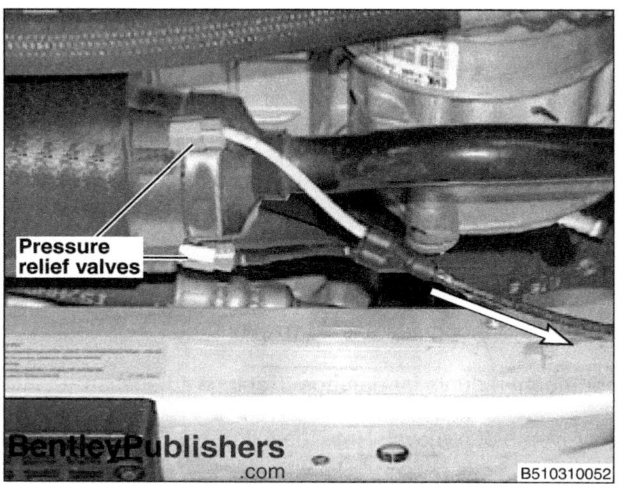

◄ Working at front stabilizer oscillating motor, pull vent line out of Y-connector (**arrow**).

> ***CAUTION—***
> - *Do not detach or disassemble pressure relief valves.*

◄ Note carefully installation positions of fluid pressure lines and brackets at oscillating motor. Then detach pressure line unions (**arrows**) from motor. Be prepared to catch dripping hydraulic fluid and dispose in a safe manner.

> ***CAUTION—***
> - *Do not allow dirt to enter hydraulic lines. Plug fluid line openings.*

To avoid confusion when installing, pressure lines are different dimensions and the unions have different threads.

BentleyPublishers.com
BentleyPublishers.com—All Rights Reserved

Front stabilizer bar, removing and installing (active stabilizer)

- Right side: If necessary, remove hydraulic line bracket mounting bolt (**arrow**) from stabilizer bar mount.
- Left side: If necessary, detach hydraulic line bracket from stabilizer bar mount.

- Remove stabilizer bar mount fasteners (**arrows**) and lower bar.

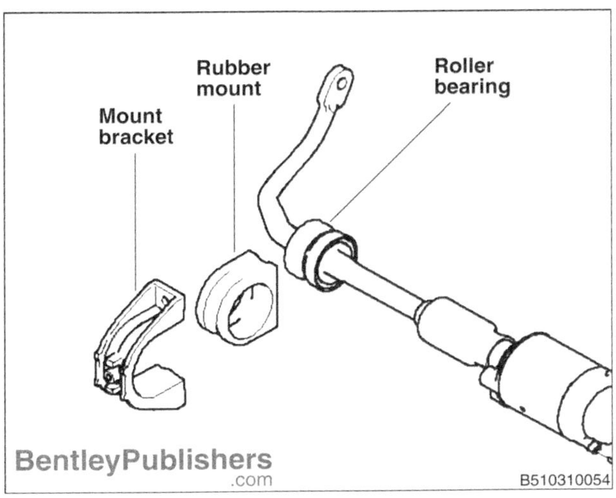

- Separate stabilizer bar mount brackets and rubber mounts from roller bearings.
  - Check roller bearings for free movement. Slight discharge of grease from bearing is acceptable. Replace stabilizer bar if a bearing binds or is otherwise damaged.
  - Replace rubber mounts if damaged. Keep rubber grease-free.
- When installing, note the following.
  - Replace self-locking fasteners.
  - To prevent stabilizer bar distortion, install left mount bracket first (without elongated holes). Then install right mount bracket (with elongated holes).
  - Use Torx tool (T30) to counterhold shaft of stabilizer bar link ball joint while tightening nut.

Tightening torques	
Stabilizer bar mount to front subframe (replace M8 fasteners with new)	30.5 Nm (22 ft-lb)
Stabilizer link to stabilizer bar (replace M10 ZNS3 nut with new)	65 Nm (48 ft-lb)

310

BentleyPublishers.com
BentleyPublishers.com—All Rights Reserved

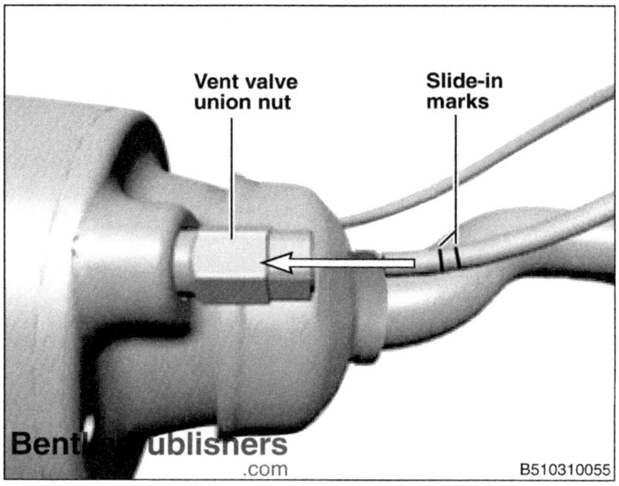

– Attach fluid pressure lines and brackets. Make sure pressure lines do not contact stabilizer bar. Minimum distance = 5 mm (¼ in).

Tightening torque	
Fluid pressure line to oscillating motor (union nut)	30 Nm (22 ft-lb)

New stabilizer bar is equipped with new pressure relief valves. Use new vent hoses and Y-connection.

- Remove plugs from pressure relief valves. Loosen valve union nut.
- Yellow vent hose is toward front of vehicle. Black hose is at rear.
- Slide vent lines into pressure relief valves up to middle of marks and tighten union nuts.

Tightening torques	
Vent hose to vent valve	5 Nm (4 ft-lb)
Vent valve to oscillating motor	14 Nm (10 ft-lb)

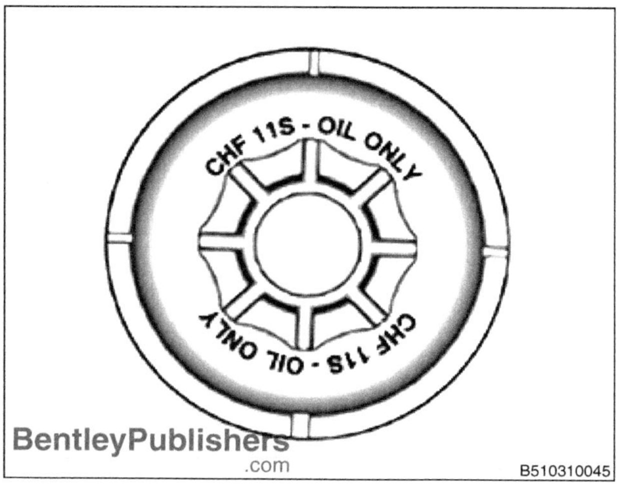

Fill hydraulic fluid reservoir (same as power steering fluid reservoir).

Active stabilizer system fluid	
Hydraulic fluid for tandem pump system	Pentosin CHF 11S

– Use BMW scan tool to bleed air out of active stabilizer system. See **Hydraulic system, bleeding** in this repair group. Check lines and connections for leaks.

– Install front end reinforcement using new fasteners. See **Front end reinforcement panel, removing and installing** in this repair group.

– Carry out commissioning procedure. See **300 Suspension, Steering and Brakes–General**.

Front stabilizer bar links, replacing

The stabilizer bar links attach the stabilizer bar to the strut assemblies.

- Rear-wheel drive: Stabilizer bar link upper ball joint attaches to separate triangular bracket which bolts to steering knuckle.
- xDrive: Stabilizer bar link upper ball joint bolts directly to strut tube.

Replace stabilizer bar links in pairs.

– Raise front of vehicle and remove front wheels.

> ***WARNING —***
> - *Make sure the vehicle is stable and well supported at all times. Use a professional automotive lift or jack stands designed for the purpose. A floor jack is not adequate support.*

BentleyPublishers.com
BentleyPublishers.com—All Rights Reserved

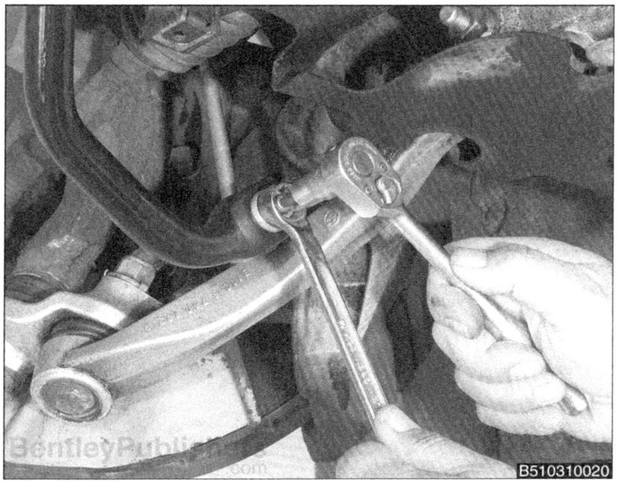

Detach stabilizer links on both sides from stabilizer bar. Use Torx tool (T30) to counterhold shaft of stabilizer bar link ball joint while loosening nut.

- Similarly, detach stabilizer bar link from steering knuckle bracket or strut tube on each side.

- Installation is reverse of removal. Replace self-locking fasteners.

Tightening torques	
Road wheel to hub	120 ± 10 Nm (89 ± 7 ft-lb)
Stabilizer link to stabilizer bar or upper link mount (replace M10 ZNS3 nut with new)	65 Nm (48 ft-lb)

BentleyPublishers.com—All Rights Reserved

BentleyPublishers.com
BentleyPublishers.com—All Rights Reserved

311 Front Axle Differential

TABLES

GENERAL

This repair group covers service and replacement of front drive axles and front differential. Front differential internal repair is not covered.

See **300 Suspension, Steering and Brakes–General** for a general description of the front suspension.

See also:

- **260 Driveshafts** for front driveshaft removal
- **310 Front Suspension**

Front differential and drive axles

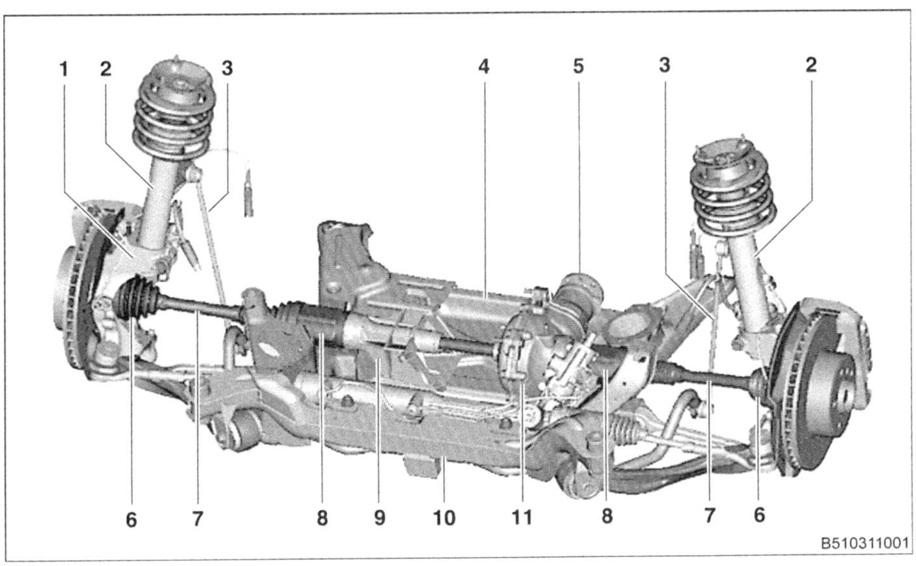

1. Steering knuckle (swivel bearing)
2. Strut assembly
3. Stabilizer bar link
4. Front end reinforcement panel
5. Front driveshaft flange
6. Outer CV joint
7. Front drive axle
8. Inner CV joint
9. Right side bearing pedestal
10. Front subframe (axle carrier)
11. Front differential

Front axle description

The front axle differential and the right side bearing pedestal bolt to each side of the oil pan. Power is transmitted to the front differential from the xDrive transfer case via the front driveshaft, then to front drive hubs through two drive axles. The right drive axle inner constant velocity (CV) joint shaft extends through the right side bearing pedestal and the oil pan into the front differential.

Warnings and Cautions

> ***WARNING—***
> - *Do not reuse self-locking fasteners. They are designed to be used only once and may fail if reused. Replace with new.*

> ***CAUTION—***
> - *Do not drive the vehicle with the front end reinforcement panel removed. See* **310 Front Suspension***.*
> - *When reinstalling the reinforcement panel after repairs, replace panel mounting fasteners with new.*
> - *Due to the chemical and corrosion characteristics of aluminum, follow these precautions when working with aluminum components:*
>   *—Do not bring into contact with battery acid.*
>   *—Do not use wire brushes with brass or iron bristles. Only use brushes with stainless steel bristles.*
>   *—Do not expose to flying sparks from grinding or cutting operations.*
>   *—Do not subject to steel welding splashes.*
>   *—Do not expose to temperatures over 80°C (176°F), even for short periods. Temperatures in painting facilities are not a problem.*

DRIVE AXLES

Front drive axles use two different types of CV joint.

- Outer CV joint: Traditional ball-and-cage design allows power to be delivered from axle to wheel hub continuously while allowing suspension and steering motion.
- Inner joint: Triple roller bearing joint minimizes the amount of vibration and noise transmitted back through vehicle drivetrain while allowing axle to move in and out to compensate for suspension travel.

To replace a CV joint or boot, remove the drive axle from the vehicle.

Front drive axle, removing and installing

– Raise car and remove front wheel.

> ***WARNING—***
> - *Make sure the vehicle is stable and well supported at all times. Use a professional automotive lift or jack stands designed for the purpose. A floor jack is not adequate support.*

BentleyPublishers.com
BentleyPublishers.com—All Rights Reserved

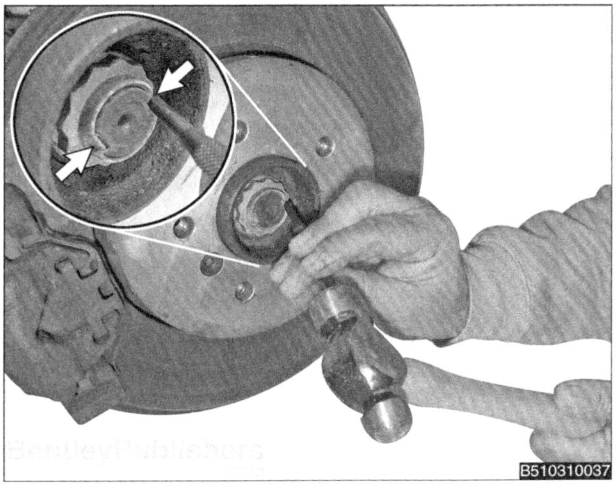

Using a suitable punch, straighten staked sides of wheel hub collar nut (**arrows**).

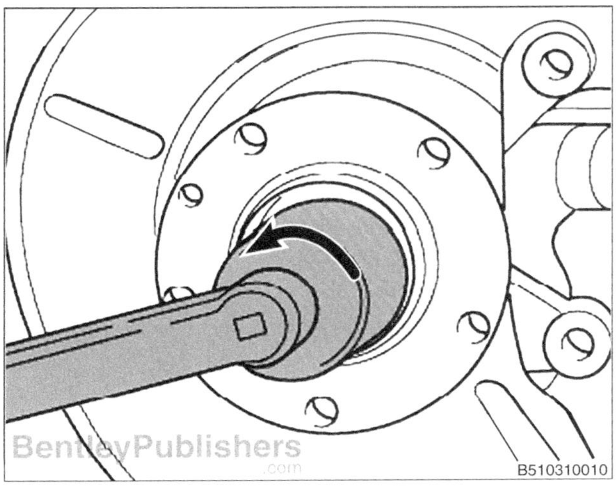

With an assistant holding down brake pedal, remove collar nut and discard.

> ***CAUTION—***
> - *The drive axle collar nut is tightened to a high torque. If necessary, reinstall road wheel and place vehicle on floor before breaking collar nut free.*

- Detach front brake caliper from steering knuckle and suspend aside using stiff wire. Do not detach brake fluid hose. See **340 Brakes**.
- Remove brake disc. See **340 Brakes**.
- Remove ABS wheel speed sensor from steering knuckle. See **340 Brakes**. Detach sensor electrical harness from bracket on strut.
- Detach control arm and tension strut from steering knuckle. See **310 Front Suspension**.
- Detach tie rod end from steering knuckle. See **320 Steering and Wheel Alignment**.
- Detach stabilizer link from strut tube. Use Torx tool (T30) to counterhold shaft of stabilizer bar link ball joint while removing nut.
- Rotate steering knuckle to one side and use BMW special tools 32 2 200 or equivalent to press drive axle stub out of wheel bearing hub. Tie up drive axle end using stiff wire.

> ***CAUTION—***
> - *In order to avoid damaging drive axle threads, do not drive out stub axle using impact tool.*

- Remove splash shield and front end reinforcement panel. See **310 Front Suspension**.

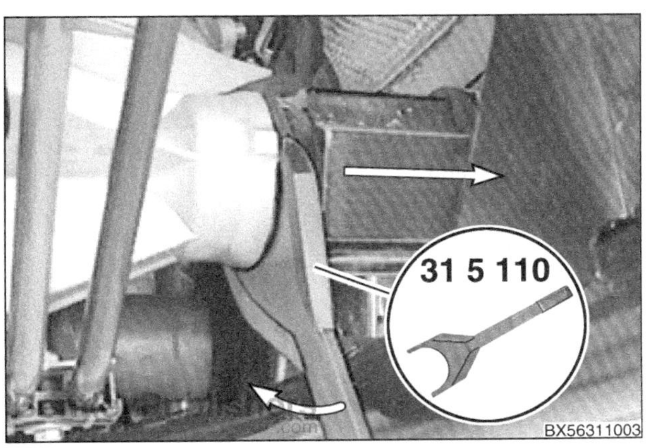

Use BMW special tool 31 5 110 or equivalent prying tool to force drive axle inner CV joint shaft out of front differential or right bearing pedestal output flange. Be prepared to catch dripping oil.

- Replace differential or bearing pedestal flange seal while axle is removed from vehicle. See **Front drive axle flange seal, replacing** in this repair group.

BentleyPublishers.com—All Rights Reserved

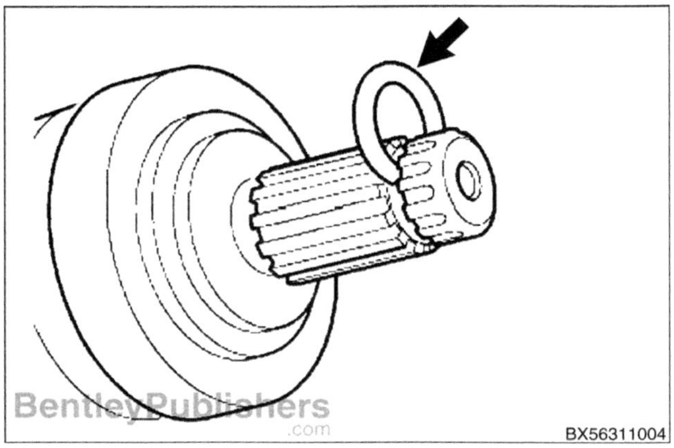

– Replace CV joint boots as necessary. See **CV joint boots, replacing** in this repair group.

◄ Replace inner CV joint locking circlip (**arrow**).

– When installing drive axle inner end:
  - Coat sealing lip of flange seal with transmission fluid.
  - Push inner CV joint shaft into differential or bearing pedestal flange until locking circlip snaps in place.

– Lightly oil drive axle spline teeth and insert drive axle into wheel hub.

> ***CAUTION—***
> *• Do not oil axle or collar nut threads.*

– Installation is reverse of removal. Note the following:
  - Replace self-locking fasteners.
  - Use Torx tool (T30) to counterhold shaft of stabilizer bar link ball joint while tightening nut.
  - Clean ABS wheel speed sensor bore hole and lubricate with Staburags® NBU 12/K or equivalent grease.
  - Oil collar nut and wheel hub contact surfaces lightly. Do not allow oil to contaminate drive axle threads. Final tightening with wheel on ground. Stake collar nut after final tightening.
  - Check front differential oil level when job is complete. See **Front differential oil level, checking** in this repair group.
  - Install front end reinforcement panel using new fasteners. See **310 Front Suspension**.

Tightening torques	
Brake caliper to steering knuckle (replace M12 x 1.5 x 43 mm ZNS bolts with new)	110 Nm (81 ft-lb)
Brake disc to wheel hub (replace M8 x 14 mm Allen bolt with new)	16 Nm (12 ft-lb)
Control arm to steering knuckle (replace M14 x 1.5 self-locking flange nut with new)	165 Nm (122 ft-lb)
Drive axle to wheel hub (replace M27 x 1.5 collar nut with new)	420 Nm (310 ft-lb)
Front end reinforcement panel to subframe (replace M10 x 35 mm 10.9 grade ZNS bolts with new): • Stage 1 • Stage 2	 56 Nm (41 ft-lb) + additional 90°
Road wheel to hub	120 ± 10 Nm (89 ± 7 ft-lb)
Stabilizer link to strut tube (replace M10 ZNS3 nut with new)	65 Nm (48 ft-lb)
Tension strut to steering knuckle ball joint (replace M14 x 1.5 self-locking flange nut with new)	165 Nm (122 ft-lb)
Tie rod end to steering knuckle (replace M14 x 1.5 ZNS3 nut with new)	165 Nm (122 ft-lb)
Wheel speed sensor to steering knuckle	8 Nm (6 ft-lb)

BentleyPublishers.com
BentleyPublishers.com—All Rights Reserved

Front drive axle flange seal, replacing

The left front drive axle is fitted to the front differential. The right front drive axle is fitted to the right bearing pedestal. See **Front drive components** in this repair group.

There is a flange seal on each side, in the differential and in the bearing pedestal. In case of flange seal oil leaks, follow this procedure to replace the seal(s).

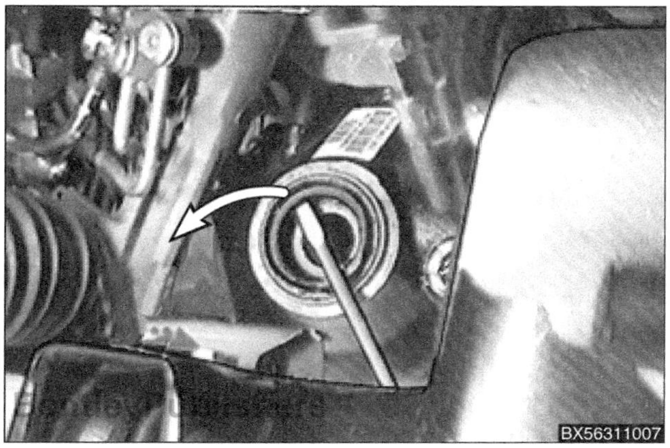

– Raise front of vehicle and support safely.

> ***WARNING —***
> - *Make sure the vehicle is stable and well supported at all times. Use a professional automotive lift or jack stands designed for the purpose. A floor jack is not adequate support.*

– Remove front drive axle. See **Front drive axle, removing and installing** in this repair group.

◄ Left side flange: Use seal removal tool or screwdriver to pry (**arrow**) flange seal out of differential.

◄ Right side flange: Use seal removal tool or screwdriver to pry (**arrow**) flange seal out of right bearing pedestal.

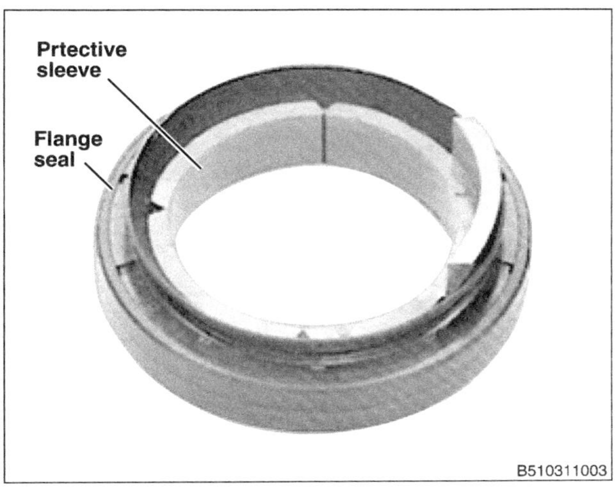

◄ Remove protective sleeve from new flange seal and save for use during drive axle installation. Seal is equipped with protective sleeve to prevent sealing lip from damage during axle installation.

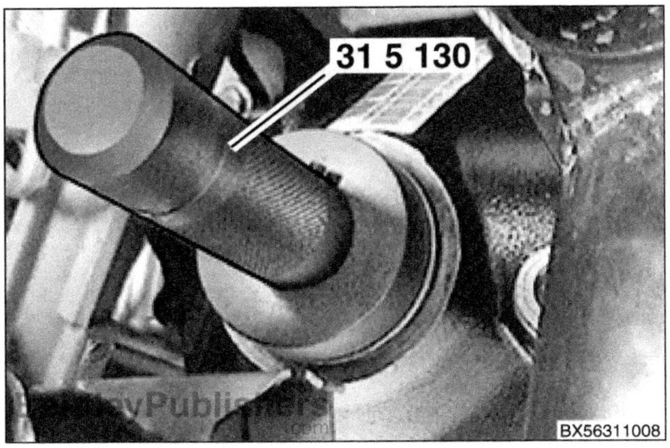

Use BMW special tool 31 3 130 or equivalent seal driver to drive flange seal into differential or bearing pedestal as far as it will go.

- Coat sealing lip of flange seal with transmission fluid.
- Insert protective sleeve into seal.

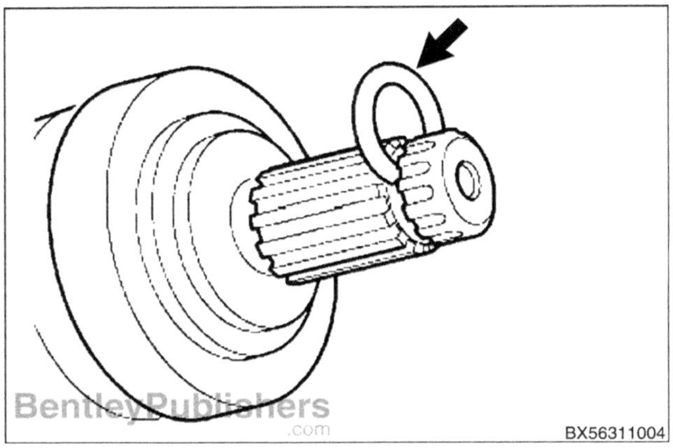

Replace inner CV joint locking circlip (**arrow**).

- Install drive axle:
  - Insert inboard end of drive axle partially into differential housing or bearing pedestal.
  - Withdraw protective sleeve from sealing lip, cut protective sleeve and remove.
  - Continue pressing drive axle in until spring clip snaps audibly into place.
- Installation is reverse of removal. Note the following:
  - Replace self-locking fasteners.
  - Reinstall drive axle stub in steering knuckle. See **Front drive axle, removing and installing** in this repair group.
  - Check front differential oil level when job is complete. See **Front differential oil level, checking** in this repair group.
  - Install front end reinforcement panel using new fasteners. See **310 Front Suspension**.

CV joint boots, replacing

When replacing CV boots, use a complete boot repair kit. The kit includes new boots, clamping bands, special lubricant and a new CV joint shaft circlip.

- Remove drive axle. See **Front drive axle, removing and installing** in this repair group.

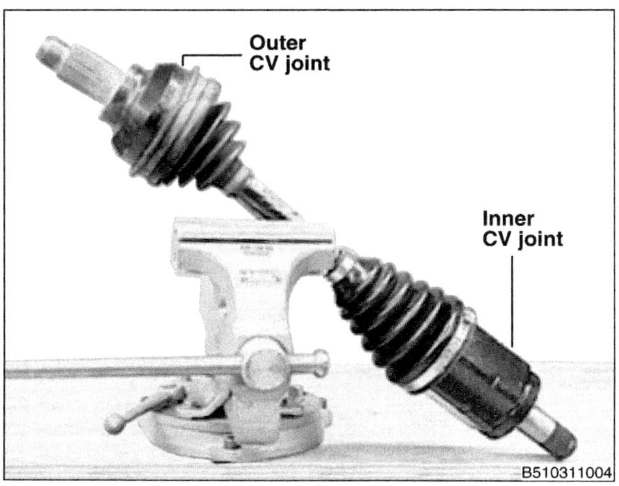

Place axle in shop vice with aluminum jaws.

BentleyPublishers.com
BentleyPublishers.com—All Rights Reserved

B510311005

◀ Release retaining clamps (**arrows**) from both ends of outer CV boot. Cut off boot and discard.

- Using a soft-faced hammer, pound outer CV joint off drive axle.

- Inspect outer CV joint carefully.
  - Look for galling, pitting and other signs of wear or physical damage.
  - Polished surfaces or visible ball tracks alone are not necessarily cause for replacement.
  - Discoloration (overheating) of balls indicates lack of lubrication. Replace joint.

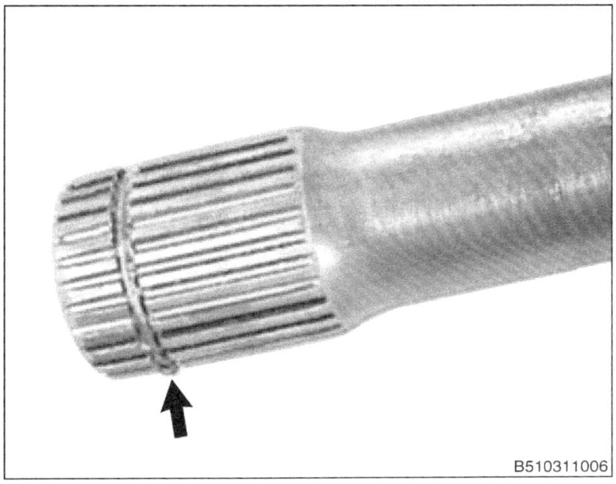

B510311006

◀ Using a flat blade screwdriver, pry spring clip (**arrow**) off drive axle splines.

- Clean old lubricant off axle splines.

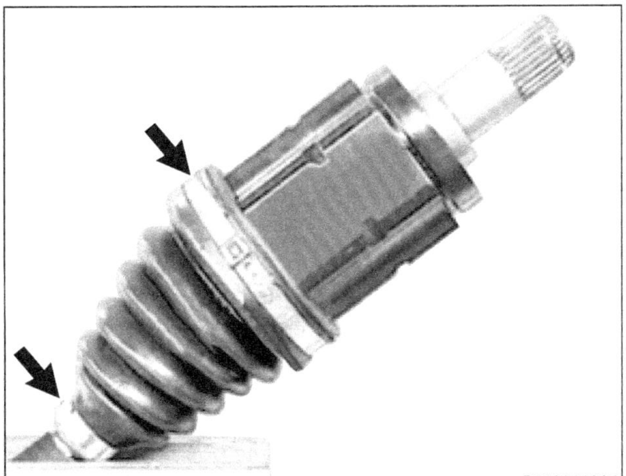

B510311007

◀ Release retaining clamps (**arrows**) from both ends of inner CV boot. Cut off boot and discard.

BentleyPublishers.com—All Rights Reserved

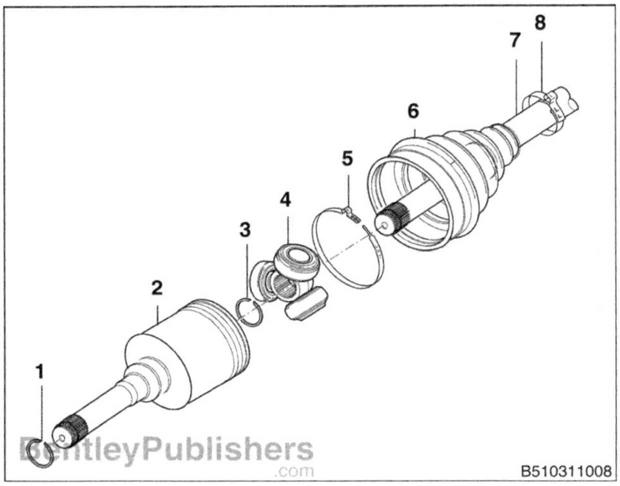

Inner CV joint (triple-roller) components

1. Locking circlip
2. Inner CV joint housing
3. Circlip
4. Triple roller bearing
5. Clamp
6. Inner CV joint boot
7. Drive axle shaft
8. Clamp

– Slide housing (**2**) off triple-roller bearing.

– Clean old lubricant off axle and triple roller bearing.

– Pack triple roller bearing and boot with fresh CV joint grease.

CV joint lubricant capacity	
Inner CV joint	80 grams (2.8 oz)

– Install new inner CV joint boot over outer axle end.

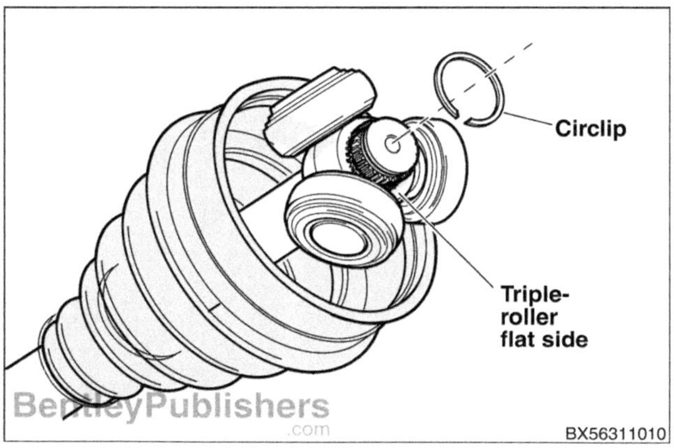

If triple-roller bearing is removed from axle shaft, note that it is installed with flat side of joint facing retaining circlip.

– Insert triple roller bearing into inner CV joint housing.

– Secure boot to housing and axle shaft using clamps supplied with boot kit. Pinch clamps tight using special clamp pliers.

– Place new clamping bands and outer CV boot over drive axle.

– Replace spring clip on splined end of drive axle.

– Apply Loctite® 270 or an equivalent heavy-duty locking compound to drive axle splines.

> ***CAUTION—***
> *• Do not let locking compound contact balls in joint. Apply only a thin coat to cover splines.*

– Pack joint with fresh CV joint grease.

CV joint lubricant capacity	
Outer CV joint	80 grams (2.8 oz)

– Tap CV joint onto splined end of drive axle until spring clip snaps audibly into place.

– Using clamp pliers, secure retaining clamp into position tightly sealing large end of boot against CV joint.

BentleyPublishers.com
BentleyPublishers.com—All Rights Reserved

Before installing small boot clamp:

- Flex CV joint as far over as it will go.
- Insert small screw-driver between boot and axle-shaft to "burp" air from boot.

— With outer boot full of grease and any air eliminated from boot, secure small end of boot on CV joint by pinching clamp with pliers.

— Remainder of installation is reverse of removal. See **Front drive axle, removing and installing** in this repair group.

— Check front differential oil level and top off if necessary. See **Front differential oil level, checking** in this repair group.

Front Differential

The front differential, bolted to the left side of the engine oil pan, is filled with lifetime oil that ordinarily does not need to be changed. If service is required, BMW recommends using only a specially formulated synthetic gear oil (SAF-XO).

Axle flange seal replacement is covered in **Front drive axle flange seal, replacing** in this repair group.

Replacement of O-ring seal between engine oil pan and front differential is covered in **Front differential, removing and installing** in this repair group.

Replacement of O-ring seal between engine oil pan and right side bearing pedestal is covered in **Right axle bearing pedestal, removing and installing** in this repair group.

Table a. Differential fluid	
Capacity	0.6 liter (0.63 US qt)
BMW specification	SAF-XO Synthetic

Front drive components

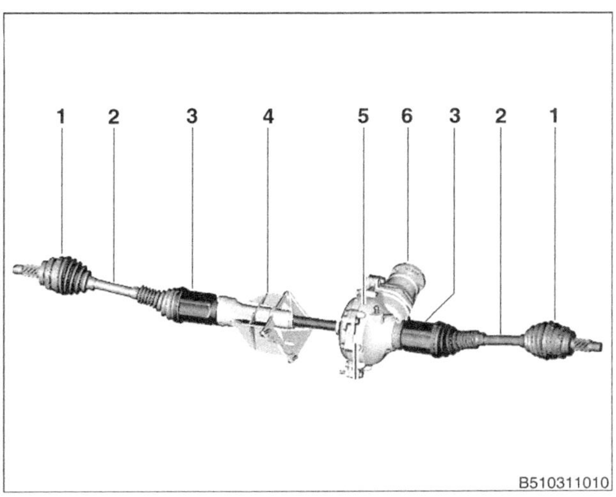

1. Outer CV joint
2. Drive axle shaft
3. Inner CV joint
4. Right side bearing pedestal
5. Front differential
6. Front driveshaft flange

311

Front differential oil level, checking

– Drive vehicle a short distance to warm differential oil.

– Raise vehicle and support safely.

> ***WARNING —***
> *• Make sure the vehicle is stable and well supported at all times. Use a professional automotive lift or jack stands designed for the purpose. A floor jack is not adequate support.*

– Remove splash shield and front end reinforcement. See **310 Front Suspension**.

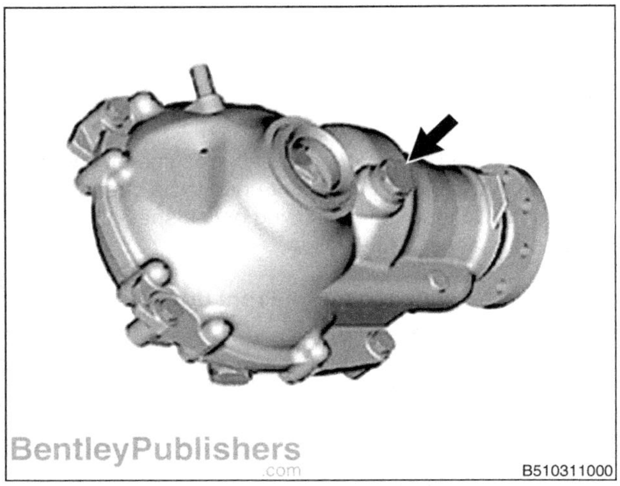

◄ With vehicle level, check oil at filler plug (**arrow**):
- Place a drain pan under oil filler plug before removing.
- Oil level is correct when it just reaches edge of filler hole.

– If necessary, top up fluid with synthetic gear oil (SAF-XO).

– Replace filler plug sealing ring.

– Install and tighten oil filler plug when oil level is correct.

Tightening torque	
Oil drain / fill plug to differential	60 Nm (44 ft-lb)

Front differential oil, draining and filling

– Drive vehicle a short distance to warm differential oil.

– Raise vehicle and safely support.

> ***WARNING —***
> *• Make sure the vehicle is stable and well supported at all times. Use a professional automotive lift or jack stands designed for the purpose. A floor jack is not adequate support.*

– Remove splash shield and front end reinforcement panel. See **310 Front Suspension**.

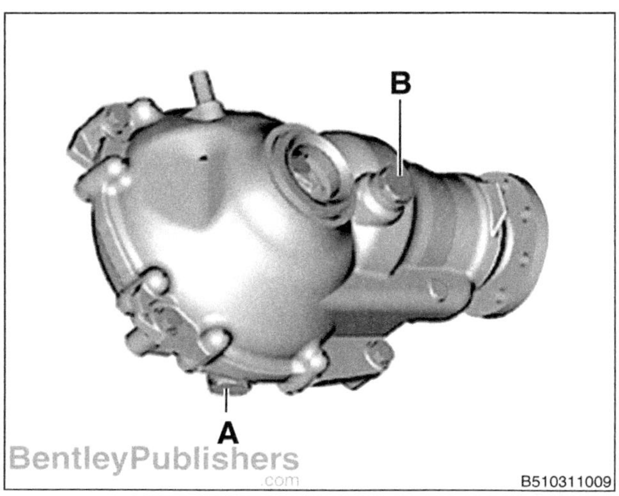

◄ With vehicle level:
- Place a drain pan under oil drain plug.
- Remove drain plug (**A**) and allow oil to drain into pan.

> ***WARNING —***
> *• Pull the loose plug away quickly to avoid being scalded by hot oil. Use gloves to protect your hands.*

– When oil flow has diminished to an occasional drip, reinstall drain plug with new sealing ring.

– Remove filler plug (**B**). Add synthetic gear oil (SAF-XO) and check level. See **Front differential oil level, checking** in this repair group.

Tightening torque	
Oil drain / fill plug to differential	60 Nm (44 ft-lb)

BentleyPublishers.com
BentleyPublishers.com—All Rights Reserved

Front differential, removing and installing

- Disconnect negative (-) battery cable.

> ***CAUTION—***
> *• Prior to disconnecting the battery, read the battery disconnection cautions in* **001 Warnings and Cautions**.

- Raise engine hood to service position. See **410 Fender, Engine Hood**.
- Remove air filter housing. See **130 Fuel Injection**.
- Remove upper engine cover (ignition coil cover). See **020 Maintenance**.
- Raise vehicle and support safely.

> ***WARNING—***
> *• Make sure the vehicle is stable and well supported at all times. Use a professional automotive lift or jack stands designed for the purpose. A floor jack is not adequate support.*

- Remove splash shield and front end reinforcement panel from under engine.
- Remove suspension side splash shields.

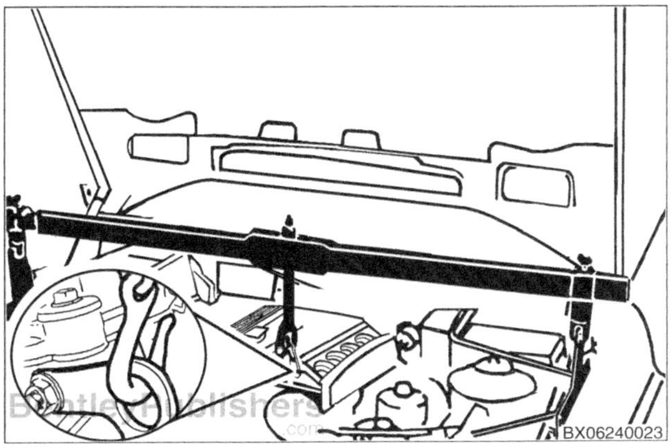

- Install engine support brace across engine bay.

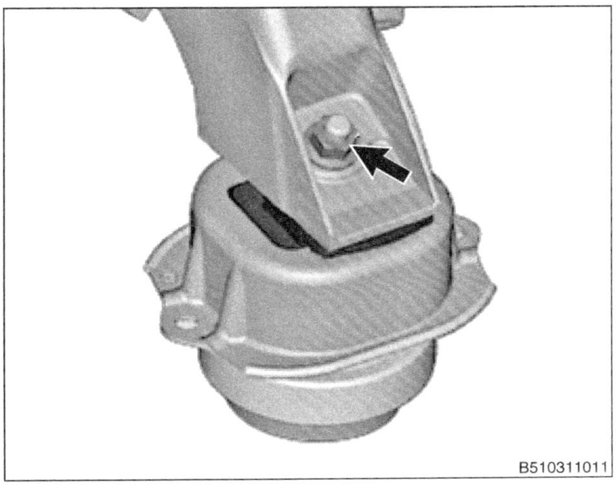

- Working underneath engine, remove right and left engine mount upper nuts (**arrow**). Raise engine slightly (¼ - ½ in) to release load on engine mounts.
- Siphon out power steering fluid.

311

BentleyPublishers.com—All Rights Reserved

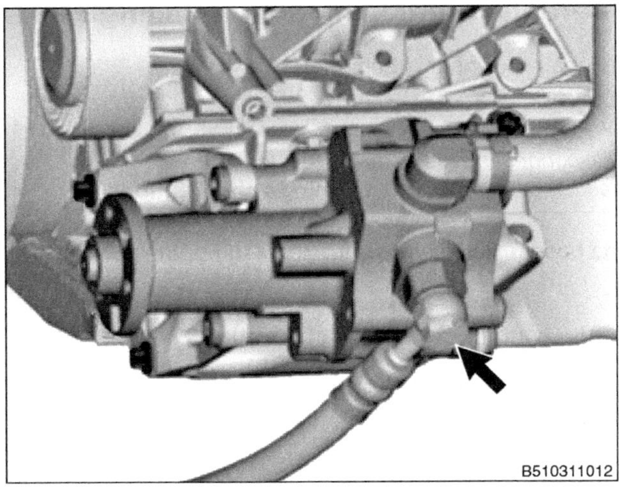

◄ Working underneath engine, remove pressure line banjo bolt (**arrow**) from power steering pump and detach pressure line. Catch dripping fluid in a shop towel.

– Detach fluid line from transmission fluid cooler.

– Drain front differential oil. See **Front differential oil, draining and filling** in this repair group.

– Drive axles:

- Remove left drive axle. See **Front drive axle, removing and installing** in this repair group.
- Pull right drive axle approx. 6 to 7 cm (2½ - 3 in) out of front differential.

– Remove front driveshaft. See **260 Driveshafts**.

◄ Detach steering spindle from steering rack:

- Move steering wheel so that front wheels point straight ahead. Vehicle with steering lock: Remove ignition key and make sure steering locks in straight ahead position.
- With front wheels pointing straight ahead, make sure steering spindle pointer and steering rack pointer line up (**arrows**). Remove lower steering spindle pinch bolt and detach steering universal joint from steering rack shaft.

> ***CAUTION—***
> - *While steering spindle is detached from steering rack, make sure steering wheel is not rotated. Rotating the wheel may damage the airbag and horn contact spring.*

◄ Support front subframe with transmission jack. Remove subframe mounting bolts (**arrows**) and lower subframe sufficiently to allow access to front differential mounting bolts.

> ***CAUTION—***
> - *Do not bend or stretch fluid lines and hoses when lowering or raising subframe.*
> - *Automatic transmission vehicle: Disconnect transmission fluid cooler lines from subframe.*

BentleyPublishers.com
BentleyPublishers.com–All Rights Reserved

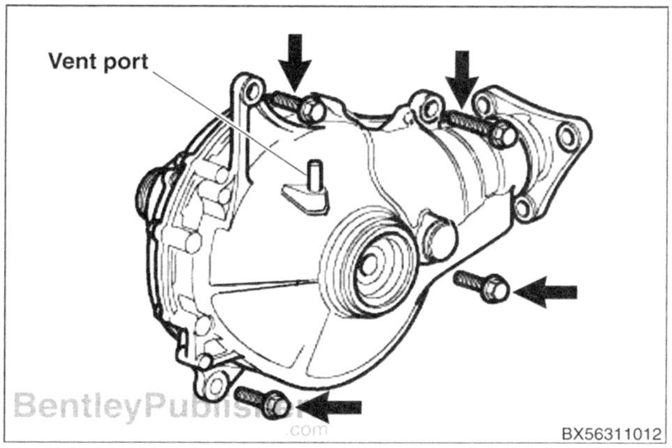

Remove bolts (**arrows**) and lower differential carefully, making sure that no lines, hoses or harnesses become snagged.

> ***CAUTION—***
> *• Support differential as last bolt is removed.*

Remove sealing O-ring (**arrow**) between oil pan and differential and replace with new.

– Install differential to oil pan.

Tightening torques	
Front differential to oil pan	65 Nm (48 ft-lb)

– Remainder of assembly is reverse of removal. Keep in mind:
- Replace drive axle flange seal before installing left drive axle.
- Replace power steering fluid line banjo bolt sealing rings.
- Fill and bleed power steering. See **320 Steering and Wheel Alignment**.
- Fill front differential with oil. See **Front differential oil level, checking** in this repair group.
- Install front end reinforcement panel using new fasteners. See **310 Front Suspension**.
- Use BMW scan tool to carry out steering angle sensor adjustment.

Tightening torques	
Engine mount to engine bracket (replace M10 10.9 grade self-locking nuts with new)	56 Nm (41 ft-lb)
Front end reinforcement panel to subframe (replace M10 x 35 mm 10.9 grade ZNS bolts with new): • Stage 1 • Stage 2	 56 Nm (41 ft-lb) + additional 90°
Front subframe to chassis (replace M12 x 1.5 10.9 grade bolts with new): • Stage 1 • Stage 2	 100 Nm (74 ft-lb) + additional 90°
Steering fluid line banjo bolt to steering pump (M14 or M16, replace sealing O-rings)	33 - 35 Nm (24 - 25 ft-lb)
Steering spindle universal joint to steering rack (pinch bolt)	22 Nm (17 ft-lb)

311

BentleyPublishers.com
BentleyPublishers.com—All Rights Reserved

Right axle bearing pedestal, removing and installing

– Raise vehicle and support safely.

> ***WARNING —***
> *• Make sure the vehicle is stable and well supported at all times. Use a professional automotive lift or jack stands designed for the purpose. A floor jack is not adequate support.*

– Remove splash shield and front end reinforcement panel. See **310 Front Suspension**.

– Remove right front drive axle. See **Front drive axle, removing and installing** in this repair group.

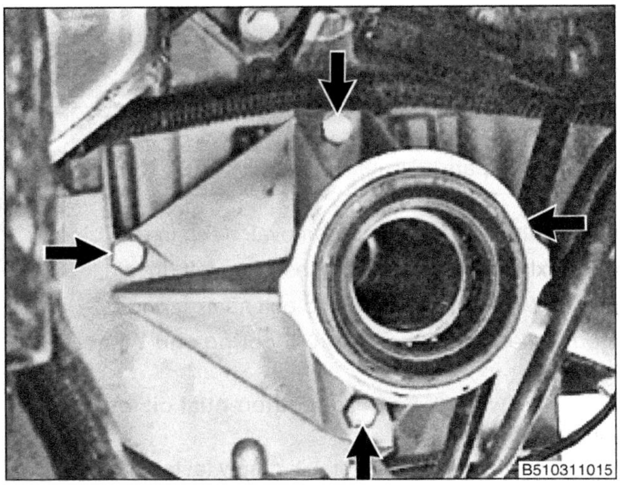

< Remove bearing pedestal mounting bolts (**arrows**) at side of oil pan. Lift off pedestal.

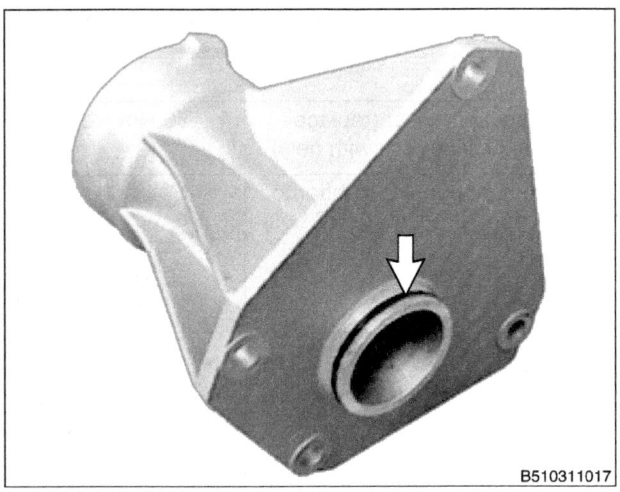

< Replace sealing O-ring (**arrow**). Coat O-ring with assembly fluid.

– Replace drive axle flange seal before installing drive axle. See **Front drive axle flange seal, replacing** in this repair group.

– Installation is reverse of removal. Remember to replace front end reinforcement panel fasteners with new.

Tightening torques	
Bearing pedestal to oil pan	21 Nm (16 ft-lb)
Front end reinforcement panel to subframe (replace M10 x 35 mm 10.9 grade ZNS bolts with new): • Stage 1 • Stage 2	 56 Nm (41 ft-lb) + additional 90°

BentleyPublishers.com
BentleyPublishers.com—All Rights Reserved

Front differential input flange seal, replacing

– Raise vehicle and support safely.

> ***WARNING —***
> • *Make sure the vehicle is stable and well supported at all times. Use a professional automotive lift or jack stands designed for the purpose. A floor jack is not adequate support.*

– Remove splash shield and front end reinforcement panel. See **310 Front Suspension**.

– Remove front driveshaft. See **260 Driveshafts**.

– Pry out input flange retaining nut lock plate.

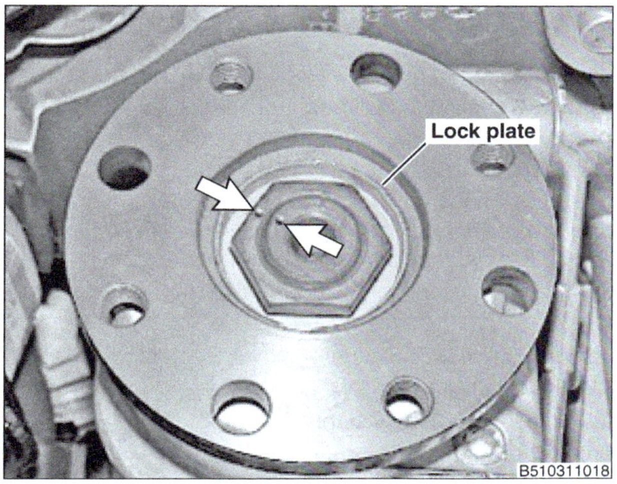

◄ Using a center punch, mark relation of input flange retaining nut to shaft (**arrows**).

◄ Counterhold input flange using BMW special tool 23 0 020 and remove nut.

> ***CAUTION—***
> • *Do not place flange holder pins in threaded holes of flange.*

– Remove input flange using hub puller. Be prepared to catch dripping oil.

BentleyPublishers.com
BentleyPublishers.com—All Rights Reserved

Front differential input flange seal, replacing

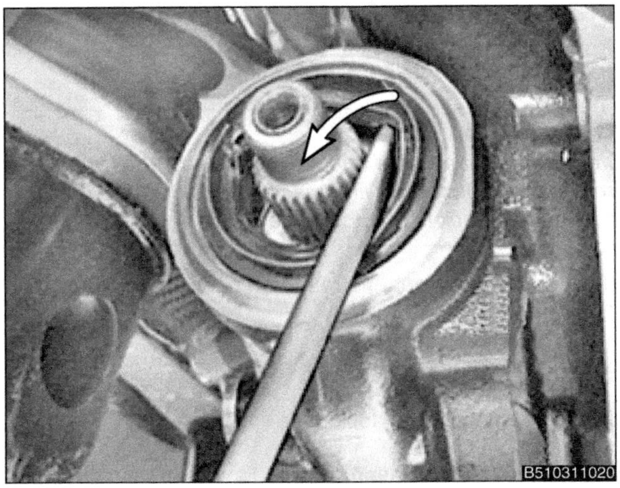

◀ Using seal puller or flat screwdriver, pry (**arrow**) seal out of differential housing.

– Coat sealing edges of new seal with transmission fluid and drive into differential housing using seal installer (BMW special tool 31 5 130 or equivalent.

– Clean input flange and install into differential housing.

◀ Install flange nut. Counterhold flange and tighten down nut until punch marks (**arrows**) align. Install new input flange retaining nut locking plate.

> ***CAUTION—***
> - *Do not torque input flange retaining nut beyond match marks. Over-torquing can damage differential internals.*
> - *Do not replace input flange or input flange locking nut.*

– Remainder of installation is reverse of removal. Keep in mind:

- Check front differential oil level and top off if necessary. See **Front differential oil level, checking** in this repair group.
- Install front end reinforcement panel using new fasteners. See **310 Front Suspension**.

Tightening torque	
Front driveshaft to differential (replace M8 ZNS bolts with new)	20 Nm (15 ft-lb) + additional 45°
Oil drain / fill plug to differential	60 Nm (44 ft-lb)

BentleyPublishers.com
BentleyPublishers.com–All Rights Reserved

320 Steering and Wheel Alignment

TABLES

GENERAL

This repair group covers steering wheel and column removal and steering system service, including wheel alignment information. For additional information, see:

- **310 Front Suspension** for active stabilizer
- **513 Interior Trim** for steering column trim
- **612 Switches** for steering column switch cluster (SZL), ignition switch and START / STOP switch
- **721 Airbag System (SRS)** for driver airbag

Bentley Publishers
BentleyPublishers.com—All Rights Reserved

Steering system

5 Series models are equipped with a power-assisted steering rack mounted to an aluminum subframe. Power boost for the steering is provided by an engine-driven hydraulic pump.

The steering wheel connects to the steering rack via an adjustable steering column which incorporates a sliding spindle to dampen vibration and noise.

Power steering fluid is supplied from the fluid reservoir to the pump and to the steering rack via rubber and metal lines. The return line from the steering rack to the reservoir loops forward and connects to the steering fluid cooler.

Steering components

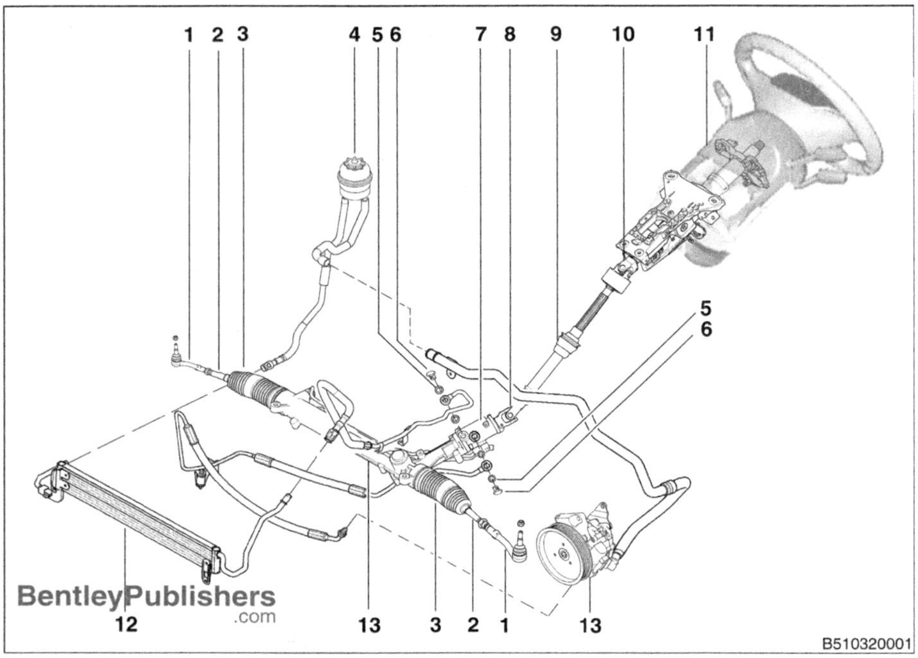

1. Tie rod end
2. Inner tie rod
3. Steering rack boot
4. Steering fluid reservoir
5. Sealing O-ring
   - Replace with new
6. Banjo bolt M14 or M16
   - Tighten to 33 - 35 Nm (24 - 26 ft-lb)
7. Steering rack gear housing with servotronic valve
8. Steering spindle lower universal joint
   - Rethread threads in pinch collar
   - Replace M8 Torx pinch bolt with new
   - Tighten to 22 Nm (16 ft-lb)
9. Lower steering spindle
10. Steering spindle upper universal joint
    - Rethread threads in pinch collar
    - Replace M8 Torx pinch bolt with new
    - Tighten to 22 Nm (16 ft-lb)
11. Steering column
12. Steering fluid cooler
13. Steering rack
14. Power steering pump.

Servotronic

The servotronic system uses an electrohydraulic valve at the steering rack to control the degree of steering assistance provided by the power steering as a function of vehicle speed. The flow of hydraulic fluid is restricted to a greater or lesser extent depending on how the servotronic valve is actuated electronically by the safety and gateway module (SGM).

The servotronic valve is actuated when terminal 15 is ON (ignition ON). The vehicle speed signal from the DSC control module is used by SGM software to modulate the servotronic valve signal.

BentleyPublishers.com
BentleyPublishers.com—All Rights Reserved

Current supply to the servotronic valve is interrupted by the faults listed below. Under these circumstances, steering assistance is limited to minimum:

- Vehicle speed signal from DSC control module incorrect; or no vehicle speed signal.
- Terminal status from CAS control module via K-CAN incorrect or missing.
- Electrical fault in servotronic valve circuit (exception: short-circuit to positive).

In case of a short-circuit to positive, full voltage is applied to the servotronic valve, activating it fully and increasing steering assistance to maximum.

Active steering

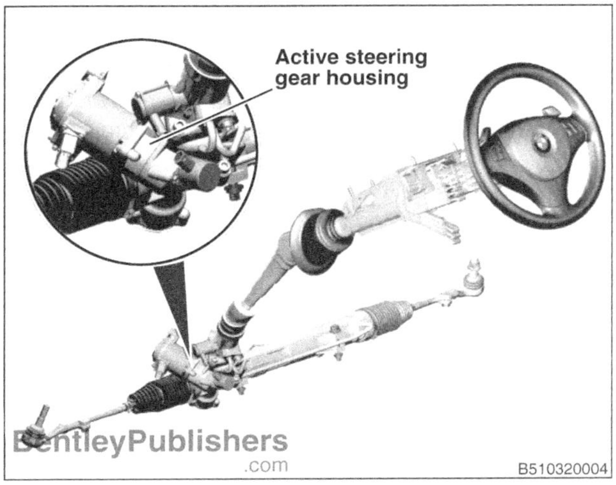

The optional active front steering system (acronyms AS, AFS or AL). provides steering power assist with variable steering ratio. Steering ratio adapts to road speed and to the steering angle requested by the driver. The steering is designed to be indirect at high speeds and direct at low speeds.

At slow speeds and when parking, vehicle maneuverability is significantly increased so that the driver does not need to grip the steering wheel excessively hard. When the vehicle is stationary, 2 turns are enough to move the steering wheel lock to lock.

At high speeds (above 75 mph / 120 kph), active steering allows a more indirect steering gear ratio than conventional steering. Unintentional steering movements are prevented in conjunction with the increased steering torque level.

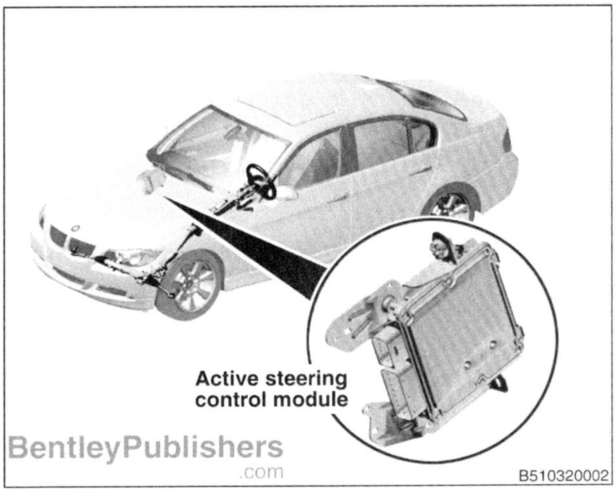

The active steering control module is in the right footwell, attached to the floor pan and protected by a kick plate.

In vehicles equipped with active steering, the Servotronic valve is activated by the active steering control module.

320

BentleyPublishers.com–All Rights Reserved

Active steering, adjusting

Use BMW scan tool to adjust (or commission) active steering system under the following circumstances:

- If front suspension or steering components are repaired, modified or aligned.
- If battery is disconnected and reconnected.
- If electrical connections on the steering column switch cluster (SZL) are disconnected and reconnected.
- If active steering or DSC electronic components are replaced, coded or programmed.

Note that steering wheel position and the position of the steered wheels are synchronized as soon as the engine is started. This ensures that the positions of the steering wheel and the road wheels match if, for example, the steering wheel was moved while the vehicle was at a standstill with the ignition switched OFF.

The synchronization procedure can cause the steering wheel or the vehicle's front wheels to move. Movements of the steering wheel or the vehicle's front wheels might be perceptible while synchronization is in progress. Synchronization also occurs while the vehicle is on the move, but the process is extremely slow and virtually imperceptible.

Power steering fluid

The power steering system is permanently filled with either ATF or Pentosin hydraulic fluid. Routinely adding fluid is not required unless the system is leaking.

> ***CAUTION—***
> - *Power steering reservoir cap is marked with the type of fluid being used— ATF or CHF. Do not mix.*

Hydraulic fluid
As recommended on reservoir cap for power steering, active steering, active stabilizer: • Automatic transmission fluid (ATF) • Pentosin CHF 11S

For filling and bleeding procedure, see **Power steering system, bleeding and filling** in this repair group.

BentleyPublishers.com
BentleyPublishers.com—All Rights Reserved

Hydraulic line quick disconnect coupling

< To disconnect quick disconnect coupling (**inset**), press plastic ring into coupling (**arrow**). Be prepared to catch dripping fluid.

– To reconnect, push coupling together until a distinct click is heard. Attempt to pull coupling apart forcefully to test for correct seating.

Warnings and Cautions

WARNING—

- *The airbag mounted in the steering wheel is an explosive device. Treat it with extreme caution. Follow the airbag removal procedure in* **721 Airbag System (SRS)**.
- *Serious injury may result if airbag system service is attempted by persons unfamiliar with the BMW SRS and its approved service procedures.*
- *Before performing any work involving airbags, disconnect the negative (-) battery cable.*
- *Prior to disconnecting the battery, read the battery disconnection cautions given in* **001 Warnings and Cautions**.
- *BMW airbags are equipped with a back-up power supply inside the airbag control module. Observe a 5 second waiting period after disconnecting the battery cable to allow the reserve power supply to discharge.*
- *Do not reuse self-locking nuts. They are designed to be used only once and may fail if reused. Replace with new.*
- *Do not install bolts and nuts coated with undercoating wax, as correct tightening torque cannot be assured. Clean the threads with solvent before installation, or install new parts.*
- *Do not attempt to weld or straighten steering components. Replace damaged parts.*
- *Power steering and hydraulic fluid (ATF or Pentosin) are injurious to skin and eyes. Wear protective gloves and eyewear.*
- *Follow local, state and federal regulations for safe disposal of power steering fluid.*

320

BentleyPublishers.com—All Rights Reserved

> ***CAUTION—***
> - *When filling or bleeding steering and hydraulic system, do not allow contamination of fluid or reservoir.*
> - *Using contaminated equipment to add fluid may introduce dirt particles into the fluid reservoir and significantly reduce the service life of system components.*
> - *To avoid electrochemical corrosion to engine components made of aluminum-magnesium alloy, do not use steel fasteners. Use aluminum fasteners only.*
> - *The end faces of aluminum fasteners are usually painted blue. For reliable identification, test fasteners for aluminum composition with magnet.*
> - *Replace aluminum bolts each time they are loosened.*
> - *Follow torque instructions, including angle of rotation specifications, when installing aluminum fasteners.*
> - *To prevent marring interior trim, work with plastic prying tools or wrap the tips of screwdrivers and pliers with tape before prying out switches or trim.*
> - *Static discharge may permanently damage an electronic component such as control module or sensor. Handle electronic components using accepted static prevention equipment and techniques. See* **600 Electrical System–General**.

TROUBLESHOOTING

Corrosion damage

Inspect tie rod ends and steering rack boots periodically for damage or tears. Replace as necessary. See **Tie rod end, replacing** and **Steering rack boot, replacing** in this repair group.

Note that rodent bite damage to rack boot is difficult to detect. Bite marks are often small and only on one side. Leakage at one rack boot can result in corrosion of rack shaft metal at both ends.

◀ If steering rack metal shows sign of corrosion, replace complete rack. See **Steering rack, removing and installing (rear-wheel-drive)** or **Steering rack, removing and installing (xDrive)** in this repair group.

BentleyPublishers.com
BentleyPublishers.com—All Rights Reserved

Active steering faults

The driver is alerted to active steering system faults by a warning icon and a Check Control warning triangle in instrument cluster.

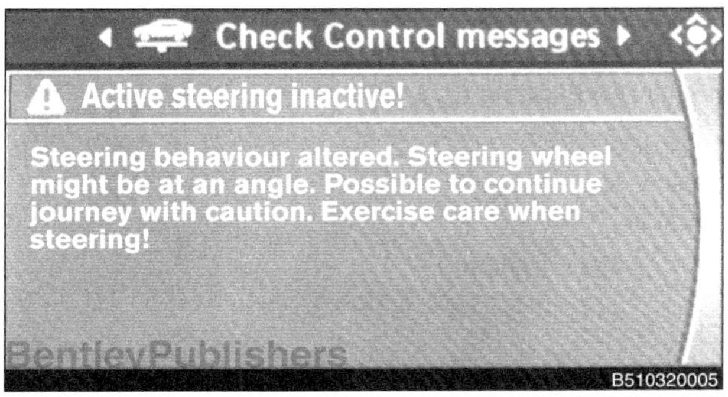

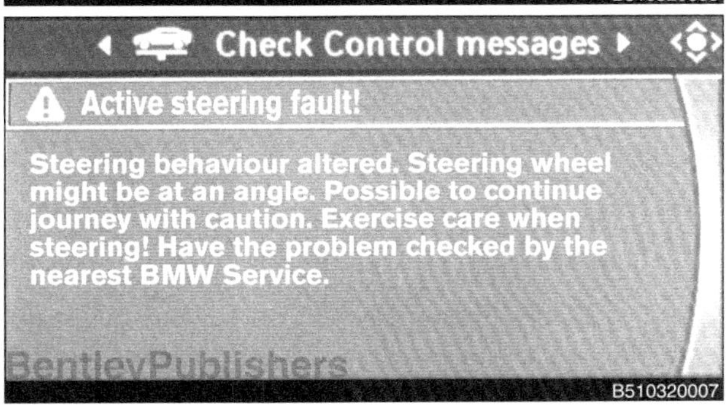

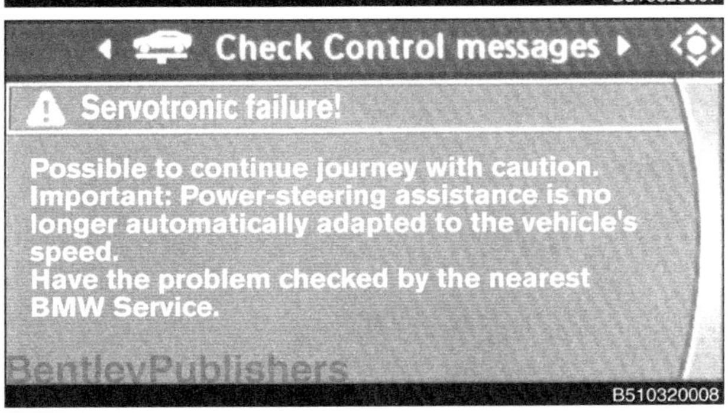

Check Control messages in center dashboard (iDrive) display details about active steering failure and instruction on how to proceed.

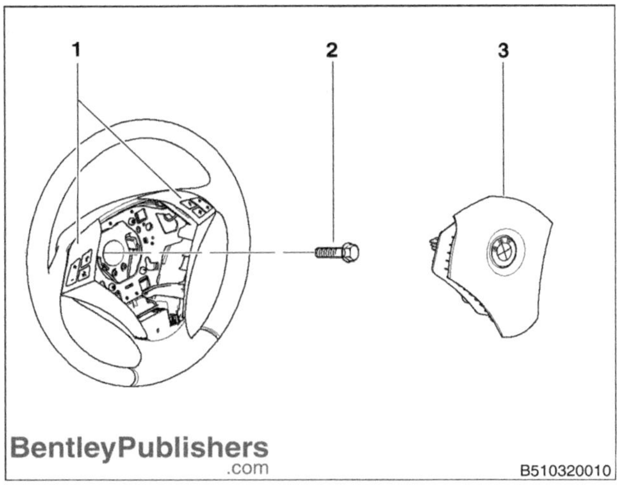

Steering Wheel

5 Series vehicles are equipped with one of two different styles of steering wheel:

Steering wheel components

1. Multifunction switches
2. M12 mounting bolt
   - Tighten to 63 Nm (46 ft-lb)
3. Airbag unit and horn button

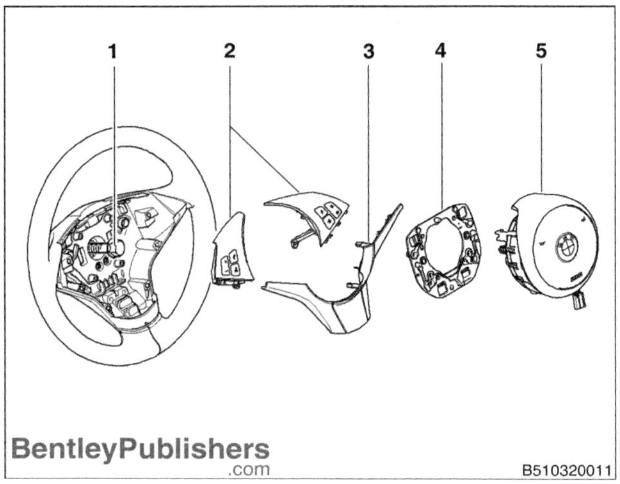

Sport steering wheel components

1. M12 mounting bolt
   - Tighten to 63 Nm (46 ft-lb)
2. Multifunction switches
3. Plastic trim
4. Mounting plate
5. Airbag unit and horn button

Steering angle sensor

The steering angle sensor is integral with the steering column switch cluster (SZL). For more information, including removal and installation of the SZL, see **612 Switches**.

To reset steering angle sensor, connect vehicle to BMW scan tool and carry out steering angle sensor adjustment system function.

Steering wheel, removing and installing

– Center steering wheel. Make sure front wheels are pointed straight ahead.

– Extend steering wheel as far back and as low as possible.

– Disconnect negative (-) cable from battery.

> ***CAUTION—***
> *• Prior to disconnecting the battery, read the battery disconnection cautions given in* **001 Warnings and Cautions**.

BentleyPublishers.com
BentleyPublishers.com—All Rights Reserved

– Remove driver airbag in steering wheel. See **721 Airbag System (SRS)**.

> ***WARNING —***
> - *The steering wheel mounted airbag is an explosive device. Treat with extreme caution. Follow the airbag removal procedure in* **721 Airbag System (SRS)**.
> - *Airbag removal may set SRS fault codes. The SRS indicator light remains ON until problems are corrected and fault memory cleared.*

< Remove steering wheel center bolt (**arrow**).

> ***CAUTION—***
> - *The SRS contact reel (also known as airbag contact ring / slip ring) is integral with the steering column switch block. The contact reel is a wound coil of wire that ensures continuous electrical contact for the airbag and steering wheel electrical components. When the steering wheel is removed the contact reel is locked and its position must not be altered.*

< The steering column and steering wheel are match marked at the factory. Check for proper alignment of steering column marks (**arrows**) before removing steering wheel.

– Remove steering wheel.

– When installing steering wheel:
  - Align steering wheel and column match marks.
  - Align steering wheel to alignment pins located on steering column switch block.
  - Install steering column center bolt. Do not over-torque.

Tightening torques	
Steering wheel to steering shaft (M12)	63 Nm (47 ft-lb)

– Carefully install airbag. See **721 Airbag System (SRS)**.

BentleyPublishers.com
BentleyPublishers.com—All Rights Reserved

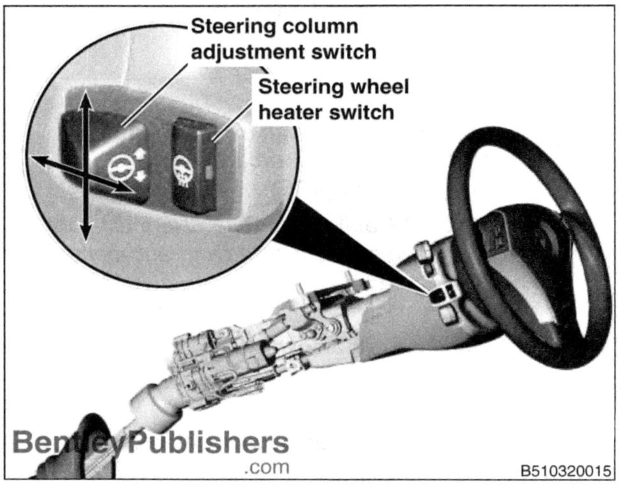

STEERING COLUMN

Adjustable steering column

The steering column adjustment switch is used to control two motors in the steering column, one for tilt and the other for extension.

Column position is memorized and stored, along with driver seat and outside rear view mirror positions, in the seat memory module.

Steering column components

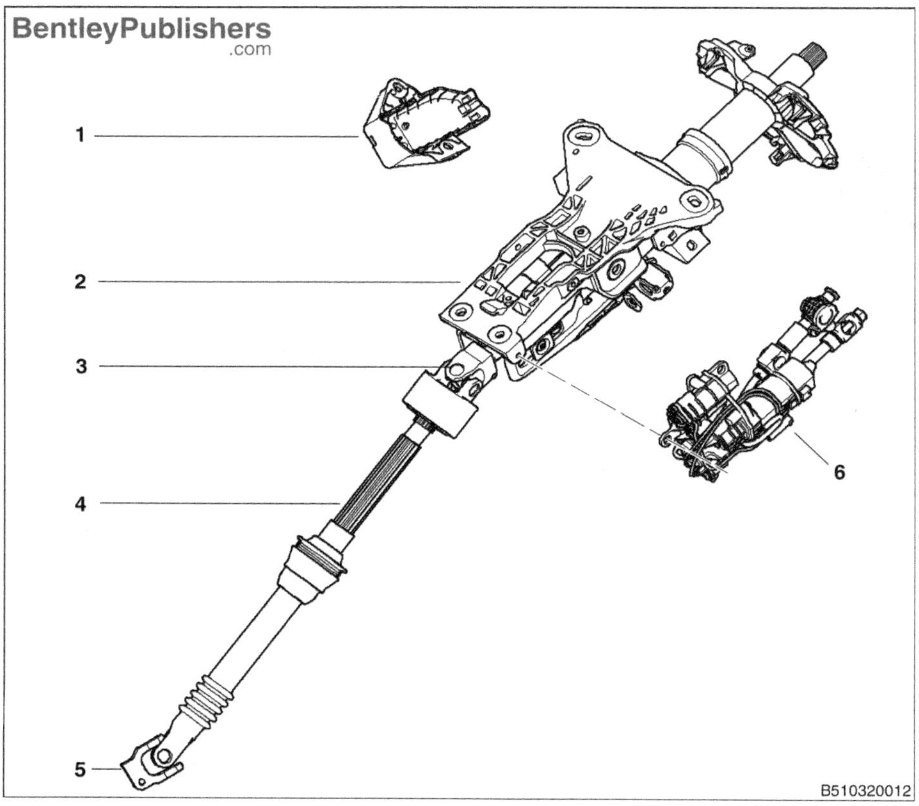

1. Electric steering lock or ELV (2006 only)
2. Steering column assembly
3. Upper steering shaft universal joint
4. Steering shaft spindle
5. Lower steering shaft universal joint
6. Steering column adjustment motors

BentleyPublishers.com
BentleyPublishers.com—All Rights Reserved

Steering column, removing and installing

- Center steering wheel. Make sure front wheels are pointed straight ahead. Extend steering wheel as far back and as low as possible.
- 2004 - 2005 model: Remove ignition key.
- Disconnect negative (-) battery cable and cover battery terminal to keep cable from accidentally contacting terminal.

> ***CAUTION—***
> - *Prior to disconnecting the battery, read the battery disconnection cautions given in* **001 Warnings and Cautions**.

- Remove driver airbag. See **721 Airbag System (SRS)**.

> ***WARNING—***
> - *The steering wheel mounted airbag is an explosive device. Treat with extreme caution. Follow the airbag removal procedure in* **721 Airbag System (SRS)**.

- Remove steering wheel. See **Steering wheel, removing and installing** in this repair group.

> ***CAUTION—***
> - *Do not move front wheels from straight-ahead position while steering wheel is off.*
> - *Do not turn or twist the airbag and horn contact spring.*

320

- Remove steering column trim, left lower dashboard trim and left dashboard panel. See **513 Interior Trim**.
- Remove airbag and horn contact spring and steering column switch cluster (SZL). See **612 Switches**.

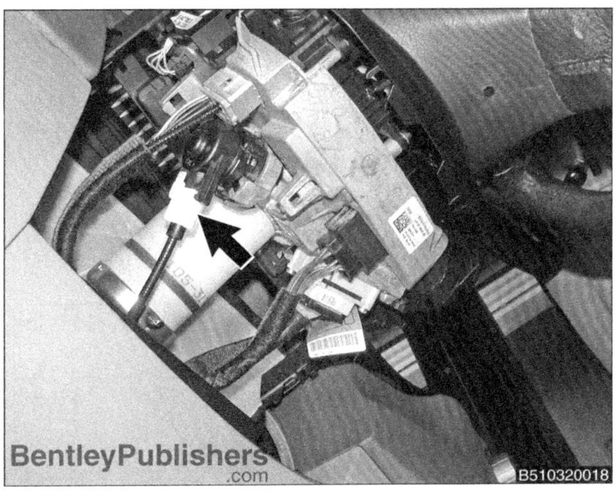

< 2004 - 2005 model with automatic transmission: Detach ignition interlock cable end (**arrow**) from steering lock.

- Detach wiring harnesses and cables from steering column and place to one side.

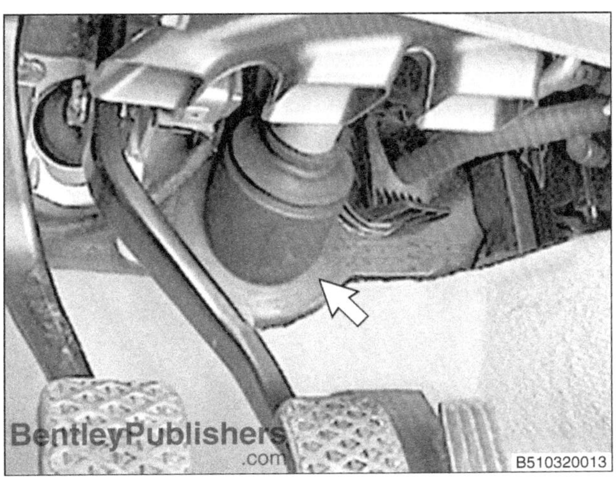

< Working in left footwell, slide sleeve (**arrow**) for steering spindle upwards on steering column.

BentleyPublishers
BentleyPublishers.com—All Rights Reserved

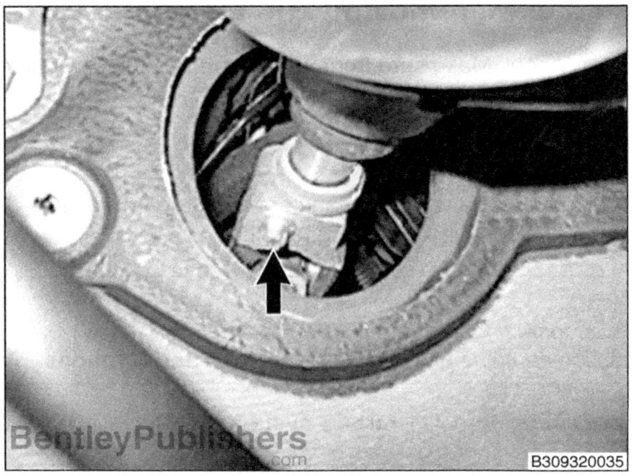

◀ Remove steering shaft upper universal joint pinch bolt (**arrow**).

◀ Working underneath steering column:

- Mark installation position of steering column.
- Loosen and remove steering column fasteners (**arrows**). Be sure to have an assistant support column as last bolts are removed.
- Guide column off steering spindle universal joint and remove.

– Installation is reverse of removal. Remember to:

- Recut threads in spindle universal joint pinch collar.
- Slide universal joint splines over column shaft. Splines fit only one way. Install new Torx pinch bolt. Pinch bolt rests in groove of steering column shaft.
- Fit spindle sleeve back into opening in floorboard. Align marks on sleeve and floorboard.
- Use new bolts to mount column to dashboard and dashboard brace. Match position marks made previously.

Tightening torques	
Steering column to dashboard or dashboard brace (replace M8 bolt with new)	21 Nm (15 ft-lb)
Steering shaft spindle universal joint pinch bolt (recut pinch collar threads, replace M8 Torx bolt with new)	22 Nm (16 ft-lb)

– Check steering system for freedom of movement through entire steering column adjustment range.

– Carry out steering angle sensor adjustment using BMW scan tool. If replacing steering angle sensor, code using BMW scan tool before performing steering angle offset check.

– If airbag warning light illuminates, use BMW scan tool to check and clear fault codes.

– Test drive vehicle to check directional stability.

BentleyPublishers.com
BentleyPublishers.com—All Rights Reserved

Steering shaft spindle, removing and installing

– Raise front of car and support safely.

> ***WARNING —***
> • *Make sure the car is firmly supported on jack stands designed for the purpose. Place jack stands underneath structural chassis points, not underneath suspension parts.*

– Remove splash shield and front end reinforcement panel from underneath engine. Also remove left side splash shield at front suspension.

◄ Remove steering spindle lower pinch bolt and slide universal joint off steering rack shaft (**arrow**).

> ***CAUTION—***
> • *While steering spindle is detached from steering rack, make sure steering wheel is not rotated. Rotating the wheel may damage the airbag and horn contact spring.*

– Remove left lower dashboard trim (pedal cluster trim). See **513 Interior Trim**.

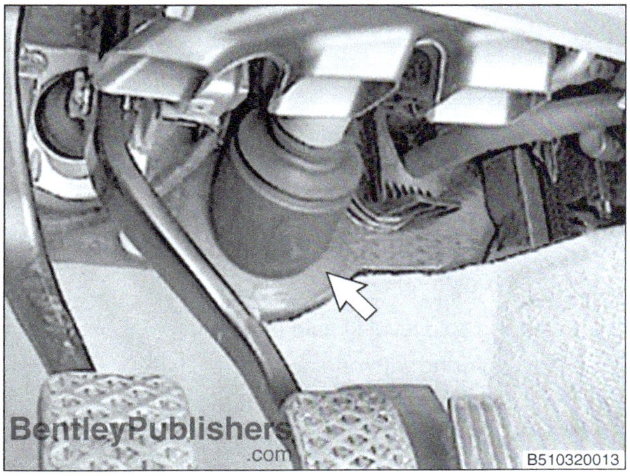

◄ Working in left footwell, slide sleeve (**arrow**) for steering spindle upwards on steering column.

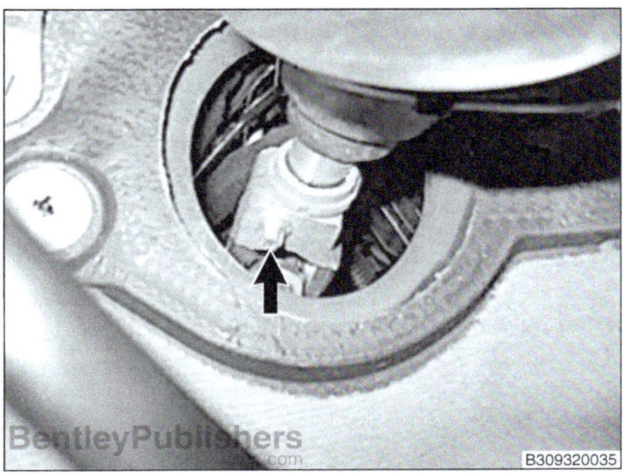

◄ Remove steering shaft upper universal joint pinch bolt (**arrow**) and slide universal down and off steering column shaft.

– Installation is reverse of removal. Remember to:
- Rethread threads in spindle universal joint pinch collars.
- Slide universal joint splines over steering column and steering rack shafts. Splines fit only one way. Install new Torx pinch bolts. Pinch bolts rest in shaft grooves.
- Fit spindle sleeve back into opening in floorboard. Align marks on sleeve and floorboard.

Tightening torque	
Steering shaft spindle universal joint pinch bolt (recut pinch collar threads, replace M8 Torx bolt with new)	22 Nm (16 ft-lb)

– Check steering system for freedom of movement through entire steering column adjustment range. Check directional stability of vehicle on test drive.

– Carry out steering angle sensor adjustment using BMW scan tool.

BentleyPublishers
BentleyPublishers.com – All Rights Reserved

Power Steering Pump

Power assist is provided by a belt-driven pump at the lower left front of the engine. The power steering fluid reservoir is located in the engine compartment:

- 6-cylinder model: Left side of engine compartment
- V8 models: Right side of engine compartment

Power steering system, bleeding and filling

The power steering system is permanently filled with either CHF (Pentosin hydraulic fluid) or ATF. Routinely adding fluid is not required unless the system is leaking.

> ***CAUTION—***
> - *Power steering reservoir cap is marked with the type of fluid being used— CHF or ATF. Do not mix.*

To check power steering fluid level in fluid reservoir:

- Park vehicle on level ground with engine OFF.
- Level is correct if it is between **MIN** and **MAX** marks on dipstick.
- If level is below **MIN** mark, add fluid to reservoir to bring level up.

> ***CAUTION—***
> - *Correct power steering fluid level can be as much as 10 mm (0.4 in) higher than **MAX** mark when engine is at normal operating temperature. Do not draw off fluid to **MAX** mark when checking fluid with engine warmed up.*

– To bleed power steering system, start engine. Turn steering wheel twice to left lock and right lock.

– Recheck fluid level with engine off. Fill to MAX.

Power steering fluid application	
Black label (check cap)	ATF Dexron III
Green label (check cap)	Pentosin CHF 11 S

– Hand-tighten reservoir cap.

BentleyPublishers.com
BentleyPublishers.com—All Rights Reserved

Power steering pump, removing and installing (M54 engine)

- Drain power steering fluid reservoir using clean syringe. Do not reuse fluid.
- Mark engine accessory belt direction of rotation, then remove belt. See **020 Maintenance**.
- Raise front of car and support safely.

> ***WARNING*** *—*
> - *Make sure the car is firmly supported on jack stands designed for the purpose. Place jack stands underneath structural chassis points, not underneath suspension parts.*

- Remove splash shield underneath front of engine. See **020 Maintenance**.
- Remove power steering pump pulley.
- Remove alternator idler pulley.

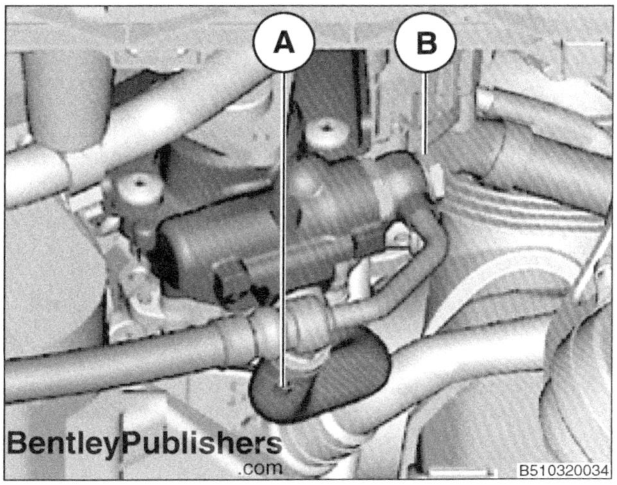

◀ Working at power steering pump, detach fluid lines. Be prepared to catch dripping fluid. Plug open fluid ports.
- Cut supply hose clamp (**A**) and separate hose from pump inlet.
- Remove banjo bolt (**B**) and remove pressure line.

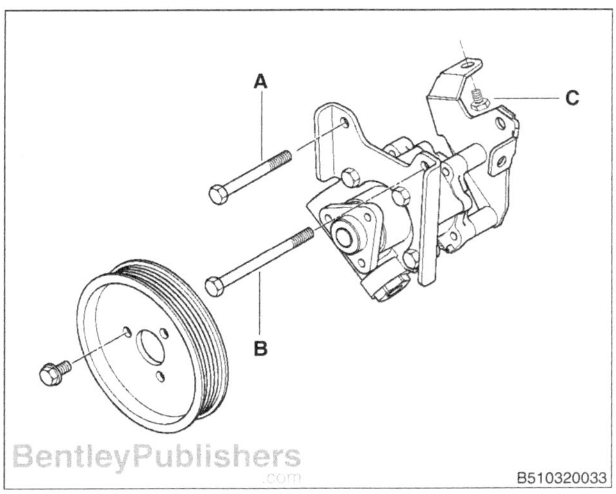

◀ Remove pump mounting bolts and lower pump.
- **A** = M8 x 70 mm ZNS bolt
- **B** = M8 x 100 mm ZNS bolt
- **C** = M8 x 22 mm ZNS bolt

- Installation is reverse of removal. Keep in mind the following:
  - Use new O-ring seals at pressure line banjo bolt. Route pressure line tension-free and with sufficient distance to adjoining components.
  - Match markings on supply line and pump connection. Use new hose clamp at connection.
  - Lettering on pump pulley faces forward.

Tightening torques	
Pressure line to pump (M16 banjo bolt, use new O-ring seals)	35 Nm (26 ft-lb)
Pulley to power steering pump (M8 x 14 mm 10.9 grade bolts)	27 Nm (20 ft-lb)
Power steering pump to engine block (M8 ZNS bolts)	21 Nm (16 ft-lb)

- Fill and bleed steering system with correct fluid. See **Power steering system, bleeding and filling** in this repair group.
- Run engine and check fluid connections for leaks.

BentleyPublishers
BentleyPublishers.com—All Rights Reserved

Power steering pump, removing and installing (N52 engine)

– Drain power steering fluid reservoir using clean syringe. Do not reuse fluid.

– Raise front of car and support safely.

> ***WARNING —***
> • *Make sure the car is firmly supported on jack stands designed for the purpose. Place jack stands underneath structural chassis points, not underneath suspension parts.*

– Remove splash shield underneath front of engine. See **020 Maintenance**.

– Remove air filter housing. See **130 Fuel Injection**.

◄ Working from above on left side of engine:

- Cut supply hose clamp (**A**) and separate hose from pump inlet. Be prepared to catch dripping fluid. Plug open fluid ports.
- Remove steering pump rear bracket mounting bolts (**B**). Discard aluminum bolts.

– Mark engine accessory belt direction of rotation, then remove belt. See **020 Maintenance**.

– Working underneath front of vehicle, remove power steering pump pulley.

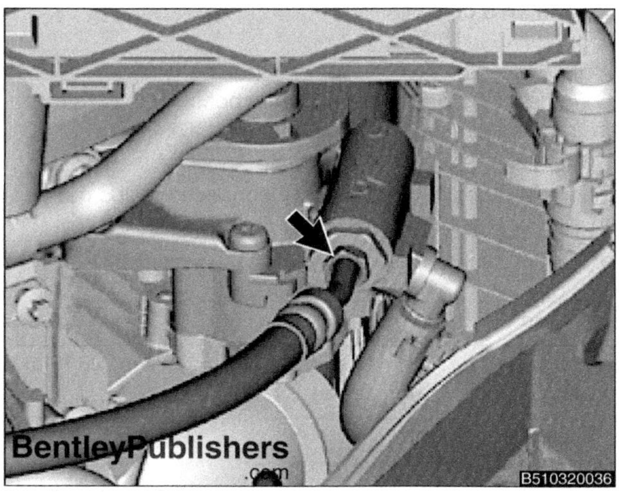

◄ Loosen union nut (**arrow**) and detach pressure line from pump. Be prepared to catch dripping fluid. Plug open fluid ports.

BentleyPublishers.com
BentleyPublishers.com—All Rights Reserved

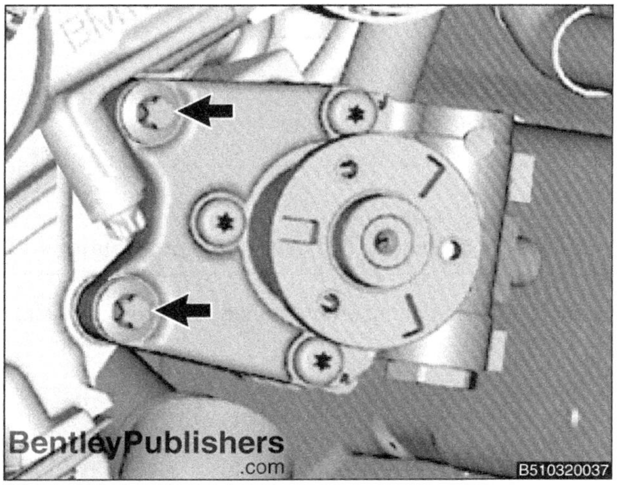

◀ Remove pump mounting bolts (**arrows**) and lower pump. Discard aluminum bolts.

– When installing pump, replace aluminium bolts with new, tighten finger tight and observe the following tightening sequence:
  1. Tighten rear bracket fasteners to 2 Nm (18 in-lb).
  2. Tighten front fasteners to 2 Nm (18 in-lb).
  3. Tighten front fasteners to final specification.
  4. Loosen rear bracket fasteners and make sure bracket is flush with engine block.
  5. Tighten rear bracket fasteners to final specification.

Tightening torque	
Power steering pump to block (replace M10 x 33 mm aluminum bolts with new)	20 Nm (15 ft-lb) additional 90°

– Remainder of installation is reverse of removal. Keep in mind the following:
  - Route fluid lines tension-free and with sufficient distance to adjoining components.
  - Match markings on supply line and pump connection. Use new hose clamp at connection.
  - Lettering on pump pulley faces forward.

Tightening torques	
Pressure line to pump	36 Nm (27 ft-lb)
Pulley to power steering pump (M8 8.8 grade bolts)	21 Nm (20 ft-lb)

– Fill and bleed steering system with correct fluid. See **Power steering system, bleeding and filling** in this repair group.

– Run engine and check fluid connections for leaks.

BentleyPublishers.com
BentleyPublishers.com–All Rights Reserved

Power steering pump, removing and installing (N54 engine)

– Disconnect negative (-) battery cable and cover battery terminal to keep cable from accidentally contacting terminal.

> ***CAUTION—***
> *• Prior to disconnecting the battery, read the battery disconnection cautions given in* **001 Warnings and Cautions**.

– Mark engine accessory belt direction of rotation, then remove belt. See **020 Maintenance**.

– Drain power steering fluid reservoir using clean syringe. Do not reuse fluid.

– Raise front of car and remove front wheels.

> ***WARNING—***
> *• Make sure the car is firmly supported on jack stands designed for the purpose. Place jack stands underneath structural chassis points, not underneath suspension parts.*

– Remove splash shield and front end reinforcement panel from underneath engine. Also remove side splash shields at front suspension.

– Install engine support brace across engine bay and support weight of engine.

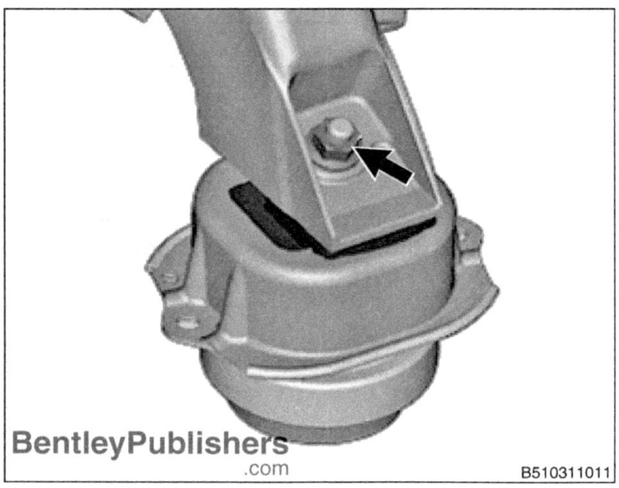

◄ Working underneath engine, remove right and left engine mount upper nuts (**arrow**). Raise engine slightly (¼ - ½ in) to release load on engine mounts.

– Detach both tie rod ends from steering knuckles. See **Tie rod end, replacing** in this repair group.

– xDrive: Detach left and right control arms from steering knuckles. Detach left and right front stabilizer bar links from stabilizer bar. See **310 Front Suspension**.

◄ Detach steering spindle from steering rack:

- Move steering wheel so that front wheels point straight ahead. Vehicle with steering lock: Remove ignition key and make sure steering locks in straight ahead position.
- With front wheels pointing straight ahead, make sure steering spindle pointer and steering rack pointer line up (**arrows**). Remove lower steering spindle pinch bolt and detach steering universal joint from steering rack shaft.

> ***CAUTION—***
> *• While steering spindle is detached from steering rack, make sure steering wheel is not rotated. Rotating the wheel may damage the airbag and horn contact spring.*

BentleyPublishers.com
BentleyPublishers.com—All Rights Reserved

– If applicable, detach electrical connectors at front suspension ride height sensor, Servotronic sensor and engine mount vacuum line.

– Vehicle with active stabilizers (dynamic drive):

- Remove right front wheel housing liner. See **410 Fenders, Engine Hood**.
- Separate fluid lines to front stabilizer oscillating motor. Be prepared to catch dripping fluid. Plug open fluid lines immediately.

– Vehicle with active steering: Detach active steering gearbox electrical connectors. Detach wiring harness from front subframe.

◄ Support front subframe with transmission jack. Remove subframe mounting bolts (**arrows**) and lower subframe about 100 mm (4 in).

> ***CAUTION—***
> - *Do not bend or stretch fluid lines and hoses when lowering or raising subframe.*
> - *N54 with automatic transmission: Disconnect transmission fluid cooler lines from subframe.*

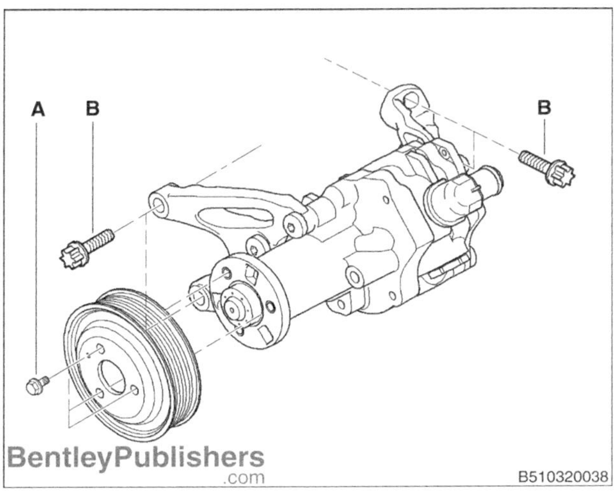

◄ Working underneath front of vehicle:

- Remove power steering pump pulley bolts (**A**) and take pulley off pump.
- Remove pump mounting bolts (**B**) and pull pump forward. Discard aluminum bolts.

BentleyPublishers.com
BentleyPublishers.com—All Rights Reserved

Power steering pump, removing and installing (N54 engine)

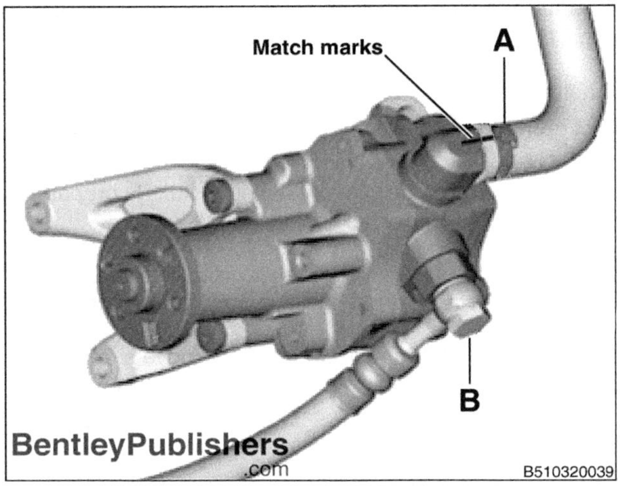

◀ Working at steering pump, detach fluid lines. Be prepared to catch dripping fluid. Plug open fluid ports.

- Cut supply hose clamp (**A**) and separate hose from pump inlet. Note match marks on supply hose and pump connection.
- Remove banjo bolt (**B**) and remove pressure line.
- Lower steering pump.

– Attach fluid lines to new pump:

- Use new O-ring seals at pressure line banjo bolt. Route pressure line tension-free and with sufficient distance to adjoining components.
- Match markings on supply line and pump connection. Use new hose clamp at connection.

Tightening torques	
Pressure line to pump (M14 banjo bolt, use new O-ring seals)	33 Nm (24 ft-lb)

– Raise pump into position on engine block. Replace aluminium bolts with new and tighten finger tight. Observe the following tightening sequence:

1. Tighten rear fasteners to 2 Nm (18 in-lb).
2. Tighten front fasteners to 2 Nm (18 in-lb).
3. Tighten front fasteners to final specification.
4. Loosen rear fasteners and make sure bracket is flush with engine block.
5. Tighten rear bracket fasteners to final specification.

Tightening torque	
Power steering pump to block (replace M10 x 35 mm aluminum bolts with new)	38 Nm (28 ft-lb)

– Reinstall pump pulley. Lettering on pump pulley faces forward.

Tightening torque	
Pulley to power steering pump (M8 x 12 mm 8.8 grade bolts)	22 Nm (16 ft-lb)

BentleyPublishers.com
BentleyPublishers.com—All Rights Reserved

– Remainder of installation is reverse of removal. Keep in mind the following:
  - Raise front subframe into place carefully, watching to make sure fluid lines and electrical harnesses are not crushed.
  - Use new fasteners as indicated in torque specifications table.

Tightening torques	
Control arm to steering knuckle (replace M14 x 1.5 self-locking flange nut with new)	165 Nm (122 ft-lb)
Engine mount to engine bracket (replace M10 10.9 grade self-locking nuts with new)	56 Nm (41 ft-lb)
Front end reinforcement panel to subframe (replace M10 x 35 mm 10.9 grade ZNS bolts with new): • Stage 1 • Stage 2	 56 Nm (41 ft-lb) + additional 90°
Front subframe to chassis (replace M12 x 1.5 10.9 grade bolts with new): • Stage 1 • Stage 2	 100 Nm (74 ft-lb) + additional 90°
Road wheel to hub	120 ± 10 Nm (89 ± 7 ft-lb)
Stabilizer link to stabilizer bar (replace M10 ZNS3 nut with new)	65 Nm (48 ft-lb)
Steering shaft spindle universal joint pinch bolt (recut pinch collar threads, replace M8 Torx bolt with new)	22 Nm (16 ft-lb)
Tie rod ball joint to steering knuckle (replace M14 x 1.5 ZNS3 nut with new)	80 Nm (59 ft-lb)

– Fill and bleed steering system with correct fluid. See **Power steering system, bleeding and filling** in this repair group.

– Run engine and check fluid connections for leaks.

– Carry out steering angle sensor adjustment using BMW scan tool.

BentleyPublishers.com
BentleyPublishers.com—All Rights Reserved

Power steering pump, removing and installing (V8 engine)

– Drain power steering fluid reservoir using clean syringe. Do not reuse fluid.

– Mark engine accessory belt direction of rotation, then remove belt. See **020 Maintenance**.

– Raise front of car and support safely.

> ***WARNING —***
> - *Make sure the car is firmly supported on jack stands designed for the purpose. Place jack stands underneath structural chassis points, not underneath suspension parts.*

– Remove splash shield underneath front of engine. See **020 Maintenance**.

– Remove power steering pump pulley.

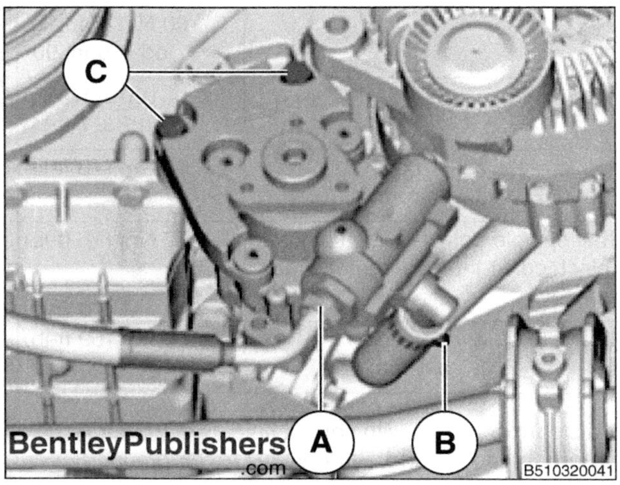

◄ Working at steering pump, detach fluid lines. Be prepared to catch dripping fluid. Plug open fluid ports.

- Cut supply hose clamp (**A**) and separate hose from pump inlet.
- Loosen union nut (**B**) and remove pressure line.
- Remove pump mounting fasteners (**C**) and lower pump.

– Installation is reverse of removal. Keep in mind the following:

- Route pressure line tension-free and with sufficient distance to adjoining components.
- Match markings on supply line and pump connection. Use new hose clamp at connection.
- Lettering on pump pulley faces forward.

Tightening torques	
Pressure line to pump	36 Nm (27 ft-lb)
Pulley to power steering pump (M8 x 14 mm 10.9 grade bolts)	27 Nm (20 ft-lb)
Power steering pump to engine block (M8 8.8 grade fasteners)	21 Nm (16 ft-lb)

– Fill and bleed steering system with correct fluid. See **Power steering system, bleeding and filling** in this repair group.

– Run engine and check fluid connections for leaks.

BentleyPublishers.com
BentleyPublishers.com—All Rights Reserved

Mechanical Steering Components

Steering rack, removing and installing (rear-wheel-drive)

– Disconnect negative (-) battery cable and cover battery terminal to keep cable from accidentally contacting terminal.

> ***CAUTION—***
> • *Prior to disconnecting the battery, read the battery disconnection cautions given in* **001 Warnings and Cautions**.

– Drain power steering fluid reservoir using clean syringe. Do not reuse fluid.

– Raise front of car and remove front wheels.

> ***WARNING —***
> • *Make sure the car is firmly supported on jack stands designed for the purpose. Place jack stands underneath structural chassis points, not underneath suspension parts.*

– Remove splash shield and front end reinforcement panel from underneath engine. Also remove side splash shields at front suspension.

320

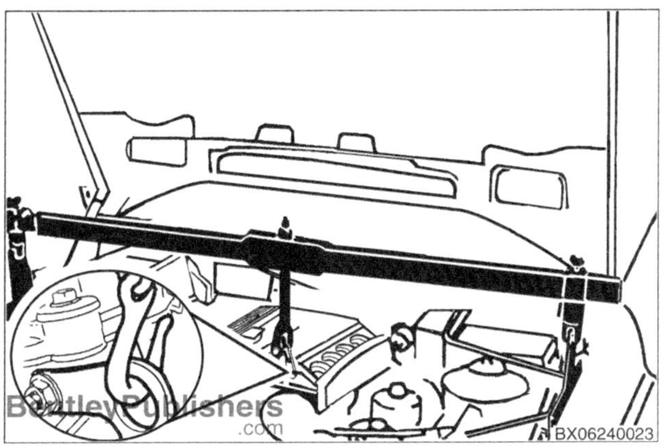

◀ Install engine support brace across engine bay.

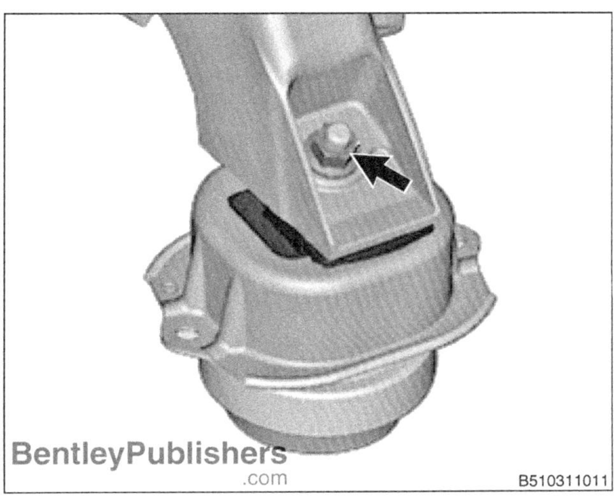

◀ Working underneath engine, remove right and left engine mount upper nuts (**arrow**). Raise engine slightly (¼ - ½ in) to release load on engine mounts.

– Detach both tie rod ends from steering knuckles. See **Tie rod end, replacing** in this repair group.

BentleyPublishers.com—All Rights Reserved

Detach steering spindle from steering rack:

- Move steering wheel so that front wheels point straight ahead. Vehicle with steering lock: Remove ignition key and make sure steering locks in straight ahead position.
- With front wheels pointing straight ahead, make sure steering spindle pointer and steering rack pointer line up (**arrows**). Remove steering spindle pinch bolt and detach steering universal joint from steering rack shaft.

CAUTION—
- *While steering spindle is detached from steering rack, make sure steering wheel is not rotated. Rotating the wheel may damage the airbag and horn contact spring.*

– If applicable, detach electrical connectors at front suspension ride height sensor, Servotronic sensor and engine mount vacuum line.

– Vehicle with active stabilizers (dynamic drive):

- Remove right front wheel housing liner. See **410 Fenders, Engine Hood**.
- Separate fluid lines to front stabilizer oscillating motor. Be prepared to catch dripping fluid. Plug open fluid lines immediately.

– Vehicle with active steering: Detach active steering gearbox electrical connectors. Detach wiring harness from front subframe.

Support front subframe with transmission jack. Remove subframe mounting bolts (**arrows**) and lower subframe about 100 mm (4 in).

CAUTION—
- *Do not bend or stretch fluid lines and hoses when lowering or raising subframe.*
- *N54 with automatic transmission: Disconnect transmission fluid cooler lines from subframe.*

– Remove left engine mounting bracket from subframe.

– If necessary, remove heat shield from power steering rack.

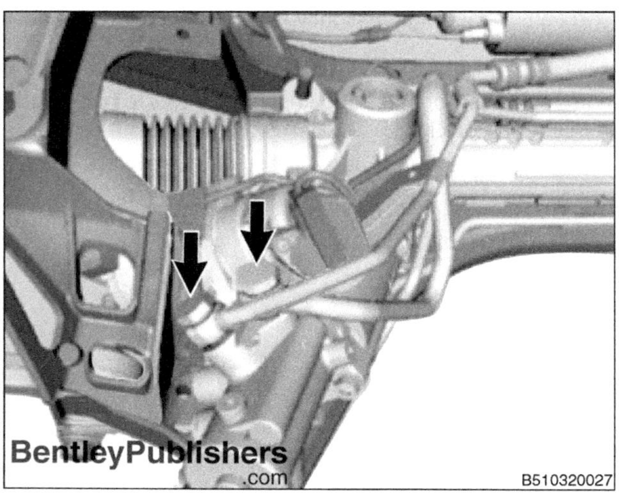

Working at steering rack, remove banjo bolts (**arrows**) and disconnect fluid lines from rack gear box.

- Be prepared to catch dripping fluid. Plug open fluid ports.
- Discard sealing O-rings.

BentleyPublishers.com
BentleyPublishers.com—All Rights Reserved

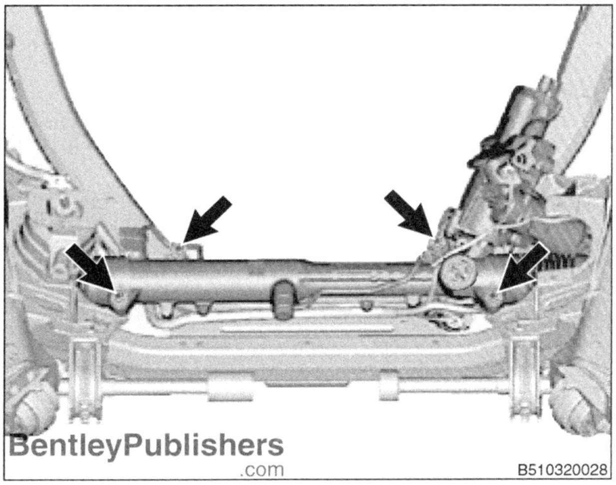

◄ Remove steering rack mounting fasteners (**arrows**).

– Vehicle with active steering: Detach electrical connectors at rack gear box.

– Slide rack to right and remove toward rear. Check for wiring, fluid lines or mechanical parts that might become snagged as rack is removed.

– When installing, use new mounting bolts and self-locking nuts. Tighten in two stages.

Tightening torque	
Steering rack to front subframe (replace M10 10.9 grade fasteners with new) • Stage 1 • Stage 2	 56 Nm (41 ft-lb) additional 90°

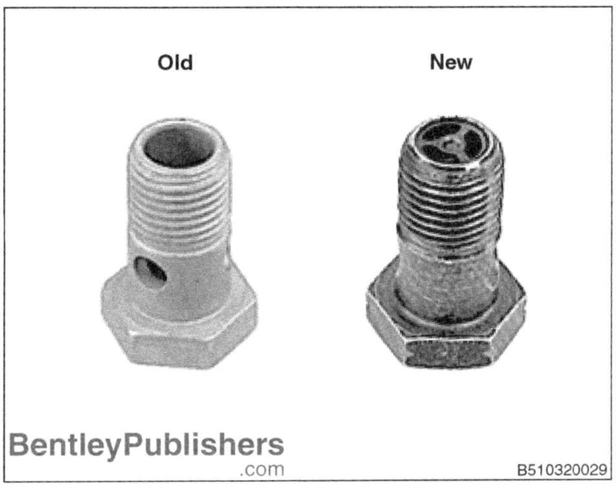

◄ When reattaching fluid lines:

- Replace old banjo bolt on pressure line with new one fitted with check valve.
- Use new sealing O-rings.

Tightening torque	
Steering fluid line banjo bolt to steering pump (M14 or M16, replace sealing O-rings)	33 - 35 Nm (24 - 25 ft-lb)

– Remainder of installation is reverse of removal. Make sure fluid lines and electrical harnesses are routed as before and connected free of stress. Keep in mind the following:

- Raise front subframe into place carefully, watching to make sure fluid lines and electrical harnesses are not crushed.
- Use new fasteners as indicated in torque specifications table.
- Fill and bleed steering and hydraulic system with correct fluid. See **Power steering system, bleeding and filling** in this repair group.
- Check fluid connections for leaks.
- Perform wheel alignment.
- Carry out steering angle sensor adjustment using BMW scan tool.
- If necessary, adjust active steering. See **Active steering, adjusting** in this repair group.
- If necessary, commission active stabilizer system. See **310 Front Suspension**.

Tightening torques	
Engine mount to engine bracket (replace M10 10.9 grade self-locking nuts with new)	56 Nm (41 ft-lb)
Front end reinforcement panel to subframe (replace M10 x 35 mm 10.9 grade ZNS bolts with new): • Stage 1 • Stage 2	 56 Nm (41 ft-lb) + additional 90°

BentleyPublishers.com
BentleyPublishers.com—All Rights Reserved

Tightening torques	
Front subframe to chassis (replace M12 x 1.5 10.9 grade bolts with new): • Stage 1 • Stage 2	 100 Nm (74 ft-lb) + additional 90°
Road wheel to hub	120 ± 10 Nm (89 ± 7 ft-lb)
Steering shaft spindle universal joint pinch bolt (recut pinch collar threads, replace M8 Torx bolt with new)	22 Nm (16 ft-lb)
Tie rod ball joint to steering knuckle (replace M14 x 1.5 ZNS3 nut with new)	80 Nm (59 ft-lb)

Steering rack, removing and installing (xDrive)

– Disconnect negative (-) battery cable and cover battery terminal to keep cable from accidentally contacting terminal.

> ***CAUTION—***
> * *Prior to disconnecting the battery, read the battery disconnection cautions given in* **001 Warnings and Cautions**.

– Drain power steering fluid reservoir using clean syringe. Do not reuse fluid.

– Raise front of car and remove front wheels.

> ***WARNING —***
> * *Make sure the car is firmly supported on jack stands designed for the purpose. Place jack stands underneath structural chassis points, not underneath suspension parts.*

– Remove splash shield and front end reinforcement panel from underneath engine. Also remove side splash shields at front suspension.

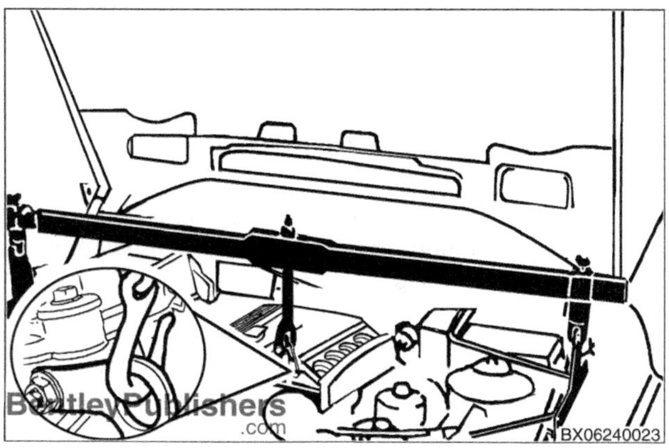

◄ Install engine support brace across engine bay.

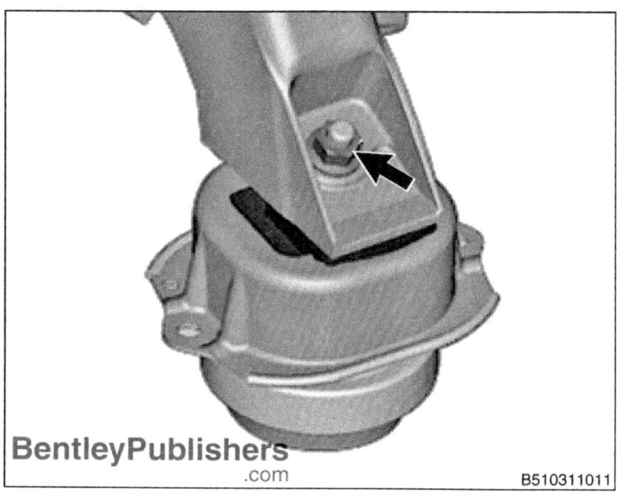

◄ Working underneath engine, remove right and left engine mount upper nuts (**arrow**). Raise engine slightly (¼ - ½ in) to release load on engine mounts.

– Detach left and right tie rod ends from steering knuckles. See **Tie rod end, replacing** in this repair group.

– Detach left and right control arms from steering knuckles. See **310 Front Suspension**.

– Detach left and right front stabilizer bar links from stabilizer bar. See **310 Front Suspension**.

BentleyPublishers.com
BentleyPublishers.com—All Rights Reserved

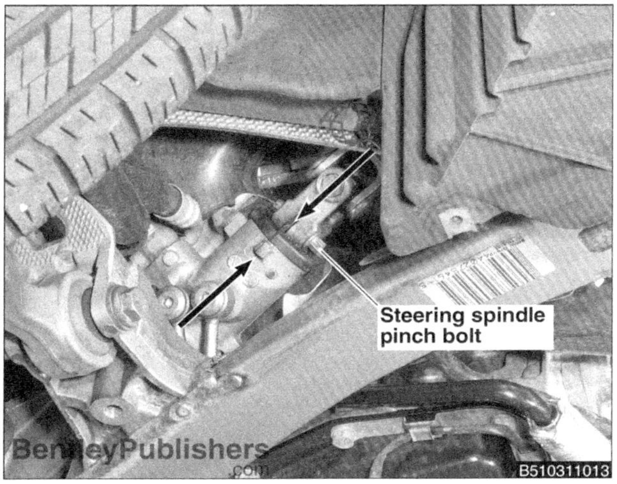

< Detach steering spindle from steering rack:

- Move steering wheel so that front wheels point straight ahead. Vehicle with steering lock: Remove ignition key and make sure steering locks in straight ahead position.
- With front wheels pointing straight ahead, make sure steering spindle pointer and steering rack pointer line up (**arrows**). Remove steering spindle pinch bolt and detach steering universal joint from steering rack shaft.

> ***CAUTION—***
> - *While steering spindle is detached from steering rack, make sure steering wheel is not rotated. Rotating the wheel may damage the airbag and horn contact spring.*

– If applicable, detach electrical connectors at front suspension ride height sensor and Servotronic sensor.

< Support front subframe with transmission jack. Remove subframe mounting bolts (**arrows**) and lower subframe until steering rack rests on front drive axle (about 100 mm or 4 in).

> ***CAUTION—***
> - *Do not bend or stretch fluid lines and hoses when lowering or raising subframe.*
> - *Make sure tension strut bushing cover is free to move.*
> - *N54 with automatic transmission: Disconnect transmission fluid cooler lines from subframe.*

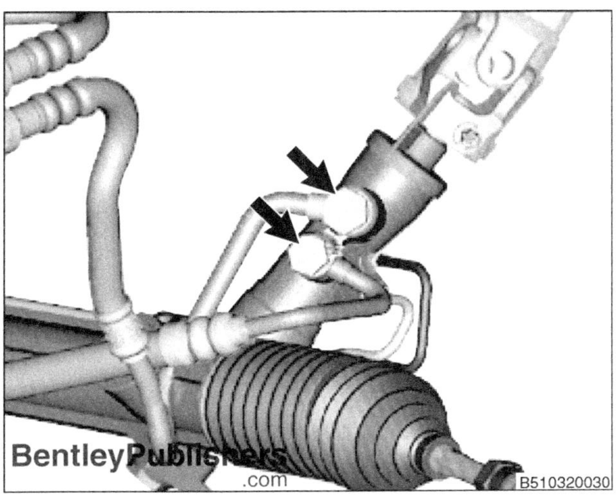

< Working at steering rack, remove banjo bolts (**arrows**) and disconnect fluid lines from rack gear box.

- Be prepared to catch dripping fluid. Plug open fluid ports.
- Discard sealing O-rings.

BentleyPublishers
BentleyPublishers.com—All Rights Reserved

- Remove steering rack mounting fasteners (**arrows**).
- Push front subframe forward and guide steering rack past drive axle.
- Lower subframe as much as possible without damaging hose and lines. Slide rack toward rear, rotate and remove from left side. Check for wiring, fluid lines or mechanical parts that might become snagged as rack is removed.
- When installing, use new mounting bolts and self-locking nuts.

Tightening torque	
Steering rack to front subframe (replace M12 10.9 grade fasteners with new)	120 Nm (89 ft-lb)

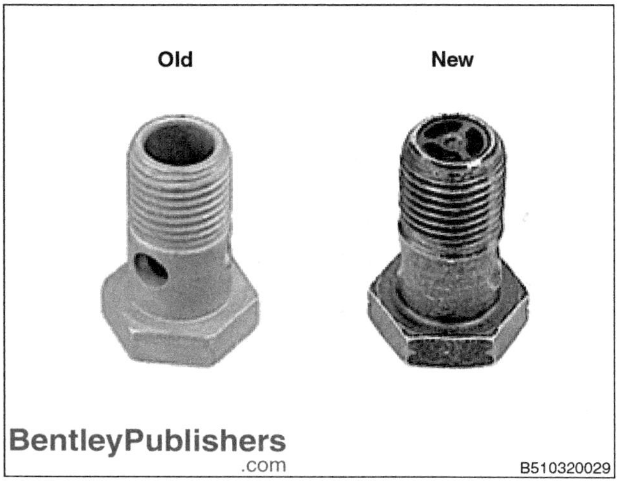

- When reattaching fluid lines:
  - Replace old banjo bolt on pressure line with new one fitted with check valve.
  - Use new sealing O-rings.

Tightening torque	
Steering fluid line banjo bolt to steering pump (M14 or M16, replace sealing O-rings)	33 - 35 Nm (24 - 25 ft-lb)

- Remainder of installation is reverse of removal. Make sure fluid lines and electrical harnesses are routed as before and connected free of stress. Keep in mind the following:
  - Raise front subframe into place carefully, watching to make sure fluid lines and electrical harnesses are not crushed.
  - Use new fasteners as indicated in torque specifications table.
  - Fill and bleed steering and hydraulic system with correct fluid. See **Power steering system, bleeding and filling** in this repair group.
  - Check fluid connections for leaks.
  - Perform wheel alignment.
  - Carry out steering angle sensor adjustment using BMW scan tool.

Tightening torques	
Control arm to steering knuckle (replace M14 x 1.5 self-locking flange nut with new)	165 Nm (122 ft-lb)
Engine mount to engine bracket (replace M10 10.9 grade self-locking nuts with new)	56 Nm (41 ft-lb)
Front end reinforcement panel to subframe (replace M10 x 35 mm 10.9 grade ZNS bolts with new): • Stage 1 • Stage 2	 56 Nm (41 ft-lb) + additional 90°
Front subframe to chassis (replace M12 x 1.5 10.9 grade bolts with new): • Stage 1 • Stage 2	 100 Nm (74 ft-lb) + additional 90°

BentleyPublishers.com
BentleyPublishers.com—All Rights Reserved

Tightening torques	
Road wheel to hub	120 ± 10 Nm (89 ± 7 ft-lb)
Stabilizer link to stabilizer bar (replace M10 ZNS3 nut with new)	65 Nm (48 ft-lb)
Steering shaft spindle universal joint pinch bolt (recut pinch collar threads, replace M8 Torx bolt with new)	22 Nm (16 ft-lb)
Tie rod ball joint to steering knuckle (replace M14 x 1.5 ZNS3 nut with new)	80 Nm (59 ft-lb)

Tie rod end, replacing

– Raise front of car. Remove road wheel.

> ***WARNING —***
> - *Make sure the car is firmly supported on jack stands designed for the purpose. Place jack stands underneath structural chassis points, not underneath suspension parts.*

– Remove splash shield underneath engine.

◄ Make reference measurement (**A**) of outer tie rod end to tie rod. Record measurement.

- Accurate measurement of tie rod end with reference to tie rod helps approximate correct wheel alignment when new parts are installed.

◄ Working at outer end of tie rod:

- Remove tie rod end nut (**A**) at steering knuckle. If necessary, counterhold tie rod ball joint shaft with Torx tool.
- Loosen tie rod adjustment lock nut (**B**).

320

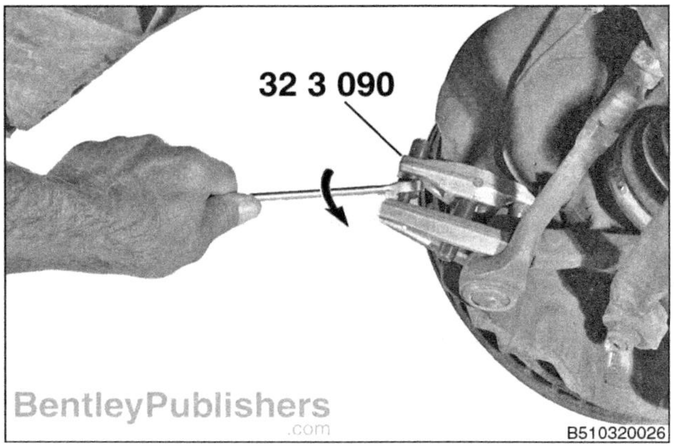

- Use BMW special tool 32 3 090 or equivalent to press tie rod end out of steering knuckle.
- Unscrew tie rod end from tie rod shaft.
- Check steering rack boot carefully for damage. Replace if necessary. See **Steering rack boot, replacing** in this repair group.
- Installation is reverse of removal.
  - Make sure threaded parts are clean.
  - Use anti-seize paste on inner tie rod threads.
  - Use new self-locking nuts where applicable.
  - Use previously made reference marks to set toe temporarily.

Tightening torques	
Lock nut to tie rod end (M16 x 1.5)	51 Nm (38 ft-lb)
Tie rod ball joint to steering knuckle (replace M14 x 1.5 ZNS3 nut with new)	80 Nm (59 ft-lb)

- After reassembling steering:
  - Have car aligned.
  - Adjust active steering (if applicable).
  - Carry out steering angle sensor adjustment using BMW scan tool.

Tie rod, replacing

- Raise front of car. Remove road wheel.

> ***WARNING —***
> - *Make sure the car is firmly supported on jack stands designed for the purpose. Place jack stands underneath structural chassis points, not underneath suspension parts.*

- Make length reference mark on tie rod, then remove outer tie rod end. See **Tie rod end, replacing**. Remove lock nut.
- Cut rack boot clamps and slide boot off. Check boot carefully for damage. Replace if necessary. See **Steering rack boot, replacing** in this repair group.

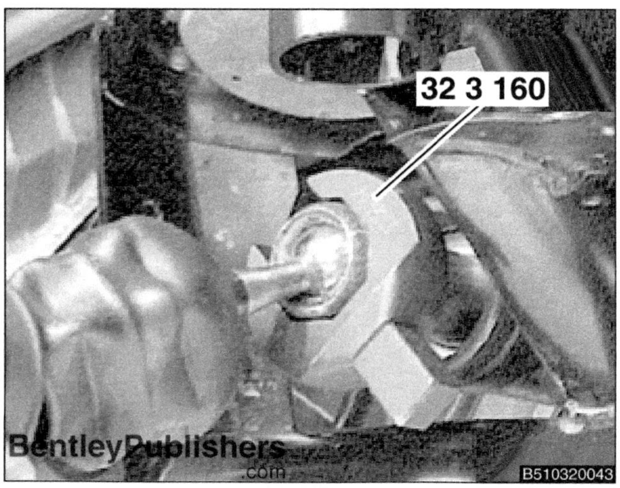

- Use BMW special tool 32 3 160 or equivalent 38 mm crowfoot wrench to loosen and remove inner tie rod shaft from steering rack.

> ***CAUTION—***
> - *To avoid damage to steering rack while removing tie rod, turn steering until end of rack is as far as possible inside rack housing.*

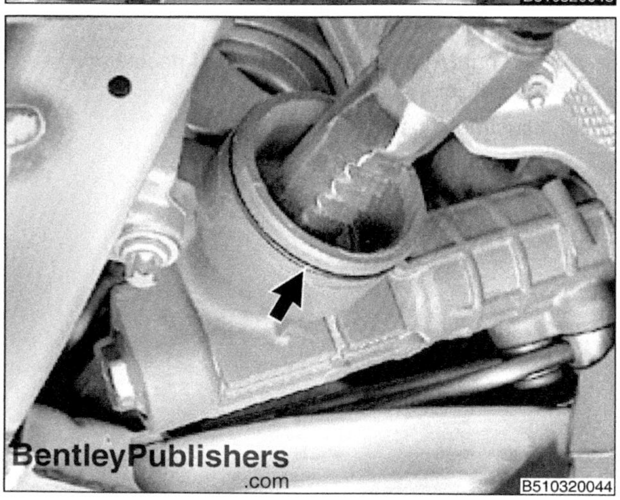

- Replace rack end sealing O-ring (**arrow**).
- When reassembling, grease tie rod taper so that rack boot support buffer or small end of rack boot slides on tie rod when tie rod is tightened, preventing rack boot from twisting.

BentleyPublishers.com
BentleyPublishers.com—All Rights Reserved

– Installation is reverse of removal, noting the following:
  - Make sure threaded parts are clean.
  - Use anti-seize paste on inner tie rod threads.
  - Use new self-locking nuts where applicable.
  - Use previously made reference marks to set toe temporarily.

Tightening torques	
Lock nut to tie rod end (M16 x 1.5)	51 Nm (38 ft-lb)
Tie rod ball joint to steering knuckle (replace M14 x 1.5 ZNS3 nut with new)	80 Nm (59 ft-lb)
Tie rod to steering rack	110 Nm (81 ft-lb)

– After reassembling steering:
  - Have car aligned.
  - Adjust active steering (if applicable).
  - Carry out steering angle sensor adjustment using BMW scan tool.

Steering rack boot, replacing

– Raise front of car. Remove road wheel.

> ***WARNING —***
> - *Make sure the car is firmly supported on jack stands designed for the purpose. Place jack stands underneath structural chassis points, not underneath suspension parts.*

– Remove splash shield underneath engine and side splash shields.

– Make length reference mark on tie rod, then remove outer tie rod end. See **Tie rod end, replacing**.

– Cut rack boot band clamps and slide off boot. Inspect boot for any sign of damage. Replace if necessary. New rack boot kit comes with new band clamps.

– Inspect exposed ends of steering rack shaft. If polished surface shows signs of corrosion, replace complete steering rack. See **Corrosion damage** in this repair group.

– Grease tie rod taper so that small end of rack boot slides on tie rod when tie rod is tightened, preventing rack boot from twisting.

– Slide new boot into position and replace band clamps. Reinstall tie rod end using previously made reference marks.

Tightening torques	
Tie rod end to tie rod (pinch bolt)	40 Nm (30 ft-lb)
Tie rod to steering knuckle (use new nut)	80 Nm (59 ft-lb)

– After reassembling steering:
  - Have car professionally aligned.
  - Carry out steering angle sensor adjustment using BMW scan tool.

Bentley Publishers.com
BentleyPublishers.com—All Rights Reserved

WHEEL ALIGNMENT

Proper handling, stability and driving ease depend upon the correct alignment of all four wheels. The front axle is aligned in relation to the rear axle, then the front wheels are aligned in relation to one another. This is known as a four-wheel or thrust-axis alignment.

BMW vehicles use a sophisticated multi-link suspension at the front and rear of the car. Proper alignment requires computerized alignment equipment.

For front and rear alignment specifications, see **Alignment specifications** in this repair group.

Preparing for alignment

The following conditions are necessary before wheel alignment:

- Correct wheels and identical tires in good condition and inflated correctly. Tire tread depth for each axle may differ left to right by max. 1 - 2 mm (0.04 - 0.08 in).
- Steering and suspension parts and bushings undamaged and showing no signs of abnormal wear.
- Wheel bearings in good condition.
- Ride height in accordance with specifications. See **Ride height** in this repair group.
- Vehicle with electronic height control (EHC): Remove EHC fuse to prevent automatic height adjustment. See **ECL Electrical Component Locations**.
- Vehicle with active steering: Remove active steering fuse to deactivate active steering. See **ECL Electrical Component Locations**.
- Vehicle in normal loaded position. See **Normal loaded position** in this repair group.

Normal loaded position

BMW defines vehicle normal loaded position as follows.

Normal loaded position	
Each front seat, seats in central position	68 Kg (150 lb)
Center of rear seat	68 Kg (150 lb)
Cargo compartment	21 Kg (46 lb)
Fuel tank	Full

Ride height

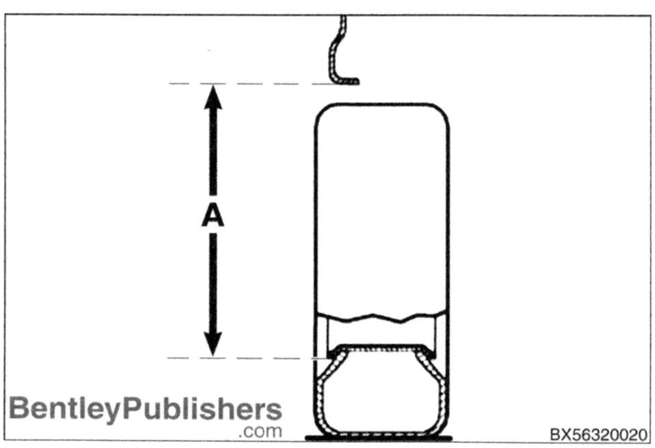

Measure ride height (**A**) from center of fender arch to bottom of wheel rim.

- Car in normal loaded position on the ground. See **Normal loaded position** in this repair group.
- Specified tires and wheels, correct tire pressure, even tire wear.
- Correct wheel bearing play.

– If ride height is outside specifications listed, install new springs. Suspension spring removal and installation is covered in **310 Front Suspension** and **330 Rear Suspension**.

BentleyPublishers.com
BentleyPublishers.com—All Rights Reserved

Table a. Front ride height (measurement A)

Wheel size	Ride height in mm (in)					
	Normal suspension	**Sport suspension**	**Rough road package**	**Long version**	**Security version**	**xDrive (E60, E61)**
16 inch	589 (23.189)	574 (22.5984)	607 (23.8976)			
17 inch	605 (23.8189)	589 (23.189)	622 (24.4882)	610 (24.0157)	605 (23.8189)	619 (24.3701)
18 inch	617 (24.2913)	602 (23.7008)	635 (25)		617 (24.2913)	632 (24.8819)
19 inch	630 (24.8031)	614 (24.5669)	648 (25.5118)			
Maximum combined deviation from specification: 10 mm (0.4 in) Maximum difference between left and right: 10 mm (0.4 in)						

Table b. Rear ride height, Sedan (E60) (measurement A)

Wheel size	Ride height in mm (in)					
	Normal suspension	**Sport suspension**	**Rough road package**	**Long version**	**Security version**	**xDrive**
16 inch	573 (22.5591)	558 (21.9685)	589 (23.189)			
17 inch	588 (23.1496)	573 (22.5591)	604 (23.7795)	593 (23.3465)	588 (23.1496)	599 (23.5827)
18 inch	601 (23.6614)	585 (23.0315)	617 (24.2913)		600 (23.622)	611 (24.0551)
19 inch	613 (24.1339)	598 (23.5433)	629 (24.7638)		612 (24.0945)	
Maximum combined deviation from specification: 10 mm (0.4 in) Maximum difference between left and right: 10 mm (0.4 in)						

Table c. Rear ride height, Sports Wagon (E61) (measurement A)

Wheel size	Ride height in mm (in)		
	Normal suspension	**Sport suspension**	**xDrive**
16 inch	588 (23.1496)	578 (22.7559)	
17 inch	603 (23.7402)	593 (23.3465)	603 (23.7402)
18 inch	615 (24.2126)	605 (23.8189)	616 (24.252)
Maximum combined deviation from specification: 10 mm (0.4 in) Maximum difference between left and right: 10 mm (0.4 in)			

Front toe, setting

Set front toe before adjusting camber. Camber and toe influence each other. Toe is the difference in the distance between the front of the wheels and the rear of the wheels. It is adjusted by altering the length of the tie rods.

Make toe adjustments with vehicle in normal loaded position. See **Normal loaded position** in this repair group.

– Center steering wheel. Make sure front wheels are pointed straight ahead.

BentleyPublishers.com
BentleyPublishers.com—All Rights Reserved

– Raise car and support safely.

> ***WARNING —***
> - *Make sure the car is firmly supported on jack stands designed for the purpose. Place jack stands underneath structural chassis points, not underneath suspension parts.*

– Remove splash shield and front end reinforcement panel from under engine.

– Remove side suspension splash shields.

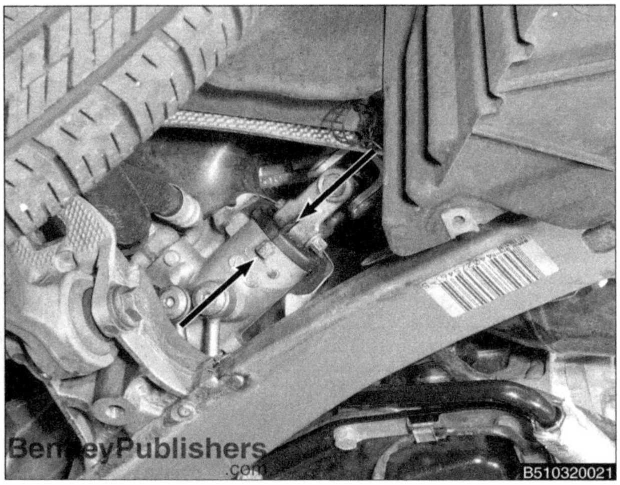

< Working underneath vehicle, make sure steering spindle pointer and steering rack pointer line up (**arrows**).

– Use steering wheel holding tool to lock steering wheel in centered position.

– Working underneath vehicle, clean threads on tie rod shaft.

< Counterhold tie rod end and loosen tie rod adjuster lock nut (**A**).
- To adjust toe, counterhold tie rod end and turn inner tie rod shaft (**curved arrow**) to change length.
- To keep steering wheel centered, adjust both tie rods equal amounts.
- Loosen or remove rack boot small end clamp (**B**) so that rack boot moves freely on tie rod shaft and does not become twisted. .

Tightening torques	
Lock nut to tie rod end (M16 x 1.5)	51 Nm (38 ft-lb)

BentleyPublishers.com
BentleyPublishers.com—All Rights Reserved

Front camber, adjusting

Set front toe before adjusting camber. Camber and toe influence each other. Camber is the deviation of each wheel from vertical. If necessary, adjust camber by removing front strut top centering pin.

> ***CAUTION—***
> - *Do not attempt to straighten out accident-caused changes in axle geometry by adjusting the camber.*

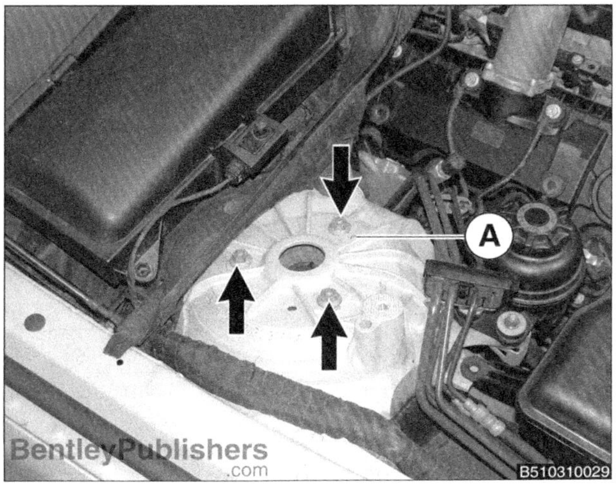

- Working at strut top bearing:
  - Remove centering pin (**A**).
  - Loosen strut top bearing mounting nuts (**arrows**) 1 - 1½ turns.

– Raise vehicle until wheels are off ground. Support vehicle safely.

> ***WARNING —***
> - *Make sure the vehicle is stable and well supported at all times. Use a professional automotive lift or jack stands designed for the purpose. A floor jack is not adequate support.*

– Working below vehicle, use compressed air to clean off debris in wheel housing near strut top bearing.

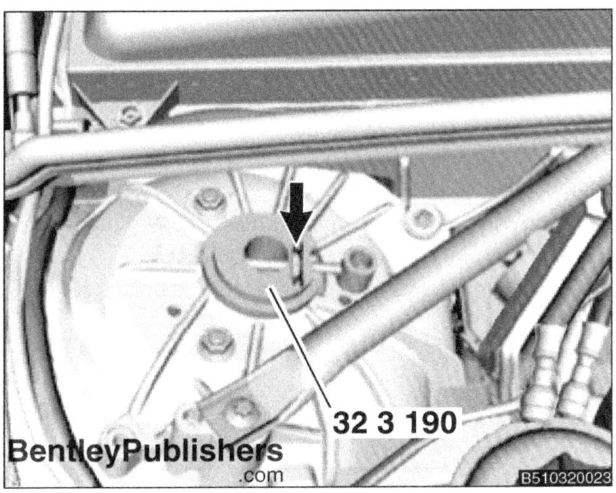

- Fit BMW special tool 32 3 190 over inboard strut top nut.
  - Turn knurled nut on tool (**arrow**) to adjust camber to specified value.
  - Replace outboard strut top mounting nuts with new, then tighten down.

– Remove special tool and replace inboard strut top mounting nut with new, then tighten down.

Tightening torque	
Strut top bearing to chassis (replace M8 shoulder nuts with new)	34 Nm (25 ft-lb)

– Check directional stability of vehicle. If necessary, reset toe. See **Front toe, setting**.

– Vehicle with active steering: Use BMW scan tool to zero steering angle sensor and commission active steering.

320

BentleyPublishers.com—All Rights Reserved

Rear camber, adjusting

Adjust rear camber before setting toe. A camber change means a toe change as well. Camber is the deviation of each wheel from vertical.

– Raise vehicle and support safely.

> ***WARNING —***
> *• Make sure the vehicle is stable and well supported at all times. Use a professional automotive lift or jack stands designed for the purpose. A floor jack is not adequate support.*

◄ Working at swing arm underneath rear suspension:

- Replace nut on swing arm eccentric bolt (**arrow**) with new. Tighten to 5 Nm (finger tight).
- Lower vehicle on alignment rack and use eccentric bolt to adjust camber (**curved arrow**).
- Tighten swing arm eccentric bolt to specification with vehicle in normal loaded position.

Tightening torque	
Swing arm eccentric fastener to rear subframe (replace nut, tighten with suspension loaded)	165 Nm (122 ft-lb)

Rear toe, setting

Adjust rear camber before setting toe. A camber change means a toe change as well. Toe is the difference in the distance between the front of the wheels and the rear of the wheels.

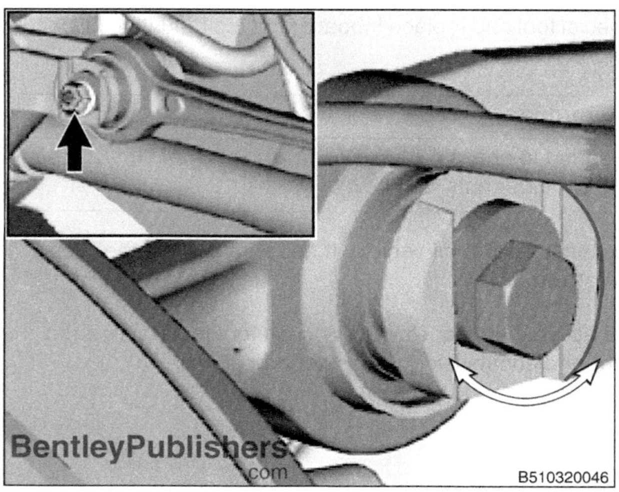

◄ Working at guide link:

- Replace nut (**arrow**) on eccentric bolt. Tighten to 5 Nm (finger tight).
- Lower vehicle on alignment rack and use eccentric bolt to set toe (**curved arrow**).
- Tighten guide link eccentric bolt to specification with vehicle in normal loaded position.

Tightening torque	
Guide link eccentric fastener to rear subframe (replace nut, tighten with suspension loaded)	65 Nm (48 ft-lb)

Caster

Front and rear caster are fixed and deviations are usually the result of worn or damaged suspension or body parts.

– Check front suspension arms and bushings for wear, damage and deformation if front caster problems are present.

– Check rear axle subframe and traction struts if rear caster problems are present.

BentleyPublishers.com
BentleyPublishers.com–All Rights Reserved

Alignment specifications

Table d. Alignment specifications, Sedan (E60)

	Specifications				
Parameter	**Standard suspension**	**Sport suspension**	**Rough road suspension**	**Long wheelbase**	**xDrive**
Front axle					
Total toe	0° 8' ± 12'	0° 8' ± 12'	0° 8' ± 12'	0° 12' ± 12'	0° 12' ± 12'
Adjustment total toe*	0° 8' ± 4'	0° 8' ± 4'	0° 8' ± 4'	0° 12' ± 4'	0° 12' ± 4'
Toe difference between right and left**	max 12'	max 12'	max 12'	max 12'	max 12'
Camber (difference between left / right, maximum 30')	-12' ± 30'	-30' ± 30'	-12' ± 30'	-05' ± 30'	-20' ± 30'
Adjustment camber*	-12' ± 25'	-30' ± 25'	-12' ± 25'	-05' ± 25'	-20' ± 25'
Toe angle difference with 20° lock on inside wheel (difference between left / right, maximum 30')	1° 41' ± 30'	1° 47' ± 30'	1° 43' ± 30'	1° 42' ± 30'	1° 53' ± 30'
Front caster (difference between left / right maximum 30')					
Front wheel displacement	0° ± 15'	0° ± 15'	0° ± 15'	0° ± 15'	0° ± 15'
Maximum wheel lock					
Inside wheel (approx.)	43° 22'	43° 01'	43° 58'	43° 22'	39° 42'
Outside wheel (approx.)	34° 00'	33° 38'	34° 22'	34° 06'	31° 38'
Rear axle					
Total toe	0° 18' ± 12'	0° 18' ± 12'	0° 18' ± 12'	0° 18' ± 12'	0° 18' ± 12'
Adjustment total toe*	0° 18' ± 4'	0° 18' ± 4'	0° 18' ± 4'	0° 18' ± 4'	0° 18' ± 4'
Camber (difference between left / right maximum 30')	-2° 00' ± 25'	-2° 00' ± 25'	-2° 00' ± 25'	-2° 00' ± 25'	-2° 00' ± 25'
Adjustment camber*	-2° 00' ± 5'	-2° 00' ± 5'	-2° 00' ± 5'	-2° 00' ± 5'	-2° 00' ± 5'
Geometrical axis deviation	0° ± 12'	0° ± 12'	0° ± 12'	0° ± 12'	0° ± 12'

* To minimize inaccuracy, use a narrower tolerance
** Front axle toe difference = criterion for steering wheel inclination

BentleyPublishers.com—All Rights Reserved

Table e. Alignment specifications, Sports Wagon (E61)

Parameter	Specifications		
	Standard suspension	**Sport suspension**	**xDrive**
Front axle			
Total toe	0° 12' ± 10'	0° 12' ± 10'	0° 12' ± 12'
Adjustment total toe*	0° 12' ± 4'	0° 12' ± 4'	0° 12' ± 4'
Toe difference between right and left**	max 12'	max 12'	max 12'
Camber (difference between left / right, maximum 30')	-12' ± 30'	-30' ± 30'	-20' ± 30'
Adjustment camber*	-12' ± 25'	-30' ± 25'	-20' ± 25'
Toe angle difference with 20° lock on inside wheel (difference between left / right, maximum 30')	1° 41' ± 30'	1° 47' ± 30'	1° 53' ± 30'
Front caster (difference between left / right maximum 30')			
Front wheel displacement	0° ± 15'	0° ± 15'	0° ± 15'
Maximum wheel lock			
Inside wheel (approx.)	43° 22'	43° 01'	39° 42'
Outside wheel (approx.)	34° 00'	33° 38'	31° 38'
Rear axle			
Total toe	0° 18' ± 12'	0° 18' ± 12'	0° 18' ± 12'
Adjustment total toe*	0° 18' ± 4'	0° 18' ± 4'	0° 18' ± 4'
Camber (difference between left / right maximum 30')	-2° 00' ± 25'	-2° 00' ± 25'	-2° 00' ± 25'
Adjustment camber*	-2° 00' ± 5'	-2° 00' ± 5'	-2° 00' ± 5'
Geometrical axis deviation	0° ± 12'	0° ± 12'	0° ± 12'

* To minimize inaccuracy, use a narrower tolerance
** Front axle toe difference = criterion for steering wheel inclination

BentleyPublishers.com
BentleyPublishers.com—All Rights Reserved

330 Rear Suspension

General

This repair group covers removal and replacement of rear suspension components.

See also:

- **300 Suspension, Steering and Brakes–General** for a general description of the 5 Series suspension.
- **310 Front Suspension** for front suspension component replacement and active stabilizer system.
- **320 Steering and Wheel Alignment** for ride height and alignment specifications.
- **331 Rear Axle Differential** for rear drive axle and differential repairs.

BentleyPublishers.com
BentleyPublishers.com–All Rights Reserved

Special service tools are required for some of the work described in this repair group. Most of these tools are specialized press jigs and pullers that may be replaced by standard pullers of various sizes. Read the procedures through before beginning any job.

Rear suspension description

The rear suspension uses a subframe as a rigid mounting platform for the differential (also called final drive) and rear suspension components. The differential is mounted to the subframe and the subframe is bolted to the body through rubber bushings which help isolate drivetrain noise and vibration.

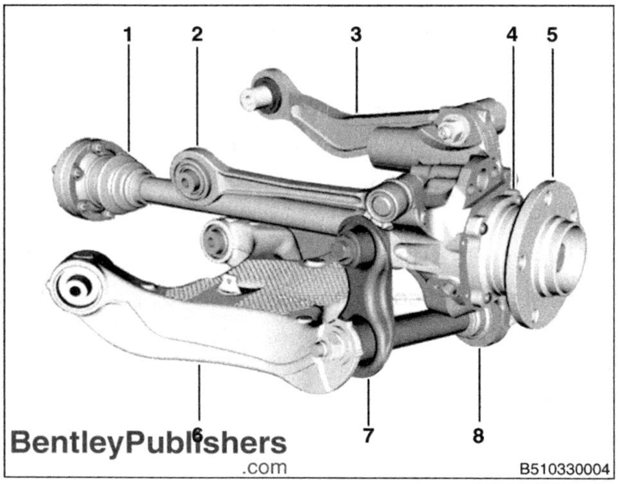

In the 5 Series multi-link rear suspension, four control arms (or links) on each side locate the rear wheels. The lower control arm, called the swing arm, is attached to the wheel bearing carrier via a ball joint at rear and an articulating integral link at front.

1. Rear drive axle
2. Guide arm (traction strut)
3. Control arm
4. Wheel bearing carrier
5. Wheel hub
6. Swing arm
7. Integral link
8. Ball joint

Shock absorber and spring applications differ on Sedan and Sports Wagon models:

- Sedan (E60) rear springs and struts are combined in a conventional MacPherson strut assembly.
- Sports Wagon (E61) gains larger and flatter cargo space and better load capacity by being equipped with separate air springs and shock absorbers. Air spring heights are determined by the electronic height control (EHC) system.

BentleyPublishers.com
BentleyPublishers.com—All Rights Reserved

Sedan rear suspension component

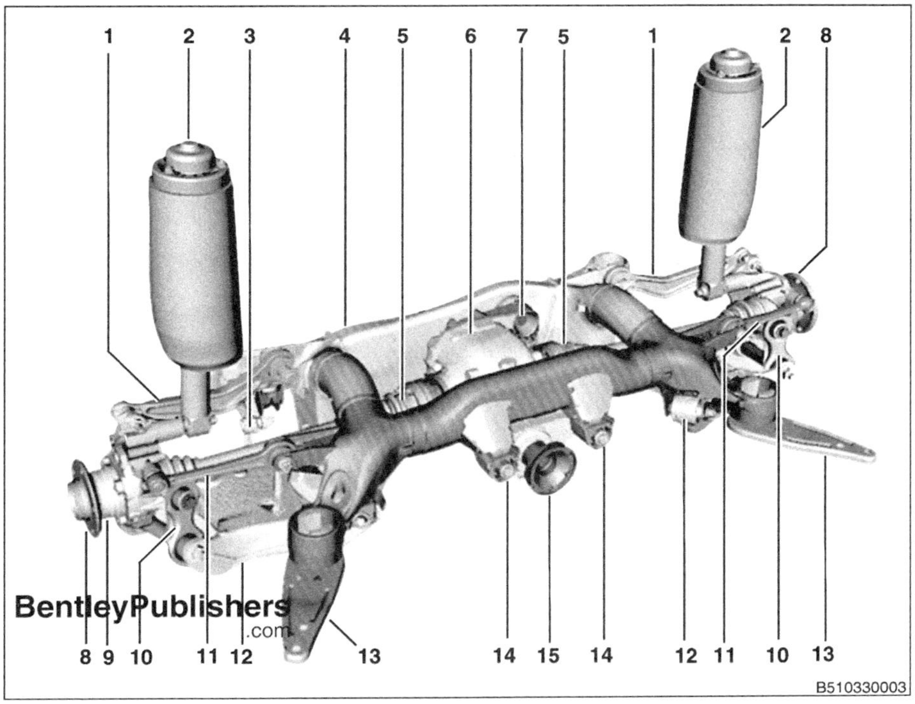

1. Control arm
2. Strut and coil spring assembly
3. Stabilizer bar link
4. Rear subframe
5. Rear drive axle
6. Rear differential
7. Differential rear mount bushing
8. Wheel hub
9. Wheel bearing carrier
10. Integral link
11. Guide arm (traction strut)
12. Swing arm
13. Thrust rod
14. Differential front mount bushings
15. Differential input flange

Sports Wagon rear suspension components

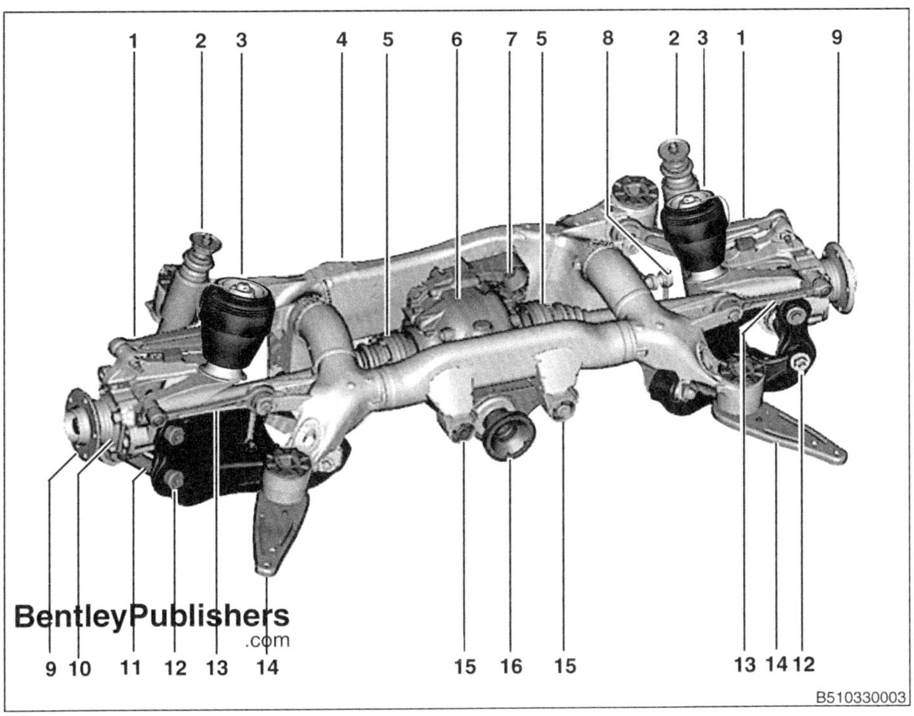

1. Control arm
2. Shock absorber
3. Air spring
4. Rear subframe
5. Rear drive axle
6. Rear differential
7. Differential rear mount bushing
8. Stabilizer bar link
9. Wheel hub
10. Wheel bearing carrier
11. Swing arm
12. Integral link
13. Guide arm (traction strut)
14. Thrust rod
15. Differential front mount bushings
16. Differential input flange

BentleyPublishers.com
BentleyPublishers.com–All Rights Reserved

Warnings and Cautions

WARNING —
- *Do not reuse self-locking fasteners.*
- *Wear protective gloves and eye wear when working with hydraulic fluid.*
- *Drain the EHC (self-leveling suspension) system before carrying out any work on the air supply compressor, air pipes or air springs. See* **EHC, disabling, deflating, refilling** *in this repair group.*

CAUTION—
- *Due to the chemical and corrosion characteristics of aluminum, follow these precautions when working with aluminum components:*
  —Do not bring into contact with battery acid.
  —Do not use wire brushes with brass or iron bristles.
  —Do not expose to flying sparks.
  —Do not subject to steel welding splashes.
  —Do not expose to temperatures over 80°C (176°F), even for short periods. Paint shop temperatures are not a problem.
- *When filling or bleeding active stabilizer bar hydraulic system, do not allow contamination of fluid or reservoir.*
- *Do not allow the vehicle to stand on its wheels when the EHC system is disabled or drained. Damage to air springs may result.*
- *When disconnecting and connecting EHC air lines, do not allow contamination to enter the lines.*

ELECTRONIC HEIGHT CONTROL (EHC)

EHC description

Sports Wagon (E61) rear suspension is equipped with air springs. Electronic control of air spring height allows the rear of the wagon to remain at a relatively stable height, thereby maintaining constant suspension geometry and handling stability. The electronic height control (EHC) system is also referred to as self-leveling suspension.

Air springs and shock absorbers are supported independently on the floor pan of the vehicle, allowing for flatter floor and larger cargo space.

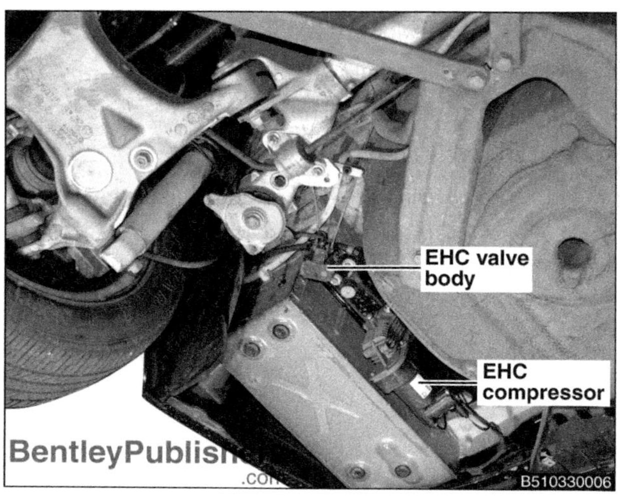

The EHC air supply system is underneath the rear of the vehicle between the spare wheel well and battery box.

BentleyPublishers.com
BentleyPublishers.com—All Rights Reserved

EHC compressor filtered air intake is at right rear of vehicle behind cargo compartment paneling.

EHC power supply

The EHC power supply fuse is in the rear fuse panel at right side of cargo compartment behind trim.

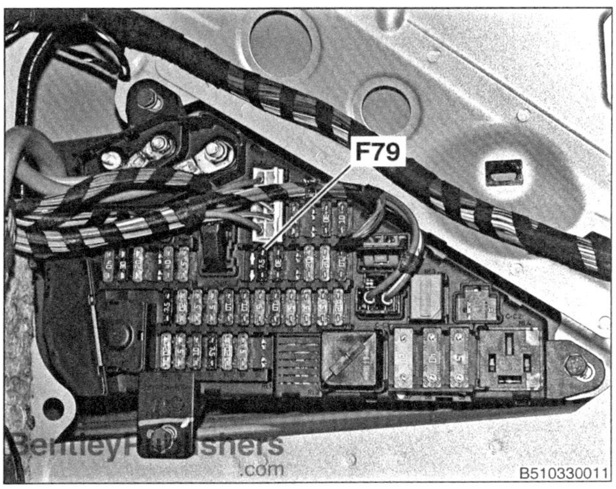

F79, 7.5 A (to 09 / 2005)

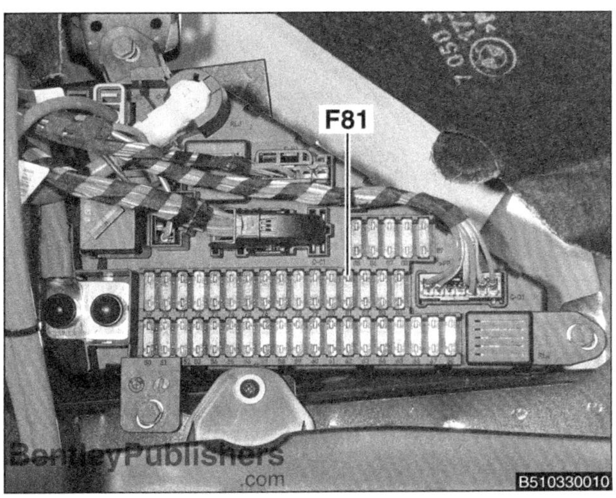

F81, 7.5 A (from 09 / 2005)

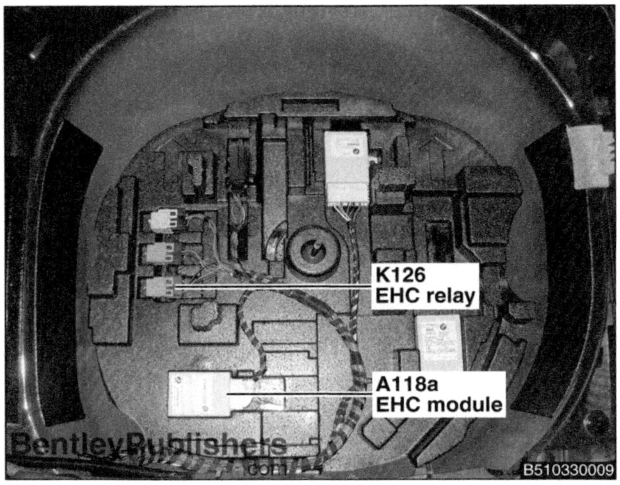

EHC control electronics

EHC power relay and control module are in the cargo compartment, underneath the spare tire.

◀ **A118a** EHC control module
K126 EHC relay

EHC, disabling, deflating, refilling

– Switch ignition OFF and remove key.

– Raise vehicle and support safely.

> ***WARNING —***
> • *Make sure the vehicle is stable and well supported at all times. Use a professional automotive lift or jack stands designed for the purpose. A floor jack is not adequate support.*

– Remove EHC power supply fuse in rear fuse panel. See **EHC power supply** in this repair group.

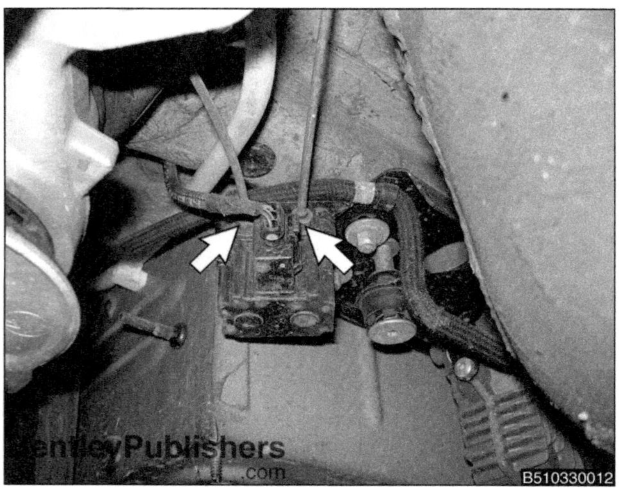

◀ Working underneath right rear of vehicle, remove plastic cover to gain access to air compressor and valve body.
- Label air lines (**arrows**), then loosen union nuts until hiss of escaping air can be heard.
- Fully release lines once sound of escaping air stops.

> ***CAUTION—***
> • *Plug open air lines and ports.*
> • *Do not lower vehicle on its wheels while air springs are depressurized.*

– When installing:
- Check union nuts for contamination. Clean if necessary.
- Start union nut by hand carefully to avoid damaging plastic threads on valve body.

Tightening torque	
EHC air line to valve body (4 mm line)	2 Nm (18 in-lb)

– Reinstall EHC power supply fuse.

– Connect BMW scan tool.

– Switch ignition ON. Use scan tool “control unit functions” menu to activate EHC and inflate rear air springs.

BentleyPublishers.com
BentleyPublishers.com—All Rights Reserved

EHC faults

The driver is alerted to EHC system faults by a warning icon and a Check Control warning triangle in instrument cluster.

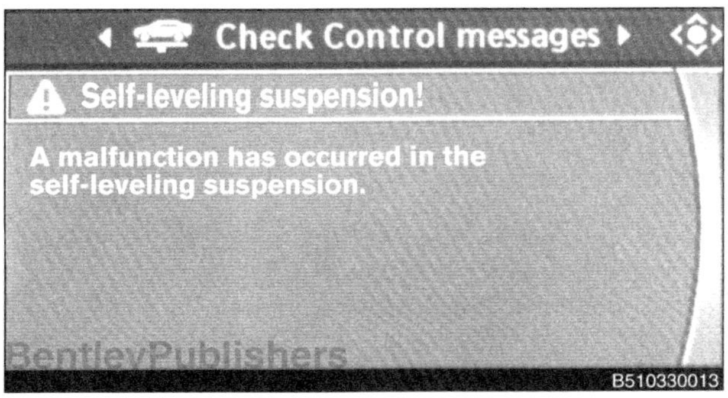

Check Control in center dashboard (iDrive) displays a self-leveling (EHC) malfunction message.

Stop and check the vehicle. If it is considerably lower at the rear than at the front, or on one side, go to the nearest BMW center or other qualified repair shop. Drive with appropriate caution in the meantime. The vehicle has reduced ground clearance and driving comfort may be noticeably reduced.

Even if the position of the vehicle is normal, make sure the vehicle is checked by a qualified repair shop.

Rear Strut Assembly (Sedan)

Sedan rear shock absorbers are MacPherson struts. The strut is a major component of the suspension and supports the spring. Most strut assembly components are available as replacement parts.

If a shock absorber or strut is defective on one side of the vehicle, replace both shock absorbers on that axle. If a coil spring is broken due to corrosion, replace both coil springs on the axle.

Rear strut, upper strut mount or coil spring replacement is a two-step procedure:

- Removal of strut assembly from vehicle.
- Disassembly and replacement of components on work bench.

Special tools are necessary for compressing the coil spring safely. Be sure to read the complete procedure before starting work.

Rear strut assembly, removing and installing (Sedan)

– Raise car and remove rear wheel.

> ***WARNING —***
> - *Make sure the car is firmly supported on jack stands designed for the purpose. Place jack stands underneath structural chassis points, not underneath suspension parts.*

– Detach stabilizer link from swing arm.

– Support rear wheel bearing carrier with suitable jack.

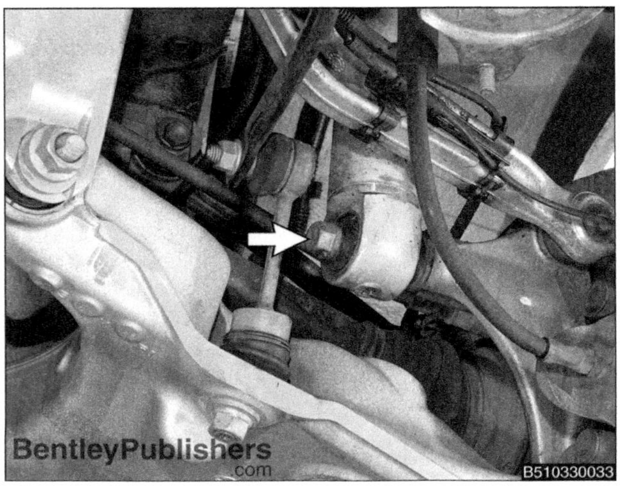

◄ Remove strut assembly lower mounting bolt (**arrow**) and detach strut base from wheel bearing carrier. Tie strut up with stiff wire to prevent it from falling out.

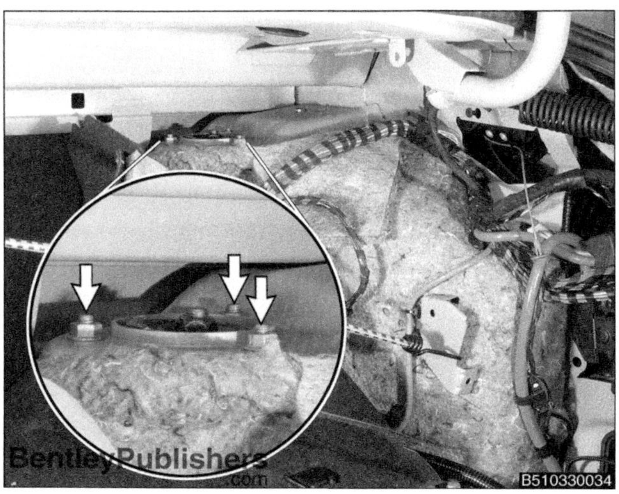

◄ Working in trunk:

- Peel back trunk side trim covering wheel well.
- Remove sound absorbing materials to gain access to strut top mount nuts (**arrows**).
- Remove nuts.

– Working in wheel arch opening underneath vehicle:

- Push wheel bearing carrier down
- Guide strut assembly down between guide arm and control arm and lift out of wheel housing.

– If replacing strut, coil spring or upper strut mount, disassemble and reassemble strut assembly on bench. See **Rear strut assembly, disassembling and assembling** in this repair group.

BentleyPublishers.com
BentleyPublishers.com—All Rights Reserved

– Installation is reverse of removal. Note the following:
- Replace upper strut mount nuts and lower strut mount bolt with new.
- Make sure that strut lower bushing mates squarely with bore in wheel bearing housing before installing lower bolt. Do not use bolt to force strut bushing into alignment.

Tightening torques	
Road wheel to hub	120 ± 10 Nm (89 ± 7 ft-lb)
Strut lower mount to wheel bearing carrier (replace M14 x 105 mm ZNS3 bolt with new)	165 Nm (122 ft-lb)
Strut upper mount to chassis (replace M8 ZNS3 self-locking nuts with new)	28 Nm (21 ft-lb)

Rear strut assembly components (Sedan)

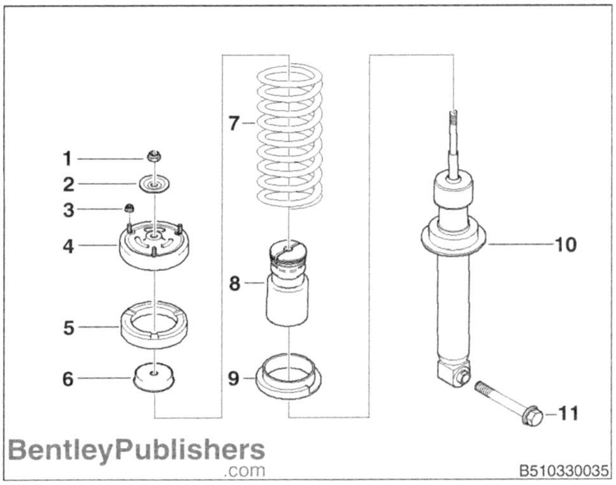

1. M12 x 1.5 self-locking nut
   - Replace with new
   - Tighten to 27 Nm (20 ft-lb)
2. Washer plate
3. M8 ZNS3 self-locking nut
   - Replace with new
   - Tighten to 28 Nm (21 ft-lb)
4. Upper strut mount
5. Upper spring pad
6. Support
7. Coil spring
8. Rubber stop
9. Lower spring pad
10. Strut with lower spring seat
11. M14 x 105 mm bolt
    - Replace with new
    - Tighten to 165 Nm (122 ft-lb)

BentleyPublishers.com
BentleyPublishers.com–All Rights Reserved

Rear strut assembly, disassembling and assembling (Sedan)

If replacing strut, upper strut mount or coil spring, first remove the strut assembly from the vehicle. See **Rear strut assembly, removing and installing (Sedan)** in this repair group.

BMW recommends the use of BMW spring compressor (special tool 31 3 340). A number of different spring compressor tools are available. Photos in this section illustrate the function of one such tool in use.

WARNING —
- *Do not attempt to disassemble the strut assembly without a spring compressor designed specifically for this job.*
- *Prior to each use, check special tool for functionality.*
- *Do not use a damaged tool.*
- *Do not make any modifications to tool.*
- *Use correct size spring retainers when compressing coil spring.*
- *When assembling BMW spring compressor (special tool 31 3 340), make sure spring retainer plates are felt and heard snapping into place. Check seating of spring retainers carefully.*

– Clamp spring compressor in shop vise.

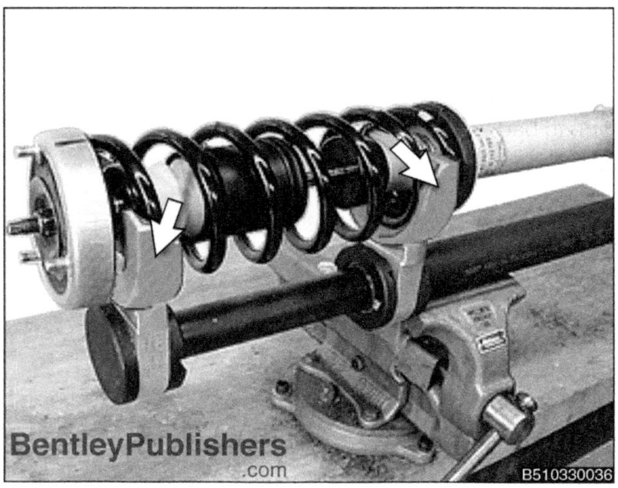

◄ Clean coil spring to remove coarse dirt, then position between compressor spring retainers (**arrows**) as shown. Compress spring.

WARNING —
- *Make sure that when tensioned, the spring coil rests completely in the spring holder recess.*
- *Do not tighten or loosen spring compressor with an impact tool.*
- *Tighten down the coil spring until stress on the upper strut mount is relieved.*
- *Only loosen upper strut nut if spring coils are completely inserted in the spring holder grooves. If necessary, loosen compressor, reposition and recompress.*

◄ Counterhold strut piston shaft with small wrench and remove upper strut nut (**arrow**).

– Remove upper strut mount and spring pad.

– If a new coil spring is being installed, relieve tension on spring compressor and remove coil spring.

– Check rubber stop and spring pads. Replace as necessary.

– Replace strut, upper strut mount or spring, as needed. If one spring is corroded or damaged, replace both springs as a matched pair.

BentleyPublishers.com
BentleyPublishers.com—All Rights Reserved

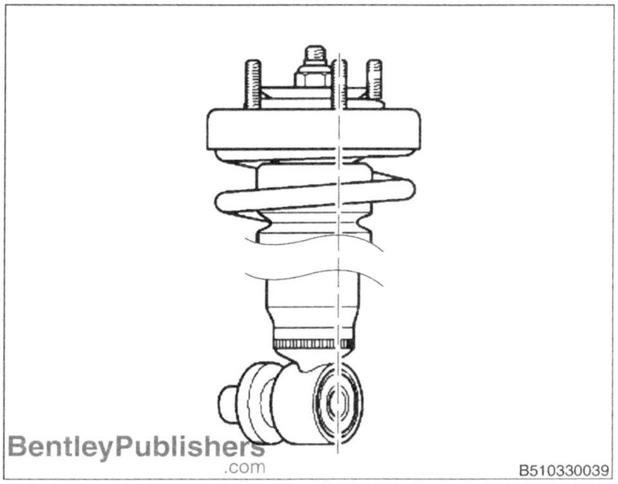

Assemble strut with coil spring and tighten down upper mount using new nut.

- Upper mount alignment: One stud on upper mount lines up with short end of strut bushing.

Tightening torque	
Strut shaft to strut top bearing (replace self-locking M12 x 1.5 nut with new)	27 Nm (20 ft-lb)

– Be sure coil spring ends line up correctly before releasing spring compressor. See following illustrations.

Upper coil spring end: Place spring end in spring pad stop (**inset**).

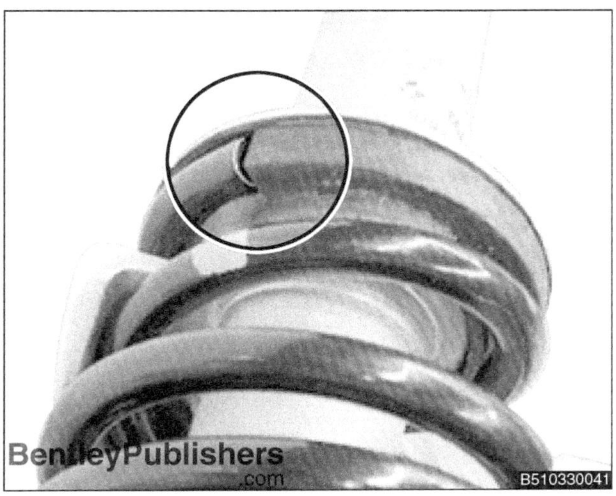

Lower coil spring end: Place spring end against spring pad stop (**inset**).

– Release spring compressor carefully and evenly, allowing spring to expand slowly.

– Reinstall strut assembly in vehicle.

Rear Shock Absorbers and Air Springs (Sports Wagon)

Sports Wagon models are equipped with rear air springs installed separately from the shock absorbers. See **Sports Wagon rear suspension components** in this repair group.

If a shock absorber is defective on one side of the vehicle, replace both shock absorbers on that axle.

Rear shock absorber components (Sports Wagon)

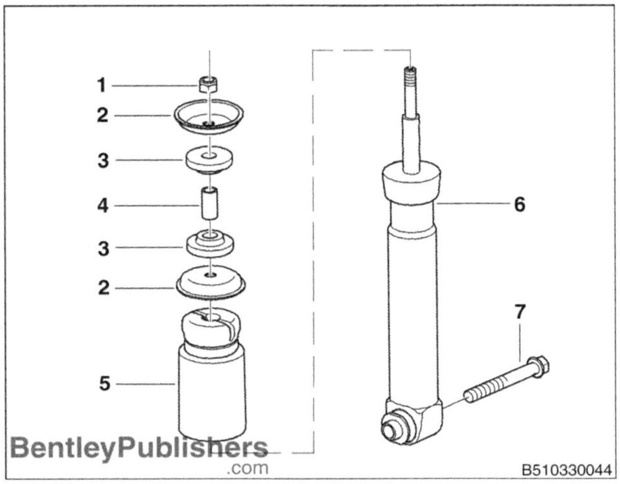

1. M12 x 1.5 self-locking nut
   - Replace with new
   - Tighten to 27 Nm (20 ft-lb)
2. Support washer
3. Support bushing
4. Spacer sleeve
5. Rubber bump stop and dust sleeve
6. Shock absorber
7. M14 x 105 mm bolt
   - Replace with new
   - Tighten to 165 Nm (122 ft-lb)

Rear shock absorber, removing and installing (Sports Wagon)

– Raise vehicle and support safely.

> ***WARNING —***
> - *Make sure the car is firmly supported on jack stands designed for the purpose. Place jack stands underneath structural chassis points, not underneath suspension parts.*

– Disable and deflate EHC system. See **EHC, disabling, deflating, refilling** in this repair group.

Remove shock absorber lower mounting bolt (**arrow**). Detach shock absorber from swing arm and tie securely to prevent it from falling out.

BentleyPublishers.com
BentleyPublishers.com—All Rights Reserved

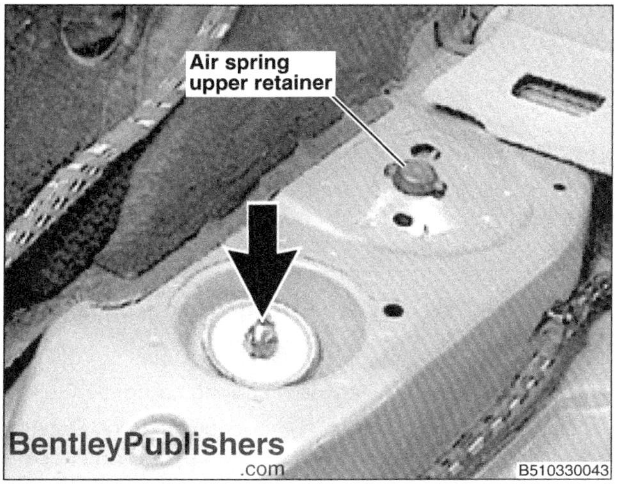

Working in cargo compartment:

- Remove floor covering and side plastic rails to gain access to shock absorber upper mountings.
- Unscrew upper shock absorber nut (**arrow**) and lift off upper support washer.

– Remove shock absorber from below vehicle.

– Inspect rubber bump stop, dust sleeve and support bushings for wear and damage. Replace as necessary.

– Transfer used and new parts to new shock absorber. Installation is reverse of removal. Note the following:

- Replace upper strut mount nut and lower strut mount bolt with new.
- Make sure that shock absorber lower bushing mates squarely with bore in swing arm before installing lower bolt. Do not use bolt to force shock absorber bushing into alignment.

Tightening torques	
Road wheel to hub	120 ± 10 Nm (89 ± 7 ft-lb)
Shock absorber lower bolt to swing arm (replace M14 x 105 mm ZNS3 bolt with new)	165 Nm (122 ft-lb)
Shock absorber upper mount to frame rail (replace M8 ZNS3 self-locking nut with new)	27 Nm (20 ft-lb)

– Reactivate and fill EHC system. See **EHC, disabling, deflating, refilling** in this repair group.

330

Air spring components (Sports Wagon)

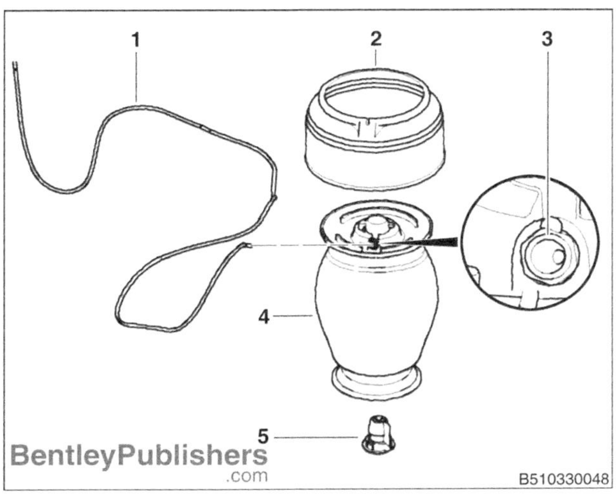

1. Air line
2. Protective boot
3. Union nut
   - Plastic air line connection
   - Tighten to 2 Nm (18 in-lb)
4. Air spring
5. Plastic retainer

BentleyPublishers.com
BentleyPublishers.com–All Rights Reserved

Air spring, removing and installing (Sports Wagon)

– Raise vehicle and remove rear wheel.

> ***WARNING —***
> *• Make sure the car is firmly supported on jack stands designed for the purpose. Place jack stands underneath structural chassis points, not underneath suspension parts.*

– Disable and deflate EHC system. See **EHC, disabling, deflating, refilling** in this repair group.

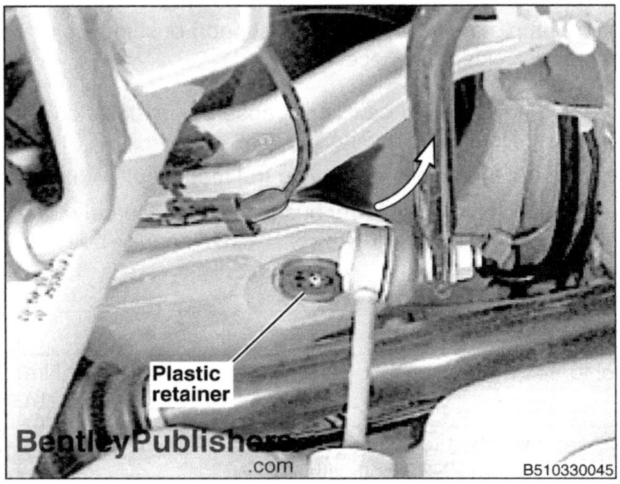

◄ Working at base of air spring above wheel bearing carrier:

- Tilt lower section of air spring (**arrow**) by hand toward front. Rotate counterclockwise 25° against resistance of lower plastic retainer to detach.
- Note that broken off upper plastic retainer on air spring is not a problem.

◄ Squeeze air spring (**arrows**) and tilt top outward.

> ***CAUTION—***
> *• Do not kink, twist or crush air line.*

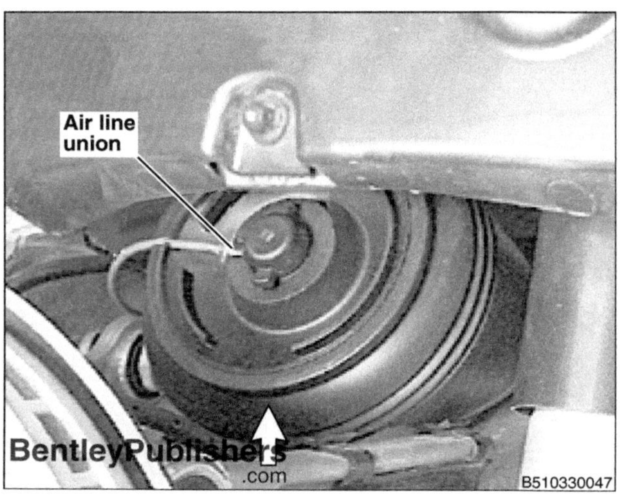

◄ Detach air line from air spring at union.

- Moisten air spring upper protective boot (**arrow**) with soapy solution, then remove air line, air line union and boot from spring.
- Plug or tape up open air ports immediately.
- Note that union nut stays on air line.
- Lift out air spring.

– Before reinstalling, check airline, union nuts and lower retainer for damage. Replace as necessary.

BentleyPublishers.com
BentleyPublishers.com—All Rights Reserved

– To reinstall, place air spring at its location and place protective boot over it. Clean air line union carefully, then screw in by hand to avoid damaging plastic threads.

Tightening torque	
Air line to air spring (4 mm line)	2 Nm (18 in-lb)

◄ Tip air spring into place and fit upper retainer in frame rail slot. Rotate spring approx. 25° to lock in place. This ensures alignment of air spring to lower retainer.

- Note that broken off upper plastic retainer on air spring is not a problem.
- To make sure air line does not slip out of protective boot groove, pull air line downward during installation.

> ***CAUTION—***
> *• Do not kink, twist or crush air line.*

– Align air spring to lower retainer and press down until retainer snaps into place. It is not necessary to replace retainer.

Tightening torque	
Road wheel to hub	120 ± 10 Nm (89 ± 7 ft-lb)

– Reactivate and fill EHC system. See **EHC, disabling, deflating, refilling** in this repair group.

Rear Suspension Arms, Subframe and Bushings

Rear suspension arm components

The diagram below illustrates Sedan rear control arms. Sports Wagon is similar.

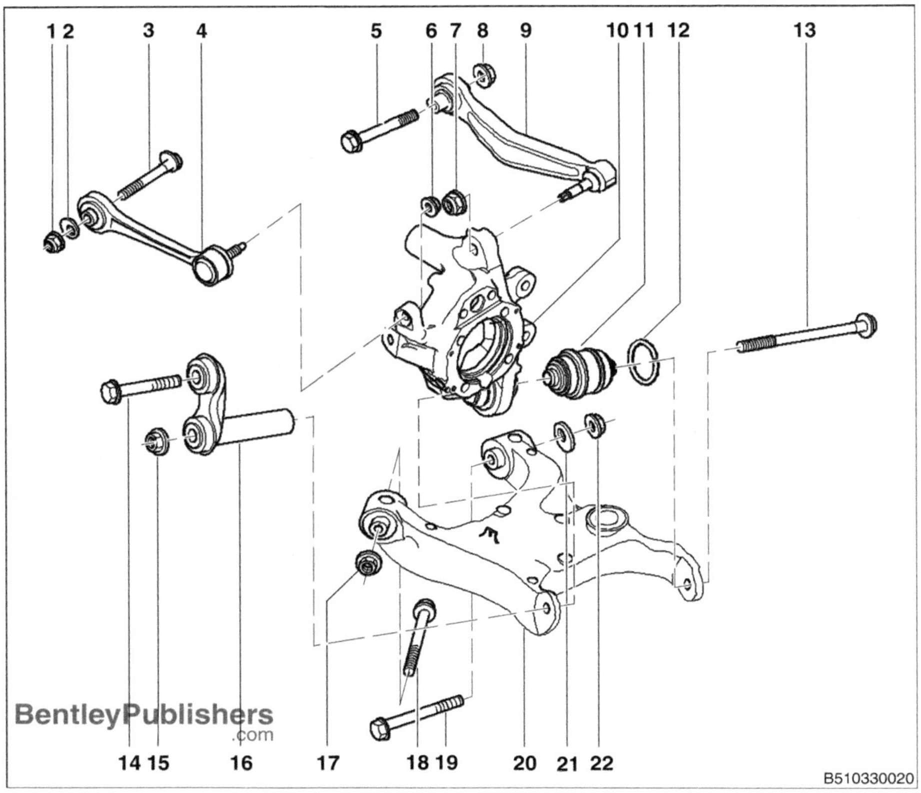

1. Self-locking M10 nut (replace)
   - Tighten with suspension loaded
   - Tighten to 65 Nm (48 ft-lb)
2. Eccentric washer
3. Eccentric M10 x 66 mm ZNS3 bolt
   - Adjusts rear toe
4. Guide arm (traction strut)
5. M12 1.5 x 92 mm ZNS3 bolt
6. Self-locking M10 nut (replace)
   - Tighten to 65 Nm (48 ft-lb)
7. Self-locking M14 x 1.5 nut (replace)
   - Tighten to 175 Nm (129 ft-lb)
8. Self-locking M12 x 1.5 nut (replace)
   - Tighten with suspension loaded
   - Tighten to 100 Nm (74 ft-lb)
9. Upper control arm
10. Wheel bearing carrier
11. Swing arm ball joint
12. Circlip
13. M16 x 275 mm ZNS3 bolt
14. M12 x 1.5 x 67 mm ZNS3 bolt
15. Self-locking M16 x 1.5 nut (replace)
   - Tighten with suspension loaded
   - Tighten to 240 Nm (177 ft-lb)
16. Integral link
17. Self-locking M12 x 1.5 nut (replace)
   - Tighten with suspension loaded
   - Tighten to 100 Nm (74 ft-lb)
18. M12 x 1.5 x 98 mm ZNS3 bolt
19. Eccentric M14 x 100 mm ZNS3 bolt
   - Adjusts rear camber
20. Swing arm
21. Eccentric washer
22. Self-locking M14 x 1.5 ZNS3 nut (replace)
   - Tighten with suspension loaded
   - Tighten to 165 Nm (122 ft-lb)

BentleyPublishers.com
BentleyPublishers.com—All Rights Reserved

Upper control arm, removing and installing

- Sports Wagon: Disable and deflate EHC system. See **EHC, disabling, deflating, refilling** in this repair group.
- Raise car and remove wheel.

> ***WARNING —***
> - *Make sure the car is firmly supported on jack stands designed for the purpose. Place jack stands underneath structural chassis points, not underneath suspension parts.*

- Release wheel speed sensor and brake pad sensor wires from brackets. Note that brake pad sensor is only on right side.

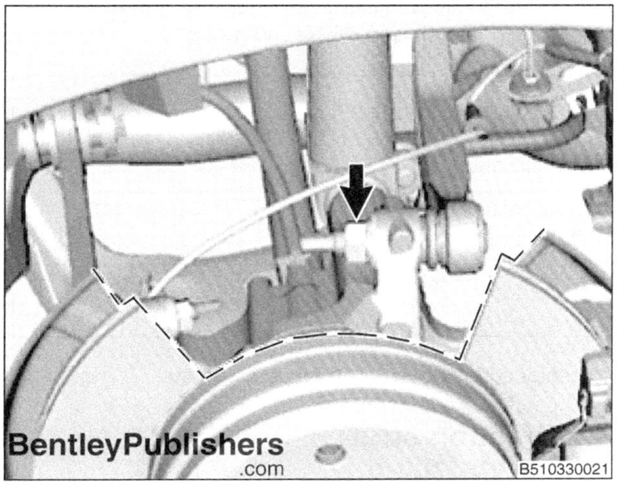

◄ Remove control arm ball joint nut (**arrow**) at wheel bearing carrier. Separate control arm from wheel bearing carrier. (Brake backing plate and disc cut away for purpose of illustration.)

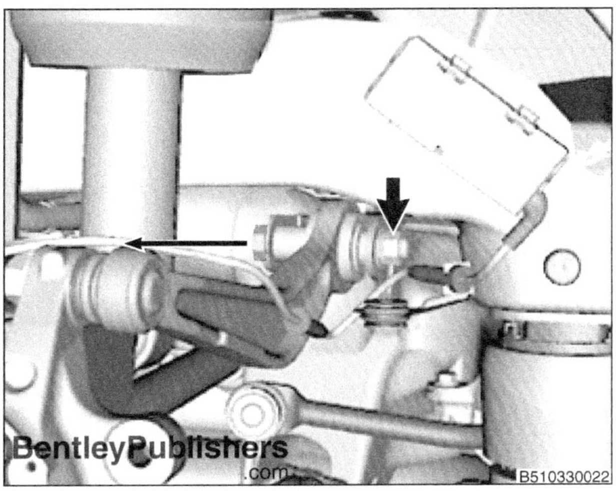

◄ Remove control arm mounting nut (**arrow**) at subframe. Slide mounting bolt out and remove control arm.

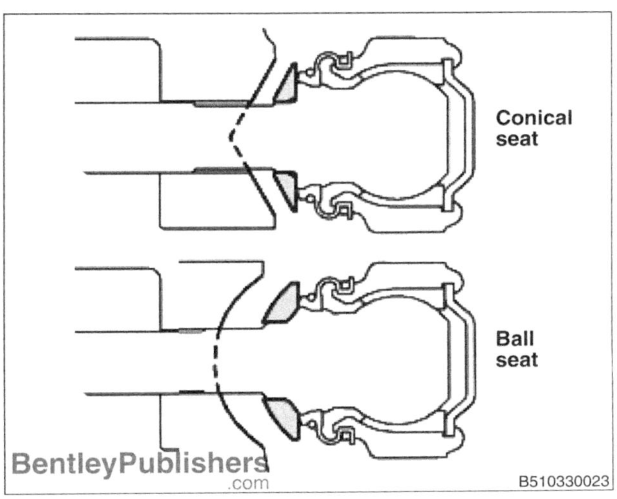

◄ If installing new control arm, be sure to match shape of wheel bearing carrier seat for ball joint.

BentleyPublishers.com
BentleyPublishers.com–All Rights Reserved

– Installation is reverse of removal. Remember to:
  - Keep control arm ball joint shaft and wheel bearing carrier bore clean and grease-free.
  - Replace self-locking fasteners.
  - Head of control arm to subframe bolt faces forward.
  - Tighten control arm to subframe fastener to final torque only after vehicle is lowered and suspension has settled.
  - Sports Wagon: Reactivate and fill EHC system. See **EHC, disabling, deflating, refilling** in this repair group.

Tightening torques	
Control arm to subframe (replace M12 x 1.5 self-locking nut with new)	100 Nm (74 ft-lb)
Control arm to wheel bearing carrier (replace M14 x 1.5 self-locking nut with new)	175 Nm (129 ft-lb)
Road wheel to hub	120 ± 10 Nm (89 ± 7 ft-lb)

– Have car professionally aligned when job is complete.

Guide arm (traction strut), removing and installing

– Raise car and remove wheel.

> ***WARNING —***
> - *Make sure the car is firmly supported on jack stands designed for the purpose. Place jack stands underneath structural chassis points, not underneath suspension parts.*

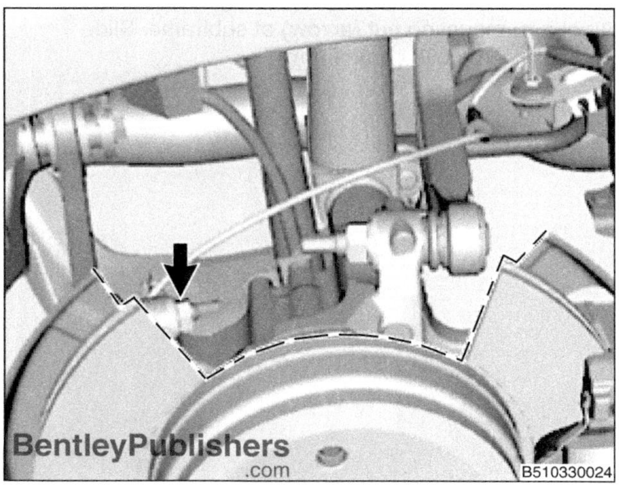

◄ Remove guide arm ball joint nut (**arrow**) at wheel bearing carrier. Separate guide arm from wheel bearing carrier. (Brake backing plate and disc cut away for purpose of illustration.)

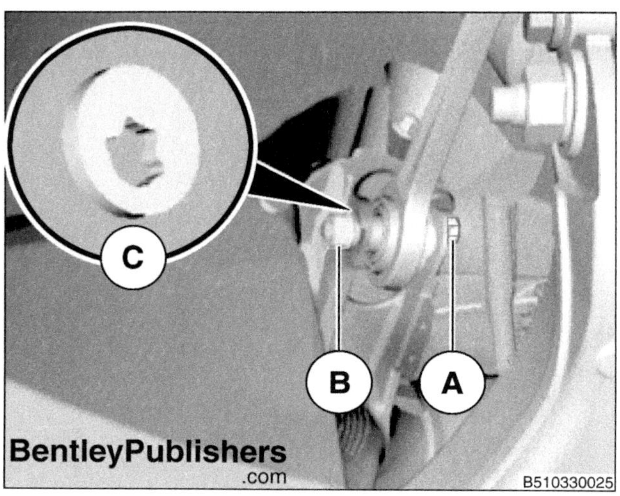

◄ Working at guide arm to subframe bolt:
  - Mark position of eccentric bolt (rear toe adjustment bolt) (**A**).
  - Remove mounting nut (**B**).
  - Remove eccentric washer (**C**) and slide out bolt.

BentleyPublishers.com
BentleyPublishers.com—All Rights Reserved

- Installation is reverse of removal. Remember to:
  - Keep guide arm ball joint shaft and wheel bearing carrier bore clean and grease-free.
  - Replace self-locking fasteners.
  - Head of guide arm to subframe eccentric (toe-adjusting) bolt faces backward.
  - Refit eccentric washer on eccentric bolt and match marks made previously.
  - Tighten guide arm to subframe fastener to final torque only after vehicle is lowered and suspension has settled.

Tightening torques	
Guide arm to subframe (replace M10 self-locking nut with new)	65 Nm (48 ft-lb)
Guide arm to wheel bearing carrier (replace M10 self-locking nut with new)	65 Nm (48 ft-lb)
Road wheel to hub	120 ± 10 Nm (89 ± 7 ft-lb)

- Have car professionally aligned when job is complete.

Swing arm, removing and installing

- Sports Wagon: Disable and deflate EHC system. See **EHC, disabling, deflating, refilling** in this repair group.

- Sports Wagon: Detach drive axle at differential output flange. See **331 Rear Axle Differential**.

- Raise car and remove wheel.

> ***WARNING—***
> - *Make sure the car is firmly supported on jack stands designed for the purpose. Place jack stands underneath structural chassis points, not underneath suspension parts.*

- Support rear wheel bearing carrier with suitable jack.

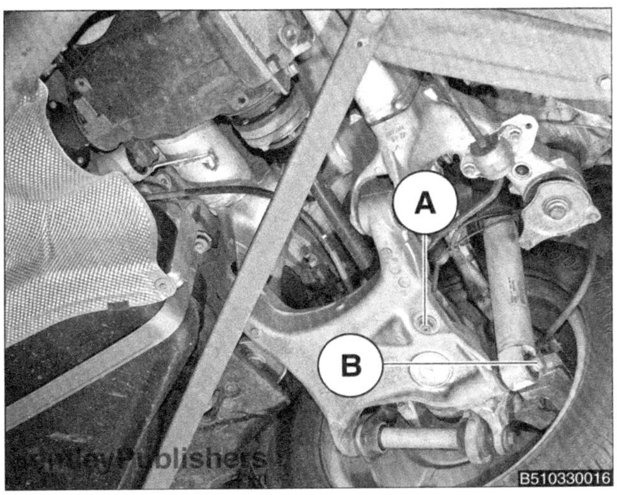

- Working underneath swing arm:
  - Disconnect control rod for suspension level sensor.
  - Remove stabilizer link nut (**A**) at swing arm and detach stabilizer link. See **Stabilizer bar link, removing and installing** in this repair group.
  - Sports Wagon: Remove shock absorber fastener (**B**) at swing arm and detach shock absorber. See **Rear shock absorber, removing and installing (Sports Wagon)** in this repair group.

- Mark (**arrows**) position of rear inboard swing arm bolt (rear camber adjusting bolt), then remove nut. Remove eccentric washer and slide out bolt.

- Remove front inboard swing arm fasteners.

- Working at integral link, remove swing arm mounting nut (**arrow**). Drive outboard mounting bolt toward rear and pull swing arm downward.

- Note orientation of swing arm bolt heads:
  - Eccentric (camber adjusting) bolt head points forward.
  - Front inboard bolt head points backward.
  - Outboard (integral link) bolt head points backward.

- Installation is reverse of removal. Remember to:
  - Replace self-locking fasteners.
  - Refit eccentric washer on camber bolt and match marks made previously.
  - Tighten swing arm and shock absorber fasteners to final torque only after vehicle is lowered and suspension has settled.
  - Sports Wagon: Reactivate and fill EHC system. See **EHC, disabling, deflating, refilling** in this repair group.

Tightening torques	
Drive axle to differential output flange (replace M10 x 43 mm ZNS3 Torx bolts with new)	80 Nm (59 ft-lb)
Integral link to swing arm (M16 x 275 mm ZNS3 bolt, replace nut with new)	240 Nm (177 ft-lb)
Road wheel to hub	120 ± 10 Nm (89 ± 7 ft-lb)
Shock absorber to swing arm (replace M14 x 1.5 x 105 mm ZNS bolts with new)	165 Nm (122 ft-lb)
Stabilizer link to swing arm (replace M12 x 1.5 ZN3 self-locking nut with new)	100 Nm (74 ft-lb)
Swing arm to subframe (M12 x 1.5 x 98 mm ZNS3 bolt, replace nut with new)	100 Nm (74 ft-lb)
Swing arm to subframe (M14 x 100 mm eccentric camber adjusting bolt, replace nut with new)	165 Nm (122 ft-lb)

- Have car professionally aligned when job is complete.

BentleyPublishers.com
BentleyPublishers.com—All Rights Reserved

Integral link, removing and installing

– Sports Wagon: Disable and deflate EHC system. See **EHC, disabling, deflating, refilling** in this repair group.

– Sports Wagon: Detach drive axle at differential output flange. See **331 Rear Axle Differential**.

– Raise car and remove wheel.

> ***WARNING—***
> • *Make sure the car is firmly supported on jack stands designed for the purpose. Place jack stands underneath structural chassis points, not underneath suspension parts.*

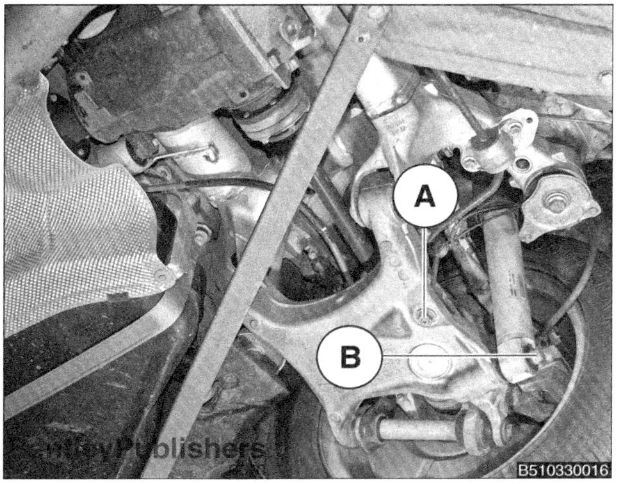

– Support rear wheel bearing carrier with suitable jack.

◄ Working underneath swing arm:

- Disconnect control rod for suspension level sensor.
- Remove stabilizer link nut (**A**) at swing arm and detach stabilizer link. See **Stabilizer bar link, removing and installing** in this repair group.
- Sports Wagon: Remove shock absorber fastener (**B**) at swing arm and detach shock absorber. See **Rear shock absorber, removing and installing (Sports Wagon)** in this repair group.

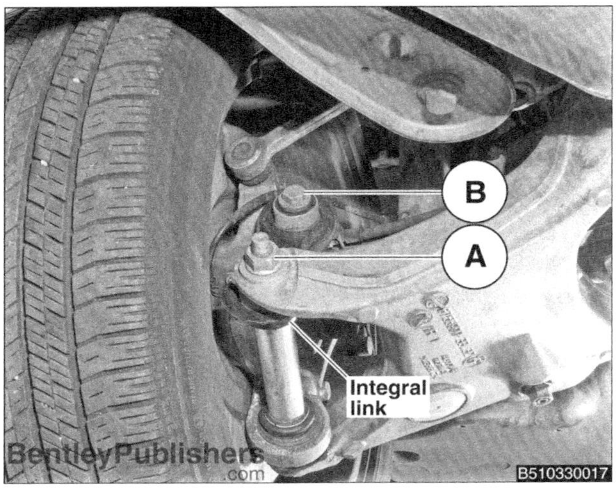

◄ Working at integral link:

- Remove swing arm mounting nut (**A**). Drive lower bolt out toward rear and pull swing arm downward.
- Remove integral link mounting bolt (**B**) and remove link.

– Installation is reverse of removal. Remember to:

- Replace self-locking fasteners.
- Tighten swing arm and shock absorber fasteners to final torque only after vehicle is lowered and suspension has settled.
- Sports Wagon: Reactivate and fill EHC system. See **EHC, disabling, deflating, refilling** in this repair group.

Tightening torques	
Drive axle to differential output flange (replace M10 x 43 mm ZNS3 Torx bolts with new)	80 Nm (59 ft-lb)
Integral link to swing arm (M16 x 275 mm ZNS3 bolt, replace nut with new)	240 Nm (177 ft-lb)
Integral link to wheel bearing carrier (M12 x 1.5 x 67 mm ZNS3 bolt)	100 Nm (74 ft-lb)
Road wheel to hub	120 ± 10 Nm (89 ± 7 ft-lb)
Shock absorber to swing arm (replace M14 x 1.5 x 105 mm ZNS bolts with new)	165 Nm (122 ft-lb)
Stabilizer link to swing arm (replace M12 x 1.5 ZN3 self-locking nut with new)	100 Nm (74 ft-lb)

– Have car professionally aligned when job is complete.

BentleyPublishers.com—All Rights Reserved

Swing arm ball joint, removing and installing

Replace ball joints in pairs.

— Sports Wagon: Disable and deflate EHC system. See **EHC, disabling, deflating, refilling** in this repair group.

— Sports Wagon: Detach drive axle at differential output flange. See **331 Rear Axle Differential**.

— Raise car and remove wheel.

> ***WARNING —***
> • *Make sure the car is firmly supported on jack stands designed for the purpose. Place jack stands underneath structural chassis points, not underneath suspension parts.*

— Remove integral link. See **Integral link, removing and installing** in this repair group.

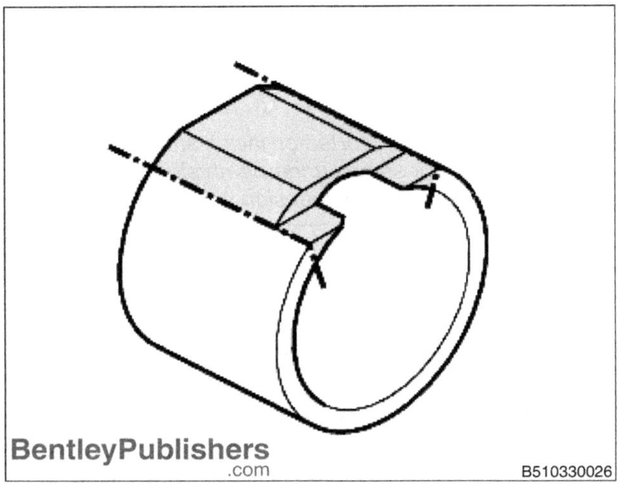

◄ Modify BMW press too 33 4 191 as illustrated. Cut area highlighted in grey and deburr edges.

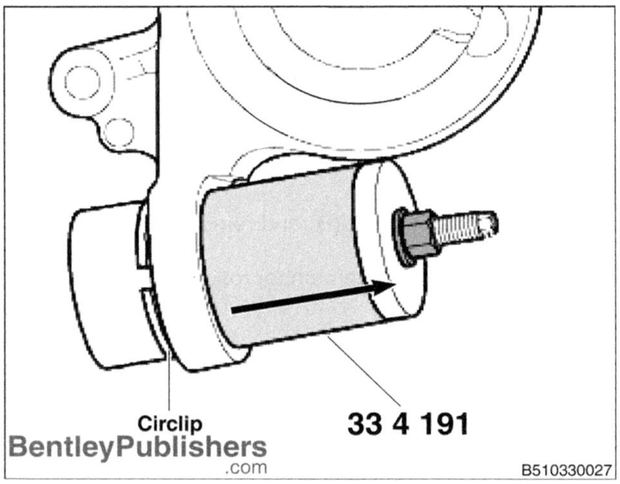

◄ Working at swing arm ball joint:
- Remove circlip.
- Use press tools to force ball joint out of wheel bearing carrier. Orient newly cut slot in special tool to point up toward wheel bearing.
- Reinstall using similar tools.
- Install new circlip.

— Installation is reverse of removal. Remember to:
- Replace self-locking fasteners.
- Tighten swing arm and shock absorber fasteners to final torque only after vehicle is lowered and suspension has settled.
- Sports Wagon: Reactivate and fill EHC system. See **EHC, disabling, deflating, refilling** in this repair group.

Tightening torques	
Drive axle to differential output flange (replace M10 x 43 mm ZNS3 Torx bolts with new)	80 Nm (59 ft-lb)
Integral link to swing arm (M16 x 275 mm ZNS3 bolt, replace nut with new)	240 Nm (177 ft-lb)
Integral link to wheel bearing carrier (M12 x 1.5 x 67 mm ZNS3 bolt)	100 Nm (74 ft-lb)
Road wheel to hub	120 ± 10 Nm (89 ± 7 ft-lb)
Shock absorber to swing arm (replace M14 x 1.5 x 105 mm ZNS bolts with new)	165 Nm (122 ft-lb)
Stabilizer link to swing arm (replace M12 x 1.5 ZN3 self-locking nut with new)	100 Nm (74 ft-lb)

— Have car professionally aligned when job is complete.

BentleyPublishers.com
BentleyPublishers.com—All Rights Reserved

Rear subframe, removing and installing

– Sports Wagon: Disable and deflate EHC system. See **EHC, disabling, deflating, refilling** in this repair group.

– Raise vehicle and support safely.

> *WARNING —*
> - *Be sure the vehicle is stable and well supported. Removal of the rear subframe may upset the balance of the vehicle. Use tall jack stands to support the vehicle at all four corners.*
> - *Alternatively, place 100 kg (220 lbs) in trunk or cargo compartment before starting work.*
> - *Once it is raised up, make sure that the vehicle cannot be raised or lowered and that it does not move off the lift locating plates.*

– Remove both rear wheels.

– Remove rear brake calipers and tie aside securely with stiff wire. Do not separate brake lines. Do not hang using brake hose. See **340 Brakes**.

– Remove both rear brake rotors. See **340 Brakes**.

– Remove both rear wheel speed sensors and unclip wiring harnesses from control arms and subframe. See **340 Brakes**.

– Sports Wagon: Remove X-brace underneath rear suspension.

– Remove complete exhaust system. See **180 Exhaust System**.

– Detach rear driveshaft and tie aside. See **260 Driveshafts**.

– Sedan: Remove both rear strut assembly lower bolts and detach assembly from wheel bearing carriers. See **Rear strut assembly, removing and installing (Sedan)** in this repair group.

– Sports Wagon: Detach bottom of air springs from subframe. See **Air spring, removing and installing (Sports Wagon)** in this repair group.

– Sports Wagon: Detach both rear shock absorbers from swing arms.

– Disconnect ride-height sensor electrical connector and unclip harness from subframe.

– Disengage both parking brake cables from brake backing plates. See **340 Brakes**.

◄ Working underneath rear of vehicle, remove fasteners (**arrows**) and remove splash shields underneath front subframe mounting bolts.

BentleyPublishers.com–All Rights Reserved

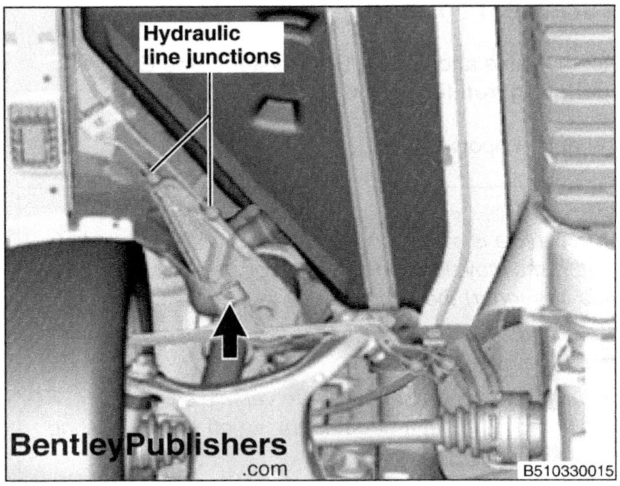

Vehicle with active stabilizer system:

- Remove hydraulic line anchor mounting bolt (**arrow**) at right thrust rod.
- Separate hydraulic lines at junctions. Be prepared to catch dripping hydraulic fluid and dispose in a safe manner.

> ***CAUTION—***
> - *Do not allow dirt to enter hydraulic lines. Plug fluid line openings.*

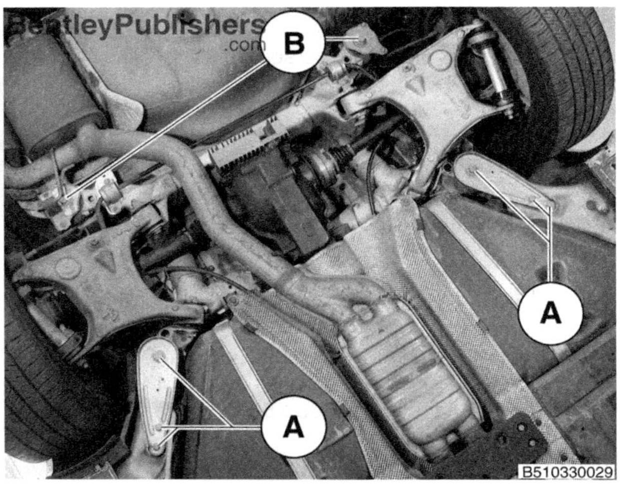

While supporting subframe with transmission jack or other suitable device:

- Remove thrust rod mounting bolts (**A**) and remove both thrust rods.
- Remove subframe rear mounting bolts (**B**) and carefully lower subframe, checking to make sure no lines, hoses or harnesses become snagged.

– To install, jack subframe into position. Install thrust rods in their original positions and tighten subframe mounting bolts.

Tightening torque	
Rear subframe to body (M14 x 1.5 x 138 mm)	165 Nm (122 ft-lb)
Thrust rod to body (M10 x 1.5 10.9 grade)	48 Nm (35 ft-lb)

– Remainder of installation is reverse of removal, noting the following.

- Tighten shock absorber fastener to final torque only after vehicle is lowered and suspension has settled.
- Vehicle with active stabilizers: Make sure hydraulic lines are correctly isolated and free of tension. Fill and bleed hydraulic system. Check for leaks once engine is running. See **310 Front Suspension**.
- Sports Wagon: Reactivate and fill EHC system. See **EHC, disabling, deflating, refilling** in this repair group.
- Bleed entire brake system before driving car. See **340 Brakes**.
- Adjust parking brake.

Tightening torques	
Brake caliper to wheel bearing carrier (M12 x 1.5 x 43 mm ZNS3 bolts)	110 Nm (81 ft-lb)
Brake rotor to wheel hub (M8 x 14 mm bolt)	16 Nm (12 ft-lb)
Hydraulic line junction union nuts	30 Nm (22 ft-lb)
Road wheel to hub	120 ± 10 Nm (89 ± 7 ft-lb)
Shock absorber to swing arm (replace M14 x 1.5 x 105 mm ZNS bolts with new)	165 Nm (122 ft-lb)

BentleyPublishers.com
BentleyPublishers.com—All Rights Reserved

Tightening torques	
Strut to wheel bearing carrier (replace M14 x 1.5 x 105 mm ZNS3 bolt with new)	165 Nm (122 ft-lb)
Wheel speed sensor to wheel bearing carrier	8 Nm (6 ft-lb)
X-brace to chassis • M10 bolts • M10 nut	 59 Nm (44 ft-lb) 41 Nm (30 ft-lb)

– Have car professionally aligned when job is complete.

Rear subframe bushings, removing and installing

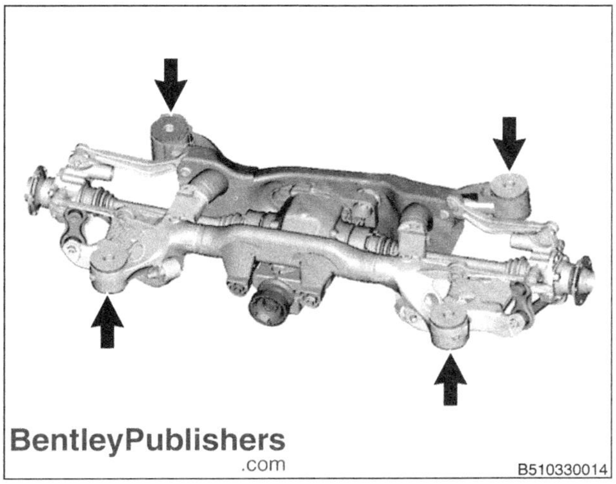

The aluminum subframe bolts to the vehicle chassis via 4 large rubber bushings (**arrows**).

– Lower rear subframe. See **Rear subframe, removing and installing** in this repair group.

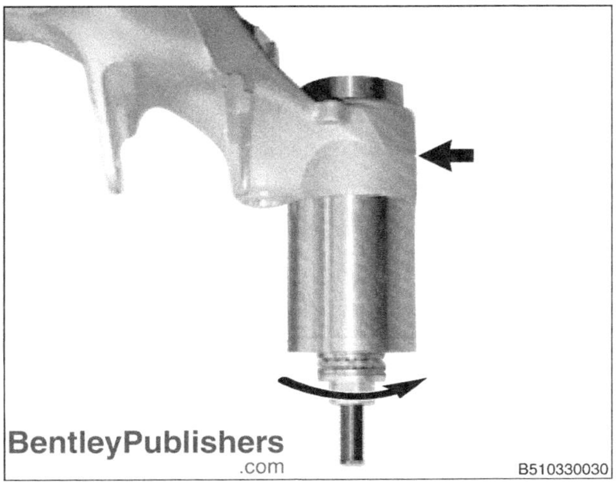

Working at subframe bushing boss (**arrow**):

- Assemble press tools for bushing removal as illustrated.
- Use hot air blower to heat subframe boss at maximum 600°C (1100°F) for approx. 2 minutes.
- Rotate press tool nut to remove bushing.

> ***WARNING —***
> *• Do point hot air jet at fuel tank.*

BentleyPublishers.com
BentleyPublishers.com—All Rights Reserved

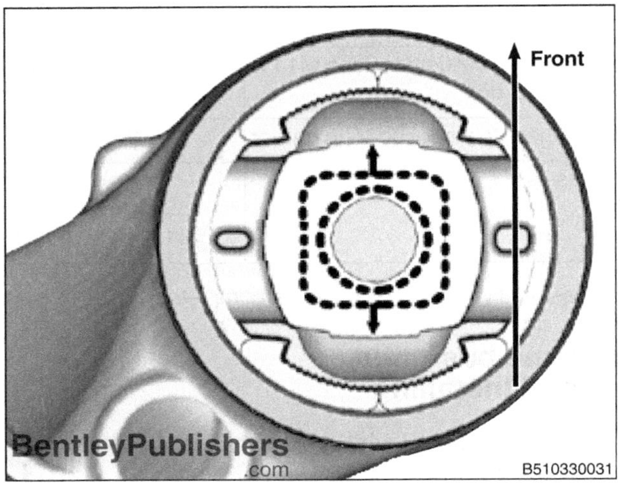

When installing:

- Orient bushing so that inner sleeve (**arrows**) point to front and rear as illustrated. Maximum deviation allowed from this orientation is 7°.
- Coat bushing with Circolight® antifriction agent.
- Press bushing in using special tools.

Rear Wheel Bearings

The rear wheel bearing is bolted to the rear wheel bearing carrier. The wheel bearing can be replaced without removing the rear wheel bearing carrier. The rear wheel hub is pressed into the rear wheel bearing. Removing the hub destroys the wheel bearing. To service the wheel bearing, remove the wheel hub and replace the wheel bearing.

Rear wheel bearing components

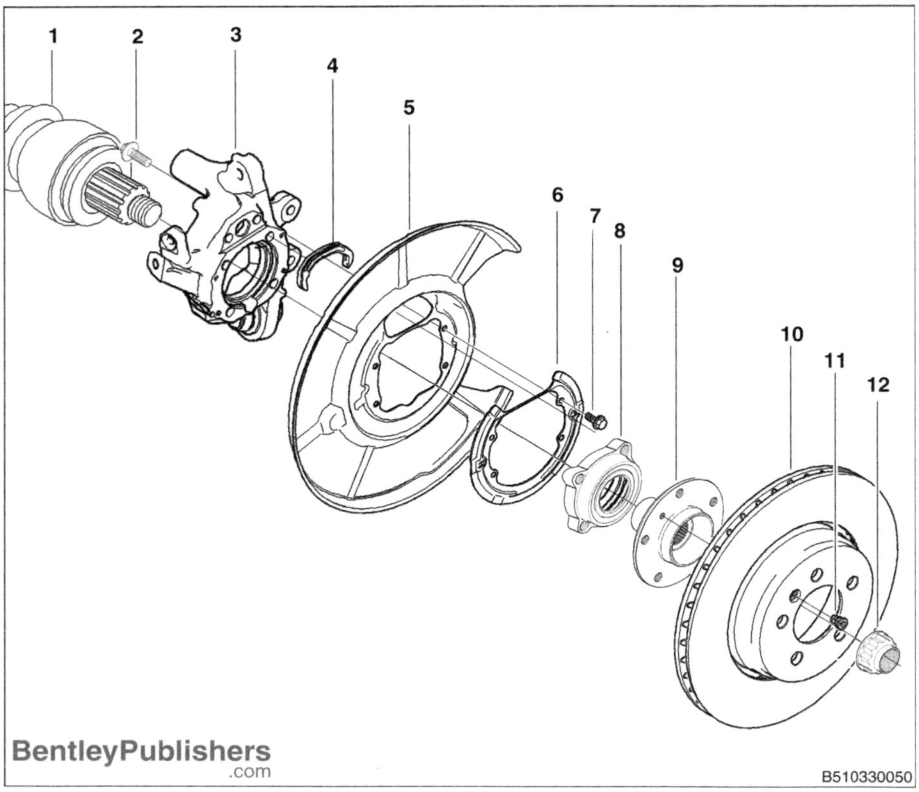

1. Drive axle
2. M10 x 1 x 34 mm Torx bolt
   - Replace with new
   - Tighten in 2 stages: 30 Nm (22 ft-lb) additional 110°
3. Wheel bearing carrier
4. Gasket
5. Backing plate
6. Reinforcement
7. M6 x 14 mm 8.8 grade bolt
8. Wheel bearing
9. Wheel hub
10. Brake rotor
11. M8 x 14 Allen bolt
    - Tighten to 16 Nm (12 ft-lb)
12. M27 collar nut
    - Replace with new
    - Tighten to 420 Nm
    - Stake to axle end

BentleyPublishers.com
BentleyPublishers.com—All Rights Reserved

Rear wheel bearing carrier, removing and installing

– Raise vehicle and remove rear wheel.

> ***WARNING —***
> • *Make sure the car is firmly supported on jack stands designed for the purpose. Place jack stands underneath structural chassis points, not underneath suspension parts.*

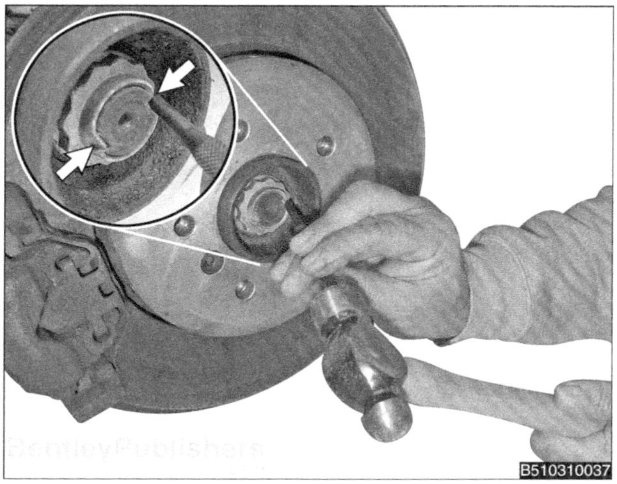

Using a suitable punch, straighten staked sides of wheel hub collar nut (**arrows**).

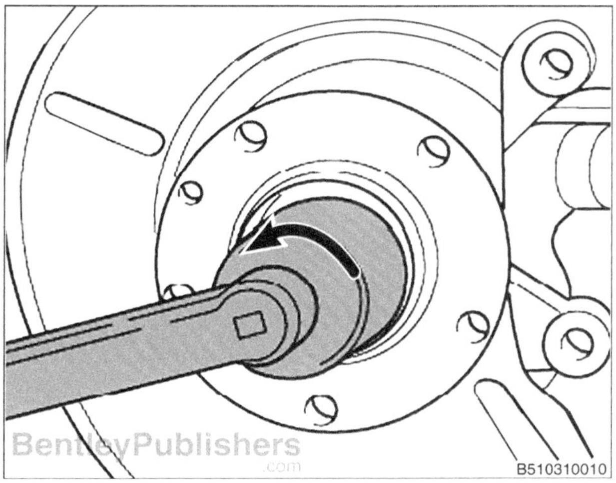

With an assistant holding down brake pedal, remove collar nut and discard.

> ***CAUTION—***
> • *The drive axle collar nut is tightened to a high torque. If necessary, reinstall road wheel and place vehicle on floor before breaking collar nut free.*

– Sports Wagon: Disable and deflate EHC system. See **EHC, disabling, deflating, refilling** in this repair group.

– Sports Wagon: Remove air spring. See **Air spring, removing and installing (Sports Wagon)** in this repair group.

– Remove brake caliper and brake rotor. Hang brake caliper aside with stiff wire. Do not disconnect brake fluid hoses. See **340 Brakes**.

– Disconnect parking brake cable from brake shoe expander. See **340 Brakes**.

– If replacing wheel bearing carrier with new: Remove rear wheel hub. See **Rear wheel hub, removing and installing** in this repair group.

– Remove ABS wheel speed sensor from wheel bearing carrier. See **340 Brakes**.

– Detach rear stabilizer link from stabilizer bar. See **Stabilizer bar link, removing and installing** in this repair group.

– Sedan: Detach strut from wheel bearing carrier. See **Rear strut assembly, removing and installing (Sedan)** in this repair group.

– Remove integral link. See **Integral link, removing and installing** in this repair group.

– Use BMW special tools 33 2 200 or equivalent to press drive axle out of wheel bearing hub.

> ***CAUTION—***
> • *In order to avoid damaging drive axle threads, do not drive out stub axle using impact tool.*

BentleyPublishers.com
BentleyPublishers.com—All Rights Reserved

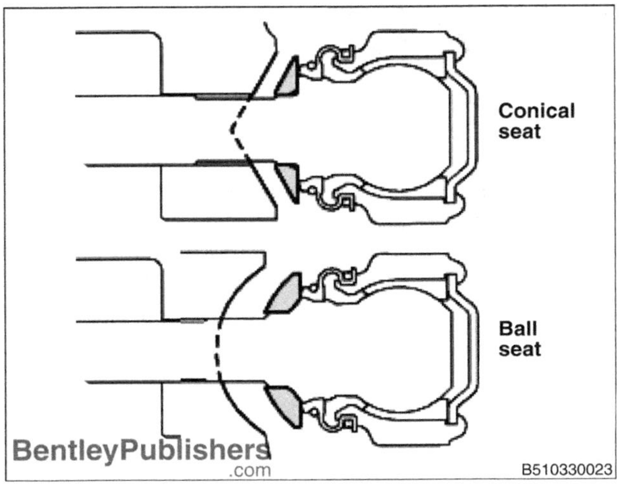

- Detach guide arm from wheel bearing carrier. See **Guide arm (traction strut), removing and installing** in this repair group.

- Detach upper control arm from wheel bearing carrier. See **Upper control arm, removing and installing** in this repair group.

- Remove wheel bearing carrier.

If installing new wheel bearing carrier, be sure to match shape of upper control arm ball joint.

- Installation is reverse of removal. Remember to:
  - Replace self-locking fasteners.
  - Tighten swing arm and shock absorber fasteners to final torque only after vehicle is lowered and suspension has settled.
  - Tighten drive axle collar nut with wheel on the ground. Stake nut to axle end.
  - Sports Wagon: Reactivate and fill EHC system. See **EHC, disabling, deflating, refilling** in this repair group.
  - Adjust parking brake.

Tightening torques	
Brake caliper to wheel bearing carrier (M12 x 1.5 x 41 mm ZNS3 bolts)	110 Nm (81 ft-lb)
Brake rotor to hub (M8 x 14 mm Allen bolt)	16 Nm (12 ft-lb)
Control arm to wheel bearing carrier (replace M14 x 1.5 self-locking nut with new)	175 Nm (129 ft-lb)
Drive axle collar nut to wheel hub (replace M27 collar nut with new, lightly oil contact face, stake nut to axle end)	420 Nm (310 ft-lb)
Guide arm to wheel bearing carrier (replace M10 self-locking nut with new)	65 Nm (48 ft-lb)
Integral link to swing arm (M16 x 275 mm ZNS3 bolt, replace nut with new)	240 Nm (177 ft-lb)
Integral link to wheel bearing carrier (M12 x 1.5 x 67 mm ZNS3 bolt)	100 Nm (74 ft-lb)
Road wheel to hub	120 ± 10 Nm (89 ± 7 ft-lb)
Shock absorber to swing arm (replace M14 x 1.5 x 105 mm ZNS bolts with new)	165 Nm (122 ft-lb)
Stabilizer bar link to stabilizer bar (M10 x 1.5 x ZNS3 nut)	65 Nm (48 ft-lb)
Wheel speed sensor to wheel bearing carrier	8 Nm (6 ft-lb)

- Have car professionally aligned when job is complete.

BentleyPublishers.com
BentleyPublishers.com—All Rights Reserved

Rear wheel hub, removing and installing

– Raise car and remove wheel.

> ***WARNING —***
> - *Make sure the vehicle is stable and well supported at all times. Use a professional automotive lift or jack stands designed for the purpose. A floor jack is not adequate support.*

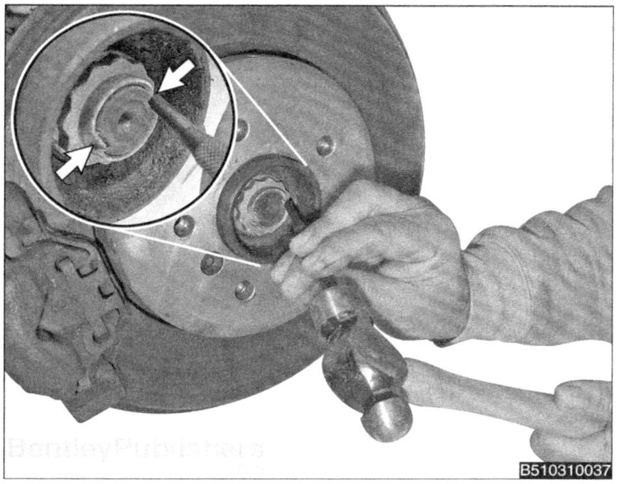

◄ Using a suitable punch, straighten staked sides of wheel hub collar nut (**arrows**).

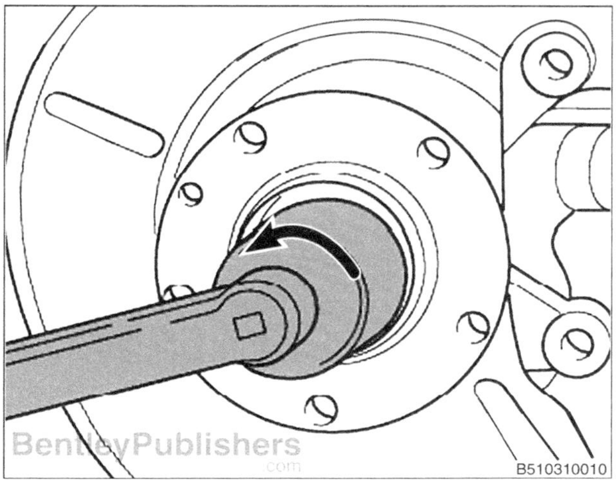

◄ With an assistant holding down brake pedal, remove collar nut and discard.

> ***CAUTION—***
> - *The drive axle collar nut is tightened to a high torque. If necessary, reinstall road wheel and place vehicle on floor before breaking collar nut free.*

– Detach front brake caliper from steering knuckle and suspend aside using stiff wire. Do not detach brake fluid hose. See **340 Brakes**.

– Remove brake rotor. See **340 Brakes**.

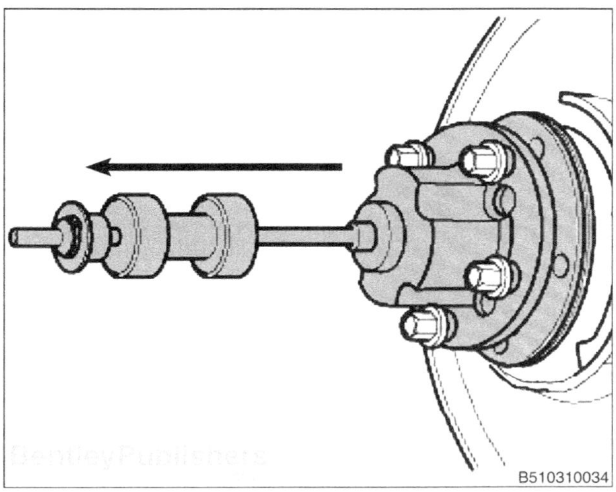

◄ Attach slide hammer hub puller to wheel hub using 5 wheel bolts. Drive hub off wheel bearing.

- Note that wheel bearing inner race is removed with hub, thus destroying bearing. Separate race from hub and discard race. Do not damage hub.

– Replace wheel bearing. See **Rear wheel bearing, replacing** in this repair group.

BentleyPublishers.com
BentleyPublishers.com–All Rights Reserved

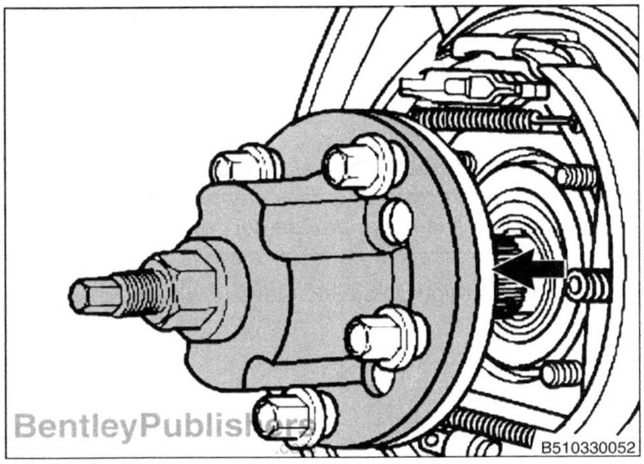

To install wheel hub:

- Oil hub (**arrow**) lightly and place over splines of output shaft.
- Use BMW special tools or equivalent to draw hub into wheel bearing.

– Remainder of installation is reverse of removal. Remember to:

- Tighten drive axle collar nut with wheel on the ground. Stake nut to axle end.
- Adjust parking brake when job is complete.

Tightening torques	
Brake caliper to wheel bearing carrier (M12 x 1.5 x 41 mm ZNS3 bolts)	110 Nm (81 ft-lb)
Brake rotor to hub (M8 x 14 mm Allen bolt)	16 Nm (12 ft-lb)
Drive axle collar nut to wheel hub (replace M27 collar nut with new, lightly oil contact face, stake nut to axle end)	420 Nm (310 ft-lb)
Road wheel to hub	120 ± 10 Nm (89 ± 7 ft-lb)

Rear wheel bearing, replacing

– Remove rear wheel hub (drive flange). See **Rear wheel hub, removing and installing** in this repair group.

> ***CAUTION—***
> - *The wheel bearing is destroyed when the drive flange is removed. Replace wheel bearing.*

– Press wheel bearing inner race off wheel hub shaft.

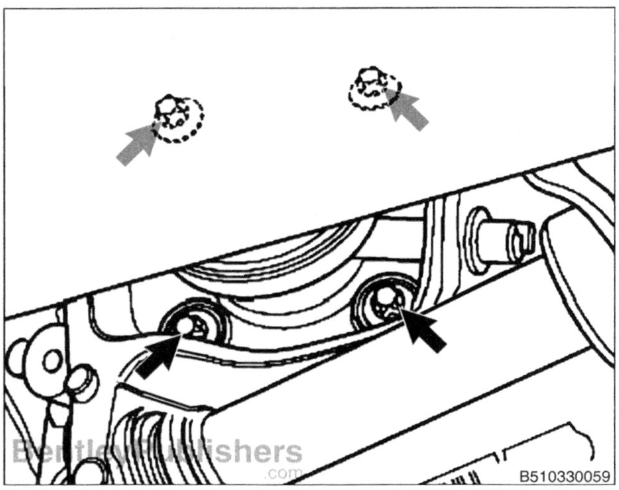

Working at back of wheel bearing carrier, remove wheel bearing fasteners and remove bearing.

BentleyPublishers.com
BentleyPublishers.com—All Rights Reserved

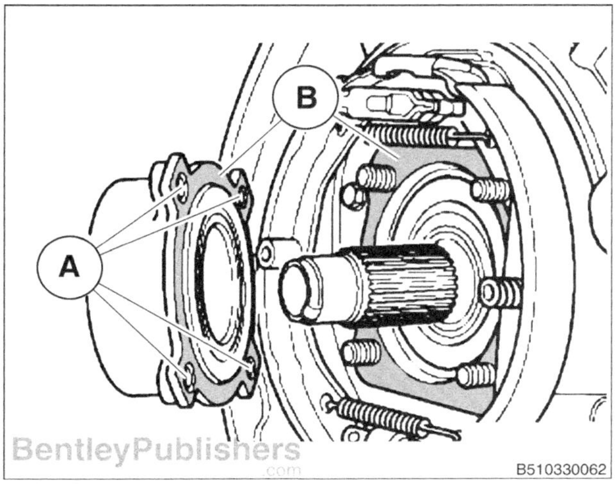

To install, make sure threads (**A**) and contact surface of wheel bearing and wheel bearing (**B**) carrier are clean and free of oil and grease. If necessary, rethread bearing bolt holes (**A**).

— Install new wheel bearing.

Tightening torque	
Wheel bearing to wheel bearing carrier (replace M10 x 1 x 34 mm Torx bolts with new): • Stage 1 • Stage 2	 30 Nm (22 ft-lb) additional 110°

— Remainder of installation is reverse of removal. Remember to:

- Tighten drive axle collar nut with wheel on the ground. Stake nut to axle end.
- Adjust parking brake when job is complete.

Tightening torques	
Brake caliper to wheel bearing carrier (M12 x 1.5 x 41 mm ZNS3 bolts)	110 Nm (81 ft-lb)
Brake rotor to hub (M8 x 14 mm Allen bolt)	16 Nm (12 ft-lb)
Drive axle collar nut to wheel hub (replace M27 collar nut with new, lightly oil contact face, stake nut to axle end)	420 Nm (310 ft-lb)
Road wheel to hub	120 ± 10 Nm (89 ± 7 ft-lb)

330

BentleyPublishers.com
BentleyPublishers.com—All Rights Reserved

Rear Stabilizer Bar

The rear stabilizer bar is mounted to the rear subframe and attached to the swing arms with stabilizer bar links.

Active stabilizer bar (dynamic drive)

5 Series vehicles may be equipped with the optional active stabilizer system (acronym ARS). The system is also referred to as dynamic drive.

With this system, front and rear stabilizer bars are split in half, with the halves connected to each other via hydraulic oscillating motors. An electronic control module operates through a valve body to control the coupling of the two stabilizer bar halves.

For additional details, including active stabilizer troubleshooting and commissioning, see **300 Suspension, Steering and Brakes–General**. Hydraulic bleeding of active stabilizer system is in **310 Front Suspension**.

Active stabilizer components, rear

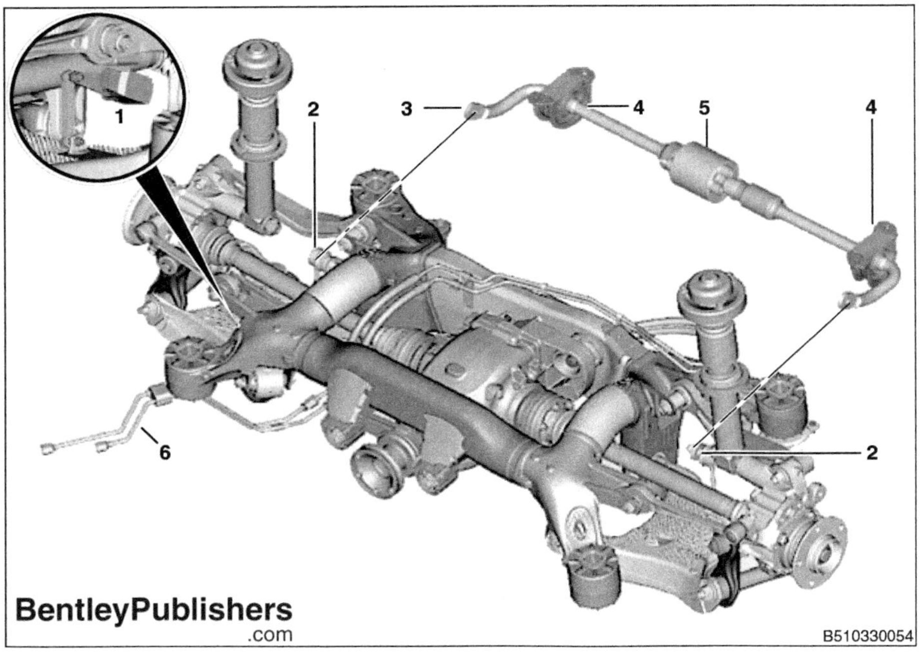

1. Rear ride height sensor
2. Stabilizer bar link
3. Stabilizer bar
4. Stabilizer bar mount
5. Rear oscillating motor
6. Oscillating motor pressure lines

BentleyPublishers.com
BentleyPublishers.com–All Rights Reserved

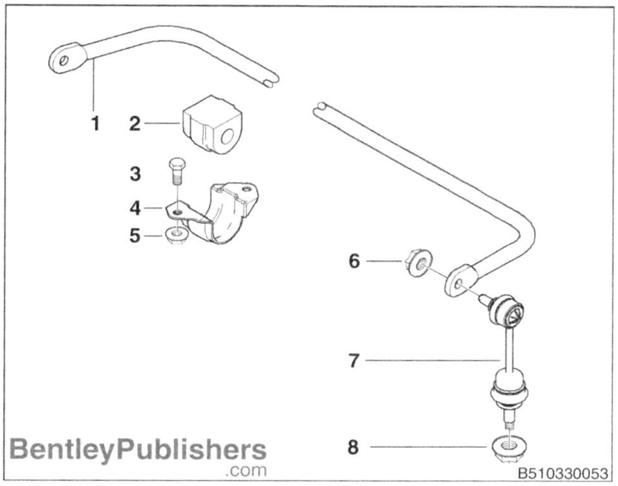

Rear stabilizer bar components (mechanical stabilizer)

1. Stabilizer bar
2. Stabilizer bar mounting bushing
3. M8 x 30 mm 10.9 grade bolt
4. Stabilizer bar mount
5. M8 ZNS3 self-locking nut
    - Replace with new
    - Tighten to 28 Nm (21 ft-lb)
6. M10 ZNS3 self-locking nut
    - Replace with new
    - Tighten to 65 Nm (48 ft-lb)
7. Stabilizer bar link
8. M12 x 1.5 ZNS3 self-locking nut
    - Replace with new
    - Tighten to 100 Nm (74 ft-lb)

Rear stabilizer bar, removing and installing (mechanical stabilizer)

The rear stabilizer is attached to the rear subframe.

— Raise vehicle and safely support.

> ***WARNING—***
> - *Make sure the car is firmly supported on jack stands designed for the purpose. Place jack stands underneath structural chassis points, not underneath suspension parts.*

— Detach stabilizer links from both ends of stabilizer bar. See **Stabilizer bar link, removing and installing** in this repair group.

— If necessary, remove vibration damper from right stabilizer bar mount.

◄ Remove mounting nuts (**arrows**) at both stabilizer bar mounts.

— Slide stabilizer bar sideways to remove.

— Check rubber mounting bushings. Replace if necessary.

— Installation is reverse of removal. Remember to replace self-locking fasteners.

Tightening torques	
Stabilizer bar mount to subframe (replace M8 ZNS3 self-locking nuts with new)	28 Nm (21 ft-lb)
Stabilizer bar link to stabilizer bar (replace M10 x 1.5 x ZNS3 self-locking nut with new)	65 Nm (48 ft-lb)

BentleyPublishers
BentleyPublishers.com—All Rights Reserved

Rear stabilizer bar, removing and installing (active stabilizer bar)

The rear stabilizer is attached to the rear subframe.

- Use BMW scan tool to read and print fault codes.
- Raise rear of vehicle and support safely.

> ***WARNING —***
> - *Make sure the vehicle is stable and well supported at all times. Use a professional automotive lift or jack stands designed for the purpose. A floor jack is not adequate support.*

- Sports Wagon: Remove X-brace underneath rear suspension.

< Remove pressure line bracket screws (**arrows**).

- Sports Wagon: If necessary, remove vibration damper from right stabilizer bar mount.

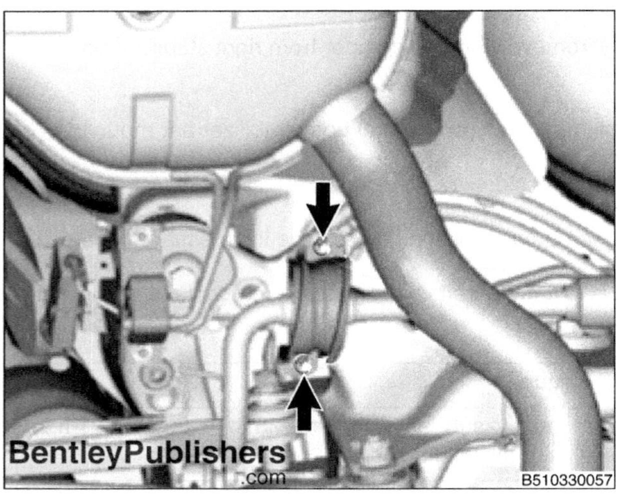

< Remove mounting nuts (**arrows**) at both stabilizer bar mounts.

BentleyPublishers.com
BentleyPublishers.com–All Rights Reserved

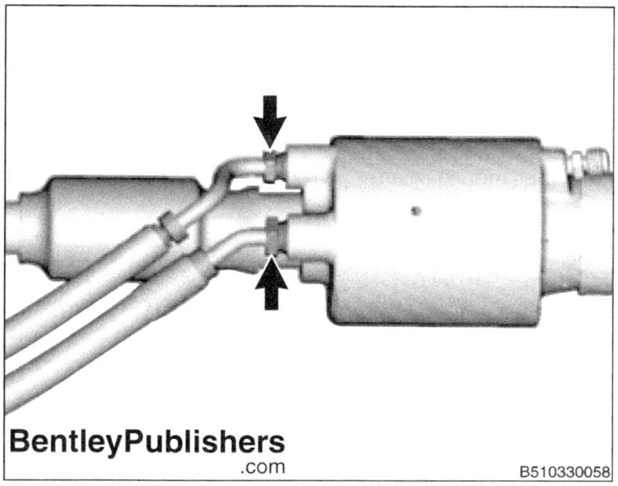

Lower stabilizer bar and note carefully installation positions of fluid pressure lines and brackets at oscillating motor. Then detach pressure line unions (**arrows**) from motor. Be prepared to catch dripping hydraulic fluid and dispose in a safe manner.

> ***CAUTION—***
> - *Do not allow dirt to enter hydraulic lines. Plug fluid line openings.*

To avoid confusion when installing, pressure lines are different dimensions and the unions have different threads.

- Detach stabilizer links from both ends of stabilizer bar. See **Stabilizer bar link, removing and installing** in this repair group.
- Slide stabilizer bar sideways to remove.

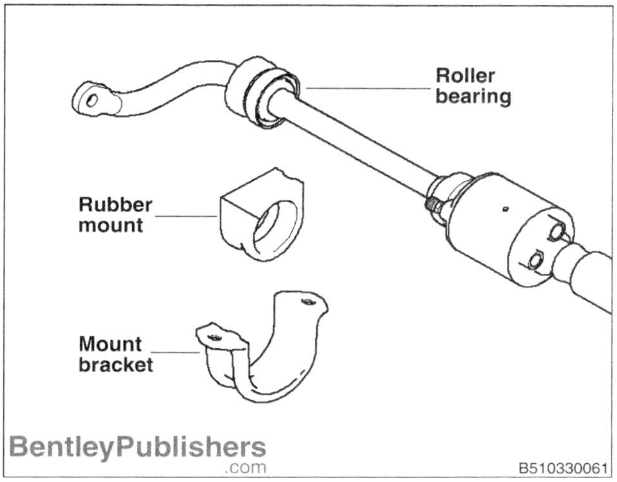

Separate stabilizer bar mount brackets and rubber mounts from roller bearings.

- Check roller bearings for free movement. Slight discharge of grease from bearing is acceptable. Replace stabilizer bar if a bearing binds or is otherwise damaged.
- Replace rubber mounts if damaged. Keep rubber grease-free.

- When reinstalling, first attach stabilizer bar to stabilizer links.
  - Use new self-locking nuts.
  - Use Torx tool (T30) to counterhold shaft of stabilizer bar link ball joint while tightening nut.

Tightening torque	
Stabilizer bar link to stabilizer bar (replace M10 x 1.5 x ZNS3 self-locking nuts with new)	65 Nm (48 ft-lb)

- Attach pressure line unions to oscillating motor finger-tight.

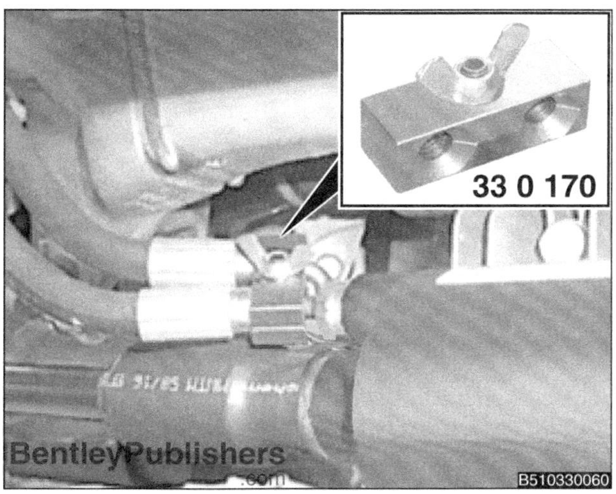

Place special tool 33 0 170 over lines and tighten wing nut. This forces pressure lines into correct alignment with motor. Tighten fluid lines. Make sure lines do not contact stabilizer bar. Minimum distance = 5 mm (¼ in).

Tightening torque	
Fluid pressure line to oscillating motor (union nut)	30 Nm (22 ft-lb)

- Reinstall stabilizer bar to subframe, noting the following.
  - To prevent stabilizer bar distortion, install left mount bracket first (without elongated holes). Then install right mount bracket (with elongated holes).
  - Replace self-locking fasteners.

Tightening torque	
Stabilizer bar mount to subframe (replace M8 ZNS3 self-locking nuts with new)	28 Nm (21 ft-lb)

BentleyPublishers.com
BentleyPublishers.com - All Rights Reserved

– Align pressure lines to subframe and tighten bracket screws.

Tightening torque	
Pressure lines to subframe	8 Nm (6 ft-lb)

– Remainder of installation is reverse of removal.

Tightening torque	
X-brace to chassis • M10 bolts • M10 nut	 59 Nm (44 ft-lb) 41 Nm (30 ft-lb)

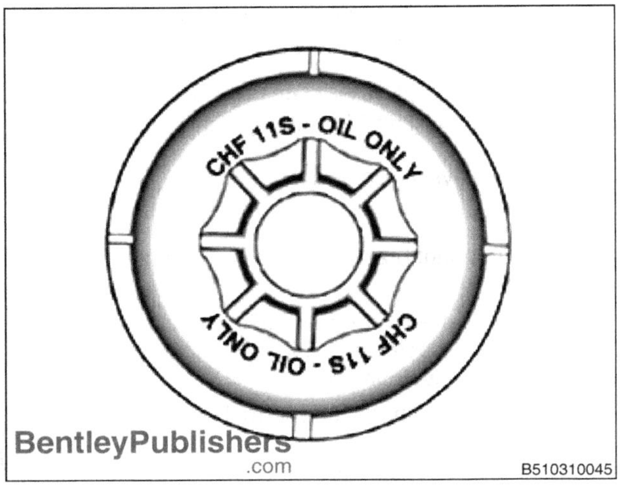

◀ Fill hydraulic fluid reservoir (same as power steering fluid reservoir).

Active stabilizer system fluid	
Hydraulic fluid for tandem pump system	Pentosin CHF 11S

– Use BMW scan tool to bleed air out of active stabilizer system. See **Hydraulic system, bleeding** in this repair group. Check lines and connections for leaks.

– Carry out commissioning procedure. See **Active stabilizer, commissioning** in **300 Suspension, Steering and Brakes–General**.

Stabilizer bar link, removing and installing

– Raise vehicle and safely support.

> ***WARNING—***
> *• Make sure the car is firmly supported on jack stands designed for the purpose. Place jack stands underneath structural chassis points, not underneath suspension parts.*

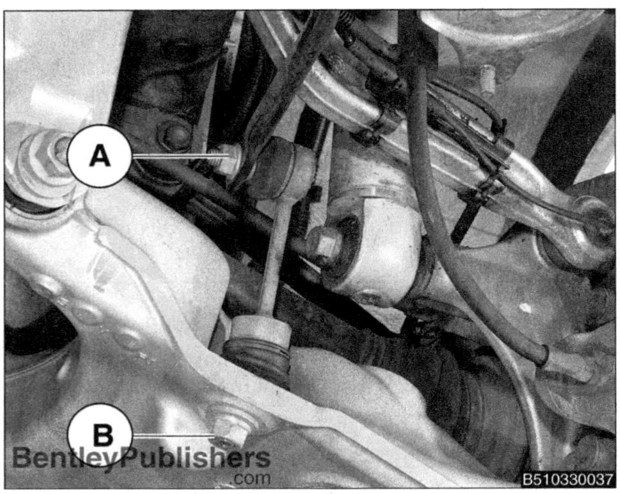

◀ Remove stabilizer link mounting nut (**A**) at stabilizer bar. Use Torx tool (T30) to counterhold shaft of stabilizer bar link ball joint while loosening nut.

– Remove stabilizer link mounting fastener (**B**) at swing arm.

– Press out stabilizer link ball joint and remove stabilizer link.

– Installation is reverse of removal.

Tightening torques	
Stabilizer bar link to subframe (replace M12 x 1.5 ZN3 self-locking nut with new)	100 Nm (74 ft-lb)
Stabilizer bar link to stabilizer bar (replace M10 x 1.5 x ZNS3 self-locking nut with new)	65 Nm (48 ft-lb)

BentleyPublishers.com
BentleyPublishers.com—All Rights Reserved

331 Rear Axle Differential

TABLE

GENERAL

This repair group covers removal and repair information for the rear differential (final drive), axle shafts, CV joints, CV joint boots, and differential seal replacement.

Internal repairs of the differential assembly are not covered in this manual.

Rear differential and rear drive axles

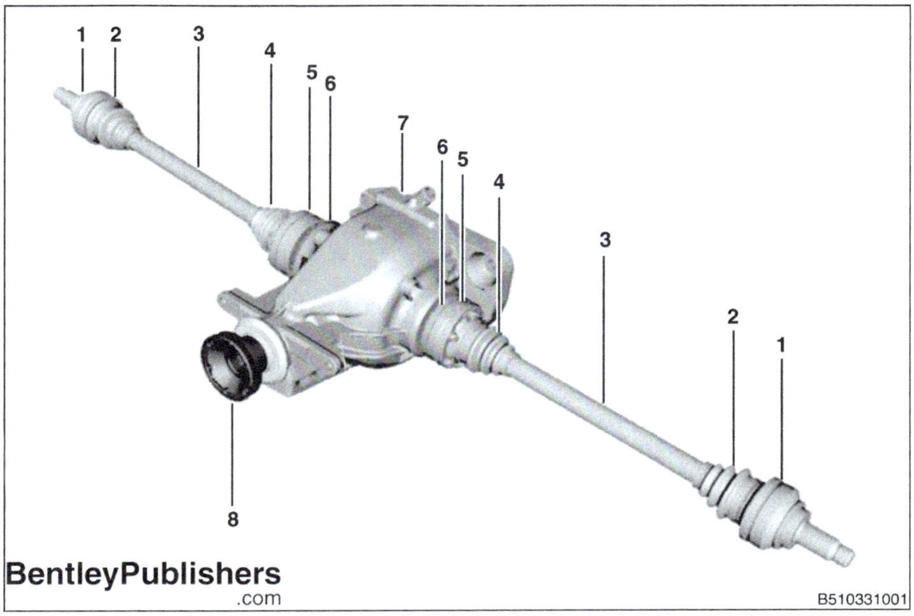

1. Outer CV joint
2. Outer CV joint boot
3. Drive axle
4. Inner CV joint boot
5. Inner CV joint
6. Differential output flange
7. Differential
8. Differential input flange

BentleyPublishers.com
BentleyPublishers.com–All Rights Reserved

Warnings and Cautions

WARNING —
- *Do not reuse self-locking fasteners.*

CAUTION—
- *Due to the chemical and corrosion characteristics of aluminum, follow these precautions when working with aluminum components:*
  —Do not bring into contact with battery acid.
  —Do not use wire brushes with brass or iron bristles.
  —Do not expose to flying sparks.
  —Do not subject to steel welding splashes.
  —Do not expose to temperatures over 80°C (176°F), even for short periods. Paint shop temperatures are not a problem.
- *Do not allow the vehicle to stand on its wheels when the EHC system is disabled or drained. Damage to air springs may result.*
- *When disconnecting and connecting EHC air lines, do not allow contamination to enter the lines.*

REAR DIFFERENTIAL

Rear differential applications

Application information and oil fill capacities for rear differentials is in **Table a**.

Table a. Rear differential applications

Engine	Transmission	Rear differential	Capacity
M54, N52	Manual, automatic	188L / 188LW	1.0 ltr (1.05 US qt)
N54	Automatic	188LW	1.0 ltr (1.05 US qt)
N54	Manual	215LW	1.2 ltr (1.27 US qt)
N62, N62 TU	Automatic	188L / 188LW	1.0 ltr (1.05 US qt)
N62, N62 TU	Manual	215K	1.4 ltr (1.48 US qt)

Differential oil, draining and filling

The 5 Series rear differential is filled with lifetime oil that ordinarily does not need to be changed. BMW recommends using only a specially formulated synthetic gear oil (SAF-XO) that is available through an authorized BMW dealer parts department. For additional information on this lubricant and any other lubricants that may be compatible, contact an authorized BMW dealer service department.

– Drive vehicle to warm differential oil.

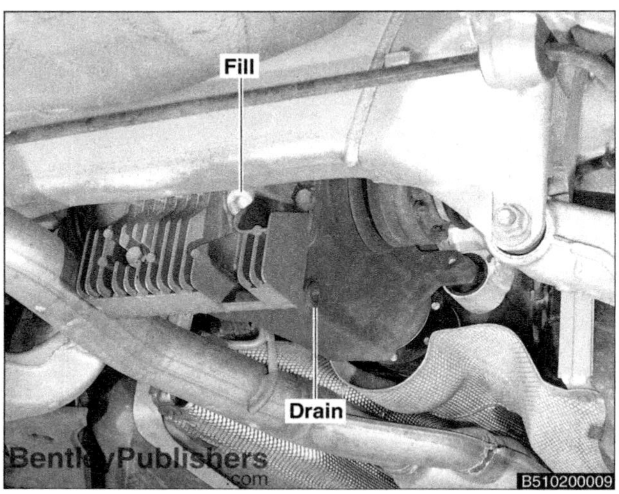

Remove differential **Drain** plug and drain oil into suitable container.

BentleyPublishers.com
BentleyPublishers.com—All Rights Reserved

– Reinstall **Drain** plug.
 • Plug with sealing O-ring: Replace O-ring.
 • Plug without sealing O-ring: Replace plug.

Tightening torque	
Drain plug to rear differential housing: • Plug with sealing O-ring (replace O-ring) • Plug without sealing O-ring (replace plug)	 65 Nm (48 ft-lb) 60 Nm (44 ft-lb)

– Remove **Fill** plug and add rear differential oil and check level. See **Differential oil level, checking** in this repair group.

– Reinstall **Fill** plug.
 • Plug with sealing O-ring: Replace O-ring.
 • Plug without sealing O-ring: Replace plug.

Tightening torque	
Fill plug to rear differential housing: • Plug with sealing O-ring (replace O-ring) • Plug without sealing O-ring (replace plug)	 65 Nm (48 ft-lb) 60 Nm (44 ft-lb)

Differential oil level, checking

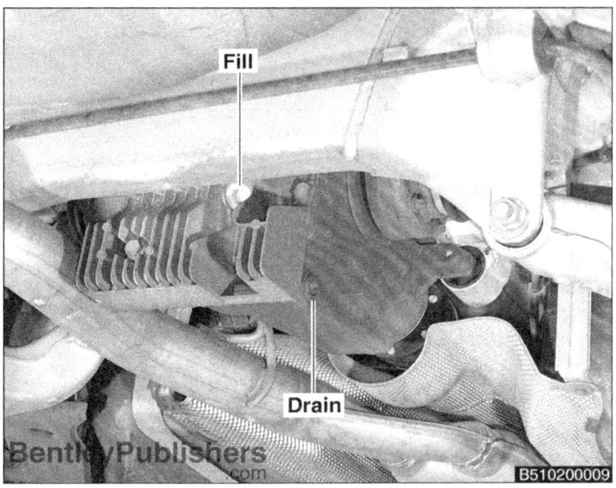

◄ Check lubricant level with vehicle on a level surface:
 • Remove differential **Fill** plug.
 • Level is correct when fluid just reaches edge of plug hole.
 • If necessary, top up fluid.
 • Plug with sealing O-ring: Replace O-ring.
 • Plug without sealing O-ring: Replace plug.
 • Install and tighten oil **Fill** plug when oil level is correct.

Tightening torque	
Fill / drain plug to rear differential housing: • Plug with sealing O-ring (replace O-ring) • Plug without sealing O-ring (replace plug)	 65 Nm (48 ft-lb) 60 Nm (44 ft-lb)

Differential, removing and installing

– Raise car and support safely.

> ***WARNING —***
> *• Make sure the car is firmly supported on jack stands designed for the purpose. Place jack stands underneath structural chassis points, not underneath suspension parts.*

– Disconnect rear driveshaft from differential. See **260 Driveshaft**.

◄ Remove drive axle CV joint bolts (**arrows**) at left and right sides. Suspend drive axles from body using stiff wire.

BentleyPublishers
BentleyPublishers.com—All Rights Reserved

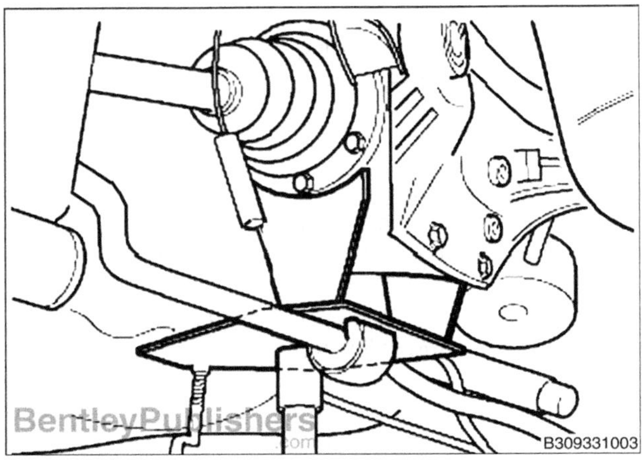

Support differential with special support or suitable transmission jack. Make sure jack does not damage output flange dust covers:

- Remove differential front mounting bolts.
- Remove differential rear mounting bolt.
- Remove vent line from differential.
- Carefully lower differential unit.

– To install, lift differential into place:

- Install mounting bolts finger tight.
- Remove differential support.
- Tighten both front mounting bolts.
- Tighten rear mounting bolt.

Tightening torques	
Differential to subframe • Front mounting bolts (M12) • Rear mounting bolt (M14)	 100 Nm (74 ft-lb) 165 Nm (122 ft-lb)

– Remainder of installation is reverse of removal. Remember to check differential oil. See **Differential oil level, checking** in this repair group.

Tightening torques	
Drive axle to differential output flange (replace M10 x 43 mm ZNS3 Torx bolts with new)	80 Nm (59 ft-lb)
Rear driveshaft CV joint to differential input (replace M10 x 52 mm ZNS bolts with new) • Stage 1 • Stage 2	 20 Nm (15 ft-lb) additional 90°

Differential input shaft seal, replacing

– Remove exhaust system. See **180 Exhaust System**.

– Remove drive shaft. See **260 Driveshaft**.

– Lift out differential input flange collar nut locking plate.

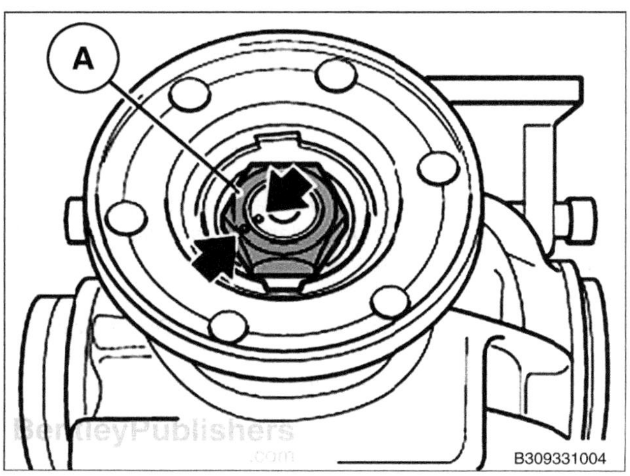

Mark position of collar nut (**A**) on input flange with a center punch (**arrows**).

BentleyPublishers.com
BentleyPublishers.com–All Rights Reserved

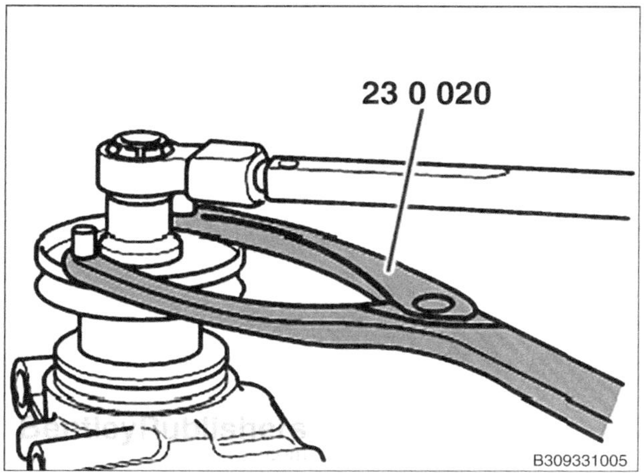

Using BMW special tool 23 0 020 or equivalent, counterhold input flange and remove collar nut. For clarity, illustration shows differential removed.

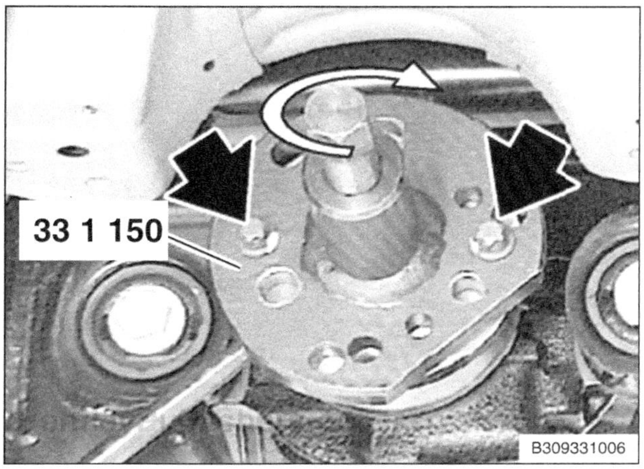

188L / 188LW rear differential: Using BMW special tool 33 1 150 or equivalent puller, remove differential input flange.

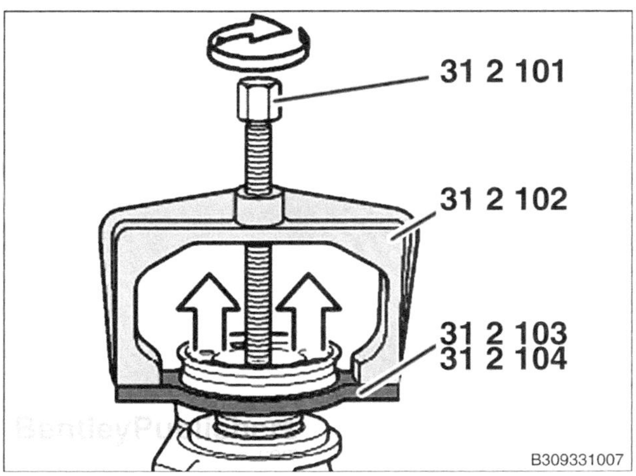

215K / 215L / 215LW rear differential: Using BMW special tools 31 2 101, 31 2 102, 31 2 103 and 31 2 104 or equivalent puller, remove differential input flange.

BentleyPublishers.com
BentleyPublishers.com—All Rights Reserved

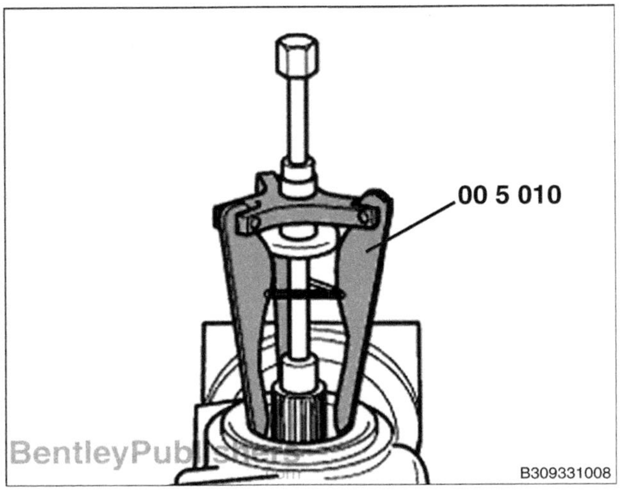

Using BMW special tool 00 5 010 or equivalent, remove seal from differential housing.

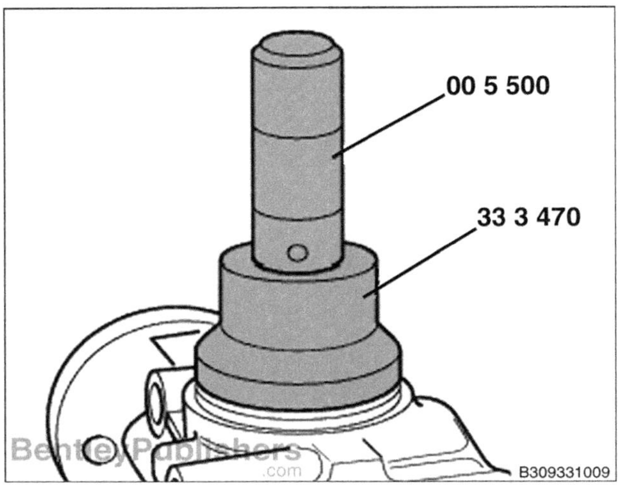

Using BMW special tools 00 5 500 and 33 3 470 or equivalent, drive new seal into place until fully seated.

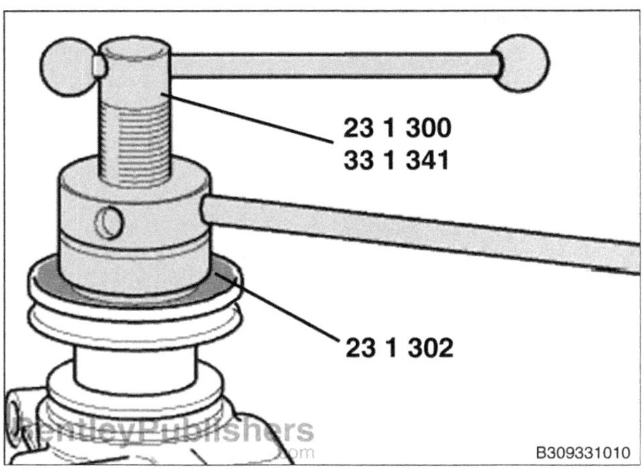

Coat drive flange sealing area and seal lip with differential oil:

- Install drive flange on pinion shaft.
- Using BMW special tools 23 1 300, 33 1 341 and 23 1 302, press drive flange on pinion shaft just far enough to allow collar nut to be installed.

BentleyPublishers.com
BentleyPublishers.com—All Rights Reserved

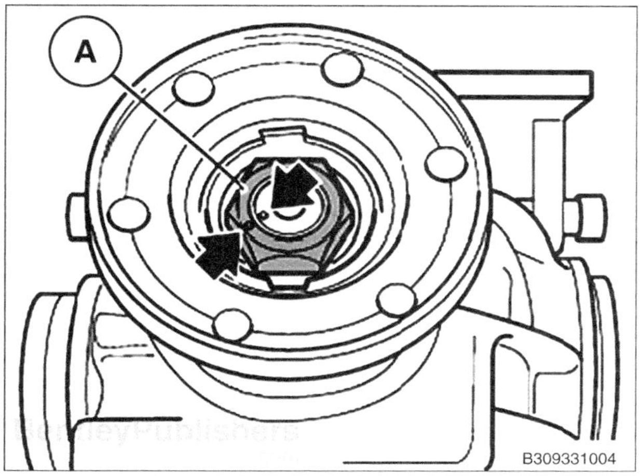

< Tighten collar nut (**A**) only until previously marked points are aligned (**arrow**).

> ***CAUTION—***
> - *Do not under any circumstances tighten collar nut beyond the marked points. Otherwise clamping sleeve inside differential housing may be damaged.*

– Using a suitable drift pin punch, install new collar nut locking plate. Stake plate in place.

– Remainder of installation is reverse of removal.

– Check differential oil level. See **Differential oil level, checking** in this repair group.

Tightening torque	
Rear driveshaft CV joint to differential input (replace M10 x 52 mm ZNS bolts with new) • Stage 1 • Stage 2	 20 Nm (15 ft-lb) additional 90°

Differential output shaft seal, replacing

– Remove exhaust system (if necessary). See **180 Exhaust System**.

< Remove drive axle CV joint bolts (**arrows**). Suspend drive axle from body using stiff wire.

< Use soft-faced hammer to drive output flange out of differential (**arrow**).

– If necessary, pry off dust cover.

BentleyPublishers.com
BentleyPublishers.com–All Rights Reserved

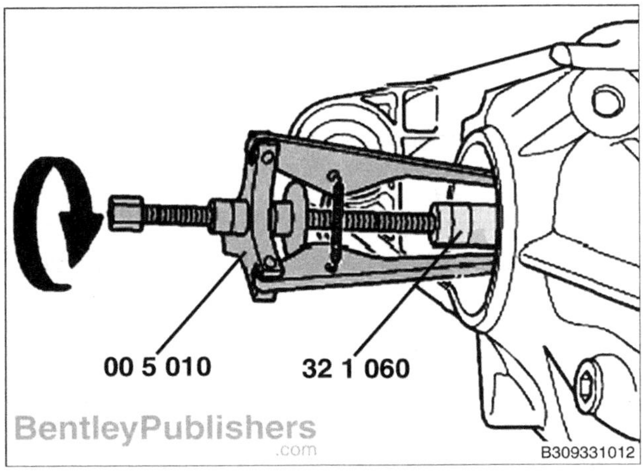

- Remove seal using BMW special tools 00 5 010 and 32 1 060 / 33 1 308 or equivalent.

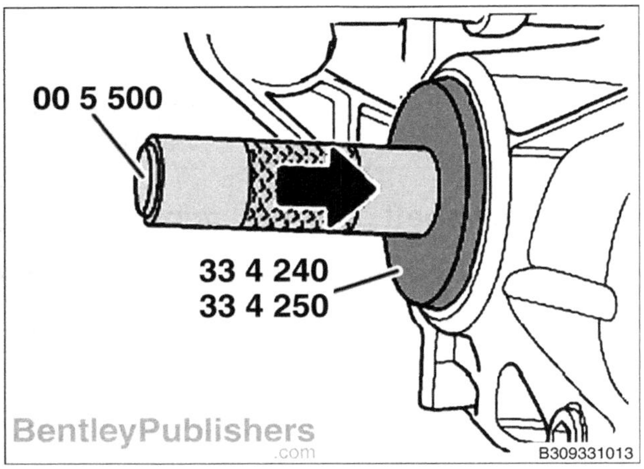

- Coat seal with differential oil and drive into place using BMW special tools 00 5 500 and 33 4 240 / 33 4 250 or equivalent.

  Tool dimensions:
  - 188L / 188LW differential: 33 4 240 (90 x 44 x 10 mm)
  - 215k / 215L / 215LW differential: 33 4 250 (100 x 50 x 10 mm)

– Replace output flange dust cover (if applicable).

- Replace snap ring (**arrow**) on differential output flange shaft:
  - Coat output flange shaft with differential oil where it contacts shaft seal.
  - Push output flange into differential until flange splines engage splines of differential gear and snap ring can be heard to snap into place.

– Remainder of installation is reverse of removal. Remember to:
  - Check differential oil level. See **Differential oil level, checking** in this repair group.

Tightening torque	
Drive axle to differential output flange (replace M10 x 43 mm ZNS3 Torx bolts with new)	80 Nm (59 ft-lb)

BentleyPublishers.com
BentleyPublishers.com—All Rights Reserved

Differential mounts, replacing

- Remove differential. See **Differential, removing and installing** in this repair group.

Rear differential mounts

- Remove differential rear rubber mount using suitable puller.
- Coat new mount with Circolight® anti-friction agent and draw mount in firmly using suitable press.

Front differential mounts

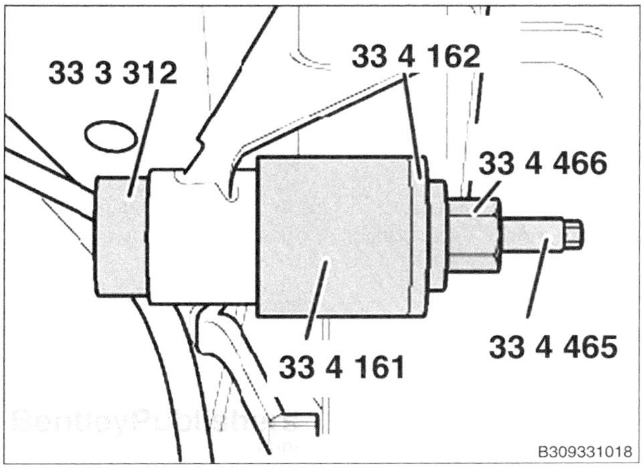

◄ Using BMW special tools 33 3 312, 33 4 161, 33 4 162, 33 4 465 and 33 4 466, remove differential front mount.

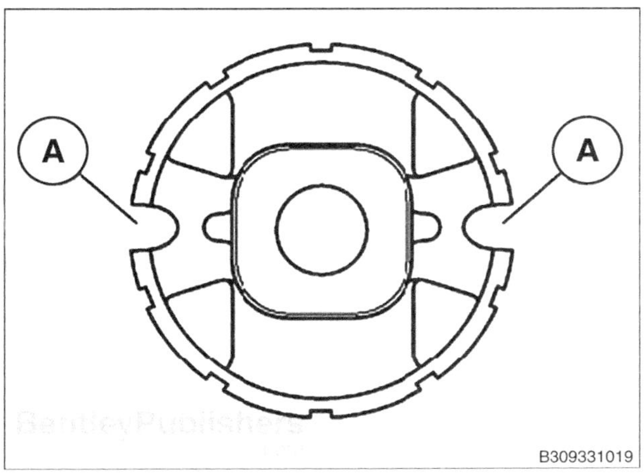

◄ Position slots (**A**) horizontally in rear subframe.

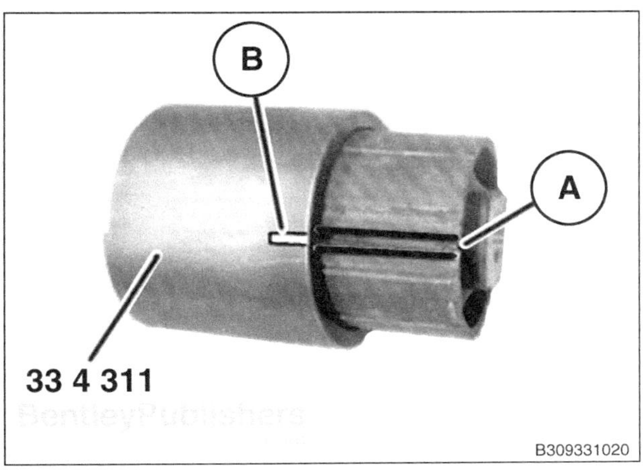

◄ Coat bearing sleeve in rear subframe and rubber mount with Circolight® anti-friction agent:

- Slide rubber mount from side with large diameter into BMW special tool 33 4 311.
- Slot **A** must line up with mark **B**.

BentleyPublishers.com
BentleyPublishers.com–All Rights Reserved

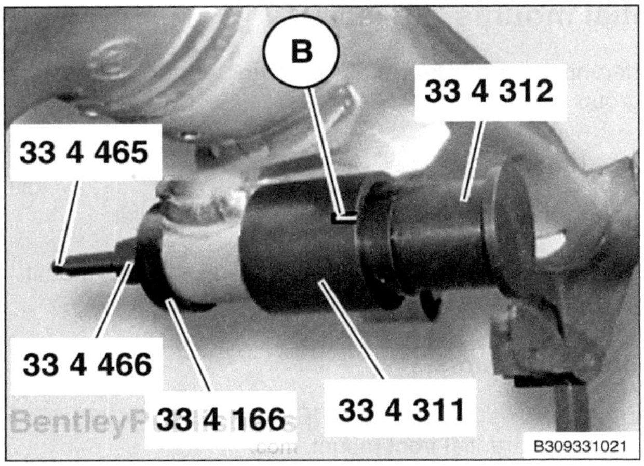

Using BMW special tools 33 4 166, 33 4 311, 33 4 312, 33 4 465 and 33 4 466, install front differential mount as far as it will go.

– Remainder of installation is reverse of removal.

Rear Drive Axles

Drive axles use constant-velocity (CV) joints on both ends. For replacement parts, only CV joint boots or complete axles are offered by BMW.

Rear drive axle assembly

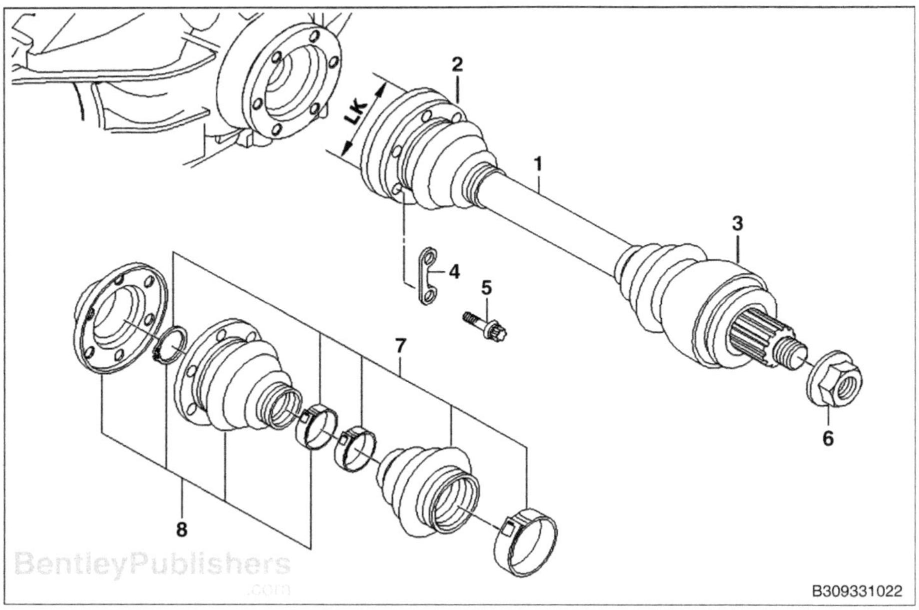

1. Drive axle
2. Inner CV joint
3. Outer CV joint
4. Reinforcement
5. Torx bolt
6. Collar nut
7. Outer CV boot kit
8. Inner CV boot kit

BentleyPublishers.com
BentleyPublishers.com—All Rights Reserved

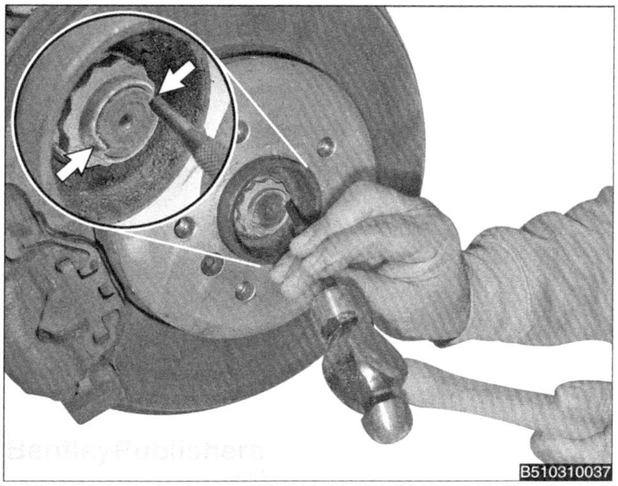

Rear drive axle, removing and installing

– Raise vehicle and remove rear wheel.

> ***WARNING—***
> • *Make sure the car is firmly supported on jack stands designed for the purpose. Place jack stands underneath structural chassis points, not underneath suspension parts.*

◀ Using a suitable punch, straighten staked sides of wheel hub collar nut (**arrows**).

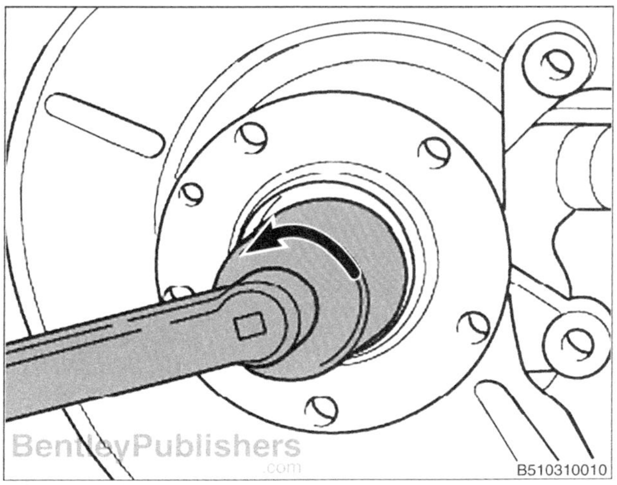

◀ With an assistant holding down brake pedal, remove collar nut and discard.

> ***CAUTION—***
> • *The drive axle collar nut is tightened to a high torque. If necessary, reinstall road wheel and place vehicle on floor before breaking collar nut free.*

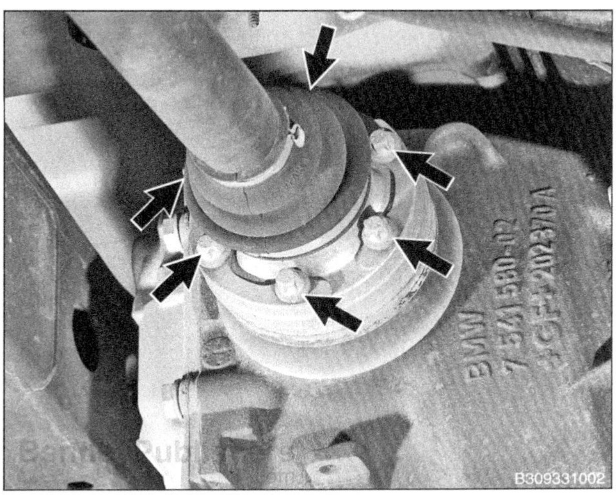

◀ Remove drive axle mounting bolts (**arrows**) from differential flange.

– Support rear wheel bearing carrier with a suitable jack.

– Using jack, lift wheel bearing carrier so that inboard side of drive axle clears differential.

BentleyPublishers.com—All Rights Reserved

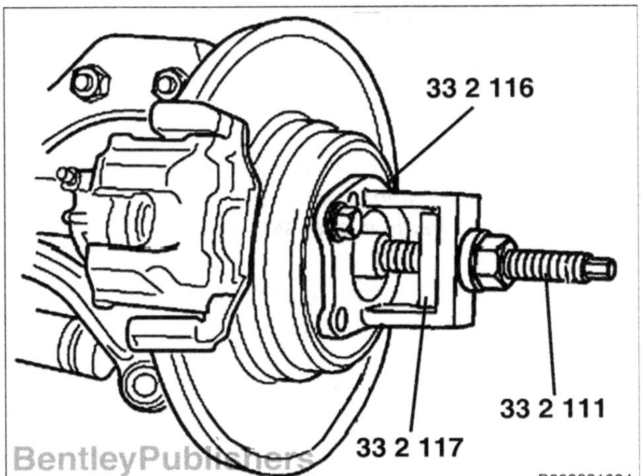

B309331024

Press drive axle out of wheel hub. Remove drive axle towards center of vehicle.

– Installation is reverse of removal. Remember to:
  - Replace CV joint bolts and axle collar nut.
  - Tighten drive axle collar nut with wheel on the ground. Stake nut to axle end.

Tightening torques	
Drive axle collar nut to wheel hub (replace M27 collar nut with new, lightly oil contact face, stake nut to axle end)	420 Nm (310 ft-lb)
Drive axle to differential output flange (replace M10 x 43 mm ZNS3 Torx bolts with new)	80 Nm (59 ft-lb)
Road wheel to hub	120 ± 10 Nm (89 ± 7 ft-lb)

CV joint boots

When replacing CV boots, use a complete boot repair kit. The kit includes new boots, clamping bands, special lubricant and a new CV joint shaft circlip.

The outer CV joint cannot be removed from the axle shaft. In order to replace the outer CV boot, it is necessary to remove the inner joint and boot first.

If the CV joints are worn or defective, a complete rebuilt axle shaft is available from an authorized BMW dealer parts department.

– Remove drive axle. See **Rear drive axle, removing and installing** in this repair group.

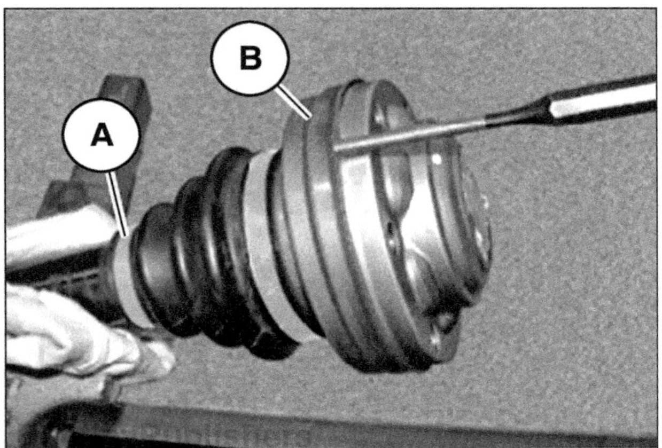

B309331026

Working with axle shaft on bench, cut off old boot clamps and remove boot (**A**) from inner joint (**B**) with suitable tool. Clean old grease off joint and shaft.

– Remove dust cover from inner CV joint.

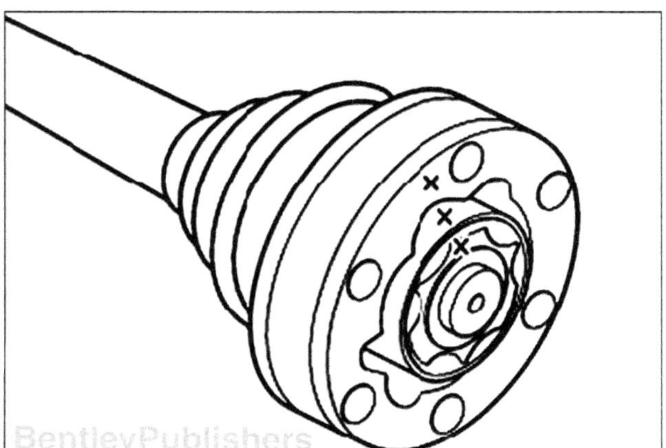
B309331025

Mark CV joint component positions in relation to each other. This aids in reassembly if CV joint comes apart.
- Ball hub
- Ball cage
- CV joint

BentleyPublishers.com
BentleyPublishers.com—All Rights Reserved

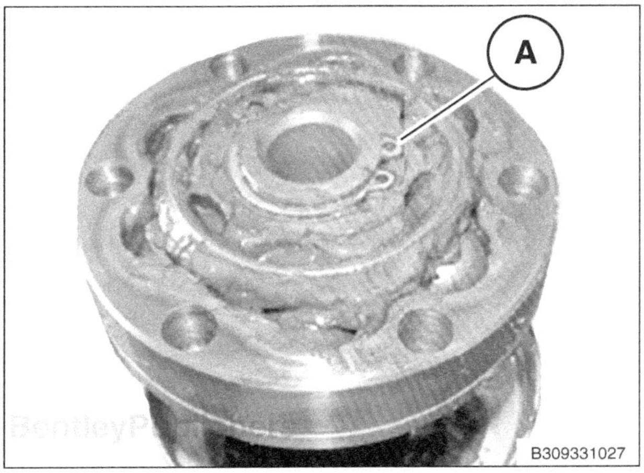

◄ Remove circlip (**A**).

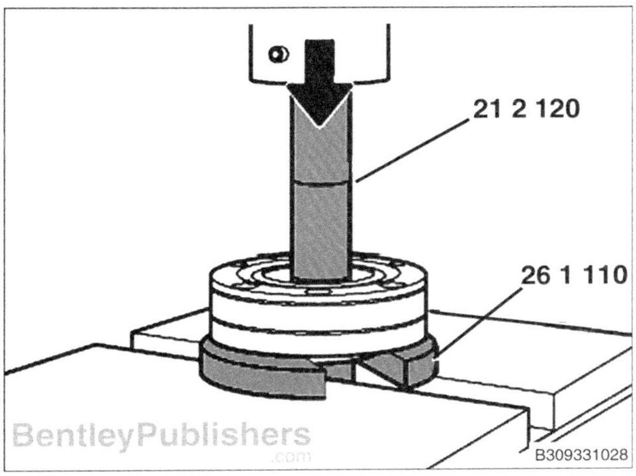

◄ Support inner hub with BMW special tool 26 1 110 and press axle shaft out of joint using BMW special tool 21 2 120.

– Clean all old lubricant off shaft splines and inner joint splines.

> ***CAUTION—***
> - *If the CV joint must be disassembled for cleaning and inspection, be sure to matchmark inner and outer race and intermediate ball cage. This allows reassembly of parts in their original positions.*

– To inspect a CV joint, clean away the grease and look for galling, pitting, and other signs of wear or physical damage.

- Polished surfaces or visible ball tracks alone are not necessarily cause for replacement.
- Discoloration due to overheating indicates lack of lubrication. Replace CV joint.

> ***CAUTION—***
> - *Do not let the ball hub pivot too far in the outer ring of the joint. The balls may fall out.*

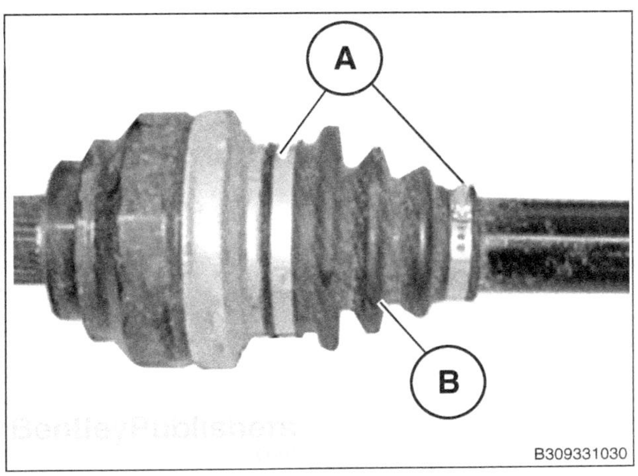

◄ Working at outer CV joint boot, cut off old boot clamps (**A**) and remove boot (**B**) from outer CV joint. Clean old grease off CV joint.

– Slide new outer CV joint boot on axle shaft. Fill boot and joint with CV grease, then slide big end of boot over joint. Install big clamp.

Before installing small boot clamp:

- Flex CV joint as far over as it will go.
- Insert small screw-driver between boot and axle-shaft to "burp" air from boot.
- Install new small clamp.

– Slide inner CV joint boot on axle shaft.

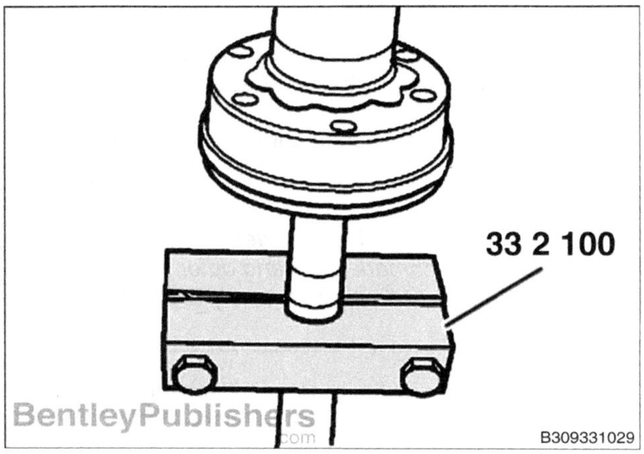

While supporting axle shaft with BMW special tool 33 2 100, press inner hub of CV joint onto shaft. Install a new circlip.

– Attach boot to CV joint and install new boot clamps. Be sure to "burp" air from boot.

– Use sealing gel to seal dust cover to CV joint prior to reinstallation.

– Installation is reverse of removal. See **Rear drive axle, removing and installing** in this repair group.

BentleyPublishers.com
BentleyPublishers.com—All Rights Reserved

340 Brakes

GENERAL

This repair group covers service for the brake system and for electronic braking and stability control (traction control) systems.

See also:
- **020 Maintenance** for basic brake system checks
- **300 Suspension, Steering and Brakes–General**
- **612 Switches** for brake light switch and steering angle sensor
- **ECL Electrical Component Locations** for fuse and component access
- **ELE Electrical Wiring Diagrams**

BentleyPublishers
BentleyPublishers.com–All Rights Reserved

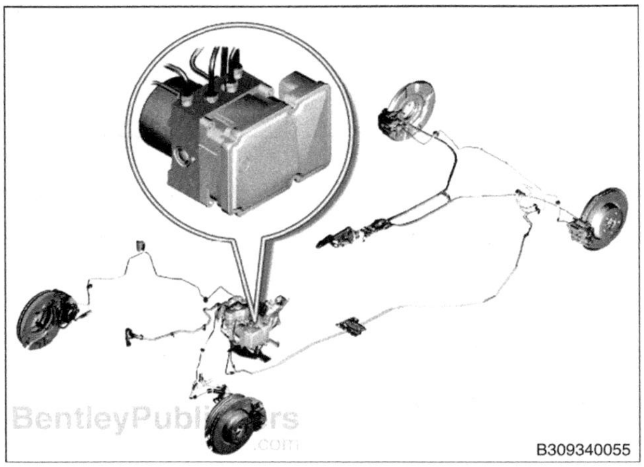

Brake system

5 Series models are equipped with vacuum power-assisted four-wheel disc brakes with integral antilock braking system (ABS). Single-piston calipers act on vented front and rear rotors. Two-stage brake pad wear sensors work along with condition based service (CBS) system to indicate when brake pads need replacement. The first stage is activated when brake pad linings reach a thickness of 6 mm (¼ in), the second is activated at a thickness of 4 mm (1/6 in). The dual drum-type parking brake system is integrated with the rear brake rotors.

Electronic braking and stability control

5 Series models are equipped with dynamic stability control (DSC). DSC is a computer controlled traction control system that uses the ABS system in conjunction with engine management controls to control wheel spin during acceleration and maintain vehicle stability while braking.

Troubleshooting

Brake performance is mainly affected by three factors:

- Level and condition of brake fluid
- Ability of brake system to create and maintain pressure
- Condition of friction components

Air in brake fluid makes the brake pedal feel spongy during braking or increases the brake pedal force required to stop. Fluid contaminated by moisture or dirt corrodes the system. Inspect the brake fluid inside the reservoir. If it is dirty or murky, or is more than two years old, replace fluid. See **Brake system, bleeding** in this repair group.

To check the function of the brake master cylinder, hold the brake pedal down hard with the engine running. If the pedal slowly falls to the floor, either the master cylinder is leaking internally or fluid is leaking externally.

Inspect the brake rotors for glazing, discoloration and scoring. Steering wheel vibration while braking at speed is often caused by warped rotors, but can also be caused by worn suspension components.

When troubleshooting, keep in mind that tire inflation, wear and temperature can all have an affect on braking. See **300 Suspension, Steering and Brakes–General** for additional suspension and brake system troubleshooting.

BentleyPublishers.com
BentleyPublishers.com—All Rights Reserved

Table a lists symptoms of brake problems, probable causes, and suggested corrective actions. Unless noted otherwise, relevant repairs are described later in this repair group.

Table a. Brake system troubleshooting

Symptom	**Probable cause**	**Repairs**
Brake squeal	Brake pad carriers dirty or corroded or pads loose (poor fit) in pad carrier	Remove brake pads and clean calipers. Use original equipment pads for proper fit and use BMW anti-squeal compound during installation.
	Brake pads heat-glazed or oil-soaked	Replace brake pads. Clean rotors. Replace leaking calipers as required.
	Wheel bearings worn (noise most pronounced when turning)	Replace worn bearings. See **310 Front Suspension** or **330 Rear Suspension**.
	Incorrectly installed brake pads, parking brake shoes, or brake parts	Check component installation. Check anti-rattle springs. Replace as necessary.
Pedal goes to floor when braking	Brake fluid loss due to system leaks	Check fluid level and inspect for signs of leakage.
	Master cylinder or traction control system faulty	Replace master cylinder. Diagnose traction control system using scan tool.
Low pedal after system bleeding	Master cylinder faulty	Replace master cylinder.
Pedal spongy or brakes work only when pedal is pumped	Air in brake fluid	Bleed system using scan tool.
	Master cylinder or traction control system faulty	Replace master cylinder. Diagnose traction control system using scan tool.
Excessive braking effort	Brake pads wet	Use light pedal pressure to dry pads while driving.
	Brake pads heat-glazed or fluid-soaked	Replace brake pads and rotors. Replace leaking calipers.
	Vacuum pump, vacuum booster, or vacuum hose connections to booster faulty	Inspect vacuum lines. Test vacuum booster and replace as required. Test brake booster check valve for one-way flow.
Brakes pulsate, chatter or grab	Warped brake rotors	Resurface or replace rotors.
	Brake pads worn	Replace brake pads.
	Brake pads heat-glazed or fluid-soaked	Clean rotors. Replace leaking calipers.
Uneven braking, car pulls to one side, rear brakes lock	Incorrect tire pressures or worn tires	Inspect tire condition. Check and correct tire pressures.
	Brake pads on one side of car heat-glazed or fluid-soaked	Replace brake pads. Clean rotors. Replace leaking calipers.
	Brake caliper or brake pads binding	Clean and recondition brakes.
	Worn suspension components	Inspect for worn or damaged suspension components. See **310 Front Suspension** or **330 Rear Suspension**.
Brakes drag, bind or overheat	Brake caliper or brake pads binding	Clean or replace caliper.
	Master cylinder or traction control system faulty	Replace master cylinder. Diagnose traction control system using scan tool.

BentleyPublishers.com
BentleyPublishers.com—All Rights Reserved

Warnings and Cautions

WARNING —
- *Make sure that the brake system is bled using a BMW scan tool or equivalent. See* **Bleeding Brakes** *in this repair group.*
- *Semi-metallic and metallic brake friction materials in brake pads or shoes produce dangerous dust. Treat all brake dust as a hazardous material. Do not create dust by grinding, sanding, or cleaning brake friction surfaces with compressed air.*
- *Brake fluid is poisonous, corrosive and dangerous to the environment. Wear safety glasses and rubber gloves when working with brake fluid. Do not siphon brake fluid with your mouth. Dispose of brake fluid properly.*
- *Do not reuse self-locking nuts, bolts or fasteners. They are designed to be used only once and may fail if reused. Replace them with new self-locking fasteners.*
- *A car with electronic stability control is still subject to normal physical laws. Avoid excessive speeds for the road conditions encountered.*

CAUTION—
- *Brake fluid damages paint. Immediately clean brake fluid spilled on painted surfaces and wash with water.*
- *Use new brake fluid from a fresh, unopened container. Brake fluid absorbs moisture from the air. This can lead to corrosion problems in the braking system and also lowers the fluid boiling point.*
- *When working on brake fluid lines:*
  -Do not mix up fluid lines at the master cylinder or hydraulic unit. Label unions before disconnecting.
  -Do not kink brake lines.
  -Plug open lines and brake fluid ports to prevent contamination.
- *Tighten brake hoses on front wheels with wheels in straight ahead position.*
- *If carrying out electric welding work, be sure to disconnect electrical harness connector from electronic control module.*
- *Do not expose electronic control modules to high sustained heat. Maximum heat exposure:*
  -95°C (203°F) for short periods of time
  -85°C (185°F) for long periods of time (approx. 2 hours)
- *Replace brake pad sensor with new if sensor is removed. Loss of spring tension in the sensor may cause it to fall off the pads.*

BentleyPublishers.com
BentleyPublishers.com—All Rights Reserved

Bleeding Brakes

> ***WARNING —***
> - *The BMW traction control system uses electronic controls and a sophisticated hydraulic unit in the brake system. Once air enters the hydraulic unit, it is very difficult to remove using traditional methods. For this reason, use BMW scan tool or equivalent to pressure bleed brakes.*
> - *When flushing brake fluid from the system, use extreme care to not let the brake fluid reservoir run dry. If air enters the hydraulic unit, use the BMW scan tool to bleed the brake system before the vehicle is driven.*

Brake bleeding is usually done for one of two reasons:

- To replace old brake fluid as part of routine maintenance
- To expel trapped air in the system that resulted from opening the brake hydraulic system during repairs.

BMW recommends completely replacing the brake fluid in the entire brake system, including the brake fluid reservoir, at least once every two years. When replacing brake fluid or bleeding brakes, use the correct specification brake fluid.

Brake fluid application	
BMW preferred fluid	Low viscosity DOT 4 brake fluid
Low viscosity DOT 4 brake fluid may be used in all 5 Series vehicles. DOT 4 and low viscosity DOT 4 brake fluid can be mixed.	

When adding or replacing brake fluid, add new brake fluid from an unopened container. It is important to bleed the entire system when any part of the hydraulic system is opened.

If you are certain no air was introduced into the master cylinder or DSC hydraulic unit, bleed the brakes at the calipers using a pressure bleeder. See **Brake calipers, bleeding** in this repair group.

However, if air enters the DSC hydraulic unit, bubbles may adhere to the edges and internal valves of the unit and these cannot be removed via conventional flushing. In that case it is necessary to bleed brakes using a BMW scan tool or equivalent. The special tool creates pulsations or vibrations which loosen and flush the bubbles.

Therefore, if you are in any doubt about introduced air into the brake system, use BMW scan tool to bleed the system or have the brakes bled by a BMW workshop. See **Brake system, bleeding** in this repair group.

When bleeding the brakes, start at the wheel farthest from the master cylinder and progress in the following order:

- right rear brake
- left rear brake
- right front brake
- left front brake

BentleyPublishers.com
BentleyPublishers.com—All Rights Reserved

Brake calipers, bleeding

– Remove left side cabin microfilter housing (at left rear engine compartment) to access brake fluid reservoir. See **020 Maintenance**.

◄ Top off brake fluid to **MAX** marked on reservoir.

– Connect pressure bleeder to reservoir. Pressurize system to approximately 2 bar (29 psi).

> ***CAUTION—***
> • *Do not exceed a pressure of 2 bar (29 psi) when pressure bleeding the brake system. Excessive pressure damages the brake fluid reservoir.*

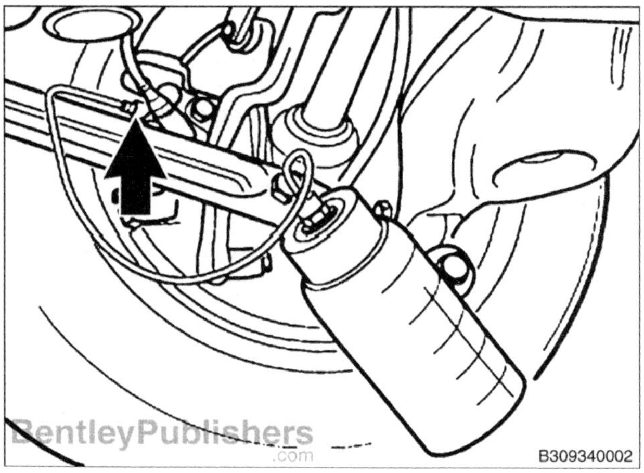

◄ Connect bleeder hose and bottle to **right rear** caliper bleeder screw (**arrow**). Have a helper hold brake pedal down.

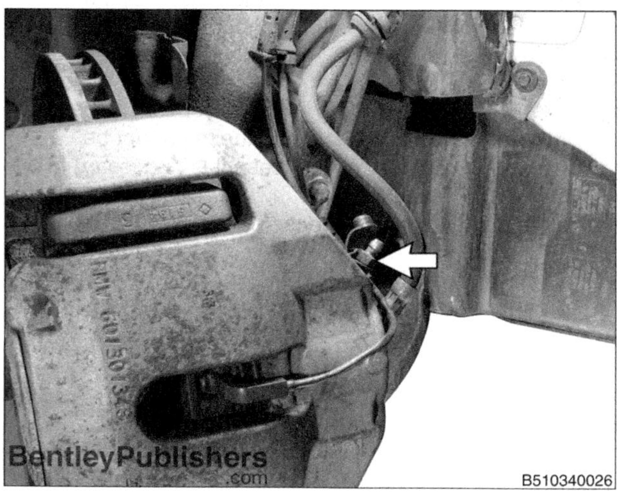

◄ Open bleeder screw (**arrow**). Have helper slowly pump brakes about 10 times with bleeder screw open, holding pedal down on last pump. When escaping fluid is free of air bubbles, close bleeder screw. (Photo illustrates front brake caliper. Rear caliper is similar.)

> ***CAUTION—***
> • *Make sure bleeder hose remains submersed in clean brake fluid whenever the bleeder valve is open.*

– Close bleeder screw and release brake pedal. Refill brake fluid reservoir and proceed to left rear wheel.

– Continue bleeding remaining wheels in the following order:
- **left rear**
- **right front**
- **left front**

Tightening torques	
Bleeder to caliper: • 9 mm screw • 11 mm screw	 7 - 11 Nm (5 - 8 ft-lb) 12 - 16 Nm (9 - 12 ft-lb)

BentleyPublishers.com
BentleyPublishers.com—All Rights Reserved

Brake system, bleeding

The procedure below requires the use of BMW scan tool.

– Remove left side cabin microfilter housing (at left rear engine compartment) to access brake fluid reservoir. See **020 Maintenance**.

◄ Top off brake fluid to **MAX** marked on reservoir.

– Connect pressure bleeder to reservoir. Pressurize system to approximately 2 bar (29 psi).

> ***CAUTION—***
> • *Do not exceed a pressure of 2 bar (29 psi) when pressure bleeding the brake system. Excessive pressure damages the brake fluid reservoir.*

– Connect BMW scan tool or equivalent to OBD II plug on left door post.

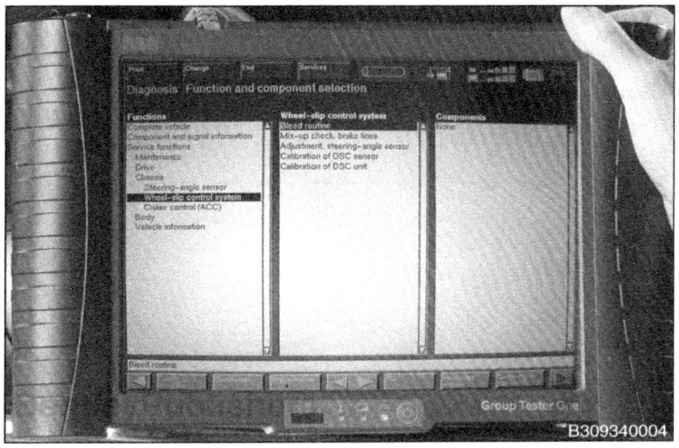

◄ Set tool to service function Bleeding ABS / DSC.

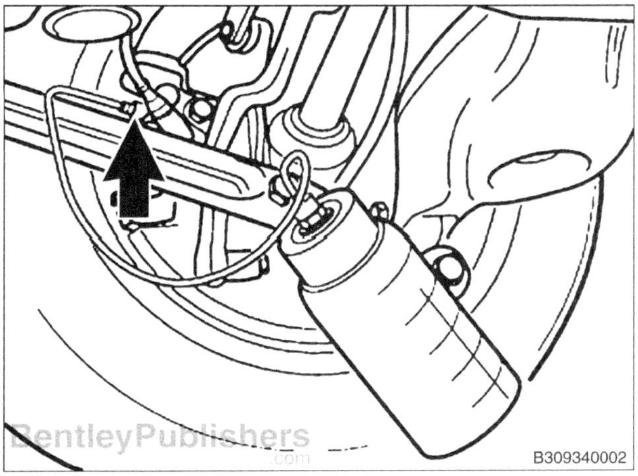

◄ Connect bleeder hose and bottle to **right rear** caliper bleeder screw (**arrow**). Have a helper hold brake pedal down.

BentleyPublishers.com
BentleyPublishers.com—All Rights Reserved

B510340026

Open bleeder screw (**arrow**). Carry out bleeding procedure using BMW scan tool on-screen instructions until clear, bubble-free fluid flows. (Photo illustrates front brake caliper. Rear caliper is similar.)

– After BMW scan tool bleeding routine has finished, press brake pedal to floor 5 times until clear, bubble-free fluid flows. Close bleeder.

> ***CAUTION—***
> • *Make sure bleeder hose remains submersed in clean brake fluid whenever the bleeder valve is open.*

– Repeat procedure at **left rear** brake.

– Repeat procedure at **right front** brake.

– Repeat procedure at **left front** brake.

– Remove brake bleeding apparatus and disconnect BMW scan tool. Top off brake fluid. Check to make sure brake fluid reservoir cap seal is intact.

Tightening torques	
Bleeder to caliper:	
• 9 mm screw	7 - 11 Nm (5 - 8 ft-lb)
• 11 mm screw	12 - 16 Nm (9 - 12 ft-lb)

BentleyPublishers.com
BentleyPublishers.com—All Rights Reserved

Brake Pads, Calipers and Rotors

Braking system components

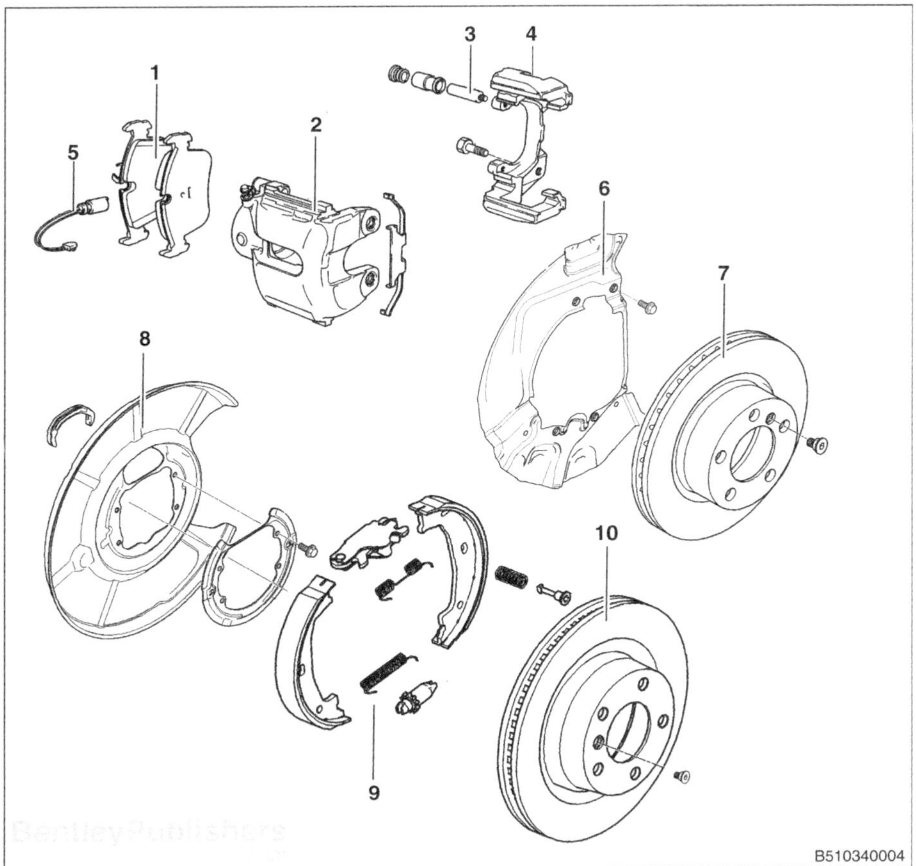

1. Brake pads
   - Wear limit: 3 mm (0.12 in).
   - Follow directional marking when installing.
2. Brake caliper
3. Brake caliper guide bolt
4. Caliper carrier
5. Brake pad wear sensor
   - On left front, right rear inner pad.
   - Replace with new pads.
   - Replace if brake pad wear warning light illuminated.
6. Front brake backing plate
7. Front brake rotor
   - See **Table c** for brake rotor specifications.
   - Clean off preservative before installing new rotor.
8. Rear brake backing plate
9. Parking brake shoes with hardware
10. Rear brake rotor
    - See **Table c** for brake rotor specifications.
    - Clean off preservative before installing new rotor.

340

Brake pads, checking and replacing

Brake pads can be replaced without disconnecting the brake fluid hose from the caliper or having to bleed the brakes. The rotors can be replaced without disassembling wheel hub and bearing. Front and rear brake pad replacement procedures are similar.

Replace pads in axle sets.

– Raise car and support safely. Remove wheels.

> ***WARNING —***
> - *Make sure the car is stable and well supported at all times. Use a professional automotive lift or jack stands designed for the purpose. A floor jack is not adequate support.*

BentleyPublishers.com
BentleyPublishers.com – All Rights Reserved

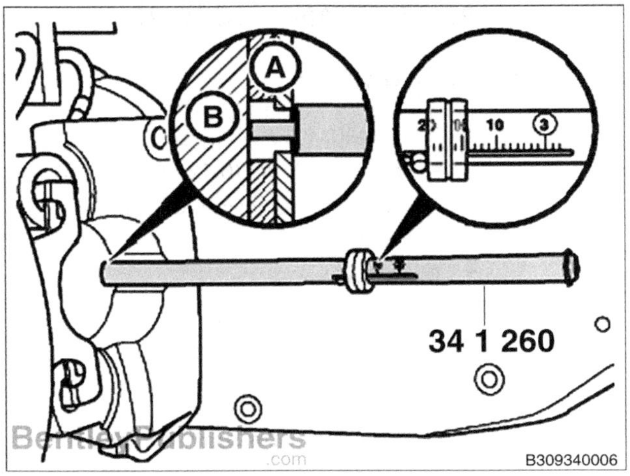

B309340006

To measure brake pad lining thickness, insert BMW special tool 34 1 260 at either right front or left rear wheel.

- Move wheel until notch for brake wear indicator can be seen through wheel opening.
- Insert tip of tool into notch so that body of tool rests on brake pad backing plate (**A**), and tip touches brake rotor (**B**).
- Replace pads if at minimum thickness. See **Table b**.

Table b. Brake pad lining minimum thickness	
Minimum pad thickness, front or rear	3 mm (0.12 in)

B510340027

Pry off brake caliper anti-rattle spring (**arrow**).

B510340028

Left front or right rear: Pull pad wear sensor off inner brake pad.

BentleyPublishers.com
BentleyPublishers.com–All Rights Reserved

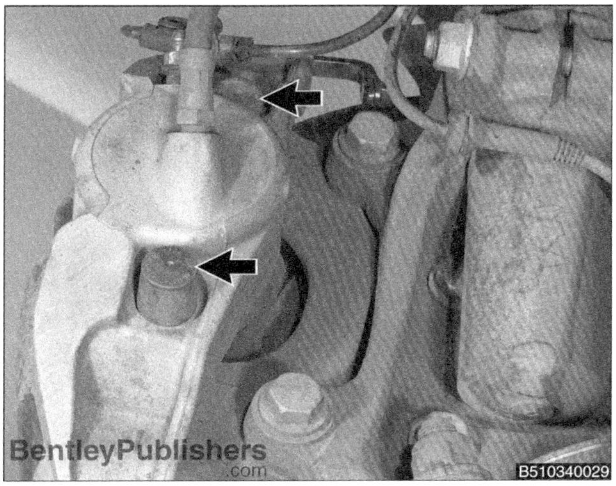

Remove plastic caps (**arrows**) from caliper guide bolts.

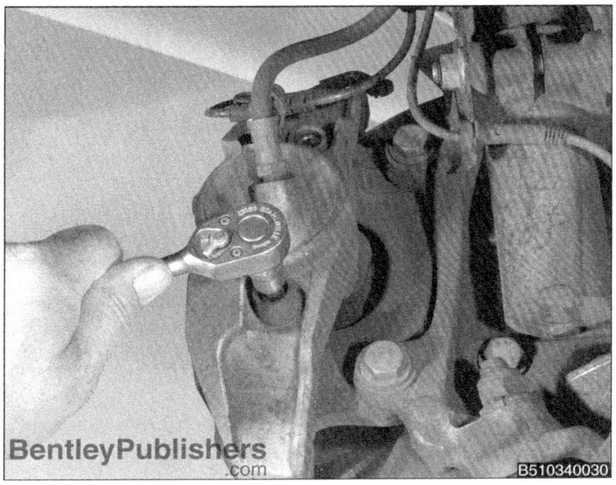

Using 7 mm Allen wrench, remove caliper guide bolts. Remove caliper from pad carrier.

- If there is a ridge on brake rotor edge, press caliper pistons back into caliper before removing caliper.

> ***CAUTION—***
> - *Before pressing in caliper pistons, siphon some fluid out of brake fluid reservoir in order to avoid paint-damaging spillover.*
> - *Do not let the brake caliper assembly hang from the brake hose. Support caliper from chassis with strong wire.*

- Remove brake pads from caliper.
- Check brake rotor thickness. Inspect for rust or damage. If necessary replace or machine brake rotors. See **Brake rotor specifications** and **Brake rotor, removing and installing** in this repair group.

Compress caliper piston using BMW special tool 34 1 050 or equivalent.

> ***CAUTION—***
> - *Before pressing in caliper pistons, siphon some fluid out of brake fluid reservoir in order to avoid paint-damaging spillover.*

- Check caliper dust boots for damage and replace if necessary.
- Clean contact surfaces between caliper and caliper carrier.

BentleyPublishers.com
BentleyPublishers.com—All Rights Reserved

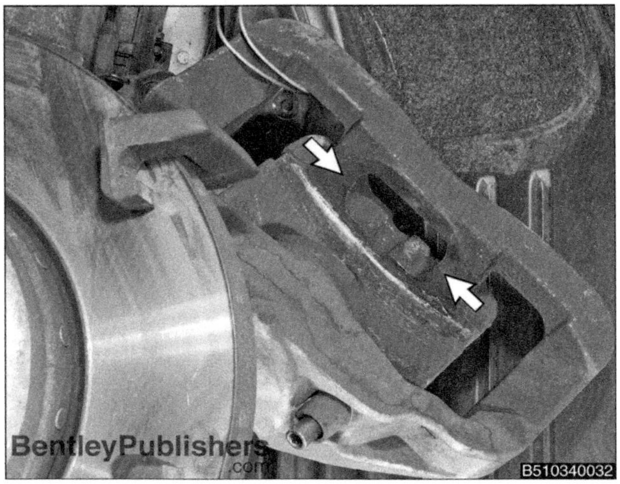

Apply thin coating of BMW anti-squeal compound to caliper contact face (**arrows**).

> ***CAUTION—***
> - *Do not let caliper rubber dust sleeve come in contact with anti-squeal compound. The compound causes rubber to swell and deteriorate.*
> - *Do not apply grease to brake pad backing plate.*

Apply thin coating of BMW anti-squeal compound to brake pad rests (**arrows**) on pad carrier.

> ***CAUTION—***
> - *In order to preserve brake caliper surface coating use brake cleaner only to clean.*

– If installing directional pads, be sure to install them as indicated by manufacturer.

– Where applicable, insert brake pad wear sensor into cutout in new pad. Push sensor into pad cutout until it locks into place.

> ***CAUTION—***
> - *If brake pad wear sensor light illuminated prior to brake pad replacement, replace wear sensor and reset CBS data.*
> - *Brake pad sensor can only be installed once, then needs replacing. See* ***020 Maintenance****.*
> - *Brake pad wear sensor is fragile. Be careful not to damage during installation.*

– Route pad wear sensor wiring through caliper opening and under bleeder dust cap.

– Remainder of installation is reverse of removal.
  - Clean brake caliper guide bolts (7 mm Allen). Replace ones which are not in perfect condition. Do not grease.
  - Before driving car, pump brake pedal several times so that brake pads contact brake rotors.
  - Check that brake fluid level is correct. Top off to MAX marking.
  - Hold ignition key for at least 30 seconds in accessory position without starting engine. This clears fault codes in memory and turns brake pad warning light OFF. If light remains ON, reset condition based service (CBS) system. See **020 Maintenance**.

Tightening torques	
Brake caliper to pad carrier (7 mm Allen)	30 ± 2 Nm (22 ± 1 ft-lb)
Road wheel to hub	120 ± 10 Nm (89 ± 7 ft-lb)

BentleyPublishers.com
BentleyPublishers.com—All Rights Reserved

Brake caliper and pad carrier, removing and installing

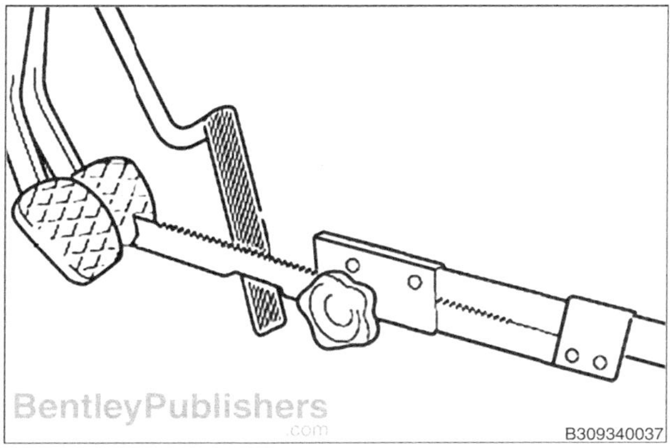

◄ Insert pedal prop and depress brake pedal slightly. This prevents brake fluid from escaping when brake lines are detached.

– Raise car and support safely. Remove wheels.

> ***WARNING —***
> - *Make sure the car is stable and well supported at all times. Use a professional automotive lift or jack stands designed for the purpose. A floor jack is not adequate support.*

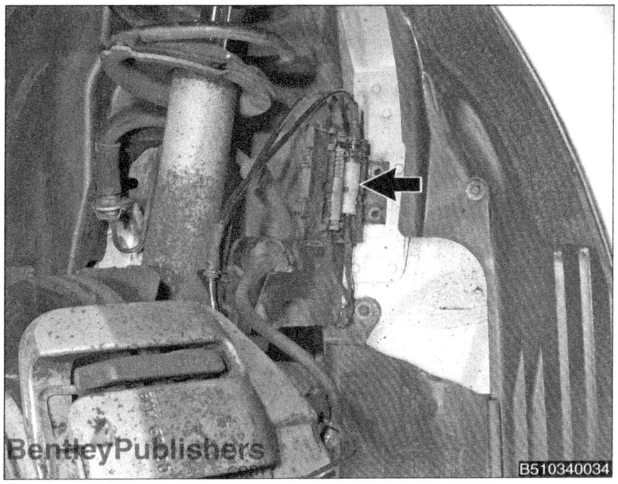

◄ Left front wheel: Working in junction box in wheel housing, separate brake pad warning sensor electrical harness (**arrow**).

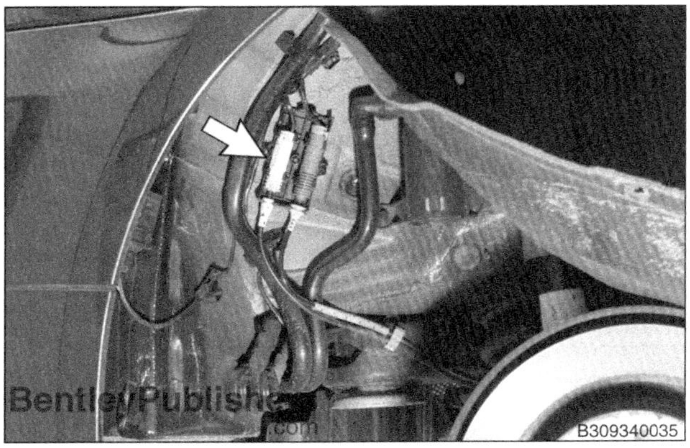

– Right rear wheel: Remove right rear wheel housing liner. See **410 Fenders, Engine Hood**.

◄ Working in junction box in wheel housing, separate brake pad warning sensor electrical harness (**arrow**).

340

BentleyPublishers.com
BentleyPublishers.com–All Rights Reserved

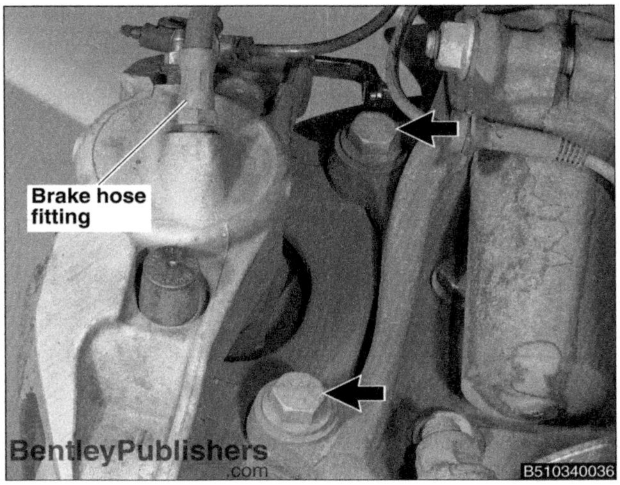

Working at brake caliper:

- Loosen brake hose fitting at caliper. Use shop towel to catch dripping brake fluid.
- Remove pad carrier and caliper mounting bolts (**arrows**) from steering knuckle (front caliper) or from wheel bearing carrier (rear caliper).

— Lift off caliper and spin assembly off hose. Plug open fluid lines and ports.

— Inspect brake caliper for signs of leakage. Check that caliper piston slides smoothly into caliper. Replace caliper if any faults are found.

— Installation is reverse of removal. Keep in mind the following:

- Tighten front caliper hoses with front wheel pointing dead ahead.
- Make sure brake hoses are not twisted or tensioned in any way.
- Make sure brake pad warning sensor harness does not chafe on wheel or other moving parts.
- Bleed entire brake system before driving car. See **Brake calipers, bleeding** in this repair group.

Tightening torques	
Brake caliper to pad carrier (7 mm Allen)	30 ± 2 Nm (22 ± 1 ft-lb)
Brake fluid hose to caliper	24 Nm (18 ft-lb)
Brake fluid hose to steel line	16 Nm (12 ft-lb)
Front brake caliper to steering knuckle (replace M12 x 1.5 x 43 mm ZNS3 bolts with new)	110 Nm (81 ft-lb)
Rear brake caliper to wheel bearing carrier (M12 x 1.5 x 41 mm ZNS3 bolts)	110 Nm (81 ft-lb)
Road wheel to hub	120 ± 10 Nm (89 ± 7 ft-lb)

Brake rotor specifications

When inspecting brake rotors and pads, replace pads only if brake rotor thickness is greater than or equal to minimum thickness (MIN TH) stamped on edge of brake rotor. This thickness is nominal thickness (when new) minus 1.6 mm (0.06 in). Otherwise, replace brake pads and rotors together.

Brake rotor dimensions vary based on model, engine size and drive configuration (rear-wheel drive or xDrive). When replacing rotors, match new parts to old parts on vehicle. If in doubt, check with BMW parts department.

Table c. Brake rotor specifications

Model	Parameter	Front: mm (in)	Rear: mm (in)
525i 525xi	Diameter	310 (12.2)	320 (12.6)
	MIN TH	22.4 (0.88)	18.4 (0.72)

BentleyPublishers.com
BentleyPublishers.com—All Rights Reserved

Table c. Brake rotor specifications			
Model	**Parameter**	**Front: mm (in)**	**Rear: mm (in)**
528i 528xi 530i 530xi 535xi	Diameter	324 (12.8)	320 (12.6)
	MIN TH	28.4 (1.11)	18.4 (0.72)
535i Sedan 545i 550i	Diameter	348 (13.7)	345 (13.6)
	MIN TH	28.4 (1.11)	22.4 (0.88)
Max. machine limit per friction ring side		0.8 (0.03)	

Brake rotor, removing and installing

Machine or replace brake rotors in pairs per axle.

– Raise car and support safely. Remove wheels.

> ***WARNING —***
> - *Make sure the car is stable and well supported at all times. Use a professional automotive lift or jack stands designed for the purpose. A floor jack is not adequate support.*

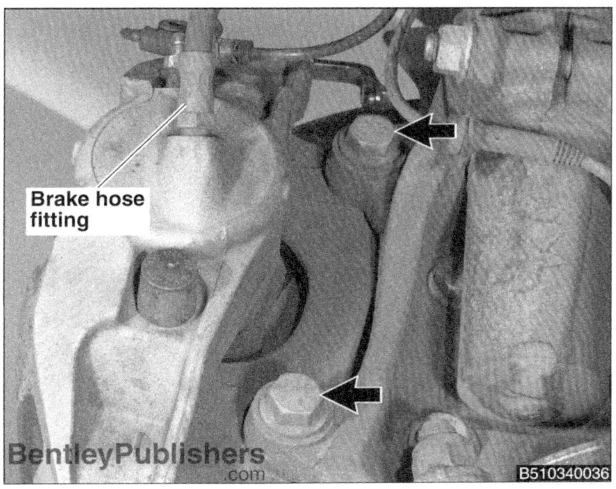

Remove brake caliper carrier bolts (**arrows**). Suspend brake caliper from chassis using stiff wire. Do not detach brake hydraulic line.

Remove brake rotor mounting screw (**arrow**) and remove rotor.

– Inspect rotor for cracks, signs of overheating and scoring.

– Minimum allowable thickness (MIN TH) is stamped on rotor hub. Measure rotor braking surface with a micrometer at eight to ten different points and use the smallest measurement. Compare to specifications in **Table c**.

– If rotor does not pass minimum thickness requirements or is damaged, replace rotor.

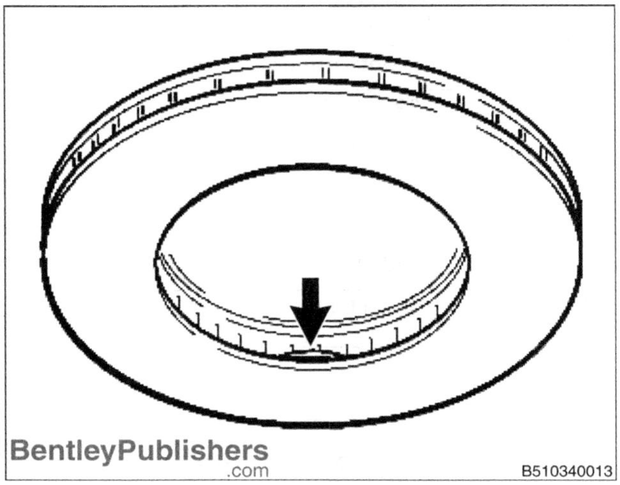

⊲ Vented brake rotors are equipped with an integrated balance weight (**arrow**). Do not remove or reposition weight.

– Clean rotor with brake cleaner before installing.

– If new rear brake rotors are installed, adjust parking brake. See **Parking brake, adjusting** in this repair group.

Tightening torques	
Brake rotor to wheel hub (replace M8 x 14 mm Allen bolt with new)	16 Nm (12 ft-lb)
Front brake caliper to steering knuckle (replace M12 x 1.5 x 43 mm ZNS3 bolts with new)	110 Nm (81 ft-lb)
Rear brake caliper to wheel bearing carrier (M12 x 1.5 x 41 mm ZNS3 bolts)	110 Nm (81 ft-lb)
Road wheel to hub	120 ± 10 Nm (89 ± 7 ft-lb)

Brake Master Cylinder

⊲ The brake master cylinder is mounted to the front of the vacuum booster on the driver side bulkhead.

1. Brake vacuum booster
2. Brake fluid reservoir
3. Brake fluid level sensor
4. Vacuum line check valve
5. Brake master cylinder
6. Vacuum line from intake manifold or auxiliary vacuum pump

Brake master cylinder, removing and installing

> ***CAUTION—***
> *• Brake fluid is highly corrosive and dangerous to the environment. Dispose of it properly.*

– Remove left side cabin microfilter housings. See **020 Maintenance**.

– Remove left lower dashboard trim. See **513 Interior Trim**.

⊲ Working at brake fluid reservoir at left rear of engine compartment:

- Detach brake fluid level sensor connector (**arrow**).
- Remove reservoir cap.
- Using a clean syringe, empty brake fluid reservoir.

– Remove brake fluid reservoir by tilting sideways out of grommets.

BentleyPublishers.com
BentleyPublishers.com–All Rights Reserved

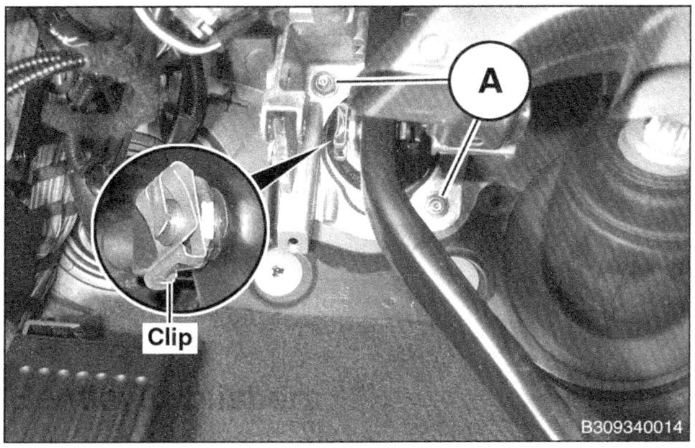

Working in passenger compartment underneath left side dashboard:

- Remove brake booster mounting fasteners (**A**).
- Remove brake booster push rod pin clip (**inset**) and detach booster pushrod from brake pedal.

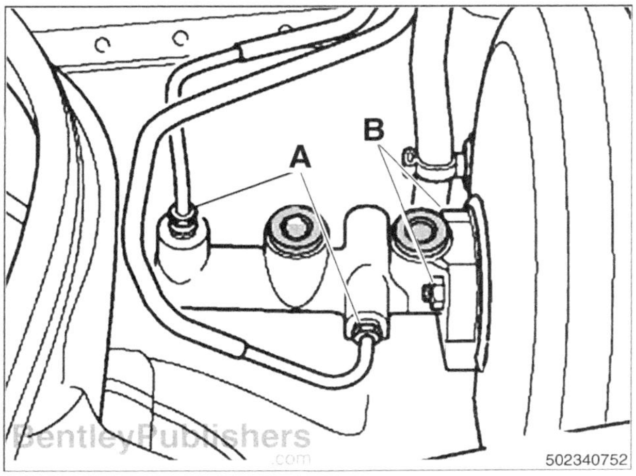

Disconnect master cylinder brake lines (**A**).

- Plug open brake lines to prevent contamination.
- Remove master cylinder mounting nuts (**B**) and remove master cylinder.

– During reinstallation, make sure fluid couplings, thread bores and mating surfaces are clean.

– Mount master cylinder to brake booster using new sealing O-ring and new self-locking nuts.

CAUTION—
- *Be sure to align master cylinder pushrod and booster pushrod.*
- *Do not over-torque master cylinder mounting nuts. This could damage the brake booster and prevent proper vacuum build-up.*
- *After 30 minutes, retighten master cylinder mounting nuts.*

Tightening torque	
Brake master cylinder to brake booster	26 Nm (19 ft-lb)

– Connect brake fluid lines to master cylinder.

Tightening torques	
Brake fluid line to master cylinder • M10 x 1 union nut • M12 x 1 union nut	 14 Nm (10 ft-lb) 17 Nm (13 ft-lb)

– Reinstall fluid reservoir using new sealing grommets.

– Fasten brake booster. After 30 minutes, retighten.

Tightening torque	
Brake booster to body panel (M8 nuts)	21 Nm (15 ft-lb)

– Remainder of installation is reverse of removal. Remember to:

- Top up with fresh brake fluid.
- Bleed entire brake system. See **Brake system, bleeding** in this repair group.

BentleyPublishers.com—All Rights Reserved

Auxiliary brake booster vacuum pump

5 Series models with N52, N54 or V8 engine use an auxiliary pump to supply vacuum to the brake booster, rather than using traditional manifold vacuum.

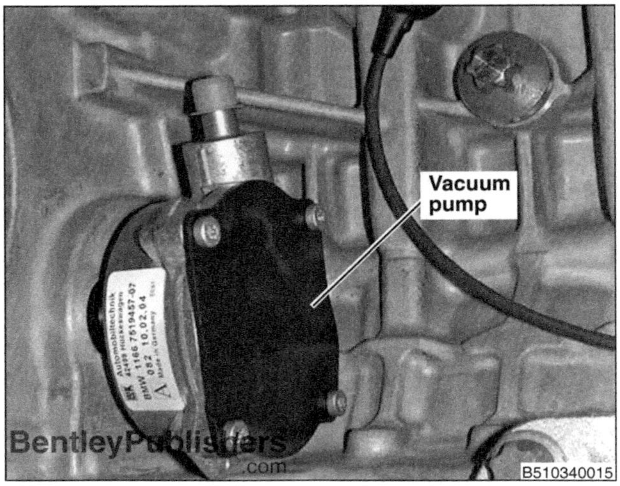

N52 engine: Vacuum pump, in front lower left of engine, is driven by oil pump chain.

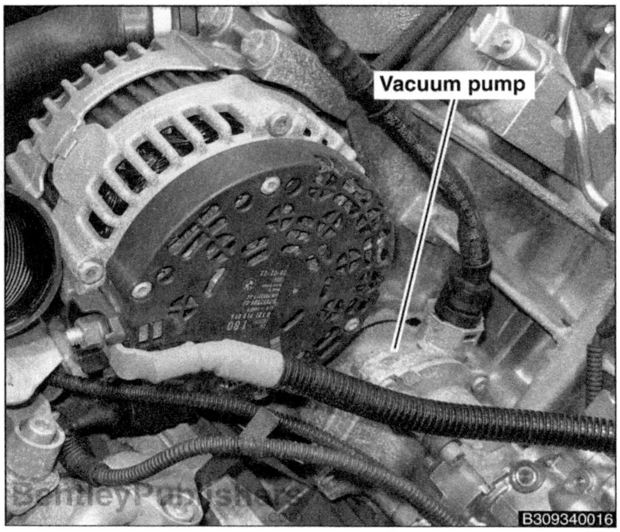

N54 engine: Vacuum pump operates in tandem with high-pressure fuel pump, both driven by oil pump chain in front lower left of engine

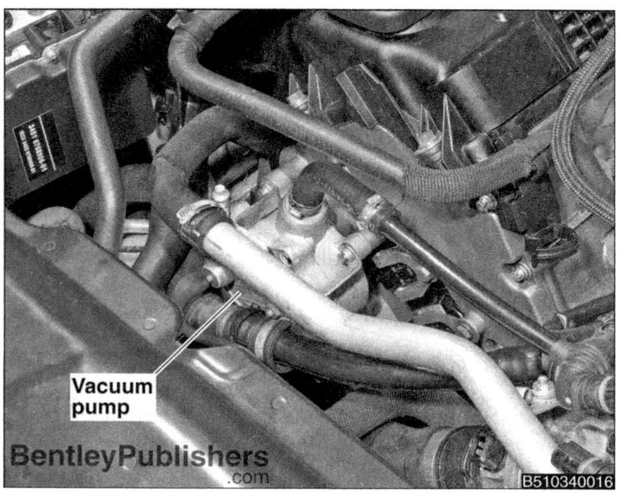

V8 engine: Vacuum pump at right front of engine is driven by right exhaust camshaft.

BentleyPublishers.com
BentleyPublishers.com—All Rights Reserved

Parking Brake

Parking brake components

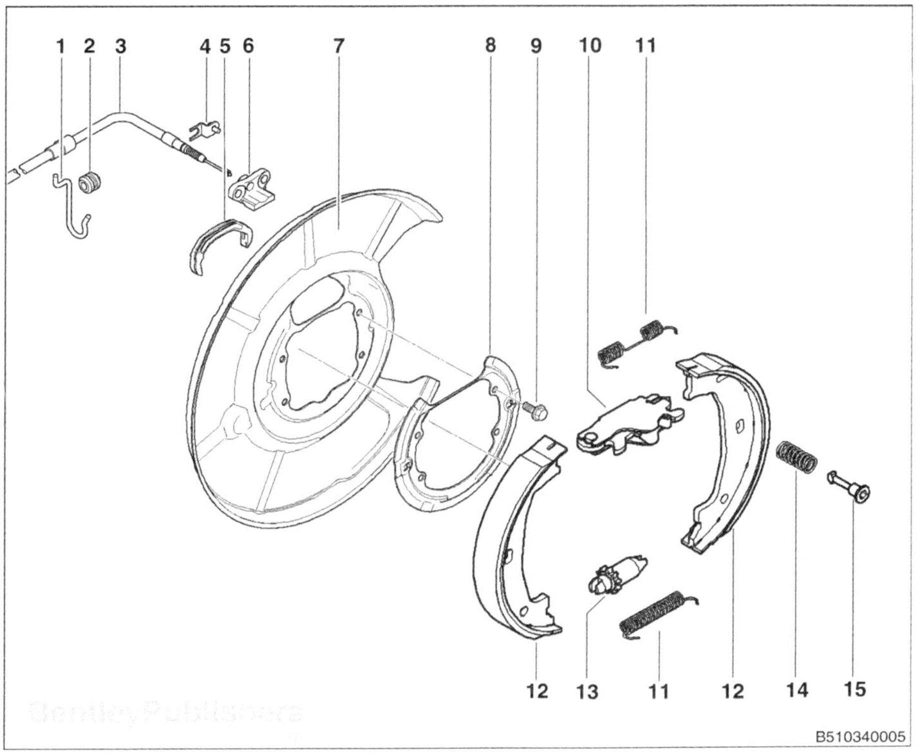

1. Bracket
2. Grommet
3. Parking brake Bowden cable
4. Bracket
5. Gasket
6. Support
7. Backing plate
8. Reinforcement
9. Bolt
10. Brake shoe expander
11. Return spring
12. Parking brake shoe
13. Brake shoe adjuster
14. Hold-down spring
15. Hold-down pin

Parking brake, adjusting

The parking brake is a brake drum system integrated into the rear brake rotors. Adjust with rear of car raised off the ground and the wheels installed. Make sure parking brake Bowden cables move freely.

Adjust parking brake under the following circumstances:

- After replacing parking brake shoes
- After replacing rear brake rotors
- Excessive stroke of parking brake handle required for actuation (more then 10 clicks)
- After replacement of brake shoe adjuster or parking brake cable(s)

– Disengage parking brake handle boot from center console.

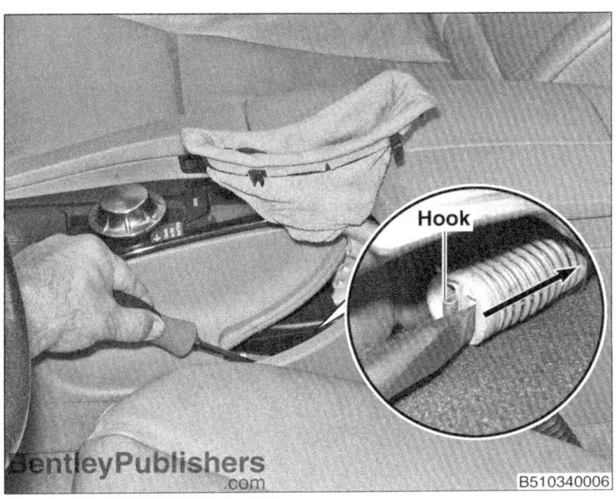

Relieve parking brake cable tension: Using a screwdriver, lock parking brake handle adjuster by pressing spring back to engage hook. **Inset** shows hook engaged.

– Raise rear of car.

> ***WARNING —***
> - *Make sure that the car is firmly supported on jack stands designed for the purpose. Place jack stands beneath structural chassis points. Do not place jack stands under suspension parts.*

BentleyPublishers.com–All Rights Reserved

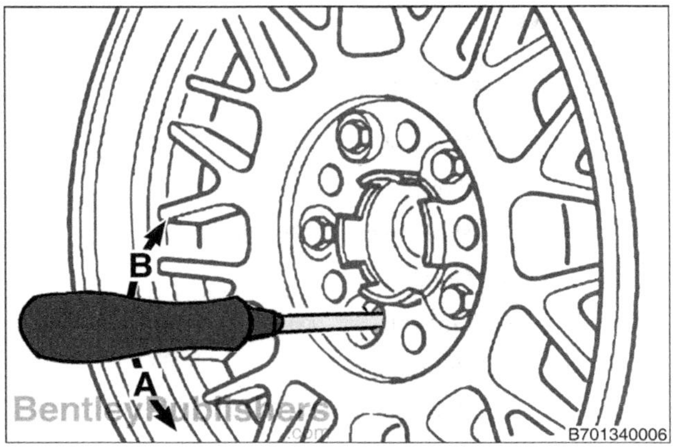

◀ Remove one rear wheel lug bolt and rotate wheel until parking brake adjuster is visible through lug bolt hole.

- Use flat-bladed screwdriver to reach into brake drum through bolt hole and turn brake adjuster.
- Left rear wheel: Turn adjuster in direction **A** to spread brake shoes.
- Right rear wheel: Turn adjuster in direction **B** to spread brake shoes.

– Turn adjuster until wheel no longer turns. Back adjuster off 8 teeth.

Parking brake adjusting (initial)	
Back off adjuster through wheel lug bolt hole	8 teeth

– Working at parking brake handle in center console, use screwdriver to release hook at brake cable spring.

– Set parking brake several times to seat brakes.

– Release lever and make sure rear wheels turn freely. If any binding is noticed, back off adjuster(s) one tooth at a time.

– Switch ignition ON. Pull up parking brake lever 1 notch and make sure that parking brake warning light switches ON. If not, adjust parking brake warning light contact switch.

– Install parking brake lever boot. Install road wheel lug bolt.

Tightening torque	
Road wheel to hub	120 ± 10 Nm (89 ± 7 ft-lb)

Parking brake shoes, removing and installing

– Disengage parking brake handle boot from center console.

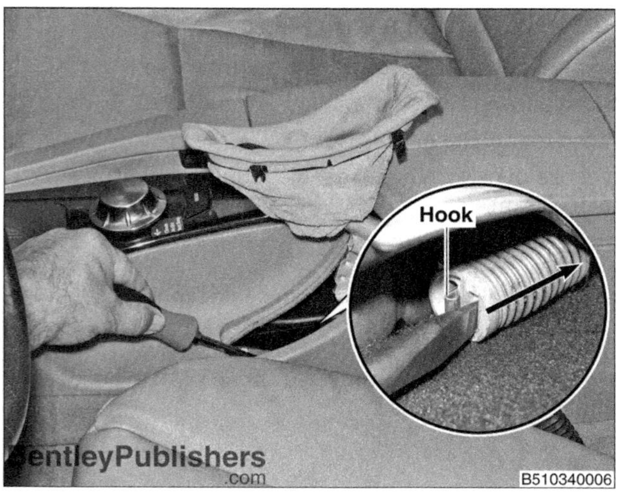

◀ Relieve parking brake cable tension: Using a screwdriver, lock parking brake handle adjuster by pressing spring back to engage hook. **Inset** shows hook engaged.

– Raise rear of car and remove wheels.

> ***WARNING*** *—*
> - *Make sure that the car is firmly supported on jack stands designed for the purpose. Place jack stands beneath structural chassis points. Do not place jack stands under suspension parts.*

– Remove brake calipers and brake discs. Hang brake calipers aside with stiff wire. Do not disconnect brake fluid hoses. See **Brake rotor, removing and installing** in this repair group.

BentleyPublishers.com
BentleyPublishers.com—All Rights Reserved

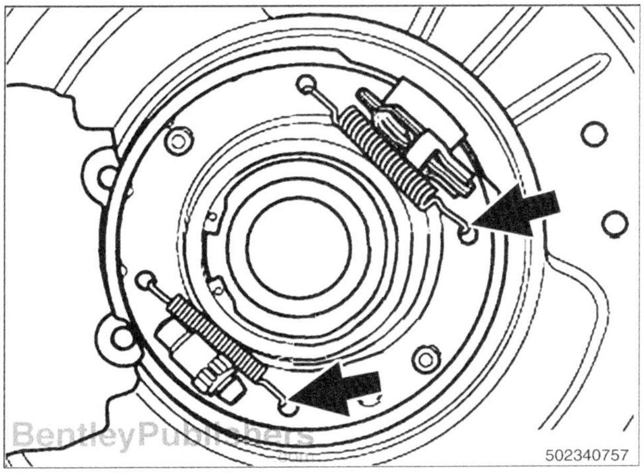

Unhook return springs (**arrows**) from brake shoes.

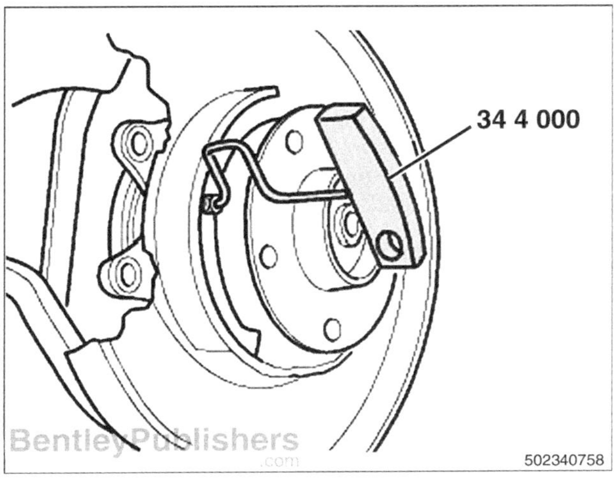

Remove brake shoe hold-down pins using BMW special tool 34 4 000 or 6 mm Allen wrench.

- Rotate hold-down pin ¼ turn to release.
- Spread shoes apart and lift out.

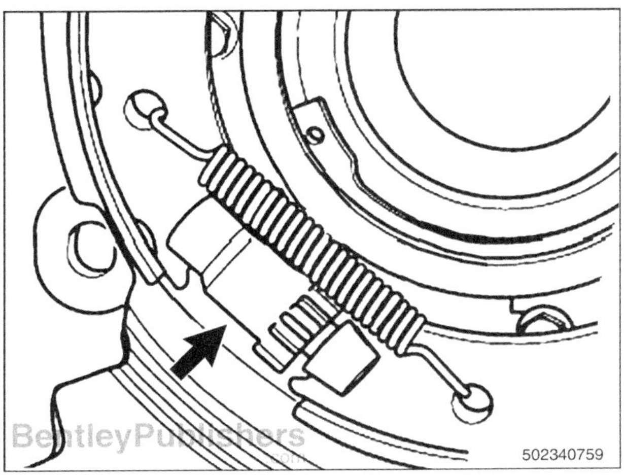

Note position of brake shoe adjuster (**arrow**). Right side is illustrated. Left is mirror image of this.

– Installation is reverse of removal. Keep in mind the following:

- Apply thin coat of grease to sliding parts and pins before reassembly.
- Be sure to adjust parking brake. See **Parking brake, adjusting** in this repair group.

Tightening torques	
Brake caliper to wheel bearing carrier (M12 x 1.5 x 41 mm ZNS3 bolts)	110 Nm (81 ft-lb)
Brake rotor to hub (M8 x 14 mm Allen bolt)	16 Nm (12 ft-lb)
Road wheel to hub	120 ± 10 Nm (89 ± 7 ft-lb)

BentleyPublishers.com
BentleyPublishers.com—All Rights Reserved

Parking brake cable, replacing

The parking brake is actuated by two separate Bowden cables. Each cable can be replaced separately.

– Raise vehicle and remove rear wheel.

> ***WARNING —***
> *• Make sure the car is firmly supported on jack stands designed for the purpose. Place jack stands beneath structural chassis points. Do not place jack stands under suspension parts.*

– Remove brake calipers and brake discs. Hang brake calipers aside with stiff wire. Do not disconnect brake fluid hoses. See **Brake rotor, removing and installing** in this repair group.

– Remove exhaust system and heat shield. See **180 Exhaust System**.

– Working inside car, gain access to base of parking brake handle by removing center console. See **513 Interior Trim**.

◀ Relieve parking brake cable tension: Using a screwdriver, lock parking brake handle adjuster by pressing spring back to engage hook. **Inset** shows hook engaged.

◀ Use screwdriver to pry plastic clip off parking brake balance bar (**arrow**).

BentleyPublishers.com
BentleyPublishers.com—All Rights Reserved

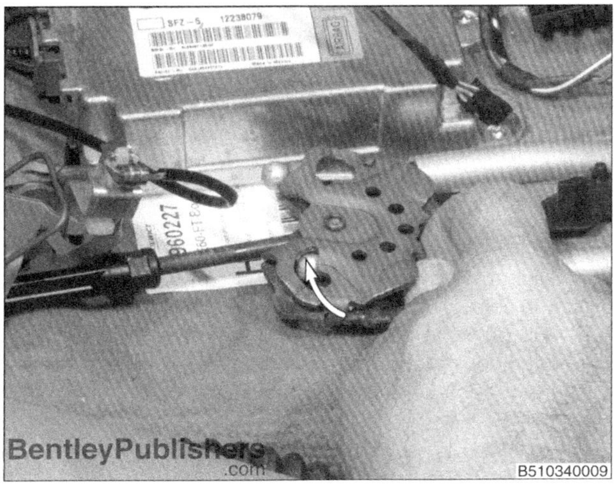

Slide Bowden cable ball end toward center of balance bar (**arrow**) and out toward rear of vehicle.

- Remove parking brake shoes. See **Parking brake shoes, removing and installing** in this repair group.

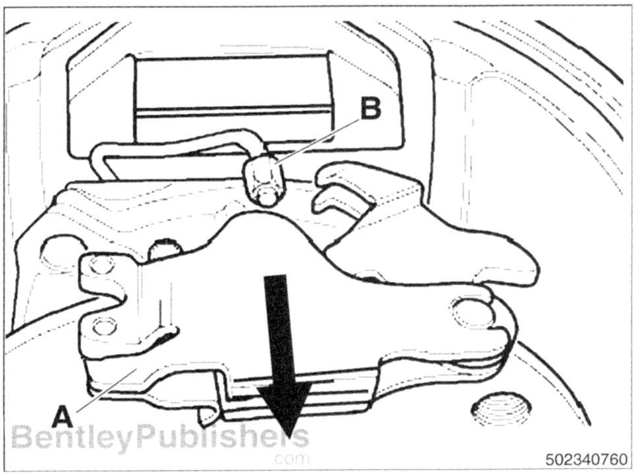

Pull shoe expander (**A**) forward and disconnect parking brake cable (**B**).

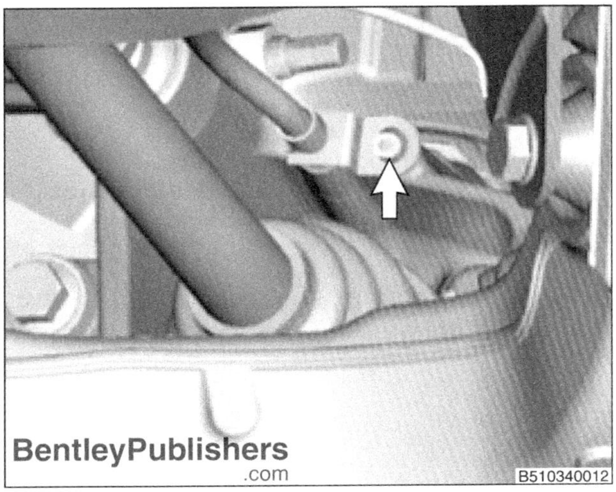

Working at wheel bearing carrier, remove fastener and cable hold down (**arrow**).

- Working underneath vehicle, detach cable from brackets. Pull cable out of brake backing plate and out of tube guide underneath vehicle.
- Installation is reverse of removal. Adjust parking brake. See **Parking brake, adjusting** in this repair group.

Tightening torques	
Brake cable bracket to wheel bearing carrier (M6 bolt)	8 Nm (6 ft-lb)
Brake caliper to wheel bearing carrier (M12 x 1.5 x 41 mm ZNS3 bolts)	110 Nm (81 ft-lb)
Brake rotor to hub (M8 x 14 mm Allen bolt)	16 Nm (12 ft-lb)
Road wheel to hub	120 ± 10 Nm (89 ± 7 ft-lb)

Antilock Brake and Stability Control

5 Series vehicles are equipped with antilock brakes (ABS) with dynamic stability control (DSC). This manual refers to these systems as ABS. DSC is specified when necessary.

Antilock brake system (ABS)

The electronically controlled antilock brake system (ABS) maintains vehicle stability and control during emergency braking by preventing wheel lock-up. ABS provides optimum deceleration and stability during adverse conditions. It automatically adjusts brake system hydraulic pressure at each wheel to prevent wheel lock-up.

Dynamic stability control (DSC)

DSC works in conjunction with ABS and the engine management system to enhance vehicle control. The main DSC function is to maintain contact between the tires and the road surface under all driving conditions. This is achieved through exact application and management of braking and drivetrain forces.

DSC is active throughout the driving range, helping to stabilize the vehicle in cornering and avoidance maneuvers by adjusting engine controls such as throttle, ignition, fuel injection and the application of brake pressure individually to the wheels.

The DSC control module uses various inputs to determine vehicle instability during braking, cornering, or reduced traction situations. Based upon these inputs the ABS / DSC control module sends outputs to the engine control module and the ABS / DSC hydraulic unit to activate torque reduction protocols and braking intervention.

The ABS / DSC control module, operating through the hydraulic unit, modulates braking force at the wheels. In addition, DSC overrides throttle opening to reduce engine torque and maintain vehicle traction. Because throttle is controlled electronically, the driver cannot increase engine power output during DSC intervention regardless of how far the accelerator pedal is pushed.

Traction control also comes into operation during deceleration. Decelerating on snowy or icy road surfaces can lead to rear wheel slip. If a rear wheel starts to drag or lock up, DSC can limit the problem by adjusting throttle, fuel injection and ignition timing.

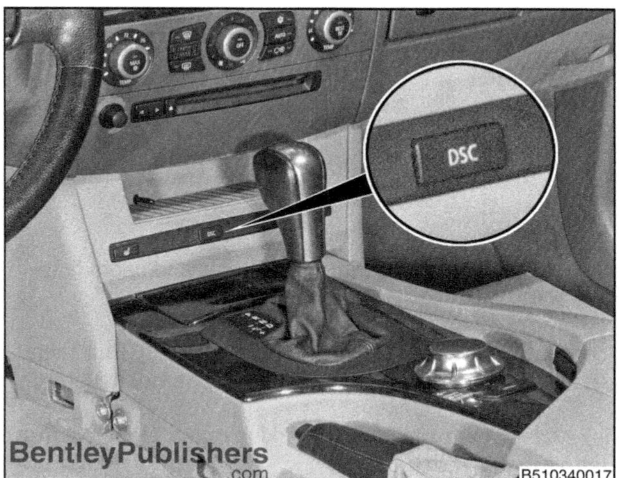

DSC can be toggled ON and OFF by a switch labeled DSC or DTC in the center console switching center. Turning DSC OFF does not disable ABS functions.

BentleyPublishers.com
BentleyPublishers.com—All Rights Reserved

Bus systems and traction control

Serial bus systems connect the engine control module (ECM), transmission control module (TCM), ABS / DSC control module, active steering control module, active stabilizer (dynamic drive) control module, longitudinal management system (LDM) and active cruise control (ACC). Bus systems allow control modules to share sensor information and other electronic data. See **600 Electrical System–General** for additional details.

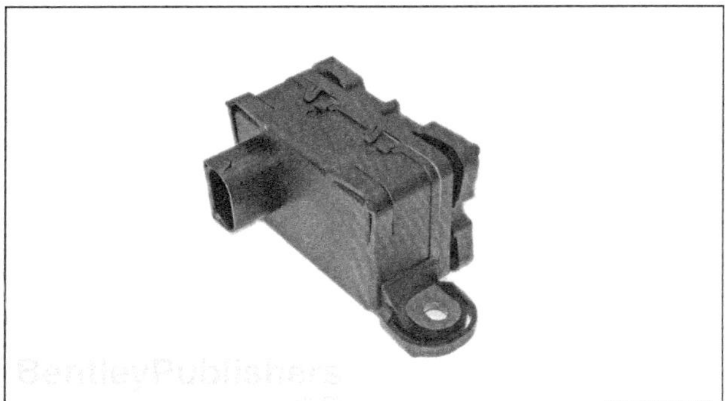

B309340047

DSC acceleration sensor

The DSC acceleration sensor utilizes two yaw rate sensor elements to detect yaw (rotation about the vertical axis) rate and lateral acceleration (G-force) rate.

For rotational acceleration (yaw), the sensor produces a reference signal and a linear voltage signal. The sensor also produces a linear voltage signal for lateral acceleration. The DSC control module uses this input to determine rotational and side forces acting on the vehicle.

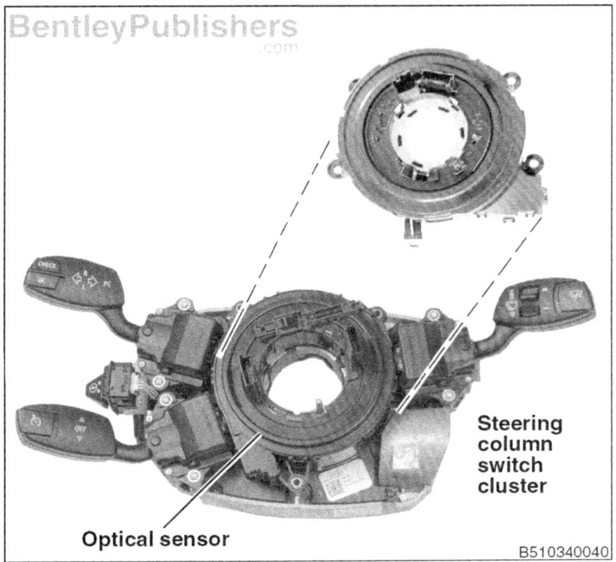

B510340040

Steering angle sensor

The steering angle sensor is located in the steering column switch cluster (SZL). It optically measures the angle of rotation of the steering wheel to determine steering angle and rate of steering.

After repairs to the steering or suspension system, use a BMW scan tool or equivalent to calibrate the sensor. Once calibration is complete, the sensor sends an identification number over the CAN-bus to the DSC control module. The ID confirms that the sensor is calibrated correctly.

DSC control module logic checks the plausibility of the steering angle sensor signal against other inputs such as wheel speeds and acceleration inputs.

If battery voltage is interrupted, steering wheel angle of rotation is recalculated by the DSC control module evaluating wheel speeds. If replacing steering angle sensor, remember to recalibrate sensor.

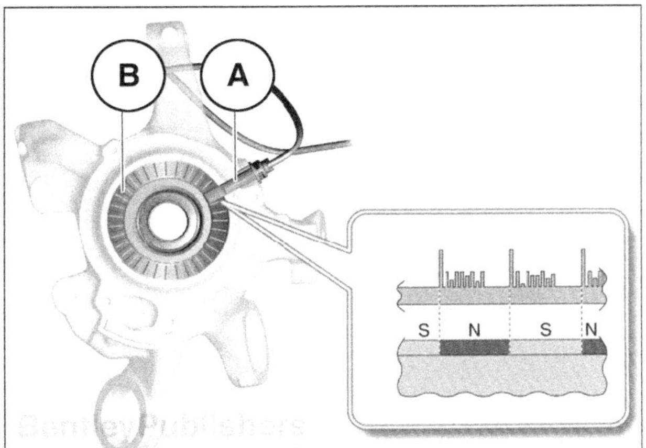

B309340049

Wheel speed sensors

A wheel speed sensor (**A**) is mounted at each wheel along with a ring of magnets (**B**) used to measure direction of wheel rotation and speed. The electrical resistance of the wheel speed sensor changes under the influence of the magnetic field. Each increment generates 2 pulses in the wheel speed sensor (96 pulses / revolution of wheel). When the vehicle is stationary the sensor sends one pulse every 0.75 second.

BentleyPublishers.com–All Rights Reserved

DSC and xDrive

xDrive (all-wheel drive system) uses information from the DSC system to control a multi-plate clutch in the transfer case. The multi-plate clutch varies torque split between front and rear wheels.

xDrive calculates the necessary torque split from information supplied by DSC:

- Accelerator pedal position
- Engine torque
- Dynamic state of vehicle (drive dynamic state)

DSC Components and Repairs

> ***CAUTION—***
> - *If the tires on the car are of different makes, the DSC system may overreact. Only fit tires of the same make and tread pattern.*
> - *In adverse conditions, such as trying to rock the car out of deep snow or other soft surface, or when snow chains are fitted, switch traction control OFF and allow the car driveline to operate conventionally.*

DSC faults

DSC systems are self-diagnosing and store fault codes (diagnostic trouble codes or DTCs) in the DSC control module. For information on how to access DTCs, see **020 Maintenance**.

ABS / DSC components are coded to each other and to the vehicle using BMW scan tool. If you do not have access to a scan tool, be sure to have newly installed ABS / DSC components coded at a qualified BMW workshop with the correct equipment.

◄ The driver is alerted to DSC faults by a warning icon and a Check Control warning triangle in instrument cluster. A gong also sounds.

BentleyPublishers.com
BentleyPublishers.com—All Rights Reserved

DSC power supply

See **ECL Electrical Component Locations** for fuse panel access information.

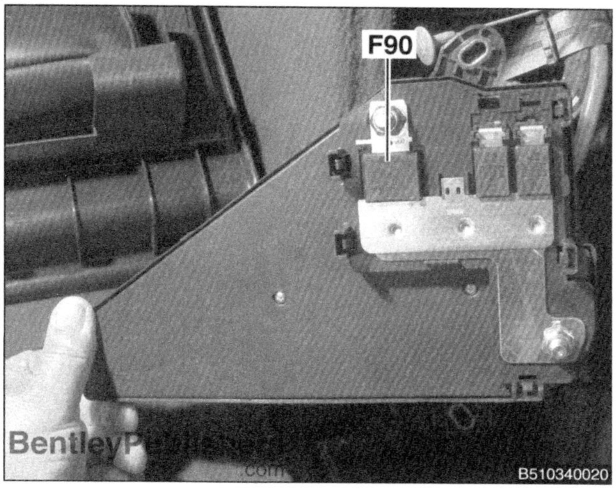

Fusible links in back of rear fuse panel, right rear trunk or cargo compartment:

- **F90** 200A: Supplies power to fuses F1 - F33

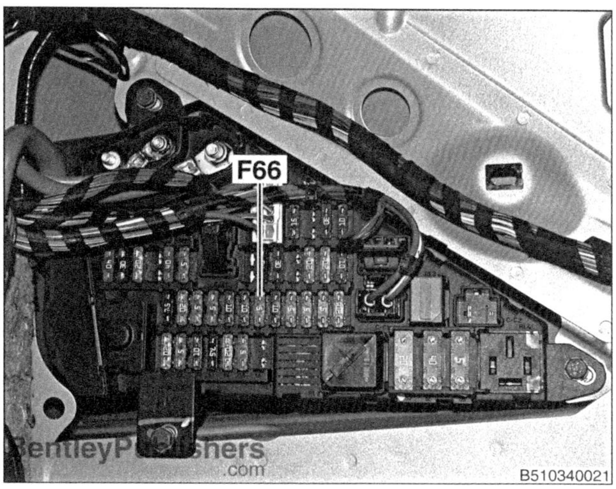

Rear fuse panel, right rear trunk or cargo compartment, to 09 / 2005:

- **F66** 5A: DSC control module

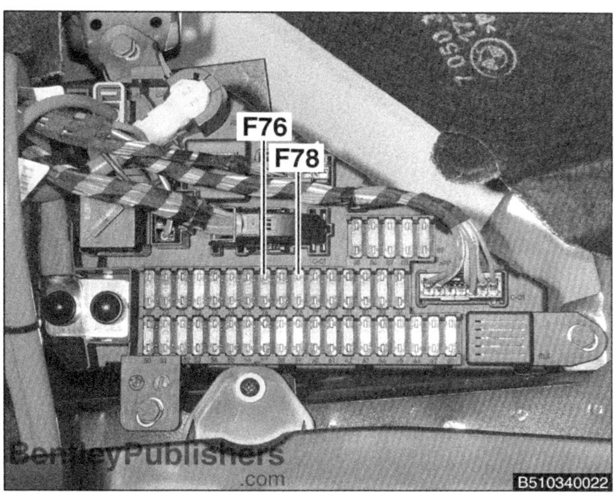

Rear fuse panel, right rear trunk or cargo compartment, from 09 / 2005:

- **F76** 10A: Active stabilizer (dynamic drive)
- **F78** 5A: DSC control module, transfer case

BentleyPublishers
BentleyPublishers.com - All Rights Reserved

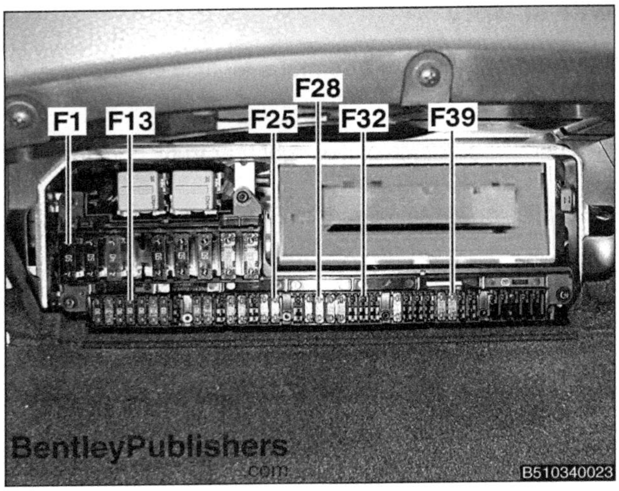

Front fuse panel, behind glove compartment:

- **F1** 50A: DSC control module
- **F13** 30A: Transfer case
- **F25** 30A: DSC control module
- **F28** 20A: Steering column switch cluster (SZL)
- **F32** 10A: Active stabilizer (dynamic drive)
- **F39** 5A: Steering column switch cluster (SZL)

DSC controls

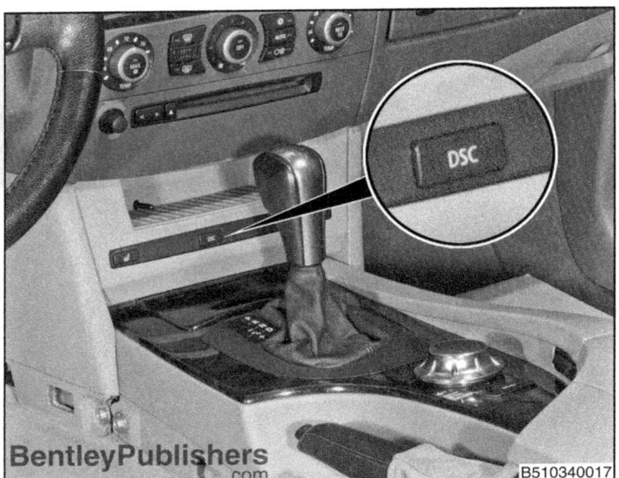

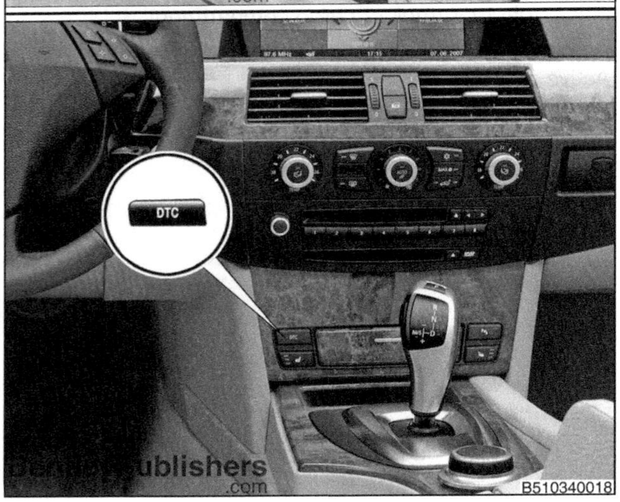

Lower center dashboard above center console:

- **DSC** (dynamic stability control) switch
- **DTC** (dynamic traction control) switch

BentleyPublishers.com
BentleyPublishers.com—All Rights Reserved

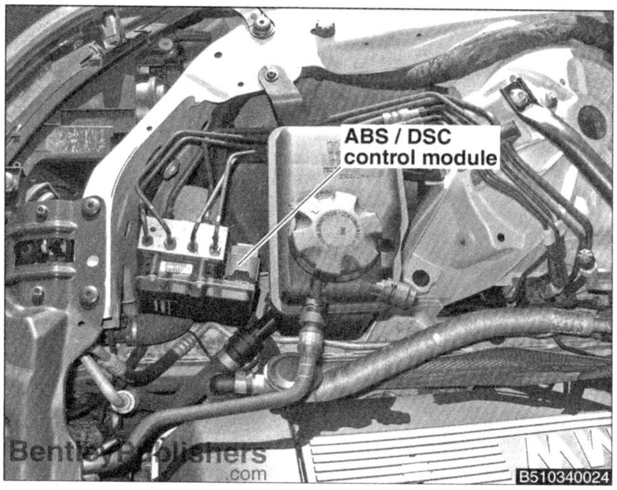

Right front engine compartment:

- **A65** ABS / DSC control module

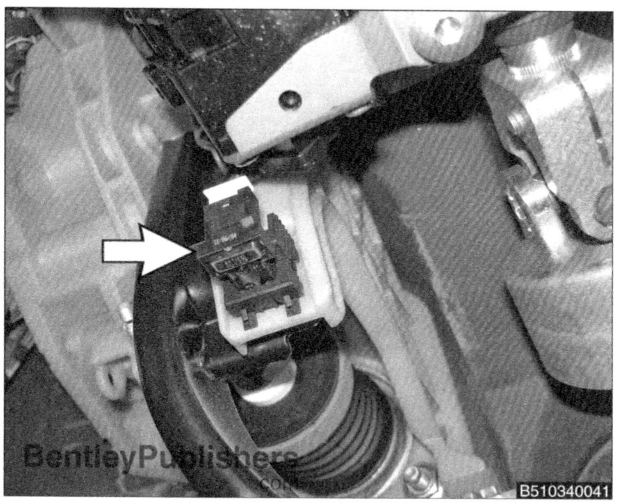

Above brake pedal, underneath left side dashboard (**arrow**):

- **S29** Brake light switch

DSC control module, removing and installing

DSC electronic control module and DSC hydraulic unit are combined in one assembly in right front engine compartment.

– Use BMW scan tool to read out and record DSC control module fault memory.

– Using a syringe, remove brake fluid from master cylinder reservoir.

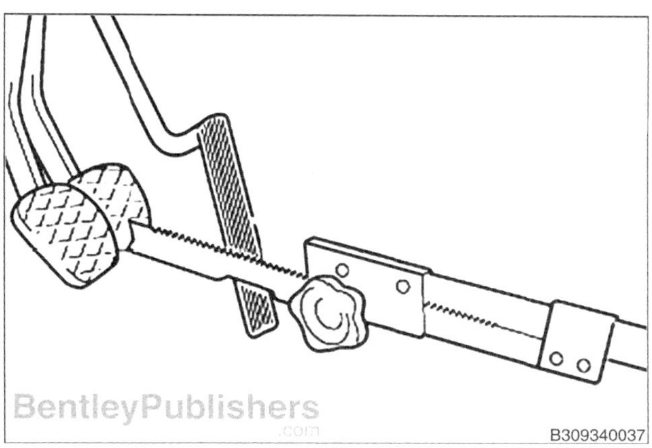

Insert pedal prop and depress brake pedal slightly. This prevents brake fluid from escaping when brake lines are detached.

BentleyPublishers.com–All Rights Reserved

Working at DSC module:

- Mark brake line positions (**arrows**) at hydraulic unit. Detach brake lines. Be prepared to catch dripping fluid. Plug open ports and lines.
- Disconnect electronic module electrical connector (**A**).
- Remove control module bracket mounting bolt (**B**). Lift module out of engine compartment.

> ***CAUTION—***
> *• Do not bend brake lines.*

– Installation is reverse of removal.

Tightening torques	
Brake fluid line to master cylinder (M10 x 1 or M12 x 1 union nut)	17 Nm (13 ft-lb)

– Bleed brake system. See **Brake system, bleeding** in this repair group.

– Use BMW scan tool to:

- Code DSC module.
- Adjust steering angle sensor.
- Adjust brake pressure sensor.
- Adjust acceleration sensors.

Wheel speed sensor, replacing

– Raise car and support safely.

> ***WARNING—***
> *• Make sure the car is stable and well supported at all times. Use a professional automotive lift or jack stands designed for the purpose. A floor jack is not adequate support.*

Front wheel speed sensor

Working in junction box in wheel housing, separate wheel speed sensor electrical harness (**arrow**). (Left front junction box illustrated. Right front contains single harness connector.)

BentleyPublishers.com
BentleyPublishers.com—All Rights Reserved

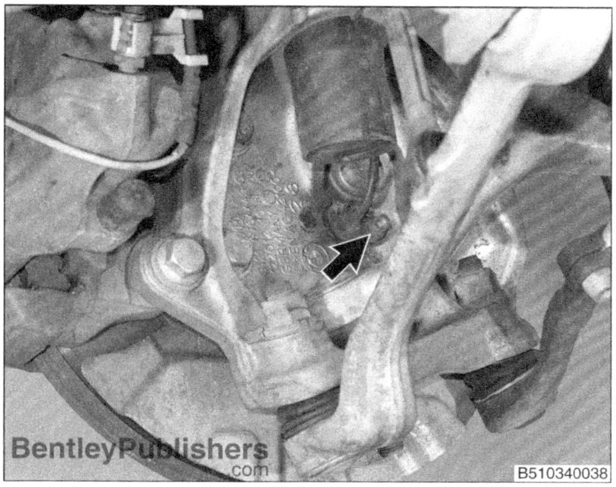

◀ Unscrew sensor mounting bolt (**arrow**) at steering knuckle and remove sensor.

– Installation is reverse of removal. Apply thin coat of Staburags®NBU 12/K or equivalent grease to speed sensor and housing.

Tightening torque	
ABS wheel speed sensor to steering knuckle	8 Nm (6 ft-lb)

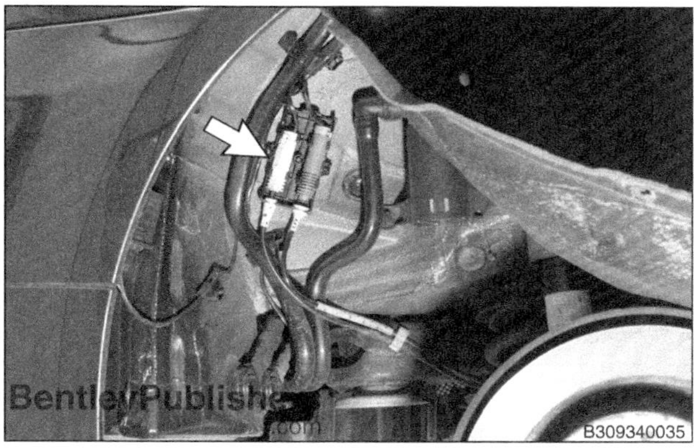

Rear wheel speed sensor

– Remove rear wheel well housing liner. See **410 Fenders, Engine Hood**.

◀ Working in junction box in wheel housing, separate wheel speed sensor electrical harness (**arrow**).

◀ Unscrew sensor mounting bolt (**arrow**) in rear wheel bearing carrier.

– Installation is reverse of removal. Apply thin coat of Staburags®NBU 12/K or equivalent grease to speed sensor and housing.

Tightening torque	
ABS wheel speed sensor to rear wheel bearing carrier	8 Nm (6 ft-lb)

Bentley Publishers.com
BentleyPublishers.com—All Rights Reserved

ABS impulse wheel

The front and rear ABS impulse wheels are integral with the inner wheel bearing seal and not available as separate parts. See **310 Front Suspension** or **330 Rear Suspension** for wheel bearing replacement procedures.

DSC acceleration sensor, removing and installing

– Remove right front seat. See **520 Seats**.

– Remove right door sill trim. See **513 Interior Trim**.

– Fold right front carpet, lay aside and remove carpet underlay.

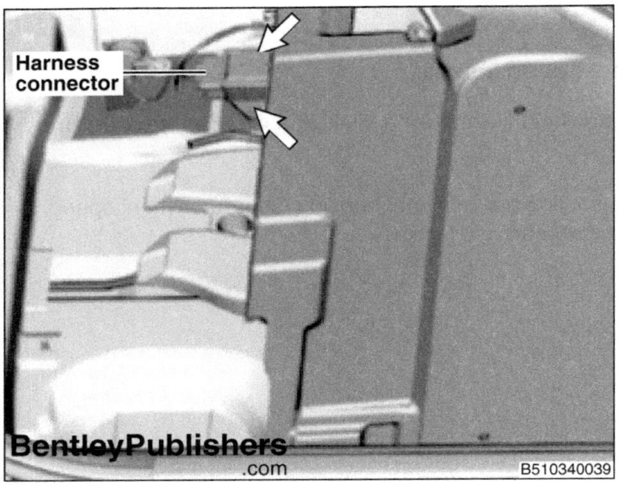

◄ Working at sensor:
- Detach electrical sensor.
- Remove mounting screws (**arrows**) and lift out sensor.

– Installation is reverse of removal.

> ***CAUTION—***
> - *Be sure to tighten the DSC acceleration sensor to specified torques. The sensor is vibration sensitive and causes DSC malfunctions if installed improperly.*

Tightening torque	
DSC acceleration sensor to body	8 Nm (6 ft-lb)

– Program using BMW scan tool.

Steering angle sensor

The steering angle sensor is integral with the steering column switch cluster (SZL). For more information, including removal and installation of the SZL, see **612 Switches**.

– To reset steering angle sensor, connect vehicle to BMW scan tool and carry out steering angle sensor adjustment system function.

BentleyPublishers.com
BentleyPublishers.com—All Rights Reserved

400 Body–General

General

This chapter covers system descriptions and general information for the repair groups in partitions **4 Body** and **5 Body Equipment**.

The 5 Series chassis was introduced in the North America market as a Sedan (E60) in fall 2003 and Sports Wagon (E61) in spring 2005. Both rear-wheel drive and all-wheel drive (xDrive) versions were offered.

BentleyPublishers
.com

B510400001

BentleyPublishers.com
BentleyPublishers.com–All Rights Reserved

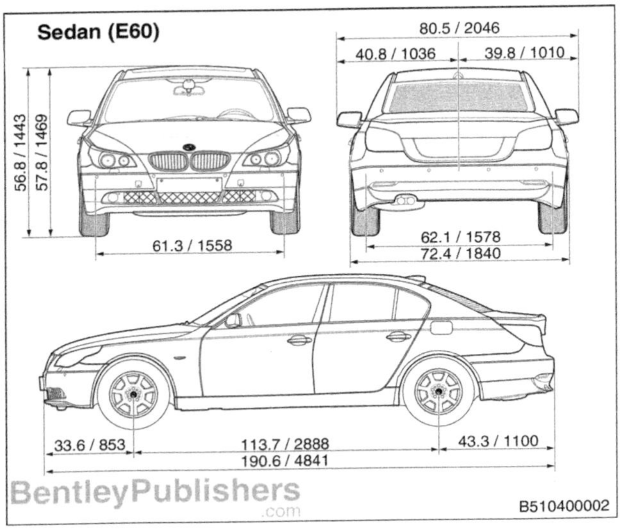

B510400002

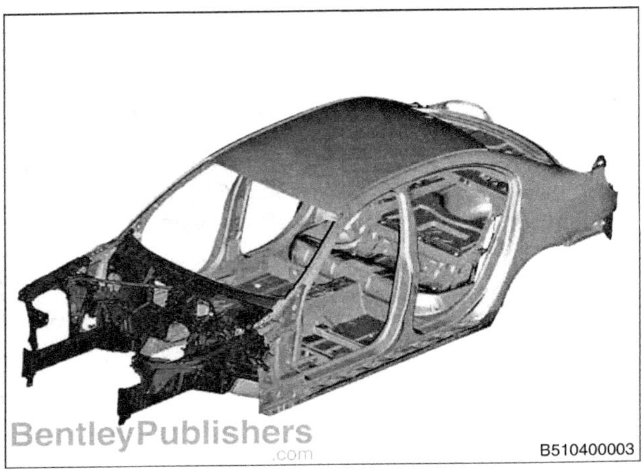

B510400003

Body Assembly

Dimensions

In the accompanying illustration, dimensions for 2004 - 2010 Sedan models are given in inches / mm.

Body dimensions have been increased from the previous generation (E39) to improve knee room in the rear seat area by 5 cm (2 in) and to increase the luggage compartment capacity, which at 520 L (18.36 ft^3) has the largest capacity of all current BMW vehicles including the E65.

Body

In order to make the Sedan (E60) and Sports Wagon (E61) significantly lighter than the previous generation 5 Series (E39), BMW designed a reduced-weight aluminum front end (GRAV). The front end consists almost entirely of aluminum while the transition to steel occurs in the front bulkhead area. Aluminum components are shown in darker shaded (front clip) section.

BentleyPublishers.com
BentleyPublishers.com–All Rights Reserved

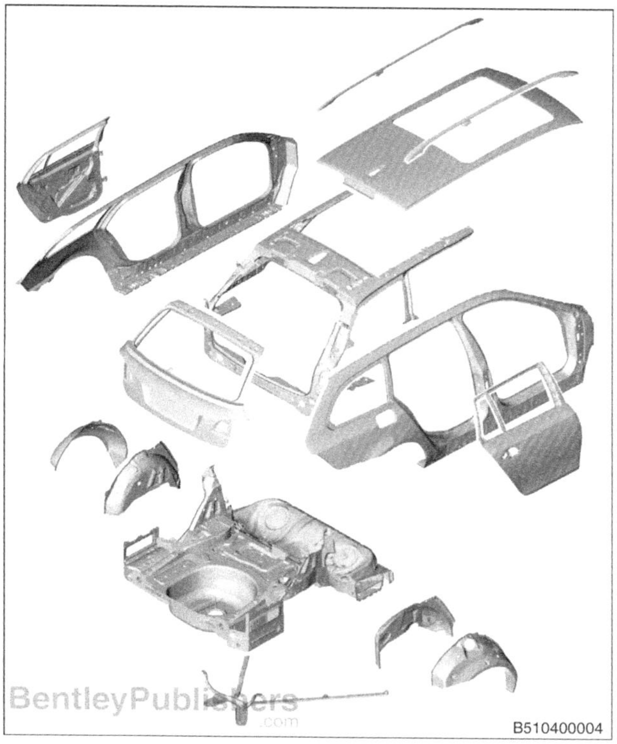

The bodies of the Sedan and Sports Wagon are identical in structure and design from the front bumper to the B-pillar. The rear doors, rear floor pans, inner halves of the rear wheel arches, rear side frames and pillars, rear roof frame and roof outer skin have been redesigned for the Sports Wagon. The rear hatch was designed similar to the E39 Sports Wagon with a separately opening rear window. To increase the body rigidity, a V-shaped tension strut is fitted on the underbody on vehicles equipped with the panoramic glass sunroof.

The Sports Wagon luggage compartment is flat with the rear seat backrests folded down. 4 lashing eyes are fitted as standard on the floor of the luggage compartment. The vehicle tool kit is located in a utility box on the left-hand frame side member. The lockable luggage compartment floor can be raised and with the aid of gas spring struts (similar to the hood) and remains in position until it is pressed down again. A storage area is provided under the luggage compartment floor. Below this is the spare wheel or a further storage area (35 liters) if the vehicle is equipped with run-on flat tires.

Safety and Security

Passenger safety

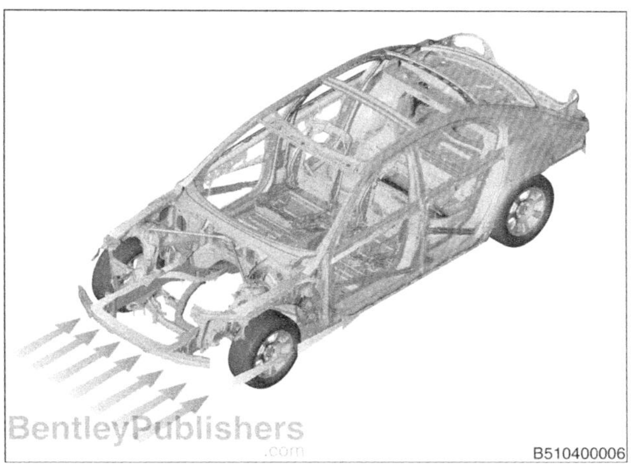

BMW passive safety is based on the principle of protecting the passenger compartment with dynamic deformation zones. One of the first considerations in designing a safe vehicle is the body and chassis design. It is vital to protect the passenger in the event of an impact. This is accomplished by maintaining the integrity of the "passenger cell" during an accident. The main consideration during an impact is to avoid transferring crash energy to the vehicle occupants. By adding energy absorbing crash elements to the vehicle, this energy transfer is reduced considerably.

BentleyPublishers
BentleyPublishers.com—All Rights Reserved

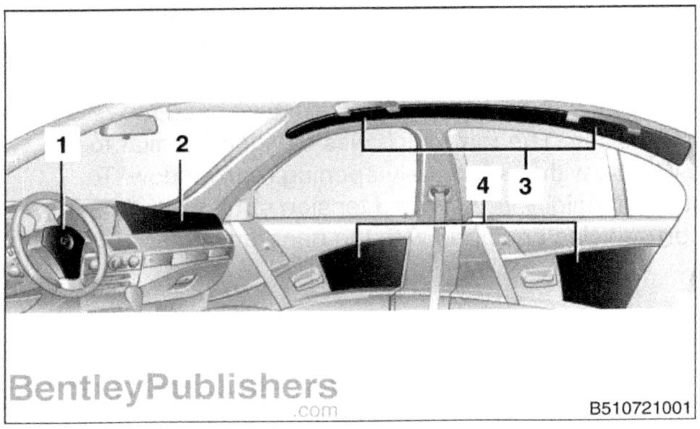

Restraint system

The passive restraint system consists of seat belts with pyrotechnic (explosive) tensioners and a minimum of 6 airbags.

1. Driver airbag
2. Passenger airbag
3. Curtain side airbag
4. Side airbag (rear optional)

For additional restraint system information, see:

- **720 Seat belts**
- **721 Airbag System (SRS)**

Door handles

In vehicle equipped with comfort access, the front door handle contains the components of the electronic outer door handle module (TAGE).

If a vehicle remote key is in the vicinity of the vehicle, a hand inserted between the door handle and the door automatically unlocks the car.

See **515 Central Locking and Anti-theft**.

Central locking

All doors locks as well as trunk lid lock, tailgate lock and fuel filler flap lock are integrated into central locking and can be operated via the following:

- Remote key
- Driver's door lock barrel (door lock)
- Central locking button
- Electronic outer door handle module (TAGE) (comfort access vehicle)

There are two different central locking systems in 5 Series models dependant on production date. See **515 Central Locking and Anti-theft** for more information.

Antitheft alarm (DWA)

When armed, the antitheft system monitors door lock contacts and trunk and engine hood locks and sounds an alarm if it detects tampering. Included in the enhanced system is a tilt sensor to protect against the vehicle being towed away, and an interior motion sensor.

For more information, see **515 Central Locking and Anti-theft**.

BentleyPublishers.com
BentleyPublishers.com—All Rights Reserved

Interior Features

Instruments, iDrive

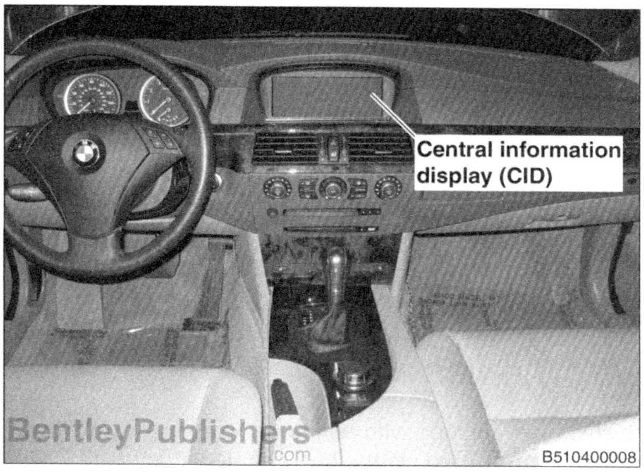

The instrument cluster and dashboard layout is similar to other BMW models with all of the controls in easy reach of the driver.

The dashboard is equipped with the iDrive central information display (CID) or car information computer (CIC). The display screen is used for a variety of vehicle control and diagnostic tasks and is also capable of displaying navigation information.

Navigation, climate control, sound system and other functions are controlled via iDrive controller.

Functions and controls essential to the driver are logically grouped and easy to reach. Instruments and controls are fully integrated into the overall design of the vehicle.

- Tilt-telescopic steering wheel has 30 mm (1.2 in) of vertical and longitudinal adjustment.
- Multifunction steering wheel contains two key pads containing controls for the sound system, telephone and cruise control.

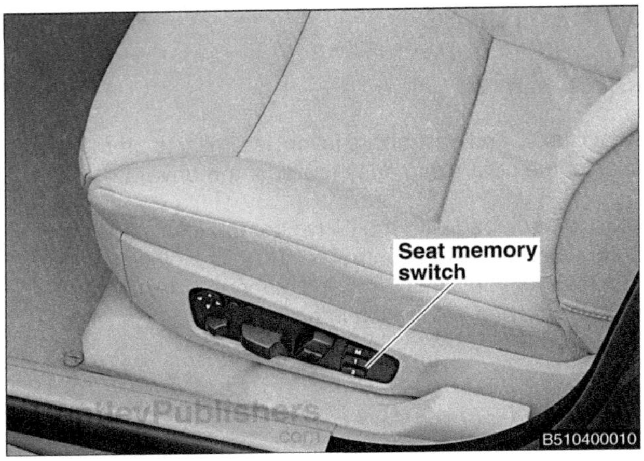

Seats

Ergonomically engineered seats are constructed from polyurethane foam containing areas or zones of different firmness. They offer superior lateral support without constricting the occupant.

In cars equipped with seat position memory, two different seat configurations can be memorized by the seat control module. Seat memory coordinates with outside mirror memory. These functions are controlled by the seat memory module.

For additional information, see **520 Seats**.

BODY ELECTRONICS

5 Series vehicles are equipped with an integrated complex of electronic modules connected via the bus system. An overview of the bus system is in **600 Electrical System–General**. For details of electrical functions, see:

- **121 Battery, Starter, Alternator**
- **512 Door Windows**
- **515 Central Locking and Anti-theft**
- **520 Seats**
- **611 Wipers and Washers**
- **620 Instruments**
- **630 Lights**

BentleyPublishers.com
BentleyPublishers.com—All Rights Reserved

410 Fenders, Engine Hood

GENERAL

This repair group covers replacement of the front fenders and removal and installation of the engine hood.

> ***CAUTION—***
> - *The body is painted at the factory after assembly. Realignment of body panels may expose unpainted metal. Paint all exposed metal once the work is complete.*

Special tools

Most body repairs can be performed using regular automotive service tools. Special BMW tools are required to set engine hood into service position.

FRONT FENDERS

Front wheel housing liner, removing and installing

– Raise front of vehicle and remove wheel.

> ***WARNING—***
> - *Make sure that the car is firmly supported on jack stands designed for the purpose. Place jack stands beneath structural chassis points. Do not place jack stands under suspension parts.*

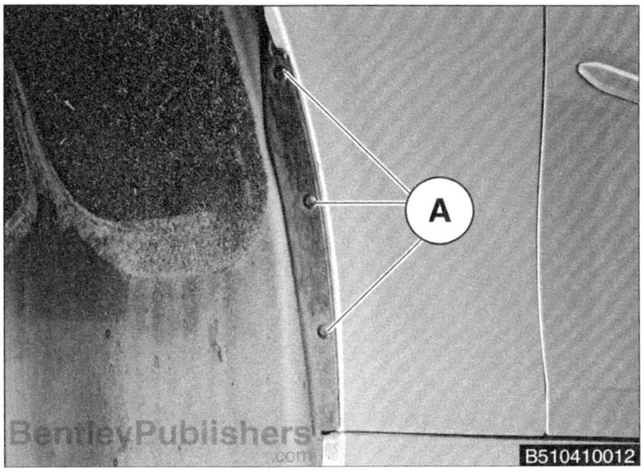

◄ Remove clips (**A**) from fender edge by driving center out with a suitable sized punch, then prying out clip with suitable plastic prying tool.

NOTE—
- *The clips are not reusable. Replace with new.*

BentleyPublishers.com—All Rights Reserved

B510410011

- Remove fasteners (**arrows**) from wheel housing liner.
- Remove liner.
- Installation is reverse of removal.

Front fender, removing and installing

- Raise hood to service position. See **Engine hood, raising to service position** in this repair group.
- Remove front wheel housing liner. See **Front wheel housing liner, removing and installing** in this repair group.

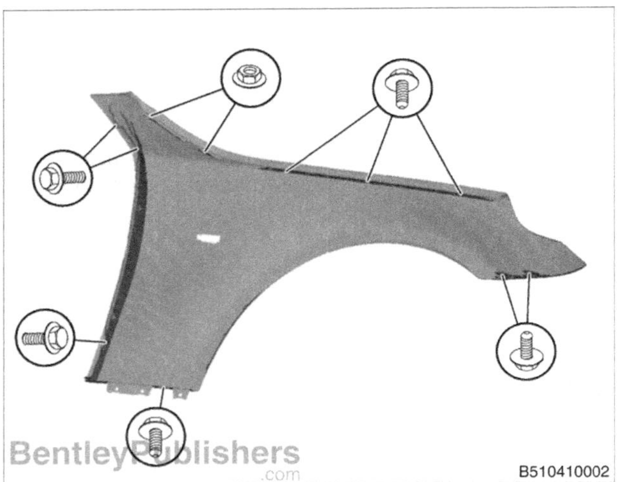

B510410002

- Fastener locations shown for Sedan and Sports Wagon models.

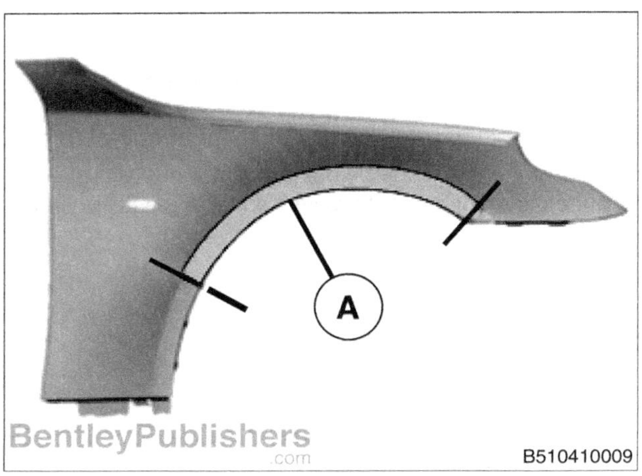

B510410009

- Apply sprayable sealing wax Terostat 9320 or equivalent to underside of fender in area **A**.

BentleyPublishers.com
BentleyPublishers.com—All Rights Reserved

Engine Hood

Engine hood latch components

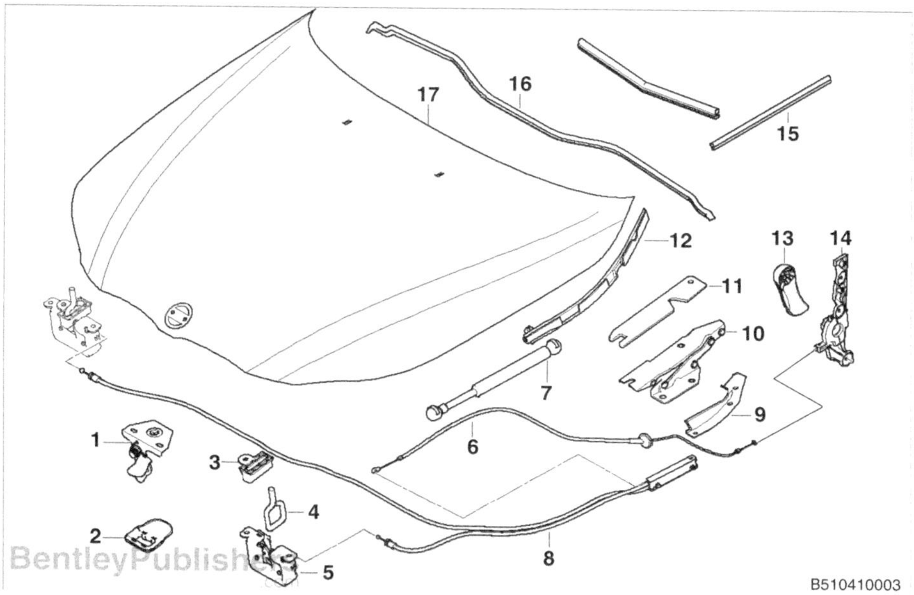

1. Hood latch
2. Hood latch end support
3. Twist protection
4. Safety bracket
5. Hood lock
6. Hood release cable, rear
7. Gas support strut
8. Hood release cable, front
9. Water gutter cover
10. Left hood hinge
11. Spacer
12. Left hood seal
13. Hood release lever
14. Combination bracket
15. Side hood seal
16. Center hood seal
17. Hood

Engine hood, aligning

When installing hood, align hinges as close to original painted surface as possible. Movement of hood on its attaching hardware may require touch up paint.

- Loosen left upper and lower hinge mounting bolts (**arrows**). Make sure hood can move relative to hinge.

- Repeat on right.

BentleyPublishers
BentleyPublishers.com—All Rights Reserved

Align hood so gap (**arrows**) to fender is as even as possible.

Tightening torque	
Hood to hood hinge (M8 x 24 mm, 8.8 grade)	18.5 Nm (13.5 ft-lb)

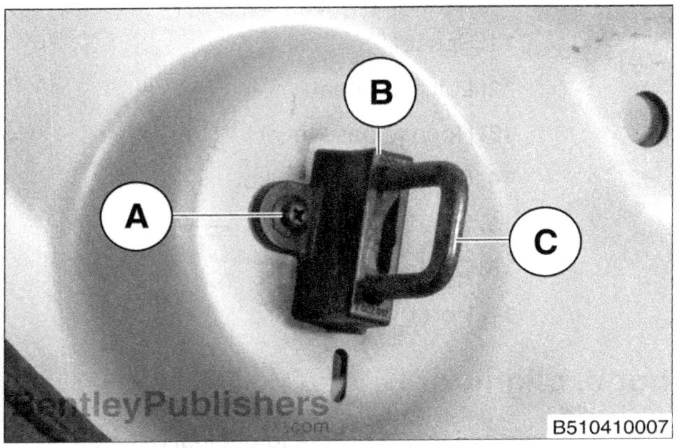

If front height adjustment is off, adjust safety brackets:

- Release screw (**A**).
- Remove cover (**B**).
- Adjust height by turning latch (**C**).

If height adjustment of engine hood lid to front fenders is off, turn front buffer stops (**arrow**) on left and right.

> ***CAUTION—***
> - *To avoid excessive wind noise, make sure hood does not protrude beyond front fenders.*

BentleyPublishers.com
BentleyPublishers.com—All Rights Reserved

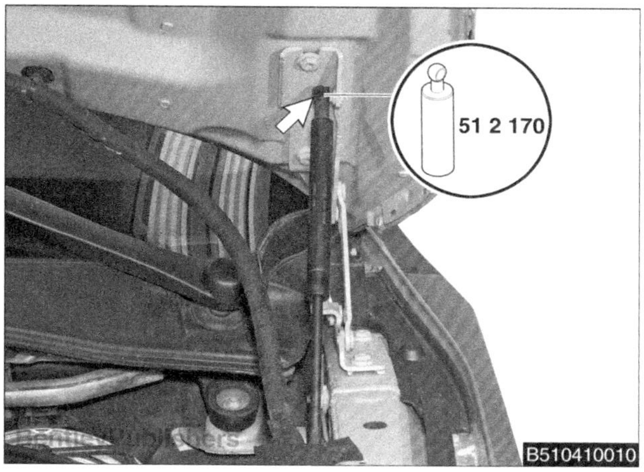

Engine hood, raising to service position

◀ Open hood. With help of an assistant, support hood and remove retaining clips (**arrow**) on upper end of left gas support strut.

– Slide BMW special tool 51 2 170 (**inset**) over housing of gas spring strut then clip ball socket of special tool on ball stud on hood.

– Repeat on right.

Engine hood, removing and installing

> ***WARNING —***
> • *The hood is heavy. Before removing the hood supports, be sure to have an assistant help support the hood.*

– Mark position of hood hinge mounting plates.

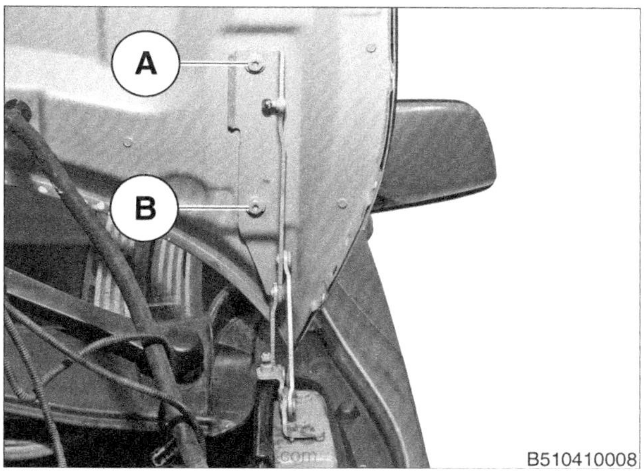

◀ Loosen upper hinge mounting bolts (**A**) and remove lower bolts (**B**). Lift hood off carefully.

– Installation is reverse of removal. Note the following:

- Repair any paint damage and paint any exposed metal.
- Check hood alignment. See **Engine hood, aligning** in this repair group.

Tightening torque	
Hood to hood hinge (M8 x 24 mm, 8.8 grade)	18.5 Nm (13.5 ft-lb)

BentleyPublishers.com
BentleyPublishers.com—All Rights Reserved

BentleyPublishers.com
BentleyPublishers.com—All Rights Reserved

411 Doors

General

This repair group covers front and rear door repair information and includes door trim panel service.

> **CAUTION—**
> - *The body is painted at the factory after assembly. Realignment of body panels may expose unpainted metal. Paint exposed metal once work is complete.*

For additional information, see:

- **512 Door Windows** for power door windows and door glass replacement
- **515 Central Locking and Anti-theft** for power door locks
- **721 Airbag system (SRS)** for side airbags

Warnings and cautions

> **CAUTION—**
> - *Prior to disconnecting the battery, read the battery disconnection cautions given in* **001 Warnings and Cautions**.
> - *To avoid damaging plastic interior trim, use a plastic prying tool or a screwdriver with the tip wrapped with masking tape.*

> **WARNING —**
> - *E60 and E61 cars are fitted with side-impact airbags in the front doors and optional side impact airbags in the rear doors. When servicing doors, always disconnect the negative (-) battery terminal. See* **721 Airbag System (SRS)** *for cautions and procedures relating to the airbag system.*

BentleyPublishers
BentleyPublishers.com–All Rights Reserved

Doors

Door check, replacing

– Close door window completely.

– Disconnect negative (-) battery cable.

> ***CAUTION—***
> - *Prior to disconnecting the battery, read the battery disconnection cautions given in* **001 Warnings and Cautions.**

– Remove interior door panel. See **Door trim panel, removing and installing** in this repair group.

◄ Remove door check seal (**C**).

– Release door check screws (**A** and **B**).

◄ Working inside door, remove door check assembly (**A**).

– Installation is reverse of removal.

Tightening torques	
Door check to body	21 Nm (15 ft-lb)
Door check to door	10 Nm (7 ft-lb)

BentleyPublishers.com
BentleyPublishers.com—All Rights Reserved

Door hinge assembly, front

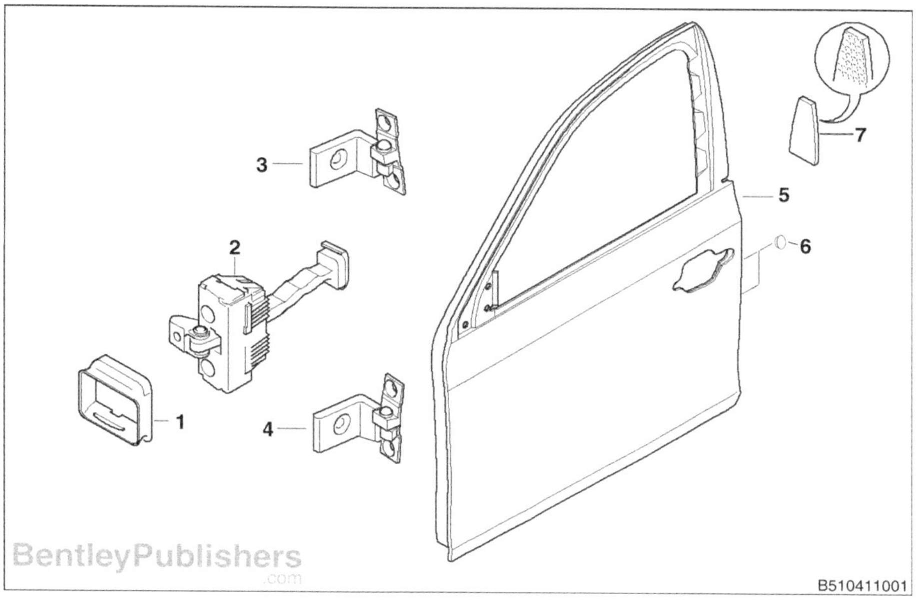

1. Door check seal
2. Door check
3. Upper door hinge
4. Lower door hinge
5. Door
6. Plug
7. Reflector

Door, adjusting

Front and rear door adjustment procedures are similar.

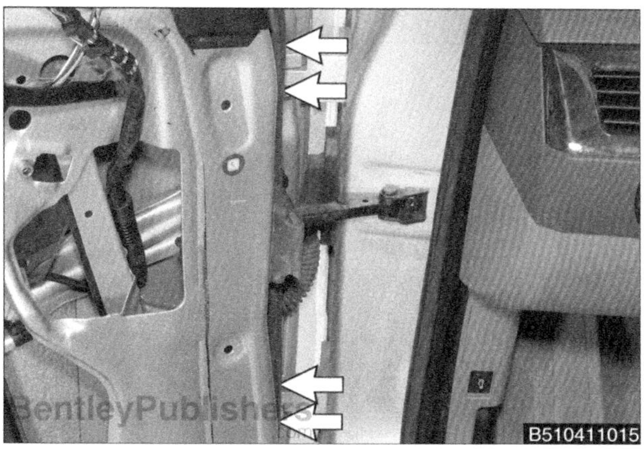

◄ If installed door is uneven or out of parallel, loosen nuts at door hinges (**arrows**) to adjust at top and bottom.

– If more adjustment is necessary, spacers (or shims) can be used to correct door alignment. Spacers are placed behind hinges.

Check that adjoining body parts are flush in terms of height (**dotted line**) and correct if necessary. After adjustment, tighten hinge nuts and screws.

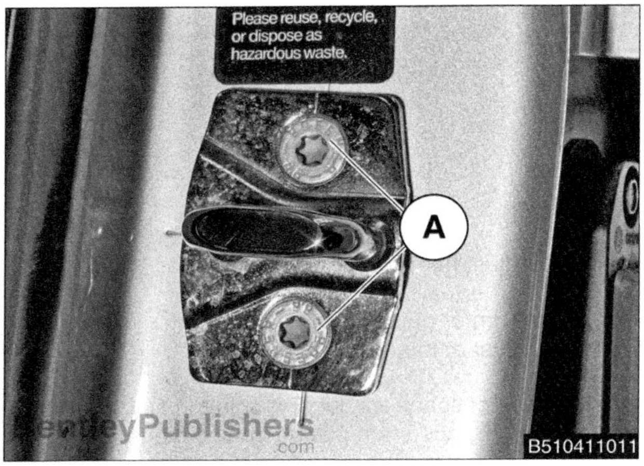

Loosen screws (**A**) to adjust lock striker.

NOTE —
- *When door is closed, lock striker must not touch or brush against door lock. Avoid scratching door.*

Tightening torques	
Door to door hinge (M8 x 28 mm)	20 Nm (15 ft-lb)
Lock striker to body	21 Nm (15 ft-lb)

Door, removing and installing

Front and rear door removal and installation are similar.

Disconnect negative (-) battery cable.

CAUTION—
- *Prior to disconnecting the battery, read the battery disconnection cautions given in* **001 Warnings and Cautions**.

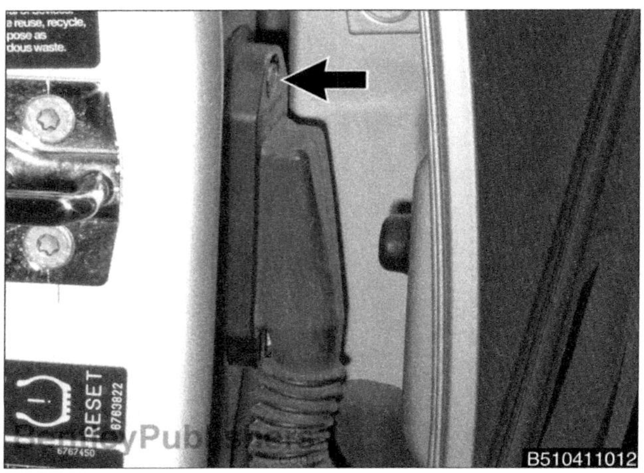

Remove harness connector mounting bolt (**arrow**) at door pillar.

BentleyPublishers.com
BentleyPublishers.com—All Rights Reserved

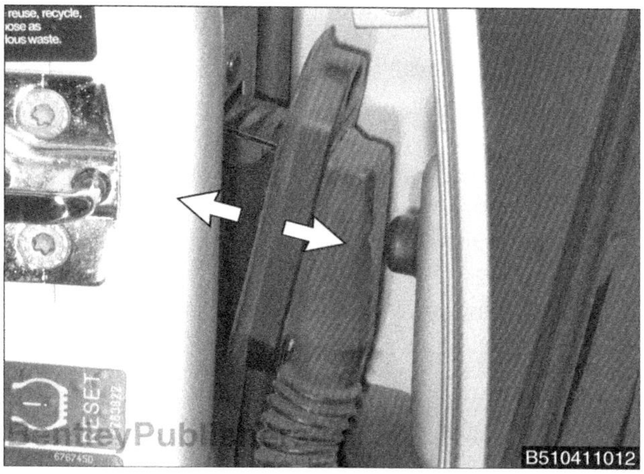

Pull plug connection from door pillar by lifting it up and out.

- Detach plug connection (**arrows**).

Remove door check mounting bolt (**A**).

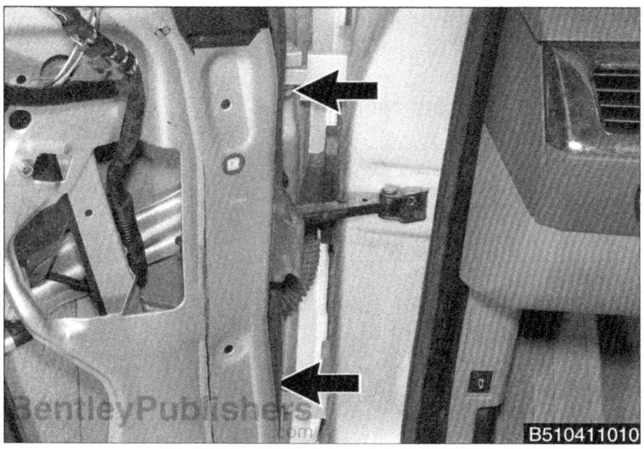

With door fully open, unscrew door hinge pin bolts (**arrows**) from top and bottom hinges.

BentleyPublishers
BentleyPublishers.com–All Rights Reserved

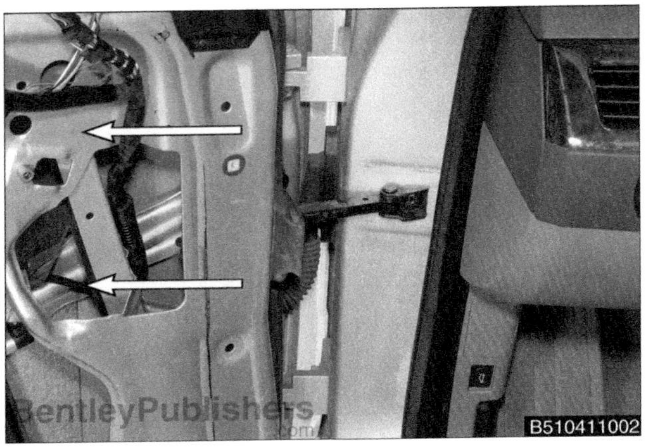

◄ Remove door by pulling out sideways.

– Installation is reverse of removal. Note the following:
  - Mount and align rear doors first (if applicable), followed by front doors.
  - Align door so that panel gaps are equal on either side. If necessary, adjust door hinges. See **Door, adjusting** in this repair group.
  - Adjust lock striker so that trailing edge of front door is slightly higher than leading edge of rear door.
  - Repair paint damage and paint exposed metal.

Tightening torques	
Door check to body	21 Nm (15 ft-lb)
Door electrical harness to body (M5 x 16 mm)	3 Nm (27 in-lb)
Door hinge pin bolts	27 Nm (20 ft-lb)

Door Panels

Sedan left front door panel is illustrated. Other door panels are similar.

Door trim panel, removing and installing

WARNING —
- *E60 and E61 cars are fitted with side-impact airbags in the front doors and optional side impact airbags in the rear doors. When servicing doors, always disconnect the negative (-) battery terminal. See* **721 Airbag System (SRS)** *for cautions and procedures relating to the airbag system.*

NOTE —
- *Vehicle through 03 / 2007 shown, later vehicles similar*

– Disconnect negative (-) battery cable.

CAUTION—
- *Prior to disconnecting the battery, read the battery disconnection cautions given in* **001 Warnings and Cautions**.

◄ Pry out footwell light (**arrow**) using plastic prying tool, disconnect from harness connector and remove.

BentleyPublishers.com
BentleyPublishers.com—All Rights Reserved

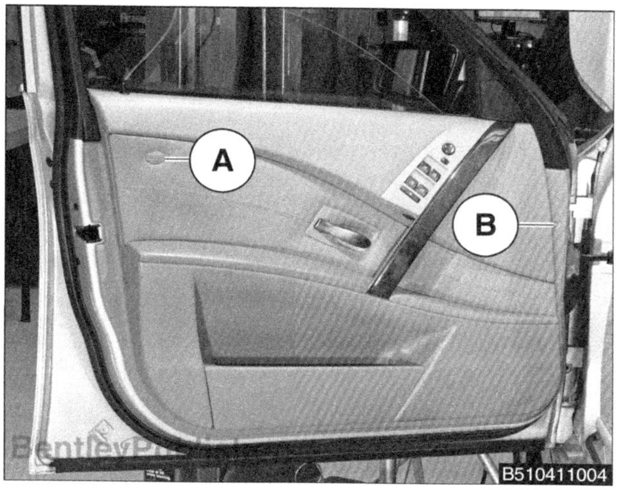

◀ Pry out airbag emblem (**A**) on door trim panel using plastic prying tool and release screw underneath.

– Remove cap (**B**) and screw underneath.

◀ Unclip door trim panel starting at bottom. Unclip door trim panel clips (**arrows**) using plastic prying tool.

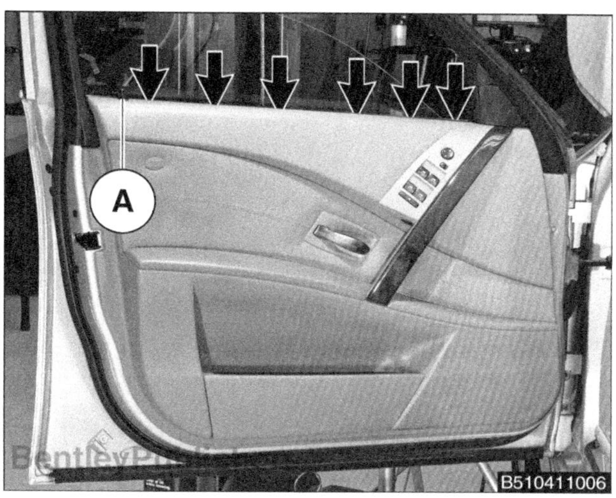

◀ Carefully unclip door trim panel from retainers (**arrows**) by pulling straight out. Pull door trim panel up out of central mounting clip and release button (**A**).

BentleyPublishers
BentleyPublishers.com–All Rights Reserved

Door trim panel, removing and installing

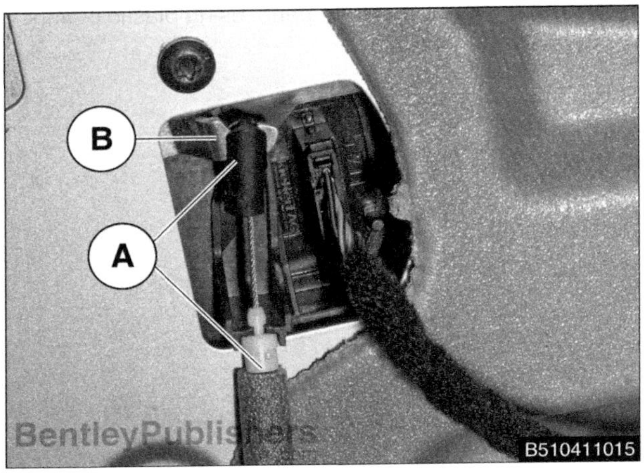

B510411015

◄ Unhook Bowden cable (**A**) from inner door opener by unlatching tab (**B**).

– Disconnect side airbag harness connector.

– Disconnect wiring harness for window switch panel and remove door trim panel.

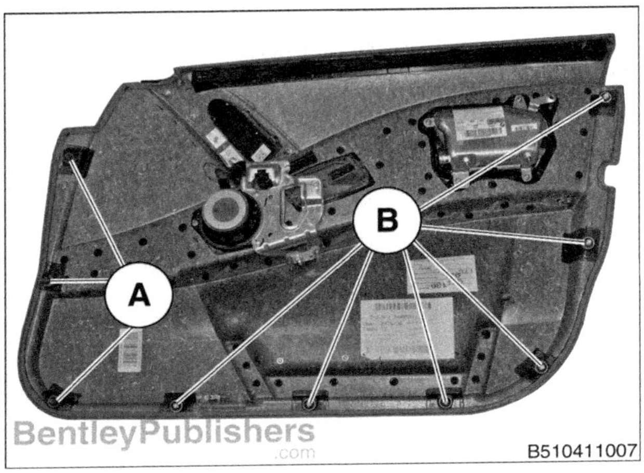

B510411007

◄ If necessary, replace any broken clips (**A** and **B**). Install clips in correct position. Clips **A** are gray and clips **B** are white.

– Attach central mounting clip to door trim panel. Remainder of installation is reverse of removal. Check door lock mechanism and window for ease of movement and proper function.

Tightening torque	
Door trim panel to door (screw under airbag emblem)	8 Nm (71 in-lb)
Door trim panel to door (screw under cap)	0.8 Nm (7 in-lb)

BentleyPublishers.com
BentleyPublishers.com—All Rights Reserved

412 Trunk lid, Tailgate

TABLE

GENERAL

This repair group covers trunk lid and tailgate removal and installation. Replacement procedure for gas-charged support struts that hold tailgate in open position is also covered.

CAUTION—
- *The body is painted at the factory after assembly. Realignment of body panels may expose unpainted metal. Paint exposed metal once work is complete.*

TRUNK LID (SEDAN)

Emergency release

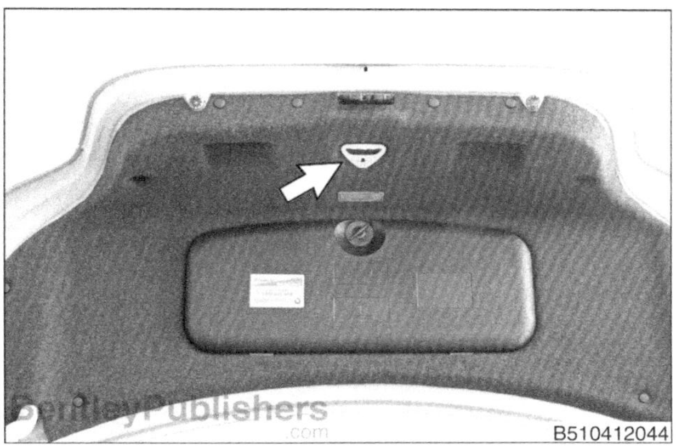

In case of emergency, pull handle (**arrow**) to open trunk lid from inside luggage compartment.

BentleyPublishers.com
BentleyPublishers.com–All Rights Reserved

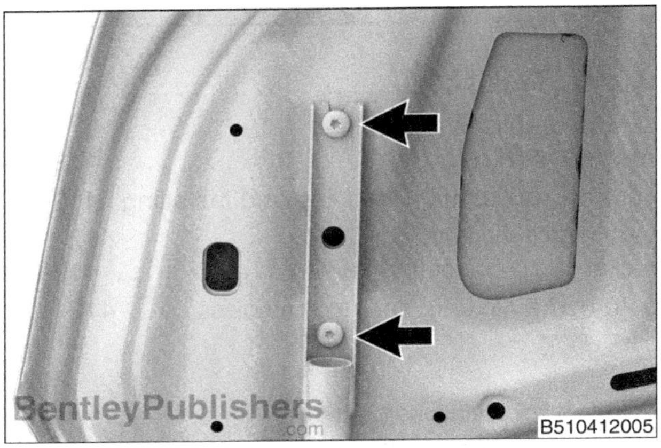

Trunk lid, removing and installing

- Raise and support trunk lid.
- Remove trunk lid trim panel. See **513 Interior Trim**.
- Mark hinge and hinge fastener locations for reinstallation.
- ◄ Unscrew left side hinge fasteners (**arrows**). Repeat on right and remove trunk lid with help of assistant.
- Installation is reverse of removal. If necessary, align trunk lid to body. See **Trunk lid, aligning** in this repair group.

Tightening torque	
Trunk lid to hinge (M8 x 19.5 mm, 8.8 grade)	18.5 Nm (13.5 ft-lb)

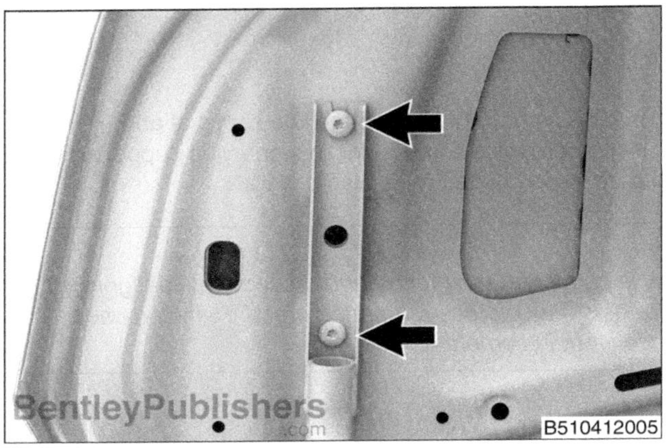

Trunk lid, aligning

- Raise and support trunk lid.
- Remove trunk lid trim panel. See **513 Interior Trim**.
- ◄ Loosen left and right hinge fasteners (**arrows**) so that trunk lid can be moved to set correct gap. Tighten bolts.
- Remove trunk lower trim panel and left side trim panel. See **513 Interior Trim**.

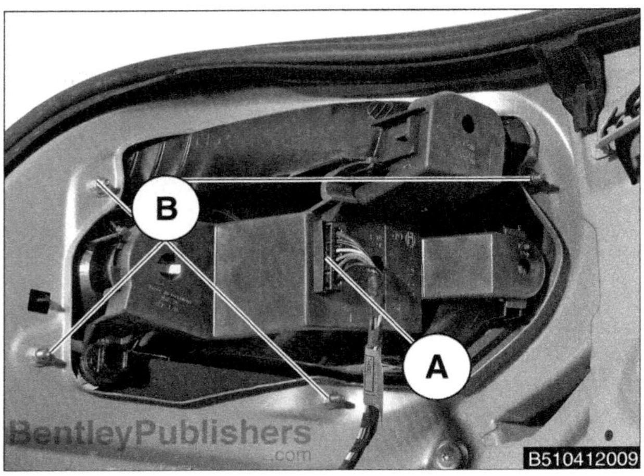

- ◄ Disconnect electrical connector (**A**).
- Release fasteners (**B**) and remove taillight.

BentleyPublishers.com
BentleyPublishers.com—All Rights Reserved

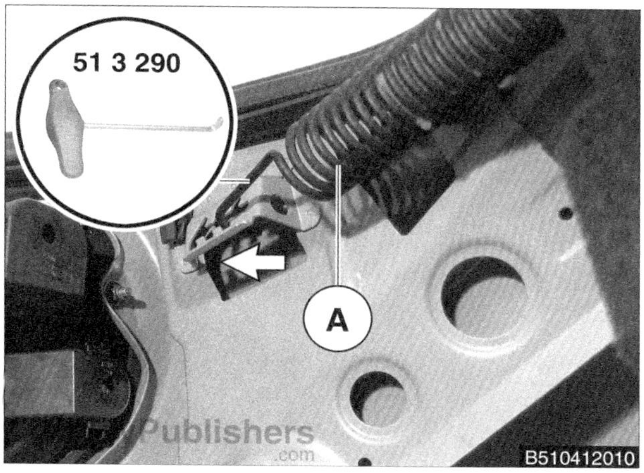

Using BMW special tool 51 3 290 (**inset**) or equivalent, pull trunk lid tension spring (**A**) toward rear to release from side panel.

> ***CAUTION—***
> *• Support trunk lid in open position.*

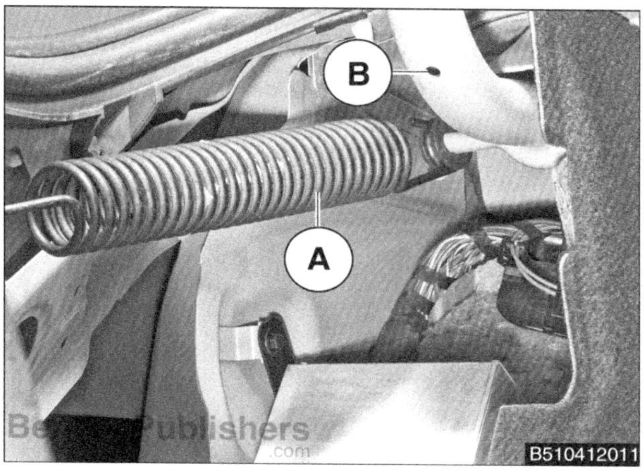

Unhook tension spring (**A**) from trunk lid hinge (**B**).

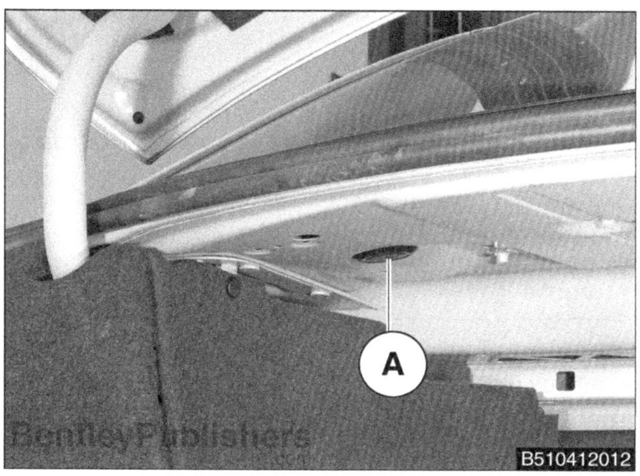

Remove cap (**A**) from opening.

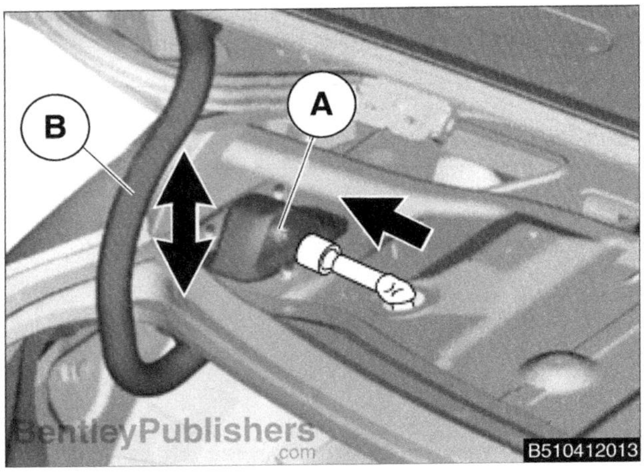

Loosen fastener (**A**) until trunk lid hinge (**B**) can be moved.

NOTE—
- *Rear parcel shelf is shown removed for purpose of clarity.*

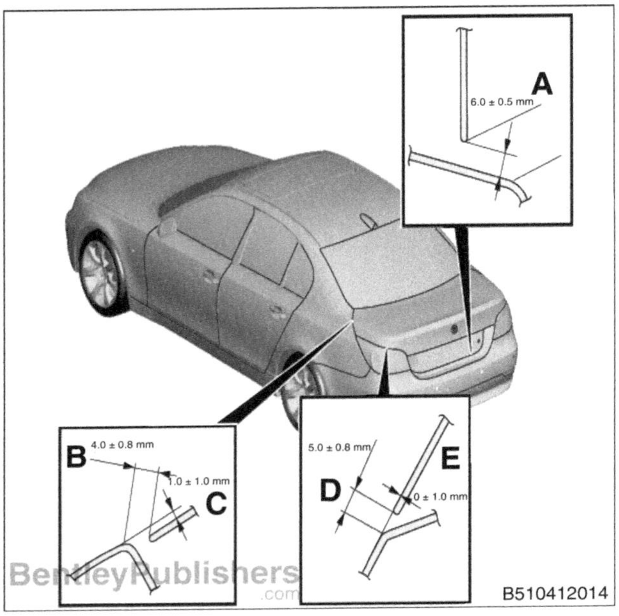

Sedan: Set gap measurements as listed below.

Table a. Trunk lid gap specifications (Sedan)
• Dimension **A** . . . . 6.0 ± 0.5 mm (0.24 ± 0.02 in)
• Dimension **B** . . . . 4.0 ± 0.8 mm (0.16 ± 0.03 in)
• Dimension **C** . . . . 1.0 ± 1.0 mm (0.04 ± 0.04 in)
• Dimension **D** . . . . 5.0 ± 0.8 mm (0.20 ± 0.03 in)
• Dimension **E** . . . . 0.0 ± 1.0 mm (0.00 ± 0.04 in)

CAUTION—
- *Before closing trunk lid, be sure that trunk lid and body do not scrape.*

– Adjust trunk lock striker. See **515 Central Locking and Anti-Theft.**

– Unscrew trunk lid detent buffers until left and right sides of lid rest on buffers with trunk lid closed.

NOTE—
- *Be sure top surface of trunk lid is even with top surfaces of fenders.*

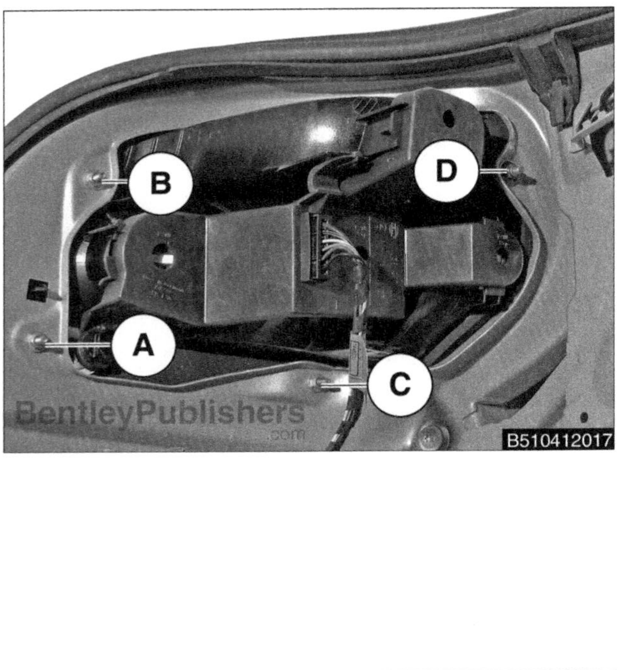

Installation is reverse of removal. Tighten taillight fasteners in the following sequence: (**A**, **B**, **C**, **D**).

Tightening torques	
Taillight to body	2.5 Nm (22 in-lb)
Trunk hinge to body (M8)	20 Nm (15 ft-lb)
Trunk lid to hinge (M8 x 19.5 mm)	18.5 Nm (13.5 ft-lb)
Trunk striker to body	10 Nm (7 ft-lb)

BentleyPublishers.com
BentleyPublishers.com—All Rights Reserved

TAILGATE (SPORTS WAGON)

Tailgate service position

- Raise and support tailgate.
- Remove tailgate gas strut or hydraulic cylinder. See **Tailgate gas strut, removing and installing** or **Tailgate hydraulic cylinder, removing and installing** in this repair group.

◄ Replace with BMW special tool 51 0 050 or equivalent.

Tailgate, removing and installing

- Open rear window.
- Unclip and remove tailgate upper rear window frame trim panel.
- Unplug electrical connectors.
- Unhook tailgate gas strut from tailgate. See **Tailgate gas strut, removing and installing** in this repair group.

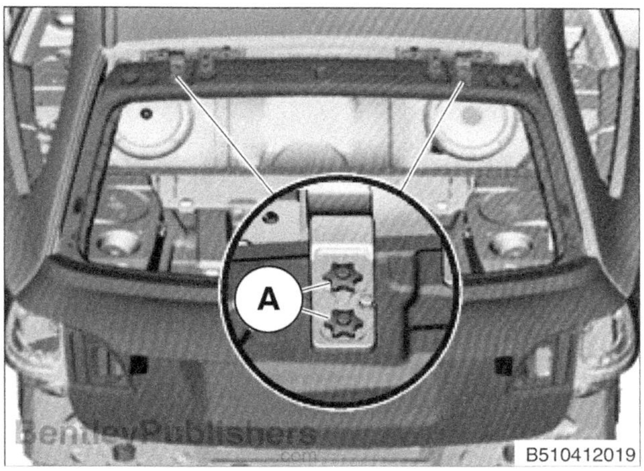

◄ Remove fasteners (**A**) and remove tailgate with help from assistant.

> ***CAUTION—***
> - *The tailgate is very heavy. Use assistance to support both sides prior to removal of hinge bolts.*

- Installation is reverse of removal. To align, see **Tailgate, aligning** in this repair group.

Tightening torque	
Tailgate to tailgate hinge (M10)	38 Nm (28 ft-lb)

Tailgate gas strut, removing and installing

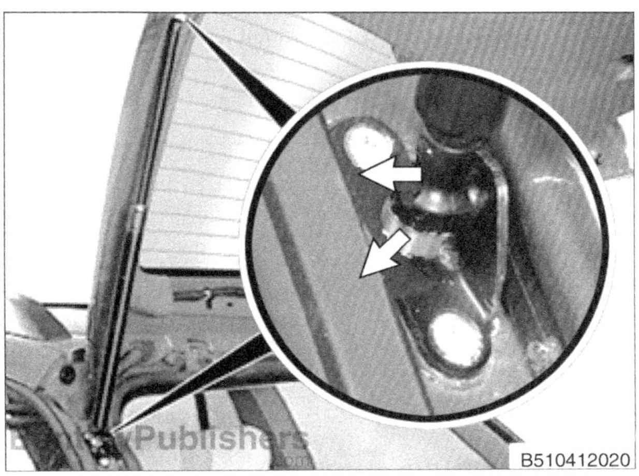

◄ Open and support tailgate. Remove retaining clip (**arrows**) on upper end of gas strut.

> ***CAUTION—***
> - *The tailgate is very heavy and will close without both pressurized struts installed. Properly support tailgate prior to removal of a strut.*

- Pull gas strut off tailgate.
- Remove retaining clip from lower end of gas strut.
- Remove gas strut.
- Installation is reverse of removal. Replace retaining clips if damaged during removal.

BentleyPublishers.com
BentleyPublishers.com–All Rights Reserved

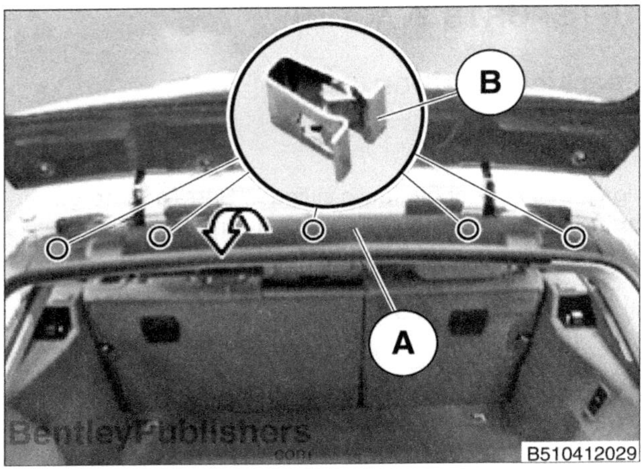

Tailgate, aligning

- Using plastic prying tool, remove upper rear window frame trim (**A**) by releasing from trim clips (**B**).

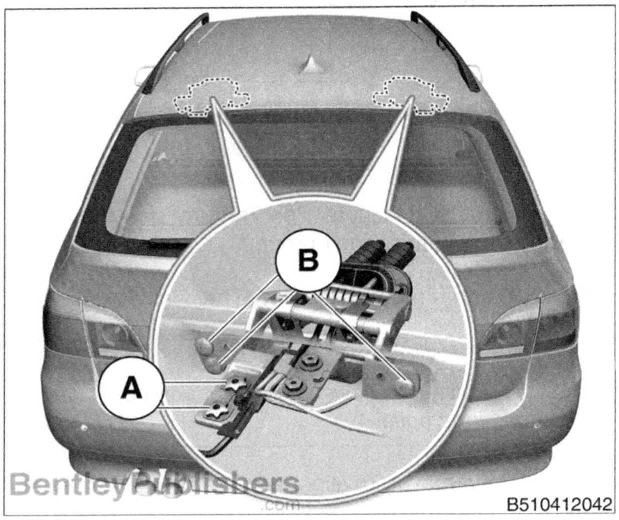

- Loosen tailgate-to-hinge fasteners (**A**) so tailgate can move relative to hinge.

– Loosen hinge-to-body fasteners (**B**) so hinge can move relative to side panel.

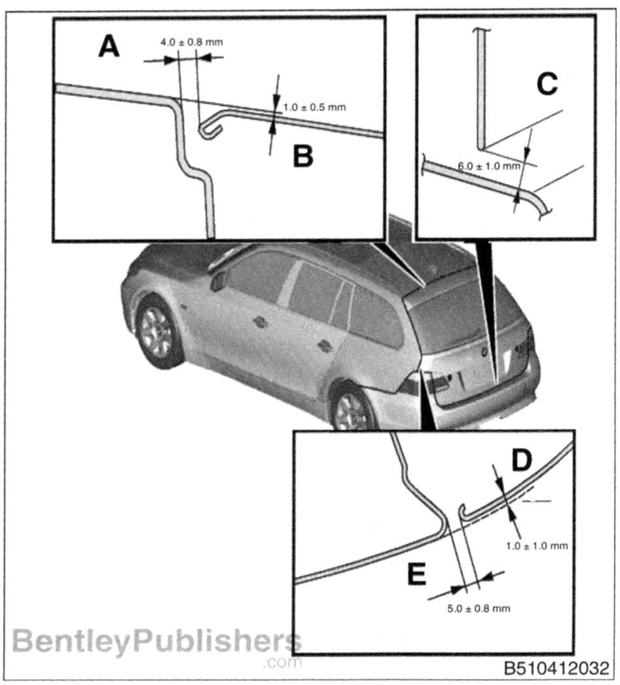

- Set gap measurements as listed below for Sports Wagon.

Table b. Tailgate gap specifications (Sports Wagon)	
• Dimension **A**	4.0 ± 0.8 mm (0.16 ± 0.03 in)
• Dimension **B**	1.0 ± 0.5 mm (0.04 ± 0.02 in)
• Dimension **C**	6.0 ± 1.0 mm (0.24 ± 0.04 in)
• Dimension **D**	1.0 ± 1.0 mm (0.04 ± 0.04 in)
• Dimension **E**	5.0 ± 0.8 mm (0.20 ± 0.03 in)

Tightening torques	
Tailgate hinge to body (M8)	20 Nm (15 ft-lb)
Tailgate to tailgate hinge (M10 nut)	38 Nm (28 ft-lb)

> ***CAUTION—***
> *• Before closing truck lid, be sure that trunk lid and body do not scrape.*

– Adjust tailgate striker assembly. See **515 Central Locking and Anti-Theft**.

BentleyPublishers.com
BentleyPublishers.com—All Rights Reserved

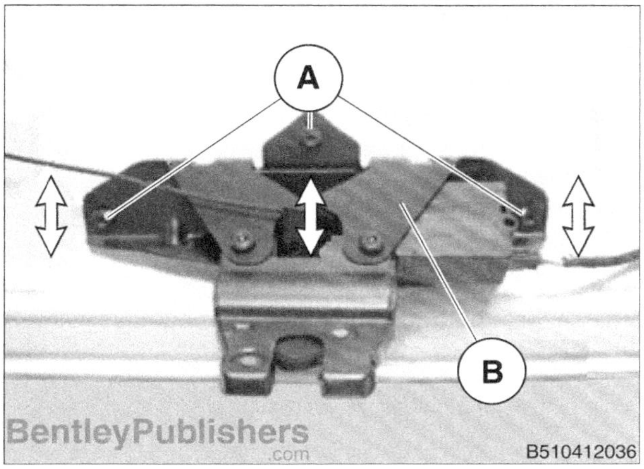

- If necessary, loosen fasteners (**A**) so that upper section of tailgate lock (**B**) can be adjusted.

- Check lock adjustment. Repeat if necessary.

Rear Spoiler (Sports Wagon)

Rear spoiler, removing and installing

Removing

- Open and raise tailgate.

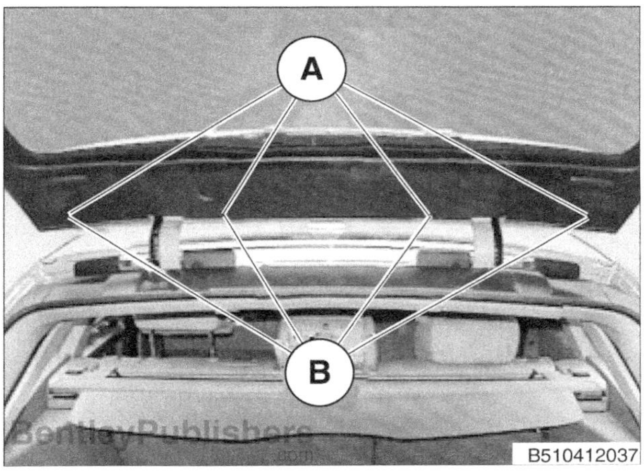

- Pry out plugs (**A**) and remove screws underneath (**B**).

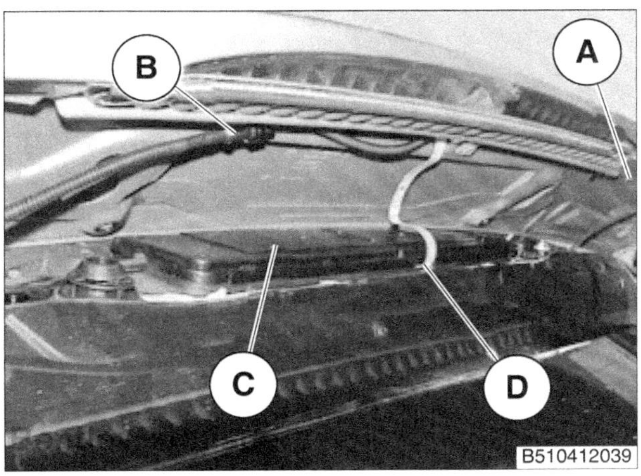

- Raise rear spoiler slightly.
- Disconnect electrical connector (**A**) and hose (**B**).
- Pull up and remove antenna amplifier cover (**C**).
- Disconnect electrical connector (**D**).
- Remove rear spoiler from tailgate and set down on clean surface.
- Installation is reverse of removal.

Tightening torque	
Rear spoiler to tailgate	8 Nm (71 in-lb)

BentleyPublishers
BentleyPublishers.com—All Rights Reserved

AUTOMATIC TAILGATE

Tailgate hydraulic cylinder, removing and installing

Removal of tailgate hydraulic cylinder and hydraulic lines requires disconnecting hydraulic lines at hydraulic unit, and removing headliner and left rear speaker. After replacement of hydraulic cylinder, check fluid level in hydraulic unit (see **Tailgate hydraulic fluid, checking and correcting level** in this repair group) and set readiness (see **Tailgate hydraulic unit, setting readiness** in this repair group).

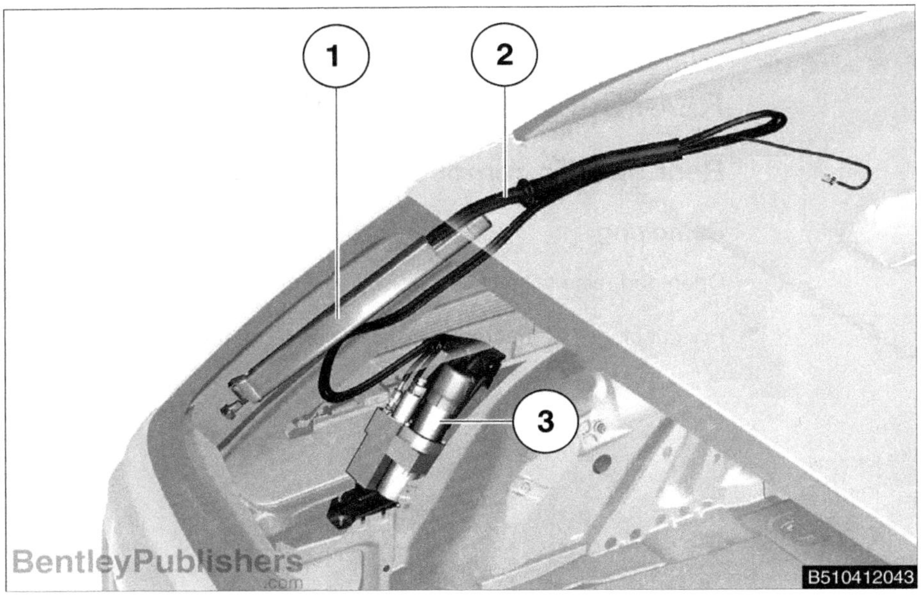

1. Hydraulic cylinder
2. Hydraulic lines
3. Hydraulic unit

Tailgate hydraulic unit, removing and installing

– Raise tailgate and put into service position. See**Tailgate service position** in this repair group.

– Remove rear left side trim panel and sound insulation.

NOTE —

• Have a cloth ready to catch any spilled hydraulic fluid.

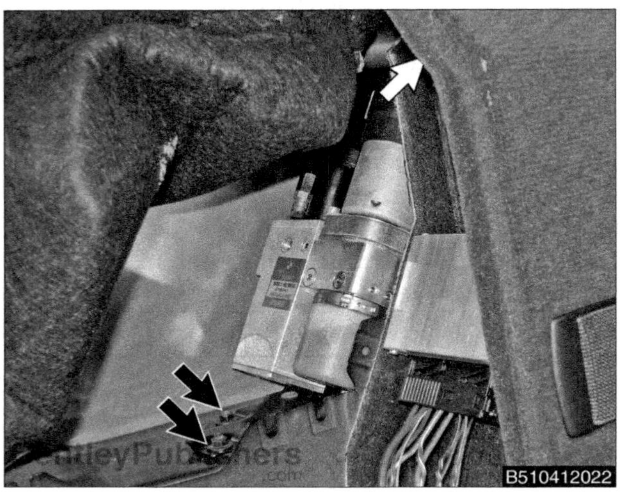

< Release fasteners (**arrows**) securing hydraulic unit bracket to side panel.

BentleyPublishers.com
BentleyPublishers.com—All Rights Reserved

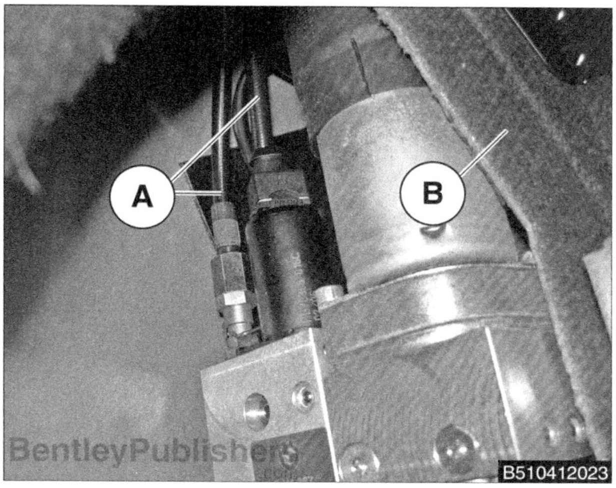

Unclip hydraulic lines (**A**) from bracket (**B**) and remove hydraulic unit.

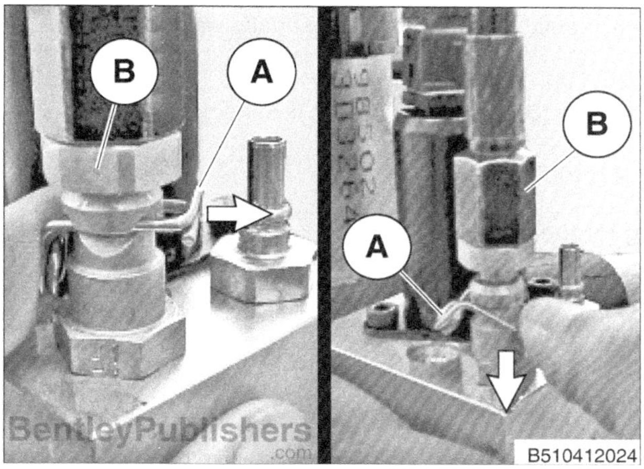

Disconnect hose coupling (**B**) from hydraulic unit by sliding retainer wire (**A**) to right, pressing downward, and pulling hose coupling up off hydraulic unit. Be sure to catch any escaping hydraulic fluid. Dispose of safely.

- Installation is reverse of removal. Check and correct fluid level if necessary. See **Tailgate hydraulic fluid, checking and correcting level** in this repair group.
- Set readiness. See **Tailgate hydraulic unit, setting readiness** in this repair group.

Tailgate hydraulic fluid, checking and correcting level

- Open and close tailgate via automatic function to bleed hydraulic system.
- Open tailgate again via automatic function.
- Remove tailgate hydraulic unit. See **Tailgate hydraulic unit, removing and installing** in this repair group.

Checking fluid level

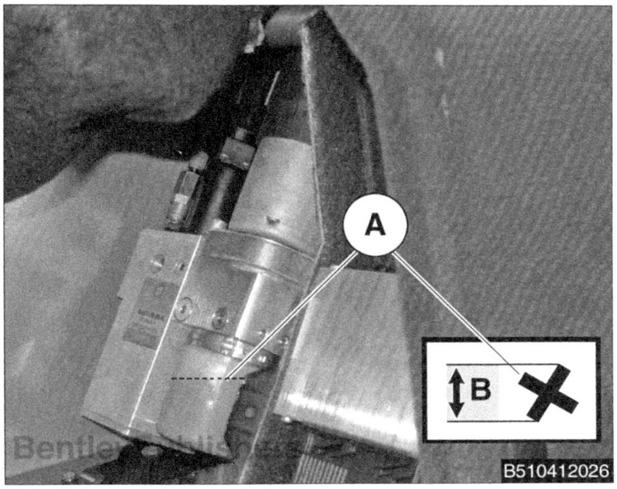

Set hydraulic unit down on a horizontal surface at angle of installation in vehicle. Fluid level (**A**) must match marking **B**.

BentleyPublishers
BentleyPublishers.com–All Rights Reserved

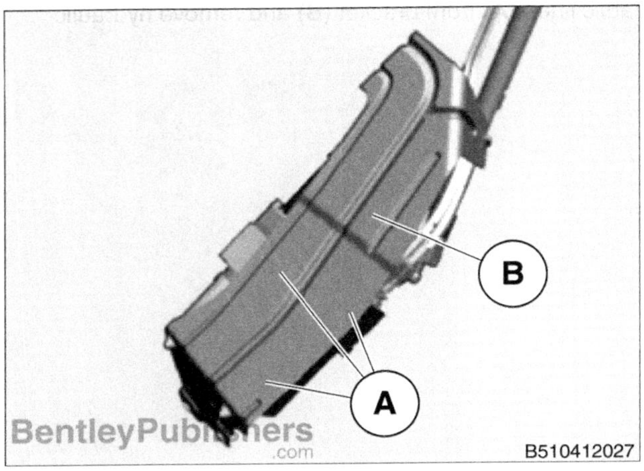

Correcting fluid level

◄ Remove fasteners (**A**) and separate hydraulic unit from bracket (**B**).

◄ Set tailgate hydraulic unit down on level surface with fluid tank (**A**) facing up.

– Open oil filler plug (**B**) and add hydraulic fluid as necessary.

– Installation is reverse of removal.

Tightening torques	
Oil filler plug to fluid tank	1 Nm (9 in-lb)
Tailgate hydraulic unit to bracket	2 Nm (18 in-lb)

Tailgate hydraulic unit, setting readiness

When a new hydraulic cylinder is installed, make sure system detects “rear lid open” deactivation point.

– Disconnect and reconnect battery.

– Press button on tailgate. Tailgate hydraulic unit runs for 13 seconds in “open” direction. Readiness is set.

BentleyPublishers.com
BentleyPublishers.com—All Rights Reserved

510 Exterior Trim, Bumpers

General

This repair group includes repair information for the outside rear view mirrors, front and rear bumper and the easily removable exterior trim parts.

The photos in this group illustrate a 2004 545i. Other models may vary.

See also:

- **411 Doors** for door panels
- **513 Interior Trim**

> ***CAUTION—***
> - *Prior to disconnecting the battery, read the battery disconnection cautions given in* **001 Warnings and Cautions**.
> - *To prevent marring the trim, work with a plastic prying tool or wrap a screwdriver tip with masking tape before prying out trim panels, switches or electrical accessories.*

510

BentleyPublishers.com
BentleyPublishers.com—All Rights Reserved

Outside Rear View Mirror

Outside mirror glass, removing and installing

To avoid breaking, move mirror glass carefully and slowly. For added protection, cover glass completely with masking tape or wear hand protection when removing.

> ***CAUTION—***
> *• Make sure mirror glass is at or above room temperature before removing. Otherwise, small plastic parts or glass may break.*

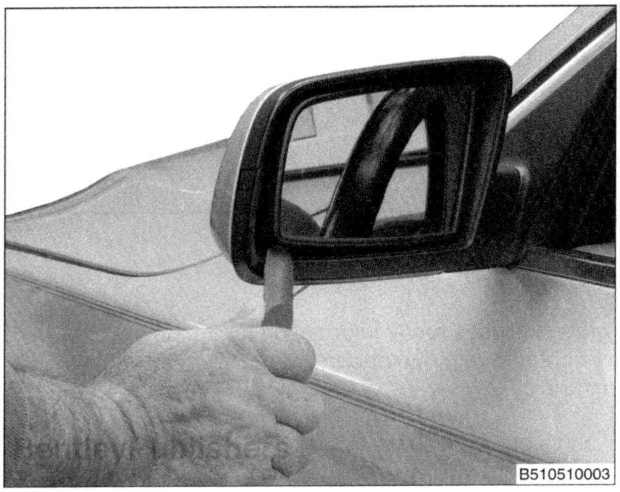

– Tilt mirror housing out.

– With mirror glass squarely positioned in housing frame, push bottom of glass fully into housing.

< Insert plastic wedge under corner of glass and gently pry glass up until glass disengages from retaining clips.

– Lift glass out and store mirror glass safely.

– When installing:
  - Position mirror glass in housing while aligning plastic retainers.
  - Carefully lever top center of glass down to engage retaining clips.
  - Check that mirror moves freely in all directions.

Outside mirror housing, removing and installing

– Remove interior door trim panel. See **411 Doors**.

– Remove door window frame cover.

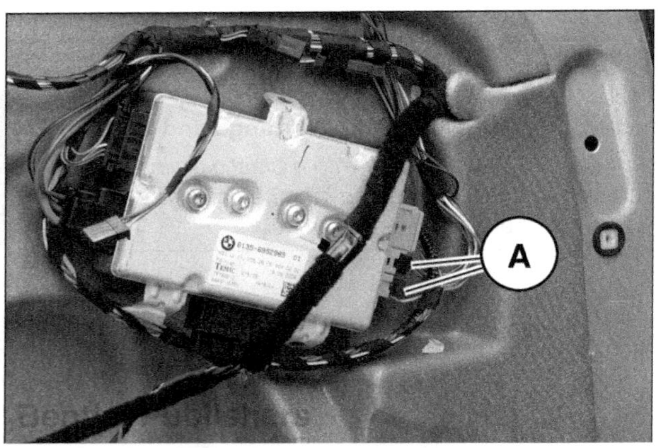

< Disconnect electrical connectors (**A**). Vehicle equipped with door module shown, others are similar.

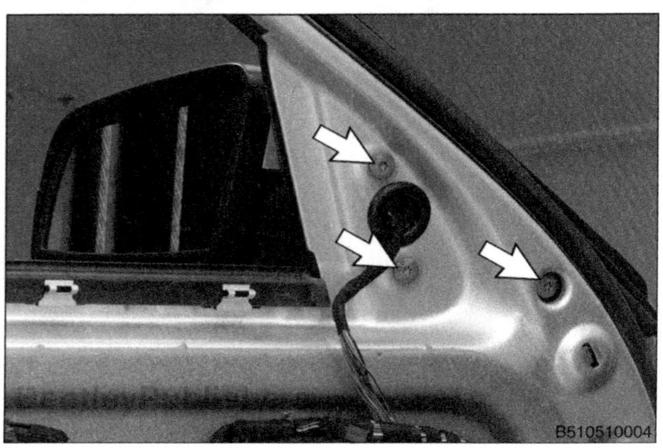

< Support mirror from outside and remove fasteners (**arrows**).

– Installation is reverse of removal.

NOTE—
• Check mirror function before reinstalling door panel trim.

Tightening torques	
Outside mirror to door • Previously installed mirror • New mirror	 6 Nm (4.5 ft-lb) 8 Nm (6 ft-lb)

BentleyPublishers.com
BentleyPublishers.com—All Rights Reserved

Bumpers

Front bumper assembly

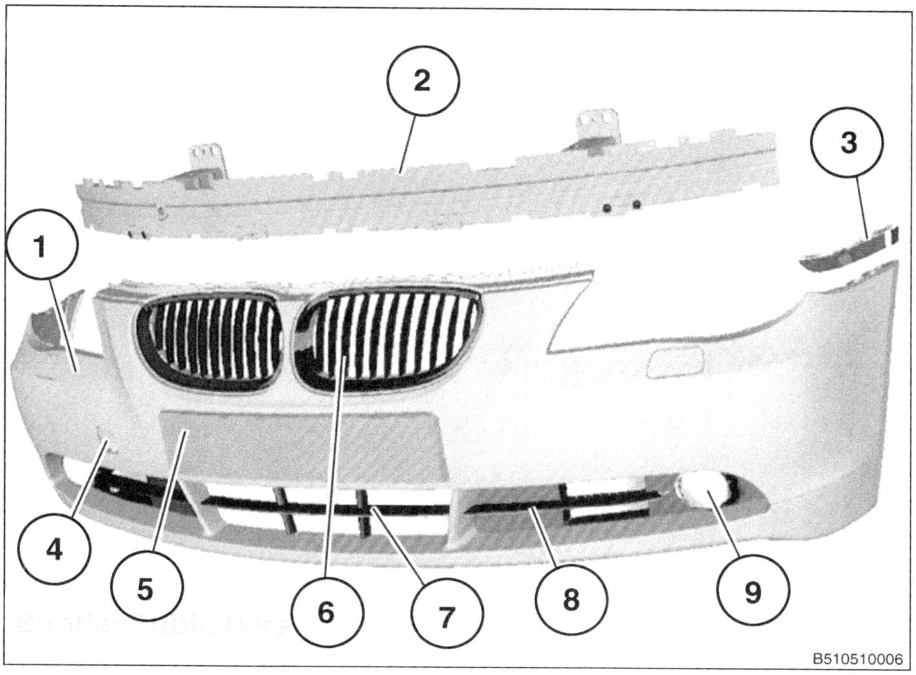

1. Front bumper cover
2. Front bumper support
3. Bracket
4. Towing eye cover
5. License plate holder
6. Radiator grill
7. Lower grill
8. Air duct trim
9. Foglight

Front bumper cover, removing and installing

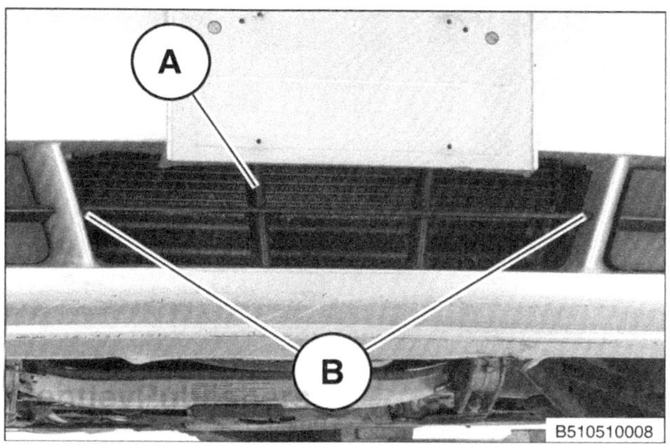

< Release lower grill catches (**B**) and remove grill (**A**).

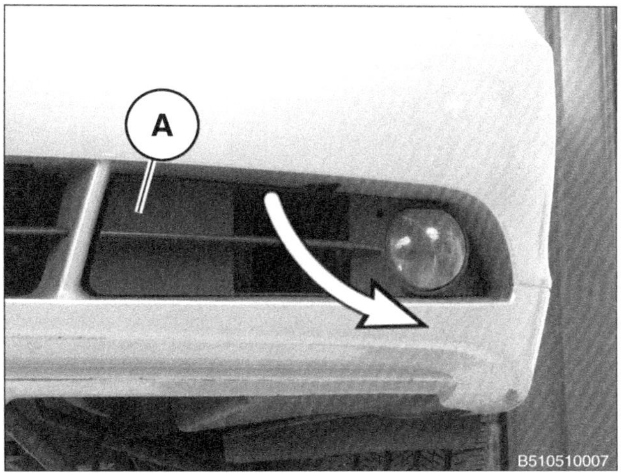

< Release air duct grill catches in direction of **arrow** and remove grill (**A**).

BentleyPublishers.com–All Rights Reserved

Working in left front wheel well, remove wheel housing trim fasteners (**A**). Repeat procedure for right side of vehicle.

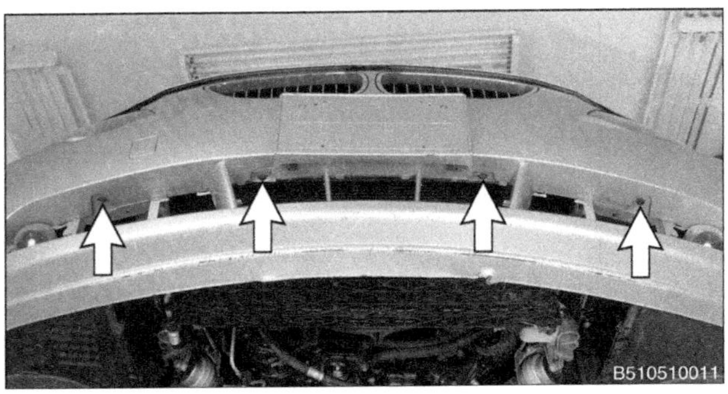

Remove bumper fasteners (**arrows**).

Lift off trim strip with seal (**A**).

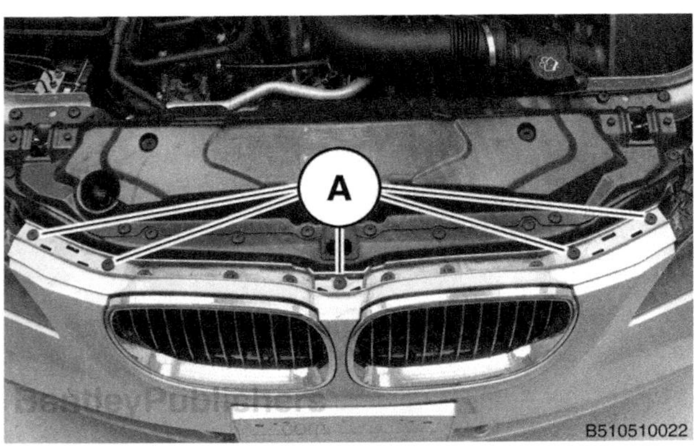

Remove bumper cover fasteners (**A**).

BentleyPublishers.com
BentleyPublishers.com–All Rights Reserved

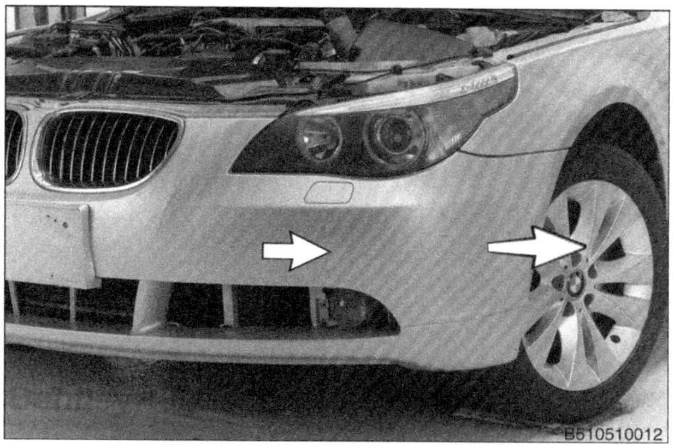

- Detach bumper cover at sides in direction of **arrows**.
- Slide bumper cover forward just enough to detach connections for headlight washers, foglights and parktronic sensors (if equipped).
- Remove bumper cover forward with help from an assistant.
- Installation is reverse of removal.

Tightening torques	
Bumper cover to front end	5 Nm (3.6 ft-lb)

Rear bumper assembly

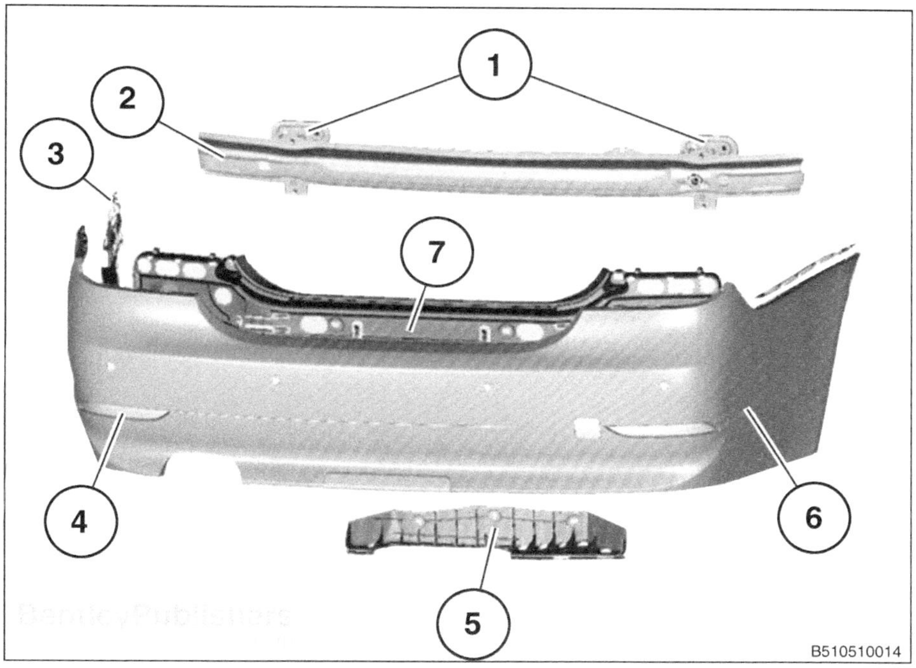

1. Impact absorber
2. Bumper support
3. Side holder
4. Rear reflector
5. Bottom guide
6. Bumper cover
7. Top guide

Rear bumper cover, removing and installing

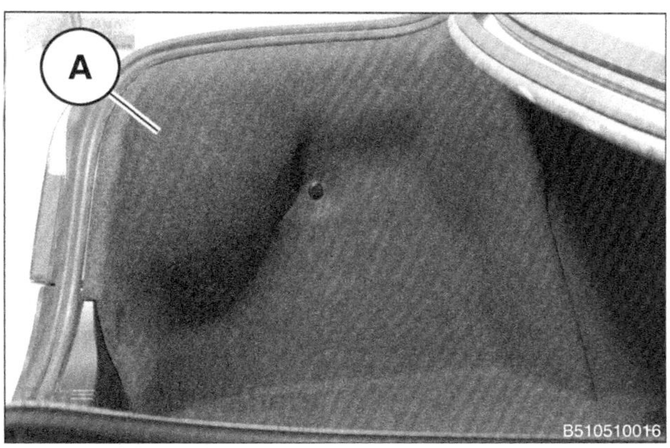

- Working in trunk, remove left and right side trim panel (**A**). Left side shown, right side is similar.

BentleyPublishers.com—All Rights Reserved

Rear bumper cover, removing and installing

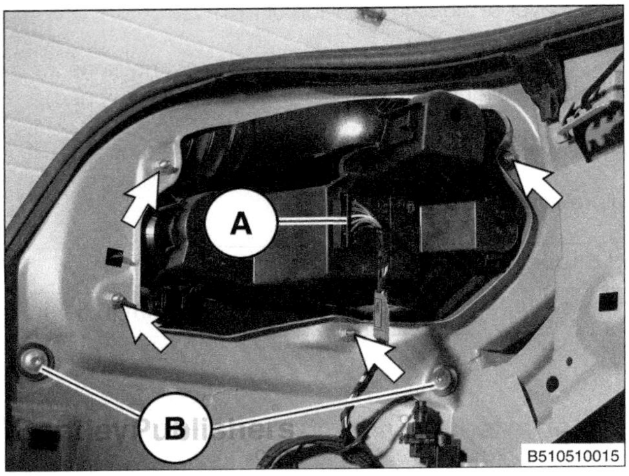

Working at left taillight assembly:

- Disconnect electrical connector (**A**).
- Remove taillight fasteners (**arrows**).
- Remove taillight assembly from body panel.
- Remove bumper cover fasteners (**B**)
- Repeat procedure for right side of vehicle.

Working in left rear wheel well:

- Remove fasteners (**A**).
- Pull rear bumper cover out of mounting bracket in direction or **arrows**.
- Repeat procedure for right side of vehicle.

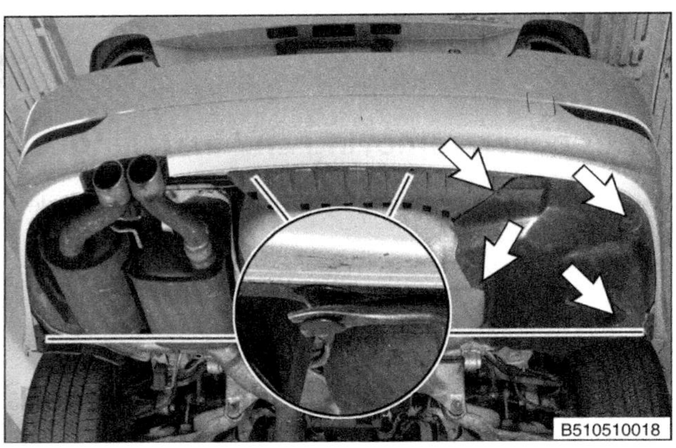

Working under rear bumper, Remove expansion rivets (**inset**). Release bumper trim fasteners (**arrows**).

BentleyPublishers.com
BentleyPublishers.com–All Rights Reserved

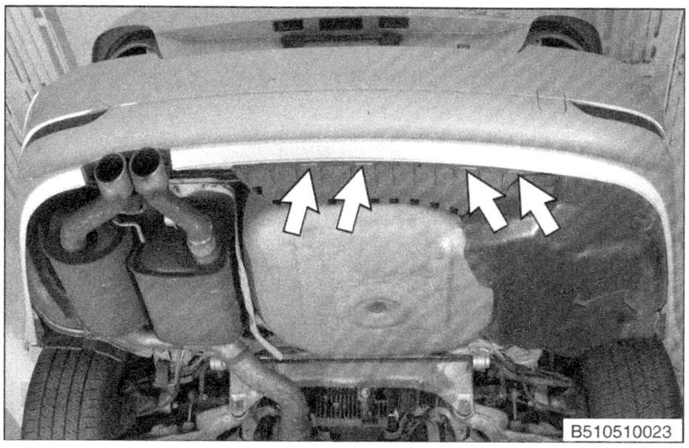

◀ Using a screwdriver, pry out bumper retaining tabs (**arrows**).

◀ Pull left and right side upper edges out in direction of **arrows**. Then remove bumper cover by pulling back and up.

– Disconnect electrical connectors for parktronic sensors, if equipped.

– Remove bumper cover with help from an assistant. Be careful not to damage bumper catches.

– Installation is reverse of removal.

Rear bumper impact absorber, replacing

– Remove rear bumper cover. See **Rear bumper cover, removing and installing** in this repair group.

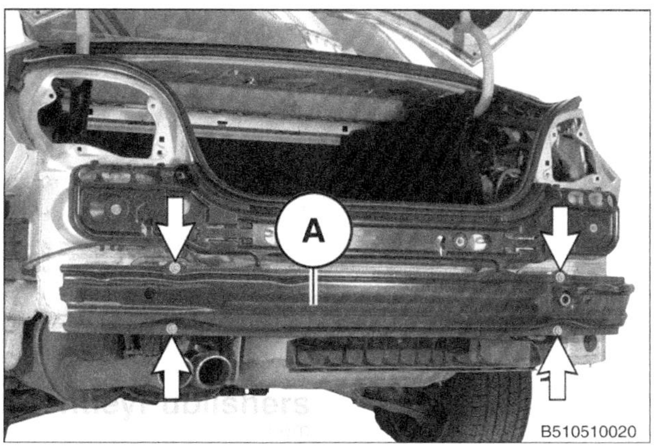

◀ Remove fasteners (**arrows**) and remove impact absorber (**A**).

– Installation is reverse of removal. Remember to replace defective panel nuts.

Tightening torques	
Impact absorber to body • M10	36 Nm (26.5 ft-lb)

BentleyPublishers
BentleyPublishers.com—All Rights Reserved

Exterior Trim

Exterior trim is attached to the body with plastic clips and fasteners that may be damaged during removal. Be sure to have necessary fasteners on hand when reinstalling exterior trim pieces.

BMW emblem, removing and installing

The procedure given below applies to both front and rear emblems.

– Wrap end of a screwdriver with tape.

> ***CAUTION—***
> *• Protect hood paint by covering area around emblem with tape.*

< Pry up emblem carefully on either side (**arrows**).

– Installation is reverse of removal.
  - Replace plastic inserts in body if damaged.
  - If emblem fits loosely, use a small amount of body molding tape or adhesive on rear of emblem before installing.

Radiator grill, removing and installing

– Remove front bumper cover. See **Front bumper cover, removing and installing** in this repair group.

< Release catches (**inset**) and remove radiator grill in direction of **arrow**.

– Once grill is removed, pry remaining tabs to separate inner grill from chrome trim ring.

– Before installation, assemble chrome trim ring and grill insert securely. Replace components if tabs or catches are broken or missing.

– Install grill and trim ring in appropriate location in hood. Press on trim ring until grill snaps into place.

– Assemble front bumper cover.

BentleyPublishers.com
BentleyPublishers.com—All Rights Reserved

512 Door Windows

GENERAL

This repair group covers door glass and window regulator repair information. The windshield, rear window and fixed rear door glass are bonded using special adhesives and tools. For bonded glass replacement, see an authorized BMW service facility or an automotive glass installer.

See also:

- **411 Doors** for door trim panel
- **515 Central Locking and Anti-theft**
- **612 Switches** for electric window switch replacement
- **721 Airbag System (SRS)** for side-impact airbag information

512

Power window features

- **To 09 / 2005**: Front door window motor activated by door module.
- **From 09 / 2005**: Front door window motors activated by body gateway module (KGM).
- Rear door window motors activated by body base module (KBM).
- Convenience operations controlled by car access system (CAS).
- One-touch operation in both directions on all four windows.
- Cable-type window regulator used for all door windows.
- Anti-trap protection monitors window motor torque.
- Convenience opening / closing from driver's lock cylinder and from remote key.
- Window operation customized by owner with Car Memory function. See **515 Central Locking and Anti-theft.**
- After ignition is switched OFF, electric windows can be operated until a door is opened, or until 16 minutes has elapsed.

BentleyPublishers.com
BentleyPublishers.com–All Rights Reserved

Power window control modules to 09 / 2005

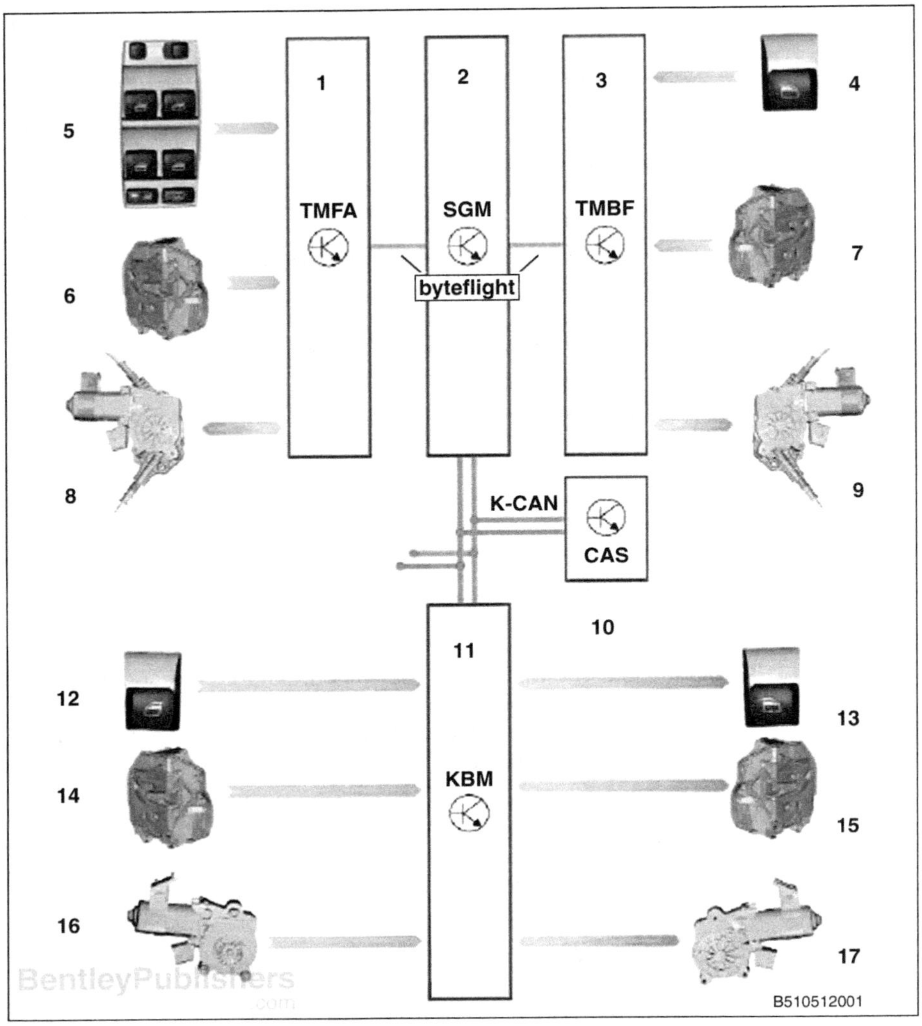

1. Left front door module (TMFA)
2. Safety gateway module (SGM)
3. Right front door module (TMBF)
4. Right front window switch
5. Left front window master switch
6. Left front door latch
7. Right front door latch
8. Left front window motor
9. Right front window motor
10. Car access system (CAS)
11. Basic body module (KBM)
12. Left rear window switch
13. Right rear window switch
14. Left rear door latch
15. Right rear door latch
16. Left rear window motor
17. Right rear window motor

Car access system (CAS) is the master control module for lowering and raising the windows.

Door modules (TMFA, TMBF) contain separate modules for door electronics and advanced safety electronics (ASE).

Door modules are the interface between:

- Left front window master switch
- Right front window switch
- Power-window motor
- Door contacts

Safety gateway module (SGM) converts and forwards messages to and from the byteflight and K-CAN.

Basic body module (KBM) receives signal from the rear door window switches via CAN bus through SGM and provides power and ground to the rear window motors.

BentleyPublishers.com
BentleyPublishers.com—All Rights Reserved

Power window control modules from 09 / 2005

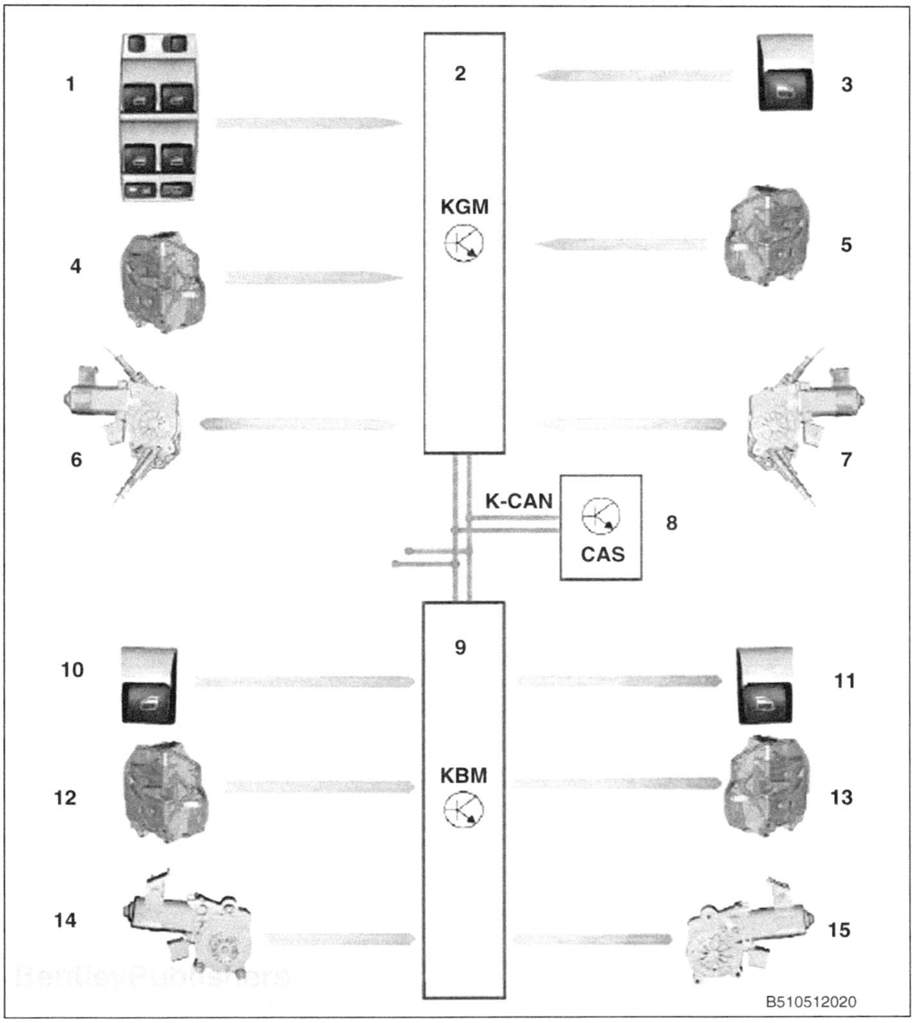

1. Left front window master switch
2. Body gateway module (KGM)
3. Right front window switch
4. Left front door latch
5. Right front door latch
6. Left front window motor
7. Right front window motor
8. Car access system (CAS)
9. Basic body module (KBM)
10. Left rear window switch
11. Right rear window switch
12. Left rear door latch
13. Right rear door latch
14. Left rear window motor
15. Right rear window motor

Car access system (CAS) is the master control module for lowering and raising the windows.

Body gateway module (KGM) receives signal from the front door window switches and provides power and ground to the rear window motors.

Basic body module (KBM) receives signal from the rear door window switches via CAN bus through SGM and provides power and ground to the rear window motors.

BentleyPublishers.com
BentleyPublishers.com–All Rights Reserved

Window switches

Front window switches

To 09 / 2005: When the power window switch for the window in the driver door or front passenger door is operated the driver window switch cluster sends a signal via the LIN-bus to left front door module (TMFA). TMFA then actuates the corresponding power window motor.

From 09 / 2005: When the power window switch for the window in the driver's door or front passenger's door is operated, the driver window switch cluster sends a signal via the LIN-bus to KGM. KGM then actuates the corresponding window motor.

To 09 / 2005: When the power window switch for the window in a rear door is operated, the driver window switch cluster sends a signal via LIN-bus to TMFA. TMFA sends the signal via byteflight to the SGM. SGM converts the signal and sends it on K-CAN to CAS. CAS evaluates the signal and sends a command to SGM on K-CAN. SGM forwards the command signal via the byteflight to right front door module (TMBF). TMBF receives the command for window movement, and the window motor is then moved.

From 09 / 2005: When the power window switch for the window in a rear door is operated, the driver window switch cluster sends a signal via LIN-bus to KGM. KGM then sends a signal to KBM and KBM actuates the corresponding window motor.

Rear window switches

When the power window switches in the rear doors are operated a signal is sent to KBM and KBM then actuates the corresponding window motor.

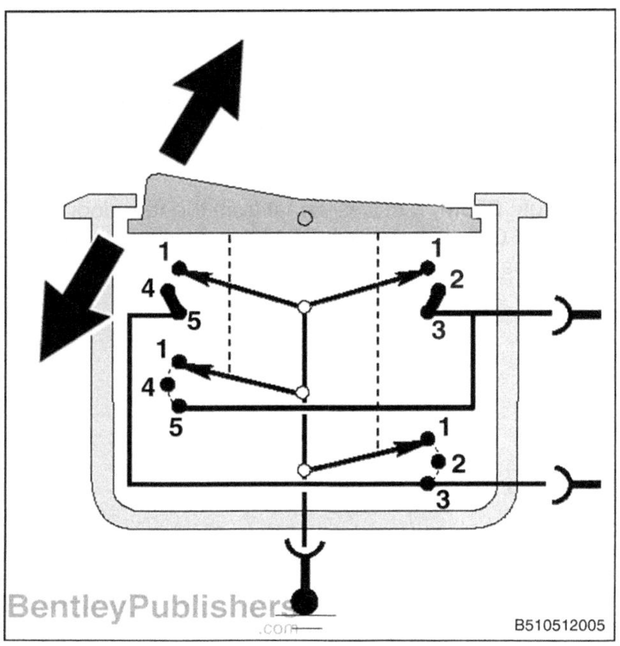

Window switch positions:

1. Off
2. Open
3. One touch open
4. Close
5. One touch close

The window toggle switch provides the corresponding control module with a coded ground signal. Holding a switch at the first detent provides a single ground signal on one wire requesting the module to open the window. When released, the ground signal is removed and the window motor stops.

Momentarily pushing the switch to the second detent and releasing provides an additional ground signal on the second wire requesting one touch mode. The corresponding module lowers the window automatically until it reaches the end position.

BentleyPublishers.com
BentleyPublishers.com—All Rights Reserved

Rear window child lockout switch

The rear window child lockout switch (**arrow**) is incorporated in the driver's window switch block.

To 09 / 2005: When activated, lock out switch provides a signal via LIN-bus to TMFA. TMFA sends the signal via byteflight to the SGM. SGM converts the signal and sends it on K-CAN to CAS. CAS prevents windows from being operated from the rear door switches.

From 09 / 2005: When activated, lock out switch provides a signal via LIN-bus to the KGM. KGM sends the signal via K-CAN to CAS. CAS prevents windows from being operated from the rear door switches.

The lockout switch signal is overridden if a multiple restraint system (MRS) crash signal is received.

Power window motors

The window motors are mounted on cable regulators. The window motor control circuit consists of two wires for operating the motor in both directions.

To 09 / 2005: The motors are activated by relays in the TMFA, TMBF and KBM. The relays provide either power or ground depending on the direction of window travel. TMFA, TMBF and KBM controls the polarity based on a request to run the window from a window switch or a convenience opening / closing signal.

From 09 / 2005: The motors are activated by relays in the KGM. he relays provide either power or ground depending on the direction of window travel. KGM controls the polarity based on a request to run the window from a window switch or a convenience opening / closing signal.

Convenience opening / closing

The CAS module provides the convenience open / close feature, providing control of the power windows (and sunroof) from outside the vehicle using the key in the driver's door lock, or with the remote key.

Anti-trap protection is active during convenience closing from the driver's door lock.

If the CAS module receives a request to operate convenience closing or opening for more than 110 seconds, the function is deactivated and a fault code is stored.

Window anti-trap protection

Anti-trapping function does not prevent an object being trapped but rather limits the trapping force to maximum 80 Nm. The power window motor is reversed when this force is exceeded.

Anti-trapping function is based on the evaluation of power window motor Hall sensor pulses. During initialization, the speed of Hall sensor pulses is learned. Speed fluctuations within a certain range triggers the anti-trapping function.

BentleyPublishers.com–All Rights Reserved

If window is not initialized or a faulty hall sensor is detected the window will only move in small increments.

Panic mode

Panic mode can be used to close a window in the event of an emergency. It is triggered by overpulling - releasing - overpulling (overpulling means pulling the switch beyond its limit stop) the power window switch. It is necessary to release and overpull the switch again in order to deactivate the anti-trapping protection function, which is still active the first time the switch is overpulled.

Overpulling the power window switch the second time within 4 seconds closes the window with maximum force.

Anti-trapping function is not active during panic mode. The window closes at the maximum closing force and does not reverse.

Warnings and Cautions

WARNING—
- *5 Series models are fitted with side-impact airbags in the front door trim panel. When servicing the door windows on cars with side-impact airbags, disconnect the negative (–) battery cable. See* **721 Airbag System (SRS)** *for cautions and procedures relating to the airbag system.*
- *Wear hand and eye protection when working with broken glass.*
- *If a window is broken, vacuum all of the glass bits out of the door cavity. Use a blunt screwdriver to clean out any remaining glass pieces from the window guide rails.*
- *When servicing the door windows, the harness connector to the window regulator should always be disconnected to prevent pinching fingers in the moving window mechanism.*

CAUTION—
- *Prior to disconnecting the battery, read the battery disconnection cautions given in* **001 Warnings and Cautions**.

BentleyPublishers.com
BentleyPublishers.com—All Rights Reserved

Window Service

Front door window, removing and installing

– Remove door panel. See **411 Doors**.

– Using BMW scan tool, print or note Car and Key Memory settings.

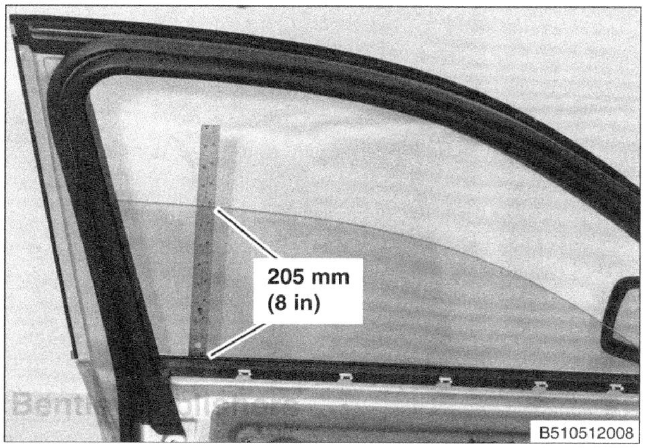

◀ Lower window to 205 mm (8 in.).

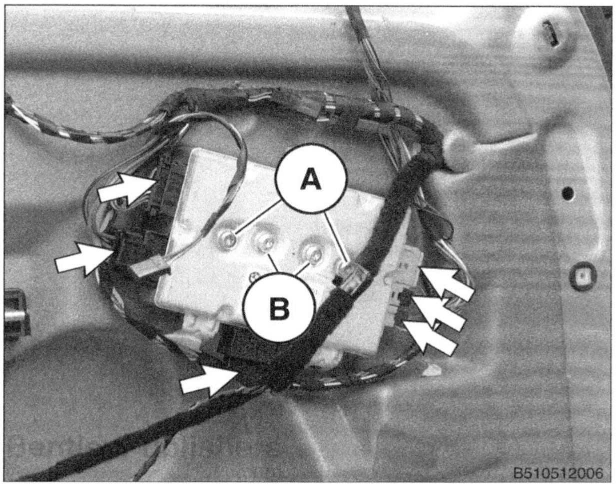

◀ **To 09 / 2005**: Detach door module electrical connectors (**arrows**), remove fasteners (**A**) and remove door module from door.

> ***CAUTION—***
> • *Do not remove center door module fasteners (**B**). Doing so will damage door module.*

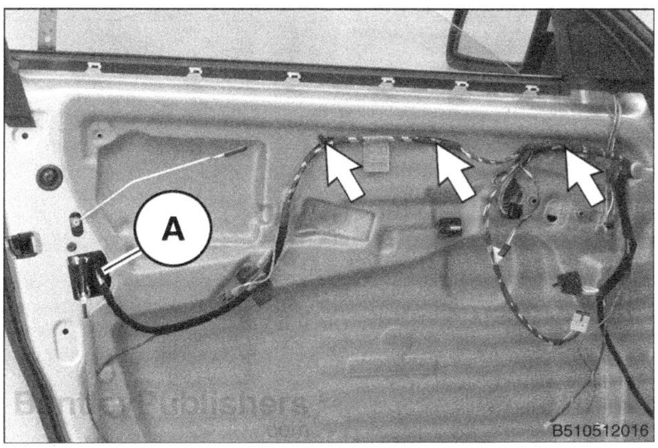

◀ Detach wire mount brackets (**arrows**). Disconnect door latch electrical connector (**A**) and lay wire harness aside.

BentleyPublishers.com
BentleyPublishers.com–All Rights Reserved

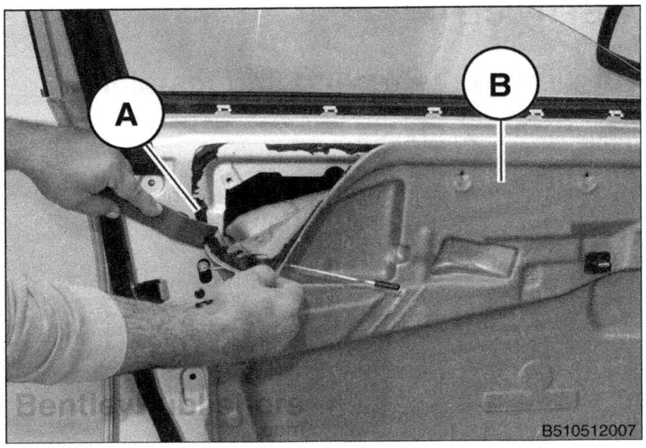

Using a sharp plastic scraper, lever sealing bead (**A**) off vapor barrier (**B**).

> ***CAUTION—***
> - *Use care when peeling back vapor barrier. Replace a damaged vapor barrier to prevent possible water leaks and insure proper operation of side impact airbags.*

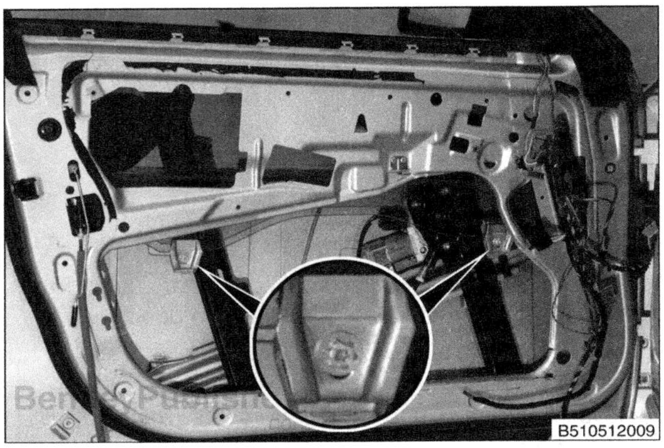

Loosen window glass mount fasteners (**inset**).

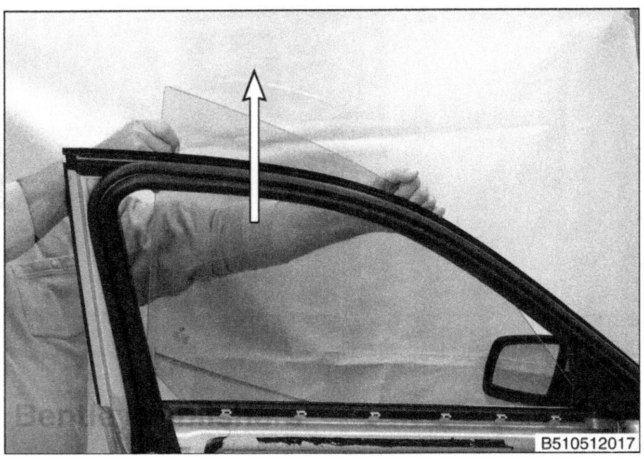

Tilt and lift glass out of door in direction of **arrow**.

– Installation is reverse of removal, noting the following:
- Adjust door window if necessary. See **Front door window, adjusting,** in this repair group.
- Initialize power window. See **Initializing windows** in this repair group.

Tightening torques	
Door window glass mount	8 Nm (5.9 ft-lb)

BentleyPublishers.com
BentleyPublishers.com—All Rights Reserved

Front door window, adjusting

- Using BMW scan tool, print or note Car and Key Memory settings.
- Fully close door window to be adjusted.
- Remove front door panel. See **411 Doors**.

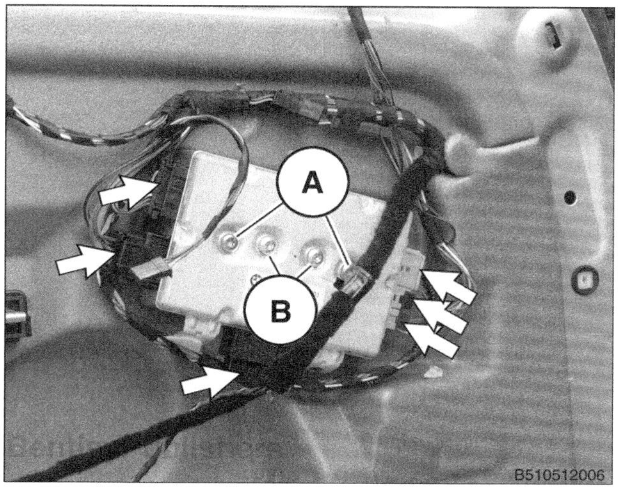

To 09/2005: Detach door module electrical connectors (**arrows**), remove fasteners (**A**) and remove door module from door.

> ***CAUTION—***
> - *Do not remove center door module fasteners (**B**). Doing so will damage door module.*

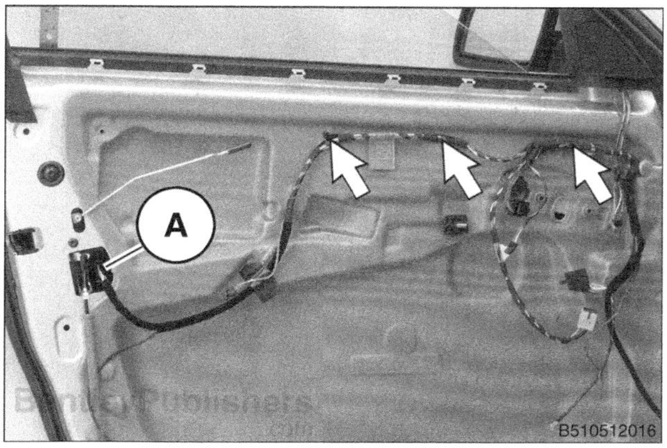

Detach wire mount brackets (**arrows**). Disconnect door latch electrical connector (**A**) and lay wire harness aside.

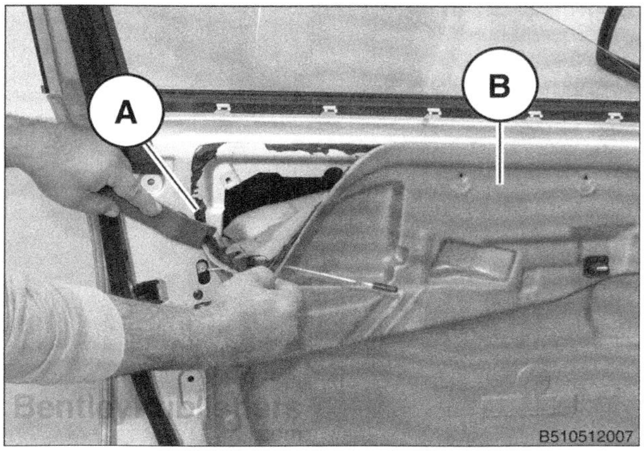

Using a sharp plastic scraper, lever sealing bead (**A**) off vapor barrier (**B**).

> ***CAUTION—***
> - *Use care when peeling back vapor barrier. Replace damaged vapor barrier to prevent possible water leaks and insure proper operation of side impact airbags.*

512

BentleyPublishers.com
BentleyPublishers.com—All Rights Reserved

Front door window, adjusting

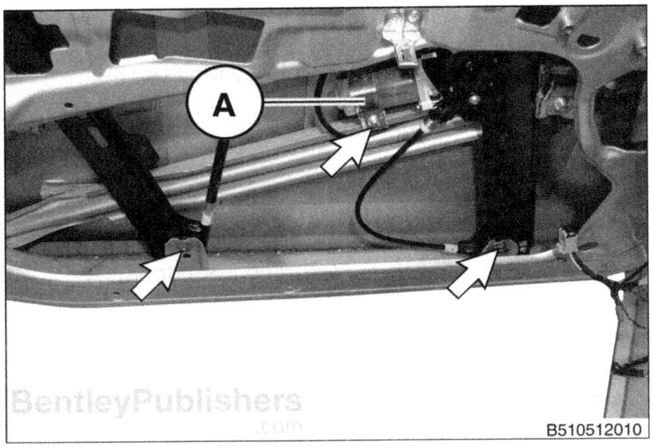

B510512010

◄ Loosen window regulator fasteners (**arrows**). Do not remove window motor mount (**A**). Disconnect power window motor electrical connector.

> ***CAUTION—***
> *• Do not loosen fasteners at top of window regulator.*

B510512019

◄ Connect fused jumper leads to power window motor terminals (**arrows**). Open window approximately 100 mm (4 in) by connecting jumper leads to an external 12 volts power source.

> ***CAUTION—***
> *• Do not activate power window motor in excess of 0.5 seconds once limit position is reached.*

NOTE—

• Reversing polarity of jumper leads reverses direction of power window motor.

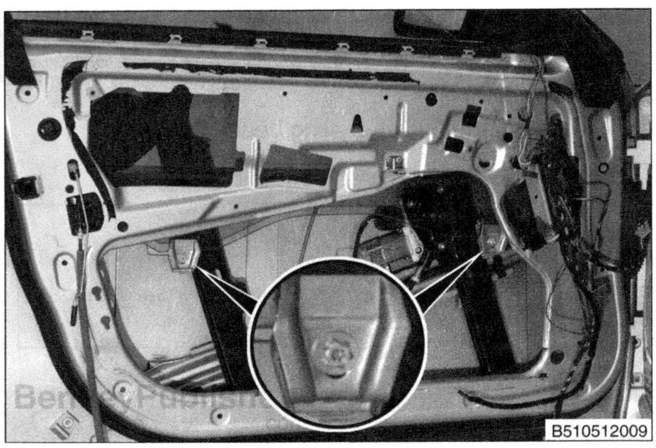
B510512009

◄ Loosen window glass fasteners (**A**).

– Close window and tighten window glass fasteners (**inset**).

Tightening torques	
Door window glass mount	8 Nm (5.9 ft-lb)

BentleyPublishers.com
BentleyPublishers.com—All Rights Reserved

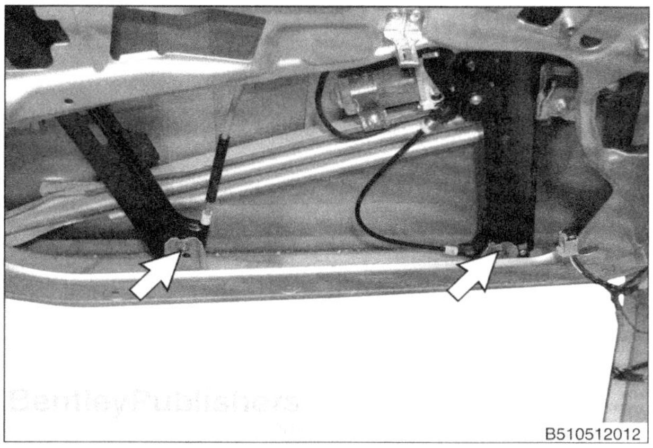

Open window approximately 40 mm (1.5 in. then tighten window regulator fasteners (**A**).

Tightening torques	
Regulator to door	8 Nm (5.9 ft-lb)

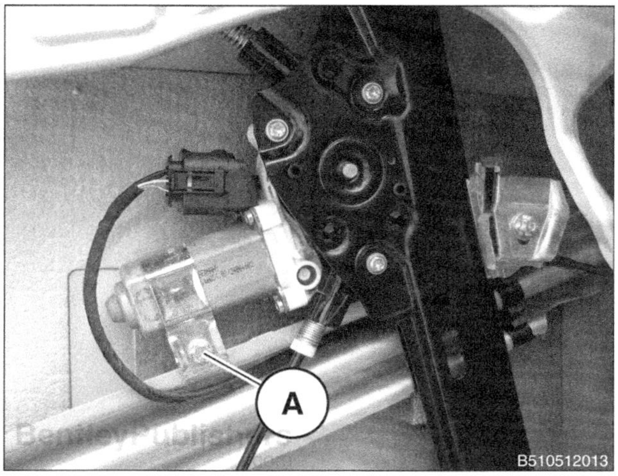

Close door window until fastener (**A**) is accessible and tighten fastener (**A**).

– Close door window completely and check adjustment.

– Once adjustment is complete:
  - Using BMW scan tool, check Car and Key Memory settings against previously made notes.
  - Initialize power window. See **Initializing windows** in this repair group.

Tightening torques	
Door window glass mount	8 Nm (5.9 ft-lb)
Window motor mount	8 Nm (5.9 ft-lb)
Regulator to door	8 Nm (5.9 ft-lb)

Window Regulator Service

The electronically-controlled window motor does not have mechanical end positions. For this reason the motor can be removed and installed with the window in any position.

Front window regulator, removing and installing

Remove window regulator and motor as one unit. If necessary, separate the two on the bench.

There are two window rails in each front window regulator, only one in the rear window regulator. Otherwise the front and rear regulators are similar.

- Remove front door panel. See **411 Doors**.
- Remove window glass. See **Front door window, removing and installing** in this repair group.

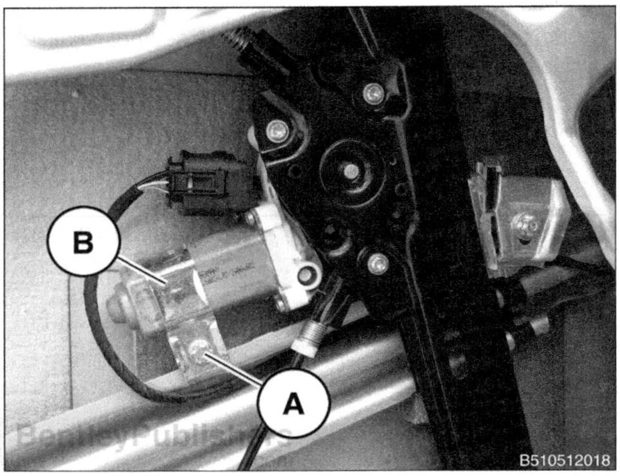

◄ Loosen window motor mount fastener (**A**) and remove mount (**B**) from window motor.

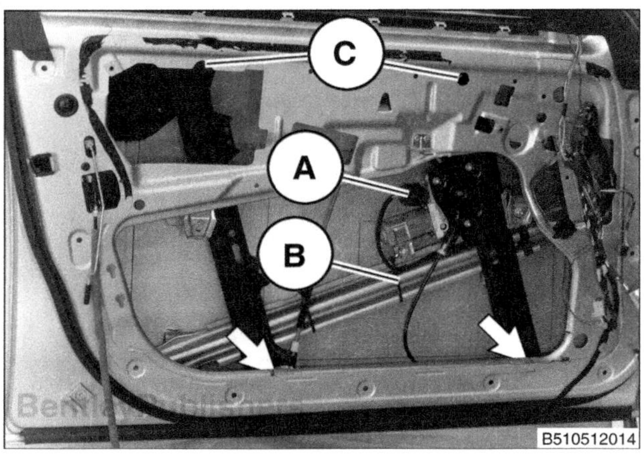

◄ Disconnect window motor electrical connector (**A**) and cut wire tie (**B**). Loosen window regulator fasteners (**arrows**) and remove window regulator fasteners (**C**).

BentleyPublishers.com
BentleyPublishers.com—All Rights Reserved

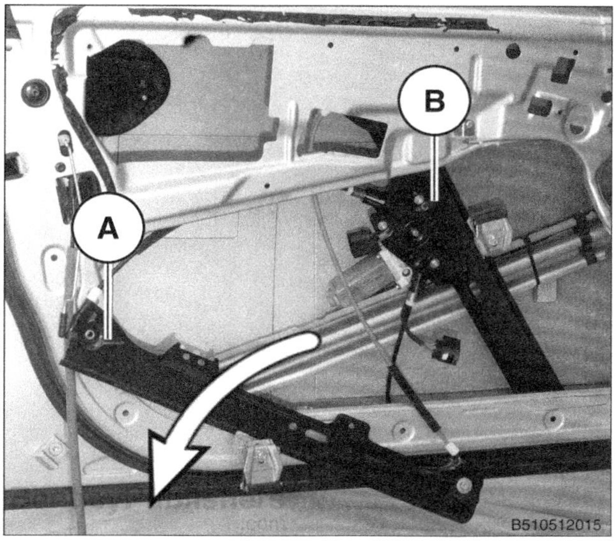

◀ Lift rear window regulator rear rail (**A**) out of door then front rail (**B**).

– Installation is reverse of removal, noting the following:
  - Route wiring harnesses to keep them away from moving window mechanism. Use new wire ties as necessary.
  - Initialize power window. See **Initializing windows** in this repair group.

Tightening torques	
Door window glass mount	8 Nm (5.9 ft-lb)
Window motor mount to redulator	8 Nm (5.9 ft-lb)
Window regulator to door	8 Nm (5.9 ft-lb)
Window motor to regulator	4 Nm (4.4 ft-lb)

Initializing Windows

Window control module learns two parameters during initialization:
- Upper end stop (normalization of the power windows).
- Power characteristic curve for opening and closing (learning the characteristic curve).

Initialization can be carried out using the control switch in each door or the power-window switch block in the driver's door.

Initialization prerequisites

- Vehicle stationary.
- Adequate battery voltage present (12.4 volts), connect vehicle power supply if necessary.
- Terminal R (radio readiness) or terminal 15 switched ON (ignition ON) See **020 Maintenance** for vehicle key states.
- All doors closed.

Performing initialization

– Fully close window by pulling the switch so that top end stop is reached (normalization).

– Open window completely so that lower end stop is reached.

– Pull control switch into one touch operation mode, then pull to close (one touch operation, close) position and hold it there until the window has reached the top end stop (learning the characteristic curve).

Initialization is successful if the window can be opened and closed without faults in one touch operation mode.

Window moves in small increments if not initialized.

BentleyPublishers.com
BentleyPublishers.com—All Rights Reserved

513 Interior Trim

GENERAL

This repair group covers interior trim removal and installation.

See also:

- **411 Doors** for interior door trim panel removal
- **520 Seats** for rear side bolster removal
- **612 Switches**
- **620 Instruments**
- **640 Heating and Air-conditioning**
- **650 Radio**
- **721 Airbag System (SRS)** for passenger airbag

BentleyPublishers.com
BentleyPublishers.com–All Rights Reserved

Trim clips

Interior trim and finish panels are clipped or screwed into place. When removing trim that is held in place with clips, be sure to have spare clips (also known as plastic rivets) on hand before beginning the job.

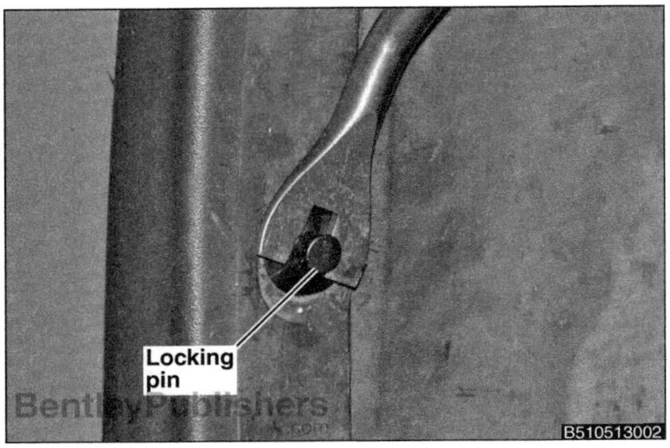

Many trim clips may be reused if removed correctly:

- Carefully pry out center locking pin.
- Lever out clip using removal tool.

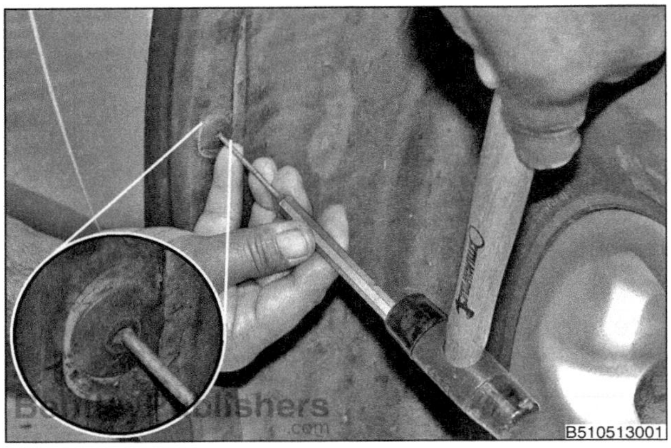

Other trim retaining clips are designed to be used only once:

- Use thin pin punch to drive in headless pin (**inset**) in center of clip.
- Lever out clip using removal tool.

Warnings and Cautions

WARNING *—*

- *Serious injury may result if airbag system service is attempted by persons unfamiliar with the BMW SRS and its approved service procedures.*
- *Before performing any work involving airbags, disconnect the negative (-) battery cable. See* **721 Airbag System (SRS)**.
- *BMW airbags are equipped with a back-up power supply inside the airbag control module. Observe a 5 second waiting period after disconnecting the battery cable to allow the reserve power supply to discharge.*

CAUTION *—*

- *When working on electrical switches or lights, disconnect the negative (-) battery cable and insulate the cable end to prevent accidental reconnection.*
- *Prior to disconnecting the battery, read the battery disconnection cautions given in* **001 Warnings and Cautions**.
- *To prevent marring the trim, work with a plastic prying tool or wrap a screwdriver tip with masking tape before prying out trim panels, switches or electrical accessories.*

BentleyPublishers.com
BentleyPublishers.com—All Rights Reserved

DASHBOARD

Right dashboard trim strip, removing and installing

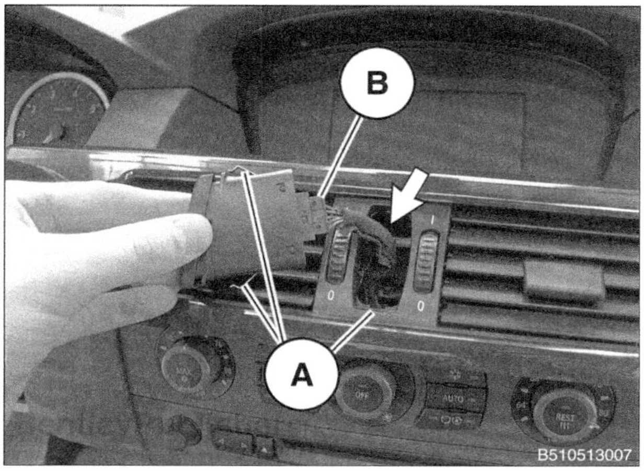

- Using a plastic prying tool, release hazard warning switch retaining tabs (**A**) and pull switch out of dashboard. Disconnect electrical connector (**B**) and remove switch.
- Remove trim strip fastener (**arrow**) if equipped.

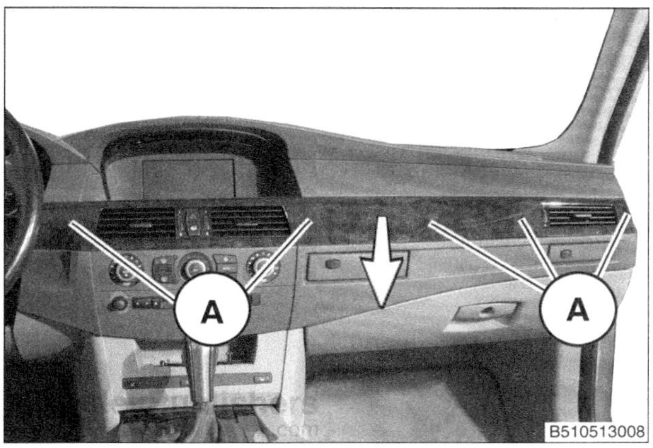

- Starting at outside of trim strip, pry in areas (**A**) using a plastic prying tool and remove in direction of **arrow**.

> ***CAUTION—***
> *• Do not twist prying tool during removal of trim panel.*

- Installation is reverse of removal. Replace missing or damaged trim panel clips.

Left dashboard trim strip, removing and installing

- Remove headlight switch. See **612 Switches**.

- Remove trim strip fasteners (**A**). Using plastic prying tool, unclip trim strip in area **B** and remove in direction of **arrow**.
- Installation is reverse of removal. Replace missing or damaged trim panel clips.

BentleyPublishers.com—All Rights Reserved

Center dashboard trim panel, removing and installing

– Remove right dashboard trim strip. See **Right dashboard trim strip, removing and installing** in this repair group.

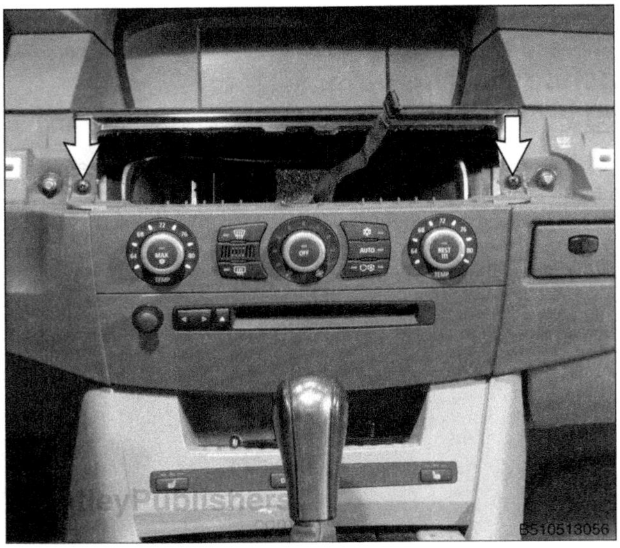

◄ Remove trim panel fasteners (**arrows**).

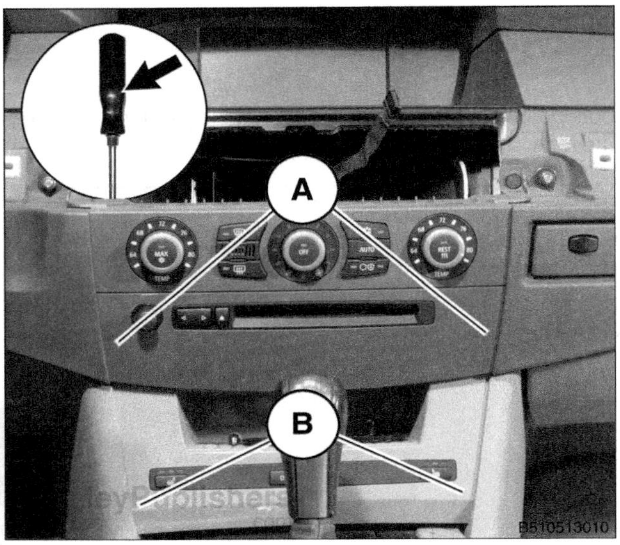

◄ Insert 180 mm (7 in) flat blade screwdriver (**inset**) behind trim panel and release clips in area **A**.

– Using a plastic prying tool, release clips in area **B** and pull back trim panel.

– Disconnect electrical connectors and remove trim panel.

– Installation is reverse of removal. Replace missing or damaged trim panel clips.

Right lower dashboard trim, removing and installing

– Move passenger seat all the way back.

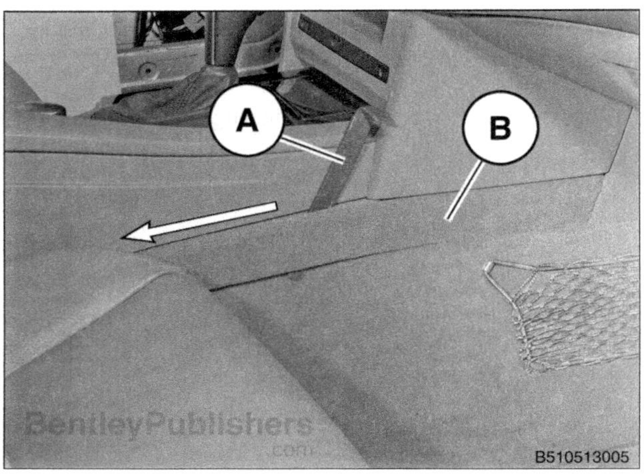

◄ Using plastic prying tool (**A**), slide trim panel (**B**) towards rear of vehicle (**arrow**) and unclip.

BentleyPublishers.com
BentleyPublishers.com—All Rights Reserved

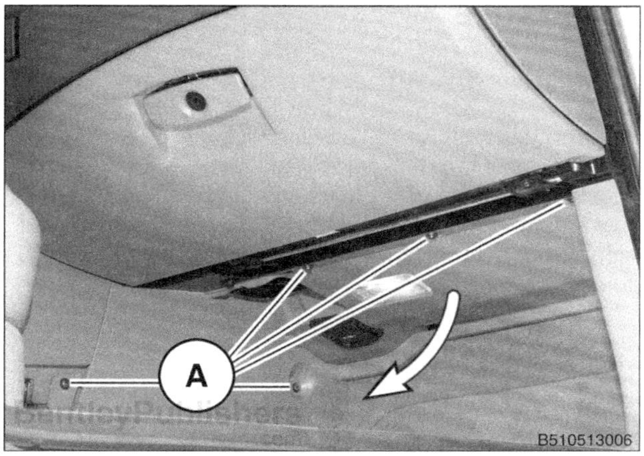

- Remove trim panel fasteners (**A**) and lower trim panel in direction of **arrow**.
- Disconnect footwell light electrical connector if equipped.
- Remove trim panel.
- Installation is reverse of removal.

Left lower dashboard trim, removing and installing

- Move driver seat all the way back.

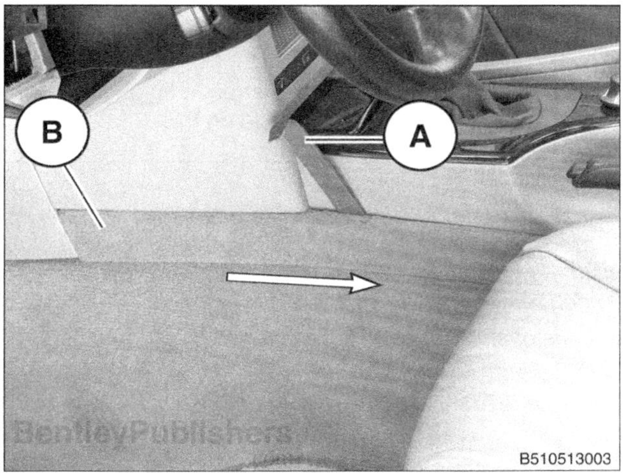

- Using plastic prying tool (**A**), slide trim panel (**B**) towards rear (**arrow**) of vehicle and unclip.

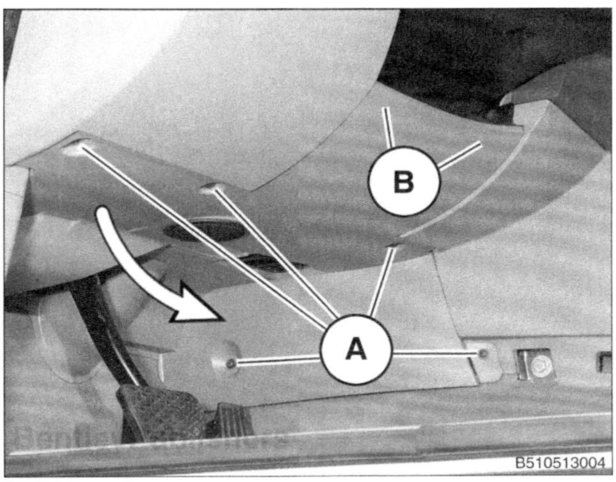

- Remove trim panel fasteners (**A**). Using plastic prying tool disconnect clips in area (**B**) and partially lower trim panel in direction of **arrow**.
- Lower trim panel and disconnect fiber optic programming connector (if equipped) and footwell light electrical connector.
- Remove trim panel.
- Installation is reverse of removal.

Center Console

Center console, removing and installing

– Remove iDrive control knob. See **612 Switches**.

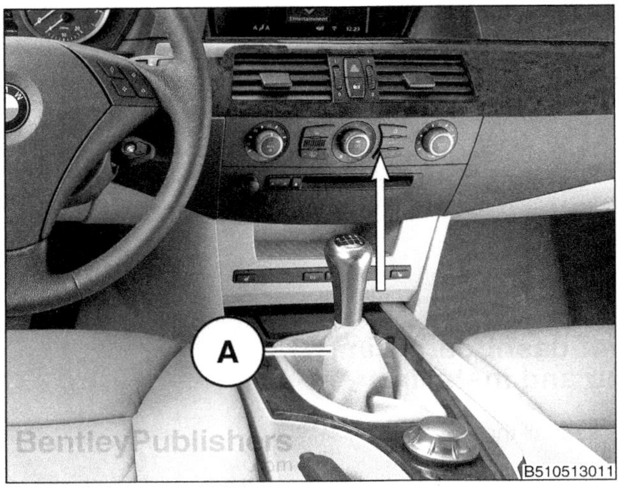

Manual transmission: Pull gearshift knob straight up (**arrow**) to remove. It may require up to 90 lbs. of force to do this. Remove shift boot by grasping boot material (**A**) and pulling up to release tabs. Be careful not to break retaining tabs.

Automatic transmission: Pull gearshift knob straight up (**arrow**) to remove. It may require up to 90 lbs. of force to do this.

> ***CAUTION—***
> - *Automatic transmissions for 2008 - 2009 models (starting with 03 / 2007 production) are equipped with gear selector switch (GWS) instead of a shift lever. The procedure given here for gearshift knob removal does not apply.*
> - *Some versions of shift boot and knob are one piece. Release boot from console before removing knob.*
> - *To prevent damage to shift knob locating tabs, do not twist while removing.*

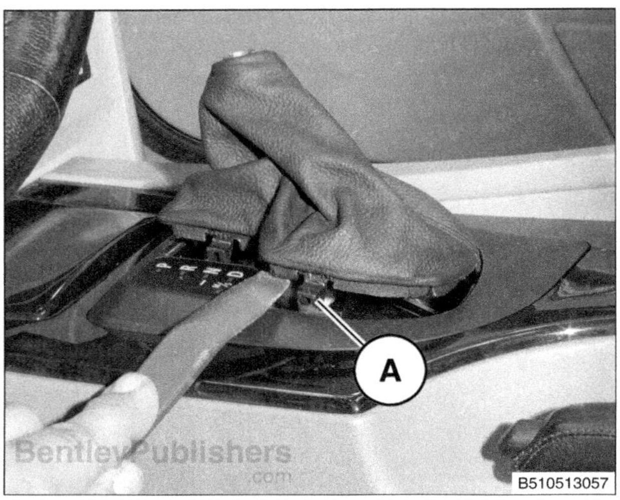

Automatic transmission: Using a plastic prying tool, lever and lift out shift boot. Be careful not to break retaining tabs (**A**).

BentleyPublishers.com
BentleyPublishers.com—All Rights Reserved

Automatic transmission: Grasp and pull out trim piece (**A**). Disconnect electrical connectors and remove in direction of **arrow**.

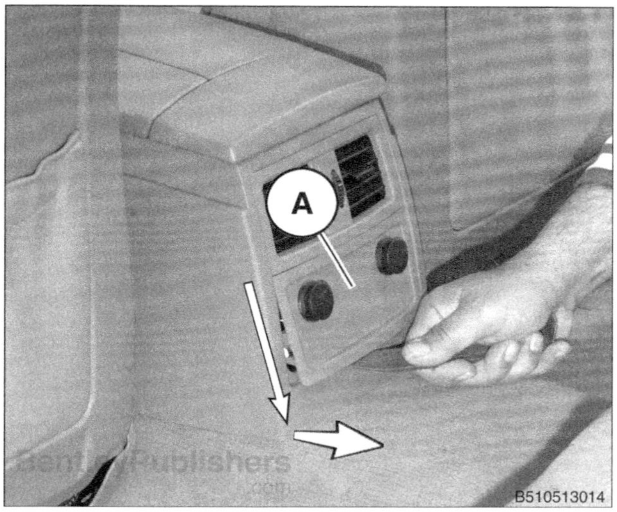

Using a plastic prying tool unclip and remove rear center console insert (**A**) in direction of **arrows**. Disconnect electrical connectors, if equipped.

Using a flat blade screwdriver, detach clips (**A**). Unclip and remove vent (**B**) in direction of **arrow**. If equipped, disconnect electrical connectors.

BentleyPublishers.com—All Rights Reserved

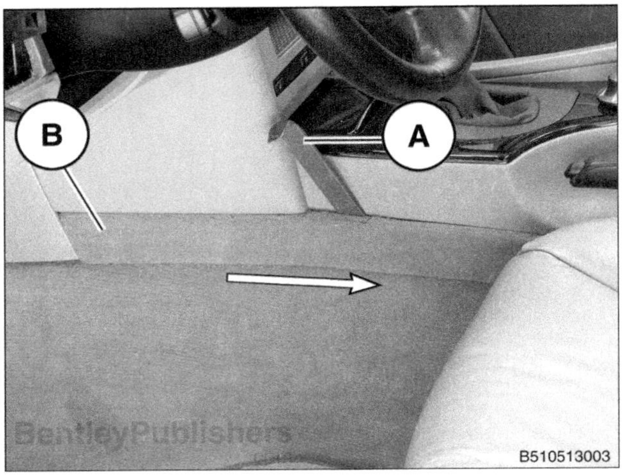

◀ Using plastic prying tool, slide trim panel (**A**) towards rear (**arrow**) of vehicle and unclip. Left side shown; right side is similar.

◀ Working at rear of center console, remove fasteners (**arrows**).

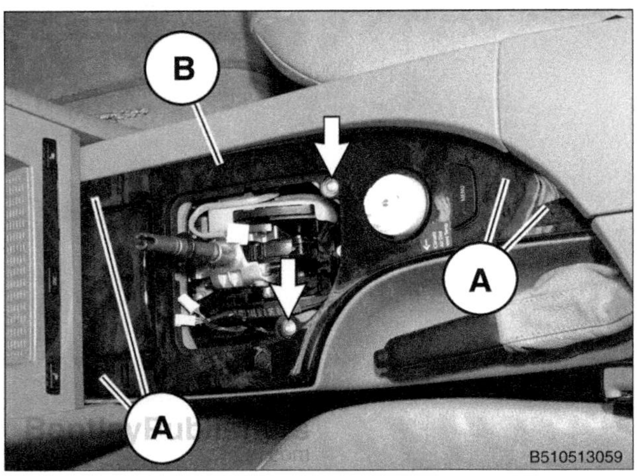

◀ Working at front of center console:

- Remove fasteners (**arrows**).
- Remove ashtray insert.
- Using a plastic prying tool, lever trim (**B**) in areas **A** up enough to disconnect iDrive menu button electrical connector.
- Remove trim (**B**).

BentleyPublishers.com
BentleyPublishers.com—All Rights Reserved

Working at side of center console, remove fastener (**arrow**). Repeat procedure for right side.

Working at front of center console, remove fasteners (**arrows**) if equipped.

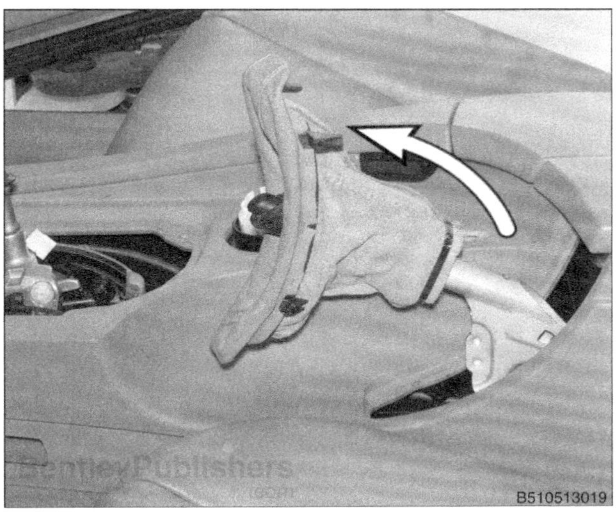

Pull parking brake handle boot (**arrow**) over handle.

BentleyPublishers.com–All Rights Reserved

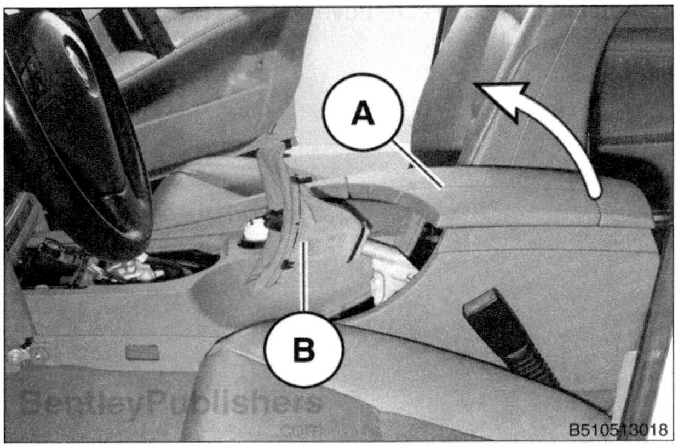

Raise center console (**A**) at rear in direction of **arrow** while feeding parking brake lever boot (**B**) through opening. Disconnect electrical connectors and remove center console (**A**).

– Installation is reverse of removal. Remember to:
  - Connect electrical connectors.
  - Connect air ducts.
  - Replace any damaged panel clips.

Glove Compartment

Glove compartment, removing and installing

– Remove door sill trim. See **Door sill trim, removing** in this repair group.

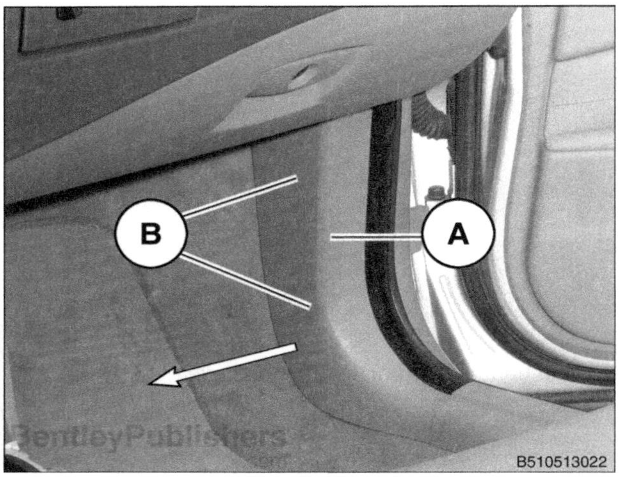

Unclip right side kick panel trim (**A**) from retaining points (**B**) in direction of **arrow** and remove.

– Remove right lower dashboard trim. See **Right lower dashboard trim, removing and installing in this repair** group.

– Open glove compartment.

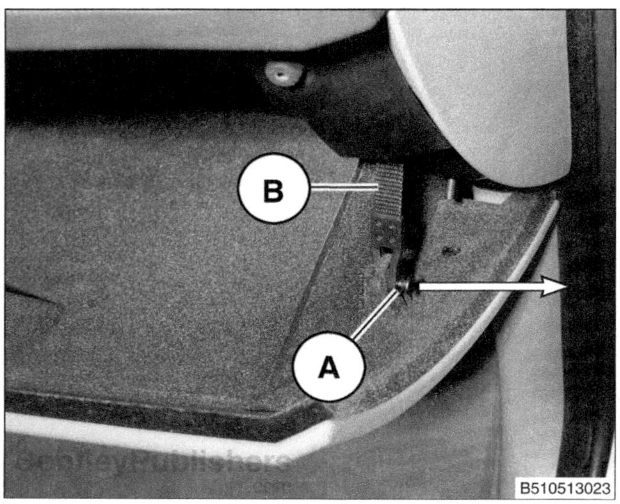

Working on right side of glove compartment, remove pin (**A**) for retaining strap (**B**) and disconnect.

BentleyPublishers.com
BentleyPublishers.com—All Rights Reserved

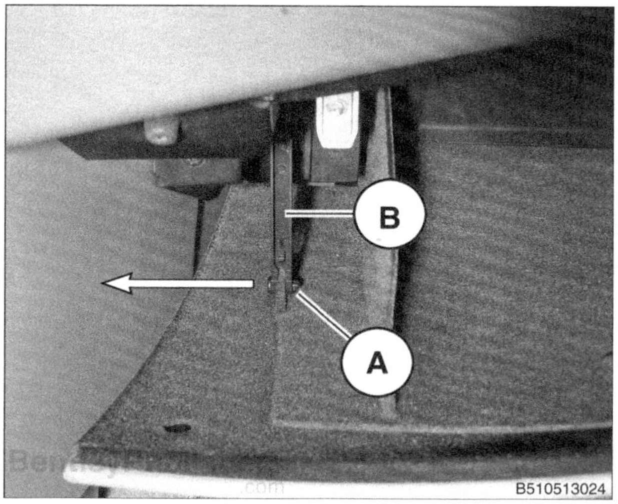

Working on left side of glove compartment, remove pin (**A**) for shock absorber (**B**) and disconnect.

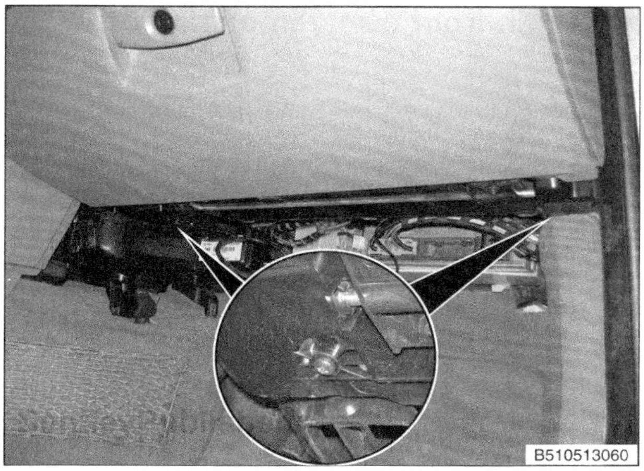

Working below glove compartment, remove fasteners (**inset**).

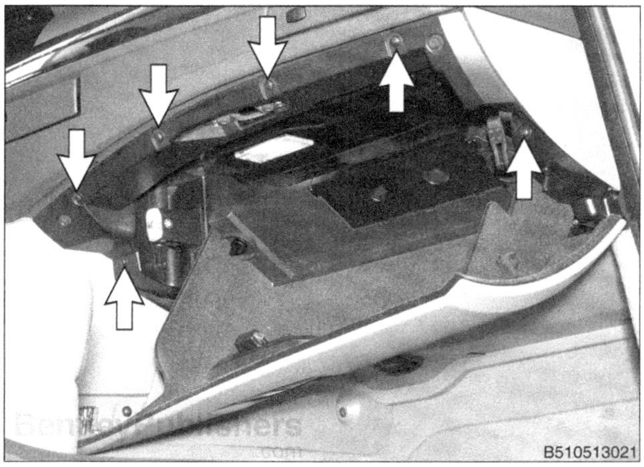

Remove fasteners (**arrows**) and pull glove compartment out slightly.

BentleyPublishers.com–All Rights Reserved

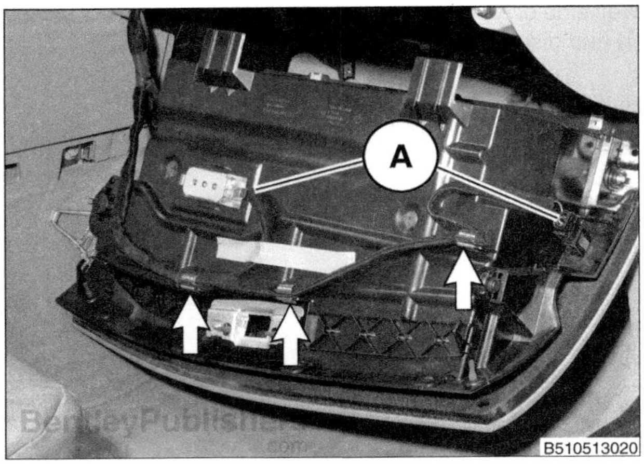

Disconnect electrical connectors (**A**), remove harness from clips (**arrows**). Remove glove compartment.

– Installation is reverse of removal. Remember to:
  - Connect electrical connectors.
  - Replace any damaged panel clips.

Covers and Trim Panels

Steering column front trim, removing and installing

– Adjust steering column to fully extended and raised position.

– Remove steering wheel. See **320 Steering and Wheel Alignment**.

Remove front trim by pulling toward rear of vehicle (**arrows**).

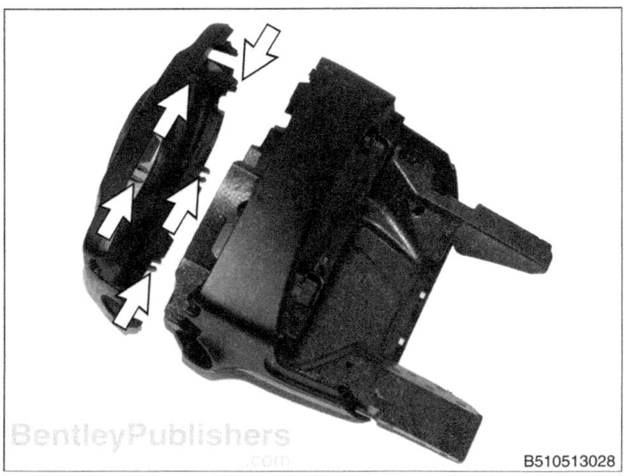

Installation is reverse of removal. Check that panel clips (**arrows**) engage properly.

BentleyPublishers.com
BentleyPublishers.com—All Rights Reserved

Steering column upper trim, removing and installing

– Remove steering column lower trim. See **Steering column lower trim, removing and installing** in this repair group.

– Adjust steering column to fully extended and lowered position.

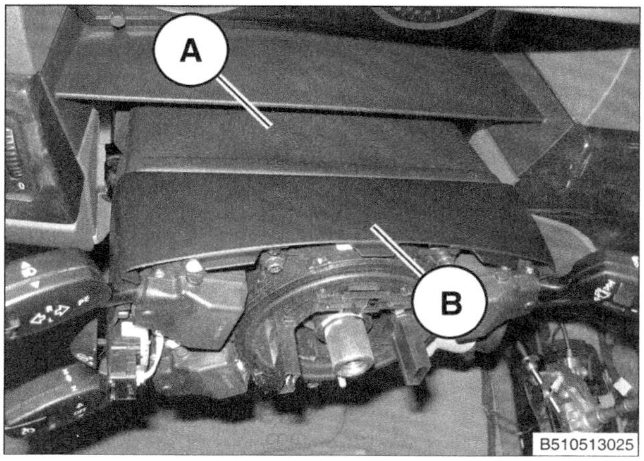

◄ Using a plastic prying tool, unclip steering column cover trim (**A**) from upper cover (**B**).

◄ Remove upper trim fasteners (**inset**).

◄ Remove by bending sides of upper trim out (**curved arrow**) and lifting up (**arrow**).

– Installation is reverse of removal.

Steering column lower trim, removing and installing

– Remove steering column front trim. See **Steering column lower trim, removing and installing** in this repair group.

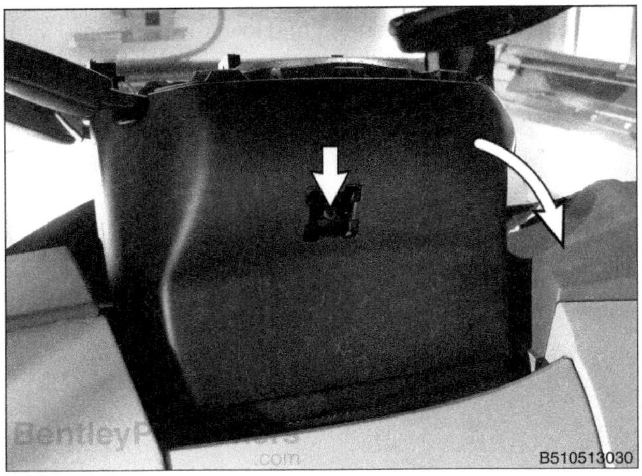

B510513030

◄ Remove trim fastener (**arrow**). Unclip and lower trim in direction of **curved arrow**.

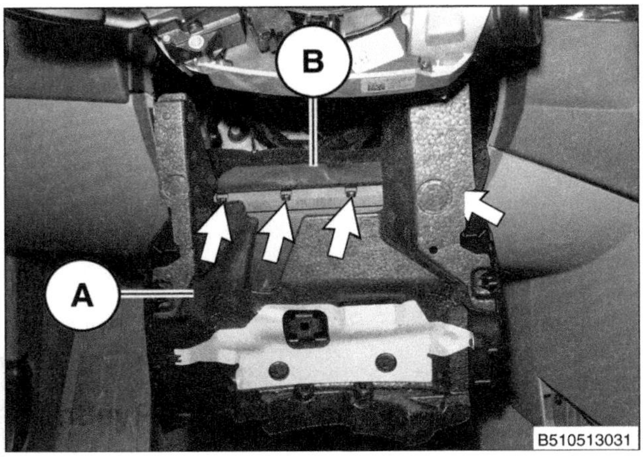

B510513031

◄ Tilt steering column trim (**A**) down and unclip (**arrows**) from cover (**B**).

– Remove lower trim.

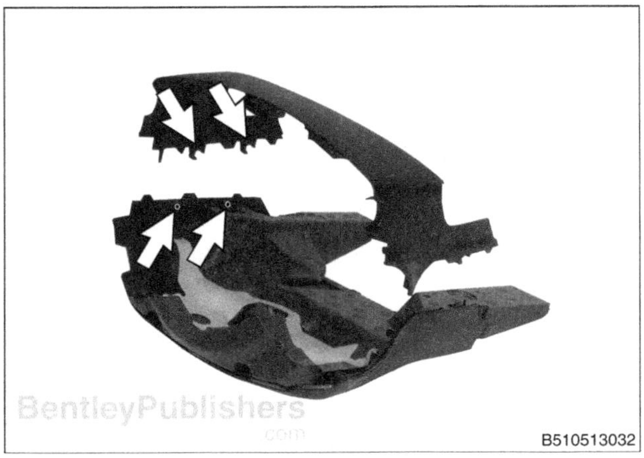

B510513032

◄ Installation is reverse of removal. Check that panel clips (**arrows**) engage properly.

BentleyPublishers.com
BentleyPublishers.com—All Rights Reserved

Door sill trim, removing

Using a plastic prying tool (**A**) unclip door sill trim clips (**arrows**) starting at rear.

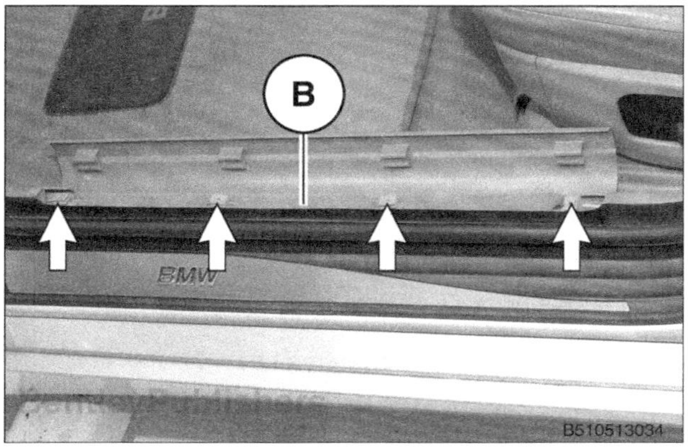

Remove panel clips (**arrows**) left in floor panel (**B**) after trim panel removal if necessary.

– Installation is reverse of removal. Replace any damaged panel clips.

A-pillar trim, removing and installing

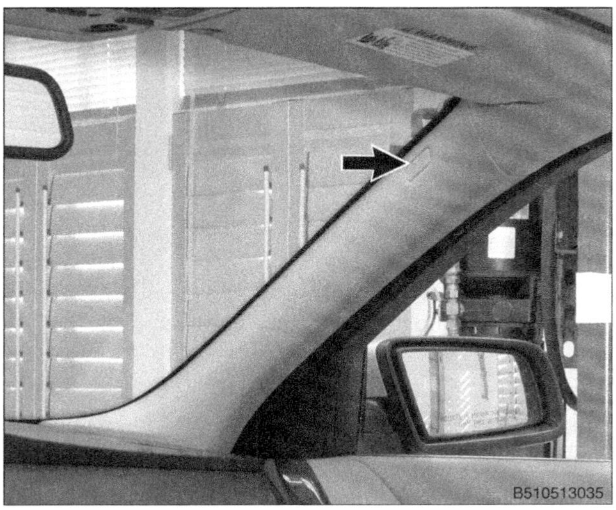

Using plastic trim tool remove cover (**arrow**) then remove fastener beneath it.

BentleyPublishers.com
BentleyPublishers.com—All Rights Reserved

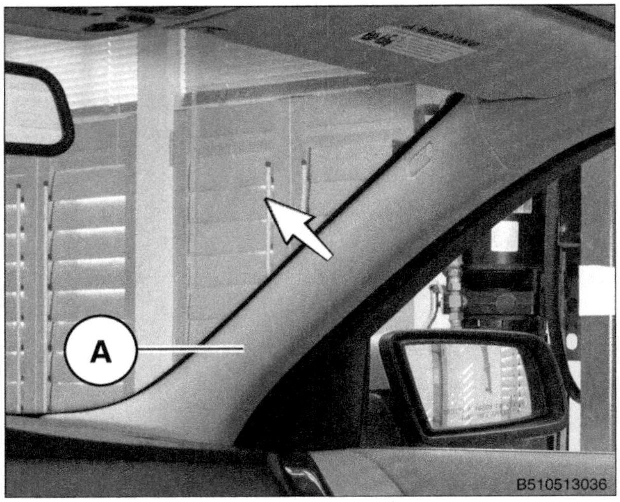

- Release door seal from A-pillar trim.
- Unclip trim from A-pillar in area **A** and remove out of dashboard in direction of **arrow**.
- Installation is reverse of removal. Replace any damaged panel clips

B-pillar trim, removing and installing

- Remove driver door sill trim. See **Door sill trim, removing and installing** in this repair group.

Lower B-pillar trim

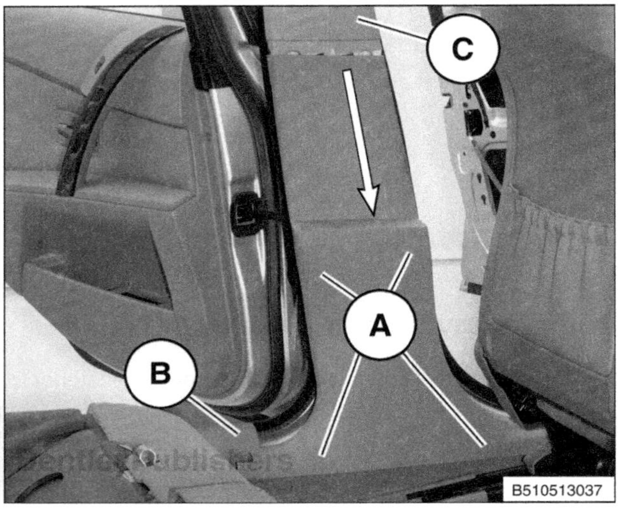

- Snap trim out of trim clips in area **A** and slide trim panel out of rear door sill trim (**B**) and door seal.
- Carefully guide trim panel out of upper trim panel (**C**).

Upper B-pillar trim

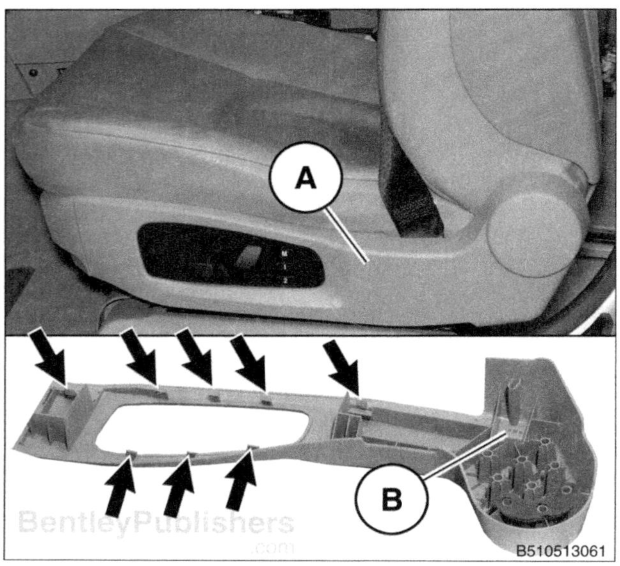

- Remove lower B-pillar trim
- Using a plastic prying tool, release retaining tabs (**arrows**) for for seat trim (**A**). Lift seat trim up to release tab **B** and remove.

BentleyPublishers.com
BentleyPublishers.com—All Rights Reserved

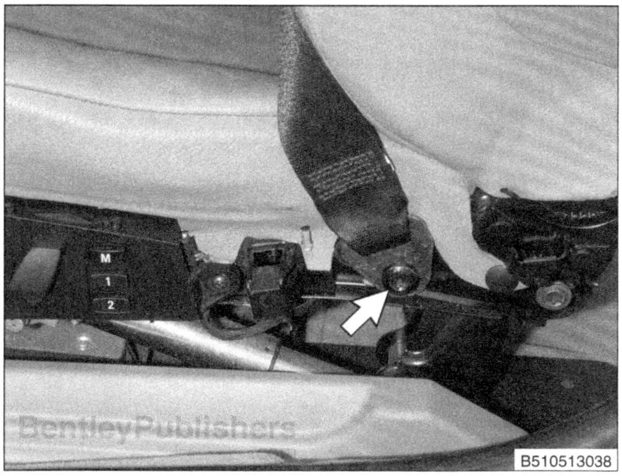

< Remove seat belt fastener (**arrow**) and discard. Disconnect seat belt and lay aside.

< Disconnect upper B-pillar trim panel trim clips (**arrows**).

< Remove trim panel in direction of **arrow**.

– Installation is reverse of removal. Remember to:

- Replace any damaged panel clips.
- Replace fastener for seat belt.

Tightening torque	
Seatbelt to front seat (Replace fastener)	36 Nm (26.5 ft-lb)

BentleyPublishers.com—All Rights Reserved

C-pillar trim, removing and installing

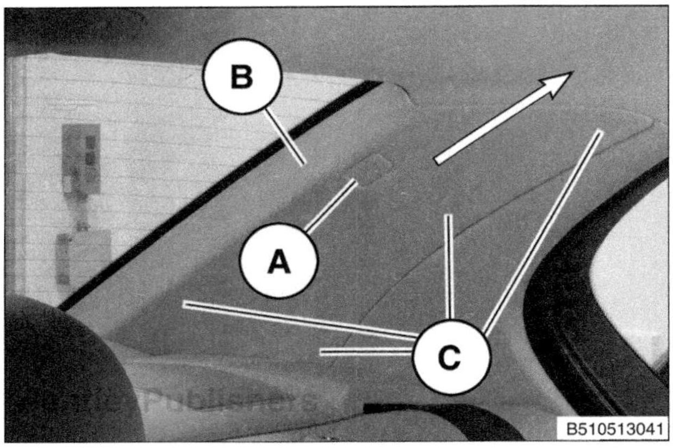

- Remove cap (**A**) and fastener underneath.
- Disconnect clips in area **C** and remove trim panel (**B**) in direction of **arrow**.
- Installation is reverse of removal. Replace any damaged panel clips.

Tightening torque	
Panel to C-pillar	3.5 Nm (2.5 ft-lb)

Trunk lid trim panel, removing and installing

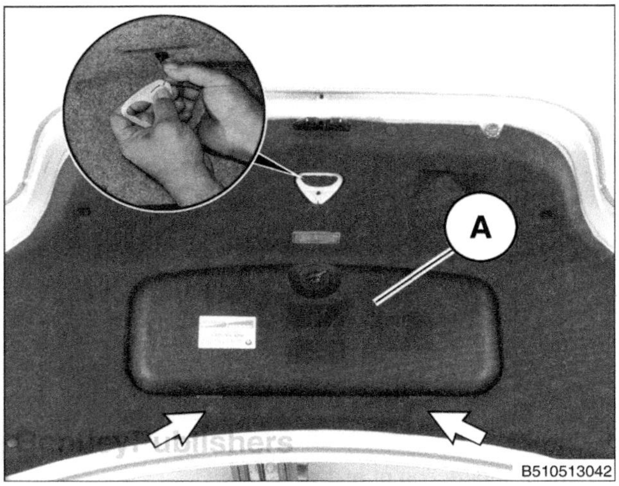

- Remove emergency trunk release handle by pushing lever and releasing cable (**inset**).

- Remove trim panel clips (**inset**) and remove trunk lid trim panel.
- Installation is reverse of removal. Remember to replace damaged trim panel clips.

BentleyPublishers.com
BentleyPublishers.com—All Rights Reserved

Trunk lower trim panel, removing and installing

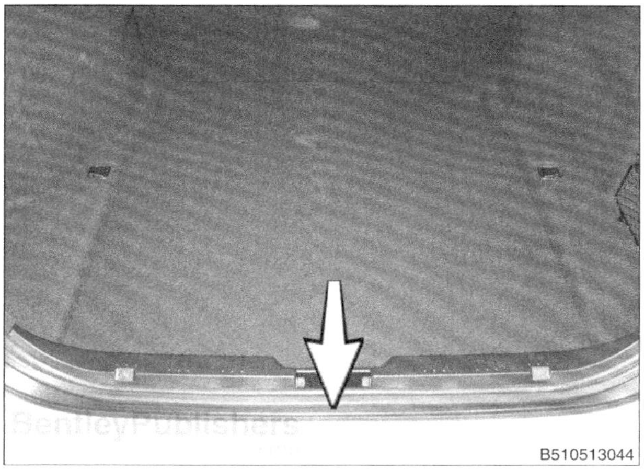

◀ Remove trunk floor trim panel in direction of **arrow**.

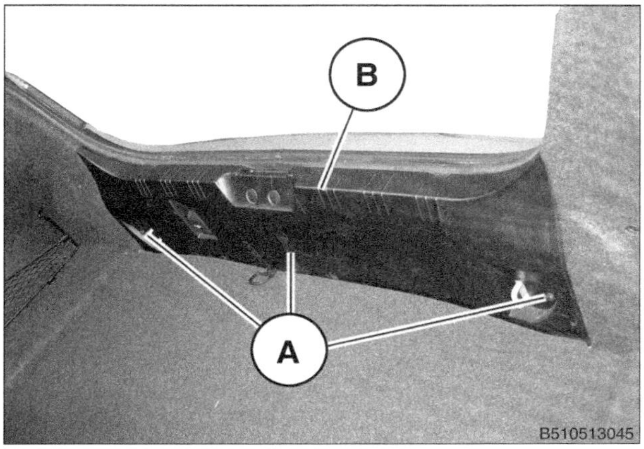

◀ Remove plastic cover and fasteners (**A**). Lift out and remove trunk rear trim panel (**B**).

– Installation is reverse of removal. Remember to replace damaged trim panel clips.

Trunk side trim panel, removing and installing

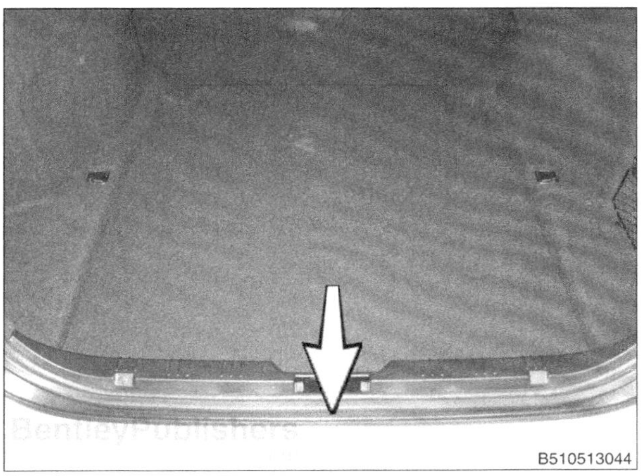

◀ Remove trunk floor trim panel in direction of **arrow**.

Right side shown, left is similar.

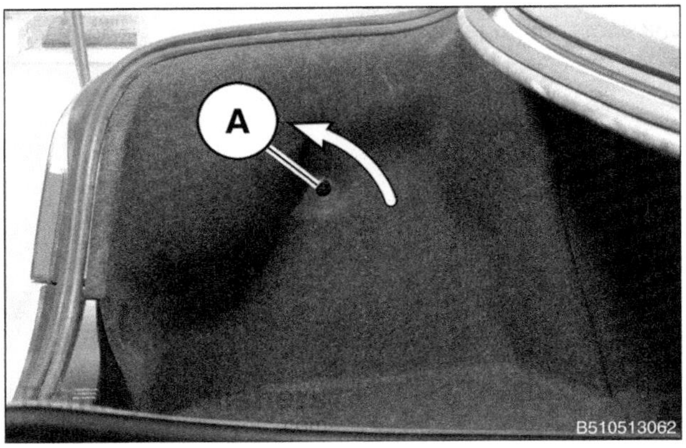

- Remove panel fastener (**A**) by rotating 90° with flat blade screwdriver and remove panel.

- Installation is reverse of removal. Remember to replace damaged trim panel clips.

Tailgate trim panel, removing and installing

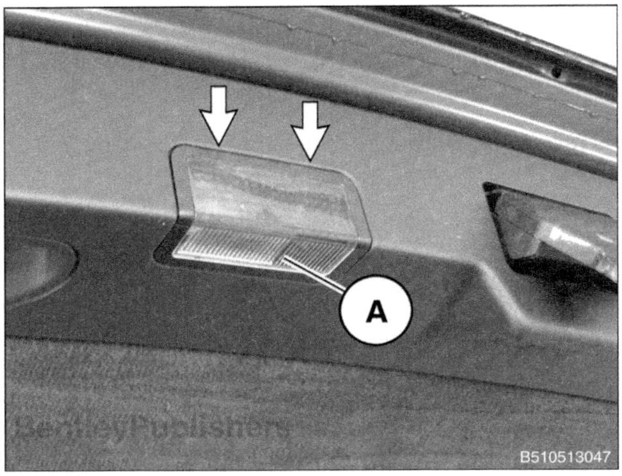

- Working at left side of tailgate: Lever tailgate light (**A**) out at recesses (**arrows**). Disconnect electrical connector and remove light. Repeat procedure for right side light if equipped.

- Models with auto tailgate: Using plastic prying tool, lever out tailgate switch (**A**). Disconnect electrical connector and remove switch.

BentleyPublishers.com
BentleyPublishers.com—All Rights Reserved

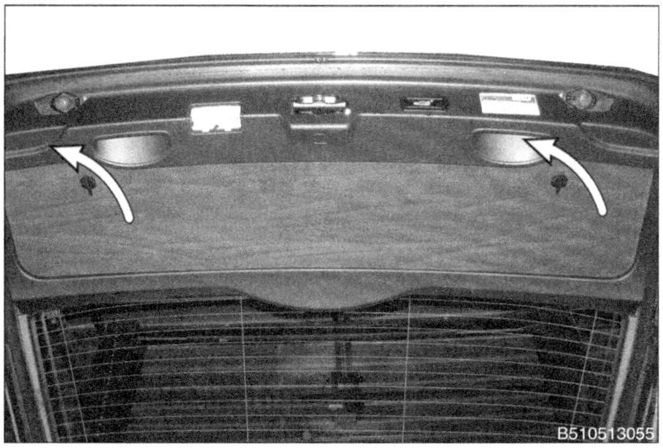

◄ Rotate retainer in direction of **arrows** and open panel.

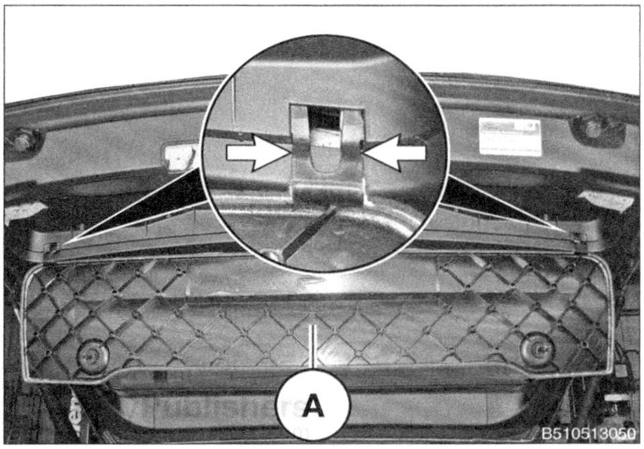

◄ Remove panel (**A**) by squeezing retaining tabs (**inset**) in direction of **arrows** and unclipping.

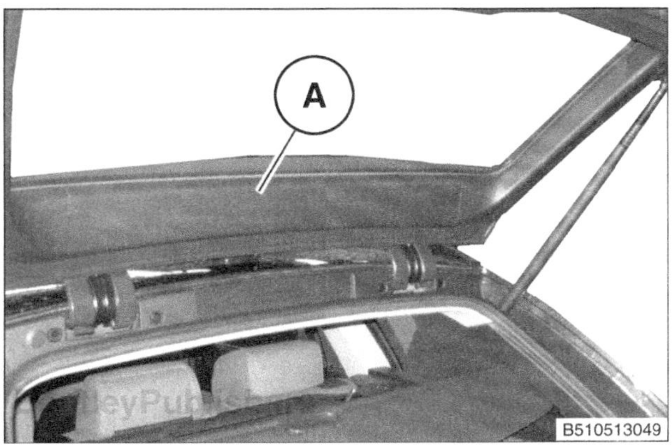

◄ Using a plastic prying tool, remove tailgate center trim panel (**A**).

Tailgate trim panel, removing and installing

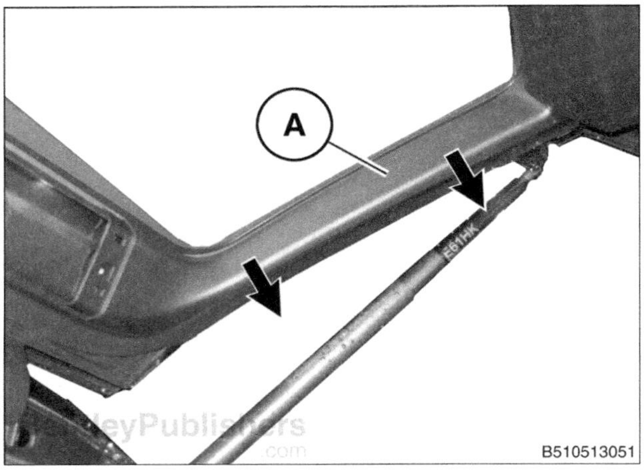

◄ Using a plastic pryingl tool, remove tailgate right side trim panel (**A**) in direction of **arrows**. Repeat procedure for left side.

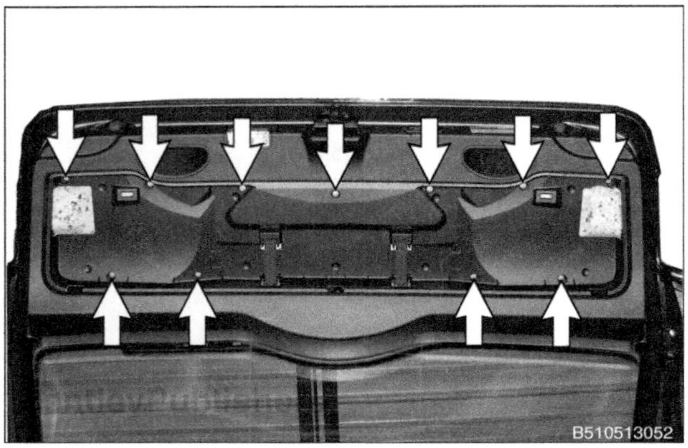

◄ Remove tailgate trim panel fasteners (**arrows**).

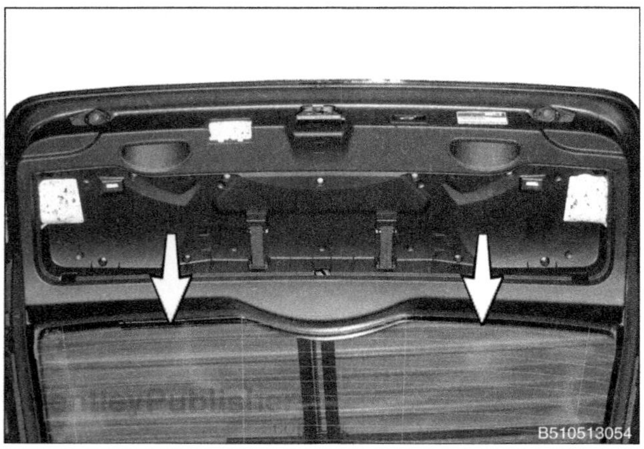

◄ Carefully remove trim panel in direction of **arrows**.

– Installation is reverse of removal. Remember to replace damaged trim panel clips.

BentleyPublishers.com
BentleyPublishers.com—All Rights Reserved

515 Central Locking and Anti-theft

General

This section covers repair information for door, trunk and tailgate locks, central locking, electronic immobilization (EWS) and anti-theft alarm (DWA).

5 Series vehicles are equipped with sophisticated self-diagnostic electrical systems. When experiencing malfunctions relating to central locking or anti-theft systems, using a BMW scan tool can usually pinpoint electrical faults quickly and safely.

See also:

- **411 Doors** for inner door trim removal
- **513 Interior Trim** for tailgate trim removal
- **600 Electrical System–General**
- **ECL Electrical Component Locations**
- **ELE Electrical Wiring Diagrams**

BentleyPublishers
BentleyPublishers.com–All Rights Reserved

Fuel filler flap emergency opening

The fuel filler flap can be unlocked manually in the event of an electrical defect. The emergency release is located in the luggage compartment on the right behind luggage compartment trim.

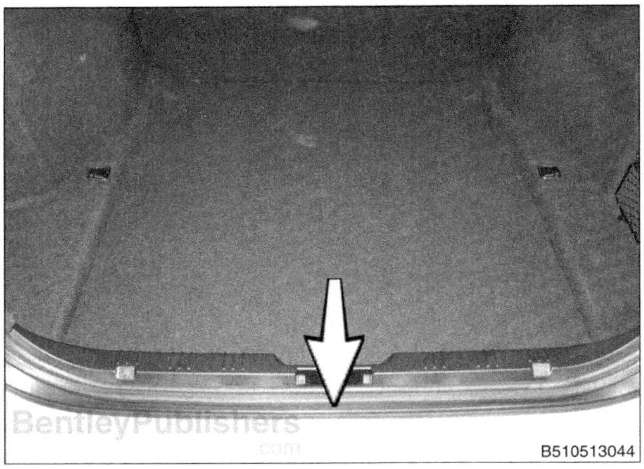

B510513044

- Remove trunk floor trim panel in direction of **arrow**.

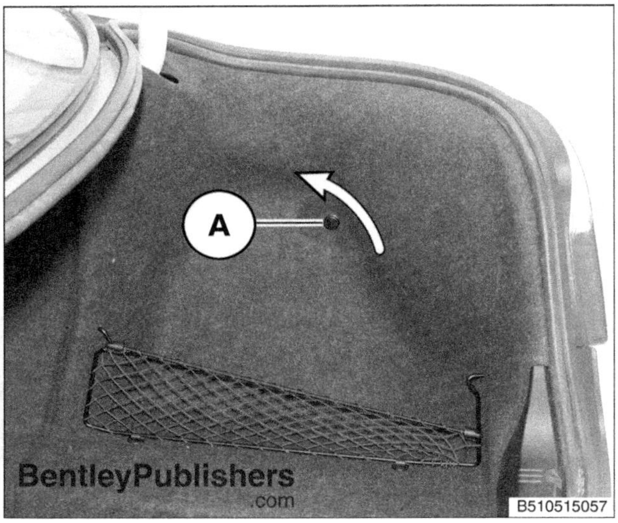

B510515057

- Remove panel fastener (**A**) by rotating 90° with flat blade screwdriver and remove panel.

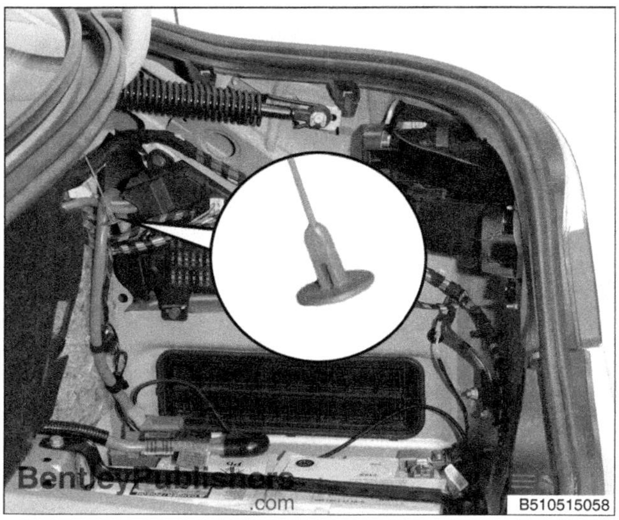

B510515058

- Pull green knob (**inset**) with fuel pump symbol to release fuel filler flap.

BentleyPublishers.com
BentleyPublishers.com—All Rights Reserved

Central Locking

Central locking in 5 Series vehicles controls the door locks, trunk lid lock and fuel filler flap lock.

Central locking is distributed over several control modules. This offers the advantage that sensors and actuators are connected directly to a control module in the vicinity of their installed location, resulting in shorter cable connections. By interlinking the control modules, it is possible to exchange sensor data between them.

There are two different central locking systems in 5 Series models dependant on production date.

To 09 / 2005: The primary control modules for central locking in the 5 Series are:

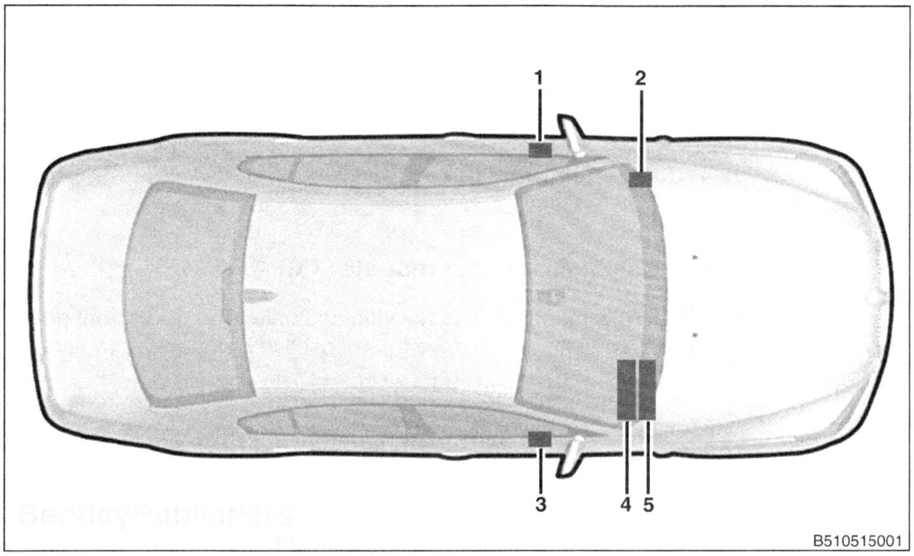

1. Driver door module (TMFA)
2. Car access system (CAS)
3. Passenger door module (TMBF)
4. Basic body module (KBM)
5. Safety gateway module (SGM)

From 09 / 2005: The primary control modules for central locking in the 5 Series are:

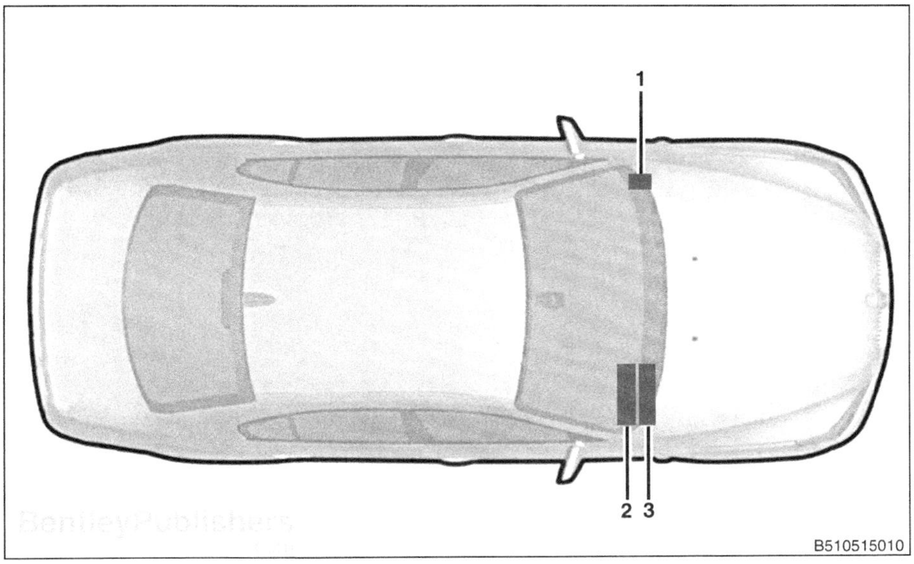

1. Car access system (CAS)
2. Basic body module (KBM)
3. Body gateway module (KGM)

BentleyPublishers.com—All Rights Reserved

Central locking can be operated via the following components:

- Remote control / remote key.
- Driver door lock barrel (door lock).
- Center-lock button.
- Electronic outer door handle module (TAGE) in connection with comfort access.

Car access system (CAS)

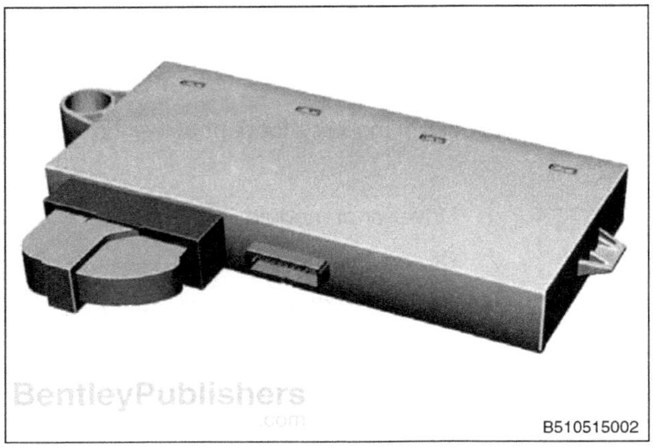
B510515002

Car access system (CAS) is the master control module for central locking, it issues the enable signal to activate a lock / unlock function. When CAS receives a lock / unlock signal from the remote control receiver, it checks whether the remote key is valid and belongs to the vehicle. If the check / authentication is successful CAS will forward the request to activate central locking via body bus system (K-CAN) to:

- **to 9 / 2005**: SGM and KBM.
- **from 9 / 2005**: KGM and KBM

NOTE—

- *For summary of CAS functions* See **600 Electrical System–General**.

To 9 / 2005: Front door module (TMFA, TMBF)

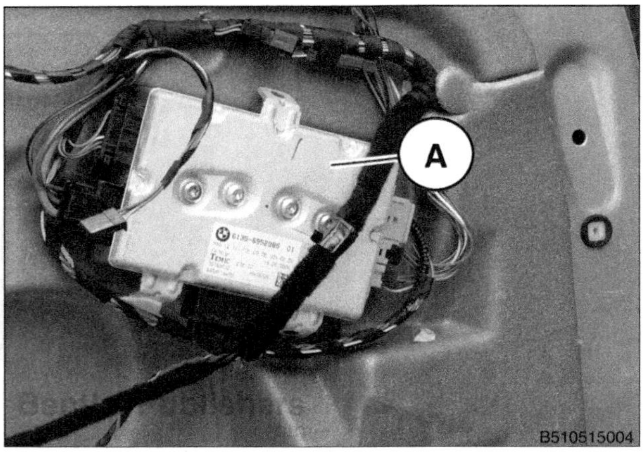

B510515004

TMFA (**A**) is responsible for locking and unlocking driver front door. It contains relays which drive the respective lock / unlock motors.

TMBF is responsible for locking and unlocking passenger front door. It contains relays which drive the respective lock / unlock motors.

To 9 / 2005: Safety gateway module (SGM)

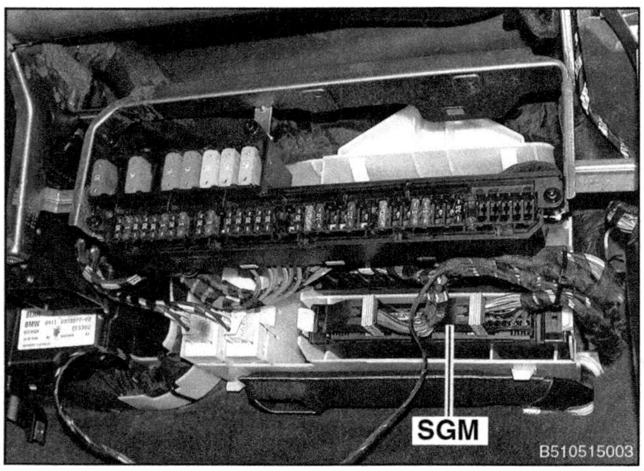

B510515003

SGM converts and forwards messages to and from byteflight and K-CAN.

BentleyPublishers.com
BentleyPublishers.com—All Rights Reserved

From 9 / 2005: Body gateway module (KGM)

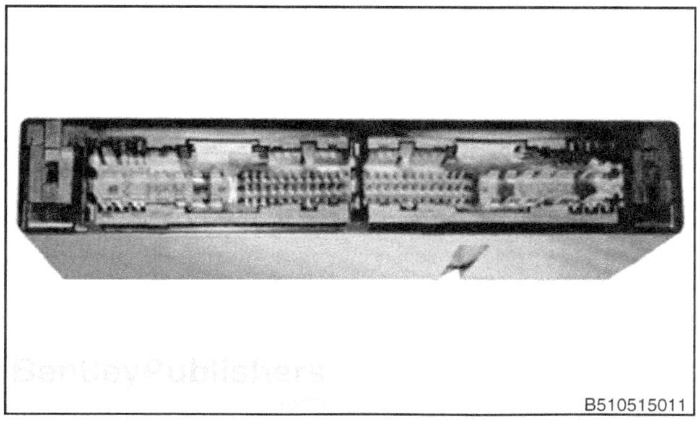

KGM is responsible for locking and unlocking the front doors. It contains relays which drive respective lock / unlock motors.

Basic body module (KBM)

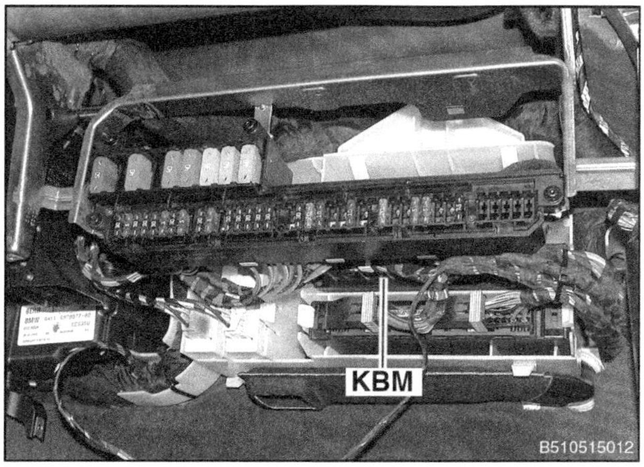

KBM is responsible for locking and unlocking passenger doors and trunk or tailgate of vehicle. It contains relays which drive the respective lock / unlock motors.

Central locking features

Central locking can be activated when driver door is closed.

Vehicle locking / unlocking is initiated by the following components:

- Remote control / remote key.
- Central locking button on center dashboard.
- Mechanical key / spare key.

Driver identification is incorporated within the remote key and is also used for the optional comfort access (CA) system. CA module activates the identification transmitter in the remote key by way of a radio signal. This makes it possible to unlock the vehicle without actively using the remote key.

The central locking system activates the following system components:

- Driver and front passenger door locks.
- Rear door locks.
- Fuel filler flap lock.
- Trunk lid or tailgate.

BentleyPublishers.com—All Rights Reserved

Unlocking / locking

When the remote key unlock button is pressed, the signal is received by the rear window antenna and forwarded to the remote control receiver. The remote control receiver forwards the signal to CAS.

The signal from the remote key is verified in CAS. If signal is valid:

- **To 9 / 2005** SGM and KBM are authorized to unlock vehicle.
- **From 9 / 2005** KGM and KBM are authorized to unlock vehicle.

To 9 / 2005: Remote key signal path

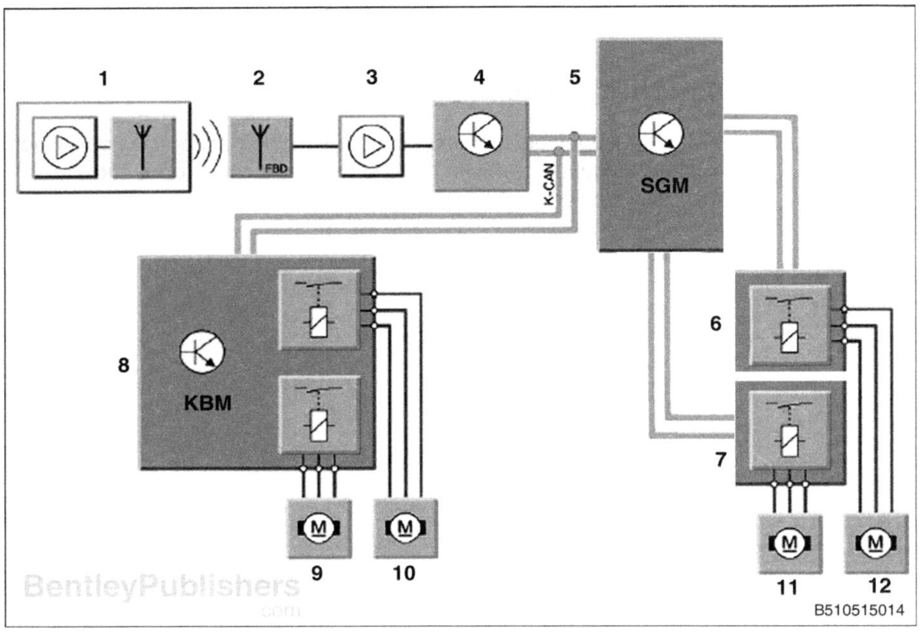

1. Remote key
2. Rear window antenna
3. Remote key receiver
4. Car access system (CAS)
5. Safety gateway module (SGM)
6. Driver door module (TMFA)
7. Passenger door module (TMBF)
8. Body basic module (KBM)
9. Lock actuator, trunk or tailgate
10. Lock actuator, passenger door, rear doors and fuel filler flap
11. Lock actuator, front passenger door
12. Lock actuator, driver door

From 9 / 2005: Remote key signal path

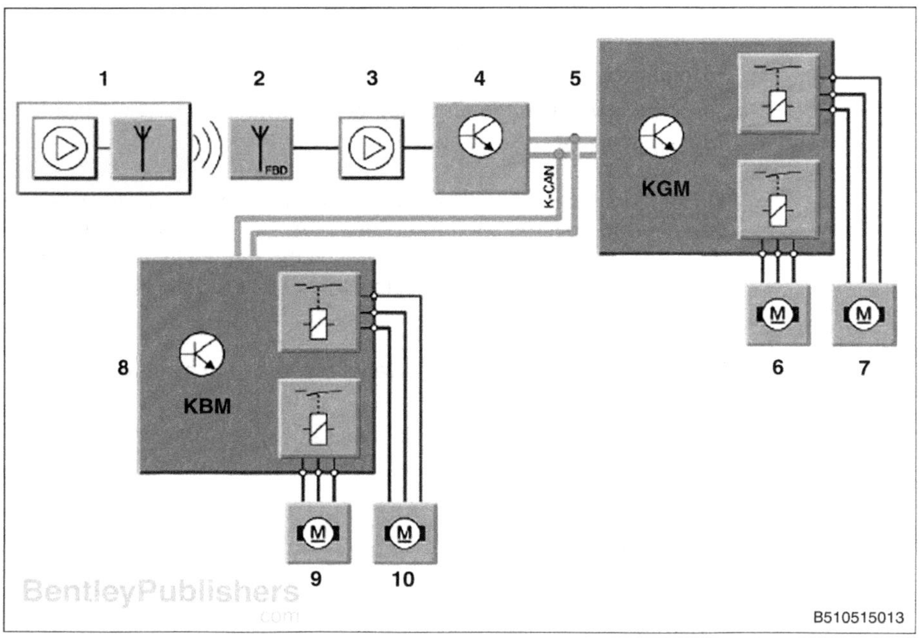

1. Remote key
2. Rear window antenna
3. Remote key receiver
4. Car access system (CAS)
5. Body gateway module (KGM)
6. Lock actuator, front passenger door
7. Lock actuator, driver door
8. Body basic module (KBM)
9. Lock actuator, trunk / tailgate
10. Lock actuator, passenger door, rear doors and fuel filler flap

BentleyPublishers.com
BentleyPublishers.com—All Rights Reserved

Mechanical key locking / unlocking

The vehicle can be locked / unlocked using a mechanical key in the driver door lock. Driver door module (TMFA) monitors the lock cylinder Hall sensor signal. When the lock is turned, a change in the Hall sensor state initiates locking / unlocking.

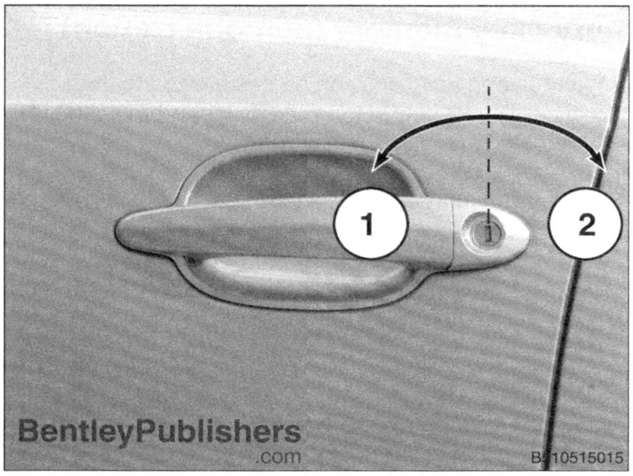

Door key positions for electrical / manual locking and unlocking are:

1. Turn once: Unlocks driver door
   Turn twice: Unlocks remaining doors, trunk, tailgate, fuel filler
   Turn and hold (door closed): Opens windows, sunroof
2. Turn once: Locks all doors, tailgate, fuel filler
   Turn and hold (door closed): Closes windows, sunroof

Selective unlocking

When unlocking with the remote key, the first unlock request only unlocks the driver's door. A second unlock request unlocks remaining doors, tailgate and fuel flap.

Central locking button

Central locking button (**arrow**) is in the center of dashboard.

- Central locking button locks doors and tailgate. The fuel filler flap remains unlocked for refueling purposes.
- If a door is opened from inside while centrally locked, the remaining doors remain locked.

The open door can be relocked, when closed, by manually locking or pressing the central locking button twice. This synchronizes the door locks.

Single lock and double lock function

Each door lock actuator incorporates two lock positions.

Single lock controls the mechanical lock mechanism when the central lock switch is pressed. The lock mechanism is fully locked at this point but can still be opened from interior by pulling interior door handle twice or by pressing central lock switch again. When single lock function is activated, the fuel filler flap is not locked.

Double lock, also known as central arrest, is activated only when vehicle is locked from the outside at the driver's door lock with a key or when CAS receives a lock request from the remote key. In this case the double lock motor is activated, mechanically offsetting an internal rod in the lock actuator. This disables the actuator from unlocking the vehicle from the interior and prevents the doors from being unlocked by any means other than an unlock request at the driver's door or via the remote key.

Vehicle accident unlocking

During a vehicle accident the airbag control module sends a signal to CAS to unlock the vehicle. The remote key and central locking button are disabled until ignition status (terminal R) is changed from ON to OFF back to ON.

Car memory / key memory (personalization)

A number of features and functions can be customized to the driver(s) preference. The identity of the vehicle user is provided by a signal from the remote key.

Car memory and key memory are two separate functions, although they are marketed as a combined feature.

Car memory

Prior to new vehicle delivery, the BMW factory scan tool is used to code driver preferences into the appropriate control modules. Thereafter these choices cannot be changed without recoding using a BMW factory scan tool or equivalent. Only one car memory setting setting per vehicle is possible.

The functions that can be set using car memory include:

- Audible confirmation (siren chirp) and visual confirmation (via turn signal lights) when alarm is armed. Siren chirp can be deactivated with BMW factory scan tool or equivalent.
- Daytime running lights ON or OFF.
- Pathway lighting allows use of the dimmer switch to turn on the headlights and interior lights for 40 seconds.
- Automatic locking of central locking system once vehicle reaches 10 mph.
- Selective unlocking. Unlocks only drivers door on first actuation of unlock button on remote key. Unlocks all doors, trunk and fuel filler door after second actuation of button on remote key.

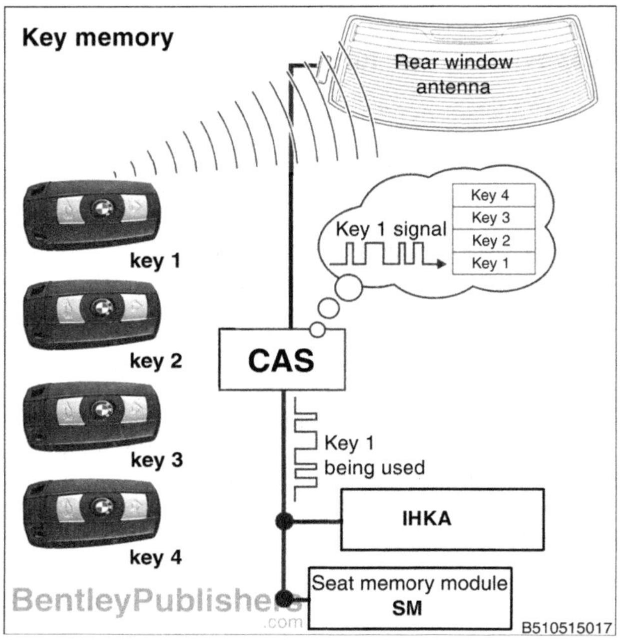

Key memory

When a remote key (identification transmitter) is used to lock or unlock the car, the user is identified by CAS. Multiple keys can be programmed with the personalization feature. The use of a personalized key then triggers car memory functions such as heating and air-conditioning settings or memory seat position adjustment.

Most key memory options can be programed using the iDrive menu. Some programming requires the use of BMW factory scan tool or equivalent.

Available key memory functions vary based on vehicle equipment. The functions that can be set include:

- Audio tone settings and radio presets.
- Language of displays and voice functions.
- Climate control temperature.
- Seat / mirror memory settings.
- Lighting preferences.

BentleyPublishers.com
BentleyPublishers.com—All Rights Reserved

Comfort access (CA)

Comfort access (CA), optional in 5 Series models, is a passive access system which allows the vehicle to be unlocked by grasping the outer door handle, provided the remote key is located within a radius of no more than approx. 0.5 - 1.5 m (1½ - 4½ ft) from the vehicle. With CA it is sufficient for the driver to carry the remote key on his or her person in order to open or start the vehicle.

Modules and other components which control and operate CA:

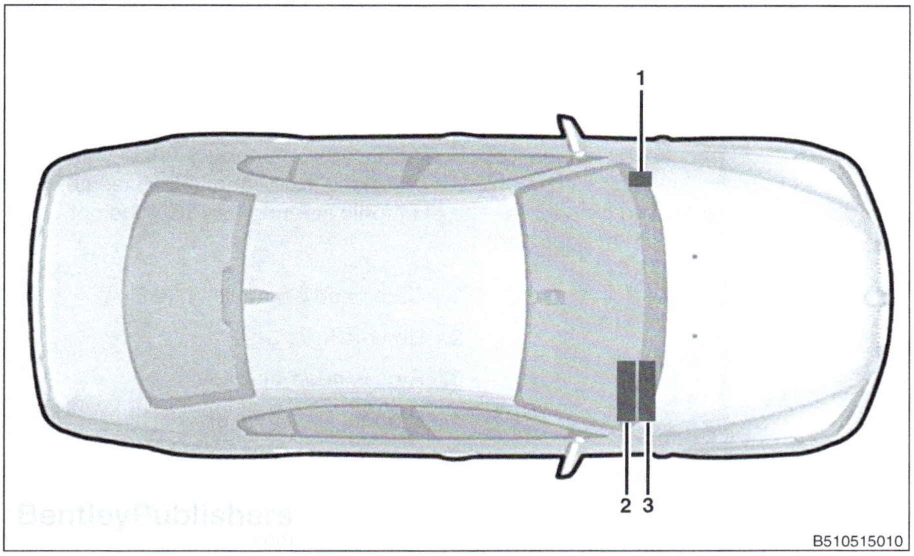

1. Car access system (CAS)
2. Basic body module (KBM)
3. Body gateway module (KGM)

When a remote key comes into range of the CA antennas, the signal is sent to CAS to be verified. If seen as valid the lock / unlock function is granted.

Comfort access antennas

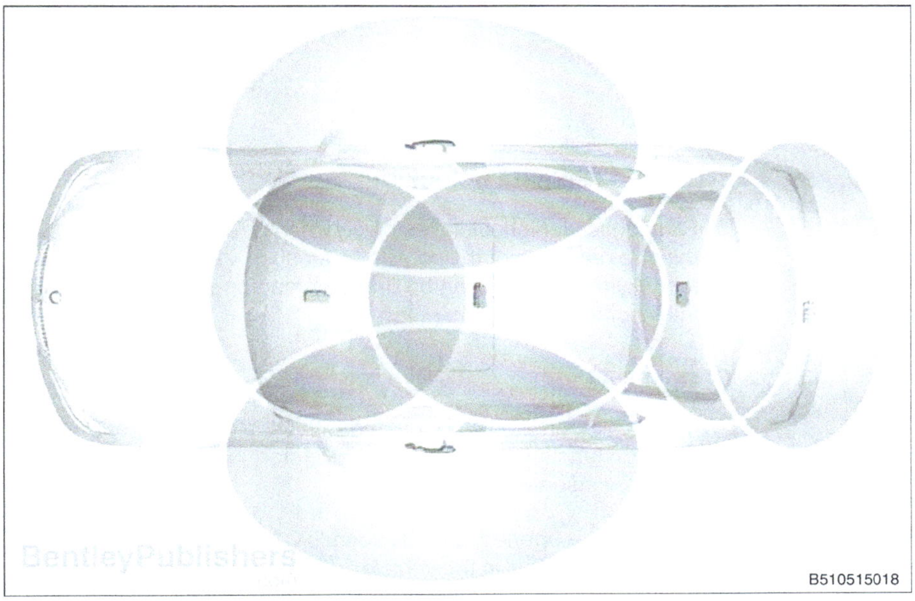

- Driver door handle
- Passenger door handle
- Rear bumper
- Luggage compartment
- Interior front
- Interior center

515

BentleyPublishers.com—All Rights Reserved

Comfort access features

Passive entry

Passive entry enables access to vehicle without operating the remote key. The vehicle recognizes the presence of a valid remote key and access is granted by grasping the outer door handle.

Unlocking

A capacitive sensor in door handle module (TAGE) recognizes that the handle has been grasped and activates the transmit antenna. The transmit antenna sends a signal to the remote key. In turn, the identification transmitter sends a 315 MHz high frequency signal to the remote control receiver, which contains the authentication request. CAS checks the remote key authentication code. Following successful authentication, CAS issues an enable signal to unlock vehicle and initiates vehicle unlocking. KGM and KBM executes the unlocking procedure. The electronic steering lock (ELV) is not yet unlocked (if equipped).

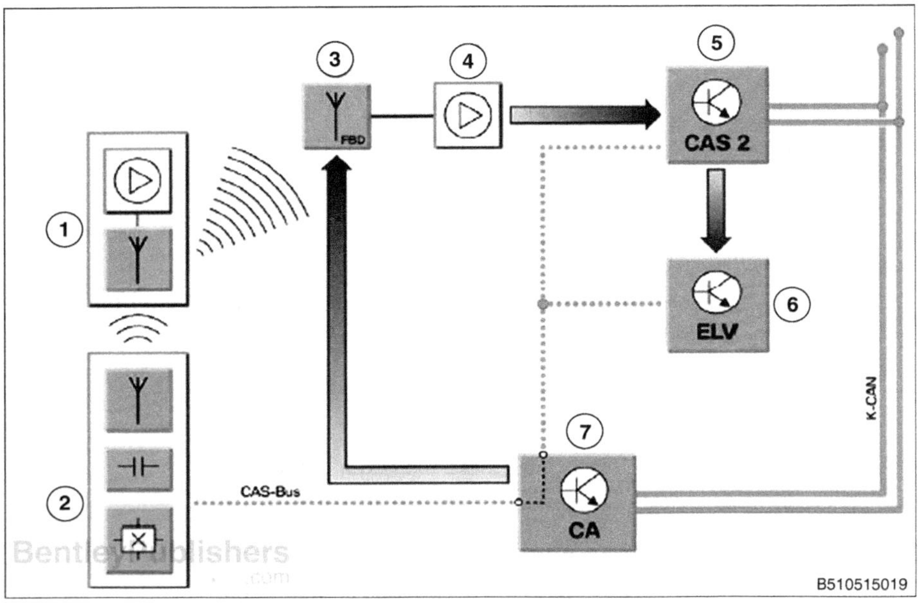

1. Door handle module (TAGE)
2. Remote key
3. Rear window antenna
4. Remote key receiver
5. Car access system (CAS)
6. Electronic steering lock (ELV)
7. Comfort access (CA) module

Trunk lid / tailgate unlocking

An authentication check takes place when a remote key is located within the rear area of the vehicle. After successful authentication, the trunk lid or tailgate can be unlocked and opened with the outer trunk lid or tailgate button.

Passive go

When a valid remote key is located within the vehicle interior, the passive go function makes it possible to start the vehicle without the remote key inserted in the key slot.

When a door is opened, after 3 seconds CAS checks to see whether a valid remote key is in the vehicle. The remote replies with data for the electronic vehicle immobilizer. CAS enables engine start by sending a signal to the ECM. If equipped, the electric steering lock (ELV) is unlocked.

BentleyPublishers.com
BentleyPublishers.com—All Rights Reserved

Passive exit

Passive exit makes it possible to lock the vehicle without using a remote key. After a door is closed, the locking procedure is started by touching the recognition area on the outer door handle. The door handle module (TAGE) sends the request to CAS to lock the vehicle.

If a valid remote key is detected in the interior of the vehicle during locking with another valid remote key from the exterior, the key left in the vehicle is deactivated until the vehicle in unlocked again.

DOOR HANDLES AND LOCKS

In the procedures that follow, door handle removal, door lock cylinder removal and door latch removal are covered for left front door. Other doors are similar although simpler, due to the lack of lock cylinder.

Inner door panel removal is covered in **411 Doors**.

Door lock cylinder, removing and installing

– Remove door latch. See **Door latch, removing and installing** in this repair group.

– Remove door handle. See **Door handle, removing and installing** in this repair group.

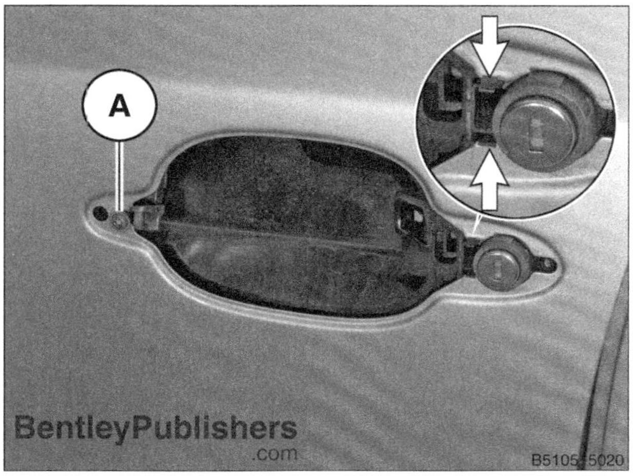

◄ Loosen carrier fastener (**A**), but do not remove. Release carrier catches (**arrows in inset**) while pressing lock cylinder into door.

– Working inside door, disconnect comfort access connector, if equipped.

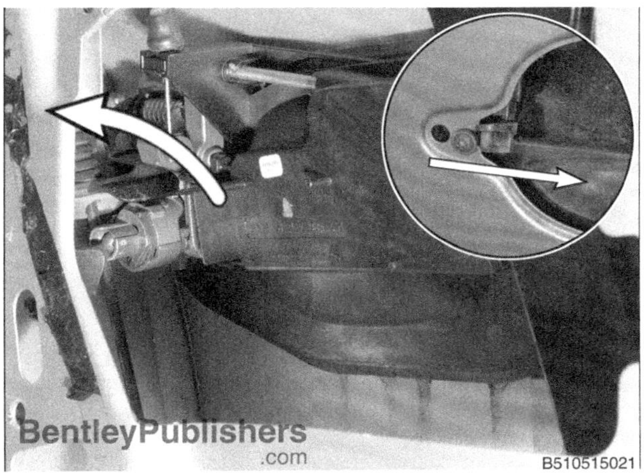

◄ Working inside door, pull carrier in direction of **arrow** while sliding fastener out of slot (**inset**).

– Remove carrier from door.

– Installation is reverse of removal.

- Replace door handle gaskets if damaged.
- Reset comfort access (CA) using BMW scan tool. See **600 Electrical System-General**.

BentleyPublishers.com—All Rights Reserved

Door handle, removing and installing

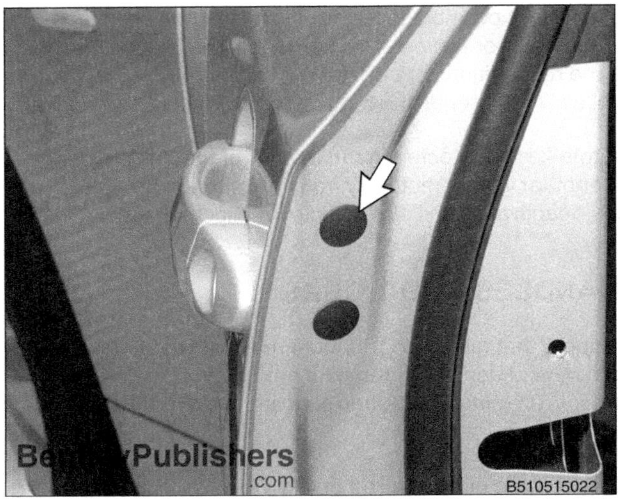

Remove rubber plug (**arrow**).

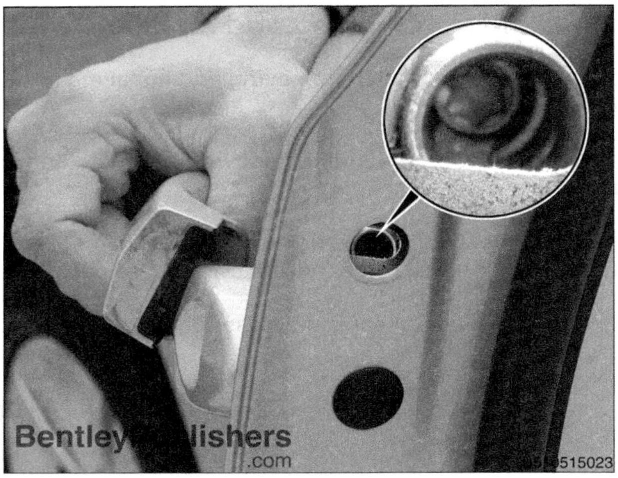

Pull outside handle up until fastener (**inset**) is accessible and hold in place.

– Turn fastener (**inset**) counterclockwise to secure handle and lock actuator in position. If lock actuator is not properly secured in position, lock actuator lever may return into latch and door panel removal will be required to reset.

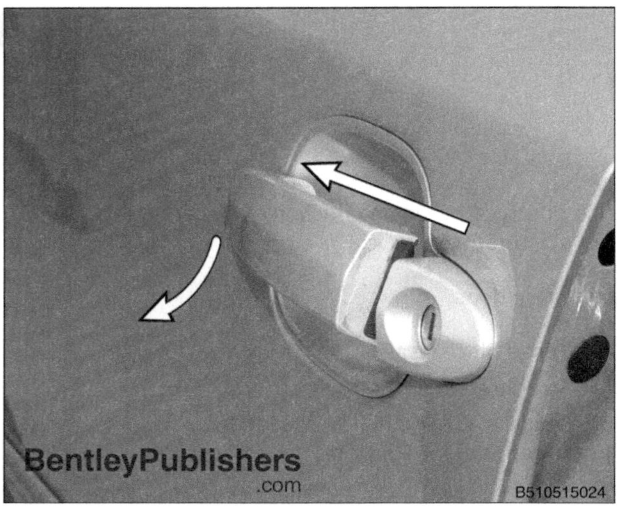

Pull outside handle out of rear lock actuator in direction of **arrow**.

– Pull outside handle downward (**curved arrow**) from front of lock actuator.

BentleyPublishers.com
BentleyPublishers.com—All Rights Reserved

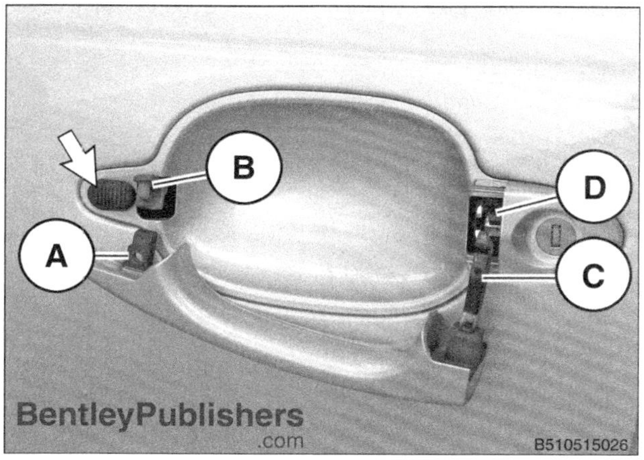

Installation is reverse of removal.

- Check that stop pad (**arrow**) is in place.
- Insert outside handle bore (**A**) correctly in mounting (**B**).
- Engage outside handle (**C**) correctly in lock actuator (**D**).
- Replace door handle gaskets if damaged.
- Reset comfort access (CA) module using BMW scan tool. See **600 Electrical System-General**.

Tightening torque	
Handle to lock actuator	3 Nm (26.5 in-lb)

Door handle trim, removing and installing

– Remove door handle. See **Door handle, removing and installing** in this repair group.

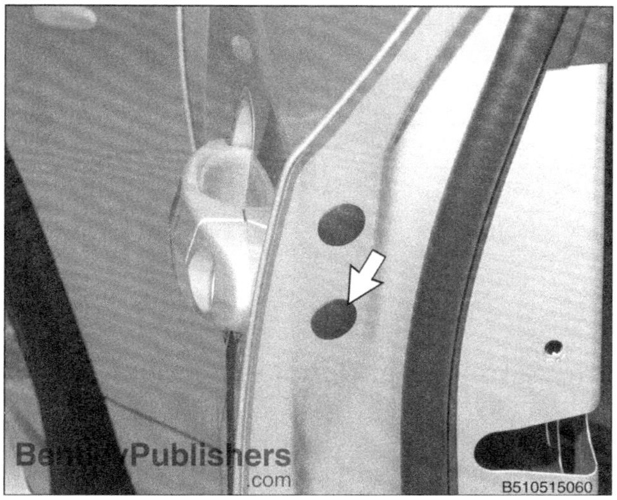

Remove rubber plug (**arrow**).

Remove trim fastener (**arrow**).

BentleyPublishers
BentleyPublishers.com–All Rights Reserved

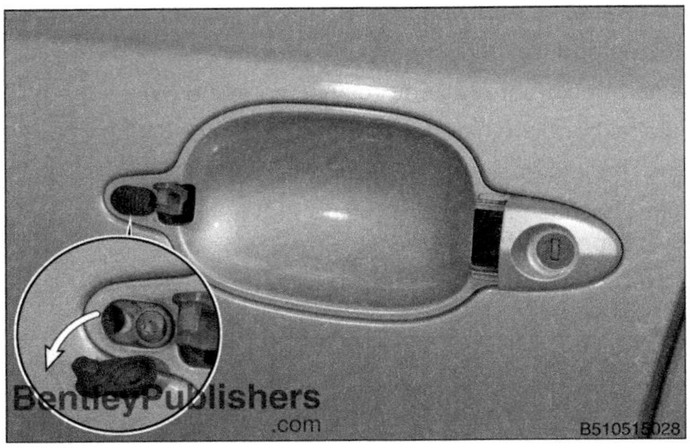

Remove door handle stop pad by pulling off in direction of **arrow**.

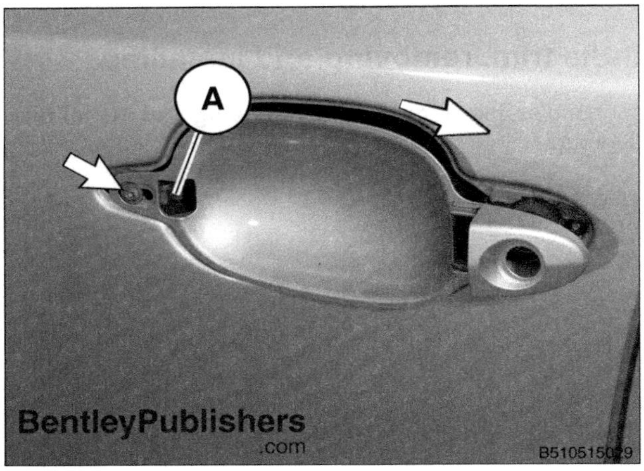

Remove fastener (**arrow**). Press handle mount (**A**) into door.

- Slide trim out, towards rear of vehicle and remove in direction of **arrow**.
- Installation is reverse of removal.
  - Replace door handle gaskets if damaged.
  - Reset comfort access (CA) using BMW scan tool. See **600 Electrical System-General**.

Tightening torque	
Cover to handle mount	2.5 Nm (22.1 in-lb)
Handle to lock actuator	3 Nm (26.5 in-lb)

Door latch, removing and installing

CAUTION—
- *Do not use silicone lubricant on door latch. Latch faults have been traced to silicone contamination.*

- Remove door trim panel. See **411 Doors**.

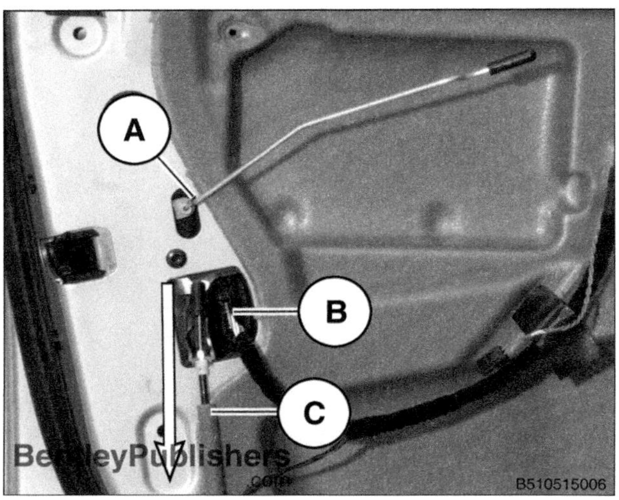

Working at corner of door at door latch assembly:
- Detach electrical connector (**A**) from latch assembly.
- Detach door lock rod (**B**) from latch assembly.
- Pull Bowden cable (**C**) down (**arrow)** to detach from latch assembly.

BentleyPublishers.com
BentleyPublishers.com—All Rights Reserved

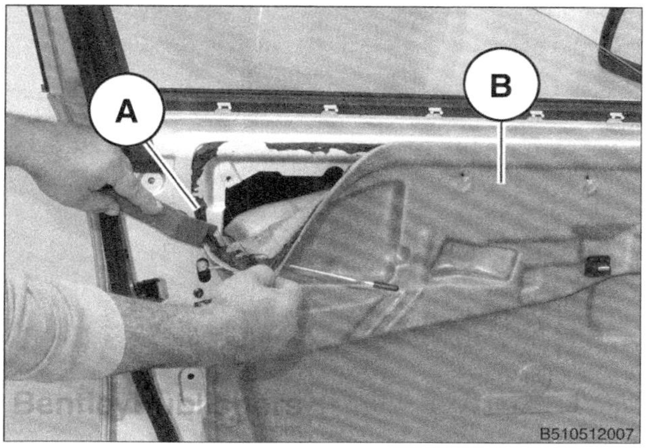

Using a sharp plastic scraper, lever sealing bead (**A**) off vapor barrier (**B**).

> ***CAUTION—***
> - *Use care when peeling back vapor barrier. A damaged vapor barrier should be replaced to prevent possible airbag malfunction and water leaks.*

- Partially remove window regulator to gain access to door latch. See **512 Door Windows**.

- Remove door lock cylinder. See **Door lock cylinder, removing and installing** in this repair group.

Working inside door, disconnect Bowden cable (**A**) from latch (**B**).

Remove latch fasteners (**arrows**).

- Remove latch assembly from inside door.

515

BentleyPublishers.com—All Rights Reserved

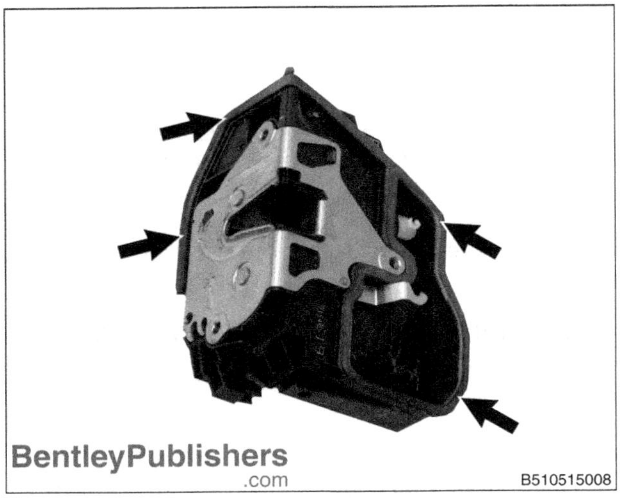

Installation is reverse of removal, keeping in mind the following:

- Do not bend door lock button metal rod when inserting through door skin opening.
- Hand tighten fasteners to properly seat door latch seal (**arrows**) before tightening.
- Replace latch mounting screws and use threadlock compound.
- Make sure seal on latch is not damaged.
- Make sure inner door cavity vapor barrier is installed intact and leak-proof.

Tightening torque	
Latch assembly to door (replace screws)	9.5 Nm (7 ft-lb)

TRUNK AND TAILGATE LOCKS

Trunk lock cylinder

– Remove trunk lid trim panel. See **513 Interior Trim**.

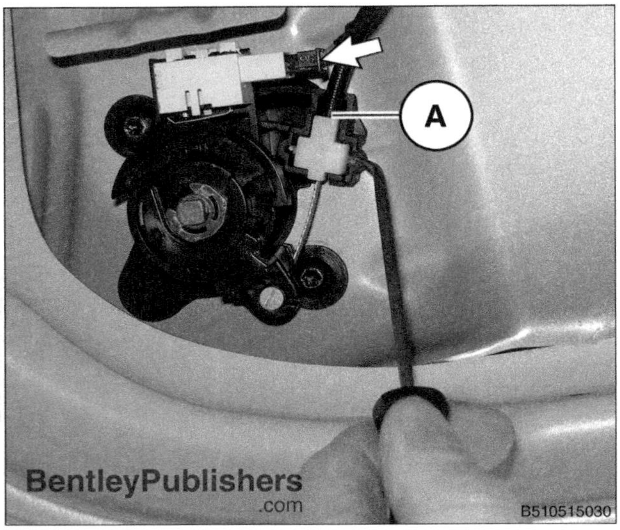

Disconnect electrical connector (**arrow**). Unclip bowden cable (**A**) from holder and remove from lock cylinder using a small pick.

Remove lock cylinder fasteners (**arrows**).

BentleyPublishers.com
BentleyPublishers.com–All Rights Reserved

Rotate lock cylinder until tabs (**arrows**) are visible through slots in trunk lid and remove lock cylinder from trunk lid.

– Installation is reverse of removal. Remember to inspect and replace lock cylinder seal if necessary.

Trunk lock assembly

– Remove trunk lid trim panel. See **513 Interior Trim**.

Disconnect electrical connector (**arrow**) from trunk lock assembly.

– Remove fasteners (**A**) and pull latch assembly out of trunk lid.

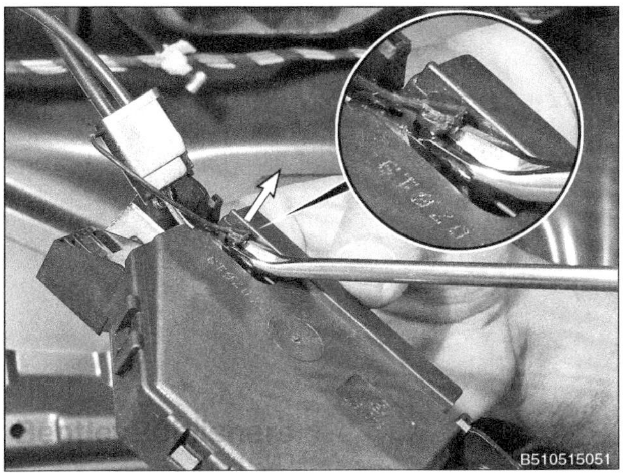

Lever out locking pin (**inset**) in direction of **arrow** and remove bowden cable from lock assembly.

– Installation is reverse of removal.

Tightening torque	
Lock assembly to trunk lid	10 Nm (7.3 ft-lb)

BentleyPublishers
BentleyPublishers.com—All Rights Reserved

Trunk lock striker, removing and installing

- Remove lower trunk trim panel. See **513 Interior**.

- Remove trunk lock striker fasteners (**arrows**) then remove striker.
- Installation is reverse of removal. Adjust striker if necessary. See **Trunk lock striker, adjusting** in this repair group.

Tightening torque	
Trunk lock striker to body	10 Nm (7.3 ft-lb)

Trunk lock striker, adjusting

- Completely screw in trunk detent buffers (**arrows**).

- Remove trunk lock striker fastener access covers (**arrows**).

BentleyPublishers.com
BentleyPublishers.com—All Rights Reserved

◀ Inspect fulcrum pads (**arrows**) for wear of damage, replace if necessary. Interior trim panel removed for clarity.

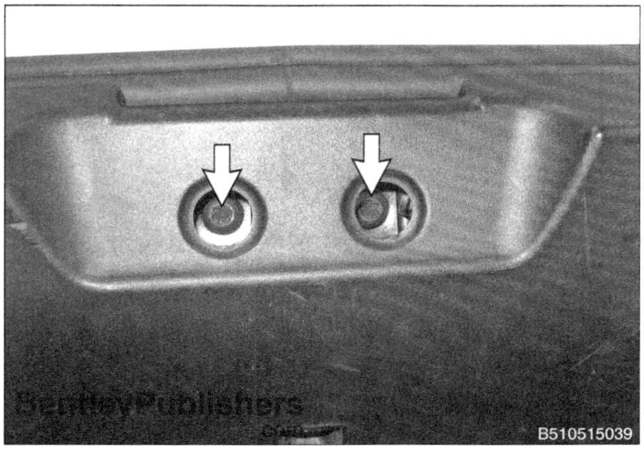

◀ Loosen trunk lock striker fasteners (**A**) slightly, just enough so it is able to move and center itself.

- Close trunk lid. This allows for the trunk lock striker to adjust itself.

- Open trunk and tighten fasteners. Check adjustment and repeat procedure if necessary.

Tightening torque	
Trunk lock striker to body	10 Nm (7.3 ft-lb)

- Reassemble trunk interior trim.

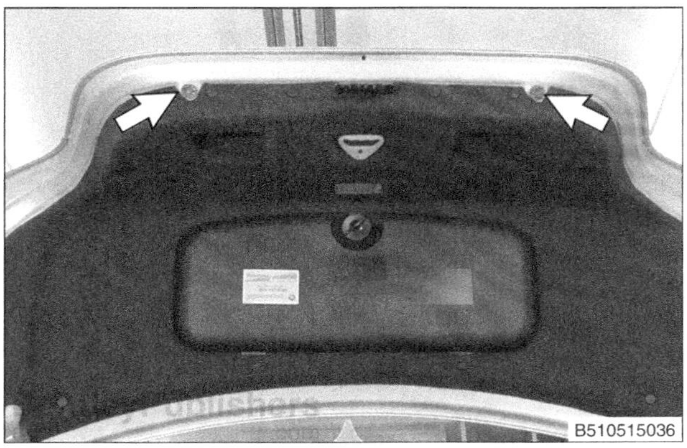

◀ Unscrew trunk lid detent buffers (**arrows**) until closed rear trunk lid left and right detent buffers make contact with body panels. Trunk lid should be even with rear body panels.

BentleyPublishers.com
BentleyPublishers.com—All Rights Reserved

Tailgate lock

To gain access to tailgate locking mechanism, remove tailgate trim. See **513 Interior Trim**.

Tailgate lock assembly, removing and installing

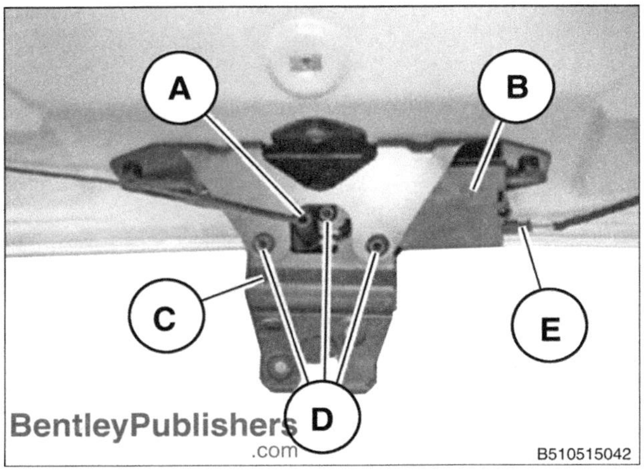

- Open upper tailgate and remove trim. See **513 Interior Trim**.
- Remove emergency release cable from latch (**A**).
- Disconnect lock actuator (**B**) electrical connector (**E**).
- Remove tailgate lock fasteners (**D**) and remove lock (**C**) from tailgate.
- Installation is reverse of removal.

Tightening torque	
Lock to tailgate lid	10 Nm (7.3 ft-lb)

Tailgate striker assembly, removing and installing

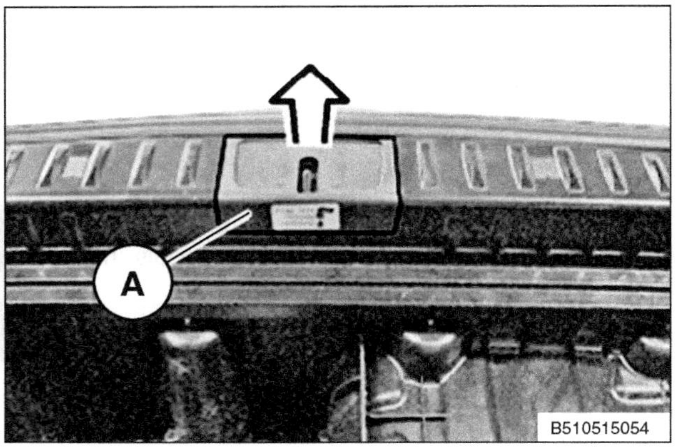

- Snap out cover (**A**) in direction of **arrow** and remove.

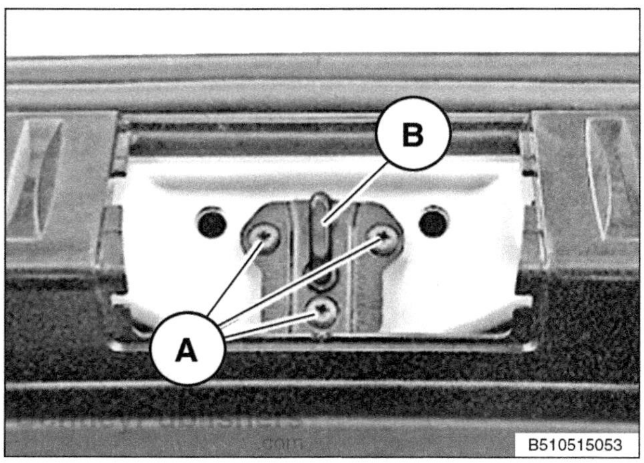

- Remove fasteners (**A**) and remove striker (**B**).
- Installation is reverse of removal.

Tightening torque	
Lock to tailgate lid	10 Nm (7.3 ft-lb)

BentleyPublishers.com
BentleyPublishers.com—All Rights Reserved

Tailgate striker assembly, adjusting

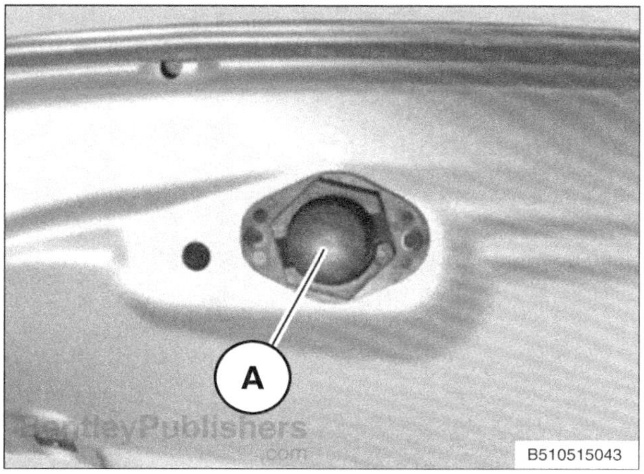

Fully screw in left and right side bump stops (**A**). See **412 Trunk Lid, Tailgate**.

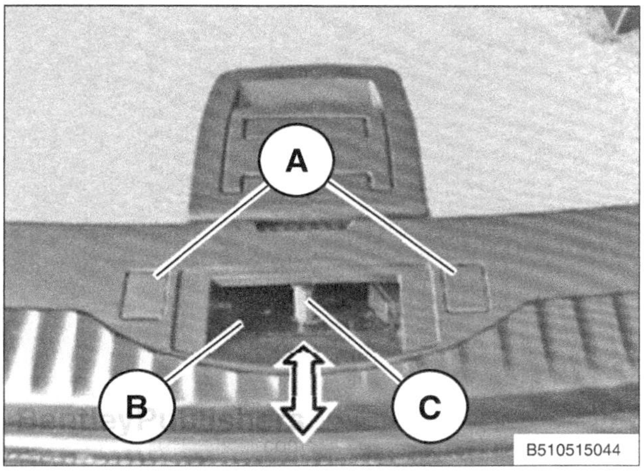

Snap out covers (**A**) and remove. Loosen fasteners beneath covers slightly, just enough so striker it is able to move and center itself.

> ***CAUTION—***
> - *Check that tailgate fasteners do not strike tailgate during adjustment.*

- Close rear lid. Tailgate will center itself.
- Open lid and tighten tailgate striker fasteners (**A**).

Tightening torque	
Striker to body	10 Nm (7.3 ft-lb)

- Install striker trim panel.

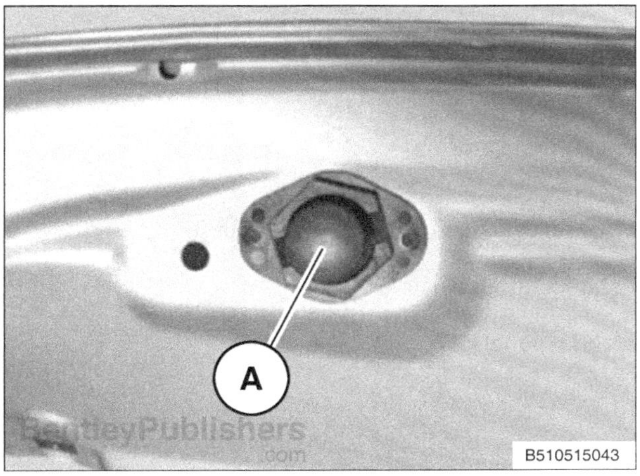

Unscrew and adjust bump stops (**A**) until tailgate is flush with side body panel.

- Check for proper adjustment and tailgate operation. Repeat adjustment if necessary.

BentleyPublishers.com—All Rights Reserved

Rear window lock assembly

The rear window (tailgate glass) lock assembly is part of the rear wiper motor. For removing and installing see **611 Wipers and Washers**.

Rear window lock, adjusting

– Check that tailgate is adjusted properly.

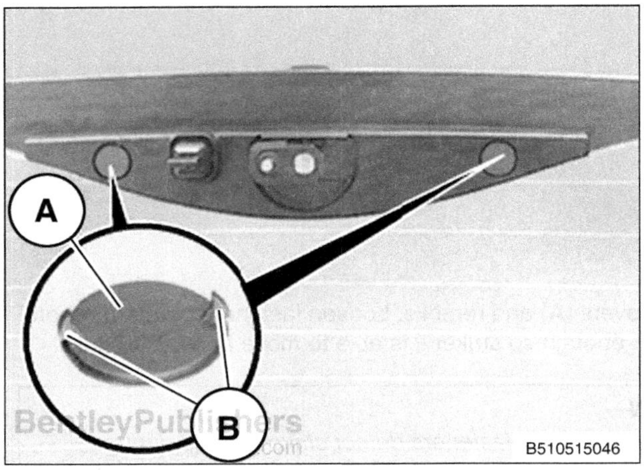

◄ Remove plastic covers (**A**). Be careful not to break catches (**B**).

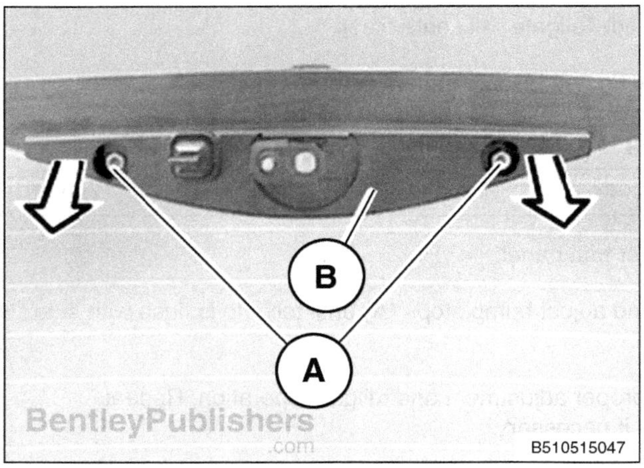

◄ Remove fasteners (**A**) then trim panel (**B**).

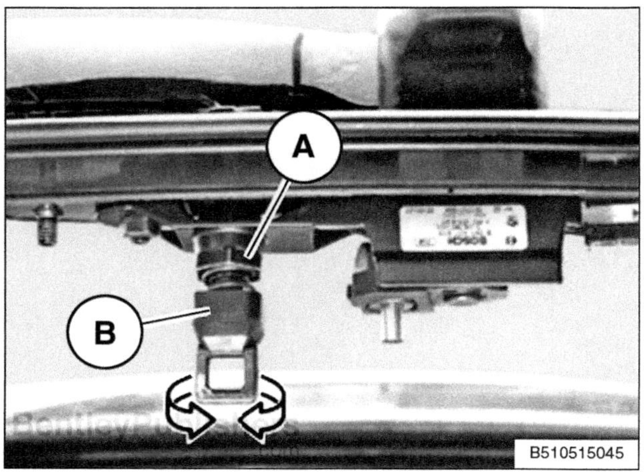

◄ Loosen nut (**A**). Adjust striker by turning in or out (**B**).

– Tighten nut and recheck adjustment.

– Install tailgate trim panels.

BentleyPublishers.com
BentleyPublishers.com—All Rights Reserved

Electronic Immobilizer (EWS)

EWS is a shared system between car access system (CAS) and digital motor electronics (DME). If an authorized key is recognized by CAS a start signal is authorized and sent to the DME. Troubleshooting is performed with a BMW factory scan tool or equivalent.

Anti-theft System (DWA)

DWA components

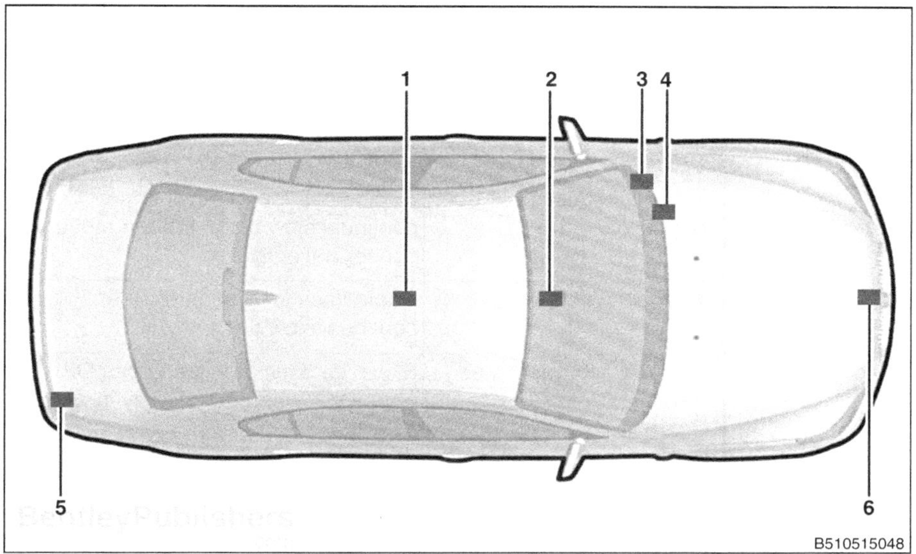

1. Ultra sonic sensor (USIS)
2. DWA LED
3. Car access system (CAS)
4. Light module (LM)
5. Tilt sensor / alarm siren (SINE)
6. Hood switch

DWA operation

When the anti-theft system (DWA) is armed:

- The doors, hood, windows and trunk lid are monitored against forcible entry.
- A tilt sensor protects the vehicle from being jacked up or towed away.
- Movement inside the vehicle is monitored.
- Battery voltage is monitored.

The system responds to unauthorized vehicle entry or attempted theft by activating the following:

- Alarm siren sounds for 30 seconds.
- Hazard warning lights and high beams flash for approx. 5 minutes.

The system is armed or disarmed from the driver door lock or the remote transmitter when the vehicle is locked or unlocked. The interior compartment monitor is activated approx. 30 seconds after the vehicle is locked.

The system indicates that it is armed by flashing the hazard warning lights once and emitting a brief chirp from the siren.

If the alarm is activated, the tailgate may still be opened using button **3** on the remote key. The alarm resets when the tailgate is closed.

Following the triggering of an alarm, the system resets and can trigger again if further tampering to the vehicle is detected.

The interior protection ultrasonic sensor (USIS) is in the DWA main control module. It monitors the passenger compartment and receives inputs from corresponding modules for alarm activation, deactivation and triggering.

DWA LED display

The DWA status LED (**arrow**) is below the interior rear view mirror. LED displays are listed in **Table a**.

Table a. DWA LED status

DWA status	LED signal
Disarmed	OFF
Armed	Continual slow flash.
Armed with tailgate or door not fully closed	Rapid flash for 10 second, then continual slow flash. Interior motion sensor not activated.
Alarm activated	Rapid flash for 5 minutes, then continual slow flash.
Disarmed after activated alarm	Blinks for 5 minutes then goes OFF, or goes OFF when remote key in inserted.

DWA alarm siren

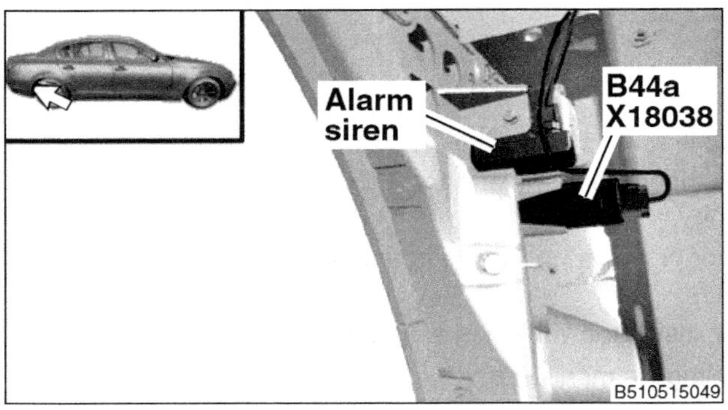

The DWA alarm siren is at the right side rear inner bumper, behind the plastic inner fenderwell.

The siren is powered by a separate rechargeable battery. This battery is recharged by the vehicle electrical system when DWA is not armed.

Tilt sensor

The tilt sensor is integrated into the alarm siren. It registers the vehicle rest position when armed and detects changes in position, for example jacking of the vehicle.

When flat-bed transporting the vehicle, lock the vehicle twice within 10 seconds to switch tilt sensor OFF.

BentleyPublishers.com
BentleyPublishers.com—All Rights Reserved

Interior protection

The interior motion ultrasonic sensor (USIS) is in the center of the headliner. Due to the design of the vehicle interior, install sensor in the correct direction to ensure proper operation of the system. The arrow on sensor points toward front of vehicle

As with the tilt sensor, when flat-bed transporting the vehicle, lock the vehicle twice within 10 seconds to switch interior motion sensor OFF.

The USIS system uses ultrasound to sense motion inside its detection cone. Every time DWA is armed, the USIS sensor adapts to whatever objects are stationary in the interior. USIS also checks for background hiss (wind noise through a partially open window) and adapts for this.

- If detected echoes and sounds are consistent, no movement is detected.
- If the echoes are altered (inconsistent), USIS determines motion in the interior compartment.
- If motion is detected, the USIS changes from a pulsed signal to a constant signal and the echo is compared again.
- If the inconsistency is still present USIS triggers the alarm.

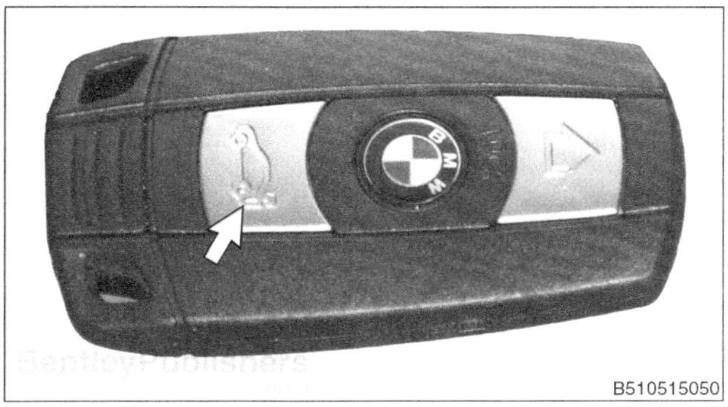

Panic mode

If the tailgate button (**arrow**) is pressed and held for more than 2 seconds, USIS activates the siren in panic mode. This function is available with either an armed or disarmed DWA.

515

Emergency disarming

Emergency disarming occurs automatically when a recognized key is inserted into the ignition lock or a remote key is inserted into the key slot. If the key is accepted by CAS, the doors are unlocked and the alarm deactivated.

BentleyPublishers.com
BentleyPublishers.com–All Rights Reserved

520 Seats

General

This repair group covers removal and installation of the front and rear seats.

See also:

- **ECL Electrical Component Locations**
- **ELE Electrical Wiring Diagrams**

Seat memory

Seat and outside rear-view mirror positions can be memorized and stored in the seat control module for two different users. The seat control module is mounted to the bottom of the seat (**inset**).

BentleyPublishers.com–All Rights Reserved

Seat memory position, setting

– With transmission range selector in P, switch ignition ON and adjust seat and mirrors to desired position.

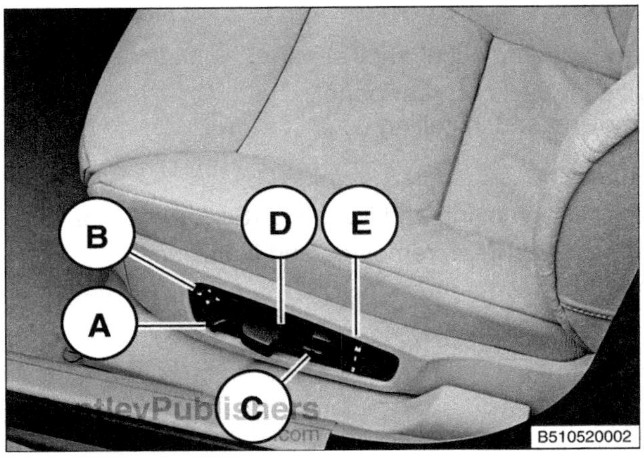

Press **M** on memory switch and press one of the two memory switches (**E**).

- The current position of the mirrors, seat and steering column is now stored in memory.

Power seat controls.

- **A** — Thigh support
- **B** — Lumbar support
- **C** — Backrest width adjustment
- **D** — Seat adjustment
- **E** — Memory buttons

Front Seats

Seat repair and component replacement is possible once the seat is removed from the vehicle. Before servicing front seats read safety regulations on handling pyrotechnic seat belt tensioners. See **720 Seat Belts** and **721 Airbag System (SRS)**.

WARNING —
- *The front seats are equipped with pyrotechnic seat belt tensioners. These tensioners are powerful devices. Handle with extreme care. Incorrect handling can trigger off the airbag or tensioner and cause injury.*
- *BMW recommends that repair or replacement work on airbag and pyrotechnic devices be carried out by a qualified BMW technician.*
- *Be sure to disconnect the battery and wait 30 seconds before attempting to work on pyrotechnic devices.*
- *During body straightening and welding with an electric arc welder, disconnect the battery and the connection to the pyrotechnic gas generators.*

BentleyPublishers.com
BentleyPublishers.com—All Rights Reserved

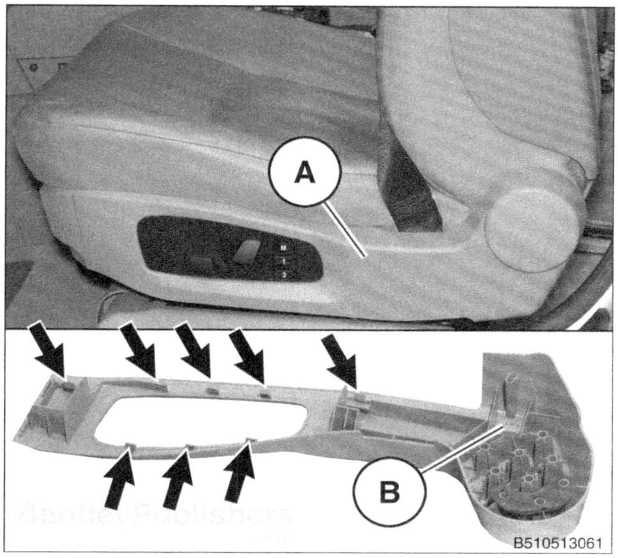

Front seat, removing and installing

< Using a plastic prying tool, release retaining tabs (**arrows**) for for seat trim (**A**). Lift seat trim up to release tab **B** and remove.

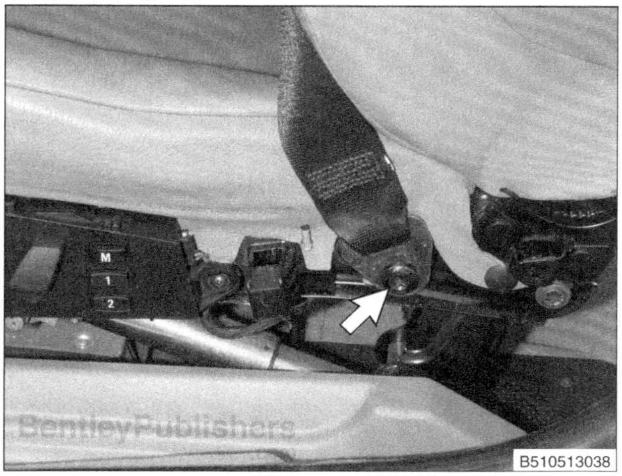

< Remove fastener (**arrow**) for seat belt and discard. Disconnect seat belt from tab and lay aside.

– Move seat as far forward as possible.

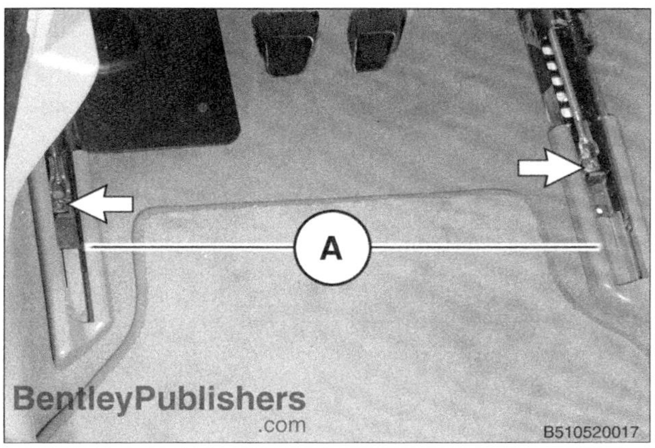

< Remove rear seat trim covers (**A**) and mounting bolts (**arrows**). Discard mounting bolts.

– Move seat as far back as possible.

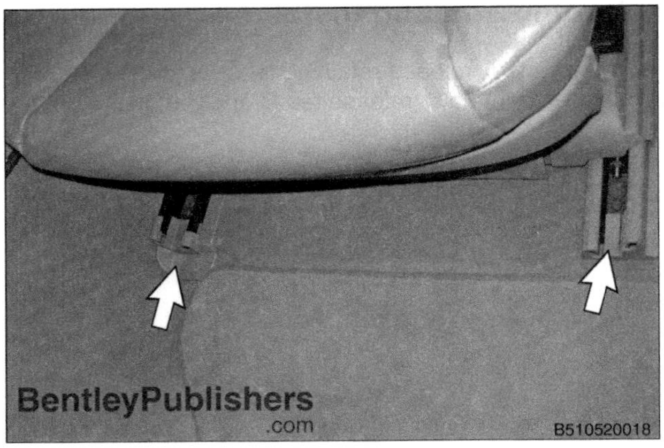

Remove front seat rail trim covers (**arrows**).

Remove front seat mounting bolts (**arrows**) and discard.

- Tilt seat back and remove front trim cover.

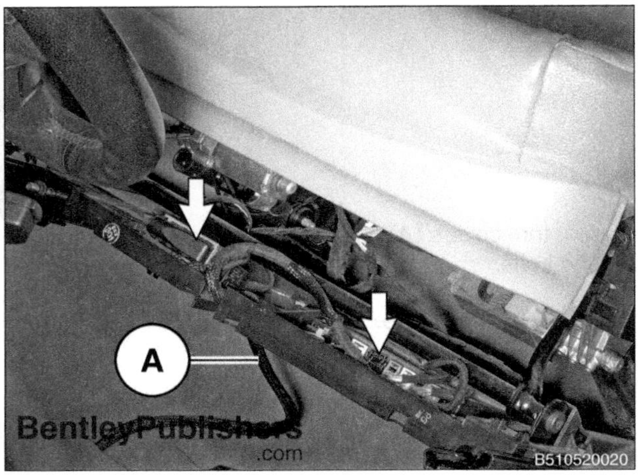

Disconnect electrical connectors (**arrows**) and unclip harness (**A**) from seat bottom.

- Cover door sill with protective covers and remove front seat from car.

> ***CAUTION—***
> - *Use a blanket to protect the seat upholstery from damage if placing the seat upside down for repairs.*

- Installation is reverse of removal. Use wire ties or equivalent means to keep seat harness wiring from fraying.

Tightening torques	
Front seat to floor (replace bolts)	44 Nm (32.4 ft-lb)
Seatbelt to front seat (replace bolt)	36 Nm (26.5 ft-lb)

BentleyPublishers.com
BentleyPublishers.com–All Rights Reserved

Front seat control switch, removing and installing

– Move seat as far forward as possible and to fully raised position.

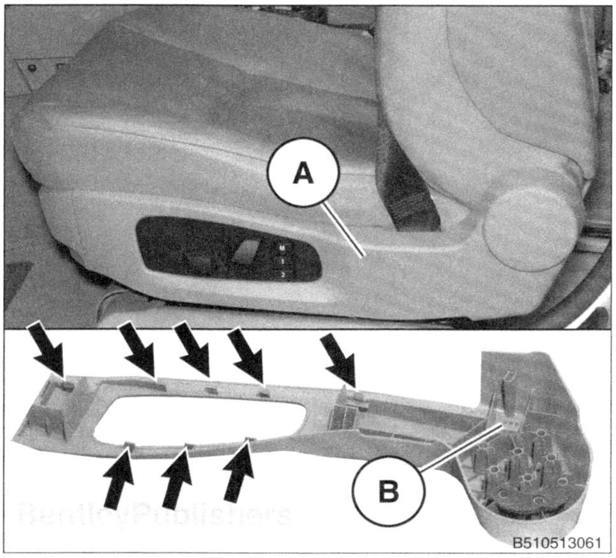

◄ Using a plastic prying tool release retaining tabs (**arrows**) for for seat trim (**A**). Lift seat trim up to release tab **B** and remove.

◄ Remove seat control switch fasteners (**arrows**).

– Disconnect seat switch electrical connector and remove seat switch.

– Installation is reverse of removal.

Passenger seat occupancy sensor

Vehicles covered by this manual are equipped with a passenger seat occupancy sensor in the passenger seat cushion. This device is part of the two-stage deployment airbag system. Access to the seat occupancy sensor is described below. 5 Series models use an OC3 mat which is integrated in the seat cushion. The seat cushion and occupancy mat are replaced as a unit.

– Using BMW scan tool, disable occupancy detection mat from restraint control module.

– Remove passenger seat. See **Front seat, removing and installing** in this repair group.

– Remove front seat backrest rear panel. See **Front seat backrest rear panel, removing and installing** in this repair group.

– Remove seat control switches. See **Front seat control switches, removing and installing** in this repair group.

Passenger seat occupancy sensor

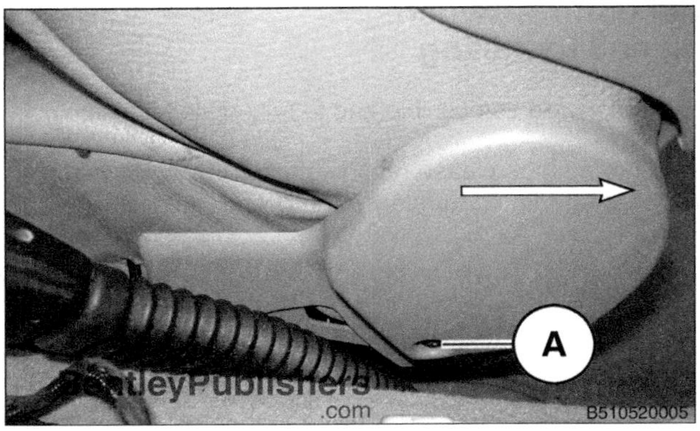

Remove fastener (**A**) and remove trim panel in direction of **arrow**.

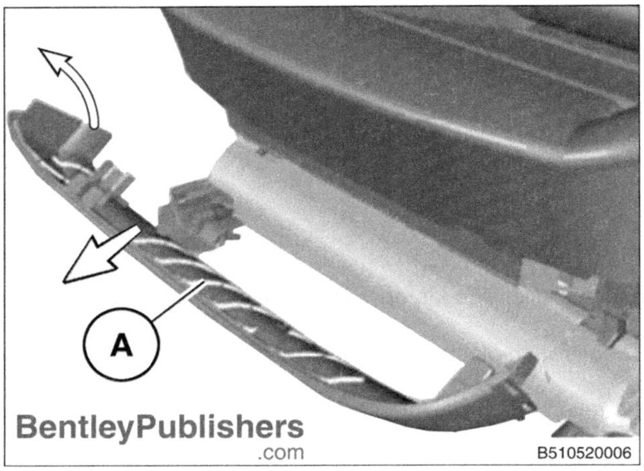

Remove seat front trim (**A**) by pressing side out (**curved arrows**) and pulling forward (**arrow**).

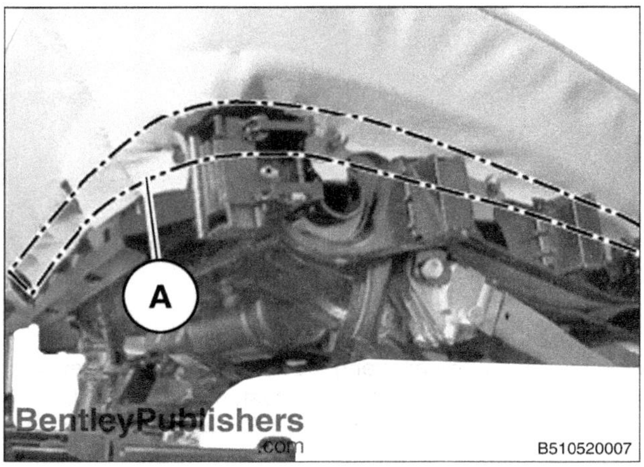

Unhook seat cover from retaining points (**A**).

BentleyPublishers.com
BentleyPublishers.com—All Rights Reserved

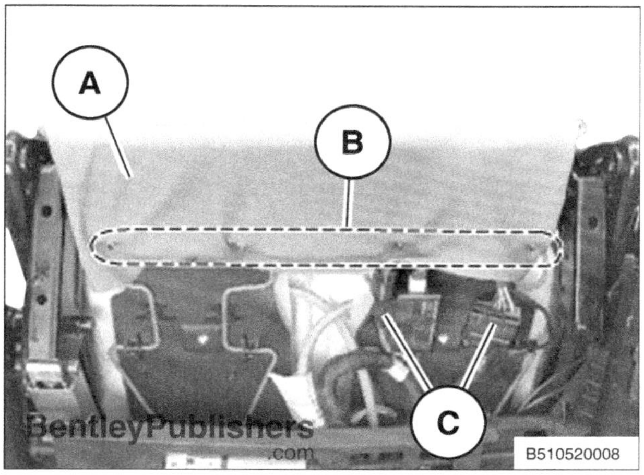

◄ Unhook seat cover (**A**) in area **B**. Disconnect electrical connectors (**C**) and remove seat cover with cushion.

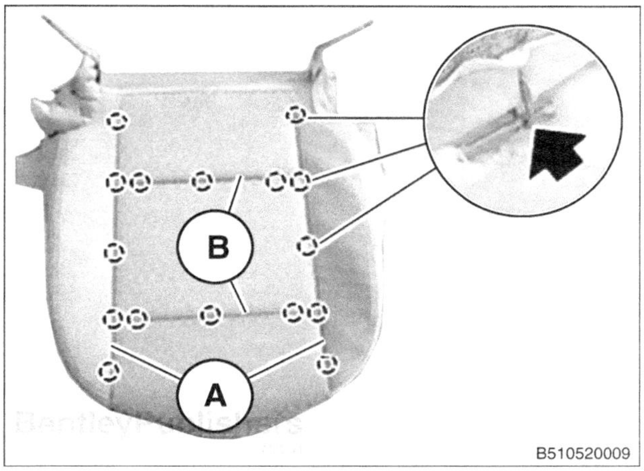

◄ Detach all retainers (**inset**) in side area from longitudinal wires (**A**). Carefully fold back seat cover and release retainers from cross-wires (**B**).

– Remove seat cover from cushion.

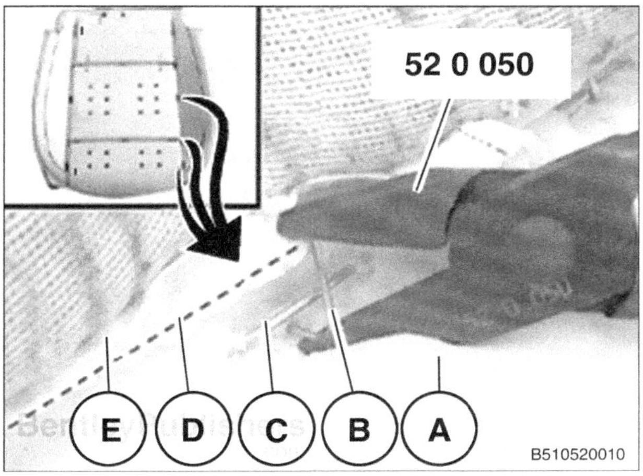

◄ Using BMW special tool 52 0 050, bend new clip (**B**) closed.

- **A** - Padding
- **B** - Clips
- **C** - Trim wire in seat cover
- **D** - Trim wire in cushion
- **E** - Seat cover

– Remainder of installation is reverse of removal. Remember to enable occupancy mat with BMW scan tool.

BentleyPublishers.com
BentleyPublishers.com—All Rights Reserved

Front seat backrest rear panel, removing and installing

The back of the front seat is accessible by removing the rigid backrest panel.

Left side seat is shown. When servicing right seat, transpose the pull and striking sides accordingly.

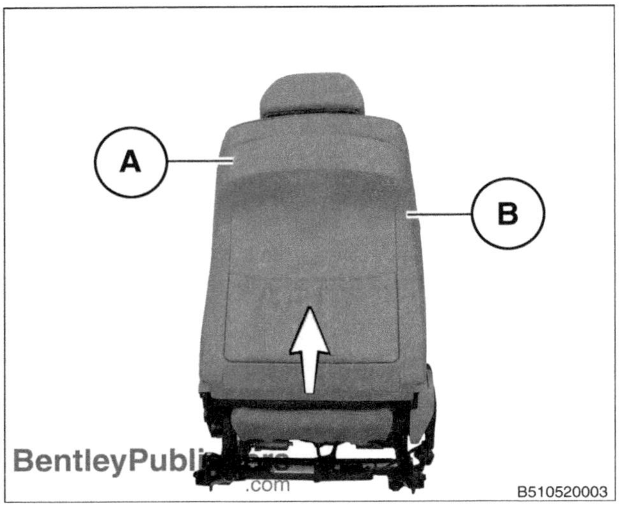

- Grip between rear panel and backrest cover (**A**) on door side until clip can be felt. Do not pull on panel.
- At same time, firmly push towards middle of seat on inner side (**B**) at top. Upper retainers release out of seat guide.
- Pull rear panel up (**arrow**) out of lower guides.
- Installation is reverse of removal. Remember to replace damaged panel clips.

REAR SEATS

Rear seat, removing and installing (Sedan)

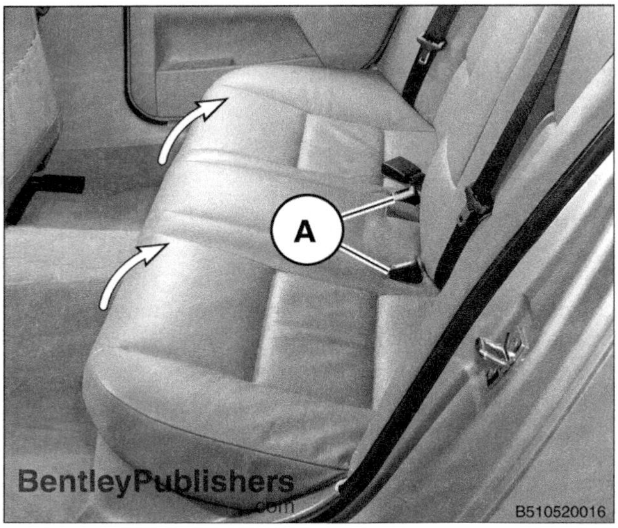

- Pull seat cushion up and away (**arrows**) from pedestal and feed seat belt buckles (**A**) through opening in seat. Lift out rear seat.

BentleyPublishers.com
BentleyPublishers.com–All Rights Reserved

Working in trunk, locate backrest locks (**inset**).

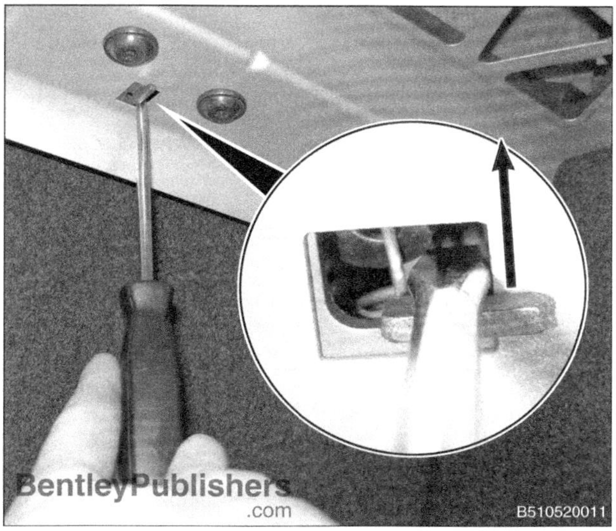

Using a screwdriver, release backrest lock (**inset**) in direction of **arrow**.

Remove seat belt buckle fasteners (**arrows**) and discard. Remove seat belt buckles.

BentleyPublishers
BentleyPublishers.com–All Rights Reserved

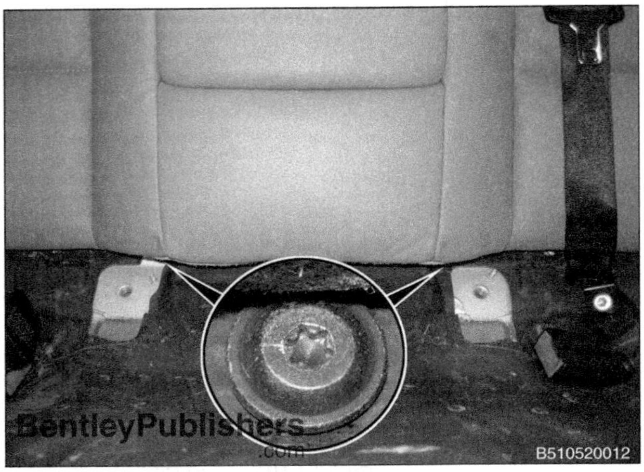

➤ Working at center of rear seat, remove backrest fasteners (**inset**).

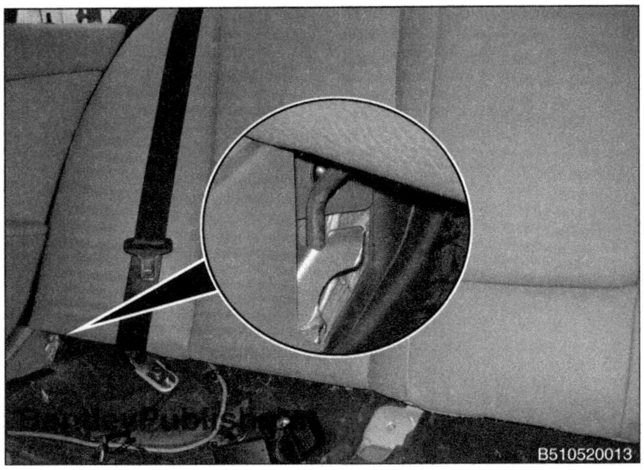

➤ Tilt backrest forward while lifting guides (**inset**) out of body panel.

– Lift backrest while feeding out seat belts and remove backrest.

– Installation is reverse of removal.

Tightening torque	
Seat belt buckle to body (replace fastener)	36 Nm (26.5 ft-lb)
Rear seat belt to body	31 Nm (22.8 ft-lb)

Rear seat headrest, removing and installing (Sedan)

– Remove rear seat backrest. See **Rear seat, removing and installing (Sedan)** in this repair group.

– Slide headrest as far up as possible.

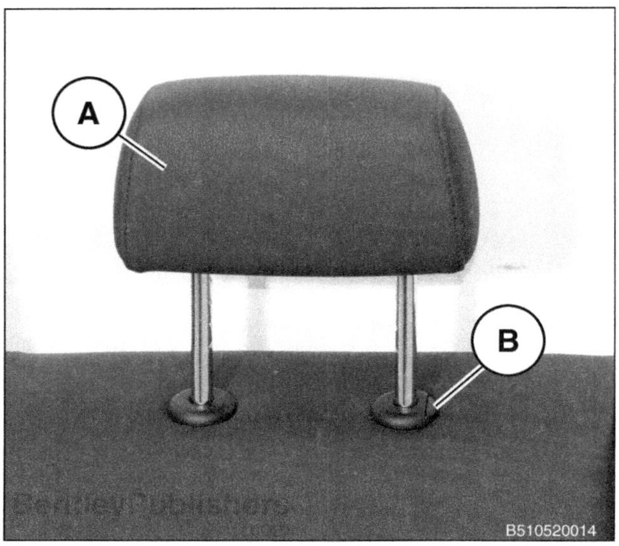

➤ Press lock (**B**) and pull headrest (**A**) upward to remove.

– Installation is reverse of removal.

BentleyPublishers.com
BentleyPublishers.com—All Rights Reserved

540 Sunroof

General

This repair group covers the slide-tilt sunroof assembly. There are two types of sunroofs available: one-piece standard sunroof and two-piece panorama sunroof. Service procedures are different for each type.

Replacement of the cable assembly or frame require that the headliner be removed. Removal and repair of the sunroof assembly and other components inside the roof cavity is beyond the scope of this manual.

Functional overview

Vent position: When the sunroof is set to the vent position (rear of panel up), the inside headliner panel automatically moves rearward. This provides cabin ventilation with less wind noise.

Anti-trap: This feature uses a Hall sensor to detect obstructions while the sunroof is closing in the automatic close function. The anti-trap feature is shut down prior to full closing (4 mm from full closed) to allow the sunroof to seat into the seal.

CAUTION—
- *The anti-trap feature is not functional when the switch is held in the manual close position.*

Unloading: The sunroof is deactivated during engine start. This function is to provide maximum starting power to the engine starter motor. The sunroof control module receives the consumer cut off signal via K-CAN.

A **water channel** is incorporated into the sunroof assembly. It automatically rises when the sunroof panel is in the vent position and prevents water from entering the interior during braking.

BentleyPublishers.com
BentleyPublishers.com—All Rights Reserved

Sunroof, emergency closing

Standard sunroof shown in illustrations.

- Using a plastic prying tool, carefully pry interior light out of headliner at **arrows**, disconnect electrical connectors and remove.
- Using a plastic prying tool, carefully pry switch panel (**A**) out of headliner, disconnect electrical connectors and remove.

- Insert 4 mm Allen into drive in sunroof motor (**arrow**). Turn Allen to manually close sunroof.

Sunroof, emergency operation mode

When an electrical fault is detected the sunroof operates in emergency operation mode, moving for 750 milliseconds at a time. Emergency function is available when CAS sends an emergency mode enabled signal via K-CAN to sunroof control module.

BentleyPublishers.com
BentleyPublishers.com—All Rights Reserved

Sunroof Adjustments

Sunroof, initializing

Initialization allows the sunroof control module to learn the end positions of the motor travel. Hall sensors send motor rotation pulses to the control module, which counts the pulses to determine panel position.

The sunroof initialization process consists of two steps:

- Normalization
- Setting characteristic curve

– Initialize sunroof in the following situations:
  - Malfunctions such as lack of one-touch function, lack of comfort opening / closing or lack of safety (anti-trapping) features.
  - If sunroof motor is removed and reinstalled.
  - If power supply to the sunroof is interrupted.
  - If repairs have been performed on the slide / tilt mechanism.

> ***WARNING —***
> *• There is no anti-trapping protection during initialization.*

Sunroof normalization

During normalization the mechanical end positions of sunroof movement are detected and stored by the control module.

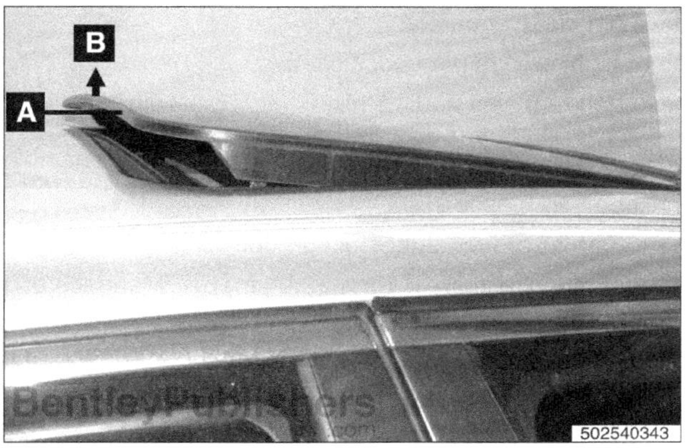

◄ Press and hold sunroof switch in raised (vent) position.
  - After reaching end position (**A**), keep switch pressed for approximately 20 seconds longer.

– Normalization is complete when sunroof in raised position (**A**) moves briefly up (**B**).

– Release sunroof switch.

Setting characteristic curve

– Models with standard sunroof: Characteristic curve is learned manually in the following steps:
  - Release switch within 5 seconds of normalization completion.
  - Press and hold switch in sunroof in raised (vent) position until sunroof has returned to closed position after a complete opening operation.
  - Characteristic curve learning is complete. Release sunroof switch.

BentleyPublishers.com—All Rights Reserved

– Models with panorama sunroof: Characteristic curve is learned manually in the following steps:
  - After completion of normalization process, press and hold switch in raised (vent) position. Hold during procedure described below.
  - The panorama glass roof stops for 5 seconds in the final raised position after normalization.
  - Sunroof headliner panel closes completely and then open completely.
  - The panorama glass then moves into closed position.
  - The panorama glass will then moves to open position and immediately back to the closed position.
  - Sunroof headliner panel closes completely.
  - Characteristic curve learning is complete. Release sunroof switch.

– All models: Check sunroof operation, including anti-trap protection.

Sunroof, adjusting

The sunroof is controlled by a set of cables that move the sunroof panel along guide rails when the motor is operated. The sunroof can be adjusted without removing it from the car.

NOTE—
- *Be sure to check drains in front corners of sunroof carrier if water is entering car through headliner.*

– Adjust sunroof panel under the following circumstances:
  - Sunroof misaligned with roof.
  - Sunroof does not close squarely.
  - Wind noise at high speeds (sunroof closed).
  - Sunroof has been removed and installed.

– For correct sunroof alignment:
  - Sunroof must be fully closed.
  - Gap must be even all around edges of sunroof.
  - Front of sunroof must be flush to 1 mm (0.04 in) below surface of roof.
  - Rear of sunroof must be flush to 1 mm (0.04 in) above surface of roof.

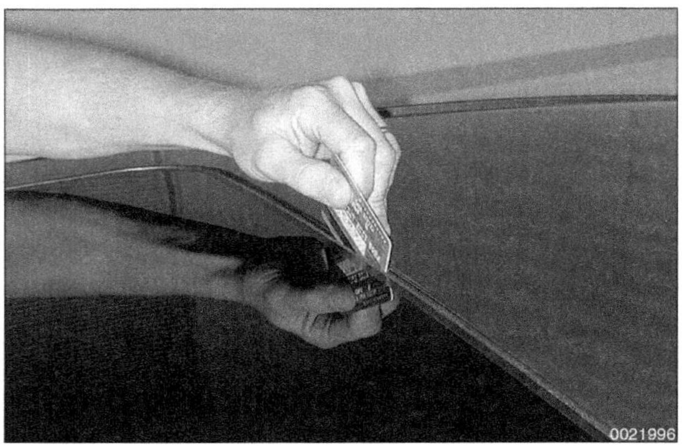

◄ Use credit card to measure gap. Card should insert through gap with equal resistance all around perimeter.

BentleyPublishers.com
BentleyPublishers.com—All Rights Reserved

Standard sunroof

Slide sunroof headliner panel back into roof cavity until approx. 60 mm (2.36 in) is exposed.

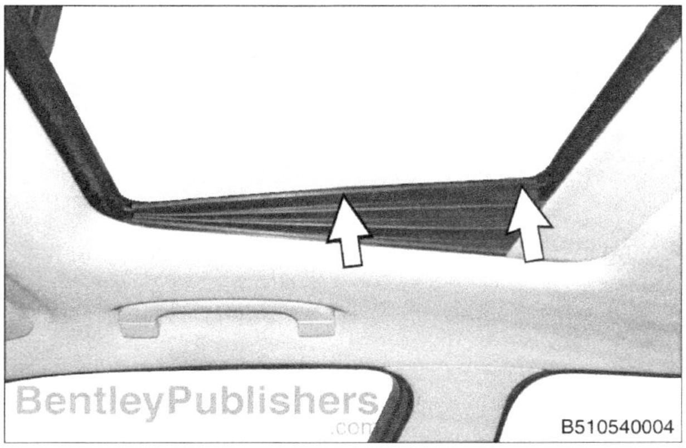

Raise glass sunroof panel to vent position to expose pleated rubber panel (**arrow**).

– Starting at rear section of pleated rubber panel, carefully pull pleated rubber panel off slotted mount on sunroof glass panel. Unlcip pleated rubber panel at front and slide forward to remove.

> ***CAUTION—***
> • *The pleated rubber panels tear easily.*

Sunroof panel is retained by three fasteners (**arrows**) on each side. To adjust sunroof height:

- Loosen fasteners until sunroof can be moved.
- Push sunroof panel up or down until correct height is achieved.
- Reinstall sunroof pleated rubber panels.

Sunroof height adjustment specifications	
Rear	Flush / 1 mm (0.04 in) higher than roof top
Front	Flush / 1 mm (0.04 in) lower than roof top

Tightening torque	
Glass panel to sunroof frame	4 Nm (35 in-lb)

BentleyPublishers.com—All Rights Reserved

Panorama sunroof

– Close glass sunroof panel to expose rubber pleated panel.

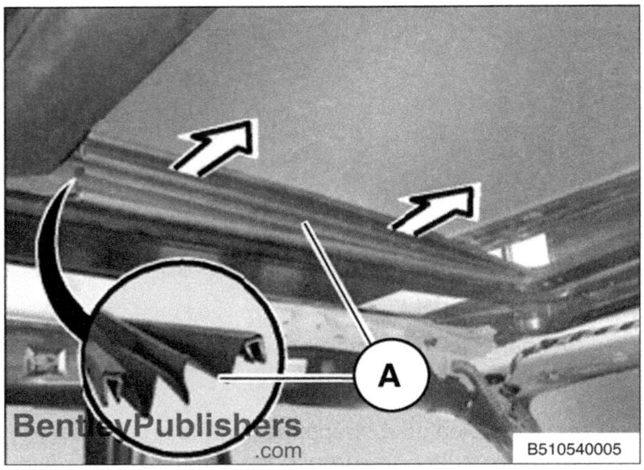

◄ Grasp and remove pleated rubber panel from guide in directions of **arrows**.

Pleated rubber panels cannot be removed without damaging them. Replace with new every time removed.

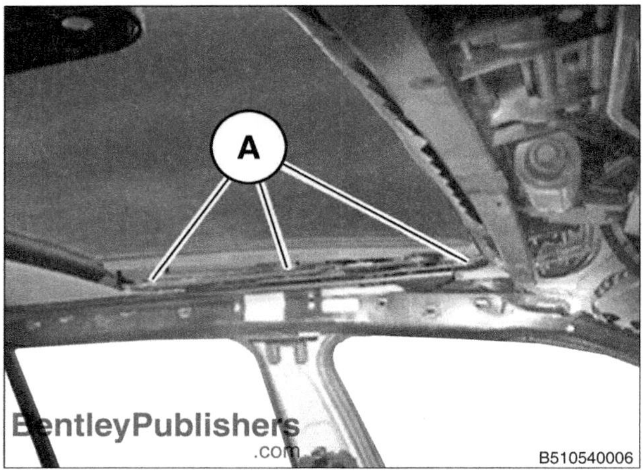

◄ Loosen mounting fasteners (**A**) on left and right sides until sunroof can be moved.

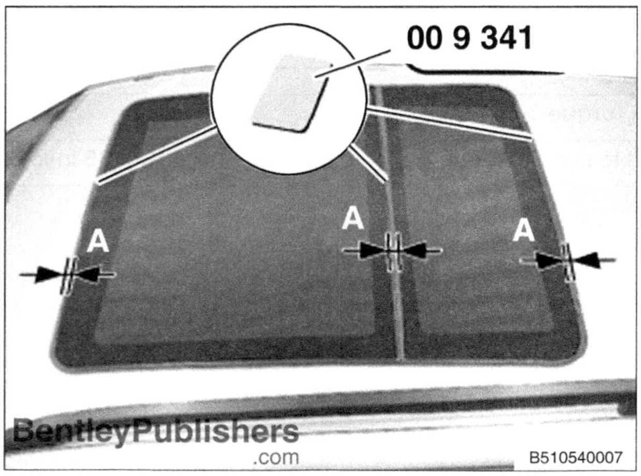

◄ Adjust panorama sunroof glass panels for consistent front and rear gaps (**A**).

- Special tool 00 9 341 is used to check gap spacing.
- **A** = 5.8 ± 0.3 mm (0.23 ± 0.012 in)

– Tighten mounting screws for glass panels.

Tightening torque	
Glass panel to sunroof frame (replace fasteners)	4 Nm (35 in-lb)

– Install new rubber pleated rubber panels.

BentleyPublishers.com
BentleyPublishers.com–All Rights Reserved

Sunroof Motor

Sunroof motor, replacing

The procedure shown below applies to models with standard sunroof. It also applies to the front sunroof motor for models with panorama sunroof.

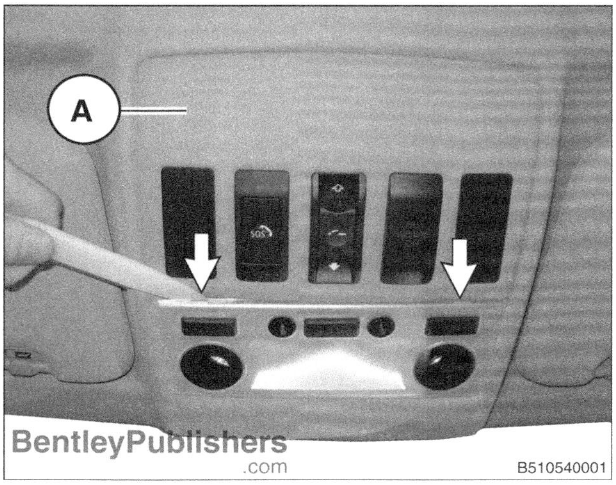

- Using a plastic prying tool, carefully pry interior light out of headliner at **arrows**, disconnect electrical connectors and remove.
- Using a plastic prying tool, carefully pry switch panel (**A**) out of headliner, disconnect electrical connectors and remove.

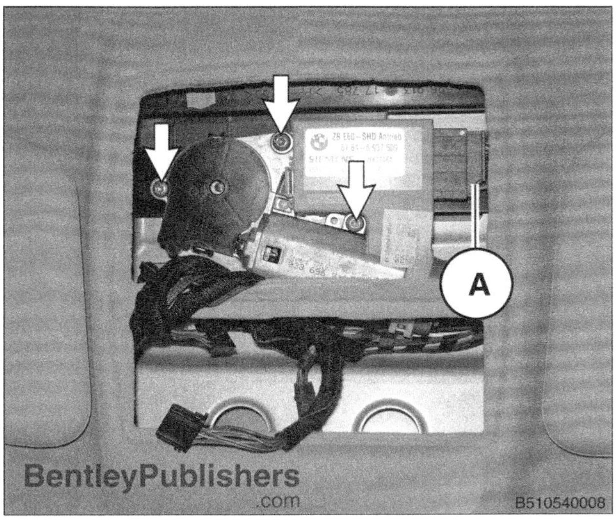

- Remove sunroof motor mounting fasteners (**arrows**) and disconnect electrical connector (**A**). Remove motor.
- Installation is reverse of removal.
  - Be sure to properly engage sunroof motor drive gear sprocket into sunroof drive gear.
  - Perform coding and programming if motor was replaced. See **600 Electrical System General**.
  - Perform initialization procedure. See **Sunroof, initializing** in this repair group.

Tightening torque	
Sunroof motor to sunroof carrier	3.5 Nm (30 in-lb)

BentleyPublishers.com
BentleyPublishers.com—All Rights Reserved

BentleyPublishers.com
BentleyPublishers.com—All Rights Reserved

WARNING

Your common sense, good judgement and general alertness are crucial to safe and successful service work. Before attempting any work on your BMW, be sure to read **001 Warnings and Cautions** *and the copyright page at the front of the manual. Review these warnings and cautions each time you prepare to work on your car. Please also read any warnings and cautions that accompany the procedures in the manual.*

A

BentleyPublishers.com
BentleyPublishers.com—All Rights Reserved

BentleyPublishers.com
BentleyPublishers.com—All Rights Reserved

WARNING

Your common sense, good judgement and general alertness are crucial to safe and successful service work. Before attempting any work on your BMW, be sure to read **001 Warnings and Cautions** *and the copyright page at the front of the manual. Review these warnings and cautions each time you prepare to work on your car. Please also read any warnings and cautions that accompany the procedures in the manual.*

BentleyPublishers.com—All Rights Reserved

E

BentleyPublishers.com
BentleyPublishers.com—All Rights Reserved

WARNING

Your common sense, good judgement and general alertness are crucial to safe and successful service work. Before attempting any work on your BMW, be sure to read **001 Warnings and Cautions** *and the copyright page at the front of the manual. Review these warnings and cautions each time you prepare to work on your car. Please also read any warnings and cautions that accompany the procedures in the manual.*

BentleyPublishers.com
BentleyPublishers.com—All Rights Reserved

BentleyPublishers.com
BentleyPublishers.com—All Rights Reserved

WARNING

Your common sense, good judgement and general alertness are crucial to safe and successful service work. Before attempting any work on your BMW, be sure to read **001 Warnings and Cautions** *and the copyright page at the front of the manual. Review these warnings and cautions each time you prepare to work on your car. Please also read any warnings and cautions that accompany the procedures in the manual.*

BentleyPublishers.com
BentleyPublishers.com—All Rights Reserved

BentleyPublishers.com
BentleyPublishers.com—All Rights Reserved

WARNING

Your common sense, good judgement and general alertness are crucial to safe and successful service work. Before attempting any work on your BMW, be sure to read **001 Warnings and Cautions** *and the copyright page at the front of the manual. Review these warnings and cautions each time you prepare to work on your car. Please also read any warnings and cautions that accompany the procedures in the manual.*

BentleyPublishers
BentleyPublishers.com–All Rights Reserved

BentleyPublishers.com
BentleyPublishers.com—All Rights Reserved

> ***WARNING***
>
> *Your common sense, good judgement and general alertness are crucial to safe and successful service work. Before attempting any work on your BMW, be sure to read* **001 Warnings and Cautions** *and the copyright page at the front of the manual. Review these warnings and cautions each time you prepare to work on your car. Please also read any warnings and cautions that accompany the procedures in the manual.*

BentleyPublishers.com
BentleyPublishers.com—All Rights Reserved

U

V

W

X

Y

BentleyPublishers.com
BentleyPublishers.com—All Rights Reserved

Selected Books and Repair Information From Bentley Publishers

Motorsports

Alex Zanardi: My Sweetest Victory
Alex Zanardi and Gianluca Gasparini
ISBN 978-0-8376-1249-2

The Unfair Advantage *Mark Donohue and Paul van Valkenburgh*
ISBN 978-0-8376-0069-7

Driving Forces: The Grand Prix Racing World Caught in the Malestrom of the Third Reich
Peter Stevenson
ISBN 978-0-8376-0217-2

Equations of Motion – Adventure, Risk and Innovation
William F. Milliken
ISBN 978-0-8376-1570-7

The Racing Driver – The Theory and Practice of Fast Driving
Denis Jenkinson
ISBN 978-0-8376-0201-1

Engineering

Physics for Gearheads *Randy Beikmann*
ISBN 978-0-8376-1615-5

Bosch Automotive Handbook
Updated 9th Edition *Robert Bosch GmbH*
ISBN 978-0-8376-1732-9

Bosch Fuel Injection and Engine Management *Charles O. Probst, SAE*
ISBN 978-0-8376-0300-1

Maximum Boost: Designing, Testing, and Installing Turbocharger Systems
Corky Bell ISBN 978-0-8376-0160-1

Supercharged! Design, Testing and Installation of Supercharger Systems
Corky Bell ISBN 978-0-8376-0168-7

Race Car Aerodynamics *Joseph Katz*
ISBN 978-0-8376-0142-7

Scientific Design of Exhaust and Intake Systems *Phillip H. Smith & John C. Morrison* ISBN 978-0-8376-0309-4

Audi Repair Manuals

Audi A4 Service Manual: 2002-2008, 1.8L Turbo, 2.0L Turbo, 3.0L, 3.2L
Bentley Publishers
ISBN 978-0-8376-1574-5

Audi A4 Service Manual: 1996-2001, 1.8L Turbo, 2.8L *Bentley Publishers*
ISBN 978-0-8376-1675-9

Audi TT Service Manual: 2000-2006, 1.8L turbo, 3.2 L *Bentley Publishers*
ISBN 978-0-8376-1625-4

Audi A6 (C5 platform) Service Manual: 1998-2004 *Bentley Publishers*
ISBN 978-0-8376-1670-4

BMW

BMW 3 Series (F30, F31, F34) Service Manual: 2012-2015 *Bentley Publishers*
ISBN 978-0-8376-1752-7

BMW 3 Series (E36) Service Manual: 1992-1998 *Bentley Publishers*
ISBN 978-0-8376-1709-1

BMW 3 Series (E30) Service Manual: 1984-1990 *Bentley Publishers*
ISBN 978-0-8376-1647-6

BMW X5 (E53) Service Manual: 2000-2006 *Bentley Publishers*
ISBN 978-0-8376-1643-8

BMW Z3 Service Manual: 1996-2002
Bentley Publishers
ISBN 978-0-8376-1617-9

BMW 7 Series (E32) Service Manual: 1988-1994 *Bentley Publishers*
ISBN 978-0-8376-1619-3

BMW 7 Series (E38) Service Manual: 1995-2001 *Bentley Publishers*
ISBN 978-0-8376-1618-6

Memoirs of a Hack Mechanic
Rob Siegel ISBN 978-0-8376-1720-6

Porsche

Porsche 911 (Type 996) Service Manual: 1999-2005 *Bentley Publishers*
ISBN 978-0-8376-1710-7

Porsche 911 Carrera Service Manual: 1984-1989 *Bentley Publishers*
ISBN 978-0-8376-1696-4

Porsche 911 SC Service Manual: 1978-1983 *Bentley Publishers*
ISBN 978-0-8376-1705-3

Porsche 911 (964): Enthusiast's Companion *Bentley Publishers*
ISBN 978-0-8376-0293-6

Porsche Boxster. Boxster S (986) Service Manual: 1997-2004
Bentley Publishers
ISBN 978-0-8376-1645-2

Porsche: Excellence Was Expected
Karl Ludvigsen ISBN 978-0-8376-0235-6

Ferdinand Porsche – Genesis of Genius
Karl Ludvigsen ISBN 978-0-8376-1557-8

Porsche – Origin of the Species
Karl Ludvigsen ISBN 978-0-8376-1331-4

Volkswagen

Volkswagen Jetta, Golf, GTI Service Manual: 1999-2005 *Bentley Publishers*
ISBN 978-0-8376-1678-0

Volkswagen Jetta Service Manual: 2005-2010 *Bentley Publishers*
ISBN 978-0-8376-1616-2

Volkswagen Passat Service Manual: 1998-2005 *Bentley Publishers*
ISBN 978-0-8376-1669-8

Volkswagen Jetta, Golf, GTI: 1993-1999, Cabrio: 1995-2002 Service Manual *Bentley Publishers*
ISBN 978-0-8376-1660-5

Volkswagen GTI, Golf, Jetta Service Manual: 1985-1992 *Bentley Publishers*
ISBN 978-0-8376-1637-7

MINI Repair Manuals

MINI Cooper Service Manual: 2002-2006 *Bentley Publishers*
ISBN 978-0-8376-1639-1

MINI Cooper Service Manual: 2007-2013 *Bentley Publishers*
ISBN 978-0-8376-1730-5

Featured Repair Manuals

BMW X3 (E83) Service Manual: 2004-2010
Bentley Publishers
ISBN 978-0-8376-1731-2

BMW 3 Series (E90, E91, E92, E93) Service Manual: 2006-2011
Bentley Publishers
ISBN 978-0-8376-1723-7

BMW 3 Series (E46) Service Manual: 1999-2005
Bentley Publishers
ISBN 978-0-8376-1657-5

BMW 5 Series (E39) Service Manual: 1997-2003
Bentley Publishers
ISBN 978-0-8376-1672-8

B BentleyPublishers®
Automotive Reference

Bentley Publishers has published service manuals and automobile books since 1950. For more information, please contact Bentley Publishers at 1734 Massachusetts Avenue, Cambridge, MA 02138 USA, or visit our web site at
`BentleyPublishers.com`